DIAGNOSTIC IMAGING

MUSCULOSKELETAL: TRAUMA

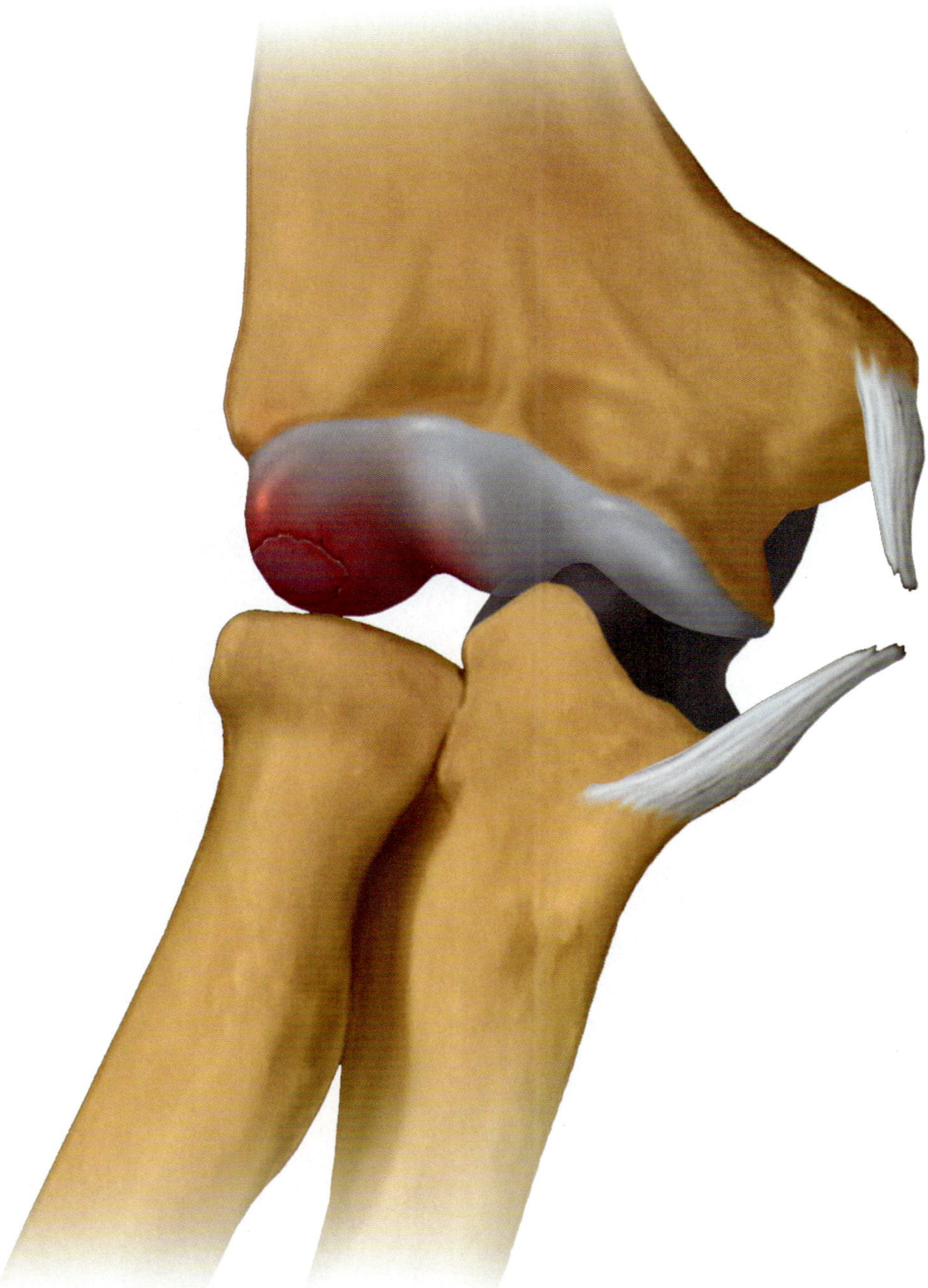

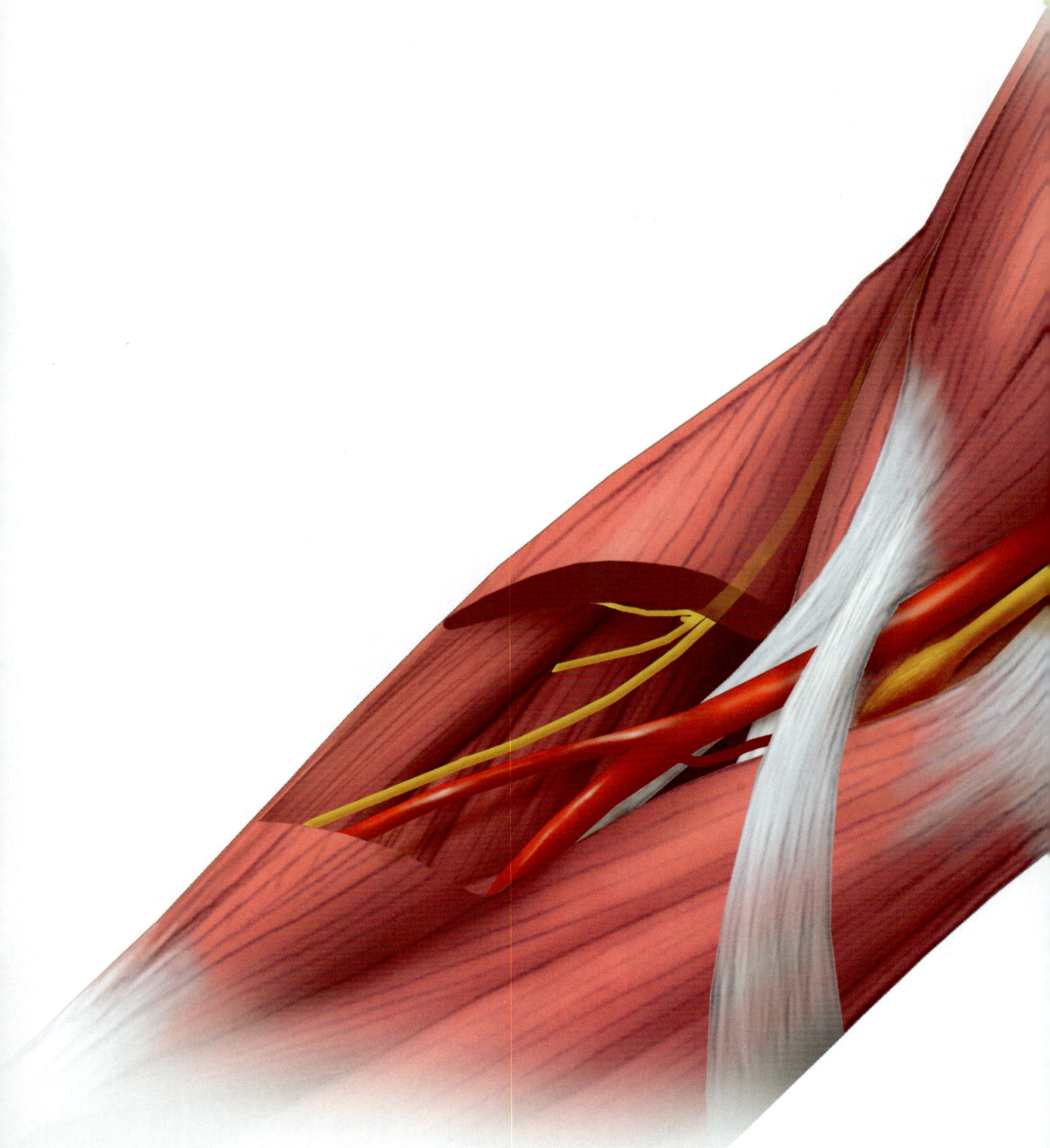

DIAGNOSTIC IMAGING
MUSCULOSKELETAL: TRAUMA

Andrew Sonin, MD
Musculoskeletal Radiology
Radiology Imaging Associates
Englewood, CO
Clinical Assistant Professor of Radiology
University of Colorado School of Medicine
Denver, CO

B.J. Manaster, MD, PhD
Professor of Radiology
University of Utah School of Medicine
Salt Lake City, UT

Carol L. Andrews, MD
Musculoskeletal Radiology
Nationalrad
Specialists in Diagnostic Imaging
Deerfield Beach, FL

Julia Crim, MD
Chief of Musculoskeletal Radiology
Professor of Radiology
University of Utah School of Medicine
Salt Lake City, UT

Michael J. Tuite, MD
Chief of Musculoskeletal Radiology
Professor of Radiology
University of Wisconsin Medical School
Madison, WI

Adam C. Zoga, MD
Director of Musculoskeletal MRI and Ambulatory Imaging
Centers
Associate Professor of Radiology
Thomas Jefferson University Hospital
Philadelphia, PA

AMIRSYS®
Names you know. Content you trust.®

First Edition

Second Printing - August 2013
Third Printing - May 2014

Printed in Canada by Friesens, Altona, Manitoba, Canada

ISBN: 978-1-931884-80-8

Notice and Disclaimer

The information in this product ("Product") is provided as a reference for use by licensed medical professionals and no others. It does not and should not be construed as any form of medical diagnosis or professional medical advice on any matter. Receipt or use of this Product, in whole or in part, does not constitute or create a doctor-patient, therapist-patient, or other healthcare professional relationship between the copyright holders and any recipient. This Product may not reflect the most current medical developments, and the copyright holders (individually and jointly), make no claims, promises, or guarantees about accuracy, completeness, or adequacy of the information contained in or linked to the Product. The Product is not a substitute for or replacement of professional medical judgment. The copyright holders, and their affiliates, authors, contributors, partners, and sponsors, individually and jointly, disclaim all liability or responsibility for any injury and/or damage to persons or property in respect to actions taken or not taken based on any and all Product information.

In the cases where drugs or other chemicals are prescribed, readers are advised to check the Product information currently provided by the manufacturer of each drug to be administered to verify the recommended dose, the method and duration of administration, and contraindications. It is the responsibility of the treating physician relying on experience and knowledge of the patient to determine dosages and the best treatment for the patient.

To the maximum extent permitted by applicable law, the copyright holders provide the Product AS IS AND WITH ALL FAULTS, AND HEREBY DISCLAIMS ALL WARRANTIES AND CONDITIONS, WHETHER EXPRESS, IMPLIED OR STATUTORY, INCLUDING BUT NOT LIMITED TO, ANY (IF ANY) IMPLIED WARRANTIES OR CONDITIONS OF MERCHANTABILITY, OF FITNESS FOR A PARTICULAR PURPOSE, OF LACK OF VIRUSES, OR ACCURACY OR COMPLETENESS OF RESPONSES, OR RESULTS, AND OF LACK OF NEGLIGENCE OR LACK OF WORKMANLIKE EFFORT. ALSO, THERE IS NO WARRANTY OR CONDITION OF TITLE, QUIET ENJOYMENT, QUIET POSSESSION, CORRESPONDENCE TO DESCRIPTION OR NON-INFRINGEMENT, WITH REGARD TO THE PRODUCT. THE ENTIRE RISK AS TO THE QUALITY OF OR ARISING OUT OF USE OR PERFORMANCE OF THE PRODUCT REMAINS WITH THE READER.

The respective copyright holders, individually and jointly, disclaim all warranties of any kind if the Product was customized, repackaged or altered in any way by any third party.

Library of Congress Cataloging-in-Publication Data

Diagnostic imaging. Musculoskeletal : trauma / Andrew Sonin ... [et al.]. -- 1st ed.
p. ; cm.
Includes bibliographical references and index.
ISBN 978-1-931884-80-8 (alk. paper)
1. Musculoskeletal system--Wounds and injuries--Imaging--Handbooks, manuals, etc. 2. Musculoskeletal system--Wounds and injuries--Imaging--Atlases. I. Sonin, Andrew.
[DNLM: 1. Musculoskeletal System--injuries--Handbooks. 2. Musculoskeletal System--radiography--Handbooks. 3. Diagnosis, Differential--Handbooks. 4. Diagnostic Imaging--methods--Handbooks. 5. Fractures, Bone--diagnosis--Handbooks. WE 39 D5386 2010]

RD680.D53 2010
617.4'7044--dc22
2010016355

With love to my book-widow Karen and my beautiful and patient children Kaitlyn, Aaron, and Evan.

Also, a special thank you to those who have provided extraordinary education, opportunity, and encouragement through my career: Jeremy J. Kaye, Lee F. Rogers, Steve W. Fitzgerald, Mark J. Murphey, and B. J. Manaster.
AS

To Steve, Tracy Joy, Katy Rose, Adeline Rose, and Elodie Charlotte.

Additionally, I would like to recognize those mentors who have profoundly influenced my career: Terry Hudson, Alex Norman, David Bragg, and Charles Oxnard. Thank you for making my academic life so rewarding.
BJM

To John, Sara, and Jenna.
CLA

With love and gratitude to Lester, Philip, and Eleanor.
JC

To my wife Monica and daughter Phoebe, whose support for this project was invaluable.
MJT

Thanks JZ; Helena and Olivia, keep reading!
ACZ

To Anna, Lucy, Tucker, Anders, and Clara.
RH

PREFACE

When I was presented with the opportunity to write *Diagnostic Imaging: Musculoskeletal Trauma* for Amirsys, it quickly became apparent that this was an opportunity to create the book that I had always wanted on my own reading room bookshelf: A comprehensive work covering all areas of orthopedic trauma and sports medicine imaging. Instead of relying on one book to reference the grading system of distal femoral fractures and another to refresh my memory on all the variants of SLAP lesions of the shoulder, this book has it all.

This book maintains the successful bulleted-text format used in other Amirsys texts, providing concise but complete information on a very wide range of topics in orthopedic trauma and sports medicine imaging. Thousands of never-before published case images provide state-of-the-art image quality throughout the full breadth of pathology discussed. Hundreds of eye-popping full-color illustrations enhance the visual nature of the book and are used to demonstrate the three-dimensional nature of pertinent anatomy and pathology, such as fracture classification systems. References are up-to-the minute, and information in the text reflects this.

Each clinical section has been written by a recognized expert in the field. The result is not only a body of information reflecting the voice of experience, but a collection of some great illustrative examples of the pathology discussed in each chapter. Rather than simply providing one "classic" example of each type of pathology, we have striven to provide examples of common and unusual variants that may be encountered in clinical practice. An introductory chapter for each anatomic region provides the author's perspective on the anatomic and imaging issues relevant to that region.

Another benefit to the expanded format of this book is the huge collection of supplemental online material found in Amirsys eBook Advantage™; even in a book of this size, some information could not be included in the printed version. Every chapter is associated with additional illustrative cases, and in many cases, significant additional text and references. Besides the supplemental material, all of the material contained in the hard copy of the text is included in the online version, which means that you will have a copy of *Diagnostic Imaging: Musculoskeletal Trauma* available to you wherever you can access the web!

We think that you will find this text an invaluable asset in your clinical practice. We have made every attempt to make this book relevant, easy to use, and visually stunning. It should appeal to both the seasoned subspecialist musculoskeletal radiologist and the junior radiology resident, and everyone in between. We hope you agree.

Andrew Sonin, MD
Musculoskeletal Radiology
Radiology Imaging Associates
Englewood, CO
Clinical Assistant Professor of Radiology
University of Colorado School of Medicine
Denver, CO

ACKNOWLEDGEMENTS

Text Editing

Kellie J. Heap
Arthur G. Gelsinger, MA
Katherine Riser, MA
Dave L. Chance, MA

Image Editing

Jeffrey J. Marmorstone
Danny C. La

Medical Editing

Janine E. Morris, MD

Illustrations

Richard Coombs, MS
Lane R. Bennion, MS
Laura C. Sesto, MA

Art Direction and Design

Laura C. Sesto, MA

Associate Editor

Ashley R. Renlund, MA

Production Lead

Melissa A. Hoopes

AMIRSYS®
Names you know. Content you trust.®

TABLE OF CONTENTS

SECTION 1 Introduction

Overview

Traumatic Injury, Special Topics

SECTION 2 Shoulder and Humerus

Introduction and Overview

Osseous Pathology

Musculotendinous Pathology

Shoulder Girdle

Rotator Cuff

Meniscal Pathology

Tendon Pathology

Other Pathology

SECTION 7
Ankle and Foot

Introduction and Overview

Osseous Injuries

Ligament Injuries

Musculotendinous Injuries

Ankle Impingement Syndromes

Miscellaneous Causes of Foot Pain

Postoperative Ankle and Foot

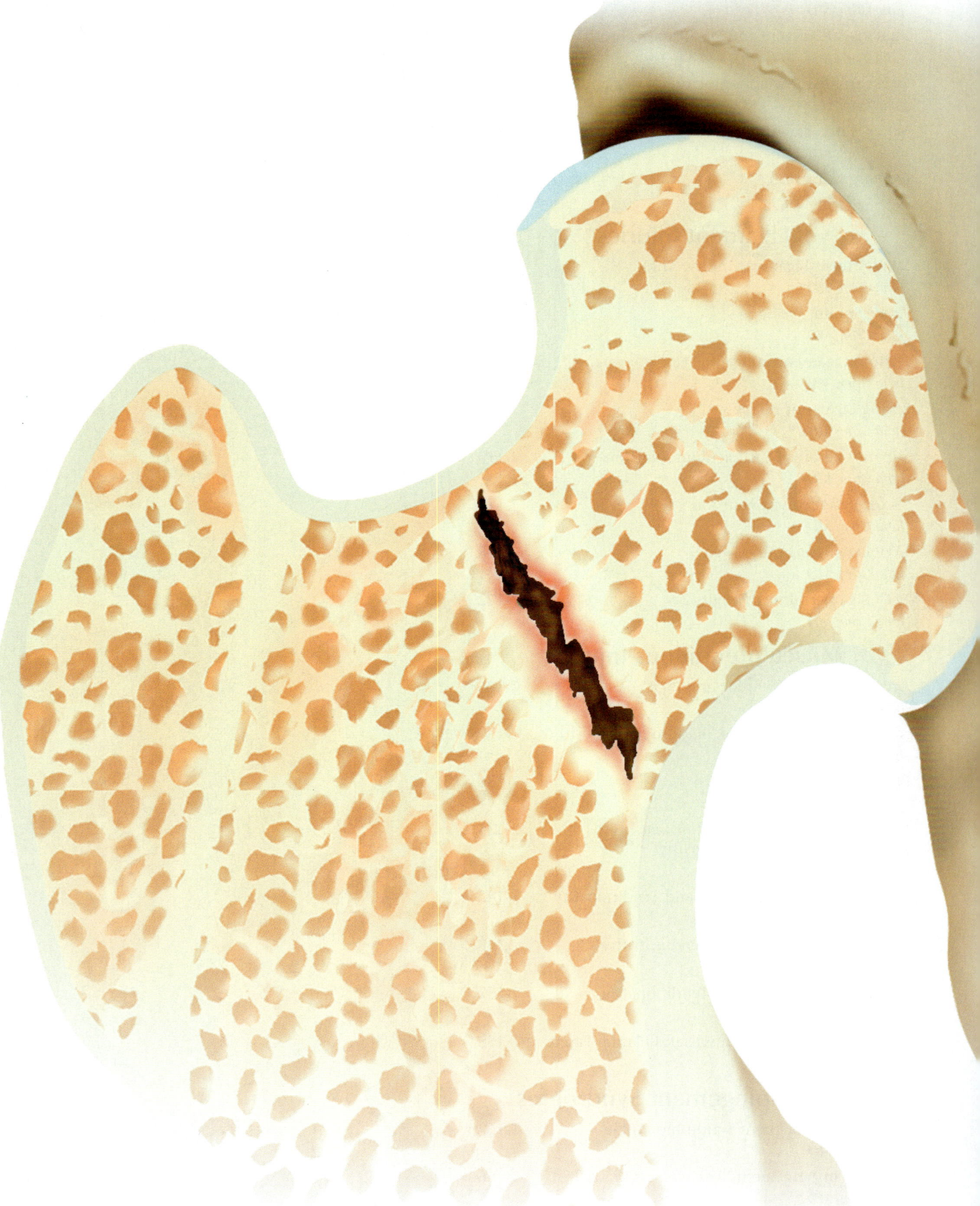

SECTION 1
Introduction

Overview

Traumatic Injury, Special Topics

INTRODUCTION TO TRAUMATIC INJURY

Purpose

This volume is designed to describe and illustrate the gamut of traumatic and sports-related injuries to the extremities encountered in the practice of radiology. Emphasis is placed on imaging patterns and the interrelation between and critical differentiation among entities, so that the reader may quickly focus on the most appropriate imaging features, descriptive language, and differential diagnosis for the entity at hand. Images and illustrations are organized in order to display the range of appearance for each entity, including potential complications and postoperative appearance. References at the end of each chapter provide the reader with additional avenues of exploration if desired.

Terminology

Anatomic terminology used in this book follows conventional medical literature guidelines. For example, the radial aspects of the elbow, forearm, and wrist are termed "lateral" and the ulnar aspects "medial." The terms "superior" and "inferior" have been used to indicate direction in the z-axis of the body instead of "cephalad" and "caudad." The term "mesial" is used to indicate direction toward the center of a limb but not necessarily the center of the body. Specific terminology used for each anatomic region is described in the introductory chapter for that section.

A variety of acronyms are found in the medical literature to refer to injury patterns, imaging findings, and operative approaches. Commonly used acronyms are presented in this book but are spelled out on their first use in each chapter.

The commonly encountered clinical entity of tendon injury is described in this book using the term "tendinopathy," as opposed to "tendinosis" or "tendinitis." This is based on the etymologic origins of these terms. The suffix "-osis" refers to "a process, condition, or state" without being terribly specific about what the process is. The suffix "-itis" refers to "inflammation of"; thus the term "tendinitis" may be appropriate in rare circumstances in which the tendon is inflamed due to acute trauma or infection. However, most cases of tendon pathology encountered in sports imaging are the result of a process of chronic repetitive microinjury and repair, resulting in a weakened tendon unit that is more prone to macroscopic tear. This process has been termed "angiofibroblastic hyperplasia" and is characterized by ingrowth of fibroblastic repair tissue and neovascularity; there is very little inflammatory component. Thus the suffix "-opathy," meaning "disease or disorder of," most accurately describes the underlying process.

Although the imaging of some areas (e.g., the shoulder) typically requires nonorthogonal imaging planes to provide anatomically appropriate images, the terms "coronal" and "sagittal" have been used to imply "oblique coronal" and "oblique sagittal" planes conventionally used in those regions. When a nonconventional obliquity is illustrated, a description of the imaging plane used is provided. The terms "CT" and "MR" have been used throughout the book to indicate "computed tomography" and "magnetic resonance" imaging techniques.

Anatomy-based Imaging Issues

For the sake of consistency, images have been presented so that they appear as "right-sided" throughout the book. Thus, all radiographs, CT and MR images, and graphic illustrations will share a uniform orientation for easier correlation between different examples and modalities. This was not done for cases involving the pelvis and chest in order to maintain anatomic relevance.

Imaging Protocols

While some specific description is provided for the acquisition of certain radiographic views, description of most CT and MR imaging techniques is given only in general terms. This was done with the understanding that imaging equipment from different manufacturers, and often different levels of equipment from the same manufacturers, have very different capabilities and use a wide range of descriptive language to provide similar imaging results. Thus, terms such as "T2-weighted fat-suppressed MR images" describe the generic appearance of anatomy that may be accomplished using a wide variety of proprietary MR techniques. In addition, the armamentarium of imaging techniques changes constantly, and new pulse sequences and hardware devices become available that may alter the method used by a particular radiologist to accomplish the same end.

In general, CT imaging of joints or extremities is best accomplished using a multislice scanner to acquire very thin slices through the affected anatomic region, with subsequent creation of anatomically appropriate reformatted images in additional planes from the acquired data. To avoid reconstruction artifact in small joints, a minimum of 6 to 8 detector rows is advisable, and more pleasing images are produced with higher detector counts (typically 32 to 64).

MR imaging of joints and extremities is best accomplished using volume coils in order to maximize signal-to-noise ratio of the images and minimize field inhomogeneity that may hinder or prevent chemical fat suppression. Joint-specific volume coils are available from most magnet manufacturers for commonly imaged anatomic regions. Optimization of images is usually a trade-off between minimizing the field-of-view to cover the area of interest and the maintenance of appropriate signal-to-noise on the images through manipulation of pulse sequence parameters. The increasing availability of multichannel extremity coils and higher strength magnets provide greater opportunity to improve both of these goals simultaneously.

The use of gadolinium arthrography can be very helpful in the detection and description of certain intraarticular pathologies. The application of this technique is described and illustrated where appropriate. This may be accomplished by direct injection of the joint (direct MR arthrography) or by intravenous injection of gadolinium followed by delayed MR imaging of the joint (indirect MR arthrography). Direct arthrography has the advantage of distending the joint but is invasive. Indirect arthrography is less invasive but provides no joint distension and may result in enhancement of tissues (such as a hyperemic but intact rotator cuff) that can confuse image interpretation.

The clinical utility of ultrasound in the evaluation of musculoskeletal injury continues to grow and is currently an area of intense research and publication. Ultrasound can provide exquisite anatomic detail of soft tissues, particularly in areas close to the body surface; because the ultrasound beam deteriorates with the depth of tissue it needs to penetrate, technical limitations are often encountered in the evaluation of deeper structures (and particularly in large patients). The technique uses no ionizing radiation and is noninvasive. However, musculoskeletal ultrasound is heavily dependent on operator skill, and a steep learning curve may be encountered as one seeks to gain expertise in this field. In geographic regions where access to MR imaging is limited, musculoskeletal ultrasound has grown to be a reliable and ubiquitous problem-solving imaging modality. In most of the United States, probably because of the relative availability of MR imaging in most communities, the use of ultrasound in the imaging diagnosis of sports injury lags behind that in other parts of the world.

Pathologic Issues

In many cases, the manifestations and etiologies of trauma and injury are very different for children than they are for adults. Description and illustration of issues specific to pediatric patients is provided where appropriate. Dedicated chapters are presented on the topics of child abuse and physeal injuries.

Orthopedic surgeons commonly use classification and grading systems to categorize injuries. These systems are usually helpful in determining appropriate therapy for a particular injury. The commonly used classification and grading systems for each injury are provided and illustrated.

Pathology-based Imaging Issues

Each chapter contains discussion of the advantages and disadvantages of particular imaging techniques in diagnosing and characterizing a particular entity. Some generalizations may be made, however.

Radiography is usually the 1st-line tool in the evaluation of acute traumatic injury to the limbs, primarily as a way to detect fractures and dislocations. Soft tissue injury is much less well delineated on radiographs, however, and the information provided regarding the soft tissue components of an injury tends to be nonspecific.

CT, because of its superior tissue density differentiation, can be more helpful in detecting and characterizing some soft tissue findings, but its clinical utility is mainly based on evaluation of osseous abnormalities. Small fractures are generally better demonstrated on CT than on radiographs because of the former's tomographic nature, which alleviates confusion due to overlapping structures. The addition of contrast arthrography to CT can be helpful in evaluating certain intraarticular structures and provides value in the detection of rotator cuff tear, labral injury in the shoulder and hip, meniscal injury in the knee, and cartilage injury throughout the extremities.

MR imaging uses a strong magnetic field and radiofrequency energy pulses to manipulate the energy state of protons within tissue. Complex machinery detects subtle differences in how different tissues respond to this energy deposition and provides exquisitely detailed information about soft tissues. For this reason, MR imaging has been the mainstay of diagnosis in acute and chronic injury to soft tissues, such as ligaments, tendons, and muscles. In addition, MR can provide detailed evaluation of intraarticular structures, often without the injection of contrast material, so that fibrocartilage (meniscus and labrum), chondral cartilage, and synovial pathology is well demonstrated. There are some specific situations in which the use of intraarticular contrast injection can increase the diagnostic accuracy of MR, such as in the postoperative patient. MR imaging also provides a sensitive method for the detection of bone marrow abnormalities, such as marrow edema, osteonecrosis, and tumor infiltration, but is not particularly sensitive for disruption of cortical bone unless a fracture is displaced.

Ultrasound provides another excellent method for studying the soft tissues of the extremities, and, as indicated above, is particularly useful in the evaluation of structures close to the skin surface. In addition, ultrasound provides real-time information regarding the motion of structures and is thus valuable in the evaluation of transient phenomena, such as tendon impingement or subluxation. Limitations of ultrasound include the inability of the beam to penetrate dense material, so that bone and metal effectively block the evaluation of anything located deeper in the body. For this reason, ultrasound is not commonly useful for the evaluation of intraarticular pathology. Ultrasound also does not perform well when encountering air collections, since sound waves travel poorly through air.

Selected References

1. Buckwalter KA: Current concepts and advances: computerized tomography in sports medicine. Sports Med Arthrosc. 17(1):13-20, 2009
2. Coris EE et al: Imaging in sports medicine: an overview. Sports Med Arthrosc. 17(1):2-12, 2009
3. Jacobson JA: Musculoskeletal ultrasound: focused impact on MRI. AJR Am J Roentgenol. 193(3):619-27, 2009
4. Krug R et al: Imaging of the musculoskeletal system in vivo using ultra-high field magnetic resonance at 7 T. Invest Radiol. 44(9):613-8, 2009
5. Renner JB: Conventional radiography in musculoskeletal imaging. Radiol Clin North Am. 47(3):357-72, 2009
6. Sanchez TR et al: MR imaging of pediatric trauma. Magn Reson Imaging Clin N Am. 17(3):439-50, v, 2009
7. Sofka CM et al: The history of clinical musculoskeletal radiology. Radiol Clin North Am. 47(3):349-56, 2009
8. Gandhi RT et al: Technical considerations and potential advantages of musculoskeletal imaging at 3.0 Tesla. Semin Musculoskelet Radiol. 12(3):185-95, 2008
9. Jaramillo D et al: Pediatric musculoskeletal MRI: basic principles to optimize success. Pediatr Radiol. 38(4):379-91, 2008
10. Meyer JS et al: Musculoskeletal MR imaging at 3 T. Magn Reson Imaging Clin N Am. 16(3):533-45, vi, 2008
11. Buckwalter KA: Optimizing imaging techniques in the postoperative patient. Semin Musculoskelet Radiol. 11(3):261-72, 2007
12. Illustrated Stedman's Medical Dictionary. 24th ed. Baltimore: Williams & Wilkins, 1982

STRESS/INSUFFICIENCY FRACTURE

Key Facts

Terminology

- Fatigue fracture, insufficiency fracture, osteoporotic fracture, stress reaction
- Stress or fatigue fracture results from abnormal repetitive stress on normal bone
- Insufficiency fracture results from normal stress on abnormal bone

Imaging

- Radiographs: Linear sclerosis ± periosteal reaction
- MR: Low signal fracture line with surrounding marrow and soft tissue edema
- Radiographs identify 28% of acute stress fractures
- MR: Marrow edema; focal cortical or medullary fracture may not be seen
  - Diffuse marrow edema ± periosteal reaction without focal fracture = stress reaction
- Bone scan: Highly sensitive; less specific than MR
  - Delayed findings after injury
- Radiograph most appropriate 1st step
- If radiographs negative → MR

Pathology

- Fatigue fracture
  - Repetitive activity, such as athletics
- Insufficiency fracture
  - Senile osteoporosis
  - Anorexia
  - Rheumatoid arthritis
  - Exogenous steroid use
  - Stress shielding from joint prosthesis

Clinical Issues

- Fatigue fracture: Young athletic adolescents or adults
- Insufficiency fracture: Common in elderly or debilitated patients
- Site usually relates to type of repetitive activity

*(**Left**) Oblique radiograph shows a subtle transverse band of increased density in the distal fibula ➡ representing an insufficiency fracture in this osteopenic patient. (**Right**) AP radiograph shows periosteal new bone formation ➡ around the distal 2nd metatarsal in a runner with foot pain for 3 weeks. A very subtle fracture line ⇨ is present, but was missed on initial interpretation. Sometimes callus is the 1st or only clue to a metatarsal stress fracture.*

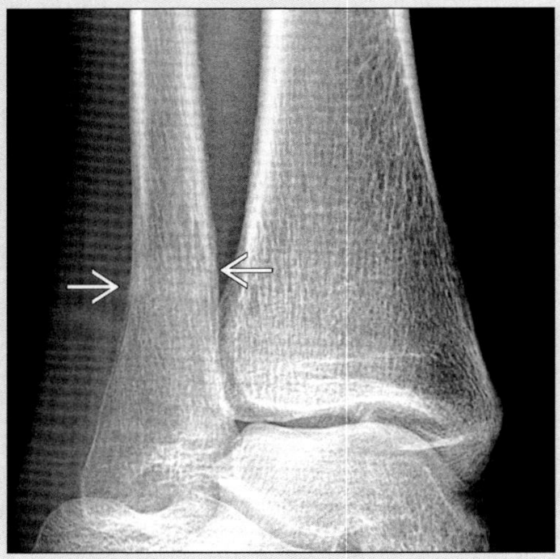

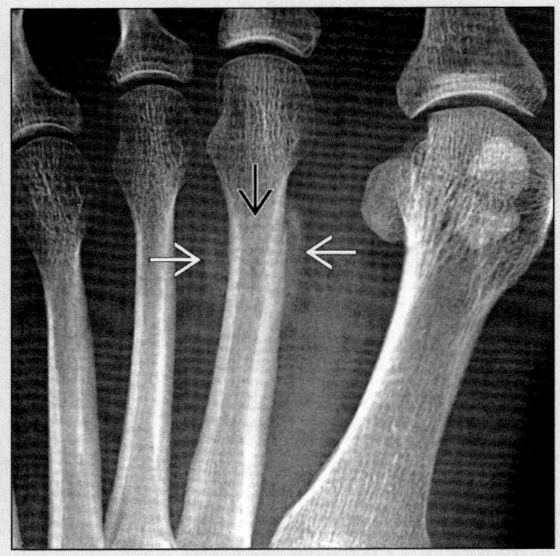

*(**Left**) Longitudinal oblique T2WI FS MR shows diffuse marrow edema throughout the 2nd metatarsal shaft indicating a stress reaction. There is fusiform cortical thickening in the midshaft ➡ indicating a more subacute or chronic time course to the injury. Radiographs were read as normal. (**Right**) Coronal T2WI FS MR shows a transverse stress fracture ➡ of the proximal lateral tibia in an adolescent long distance runner with leg pain. The posterior proximal tibia is a common place for stress injury.*

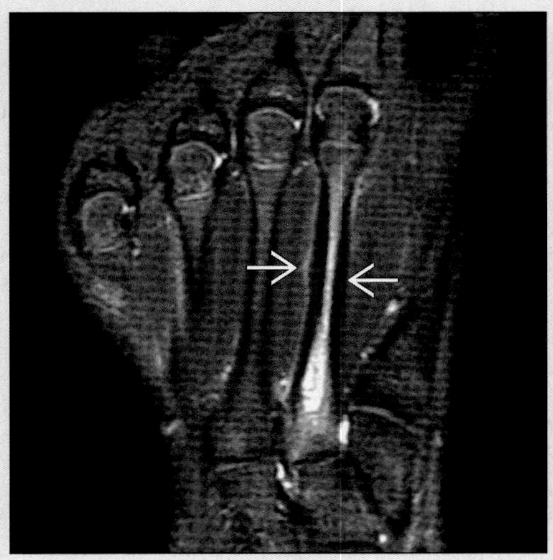

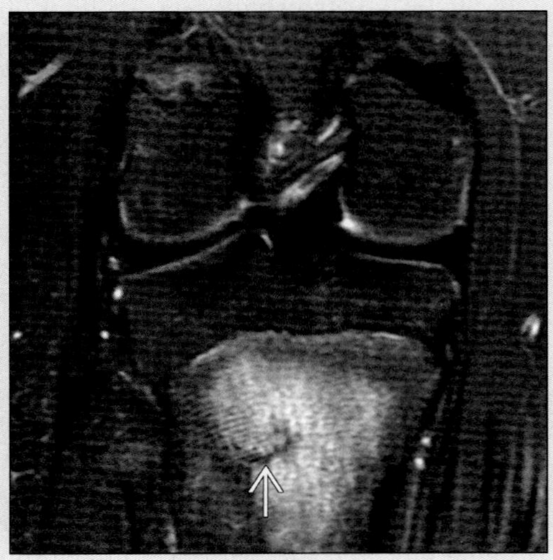

STRESS/INSUFFICIENCY FRACTURE

TERMINOLOGY

Synonyms

- Fatigue fracture, insufficiency fracture, osteoporotic fracture, stress reaction

Definitions

- Stress or fatigue fracture results from abnormal repetitive stress on normal bone
- Insufficiency fracture results from normal stress on abnormal bone

IMAGING

General Features

- Best diagnostic clue
    - Radiographs: Linear sclerosis ± periosteal reaction
    - MR: Low signal fracture line with surrounding marrow and soft tissue edema
- Location
    - Pelvis
        - Sacrum
        - Pubic rami (inferior > superior)
    - Femur
        - Basicervical
        - Medial cortex in proximal & midshaft; proximal 1/3, including subtrochanteric region, most common
        - Posterior cortex in distal shaft
    - Long bone longitudinal
        - Femur
        - Tibia
        - Fibula
    - Tibia
        - Posterior cortex
        - Incidence: Proximal 1/3 > distal 1/3 > middle 1/3
        - "Shin splints" in runners
    - Fibula
        - Shaft proximal to lateral malleolus most common
        - Proximal 1/3 less common
    - Ankle
        - Calcaneus
        - Navicular
    - Foot
        - Metatarsals (90% involve 2nd & 3rd; 4th next most common)
            - "March fracture" seen in military recruits
        - Sesamoids of great toe
    - Upper extremity and ribs very uncommon
        - Forearm fractures due to crutch use
        - Coracoid fractures in shooting ("trapshooter's shoulder")
- Size
    - May range from tiny subchondral focus to large linear fracture
- Morphology
    - May be linear, crescentic, band-like, or cloud-like

Radiographic Findings

- Early stress fracture
    - Radiographs identify 28%
    - Usually normal
- Subacute or chronic stress fracture
    - Radiographs more specific
    - Linear sclerosis on radiograph, focal periosteal reaction
    - Sclerosis usually perpendicular to major trabeculae
    - Subchondral insufficiency fracture: Focal or diffuse subcortical sclerosis

CT Findings

- Linear lucency of fracture line seen
- Adjacent sclerosis and reaction, depending on age of fracture
- Orthogonal reformats optimal for evaluating longitudinal stress fracture

MR Findings

- Early stress fracture
    - Marrow edema
        - Low signal on T1WI, high signal on T2WI MR
        - May be focal or diffuse
    - Focal cortical or medullary fracture may not be seen
        - Diffuse marrow edema ± periosteal reaction without focal fracture = stress reaction
    - Surrounding soft tissue edema
- Subacute or chronic stress fracture
    - Fracture line better seen on all sequences
    - Marrow and soft tissue edema less prominent
    - Periosteal new bone formation near fracture site

Nuclear Medicine Findings

- Early stress fracture: Tc-99m bone scan shows ill-defined increased uptake
- Subacute or chronic stress fracture: Bone scan shows more focal uptake
- Highly sensitive; less specific than MR
- High negative predictive value
- Delayed findings after injury
    - Up to 48 hours in fatigue fracture
    - Up to 72 hours in osteoporotic insufficiency fracture

Imaging Recommendations

- Best imaging tool
    - Radiograph most appropriate 1st step
    - If radiographs negative → MR

DIFFERENTIAL DIAGNOSIS

Tumor

- Periosteal bone formation may indicate surface osteosarcoma

Infection

- Osteomyelitis demonstrates diffuse or focal marrow edema and soft tissue swelling

Soft Tissue Injury

- Soft tissue edema at site of injury, but little marrow abnormality

PATHOLOGY

General Features

- Etiology
    - Bone responds to repetitive loading by remodeling
        - Cortical bone initially remodels by osteoclast-mediated resorption
        - Osteoblastic bone replacement follows resorption

STRESS/INSUFFICIENCY FRACTURE

- Stress fracture occurs in vulnerable period when resorption outpaces bone replacement
- Stress fractures evolve over time
  - Early, injury is distributed over long segment of cortex (termed stress reaction)
  - With ongoing microfractures, focal segment may weaken more rapidly
    - This segment becomes focal point for bone deformation during repeated loading
    - Microfractures progress at this site
    - Reparative process becomes more focal, related to microfracture site
- Muscle fatigue may also contribute to injury
  - If muscle is close to exhaustion, it provides less stress shielding & may increase load on bone
- Fatigue fracture
  - Repetitive activity, such as athletics
    - Running, marching, push-ups
- Insufficiency fracture
  - Senile osteoporosis
  - Anorexia
  - Rheumatoid arthritis or other inflammatory conditions
  - Exogenous steroid use
  - Ethanol abuse
  - Renal osteodystrophy
  - Radiation osteitis
  - Stress shielding from joint prosthesis

CLINICAL ISSUES

Presentation

- Most common signs/symptoms
  - Localized pain
  - Focal tenderness
  - Soft tissue swelling

Demographics

- Age
  - Fatigue fracture: Young athletic adolescents or adults
  - Insufficiency fracture: Common in elderly or debilitated patients
- Gender
  - Fatigue fracture: M > F
  - Insufficiency fracture: M < F
- Epidemiology
  - Up to 10% of patients seen in sports medicine clinics have stress fracture
    - Particularly common in runners
  - Site usually relates to type of repetitive activity
    - Runners: Pubic rami, basicervical femoral neck, tibial shaft, fibula, metatarsal shaft, sesamoid
    - Backpacking (& carrying heavy bookbags): Upper ribs, clavicle
    - Tennis: Humerus, ulna, metacarpal
    - Baseball pitching: Humerus, scapula, 1st rib, ulna
    - Baseball batting: Rib
    - Baseball catching: Patella
    - Basketball: Patella, tibia (anterior mid shaft, may heal poorly), calcaneus, sacrum, navicular
    - Javelin throwing: Ulna, humerus
    - Soccer: Tibia, pubis
    - Swimming: Tibia, metatarsals
    - Skating: Fibula, tibia, metatarsals
    - Curling: Ulna
    - Aerobics: Tibia, fibula
    - Ballet dancers: Spine (pars interarticularis), tibia, fibula, metatarsals, sesamoid
    - Cricket: Humerus, spine
    - Golf: Lower ribs, ulna, sternum, tibia, hook of hamate
    - Fencing: Pubis
    - Handball: Metacarpals
    - Water-skiing: Pars interarticularis
    - Rowing: Lower ribs
    - Trap shooting: Coracoid process
    - Volleyball: Pisiform
    - Gymnast: Fibula, pars interarticularis, humerus (periostitis at pectoralis insertion, from iron ring performance)
    - Wrestlers: Sternum
    - Weightlifters: Humerus (periostitis at pectoralis insertion, from repeated lateral curls)

Natural History & Prognosis

- If protected, will heal without consequence
- If protection is not instituted or adhered to, may progress to complete fracture

Treatment

- Protect from weight-bearing; if advanced, immobilization
- Complete fracture in osteoporotic patient may warrant surgical fixation

DIAGNOSTIC CHECKLIST

Consider

- Stress fracture in typical locations

Image Interpretation Pearls

- Radiographs often normal, particularly in early phase of injury
- Seek appropriate history if stress injury suspected
- Biopsy of stress fracture should be avoided
  - Biopsy further weakens bone → longer immobilization
  - Biopsy of fracture may be confused histologically with aggressive bone tumor

SELECTED REFERENCES

1. Brockwell J et al: Stress fractures of the foot and ankle. Sports Med Arthrosc. 17(3):149-59, 2009
2. Davis KW: Transient migratory osteoporosis or insufficiency fracture? J Clin Rheumatol. 15(3):153, 2009
3. Hodnett PA et al: MR imaging of overuse injuries of the hip. Magn Reson Imaging Clin N Am. 17(4):667-79, vi, 2009
4. Krestan C et al: Imaging of insufficiency fractures. Eur J Radiol. 71(3):398-405, 2009
5. Campbell SE et al: Imaging of stress injuries of the pelvis. Semin Musculoskelet Radiol. 12(1):62-71, 2008
6. Silva MJ: Biomechanics of osteoporotic fractures. Injury. 38 Suppl 3:S69-76, 2007
7. Pepper M et al: The pathophysiology of stress fractures. Clin Sports Med. 25(1):1-16, vii, 2006
8. Snyder RA et al: Epidemiology of stress fractures. Clin Sports Med. 25(1):37-52, viii, 2006
9. Knapp TP et al: Stress fractures: general concepts. Clin Sports Med. 16(2):339-56, 1997

STRESS/INSUFFICIENCY FRACTURE

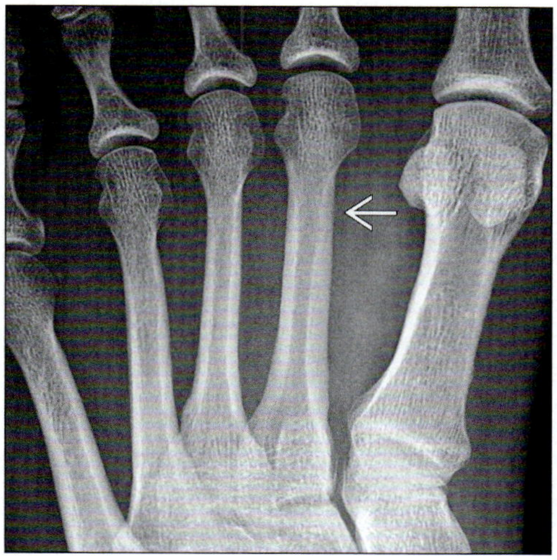

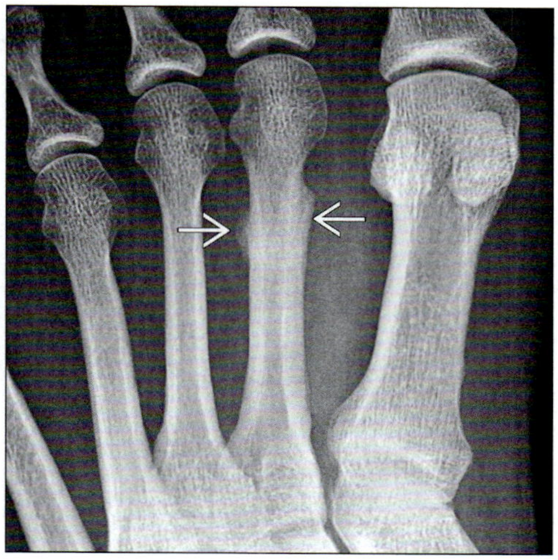

(Left) *AP radiograph in a runner complaining of foot pain for 1 week was read as normal. In retrospect, there is a very subtle focus of periosteal reaction in the distal medial shaft of the 2nd metatarsal representing a stress fracture.* ***(Right)*** *AP radiograph from the same patient 4 weeks later shows completion of the 2nd metatarsal stress fracture and mature periosteal bone around it. The patient continued to run between the 2 examinations.*

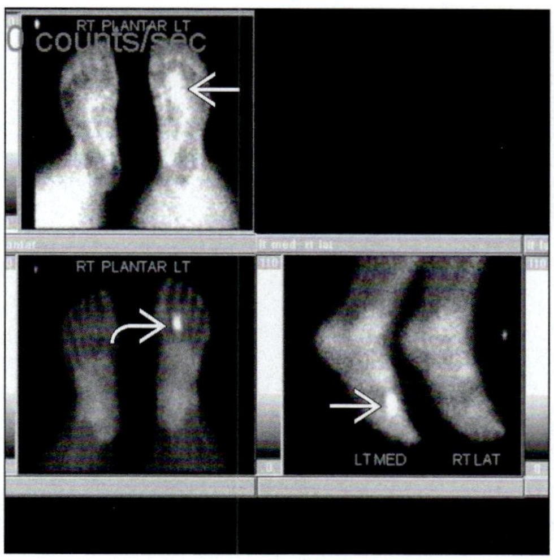

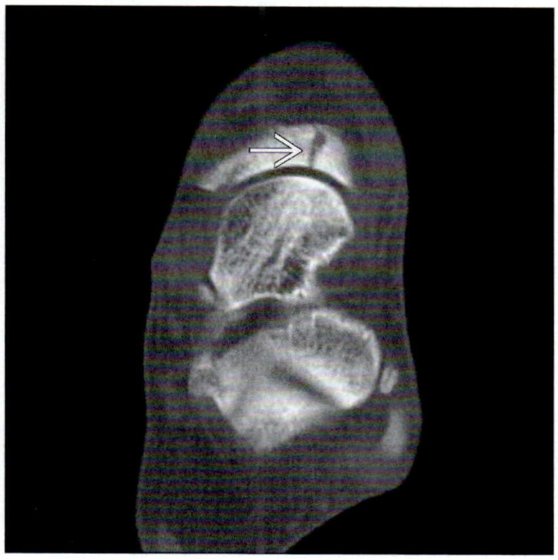

(Left) *Triple-phase bone scintigraphy of the feet in the same patient obtained at the time of the 2nd radiograph shows focal increased uptake in the distal 2nd metatarsal on blood pool and delayed-phase images. Stress injury usually shows increased uptake in all phases.* ***(Right)*** *Axial bone CT shows a sagittal stress fracture of the tarsal navicular in a runner. This fatigue injury is hard to diagnose on radiographs and is sometimes bilateral.*

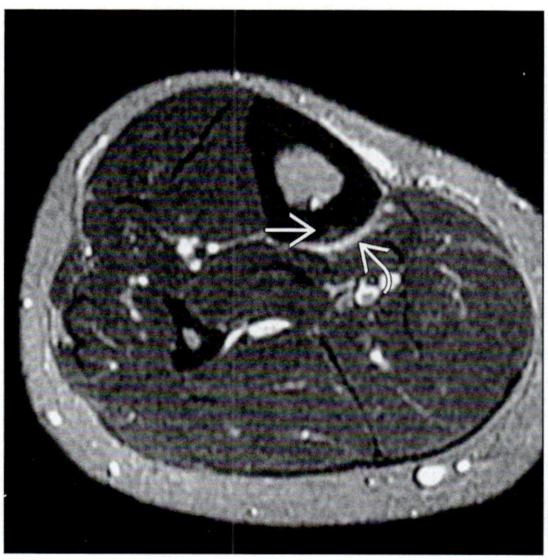

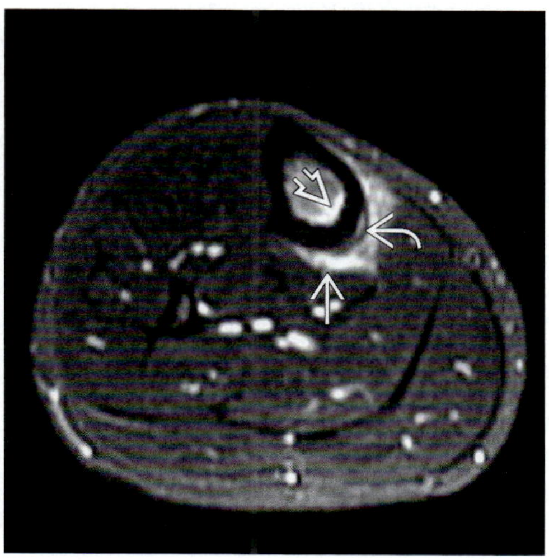

(Left) *Axial T2WI FS MR demonstrates subtle periosteal edema and a small area of increased intracortical signal in the posteromedial aspect of the tibia. This is an early stress fracture. Note the lack of marrow edema.* ***(Right)*** *Axial T2WI FS MR shows a more florid example of stress fracture of the tibia. There is periosteal and endosteal edema, and periosteal new bone formation is present. A cortical fracture line is not seen and in many cases is never found.*

STRESS/INSUFFICIENCY FRACTURE

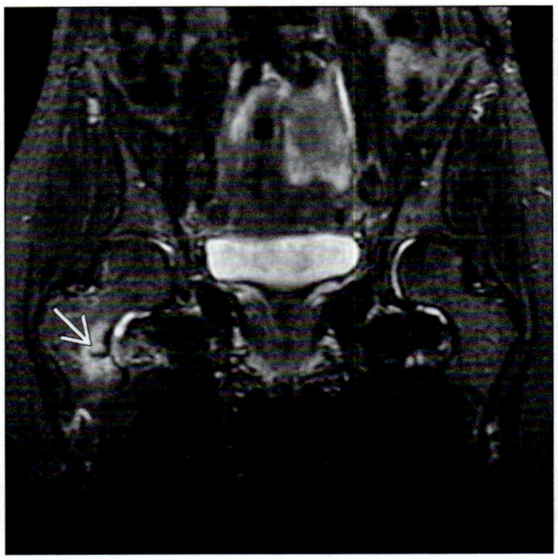
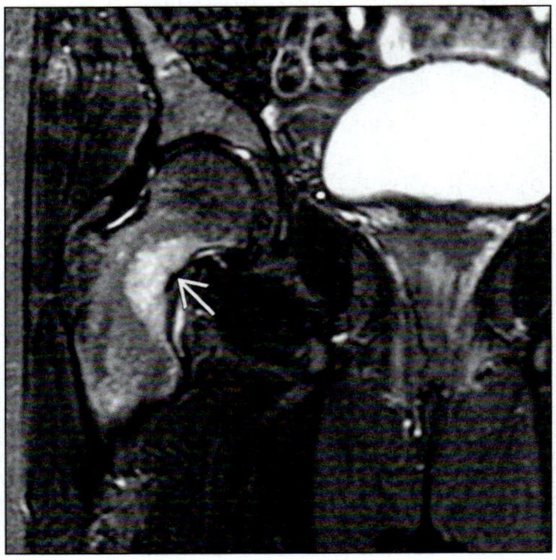

(Left) *Coronal T2WI FS MR demonstrates the typical appearance of a fatigue fracture ➡ of the basicervical femoral neck in an avid runner. These fractures invariably begin in the inferomedial cortex, and if undetected, may progress to complete fractures through the femoral neck, a devastating injury.* ***(Right)*** *Coronal T2WI FS MR shows florid marrow edema in the medial femoral neck in another runner with hip pain. History and subtle fracture line ➡ distinguish this from tumor.*

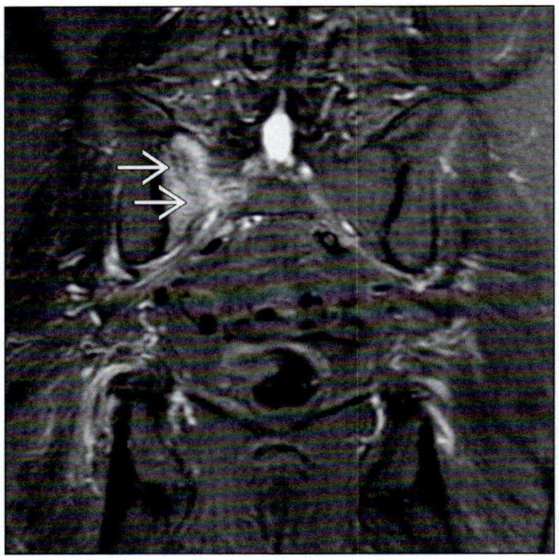
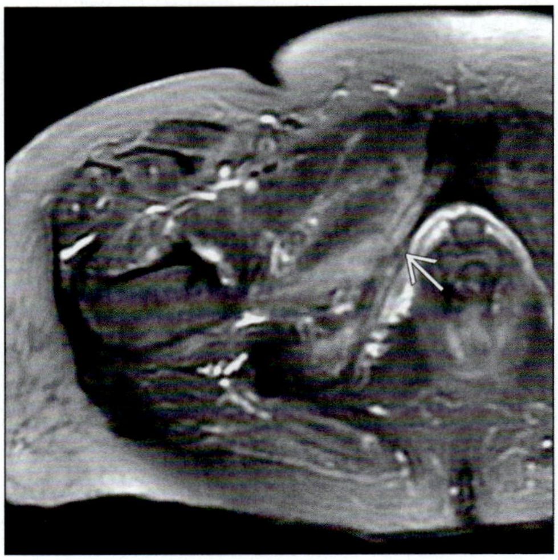

(Left) *Coronal T2WI FS MR demonstrates a vertical insufficiency fracture of the sacral ala ➡ with extensive surrounding marrow edema. This is a very common pattern of injury in the elderly due to osteoporosis, and should be actively sought in the evaluation of these patients.* ***(Right)*** *Axial T2WI FS MR from the same patient shows a fracture of the ipsilateral inferior obturator ramus ➡. Coexistence of sacral and ipsilateral obturator fractures in elderly osteoporotic fractures is typical.*

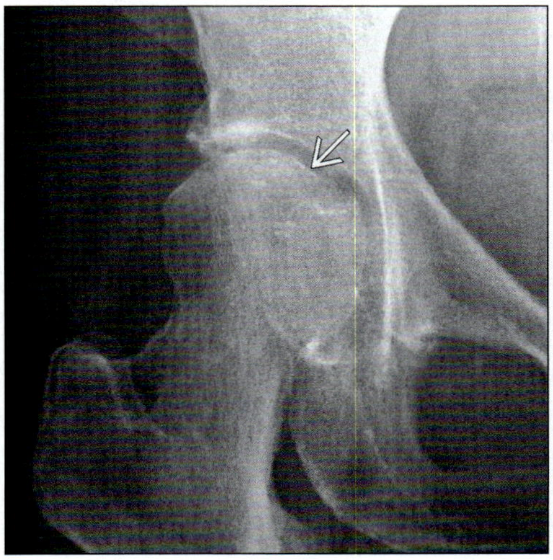
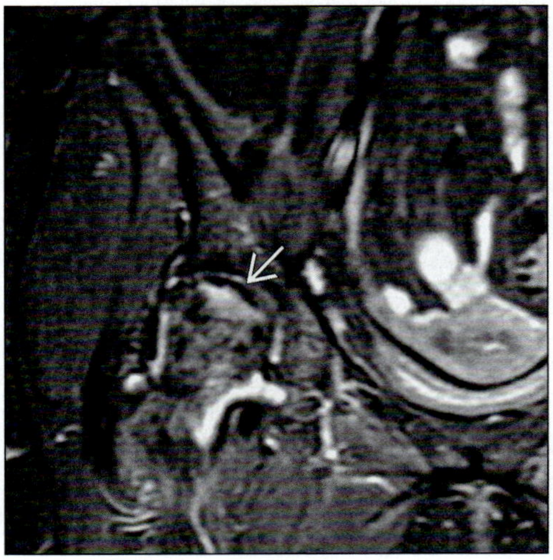

(Left) *AP radiograph shows flattening of the medial femoral head with subtle subchondral sclerosis ➡ in this elderly osteoporotic woman with hip pain.* ***(Right)*** *Coronal T2WI FS MR from the same patient shows a linear subchondral insufficiency fracture ➡ and marrow edema. The distinction between fracture and osteonecrosis is important in these cases; fracture lines usually parallel the articular surface, and the margins of osteonecrosis curve away from the surface.*

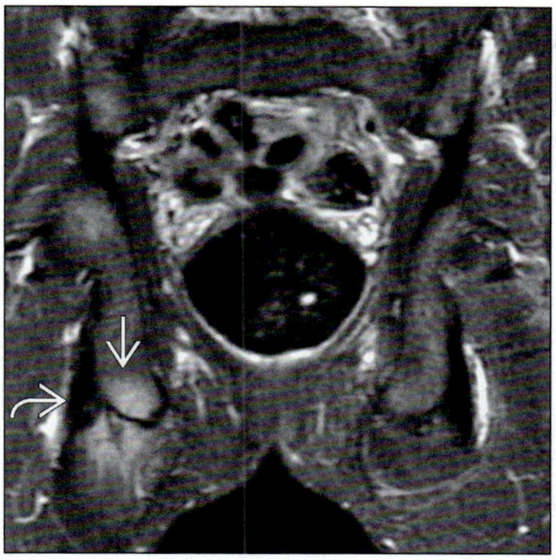

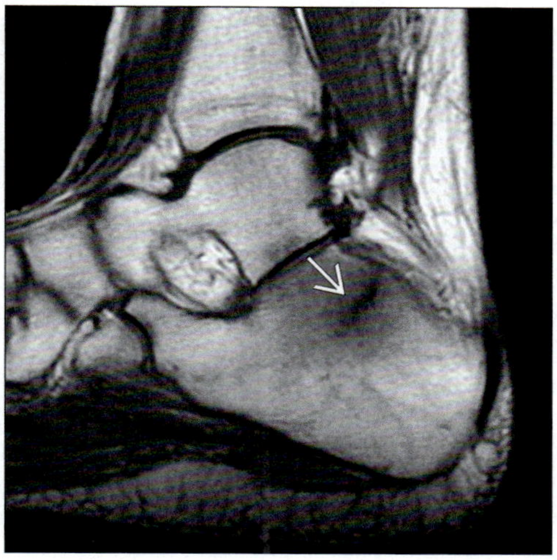

(Left) *Coronal T2WI FS MR shows asymmetric marrow edema in the right ischial tuberosity → and in the adjacent soft tissues. The hamstring origin → is intact. This is a stress injury related to hurdling (in which there are repeated sudden flexions of the hip).* ***(Right)*** *Sagittal T1WI MR in an elderly woman with foot pain after taking up walking shows a vertical fracture line → in the dorsal calcaneus. This is a typical pattern for calcaneal insufficiency fracture.*

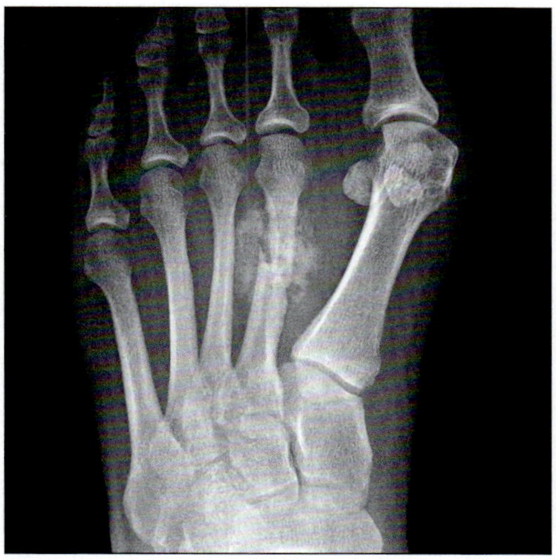

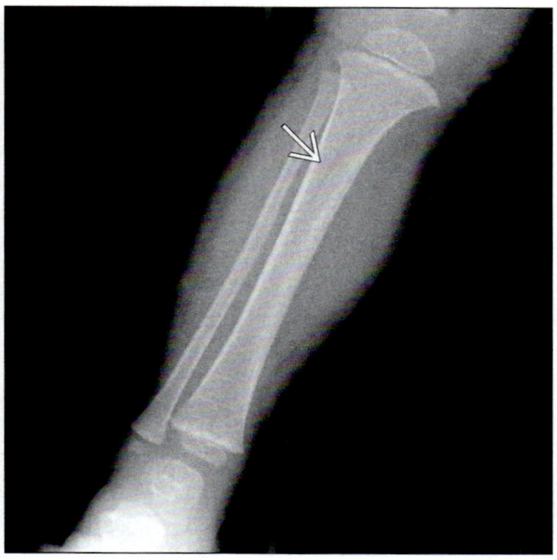

(Left) *AP radiograph demonstrates exuberant callus formation around a stress fracture of the midshaft of the 2nd metatarsal in a runner who failed to seek treatment despite several weeks of severe foot pain while running. In the absence of an obvious fracture, such callus could be mistaken for tumor osteoid.* ***(Right)*** *AP radiograph demonstrates a nondisplaced spiral fracture → of the mid-tibia in a young child. This is a toddler fracture and is commonly seen as children begin to walk.*

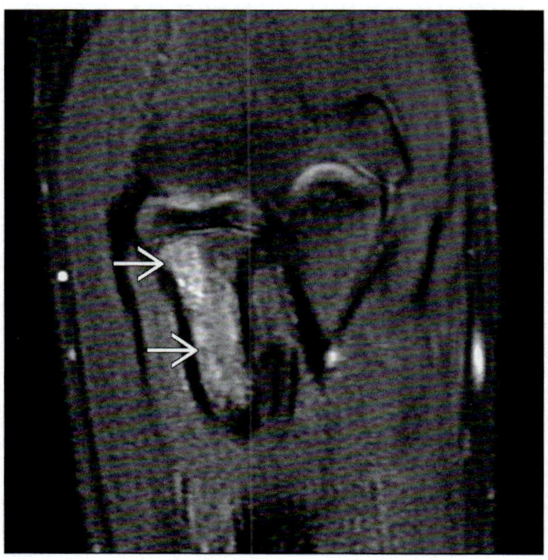

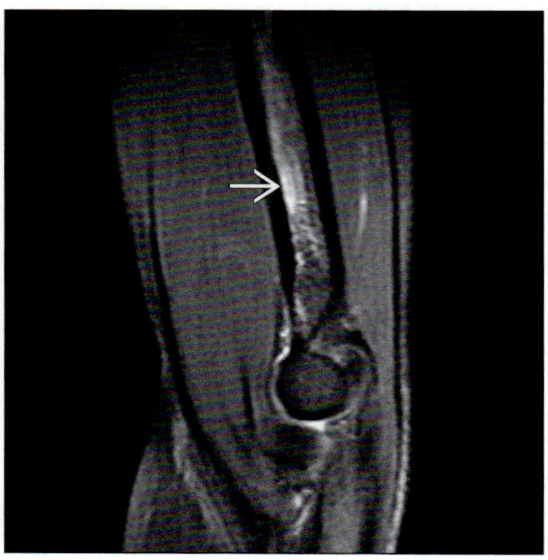

(Left) *Coronal T2WI FS MR of the anterior elbow in a high school baseball pitcher with elbow pain shows diffuse marrow edema in the proximal radius → without a discrete fracture line. This is an unusual site for stress injury, and the patient was rested for 4 weeks with close follow-up. Symptoms improved and MR returned to normal.* ***(Right)*** *Sagittal T2WI FS MR in a college tennis player shows stress edema → in the distal humeral shaft. Patient improved with rest, which is the hallmark of stress injury.*

PATHOLOGIC FRACTURE

Key Facts

Terminology
- Fracture through abnormal osseous lesion

Imaging
- Diagnostic clues
  - Fracture line extending through focal destructive lesion in bone
  - Avulsion fracture in unusual location
  - Transverse fracture common
  - Surrounding permeative change in bone
  - Endosteal scalloping, cortical destruction
  - Chondroid or osteoid matrix
  - Associated soft tissue mass
  - Additional tumor deposits
- Typical locations
  - Metastatic deposit or multiple myeloma: Spine, pelvis, femur, humerus
  - Enchondroma: Phalanges of fingers
  - Unicameral bone cyst: Long bones in children

Pathology
- Primary bone tumor
  - Benign: Unicameral bone cyst, enchondroma, fibrous cortical defect, lipoma
  - Malignant: Osteosarcoma, chondrosarcoma, fibrosarcoma, primary lymphoma of bone
- Multifocal processes: Metastases, multiple myeloma
- Postsurgical defect, metabolic disease, infection

Clinical Issues
- Tumors most frequently metastatic to bone in adults: Breast, prostate, lung

Diagnostic Checklist
- Consider pathologic fracture when fracture in unusual location, transverse orientation, or in absence of appropriate trauma history

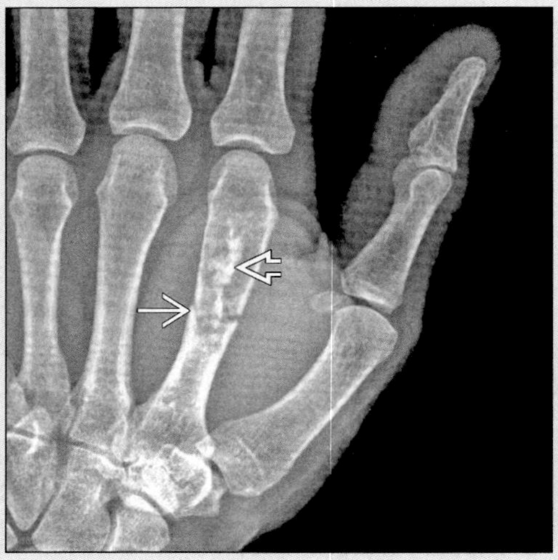

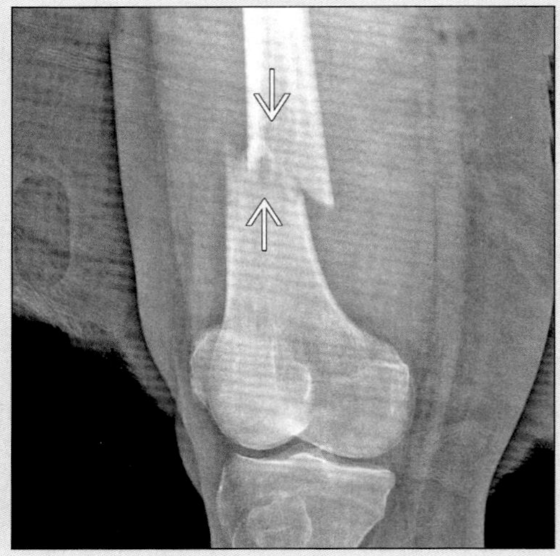

(Left) *AP radiograph shows a transverse fracture ➡ through an expansile lytic lesion of the 2nd metacarpal. Chondroid matrix ➡ is seen in the center of the lesion, confirming an underlying enchondroma. Many enchondromas of the hand and fingers have no visible matrix.* ***(Right)*** *Oblique radiograph of the distal femur shows a fracture of the distal femoral shaft through an area of lucency ➡ in the bone that raises the suspicion of a pathologic fracture. This was a breast cancer metastasis.*

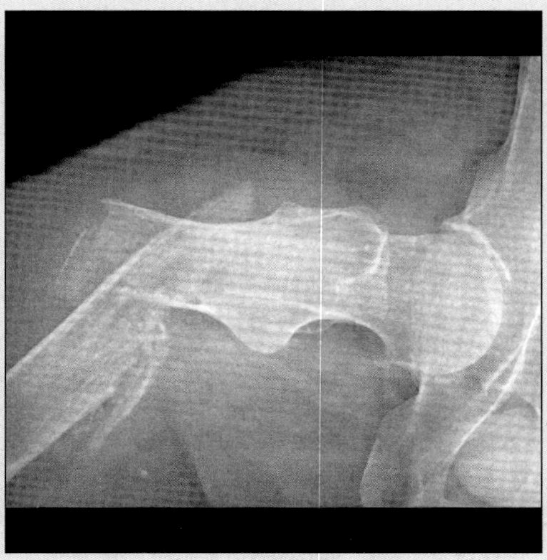

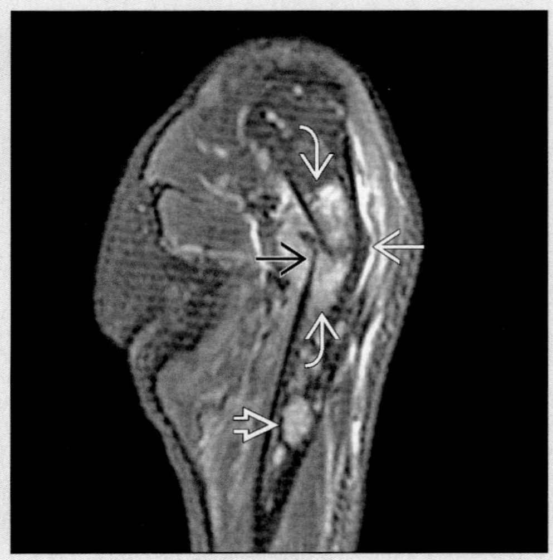

(Left) *Frog leg lateral radiograph shows a comminuted subtrochanteric fracture of the femur. This is an unusual location for traumatic fracture and should raise the suspicion of an underlying lesion.* ***(Right)*** *Sagittal T2WI FS MR shows a pathologic fracture ➡ through an osseous metastasis in the humerus. Additional metastatic deposits ➡ are present more distally. Clues to an underlying lesion are well-defined borders ➡ and scalloping or destruction of the cortex ➡.*

PATHOLOGIC FRACTURE

TERMINOLOGY

Definitions

- Fracture through abnormal osseous lesion
  - Tumor
  - Infection
  - Postsurgical
  - Metabolic
  - Congenital

IMAGING

General Features

- Best diagnostic clue
  - Fracture line extending through focal destructive lesion in bone
  - Avulsion fracture in unusual location
- Location
  - Metastatic deposit or multiple myeloma: Typically femur or humerus
    - Avulsion of lesser trochanter or ischial tuberosity in adult
    - Subtrochanteric femoral shaft
  - Enchondroma: Finger phalanges
  - Unicameral bone cyst: Long bones in children
- Size
  - Underlying lesion may range from small cortical lesion to large destructive process
  - Injury may range from subtle cortical interruption to comminuted fracture complex
- Morphology
  - Transverse fracture common
    - Oblique or spiral patterns usually seen in trauma of long bones

Radiographic Findings

- Surrounding permeative change in bone
- Endosteal scalloping
- Chondroid or osteoid matrix
- Associated soft tissue mass

CT Findings

- Cortical destruction
- Tumor matrix
- Aggressive periosteal reaction
- Endosteal scalloping
- Associated soft tissue mass

MR Findings

- Factors suggesting pathologic fracture
  - Well-defined mass-like ↓ T1 marrow signal in pathologic fracture (83% vs. 7% in nonpathologic fractures)
  - Abnormal muscle signal in pathologic fracture (83% vs. 48%)
  - Soft tissue mass in pathologic fracture (67% vs. 0%)
  - Endosteal scalloping in pathologic fracture (58% vs. 0%)
  - May see additional tumor deposits in metastatic disease or multiple myeloma
- Fracture-related hematoma may confuse issue of soft tissue mass
  - MR + C may be confusing; post-traumatic hemorrhage and neovascularity may mimic underlying tumor
- In- and out-of-phase T1WI may be useful in differentiating marrow edema from infiltrative process
  - Marrow edema: Signal drop-out on out-of-phase images since marrow fat is retained
  - Infiltrative process: No signal drop-out since marrow fat is replaced
- Diffusion-weighted imaging
  - Tumor infiltration of marrow → decreased diffusion

Imaging Recommendations

- Best imaging tool
  - Radiograph + high level of suspicion as initial modality
  - MR to identify underlying focal lesion
- Protocol advice
  - Radiographs: Multiple views to characterize lesion
  - MR: T2WI without FS to evaluate possible underlying mass and differentiate from hematoma

DIFFERENTIAL DIAGNOSIS

Insufficiency Fracture

- Fracture through osteoporotic bone
- Usually no focal underlying lesion

Fatigue Fracture

- Injury due to excessive repetitive force
- Normal underlying bone

Surgical Defect

- Surgical history important
  - Previous biopsy or osteotomy

PATHOLOGY

General Features

- Etiology
  - Primary bone tumor
    - Benign: Unicameral bone cyst, enchondroma, fibrous cortical defect, lipoma
    - Malignant: Osteosarcoma, chondrosarcoma, fibrosarcoma, primary lymphoma of bone
  - Multifocal process
    - Metastatic disease, multiple myeloma, lymphoma
  - Postsurgical defect
    - Biopsy site, bone graft harvest, osteotomy
  - Metabolic: Fragile bone in renal osteodystrophy
  - Radiation osteitis: Fragile bone at risk for fx
  - Bone infarct
  - Arthritis: Degenerative or inflammatory cyst
  - Congenital: Osteogenesis imperfecta
  - Infection
    - Bone abscess
    - Osteomyelitis: Fx rare; consider underlying squamous cell carcinoma
- Associated abnormalities
  - Adjacent soft tissue mass
  - Osseous lesions in other locations

Gross Pathologic & Surgical Features

- Malignant underlying lesion

PATHOLOGIC FRACTURE

- Tumor may spread along fracture planes
- Tumor may enter bloodstream through disrupted blood vessels

CLINICAL ISSUES

Presentation

- Most common signs/symptoms
  - Acute pain, often without significant trauma
  - Swelling and pain in area often predates trauma

Demographics

- Age
  - Any, but older adults are more likely to have osseous metastatic disease
- Epidemiology
  - Tumors most frequently metastatic to bone in adults
    - Breast
    - Prostate
    - Lung
  - Common benign tumors with pathologic fracture
    - Unicameral bone cyst
    - Enchondroma (especially in fingers)

Natural History & Prognosis

- Tumor may spread locally due to fracture
- Tumor may spread hematogenously due to entry into disrupted blood vessels

Treatment

- Benign lesion
  - Curettage
  - Bone graft or cement
  - Hardware for support as necessary
  - Some benign lesions thought by some to react to fracture with healing response: Simple bone cyst & nonossifying fibroma (fibroxanthoma) may heal with only external support (casting)
  - Fibrous dysplasia, though benign, generally is not cured by curettage; only complications of this disease are treated
- Malignant lesion (primary)
  - Reduction & support of fracture
    - Note: Must identify fractures as suspicious for pathologic lesion; intramedullary rodding must be avoided so as to not spread tumor locally
  - Biopsy; if positive for tumor, staging workup
- Malignant lesion (metastatic)
  - For a weight-bearing long bone, IM rodding considered superior to ORIF
  - Radiation may be considered if lesion is responsive; may help alleviate pain but may retard osseous healing
  - Cement or bone graft may be used to fill a significant defect
  - Radiofrequency ablation may be used in selected lesions for pain control

DIAGNOSTIC CHECKLIST

Consider

- Pathologic fracture when fracture in unusual location, transverse orientation without substantial trauma, or any fracture pattern in absence of appropriate trauma history

Image Interpretation Pearls

- Look for other focal lesions to confirm presence of metastatic disease
- Underlying lesion may be difficult to characterize due to trauma, hematoma, healing response, or surgery

Reporting Tips

- Suggest metastatic work-up in appropriate setting

SELECTED REFERENCES

1. Kim MS et al: Prognostic effect of pathologic fracture in localized osteosarcoma: a cohort/case controlled study at a single institute. J Surg Oncol. 100(3):233-9, 2009
2. Bramer JA et al: Do pathological fractures influence survival and local recurrence rate in bony sarcomas? Eur J Cancer. 43(13):1944-51, 2007
3. Shenoy R et al: Tumours of the hand presenting as pathological fractures. Acta Orthop Belg. 73(2):192-5, 2007
4. Erly WK et al: The utility of in-phase/opposed-phase imaging in differentiating malignancy from acute benign compression fractures of the spine. AJNR Am J Neuroradiol. 27(6):1183-8, 2006
5. James SL et al: Atraumatic avulsion of the lesser trochanter as an indicator of tumour infiltration. Eur Radiol. 16(2):512-4, 2006
6. Snyder BD et al: Predicting fracture through benign skeletal lesions with quantitative computed tomography. J Bone Joint Surg Am. 88(1):55-70, 2006
7. Fayad LM et al: Distinction of long bone stress fractures from pathologic fractures on cross-sectional imaging: how successful are we? AJR Am J Roentgenol. 185(4):915-24, 2005
8. Bacci G: Pathologic fracture in osteosarcoma. J Bone Joint Surg Am. 85-A(9):1848; author reply 1848-9, 2003
9. McVey MJ et al: Pathologic fracture of metacarpal enchondroma: case study and differential diagnosis. J Manipulative Physiol Ther. 25(5):340-4, 2002
10. Spuentrup E et al: Diffusion-weighted MR imaging for differentiation of benign fracture edema and tumor infiltration of the vertebral body. AJR Am J Roentgenol. 176(2):351-8, 2001
11. Wai EK et al: Pathologic fractures of the proximal femur secondary to benign bone tumors. Clin Orthop Relat Res. (393):279-86, 2001
12. Lingg GM et al: Insufficiency and stress fractures of the long bones occurring in patients with rheumatoid arthritis and other inflammatory diseases, with a contribution on the possibilities of computed tomography. Eur J Radiol. 26(1):54-63, 1997
13. Wall JE et al: Fractures in children treated with radiotherapy for soft tissue sarcoma. Orthopedics. 19(8):657-64, 1996
14. Dijstra S et al: Impending and actual pathological fractures in patients with bone metastases of the long bones. A retrospective study of 233 surgically treated fractures. Eur J Surg. 160(10):535-42, 1994
15. Walker LG et al: Pathologic fracture of the proximal ulna through a bone graft donor site. J Hand Surg Am. 15(5):781-4, 1990
16. Harrington KD: Impending pathologic fractures from metastatic malignancy: evaluation and management. Instr Course Lect. 35:357-81, 1986
17. Bertin KC et al: Isolated fracture of the lesser trochanter in adults: an initial manifestation of metastatic malignant disease. J Bone Joint Surg Am. 66(5):770-3, 1984
18. Bohrer SP: Fracture complicating bone infarcts and-or osteomyelitis in sickle-cell disease. Clin Radiol. 22(1):83-8, 1971

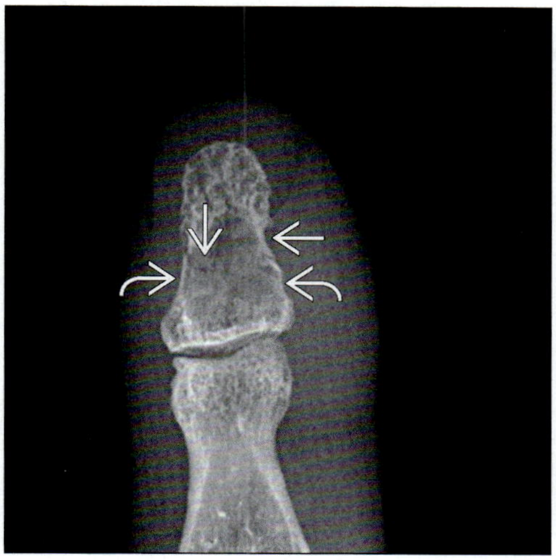

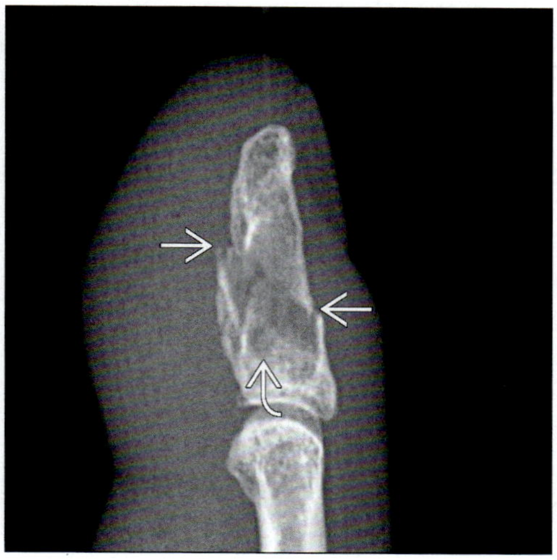

(Left) *AP radiograph shows a comminuted fracture ➔ through an expansile lucent lesion in the terminal phalanx of a finger in an adult. There is surrounding soft tissue swelling. The patient suffered mild trauma. Note the cortical thinning around the lesion ↪.* ***(Right)*** *Lateral radiograph from the same patient shows the fracture ➔ and the expansile nature of the underlying lesion ↪. This is a pathologic fracture through an enchondroma, the most common such lesion of the hand and fingers.*

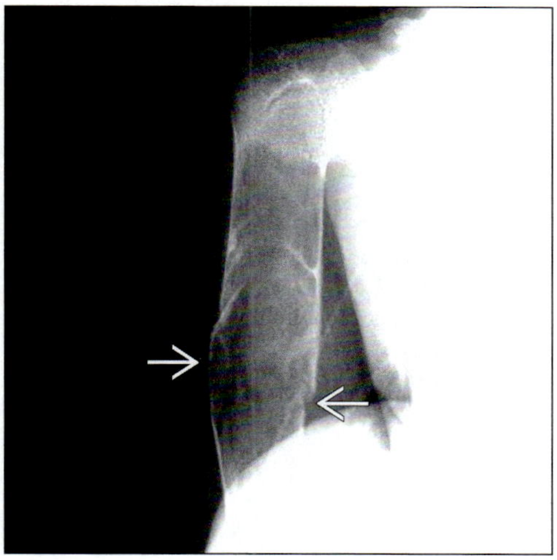

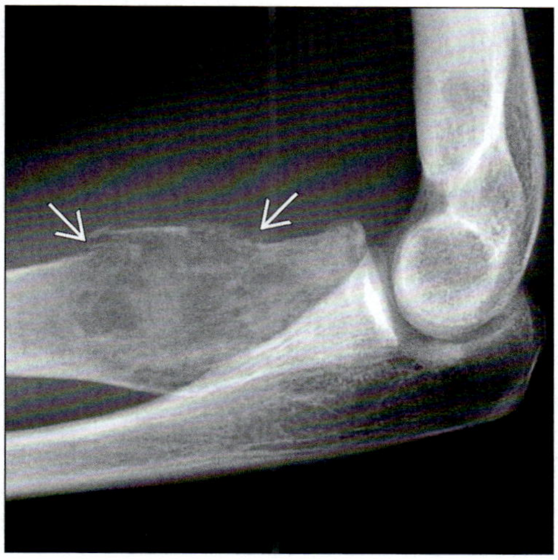

(Left) *AP radiograph from a child demonstrates a pathologic fracture ➔ through a large expansile lesion in the proximal humerus after low-grade trauma. This was a unicameral bone cyst, the most common such lesion of large long bones in children.* ***(Right)*** *Lateral radiograph shows a pathologic fracture ➔ of an expansile lytic lesion of the proximal radius in an adult. There is hazy matrix in the lesion; this was biopsy-proven fibrous dysplasia.*

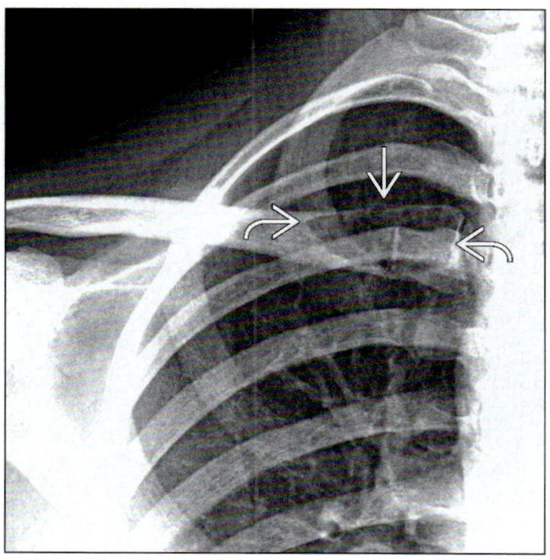

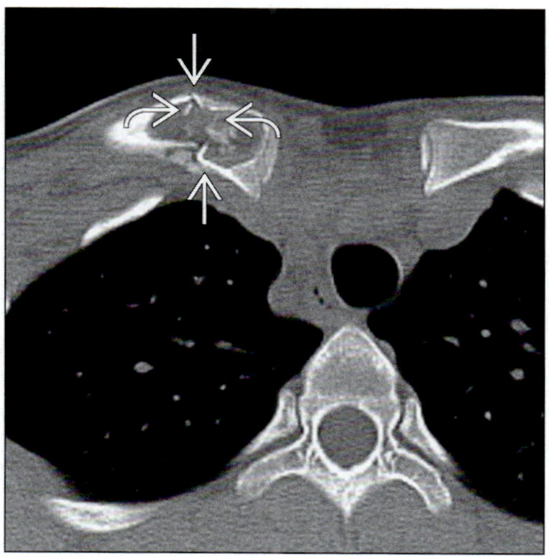

(Left) *AP radiograph from a 15-year-old boy with anterior chest pain after mild trauma demonstrates a subtle cortical fracture ➔ through a well-circumscribed expansile lucent lesion ↪ in the medial clavicle.* ***(Right)*** *Axial bone CT from the same patient shows the fracture ➔ extending through the lesion. Several small cortical fragments ↪ are seen within the lesion; this is the "fallen fragment" sign and indicates a cystic lesion. This was a unicameral bone cyst.*

PATHOLOGIC FRACTURE

*(**Left**) AP radiograph shows a pathologic fracture through a well-defined lytic lesion in the lateral femoral condyle. The lesion extends to the articular surface of the femur. No matrix is seen in the lesion. (**Right**) Axial noncontrast CT shows expansion of the lateral femoral condylar cortex by the mass, with multiple cortical fractures. Density of the lesion is similar to skeletal muscle. This was a biopsy-proven giant cell tumor.*

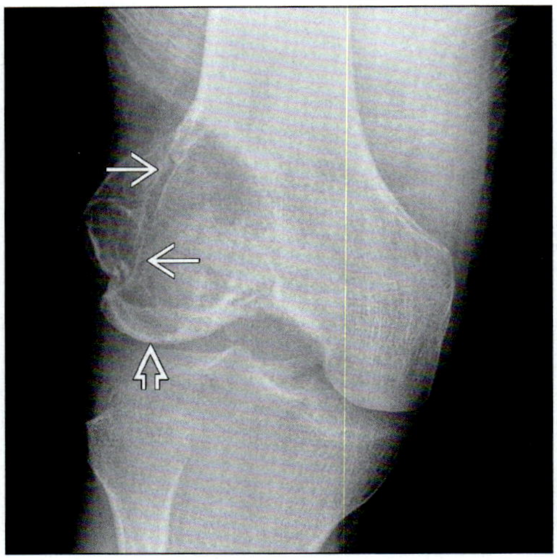

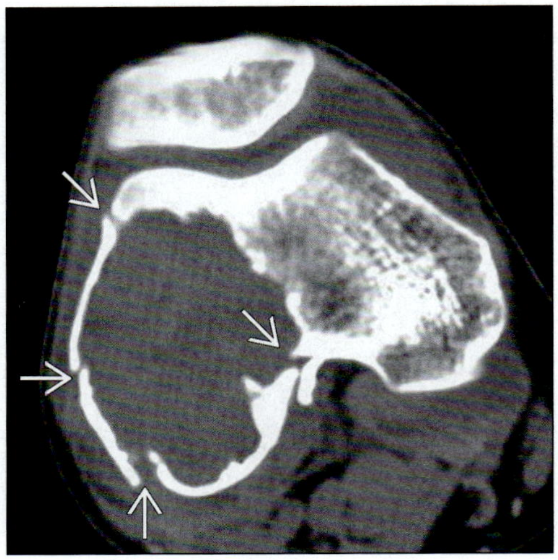

*(**Left**) AP radiograph shows an expansile lytic lesion in the base of the superior obturator ramus with a tiny cortical fracture line superiorly. Patient had sudden onset of pain when walking and had no history of malignancy. (**Right**) Coronal T1WI MR from the same patient shows the fracture line and the expansile lesion. The lesion followed fat signal on all sequences. This is a pathologic fracture through a benign intraosseous lipoma.*

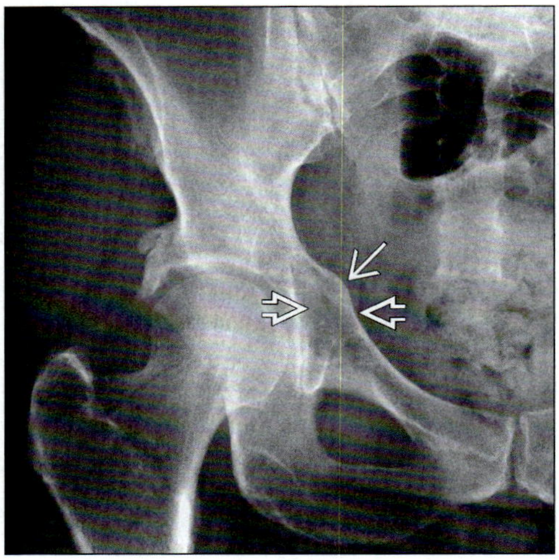

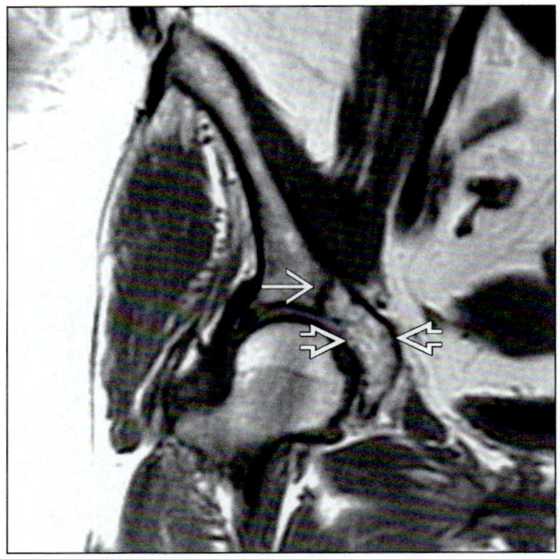

*(**Left**) Sagittal T1WI MR shows a fracture in the calcaneus extending into a well-defined but irregularly shaped focus of fat signal surrounded by serpentine low signal. This is a pathologic fracture through a bone infarct. Another infarct is seen in the talar dome in this patient with lupus. (**Right**) Lateral radiograph of the leg shows a transverse fracture of the tibia. Cortex of the tibia is diffusely thick, and the bone is expanded; this is a pathologic fracture in Paget disease.*

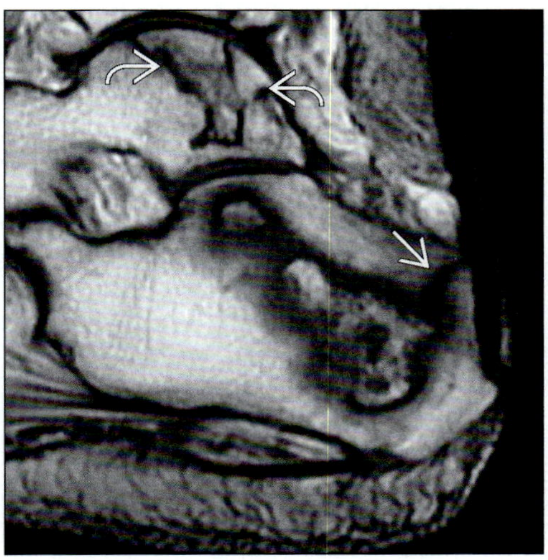

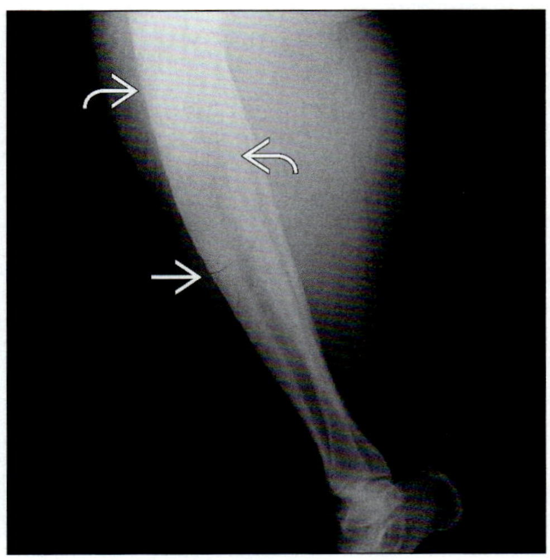

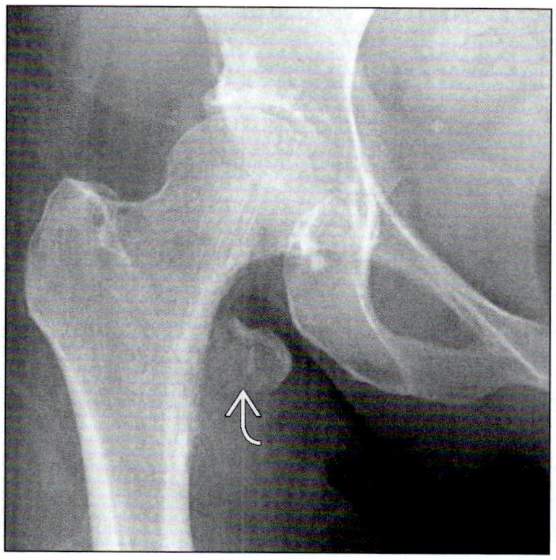

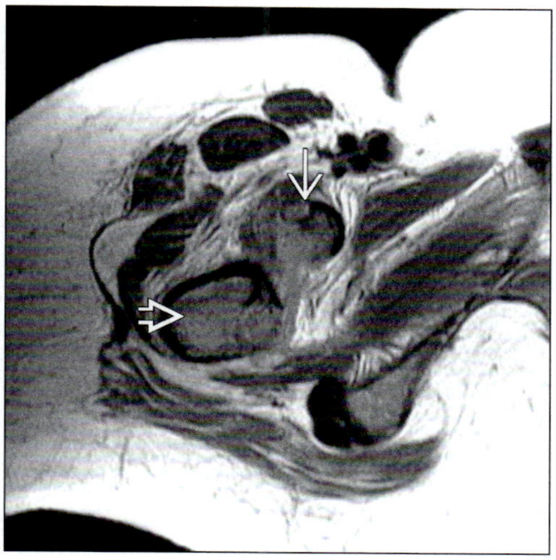

(Left) *AP radiograph from a 66-year-old woman with a history of breast cancer shows avulsion of the lesser trochanter of the femur. The avulsed edges of bone are irregular and poorly defined. Lesser trochanter avulsion in an adult should lead to a high level of suspicion for underlying pathologic lesion.* ***(Right)*** *Axial T1WI MR from the same patient shows the displaced lesser trochanter. Both the trochanter and the femur demonstrate abnormally low marrow signal intensity.*

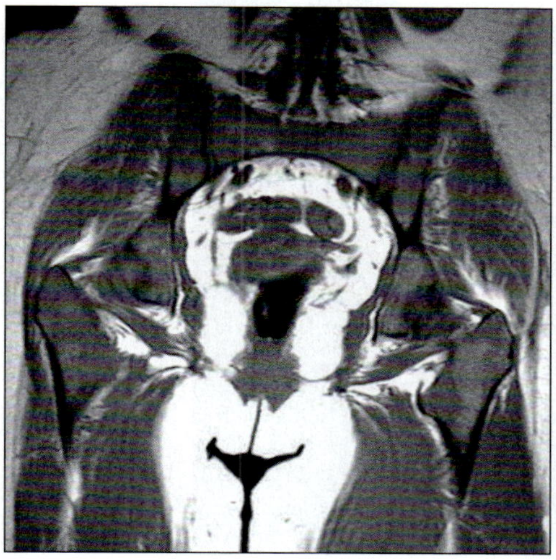

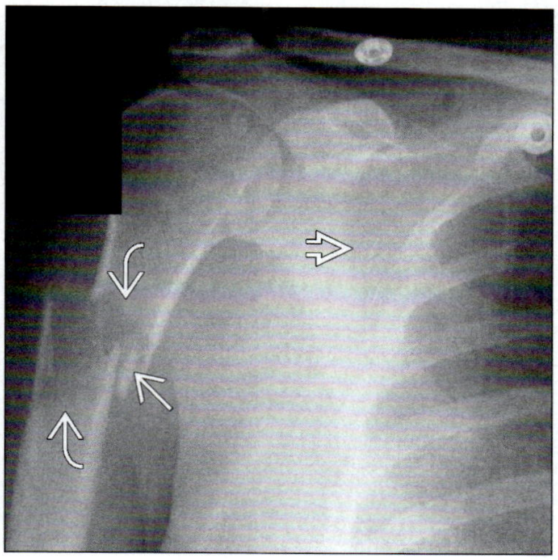

(Left) *Coronal T1WI MR of the whole pelvis from the same patient shows diffusely low signal intensity throughout the marrow of the pelvis and proximal femora consistent with diffuse osseous metastatic disease.* ***(Right)*** *AP radiograph shows an oblique fracture of the proximal humerus through an ill-defined lucent lesion. Surgical clips in the axilla indicate prior mastectomy for breast cancer. Biopsy proved this to be a breast cancer metastasis.*

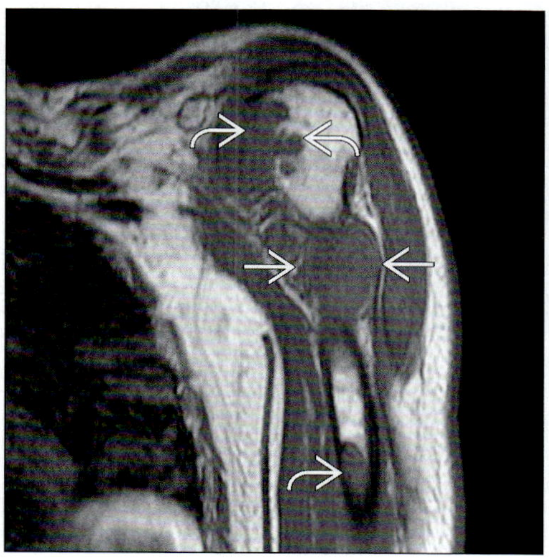

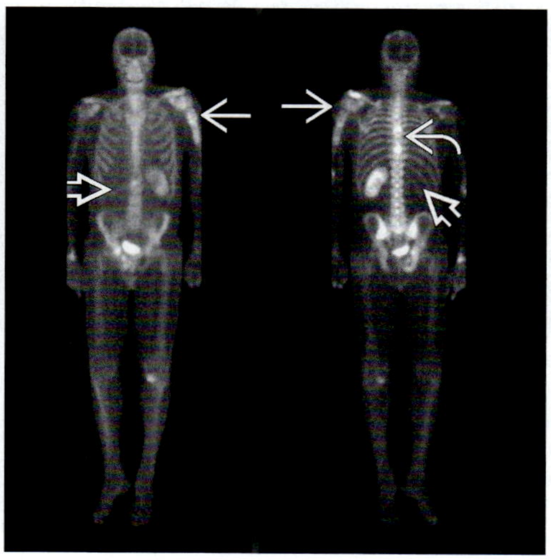

(Left) *Coronal T1WI MR from a different patient shows an expansile mass at the site of fracture, confirming its pathologic nature. Several other destructive osseous lesions are seen in the humerus.* ***(Right)*** *AP and PA views from a whole-body bone scan of the same patient shows multiple focal areas of increased uptake in the humerus and additional lesions in the thoracic spine. Absent kidney indicates prior nephrectomy for renal cell carcinoma.*

PHYSEAL INJURY (SALTER-HARRIS)

Key Facts

Terminology

- Salter-Harris fracture, Salter fracture
- Fracture affecting physis in skeletally immature patient

Imaging

- Fracture extends into, across, or through physis
- Linear fracture in metaphysis ± epiphysis
- MR
  - ↑ T2 signal in physis
  - ± chondral injury
  - Soft tissue swelling
  - Joint effusion
  - Detection of premature post-traumatic physeal closure
- Opposite-side comparison radiographs may help confirm subtle Salter-Harris type I fracture

Pathology

- Type I: Involves only physis
- Type II: Involves physis and metaphysis
- Type III: Involves physis and epiphysis
- Type IV: Involves physis, metaphysis, & epiphysis
- Type V: Crush fracture involving all or part of physis (rare)

Clinical Issues

- Average age: F = 11 years, M = 12 years
- 18% of childhood fractures involve physeal injury
- Prognosis worse in lower extremities regardless of Salter-Harris classification
- Most substantial complications occur at knee and ankle
  - Premature epiphyseal closure
  - Limb shortening or angulation
  - Joint incongruity

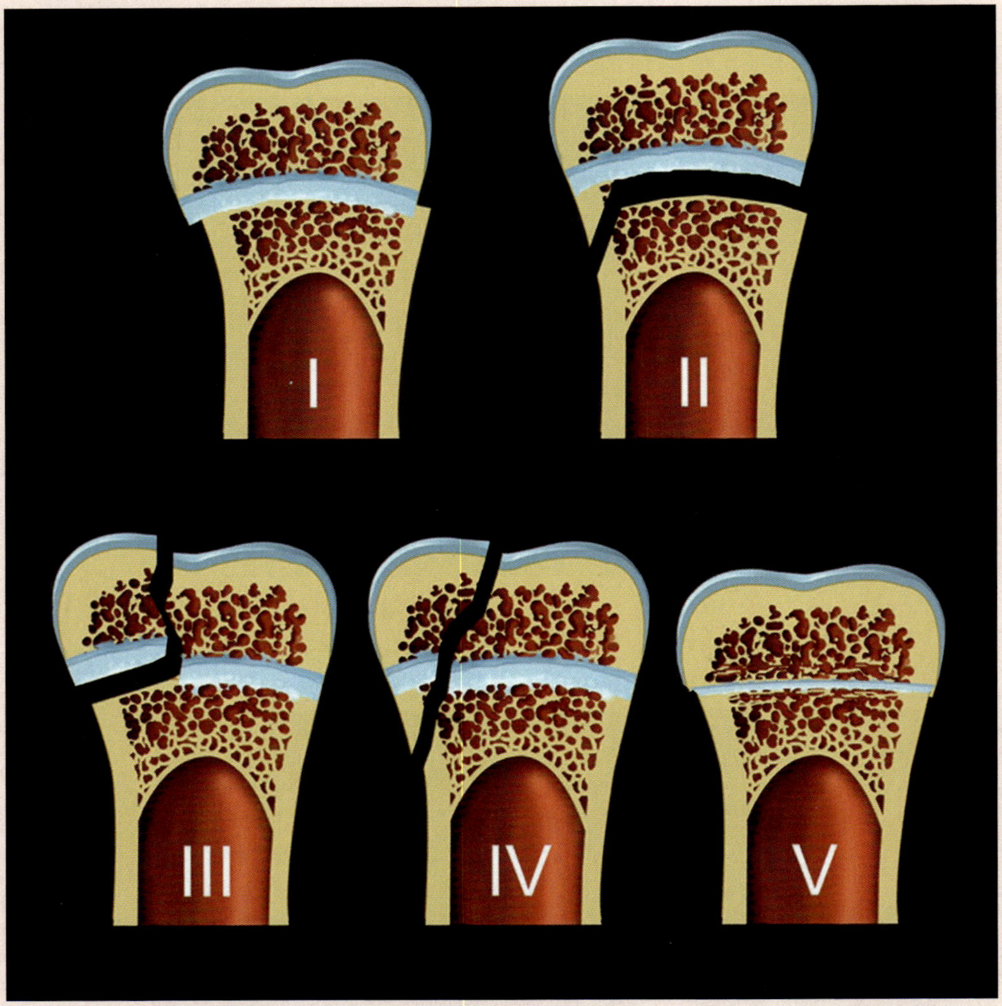

Graphic representation of the Salter-Harris classification of injuries involving the growth plate (physis) of long bones in children shows the 5 most common types of injury. Type I injury involves a transverse shearing injury through the physis ± displacement or widening. Type II injury involves a metaphyseal fracture extending into the physis. Type III injury involves an epiphyseal fracture (often involving the articular surface) extending into the physis. Type IV injury involves a fracture of the metaphysis crossing the physis into the epiphysis. Type V, the rarest and most severe type, involves a crush injury to the physis, which destroys the layer of growing cells and halts growth.

PHYSEAL INJURY (SALTER-HARRIS)

TERMINOLOGY

Abbreviations

- Salter-Harris (SH) fracture, Salter fracture

Definitions

- Fracture affecting physis in skeletally immature patient

IMAGING

General Features

- Location
  - Upper extremity
    - Distal radius: 28%
    - Fingers: 26%
    - Distal humerus: 7%
    - Proximal radius: 5%
    - Distal ulna: 5%
    - Metacarpals: 4%
    - Proximal humerus: 2%
    - Clavicle: 1%
    - Proximal ulna: 1%
  - Lower extremity
    - Distal tibia: 9%
    - Toes: 7%
    - Distal fibula: 3%
    - Metatarsals: 1%
    - Proximal tibia: 1%
    - Distal femur: 1%
- Morphology
  - Fracture extends into, across, or through physis

Radiographic Findings

- Radiography
  - Type I-IV
    - Adjacent soft tissue swelling
    - Joint effusion
  - Type I
    - Widening of part or all of physis ± displacement
    - May appear normal on radiograph; focal soft tissue swelling may be only hint of injury
  - Type II: Lucent fracture line(s) extending through metaphysis and into physis
  - Type III: Lucent fracture line(s) extending through epiphysis and into physis
  - Type IV: Lucent fracture line(s) extending through metaphysis, across physis, and into epiphysis
  - Type V: Narrowing of physis
  - Post-treatment
    - Persistent physeal widening > 3 mm post-reduction → periosteal entrapment in fracture likely

CT Findings

- Used to evaluate anatomic extent and degree of displacement of the more complex Salter-Harris injuries
  - Most commonly used with triplane fractures of distal tibia (Salter-Harris IV)
- Used to evaluate focal bony bridging across physis during healing process (most common in SH IV or V)

MR Findings

- ↑ T2 signal in physis
  - Comparison to other physes in field of view can be helpful
- Linear fracture in metaphysis ± epiphysis
  - Usually ↓ signal on T1 and T2
  - Surrounded by marrow edema
- ± chondral injury
- Soft tissue swelling
- Joint effusion
- Detection of premature post-traumatic physeal closure
  - Quantification of closure as percentage of total ara of physis

Ultrasonographic Findings

- Evaluation of physeal birth injuries presenting as "pseudodislocation" in long bones with unformed ossification centers
  - Proximal and distal humerus, proximal femur most common

Nuclear Medicine Findings

- Bone scan
  - Focal increased radiotracer uptake
  - Pinhole collimation → increased spatial resolution for small parts

Imaging Recommendations

- Protocol advice
  - Opposite-side comparison radiographs may help confirm subtle Salter-Harris type I fracture

PATHOLOGY

General Features

- Etiology
  - Structure of normal epiphysis
    - Germinal zone closest to epiphyseal ossification center: Small active chondrocytes emerge from resting chondrocytes
    - Proliferative zone: Flattened chondrocytes arranged in columns
    - Hypertrophic zone: Swollen chondrocytes arranged in columns
    - Provisional calcification zone: Chondrocytes die, cartilage matrix calcifies, osteoclasts form osteoid
    - Perichondral ring: Layer of cartilaginous tissue contiguous with adjacent periosteum of metaphysis and epiphysis
    - Surface of metaphyseal and epiphyseal interface with physis is irregular or corrugated
      - Consists of small bony projections, undulations, knobs and ridges (mammillary processes)
    - Metaphysis and epiphysis receive most arterial supply from separate sources
      - Fracture through physis does not interfere with blood supply of either epiphysis or metaphysis
      - Exception: Femoral capital and radial head epiphyses, which are intraarticular
  - Joint capsule and ligamentous structures are stronger than physis in children
    - Physeal fractures in children are mechanical equivalent of ligamentous injuries in adults
  - Structure of physeal fracture
    - Damage due to shear, grinding, compression force

PHYSEAL INJURY (SALTER-HARRIS)

- Fracture plane undulates within proliferative, hypertrophic, and provisional calcification zones
- Fibrin appears within cleavage, cartilaginous cells continue to grow, epiphyseal plate thickens as cellular columns lengthen
- Fibrin gone and normal growth pattern restored in about 21 days

Staging, Grading, & Classification

- Type I: Involves only physis
- Type II: Involves physis and metaphysis
- Type III: Involves physis and epiphysis
- Type IV: Involves physis, metaphysis, & epiphysis
- Type V: Crush fracture involving all or part of physis (rare)
  - Usually 1st recognized when cone epiphyses or partial epiphyseal arrest becomes apparent later
- Type VI-IX, as described by Ogden (rare)
  - Type VI: Perichondral ring injury
  - Type VII: Intraepiphyseal fracture not involving physis
  - Type VIII: Metaphyseal fracture not involving physis directly but → ischemic growth disturbance
  - Type IX: Periosteal injury → disturbed diaphyseal growth

CLINICAL ISSUES

Presentation

- Most common signs/symptoms
  - Pain, swelling, point tenderness, limited range of motion, inability to bear weight

Demographics

- Age
  - Average
    - F: 11 years
    - M: 12 years
  - Ages 16 and 17: Physeal fractures more common in males; physes have closed in females
- Gender
  - M > F
- Epidemiology
  - 18% of childhood fractures involve physeal injury
  - Relative incidence of Salter-Harris fractures
    - Type I: 8.5%
    - Type II: 73% (most common)
    - Type III: 6.5%
    - Type IV: 12%
    - Type V: < 1% (rarest)
  - Exceptions to general relative incidence
    - Distal humerus: Almost all fractures are type IV
    - Distal tibia: Types II, III, and IV equally common

Natural History & Prognosis

- Most uncomplicated type I-III fractures do well with conservative or surgical therapy
- Prognosis worse in lower extremities regardless of Salter-Harris classification
- Complications
  - Overall complication rate: ~ 14%
  - Most substantial complications occur at knee and ankle
  - Premature epiphyseal closure
    - Limb shortening (early fusion across entire physis) or angulation (early fusion across small portion of physis, continued growth in remaining portions)
    - Premature epiphyseal closure in lower tibial fractures
      - 27% overall rate
      - 21% in triplane (SH IV) fractures, rare in Tillaux (SH III) fractures
  - Joint incongruity
  - Persistently trapped periosteum
    - May prevent healing

Treatment

- Casting for low Salter-Harris categories
- Open reduction and internal fixation often required with higher categories

DIAGNOSTIC CHECKLIST

Consider

- Follow knee and ankle fractures for at least a year or until skeletal maturity for early detection of premature closure of epiphyses

Image Interpretation Pearls

- Subtle type I injury may require comparison to contralateral extremity for diagnosis

SELECTED REFERENCES

1. Basener CJ et al: Growth disturbance after distal femoral growth plate fractures in children: a meta-analysis. J Orthop Trauma. 23(9):663-7, 2009
2. Leary JT et al: Physeal fractures of the distal tibia: predictive factors of premature physeal closure and growth arrest. J Pediatr Orthop. 29(4):356-61, 2009
3. Mubarak SJ et al: Classification of proximal tibial fractures in children. J Child Orthop. 3(3):191-7, 2009
4. Shi DP et al: Epiphyseal and physeal injury: comparison of conventional radiography and magnetic resonance imaging. Clin Imaging. 33(5):379-83, 2009
5. Arkader A et al: Predicting the outcome of physeal fractures of the distal femur. J Pediatr Orthop. 27(6):703-8, 2007
6. Schnetzler KA et al: The pediatric triplane ankle fracture. J Am Acad Orthop Surg. 15(12):738-47, 2007
7. Ilharreborde B et al: Long-term prognosis of Salter-Harris type 2 injuries of the distal femoral physis. J Pediatr Orthop B. 15(6):433-8, 2006
8. Craig JG et al: The distal femoral and proximal tibial growth plates: MR imaging, three-dimensional modeling and estimation of area and volume. Skeletal Radiol. 33(6):337-44, 2004
9. Swischuk LE et al: Frequently missed fractures in children (value of comparative views). Emerg Radiol. 11(1):22-8, 2004
10. Barmada A et al: Premature physeal closure following distal tibia physeal fractures: a new radiographic predictor. J Pediatr Orthop. 23(6):733-9, 2003
11. Ecklund K et al: Patterns of premature physeal arrest: MR imaging of 111 children. AJR Am J Roentgenol. 178(4):967-72, 2002
12. Koury SI et al: Recognition and management of Tillaux fractures in adolescents. Pediatr Emerg Care. 15(1):37-9, 1999
13. Carey J et al: MRI of pediatric growth plate injury: correlation with plain film radiographs and clinical outcome. Skeletal Radiol. 27(5):250-5, 1998
14. Rogers LF et al: Imaging of epiphyseal injuries. Radiology. 191(2):297-308, 1994

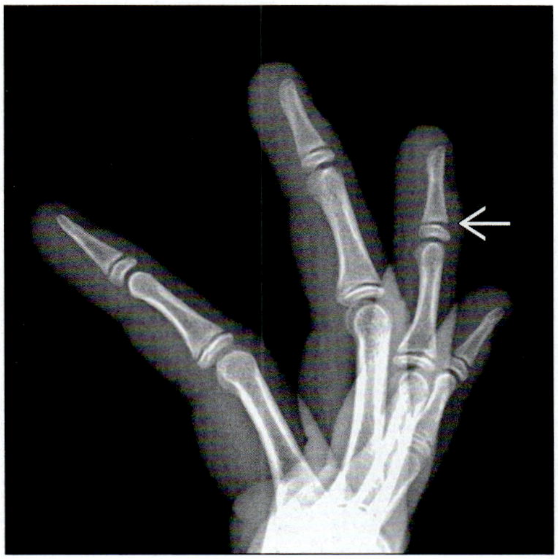

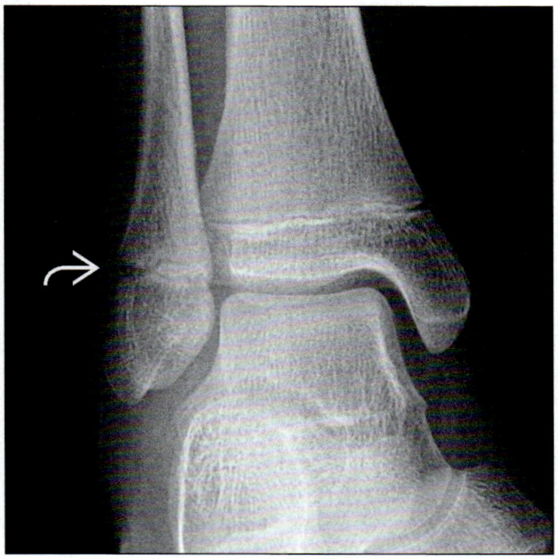

(Left) *Lateral radiograph of the fingers demonstrates subtle dorsal widening of the physis of the distal phalanx of the 4th finger compared to the other digits, with overlying soft tissue swelling. This is a Salter I injury.* ***(Right)*** *Oblique radiograph of the ankle in a child with lateral ankle pain after a fall demonstrates lateral widening of the distal fibular physis consistent with a Salter I injury. The fibular physis should be no wider than the lateral aspect of the tibial physis.*

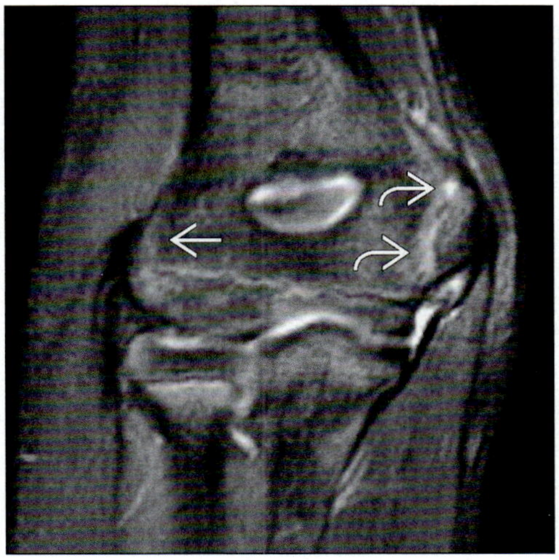

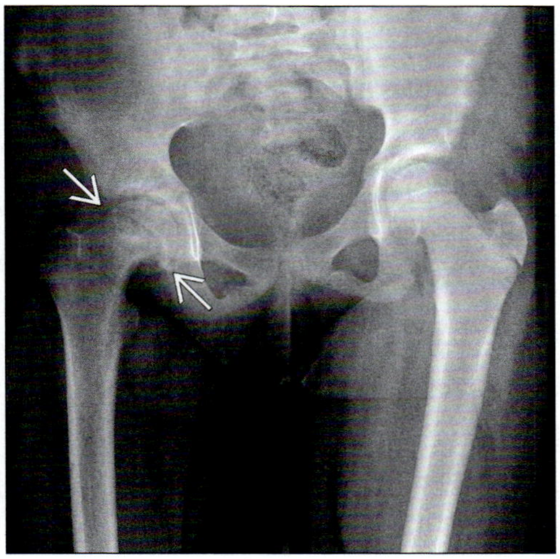

(Left) *Coronal T2WI FS MR of the elbow demonstrates elevated signal in the physis of the medial humeral epicondyle in a little league baseball pitcher with medial elbow pain. This represents little leaguer's elbow and can lead to displacement of the epiphysis. Note the closed lateral epicondylar epiphysis.* ***(Right)*** *AP radiograph shows slipped capital femoral epiphysis of the right hip, a Salter I injury. The left hip is normal.*

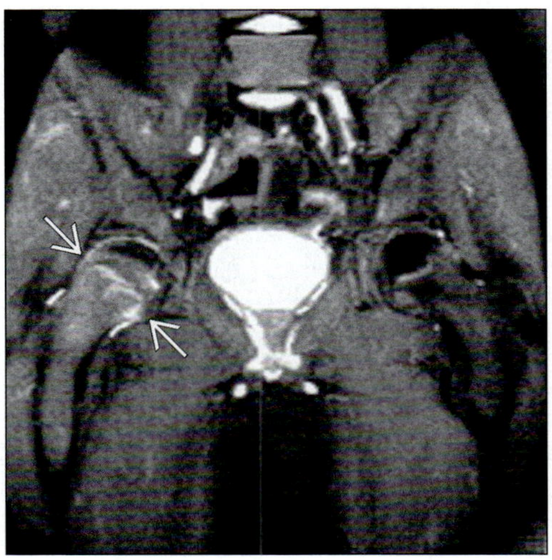

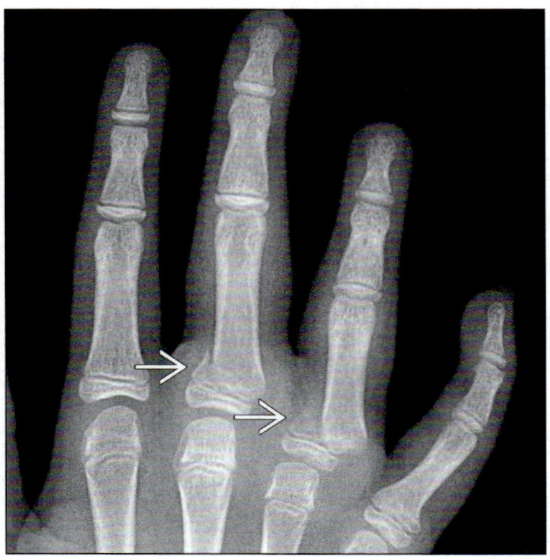

(Left) *Coronal T2WI FS MR demonstrates elevated signal in the proximal right femoral physis with surrounding marrow edema consistent with a Salter I injury. There was no radiographic evidence of slipped capital femoral epiphysis (SCFE) in this patient. Patient was treated surgically based on the MR findings.* ***(Right)*** *AP radiograph demonstrates fractures of the metaphyses extending into the growth plates of the proximal phalanges of the 3rd and 4th fingers consistent with Salter II fractures.*

(Left) *Coronal reformat bone CT shows a nondisplaced metaphyseal fracture → of the distal lateral femur entering the physis, consistent with a Salter II fracture. The fracture was not visible on routine radiographs. MR is more commonly used in this clinical setting due to its lack of ionizing radiation.* ***(Right)*** *AP radiograph shows marked displacement of fractures of the distal radius and ulna. Involvement of the radial growth plate is difficult to assess on this view.*

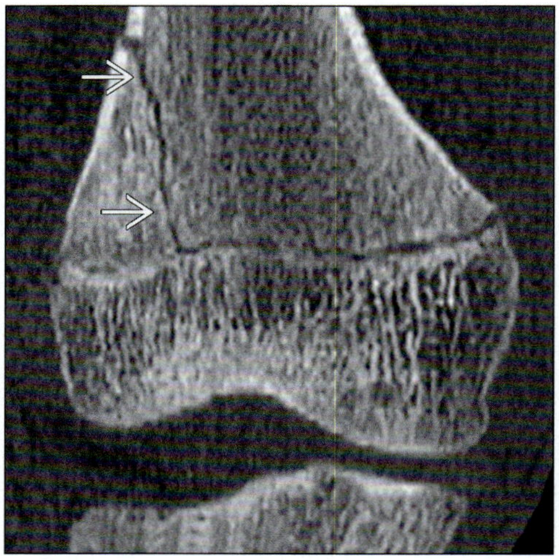

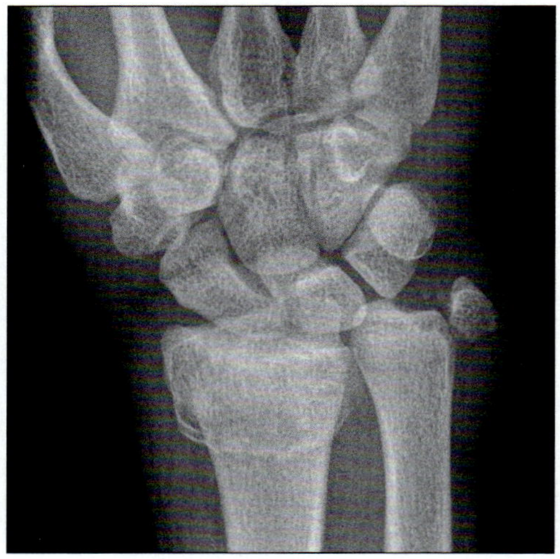

(Left) *Lateral radiograph from the same patient confirms extension of the radial fracture through the physis → and shows a triangular metaphyseal fragment → attached to the displaced epiphysis, consistent with a Salter II injury.* ***(Right)*** *AP radiograph shows a displaced fracture of the distal lateral tibial epiphysis → extending through the growth plate but not into the metaphysis. This is a juvenile Tillaux fracture, a type of Salter III injury.*

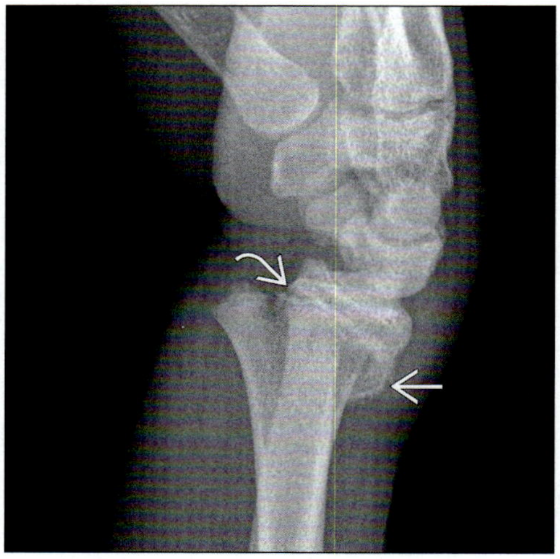

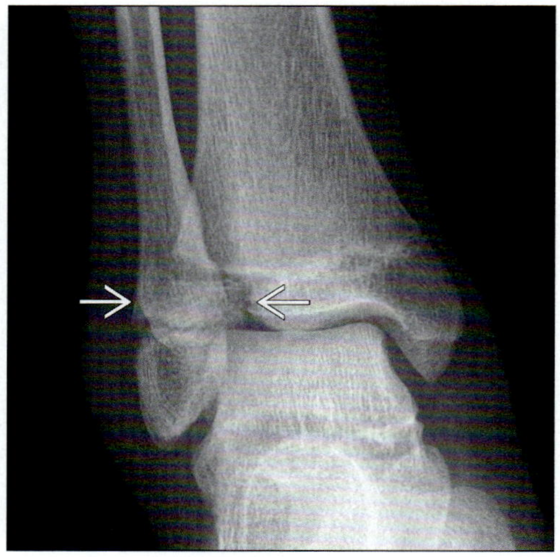

(Left) *Lateral radiograph from the same patient shows anterior displacement of the epiphyseal fracture fragment →. This fracture is typical in 12-15 year olds due to partial closure of medial distal tibial physis and patency of the lateral physis, restricting fracture propagation to the lateral tibia.* ***(Right)*** *Coronal reformat bone CT from the same patient shows the closed medial tibial physis → and patency of the lateral physis →. The vertical fracture → enters the articular surface.*

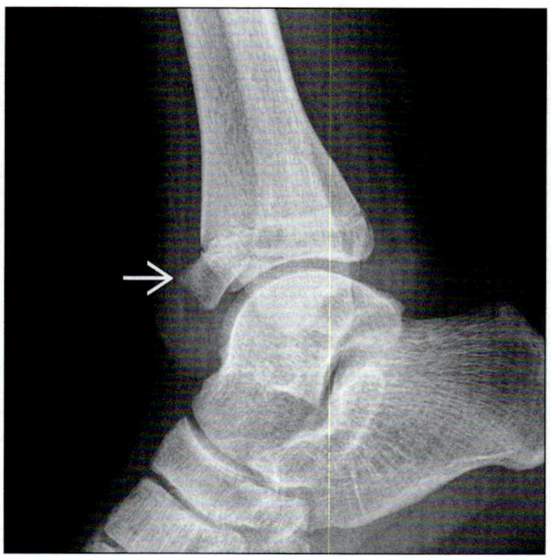

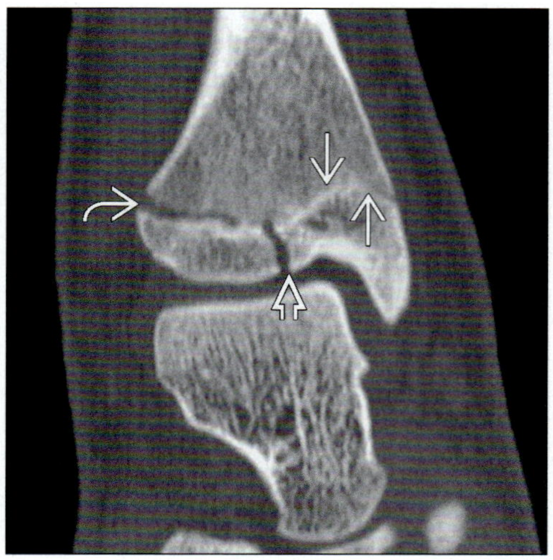

PHYSEAL INJURY (SALTER-HARRIS)

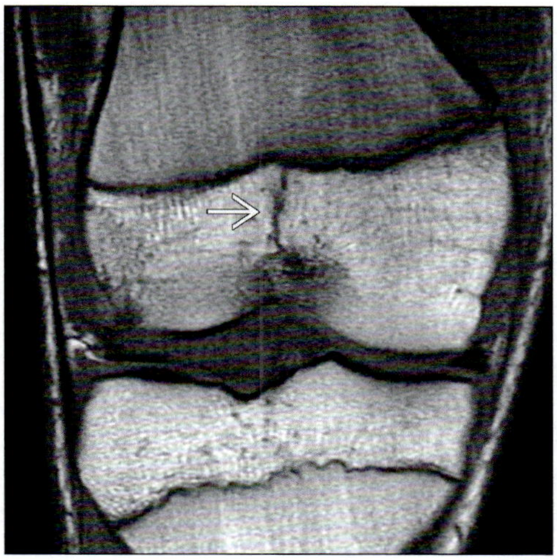

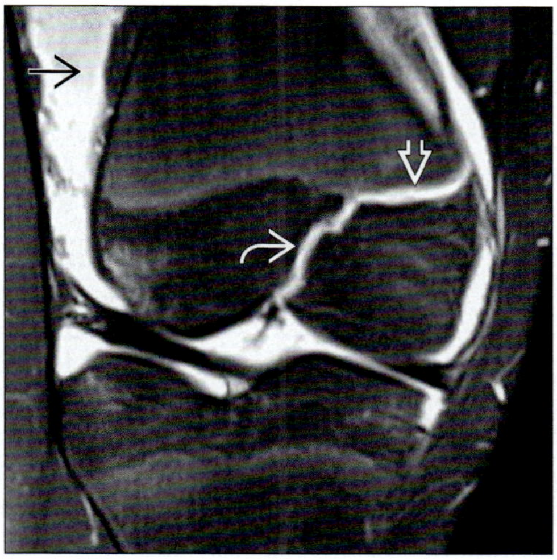

(Left) *Coronal T1WI MR shows a nondisplaced vertical fracture of the distal femoral epiphysis → extending into the intercondylar notch consistent with a Salter III injury. This was not visible on radiographs.* ***(Right)*** *Coronal T2WI FS MR shows a displaced Salter III fracture of the distal femur extending vertically through the roof of the intercondylar notch ↪ and medially through the physis ⇨. A large lipohemarthrosis → is present.*

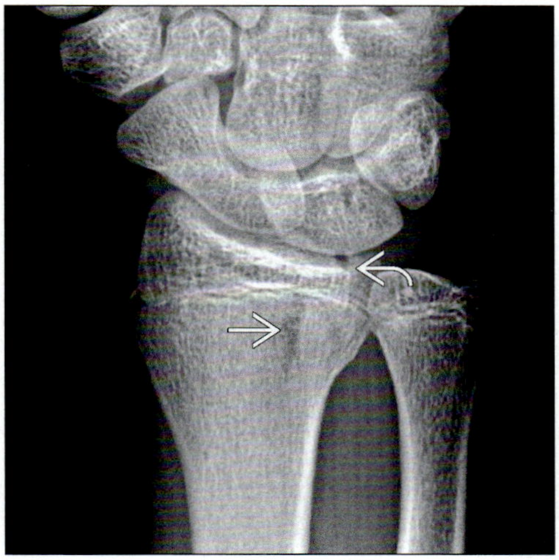

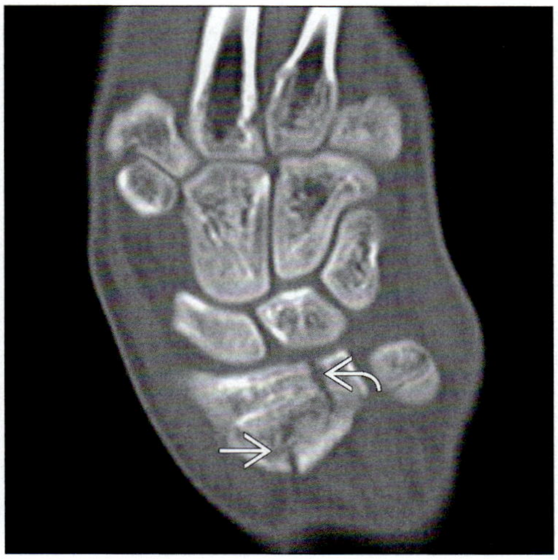

(Left) *AP radiograph shows subtle linear lucency in the distal radial metaphysis → and another in the distal radial epiphysis ↪ more medially. This suggests a Salter IV injury. The width of the growth plate is normal.* ***(Right)*** *Coronal reformat bone CT from the same patient confirms the fracture lines in the radial metaphysis → and epiphysis ↪. Confirmation of involvement of the radial articular surface is important for treatment and prognosis.*

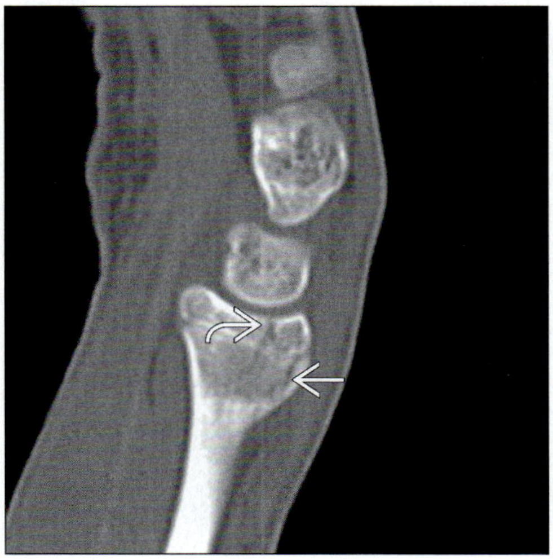

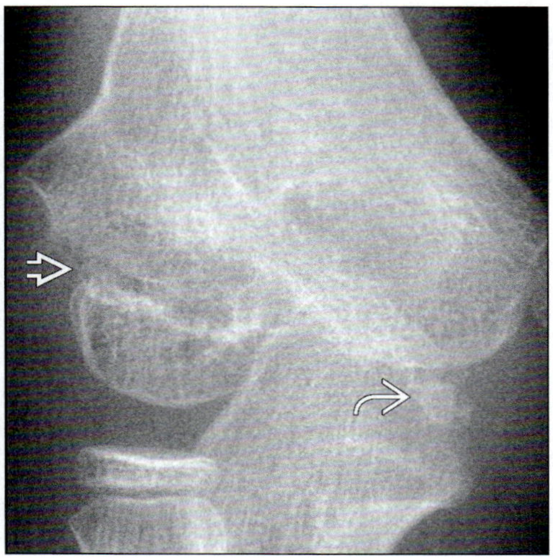

(Left) *Sagittal reformat bone CT from the same patient shows the metaphyseal → and epiphyseal ↪ components of the Salter IV injury.* ***(Right)*** *AP radiograph demonstrates a malunion of a Salter IV lateral humeral condylar fracture. The capitellum maintains its normal relationship to the metaphyseal fragment ⇨. That whole segment is proximally and laterally displaced several millimeters and has healed in this position. The trochlear ossification center is seen in its normal position ↪.*

CHILD ABUSE: THE EXTREMITIES

Key Facts

Terminology

- Nonaccidental trauma, trauma X, battered child syndrome, shaken baby syndrome

Imaging

- Lower extremity fracture in infant prior to walking age
- Metaphyseal corner fracture
  - Most common in proximal humerus, distal femur, and proximal and distal tibia
  - Triangular bone fragment at edge of metaphysis close to growth plate
- Bucket-handle fracture
  - Similar to corner fracture but involves more of metaphyseal circumference
- Subperiosteal new bone formation
  - Appears 5-14 days after trauma
- Skeletal survey
  - Healing fractures of differing ages
  - Metaphyseal corner, posterior rib, scapular, sternal fractures; thoracolumbar compression fractures; spinous process avulsion fractures

Top Differential Diagnoses

- Rickets
- Leukemia
- Osteogenesis imperfecta

Clinical Issues

- Discordance between stated history & injury on imaging (history of minimal or no trauma)
- Most < 1 year
- Almost all < 6 years
- Recognition of abuse is key
- Notification to local Child Protection Agency mandated

__(Left)__ AP radiograph shows a metaphyseal corner fracture ➔ of the proximal humerus in a small child. This is a hallmark finding suggesting child abuse, as fractures in this location rarely are due to accidental trauma. Corner fractures strongly suggest a violent twisting injury due to abuse by another individual. __(Right)__ AP radiograph shows a bucket-handle fracture of the distal radius. A circumferential fragment ➔ has separated from the metaphysis. A small corner fracture ➔ is also present.

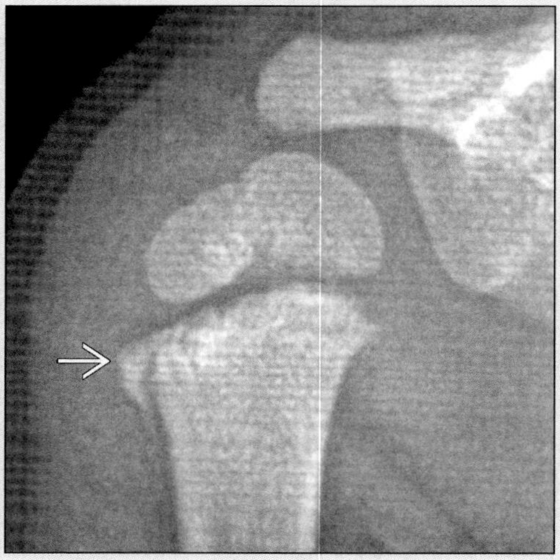

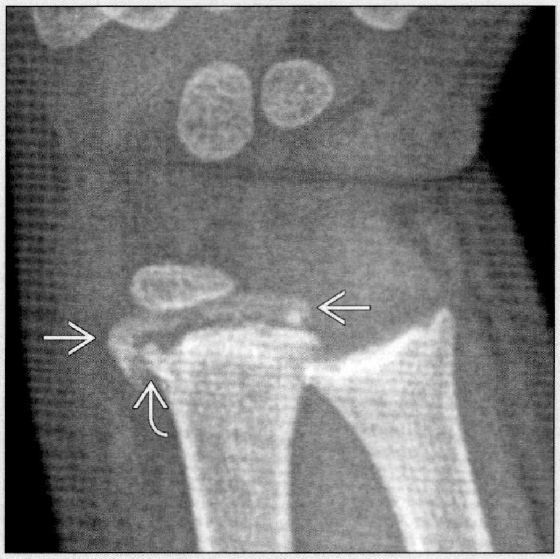

__(Left)__ Anteroposterior radiograph from a young child who would not walk shows a buckle fracture of the 5th metatarsal ➔. __(Right)__ Anteroposterior projection from a radionuclide bone scan in the same patient discloses varying increased uptake in both tibial diaphyses ➔ and in the foot ➔. The presence of multiple fractures, particularly with varying degrees of activity (and therefore likely of different ages) is highly suggestive of child abuse.

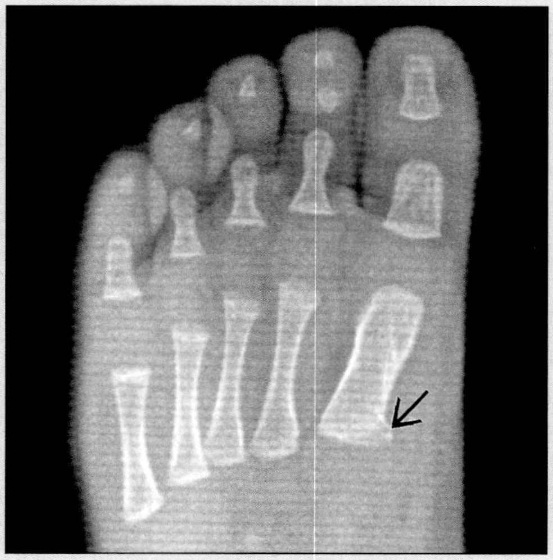

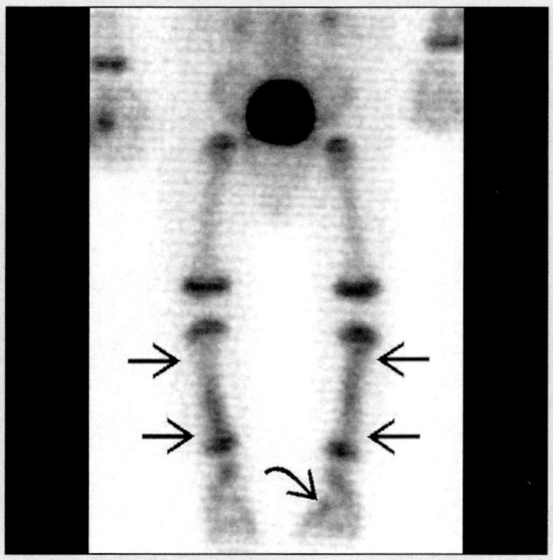

CHILD ABUSE: THE EXTREMITIES

TERMINOLOGY

Synonyms

- Nonaccidental trauma (NAT)
- Trauma X
- Battered child syndrome
- Shaken baby syndrome

Definitions

- Skeletal injury to a child due to violent act of another individual
  - Other organ injuries often seen as well

IMAGING

General Features

- Best diagnostic clue
  - Metaphyseal corner fracture
- Location
  - Metaphyses of long bones
    - Most common locations
      - Proximal humerus
      - Distal femur
      - Distal tibia
  - Finger fracture
  - Lower extremity fracture in infant prior to walking age
- Size
  - May be quite small
- Morphology
  - Plane of view may determine whether a particular fracture has corner or bucket-handle fracture appearance
    - If x-ray beam tangential to physis → corner fracture
    - If x-ray beam oblique to physis → bucket-handle fracture

Radiographic Findings

- Metaphyseal corner fracture
  - Triangular bone fragment at edge of metaphysis close to growth plate
- Bucket-handle fracture
  - Similar to corner fracture but involves more of metaphyseal circumference
  - Crescentic or annular rim of bone from metaphysis at periosteal collar
    - Site of cortical bone resorption during growth remodeling
- Subperiosteal new bone formation
  - Response to periosteal hemorrhage
  - Appears 5-14 days after trauma
  - Can be a normal finding in long bones of infants (physiologic)
    - Usually thin, regular, and bilaterally symmetric
    - Physiologic periosteal reaction peaks at 3-4 months, resolves by 6 months of age
- Skeletal survey
  - Healing fractures of differing ages
  - Metaphyseal corner fractures
  - Posterior rib fractures
  - Scapular fractures
  - Sternal fractures
  - Thoracolumbar spine compression fractures
  - Spinous process avulsion fractures

CT Findings

- Similar to radiographs but more sensitive for subtle injury
  - Early periosteal new bone formation

MR Findings

- Marrow edema or growth plate injury in young child
- Associated soft tissue injury
  - Muscle edema
  - Hematoma
  - Subcutaneous contusion

Nuclear Medicine Findings

- Bone scan
  - Fractures of different ages suggest child abuse
  - Sensitive but relatively nonspecific

Imaging Recommendations

- Best imaging tool
  - Radiographs
- Protocol advice
  - Skeletal survey
  - Bone scan if > 2 years or if skeletal survey equivocal & clinical suspicion high

DIFFERENTIAL DIAGNOSIS

Metabolic Bone Disease

- Rickets
  - Metaphyseal irregularity
  - Subperiosteal new bone formation
  - May fracture

Neoplasm

- Leukemia
  - Metaphyseal bands of lucency
  - Subperiosteal new bone formation

Central Nervous System Abnormality

- Myelomeningocele
  - Deceased sensation in lower extremities → increased fracture risk

Congenital

- Osteogenesis imperfecta
- Spondyloepiphyseal dysplasia
- Metaphyseal chondrodysplasia

PATHOLOGY

General Features

- Etiology
  - Force applied to child directly by larger person
    - Striking, dropping, twisting, pulling, pushing, kicking, biting
- Associated abnormalities
  - Central nervous system (CNS) trauma
    - Parenchymal hemorrhage in brain
    - Subdural or subarachnoid bleed
    - Skull fracture
    - Spine fracture
  - Visceral organ injury
    - Liver

- Spleen
- Pancreas
- Duodenum

Gross Pathologic & Surgical Features

- Immature bones weakest at metaphyseal-physeal junction
  - Pulling or torsional forces on limbs result in bone failure at this location

CLINICAL ISSUES

Presentation

- Most common signs/symptoms
  - Discordance between stated history & injury on imaging
    - History of minimal or no trauma
- Other signs/symptoms
  - Retinal hemorrhage in shaken baby
  - Apnea
  - Seizures
  - Fussiness
  - Bruises
  - Cigarette or scalding burns
- Clinical profile
  - Poor feeding
  - Small child for age

Demographics

- Age
  - Most < 1 year
  - Almost all < 6 years
- Gender
  - M > F (up to 2:1)
- Epidemiology
  - 30% of fractures in infants are from abuse
  - 17-25:100,000 annual incidence
  - Most common cause of traumatic death in infancy
    - 1,200 deaths per year in USA
    - Estimated > 1,000,000 children seriously injured/year in USA
  - Risk factors
    - < 1 year, prematurity, twin, male, physical handicap, stepchild
    - Young parents; ↓ socioeconomic status, 1/3 of perpetrators under influence of alcohol or drugs

Treatment

- Recognition of abuse is key
- Notification to local Child Protection Agency mandated
- Multidisciplinary child abuse & neglect team intervention

DIAGNOSTIC CHECKLIST

Consider

- Associated CNS or visceral injury
- Abuse of siblings

Image Interpretation Pearls

- Discordance of degree of injury and severity of injury history should be alert of abuse

Reporting Tips

- Better to overcall and question abuse than to miss a case of abuse that may result in death
- Document concern for child abuse and the individual in Child Protection to whom it was reported

SELECTED REFERENCES

1. Ravichandiran N et al: Delayed identification of pediatric abuse-related fractures. Pediatrics. 125(1):60-6, 2010
2. Hansoti B et al: Limb fractures and nonaccidental injury in children less than 24 months of age. Eur J Emerg Med. 15(2):63-6, 2008
3. Peters ML et al: The presence of bruising associated with fractures. Arch Pediatr Adolesc Med. 162(9):877-81, 2008
4. Coffey C et al: The risk of child abuse in infants and toddlers with lower extremity injuries. J Pediatr Surg. 40(1):120-3, 2005
5. Grayev AM et al: Metaphyseal fractures mimicking abuse during treatment for clubfoot. Pediatr Radiol. 31(8):559-63, 2001
6. Nimkin K et al: Fractures of the hands and feet in child abuse: imaging and pathologic features. Radiology. 203(1):233-6, 1997
7. Kleinman PK et al: A regional approach to the classic metaphyseal lesion in abused infants: the proximal humerus. AJR Am J Roentgenol. 167(6):1399-403, 1996
8. dos Santos LM et al: Soft tissue swelling with fractures: abuse versus nonintentional. Pediatr Emerg Care. 11(4):215-6, 1995
9. Conway JJ et al: The role of bone scintigraphy in detecting child abuse. Semin Nucl Med. 23(4):321-33, 1993
10. Stewart G et al: Trauma in infants less than three months of age. Pediatr Emerg Care. 9(4):199-201, 1993
11. Oestreich AE et al: The periphysis and its effect on the metaphysis: I. Definition and normal radiographic pattern. Skeletal Radiol. 21(5):283-6, 1992
12. McClelland CQ et al: Fractures in the first year of life. A diagnostic dilemma. Am J Dis Child. 136(1):26-9, 1982
13. Paterson CR: Vitamin D deficiency rickets simulating child abuse. J Pediatr Orthop. 1(4):423-5, 1981
14. Jaffe AC et al: Multiple metatarsal fractures in child abuse. Pediatrics. 60(4 Pt 2):642-3, 1977
15. Wilkinson RH et al: Pediatric skeletal trauma. Curr Probl Diagn Radiol. 6(2):1-37, 1976

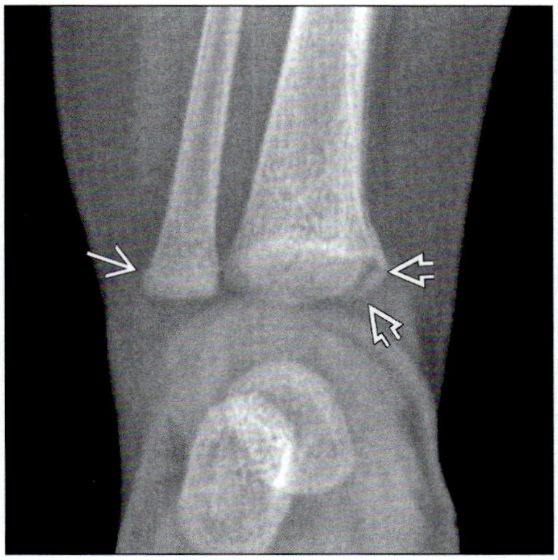

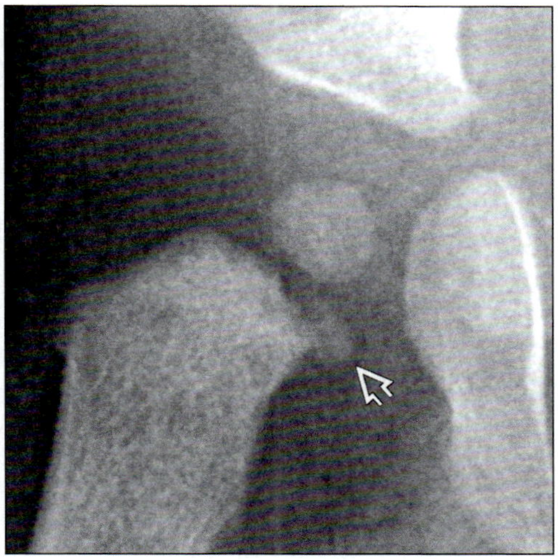

(Left) *AP radiograph demonstrates a tibial fracture → displaying features of a corner fracture (so termed for its location) and a bucket-handle fracture. A small corner fracture of the fibula → is also present.* ***(Right)*** *Anteroposterior radiograph of the hip in an infant shows a metaphyseal corner fracture of the proximal femur →. This child was not yet walking. Fractures in the lower extremity in children not yet walking are highly suggestive of abuse.*

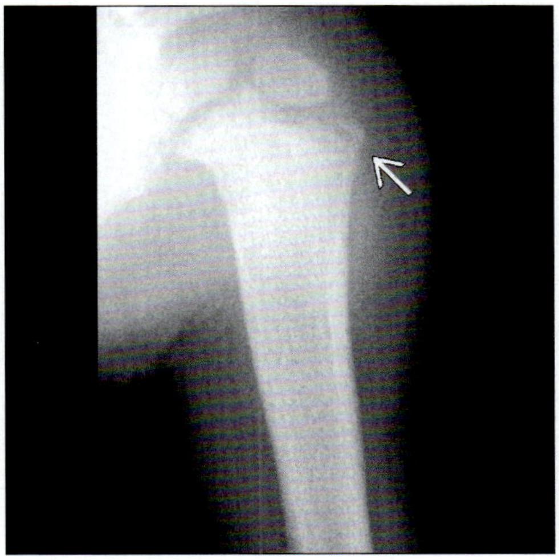

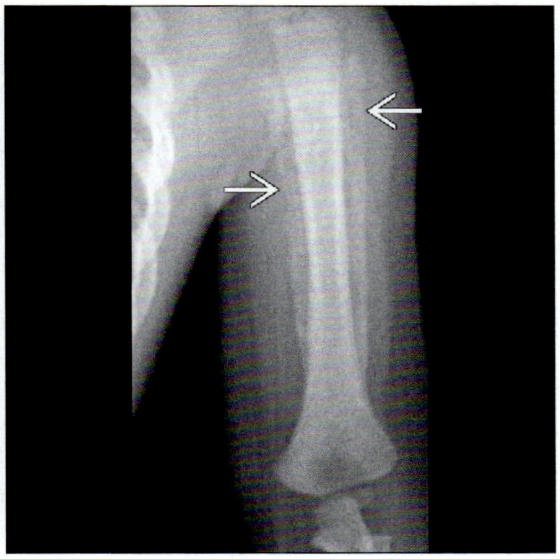

(Left) *Anteroposterior radiograph demonstrates a metaphyseal corner fracture →, which is highly suggestive of child abuse.* ***(Right)*** *Anteroposterior radiograph from the same child obtained 2 weeks later shows dense periosteal reaction → related to subperiosteal bleeding from the original metaphyseal fracture. This degree of periosteal reaction is far greater than physiologic periosteal bone formation and was not symmetrical with the contralateral limb.*

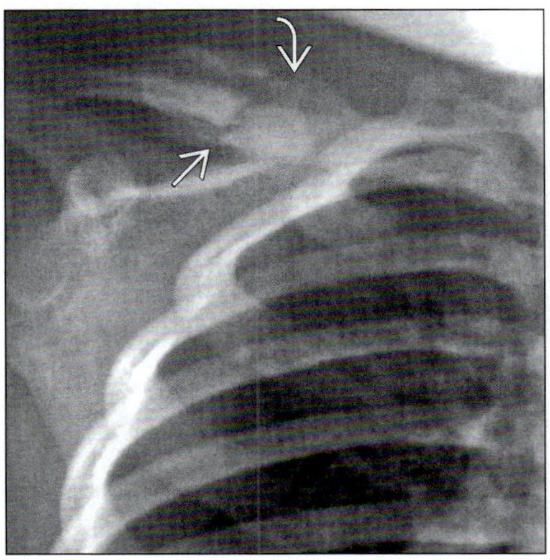

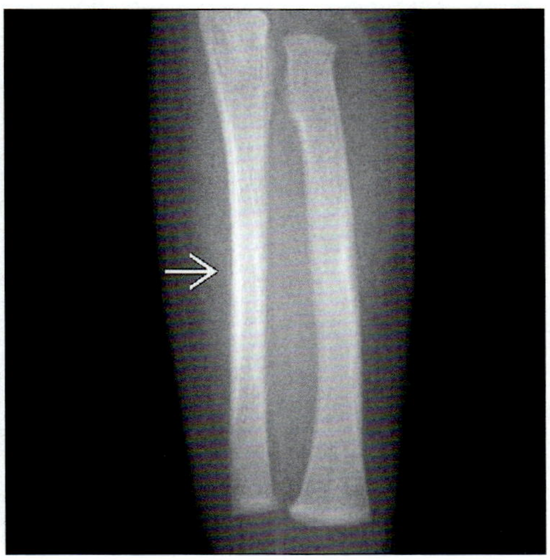

(Left) *Anteroposterior radiograph of the contralateral clavicle from the same child shows a subacute midshaft fracture →. Note the thick periosteal bone formation around the fracture →.* ***(Right)*** *Anteroposterior radiograph from the same child shows periosteal reaction involving the ulnar shaft →. No history of significant recent trauma was elicited. Regardless of history, multiple fractures of different ages are highly suggestive of child abuse.*

MUSCLE INJURY

Key Facts

Terminology

- Muscle strain, muscle tear, intramuscular injury, myotendinous injury, delayed onset muscle soreness (DOMS)
- Injury occurring within muscle belly or at myotendinous junction (MTJ)

Imaging

- Muscle contusion
  - Focal edema in muscle at site of injury
- DOMS: Diffuse edema in group of muscles 48-72 hours after exercise
- Muscle strain
  - Acute: Feathery ↑ SI on T2WI coursing along muscle fibers
- Intramuscular hematoma
  - Focal mass with variable signal intensity
- Partial or complete tear
  - Disruption of some or all muscle fibers
  - Most common at musculotendinous junction
- CT: May see subtle increased attenuation in muscle in setting of acute microhemorrhage
- US: Hypoechoic area or focal disruption within muscle

Pathology

- Unlike tendon or ligament, muscle has potential to repair itself at least to some degree

Clinical Issues

- Low-grade strain heals without sequelae
- Partial tear may → permanent defect and persistent weakness
- Complete tear → chronic atrophy of muscle
- Myositis ossificans may be chronic result of severe injury

__(Left)__ Coronal T2WI FS MR shows diffuse muscle edema → manifested by increased signal in the supraspinatus in a weightlifter with shoulder pain 2 days after exertion. This is an example of delayed onset muscle soreness (DOMS), a self-limited condition. __(Right)__ Axial T2WI MR shows feathery edema → in the soleus muscle around the myotendinous junction (MTJ) after an acute injury in a runner. The MTJ is the most common location for muscle strain injury or tear.

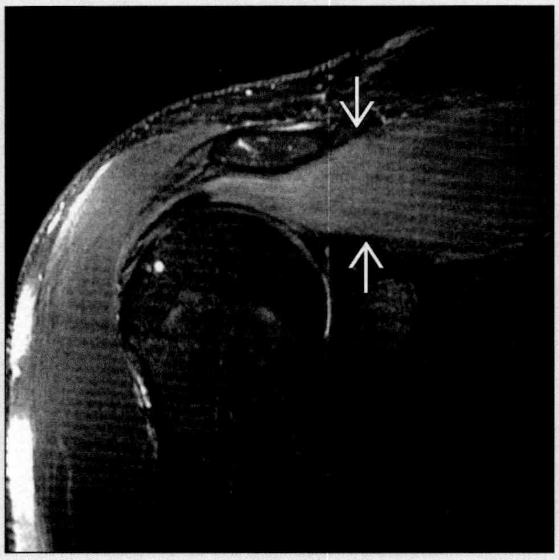

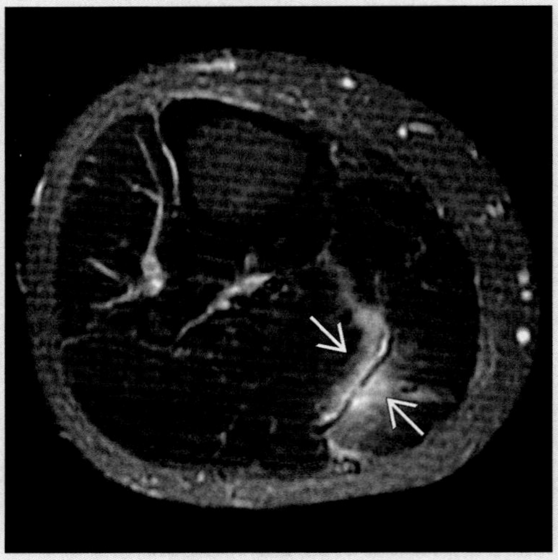

__(Left)__ Sagittal T2WI FS MR shows a large fluid collection → with surrounding edema in the vastus intermedius muscle due to a direct blow injury involving a motor vehicle collision. Muscle contusion may range from mild edema to large hematoma formation. __(Right)__ Axial T1WI MR from the same patient 3 months later shows an intramuscular hematoma → within the vastus intermedius. There is high signal in the rim of the collection indicating methemoglobin; the central portion ⇨ is liquefied.

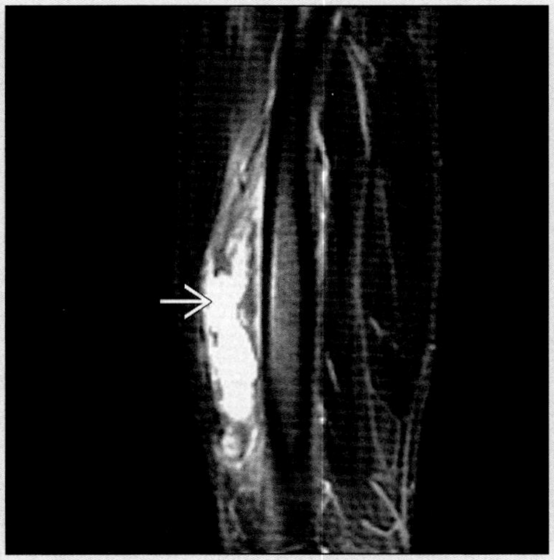

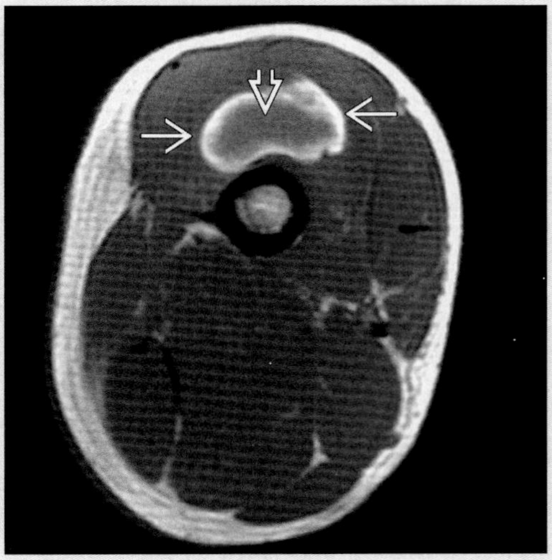

MUSCLE INJURY

TERMINOLOGY

Synonyms

- Muscle strain, muscle tear, intramuscular injury, myotendinous injury, delayed onset muscle soreness (DOMS)

Definitions

- Injury occurring within muscle belly or at myotendinous junction (MTJ)

IMAGING

General Features

- Best diagnostic clue
  - Increased T2 signal intensity (SI) within muscle after acute injury
- Location
  - Muscle belly or myotendinous junction
- Morphology
  - Ranges from interstitial edema to intramuscular hematoma to complete disruption of myotendinous unit (MTU)
    - Intramuscular injury: Feathery edema following course of muscle fibers
    - Intramuscular hematoma: Round or ovoid mass

CT Findings

- May see subtle increased attenuation in muscle in setting of acute microhemorrhage
- Acute hematoma: High-density mass
  - Becomes lower attenuation as it becomes more chronic
  - Eventually may liquefy → fluid attenuation
- Fat atrophy in muscle in chronic tear

MR Findings

- Muscle contusion
  - Focal edema in muscle at site of injury
- DOMS: Diffuse edema in group of muscles 48-72 hours after exercise
- Muscle strain
  - Acute: Feathery ↑ SI on T2WI coursing along fibers
  - Severe edema may → compartment syndrome
  - Chronic: Foci of hemosiderin deposition show ↓ SI on all sequences
    - Best seen on gradient echo
  - Chronic strain may weaken superficial fascia → transfascial muscle herniation
- Intramuscular hematoma
  - Focal mass with variable signal intensity
    - Signal characteristics depend on age of blood products
    - May enhance acutely or even subacutely; chronic hematoma generally has liquified non-enhancing center
- Partial or complete tear
  - Disruption of some or all muscle fibers
  - Most common at MTJ
  - May be associated with intramuscular hematoma formation

Ultrasonographic Findings

- Muscle strain: Mildly decreased echogenicity within affected muscle
- Intramuscular hematoma: Heterogeneous avascular mass within muscle
- Partial tear: Hypoechoic area within muscle
- Complete tear: Focal disruption of MTU, usually at MTJ

Imaging Recommendations

- Best imaging tool
  - MR
- Protocol advice
  - Gradient echo MR + FS most sensitive to subtle areas of acute interstitial hemorrhage and chronic hemosiderin

PATHOLOGY

General Features

- Unlike tendon or ligament, muscle can repair itself

Staging, Grading, & Classification

- MR grading system for muscle injury
  - Grade I: Strain
  - Grade II: Partial tear
  - Grade III: Complete tear

CLINICAL ISSUES

Presentation

- Most common signs/symptoms
  - Acute pain and swelling after injury
  - Weakness in affected muscle(s)

Natural History & Prognosis

- Low-grade strain heals without sequelae
- Partial tear may → permanent defect and persistent weakness
- Complete tear → chronic atrophy of muscle
- Severe muscle edema can → compartment syndrome ± myonecrosis
- Chronic muscle strain may → transfascial muscle herniation
- Myositis ossificans may be chronic result of severe injury

Treatment

- Complete tear at MTJ can be surgically repaired
- Intramuscular tear more difficult to repair

DIAGNOSTIC CHECKLIST

Reporting Tips

- Describe location of tear (muscle belly vs. MTJ)
- Describe degree of atrophy associated

SELECTED REFERENCES

1. Shelly MJ et al: MR imaging of muscle injury. Magn Reson Imaging Clin N Am. 17(4):757-73, vii, 2009
2. Koh ES et al: Ultrasound of skeletal muscle injury. Semin Musculoskelet Radiol. 11(2):162-73, 2007
3. Boutin RD et al: Imaging of sports-related muscle injuries. Radiol Clin North Am. 40(2):333-62, vii, 2002
4. May DA et al: Abnormal signal intensity in skeletal muscle at MR imaging: patterns, pearls, and pitfalls. Radiographics. 20 Spec No:S295-315, 2000

(Left) *Axial PDWI FS MR shows feathery edema ➔ in the posterior fibers of the deltoid muscle in a 27-year-old professional tennis player. Findings are consistent with muscle strain or delayed onset muscle soreness (DOMS). There is no focal rupture.* ***(Right)*** *Sagittal T2WI FS MR from the same patient shows the sharp demarcation between the edematous deltoid muscle ➔ and the normal infraspinatus ➔ and teres minor ➔ muscles.*

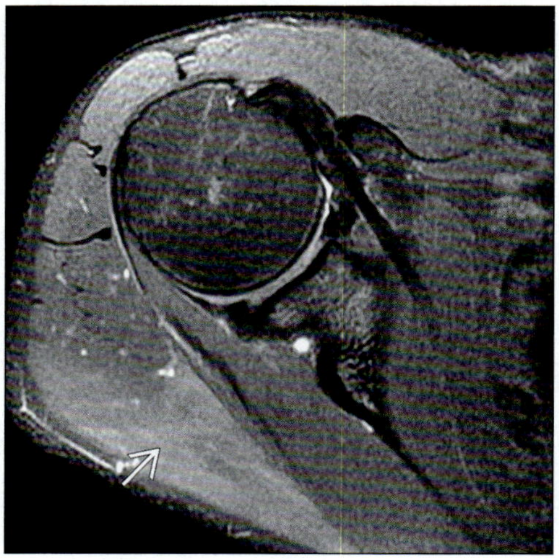

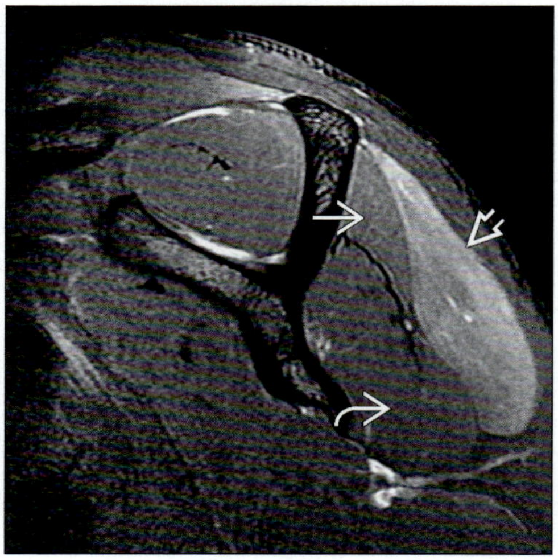

(Left) *Axial T2*-weighted gradient echo image from a patient with chronic thigh injury shows focal hemosiderin deposition creating signal void ➔ at the site of previous myotendinous junction tear in the rectus femoris.* ***(Right)*** *Axial T2WI FS MR shows an acute intramuscular tear in the gluteus maximus ➔ after a severe twisting injury. Feathery edema ➔ extends along muscle fiber planes more laterally without disrupting the architecture of the muscle.*

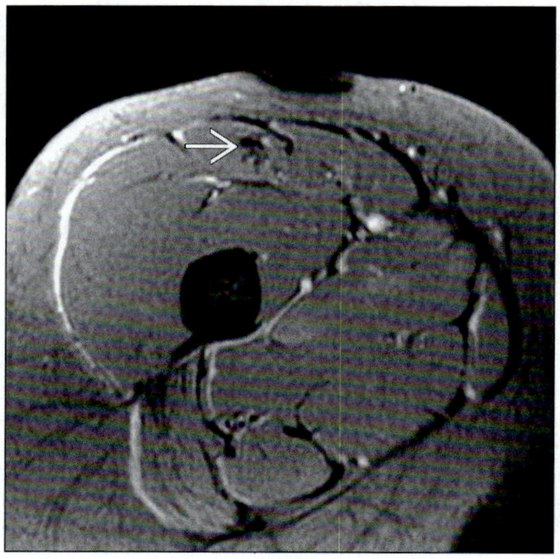

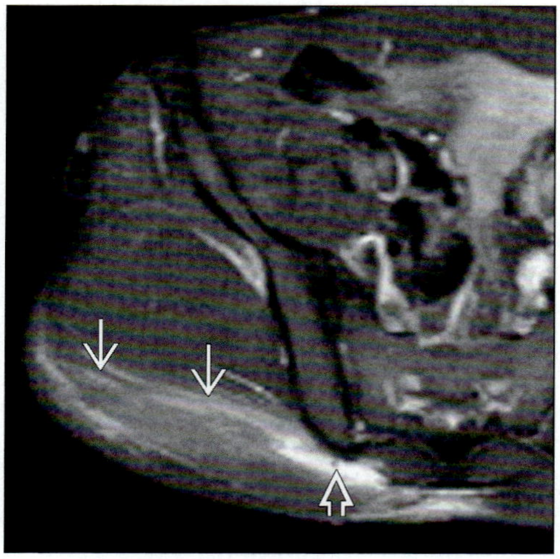

(Left) *Axial T2WI MR shows an acute low-grade tear of the myotendinous junction of the soleus in a tennis player. The tendon ➔ is intact. There is feathery edema ➔ in the muscle fibers near the MTJ without a discrete fiber defect or hematoma.* ***(Right)*** *Axial T1WI C+ FS MR from the same patient shows significant enhancement of the injured muscle ➔. Enhancement is usually present after acute injury and should not be mistaken for underlying pathology.*

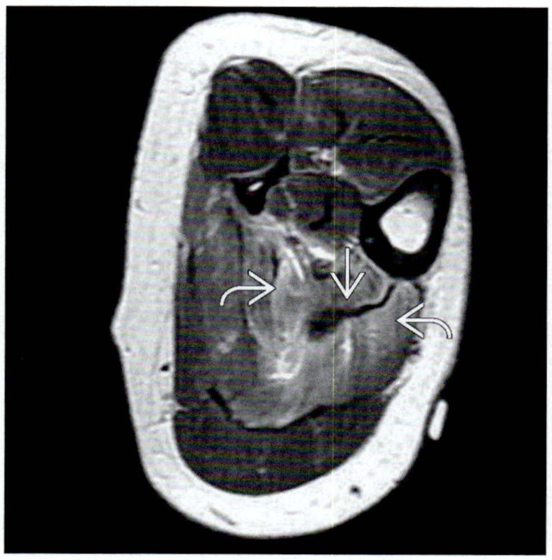

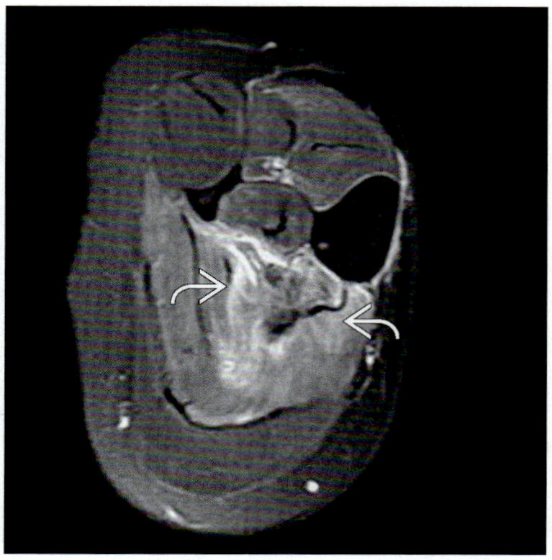

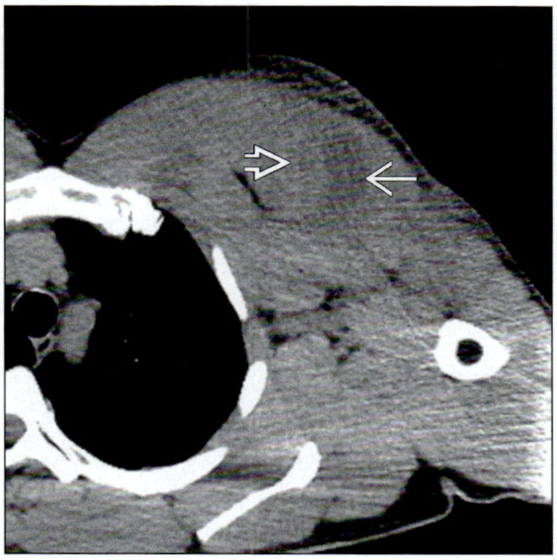

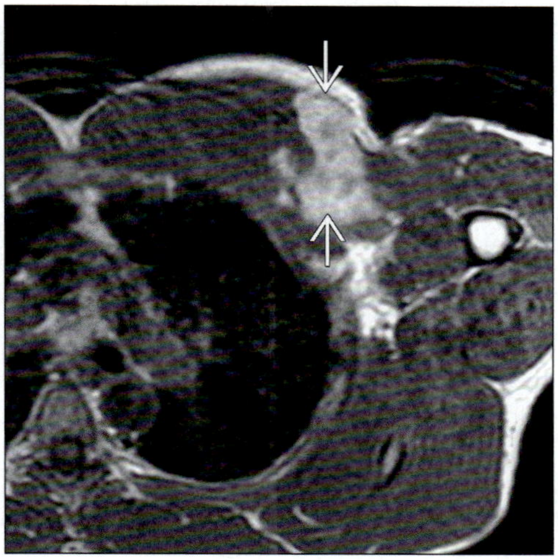

(Left) *Axial noncontrast CT in a weightlifter with acute anterior chest pain and swelling after doing bench presses shows marked swelling and heterogeneous density in the pectoralis major. Areas of high density* ⇨ *indicate intramuscular microhemorrhage; low density* ➔ *indicates fluid.* ***(Right)*** *Axial T1WI MR from the same patient shows high signal hematoma* ➔ *filling the defect. At surgery, this was a complete tear of the muscle belly of the pectoralis major.*

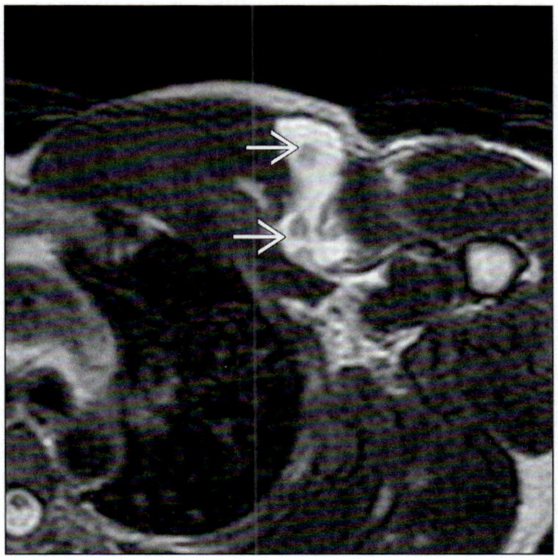

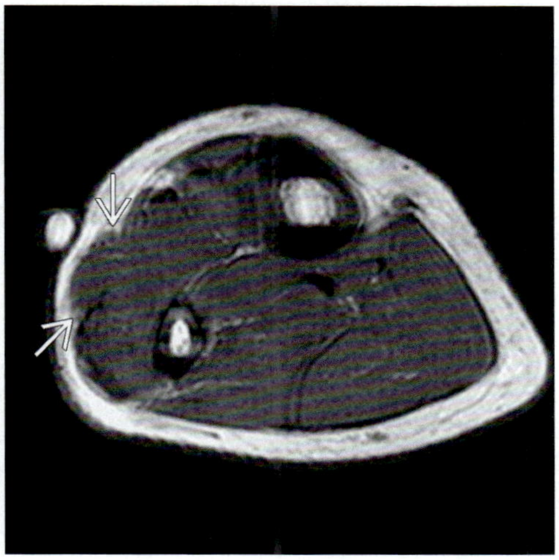

(Left) *Axial T2WI MR from the same patient shows high signal in the hematoma with areas of lower signal* ➔ *likely representing blood products of different ages.* ***(Right)*** *Axial T1WI MR from the leg of a long-distance runner with anterior leg pain while running shows a focal transfascial herniation* ➔ *of the extensor digitorum longus muscle. The anterior compartment of the leg is the most common location for this injury.*

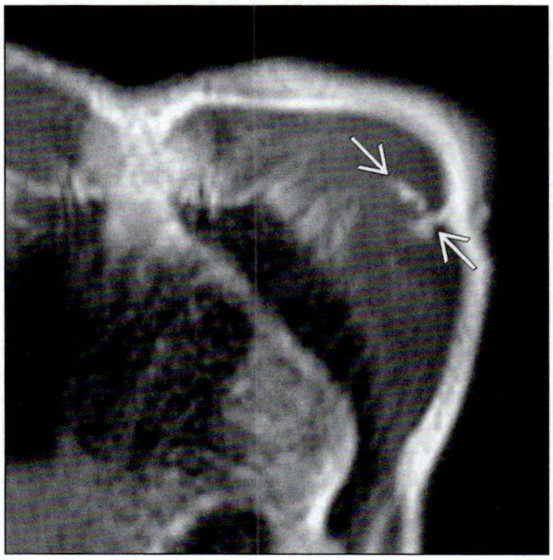

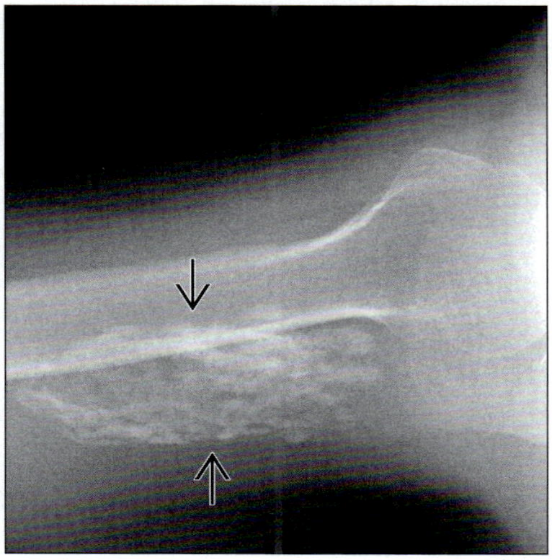

(Left) *Coronal T1WI MR in a patient who suffered a previous partial tear of the pectoralis major muscle and complained of a focal defect in the anterior chest shows a fat collection* ➔ *in the pectoralis near the myotendinous junction. This is post-traumatic muscle atrophy, an entity that can manifest as a clinically palpable defect.* ***(Right)*** *Axillary lateral radiograph of the shoulder demonstrates well-defined ossification* ⇨ *outlining a portion of the triceps muscle. This is myositis ossificans.*

HEMATOMA

Key Facts

Terminology

- Hematoma: Localized collection of blood (usually clotted) outside blood vessel but contained within a space

Imaging

- Usually isodense to muscle on noncontrast CT but can be higher density in case of acute hemorrhage
- Contrast extravasation into hematoma on contrast-enhanced CT suggests active bleeding
- Fluid-cellular layer (a.k.a. hematocrit level)
- Typically no abnormal enhancement in acute stage
- MR appearance is highly complex and variable
  - Much less predictable than intracranial hematomas due to larger and more complex/loculated nature
  - Chronic hematoma may appear infiltrative, with adjacent edema/inhomogeneous enhancement

Top Differential Diagnoses

- Solid tumor
- Muscle infarct
- Ganglion cyst

Pathology

- Sports activity
- Traumatic
- Nontraumatic
  - Anticoagulation therapy
  - Clotting disorder
  - Intratumoral bleeding

Diagnostic Checklist

- History of anticoagulation or trauma
- Unexplained spontaneous bleed → must exclude underlying neoplasm

__(Left)__ Axial T1WI MR shows a crescentic focus of increased signal within the vastus intermedius muscle in the mid-thigh in a football player with thigh pain after being tackled. High signal within a mass on T1WI usually indicates blood products, lipid, or proteinaceous fluid. This is an intramuscular hematoma. __(Right)__ Axial T1WI MR of the mid-arm shows a mass in the biceps muscle with a rim of hyperintense signal indicating a subacute hematoma. The central portion remains hypointense.

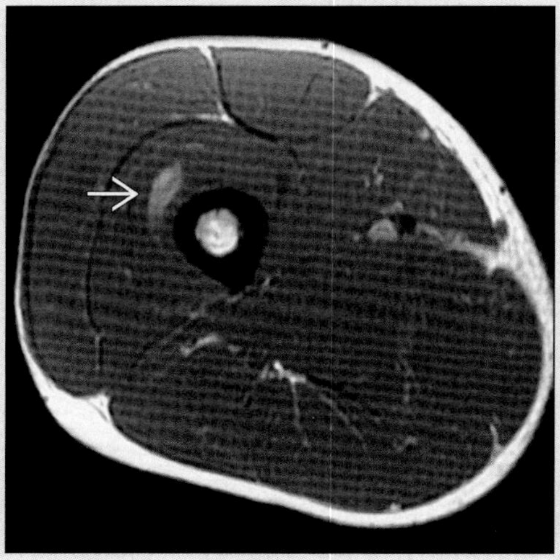

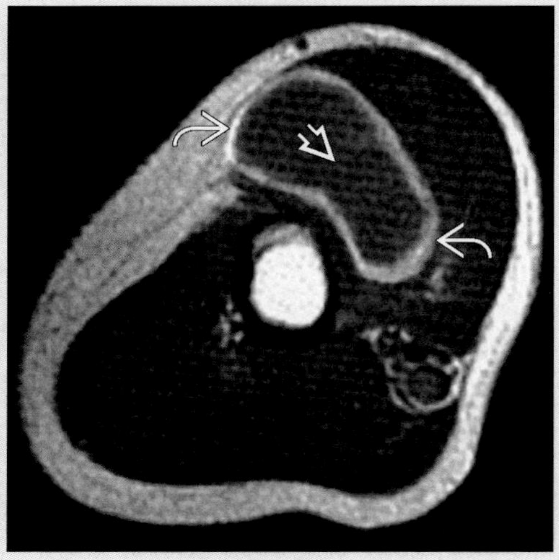

__(Left)__ Axial PDWI MR shows a large subcutaneous hematoma lateral to the greater femoral trochanter in a patient status post recent motor vehicle accident. Note that there are areas of varying signal intensity within the mass, which is typical for hematoma. __(Right)__ Sagittal T2WI MR shows a large low signal hematoma interposed between the patella and the torn and retracted quadriceps tendon in this diabetic patient.

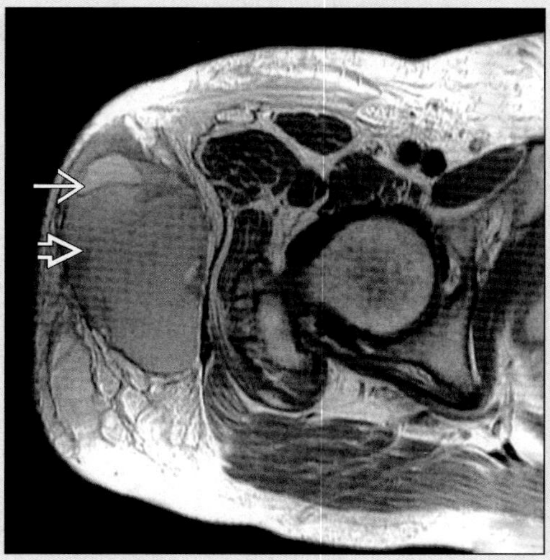

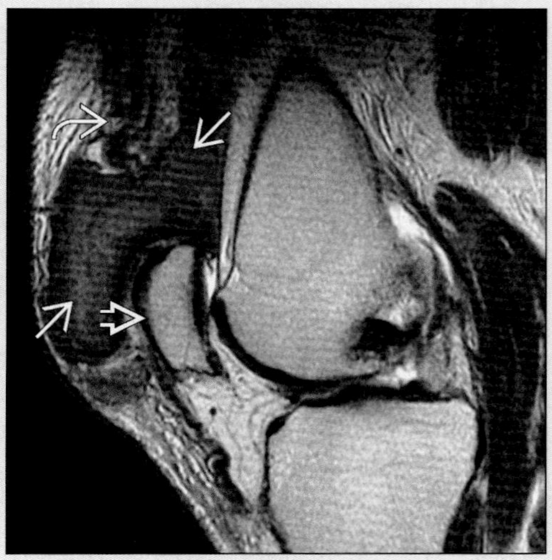

HEMATOMA

TERMINOLOGY

Definitions

- Hematoma: Localized collection of blood (usually clotted) outside blood vessel but contained within a space
- Interstitial hemorrhage: Blood interspersed within muscle

IMAGING

General Features

- Best diagnostic clue
  - Soft tissue mass with MR characteristics of blood products
    - Usually ↑ T2 signal
    - Often ↑ T1 signal
- Location
  - Intra- or extramuscular, depending on mechanism
- Size
  - Variable
- Morphology
  - Often extends longitudinally within fascial planes

Radiographic Findings

- Radiographs may show focal soft tissue mass
- Other evidence of trauma (e.g., joint effusion or fracture)

CT Findings

- CT shows asymmetric enlargement of involved muscle or tissue plane
  - Usually isodense to muscle on noncontrast CT but can be higher density in case of acute hemorrhage
  - Evolution to seroma will appear hypodense
- Contrast extravasation into hematoma on contrast-enhanced CT suggests active bleeding

MR Findings

- T1WI
  - Initially iso- or hypointense to muscle (deoxyhemoglobin)
    - Then hyperintense (methemoglobin)
    - Eventually hypointense (hemosiderin)
  - Chronic seroma appears hypointense (fluid)
  - Development of fibrosis or heterotopic ossification → foci of low signal
  - Associated areas of intraparenchymal muscle hemorrhage would show high signal on both T1WI and T2WI
- T2WI
  - Initially hyperintense to muscle (oxyhemoglobin)
  - Then iso- or hypointense (deoxyhemoglobin)
  - Then hyperintense (as intracellular methemoglobin becomes extracellular with RBC breakdown)
  - Eventually hypointense (hemosiderin)
  - Chronic seroma appears hyperintense (fluid)
  - Reactive edema (↑ T2 signal) surrounding hematoma may be seen at any stage
  - Fluid-cellular layer (a.k.a. hematocrit level)
    - Usually in patients who are anticoagulated or have intratumoral bleeding
- T1WI C+
  - Typically no abnormal enhancement in acute stage
  - Chronic hematomas often incite adjacent soft tissue reaction
    - Inhomogeneous enhancement in multiple muscle groups &/or neurovascular bundles
    - Infiltrative appearance can make chronic hematoma indistinguishable from aggressive tumor
- MR appearance is highly complex and variable
  - Varies by differences in individual MR unit field strengths and different patterns of blood breakdown
  - Much less predictable than intracranial hematomas due to larger and more complex/loculated nature

Ultrasonographic Findings

- Grayscale ultrasound
  - Asymmetric muscle enlargement
    - Hematoma acutely anechoic, with irregular echoic areas developing later
- Color Doppler
  - No internal color flow

Angiographic Findings

- Hematoma: No internal vascularization

Imaging Recommendations

- Best imaging tool
  - MR
- Protocol advice
  - Include T2WI without FS to better assess signal characteristics
  - Consider MR + C to assess internal vascularity or possible intratumoral bleed

DIFFERENTIAL DIAGNOSIS

Solid Tumor

- Neoplasm may have similar signal characteristics
- Neoplasms may bleed → hematoma
- Most neoplasms have internal vascularity

Muscle Infarction

- History of diabetes
- Internal architecture similar to normal muscle

Ganglion Cyst

- Fluid signal
- No significant surrounding edema

PATHOLOGY

General Features

- Etiology
  - Sports activity
    - Muscle tear (strain), often at myotendinous junction
  - Traumatic
    - Blunt or penetrating trauma
  - Nontraumatic
    - Anticoagulation therapy
    - Clotting disorder
    - Intratumoral bleeding
      - Neoplasms with ↑ vascularity may spontaneously bleed
- Associated abnormalities

  - Muscle or tendon tear
  - Surrounding muscular edema/inflammation
  - Fracture

Gross Pathologic & Surgical Features

- Hematomas progress through various stages of blood breakdown: Oxyhemoglobin, deoxyhemoglobin, intracellular methemoglobin, extracellular methemoglobin, hemosiderin
- Appearance often heterogeneous due to blood products of different stages

CLINICAL ISSUES

Presentation

- Most common signs/symptoms
  - Pain and focal swelling
    - Usually sudden onset
  - Signs/symptoms related to blunt or penetrating trauma
- Other signs/symptoms
  - Swelling, ecchymoses, or other evidence of trauma

Demographics

- Age
  - Nontraumatic: Elderly, usually related to anticoagulation therapy
- Gender
  - Traumatic or activity-related: M > F
  - Nontraumatic: M = F
- Epidemiology
  - Predisposing conditions
    - Iatrogenic
      - Anticoagulation therapy
    - Congenital
      - Hemophilia
      - Henoch-Schönlein purpura, von Willebrand disease, and other bleeding diatheses

Natural History & Prognosis

- Without treatment (conservative or surgical), rebleeding will often occur
- Can lead to compartment syndrome
  - Venous congestion → neurovascular compromise → muscle degeneration/necrosis

Treatment

- Conservative treatment (i.e., activity restriction) if asymptomatic
- Surgical evacuation if symptomatic (weakness, hyperesthesia, pain on passive flexion)

DIAGNOSTIC CHECKLIST

Consider

- History of anticoagulation or trauma
- Tumor may mimic hematoma and hematoma may mimic tumor
  - Intratumoral bleed may appear as typical hematoma
    - Unexplained spontaneous bleed → must exclude underlying neoplasm
    - Watch for mass adjacent to hematoma
  - Chronic hematoma may appear mass-like
    - Inhomogeneous, with enhancement (though rarely centrally)

Image Interpretation Pearls

- Hematoma is avascular
  - MR + C, CECT, or Doppler ultrasound to detect internal flow

Reporting Tips

- Try to determine acuity of bleed if possible

SELECTED REFERENCES

1. Kontogeorgakos VA et al: Extremity soft tissue sarcomas presented as hematomas. Arch Orthop Trauma Surg. Epub ahead of print, 2009
2. McKenzie G et al: Pictorial review: Non-neoplastic soft-tissue masses. Br J Radiol. 82(981):775-85, 2009
3. Shelly MJ et al: MR imaging of muscle injury. Magn Reson Imaging Clin N Am. 17(4):757-73, vii, 2009
4. Taïeb S et al: Soft tissue sarcomas or intramuscular haematomas? Eur J Radiol. 72(1):44-9, 2009
5. Wu JS et al: Soft-tissue tumors and tumorlike lesions: a systematic imaging approach. Radiology. 253(2):297-316, 2009
6. Nett MP et al: Magnetic resonance imaging of acute "wiiitis" of the upper extremity. Skeletal Radiol. 37(5):481-3, 2008
7. Theodorou SJ et al: Imaging findings of complications affecting the upper extremity in intravenous drug users: featured cases. Emerg Radiol. 15(4):227-39, 2008
8. Allen DJ et al: Primary malignancies mistaken for pseudotumours in haemophilic patients. Haemophilia. 13(4):383-6, 2007
9. Papp DF et al: Magnetic resonance imaging of soft-tissue tumors: determinate and indeterminate lesions. J Bone Joint Surg Am. 89 Suppl 3:103-15, 2007
10. Liu PT et al: Chronic expanding hematoma of the thigh simulating neoplasm on gadolinium-enhanced MRI. Skeletal Radiol. 35(4):254-7, 2006
11. Niimi R et al: Soft-tissue sarcoma mimicking large haematoma: a report of two cases and review of the literature. J Orthop Surg (Hong Kong). 14(1):90-5, 2006
12. Suzuki T et al: Arterial injury associated with acute compartment syndrome of the thigh following blunt trauma. Injury. 36(1):151-9, 2005
13. Connell DA et al: Longitudinal study comparing sonographic and MRI assessments of acute and healing hamstring injuries. AJR Am J Roentgenol. 183(4):975-84, 2004
14. Gomez P et al: High-grade sarcomas mimicking traumatic intramuscular hematomas: a report of three cases. Iowa Orthop J. 24:106-10, 2004
15. Saotome K et al: Enlarging intramuscular hematoma and fibrinolytic parameters. J Orthop Sci. 8(2):132-6, 2003
16. Bush CH: The magnetic resonance imaging of musculoskeletal hemorrhage. Skeletal Radiol. 29(1):1-9, 2000
17. Aoki T et al: The radiological findings in chronic expanding hematoma. Skeletal Radiol. 28(7):396-401, 1999
18. Lin WY et al: Subcutaneous hematoma detected on Tc-99m pyrophosphate muscle imaging. Clin Nucl Med. 23(1):56, 1998
19. Jaovisidha S et al: Hemophilic pseudotumor: spectrum of MR findings. Skeletal Radiol. 26(8):468-74, 1997
20. Takebayashi S et al: Sonographic findings in muscle strain injury: clinical and MR imaging correlation. J Ultrasound Med. 14(12):899-905, 1995
21. Mittlmeier T et al: Compartment syndrome of the foot after intraarticular calcaneal fracture. Clin Orthop Relat Res. (269):241-8, 1991
22. De Smet AA et al: Magnetic resonance imaging of muscle tears. Skeletal Radiol. 19(4):283-6, 1990

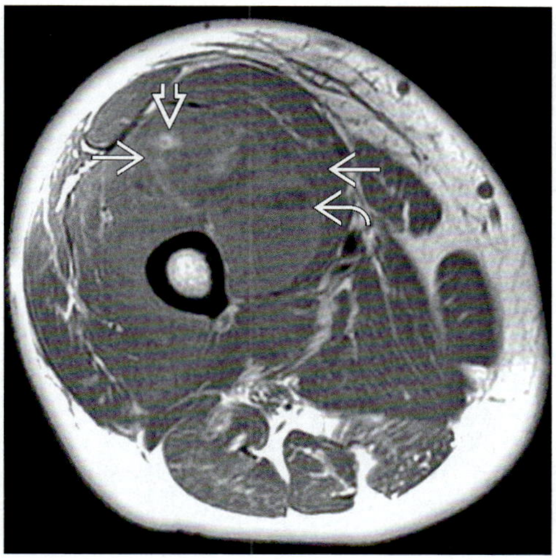

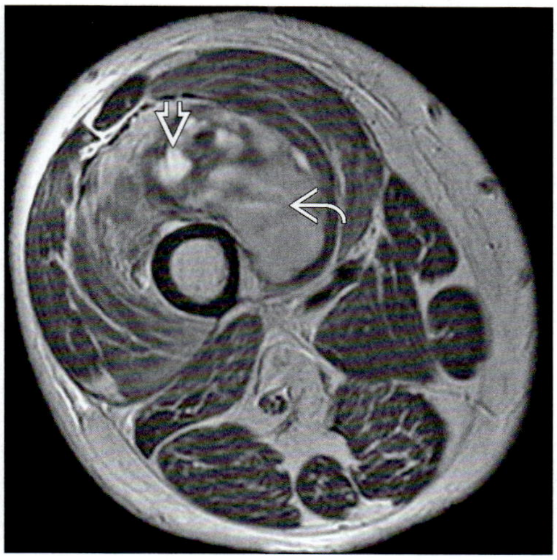

(Left) *Axial T1WI MR shows a large, complex mass → in the vastus intermedius muscle in the mid-thigh in a patient on warfarin (Coumadin). Some portions of the mass are hyperintense to muscle →, indicating blood or fat. There is a fluid-fluid level → posteriorly.* ***(Right)*** *Axial T2WI MR from the same patient shows marked heterogeneity of signal in the mass, with areas anteriorly that appear cystic → and the fluid level more posteriorly →. Areas of high signal on T1 and T2 images represent blood.*

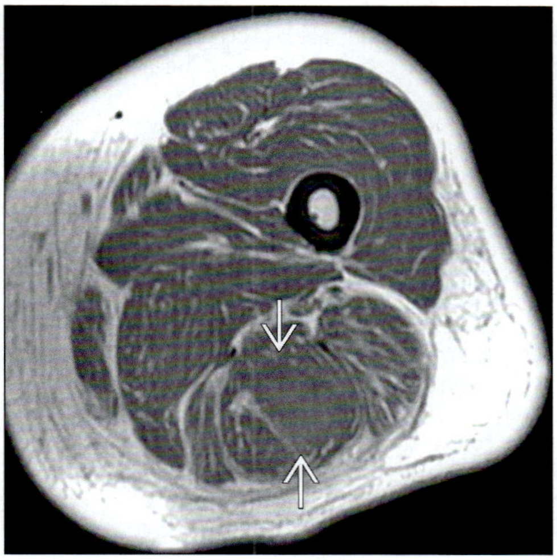

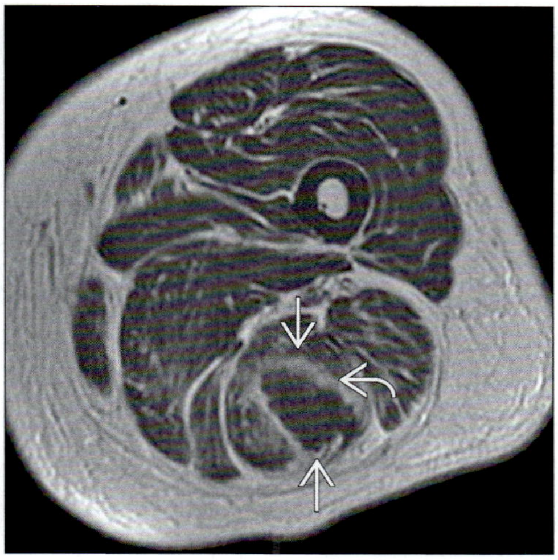

(Left) *Axial T1WI MR shows a vague, mass-like area of uniform tissue → isointense to skeletal muscle in the hamstrings of the posterior thigh.* ***(Right)*** *Axial T2WI MR from the same patient shows the mass → more clearly. The central portion demonstrates hypointense signal relative to normal muscle; there is a thin rim of hyperintense signal → around the periphery. Dark signal on T1- and T2-weighted sequences usually indicates an acute hematoma.*

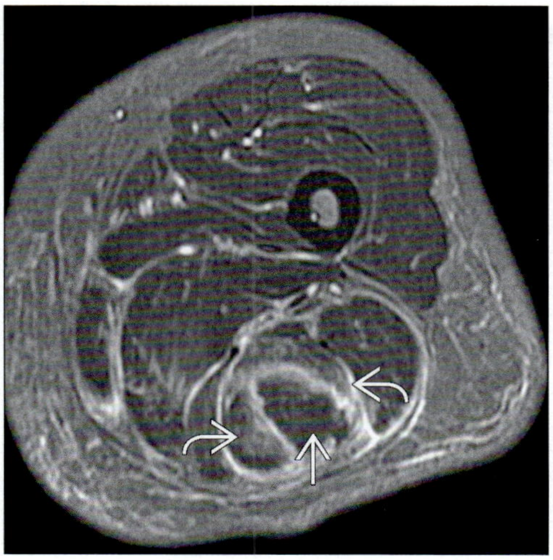

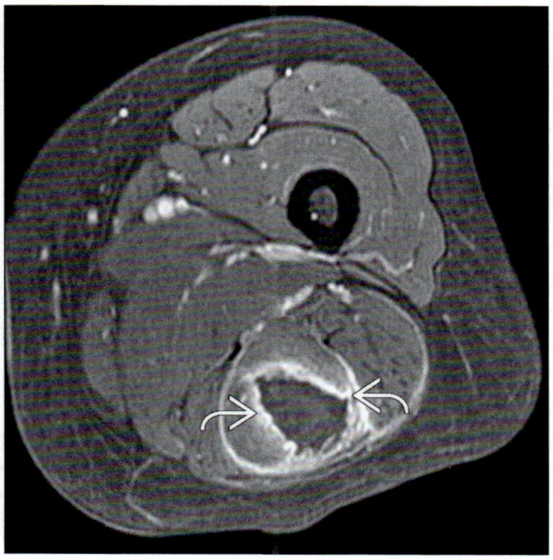

(Left) *Axial T2WI FS MR from the same patient demonstrates persistent low signal in the center of the mass → with reactive edema in the surrounding muscles →.* ***(Right)*** *Axial T1WI C+ FS MR from the same patient shows intense enhancement → of the reactive/post-traumatic tissue around the hematoma. The hematoma itself remains low signal; central enhancement should trigger concern for an underlying neoplasm, as bland hematoma does not enhance.*

HEMATOMA

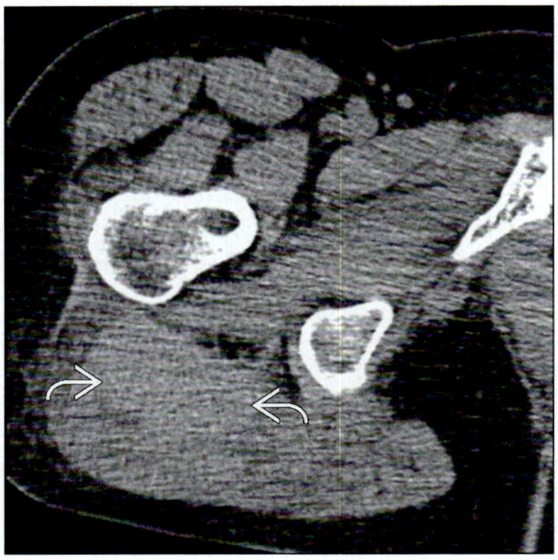
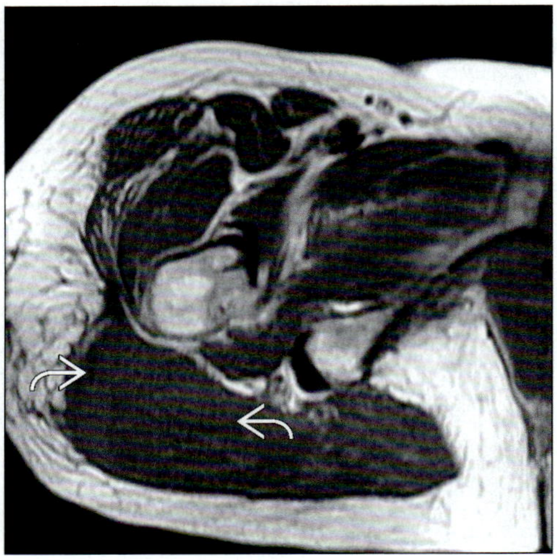

(Left) *Axial noncontrast CT demonstrates a mass-like area of slight hyperdensity in the gluteus maximus in an elderly patient status post fall on the hip. CT can demonstrates acute hematoma if careful attention is paid to the subtle increased attenuation usually present in the blood collection. As the hematoma ages, its density decreases.* ***(Right)*** *Axial T1WI MR from the same patient on the same day shows the hematoma has signal intensity similar to that of normal muscle.*

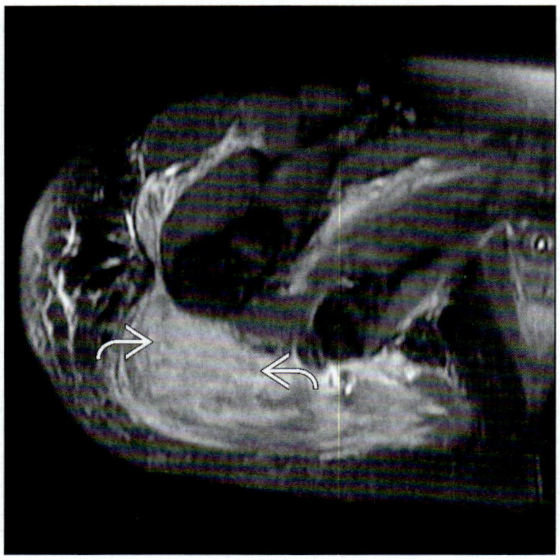
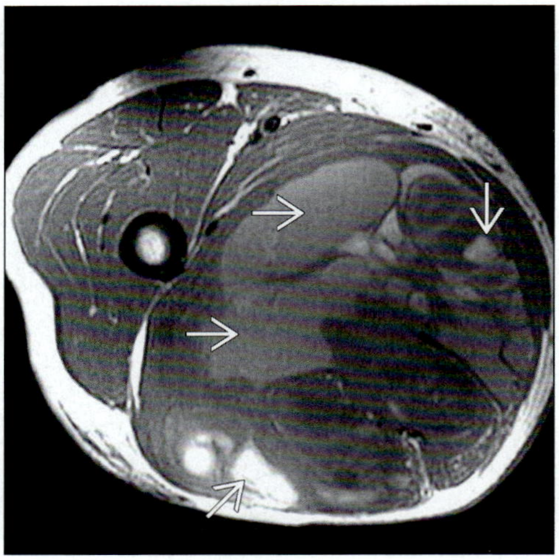

(Left) *Axial T2WI FS MR from the same patient shows high signal in the hematoma. In hyperacute cases such as this, blood products usually demonstrate signal characteristics similar to those of water.* ***(Right)*** *Axial T1WI MR from a 68-year-old man with sudden onset of pain and swelling in the posterior thigh demonstrates a large heterogeneous mass. Areas of high signal within the mass suggest blood products or fat content.*

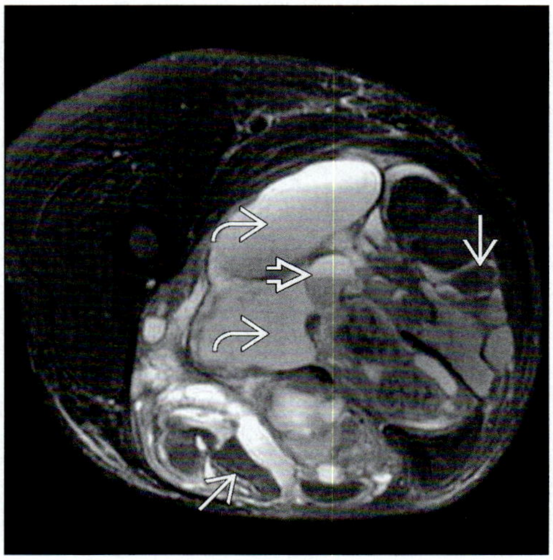
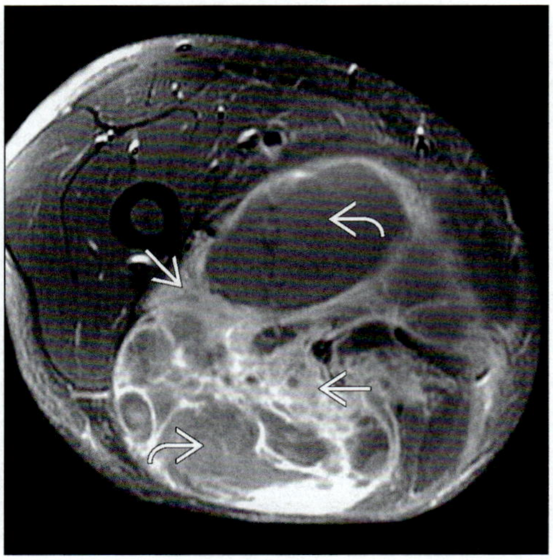

(Left) *Axial T2WI FS MR from the same patient shows low signal in some of the high signal T1 regions, which could be blood or fat. Persistent high signal in other areas more likely represents blood. There are also fluid-fluid levels.* ***(Right)*** *Axial T1WI C+ FS MR from the same patient shows extensive irregular enhancement throughout the mass between areas of nonenhancing blood. This was a high-grade sarcoma that spontaneously bled. Bland hematomas do not enhance.*

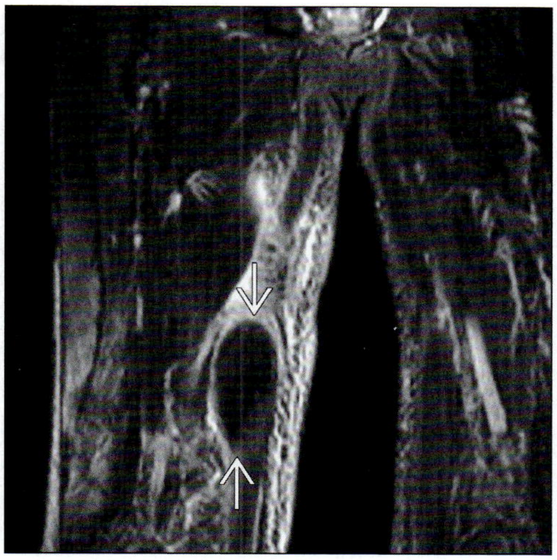

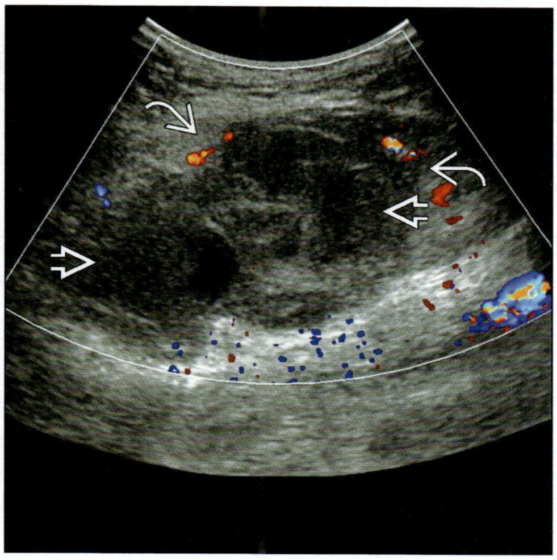

(Left) *Coronal T2WI FS MR of the upper medial thigh in a patient with swelling and pain here after trauma shows a rounded mass → in the medial subcutaneous tissues surrounded by soft tissue edema. The low signal within the mass suggests hematoma.* ***(Right)*** *Longitudinal oblique ultrasound from the same patient using color Doppler technique shows mild hyperemia → but no flow within the collection itself →. Simple hematomas should demonstrate no significant internal blood flow.*

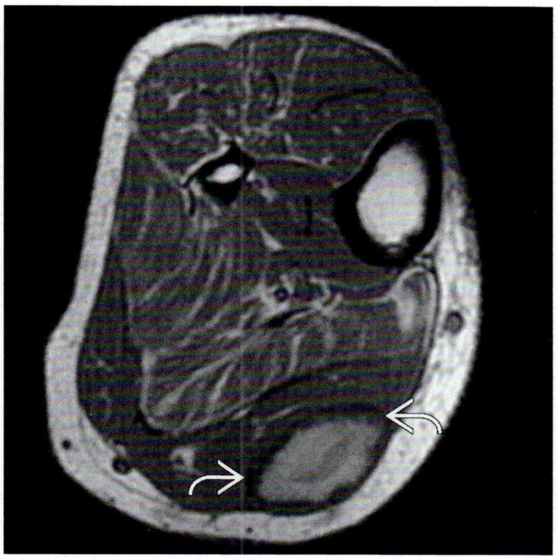

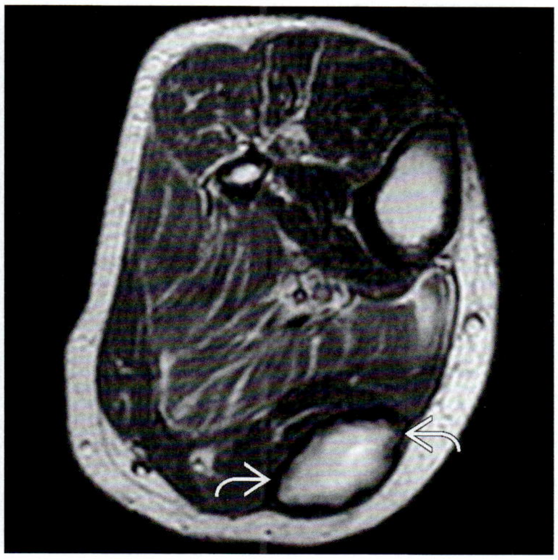

(Left) *Axial T1WI MR shows an ovoid mass in the gastrocnemius muscle with high signal centrally and a low signal rim →. This appearance is virtually pathognomonic for subacute or chronic hematoma; the low signal rim indicates hemosiderin formation.* ***(Right)*** *Axial T2WI MR from the same patient shows persistent high signal in the central aspect of the mass and persistent low signal in the rim →, consistent with methemoglobin centrally and hemosiderin peripherally.*

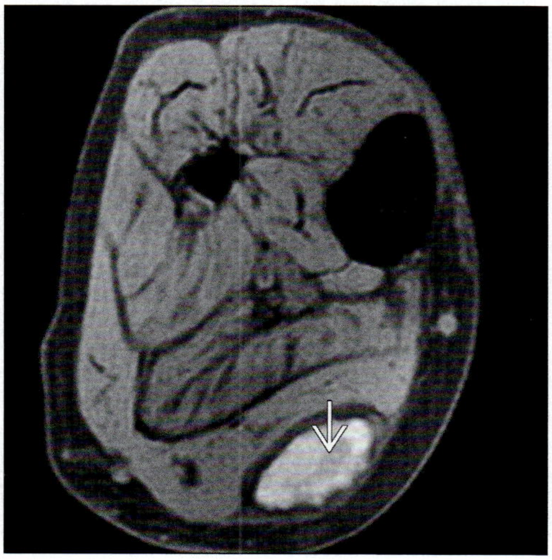

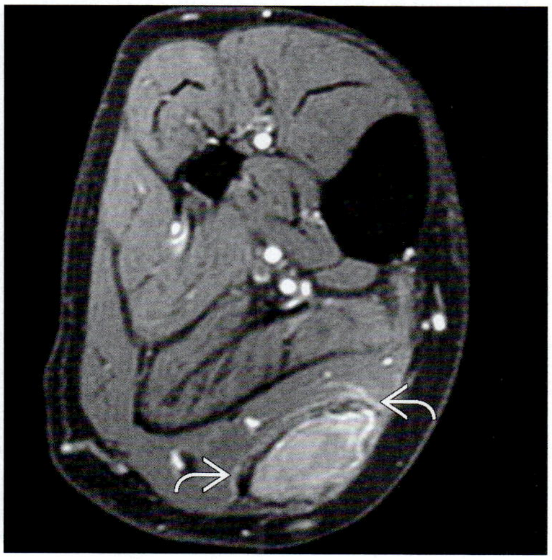

(Left) *Axial T1WI FS MR from the same patient shows the high signal methemoglobin in the central aspect of the hematoma →. In some cases, subtle elevated T1 signal may be better seen on T1-weighted images with fat saturation.* ***(Right)*** *Axial T1WI C+ FS MR from the same patient shows minimal enhancement of the rim of the hematoma →. Central enhancement is difficult to exclude because of the inherent high signal in the mass.*

FOREIGN BODY

Key Facts

Terminology

- Foreign body (FB)

Imaging

- Object (e.g., bullet, nail, staple) may be recognizable based on radiographic appearance
- Radiographs will detect 80% of all foreign bodies
- Only 15% of wooden foreign bodies are visible on radiography
  - Wood is initially low density due to air content
  - Chronically retained wood becomes hyperdense due to absorption of blood products and exudate
- T1WI C+ FS MR useful in evaluating FB complications
  - FB more conspicuous as central defect in enhancing reactive tissue
  - Can be useful in identifying small FBs on MR; FB may be easily overlooked on routine MR
- High frequency US transducers (↑ resolution, ↓ penetration) > sensitivity for FBs
- All foreign bodies are initially hyperechoic
- Wooden foreign bodies become less echogenic over time
- Plain radiographs in multiple views to localize FB
- Ultrasound next step if negative radiograph

Top Differential Diagnoses

- Film-screen artifact
- Surgically implanted object

Clinical Issues

- 38% of FBs missed by initial treating physician
- FB may be left in place if it is inert, asymptomatic, not a threat to function, ± difficult to remove due to small size
- Should be removed if reactive, large, concerning to patient, or near vital structures

*(**Left**) Lateral radiograph shows a large metal nail ➡ overlying the talus. A single radiographic projection cannot determine the anatomic location of a foreign body in 3D tissue; at least 1 more view is required to localize the object. (**Right**) AP radiograph from the same patient shows the nail → directly superimposed over the talus. The information on both images together confirms that the nail is embedded in the talus and not simply in the adjacent soft tissues.*

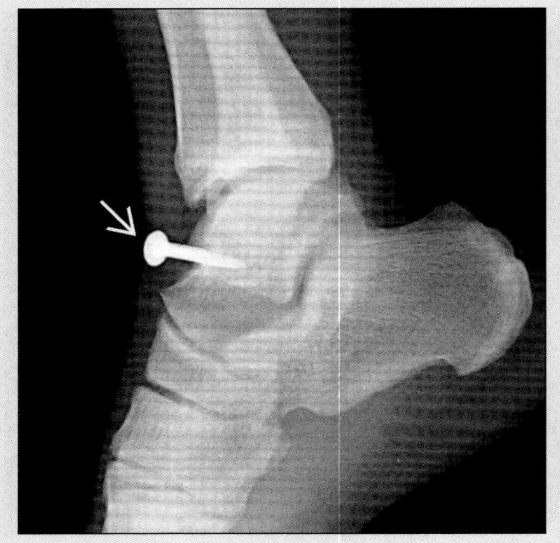

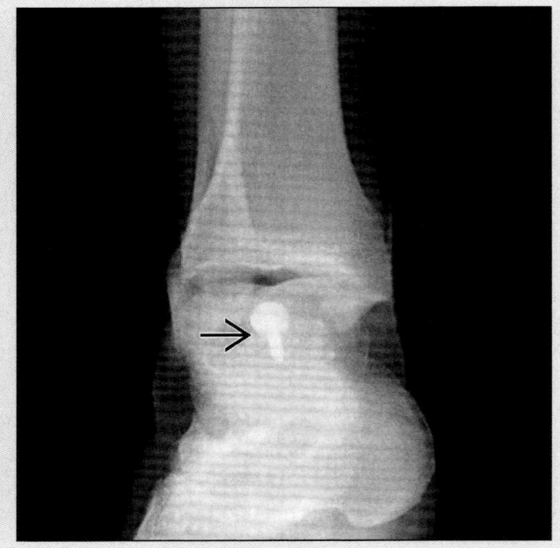

*(**Left**) Coronal T1WI C+ FS MR of the forefoot shows a rectangular focus of signal void ⇩ in the plantar soft tissues surrounded by mildly enhancing reactive tissue. This is a piece of wood. Fresh wood contains air and often appears as signal void on MR images. (**Right**) Sagittal MR arthrogram T1WI of the wrist after gadolinium arthrogram → shows a large round signal void ↪ creating marked magnetic field distortion. This was a tiny metal fragment in the patient's hand from previous trauma.*

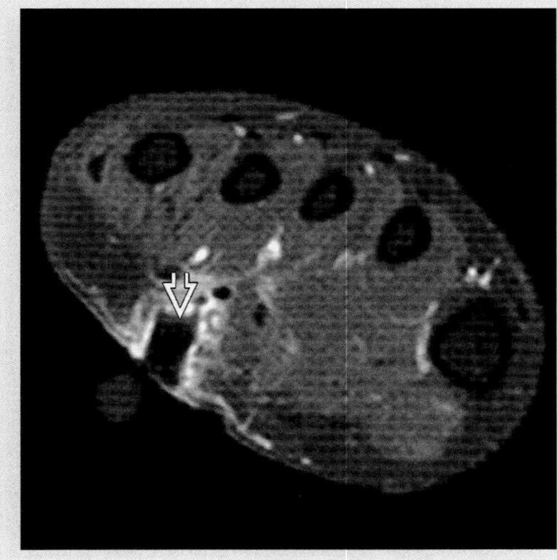

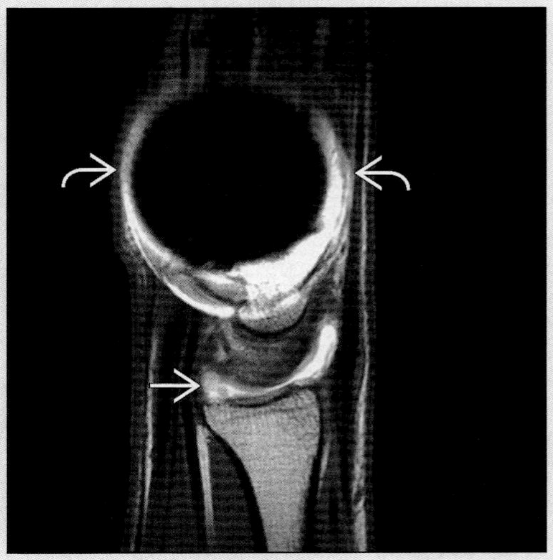

FOREIGN BODY

TERMINOLOGY

Abbreviations

- Foreign body (FB)

Definitions

- Object of external origin that is retained within tissues of body

IMAGING

General Features

- Best diagnostic clue
  - Radiodensity on plain film with contours more characteristic of artificial than of natural material
- Location
  - Most common in hands but can present anywhere
- Morphology
  - Object (e.g., bullet, nail, staple) may be recognizable based on radiographic appearance

Radiographic Findings

- Radiographs will detect 80% of all foreign bodies
- Metal and glass (even nonleaded) are radiopaque relative to soft tissues
  - It is a myth that only leaded glass can be detected by radiography
- Wood and plastic are often isodense to muscle/water and not detected by plain film
  - Only 15% of wooden foreign bodies are visible on radiography
  - Large piece of porous wood may contain sufficient air to create lucent defect in soft tissues

CT Findings

- Metal and glass are hyperdense
  - Metal may cause significant streak artifacts
- Wood is initially low density due to air content
  - Chronically retained wood becomes hyperdense due to absorption of blood products and exudate

MR Findings

- Usually not necessary in acute setting
  - Useful for planning surgical approach
- T1WI FS C+ MR useful in evaluating FB complications
  - Inflammatory reaction
    - FB more conspicuous as central defect in enhancing reactive tissue
    - Can be useful in identifying small FBs on MR
  - Abscess
  - Osteomyelitis
  - Granuloma
  - Scar formation
- Metal FB → susceptibility artifact
- Wood and glass will create geometric areas of signal void but no susceptibility artifact
  - Wood is hypointense to skeletal muscle on both T1WI and T2WI sequences
- Should **not** be used in evaluating metallic FBs in critical locations, such as eye or CNS
  - Ferromagnetic materials may be displaced by magnetic field with catastrophic results

Ultrasonographic Findings

- High frequency transducers (↑ resolution, ↓ penetration) are more sensitive than those of lower frequency
  - Operator competence is limiting factor in clinical setting
- All foreign bodies are initially hyperechoic
  - Wooden foreign bodies become less echogenic over time
    - 87% sensitivity and 97% specificity for wooden foreign bodies as small as 2.5 mm
- ↑ conspicuity due to hypoechoic halo of granulation tissue, edema, or hemorrhage
  - Halo usually develops > 24 hours after FB entry
- Sonographic artifacts can aid in identification
  - Objects with rough surface (i.e., wood) have "clean" posterior acoustic shadowing
  - Objects with smooth surface (i.e., glass, metal) have "dirty" shadowing and reverberation artifacts
  - Plastic foreign bodies are less echogenic but have marked posterior shadowing

Imaging Recommendations

- Best imaging tool
  - Plain radiographs in multiple views to localize FB
  - Ultrasound next step if negative radiograph
    - Intraoperative sonographic guidance can also be used during surgical removal
- Protocol advice
  - "Soft tissue" technique radiographs (i.e., lower kV and higher mA) may be more sensitive for small nonmetallic FBs

DIFFERENTIAL DIAGNOSIS

Film-Screen Artifact

- Will not triangulate to constant anatomic location
- Less of an issue with modern DR systems

Surgically Implanted Object

- Surgical suture, clip, or staple

PATHOLOGY

General Features

- Etiology
  - Extremity foreign bodies are almost exclusively due to penetrating trauma
  - Paint and grease guns cause numerous digital injuries annually
    - Index finger is most commonly involved since it is placed over nozzle opening
    - Pressure injected contents dissect a path through tissue planes
      - Path commonly follows flexor tendon sheaths into palm
    - Oil or paint thinner causes radiolucent stripe
    - Paint is radiopaque if it contains lead or other heavy metals
- Associated abnormalities
  - Fractures, vascular injury/hemorrhage, or nerve injury from penetrating trauma

FOREIGN BODY

Gross Pathologic & Surgical Features

- Wood: 36%
- Glass: 23%
- Metal: 20%
- Other (plastic, gravel, etc.): 21%

CLINICAL ISSUES

Presentation

- Most common signs/symptoms
  - Pain &/or mass
  - Occasionally asymptomatic
    - Incidental finding on study performed for another reason
- ± history of penetrating trauma given by patient
- Missed FBs are 1 of leading causes of malpractice claims against ED physicians
  - 38% of FBs missed by initial treating physician
  - 13% present with complication
- Average time for FB retention is 7 months
  - 43% removed within 1 week
  - 11% retained longer than 1 year

Demographics

- Epidemiology
  - 40% of FBs in hand (most common site) are work-related

Natural History & Prognosis

- Can travel locally
- Complications
  - Infection
  - Granuloma formation
  - Neurovascular injury due to local inflammatory reaction
  - Can enter blood vessel and travel to heart ± pulmonary circulation

Treatment

- FB may be left in place if it is inert, asymptomatic, not a threat to function, ± difficult to remove due to small size
- Should be removed if reactive, large, concerning to patient, or near vital structures

DIAGNOSTIC CHECKLIST

Consider

- Old episode of penetrating trauma

Image Interpretation Pearls

- Triangulate location of FB on at least 2 radiographic views

Reporting Tips

- Measure size of object
- Describe approximate depth from skin surface

SELECTED REFERENCES

1. Aras MH et al: Comparison of the sensitivity for detecting foreign bodies among conventional plain radiography, computed tomography and ultrasonography. Dentomaxillofac Radiol. 39(2):72-8, 2010
2. Ando A et al: Imaging features of foreign body granuloma in the lower extremities mimicking a soft tissue neoplasm. Ups J Med Sci. 114(1):46-51, 2009
3. Creel SA et al: Sonographic surface localization of subcutaneous foreign bodies and masses. J Clin Ultrasound. 37(3):158-60, 2009
4. Mamede RC et al: Do foreign bodies migrate through the body towards the heart? Braz J Otorhinolaryngol. 75(2):195-9, 2009
5. Abu Hassan FO: Retained toothpick causing pseudotumor of the first metatarsal: a case report and literature review. Foot Ankle Surg. 14(1):32-5, 2008
6. Ozsarac M et al: Glass foreign body in soft tissue: possibility of high morbidity due to delayed migration. J Emerg Med. Epub ahead of print, 2008
7. Visvanathan V et al: Ultrasound-assisted removal of an airgun pellet from the hand. J Plast Reconstr Aesthet Surg. 61(2):234-5, 2008
8. Tagliafico A et al: Wrist and hand ultrasound. Semin Musculoskelet Radiol. 11(2):95-104, 2007
9. Muehleman C et al: Multiple-image radiography for human soft tissue. J Anat. 208(1):115-24, 2006
10. Turkcuer I et al: Do we really need plain and soft-tissue radiographies to detect radiolucent foreign bodies in the ED? Am J Emerg Med. 24(7):763-8, 2006
11. Crankson S et al: Ultrasound in the diagnosis and treatment of wooden foreign bodies in the foot. Ann Saudi Med. 24(6):480-1, 2004
12. Dumarey A et al: Large wooden foreign body in the hand: recognition of occult fragments with ultrasound. Emerg Radiol. 10(6):337-9, 2004
13. Hung YT et al: Ultrasound for the detection of vegetative foreign body in hand--a case report. Hand Surg. 9(1):83-7, 2004
14. Dean AJ et al: Technique for emergency medicine bedside ultrasound identification of a radiolucent foreign body. J Emerg Med. 24(3):303-8, 2003
15. Graham DD Jr: Ultrasound in the emergency department: detection of wooden foreign bodies in the soft tissues. J Emerg Med. 22(1):75-9, 2002
16. Peterson JJ et al: Wooden foreign bodies: imaging appearance. AJR Am J Roentgenol. 178(3):557-62, 2002
17. Horton LK et al: Sonography and radiography of soft-tissue foreign bodies. AJR Am J Roentgenol. 176(5):1155-9, 2001
18. Jacobson JA et al: Wooden foreign bodies in soft tissue: detection at US. Radiology. 206(1):45-8, 1998
19. Kornreich L et al: Preoperative localization of a foreign body by magnetic resonance imaging. Eur J Radiol. 26(3):309-11, 1998
20. Leung A et al: Intraoperative sonography-guided removal of radiolucent foreign bodies. J Pediatr Orthop. 18(2):259-61, 1998
21. Grechenig W et al: [Ultrasound imaging of foreign bodies--an experimental study] Biomed Tech (Berl). 41(11):308-15, 1996
22. Manthey DE et al: Ultrasound versus radiography in the detection of soft-tissue foreign bodies. Ann Emerg Med. 28(1):7-9, 1996
23. Gooding GA: Foreign bodies. Clin Diagn Ultrasound. 30:99-111, 1995
24. Monu JU et al: Soft-tissue masses caused by long-standing foreign bodies in the extremities: MR imaging findings. AJR Am J Roentgenol. 165(2):395-7, 1995
25. Dupuy DE et al: MR imaging of a patient with a ferromagnetic foreign body. AJR Am J Roentgenol. 160(4):893, 1993
26. Donaldson JS: Radiographic imaging of foreign bodies in the hand. Hand Clin. 7(1):125-34, 1991
27. Bodne D et al: Imaging foreign glass and wooden bodies of the extremities with CT and MR. J Comput Assist Tomogr. 12(4):608-11, 1988

FOREIGN BODY

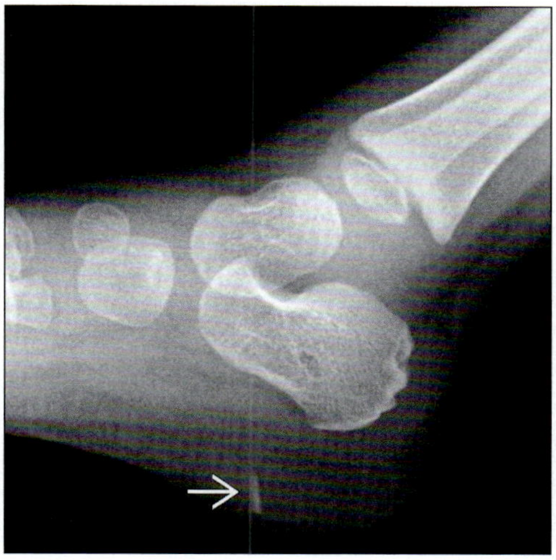

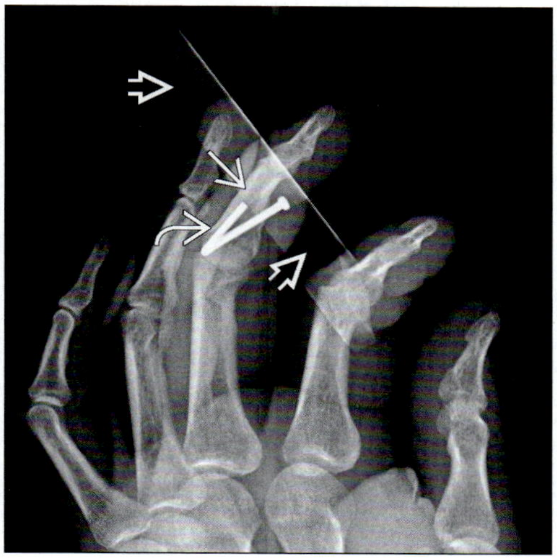

(Left) *Lateral radiograph shows an elongate foreign body in the plantar soft tissues of the heel of a child who stepped on glass. Glass is usually well seen on radiographs using soft tissue technique.* ***(Right)*** *Oblique radiograph shows a bent metal nail passing through a wood block and into the middle phalanx of the 3rd finger. Note the very low density of the wood, which contains a significant amount of air; small wooden bodies may be difficult to detect on routine radiographs.*

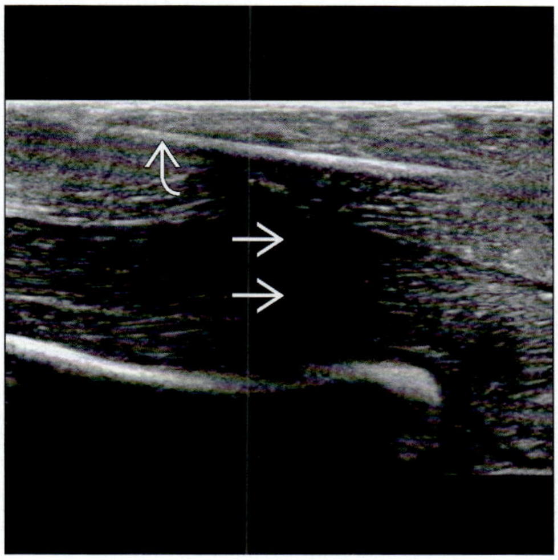

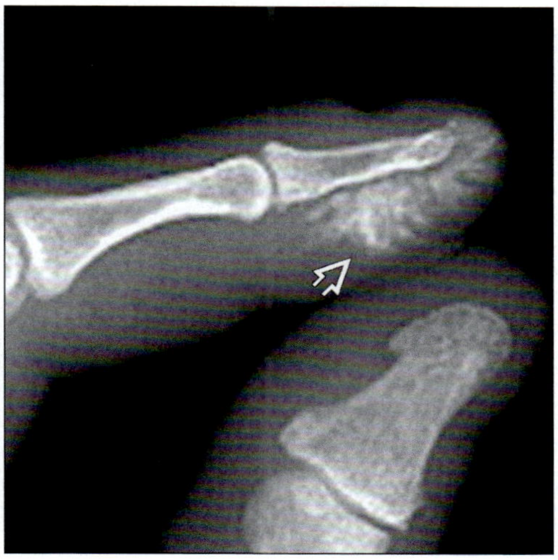

(Left) *Ultrasound shows a wooden toothpick embedded in the muscle as a hyperechoic focus with complete posterior shadowing. Note that a linear, high frequency (12 MHz) transducer was used.* ***(Right)*** *Lateral radiograph shows a radiodense foreign material in the volar soft tissues of the index finger. Paint gun injuries in which there is skin penetration have this typical appearance due to the lead in the paint. The pain may dissect along tissue planes over large distances.*

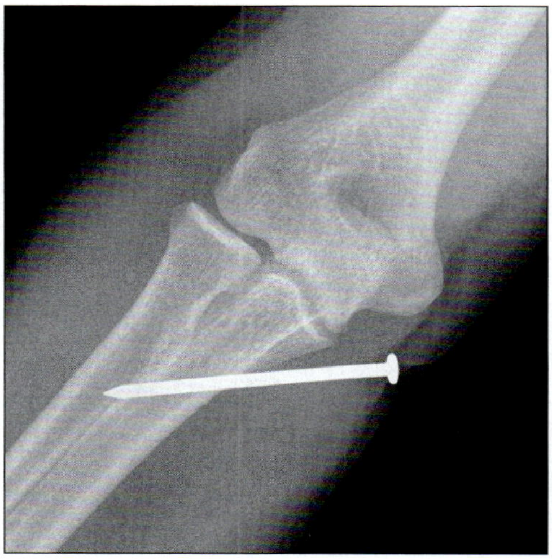

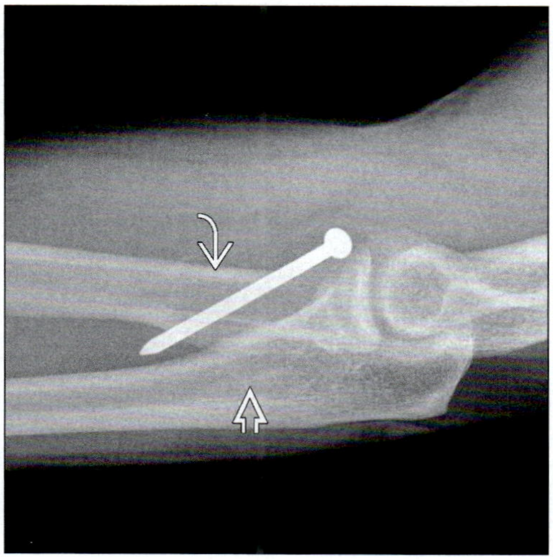

(Left) *AP radiograph of the elbow demonstrates a large metal nail overlying the proximal medial forearm. Its depth and relationship to the forearm bones cannot be accurately determined in this plane alone.* ***(Right)*** *Lateral radiograph from the same patient shows that the nail passes anterior to the ulna and posterior to the radius, missing both bones and traveling between them. Surgical extraction of foreign bodies is more easily planned with specific knowledge of their anatomic location.*

FOREIGN BODY

*(**Left**) Lateral radiograph shows a metal needle fragment ➔ overlying the calcaneus. Based on this single projection, the needle may be within the bone or medial or lateral to it. (**Right**) Axial oblique radiograph from the same patient was performed to localize the metal fragment. The needle ➔ can be seen projecting lateral to the lateral border of the calcaneus ➔ and is therefore in the adjacent soft tissues. Ultrasound is also useful in this setting.*

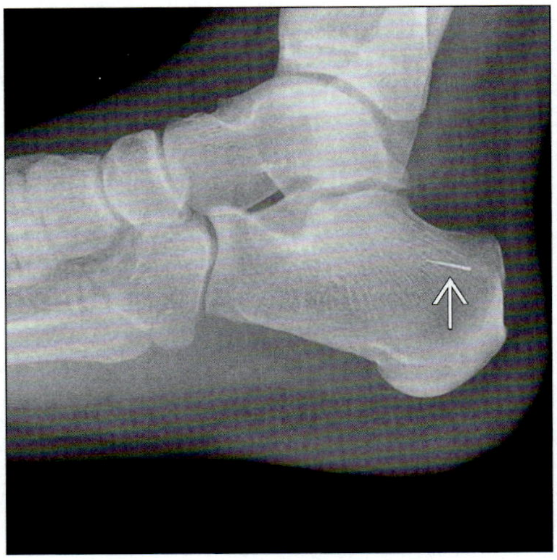

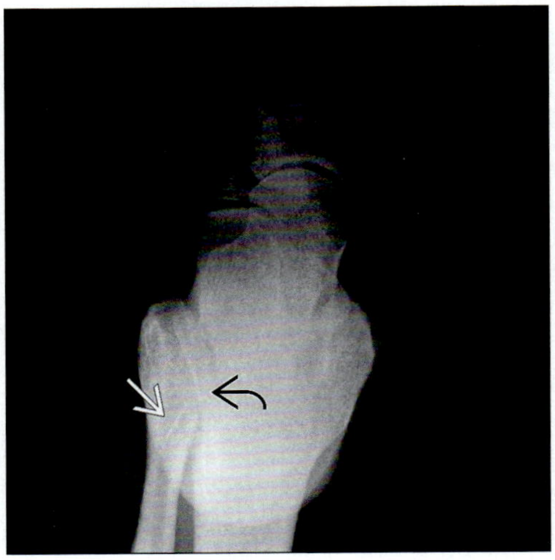

*(**Left**) Sagittal T2W FS MR serial images from lateral to medial in the foot of an 8-year-old girl who stepped on a toothpick several weeks ago shows an elongate focus of signal void ➔ representing the wooden toothpick surrounded by abscess fluid. Marrow edema ➔ in the 3rd metatarsal represents osteomyelitis. (**Right**) Longitudinal oblique T2WI FS MR serial images from the same patient show the oblique course of wood fragment ➔ through the foot and osteomyelitis of the 3rd metatarsal ➔.*

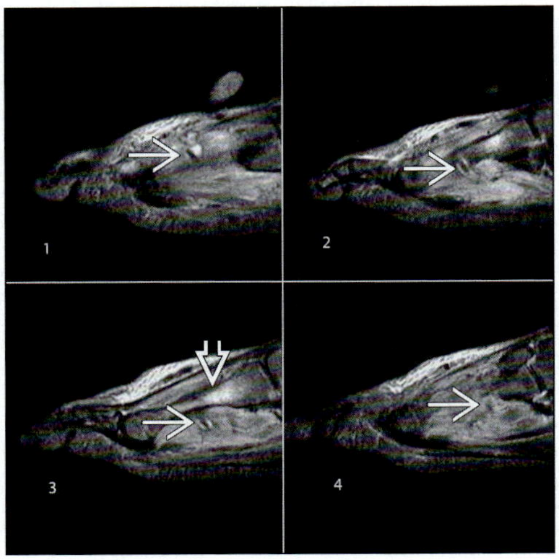

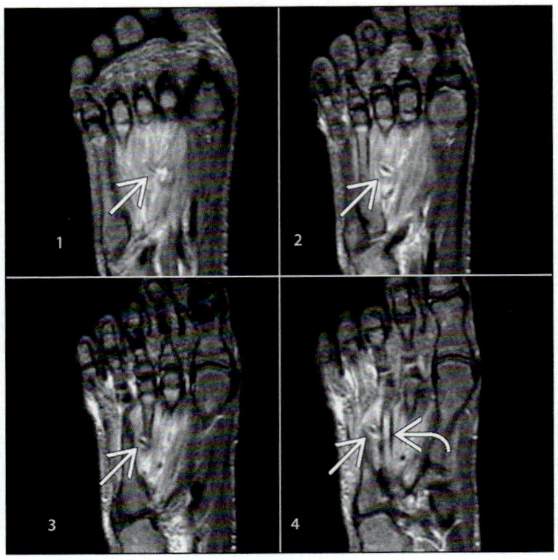

*(**Left**) AP radiograph demonstrates a metal staple in the tip of the finger. Some foreign bodies are easily characterized by their shape on radiographs; others may require multiple projections to assess their 3D nature. (**Right**) AP radiograph during surgery for placement of a prosthetic hip demonstrates a broken drill bit ➔ embedded in the bone during the procedure. Intraoperative radiographs are often used to identify and localize retained surgical instruments and materials.*

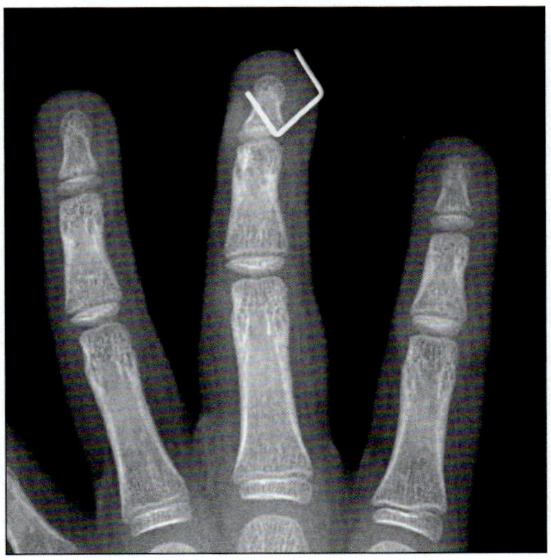

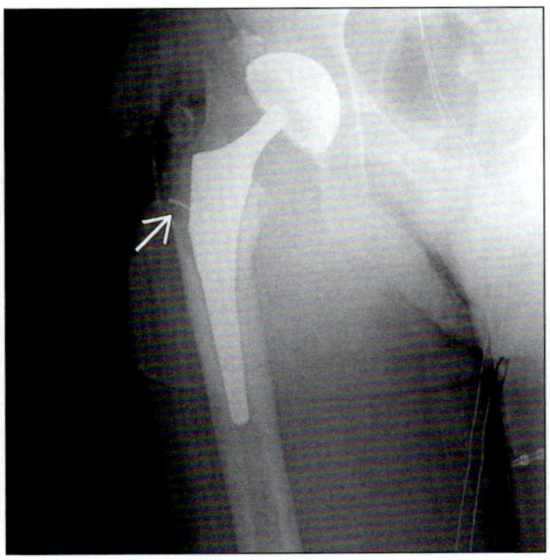

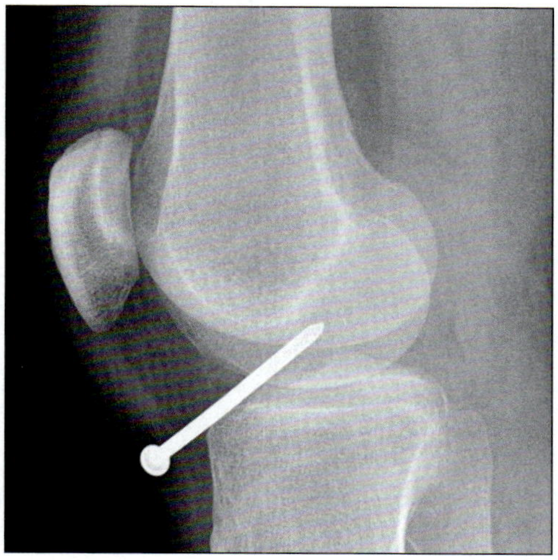
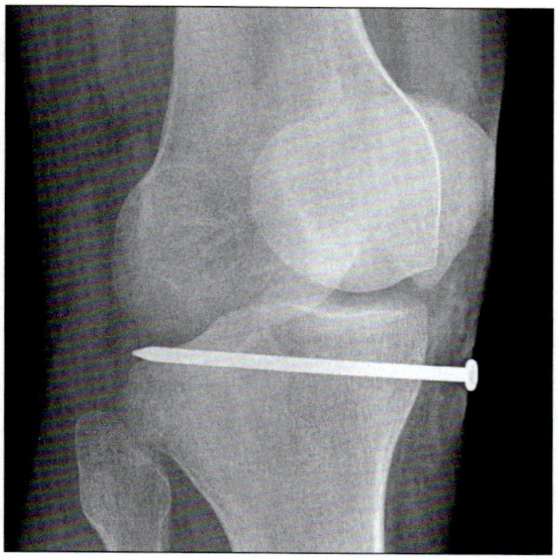

***(Left)** Lateral radiograph shows a large metal nail projecting over the knee. From this single projection, it is impossible to discern whether the nail penetrates into the femoral condyles or is in the knee joint. **(Right)** Oblique radiograph from the same patient suggests that the nail may be intraarticular, but the lack of overlap of the nail with the femoral condyles excludes the possibility of penetration into the distal femur.*

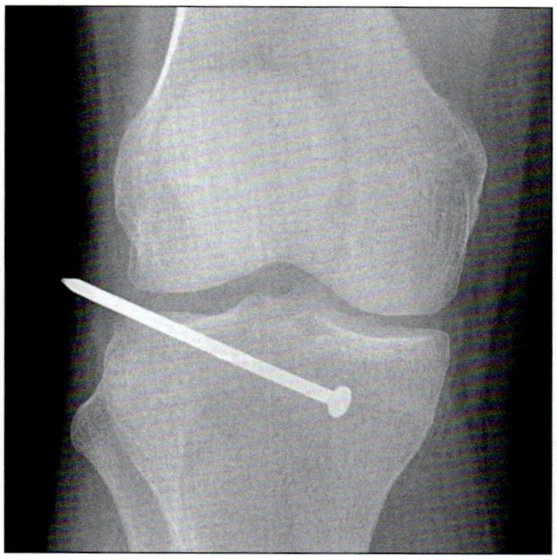
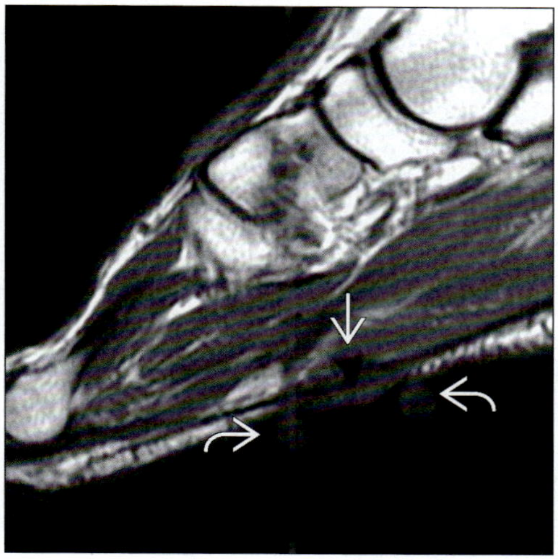

***(Left)** AP radiograph from the same patient shows that the nail's path is through extraarticular lateral soft tissues. The lack of intraarticular gas on this study is also a clue that there is no entry of the object into the joint. **(Right)** Sagittal T1WI MR from a patient who stepped on glass shows a triangular signal void ➔ between the 2 skin markers ↷. Radiographs showed no foreign body. It can be difficult to identify foreign material on T1WI without the benefit of intravenous contrast.*

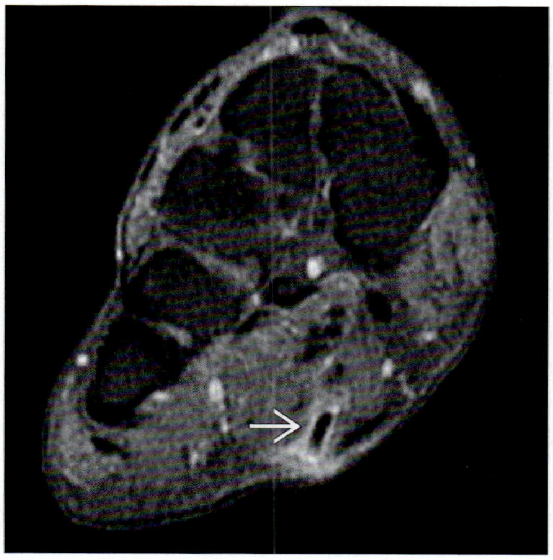
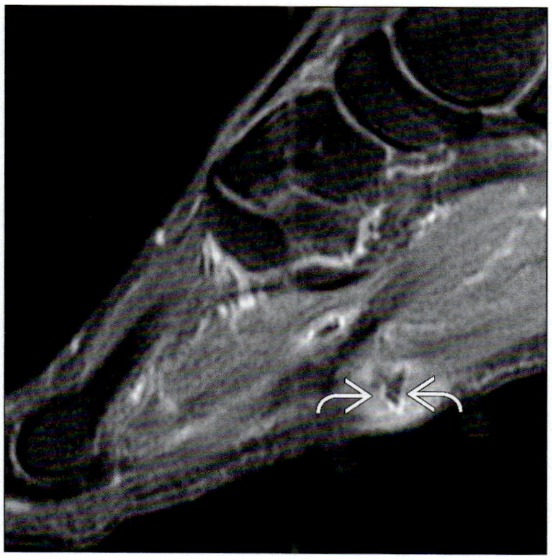

***(Left)** Coronal PDWI FS MR from the same patient demonstrates a rectangular signal void ➔ in the plantar soft tissues of the foot surrounded by mild reactive edema. Soft tissue reaction around an embedded object may help identify the object on fluid-sensitive sequences, but may not be present for several days after injury. **(Right)** Sagittal T1 C+ FS MR from the same patient shows post-contrast enhancement of reactive granulation tissue ↷ around the fragment, making it much more conspicuous.*

SECTION 2
Shoulder and Humerus

Introduction and Overview

Osseous Pathology

Musculotendinous Pathology

Shoulder Girdle

Rotator Cuff

Proximal Biceps Tendon

Ligamentous Pathology

Instability and Labrum

Superior Labrum

Combined Cuff and Labral Lesions

Neurogenic Pathology

SHOULDER AND HUMERUS OVERVIEW

Overview

The shoulder is the 2nd most commonly imaged peripheral joint by MR, exceeded only by the knee in most practices. Shoulder MR imaging is ordered regularly both because shoulder pain is prevalent (representing the 3rd most common complaint by patients after headache and back pain) and because shoulder pathology is often undetectable on radiographs but well demonstrated on MR images. This section will review the MR appearance of trauma to the rotator cuff, biceps tendon, glenoid labrum, and articular cartilage. In addition, the imaging findings of fractures and dislocations of the shoulder girdle will be presented.

Terminology and Conventions

The use of acronyms to describe labroligamentous pathology in the shoulder has become prevalent; memorizing the constantly expanding lexicon assigned to these lesions can be challenging even to the seasoned shoulder imaging expert. While terms such as "ALPSA" (anterior labroligamentous periosteal sleeve avulsion) and "GLAD" (glenoid labrum articular disruption) are described and illustrated, more straightforward terms have been applied where applicable.

As elsewhere in this text, the term "tendinopathy" is preferred to "tendinosis" or "tendinitis" to describe tendinous microtrauma and degeneration. Although the surgical literature still refers to rotator cuff hydroxyapatite deposition disease as "calcific tendinitis," the more pathologically accurate term "calcific tendinopathy" is used. In addition, there is some controversy in the literature over the terms "intrasubstance" or "interstitial" tear when describing certain appearances of rotator cuff tear. These lesions are discussed both in the "Partial Thickness Tear" and "Tendinopathy" chapters because of ongoing controversy regarding the clinical significance of these tears relative to those that involve the surface of the rotator cuff tendons.

The terminology describing rotator cuff tears is often unique. Injuries to almost all other tendons in the body are described in terms of a 3-level system: Grade 1 (strain), grade 2 (partial tear), and grade 3 (complete tear). Because the supraspinatus, infraspinatus, and teres minor tendons are flat and fuse with each other, injury to the tendons of the rotator cuff are instead described in terms of "fraying" (irregularity of the tendon surface), "partial thickness tear," "full thickness tear," and "large" or "massive" full thickness tear (referring to those involving the entire width of one or more of the tendons).

Because the location of a glenoid labral tear often determines the symptoms and prognosis, labral tear classification schemes are typically based on location. Categories of labral pathologies are described as being in 1 of 3 regions: Anterior, posterior, or superior, or as bridging more than 1 region (e.g., "posterosuperior"). Surgeons describe labral tear location using a clock-face scheme in which 12:00 refers to the position of the biceps anchor, 3:00 is defined as the anterior mid-glenoid, and 6:00 corresponds to the inferior margin of the glenoid. Some authors reverse this, however, depending on whether the right or left shoulder is involved.

Although MR and CT imaging sequences are generally prescribed in oblique planes reflecting the angle of the scapula relative to the body, it is conventional to refer to these oblique imaging planes as "coronal" and "sagittal" in discussion of the shoulder.

Anatomic Considerations

The glenohumeral articulation is the most mobile joint in the body. It is a ball-and-socket joint but with a shallow socket (the glenoid fossa) allowing a wide range of motion. The coracoacromial arch provides a superior osseous restraint to humeral motion superiorly, but the shoulder is vulnerable to large anterior, posterior, or inferiorly directed forces. Glenohumeral stability is maintained by the

- Glenoid labrum, which deepens the glenoid fossa and is the attachment site of the glenohumeral ligaments
- Glenohumeral ligaments, which are focal thickenings of the joint capsule
- Muscles and tendons that cross the joint and act as dynamic stabilizers to maintain congruity of the humeral head in the glenoid fossa
- Adhesion-cohesion forces (the "suction cup effect")

The glenoid labrum is a triangular structure in cross section that extends around the periphery of the osseous glenoid rim. The labrum is made of fibrous tissue except for a small fibrocartilage transition zone at its attachment to the hyaline cartilage of the glenoid. Because the labrum is fibrous, its shape can be variable on MR images and can have a triangular, rounded, or notched appearance. Although the labrum acts as a bumper to deepen the glenoid socket, its main role is as a continuation of the joint capsule and the insertion site of the glenohumeral ligaments and long head of the biceps tendon.

The main muscles of clinical concern in the shoulder are the muscles of the rotator cuff: The supraspinatus, infraspinatus, teres minor, and subscapularis. The supraspinatus tendon inserts on the superior facet of the greater humeral tuberosity, while the infraspinatus and teres minor tendons insert along the posterior aspect of the tuberosity. The subscapularis tendon inserts on the lesser humeral tuberosity. These tendons form a "cuff" of continuous tissue in their distal 1.5 cm that act to abduct and internally and externally rotate the humeral head and serve as active stabilizers of the glenohumeral joint.

The biceps long head tendon also crosses the glenohumeral joint and acts as a dynamic stabilizer of the joint as well as a forward flexor. The tendon ascends in the bicipital groove between the lesser tuberosity and the anterior aspect of the greater tuberosity before exiting the groove and curving medially within the biceps pulley sling. The sling is composed of the fused lateral portions of the coracohumeral and superior glenohumeral ligaments. When the tendon exits the pulley sling it is oriented horizontally and becomes intraarticular. The tendon has a variable insertion onto the superior labrum and supraglenoid tubercle, but it may insert directly onto the joint capsule, or may be absent.

The intraarticular portion of the humeral head is predominantly covered by hyaline cartilage, except for the "bare area" posteriorly, which can mimic a Hill-Sachs lesion on more caudal axial CT or MR sections. The glenoid fossa has a "central bare spot" that should not be confused with a pathologic focal chondral defect.

Pathologic Considerations

Shoulder injuries can occur from a variety of causes, including acute trauma, overuse, or age-related injuries. Fractures around the shoulder are common after trauma and should be carefully searched for on trauma chest radiographs. Dislocation is frequent both at the glenohumeral joint and at the acromioclavicular joint, the latter often called a "separated shoulder." Overuse injuries can cause internal impingement pain and decreased throwing velocity in overhead throwing athletes. Rotator cuff tears are usually the end result of weakening of the tendon due to age-related tendinopathy, but they occasionally occur after severe trauma.

Several commonly encountered normal variants occur around the shoulder, including the os acromiale and sublabral foramen, that are usually asymptomatic. Conversely, rotator cuff tears may be clinically asymptomatic; the tendon has few nerve fibers, and most of the pain associated with cuff pathology likely arises from synovitis or mechanical catching of redundant synovium by the tear.

Imaging Considerations

A variety of radiographic views are used for imaging the shoulder. The standard series often includes an anteroposterior (AP) internal rotation, Neer AP (or Grashey) view with external rotation, and the axillary view. Some prefer an AP external rotation view, while others obtain a Zanca AP view, which has 10-20° of cephalad tilt to better visualize the acromioclavicular joint. A scapular-Y view may be obtained for certain fractures or in patients who cannot abduct the arm for an axillary view. The arch view is useful for determining the shape of the anterior acromion. The West Point view profiles the anteroinferior glenoid rim and is therefore more sensitive for detection of Bankart fractures. Multiple additional specialty views, including the Stryker notch and Rockwood views, may be helpful for profiling specific shoulder pathology.

CT scans are helpful for characterizing complex scapular and proximal humeral fractures, especially for assessing involvement of the articular surfaces. Ultrasound is increasingly used in the shoulder to evaluate the rotator cuff and biceps tendons and to guide therapeutic injections. Most practices no longer perform shoulder arthrography except as a precursor to MR or CT arthrography. Although full thickness rotator cuff tears can be diagnosed with conventional arthrography, tear size and morphology are better evaluated with an accompanying MR or CT exam.

MR imaging is performed with the patient prone and the humerus comfortably externally rotated. The main challenges associated with shoulder MR are motion artifact and poor signal-to-noise ratio. The anatomy of the shoulder prevents shoulder coils from either rigidly immobilizing the joint (to prevent motion artifact from breathing) or encircling the joint (to provide uniform signal intensity across the imaging volume). Image quality is generally better on 3 Tesla scanners and with the more rigid 8 channel or higher shoulder coils.

Imaging Protocols

No single preferred MR protocol exists for imaging the shoulder, but most radiologists agree with several basic principles. Because the scapula is angled 30° anteriorly relative to the body, the standard imaging planes are oriented relative to the glenoid articular surface and the supraspinatus tendon. The oblique coronal plane is oriented perpendicular to the glenoid fossa (or parallel to the long axis of the supraspinatus tendon). The oblique sagittal plane is oriented parallel to the glenoid face (or orthogonal to the oblique coronal plane). Axial images are usually acquired without anatomic angulation.

Two other slice positioning techniques can sometimes be helpful and deserve mention. Because the scapula is tilted anteriorly, angling the axial images perpendicular to the true vertical axis of the glenoid fossa better profiles the 3:00 and 9:00 positions of the labrum and reduces the partial volume averaging of anteroinferior and posterosuperior labral tears. Adding angled oblique sagittal images perpendicular to the distal-most fibers of the supraspinatus tendon can show better the tendon tears that commonly occur in this region.

Because labral tears can be difficult to see with conventional MR technique, direct MR arthrography is often used in the setting of suspected instability lesions. MR arthrography has higher accuracy for delineation of labroligamentous pathology compared to routine shoulder MR imaging due to joint distension and increased contrast-to-noise afforded by the injectate. Abduction external rotation (ABER) images can further improve the accuracy of MR arthrography for certain pathology. The ABER position relaxes the posterosuperior rotator cuff so small internal impingement tears and delamination tears are more apparent. ABER images are perpendicular to the anteroinferior and posterosuperior labrum so there is less partial averaging of these structures, and Bankart-type tears (especially nondisplaced Perthes lesions) and posterosuperior internal impingement labral tears are more easily seen. Finally, the ABER position pulls the anterior band inferior glenohumeral ligament taut, which can distract and therefore make more apparent small Bankart tears. Indirect MR arthrography can also improve sensitivity over conventional MR for cuff and labral pathology.

Multiple MR pulse sequences have been advocated for imaging of the shoulder. High accuracy for cuff tears has been reported with fat-suppressed fast spin-echo T2-weighted sequences in the oblique coronal and sagittal planes, often with a TE of 45-55 msec at 1.5 Tesla. It is also helpful to obtain at least one T1- or proton density-weighted sequence without fat suppression to evaluate for fatty muscle atrophy. The sequences used for axial images are more variable, with some preferring a fat-suppressed intermediate-weighted sequence (TE = 30-40 msec at 1.5T), while others obtain T2*-weighted gradient echo &/or proton density-weighted sequences. The choice of sequences generally depends on personal preference and machine capabilities.

Selected References

1. Cowderoy GA et al: Overuse and impingement syndromes of the shoulder in the athlete. Magn Reson Imaging Clin N Am. 17(4):577-93, v, 2009
2. Murray PJ et al: Clinical update: MR imaging of the shoulder. Sports Med Arthrosc. 17(1):40-8, 2009
3. Steinbach LS: MRI of shoulder instability. Eur J Radiol. 68(1):57-71, 2008

SHOULDER AND HUMERUS OVERVIEW

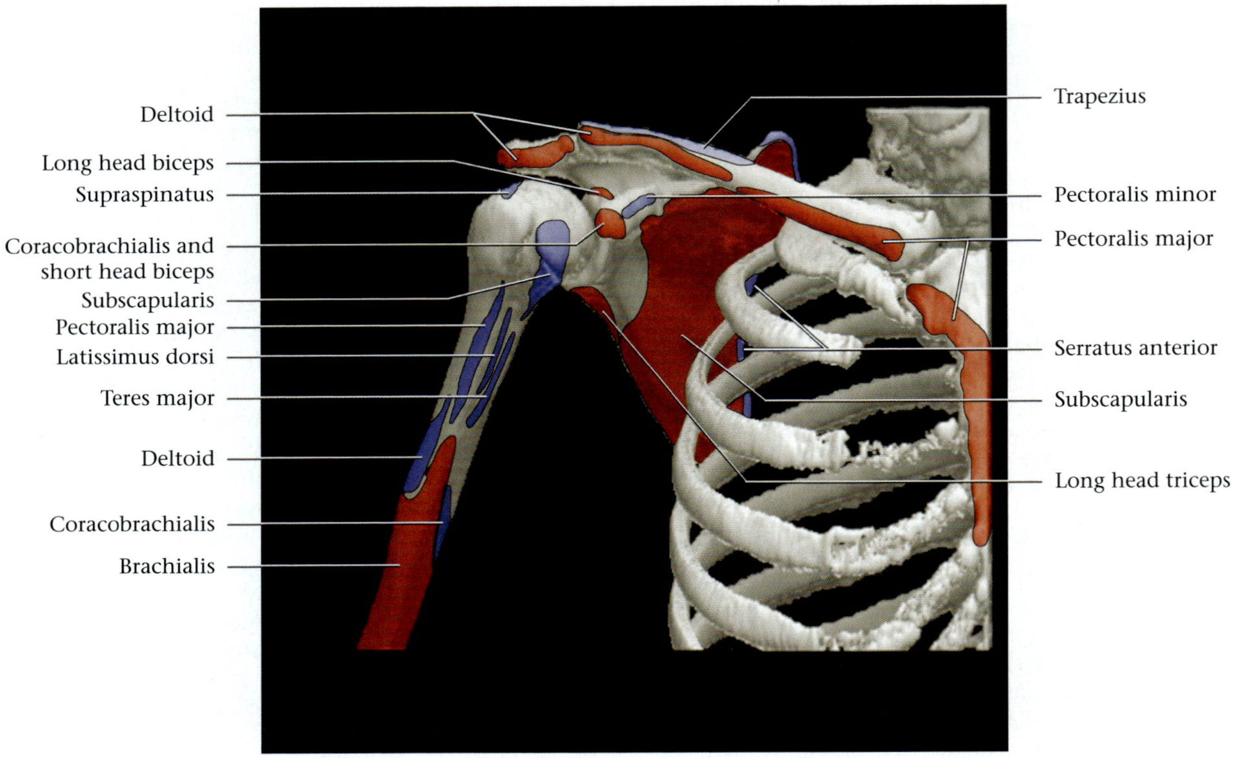

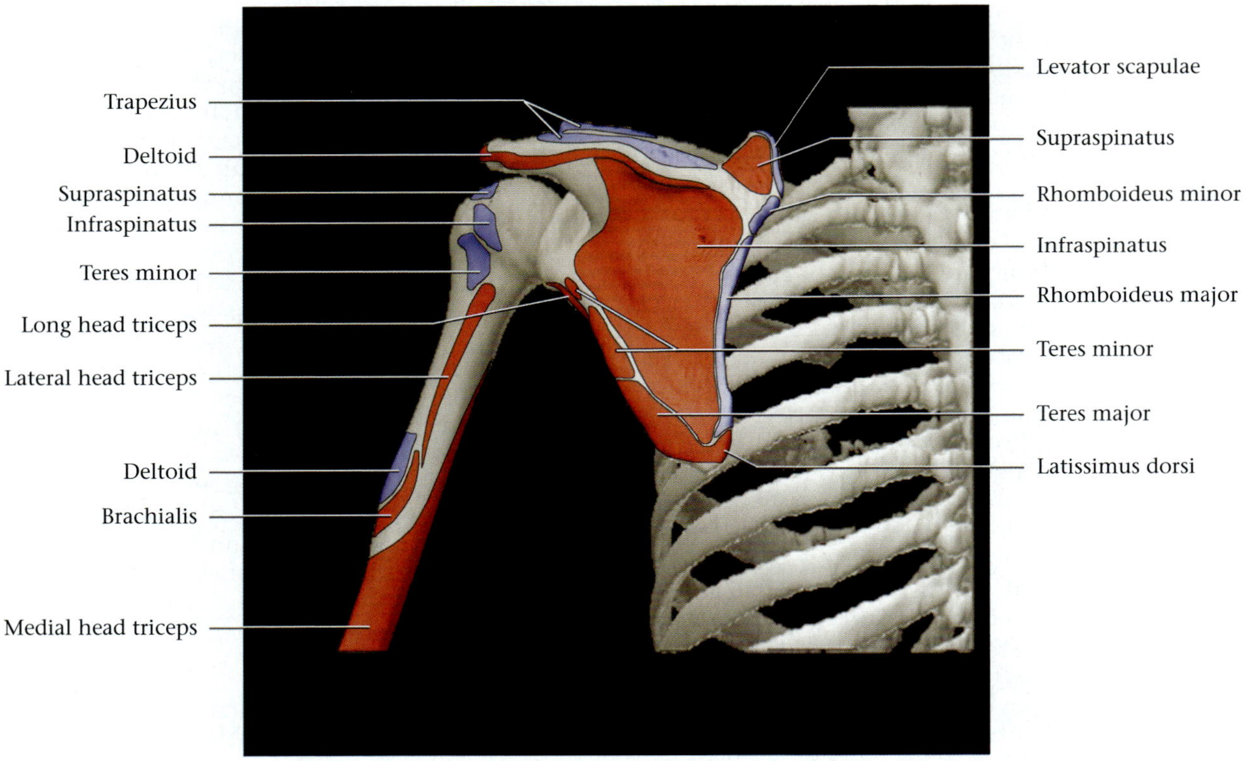

(Top) *Anterior view of the shoulder from a 3D CT image shows the muscle origins in red and the insertions in blue. Note that the glenoid fossa is shallow to allow a wide range of motion at the shoulder joint. The muscles that cross the glenohumeral joint act as dynamic stabilizers to help prevent dislocation. These include the rotator cuff muscles (supraspinatus, infraspinatus, teres minor, and subscapularis), as well as other muscles around the shoulder, including the long head biceps and pectoralis major.* ***(Bottom)*** *Posterior view of the shoulder from a 3D CT shows the muscle origins in red and the insertions in blue. Notice the supraspinatus muscle origin from the posterior scapula above (and partly hidden by) the scapular spine and its insertion onto the superior facet of the greater tuberosity. The infraspinatus and teres minor tendons insert more inferiorly on the posterior portion of the greater tuberosity.*

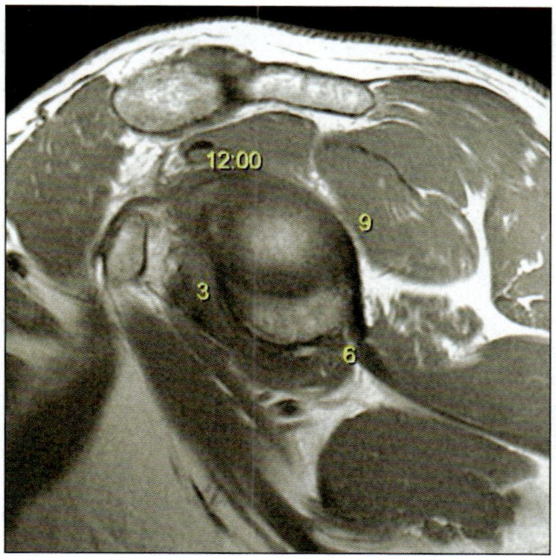

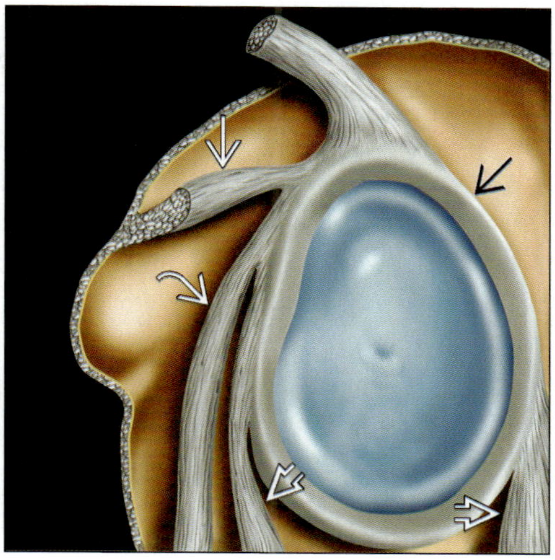

(Left) *Oblique sagittal T1WI MR shows the "clock face" classification used to localize labral tears. In this system, 12 o'clock is the superior insertion site of the biceps anchor, and 3 o'clock is defined as the anterior mid-glenoid. We will mostly use a description to localize tears; for example, a tear from 4-5 o'clock will be called anteroinferior.* ***(Right)*** *Sagittal graphic shows the glenoid labrum → and the origins of the superior →, middle →, and inferior → glenohumeral ligaments.*

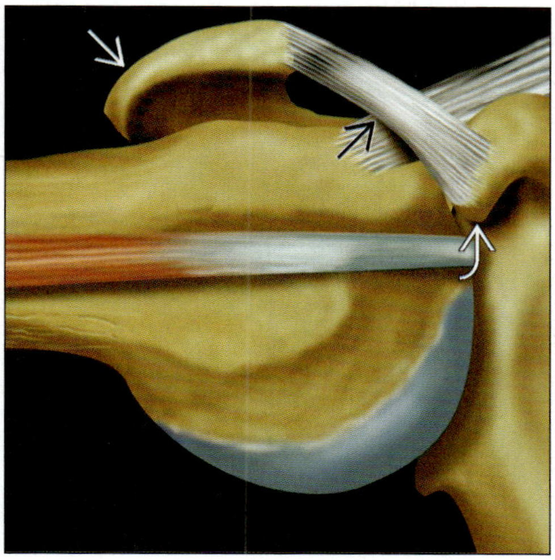

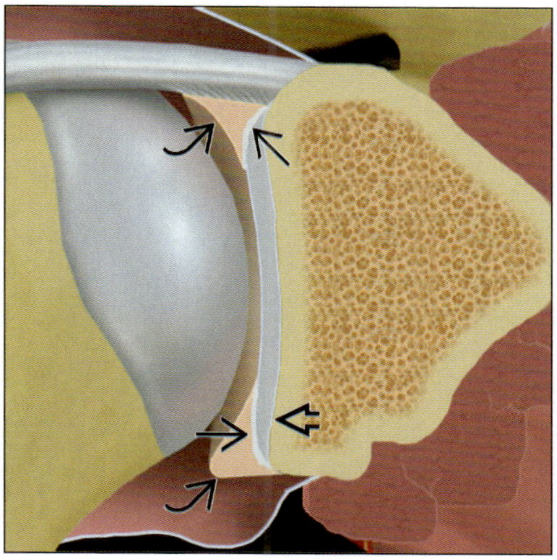

(Left) *AP graphic with the arm abducted 90° shows the coracoacromial arch made up of the acromion process →, the coracoacromial ligament →, and the coracoid process →. The coracoacromial arch protects the top of the shoulder.* ***(Right)*** *Coronal graphic shows the labrum → and the fibrocartilage transition zone → between the fibrous labrum and the glenoid hyaline cartilage →. The labrum functions mainly as the insertion of the glenohumeral ligaments and the long head biceps tendon.*

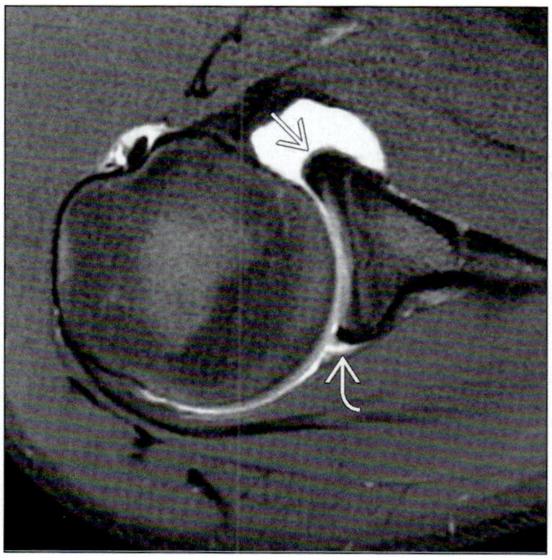

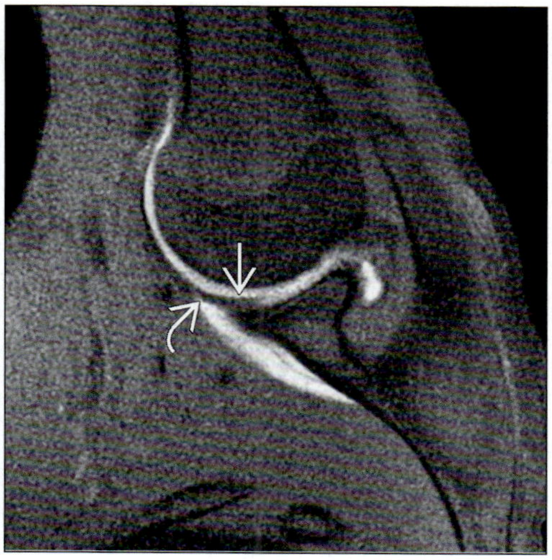

(Left) *Axial MR arthrogram T1WI FS shows a normal anterior labrum →, which appears rounded on this image. The patient has a posterior labral tear →.* ***(Right)*** *ABER MR arthrogram T1WI FS from the same patient shows that the anterior labrum in the same region is now triangular → in shape. With the arm in abduction and external rotation, the anterior band of the inferior glenohumeral ligament → is pulled taut and the true insertion of the ligament into the tip of the labrum is better seen.*

SHOULDER AND HUMERUS OVERVIEW

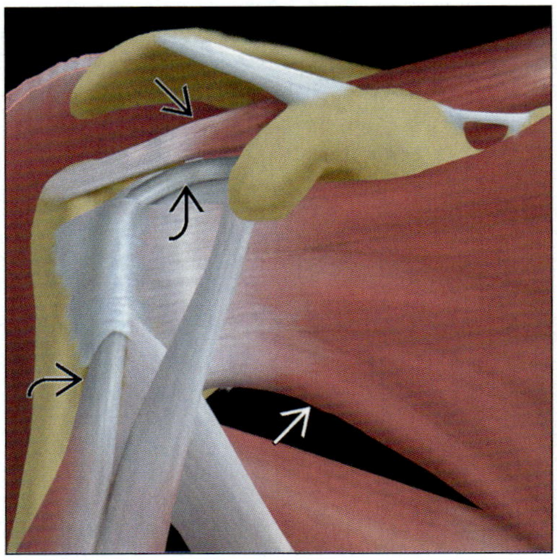

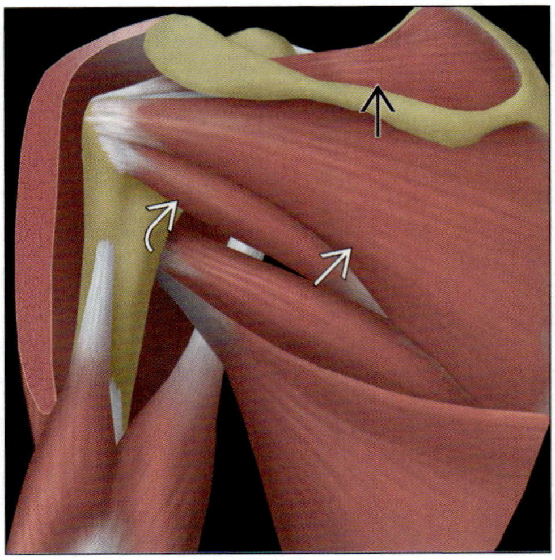

(Left) *AP graphic shows the supraspinatus → and subscapularis → muscles. The long head biceps tendon → ascends in the bicipital groove and then curves medially within the biceps pulley sling (cut away) before running horizontally through the joint to attach onto the superior labrum and supraglenoid tubercle.* ***(Right)*** *PA graphic with the deltoid muscle partially cut away shows the supraspinatus →, infraspinatus →, and teres minor → muscles attaching onto the greater tuberosity.*

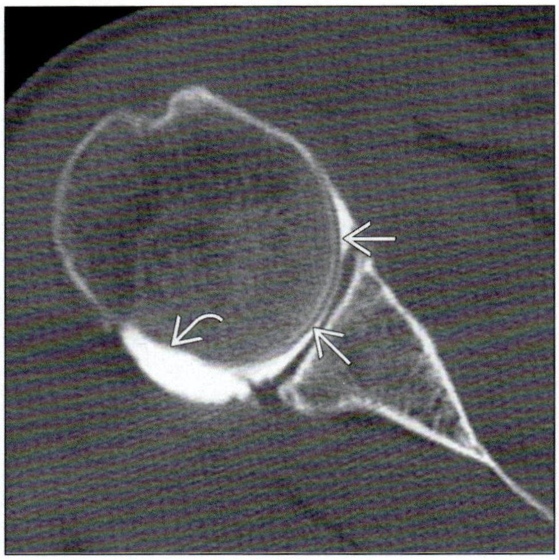

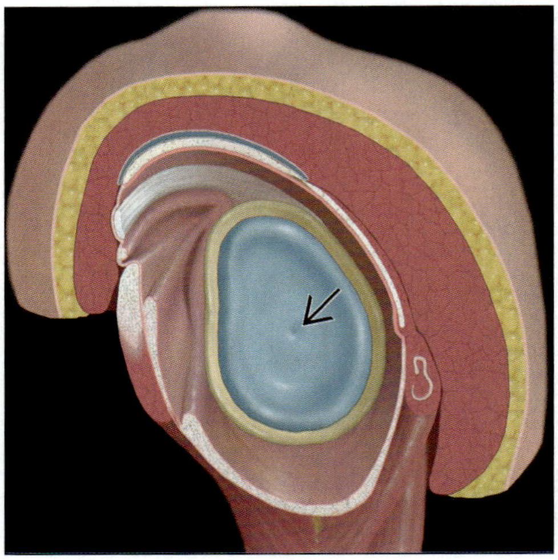

(Left) *Axial CT arthrogram shows the hyaline cartilage covering the articular surface of the humeral head → and the intraarticular "bare area" →. The bare area can sometimes mimic a Hill-Sachs lesion on more inferior slices; a true Hill-Sachs lesion will be seen on slices through the superior aspect of the humeral head.* ***(Right)*** *Sagittal graphic shows the "central bare area" of the glenoid fossa →. The central bare area should not be confused with a pathologic focal chondral defect.*

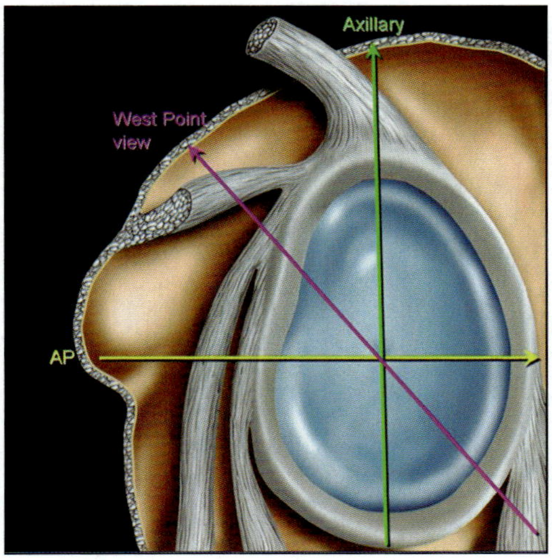

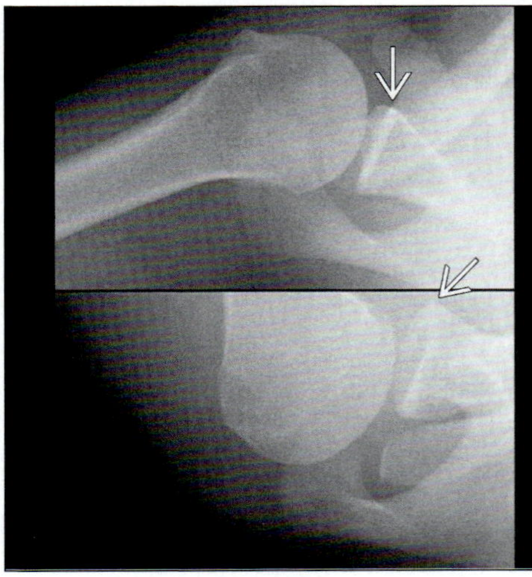

(Left) *Sagittal graphic shows the direction of the x-ray beam for 3 radiographic views of the shoulder. If the AP view (yellow) is taken with the patient turned 30° to profile the glenoid fossa, it is called a Neer AP or Grashey view. The axillary view (green) is best for showing anterior or posterior glenohumeral joint dislocations. The West Point view (purple) is preferred for showing small Bankart fractures.* ***(Right)*** *Axillary (top) and West Point (bottom) views show a normal anterior glenoid →.*

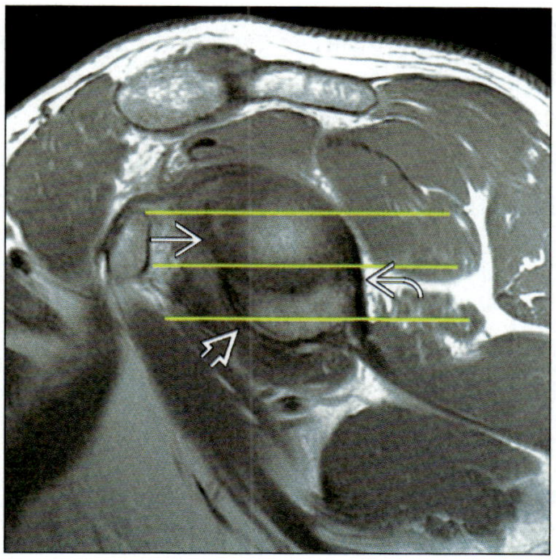

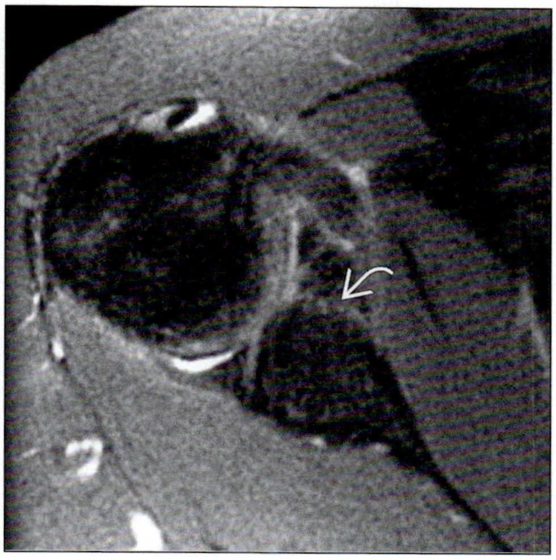

(Left) *Sagittal T1WI MR shows 3 localizer lines for a standard axial sequence of the shoulder. Note the anterior tilt of the scapula so that the resulting images are most transverse through the anterosuperior ➔ and posteroinferior ➔ labrum. There will be some partial averaging of the anteroinferior labrum ➔, which courses obliquely through the sections.* ***(Right)*** *Axial PDWI FS MR through the inferior glenoid fossa shows a partially healed Bankart fracture ➔.*

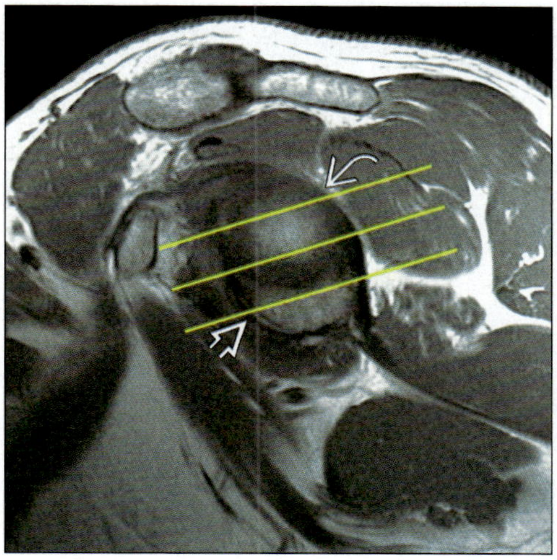

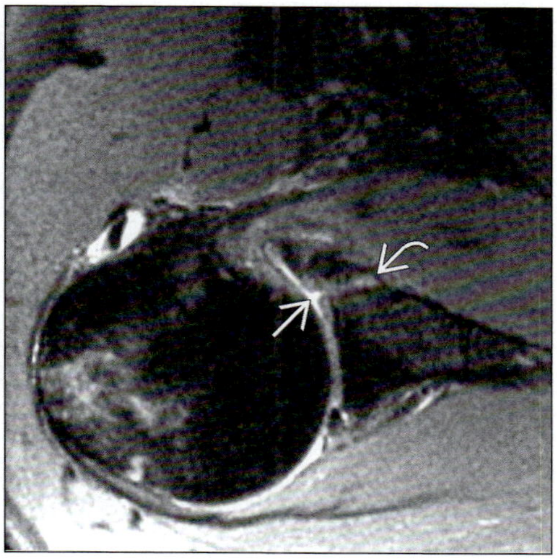

(Left) *Sagittal T1WI MR shows localizer lines for angled axial images orthogonal to the long axis of the glenoid. The resulting images have less partial averaging of tears of the anteroinferior ➔ and posterosuperior ➔ labrum.* ***(Right)*** *Angled axial PDWI FS MR from the same patient as the previous axial image shows the old Bankart fracture ➔. There is less elongation of the glenoid, and the section is more perpendicular to the Bankart fracture line. A small chondral defect ➔ is now seen.*

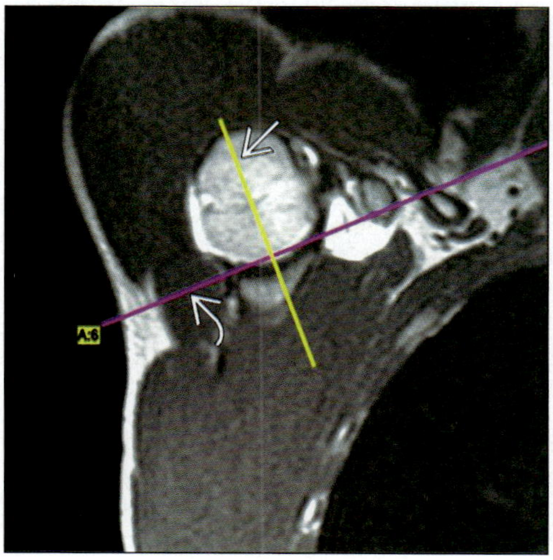

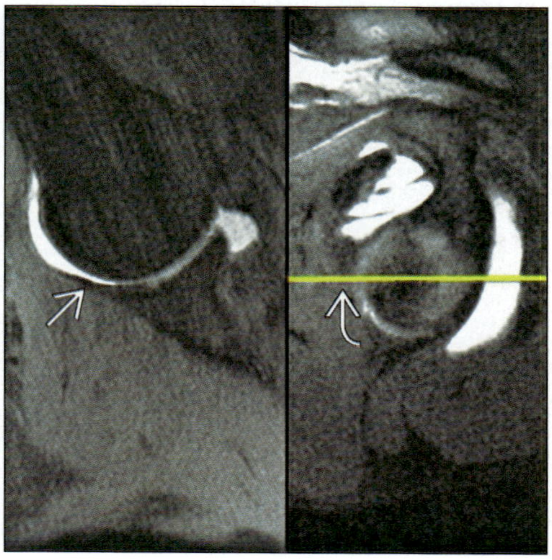

(Left) *Coronal T1WI MR with the arm in ABER position shows the localization of an axial ABER image ➔. A 2nd localizer line ➔ is for an oblique sagittal-like image.* ***(Right)*** *ABER MR arthrogram T1WI FS screen capture from the same patient shows the normal anterior labrum on the axial ABER image at left ➔. On the right is the oblique sagittal-like image showing the yellow localizer line ➔ for the axial ABER image transverse to the anteroinferior labrum.*

STERNOCLAVICULAR TRAUMA

Key Facts

Terminology

- Sternoclavicular (SC) dislocation, subluxation, sprain
- SC ligament injury with various degrees/directions of medial clavicle displacement from SC joint

Imaging

- Abnormal position of medial clavicle in relation to manubrium
- Anterior (presternal) dislocation (AD): Medial clavicle anterior or anterosuperior to manubrium
- Posterior (retrosternal) dislocation (PD): Medial clavicle posterior or posterosuperior to manubrium
- CT best at revealing position of medial clavicle and associated injuries

Pathology

- Usually result of profound direct or indirect force
- Anterior dislocation
  - Atraumatic: Ligamentous laxity in young adults and teenagers when arm raised over head
  - Traumatic: Indirect; anterior force applied to shoulder that is posterior to SC joint
- Posterior dislocation
  - Direct: Blow to medial clavicle

Clinical Issues

- Posterior dislocation: Potential life-threatening emergency
- If < 22 years of age, apparent SC dislocation is usually a physeal fracture

Diagnostic Checklist

- AD are far more common than PD; both are rare
- PD can result in damage to superior mediastinal structures that can be life-threatening
- CT is preferred diagnostic modality
- Radiographs unreliable for dislocation direction

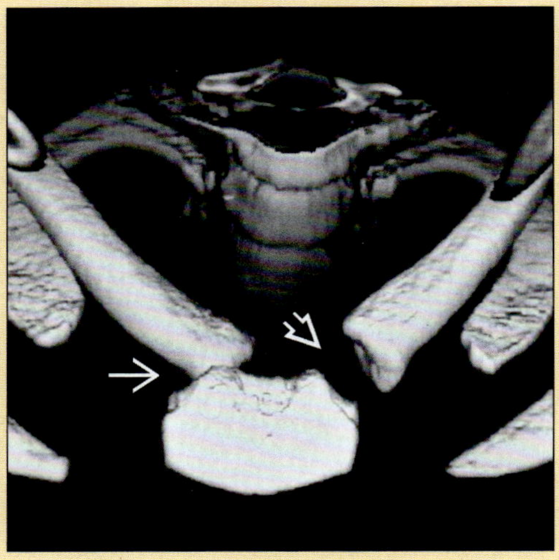

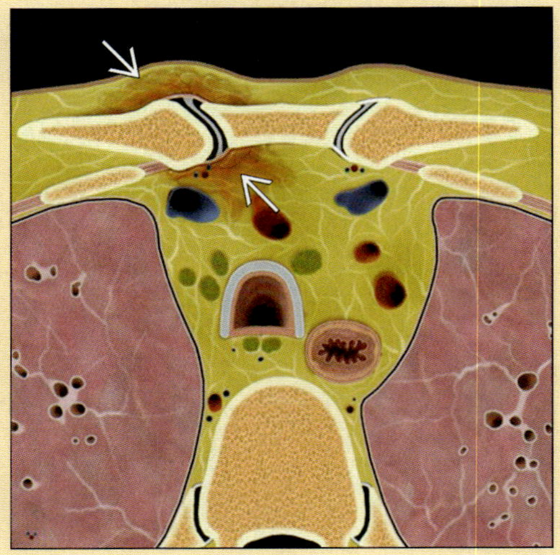

(Left) *Frontal bone CT 3D reformat shows the right posterior SC joint dislocation. This image shows nicely the normal arrangement of the left SC joint. The clavicle is normally slightly superiorly placed relative to the manubrium because of the interposed triangular-shaped disc.* ***(Right)*** *Axial graphic shows grade I SC injury with stretched SC ligaments and adjacent hemorrhage and edema. The SC joint is radiographically normal in grade I injuries, and the ligaments are macroscopically intact.*

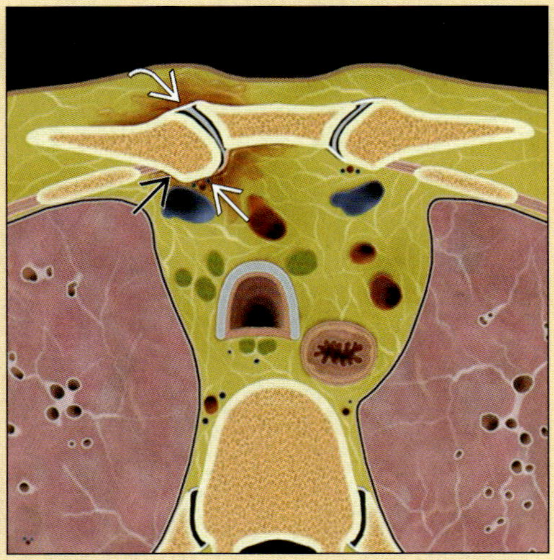

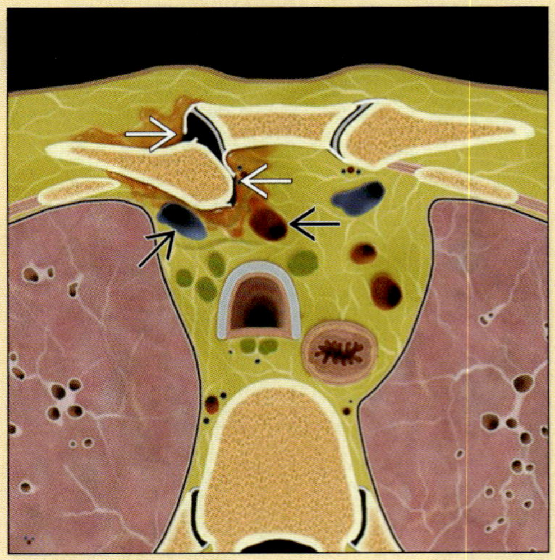

(Left) *Axial graphic shows grade II SC injury with (posterior) subluxation of the medial clavicle, torn posterior SC ligament, and stretched but intact anterior capsule.* ***(Right)*** *Axial graphic shows grade III posterior dislocation of the medial clavicle. The sternoclavicular ligaments and capsule are torn. The costoclavicular ligaments are also torn in grade III SC injuries. Traumatic posterior dislocations can also injure the superior mediastinal great vessels.*

STERNOCLAVICULAR TRAUMA

TERMINOLOGY

Abbreviations

- Sternoclavicular (SC) dislocation, subluxation, sprain

Definitions

- SC ligament injury with various degrees/directions of medial clavicle displacement from SC joint

IMAGING

General Features

- Best diagnostic clue
  - Abnormal position of medial clavicle in relation to manubrium
- Location
  - Anterior (presternal) dislocation (AD): Medial clavicle anterior or anterosuperior to manubrium
  - Posterior (retrosternal) dislocation (PD): Medial clavicle posterior or posterosuperior to manubrium
- Size
  - Symptomatic sprain to complete SC dislocation

Radiographic Findings

- Often misleadingly normal
  - Normal SC joints can appear subluxated on radiographs due to the oblique angle of the joint
  - Overlapping thoracic structures obscure the joint
- Routine: PA, oblique, lateral views of SC joints
  - PA: Difference in relative craniocaudal positions of medial clavicles > 50% width of clavicle head
- Special views
  - Rockwood (serendipity) view: Beam directed cephalad through manubrium in supine patient
    - AD: Medial clavicle above horizontal plane
    - PD: Medial clavicle below horizontal plane
  - Hobbs view: Patient seated, leaning over table with beam aimed through cervical spine
    - Approximates 90° lateral view of SC joint
- Stress maneuver
  - Classic: Ipsilateral arm across chest and pulling against contralateral elbow
  - Can also use with CT

CT Findings

- CECT
  - Identifies vascular and soft tissue injury
  - Easily distinguishes direction of dislocation
  - Useful for surgical planning
  - Can obtain quickly in trauma setting

MR Findings

- T1WI
  - Identify ligament disruption
- T2WI
  - Increased signal ligament tears, effusion, ST edema
- MRA
  - Identify vascular injury with PD
- Superior to CT in identifying articular cartilage and disc injury, joint effusions, ligament injury

Ultrasonographic Findings

- Quick preoperative screen for dislocation
- Color Doppler allows quick assessment of vessels
- Confirm reduction before leaving operating room
- Correct diagnosis of SC joint malposition found in 89% of cases for US vs. 13% for radiographs
- Limited number of individuals with expertise in use

Imaging Recommendations

- Best imaging tool
  - CT best at revealing position of medial clavicle and associated injuries
- Protocol advice
  - Radiographs: Special views ± stress maneuvers
  - CT
    - Thin slices (≤ 1 mm) to include both SC joints + medial half of both clavicles
    - Coronal reformats useful; angle to plane of joint
    - Neutral position or with stress maneuver
    - Intravenous contrast: ↑ conspicuity of vessels
  - MR
    - Coronal, sagittal, axial, and oblique axial planes
    - T1, T2, and STIR or T2 fat-suppressed

DIFFERENTIAL DIAGNOSIS

Physeal Fractures

- Medial clavicle physis closes between 22-26 years
- If < 22 years, displaced clavicle is usually a physeal fracture and not a true dislocation
- Ligaments remain intact

Osteomyelitis

- Uncommon: Usually occurs after surgery or radiation
- Other clinical scenarios
  - IV drug abuse
  - Endocarditis
  - Adjacent mediastinal infection
- Abscess & tissue destruction

Neoplasia

- Metastatic disease or primary neoplasm (multiple myeloma, lymphoma, or chondrosarcoma)
- Bone destruction, tumor matrix, associated soft tissue mass

Arthritis

- Particularly rheumatoid arthritis, scleroderma, ankylosing spondylitis, SAPHO syndrome, other collagen-vascular diseases
- Joint space narrowing, erosions, osteophyte formation, sclerosis, & capsular thickening

PATHOLOGY

General Features

- Etiology
  - Usually result of profound direct or indirect force
  - Anterior dislocation
    - Atraumatic: Ligamentous laxity in young adults and teenagers when arm raised over head
    - Traumatic: Indirect; anterior force applied to shoulder that is posterior to SC joint
  - Posterior dislocation
    - Direct: Blow to medial clavicle
    - Indirect: Posterolaterally applied force to shoulder that is anterior to SC joint
  - Most result from motor vehicle accident (40%) or sports-related trauma (21%)

STERNOCLAVICULAR TRAUMA

- Associated abnormalities
  - AD: Pneumothorax, hemothorax, pulmonary contusion, rib fractures if due to significant trauma
  - PD: Great vessel, trachea, esophagus, and nerve injury of superior mediastinum

Staging, Grading, & Classification

- Severity grades of SC joint injury
  - Grade I: Sprain = incomplete tear or stretching of SC and costoclavicular ligaments
  - Grade II: SC joint subluxation = complete tear of SC ligament + partial tear of capsule
  - Grade III: SC dislocation = complete tear of SC and costoclavicular ligaments + complete tear of capsule

Gross Pathologic & Surgical Features

- Joint capsule disruption
- Disruption of posterior SC ligament
- Anterior SC ligament is much weaker than posterior

CLINICAL ISSUES

Presentation

- Most common signs/symptoms
  - Pain, swelling, and deformity at SC joint
- Other signs/symptoms
  - Chest + shoulder pain exacerbated by arm movement or assuming supine position
  - Affected arm supported across chest by opposite arm
  - Head tilted toward affected side
  - Affected shoulder: Shortened + thrust forward
- Clinical profile
  - Anterior dislocation
    - Less painful than PD
    - Prominent medial clavicle accentuated by shoulder motion
  - Posterior dislocation
    - Potential life-threatening emergency
    - Dyspnea, dysphagia, venous congestion, paresthesias
    - Sulcus-type deformity near sternum: Soft tissue swelling can obscure → false appearance of AD
    - Physical findings more subtle than with AD
  - Atraumatic dislocation: Only mild symptoms

Demographics

- Age
  - Young adults
  - Older patients may develop AD without clear history of trauma → painless mass over medial clavicle
- Gender
  - M > F
  - Atraumatic more common in young girls
- Epidemiology
  - SC dislocations are rare
    - 3% of shoulder girdle injuries
    - < 1% of all dislocations
    - 0.6% of all trauma patients
  - 90-95% of SC dislocations are anterior
  - Often associated with significant injuries
    - > 2/3 of AD associated with serious injuries
    - 25% of PD have superior mediastinal injury
    - Associated injuries not as serious for AD as PD

Natural History & Prognosis

- Most regain adequate upper extremity function
- AD: Symptoms usually subside rapidly
- PD: Significant disability & (rarely) death

Treatment

- Grade I: Ice for 12-24 hours, heat, immobilize for 3-4 days
- Grade II: Ice for 12 hours, heat for 12-24 hours, reduce subluxation, immobilize for 3-6 weeks
- Anterior dislocation (grade III)
  - Usually closed reduction under conscious sedation
  - Immobilization for 4-6 weeks following reduction
  - Recurrent painful dislocations may require surgery
- Posterior dislocation (grade III)
  - Closed reduction preferred < 24 hours after injury; often done in OR under general anesthesia
  - Surgery if closed reduction fails or joint unstable
  - Immobilization for 4-6 weeks following reduction
  - Should have cardiothoracic surgery available
- Atraumatic: Benign course → treatment not required
- Complications (nonoperative management)
  - AD: Cosmetic deformity, degenerative changes, persistent pain, weakness, instability with activity
  - PD: Great vessel, tracheal, esophageal, or brachial plexus injury

DIAGNOSTIC CHECKLIST

Consider

- SC dislocations are rare
- AD are far more common than PD
- Superior mediastinal structure damage in PD

Image Interpretation Pearls

- CT is preferred diagnostic modality
- Radiographs unreliable for dislocation direction

SELECTED REFERENCES

1. Ernberg LA et al: Radiographic evaluation of the acromioclavicular and sternoclavicular joints. Clin Sports Med. 22(2):255-75, 2003
2. Aslam M et al: Pictorial review: MRI of the sternum and sternoclavicular joints. Br J Radiol. 75(895):627-34, 2002
3. Salgado RA et al: Post-traumatic posterior sternoclavicular dislocation: case report and review of the literature. Emerg Radiol. 9(6):323-5, 2002
4. McCulloch P et al: Radiographic clues for high-energy trauma: three cases of sternoclavicular dislocation. AJR Am J Roentgenol. 176(6):1534, 2001
5. Ferrera PC et al: Sternoclavicular joint injuries. Am J Emerg Med. 18(1):58-61, 2000
6. Destouet JM et al: Computed tomography of the sternoclavicular joint and sternum. Radiology. 138(1):123-8, 1981

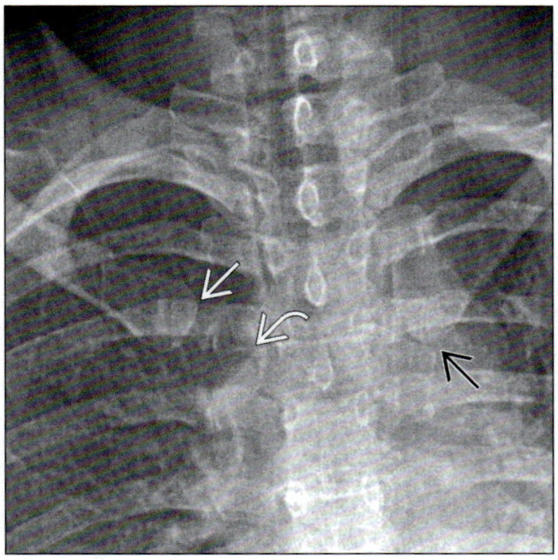

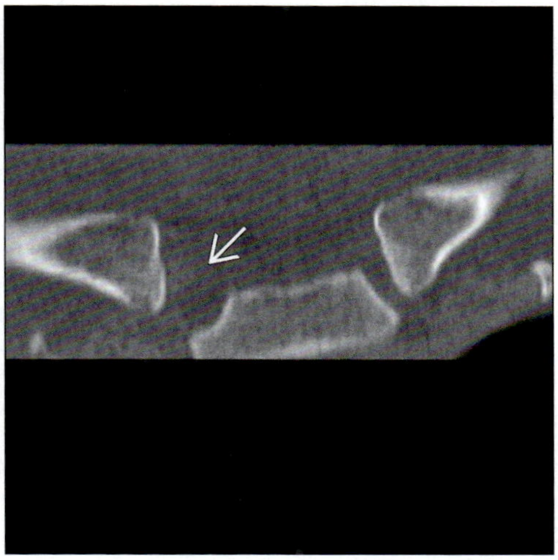

*(**Left**) AP radiograph shows widening between the right clavicle and sternal (SC) joint articular surface. The left SC joint is normal. This is a grade III dislocation with torn SC ligaments and joint capsule and torn costoclavicular ligaments. It is important to look at the SC joint on trauma chest radiographs because of the risk of vascular injury. (**Right**) Coronal bone CT reformat in the same patient shows the widened right SC joint. There is minimal anterior or posterior displacement.*

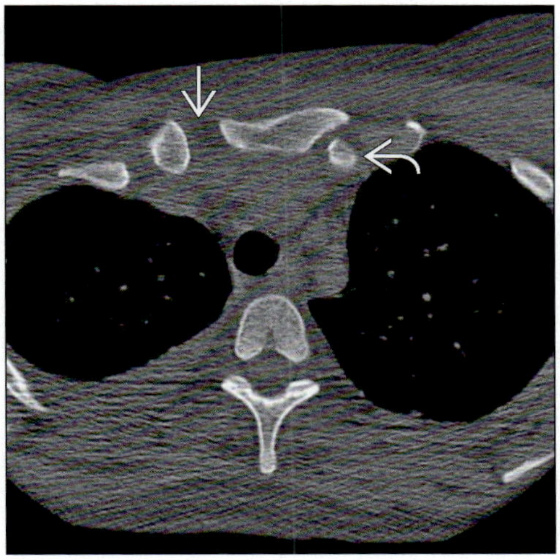

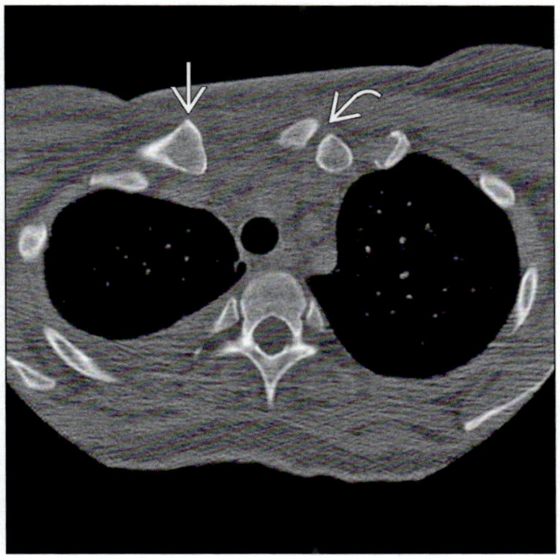

*(**Left**) Axial bone CT in the same patient shows the slight anterior and lateral dislocation of the right medial clavicle. Note the normal position of the left clavicular head. (**Right**) Axial bone CT in the same patient shows the right medial clavicle dislocated laterally and slightly anteriorly. Axial CT images are helpful to exclude posterior dislocations, which are associated with vascular injury. Note the normal left SC joint.*

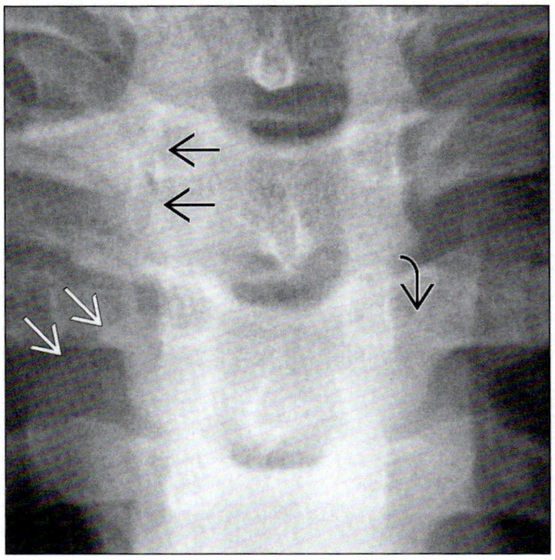

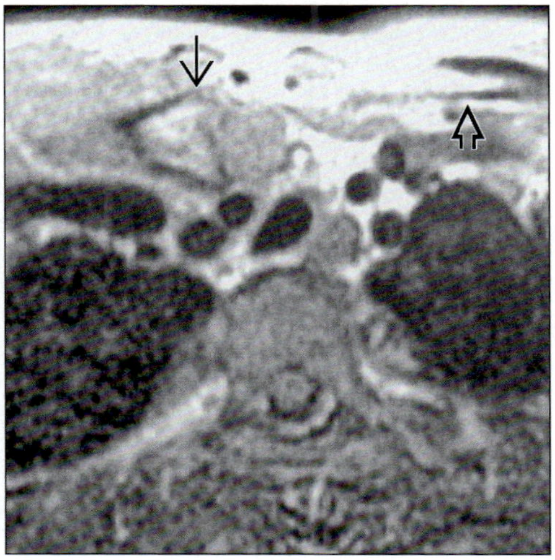

*(**Left**) Anteroposterior radiograph shows a superior dislocation of the right clavicular articular surface relative to its site of articulation on the manubrium. The contralateral normal SC joint is seen. (**Right**) Axial T1WI MR in the same patient shows the posterior & superior dislocation of the right clavicle relative to the left. The superiorly displaced head of the right clavicle is seen in the same plane as the left mid clavicle. There is no significant compression of major vessels.*

CLAVICLE FRACTURE

Key Facts

Imaging

- Linear fracture line, usually in middle 3rd
- Anterior and inferior displacement of lateral fragment
- Radiographs: Anteroposterior and 45° cephalic tilt anteroposterior view

Top Differential Diagnoses

- Acromioclavicular (AC) dislocation: Pain and point tenderness over AC joint following direct trauma onto point of shoulder
- Post-traumatic osteolysis of distal clavicle: Shoulder pain within variable time period (weeks to years) following episode of often minor trauma or repetitive trauma in weightlifters

Pathology

- Direct fall onto shoulder (most common), direct blow to clavicle, or fall on outstretched hand
- Associations
  - Rib fractures
  - Pneumothorax or hemothorax
  - Acromioclavicular disruptions
  - Subclavian vessel or brachial plexus injury

Clinical Issues

- Common, accounting for 5% of all fractures
- 50% occur in children under the age of 10
- Majority of fractures heal without difficulty
- Nonunion is rare (1-4% of cases); more likely to occur with unstable distal clavicle or poorly immobilized fracture
- Post-traumatic osteoarthritis is common in type III distal clavicle fractures

Diagnostic Checklist

- Associated injuries, especially with high mechanism trauma

***(Left)** AP radiograph shows a comminuted fracture in the middle 3rd of the clavicle ➔, which is involved in 80% of clavicle fractures. Displaced fractures, as in this case, are at risk for nonunion. **(Right)** AP radiograph with cephalad angulation shows a fracture in the lateral 3rd of the clavicle ➔. In this type I lateral 3rd fracture, the fracture is interligamentous and the coracoclavicular ligaments are intact. These are therefore minimally displaced and usually heal well.*

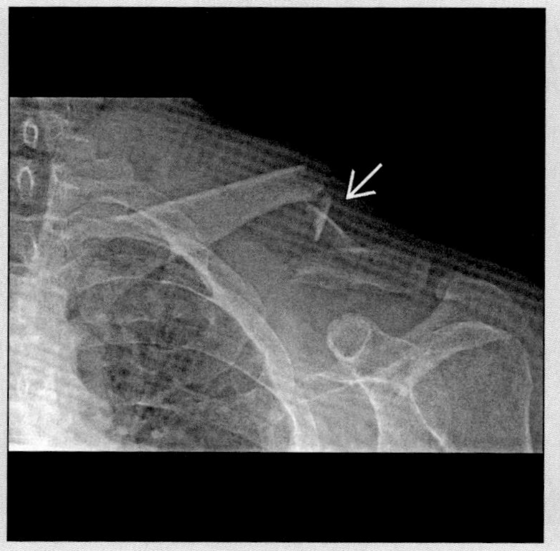

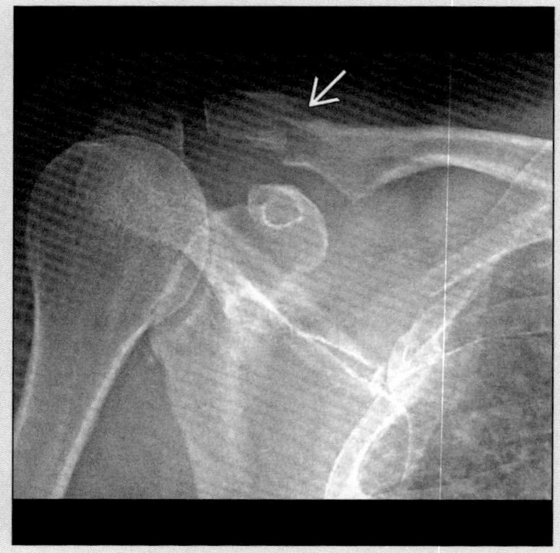

***(Left)** AP radiograph shows a fracture in the lateral 3rd of the clavicle ➔. In this type II lateral 3rd fracture, the medial clavicle is detached from the coracoclavicular ligaments and thus elevated. These fractures have a higher risk of nonunion than the middle 3rd or type I fractures. **(Right)** AP radiograph with cephalad angulation shows a fracture in the medial 3rd of the clavicle ➔. These account for only 5% of clavicle fractures, and are often due to severe impact to the lateral shoulder.*

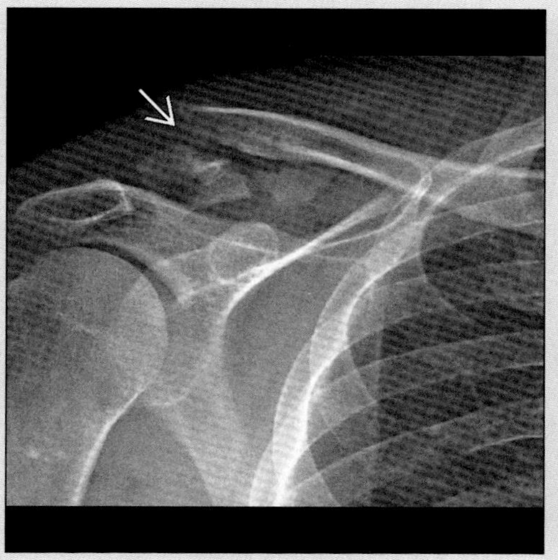

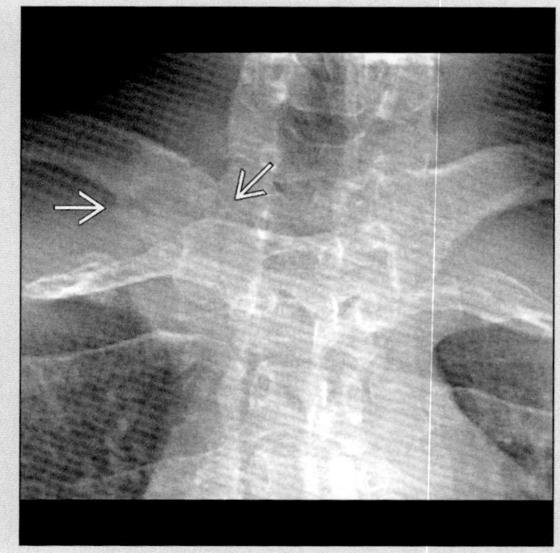

CLAVICLE FRACTURE

TERMINOLOGY

Definitions

- Displaced and nondisplaced fractures of clavicle

IMAGING

General Features

- Best diagnostic clue
  - Linear fracture line, usually in middle 3rd
  - Anterior and inferior displacement of lateral fragment
- Size
  - Varies from nondisplaced cortical break to displaced &/or comminuted fracture

Radiographic Findings

- Linear lucent fracture line ± displacement

CT Findings

- Can aid in identifying intraarticular involvement and nondisplaced medial fractures

MR Findings

- T2WI
  - Evaluate for ligament injury to acromioclavicular (AC) or sternoclavicular (SC) joints

Imaging Recommendations

- Best imaging tool
  - Radiograph: Anteroposterior and 45° cephalic tilt anteroposterior view

DIFFERENTIAL DIAGNOSIS

Clinical

- AC dislocation: Pain and point tenderness over AC joint following direct trauma onto point of shoulder
- SC joint dislocation: Pain over SC joint ± deformity

Radiographic

- Congenital pseudoarthrosis of clavicle
- Post-traumatic osteolysis of distal clavicle
  - Distal aspect of clavicle
  - Follows episode of often minor trauma by weeks/years
  - Repetitive trauma in weightlifters

PATHOLOGY

General Features

- Etiology
  - Direct fall onto shoulder (most common), direct blow to clavicle, or fall on outstretched hand
- Associated abnormalities
  - Rib fractures
  - Pneumothorax or hemothorax
  - Acromioclavicular or sternoclavicular disruptions
  - Subclavian vessel or brachial plexus injury

Staging, Grading, & Classification

- Allman classification
  - Group 1: Middle 3rd (80%)
  - Group 2: Distal 3rd (15%)
  - Group 3: Medial 3rd (5%)
- Neer classification of distal clavicle fractures
  - Type I: Between AC joint and CC ligament attachment, minimal displacement, and intact ligaments
  - Type II: At level of CC ligament, displaced with detachment of ligaments from proximal fragment
  - Type III: At articular surface of AC joint

CLINICAL ISSUES

Presentation

- Most common signs/symptoms
  - Pain, swelling, and palpable deformity

Demographics

- Epidemiology
  - Common: 5% of all fractures
  - 50% occur in children under age of 10
  - M:F = 2.1:1

Natural History & Prognosis

- Majority of fractures heal without difficulty
- Nonunion is rare (1-4% of cases); more likely to occur with unstable distal clavicle or poorly immobilized fx
- Post-traumatic osteoarthritis is common in type III distal clavicle fractures

Treatment

- Conservative: Protected immobilization
- Surgical: ORIF with plate/screw or IM rod placement
  - Indications: Painful nonunion or unacceptable cosmetic deformity
- Complications: Neurovascular symptoms from compression of subclavian vessels or brachial plexus; malunion with shortening may produce aesthetically unappealing deformity

DIAGNOSTIC CHECKLIST

Consider

- Associated injuries, especially with high mechanism trauma
- SC joint injury
- Nerve or vascular damage

SELECTED REFERENCES

1. Khan LA et al: Fractures of the clavicle. J Bone Joint Surg Am. 91(2):447-60, 2009
2. Kim W et al: Management of acute clavicle fractures. Orthop Clin North Am. 39(4):491-505, vii, 2008
3. Throckmorton T et al: Fractures of the medial end of the clavicle. J Shoulder Elbow Surg. 16(1):49-54, 2007
4. Jackson WF et al: Fractures of the lateral third of the clavicle: an anatomic approach to treatment. J Trauma. 61(1):222-5, 2006
5. Hanby CK et al: Medial clavicle physis fracture with posterior displacement and vascular compromise: the value of three-dimensional computed tomography and duplex ultrasound. Orthopedics. 26(1):81-4, 2003

ACROMIOCLAVICULAR JOINT INJURY

Key Facts

Terminology

- AC (acromioclavicular) separation
- AC dislocation, shoulder separation
- Subluxation or dislocation at AC joint, or sprain of AC ligament

Imaging

- Widening of AC joint with variable displacement of distal clavicle
  - > 6 mm of widening
- Type III: Elevated lateral clavicle
  - CC distance > 14 mm
  - CC distance > 50% wider compared to uninjured side
  - AC joint is usually wide as well
- Comparison with uninjured side can be helpful
- Can take AP view of both AC joints on single cassette

Pathology

- Direct force onto point of shoulder
  - Usually from fall
- Type I: AC ligaments sprained but functionally intact; CC ligaments spared
- Type II: AC ligaments completely disrupted; CC ligaments sprained but functionally intact
- Type III: AC and CC ligaments completely disrupted with superior displacement of clavicle
- Type IV: Complete AC and CC ligament disruption + posterior displacement of clavicle into trapezius
- Type V: Severe form of type III, complete AC and CC ligament disruption with marked superior displacement of clavicle
- Type VI: Inferior dislocation of distal clavicle

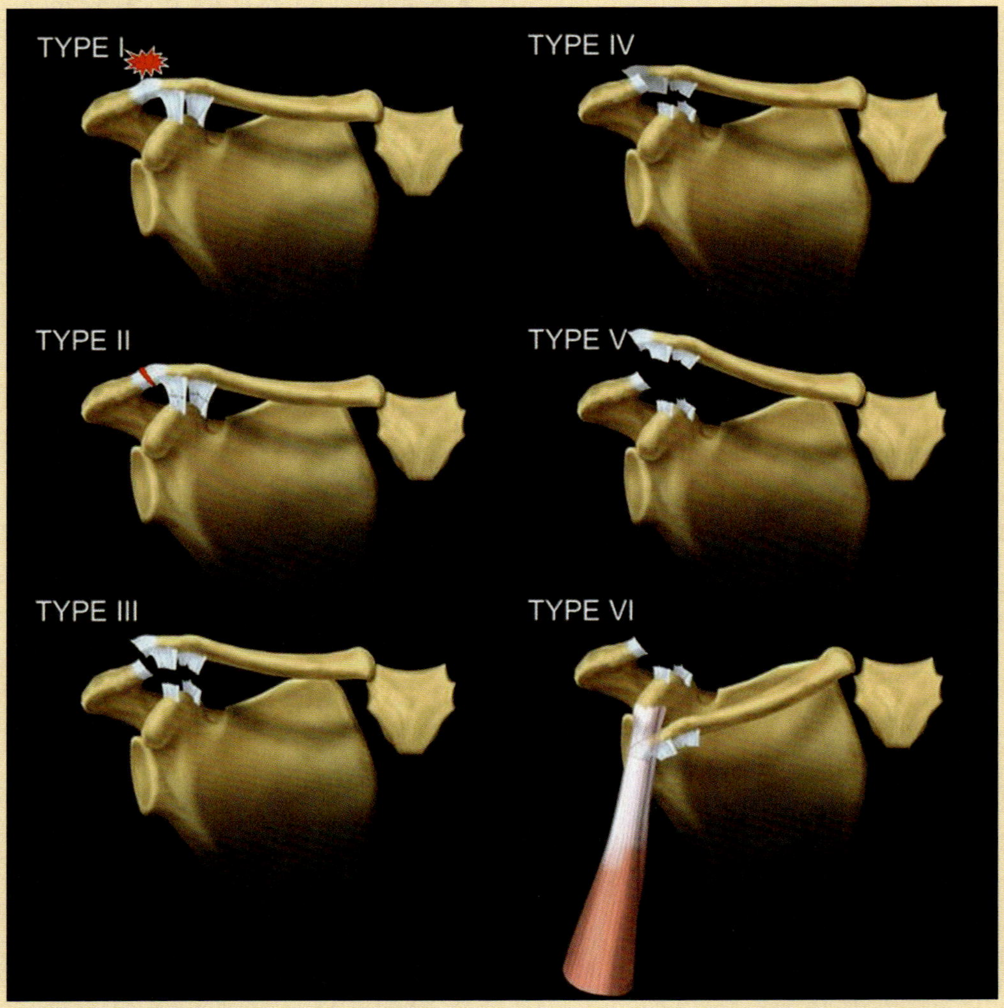

Graphic shows AP views of the shoulder demonstrating the Allman and Rockwood classification of AC joint injuries. Type I shows sprain of the AC joint ligament (normal radiograph). In type II the AC ligament is torn so the AC joint is wide, but the CC ligaments are intact so the clavicle is not elevated. Type III has torn AC and CC ligaments, and the distal clavicle is elevated. In type IV the distal clavicle is posteriorly dislocated, seen best on an axillary view radiograph. Type V is a severe superior dislocation of the distal clavicle, which then lies subcutaneously. Type VI is an inferior dislocation of the distal clavicle, which can be either subcoracoid (pictured) or subacromial.

ACROMIOCLAVICULAR JOINT INJURY

TERMINOLOGY

Abbreviations
- Acromioclavicular (AC) separation

Synonyms
- AC dislocation, shoulder separation

Definitions
- Subluxation or dislocation at AC joint or sprain of AC ligament
  - Distal clavicle displaced relative to acromion

IMAGING

General Features
- Best diagnostic clue
  - Widening of AC joint with variable displacement of distal clavicle
- Size
  - > 6 mm of widening

Radiographic Findings
- Normal in type I AC injury
- AP view: Increased AC distance in type II
  - > 6 mm
  - Measure at narrowest
  - Borderline cases: Compare with other shoulder
  - Do not let patient support injured arm
    - Supporting arm can reduce displacement
- Type III: Elevated lateral clavicle
  - 20% of normal shoulders have slight elevation
    - May need radiograph of opposite shoulder
  - AC joint is usually wide as well
  - Increased coracoclavicular (CC) distance
    - > 50% wider compared to uninjured side
    - > 14 mm
  - If CC distance is normal, consider coracoid fracture
    - Coracoid fx is seen best on axillary or Stryker notch view
- Higher grade injuries
  - Marked superior, posterior, or inferior dislocation of distal clavicle relative to acromion
- Axillary view: Shows posterior dislocation in type IV
- Zanca view
  - 10-15° cephalad angulation
  - Projects AC joint above acromion process and spine
  - Sometimes shows AC joint better
- Stress views
  - 10-15 lb suspended by straps from both wrists
  - Tell patient to relax both shoulders and let arms be pulled down
  - AP view of both AC joints

CT Findings
- Increased AC ± CC distances and variable distal clavicle displacement
- Helpful when patients cannot position for axillary view
  - Presurgical planning for type IV and VI dislocations

MR Findings
- T1WI
  - Widened AC joint or elevated distal clavicle
  - AC and CC ligament disruption
- T2WI FS
  - Soft tissue edema around AC joint
  - Joint effusion
  - Marrow edema adjacent to joint

Imaging Recommendations
- Best imaging tool
  - AP radiograph
    - Comparison with uninjured side can be helpful
      - Can take AP view of AC joints on single cassette
    - Adding wrist weights rarely recommended
      - Shows true severity of injury, but often unimportant
      - No weights shows alignment during healing
      - If near anatomic without weights, usually heals well
  - Axillary view to exclude type IV acromioclavicular separation
- Protocol advice
  - AP radiograph of both AC joints and clavicles on single cassette
  - Axillary view of injured shoulder

Ultrasonographic Findings
- Grayscale ultrasound
  - Wide AC joint > 6 mm
  - Tender to localized palpation with transducer
- Power Doppler
  - Increased blood flow around injured joint

DIFFERENTIAL DIAGNOSIS

Post-Traumatic Distal Clavicle Osteolysis
- Painful AC joint and distal clavicle
- May be from stress fx of distal clavicle
- 2 mechanisms
  - Continued activity after type I injury
  - Weightlifting; repetitive trauma

Lateral Clavicle Fracture
- Pain and swelling following acute trauma

Normal Pediatric/Adolescent AC
- Incomplete ossification of lateral clavicle and medial acromion
- Lateral clavicle epiphysis ossifies/fuses @ ~ 19 years old
- Compare to normal side: < 25% difference

Septic AC Joint
- Indistinct subchondral bone plate

Scapulothoracic Dissociation
- Severe AC separation with brachial plexus injury
- May also have sternoclavicular diastasis or clavicle fracture

PATHOLOGY

General Features
- Etiology
  - Direct force onto point of shoulder
    - Usually from fall
    - Most common mechanism of injury
  - Fall onto outstretched hand
  - Indirect upward force/downward pull of arm

ACROMIOCLAVICULAR JOINT INJURY

 - Inferior dislocation: Severe direct force onto lateral clavicle
- Associated abnormalities
  - Acromial, clavicular, and rib fractures

Staging, Grading, & Classification

- Type I: AC ligaments sprained but functionally intact; CC ligaments spared
  - Normal radiograph
  - Tender, painful AC joint
- Type II: AC ligaments completely disrupted; CC ligaments sprained but functionally intact
  - Wide AC joint
  - AC subluxation
- Type III: AC and CC ligaments completely disrupted with superior displacement of clavicle
  - Mild to moderate superior displacement of clavicle
  - AC dislocation
- Type IV: Complete AC and CC ligament disruption with posterior displacement of clavicle into trapezius muscle
  - AP radiograph can appear normal
  - Axillary view shows posterior dislocation of clavicle
- Type V: Similar to, but more severe than type III
  - Complete AC and CC ligament disruption with marked superior displacement of clavicle
  - CC distance usually 2-3x > normal
  - Distal clavicle lies subcutaneously
  - Torn deltoid and trapezius muscle attachments
  - Usually markedly drooped shoulder
- Type VI: Inferior dislocation of distal clavicle
  - AC ligament disruption ± CC ligament disruption with inferior displacement of clavicle
  - Shoulder has "flat" appearance
  - 2 types
    - Subcoracoid
      - Requires severe trauma; often other associated injuries
    - Subacromial
      - Decreased coracoclavicular distance

Gross Pathologic & Surgical Features

- Torn AC ligament
- Torn CC ligaments in higher grade injuries
  - Conoid ligament
    - Extends from coracoid process to conoid tubercle of clavicle
  - Trapezoid ligament
    - Extends from lateral superior coracoid process to trapezoid line on clavicle

CLINICAL ISSUES

Presentation

- Most common signs/symptoms
  - Pain and tenderness over AC joint
    - Type I: Minimal pain with arm movement
  - Type II
    - Distal clavicle slightly cephalad of acromion
  - Type III
    - Distal clavicle elevated, shoulder depressed
    - Arm held adducted

Demographics

- Age
  - Most common in those 15-44 years
  - Rare < 12 years
- Gender
  - M > F = 5-10:1
- Epidemiology
  - Common injury in hockey and rugby

Natural History & Prognosis

- Often have residual clicking and mild pain
- Weakness and instability can also persist

Treatment

- Conservative
  - Type I and II; most type III
  - Ice, analgesics, and shoulder rest in sling (1-4 weeks)
  - Heavy lifting/sports restriction (6-12 weeks)
- Surgical (rare)
  - Type IV-VI and type III in athletes/heavy laborers
  - AC resection (Mumford procedure)
  - Repair of AC and CC ligaments
  - Dynamic muscle transfers
- Complications
  - Acromioclavicular osteoarthritis
  - Post-traumatic osteolysis of distal clavicle
  - Coracoclavicular ossification
  - Septic AC joint: Especially after type V with skin injury

DIAGNOSTIC CHECKLIST

Consider

- Associated fractures, especially in high-grade dislocations

Image Interpretation Pearls

- Slight elevation of lateral clavicle can be normal finding, but AC distance should be < 7 mm
- Weight-bearing views unnecessary, do not alter treatment

SELECTED REFERENCES

1. Alyas F et al: MR imaging appearances of acromioclavicular joint dislocation. Radiographics. 28(2):463-79; quiz 619, 2008
2. Bishop JY et al: Treatment of the acute traumatic acromioclavicular separation. Sports Med Arthrosc. 14(4):237-45, 2006
3. Barnes CJ et al: Magnetic resonance imaging of the coracoclavicular ligaments: its role in defining pathoanatomy at the acromioclavicular joint. J Surg Orthop Adv. 13(2):69-75, 2004
4. Antonio GE et al: Pictorial essay. MR imaging appearance and classification of acromioclavicular joint injury. AJR Am J Roentgenol. 180(4):1103-10, 2003
5. Ernberg LA et al: Radiographic evaluation of the acromioclavicular and sternoclavicular joints. Clin Sports Med. 22(2):255-75, 2003

ACROMIOCLAVICULAR JOINT INJURY

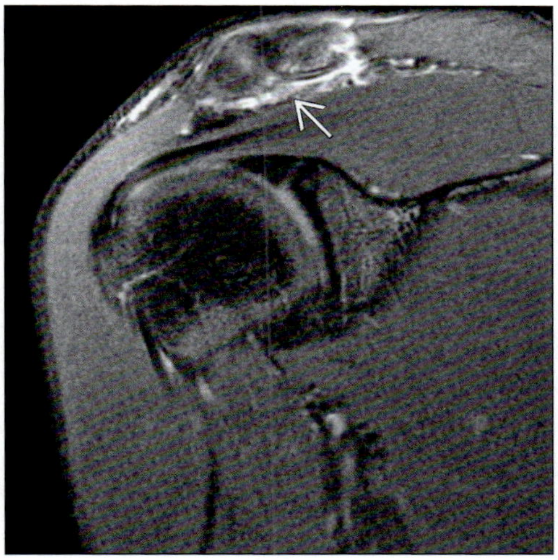

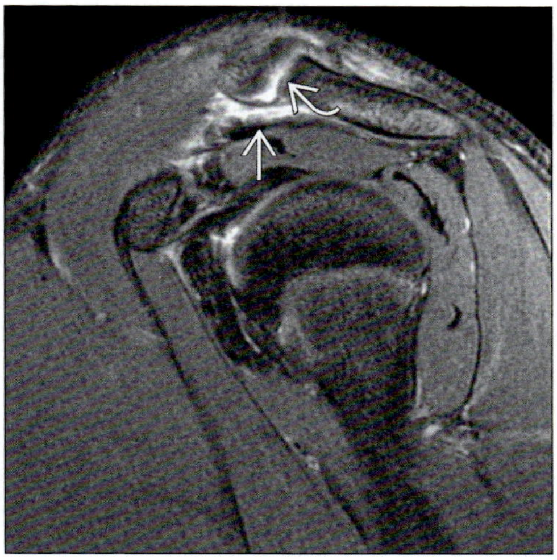

(Left) *Coronal oblique T2WI FS MR in a patient with a type I AC separation after a fall onto the point of the shoulder. There is edema around the AC joint ➔ without elevation of the lateral clavicle.* ***(Right)*** *Oblique sagittal T2WI FS MR in the same patient also shows the edema around the AC joint ➔ and an AC joint effusion ↪. In a patient with pain over the AC joint after trauma without elevation of the distal clavicle, these findings are consistent with a type I AC separation.*

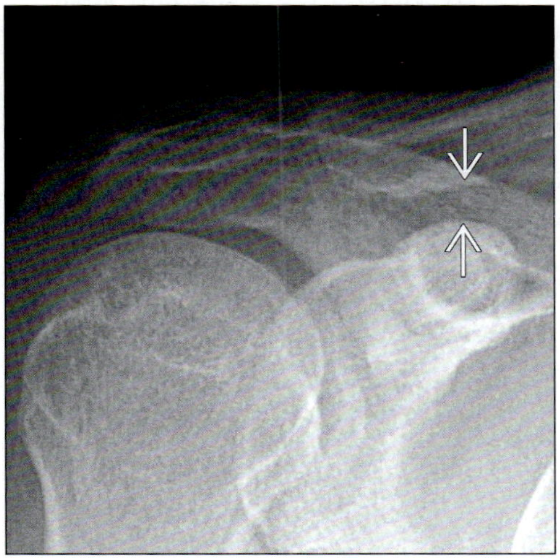

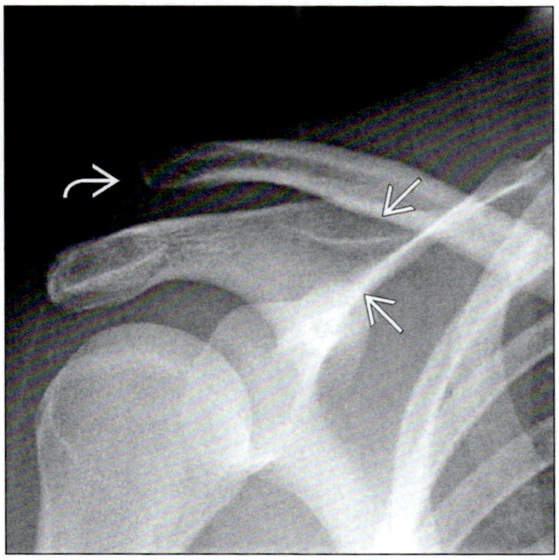

(Left) *AP radiograph in a young man with pain after a fall. The AC joint width measures 12.4 mm; the upper limit of normal is 6 mm. The coracoclavicular distance ➔ is normal, so this is a type II AC separation.* ***(Right)*** *AP radiograph in a patient with a type III AC separation. There is elevation of the distal clavicle ↪ relative to the acromion, as well as widening between the clavicle and the coracoid process ➔ by greater than 50% compared with the normal side (not shown).*

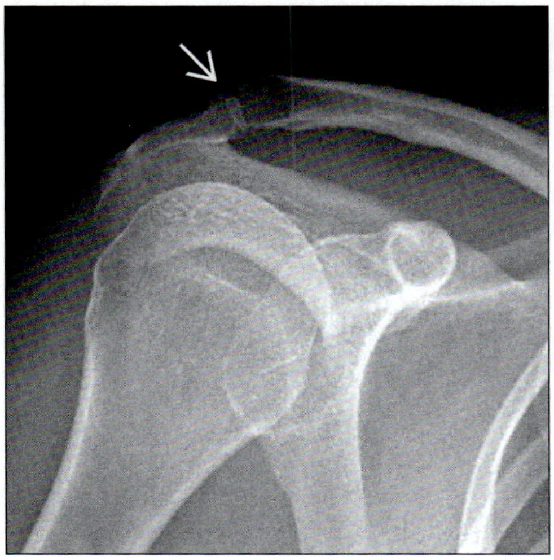

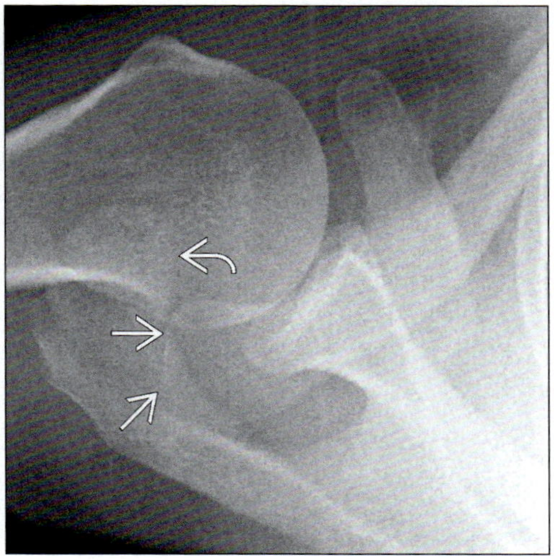

(Left) *AP radiograph in a patient with shoulder pain after trauma. There is loss of congruency at the acromioclavicular joint ➔ but no widening or elevation of the lateral clavicle.* ***(Right)*** *Axillary radiograph in the same patient shows posterior dislocation of the clavicle ➔ relative to the acromial articular surface ↪. This is a type IV acromioclavicular separation. Surgery is often required to reduce the clavicle out of the trapezius muscle.*

POST-TRAUMATIC OSTEOLYSIS DISTAL CLAVICLE

Key Facts

Terminology

- a.k.a. atraumatic osteolysis of distal clavicle (AODC), distal clavicular osteolysis (DCO), stress-induced osteolysis, clavicular osteolysis
- Resorption of distal clavicle from repetitive heavy lifting or after acute AC joint trauma

Imaging

- Radiograph
  - Early: May be normal or have evidence of prior low-grade AC separation
  - Focal loss of distal clavicle subchondral bone plate
  - Later: Resorption of 0.5-3 cm of distal clavicle
  - Subchondral cysts
  - Can have erosion or cupping of acromion
- MR
  - Marrow edema in distal clavicle
  - Linear subchondral fracture line
  - Distended capsule of AC joint

Top Differential Diagnoses

- MR distal clavicle edema
  - Common: 12% of shoulder MRs
  - May be from DJD or early stress reaction
- Rheumatoid arthritis or infection
  - Acromion affected as well as clavicle

Pathology

- Subchondral traumatic or fatigue fracture of lateral clavicle

Clinical Issues

- Pain over AC joint

Diagnostic Checklist

- Relative sparing of medial clavicle
- MR: Look for subchondral fracture line

__(Left)__ AP radiograph of a 20-year-old college softball player with persistent shoulder pain 3 months after a collision with the catcher. There is resorption of the lateral clavicle ➡ consistent with post-traumatic osteolysis. There is also some resorption of the medial acromion ➡, less common in this condition and probably from hyperemia/synovitis. __(Right)__ AP radiograph in a 36-year-old recreational weight lifter. There are areas of subtle subchondral bone plate indistinctness ➡.

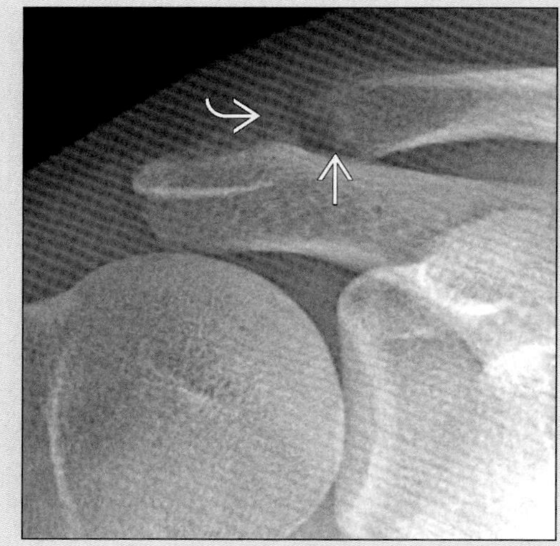

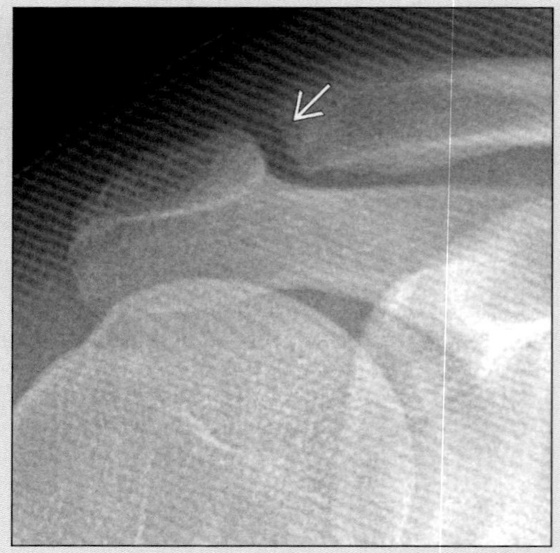

__(Left)__ Axial T2WI FS MR through the distal clavicle in the same patient shows bone marrow edema ➡ and a hypointense subchondral fx line ➡. __(Right)__ Coronal oblique T2WI FS MR in the same patient shows the distal clavicle bone marrow edema ➡ and the subchondral fracture line ➡. The subchondral fracture in this patient is probably a fatigue-type stress fracture. Post-traumatic osteolysis can also occur after a type I or II AC separation.

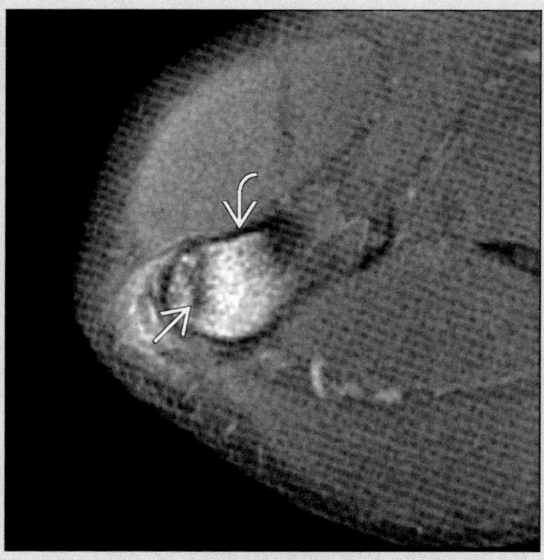

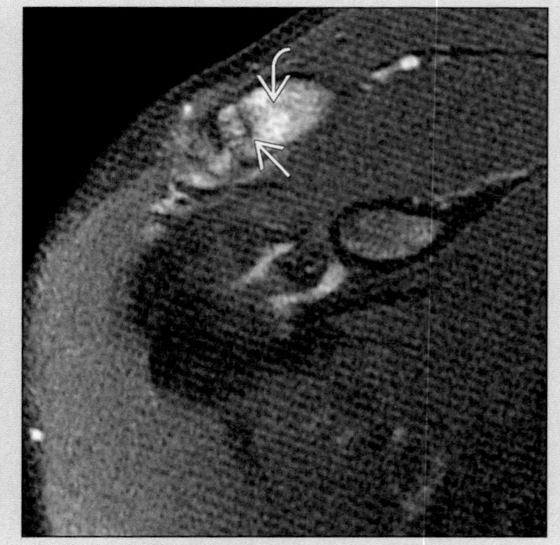

POST-TRAUMATIC OSTEOLYSIS DISTAL CLAVICLE

TERMINOLOGY

Abbreviations
- Atraumatic osteolysis of distal clavicle (AODC), distal clavicular osteolysis (DCO)

Synonyms
- Stress-induced osteolysis, clavicular osteolysis

Definitions
- Resorption of distal clavicle from repetitive heavy lifting or after acute AC joint trauma

IMAGING

General Features
- Best diagnostic clue
  - Osteolysis of distal clavicle with sparing of medial acromion

Radiographic Findings
- Radiography
  - Early: May be normal or have evidence of prior low-grade AC separation
    - Widened AC joint
  - Focal loss of distal clavicle subchondral bone plate
  - Later: Resorption of 0.5-3 cm of distal clavicle
  - Subchondral cysts
  - Can have erosion or cupping of acromion
  - Dystrophic calcification

MR Findings
- T1WI
  - Low signal in distal clavicle
  - Linear hypointense subchondral fracture line
  - Irregularity of subchondral bone plate
- T2WI FS
  - Marrow edema in distal clavicle
    - Medial acromion edema may be present but less common
  - Linear hypointense fx line with surrounding edema
  - Distended capsule of AC joint

Imaging Recommendations
- Best imaging tool
  - MR to show edema and subchondral fx

DIFFERENTIAL DIAGNOSIS

MR Distal Clavicle Edema
- Common: 12% of shoulder MRs
- May be from osteoarthritis or early stress reaction

Septic AC Joint
- Resorption of medial acromion also
- May occur after type V AC separation

Rheumatoid Arthritis
- Associated hand/wrist erosions
- Erosions at medial acromion as well as clavicle

Hyperparathyroidism
- Associated subligamentous & subperiosteal resorption

PATHOLOGY

General Features
- Subchondral traumatic or fatigue fracture of lateral clavicle
- Osteolysis
  - Bone resorption may be due to mild instability in this non-weight-bearing joint
- Other possible etiologies
  - Post-traumatic synovitis
  - Autonomic dysfunction
  - Ischemic necrosis

Gross Pathologic & Surgical Features
- Synovitis
- Degeneration and fissures in cartilage

Microscopic Features
- Abundant osteoblasts and microfractures
- Osteoclast-like giant cells

CLINICAL ISSUES

Presentation
- Most common signs/symptoms
  - Pain over AC joint
- Other signs/symptoms
  - Diminished lifting strength
  - Local crepitation
  - Restricted mobility

Natural History & Prognosis
- Lysis begins 2-3 weeks after injury
- Reparative changes can take 4-6 months
- Subchondral bone reconstitutes with mild deformity

Treatment
- Rest/immobilization
  - Can shorten recovery time

DIAGNOSTIC CHECKLIST

Consider
- Rheumatoid arthritis or infection

Image Interpretation Pearls
- Relative sparing of adjacent acromion
- MR: Look for subchondral fracture line

SELECTED REFERENCES

1. Kassarjian A et al: Distal clavicular osteolysis: MR evidence for subchondral fracture. Skeletal Radiol. 36(1):17-22, 2007
2. Yu YS et al: MR observations of postraumatic osteolysis of the distal clavicle after traumatic separation of the acromioclavicular joint. J Comput Assist Tomogr. 24(1):159-64, 2000
3. de la Puente R et al: Post-traumatic and stress-induced osteolysis of the distal clavicle: MR imaging findings in 17 patients. Skeletal Radiol. 28(4):202-8, 1999
4. Patten RM: Atraumatic osteolysis of the distal clavicle: MR findings. J Comput Assist Tomogr. 19(1):92-5, 1995

SCAPULAR TRAUMA

Key Facts

Imaging

- Linear fracture line on AP, scapular-Y, and axillary radiographs
- Medialization of glenohumeral joint
- Scapular body fx
  - 50-85% of scapula fxs
- Scapular neck fx
  - 10-60% of scapula fxs
- Scapula fxs often identifiable on trauma CXR
  - 80% of patients with scapula fx had other fxs on CXR
- Bone CT reformats: Oblique coronal and oblique sagittal relative to glenoid

Top Differential Diagnoses

- Clinical
  - Proximal humerus fracture
- Radiographic
  - Os acromiale
  - Unfused apophyses
  - Glenoid hypoplasia/dysplasia

Pathology

- Usually direct high-energy trauma
  - MVA in 50%, motorcycle accident in 20%
- Significant associated injuries in 80-90%
  - Pneumothorax in 15-50%
  - Rib fractures in 15-50%
  - Pulmonary contusion in 15-50%
  - Clavicle fx in 25-35%
  - Head trauma in 40%

Clinical Issues

- Surgical indications
  - Displaced intraarticular fracture
  - Severely comminuted

__(Left)__ AP radiograph of a patient with shoulder pain after an MVA. There is a severely comminuted fracture of the left scapula involving the entire body of the scapula, as well as the scapular spine. The glenoid articular surface is not well seen inferiorly; a CT is typically done to evaluate intraarticular extent. __(Right)__ Scapular-Y radiograph in the same patient shows the posterior rotation of a large scapula body fragment and the fracture at the base of the scapular spine.

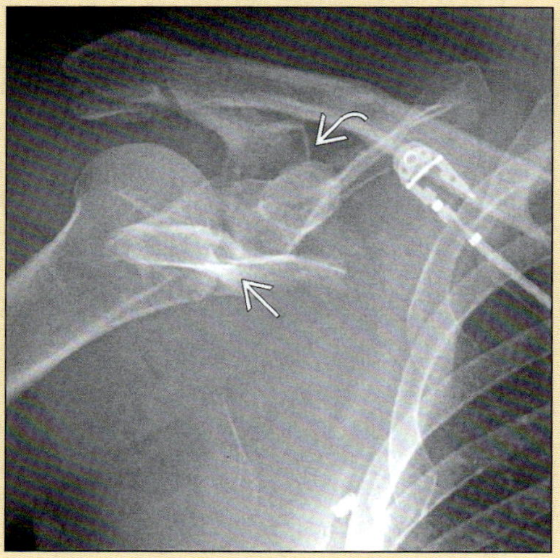

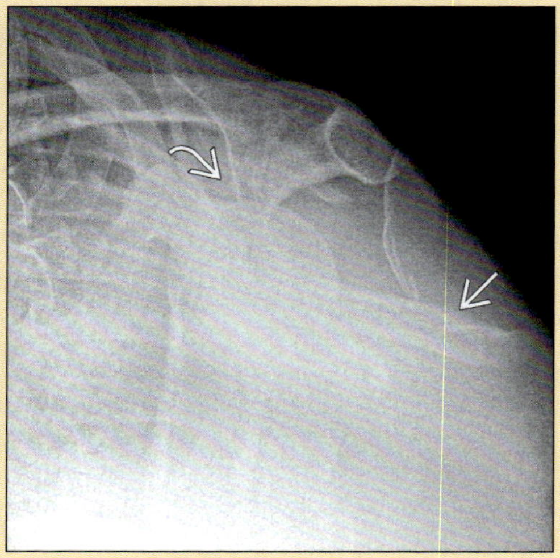

__(Left)__ 3D bone CT of the same patient shows the fracture of the scapular body and the fracture of the base of the scapular spine. Rotating the 3D images helps the orthopedist in surgical planning. __(Right)__ Axial bone CT of the same patient shows the large posteriorly rotated scapular body fragment as well as the extension to the glenoid articular surface. There are several additional displaced smaller fragments. This patient required surgical reduction and plate fixation.

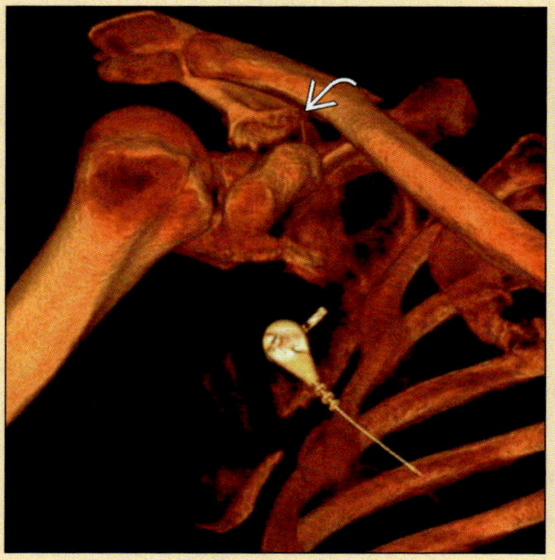

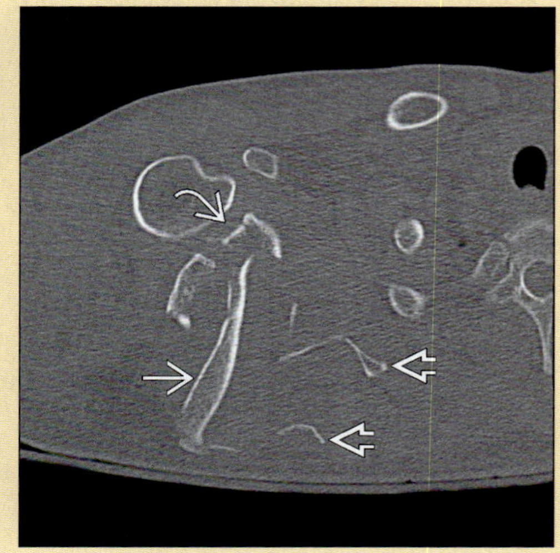

SCAPULAR TRAUMA

TERMINOLOGY

Synonyms

- Glenoid fx, acromion fx, coracoid fx, body fx

Definitions

- Fracture of scapula due to direct trauma or avulsion
  - Also dislocation fxs of glenoid rim

IMAGING

General Features

- Best diagnostic clue
  - Linear fracture line on AP, scapular-Y, and axillary radiographs
  - Medialization of glenohumeral joint
    - Seen best on true AP view of scapula (30° mediolateral AP projection)
- Location
  - Scapular body fx
    - 50-85% of scapula fxs
    - Usually from severe direct trauma
  - Scapular neck fx
    - 10-60% of scapula fxs
    - May have inferior displacement of lateral (glenoid fossa) fragment
  - Glenoid fossa (intraarticular) fx
    - Mostly 2° shoulder dislocation
    - Larger fxs may require surgery
  - Acromion and coracoid fxs
    - Direct trauma or avulsion fxs
  - Spinous process fx
    - Usually from direct trauma

Radiographic Findings

- Scapula fxs often identifiable on review of trauma CXR
  - On prospective reads, only 57% of scapular fxs mentioned in dictation
  - 80% of patients with scapula fx had other fxs on CXR
    - Mostly humerus, clavicle, rib fxs
- Linear lucency through scapula
- Scapula series
  - True AP of scapula
  - Scapular-Y view
  - Axillary view helpful for scapula neck and glenoid rim fxs

CT Findings

- Bone CT
  - More sensitive for nondisplaced fxs
  - Better demonstrates intraarticular extension
    - Quantitate gap &/or step-off of articular surface, although severe OA rare
  - 3D images assist in surgical planning

MR Findings

- T1WI
  - Decreased signal fx line and marrow edema
- T2WI FS
  - Decreased signal intensity fracture line
  - Increased signal marrow edema
  - Increased signal intensity of surrounding musculature from injury
- MR shows associated ligamentous, muscular, or brachial plexus injury

Imaging Recommendations

- Best imaging tool
  - Initial evaluation: True AP and scapular-Y views of shoulder
  - Best imaging: Bone CT with multiplanar reformats
- Protocol advice
  - Bone CT reformats: Oblique coronal and oblique sagittal relative to glenoid
    - 3D images for complicated fxs

DIFFERENTIAL DIAGNOSIS

Clinical

- Proximal humerus fracture
  - Similar mechanism of injury and presentation
  - More swelling and ecchymosis
- Glenohumeral dislocation
- Glenoid osteochondral injury

Radiographic

- Os acromiale
  - 2% of shoulders, 66% bilateral
  - Anatomic variant consisting of fibrocartilaginous union of acromion to scapular spine
  - May contribute to shoulder impingement and rotator cuff tear
- Glenoid hypoplasia
  - Normal variant
  - Associated with multidirectional instability and posterior instability/labral tears
- Unfused apophyses
  - Acromion physis fuses at 25-30 years
  - Coracoid base ossification center fuses at 14-16 years
    - Physis extends across upper 1/3 of glenoid
  - Coracoid tip and crescent-shaped medial base centers fuse at 18-25 years
  - Glenoid lower pole fuses at 20-25 years
  - Inferior angle and medial border fuse at 20-25 years

PATHOLOGY

General Features

- Etiology
  - Usually direct high-energy trauma
    - MVA in 50%, motorcycle accident in 20%
    - Scapula is surrounded by muscle and articulates with clavicle and humerus, both of which are likely to fracture before scapula
  - Glenohumeral dislocation may lead to glenoid rim fracture
- Associated abnormalities
  - Significant associated injuries in 80-90%
    - Pneumothorax in 15-50%
      - May be delayed in onset by several days
    - Rib fractures in 15-50%
    - Pulmonary contusion in 15-50%
    - Clavicle fx in 25-35%
      - Usually with glenoid fx
    - Brachial plexus injury in 5-10%
      - Poor prognosis
      - Arterial injury in 50%

SCAPULAR TRAUMA

- Head trauma in 40%
- Arterial injury in 10%
- Suprascapular nerve injury
  - Fracture extension into suprascapular or spinoglenoid notch
- Mortality rate 10-15% in patients with scapular fx
  - Usually from other injuries or pneumonia complications

Staging, Grading, & Classification

- Zdrakovic and Damholt system
  - Type I: Fractures of body
  - Type II: Fractures of coracoid and acromion
  - Type III: Glenoid and neck fxs
    - More difficult to treat
    - 6% of scapula fxs
- Thompson system
  - Type I: Coracoid, acromion, and small body fxs
  - Type II: Glenoid and neck fxs
  - Type III: Major scapular body fxs
  - Types II and III more likely to have associated injuries
- Kuhn classification of acromion fractures
  - Type I: Minimally displaced
  - Type II: Displaced, but no loss of subacromial space
  - Type III: Displaced with associated loss of subacromial space
- Glenoid neck fractures
  - Stable: No clavicle fracture or AC joint separation
  - Unstable: Associated clavicle fracture or AC dislocation
- Coracoid process fractures
  - Proximal to coracoclavicular ligaments
  - Distal to coracoclavicular ligaments
- Intraarticular glenoid fractures (Ideberg classification)
  - Type I: Rim fx glenoid
    - Ia: Anterior
    - Ib: Posterior
  - Type II: Transverse glenoid fracture, extending to axillary border of scapula, inferior fragment displaced with humeral head
  - Type III: Oblique glenoid fracture, exiting at superior border of scapula
    - Often associated AC dislocation or fx
  - Type IV: Horizontal (transverse) fracture crossing scapula from glenoid, exiting through medial border of scapular body
  - Type V: Type IV plus 2nd fx separating inferior glenoid fragment
  - Type VI: Severe comminuted fx of glenoid articular surface

CLINICAL ISSUES

Presentation

- Most common signs/symptoms
  - Pain and tenderness over shoulder and scapula
  - Shoulder may appear flattened
  - Prominent acromion
  - Arm is held adducted; abduction is very painful
  - Less ecchymosis than upper humerus fx

Demographics

- Age
  - Mean: 35-45 years old
- Gender
  - M > F
- Epidemiology
  - Approximately 0.5% of all fractures
  - 3-5% of shoulder girdle fxs
  - Usually high-energy trauma
    - Motor vehicle accident in 50%
    - Motorcycle accident in 20%

Natural History & Prognosis

- Usually good result with conservative treatment
  - May have decreased ROM

Treatment

- Conservative
  - 90% of scapula fractures are minimally displaced or nondisplaced
  - Goal to restore function, minimize long-term stiffness
- Surgical indications
  - Displaced intraarticular fracture
    - Especially if humeral head displaced with fragment
  - Severely displaced extraarticular fracture
    - Severely comminuted
    - Significant displacement or angulation of scapular neck fx
    - Significantly displaced scapular spine, coracoid, and acromial fxs
    - "Floating shoulder" with clavicle fracture

DIAGNOSTIC CHECKLIST

Consider

- CT of scapula if diagnosis unsure

Image Interpretation Pearls

- Carefully inspect scapula on trauma chest radiograph

Reporting Tips

- Comment on glenoid articular surface involvement

SELECTED REFERENCES

1. Zlowodzki M et al: Treatment of scapula fractures: systematic review of 520 fractures in 22 case series. J Orthop Trauma. 20(3):230-3, 2006
2. Cole PA: Scapula fractures. Orthop Clin North Am. 33(1):1-18, vii, 2002
3. Resnick D: Diagnosis of Bone and Joint Disorders. 4th ed. Philadelphia: W.B. Saunders Company. 2817-20, 2824-7, 2002
4. Bucholz RW et al: Rockwood and Green's Fractures in Adults. 5th ed. Philadelphia: Lippincott Williams and Wilkins. 1079-1092, 2001
5. Goss TP: The scapula: coracoid, acromial, and avulsion fractures. Am J Orthop. 25(2):106-15, 1996

SCAPULAR TRAUMA

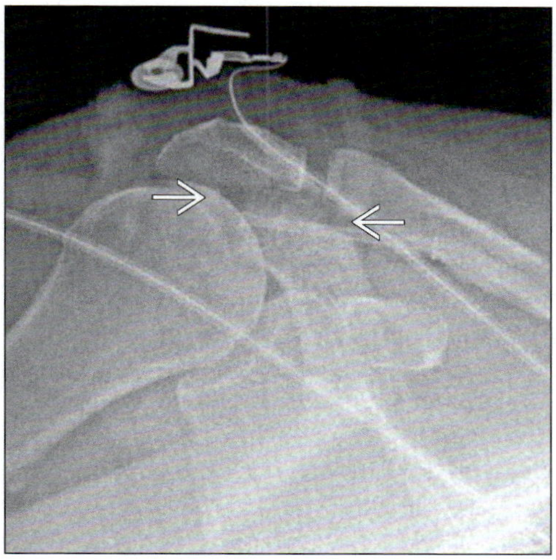

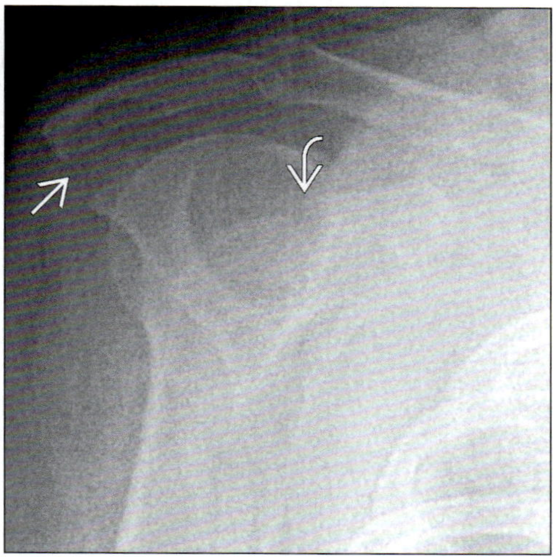

(Left) *AP radiograph of a patient after an MVA. There is a lucency at the base of the acromion process ➔, consistent with a fracture.* ***(Right)*** *Scapular-Y radiograph of the same patient confirms the lucency at the junction of the acromion process and scapular spine ➔. Most acromion process fractures can be treated conservatively. There is also a lucency at the base of the coracoid process ➔. The patient could not tolerate positioning for an axillary view, so CT was performed.*

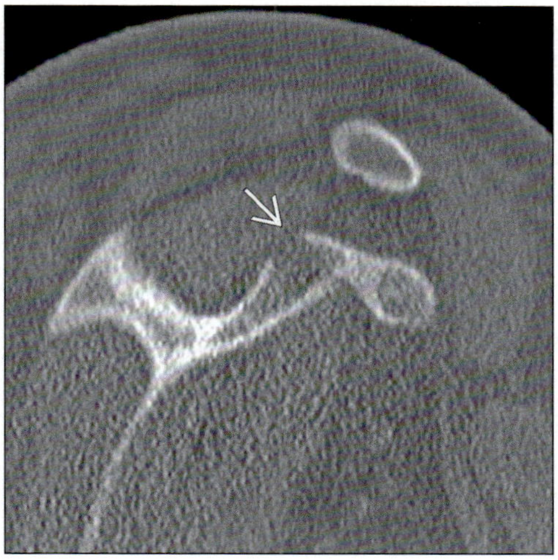

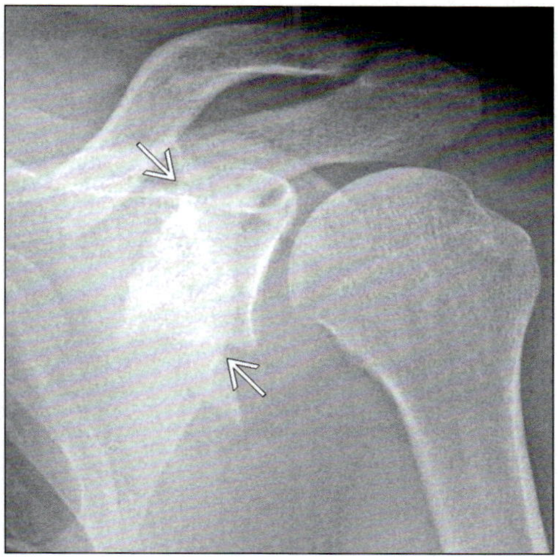

(Left) *Sagittal reformat bone CT in the same patient confirms the fracture at the base of the coracoid process ➔ as well as the acromion fx (not shown). Minimally displaced or angulated coracoid process fractures are usually treated conservatively.* ***(Right)*** *AP radiograph in a patient with a "flattened" appearing shoulder after an MVA. There is an impacted fracture of the scapular neck ➔ and no apparent involvement of the articular surface, which could be confirmed with a CT.*

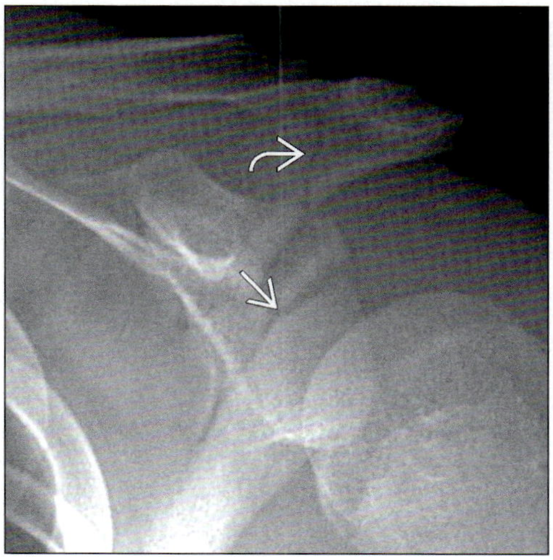

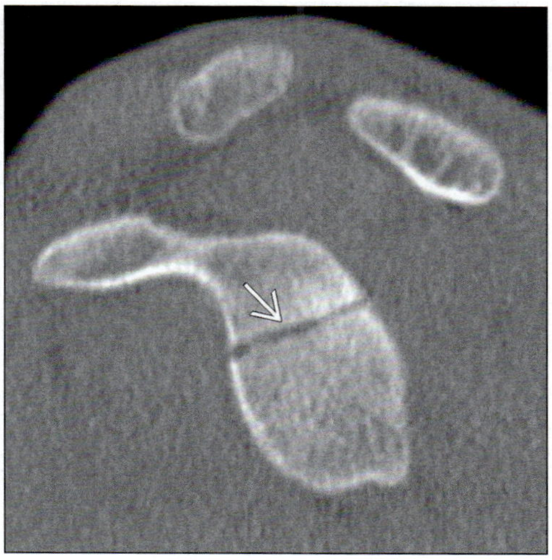

(Left) *AP radiograph in a patient complaining of shoulder pain 2 days after an MVA. There is a lucency through the glenoid articular surface ➔ consistent with a type II glenoid fracture. There is also a fracture of the acromion process ➔. The fractures were not well seen on the CXRs.* ***(Right)*** *Sagittal reformat bone CT of the same patient confirms the minimal displacement of glenoid articular surface ➔. Minimally displaced type II glenoid fractures can usually be treated conservatively.*

ANTERIOR GLENOHUMERAL DISLOCATION

Key Facts

Imaging

- Humeral head displaced anteriorly, inferiorly, and medially
  - Below coracoid process
  - May be "locked": Humeral head impaled on glenoid rim
  - Bankart and Hill-Sachs fxs
- Anterior instability or postreduction
  - 35% of fractures visible only on postreduction
- CT postreduction: More sensitive for showing fxs
- MR postreduction
  - Bankart lesion: Anteroinferior labral tear
  - Posterolateral humeral head bone contusion or Hill Sachs impaction fracture
- MR arthrogram
  - More sensitive than conventional MR for Bankart labral tear

Pathology

- Glenohumeral joint is most commonly dislocated joint in body
- 95% of glenohumeral dislocations are anterior
- Hill-Sachs impaction fx in 80%
- Bankart fx
  - Can go onto nonunion or fragment can be resorbed
  - Glenoid rim defect often enlarges with repeated dislocation
- Bankart lesion labral tear and variants
  - Occurs in 75% of patients < 40 years old
- Older patients
  - Anterior band IGHL tear
  - Rotator cuff tear
  - Greater tuberosity avulsion fx

*(**Left**) AP radiograph in a patient with shoulder pain after trauma shows inferior and medial displacement of the humeral head → relative to the glenoid fossa ⇨. The humeral head in anterior dislocation is usually pulled medially by muscles and displaced inferiorly under the coracoid process. (**Right**) Axillary lateral radiograph of the same patient shows the humeral head displaced anterior ➡, as well as medial to the glenoid fossa ⇨. 95% of all glenohumeral dislocations are anterior.*

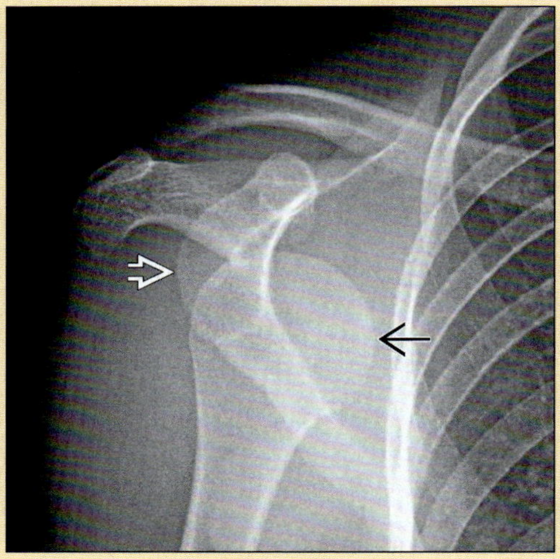

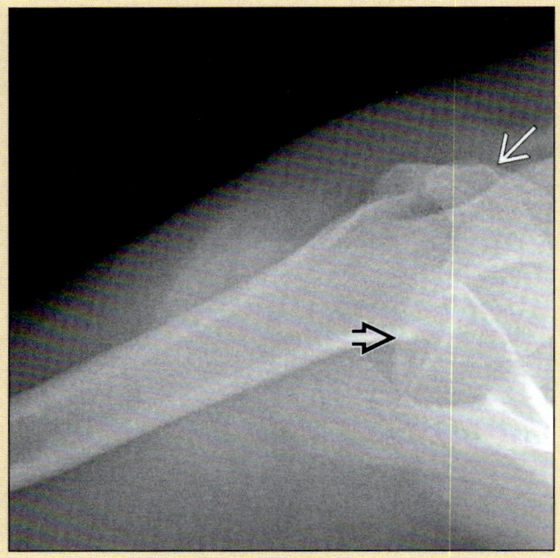

*(**Left**) Lateral graphic of the glenoid fossa shows a fracture → of the anteroinferior glenoid rim. This is a Bankart fracture and is seen in approximately 15% of patients after an initial anterior dislocation. These can go onto nonunion or resorption of the fragment if repeated anterior dislocations. (**Right**) Axillary lateral radiograph shows a Bankart fracture → of the anterior glenoid rim. A Bankart fracture indicates that the patient has had a prior anterior dislocation.*

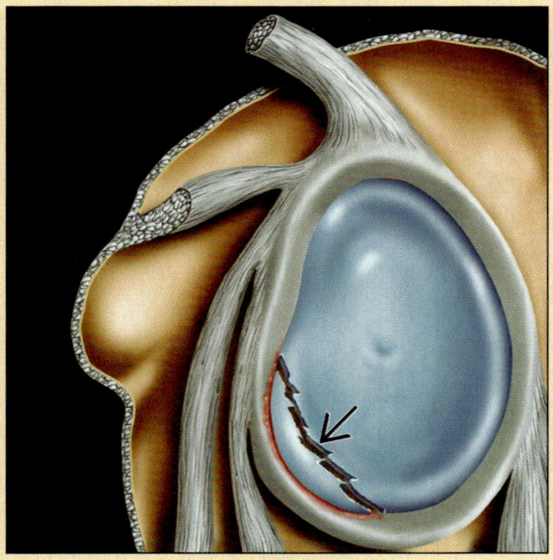

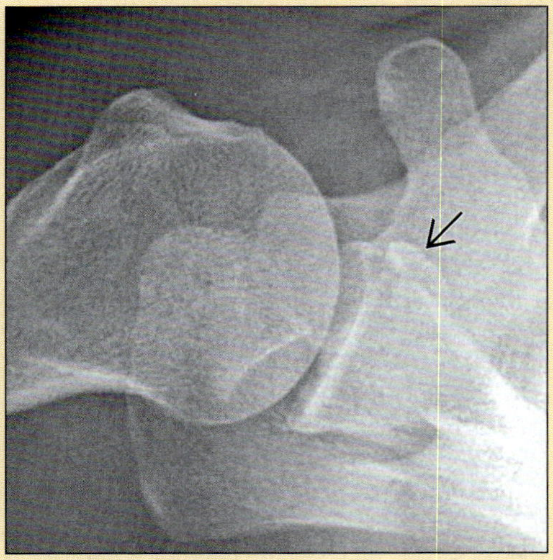

ANTERIOR GLENOHUMERAL DISLOCATION

TERMINOLOGY

Synonyms

- Shoulder dislocation

Definitions

- Anterior displacement of humeral head with loss of articulation with glenoid

IMAGING

General Features

- Best diagnostic clue
  - Anterior, inferior, and medial displacement of humeral head
    - Hill-Sachs and Bankart fractures
- Location
  - Humeral head below coracoid process

Radiographic Findings

- AP view
  - Humeral head displaced medial and caudal to glenoid fossa
  - Typically inferior to coracoid process
  - May be "locked" with anterior glenoid rim sitting in impaction fx of humeral head (Hill-Sachs lesion)
  - Older patients
    - Greater tuberosity fx
    - Rotator cuff tear
- Axillary view
  - Humeral head displaced anteriorly
  - May have fx of anterior glenoid rim (Bankart fx)
    - Bankart fx in 15% after initial dislocation
- Scapular-Y view
  - Some cannot position arm for axillary view, so scapular-Y view good for showing direction of dislocation
  - Humeral head displaced anteriorly
- Anterior instability on postreduction radiographs
  - 35% of fractures after dislocation only visible on postreduction radiographs
- AP with internal rotation
  - Hill-Sachs lesion
    - Posterolateral impaction fracture
    - Present in 80% after initial dislocation
    - Often obscured on external rotation or axillary view
- West Point view
  - Angled axillary view
  - Profiles 4-5 o'clock portion of glenoid rim
  - More sensitive for small Bankart fx than axillary view

CT Findings

- Postreduction: More sensitive for showing Hill-Sachs and Bankart fxs than radiograph
  - For large Hill-Sachs and Bankart fxs: Helps in planning bone graft if recurrent dislocation
- CT arthrography
  - Useful for diagnosing labral tears, especially if metal from prior surgery (concern for MR artifact)

MR Findings

- Postreduction
- Axial PD/intermediate FS
  - Anteroinferior
    - Bankart lesion: Anteroinferior labral tear
    - Bankart fx
    - Torn anterior band inferior glenohumeral ligament
  - Posterolateral humeral head
    - Bone contusion
    - Hill-Sachs impaction fx
- MR arthrogram
  - More sensitive than conventional MR for Bankart labral tear
  - ABER position oblique axial
    - Most sensitive for detecting Bankart labral tear

Imaging Recommendations

- Best imaging tool
  - Radiograph: AP and axillary view
    - AP with internal rotation after reduction to see Hill-Sachs lesion
- Protocol advice
  - Postreduction: MR arthrogram + ABER

DIFFERENTIAL DIAGNOSIS

Posterior Dislocation

- Usually less displacement on AP radiograph
- Confirmed with axillary or scapular-Y view

Inferior Humeral Head Subluxation

- Subluxed and not dislocated
- Due to muscle weakness, usually after humeral head fx

Multidirectional Instability

- Patient complains of instability
- No prior dislocation, no labral tear
- Familial, or in athletes whose shoulders are repeatedly put through wide range of motion (e.g., swimmers)

PATHOLOGY

General Features

- Etiology
  - Glenohumeral joint is most commonly dislocated joint in body
    - 45% of all dislocations
  - Glenoid fossa is shallow so as to allow wide range of motion
    - Relatively easy to dislocate and tear soft tissue stabilizers
  - Shoulder stability from
    - Dynamic stabilizers
      - Rotator cuff muscles
      - Pectoralis major, teres major
    - Static stabilizers
      - Labrum, joint capsule, glenohumeral ligaments
    - Adhesion-cohesion
    - Suction cup effect
  - 95% of glenohumeral dislocations are anterior
    - Mechanism
      - Arm in abduction and external rotation, anterior blow to distal arm
      - Humeral head levers out anteriorly as humeral neck contacts acromion

ANTERIOR GLENOHUMERAL DISLOCATION

    - Direct blow to back of shoulder or violent arm traction
- Associated abnormalities
  - Hill-Sachs impaction fx in 80%
    - Stable: < 20% of articular surface involved
    - Intermediate: 20-40% of articular surface involved
    - Unstable (likely to re-dislocate): > 40% of articular surface involved
    - Engaging: If glenoid rim drops into Hill-Sachs depression during functional movement
  - Bankart fx
    - Can go on to nonunion or fragment can be resorbed
    - Glenoid rim defect often enlarges with repeated dislocation
    - If glenoid rim defect > 20-25% of glenoid fossa, may need bone graft
  - Bankart lesion (anterior labral tear) and variants
    - Occurs in 75% of patients < 40 years old after anterior dislocation
    - Bankart lesion: Detached labral fragment
    - Variants
      - Partial Bankart labral tear
      - Anterior labrum periosteal sleeve avulsion (ALPSA)
      - Perthes lesion
  - Glenolabral articular disruption (GLAD) lesion
  - Bony humeral avulsion of IGHL (BHAGL)
  - Older patients
    - Anterior band inferior glenohumeral ligament (IGHL) tear
      - Near labral attachment in 40%
      - Intrasubstance in 35%
      - Humeral avulsion of IGHL (HAGL) in 25%
    - Rotator cuff tear
      - > 40 years old: 30%
      - > 60 years old: 80%
    - Greater tuberosity avulsion fx
      - 4x more common in those > 30 years old
    - Subscapularis tendon tear or lesser tuberosity avulsion fx
  - Nerve and vascular injuries: Uncommon
    - Axillary nerve: Lose sensation in lateral shoulder
  - Paralabral cyst

Staging, Grading, & Classification

- Type 1: Partial Bankart labral tear
- Type 2: Detached labrum (true Bankart tear)
- Type 3: Severe detachment with attenuated labrum
- Type 4: Bankart glenoid rim fx

CLINICAL ISSUES

Presentation

- Most common signs/symptoms
  - Pain
  - Shoulder feels like it "popped out"
- Other signs/symptoms
  - Sulcus just beneath acromion
  - Prominent low-lying humeral head
  - Arm held in slight abduction and external rotation

Demographics

- Age
  - Peak: 15-25 years old
- Gender
  - M > F

Natural History & Prognosis

- Recurrent instability
  - < 20 years old: 90% recurrence
    - Often require Bankart repair
  - > 40 years old: 10-15% recurrence
- Chronic pain common

Treatment

- Closed reduction
  - Perform promptly to avoid muscle spasm and enlargement of Hill-Sachs fx
- Strengthen muscles that act as dynamic stabilizers
- Teach patient to avoid positions where dislocation occurs
- Surgery
  - Rarely required for reduction
  - Common for instability in young people
    - Bankart labral tear repair
    - Rare: Bone graft to fill anteroinferior glenoid rim defect or Hill-Sachs impaction fx

DIAGNOSTIC CHECKLIST

Consider

- Presence of associated fractures

Image Interpretation Pearls

- If shoulder dislocated on AP view, 95% are anterior

Reporting Tips

- Mention Hill-Sachs or Bankart fx

SELECTED REFERENCES

1. Connolly S et al: Irreducible anterior dislocation of the shoulder due to soft tissue interposition of subscapularis tendon. Skeletal Radiol. 37(1):63-5, 2008
2. Diederichs G et al: CT-based patient-specific modeling of glenoid rim defects: a feasibility study. AJR Am J Roentgenol. 191(5):1406-11, 2008
3. Griffith JF et al: Prevalence, pattern, and spectrum of glenoid bone loss in anterior shoulder dislocation: CT analysis of 218 patients. AJR Am J Roentgenol. 190(5):1247-54, 2008
4. Antonio GE et al: First-time shoulder dislocation: High prevalence of labral injury and age-related differences revealed by MR arthrography. J Magn Reson Imaging. 26(4):983-91, 2007
5. Griffith JF et al: CT compared with arthroscopy in quantifying glenoid bone loss. AJR Am J Roentgenol. 189(6):1490-3, 2007
6. Widjaja AB et al: Correlation between Bankart and Hill-Sachs lesions in anterior shoulder dislocation. ANZ J Surg. 76(6):436-8, 2006
7. Griffith JF et al: Anterior shoulder dislocation: quantification of glenoid bone loss with CT. AJR Am J Roentgenol. 180(5):1423-30, 2003
8. Ito H et al: Radiographic evaluation of the Hill-Sachs lesion in patients with recurrent anterior shoulder instability. J Shoulder Elbow Surg. 9(6):495-7, 2000

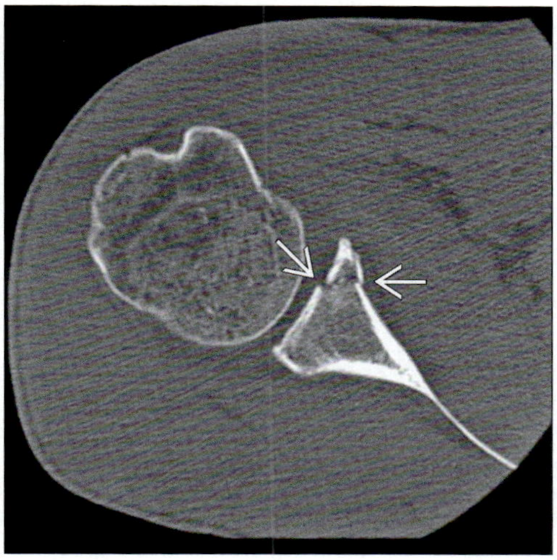

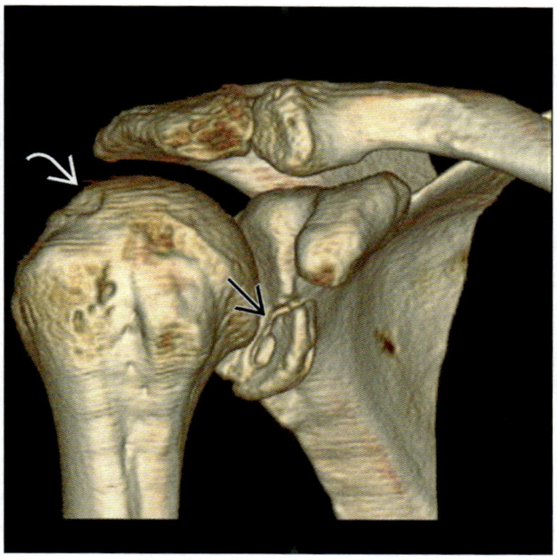

(Left) *Axial bone CT shows a patient who had a prior dislocation reduced and now with anterior instability and pain. There is a Bankart fracture of the anterior glenoid rim ➔. CT can evaluate a Bankart fracture not well seen on radiograph and help with preoperative planning.* ***(Right)*** *AP view 3D bone CT of the same patient shows the Bankart fracture of the anteroinferior glenoid rim ➔. The patient also has a small Hill-Sachs impaction fracture ➔ of the posterolateral superior humeral head.*

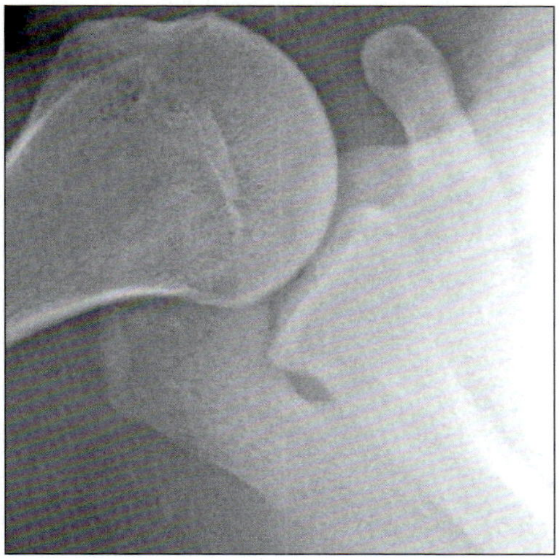

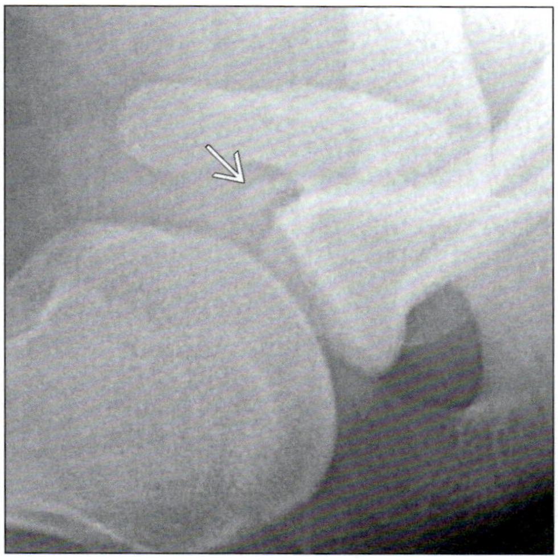

(Left) *Axillary lateral radiograph shows a patient with a prior dislocation that was reduced, complaining of continued pain and instability appears normal, as did the AP view (not shown).* ***(Right)*** *West Point radiograph of the same patient shows a bone fragment ➔ adjacent to the anteroinferior glenoid, a Bankart fracture. The West Point view, which profiles the anteroinferior glenoid rim, can be used like CT to see small Bankart fractures not easily seen on AP or axillary radiographs.*

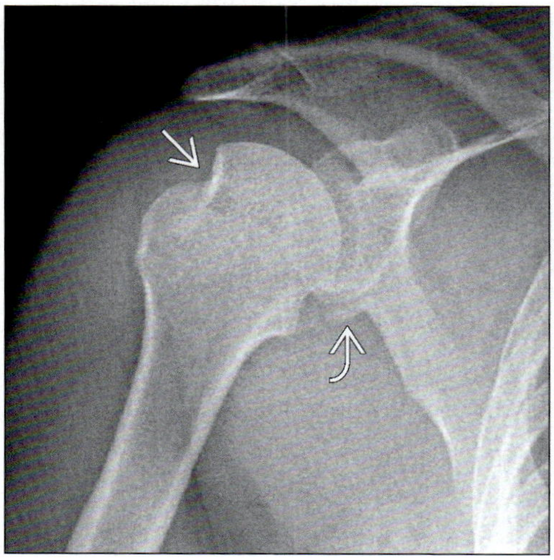

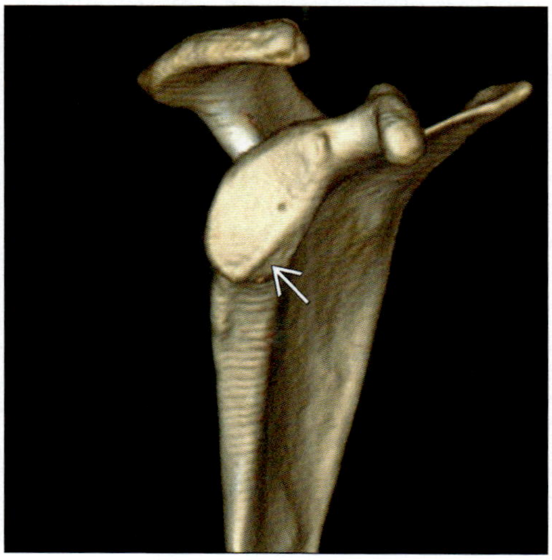

(Left) *AP radiograph of a patient with a prior dislocation that was reduced shows a Hill-Sachs fracture of the posterolateral humeral head ➔ and a small Bankart fracture ➔.* ***(Right)*** *Lateral 3D bone CT of a patient with multiple anterior dislocations and increasing instability shows a defect in the anteroinferior glenoid rim ➔. Resorption of the Bankart fragment and mechanical wear of the glenoid rim can occur from repeated dislocations.*

POSTERIOR GLENOHUMERAL DISLOCATION

Key Facts

Terminology

- Posterior displacement of humeral head with loss of articulation with glenoid fossa
  - Less common than posterior subluxation, or posterior instability without acute traumatic event

Imaging

- Arm in fixed internal rotation
- Usually dislocates straight posterior (subacromial)
- "Trough" sign (reverse Hill-Sachs)
  - Vertical linear sclerosis of medial humeral head
  - 75% of posterior dislocations
  - Due to anteromedial humeral head impaction fracture on posterior glenoid rim
- "Lightbulb" sign (fixed internal rotation)
  - Lesser tuberosity medial and greater tuberosity lateral so proximal humerus shaped like light bulb
- "Rim" sign (shoulder joint width > 6 mm)
- Loss of normal "half-moon overlap"
  - Due to lateral displacement of dislocated humeral head
- Axillary view: Posterior displaced humeral head

Pathology

- 2-4% of glenohumeral dislocations
- Mechanism
  - Epileptic seizure: Most common
  - Fall on outstretched hand or blow to flexed, adducted, internal rotated shoulder
  - Increased risk if hypoplastic glenoid

Diagnostic Checklist

- Consider posterior dislocation if shoulder fixed in internal rotation post trauma
- Need axillary or scapular-Y view
  - Posterior dislocation often missed on AP view

*(**Left**) AP radiograph of a patient with shoulder pain after a motorcycle accident. Note the wide glenohumeral joint ➡, the so-called "rim" sign, and loss of the normal "half-moon overlap." The humerus is also internally rotated causing the "light bulb" sign where the lesser tuberosity ⮫ projects medially. (**Right**) Axillary lateral radiograph of the same patient shows that the humeral head is "perched" on the posterior rim of the glenoid ➡. Most posterior dislocations are straight posterior.*

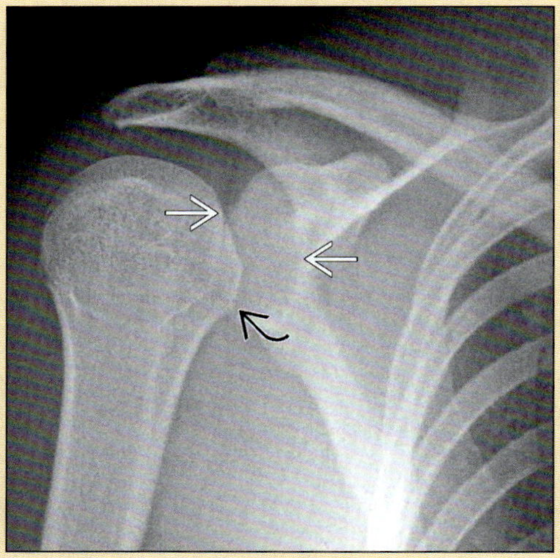

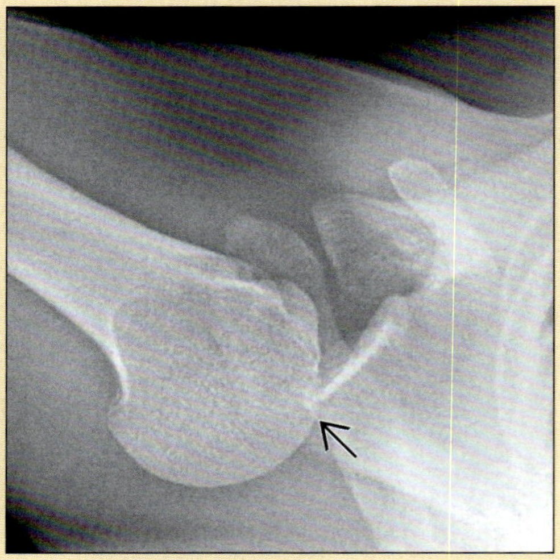

*(**Left**) Axial graphic shows a common appearance after reducing a posterior dislocation. Note the reverse Bankart fx ➡, reverse Hill-Sachs fx ⮫, and injury to the periosteum ⇨. (**Right**) Axillary radiograph of a football player who felt that his shoulder was "going out posteriorly" after a hard fall onto an adducted arm. There is a bony fragment off the posterior glenoid ➡ with an adjacent donor site in the glenoid rim ⮫. A reverse Bankart fracture was confirmed at surgery.*

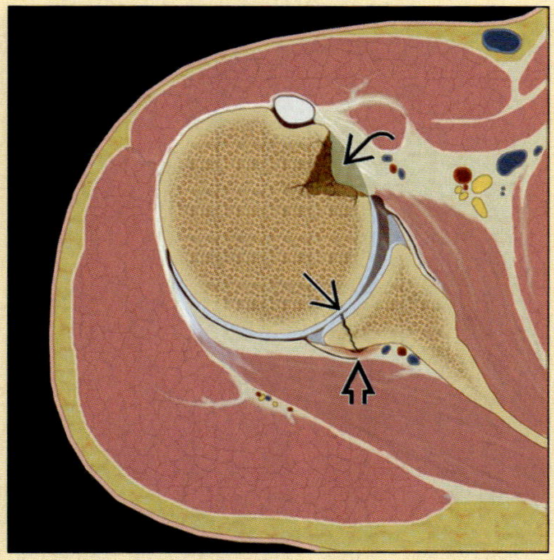

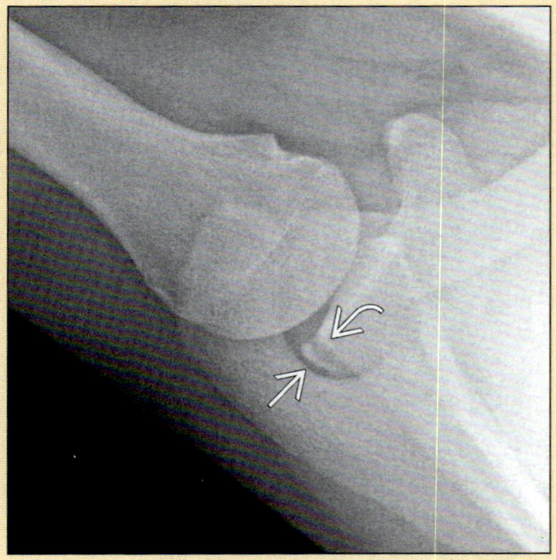

POSTERIOR GLENOHUMERAL DISLOCATION

TERMINOLOGY

Synonyms

- Posterior shoulder dislocation

Definitions

- Posterior displacement of humeral head with loss of articulation with glenoid fossa
  - Less common than posterior subluxation or posterior instability without acute traumatic event

IMAGING

General Features

- Best diagnostic clue
  - Incongruent glenohumeral joint
    - Arm in fixed internal rotation
- Location
  - Usually dislocates straight posterior (subacromial)
    - Rarely: Subspinous or subglenoid
- Size
  - Reverse Hill-Sachs impaction fracture variable in size
    - Best predictor of likelihood of recurrent dislocation

Radiographic Findings

- AP view
  - Joint appears incongruent: Narrow or wide
    - Narrow if "locked"
      - Posterior glenoid rim impaled into reverse Hill-Sachs impaction fracture
    - Wide if "perched"
      - Humeral head resting on posterior glenoid rim
      - No significant reverse Hill-Sachs or reverse Bankart fracture
  - Findings often subtle: Little displacement on AP view
    - 50% missed on initial interpretation
    - Chronic posterior dislocation not uncommon
  - "Trough" sign
    - Vertical linear sclerosis medial humeral head
    - Present in 75% of posterior dislocations
    - Due to reverse Hill-Sachs impaction fracture
      - Impacted bone along fx margin parallel to radiograph beam
  - "Lightbulb" sign
    - Lesser tuberosity medial and greater tuberosity lateral: Proximal humerus shaped like light bulb
    - Due to marked internal rotation
  - "Rim" sign
    - Widened glenohumeral joint
    - > 6 mm, seen in 2/3 of posterior dislocations
  - Loss of normal "half-moon overlap"
    - Due to lateral displacement of dislocated humeral head
- Axillary view
  - Posteriorly displaced humeral head
  - Best view for confirming posterior dislocation
  - Often "locked": Humeral head stuck on posterior glenoid rim
    - Posterior glenoid rim sitting in reverse Hill-Sachs defect
  - Posterior Bankart fracture
    - Fracture of posterior glenoid rim
    - 30% of posterior dislocations
- Scapular-Y view
  - Posterior displaced humeral head
  - Obtain if patient cannot abduct for axillary view
- **Posterior instability** or **post-reduction**
  - AP and axillary views: Fractures better seen
  - Posterior instability: May also be seen in patients without prior posterior dislocation

CT Findings

- Prior to open reduction: Preoperative planning
  - Size of reverse Hill-Sachs lesion
  - Presence of reverse Bankart fx
- Post-reduction: Assess fxs not well seen on radiograph
  - Recurrent dislocations may cause posterior glenoid rim defect
- CT arthrography
  - Useful for diagnosing labral tears, especially if metal from prior surgery (concern for MR artifact)

MR Findings

- **Post-reduction**
- Axial PD/intermediate FS
  - Glenoid
    - Posterior Bankart lesion: Posterior labral tear
      - Also common in athletes with instability but no prior posterior dislocation
    - Posterior Bankart fx
  - Humeral head
    - Anteromedial humeral head bone contusion
    - Reverse Hill-Sachs impaction fracture
- MR arthrogram
  - More sensitive than conventional MR for reverse Bankart labral tear

Imaging Recommendations

- Best imaging tool
  - Axillary or scapular-Y view
    - Posterior dislocation easy to overlook on AP radiograph
- Protocol advice
  - Post-reduction: MR arthrogram to evaluate posterior labrum

DIFFERENTIAL DIAGNOSIS

Hypoplastic (Dysplastic) Glenoid

- Congenital flat/convex/retroverted posterior glenoid fossa
- Hypertrophy of posterior labrum
- Posterior labral tear, instability common
- Increased risk of posterior dislocation

Multidirectional Instability, Shoulder

- No prior dislocation
- No posterior labral tear
- In athletes where shoulder is repeatedly put thru wide range of motion, such as swimmers
- May be familial

Bennett Lesion

- Throwing athletes
- Posterior glenoid rim intact

POSTERIOR GLENOHUMERAL DISLOCATION

Adhesive Capsulitis, Shoulder
- In the clinical differential diagnosis
- Often after trauma
- No reverse Hill-Sachs lesion

PATHOLOGY

General Features
- Etiology
  - 2-4% of glenohumeral dislocations
  - Epileptic seizure (most common)
    - Strong contraction of posterior internal rotators
      - Latissimus dorsi and teres major
    - Anterior internal rotators also much stronger than external rotators
    - Can be bilateral
      - If patient has bilateral posterior dislocation and does not know why, think 1st onset of epilepsy
  - Severe electric shock
  - Fall on outstretched hand or blow to anterior shoulder
    - Occurs usually when shoulder flexed, adducted, internal rotated
  - Severe direct blow to anterior shoulder
  - American football linemen ("stiffarming")
  - Increased risk if hypoplastic or retroverted glenoid
  - Children: Posterior dislocation rare
    - Birth-related paralysis
    - Congenital severely hypoplastic glenoid
- Associated abnormalities
  - Fractures
    - Reverse Hill-Sachs (McLaughlin lesion, "trough" sign) at anteromedial humeral head
    - Reverse Bankart fracture
  - Soft tissue injuries
    - Reverse Bankart lesion (labral tear)
    - Posterior labrocapsular periosteal sleeve avulsion (POLPSA)
    - Tear of posterior band of inferior glenohumeral ligament (IGHL)
      - Midsubstance tear
      - Posterior humeral avulsion of the glenohumeral ligament (PHAGL)
    - Teres minor muscle/tendon tear/strain
  - Lesser tuberosity fracture
  - Neurovascular injuries: Rare
  - Humeral neck fractures may have associated glenohumeral dislocation

Staging, Grading, & Classification
- Humeral head defect: Percentage of articular surface involved
  - Small: < 20%
  - Medium: 20-40%
  - Large: > 40%

CLINICAL ISSUES

Presentation
- Most common signs/symptoms
  - Shoulder pain, especially post-ictal
  - Limited range of motion
- Other signs/symptoms
  - Arm in fixed internal rotation
  - Prominent coracoid process, flattened anterior shoulder, prominent rounded posterior shoulder

Demographics
- Age
  - Average: 50 years old
    - Older than anterior dislocation

Natural History & Prognosis
- High rate of missed dislocation
  - "Chronic posterior dislocation": Locked > 3 weeks
- When locked, may require open reduction
- Chronic pain common

Treatment
- Nonoperative: If reverse Hill-Sachs defect < 20%
  - Strengthen external rotators
  - Avoid dislocation position (adduction and internal rotation)
- Surgery: Instability or recurrent posterior dislocations
  - Posterior capsular shift
  - Reverse Bankart lesion: Labral repair
  - Reverse Hill-Sachs lesion
    - < 20%: May need partial subscapularis tendon transfer into defect
    - 20-40%: Lesser tuberosity transfer into defect
      - "McLaughlin procedure"
    - > 40%: Allograft into defect, or hemiarthroplasty
  - Reverse Bankart fracture: If large may need screw fixation
  - Large posterior rim defect: Glenoid rim bone graft

DIAGNOSTIC CHECKLIST

Consider
- Posterior dislocation if fixed internal rotation post trauma

Image Interpretation Pearls
- Need axillary or scapular-Y view
  - Posterior dislocation often missed on AP view

Reporting Tips
- Mention if reverse Hill-Sachs or reverse Bankart fx

SELECTED REFERENCES

1. Harish S et al: Imaging findings in posterior instability of the shoulder. Skeletal Radiol. 37(8):693-707, 2008
2. Kowalsky MS et al: Traumatic posterior glenohumeral dislocation: classification, pathoanatomy, diagnosis, and treatment. Orthop Clin North Am. 39(4):519-33, viii, 2008
3. Saupe N et al: Acute traumatic posterior shoulder dislocation: MR findings. Radiology. 248(1):185-93, 2008
4. Chung CB et al: Humeral avulsion of the posterior band of the inferior glenohumeral ligament: MR arthrography and clinical correlation in 17 patients. AJR Am J Roentgenol. 183(2):355-9, 2004
5. Cicak N: Posterior dislocation of the shoulder. J Bone Joint Surg Br. 86(3):324-32, 2004
6. Tung GA et al: MR arthrography of the posterior labrocapsular complex: relationship with glenohumeral joint alignment and clinical posterior instability. AJR Am J Roentgenol. 180(2):369-75, 2003

POSTERIOR GLENOHUMERAL DISLOCATION

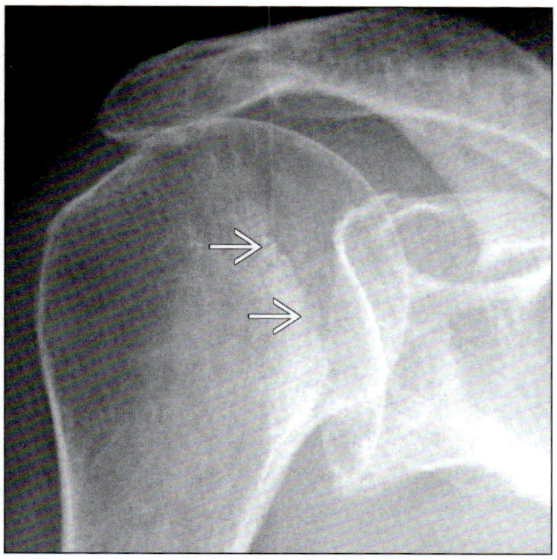

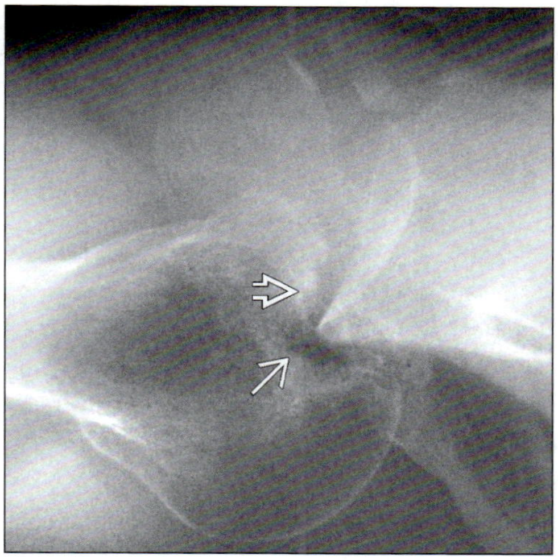

(Left) *Neer AP radiograph shows a lack of congruence of the glenohumeral joint with increased overlap. There is also a "trough" sign ➔ indicating a reverse Hill-Sachs impaction fracture of the anteromedial humeral head.* ***(Right)*** *Axillary radiograph of the same patient confirms the locked or chronic posterior dislocation. The "trough" sign on the AP radiograph is made by the sagittally oriented impacted bone ➔ that forms the lateral margin of the reverse Hill-Sachs fracture ➔.*

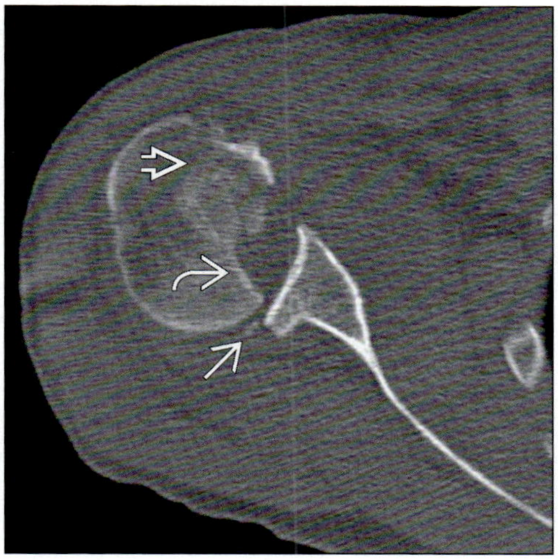

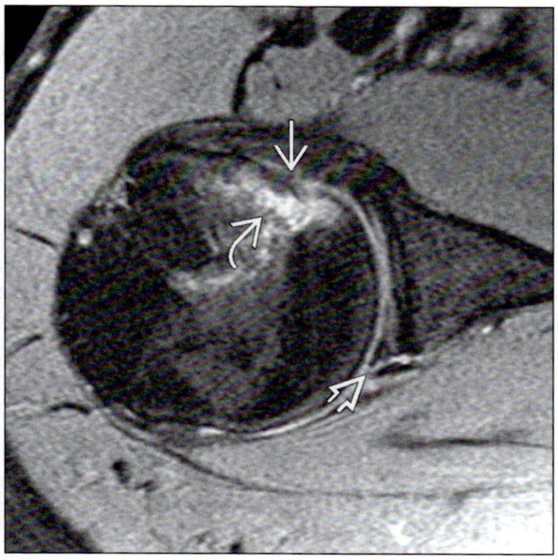

(Left) *Axial bone CT after reduction of a posterior dislocation shows a reverse Bankart fracture ➔, reverse Hill-Sachs impaction fracture ➔, and a fracture of the lesser tuberosity ➔. Lesser tuberosity fxs are common with posterior dislocation.* ***(Right)*** *Axial PDWI FS MR post reduction after a posterior dislocation shows a small reverse Hill-Sachs impaction fracture ➔ with associated bone bruise ➔. There is also a tear of the posterior labrum ➔, a reverse Bankart lesion.*

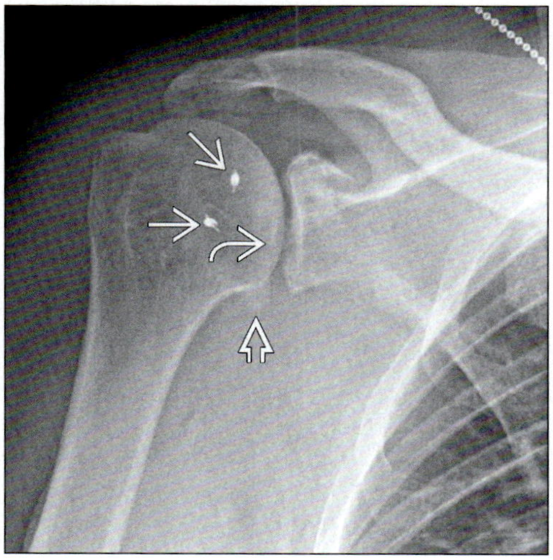

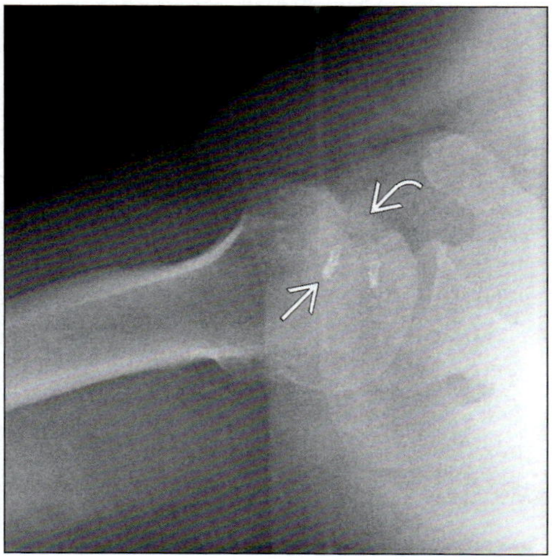

(Left) *AP radiograph following surgery for recurrent posterior instability after multiple posterior dislocations. There are 2 suture anchors ➔ projecting over the humeral head. There is also a "trough" sign ➔ and old callus ➔ from a reverse Hill-Sachs fracture.* ***(Right)*** *Axillary radiograph of the same patient again shows that the 2 suture anchors ➔ are in the reverse Hill-Sachs impaction fracture ➔. The patient had a subscapularis tendon transfer to prevent recurrent dislocations.*

INFERIOR GLENOHUMERAL DISLOCATION AND LUXATIO ERECTA

Key Facts

Terminology
- Inferior shoulder dislocation, luxatio erecta humeri

Imaging
- Humeral head located inferior to glenoid fossa
- Almost all are luxatio erecta with arm fixed in abduction
- AP radiograph
  - Pathognomonic: Arm abducted with inferiorly located humeral head

Pathology
- Headfirst fall onto fully abducted outstretched arm
  - Hyperabduction with humeral head/neck leveraged out by lateral acromion
  - Axial force drives humeral head inferiorly
- Torn inferior capsule
- Greater tuberosity fx or rotator cuff tear
- Brachial plexus symptoms prior to relocation
- Also
  - Fx acromion, clavicle, coracoid, glenoid rim
  - Persistent brachial plexus injury
  - Venous thrombosis
  - Axillary artery injury
- Humeral head protrudes through rent in inferior capsule
- May require surgical reduction if inferior capsule hole is too tight to reduce

Clinical Issues
- Arm held between 110-160° abduction
- Shoulder pain
- Sensory deficit: 60-100%
  - Most resolve after reduction
  - Usually small patch over deltoid
- Immediate reduction recommended

(Left) AP bone CT 3D image shows inferior dislocation of the humeral head relative to the glenoid fossa. The arm is abducted, diagnostic for luxatio erecta. There is also a fracture of the greater tuberosity. (Right) AP radiograph in the same patient shows the inferior dislocation of the humeral head relative to the glenoid fossa. The humerus is parallel to the spine of the scapula, the common position in luxatio erecta. Again note the greater tuberosity fragment.

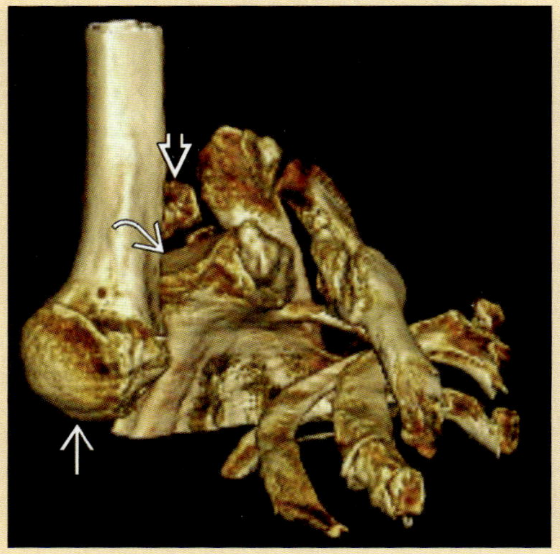

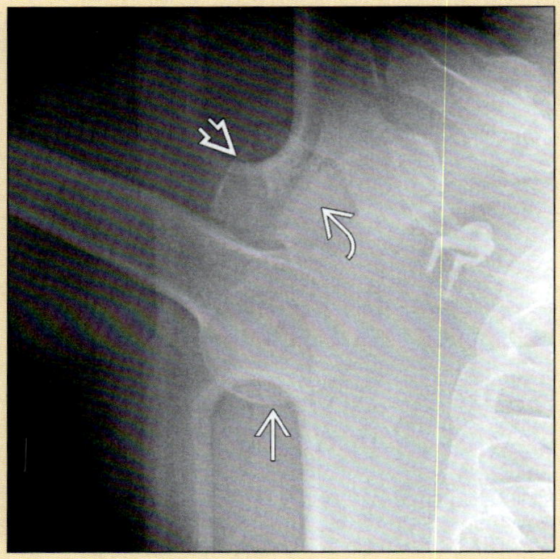

(Left) AP radiograph in a patient with bilateral luxatio erecta shows both humeral heads dislocated inferiorly and both arms abducted so that the humeri are parallel to the spines of the scapula. A greater tuberosity fracture is seen in the right shoulder. (Right) Axillary radiograph of the left shoulder in the same patient shows the humeral head to be magnified due to inferior (and slight anterior) dislocation, putting it further from the imaging plate.

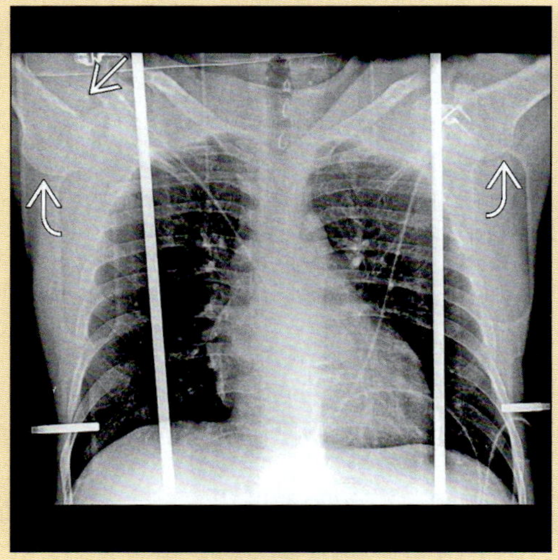

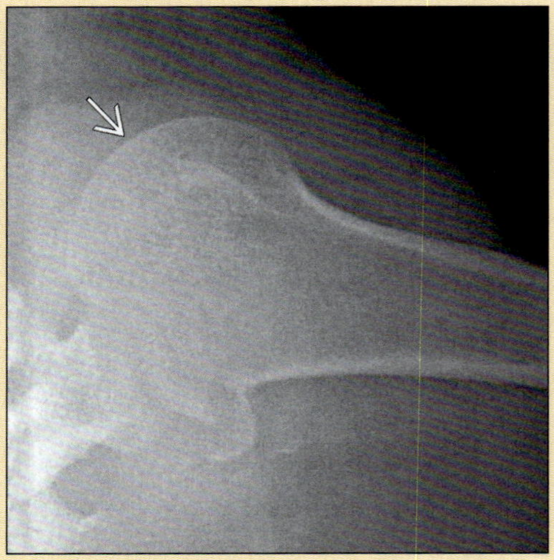

INFERIOR GLENOHUMERAL DISLOCATION AND LUXATIO ERECTA

TERMINOLOGY

Synonyms

- Inferior shoulder dislocation, luxatio erecta humeri

Definitions

- Inferior displacement of humeral head with loss of articulation with glenoid fossa

IMAGING

General Features

- Best diagnostic clue
  - Humeral head located inferior to glenoid fossa
    - Almost all are luxatio erecta with arm fixed in abduction

Radiographic Findings

- AP view
  - Humeral head inferior to glenoid
  - Arm in abduction
  - May have displaced greater tuberosity fracture

MR Findings

- Post-reduction
  - Torn inferior capsule
  - Rotator cuff tear or greater tuberosity fracture

Imaging Recommendations

- Best imaging tool
  - AP radiograph
    - Pathognomonic: Arm abducted with inferiorly located humeral head

DIFFERENTIAL DIAGNOSIS

Inferior Humeral Head Subluxation

- Arm not in fixed abduction
- Subluxed only (still contact between joint surfaces)
- Due to muscle weakness (stroke or surgical neck fx)

Anterior Dislocation

- Humeral head may lie somewhat inferiorly
- Arm not in fixed abduction

PATHOLOGY

General Features

- Etiology
  - Headfirst fall onto fully abducted outstretched arm
    - Hyperabduction with humeral head/neck leveraged out by lateral acromion
    - Axial force drives humeral head inferiorly
- Associated abnormalities
  - Torn inferior capsule
  - Greater tuberosity fx or rotator cuff tear
  - Brachial plexus symptoms prior to relocation
  - Also
    - Fx acromion, clavicle, coracoid, glenoid rim
    - Persistent brachial plexus injury
    - Venous thrombosis
    - Axillary artery injury

Staging, Grading, & Classification

- Closed: Skin intact
- Open: Protrudes through axillary skin

Gross Pathologic & Surgical Features

- Humeral head protrudes through rent in inferior capsule
- May require surgical reduction if inferior capsule hole is too tight to reduce

CLINICAL ISSUES

Presentation

- Most common signs/symptoms
  - Arm held between 110-160° abduction
  - Shoulder pain
  - Sensory deficit: 60-100%
    - Most resolve after reduction
    - Usually small patch over deltoid
- Other signs/symptoms
  - Arm in external rotation
  - Decreased pulses due to axillary artery injury

Demographics

- Epidemiology
  - 0.5-1% of shoulder dislocations

Natural History & Prognosis

- Inferior capsule heals on its own
- Arthroscopy if torn rotator cuff

Treatment

- Immediate reduction
  - Reduces chance of long-term neurological injury
- May require surgical reduction if inferior capsule hole is too tight to reduce

DIAGNOSTIC CHECKLIST

Consider

- Luxatio erecta if arm abducted
- Remember association with greater tuberosity fracture &/or rotator cuff tear

Image Interpretation Pearls

- Humerus parallel to scapular spine

Reporting Tips

- Mention if any fractures

SELECTED REFERENCES

1. Fung DA et al: Asymmetric bilateral shoulder dislocation involving a luxatio erecta dislocation. Am J Orthop. 37(5):E97-8, 2008
2. Begaz T et al: Luxatio erecta: inferior humeral dislocation. J Emerg Med. 31(3):303-4, 2006
3. Sewecke JJ et al: Bilateral luxatio erecta: a case report and review of the literature. Am J Orthop. 35(12):578-80, 2006
4. Yamamoto T et al: Luxatio erecta (inferior dislocation of the shoulder): a report of 5 cases and a review of the literature. Am J Orthop. 32(12):601-3, 2003
5. Downey EF Jr et al: Unusual dislocations of the shoulder. AJR Am J Roentgenol. 140(6):1207-10, 1983

GREATER TUBEROSITY FRACTURE

Key Facts

Terminology

- Avulsion fx through base of greater tuberosity
  - Isolated: Usually after fall
  - Associated with other injuries
  - Humeral head/neck fx
  - After glenohumeral dislocation
- Appearance varies
  - Radiographically occult, nondisplaced, seen only on CT or MR
  - Markedly displaced

Imaging

- AP view
  - Linear fx line extending from lateral humeral head cortex to medial supraspinatus tendon footprint
- MR: Linear low signal fx line, base of greater tuberosity
  - T2WI: Marrow edema in greater tuberosity
- US: Step-off in echogenic lateral cortex of greater tuberosity

Top Differential Diagnoses

- Humeral head/neck fracture
- Extended Hill-Sachs lesion
- Enthesopathic cysts
- MR: T2WI "edema" from poor fat suppression
- Extended Hill-Sachs Lesion

Pathology

- Direct impact of greater tuberosity on acromion process
- Avulsion by rotator cuff tendons

*(**Left**) AP graphic shows a fracture of the greater tuberosity ↪. Greater tuberosity fractures can occur either from a fall with direct impaction of the tuberosity onto the acromion process or from an anterior subluxation/ dislocation. (**Right**) AP external rotation radiograph in a patient with shoulder pain after trauma shows a fracture through the base of the greater tuberosity ➔. Fractures with less than 5 mm superior displacement are usually treated conservatively.*

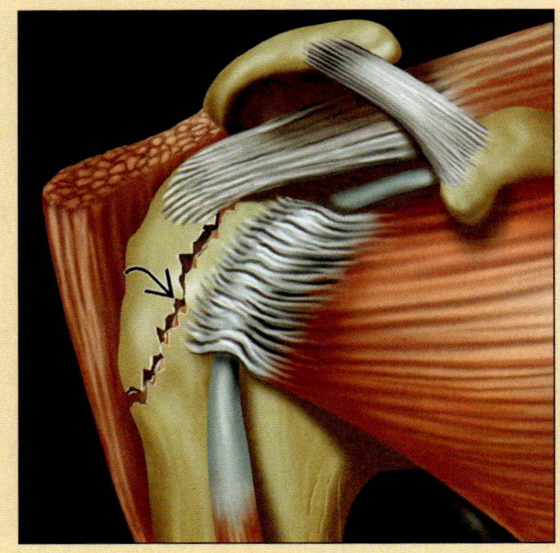

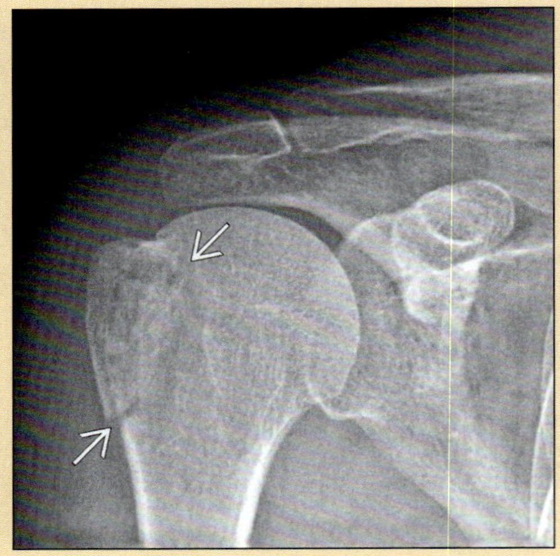

*(**Left**) AP radiograph of a patient with shoulder pain after a fall shows a nondisplaced fx of the greater tuberosity ➔, which was not diagnosed on original interpretation of the radiograph. (**Right**) Coronal oblique T2WI FS MR of the same patient performed 1 week later for continued shoulder pain. There is a high-signal fx line ➔ now with slight displacement of the greater tuberosity ⇨. There is also a possible small articular surface partial RCT ↪. The patient did well with rest.*

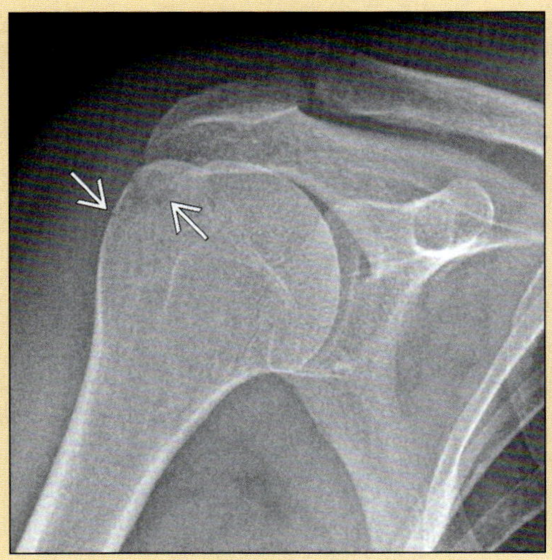

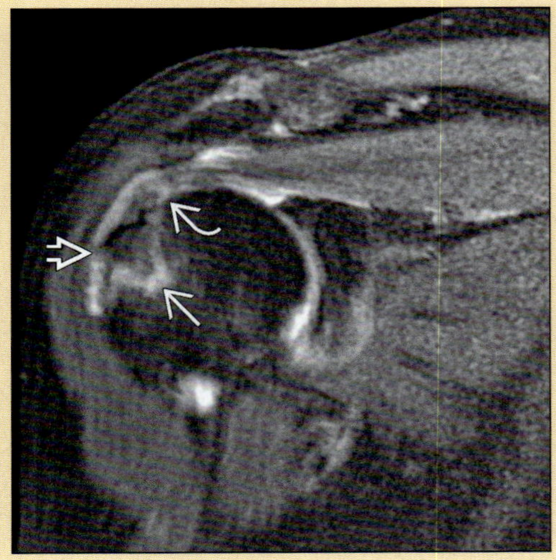

GREATER TUBEROSITY FRACTURE

TERMINOLOGY

Synonyms

- Rotator cuff avulsion fracture

Definitions

- Avulsion fx through base of greater tuberosity (GT)
  - Isolated: Usually after fall
  - Associated with other injuries
    - Humeral head/neck fracture
    - After glenohumeral dislocation

IMAGING

General Features

- Best diagnostic clue
  - Disruption of lateral cortex
- Size
  - May be nondisplaced and radiographically occult

Radiographic Findings

- AP view
  - Linear fx line extending from lateral humeral head cortex to medial supraspinatus tendon footprint

CT Findings

- Same as radiograph
- Subtle disruption of trabecula

MR Findings

- T1WI
  - Linear low signal fracture line at base of greater tuberosity
- T2WI FS
  - Marrow edema in greater tuberosity

Ultrasonographic Findings

- Step-off in lateral cortex of GT

Imaging Recommendations

- Best imaging tool
  - Nondisplaced fx: MR

DIFFERENTIAL DIAGNOSIS

Humeral Head/Neck Fracture

- Surgical neck fracture may be nondisplaced

Extended Hill-Sachs Lesion

- After anterior dislocation

Enthesopathic Cysts

- Usually rounded rather than linear

GT "Edema" from Poor Fat Saturation

- No fx line, high signal on T1 MR also

Rotator Cuff Tear

- Can be difficult to distinguish from GT fx clinically
- May have both RCT and greater tuberosity fx

PATHOLOGY

General Features

- Etiology
  - Direct impact of GT on acromion process
  - Avulsion by rotator cuff tendons
    - Glenohumeral joint subluxation/dislocation
- Associated abnormalities
  - Humeral head/neck fx
  - Dislocation lesions

Staging, Grading, & Classification

- Non-/minimally displaced
  - < 5 mm superior and < 10 mm posterior displacement
- Displaced

CLINICAL ISSUES

Presentation

- Most common signs/symptoms
  - Shoulder pain after fall or anterior/inferior dislocation
  - Decreased range of motion

Demographics

- Age
  - < 40 years; M > F
    - Nondisplaced
    - Supraspinatus tendon stronger than bone
    - Most common shoulder fx in skiers
  - > 50 years; F > M
    - Osteoporotic
- Epidemiology
  - 10% of anterior dislocations
  - GT involved in 40% of proximal humerus fx

Natural History & Prognosis

- Most heal with conservative treatment and no sequelae

Treatment

- Nondisplaced: Conservative
  - If continued pain, may require subacromial decompression
    - MR to evaluate for rotator cuff tear
- Displaced: Surgical reduction and screw fixation

DIAGNOSTIC CHECKLIST

Consider

- Anterior dislocation if isolated GT fx on radiograph

Image Interpretation Pearls

- May initially detect on MR; "cuff" pain with normal radiograph

SELECTED REFERENCES

1. Mason BJ et al: Occult fractures of the greater tuberosity of the humerus: radiographic and MR imaging findings. AJR Am J Roentgenol. 172(2):469-73, 1999
2. Zanetti M et al: MR imaging for traumatic tears of the rotator cuff: high prevalence of greater tuberosity fractures and subscapularis tendon tears. AJR Am J Roentgenol. 172(2):463-7, 1999
3. Reinus WR et al: Fractures of the greater tuberosity presenting as rotator cuff abnormality: magnetic resonance demonstration. J Trauma. 44(4):670-5, 1998

OSTEOCHONDRAL INJURY, SHOULDER

Key Facts

Terminology

- Abbreviations: OCI, OC lesion, OCD
- Synonyms: (Osteo)chondral defect or lesion, osteochondritis dissecans
- Definitions: Traumatic injury involving articular surface of humeral head or glenoid fossa
- Isolated lesions are rare
- More commonly associated with other injuries
  - Bankart fx, large Hill-Sachs fx
  - Reverse Bankart fx, reverse Hill-Sachs fx
  - Glenolabral articular disruption (GLAD) lesion
  - Glenoid articular disruption or divot (GARD) lesion

Imaging

- Isolated OC lesion: Radiograph usually normal
- MR, fluid-sensitive sequence
  - High signal fluid in focal chondral defect
  - ± focal subchondral edema
- MR arthrogram
  - Most sensitive for chondral defect, intraarticular chondral body, labral tears

Top Differential Diagnoses

- Central bare area of glenoid
  - Normal variant
- Osteoarthritis
  - Osteophytes, subchondral sclerosis
- Osteonecrosis
  - Serpentine "double line" sign

Diagnostic Checklist

- Consider instability if lesion near glenoid rim
- If chondral defect present, seek intraarticular body

(Left) *Axial T1WI FS MR arthrogram image in a 31-year-old man with shoulder pain after a fall. There is a chondral defect of the glenoid ➔ with a cartilage flap ➔. This patient also had loose bodies in the axillary recess (not shown), which were removed, and a microfracture of the grade 4 cartilage defect.* ***(Right)*** *Coronal graphic shows a grade 4 focal chondral defect of the humeral head ➔. The cartilage-bone interface is the weakest point, thus many acute trauma defects are grade 4.*

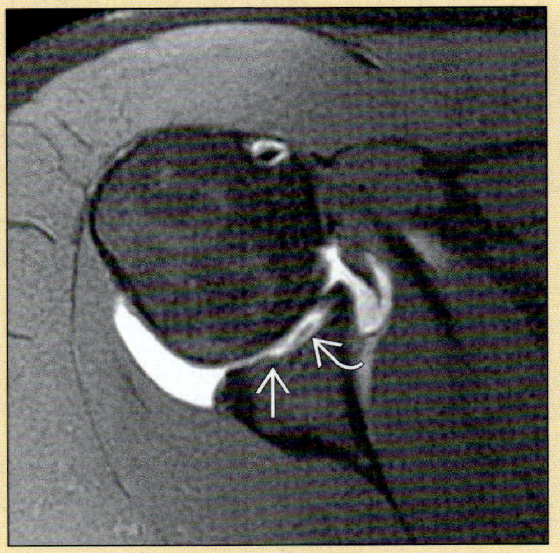

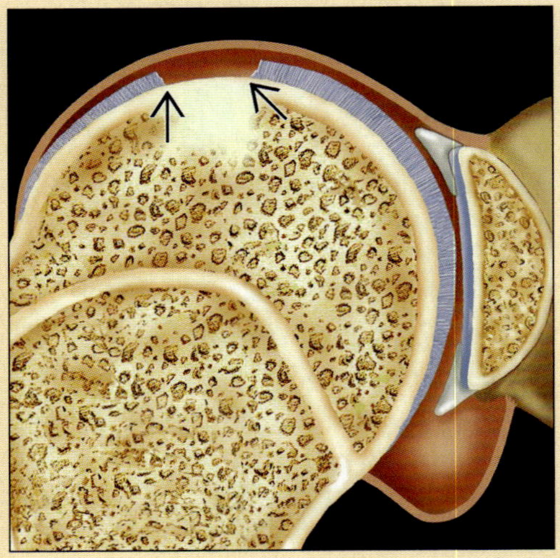

(Left) *Coronal oblique T2WI FS MR of the shoulder shows the full thickness chondral defect ➔ of the humeral head. Fluid extends through the defect and abuts the cortex of the humeral head ➔.* ***(Right)*** *Sagittal T2WI MR in the same patient shows a well-defined full thickness chondral defect ➔ in the superior aspect of the humeral head. The defect is seen as fluid signal abutting the cortex of the humerus ➔ between sharply marginated areas of normal articular cortex ➔.*

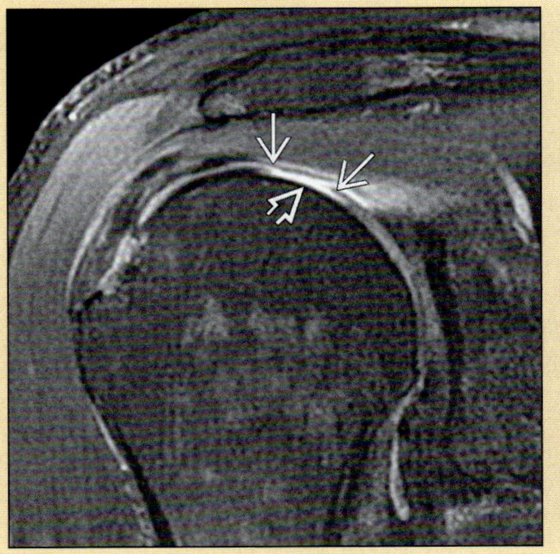

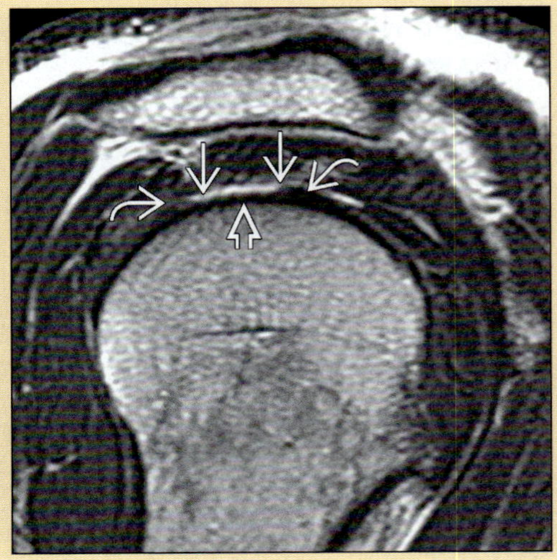

OSTEOCHONDRAL INJURY, SHOULDER

TERMINOLOGY

Abbreviations

- Osteochondral injuries (OCI), osteochondral (OC) lesion, OCD

Synonyms

- (Osteo)chondral defect, osteochondritis dissecans

Definitions

- Traumatic injury involving articular surface of humeral head or glenoid fossa
  - Isolated lesions are rare
  - More commonly associated with other injuries
    - Bankart fx, large Hill-Sachs fx
      - Anterior dislocation, often capsule injury
    - Reverse Bankart fx, reverse Hill-Sachs fx
      - Posterior dislocation
    - Glenolabral articular disruption (GLAD) lesion
      - Chondral injury and superficial tear anteroinferior labrum
    - Glenoid articular disruption or divot (GARD)
      - Osteochondral defect and labral tear
      - Can be anterior or posterior
    - Internal impingement humeral head lesions
      - Posterior greater tuberosity

IMAGING

General Features

- Best diagnostic clue
  - MR: Focal chondral defect
    - ± underlying osseous changes
- Location
  - Humeral head or glenoid fossa

Radiographic Findings

- Isolated OC lesion: Radiograph usually normal

CT Findings

- May see subchondral cyst or sclerotic area
- CT arthrography
  - Contrast fills focal chondral defect
  - Identify radiolucent chondral loose body
  - ± associated labral tears

MR Findings

- PD/intermediate FS
  - High signal fluid in focal chondral defect
  - ± focal subchondral edema
  - ± loose bodies
- MR arthrogram
  - Most sensitive for chondral defect, intraarticular chondral body, labral tears

Imaging Recommendations

- Best imaging tool
  - MR arthrography

DIFFERENTIAL DIAGNOSIS

Central Bare Area of Glenoid

- Normal variant

Osteoarthritis

- No specific injury
- Also have osteophytes, subchondral sclerosis

Osteonecrosis

- Serpentine "double line" sign

PATHOLOGY

General Features

- Etiology
  - Shear injury to articular surface

Staging, Grading, & Classification

- Grade 1: No defect (softening at surgery)
- Grade 2: < 50% chondral defect
- Grade 3: > 50% chondral defect
- Grade 4: Full thickness chondral defect

CLINICAL ISSUES

Presentation

- Most common signs/symptoms
  - Pain
  - Locking if loose body

Demographics

- Age
  - Peak: 15-40 years; active patients
- Gender
  - M > F

Natural History & Prognosis

- If persistent pain, usually requires surgery

Treatment

- Debridement and lavage
- If large and full thickness: Microfracture, mosaicplasty

DIAGNOSTIC CHECKLIST

Consider

- Instability if lesion near glenoid rim

Image Interpretation Pearls

- If chondral defect present, look for intraarticular body

SELECTED REFERENCES

1. Kircher J et al: Osteochondral autologous transplantation for the treatment of full-thickness cartilage defects of the shoulder: results at nine years. J Bone Joint Surg Br. 91(4):499-503, 2009
2. Wiener E et al: Delayed gadolinium-enhanced MRI of cartilage (dGEMRIC) of cadaveric shoulders: comparison of contrast dynamics in hyaline and fibrous cartilage after intraarticular gadolinium injection. Acta Radiol. 50(1):86-92, 2009
3. Ly JQ et al: Bare area of the glenoid: magnetic resonance appearance with arthroscopic correlation. J Comput Assist Tomogr. 28(2):229-32, 2004
4. Sanders TG et al: The glenolabral articular disruption lesion: MR arthrography with arthroscopic correlation. AJR Am J Roentgenol. 172(1):171-5, 1999

HUMERAL HEAD/NECK FRACTURE

Key Facts

Terminology

- Surgical neck fx, proximal humerus fx
- Fracture of humeral head or neck
  - Usually involves surgical neck
  - May also have fx of greater tuberosity, lesser tuberosity, or articular surface

Imaging

- AP, axillary, and scapular-Y views
  - Often need all 3 views to correctly classify fx
- CT: Assists in defining degree of displacement &/or angulation of fragments
- MR: Evaluate for torn rotator cuff tendons or ligaments
  - Help identify nondisplaced greater tuberosity fxs

Pathology

- Fall on outstretched arm with severe abduction of shoulder
- Direct or indirect high-energy trauma
- Associated injuries
  - Axillary n. injury: Most commonly injured nerve
  - Injury to axillary vessels
  - Entrapped long head biceps t. between fracture fragments
  - Humeral head osteonecrosis
- Neer classification
  - 1 part: No fragments displaced > 1 cm or angulated over 45°
  - 2 part: 1 fragment displaced > 1 cm or angulated over 45°
  - 3 part: 2 fragments displaced > 1 cm or angulated over 45°
  - 4 part: All 3 fragments displaced > 1 cm or angulated over 45° relative to others

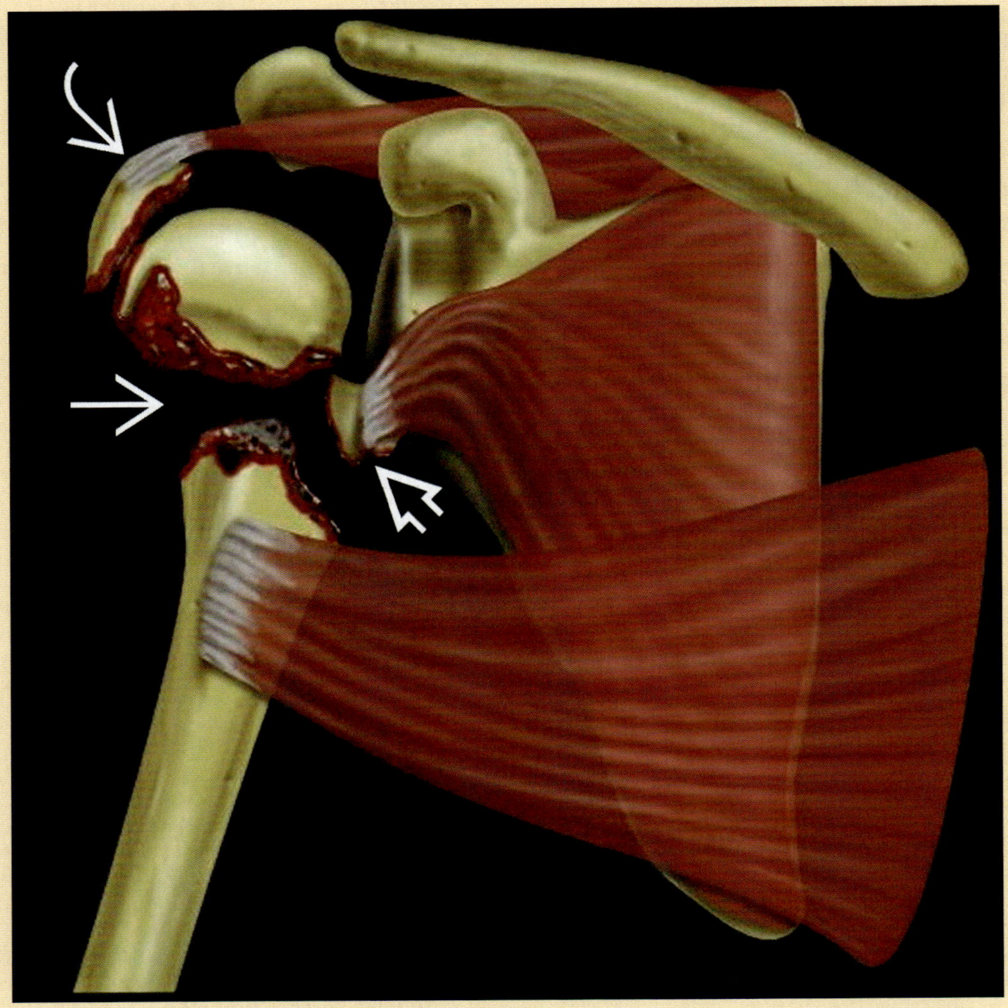

AP graphic of a Neer 4 part fracture. There are fractures involving the surgical neck ➔, as well as the greater tuberosity ➔ and lesser tuberosity ➔. The Neer classification of humeral head/neck fractures does not classify by the number of fx lines but the number of displaced or angulated fragments. To be considered displaced, the distance between the fragments must be at least 1 cm. To be angulated, fragment rotation must be at least 45°. In this drawing, there is > 1 cm displacement at all 3 fractures, and > 45° angulation between the rotated humeral head fragment and the shaft. 80% of humeral head/neck fxs are 1 part, with minimal displacement and angulation.

HUMERAL HEAD/NECK FRACTURE

TERMINOLOGY

Synonyms

- Surgical neck fx, proximal humerus fx

Definitions

- Fracture of humeral head or neck
  - Usually involves surgical neck
  - May also have fx of greater tuberosity, lesser tuberosity, or articular surface

IMAGING

General Features

- Best diagnostic clue
  - Oblique or transverse fracture line just below tuberosities
    - Often additional fracture of greater &/or lesser tuberosities
    - May be displaced or angulated
- Location
  - Surgical neck of humerus
  - Humeral head
  - Lesser tuberosity
  - Greater tuberosity
- Size
  - Varies from nondisplaced fracture to severe comminution and displacement of fragments
- Morphology
  - Fractures described according to separate fragments and degree of displacement and angulation

Radiographic Findings

- AP, axillary, and scapular-Y views
  - Often need all 3 views to correctly classify fx
  - Evaluate displacement and angulation
- Radiolucent fracture line extending through surgical neck/humeral head or tuberosity
  - Fracture lines may be difficult to see due to osteopenia, obesity, overlapping fragments, or nondisplacement
  - Impacted fractures may be seen as sclerotic lines
  - May be difficult to distinguish comminuted surgical neck fragment from tuberosity fracture
- Lipohemarthrosis may be demonstrated by upright or decubitus projections

CT Findings

- Bone CT
  - Assists in defining degree of displacement &/or angulation of fragments
    - Helpful in identifying glenohumeral dislocation
    - May demonstrate associated glenoid fracture
  - More sensitive for nondisplaced fxs
  - May demonstrate lipohemarthrosis
  - Aids in surgical planning

MR Findings

- T2WI FS
  - Evaluate for torn rotator cuff tendons or ligaments
- Helpful to identify nondisplaced greater tuberosity fractures

Imaging Recommendations

- Best imaging tool
  - AP, scapular-Y view, and axillary views of shoulder
- Protocol advice
  - CT to better characterize fx alignment

DIFFERENTIAL DIAGNOSIS

Glenohumeral Dislocation

- Loss of articulation of humeral head with glenoid
  - May be associated with humeral head/neck fracture
  - Older patients may have greater tuberosity fx
- Less soft tissue swelling

Glenoid Rim Fracture

- Usually after glenohumeral dislocation

Unfused Growth Plate

- May mimic fracture in skeletally immature patient

PATHOLOGY

General Features

- Etiology
  - Fall on outstretched arm with severe abduction of shoulder
  - Direct or indirect high-energy trauma
- Associated abnormalities
  - Axillary nerve injury: Most commonly injured nerve
    - Less common: Suprascapular nerve, musculocutaneous nerve, brachial plexus
  - Injury to axillary vessels
  - Entrapped long head biceps tendon between fracture fragments
    - Prevents fx reduction
  - Humeral head osteonecrosis
    - Especially anatomic neck fxs

Staging, Grading, & Classification

- Neer classification
  - 4 potential fragments: Head, shaft, greater and lesser tuberosities
  - Classified not by number of fx lines, but by how many displaced/angulated fragments
  - 1 part: No fragments displaced > 1 cm or angulated over 45°
    - Fragments are in continuity, held together by soft tissue
  - 2 part: 1 fragment displaced > 1 cm or angulated over 45°
  - 3 part: 2 fragments displaced > 1 cm or angulated over 45°
  - 4 part: All 3 fragments displaced > 1 cm or angulated over 45° relative to others
    - "Classic 4 part": Fracture/dislocation
    - "Valgus-impacted 4 part fx": Humeral head rotated to face superiorly
- Articular surface involvement also should be noted
  - Impaction fracture of humeral head
    - Grading: Percentage of articular surface involved
  - Split fracture of humeral head
- 80% of humeral neck fxs are nondisplaced (1 part)

Gross Pathologic & Surgical Features

- Fx fragments are displaced by tendinous insertions

HUMERAL HEAD/NECK FRACTURE

- Humeral shaft displaced medially by pectoralis major muscle
- Greater tuberosity displaced superiorly by supraspinatus, posteriorly by infraspinatus
- Lesser tuberosity displaced medially by subscapularis

CLINICAL ISSUES

Presentation

- Most common signs/symptoms
  - Pain, swelling
  - Swelling may worsen for several days
- Other signs/symptoms
  - Deformity if there is displacement of fragments
  - Inferior displacement of humerus due to muscle atony

Demographics

- Age
  - Older patients
    - Related to osteopenia
    - Surgical neck fx usually comminuted
  - Younger patients
    - High-energy trauma
    - Variable fracture pattern
      - May be tuberosity or humeral head fracture only
- Gender
  - Elderly: M < F
    - Osteoporosis
- Epidemiology
  - 2-3% of upper extremity fractures
  - 1% of all fractures

Natural History & Prognosis

- Nondisplaced fractures may heal with conservative management
  - Some degree of angulation may be tolerated due to compensation by range of motion at shoulder
  - Bone contact necessary for healing to occur
- Fractures with significant displacement or rotation of fragments may require surgical management

Treatment

- Options, risks, complications
  - Osteonecrosis: After displaced fractures or uncommon anatomic neck fracture
  - Osteoarthritis if intraarticular fracture
  - Heterotopic bone formation
  - Traumatic rotator cuff tear
- Immediate surgery for neurovascular compromise
- Minimally displaced fracture
  - Gradual mobilization and physical therapy
  - Elderly and sedentary patients: > 1 cm of displacement may occasionally be treated as "nondisplaced" as long as there is some bony contact of fragments
  - Surgery for neurovascular compromise, open fracture
- 2 part surgical neck fracture
  - Closed reduction tried 1st
  - Surgery if unable to reduce anterior angulation to < 45° or if displaced by entrapped biceps tendon
- 2 part greater tuberosity fracture
  - Surgery for > 5 mm of superior displacement or > 10 mm of posterior displacement
- 2 part lesser tuberosity fracture
  - Closed reduction if fragment does not block internal rotation
- 3 and 4 part fractures
  - Usually surgical reduction and fixation
- Guidelines may be adjusted depending on patient factors

DIAGNOSTIC CHECKLIST

Consider

- Injury to axillary vessels
- Injury to brachial plexus or axillary nerve
- Associated injury or fracture of chest wall, neck, head

Image Interpretation Pearls

- AP, axillary, and scapular-Y views required for full evaluation of proximal humerus fractures
  - Watch for overlapping fragments and impaction
  - Tuberosity fractures often missed or misdiagnosed on initial radiographs
- CT may be helpful in characterizing complex proximal humerus fractures
  - Tuberosity fractures typically well demonstrated by CT
  - Rotation of fragments well demonstrated by CT

SELECTED REFERENCES

1. Shrader MW et al: Understanding proximal humerus fractures: image analysis, classification, and treatment. J Shoulder Elbow Surg. 14(5):497-505, 2005
2. Bucholz RW et al: Rockwood and Green's Fractures in Adults. Vol 1. 5th ed. Philadelphia: Lippincott Williams and Wilkins. 997-1023, 2001
3. Blake R et al: Emergency department evaluation and treatment of the shoulder and humerus. Emerg Med Clin North Am. 17(4):859-76, vi, 1999
4. Bernstein J et al: Evaluation of the Neer system of classification of proximal humeral fractures with computerized tomographic scans and plain radiographs. J Bone Joint Surg Am. 78(9):1371-5, 1996
5. Kilcoyne RF et al: The Neer classification of displaced proximal humeral fractures: spectrum of findings on plain radiographs and CT scans. AJR Am J Roentgenol. 154(5):1029-33, 1990

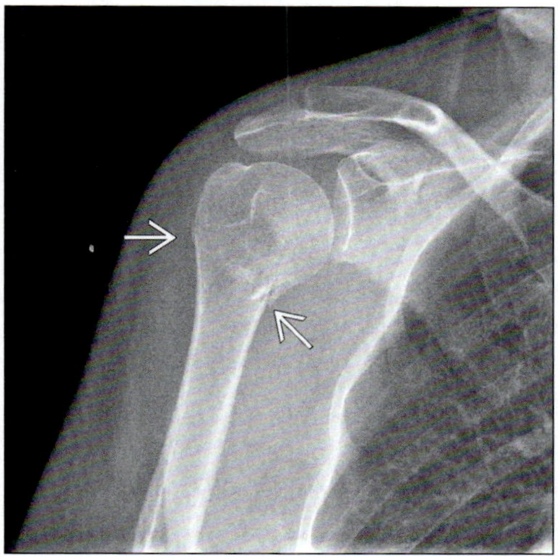

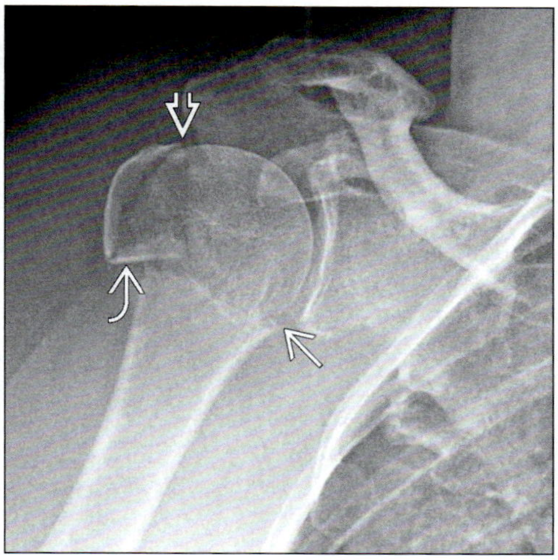

(Left) *AP radiograph shows a minimally displaced fx of the surgical neck. Because there is less than 1 cm of displacement and less than 45° of angulation, this is considered a Neer 1 part fx.* ***(Right)*** *AP radiograph shows a minimally displaced surgical neck fx. There is a greater tuberosity fx with rotation of an inferior fragment greater than 45°, so this is a Neer 2 part fx. Because of the minimal displacement, the patient was treated conservatively.*

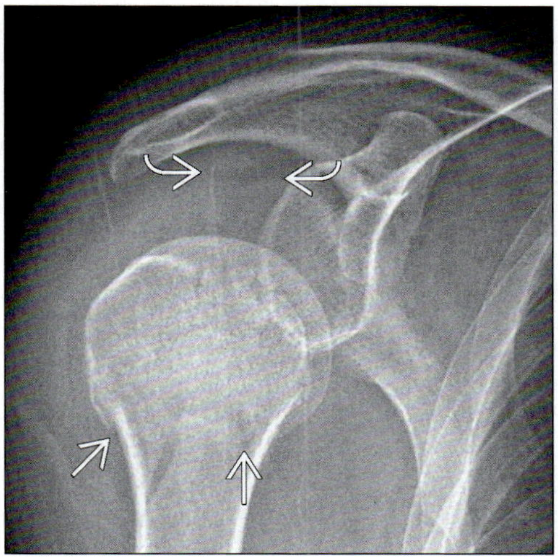

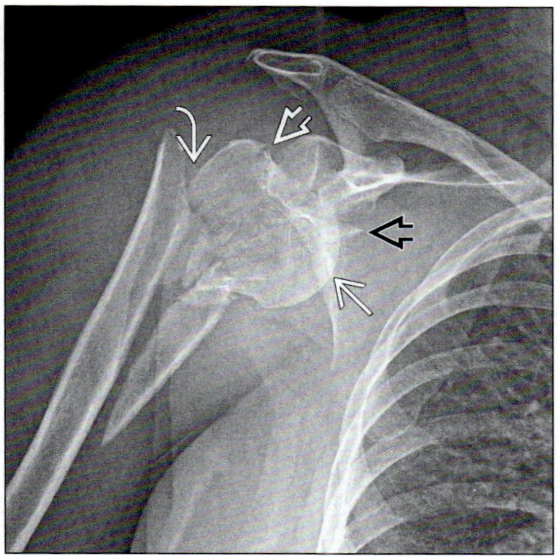

(Left) *AP radiograph shows a Neer 1 part humerus fx. There is a lipohemarthrosis indicating intraarticular involvement and decreased muscle tone causing mild inferior subluxation.* ***(Right)*** *AP radiograph shows an anterior dislocation of the humeral head and displaced fractures of the humeral neck and lesser tuberosity. This is a Neer 3 part fx/dislocation. There is also a minimally displaced fracture of the greater tuberosity. The patient underwent hemiarthroplasty.*

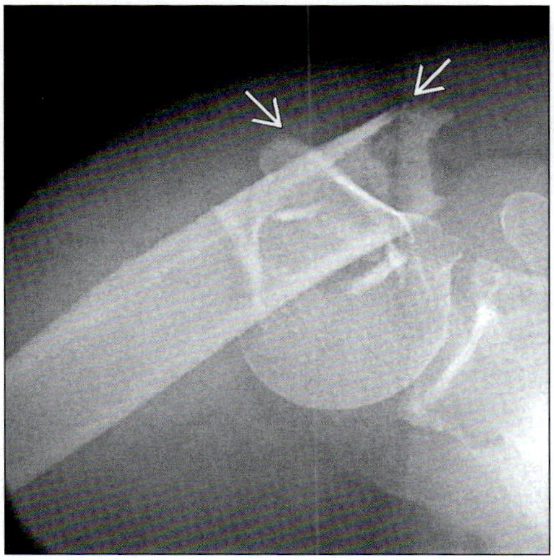

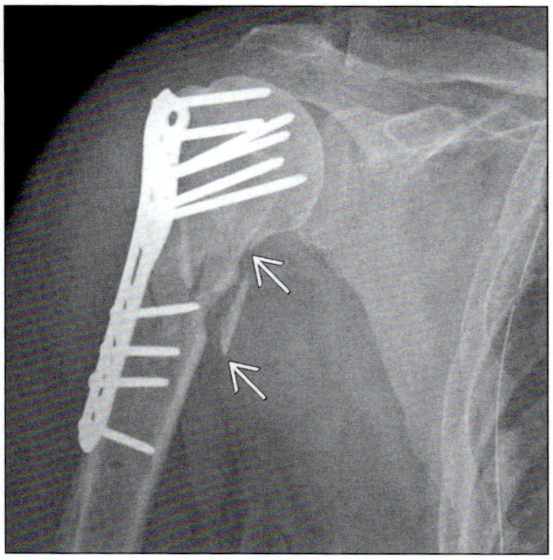

(Left) *Axillary radiograph in a 67-year-old man after fall from a ladder. There is over 1 cm of displacement and ~ 90° angulation at a humeral neck fracture, making this is a Neer 2 part fx. Attempts at closed reduction were unsuccessful with persistent angulation.* ***(Right)*** *AP radiograph of the same patient shows interval reduction and plate and screw fixation at the comminuted surgical neck fracture. Surgical fixation prevents malunion, which is associated with continued pain and weakness.*

LITTLE LEAGUER'S SHOULDER

Key Facts

Terminology
- Proximal humerus Salter I stress fracture, proximal humeral epiphysiolysis, osteochondrosis of proximal epiphysis, slipped capital humeral epiphysis
- Rotational stress fracture of proximal humeral physis

Imaging
- Widened proximal humeral lateral physis
  - MR: Marrow edema adjacent to widened physis
- May need opposite shoulder for comparison
- Chronic: Also sclerosis, cysts, demineralization, fragmentation
- Healing: Periosteal reaction
- Widened proximal humeral physis
  - Widening is greatest laterally
- T2 high signal in wide physis

Pathology
- Overhead throwing in pediatric/adolescent athletes
- Baseball pitchers
- Age range: 12-16 years old, average: 13 years old
- Increased incidence: High pitch count, throwing curveballs, pitching year round
- Usually baseball pitchers but also described in cricket bowlers, volleyball and badminton players

Diagnostic Checklist
- Widened physis in dominant arm
- Other shoulder injuries seen in overhead throwing athletes
- Pathologic internal impingement
- Acromial physis stress injury
- Rotator cuff strain

*(**Left**) AP radiograph of the dominant arm in a 12-year-old little league baseball pitcher with pain from throwing. The radiograph shows widening of the physis →. Comparison with the other arm can sometimes be helpful in more subtle cases of physeal widening. (**Right**) Coronal oblique T1WI MR of the shoulder in the same patient also shows the marked widening of the physis with intermediate signal →. Prominent red marrow in the humeral metaphysis ⇨ is normal at this age.*

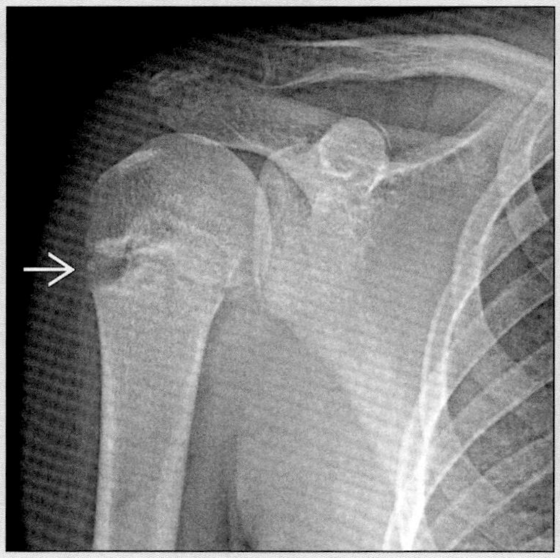

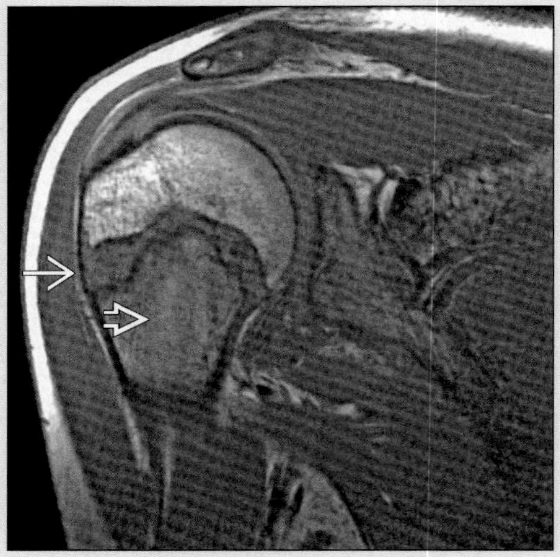

*(**Left**) Coronal oblique T2WI FS MR of the same patient shows heterogeneous increased signal at the physis → with mild periostitis ↷. Note that the marked widening of the physis is more apparent on the radiograph and T1WI. (**Right**) Sagittal oblique T2WI FS MR in the same patient shows the increased signal at the physis →. Little leaguer's shoulder, or a Salter I stress fractures of the proximal humerus, almost always occurs in the throwing arm of young baseball pitchers.*

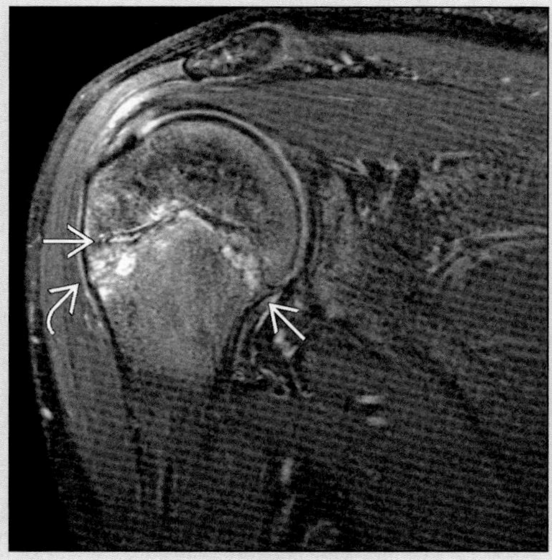

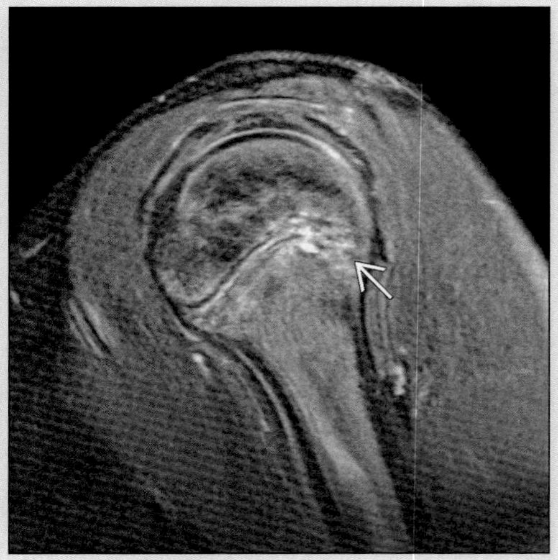

LITTLE LEAGUER'S SHOULDER

TERMINOLOGY

Synonyms

- Proximal humerus Salter I stress fracture, proximal humeral epiphysiolysis, osteochondrosis of proximal epiphysis, slipped capital humeral epiphysis

Definitions

- Rotational stress fracture of proximal humeral physis, typically in 12-16 year old

IMAGING

General Features

- Best diagnostic clue
  - Widened proximal humeral lateral physis
    - MR: Marrow edema adjacent to widened physis
- Location
  - Proximal humerus physis

Radiographic Findings

- Radiography
  - Widened proximal humeral physis
    - Widening is most prominent laterally
  - May need opposite shoulder for comparison
  - Chronic: Also sclerosis, cysts, demineralization, fragmentation
  - Healing: Periosteal reaction

MR Findings

- Widened proximal humeral physis, especially laterally
- ↑ T2 signal in widened physis
- Adjacent marrow edema in epiphysis and metaphysis
- Periosteal reaction: Linear ↑ T2 signal typically adjacent to lateral outer cortex

Imaging Recommendations

- Best imaging tool
  - AP and axillary views of shoulder
- Protocol advice
  - Shoulder MR if radiograph is negative
  - Widening at physis best seen on T1

DIFFERENTIAL DIAGNOSIS

Acute Salter I Fracture

- Sudden shoulder pain after acute injury
- May be displaced (and not simply widened physis)
- "Slipped capital humeral epiphysis" in gymnasts

Infection

- Irregularity mainly on metaphyseal side
- Fever, ↑ WBC, ↑ ESR

Rickets, Renal Osteodystrophy

- Bilateral
- Not focally painful

PATHOLOGY

General Features

- Etiology
  - Overhead throwing pediatric/adolescent athletes
    - Usually baseball pitchers but also described in cricket bowlers, volleyball and badminton players

Staging, Grading, & Classification

- Early
  - Widened physis
  - T2WI MR: Marrow edema, periostitis
- Chronic
  - Radiographic periosteal reaction, sclerosis, fragmentation

CLINICAL ISSUES

Presentation

- Most common signs/symptoms
  - Shoulder pain when throwing baseball or other overhand throwing
  - Tenderness to palpation over proximal humerus
- Other signs/symptoms
  - Pain with external rotation of humerus against resistance

Demographics

- Age
  - Range: 12-16 years old, average: 13 years old
- Gender
  - Usually males

Natural History & Prognosis

- Pain usually resolves with rest
- Widened physis can persist even when patient is asymptomatic and has returned to throwing

Treatment

- Rest
  - 6 weeks, longer if still symptomatic
  - Some recommend no pitching until physis closes
- Gradual addition of light throwing
- Typically 3 months until asymptomatic pitching

DIAGNOSTIC CHECKLIST

Consider

- Other shoulder injuries seen in overhead throwing athletes
  - Pathologic internal impingement
  - Rotator cuff strain

SELECTED REFERENCES

1. Hatem SF et al: MRI of Little Leaguer's shoulder. Skeletal Radiol. 35(2):103-6, 2006
2. Song JC et al: MRI findings in Little Leaguer's shoulder. Skeletal Radiol. 35(2):107-9, 2006
3. Fleming JL et al: Little Leaguer's shoulder. Skeletal Radiol. 33(6):352-4, 2004
4. Carson WG Jr et al: Little Leaguer's shoulder. A report of 23 cases. Am J Sports Med. 26(4):575-80, 1998

TUG LESION, HUMERUS

Key Facts

Terminology

- Chronic avulsive injury, proliferative periostitis, metaphyseal fibrous defect, cortical irregularity syndrome, periosteal (juxtacortical) desmoid
- Reactive fibroosseous mass at bone attachment of major muscle
  - Pectoralis major and deltoid insertions
  - Humerus less common than distal femur "cortical desmoid"

Imaging

- Radiograph: Focal cortical thickening with hyperlucent areas
- MR
  - Irregular cortical thickening
  - T2 high signal deep to periosteum or within cortex
  - No soft tissue mass

Top Differential Diagnoses

- Normal deltoid tubercle
  - Deltoid tubercle/ridge can be slightly irregular
- Enthesopathy
  - Older patients
- Healing fracture
  - Linear cortical fx line usually visible
- Chronic osteomyelitis
  - Clinical signs of infection: ± fever, ↑ WBC, ↑ ESR
- Melorheostosis
  - "Dripping wax" appearance
- Osteoid osteoma
  - Lytic nidus, surrounding dense new bone formation
- Surface osteo-/chondrosarcoma
  - Aggressive periosteal reaction
  - Soft tissue mass on MR

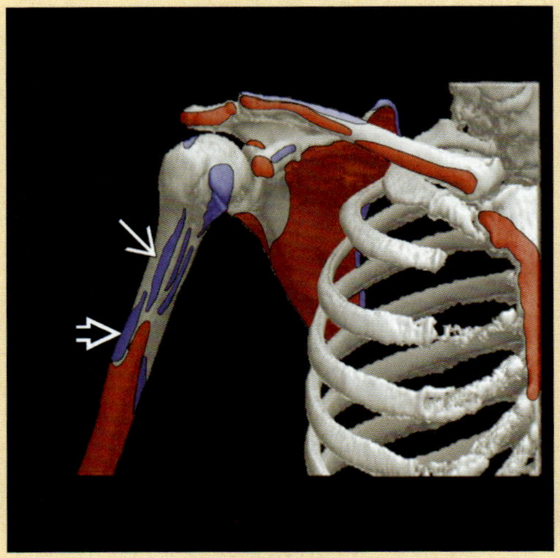
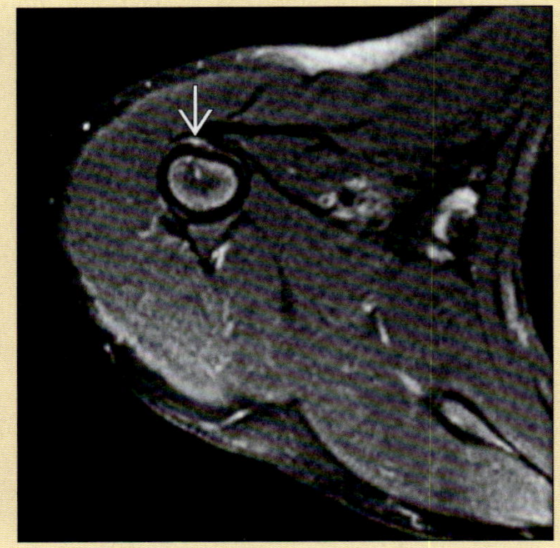

***(Left)** AP graphic shows the tendon attachment sites of the pectoralis major ➡ and deltoid muscles ⇨. These are the most common sites for tug lesions in the humerus. **(Right)** Axial T2WI FS MR in a 12-year-old girl with anterior mid-upper arm pain. There is high signal along the anterior humeral cortex ➡ in the region of the pectoralis tendon insertion. Although there was no cortical irregularity in this patient, some patients will present with cortical thickening and areas of hyperlucency.*

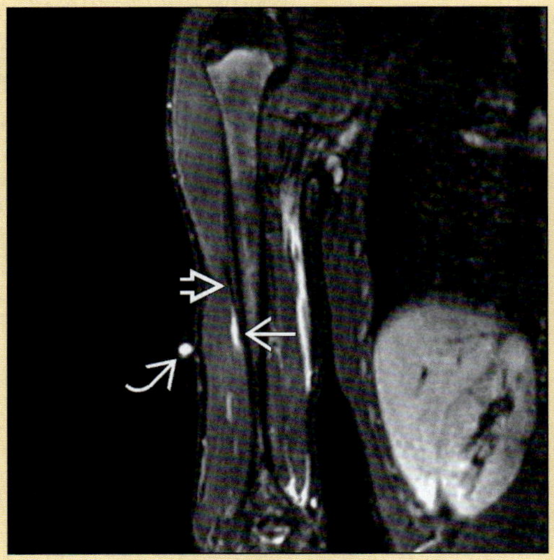
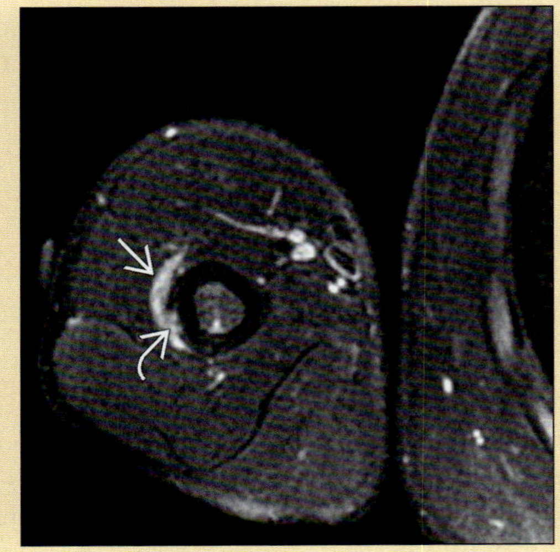

***(Left)** Coronal T2WI FS MR in a 14-year-old boy with continuing focal arm pain ↗ 2 months after a football injury. There is high signal ➡ near the insertion of the tendon of the deltoid muscle ⇨. **(Right)** Axial T2WI FS MR in the same patient shows the high signal hematoma ➡ and cortical irregularity ↗ at the deltoid insertion site. The patient had been playing sports continually and was instructed to rest the arm. The arm pain gradually resolved with rest.*

TUG LESION, HUMERUS

TERMINOLOGY

Synonyms

- Chronic avulsive injury, proliferative periostitis, metaphyseal fibrous defect, cortical irregularity syndrome, periosteal (juxtacortical) desmoid

Definitions

- Reactive fibroosseous mass at bone attachment of major muscle
  - Humerus: Pectoralis major and deltoid insertions
  - Humerus less common than distal femur "cortical desmoid"

IMAGING

General Features

- Best diagnostic clue
  - Irregular cortical thickening with focal lucent areas at insertion of pectoralis major or deltoid
- Location
  - Pectoralis major insertion
    - "Ringman's lesion" in gymnasts
  - Deltoid insertion
- Size
  - 1-3 cm

Radiographic Findings

- Focal cortical thickening with hyperlucent areas

CT Findings

- Focal periosteal reaction, may contain some noncalcified areas

MR Findings

- T1WI
  - Irregular cortical thickening
  - Tendon usually normal
- T2WI FS
  - High signal deep to periosteum or within cortex
  - May have adjacent soft tissue edema
  - No soft tissue mass

Nuclear Medicine Findings

- Bone scan
  - Focal increased activity at muscle insertion site

Imaging Recommendations

- Best imaging tool
  - MR for cases that are equivocal on radiograph

DIFFERENTIAL DIAGNOSIS

Normal Deltoid Tubercle

- Deltoid tubercle/ridge can be slightly irregular

Enthesopathy

- Occurs at muscle attachments also
- Older patients
- Dense/solid focal cortical thickening

Healing Fracture

- Linear cortical fx line usually visible

Chronic Osteomyelitis

- Brodie abscess, sinus tract
- Clinical signs of infection: ± fever, ↑ WBC, ↑ ESR

Melorheostosis

- "Dripping wax" appearance
- Usually more extensive

Osteoid Osteoma

- Lytic nidus, surrounding dense new bone formation

Surface Osteo-/Chondrosarcoma

- Aggressive periosteal reaction
- Soft tissue mass on MR

PATHOLOGY

General Features

- Etiology
  - Avulsive injury
    - Chronic repetitive injury
    - Acute traumatic event with re-injury

CLINICAL ISSUES

Presentation

- Most common signs/symptoms
  - Noted incidentally on radiograph in young athlete
  - May present with focal pain and tenderness

Demographics

- Age
  - 15-20 years old
- Gender
  - M > F

Natural History & Prognosis

- Resolves with rest
- Mild cortical thickening may persist

Treatment

- Refrain from aggravating sports/activities until asymptomatic

DIAGNOSTIC CHECKLIST

Image Interpretation Pearls

- Do not recommend biopsy
  - Histology can be misinterpreted as osteosarcoma

SELECTED REFERENCES

1. Donnelly LF et al: Chronic avulsive injuries of childhood. Skeletal Radiol. 28(3):138-44, 1999
2. Donnelly LF et al: Chronic avulsive injury of the deltoid insertion in adolescents: imaging findings in three cases. Radiology. 211(1):233-6, 1999
3. Fulton MN et al: Cortical desmoid-like lesion of the proximal humerus and its occurrence in gymnasts (ringman's shoulder lesion). Am J Sports Med. 7(1):57-61, 1979

HUMERAL SHAFT FRACTURE

Key Facts

Terminology

- Midshaft humerus fx, humerus diaphysis fx
- Humerus fracture occurring between upper pectoralis major insertion and supracondylar ridge
  - Fracture extends below supracondylar ridge
- Humeral neck fracture
  - Fracture extends above pectoralis major insertion

Imaging

- Transverse lucency through humeral shaft
- Usually varus angulation
  - Proximal fragment abducted by deltoid
  - Distal fragment pulled proximally by biceps and brachialis muscles
- 70% in middle 1/3
- Transverse (61%), oblique (18%), or spiral (17%)

Top Differential Diagnoses

- Nutrient canal
  - Straight with sclerotic margins
- Supracondylar or elbow fracture

Clinical Issues

- Radial nerve injury in 16%
- Moderate rate of nonunion, especially open or transverse noncomminuted
  - Nonunion in 6% of open, < 2% of closed fxs
- Treatment
  - Long arm brace, coaptation splint, or cast brace
  - Surgery if reduced angulation cannot be maintained, gap remains at fx site, or open fracture

Diagnostic Checklist

- Pathologic fx if underlying lytic lesion
- Mention if varus angulation > 15°, or persistent fx gap

*(**Left**) Lateral & AP radiographs of the right humerus in a 62-year-old woman with pain & deformity after a fall from standing height. Note the midshaft humeral fx ➔ with shortening, anterior displacement, and varus angulation. The patient was unable to extend the wrist. Due to pain, she underwent closed reduction under anesthesia. One month later her radial nerve palsy persists. (**Right**) AP radiograph shows a comminuted spiral fx ➔ of the humerus. The fx healed well in a fracture brace.*

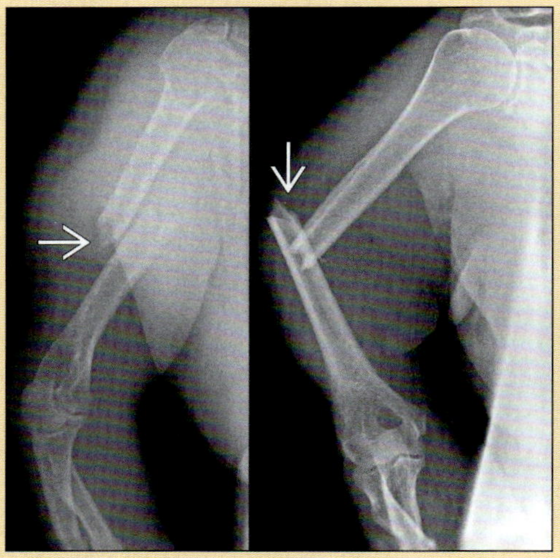

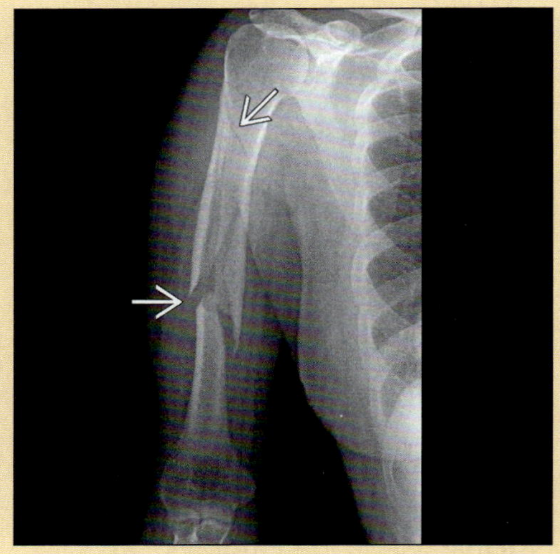

*(**Left**) AP radiographs of a patient healing with malunion. 3 months after the fracture (left) there is callus ➔ but 30° of varus angulation. The patient then had surgical reduction (right) with plate fixation ➔. (**Right**) Posterior view of the right arm shows the close proximity of the radial nerve ➔ to the humeral shaft. 16% of patients with humeral shaft fractures have a radial nerve palsy, which presents as "wrist drop." Most improve after reduction of the displacement and angulation.*

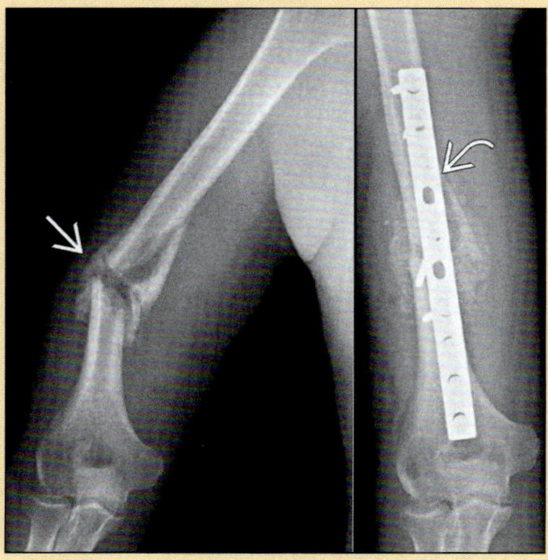

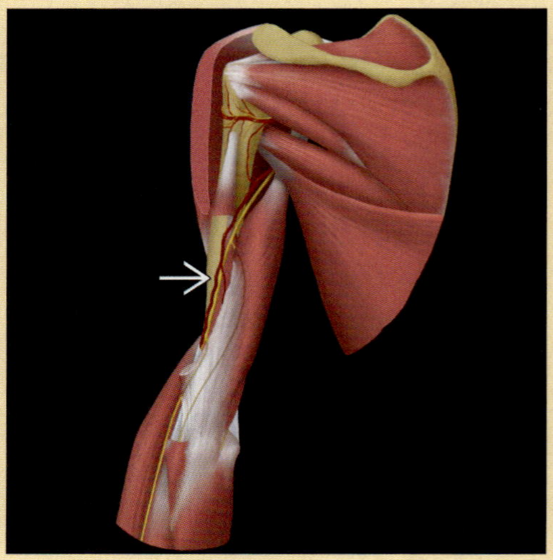

HUMERAL SHAFT FRACTURE

TERMINOLOGY

Synonyms

- Midshaft humerus fx, humerus diaphysis fx

Definitions

- Humerus fracture occurring between upper pectoralis major insertion and supracondylar ridge

IMAGING

General Features

- Best diagnostic clue
  - Transverse lucency through humeral shaft
    - Usually varus angulation
      - Proximal fragment abducted by deltoid
      - Distal fragment pulled proximally by biceps and brachialis muscles
- Location
  - 70% in middle 1/3
- Morphology
  - Transverse (61%), oblique (18%), or spiral (17%)

Radiographic Findings

- Transverse lucency on AP or lateral radiographs
  - Often varus angulation

Imaging Recommendations

- Best imaging tool
  - AP and lateral radiographs out of splint

DIFFERENTIAL DIAGNOSIS

Nutrient Canal

- Straight with sclerotic margins

Supracondylar or Elbow Fracture

- Fracture extends below supracondylar ridge

Humeral Neck Fracture

- Fracture extends above pectoralis major insertion

PATHOLOGY

General Features

- Etiology
  - Fall, motor vehicle accident, gunshot injury
    - Bending forces: Angulated transverse fx
    - Torsional force: Spiral fx
  - Rarely: Overhead throwing in nonconditioned individuals

Staging, Grading, & Classification

- AO-ASIF
  - Type A: No comminution
  - Type B: Butterfly fragment
  - Type C: Comminuted

Gross Pathologic & Surgical Features

- Usually varus angulation
  - Closed reduction: Attempt slight valgus

CLINICAL ISSUES

Presentation

- Most common signs/symptoms
  - Pain, deformity, swelling
  - Arm is shortened
- Other signs/symptoms
  - Can present as adhesive capsulitis
  - Radial nerve injury in 16%
    - Nerve runs in radial sulcus along posterior shaft
      - Particularly fractures at junction of mid and distal 1/3 where it pierces lateral intermuscular septum
    - "Wrist drop" deformity
    - Palsy is usually reversible with reduction
  - Brachial artery injury in up to 3%

Demographics

- Age
  - Average: 54.8 years old
- Epidemiology
  - 1.2% of all fxs in adults

Natural History & Prognosis

- Moderate rate of nonunion, especially open or transverse noncomminuted
  - Nonunion in 6% of open, < 2% of closed fxs

Treatment

- Long arm brace, coaptation splint, or cast brace
- Surgery if
  - Cannot maintain reduced angulation
    - Desire < 10-20° anterior and < 10-30° varus angulation
  - Gap remains at fx site
  - Displaced or segmental fx
  - Open fracture
  - Prolonged recumbency
  - Ipsilateral forearm fx
  - Bilateral humeral shaft fxs (to allow self care)
  - Noncompliant patient

DIAGNOSTIC CHECKLIST

Consider

- Pathologic fx if underlying lytic lesion

Reporting Tips

- Mention if varus angulation > 15°, or persistent fx gap

SELECTED REFERENCES

1. Elton SG et al: Management of radial nerve injury associated with humeral shaft fractures: an evidence-based approach. J Reconstr Microsurg. 24(8):569-73, 2008
2. Ekholm R et al: Fractures of the shaft of the humerus. An epidemiological study of 401 fractures. J Bone Joint Surg Br. 88(11):1469-73, 2006
3. Shao YC et al: Radial nerve palsy associated with fractures of the shaft of the humerus: a systematic review. J Bone Joint Surg Br. 87(12):1647-52, 2005
4. Young CF: Injuries of the humeral shaft. Br J Hosp Med (Lond). 66(11):M74-5, 2005

PECTORALIS INJURY

Key Facts

Terminology
- Synonym: Pectoralis major tear

Imaging
- Location
  - Tendon-bone enthesis at humerus
  - Myotendinous junction tears: 2-6 cm from insertion
  - Intramuscular tears
- Normal tendon
  - Appears to arise from posterior aspect of muscle
  - Thin (3-5 mm) AP
  - 4-6 cm cranial caudal extent
  - Lies immediately anterior to biceps and coracobrachialis muscles
  - Inserts onto lateral lip of inferior bicipital groove
- MR
  - High signal at tendon disruption
  - Edema or hemorrhage in soft tissues around tear
- Ultrasound
  - Tendon fiber disruption + hypoechoic hematoma
- Protocol tips
  - Scan humerus in slight external rotation
  - Axial T1WI and T2WI FS from quadrilateral space → superior aspect of deltoid tubercle
  - Oblique coronal MR images parallel to long axis of tendon

Clinical Issues
- Swelling, pain, ecchymosis
- Weakness with adduction
- Audible "pop," palpable defect
- Muscle asymmetry with contralateral muscle
- Etiology
  - Weightlifting (bench pressing), American football

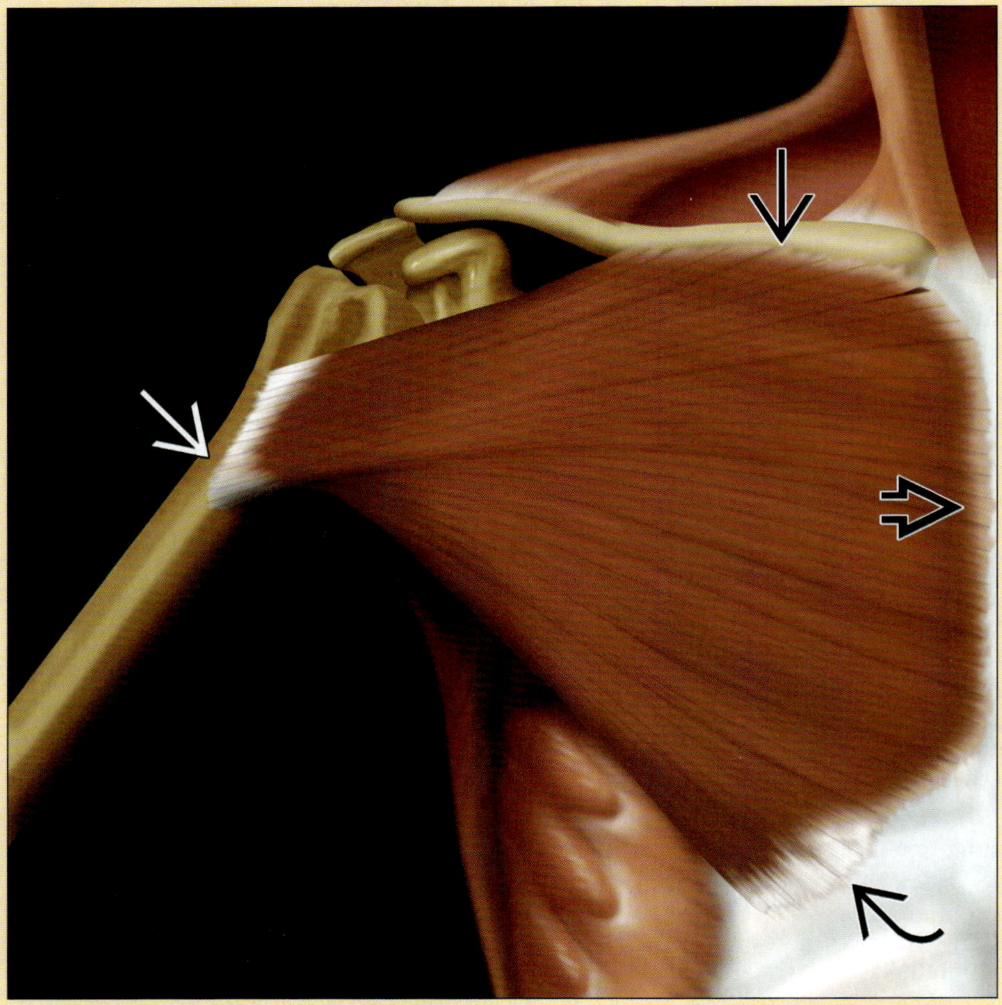

AP graphic of a normal pectoralis major muscle and tendon. The pectoralis major origin has 2 heads, clavicular and sternal. The more superior clavicular head arises from the medial 2/3 of the clavicle ➙. The sternal head is divided into a larger, more superior main sternal portion ⇨, and a smaller abdominal portion ↗. The main sternal head arises from the sternum and ribs 1-5, while the abdominal portion arises from ribs 5 and 6 as well as the fascia of the external oblique and transversalis muscles. Three lamina are thus formed, which then twist > 90° so that the clavicle fibers insert on the humerus inferior and superficial ➡, and the abdominal fibers are superior and deep.

PECTORALIS INJURY

TERMINOLOGY

Synonyms

- Pectoralis major tear

Definitions

- Tear of pectoralis major tendon or muscle

IMAGING

General Features

- Best diagnostic clue
  - MR: Hemorrhage and edema at tear
- Location
  - Tendon-bone enthesis at humerus
  - Myotendinous junction tears: 2-6 cm from insertion
  - Intramuscular tears
- Normal tendon
  - Thin (3-5 mm) AP
  - 4-6 cm cranial caudal extent

MR Findings

- T1WI
  - Gap in course of low signal pectoralis tendon
- T2WI FS
  - High signal at tendon disruption
  - Edema or hemorrhage in soft tissues around tear
    - May involve either or both clavicular head and sternal head portions

Ultrasonographic Findings

- Tendon fiber disruption with hypoechoic hematoma

Imaging Recommendations

- Best imaging tool
  - MR: Axial T1WI and T2WI FS
- Protocol advice
  - Scan humerus in slight external rotation
  - Axial T1WI and T2WI FS
    - From quadrilateral space → superior aspect of deltoid tubercle
  - Oblique coronal MR images parallel to long axis of tendon

DIFFERENTIAL DIAGNOSIS

Rotator Cuff Tear

- Pain or weakness with abduction

Glenoid Labral Tear

- Common in similar demographic group

Biceps Tendon Injury

- Can mimic pectoralis major tears at tendon-bone insertion

PATHOLOGY

General Features

- Etiology
  - Excessive tension on contracted muscle
  - Intramuscular tear: Direct blow

Staging, Grading, & Classification

- Prognosis and treatment based on
  - Grade
    - Low grade: < 50% partial tear
    - High grade: > 50% partial tear
    - Complete tears
    - Note: High-grade partial tears at myotendinous junction most common in surgical practice
      - Intramuscular and myotendinous tears typically partial
  - Location
    - Proximal intramuscular tears
    - Myotendinous junction
    - Tendon-bone interface at humerus
  - Pectoralis head involved
    - Clavicular head vs. sternal head portions
  - Acute or chronic
- Tietjen classification
  - I: Sprain or contusion
  - II: Partial tears
  - III: Complete tears
    - IIIA: Sternoclavicular origin
    - IIIB: Muscle belly
    - IIIC: Musculotendinous junction
    - IIID: Humeral insertion

CLINICAL ISSUES

Presentation

- Most common signs/symptoms
  - Swelling, pain, ecchymosis
- Other signs/symptoms
  - Weakness with adduction
  - Audible "pop," palpable defect

Demographics

- Epidemiology
  - Weightlifting (bench pressing), American football
- Male, 20-40 years old

Natural History & Prognosis

- Proximal and myotendinous partial tears usually heal
- Complete tears: Weakness, cosmetic defect

Treatment

- Small partial tears: Conservative Rx
- Humeral avulsion: Early surgical repair
  - Prevent tendon retraction
  - Chronic tears: Worse functional outcome if surgery > 6-8 weeks after injury

SELECTED REFERENCES

1. Zvijac JE et al: Pectoralis major tears: correlation of magnetic resonance imaging and treatment strategies. Am J Sports Med. 34(2):289-94, 2006
2. Carrino JA et al: Pectoralis major muscle and tendon tears: diagnosis and grading using magnetic resonance imaging. Skeletal Radiol. 29(6):305-13, 2000
3. Lee J et al: MR imaging assessment of the pectoralis major myotendinous unit: an MR imaging-anatomic correlative study with surgical correlation. AJR Am J Roentgenol. 174(5):1371-5, 2000
4. Connell DA et al: Injuries of the pectoralis major muscle: evaluation with MR imaging. Radiology. 210(3):785-91, 1999

PECTORALIS INJURY

(Left) **AP graphic shows the muscle origins and insertions around the shoulder. The pectoralis major origin is from the clavicle and sternum . The tendon inserts on the lateral lip of the inferior bicipital groove .** ***(Right)*** *AP graphic of a myotendinous junction partial tear of the pectoralis major muscle. There is a tear mainly of the clavicle head portion of the muscle in the myotendinous region, with sparing of the sternal head portion . The deltoid muscle is also shown .*

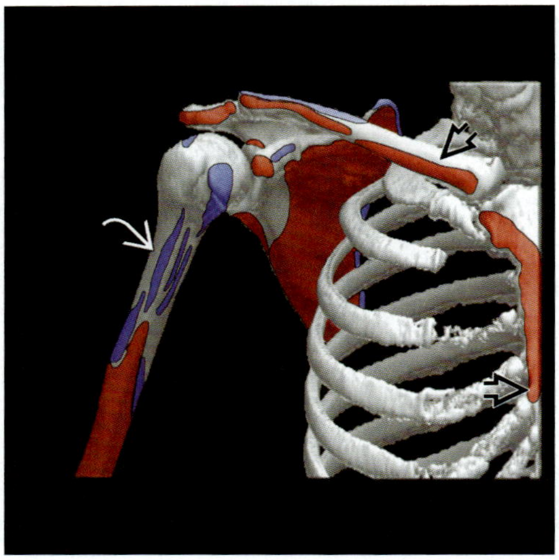

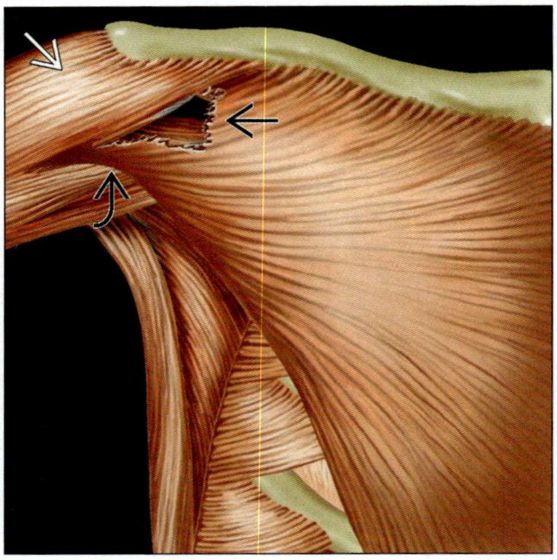

(Left) *Axial T1WI MR of a 58-year-old man with pain and tenderness over the pectoralis region after a fall with hyperextension injury 5 months prior. The tendon near the humeral insertion is intact.* ***(Right)*** *Axial T2WI FS MR in the same patient at a more superior level shows intact fibers in the upper myotendinous region. The patient therefore does not have a complete tear of the pectoralis major at either of the 2 most common sites, the humeral insertion or the myotendinous junction.*

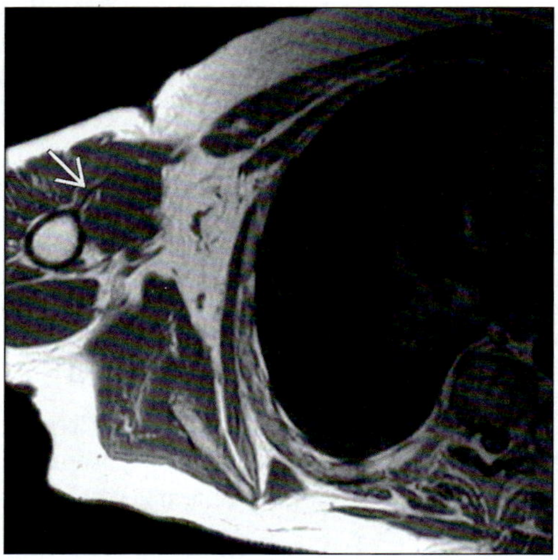

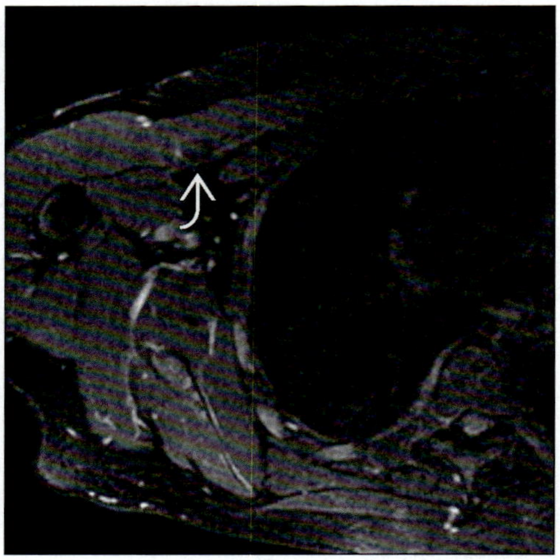

(Left) *Axial T2WI FS MR in the same patient more inferiorly following the tendon laterally shows focal edema at the pectoralis major myotendinous junction.* ***(Right)*** *Axial T2WI FS MR in the same patient, on a slice more inferiorly at the level of the tendon attachment , shows edema and fluid near the myotendinous junction . The patient had preserved strength with adduction testing and was treated conservatively for a presumed low-grade (< 50%) partial pectoralis major tear.*

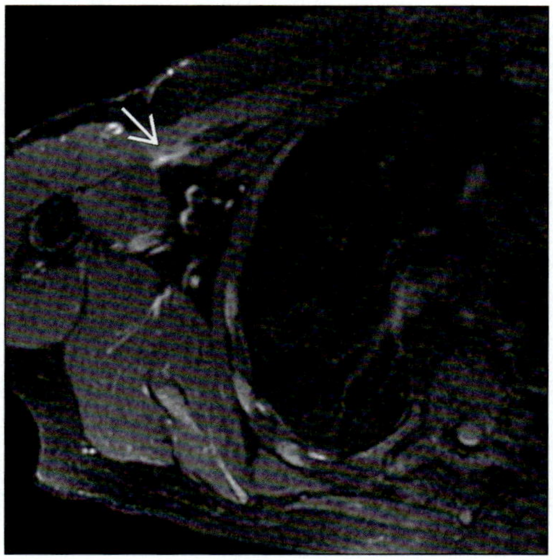

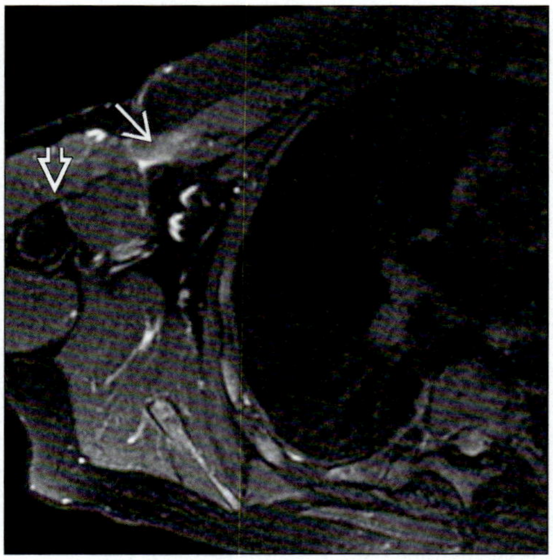

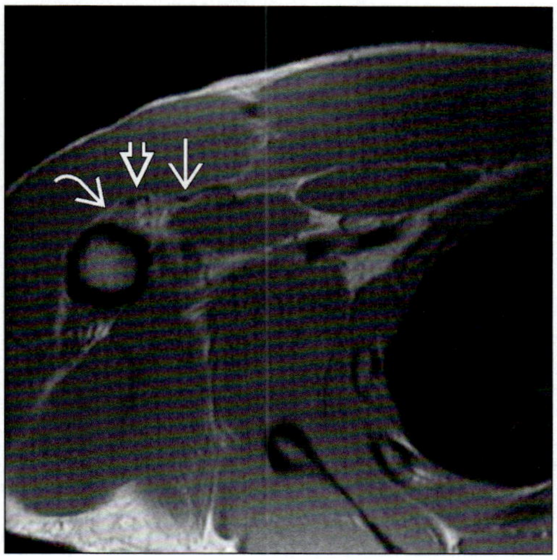

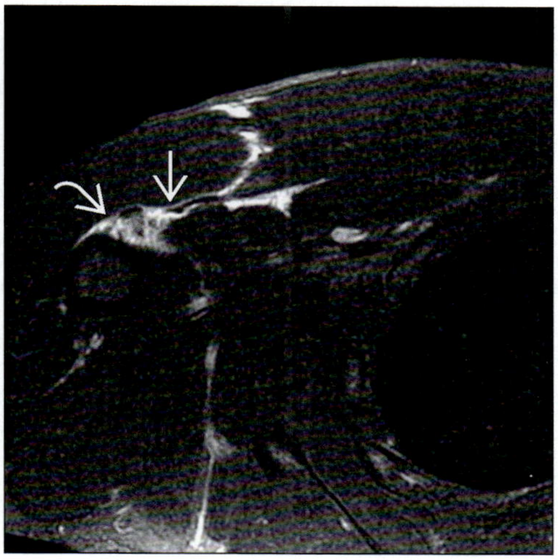

(Left) *Axial T1WI MR in a 22-year-old man with severe pain while doing a heavy bench press. The pectoralis major tendon is not seen at the humeral insertion site and is retracted medially. Deep to the biceps tendon there is high signal at the pectoralis insertion site.* ***(Right)*** *Axial T2WI FS MR in the same patient at the same level shows high signal hemorrhage near the insertion of the tendon and medial retraction of the tendon edge.*

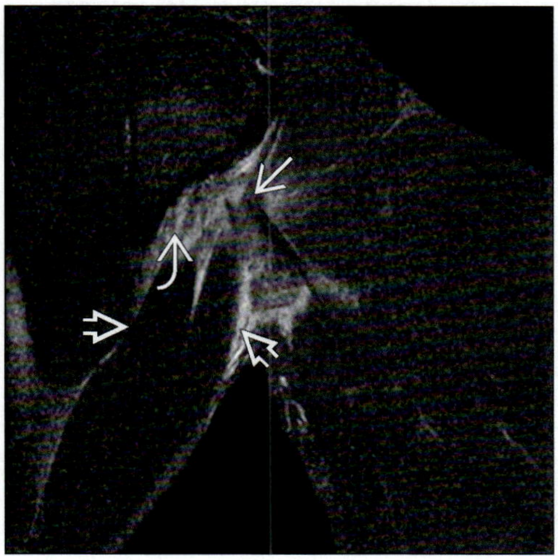

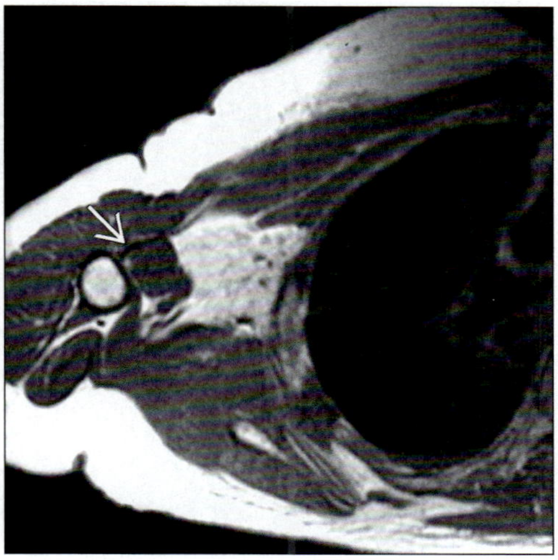

(Left) *Oblique coronal T2WI FS MR in the same patient, parallel to the course of the pectoralis major, shows retraction of the tendon edge and edema and hemorrhage near the tendon insertion into the humerus. The 2 heads of the biceps muscle are intact.* ***(Right)*** *Axial T1WI MR in a 43-year-old woman with anterior chest muscle pain after an injury sustained while diving into a pool. The pectoralis major tendon is normal at the insertion.*

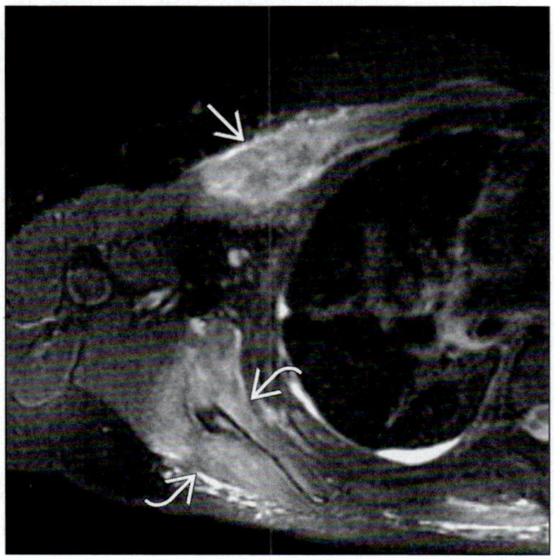

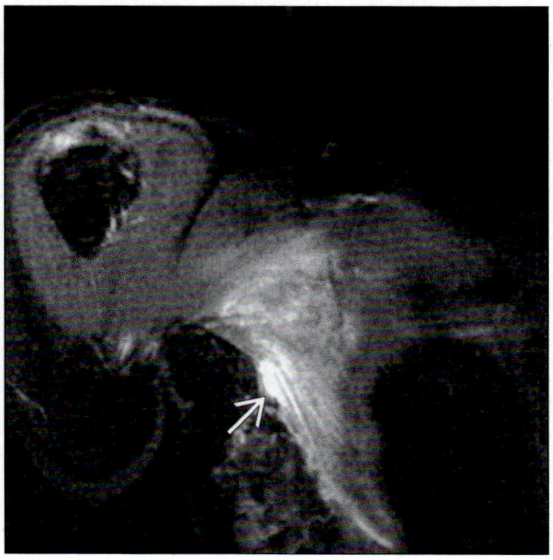

(Left) *Axial T2WI FS MR in the same patient shows high signal in the pectoralis major muscle, consistent with an intramuscular partial tear. There is also edema from injury to the teres minor and subscapularis muscles.* ***(Right)*** *Coronal oblique T2WI FS MR in the same patient, parallel to the orientation of the pectoralis major muscle, again shows the edema in the pectoralis major muscle, consistent with an intramuscular partial tear.*

DELTOID MUSCLE INJURY

Key Facts

Terminology
- Deltoid muscle tear, deltoid dehiscence, deltoid detachment
- Spontaneous or postsurgical tear of deltoid muscle

Imaging
- MR
  - Deltoid defect with focal area of T2 fluid signal intensity
- Spontaneous tear
  - Proximal musculotendinous junction
  - Middle 1/3 of muscle
  - May have subcutaneous fluid collection or edema or intradeltoid cyst
- Postsurgical
  - Tear near acromion process if prior lateral acromioplasty
  - Tear can also occur where surgical incision/repair of deltoid was performed
  - Metal susceptibility artifacts from arthroscopy instrument debris may be present

Top Differential Diagnoses
- Deltoid muscle strain
  - No focal fluid collection as seen in tear
- Denervation edema

Clinical Issues
- Pain
- Weakness
- Palpable defect
- Chronic full thickness cuff tear or prior rotator cuff surgery
- Conservative treatment

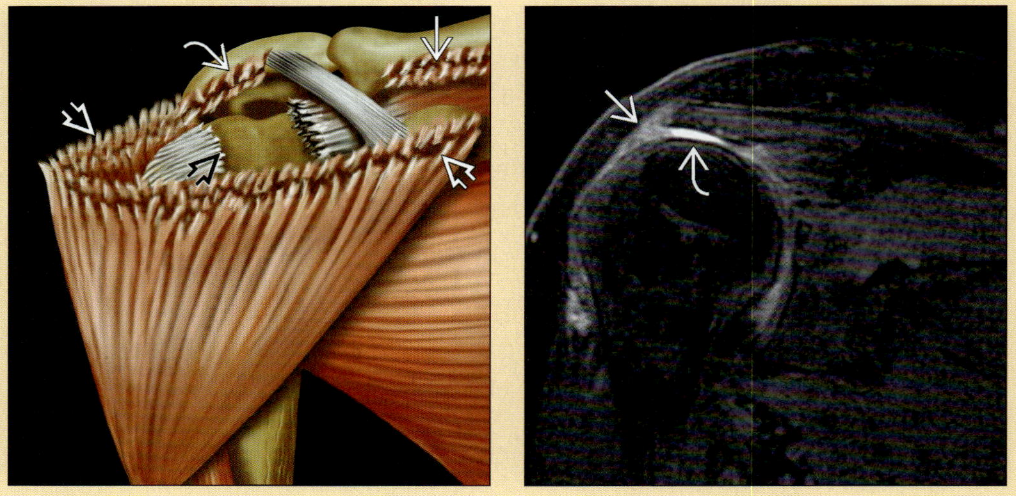

(Left) AP graphic of a large deltoid muscle tear. The tear involves both the acromial and clavicular regions of the muscle. There is also a full thickness rotator cuff tear, which is usually present with spontaneous deltoid tears. (Right) Coronal oblique T2WI FS MR shows a full thickness tear of the deltoid muscle with increased signal tissue in the defect. There is also a large full thickness supraspinatus tendon tear with a high-riding humeral head.

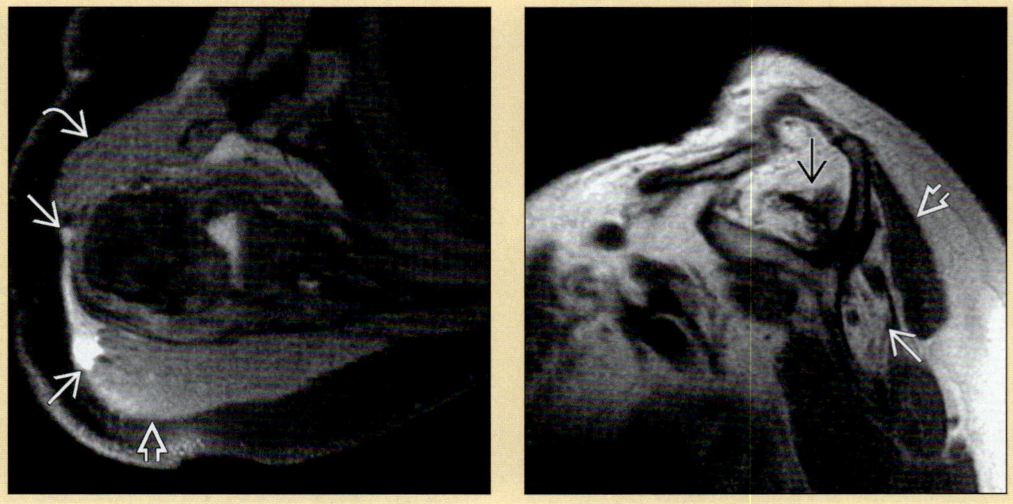

(Left) Axial T2WI FS MR in the same patient shows the full thickness tear of the deltoid muscle with fluid adjacent to the subcutaneous fat. There is sparing of the anterior and posterior portions of the deltoid. (Right) Sagittal T1WI MR in the same patient shows atrophy and fatty replacement of the supraspinatus and infraspinatus muscles. Most deltoid tears also have a chronic cuff tear. There is no significant atrophy of the posterior deltoid muscle.

DELTOID MUSCLE INJURY

TERMINOLOGY

Synonyms

- Deltoid muscle tear, dehiscence, detachment

Definitions

- Spontaneous or postsurgical deltoid tear

IMAGING

General Features

- Best diagnostic clue
  - Deltoid defect with focal T2 fluid signal intensity
  - Postsurgical tear: Metal debris artifact
- Location
  - Myotendinous junction or origin of deltoid

MR Findings

- Defect in deltoid muscle with focal area of fluid signal intensity
  - Spontaneous tear
    - Proximal musculotendinous junction
    - Middle 1/3 of muscle
    - Subcutaneous fluid collection or edema
  - Postsurgical
    - Tear near acromion process if prior lateral acromioplasty
    - Tear can also occur where surgical incision/repair of deltoid was performed
    - Metal susceptibility artifacts from arthroscopy instrument debris may be present

Ultrasonographic Findings

- Anechoic region in deltoid muscle

Imaging Recommendations

- Best imaging tool
  - Conventional MR
- Protocol advice
  - Standard MR planes with fat-suppressed T2WI

DIFFERENTIAL DIAGNOSIS

Deltoid Muscle Strain

- Focal muscle edema at myotendinous junction
- No focal fluid collection as seen in tear
- "Wii-itis" from overplaying bowling video game
  - Triceps edema

Deltoid Contusion

- Focal muscle edema from direct blow to muscle

Deltoid Contracture

- Fibrotic cord extending from acromion to deltoid tuberosity
- Due to repeated intramuscular injections

Atrophy and Denervation Edema

- Age-related atrophy
  - Deltoid often more severe than supraspinatus
- Postoperative atrophy
  - More common in classic open rotator cuff repairs
- Denervation atrophy (preceded by muscle edema)
  - Quadrilateral space syndrome
  - Parsonage-Turner
  - Axillary nerve injury from dislocation or proximal humerus fracture

Deltoid Lipoma

- T1WI: Well-defined homogeneous high signal mass
- Deltoid muscle is common site for lipoma

PATHOLOGY

Staging, Grading, & Classification

- Partial thickness deltoid tear
  - Usually involves undersurface of tendon
- Full thickness deltoid tear
  - Spontaneous tear: Middle 1/3 of muscle
  - Postsurgical: Adjacent to lateral acromioplasty
- Complete deltoid detachment

CLINICAL ISSUES

Presentation

- Most common signs/symptoms
  - Lateral shoulder and arm pain
- Other signs/symptoms
  - Weakness
  - Palpable defect

Demographics

- Age
  - Spontaneous tear in 70-75 year olds
- Gender
  - M:F = 1:2
- Epidemiology
  - Massive cuff tear or prior cuff surgery
  - Uncommon: 0.3% of shoulder MR scans

Natural History & Prognosis

- Tears may heal with scar tissue

Treatment

- Conservative
- Shoulder arthroplasty for chronic FT RCT
  - Deltoid tear may be contraindication

DIAGNOSTIC CHECKLIST

Consider

- Deltoid tear if massive rotator cuff tear

Image Interpretation Pearls

- Subcutaneous fluid or intramuscular cyst

SELECTED REFERENCES

1. Ashry R et al: Muscle atrophy as a consequence of rotator cuff tears: should we compare the muscles of the rotator cuff with those of the deltoid? Skeletal Radiol. 36(9):841-5, 2007
2. Ilaslan H et al: Deltoid muscle and tendon tears in patients with chronic rotator cuff tears. Skeletal Radiol. 36(6):503-7, 2007
3. Bianchi S et al: Imaging findings of spontaneous detachment of the deltoid muscle as a complication of massive rotator cuff tear. Skeletal Radiol. 35(6):410-5, 2006

DELTOID MUSCLE INJURY

*(**Left**) Coronal oblique T2WI in a patient with a prior open rotator cuff debridement and partial subacromial decompression, and ongoing pain at the incision site. There is focal thinning and increased signal in the deltoid muscle. (**Right**) Axial fat-suppressed intermediate-weighted image in the same patient shows the postsurgical scar in the deltoid with metal susceptibility artifact. Although the patient had pain and a small post-op defect, no large deltoid detachment was seen.*

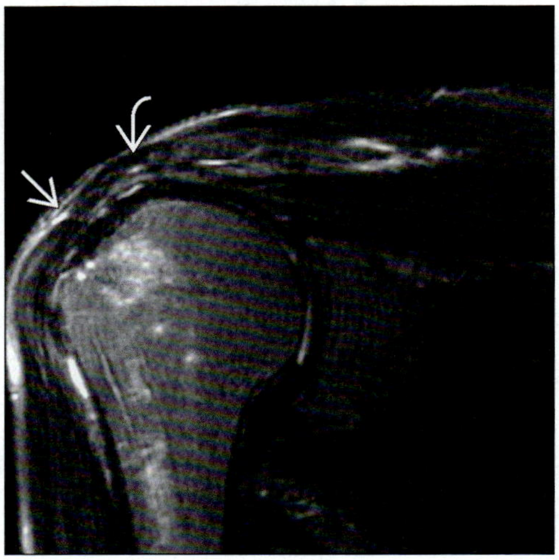

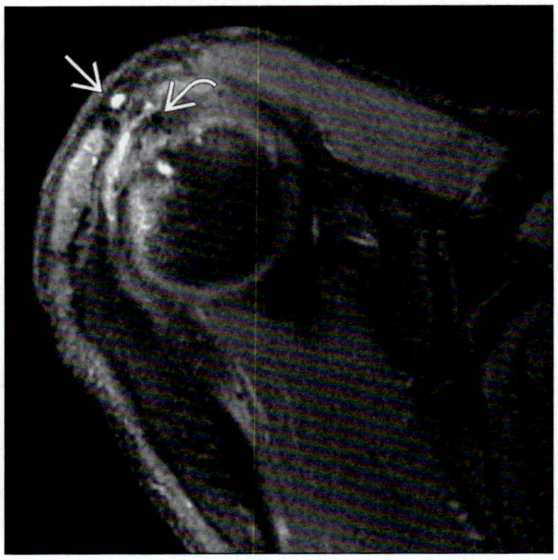

*(**Left**) Neer AP radiograph of a patient with a prior lateral acromioplasty and rotator cuff debridement, who now has recurrent shoulder pain. (**Right**) Coronal oblique T2WI FS MR in the same patient shows extensive metal susceptibility artifacts and a defect in the proximal deltoid from old dehiscence. Lateral acromioplasty often weakens the deltoid attachment. The patient also had articular and bursal surface partial thickness supraspinatus tendon tears at subsequent arthroscopy.*

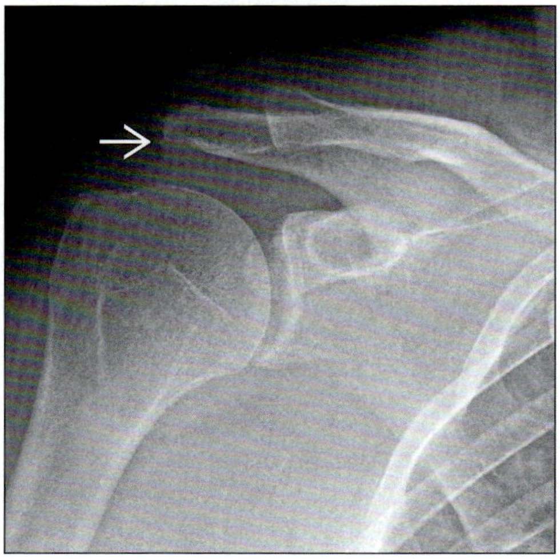

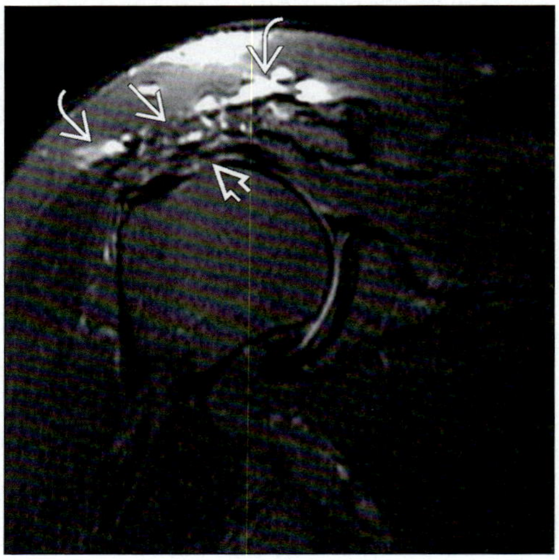

*(**Left**) Oblique sagittal T1WI in the same patient shows the metal susceptibility artifact and the focal defect in the deltoid muscle. There is also focal thinning of the supraspinatus tendon consistent with the recurrent tear seen at revision arthroscopy. (**Right**) Adjacent oblique sagittal T1WI MR in the same patient also shows the defect in the deltoid muscle. Because there was adequate scar at the old deltoid dehiscence site, no deltoid repair was done at the revision surgery.*

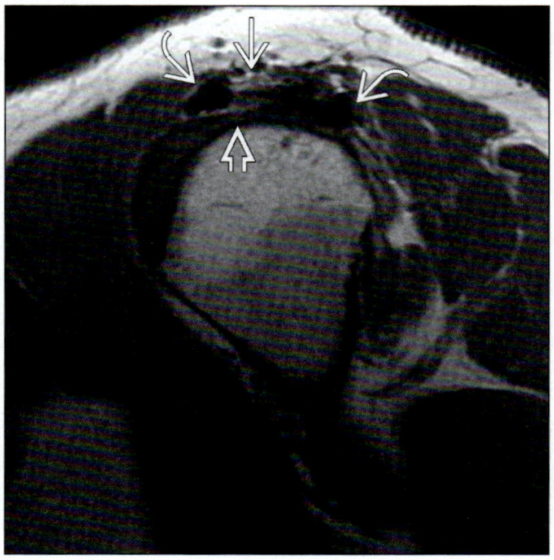

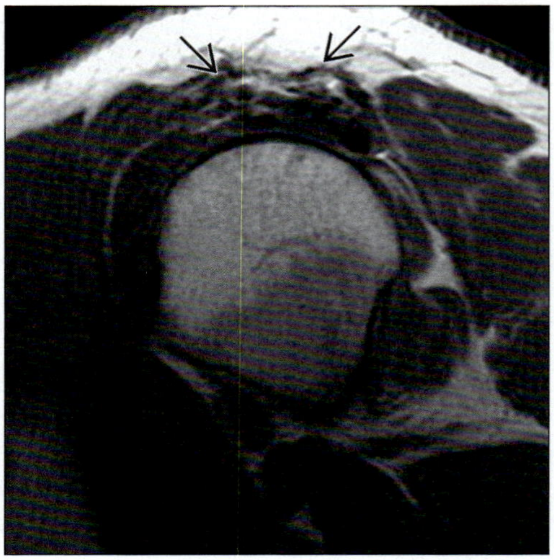

DELTOID MUSCLE INJURY

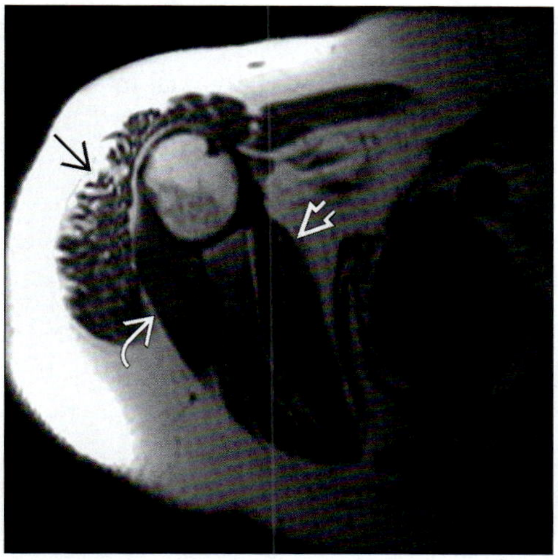

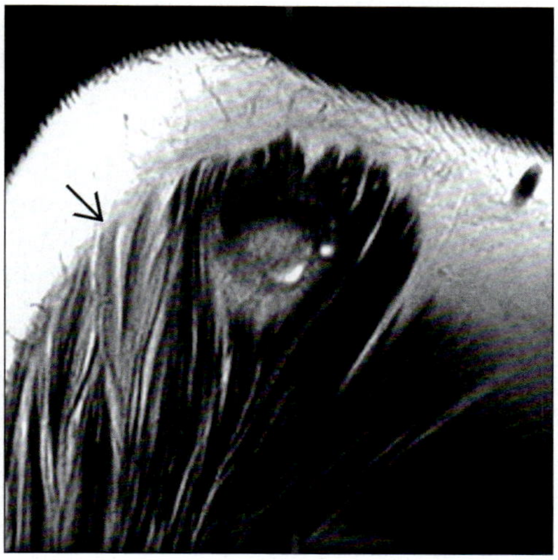

(Left) *Axial T1WI MR shows mild fatty replacement atrophy → of the deltoid muscle in a 48-year-old diabetic patient. The infraspinatus → and subscapularis → muscles are normal.* ***(Right)*** *Sagittal T2WI MR without fat suppression shows mild fatty atrophy of the deltoid muscle →. With normal aging, the deltoid often shows slightly greater fatty atrophy than the supraspinatus muscle. The deltoid can therefore be a conservative internal reference for rotator cuff muscle atrophy.*

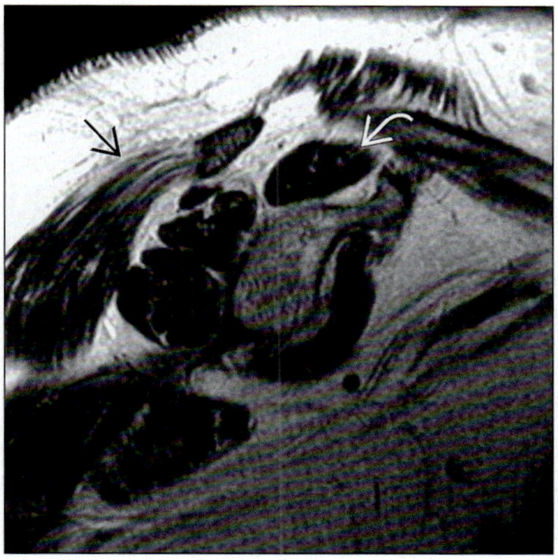

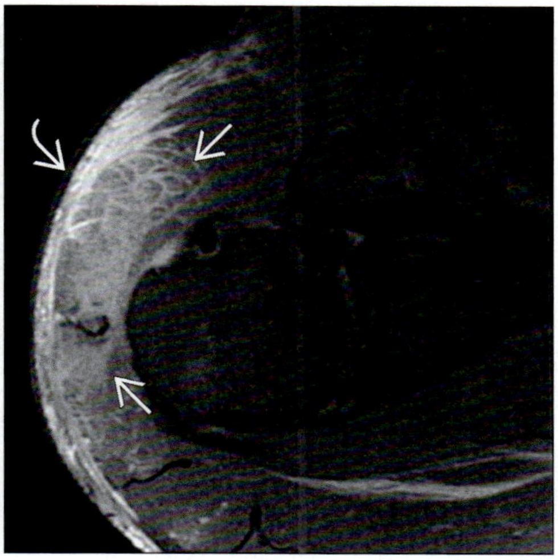

(Left) *Sagittal T2WI without fat suppression in the same patient demonstrates mild fatty atrophy of the deltoid muscle →, increased compared to the supraspinatus muscle →. The patient did not have a rotator cuff tear. The deltoid atrophy was felt to be due to chronic shoulder pain and limited activity.* ***(Right)*** *Axial T2WI FS MR after a direct blow to the shoulder shows high signal in both the deltoid muscle → and overlying subcutaneous fat → consistent with a contusion.*

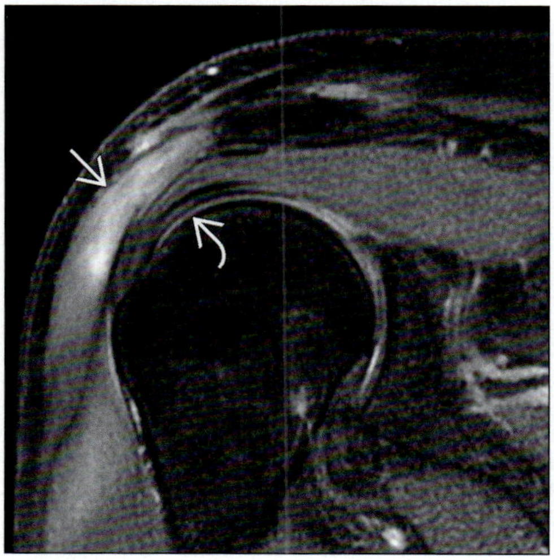

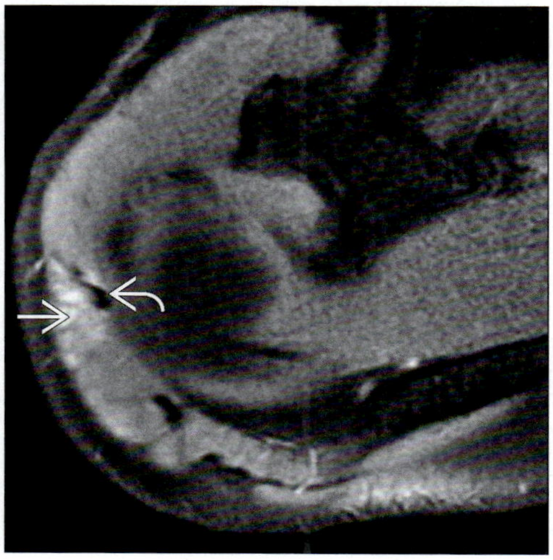

(Left) *Oblique coronal T2WI in a patient with shoulder pain after an MVA shows high signal in the deltoid muscle →, with sparing of the supraspinatus tendon →. The patient did not have bruising over the area of deltoid edema.* ***(Right)*** *Oblique axial image in the same patient shows that the area of deltoid muscle injury → is adjacent to 1 of the deltoid tendons →. Injury at the myotendinous junction is common in muscle strains or partial tears from overstretch injury.*

PROXIMAL TRICEPS INJURY

Key Facts

Terminology
- Strain or tear of proximal triceps muscle

Imaging
- MR
  - T2 high signal within proximal triceps muscle
  - Intramuscular; usually at proximal myotendinous junction
  - Perimuscular T2 high signal edema
  - May involve single or multiple heads of muscle
  - T1: Usually normal; may have areas of high signal from extracellular methemoglobin
- Radiographs
  - Usually normal
  - Old injuries may show myositis ossificans
- Ultrasound
  - Hypoechoic region within triceps muscle
  - Focally tender to palpation with transducer
- Imaging recommendations
  - MR with fat-suppressed T2WI
  - Sagittal T2WI shows proximal muscle long axis best
  - If concern for triceps injury, obtain axial images distal to humeral head

Top Differential Diagnoses
- Neurogenic muscle edema
- Distal triceps tear
- Triceps muscle flap

Pathology
- Grade 1: Strain
- Grade 2: Partial muscle tear
- Grade 3: Complete rupture of 1 of 3 heads

Clinical Issues
- Pain posterior upper arm

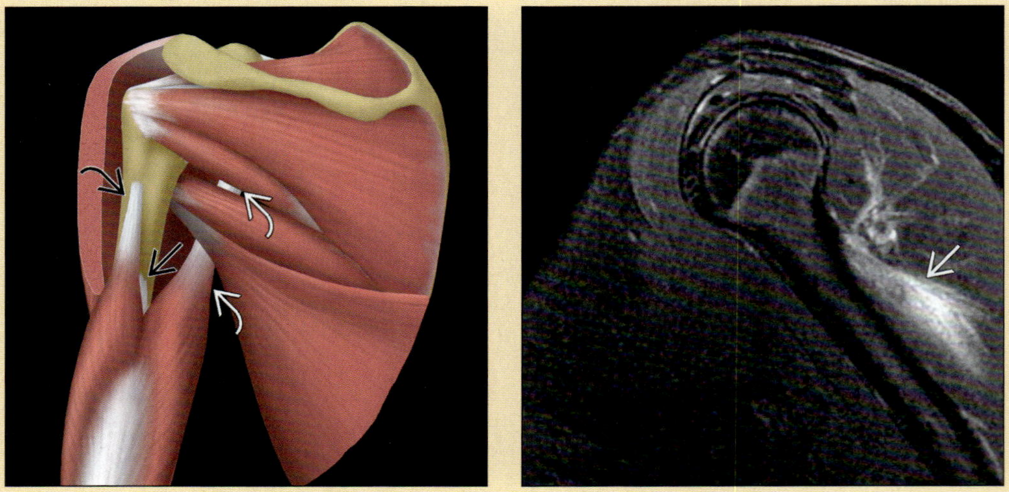

(Left) *Graphic shows posterior view of the 3 heads of the proximal triceps muscle, including the long head ➔ arising from the inferior glenoid, the lateral head ➔, and the medial head ➔, which originates deep and distal to the others.* ***(Right)*** *Oblique sagittal T2WI FS MR of a 16-year-old swimmer with posterior shoulder pain after an injury shows edema within the proximal portion of the lateral head of the triceps muscle ➔ as well as perimuscular edema consistent with a muscle strain.*

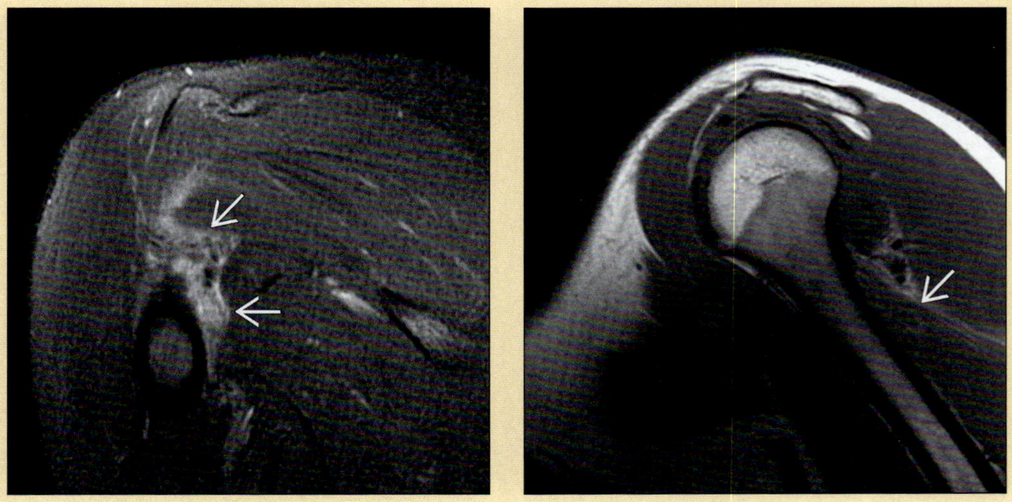

(Left) *Coronal oblique T2WI FS MR in a 16 year old with pain in the posterior shoulder and upper arm after an injury. There is high signal ➔ adjacent to the posterior proximal humeral diaphysis near the origin of the lateral head of the triceps muscle.* ***(Right)*** *Oblique sagittal T1WI MR in the same patient demonstrates the origin of the lateral head of the triceps muscle ➔. Proximal triceps muscle injuries are uncommon. Grade 1 muscle strain injuries like this can be treated conservatively.*

PROXIMAL TRICEPS INJURY

TERMINOLOGY

Synonyms

- Triceps brachii injury

Definitions

- Strain or tear of proximal triceps muscle

IMAGING

General Features

- Best diagnostic clue
  - MR T2 high signal within proximal triceps muscle
- Location
  - Intramuscular, usually at proximal myotendinous junction

Radiographic Findings

- Usually normal
- Old injuries may show heterotopic ossification

MR Findings

- T1WI
  - Usually normal
  - May have areas of high signal from extracellular methemoglobin
- T2WI FS
  - Intramuscular high signal
  - Perimuscular high signal edema
  - May involve 1 or multiple heads of muscle

Ultrasonographic Findings

- Hypoechoic region within triceps muscle
  - Focally tender to palpation with transducer

Imaging Recommendations

- Best imaging tool
  - MR with fat-suppressed T2WI
- Protocol advice
  - Sagittal T2WI is most important plane/sequence
  - If concern for triceps injury, extend axial images distal to humeral head
    - More distal than usual imaging for shoulder injury

DIFFERENTIAL DIAGNOSIS

Neurogenic Muscle Edema

- Motor nerve compression
- Neuritis

Distal Triceps Tear

- Avulsion near olecranon process
- Much more common than proximal triceps injuries
  - Nonetheless, rare injury
- Weakness with elbow extension; similar presentation as proximal triceps injury
  - Pain near elbow joint (distal to expected site of pain for proximal triceps injury)

Triceps Long Head Flap

- Used for repair of large rotator cuff tears
- Watch for evidence of surgery to triceps muscle
  - Metal artifact
  - Abnormal position of long head triceps muscle

PATHOLOGY

General Features

- Etiology
  - Athletic trauma
    - Weightlifters
    - American football linemen
  - Anabolic steroid users
  - Systemic diseases affecting tendinous integrity
    - Common diseases include
      - Hyperparathyroidism
      - Renal osteodystrophy
      - Rheumatoid arthritis
    - Usually affects distal rather than proximal triceps

Staging, Grading, & Classification

- Grade 1: Strain
  - No macroscopic muscle tear
  - Microscopic tear of collagen fibers
- Grade 2: Partial muscle tear
  - Tear through a portion of 1 of the heads of triceps muscle
  - At least some muscle fibers of each head still intact
- Grade 3: Complete rupture of 1 of 3 heads
  - Complete muscle tear of all 3 heads very uncommon

CLINICAL ISSUES

Presentation

- Most common signs/symptoms
  - Pain in posterior upper arm
  - Weakness or loss of elbow extension
- Other signs/symptoms
  - Bruising of posterior upper arm

Demographics

- Age
  - 20-40 years of age
- Gender
  - Male > female

Natural History & Prognosis

- Partial muscle tears heal with rest
- Complete tears require surgery

Treatment

- Surgery for complete ruptures
  - Most do well with repair

DIAGNOSTIC CHECKLIST

Reporting Tips

- Mention which of the 3 muscle heads are torn as well as extent of tear

SELECTED REFERENCES

1. Singh RK et al: Complete rupture of the triceps brachii muscle. Br J Sports Med. 36(6):467-9, 2002
2. Holleb PD et al: Triceps brachii injuries. Sports Med. 10(4):273-6, 1990
3. Aso K et al: Muscle belly tear of the triceps. Am J Sports Med. 12(6):485-7, 1984

ROTATOR CUFF IMPINGEMENT

Key Facts

Terminology

- Painful compression of supraspinatus tendon and subacromial bursa between coracoacromial (CA) arch and humeral head
  - Older individuals
- Synonyms: External impingement, extrinsic impingement, subacromial impingement syndrome (SIS/SAIS), anterior subacromial impingement syndrome (ASIS), supraspinatus impingement, outlet impingement

Imaging

- Hooked anterior acromion on arch view
- Acromioclavicular joint inferior osteophytes
  - ≥ 2 mm directed inferiorly
- Arch, outlet, or coracoacromial arch view
  - Hooked or downturned anterior acromion
  - Decreased acromiohumeral distance anteriorly compared to posteriorly
- Os acromiale
- Large lateral extension of acromion
  - ± lateral acromial tilt
- MR: Oblique coronal images
  - Tendinopathy: Increased signal and focal swelling of supraspinatus tendon
  - Subacromial/subdeltoid bursitis

Top Differential Diagnoses

- Calcific tendinopathy
  - Supraspinatus (SS) tendon or subacromial (SA) bursa calcification on radiograph

Diagnostic Checklist

- Do not attempt to diagnose rotator cuff impingement by MR; can only be diagnosed clinically

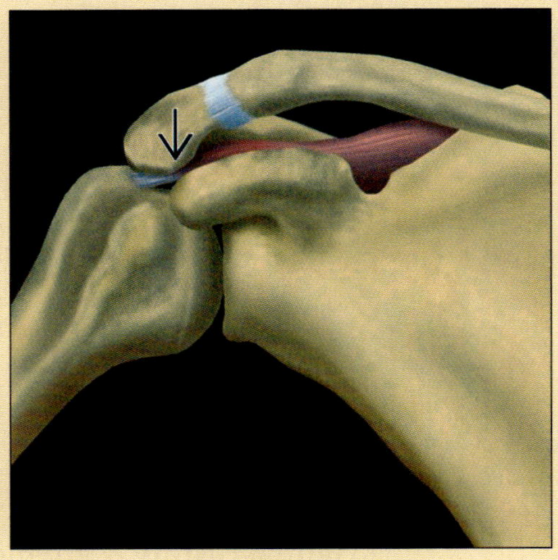
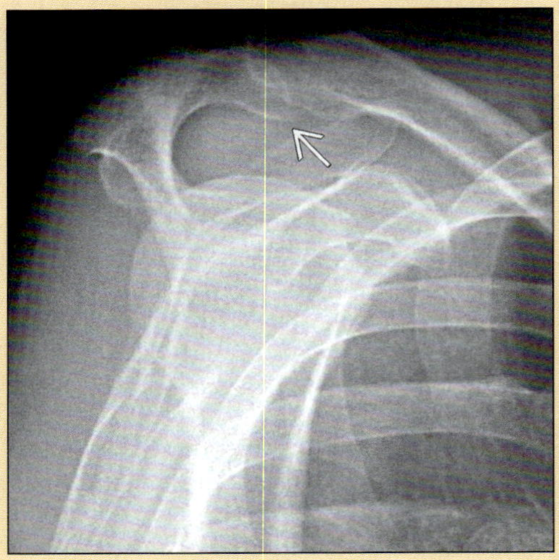

(Left) *AP graphic shows a downturned, or type III anterior acromion ➔, compressing the anterior supraspinatus tendon. The acromion also compresses the richly innervated subacromial bursa and joint capsule.* ***(Right)*** *Arch view of a patient with a hooked, or type III acromion ➡. The anterior acromion is downturned, which may indent the superior surface of the supraspinatus tendon. An anterior subacromial resection will often relieve impingement pain and allow healing of a repaired rotator cuff.*

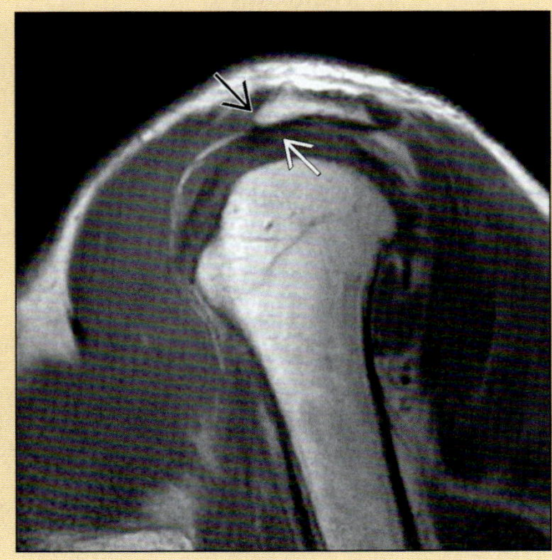
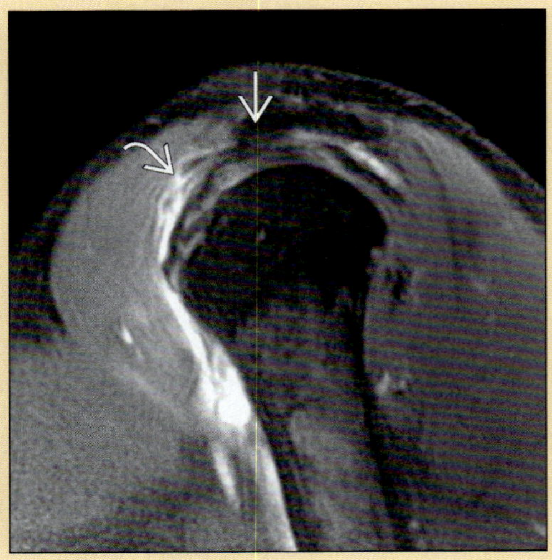

(Left) *Sagittal T1WI MR shows a type III acromion ➔ compressing the subacromial bursa and indenting the supraspinatus tendon ➡. The compression typically worsens with flexion and abduction.* ***(Right)*** *Sagittal T2WI FS MR in the same patient shows the anterior acromion ➡ indenting the supraspinatus tendon. There is subacromial bursitis ⤻, which can be from repetitive compression of the bursa by the anterior acromion during flexion and abduction.*

ROTATOR CUFF IMPINGEMENT

TERMINOLOGY

Abbreviations

- Subacromial impingement syndrome (SIS/SAIS)
- Anterior subacromial impingement syndrome (ASIS)

Synonyms

- External impingement, extrinsic impingement, supraspinatus impingement, outlet impingement, classic impingement

Definitions

- Painful compression of supraspinatus tendon and subacromial bursa between coracoacromial arch and humeral head
  - Coracoacromial arch
    - Anterior acromion process
    - Coracoacromial (CA) ligament
    - Acromioclavicular (AC) joint

IMAGING

General Features

- Best diagnostic clue
  - Hooked or type III anterior acromion on arch view
  - AC joint inferior osteophytes
  - Os acromiale
- Size
  - AC joint inferior osteophytes ≥ 2 mm
  - Increased lateral extension of acromion
    - Graded by measuring horizontal distance from glenoid to lateral aspect of acromion process, and glenoid to lateral cortex of humeral head
    - Glenoid → acromion/glenoid → humerus ratio > 0.7 associated with rotator cuff tears
- Morphology
  - Type III hooked anterior acromion

Radiographic Findings

- AP view
  - AC joint osteophytes
    - ≥ 2 mm directed inferiorly
  - Decreased acromiohumeral distance
    - < 7 mm
    - Chronic full thickness rotator cuff tear (RCT) ± fatty atrophy of cuff
    - May be seen in congenital subacromial (SA) stenosis
      - No superior migration of humeral head
      - Impingement pain ± RCT
      - Many will go on to develop RCT
- Zanca view: 15-20° cranial angulation
  - Shows inferior surface of AC joint better
- Arch view or outlet view or CA arch view
  - Used to grade anterior acromion process
    - Type I: Flat
    - Type II: Normal curved
    - Type III: Hooked or downturned
      - Congenital, not age-related
      - Associated with impingement pain and rotator cuff tears
  - Anterior acromion: CA enthesophyte
    - Age-related, reactive
    - Type III acromion and enthesophyte can coexist
      - Both contribute to impingement
  - Decreased acromiohumeral distance
    - Anteriorly > posteriorly
- Axillary view: Os acromiale

CT Findings

- Oblique sagittal reformatted images
  - Hooked anterior acromion
- Oblique coronal reformatted images
  - AC inferior osteophytes
- CT arthrography
  - RCT

MR Findings

- T1WI
  - Oblique sagittal
    - Hooked acromion
    - Thick CA ligament
    - Anterior acromion or AC joint indents supraspinatus (SS)
    - Os acromiale
      - Step-off deformity at junction with acromion process
  - Oblique coronal
    - Tendinopathy: Increased signal and focal swelling of SS tendon
- T2WI FS
  - Thick synovitic SA/subdeltoid (SD) bursa
  - Os acromiale
    - Pseudarthrosis type with fluid in synchondrosis
  - Tendinopathy
    - Abnormal increased signal within SS tendon
    - Focal swelling of SS tendon
  - Rotator cuff tear
    - High signal extending to tendon surface

Ultrasonographic Findings

- Grayscale ultrasound
  - CA arch indents SS during flexion/abduction
    - Reproduction of pain with compression of bursa
  - Tendinopathy
    - Focal swelling and hypoechoic changes of SS tendon
  - Rotator cuff tear
    - Focal hypo- or anechoic region extending to SS tendon surface

Imaging Recommendations

- Best imaging tool
  - MR to diagnose or exclude RCT
- Protocol advice
  - FS T2WI in 3 planes
  - Non-fat-suppressed T1 or PD in at least 1 plane

DIFFERENTIAL DIAGNOSIS

Intrinsic Causes of Tendinopathy/RCT

- Patients can have impingement-type pain without extrinsic causes
  - SA/SD bursitis
- Intrinsic causes of tendinopathy/cuff tear
  - Ischemic changes/poor healing in hypovascular critical zone of RC
  - Myxoid degeneration
  - Repetitive microtrauma

ROTATOR CUFF IMPINGEMENT

Internal Impingement
- Young overhead throwing athletes
- Pain in late cocking phase of overhead throwing

Secondary Impingement
- Classic impingement pain but due to instability
- No abnormality of CA arch

Subcoracoid Impingement
- Impingement of subscapularis (SC) tendon

AC Joint Osteoarthritis
- Tender over AC joint

Calcific Tendinopathy
- M < F
- Age: 5th and 6th decade
- SS tendon or SA bursa calcification on radiograph

Acute Traumatic Cuff Tear
- May have preexisting impingement syndrome

Spinoglenoid (SG)/Suprascapular Notch Ganglion
- Paralabral cyst compresses suprascapular nerve

Adhesive Capsulitis
- Decreased range of motion (ROM)
- Thickened axillary recess capsule
- Scarring in anterior rotator interval

PATHOLOGY

General Features
- Etiology
  - Repetitive indenting and compression of RC and SA/SD bursa by CA arch structures

Staging, Grading, & Classification
- Neer impingement stages
  - Stage I: Edema and hemorrhage
    - Patients ≤ 25 years
  - Stage II: Fibrosis and thickening of SA/SD bursa with tendinopathy
    - Patients 25-40 years
    - May resolve temporarily with rest, but residual damage to RC and bursa
  - Stage III: Rotator cuff tear
    - Patients > 40 years old
    - Operative RC repair and acromioplasty often necessary
- Bigliani anterior acromial shape
  - Type I: Flat
    - Uncommon
  - Type II: Curved undersurface
    - Most common
  - Type III: Hooked acromion
    - Downturned anterior acromion
    - Associated with impingement syndrome
  - Type IV: Convex undersurface is rare

CLINICAL ISSUES

Presentation
- Most common signs/symptoms
  - Symptom: Pain during activities with arm in 90° flexion/abduction
  - Signs
    - Neer impingement sign
      - Pain with 90-140° passive flexion
    - Hawkins-Kennedy impingement sign
      - Pain with passive internal rotation while in 90° flexion
    - Painful arc sign
      - Pain during active elevation between 60-120° in plane of scapula

Demographics
- Age
  - > 40 years
- Gender
  - M > F

Natural History & Prognosis
- Younger patients improve with physical therapy and cessation of painful activities
- Older patients may require surgery

Treatment
- Initial treatment is physical therapy
  - Strengthen RC muscles
    - Reestablish correct scapular and glenohumeral kinematics
    - Strengthen humeral head depressors (infraspinatus [IS], teres minor [TM], and SC) and pectoralis major muscles
- SA injection with anesthetic and steroid
  - Neer impingement test: Injection of 10 mL lidocaine into SA/SD bursa
    - Positive test: Pain with overhead motion resolves after injection
    - Good predictor of successful pain relief after decompression surgery
- Surgery: Impingement without RC tear
  - Anterior SA decompression
  - AC joint resection (Mumford procedure)
  - Debridement of synovitis, especially in SA/SD bursa
- Surgery: Impingement with RC tear
  - Partial thickness RC tear: Debridement
  - Full thickness RC tear: Surgical repair

DIAGNOSTIC CHECKLIST

Consider
- If recent trauma, look for nondisplaced greater tuberosity fracture

Image Interpretation Pearls
- Muscular individuals: CA arch can indent RC without impingement symptoms

Reporting Tips
- Do not attempt to diagnose rotator cuff impingement by MR; can only be diagnosed clinically
- Mention MR findings associated with impingement
  - Tendinopathy, CA arch indenting cuff

SELECTED REFERENCES

1. Campbell RS et al: External impingement of the shoulder. Semin Musculoskelet Radiol. 12(2):107-26, 2008

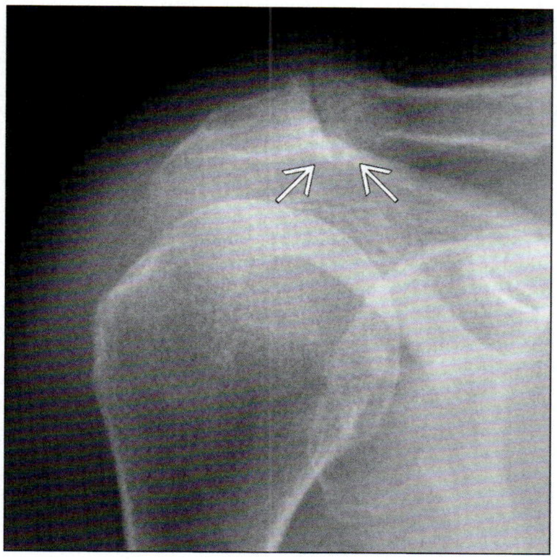

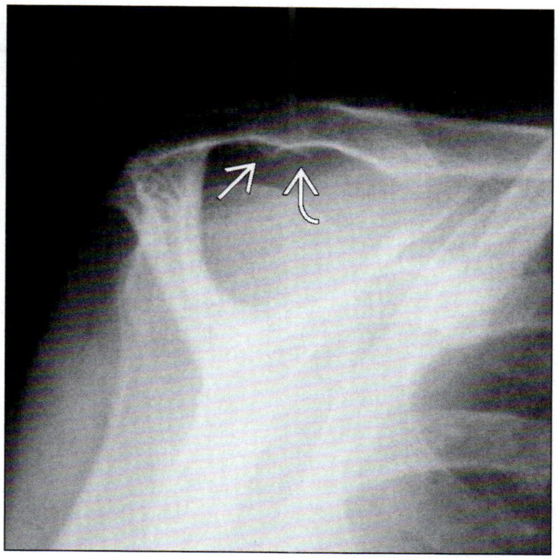

(Left) *Anteroposterior radiograph shows inferiorly directed acromioclavicular joint osteophytes . The osteophytes can indent the superior surface of the supraspinatus tendon. AC joint resection often relieves impingement pain and allows healing of a repaired rotator cuff.* ***(Right)*** *Arch view of the same patient with acromioclavicular joint DJD. There are inferiorly-directed osteophytes at the acromioclavicular joint as well as a coracoacromial enthesophyte .*

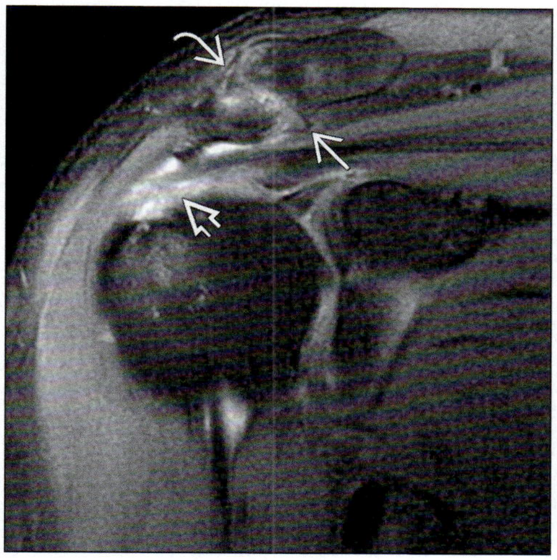

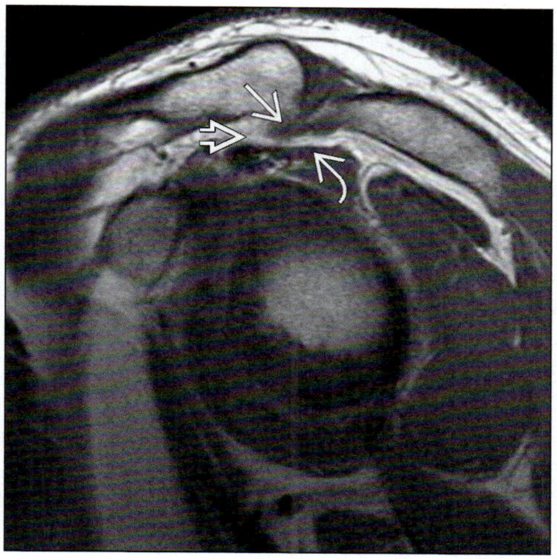

(Left) *Coronal oblique T2WI FS MR shows AC DJD with synovial hypertrophy and inferiorly directed osteophytes . There is a partial tear of the supraspinatus tendon .* ***(Right)*** *Sagittal T1WI MR in the same patient shows inferiorly directed AC joint osteophytes . There is slight indentation of the superior surface of the supraspinatus tendon , although there is peribursal fat around the subacromial bursa . Inferior AC osteophytes are associated with rotator cuff tears.*

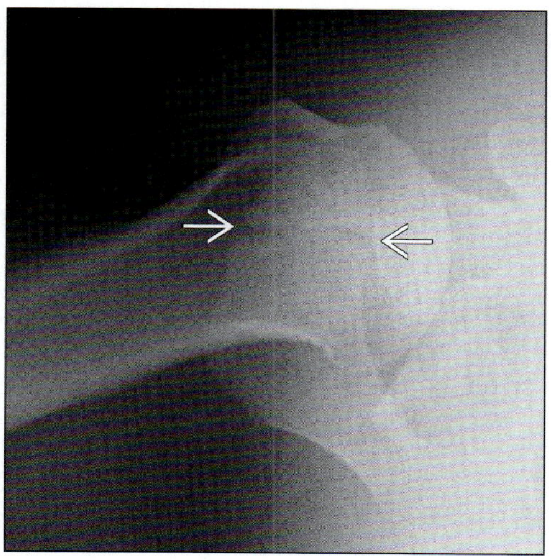

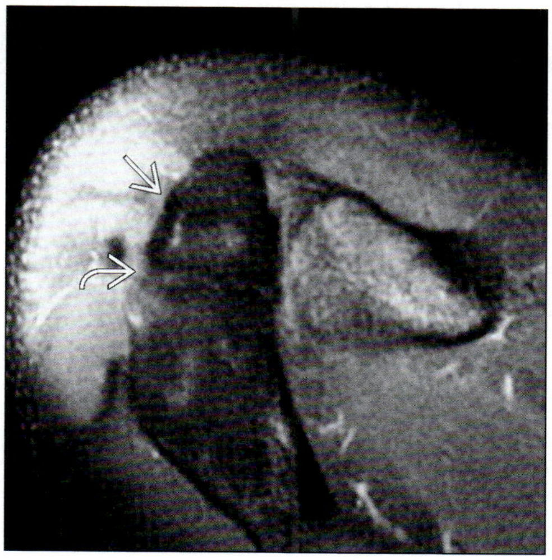

(Left) *Axillary view shows an os acromiale with lucency between the os and adjacent acromion process . A mobile os acromiale can repetitively impact down on the subacromial bursa and rotator cuff during abduction and may require resection or fixation.* ***(Right)*** *Axial PDWI FS MR in the same patient shows an os acromiale with a fibrocartilaginous union . This patient does not have fluid or a pseudarthrosis at the junction of the os with the acromion process, indicating minimal motion.*

ROTATOR CUFF IMPINGEMENT

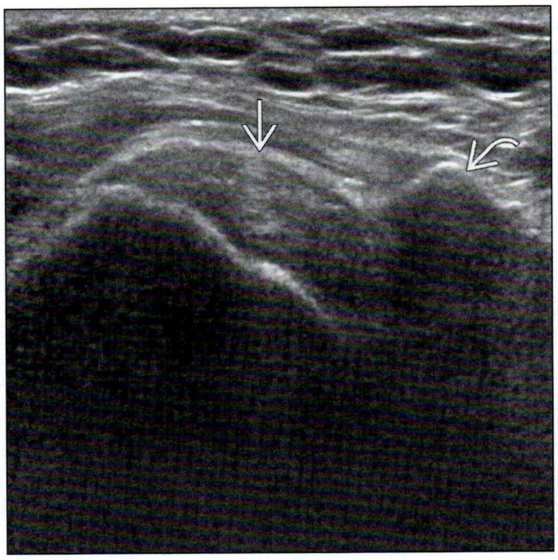

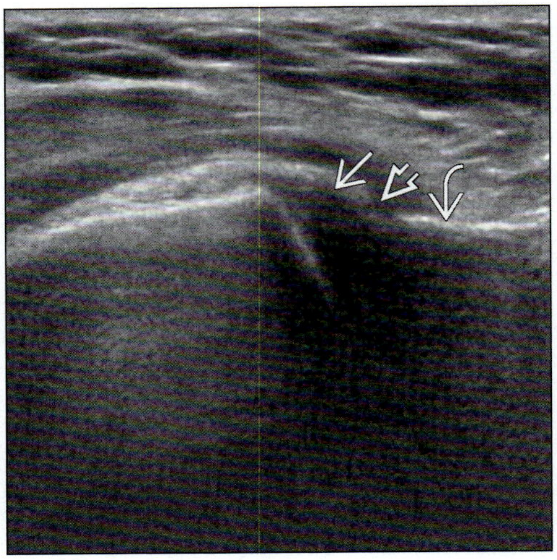

(Left) *Longitudinal ultrasound of the supraspinatus tendon in neutral position shows the cuff and acromion process. The tendon is slightly thickened, but no cuff tear is seen.* ***(Right)*** *Longitudinal ultrasound in the same patient with the arm abducted and flexed and the patient reproducing his impingement pain. The anterior acromion indents the rotator cuff and compresses the subacromial bursa. This is the ultrasound correlate of a clinical impingement test.*

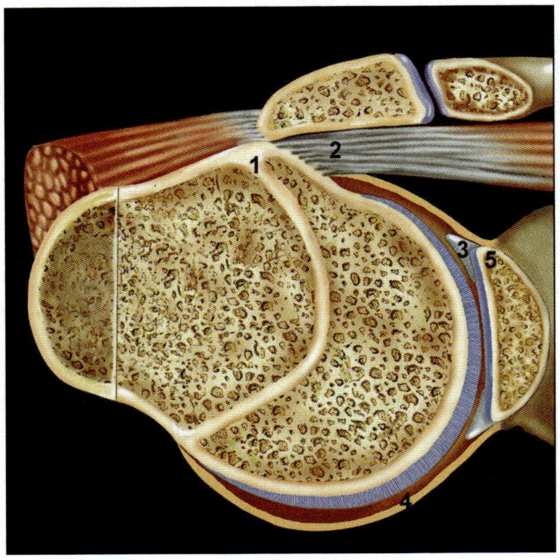

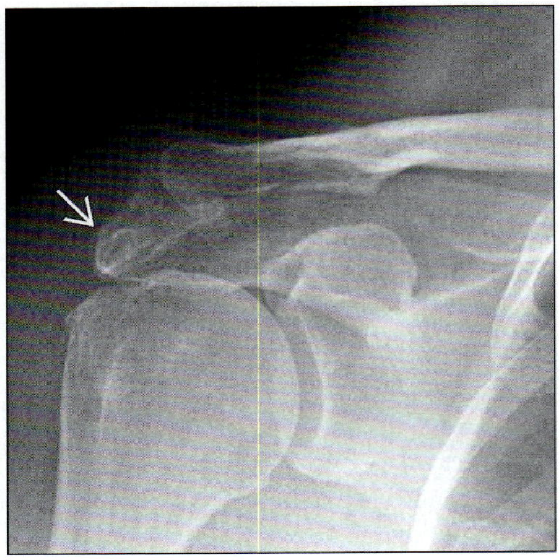

(Left) *Coronal graphic shows contact between the anterolateral acromion and the lateral rotator cuff (2) and greater tuberosity (1). Note the labrum (3), capsule (4), and glenoid (5). Contact between the acromion and the greater tuberosity is less common than with the supraspinatus tendon.* ***(Right)*** *Neer AP radiograph shows severe lateral downsloping of the lateral acromion. True severe lateral downsloping as a cause of impingement pain is less common than a hooked anterior acromion.*

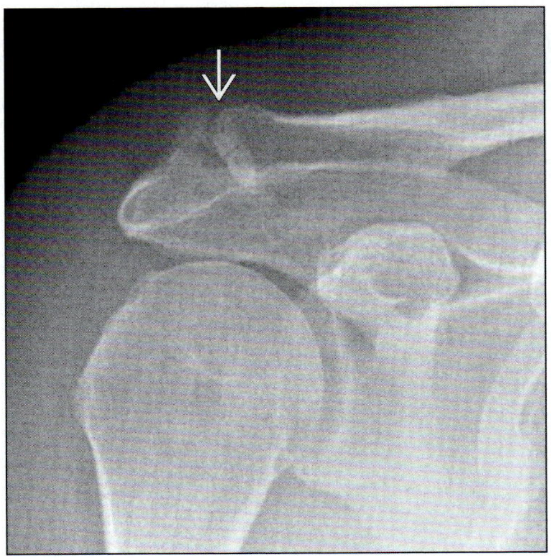

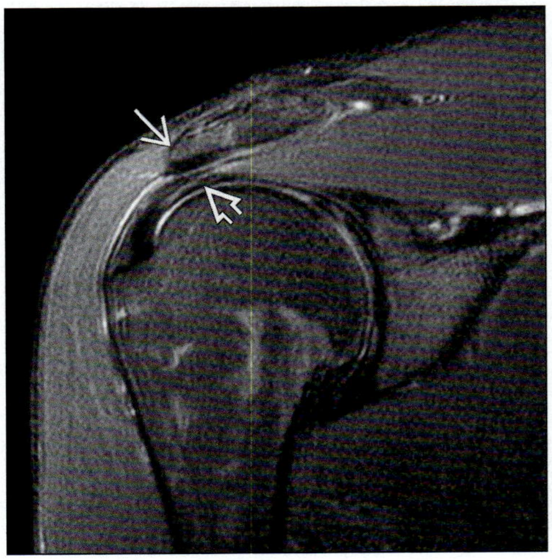

(Left) *AP internal rotation radiograph in the same patient shows AC DJD. The lateral downsloping of the more anterior part of the acromion is not as well seen.* ***(Right)*** *Coronal oblique T2WI FSE MR of a patient with marked downsloped lateral acromion indenting the rotator cuff. The compression of the bursa and cuff is typically worse with abduction. Decompression of the lateral acromion is done carefully so as not to weaken the origin of the deltoid muscle.*

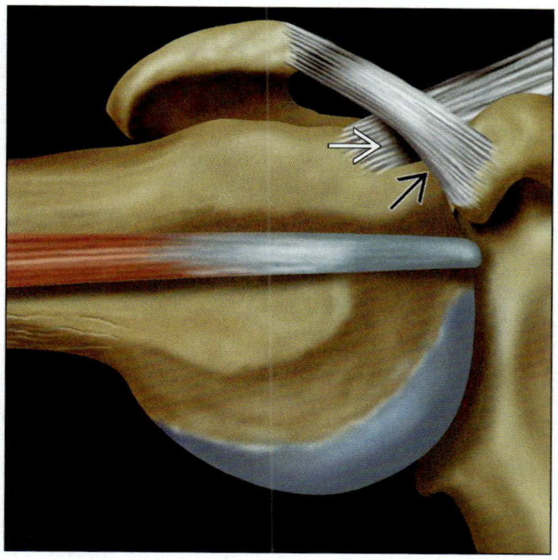
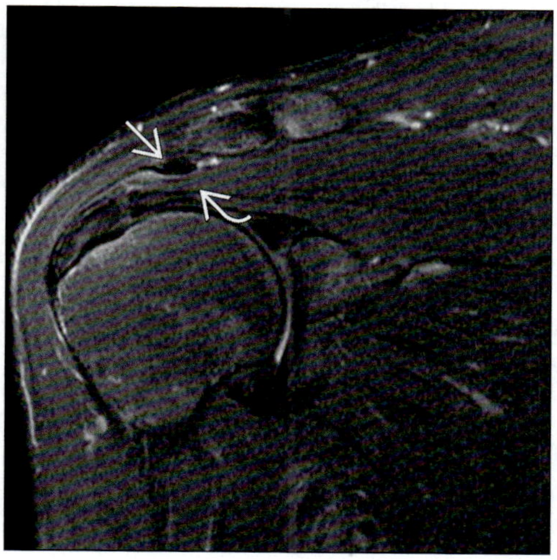

(Left) AP graphic with the arm abducted and slightly forward flexed shows contact between the anterior supraspinatus tendon ➡ and the coracoacromial ligament ⇨. Abrasion of the cuff occurs mainly when the ligament is thickened. ***(Right)*** Coronal oblique T2WI FS MR shows a thick coracoacromial ligament ➡ near its origin from the anterior acromion process. The thick ligament is indenting the supraspinatus tendon ↪. Part of the treatment of impingement pain is resection of the ligament.

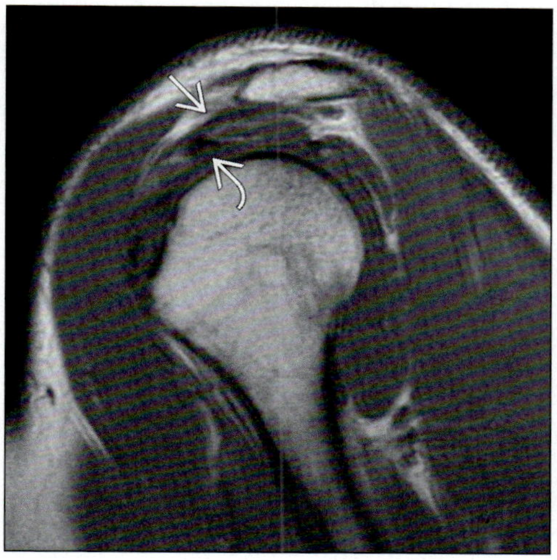
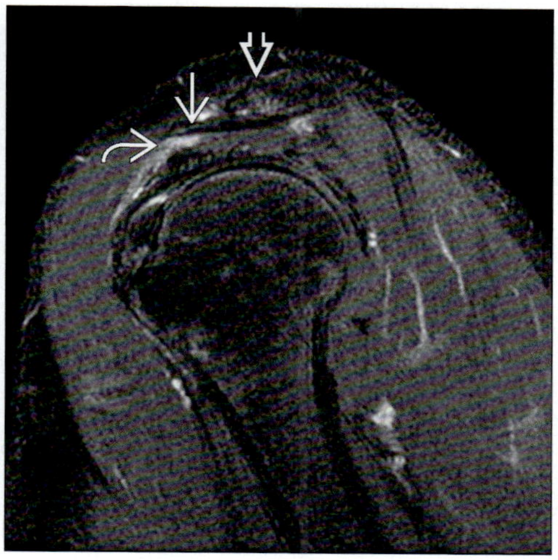

(Left) Sagittal T1WI MR in the same patient shows the thick coracoacromial ligament ➡ contacting the anterior supraspinatus tendon ↪. The coracoacromial ligament forms the anterior part of the coracoacromial arch. ***(Right)*** A slightly more lateral sagittal T2WI FS MR in the same patient also shows the thick coracoacromial ligament ➡ arising from the acromion process ⇨. There is mild subacromial bursitis adjacent to the ligament ↪.

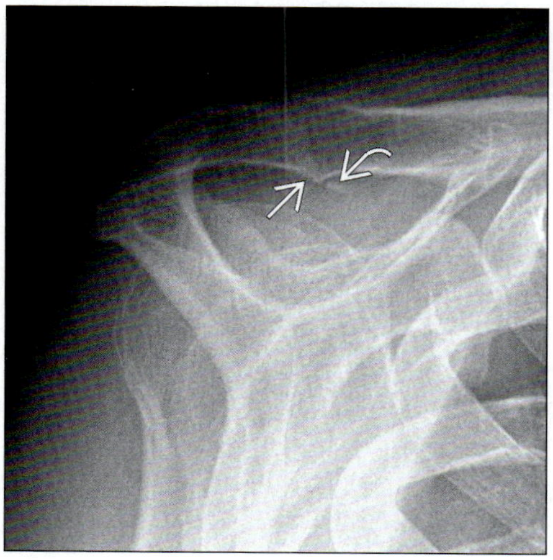
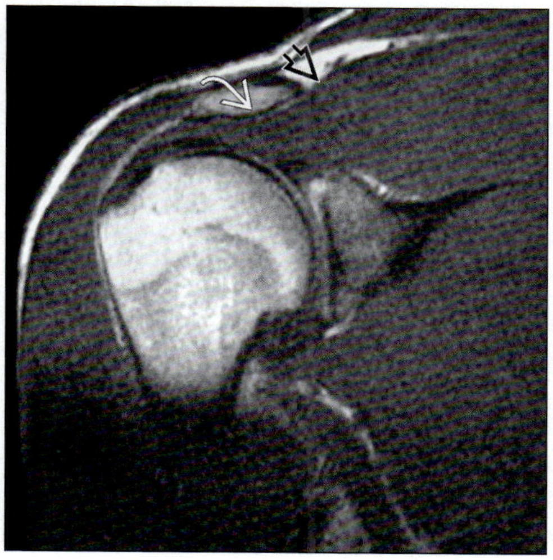

(Left) Scapular-Y lateral x-ray shows a coracoacromial ligament enthesophyte ➡ and ligamentous ossification ↪. Even when the anterior acromion is not "hooked," a rigid enthesophyte can be an osseous cause of impingement pain and is usually resected during decompression surgery. ***(Right)*** Coronal oblique T1WI MR in a young bodybuilder with no impingement symptoms. Note the asymptomatic indentation of the well-developed supraspinatus muscle ⇨ by the acromion process ↪.

OS ACROMIALE

Key Facts

Imaging

- Radiograph
  - Linear lucency with sclerotic margins across acromion process
  - Lucency is usually perpendicular to AC joint
  - Seen best on axillary view
- MR
  - Step-off associated with rotator cuff tears
  - "Stable" os acromiale: Low SI line, all sequences
  - May have marrow edema adjacent to synchondrosis
  - Controversial if marrow edema associated with impingement or rotator cuff tears
  - Pseudoarthrosis or nonunion if high signal fluid at synchondrosis
  - Os acromiale usually seen best on axial image
  - "Double AC joint" on oblique sagittal image

Top Differential Diagnoses

- Normal apophysis in young adult (up to 20-25 years of age)
- Acromial fracture
- Clinical: Outlet impingement syndrome
- Clinical: Acromioclavicular joint injury or arthritis

Pathology

- If persistent os acromiale, may be unstable or stable
  - XR cannot distinguish stable from unstable
- Controversial if unstable os associated with impingement or rotator cuff tears
  - Theory: Dynamic impingement caused by deltoid pulling os acromiale down and impacting supraspinatus

Clinical Issues

- Painful acromion if unstable nonunion

*(**Left**) Graphic of the top of the shoulder shows 3 types of persistent synchondroses of the acromion. The most anterior is between the mesoacromion → and preacromion ↷. The middle is between the mesoacromion and meta-acromion ⇨. The most posteromedial is between the meta-acromion and the acromial base. (**Right**) Axillary radiograph of a 59-year-old man shows a type A os acromiale. Note the lucency across the acromion → at the posterior margin of the articulation with the clavicle.*

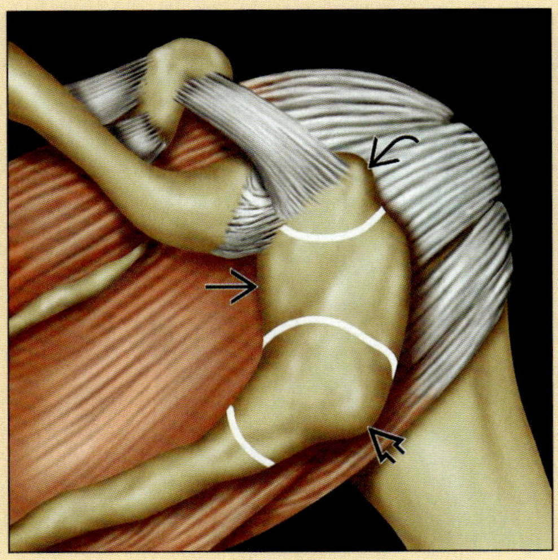

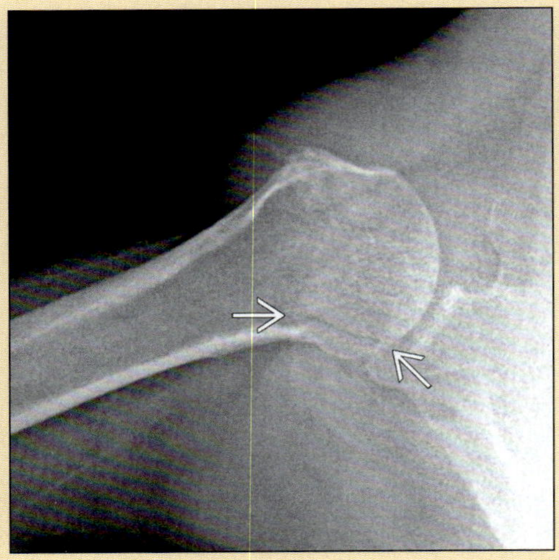

*(**Left**) Arch view of the shoulder in the same patient shows lucency across the acromion process →, which is separate from the acromioclavicular joint ↷. This type A os acromiale is the most common, and the synchondrosis is usually at the posterior margin of the acromioclavicular joint. (**Right**) AP radiograph in the same patient. The os acromiale and its synchondrosis lucency are difficult to identify on an AP radiograph, so it is important to obtain an axillary view prior to cuff surgery.*

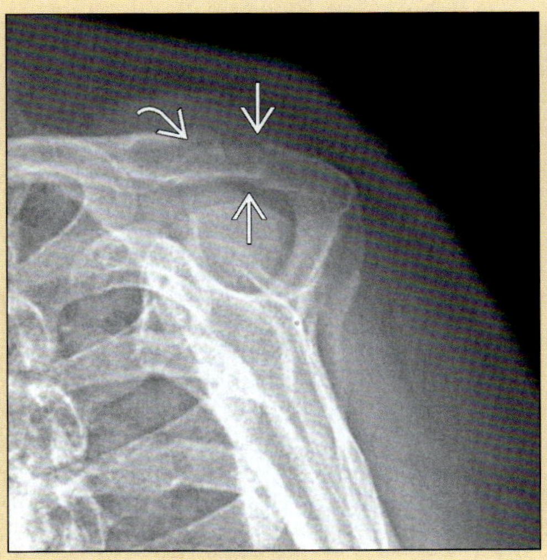

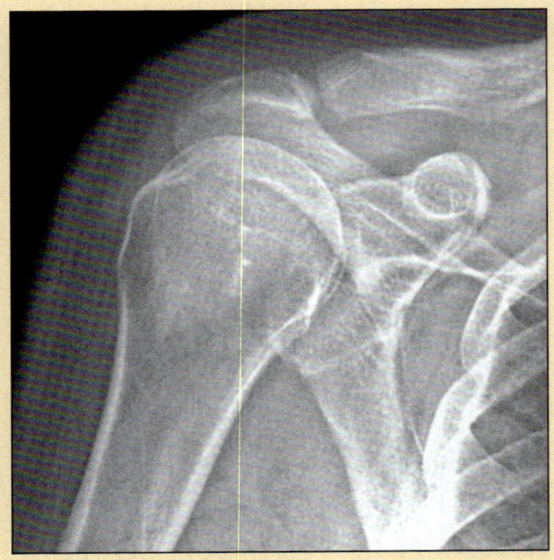

OS ACROMIALE

TERMINOLOGY

Synonyms

- Unfused acromial apophysis

Definitions

- Failure of osseous fusion between acromial apophysis and base of acromion process

IMAGING

General Features

- Best diagnostic clue
  - Separate articulation in anterior acromion on axillary view
- Location
  - Acromion process

Radiographic Findings

- Linear lucency with sclerotic margins across acromion process
  - Lucency is usually perpendicular to AC joint
  - Seen best on axillary view
- Lucency may be at 1 of 3 locations on acromion process
- Arch or scapular-Y view: Additional articulation in anterior acromion separate from AC joint
- May have osteophytes at synchondrosis
- Contralateral views obtained in teenager to differentiate normal open apophysis from os acromiale often not helpful
  - Os acromiale is bilateral in 1-2/3 of patients

CT Findings

- Linear lucency across acromion process
- Easier to identify than on radiographs
- Sclerotic margins at articulation
  - May have osteophytes and subchondral cysts

MR Findings

- T1WI
  - Low signal line across acromion process
    - Usually just posterior to AC joint
    - May have osteophytes along articulation
  - Sagittal view: Look for step-off between os and acromial base
    - Associated with rotator cuff tears
- T2WI FS
  - "Stable" os acromiale will have low signal line only
  - May have marrow edema adjacent to synchondrosis
    - Not associated with impingement or rotator cuff tears
  - Pseudoarthrosis or nonunion if there is high signal fluid at articulation
    - Controversial if associated with impingement or RCT
- Os acromiale usually seen best on axial image
- "Double AC joint" on oblique sagittal image
- Often difficult to identify on oblique coronal image

Ultrasonographic Findings

- Linear hypoechoic region on sagittal longitudinal view of acromion process

Imaging Recommendations

- Best imaging tool
  - Axillary view
- Protocol advice
  - On MR, included axial images superiorly to include acromion process

DIFFERENTIAL DIAGNOSIS

Normal Apophysis in Young Adult

- Acromial ossification centers appear age 15 years
- Should fuse by age 25 years
- Can have physeal stress fracture
  - Will be tender over physis

Acromial Fracture

- No sclerotic margins at lucent line
- Fracture line will typically have oblique orientation

Clinical: Outlet Impingement Syndrome

- Similar clinical exam, including positive Neer impingement test
- Pain relief with subacromial anesthetic injection

Clinical: Acromioclavicular Joint Injury or Arthritis

- Maximum pain occurs with palpation of AC joint

PATHOLOGY

General Features

- 3 potential ossification centers of acromion
  - Preacromion: Most distal
  - Mesoacromion: Mid
  - Meta-acromion: Proximal, adjacent to base of acromion
    - Encompasses posterolateral angle of acromion
- If persistent os acromiale, may be unstable or stable
  - Cannot distinguish stable from unstable by radiographs
  - MR: Unstable will have high signal at junction
- Controversial if unstable os associated with impingement or rotator cuff tears
  - Theory: Dynamic impingement caused by deltoid pulling os acromiale down and impacting supraspinatus
  - Inferior osteophytes at pseudoarthrosis may contact and abrade supraspinatus tendon
  - Unstable os may be associated with DJD of AC joint

Staging, Grading, & Classification

- Type A: Between mesoacromion and meta-acromion
  - Most common
- Type B: Between meso- and preacromion
  - Small, often < 4 mm
  - If painful can be resected with few complications
- Type C: Between meta-acromion and base of acromion process
- Type D: Between both pre- and mesoacromion, and meso and meta-acromion
  - Bipartite os acromiale
- Type E: Between pre- and mesoacromioin, and meta-acromion and base of acromion

OS ACROMIALE

- Type F: Between meso- and meta-acromion, and meta-acromion with base of acromion
- Type G: Unfused between all 3 ossification centers
  - Tripartite os acromiale

CLINICAL ISSUES

Presentation

- Most common signs/symptoms
  - Painful acromion if unstable nonunion
    - Tender to palpation of anterior acromion or articulation
    - Similar symptoms occur in adolescents who have stress injury to ossification center physis
  - Impingement or AC joint pain
    - Controversial if caused by unstable os acromiale
- Other signs/symptoms
  - Weakness with forward flexion/arm elevation
  - Night pain
  - Pain with overhead activity
  - Limited range of motion
  - Clicking with activity

Demographics

- Gender
  - M:F = 2:1
- Ethnicity
  - 2x as common in African-Americans compared with Caucasians
- 1-15% incidence
  - Most concur ~ 7%
- Bilateral in 33-62%

Natural History & Prognosis

- Controversial if cause of impingement or rotator cuff tears
- Opinion 1: No association with cuff tears
  - Oulette et al Skeletal Radiol 2007
    - Arthroscopy correlation in 19 patients with os acromiale and 12 controls
    - No difference in prevalence of rotator cuff tears
    - Step-off at synchondrosis: Trend toward association with RCT
    - Similar results in Boehm JBJS Br 2003
- Opinion 2: Unstable os associated with impingement and cuff tears
  - Mostly small anecdotal studies
  - Hutchinson et al Arthroscopy 1993
    - 3 patients with impingement syndrome and os acromiale, poor results with subacromial decompression only
  - Jerosch et al Unfallchirurg 1994
    - 12 patients with os acromiale did worse postoperatively than 110 patients without

Treatment

- Painful os acromiale
  - Injection into articulation
  - Resection, particularly if small os acromiale
- Impingement surgery with unstable os
  - Os acromiale excision (particularly type B or < 4 mm)
    - Best results if done arthroscopically to preserve as much of deltoid attachment as possible
    - Poor results for large os fragments
      - Weakens deltoid muscle attachment
  - Fusion of os acromiale
    - Best technique if large fragment
    - Screws, wires, and bone graft
    - Important to stabilize os acromiale if doing subacromial decompression
  - Most important: Recognize os acromiale prior to impingement/RCT surgery
    - Alters surgical procedure

DIAGNOSTIC CHECKLIST

Consider

- Os acromiale if lucent line on radiograph and patient > 25 years old

Reporting Tips

- Fluid at articulation = nonunion or pseudoarticulation

SELECTED REFERENCES

1. Smith J et al: Role of sonography in the evaluation of unstable os acromiale. J Ultrasound Med. 27(10):1521-6, 2008
2. Burbank KM et al: Incidence of os acromiale in patients with shoulder pain. Am J Orthop. 36(3):153-5, 2007
3. Ouellette H et al: Re-examining the association of os acromiale with supraspinatus and infraspinatus tears. Skeletal Radiol. 36(9):835-9, 2007
4. Sahajpal D et al: Surgical management of os acromiale: a case report and review of the literature. Bull NYU Hosp Jt Dis. 65(4):312-6, 2007
5. Abboud JA et al: Surgical treatment of os acromiale with and without associated rotator cuff tears. J Shoulder Elbow Surg. 15(3):265-70, 2006
6. Kurtz CA et al: Symptomatic os acromiale. J Am Acad Orthop Surg. 14(1):12-9, 2006
7. Pagnani MJ et al: Painful os acromiale (or unfused acromial apophysis) in athletes. J Shoulder Elbow Surg. 15(4):432-5, 2006
8. Boehm TD et al: Rotator cuff tears associated with os acromiale. Acta Orthop. 76(2):241-4, 2005
9. Youm T et al: Os acromiale: evaluation and treatment. Am J Orthop. 2005 Jun;34(6):277-83. Erratum in: Am J Orthop. 34(8):382, 2005
10. Lee DH et al: The double-density sign: a radiographic finding suggestive of an os acromiale. J Bone Joint Surg Am. 86-A(12):2666-70, 2004
11. Boehm TD et al: Os acromiale associated with tear of the rotator cuff treated operatively. Review of 33 patients. J Bone Joint Surg Br. 85(4):545-9, 2003
12. Boehm TD et al: Ultrasonographic appearance of os acromiale. Ultraschall Med. 24(3):180-3, 2003
13. Uri DS et al: Os acromiale: evaluation of markers for identification on sagittal and coronal oblique MR images. Skeletal Radiol. 26(1):31-4, 1997
14. Jerosch J et al: [Arthroscopic subacromial decompression--indications in os acromiale?.] Unfallchirurg. 97(2):69-73, 1994
15. Park JG et al: Os acromiale associated with rotator cuff impingement: MR imaging of the shoulder. Radiology. 193(1):255-7, 1994
16. Hutchinson MR et al: Arthroscopic decompression of shoulder impingement secondary to Os acromiale. Arthroscopy. 9(1):28-32, 1993

OS ACROMIALE

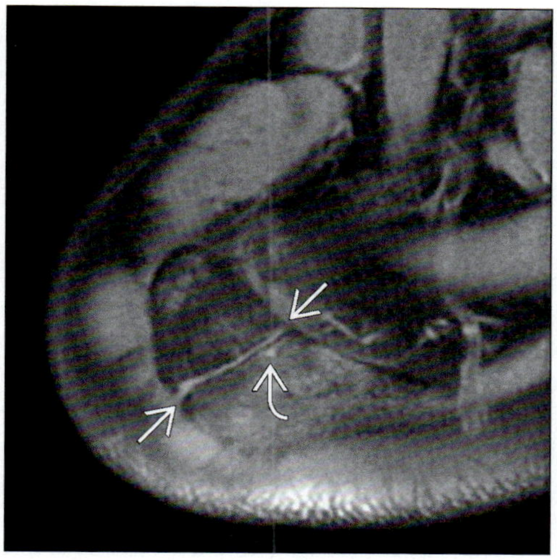

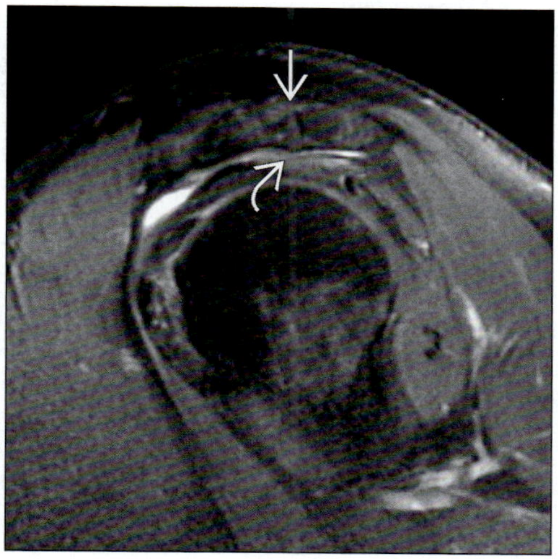

(Left) *Axial PDWI FS MR, in the same patient as shown in prior radiograph, shows intermediate signal at the synchondrosis* → *with no fluid signal intensity. There is a small subchondral cyst* ↪. ***(Right)*** *Oblique sagittal T2WI FS MR in the same patient shows no high signal fluid at the synchondrosis* →. *There is a small osteophyte at the inferior margin of the articulation* ↪, *but no step-off. This is the most common location within the acromion for an os acromiale.*

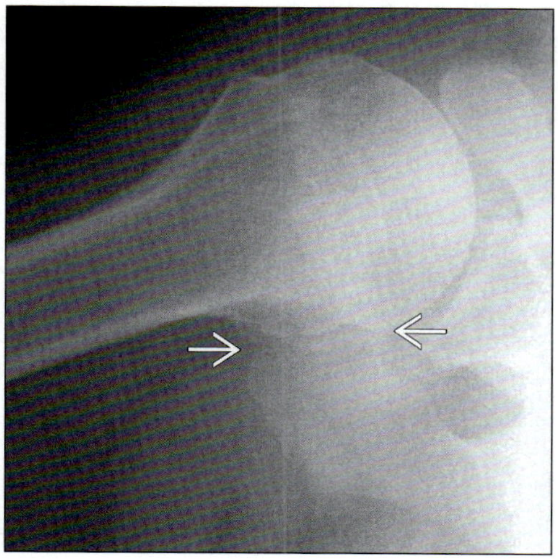

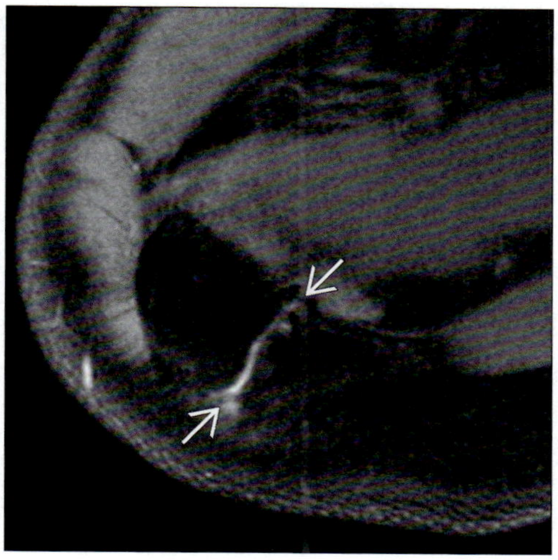

(Left) *Axillary lateral radiograph of a 66-year-old man with pain and tenderness over the acromion process. There is a lucency across the acromion process* → *consistent with a type A os acromiale.* ***(Right)*** *Axial T2WI FS MR in the same patient shows high signal fluid* → *in the linear defect across the acromion process, consistent with a pseudoarthrosis and unstable os acromiale. The patient had good pain relief after an injection into the subdeltoid bursa.*

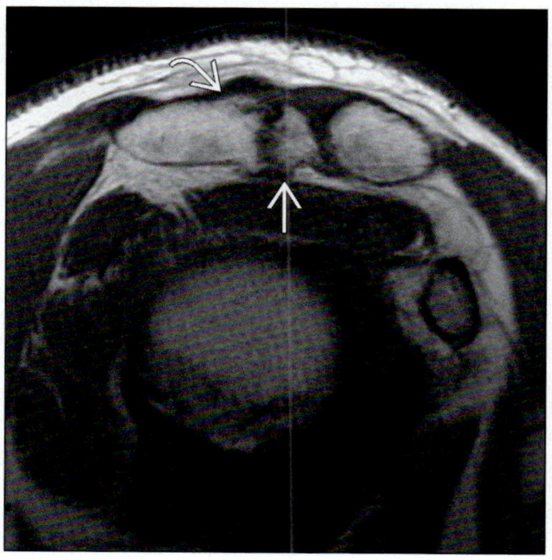

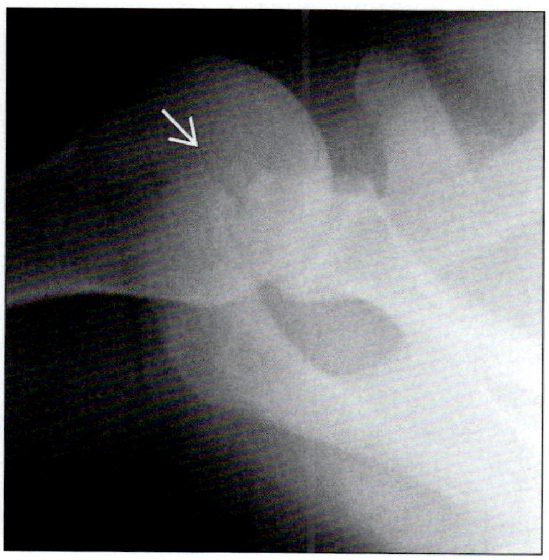

(Left) *Oblique sagittal T1WI MR in the same patient shows inferior osteophytes at the margins of the synchondrosis* →. *Note the adjacent acromioclavicular joint* ↪ *in this patient with a "double AC joint" sign.* ***(Right)*** *Axillary radiograph in a patient with a type B os acromiale* →. *This is the 2nd most common type of os acromiale. The os acromiale fragment is usually resected in patients who undergo anterior subacromial decompression for impingement or rotator cuff pain.*

ROTATOR CUFF TENDINOPATHY

Key Facts

Terminology
- Tendinosis, tendinitis, rotator cuff impingement syndrome, painful shoulder syndrome

Imaging
- MR
  - High signal intensity within rotator cuff tendon
  - Increased superior-to-inferior thickness of supraspinatus tendon
- Ultrasound
  - Hypoechoic region within center of tendon

Top Differential Diagnoses
- Intratendinous partial cuff tear/fissure/cleft
  - Some include intratendinous tears in spectrum of rotator cuff tendinopathy
- Calcific tendinopathy
  - Low signal focal areas corresponding to calcium hydroxyapatite
- Angular anisotropy
  - "Magic angle" artifact
- Normal tendon overlap
  - Oblique sagittal image confirms this to be partial averaging of connective tissue between tendons
- Rotator cuff tear

Pathology
- Intrinsic causes
  - Aging/degeneration from chronic overuse
- Extrinsic causes
  - Rotator cuff impingement
- Histology
  - Fissures in tendon fibers, mucoid and eosinophilic degeneration

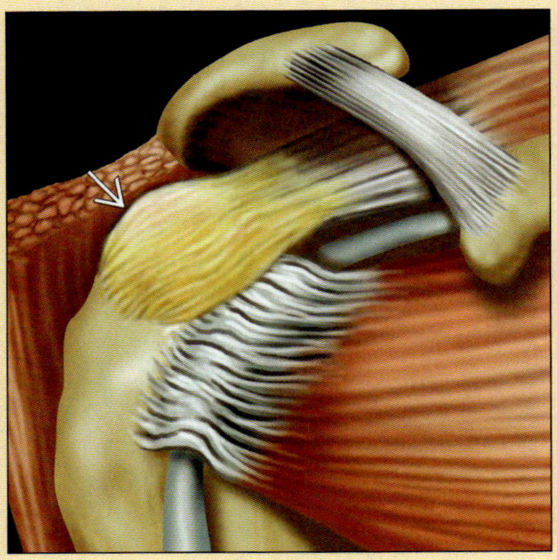
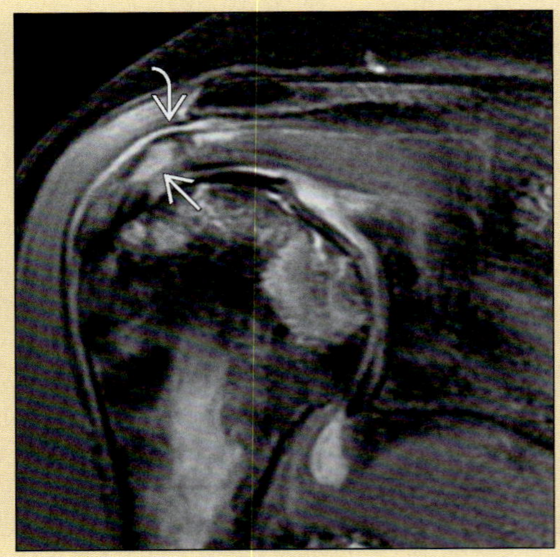

(Left) *AP graphic of the anterior shoulder. There is focal swelling and degeneration ➔ in the lateral supraspinatus tendon at an area of tendinopathy. Tendinopathy is common in the supraspinatus tendon in older patients.* ***(Right)*** *Coronal oblique T2WI FS MR shows severe tendinopathy. There is abnormal high signal within the substance of the supraspinatus tendon ➔, with focal swelling shown by the bulge in the superior surface of the tendon ➔. The surfaces of the tendon are intact.*

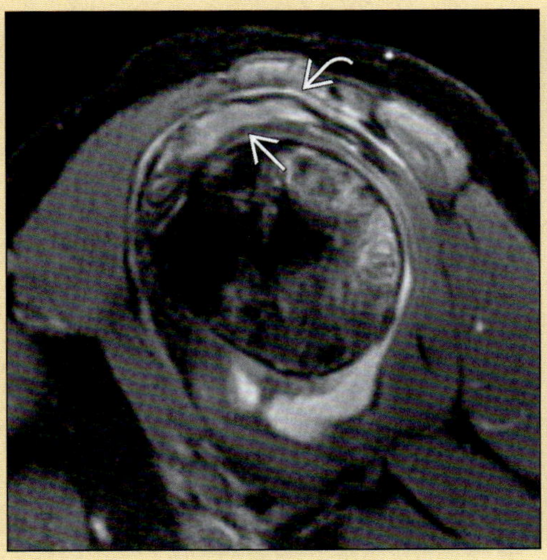
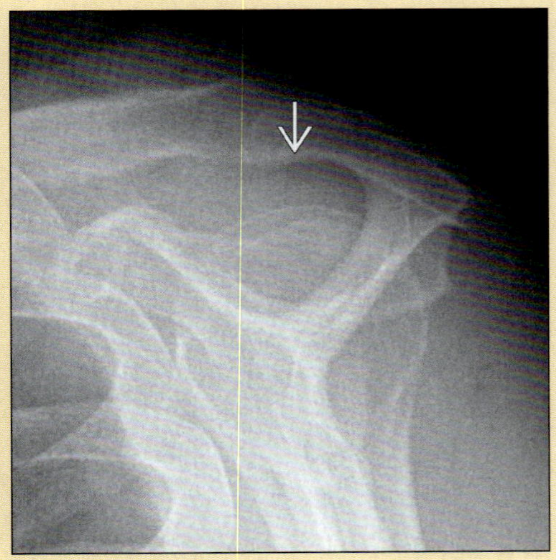

(Left) *Oblique sagittal T2WI FS MR in the same patient again shows the intrasubstance increased signal ➔ with associated swelling of the tendon ➔. No partial or full thickness supraspinatus tendon tear was visible at arthroscopy.* ***(Right)*** *Arch view radiograph in the same patient shows a type III anterior acromion process ➔. The patient had impingement symptoms that did not improve with physical therapy, and underwent an arthroscopic anterior subacromial decompression.*

ROTATOR CUFF TENDINOPATHY

TERMINOLOGY

Synonyms

- Tendinosis, tendinitis, rotator cuff impingement syndrome, painful shoulder syndrome
  - Tendinitis is not preferred term because there are no inflammatory cells

Definitions

- Collagenous degeneration of rotator cuff tendons, usually supraspinatus tendon

IMAGING

General Features

- Best diagnostic clue
  - Thickened and increased signal intensity of tendon on T2WI MR
- Location
  - Predominantly supraspinatus tendon
  - Can occur in subscapularis, especially if subcoracoid impingement
- Size
  - Focal tendon thickness > 8 mm

Radiographic Findings

- Rotator cuff impingement bone findings
  - Acromioclavicular joint osteoarthritis with inferior osteophytes
  - Hooked anterior acromion
  - Enthesopathy of greater tuberosity
    - Cysts or sclerosis

MR Findings

- T2WI FS
  - High signal intensity within rotator cuff tendon
    - Fluid signal intensity represents intratendinous or intrasubstance partial tear/cleft/fissure
  - Increased superior-to-inferior thickness of supraspinatus tendon
  - ± subacromial/subdeltoid fluid
  - Beware: Do not overcall tendinopathy
    - Normal intrasubstance layers of supraspinatus tendon (SST) can be intermediate signal
      - Signal intensity is lower than in tendinopathy
      - Cuff will not be focally swollen
    - If high signal extends to 1 surface of tendon, then it is an articular or bursal surface partial thickness cuff tear
    - Hyperintensity extending from articular to bursal surface indicates full thickness cuff tear
- PD/intermediate
  - High signal within substance of tendon
  - Increased superior-to-inferior thickness of tendon
    - Compare with normal signal intensity areas of tendon so as not to overcall "magic angle" artifact
  - Acromioclavicular joint osteophytes or synovial/ capsule hypertrophy
  - Oblique sagittal image: Hooked anterior acromion process
- MR arthrography
  - No defect in surface of tendon

Ultrasonographic Findings

- Hypoechoic region within center of tendon
  - Preserved echogenicity of articular and bursal surfaces of tendon

Imaging Recommendations

- Best imaging tool
  - Coronal and sagittal fat-suppressed T2WI MR
- Protocol advice
  - High-resolution images in order to distinguish tendinopathy from partial thickness cuff tear

DIFFERENTIAL DIAGNOSIS

Intratendinous Partial Cuff Tear/Fissure/Cleft

- Intratendinous or intrasubstance signal is fluid intensity
- Some include intratendinous tears in spectrum of rotator cuff tendinopathy

Calcific Tendinopathy

- Focal low signal area corresponding to calcium hydroxyapatite
- Hyperintense area around calcification
- May eventually rupture into subacromial/subdeltoid bursa

Angular Anisotropy

- "Magic angle" artifact
- Affected longitudinally oriented collagen fibers are oriented at 55° to main magnetic field B_0
- Occurs in distal curving portion of supraspinatus tendon
- More apparent on short TE images, less so on T2WI

Normal Tendon Overlap

- Intermediate signal on oblique coronal images where supraspinatus and infraspinatus tendons overlap
- Oblique sagittal image confirms this to be partial averaging of connective tissue between tendons

Rotator Cuff Tear

- High signal in cuff tears extends to surface of tendon in tears visible on inspection by arthroscopist
  - Classified as articular or bursal partial thickness or full thickness tears
- Cuff tears and tendinopathy often coexist near each other

PATHOLOGY

General Features

- Etiology
  - Intrinsic causes
    - Aging/degeneration from chronic overuse
    - Repetitive microtrauma with inadequate healing, reparative type III collagen
    - Decreased vascularity, poor healing
    - Mucoid and eosinophilic degeneration
  - Extrinsic causes
    - Coracoacromial arch causing rotator cuff impingement
      - Hooked anterior acromion
      - Acromioclavicular joint DJD
      - Os acromiale
    - Weak humeral head depressors

ROTATOR CUFF TENDINOPATHY

- Pectoralis major, teres minor and major, latissimus dorsi muscles weaken with aging
- Deltoid muscle weakly opposed, so supraspinatus tendon repetitively impacts undersurface of acromion

Staging, Grading, & Classification

- Neer impingement stages
  - Stage I
    - Reversible edema and hemorrhage
    - Active patients, usually < 25 years old
  - Stage II
    - Tendinopathy and fibrosis
    - 25-40 years old
  - Stage III
    - Enthesopathy and rotator cuff tear
    - > 40 years old

Gross Pathologic & Surgical Features

- Swollen bumpy surface of tendon
  - Discoloration: Yellow or pink from inflammation
- No disruption of tendon fibers of surface of cuff
- Histology
  - Fissures and splits in tendon fibers
  - Increased type 3 collagen, which has smaller fiber diameter
  - Degradation of mucopolysaccharide, increased glycosaminoglycan and proteoglycan
  - Mucoid and eosinophilic degeneration, fibrillary degeneration
  - Loss of cellularity with apoptosis (cell death)
  - Angiofibroblastic hyperplasia

CLINICAL ISSUES

Presentation

- Most common signs/symptoms
  - Pain, especially at night
  - Weakness
  - Decreased shoulder range of motion

Demographics

- Age
  - > 40 years old (average: 55 years old)
  - Also seen in younger athletes (i.e., swimmers)
- Gender
  - M = F

Natural History & Prognosis

- Controversial
  - Some believe tendinopathy may progress to partial or full thickness rotator cuff tear

Treatment

- NSAIDs
- Physical therapy
  - Improve range of motion
  - Strengthen shoulder depressor muscles
  - Promote healing of rotator cuff
- Subacromial injection
  - Corticosteroid and local anesthetic
  - Relieves pain, reduces thickness of inflamed subacromial/subdeltoid bursa
- Surgery
  - Decrease osseous impingement by coracoacromial arch
    - Anterior subacromial decompression
      - Convert hooked acromion to flat undersurface of anterior acromion
    - Acromioclavicular joint resection
      - If inferior osteophytes > 2 mm
  - For severe tendinopathy, may incise cuff and debride area of tendinopathy
    - Especially if there is intrasubstance cleft or fissure

DIAGNOSTIC CHECKLIST

Image Interpretation Pearls

- Only call tendinopathy if swollen and abnormal increased signal on T2WI

Reporting Tips

- Determine whether surgeon prefers term "intrasubstance partial cuff tear" or "intrasubstance cleft/fissure"

SELECTED REFERENCES

1. Lewis JS: Rotator cuff tendinopathy/subacromial impingement syndrome: is it time for a new method of assessment? Br J Sports Med. 43(4):259-64, 2009
2. Mayerhoefer ME et al: Shoulder impingement: relationship of clinical symptoms and imaging criteria. Clin J Sport Med. 19(2):83-9, 2009
3. Teefey SA et al: Shoulder Ultrasound vs MRI for rotator cuff pathology. PM R. 1(5):490-5, 2009
4. Fritz LB et al: Cystic changes at supraspinatus and infraspinatus tendon insertion sites: association with age and rotator cuff disorders in 238 patients. Radiology. 244(1):239-48, 2007
5. Hambly N et al: Rotator cuff impingement: correlation between findings on MRI and outcome after fluoroscopically guided subacromial bursography and steroid injection. AJR Am J Roentgenol. 189(5):1179-84, 2007
6. Sein ML et al: Reliability of MRI assessment of supraspinatus tendinopathy. Br J Sports Med. 41(8):e9, 2007
7. Vanhoenacker FM et al: MR arthrography of the rotator cuff. JBR-BTR. 90(5):338-44, 2007
8. Uri DS: MR imaging of shoulder impingement and rotator cuff disease. Radiol Clin North Am. 35(1):77-96, 1997
9. Williams GR Jr et al: Anatomic, histologic, and magnetic resonance imaging abnormalities of the shoulder. Clin Orthop Relat Res. (330):66-74, 1996
10. Timins ME et al: Increased signal in the normal supraspinatus tendon on MR imaging: diagnostic pitfall caused by the magic-angle effect. AJR Am J Roentgenol. 165(1):109-14, 1995
11. Vahlensieck M et al: Two segments of the supraspinous muscle: cause of high signal intensity at MR imaging? Radiology. 186(2):449-54, 1993
12. Fu FH et al: Shoulder impingement syndrome. A critical review. Clin Orthop Relat Res. (269):162-73, 1991
13. Kjellin I et al: Alterations in the supraspinatus tendon at MR imaging: correlation with histopathologic findings in cadavers. Radiology. 181(3):837-41, 1991

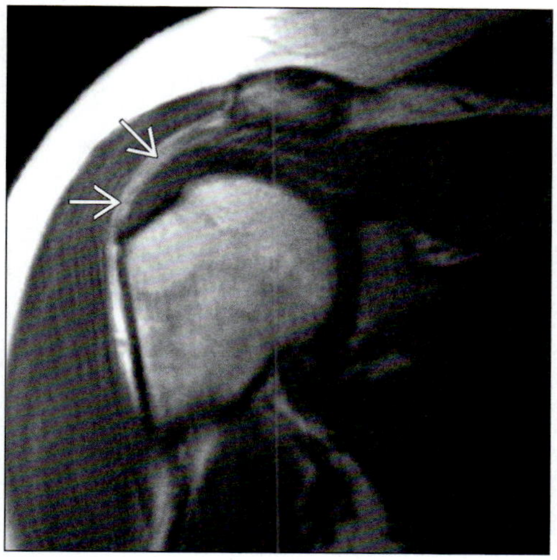

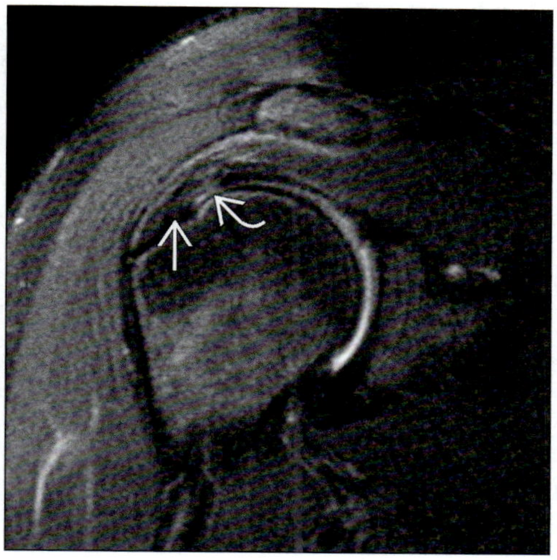

(Left) *Coronal oblique T1WI MR shows mild diffuse increased signal within the lateral supraspinatus tendon* ➔. ***(Right)*** *Coronal oblique T2WI FS MR in the same patient now shows only a small focal area of possible tendinopathy* ➔. *The remainder of the cuff is normal, with areas of intermediate intratendinous signal* ➔ *associated with layers with more connective tissue and less densely packed collagen fibers. The increased signal on the T1WI is probably "magic angle" artifact.*

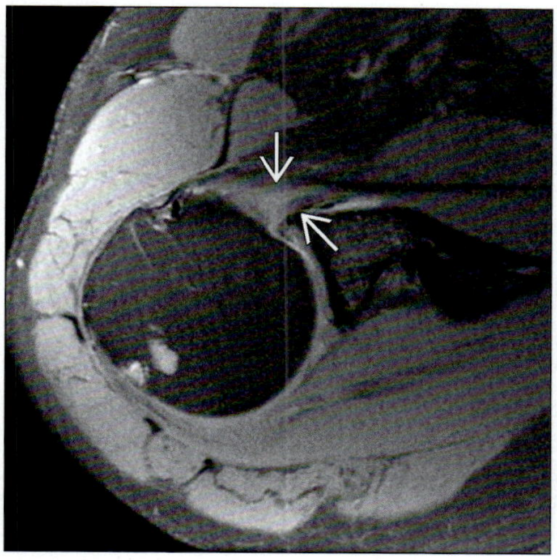

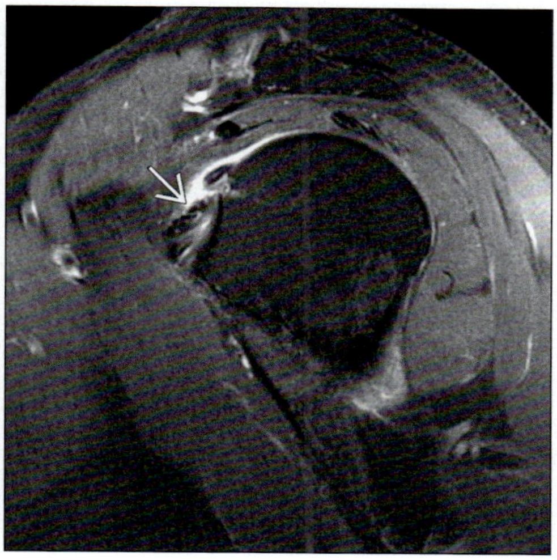

(Left) *Axial intermediate-weighted FS MR in a 60-year-old physically active man (chopping wood, etc.) shows swelling and increased signal in the lateral subscapularis tendon* ➔. ***(Right)*** *Oblique sagittal T2WI FS MR in the same patient shows the increased signal within the lateral subscapularis tendon* ➔ *near the insertion onto the lesser tuberosity. Although this patient's coracohumeral interval was normal, subscapularis tendinopathy can be associated with subcoracoid impingement syndrome.*

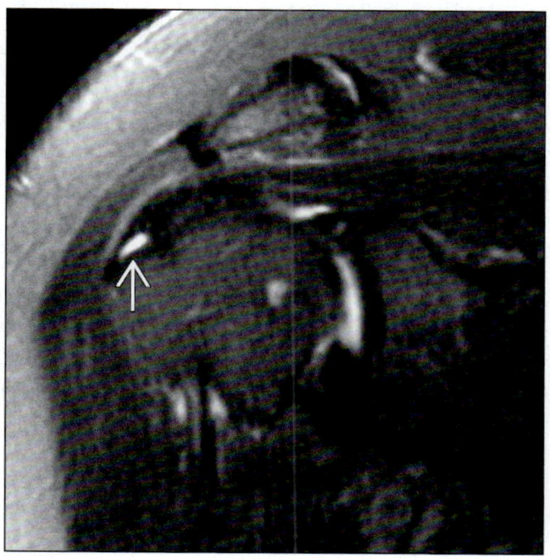

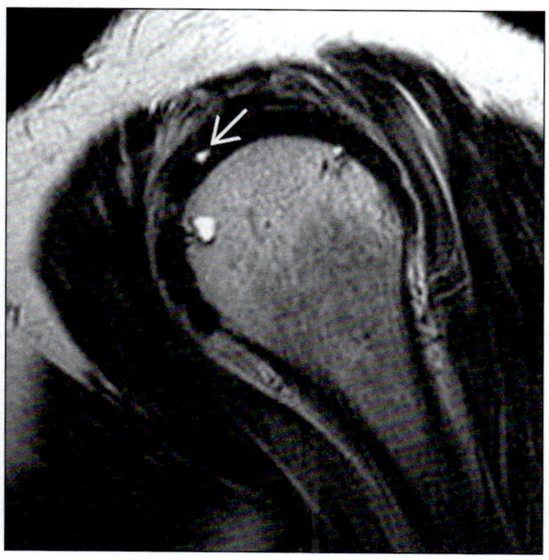

(Left) *Coronal oblique T2WI FS MR shows a focal area of fluid signal intensity in the lateral supraspinatus tendon* ➔. *The articular and bursal surfaces of the tendon are intact.* ***(Right)*** *Oblique sagittal T2WI MR in the same patient shows the intrasubstance fissure within the cuff* ➔. *Some prefer to term this an "intrasubstance partial thickness cuff tear." If severe, surgeons may incise the cuff and debride the abnormal area. This patient's symptoms resolved with physical therapy.*

ROTATOR CUFF PARTIAL THICKNESS TEAR

Key Facts

Terminology

- Tear involving only a portion of rotator cuff tendon thickness so that tear depth is less than full thickness of tendon

Imaging

- 95% supraspinatus
- Articular:bursal surface tears = 3:1
  - Usually in lateral 1.5 cm of supraspinatus tendon
- Other partial tear types
  - Intratendinous/intrasubstance partial tear
  - Rim rent tear
  - Posterior cuff articular surface partial tears in throwers with internal impingement
- MR
  - Focal increased T2WI signal extending to surface of rotator cuff but not extending across entire thickness of cuff
  - Often fluid signal intensity; this may be obscured by granulation tissue/fibrosis within the tear
  - May appear as thinned tendon
- MR arthrography
  - T1WI FS: High signal contrast extends into articular surface tear
  - T2 images also obtained to show bursal surface tears, intratendinous fissures
- Ultrasound
  - Focal hypoechoic or anechoic area involving surface of tendon but not extending across entire thickness of tendon

Diagnostic Checklist

- Report size, depth, and articular or bursal surface involved by tear

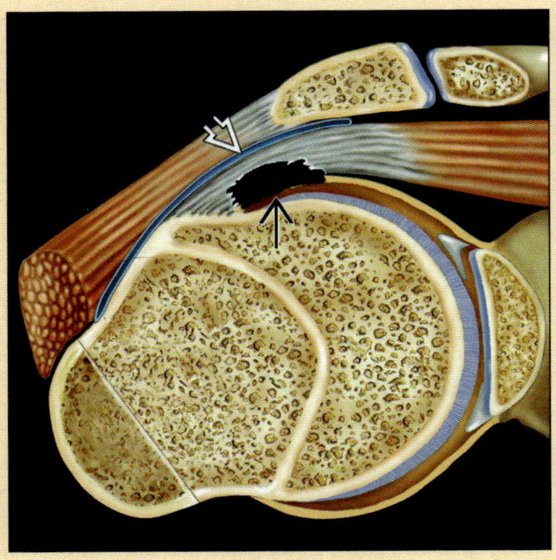
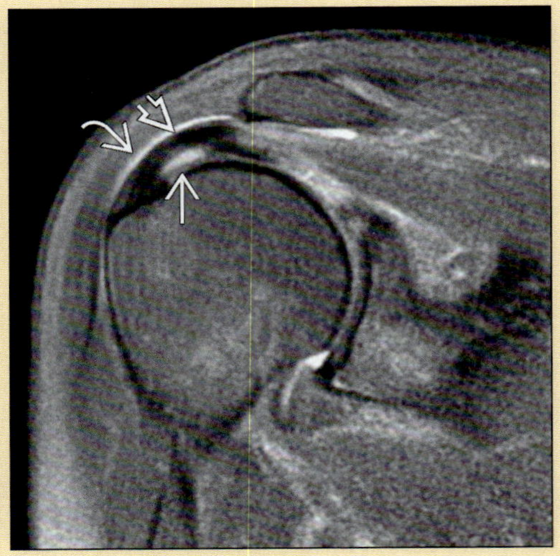

(Left) *AP graphic shows an articular surface partial thickness rotator cuff tear ➔. The tear extends only part-way across the tendon and disrupts the undersurface of the tendon. The bursal-side fibers of the cuff are intact and the subacromial/subdeltoid bursa ➔ is uninvolved.* ***(Right)*** *Coronal oblique T2WI FS MR shows high signal within a tear involving the articular surface of the supraspinatus tendon ➔. The bursal surface of the cuff ➔ is intact, and there is a small SA/SD bursal effusion ➔.*

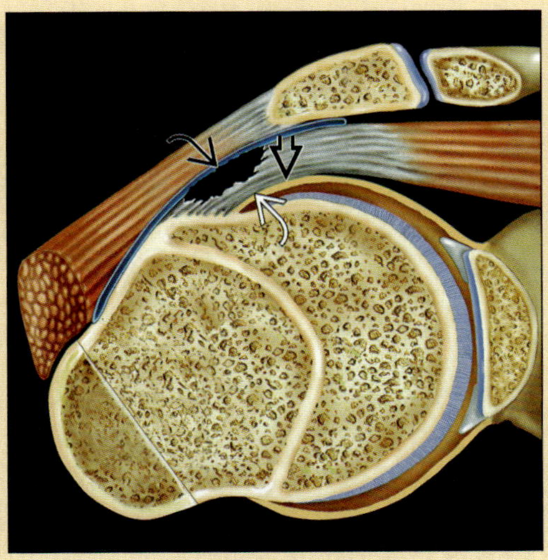
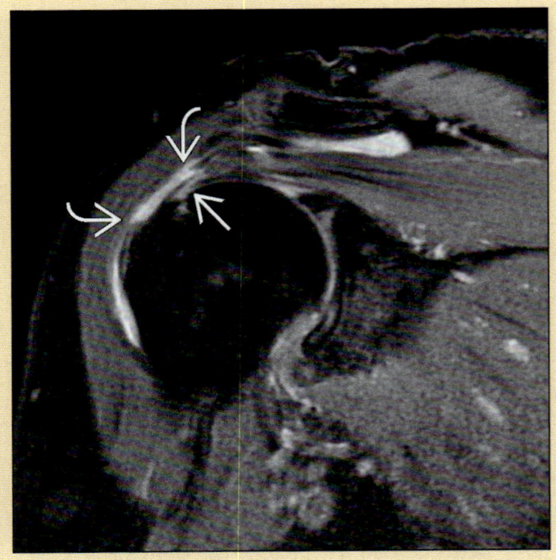

(Left) *AP graphic of a bursal surface partial thickness rotator cuff tear ➔. The articular-side fibers of the cuff ➔ and the joint capsule ➔ are intact.* ***(Right)*** *Coronal oblique T2WI FS MR shows high signal in a tear extending to the bursal surface of the supraspinatus tendon ➔. The articular surface of the cuff ➔ was intact at surgery, although the original MR reading questioned a small articular surface tear. Bursal partial tears are less common than articular partial tears.*

ROTATOR CUFF PARTIAL THICKNESS TEAR

TERMINOLOGY

Abbreviations

- Partial (thickness) rotator cuff tear (PRCT/PTRCT)

Synonyms

- Partial tear, articular/undersurface tear, bursal/superior surface tear, intratendinous/intrasubstance/interstitial tear

Definitions

- Tear involving only a portion of rotator cuff tendon thickness so that tear depth is less than full thickness of tendon

IMAGING

General Features

- Best diagnostic clue
  - MR T2 high signal within tendon that involves only a portion of tendon thickness
- Location
  - 95% supraspinatus
  - Articular:bursal surface tears = 3:1
    - 30-50% of bursal surface partial tears also have articular surface partial thickness tear
  - Usually in lateral 1.5 cm of supraspinatus tendon
  - Other partial tear types
    - Intratendinous/intrasubstance/interstitial partial tear
      - Controversial if these should be included with articular & bursal surface partial tears
      - Some refer to these as intrasubstance cleft/ fissure, a severe form of tendinopathy
      - There is debate if these are a cause of pain
      - Many orthopedists do not consider these a surgical lesion, and will not incise intact surface of cuff to repair them
    - Rim rent tear
      - Articular surface partial supraspinatus tendon tear at insertion onto greater tuberosity
      - Also called PASTA (partial articular surface tendon avulsion) lesion
    - Posterior cuff articular surface partial thickness tear
      - Seen in throwers with internal impingement
- Size
  - Most 1-3 cm diameter

Radiographic Findings

- Bone findings associated with PTRCT
  - Type III acromion
  - Acromioclavicular joint osteophytes
  - ± os acromiale
- Arthrography
  - Contrast extends into substance of rotator cuff tendon
    - Shoulder joint injection shows articular surface partial thickness tears only
  - No opacification of subacromial/subdeltoid bursa
  - Bursal surface partial thickness tears not seen
    - Bursal tears can be seen with subacromial/ subdeltoid bursography

CT Findings

- CT arthrography
  - Contrast seen extending into substance of rotator cuff tendon
    - Articular surface partial thickness tears only
  - Useful in patients with contraindication to MR
  - 95% accuracy for articular partial tears

MR Findings

- T1WI
  - Focal increased signal involving only a portion of thickness of rotator cuff
    - T1/PD images of partial tears have similar findings as tendinopathy
  - Bone findings associated with rotator cuff tear (RCT): Type III acromion, etc.
- T2WI FS
  - Focal increased signal within tendon that does not extend across entire thickness of cuff
    - Often fluid signal intensity
    - Tear can be obscured by granulation tissue/fibrosis within tear
  - Thinned tendon
    - < 5-7 mm thickness
    - Tendon appears only to be thinned when margins of tear are smooth and not abrupt
  - Conventional MR: 65-70% sensitivity, 90-95% specificity
  - Secondary signs
    - Enthesopathic cysts of anterior greater tuberosity
      - Posterior humeral head cysts not associated with cuff tears
    - Intramuscular cyst
      - 80-95% of patients with intramuscular cyst have RCT
      - Most are partial thickness tears
- MR arthrography
  - T1WI FS: High signal contrast extends into articular surface tear
    - T2 images also obtained to show bursal surface tears and intratendinous fissures
  - 95% accuracy for articular surface partial tears
  - Abduction external rotation (ABER) images
    - More sensitive for
      - Internal impingement posterior cuff articular surface partial thickness tears
      - Delaminating/horizontal tears
- Indirect MR arthrography (intravenous contrast injection with delayed MR imaging)
  - 80-90% sensitivity for partial tears, especially with ABER images
  - Advantages
    - No intraarticular injection, so no need for fluoro or ultrasound guidance
    - Granulation tear within a tear enhances
    - Enhancement of subacromial (SA)/subdeltoid (SD) bursal fluid, so bursal surface tears better seen
  - Disadvantage
    - Tendinosis can enhance similar to tear

Ultrasonographic Findings

- Focal hypoechoic or anechoic area involving surface of tendon but not extending across entire thickness of tendon

ROTATOR CUFF PARTIAL THICKNESS TEAR

- Alternative to conventional MR
  - Dynamic exam
    - Compression test: Press down with transducer; if deltoid indents into cuff defect → "deltoid herniation" sign
  - Less expensive
  - Especially if patient has contraindication to MR: Pacemaker, etc.
  - 65% sensitivity, 90-95% specificity

Imaging Recommendations

- Best imaging tool
  - Direct MR arthrography
- Protocol advice
  - Small field-of-view, high resolution T1W FS and T2W FS images
  - High quality, high channel shoulder coil

DIFFERENTIAL DIAGNOSIS

Radiographic

- Rotator cuff full thickness tear
  - High T2 signal extends from articular to bursal surface of cuff
- Rotator cuff tendinopathy
  - Intermediate to high signal on T2 images confined to midsubstance of cuff
- Rotator cuff calcific tendinitis
  - Low signal calcium with surrounding T2 high signal edema within cuff
  - Late in disease course will often rupture through bursal surface of cuff; may develop calcific bursitis

Clinical

- Adhesive capsulitis, shoulder
- Rotator cuff impingement
- Often difficult to diagnose a partial tear clinically; MR or ultrasound may be needed to exclude rotator cuff tear

PATHOLOGY

General Features

- Etiology
  - Intrinsic: Overuse, age-related degeneration
    - Articular surface fibers thinner, stiffer, and less well vascularized than bursal surface
  - Extrinsic: Impingement structures (type III acromion, etc.), humeral head depressor muscle weakness
  - Acute trauma

Staging, Grading, & Classification

- Ellman classification: Depth of tear
  - Grade 1: < 3 mm
  - Grade 2: 3-6 mm
  - Grade 3: > 6 mm
- Snyder classification: Side involved and defect size
  - A1: Articular surface fraying
  - A2: Articular, < 2 cm
  - A3: Articular, 2-3 cm
  - A4: Articular, > 3 cm
  - B1: Bursal surface fraying
  - B2: Bursal, < 2 cm
  - B3: Bursal, 2-3 cm
  - B4: Bursal, > 3 cm

CLINICAL ISSUES

Presentation

- Most common signs/symptoms
  - Pain, especially at night
  - Positive impingement tests

Demographics

- Age
  - Peak: 40-50 years old
- Gender
  - Male slightly > female
- Epidemiology
  - > 40 years: PTRCT in 35%
    - May be asymptomatic

Natural History & Prognosis

- Most progress to full thickness tear within 2 years
- May stabilize and become asymptomatic with physical therapy

Treatment

- Physical therapy
- Surgery
  - Small shallow tears: Debridement
  - Large deep partial thickness tears: Cuff repair

DIAGNOSTIC CHECKLIST

Consider

- Angled oblique sagittal images perpendicular to lateral supraspinatus tendon
  - Less volume averaging artifact than standard oblique sagittal images

Image Interpretation Pearls

- Look especially for articular surface tears → more common than bursal surface tears

Reporting Tips

- Report size, depth, and articular or bursal surface involved by tear
  - Helps in surgical planning

SELECTED REFERENCES

1. de Jesus JO et al: Accuracy of MRI, MR arthrography, and ultrasound in the diagnosis of rotator cuff tears: a meta-analysis. AJR Am J Roentgenol. 192(6):1701-7, 2009
2. Manvar AM et al: Clinical significance of intramuscular cysts in the rotator cuff and their relationship to full- and partial-thickness rotator cuff tears. AJR Am J Roentgenol. 192(3):719-24, 2009
3. Vlychou M et al: Symptomatic partial rotator cuff tears: diagnostic performance of ultrasound and magnetic resonance imaging with surgical correlation. Acta Radiol. 50(1):101-5, 2009
4. Vinson EN et al: Rim-rent tear of the rotator cuff: a common and easily overlooked partial tear. AJR Am J Roentgenol. 189(4):943-6, 2007
5. Meister K et al: MR arthrography of partial thickness tears of the undersurface of the rotator cuff: an arthroscopic correlation. Skeletal Radiol. 33(3):136-41, 2004

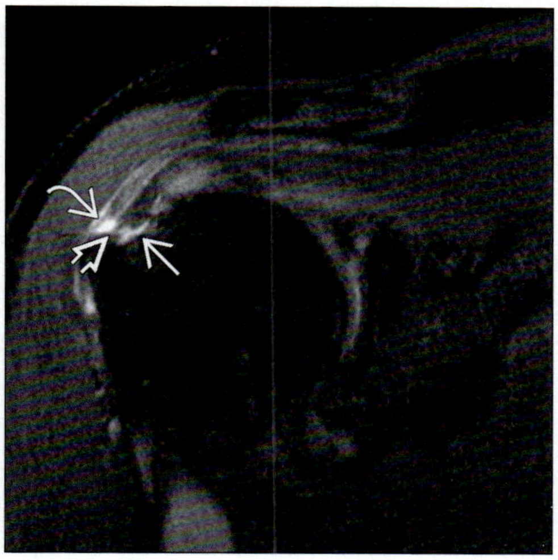

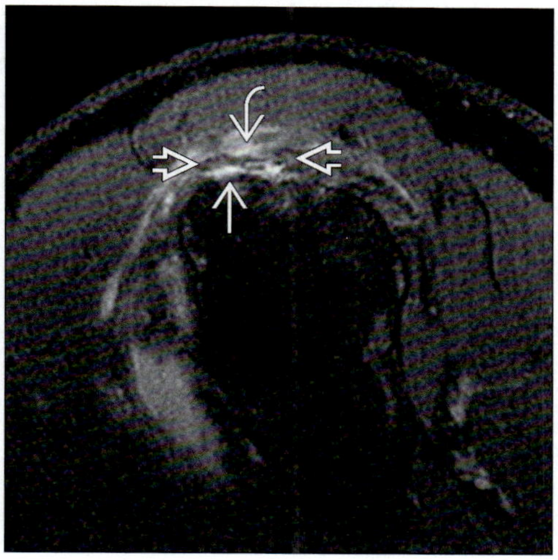

(Left) *Coronal oblique T2WI FS MR in a patient with articular and bursal partial tears of the supraspinatus tendon. There is high signal involving both the articular and bursal surfaces of the tendon, with a thin strip of intact cuff midsubstance.* ***(Right)*** *Oblique sagittal T2WI FS MR in the same patient shows articular and bursal partial tears and intact midsubstance tissue. Half of all bursal partial tears are associated with articular surface tears.*

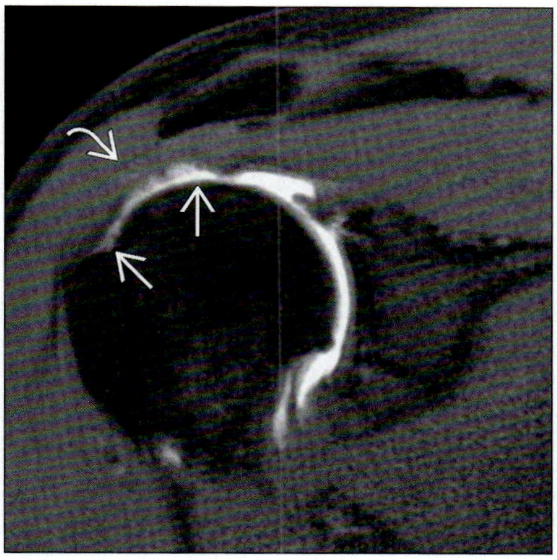

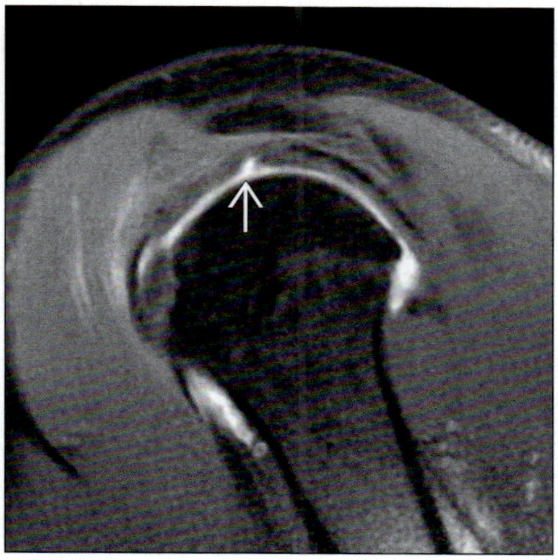

(Left) *Coronal oblique MR arthrogram T1WI FS shows a broad partial thickness tear involving the articular surface of the supraspinatus tendon, with no contrast in the SA/SD bursa.* ***(Right)*** *Oblique sagittal MR arthrogram T1WI FS in the same patient shows that the articular surface partial thickness rotator cuff tear has a small anteroposterior width. Direct MR arthrography has a 95% accuracy for articular surface partial thickness cuff tears.*

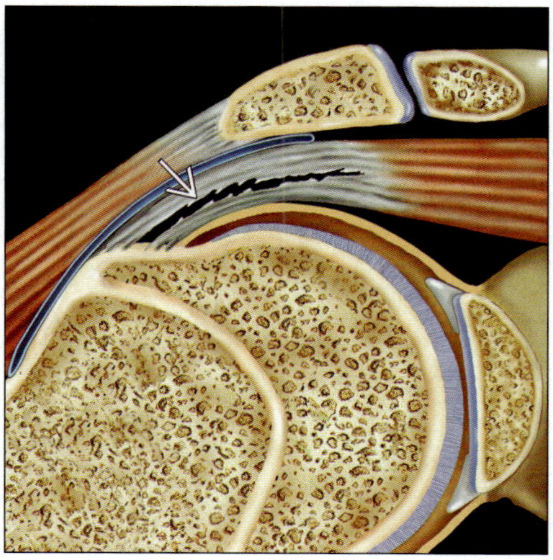

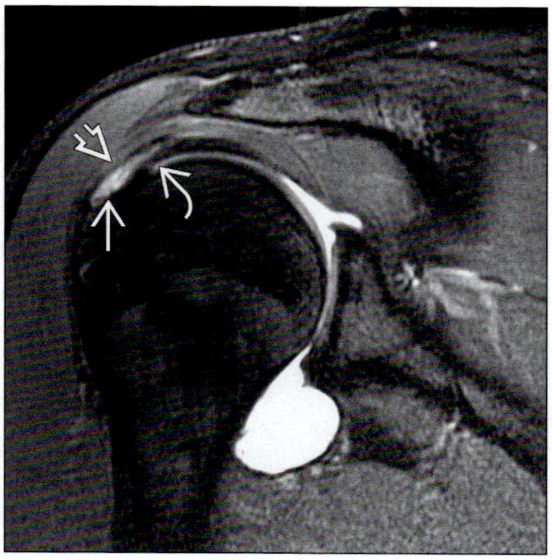

(Left) *AP graphic of the shoulder shows an intrasubstance or interstitial fissure/tear of the rotator cuff. Some authors call these intrasubstance partial thickness tears, while others consider these severe tendinopathy.* ***(Right)*** *Coronal oblique MR arthrogram T2WI FS in a patient with an intrasubstance fissure/tear. The articular and bursal surface fibers of the cuff are intact, so these interstitial (intratendinous) tears are not seen at arthroscopy or bursoscopy.*

ROTATOR CUFF PARTIAL THICKNESS TEAR

*(**Left**) Coronal oblique T2WI FS MR shows a rim rent tear →, an articular surface partial tendon avulsion at the insertion. This patient also has an intramuscular cyst →, which are associated with partial thickness tears. (**Right**) Longitudinal oblique ultrasound in a patient with a contraindication to having an MR scan. There is a partial cuff tear → at the insertion onto the tuberosity →, with calipers measuring the tear size. The bursal surface of the cuff → is intact.*

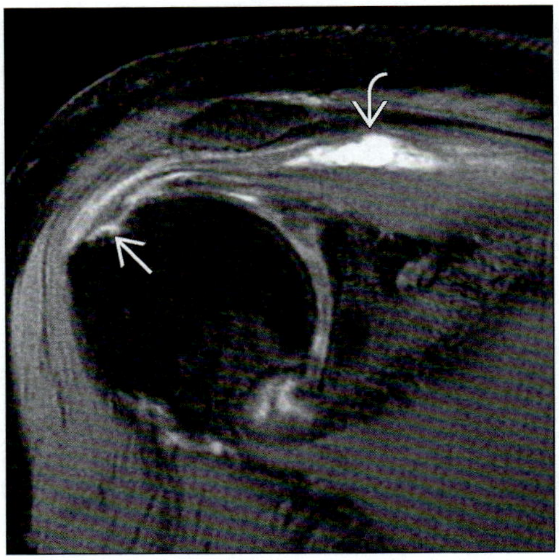

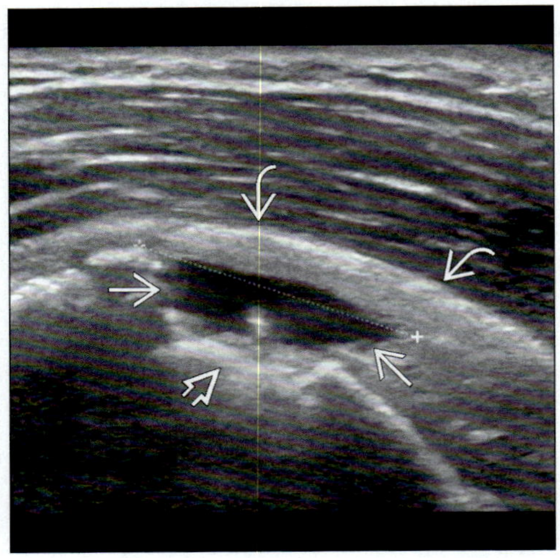

*(**Left**) Coronal oblique T2WI FS MR in a patient with shoulder pain shows a bursal surface partial thickness tear →. The articular surface of the cuff is intact →. (**Right**) Oblique sagittal T2WI FS MR in the same patient also shows the bursal surface partial thickness tear →. The tear, as well as the intact articular surface of the cuff, are blurry due to partial-averaging that occurs because the lateral cuff is curving downward within the oblique sagittal slice.*

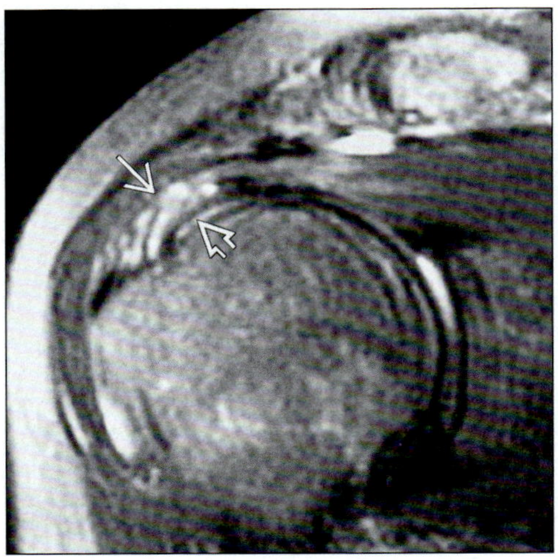

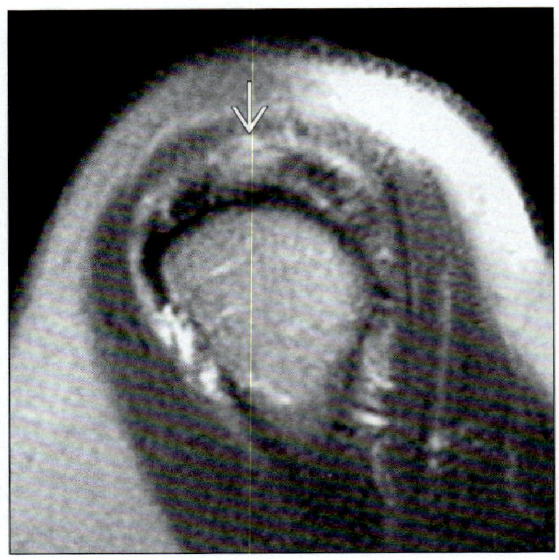

*(**Left**) Coronal oblique T2WI FS MR in the same patient shows localizer lines → for angled oblique sagittal images perpendicular to the lateral supraspinatus tendon, where most cuff tears occur. (**Right**) Angled oblique sagittal T2WI FS MR in the same patient shows the bursal surface partial tear →. Note the better detail of the tear compared with the standard oblique sagittal image, as well as the sharp delineation of the hyaline cartilage → and intact articular surface of the cuff →.*

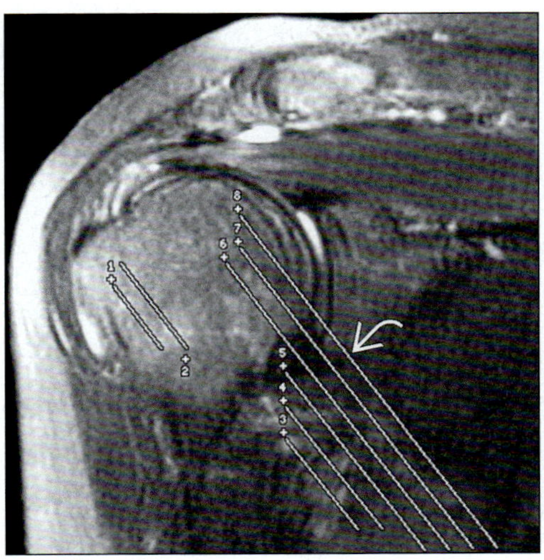

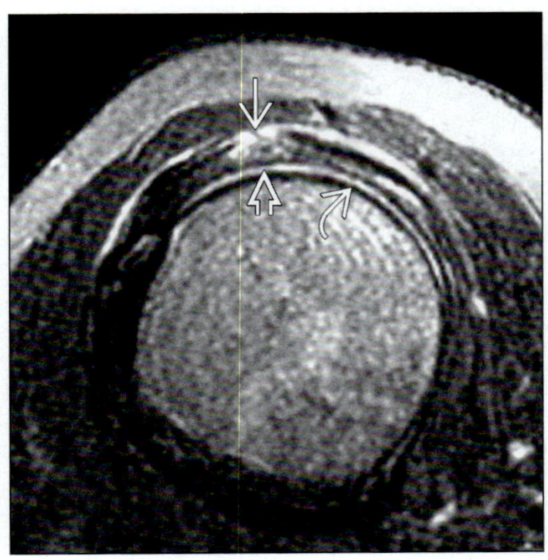

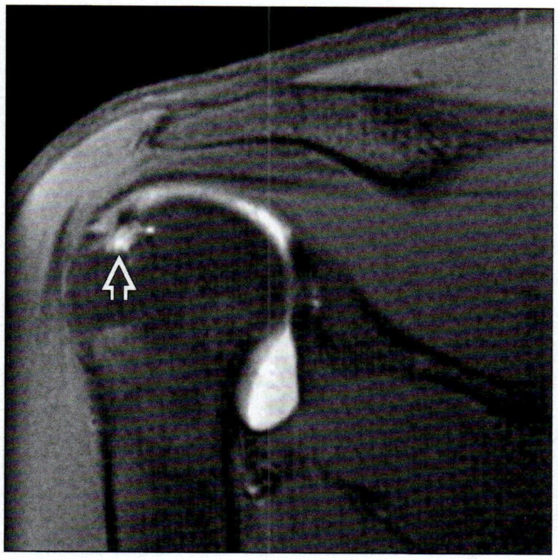

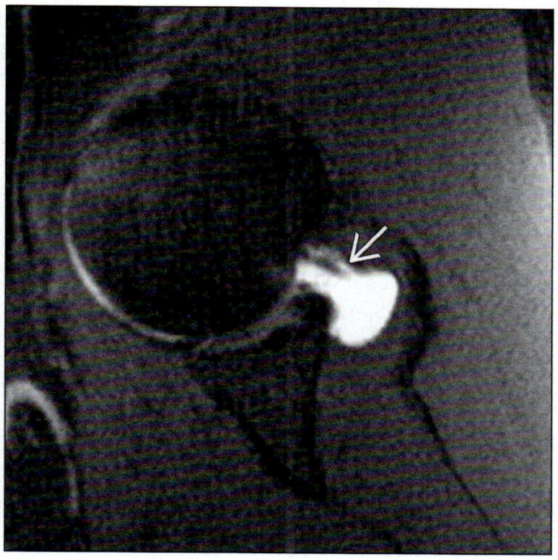

(Left) *Coronal oblique MR arthrogram T1WI FS in a young athlete with shoulder pain. There is an enthesopathic cyst in the posterior greater tuberosity, but no definite cuff tear is seen.* ***(Right)*** *ABER MR arthrogram T1WI FS in the same patient shows an articular surface partial thickness cuff tear of the superior aspect of the infraspinatus tendon. In overhead throwing athletes, direct MR arthrogram ABER images are more sensitive for posterior cuff tears than standard images.*

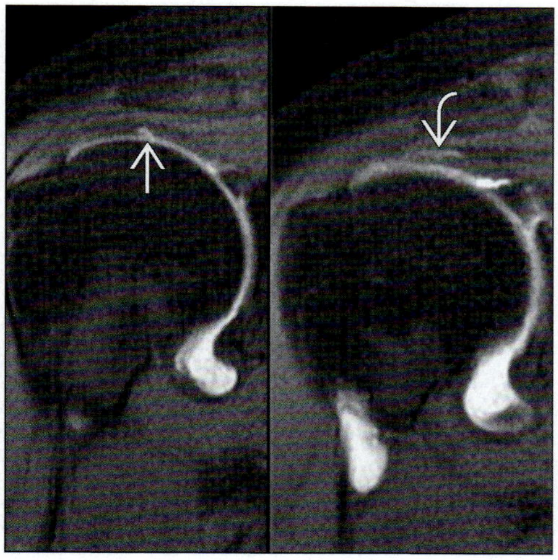

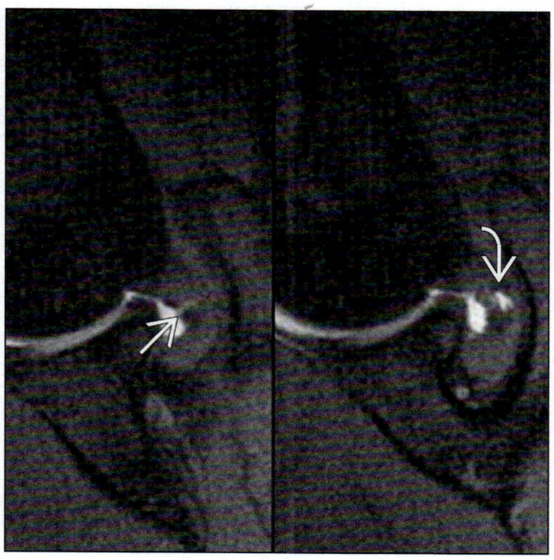

(Left) *Coronal oblique MR arthrogram T1WI FS adjacent images. There is a linear articular surface partial tear of the supraspinatus tendon. On the adjacent image, there is faint linear signal within the tendon.* ***(Right)*** *ABER MR arthrogram T1WI FS adjacent images in the same patient. The articular surface partial tear is seen, as well as better visualization of the deep extension of the tear. ABER images often show the delamination extent of tears better than standard images.*

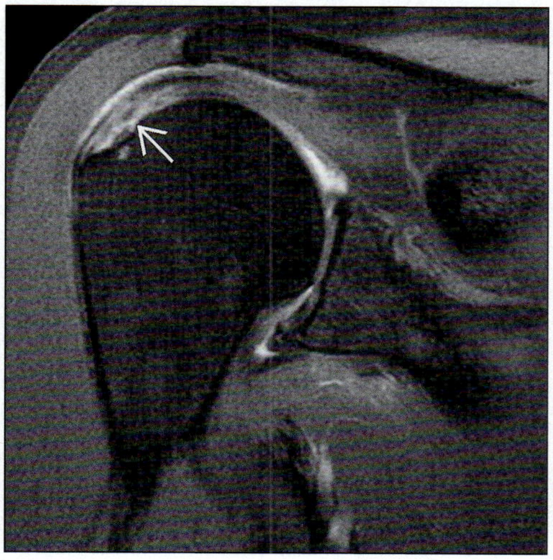

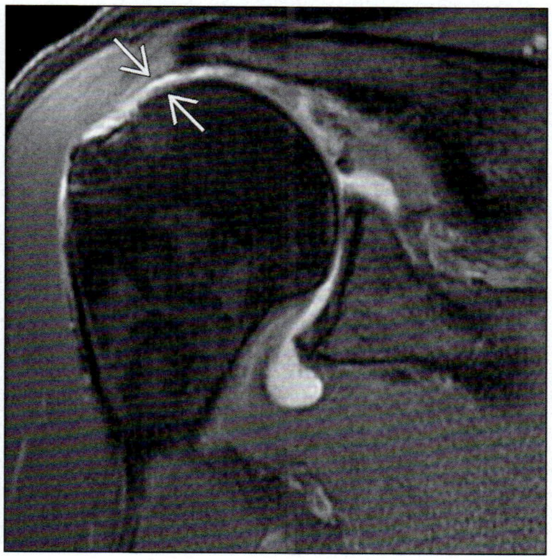

(Left) *Coronal oblique T2WI FS MR shows high signal disrupting the articular surface of the cuff, indicating a partial thickness rotator cuff tear. The patient briefly tried physical therapy but was not compliant.* ***(Right)*** *Coronal oblique T2WI FS MR in the same patient 1 year later, performed because of continuing shoulder pain. There is now a full thickness rotator cuff tear. Although some patients improve with physical therapy, many partial thickness tears will progress to a FTRCT.*

ROTATOR CUFF FULL THICKNESS TEAR

Key Facts

Terminology

- Full thickness (superior to inferior) rotator cuff or supraspinatus tendon tear (FTRCT/FTSSTT)
  - Tear extending from articular to bursal surface of rotator cuff tendon
  - Tear does not need to involve complete tendon, or lead to retraction of entire tendon
- If also full-width (anterior to posterior) tendon tear: Complete cuff tear

Imaging

- 95% of rotator cuff tears involve supraspinatus tendon
  - Usually involves lateral 1.5 cm of supraspinatus tendon
- Radiographs: Decreased acromiohumeral distance (< 7 mm)
  - Indicates chronic full thickness tear
- MR
  - High T2 signal extending from articular to bursal surface of tendon
  - No tendon between humeral head and acromion
- Ultrasound
  - Hypo- or anechoic region in cuff extending from humeral head to subacromial/subdeltoid bursa

Top Differential Diagnoses

- Angular anisotropy ("magic angle")
- Partial thickness cuff tear
  - Intact articular or bursal surface of cuff
- Severe tendinopathy
- Rotator cuff impingement

Clinical Issues

- Pain, especially at night
- Pain and weakness with abduction

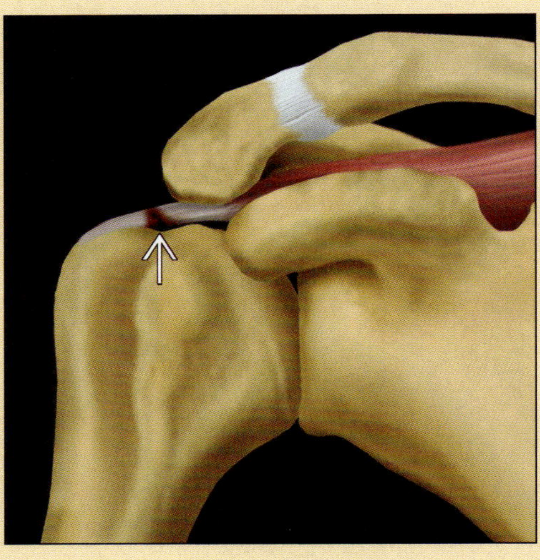

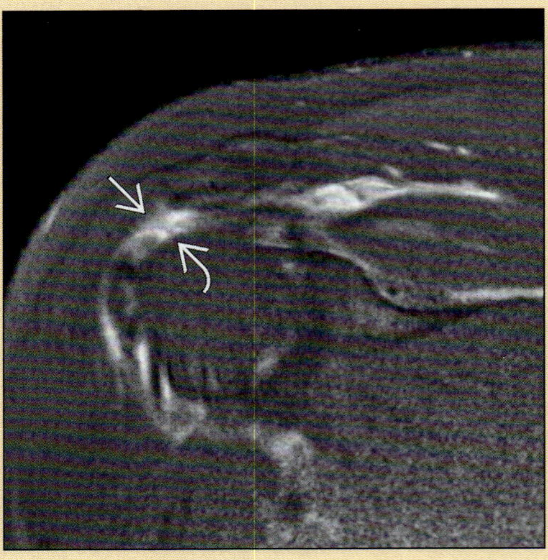

***(Left)** AP graphic shows a small full thickness rotator cuff tear extending from the articular to the bursal surface of the tendon. The supraspinatus tendon is the most common cuff tendon to tear. **(Right)** Coronal oblique T2WI FS MR shows high signal extending from the articular surface to the bursal surface of the supraspinatus tendon, consistent with a full thickness rotator cuff tear. The anteriolateral "critical zone" of the supraspinatus tendon is a common location for cuff tears.*

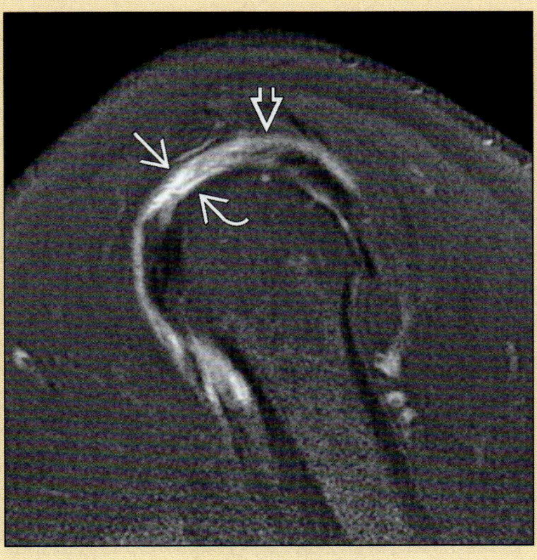

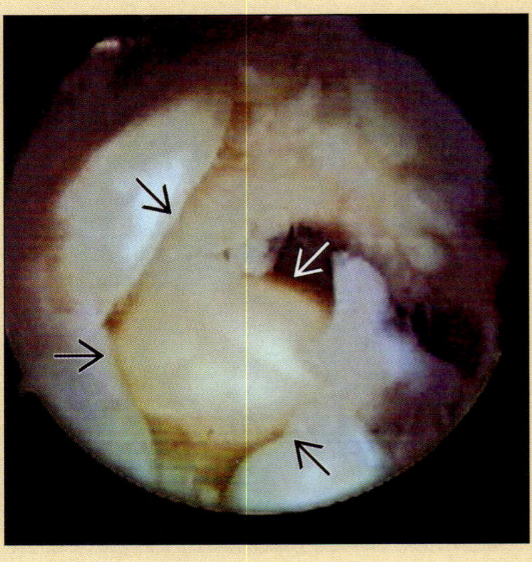

***(Left)** Oblique sagittal T2WI FS MR in the same patient shows the high signal extending from the articular to the bursal surface of the cuff. Note that the more posterior portion of the supraspinatus tendon is intact, making this a full thickness rotator cuff tear, but not a complete or full-width tear. **(Right)** Arthroscopic photograph in the same patient, looking from the subdeltoid bursa through the rotator cuff tear down onto the humeral head.*

ROTATOR CUFF FULL THICKNESS TEAR

TERMINOLOGY

Abbreviations

- Rotator cuff tear (RCT)
- Full thickness rotator cuff or supraspinatus tendon tear (FTRCT, FTSSTT)

Definitions

- Tear extending from articular to bursal surface of rotator cuff tendon
  - Tear does not need to involve complete tendon, or lead to retraction of entire tendon
- If also full-width tendon tear: Complete cuff tear

IMAGING

General Features

- Best diagnostic clue
  - High signal on T2WI MR involving entire thickness of tendon
- Location
  - 95% of FTRCT occur in supraspinatus tendon (SST)
    - Usually in lateral 1.5 cm of SST
      - "Critical zone" is relatively hypovascular
      - Lateral to rotator crescent (band-like thickening of tendon)
      - FTRCT can also be tendon avulsion at insertion onto greater tuberosity
    - Most infraspinatus full thickness tears also have supraspinatus tendon tear
- Size
  - Can be small (< 1 cm in diameter) to massive (> 5 cm)

Radiographic Findings

- Decreased acromiohumeral distance (< 7 mm)
  - Indicates chronic full thickness tear
- Rotator cuff arthropathy
  - Seen in chronic full thickness rotator cuff tears
  - Wear of undersurface of acromion process
    - Sclerosis and increased concavity of undersurface of acromion process
  - Glenohumeral joint osteoarthritis
- Associated osseous findings: Type III acromion
- Arthrography
  - Contrast injected into glenohumeral joint communicates with subacromial/subdeltoid (SA/SD) bursa
  - Geyser sign: FTRCT with contrast extending from SA/SD bursa into acromioclavicular joint

CT Findings

- CT arthrography
  - Contrast visualized in cuff defect
  - Contrast opacifies SA/SD bursa
  - Useful in patients with contraindication to MR
  - 95-100% accuracy for FTRCT
  - Preferred to ultrasound by some orthopedists when MR not available because images appear similar to MR

MR Findings

- T1WI
  - Defect in rotator cuff tendons with ↑ signal
  - Atrophy of involved muscle(s) in chronic complete FTRCT
    - Occupation ratio: Supraspinatus muscle cross-sectional area divided by supraspinatus fossa area on oblique sagittal image
      - > 0.6: Normal or mild atrophy
      - 0.4-0.6: Moderate atrophy
      - < 0.4: Severe atrophy
    - Muscle fat content: Goutallier grading system
      - Grade 0: No fat streaks
      - Grade 1: Few fatty streaks
      - Grade 2: Equal amounts of muscle and fat
      - Grade 3: More fat than muscle
- T2WI FS
  - 2 main signs
    - High signal extending from articular to bursal surface of tendon
      - Often fluid signal intensity
      - Signal may be less than fluid due to frayed edges of tendon, etc.
    - No tendon between humeral head and acromion
      - Seen in 10% of FTRCTs
      - Mostly chronic tears with no effusion
      - Cause of false-negative MR interpretation
  - 90-95% accuracy for FTRCT
  - Sagittal image: "Bald spot" sign is high signal defect within crescent of low signal cuff
  - Chronic full-width (complete) FTSSTT
    - Muscle atrophy
    - Retraction of tendon edge
- T1WI C+ FS
  - Direct MR arthrography
    - High signal-to-noise T1WI with dilute gadolinium produce high-quality images
    - Contrast injected into glenohumeral joint extends through tear into SA/SD bursa
      - Contrast within SA/SD bursa helps identify small FTRCTs
    - 95-100% accuracy for full thickness rotator cuff tears
  - Indirect MR arthrography
    - Intravenous gadolinium injection followed by delayed MR imaging
      - No intraarticular injection, so do not need fluoroscopy or US
    - More helpful for partial thickness tears where conventional MR accuracy is lower
    - SA/SD bursa enhances, so high signal within bursa cannot be used sign of FTRCT
    - > 95% accuracy for FTRCT
      - Abduction and external rotation (ABER) images increase accuracy to 100%

Ultrasonographic Findings

- Hypo- or anechoic region in cuff extending from humeral head to subacromial/subdeltoid bursa
  - "Uncovered cartilage" sign: Anechoic fluid adjacent to isoechoic hyaline cartilage
- Loss of convexity of superior surface of cuff
- Pros: Hands-on dynamic exam, less expensive and tolerated better than MR, high spatial resolution
- Cons: Operator dependent, steep learning curve, poorer contrast resolution than MR
- 90-95% accuracy for FTRCT

ROTATOR CUFF FULL THICKNESS TEAR

Imaging Recommendations

- Best imaging tool
  - Oblique coronal fat-suppressed T2WI
    - 90% of RCTs seen best with this sequence
    - Correlate with sagittal imaging, especially to evaluate supraspinatus/infraspinatus overlap region
  - Conventional MR: Noninvasive, high accuracy
- Protocol advice
  - Include oblique sagittal images
    - 10% of RCTs seen better

DIFFERENTIAL DIAGNOSIS

Imaging

- Angular anisotropy ("magic angle" phenomenon)
  - Band of increased signal extending across lateral cuff on short TE sequences
  - Occurs where cuff fibers oriented 55° to B_0
  - Becomes normal on T2WI
- Rotator cuff partial thickness tear
  - Intact articular or bursal surface of cuff
- Rotator cuff tendinopathy
  - Some intact low signal fibers can be seen extending through area of increased signal

Clinical

- Adhesive capsulitis
- Glenohumeral joint osteoarthritis
- Rotator cuff impingement
- Clinical exam often difficult: MR or ultrasound may be needed to exclude rotator cuff tear

PATHOLOGY

General Features

- Etiology
  - Most FTRCTs occur in patients with preexisting degeneration or partial thickness rotator cuff tear
  - Intrinsic cuff failure
    - Overuse and age-related degeneration from mechanical damage, inflammatory mediators, altered expression of degradative enzymes
    - Decreased vascularity with age
  - Extrinsic cuff impingement: Can be 1° impingement or 2° to abnormal glenohumeral motion
    - Primary impingement
      - Osseous impingement: Type III acromion, inferior AC joint osteophytes, ± os acromiale
      - Thickened coracoacromial ligament
    - Secondary impingement
      - Weak humeral head depressors allow elevation of humeral head with impingement
  - Instability: Multidirectional or microinstability
  - Acute cuff trauma
    - Older patients secondary to anterior dislocation
    - FTRCT occurs rarely following trauma in younger patients with otherwise healthy rotator cuff

Staging, Grading, & Classification

- Maximum diameter of full thickness defect (involves 1 or more tendons of RCT)
  - Small: < 1 cm
  - Medium: 1-3 cm
  - Large: > 3 cm but not complete
  - Complete or massive: Entire tendon complex torn
- Measure in 2 dimensions (use coronal for medial-lateral measurement of retraction and sagittal for anterior-posterior tear dimension)

Gross Pathologic & Surgical Features

- Arthroscopist can look from glenohumeral joint into subacromial/subdeltoid bursa

CLINICAL ISSUES

Presentation

- Most common signs/symptoms
  - Pain, especially at night
  - Pain and weakness with abduction
  - Positive impingement test
  - Complete FTSSTT: Unable to abduct initial 10°

Demographics

- Age
  - > 40 years old
- Gender
  - Slight male predominance
- Epidemiology
  - > 60 years: FTRCT in 25%
    - Not all are symptomatic
  - < 10% of FTRCT are acute at presentation

Natural History & Prognosis

- Pain may improve with physical therapy
- Some patients progress to cuff arthropathy

Treatment

- Small-medium FTRCT: Cuff repair
- Complete FTRCT
  - Nonrepairable if significant muscle atrophy or retraction of tendon edge medial to glenoid
  - Debridement may improve symptoms
  - If cuff arthropathy, may need arthroplasty

DIAGNOSTIC CHECKLIST

Consider

- FTRCT in patients > 40 years old after anterior dislocation

Image Interpretation Pearls

- Look for high-riding humeral head on MR
  - Can overlook FTRCT on MR if no fluid

Reporting Tips

- Give diameter of FTRCT in 2 dimensions
- Describe severity of fatty atrophy of muscle

SELECTED REFERENCES

1. de Jesus JO et al: Accuracy of MRI, MR arthrography, and ultrasound in the diagnosis of rotator cuff tears: a meta-analysis. AJR Am J Roentgenol. 192(6):1701-7, 2009
2. Jung JY et al: Comparison study of indirect MR arthrography and direct MR arthrography of the shoulder. Skeletal Radiol. 38(7):659-67, 2009
3. Magee T et al: 3.0-T MRI of the supraspinatus tendon. AJR Am J Roentgenol. 187(4):881-6, 2006

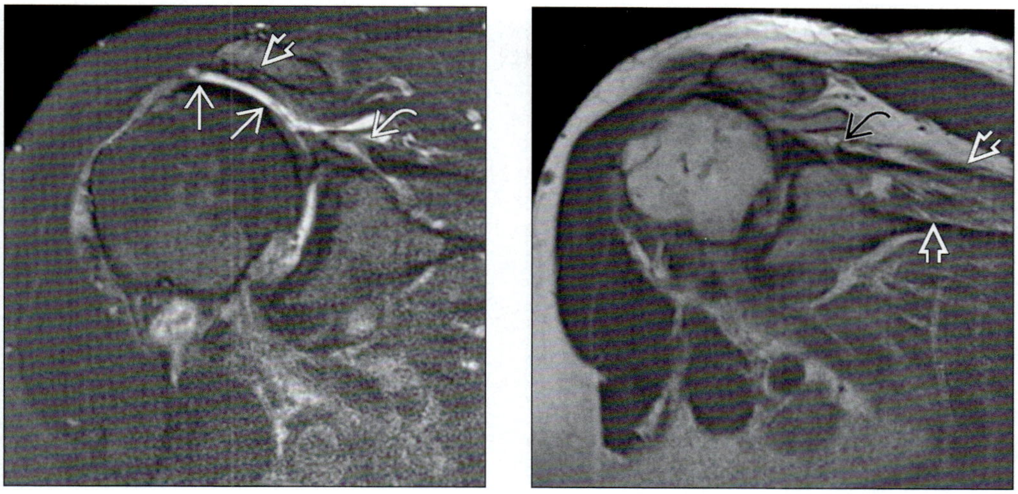

(Left) *Coronal oblique T2WI FS MR shows a chronic massive full thickness RCT with retraction of the tendon edge. There is chronic wear of the undersurface of the acromion from the high-riding humeral head, indicating cuff arthropathy.* ***(Right)*** *Coronal oblique T1WI MR in the same patient confirms that the retracted supraspinatus tendon edge is medial to the glenoid fossa, making this an unrepairable tear. There is also atrophy of the supraspinatus muscle.*

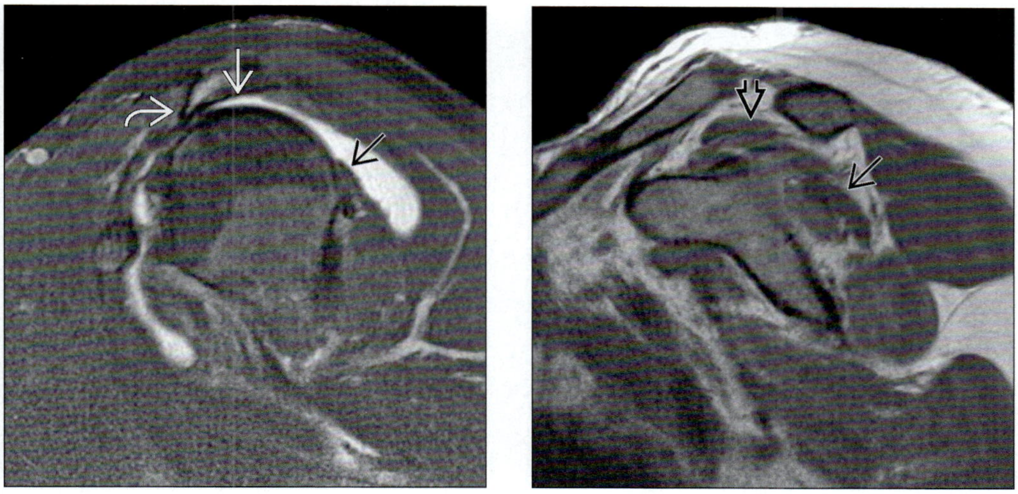

(Left) *Oblique sagittal T2WI FS MR in the same patient shows the full thickness tear of the supraspinatus tendon extending to involve most of the infraspinatus tendon. Most infraspinatus tendon tears are associated with large supraspinatus tears. There is an anterior acromial enthesophyte.* ***(Right)*** *Oblique sagittal T1WI MR more medially through the scapular neck in the same patient shows the atrophy of the supraspinatus and infraspinatus muscles.*

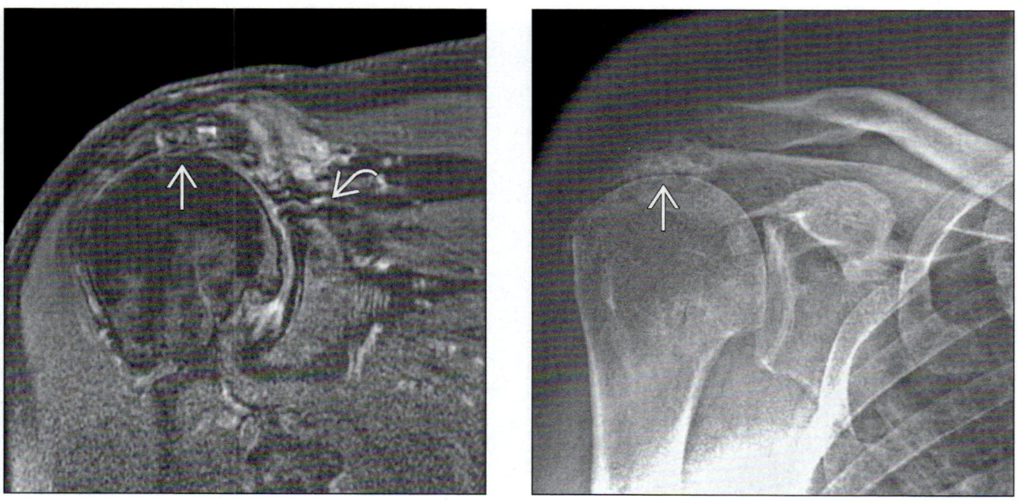

(Left) *Coronal oblique T2WI FS MR in a patient with a chronic FTRCT and minimal joint fluid. There is no supraspinatus tendon between the humeral head and the undersurface of the acromion process. The retracted tendon edge is medial to the glenoid fossa. In patients without a joint effusion, it is important to look for an intact tendon so as not to miss a FTRCT.* ***(Right)*** *Neer AP radiograph in the same patient shows the high-riding humeral head of a chronic FTRCT.*

ROTATOR CUFF FULL THICKNESS TEAR

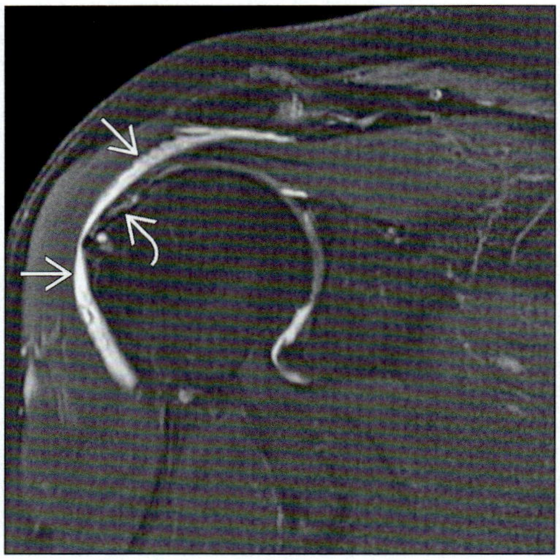

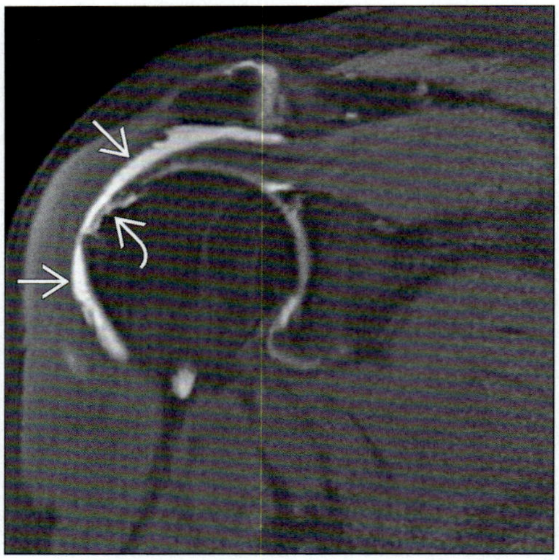

(Left) *Coronal oblique T2WI FS MR shows an apparent deep articular surface partial thickness tear ➔ of the supraspinatus tendon. There is also a large subacromial/ subdeltoid bursa effusion ➔.* ***(Right)*** *Coronal oblique MR arthrogram T1WI FS in the same patient shows that the SA/SD bursal fluid is high signal gadolinium contrast ➔, indicating a full thickness rotator cuff tear. Contrast extending across the entire cuff at the tear ➔ is more easily seen on the T1-weighted image.*

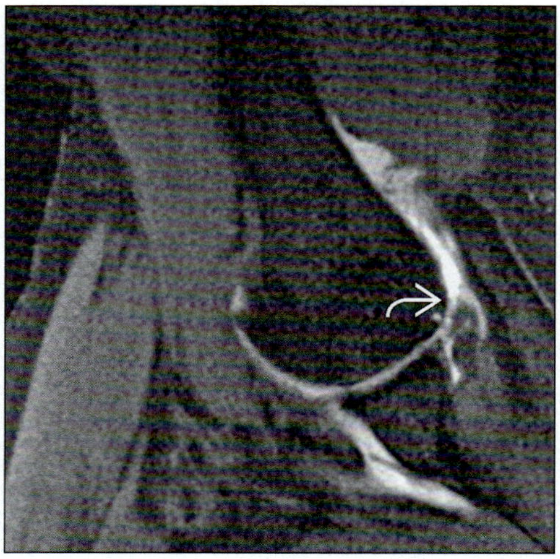

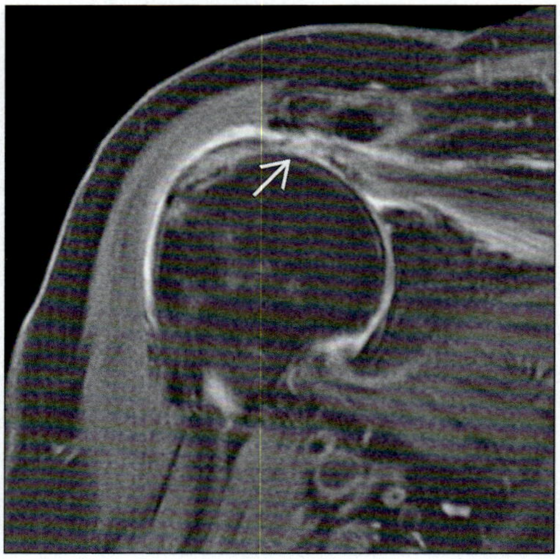

(Left) *ABER MR arthrogram T1WI FS in the same patient confirms the full thickness cuff tear ➔. Abduction external rotation images are mostly used to better visualize Bankart tears and internal impingement lesions. However, by relaxing the tension on the cuff it can also show small tears.* ***(Right)*** *Coronal oblique T2WI FS MR shows a full thickness rotator cuff tear at the myotendinous junction of the supraspinatus tendon ➔. Tears in this location are less common than more lateral tears.*

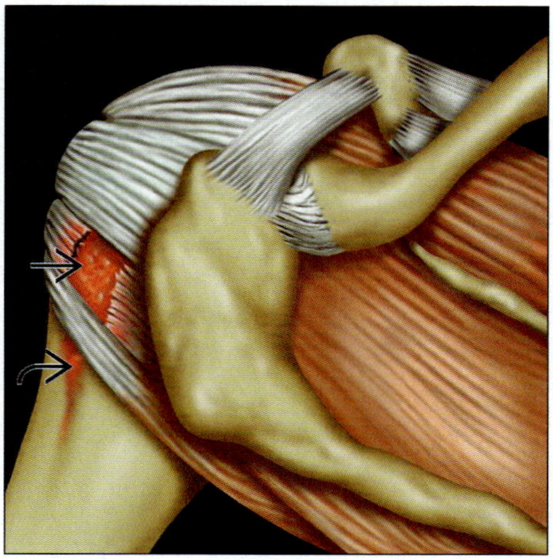

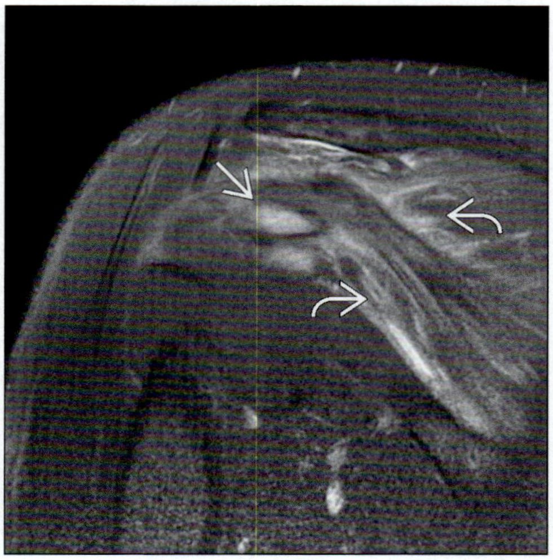

(Left) *PA graphic of the shoulder shows an isolated tear of the posterior cuff (infraspinatus or teres minor tendons) ➔. There is hemorrhage at the tear site extending down the posterior humerus ➔.* ***(Right)*** *Coronal oblique T2WI FS MR shows a full thickness tear of the infraspinatus tendon ➔ with intramuscular edema ➔. Most infraspinatus tendon tears are in the superior aspect of the tendon and result from posterior extension of a massive supraspinatus tendon tear.*

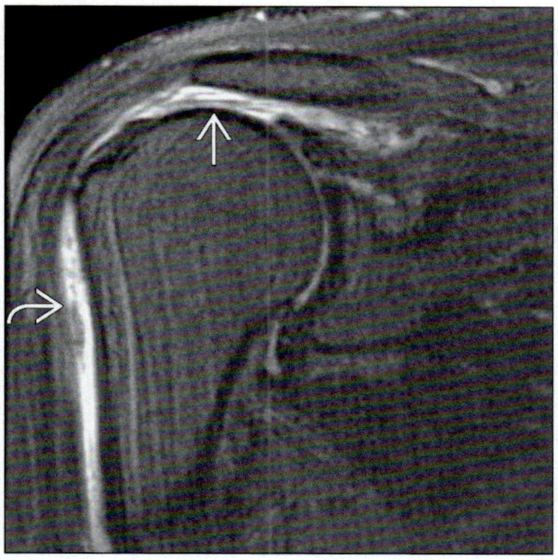

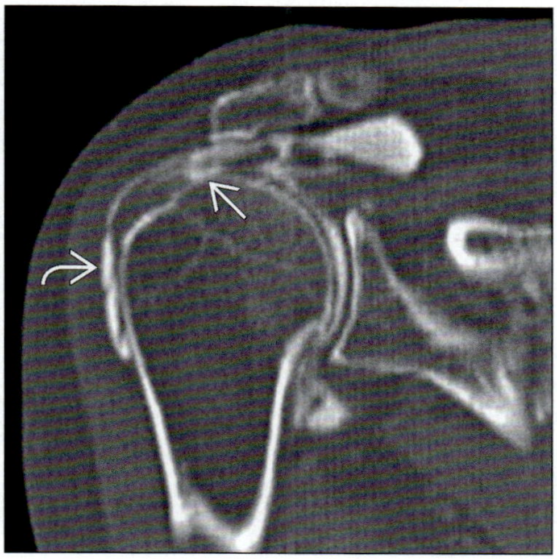

(Left) *Coronal oblique T2WI FS MR in a 73-year-old man with shoulder pain and weakness after a fall. There is fluid and hemorrhage in an acute full thickness rotator cuff tear extending into the SA/SD bursa. Most acute tears are in tendons with preexisting degeneration.* ***(Right)*** *Coronal oblique CT arthrogram shows contrast extending through a defect in the supraspinatus tendon and filling of the subacromial/subdeltoid bursa, indicating a full thickness rotator cuff tear.*

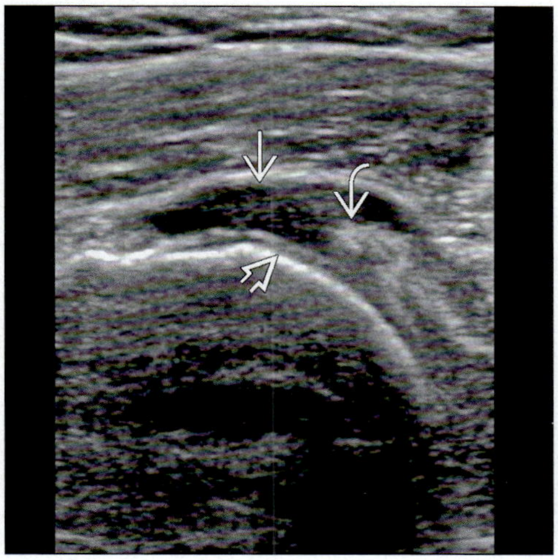

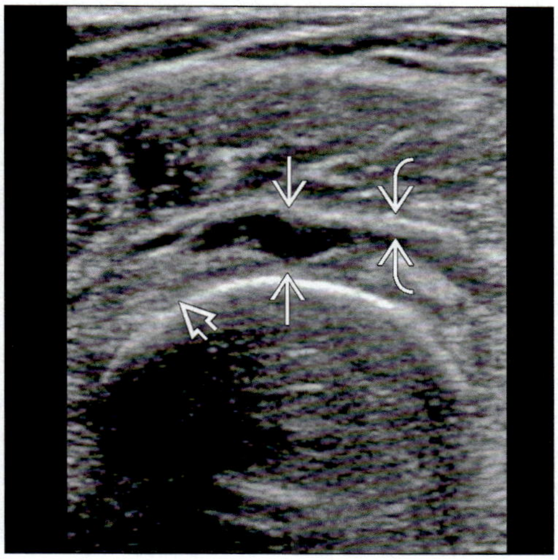

(Left) *Longitudinal coronal oblique ultrasound shows anechoic fluid and the frayed tendon edge of a full thickness rotator cuff tear. Note the "uncovered cartilage" sign.* ***(Right)*** *Transverse sagittal oblique ultrasound in the same patient shows the fluid within the full thickness rotator cuff tear, with some intact supraspinatus tendon adjacent to the tear. Note the loss of convexity of the hyperechoic capsule of the subdeltoid bursa.*

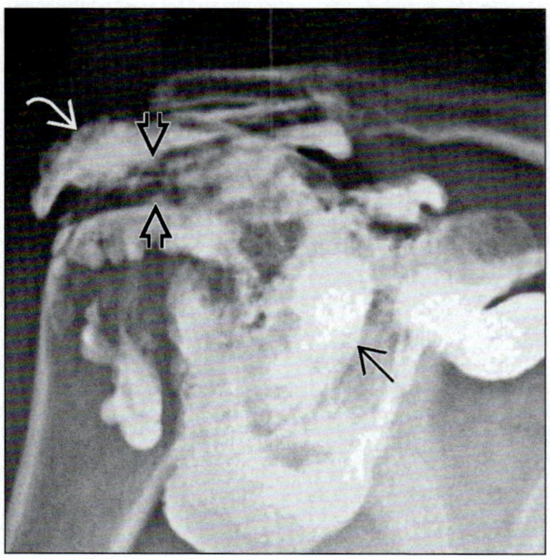

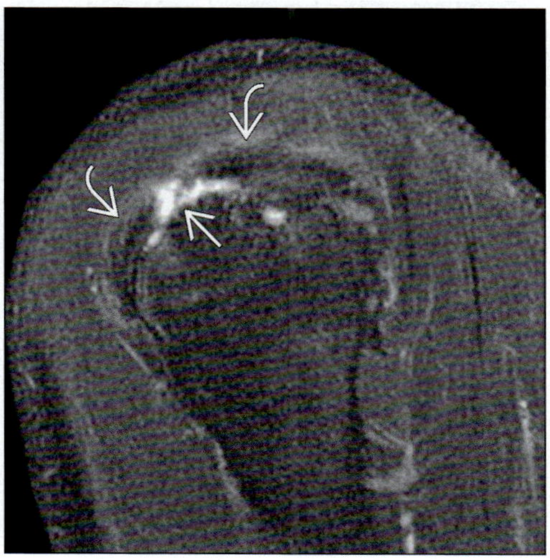

(Left) *AP arthrogram in a patient with a full thickness rotator cuff tear demonstrates that there is contrast extending from the glenohumeral joint into the SA/SD bursa. The tear through the rotator cuff tendon is better seen if followed by a post-arthrogram CT.* ***(Right)*** *Oblique sagittal T2WI FS MR in a patient with a FTRCT shows the "bald spot" sign. There is intact low signal tendon anterior and posterior to the tear, mimicking hair around a central bald spot.*

ROTATOR INTERVAL TEAR

Key Facts

Terminology
- Rotator cuff interval (RCI), rotator interval (RI)
- Tear of coracohumeral ligament (CHL), superior glenohumeral ligament (SGHL), or joint capsule between supraspinatus and subscapularis tendons
  - CHL and SGHL can instead be stretched and lax after injury; may have similar instability symptoms as RCI tear

Imaging
- Rotator interval
  - Triangle with base at coracoid process and lateral apex at transverse humeral ligament
  - Superior border: Anterior supraspinatus tendon
  - Inferior border: Cranial edge of subscapularis tendon
- Tear on MR
  - T2 high signal in RCI
  - Disrupted fibers of CHL
  - Chronic tears: CHL and SGHL thickened and scarred
  - "Hidden lesion": Biceps subluxed at entrance to groove with tear of CHL/SGHL biceps pulley in lateral rotator interval
- MR arthrography
  - Tear: Contrast extends to base of coracoid process

Top Differential Diagnoses
- Adhesive capsulitis
  - Thick CHL with fibrosis in surrounding fat

Pathology
- Type 1 RCI lesion: Fibrosis = adhesive capsulitis
- Type 2 RCI lesion: Tear

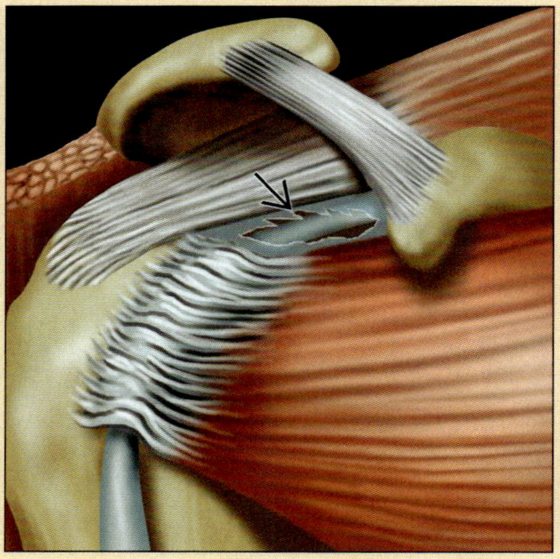
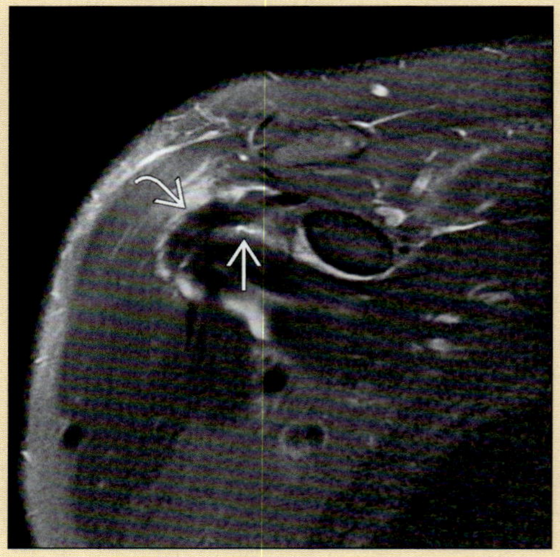

(Left) *AP graphic of a tear of the RCI capsule and ligaments ➡. The RCI is a triangle with its base at the coracoid and lateral apex at the transverse humeral ligament, bounded on top by the supraspinatus and inferiorly by the subscapularis tendon. The biceps tendon passes through the RCI.* ***(Right)*** *Coronal oblique T2WI FS MR through the anterior shoulder shows a small tear ➡ of the RCI capsule medial to the biceps tendon ➡. RCI tears are seen best on oblique sagittal images.*

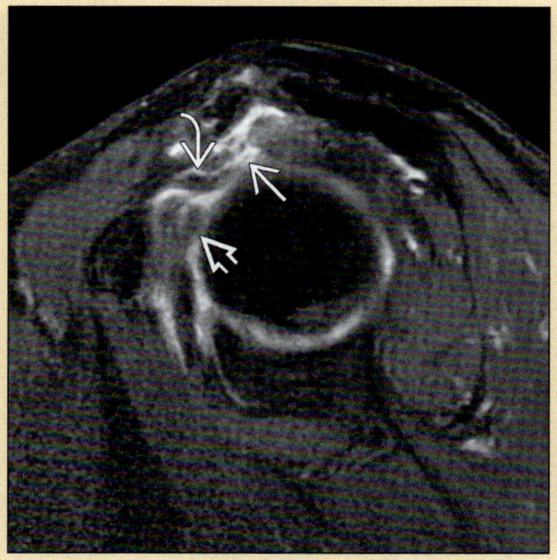
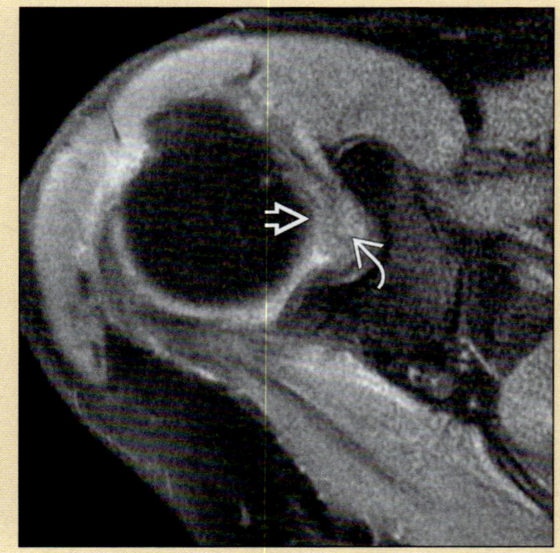

(Left) *Oblique sagittal T2WI FS MR shows edema in the RCI ➡ and a thickened partially torn coracohumeral ligament ➡. There is tendinopathy in a dislocated biceps tendon ➡.* ***(Right)*** *Axial PDWI FS MR in the same patient shows a torn SGHL ➡, while the biceps ➡ is not well seen due to severe tendinopathy. The SGHL and CHL are the 2 main ligaments in the RCI and form the biceps pulley; when they tear, the biceps dislocates deep to the subscapularis tendon.*

ROTATOR INTERVAL TEAR

TERMINOLOGY

Abbreviations

- Rotator cuff interval (RCI), rotator interval (RI)

Synonyms

- "Hidden" lesion if subluxed biceps due to RCI tear

Definitions

- Tear of coracohumeral ligament (CHL), superior glenohumeral ligament (SGHL), or joint capsule between supraspinatus and subscapularis tendons
  - CHL and SGHL can instead be stretched and lax after injury
- Normal rotator interval
  - CHL and SGHL fuse laterally & form biceps pulley
  - RCI prevents excessive flexion and external rotation
  - RCI is tunnel through which the biceps passes

IMAGING

General Features

- Best diagnostic clue
  - MR T2 high signal in RCI
- Location
  - RCI = triangle with base at coracoid process and lateral apex at transverse humeral ligament
    - Superior border: Anterior supraspinatus tendon
    - Inferior border: Cranial edge of subscapularis
- Size
  - Small isolated tears
  - Large tears associated with anterior supraspinatus tendon tear and cranial subscapularis tear with torn/ dislocated biceps tendon

MR Findings

- T1WI
  - Intermediate signal surrounding indistinct CHL
- T2WI FS
  - High signal in RCI
  - Disrupted fibers of CHL
  - Chronic tears: CHL and SGHL thickened and scarred
  - "Hidden lesion": Biceps subluxed
    - Torn biceps pulley in lateral RCI
    - Lesion is "hidden" to arthroscopist due to intact joint capsule
- MR arthrography
  - Normal: CHL and anterior SGHL well seen
  - Tear: Contrast extends to base of coracoid process
  - Acute tears: Contrast extravasates through RCI

Imaging Recommendations

- Best imaging tool
  - Direct MR arthrography
- Protocol advice
  - If RCI tear suspected, do not use interval approach for arthrogram injection

DIFFERENTIAL DIAGNOSIS

Congenital RCI Defect or Absent CHL or SGHL

- Uncommon

Adhesive Capsulitis

- Thick CHL with fibrosis in surrounding fat

Subcoracoid Impingement

- ↓ coracohumeral interval, lesser tuberosity cysts
- Can also have RCI tear

Microinstability

- Superior labral tear with mild instability symptoms
  - If labral tear involves anterosuperior labrum, can also have rotator interval tear

Superior Labrum Anterior Cuff (SLAC) Lesion

- Tear of anterior SST and anterosuperior labrum
- Can also have RCI tear

PATHOLOGY

General Features

- Etiology
  - Anterior dislocation
    - Interval torn in up to 50% of anterior dislocations
  - Acute trauma to anterosuperior capsule
  - Overhead throwers/laborers: Chronic tears
- Associated abnormalities
  - Biceps tendon subluxation
    - Can be intermittent and not seen on routine MR
  - Anterior supraspinatus tendon tear
  - Cranial subscapularis tendon tear
    - 45% have RCI tear
  - Anterior and anterosuperior labral tear

Staging, Grading, & Classification

- Type 1 RCI lesion: Fibrosis = adhesive capsulitis
- Type 2 RCI lesion: Tear

CLINICAL ISSUES

Presentation

- Most common signs/symptoms
  - Anterior shoulder pain
  - Mild instability
- Other signs/symptoms
  - "Sulcus" sign: Inferior translation of humeral head

Natural History & Prognosis

- Failure to repair torn RCI is cause of persistent pain and instability after instability surgery

Treatment

- Surgical closure of defect

DIAGNOSTIC CHECKLIST

Image Interpretation Pearls

- Evaluate RCI on oblique sagittal images

SELECTED REFERENCES

1. Lee JC et al: MRI of the rotator interval of the shoulder. Clin Radiol. 62(5):416-23, 2007
2. Vinson EN et al: Magnetic resonance imaging findings associated with surgically proven rotator interval lesions. Skeletal Radiol. 36(5):405-10, 2007
3. Morag Y et al: MR arthrography of rotator interval, long head of the biceps brachii, and biceps pulley of the shoulder. Radiology. 235(1):21-30, 2005

ROTATOR INTERVAL TEAR

*(**Left**) AP graphic with 3 sequential images of the RCI. The coracohumeral ligament ➔ merges with the superior glenohumeral ligament ➔ to form a "T" on the middle image, and the biceps pulley ➔ laterally. (**Right**) Oblique sagittal MR arthrogram T2WI FS of a normal RCI shows the CHL ➔ reinforcing the capsule of the RCI. The SGHL ➔ is merging with the CHL to form a "T," the beginning of the biceps pulley that will stabilize the biceps tendon ➔ as it turns to enter the groove.*

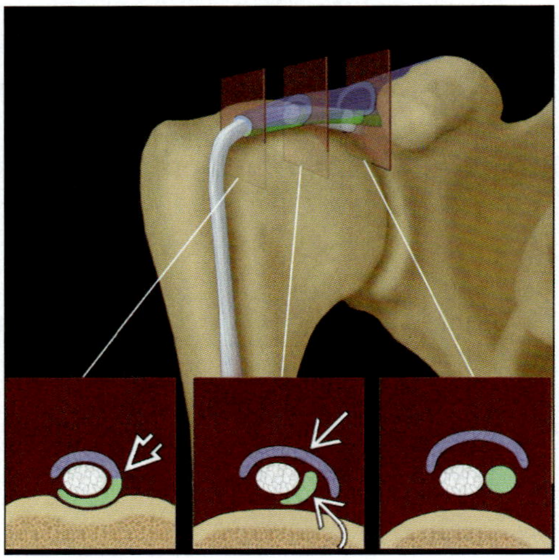

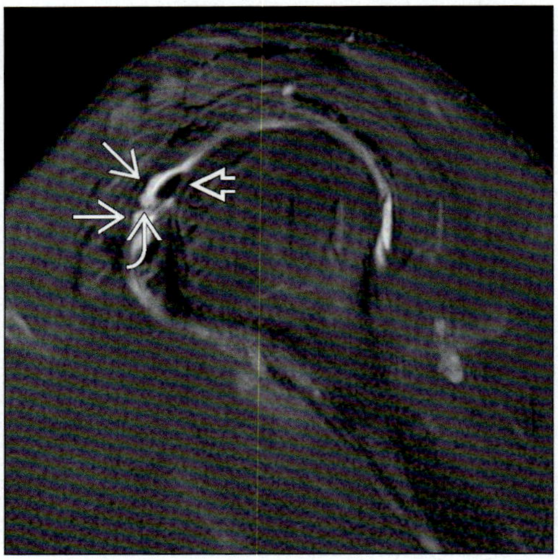

*(**Left**) Axial MR arthrogram PDWI FS of a normal RCI shows an intact superior glenohumeral ligament ➔ arising from the anterosuperior labrum ➔ just inferior to the biceps tendon and curving laterally to merge with the coracohumeral ligament. (**Right**) Oblique sagittal MR arthrogram T2WI FS in the same patient shows the CHL ➔ merging with the lateral portion of the SGHL ➔ to form a sling (biceps pulley) around the biceps tendon ➔ in the lateral aspect of the RCI.*

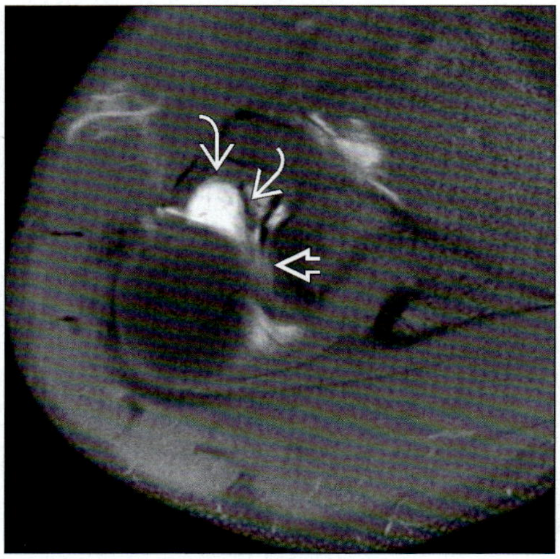

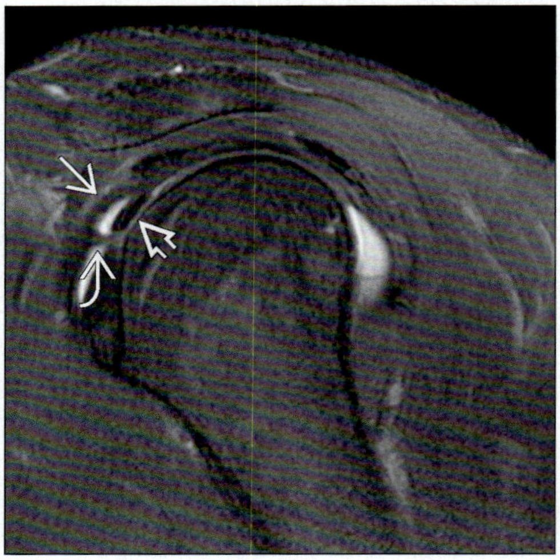

*(**Left**) Axial T2WI FS MR of a RCI tear shows high signal in the expected location of the lateral SGHL ➔ and subluxation of the biceps tendon ➔. The patient also has a rotator cuff tear ➔. (**Right**) Oblique sagittal T2WI FS MR in the same patient shows that the CHL is slightly lax medially ➔ and torn near its coracoid insertion ➔. The more superficial coracoacromial ligament ➔ is intact. There is mild tendinopathy of the subluxed biceps ➔. The rotator cuff is torn ➔.*

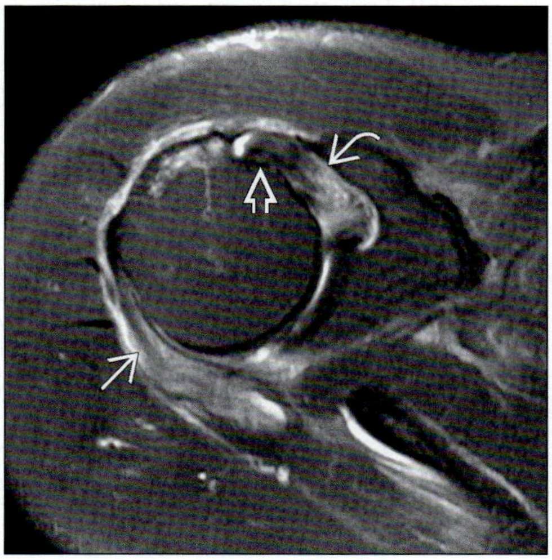

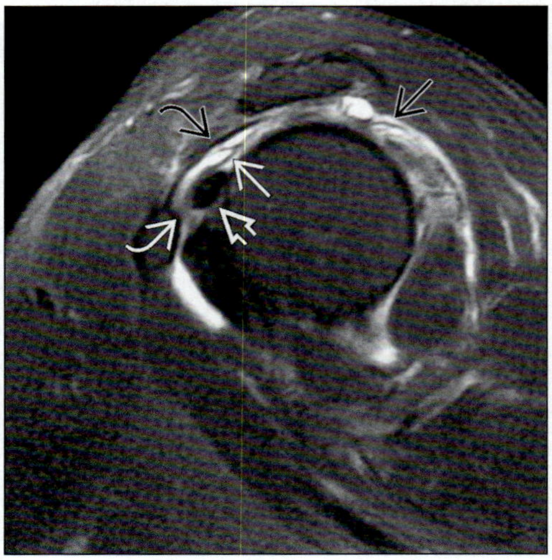

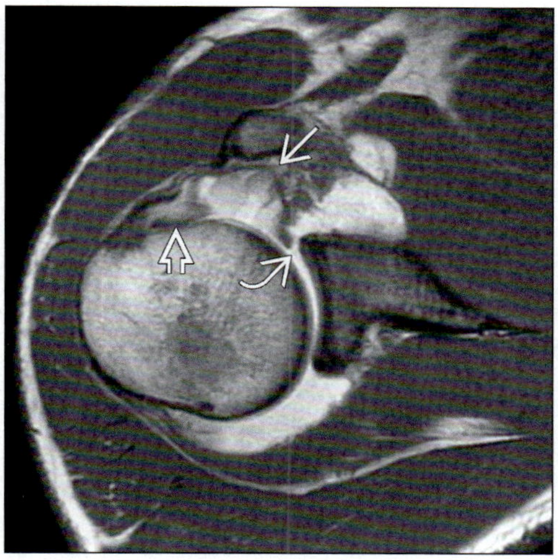

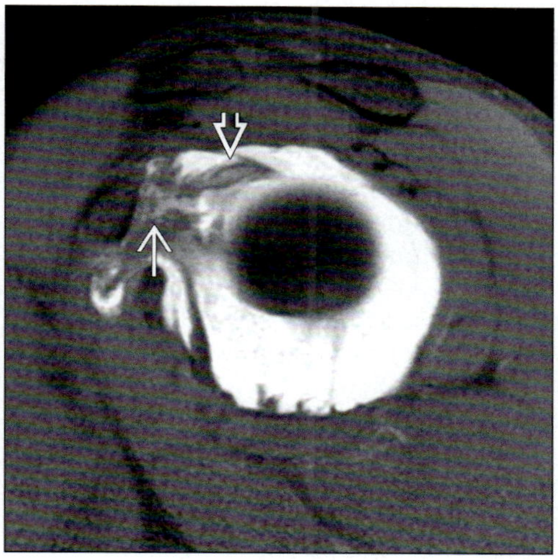

(Left) *Axial MR arthrogram T1WI shows a torn rotator interval capsule with subluxed and partially torn biceps. The patient had an anterior dislocation and labral tear, which are associated with RCI tears.* ***(Right)*** *Oblique sagittal MR arthrogram T1WI FS in the same patient again shows the torn rotator interval capsule and partial biceps tear. Contrast extending to the posterior aspect of the coracoid process is a sign of RCI tear on direct MR arthrography.*

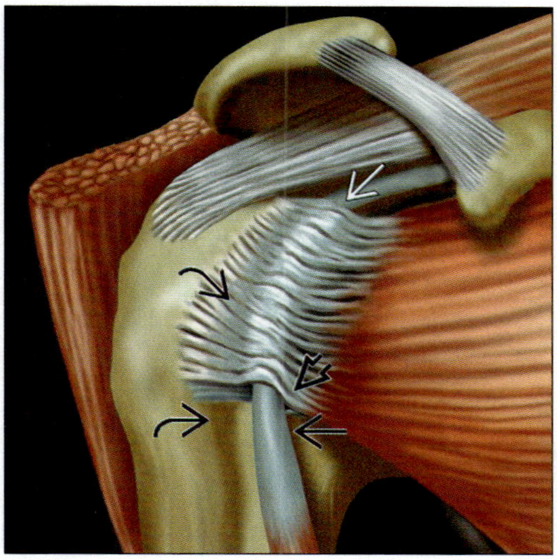

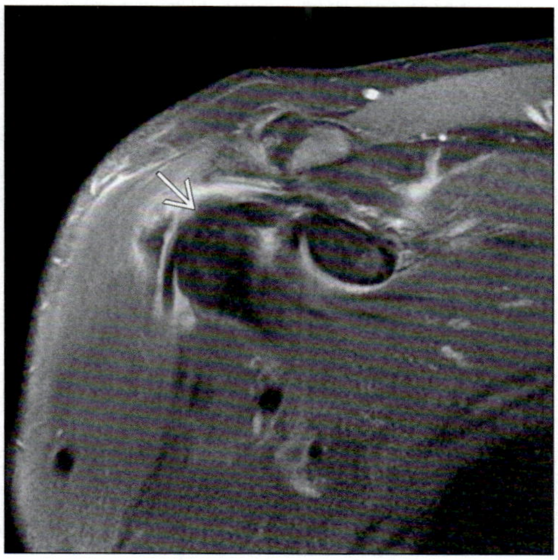

(Left) *AP graphic shows a "hidden lesion" with slight medial subluxation of the biceps tendon out of the bicipital groove and deep to the subscapularis tendon. The lesion is from a partial tear of the subscapularis at the insertion and a biceps pulley tear in the lateral RCI. The joint capsule is intact so the lesion is "hidden" at arthroscopy.* ***(Right)*** *Coronal oblique PDWI FS MR shows slight subluxation of the biceps tendon in a patient with a "hidden lesion" and RCI tear.*

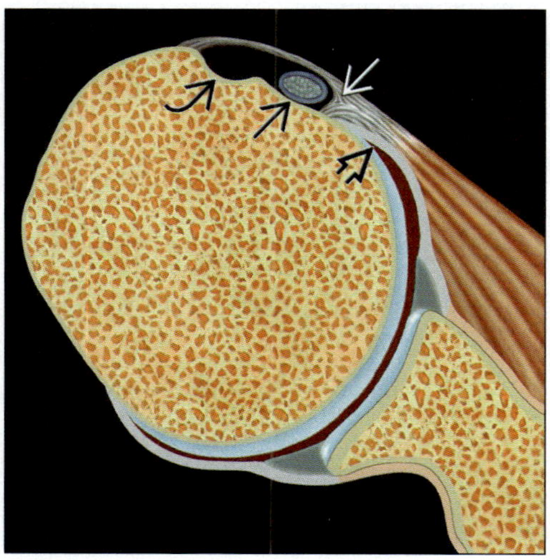

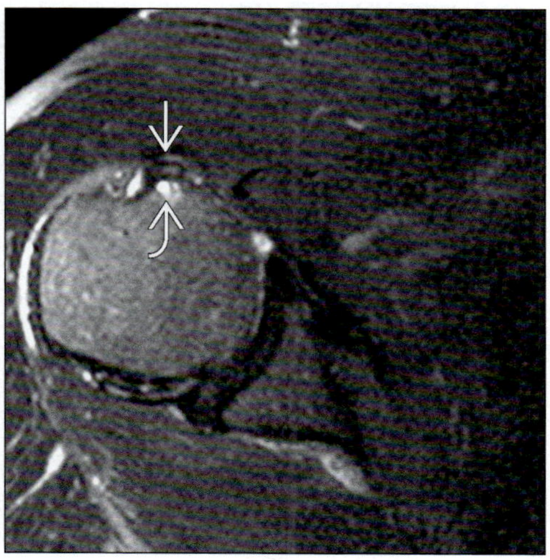

(Left) *Axial graphic of a "hidden lesion." There is medial subluxation of the biceps tendon out of the bicipital groove from a partial tear of the subscapularis tendon at its insertion and an intact joint capsule.* ***(Right)*** *Axial T2WI FS MR shows subluxation of the biceps tendon, which is perched on the lesser tuberosity. If the capsule and medial fibers of the subscapularis are intact, this is a "hidden lesion." There are also cysts in the lesser tuberosity.*

SUBSCAPULARIS TEAR

Key Facts

Terminology

- Subscapularis (SSC) tears
- Torn fibers of subscapularis tendon
  - Typically in cranial portion of tendon and 1-2 cm from lesser tuberosity

Imaging

- MR
  - High T2/fluid signal defect in tendon
  - Partial tears: Usually cranial portion near lesser tuberosity
  - Typically involves deep articular surface of tendon
  - Complete tears: No low signal tendon fibers inserting on lesser tuberosity ("naked humerus" sign: Oblique coronal anterior image with fluid covering anterior lesser tuberosity)
  - Tears seen best on axial and oblique sagittal images

Pathology

- Etiology: Trauma
  - Fall can cause isolated complete tears
  - Glenohumeral dislocation: Especially > 40 years
- Etiology: Extension of anterior supraspinatus tears
  - Common
  - Also have anterior interval tear
  - May have tear or dislocation of long head biceps tendon
- Etiology: Subcoracoid or anterosuperior impingement

Clinical Issues

- Pain, especially at night
- Weakness with internal rotation
- Anterior shoulder tenderness

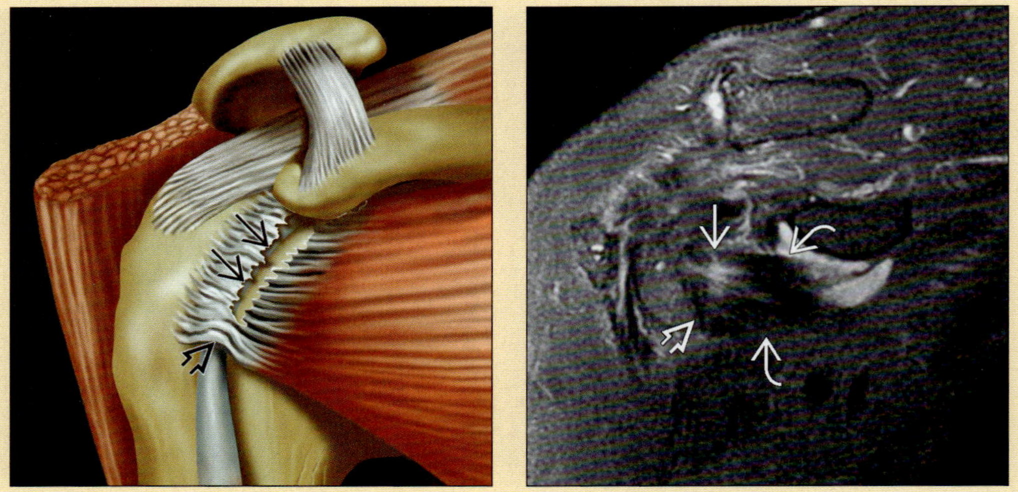

*(**Left**) AP graphic shows a partial tear of the subscapularis tendon → with some intact caudal fibers →. Most subscapularis tendon tears involve the cranial portion of the tendon and occur laterally near the insertion onto the lesser tuberosity. (**Right**) Coronal oblique T2WI FS MR through the anterior shoulder shows the subscapularis tendon → with high signal in the cranial portion → consistent with a partial tear. There are some intact fibers in the caudal portion → of the tendon.*

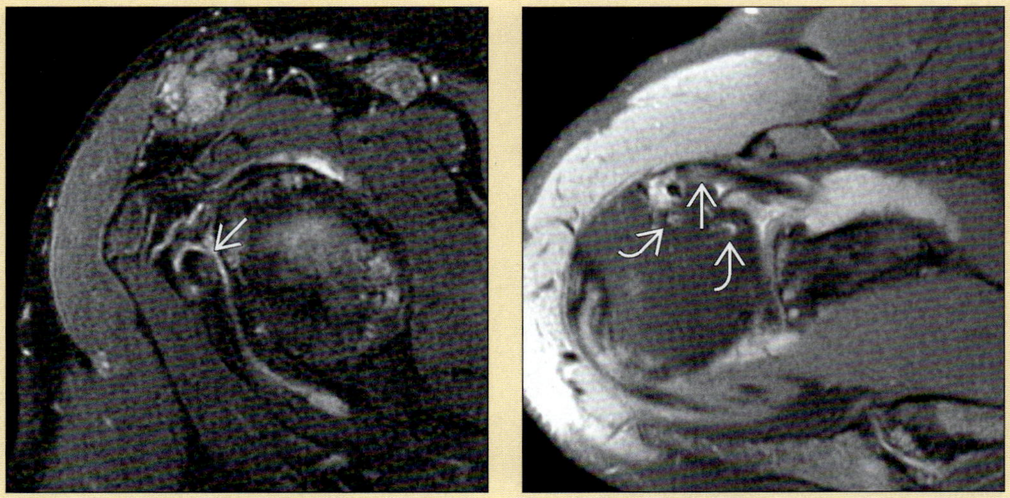

*(**Left**) Oblique sagittal T2WI FS MR shows partial subscapularis tear with high signal extending to the tendon surface in the cranial portion →. (**Right**) Axial PDWI FS MR in the same patient shows increased signal → involving the articular surface of the subscapularis tendon from a partial tear. Most subscapularis partial tears involve the articular surface of the tendon and are best seen on axial and oblique sagittal images. Note the lesser tuberosity cysts → in this chronic tear.*

SUBSCAPULARIS TEAR

TERMINOLOGY

Abbreviations
- Subscapularis (SSC) tears

Synonyms
- Subscapularis rupture, anterior cuff tear

Definitions
- Torn fibers of subscapularis tendon
  - Typically in cranial portion of tendon and 1-2 cm from lesser tuberosity
- Normal subscapularis function
  - Internal rotation
  - Adduction
  - Reinforcement of anterior shoulder capsule

IMAGING

General Features
- Best diagnostic clue
  - Focal disruption of subscapularis tendon
- Location
  - Most commonly in cranial 1/3 of tendon
  - Usually lateral near lesser tuberosity insertion
- Size
  - Most are small (≤ 1 cm) partial tears
  - Caudal portion only involved with large tears
- Morphology
  - Normal tendon forms from 4-6 collagen bundles

Radiographic Findings
- Lesser tuberosity cysts: Chronic subscapularis tear

CT Findings
- Lesser tuberosity cysts in chronic tears
- CT arthrography: Partial tears have gap in cranial lateral portion of SSC tendon

MR Findings
- T1WI
  - Partial tear: Intermediate signal in lateral tendon
  - Complete tear: Loss of low signal tendon fibers
    - Chronic tears: Atrophy of SSC muscle
    - Patients with muscle atrophy > 50% usually do poorly after attempted surgical repair
- T2WI FS
  - High signal defect in tendon
    - Partial tears: Usually cranial portion near lesser tuberosity
      - Deep > superficial tendon surface involvement
    - Complete tears: No low signal tendon fibers inserting on lesser tuberosity
      - "Naked humerus" sign: Oblique coronal anterior image with fluid covering anterior lesser tuberosity
  - Tears seen best on axial and oblique sagittal images
  - Secondary signs
    - Long head biceps tendon tear
    - Long head biceps subluxation or dislocation
    - Tear of anterior supraspinatus tendon
  - Conventional MR: Sensitivity 30% in some studies
- Direct MR arthrography
  - Contrast extending into defect in subscapularis tendon
  - Additional signs
    - Leakage of contrast onto lesser tuberosity
      - Scar tissue may prevent this
    - Fatty degeneration of subscapularis muscle
      - Usually superior portion of muscle
    - Biceps subluxation in 20-50%
  - Oblique sagittal images > axial images for specificity and accurate grading of SSC tears
  - Sensitivity = 90%, specificity = 80%

Ultrasonographic Findings
- Hypoechoic defect in subscapularis tendon

Imaging Recommendations
- Best imaging tool
  - MR with axial and oblique sagittal images
- Protocol advice
  - Fat-suppressed T2-weighted images

DIFFERENTIAL DIAGNOSIS

Subscapularis Tendinopathy
- Thickened tendon with intratendinous increased T2WI signal
- No extension of high signal to surface of tendon

Biceps Tear or Dislocation/Subluxation
- Also have pain in anterior shoulder
- May be associated with cranial SSC tears
- Can have biceps dislocation and biceps pulley tears without subscapularis tendon tear
  - Biceps pulley = coracohumeral & superior glenohumeral ligaments that stabilize biceps tendon within joint and at entrance to bicipital groove

Subscapularis Recess Effusion
- Can have partial averaging of fluid with tendon, especially on oblique coronal images
- Review of axial and oblique sagittal images will show if SSC tendon is actually torn

Anterosuperior Impingement
- Patients typically have more than 1 of the following
  - Partial subscapularis tear
  - Superior glenohumeral ligament tear
  - Anterior articular surface partial thickness supraspinatus tendon tear
  - Long head biceps tendon tear and subluxation/dislocation

Rotator Interval Tear
- Can have partial subscapularis tendon tear along with coracohumeral ligament tear and supraspinatus tendon tear

PATHOLOGY

General Features
- Etiology
  - Trauma
    - Less common than from extension of supraspinatus tear
    - Fall can cause isolated complete tears
    - Glenohumeral dislocation: Especially > 40 years old

SUBSCAPULARIS TEAR

- Anterior dislocation: Subscapularis tendon strain common
- Subscapularis tendon often thin and somewhat lax in patients with instability after anterior dislocation
- Posterior dislocation: Subscapularis tear in 14%
- Aggressive muscle contraction
  - Hard throwing
- Extension of anterior supraspinatus tears
  - Common
  - Also have anterior interval tear
  - May have tear or dislocation of long head biceps tendon
  - 30% with full thickness rotator cuff tear of supraspinatus tendon also have tear of subscapularis tendon
- Subcoracoid or anterosuperior impingement
  - Subcoracoid impingement
    - Pain with adduction, forward flexion, and internal rotation
    - Consider if coracohumeral distance < 11 mm on axial image
  - Anterosuperior impingement
    - Pain from lesser tuberosity contacting anterosuperior labrum with flexion & internal rotation
    - Tear of biceps pulley leads to medial and superior migration of humeral head
    - Also have tears of biceps tendon, anterior supraspinatus tendon, and superior glenohumeral ligament
- Post-tenotomy rupture
  - Subscapularis tendon often surgically divided during open anterior labral repair
  - Tendon can rupture at repaired tenotomy site

Staging, Grading, & Classification

- Pfirrmann grading system
  - Grade 1: Partial tear < 1/4 of craniocaudal diameter
  - Grade 2: Partial tear ≥ 1/4 of craniocaudal diameter
  - Grade 3: Complete detachment
- Fatty atrophy > 50% of muscle: Poor prognosis for repair

Gross Pathologic & Surgical Features

- Lax anterior capsule at arthroscopy
- Scar tissue can form at tear and resynovialize, so tear may not be obvious at arthroscopy

CLINICAL ISSUES

Presentation

- Most common signs/symptoms
  - Pain, especially at night
  - Weakness with internal rotation
  - Anterior shoulder tenderness
  - Usually gradual onset of symptoms and delayed presentation
- Other signs/symptoms
  - Lift-off test: Unable to lift hand against resistance from small of back
  - Belly-off sign: Unable to keep palm of hand on abdomen when positioned passively
  - Internal rotation lag sign: Arm maximally internally rotated behind small of back; patient unable to hold position

Demographics

- Age
  - Mean: 50 years old
- Gender
  - M:F = 2-3:1
- Epidemiology
  - 2-3.5% of rotator cuff tears

Natural History & Prognosis

- Partial tears can progress to larger tears
- Complete tears may form scar tissue, regain some function

Treatment

- Small partial tears: Debridement
- Large or complete tears: Surgical repair
  - If significant atrophy, may require tendon/muscle transfer

DIAGNOSTIC CHECKLIST

Consider

- Subscapularis tear if large anterior supraspinatus tear or subluxed/dislocated biceps tendon

Image Interpretation Pearls

- Evaluate cranial portion of SSC tendon
  - Caudal 1/2 of tendon is usually normal

SELECTED REFERENCES

1. Saupe N et al: Acute traumatic posterior shoulder dislocation: MR findings. Radiology. 248(1):185-93, 2008
2. Studler U et al: Abnormalities of the lesser tuberosity on radiography and MRI: association with subscapularis tendon lesions. AJR Am J Roentgenol. 191(1):100-6, 2008
3. Arrigoni P et al: Calcific tendonitis of the subscapularis tendon causing subcoracoid stenosis and coracoid impingement. Arthroscopy. 22(10):1139, 2006
4. Bergin D et al: Abnormalities on MRI of the subscapularis tendon in the presence of a full-thickness supraspinatus tendon tear. AJR Am J Roentgenol. 186(2):454-9, 2006
5. Lyons RP et al: Subscapularis tendon tears. J Am Acad Orthop Surg. 13(5):353-63, 2005
6. Richards DP et al: Relation between narrowed coracohumeral distance and subscapularis tears. Arthroscopy. 21(10):1223-8, 2005
7. Tung GA et al: Subscapularis tendon tear: primary and associated signs on MRI. J Comput Assist Tomogr. 25(3):417-24, 2001
8. Li XX et al: MR evaluation of subscapularis tears. J Comput Assist Tomogr. 23(5):713-7, 1999
9. Pfirrmann CW et al: Subscapularis tendon tears: detection and grading at MR arthrography. Radiology. 213(3):709-14, 1999

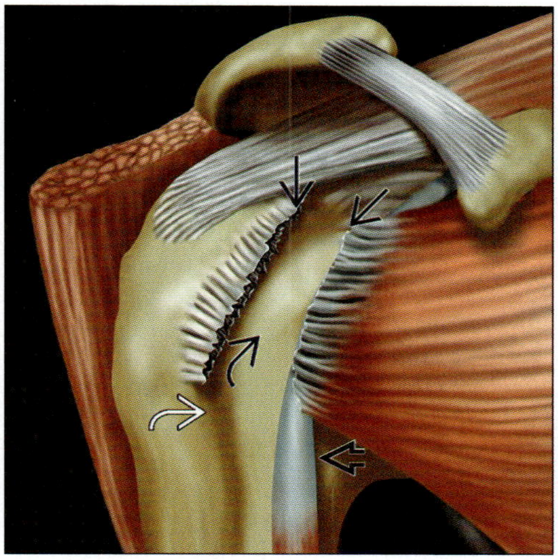

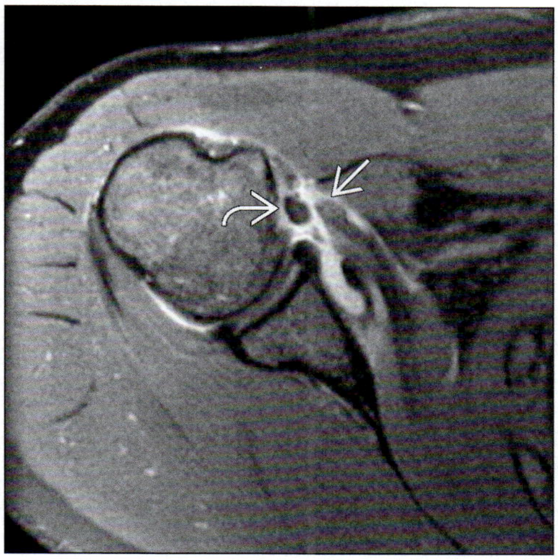

(Left) *AP graphic of a complete tear of the subscapularis tendon ➙. The tendon is avulsed from the lesser tuberosity ➙, and there is medial dislocation of the biceps tendon ➙ out of the bicipital groove ➙.* ***(Right)*** *Axial PDWI FS MR shows a complete tear ➙ of the subscapularis tendon. The long head biceps tendon is dislocated medially ➙ and has increased signal consistent with tendinopathy. Biceps abnormalities are present in about 1/2 of patients with subscapularis tears.*

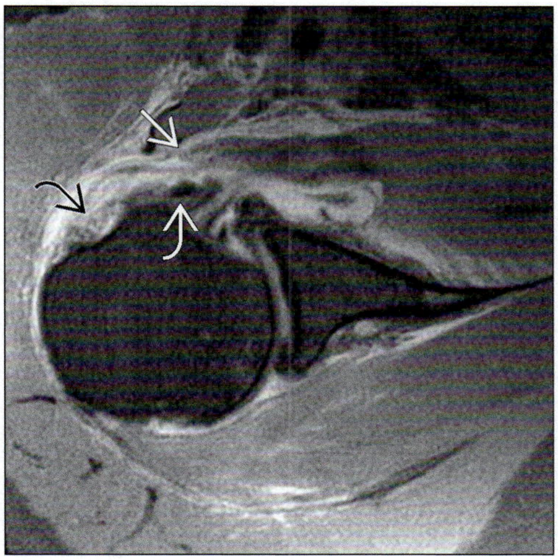

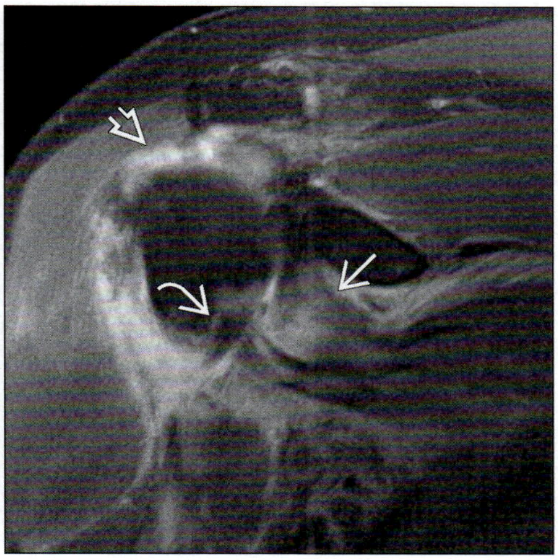

(Left) *Axial PDWI FS MR shows a complete subscapularis tendon tear ➙. There is also medial dislocation of the long head biceps tendon ➙ and an empty bicipital groove ➙.* ***(Right)*** *Coronal oblique T2WI FS MR in the same patient shows a full thickness tear ➙ of the supraspinatus tendon, subscapularis tear ➙, and dislocated biceps tendon ➙. 2/3 of patients with subscapularis tear also have a supraspinatus tear, usually involving the anterior portion of the tendon.*

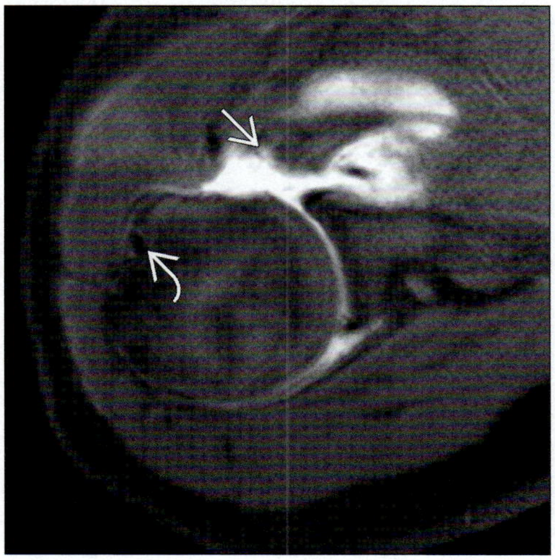

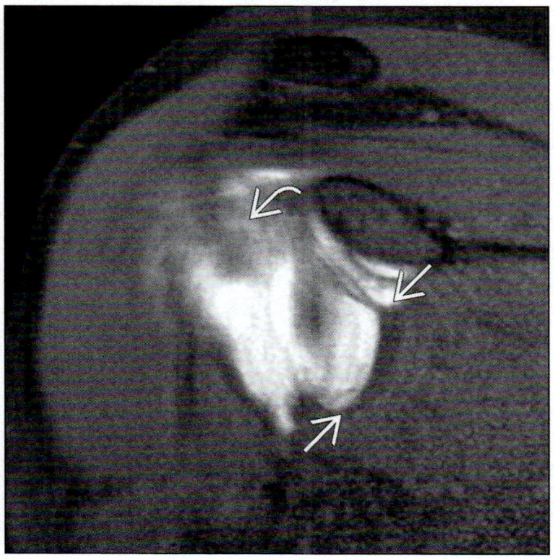

(Left) *Axial MR arthrogram T1WI FS shows a complete tear of the subscapularis tendon ➙ in a patient with a prior anterior shoulder dislocation. The long head biceps tendon ➙ is located in normal position in the bicipital groove.* ***(Right)*** *Coronal oblique MR arthrogram T1WI FS through the anterior shoulder in the same patient shows a "naked humerus" sign with fluid adjacent to the lesser tuberosity ➙. The torn edge of the subscapularis tendon ➙ is retracted medially.*

ROTATOR CUFF POSTOPERATIVE REPAIR

Key Facts

Terminology

- Debridement or repair of rotator cuff tear

Imaging

- Radiograph: Decompression ostectomy changes, suture anchors in greater tuberosity
- MR: Normal postoperative appearance
  - Metal artifact
  - Polymer or bioabsorbable tacks & anchors
  - Only 10% of post-op tendons appear normal
  - Residual cuff defect < 1 cm common
  - SA/SD effusion common in asymptomatic patients
  - Shoulder effusion in 90%
- Recurrent tear
  - Absence of tendon
  - Fluid signal in > 1 cm defect
- Direct MR arthrography
  - 90% accuracy for full thickness tears
  - Specificity for recurrent partial tear = 25%
- Other complications
  - Loose suture anchor or tack
  - Deltoid dehiscence
  - Inadequate decompression
  - Adhesions from scar/fibrosis

Clinical Issues

- 25% of patients are symptomatic after rotator cuff surgery
- Causes of pain other than cuff re-tear: Hematoma, adhesions, etc.

Diagnostic Checklist

- Use term "defect" rather than "recurrent cuff tear" if tendon discontinuity with fluid signal measures < 1 cm diameter

*(**Left**) Lateral graphic shows an open repair of a full thickness supraspinatus tendon tear. A trough is made in the lateral greater tuberosity →, and the torn edge of the cuff ↪ is sutured into the trough. (**Right**) Oblique sagittal T2WI FS MR in a patient with a prior mini-open rotator cuff repair. There is residual high signal at the trough site ➔ and micrometallic artifact ➔ from the burr used for decompression. No fluid-signal defect to indicate a re-tear is seen on this image.*

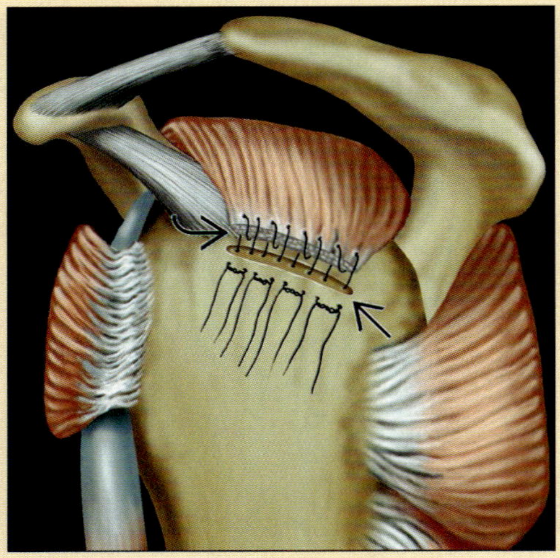

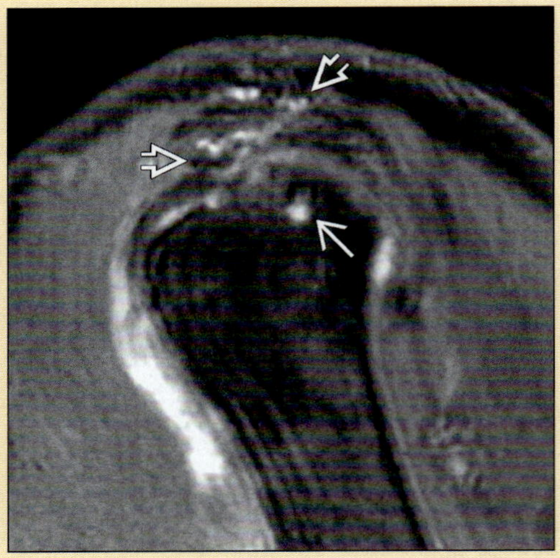

*(**Left**) Coronal oblique T2WI FS MR in the same patient shows normal postoperative appearance of the repaired cuff, with minimal increased signal ➔ at the repair site. Also seen is signal elevation in the trough ↪ and micrometallic artifact ➔. (**Right**) Coronal oblique T2W FS MR adjacent images in a patient with a prior arthroscopic rotator cuff repair. The bioabsorbable tack ➔ is in normal position, and there is no re-tear of the supraspinatus tendon ↪. There is a SA/SD effusion ➔.*

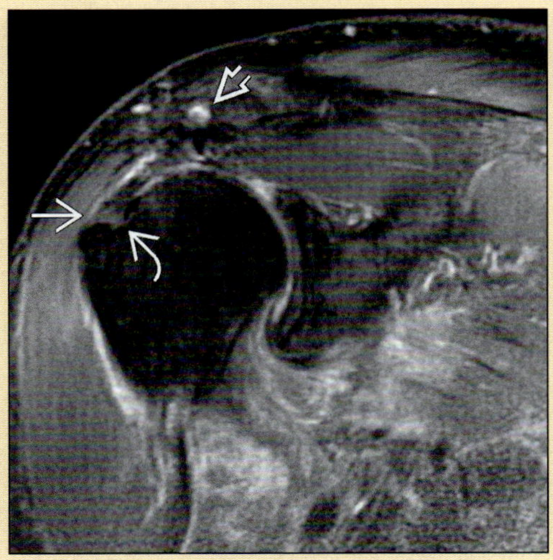

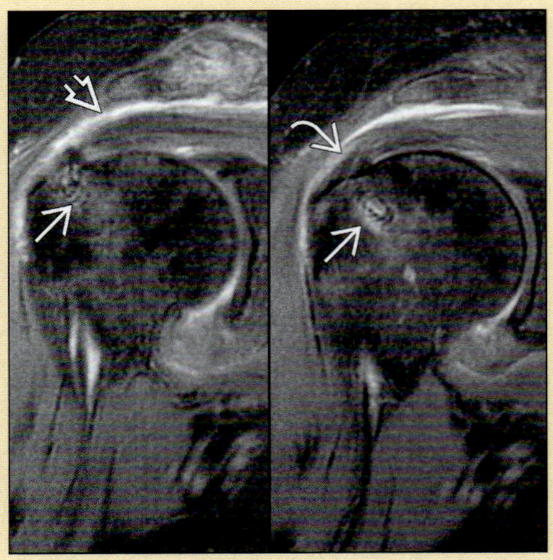

ROTATOR CUFF POSTOPERATIVE REPAIR

TERMINOLOGY

Abbreviations

- Anterior subacromial decompression (ASAD)
- Full thickness rotator cuff tear (FTRCT)

Synonyms

- Cuff debridement used for partial tears

Definitions

- Debridement or repair of rotator cuff tear
  - Usually with coracoacromial arch decompression
  - Arthroscopic, mini-open, or open repairs
- Standard surgical procedures
  - Debridement of shallow partial thickness tears (< 1/2 to 2/3 thickness of cuff)
  - Repair of full thickness or deep partial thickness tears
    - Bone trough (transosseous)
    - Suture anchors
    - Bioabsorbable tacks
  - Anterior subacromial decompression (acromioplasty)
  - Acromioclavicular joint resection (Mumford procedure)
  - Coracoacromial ligament resection
- Postoperative integrity of cuff does not necessarily equate to lack of symptoms

IMAGING

General Features

- Best diagnostic clue
  - Radiograph: Decompression ostectomy changes, suture anchors in greater tuberosity
  - MR: Micrometallic artifacts
- Location
  - Supraspinatus tendon most common
- Size
  - Cuff defects < 1 cm common post-op

Radiographic Findings

- Decompression of coracoacromial arch structures
  - Anterior subacromial decompression
  - Acromioclavicular joint resection (Mumford procedure)
- Suture anchors in greater tuberosity
  - Presume loosening if anchor outside of bone
- Trough in greater tuberosity if mini-open repair
- Mild superior subluxation of humeral head due to
  - Capsular tightening
  - Scarring
  - Cuff atrophy
  - Bursectomy

CT Findings

- CT arthrography
  - Alternative to MR if concern for significant metal artifact
  - Contrast in subacromial/subdeltoid (SA/SD) bursa common post-op
    - Surgeon often does not obtain watertight seal
  - Re-tear if > 1 cm defect in cuff

MR Findings

- T1WI
  - Expect increased signal within repaired cuff
  - Trough in lateral greater tuberosity if mini-open repair
- T2WI FS
  - **Normal postoperative appearance**
    - Metal artifact
      - Micrometallic artifact from burr used for decompression
      - Metallic suture anchors
    - Polymer or bioabsorbable tacks & anchors
  - Only 10% of post-op tendons appear normal
    - Increased T2 signal in normal post-op tendon in > 50%
      - Intratendinous high signal persists for long time post-op, mimics residual/recurrent tear
      - High signal granulation tissue around sutures
    - Residual cuff defect < 1 cm common
    - Cuff defects in 20% of asymptomatic and 50% of symptomatic patients
    - SA/SD effusion common in asymptomatic patients (67%)
    - Shoulder joint effusion in 90%
    - Loss of peribursal fat
    - Low signal intensity marrow in remaining anterior acromion from marrow fibrosis
    - Coracoacromial arch decompression
      - Flattened undersurface anterior acromion
      - Wide AC joint
  - **Recurrent tear**
    - Can be difficult to accurately diagnose re-tear with MR, especially partial tear
    - Cardinal signs
      - Absence of tendon
      - Fluid signal in > 1 cm defect
    - 80-90% accuracy for recurrent FTRCT
    - Low specificity for recurrent partial thickness tears (25%)
      - Normal postoperative cuff thinning can mimic partial tear
    - Poor accuracy for correctly determining size of recurrent tears
      - MR underestimates re-tear size
    - Helpful to know size of original tear, and what type of surgery performed
  - **Other complications**
    - Loose suture anchor or tack
      - Loose hardware can cause pain, re-tear
    - Deltoid dehiscence
      - Can occur after open repair
    - Inadequate decompression
      - Continued hooked anterior acromion, inferior osteophytes
    - Acromion fracture
    - Teres minor and deltoid atrophy if injury to anterior branch of axillary nerve
    - Infection, hematoma
    - Adhesions from scar/fibrosis
- T2* GRE
  - Metal artifact worse
    - Intravoxel dephasing T2* effect
- **Direct MR arthrography**
  - Best evaluation of post-op tendon

ROTATOR CUFF POSTOPERATIVE REPAIR

- 90% accuracy for full thickness tears
- Contrast in SA/SD: < 50% specificity and only 92% sensitivity for FTRCT
  - Repairs often not watertight but functional
  - Poor quality scar tissue can form at re-tear site, preventing opacification of SA/SD bursa
- Partial tears: 50% sensitivity
  - Difficult to distinguish thinned cuff from partial tear

Ultrasonographic Findings

- Anechoic region replacing isoechoic tendon
- 80-90% accuracy for recurrent rotator cuff tear

Imaging Recommendations

- Best imaging tool
  - Direct MR arthrography
- Protocol advice
  - Avoid gradient echo images

DIFFERENTIAL DIAGNOSIS

Normal Postoperative Cuff with High Signal ± Small Communicating Defect

- Fluid signal defect in cuff should be < 1 cm to be considered normal postoperative appearance

Adhesive Capsulitis

- Often from inadequate postoperative physical therapy

Secondary Impingement from Instability

- Capsulorrhaphy required to resolve impingement symptoms

Biceps Tendinopathy/Dislocation

- Biceps pathology often associated with cuff tears
- May not be adequately addressed at time of cuff repair

Glenohumeral Joint Osteoarthritis

- Associated with chronic full thickness tears

PATHOLOGY

General Features

- Etiology
  - Re-tear
    - Re-injury of incompletely healed tendon
    - Failure of sutures or loose surgical tacks

Staging, Grading, & Classification

- Re-tear: Bateman grade
  - Grade 1 (small): < 1 cm (can be normal postoperative defect)
  - Grade 2 (medium): 1-3 cm
  - Grade 3 (large): 3-5 cm
  - Grade 4 (massive): > 5 cm

Gross Pathologic & Surgical Features

- Tendon heals with fibrous tissue

CLINICAL ISSUES

Presentation

- Most common signs/symptoms
  - 25% of patients are symptomatic after rotator cuff surgery
    - May not have re-tear of cuff: Hematoma, adhesions, etc.
    - Re-tear more likely if large initial tear and muscle atrophy or fatty replacement (Goutallier stage > 1)
  - **Re-tear**
    - Pain, especially at night
    - Decreased abduction strength

Natural History & Prognosis

- Poor prognosis for large re-tears, especially if associated with muscle atrophy and fatty replacement

Treatment

- Post-op pain: Physical therapy
- Revision repair for re-tear if no significant atrophy or retraction, and solid tendon edges without fragmentation

DIAGNOSTIC CHECKLIST

Image Interpretation Pearls

- Use term "defect" rather than "recurrent cuff tear" if tendon discontinuity with fluid signal measures < 1 cm diameter

SELECTED REFERENCES

1. McMenamin D et al: Imaging of the shoulder after surgery. Eur J Radiol. 68(1):106-19, 2008
2. Liem D et al: Magnetic resonance imaging of arthroscopic supraspinatus tendon repair. J Bone Joint Surg Am. 89(8):1770-6, 2007
3. Potter HG et al: Imaging of the rotator cuff following repair: human and animal models. J Shoulder Elbow Surg. 16(5 Suppl):S134-9, 2007
4. Duc SR et al: Diagnostic performance of MR arthrography after rotator cuff repair. AJR Am J Roentgenol. 186(1):237-41, 2006
5. Mellado JM et al: MR assessment of the repaired rotator cuff: prevalence, size, location, and clinical relevance of tendon rerupture. Eur Radiol. 16(10):2186-96, 2006
6. Wu J et al: MRI of the postoperative shoulder. Clin Sports Med. 25(3):445-64, vi, 2006
7. Mellado JM et al: Surgically repaired massive rotator cuff tears: MRI of tendon integrity, muscle fatty degeneration, and muscle atrophy correlated with intraoperative and clinical findings. AJR Am J Roentgenol. 184(5):1456-63, 2005
8. Mohana-Borges AV et al: MR imaging and MR arthrography of the postoperative shoulder: spectrum of normal and abnormal findings. Radiographics. 24(1):69-85, 2004
9. Zlatkin MB: MRI of the postoperative shoulder. Skeletal Radiol. 31(2):63-80, 2002
10. Zanetti M et al: MR imaging after rotator cuff repair: full-thickness defects and bursitis-like subacromial abnormalities in asymptomatic subjects. Skeletal Radiol. 29(6):314-9, 2000
11. Spielmann AL et al: Shoulder after rotator cuff repair: MR imaging findings in asymptomatic individuals--initial experience. Radiology. 213(3):705-8, 1999
12. Gusmer PB et al: MR imaging of the shoulder after rotator cuff repair. AJR Am J Roentgenol. 168(2):559-63, 1997
13. Magee TH et al: MR imaging of the shoulder after surgery. AJR Am J Roentgenol. 168(4):925-8, 1997

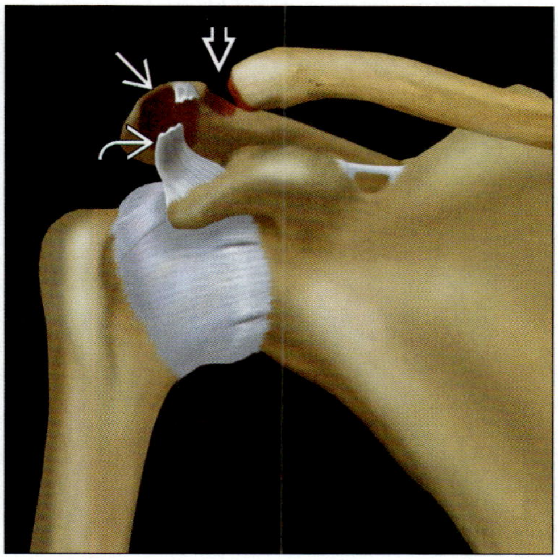

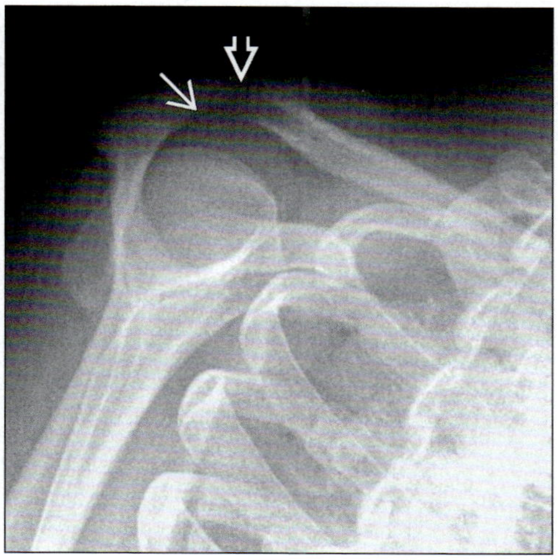

(Left) *AP graphic shows coracoacromial arch decompression, often done at the same time as cuff debridement/repair. There is an anterior acromioplasty ➔, acromioclavicular joint resection ➔, and lysis of the coracoacromial ligament ➔.* ***(Right)*** *Arch view shows an anterior acromioplasty ➔ and acromioclavicular joint resection ➔. Flattening the acromial undersurface and removing inferior AC joint osteophytes helps the postoperative cuff heal as well as reduce impingement pain.*

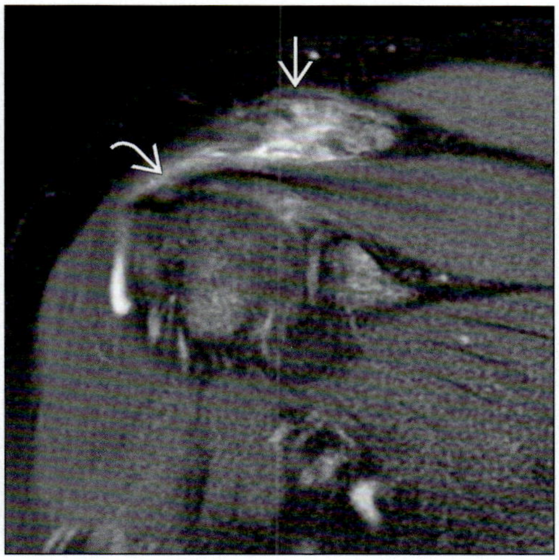

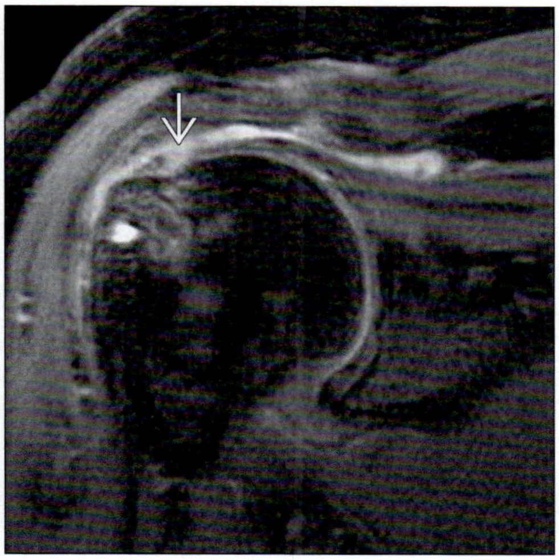

(Left) *Coronal oblique T2WI FS MR of the same patient, 2 months after anterior subacromial decompression and AC joint resection. There is increased signal fibrovascular tissue at the resected AC joint ➔ but an intact rotator cuff ➔.* ***(Right)*** *Coronal oblique T2WI FS MR shows normal healing tendon at 2nd look arthroscopy. There is high, but not fluid, signal in the supraspinatus tendon ➔. T2 high signal in the tendon can represent normal healing fibrous/granulation tissue.*

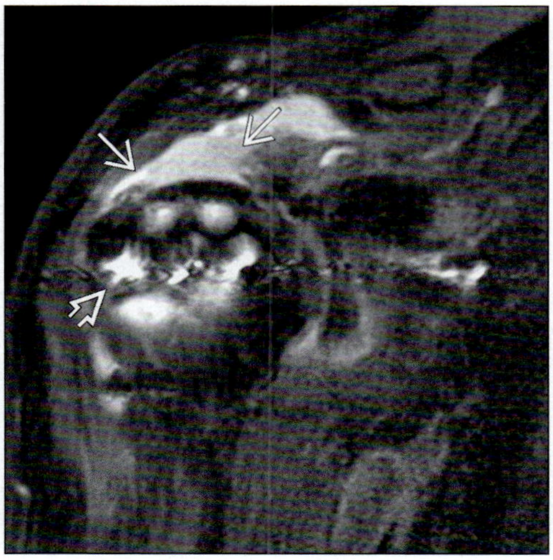

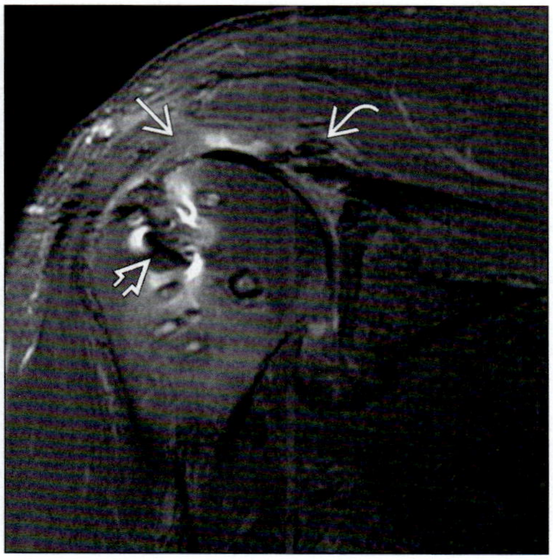

(Left) *Coronal oblique T2WI FS MR of cuff re-tear shows fluid defect > 1 cm ➔ in the supraspinatus tendon. There is artifact ➔ from 3 metal suture anchors in the humeral head.* ***(Right)*** *Coronal oblique T2WI FS MR of a re-tear shows > 1 cm defect in supraspinatus tendon ➔. The tendon edge is retracted but remains lateral to the glenoid rim ➔. Polymer tacks and metal artifact from suture anchors ➔ are seen. Fluid gap and absent tendon are the 2 main signs of a cuff re-tear.*

ROTATOR CUFF POSTOPERATIVE REPAIR

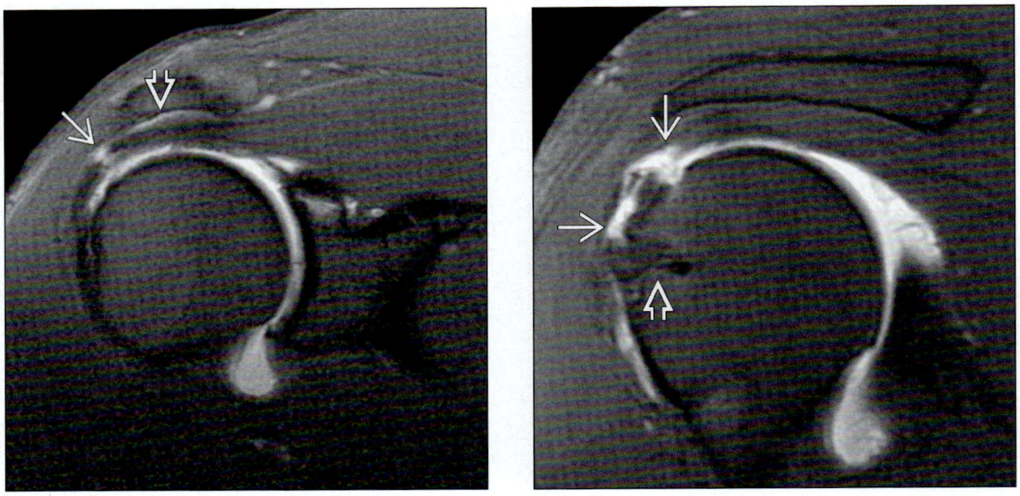

(Left) *Oblique coronal MR arthrogram T1WI FS after cuff repair shows a subcentimeter full thickness defect ➔ in the supraspinatus tendon, with a small amount of contrast in the SA bursa ➔. Such small defects are often present after surgery and do not necessarily warrant surgical repair.* ***(Right)*** *Coronal oblique MR arthrogram T1WI FS shows a 2 cm high-grade partial thickness re-tear ➔ after cuff repair ➔. There is no extension of contrast into the SA/SD bursa. Surgical repair was performed.*

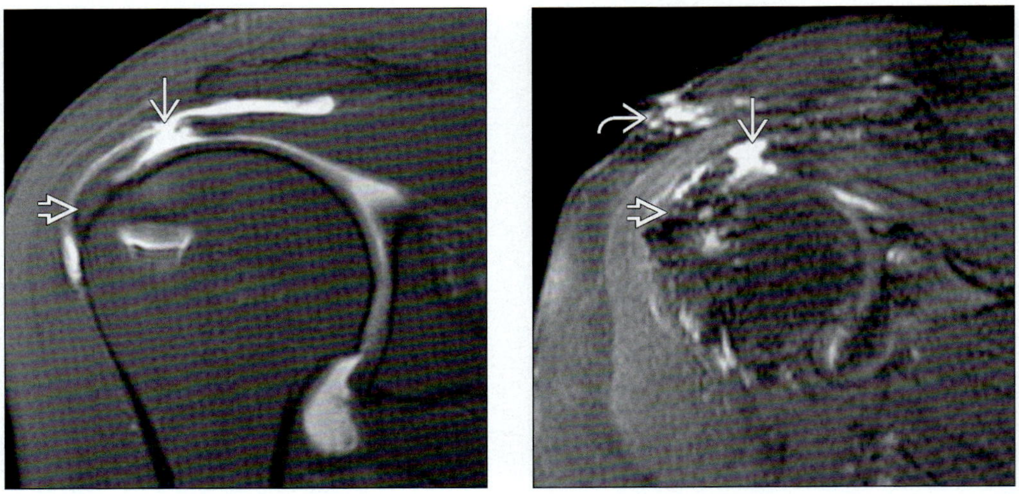

(Left) *Coronal MR arthrogram T1WI FS shows a small critical zone tear ➔ of the supraspinatus tendon after rotator cuff repair performed more distally ➔. Re-tear of the cuff at a location different than the operative site is not uncommon.* ***(Right)*** *Coronal oblique T2WI FS MR from a different patient shows a critical zone tear ➔ after previous rotator cuff repair at the insertional footprint on the humeral head ➔. Micrometallic artifact from previous acromioplasty ➔ is also seen.*

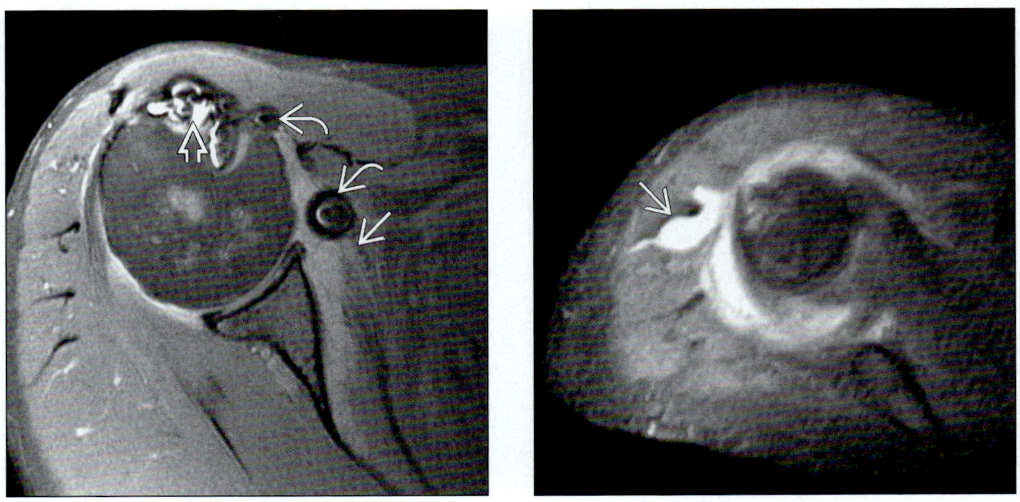

(Left) *Axial PDWI FS MR after repair of the supraspinatus and subscapularis shows a complete re-tear of the subscapularis tendon ➔ with retraction and atrophy of the muscle belly. Artifact from the previous repair is seen along the course of the subscapularis ➔ and in the humeral head ➔.* ***(Right)*** *Axial MR arthrogram T1WI FS several weeks after cuff repair shows dissection of injected contrast into the surgical cleft of the deltoid incision ➔. The patient complained of pain in this area.*

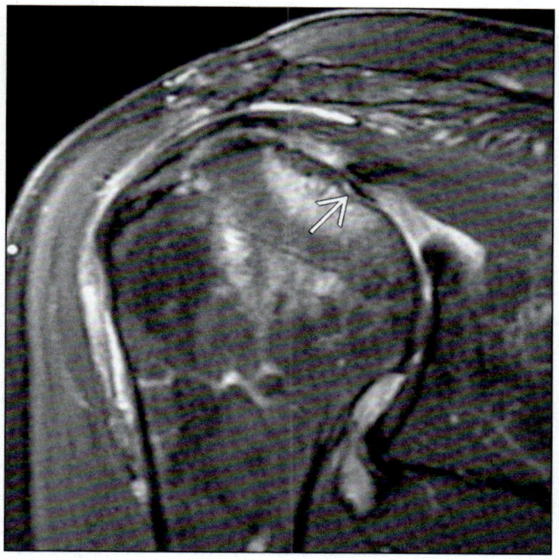
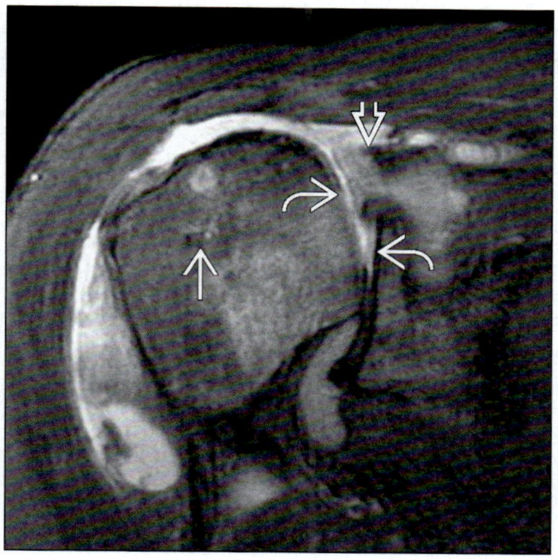

(Left) *Coronal T2WI FS MR 3 weeks after cuff repair shows a subchondral fracture → of the humeral head, presumably on the basis of intraoperative trauma.* ***(Right)*** *Coronal T2WI FS MR 6 months after rotator cuff repair shows a pin track →. There is a massive re-tear of the retracted cuff → and a joint effusion extends into the SA/SD bursa. Accelerated osteoarthritis has resulted in compete cartilage loss on both sides of the shoulder joint →. Shoulder aspiration showed no infection.*

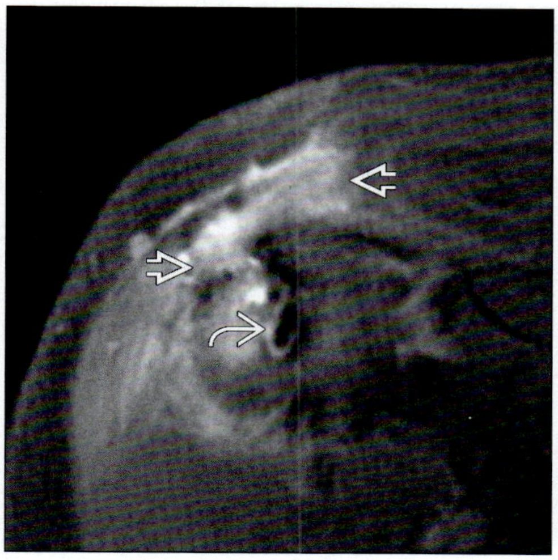
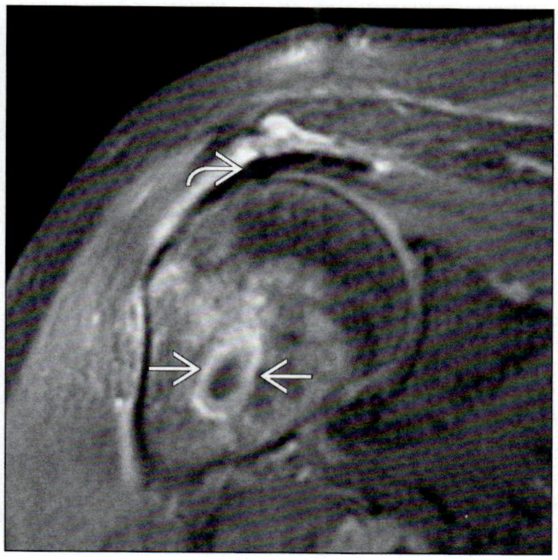

(Left) *Coronal T2WI FS MR in a 54-year-old man after rotator cuff repair shows artifact → from cuff fixation and a large high-signal gap → in the expected course of the repaired supraspinatus tendon.* ***(Right)*** *Coronal T2WI FS MR more anteriorly in the same patient shows an intraosseous abscess → in the humerus surrounded by mild marrow edema. The infraspinatus portion of the cuff → remains intact. The patient had a sinus tract draining to the superior skin surface.*

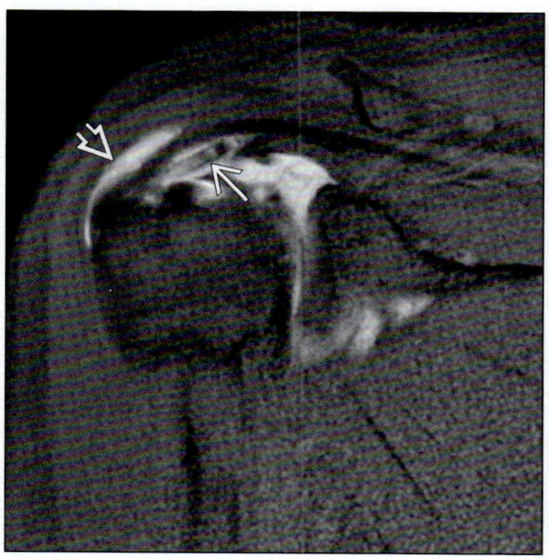
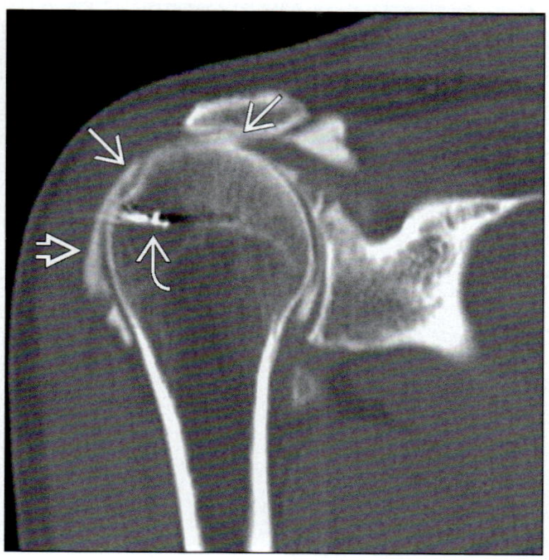

(Left) *Coronal oblique MR arthrogram T1WI FS image in a patient with a loose polymer tack → shows contrast in the SA/SD bursa → from a recurrent FTRCT. Tear was seen on a more anterior image at the site where the tack failed.* ***(Right)*** *Coronal oblique CT arthrogram shows a large defect in the supraspinatus tendon → with contrast in the subacromial/subdeltoid bursa →, indicating a re-tear of the rotator cuff. A metallic suture anchor is present in the greater tuberosity →.*

CALCIFIC ROTATOR CUFF TENDINOPATHY

Key Facts

Terminology

- Calcium hydroxyapatite deposition (Ca HAD)
- Calcific "tendinitis" and bursitis, calcium HAD crystal disease, calcifying tendinitis

Imaging

- Supraspinatus tendon involved in 80%
- MR
  - Slight "blooming" artifact around low signal calcification
  - Tendinopathy: ± band of edema around calcification
  - Calcific bursitis: Intrabursal low signal Ca, effusion, peribursal inflammation
- Ultrasound
  - Echogenic foci within tendon or bursa
  - Best modality for image-guided lavage

Top Differential Diagnoses

- Hyperparathyroidism
- Tumoral (idiopathic) calcinosis
- Pyrophosphate arthropathy

Pathology

- Fluffy calcification: Acute, painful
- Solid & well-defined Ca HAD: Chronic, often asymptomatic

Clinical Issues

- Pain worse at night
- Radiates to deltoid insertion
- Treatment
  - Ultrasound-guided aspiration/lavage
  - Corticosteroid in subacromial/subdeltoid (SA/SD) bursa

*(**Left**) AP graphic shows a calcification → within the subacromial/subdeltoid (SA/SD) bursa ↪. The most common form of tendon calcification around the shoulder is calcium hydroxyapatite, and the supraspinatus tendon and adjacent SA/SD bursa are the most common sites of deposition. (**Right**) AP radiograph shows calcification in the supraspinatus tendon ➡ in a patient with shoulder pain. The calcification is somewhat fluffy appearing, which is common in painful calcific tendinopathy.*

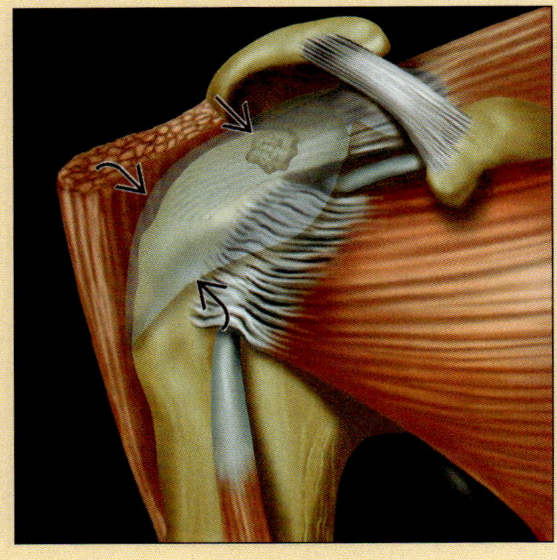

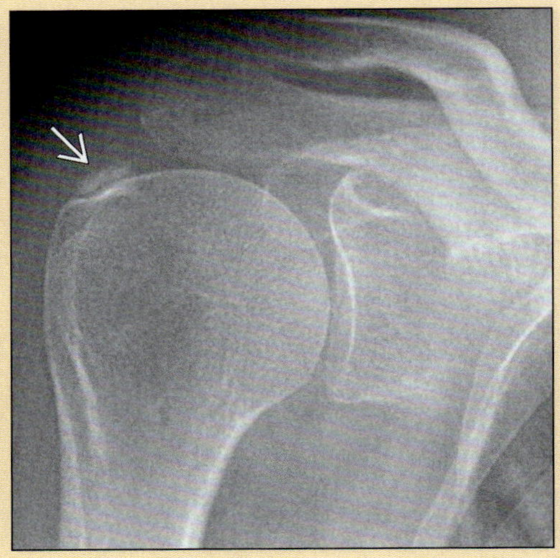

*(**Left**) Coronal oblique T2WI FS MR in the same patient shows the markedly low signal calcium within the supraspinatus tendon ➡. There is high signal synovitis or fluid in the SA/SD bursa consistent with bursitis ↪, which is often present in the painful phase of calcific tendinopathy. (**Right**) Adjacent coronal oblique T2WI FS MR in the same patient shows low signal calcium within the tendon ➡. The low signal often appears slightly larger on MR than radiograph due to "blooming" artifact.*

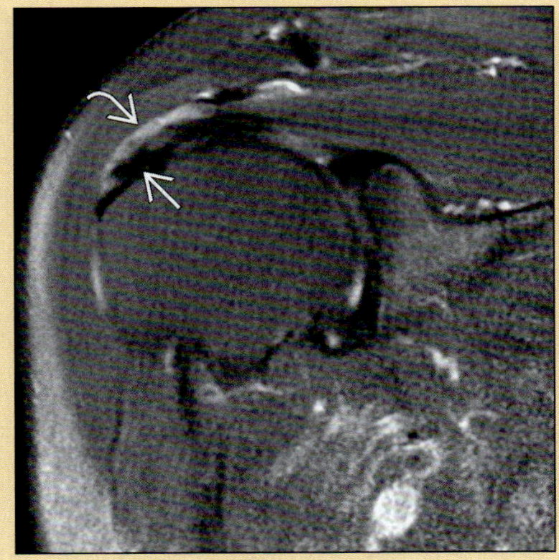

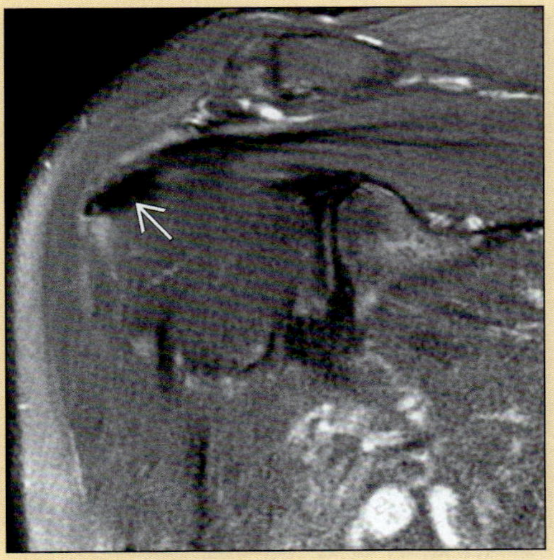

CALCIFIC ROTATOR CUFF TENDINOPATHY

TERMINOLOGY

Abbreviations

- Calcium hydroxyapatite deposition (Ca HAD)
- Calcium pyrophosphate dihydrate (CPPD)

Synonyms

- Calcific "tendinitis" and bursitis, calcifying tendinitis, calcific tendinosis

Definitions

- Ca HAD in shoulder tendons and bursa

IMAGING

General Features

- Best diagnostic clue
  - Calcification cranial to humeral head
- Location
  - Supraspinatus tendon: 80%
  - Other cuff tendons, subdeltoid bursa

Radiographic Findings

- Calcification within rotator cuff tendon
- ± sclerosis or erosions of tuberosity
- "Milwaukee shoulder": HAD and CPPD with glenohumeral joint destruction

MR Findings

- T1WI
  - ↓ signal with slight "blooming" artifact
- T2WI FS
  - Tendinopathy
    - Low signal calcium deposit
    - Band of ↑ signal edema around deposit
  - Bursitis
    - Low signal intrabursal calcium
    - SA/SD bursa effusion, peribursal inflammation
  - 70% accuracy

Ultrasonographic Findings

- Echogenic foci within tendon or bursa
- Best modality for image-guided lavage

Imaging Recommendations

- Best imaging tool
  - Radiographs
- Protocol advice
  - Multiple views to localize calcification

DIFFERENTIAL DIAGNOSIS

Hyperparathyroidism

- Bone resorption (subperiosteal, acroosteolysis, etc.)

Loose Bodies

- Glenohumeral degenerative joint disease

Tumoral (Idiopathic) Calcinosis

- Painless, 15-30 years old

PATHOLOGY

General Features

- Etiology
  - Ischemia
  - Repetitive microtrauma
  - Metabolic causes: Familial, renal failure

Staging, Grading, & Classification

- Uhthoff & Loehr
  - Precalcific stage: Fibrocartilaginous changes
  - Formative phase: Fluffy calcification
  - Resting phase: May have mechanical symptoms
  - Resorptive phase: Fluffy, inflamed, & painful
  - Postcalcific stage: Tendon heals

Gross Pathologic & Surgical Features

- Whitish deposit in hyperemic tendon & bursa

CLINICAL ISSUES

Presentation

- Most common signs/symptoms
  - Shoulder pain worse at night
    - Radiates to deltoid insertion
  - Tender to palpation
  - Chronic: Restricted range of motion

Demographics

- Age
  - 40-70 years old
- Gender
  - M < F
- Epidemiology
  - Present in 3-7% of population, 1/3 symptomatic
  - Bilateral in 23%

Natural History & Prognosis

- Self limited, painful during resorption
- Intratendinous HAD may rupture into SA/SD bursa

Treatment

- Ultrasound-guided aspiration/lavage
  - Follow with corticosteroid into SA/SD bursa
- Extracorporeal shock wave therapy
- Arthroscopic removal of HAD

DIAGNOSTIC CHECKLIST

Image Interpretation Pearls

- Easy to miss on MR; radiographs required

Reporting Tips

- Identify which tendon, or if bursa is involved

SELECTED REFERENCES

1. Serafini G et al: Rotator cuff calcific tendonitis: short-term and 10-year outcomes after two-needle us-guided percutaneous treatment--nonrandomized controlled trial. Radiology. 252(1):157-64, 2009
2. Siegal DS et al: Calcific tendinitis: a pictorial review. Can Assoc Radiol J. 60(5):263-72, 2009
3. del Cura JL et al: Sonographically guided percutaneous needle lavage in calcific tendinitis of the shoulder: short- and long-term results. AJR Am J Roentgenol. 189(3):W128-34, 2007

CALCIFIC ROTATOR CUFF TENDINOPATHY

(Left) AP radiograph shows calcification lateral and inferior to the greater tuberosity, consistent with HAD within the SA/SD bursa. Bursal calcification often occurs when calcific tendinopathy ruptures into the bursa during the painful active phase. **(Right)** Coronal oblique T2WI FS MR in the same patient shows the calcification within the bursa. Although the calcification is fairly solid appearing on x-ray, the presence of edema and fluid in the bursa indicates it may be causing symptoms.

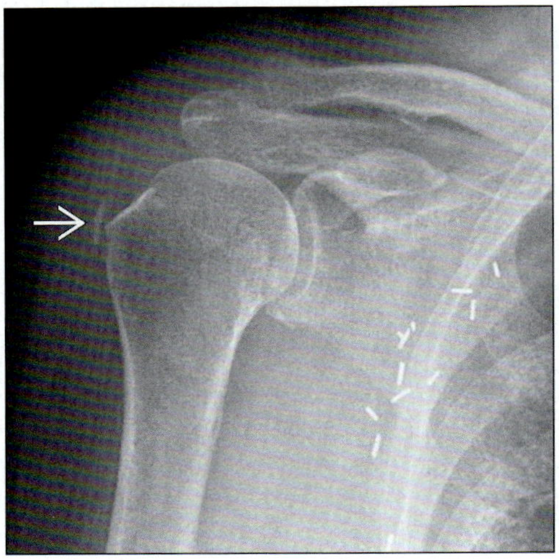

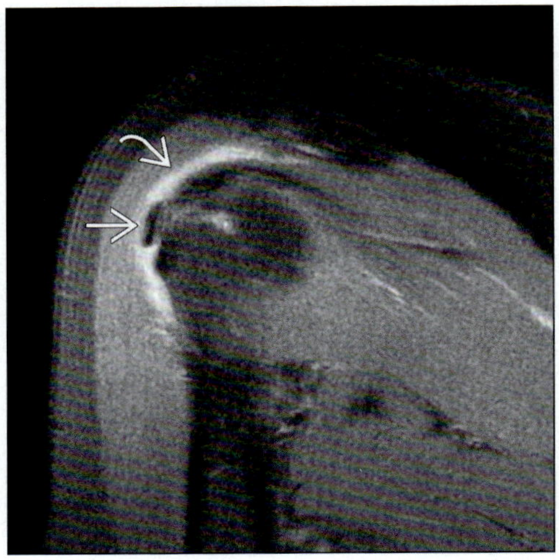

(Left) Axillary lateral radiograph in a patient with anterior shoulder pain shows a slightly fluffy calcification medial to the lesser tuberosity. **(Right)** Axial PDWI FS MR in the same patient shows the low signal calcification within the subscapularis tendon, with adjacent tendinopathy and mild bursitis. Calcific tendinopathy can involve any of the rotator cuff tendons and, when symptomatic, is often associated with adjacent synovitis or bursitis.

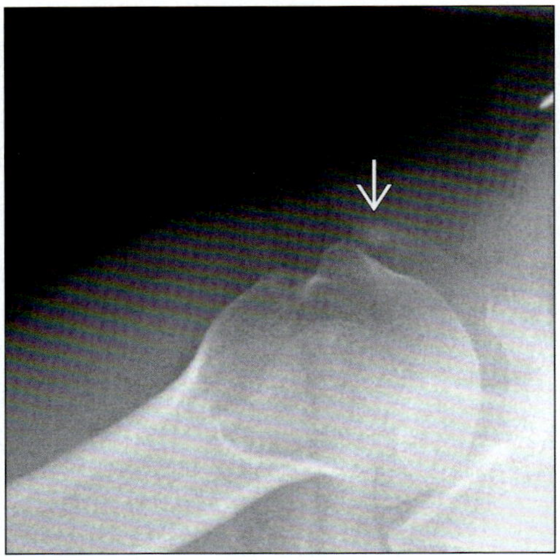

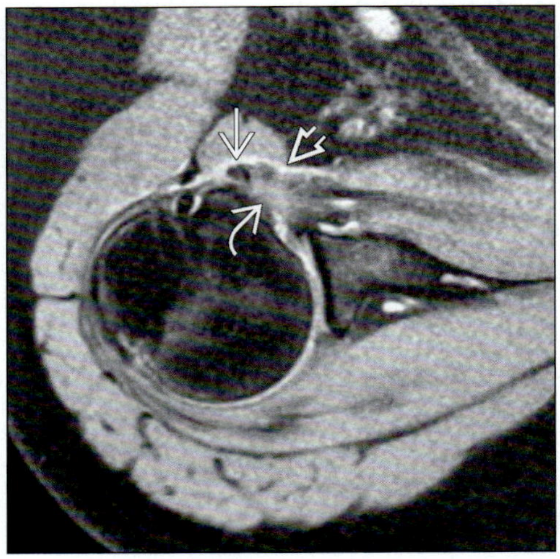

(Left) AP graphic of severe calcific bursitis shows multiple calcifications within the SA/SD bursa, as well as in the peribursal space. During the active phase, calcium HAD crystals can rupture out of the bursa into the peribursal soft tissue. **(Right)** AP radiograph shows calcification both in the supraspinatus tendon and the subdeltoid bursa. Just as calcific bursitis can rupture into the soft tissue, calcific tendinopathy often ruptures into the bursa during the painful active phase.

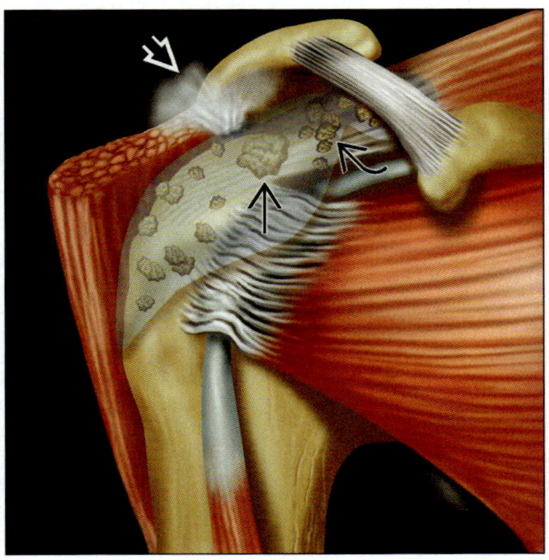

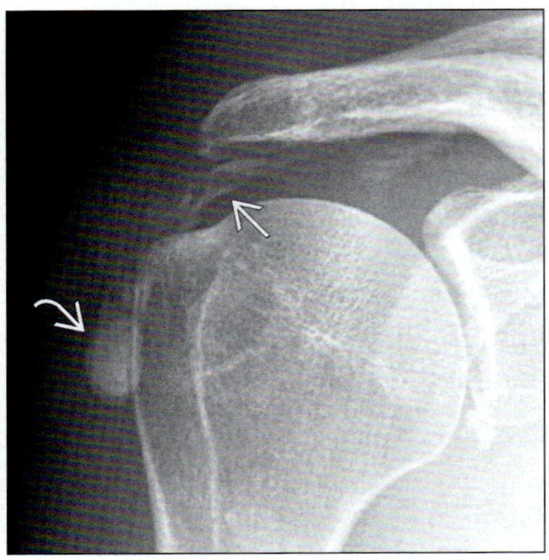

CALCIFIC ROTATOR CUFF TENDINOPATHY

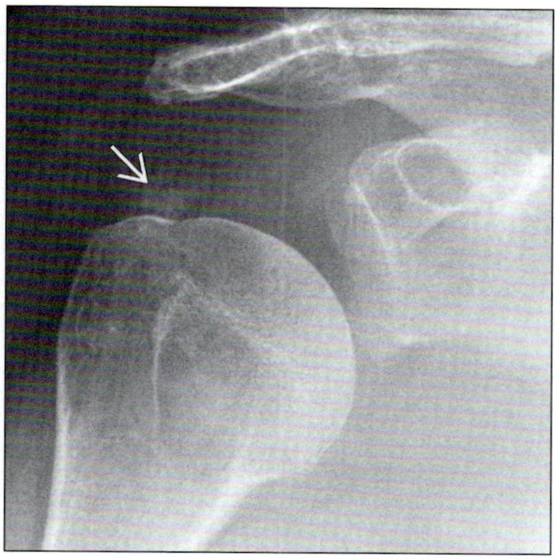

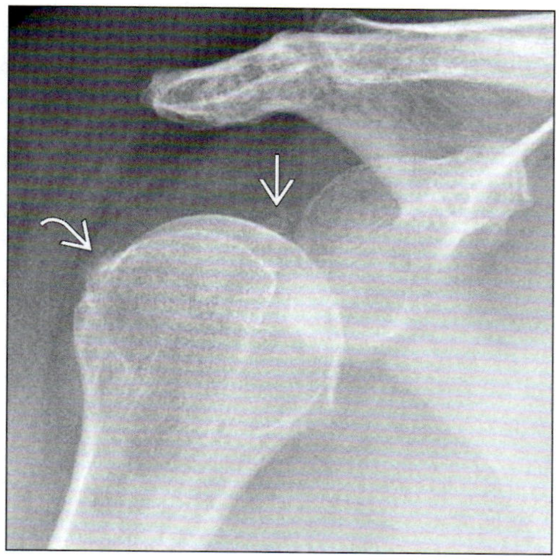

(Left) *AP radiograph shows fluffy calcification* → *probably within the supraspinatus tendon, concerning for calcific tendinopathy.* ***(Right)*** *AP internal rotation radiograph in the same patient shows the calcification moving with the supraspinatus tendon* →*, confirming that this is calcific tendinopathy. Bursal calcifications typically do not change position with internal-external rotation. The patient also has calcific tendinopathy of the infraspinatus tendon* ↷.

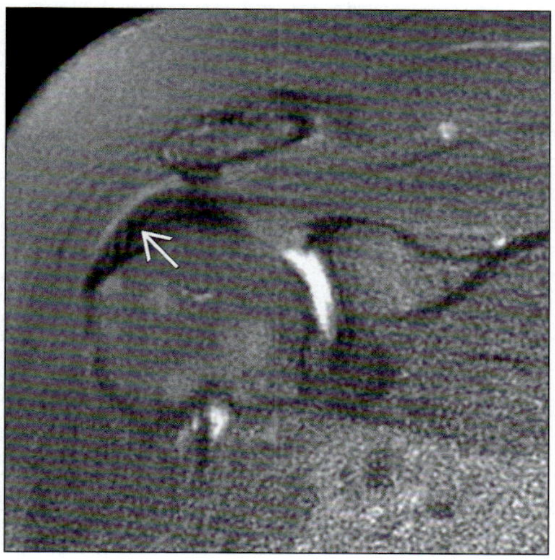

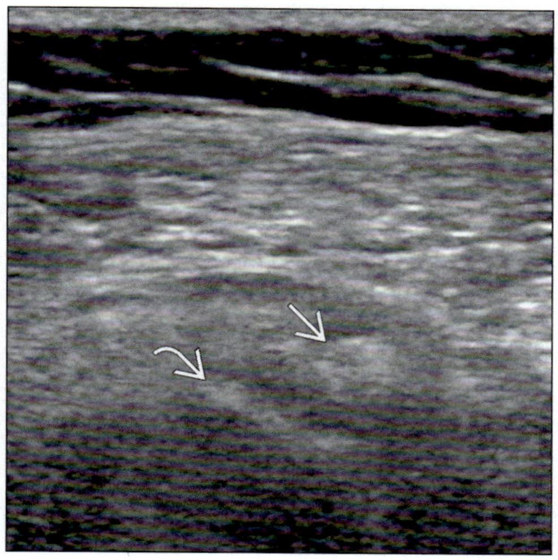

(Left) *Coronal oblique T2WI FS MR in the same patient shows a faint area of markedly low signal within the supraspinatus tendon* → *corresponding to the area of calcification on x-ray.* ***(Right)*** *Longitudinal oblique ultrasound in the same patient shows the echogenic calcification* → *within the supraspinatus tendon near the superior facet of the greater tuberosity* ↷*. Calcific tendinopathy is often easier to identify on ultrasound than on MR and, when very dense, can cause shadowing.*

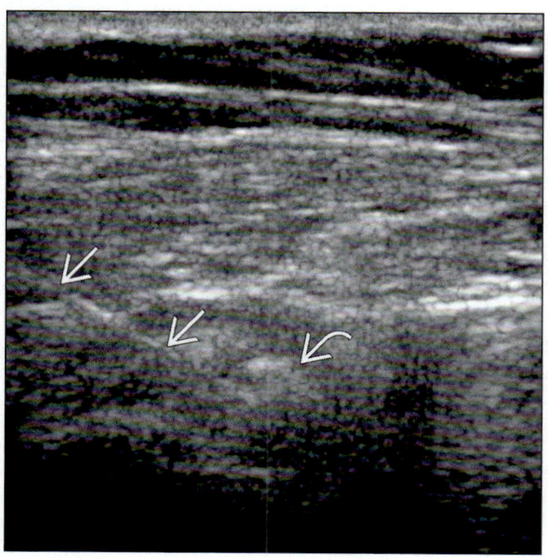

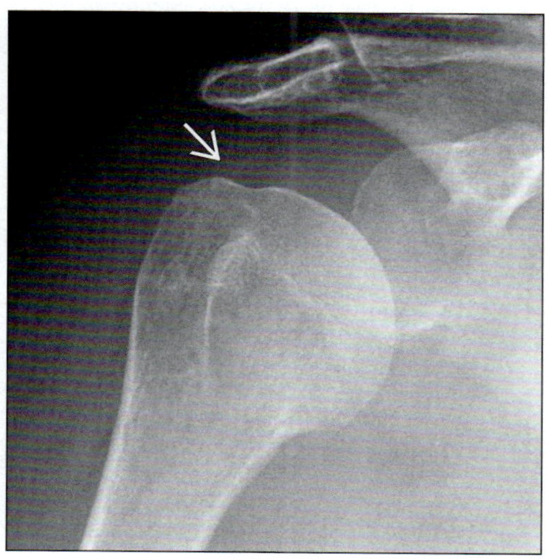

(Left) *Longitudinal oblique ultrasound in the same patient shows a 20-gauge needle* → *with its tip in the calcification* ↷*. Pulse lavage was done with 1% lidocaine to break-up and aspirate the calcium, followed by a corticosteroid injection into the SA/SD bursa.* ***(Right)*** *AP radiograph 3 months later shows decreased calcification within the supraspinatus tendon* →*. The patient was symptom free. Ultrasound-guided pulsed lavage is effective in the treatment of painful calcific tendinopathy.*

BICEPS TENDINOPATHY, SHOULDER

Key Facts

Terminology
- Long head biceps tendinosis, or "tendinitis"
- Microscopic tears within BT
- Often associated with tenosynovitis when painful

Imaging
- **MR**
  - Increased signal within BT
  - Focal swelling
  - High signal that completely surrounds tendon
  - Indistinct surface of tendon
  - "Hourglass" BT: Focal swelling at groove entrance; "trigger shoulder": Limited arm elevation
- Evaluate BT on oblique sagittal (intraarticular) and axial (intertubercular) images
- **Ultrasound**
  - Hypoechoic areas within tendon
  - Tenderness to palpation with transducer

Top Differential Diagnoses
- Biceps partial tear
  - Can often be difficult to distinguish from tendinopathy
  - High signal in BT extends to tendon surface
- "Magic angle"
  - Signal intensity decreases on T2-weighted images

Pathology
- Impingement syndrome or RCT: 95%
- Primary biceps tenosynovitis: 5%

Clinical Issues
- Anterior shoulder pain extending down arm
- Pain with lifting or pulling
- Tenderness to palpation along bicipital groove

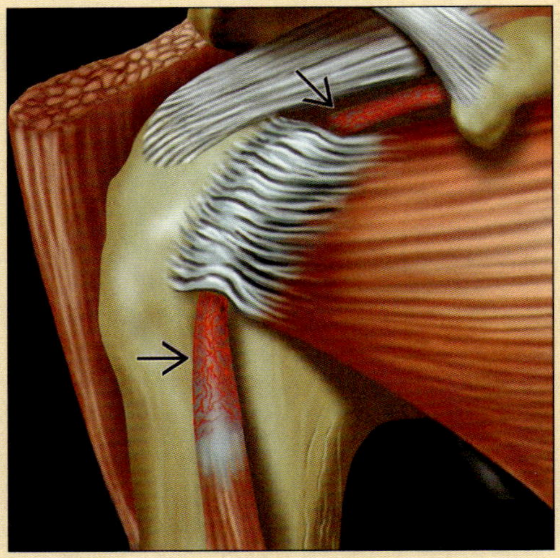

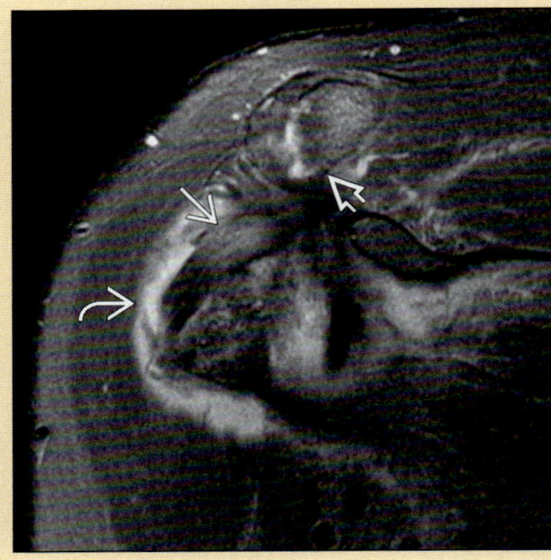

*(**Left**) AP graphic shows BT tenosynovitis with engorged blood vessels → surrounding the tendon. Tendinopathy with mucoid change in the tendon is often present as well. (**Right**) Coronal oblique T2WI FS MR shows increased signal in a focally enlarged horizontal portion of the BT →. The surface of the tendon remains low signal, indicating that this is tendinopathy and not a partial biceps tendon tear. There is also AC joint DJD ⇨ and an effusion ↪ from a rotator cuff tear (not shown).*

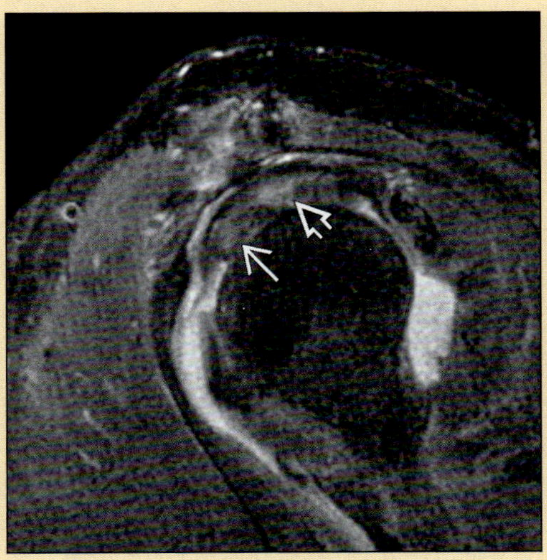

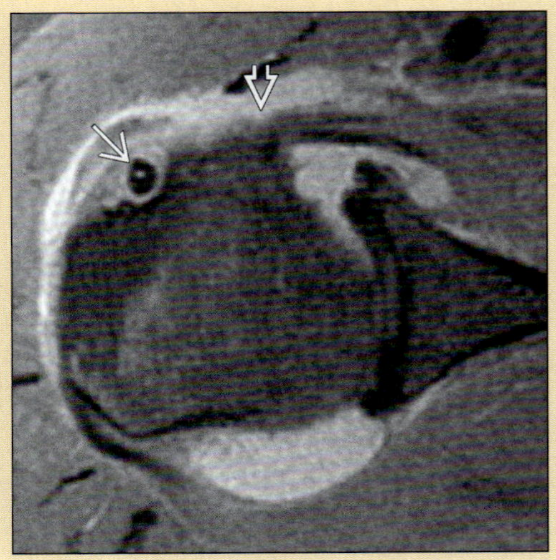

*(**Left**) Oblique sagittal T2WI FS MR in the same patient shows increased signal within the enlarged biceps tendon →. There is also high signal in the anterior supraspinatus tendon ⇨, which is associated with biceps tendinopathy. (**Right**) Axial PDWI FS MR in the same patient shows less severe high signal tendinopathy → in the intertubercular portion of the biceps tendon. There is partial tear of the cranial subscapularis tendon ⇨, which is associated with biceps tendinopathy.*

BICEPS TENDINOPATHY, SHOULDER

TERMINOLOGY

Abbreviations
- Biceps tendon (BT)

Synonyms
- Long head biceps tendinosis, or "tendonitis"

Definitions
- Mucoid change within BT
- 2 types appear similar on MR
  - Painful tendinopathy
    - Usually associated with tenosynovitis
  - Degenerative tendinopathy
    - Often asymptomatic
    - May also have lipoid change within tendon

IMAGING

General Features
- Best diagnostic clue
  - Thickened tendon + increased BT signal
- Location
  - Intraarticular &/or intertubercular portions

MR Findings
- T1WI
  - ↑ diameter and signal within BT
- T2WI FS
  - Increased signal within BT
  - Diffuse or focal swelling
  - High signal around tendon > > joint effusion
    - Tenosynovitis if enhances with IV contrast
  - Indistinct surface of tendon
  - "Hourglass" BT: Focal swelling at groove entrance
    - "Trigger shoulder": Limited arm elevation
- MR arthrography
  - Better shows contour irregularity and caliber changes
  - 80-85% accuracy
- Indirect MR arthrography
  - Enhancing synovitis

Ultrasonographic Findings
- Hypoechoic areas within tendon
- Tenderness to palpation with transducer

Imaging Recommendations
- Best imaging tool
  - T2WI MR ± FS
- Protocol advice
  - Evaluate BT on oblique sagittal (intraarticular) and axial (intertubercular) images

DIFFERENTIAL DIAGNOSIS

Biceps Tendon Tear
- High signal in BT extends to tendon surface
- Often difficult to distinguish from tendinopathy

Rotator Cuff Impingement
- 95% of patients with biceps tendinopathy

"Magic Angle" Artifact
- Signal intensity decreases on T2WI

PATHOLOGY

General Features
- Etiology
  - Impingement syndrome or RCT: 95%
  - Primary biceps tenosynovitis: 5%
    - Osteophytes along bicipital groove, or congential supratubercular ridge
- Associated abnormalities
  - Subcoracoid impingement syndrome
  - Superior labrum anterior cuff (SLAC) lesion
  - Subscapularis tendon tear
  - Biceps dislocation

Staging, Grading, & Classification
- Impingement tendinopathy
  - Involves intraarticular portion of BT
- Subluxation tendinopathy
- Attrition tendinopathy
  - Narrow bicipital groove

Gross Pathologic & Surgical Features
- Thickened, indurated tendon

CLINICAL ISSUES

Presentation
- Most common signs/symptoms
  - Anterior shoulder pain extending down arm
  - Pain with lifting or pulling
  - Tenderness to palpation along bicipital groove
- Other signs/symptoms
  - Positive clinical tests
    - Yergason test: Pain with resisted supination
    - Speed test: Biceps resistance test

Demographics
- Age
  - 40-60 years old
- Epidemiology
  - 10-25% of shoulder MRs

Natural History & Prognosis
- Mild: Often subsides with conservative therapy
- Severe: Tenotomy (resect) or tenodesis (cut and reattach)

Treatment
- Rest and NSAIDs
- Ultrasound-guided biceps tendon sheath injection
- Surgery: Acromioplasty to treat impingement
  - Severe tendinopathy: Tenotomy or tenodesis

DIAGNOSTIC CHECKLIST

Consider
- BT tendinopathy if anterosuperior rotator cuff tear

SELECTED REFERENCES

1. Buck FM et al: Degeneration of the long biceps tendon: comparison of MRI with gross anatomy and histology. AJR Am J Roentgenol. 193(5):1367-75, 2009
2. Beltran J et al: Shoulder: labrum and bicipital tendon. Top Magn Reson Imaging. 14(1):35-49, 2003

(Left) *Axial PDWI FS MR shows increased signal within the biceps tendon, which is also dislocated out of the bicipital groove.* ***(Right)*** *Oblique sagittal T2WI FS MR in the same patient shows the medial dislocation of the biceps tendon, with intratendinous signal indicating tendinopathy. There is an anterior supraspinatus partial thickness tendon tear and a cranial subscapularis partial tear. Anterior cuff tears are associated with biceps tendinopathy and dislocation.*

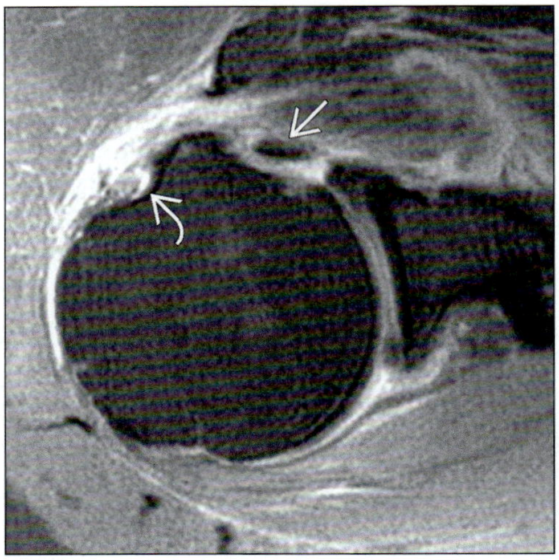

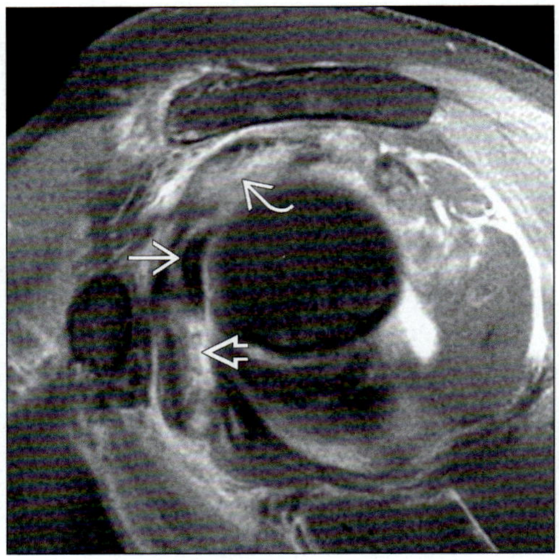

(Left) *Coronal oblique T2WI FS MR shows focal enlargement of the intraarticular portion of the BT, which then narrows and becomes low signal entering the groove. This is the "hourglass" type of biceps tendinopathy, which can cause a "trigger shoulder" with limited arm elevation.* ***(Right)*** *Coronal oblique T2WI FS MR in the same patient shows increased signal within the BT where there is (milder) tendinopathy in the intertubercular portion below the entrance to the groove.*

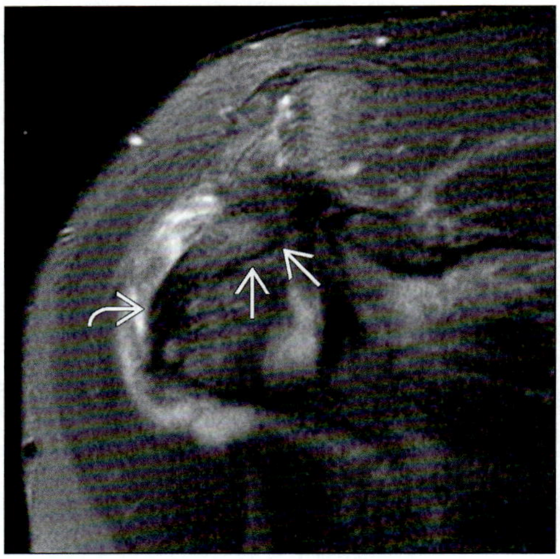

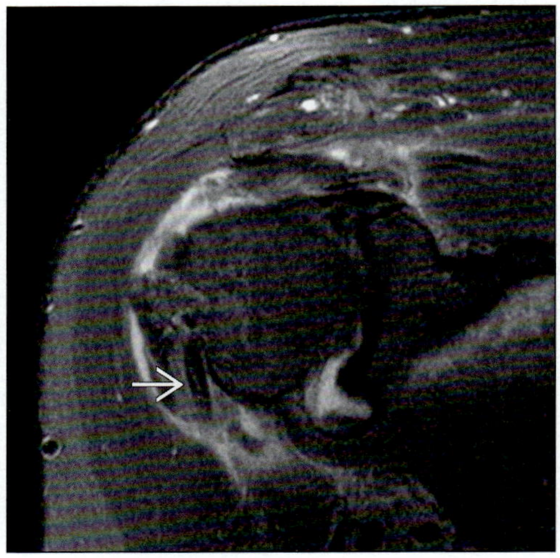

(Left) *Coronal oblique T1WI MR shows increased signal in the biceps tendon as it enters the bicipital groove and is oriented at 55° to B0. A T2WI is needed to determine if this is due to "magic angle" artifact, or represents severe tendinopathy.* ***(Right)*** *Coronal oblique T2WI FS MR in the same patient shows that the biceps tendon contains low signal in this region and therefore the increased T1WI signal is due mostly to "magic angle" artifact. There is mild tendinopathy.*

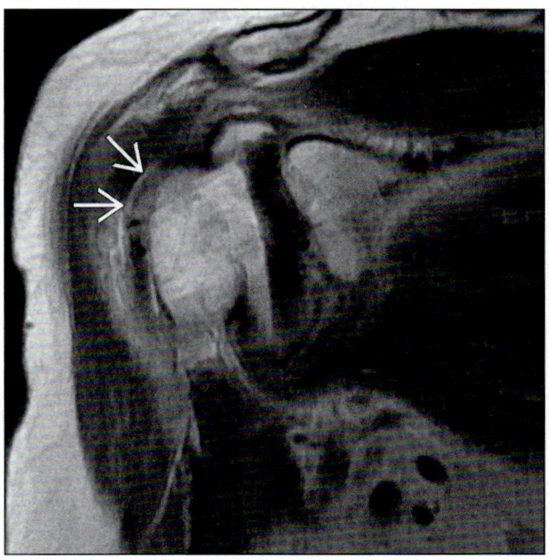

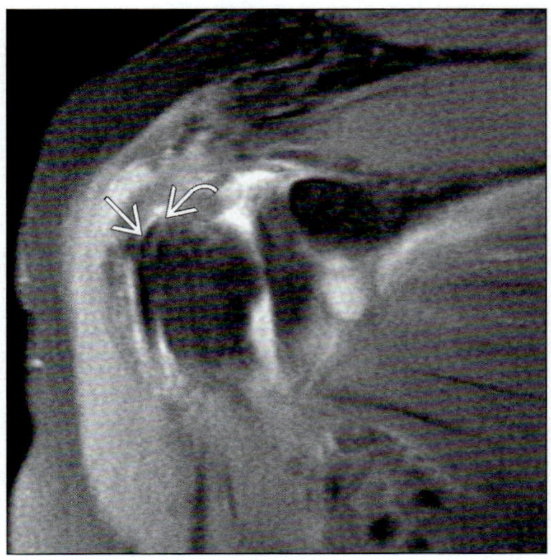

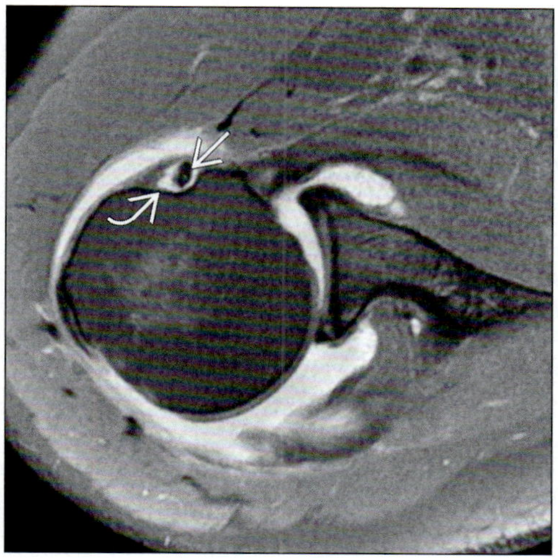

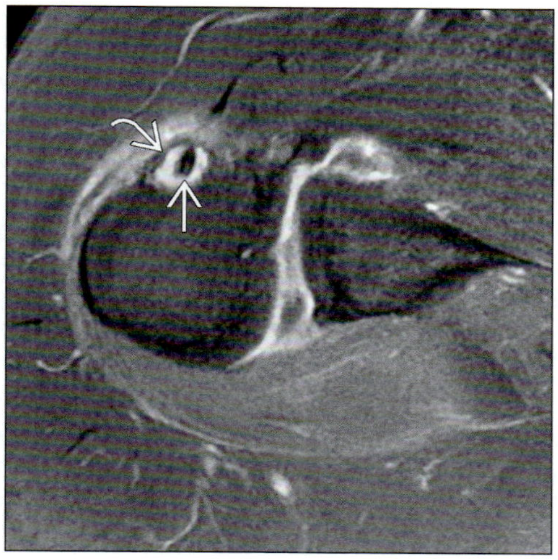

(Left) *Axial PDWI FS MR shows high signal around the BT consistent with effusion or tenosynovitis. There is also increased signal within the BT, indicating tendinopathy.* ***(Right)*** *Axial T1WI C+ FS MR in the same patient shows that the synovium enhances, indicating tenosynovitis. The high signal in the BT comes close to involving the surface of the tendon in this patient with tendinopathy; it can be difficult to distinguish tendinopathy from a partial tear.*

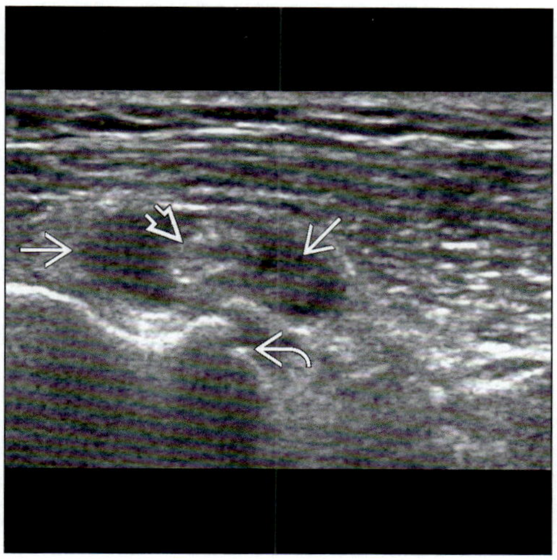

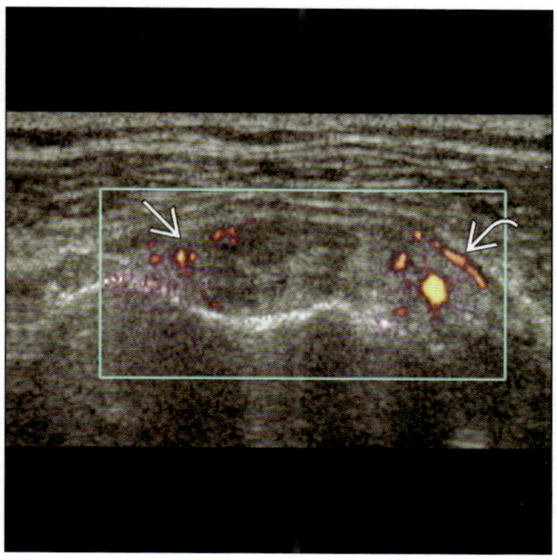

(Left) *Transverse ultrasound shows an enlarged diameter and hypoechoic changes in the BT. There is either surrounding hypoechoic effusion or tenosynovitis, and the biceps is subluxed onto the lesser tuberosity.* ***(Right)*** *Transverse power Doppler ultrasound in the same patient shows increased blood flow in the BT sheath, indicating that the hypoechoic area around the tendon is tenosynovitis. There is also increased vascularity at the adjacent subscapularis tendon.*

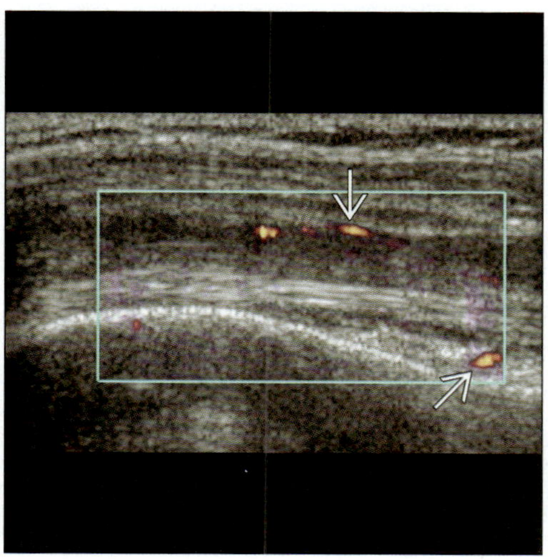

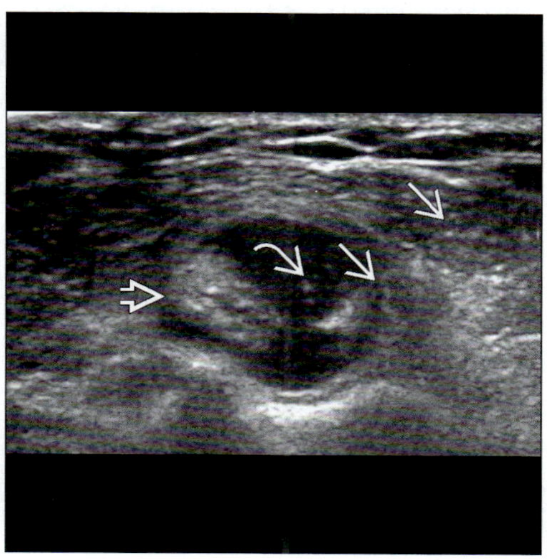

(Left) *Longitudinal power Doppler ultrasound in the same patient shows the increased blood flow in the biceps tendon sheath from tenosynovitis. Biceps tendinopathy is often associated with tenosynovitis in patients with pain.* ***(Right)*** *Transverse ultrasound in the same patient during selective biceps tendon sheath injection shows the needle placed into the BT sheath without piercing the tendon. Hyperechoic corticosteroid and anesthetic are seen during the injection.*

BICEPS TENDON TEAR, INTRAARTICULAR

Key Facts

Terminology

- Long head BT partial tear, complete tear, or rupture

Imaging

- Intraarticular (horizontal) portion more common than intertubercular (vertical) portion
- Type IV SLAP tear
    - Partial tear at superior labrum/supraglenoid tubercle origin
- Tears can occur at entrance to bicipital groove
    - Small tears here can be an isolated painful lesion
- **MR**
    - High T2 SI in BT extending to tendon surface
- Rupture
    - Absent segment of BT
    - Empty bicipital groove
    - Scar tissue can fill groove and mimic intact BT
    - BT may be fixated within groove despite intraarticular rupture, usually with chronic tears
- Type IV SLAP tear
    - High T2 signal in BT at proximal origin
    - Abnormal high T2 signal in superior labrum

Top Differential Diagnoses

- Normal BT variants
    - Capsular insertion, bifid, or congenital absent BT
- Biceps tendinopathy
    - High signal does not extend to tendon surface
- Prior biceps tenotomy/tenodesis

Clinical Issues

- Pain in bicipital groove
- "Rupture salvatrice": Pain relief after rupture of chronic BT partial tear
- Rupture: May hear a "pop"

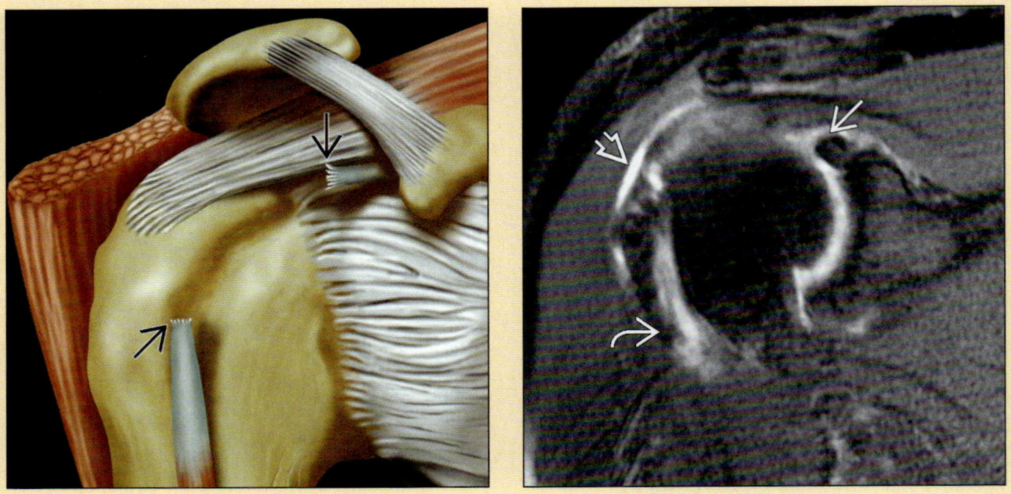

(Left) *AP graphic shows a complete tear of the long head of the biceps tendon →, with retraction of the distal segment.* ***(Right)*** *Coronal oblique MR arthrogram T1WI FS shows a complete tear of the biceps tendon in a 66-year-old man who felt a "pop" while lifting a heavy object. There is only a small stump at the superior labrum →, while the distal portion is in the lower bicipital groove →. There is also contrast in the subacromial/subdeltoid bursa → from a tear of the rotator interval.*

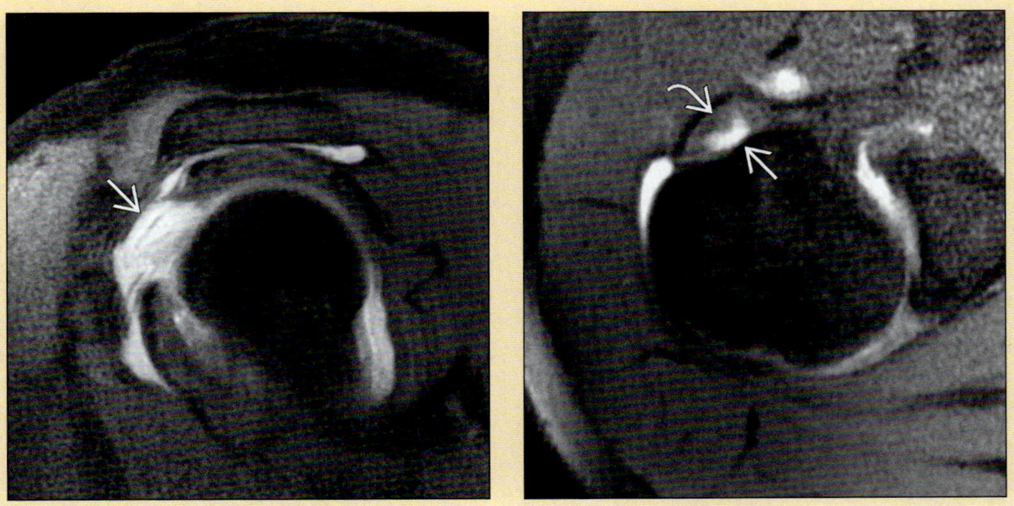

(Left) *Oblique sagittal MR arthrogram T2WI FS in the same patient shows fluid in the torn rotator interval → and absence of the biceps tendon. Biceps tendon ruptures may have an associated tear of the biceps pulley and interval.* ***(Right)*** *Axial MR arthrogram PDWI FS in the same patient shows fluid in the bicipital groove → with some capsular thickening laterally → but an absent long head biceps tendon. The distal stump of the tendon was visible on a more caudal section.*

BICEPS TENDON TEAR, INTRAARTICULAR

TERMINOLOGY

Abbreviations

- Biceps tendon (BT)
- **S**uperior **L**abral tear with **A**nterior and **P**osterior extension (SLAP)

Synonyms

- Long head BT partial tear or complete tear/rupture

Definitions

- Macroscopic tear of BT fibers in shoulder

IMAGING

General Features

- Best diagnostic clue
  - Tear or gap in BT
- Location
  - Intraarticular (horizontal) portion more common than intertubercular (vertical) portion
  - Type IV SLAP tear
    - Partial tear at superior labrum/tubercle
  - Tears can occur at entrance to bicipital groove
    - Small, can be difficult to see on MR
- Size
  - Thinned tendon = chronic partial tear

MR Findings

- T2WI FS
  - High signal in BT extends to tendon surface
  - Rupture
    - Absent segment of BT
    - Empty bicipital groove
    - Scar tissue can fill groove and mimic intact BT
    - BT may be fixated within groove despite intraarticular rupture, usually with chronic tears
  - Type IV SLAP tear
    - High signal in BT at proximal origin
    - Abnormal high signal in superior labrum
  - Accuracy: 80%
- MR arthrography
  - 90-95% accuracy for rupture

Imaging Recommendations

- Best imaging tool
  - T2WI FS MR
- Protocol advice
  - Evaluate BT on oblique sagittal (intraarticular) and axial (intertubercular) images

DIFFERENTIAL DIAGNOSIS

Normal Biceps Tendon Variants

- Bifid or congenitally absent BT

Biceps Tendinopathy

- High signal does not extend to tendon surface
- Can be difficult to distinguish from BT tear

Prior Biceps Tenotomy/Tenodesis

- Micrometallic artifact

PATHOLOGY

General Features

- Etiology
  - Anterior supraspinatus tears that extend across anterior interval → subscapularis tendon
    - 25% of supraspinatus (SST) tears have biceps tear
  - Chronic overuse with repetitive microtrauma

Staging, Grading, & Classification

- Snyder grade
  - Grade 1: < 25% fraying
  - Grade 2: 25-50% partial tear
  - Grade 3: > 50% partial tear
  - Grade 4: Rupture

Gross Pathologic & Surgical Features

- Distal tendon may retract into upper arm

CLINICAL ISSUES

Presentation

- Most common signs/symptoms
  - Pain in bicipital groove
  - "Rupture salvatrice": Pain relief after rupture of chronic BT partial tear
    - If continued pain, usually from RCT
  - Acute tears in younger patients are painful
- Other signs/symptoms
  - May hear a "pop"
  - "Popeye arm": Bulge from retracted muscle belly
    - Less obvious than with distal biceps rupture at elbow

Demographics

- Age
  - 50-60 years old
- Gender
  - M > F

Natural History & Prognosis

- BT may be flattened prior to tearing

Treatment

- Shallow tears: Debridement
- Deep partial tears: Tenodesis or tenotomy
- Ruptures: Debridement of stump edges

DIAGNOSTIC CHECKLIST

Consider

- BT tear if anterosuperior rotator cuff tear

Reporting Tips

- Describe location of tear

SELECTED REFERENCES

1. Gaskin CM et al: Focal partial tears of the long head of the biceps brachii tendon at the entrance to the bicipital groove: MR imaging findings, surgical correlation, and clinical significance. Skeletal Radiol. 38(10):959-65, 2009
2. Beall DP et al: Association of biceps tendon tears with rotator cuff abnormalities: degree of correlation with tears of the anterior and superior portions of the rotator cuff. AJR Am J Roentgenol. 180(3):633-9, 2003

BICEPS TENDON TEAR, INTRAARTICULAR

(Left) Axial PDWI FS MR shows high signal within the biceps tendon extending to the tendon surface, indicating a longitudinal partial tear → of the intertubercular portion of the long head of the BT. (Right) Oblique sagittal T2WI FS MR in the same patient shows increased signal in the horizontal, or intraarticular, portion of the BT →. This portion of the BT was torn at surgery. It can sometimes be difficult to distinguish partial tears from severe tendinopathy of the BT.

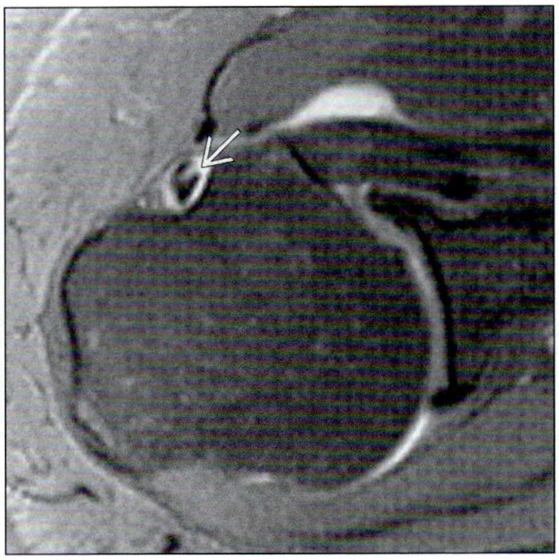

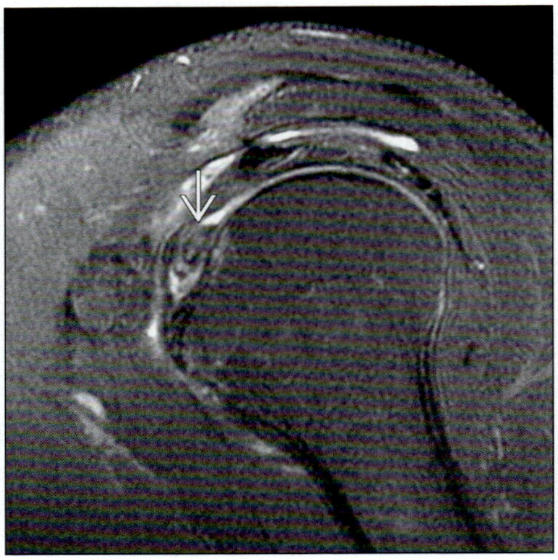

(Left) Axial PDWI FS MR shows a deep longitudinal partial tear → of the long head of the biceps tendon in the bicipital groove. The biceps tendon is also enlarged. (Right) Axial PDWI FS more cranial MR in the same patient shows that the biceps tendon has a "longitudinal split" partial tear into 2 strands. The lateral strand is still in the bicipital groove →, while the more medial segment is dislocated ↪. The dislocated strand indicates a tear of the biceps pulley.

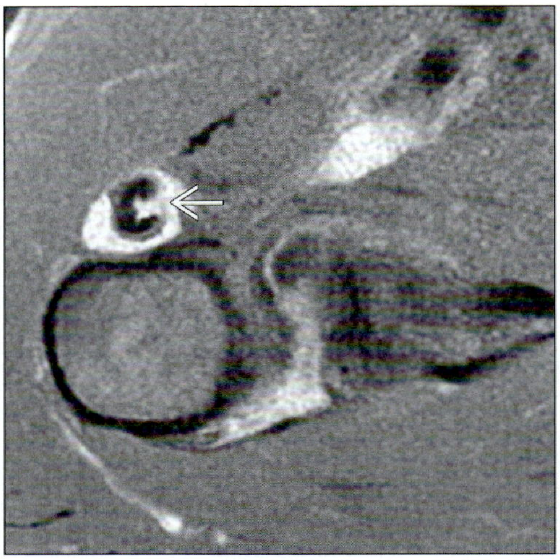

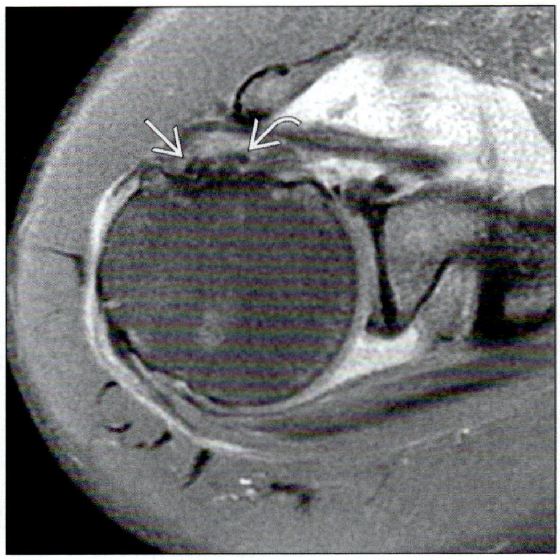

(Left) Lateral graphic of a type IV superior labrum anterior-posterior (SLAP) tear. In a type IV SLAP tear, the tear of the superior labrum → extends up into the biceps tendon ↪. (Right) Coronal oblique and sagittal T2W FS MR images in a patient with a type IV SLAP tear. There is abnormal high signal in the superior labrum → from a SLAP tear. On the sagittal image, there is also linear high signal in the BT ↪ from a partial tear near the origin from the superior labrum.

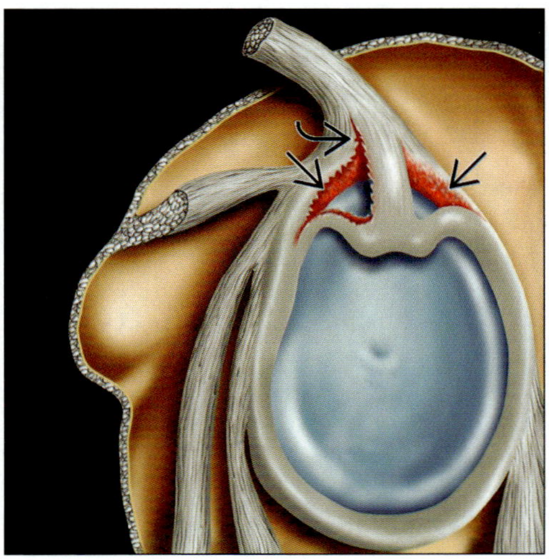

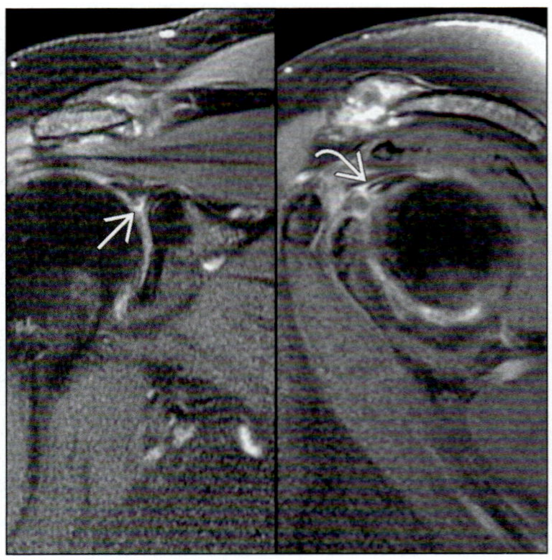

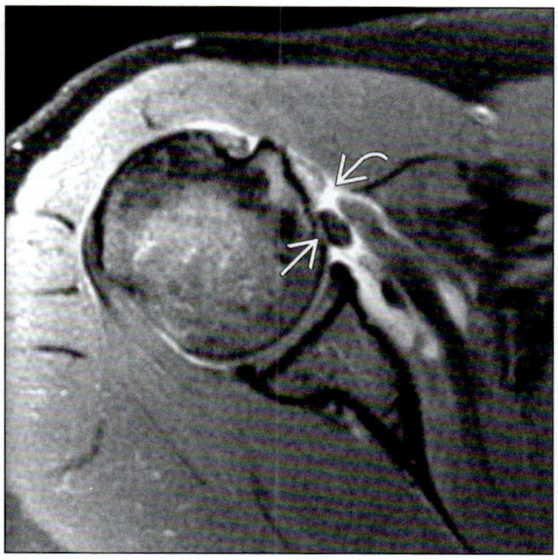

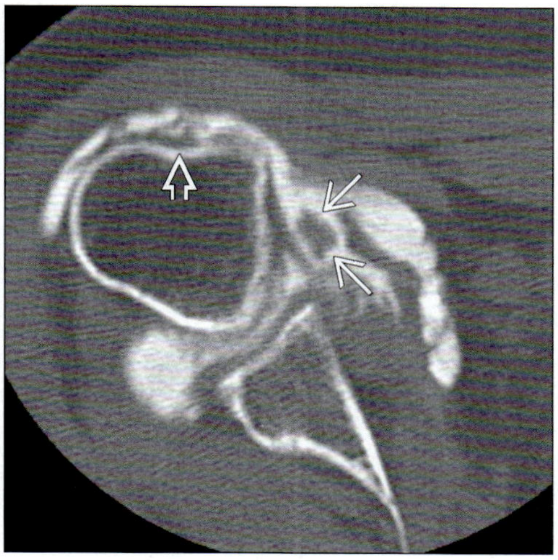

(Left) *Axial PDWI FS MR shows a dislocated long head of the biceps tendon with slight surface irregularity anteriorly from a 10% partial tear. Many dislocated biceps tendons also have a partial tear. There is also a tear of the subscapularis tendon.* ***(Right)*** *Axial CT arthrogram shows a dislocated long head biceps tendon with 10% partial tearing. The tendon is also enlarged from tendinopathy, and there is debris or synovitis in an otherwise empty bicipital groove.*

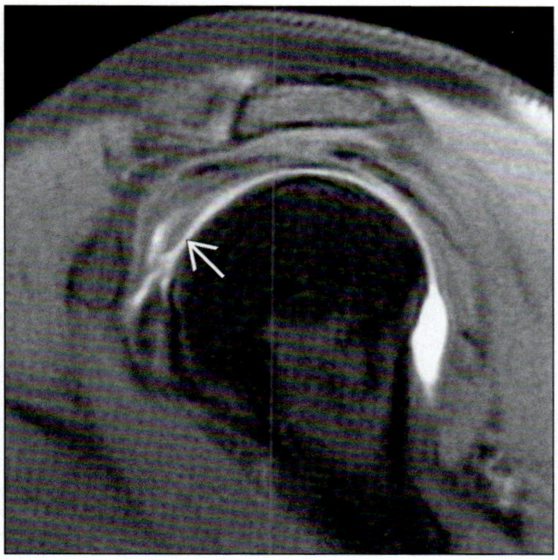

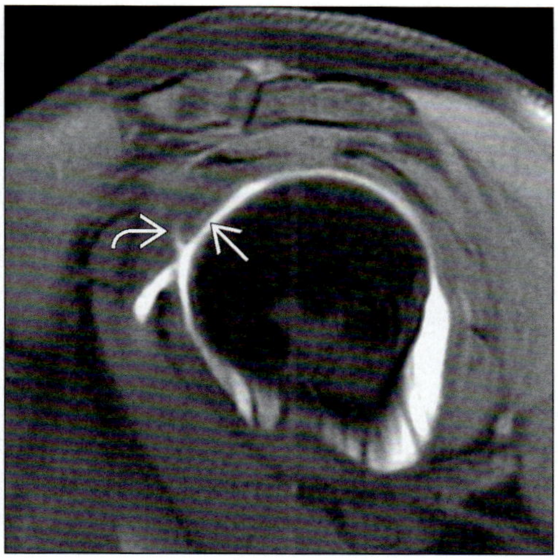

(Left) *T1WI FS of a normal variant intracapsular origin of the biceps tendon shows the biceps tendon to be well-defined and adjacent to the capsule in the anterior interval.* ***(Right)*** *On a more medial oblique sagittal MR arthrogram T1WI FS in the same patient, there is less fluid between the biceps tendon and the joint capsule. Fluid usually completely surrounds the intraarticular BT on MR arthrogram images.*

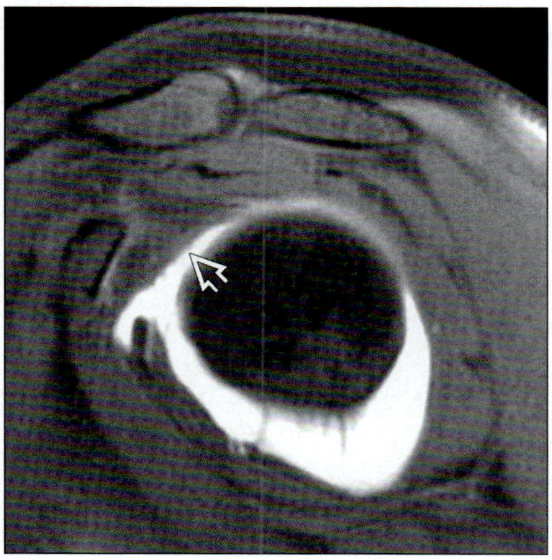

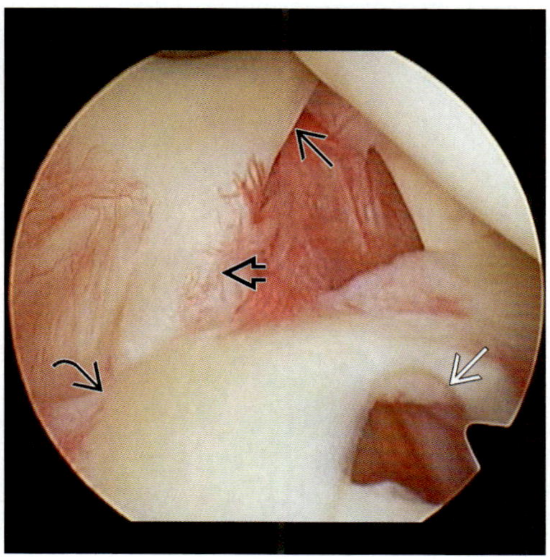

(Left) *On the next more medial oblique sagittal MR arthrogram T1WI FS in the same patient, the BT is now indistinguishable from the joint capsule in the anterior interval. This normal variant can be confused with a biceps tendon rupture.* ***(Right)*** *Lateral arthroscopic photograph of the same patient shows the biceps tendon fusing with the joint capsule prior to inserting on the superior labrum and supraglenoid tubercle. There is also a sublabral foramen normal variant.*

BICEPS TENDON DISLOCATION

Key Facts

Terminology
- Biceps tendon (BT) subluxation, BT instability
- Displacement of BT out of bicipital groove
- Torn biceps pulley (coracohumeral ligament and superior glenohumeral ligament)
  - ± subscapularis or transverse ligament tear

Imaging
- MR: Subluxation
  - BT perched on lateral edge of lesser tuberosity
  - Mild displacement with chronic cuff impingement or "hidden lesion" (partial subscapularis tendon tear with intact joint capsule)
- MR: Dislocation
  - BT medially displaced completely out of bicipital groove
  - "Empty bicipital groove" sign
- Intraarticular dislocation
  - Subscapularis avulsed from lesser tuberosity
- Dislocation anterior to subscapularis
  - Transverse humeral ligament torn

Top Differential Diagnoses
- Biceps rupture
- Intraarticular loose body
- Displaced anterior labral tear

Pathology
- Rotator cuff tear with biceps pulley tear

Clinical Issues
- Tenotomy (tendon resected) or tenodesis (tendon cut and reattached in groove)

Diagnostic Checklist
- Identify "hidden lesion" if present

__(Left)__ AP graphic shows a tear of the subscapularis tendon with medial dislocation of the long head of the biceps tendon . Biceps dislocation is associated with subscapularis tendon tears, although most subscapularis tears are partial tears involving the cranial portion of the tendon. __(Right)__ Coronal oblique T2WI FS MR shows medial dislocation of the biceps tendon out of the bicipital groove . There is also a tear of the subscapularis tendon .

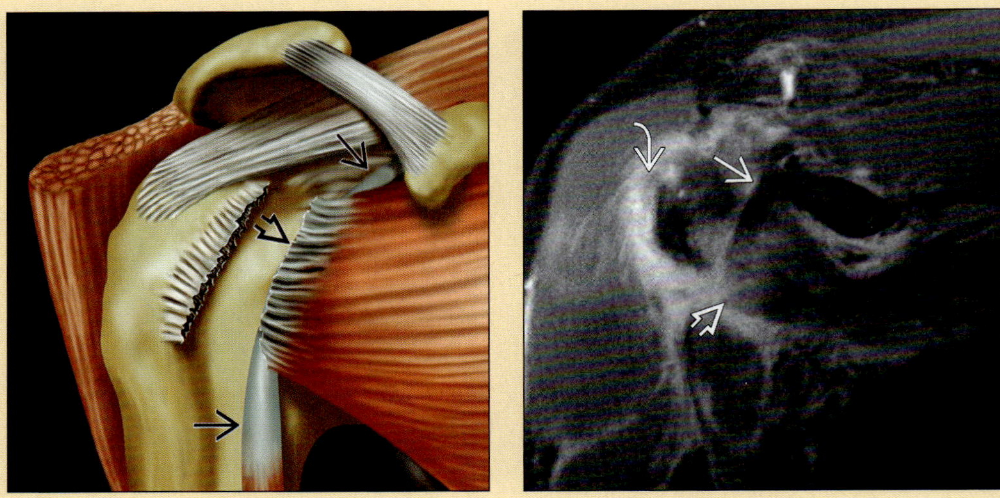

__(Left)__ Axial PDWI FS MR from the same patient shows the biceps tendon medial to the lesser tuberosity with an empty bicipital groove , indicating a biceps dislocation. Also seen is the tear of the subscapularis tendon . __(Right)__ Oblique sagittal T2WI FS MR from the same patient shows the biceps tendon turning inferiorly while still medial to the lesser tuberosity. The subscapularis tendon is irregular and completely torn more laterally.

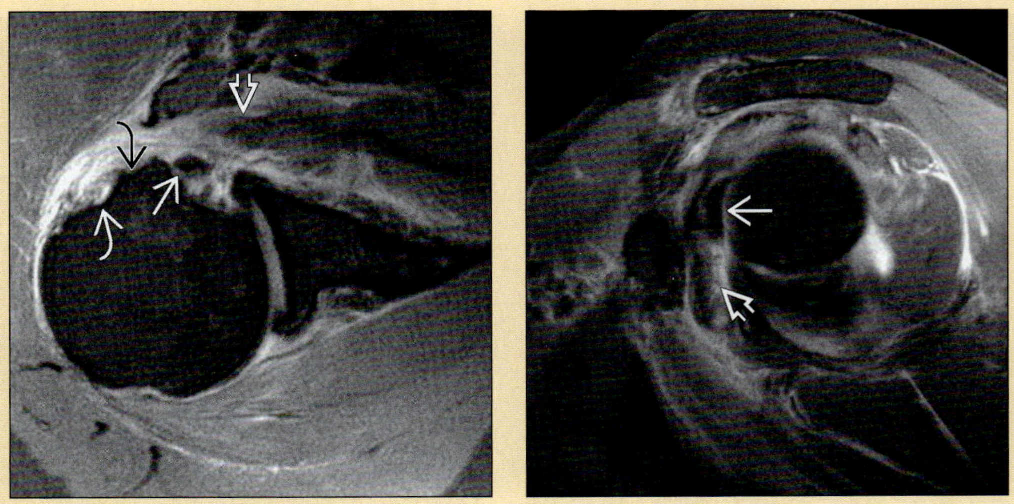

BICEPS TENDON DISLOCATION

TERMINOLOGY

Abbreviations
- Biceps tendon (BT)

Synonyms
- BT subluxation, BT instability

Definitions
- Displacement of BT out of bicipital groove
  - ± subscapularis or transverse ligament tear
- Biceps pulley: Coracohumeral ligament (CHL) and superior glenohumeral ligament (SGHL)

IMAGING

General Features
- Best diagnostic clue
  - Displacement of BT on axial MR images
- Size
  - Dislocation
    - BT displaced completely out of bicipital groove
    - Usually does not reduce
  - Subluxation
    - Mild displacement with chronic cuff tear

MR Findings
- T2WI FS
  - Biceps tendon displaced medially
  - Subluxation
    - BT perched on lateral edge of lesser tuberosity
    - Flattened BT
    - "Hidden lesion": Partial subscapularis tendon tear with intact joint capsule
      - Important to recognize, as this lesion not seen at arthroscopy
  - Dislocation
    - BT displaced completely out of bicipital groove
    - Empty bicipital groove in 80-90%
    - False-negative: Scar can fill groove and mimic an intact biceps tendon
    - **Intraarticular dislocation**
      - Subscapularis avulsed from lesser tuberosity
      - Can mimic labral tear if dislocates medially adjacent to anterior labrum
    - **Dislocation anterior to subscapularis**
      - Transverse humeral ligament torn
      - ± tear of superficial subscapularis fibers
  - Tear of biceps pulley (CHL and SGHL)

Ultrasonographic Findings
- No biceps seen in groove
- Biceps tendon identified medial to groove

Imaging Recommendations
- Best imaging tool
  - Axial MR images

DIFFERENTIAL DIAGNOSIS

Biceps Tendon Rupture
- Displaced BT not seen

Intraarticular Loose Body
- Can mimic intraarticular dislocated biceps tendon

Displaced Anterior Labral Tear
- BT seen in groove

PATHOLOGY

General Features
- Etiology
  - Rotator cuff tear with biceps pulley tear
  - Hypoplastic bicipital groove and repeated trauma
    - Younger patients
- Associated abnormalities
  - Biceps tendon partial tear
  - Subscapularis tendon tear
  - Subcoracoid impingement
  - Lesser tuberosity fracture

Staging, Grading, & Classification
- Position relative to subscapularis tendon
  - Posterior to subscapularis
    - Subscapularis detached from lesser tuberosity
    - Subscapularis not retracted if transverse humeral ligament is intact
  - Anterior to subscapularis
    - Torn transverse humeral ligament ± variable portion of anterior subscapularis tendon fibers

CLINICAL ISSUES

Presentation
- Most common signs/symptoms
  - Pain & tenderness in bicipital groove
  - Pain with lifting or pulling
- Other signs/symptoms
  - "Pop" or "snap" during shoulder range of motion

Natural History & Prognosis
- Continued pain if untreated

Treatment
- Tenotomy (tendon resected) or tenodesis (tendon cut and reattached in groove)
- Rarely, reduction and surgical repair of biceps pulley

DIAGNOSTIC CHECKLIST

Reporting Tips
- State if BT anterior or posterior to subscapularis
- Identify "hidden lesion" if present

SELECTED REFERENCES

1. Tonino PM et al: Complex shoulder disorders: evaluation and treatment. J Am Acad Orthop Surg. 17(3):125-36, 2009
2. Lin A et al: Magnetic resonance imaging of the long head of the biceps tendon: benefit of coplanar image. J Comput Assist Tomogr. 31(4):642-8, 2007
3. Morag Y et al: MR arthrography of rotator interval, long head of the biceps brachii, and biceps pulley of the shoulder. Radiology. 235(1):21-30, 2005
4. Spritzer CE et al: Assessment of instability of the long head of the biceps tendon by MRI. Skeletal Radiol. 30(4):199-207, 2001

BICEPS TENDON DISLOCATION

*(**Left**) Axial PDWI FS MR shows dislocation of the biceps tendon anterior to the subscapularis tendon. The biceps pulley is torn, allowing dislocation. The transverse ligament is torn, allowing the biceps to dislocate out of the groove and anterior to the subscapularis. Some of the anterior fibers of the subscapularis are torn as well. (**Right**) Coronal oblique T2WI FS MR from the same patient shows the dislocated biceps tendon passing anterior to the subscapularis tendon.*

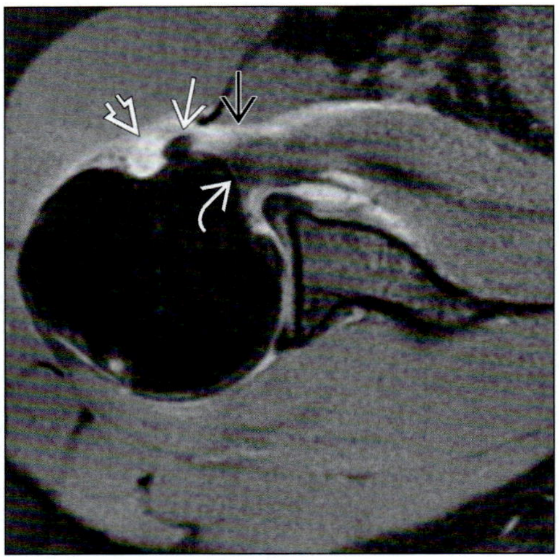

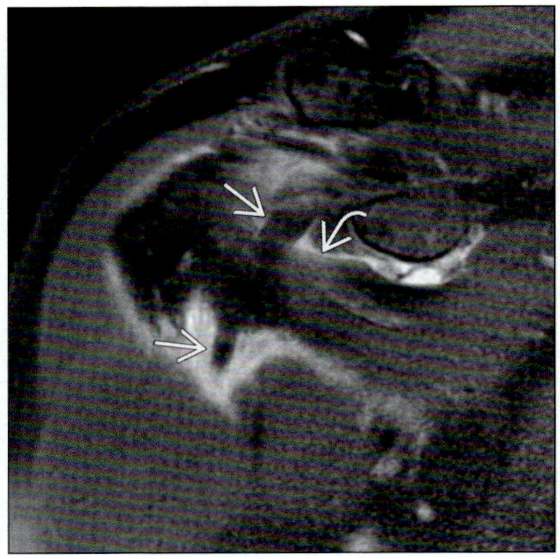

*(**Left**) Axial PDWI FS MR shows a dislocated biceps tendon deep to the subscapularis tendon. The subscapularis tendon is avulsed from the edematous lesser tuberosity, but note some intact fibers contiguous with the transverse ligament, preventing retraction of the subscapularis. (**Right**) Axial PDWI FS MR shows a medially dislocated biceps tendon adjacent to the anterior labrum. The biceps should not be mistaken for the middle glenohumeral ligament or for a labral tear.*

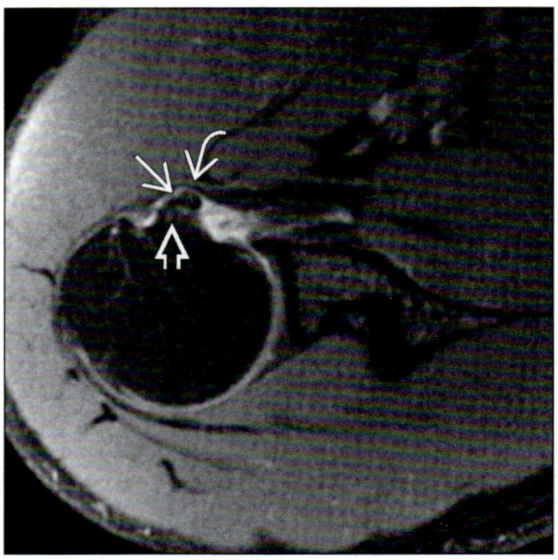

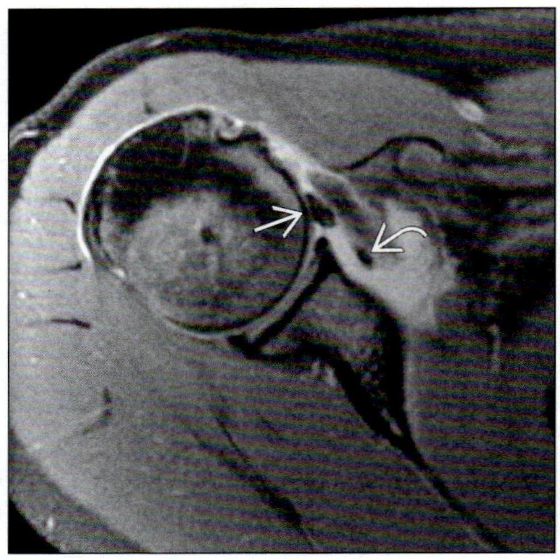

*(**Left**) Sagittal graphic shows the coracohumeral ligament within the rotator interval and enveloping the anterior edge of the supraspinatus tendon. The inferior aspect of the ligament is beginning to wrap under the biceps tendon and will soon fuse with the superior glenohumeral ligament to form the biceps pulley sling, which keeps the biceps tendon within the groove. (**Right**) Sagittal T2WI FS MR shows a tear of the biceps pulley in a patient with biceps tendon subluxation.*

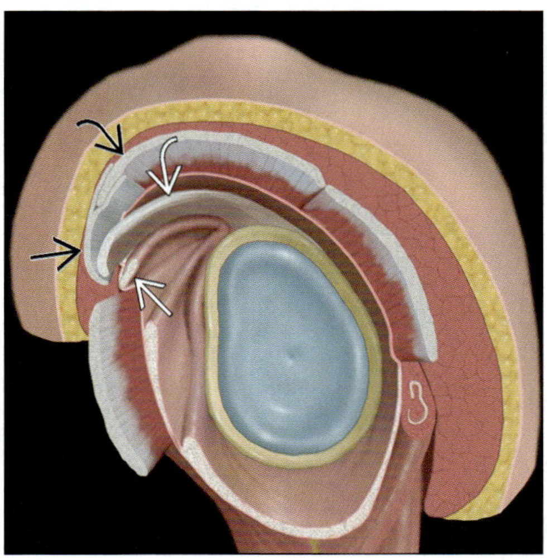

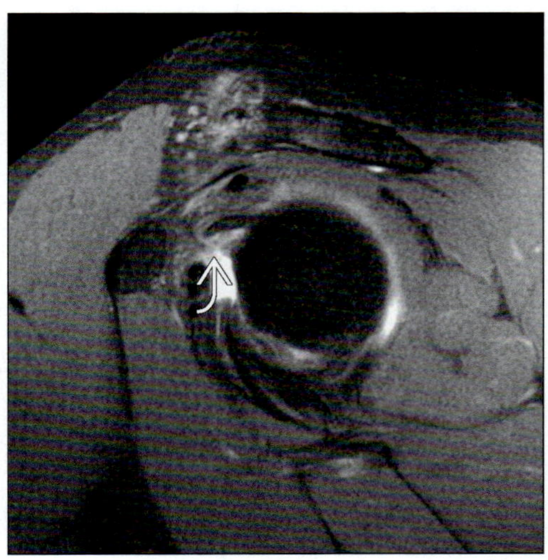

BICEPS TENDON DISLOCATION

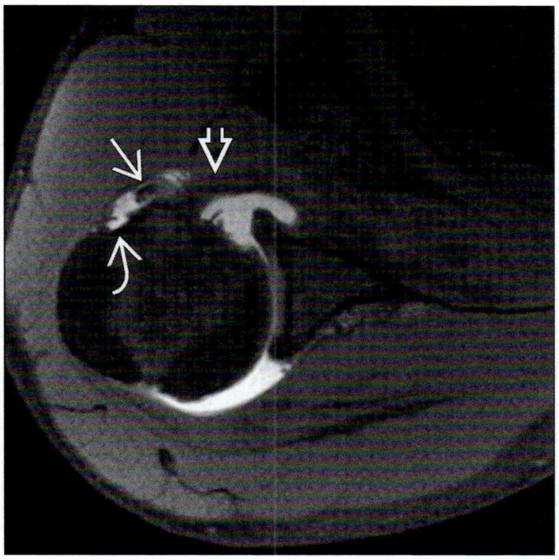

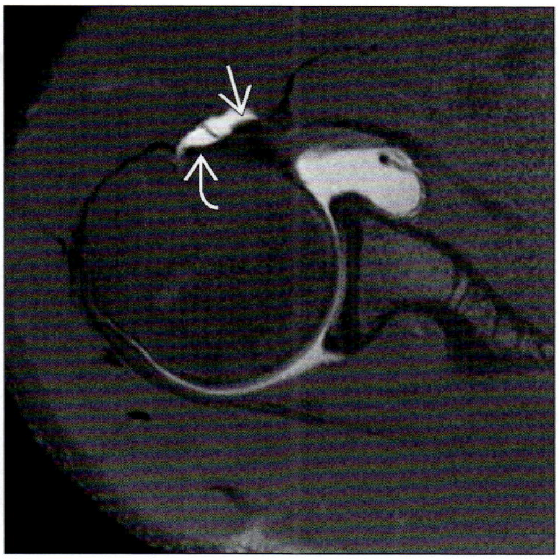

(Left) *Axial MR arthrogram T1WI FS shows the biceps tendon dislocated medially, anterior to the subscapularis tendon, with an "empty groove" sign. This indicates that the transverse ligament and biceps pulley are torn.* ***(Right)*** *Axial MR arthrogram T1WI FS in an 18 year old with shoulder pain while throwing shows a shallow bicipital groove. The biceps tendon is dislocated medially, anterior to the lesser tuberosity. A shallow groove is a risk factor for dislocation.*

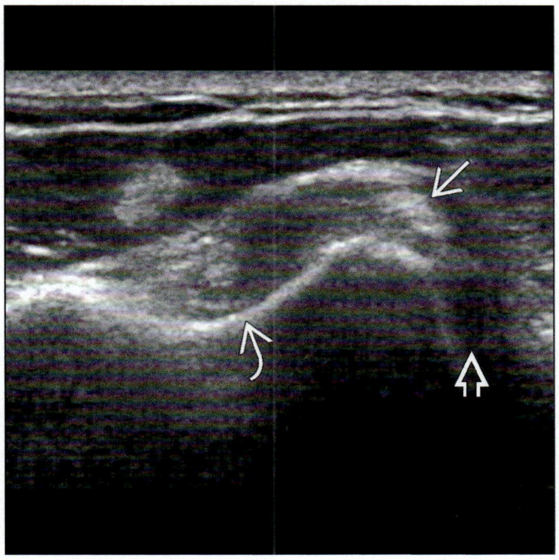

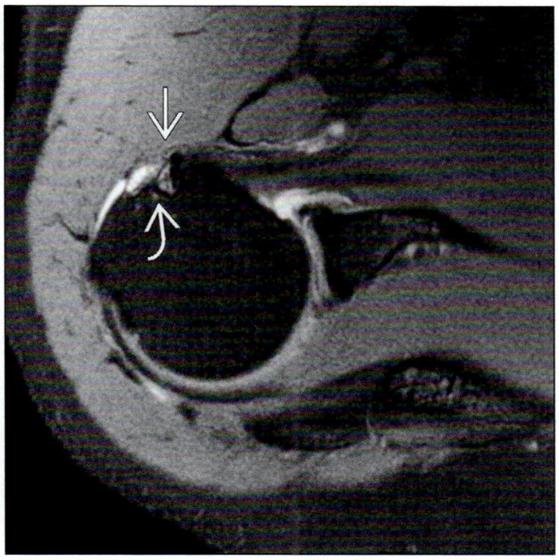

(Left) *Transverse ultrasound shows medial dislocation of the biceps tendon and an "empty groove" sign with synovitis and hypoechoic fluid within the bicipital groove. The subscapularis tendon is also torn.* ***(Right)*** *Axial PDWI FS MR shows subluxation of the biceps tendon partly out of the bicipital groove. A common cause of mild biceps subluxation is a tear of the biceps pulley associated with tears of the anterior supraspinatus and cranial subscapularis tendons.*

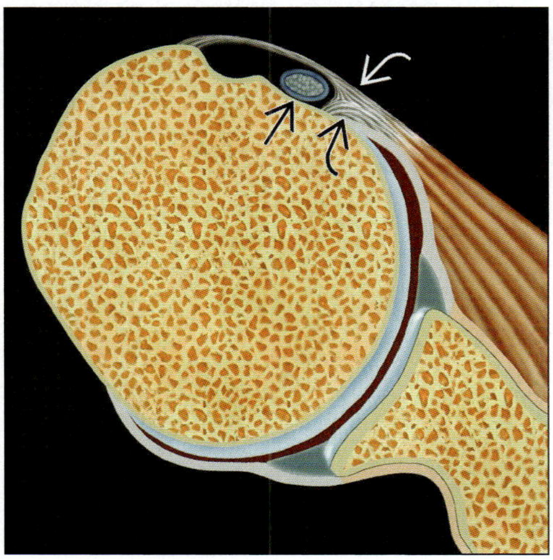

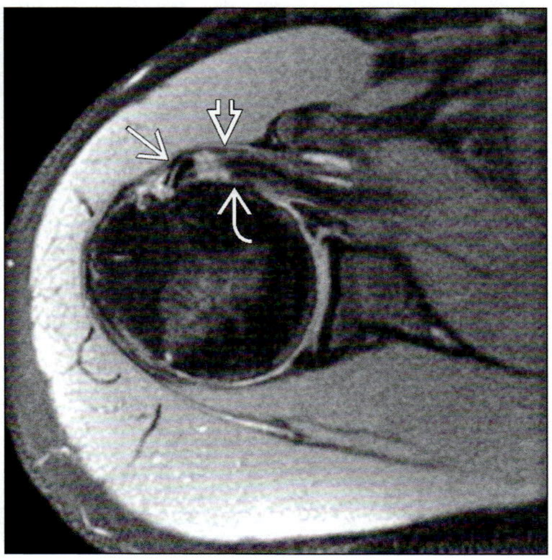

(Left) *Axial graphic shows a "hidden lesion" with mild dislocation of the biceps tendon into the subscapularis tendon from partial tearing of the lateral subscapularis tendon insertion. Because the joint capsule is intact, the biceps dislocation is not seen at arthroscopy.* ***(Right)*** *Axial PDWI FS MR shows a "hidden lesion." The biceps tendon is subluxed into the subscapularis tendon, but the deep and superficial fibers of the subscapularis tendon are intact.*

NORMAL LABRAL VARIANTS

Key Facts

Terminology

- Sublabral foramen
  - Unattached anterosuperior labrum
- Buford complex
  - Absent anterosuperior labrum with thick, cord-like middle glenohumeral ligament
- Superior recess
  - Partially unattached superior labrum

Imaging

- Anterosuperior labrum: SLF and Buford complex
  - High signal between labrum and glenoid rim (SLF), or absent labrum (Buford complex)
- Superior labrum: Superior recess
  - High signal curves medially at labral-chondral junction

Top Differential Diagnoses

- Microinstability, shoulder
- SLAP lesions V-X
- Bankart lesion

Pathology

- Sublabral foramen
  - No hemorrhage or fraying at unattached labrum
- Buford complex
  - No hemorrhage at glenoid rim, thick cord-like MGHL
- Superior recess
  - Stable biceps anchor
  - Recess is synovial lined without hemorrhage

Diagnostic Checklist

- Consider variants if labral "abnormality" between 12-3 o'clock

*(**Left**) Lateral graphic shows a sublabral foramen →. There is a congenitally unattached labrum → extending from just inferior to the origin of the middle glenohumeral ligament → to just above the origin of the anterior band of the inferior glenohumeral ligament →. A SLF is seen in about 10% of people. (**Right**) Lateral arthroscopic photograph shows a sublabral foramen → with unattached labrum →. Note the middle glenohumeral ligament →, glenoid rim →, and humeral head →.*

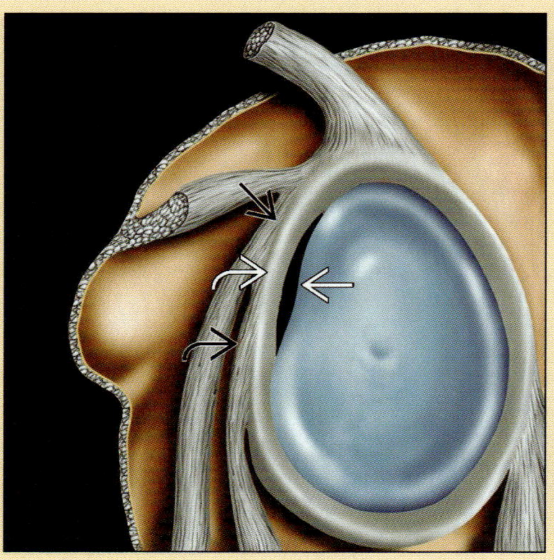

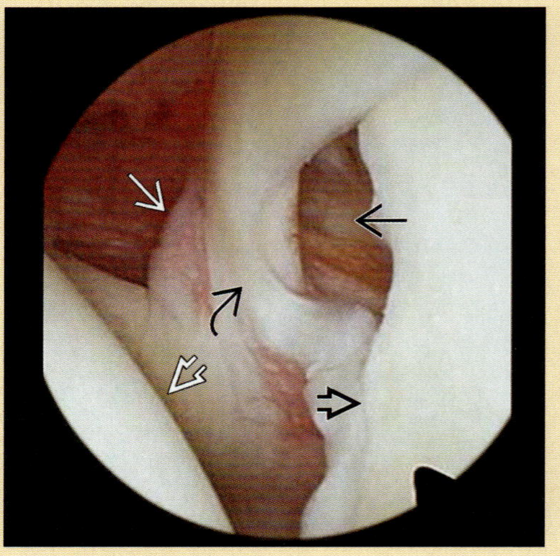

*(**Left**) AP graphic shows a superior recess →. Superior labrum is partly unattached, although there is secure attachment to the glenoid rim superomedially and a stable biceps anchor →. A superior recess of some depth is seen in 75% of people. (**Right**) Lateral arthroscopic photograph shows a superior recess → between the superior labrum → and the glenoid rim →. The biceps anchor → was stable when examined during arthroscopy. The recess is synovial lined and is a blind-ending pouch.*

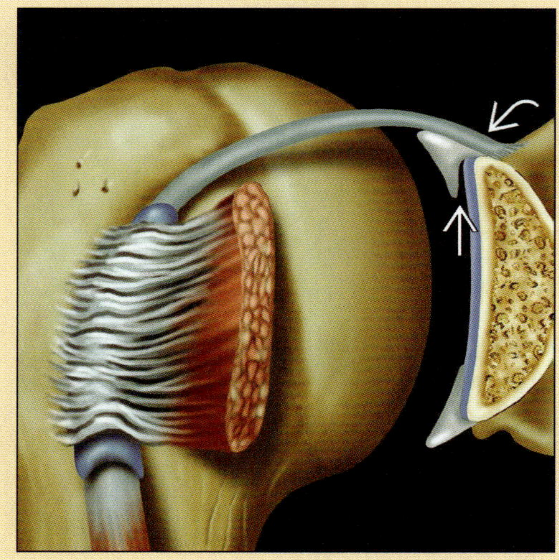

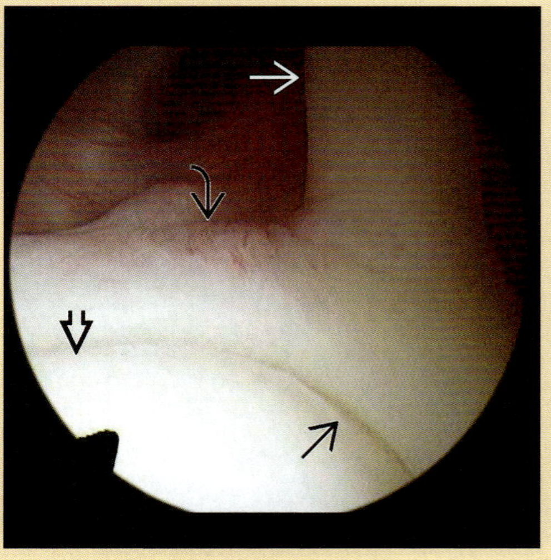

NORMAL LABRAL VARIANTS

TERMINOLOGY

Synonyms

- Sublabral foramen (SLF) or sublabral hole, Buford complex, superior recess

Definitions

- Sublabral foramen
  - Unattached anterosuperior labrum
- Buford complex
  - Absent anterosuperior labrum with thick, cord-like middle glenohumeral ligament
- Superior recess
  - Partially unattached superior labrum

IMAGING

General Features

- Best diagnostic clue
  - Anterosuperior labrum: SLF & Buford complex more common than isolated labral tears
  - Superior labrum: ↑ MR signal curves medially
- Location
  - Anterosuperior labrum 1-3 o'clock: SLF & Buford
  - Superior labrum 12-1 o'clock: Superior recess

MR Findings

- Anterosuperior labrum: SLF and Buford complex
  - High signal between labrum and glenoid rim (SLF), or absent labrum (Buford complex)
  - Between origins of middle glenohumeral ligament (MGHL) and anterior band inferior glenohumeral ligament (IGHL)
  - Does not extend to involve superior labrum or anteroinferior labrum
  - SLF: Unattached but otherwise intact labrum
    - Extends posterior to biceps anchor in 55%
  - Buford complex: Thick diameter of MGHL
- Superior labrum: Superior recess
  - High signal curves medially at labral-chondral junction
  - High signal only within a recess: Does not extend medially across entire labral attachment
  - Distinguish from SLAP tear
    - High signal curves laterally
    - "Double Oreo" sign: 2 high signal lines, 1 is superior recess and other is superior labrum anterior-posterior (SLAP) tear
    - High signal too wide on oblique coronal images of SLAP
      - > 2 mm on conventional MR
      - > 3 mm on direct MR arthrography
    - Labrum detached: Type III SLAP tear
    - Signal posterior to biceps anchor in SLAP
      - Only 75% specificity at conventional MR, 55% at MR arthrography

Imaging Recommendations

- Best imaging tool
  - MR arthrography

DIFFERENTIAL DIAGNOSIS

Microinstability, Shoulder

- Can have isolated tear of anterosuperior labrum

SLAP Lesions V-X

- Type V SLAP tear extends into anterior labrum

Bankart Lesion

- Bankart tears can extend into anterosuperior labrum

PATHOLOGY

General Features

- Etiology
  - Congenital variants

Staging, Grading, & Classification

- Superior labrum
  - Type 1: No recess, labrum firmly attached
  - Type 2: Small superior recess
  - Type 3: Deep superior recess

Gross Pathologic & Surgical Features

- Sublabral foramen & Buford complex
  - No hemorrhage or fraying
- Superior recess
  - Stable biceps anchor
  - Recess is synovial lined without hemorrhage

CLINICAL ISSUES

Demographics

- Gender
  - M = F
- Epidemiology
  - SLF: 10%
  - Buford complex: 1-2%
  - Superior recess: 75%

Natural History & Prognosis

- Asymptomatic

Treatment

- None
  - "Fixing" a SLF or superior recess → restricted ROM

DIAGNOSTIC CHECKLIST

Consider

- Variant if labral "abnormality" between 12-3 o'clock

SELECTED REFERENCES

1. Shortt CP et al: Association of glenoid morphology and anterosuperior labral variation. J Comput Assist Tomogr. 33(4):584-6, 2009
2. Waldt S et al: Variants of the superior labrum and labro-bicipital complex: a comparative study of shoulder specimens using MR arthrography, multi-slice CT arthrography and anatomical dissection. Eur Radiol. 16(2):451-8, 2006
3. Tuite MJ et al: Sublabral foramen and buford complex: inferior extent of the unattached or absent labrum in 50 patients. Radiology. 223(1):137-42, 2002

NORMAL LABRAL VARIANTS

*(**Left**) Oblique sagittal MR arthrogram T1WI FS in a patient with a sublabral foramen shows the anterosuperior labrum unattached to the glenoid rim, and there is a thick MGHL. (**Right**) Axial MR arthrogram PDWI FS in the same patient shows the unattached anterosuperior labrum and the middle glenohumeral ligament beginning to fuse with the anterior capsule. In an SLF, the labrum is separated from the glenoid rim at the labral-chondral junction and is otherwise intact.*

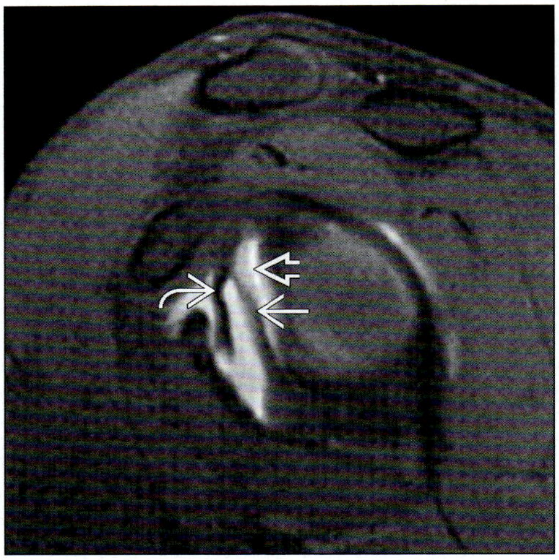

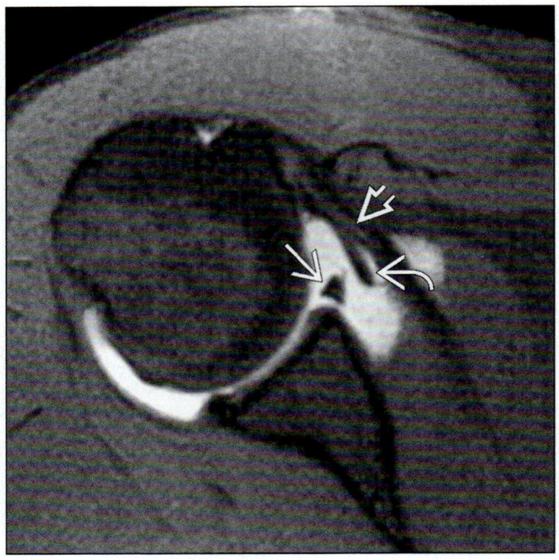

*(**Left**) Sagittal graphic of a Buford complex shows congenital absence of the anterosuperior labrum between the origins of the middle glenohumeral ligament and anterior band inferior glenohumeral ligament. There is also a thick, cord-like MGHL. The Buford complex is seen in 1-2% of people and is not unstable. (**Right**) Sagittal MR arthrogram T1WI FS in a patient with a Buford complex shows absence of the anterosuperior labrum and a thick, cord-like MGHL.*

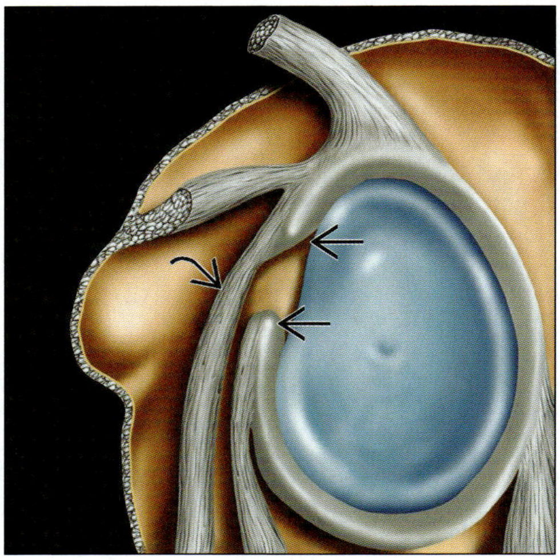

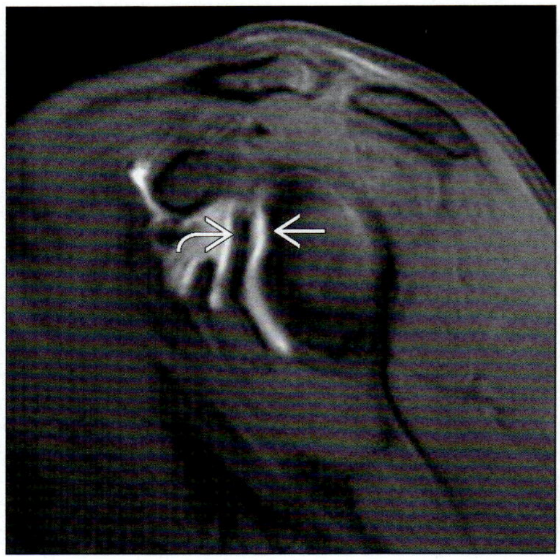

*(**Left**) Axial MR arthrogram T1WI FS in the same patient shows a bare anterosuperior glenoid rim and a nearby low signal structure. The differential diagnosis on this slice is a normal variant vs. a torn labrum. (**Right**) Axial MR arthrogram T1WI FS more inferiorly in the same patient shows reappearance of normal labrum with continuation of the nearby thick MGHL confirming this as a Buford complex. More inferiorly (not shown), the MGHL fuses with the anterior capsule.*

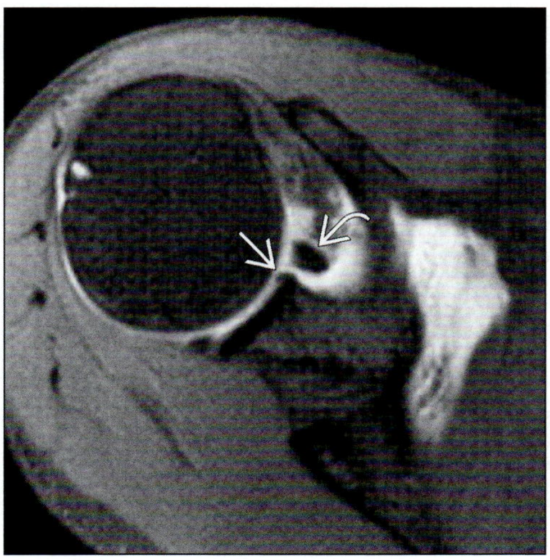

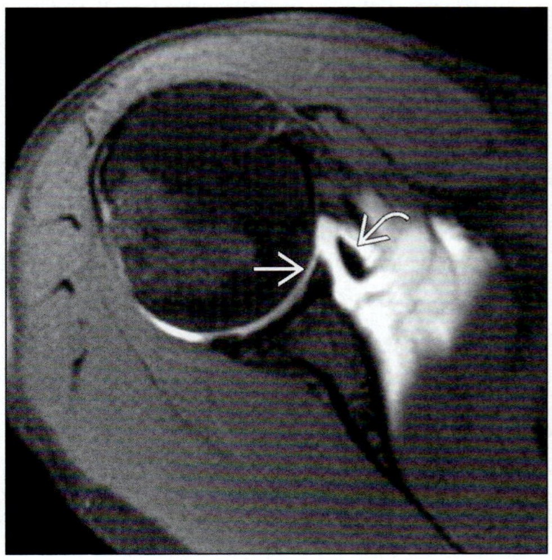

NORMAL LABRAL VARIANTS

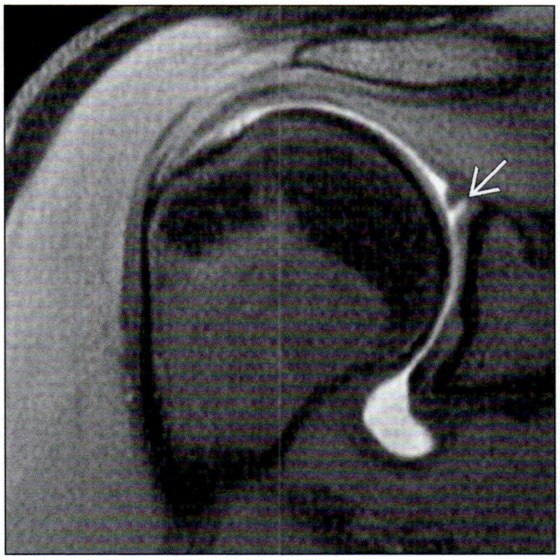

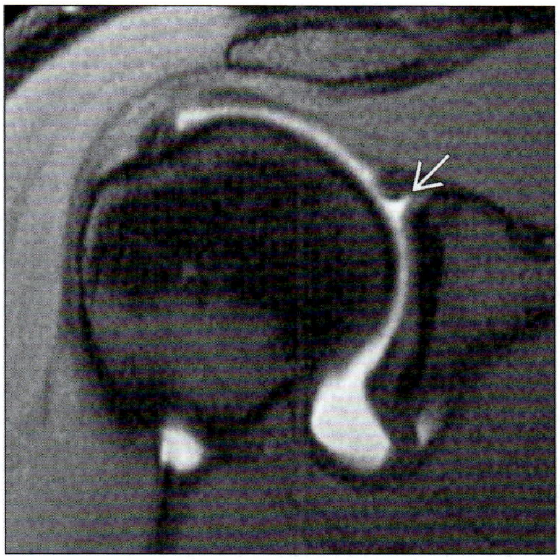

(Left) *Coronal oblique MR arthrogram T1WI FS in a patient with an arthroscopically proven normal variant superior recess shows contrast extending between the superior labrum and adjacent glenoid articular cartilage.* ***(Right)*** *Coronal oblique MR arthrogram T1WI FS in the same patient again shows the superior recess. The high signal should curve medially at the labral-chondral junction and should be a blind-ending pouch. The superior recess can be of variable depth.*

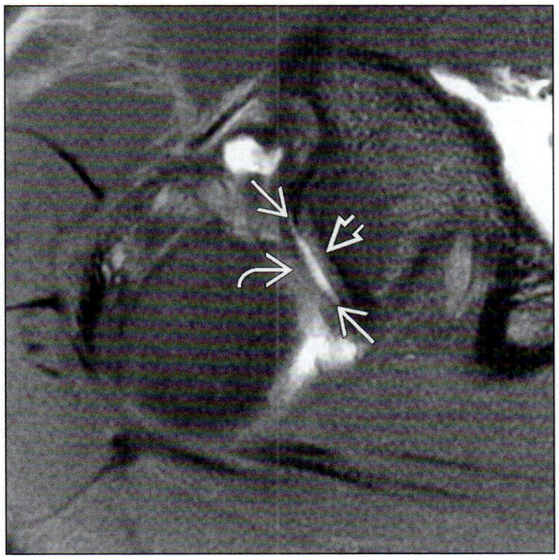

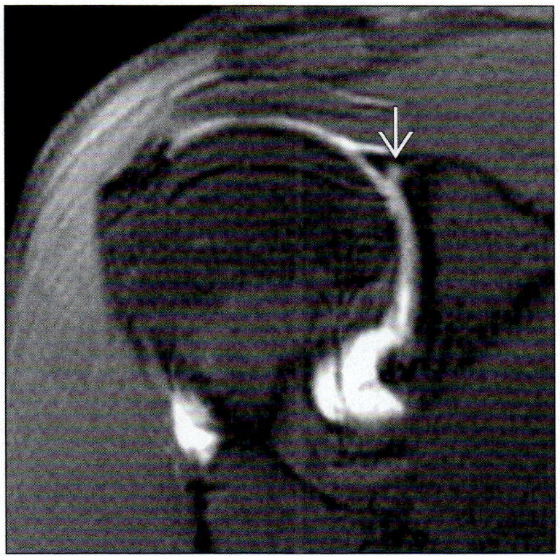

(Left) *Axial MR arthrogram PDWI FS in a patient with a superior recess shows contrast in the recess between the superior labrum and the glenoid articular cartilage. On MR arthrography, the superior recess can extend posterior to the biceps anchor in 55% of cases.* ***(Right)*** *Coronal oblique MR arthrogram T2WI FS of a superior recess shows slight widening (< 3 mm) within the recess. Most superior recesses are never wider than 3 mm on MR arthrography.*

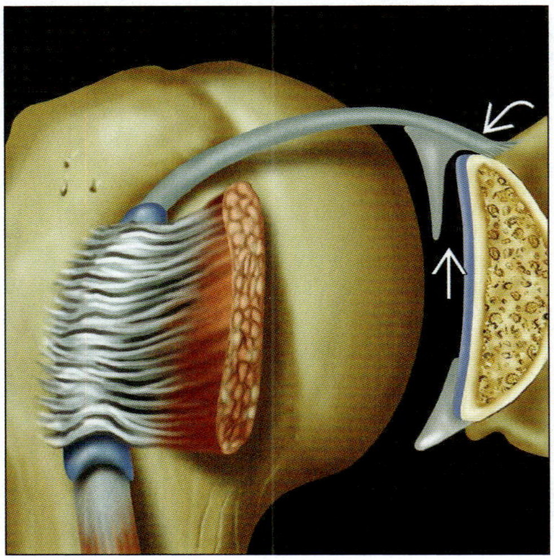

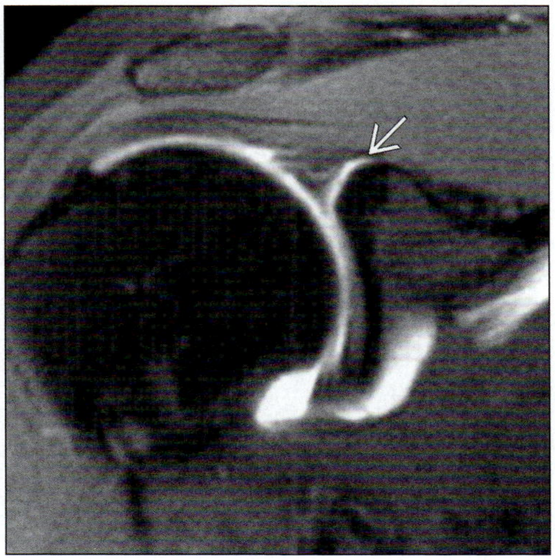

(Left) *AP graphic of a type 3 superior labral attachment with a deep superior recess. Note that the recess is a blind-ending pouch and that the labrum is not completely detached, and the biceps anchor has a stable attachment to the labrum and supraglenoid tubercle. The deepest reported recess measured 1.4 cm.* ***(Right)*** *Coronal oblique MR arthrogram T1WI FS shows an arthroscopically proven deep superior recess. The recess was fully lined with synovium, and the biceps anchor was stable.*

ADHESIVE CAPSULITIS, SHOULDER

Key Facts

Terminology
- Frozen shoulder, scapulohumeral periarthritis
- ↓ ROM from tight, fibrotic capsule
  - Predilection for rotator interval, inferior capsule, biceps anchor

Imaging
- Fibrosis in rotator interval, thick coracohumeral ligament
- Thick capsule/synovium at rotator interval and in axillary recess
- Decreased joint capacity at arthrography
- **MR**: Can be normal, or
  - Low signal fibrosis in anterior interval
  - "Subcoracoid triangle" sign
- MR arthrogram
  - < 8-10 mL joint capacity before contrast extravasation at arthrography
  - Thick coracohumeral ligament ≥ 4 mm
  - Thick anterior interval joint capsule ≥ 7 mm
  - ↓ volume of axillary recess < 0.7 mL
  - Thick axillary recess joint capsule > 3 mm

Top Differential Diagnoses
- Calcific tendinopathy
- Glenohumeral osteoarthritis
- Rotator cuff impingement

Pathology
- Etiology: Idiopathic most common

Clinical Issues
- Shoulder pain
- Decreased range of motion
- Treatment
  - Physical therapy
  - Anesthetic and corticosteroid joint injection

(Left) AP graphic of adhesive capsulitis shows thickening of the axillary recess joint capsule ➔ with synovitis ➔. The axillary recess and rotator interval are the 2 main sites affected by adhesive capsulitis. (Right) Coronal oblique T2WI FS MR shows a patient with adhesive capsulitis. There is thickening and high signal of the axillary recess capsule and synovium ➔. The axillary recess is also small from capsular fibrosis, seen better on MR arthrography.

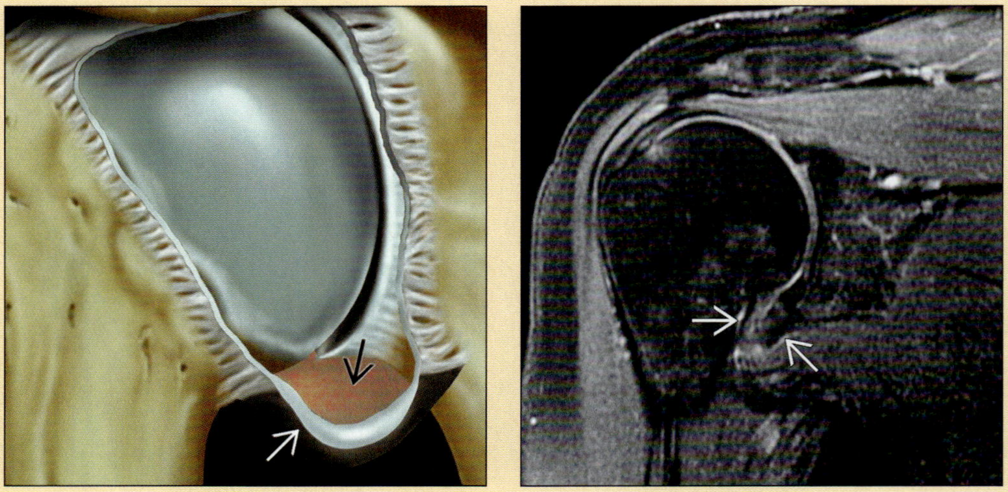

(Left) Oblique sagittal T1WI MR from the same patient shows fibrosis in the rotator interval causing thickening of the coracohumeral ligament ➔. There is also thickening of the adjacent joint capsule ➔. Adhesive capsulitis often affects the rotator interval. (Right) Coronal oblique MR arthrogram T2WI FS shows decreased fluid volume in the axillary recess ➔ with a thickened capsule ➔. Some of the injected contrast has extravasated from the subscapularis recess ➔.

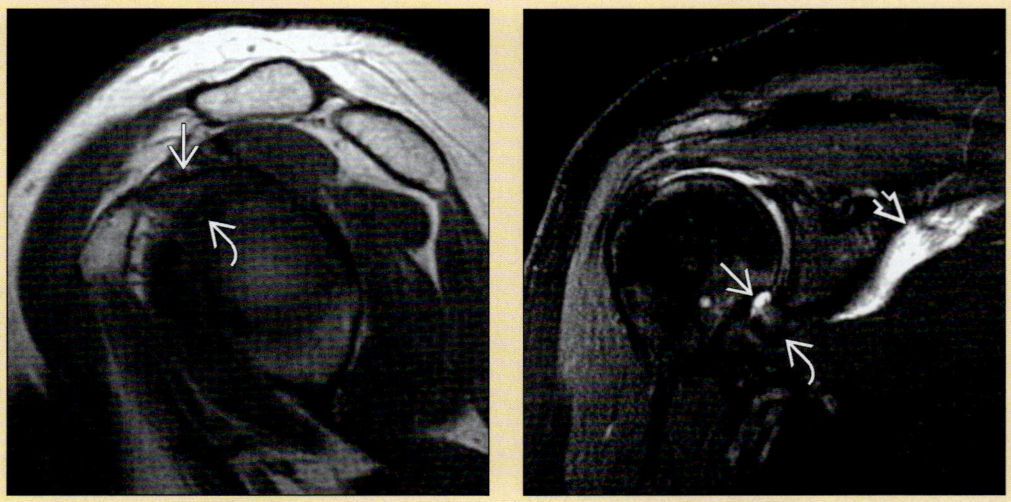

ADHESIVE CAPSULITIS, SHOULDER

TERMINOLOGY

Abbreviations
- Adhesive capsulitis (AC)

Synonyms
- Frozen shoulder, scapulohumeral periarthritis

Definitions
- ↓ range of motion (ROM) from tight, fibrotic capsule

IMAGING

General Features
- Best diagnostic clue
  - Fibrosis in rotator interval
  - Thick capsule/synovium in rotator interval and axillary recess
  - Decreased joint capacity at arthrography
- Location
  - Predilection for rotator interval, inferior capsule, biceps anchor

Radiographic Findings
- Arthrography
  - < 8-10 mL joint capacity before contrast extravasation (controversial)
    - False-negative if full thickness rotator cuff tear

MR Findings
- T1WI
  - Low signal fibrosis in rotator interval
  - Thick coracohumeral ligament (CHL)
  - Thick capsule in rotator interval and axillary recess
  - "Subcoracoid triangle" sign: Loss of fat in triangle between coracohumeral ligament, coracoid process, and joint capsule
- T2WI FS
  - ↑ signal in rotator interval & joint capsule
  - Pericapsular edema
- T1WI C+ FS
  - Enhancing, thickened capsule in rotator interval
- MR arthrography
  - Thick coracohumeral ligament ≥ 4 mm
  - Thick rotator interval joint capsule ≥ 7 mm
  - < 8-10 mL joint capacity
  - ↓ volume of axillary recess < 0.7 mL
  - Axillary recess joint capsule/synovium > 3-4 mm

Imaging Recommendations
- Best imaging tool
  - MR arthrography
- Protocol advice
  - Include non-fat-suppressed oblique sagittal image

DIFFERENTIAL DIAGNOSIS

Calcific Tendinopathy, Shoulder
- Calcification on radiograph

Osteoarthritis, Shoulder
- Inferior humeral head osteophyte

Rotator Cuff Impingement
- Pain with impingement tests

PATHOLOGY

General Features
- Etiology
  - Idiopathic: Common
  - Also: Diabetes, hypothyroidism, severe trauma
- Associated abnormalities
  - Rotator cuff tendinopathy and tear (RCT)
    - Shoulder capacity normal if full thickness RCT

Staging, Grading, & Classification
- Stage 1: Symptoms 0-3 months, limited ROM, pain
- Stage 2: Symptoms 3-9 months, "freezing" stage, severely restricted ROM, pain
  - Most common stage at presentation
- Stage 3: Symptoms 9-15 months, "frozen" stage, severe stiffness, minimal pain
- Stage 4: Symptoms 15-24 months, "thawing" stage, improving ROM, minimal pain

Gross Pathologic & Surgical Features
- Synovial inflammation
- Capsular fibrosis

CLINICAL ISSUES

Presentation
- Most common signs/symptoms
  - Shoulder pain
  - Decreased active and passive range of motion, especially forward flexion and external rotation

Demographics
- Age
  - 40-60 years old
- Gender
  - F > > M

Natural History & Prognosis
- Self-limited disease, may take up to 2 years to resolve

Treatment
- Physical therapy
- Anesthetic and corticosteroid joint injection
- Distension arthrography & manipulation
  - Often too painful to perform without sedation
- Manipulation under anesthesia, capsulotomy

DIAGNOSTIC CHECKLIST

Image Interpretation Pearls
- Evaluate rotator interval on sagittal short TE, non-fat-suppressed images

SELECTED REFERENCES

1. Kim KC et al: Adhesive capsulitis of the shoulder: dimensions of the rotator interval measured with magnetic resonance arthrography. J Shoulder Elbow Surg. 18(3):437-42, 2009
2. Mengiardi B et al: Frozen shoulder: MR arthrographic findings. Radiology. 233(2):486-92, 2004
3. Lee MH et al: Adhesive capsulitis of the shoulder: diagnosis using magnetic resonance arthrography, with arthroscopic findings as the standard. J Comput Assist Tomogr. 27(6):901-6, 2003

(Left) *Oblique sagittal T1WI MR in a patient with adhesive capsulitis shows mild thickening and ↑ signal in the coracohumeral ligament →. A coracohumeral ligament thickness of ≥ 4 mm is 60% sensitive and 95% specific for adhesive capsulitis.* ***(Right)*** *Oblique sagittal T2WI FS MR in the same patient shows mild edema around the coracohumeral ligament → and the rotator interval capsule →. Edema in the rotator interval is common in the early stages of adhesive capsulitis.*

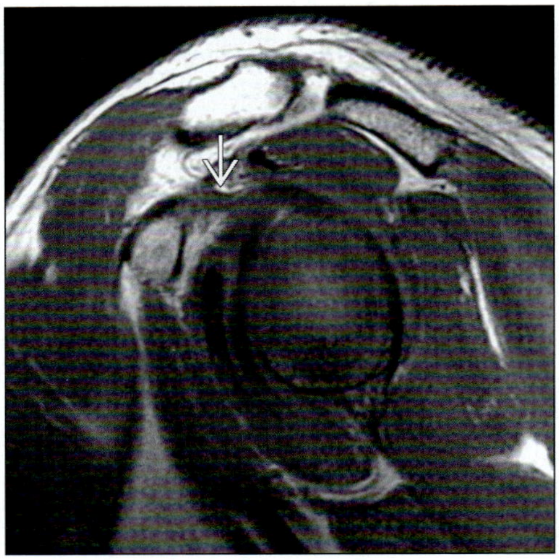

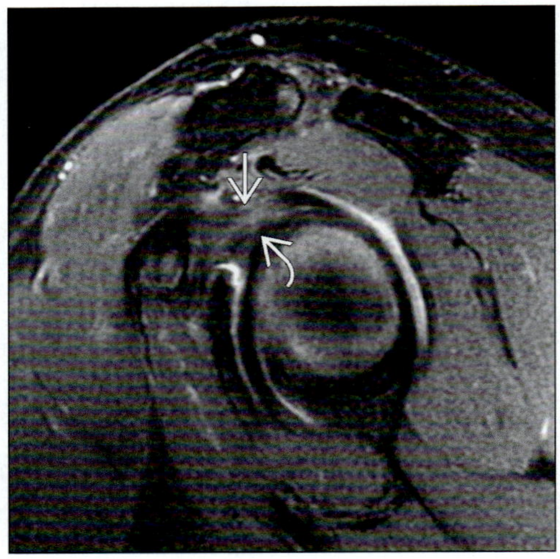

(Left) *Oblique sagittal T1WI MR in a patient with adhesive capsulitis shows moderate thickening of the coracohumeral ligament →, as well as the rotator interval joint capsule →. Thickening of the CHL often begins dorsally at the joint capsule and progresses toward the coracoid.* ***(Right)*** *Oblique sagittal T1WI MR shows a patient with AC and severe involvement of the rotator interval, with fibrosis around the coracohumeral ligament → and in the subcoracoid fat →.*

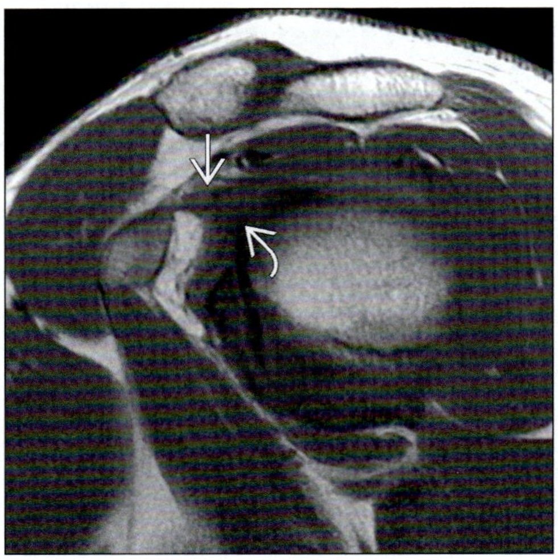

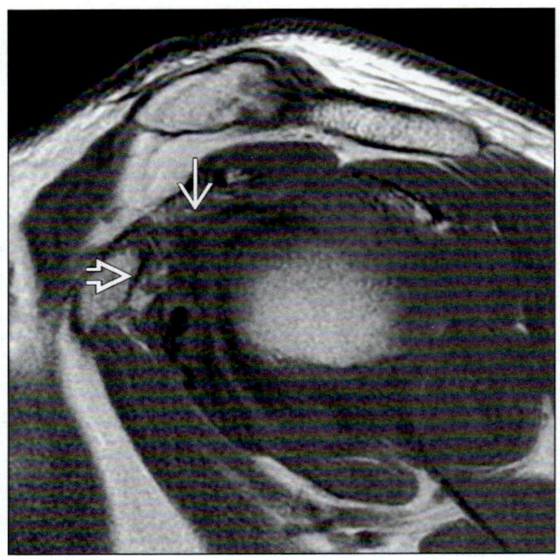

(Left) *Oblique sagittal T1WI MR of a patient with severe adhesive capsulitis and a "subcoracoid triangle" sign shows complete loss of the subcoracoid fat →. This is often seen late in the disease when there is marked shoulder stiffness.* ***(Right)*** *Coronal oblique MR arthrogram T2WI FS in a patient with adhesive capsulitis shows thickening of the capsule → and a small volume → in the axillary recess. There is also extravasation of contrast from the glenohumeral joint →.*

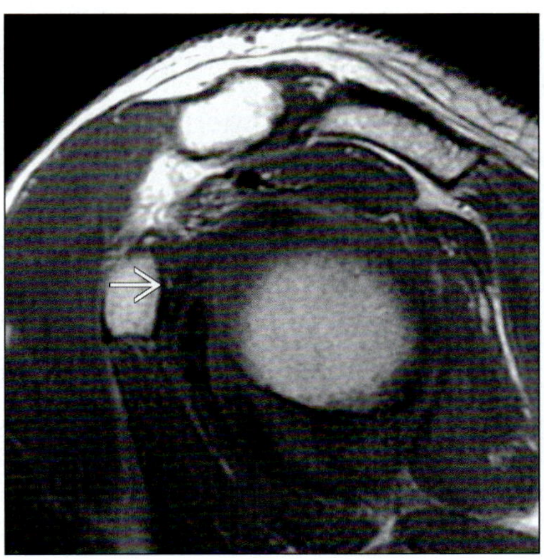

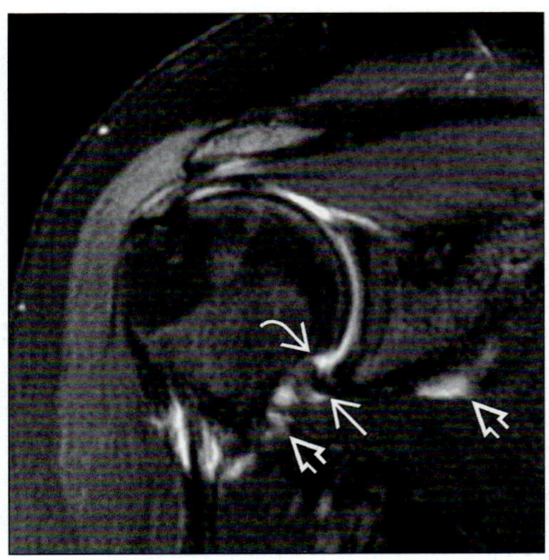

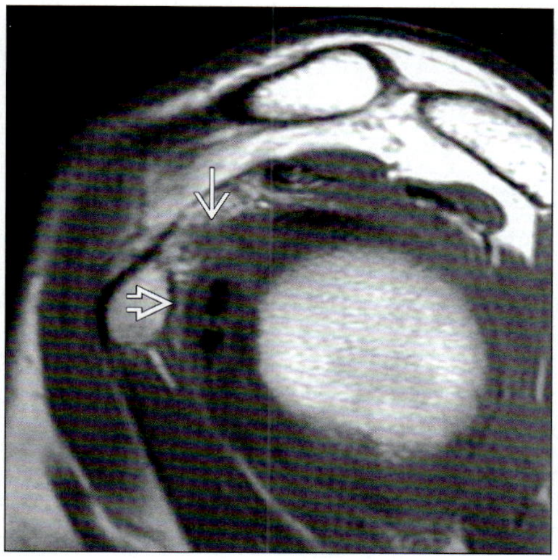

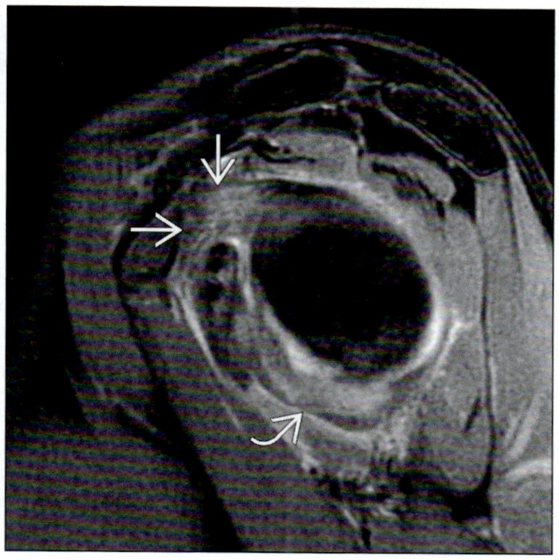

(Left) *Oblique sagittal T1WI MR of a patient with adhesive capsulitis shows fibrosis around the coracohumeral ligament and in the subcoracoid fat in the rotator interval.* ***(Right)*** *Oblique sagittal T2WI FS MR in the same patient shows mild edema in the rotator interval. There is also pericapsular edema in the axillary recess. These are the 2 most common sites for adhesive capsulitis, and pericapsular edema is often seen in the 1st 9 months of the disease.*

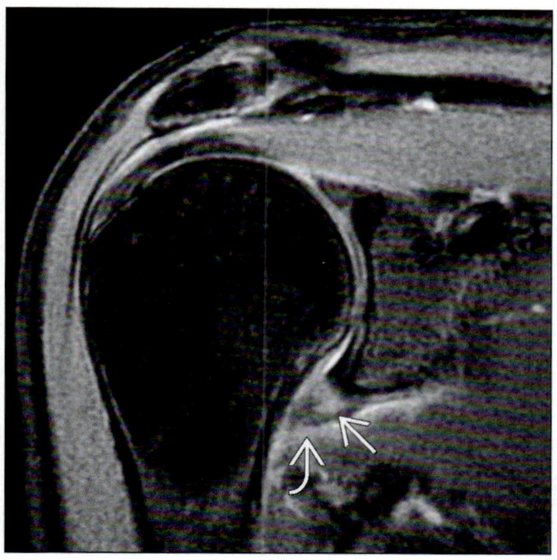

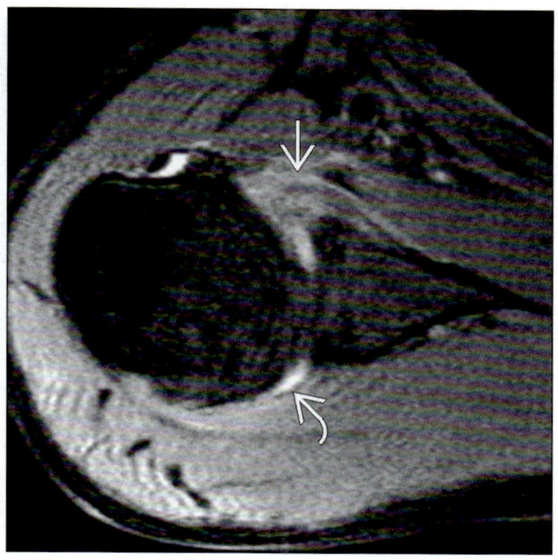

(Left) *Coronal oblique T2WI FS MR in the same patient shows capsular thickening and edema and pericapsular edema in the axillary recess. The normal axillary recess capsule can appear thick on conventional MR images in adduction but should be low signal on all sequences.* ***(Right)*** *Axial PDWI FS MR in the same patient shows increased signal in the anteroinferior capsule, with pericapsular edema. Note the relative sparing of the posterior capsule.*

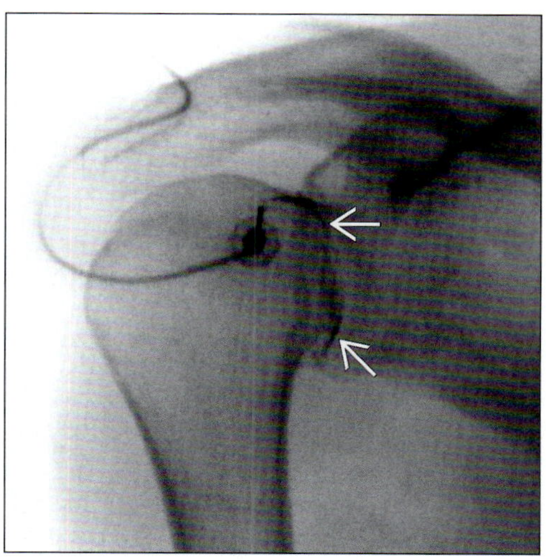

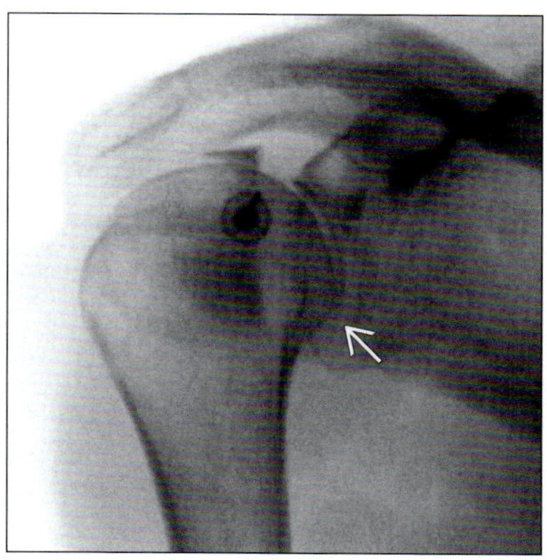

(Left) *AP arthrogram in the same patient shows contrast injected into the glenohumeral joint, confirming an intraarticular location.* ***(Right)*** *AP arthrogram in the same patient after the injection of anesthetic and corticosteroid shows dilution of the contrast and a tight axillary recess. Intraarticular injection of corticosteroid can improve pain and shorten the time course of adhesive capsulitis, and injection is less invasive than arthroscopic capsulotomy.*

BANKART LESION

Key Facts

Terminology
- Detached anteroinferior labrum after anterior dislocation

Imaging
- Anteroinferior labrum and associated ABIGL detached from glenoid rim
- Torn scapular periosteum
- High T2 signal completely across anteroinferior labrum on axial images
  - Tear usually at labral-cartilage junction
- Absent/irregular anteroinferior labrum
  - Chronic Bankart lesion
- Direct MR arthrography
  - Contrast extends completely across anteroinferior labrum in acute lesions
  - Higher sensitivity than conventional MR
  - ABER images: Improved sensitivity for anterior labral tears

Top Differential Diagnoses
- Bankart fracture
- Bankart variation lesions
- Normal labral variants

Clinical Issues
- Prior anterior dislocation
- Shoulder pain
- Anterior instability and apprehension, particularly with abduction and external rotation
- < 40 years old
- M > F

Diagnostic Checklist
- Look for Hill-Sachs fracture or bone contusion

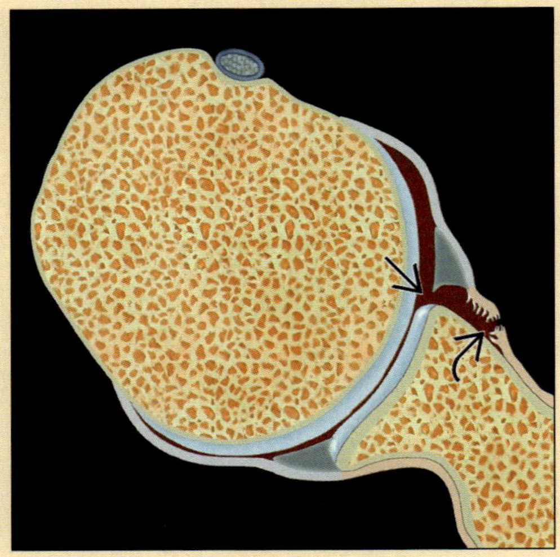

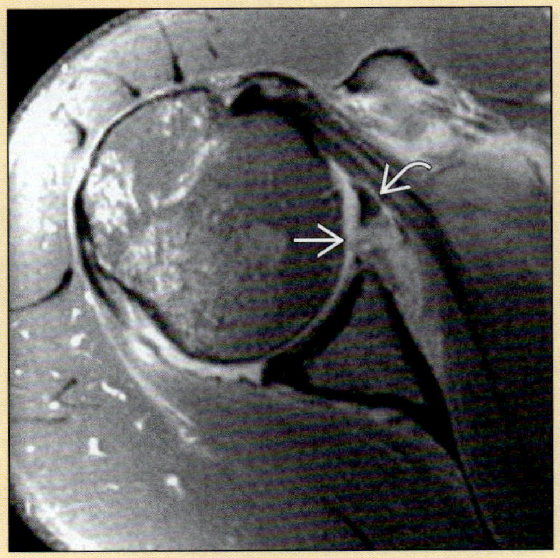

(Left) *Axial graphic of a classic Bankart lesion. There is a complete tear through the anteroinferior labrum → and torn scapular periosteum ↪, causing labral detachment. The ABIGL, the main anterior stabilizer, attaches to the labrum and therefore becomes ineffective. The labrum-articular cartilage junction is the most common site to tear in Bankart lesions.* ***(Right)*** *Axial PDWI FS MR of a Bankart lesion shows a tear through the anterior labrum → causing a detached segment of labrum ↪.*

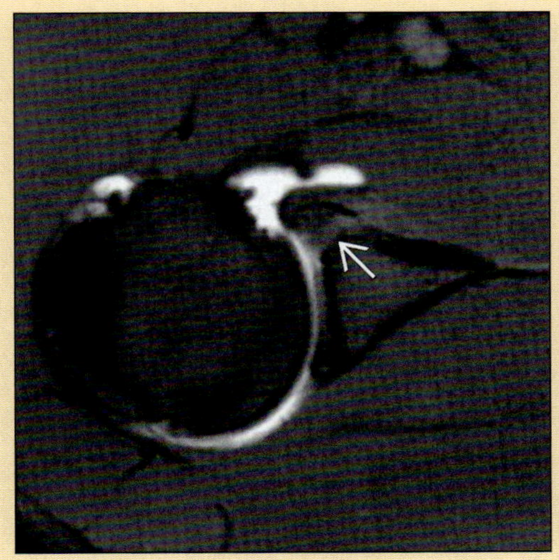

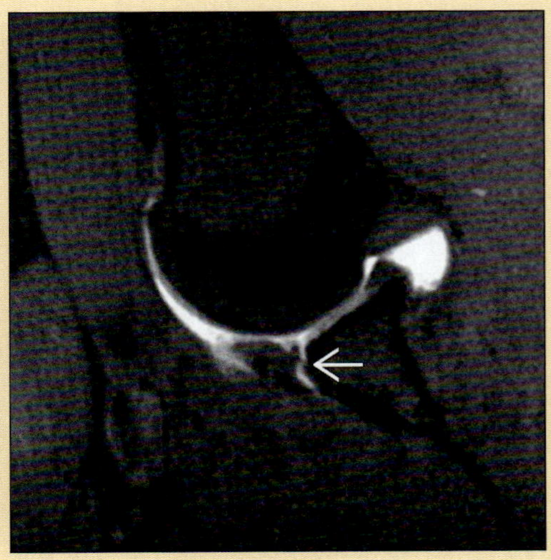

(Left) *Axial MR arthrogram PDW FS image shows high signal between the labrum and glenoid rim → indicating a Bankart lesion. The signal is not as bright as gadolinium, probably from fibrovascular tissue within the tear site.* ***(Right)*** *ABER MR arthrogram T1W FS image in the same patient now shows gadolinium signal between the detached labrum and the glenoid rim → confirming the Bankart lesion. ABER images, which cause traction on the anteroinferior labrum, can make Bankart lesions more visible.*

BANKART LESION

TERMINOLOGY

Abbreviations

- Anterior band inferior glenohumeral ligament (ABIGL)

Synonyms

- Anteroinferior labral tear

Definitions

- Detached anteroinferior labrum after anterior shoulder dislocation

IMAGING

General Features

- Best diagnostic clue
  - High T2 signal through labrum on MR images
    - History of anterior dislocation
- Location
  - Anteroinferior labrum between 3 and 6 o'clock
- Size
  - Small or large tears that involve anteroinferior labrum, extending superiorly
- Morphology
  - Anteroinferior labrum and ABIGL detached from glenoid rim
  - Torn scapular periosteum

Radiographic Findings

- Anterior dislocation
- Hill-Sachs impaction fracture
  - Best seen on internal rotation view
  - Seen in 80% after anterior dislocation
- No anterior glenoid rim fracture (if fractured, considered a Bankart fracture rather than Bankart lesion)

CT Findings

- Hill-Sachs impaction fracture
- Arthrography
  - Contrast extends completely across anteroinferior labrum
  - Absent anteroinferior labrum
  - Sensitivity: 95%

MR Findings

- PD/intermediate FS
  - High signal completely across anteroinferior labrum on axial images
    - Tear usually at labral-cartilage junction
    - Tear may extend into anterosuperior labrum or inferior labrum on oblique coronal images
    - Recent dislocation: Edema around anterior capsule + anterior glenoid bone marrow edema
  - Absent anteroinferior labrum
    - Chronic Bankart lesion
    - Increased signal adjacent to glenoid rim
    - Displaced labrum resorbed or scarred to capsule
  - Hill-Sachs bone contusion or impaction fracture
  - Sensitivity: 70%
    - Higher for acute Bankart lesions, lower for chronic instability
    - 90% sensitivity with 3 Tesla MR
- Optimal pulse sequence controversial
  - Short TE sequences (T2* gradient echo, PD)
    - Pro: More Bankart lesions visible
      - Gradient echo: Thin section volume acquisition more sensitive for tears
    - Con: Increased signal in normal labrum
      - Myxoid change
      - "Magic angle" artifact
  - T2WI
    - Pro: High signal in labrum is always abnormal
    - Con: Low sensitivity
      - Some labral tears fill with fibrovascular tissue (↓ T2 signal)
- Indirect MR arthrography
  - Pro: Intravenous injection does not require image guidance
  - Con: No distension to outline labrum, normal labrum enhances
  - Sensitivity: > 90%
- Direct MR arthrography
  - Contrast extends completely across anteroinferior labrum
  - Absent anteroinferior labrum
  - T1WI: High signal-to-noise ratio
  - Abduction external rotation (ABER) images show tear better
  - Sensitivity: 95%

Imaging Recommendations

- Best imaging tool
  - Direct MR arthrography with ABER positioning
- Protocol advice
  - Angled axial images perpendicular to anteroinferior labrum, or ABER position images may show tear better

DIFFERENTIAL DIAGNOSIS

Bankart Variation Lesions

- Partial Bankart tear: Labrum not completely detached
- Also includes Perthes, ALPSA, GLAD/GARD, and ABIGL/capsule tears

Perthes Lesion

- Scapular periosteum intact, labrum not displaced

ALPSA Lesion

- Scapular periosteum intact, labrum displaced
  - Chronic ALPSA lesion: Synovializes over labrum, which is scarred down to scapular neck

GLAD/GARD Lesion

- Anterior glenoid rim chondral lesion and labral tear

IGL Tears

- Common in older patients after anterior dislocation

HAGL Lesion

- ABIGL disrupted at humeral insertion

Normal Labral Variants

- Sublabral foramen and Buford complex occur in anterosuperior labrum

PATHOLOGY

General Features

- Etiology

BANKART LESION

- Anterior glenohumeral dislocation
  - Injury to anterior labrocapsular structures
    - Bankart lesion labral tear
    - Bankart fracture of anteroinferior glenoid rim
    - Bankart variation lesions
    - HAGL lesion
- Associated abnormalities
  - Type V SLAP tear: SLAP tear extends into Bankart lesion
  - Floating ABIGL: Bankart lesion and HAGL lesion
  - Extensive labral tear: > 180° can involve anteroinferior labrum

Gross Pathologic & Surgical Features

- Detachment of labrum and associated ABIGL, hemorrhage

CLINICAL ISSUES

Presentation

- Most common signs/symptoms
  - Prior anterior dislocation
  - Shoulder pain
  - Anterior instability and apprehension, particularly with abduction and external rotation
- Other signs/symptoms
  - Drawer test
    - Sitting with forearm on lap, humerus pushed against glenoid and forward; positive sign with anterior displacement
  - Sulcus test
    - Sitting with arm relaxed, humerus pushed down (caudally) and sulcus forms with inferior displacement of humeral head
  - Fulcrum test
    - Supine with arm abducted 90° and extended while other hand under shoulder as a fulcrum; positive if anterior subluxation
  - Crank or apprehension test
    - Sitting with arm abducted 90°, pull back on wrist (external rotation) while other hand on back of shoulder; positive if patient is apprehensive, feeling it is going to dislocate

Demographics

- Age
  - < 40 years old
- Gender
  - M > F
- Epidemiology
  - Glenohumeral dislocation: Common; 50% of all joint dislocations
  - Anterior dislocation: 95% of all glenohumeral dislocations
  - Bankart lesion occurs in 75% of anterior dislocations in patients < 40 years old

Natural History & Prognosis

- Most Bankart labral tears do not heal
- Recurrent dislocation in 50%
  - 50% of recurrent dislocators require surgery

Treatment

- Conservative
  - Arm sling to protect stability
  - At 3 weeks, begin rotator cuff strengthening
- Bankart repair, open or arthroscopic
  - Capsulolabral reconstruction
  - Refixation of labrum back to glenoid rim

DIAGNOSTIC CHECKLIST

Consider

- Bankart lesion should be sought in presence of superior humeral head posterolateral bone contusion (Hill-Sachs)

Reporting Tips

- Extent of tear
- Size of Hill-Sachs defect

SELECTED REFERENCES

1. Dietrich TJ et al: Articular cartilage and labral lesions of the glenohumeral joint: diagnostic performance of 3D water-excitation true FISP MR arthrography. Skeletal Radiol. 39(5):473-80, 2010
2. Jung JY et al: Comparison study of indirect MR arthrography and direct MR arthrography of the shoulder. Skeletal Radiol. 38(7):659-67, 2009
3. Schreinemachers SA et al: Is a single direct MR arthrography series in ABER position as accurate in detecting anteroinferior labroligamentous lesions as conventional MR arthography? Skeletal Radiol. 38(7):675-83, 2009
4. Song HT et al: The usefulness of virtual MR arthroscopy as an adjunct to conventional MR arthrography in detecting anterior labral lesions of the shoulder. AJR Am J Roentgenol. 192(4):W149-55, 2009
5. Lee MJ et al: Gradient-recalled echo sequences in direct shoulder MR arthrography for evaluating the labrum. Skeletal Radiol. 37(1):19-25, 2008
6. Antonio GE et al: First-time shoulder dislocation: High prevalence of labral injury and age-related differences revealed by MR arthrography. J Magn Reson Imaging. 26(4):983-91, 2007
7. Van der Woude HJ et al: MR arthrography in glenohumeral instability. JBR-BTR. 90(5):377-83, 2007
8. Robinson G et al: Normal anatomy and common labral lesions at MR arthrography of the shoulder. Clin Radiol. 61(10):805-21, 2006
9. Song HT et al: Anterior-inferior labral lesions of recurrent shoulder dislocation evaluated by MR arthrography in an adduction internal rotation (ADIR) position. J Magn Reson Imaging. 23(1):29-35, 2006
10. Ito Y et al: Computerized assessment of Bankart lesions under tension with magnetic resonance arthrography. J Shoulder Elbow Surg. 14(3):247-51, 2005
11. Takase K et al: Intraarticular lesions in traumatic anterior shoulder instability: a study based on the results of diagnostic imaging. Acta Orthop. 76(6):854-7, 2005
12. Waldt S et al: Anterior shoulder instability: accuracy of MR arthrography in the classification of anteroinferior labroligamentous injuries. Radiology. 237(2):578-83, 2005
13. Applegate GR et al: Chronic labral tears: value of magnetic resonance arthrography in evaluating the glenoid labrum and labral-bicipital complex. Arthroscopy. 20(9):959-63, 2004
14. McCauley TR: MR imaging of the glenoid labrum. Magn Reson Imaging Clin N Am. 12(1):97-109, vi-vii, 2004
15. Rafii M: Non-contrast MR imaging of the glenohumeral joint. Part II. Glenohumeral instability and labrum tears. Skeletal Radiol. 33(11):617-26, 2004

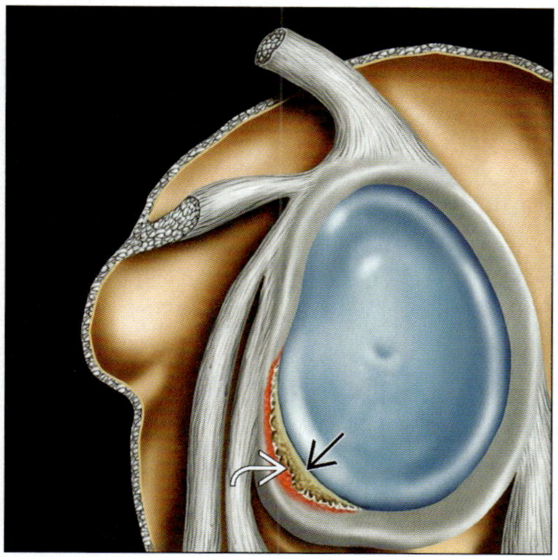

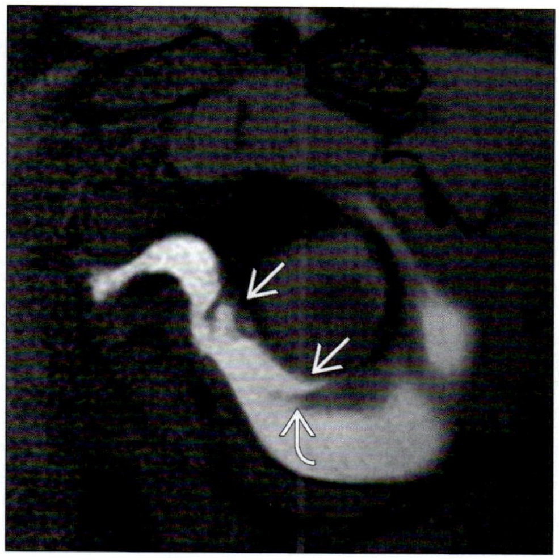

(Left) *Sagittal graphic shows a Bankart lesion → with detachment of the anteroinferior labrum. There is hemorrhage (depicted in red) at the tear site ➔ indicating a recent injury. The most common site for a Bankart lesion is at the labral-articular cartilage junction at the glenoid rim.* ***(Right)*** *Oblique sagittal MR arthrogram T1W FS image shows a Bankart lesion ➔ in a patient with a prior anterior dislocation. A portion of the detached anterior labrum ➔ is seen.*

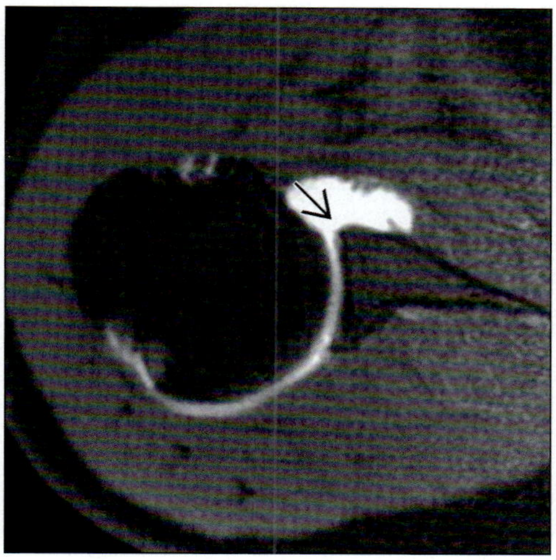

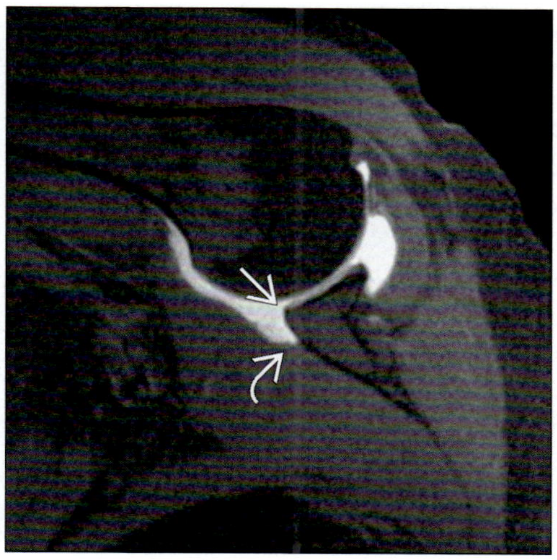

(Left) *Axial MR arthrogram T1W FS image of a patient with chronic anterior instability and a recent dislocation episode shows absent anterior labrum with a bare glenoid rim →. The labrum may be displaced or resorbed in patients with chronic tears.* ***(Right)*** *ABER MR arthrogram T1W FS image in the same patient shows the chronic Bankart lesion with bare glenoid rim ➔. The anterior band IGL is not seen, and there is medial stripping of the anterior capsule attachment to the scapula neck ➔.*

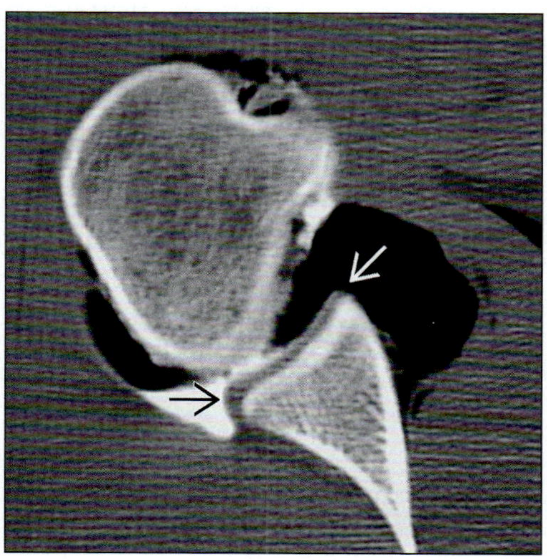

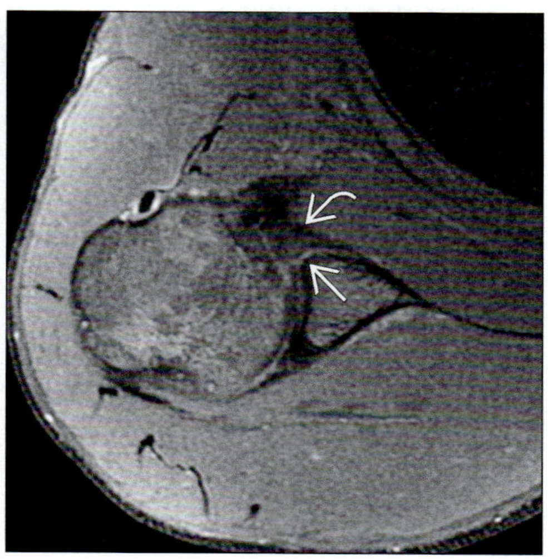

(Left) *Axial bone CT double contrast arthrogram of a chronic Bankart lesion shows an absent anterior labrum with bare glenoid rim ➔. The posterior labrum is normal →.* ***(Right)*** *Axial PDWI FS MR of an old Bankart lesion in a patient with anterior instability shows increased signal ➔ between the anterior labrum ➔ and the glenoid rim. Bankart tears can be subtle on conventional MR images in the absence of a joint effusion or recent dislocation.*

BANKART VARIATION LESIONS

Key Facts

Terminology

- Anteroinferior labral tear without complete detachment (as seen in true Bankart lesion)
  - Partial Bankart labral tear
  - Perthes lesion (labral tear with intact periosteum)
  - Also: ALPSA lesion, GLAD/GARD lesion

Imaging

- **MR**: Increased signal extends partly through labrum
  - Partial Bankart tear: Some intact labral tissue prevents complete detachment of labrum
  - Perthes lesion: Labrum is torn from glenoid rim but held in place by intact periosteum
  - Hill-Sachs bone contusion or impaction fracture
  - Sensitivity on routine MR < 70%
- **Direct MR arthrogram**: Contrast extends into labral tear
  - Sensitivity: 50-70% on axial images
- ABER images: 10-20% ↑ sensitivity over routine axial MR arthrogram images
  - Traction on anterior labrum widens tear
  - Sections are perpendicular to anteroinferior labrum so less partial averaging

Top Differential Diagnoses

- Bankart lesion
  - Complete anterior labral tear with detachment
- IGL tears
  - Tear is more lateral, involving anterior capsule and anterior band IGL

Clinical Issues

- Anterior instability and apprehension
- Shoulder pain
  - Symptoms may be less severe than with true Bankart labral tear

__(Left)__ Lateral graphic shows a partial Bankart tear of the anteroinferior labrum ⇒. Partial Bankart tears are considered a Bankart variation lesion, because a true Bankart lesion is a complete tear with detachment of the anteroinferior labrum. __(Right)__ Axial MR arthrogram T1WI FS shows contrast extending partly into the anteroinferior labrum ➡ from a partial Bankart lesion. These lesions can result from severe anterior subluxation and may be associated with instability.

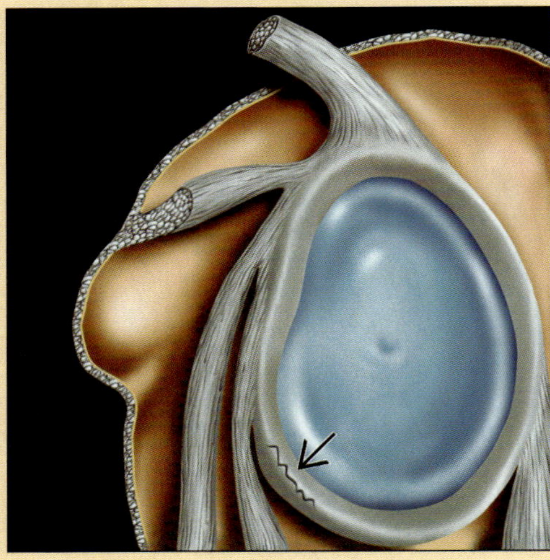

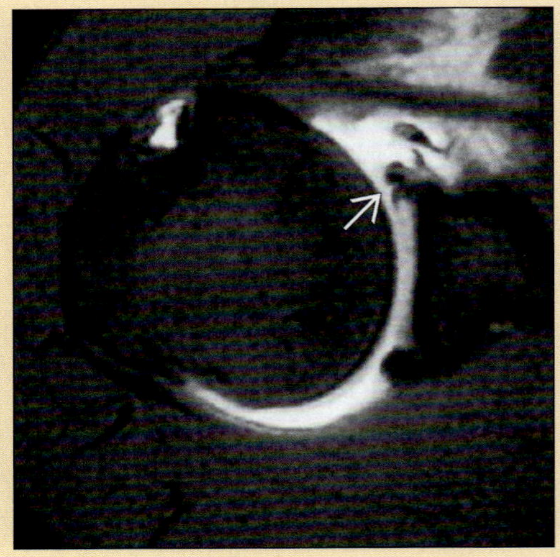

__(Left)__ Axial graphic shows a complete labral tear ⇒ but with a periosteal sleeve ➡ still attached, making this a Bankart variation lesion. If the labrum is not displaced from the glenoid rim, it is a Perthes lesion. __(Right)__ Axial MR arthrogram T1WI FS shows a Perthes lesion with contrast extending across the labrum ➡ and an intact periosteal sleeve ➡ connecting the labrum to the anterior scapular neck. ABER images may be needed to distinguish a partial Bankart from a Perthes lesion.

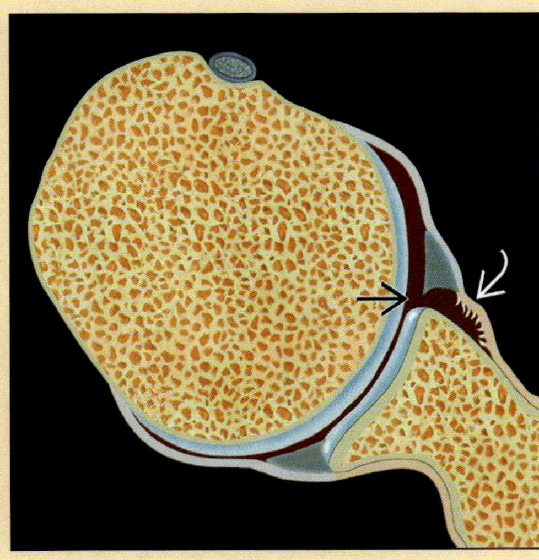

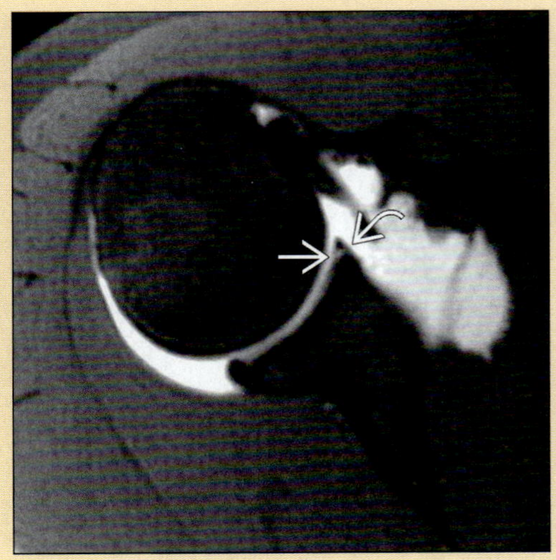

BANKART VARIATION LESIONS

TERMINOLOGY

Synonyms

- Perthes lesion
- Partial Bankart labral tear or lesion
- Also: ALPSA (**a**nterior **l**abrum **p**eriosteal **s**leeve **a**vulsion) lesion; GLAD/GARD (**g**lenoid **l**abrum **a**rticular **d**isruption/**g**lenoid **a**rticular **r**im **d**isruption) lesions may also be Bankart variation lesions

Definitions

- Anteroinferior labral tear without complete detachment (as is seen in true Bankart lesion)

IMAGING

General Features

- Best diagnostic clue
  - High signal extending partly through labrum for partial Bankart tear; periosteum intact
- Location
  - Anteroinferior labrum

CT Findings

- CT arthrography
  - Contrast extends partly across labrum

MR Findings

- PD/intermediate FS
  - Increased signal extends partly through labrum
  - **Partial Bankart tear**: Some intact labral tissue prevents complete detachment of labrum
  - **Perthes lesion**: Labrum is torn from glenoid rim but held in place by intact periosteum
  - Hill-Sachs bone contusion or impaction fracture
  - Sensitivity on routine MR < 70%
    - Low sensitivity if no effusion or recent dislocation episode
- Direct MR arthrography
  - Contrast extends into labral tear
  - Minimal displacement of torn portion of labrum on routine axial adduction images
  - Sensitivity: 50-70% on axial images of direct MR arthrogram
  - ABER (**ab**duction/**e**xternal **r**otation) images: 10-20% ↑ sensitivity over routine axial MR arthrogram images
    - Traction on anterior labrum widens tear
    - Sections are perpendicular to anteroinferior labrum so less partial averaging

Imaging Recommendations

- Best imaging tool
  - Direct MR arthrography
- Protocol advice
  - Add ABER images to increase sensitivity

DIFFERENTIAL DIAGNOSIS

Bankart Lesion

- Complete anterior labral tear with detachment

Inferior Glenohumeral Ligament Tears

- Tear is more lateral, involving anterior capsule and anterior band IGL

PATHOLOGY

General Features

- Etiology
  - Anterior glenohumeral dislocation
- Associated abnormalities
  - Hill-Sachs bone contusion or impaction fracture

Staging, Grading, & Classification

- Partial Bankart tear: Tear extends only partly through labrum
- Perthes lesion: Tear extends entirely through labrum, but periosteum intact
  - Labrum not displaced
- ALPSA lesion: Tear extends entirely through labrum with intact periosteum
  - Labrum displaced medially
- GLAD/GARD lesion: Labral tear + chondral lesion at glenoid rim

Gross Pathologic & Surgical Features

- Hemorrhage at partial tear site

CLINICAL ISSUES

Presentation

- Most common signs/symptoms
  - Shoulder pain
  - Anterior instability and apprehension
    - Symptoms may be less severe than with true Bankart labral tear

Demographics

- Age
  - < 40 years old
- Gender
  - M > F

Natural History & Prognosis

- Often continued instability and pain

Treatment

- If continued instability: Labrocapsular repair

DIAGNOSTIC CHECKLIST

Consider

- Bankart variation considered if high T2 signal extends partly across anteroinferior labrum or periosteum intact

Image Interpretation Pearls

- ABER images significantly ↑ sensitivity for partial Bankart tear and Perthes lesion; consider using in cases of instability

SELECTED REFERENCES

1. Schreinemachers SA et al: Is a single direct MR arthrography series in ABER position as accurate in detecting anteroinferior labroligamentous lesions as conventional MR arthography? Skeletal Radiol. 38(7):675-83, 2009
2. Song HT et al: The usefulness of virtual MR arthroscopy as an adjunct to conventional MR arthrography in detecting anterior labral lesions of the shoulder. AJR Am J Roentgenol. 192(4):W149-55, 2009

*(**Left**) Axial MR arthrogram T1WI FS of a patient with a partial Bankart lesion. Contrast extends partly across the attachment of the anteroinferior labrum to the glenoid rim with intact labral tissue more anteriorly. (**Right**) Axial MR arthrogram T1W FS adjacent image in the same patient again shows contrast extending partly between the anteroinferior labrum and the adjacent glenoid rim. Although partial Bankart lesions have some intact labral tissue, they can cause instability.*

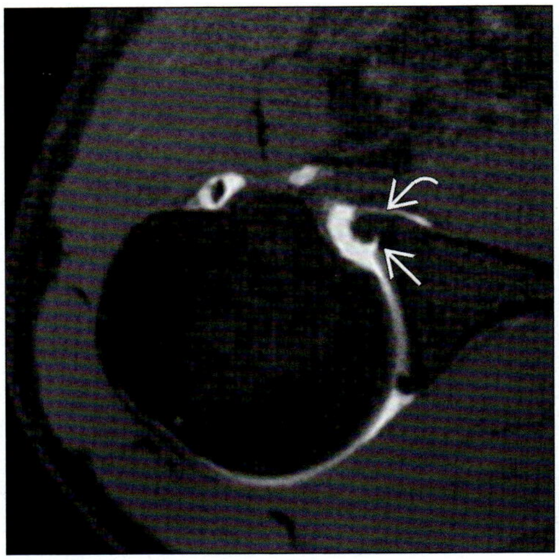

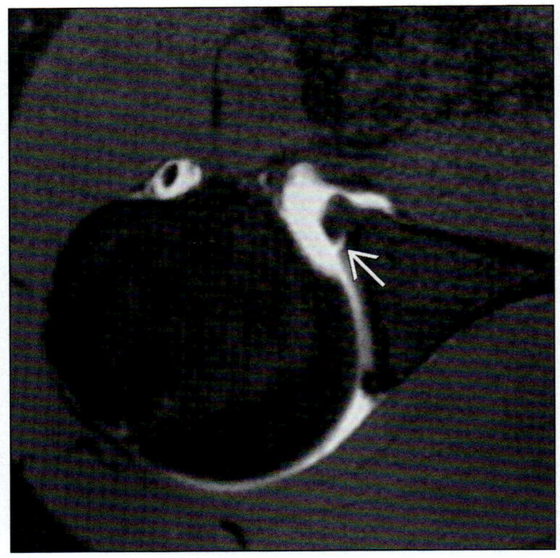

*(**Left**) Axial MR arthrogram T1WI FS of a patient with a partial Bankart tear shows a small indentation in the anterior labrum. Note the attachment of the anterior band IGL. (**Right**) Axial MR arthrogram T1W FS adjacent image in the same patient again shows a small indentation in the anterior labrum. This could be a fold in the labrum, but it is at the level of the attachment of the ABIGL. In a patient with instability, a minor irregularity such as this may indicate partial Bankart lesion.*

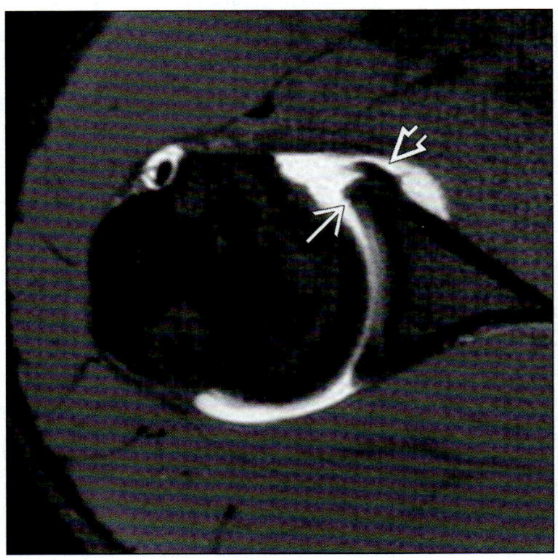

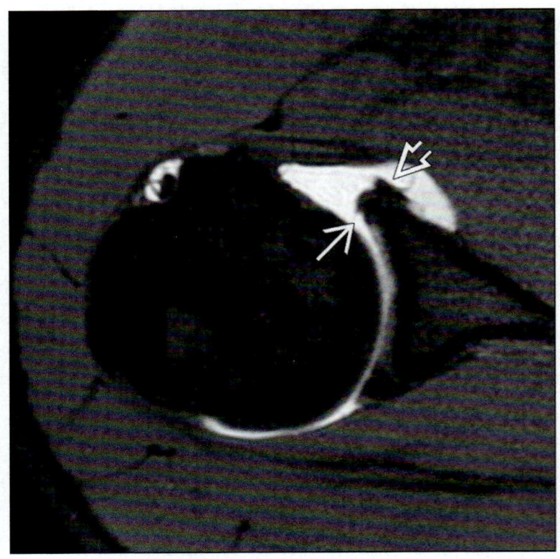

*(**Left**) ABER MR arthrogram T1WI FS in the same patient shows a partial Bankart tear. Traction on the anterior band of the IGL confirms that there is still some intact peripheral labral tissue. (**Right**) ABER MR arthrogram T1W FS adjacent image in the same patient again shows the partial Bankart lesion. ABER images often show the tear better and more accurately demonstrate the true depth of a partial Bankart tear. Partial Bankart tears can cause instability and apprehension.*

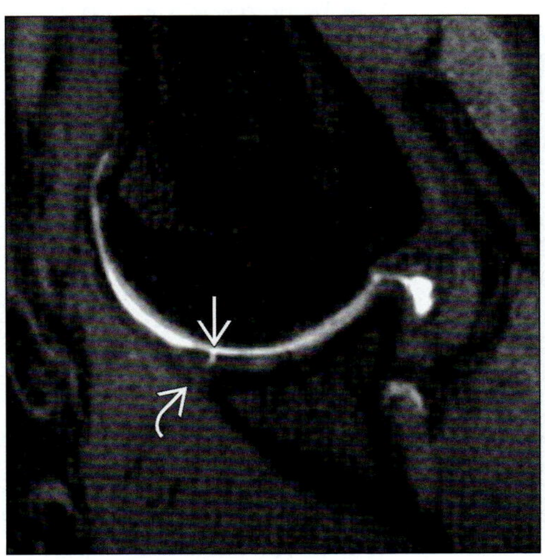

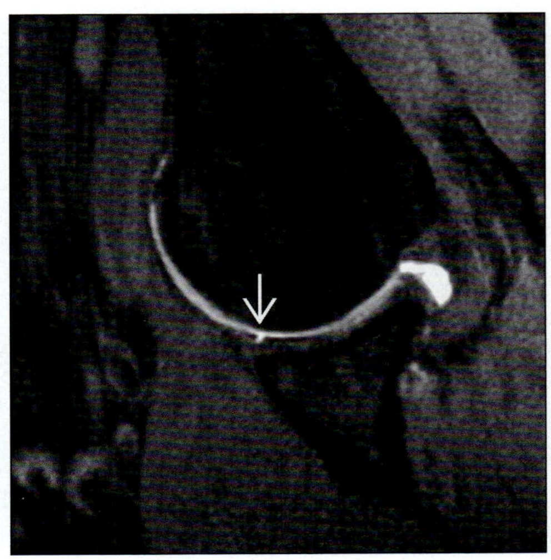

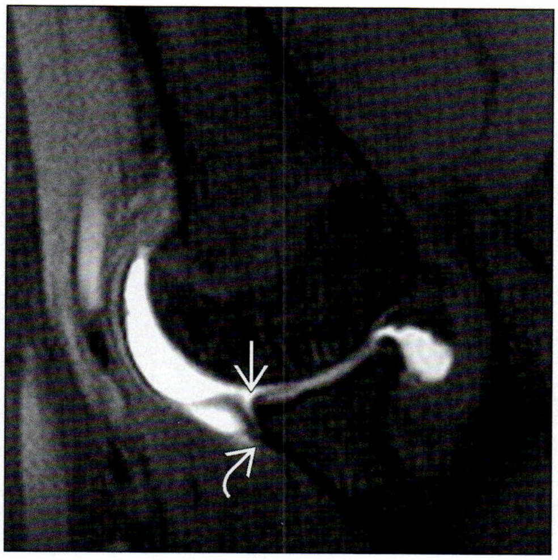

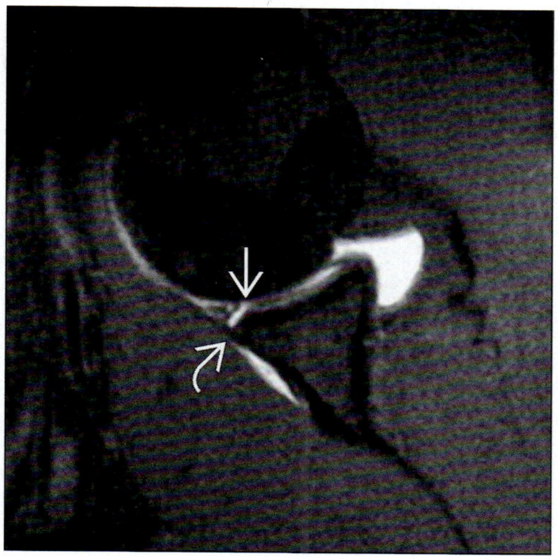

(Left) *ABER MR arthrogram T1WI FS shows a Perthes lesion. Contrast extends across the anteroinferior labrum, which is attached by a periosteal sleeve and is not displaced medially from the glenoid rim.* ***(Right)*** *ABER MR arthrogram T1WI FS demonstrates a Perthes lesion. The labrum is torn from the glenoid but is not displaced and is still attached to the scapula by a periosteal sleeve. The periosteum is not stripped from the scapular neck as on the prior example.*

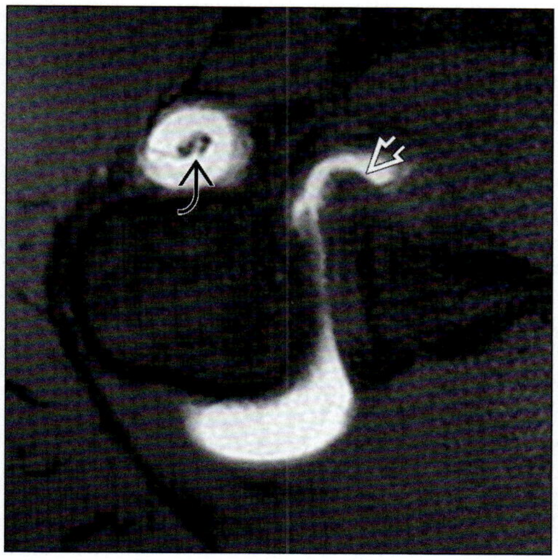

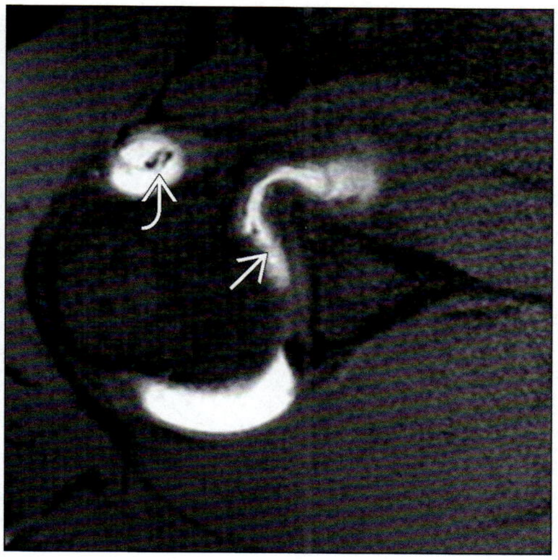

(Left) *Axial MR arthrogram T1WI FS in a patient with a Perthes lesion. On this inferior slice, no definite tear is seen of the anteroinferior labrum. There is a partial tear of the biceps tendon.* ***(Right)*** *Axial MR arthrogram T1WI FS more superiorly in the same patient shows a possible partial tear of the anteroinferior labrum. The patient had apprehension and mild anterior instability on clinical examination. Again seen is the partial tear of the biceps tendon.*

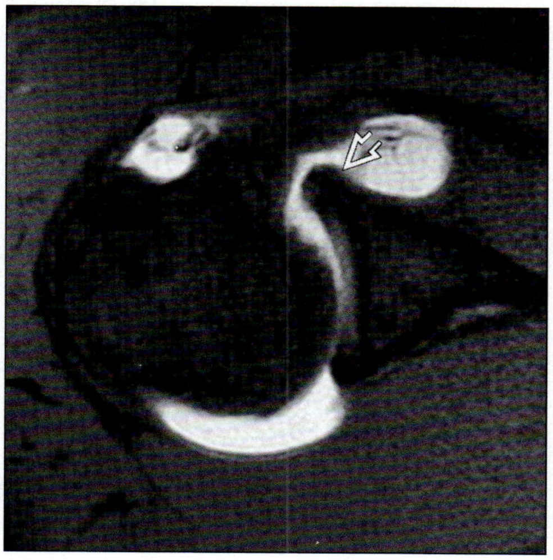

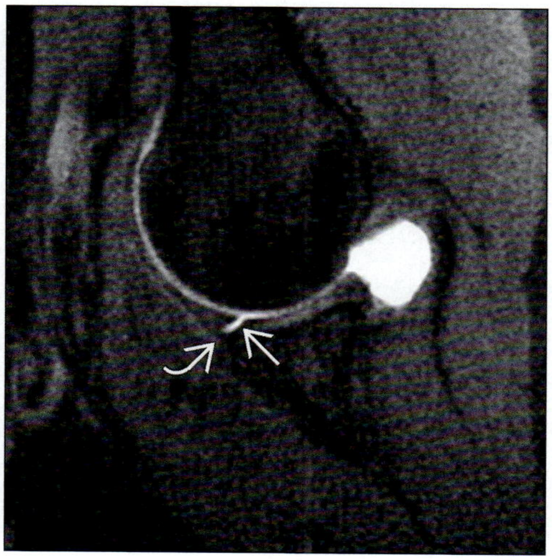

(Left) *Axial MR arthrogram T1WI FS slightly more superiorly in the same patient again shows no definite tear of the labrum.* ***(Right)*** *ABER MR arthrogram T1WI FS in the same patient shows a tear of the anteroinferior labrum with the labrum still attached to the scapula. At arthroscopy, the labrum was not displaced and attached only by scapular periosteum, so this was called a Perthes lesion. ABER images improve the sensitivity for small Bankart variation lesions by 10-20%.*

ALPSA LESION

Key Facts

Terminology

- Anterior **l**abrum **p**eriosteal **s**leeve **a**vulsion (ALPSA)
- Torn anterior labrum medially displaced and attached to scapula only by periosteum
- Bankart variation lesion
  - These also include Perthes lesion, partial Bankart tear, GLAD lesion

Imaging

- MR: Labrum displaced from glenoid rim with low T2 signal attachment of periosteum to scapular neck
  - Acute ALPSA: Hemorrhage and edema of anterior capsuloligamentous tissues
  - Chronic ALPSA: Focal low signal mass (displaced labrum) adhered to scapular neck
- Better seen on direct MR arthrogram

Top Differential Diagnoses

- Bankart lesion
- Perthes lesion
- GLAD lesion
- IGL tears

Clinical Issues

- Shoulder pain
- Recurrent anterior dislocations
  - Patients often have more dislocation episodes than with a classic Bankart lesion
- Most frequently < 25 years old
- M > F
- ALPSA seen in 20-25% of anterior dislocations
- Treatment: Anterior labrocapsular repair
  - Chronic ALPSA: Labrum 1st dissected from scapula and then labrocapsular repair

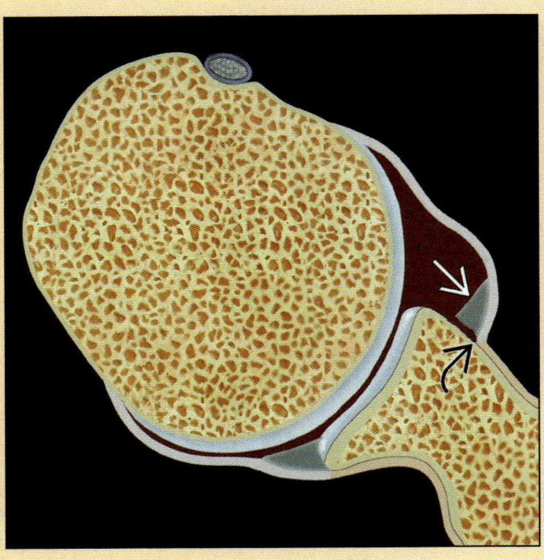

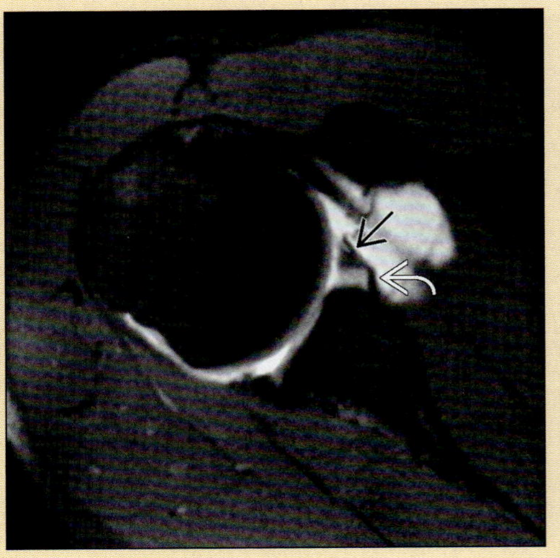

(Left) *Axial graphic of an ALPSA lesion shows the anterior labrum ➡ torn from the glenoid rim but still attached to the scapular neck by a periosteal sleeve ➡.* ***(Right)*** *Axial MR arthrogram T1WI FS of an ALPSA lesion shows the labrum ➡ avulsed from the glenoid rim but still attached to the scapular neck by a periosteal sleeve ➡. ALPSA lesions occur in 20-25% of anterior dislocations and are typically seen in patients under 25 years old who have multiple anterior dislocation episodes.*

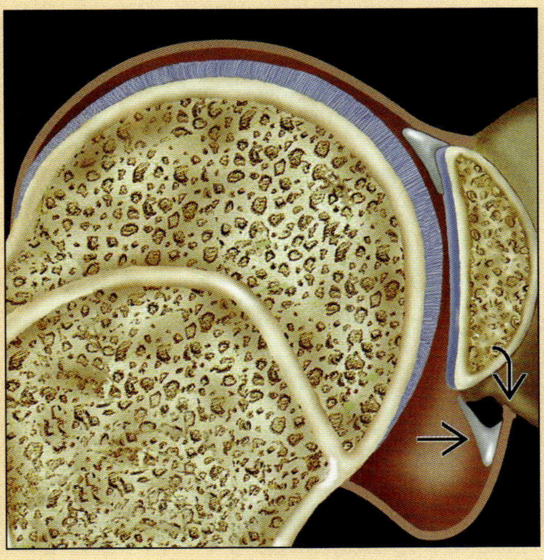

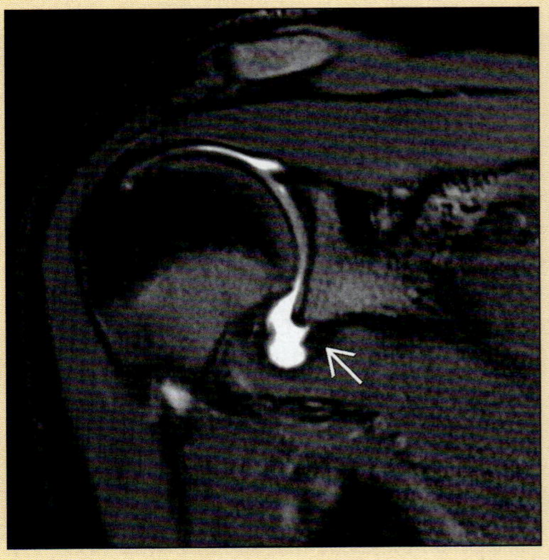

(Left) *Coronal oblique graphic of an ALPSA lesion shows the torn labrum displaced medially and rotated inferiorly ➡ but still attached to the glenoid rim by a periosteal sleeve ➡.* ***(Right)*** *Coronal oblique MR arthrogram T2WI FS of an ALPSA lesion shows the displaced labrum inferior to the scapular neck ➡. If the labral injury is old and the displaced fragment becomes adhered to the scapular neck and covered by synovium and fibrous tissue, it is called a chronic ALPSA lesion.*

ALPSA LESION

TERMINOLOGY

Abbreviations

- Anterior **l**abrum **p**eriosteal **s**leeve **a**vulsion (ALPSA)

Synonyms

- Bankart variation lesion
  - Also includes Perthes lesion, partial Bankart tear, **g**leno**l**abral **a**rticular **d**isruption (GLAD) lesion

Definitions

- Torn anterior labrum medially displaced and attached to scapula only by periosteum

IMAGING

General Features

- Best diagnostic clue
  - Medially displaced labrum with low signal attachment to scapular neck on MR
- Location
  - Anterior labrum

Radiographic Findings

- ± Hill-Sachs impaction fracture
- ± anterior glenoid rim deficiency

MR Findings

- PD/intermediate FS
  - Labrum displaced from glenoid rim with low T2 signal attachment to scapular neck (periosteum)
    - Acute ALPSA: Hemorrhage and edema of anterior capsuloligamentous tissues
    - Chronic ALPSA: Focal low signal mass (displaced labrum) adhered to scapular neck
- MR arthrography
  - ALPSA: Labrum surrounded by contrast except for periosteal attachment to scapula
  - Chronic ALPSA
    - Contrast adjacent to bare glenoid rim
    - Small extraarticular mass (displaced labrum) adhered to scapular neck
  - ABER (abduction external rotation) images may show tear better

Imaging Recommendations

- Best imaging tool
  - Direct MR arthrogram

DIFFERENTIAL DIAGNOSIS

Perthes Lesion

- Similar to ALPSA lesion: Labrum torn but attached by periosteal sleeve
- Labrum not medially displaced

Bankart Lesion

- Periosteum ruptured
- Labrum may be resorbed or adhered to capsule

GLAD Lesion

- Labral tear with glenoid chondral lesion

Inferior Glenohumeral Ligament Tears

- Tear of the capsuloligamentous complex lateral to labrum

PATHOLOGY

General Features

- Etiology
  - Anterior glenohumeral dislocation
- Associated abnormalities
  - Hill-Sachs lesion
  - May have Bankart fracture or deficient glenoid rim

Staging, Grading, & Classification

- ALPSA (acute)
  - Labrum displaced medially, anterior to scapula
- Chronic ALPSA
  - Labrum and periosteal sleeve fibrosed to scapula
  - Labrum covered by synovium & fibrous tissue

Gross Pathologic & Surgical Features

- Labrum medially displaced but attached to scapular neck by periosteum
  - Inferiorly rotated

CLINICAL ISSUES

Presentation

- Most common signs/symptoms
  - Shoulder pain
  - Recurrent anterior dislocations
    - Patients often have more dislocation episodes than with a classic Bankart lesion

Demographics

- Age
  - < 25 years old
- Gender
  - M > F
- Epidemiology
  - 20-25% of anterior dislocations

Natural History & Prognosis

- Instability and recurrent dislocations
- Medially displaced labrum becomes covered by synovialized fibrous tissue: Chronic ALPSA

Treatment

- Anterior labrocapsular repair
- Chronic ALPSA: Labrum dissected from scapula and then labrocapsular repair

DIAGNOSTIC CHECKLIST

Consider

- Chronic ALPSA if bare glenoid rim but labrum not seen in joint
- Medial labral displacement differentiates from Perthes

SELECTED REFERENCES

1. Schreinemachers SA et al: Is a single direct MR arthrography series in ABER position as accurate in detecting anteroinferior labroligamentous lesions as conventional MR arthography? Skeletal Radiol. 38(7):675-83, 2009
2. Ozbaydar M et al: Results of arthroscopic capsulolabral repair: Bankart lesion versus anterior labroligamentous periosteal sleeve avulsion lesion. Arthroscopy. 24(11):1277-83, 2008

ALPSA LESION

*(**Left**) Axial MR arthrogram T1WI FS of a patient with prior dislocations and an ALPSA lesion shows that the labrum is torn from the glenoid rim and displaced medially ➡ but is still attached to the scapular neck by a periosteal sleeve ➡. (**Right**) Axial MR arthrogram T1W FS more cranial image in the same patient shows the avulsed labrum ➡ surrounded by synovitis ➡. The periosteal sleeve ➡ is somewhat wavy. A Hill-Sachs impaction fracture is visible in the posterior humeral head ➡.*

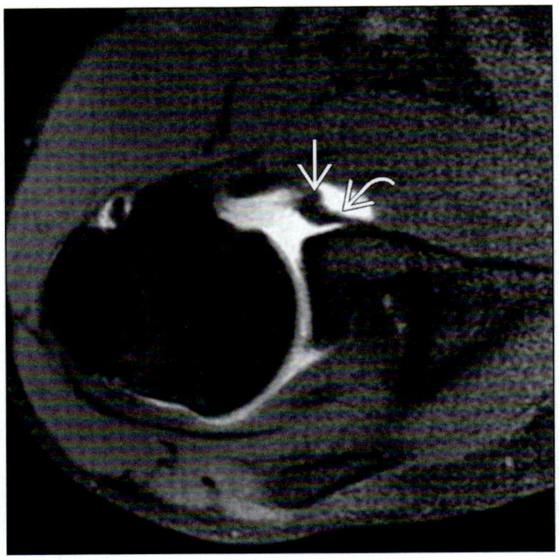

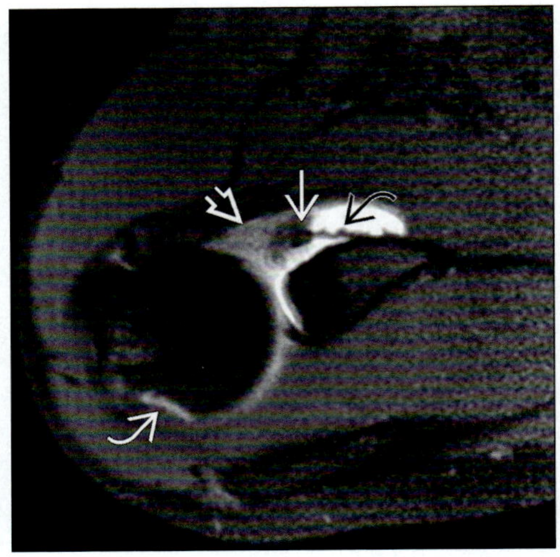

*(**Left**) Axial MR arthrogram T1WI FS of a patient with an ALPSA lesion shows a bare anterior glenoid rim and a medially displaced labral fragment ➡ attached to the scapular neck by periosteum ➡. (**Right**) Axial MR arthrogram T1W FS image more inferiorly in the same patient again shows the torn and displaced labrum ➡ with a periosteal sleeve ➡. There is a blunted glenoid rim ➡ from either a Bankart fracture or mechanical erosion from multiple previous dislocation episodes.*

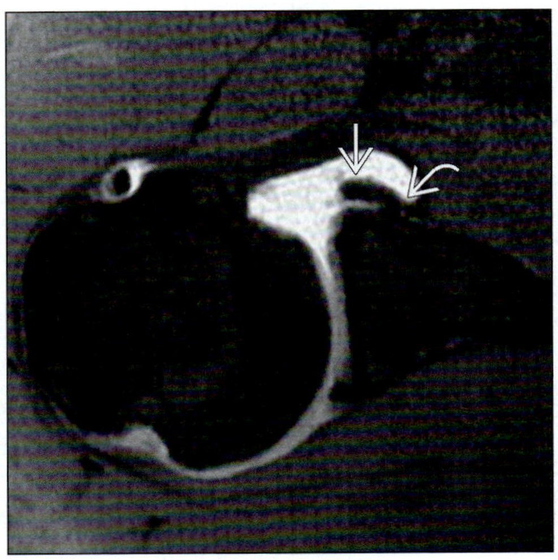

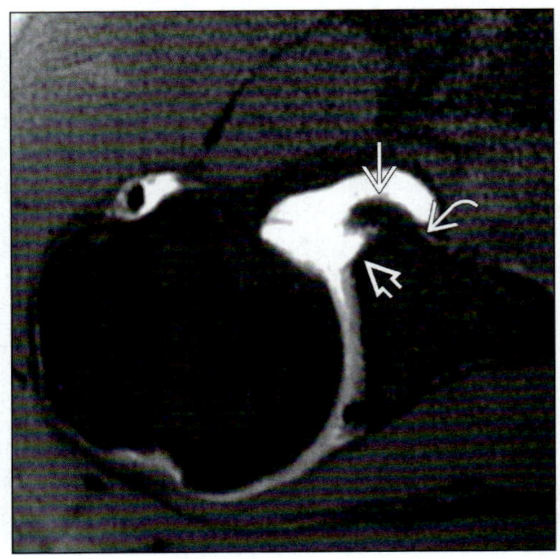

*(**Left**) ABER MR arthrogram T1W FS image in the same patient shows the displaced labral fragment ➡ attached to the scapular neck by the periosteal sleeve ➡. (**Right**) ABER MR arthrogram T1W FS adjacent image in the same patient again shows the torn labrum ➡ and periosteal sleeve ➡. The anterior band IGL ➡ is still intact, but it is attached to an unstable anteroinferior labrum so patient has instability. ALPSA lesions are easiest to diagnose on direct MR arthrography.*

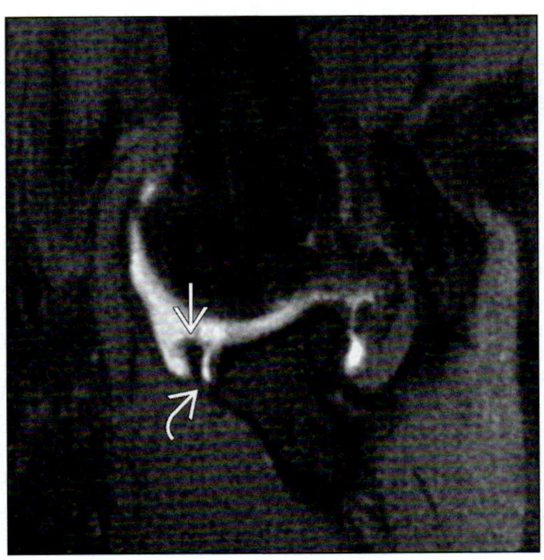

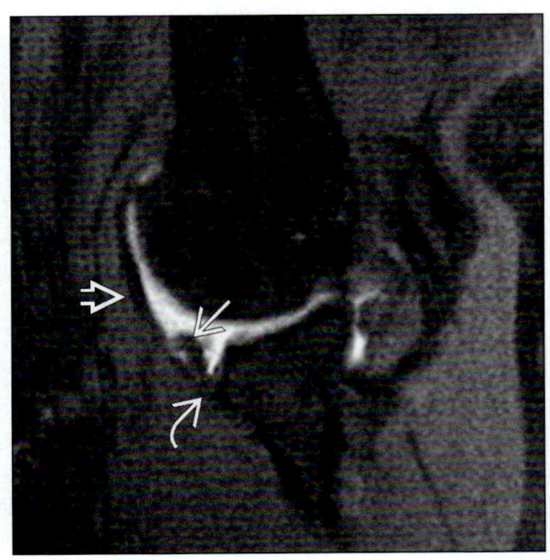

ALPSA LESION

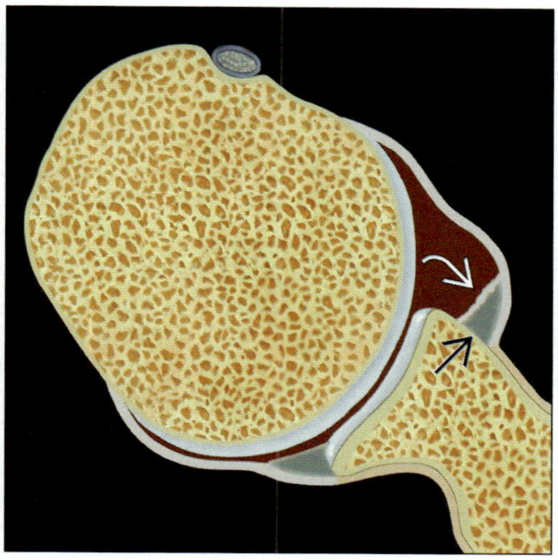

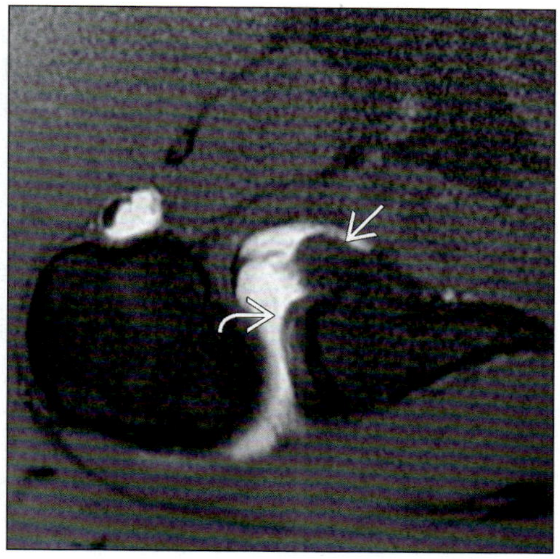

(Left) *Axial graphic of a chronic ALPSA lesion shows the medially displaced labrum → adhered to the scapular neck and covered by synovial tissue →.* ***(Right)*** *Axial MR arthrogram T1WI FS of a chronic ALPSA lesion shows the torn anteroinferior labrum displaced medially →, adjacent to the scapular neck. The labrum also rotates inferiorly, so it appears large on this axial section. There is only a small amount of residual labrum attached to the glenoid rim →.*

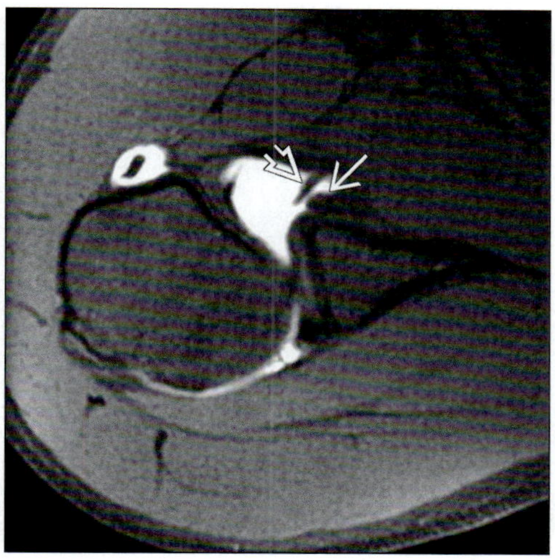

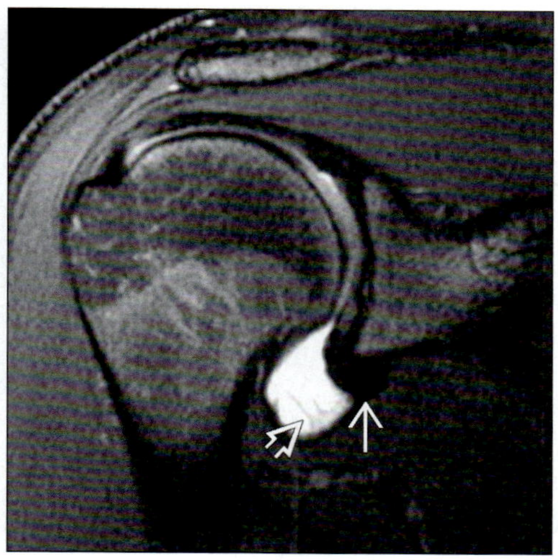

(Left) *Axial MR arthrogram T1WI FS of a patient with a chronic ALPSA lesion shows the labrum adhered to the scapular neck →. A portion of the middle glenohumeral ligament is also seen → and should not be confused for a Bankart lesion.* ***(Right)*** *Coronal oblique MR arthrogram T2WI FS in the same patient shows the chronic ALPSA lesion → rotated inferiorly and attached to the anteroinferior scapular neck. The labrum is covered by synovium, and there is mild adjacent synovitis →.*

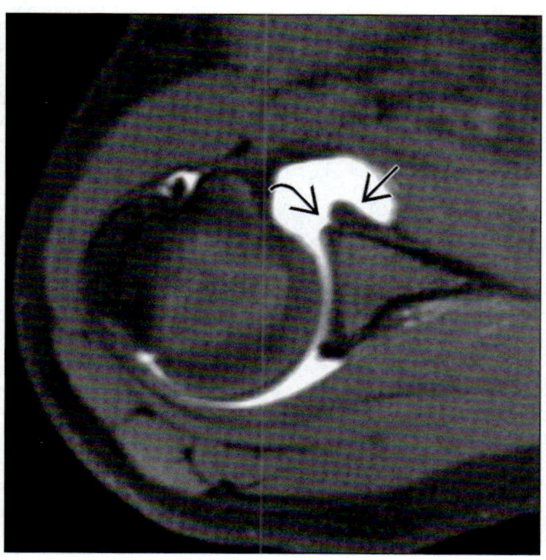

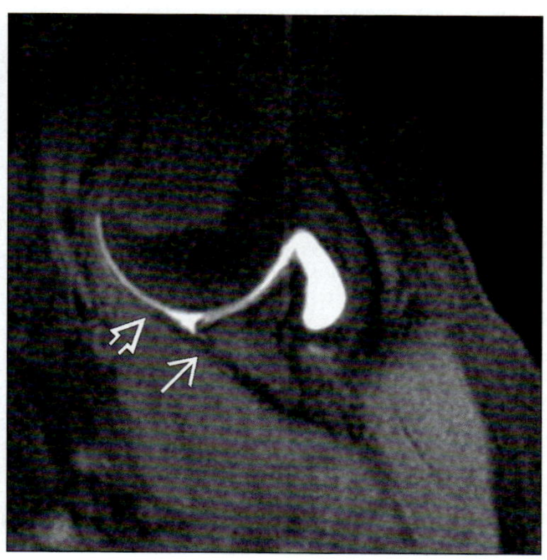

(Left) *Axial MR arthrogram T1WI FS of a chronic ALPSA lesion shows the anteroinferior labrum displaced medially → and covered with synovium. Diagnosing an ALPSA lesion on MR helps the surgeon to know to look for the crease → between the glenoid rim and the synovium covering the labrum, which is where the incision is made in order to dissect out the labrum in preparation for repair.* ***(Right)*** *ABER MR arthrogram T1WI FS in the same patient shows the IGL → attached to the ALPSA lesion →.*

GLAD/GARD LESION

Key Facts

Terminology
- Glenolabral articular disruption (GLAD)
  - Anteroinferior chondral lesion with superficial tear of labrum and no instability on clinical exam
  - This is original definition; some now use "GLAD" for any labral tear with adjacent chondral lesion
- Glenoid articular rim disruption (GARD) (or Divot)
  - Chondral or osteochondral lesion adjacent to labral tear
  - Unstable on clinical exam

Imaging
- Articular cartilage flap or defect near glenoid rim
- Labral tear adjacent to chondral lesion
- Small osteochondral lesion (rare)
- **GLAD** lesion
  - Also has partial tear of anterior labrum
- **GARD** lesion
  - Also has adjacent Bankart, reverse Bankart, or other Bankart variation lesion

Top Differential Diagnoses
- Bankart fracture
- Glenoid OCD
- Perthes lesion

Pathology
- GLAD lesion
  - Impaction injury with arm in abduction and external rotation
  - No dislocation usually associated
- GARD lesion
  - Glenohumeral dislocation or subluxation

Clinical Issues
- Shoulder pain

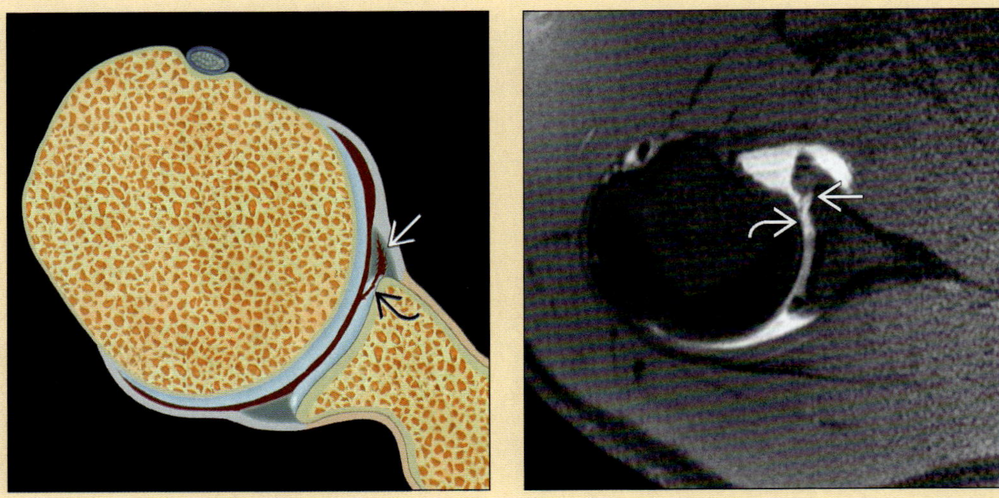

***(Left)** Axial graphic of a GLAD lesion shows a partial tear of the labrum ➔ and a flap tear of the articular cartilage ⇨. The classic GLAD lesion presents with shoulder pain but without instability and makes up ~ 2% of labroligamentous injuries. **(Right)** Axial MR arthrogram T1WI FS of a 20-year-old college football player who fell on his elbow with his arm abducted and externally rotated shows a chondral defect ➔ and partial tear of the anterior labrum ➔, indicating a GLAD lesion.*

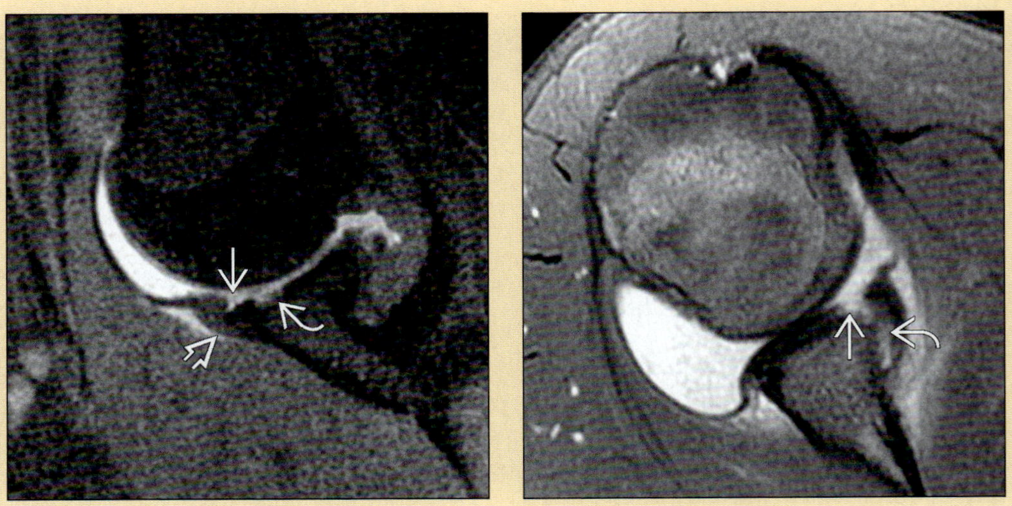

***(Left)** ABER MR arthrogram T1WI FS of a GLAD lesion in the same patient again shows the chondral lesion ➔. With ABER position, the partial tear of the labrum is better seen ➔. The labral attachment to the glenoid ⇨ was otherwise stable. **(Right)** Axial MR arthrogram PDWI FS shows a GARD lesion with a displaced tear of the anterior labrum and a peel-back lesion of the anterior periosteum ➔ associated with a defect in the adjacent anterior glenoid cartilage ➔.*

GLAD/GARD LESION

TERMINOLOGY

Abbreviations

- Glenolabral articular disruption (GLAD)
- Glenoid articular rim disruption (GARD) (or Divot)

Synonyms

- Bankart variation lesion

Definitions

- GLAD lesion: Anteroinferior chondral lesion with superficial (small partial) tear of labrum and no instability on clinical exam
  - This is original definition; some now use "GLAD" for any labral tear with adjacent chondral lesion
- GARD lesion: Chondral or osteochondral lesion adjacent to large instability-causing labral tear
  - Unstable on clinical exam

IMAGING

General Features

- Best diagnostic clue
  - Glenoid rim chondral defect with labral tear on MR
- Location
  - GLAD lesion: Anteroinferior glenoid rim
  - GARD lesion: Anterior or posterior glenoid rim

Radiographic Findings

- GARD lesion: ± Hill-Sachs impaction fracture

MR Findings

- PD/intermediate FS
  - Articular cartilage lesion near glenoid rim
  - Labral tear adjacent to chondral lesion
  - Small osteochondral lesion (rare)
- **GLAD** lesion
  - Partial tear of anterior labrum
  - No Hill-Sachs bone contusion or fracture
- **GARD** lesion
  - Also has an adjacent Bankart, reverse Bankart, or other Bankart variation lesion
- Direct MR arthrography
  - High signal contrast in chondral lesion and labral tear
  - > 95% sensitivity

Imaging Recommendations

- Best imaging tool
  - Direct MR arthrography
- Protocol advice
  - Conventional MR: Include cartilage-sensitive axial sequence

DIFFERENTIAL DIAGNOSIS

Bankart Fracture

- Complete osteochondral fracture of the anterior glenoid rim with separate bone fragment
- Osteochondral defect seen in some GARD lesions only involves the articular surface and does not extend across the glenoid osseous margin

Glenoid Osteochondral Fracture

- No labral tear, usually not near glenoid rim

PATHOLOGY

General Features

- Etiology
  - GLAD lesion
    - Impaction injury with arm in abduction and external rotation
    - No dislocation usually associated
  - GARD lesion
    - Glenohumeral dislocation or subluxation

Staging, Grading, & Classification

- GLAD lesion
  - Chondral defect
  - Chondral flap tear
  - ± chondral intraarticular body
- GARD lesion
  - Chondral lesion
  - Osteochondral lesion
  - Labral tear: Bankart, reverse Bankart, etc.

Gross Pathologic & Surgical Features

- Labral tear with adjacent chondral lesion
  - GLAD lesion: IGL intact
  - GARD lesion: Labroligamentous tear

CLINICAL ISSUES

Presentation

- Most common signs/symptoms
  - Shoulder pain
  - **GLAD** lesion
    - No clinical instability
      - May have mild subluxation when examined under anesthesia
    - Uncommon: Apprehension sign
  - **GARD** lesion
    - Instability and apprehension

Demographics

- Age
  - 20-40 years old
- Gender
  - M > F

Natural History & Prognosis

- Some patients improve with physical therapy
- Many require surgery

Treatment

- Debridement of chondral lesion
- GLAD: ± debridement of labral lesion
- GARD: Labroligamentous repair

SELECTED REFERENCES

1. Antonio GE et al: First-time shoulder dislocation: High prevalence of labral injury and age-related differences revealed by MR arthrography. J Magn Reson Imaging. 26(4):983-91, 2007
2. Waldt S et al: Anterior shoulder instability: accuracy of MR arthrography in the classification of anteroinferior labroligamentous injuries. Radiology. 237(2):578-83, 2005
3. Sanders TG et al: The glenolabral articular disruption lesion: MR arthrography with arthroscopic correlation. AJR Am J Roentgenol. 172(1):171-5, 1999

HAGL LESION

Key Facts

Terminology
- Humeral avulsion of glenohumeral ligament (HAGL)

Imaging
- Disrupted low signal IGL fibers near humeral insertion
- Periligamentous edema and hemorrhage
- "J" sign: Torn lateral IGL lays inferiorly on coronal image
- Direct MR arthrography
  - Contrast extravasation through tear: Acute or large nonhealing tear

Top Differential Diagnoses
- Medial IGL tears
- Bankart lesion
- Subscapularis tears

Pathology
- **Classification**
  - HAGL with torn insertion of ABIGL: Common, associated with instability
  - Axillary pouch HAGL with intact ABIGL
  - BHAGL: Bone avulsion of IGL insertion on humerus
  - Floating IGL: HAGL and Bankart lesion

Clinical Issues
- Shoulder pain
- Apprehension in abduction and external rotation
- Small tears without instability may heal
- Large tears usually fail to heal

Diagnostic Checklist
- Consider HAGL if anterior dislocation and no labral tear seen

__(Left)__ AP graphic of a HAGL lesion shows the IGL tear near the humeral insertion. The tear involves the anterior band → and axillary pouch portions of the IGL →. __(Right)__ Coronal oblique MR arthrogram T2WI FS shows the "J" sign of a HAGL lesion. The IGL → is torn from its humeral attachment causing it to droop and form a J-shaped appearance. There is also extravasation of joint fluid → through the persistent ligament tear in this patient with instability after a prior dislocation.

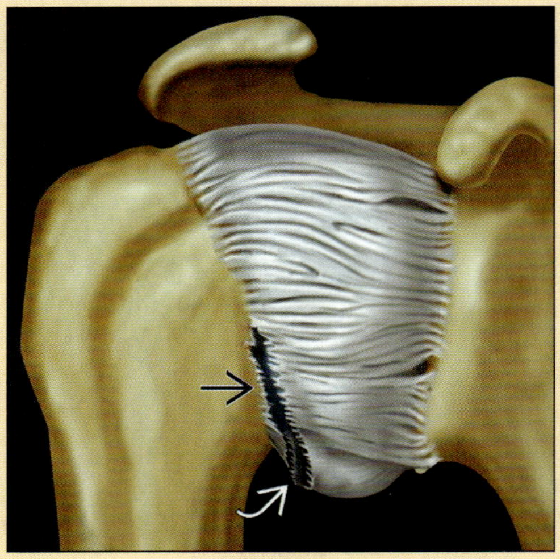

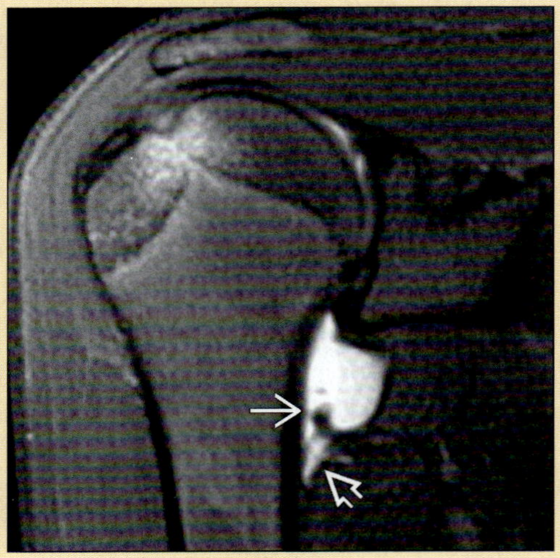

__(Left)__ ABER MR arthrogram T1WI FS in the same patient shows the HAGL lesion also involving the anterior band portion of the IGL. The ABIGL is torn from its insertion →, and there is extravasation of contrast →. HAGL lesions with instability often require surgery. __(Right)__ Axial PDWI FS MR shows a recent HAGL lesion with a tear in the ABIGL at its humeral insertion →. The medial portion of the ABIGL is wavy →, and there is edema and hemorrhage in the pericapsular soft tissues →.

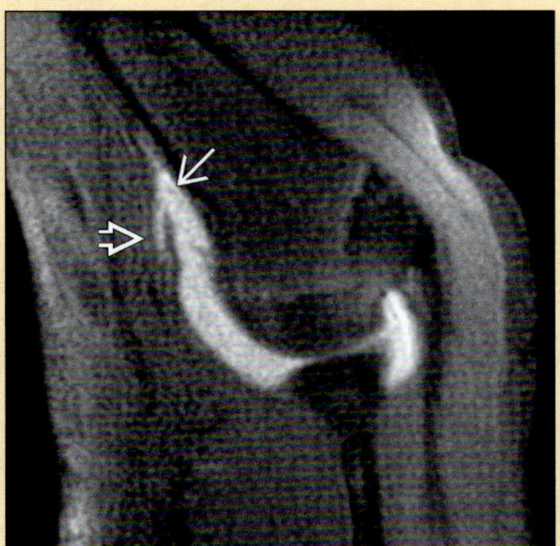

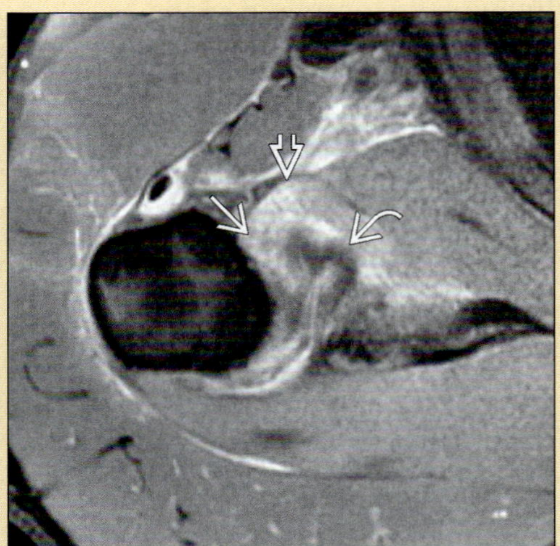

HAGL LESION

TERMINOLOGY

Abbreviations

- Humeral avulsion of glenohumeral ligament (HAGL)

Synonyms

- Lateral inferior glenohumeral ligament (IGL) tear
- If avulsion fracture: Bony humeral avulsion of glenohumeral ligament (BHAGL)

Definitions

- Avulsion of IGL from humeral insertion

IMAGING

General Features

- Best diagnostic clue
  - Disrupted IGL fibers near humeral insertion on oblique coronal images

Radiographic Findings

- Usually normal
- BHAGL: Small bone avulsion fragment near medial humeral metaphysis

CT Findings

- CT arthrography
  - Extravasation of contrast through IGL tear near humeral insertion

MR Findings

- T2WI FS
  - Disrupted low signal IGL fibers near humeral insertion
  - Periligamentous edema and hemorrhage
  - "J" sign: Torn lateral IGL droops inferiorly on coronal images
    - U-shaped appearance of normal intact IGL becomes J-shaped
  - HAGL involving anterior band IGL (ABIGL) seen best on axial images
  - Hill-Sachs fractures and bone contusions less common than with Bankart lesions
- Direct MR arthrography
  - Contrast extravasation through tear: Acute or large nonhealing tear

Imaging Recommendations

- Best imaging tool
  - MR
- Protocol advice
  - Oblique coronal fat-suppressed T2-weighted images

DIFFERENTIAL DIAGNOSIS

Medial IGL Tears

- Midsubstance or glenoid avulsion IGL tears (GAGL) occur more medially

Bankart Lesion

- Anteroinferior labrum torn
- Associated Hill-Sachs impaction fracture more common

Subscapularis Tears

- Can also occur after anterior dislocation
- Edema & hemorrhage in muscle more anteriorly

PATHOLOGY

General Features

- Etiology
  - Anterior glenohumeral dislocation
- Associated abnormalities
  - Rotator cuff tear
    - Older patients

Staging, Grading, & Classification

- HAGL with torn insertion of ABIGL
  - Common
  - Instability after anterior dislocation
- Axillary pouch HAGL with intact ABIGL
- BHAGL: Bone avulsion of IGL insertion on humerus
- Floating IGL: HAGL and Bankart lesion
  - Results in "floating" ligament with no attachment

CLINICAL ISSUES

Presentation

- Most common signs/symptoms
  - Shoulder pain
  - Apprehension in abduction and external rotation
- Other signs/symptoms
  - Decreased external rotation
  - Crepitus

Demographics

- Age
  - 20-50 years old
    - Slightly higher mean age than Bankart lesions
- Gender
  - M > F
- Epidemiology
  - 4% of patients with anterior instability

Natural History & Prognosis

- Small tears without instability may heal
- Large tears usually fail to heal

Treatment

- Small tears: Protect shoulder, then physical therapy
- Large tears or instability: Surgical capsuloligamentous repair

DIAGNOSTIC CHECKLIST

Consider

- HAGL if anterior dislocation and no labral tear seen

SELECTED REFERENCES

1. Melvin JS et al: MRI of HAGL lesions: four arthroscopically confirmed cases of false-positive diagnosis. AJR Am J Roentgenol. 191(3):730-4, 2008
2. Bui-Mansfield LT et al: Humeral avulsion of the glenohumeral ligaments: the HAGL lesion. Am J Sports Med. 35(11):1960-6, 2007
3. Rhee YG et al: Anterior shoulder instability with humeral avulsion of the glenohumeral ligament lesion. J Shoulder Elbow Surg. 16(2):188-92, 2007

INFERIOR GLENOHUMERAL LIGAMENT INJURY

Key Facts

Terminology
- Inferior glenohumeral ligament (IGL)
- Tear of IGL, usually after glenohumeral dislocation
- Glenoid avulsion of glenohumeral ligament (GAGL)
  - IGL tear near the glenoid rim/labrum
  - More common than midsubstance IGL tears
- Anterior band of IGL (ABIGL)

Imaging
- T2 high signal edema in and around IGL
- Thickened and disrupted low signal IGL fibers
- Uncommon: Hill-Sachs fracture or bone contusion
- **Direct MR arthrography**
  - Extravasation of contrast through IGL defect
  - Old healed IGL tear: Focal thinning or thickening of IGL with laxity

Top Differential Diagnoses
- Bankart lesion
  - Labroligamentous complex tear involves the labrum
- HAGL lesion
  - IGL near humeral insertion

Pathology
- Grade 1: Sprained IGL
  - 15-30% after anterior dislocation
  - May heal with laxity
- Grade 2: Partial tear
- Grade 3: Complete tear of entire IGL (rare)

Clinical Issues
- IGL tears generally in patients > 40 years old
- IGL tear with rotator cuff tear common after dislocation in patients > 50 years old

__(Left)__ AP graphic shows a torn IGL involving the axillary pouch ➔ and ABIGL ➩. Tears involving the ABIGL often cause instability, and if large, may need to be surgically repaired. __(Right)__ Coronal oblique T2WI FS MR of a patient with a recent anterior dislocation shows a torn IGL with surrounding edema ➔. The axillary pouch portion of the IGL can be torn after either an anterior or inferior (luxatio erecti) dislocation. A portion of normal IGL ➔ is also seen.

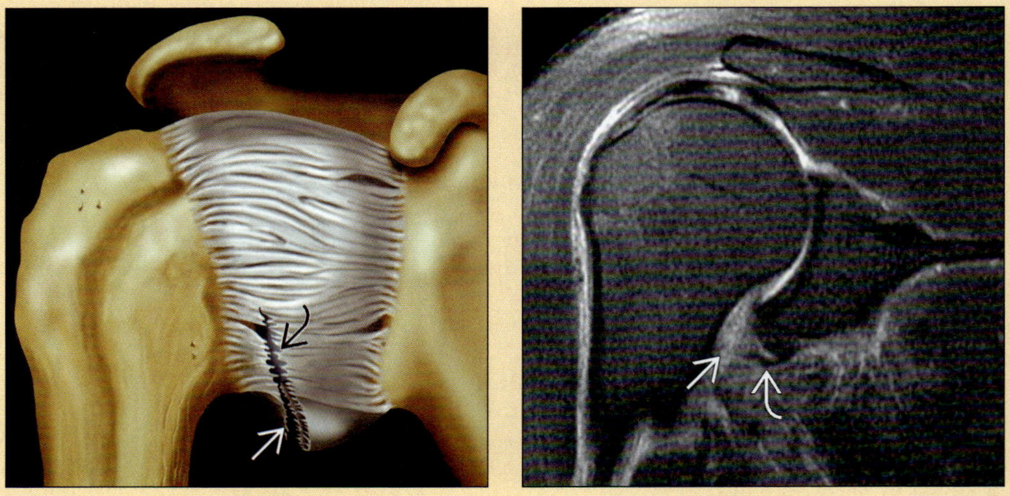

__(Left)__ Axial MR arthrogram T1WI FS of a patient with a torn IGL after an anterior shoulder dislocation. The ABIGL is torn and frayed ➔, and there is contrast extravasation into the anterior soft tissues ➔. __(Right)__ ABER MR arthrogram T1WI FS of the same patient shows the tear in the ABIGL ➔ and the extravasated contrast ➔. A portion of intact ABIGL is also seen ➔. IGL tears after dislocation are more common in older patients. Small tears often heal without surgery.

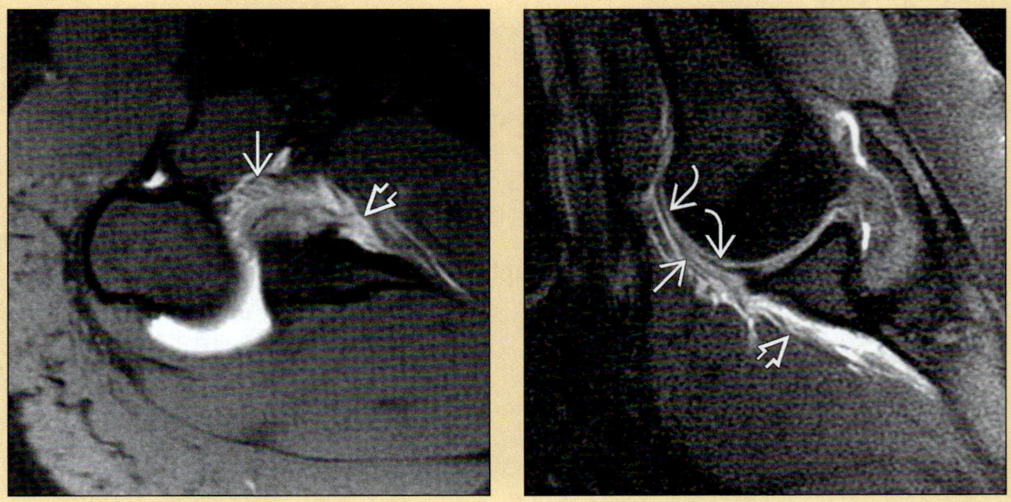

INFERIOR GLENOHUMERAL LIGAMENT INJURY

TERMINOLOGY

Abbreviations

- Inferior glenohumeral ligament (IGL)
- Anterior band of IGL (ABIGL)

Synonyms

- Glenoid avulsion of glenohumeral ligament (GAGL)
  - IGL tear near the glenoid rim/labrum
- Humeral avulsion of glenohumeral ligament (HAGL)
  - IGL tear at humeral insertion
- Both more common than midsubstance IGL tears

Definitions

- Tear of IGL, usually after glenohumeral dislocation

IMAGING

General Features

- Best diagnostic clue
  - Contrast extravasation through IGL at MR arthrography
- Location
  - Inferior shoulder
- Size
  - Small (common, usually heal) to large

MR Findings

- T2WI FS
  - T2 high signal edema in and around the IGL
  - Thickened and disrupted low signal IGL fibers
  - Uncommon: Hill-Sachs fracture or bone contusion
- Direct MR arthrography
  - Extravasation of contrast through IGL defect
  - ABER (abduction external rotation) may show ABIGL tear better
  - Old healed IGL tear: Focal thinning or thickening of IGL with laxity

Imaging Recommendations

- Best imaging tool
  - Direct MR arthrography
- Protocol advice
  - Include T2WI to look for ligament and periligamentous edema

DIFFERENTIAL DIAGNOSIS

Bankart Lesion

- Labroligamentous complex tear involves the labrum

HAGL Lesion

- IGL tear near humeral insertion

PATHOLOGY

General Features

- Etiology
  - Anterior dislocation
    - Anterior band IGL
    - Variable extension into axillary pouch portion of IGL
  - Inferior dislocation
    - Axillary pouch IGL tear
- Associated abnormalities
  - Stretched/torn anterior capsule
  - Rotator cuff tear in older patients

Staging, Grading, & Classification

- Grade 1: Sprained IGL
  - Common (15-30%) after anterior dislocation
  - May heal with laxity
- Grade 2: Partial tear
- Grade 3: Complete tear of IGL, including anterior and posterior bands
  - Extremely rare

Gross Pathologic & Surgical Features

- Tear through IGL with hemorrhage
- After anterior dislocation: Torn ABIGL with dislocatable humeral head

CLINICAL ISSUES

Presentation

- Most common signs/symptoms
  - Recent dislocation
  - Shoulder pain
  - Instability, apprehension: Chronic tear

Demographics

- Age
  - > 40 years old
    - IGL tear with rotator cuff tear common after dislocation in patients > 50 years old
- Gender
  - M > F
- Epidemiology
  - 1-4% of patients with recurrent anterior instability

Natural History & Prognosis

- Small tears usually heal with conservative therapy
- Large tears may become chronic and cause instability

Treatment

- Protect shoulder, then physical therapy
- Large chronic tears: Repair and capsulorrhaphy

DIAGNOSTIC CHECKLIST

Consider

- IGL tear in older patients after dislocation

SELECTED REFERENCES

1. Krug DK et al: MRI findings associated with luxatio erecta humeri. Skeletal Radiol. 39(1):27-33, 2010
2. Ng AW et al: Assessment of capsular laxity in patients with recurrent anterior shoulder dislocation using MRI. AJR Am J Roentgenol. 192(6):1690-5, 2009
3. Bencardino JT et al: MR imaging of the glenohumeral ligaments. Radiol Clin North Am. 44(4):489-502, vii, 2006
4. Mizuno N et al: Recurrent anterior shoulder dislocation caused by a midsubstance complete capsular tear. J Bone Joint Surg Am. 87(12):2717-23, 2005
5. Takubo Y et al: Magnetic resonance imaging evaluation of the inferior glenohumeral ligament: non-arthrographic imaging in abduction and external rotation. J Shoulder Elbow Surg. 14(5):511-5, 2005

BENNETT LESION

Key Facts

Terminology

- Posterior glenoid exostosis, thrower's exostosis
- Ossification at glenoid attachment of posterior band inferior glenohumeral ligament (PBIGL)

Imaging

- Radiograph: Crescentic ossification adjacent to posterior glenoid rim in a baseball pitcher
- MR: Low signal ossification arising from posteroinferior glenoid rim
- Surrounding heterogeneous signal fibrosis and thickened posterior capsule
- Acute injury: Edema around ossification
- ± posterior labral tear
  - Adjacent to Bennett lesion
  - Posterosuperior labral tear of internal impingement

Top Differential Diagnoses

- Reverse Bankart fracture
- Posterior glenoid osteophyte
- Calcific tendinopathy

Pathology

- Progression in overhead throwers
  - Thickening of PBIGL and capsule
  - Ossification at enthesis
- Associated abnormalities
  - Internal impingement labral tear and rotator cuff tear
  - Posterior labral tear

Clinical Issues

- Present in 22% of asymptomatic and 33% of symptomatic Major League pitchers

*(**Left**) PA graphic shows a Bennett lesion ➔ with surrounding capsular edema ➔. This lesion is thought to be heterotopic ossification or a traction enthesophyte at the attachment of the posterior band of the inferior glenohumeral ligament ➔. The lesion is seen mainly in baseball pitchers. (**Right**) Axial CT arthrogram of a baseball pitcher shows a Bennett lesion at the posteroinferior glenoid rim ➔. The athlete had internal impingement pain. The posterior labrum ➔ is intact.*

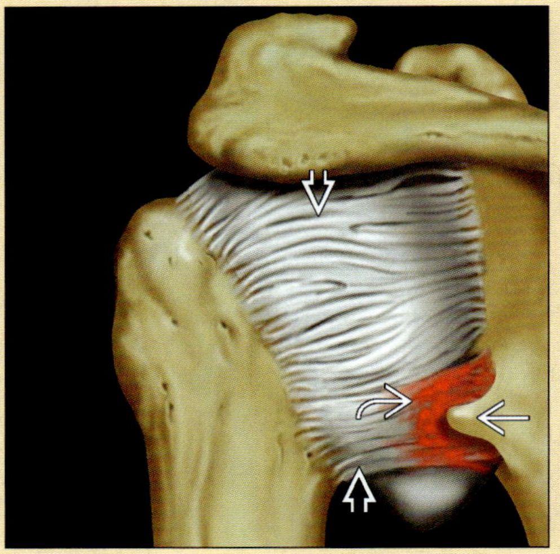

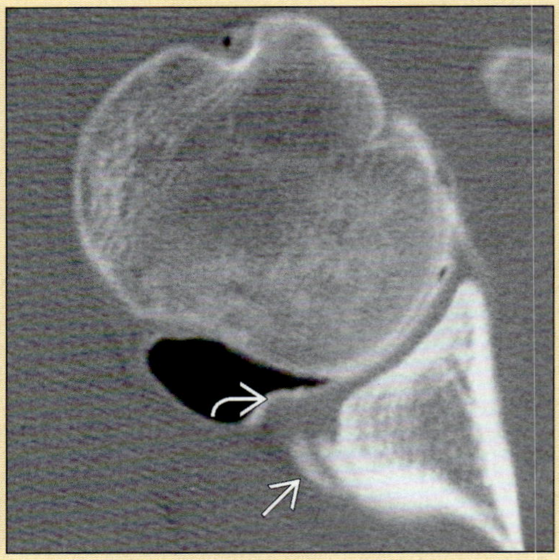

*(**Left**) Axillary lateral radiograph shows ossification adjacent to the posterior glenoid rim ➔ in a baseball pitcher. (**Right**) Axial MR arthrogram T1WI and fat-suppressed T1WI from the same patient show heterogeneous signal adjacent to the posterior glenoid ➔ representing ossification. An area of high signal on the T1WI ➔ most likely represents marrow fat. The patient also had internal impingement with labral and cuff tears, and the Bennett lesion was not felt to be the cause of his pain.*

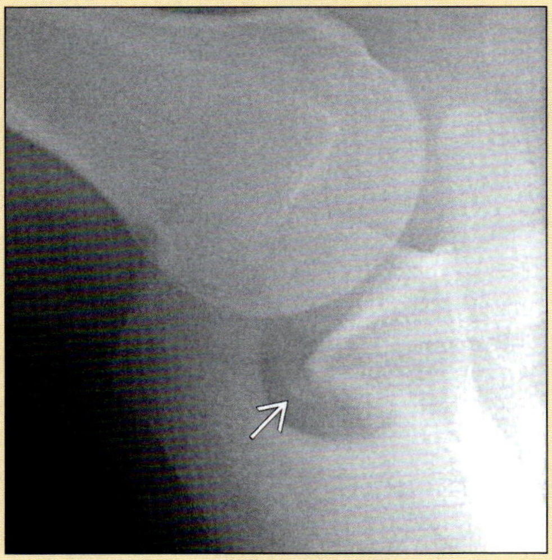

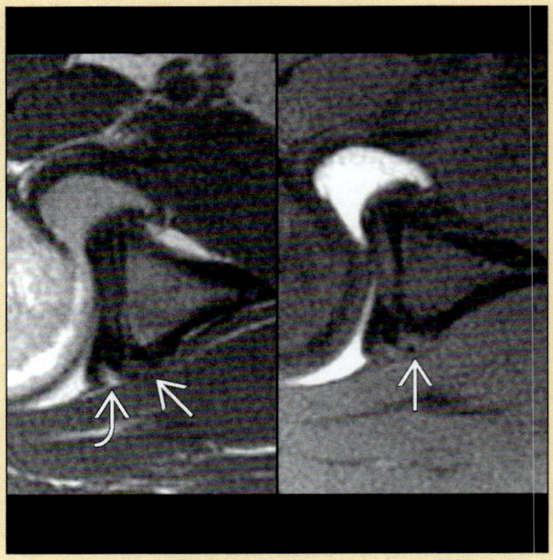

BENNETT LESION

TERMINOLOGY

Synonyms

- Posterior glenoid exostosis, thrower's exostosis

Definitions

- Ossification at glenoid attachment of posterior band inferior glenohumeral ligament (PBIGL)

IMAGING

General Features

- Best diagnostic clue
  - Ossification adjacent to posterior glenoid rim in a baseball pitcher
- Location
  - Posterior inferior glenoid rim
  - Dominant throwing arm
- Morphology
  - Crescentic

Radiographic Findings

- Crescentic ossification adjacent to posteroinferior glenoid rim
- Seen best on axillary view or angled AP view (Bennett view: 5° cranial tilt)

CT Findings

- Crescentic ossification adjacent to posterior glenoid

MR Findings

- PD/intermediate FS
  - Low signal ossification arising from posteroinferior glenoid rim
    - May blend in with low signal capsule
  - Surrounding heterogeneous signal fibrosis and thickened capsule
  - Acute injury: High signal edema around ossification
  - ± linear high signal in posterior labral tear
    - Adjacent to Bennett lesion
    - Posterosuperior labral tear of internal impingement

Imaging Recommendations

- Best imaging tool
  - CT best to determine size of ossification
  - MR best for labral tears

DIFFERENTIAL DIAGNOSIS

Reverse Bankart Fracture

- Also has donor site in posterior glenoid rim

Posterior Glenoid Osteophyte

- Usually seen with inferior humeral head osteophyte

Calcific Tendinopathy

- Calcification (not ossification)
- Typically in tendon or bursa, not capsule or glenohumeral ligament

PATHOLOGY

General Features

- Etiology
  - Heterotopic ossification or traction enthesophyte in PBIGL or posterior capsule from repetitive overhead throwing
    - High forces during deceleration phase of throwing
- Associated abnormalities
  - Internal impingement
    - Thickened PBIGL with glenohumeral internal rotation deficit (GIRD)
    - Posterosuperior labral tear or fraying
    - Posterior rotator cuff tear
  - Posterior labral tear

Staging, Grading, & Classification

- Progression in overhead throwers
  - Thickening of PBIGL and capsule
  - Ossification at enthesis

Gross Pathologic & Surgical Features

- Reactive new bone formation

CLINICAL ISSUES

Presentation

- Most common signs/symptoms
  - Bennett lesion itself usually asymptomatic
  - May have posterior shoulder pain or internal impingement symptoms

Demographics

- Age
  - 20-40 years old
- Gender
  - M > F
- Epidemiology
  - Present in 22% of asymptomatic and 33% of symptomatic Major League pitchers

Treatment

- Treat cause of shoulder pain
  - Internal impingement: Shoulder stretches to improve internal rotation
- Excise Bennett lesion if large (> 100 mm^2)

DIAGNOSTIC CHECKLIST

Consider

- Bennett lesion when imaging baseball pitchers

SELECTED REFERENCES

1. Ouellette H et al: Spectrum of shoulder injuries in the baseball pitcher. Skeletal Radiol. 37(6):491-8, 2008
2. Nakagawa S et al: Superior Bennett lesion: a bone fragment at the posterosuperior glenoid rim in 5 athletes. Arthroscopy. 23(10):1135, 2007
3. Wright RW et al: Prevalence of the Bennett lesion of the shoulder in major league pitchers. Am J Sports Med. 32(1):121-4, 2004
4. De Maeseneer M et al: The Bennett lesion of the shoulder. J Comput Assist Tomogr. 22(1):31-4, 1998

POSTERIOR LABRAL TEAR, SHOULDER

Key Facts

Terminology
- Reverse Bankart lesion, posterior Bankart tear

Imaging
- High signal extending to surface of posterior labrum
- Conventional MR: Sensitivity ~ 70% (3Tesla: 86%)
- Direct MR arthrography: Sensitivity = 90%

Top Differential Diagnoses
- Reverse Bankart fracture
- Combined Bankart and reverse Bankart tears
- Type VIII SLAP lesion
- Reverse humeral avulsion of glenohumeral ligament (RHAGL) lesion
- Reverse glenoid avulsion of glenohumeral ligament (RGAGL) lesion
- Bennett lesion

Pathology
- Repetitive trauma
- Posterior dislocation
- Hypoplastic glenoid, glenoid retroversion
- FOOSH: Posterosuperior labral tear
- Classification
  - Reverse Bankart tear
  - Partial reverse Bankart tear
  - Posterior labrocapsular periosteal sleeve avulsion (POLPSA)
  - Reverse glenolabral articular disruption (RGLAD) or glenoid articular rim defect (RGARD) lesion
  - Posterosuperior tear with paralabral ganglion cyst

Clinical Issues
- Posterior tears less associated with dislocation or unidirectional instability than anterior tears

*(**Left**) Axial graphic shows a reverse Bankart lesion. There is a torn posterior labrum → and periosteum ↝ with detachment of the labroligamentous complex. This is the classic lesion after a posterior dislocation and often results in posterior instability. (**Right**) Axial MR arthrogram PDWI FS shows a reverse Bankart tear with detached posterior labrum →. There is also posterior glenoid rim deficiency ⇨, which can be either congenital hypoplasia or mechanical erosion from prior dislocations.*

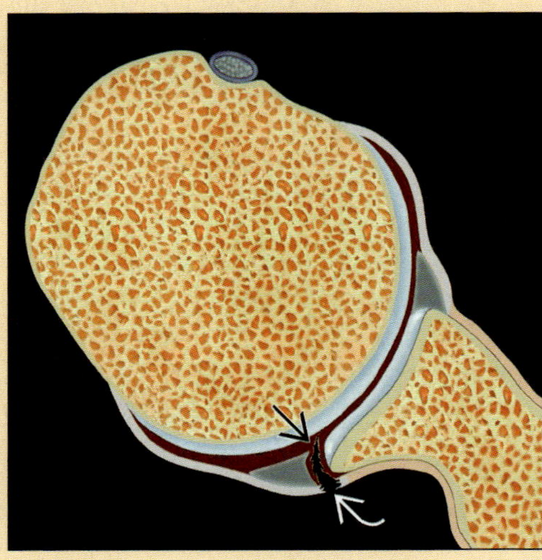

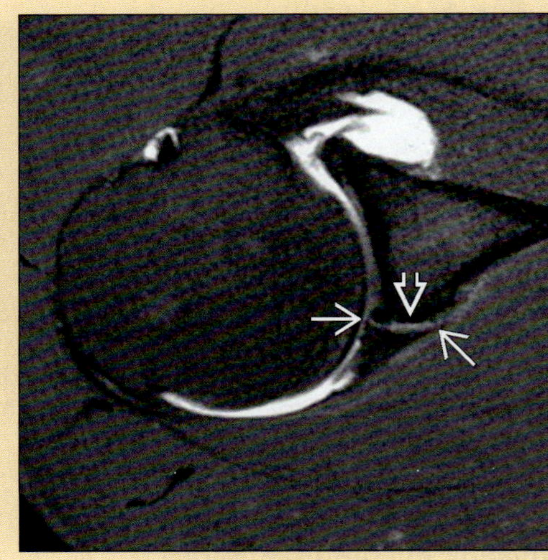

*(**Left**) Lateral graphic shows a reverse Bankart lesion extending from 8-10 o'clock →. Posterior dislocations can occur with trauma, seizure or electrical shock, and can be associated with a reverse Hill-Sachs lesion ("trough" sign) of the anteromedial humeral head. (**Right**) Sagittal MR arthrogram T2WI FS in the same patient shows the posterior labral detachment →. The posterior glenoid rim ⇨ does not have the normal convex shape, possibly worn down from recurrent dislocations.*

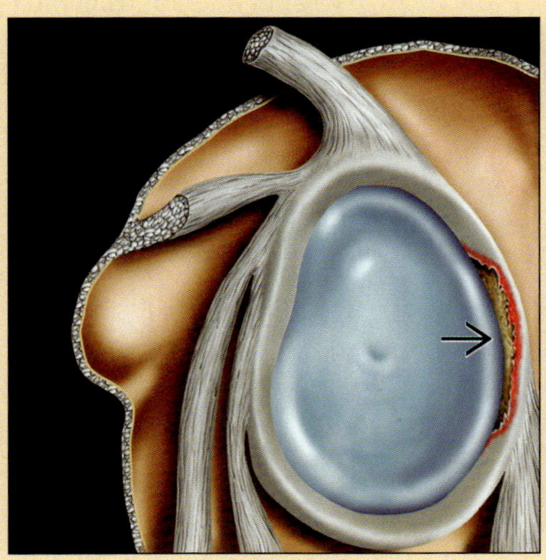

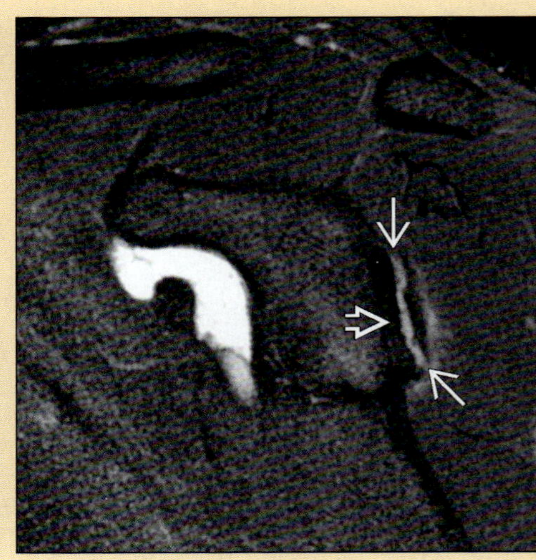

POSTERIOR LABRAL TEAR, SHOULDER

TERMINOLOGY

Synonyms

- Reverse Bankart lesion, posterior Bankart tear

Definitions

- Tear of posterior glenoid labrum

IMAGING

General Features

- Best diagnostic clue
  - Linear high signal extending to surface of labrum
- Location
  - Posterior glenoid labrum

Radiographic Findings

- Reverse Hill-Sachs impaction fracture ("trough" sign)
  - After posterior shoulder dislocation

MR Findings

- PD/intermediate FS
  - High signal extending to surface of posterior labrum
  - Sensitivity ~ 70%
    - Higher with 3T (86%)
- MR arthrography
  - High signal contrast extends into labrum
  - Sensitivity: 90%

Imaging Recommendations

- Best imaging tool
  - Direct MR arthrography

DIFFERENTIAL DIAGNOSIS

Reverse Bankart Fracture

- Fracture of posterior glenoid rim after dislocation

Extended Labral Tears

- Combined Bankart and reverse Bankart tears

Type VIII SLAP Lesion

- Superior labrum anterior-posterior (SLAP) tear extending into posterior labrum

Posterior Capsuloligamentous Tear

- Reverse humeral avulsion of glenohumeral ligament (RHAGL) lesion
- Reverse glenoid avulsion of glenohumeral ligament (RGAGL) lesion

PATHOLOGY

General Features

- Etiology
  - Repetitive trauma
    - American football linemen
  - Posterior dislocation: 50% of patients with posterior labral tear have history of prior trauma
    - Causes: Trauma, seizure, electrical shock
  - Hypoplastic glenoid, glenoid retroversion ≥ 8°
  - Fall on outstretched hand (FOOSH): Posterosuperior tear
- Associated abnormalities
  - Rotator cuff tear: After dislocation
  - Paralabral cyst: Posterosuperior labral tear
    - Spinoglenoid notch ganglion cyst
    - Cyst can also be seen with posterior SLAP tear
  - RHAGL + posterior labral tear = floating posterior band inferior glenohumeral ligament (FPIGHL)

Staging, Grading, & Classification

- Reverse Bankart tear
  - Posterior labral detachment with torn periosteum
- Partial reverse Bankart tear
  - Tear involves a portion of width of labrum
- Posterior labrocapsular periosteal sleeve avulsion (POLPSA) lesion
  - Posterior labral tear with intact periosteum
  - Posterior version of anterior labrum periosteal sleeve avulsion (ALPSA) lesion
- Reverse glenolabral articular disruption (RGLAD) or reverse glenoid articular rim defect (RGARD) lesion
  - Posterior labral tear and adjacent chondral lesion
- Kim lesion
  - Incomplete avulsion of deep portion of posterior labrum with separate shallow labral tear
- Posterosuperior labral tear
  - Often have spinoglenoid notch ganglion cyst

Gross Pathologic & Surgical Features

- Hemorrhage at posterior labral tear

CLINICAL ISSUES

Presentation

- Most common signs/symptoms
  - Shoulder pain
  - Clicking
  - Instability: Posterior or multidirectional
- Clinical profile
  - Posterior tears less associated with dislocation or unidirectional instability than anterior tears

Demographics

- Age
  - 20-30 years old
- Gender
  - M > F
- Epidemiology
  - 2-4% of glenohumeral dislocations are posterior
  - 5% of shoulder instability is posterior

Natural History & Prognosis

- Posterior instability if reverse Bankart fails to heal

Treatment

- Posterior labroligamentous repair

DIAGNOSTIC CHECKLIST

Consider

- Posterior labral tear if hypoplastic glenoid

SELECTED REFERENCES

1. Harish S et al: Imaging findings in posterior instability of the shoulder. Skeletal Radiol. 37(8):693-707, 2008
2. Saupe N et al: Acute traumatic posterior shoulder dislocation: MR findings. Radiology. 248(1):185-93, 2008

POSTERIOR LABRAL TEAR, SHOULDER

*(**Left**) Axial PDWI FS MR shows a posterior labral tear. The patient had a posterior shoulder dislocation while swinging a softball bat 2 weeks prior, which reduced by itself after 2 minutes. He also has a small reverse Hill-Sachs lesion with adjacent bone contusion. (**Right**) Axial MR arthrogram T1WI FS shows a partial reverse Bankart tear. Partial posterior labral tears often cause shoulder pain and clicking but can also be associated with instability symptoms.*

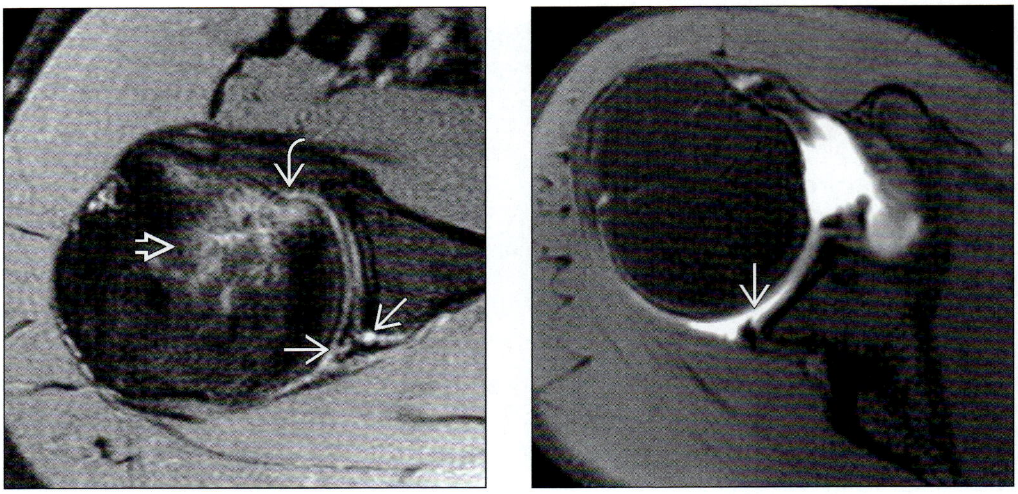

*(**Left**) Axial MR arthrogram T1WI FS shows a posterior labral tear in a patient with a prior posterior dislocation. The patient also has an adjacent chondral lesion. This combination of a labral tear and chondral lesion represents either a reverse glenolabral articular defect (GLAD) or reverse glenoid articular rim defect (GARD) lesion. (**Right**) Sagittal MR arthrogram T1WI FS in the same patient again shows the chondral lesion in this patient with a reverse GLAD lesion.*

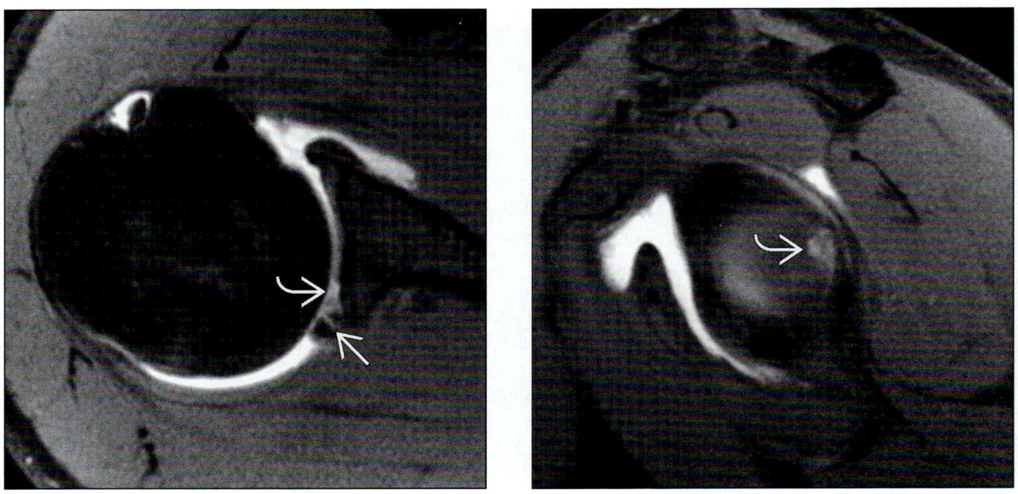

*(**Left**) Axial PDWI FS MR shows a posterior labrocapsular periosteal sleeve avulsion, or POLPSA lesion. The posterior labrum is torn but still attached to the scapular neck by periosteum. The POLPSA lesion is the posterior version of the ALPSA lesion. (**Right**) Axial PDWI FS MR shows a Kim lesion, an incomplete avulsion of the deep portion of the posterior labrum with intact attachment of the superficial labrum. A Kim lesion usually also includes a separate shallow labral tear.*

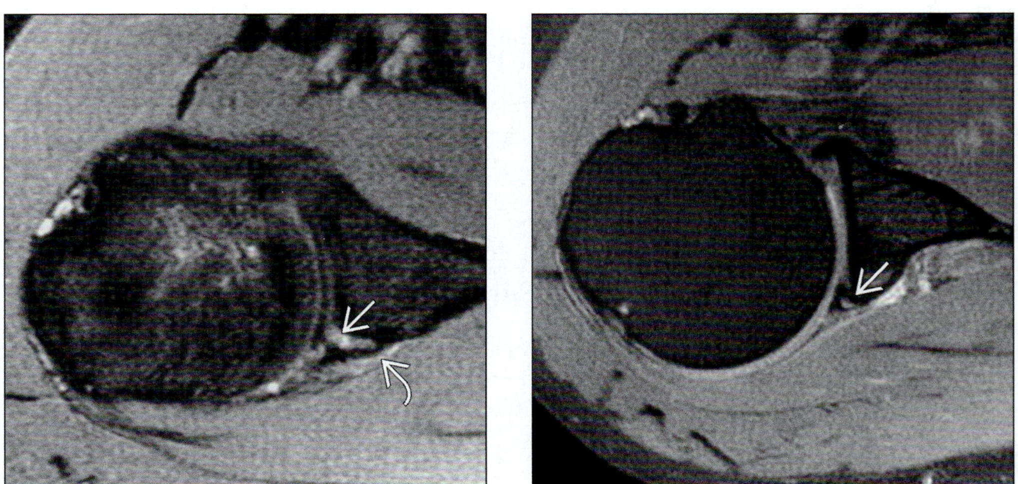

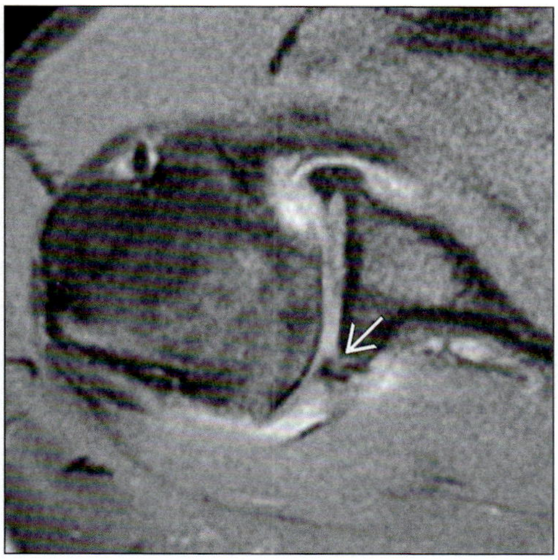

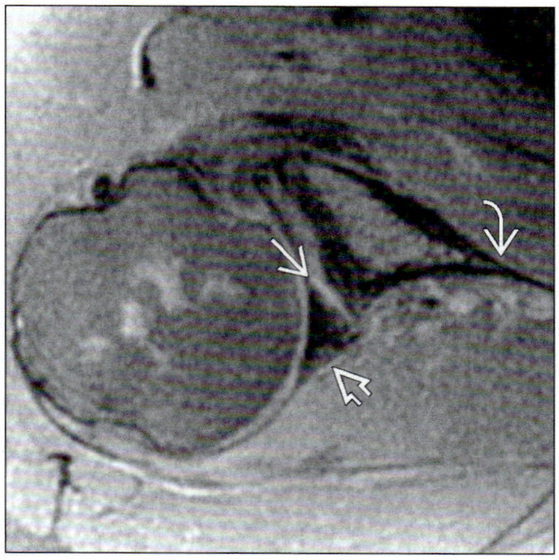

(Left) *Axial PDWI FS MR shows a posterior labral tear →. The patient had gradually increasing pain with exercise but no prior dislocation. There is a mildly hypoplastic glenoid, which is associated with atraumatic posterior labral tears.* ***(Right)*** *Axial PDWI FS MR shows a posterior labral tear →. This patient has a severely hypoplastic glenoid, with associated hypertrophy of the posterior labrum →. The patient also has glenoid retroversion relative to the plane of the scapula →.*

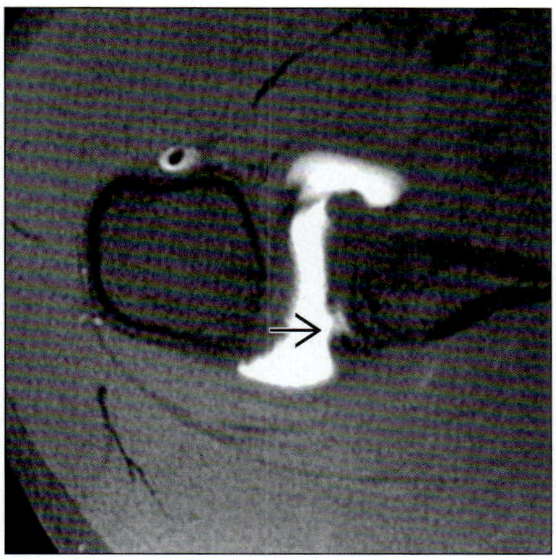

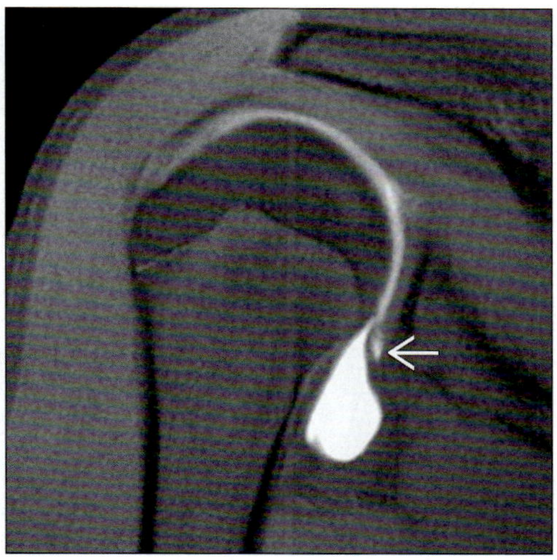

(Left) *Axial MR arthrogram T1WI FS shows a posterior labral tear → on an axial section through the inferior glenoid fossa. The patient fell headfirst with his arm fully extended while snowboarding and dislocated his shoulder.* ***(Right)*** *Coronal oblique MR arthrogram T1WI FS in the same patient also shows the tear of the posteroinferior labrum →. The location on the glenoid rim of posterior labral tears depends on the etiology and the direction of the force applied to the labrum.*

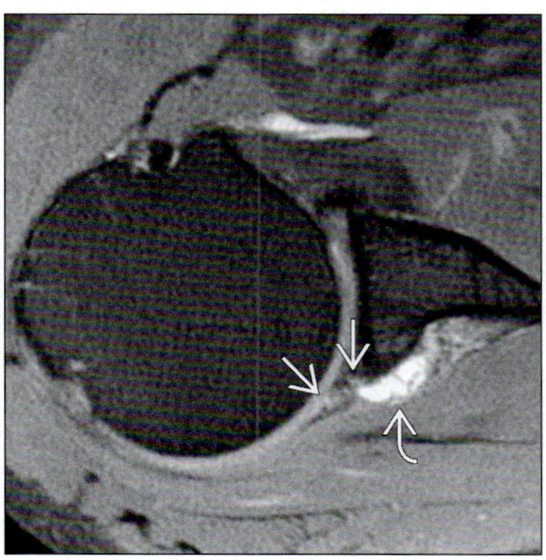

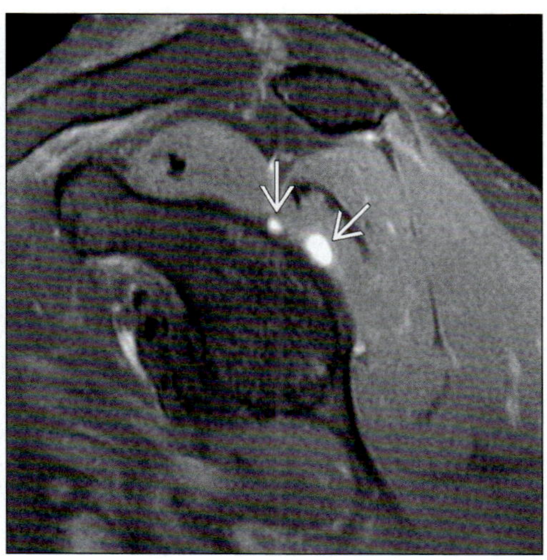

(Left) *Axial PDWI FS MR shows a posterosuperior labral tear →. There is also a paralabral cyst → adjacent to the scapular neck. Posterosuperior labral tears are associated with paralabral cysts in the spinoglenoid notch.* ***(Right)*** *Sagittal MR arthrogram T1WI FS in the same patient shows the multilobulated paralabral cyst → adjacent to the posterosuperior glenoid. This is the most common site for paralabral cysts. When large, these can cause pain and infraspinatus muscle atrophy.*

EXTENDED LABRAL TEARS

Key Facts

Terminology
- 180-360° tear, triple labral lesion, combined labral lesion
- Labral tears that involve > 1 of 3 regions (anterior, posterior, superior) around glenoid rim

Imaging
- Increased signal extending to surface of labrum in > 1 region
- MR arthrography
  - Improved sensitivity over conventional MR for entire extent of tear
  - ABER images show anteroinferior and posterosuperior tears best

Top Differential Diagnoses
- Bankart lesion
- SLAP tears
- Posterior labral tear
- Normal labral variants

Pathology
- Anterior dislocation or subluxation
- Often related to weight training
- ± hypoplastic glenoid
- Types
  - Superior & anterior (type V SLAP)
  - Anterior & posterior (Bankart-reverse Bankart)
  - Anterior, superior, & posterior tear (type IX SLAP)
  - Superior & posterior (type VIII SLAP)

Clinical Issues
- Shoulder pain
- Instability: Anterior or multidirectional
  - 10-25% of patients with instability

(Left) Sagittal graphic shows an extended labral tear with a Bankart portion ➔ and a reverse Bankart tear ➔. Most patients with this 180° tear present with anterior instability and pain after an anterior dislocation or subluxation. (Right) Axial MR arthrogram PDWI FS in a high school football player after a self-reducing anterior dislocation shows a tear involving the anterior ➔ and posterior ➔ labrum. The posterior labrum tears due to traction by the posterior capsule and ligaments.

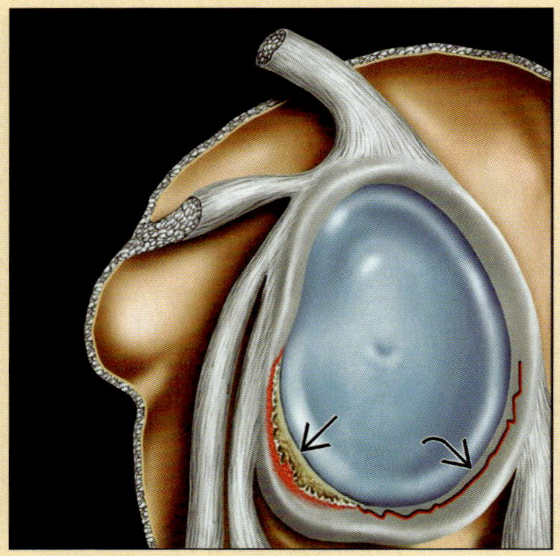

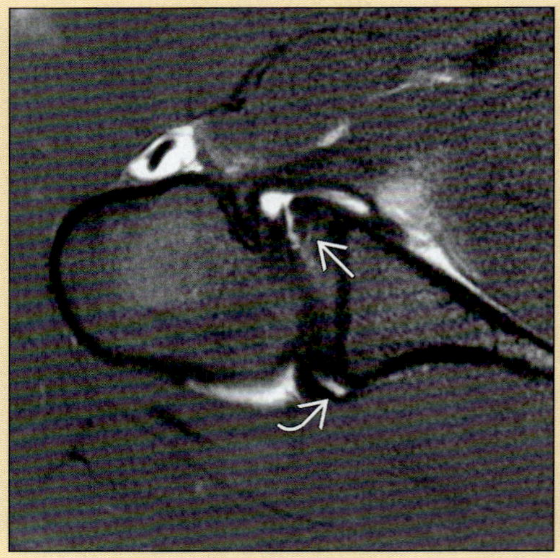

(Left) Coronal oblique MR arthrogram T2WI FS in the same patient shows the tear extending across the inferior labrum ➔, involving the caudal 180° of the labrum. (Right) Coronal (left) and axial (right) MR arthrogram T2W FS images of an 18-year-old college football player with pain and anterior instability after a subluxation episode shows a 360° labral tear. The tear involves the superior ➔, inferior ➔, anterior ➔, and posterior ➔ labrum. These are also called type IX SLAP tears.

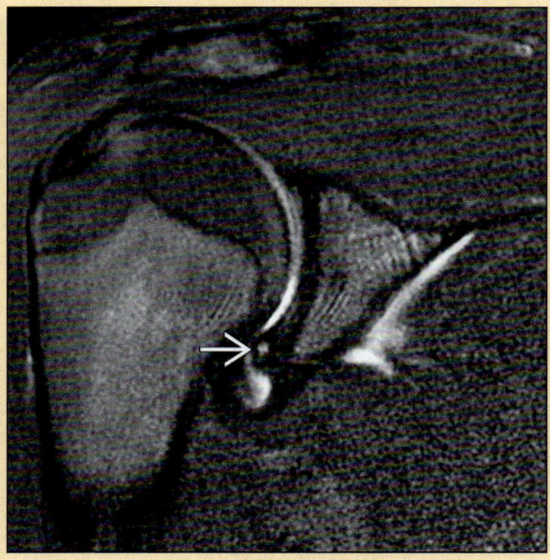

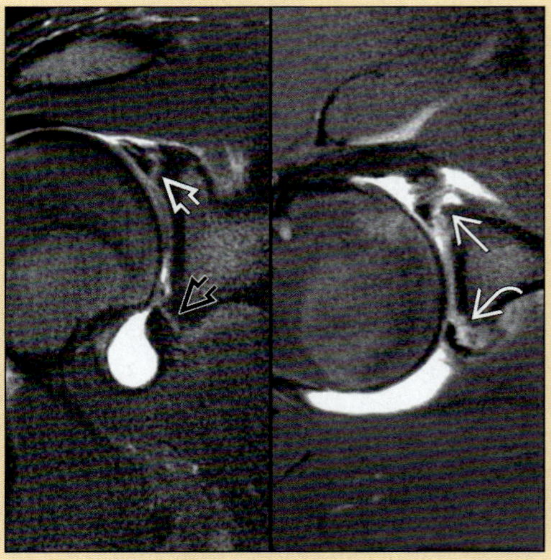

EXTENDED LABRAL TEARS

TERMINOLOGY

Abbreviations

- Superior labrum anterior-posterior (SLAP) tear

Synonyms

- 180-360° tear, triple labral lesion, combined labral lesion
- Types
  - Bankart-reverse Bankart tear
  - Type V SLAP tear (SLAP → Bankart) or type VIII SLAP tear (SLAP → posterior labrum)
  - Type IX SLAP tear (Bankart → SLAP → posterior labral tear)

Definitions

- Labral tear that involves > 1 of 3 main regions (anterior, posterior, and superior)

IMAGING

General Features

- Best diagnostic clue
  - Labral tear involving multiple regions on MR
- Size
  - 180-360° involvement of labrum

Radiographic Findings

- ± Hill-Sachs fracture
  - Reverse Hill-Sachs rare

MR Findings

- PD/intermediate FS
  - Increased signal extending to surface of labrum
  - Involves > 1 region of glenoid rim
  - Sensitivity: 75% for labral pathology but low sensitivity for entire extent of tear
- MR arthrography
  - Increased signal contrast extends into labral tear involving multiple segments of labrum
  - Improved sensitivity for entire extent of tear

Imaging Recommendations

- Best imaging tool
  - Direct MR arthrography
- Protocol advice
  - Abduction external rotation (ABER) images show anteroinferior and posterosuperior tears better

DIFFERENTIAL DIAGNOSIS

Bankart Lesion

- Large Bankart tears can extend into inferior or anterosuperior labrum

SLAP Tear

- Posterior involvement in type II SLAP tears smaller than type VIII SLAP tear

Posterior Labral Tear

- Posterior instability and apprehension more common in isolated than in posterior extended labral tears

Normal Labral Variants

- Can mimic labral tear

PATHOLOGY

General Features

- Etiology
  - Anterior dislocation or subluxation
  - Often related to weight training
- Associated abnormalities
  - ± hypoplastic glenoid

Staging, Grading, & Classification

- Superior & anterior (type V SLAP)
  - Most common extended labral tear
  - 10-14% of labral surgeries
- Anterior & posterior (Bankart-reverse Bankart)
  - 5-8% of labral surgeries
- Anterior, superior, & posterior tear (type IX SLAP)
  - 2-5% of labral surgeries
  - 360° tear, or short intact labral segment at 6 o'clock
- Superior & posterior (type VIII SLAP)
  - 2-3% of labral surgeries

Gross Pathologic & Surgical Features

- Labral tear or detachment with hemorrhage involving > 1 region of labrum

CLINICAL ISSUES

Presentation

- Most common signs/symptoms
  - Shoulder pain
  - Instability: Anterior or multidirectional most common
    - ± posterior laxity
  - Anterior apprehension

Demographics

- Age
  - 18-35 years old
- Gender
  - M > > F
- Epidemiology
  - 10-25% of patients with instability

Natural History & Prognosis

- Continued pain and instability

Treatment

- Combined labral repair

DIAGNOSTIC CHECKLIST

Consider

- Extended tear if labral tear seen in several regions

SELECTED REFERENCES

1. Chang D et al: SLAP lesions: anatomy, clinical presentation, MR imaging diagnosis and characterization. Eur J Radiol. 68(1):72-87, 2008
2. Lindauer KR et al: MR imaging appearance of 180-360 degrees labral tears of the shoulder. Skeletal Radiol. 34(2):74-9, 2005
3. Lo IK et al: Triple labral lesions: pathology and surgical repair technique-report of seven cases. Arthroscopy. 21(2):186-93, 2005

MULTIDIRECTIONAL INSTABILITY, SHOULDER

Key Facts

Terminology

- Multidirectional instability (MDI)
- Recurrent atraumatic instability
- AMBRII
  - **A**traumatic
  - **M**ultidirectional
  - **B**ilateral
  - **R**ehabilitation is usual treatment
  - **I**nferior capsule: If surgery, capsule is tightened
  - **I**nterval (rotator) is also tightened/reconstructed if surgery
- Pain from excessive anterior, posterior, and inferior humeral head translation without labral tear

Imaging

- MR
  - No tear of labrum or capsule
  - No Hill-Sachs or reverse Hill-Sachs fracture
  - ± hypoplastic glenoid fossa (radius of curvature > 100 mm)
  - ± increased glenoid retroversion (> 6°)
- MR arthrogram
  - Lax anterior and posterior capsule and inferior glenohumeral ligament

Top Differential Diagnoses

- Extended labral tears
- SLAP lesions V-X
- Microinstability, shoulder
- Rotator interval tears

Clinical Issues

- Shoulder pain or discomfort
- Sensation that shoulder subluxes or dislocates

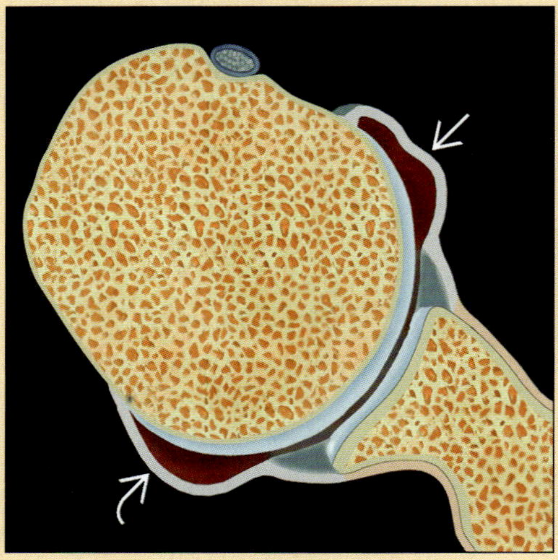

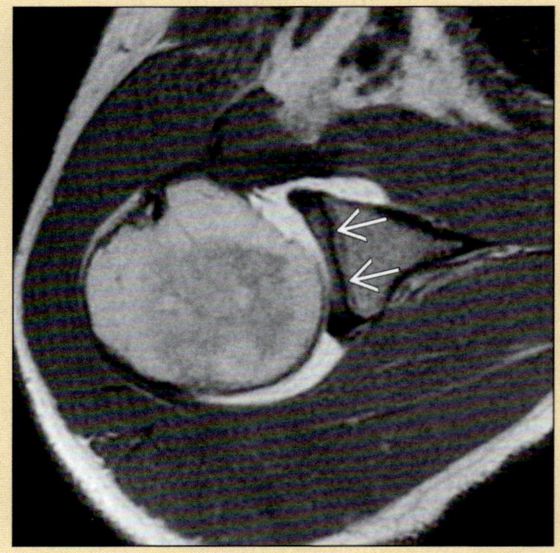

***(Left)** Axial graphic shows a lax anterior and posterior capsule without a labral tear, as seen in patients with multidirectional instability. Multidirectional instability often runs in families. **(Right)** Axial MR arthrogram T1WI of a patient with multidirectional instability (MDI) symptoms shows decreased glenoid fossa concavity and mild glenoid retroversion. These 2 findings, as well as a lax joint capsule from deficient collagen, are common in patients with MDI.*

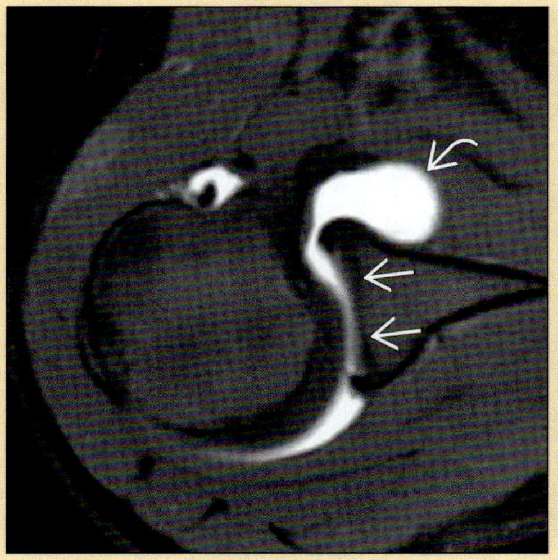

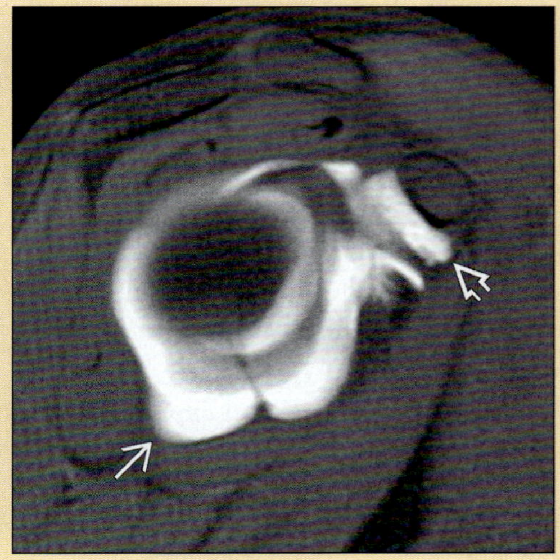

***(Left)** Axial MR arthrogram T1WI FS in a patient with MDI demonstrates a lax anterior joint capsule and a mildly hypoplastic glenoid fossa. MDI is more common in female swimmers and gymnasts. **(Right)** Oblique sagittal MR arthrogram T1WI FS in the same patient shows the lax anterior, posterior, and inferior capsule. There is also laxity of the rotator interval capsule. Exercises to strengthen the rotator cuff and restore optimal neuromuscular control can relieve symptoms.*

2

MULTIDIRECTIONAL INSTABILITY, SHOULDER

TERMINOLOGY

Abbreviations

- Multidirectional instability (MDI)
- AMBRII
  - **A**traumatic
  - **M**ultidirectional
  - **B**ilateral
  - **R**ehabilitation is usual treatment
  - **I**nferior capsule: If surgery, capsule is tightened
  - **I**nterval (rotator) is also tightened/reconstructed if surgery

Synonyms

- Recurrent atraumatic instability

Definitions

- Pain from excessive anterior, posterior, and inferior humeral head translation without labral tear

IMAGING

General Features

- Best diagnostic clue
  - Lax capsule and ligaments on MR arthrography
- Location
  - Anterior and posterior capsule

MR Findings

- No tear of labrum or capsule
- No Hill-Sachs or reverse Hill-Sachs fracture
- ± hypoplastic glenoid (radius of curvature > 100 mm)
- ± increased glenoid retroversion (> 6° from neutral)
- MR arthrogram
  - Lax anterior, posterior capsule and IGHL

Imaging Recommendations

- Best imaging tool
  - Direct MR arthrogram

DIFFERENTIAL DIAGNOSIS

Extended Labral Tears

- Traumatic tears that involve > 180° of labrum
- May have multidirectional instability on examination

SLAP Lesions V-X

- Type IX SLAP tear extends down to involve anterior and posterior labrum
- May present clinically similar to MDI

Microinstability, Shoulder

- Usually pain without significant instability symptoms

Rotator Interval Tears

- Some patients have multidirectional instability

PATHOLOGY

General Features

- Etiology
  - Congenital/familial
    - Often bilateral
  - Repetitive stretching of joint capsule combined with imbalanced dynamic stabilizers, e.g., swimmers
  - Psychiatric
    - Subjective symptoms are physical manifestation of psychiatric disorder
    - Stabilization surgery does not cure symptoms
- Genetics
  - Family history of MDI in 25%
    - Abnormal collagen production
      - Decreased cross-linking of collagen protein
      - Increased immature collagen fibers
  - ± Ehlers-Danlos
- Associated abnormalities
  - Generalized joint laxity

Gross Pathologic & Surgical Features

- Lax but otherwise intact anterior and posterior capsule
- Defect in anterior rotator interval

CLINICAL ISSUES

Presentation

- Most common signs/symptoms
  - Pain or discomfort with activities of daily living
  - Sensation that shoulder subluxes or dislocates

Demographics

- Age
  - < 30 years old
- Gender
  - F > M

Natural History & Prognosis

- Symptoms may improve with exercise

Treatment

- Rehabilitation
  - Strengthening and rebalancing of dynamic stabilizers
  - Restoration of optimal neuromuscular control
- Surgery
  - Capsulorrhaphy/capsular shift

DIAGNOSTIC CHECKLIST

Consider

- MDI in patients with instability and normal MR

Reporting Tips

- Mention whether glenoid is hypoplastic or retroverted

SELECTED REFERENCES

1. von Eisenhart-Rothe R et al: Simultaneous 3D assessment of glenohumeral shape, humeral head centering, and scapular positioning in atraumatic shoulder instability: a magnetic resonance-based in vivo analysis. Am J Sports Med. Epub ahead of print, 2009
2. Provencher MT et al: An analysis of the rotator interval in patients with anterior, posterior, and multidirectional shoulder instability. Arthroscopy. 24(8):921-9, 2008
3. Kim SH et al: Loss of chondrolabral containment of the glenohumeral joint in atraumatic posteroinferior multidirectional instability. J Bone Joint Surg Am. 87(1):92-8, 2005
4. An YH et al: Multidirectional instability of the glenohumeral joint. Orthop Clin North Am. 31(2):275-85, 2000

POSTOPERATIVE LABRAL AND INSTABILITY REPAIR, SHOULDER

Key Facts

Terminology

- Synonyms: Bankart repair, superior labrum anterior-posterior (SLAP) repair, reverse Bankart repair
- Definition: Repair of torn labrum &/or inferior glenohumeral ligament (arthroscopic or open)
  - Repair of labral tear with sutures tied down to anchors or bioabsorbable screws/tacks
  - Debridement of frayed labrum
  - Capsulorrhaphy (sutures, laser, or radiofrequency [RF]) to tighten capsule &/or inferior glenohumeral ligament
  - Bone grafts for large Hill-Sachs lesions or severe glenoid rim deficiency

Imaging

- Normal postoperative appearance
  - Metallic artifact around suture anchors or bioabsorbable screws/tacks
  - Bone graft to fill large Hill-Sachs defects or severe glenoid rim deficiency
- Recurrent tear
  - Linear high signal extending to surface of labrum
  - Insufficient labrocapsular buttress at glenoid rim
- Other causes of recurrent symptoms
  - Loose anchor or tack
  - Lax capsule
  - Subscapularis tendon tear
  - Chondrolysis

Clinical Issues

- Recurrent labral tear
  - Shoulder pain
  - Instability/subluxation
  - Clicking
  - Limited range of motion
- Recurrent/residual symptoms in 20%

__(Left)__ PA arthroscopic photograph shows a labrocapsular fixation suture anchor about to be deployed. The suture anchor → is visible within the cannula →, which is being seated on the glenoid rim →. After the anchor is placed, sutures are tied to the labrum, which is then reapproximated with the glenoid rim to reestablish its normal anatomic position. __(Right)__ Neer AP radiograph of a patient with a previous Bankart repair shows 3 suture anchors → in the anterior glenoid.

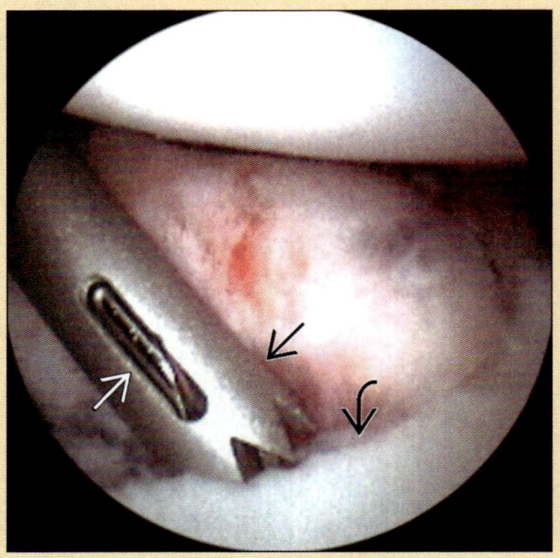

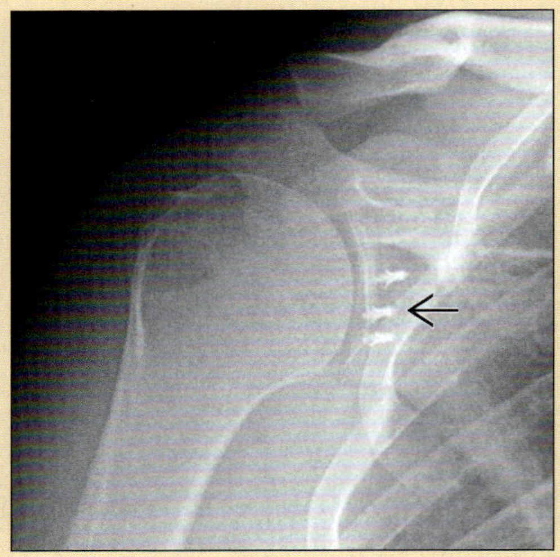

__(Left)__ Axial MR arthrogram T1WI FS in the same patient 14 months after surgery shows artifact around 1 of the suture anchors →. The anterior labrum → is healed back in its normal position on the glenoid rim, reestablishing a "bumper" and a stable insertion for the glenohumeral ligament. This MR was performed for unrelated symptoms. __(Right)__ PA arthroscopic photograph in the same patient 1 month after the MR shows synovialization over the barely visible sutures → at the Bankart repair site.

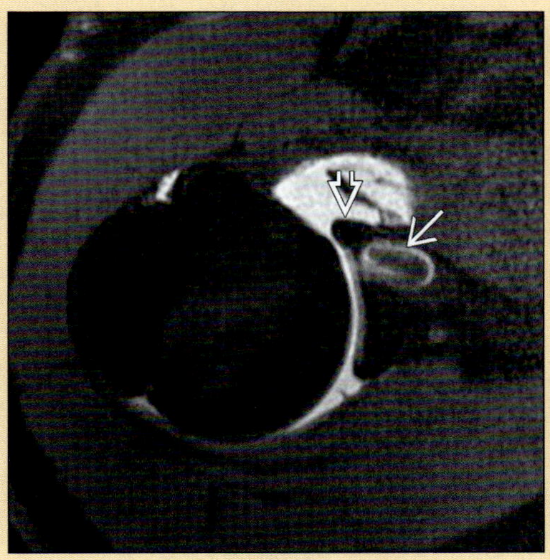

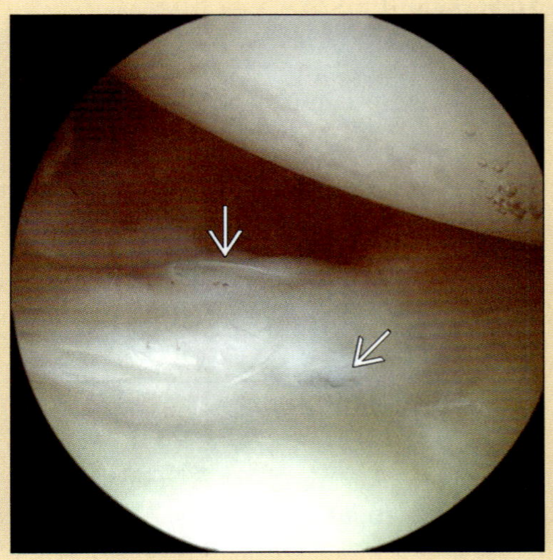

POSTOPERATIVE LABRAL AND INSTABILITY REPAIR, SHOULDER

TERMINOLOGY

Synonyms

- Bankart repair, superior labrum anterior-posterior (SLAP) repair, reverse Bankart repair

Definitions

- Repair of torn labrum &/or inferior glenohumeral ligament
  - May be performed arthroscopically or as open procedure
- Standard procedures
  - Debridement of frayed labrum
  - Labral tear repair using sutures tied to metal anchors or bioabsorbable screws/tacks
  - Capsulorrhaphy (sutures, laser, or radiofrequency [RF]) to tighten capsule and inferior glenohumeral ligament
  - Bone grafts for large Hill-Sachs lesions or severe glenoid rim deficiency

IMAGING

General Features

- Best diagnostic clue
  - Metallic artifact or bioabsorbable screws/tacks
- Location
  - Glenoid rim or glenohumeral ligament

Radiographic Findings

- Suture anchors in glenoid rim
- Older procedures: Large screw from coracoid process transfer (Bristow procedure, etc.)

MR Findings

- PD/intermediate FS
  - Normal postoperative appearance
    - Labral repair
      - Micrometallic artifact in soft tissues
      - Metal artifact around suture anchors
      - Bioabsorbable screws/tacks
      - Labroligamentous complex located in normal position on glenoid rim
    - Capsulorraphy
      - Micrometallic artifact in anterior capsule
    - Bone graft
      - Filling large Hill-Sachs defects or severe glenoid rim deficiency
  - Recurrent tear
    - Linear high signal extending to labral surface
    - Insufficient labrocapsular buttress at glenoid rim
      - Labrum positioned medially on scapular neck
    - 75% accuracy for recurrent tear
  - Other causes of recurrent symptoms
    - Loose anchor or tack
    - Lax capsule
      - Incomplete tightening of anterior capsule/inferior glenohumeral ligament
    - Subscapularis tendon tear
      - Breakdown of reanastomosis after open repair
    - Chondrolysis
      - May be associated with high volume of intraarticular bupivacaine in postoperative period
- T2* GRE
  - Metal artifact worse
    - Due to intravoxel dephasing T2* effect
- Direct MR arthrography
  - Divot between anterior labrocapsular complex and glenoid rim, especially on abduction external rotation (ABER) images
  - Extracapsular leak
  - Labrum positioned medially on scapular neck
  - 92% accuracy for recurrent labral tear

Imaging Recommendations

- Best imaging tool
  - Direct MR arthrography
- Protocol advice
  - Avoid gradient echo images due to metal artifact

DIFFERENTIAL DIAGNOSIS

Residual Cleft in Otherwise Stable Labrum

- Small labral fold/cleft may be present after repair

Suture Anchor Metallic Artifact Mimicking a Tear

- ↑ SI metal artifact may superimpose over labrum

PATHOLOGY

General Features

- Associated abnormalities
  - Rotator cuff tear in 25-33% of patients with recurrent instability

CLINICAL ISSUES

Presentation

- Most common signs/symptoms
  - Recurrent labral tear
    - Shoulder pain
    - Instability/subluxation
    - Clicking
    - Limited range of motion

Demographics

- Age
  - 20-40 years old
- Gender
  - M > F

Natural History & Prognosis

- Most patients have few or no symptoms after labroligamentous surgery
- 20% have recurrent or residual symptoms

Treatment

- Recurrent instability
  - Revision labral repair
  - Bone graft for large osseous defects

SELECTED REFERENCES

1. van der Woude HJ: Magnetic resonance imaging after surgery for shoulder instability. JBR-BTR. 92(1):43-8, 2009
2. Probyn LJ et al: Recurrent symptoms after shoulder instability repair: direct MR arthrographic assessment--correlation with second-look surgical evaluation. Radiology. 245(3):814-23, 2007

*(**Left**) Coronal oblique MR arthrogram T2WI FS after a SLAP repair shows a bioabsorbable tack → in the superior glenoid neck. The superior labrum is normal ↪ with no recurrent SLAP tear. The MR was done for unrelated symptoms. (**Right**) Axial MR arthrogram T1WI FS shows a small cleft → at the Bankart repair site. Small clefts can be seen in stable healed Bankart repair sites and should be mentioned. Artifacts from metal anchors can also create high signal lines across the labrum.*

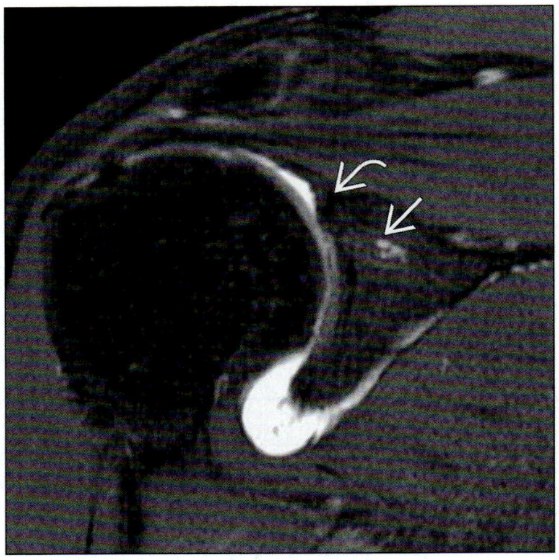

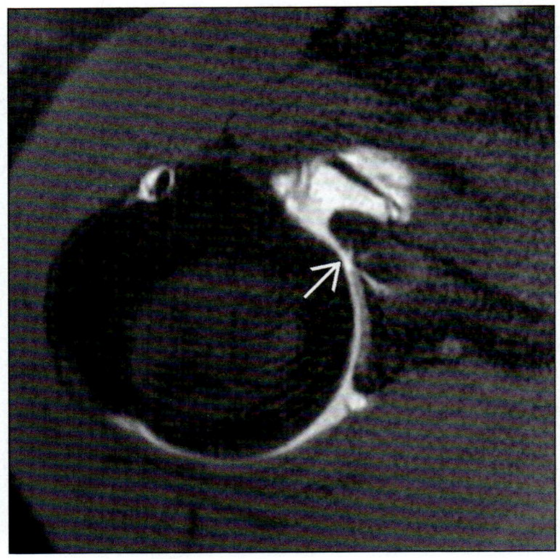

*(**Left**) Axial MR arthrogram T1WI FS shows a recurrent Bankart tear. The anterior labrum → has retorn from the glenoid rim and is displaced medially on the scapular neck. There is artifact from 1 of the Bankart repair suture anchors ↪. Bankart retears often displace medially toward the scapular neck. (**Right**) Axial MR arthrogram T1WI FS shows a re-torn anterior labrum → after prior Bankart repair. There is metal artifact ↪ in the labroligamentous structures from the prior surgery.*

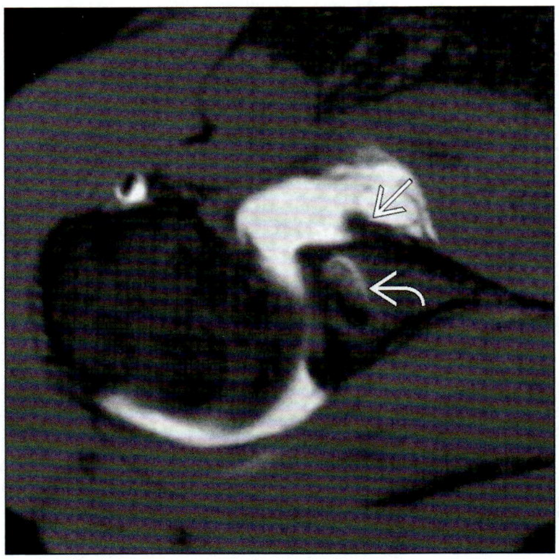

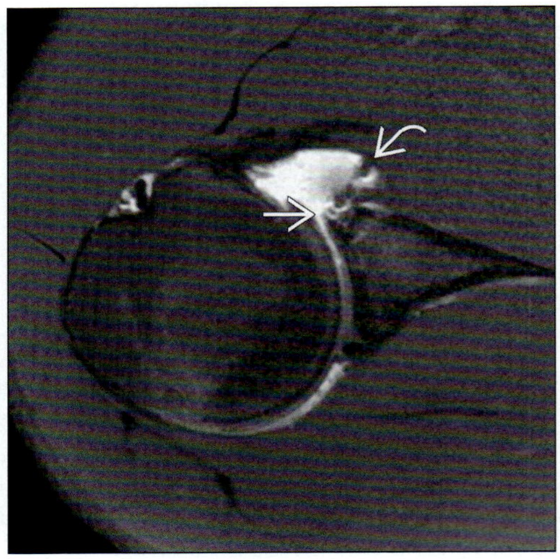

*(**Left**) Axial MR arthrogram T1WI FS shows a bioabsorbable screw → from a prior SLAP repair. There is a retear of the superior labrum with high signal ↪ between the superior labrum and the glenoid rim. (**Right**) Axial MR arthrogram T1WI shows a complete tear of the subscapularis tendon →. This patient had a prior open Bankart repair using suture anchors ↪, and the subscapularis tendon was divided and then reattached. Subscapularis ruptures are a complication of open Bankart repair.*

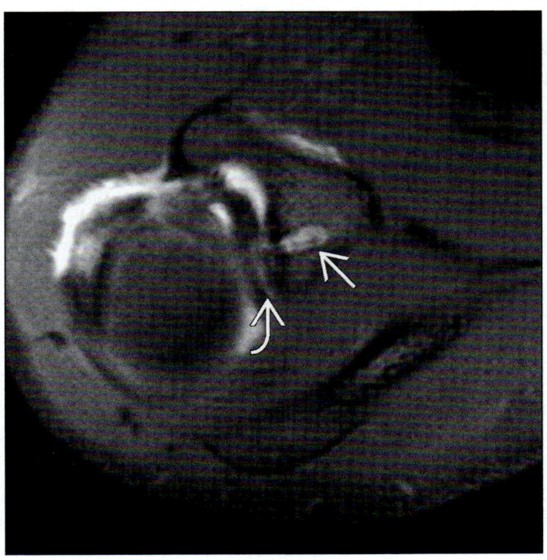

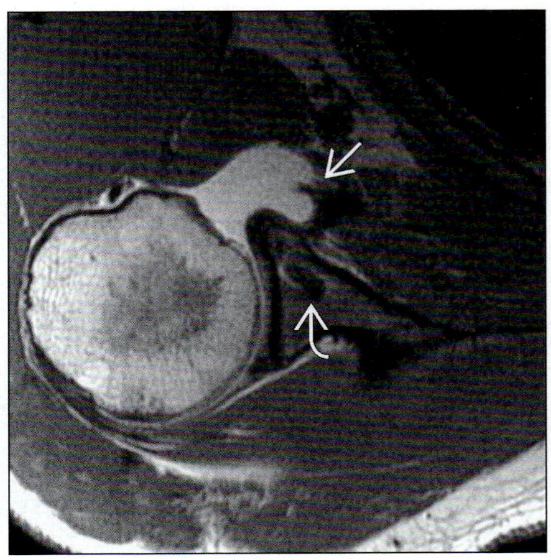

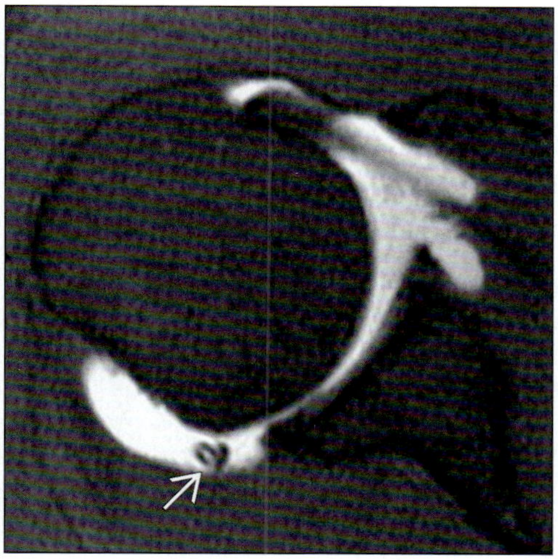

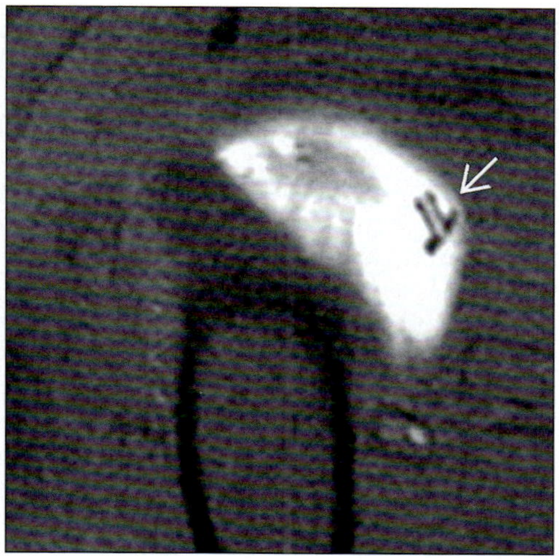

(Left) *Axial MR arthrogram T1WI FS shows a loose body in the posterior recess of the joint ➔. The body has an unusual high signal center. The patient had undergone prior labral repair with a bioabsorbable tack and has new symptoms.* ***(Right)*** *Coronal oblique MR arthrogram T1WI FS in the same patient again shows the loose body in the posterior aspect of the joint ➔. The body clearly has the appearance of a tack with a hollow fluid-filled center. Loose tacks can cause pain and locking.*

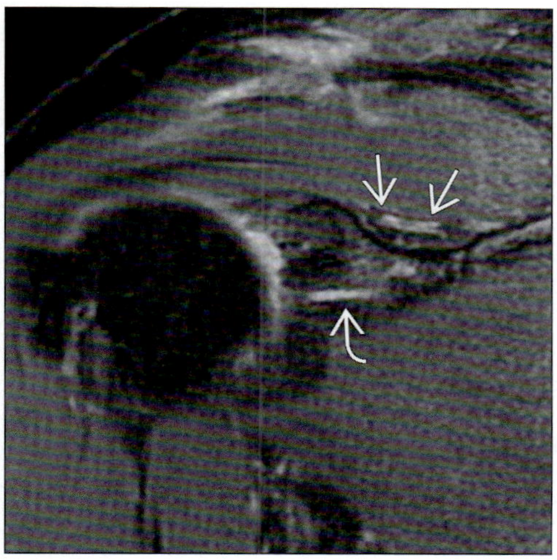

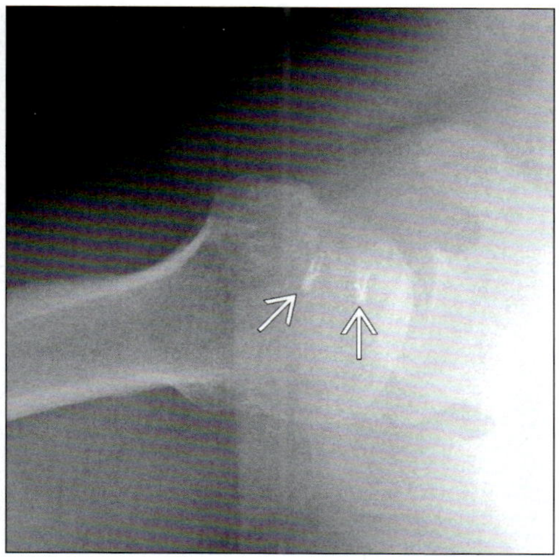

(Left) *Coronal T2WI FS MR of a patient with a prior SLAP repair shows a loose bioabsorbable tack ➔ in the suprascapular notch. Another tack is seen seated within the glenoid neck ➔.* ***(Right)*** *Axillary lateral radiograph of the same patient after a McLaughlin procedure shows 2 suture anchors ➔ in the reverse Hill-Sachs defect. In this procedure, the subscapularis is removed from its normal insertion, pulled taut, and sutured into the defect to help prevent recurrent posterior dislocation.*

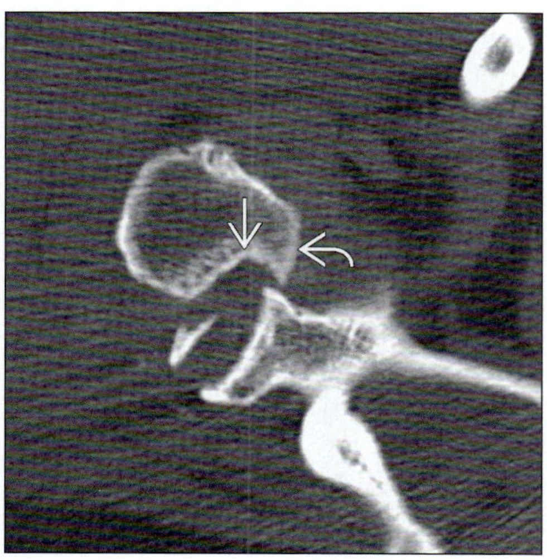

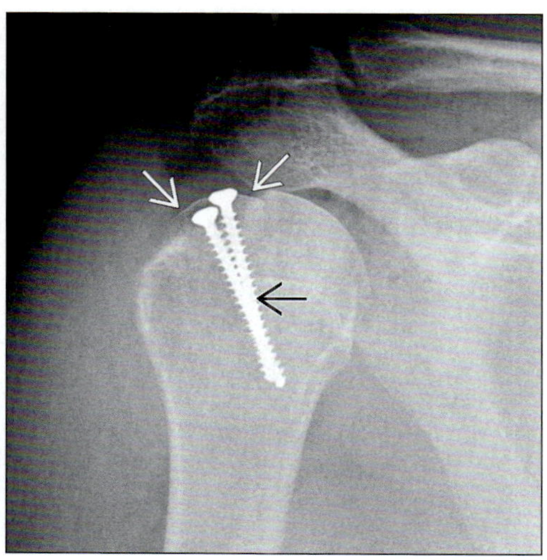

(Left) *Axial bone CT shows an anterior dislocation of the humeral head ➔ with a large Hill-Sachs lesion ➔. The patient had recurrent anterior dislocations that were difficult to reduce due to the lesion.* ***(Right)*** *AP internal rotation radiograph of the same patient after surgery shows a bone graft ➔ in the Hill-Sachs lesion with 2 fixation screws ➔. A bone graft is also sometimes required to correct glenoid rim deficiency in patients with recurrent dislocations.*

SLAP TEAR

Key Facts

Terminology

- Superior labrum anterior-posterior (SLAP) tear
- Synonym: Type I-IV SLAP tear
- Tear of superior labrum that extends anterior to posterior

Imaging

- Type I: Irregular fraying of tip of superior labrum
- Type II: Linear increased signal along long axis of superior labrum that extends to surface of labrum
  - Lateral curving high signal
  - "Double Oreo" sign: 2 high T2 signal lines (normal variant superior recess plus SLAP tear)
  - Abnormally wide high signal on coronal images
- Type III: Linear increased signal extends across entire labrum so that superior labrum is detached
- Type IV: Linear increased labral signal + surfacing signal in biceps origin

Top Differential Diagnoses

- Normal labral variants
  - Superior recess in 75% of people
  - High T2 signal curves medially at labral-chondral junction
- Normal recess between origin of biceps and superior labrum
- Biceps tendon tear, intraarticular
- Microinstability, shoulder

Clinical Issues

- Common symptoms: Shoulder pain, clicking or popping
- Positive O'Brien test
- Epidemiology: 25-30% of shoulder arthroscopies
  - 75% of these are type I

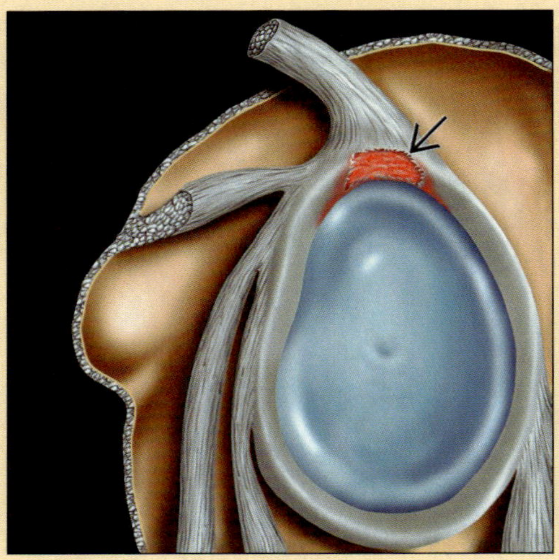

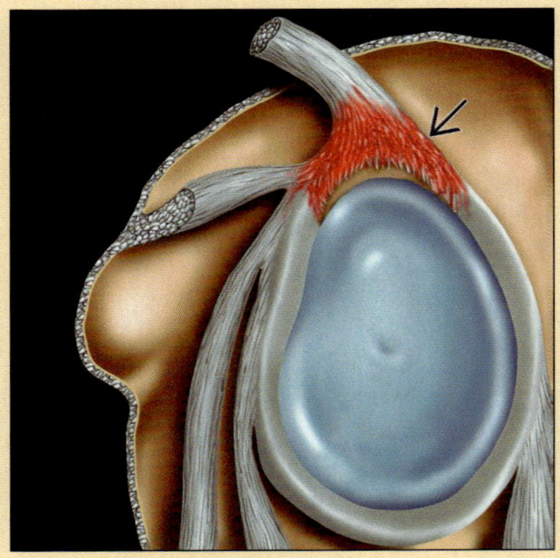

(Left) *Lateral graphic shows a type I SLAP tear. There is superficial fraying of the superior labrum → without a deep tear. Type I SLAP tears are common (20% of all shoulder arthroscopies), and the biceps anchor is stable.* ***(Right)*** *Lateral graphic shows a type II SLAP tear. There is a longitudinal tear of the superior labrum →. Type II SLAP tears are the most common subtype of the significant SLAP tears (types II-X) and are usually symptomatic and have an unstable biceps anchor.*

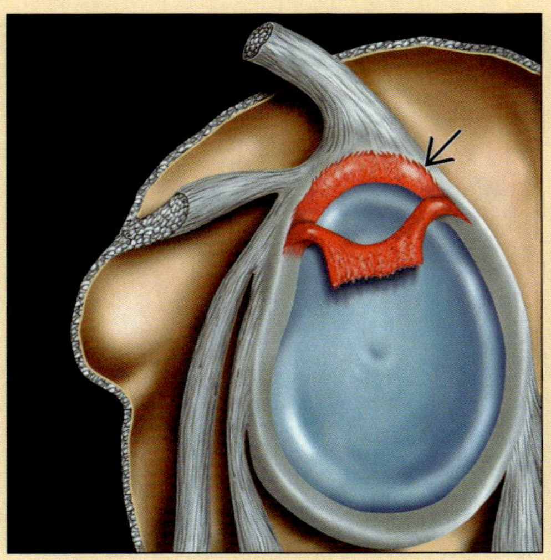

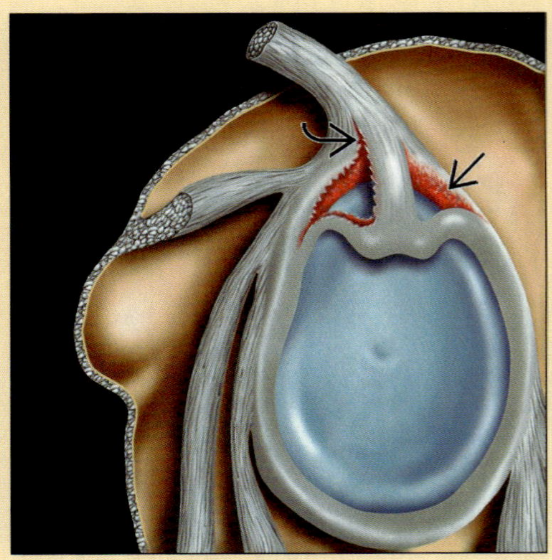

(Left) *Lateral graphic shows a type III SLAP tear. There is full width tear causing detachment of a segment of the superior labrum →. Although this appears as a bucket-handle tear at arthroscopy, these tears are typically nondisplaced at MR.* ***(Right)*** *Lateral graphic shows a type IV SLAP tear. There is a longitudinal tear of the superior labrum →, which extends into the biceps root ↪. Type IV SLAP tears are an "extended SLAP" tear but were 1 of the 4 original subtypes described, so are included here.*

SLAP TEAR

TERMINOLOGY

Abbreviations

- Superior labrum anterior-posterior (SLAP) tear

Synonyms

- Type I-IV SLAP tear

Definitions

- Longitudinal tear of superior labrum that extends anterior to posterior
  - Fraying of superior labrum = type I
  - Extended SLAP tear: Type V-X SLAP tears are a type II or III SLAP tear with extension into adjacent labrum or ligament

IMAGING

General Features

- Best diagnostic clue
  - Type II-IV: High T2 signal curving laterally in superior labrum on oblique coronal images
- Location
  - Superior labrum, 11 o'clock to 1 o'clock
- Size
  - Fraying (type I), to small type II with stable biceps, to extensive type II-IV

Radiographic Findings

- Arthrography
  - Grashey (Neer AP) view arthrography: 50% sensitivity

CT Findings

- CT arthrography
  - High attenuation fluid curves laterally in superior labrum
  - 90% sensitivity

MR Findings

- T2WI FS
  - Conventional MR: 70% sensitivity for type II-IV SLAP tears
  - Type I: Irregular fraying of tip of superior labrum
  - Type II: Linear increased signal along long axis of superior labrum that extends to surface of labrum
    - Usually extends to inferior articular surface of labrum
    - Tear can either be at labral-cartilage junction or in body of labrum
  - Type III: Linear increased signal extends across entire labrum so that superior labrum is detached
    - Labral fragment usually not displaced from glenoid on conventional MR images
  - Type IV: Linear increased labral signal + surfacing signal in biceps tendon origin
    - SLAP tear seen best on oblique coronal images
    - Biceps origin tear seen best on oblique sagittal images
      - Some biceps tears appear only as enlarged tendon with intratendinous signal
      - Biceps may instead be completely torn and avulsed from labral attachment
- Indirect MR arthrography
  - Higher sensitivity (85-90%) than conventional MR
  - Lower specificity due to enhancement of age-related labral myxoid change adjacent to superior recess
- Direct MR arthrography
  - High signal contrast curving laterally in labrum
  - Abnormally wide high signal contrast within labrum
    - > 3 mm in arthrogram
  - Anteroposterior linear high signal on axial section through superior labrum
    - 40% specificity
  - Contrast extending into labrum posterior to biceps anchor
    - 50% specificity
  - Sensitivity: 90-95%
    - Increased by adding arm traction in external rotation
- **Signs to differentiate from normal variant superior recess**
  - Lateral curving high T2 signal in tear
    - Most useful sign
  - "Double Oreo" sign: 2 high T2 signal lines
    - 1 line is normal variant superior recess, other is SLAP tear
  - Abnormally wide high T2 SI on coronal images
    - > 2 mm suggests tear rather than recess
    - Focal widening of linear high SI within labrum
  - High T2 signal extends posterior to biceps anchor
    - 70% specificity for tear
    - Controversial: Some believe superior recess can extend posterior to biceps
  - Extension into anterosuperior labrum
    - 40% of SLAP tears
    - 90% specificity (10% of people have sublabral foramen)

Imaging Recommendations

- Best imaging tool
  - Direct MR arthrography

DIFFERENTIAL DIAGNOSIS

Normal Labral Variants

- Superior recess in 75% of people
  - 10% also have sublabral foramen in anterosuperior labrum
- High SI curves medially at labral-chondral junction
- Synovial-lined blind-ending pouch
- Stable biceps anchor

Normal Recess between Origin of Biceps and Superior Labrum

- Seen on single anterior coronal section

Biceps Tendon Tear, Intraarticular

- Biceps tear at its anchor can mimic type IV SLAP lesion if there is also superior recess

Microinstability, Shoulder

- Present in some patients with SLAP tears
- Also seen in patients with isolated superior or middle glenohumeral ligament tears

SLAP Lesions V-X

- Extensive tear also involves adjacent structures
- Considered a different lesion category than SLAP I-IV

SLAP TEAR

PATHOLOGY

General Features

- Etiology
  - Fall on outstretched arm
    - Humeral head impacts against superior glenoid and labrum
  - Acute biceps traction injury (forced extension with flexed forearm)
    - Example: Carrying a couch when other person drops his/her end
  - Repetitive biceps traction injury
    - Overhead throwing athletes
- Associated abnormalities
  - Other labral or ligamentous tears (SLAP V-X)
    - Type V: Bankart lesion with SLAP tear from anterior dislocation
  - Rotator cuff tears in 40%
  - Spinoglenoid notch paralabral cyst associated with traumatic posterior SLAP tear
  - Posterosuperior cuff tear seen with posterior SLAP tear in throwers (internal impingement)

Staging, Grading, & Classification

- Snyder classification has 4 types
- Type I: Fraying of superior labrum
  - Common
  - No deep longitudinal tear of labrum
  - Usually degenerative
    - May also have myxoid degeneration within substance of labrum
  - Does not require surgery
  - Biceps anchor is stable
- Type II: Partial tear of superior labrum anterior to posterior
  - Most common significant SLAP tear
  - Attachment of long head of biceps tendon often unstable
  - Superior labrum not detached from glenoid rim
  - **Subtypes**
    - Type IIA: Anterior SLAP tear
      - Subtype: SLAC (**s**uperior **l**abrum, **a**nterior **c**uff) = type IIA SLAP and articular side anterior cuff partial tear
    - Type IIB: Posterior SLAP tear
      - Acute trauma: ± spinoglenoid notch paralabral cyst
      - Overhead throwing athletes: "Peel-back" SLAP tear; may also have posterior cuff tear
    - Type IIC: Extends equally anteriorly and posteriorly
      - Classic, most common type II subtype
- Type III: Full thickness tear with detachment of superior labral fragment
  - Labral fragment forms "bucket handle" when it separates from superior glenoid rim during arthroscopy
- Type IV: Tear extends into biceps anchor

Gross Pathologic & Surgical Features

- Snyder type I: Fraying at tip of labrum
- Type II-IV: Linear longitudinal tear of superior labrum with hemorrhage and unstable biceps anchor

CLINICAL ISSUES

Presentation

- Most common signs/symptoms
  - Shoulder pain
  - Clicking or popping
  - Biceps weakness/pain
  - Positive O'Brien test: Pain with resisted elevation with elbow extended; arm flexed 90°, adducted 15°, and internally rotated
- Other signs/symptoms
  - Microinstability

Demographics

- Age
  - Type I: > 50 years old
  - Type II: 20-60 years old
  - Type III-IV: 20-40 years old
- Gender
  - M > F
- Epidemiology
  - 25-30% of shoulder arthroscopies
    - 75% of these are type I

Natural History & Prognosis

- Type I: Common age-related finding
- Type II-IV: May cause continued pain

Treatment

- Type I: Conservative or labral debridement
- Type II-IV
  - Small tears: Labral debridement
  - Large tears: Labral repair with reattachment to superior glenoid rim

DIAGNOSTIC CHECKLIST

Consider

- Superior recess if thin labral signal curves medially at labral-cartilage junction
- Sometimes impossible to distinguish type II SLAP from superior recess on MR

Image Interpretation Pearls

- Look for superior labral linear signal that is vertical or curves laterally

SELECTED REFERENCES

1. Borrero CG et al: Magnetic resonance appearance of posterosuperior labral peel back during humeral abduction and external rotation. Skeletal Radiol. 39(1):19-26, 2010
2. Holzapfel K et al: Inter- and intraobserver variability of MR arthrography in the detection and classification of superior labral anterior posterior (SLAP) lesions: evaluation in 78 cases with arthroscopic correlation. Eur Radiol. 20(3):666-73, 2010
3. Lin E: Magnetic resonance arthrography of superior labrum anterior-posterior lesions: a practical approach to interpretation. Curr Probl Diagn Radiol. 38(2):91-7, 2009
4. Chang D et al: SLAP lesions: anatomy, clinical presentation, MR imaging diagnosis and characterization. Eur J Radiol. 68(1):72-87, 2008

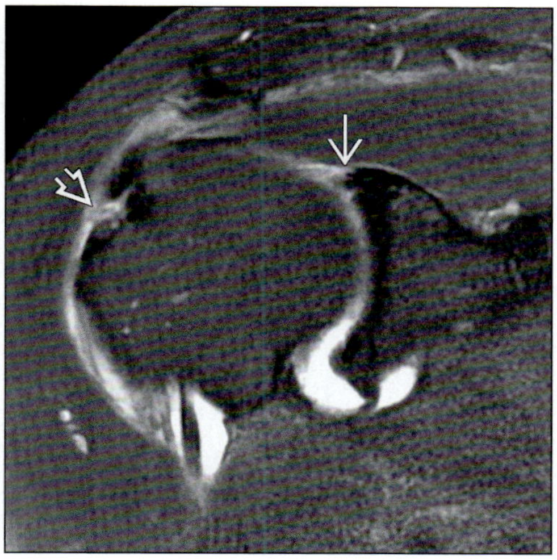

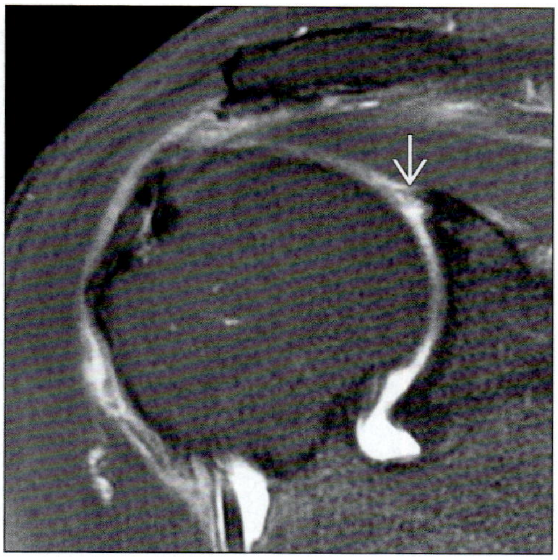

(Left) Coronal oblique MR arthrogram T1WI FS shows a type I SLAP tear with fraying of the superior labrum. There is also a rotator cuff tear. *(Right)* Coronal oblique MR arthrogram T1WI FS more posteriorly in the same patient shows the type I SLAP tear. The patient had degenerative fraying of the anterior & posterior labrum when surgery was performed to repair the rotator cuff tear. Type I SLAP tears are commonly seen in older patients at surgery performed for a rotator cuff tear.

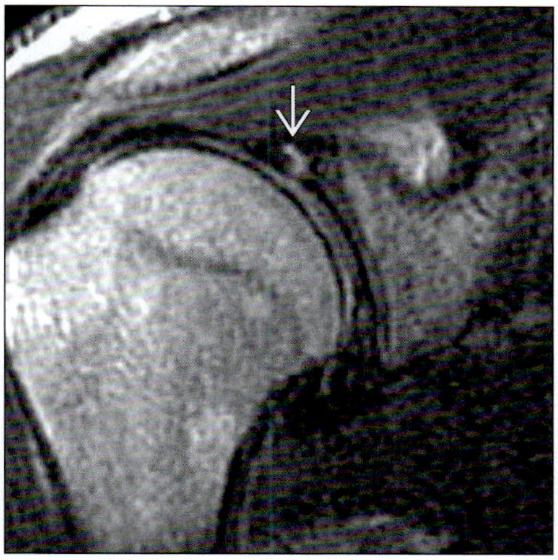

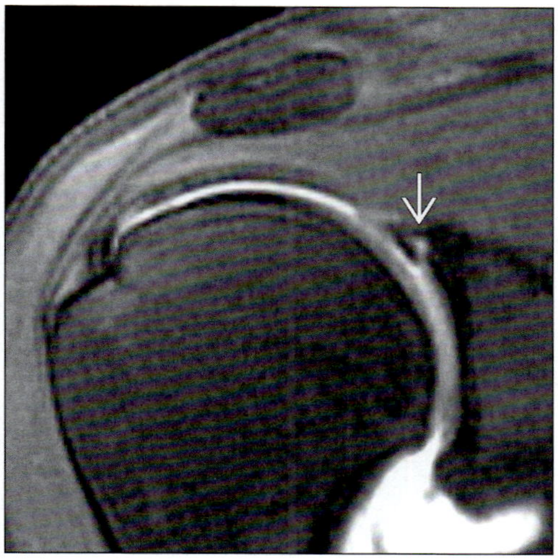

(Left) Coronal oblique T2WI FS MR shows a type II SLAP tear. There is laterally curved high signal in the superior labrum. A normal variant superior recess will instead curve medially at the labral-chondral junction. *(Right)* Coronal oblique MR arthrogram T1WI FS shows a type II SLAP tear. The contrast in the tear is mainly oriented vertically and curves laterally near the top of the tear, instead of curving medially as would be seen with a normal variant superior recess.

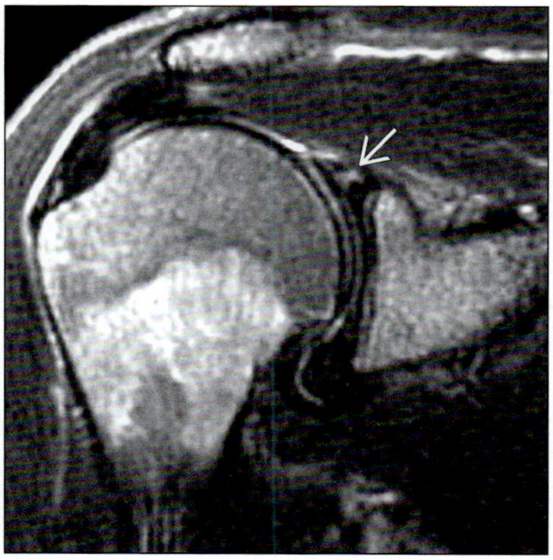

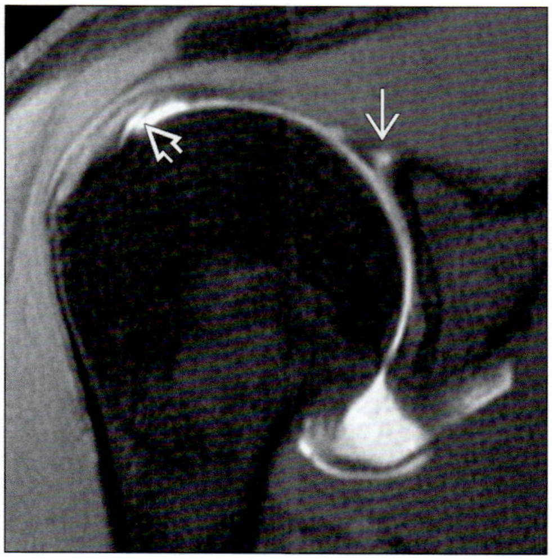

(Left) Coronal oblique T2WI FS MR shows a type II SLAP tear. Another sign of a SLAP tear is abnormally wide high signal (> 2 mm on MR) in the superior labrum. *(Right)* Coronal oblique MR arthrogram T1WI FS shows a type II SLAP tear. Contrast within the superior labrum is abnormally wide (> 3 mm on MR arthrography) and curves laterally near the superior aspect of the tear. This patient also has an articular surface partial thickness rotator cuff tear.

SLAP TEAR

(Left) *Coronal oblique MR arthrogram T2WI FS shows a "double Oreo" sign of a SLAP tear. There is 1 high signal line from a superior recess → and a 2nd line more laterally from the SLAP tear → with a thin area of intact labrum in between.* ***(Right)*** *Axial MR arthrogram T1WI FS shows a SLAP tear → visible through the superior labrum. Although common in SLAP tears, this finding of high signal at this site on a high axial image only has a 40% specificity and can be seen in patients with a prominent superior recess.*

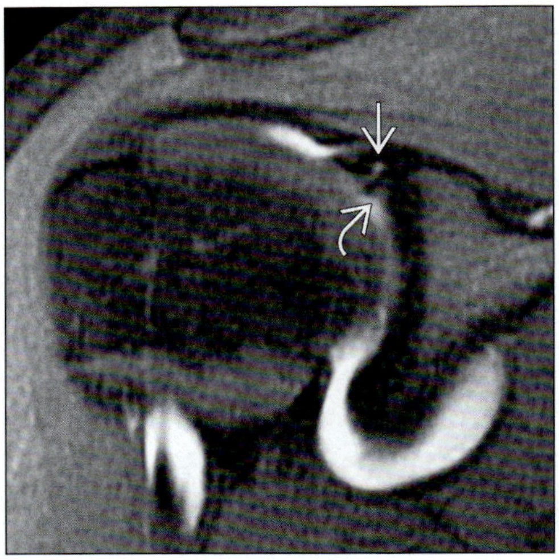

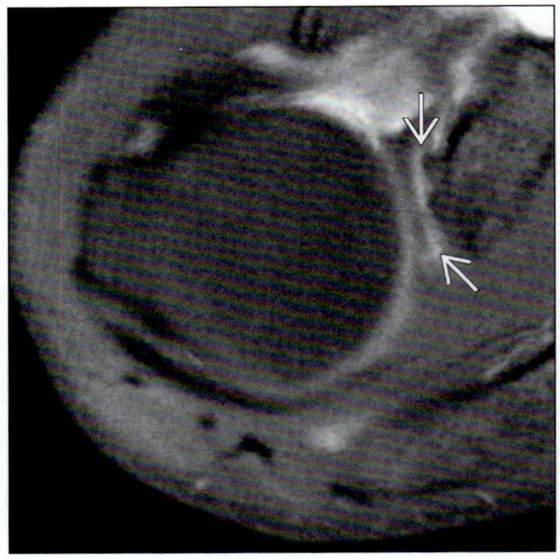

(Left) *Coronal MR arthrogram T1WI FS shows extension of a SLAP tear → into the anterosuperior labrum.* ***(Right)*** *Axial MR arthrogram T1WI FS, same patient, shows SLAP tear extending into the anterosuperior labrum →. This extension to the anterosuperior labrum has a 90% specificity, because only 10% of patients with a superior recess also have a sublabral foramen. 40% of SLAP tears show tear at this site, some of which are extended SLAP tears, such as a type V SLAP that extends down into a Bankart tear.*

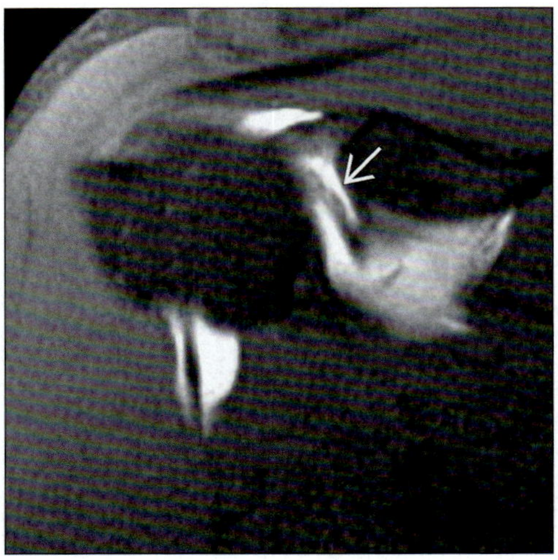

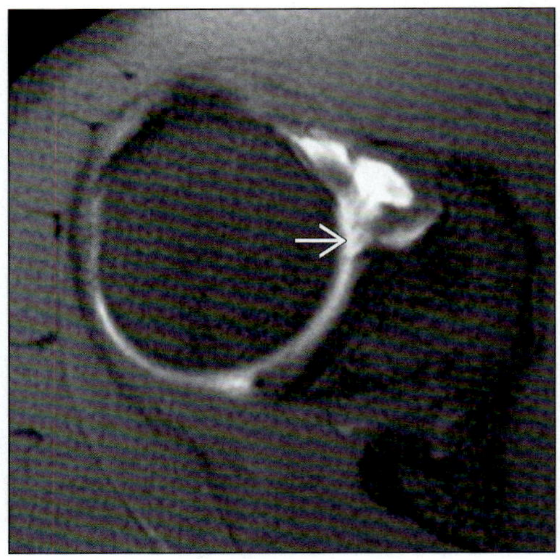

(Left) *Coronal oblique MR arthrogram T1WI FS shows extension of abnormal signal → posterior to the biceps. The tear extends across the width of the labrum, so this is a type III SLAP tear.* ***(Right)*** *Axial MR arthrogram T1WI FS from the same patient shows extension of the SLAP tear into the posterosuperior labrum →. Contrast within the labrum posterior to the biceps anchor only has a 50% specificity for a SLAP tear. In this case the signal is far posterior and > 3 mm in width, confirming a SLAP tear.*

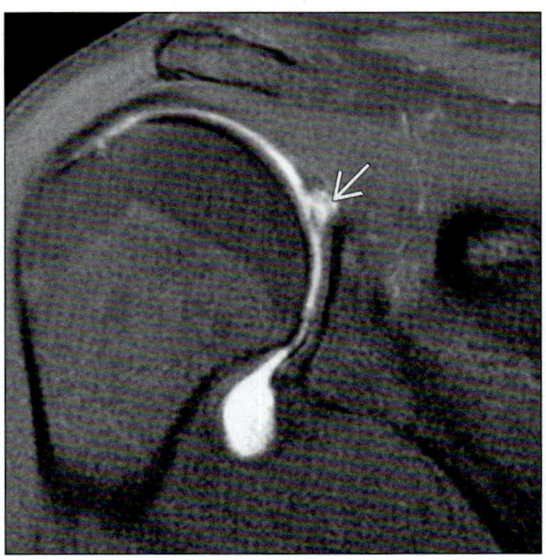

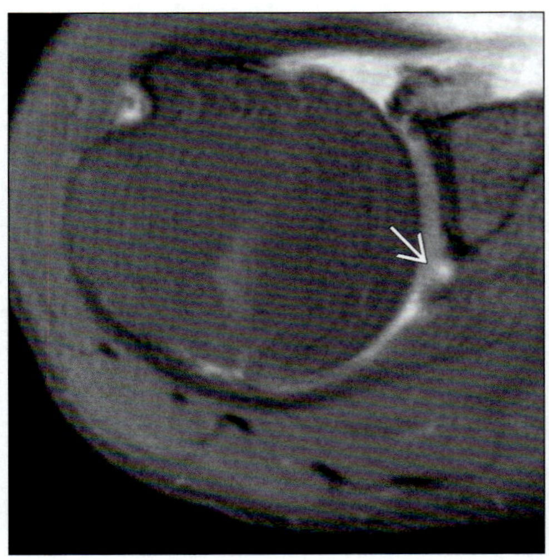

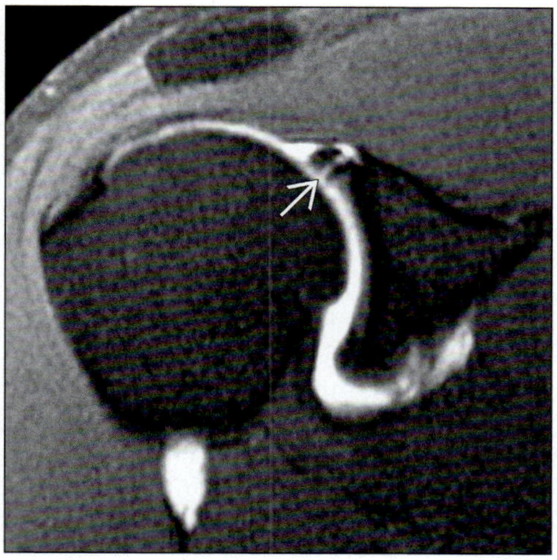

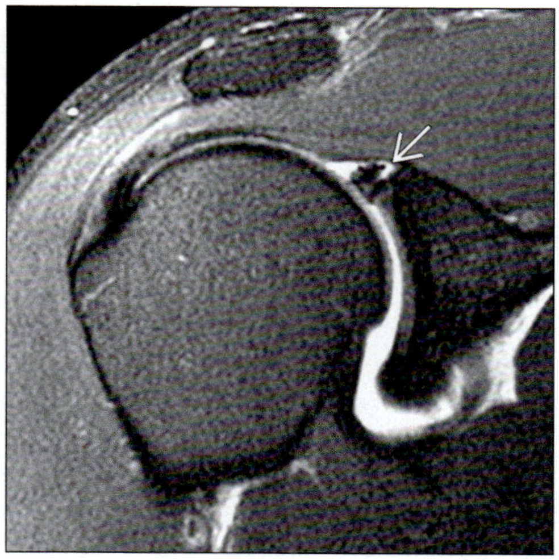

(Left) *Coronal oblique MR arthrogram T1WI FS shows a type III SLAP tear →. Contrast extends across the entire width of the superior labrum so that a labral segment is detached. Type III SLAP tears occur almost exclusively in patients 20-40 years old after trauma.* ***(Right)*** *Coronal oblique MR arthrogram T2WI FS in the same patient again shows the type III SLAP tear →. The hallmark of a type III SLAP tear at arthroscopy is a bucket-handle tear of the superior labrum.*

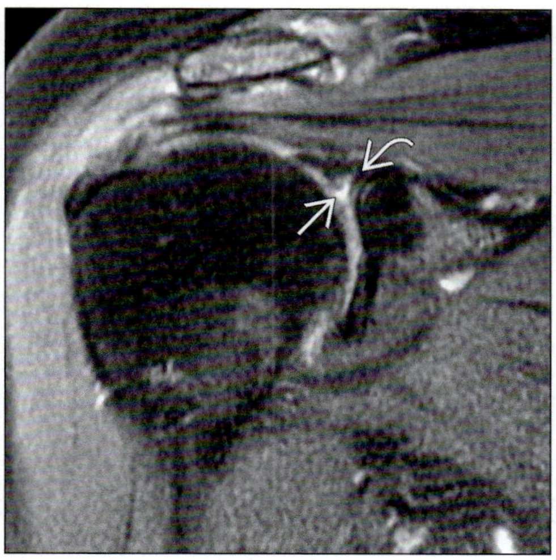

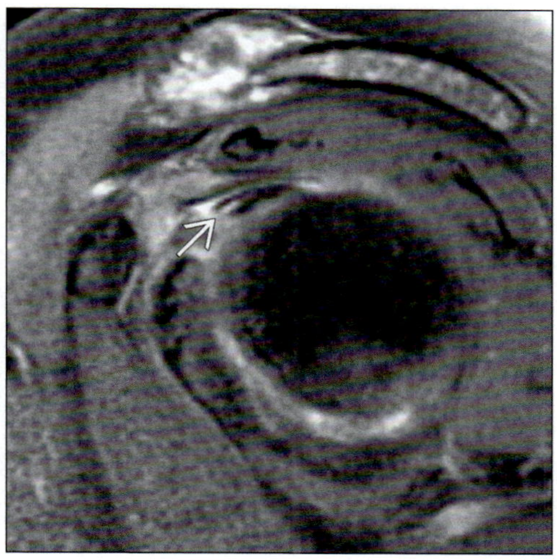

(Left) *Coronal oblique T2WI FS MR shows a type IV SLAP tear. There is high signal in the tear → with a thin strip of intact labrum ↷ adjacent to the glenoid cartilage. Although the signal in this patient curves medially, a superior recess occurs at the labral-chondral junction.* ***(Right)*** *Sagittal T2WI FS MR in the same patient shows a biceps tear → also. The combination of a longitudinal tear of the superior labrum with extension into the biceps anchor is a type IV SLAP tear.*

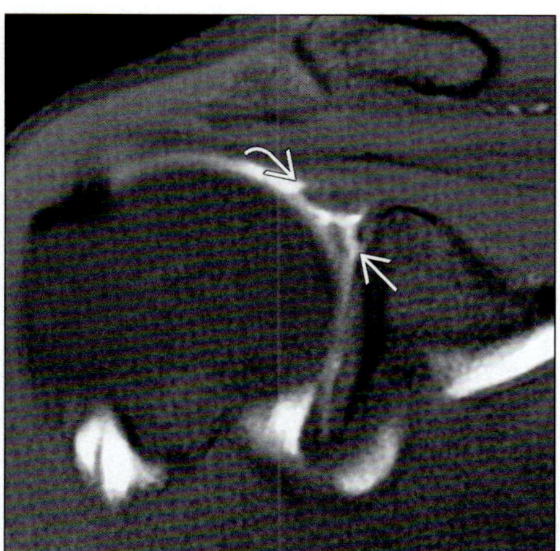

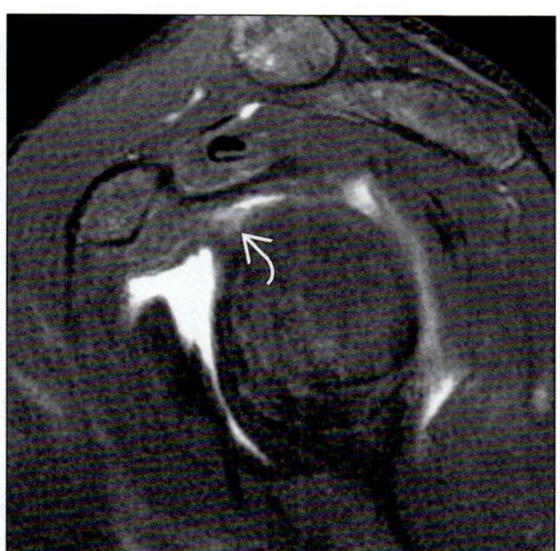

(Left) *Coronal oblique MR arthrogram T1WI FS shows a type IV SLAP tear. There is a tear of the superior labrum → and a stump where the biceps tendon is completely torn ↷.* ***(Right)*** *Oblique sagittal MR arthrogram T1WI FS in the same patient shows the complete tear of the biceps tendon ↷ near its origin from the superior labrum and supraglenoid tubercle. Type IV SLAP tears may have either a partial tear of the long head of the biceps tendon or a complete tear of the tendon root.*

EXTENDED SLAP TEAR

Key Facts

Terminology

- SLAP: Superior labrum anterior-posterior tear
- Synonym: Type V-X SLAP tear
- Tear of superior labrum anterior to posterior, which extends to involve other portions of labrum or ligaments

Imaging

- 2 components
  - SLAP tear: Laterally curved or abnormally wide signal, "double Oreo" sign
  - Extended tear: High signal extends into adjacent structure

Top Differential Diagnoses

- Normal labral variants
- Bankart lesion
- Posterior labral tear

Pathology

- Etiology
  - Type V and IX: Anterior dislocation
  - Type VI, VII, and X: Fall on outstretched arm
  - Type VIII: Posterior dislocation
- Types
  - Type V: SLAP tear extending into Bankart tear
  - Type VI: Flap tear where type III SLAP tear has 1 limb of the "bucket handle" detached
  - Type VII: SLAP tear that extends down into middle glenohumeral ligament
  - Type VIII: SLAP tear extends into reverse Bankart tear
  - Type IX: SLAP, Bankart, and reverse Bankart tears (360° or near 360° tear)
  - Type X: SLAP tear extends into rotator interval and superior glenohumeral ligament

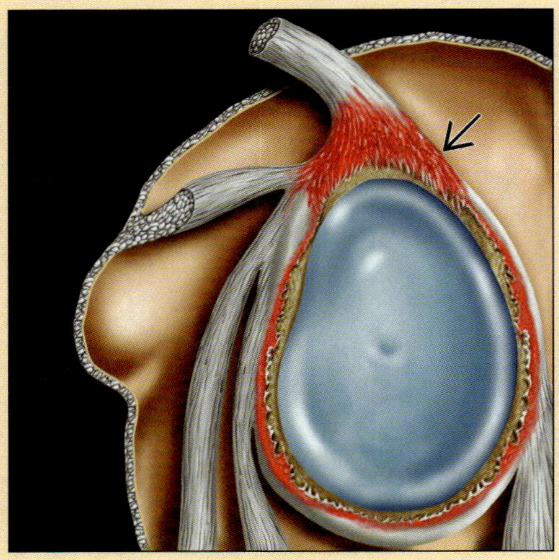

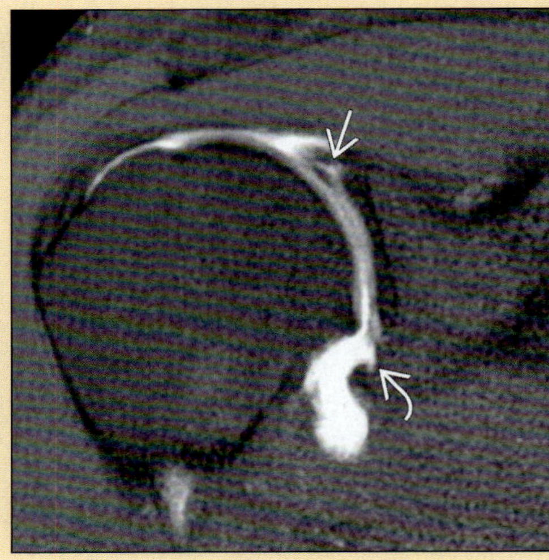

(Left) *Lateral graphic shows a type IX SLAP tear, also called a "360° labral tear" or "triple labral lesion." The tear not only involves the superior labrum → but extends around the entire labrum. These tears usually occur after an anterior dislocation and result from the "circle principle" where the entire labrum is partially avulsed due to traction by the capsule and ligaments.* ***(Right)*** *Coronal oblique MR arthrogram T1WI FS shows a SLAP tear → and a tear of the inferior labrum ↷.*

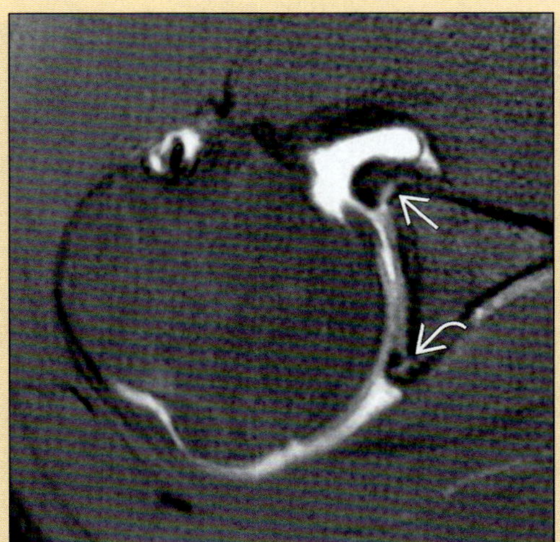

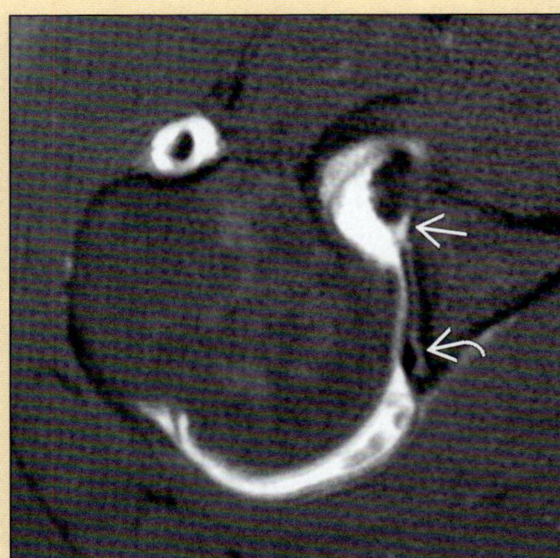

(Left) *Axial MR arthrogram T1WI FS in the same patient shows the tear involving the anterior → and posterior labrum ↷.* ***(Right)*** *Axial MR arthrogram T1WI FS near the inferior glenoid rim in the same patient shows the tear extending across the entire inferior labrum from anterior → to posterior ↷. At surgery the entire circumference of the labrum was torn. This patient had a prior anterior dislocation and complained of shoulder pain and anterior instability.*

EXTENDED SLAP TEAR

TERMINOLOGY

Abbreviations

- Superior labrum anterior-posterior (SLAP) tear

Synonyms

- Type V-X SLAP tear
  - Subtype: Triple labral lesion, 360° labral tear

Definitions

- Tear of superior labrum anterior to posterior, which involves other portions of labrum or ligaments

IMAGING

General Features

- Best diagnostic clue
  - High signal curving laterally in superior labrum, with tear extending into adjacent structure
- Location
  - Superior labrum + additional structure
- Size
  - Extensive

MR Findings

- T2WI FS
  - SLAP tear: Longitudinal linear increased signal that extends to surface of labrum
    - Lateral curving high signal
    - "Double Oreo" sign: 2 high signal lines
    - Abnormally wide high signal on coronal images
  - Extended tear: SLAP + other pathology
    - Type V: + abnormal signal in anterior labrum
    - Type VI: Flap tear with detached labrum torn free at 1 end
    - Type VII: + longitudinal high signal tear of middle glenohumeral ligament
    - Type VIII: + high signal posterior labral tear extending down to at least 9 o'clock (presuming 3 o'clock is universal anterior)
    - Type IX: + anterior and posterior labral tears
    - Type X: + high signal tear of rotator interval and superior glenohumeral ligament
- MR arthrography
  - SLAP tear
    - Laterally curved high signal contrast
    - Abnormally wide high signal contrast cleft
  - Extended tear
    - Contrast extends into tear of adjacent structure

Imaging Recommendations

- Best imaging tool
  - Direct MR arthrography

DIFFERENTIAL DIAGNOSIS

Normal Labral Variants

- Superior recess common in patients who also have non-SLAP labroligamentous tear

Bankart Lesion

- May have Bankart tear and superior recess

Posterior Labral Tear

- May have posterior labral tear and superior recess

PATHOLOGY

General Features

- Etiology
  - Type V & IX: Anterior dislocation
    - Superior labrum also tears from traction by capsule and ligaments
  - Type VI, VII, and X: Fall on outstretched arm
  - Type VIII: Posterior dislocation

Staging, Grading, & Classification

- Type V: SLAP tear extending into Bankart tear
- Type VI: Flap tear where type III SLAP tear has 1 limb of "bucket handle" detached
- Type VII: SLAP tear that extends down into middle glenohumeral ligament
- Type VIII: SLAP tear extends into reverse Bankart tear
- Type IX: SLAP, Bankart, and reverse Bankart tears
  - 360° or near 360° tear
- Type X: SLAP tear extends into tear of rotator interval and superior glenohumeral ligament

Gross Pathologic & Surgical Features

- Extensive labral tear with hemorrhage

CLINICAL ISSUES

Presentation

- Most common signs/symptoms
  - Shoulder pain
  - Popping, clicking
  - Anterior instability: Type V & IX SLAP tears

Demographics

- Age
  - 20-40 years old
- Gender
  - M > F

Natural History & Prognosis

- Continued pain

Treatment

- Extensive labroligamentous repair

DIAGNOSTIC CHECKLIST

Consider

- Extended SLAP tear if SLAP tear and injury to adjacent structure

SELECTED REFERENCES

1. Borrero CG et al: Magnetic resonance appearance of posterosuperior labral peel back during humeral abduction and external rotation. Skeletal Radiol. 39(1):19-26, 2010
2. Holzapfel K et al: Inter- and intraobserver variability of MR arthrography in the detection and classification of superior labral anterior posterior (SLAP) lesions: evaluation in 78 cases with arthroscopic correlation. Eur Radiol. 20(3):666-73, 2010
3. Lin E: Magnetic resonance arthrography of superior labrum anterior-posterior lesions: a practical approach to interpretation. Curr Probl Diagn Radiol. 38(2):91-7, 2009
4. Mohana-Borges AV et al: Superior labral anteroposterior tear: classification and diagnosis on MRI and MR arthrography. AJR Am J Roentgenol. 181(6):1449-62, 2003

EXTENDED SLAP TEAR

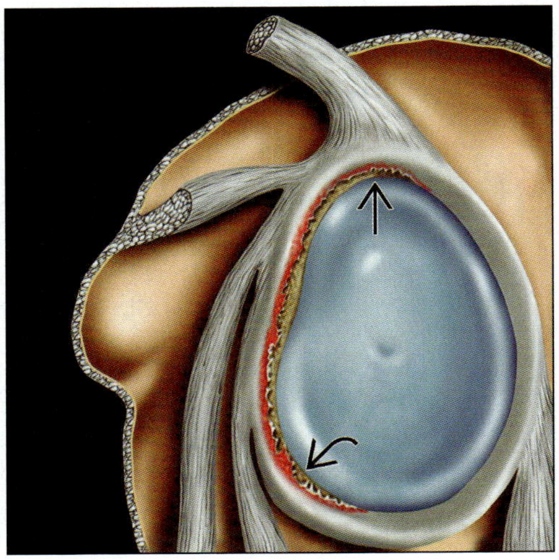

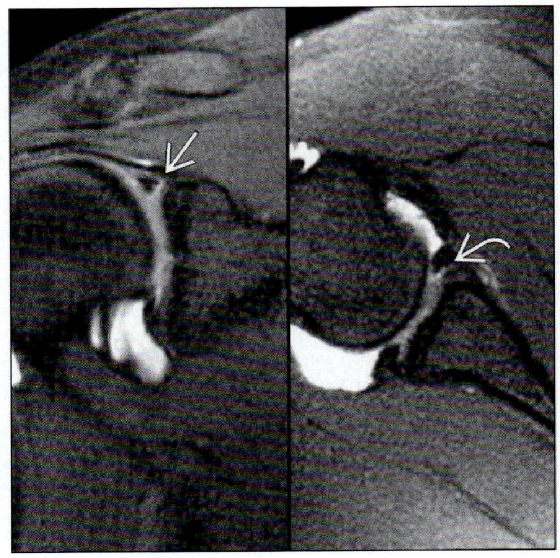

(Left) *Lateral graphic shows a type V SLAP tear. There is a tear of the superior labrum* → *, which extends into a Bankart tear* ↪ *. This is the most common type of extended SLAP tear, seen in about 20% of patients with a Bankart tear and 15% of patients with a type II-X SLAP tear.* ***(Right)*** *Coronal (left) and axial (right) MR arthrogram T1W FS images show a type V SLAP tear. The SLAP tear* → *extends down into a Bankart tear* ↪ *. This patient had anterior instability after an anterior dislocation.*

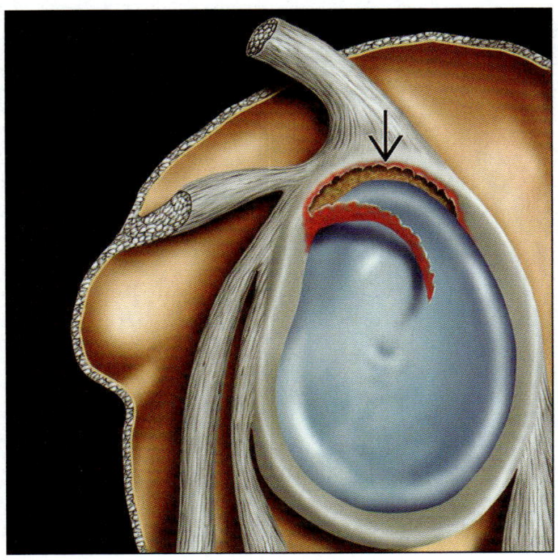

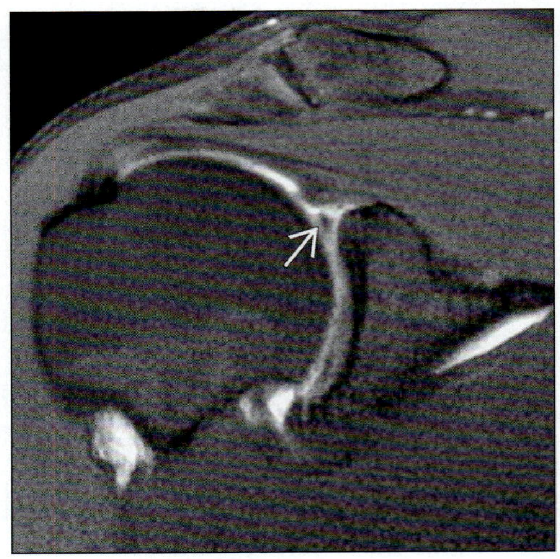

(Left) *Lateral graphic shows a type VI SLAP tear* → *. This is a flap tear of the superior labrum in which the full-width SLAP tear is detached at 1 end. A type VI SLAP tear is not truly an "extended SLAP" tear involving an adjacent structure, but it was described after the original type I-IV SLAP tear classification and is therefore included here.* ***(Right)*** *Coronal MR arthrogram T1WI FS shows a detached superior labrum* → *. This is a type VI SLAP tear since the labral segment is not attached at 1 end.*

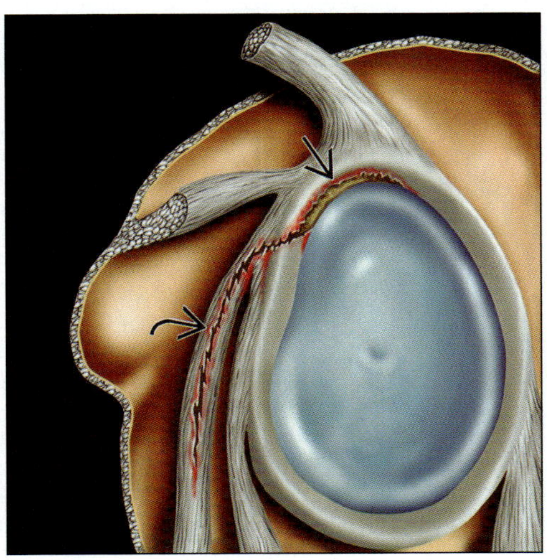

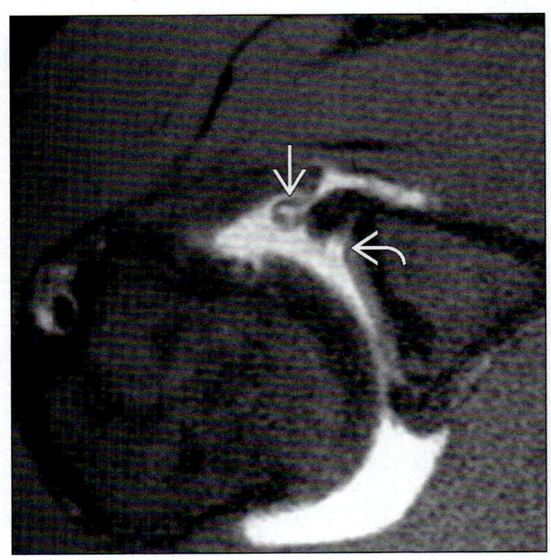

(Left) *Lateral graphic shows a type VII SLAP tear. This is a SLAP tear* → *that extends down into the middle glenohumeral ligament (MGL)* ↪ *. The MGL arises from the anterosuperior labrum inferior to the superior glenohumeral ligament and biceps and then inserts on the anterior capsule.* ***(Right)*** *Axial MR arthrogram T1WI FS shows a tear of the MGL* → *, as well as extension of the SLAP tear into the anterosuperior labrum* ↪ *. Type VII SLAP tears can result from a fall onto an outstretched hand.*

EXTENDED SLAP TEAR

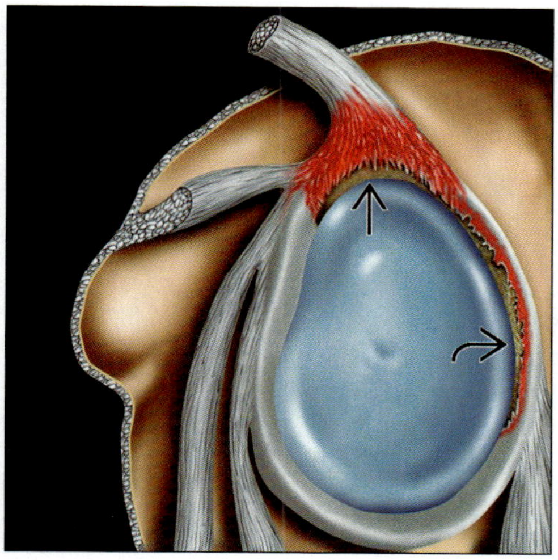

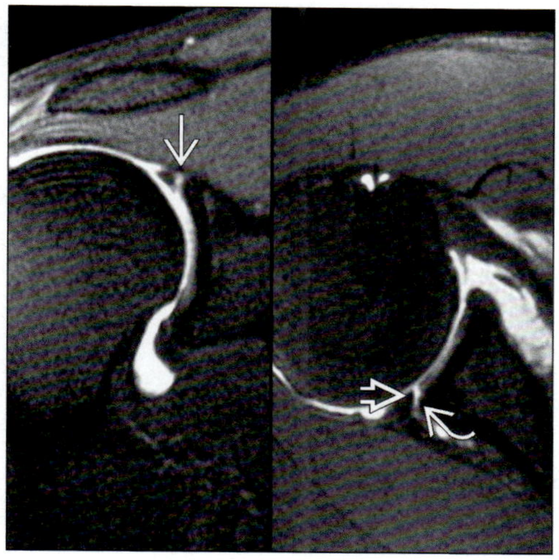

(Left) Lateral graphic shows a type VIII SLAP tear. The SLAP tear extends into a reverse Bankart tear. Type VIII SLAP tears are about 5% of all type II-X SLAP tears and usually result from a posterior dislocation. *(Right)* Coronal (left) and axial (right) MR arthrogram T1W FS images show a SLAP tear extending into a posterior labral tear. This patient presented with instability after a prior posterior dislocation. There is also a small chondral lesion adjacent to the reverse Bankart tear.

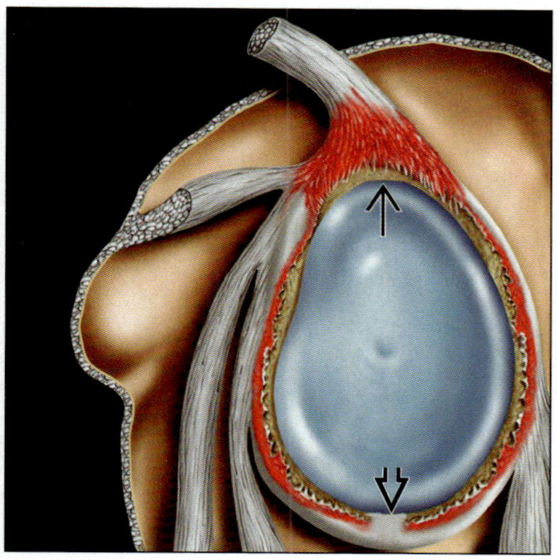

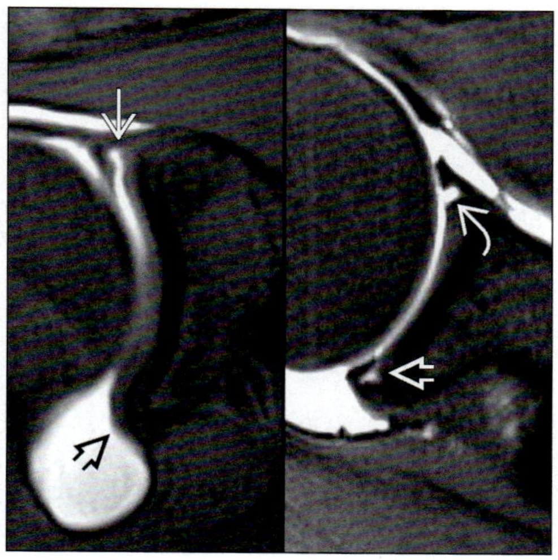

(Left) Lateral graphic shows a type IX SLAP tear. The SLAP tear extends into a Bankart and reverse Bankart tear. Type IX SLAP tears can either spare a short segment of the inferior labrum (350° tear), or involve the entire circumference. These tears are seen in about 5% of all type II-X SLAP tears. *(Right)* Coronal (left) and axial (right) MR arthrogram T1W FS images show a type IX SLAP tear extending into the anterior and posterior labrum but sparing the inferior labrum.

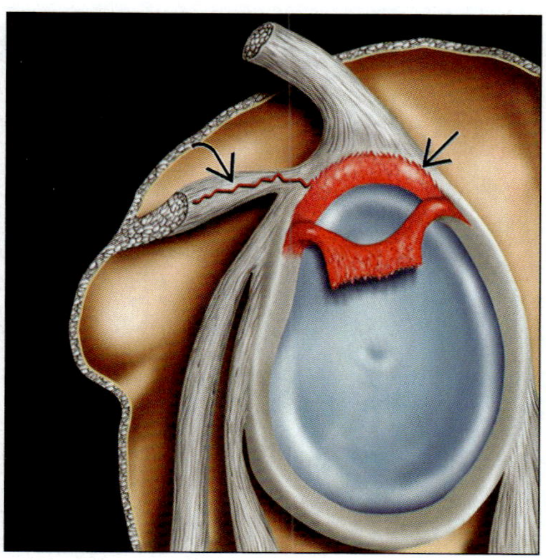

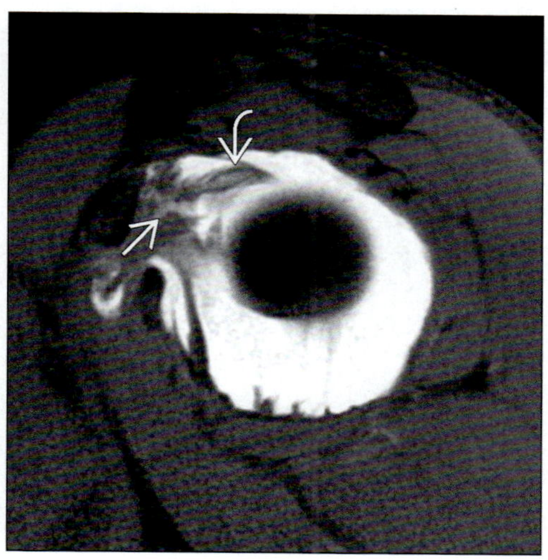

(Left) Lateral graphic shows a type X SLAP tear. The SLAP tear extends out into the superior glenohumeral ligament. These SLAP tears usually also have a tear of the rotator interval, which may be associated with a dislocated biceps tendon. *(Right)* Sagittal MR arthrogram T1WI FS shows a partial tear of the superior glenohumeral ligament and rotator interval. This is a type X SLAP tear if the superior labrum is also torn. This patient also has a partial tear of the biceps tendon.

INTERNAL IMPINGEMENT, SHOULDER

Key Facts

Terminology

- Pathologic internal impingement, posterosuperior glenoid impingement
- Pain with abduction and external rotation, decreased throwing velocity
  - Clinical diagnosis
  - Typically in overhead throwing athletes (e.g., baseball pitchers)

Imaging

- Rotator cuff: Posterior supraspinatus &/or cranial fibers of infraspinatus
  - Articular surface fraying or partial thickness tear
- Labrum: Posterosuperior labrum
  - Posterosuperior labral tear ± posterior SLAP tear
- Humeral head
  - Greater tuberosity cysts or notch
- Most sensitive imaging test: Direct MR arthrography
  - Include ABER images to better profile posterosuperior rotator cuff and labrum

Pathology

- Repetitive contact between greater tuberosity and posterosuperior glenoid rim
- Excessive external rotation from stretched lax anterior capsule
- Glenohumeral internal rotation deficit (GIRD)
- Scapular dyskinesis and fatigue of dynamic shoulder stabilizers

Clinical Issues

- Posterior shoulder pain
  - Worse with abduction and external rotation
- Decreased throwing velocity ("dead arm")

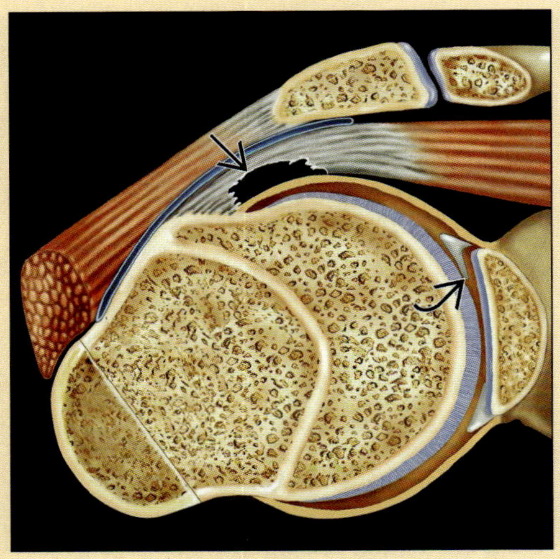

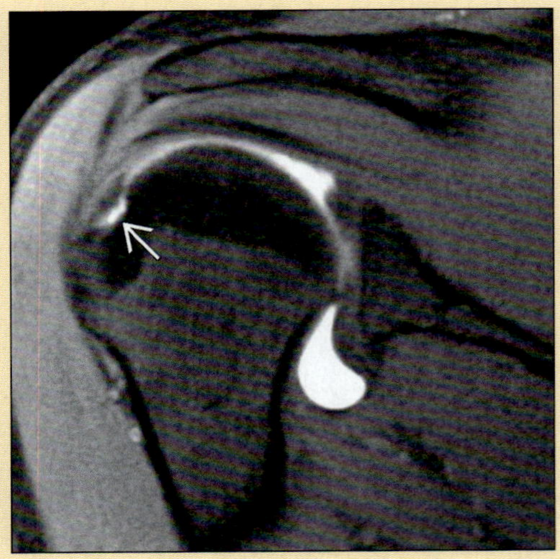

***(Left)** Oblique coronal graphic of the posterior shoulder shows an articular surface, partial thickness rotator cuff tear → and labral tear ↗. Tears in the posterosuperior cuff and labrum are common in overhead throwing athletes with severe internal impingement. On imaging, patients with clinical internal impingement may have either fraying or tears of the posterosuperior cuff and labrum. **(Right)** Coronal oblique MR arthrogram T1WI FS shows a partial tear of the posterior rotator cuff ➔.*

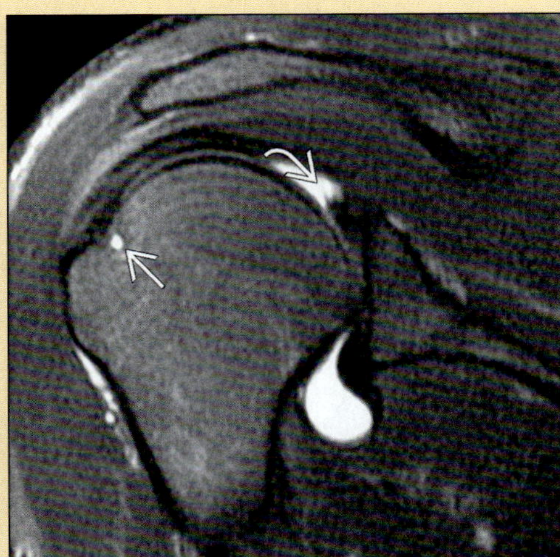

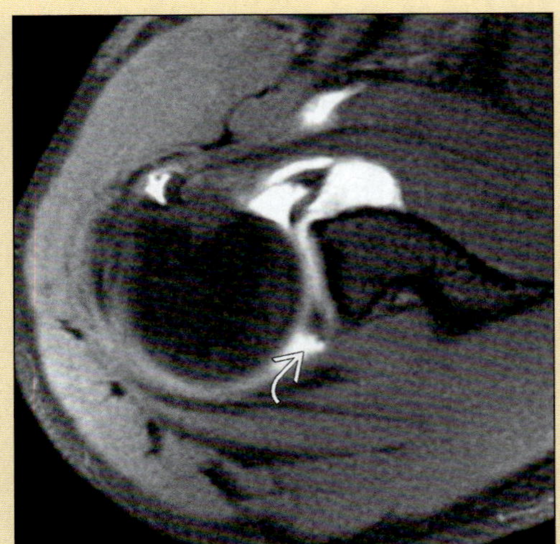

***(Left)** Coronal oblique MR arthrogram T2WI FS from the same patient shows a small bone cyst ➔ just medial to the supraspinatus tendon, as well as fraying of the labrum ↗. **(Right)** Axial MR arthrogram T1WI FS from the same patient shows fraying of the labrum ↗. A tear of the posterior supraspinatus or anterior infraspinatus tendon and fraying of the posterosuperior labrum is likely due to repetitive impaction of the posterior greater tuberosity on the posterosuperior glenoid rim.*

INTERNAL IMPINGEMENT, SHOULDER

TERMINOLOGY

Synonyms

- Pathologic internal impingement, posterosuperior glenoid impingement, posterior glenoid impingement

Definitions

- Pain with abduction and external rotation, decreased throwing velocity
  - Clinical diagnosis
  - Typically in overhead throwing athletes, e.g., professional baseball pitchers
  - Common features
    - Posterosuperior labral tear ± posterior superior labrum anterior-posterior (SLAP) tear
    - Articular surface partial thickness tear of posterior rotator cuff
    - Thick posterior capsule and posterior band of inferior glenohumeral ligament (PBIGL)
    - Lax, stretched anterior capsule

IMAGING

General Features

- Best diagnostic clue
  - Posterosuperior labral and rotator cuff tear in overhead throwing athlete
- Location
  - Posterosuperior labrum
  - Posterior supraspinatus &/or anterior aspect of infraspinatus tendon
- Size
  - Small (< 1 cm)

Radiographic Findings

- Young patient with posterior greater tuberosity cysts &/or sclerosis
- Greater tuberosity notch: "Pseudo Hill-Sachs" lesion
  - Posterior superior greater tuberosity impaction lesion
- ± Bennett lesion (ossification of posterior shoulder capsule)

MR Findings

- T2WI FS
  - Rotator cuff: Posterior supraspinatus &/or superior infraspinatus tendon
    - Articular surface fraying
    - High signal involving articular surface: Partial thickness tear
      - Tears are small and may be difficult to see with conventional MR imaging
  - Posterosuperior labrum
    - Fraying of free edge
    - High signal extending to surface of labrum: Posterosuperior labral tear ± SLAP lesion
      - Despite location, rarely have a paralabral cyst
  - Humeral head
    - Posterior greater tuberosity high signal cysts
      - Often reversible after treatment, may represent dilated vascular channels
    - Greater tuberosity notch
      - "Pseudo Hill-Sachs" lesion: More superior (near insertion of supraspinatus tendon) and < 1 cm, which is smaller than true Hill-Sachs
      - Osteochondral impaction lesion of posterior humeral head
      - Seen in 50% of baseball pitchers with internal impingement
  - Glenoid
    - Rounding of posterosuperior glenoid rim from mechanical attrition
    - Posterior capsule and PBIGL
      - Thickened near glenoid insertion
      - Bennett lesion
  - T2WI sensitivity in patients with internal impingement
    - Rotator cuff lesions = 27%
    - Labral lesions = 50%
- PD/intermediate FS
  - Axial images: Lax anterior capsule
- MR arthrography
  - High signal contrast extending into rotator cuff or labral tear
  - Abduction external rotation (ABER) images
    - Relaxes posterior rotator cuff, so tears more visible
    - ABER images are orthogonal to posterosuperior labrum
      - Less partial averaging
      - Labral tears more visible
  - MR arthrography sensitivity in patients with internal impingement
    - Rotator cuff lesions = 54%
    - Labral lesions = 50%

Imaging Recommendations

- Best imaging tool
  - Direct MR arthrography
- Protocol advice
  - Include ABER images to better profile posterosuperior rotator cuff and labrum

DIFFERENTIAL DIAGNOSIS

SLAP Tears

- Pain with anterosuperior, or directly superior, labrum provocative tests

Rotator Cuff Impingement

- Pain is usually anterior, occurs with forward flexion and in older patients

Rotator Cuff Partial Thickness Tear

- Impingement-related cuff tears are usually in anterior 1/2 of supraspinatus tendon

Microinstability, Shoulder

- Anterior rotator cuff tear ± SLAP tear

PATHOLOGY

General Features

- Etiology
  - Repetitive contact between greater tuberosity and posterosuperior glenoid rim
    - Some contact occurs normally, but in athletes it is more forceful and repetitive
  - Excessive external rotation from stretched lax anterior capsule

- Longitudinal twisting of long head biceps tendon at labral insertion; may cause "peel-back" SLAP tear
- Longitudinal twisting of rotator cuff; may cause shearing cuff tear
  - Glenohumeral internal rotation deficit (GIRD)
    - Deceleration phase of throwing causes repetitive injury and fibrosis of PBIGL at glenoid
    - Stiff and contracted PBIGL causes abnormal posterosuperior shift of humeral head in late cocking phase of throwing
      - Increased impaction force between greater tuberosity and glenoid rim
      - Worsens longitudinal twisting of long head biceps tendon labral attachment and of rotator cuff
    - Posterosuperior shift of center of rotation of humeral head reduces cam effect on anterior capsule
      - Allows increased external rotation
  - Scapular dyskinesis and fatigue of dynamic shoulder stabilizers
    - Can increase impaction forces in abduction and external rotation
- Associated abnormalities
  - Glenohumeral internal rotation deficit (GIRD)
  - Microinstability
  - Bennett lesion

Staging, Grading, & Classification

- Progression of pathology
  - Lax anterior capsule and GIRD
  - Posterior rotator cuff fraying &/or posterosuperior labral fraying
  - Posterior rotator cuff tear and SLAP tear
- Jobe clinical stages
  - Stage I: Stiffness, pain in late cocking phase of throwing
  - Stage II: Constant posterior shoulder pain
  - Stage III: No improvement with rest and conservative treatment

Gross Pathologic & Surgical Features

- Rotator cuff and labrum: Indurated, inflamed articular surface with fraying/tear

CLINICAL ISSUES

Presentation

- Most common signs/symptoms
  - Posterior shoulder pain
    - Worse with abduction and external rotation
  - Decreased throwing velocity
- Other signs/symptoms
  - Posterosuperior joint line tenderness
  - Increased external rotation
  - Decreased internal rotation

Demographics

- Age
  - 18-35 years old
- Gender
  - M > F
- Overhead throwers, occupations with repetitive abduction and external rotation

Natural History & Prognosis

- Increasing pain and decreasing throwing velocity: "Dead arm"

Treatment

- GIRD: Posterior capsule stretching exercises
  - If persists, arthroscopic selective posteroinferior capsulotomy
- Rotator cuff and posterior SLAP tears: Surgical debridement or repair

DIAGNOSTIC CHECKLIST

Consider

- Internal impingement in young overhead throwing athletes with cuff and labral tears

Image Interpretation Pearls

- Inspect posterosuperior rotator cuff and labrum closely

SELECTED REFERENCES

1. Muraki T et al: Effect of posteroinferior capsule tightness on contact pressure and area beneath the coracoacromial arch during pitching motion. Am J Sports Med. 38(3):600-7, 2010
2. Tyler TF et al: Correction of posterior shoulder tightness is associated with symptom resolution in patients with internal impingement. Am J Sports Med. 38(1):114-9, 2010
3. Drakos MC et al: Internal impingement of the shoulder in the overhead athlete. J Bone Joint Surg Am. 91(11):2719-28, 2009
4. Heyworth BE et al: Internal impingement of the shoulder. Am J Sports Med. 37(5):1024-37, 2009
5. Mihata T et al: Excessive glenohumeral horizontal abduction as occurs during the late cocking phase of the throwing motion can be critical for internal impingement. Am J Sports Med. Epub ahead of print, 2009
6. Grainger AJ: Internal impingement syndromes of the shoulder. Semin Musculoskelet Radiol. 12(2):127-35, 2008
7. Ouellette H et al: Spectrum of shoulder injuries in the baseball pitcher. Skeletal Radiol. 37(6):491-8, 2008
8. Ouellette H et al: Imaging of the overhead throwing athlete. Semin Musculoskelet Radiol. 2005 Dec;9(4):316-33. Review. Erratum in: Semin Musculoskelet Radiol. 11(1):91, 2007
9. Tuite MJ et al: Shoulder MR arthrography of the posterior labrocapsular complex in overhead throwers with pathologic internal impingement and internal rotation deficit. Skeletal Radiol. 36(6):495-502, 2007
10. Burkhart SS: Internal impingement of the shoulder. Instr Course Lect. 55:29-34, 2006
11. Laudner KG et al: Scapular dysfunction in throwers with pathologic internal impingement. J Orthop Sports Phys Ther. 36(7):485-94, 2006
12. Giaroli EL et al: MRI of internal impingement of the shoulder. AJR Am J Roentgenol. 185(4):925-9, 2005
13. Kaplan LD et al: Internal impingement: findings on magnetic resonance imaging and arthroscopic evaluation. Arthroscopy. 20(7):701-4, 2004
14. Tirman PF et al: Shoulder imaging in athletes. Semin Musculoskelet Radiol. 8(1):29-40, 2004
15. Tirman PF et al: Posterosuperior glenoid impingement of the shoulder: findings at MR imaging and MR arthrography with arthroscopic correlation. Radiology. 193(2):431-6, 1994

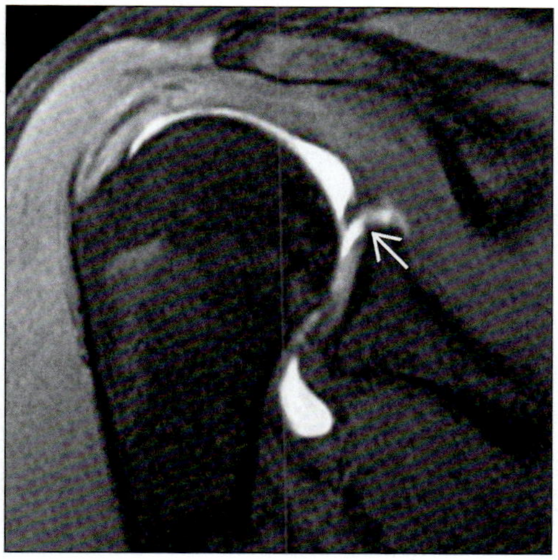

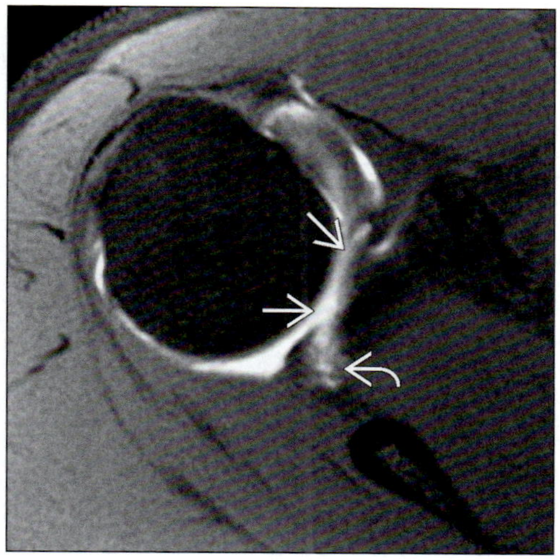

(Left) *Coronal oblique MR arthrogram T1WI FS from a 34-year-old baseball pitcher shows a posterior SLAP tear* ➡. ***(Right)*** *Axial MR arthrogram T1WI FS from the same patient shows the SLAP tear* ➡ *extending posteriorly* ↪. *Such SLAP tears are likely due to "peel-back" mechanism from excessive twisting of the long head biceps tendon insertion on the labrum during hyperexternal rotation in the late cocking phase of throwing, causing the labrum to be "peeled" off the superior glenoid rim.*

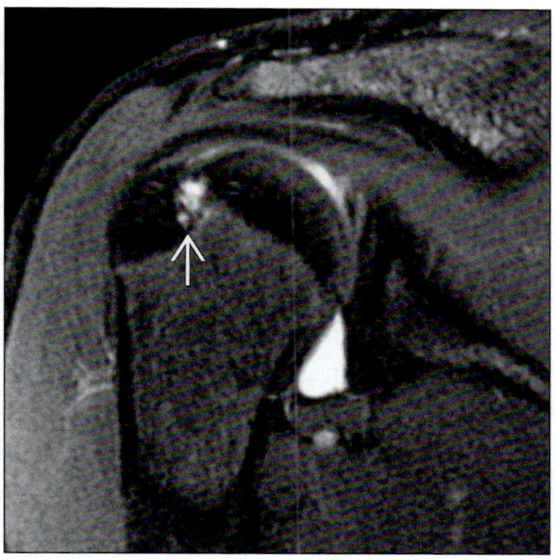

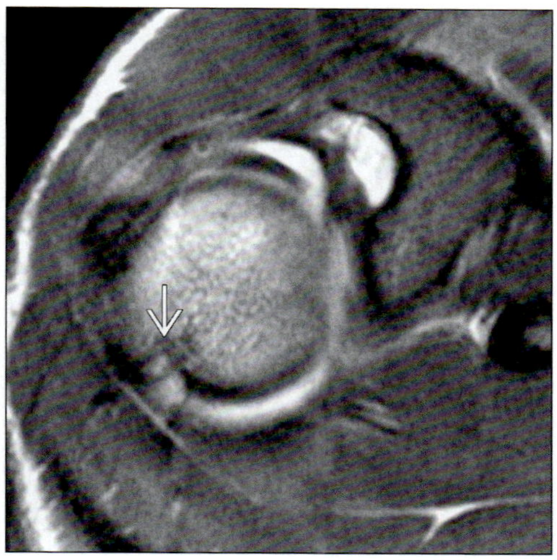

(Left) *Coronal oblique MR arthrogram T2WI FS shows cysts* ➡ *just medial to the insertion of the posterior supraspinatus tendon. Cysts in this region in a young throwing athlete are a common finding in internal impingement, and one should look for posterosuperior cuff and labral tears.* ***(Right)*** *Axial MR arthrogram T1WI shows a greater tuberosity notch* ➡ *or "pseudo Hill-Sachs" lesion. These are due to repetitive impaction of the humeral head on the posterosuperior glenoid rim during throwing.*

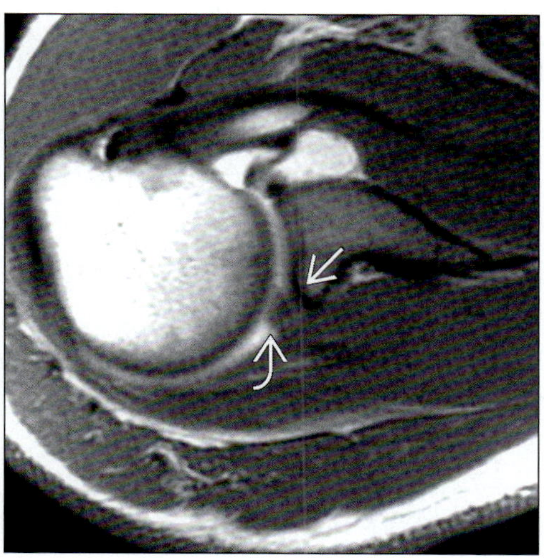

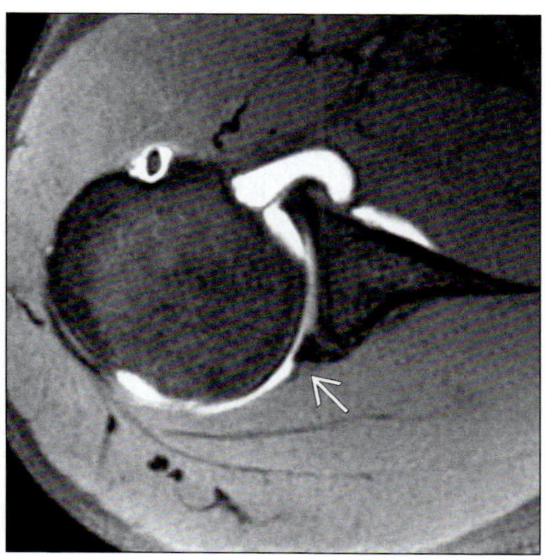

(Left) *Axial MR arthrogram T1WI shows flattening of the posterosuperior glenoid rim* ➡ *due to repetitive impaction by the greater tuberosity during throwing. There is also labral fraying* ↪. ***(Right)*** *Axial MR arthrogram T1WI FS from a patient with internal impingement and GIRD shows thickening of the posterior capsule and PBIGL near its glenoid insertion* ➡ *with loss of the normal recess. These findings are more common in throwers with GIRD and may be a precursor to a Bennett lesion.*

INTERNAL IMPINGEMENT, SHOULDER

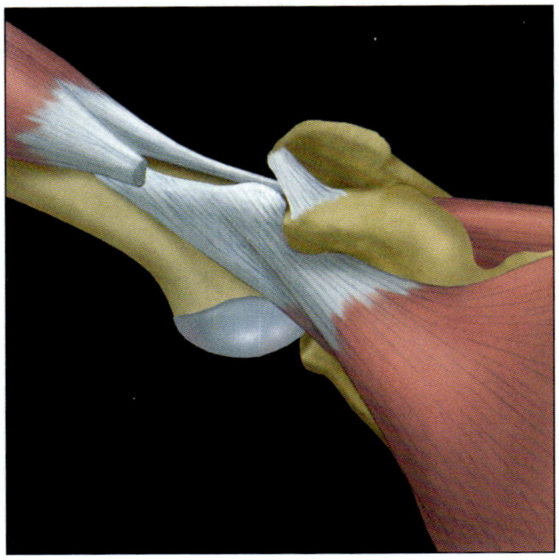

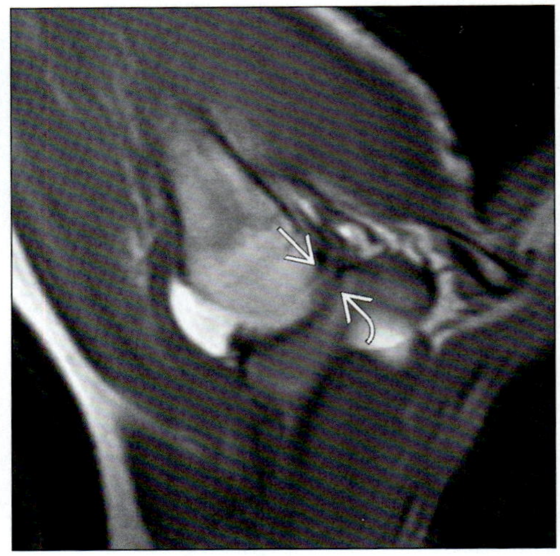

(Left) *AP graphic shows the arm position for abduction external rotation (ABER) images.* ***(Right)*** *Coronal T1WI MR localizer shows the arm in the ABER position. There is contact between the greater tuberosity and the posterosuperior glenoid rim. This contact occurs normally and only leads to internal impingement when there is repetitive forceful contact. ABER images are helpful not because they reproduce the symptomatic position, but because they show the cuff & labral tears well.*

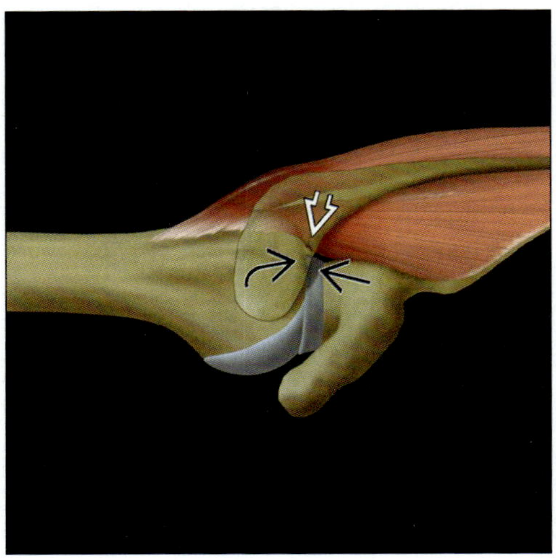

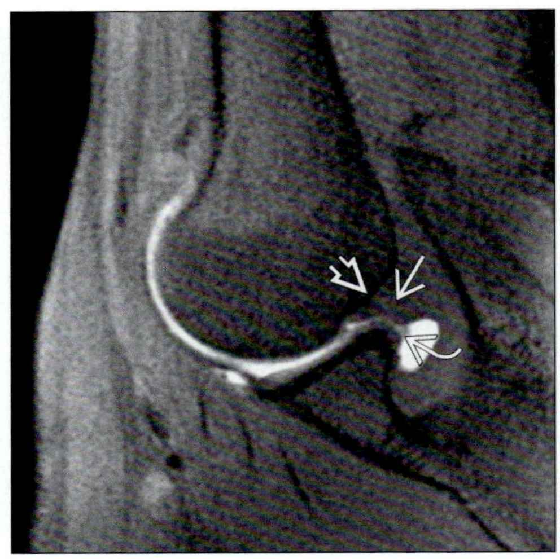

(Left) *Superior view of the ABER position shows contact between the greater tuberosity & posterosuperior labrum with labral fraying and a small cuff tear.* ***(Right)*** *ABER MR arthrogram T1WI FS in a patient with internal impingement shows the posterior supraspinatus tendon & posterosuperior labrum pinched between the greater tuberosity & glenoid rim. ABER images are orthogonal to the posterosuperior glenoid rim & thus better demonstrate cuff & labral lesions in this region.*

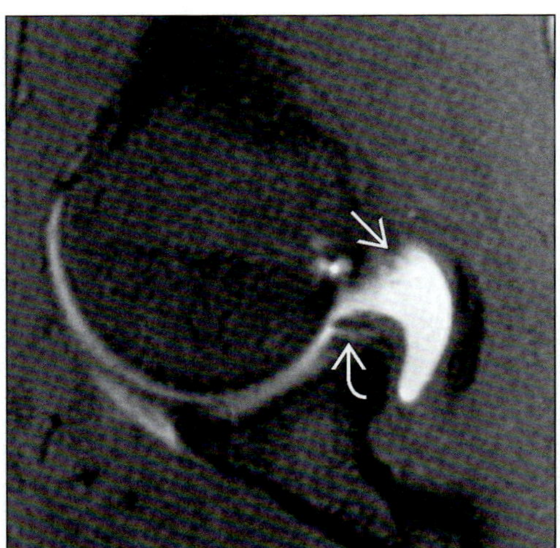

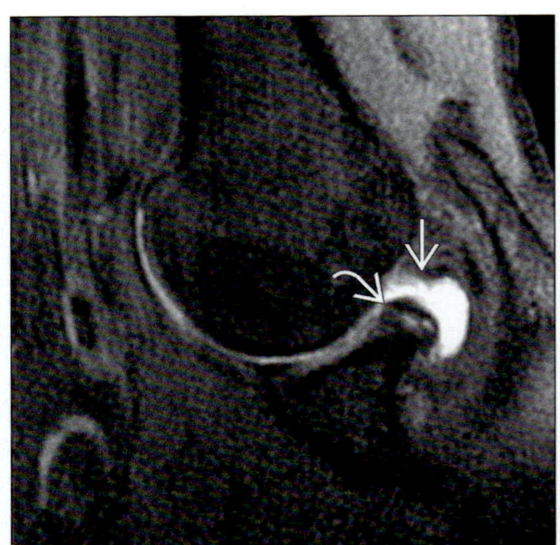

(Left) *ABER MR arthrogram T1WI FS from an 18-year-old baseball pitcher with internal impingement pain and decreased throwing velocity shows fraying of the anterior infraspinatus tendon and a tear of the posterosuperior labrum.* ***(Right)*** *ABER MR arthrogram T1WI FS from a 24-year-old volleyball player with internal impingement symptoms shows a shallow articular surface partial thickness tear of the posterior supraspinatus tendon and a posterior SLAP tear.*

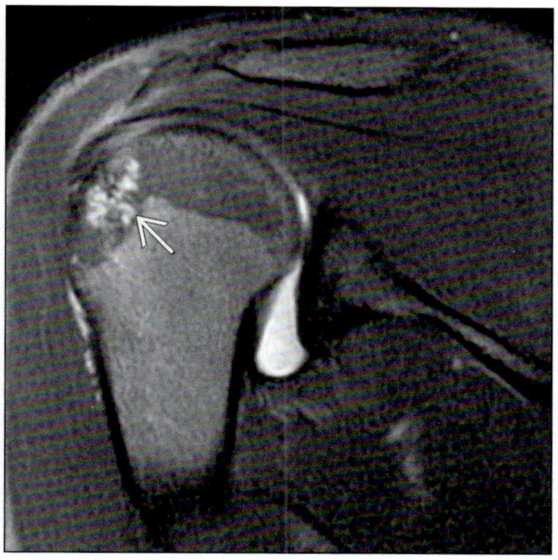

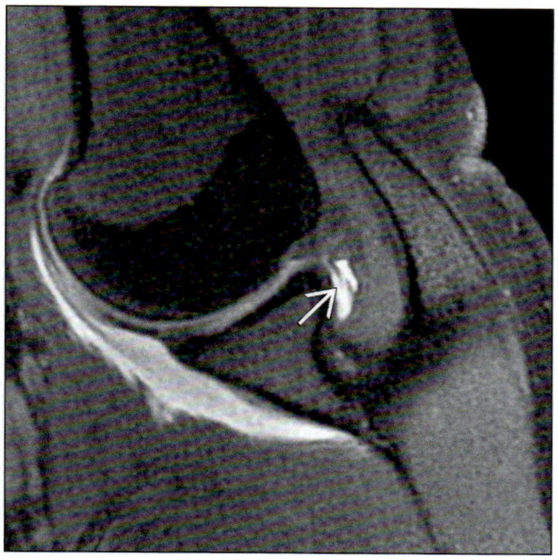

(Left) *Coronal oblique MR arthrogram T2WI FS shows abundant posterior greater tuberosity cysts → in a 21-year-old college volleyball player with internal impingement symptoms. No definite posterior cuff tear was seen on the adducted images.* ***(Right)*** *ABER MR arthrogram T1WI FS from the same patient shows a small flap tear of the posterosuperior rotator cuff →. ABER MR arthrogram images may show posterosuperior rotator cuff and labral tears not seen on standard adduction images.*

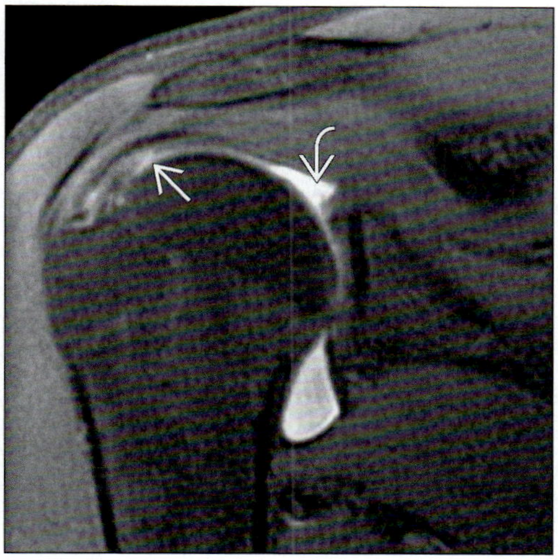

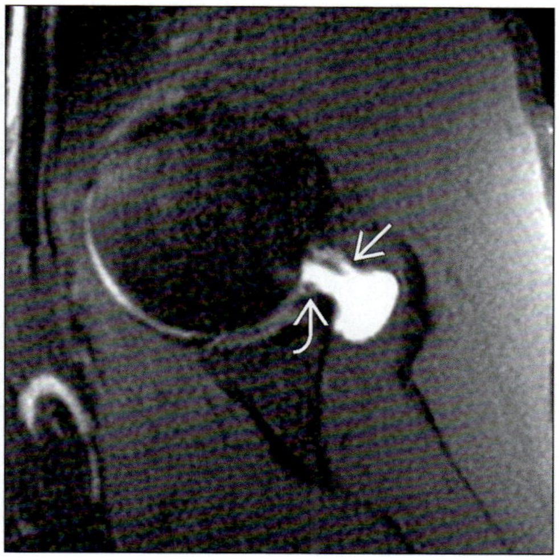

(Left) *Coronal oblique MR arthrogram T1WI FS in a 29 year old who experienced work-related shoulder pain when his arm was overhead shows a possible articular surface partial thickness tear of the posterosuperior rotator cuff →. There is fraying of the posterosuperior labrum ↗.* ***(Right)*** *ABER MR arthrogram T1WI FS from the same patient better shows the articular surface partial thickness tear of the posterosuperior cuff →. The fraying of the posterosuperior labrum is also well seen ↗.*

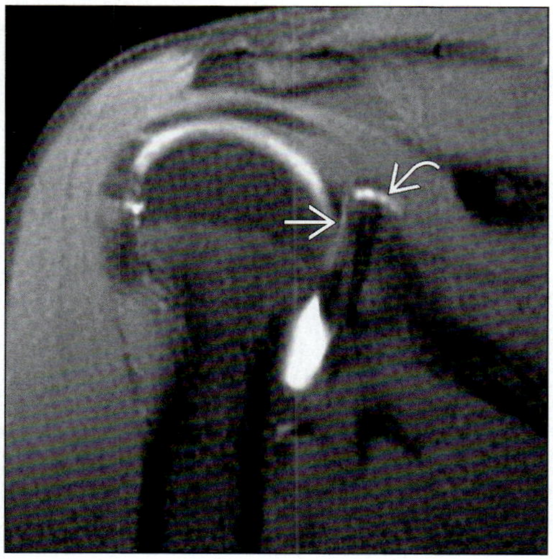

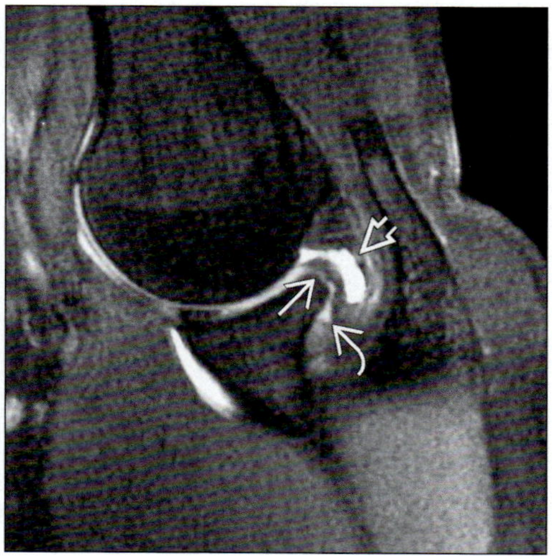

(Left) *Coronal oblique MR arthrogram T1WI FS in a 24-year-old college baseball pitcher with clinical internal impingement shows a posterior SLAP tear → and a paralabral cyst ↗.* ***(Right)*** *ABER MR arthrogram T1WI FS from the same patient shows the posterosuperior labral tear → with a small paralabral cyst ↗. There is fraying of the articular surface of the rotator cuff →. Fraying of the posterior supraspinatus or anterior infraspinatus tendon is common in internal impingement.*

MICROINSTABILITY, SHOULDER

Key Facts

Terminology
- Sometimes used interchangeably: Superior labrum/ anterior cuff (SLAC) lesion, superior glenohumeral ligament (SGL)/middle glenohumeral ligament (MGL) tear
- Clinical diagnosis based on shoulder pain and sensation of slippage
  - No true instability on examination
- Term covers some of the same injuries as seen with rotator interval tears & biceps lesions

Imaging
- Anterosuperior labral tear
- Tear of MGL or SGL

Top Differential Diagnoses
- Anterosuperior normal labral variants
- SLAP lesions
- Rotator interval tears

Pathology
- Acute trauma
- Repetitive stress, such as overhead throwing or swimming

Clinical Issues
- Shoulder pain that mimics rotator cuff tendinopathy
- Subjective sensation of slippage
- Shoulder muscles easily fatigue
- No true instability on examination

Diagnostic Checklist
- Consider anterosuperior normal labral variant (common)

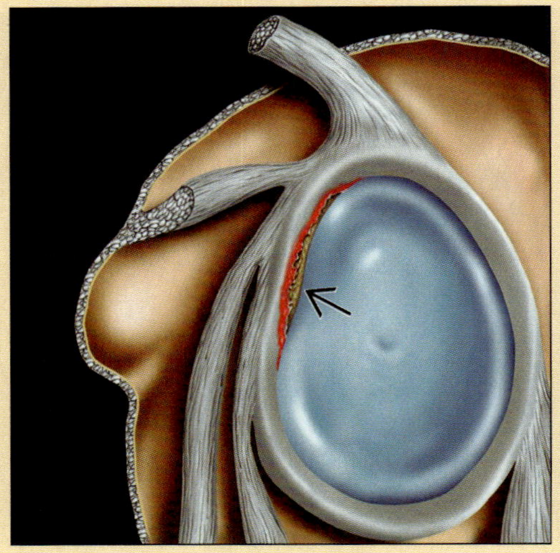

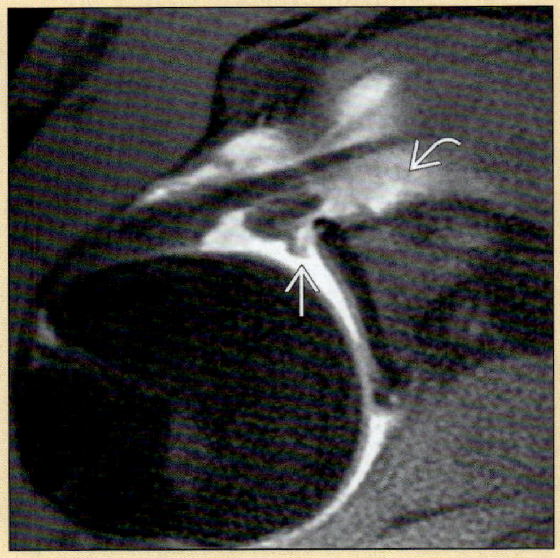

(Left) *Lateral graphic shows a tear of the anterosuperior labrum ➔ with irregular margins and hemorrhage. These tears often present with microinstability and need to be distinguished on MR from the more common anterosuperior normal labral variants.* ***(Right)*** *Axial MR arthrogram T1WI FS in a 21 year old with microinstability after acute trauma shows a tear of the anterosuperior labrum ➔, as well as injury to the capsule with contrast extravasation into the pericapsular soft tissues ➔.*

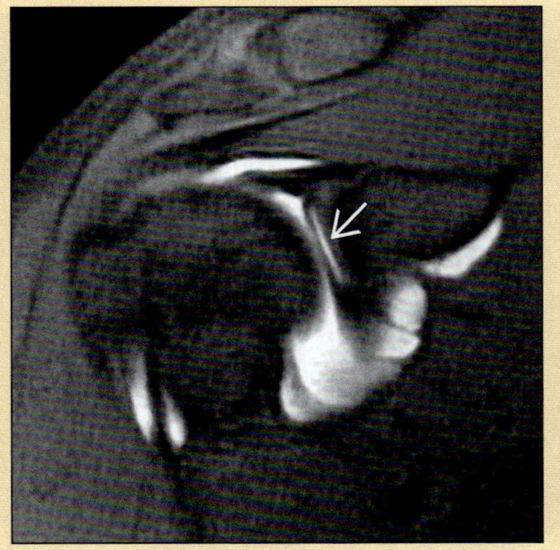

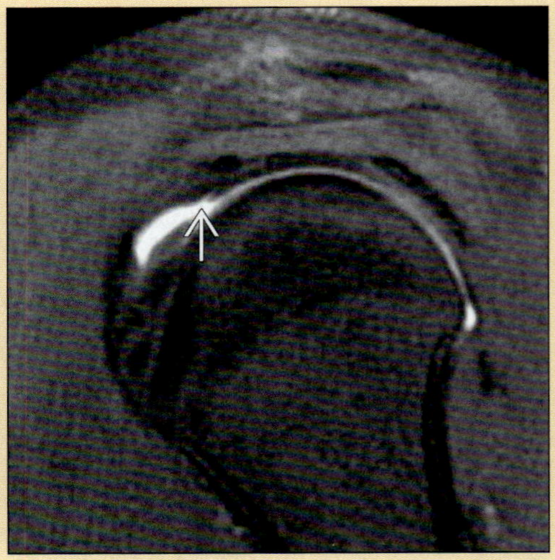

(Left) *Coronal oblique MR arthrogram T1WI FS from a 17-year-old baseball pitcher with microinstability symptoms shows undercutting of the labrum indicating either a sublabral foramen or an anterosuperior labral tear ➔.* ***(Right)*** *Sagittal MR arthrogram T1WI FS from the same patient shows a small articular surface partial thickness tear of the leading edge of the supraspinatus tendon ➔. At surgery the labrum was also torn; this combination constitutes a superior labrum, anterior cuff (SLAC) lesion.*

MICROINSTABILITY, SHOULDER

TERMINOLOGY

Abbreviations

- AIOS
  - Acquired, often in overhead athletes
  - Instability (micro) on examination
  - Overstress etiology
  - Surgery is treatment of choice

Synonyms

- Microtraumatic instability
- Sometimes used interchangeably: Superior labrum/ anterior cuff (SLAC) lesion, superior glenohumeral ligament (SGL)/middle glenohumeral ligament (MGL) tear

Definitions

- Clinical diagnosis based on shoulder pain and sensation of slippage
  - No true instability on examination
- Original definition: Anterosuperior labral, SGL, or MGL tear
  - Confusing term: Some authors include any labral tear in upper half of glenoid, including superior labrum anterior-posterior (SLAP) tears; others include internal impingement pseudolaxity from stretched anterior capsule ± posterior SLAP tear
- Term covers some of the same injuries as seen with rotator interval tears & biceps lesions

IMAGING

General Features

- Best diagnostic clue
  - Tear of anterosuperior labrum, MGL, or SGL
- Location
  - Anterosuperior labrum

MR Findings

- PD/intermediate FS
  - Anterosuperior labral tear
    - Adjacent edema in rotator interval
    - ± anterior supraspinatus tear (SLAC lesion)
  - Unattached or absent anterosuperior labrum usually a normal variant
  - High signal with disrupted fibers at labral attachment of MGL or SGL
- MR arthrography
  - High signal contrast extends into anterosuperior labrum or labral attachment of MGL or SGL

Imaging Recommendations

- Best imaging tool
  - Direct MR arthrography

DIFFERENTIAL DIAGNOSIS

Anterosuperior Normal Labral Variants

- Sublabral foramen and Buford complex mimic microinstability lesion on MR
- Present in 10-15% of shoulders

SLAP Lesions

- Some include SLAP tears with microinstability
- SLAP VII tear (extends into MGL) meets classic definition of microinstability lesion

Rotator Interval Tears

- Can be associated with anterosuperior labral tear (type X SLAP tear)

PATHOLOGY

General Features

- Etiology
  - Acute trauma
  - Repetitive stress, such as overhead throwing or swimming
- Associated abnormalities
  - Anterior supraspinatus tendon partial thickness tear
    - SLAC lesion
  - Rotator interval tear
  - Biceps pulley lesion with biceps subluxation
  - Anterosuperior chondral fissuring

CLINICAL ISSUES

Presentation

- Most common signs/symptoms
  - Shoulder pain that mimics cuff tendinopathy
  - Subjective sensation of slippage
  - Shoulder muscles easily fatigued
- Other signs/symptoms
  - No true instability on examination

Demographics

- Age
  - Young adults
- Gender
  - M > F
- Epidemiology
  - < 5% of patients with shoulder pain

Natural History & Prognosis

- May resolve with physical therapy, cessation of repetitive overhead activity

Treatment

- Continued symptoms: Labral debridement or repair imbricating anterosuperior capsule and rotator interval

DIAGNOSTIC CHECKLIST

Consider

- Anterosuperior normal labral variant if unattached/ absent anterosuperior labrum
  - Variants much more common than isolated anterosuperior labral tear

SELECTED REFERENCES

1. Drakos MC et al: Internal impingement of the shoulder in the overhead athlete. J Bone Joint Surg Am. 91(11):2719-28, 2009
2. Steinbach LS: MRI of shoulder instability. Eur J Radiol. 68(1):57-71, 2008
3. Savoie FH 3rd et al: Anterior superior instability with rotator cuff tearing: SLAC lesion. Orthop Clin North Am. 32(3):457-61, ix, 2001

SUPRASCAPULAR AND SPINOGLENOID NOTCH CYSTS

Key Facts

Terminology
- Cystic mass in spinoglenoid notch ± extension into suprascapular notch
- Most common location for paralabral cysts

Imaging
- Well-defined high T2 signal mass extending medially from posterior labrum
  - Typically begins in spinoglenoid notch
  - Often multiloculated
  - May extend into suprascapular notch
- Posterosuperior labral tear
- Infraspinatus ± supraspinatus muscle signal changes
  - High signal denervation edema
  - Infraspinatus affected 1st
  - Supraspinatus involved if suprascapular notch cyst
  - Fat atrophy: Chronic paralabral cyst
- Direct MR arthrogram: Include T2WI

Top Differential Diagnoses
- Rotator cuff atrophy syndromes
- Spinoglenoid notch venous varices
- Anterior or inferior paralabral cyst
  - Less common locations for cyst
- Juxtaarticular myxoma or other tumor

Pathology
- Posterosuperior labral tear or posterior SLAP tear
- Compresses suprascapular nerve
  - Shoulder pain
  - Denervation edema of infraspinatus ± supraspinatus muscles

Clinical Issues
- External rotation weakness with infraspinatus denervation

*(**Left**) Axial PDWI FS MR shows a small paralabral cyst ➡ in the spinoglenoid notch and an adjacent labral tear ⮕. Paralabral cysts result from joint fluid leaking through a labral tear, and the spinoglenoid notch near the posterosuperior labrum is the most common location. (**Right**) Sagittal T2WI FS MR in the same patient shows the paralabral cyst ➡ in the "low pressure" region between the supraspinatus and infraspinatus muscles and also between the scapular spine and the glenoid rim.*

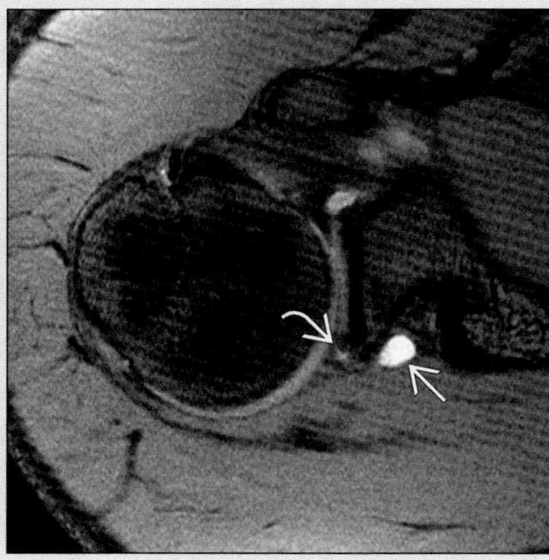

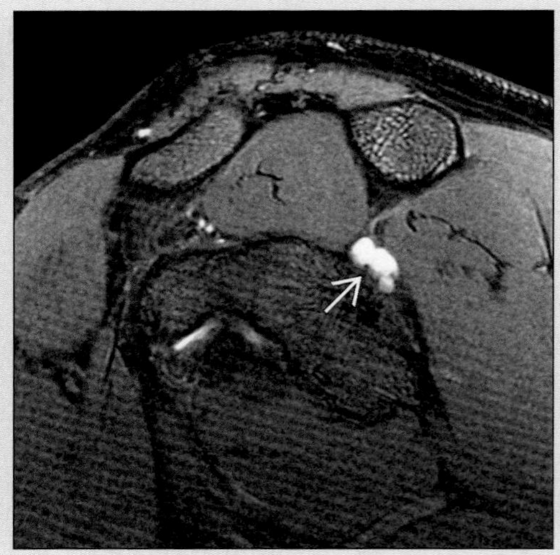

*(**Left**) Axial PDWI FS MR shows a paralabral cyst ➡ in the spinoglenoid notch. This patient also has mild elevated signal in the infraspinatus muscle ⮕ relative to the deltoid → and subscapularis ⇨ muscles. (**Right**) Coronal T2WI FS MR in the same patient shows a portion of the paralabral cyst ➡ and the edema in the infraspinatus muscle ⮕ relative to the teres minor muscle ⇨. Infraspinatus muscle edema is not uncommon and represents suprascapular nerve compression.*

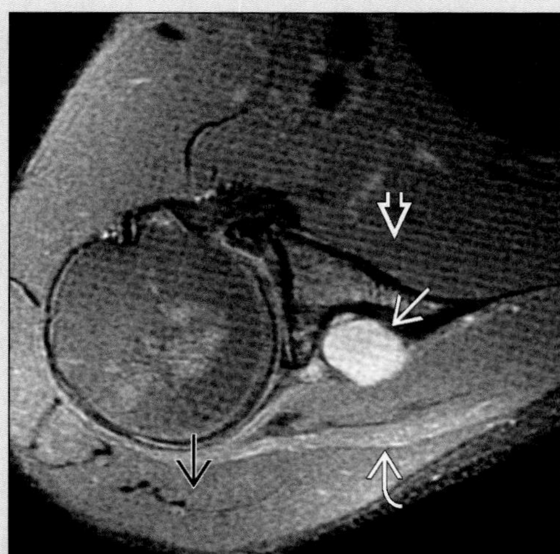

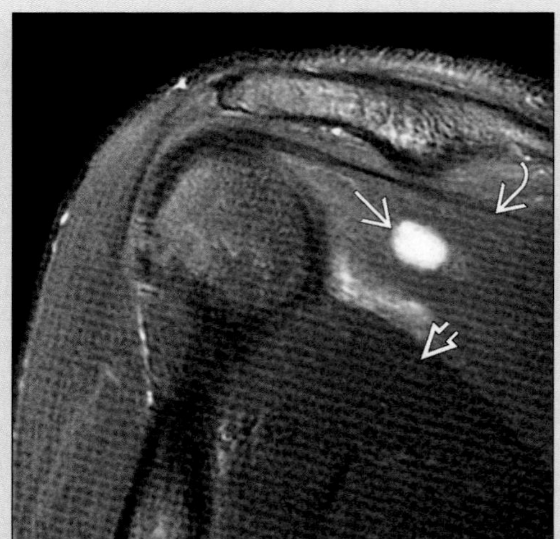

SUPRASCAPULAR AND SPINOGLENOID NOTCH CYSTS

TERMINOLOGY

Synonyms
- Paralabral cyst, glenoid labral cyst, periglenoidal ganglion cyst

Definitions
- Cystic mass in spinoglenoid notch ± extension into suprascapular notch

IMAGING

General Features
- Best diagnostic clue
  - Cyst adjacent to posterosuperior scapular neck
- Location
  - Spinoglenoid notch ± suprascapular notch

MR Findings
- T1WI
  - Decreased signal intensity paralabral mass
  - Fat atrophy of infraspinatus ± supraspinatus muscle
    - Chronic compression of suprascapular nerve
- PD/intermediate FS
  - Well-defined high signal mass extending medially from posterior labrum
    - Typically begins in spinoglenoid notch
    - Often multiloculated
    - Large cysts may → suprascapular notch
  - High signal in posterosuperior labral tear
    - Not always seen (tear can heal)
  - Infraspinatus ± supraspinatus muscle signal changes: Suprascapular nerve compression
    - High signal denervation edema
      - Infraspinatus muscle affected 1st
      - Supraspinatus if suprascapular notch cyst
    - Fat atrophy: Chronic paralabral cyst
- MR arthrography
  - High signal contrast extending into labrum
    - Posterosuperior labrum or posterior SLAP tear
  - Contrast may not communicate with, or only opacify a portion of, cyst

Ultrasonographic Findings
- Anechoic or hypoechoic paralabral mass

Imaging Recommendations
- Best imaging tool
  - PD/T2WI FS MR
- Protocol advice
  - Include T2WI on direct MR arthrograms
  - Include non-fat saturated T1 to look for atrophy

DIFFERENTIAL DIAGNOSIS

Rotator Cuff Atrophy Syndromes
- Suprascapular nerve entrapment from other causes (fibrous band, etc.)
- Parsonage-Turner viral neuritis
  - 90% involve supraspinatus and infraspinatus
  - 2/3 involve other muscles (most often deltoid)

Spinoglenoid Notch Venous Varices
- Can cause distal suprascapular nerve compression and infraspinatus atrophy

Anterior or Inferior Paralabral Cyst
- Less common location for paralabral cysts

Juxtaarticular Myxoma or Other Tumor
- Some T2 high intensity masses can mimic a cyst

PATHOLOGY

General Features
- Etiology
  - Posterosuperior labral tear allows synovial fluid to leak into paralabral soft tissues
- Associated abnormalities
  - Posterosuperior labral tear or posterior SLAP tear
  - Compresses suprascapular nerve
    - Shoulder pain
    - Denervation edema of infra- ± supraspinatus

Staging, Grading, & Classification
- Pathology is progressive
  - Spinoglenoid notch cyst, no infraspinatus edema
  - Spinoglenoid cyst + infraspinatus muscle edema
    - Compresses suprascapular nerve distal to branches to supraspinatus muscle
  - Spinoglenoid → suprascapular notch cyst
  - Spinoglenoid → suprascapular notch cyst with infraspinatus and supraspinatus edema
    - Compresses suprascapular nerve proximal to branches to both muscles
  - Paralabral cyst with muscle fat atrophy

CLINICAL ISSUES

Presentation
- Most common signs/symptoms
  - Shoulder pain
  - External rotation weakness from infraspinatus denervation

Demographics
- Epidemiology
  - Paralabral cyst in 3-5% of labral tears

Natural History & Prognosis
- Asymptomatic small cysts: Conservative treatment
- Large cysts: Muscle denervation may become chronic

Treatment
- Arthroscopic decompression cyst, repair labral tear
- Ultrasound-guided percutaneous dry needling/aspiration/injection

DIAGNOSTIC CHECKLIST

Image Interpretation Pearls
- Direct MR arthrography: Always include T2WI

SELECTED REFERENCES

1. Park JH et al: Effective arthroscopic treatment of large, anteriorly extended spinoglenoid cysts. Arch Orthop Trauma Surg. 130(3):423-6, 2010
2. Mellado JM et al: Fluid collections and juxta-articular cystic lesions of the shoulder: spectrum of MRI findings. Eur Radiol. 12(3):650-9, 2002

SUPRASCAPULAR AND SPINOGLENOID NOTCH CYSTS

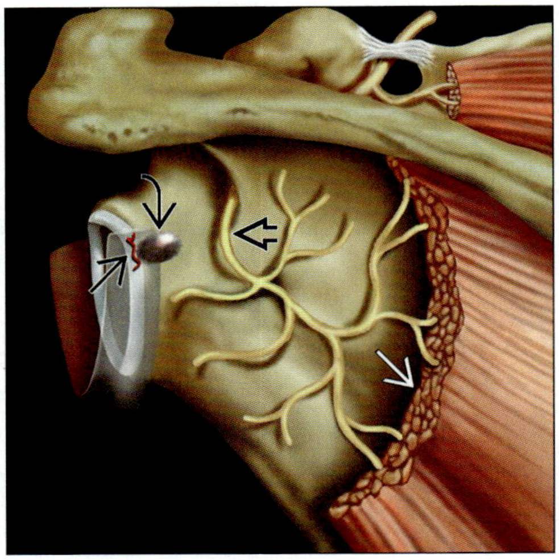

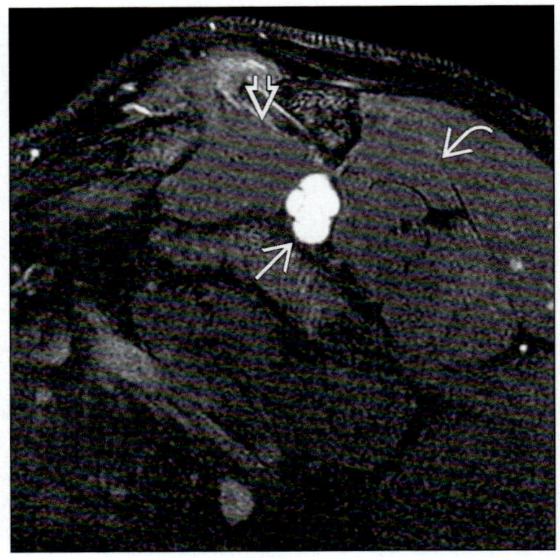

(Left) *PA graphic shows a posterosuperior labral tear → and small paralabral cyst ↷. With the lateral portion of the infraspinatus muscle cut away, the unaffected suprascapular nerve ⇨ can be seen innervating the infraspinatus muscle ➡.* ***(Right)*** *Sagittal T2WI FS MR shows a paralabral cyst in the spinoglenoid notch ➡. There is no signal abnormality of the infraspinatus ➡ or supraspinatus ➡ muscles, indicating that the motor branches of the suprascapular nerve are not involved.*

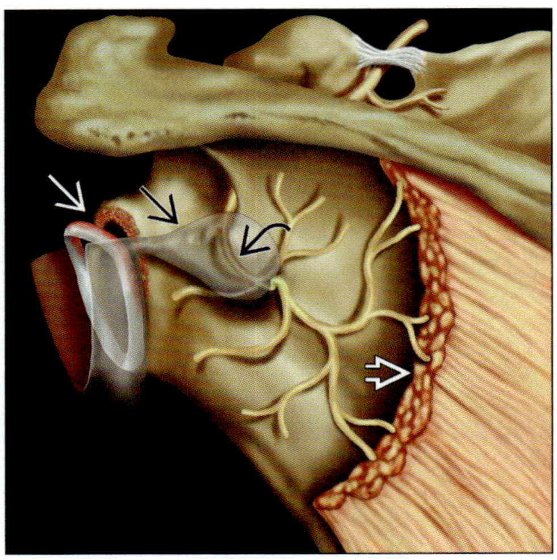

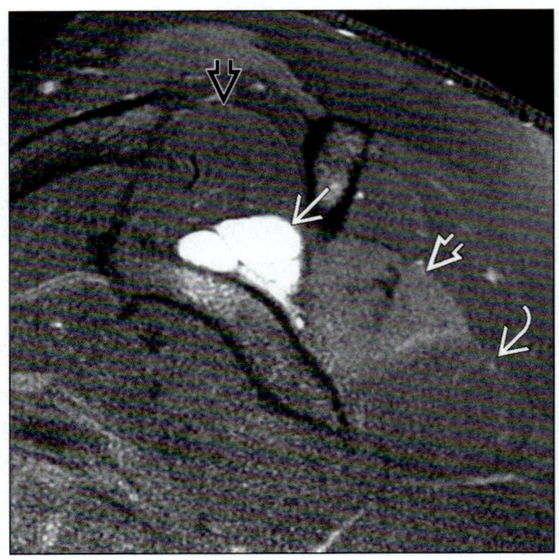

(Left) *PA graphic shows a larger paralabral cyst → arising from a superior labrum anteroposterior (SLAP) tear ➡. The cyst compresses the distal suprascapular nerve ↷ causing denervation edema ➡ of the infraspinatus muscle.* ***(Right)*** *Sagittal T2WI FS MR shows a large paralabral cyst ➡ with mild denervation edema of the infraspinatus muscle ➡; compare with the signal intensity of the teres minor ➡ and supraspinatus ⇨ muscles. Patients may have external rotation weakness.*

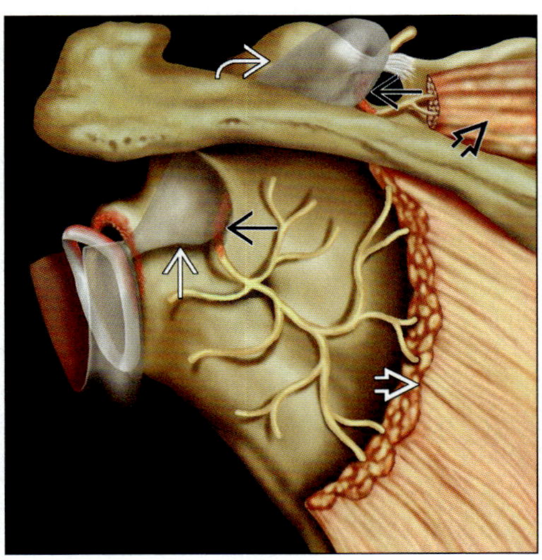

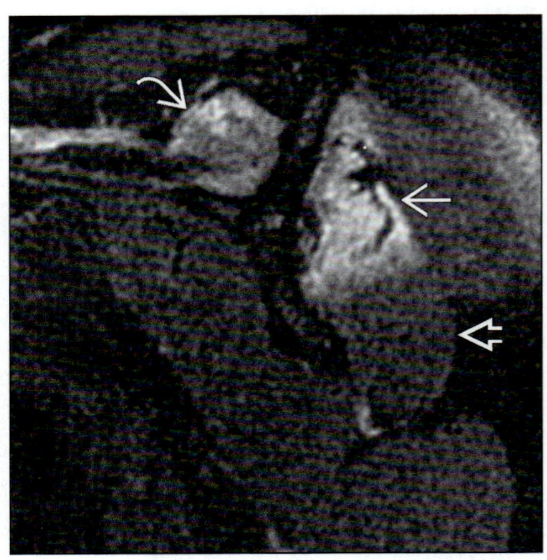

(Left) *Posterior view graphic shows a very large paralabral cyst arising from the posterior aspect of a large SLAP tear and extending from the spinoglenoid notch ➡ into the suprascapular notch ➡. There is compression of the suprascapular nerve → with denervation edema of both the supraspinatus ⇨ and infraspinatus ➡ muscles.* ***(Right)*** *Sagittal T2WI FS MR shows denervation edema of both the infraspinatus ➡ and supraspinatus ➡ muscles, with sparing of the teres minor muscle ➡.*

SUPRASCAPULAR AND SPINOGLENOID NOTCH CYSTS

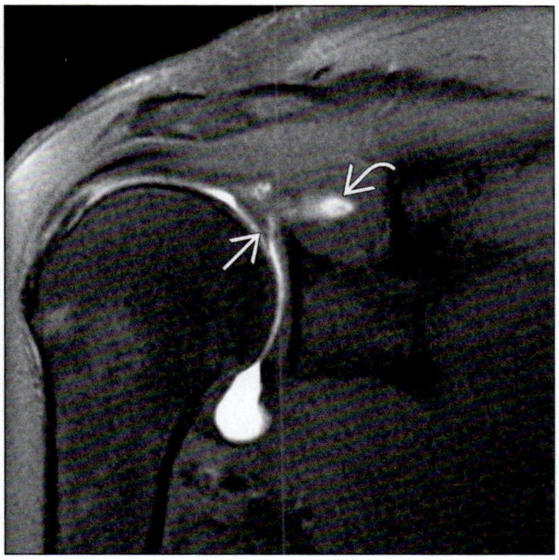

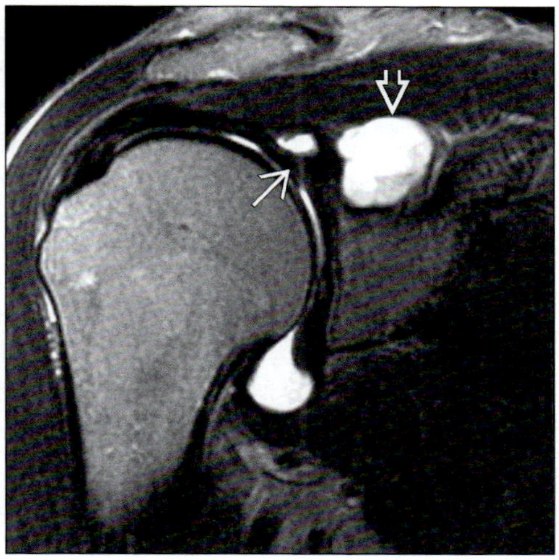

(Left) *Coronal oblique MR arthrogram T1WI FS shows contrast in a posterior SLAP tear that appears to extravasate out of the joint into the spinoglenoid notch region.* ***(Right)*** *Coronal oblique MR arthrogram T2WI FS in the same patient again shows the SLAP tear, as well as a spinoglenoid notch paralabral cyst. Because a paralabral cyst may not fill immediately with the injection of intraarticular contrast, a T2WI should always be included as part of a direct MR arthrogram.*

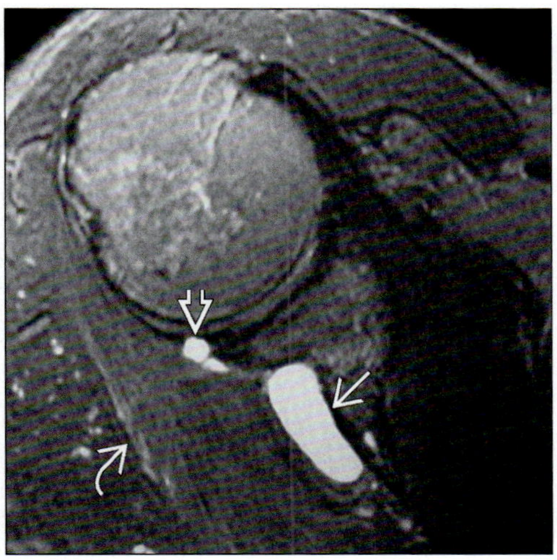

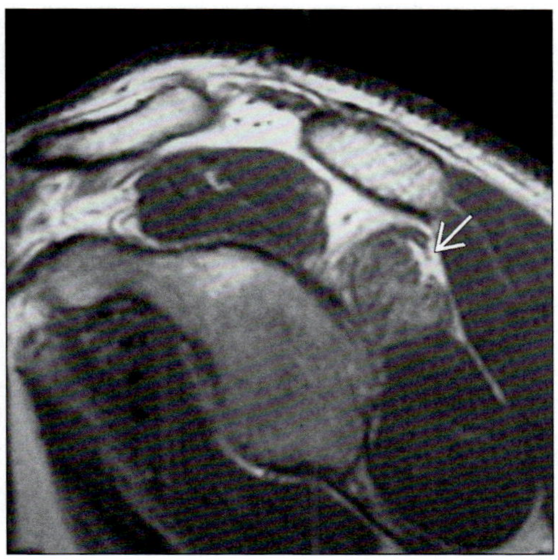

(Left) *Axial PDWI FS MR shows a paralabral cyst, posterior labral tear, and mild infraspinatus edema. The patient had shoulder pain for over a year and finally agreed to surgery a few months after the MR.* ***(Right)*** *Sagittal T1WI MR in the same patient, performed because of continued external rotation weakness months after surgery, shows fat atrophy of the infraspinatus muscle. The paralabral cyst had not recurred, so this was presumed to be due to permanent nerve damage.*

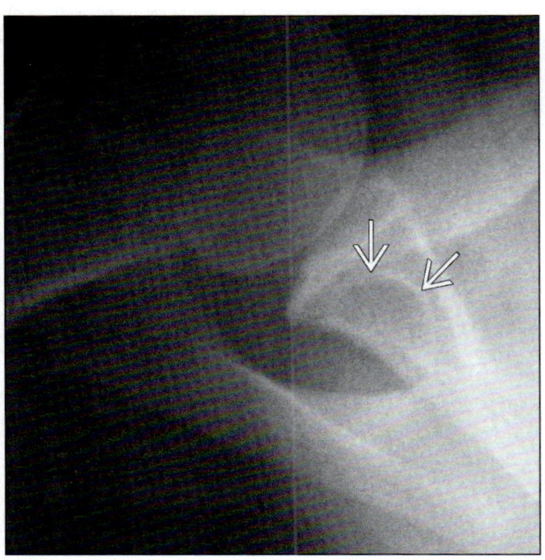

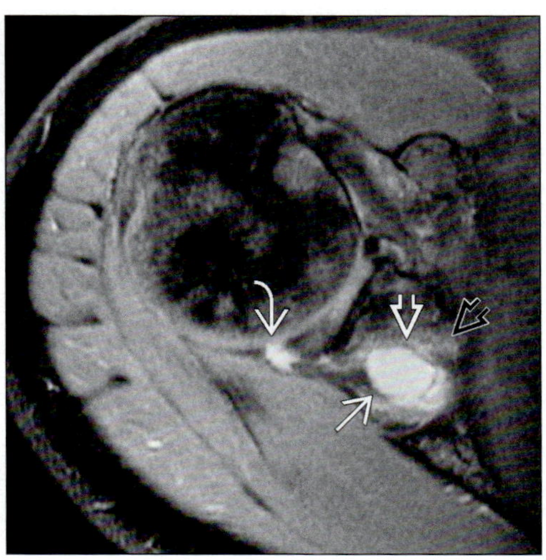

(Left) *Axillary lateral radiograph of a patient with several years of shoulder pain shows a lucency in the posterior scapular neck.* ***(Right)*** *Axial T2WI FS MR in the same patient shows a paralabral cyst and a posterosuperior labral tear. The cyst causes a pressure erosion of the posterior scapular neck with mild marrow edema, corresponding to the erosion seen on the radiograph. Bone erosion is uncommon with paralabral cysts and indicates a chronic cyst.*

ROTATOR CUFF DENERVATION SYNDROMES

Key Facts

Terminology
- Parsonage-Turner syndrome, acute brachial neuritis, quadrilateral space syndrome, neurogenic edema, denervation edema
- Edema within rotator cuff muscle(s) from motor nerve abnormality

Imaging
- Diffuse T2WI high signal in ≥ 1 rotator cuff muscle
- Increased T1 signal fat within muscle
- Decreased muscle bulk

Top Differential Diagnoses
- Rotator cuff full thickness tear
- Rotator cuff muscle injury
- Intervertebral disc herniation, cervical
- Inflammatory myopathy
- Clinical: Rotator cuff impingement
- Clinical: Calcific tendinopathy

Pathology
- Viral neuritis: Parsonage-Turner syndrome
- Fibrous bands
  - Quadrilateral space syndrome
  - Suprascapular nerve compression
- Paralabral cyst
  - Infraspinatus ± supraspinatus atrophy
- Nerve injury
  - Prior fracture, dislocation, or surgery
  - Overhead throwing athletes
  - Rotator cuff tear with teres minor nerve irritation
- Cervical radiculopathy, mass compressing brachial plexus

Clinical Issues
- Parsonage-Turner syndrome: Sudden onset severe shoulder pain and weakness

*(**Left**) PA graphic shows atrophy in the supraspinatus →, infraspinatus →, & deltoid → muscles, with sparing of the teres minor →. These are the muscles most commonly involved in Parsonage-Turner syndrome, and indicates neuritis of the suprascapular nerve and 6th cervical portion of the axillary nerve. (**Right**) Sagittal T2WI FS MR shows edema of the supraspinatus →, infraspinatus →, and deltoid muscles → in a patient with Parsonage-Turner syndrome. The teres minor → is spared.*

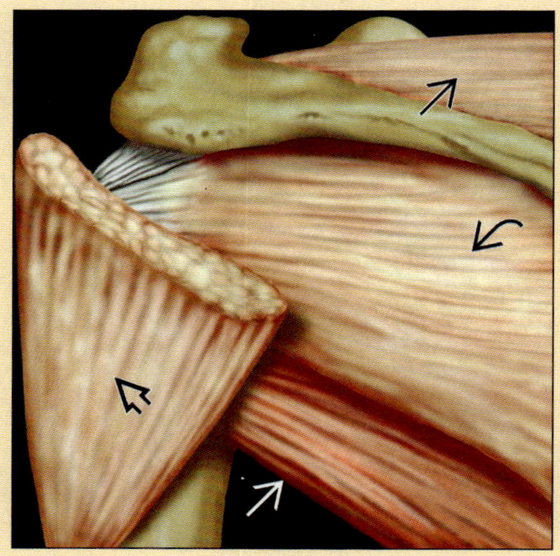

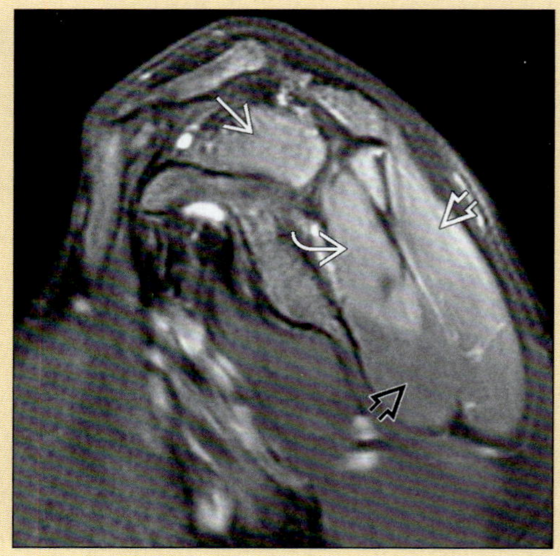

*(**Left**) PA graphic of quadrilateral space syndrome shows an indurated axillary nerve → with teres minor muscle denervation atrophy →. The quadrilateral space → is bounded by the humerus, teres minor, teres major →, and long head of the triceps → muscles. The syndrome is usually from fibrous bands within the space compressing the axillary nerve. (**Right**) Sagittal T2WI FS MR of a patient with quadrilateral space syndrome shows edema of the deltoid → and teres minor → muscles.*

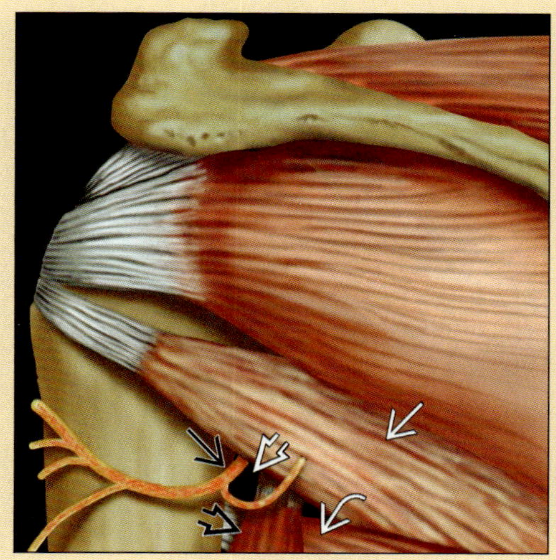

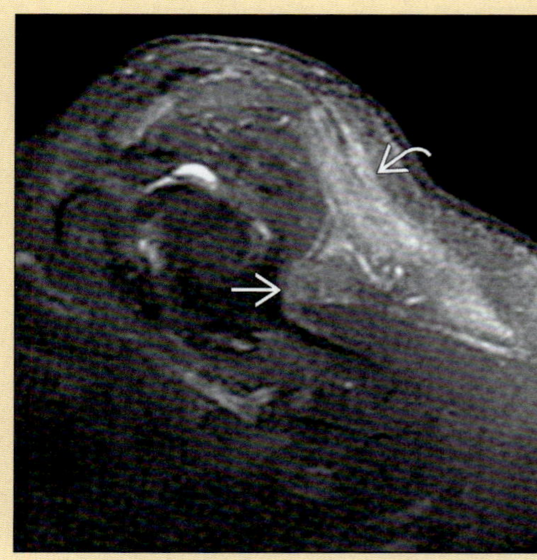

ROTATOR CUFF DENERVATION SYNDROMES

TERMINOLOGY

Synonyms

- Forms of rotator cuff denervation syndrome: Parsonage-Turner syndrome, acute brachial neuritis, quadrilateral space syndrome, neurogenic edema, denervation edema

Definitions

- Edema within rotator cuff muscle(s) from motor nerve abnormality
  - May also involve deltoid muscle
- Quadrilateral space syndrome
  - Axillary nerve compression/injury due to fibrous band or paralabral cyst
  - Teres minor and deltoid muscle denervation edema or atrophy
  - Compression of posterior humeral circumflex artery
- Parsonage-Turner syndrome (viral neuritis)
  - Usually involves suprascapular nerve
  - ~ 90% have denervation edema of supraspinatus and infraspinatus muscles
  - 2/3 have ≥ 3 muscles involved
    - Deltoid is 3rd most commonly involved muscle
    - Can have deltoid edema without teres minor edema if neuritis only of 6th cervical nerve portion of axillary nerve

IMAGING

General Features

- Best diagnostic clue
  - Diffuse T2WI high signal in 1 or more rotator cuff muscles
- Location
  - Suprascapular nerve
    - Supraspinatus muscle
    - Infraspinatus muscle
  - Axillary nerve
    - Teres minor muscle (5th cervical nerve)
    - Deltoid muscle (5th & 6th cervical nerves)
  - Subscapular nerve
    - Subscapularis muscle

MR Findings

- T1WI
  - Chronic denervation
    - Decreased muscle bulk
    - Increased T1 signal fat within muscle
      - No associated full thickness rotator cuff tear
    - Parsonage-Turner syndrome can demonstrate muscle atrophy without fatty replacement
- T2WI FS
  - Acute/subacute
    - Diffusely increased signal within 1 or more rotator cuff muscles
    - No tear of tendon(s) of involved muscle(s)
    - Parsonage-Turner syndrome: May involve other muscles (deltoid, latissimus dorsi, etc.)
  - Chronic
    - Decreased muscle bulk
    - Little to no muscle edema

Ultrasonographic Findings

- Increased echogenicity of 1 or more rotator cuff muscles
- Decreased muscle bulk

Imaging Recommendations

- Best imaging tool
  - T2WI MR

DIFFERENTIAL DIAGNOSIS

Rotator Cuff Full Thickness Tear

- Chronic complete rotator cuff tears often have fatty muscle atrophy

Rotator Cuff Muscle Injury

- Muscle strain presents as edema on MR
- Usually does not involve entirety of a muscle

Intervertebral Disc Herniation, Cervical

- Can also cause vague shoulder pain and denervation muscle edema

Inflammatory Myopathy

- Associated with elevated inflammatory markers

Clinical: Rotator Cuff Impingement

- Hooked anterior acromion process, acromioclavicular joint osteophytes, rotator cuff tendinopathy

Clinical: Rotator Cuff Partial Thickness Tear

- No denervation muscle edema
- T2WI MR high signal involving a portion of rotator cuff tendon

Clinical: Calcific Tendinopathy

- Calcification within rotator cuff tendon or bursa

PATHOLOGY

General Features

- Etiology
  - Viral neuritis: Parsonage-Turner syndrome
  - Fibrous bands
    - Quadrilateral space syndrome
    - Suprascapular nerve compression
      - Hypertrophied or calcified superior transverse (suprascapular) ligament at suprascapular notch or spinoglenoid ligament
  - Paralabral cyst
    - Arises from glenoid labral tear
      - Usually posterosuperior or posterior
    - Located in suprascapular notch → impingement of suprascapular nerve
      - Affects supraspinatus and infraspinatus muscles
    - Located in spinoglenoid notch → impingement of infraspinatus nerve
      - Affects infraspinatus muscle only
    - Inferior paralabral cyst → axillary nerve compression (quadrilateral space syndrome)
      - Affects teres minor ± deltoid muscles
  - Nerve injury (stretching/traction injury)
    - Prior glenohumeral dislocation or humeral neck fracture
    - Prior shoulder surgery

- Teres minor branch of axillary nerve is close to inferior capsule and easily injured
- Overhead throwing athletes
  - Chronic irritation or traction on either infraspinatus or axillary nerve
- 2-tendon rotator cuff tears, causing humeral de-centering and irritation of teres minor branch nerve
  - May be cause of "idiopathic" isolated teres minor atrophy
  - Teres minor nerve passes close to inferior capsule
  - Subluxation of humeral head may cause chronic irritation of nerve
  - Perineural thickening and myxoid globular degeneration of teres minor branch nerve epineurium common in elderly cadavers
- Cervical radiculopathy, spinal cord injury, brachial plexus or periscapular mass
  - Metastatic breast cancer most common to affect brachial plexus

Staging, Grading, & Classification

- Hyperacute denervation (< 1 week)
  - Muscle signal intensity is normal
- Acute/subacute denervation (1 week to several months)
  - T2WI ↑ signal in muscle
- Chronic denervation (> several months)
  - Muscle atrophy and fatty infiltration

Gross Pathologic & Surgical Features

- Increased extracellular fluid and capillary blood volume in muscle
- Neuropraxic injury to nerve myelin sheath

CLINICAL ISSUES

Presentation

- Most common signs/symptoms
  - Parsonage-Turner syndrome
    - Sudden onset severe shoulder pain
    - Abduction and external rotation weakness
    - Bilateral in 8-33%
  - Quadrilateral space syndrome
    - Pain & paresthesias in upper extremity
      - Aggravated by forward flexion, abduction, & external rotation
    - Tenderness to palpation over quadrilateral space
  - Isolated teres minor atrophy
    - Usually pain from other shoulder pathology
    - ± external rotation weakness (infraspinatus is stronger external rotator)

Demographics

- Gender
  - Parsonage-Turner syndrome: M > F
  - Quadrilateral space syndrome: M > F
- Epidemiology
  - Parsonage-Turner syndrome incidence: 1% of shoulder MR scans, 1.6 cases per 100,000 people
  - Isolated teres minor atrophy: 1-5% of shoulder MR scans
  - Infraspinatus atrophy without paralabral cyst: 4.4% of professional baseball pitchers
  - Glenohumeral dislocation: 10-15% have a post-reduction period of axillary nerve injury & pain

Natural History & Prognosis

- Parsonage-Turner syndrome: Pain resolves in weeks, but weakness takes months to several years to resolve

Treatment

- Fibrous bands: Surgical resection
- Paralabral cyst: Repair of labral tear + cyst excision

DIAGNOSTIC CHECKLIST

Consider

- Parsonage-Turner syndrome if supraspinatus and infraspinatus muscle edema without paralabral cyst

Image Interpretation Pearls

- Distribution of muscle edema ± atrophy indicates nerve(s) affected

SELECTED REFERENCES

1. Friend J et al: Teres minor innervation in the context of isolated muscle atrophy. Surg Radiol Anat. 32(3):243-9, 2010
2. Cummins CA et al: Peripheral nerve injuries in baseball players. Phys Med Rehabil Clin N Am. 20(1):175-93, x, 2009
3. Scalf RE et al: MRI findings of 26 patients with Parsonage-Turner syndrome. AJR Am J Roentgenol. 189(1):W39-44, 2007
4. Amin MF et al: An unusual cause of the quadrilateral space impingement syndrome by a bone spike. Skeletal Radiol. 35(12):956-8, 2006
5. Bencardino JT et al: Entrapment neuropathies of the shoulder and elbow in the athlete. Clin Sports Med. 25(3):465-87, vi-vii, 2006
6. Brestas PS et al: Ultrasound findings of teres minor denervation in suspected quadrilateral space syndrome. J Clin Ultrasound. 34(7):343-7, 2006
7. Gaskin CM et al: Parsonage-Turner syndrome: MR imaging findings and clinical information of 27 patients. Radiology. 240(2):501-7, 2006
8. Wilson L et al: Isolated teres minor atrophy: manifestation of quadrilateral space syndrome or traction injury to the axillary nerve? Orthopedics. 29(5):447-50, 2006
9. Cothran RL Jr et al: Quadrilateral space syndrome: incidence of imaging findings in a population referred for MRI of the shoulder. AJR Am J Roentgenol. 184(3):989-92, 2005
10. Elsayes KM et al: Value of magnetic resonance imaging for muscle denervation syndromes of the shoulder girdle. J Comput Assist Tomogr. 29(3):326-9, 2005
11. Ryan M et al: Whole body magnetic resonance imaging in the diagnosis of Parsonage Turner syndrome. Acta Radiol. 45(5):534-9, 2004
12. Sofka CM et al: Teres minor denervation on routine magnetic resonance imaging of the shoulder. Skeletal Radiol. 33(9):514-8, 2004
13. Robinson P et al: Quadrilateral space syndrome caused by glenoid labral cyst. AJR Am J Roentgenol. 175(4):1103-5, 2000
14. Bredella MA et al: Denervation syndromes of the shoulder girdle: MR imaging with electrophysiologic correlation. Skeletal Radiol. 28(10):567-72, 1999
15. Sanders TG et al: Paralabral cyst: an unusual cause of quadrilateral space syndrome. Arthroscopy. 15(6):632-7, 1999

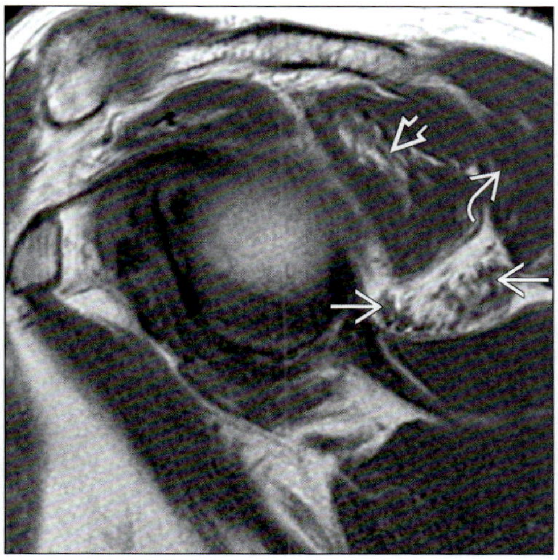

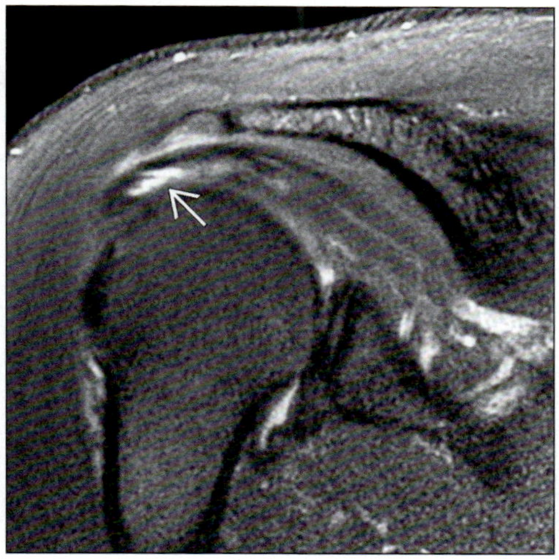

(Left) *Sagittal T1WI MR shows marked fatty atrophy of the teres minor muscle with sparing of the deltoid. This finding is common in older patients with rotator cuff tears and is presumably due to chronic irritation of the teres minor nerve from humeral head de-centering. Note mild infraspinatus atrophy.* ***(Right)*** *Coronal T2WI FS MR from the same patient shows an articular surface partial thickness tear of the infraspinatus tendon. The patient also had a supraspinatus tendon tear.*

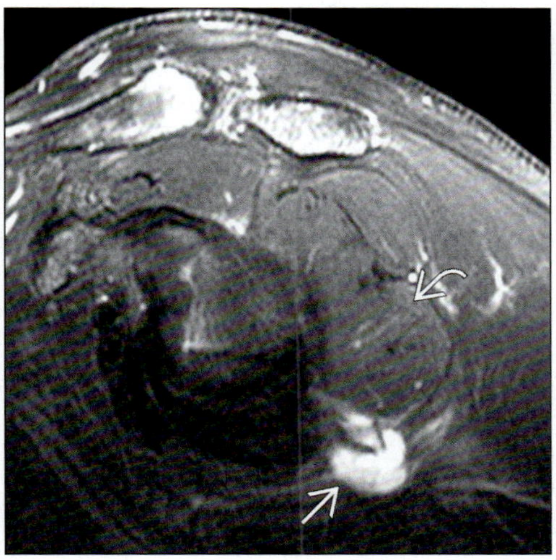

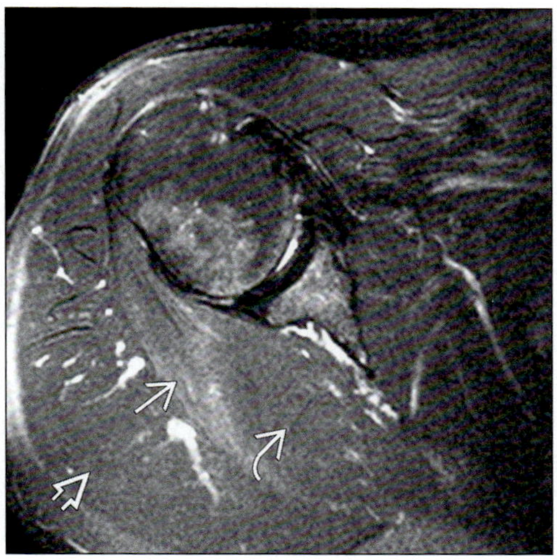

(Left) *Sagittal T2WI FS MR shows an inferior paralabral cyst in a patient with an inferior labral tear. There is mild edema of the teres minor muscle.* ***(Right)*** *Axial PDWI FS MR from the same patient shows the denervation edema of the teres minor muscle with sparing of the infraspinatus & deltoid muscles. Quadrilateral space paralabral cysts are less common than spinoglenoid notch cysts, but they can cause teres minor denervation edema from compression of the teres minor nerve.*

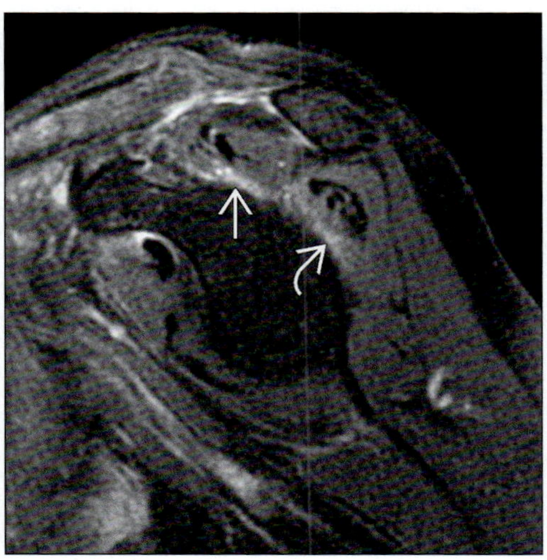

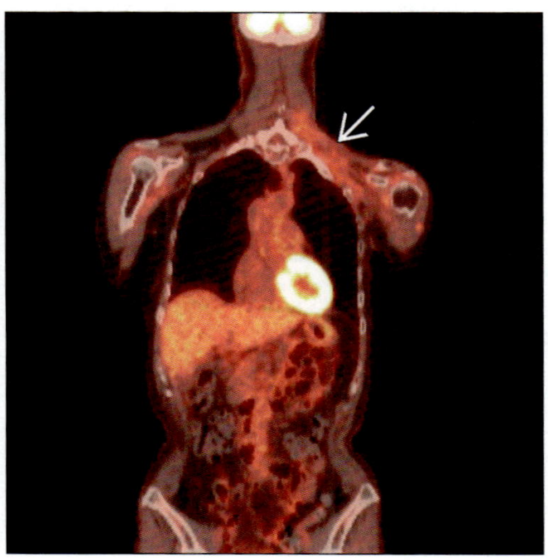

(Left) *Sagittal T2WI FS MR shows mild edema of the supraspinatus and infraspinatus muscles. The patient did not have a suprascapular paralabral cyst nor the acute onset of pain characteristic of Parsonage-Turner syndrome.* ***(Right)*** *Coronal reformat fused PET/CT image from the same patient shows increased metabolic activity around the brachial plexus. The patient had breast cancer and was found to have supraclavicular metastases encasing the brachial plexus.*

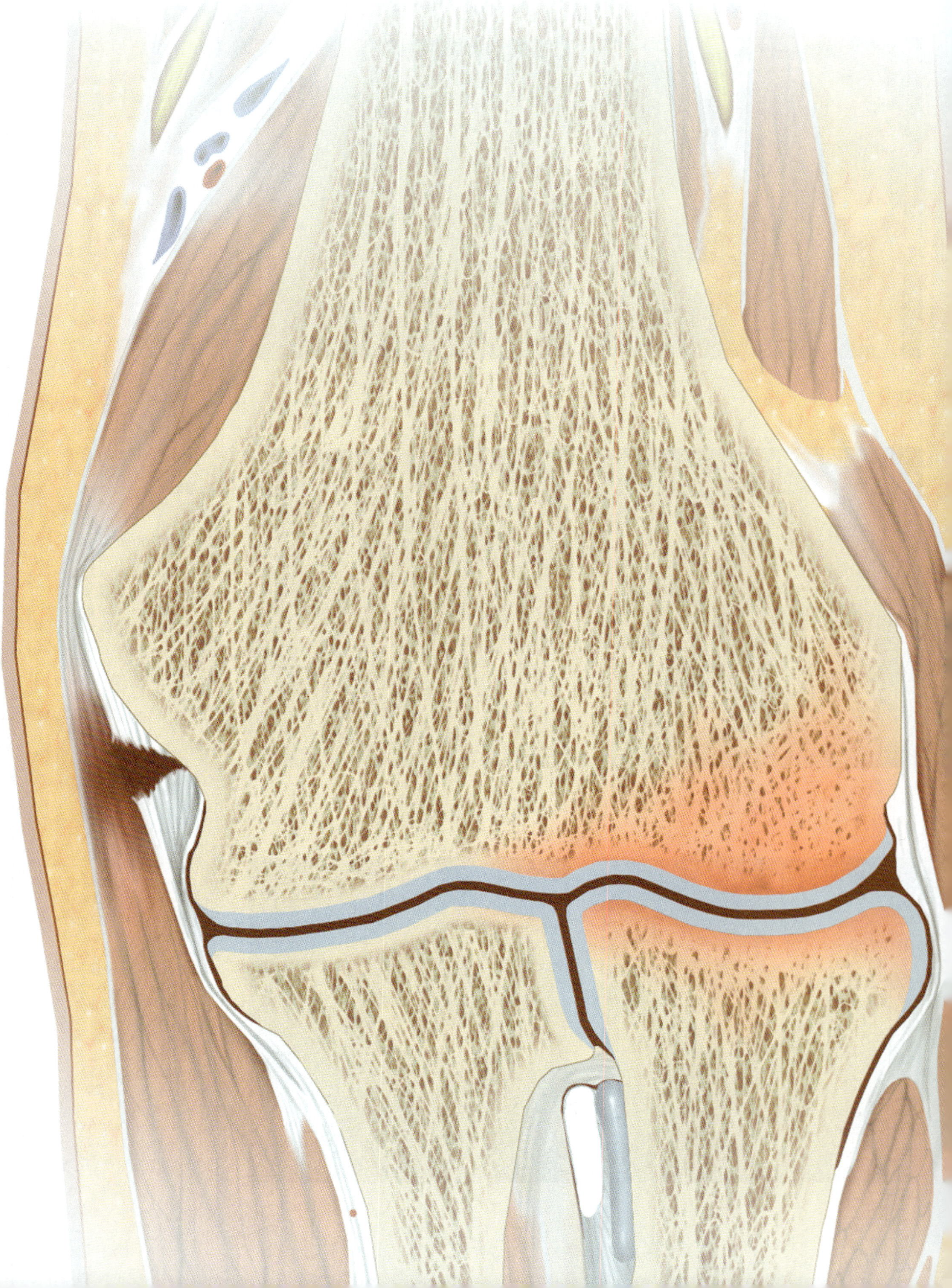

SECTION 3
Elbow

Introduction and Overview

Osseous Pathology

Ligamentous Pathology

Tendinous Pathology

Other Pathology

Neuropathies

Overview

Many radiologists approach advanced imaging of the elbow with some degree of trepidation. This is more due to a lack of familiarity with the relevant anatomy and pathology encountered in the elbow region than to any related complexity of diagnosis. The elbow is, in fact, a fairly straightforward articulation to master for the practicing radiologist. It is the rarity of elbow MR imaging referrals (relative to the more commonly encountered cases involving the knee, shoulder, and ankle) that often limits practical experience for the individual not associated with a surgeon with a particular interest in elbow pathology. This section explores the commonly (and some of the not-so-commonly) encountered pathologic conditions for which advanced imaging lends advantage.

Terminology and Conventions

It is common anatomic practice for the radial aspect of the elbow to be considered lateral, and the ulnar aspect medial. The terms "coronal" and "sagittal" used in these chapters refer to the anatomic planes as defined by the inherent anatomy of the elbow as described in Imaging Considerations below, and thus will frequently be oblique planes (with reference to the machine axis) for any given patient. As in other chapters, degenerative changes within a tendon are referred to as "tendinopathy" and not as "tendinitis" or "tendinosis" in an effort to stay true to the appropriate etymologic meanings of these terms. As elsewhere in the book, all images are presented as right-sided in order to preserve uniformity.

Anatomic Considerations

The elbow is, at its simplest, a hinge joint at the ulnohumeral articulation. There is also a rotational component at the radiocapitellar and proximal radioulnar joints to allow pronation and supination of the forearm. Capsular thickenings of the joint medially and laterally comprise the ulnar collateral ligament complex and the lateral collateral ligament complex, respectively. Tendinous insertions of the muscles of elbow flexion (biceps brachii and brachialis) and extension (triceps brachii) occur on the bones immediately adjacent to the elbow. Origins of the major flexor and extensor tendons of the wrist and fingers arise from the medial and lateral humeral epicondyles, respectively. Three major nerves pass through the elbow region and can be injured here: The ulnar nerve is particularly vulnerable because of its very superficial location in the cubital tunnel, but the median and radial nerves can be affected by a variety of congenital and acquired conditions as they pass through their respective courses in the elbow.

Several imaging pitfalls in the evaluation of the elbow, particularly by MR imaging, have been described:

- Pseudodefect of the capitellum: This is a normal groove between the capitellum and the lateral humeral condyle. On sagittal MR images, this may mimic an osteochondral defect.
- Transverse trochlear ridge: This is a normal bony ridge extending transversely across the trochlear notch of the ulna at the junction of the coronoid and olecranon processes. Because it has no overlying cartilage, it may mimic a chondral defect or osteophyte on sagittal MR images.
- Trochlear groove: These are medial and lateral indentations in the mid-aspect of the trochlear notch of the ulna near the site of the transverse trochlear ridge. Sagittal MR images through these depressions in the articular surface may mimic a fracture.

Pathologic Considerations

Injury to the elbow may be due to a discrete traumatic episode, which more often results in osseous injury or ligament tear, or to chronic repetitive injury, which may result in a variety of pathology involving bones, ligaments, or tendons. Injury patterns also vary depending on the age of the patient; when relevant, issues specific to the pediatric patient have been identified. Fractures of the distal humerus, proximal ulna, and proximal radius are usually the result of a discrete episode of trauma, and may be associated with significant soft tissue injury around the elbow joint, as well as other traumatic lesions elsewhere in the body. Grading systems for these injuries tend to focus on criteria that help determine the method of treatment, and are described in the appropriate chapters.

Ligamentous injury can be due to either acute trauma or repetitive injury, and may manifest as chronic pain or with sudden onset of symptoms during activity. Ulnar collateral ligament tear is the most common and significant elbow ligament injury, and is often seen in the setting of overhead throwing athletes. One type of repetitive injury in particular has been paid special attention: Athletes who perform overhead throwing motions are prone to develop a particular pattern of injury due to valgus stress placed on the elbow during these activities, resulting in impaction injuries in the lateral aspect of the elbow and distraction injuries in the medial aspect. This pattern is important to recognize; when caught early in its course, appropriate treatment can make the difference between return to play and a career-ending injury.

Imaging Considerations

Radiographic evaluation of the elbow should include, at a minimum, anteroposterior and lateral views. In the setting of trauma, an oblique view should be added. In cases where a subtle radial head fracture is questioned, a dedicated radial head/capitellum view is available; this involves 90° of elbow flexion and a 45° tube angulation toward the radial head to best demonstrate the radial head and neck.

Ultrasound evaluation of the elbow has been shown to be effective for tendinous pathology. Injury to the biceps, triceps, flexor, and extensor tendons can be accurately evaluated using ultrasound imaging. Recently, ultrasound has been reported to be useful in the evaluation of the ulnar collateral ligament.

MR imaging of the elbow is far less straightforward. Two positioning strategies are available. The preferred method puts greater demands on the patient (in terms of positioning and comfort) and the MR imaging technologist (in terms of correctly positioning the patient and performing the study), but results in more clinically satisfying images. This usually involves placing the extended elbow in a volume coil (such as a clamshell phased-array knee coil), which serves as a rigid frame in which to steady the extremity, but may cause uncomfortable compression of tissues. The use of such a coil often requires that the patient be scanned with his/her arm raised over the head (the "Superman" position), as the coil must be placed in or near the magnetic isocenter. Most patients prefer to accomplish this while lying prone as opposed to supine. Generous padding with cushions and pillows may mitigate some of the positional discomfort, but the technologist should warn the patient that the extremity may become painful or even numb during the imaging process as a result of the position in which it is kept. A tailored approach to imaging, in which the most clinically desired sequences for the patient's symptoms are acquired 1st, is advised.

The 2nd strategy involves letting the patient lie prone, arm at the side (with elbow extended), and using a flexible surface coil to image the elbow. While more comfortable for the patient and generally easier for the technologist, this method may provide clinically less desirable images due to limited signal-to-noise, poor fat suppression due to inhomogeneity of the magnetic field at its periphery, surface coil artifact (brighter tissue signal near the coil with rapid drop-off toward the deep portion of the extremity), and artifact from motion of the adjacent torso (respiration and other movements). In fact, the arm-at-the-side position may not be possible to accomplish at all in some larger patients due to restrictions of the magnet bore caliber.

A 3rd option is to acquire images of the elbow in an open-sided magnet so that the elbow remains in the center of the magnetic field, but the arm can be held only slightly abducted to the side. Field strength of such magnets is usually limited, and coil selection is more limited.

Whichever method is employed for elbow imaging, the forearm will likely be held in some degree of pronation. While true anatomic position implies supination of the forearm, in clinical practice it is unfeasible to expect patients to hold their arm in a supinated position for 20 to 30 minutes. A moderate degree of pronation is acceptable and is to be expected. This generally causes no diagnostic challenge, but one will notice that the radial tuberosity is rotated medially, with associated curving of the distal biceps tendon.

Once appropriate positioning has been achieved and scout images acquired, it is important that true anatomic planes of imaging be produced. For coronal images, slices should parallel a line drawn on a true axial image through the widest point of the medial and lateral humeral epicondyles, respectively. Sagittal images are prescribed 90° to the coronals. Educating the MR technologists working in your practice regarding these positioning and scan prescription guidelines will prove invaluable, and will likely significantly curtail the incidence of suboptimal scans and patient callbacks that you experience.

The use of MR arthrography has been advocated for the detection of certain pathology, chiefly undersurface partial tears of the ulnar collateral ligament and chondral defects. With modern MR imaging equipment and appropriate sequence selection, it has been my personal experience that MR arthrography adds much to the diagnosis and characterization of these lesions. Some authors have advocated the use of indirect arthrography (in which gadolinium contrast is injected intravenously, and delayed imaging of the elbow is performed after exercising the extremity to increase synovial blood flow and thus diffusion of intraarticular contrast into the joint fluid) as an alternative to direct puncture arthrography. Similarly, the addition of the **FABS** (elbow **F**lexed, shoulder **AB**ducted with **S**upination of forearm) position for evaluation of the distal biceps tendon has been described; an example is seen in the biceps tendon injury chapter. Intravenous contrast administration is usually reserved for the assessment of inflammatory or neoplastic pathology (synovitis, soft tissue masses, or infection).

Imaging Protocols

As with other joints discussed in this book, personal preferences regarding sequence selection vary widely among individuals who perform and interpret MR imaging of the elbow. Rather than take a purely prescriptive approach, I think it best to state general guidelines for this purpose. Certainly at least 1 sequence in each anatomic plane (axial, coronal, and sagittal) should be acquired. If an isotropic volume acquisition is used and diagnostic reformatted images can be created in multiple planes, then of course the 3-plane rule may not apply. I think it also best to acquire at least one T1-weighted sequence without fat suppression as a way to properly characterize tissue signal. The workhorse for most radiologists tends to be a fat-suppressed proton-density-(intermediate-)weighted or fat-suppressed T2-weighted (or STIR) MR sequence in some or all planes, as a way of studying both anatomy and tissue characteristics. Proton density-weighted sequences without chemical fat suppression provide excellent anatomic detail with high signal-to-noise, but are not particularly useful for tissue characterization. In general, one tends to develop a protocol that provides an acceptable combination of speed, variety, and reliability, and become comfortable with it.

Selected References

1. Chew ML et al: Disorders of the distal biceps brachii tendon. Radiographics. 25(5):1227-37, 2005
2. Ly JQ et al: MR imaging of the elbow: a spectrum of common pathologic conditions. Clin Imaging. 29(4):278-82, 2005
3. Kijowski R et al: Magnetic resonance imaging of the elbow. Part I: normal anatomy, imaging technique, and osseous abnormalities. Skeletal Radiol. 33(12):685-97, 2004
4. Miller TT et al: Sonography of injury of the ulnar collateral ligament of the elbow-initial experience. Skeletal Radiol. 33(7):386-91, 2004
5. Jacobson JA et al: US of the anterior bundle of the ulnar collateral ligament: findings in five cadaver elbows with MR arthrographic and anatomic comparison--initial observations. Radiology. 227(2):561-6, 2003
6. De Smet AA et al: Dynamic sonography with valgus stress to assess elbow ulnar collateral ligament injury in baseball pitchers. Skeletal Radiol. 31(11):671-6, 2002
7. Steinbach LS et al: Special focus session. MR arthrography. Radiographics. 22(5):1223-46, 2002
8. Rosenberg ZS et al: MR imaging of normal variants and interpretation pitfalls of the elbow. Magn Reson Imaging Clin N Am. 5(3):481-99, 1997
9. Palmer WE: MR arthrography: is it worthwhile? Top Magn Reson Imaging. 8(1):24-43, 1996
10. Sonin AH et al: MR imaging of sports injuries in the adult elbow: a tailored approach. AJR Am J Roentgenol. 167(2):325-31, 1996
11. Timmerman LA et al: Undersurface tear of the ulnar collateral ligament in baseball players. A newly recognized lesion. Am J Sports Med. 22(1):33-6, 1994

(Left) *Radial head/capitellar view oblique lateral radiograph shows osteochondral fracture of capitellum, and displays radial head and neck to better advantage than routine lateral view. This view is accomplished by angling the tube 45° toward the radial head off a standard lateral view.* ***(Right)*** *Axial T2WI FS MR performed with patient's arm at side shows moderate respiratory motion artifact in the phase-encoding direction, and heterogeneous signal intensity throughout the image.*

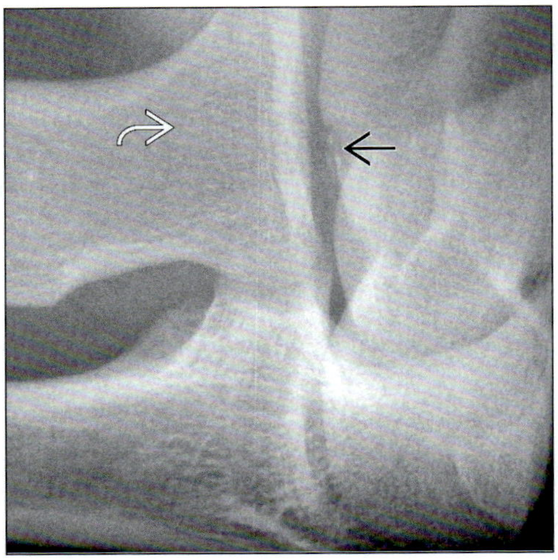

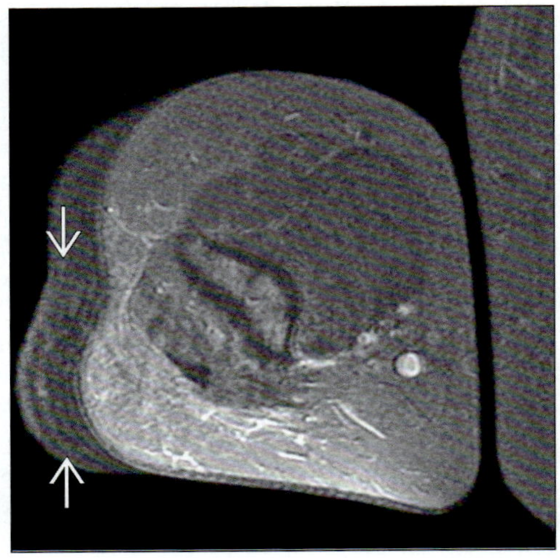

(Left) *Axial T1WI MR shows correct positioning of orthogonal imaging planes using a true axial image. The coronal plane of the elbow is defined by a line drawn through the widest parts of the humeral condyles. The sagittal plane is defined by a line drawn 90° to the coronal plane.* ***(Right)*** *Coronal graphic of elbow joint shows the extensor and flexor tendon origins and the radial and ulnar collateral ligaments. The UCL attaches to the sublime tubercle of the ulna.*

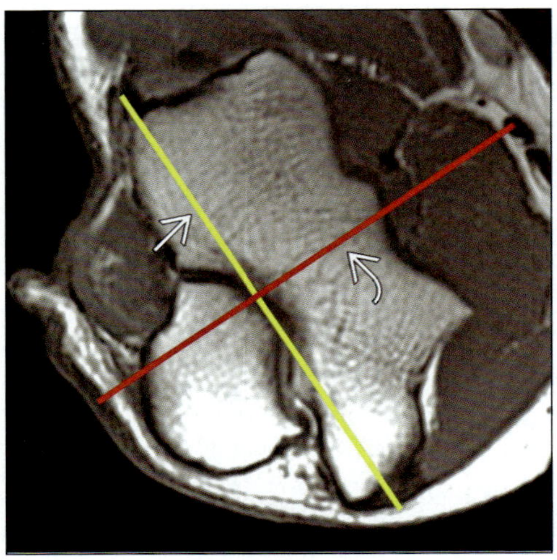

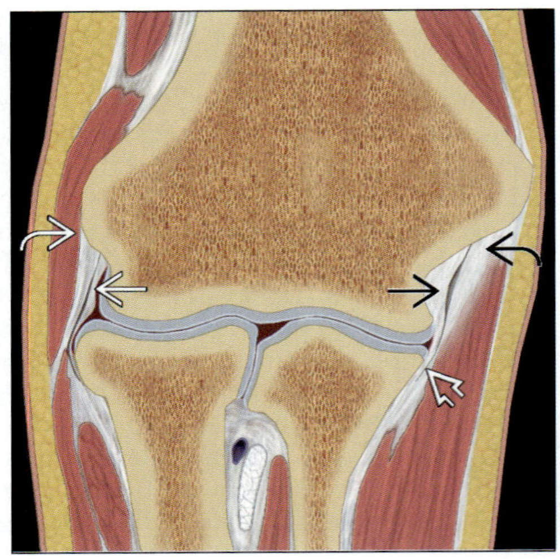

(Left) *Coronal T2WI FS MR of the elbow in a 17-year-old boy shows a normal radial collateral ligament just deep to the common extensor tendon. The annular ligament is seen medial to the radial head.* ***(Right)*** *Coronal T2WI FS MR from the same patient slightly more anteriorly shows the anterior band of the ulnar collateral ligament running from the medial epicondyle to the sublime tubercle of the ulna. The common flexor/pronator tendon is more superficial.*

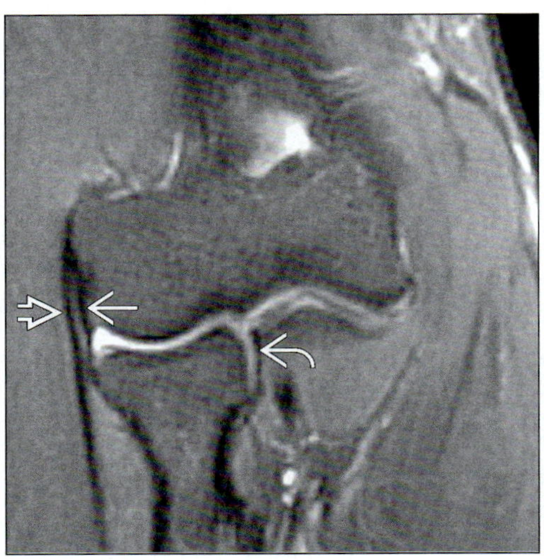

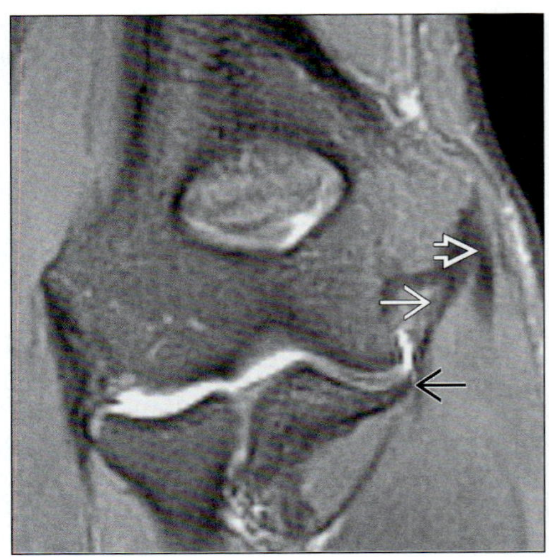

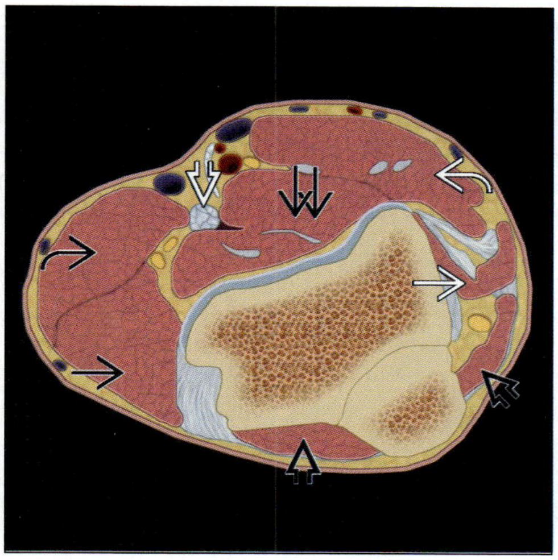

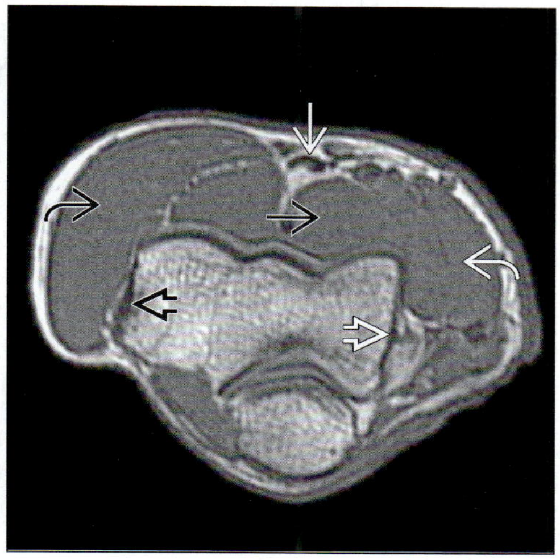

*(**Left**) Axial graphic through the distal humerus shows the extensor carpi radialis longus and brachialis muscles laterally, common flexor and pronator teres muscles medially, brachialis muscle and biceps tendon anteriorly, and triceps muscle posteriorly. (**Right**) Axial T1WI MR shows the medial and lateral humeral condyles, biceps tendon, brachialis muscle, pronator teres muscle, and brachioradialis muscle.*

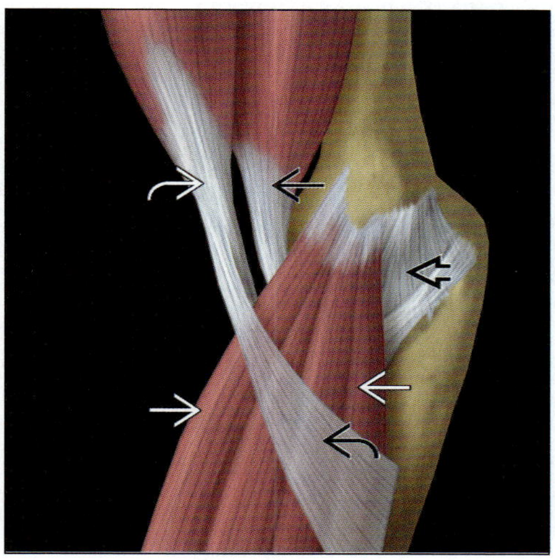

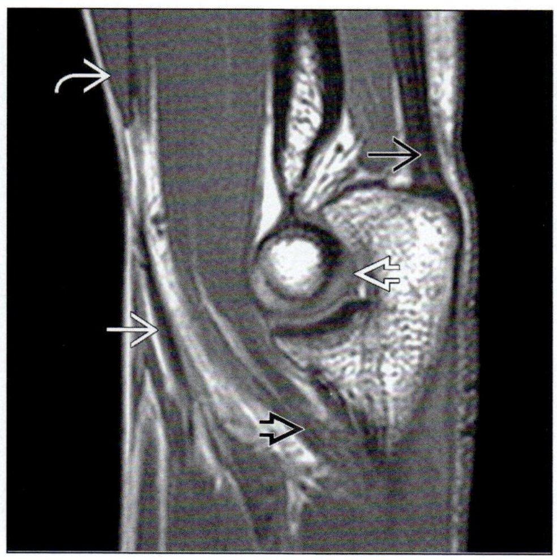

*(**Left**) Sagittal graphic seen from medial perspective shows the brachialis and biceps tendons, with the lacertus fibrosus extending medially to the aponeurosis of the common flexor muscles. Ulnar collateral ligament is also seen. (**Right**) Sagittal T1WI MR shows the ulnohumeral joint, triceps tendon, biceps muscle and tendon, and the brachialis insertion onto the proximal ulna.*

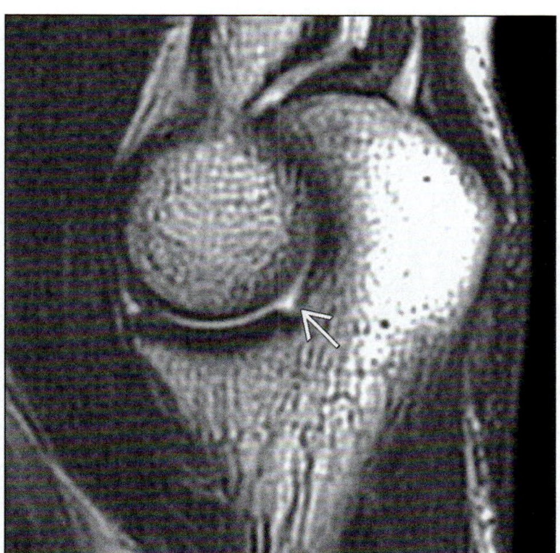

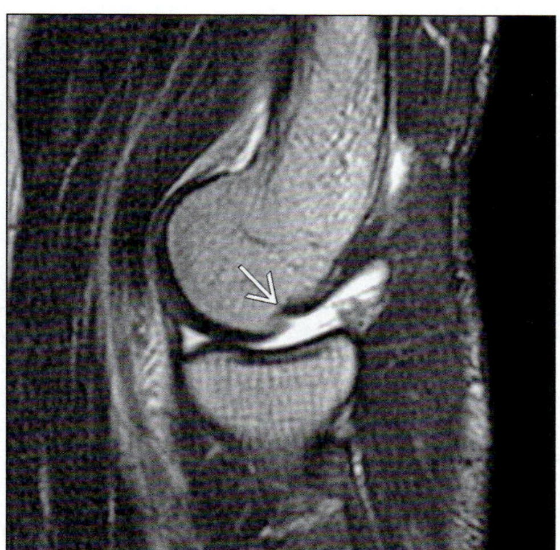

*(**Left**) Sagittal T2WI MR shows a transverse trochlear groove in the medial aspect of the trochlear articular surface, a normal indentation found in the medial and lateral aspects of the trochlear fossa that does not have overlying cartilage. This should not be mistaken for a chondral fracture. (**Right**) Sagittal T2WI MR shows the "pseudodefect of the capitellum," a normal notch posterolateral to the capitellum that should not be mistaken for an osteochondral defect.*

DISTAL HUMERAL FRACTURE

Key Facts

Terminology

- May encompass a variety of fracture subtypes
  - Supracondylar, transcondylar, and intercondylar fractures

Imaging

- Radiograph: Positive "fat pad" sign (distension of periarticular fat pads) at elbow
  - Due to hemarthrosis
- Depending on subtype, may involve distal humeral shaft, or 1 or both humeral condyles
  - Supracondylar fractures: Extracapsular
  - Condylar, transcondylar, & intercondylar fractures: Intraarticular
- CT defines fracture best
  - Extent of fracture complex
  - Involvement of articular surface of humerus

Top Differential Diagnoses

- Elbow dislocation
- Floating elbow: Fractures above and below elbow joint isolating joint

Clinical Issues

- Nondisplaced extraarticular fractures usually treated with casting & immobilization
- Unstable extraarticular fractures treated surgically
- Articular disruption treated surgically

Diagnostic Checklist

- Distension of periarticular fat pads at elbow suggests intracapsular extension of fracture complex
- Involvement of articular surface must be noted and characterized
- Try to discern extracapsular vs. intracapsular injuries with respect to elbow

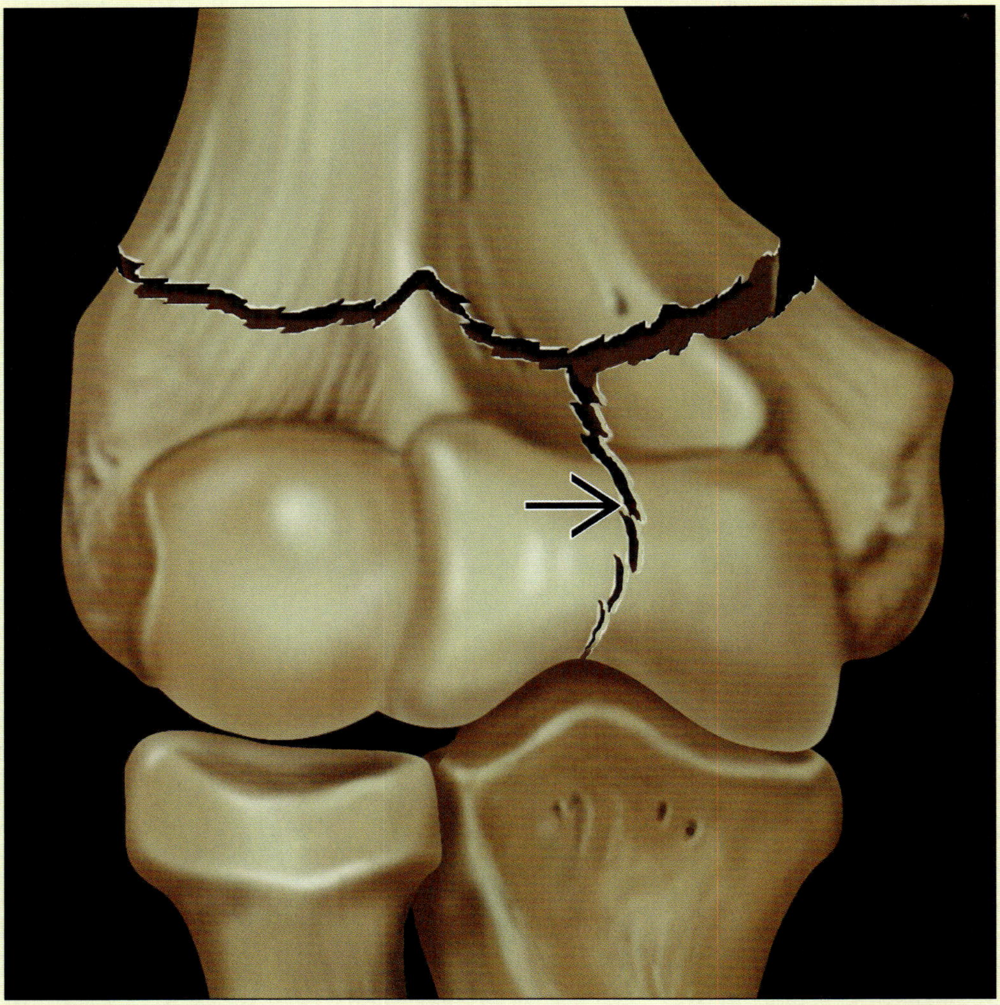

Graphic representation of T-shaped distal humeral fracture. This fracture has an intercondylar component ⇒ and separates both medial and lateral condyles from the humeral shaft. The AO classification system is used to characterize fractures involving the distal humerus. Under this system, type A fractures are extraarticular, type B fractures involve either the medial or lateral condyle (but not both), and type C fractures involve both condyles. In each type there are subtypes depending on morphology and complexity of the fracture complex. This is a C1.3 fracture by the Arbeitsgemeinschaft fur Osteosynthesefragen (AO) classification system.

DISTAL HUMERAL FRACTURE

TERMINOLOGY

Synonyms

- May encompass a variety of fracture subtypes
  - Supracondylar, transcondylar, and intercondylar fractures

Definitions

- Any fracture involving distal humeral shaft &/or 1 or both humeral condyles

IMAGING

General Features

- Best diagnostic clue
  - Positive "fat pad" sign (distension of periarticular fat pads) at elbow
    - Due to hemarthrosis
  - Lucent fracture line seen in distal humerus on radiographs
- Location
  - Depending on subtype, may involve distal humeral shaft, or 1 or both humeral condyles
  - Supracondylar fractures: Extracapsular
  - Condylar, transcondylar, and intercondylar fractures: Intraarticular
- Size
  - May range from small transverse fractures of distal humeral shaft to complex comminuted injuries involving shaft and both condyles
- Morphology
  - May involve distal shaft only, or 1 or both condyles
    - May be T- or Y-shaped
    - Articular surface of medial or lateral humeral condyles often involved
    - Displacement of capitellar or trochlear articular surfaces must be characterized

Radiographic Findings

- Radiography
  - Linear lucency or lucencies extending into or through 1 or both humeral condyles
  - Angular deformity of distal humerus
  - May see disruption of articular surfaces of medial &/ or lateral condyles

CT Findings

- Better defines complex fracture for surgical planning
  - Fracture line(s) extending through distal humeral shaft &/or humeral condyles
  - Involvement of condylar articular surfaces
  - Intraarticular fragments
- Elbow hemarthrosis if fracture is acute

MR Findings

- Marrow edema and fracture line(s) in distal humerus
- May demonstrate articular cartilage disruption
- Osseous or cartilaginous bodies in joint
- Hemarthrosis in elbow joint
- Elbow ligaments usually intact
- Entrapment or enhancement of damaged radial or median nerve

Imaging Recommendations

- Best imaging tool
  - CT: Best defines extent of fracture complex and involvement of humeral articular surface
- Protocol advice
  - Multidetector CT
  - Thin section axial images
  - Coronal and sagittal reformatted images
    - True axial of joint serves as reference for reformats
    - Coronal reformatted images based on line drawn through points of both humeral condyles on true axial image
    - Sagittal reformatted images perpendicular to coronals as above

DIFFERENTIAL DIAGNOSIS

Elbow Dislocation

- Humerus intact with displacement of radius and ulna
- Most dislocations involve posterior and lateral displacement of forearm bones relative to humerus

"Supracondylar" Fracture of Childhood

- Generally seen in pediatric population
- Generally transcondylar (intraarticular) rather than true supracondylar fracture
- Does not involve articular surface
- May be subtle on radiography

Floating Elbow

- Fractures of distal humerus and proximal forearm isolate elbow joint

PATHOLOGY

General Features

- Etiology
  - Intercondylar fractures usually due to fall on outstretched arm (FOOSH)
  - May be due to direct blow
  - May involve varus or valgus stress
  - May involve hyperflexion or hyperextension
- Associated abnormalities
  - Radial nerve injury can be seen in displaced spiral fracture of distal humeral shaft
    - Wrist drop deformity
  - Median nerve injury may be seen due to traction injury from displaced supracondylar fragment
    - "Hand of benediction" deformity
- Epidemiology
  - Typically seen in elderly
  - Distal humeral fractures account for 0.5% of all adult fractures

Staging, Grading, & Classification

- AO classification
  - A1: Extraarticular fracture, apophyseal avulsion
    - A1.1 lateral epicondyle
    - A1.2 medial epicondyle, nonincarcerated
    - A1.3 medial epicondyle, incarcerated
  - A2: Extraarticular fracture, metaphyseal simple
    - A2.1 oblique downwards and inwards
    - A2.2 oblique downwards and outwards
    - A2.3 transverse
  - A3: Extraarticular fracture, metaphyseal multifragmentary

DISTAL HUMERAL FRACTURE

- A3: Extraarticular fracture, metaphyseal multifragmentary
  - A3.1 with intact wedge
  - A3.2 with fragmented wedge
  - A3.3 complex
- B1: Partial articular fracture, lateral sagittal
  - B1.1 capitellum
  - B1.2 transtrochlear simple
  - B1.3 transtrochlear multifragmentary
- B2: Partial articular fracture, medial sagittal
  - B2.1 transtrochlear simple, through medial side (Milch I)
  - B2.2 transtrochlear simple, through groove
  - B2.3 transtrochlear multifragmentary
- B3: Partial articular fracture, frontal
  - B3.1 capitellum
  - B3.2 trochlea
  - B3.3 capitellum and trochlea
- C1: Complete articular fracture, articular simple, metaphyseal simple
  - C1.1 with slight displacement
  - C1.2 with marked displacement
  - C1.3 T-shaped epiphyseal
- C2: Complete articular fracture, articular simple, metaphyseal multifragmentary
  - C2.1 with intact wedge
  - C2.2 with fragmented wedge
  - C2.3 complex
- C3: Complete articular fracture, multifragmentary
  - C3.1 metaphyseal simple
  - C3.2 metaphyseal wedge
  - C3.3 metaphyseal complex

CLINICAL ISSUES

Presentation

- Most common signs/symptoms
  - Localized pain and swelling
  - Deformity in distal arm or elbow
- Other signs/symptoms
  - Rarely associated with radial nerve injury if spiral fracture of distal shaft with displacement

Demographics

- Age
  - T- & Y-shaped fractures more common in elderly
  - Supracondylar and transcondylar fractures more common in pediatric patients
- Gender
  - Intercondylar fractures more common in women, possibly due to decreased bone mass

Natural History & Prognosis

- Nondisplaced extracapsular fractures heal well with conservative management
- Angulated or displaced fractures treated surgically
- Inadequate reduction of articular surfaces → elbow arthritis
- If elbow range of motion not restored early → contractures

Treatment

- Nondisplaced extraarticular fractures usually treated with casting and immobilization
- Unstable extraarticular fractures treated surgically
- Disruption of articular surface of distal humerus requires meticulous repositioning to avoid post-traumatic arthritis

DIAGNOSTIC CHECKLIST

Consider

- Presence of other injuries at shoulder, forearm, elbow, and wrist
- Possibility of floating elbow (must exclude proximal forearm fractures)

Image Interpretation Pearls

- Distension of periarticular fat pads at elbow suggests intracapsular extension of fracture complex
- Subtle intraarticular extension may be missed on radiographs; consider CT
  - CT may upgrade classification & alter decisions regarding surgery

Reporting Tips

- Involvement of articular surface must be noted and characterized
- Try to discern extracapsular vs. intracapsular injuries with respect to elbow

SELECTED REFERENCES

1. Ashwood N et al: Transarticular shear fractures of the distal humerus. J Shoulder Elbow Surg. 19(1):46-52, 2010
2. Galano GJ et al: Current treatment strategies for bicolumnar distal humerus fractures. J Am Acad Orthop Surg. 18(1):20-30, 2010
3. http://www.ao-asif.ch/wps/portal. Accessed March 23, 2009
4. Ruan HJ et al: Incidence, management, and prognosis of early ulnar nerve dysfunction in type C fractures of distal humerus. J Trauma. 67(6):1397-401, 2009
5. Davies MB et al: A clinically applicable fracture classification for distal humeral fractures. J Shoulder Elbow Surg. 15(5):602-8, 2006
6. Doornberg J et al: Two and three-dimensional computed tomography for the classification and management of distal humeral fractures. Evaluation of reliability and diagnostic accuracy. J Bone Joint Surg Am. 88(8):1795-801, 2006
7. McCarty LP et al: Management of distal humerus fractures. Am J Orthop. 34(9):430-8, 2005
8. O'Driscoll SW et al: Management of the smashed distal humerus. Orthop Clin North Am. 33(1):19-33, vii, 2002
9. Ring D et al: Fractures of the distal humerus. Orthop Clin North Am. 31(1):103-13, 2000
10. Hastings H 2nd et al: Fixation of complex elbow fractures, Part I. General overview and distal humerus fractures. Hand Clin. 13(4):703-19, 1997
11. McKee MD et al: A contemporary approach to the management of complex fractures of the distal humerus and their sequelae. Hand Clin. 10(3):479-94, 1994
12. Jupiter JB et al: Fractures of the distal humerus. Orthopedics. 15(7):825-33, 1992
13. Wilkins KE: The operative management of supracondylar fractures. Orthop Clin North Am. 21(2):269-89, 1990

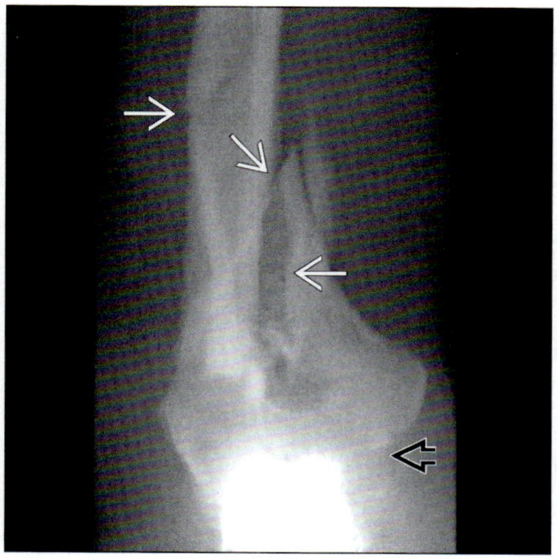

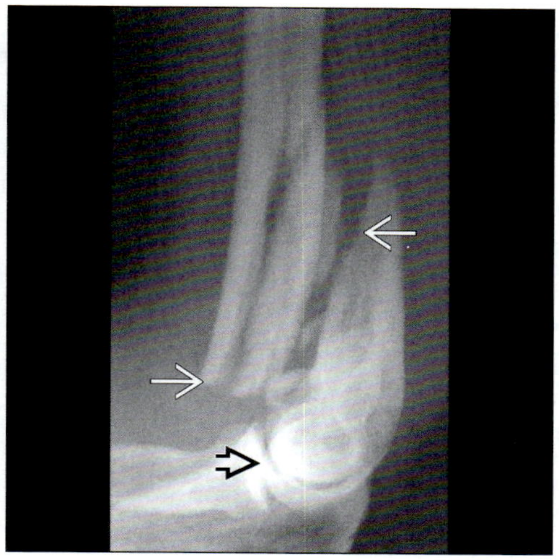

(Left) *Supracondylar fracture of the humerus in an adult. AP radiograph of the elbow demonstrates a comminuted supracondylar fracture of the humerus →. The articular surfaces of the elbow joint are intact ⇨. Sparing of the articular surfaces and its complexity makes this an A3.3 lesion by the AO classification system.* ***(Right)*** *Lateral radiograph of same patient demonstrates a comminuted supracondylar fracture of the humerus →. The articular surfaces of the elbow joint are intact ⇨.*

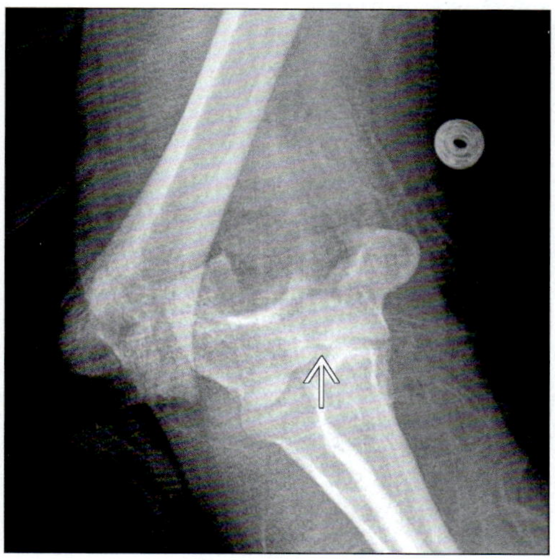

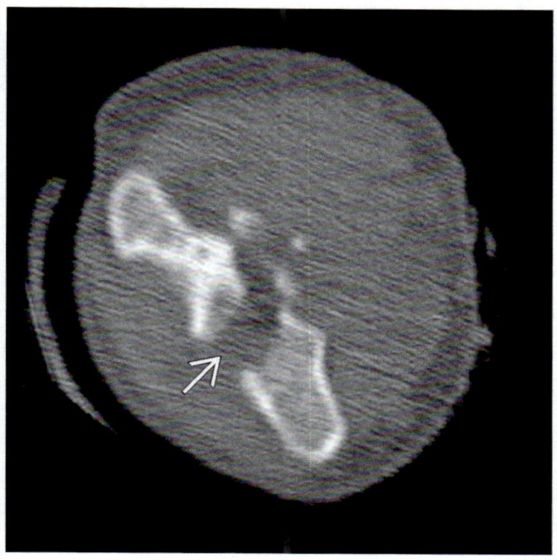

(Left) *AP radiograph of elbow in a 41-year-old man shows marked displacement of humeral condyles relative to humerus. Extension to articular surface → is difficult to see. If not appreciated, this would receive AO classification of A2.3 and in some cases could be treated with casting.* ***(Right)*** *Axial bone CT from same patient better shows intercondylar portion of the fracture →. Metaphyseal comminution is also better seen. This changes AO classification to C2.1, which requires open reduction.*

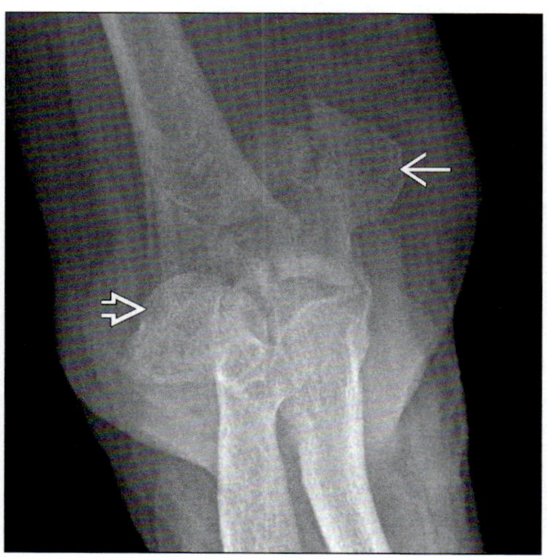

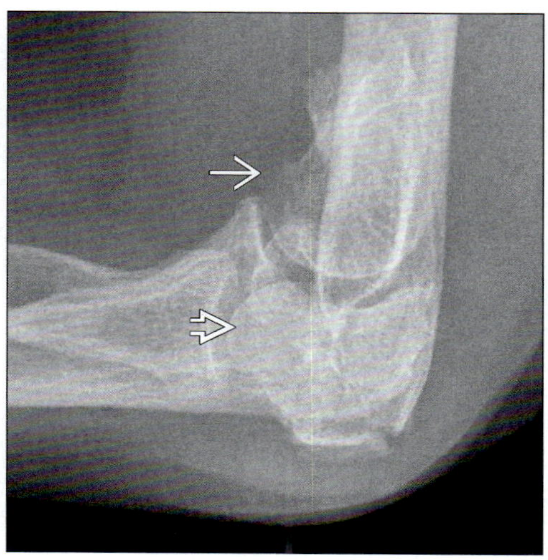

(Left) *Oblique radiograph of elbow in a 58-year-old woman shows a badly comminuted intercondylar fx of distal humerus. Note fragmentation and displacement of trochlea → and rotation and displacement of capitellum ⇨.* ***(Right)*** *Lateral radiograph in same patient shows marked superior displacement of the trochlea → and 90° rotation of capitellum ⇨. This is an AO C3.3 lesion. Fragment rotation and displacement are often better judged on radiographs than on CT.*

TRANSCONDYLAR FRACTURE, ELBOW

Key Facts

Terminology

- "Supracondylar" fracture of childhood
- Fracture of distal humerus extending across both medial and lateral condyles
- Does not involve articular surface of humerus
- Intracapsular in location

Imaging

- Radiographs usually sufficient
- Usually single large distal humeral fragment
  - Contains both humeral condyles
- Does not enter humeral articular surface
- In children, disruption of anterior humeral line
- True lateral view to assess anterior humeral line sign
- Fx line not visible in up to 25% of pediatric cases

Top Differential Diagnoses

- True supracondylar fracture
- Intercondylar fracture
- Medial condylar fracture
- Lateral condylar fracture
- Elbow dislocation

Pathology

- Extension type: More common in children & elderly
- Flexion type: Significant displacement

Clinical Issues

- Accounts for 60% of elbow fractures in children
- Due to fall on outstretched hand (FOOSH)
- Hyperextension mechanism
- Closed reduction for nondisplaced fractures
- Temporary nerve impairment in 10-16%
- Common in age < 10 years

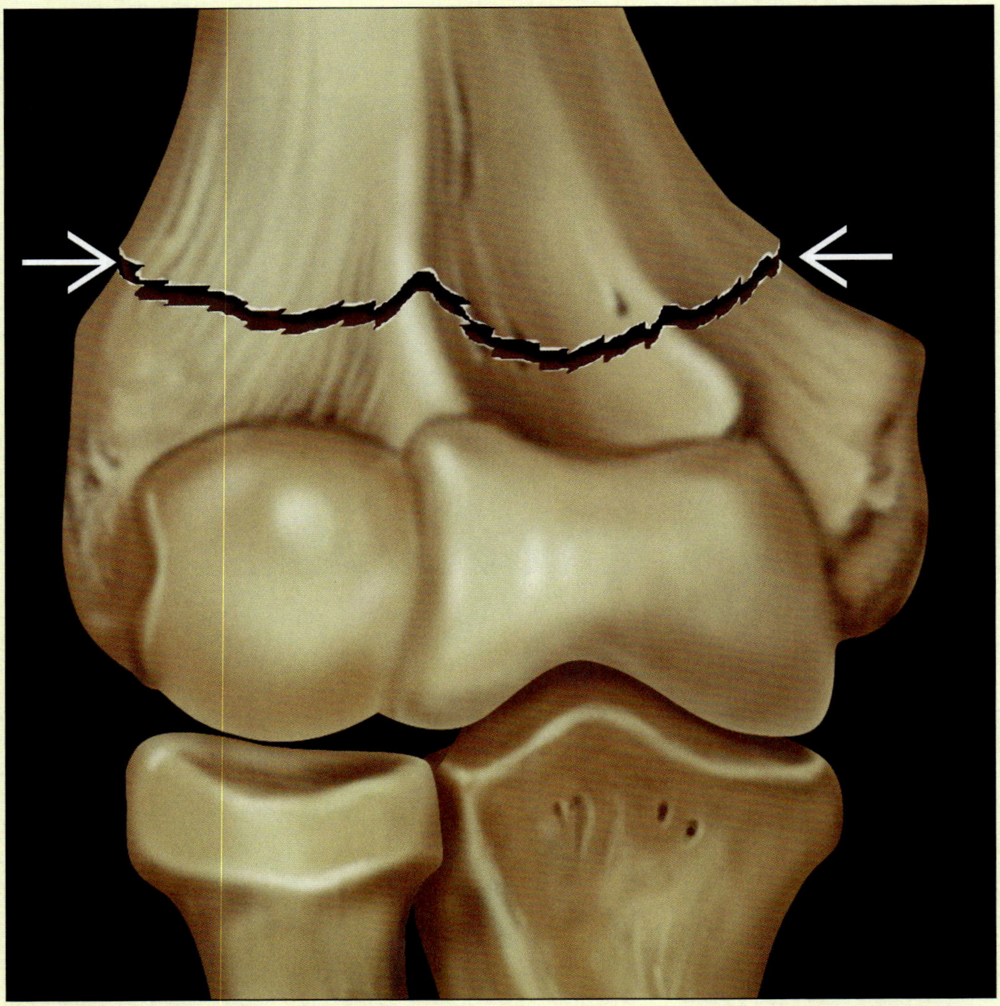

Graphic representation of a transcondylar fracture of the humerus. The fracture ➡ propagates in a horizontal plane across the proximal aspects of the humeral condyles. There is no extension to the articular surface. The transcondylar fracture is the most common elbow fracture in children and is often referred to as "supracondylar." The intraarticular location of the fracture leads to an associated elbow hemarthrosis in the setting of acute trauma, a finding that can serve as an important diagnostic clue on radiographs of a minimally displaced fracture obtained at the time of injury. A true supracondylar fracture would be extraarticular and thus would not cause an effusion.

TRANSCONDYLAR FRACTURE, ELBOW

TERMINOLOGY

Synonyms

- "Supracondylar" fracture of childhood
  - Most really transcondylar
- Gartland fracture

Definitions

- Fracture of distal humerus extending across both medial and lateral condyles
- Orientation: Horizontal
- Does not involve articular surface of humerus
- Extends into joint capsule
- Mechanism: Extension force on outstretched arm

IMAGING

General Features

- Best diagnostic clue
  - Transverse fracture line across both humeral condyles
  - Acute setting: Positive "fat pad" sign (elbow effusion)
  - Pediatric: Disruption of anterior humeral line
- Location
  - Distal humerus
    - Intracapsular portion (within elbow joint)
    - Extends across thin bone between condyles
    - Spares articular surfaces
- Size
  - Single large distal humeral fragment
    - Contains both humeral condyles
- Morphology
  - Horizontal fracture line extending across medial and lateral humeral condyles
  - Extends into elbow joint capsule
  - Distal fragment usually displaced posteriorly and medially, if at all
  - Adults: Complete fracture
  - Pediatrics: May be incomplete
    - Rare separation of distal humeral epiphysis in children

Radiographic Findings

- Horizontal fracture line traversing medial and lateral humeral condyles
  - Fx line not visible in up to 25% of pediatric cases
  - Fx line sometimes only visible on oblique view
- Does not enter humeral articular surface
- May be displaced
  - Usually posteromedially
- Positive "fat pad" sign
- Disruption of anterior humeral line in children
  - Line drawn on lateral view along anterior cortex of distal humeral shaft
  - Line should intersect middle 1/3 of capitellum
  - If line extends into anterior 1/3 of capitellum or more anteriorly → fracture

CT Findings

- Fracture line located within elbow joint capsule
- Fracture may be difficult to see on axial images
- Coronal and sagittal reformats show transverse fracture propagation
- Significant anterior displacement of condyles in flexion type
- No extension into articular surface of humerus
- Joint effusion

MR Findings

- Fracture line traverses medial and lateral condyles
- No entry into articular surface of distal humerus
- Involvement of distal humeral physis rare
  - High signal on T2WI FS in growth plate
  - Widening of physis
- Surrounding marrow edema and soft tissue swelling
- Intracapsular location of fracture
- Significant anterior displacement of condyles in flexion type
- Elbow joint effusion

Imaging Recommendations

- Best imaging tool
  - Radiographs
- Protocol advice
  - Radiograph
    - AP, lateral, and oblique views
    - True lateral view necessary to assess anterior humeral line sign
  - CT
    - Thin section true axial acquisition
    - Coronal and sagittal reformats necessary
    - Sagittal reformats useful to show degree of posterior displacement
  - MR
    - Coronal and sagittal T1WI and T2WI FS best demonstrate fracture and effusion
    - Angle images relative to humerus, not forearm
    - Coronal plane defined by line drawn through widest parts of condyles on true axial

DIFFERENTIAL DIAGNOSIS

True Supracondylar Fracture

- Extracapsular
- Displaced
- Older adults

Intercondylar Fracture

- Fracture line extends between humeral condyles
- Fracture enters articular surface

Medial Condylar Fracture

- Fracture does not involve lateral condyle
- Fracture enters articular surface

Lateral Condylar Fracture

- Fracture does not involve medial condyle
- Fracture enters articular surface

Elbow Dislocation

- Ulna displaced relative to distal humerus
- Radiocapitellar joint usually disrupted
- May have associated distal humeral fracture

TRANSCONDYLAR FRACTURE, ELBOW

PATHOLOGY

General Features

- Associated abnormalities
  - Other elbow fractures
    - Olecranon
    - Medial condyle
  - Distal radial fracture (5-6% incidence)
  - Traction injury to neurovascular structures
    - Brachial artery (0.5% incidence)
    - Anterior interosseous branch of median nerve (4% incidence)

Staging, Grading, & Classification

- Extension type
  - More common in children and elderly
  - Pediatric
    - Posterior and medial displacement
  - Elderly
    - Usually nondisplaced
- Flexion type
  - Significant displacement
  - Coronoid process trapped between displaced condyles and humeral shaft
  - May be comminuted
- Most are metaphyseal, not involving the articular surface (A2 distal humeral fractures by AO classification)

CLINICAL ISSUES

Presentation

- Most common signs/symptoms
  - Pain and swelling of elbow
  - Deformity if displaced
- Other signs/symptoms
  - Neurovascular complaints
    - Usually in displaced fractures
    - Temporary nerve impairment (10-16% incidence)

Demographics

- Age
  - Bimodal
    - Children
      - Common < 10 years old
      - Peak: 6 years old
    - Adults
      - Usually elderly
- Gender
  - M > F
- Epidemiology
  - 60% of elbow fractures in children
  - Due to fall on outstretched hand (FOOSH) + hyperextension

Natural History & Prognosis

- Good prognosis with appropriate fragment reduction and stabilization
- Complications
  - Failure of reduction
  - Cubitus varus if not appropriately reduced
  - Median nerve injury
    - May be traumatic or iatrogenic
    - Up to 5% of cases have chronic median neuropathy
  - Volkmann contracture
    - Ischemic flexion contracture of hand
    - Due to unrecognized vascular injury
    - Rare (about 0.5% incidence)

Treatment

- Closed reduction for nondisplaced fractures
- Open reduction and internal fixation for displaced fractures

DIAGNOSTIC CHECKLIST

Consider

- Floating elbow
  - Fractures of distal humerus and proximal forearm isolates elbow
- Pathologic fracture
  - Underlying destructive lesion causes irregular fracture edges and lucency
- Intercondylar fracture
  - Fracture line extends between condyles and into articular surface
- Isolated medial or lateral condylar fracture
  - Fracture only involves 1 condyle
  - Fracture line extends to articular surface

Image Interpretation Pearls

- Anterior humeral line disruption may be only finding in young children
- Presence of joint effusion indicates fracture extension into elbow joint capsule
- "Fat pad" sign may be absent if joint capsule disrupted
  - Usually seen in severely displaced fractures

Reporting Tips

- Describe any fragment displacement or angulation
- True lateral XR critical for evaluation of anterior humeral line
- Consider child abuse in nonambulatory children

SELECTED REFERENCES

1. Mallo G et al: Use of the Gartland classification system for treatment of pediatric supracondylar humerus fractures. Orthopedics. 33(1):19, 2010
2. Herman MJ et al: Relationship of the anterior humeral line to the capitellar ossific nucleus: variability with age. J Bone Joint Surg Am. 91(9):2188-93, 2009
3. Mangat KS et al: The 'pulseless pink' hand after supracondylar fracture of the humerus in children: the predictive value of nerve palsy. J Bone Joint Surg Br. 91(11):1521-5, 2009
4. Omid R et al: Supracondylar humeral fractures in children. J Bone Joint Surg Am. 90(5):1121-32, 2008
5. Shrader MW: Pediatric supracondylar fractures and pediatric physeal elbow fractures. Orthop Clin North Am. 39(2):163-71, v, 2008
6. Baratz M et al: Pediatric supracondylar humerus fractures. Hand Clin. 22(1):69-75, 2006
7. Hausman M et al: Treatment of distal humerus fractures in the elderly. Clin Orthop Relat Res. (425):55-63, 2004
8. O'Driscoll SW: Supracondylar fractures of the elbow: open reduction, internal fixation. Hand Clin. 20(4):465-74, 2004

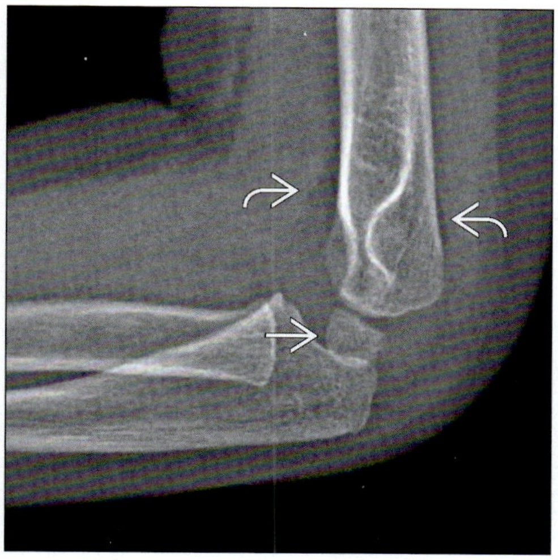

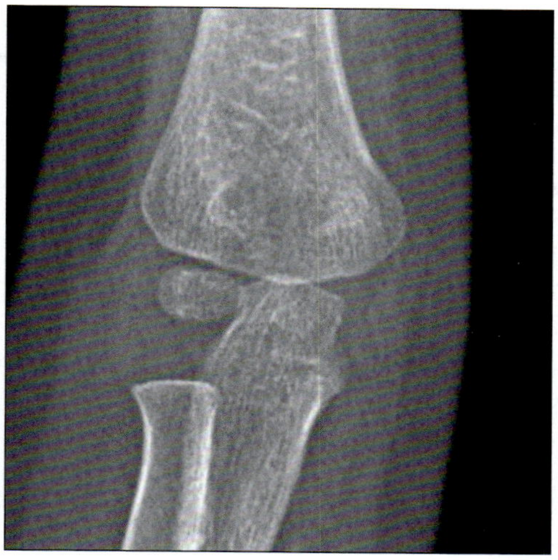

(Left) *Lateral radiograph shows a "supracondylar" humeral fracture in a 4-year-old boy after a fall on outstretched arm. Note the posterior displacement of capitellum ➡ relative to line along anterior cortex of the distal humerus. Positive "fat pad" sign ➡ confirms intraarticular nature of fracture. Thus, this is really an intercondylar fracture.* ***(Right)*** *AP radiograph from the same patient shows no fracture line, typical in this age group unless there is more significant displacement of the fracture.*

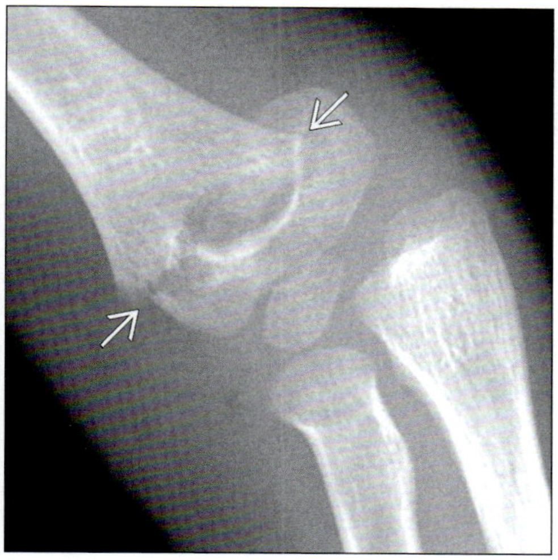

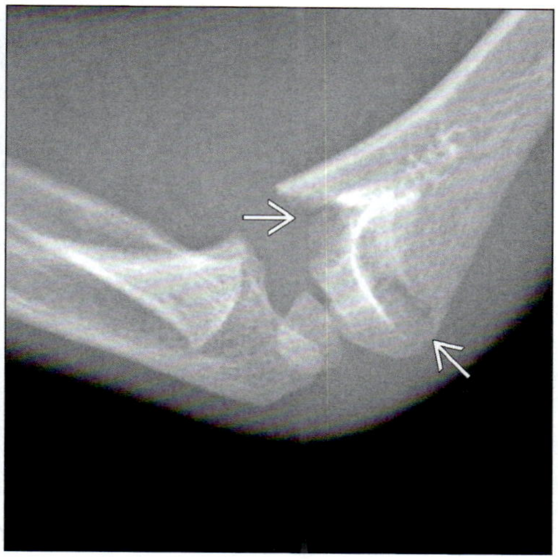

(Left) *AP radiograph demonstrates a displaced transverse fracture ➡ extending through the distal humeral metaphysis and across both condyles.* ***(Right)*** *Lateral radiograph from the same patient demonstrates a displaced transverse fracture ➡ extending through the distal humeral metaphysis and across both condyles. Elevation of the periarticular fat pads is not seen in this case, probably because of disruption of the joint capsule.*

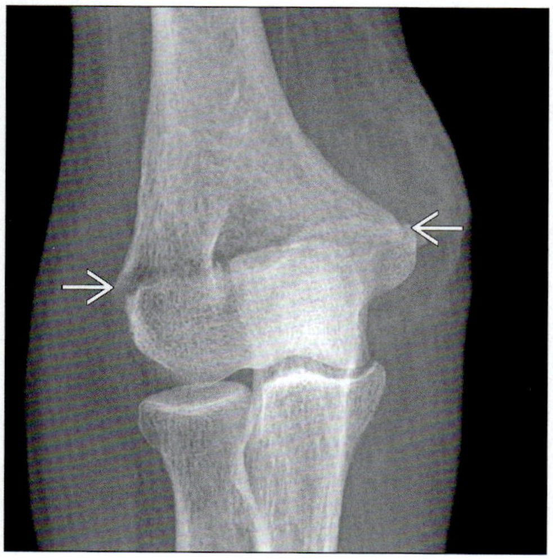

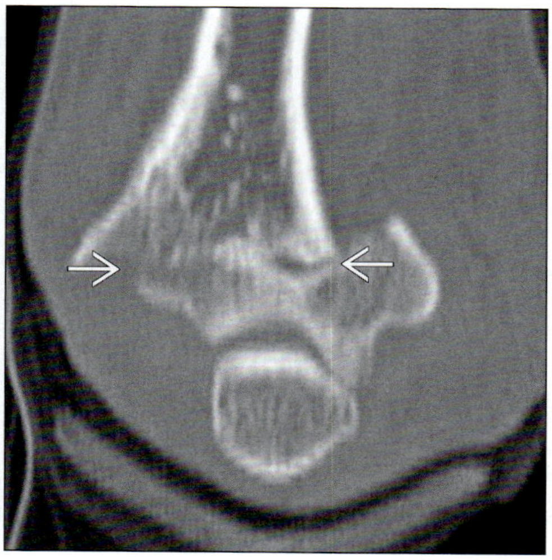

(Left) *Transcondylar humeral fracture in a 63-year-old woman after a fall on outstretched hand. The fracture line ➡ traverses both condyles but does not enter articular surface. There is subtle medial displacement of distal fragment, a common finding in these injuries.* ***(Right)*** *Coronal reformat NECT from the same patient confirms transcondylar path of fracture ➡ and emphasizes medial displacement of distal fragment. CT also confirms that the articular surface of the humerus is spared.*

LATERAL CONDYLAR FRACTURE, ELBOW

Key Facts

Terminology
- Any intraarticular fracture of lateral humeral condyle

Imaging
- Radiographs in setting of acute trauma
- Linear fracture line extending through lateral humeral condyle and into elbow joint
- CT: Useful in adult fractures
  - Extent of fracture
  - Relationship and position of various fragments
- MR: Extent of injury through unossified distal humeral epiphysis in children

Pathology
- Milch classification
  - Predicts elbow stability based on involvement of lateral trochlear ridge
  - Milch II more common; fracture line extends into joint at or medial to lateral trochlear ridge
- Jakob classification
  - Based on fragment displacement

Clinical Issues
- More common in children, peak age 6 years
- Can occur in skeletally mature patients
- Left untreated, residual pain and deformity likely, especially with Milch II injury
- 2nd most common pediatric elbow fracture

Diagnostic Checklist
- Involvement of articular cartilage in children
- Involvement of lateral trochlear ridge to differentiate stable from unstable injury
- True extent of involvement of articular cartilage only well seen on MR

*(**Left**) Graphic of a Milch type II lateral condylar fracture demonstrates extension of the fracture from the lateral condyle above the growth plate, extending through the unossified epiphysis, and into the articular surface medial to the lateral trochlear ridge. (**Right**) Coronal T2WI FS MR shows a lateral condylar fracture extending into the medial aspect of the epiphysis and disrupting the articular surface medial to the lateral trochlear ridge in this unstable Milch II injury.*

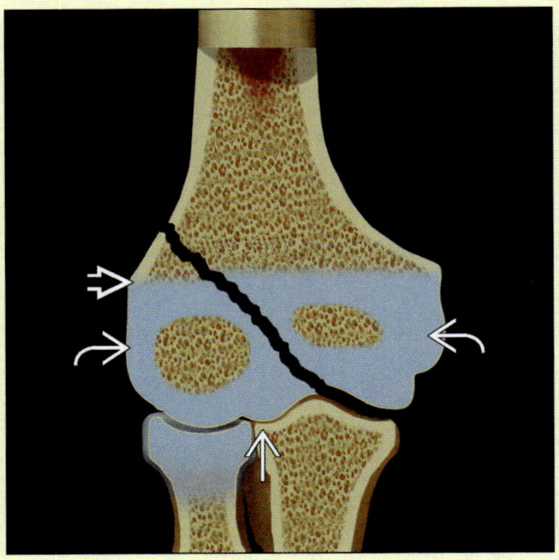

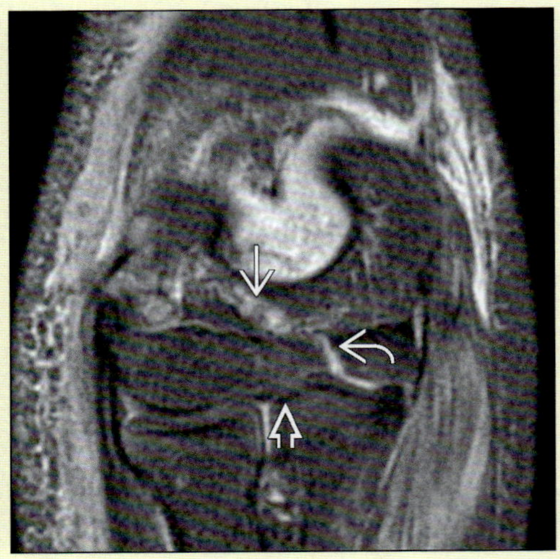

*(**Left**) Graphic of a Milch I lateral humeral condylar fracture shows the fracture entering the articular surface lateral to the lateral trochlear ridge. This injury is less likely to result in instability than a Milch II injury. (**Right**) Lateral radiograph shows a displaced lateral condylar fracture, which extends into the articular surface of the humerus lateral to the trochlear ridge. This is a Milch I fracture, although its displacement requires open reduction.*

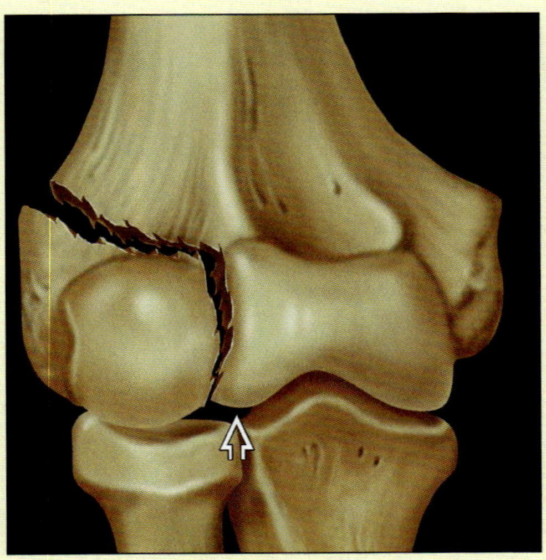

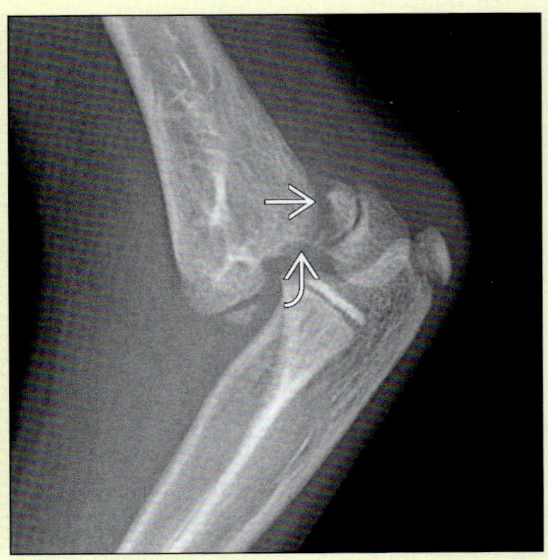

LATERAL CONDYLAR FRACTURE, ELBOW

TERMINOLOGY

Definitions

- Any fracture of lateral humeral condyle that extends intraarticularly
  - Excludes fractures only involving extraarticular portion of condyle
  - Excludes fractures involving both condyles
  - Much more common in children

IMAGING

General Features

- Best diagnostic clue
  - Linear fracture line extending through lateral humeral condyle and into elbow joint
- Location
  - Lateral condyle of distal humerus
- Size
  - May result in fragment ranging from small to large
- Morphology
  - Generally oblique in orientation
  - Extends into articular surface of distal humerus
    - Lateral to lateral trochlear ridge → stable injury
    - Medial to lateral trochlear ridge → unstable injury
    - Children → propagation through distal humeral epiphyseal cartilage (Salter-Harris type IV injury)

Radiographic Findings

- Radiography
  - Adults
    - Linear fx line extending through lateral humeral condyle
    - Best seen on AP view
    - Acute → positive "fat pad" sign
  - Children
    - Often only see small sliver of bone separated from lateral humeral condyle
    - Size of fragment on XR may grossly underestimate true extent of injury
      - May be Salter-Harris type II or IV
      - Cannot see fracture extent through unossified epiphysis
    - Acute → positive "fat pad" sign

CT Findings

- More useful in adult vs. pediatric patients
- Oblique linear lucency in lateral condyle extending to articular surface
- Entry of fracture into joint
  - Lateral to lateral trochlear ridge
  - Medial to lateral trochlear ridge

MR Findings

- Children: MR needed to see fracture extent in unossified distal humeral epiphysis
  - Osseous extent on radiographs may grossly underestimate true extent of injury
  - Defines involvement of growth plate and resultant Salter classification

Imaging Recommendations

- Best imaging tool
  - Radiographs in setting of acute trauma
    - 3 views
  - CT: Extent of fracture and relationship and position of various fragments in adults
    - Useful for characterizing the following
      - Precise fracture planes
      - Offset of articular surface
      - Displacement and rotation of fragments
  - MR: Extent of injury through unossified distal humeral epiphysis in children
    - Extension of fracture into or through distal articular cartilage
    - Involvement of growth plate (Salter II or IV)
    - Location of fracture relative to lateral trochlear ridge
- Protocol advice
  - Start with standard elbow radiographs
    - AP and lateral views
    - Oblique view helpful for nondisplaced fractures
  - CT
    - Most useful in adults
    - Thin sections acquired in true axial plane
    - Reformatted in true sagittal and true coronal planes
  - MR
    - Most useful in children
    - Coronal and sagittal T2WI FS to show
      - Fx involvement in unossified distal humeral epiphysis
      - Complex fracture planes

DIFFERENTIAL DIAGNOSIS

Intercondylar Fracture

- Fracture line usually extends between condyles and into humeral shaft in "Y" or "T" configuration
- More common in adults

Transcondylar Fracture, Elbow

- Transverse fracture through upper portions of both condyles
- Does not enter articular surface
- More common in children

Supracondylar Fracture

- Transverse fracture of distal humeral shaft sparing condyles
- Extraarticular
- Unusual injury

Elbow Dislocation

- Ulnohumeral and radiocapitellar alignment disrupted
- May be associated with other fractures

Capitellum Fracture

- Usually in adults
- Separation and rotation of articular surface of lateral humeral condyle

Lateral Epicondylar Avulsion

- Extraarticular injury due to varus stress
- Fragment may be quite small

PATHOLOGY

General Features

- Etiology

LATERAL CONDYLAR FRACTURE, ELBOW

- Usually due to varus force on elbow
- Can be due to fall on outstretched hand (FOOSH)
  - Drives proximal radius into lateral condyle

Staging, Grading, & Classification

- Milch classification
  - Predicts elbow stability based on involvement of lateral trochlear ridge
  - Milch I
    - Less common type
    - Fracture extends into elbow joint lateral to lateral trochlear ridge
    - Mediolateral elbow stability preserved
    - Salter-Harris type IV in child
  - Milch II
    - More common type
    - Fracture extends into elbow joint at or medial to lateral trochlear ridge
    - Mediolateral elbow instability
    - Salter-Harris type II (unusual) or IV in child
- Jakob classification
  - Based on fragment displacement
  - Type I
    - Not significantly displaced (< 2 mm)
    - Articular surface maintains normal alignment
  - Type II
    - 2-4 mm hinged displacement of metaphyseal fragment
    - Articular surface maintains normal alignment
  - Type III
    - Completely displaced
    - Fragment rotated
    - Unstable elbow joint

CLINICAL ISSUES

Demographics

- Age
  - More common in children
  - Range: 4-10 years old
  - Peak: 6 years old
  - Can occur in skeletally mature patients
- Epidemiology
  - 2nd most common pediatric elbow fracture
  - 17% of elbow fractures in children
  - < 5% of adult elbow fractures

Natural History & Prognosis

- Untreated → residual pain and deformity likely, especially with Milch II injury
- Complications
  - Osteoarthritis
  - Cubitus varus
  - Delayed union
  - Nonunion
  - Overgrowth of lateral condyle
    - Due to hyperemia
    - 40% of lateral condylar fractures
  - Osteonecrosis of lateral condyle
    - Very rare
  - Cubitus valgus
    - Due to lateral physeal growth arrest

Treatment

- Depends on displacement of fracture
  - Jakob I
    - Long arm cast
    - 90° elbow flexion
    - If delayed union or malalignment → surgery
  - Jakob II
    - Percutaneous pinning if possible
    - Open fixation
  - Jakob III
    - Open reduction and fixation

DIAGNOSTIC CHECKLIST

Consider

- Involvement of nonossified physis and epiphysis in children
  - Salter-Harris II or IV
- Involvement of lateral trochlear ridge to differentiate stable from unstable injury
  - Fracture lateral to ridge → more stable injury
  - Fracture medial to ridge → unstable injury

Image Interpretation Pearls

- In young children, radiographs may be misleading
  - True extent in injury may involve articular cartilage only well seen on MR
- Milch classification depends on fracture location
  - Lateral to lateral trochlear ridge → stable
  - Medial to lateral trochlear ridge → unstable elbow

Reporting Tips

- Must comment on presence or absence of involvement of lateral trochlear ridge
  - Usually requires MR in children
- Salter-Harris fracture type (II vs. IV) in children
- Degree of displacement and rotation of distal fragment

SELECTED REFERENCES

1. Orthopaedia Collaborative Knowledgebase. "Pediatric lateral humeral condyle fracture." http://www.orthopaedia.com/display/Main/Pediatric+lateral+humeral+condyle+fracture: Accessed March 15, 2009
2. Shrader MW: Pediatric supracondylar fractures and pediatric physeal elbow fractures. Orthop Clin North Am. 39(2):163-71, v, 2008
3. Sullivan JA: Fractures of the lateral condyle of the humerus. J Am Acad Orthop Surg. 14(1):58-62, 2006
4. Murnaghan JM et al: Fractured lateral epicondyle with associated elbow dislocation. Int J Clin Pract. 56(6):475-7, 2002
5. Jakob R et al: Observations concerning fractures of the lateral humeral condyle in children. J Bone Joint Surg Br. 57(4):430-6, 1975

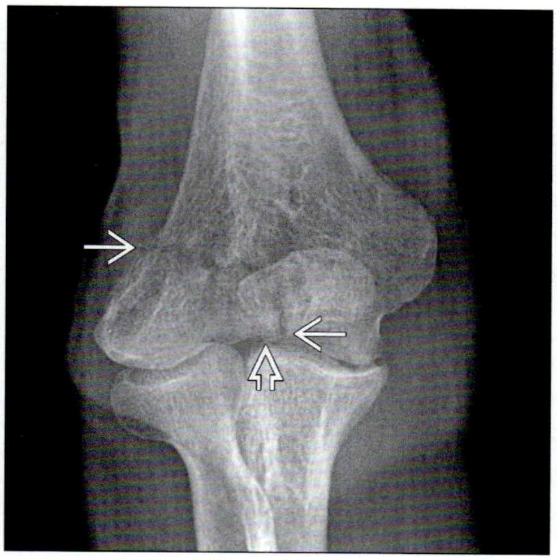

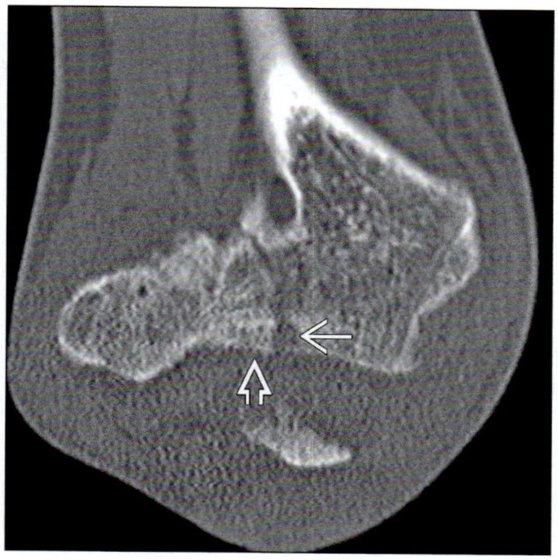

(Left) *Anteroposterior radiograph of Milch II lateral condylar fx in an adult shows a fracture line* → *entering the elbow joint medial to the lateral trochlear ridge* ⇨*. Milch II fractures are more common than type I.* ***(Right)*** *Coronal reformat bone CT in the same patient shows entry point of fx* → *to better advantage. The fracture location lateral to the lateral trochlear ridge* ⇨ *can be better delineated, thus confirming it as a Milch type II lesion and therefore prone to elbow instability.*

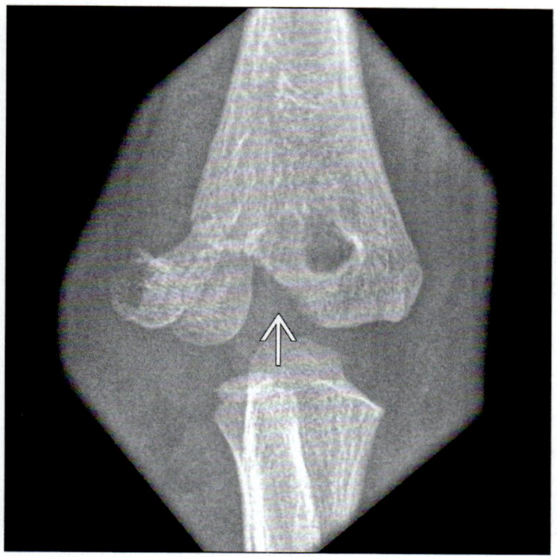

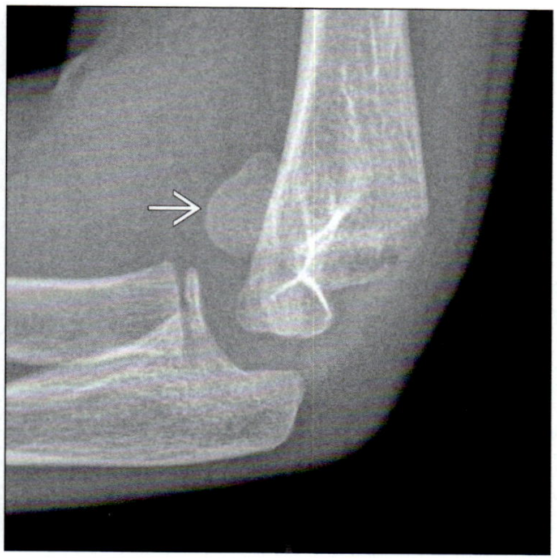

(Left) *AP radiograph shows a displaced & rotated lateral humeral condylar fx in an 8 year old extending medial to lateral trochlear ridge* →*, making it a Milch II injury. Radiocapitellar alignment is disrupted, but Salter-Harris designation cannot be made without MR.* ***(Right)*** *Lateral radiograph in the same patient shows anterior rotation of lateral condyle* →*. Fractures with this degree of displacement require open reduction and pinning. Absence of positive fat pads is probably due to joint capsule disruption.*

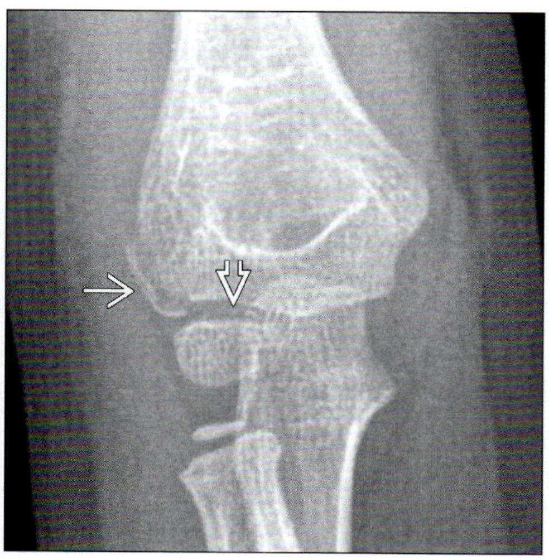

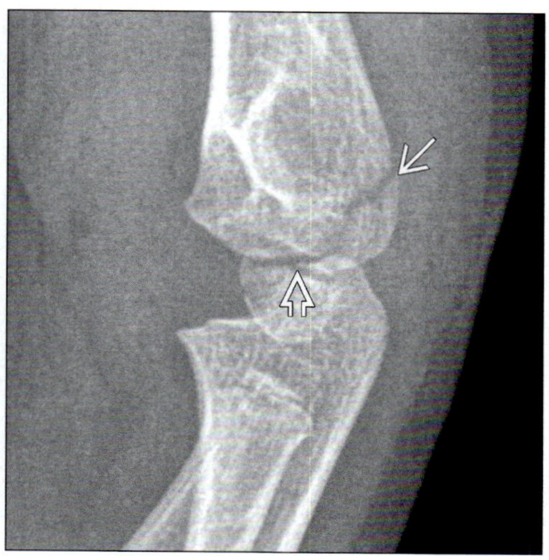

(Left) *Anteroposterior radiograph of a lateral condylar fracture in 2-year-old child shows separated osseous fragment* →*. Subtle fracture through growth plate* ⇨ *proves this is a Salter-Harris injury.* ***(Right)*** *Lateral radiograph of the same patient shows slight separation of the metaphyseal portion of the lateral condylar fracture* →*. The growth plate is disrupted* ⇨*.*

MEDIAL CONDYLAR FRACTURE, ELBOW

Key Facts

Terminology
- Intraarticular fracture through medial humeral condyle entering distal articular surface of humerus
  - Excludes fractures that only involve extraarticular portion of condyle
  - Excludes fractures involving both condyles

Imaging
- Linear fracture line extending through medial humeral condyle and into elbow joint
- Acute trauma: Radiographs
- CT: Extent of fracture and relationship and position of various fragments in adults
- MR: Extent of injury through unossified distal humeral epiphysis in children

Pathology
- Milch classification
  - Predicts elbow stability based on involvement of lateral trochlear ridge
  - If fracture extends through or lateral to trochlear ridge elbow considered unstable
- Kilfoyle classification
  - Based on displacement of medial condylar fragments

Clinical Issues
- Peak age of incidence: 7-14 years
- < 1% of pediatric elbow fractures

Diagnostic Checklist
- Involvement of lateral trochlear ridge to differentiate stable from unstable injury
- True extent of injury may involve articular cartilage only well seen on MR

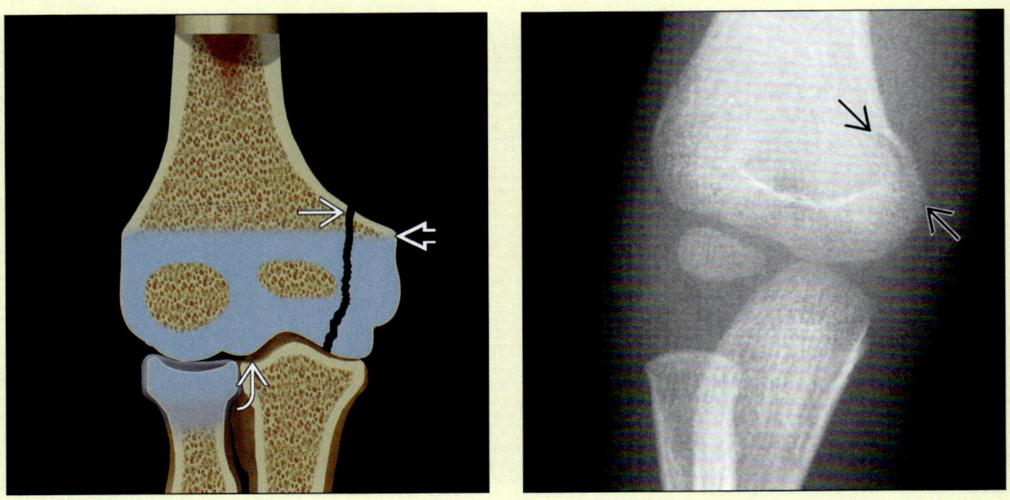

*(**Left**) Graphic illustration of a Milch type I medial humeral condylar fracture demonstrates extension of the fracture line through the ossified medial condyle ➔, across the physis ⇨, and into the articular surface. The fracture exits medial to the lateral trochlear ridge ⮫. (**Right**) AP radiograph demonstrates a subtle fracture line extending through the medial aspect of the medial humeral condyle →. This is a Milch I injury.*

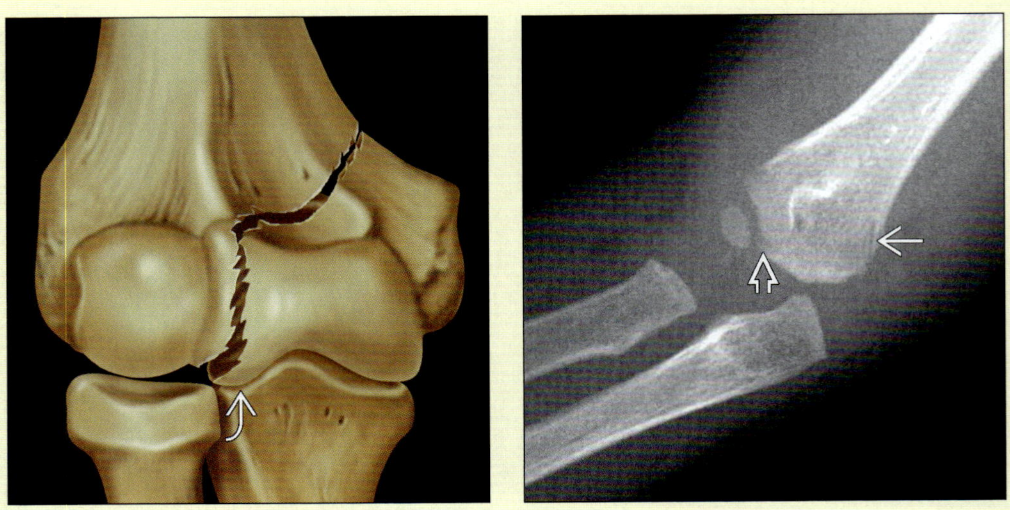

*(**Left**) Graphic of a Milch type II medial humeral condylar fracture shows the fracture entering the articular surface lateral to the lateral trochlear ridge ⮫. Type II fractures are more prone to instability. (**Right**) Anteroposterior radiograph of a medial condylar fracture in a 1 year old shows a fracture line ➔ extending into lateral articular surface of humerus ⇨. The exact site of extension into the joint surface cannot be discerned without MR in a young child. This is a Milch II injury.*

MEDIAL CONDYLAR FRACTURE, ELBOW

TERMINOLOGY

Definitions

- Intraarticular fracture through medial humeral condyle → distal articular surface or physis
  - Excludes fractures that only involve extraarticular portion of condyle
  - Excludes fractures involving both condyles

IMAGING

General Features

- Best diagnostic clue
  - Linear fracture line extending through medial humeral condyle → elbow joint
- Location
  - Medial condyle of distal humerus
- Size
  - May result in fragments ranging from small to large
  - Small fragments arising from extraarticular epicondyle are different injury
- Morphology
  - Generally oblique in orientation
  - Extends into articular surface of distal humerus
    - Medial to lateral trochlear ridge
    - Lateral to lateral trochlear ridge
    - Children: Extension through distal humeral epiphyseal cartilage (Salter-Harris type IV injury)
      - Some cases do not involve epiphysis (Salter-Harris II)

Radiographic Findings

- Radiography
  - Adults
    - Linear fracture line extending through medial humeral condyle
    - Best seen on AP view
    - Acute injury → positive "fat pad" sign
  - Pediatric
    - Often only see small sliver of bone separated from medial humeral condyle
    - Size of fragment on XR may be < < true extent of injury
    - Acute injury → positive "fat pad" sign

CT Findings

- More useful in adults than in children
- Oblique linear lucency extending from distal humeral articular surface → medial humeral condyle
- Involvement of lateral wall of trochlea important feature in determining stability of elbow
  - Medial to lateral trochlear ridge → stable
  - Lateral to lateral trochlear ridge → unstable

MR Findings

- Fracture line in lateral humeral condyle: ↓ signal on all MR sequences
- Surrounding marrow edema shows ↓ signal on T1WI and ↑ signal on T2WI
- MR crucial in defining extent of injury in children with unossified distal humeral epiphysis
  - Osseous extent on radiographs may grossly underestimate true extent of injury
  - Defines involvement of growth plate

Imaging Recommendations

- Best imaging tool
  - Acute trauma: Radiographs
    - 3 views
  - CT: Extent of fracture and relationship and position of various fragments in adults
  - MR: Extent of injury through unossified distal humeral epiphysis in children
- Protocol advice
  - Start with standard elbow radiographs
    - AP and lateral views mandatory
    - Oblique view helpful for nondisplaced fractures
  - CT
    - Most useful in adults
    - Useful for characterizing
      - Precise fracture planes
      - Location of fracture relative to lateral trochlear ridge
      - Offset of articular surface
      - Displacement of fragments
      - Rotation of fragments
    - Thin sections acquired in true axial plane
    - Reformatted in true sagittal and true coronal planes
  - MR
    - Most useful in children
    - Coronal and sagittal fat-suppressed T2-weighted images best demonstrate complex fracture planes
      - Involvement of lateral trochlear ridge
    - ↑ T2 signal in growth plate indicates Salter-Harris injury
    - Extension of fracture into or through distal articular cartilage

DIFFERENTIAL DIAGNOSIS

Intercondylar Fracture

- Fracture line usually extends between condyles and into humeral shaft in "Y" or "T" configuration
- More common in adults

Bicondylar Fracture

- Usually "Y" configuration, separates both condyles from humeral shaft
- More common in adults

Transcondylar Fracture, Elbow

- Transverse fracture of both humeral condyles sparing articular surface
- More common in young children

Elbow Dislocation

- Ulnohumeral and radiocapitellar alignment disrupted
- May be associated with other fractures

Valgus Stress Mechanism/Little Leaguer's Elbow

- Extraarticular injury due to valgus stress
- Fragment may be quite small

PATHOLOGY

General Features

- Etiology

MEDIAL CONDYLAR FRACTURE, ELBOW

  - Can be due to fall on outstretched hand (FOOSH) with elbow in valgus
  - Fall onto olecranon, driving olecranon into medial condyle
  - Avulsion of condyle due to violent contraction of flexor and pronator muscles (uncommon)

Staging, Grading, & Classification

- Milch classification
  - Predicts elbow stability based on involvement of lateral trochlear ridge
  - Milch I
    - Fracture extends into elbow joint medial to lateral trochlear ridge
    - Mediolateral elbow stability preserved
    - Salter-Harris type IV
    - More common type than Milch II
  - Milch II
    - Fracture extends into elbow joint at or lateral to lateral trochlear ridge
    - Mediolateral elbow instability
    - Salter-Harris type II or IV
    - Less common type than Milch I
- Kilfoyle classification
  - Based on displacement of medial condylar fragments
  - Type I
    - Nondisplaced fracture through medial condylar metaphysis down to physis
    - Salter-Harris type II injury
    - Does not involve distal humeral epiphysis or extend to elbow joint
  - Type II
    - Nondisplaced fracture through medial condylar metaphysis and epiphysis
    - Salter-Harris type IV injury
    - Extends into elbow joint
  - Type III
    - Displacement of fracture fragment
    - Fragment may also be rotated

CLINICAL ISSUES

Presentation

- Most common signs/symptoms
  - Pain and swelling around elbow, particularly in region of medial humeral condyle

Demographics

- Age
  - Peak incidence: 7-14 years
  - Can occur in adults
- Epidemiology
  - Very rare: < 1% of pediatric elbow fractures

Natural History & Prognosis

- Untreated → residual pain and deformity likely
  - Especially Milch II injury
- Complications
  - Osteoarthritis
  - Cubitus varus
  - Delayed union
  - Nonunion
  - Overgrowth of medial condyle

Treatment

- Depends on displacement of fracture
  - Nondisplaced fractures → conservative therapy
    - Closed casting
    - CT to assess fragment alignment while in cast
  - Displaced fractures (Kilfoyle III) must be reduced
    - Open reduction and internal fixation (ORIF)
    - Accurate fragment reduction is critical to achieving good outcome

DIAGNOSTIC CHECKLIST

Consider

- Involvement of articular cartilage in children
  - Salter IV
- Involvement of lateral trochlear ridge to differentiate stable from unstable injury
  - Fracture medial to ridge → usually more stable injury
  - Fracture lateral to ridge → usually unstable injury

Image Interpretation Pearls

- In young children, radiographs may be misleading
  - True extent of injury may involve articular cartilage only well seen on MR
- Milch classification depends on fracture location
  - Medial to lateral trochlear ridge → stable elbow
  - Lateral to lateral trochlear ridge → unstable elbow

Reporting Tips

- Must comment on presence or absence of involvement of lateral trochlear ridge
- Degree of displacement and rotation of distal fragment
- Salter-Harris fracture type (II vs. IV)

SELECTED REFERENCES

1. Leet AI et al: Medial condyle fractures of the humerus in children. J Pediatr Orthop. 22(1):2-7, 2002
2. Nagi ON et al: Fractures of the medial humeral condyle in adults. Singapore Med J. 41(7):347-51, 2000
3. Papavasiliou V et al: Fractures of the medial condyle of the humerus in childhood. J Pediatr Orthop. 7(4):421-3, 1987
4. Bensahel H et al: Fractures of the medial condyle of the humerus in children. J Pediatr Orthop. 6(4):430-3, 1986
5. Ghawabi MH: Fracture of the medial condyle of the humerus. J Bone Joint Surg Am. 57(5):677-80, 1975
6. Kilfoyle RM: Fractures of the medial condyle and epicondyle of the elbow in children. Clin Orthop Relat Res. 41:43-50, 1965

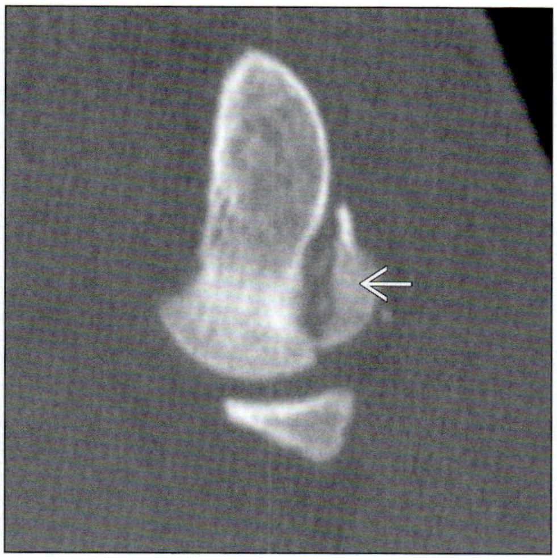

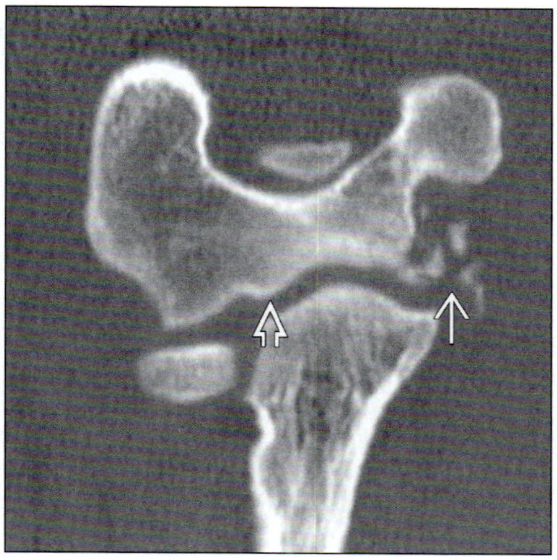

(Left) *Sagittal reformat bone CT through the medial condyle shows comminuted fracture of the medial condyle in an adult. Note the fracture fragments arising posteromedially from the medial humeral condyle.* ***(Right)*** *Coronal reformat bone CT in the same patient shows the fracture fragments arising posteromedially from the medial humeral condyle. The fracture enters the articular surface medial to the lateral trochlear ridge, and is thus a Milch I injury.*

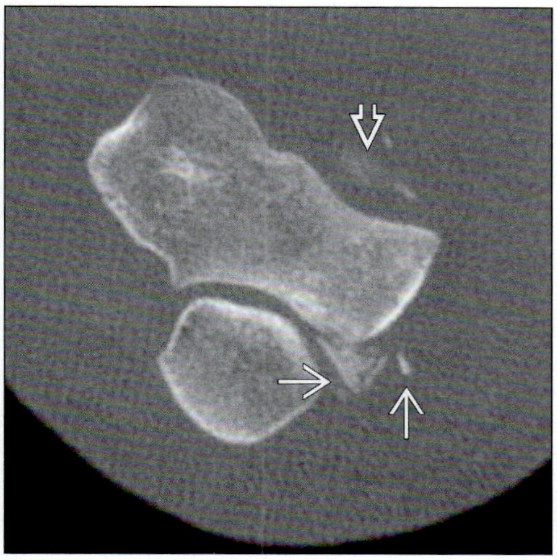

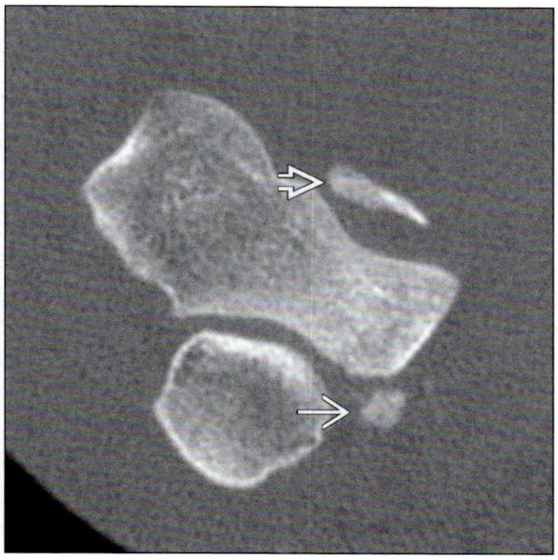

(Left) *Axial NECT in an adult shows comminuted fracture of the medial epicondyle. Note the fracture fragments arising posteromedially from the medial humeral condyle, as well as a coronoid process fracture.* ***(Right)*** *Contiguous axial NECT in the same patient shows the fracture fragments arising posteromedially from the medial humeral condyle, as well as a coronoid process fracture.*

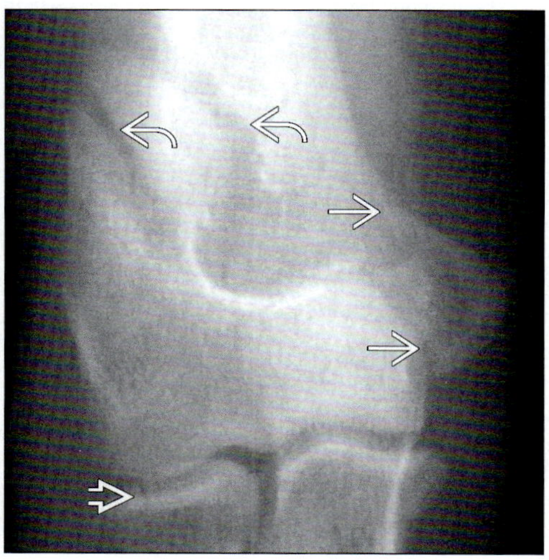

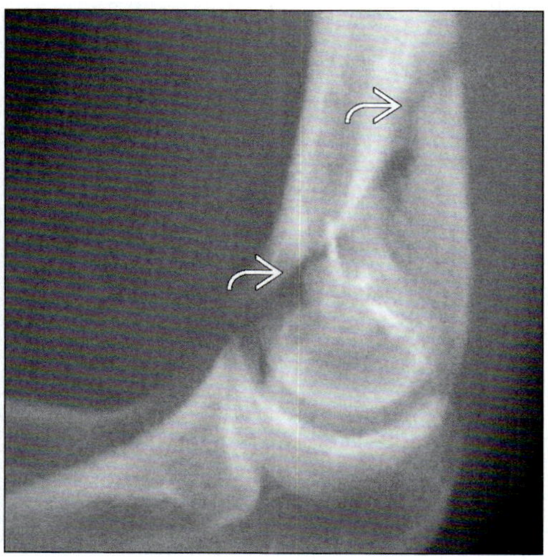

(Left) *AP radiograph of the elbow demonstrates a complex fracture of the distal humerus. A component of the fracture extends vertically through the medial aspect of the medial humeral condyle without significant displacement. A nondisplaced radial head fracture is also present.* ***(Right)*** *Lateral radiograph in the same patient shows the complex oblique nature of the fracture as it extends proximal to distal and posterior to anterior.*

CAPITELLUM FRACTURE

Key Facts

Terminology
- Fracture of articular surface of lateral condyle of humerus
- Capitellar osteochondritis
  - Old term for osteochondral fracture of capitellum
- Panner disease
  - Old term for osteochondral fractures of capitellum in children

Imaging
- Semicircular bone fragment
  - Displaced anterior to condyle
  - Rotated
- Lateral radiograph for type 1 injury
- AP view may look normal
- MR for chondral or osteochondral injury
  - Chondral injury may only be visible on T2WI
  - Marrow edema on T1WI and T2WI FS
  - Intraarticular bodies

Pathology
- Shearing forces across capitellum due to impact against radial head
- Trauma with elbow in flexion → anterior fracture
- Trauma with elbow in extension → posterior fracture

Clinical Issues
- Type 1 lesions account for < 1% of elbow fractures
- Osteochondral lesions
  - Valgus stress injury in throwers
  - Associated medial-sided soft tissue injury

Diagnostic Checklist
- Carefully search for intraarticular bodies
- Describe size and location of articular surface defect
- Describe any associated medial compartment injury

__(Left)__ Sagittal graphic representation of type 1 capitellar fracture. Note the coronal plane of the fracture ➡, with anterior displacement of the fragment. Rotation of the fragment is a common associated finding. __(Right)__ Sagittal reformat bone CT demonstrates a type I capitellar fracture ➡ with anterior displacement and rotation of the subtended fragment ⇨. A portion of the coronoid ↪ is also partially imaged; this was not fractured.

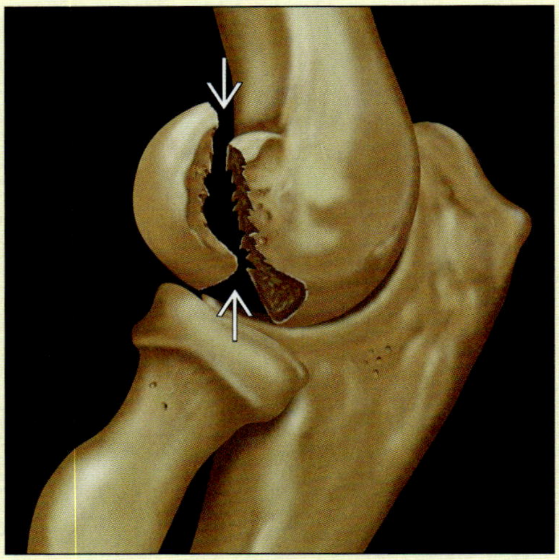
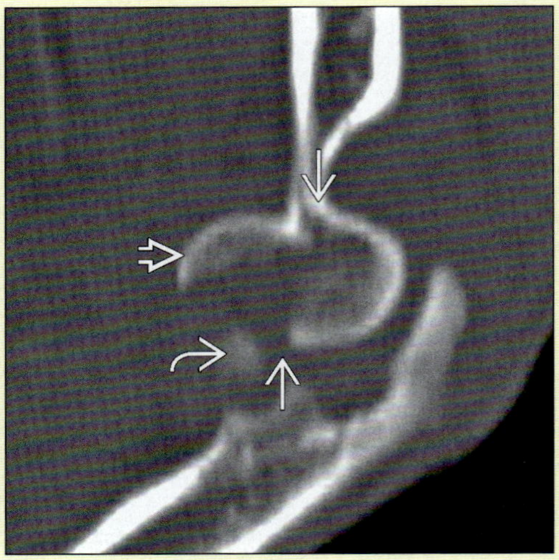

__(Left)__ Sagittal graphic of an osteochondral fracture ➡ of the anterior articular surface of the capitellum. This injury typically occurs in the setting of repetitive valgus stress, as seen in overhead throwing athletes. __(Right)__ Sagittal T2WI FS MR shows an osteochondral fracture ➡ in the anterior capitellum due to a valgus stress injury in a little league pitcher. A joint effusion ↪ is present. An osteochondral fragment was present more medially (not shown).

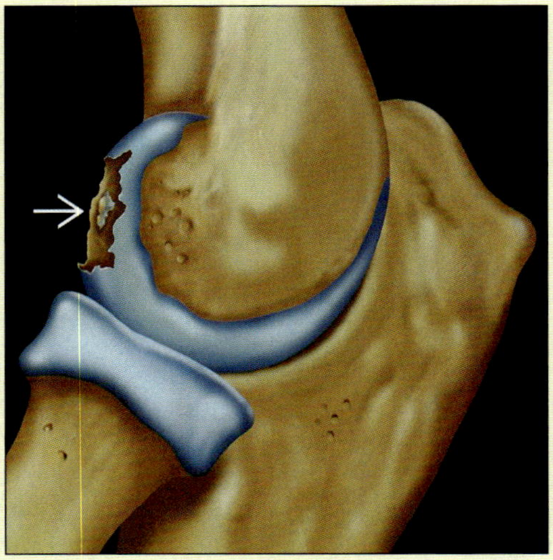
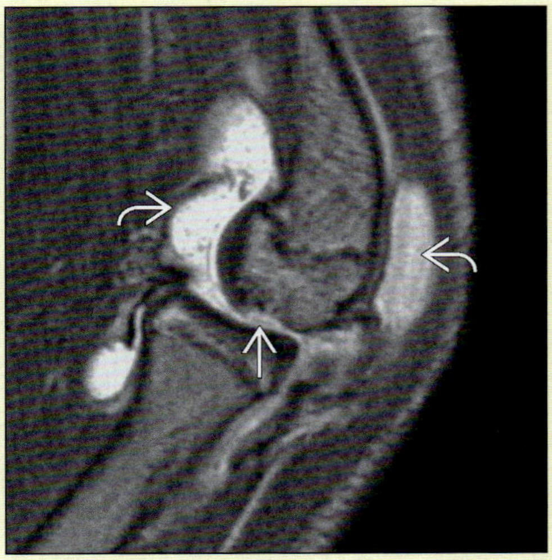

CAPITELLUM FRACTURE

TERMINOLOGY

Synonyms

- Capitulum fracture
- Capitellar osteochondritis
  - Old term for describing osteochondral fracture of capitellum in childhood
- Osteochondritis dissecans
  - Old term for osteochondral fracture
- Panner disease
  - Historically considered osteonecrosis of capitellum
  - Likely etiology: Osteochondral fx of capitellum
  - Young children
  - Radiographic irregularity of capitellum
  - Usually remodels by adulthood
- Hahn-Steinthal fracture
- Kocher-Lorenz fracture
- Brogerg-Morrey fracture
- McKee fracture
- Osborne-Cotterill lesion

Definitions

- Fracture limited to articular surface of lateral condyle of humerus
- May be osseous, osteochondral, or purely chondral

IMAGING

General Features

- Best diagnostic clue
  - Displaced and rotated semicircular bone fragment anterior to distal humerus on lateral radiograph
- Location
  - Articular surface of lateral humeral condyle (capitellum)
- Size
  - Fragment size may vary
    - Type 1: Large rotated osseous fragment
    - Type 2: Osteochondral fragment may be quite small
    - Types 3 or 4: Multiple fragments
    - Valgus stress injury may be purely chondral
- Morphology
  - Various types described
    - Large fragment separated by coronal fracture
    - May be thinner osteochondral shear fragment
    - May be comminuted
    - Focal chondral or osteochondral impaction fragment
      - Usually anterior
      - Posterior in hyperextension injury

Radiographic Findings

- AP view often normal
- Lateral view → rotated and displaced osseous fragment
- Positive "fat pad" sign, indicating effusion
- ± intraarticular fragments

CT Findings

- Fracture line in type 1 injury in coronal plane
- Fracture may extend into lateral edge of trochlea
- Osteochondral lesions well seen
- Pure chondral lesions not seen
  - CT arthrogram can help
- Loose bodies
- Joint effusion

MR Findings

- Chondral injury may only be visible on T2WI
- Marrow edema on T1WI and FS T2WI
- Joint effusion
- Intraarticular bodies
  - Usually in olecranon fossa
  - Can be in coronoid fossa
- Possible medial-sided distraction injury
  - Ulnar collateral ligament
    - Tear
    - Edema
  - Flexor/pronator tendon
    - Tear
    - Edema

Imaging Recommendations

- Best imaging tool
  - Lateral radiograph for type 1 injury
  - MR for valgus stress pattern injury
    - Often involves smaller fragment(s)
    - Sometimes purely chondral
- Protocol advice
  - MR
    - Sagittal T1WI and T2WI FS best for
      - Fracture plane
      - Fragment size and location
      - Status of articular cartilage
    - Coronal T2WI FS best for
      - Integrity of ulnar collateral ligament
      - Status of flexor/pronator tendon complex
    - Axial T2WI best for
      - Location of loose bodies
      - Status of flexor/pronator tendon complex

DIFFERENTIAL DIAGNOSIS

Pseudodefect of Capitellum

- Normal notch in posterior lateral capitellum
- Best seen on sagittal images
- Not associated with marrow edema or symptomatology

PATHOLOGY

General Features

- Etiology
  - Acute traumatic type
    - Trauma with elbow in flexion → anterior fracture
    - Trauma with elbow in extension → posterior fracture
    - Type 1 (Hahn-Steinthal type)
      - Shearing forces across capitellum
      - Due to impaction on radial head
    - Type 2 (Kocher-Lorenz type)
      - Valgus impaction of radiocapitellar joint
  - Repetitive traumatic type
    - Valgus stress mechanism of injury
    - Overhead throwing athletes
    - Can be hyperextension injury

CAPITELLUM FRACTURE

- Posterior fracture
- Gymnastics
- Volleyball
- Associated abnormalities
  - Radial head fracture
  - Ulnar-sided injury
    - Ulnar collateral ligament tear
    - Flexor/pronator tendon injury
  - Intraarticular bodies

Staging, Grading, & Classification

- Conventional system
  - Type 1: Hahn-Steinthal fracture
    - Coronal fracture plane
    - Large osseous fragment
    - Usually includes portion of trochlear lip
    - Fragment anteriorly displaced into coronoid fossa and rotated
  - Type 2: Kocher-Lorenz fracture
    - Thinner fracture depth
    - Rarely completely chondral
    - Posterior fragment displacement
  - Type 3: Broberg-Morrey fracture
    - Comminuted fracture
  - Type 4: McKee fracture
    - Comminuted fracture extending into trochlea
- Bryan and Morrey system
  - Type 1: Complete osteochondral fracture
  - Type 2: Superficial osteochondral fragment
  - Type 3: Comminuted fracture
  - Type 4: Coronal shear fracture involving capitellum and portion of trochlea
- Osborne-Cotterill lesion: Impaction fracture in posterolateral corner of capitellum
  - Related to rotatory instability of elbow
- Panner disease
  - Impaction fracture of capitellar osteochondral surface
  - Occurs in younger children (5-11 years)
  - ± residual deformity

CLINICAL ISSUES

Presentation

- Most common signs/symptoms
  - Lateral elbow pain and swelling
  - Restricted motion due to intraarticular fragment
- Other signs/symptoms
  - Medial elbow pain if associated with ulnar collateral ligament injury

Demographics

- Gender
  - Type 1: M < F
    - Likely due to ↑ carrying angle at elbow (in females)
  - Osteochondral lesions: M > F
    - Overhead throwing injuries
- Epidemiology
  - Type 1 fractures account for < 1% of elbow fractures
  - Valgus stress injuries
    - Throwing athletes
      - Injury occurs in valgus and elbow flexion
      - Defect usually anterior in capitellum
    - Gymnasts
      - Injury occurs in valgus and elbow extension
      - Defect usually posterior in capitellum

Natural History & Prognosis

- Untreated → poor function with restricted motion and early arthritis

Treatment

- Large fragments are surgically reduced with pins or screws
- Smaller fragments may simply be removed

DIAGNOSTIC CHECKLIST

Consider

- Radial head fracture
- Concomitant injury to medial structures as part of valgus stress injury
  - Ulnar collateral ligament injury
  - Flexor/pronator tendon injury
- Presence of intraarticular bodies

Image Interpretation Pearls

- Fracture may not be seen well on AP radiograph
- Assess lateral radiograph for rotated or missing capitellar articular surface
- May miss small osteochondral or purely chondral lesions entirely on radiograph

Reporting Tips

- Carefully search for intraarticular bodies
- Describe size and location of articular surface defect
- Describe any associated medial compartment injury

SELECTED REFERENCES

1. Cottalorda J et al: The often-missed Kocher-Lorenz elbow fracture. Rev Chir Orthop Traumatol. 95(7):551-4, 2009
2. Guitton TG et al: Fractures of the capitellum and trochlea. J Bone Joint Surg Am. 91(2):390-7, 2009
3. Singh AP et al: Fractures of capitellum: a review of 14 cases treated by open reduction and internal fixation with Herbert screws. Int Orthop. Epub ahead of print, 2009
4. Jeon IH et al: Osborne-cotterill lesion: an osseous defect of the capitellum associated with instability of the elbow. AJR Am J Roentgenol. 191(3):727-9, 2008
5. Cheung EV: Fractures of the capitellum. Hand Clin. 23(4):481-6, vii, 2007
6. Sen MK et al: Open reduction and internal fixation of coronal fractures of the capitellum. J Hand Surg [Am]. 32(9):1462-5, 2007
7. Faber KJ: Coronal shear fractures of the distal humerus: the capitellum and trochlea. Hand Clin. 20(4):455-64, 2004
8. Kobayashi K et al: Lateral compression injuries in the pediatric elbow: Panner's disease and osteochondritis dissecans of the capitellum. J Am Acad Orthop Surg. 12(4):246-54, 2004
9. Rizzo M et al: Fractures of the elbow's lateral column radial head and capitellum. Hand Clin. 18(1):21-42, 2002
10. McKee MD et al: Coronal shear fractures of the distal end of the humerus. J Bone Joint Surg Am. 78(1):49-54, 1996
11. Rosenberg ZS et al: Pseudodefect of the capitellum: potential MR imaging pitfall. Radiology. 191(3):821-3, 1994
12. Bryan RS et al. Fractures of the distal humerus. In Morrey BF (ed.): The Elbow and Its Disorders. Philadelphia: WB Saunders. 302-39, 1985

CAPITELLUM FRACTURE

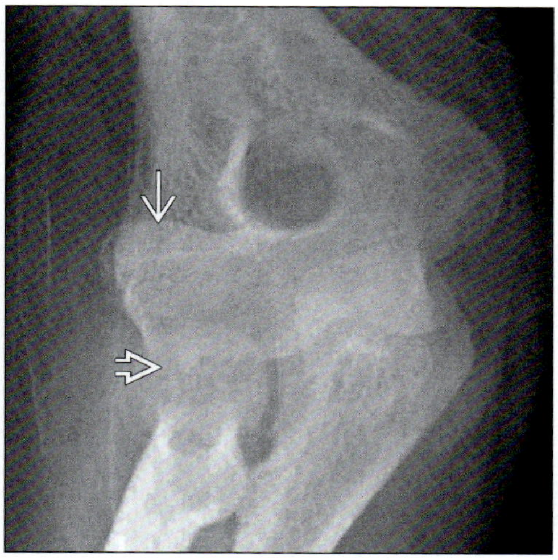

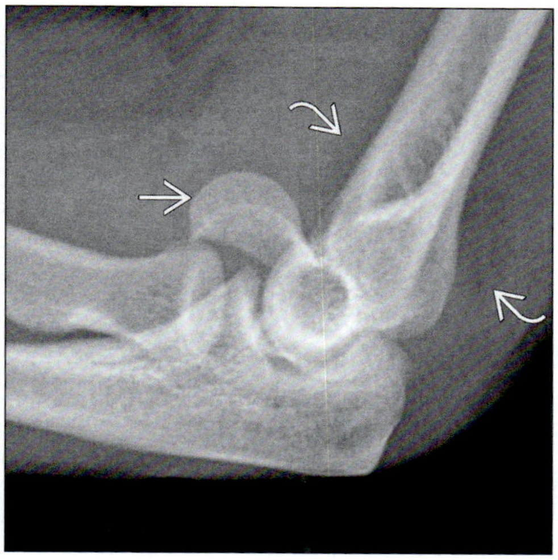

(Left) *AP radiograph in a patient with a rotated type I capitellar fracture demonstrates anterosuperior rotation of the capitellar articular surface →. Findings on the AP view can be quite subtle; poor delineation of the normal articular surface of the capitellum → is often the only finding.* ***(Right)*** *Lateral radiograph from the same patient demonstrates 90° anterosuperior rotation of capitellar articular surface →. Note distended periarticular fat pads → indicating hemarthrosis from acute injury.*

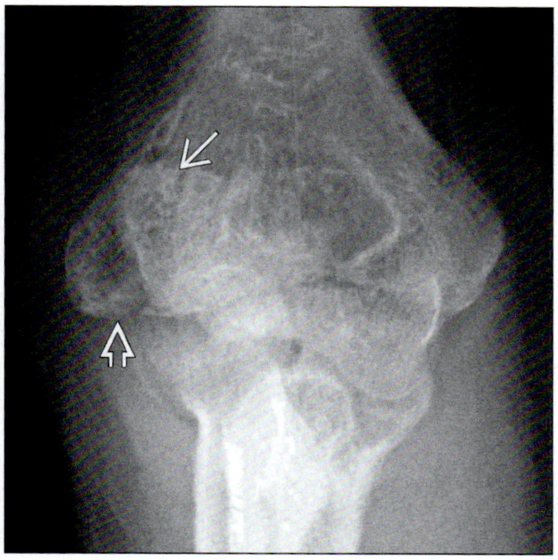

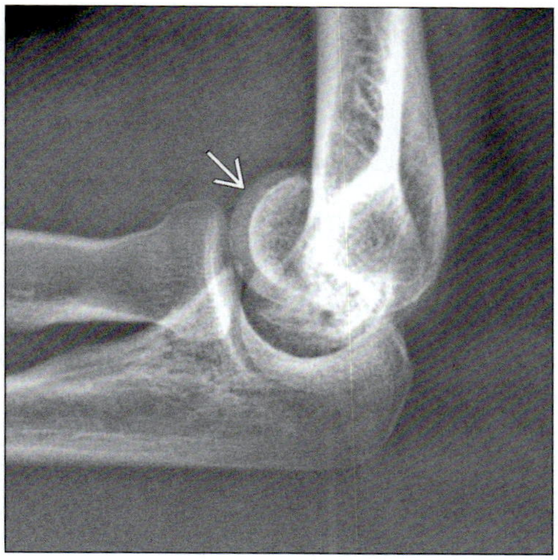

(Left) *Anteroposterior radiograph of the elbow from a patient with a slightly rotated type I fracture of the capitellum demonstrates displacement of the capitellum →. The finding is subtle on the AP view, with only mild irregularity of the capitellar border → providing a clue to the diagnosis.* ***(Right)*** *Lateral radiograph from the same patient demonstrates rotation of the capitellum →. Note that in this case, radiocapitellar alignment is only slightly abnormal in the lateral view.*

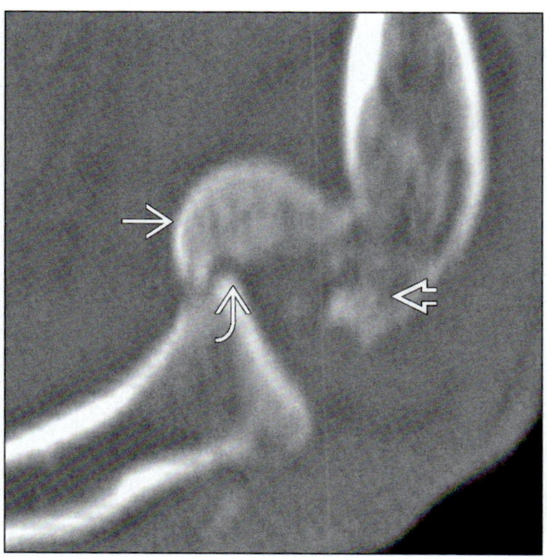

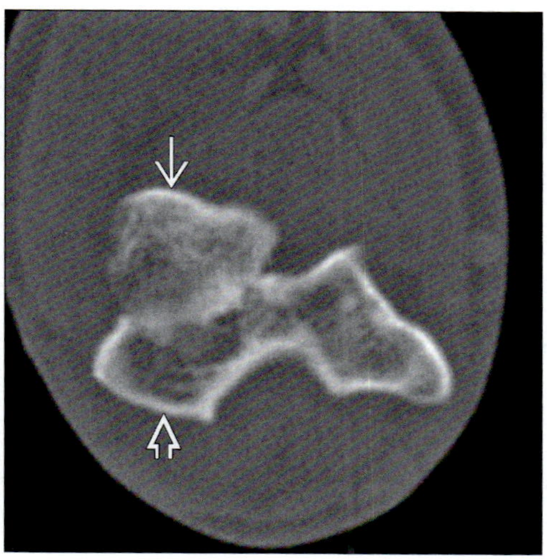

(Left) *Sagittal reformat bone CT shows a rotated capitellar fragment → anterior to the lateral humeral condyle → in this patient with a rotated type I capitellar fracture. Note the impaction of the capitellum on the superior radial head →.* ***(Right)*** *Axial bone CT from the same patient shows a rotated capitellar fragment → anterior to the lateral humeral condyle →.*

CAPITELLUM FRACTURE

(Left) *Anteroposterior radiograph shows a slightly displaced radial head fracture in a patient with a type II capitellar fracture. The capitellum appears normal.* ***(Right)*** *Lateral radiograph from the same patient shows curvilinear osteochondral fracture fragment arising from the capitellar articular surface. Because of the coronal plane of the fracture, it is not seen on the AP view. Note the distended periarticular fat pads indicating hemarthrosis.*

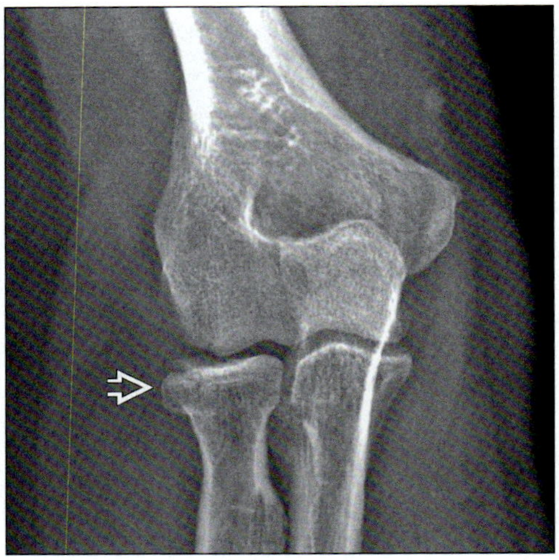

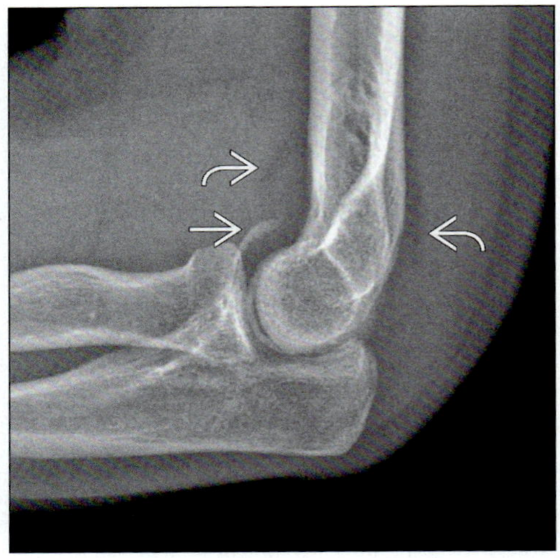

(Left) *AP radiograph shows posterolateral elbow dislocation. Note the rotated & laterally displaced capitellar fragment maintaining alignment with the radial head. This patient was proven to have comminuted (type 3) capitellar fracture associated with elbow dislocation* ***(Right)*** *Axial bone CT through the ulnohumeral joint in the same patient shows marked lateral displacement of the comminuted capitellar fragment relative to the lateral humeral condyle. Olecranon & coronoid are seen.*

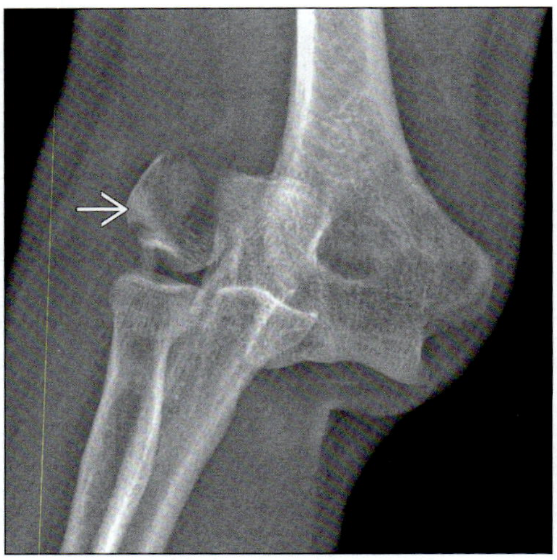

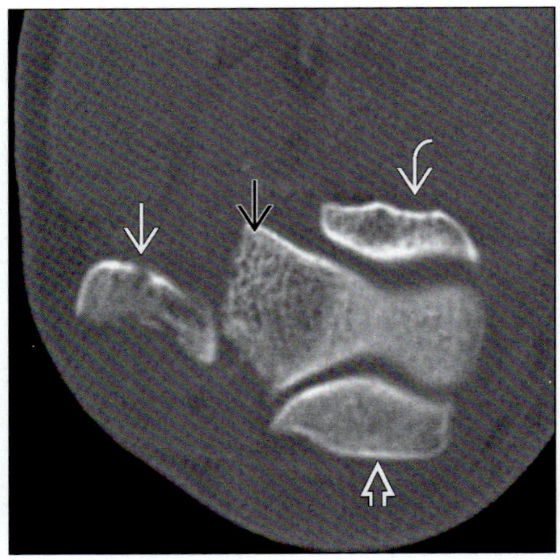

(Left) *AP radiograph of the elbow in a patient with chronic changes in the capitellum related to childhood Panner disease demonstrates flattening and osteophytosis of capitellum, mild osteophytosis of the radial head, & several intraarticular bodies.* ***(Right)*** *Sagittal T1WI MR in the same patient demonstrates degenerative changes in the capitellum, with diffuse chondral loss, subchondral sclerosis, & osteophytosis. Note also the radial head osteophytes & intraarticular body.*

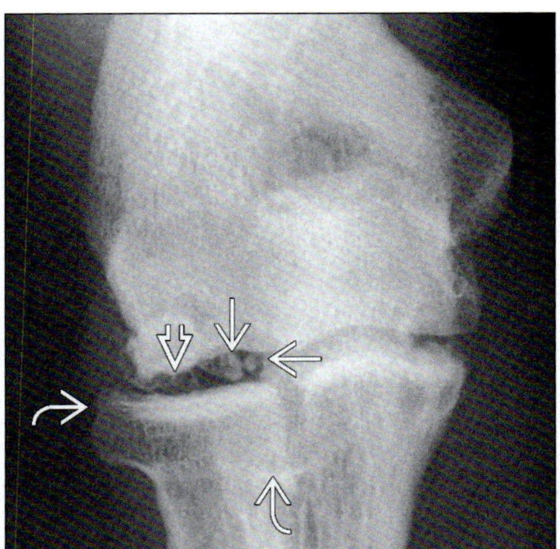

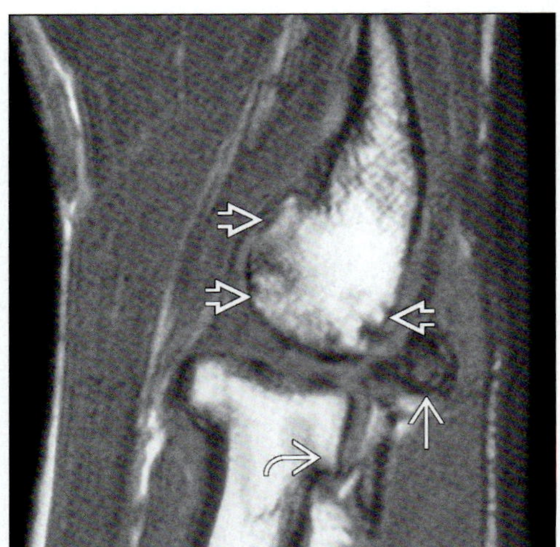

CAPITELLUM FRACTURE

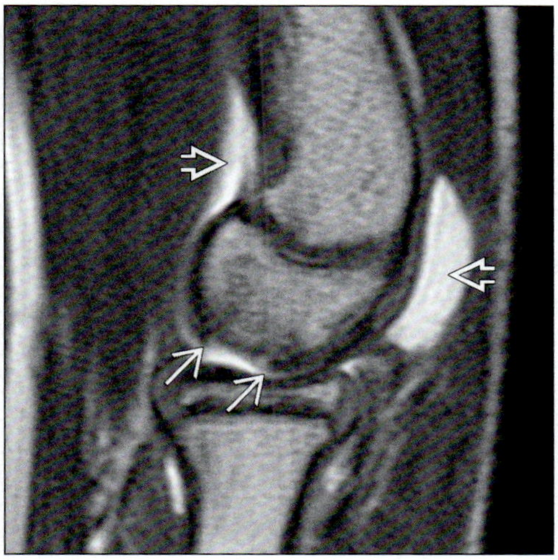

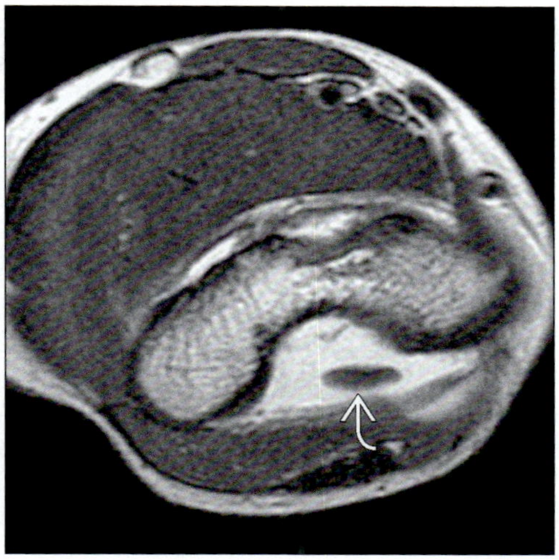

(Left) *Sagittal T2WI MR in a 12-year-old pitcher shows acute osteochondral fracture of the capitellum with joint effusion & chondral fragment in the olecranon fossa. This image demonstrates a well-defined chondral defect in the anterior articular surface of the capitellum ➡. Note the moderate joint effusion ➡.* ***(Right)*** *Axial T2WI MR in the same patient shows a curvilinear chondral fragment ➡ in the olecranon fossa surrounded by joint fluid. The fragment is likely the cause of restricted motion.*

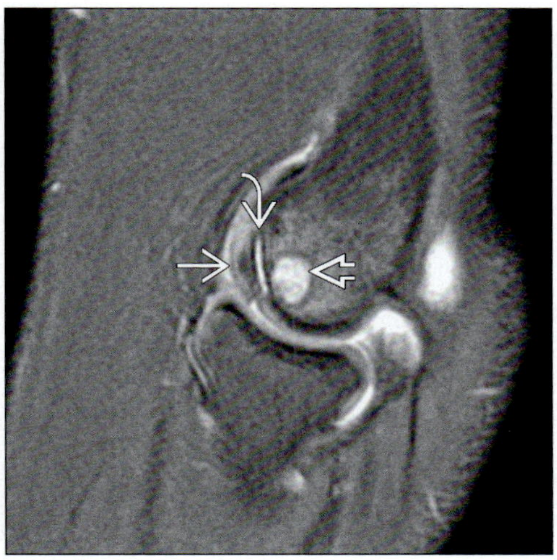

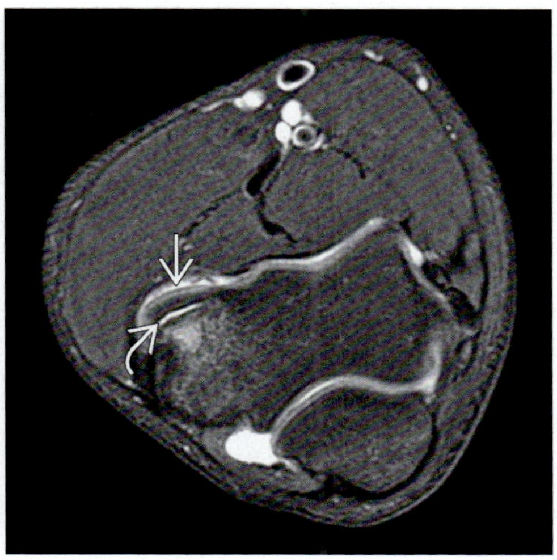

(Left) *Sagittal T2WI FS MR of the elbow of a 14-year-old baseball pitcher shows an osteochondral fracture ➡ in the anterior articular surface of capitellum. Fluid outlining the fragment ➡ indicates its instability. The subchondral cyst ➡ indicates chronicity; this is an acute-on-chronic injury.* ***(Right)*** *Axial T2WI FS MR from the same patient shows a separated fragment ➡ with underlying fluid ➡. Unstable lesions such as this must be surgically stabilized.*

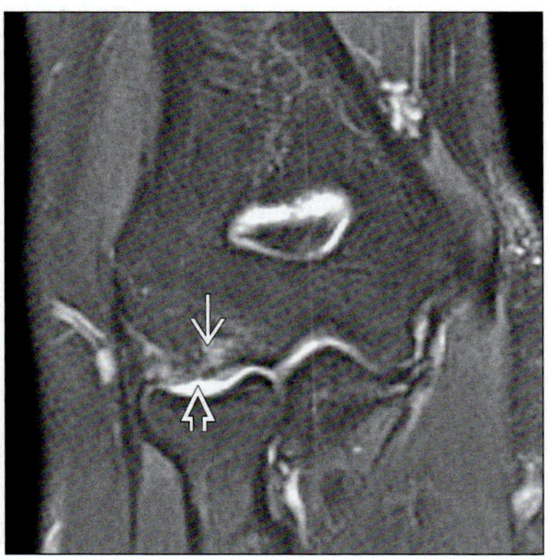

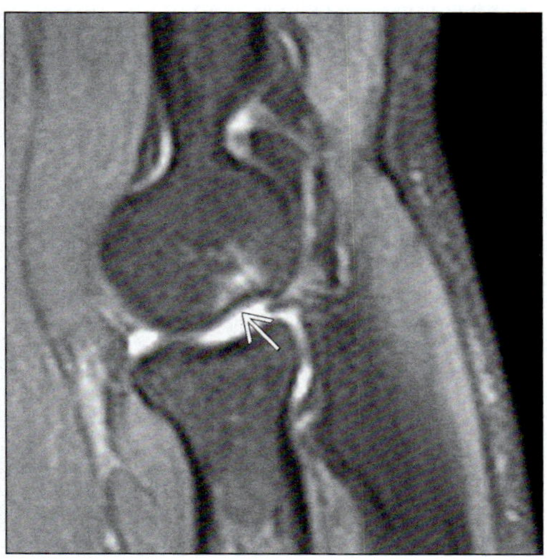

(Left) *Coronal T2WI FS MR shows the elbow of a 21-year-old college volleyball player after a hyperextension injury. Subchondral marrow edema in the capitellum ➡ and irregular chondral defect ➡ are located in a more posterior position than is typical for throwing injury.* ***(Right)*** *Sagittal PDWI FS MR from the same patient shows the posterior location of the capitellar injury ➡. Posterior capitellar contusion or fracture is typical of hyperextension mechanism of injury.*

ELBOW DISLOCATION

Key Facts

Terminology
- Ulnohumeral dislocation
- Ulnar dislocation is usually associated with radiocapitellar joint (RCJ) dislocation

Imaging
- 90% posterior or posterolateral
- Anterior or medial dislocation in < 10%
- Radial head dislocation most common in children
- Repeat radiographs after reduction to evaluate for displaced or trapped fragments

Top Differential Diagnoses
- Pediatric
  - Complete separation of distal humeral epiphysis
  - Congenital radial head dislocation
- Adult
  - Monteggia fracture

Pathology
- Fall on outstretched hand with elbow in hyperextension
- 20-56% of elbow dislocations associated with fractures
  - Medial humeral condyle most common
  - Radial head and neck 2nd most common
  - Coronoid process, especially adults
- Ligament disruptions
  - LUCL tear most common

Clinical Issues
- 10-50% of cases are sports related
- 2nd most common dislocation in adults
- Most frequent dislocation in children

__(Left)__ Sagittal graphic shows stage II lateral instability known as the perched elbow. The humeral trochlea ➡ is perched on, but not anterior to, the coronoid ➡. This is an unusual finding, and usually results in significant ligamentous instability. __(Right)__ Lateral radiograph shows a perched elbow nearly dislocated. As in the graphic, the trochlea ➡ is perched on the coronoid process ➡. The radiocapitellar joint is dislocated ➡.

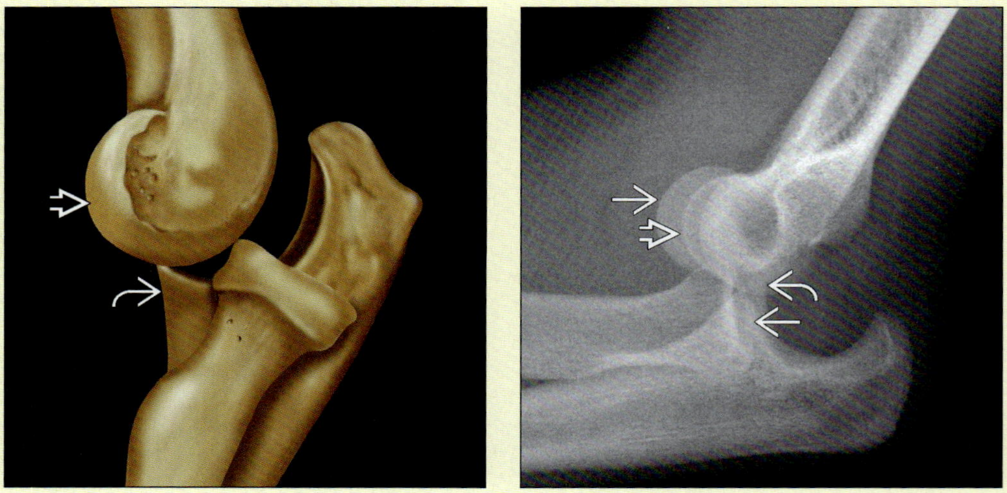

__(Left)__ Sagittal graphic shows bony injury with posterior dislocation, including a fracture of the anterior aspect of the radial head ➡ and a posterior capitellar fracture ➡. __(Right)__ Sagittal T2WI FS MR demonstrates corresponding findings in a patient after closed reduction of a posterior elbow dislocation. Fracture of the anterior aspect of the radial head ➡ and in the posterior aspect of the capitellum ➡ is a typical pattern.

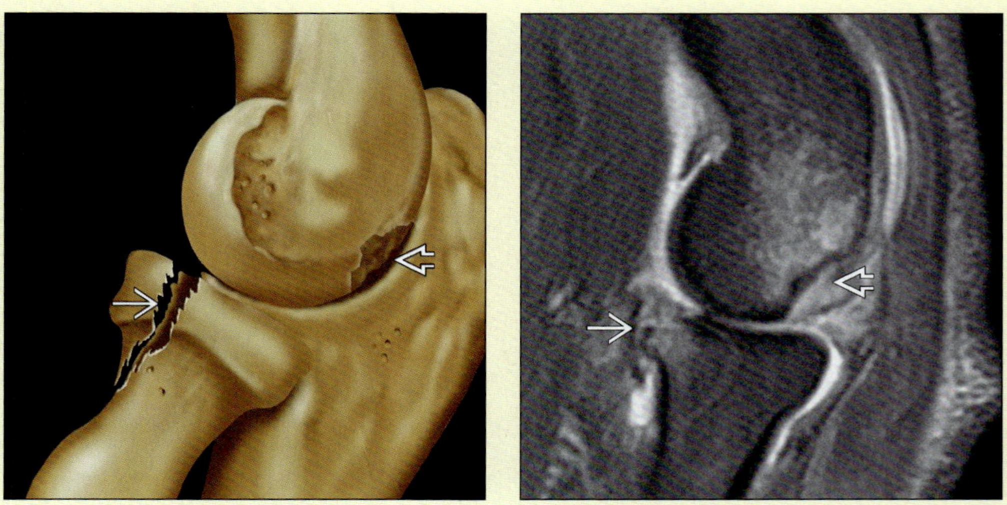

ELBOW DISLOCATION

TERMINOLOGY

Synonyms
- Ulnohumeral dislocation

Definitions
- Ulna dislocated relative to humerus
  - Ulnar dislocation is usually associated with radiocapitellar joint (RCJ) dislocation
  - Isolated dislocation of ulna is unusual
- Isolated RCJ dislocation is not true elbow dislocation
  - Can occur in children
    - Nursemaid elbow = subluxation of RCJ
  - Rare in adults
    - Often associated with proximal ulnar fracture (Monteggia injury)

IMAGING

General Features
- Best diagnostic clue
  - Loss of articulation between ulna and trochlea
    - Often associated with disruption of RCJ
  - Empty semilunar notch of olecranon seen on lateral view
- Morphology
  - Posterior, anterior, posterolateral, or posteromedial dislocation of ulna relative to humerus
    - 90% posterior or posterolateral
    - Anterior or medial dislocation in < 10%
    - Divergent dislocation (radius and ulna in opposite directions) rare
    - ± impaction or avulsion fractures
    - ± ligamentous disruptions
  - Radial head dislocation most common in children
    - Isolated RCJ dislocation unusual
    - Usually associated with elbow dislocation or ulnar fracture

Radiographic Findings
- Complete loss of articulation between trochlea and ulna
  - Usually posterior
- Associated radial head dislocation
- Concomitant fractures

CT Findings
- Bone CT
  - Post-reduction
    - Search for entrapped fracture fragments and donor sites
    - Small osseous fragments better seen on CT than MR

MR Findings
- Useful for characterized associated ligamentous injury
  - Lateral ulnar collateral ligament (LUCL)
  - Radial collateral ligament (RCL)
  - Ulnar collateral ligament (UCL)
- Can identify osteochondral defects
  - More common if reduction delayed

Imaging Recommendations
- Best imaging tool
  - Radiographs: Anteroposterior, lateral, & oblique
  - CT for entrapped fragments and subtle fractures
  - MR for ligamentous instability
- Protocol advice
  - Repeat radiographs after reduction to evaluate for displaced or trapped fragments

DIFFERENTIAL DIAGNOSIS

Pediatric
- Complete separation of distal humeral epiphysis
  - Normal relationship between ulna and humeral epiphysis is maintained
  - Medial displacement of forearm bones relative to humerus
- Congenital radial head dislocation

Adult
- Monteggia fracture
  - RCJ dislocation + angulated fracture of proximal ulnar diaphysis

PATHOLOGY

General Features
- Etiology
  - Fall on outstretched hand with elbow in hyperextension
  - Valgus carrying angle of elbow results in
    - Axial compression laterally
    - Distraction medially
  - Anterior dislocation may result from direct blow to posterior aspect of flexed elbow
- Associated abnormalities
  - 20-56% of elbow dislocations associated with fractures
    - Medial humeral condyle most common
      - May be avulsion of medial epicondyle, especially in children
      - Entrapped fragment can prevent reduction
    - Radial head and neck 2nd most common
      - ~ 10% of cases
      - "Terrible triad" = coronoid process fracture + radial head fracture + posterior dislocation
    - Coronoid process
      - Most frequent etiology of coronoid process fracture
      - Most likely source of entrapped fragment in adults
    - Lateral humeral condyle
    - Olecranon
    - Osteochondral injury
    - Distal radial fracture (Colles, etc.)
  - Ligament disruptions
    - LUCL tear
      - Most common
      - Results in posterolateral instability of elbow
    - UCL tear
    - RCL tear
  - Neurovascular injury
    - Median nerve entrapment
      - More common in children
      - Median nerve palsy
      - Usually associated with fx of medial epicondyle

ELBOW DISLOCATION

      - Median nerve entrapped in elbow after closed reduction
      - Causes radiographic depression in medial aspect of distal humeral metaphysis
    - Brachial artery injury
      - Rare
      - More likely with open injury
      - Associated with medial epicondylar avulsion

Staging, Grading, & Classification

- O'Driscoll stages of posterolateral elbow instability
  - Stage I
    - Disruption of LUCL
    - Posterolateral subluxation of ulna
  - Stage II
    - Further injury to lateral collateral ligament complex
    - Coronoid perched on trochlea
  - Stage III
    - Complete posterior dislocation + UCL disruption

CLINICAL ISSUES

Presentation

- Most common signs/symptoms
  - Pain, swelling, and deformity of elbow
    - Posterior dislocation: Elbow flexion, exaggerated olecranon prominence
    - Anterior dislocation: Elbow in full extension; forearm in supination

Demographics

- Age
  - Active young patients; mean: 30 years
- Gender
  - M:F = 2-2.5:1
- Epidemiology
  - Incidence: 6/100,000 individuals
  - 10-50% of cases are sports related
  - 2nd most common dislocation in adults
  - Most frequent dislocation in children

Natural History & Prognosis

- If uncomplicated, 60% have late symptoms of pain
- Redislocation
  - Unlike shoulder, elbow dislocation does not predispose to future dislocations
  - Associated with poor outcome

Treatment

- Early reduction needed to reduce risk of
  - Neurovascular compromise
  - Osteochondral injury
- Conservative
  - Reduction and bracing in pronation
  - Protected mobilization
- Surgical
  - Repair of ligament and tendon disruptions
  - Preservation of radial head if possible
    - Excessive comminution requires excision
  - Excision of interposed soft tissue or osteochondral fragments
- Complications
  - Immediate
    - Failed reduction due to entrapped fragments
      - Children: Medial humeral epicondyle
  - Late
    - Posterolateral instability
      - Disruption of lateral ulnar collateral ligament
      - Association with coronoid fracture
    - Heterotopic ossification: 3-5%
      - Usually located anteriorly
      - Can appear within 3-4 weeks
      - Much more common if concomitant head injury
    - Flexion contracture
    - Neuropraxias: Up to 20%
      - Median nerve
      - Ulnar nerve

DIAGNOSTIC CHECKLIST

Consider

- Associated fractures, both pre- and post-reduction
  - Evaluate coronoid process fractures carefully for size and location
  - CT may be helpful in locating these fragments if reduction is not satisfactory
- Instability
  - MR in outpatient setting can identify disruption of ligaments and tendons
- Nerve or vascular damage

Image Interpretation Pearls

- Search for intraarticular fragments that may prevent complete reduction

Reporting Tips

- Describe direction of dislocation ± radiocapitellar disarticulation

SELECTED REFERENCES

1. de Haan J et al: Simple elbow dislocations: a systematic review of the literature. Arch Orthop Trauma Surg. 130(2):241-9, 2010
2. Guitton TG et al: Anterior olecranon fracture-dislocations of the elbow in children. A report of four cases. J Bone Joint Surg Am. 91(6):1487-90, 2009
3. Morrey BF: Current concepts in the management of complex elbow trauma. Surgeon. 7(3):151-61, 2009
4. Seijas R et al: Terrible triad of the elbow. J Orthop Surg (Hong Kong). 17(3):335-9, 2009
5. Dipaola M et al: Complex elbow instability. Hand Clin. 24(1):39-52, 2008
6. Hobgood ER et al: Acute dislocations of the adult elbow. Hand Clin. 24(1):1-7, 2008
7. Kuhn MA et al: Acute elbow dislocations. Orthop Clin North Am. 39(2):155-61, v, 2008
8. Lyons RP et al: Chronically unreduced elbow dislocations. Hand Clin. 24(1):91-103, 2008
9. Martin BD et al: Complications related to simple dislocations of the elbow. Hand Clin. 24(1):9-25, 2008
10. Tashjian RZ et al: Complex elbow instability. J Am Acad Orthop Surg. 14(5):278-86, 2006
11. Chung CB et al: Magnetic resonance imaging of elbow instability. Semin Musculoskelet Radiol. 9(1):67-76, 2005
12. Pacelli LL et al: Elbow instability: the orthopedic approach. Semin Musculoskelet Radiol. 9(1):56-66, 2005
13. Saati AZ et al: Fracture-dislocation of the elbow: diagnosis, treatment, and prognosis. Hand Clin. 20(4):405-14, 2004

ELBOW DISLOCATION

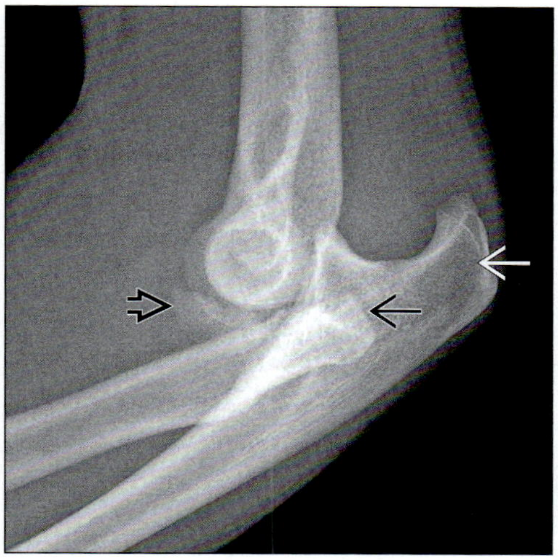
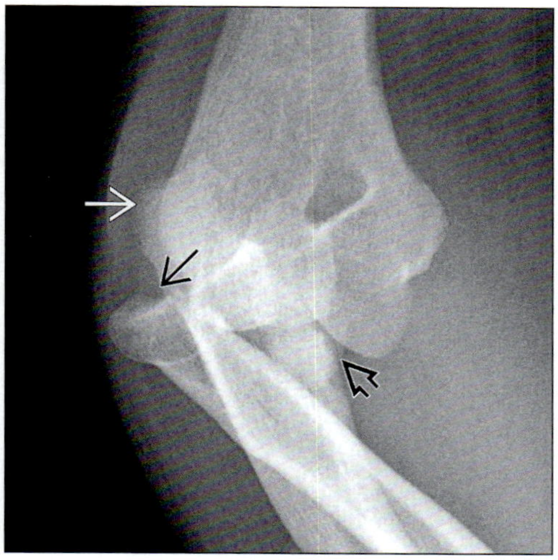

(Left) *Lateral radiograph shows a posterior dislocation of the elbow, including the olecranon and radial head. A large radial head fragment remains located anterior to the capitellum.* ***(Right)*** *Anteroposterior radiograph shows the lateral direction of a typical dislocation. Dislocated radial head with a large defect, dislocated ulna, and the radial head fracture fragment are seen. About 90% of elbow dislocations are posterolateral in direction.*

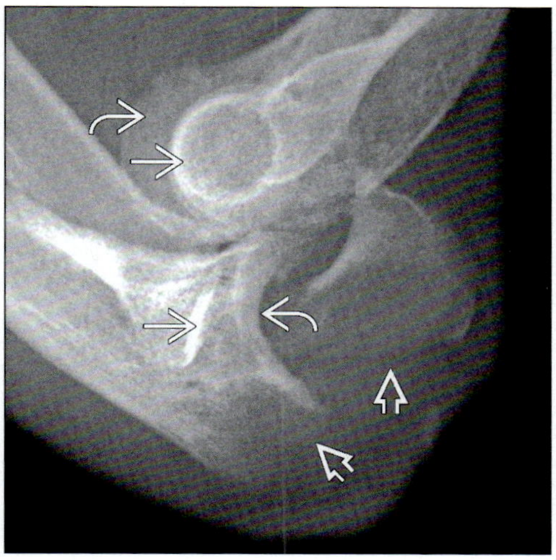
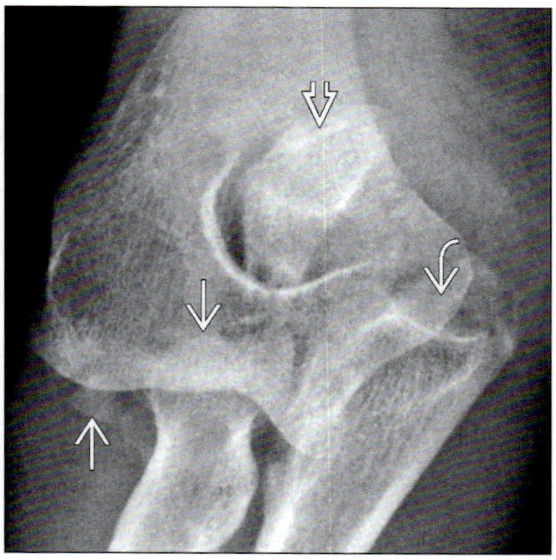

(Left) *Lateral radiograph shows a "fracture-dislocation" of the elbow. Note the posterior dislocation of the elbow joint, dislocation of the radiocapitellar joint, and a displaced fracture of the olecranon.* ***(Right)*** *AP radiograph from the same patient shows that the dislocation is posteromedial. Radial head fracture is also seen. The displaced olecranon fragment is seen overlying the olecranon fossa while the ulnar shaft is medially displaced.*

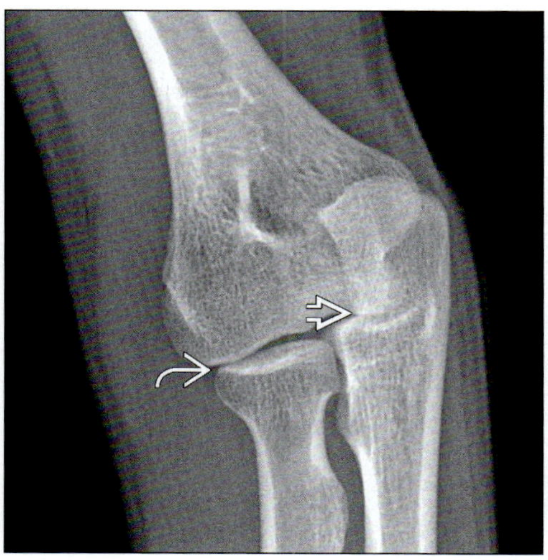
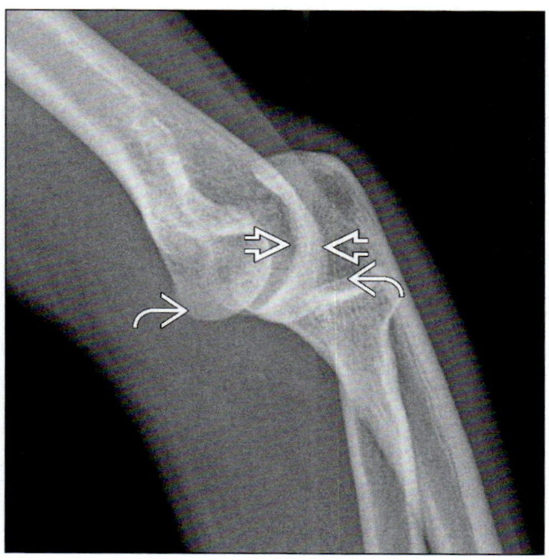

(Left) *AP radiograph shows subtle malalignment of the radiocapitellar and elbow joints. No fracture is seen.* ***(Right)*** *Lateral radiograph from the same patient demonstrates the lack of congruity of the radiocapitellar and elbow joints. This is a very subtle posteromedial dislocation and could be difficult to identify on the AP view alone. The lateral view is necessary to confirm the direction of dislocation, and is helpful in identifying fractures associated with the injury.*

ELBOW DISLOCATION

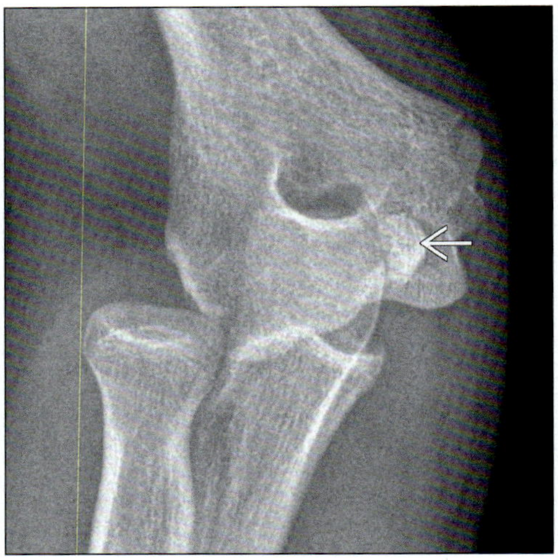

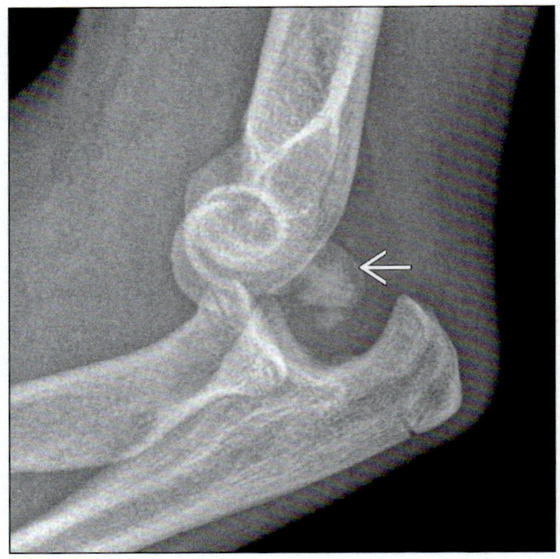

(Left) *AP radiograph of a 14-year-old athlete after elbow injury demonstrates posterolateral dislocation of the elbow and radiocapitellar joints. There is a fragment in the medial aspect of the joint without an obvious donor site.* ***(Right)*** *Lateral radiograph from the same patient shows more clearly the dislocation and entrapped fragment. The well-defined edges of the fragment and the age of the patient should be clues to a possible avulsed medial humeral epicondyle.*

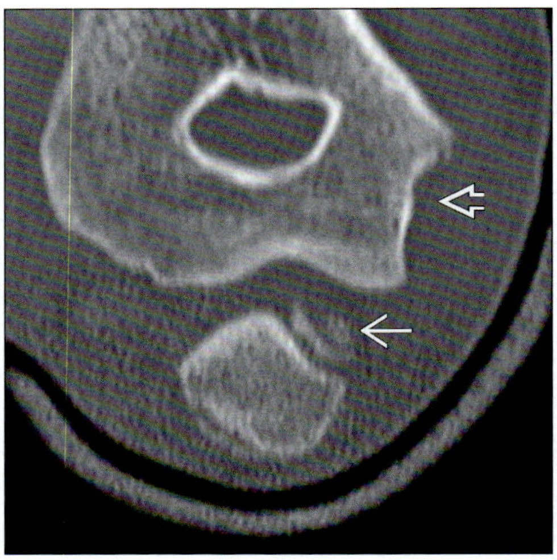

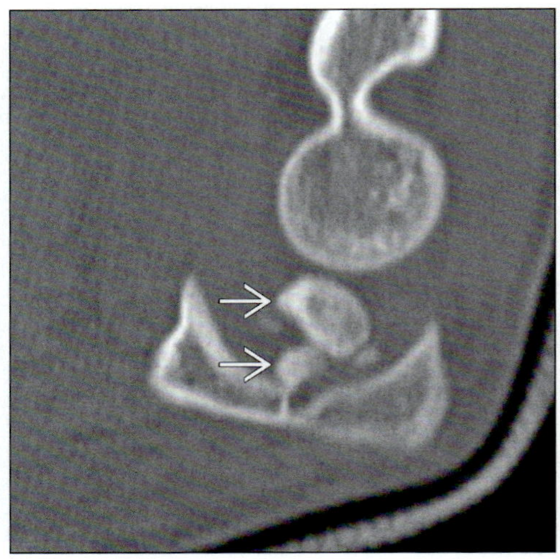

(Left) *Coronal reformat bone CT from the same patient shows the entrapped fragment and the donor site of the medial epicondyle. The smoothly corticated edge of the condyle is expected; the lack of the expected epicondylar ossification center adjacent to it must be appreciated to make this diagnosis.* ***(Right)*** *Sagittal reformat bone CT from the same patient shows the well-corticated nature of the entrapped medial epicondylar fragments, which help confirm their origin.*

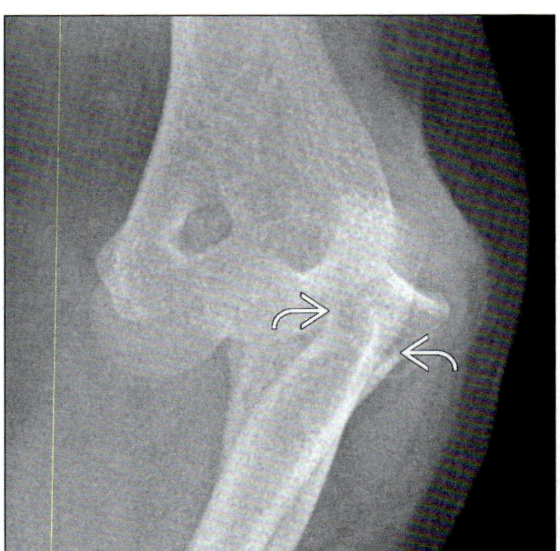

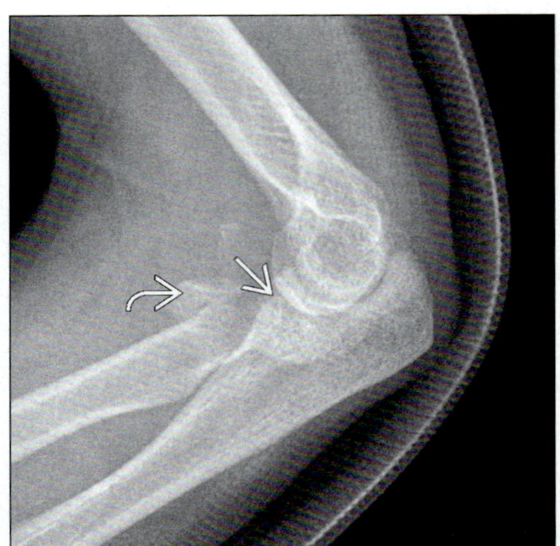

(Left) *AP radiograph of a posterolateral elbow dislocation shows irregular margins of the radial head and neck, suggesting an associated fracture here.* ***(Right)*** *Lateral radiograph from the same patient after reduction confirms a comminuted radial head and neck fracture and a coronoid fracture. The combination of elbow dislocation with radial head and coronoid fractures has been termed the "terrible triad of the elbow" because of its poor prognosis.*

ELBOW DISLOCATION

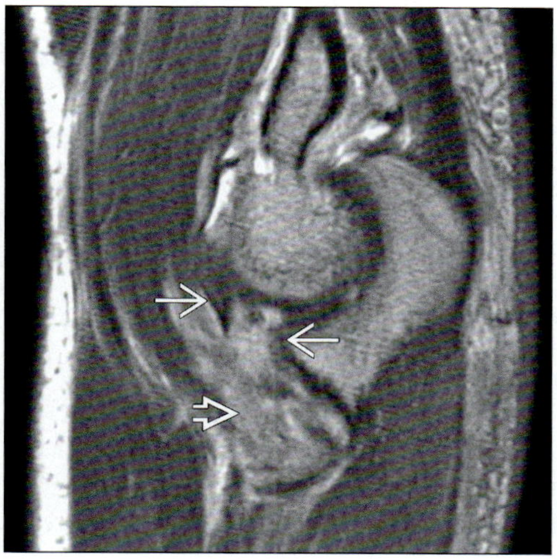

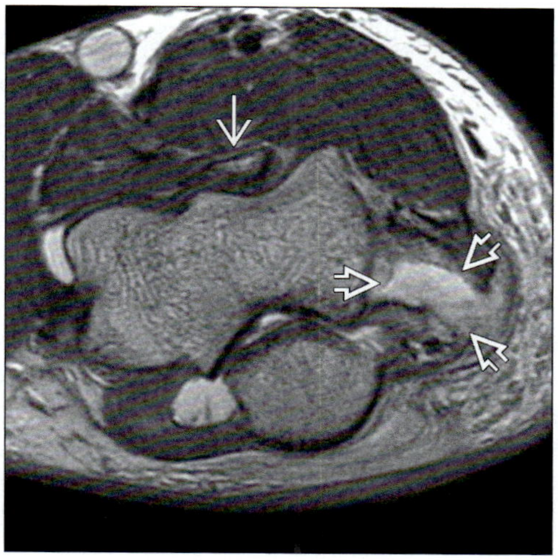

(Left) *Sagittal T2WI MR of the elbow after closed reduction of an elbow dislocation shows avulsion of the coronoid process and extensive edema in the brachialis muscle.* ***(Right)*** *Axial T2WI MR from the same patient shows the coronoid fragment anterior to the trochlea. Marked tissue disruption and fluid are present medial to the medial humeral epicondyle, indicating high-grade injury to the ulnar collateral ligament and the flexor/pronator tendon origin.*

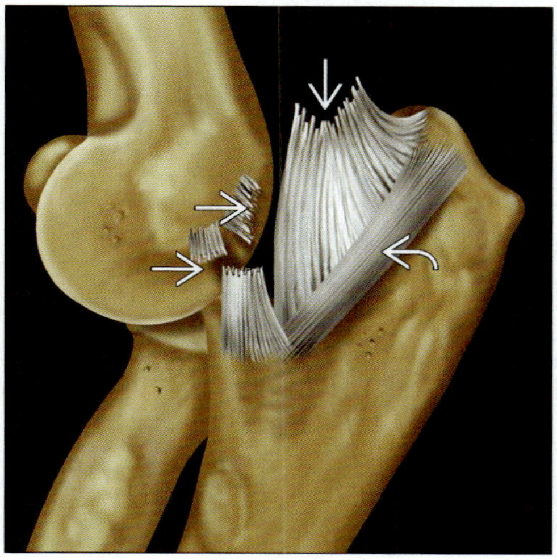

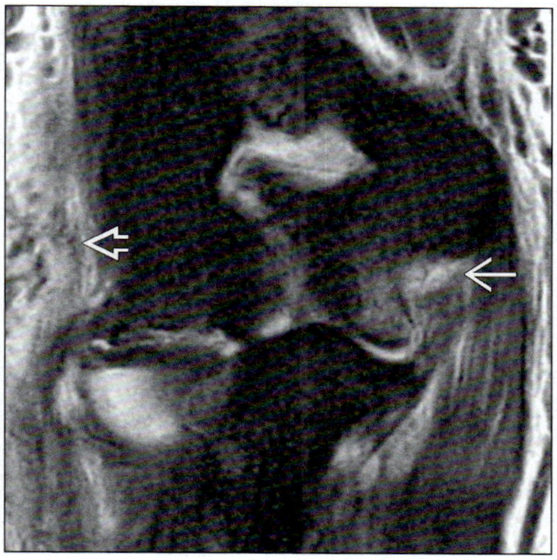

(Left) *Sagittal graphic shows posterior dislocation of the olecranon and radius with disruption of the ulnar collateral ligament complex. The transverse component remains intact.* ***(Right)*** *Coronal T2WI FS MR of the elbow after dislocation shows disruption of the proximal aspect of the anterior band of the ulnar collateral ligament. Edema in the lateral soft tissues is related to disruption of the radial and lateral ulnar collateral ligaments (not shown).*

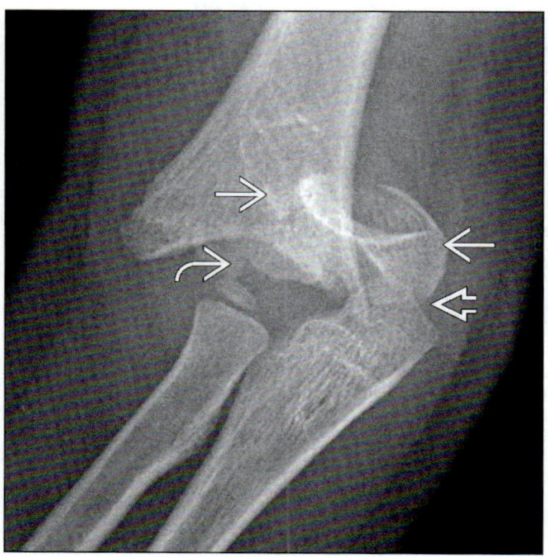

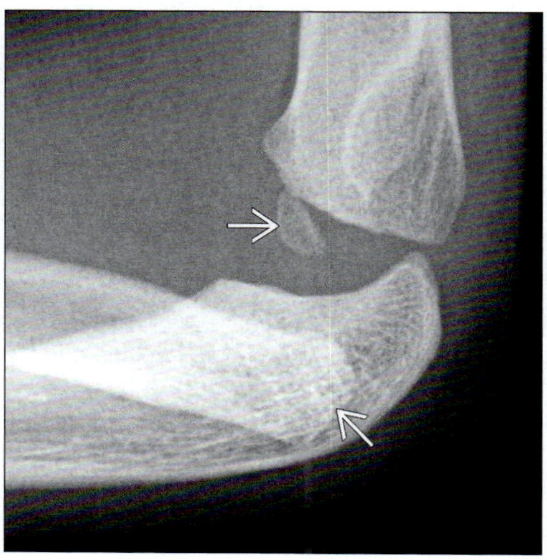

(Left) *AP radiograph of a child shows complete displacement of the distal humeral epiphysis from the distal humeral shaft. The key to distinguishing this from an elbow dislocation is noting the preservation of normal alignment of the radiocapitellar and ulnotrochlear joints.* ***(Right)*** *Lateral radiograph of a 6 year old with no antecedent trauma who complained of limited pronation of the elbow shows a congenital radial head dislocation. This finding was bilateral.*

MONTEGGIA INJURY

Key Facts

Terminology

- Monteggia fracture/dislocation = fracture of ulnar shaft + dislocation of radiocapitellar joint (RCJ)

Imaging

- Ulnar fracture usually in proximal diaphysis
- Radial head dislocated from capitellum
- Ulna angulates in same direction as RCJ dislocation
- Separate radiograph series of elbow and forearm required for complete and accurate evaluation

Pathology

- Forearm functions as ring
- Fracture or displacement of 1 requires either
  - Fx and displacement or dislocation of other, or
  - Disruption of interosseous membrane
- Classification of Monteggia fracture/dislocation: Bado
- Bado type 1
  - Anterior dislocation of RCJ
  - 65% of cases
- Bado type 2
  - Posterior or posterolateral RCJ dislocation
  - 18% of cases
- Bado type 3
  - Lateral RCJ dislocation
  - Children ages 6-9 years old
- Bado type 4
  - Type 1 + radial shaft fracture

Clinical Issues

- 1-2% of forearm fractures
- Adults > > children
- Most cases require open reduction/internal fixation
- In children, closed reduction with forearm in supination usually sufficient

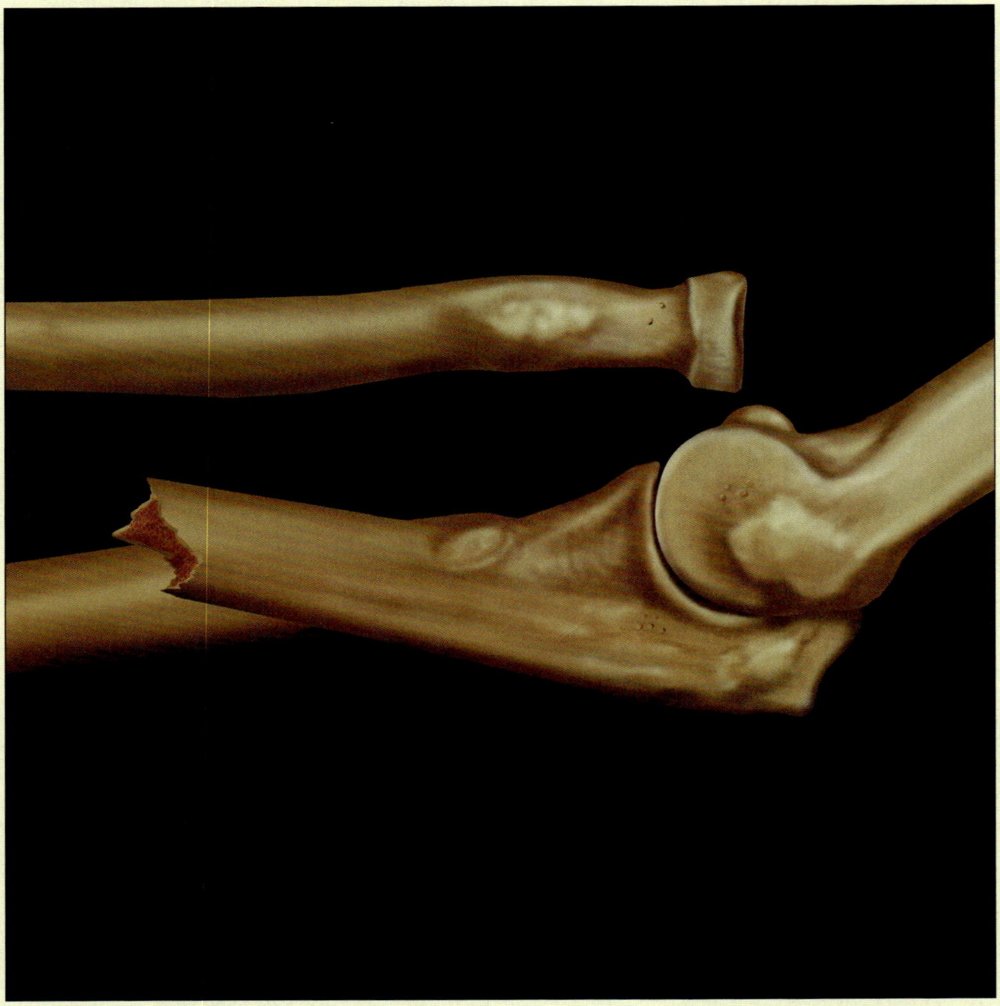

Graphic representation of a Monteggia injury. Originally described as an anterior dislocation of the radiocapitellar joint combined with an angulated fracture of the proximal ulna (as depicted here), the term has been broadened to include any radiocapitellar dislocation with a concomitant angulated ulnar fracture. One constant feature of these injuries is that the ulnar fracture is angulated in the same direction as the radial displacement. There is also dislocation of the proximal radioulnar articulation. Most of these injuries occur as a result of a fall on an outstretched hand with forced pronation of the forearm.

MONTEGGIA INJURY

TERMINOLOGY

Definitions

- Fracture of ulnar shaft + dislocation of radiocapitellar joint (RCJ)
- 1st described by Monteggia in 1814
- Originally specifically described anterior RCJ dislocation
  - Now includes any direction of dislocation

IMAGING

General Features

- Best diagnostic clue
  - Dislocated RCJ associated with ulnar shaft fracture
- Location
  - Ulnar fracture usually in proximal diaphysis
  - May be anywhere between coronoid process and midshaft
- Morphology
  - Ulnar fracture may be simple or very comminuted
    - Usually transverse or slightly oblique
    - Often with butterfly fragment
  - Degree of ulnar fracture angulation varies with degree of RCJ displacement

Radiographic Findings

- RCJ dislocation
  - Usually anterior
  - Can be posterior or lateral
- Angulated ulnar fracture
  - Proximal to mid-shaft of diaphysis
  - Angulation in same direction as RCJ dislocation

Imaging Recommendations

- Best imaging tool
  - Radiographs of elbow and forearm
- Protocol advice
  - Separate radiograph series of elbow and forearm
  - Multiple views of elbow to characterize complex injury

DIFFERENTIAL DIAGNOSIS

Isolated Ulnar Shaft Fracture

- Due to direct blow injury
- Usually not very angulated
- Overlying soft tissue injury

Isolated Radiocapitellar Joint Dislocation

- May be congenital

Anterior Elbow Dislocation

- Ulnar fracture located proximal to coronoid
- Coronoid and ulnar shaft usually displaced anteriorly

PATHOLOGY

General Features

- Etiology
  - Forearm functions as ring
    - Strong interosseous membrane binds radius to ulna
    - Fracture or displacement of 1 requires either
      - Fracture and displacement or dislocation of other, or
      - Disruption of interosseous membrane
  - Fall on outstretched hand with forced pronation
    - Probably accounts for majority of type 1 injuries
  - Fall with elbow flexed
    - May explain type 2 and 3 lesions
  - Direct blow to forearm
    - Could account for injuries with comminuted ulna
- Associated abnormalities
  - Annular ligament tear
    - Disruption of radiocapitellar joint can cause this
  - Injury to interosseous membrane of forearm

Staging, Grading, & Classification

- Bado classification
  - Type 1
    - Anterior dislocation of RCJ
    - Ulnar shaft fracture with apex anterior angulation
    - 65% of cases
  - Type 2
    - Posterior or posterolateral RCJ dislocation
    - Ulnar shaft fracture with apex posterior angulation
    - 18% of cases
  - Type 3
    - Lateral or anterolateral RCJ dislocation
    - Proximal ulnar shaft fracture
    - Children ages 5-9 years old
    - 16% of cases
  - Type 4
    - Anterior dislocation of RCJ
    - Ulnar shaft fracture
    - Fracture of proximal 1/3 of radial shaft
    - 1-5% of cases

CLINICAL ISSUES

Presentation

- Most common signs/symptoms
  - Pain and swelling of elbow and proximal forearm
  - Deformity at site of ulnar fracture if angulated
  - Restricted motion of RCJ
    - Pronation/supination
    - Elbow flexion/extension
- Other signs/symptoms
  - Posterior interosseous nerve (PIN) palsy
    - May present in delayed fashion if RCJ not reduced
    - Most common in Bado type 2 injury
    - Symptoms of PIN palsy
      - Pain with resisted supination of the forearm and frequently with resisted pronation
      - Active supination from a pronated position along with wrist flexion may reproduce the symptoms
      - Most have pain with resisted extension of extension of middle finger

Demographics

- Age
  - Adults > > children
- Gender
  - M > F
- Epidemiology

MONTEGGIA INJURY

Type 1 injuries are the most common. The key features are anterior dislocation of the radiocapitellar joint and proximal ulnar shaft fracture with apex anterior angulation.

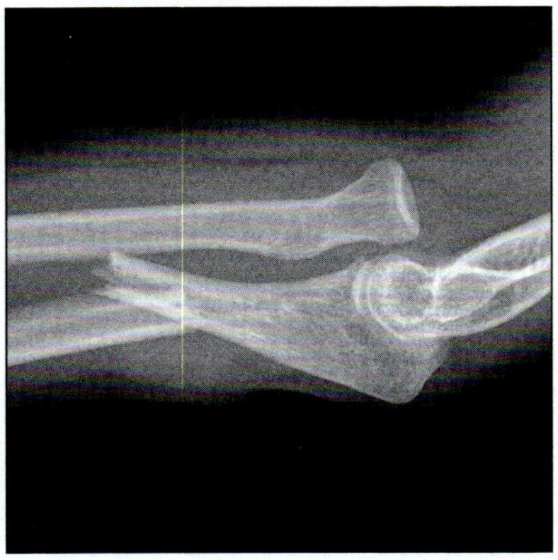

Type 2 injuries comprise about 16% of Monteggia injuries and are characterized by posterior dislocation of the radiocapitellar joint and a posteriorly angulated ulnar shaft fracture.

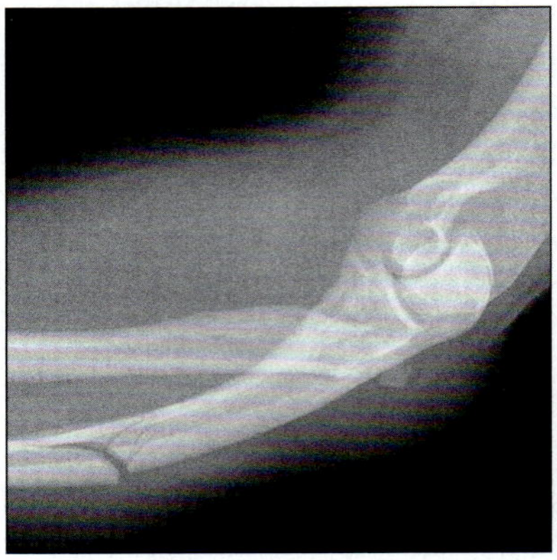

Type 3 Monteggia injuries are less common. These include lateral dislocation of the radiocapitellar joint and laterally angulated fracture of the ulnar shaft.

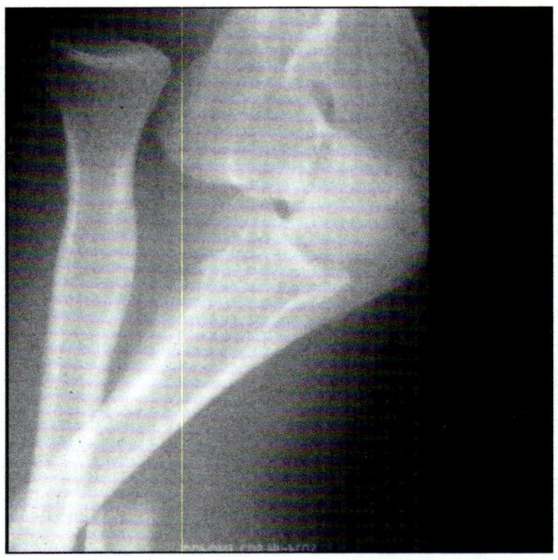

Type 4 injuries are the least common, comprising about 5% of all Monteggia injuries. These are characterized by radial shaft fracture in addition to radiocapitellar joint dislocation and ulnar shaft fracture.

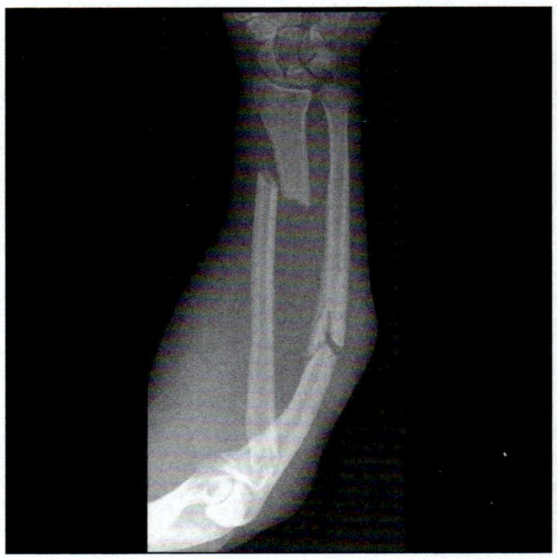

The Bado classification describes 4 types of Monteggia injuries. Classification depends on the direction of dislocation of the radiocapitellar joint, and in the case of type 4 injury, the presence of an additional radial shaft fracture. The associated ulnar fracture demonstrates apex angulation in the same direction as the radiocapitellar dislocation.

- 1-2% of forearm fractures

Natural History & Prognosis

- PIN or radial nerve palsy
  - Usually resolves spontaneously with timely RCJ reduction
- Nonunion of ulnar fracture
- Radiohumeral ankylosis
- Radioulnar synostosis
- Recurrent RCJ dislocation
- Heterotopic ossification

Treatment

- Most adult cases require open reduction and internal fixation
  - Plate and screw fixation of ulnar fracture
  - Closed reduction of RCJ may be attempted
    - Casted in 90° of flexion with supination
- In children, closed reduction with forearm in supination usually sufficient

DIAGNOSTIC CHECKLIST

Consider

- Anterior fracture/dislocation of elbow
  - Ulnar fracture will be proximal to coronoid

Image Interpretation Pearls

- Radiocapitellar line disruption
  - Line drawn through center of radial shaft and head should intersect capitellum
  - True in any projections
- RCJ dislocation occurs in direction of angulation of ulnar fracture
- May need to include midshaft of ulna to see ulnar fracture

Reporting Tips

- Describe direction of RCJ dislocation
- Describe location of ulnar fracture
- Describe direction of angulation of ulnar fracture

SELECTED REFERENCES

1. Cheung EV et al: Monteggia fracture-dislocation associated with proximal and distal radioulnar joint instability. A case report. J Bone Joint Surg Am. 91(4):950-4, 2009
2. Guitton TG et al: Long-term evaluation of surgically treated anterior monteggia fractures in skeletally mature patients. J Hand Surg Am. 34(9):1618-24, 2009
3. Eathiraju S et al: Monteggia fracture-dislocations. Hand Clin. 23(2):165-77, v, 2007
4. Wilkins KE: Changes in the management of monteggia fractures. J Pediatr Orthop. 22(4):548-54, 2002
5. Perron AD et al: Orthopedic pitfalls in the ED: Galeazzi and Monteggia fracture-dislocation. Am J Emerg Med. 19(3):225-8, 2001
6. Kay RM et al: The pediatric Monteggia fracture. Am J Orthop. 27(9):606-9, 1998
7. Ring D et al: Monteggia fractures in children and adults. J Am Acad Orthop Surg. 6(4):215-24, 1998

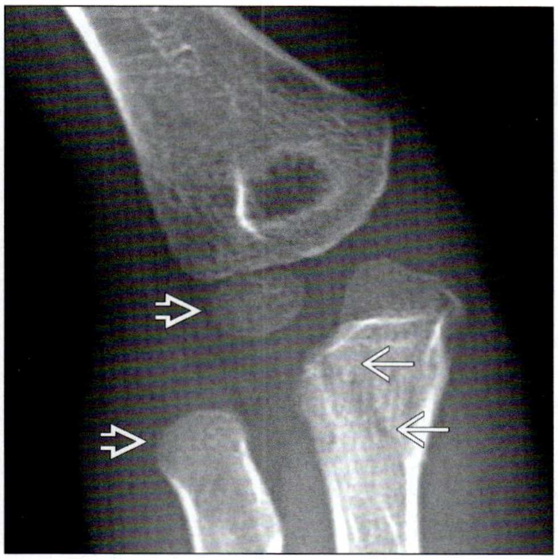
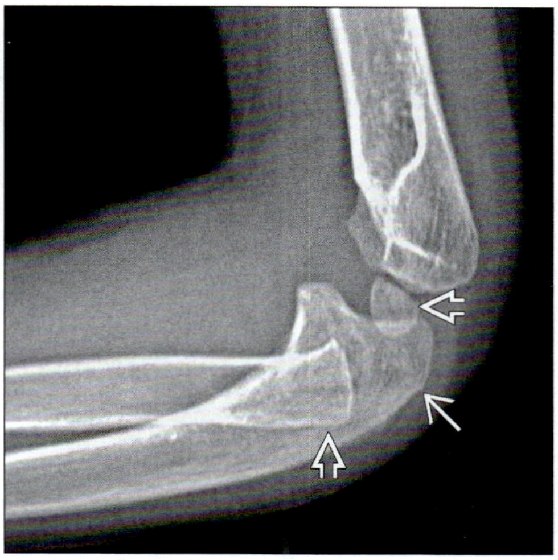

(Left) *AP radiograph from a 4 year old after a fall on an outstretched hand shows lateral dislocation of the radiocapitellar joint. A line extending through the proximal radial shaft should intersect the capitellum in all planes; here, it does not. There is also a proximal ulnar fracture.* ***(Right)*** *Lateral radiograph from the same patient shows posterior dislocation of the radiocapitellar joint. Posterior angulation at the site of the ulnar fracture is subtle. This is a type 2 injury.*

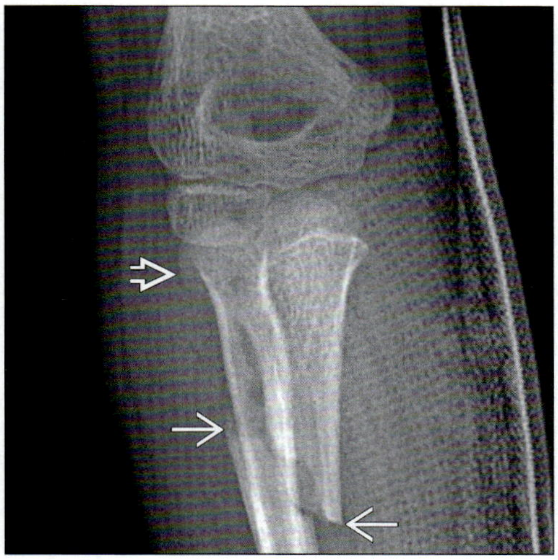
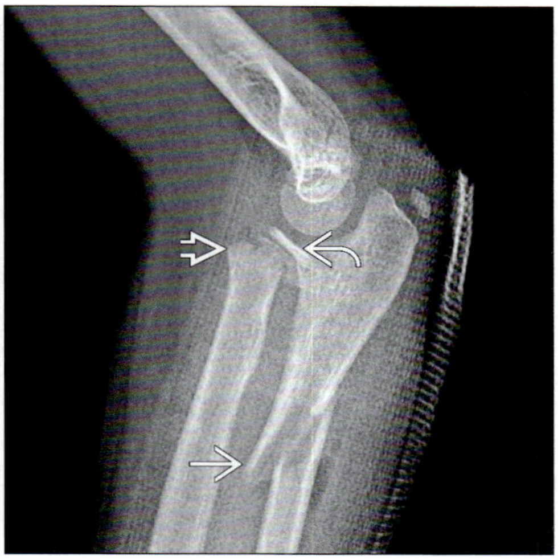

(Left) *AP radiograph from an 8 year old shows a displaced fracture of the proximal ulnar shaft. The radial head and neck are deformed, but alignment of the radiocapitellar joint is normal in this projection.* ***(Right)*** *Lateral radiograph from the same patient shows anterior angulation of the ulnar fracture. There is a fracture of the radial neck with anterior displacement, but the radial head aligns with the capitellum. This is functionally the same as a type 1 Monteggia injury.*

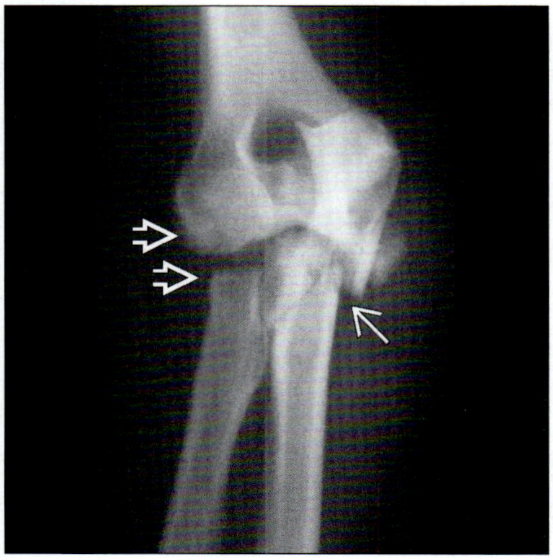
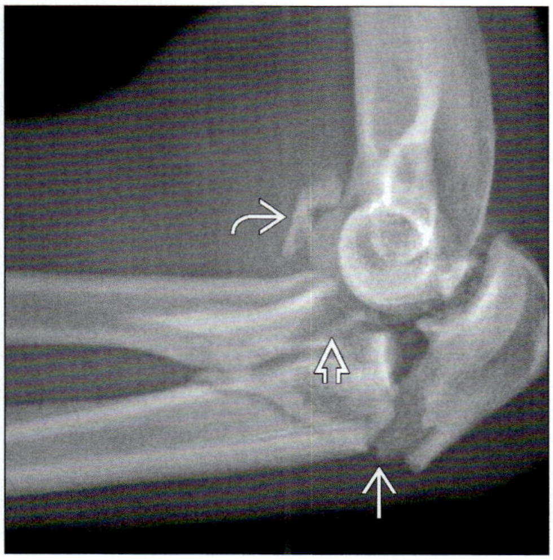

(Left) *AP radiograph of the elbow shows a comminuted fracture of the proximal ulna and a displaced and comminuted fracture of the radial head and neck.* ***(Right)*** *Lateral radiograph from the same patient shows the ulnar fracture angulated posteriorly. The fracture is located proximal to the coronoid (which is also fractured and displaced). The comminuted radial head and neck are posteriorly displaced. This is a fracture-dislocation of the elbow and not a Monteggia injury.*

RADIAL HEAD/NECK FRACTURE

Key Facts

Imaging

- Positive "fat pad" sign on lateral radiograph
- 50% of radial head and neck fxs are nondisplaced
- Radial head fx: Vertical fx line in articular cortex
- Radial neck fx: Buckle or angulation at radial head/neck junction
- Essex-Lopresti fracture
  - Severely comminuted radial head fracture
  - Proximal migration of radial shaft
  - Disruption of distal radioulnar joint (DRUJ), usually with dorsal subluxation/dislocation
- Radiographs for initial diagnosis
  - Angled radial head view may be useful

Pathology

- Impaction injury due to axial overloading of lateral elbow in fall on outstretched hand (FOOSH)
- Mason Johnston classification
  - Type 1 = displacement < 2 mm
  - Type 2 = displacement > 2 mm
  - Type 3 = comminution
  - Type 4 = comminution + dislocation

Clinical Issues

- 50% of adult elbow fractures
- 15% of pediatric elbow fractures
- Usually occur in radial neck
- Essex-Lopresti rare

Diagnostic Checklist

- Shortening of radius + severe radial head comminution should prompt evaluation of wrist
- Presence of other fractures
  - Elbow
  - Wrist

*(**Left**) Coronal graphic shows a comminuted radial head fracture. Radial head fractures are characterized using the the Mason classification system, in which type 1 injuries are displaced 2 mm or less, type 2 injuries are depressed more than 2 mm, grade 3 injuries are comminuted, and grade 4 injuries are associated with elbow dislocation. (**Right**) AP radiograph shows a minimally impacted and comminuted radial head fracture ➔ with propagation into the the head/neck junction ↪ as seen in the illustration.*

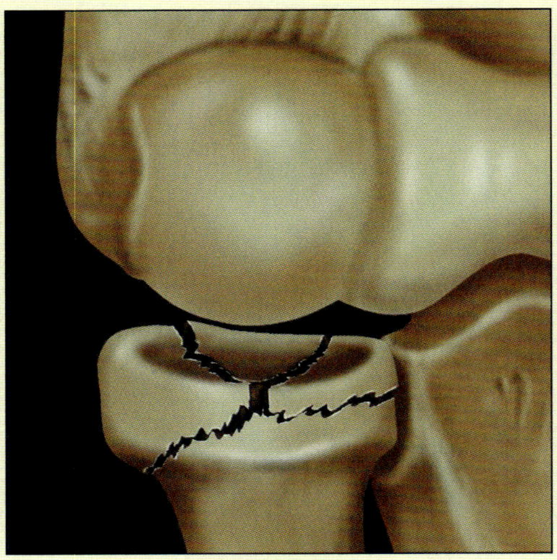

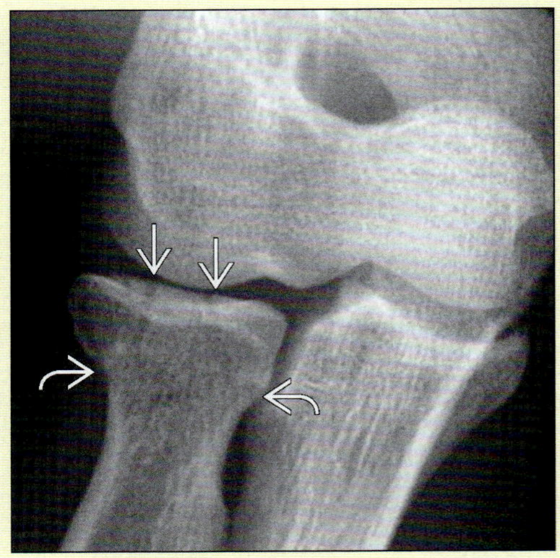

*(**Left**) AP radiograph demonstrates a comminuted radial head fracture with significant displacement of the subtended fracture fragment ➔. The articular surface of the displaced fragment is rotated 90° relative to its normal position. (**Right**) Lateral radiograph of the same patient shows that the radial head fragment ➔ is displaced posteriorly. There is distension of the periarticular fat pads ↪ indicating a hemarthrosis in this acute Mason 3 injury.*

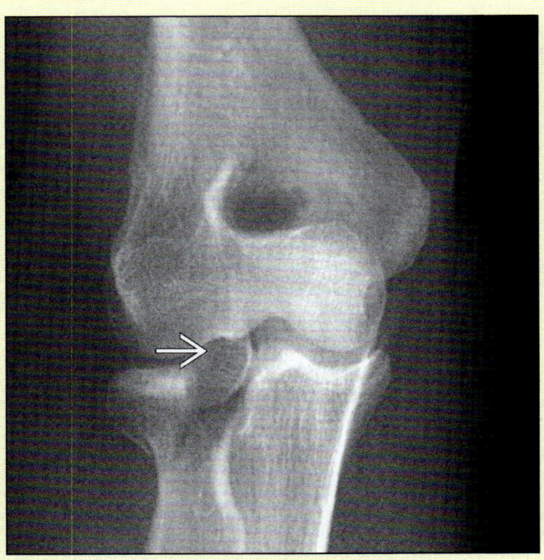

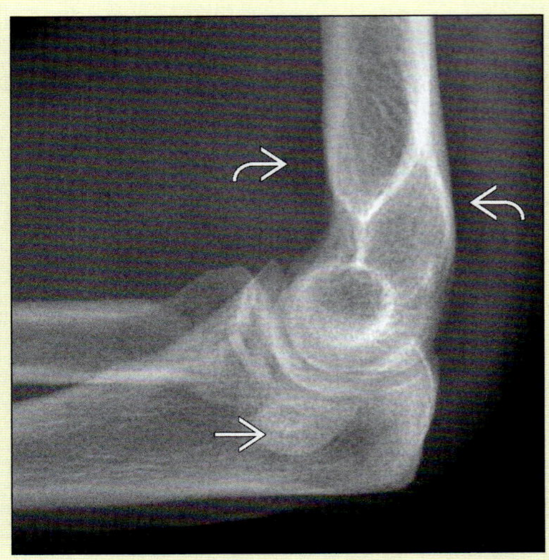

RADIAL HEAD/NECK FRACTURE

TERMINOLOGY

Synonyms
- Lateral column injury of elbow

IMAGING

General Features
- Best diagnostic clue
  - Positive "fat pad" sign on lateral radiograph
  - Radial head fracture: Vertical fracture line in articular cortex
  - Radial neck fracture: Buckle or angulation at radial head/neck junction
- Location
  - Radial head &/or neck
  - Anterolateral aspect of head more vulnerable because of lack of subchondral bone
- Size
  - Ranges from small nondisplaced to highly comminuted and significantly displaced
  - 50% of radial head and neck fxs are nondisplaced
- Morphology
  - Radial head fracture
    - Fracture line usually vertically oriented
    - Usually in lateral aspect of radial head
  - Radial neck fracture
    - Fracture line usually axially oriented
    - Impaction, particularly anteriorly
    - 20% are displaced
    - 20% are comminuted
  - Essex-Lopresti fracture
    - Severely comminuted radial head fracture
    - Proximal migration of radial shaft
    - Disruption of interosseous membrane of forearm
    - Disruption of distal radioulnar joint (DRUJ), usually with dorsal subluxation/dislocation

Radiographic Findings
- Positive "fat pad" sign
  - Anterior bowing of anterior fat pad
  - Radiographic presence of posterior fat pad
- Positive supinator "fat pad" sign
  - Blurring or loss of normal margin of supinator muscle anterior to radial head on lateral radiograph
- Radial head fracture
  - Linear fracture line
  - Abrupt step-off of cortical margin
  - "Double cortex" sign
  - Best seen on lateral view
    - Often need radial head view to confirm
- Radial neck fracture
  - Uncommon to see actual fracture line
  - Angulation at junction of radial head and neck
  - Best seen on lateral view
  - May see linear sclerosis in radial neck on AP view
- If highly comminuted or due to elbow dislocation, other fractures of adjacent bones (coronoid process, capitellum)
- Essex-LoPresti
  - Severely comminuted radial head fracture
  - Shortened radial shaft
  - Ulnar positive variance
  - DRUJ disruption
  - ± ulnar styloid fracture

CT Findings
- Can help identify subtle radial head fracture
- Characterize degree of comminution and displacement
- Evaluate stability of DRUJ in Essex-Lopresti fracture

MR Findings
- Fracture line in radial head or neck
- Marrow edema surrounding fracture
- Elbow effusion
- ± edema or disruption of ulnar collateral ligament (UCL) of elbow
- ± radiocapitellar joint disruption
- ± annular ligament tear

Imaging Recommendations
- Best imaging tool
  - Radiographs for initial diagnosis
    - Lateral view most helpful
    - Radial head view often useful
  - CT for precise mapping of fragments
  - MR for evaluation of UCL or annular ligament
- Protocol advice
  - Angled radial head view (isolates head from overlap) may be useful additional view
  - Include views of forearm and wrist if
    - Radial head severely comminuted
    - Radial shaft significantly shortened

DIFFERENTIAL DIAGNOSIS

Capitellum Fracture
- Displaced fragment located proximal to radial head
- Radial head fragments rarely migrate proximally

Osteoarthritis, Elbow
- Ring osteophyte at neck can simulate impacted fx

Elbow Dislocation
- Associated olecranon fx and ligament disruptions

Floating Elbow
- Distal humeral fracture + proximal forearm fractures

PATHOLOGY

General Features
- Etiology
  - Impaction injury due to axial overloading of lateral elbow in fall on outstretched hand (FOOSH)
  - Valgus carrying angle of elbow predisposes to lateral-sided injury
- Associated abnormalities
  - "Terrible triad of elbow"
    - Radial head fracture
    - Coronoid process fracture
    - Elbow dislocation
  - Capitellum fracture
    - Displaced fragment lies proximal to radial head
  - Distal radial fracture (Colles, etc.)
  - Scaphoid fracture
  - Triceps tendon rupture

RADIAL HEAD/NECK FRACTURE

- Dislocation of DRUJ (Essex-Lopresti)
- Injury to ulnar collateral ligament of elbow (valgus mechanism)

Staging, Grading, & Classification

- Mason Johnston classification radial head fracture
  - Type 1 = displacement < 2 mm
  - Type 2 = displacement > 2 mm
    - Segmental, involving < 30% of radial head
  - Type 3 = comminution
  - Type 4 = comminution + dislocation

CLINICAL ISSUES

Presentation

- Most common signs/symptoms
  - Point tenderness over radial head
  - Elbow pain and loss of function
- Other signs/symptoms
  - Wrist tenderness if associated wrist fracture or DRUJ disruption
- Clinical profile
  - FOOSH

Demographics

- Age
  - Radial head: Adults
  - Radial neck: Adults and children
- Gender
  - M > F
- Epidemiology
  - 50% of adult elbow fractures
    - Most common elbow fracture in adults
  - 15% of pediatric elbow fractures
    - Usually occur in radial neck
  - Essex-Lopresti rare

Natural History & Prognosis

- Early complications
  - Compartment syndrome
  - Neurovascular injury: Posterior interosseous nerve (uncommon)
  - Infection
- Late complications
  - Hardware failure
  - Nonunion/malunion
  - Infection
  - Synostosis
  - Persistent pain and stiffness

Treatment

- Radial neck fracture
  - Almost always treated conservatively unless severely comminuted or displaced
- Mason 1 fracture
  - Generally treated conservatively
    - Casting with early motion
- Mason 2 fracture
  - "Rule of 3s": Treated conservatively if
    - < 30% of articular surface involved
    - < 30° angulation of fracture
    - Displacement < 3 mm
- Mason 3 or 4 fracture
  - Generally requires ORIF
  - Resection of 1 or more fragments may be required
  - Resection of radial head and prosthesis placement if valgus instability persists
    - Usually required if comminution extends distal to annular ligament
- Essex-Lopresti fracture
  - ORIF of radial head fracture
  - If severely comminuted, radial head resection may be necessary
  - Stabilization of DRUJ

DIAGNOSTIC CHECKLIST

Consider

- Extent of comminution
- Presence of other fractures
  - Elbow
  - Wrist
- Presence of valgus instability
  - UCL injury
- Essex-Lopresti fracture

Image Interpretation Pearls

- Shortening of radius + severe radial head comminution should prompt evaluation of wrist
  - Additional fractures
  - Disruption of DRUJ (Essex-Lopresti)

Reporting Tips

- Describe displacement and comminution of radial head fracture if present
  - > 2 mm displacement = Mason 2 or greater
  - Comminuted = Mason 3

SELECTED REFERENCES

1. Dodds SD et al: Essex-lopresti injuries. Hand Clin. 24(1):125-37, 2008
2. Rosenblatt Y et al: Current recommendations for the treatment of radial head fractures. Orthop Clin North Am. 39(2):173-85, vi, 2008
3. Jackson JD et al: Radial head fractures. Hand Clin. 23(2):185-93, vi, 2007
4. Tejwani NC et al: Fractures of the radial head and neck: current concepts in management. J Am Acad Orthop Surg. 15(7):380-7, 2007
5. Auyeung J et al: The Essex-Lopresti lesion: a variant with a bony distal radioulnar joint injury. J Hand Surg [Br]. 31(2):206-7, 2006
6. Bano KY et al: Radial head fractures--advanced techniques in surgical management and rehabilitation. J Hand Ther. 19(2):114-35, 2006
7. Cunningham PM: MR imaging of trauma: elbow and wrist. Semin Musculoskelet Radiol. 10(4):284-92, 2006
8. Roidis NT et al: Current concepts and controversies in the management of radial head fractures. Orthopedics. 29(10):904-16; quiz 917-8, 2006

RADIAL HEAD/NECK FRACTURE

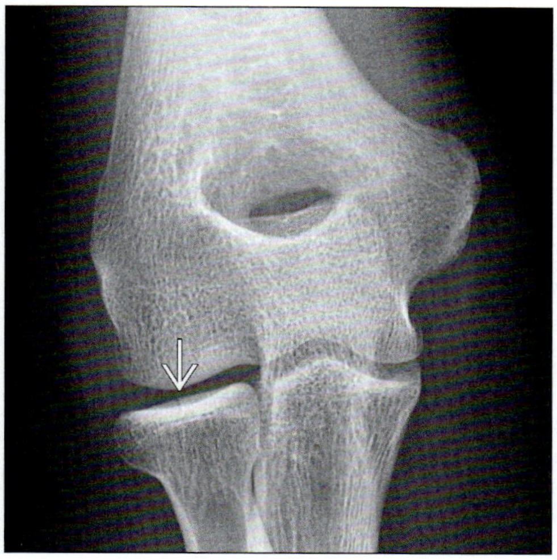

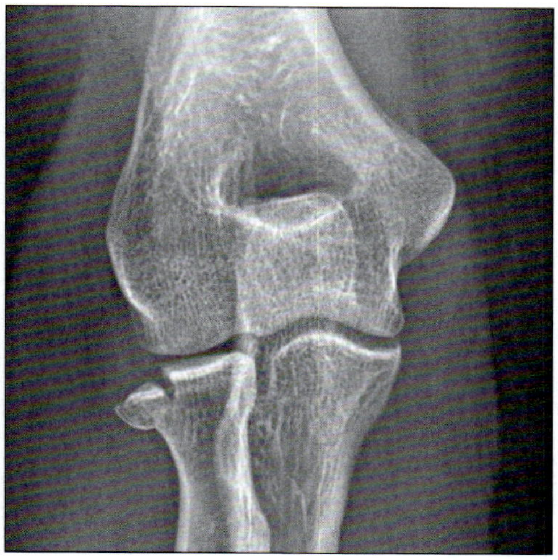

(Left) *AP radiograph of the elbow demonstrates a subtle nondisplaced articular surface fracture of the radial head ➔. Fractures displaced less than 2 mm are classified as Mason 1 injuries.* ***(Right)*** *Oblique radiograph of the elbow shows a fracture of the radial head with 4 mm of displacement. Fractures displaced more than 2 mm and involving less than 30% of the articular surface are categorized as Mason 2 injuries.*

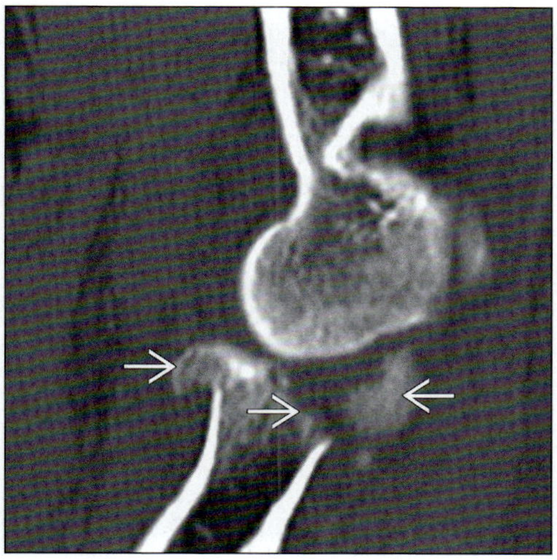

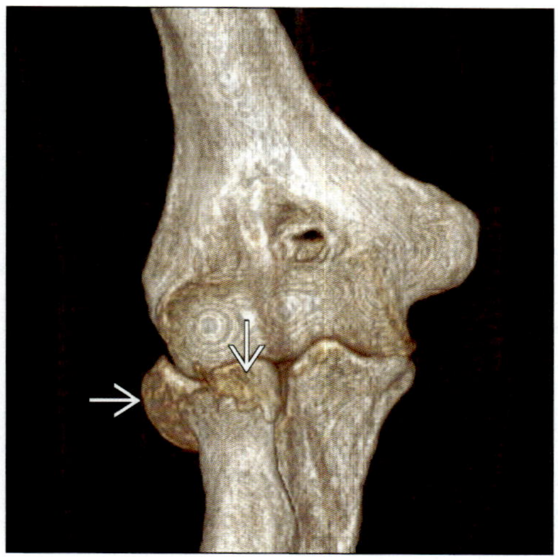

(Left) *Sagittal reformat bone CT through the radiocapitellar joint demonstrates comminution of the radial head ➔, consistent with a Mason 3 fracture pattern.* ***(Right)*** *Surface-rendered 3-D CT reformat of the same patient shows the impaction and comminution of the radial head ➔. Although such reconstructions often provide no incremental benefit to the interpreting radiologist, orthopedic surgeons often find them helpful in planning.*

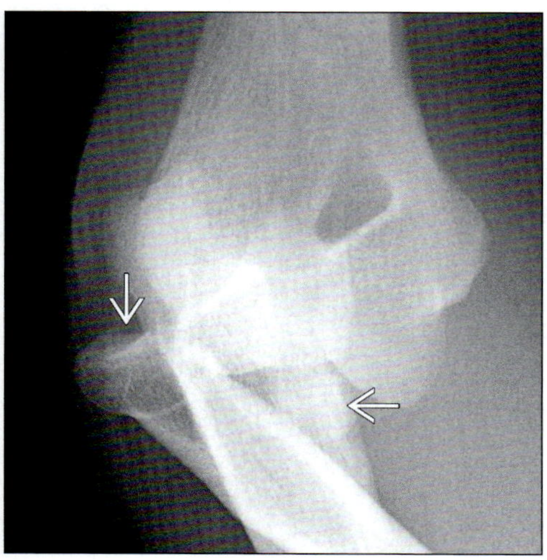

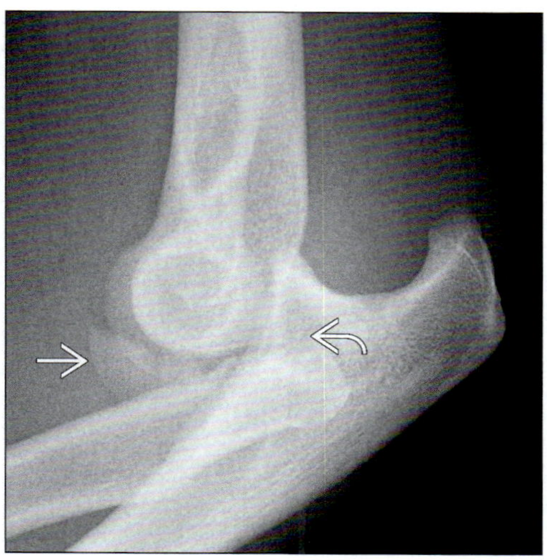

(Left) *AP radiograph demonstrates posterolateral dislocation of the elbow and radiocapitellar joint. There is a displaced fracture of the radial head ➔.* ***(Right)*** *Lateral radiograph of the same patient confirms posterior and lateral dislocation. The radial head fragment ➔ is better seen. The defect in the radial head ↰ is also well seen. This combination of elbow dislocation and radial head fracture constitutes a Mason 4 injury.*

RADIAL HEAD/NECK FRACTURE

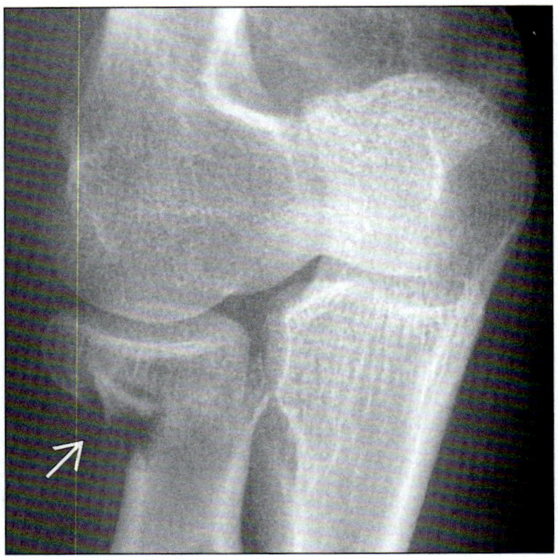

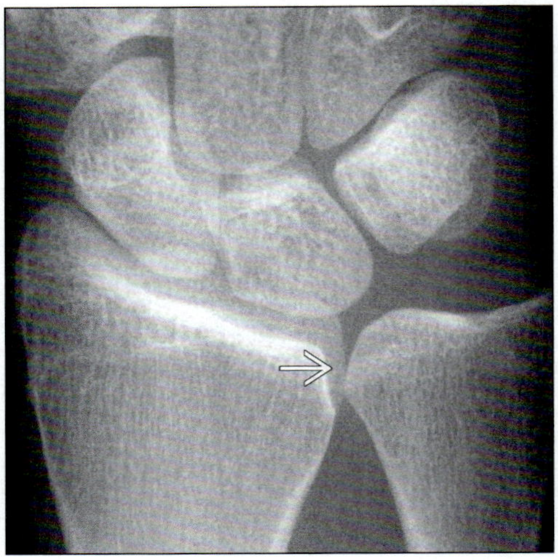

(Left) *Oblique radiograph shows a comminuted radial head/neck fracture showing displacement as well as some impaction ➔, revealing that this patient had high-energy trauma. This should alert the observer to search for a more substantial injury, such as MCL tear, coronoid process fracture, or Essex-Lopresti injury.* ***(Right)*** *PA radiograph of the wrist in the same patient confirms Essex-Lopresti injury with relative ulnar positive variance ➔ and associated instability at the distal radioulnar joint.*

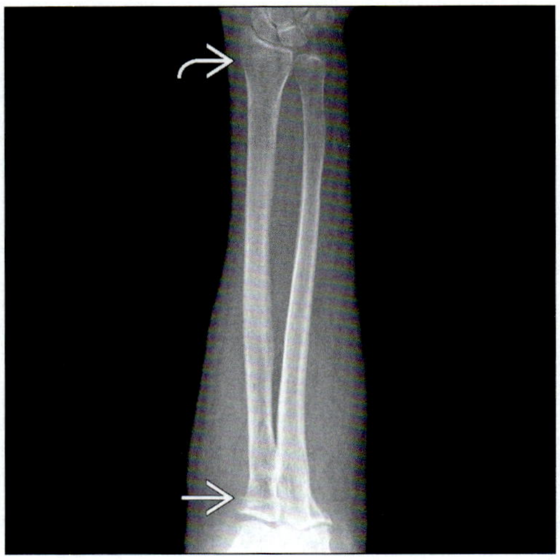

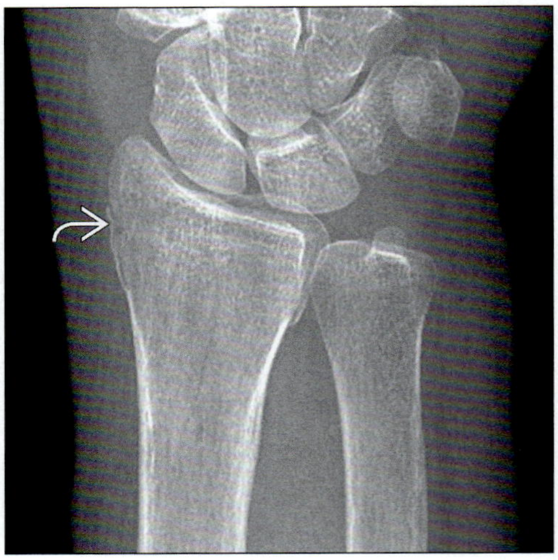

(Left) *AP radiograph of the forearm shows an impacted radial neck fracture ➔. The patient also complained of wrist pain; a nondisplaced distal radial fracture is also present ➔.* ***(Right)*** *Magnified AP view of the wrist in the same patient better shows the distal radial fracture ➔ with entry into the articular surface. Axial forces that cause radial head and neck fractures can sometimes cause concurrent wrist injuries; these should be carefully sought.*

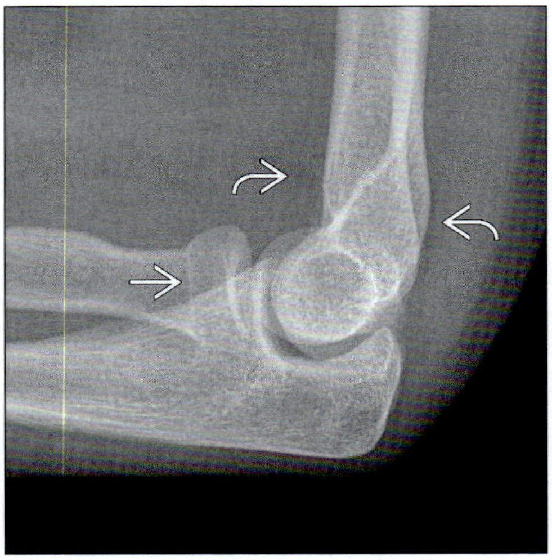

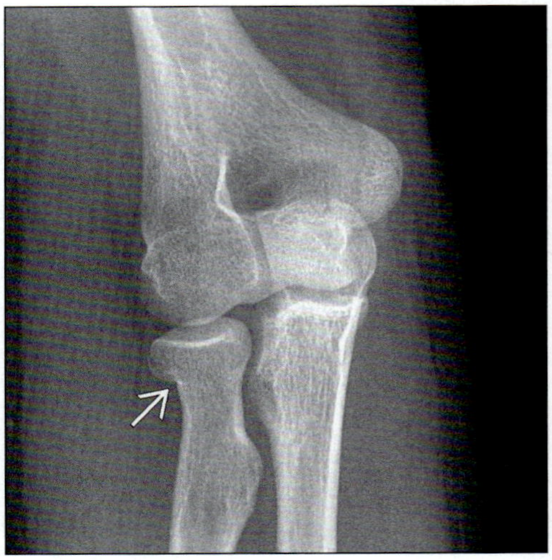

(Left) *Lateral radiograph of the elbow in the setting of acute trauma demonstrates distended periarticular fat pads ➔, attesting to a hemarthrosis. Note the sclerotic band of density across the radial neck ➔ indicating a subtle impaction fracture. Because this is a young patient, osteophytes should not be included in the differential diagnosis.* ***(Right)*** *Oblique radiograph of the same patient shows an impaction buckle of the lateral radial neck cortex ➔ confirming the fracture.*

RADIAL HEAD/NECK FRACTURE

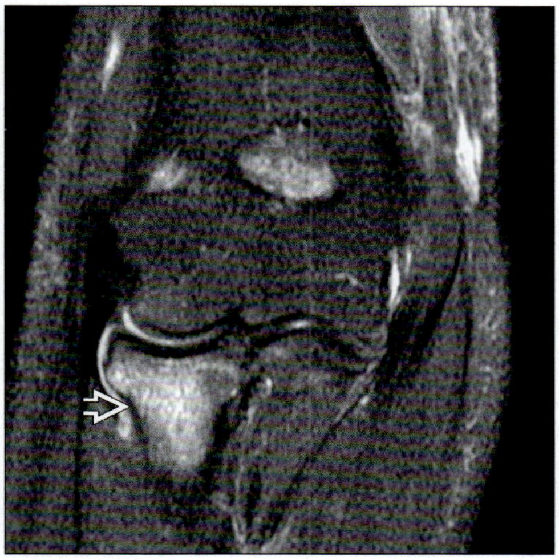

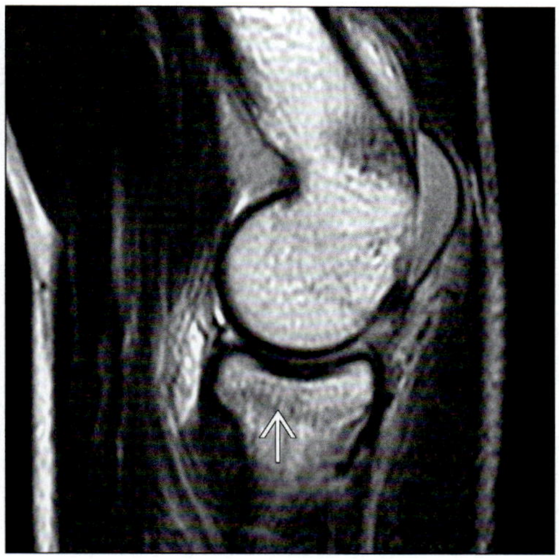

(Left) *This patient with elbow pain experienced trauma, but radiographs (not shown) showed no fracture. Coronal T2WI FS MR demonstrates marrow edema in the radial neck ⇨. In this setting, the patient should be treated for a nondisplaced radial neck fracture.* ***(Right)*** *Sagittal T2WI MR from the same patient shows a subtle fracture line ➔ in the radial neck with no cortical discontinuity or buckling. Such radiographically occult fractures may benefit from timely MR to guide therapy.*

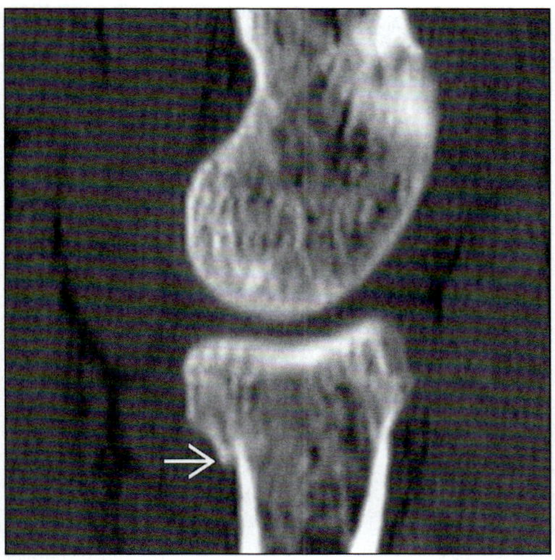

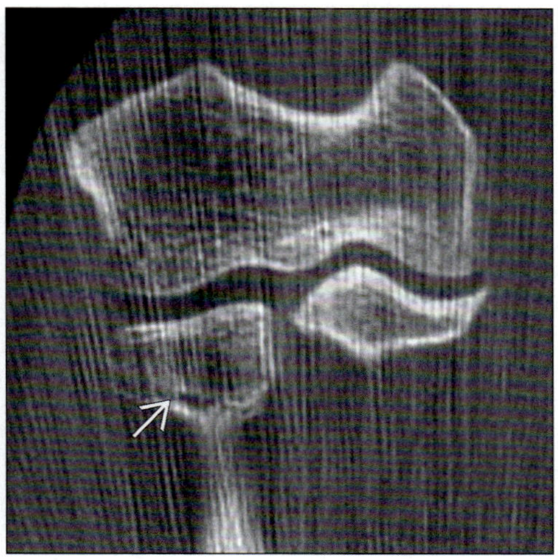

(Left) *Sagittal reformat bone CT in a patient whose radiographs were interpreted as normal after elbow trauma shows a subtle stepoff in the anterior cortex of the radial neck ➔.* ***(Right)*** *Coronal reformat bone CT in the same patient shows the fracture line ➔ propagating across the radial neck. Even in retrospect, this fracture could not be identified on radiograph. Though CT is best known for identifying and characterizing radial head fractures, sometimes neck fractures are discovered.*

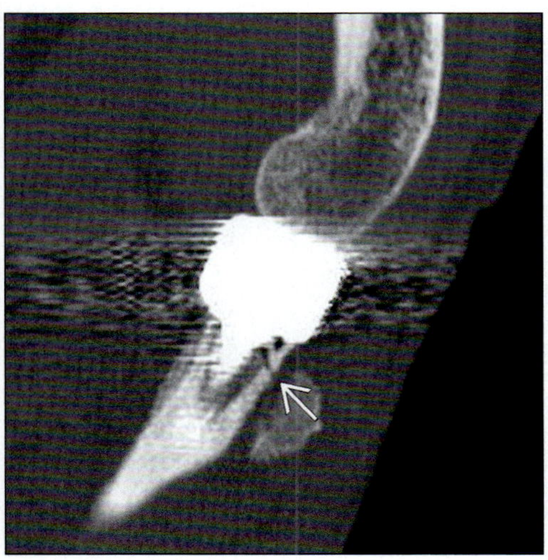

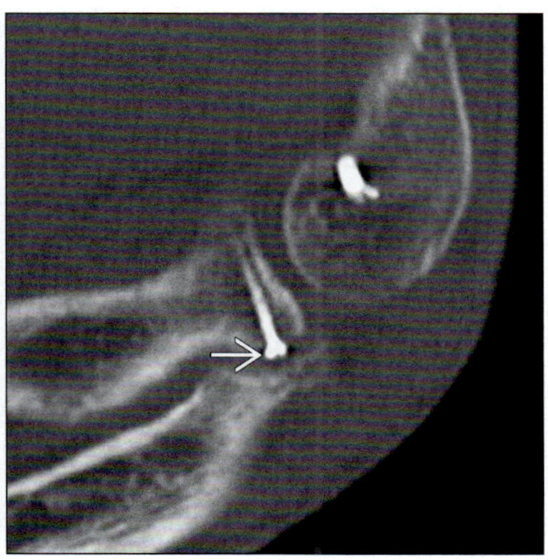

(Left) *Sagittal reformat bone CT in a patient with a radial head prosthesis demonstrates a new fracture ➔ through the posterior cortex of the radius at the level of the prosthesis stem. The fracture could not be seen on radiographs.* ***(Right)*** *Sagittal reformat bone CT in a patient who has undergone screw repair of radial and capitellar fractures shows protrusion of the radial head screw ➔ into the proximal radioulnar joint. The patient complained of pain on supination/pronation.*

OLECRANON FRACTURE

Key Facts

Terminology
- Fracture through olecranon process of ulna due to
  - Direct blow
  - Distraction injury at triceps insertion

Imaging
- Can extend medially to involve sublime tubercle
  - Attachment site of ulnar collateral ligament anterior band
  - Associated with increased instability
- Usually seen best on lateral projection

Top Differential Diagnoses
- Triceps rupture without osseous component (rare)
- Trochlear notch
  - Normal anatomic groove
- Normal ossification center
  - Well-marginated sclerotic edges
  - Fuses at 14-18 years
- Osteotomy for surgical access to elbow joint
  - Straight, transverse olecranon lucency with fixation

Pathology
- AO classification
  - Type A: Extraarticular fracture
  - Type B: Intraarticular fracture
  - Type C: Intraarticular fractures of both olecranon and radial head

Clinical Issues
- 20% of adult elbow fractures
- 6% of pediatric elbow fractures
- Nondisplaced fractures treated conservatively
- Surgery for displacement ≥ 2 mm

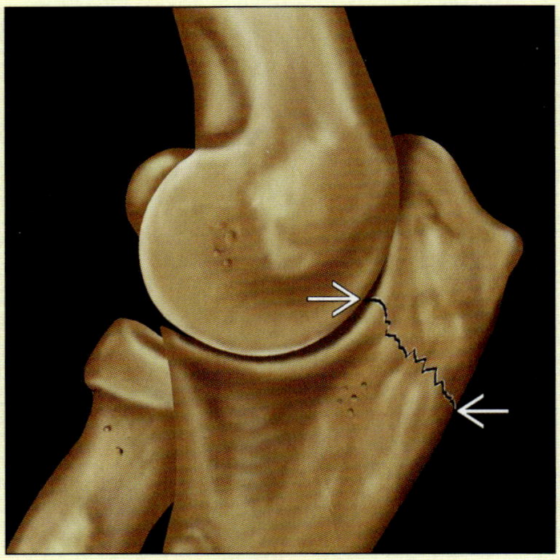
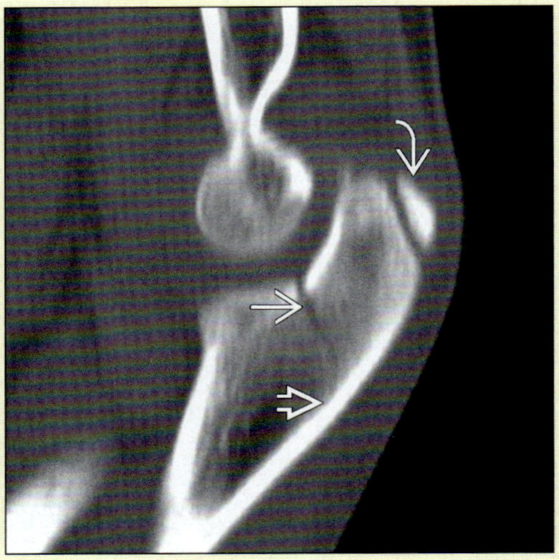

(Left) *Graphic representation of nondisplaced olecranon fracture ➔ through the midportion of the olecranon. Such fractures are frequently distracted due to traction from the attached triceps tendon.* ***(Right)*** *Sagittal reformat bone CT in a 14-year-old boy after a fall on the elbow shows a nondisplaced intraarticular fracture of the olecranon ➔. The fracture is not displaced in this case because it is an incomplete fracture ➔. The unfused apophysis ➔ is also seen.*

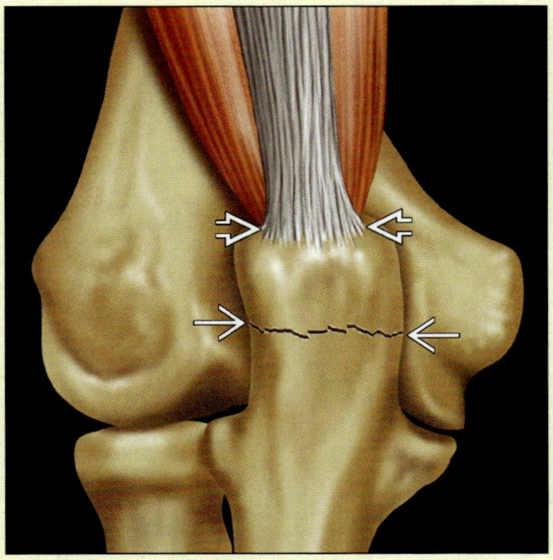
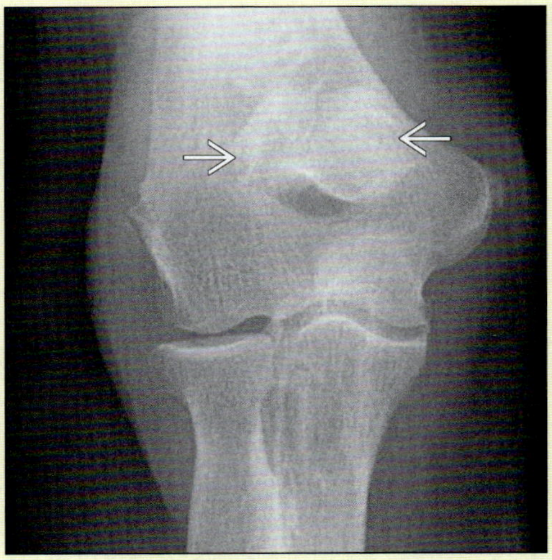

(Left) *Coronal graphic representation of a nondisplaced fracture of the olecranon ➔ seen from posterior. Note the broad attachment of the triceps tendon ➔.* ***(Right)*** *Anteroposterior radiograph demonstrates an acute olecranon fracture with significant diastasis of the subtended fragment ➔. Complete fractures, such as this, are more often distracted due to the tensile forces imparted by the triceps tendon on the dissociated fragment.*

OLECRANON FRACTURE

TERMINOLOGY

Definitions

- Fracture through olecranon process of ulna due to
  - Direct blow
  - Distraction injury at triceps insertion

IMAGING

General Features

- Best diagnostic clue
  - Linear fracture through olecranon on lateral elbow radiograph
- Location
  - Varies from base of olecranon to tip
  - Most are intraarticular
- Size
  - Fragment sizes depend on fracture location
- Morphology
  - Fracture line generally in axial plane
  - May be transverse, oblique, or comminuted
  - Can extend medially to involve sublime tubercle
    - Attachment site of ulnar collateral ligament anterior band
    - Associated with increased instability

Radiographic Findings

- May range from subtle nondisplaced fracture line to marked comminution and displacement
- Usually seen best on lateral projection
- Small markedly displaced fragment suggests triceps avulsion
- May extend to include coronoid process

MR Findings

- Fracture line best seen on sagittal and coronal T1WI and T2WI FS
- Stress injury may just show marrow edema without fracture line
- Small fractures at tip may be missed on MR

Imaging Recommendations

- Best imaging tool
  - Lateral radiograph
  - MR to assess integrity of triceps
- Protocol advice
  - Radiographs: Lateral view to assess size and position of fragment(s)
  - CT: Sagittal reformats to assess size of articular surface gap
  - MR: Sagittal and axial T2WI FS for assessment of fracture and triceps integrity

DIFFERENTIAL DIAGNOSIS

Triceps Rupture without Osseous Component (Rare)

- Pain and swelling near tip of olecranon
- Inability to extend elbow
- No osseous abnormality on radiograph

Trochlear Notch

- Normal anatomic groove

Normal Ossification Center

- Well-marginated sclerotic edges
- Fuses at 14-18 years
- Os supratrochlear: Failure of fusion

Patella Cubiti

- Elongate sesamoid in distal lateral triceps
- Smoothly corticated
- Synonym: Bipartite olecranon

Monteggia Injury

- Proximal ulnar shaft fracture (extraarticular)
- Radiocapitellar dislocation

Olecranon Bursitis

- Presents with pain over olecranon
- No osseous abnormality on radiograph

Osteotomy

- Transverse osteotomy through olecranon is common open surgical approach to elbow
- Straight, transverse olecranon lucency with fixation (screw or figure-of-8 tension band)

PATHOLOGY

General Features

- Etiology
  - Direct blow injury
    - Fall on olecranon
    - Direct impact by object
  - Distraction
    - Fall on flexed elbow
    - Elbow extension against resistance
  - Hyperextension
    - Olecranon impacts posterior trochlear groove
  - Valgus stress
    - Stress injury in overhead throwers
    - Tends to involve medial aspect

Staging, Grading, & Classification

- AO classification
  - Type A: Extraarticular fracture
  - Type B: Intraarticular fracture
  - Type C: Intraarticular fractures of both olecranon and radial head
- Schatzker classification
  - Type A: Simple transverse fracture
  - Type B: Transverse impacted fracture
  - Type C: Oblique fracture
  - Type D: Comminuted fracture
  - Type E: More distal fracture (extraarticular)
  - Type F: Fracture/dislocation

CLINICAL ISSUES

Demographics

- Age
  - Adults > children
- Gender
  - M > F
- Epidemiology
  - 20% of adult elbow fractures
  - 6% of pediatric elbow fractures

OLECRANON FRACTURE

Minimally displaced olecranon fracture with no evidence of entry into coronoid process. The radiocapitellar and elbow joints are maintained.

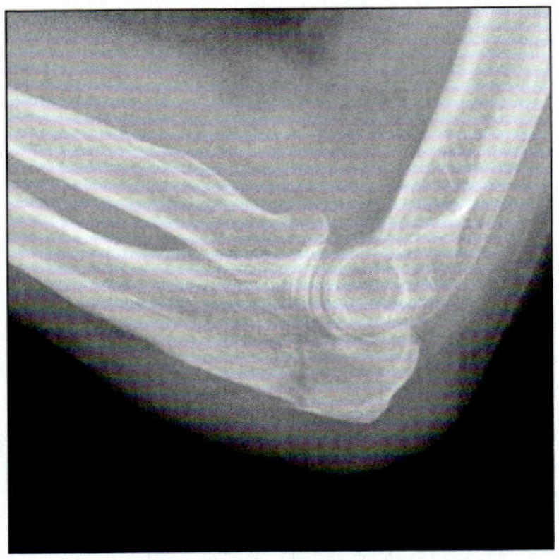

Same olecranon fracture reduction after pin and wire (tension band) fixation. For simple olecranon fractures, this is an effective treatment.

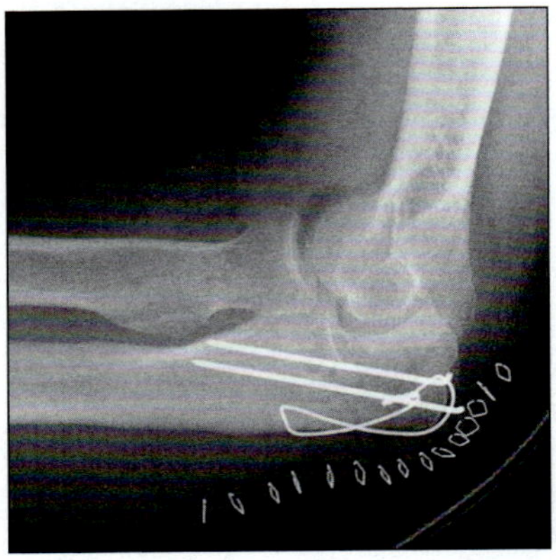

Complex distal humeral fracture. The olecranon is intact. Fixation of the humeral fracture requires olecranon osteotomy to expose the humerus.

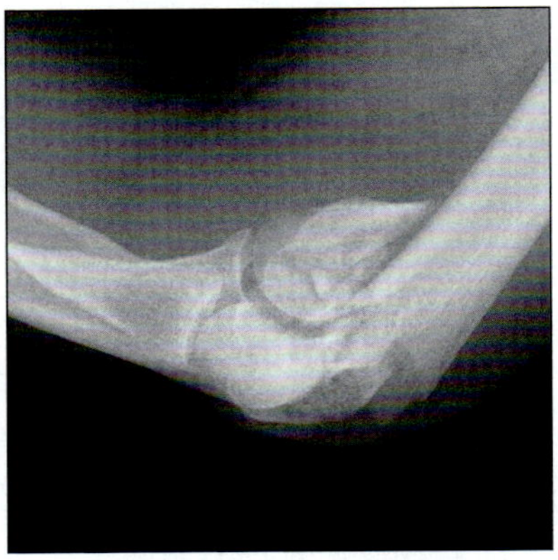

Surgical reduction of the bicondylar humeral fracture has been performed. The osteotomy has been reduced with plate and screw fixation.

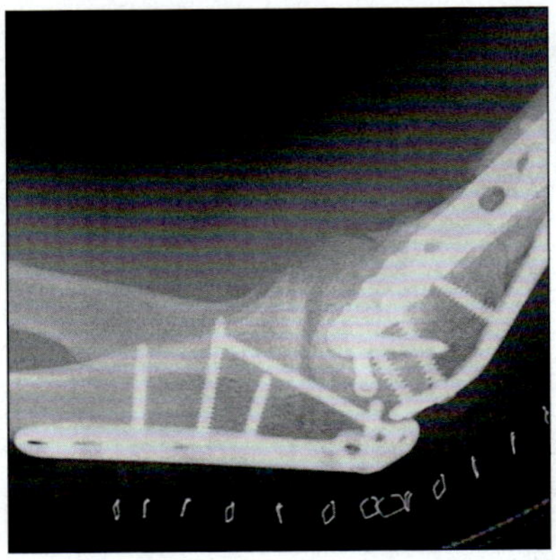

The most common types of olecranon fracture fixation are tension band wiring (row 1) and plate and screw fixation (row 2). Plate and screw fixation is usually reserved for more complex fractures, especially those extending into the coronoid process.

Natural History & Prognosis

- Complications
  - Decreased range of motion
  - Osteoarthritis
  - Hardware complications
    - Up to 80% in tension band fixation
  - Heterotopic ossification (13-14%)
  - Nonunion (< 5%)
  - Ulnar nerve injury (10%)
  - Infection (0-6%)

Treatment

- Nondisplaced fractures treated conservatively
- Surgery for displacement ≥ 2 mm
  - Fragment excision and triceps advancement
    - Elderly, osteoporotic patients
    - Inability to reattach fragment primarily
  - Tension band wiring
    - Most secure and simplest method
  - Plate fixation
    - Comminuted fractures
    - Fractures involving coronoid
    - Monteggia injuries
  - Intramedullary screw fixation
    - Often combined with tension band

DIAGNOSTIC CHECKLIST

Consider

- Triceps disruption
- Monteggia injury
- Other associated fractures

Image Interpretation Pearls

- Findings on AP view may be subtle or absent

Reporting Tips

- Describe size and displacement of fragment(s)
- Assess for other fractures

SELECTED REFERENCES

1. Huang TW et al: Tension band wiring for olecranon fractures: relative stability of Kirschner wires in various configurations. J Trauma. 68(1):173-6, 2010
2. Buijze G et al: Clinical evaluation of locking compression plate fixation for comminuted olecranon fractures. J Bone Joint Surg Am. 91(10):2416-20, 2009
3. Iannuzzi N et al: Excision and advancement in the treatment of comminuted olecranon fractures. J Orthop Trauma. 23(3):226-8, 2009
4. Kloen P et al: Treatment of proximal ulna and olecranon fractures by dorsal plating. Oper Orthop Traumatol. 21(6):571-85, 2009
5. Newman SD et al: Olecranon fractures. Injury. 40(6):575-81, 2009
6. Veillette CJ et al: Olecranon fractures. Orthop Clin North Am. 39(2):229-36, vii, 2008
7. McKay PL et al: Fractures of the proximal ulna olecranon and coronoid fractures. Hand Clin. 18(1):43-53, 2002
8. Nork SE et al: Surgical treatment of olecranon fractures. Am J Orthop. 30(7):577-86, 2001
9. Hak DJ et al: Olecranon fractures: treatment options. J Am Acad Orthop Surg. 8(4):266-75, 2000

OLECRANON FRACTURE

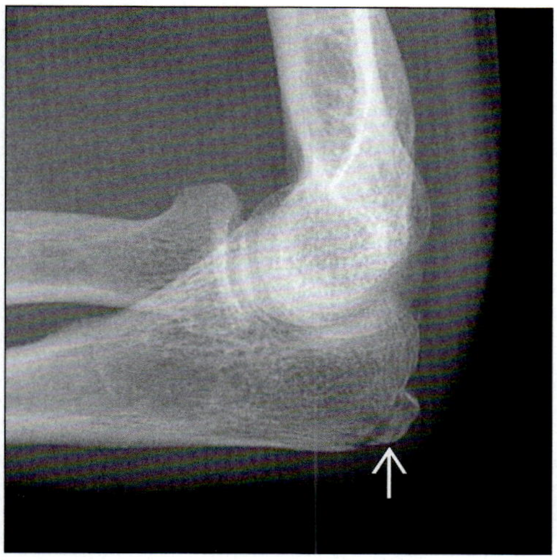
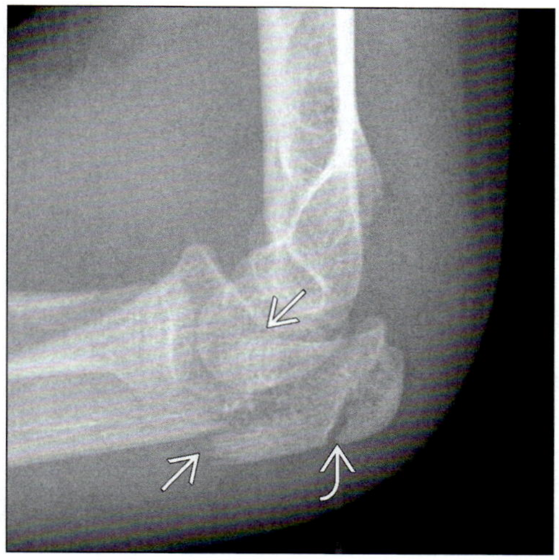

(Left) *Lateral radiograph of the elbow demonstrates a small fracture fragment ➔ at the tip of the olecranon. Edges of the fragment are irregular, distinguishing this injury from an ununited apophyseal ossification center.* ***(Right)*** *Lateral radiograph in a child after a fall demonstrates an oblique fracture through the midportion of the olecranon ➔ with mild displacement. By contrast, the physis of the olecranon ↗ demonstrates undulating, well-corticated margins.*

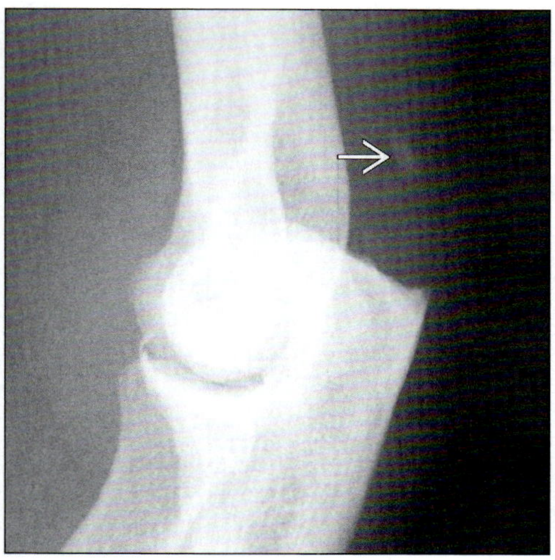
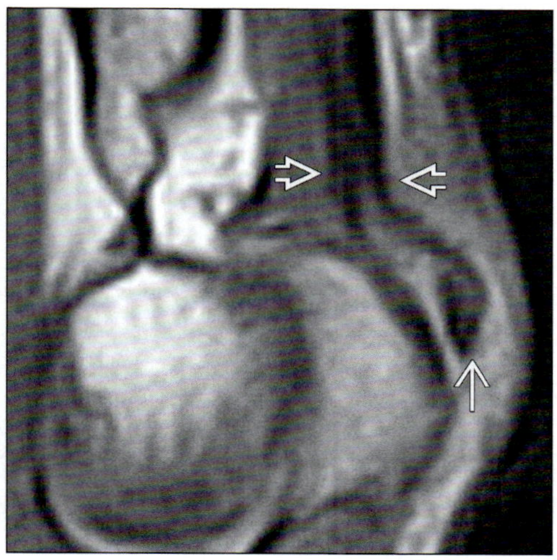

(Left) *Lateral radiograph of the elbow after a fall on an outstretched hand with a flexed elbow demonstrates a displaced fragment ➔ arising from the tip of the olecranon. The degree of displacement suggests a significant injury to the triceps.* ***(Right)*** *Sagittal PDWI MR in a different patient shows partial avulsion of the triceps tendon ⇨ with a small avulsed osseous fragment ➔. MR imaging can be helpful in gauging the degree of triceps injury.*

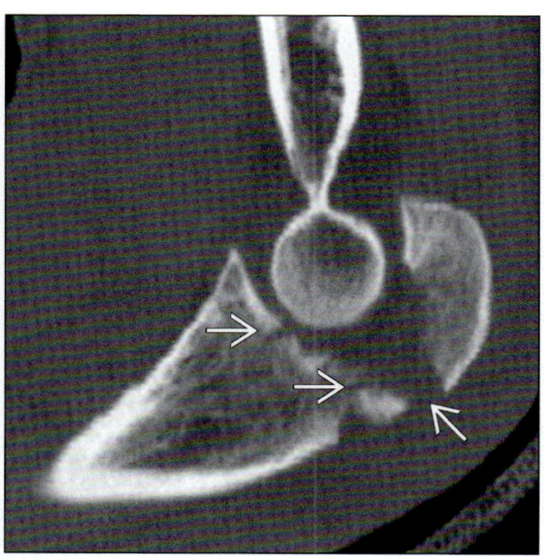
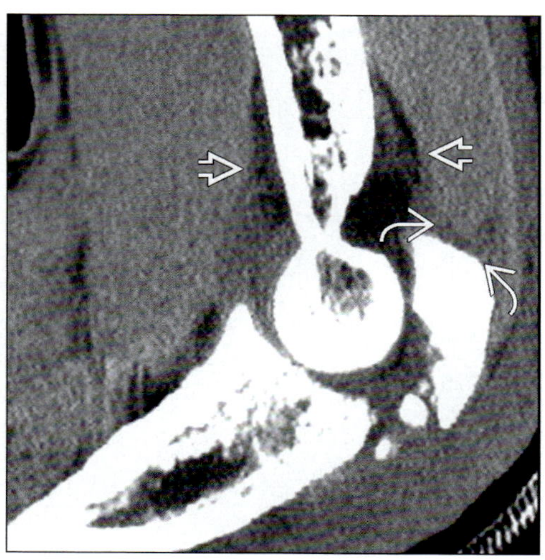

(Left) *Sagittal reformat bone CT shows comminution of the olecranon after a fall onto the elbow. Multiple sites of entry into the articular surface are present ➔.* ***(Right)*** *Sagittal reformat bone CT in soft tissue windows from the same patient shows the attachment of the triceps tendon ↗ and displacement of the periarticular fat pads ⇨. While CT can help judge the integrity of the triceps unit, MR imaging is more accurate.*

OLECRANON FRACTURE

(Left) *Lateral radiograph of the elbow in a 73-year-old woman after minor trauma to the elbow shows a distracted fracture of the olecranon. The margin of the displaced fragment is poorly defined ➔, and the entire fragment and the end of the native ulna are somewhat lucent ➔, suggesting an underlying lesion. This was a pathologic fracture through a myelomatous deposit.* **(Right)** *AP radiograph of the same patient shows medially displaced fragments ➔ from the expansile olecranon mass.*

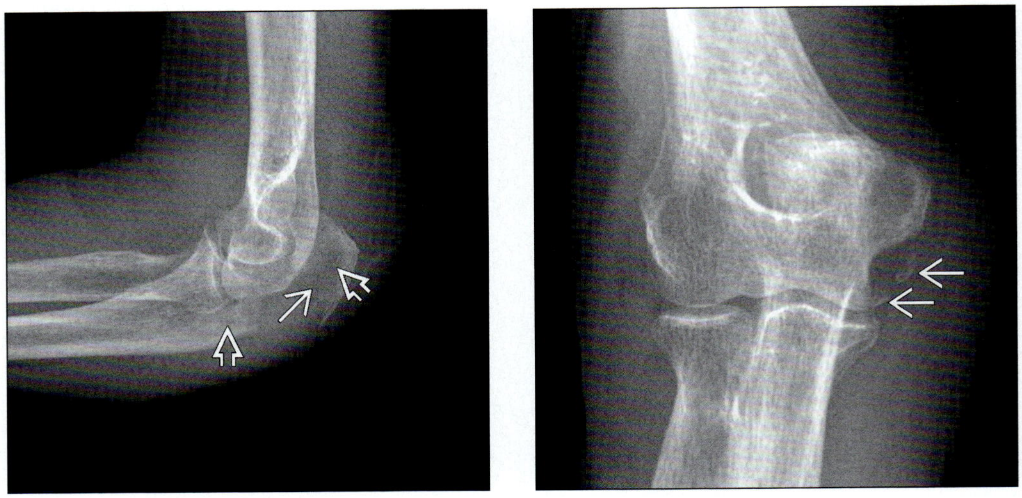

(Left) *Lateral radiograph of the elbow shows a chronic nonunion of an olecranon fracture. There is a large well-corticated osseous fragment ➔ posterior to the distal humerus, representing an ununited olecranon fracture fragment. A 2nd corticated fragment ➔ is less displaced, but also ununited.* **(Right)** *Anteroposterior radiograph in the same patient shows the major ➔ and minor ➔ fragments from the chronic nonunion of the olecranon.*

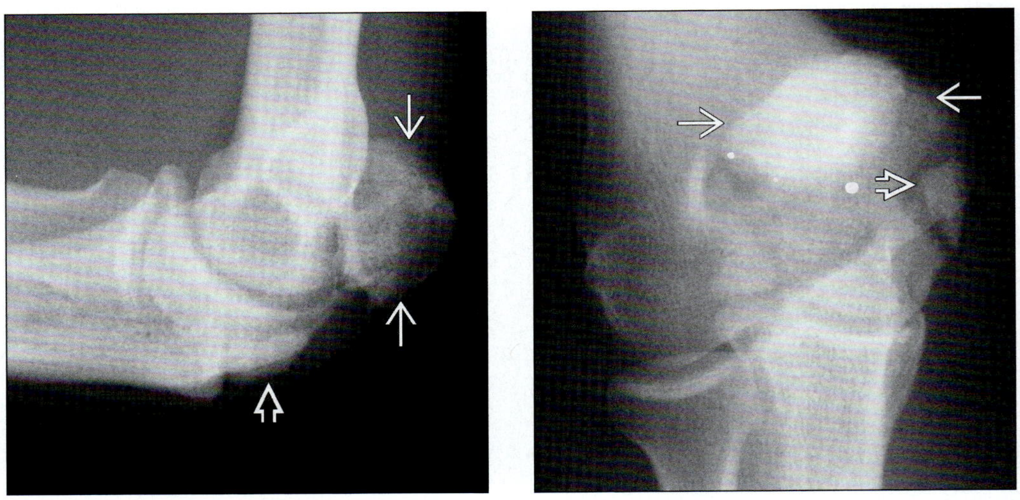

(Left) *Lateral radiograph of the elbow shows posterior dislocation of the elbow. The displaced fragment ➔ likely represents the entire olecranon apophysis, and is rotated and distracted away from the native ulna ➔. The radiocapitellar joint is dislocated as well ➔.* **(Right)** *Oblique radiograph from the same patient shows the displaced olecranon ossification center ➔ separated from the native ulna ➔. This is the equivalent of a fracture-dislocation (Schatzker type F).*

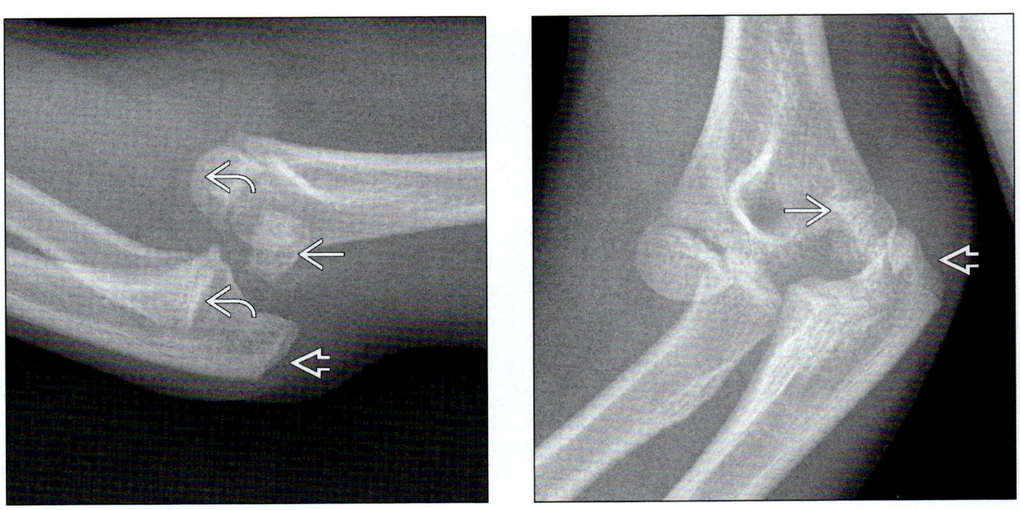

OLECRANON FRACTURE

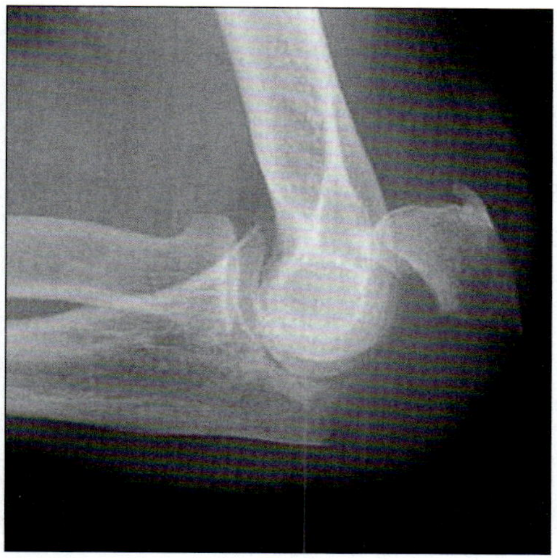

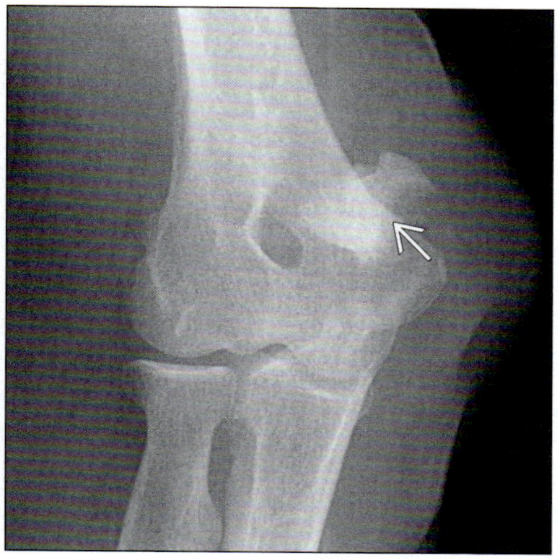

(Left) *Lateral radiograph of the elbow demonstrates an acute olecranon fracture with very significant diastasis. This results from the tensile stress of sudden contraction of a large muscle (triceps) on its insertion on a small bone such as the olecranon.* ***(Right)*** *Oblique radiograph in the same patient shows the displacement and rotation of the distracted fragment. Note the ill-defined edges of the fracture fragment ➡, indicating an acute time course for the injury.*

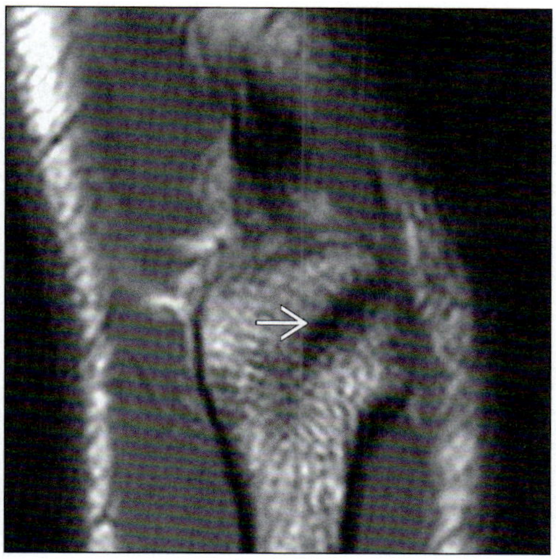

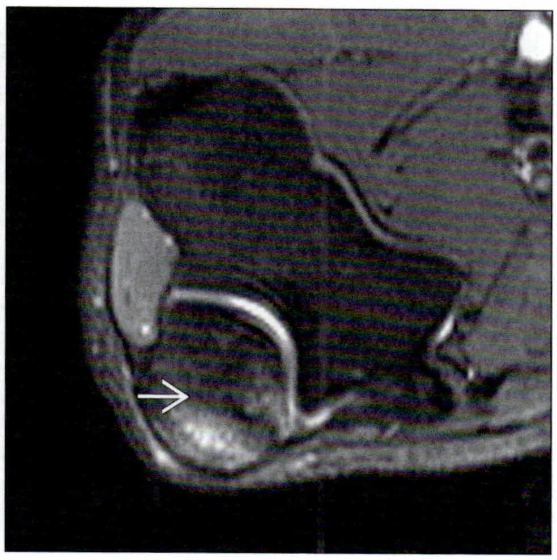

(Left) *Coronal T1WI MR shows a linear sclerotic fracture line ➡ in the medial aspect of the olecranon in a college baseball pitcher with posterior elbow pain.* ***(Right)*** *Axial PDWI FS MR in the same patient shows the fracture ➡ surrounded by high signal marrow edema. Injury to this area of the olecranon in overhead throwers is not uncommon and should be considered in the differential for elbow pain in these patients.*

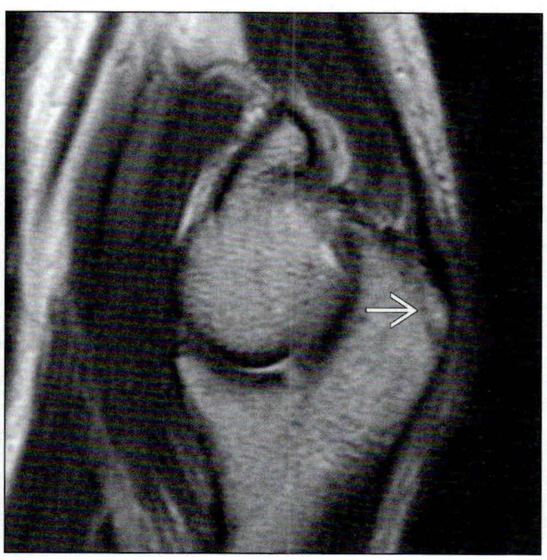

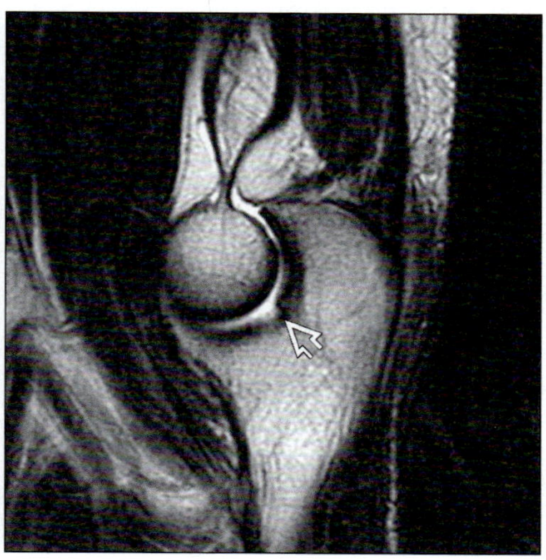

(Left) *Sagittal T2WI shows a subtle nondisplaced fracture ➡ at the tip of the olecranon. The fracture could not be seen on radiographs. Fat-suppressed T2WI (not shown) demonstrated surrounding marrow edema.* ***(Right)*** *Sagittal T2WI MR shows the normal trochlear notch ➡ in the mid-articular surface of the olecranon. This is a "bare area" in the articular cartilage and should not be mistaken for a fracture. There is no associated marrow edema or joint effusion.*

CORONOID FRACTURE

Key Facts

Terminology
- Fracture of coronoid process of ulna
- Usually associated with anterior elbow dislocation
- "Terrible triad" injury
  - Coronoid process fracture
  - Radial head fracture
  - Elbow dislocation

Imaging
- Fragment may range from tiny to whole coronoid process
- Best seen on lateral view
- Look for associated elbow ligament disruption if dislocation

Pathology
- Rare isolated fracture
- Usually due to posterior elbow dislocation
- Regan/Morrey classification
  - Type I: Fracture of tip of coronoid
  - Type II: Fracture of about 50% of coronoid
  - Type III: Fracture of entire coronoid
  - Type IV: Type II or III + medial extension
- Avulsion of brachialis tendon insertion
- Direct impaction against trochlea without dislocation

Clinical Issues
- Depends on Regan/Morrey type

Diagnostic Checklist
- Elbow dislocation as precursor
- Signs of associated elbow instability
- Radial head fracture ("terrible triad")
- Describe size of fragment (Regan/Morrey type)
- Look for associated ligamentous injuries

*(**Left**) Graphic representation of a coronoid fracture shows displacement of the entire coronoid process ➡ from the ulna ➡. This is a Regan/Morrey type III lesion. (**Right**) Lateral radiograph of the elbow demonstrates a displaced fracture of the coronoid ➡. The fracture fragment is comminuted and includes virtually the entire coronoid process; this is a Regan/Morrey type IV fracture, with medial extension including the sublime tubercle shown on subsequent CT.*

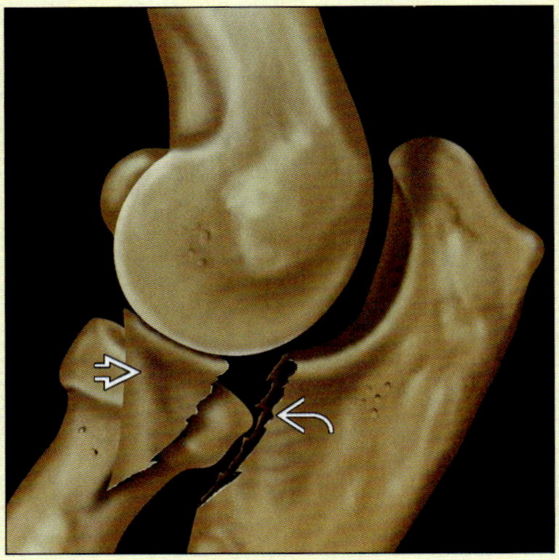

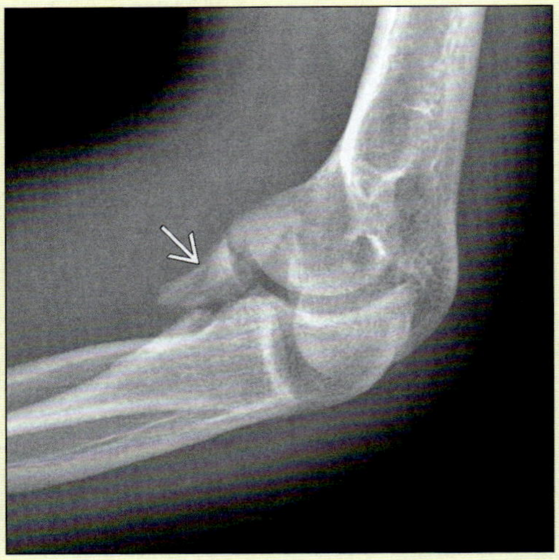

*(**Left**) Lateral radiograph shows an isolated fracture of the coronoid process ➡. There was no dislocation. Isolated coronoid fractures are rare. Displacement of the periarticular fat pads ➡ are a clue to the presence of an intraarticular fracture. This is a Regan/Morrey type I lesion. (**Right**) Lateral radiograph after posterolateral elbow dislocation shows 2 displaced triangular bone fragments ➡ arising from the coronoid process. Such fragments are often only seen on post-reduction views.*

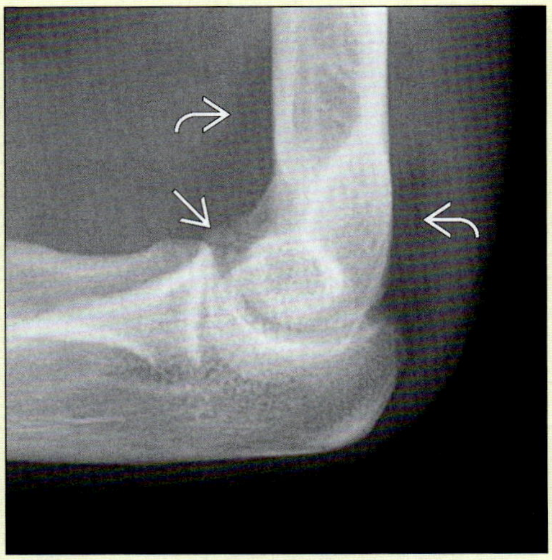

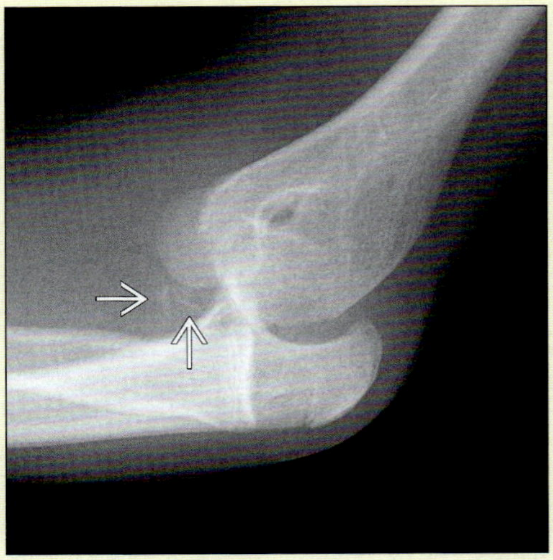

CORONOID FRACTURE

TERMINOLOGY

Definitions

- Fracture of coronoid process of ulna
  - Usually associated with anterior elbow dislocation
- "Terrible triad" injury
  - Coronoid process fracture
  - Radial head fracture
  - Elbow dislocation

IMAGING

General Features

- Best diagnostic clue
  - Separation of portion of coronoid process on lateral radiograph
- Location
  - Proximal ulna anterior to ulnotrochlear joint
- Size
  - Fragment may range from tiny to whole process
- Morphology
  - Usually triangular fragment
  - Fracture in coronal plane

Radiographic Findings

- Best seen on lateral view
- Osseous fragment adjacent to tip of coronoid process

CT Findings

- 1 or more small corticated fragments arising from coronoid

MR Findings

- Look for associated elbow ligament disruption if dislocation
  - Anterior band ulnar collateral ligament
  - Anterior capsular disruption
- Associated brachialis tendon injury

Imaging Recommendations

- Best imaging tool
  - Lateral radiograph
- Protocol advice
  - Some fractures best seen on oblique view
  - Post-reduction views best in dislocation cases

DIFFERENTIAL DIAGNOSIS

Intraarticular Body in Coronoid Fossa

- Coronoid maintains normal size and shape

Capitellar Fracture

- Curvilinear fragment, not triangular

PATHOLOGY

General Features

- Etiology
  - Usually due to posterior elbow dislocation
    - Fall on outstretched arm
    - Coronoid impacted against trochlea
  - Rare isolated fracture
    - Avulsion of brachialis tendon insertion
    - Direct impaction against trochlea without dislocation

Staging, Grading, & Classification

- Regan/Morrey classification
  - Type I: Fracture of tip of coronoid
    - Due to posterior elbow dislocation
    - Shearing mechanism
  - Type II: Fracture of about 50% of coronoid
  - Type III: Fracture of entire coronoid
  - Type IV: Type II or III + medial extension
    - May involve sublime tubercle (attachment of ulnar collateral ligament)

CLINICAL ISSUES

Natural History & Prognosis

- Usually due to posterior elbow dislocation
- Isolated coronoid fractures rare
- Complications
  - Stiffness
  - Instability
  - Heterotopic ossification
  - Fragments may become free in joint
    - Restricted elbow motion

Treatment

- Depends on Regan/Morrey type
  - Type I treated conservatively
  - Types II: Must determine degree of elbow instability
    - Assess at > 60° extension
    - Surgery required if unstable
  - Type III or IV: Primary repair of elbow ligaments
    - Coronoid may be reduced surgically
- Associated intraarticular fragments may be removed

DIAGNOSTIC CHECKLIST

Consider

- Elbow dislocation as precursor
- Signs of associated elbow instability
- Radial head fracture ("terrible triad")

Image Interpretation Pearls

- Study post-reduction radiographs for small fragments

Reporting Tips

- Describe size of fragment (Regan/Morrey classification)

SELECTED REFERENCES

1. Steinmann SP: Coronoid process fracture. J Am Acad Orthop Surg. 16(9):519-29, 2008
2. Wells J et al: Coronoid fractures of the elbow. Clin Med Res. 6(1):40-4, 2008
3. Ring D: Fractures of the coronoid process of the ulna. J Hand Surg Am. 31(10):1679-89, 2006
4. Cohen MS: Fractures of the coronoid process. Hand Clin. 20(4):443-53, 2004

FOREARM FRACTURES

Key Facts

Terminology
- "Both bone" fracture = usually midshaft fractures of radius and ulna
- Nightstick fracture = nondisplaced fracture of ulnar shaft

Imaging
- Displaced fracture of 1 bone usually results in either
  - Fracture of other bone
  - Dislocation of other bone
- Direct blow injury can result in nondisplaced fracture of only 1 bone
- Best imaging tool: Radiographs, orthogonal views

Top Differential Diagnoses
- Galeazzi injury
- Monteggia injury
- Essex-Lopresti injury

Pathology
- Axial load transmitted to radius and ulna, resulting in either
  - Paired fractures of both bones at similar location
  - Fracture of 1 and dislocation of other
- Bowing (plastic) deformity
  - Almost exclusively in children
  - May occur 2° to compression between rollers

Clinical Issues
- Common: 23% of all fractures in children
- Forearm fractures are 10x more frequent than carpal injuries in both adults and children

Diagnostic Checklist
- Look for associated injuries if only single fracture seen
- Compartment syndrome may be associated

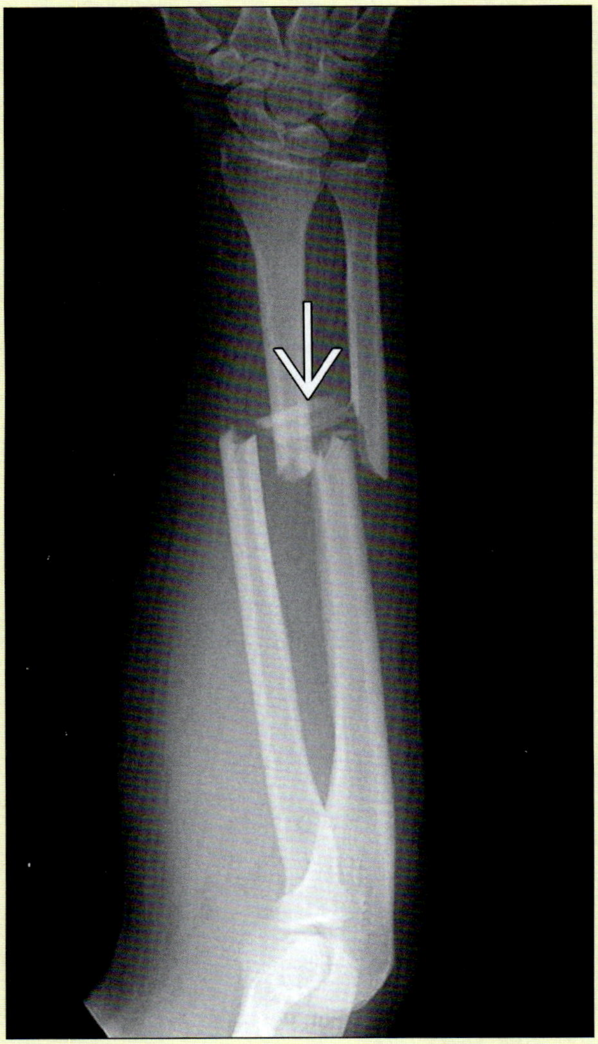

AP radiograph of a typical "both bone" forearm fracture shows midshaft injuries to both the radius and ulna, with a rotated cortical fragment ➔ arising from the radius. Note that the medial displacement of both bones is similar.

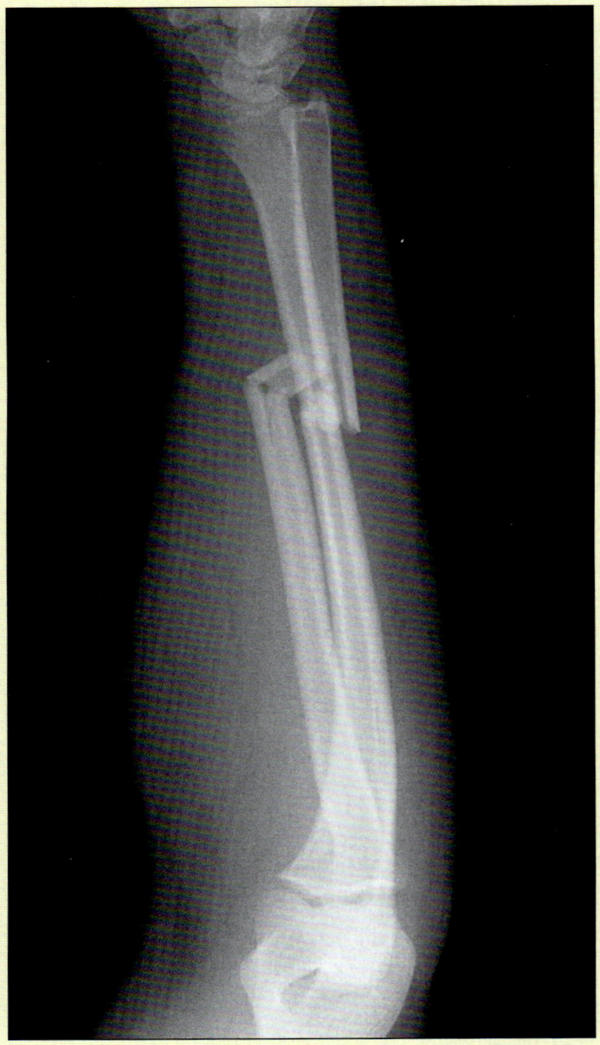

Lateral radiograph in the same patient shows that the fractures are also displaced dorsally to the same degree. Because the strong interosseous ligament remains intact, the fragments move as a single unit.

FOREARM FRACTURES

TERMINOLOGY

Synonyms

- "Both bone" fracture = usually midshaft fracture of radius and ulna
- Nightstick fracture = nondisplaced fracture of ulnar shaft
- Plastic deformity = bowing fracture

Definitions

- Fracture, complete or incomplete, of radius &/or ulna
- Fracture/dislocation: Generally fracture of 1 forearm bone with associated dislocation of other, either at elbow or wrist

IMAGING

General Features

- Best diagnostic clue
  - Fractures or dislocation of 1 or both forearm bones
- Location
  - Anywhere along forearm, from radiocapitellar joint to distal radioulnar joint (DRUJ)
- Morphology
  - Radius and ulna firmly attached by strong interosseous ligament
    - Displaced fracture of 1 bone usually results in either
      - Fracture of other bone
      - Dislocation of other bone
  - Direct blow injury can result in nondisplaced fracture of only 1 bone
  - Both bone fracture: Generally middle to distal 1/3 of forearm
  - Nightstick fracture: Isolated ulna fracture, generally middle to distal 1/3
  - Bowing fracture: Ulna more common than radius; usually complete fracture in adjacent forearm bone

Radiographic Findings

- Fracture lines, usually in midshaft locations
- Focal disruption of cortex in incomplete fracture
- Bowing of either bone

CT Findings

- May be used to confirm subtle DRUJ dislocation

Imaging Recommendations

- Best imaging tool
  - Radiographs, orthogonal views

DIFFERENTIAL DIAGNOSIS

Galeazzi Injury

- Fracture of radial shaft and dislocation of DRUJ

Monteggia Injury

- Fracture of ulnar shaft and dislocation of radiocapitellar joint

Essex-Lopresti Injury

- Comminuted fracture of radial head (or simple fracture and severe ligamentous disruption) and dislocation of DRUJ

PATHOLOGY

General Features

- Etiology
  - Fall on outstretched hand (FOOSH)
  - Axial load transmitted to radius and ulna, resulting in either
    - Paired fractures of both bones at similar location
    - Fracture of 1 and dislocation of other
  - Direct blow
    - Ulna
      - More common
      - "Nightstick injury" occurs when forearm is held up to protect face from blows
    - Radius
      - Much less common
  - Bowing (plastic) deformity
    - Almost exclusively in children
    - Rarely in adults

CLINICAL ISSUES

Demographics

- Age
  - Children aged 1-10; adults
- Epidemiology
  - Common: 23% of all fractures in children
  - Forearm fractures 10x more frequent than carpal injuries in both adults and children

Natural History & Prognosis

- Incomplete fracture heals in anatomic position
- Children's complete fracture heals with remodeling
- Plastic fracture does not remodel

Treatment

- Reduction; "both bone" fracture in adult generally requires ORIF
- Reduction and stabilization of adjacent unstable joint

DIAGNOSTIC CHECKLIST

Consider

- Look for associated injuries if only single fracture seen
- Compartment syndrome may be associated

Image Interpretation Pearls

- Study elbow and wrist carefully for associated fractures or dislocation

SELECTED REFERENCES

1. Kropman RH et al: Treatment of impacted greenstick forearm fractures in children using bandage or cast therapy: a prospective randomized trial. J Trauma. 68(2):425-8, 2010
2. Bae DS: Pediatric distal radius and forearm fractures. J Hand Surg [Am]. 33(10):1911-23, 2008
3. Moss JP et al: Diaphyseal fractures of the radius and ulna in adults. Hand Clin. 23(2):143-51, v, 2007
4. Patel RV et al: Forearm fractures. Br J Hosp Med (Lond). 68(5):M90-1, 2007
5. Sclamberg J et al: Acute plastic bowing deformation of the forearm in an adult. AJR Am J Roentgenol. 170(5):1259-60, 1998

FOREARM FRACTURES

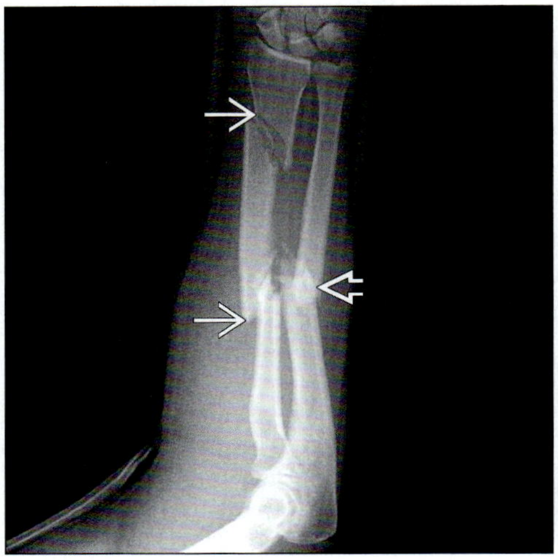

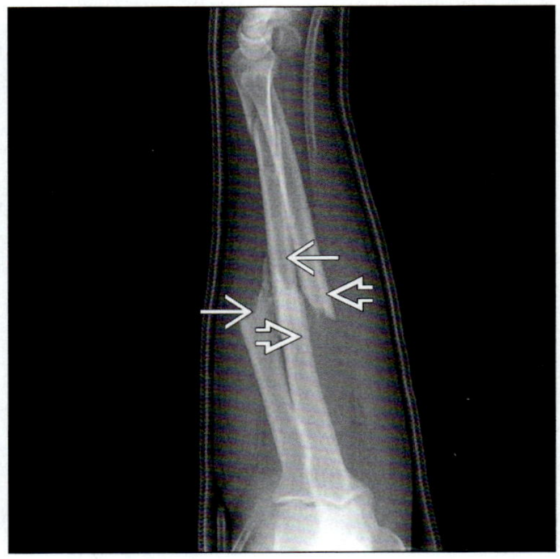

(Left) *AP radiograph demonstrates a segmental fracture of the radius* → *and a comminuted midshaft fracture of the ulna* ⇨. *Note that while the distal forearm and wrist are viewed in AP projection, the proximal forearm and elbow are in lateral projection, indicating 90° rotation at the fracture sites.* ***(Right)*** *Lateral radiograph in the same patient demonstrates 1 shaft width medial displacement of the radial* → *and ulnar* ⇨ *fractures.*

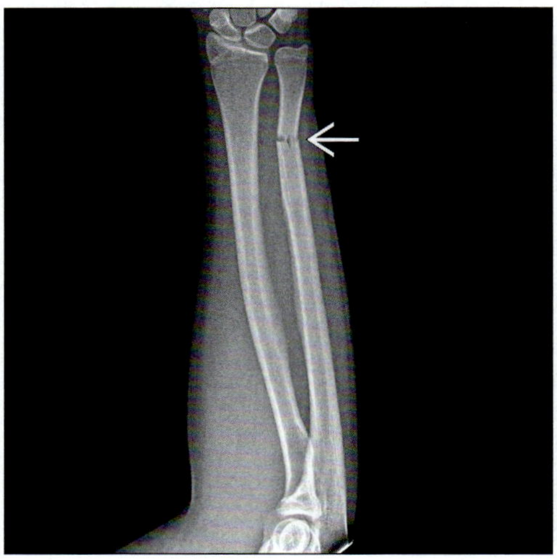

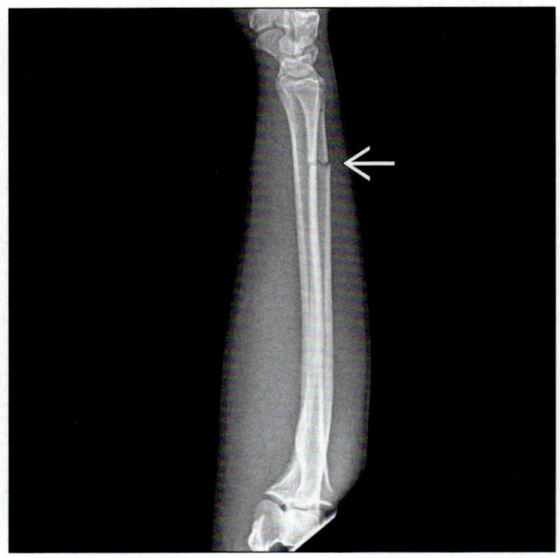

(Left) *AP radiograph demonstrates a distal ulnar fracture* → *with mild apex lateral angulation. The radius is intact. When a fracture occurs in a forearm bone due to a direct blow injury, such as this 1, the other bone often is intact.* ***(Right)*** *Lateral radiograph in the same patient confirms that although there is a distal ulnar shaft fracture* →*, the radius is intact. This is called a nightstick injury because it often results from a blow to the forearm while protecting the face or head.*

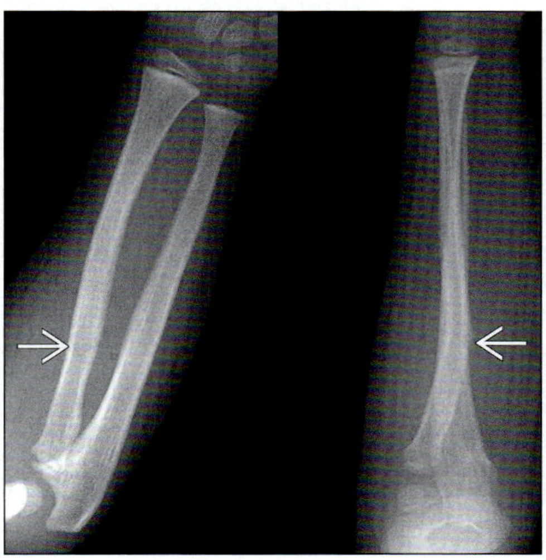

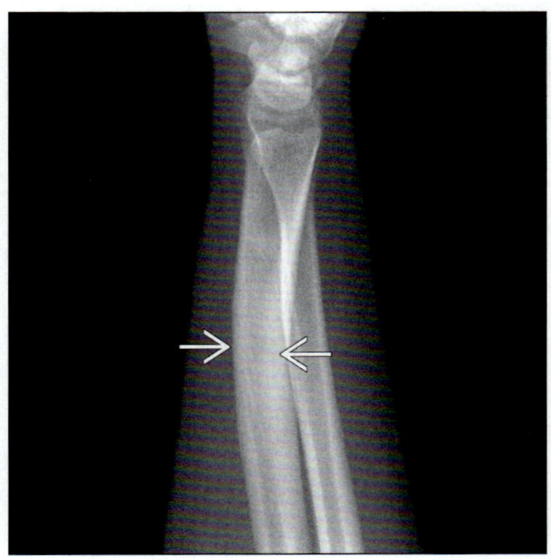

(Left) *AP and lateral radiographs demonstrate an angulated fracture of the proximal radius* →. *The ulna is intact and there is no diastasis of DRUJ. Isolated radial shaft fractures are uncommon.* ***(Right)*** *Lateral radiograph shows an apex volar bowing deformity and cortical thickening* → *in the distal radial shaft, indicating a healed plastic deformation fracture. With significant bowing of 1 forearm bone, the other bone is usually bowed or fractured, or the DRUJ is disrupted.*

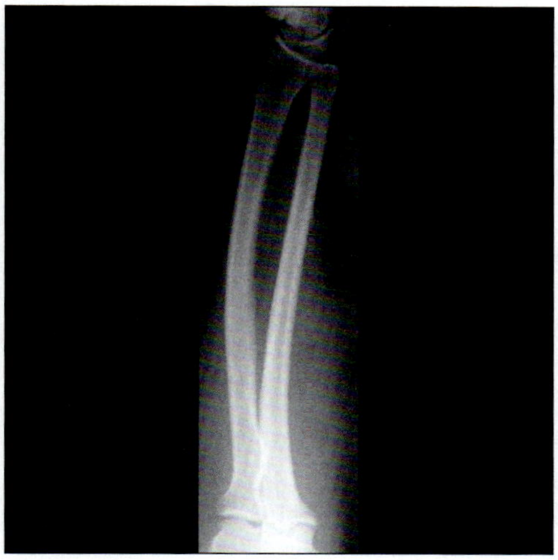

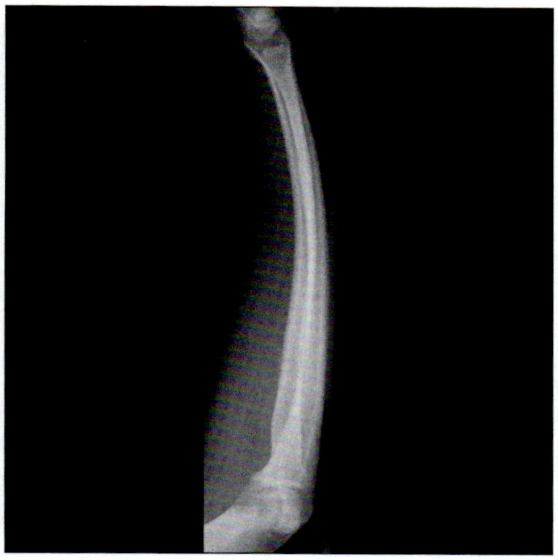

(Left) *AP radiograph in a 19-year-old man who caught his forearm in a roller press shows lateral bowing deformities in the radial and ulnar shafts. No discrete fracture line is visible, and the elbow and wrist are intact.* **(Right)** *Lateral radiograph in the same patient at the time of injury shows obvious dorsal bowing of the radial and ulnar shafts. There is no discrete fracture line.*

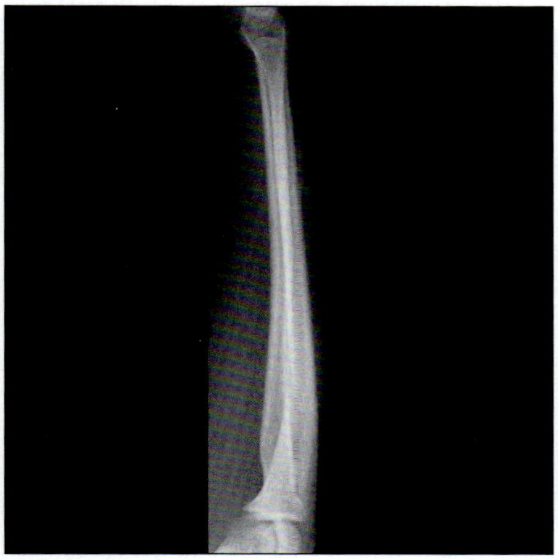

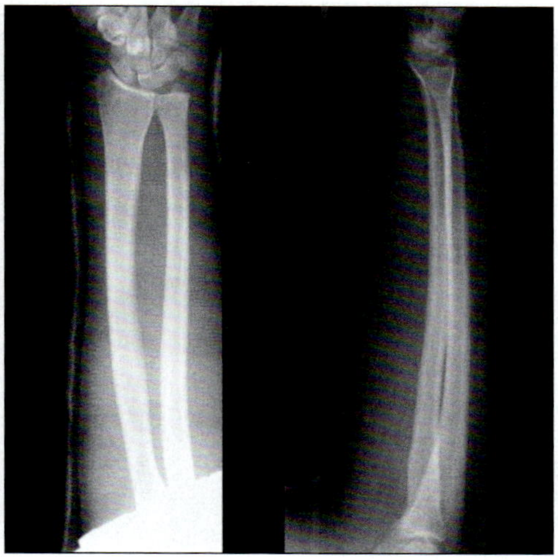

(Left) *Lateral radiograph of the opposite extremity in the same patient obtained at the time of injury shows normal morphology.* **(Right)** *Anteroposterior and lateral radiographs of the affected forearm in the same patient after closed reduction shows return to normal morphology. Reducing these fractures usually involves steady pressure on the bones placed over a fulcrum to reverse the initial injury (as in this case). In some cases the fractures must be completed to linear fractures to allow fixation.*

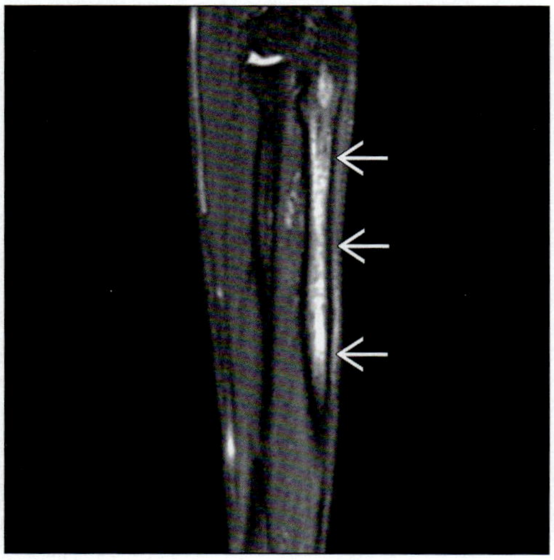

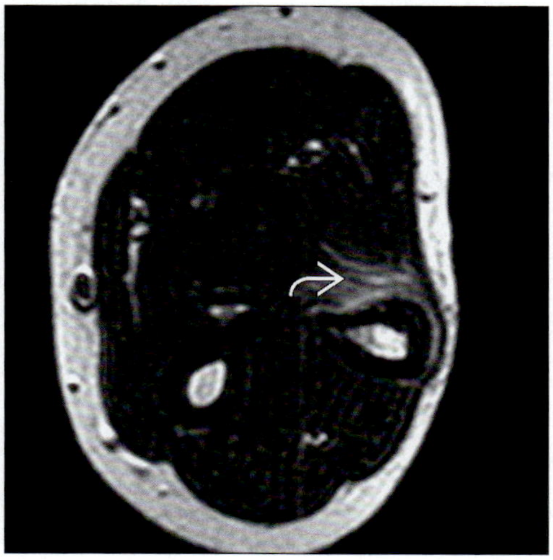

(Left) *Coronal T2WI FS MR of a 15-year-old girl on a rifle drill team with forearm pain demonstrates diffuse elevated signal throughout the ulnar shaft ➔. A discrete fracture line is not present. This is a stress injury and resolved with cessation of the activity.* **(Right)** *Axial T2WI MR in the same patient shows extensive periosteal elevation ↪ anterior to the edematous ulnar shaft. As in other parts of the body, stress fractures in the forearm are seen with repetitive motion injury.*

RADIAL COLLATERAL LIGAMENT INJURY

Key Facts

Terminology
- Radial collateral ligament (RCL)

Imaging
- Lateral elbow joint line between lateral humeral condyle and radial head
- Deep to common extensor tendons
- Best seen on coronal images
- FS PDWI or T2WI best
- Thickening, thinning, abnormal signal, or disruption in RCL

Top Differential Diagnoses
- Extensor tendon disruption
- Lateral ulnar collateral ligament injury
- Annular ligament injury
- Elbow dislocation

Pathology
- Traumatic
  - Varus stress
  - Posterolateral rotation
  - Elbow dislocation
- Iatrogenic: Related to extensor tendon debridement

Clinical Issues
- Conservative for isolated RCL injury
- Surgical repair required for
  - Higher grade PLRI
  - Elbow dislocation with multiple ligament &/or tendon injuries

Diagnostic Checklist
- Posterolateral rotatory instability
- Extensor tendon disruption
- If RCL abnormal, suspect LUCL injury as well

*(**Left**) Graphic of radial collateral ligament (RCL) disruption shows discontinuity of the RCL → in its midportion. Most tears of the ligament occur between the humeral attachment and the midportion of the ligament. The lateral ulnar collateral ligament (LUCL) → is intact. (**Right**) Coronal T2WI FS MR shows partial tears of the radial collateral ligament → and lateral ulnar collateral ligament →. The common extensor tendon is also partially torn →.*

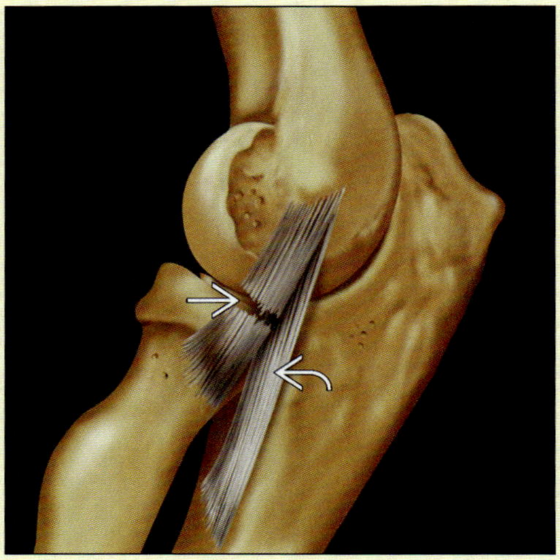

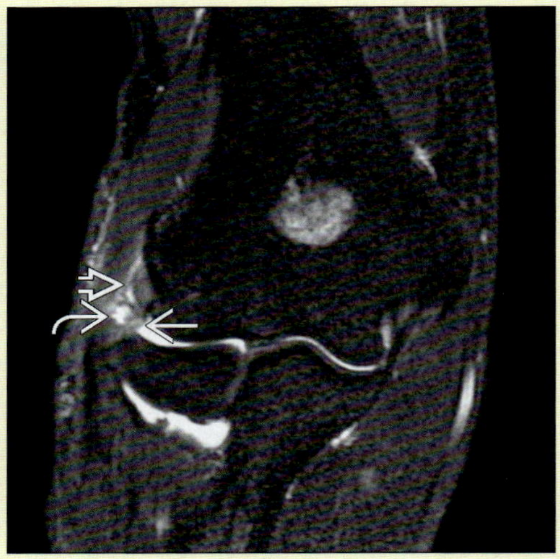

*(**Left**) Coronal T2WI FS MR obtained on a low-field magnet shows sprain (partial tear) of the RCL → with increased signal in the ligament fibers but no visible disruption. The LUCL → is also sprained. The common extensor tendon → is intact. (**Right**) Coronal T2WI FS MR after elbow dislocation shows disruption of the extensor tendon →, LUCL →, and RCL →. This combination of injuries is typical after dislocation.*

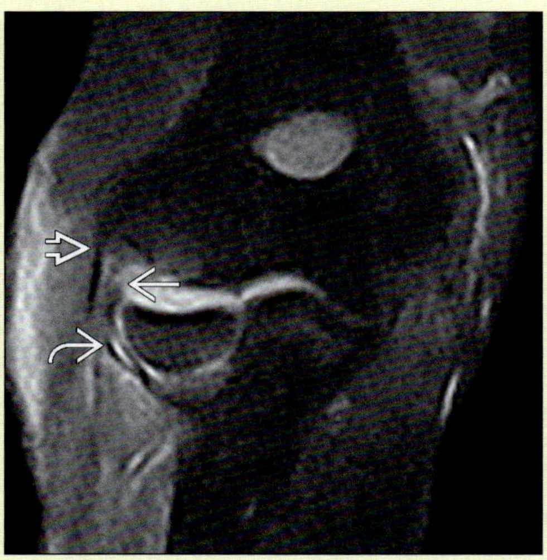

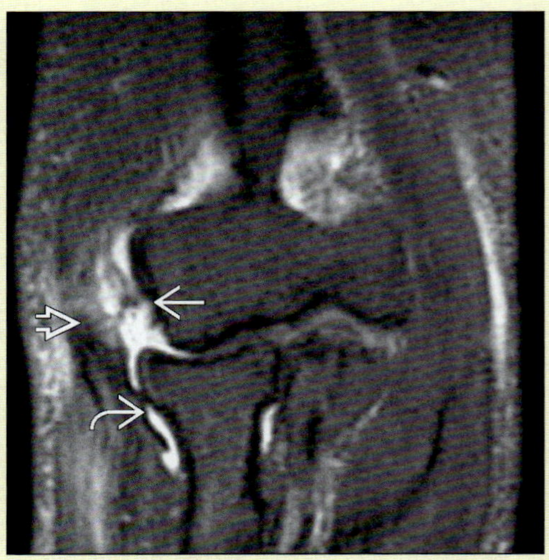

RADIAL COLLATERAL LIGAMENT INJURY

TERMINOLOGY

Abbreviations

- Radial collateral ligament (RCL) injury

Synonyms

- Lateral collateral ligament injury

Definitions

- Sprain (partial tear) or disruption of radial collateral ligament

IMAGING

General Features

- Best diagnostic clue
  - Discontinuity of fibers of radial collateral ligament on coronal MR images
- Location
  - Lateral elbow joint line between lateral humeral condyle and radial head
  - Deep to common extensor tendon
- Morphology
  - Expansion of fibrous tissue extending from lateral humeral epicondyle to radial head
  - Proximal aspect blends with proximal aspect of lateral ulnar collateral ligament (LUCL)
  - Ligament becomes more important with increasing elbow flexion

MR Findings

- Best seen on coronal images
  - Fat-suppressed intermediate or T2WI best
  - Thickening, thinning, abnormal signal, or disruption in RCL
  - May be associated with abnormalities of LUCL or extensor tendons

Imaging Recommendations

- Best imaging tool
  - MR
- Protocol advice
  - Thin section coronal T2WI FS MR best depicts normal or abnormal RCL

DIFFERENTIAL DIAGNOSIS

Common Extensor Tendon Injury, Elbow

- Often coexists with RCL and LUCL injury when due to severe varus force

Lateral Ulnar Collateral Ligament Injury

- Often associated with RCL injury

Annular Ligament Injury

- Often injured in conjunction with RCL and LUCL injury

Elbow Dislocation

- RCL, LUCL, ± UCL injury usually associated

PATHOLOGY

General Features

- Etiology
  - Traumatic
    - Varus stress
    - Posterolateral rotation
    - Elbow dislocation
  - Iatrogenic
    - Usually related to extensor tendon debridement

Staging, Grading, & Classification

- Part of staging of posterolateral rotatory instability (PLRI) of elbow (O'Driscoll staging)
  - Stage I: Lateral ulnar collateral ligament (LUCL) tear
  - Stage II: Perched elbow, tears of LUCL and RCL
  - Stage III: Complete elbow dislocation

CLINICAL ISSUES

Presentation

- Most common signs/symptoms
  - Pain and swelling over lateral elbow
  - ± varus instability
- Other signs/symptoms
  - May be associated with posterolateral rotatory instability of elbow
    - LUCL tear
    - Perched elbow
    - Elbow dislocation ± ulnar collateral ligament tear
  - May be associated with frank elbow dislocation
  - Often associated with clinical "lateral epicondylitis" (extensor tendinopathy)

Natural History & Prognosis

- Isolated RCL injury does well
- When part of PLRI → associated with chronic elbow instability

Treatment

- Conservative for isolated RCL injury
- Surgical repair required for
  - Higher grade PLRI
  - Elbow dislocation with multiple ligament &/or tendon injuries

DIAGNOSTIC CHECKLIST

Consider

- Posterolateral rotatory instability
- Extensor tendon disruption
- Iatrogenic injury

Image Interpretation Pearls

- If RCL abnormal, suspect LUCL injury as well

Reporting Tips

- Describe associated LUCL, extensor tendon injuries

SELECTED REFERENCES

1. Fern SE et al: Complex varus elbow instability: a terrible triad model. J Shoulder Elbow Surg. 18(2):269-74, 2009
2. Charalambous CP et al: Posterolateral rotatory instability of the elbow. J Bone Joint Surg Br. 90(3):272-9, 2008
3. Kijowski R et al: Magnetic resonance imaging of the elbow. Part II: Abnormalities of the ligaments, tendons, and nerves. Skeletal Radiol. 34(1):1-18, 2005
4. Mehta JA et al: Posterolateral rotatory instability of the elbow. J Am Acad Orthop Surg. 12(6):405-15, 2004

LATERAL ULNAR COLLATERAL LIGAMENT INJURY

Key Facts

Terminology
- Posterolateral rotatory instability of elbow (PLRI)

Imaging
- Ligament runs from lateral humeral condyle to lateral aspect of proximal ulna
- Forms sling around radial head
- Provides posterolateral support to elbow
- Thin section coronal T2WI FS MR sequences most helpful
- Edema or disruption of fibers of LUCL
- Proximal fibers near humeral attachment at greatest risk

Top Differential Diagnoses
- Radial collateral ligament injury
- Annular ligament injury
- Extensor tendon injury
- Elbow dislocation

Pathology
- Trauma, especially fall on outstretched hand
- Iatrogenic: Disrupted during debridement of extensor tendon
- Posterolateral rotatory instability
  - Stage 1: LUCL tear
  - Stage 2: LUCL and RCL tear
  - Stage 3a: Stage 2 + posterior UCL tear
  - Stage 3b: Stage 3a + complete UCL tear
  - Stage 3c: Complete stripping of ligaments and tendons from distal humerus

Diagnostic Checklist
- Coexistent injury to common extensor tendon
- LUCL almost always ruptured at or near humeral attachment

*(**Left**) Coronal T2WI FS MR demonstrates the normal proximal fibers of the lateral ulnar collateral ligament (LUCL) →. This structure often blends in with the adjacent radial collateral ligament (RCL) → and with the more superficially located common extensor tendon → at the lateral humeral epicondyle. (**Right**) Contiguous coronal T2WI FS MR in the same patient shows the LUCL → outlined by joint fluid as it runs posterior to the radial head → and inserts on the ulna →.*

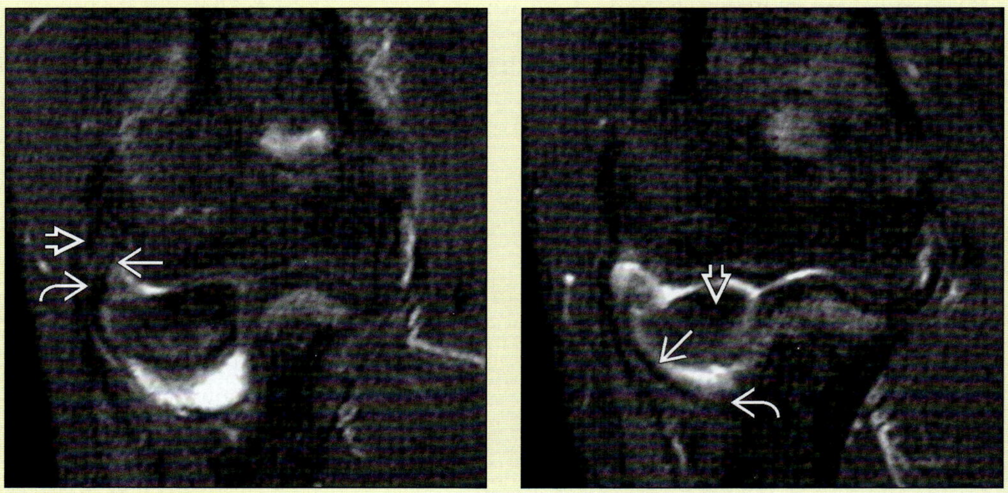

*(**Left**) Graphic representation of the lateral ulnar collateral ligament shows its origin on the lateral humeral condyle →, its passage posterior to the radius →, and its distal attachment to the crista supinator of the ulna →. The radial collateral ligament → originates nearby and attaches to the radius. (**Right**) Coronal T2WI FS MR shows disruption of the humeral attachment of the LUCL → as well as the common extensor tendon →. These 2 structures are often injured together.*

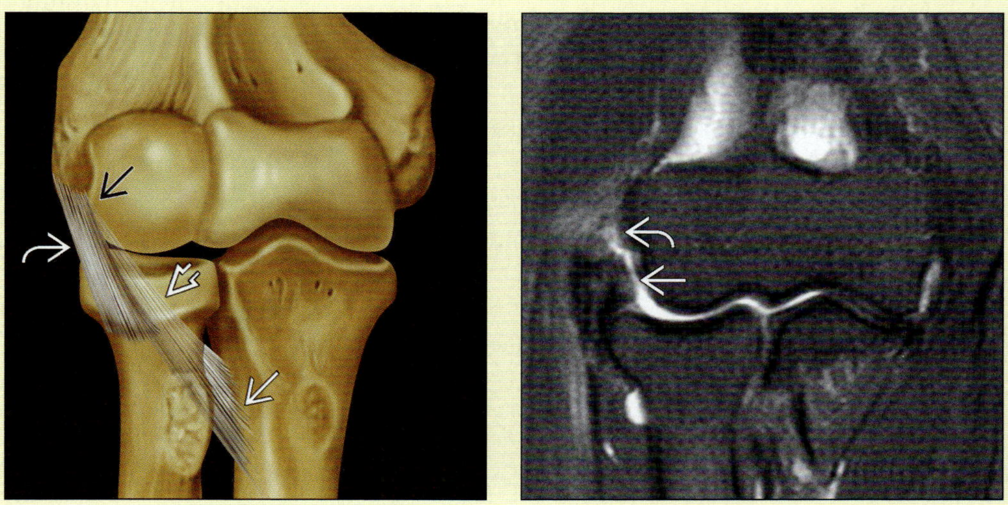

LATERAL ULNAR COLLATERAL LIGAMENT INJURY

TERMINOLOGY

Abbreviations
- Lateral ulnar collateral ligament (LUCL)

Synonyms
- Posterolateral rotatory instability of elbow (PLRI)

IMAGING

General Features
- Best diagnostic clue
  - Disruption of ligament fibers on MR
- Location
  - Ligament runs from lateral humeral epicondyle to crista supinator of proximal ulna
    - Forms sling posterior to radial head
    - Provides posterolateral support to elbow

Radiographic Findings
- Radiographs usually nondiagnostic

MR Findings
- Edema or disruption of fibers of LUCL
  - Proximal fibers near humeral attachment at greatest risk
- Often associated injury or disruption of radial collateral ligament (RCL)
- ± injury to common extensor tendon origin
- In cases of elbow dislocation → ulnar collateral ligament (UCL) injury

Ultrasonographic Findings
- Edema or disruption in lateral ligament complex of elbow

Imaging Recommendations
- Best imaging tool
  - MR
- Protocol advice
  - Thin section coronal T2WI FS MR sequences most helpful

DIFFERENTIAL DIAGNOSIS

Radial Collateral Ligament Injury
- Often injured in conjunction with LUCL

Annular Ligament Injury
- Generally in small children

Extensor Tendon Injury
- Often injured in conjunction with LUCL
  - Especially with elbow dislocation

Elbow Dislocation
- LUCL + RCL + UCL disruption

PATHOLOGY

General Features
- Etiology
  - Trauma, especially fall on outstretched hand
  - Iatrogenic: Disrupted during debridement of extensor tendon

Staging, Grading, & Classification
- Posterolateral rotatory instability
  - Stage 1
    - LUCL tear
    - Rotatory instability of elbow
    - Radial head subluxates posterior to capitellum
  - Stage 2
    - LUCL and RCL tear
      - Posterolateral dislocation/perched elbow
      - Varus instability of elbow
  - Stage 3a
    - Stage 2 + posterior UCL tear
    - Often with radial head and coronoid fracture
  - Stage 3b
    - Stage 3a + entire UCL torn
    - Rotatory, varus, and valgus elbow instability
  - Stage 3c
    - Stripping of entire distal humerus of ligamentous and tendinous attachments

CLINICAL ISSUES

Presentation
- Most common signs/symptoms
  - Posterolateral rotatory instability of elbow
    - Clicking, snapping, or giving way on elbow extension
    - Clinical test
      - Elbow flexed 20°
      - Forearm supination + applied valgus and axial force → posterolateral subluxation + apprehension

Treatment
- Surgical repair of ligament

DIAGNOSTIC CHECKLIST

Consider
- Coexistent injury to common extensor tendon

Image Interpretation Pearls
- LUCL almost always ruptured at or near humeral attachment

SELECTED REFERENCES

1. Stewart B et al: Sonography of the lateral ulnar collateral ligament of the elbow: study of cadavers and healthy volunteers. AJR Am J Roentgenol. 193(6):1615-9, 2009
2. Cheung EV: Chronic lateral elbow instability. Orthop Clin North Am. 39(2):221-8, vi-vii, 2008
3. Tran N et al: Ultrasonography of the elbow. Semin Musculoskelet Radiol. 11(2):105-16, 2007
4. Mehta JA et al: Posterolateral rotatory instability of the elbow. J Am Acad Orthop Surg. 12(6):405-15, 2004
5. Potter HG et al: Posterolateral rotatory instability of the elbow: usefulness of MR imaging in diagnosis. Radiology. 204(1):185-9, 1997
6. O'Driscoll SW: Elbow instability. Hand Clin. 10(3):405-15, 1994

ANNULAR LIGAMENT INJURY

Key Facts

Terminology

- Nursemaid's elbow: Subluxation or dislocation of radial head out of annular ligament sling
- Adult injury: Annular ligament disrupted as part of larger injury to lateral collateral ligament complex

Imaging

- Nursemaid's elbow: Clinical diagnosis
- MR for suspected disruption of lateral ligaments
  - Annular ligament best seen with joint distention
  - Thin section coronal T2WI FS MR best for determining integrity of collateral ligaments

Pathology

- Adult injury
  - Annular ligament disrupted as part of lateral ligament complex
  - Varus elbow stress
  - Elbow dislocation
  - Posterolateral instability
- Nursemaid's elbow
  - Radial head slips out from under annular ligament
  - Distal fibers of annular ligament weaker in children than adults
  - Fibers of annular ligament interpose between radial head and capitellum

Clinical Issues

- Nursemaid's elbow
  - Young child suddenly refuses to use arm
  - History of axial traction by pull on hand or wrist may be elicited
  - Typically from age 1-4
  - Complete recovery typical

(Left) *Graphic representation of the elbow ligaments viewed anteriorly demonstrates the annular ligament ➡ encircling the radial head. Other ligaments illustrated include the radial collateral ➡, lateral ulnar collateral ➡, ulnar collateral ⇨, and the oblique cord ⇨. The latter is part of the proximal radioulnar joint.* ***(Right)*** *Coronal MR arthrogram T1WI FS in a 13-year-old boy shows the normal annular ligament ➡ displaced slightly away from the radial head.*

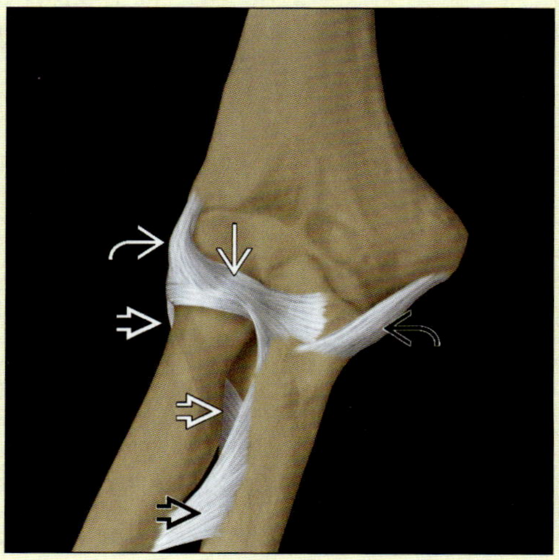

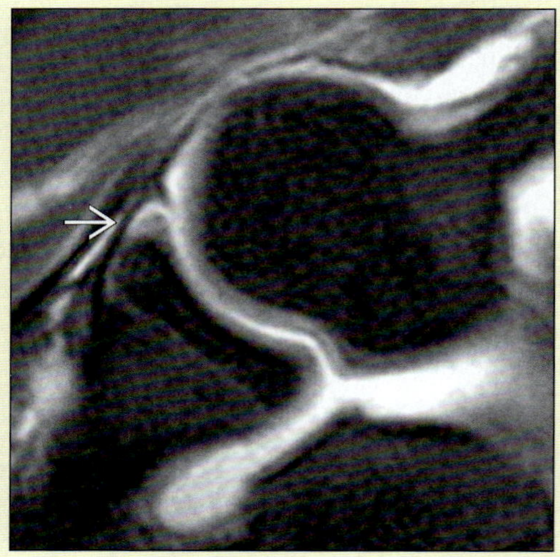

(Left) *Axial T2WI MR in the same patient shows the circumferential nature of the annular ligament ➡ at the level of the radial head.* ***(Right)*** *Axial T2WI MR shows annular ligament disruption ➡ as part of a more diffuse injury to the lateral collateral ligament complex of the elbow. In general, the relevance of injury to the annular ligament is exceeded by that affecting the radial collateral and lateral ulnar collateral ligaments.*

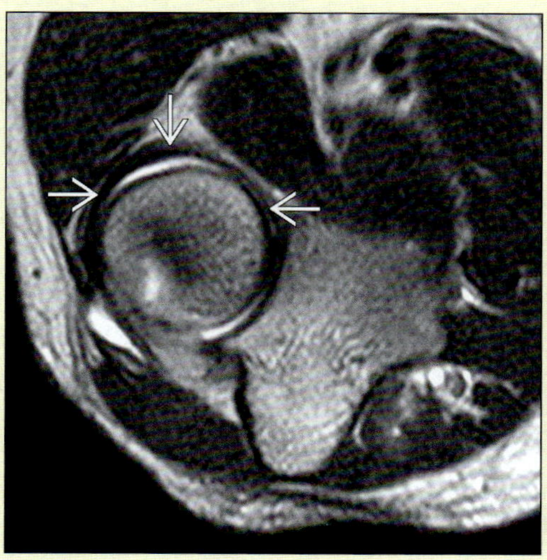

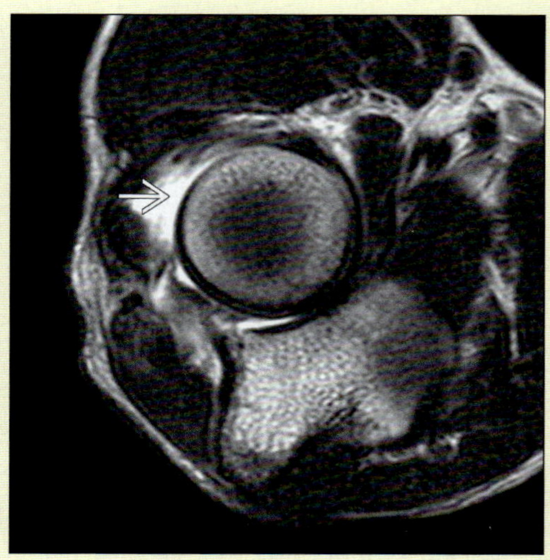

ANNULAR LIGAMENT INJURY

TERMINOLOGY

Definitions

- Nursemaid's elbow: Subluxation or dislocation of radial head out of annular ligament sling
- Adult injury: Annular ligament disrupted as part of larger injury to lateral collateral ligament complex

IMAGING

General Features

- Best diagnostic clue
  - Nursemaid's elbow: Clinical diagnosis
  - Adult injury: Elbow dislocation
- Location
  - Focal thickening of elbow capsule around radial head anteriorly, laterally, and posteriorly
- Morphology
  - Nursemaid's elbow: Interposition of part of annular ligament between radial head and capitellum

Radiographic Findings

- Nursemaid's elbow: None; radiographs will be normal

MR Findings

- Adult injury
  - Annular ligament disruption usually seen in conjunction with radial collateral and lateral ulnar collateral ligament disruption
  - Discontinuity or nonvisualization of annular ligament on coronal images
    - Coronal T2WI FS MR best for depicting ligamentous pathology

Imaging Recommendations

- Best imaging tool
  - No imaging required for nursemaid's elbow
  - MR for suspected disruption of lateral ligaments
- Protocol advice
  - Thin section coronal T2WI FS MR best for determining integrity of collateral ligaments
  - Annular ligament best seen with joint distention

PATHOLOGY

General Features

- Etiology
  - Nursemaid's elbow
    - Radial head slips out from under annular ligament
    - Distal fibers of annular ligament weaker in children than adults
    - Fibers of annular ligament interpose between radial head and capitellum
  - Adult injury
    - Annular ligament disrupted as part of lateral ligament complex
    - Varus elbow stress
    - Elbow dislocation
    - Posterolateral instability

CLINICAL ISSUES

Presentation

- Most common signs/symptoms
  - Nursemaid's elbow
    - History
      - Young child suddenly refuses to use arm
      - History of axial traction by pull on hand or wrist may be elicited
    - Exam
      - Anxiety > pain
      - Child protects arm
      - Elbow flexed 10-20°
      - Forearm slightly pronated
      - May have tenderness over radial head
  - Adult injury
    - Usually related to elbow dislocation
    - Pain and swelling over lateral elbow
    - ± varus deformity

Demographics

- Age
  - Nursemaid's elbow
    - Typically from age 1-4
    - Reported from 4 months to 15 years
- Gender
  - Nursemaid's elbow: Slight female predilection

Natural History & Prognosis

- Nursemaid's elbow: Complete recovery typical

DIAGNOSTIC CHECKLIST

Consider

- Nursemaid's elbow in young child with elbow pain and normal, pronated elbow on radiograph

Image Interpretation Pearls

- Specifically question possibility of nursemaid's elbow in appropriate clinical setting

SELECTED REFERENCES

1. Sanal HT et al: Annular ligament of the elbow: MR arthrography appearance with anatomic and histologic correlation. AJR Am J Roentgenol. 193(2):W122-6, 2009
2. Tan JW et al: Pathology of the annular ligament in paediatric Monteggia fractures. Injury. 39(4):451-5, 2008
3. Galik K et al: The effect of the annular ligament on kinematics of the radial head. J Hand Surg [Am]. 32(8):1218-24, 2007
4. Pearson BV et al: Nursemaid's elbow in a 31-year-old female. Am J Emerg Med. 25(2):222-3, 2007
5. Toupin P et al: Radial head subluxation: how long do children wait in the emergency department before reduction?. CJEM. 9(5):333-7, 2007
6. Bozkurt M et al: The annular ligament: an anatomical study. Am J Sports Med. 33(1):114-8, 2005
7. Choung W et al: Acute annular ligament interposition into the radiocapitellar joint in children (nursemaid's elbow). J Pediatr Orthop. 15(4):454-6, 1995
8. Jones J et al: "Irreducible" nursemaid's elbow. Am J Emerg Med. 13(4):491, 1995
9. Frumkin K: Nursemaid's elbow: a radiographic demonstration. Ann Emerg Med. 14(7):690-3, 1985

ULNAR COLLATERAL LIGAMENT INJURY

Key Facts

Terminology
- Medial collateral ligament injury

Imaging
- Anterior band: Most clinically relevant
- Edema around ± within ligament on T2WI
- Best seen on coronal T2WI or PDWI FS
- UCL partial tear
  - Undersurface tears more common
  - Often at junction with sublime tubercle of ulna
  - MR arthrogram: "T" sign of contrast extending in this recess on coronal T1WI FS
- UCL complete tear
  - Focal or diffuse fiber disruption of anterior band of UCL
  - Usually proximal fibers

Pathology
- Valgus stress to elbow
- Often repetitive activity, such as overhand throwing
- Grade I: Normal ligament surrounded by soft tissue swelling
- Grade II: Abnormal signal in ligament without complete disruption
- Grade III: Complete disruption of ligament

Clinical Issues
- Repaired UCL tear
- Most throwers return to pre-injury level of performance

Diagnostic Checklist
- Proximal aspect of UCL near humeral attachment often demonstrates normal heterogeneity due to splaying of fibers

__(Left)__ Graphic shows a proximal tear of ulnar collateral ligament with fiber disruption in the anterior band ➔ just distal to the medial humeral epicondylar origin of the ligament. Tear extends to the posterior band ⮫ as well. The transverse band ⇨ is intact. __(Right)__ Coronal PDWI MR in a 13-year-old baseball pitcher with medial elbow pain shows disruption of the proximal fibers of the anterior band of the ulnar collateral ligament ➔. This is the typical site of ligament disruption.

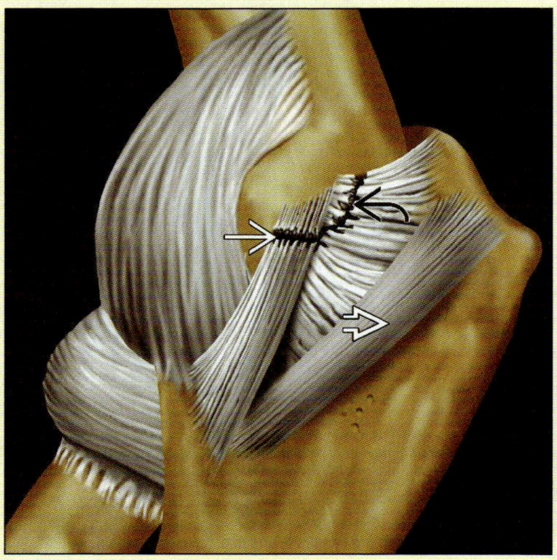

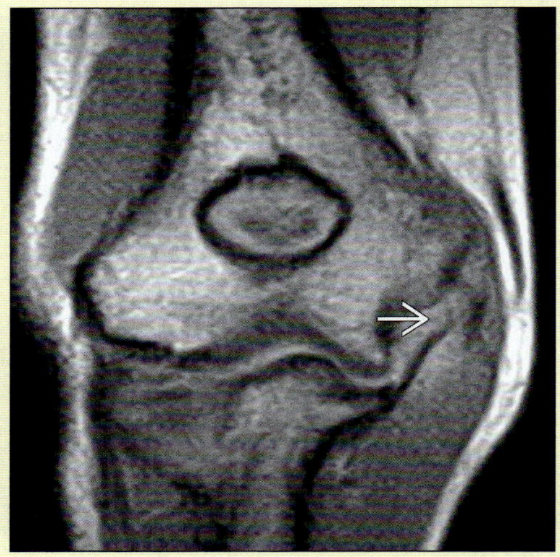

__(Left)__ Graphic shows a distal tear of the anterior band of the ulnar collateral ligament ➔ just proximal to the ligament insertion on the sublime tubercle of the coronoid process of the ulna ⮫. __(Right)__ Coronal T2WI FS MR in a 22-year-old javelin thrower with medial elbow pain shows disruption of the distal fibers of the anterior band of the ulnar collateral ligament ➔. Distal avulsions are less common than proximal injuries.

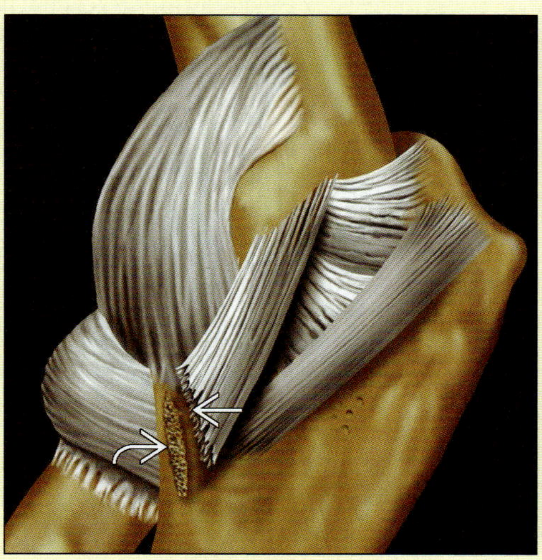

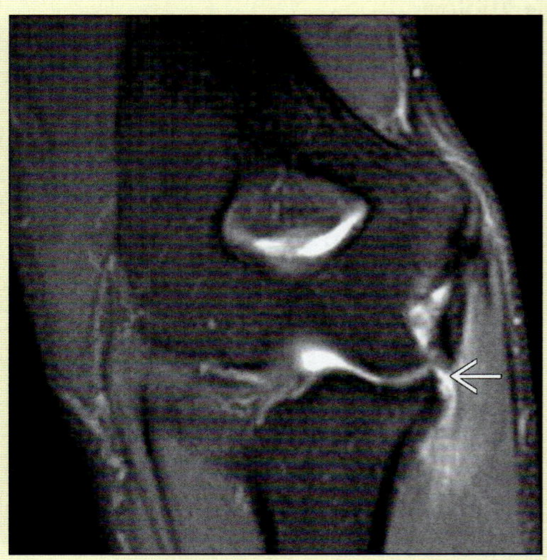

ULNAR COLLATERAL LIGAMENT INJURY

TERMINOLOGY

Abbreviations

- Ulnar collateral ligament (UCL) tear

Synonyms

- Medial collateral ligament injury

Definitions

- Sprain, tear, or avulsion of ulnar collateral ligament of elbow

IMAGING

General Features

- Best diagnostic clue
  - Disruption of fibers of anterior band of ulnar collateral ligament
- Location
  - Medial elbow at joint line
- Morphology
  - Ligament extends from medial humeral epicondyle to ulna
  - Comprised of 3 bundles
    - Anterior band: Most clinically relevant
      - Primary stabilizer to valgus force in elbow
      - Proximal attachment: Inferior aspect of medial humeral epicondyle
      - Distal attachment: Sublime tubercle of coronoid process of ulna
    - Posterior band
      - Probable thickening of joint capsule
      - Extends from medial humeral epicondyle to border of semilunar notch of ulna
    - Transverse band
      - Oriented perpendicular to long axis of anterior and posterior bands
      - No known stabilizing role

Radiographic Findings

- In setting of chronic injury
  - Hypertrophy of sublime tubercle
  - Hypertrophy of medial humeral epicondyle
  - Hypertrophy of entire upper extremity reported in high-level baseball pitchers
  - Ligamentous calcification
- Avulsion flake at sublime tubercle
- Avulsion of medial humeral epicondylar fragment

CT Findings

- Can be useful to characterize small avulsion fragments if present
- More sensitive than radiographs for ligament calcification
- Subchondral changes and osteophytes in capitellum if chronic

MR Findings

- UCL sprain
  - Edema around ± within ligament on T2WI
  - Best seen on coronal T2WI FS or PDWI
  - Fiber continuity often best documented on PDWI without FS
- UCL partial tear
  - Undersurface tears more common
  - Often at junction with sublime tubercle of ulna
  - Fluid tracks between distal ligament fibers and medial cortex of sublime tubercle on coronal T2WI
    - MR arthrogram: "T" sign of contrast extending in this recess on coronal T1WI FS
- UCL complete tear
  - Focal or diffuse fiber disruption of anterior band of UCL
  - Usually proximal fibers
  - Occasionally distal ligament
  - Best seen on coronal PDWI FS
  - Disruption of medial joint capsule with extravasation of joint fluid
  - MR arthrogram: Complete disruption of UCL with contrast extravasation
  - T1WI may exaggerate abnormal signal in ligament and make sprain look like tear
- UCL avulsion
  - Rare injury
  - May occur at sublime tubercle or medial epicondyle
  - Bone fragment usually very small and difficult to see on MR

Ultrasonographic Findings

- Loss of normal ligament anisotropy
- Focal disruption of UCL fibers

Imaging Recommendations

- Best imaging tool
  - MR
  - Some advocate MR arthrography for undersurface partial tears
- Protocol advice
  - Elbow best imaged in extension
  - Thin T2WI FS or PDWI FS in coronal plane show anterior band best
  - Tailor exam by acquiring most important sequences 1st as patient is often uncomfortable

DIFFERENTIAL DIAGNOSIS

Common Flexor/Pronator Tendon Injury, Elbow

- Located superficial to UCL
- Enlargement ± abnormal signal in origin of tendon at medial humeral epicondyle superior to UCL origin
- Flexor injury can accompany UCL injury in severe acute setting

Valgus Stress Mechanism/Little Leaguer's Elbow

- Avulsion of medial humeral epicondyle ossification center in skeletally immature thrower
- Attachment site of UCL and flexor/pronator tendon
- May displace into joint space

Elbow Dislocation

- Usually associated with tears of UCL and lateral collateral ligament complex
- Associated fractures common
  - Radial head
  - Coronoid process of ulna more anterior than sublime tubercle

Traction Spur of Coronoid Process

- Hypertrophy of sublime tubercle attachment of UCL
- Calcification of distal ligament fibers

ULNAR COLLATERAL LIGAMENT INJURY

- Connotes chronic injury
- Does not imply ligament rupture

PATHOLOGY

General Features

- Etiology
  - Valgus stress to elbow
    - Often repetitive activity, such as overhand throwing
    - Risk factor: Throwing breaking pitches
  - Elbow dislocation

Staging, Grading, & Classification

- Grade I: Normal ligament surrounded by soft tissue swelling
- Grade II: Abnormal signal in ligament without complete disruption
- Grade III: Complete disruption of ligament

Gross Pathologic & Surgical Features

- Tears usually occur in proximal fibers near humeral insertion
- Distal tear or distal avulsion can occur

CLINICAL ISSUES

Presentation

- Most common signs/symptoms
  - Medial elbow pain and swelling
    - Worse during throwing motion
  - Acute "pop" of ligament disruption sometimes felt
  - Inability to maintain usual throwing velocity after injury
  - Laxity of ligament on forced valgus stress
- Other signs/symptoms
  - May have lateral elbow pain if concomitant capitellar injury

Demographics

- Age
  - Teens and young adults
- Gender
  - M > F

Natural History & Prognosis

- Usually acute injury superimposed on chronic repetitive valgus stress pattern in thrower
- Laxity may be present prior to acute event
- Low-grade injury
  - Complete recovery while abstaining from overhand throwing
  - May recur if activity resumed
- Untreated high-grade injury
  - Partial tear often progresses to complete if not rested
  - Progressive valgus instability
  - Ulnar neuropathy
  - Accelerated osteoarthritis
  - Chronic pain
- Repaired UCL tear
  - Requires months of postoperative rehabilitation
  - Most throwers return to pre-injury level of performance

Treatment

- Low-grade sprain: Conservative therapy
  - Cessation of offending activity (throwing, etc.)
  - NSAIDs, ice
- High-grade partial or complete tear
  - Surgical reconstruction of ligament
    - Tendon graft (palmaris longus, etc.): "Tommy John procedure"
    - Primary repair of ligament

DIAGNOSTIC CHECKLIST

Consider

- Associated impaction injury to lateral aspect of elbow
- Concomitant injury to flexor/pronator tendon
- Little leaguer's elbow in appropriate population

Image Interpretation Pearls

- Proximal aspect of UCL near humeral attachment often demonstrates normal heterogeneity due to splaying of fibers
- "T" sign may be difficult to discern without large effusion or MR arthrogram

Reporting Tips

- Describe type and location of tear
  - Partial vs. complete
  - Proximal
  - Midsubstance
  - Distal
  - Avulsion

SELECTED REFERENCES

1. Meyers A et al: Ulnar collateral ligament reconstruction. Hand Clin. 24(1):53-67, 2008
2. Schwartz ML et al: Avulsion of the medial epicondyle after ulnar collateral ligament reconstruction: imaging of a rare throwing injury. AJR Am J Roentgenol. 190(3):595-8, 2008
3. Vitale MA et al: The outcome of elbow ulnar collateral ligament reconstruction in overhead athletes: a systematic review. Am J Sports Med. 36(6):1193-205, 2008
4. Ouellette H et al: Imaging of the overhead throwing athlete. Semin Musculoskelet Radiol. 2005 Dec;9(4):316-33. Review. Erratum in: Semin Musculoskelet Radiol. 11(1):91, 2007
5. Tran N et al: Ultrasonography of the elbow. Semin Musculoskelet Radiol. 11(2):105-16, 2007
6. Safran M et al: Ulnar collateral ligament of the elbow. Arthroscopy. 21(11):1381-95, 2005
7. Munshi M et al: Anterior bundle of ulnar collateral ligament: evaluation of anatomic relationships by using MR imaging, MR arthrography, and gross anatomic and histologic analysis. Radiology. 231(3):797-803, 2004
8. De Smet AA et al: Dynamic sonography with valgus stress to assess elbow ulnar collateral ligament injury in baseball pitchers. Skeletal Radiol. 31(11):671-6, 2002
9. Salvo JP et al: Avulsion fracture of the ulnar sublime tubercle in overhead throwing athletes. Am J Sports Med. 30(3):426-31, 2002
10. Glajchen N et al: Avulsion fracture of the sublime tubercle of the ulna: a newly recognized injury in the throwing athlete. AJR Am J Roentgenol. 170(3):627-8, 1998
11. Nakanishi K et al: MR arthrography of elbow: evaluation of the ulnar collateral ligament of elbow. Skeletal Radiol. 25(7):629-34, 1996

ULNAR COLLATERAL LIGAMENT INJURY

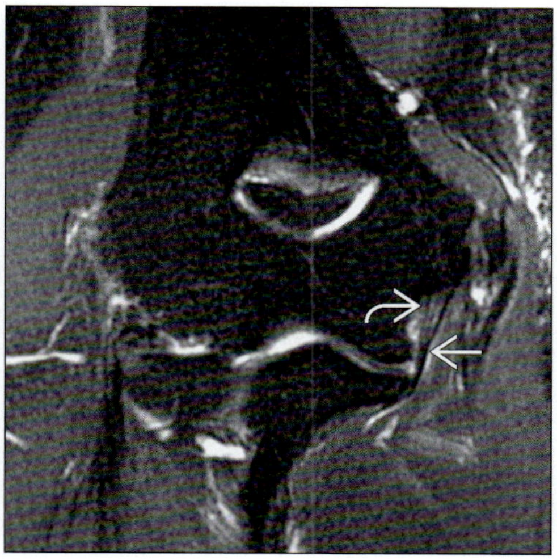

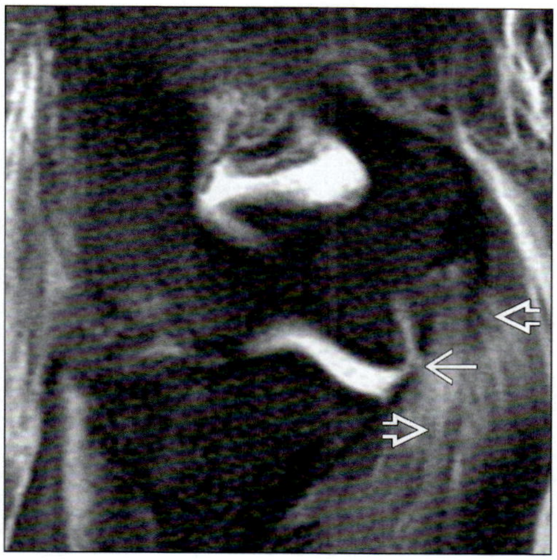

(Left) *Coronal T2WI FS MR of a normal ulnar collateral ligament anterior band, shows slightly increased signal in the proximal portion of the ligament near its humeral attachment, which should not be mistaken for a tear.* ***(Right)*** *Coronal T2WI FS MR in a baseball pitcher shows midsubstance disruption of the anterior band of the ulnar collateral ligament. There is edema surrounding the ligament and in the common flexor/ pronator myotendinous unit.*

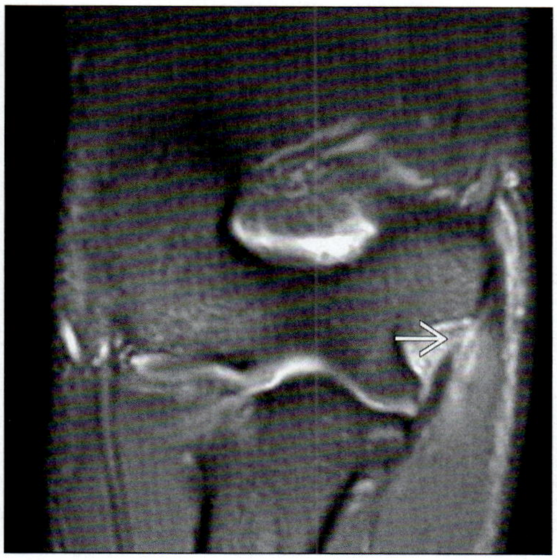

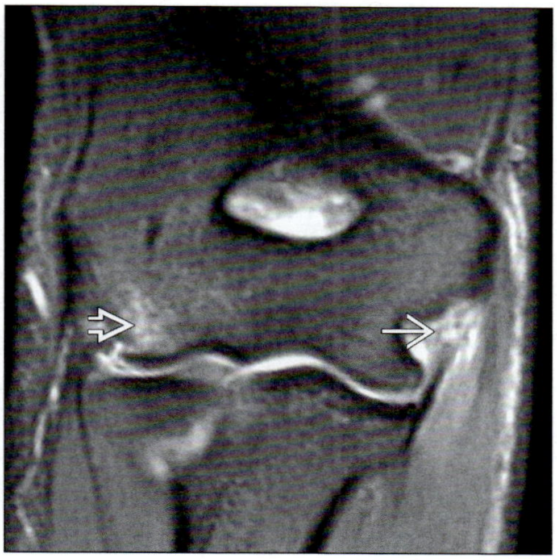

(Left) *Coronal PDWI FS MR in a 20-year-old college football quarterback shows edema around the proximal UCL with fiber disruption at its humeral insertion.* ***(Right)*** *Coronal PDWI FS MR in the same patient slightly more anteriorly shows the proximal disruption of the UCL and a subchondral contusion of the capitellum indicating impaction injury in the lateral aspect of the elbow. This injury pattern is typical in throwers.*

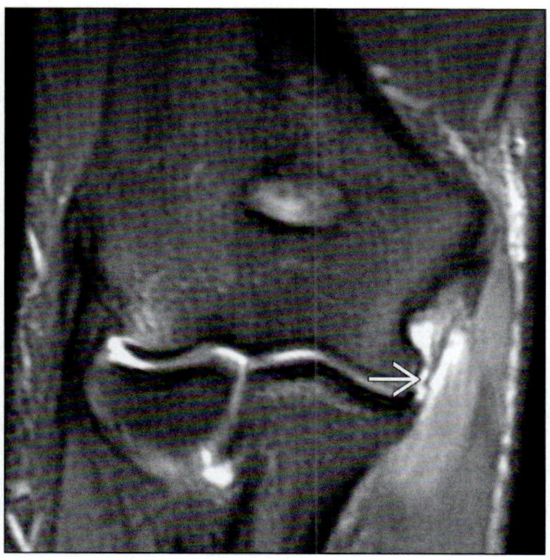

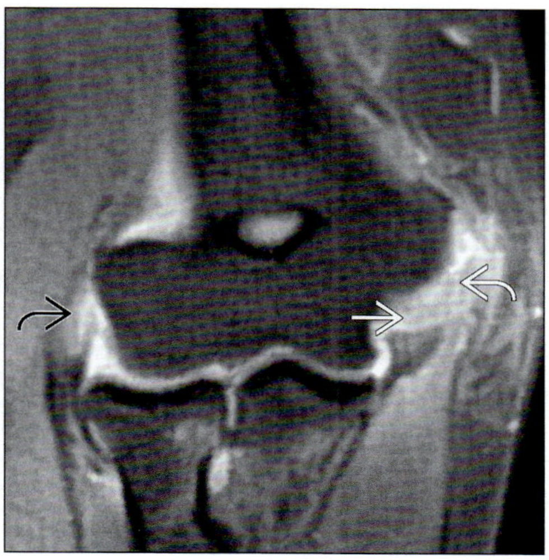

(Left) *Coronal PDWI FS MR in the same patient more anteriorly shows irregularity of the distal UCL fibers, likely a more chronic finding. Acute-on-chronic injury patterns are typical in high-level throwing athletes with UCL tears.* ***(Right)*** *Coronal T2WI FS MR in a 17-year-old baseball pitcher shows a disrupted and displaced tear of the UCL. There is also edema in the proximal flexor/ pronator tendon complex and in the proximal extensor tendons.*

ULNAR COLLATERAL LIGAMENT INJURY

*(**Left**) Coronal T2WI FS MR illustrates a distal avulsion of the anterior band of the UCL ➡ in a throwing athlete. Edema at the site confirms the acuity of the injury. (**Right**) Coronal T1WI MR in the same patient shows a tiny avulsion fragment ➡ arising from the sublime tubercle of the ulna ➡. While PD or T2WI are generally more accurate gauges of ligament disruption, small avulsion fragments are frequently seen better on T1WI due to marrow signal in the fragment.*

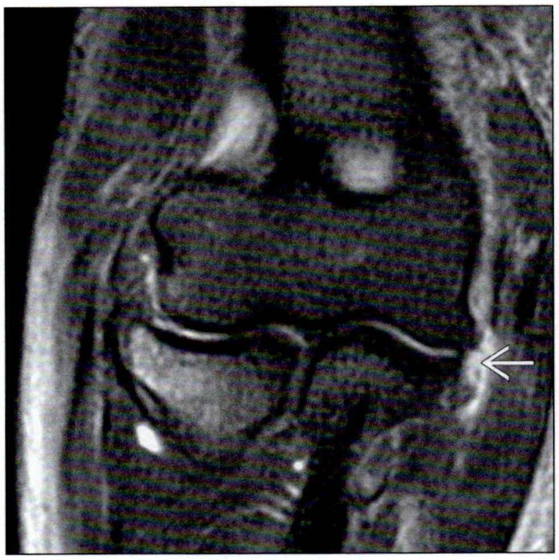

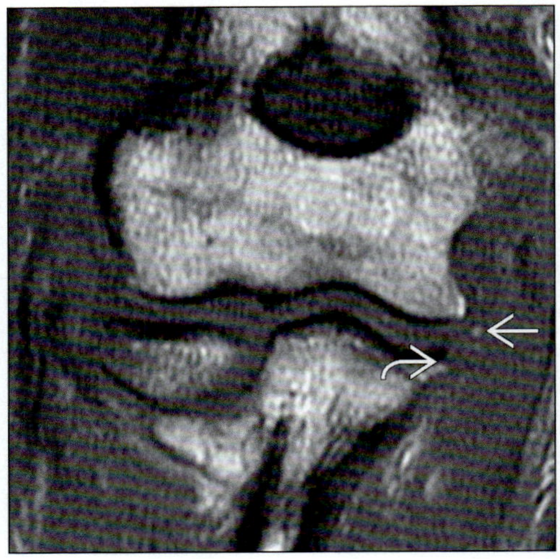

*(**Left**) Coronal T2WI FS MR after elbow dislocation shows diffuse abnormal signal in the anterior band of the UCL ➡ making precise demarcation of the site of injury difficult. Tears of the lateral ligament complex and extensor tendons ➡ are also present. (**Right**) Coronal T2WI FS MR in a 16 year old after elbow dislocation shows complete maceration of the UCL ➡ with no demonstrable fibers in the midportion of the ligament. Only the distal stump ➡ is visible.*

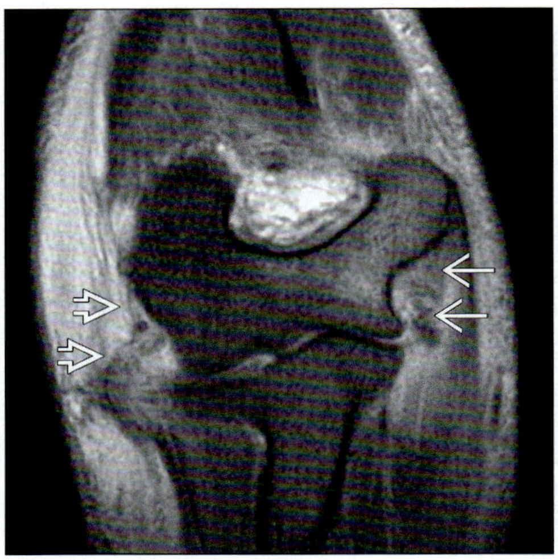

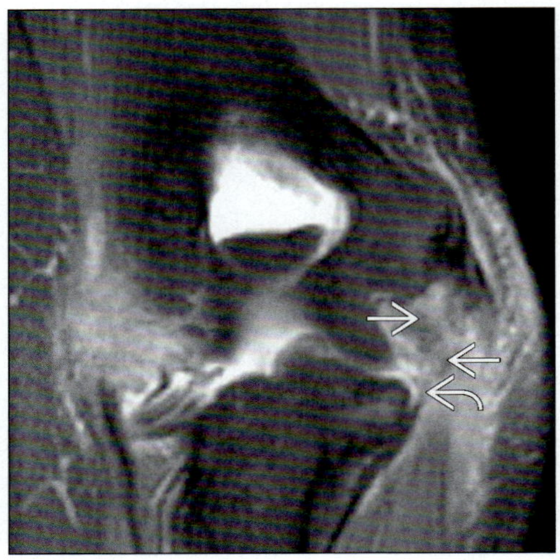

*(**Left**) Coronal T2WI FS MR shows a discrete disruption of the anterior band of the UCL ➡ at its attachment to the medial humeral epicondyle. (**Right**) Coronal T1WI MR in the same patient at the same slice location shows that the disruption of the UCL is due to avulsion of a small bone fragment ➡ from the humerus. Avulsions are often better recognized on T1WI due to the better distinction of marrow signal. Radiographs are useful adjuncts in such cases.*

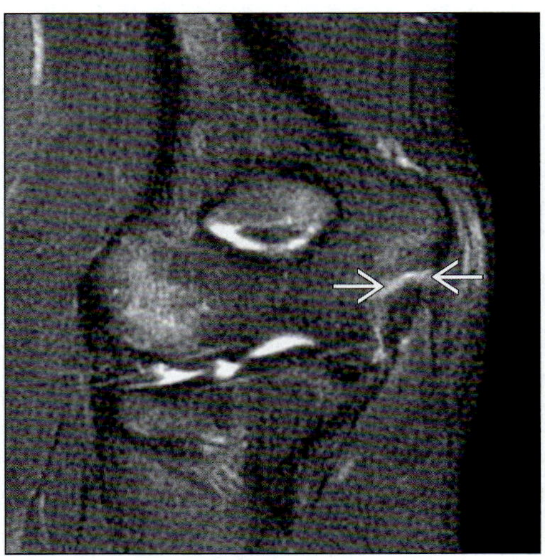

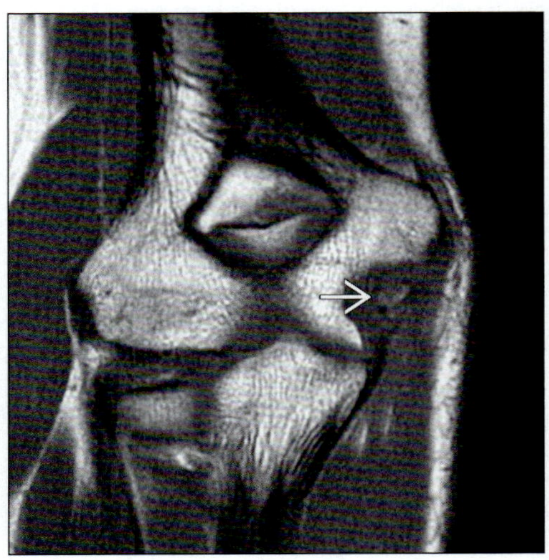

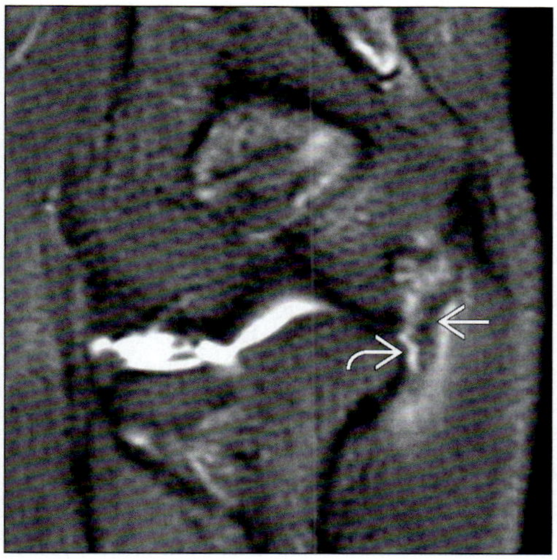

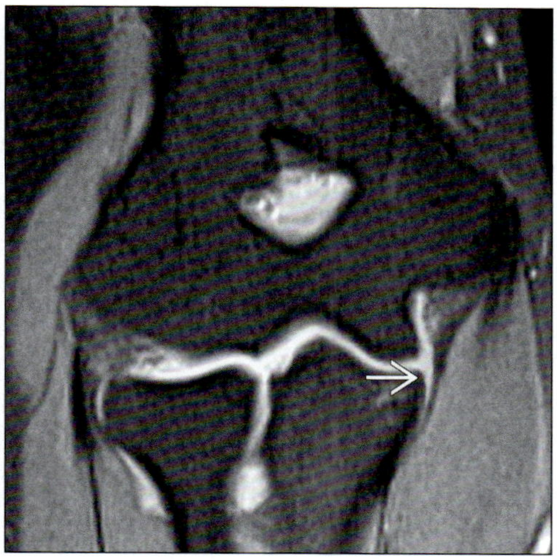

(Left) *Coronal T2WI FS MR shows a high-grade partial tear of the UCL. The entire ligament is edematous* → *. There is stripping of the deep distal fibers from their attachment to the sublime tubercle* ↷ *, comprising the "T" sign.* ***(Right)*** *Coronal MR arthrogram T1WI FS shows the "T" sign of a partial tear of the UCL* → *with contrast tracking into the space between the sublime tubercle and the torn deep fibers of the anterior band.*

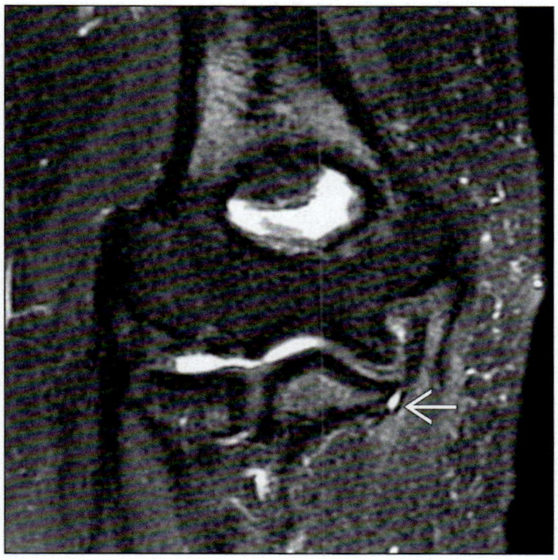

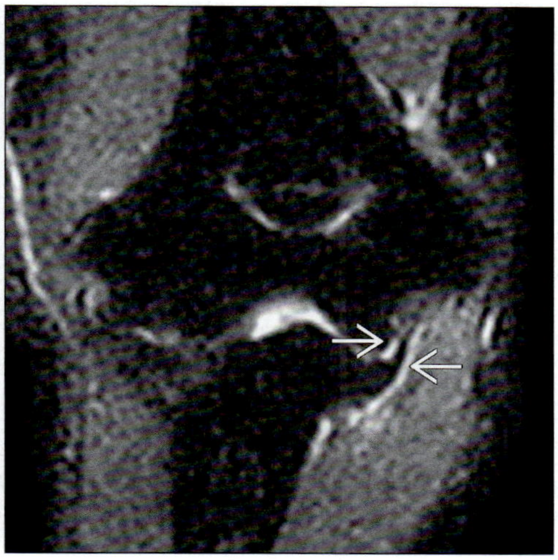

(Left) *Coronal T2WI FS MR in a patient with previous UCL partial tear shows fluid collecting in the periosteal recess medial to the sublime tubercle* → *where there was previous stripping of the deep fibers of the UCL.* ***(Right)*** *Coronal T2WI FS MR of a low-grade sprain of the UCL demonstrates edema around the anterior band of the ligament* → *. Ligamentous fibers demonstrate normal signal. This is a grade I sprain.*

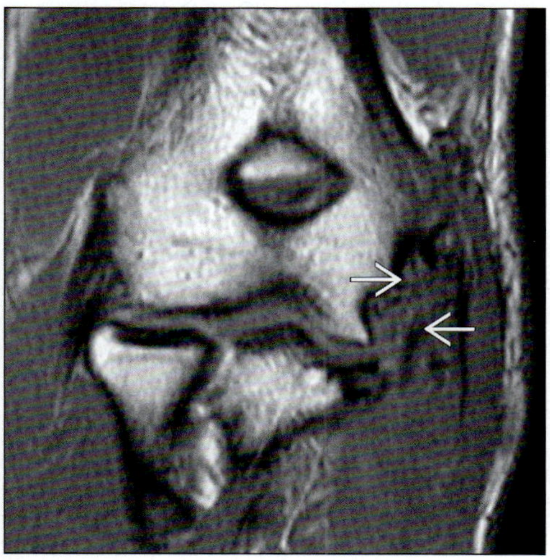

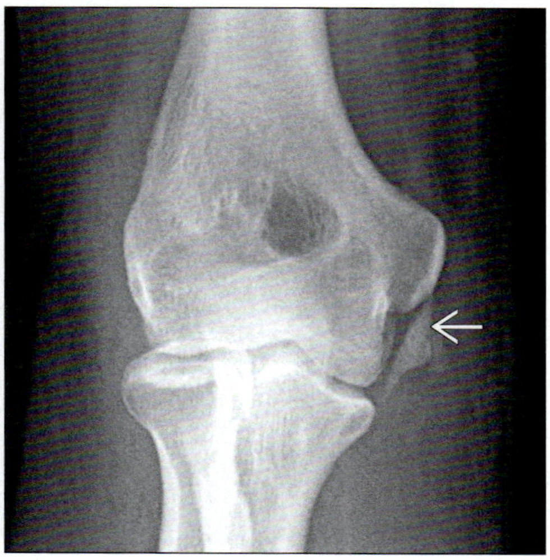

(Left) *Coronal T1WI MR after reconstruction of the UCL with primary reattachment demonstrates extensive postsurgical change* → *along the course of the anterior band. Discrete ligamentous fibers cannot be identified. This patient was stable to valgus stress.* ***(Right)*** *AP radiograph in a young adult 2 years after valgus stress injury demonstrates marked calcification of the UCL* → *. Such dystrophic calcifications can occur after 1 injury or with repetitive trauma. This case is severe.*

VALGUS STRESS MECHANISM/LITTLE LEAGUER'S ELBOW

Key Facts

Terminology
- Thrower's elbow
- Valgus stress mechanism
  - Compression of lateral structures
  - Distraction of medial structures
- Little leaguer's elbow
  - Specific pattern of valgus stress injury to elbow in adolescent throwers
  - Avulsion of medial humeral epicondyle
  - ± intraarticular entrapment of epicondyle
  - Often associated with capitellar chondral/ osteochondral injury

Imaging
- Little leaguer's elbow
  - Fragmentation of medial epicondylar apophysis
  - Displacement of medial epicondylar apophysis
- Medial distraction
  - UCL sprain or tear
  - Flexor/pronator injury
  - Medial epicondylar injury
- Lateral compression
  - Usually in anterior articular surface (elbow flexed during injury)
  - Capitellar injury
  - Chondral tear or defect
  - Chondral or osteochondral fragments in joint
- Ulnar neuropathy

Clinical Issues
- Little leaguer's elbow: 6-14 years old
- Continuing repetitive activity → worsening injury → hypertrophic and degenerative changes
- Stable (osteo)chondral capitellar fragment may become unstable or detach

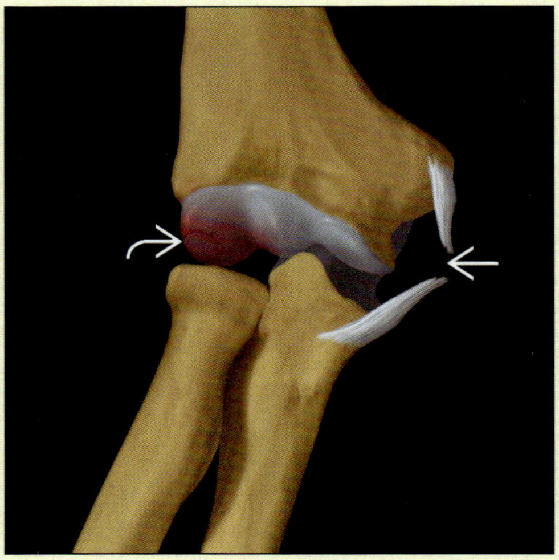
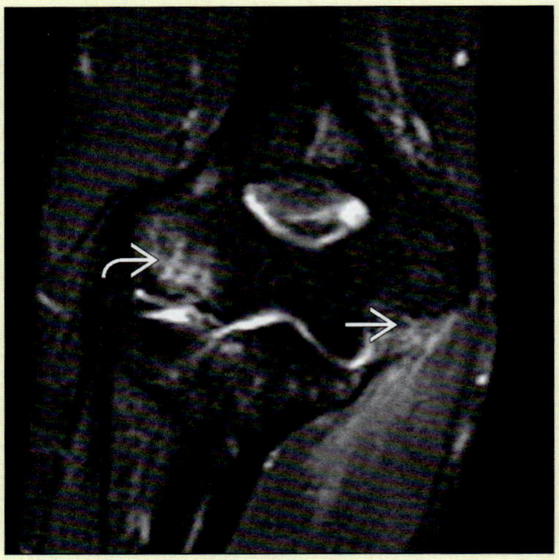

(Left) *Graphic of valgus stress pattern of elbow injury shows distraction force medially, resulting in disruption of the anterior band of the ulnar collateral ligament ➡, and impaction force laterally, resulting in chondral fracture of the capitellum ➡. In throwers, the capitellar injury is usually anterior due to flexion of the elbow during pitching.* ***(Right)*** *Coronal T2WI FS MR slightly more anteriorly in the same patient shows partial tear of the proximal UCL fibers ➡ and capitellar edema ➡.*

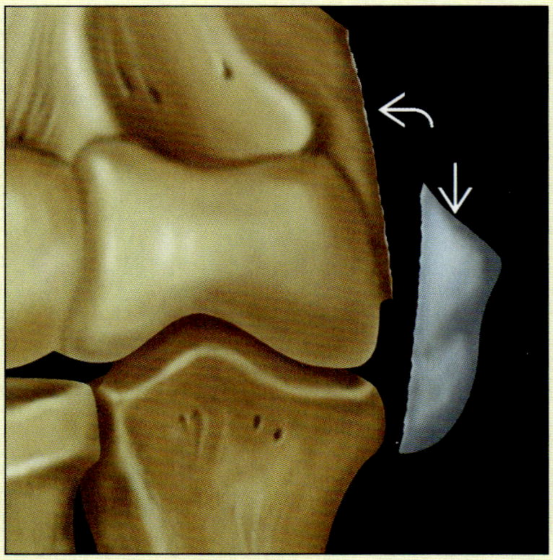
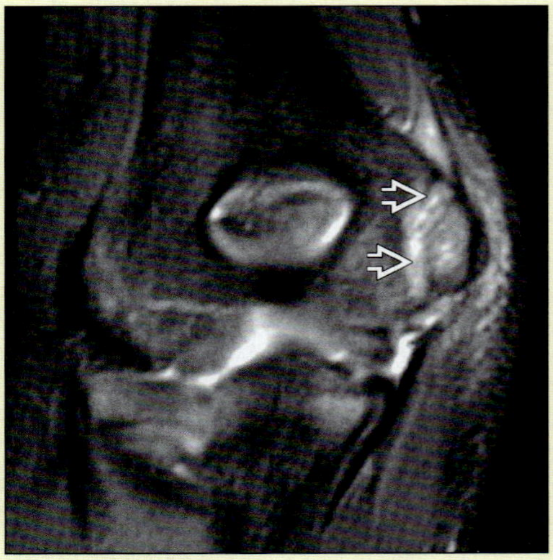

(Left) *Graphic of avulsion of the medial humeral epicondylar ossification center (little leaguer's elbow) shows inferior and medial displacement of the apophysis ➡ from its usual location ➡. The ulnar collateral ligament (UCL) and flexor/pronator tendon complex both attach to the apophysis.* ***(Right)*** *Coronal PDWI FS MR in a 13 year old after a fall shows edema in the physeal plate of the medial humeral epicondyle ➡, consistent with an acute nondisplaced avulsive injury.*

VALGUS STRESS MECHANISM/LITTLE LEAGUER'S ELBOW

TERMINOLOGY

Synonyms

- Thrower's elbow
- Little leaguer's elbow: Medial epicondylar apophysitis, medial epicondylar avulsion

Definitions

- Valgus stress mechanism
  - Pattern of injury seen in elbow due to repetitive valgus elbow stress, including
    - Compression of lateral structures
    - Distraction of medial structures
- Little leaguer's elbow
  - Specific pattern of valgus stress injury to elbow in adolescent throwers
  - Avulsion of medial humeral epicondyle
  - ± intraarticular entrapment of epicondylar fragment
  - Often associated with capitellar chondral/ osteochondral injury

IMAGING

General Features

- Best diagnostic clue
  - Combination of medial distraction and lateral compression injuries on elbow MR
- Location
  - Distraction injury
    - Usually involves anterior band of ulnar collateral ligament (UCL)
    - Occasionally see traction edema at sublime tubercle of ulna
    - Can affect flexor/pronator tendon bundle
      - Partial or complete avulsion of medial epicondylar tendon attachment in skeletally immature patients
  - Compression injury
    - Articular surface of capitellum
      - Usually anterior, as injury generally occurs with elbow in flexion
    - Medial side of olecranon

Radiographic Findings

- Little leaguer's elbow
  - Fragmentation of medial humeral epicondylar apophysis
  - Widening of medial humeral epicondylar growth plate
  - Overgrowth of medial epicondyle
  - Displacement of medial epicondylar ossification center
    - Can become entrapped in medial elbow joint
- Chronic changes
  - Hypertrophy of extremity
  - Degenerative changes

MR Findings

- **Medial distraction**
  - UCL sprain or tear
    - Edema in &/or around anterior band of ulnar collateral ligament
    - Disruption of anterior band fibers
    - Best seen on coronal PD and T2WI FS MR
  - Edema in sublime tubercle of ulna
    - Best seen on coronal T2WI FS MR
  - Flexor/pronator injury
    - Edema ± fiber disruption in or around common tendon origin
    - Edema on axial and coronal PD FS or T2WI MR
    - Fiber disruption on axial and coronal PDWI MR
    - Muscle tear on axial and coronal T2WI FS MR
  - Medial epicondylar injury
    - Marrow edema in epicondyle ± physis
      - ↑ signal on axial and coronal T2WI FS MR
      - ↓ signal on T1WI MR
    - Displacement of epicondyle
- **Lateral compression**
  - Capitellar injury
    - Usually in anterior articular surface (elbow flexed during injury)
    - Injury may be in posterior capitellum in hyperextension injury (gymnasts)
    - Subchondral marrow edema or fracture
    - Chondral tear or defect
      - Stable: Fragment in situ with no fluid interposed between it and native bone
      - Unstable: Fragment in situ or slightly displaced with fluid interposed between it and native bone
    - Osteochondral fracture
    - Best seen on sagittal and coronal PD FS and T2WI MR
  - Chondral or osteochondral fragments in joint
  - Olecranon stress or compression fracture
    - Marrow edema or linear fracture line in proximal lateral aspect of olecranon
- **Ulnar neuropathy**
  - Increased size ± signal of ulnar nerve in cubital tunnel

Imaging Recommendations

- Best imaging tool
  - MR
- Protocol advice
  - Try to maintain elbow extension for imaging
  - Patients often uncomfortable; acquire most desired sequences 1st (tailored exam)
    - Coronal T2WI FS good choice for 1st sequence in this clinical setting

DIFFERENTIAL DIAGNOSIS

Flexor/Pronator Tendon Injury

- Isolated injury to tendons without UCL injury or medial apophyseal avulsion

Extensor Tendinopathy

- Isolated injury to tendons in lateral aspect of elbow

Osteoarthritis

- Osteophytes in radiocapitellar and ulnohumeral joints

PATHOLOGY

General Features

- Etiology

VALGUS STRESS MECHANISM/LITTLE LEAGUER'S ELBOW

- Usually due to repetitive valgus stress at elbow, especially with elbow flexion
- Most commonly seen in overhead throwers

Staging, Grading, & Classification

- Medial distraction
  - Anterior band of UCL
    - Grade I: Edema around normal ligament
    - Grade II: Abnormal signal within ligament without complete disruption
      - Includes partial tear
    - Grade III: Complete disruption of ligament
- Lateral compression
  - Articular surface of capitellum
    - Subchondral impaction contusion
    - Chondral injury
      - Flap tear
      - Chondral fracture
    - Osteochondral fracture
      - Stable
      - Unstable
      - Ex-vacuo

CLINICAL ISSUES

Presentation

- Most common signs/symptoms
  - Medial ± lateral elbow pain
  - Medial instability if UCL ruptured
  - Elbow swelling
  - Pain on palpation over medial humeral epicondyle
- Other signs/symptoms
  - Restricted motion at elbow if entrapped epicondylar fragment
  - Ulnar neuropathy
    - Commonly associated with valgus stress injury
    - Repetitive traction injury to ulnar nerve in cubital tunnel
- Clinical profile
  - Typically overhead throwing athlete
    - Baseball pitcher
    - Volleyball spiker
    - Football quarterback
    - Javelin thrower
  - Gymnasts

Demographics

- Age
  - Generally young adolescents and young adults
  - Little leaguer's elbow: 6-14 years
- Gender
  - M > > F (related to activity)

Natural History & Prognosis

- Symptoms may increase gradually or present suddenly
- Symptoms usually progress as activity continues
- Strain or partial tear of UCL may convert to complete tear suddenly ("pop")
- Stable (osteo)chondral capitellar fragment may become unstable or detach
- Continuing repetitive activity → worsening injury → hypertrophic and degenerative changes
  - Traction spurs at medial humeral epicondyle and coronoid
  - Irregularity of capitellar articular surface
  - Diffuse osteoarthritis of elbow

Treatment

- Low-grade UCL sprain ± mild capitellar contusion
  - Conservative treatment
  - Cessation of offending activity most important
- High-grade UCL injury
  - Ligament repair often necessary
- Unstable osteochondral injury to capitellum
  - Arthroscopic debridement
- Little leaguer's elbow
  - Nondisplaced
    - Conservative therapy
  - Displaced/entrapped fragment
    - Open reduction and internal fixation
- Ulnar neuropathy
  - Generally self-limited; improves with cessation of activity
  - Chronic: May require ulnar nerve transposition surgery

DIAGNOSTIC CHECKLIST

Consider

- Medial traction injuries and lateral compression injuries often coexist
- Ulnar neuropathy

Image Interpretation Pearls

- Capitellar (osteo)chondral lesions often best seen on sagittal images
- UCL injury best seen on coronal images
- Chondral defect in capitellum → search for intraarticular bodies

Reporting Tips

- Describe size and location of capitellar injury
  - Categorize as subchondral, chondral, or osteochondral
  - Categorize as stable or unstable

SELECTED REFERENCES

1. Lynch JR et al: Medial collateral ligament injury in the overhand-throwing athlete. J Hand Surg Am. 33(3):430-7, 2008
2. Magra M et al: A review of epidemiology of paediatric elbow injuries in sports. Sports Med. 37(8):717-35, 2007
3. Benjamin HJ et al: Little league elbow. Clin J Sport Med. 15(1):37-40, 2005
4. Lee HH et al: Operative treatment of displaced medial epicondyle fractures in children and adolescents. J Shoulder Elbow Surg. 14(2):178-85, 2005
5. Ahmad CS et al: Valgus extension overload syndrome and stress injury of the olecranon. Clin Sports Med. 23(4):665-76, x, 2004
6. Hang DW et al: A clinical and roentgenographic study of Little League elbow. Am J Sports Med. 32(1):79-84, 2004
7. Cain EL Jr et al: Elbow injuries in throwing athletes: a current concepts review. Am J Sports Med. 31(4):621-35, 2003
8. Klingele KE et al: Little league elbow: valgus overload injury in the paediatric athlete. Sports Med. 32(15):1005-15, 2002

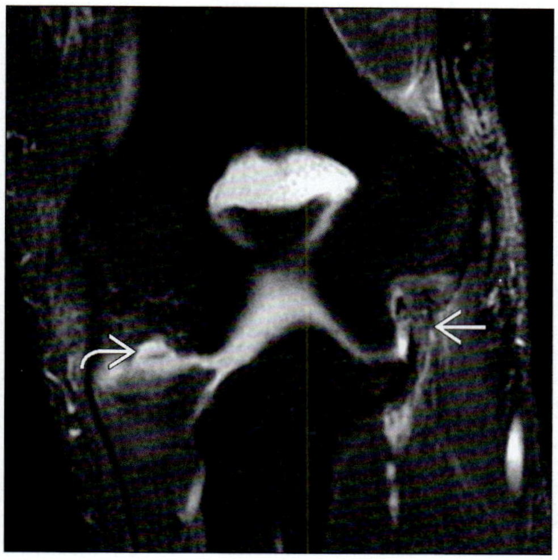

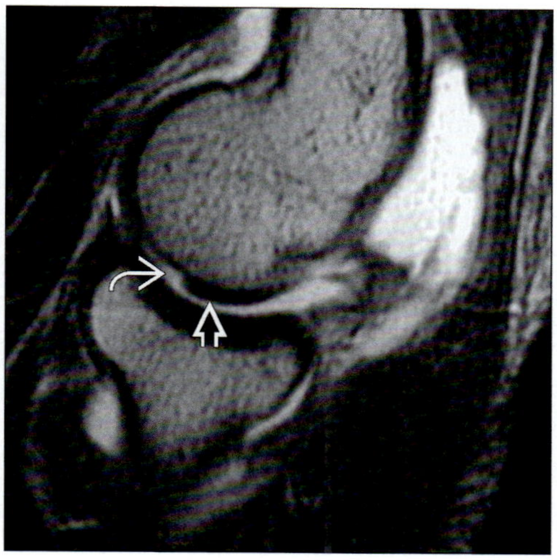

(Left) *Coronal T2WI FS MR shows a partial tear of the anterior band of the UCL with abnormal signal within and around the ligament. There is a chondral defect in the capitellum with a small fragment inside.* ***(Right)*** *Sagittal T2WI MR in the same patient demonstrates the chondral defect in the anterior articular surface of the capitellum. Normal articular cartilage is seen more posteriorly.*

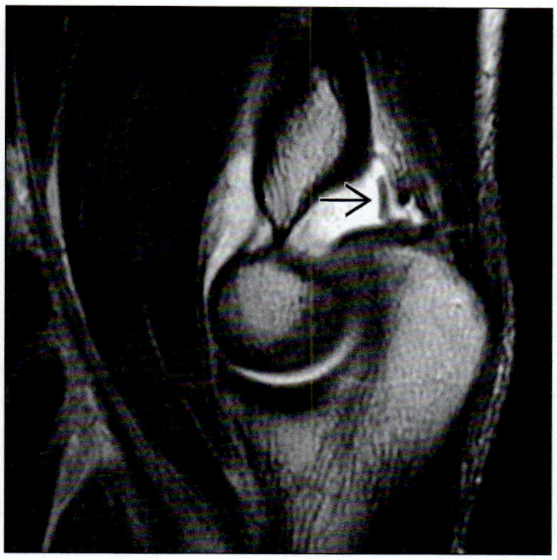

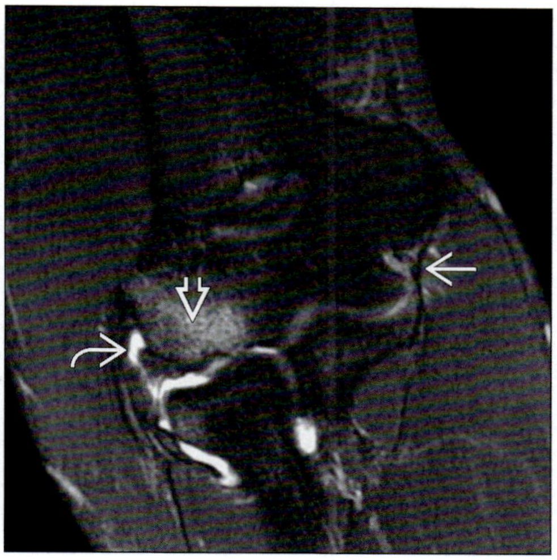

(Left) *Sagittal T2WI MR more medially in the same patient shows a chondral fragment in the olecranon fossa of the joint. The presence of a chondral or osteochondral defect of the capitellum in the setting of acute trauma should trigger a search for such bodies.* ***(Right)*** *Coronal T2WI FS MR in a college pitcher shows mild edema around the UCL and subchondral marrow edema in the anterior capitellum. There is also a partial tear of the common extensor tendon.*

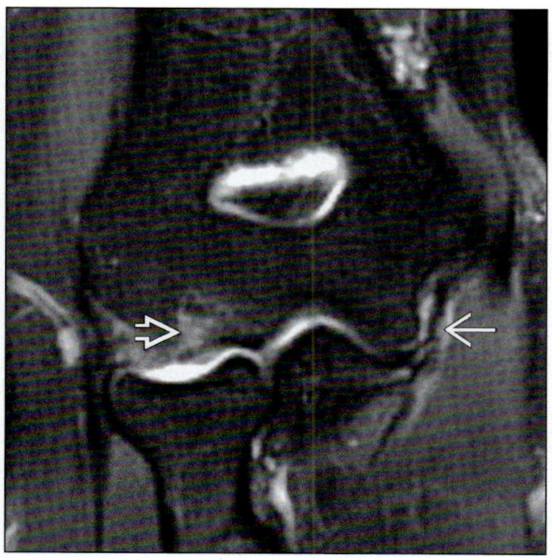

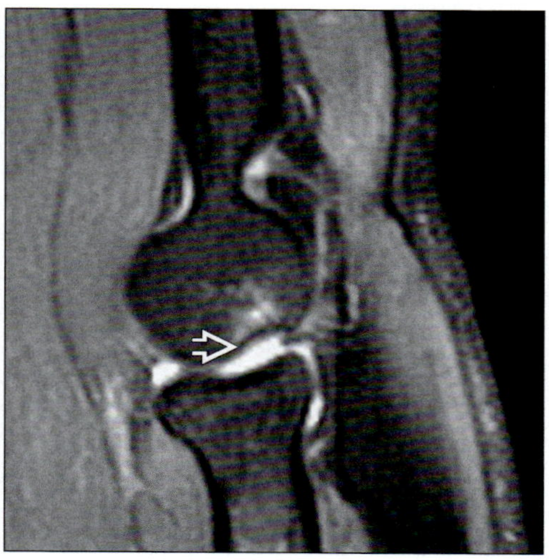

(Left) *Coronal T2WI FS MR in a college volleyball player who felt sudden elbow pain after spiking shows mild edema around the UCL, consistent with a low-grade sprain, and subchondral edema in the capitellum.* ***(Right)*** *Sagittal PDWI FS MR in the same patient shows an impaction defect in the posterior capitellum with marrow edema. The posterior location of this injury relates to the patient striking the ball with the elbow in extension.*

VALGUS STRESS MECHANISM/LITTLE LEAGUER'S ELBOW

(Left) Coronal T2WI FS MR after traumatic valgus stress injury to the elbow shows disruption of the midportion of the UCL ➡, tear of the flexor/pronator tendon ➡, and subchondral contusion in the capitellum ➡. *(Right)* Axial T2WI FS MR in the same patient shows the flexor/pronator tendon tear ➡ and a chondral fragment ➡ posterior to the capitellum arising from an acute chondral fracture in the capitellum (not shown).

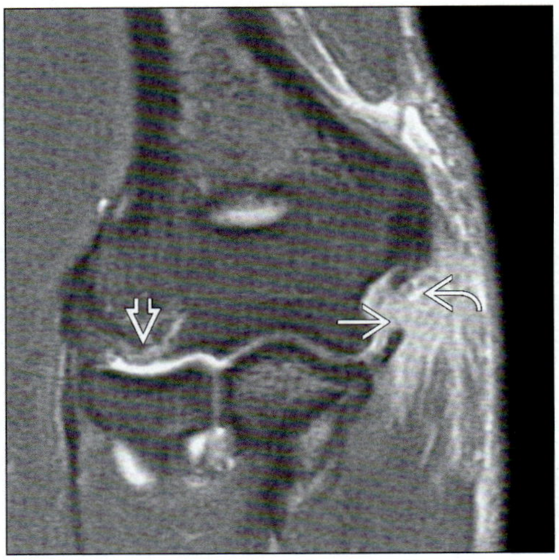

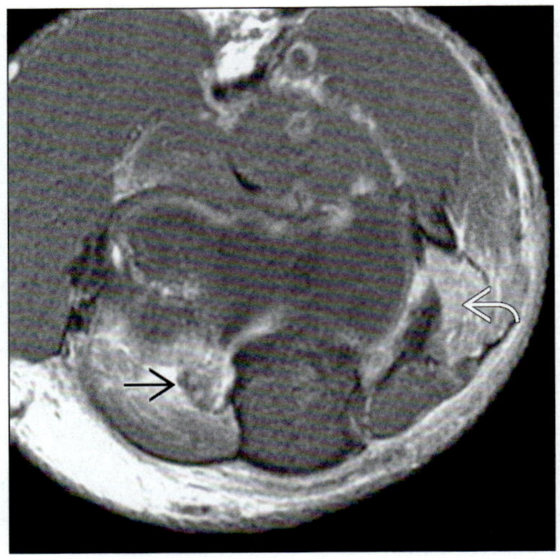

(Left) AP radiograph in a 13-year-old baseball pitcher, with sudden medial elbow pain and limited elbow motion while pitching, shows displacement of the medial humeral epicondylar ossification center ➡ into the joint. Its normal position ➡ is vacant. A trapped medial epicondylar fragment such as this usually requires open reduction. *(Right)* AP radiograph after attempted closed reduction in the same patient shows the epicondylar center out of the joint but still displaced ➡.

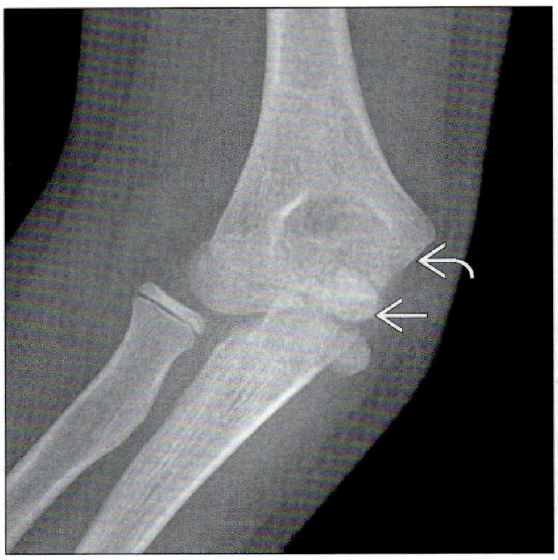

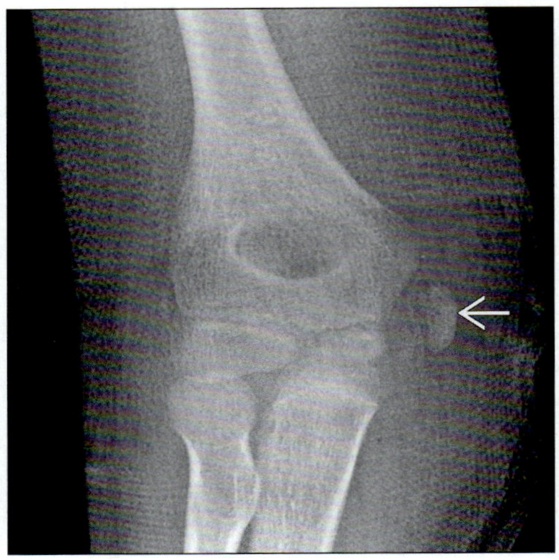

(Left) AP radiograph in a 16-year-old baseball pitcher with chronic elbow pain shows a nonunion of an old medial epicondylar avulsion injury ➡. The patient had mild medial ligamentous laxity on physical examination. *(Right)* Coronal T2WI FS MR in a 13-year-old little league pitcher with medial elbow pain shows partial avulsion of the medial humeral epicondyle ➡ with fluid interposed between the native humerus and the fragment. Epicondylar avulsions can consist of only fragments such as this.

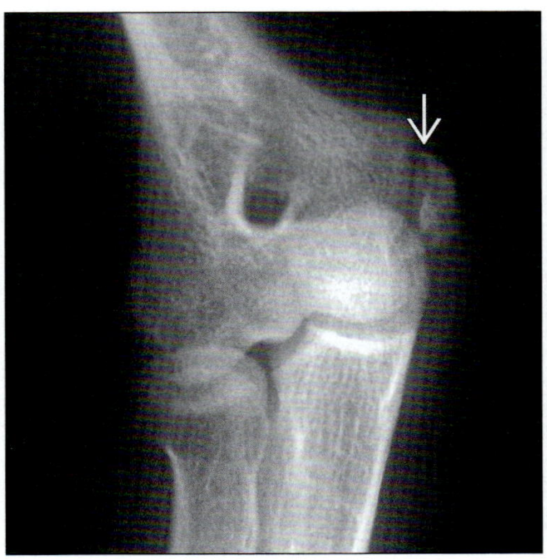

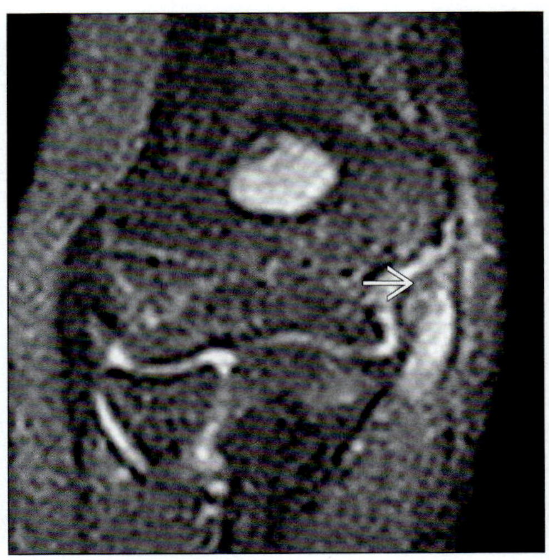

VALGUS STRESS MECHANISM/LITTLE LEAGUER'S ELBOW

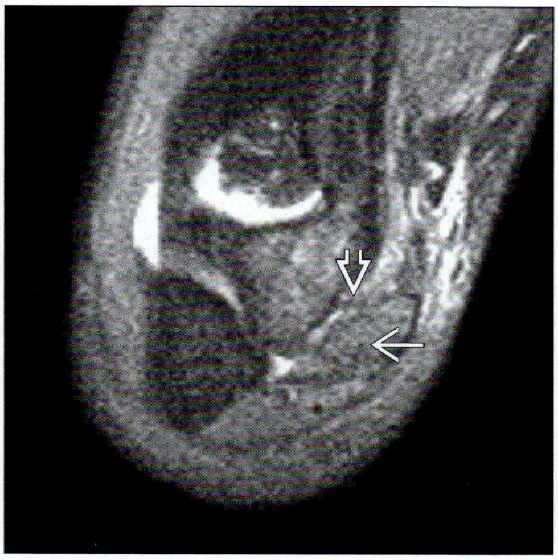

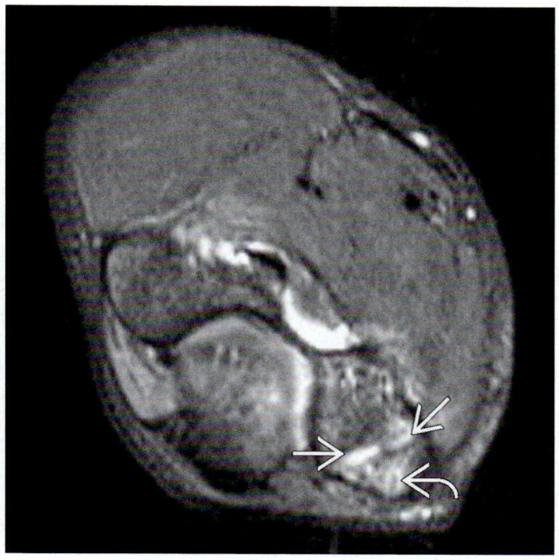

(Left) *Coronal T2WI FS MR in a 9 year old with medial elbow pain after pitching little league baseball shows partial displacement and rotation of the medial epicondylar ossification center ➡ with edema in the fragment and in the physeal space ⇨.* ***(Right)*** *Axial T2WI FS MR in a 14-year-old pitcher shows fluid signal in the physis ➡ of the medial humeral epicondyle. There is edema in the apophysis ↪ as well.*

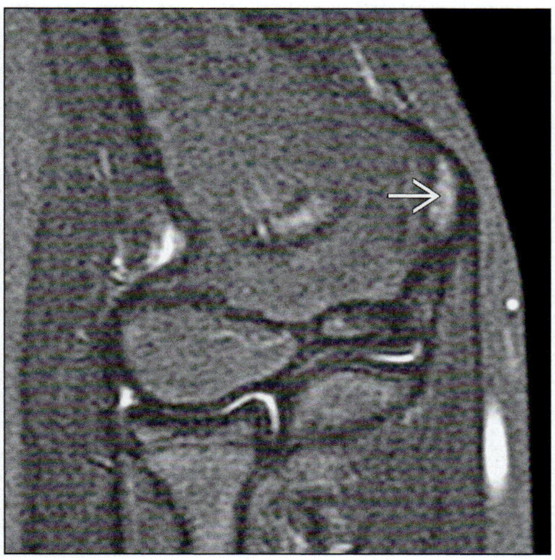

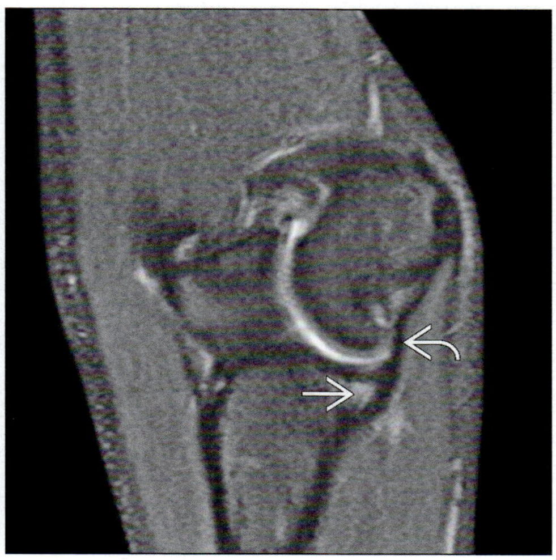

(Left) *Coronal T2WI FS MR in a 13-year-old thrower with medial elbow pain shows subtle marrow edema in a nondisplaced medial epicondyle ➡. As in other anatomic sites, ligament disruption is less common in skeletally immature patients than apophyseal injury.* ***(Right)*** *Coronal T2WI FS MR in a high school pitcher with medial elbow pain shows marrow edema in the sublime tubercle ➡ with an intact UCL ↪. This is an unusual pattern of injury in the spectrum of valgus stress.*

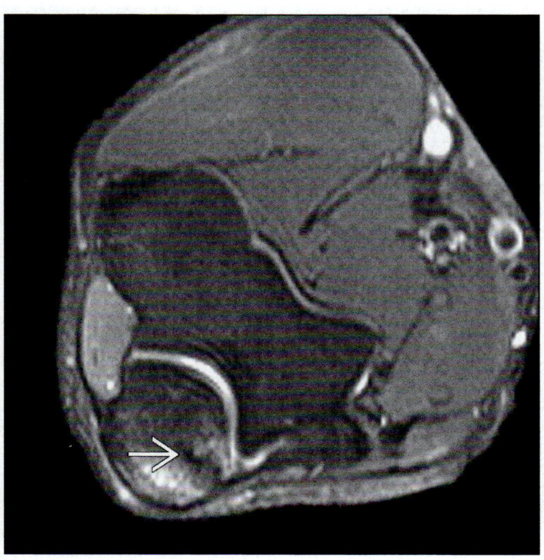

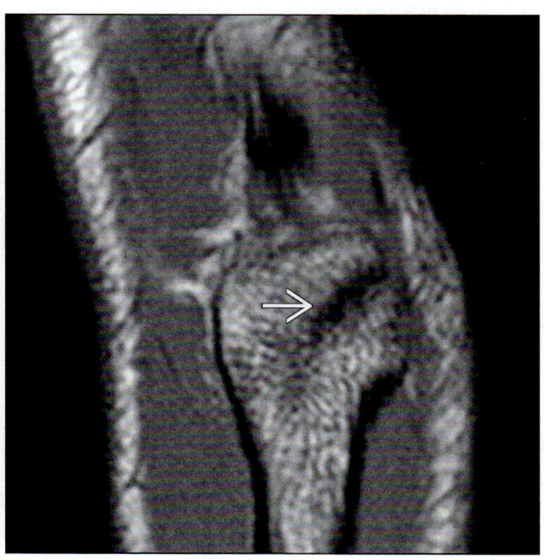

(Left) *Axial PDWI FS MR in a college baseball pitcher with posterior elbow pain shows a linear fracture ➡ with surrounding marrow edema in the proximal medial aspect of the olecranon. Other structures were normal.* ***(Right)*** *Coronal T1WI MR in the same patient shows the linear olecranon fracture ➡, which was the cause of the patient's symptoms. Such injuries are likely due to repetitive impaction of the medial olecranon against the trochlea.*

BICEPS TENDON INJURY, ELBOW

Key Facts

Terminology
- Partial or complete tear or avulsion of distal attachment of biceps brachii tendon

Imaging
- Discontinuity of biceps tendon
- Axial fat-suppressed intermediate or T2-weighted sequences most accurate in assessing
  - Location of disruption
  - Extent of retraction
  - Percentage of tendon disrupted (if partial)
- Usually at or near tendinous insertion on radial tuberosity
- MR
  - Sagittal images can be misleading
  - Long axis of biceps is oblique to true sagittal plane

Top Differential Diagnoses
- Cephalic vein
- Brachialis muscle and tendon
- Partial tear mistaken for complete tear

Clinical Issues
- Surgical reattachment of avulsed/torn tendon
- May progress to fibrosis if left untreated
- Attempting to lift extremely heavy object
- "Popping" sensation in antecubital fossa at time of acute injury
- "Popeye" sign of retracted muscle belly in mid-arm causing visible bulge
- Palpable defect in tendon at antecubital fossa

Diagnostic Checklist
- Location of tear
- Presence and degree of retraction
- Do not rely on sagittal images

*(**Left**) Graphic illustration of a partial tear of the distal biceps tendon shows disruption of the tendon fibers → at its attachment to the radial tuberosity →. Lacertus fibrosus → is the bicipital aponeurosis extension to the medial forearm. (**Right**) Sagittal T2WI MR shows disruption of the biceps tendon → with retraction into the arm. The brachialis muscle → lies deep to the biceps.*

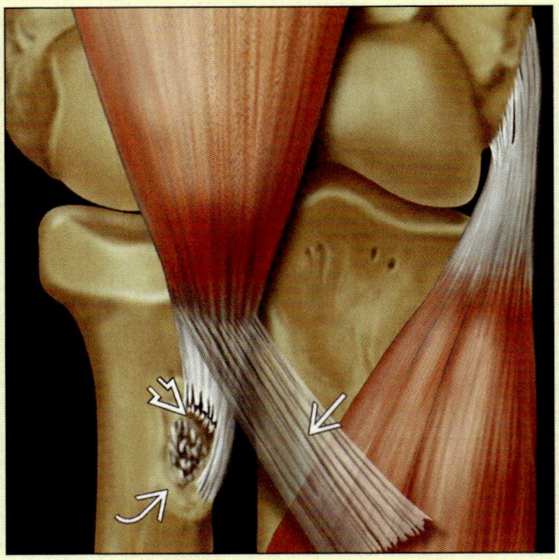

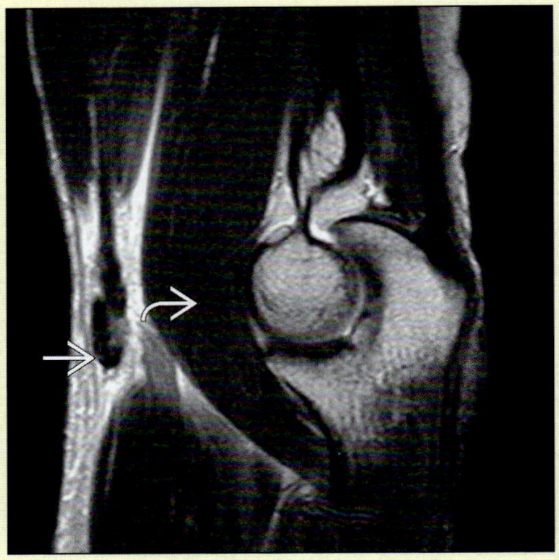

*(**Left**) Axial graphic shows the torn biceps tendon → in its distal course anterior to the radius →. Vessels and nerves → fill the fat anterior to the tendon. (**Right**) Axial T2WI MR shows complete disruption of the biceps tendon. The radial tuberosity → is bare; the soft tissues anterior to it are edematous. In the expected course of the biceps tendon → there is only edematous fat. The vessel → in the antecubital fossa should not be mistaken for the biceps tendon.*

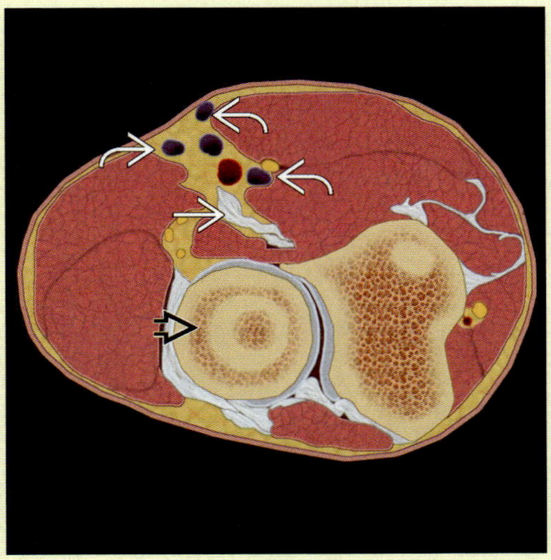

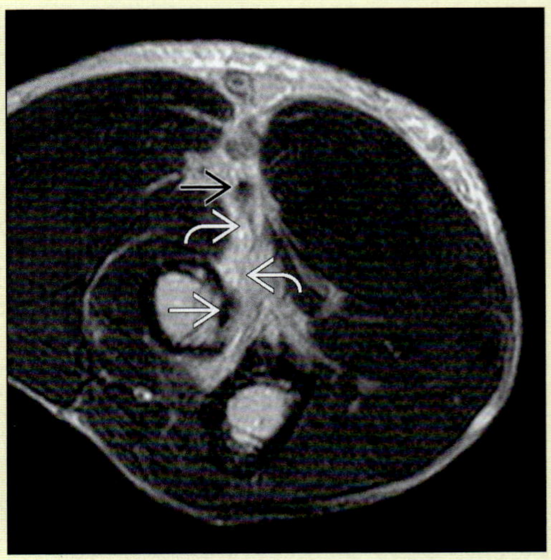

BICEPS TENDON INJURY, ELBOW

TERMINOLOGY

Synonyms

- Biceps tear, biceps rupture

Definitions

- Partial or complete tear or avulsion of distal attachment of biceps brachii tendon
- Significant degeneration of biceps tendon resulting in macroscopic change visible on MR
- Injury to biceps brachii muscle belly or myotendinous junction, usually due to acute trauma

IMAGING

General Features

- Best diagnostic clue
  - Discontinuity of some or all fibers of biceps tendon
- Location
  - Usually at or near tendinous insertion on radial tuberosity
  - Less often at distal myotendinous junction of biceps
  - Much more common in dominant arm
- Size
  - Distal biceps tears may range from nondisplaced to markedly retracted
- Morphology
  - Common distal tendon of both long and short heads inserts on radial tuberosity
  - Lacertus fibrosus: Aponeurotic sheath extending from myotendinous junction medially to fascia of anteromedial forearm
  - Biceps tendon has no tendon sheath in antecubital fossa

MR Findings

- Discontinuity of tendon fibers
- Retraction of proximal tendon and myotendinous junction
- Soft tissue swelling ± fluid in antecubital fossa along course of tendon
  - Lack of real tendon sheath → ill-defined peritendinous fluid instead of well-defined peritendinous fluid collection
- Disruption of lacertus fibrosus

Ultrasonographic Findings

- Discontinuity of biceps tendon in antecubital fossa
- Can be useful in determining extent of retraction and mobility of tendon

Imaging Recommendations

- Best imaging tool
  - MR
- Protocol advice
  - Axial fat-suppressed intermediate or T2-weighted sequences most accurate in assessing
    - Location of disruption
    - Extent of retraction
    - Percentage of tendon disrupted (if partial)
  - Axial images must extend at least 5 cm below elbow joint to include entire radial tuberosity
  - Sagittal images can be misleading
    - Long axis of biceps is oblique to true sagittal plane
    - Brachialis tendon and muscle can be mistaken for biceps
    - Cephalic vein can be mistaken for biceps
  - **Positioning issues**
    - Elbow must be straight to avoid redundancy and distortion of biceps
    - Pronation/supination not as critical
      - Pronation more comfortable, results in posterior rotation of radial tuberosity
      - Supination harder for patient to maintain, puts radial tuberosity in more anatomic position
    - FABS position (elbow **f**lexed, arm **ab**ducted, wrist **s**upinated) imaging
      - Patient prone, arm over head, elbow flexed
      - Slices prescribed parallel to long axis of tendon, coronal to arm
      - Shows distal biceps tendon in its full length, including radial attachment
      - May be useful adjunct in questionable partial tears

DIFFERENTIAL DIAGNOSIS

Cephalic Vein

- Runs longitudinally through antecubital fossa, may mimic biceps on sagittal images
- Ramifications of vein often provide clue to identity

Brachialis Muscle and Tendon

- Located deep to biceps tendon
- Myotendinous junction located more distal than that of biceps
- Inserts on ulna

Partial Tear Mistaken for Complete Tear

- Axial images provide data on percentage of tendon involved in tear

Biceps Tendinopathy vs. Partial Tear

- Distinction between partial tear and tendinopathy can be subjective
  - Pathologically, tendinopathy often includes areas of microtear of tendon
  - Different MR pulse sequences may suggest different diagnostic likelihood
    - Fluid better seen on T2WI within tendon suggests partial tear
    - On short TE MR sequences this fluid is hard to distinguish from mucinous tendon degeneration, thus making distinction between partial tear and tendon degeneration difficult or impossible

Bicipitoradial Bursitis

- Focal fluid in bursa that forms between distal tendon and radial neck
- Appears different with forearm in supination vs. pronation

PATHOLOGY

Staging, Grading, & Classification

- **Complete tear**
  - Discontinuity of tendon usually at or immediately proximal to insertion on radial tuberosity
  - May be displaced or nondisplaced

BICEPS TENDON INJURY, ELBOW

- Integrity of lacertus fibrosus serves as indicator of lack of significant retraction
- **Partial tear**
  - Tendinous fiber disruption without complete dissociation of myotendinous unit
  - May be difficult to differentiate from tendinopathy on MR
- **Tendinopathy**
  - Mucinous degeneration of tendon substance ± microtears without macroscopic tendon defect
  - Often predisposes to more significant disruption
- **Chronic tear**
  - Complete disruption of distal tendon and subsequent fibrosis of tendon to adjacent structures
    - Brachialis
    - Superficial fascia
  - May be difficult to repair due to decreased tendon mobility

CLINICAL ISSUES

Presentation

- Most common signs/symptoms
  - Pain and swelling in antecubital fossa
  - "Popping" sensation in antecubital fossa at time of acute injury
  - Limited elbow flexion with weakness (30% loss)
    - Some flexion still possible due to actions of brachialis, forearm flexors, and extensors
  - "Popeye" sign of retracted muscle belly in mid-arm causing visible bulge
  - Palpable defect in tendon at antecubital fossa
    - May be hard to detect in acute setting due to hematoma formation
- Other signs/symptoms
  - Limited supination (40% loss)
- Clinical profile
  - Classic history: Weightlifter feels "pop" in antecubital fossa while lifting heavy weight

Demographics

- Age
  - Acute traumatic rupture can occur at any age
  - Ruptures related to chronic tendinopathy usually in 4th to 6th decades
- Gender
  - M > > F
    - Likely due to activity profiles
- Epidemiology
  - Most common tendon rupture in elbow
  - More common in dominant arm
  - Distal biceps tears account for about 5% of all biceps ruptures

Natural History & Prognosis

- Usually due to eccentric contraction of biceps against heavy load
  - "Negatives" while doing curls
  - Attempting to lift extremely heavy object
- Most full-thickness tears are preceded by chronic tendinopathy
- May progress to fibrosis if left untreated
  - May scar down onto brachialis

Treatment

- Low-grade tear or tendinopathy treated conservatively
  - Rest, ice, antiinflammatory therapy, physical therapy
- Surgical reattachment of avulsed/torn tendon
  - Degree of tendon retraction at imaging influences type and size of incision
  - Integrity of lacertus fibrosus may influence operative approach

DIAGNOSTIC CHECKLIST

Consider

- Location of tear
- Presence and degree of retraction
- Status of lacertus fibrosus

Image Interpretation Pearls

- Follow course of tendon on sequential axial images to find site of disruption
- Do not rely on sagittal images

Reporting Tips

- Describe size of residual stump of tendon on radial tuberosity, if any
- Measure distance of tendon retraction

SELECTED REFERENCES

1. Rineer CA et al: Elbow tendinopathy and tendon ruptures: epicondylitis, biceps and triceps ruptures. J Hand Surg Am. 34(3):566-76, 2009
2. Bain GI et al: Treatment of partial distal biceps tendon tears. Sports Med Arthrosc. 16(3):154-61, 2008
3. Chavan PR et al: Repair of the ruptured distal biceps tendon: a systematic review. Am J Sports Med. 36(8):1618-24, 2008
4. Cohen MS: Complications of distal biceps tendon repairs. Sports Med Arthrosc. 16(3):148-53, 2008
5. Hamer MJ et al: Operative treatment of chronic distal biceps tendon ruptures. Sports Med Arthrosc. 16(3):143-7, 2008
6. Mazzocca AD et al: Distal biceps rupture. Orthop Clin North Am. 39(2):237-49, vii, 2008
7. Tran N et al: Ultrasonography of the elbow. Semin Musculoskelet Radiol. 11(2):105-16, 2007
8. Kijowski R et al: Magnetic resonance imaging of the elbow. Part II: Abnormalities of the ligaments, tendons, and nerves. Skeletal Radiol. 34(1):1-18, 2005

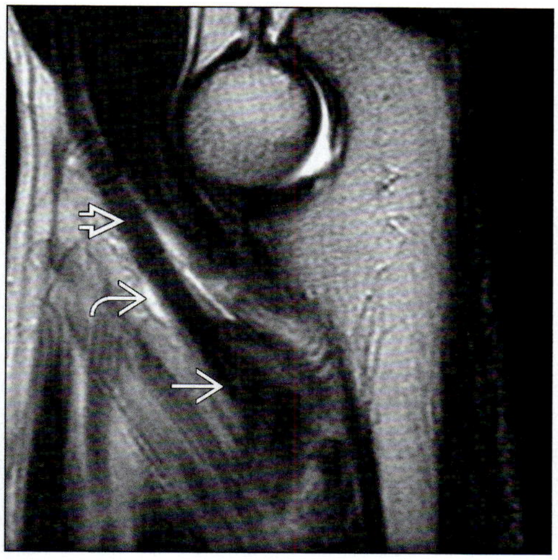
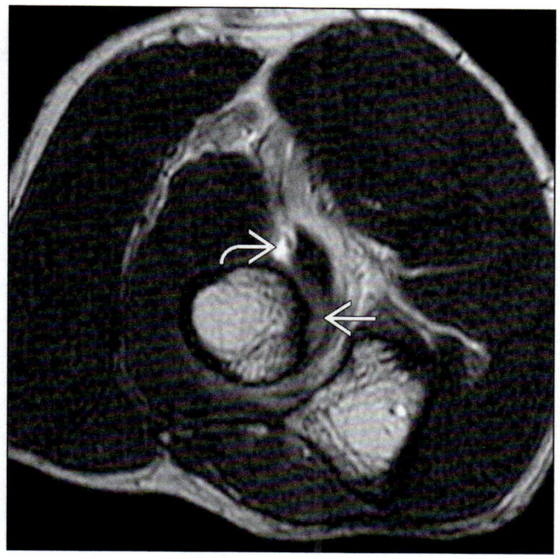

(Left) *Sagittal T2WI MR of biceps tendinopathy shows the intact biceps tendon ⇨ with thickening of the distal aspect of the tendon → and mild peritendinous fluid ↪.* ***(Right)*** *Axial T2WI MR in the same patient better shows the thickened and inhomogeneous distal biceps tendon → without demonstrable fiber interruption. Peritendinous fluid ↪ is present. The distinction between tendinopathy and partial tear can be difficult.*

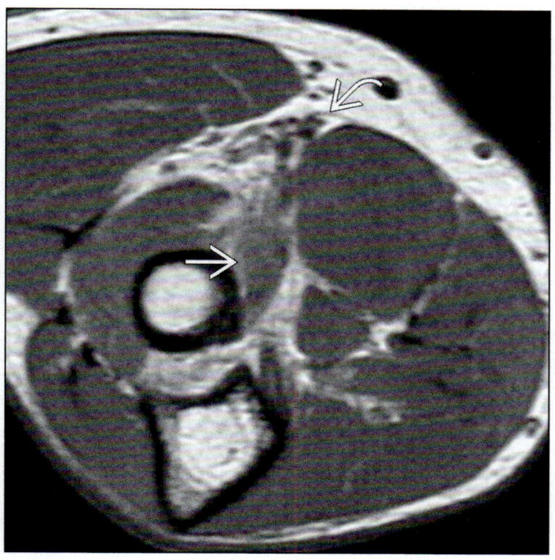
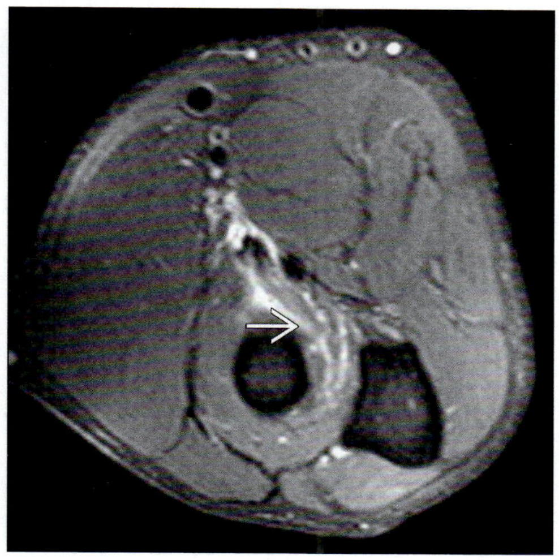

(Left) *Axial PDWI MR at the level of the radial tuberosity shows a low-grade partial tear of the tendon at its attachment →. A small portion of the lacertus fibrosus ↪ is seen on this image.* ***(Right)*** *Axial PDWI FS MR shows a higher grade partial tear of the tendon → with only a few fibers remaining intact. Cross-sectional (axial) images provide a more accurate means of determining the percentage of tendon fibers that remain intact.*

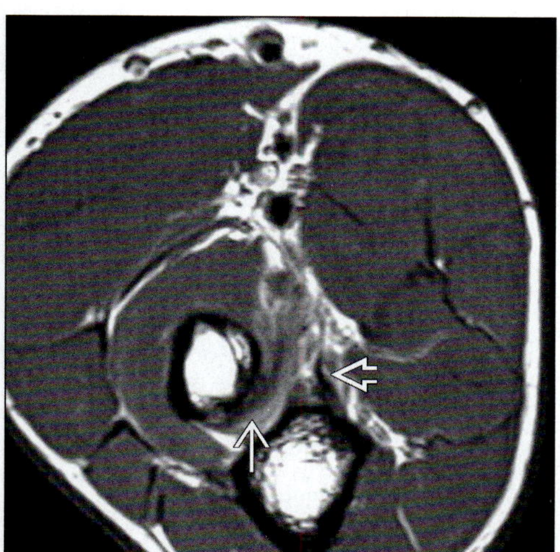
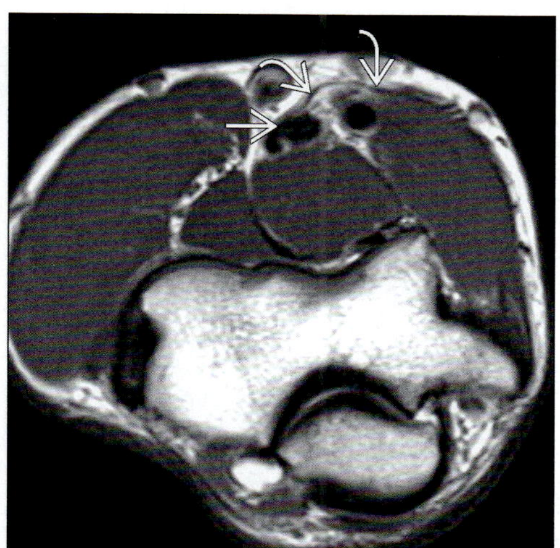

(Left) *Axial PDWI MR shows a high-grade partial tear of the distal biceps tendon → and the normal brachialis tendon ⇨ attaching to the ulna.* ***(Right)*** *Axial PDWI MR acquired more superiorly in the same patient demonstrates the biceps tendon → and the intact lacertus fibrosus ↪ extending to the fascia superficial to the flexor/pronator muscles.*

BICEPS TENDON INJURY, ELBOW

(Left) Axial PDWI MR shows a complete tear of the biceps tendon at the radial tuberosity →. There are no recognizable tendon fibers in the antecubital fossa. Edema in the fat plane normally occupied by the tendon ↪ implies a recent injury. **(Right)** Axial PDWI MR slightly more superiorly in the same patient shows disruption of the lacertus fibrosus ↪ and diffuse edema in the antecubital fossa. The biceps is absent from its usual location →.

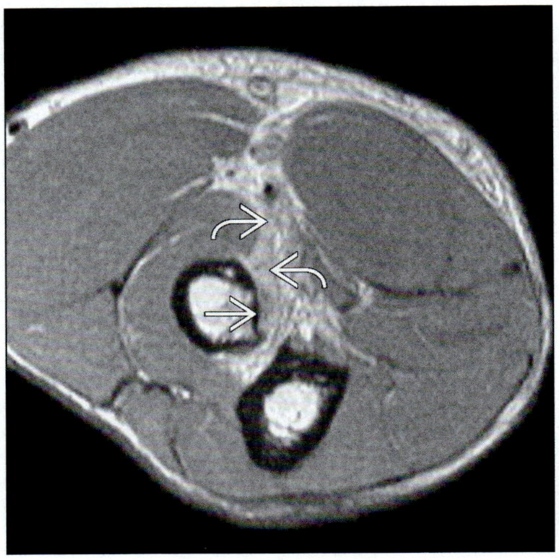

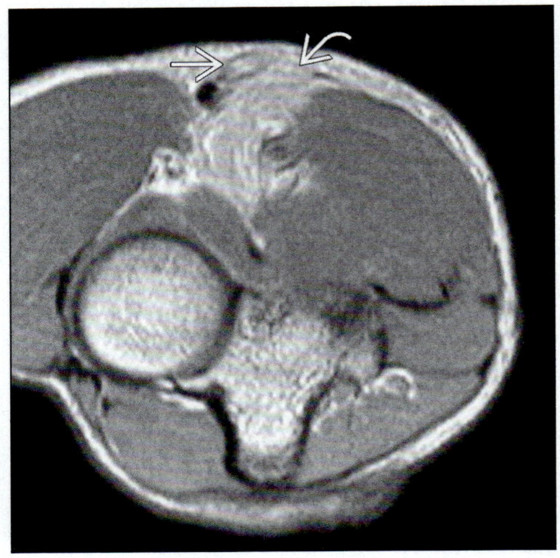

(Left) Sagittal T2WI FS MR of the arm shows linear edema in the mid-muscle belly of the biceps brachii → indicating a muscle strain injury. The distal tendon was intact. The elbow can be seen at the inferior margin of the image ⇨. **(Right)** Axial T2WI FS MR in the same patient shows the muscle edema → in the midportion of the biceps muscle belly. Such injuries are usually due to acute overuse mechanisms, such as in weightlifting.

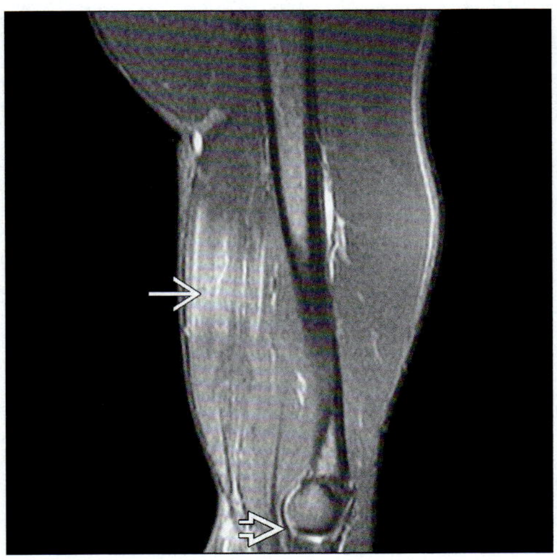

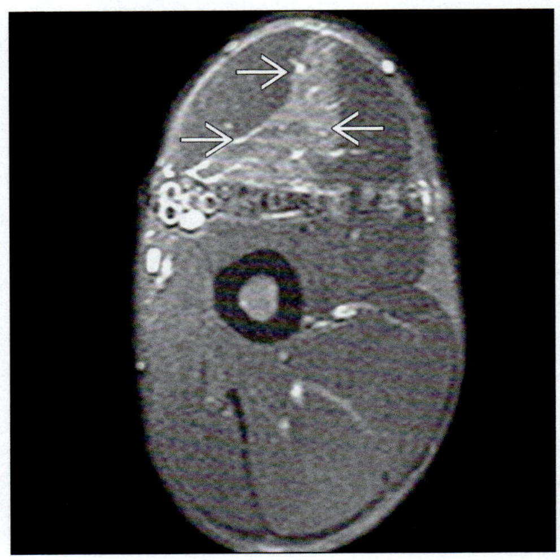

(Left) Sagittal MR arthrogram T1WI FS scout for prescribing a FABS (elbow flexion, arm abduction, wrist supination) shows the scan plane along the long axis of the humerus (and therefore, the biceps tendon) in a roughly coronal plane. **(Right)** Coronal MR arthrogram T1WI FS resulting from FABS positioning shows the intact biceps tendon → attaching to the radial tuberosity ↪. The advantage of FABS imaging is the greater length of tendon seen on 1 image.

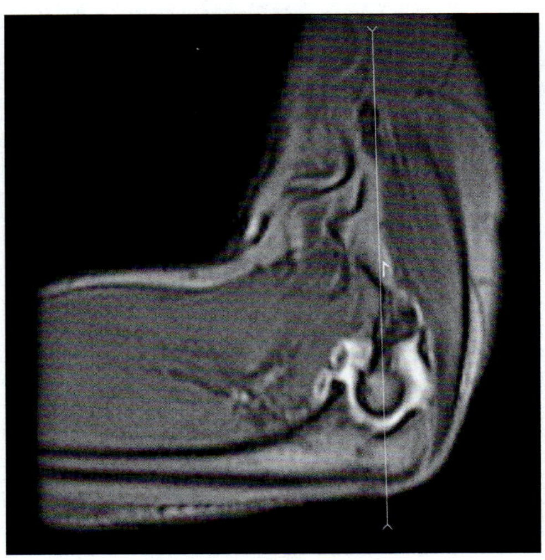

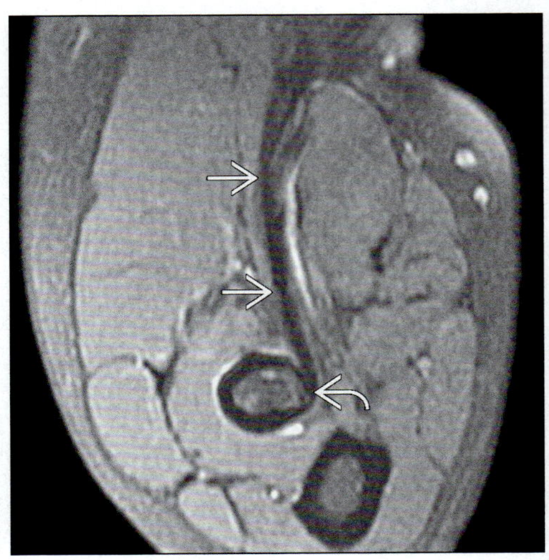

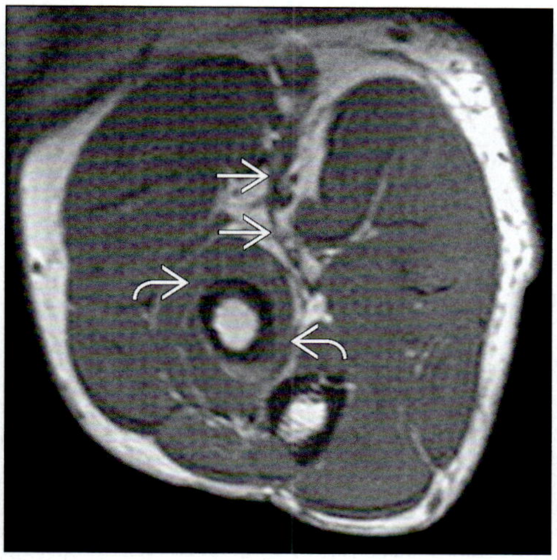

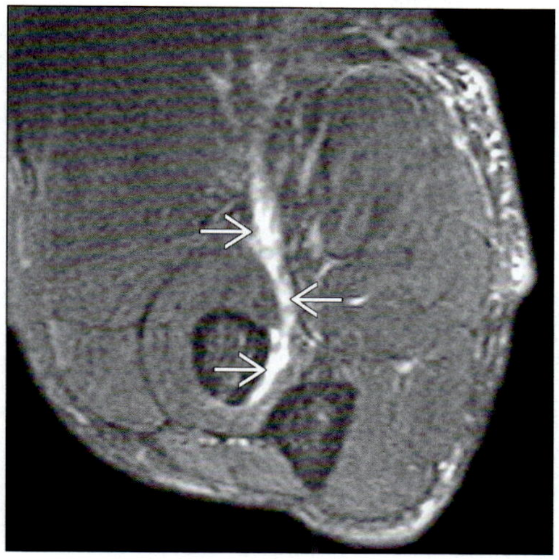

(Left) *Coronal T1WI MR obtained in the FABS position shows an absent biceps tendon in its expected location. The supinator muscle surrounds the radial neck at this level.* ***(Right)*** *Coronal T2WI FS MR FABS image in the same patient better shows the abnormal signal along the course of the torn biceps tendon. Lack of familiarity with anatomic relationships using this imaging plane can cause confusion for the inexperienced reader.*

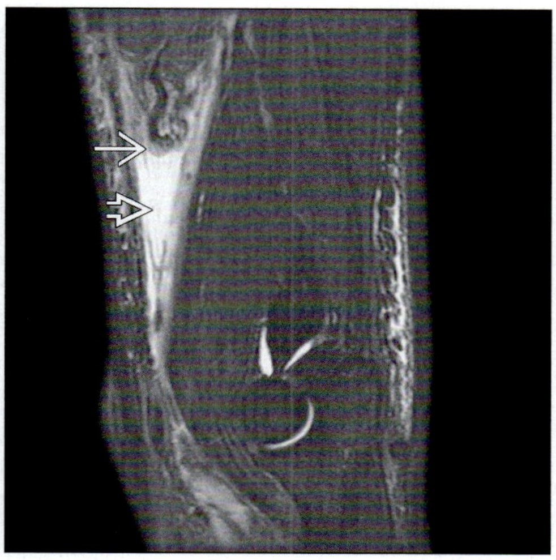

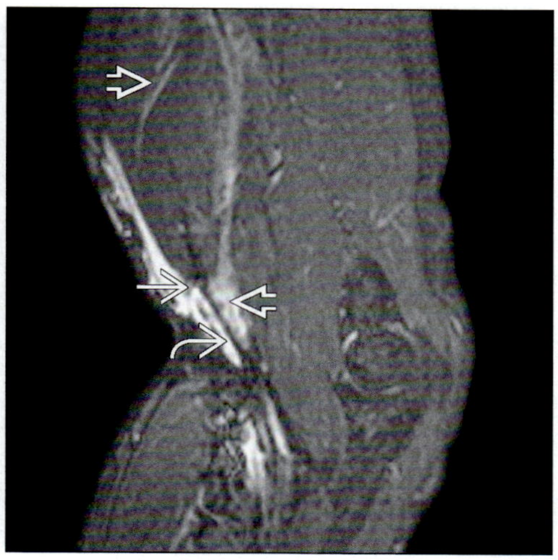

(Left) *Sagittal T2WI FS MR shows a complete tear of the biceps tendon with 14 centimeters of retraction of the tendon end and fluid in the vacated space. The distance of the tendon retraction is important for the surgeon to know prior to surgery for planning purposes.* ***(Right)*** *Sagittal T2WI FS MR shows a partial tear of the biceps at the myotendinous junction with edema around the junction and in the muscle belly. The tendon is intact.*

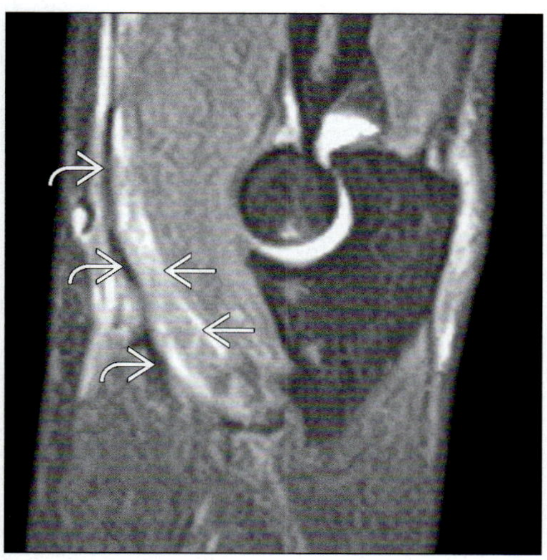

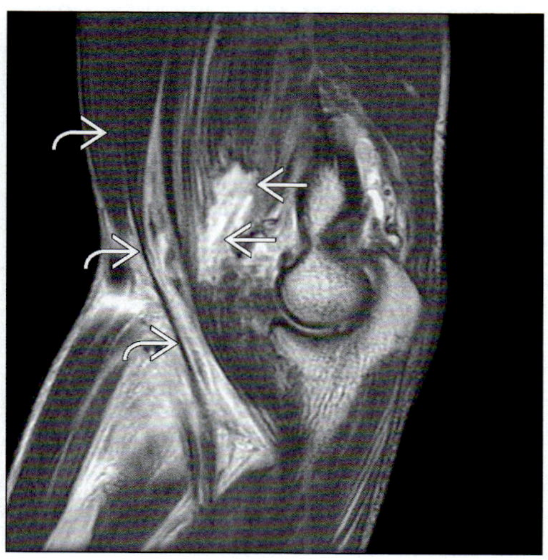

(Left) *Sagittal PDWI FS MR shows a high-grade injury of the biceps tendon without significant retraction. The linear signal void of the cephalic vein more anteriorly should not be mistaken for an intact tendon.* ***(Right)*** *Sagittal T2WI MR shows a tear of the brachialis in the distal muscle belly with surrounding edema. The biceps is located more anteriorly and is intact. Axial images (not shown) are a useful adjunct in assessing this anatomy.*

BICIPITORADIAL BURSITIS

Key Facts

Terminology

- Inflammation of bicipitoradial bursa

Imaging

- Fluid signal mass around distal biceps tendon or between biceps and radius on MR
- Flattened synovial sac not visualized on MR unless distended with fluid
- Variable amounts of internal debris
- Fluid signal collection around or deep to distal biceps tendon
- Well-defined margins
- May change shape depending on whether forearm is supinated or pronated
- Best characterized on axial images
- Combination of T1WI and T2WI to confirm fluid nature of collection

Top Differential Diagnoses

- Biceps tendinopathy or partial tear
- Biceps tear
- Antecubital mass

Pathology

- Repetitive motion
  - Especially pronation/supination of forearm
- Inflammatory arthropathy
- Synovial osteochondromatosis
  - Rheumatoid
  - Gout
- Subtendinous bursa with synovial lining

Diagnostic Checklist

- Concomitant biceps tendinopathy or partial tear

*(**Left**) Coronal graphic shows a distended bicipitoradial bursa adjacent to the distal biceps tendon. A partial tear of the biceps is present. The shape of the bursa often changes with pronation/supination. (**Right**) Axial T2WI MR shows a well-defined fluid collection anterior and posterior to the distal biceps tendon. The distal biceps tendon does not have a synovial sheath; fluid seen adjacent this portion of the tendon generally resides in an inflamed bicipitoradial bursa.*

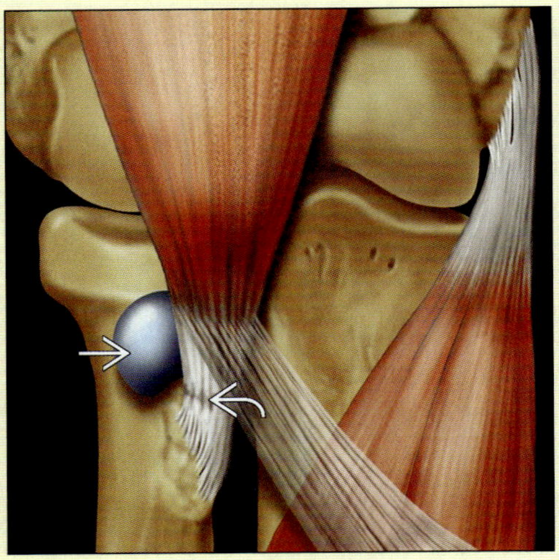

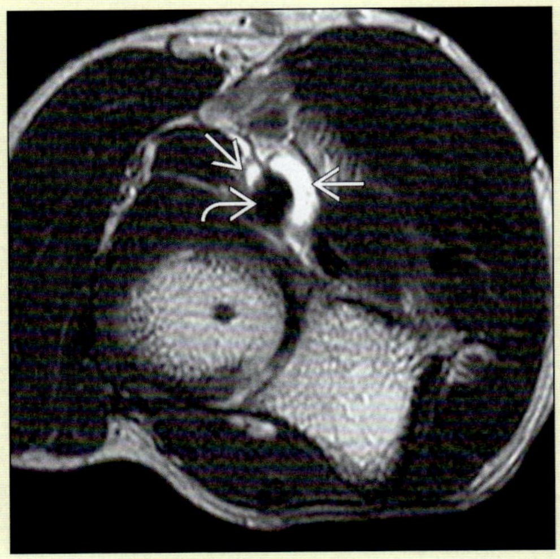

*(**Left**) Sagittal T2WI FS MR of bicipitoradial bursitis in a 34-year-old waitress presenting with an antecubital mass shows a large fluid signal collection anterior to the biceps tendon. Debris present within the inflamed bursa probably represents fibrinous and hemorrhagic material. (**Right**) Axial PDWI MR from the same patient shows a well-defined fluid collection anterior to the biceps tendon. The brachialis muscle lies deep to the biceps.*

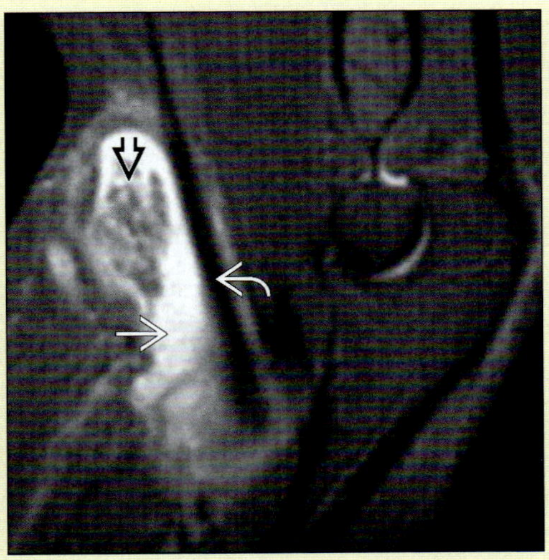

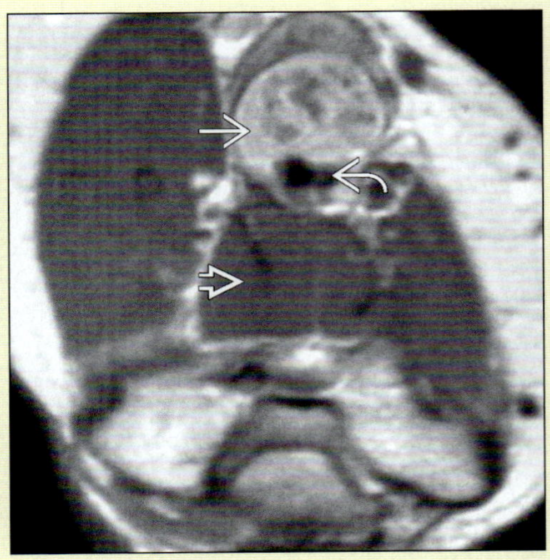

BICIPITORADIAL BURSITIS

TERMINOLOGY

Synonyms

- Cubital bursitis, antecubital bursitis, distal bicipital radial bursitis, radiobicipital bursitis

Definitions

- Inflammation of bicipitoradial bursa

IMAGING

General Features

- Best diagnostic clue
  - Fluid signal mass around distal biceps tendon or between biceps and radius on MR
- Location
  - Interposed between distal biceps tendon and anterior aspect of radial tuberosity
- Morphology
  - Flattened synovial sac not visualized on MR unless distended with fluid
  - Variable amounts of internal debris

MR Findings

- Fluid signal collection around or deep to distal biceps tendon
- Well-defined margins
- May change shape depending on whether forearm is supinated or pronated
- Often contains debris in center of bursa
- Wall may enhance with contrast; contents will not

Ultrasonographic Findings

- Typically hypo- or anechoic mass with through transmission

Imaging Recommendations

- Best imaging tool
  - MR
- Protocol advice
  - Best characterized on axial images
    - Combination of T1WI and T2WI to confirm fluid nature of collection
    - T2WI useful for characterizing debris in bursa

DIFFERENTIAL DIAGNOSIS

Biceps Partial Tear or Tendinopathy

- Abnormal signal within distal tendon
- May coexist with bicipitoradial bursitis

Biceps Tear

- Fluid may fill gap between torn tendon and radial tuberosity

Antecubital Mass

- Neoplastic
  - Synovial sarcoma
  - Peripheral nerve sheath tumor
- Nonneoplastic
  - Ganglion cyst
  - Hematoma

PATHOLOGY

General Features

- Etiology
  - Repetitive motion
    - Especially pronation/supination of forearm
  - Infection
  - Inflammatory arthropathy
    - Rheumatoid
    - Gout
  - Synovial osteochondromatosis
- Associated abnormalities
  - Associated with biceps tendinopathy/tear
- Distal biceps tendon has no synovial lining
- Fluid seen around distal biceps generally represents bursal fluid

Gross Pathologic & Surgical Features

- Subtendinous bursa with synovial lining

CLINICAL ISSUES

Presentation

- Most common signs/symptoms
  - Antecubital mass (if large) ± pain associated with supination
  - Often mistaken for neoplasm
  - Acute bursitis
    - Pain
    - Tenderness
    - Limited motion

Demographics

- Epidemiology
  - Relatively uncommon

Natural History & Prognosis

- May present insidiously: Due to repetitive activity
- Conservative treatment leads to resolution

Treatment

- Acute bursitis
  - Rest
  - Immobilization
  - NSAIDs

DIAGNOSTIC CHECKLIST

Consider

- Concomitant biceps tendinopathy or partial tear

Image Interpretation Pearls

- FS T2WI most sensitive for fluid collections
- Axial imaging plane recommended

SELECTED REFERENCES

1. Sofka CM et al: Sonography of cubital bursitis. AJR Am J Roentgenol. 183(1):51-3, 2004
2. Bond JR et al: Radiologic case study. Partial tear of the distal biceps tendon with mass-like bicipitoradial bursitis and associated hyperostosis of the radial tuberosity. Orthopedics. 26(4):376, 448-50, 2003
3. Yamamoto T et al: Bicipital radial bursitis: CT and MR appearance. Comput Med Imaging Graph. 25(6):531-3, 2001

BRACHIALIS INJURY

Key Facts

Terminology

- Muscle strain or partial or complete tear of brachialis muscle &/or tendon

Imaging

- Origin: Mid-humeral shaft anteriorly
- Insertion: Anterior tuberosity of coronoid process of ulna
- Lies deep to biceps brachii in distal arm and antecubital fossa
- Elevated T2 signal in muscle belly indicates muscle strain ± fiber tear
- Abnormal signal in and around tendon without complete disruption generally seen in partial tear
- Sagittal and axial fat-suppressed PD or T2-weighted MR images most accurate for evaluation of brachialis
- Extension of elbow necessary to avoid distortion of anatomy in anterior elbow

Top Differential Diagnoses

- Biceps brachii injury
- Supinator injury
- Flexor/pronator muscle injury

Clinical Issues

- May present as antecubital mass
- Most brachialis injuries are strains of muscle or myotendinous junction and improve with rest
- Complete tear may result in decreased strength on elbow flexion if left untreated

Diagnostic Checklist

- Concomitant biceps brachii injury
- Normal distal brachialis tendon often appears attenuated and heterogeneous
- Describe percentage of tendon disrupted in cross-section (on axial MR images)

__(Left)__ Graphic depicting the relationship of the brachialis muscle to the biceps brachii shows the brachialis muscle belly lying deep to the biceps in the arm. The myotendinous junction of the biceps lies more cephalad than that of the brachialis. The lacertus fibrosus of the biceps is shown attaching near the flexor muscles. __(Right)__ Axial T2WI FS MR shows disruption of the distal brachialis tendon just proximal to the anterior tuberosity of the ulna. The distal biceps is intact.

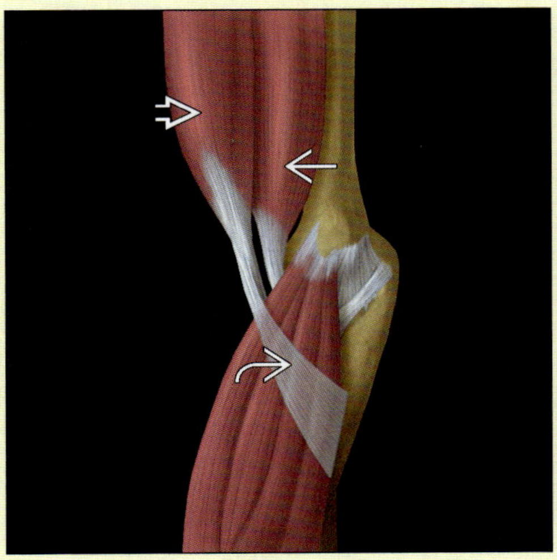

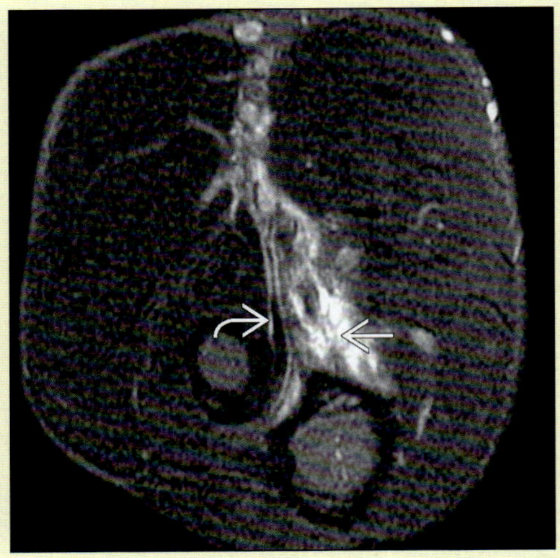

__(Left)__ Axial PDWI MR after elbow dislocation shows an intramuscular tear of the brachialis muscle belly with extensive surrounding edema. The biceps myotendinous junction is intact. __(Right)__ Sagittal T2WI MR in the same patient shows the relationship of the torn brachialis muscle to the intact biceps muscle and tendon. Because of its thicker attachment, the brachialis tendon can sometimes be assessed on sagittal images.

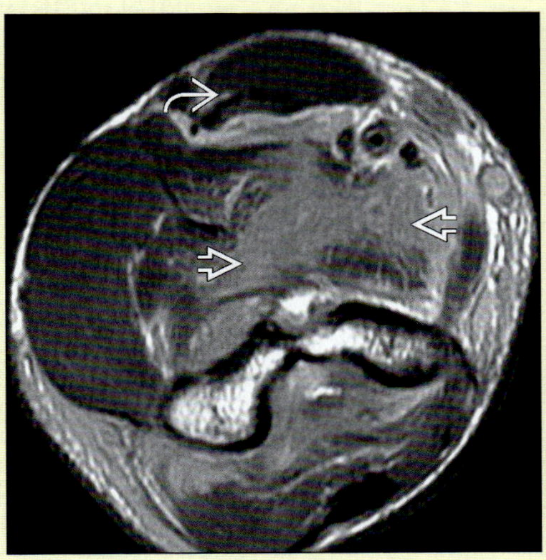

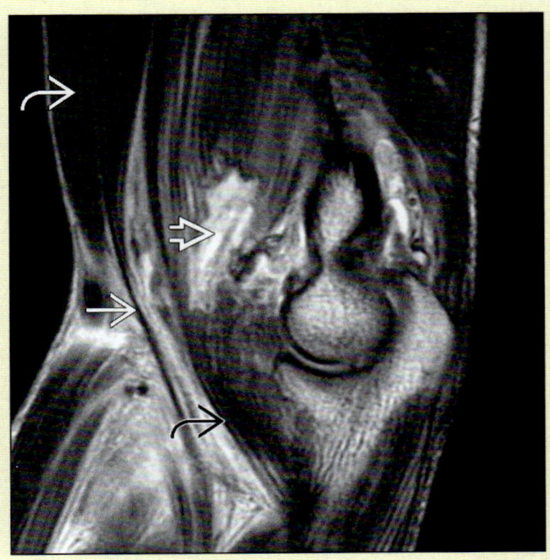

BRACHIALIS INJURY

TERMINOLOGY

Definitions

- Muscle strain, partial tear, or complete tear of brachialis muscle &/or tendon

IMAGING

General Features

- Best diagnostic clue
  - Abnormal signal in or discontinuity of brachialis myotendinous unit
- Location
  - Origin: Mid-humeral shaft anteriorly
  - Insertion: Anterior tuberosity of coronoid process of ulna
  - Lies deep to biceps brachii in distal arm and antecubital fossa
- Morphology
  - Innervated by musculocutaneous nerve

MR Findings

- Elevated T2 signal in muscle belly indicates muscle strain ± fiber tear
- Abnormal signal in and around tendon without complete disruption generally seen in partial tear
- Complete disruption of tendon, usually with fluid filling gap

Ultrasonographic Findings

- Partial or complete tear seen as hypoechoic defect in distal brachialis tendon

Imaging Recommendations

- Best imaging tool
  - MR
- Protocol advice
  - Sagittal and axial fat-suppressed PD or T2-weighted MR images most accurate for evaluation of brachialis
  - Extension of elbow necessary to avoid distortion of anatomy in anterior elbow
  - Pronation/supination of forearm less critical

DIFFERENTIAL DIAGNOSIS

Biceps Brachii Injury

- Biceps is thin tendon anterior to brachialis muscle at level of elbow joint
- Biceps inserts on radius, brachialis on ulna
- Both may result in weakened elbow flexion

Supinator Injury

- Surrounds proximal radius lateral to distal brachialis
- Injury limits supination but not flexion of elbow

Flexor/Pronator Muscle Injury

- Located more medial than brachialis
- Results in weakened wrist and finger flexion ± forearm pronation

PATHOLOGY

Staging, Grading, & Classification

- As with other myotendinous structures
- Grade 1: Abnormal signal around otherwise normal tendon
- Grade 2: Abnormal signal within tendon, but not completely ruptured
- Grade 3: Complete rupture of myotendinous unit

CLINICAL ISSUES

Presentation

- Most common signs/symptoms
  - Pain in antecubital fossa
  - Weakness on elbow flexion
- Other signs/symptoms
  - May present as antecubital mass

Natural History & Prognosis

- Most brachialis injuries are strains of muscle or myotendinous junction and improve with rest
- Complete tear may result in decreased strength on elbow flexion if left untreated
- Cosmetic deformity due to muscle atrophy
- Myositis ossificans

Treatment

- Conservative treatment for strain injury or low-grade partial tear
- Surgical repair of complete rupture

DIAGNOSTIC CHECKLIST

Consider

- Concomitant biceps brachii injury

Image Interpretation Pearls

- Normal distal brachialis tendon often appears attenuated and heterogeneous

Reporting Tips

- Describe percentage of tendon disrupted in cross-section (on axial MR images)

SELECTED REFERENCES

1. Sanal HT et al: Distal attachment of the brachialis muscle: anatomic and MRI study in cadavers. AJR Am J Roentgenol. 192(2):468-72, 2009
2. Nishida Y et al: Brachialis muscle tear mimicking an intramuscular tumor: a report of two cases. J Hand Surg Am. 32(8):1237-41, 2007
3. Holtzhausen LM et al: Elbow, forearm, wrist, and hand injuries among sport rock climbers. Clin J Sport Med. 6(3):196-203, 1996
4. Falchook FS et al: Rupture of the distal biceps tendon: evaluation with MR imaging. Radiology. 190(3):659-63, 1994
5. Fitzgerald SW et al: Distal biceps tendon injury: MR imaging diagnosis. Radiology. 191(1):203-6, 1994

TRICEPS TENDON INJURY, ELBOW

Key Facts

Terminology

- Tendinopathy, partial or complete tear, or avulsion of distal triceps tendon

Imaging

- Typically at or very close to insertion on olecranon
- Sagittal and axial T2WI FS MR most accurate for presence and size of focal tear
- Axial PDWI MR accurate for assessing tendinopathy based on size and signal characteristics
- Partial tendon tear
  - Focal discontinuity of fibers in triceps tendon
- Complete tendon tear
  - Discontinuity of triceps tendon
- Tendinopathy
  - Increased intratendinous signal on T1WI and PDWI MR

Clinical Issues

- Severe trauma to normal tendon
- Less severe trauma to abnormal tendon
  - Anabolic steroid use
  - Diabetes mellitus
  - Chronic renal failure
  - Connective tissue disease
- Low-grade injury
  - Conservative management
- High-grade partial tear, complete tear, or avulsion
  - Surgical repair

Diagnostic Checklist

- Underlying tendon pathology
- Provide location and size of tear and degree of retraction

*(**Left**) PA graphic of complete tear of the distal triceps tendon ➔ just above its insertion on the olecranon ➔. Complete rupture of this strong tendon is usually due either to massive trauma or to an underlying condition that weakens the tendon. (**Right**) Coronal T2WI FS MR of a complete rupture of the triceps tendon at its insertion on the olecranon shows the fluid-filled gap ➔ between the tip of the olecranon ➔ and the thickened and irregular tendon ➔.*

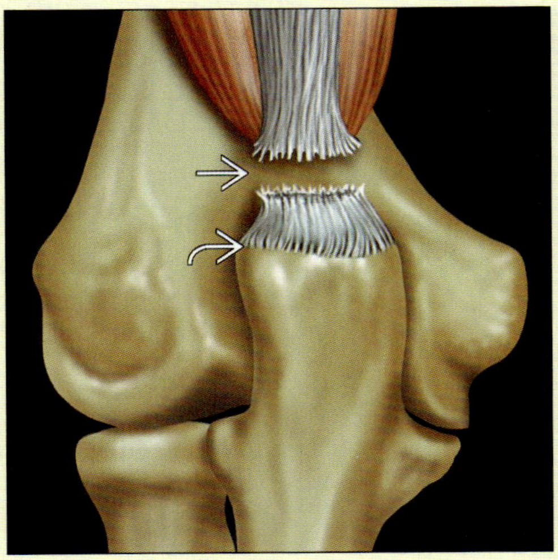

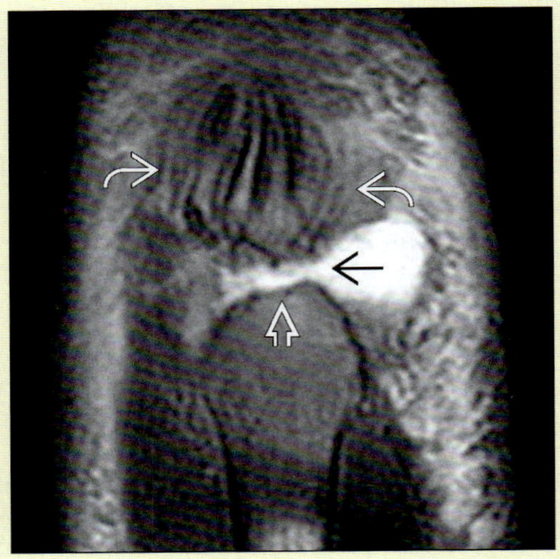

*(**Left**) PA graphic shows an olecranon fracture ➔ that could result in avulsion of the triceps tendon ➔. Such fractures, if complete, can separate from the ulna and are functionally equivalent to disruption of the triceps tendon. (**Right**) Sagittal gradient echo MR shows a displaced avulsion fracture of the tip of the olecranon ➔. The majority of the triceps tendon remains attached to the displaced fragment ➔. A small anterior remnant ➔ remains intact.*

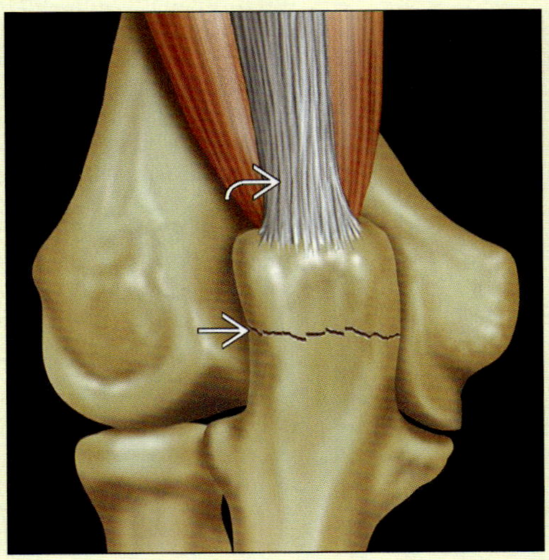

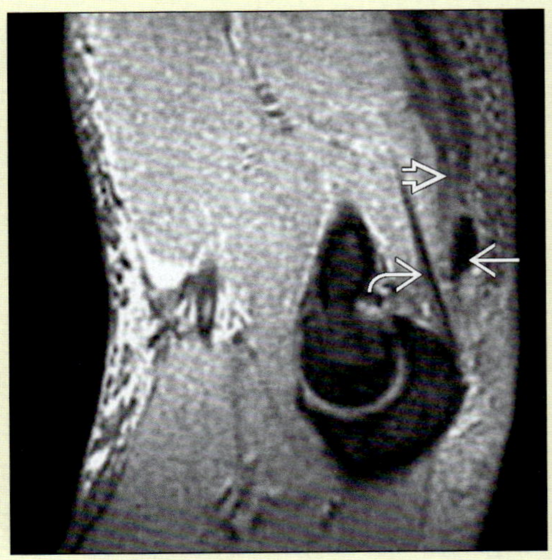

TRICEPS TENDON INJURY, ELBOW

TERMINOLOGY

Definitions

- Tendinopathy, partial or complete tear, or avulsion of distal triceps tendon

IMAGING

General Features

- Best diagnostic clue
  - Abnormal signal &/or morphology in distal triceps tendon
- Location
  - Myotendinous junction
  - Triceps tendon
    - Typically at or very close to insertion on olecranon
  - Tip of olecranon
- Size
  - May vary from microtear (tendinopathy) to displaced complete tendon tear

Radiographic Findings

- Tendon injury: Soft tissue swelling in distal posterior arm
- Tendon avulsion: Fracture of tip of olecranon with variable displacement

CT Findings

- May be useful in identifying small osseous avulsion fragment

MR Findings

- Tendinopathy
  - Tendon usually thickened
  - Increased intratendinous signal on T1WI and PDWI MR
  - May have associated subcortical marrow edema in olecranon
  - May have associated peritendinous soft tissue swelling
- Partial tendon tear
  - Focal discontinuity of fibers in triceps tendon
  - Tear usually ↑ signal on T2WI MR
  - Often associated tendinopathy in surrounding tendon
- Complete tendon tear
  - Discontinuity of triceps tendon
  - Variable retraction of proximal aspect of tendon
  - ± associated hematoma
- Osseous avulsion
  - Some or all of tendinous insertional footprint on olecranon detached
  - Variable degree of retraction/displacement

Imaging Recommendations

- Best imaging tool
  - MR
- Protocol advice
  - Sagittal and axial T2WI FS MR most accurate for presence and size of focal tear
  - Axial PDWI MR accurate for assessing tendinopathy based on size and signal characteristics

DIFFERENTIAL DIAGNOSIS

Olecranon Fracture

- Triceps tendon intact, may be attached to olecranon fragment

Olecranon Bursitis

- Fluid collection posterior to olecranon

Gout

- Thickened and irregular triceps tendon

Rheumatoid Arthritis

- Thickened and irregular triceps

Infection

- Primarily affects bone or joint

CLINICAL ISSUES

Presentation

- Most common signs/symptoms
  - Pain and swelling in distal posterior arm and elbow

Natural History & Prognosis

- Severe trauma to normal tendon
- Less severe trauma to abnormal tendon
  - Anabolic steroid use
  - Diabetes mellitus
  - Chronic renal failure
  - Gout
  - Rheumatoid arthritis

Treatment

- Low-grade injury: Conservative management
- High-grade partial tear, complete tear, or avulsion
  - Surgical repair

DIAGNOSTIC CHECKLIST

Consider

- Underlying tendon pathology

Image Interpretation Pearls

- Axials best for determining percentage of tendon torn
- Sagittals best for measuring degree of retraction in complete tear or avulsion

Reporting Tips

- Provide location and size of tear and degree of retraction

SELECTED REFERENCES

1. Keener JD et al: Insertional anatomy of the triceps brachii tendon. J Shoulder Elbow Surg. Epub ahead of print, 2010
2. Belentani C et al: Triceps brachii tendon: anatomic-MR imaging study in cadavers with histologic correlation. Skeletal Radiol. 38(2):171-5, 2009
3. Rineer CA et al: Elbow tendinopathy and tendon ruptures: epicondylitis, biceps and triceps ruptures. J Hand Surg Am. 34(3):566-76, 2009
4. Madsen M et al: Surgical anatomy of the triceps brachii tendon: anatomical study and clinical correlation. Am J Sports Med. 34(11):1839-43, 2006
5. van Riet RP et al: Surgical treatment of distal triceps ruptures. J Bone Joint Surg Am. 85-A(10):1961-7, 2003

TRICEPS TENDON INJURY, ELBOW

__(Left)__ Sagittal T2WI MR shows mild elevation of signal in the distal posterior fibers of the triceps tendon → without focal fiber disruption. These findings are consistent with tendinopathy in this weightlifter with posterior elbow pain. __(Right)__ Axial PDWI MR in the same patient at the level of the triceps tendon insertion shows elevated signal → in the posterior aspect of the tendon. This is a typical location for triceps tendinopathy. Lateral tendon fibers ↪ are normal.

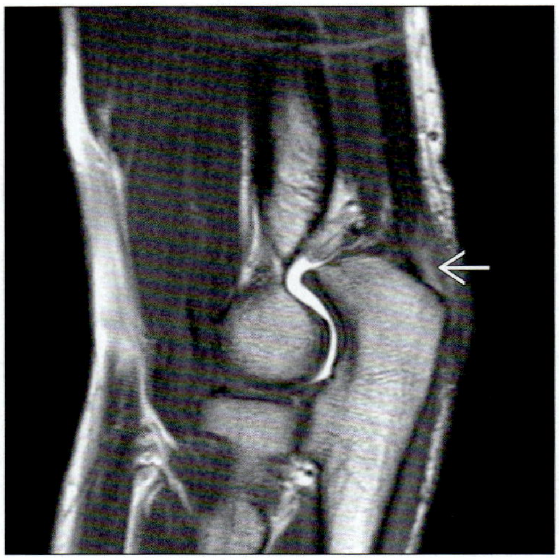

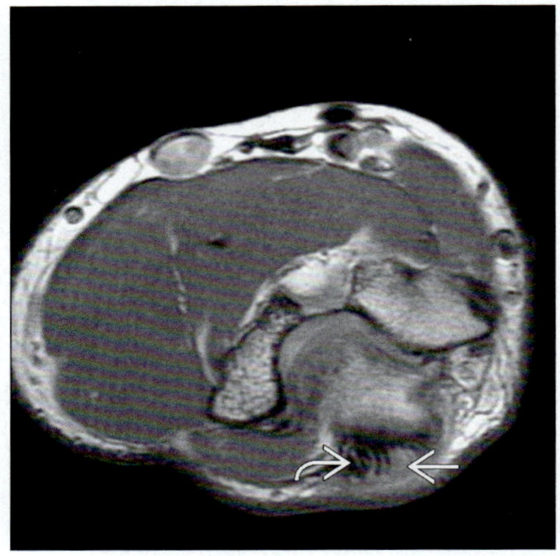

__(Left)__ Sagittal T2WI FS MR in a 22-year-old college football lineman with elbow pain shows linear increased signal → in the distal triceps, consistent with small areas of delamination. Such injuries are often due to resisted extension of the arm. __(Right)__ Coronal T2WI FS MR in the same patient shows the interstitial nature of the abnormal tendon signal → in the distal triceps. These are likely acute fiber tears in an otherwise normal tendon.

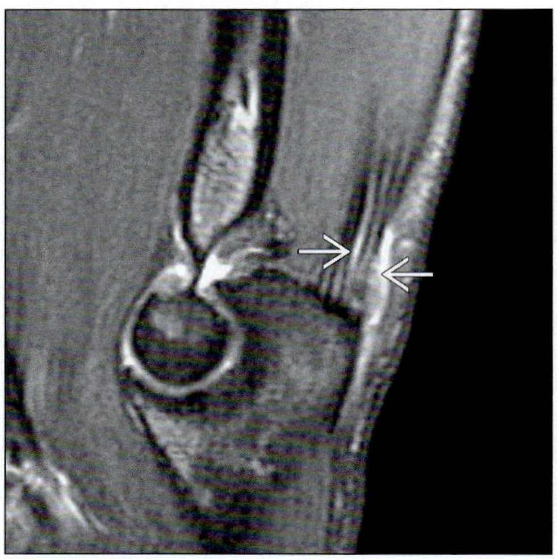

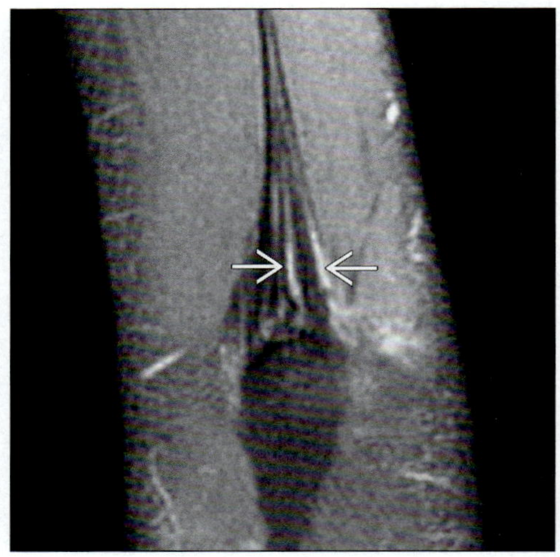

__(Left)__ Sagittal T2WI MR demonstrates a partial tear of the triceps tendon. The superficial fibers → are disrupted and retracted from the olecranon; the deeper fibers ↪ remain intact, but are irregular and thickened, suggesting underlying tendinopathy. __(Right)__ Axial T2WI MR in the same patient just above the olecranon shows focal disruption of the posterior lateral aspect of the triceps → with surrounding edema. This tear involves 20% of the tendon area.

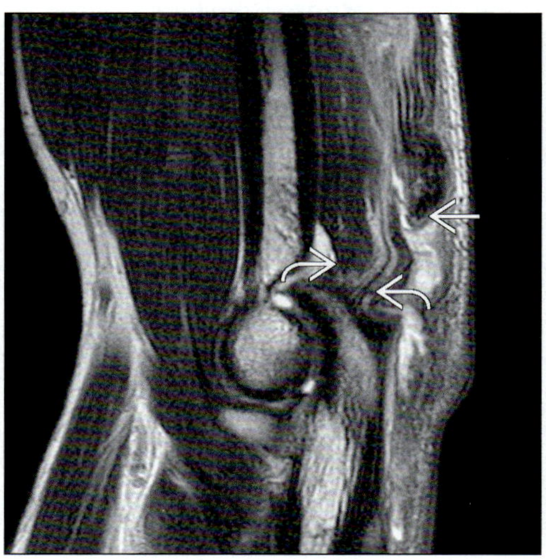

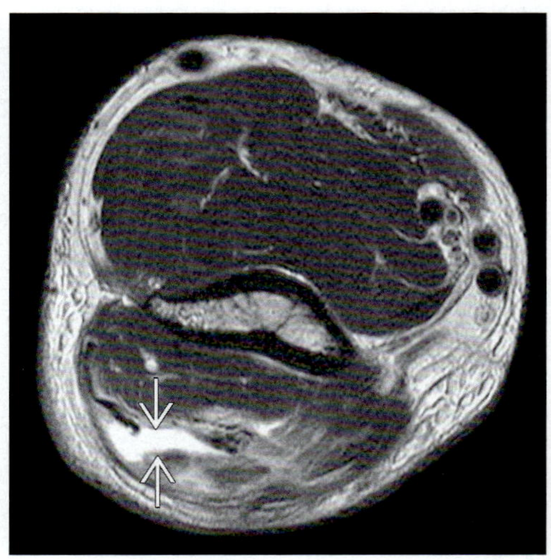

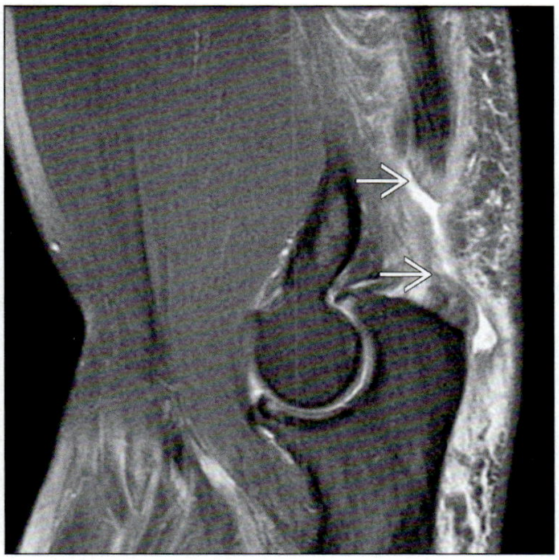
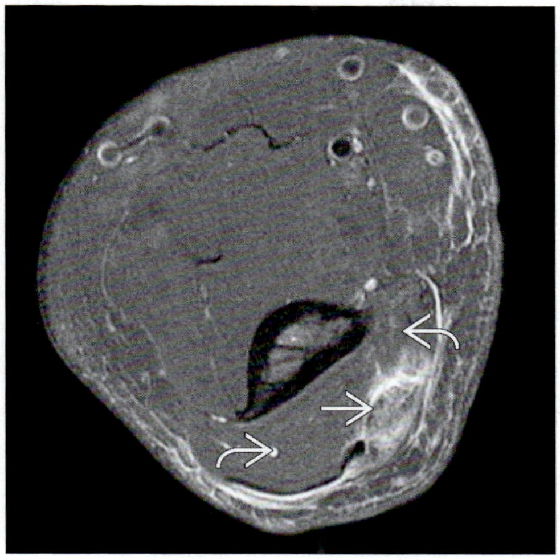

(Left) *Sagittal T2WI FS MR shows a complete rupture of the distal triceps tendon* → *with interposed fluid and extensive surrounding soft tissue swelling. It is important to describe both the site of tear relative to the olecranon and the interfragment gap (between the arrows).* ***(Right)*** *Axial T2WI FS MR in the same patient shows the complete rupture of the triceps tendon* →*. The distal muscle fibers more anterior to the tendon* ↪ *remain intact.*

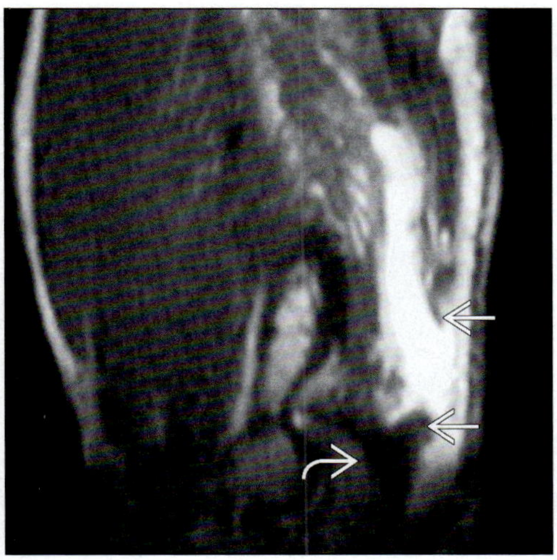
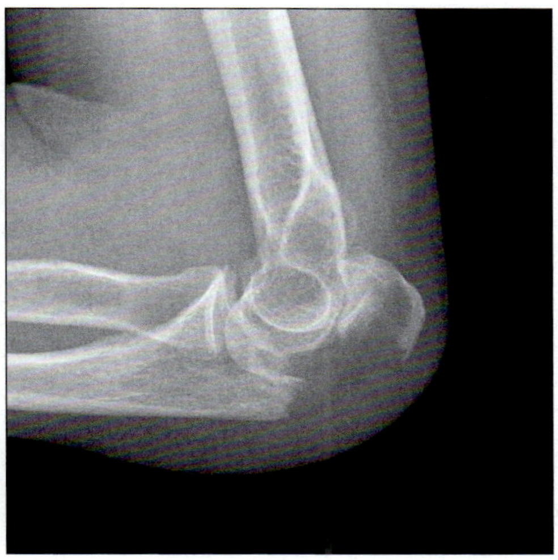

(Left) *Sagittal T2WI MR shows a complete disruption of the triceps tendon more medially, with a large gap* → *between the ends of the tendon filled with fluid/ hematoma. The olecranon* ↪ *is poorly seen because of poor coil placement.* ***(Right)*** *Lateral radiograph demonstrates a markedly displaced olecranon fracture. When the entire olecranon is displaced, the triceps tendon is attached to the displaced fragment, resulting in the functional equivalent of a complete triceps tear.*

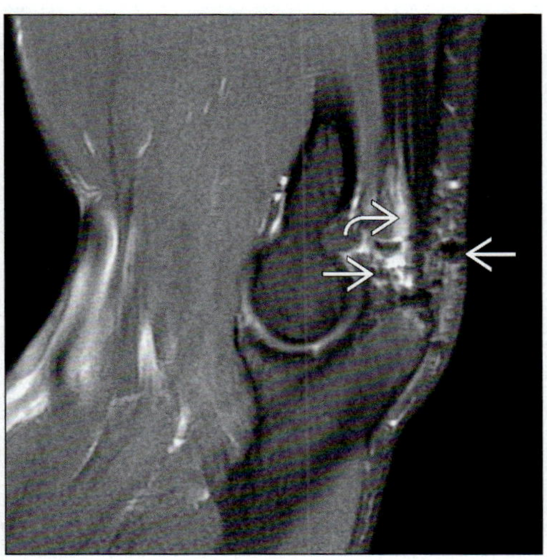
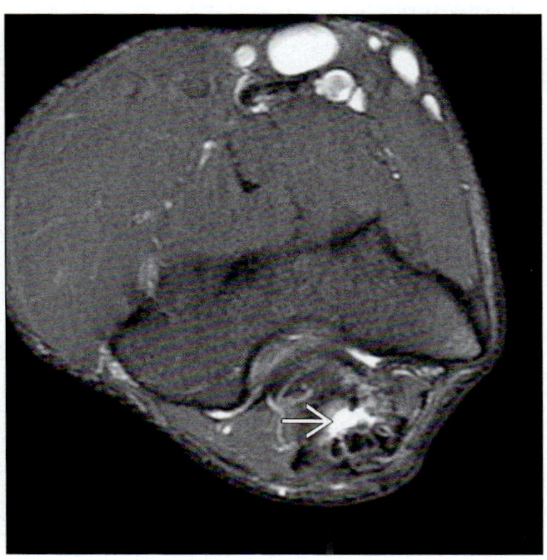

(Left) *Sagittal PDWI FS MR in a weightlifter with elbow pain after triceps tendon debridement surgery shows postoperative susceptibility artifacts* → *from prior repair. There is a focus of elevated signal* ↪ *in the anterior fibers consistent with a partial re-tear of the tendon.* ***(Right)*** *Axial PDWI FS MR in the same patient shows the focal fiber disruption* → *in the anterior aspect of the triceps, comprising about 30% of the tendon area.*

COMMON EXTENSOR TENDON INJURY, ELBOW

Key Facts

Terminology
- Extensor tendinopathy, extensor tendinitis, lateral epicondylitis, tennis elbow
- Acute or chronic injury to origin of common extensor tendon at its attachment to lateral humeral epicondyle

Imaging
- Increased size &/or signal in common extensor tendon origin
- Focal fiber disruption appears as fluid signal within tendon substance on T2WI ± fat suppression
- May range from microscopic tears to complete disruption of tendon units
- Chronic traction changes in cortex of lateral epicondyle

Pathology
- "Tendinitis" is misnomer; rarely has major inflammatory component
- Angiofibroblastic hyperplasia
- Repaired tissue not as strong as normal tendon

Clinical Issues
- Lateral elbow pain
- More severe with elbow extension
- Symptoms improve with rest, worsen with activity
- May progress to high-grade tear if not addressed

Diagnostic Checklist
- Describe approximate percentage of tendon cross-section torn
- Look for associated ligamentous injury

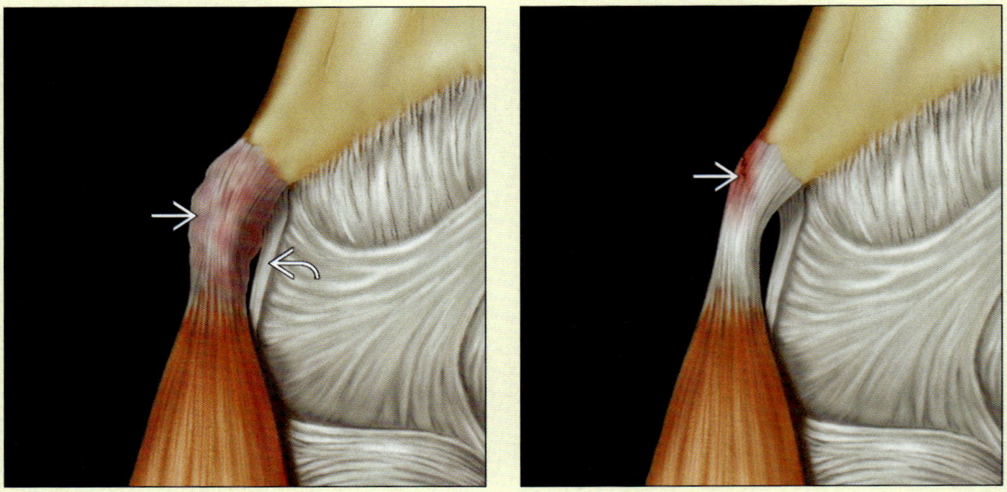

***(Left)** Graphic illustration of extensor tendinopathy of the elbow shows irregularity and thickening of the common extensor tendon → at its origin on the lateral humeral epicondyle. The radial collateral ligament ↶ lies deep to the tendon. **(Right)** Graphic illustration of a partial tear of the common extensor tendon shows focal fiber disruption → in the proximal lateral aspect of the tendon.*

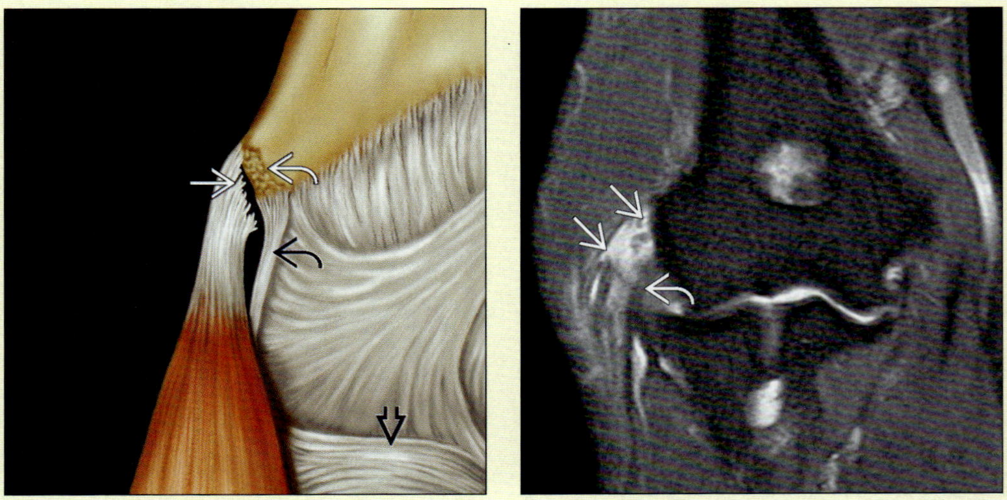

***(Left)** Graphic illustration of a partial tear of the common extensor tendon at its origin shows avulsion of the proximal tendon fibers → at the cortical margin of the lateral humeral epicondyle ↶. Partial tears usually begin in the extensor carpi radialis brevis unit. The radial collateral ligament ↶ lies deep to the tendon (RCL) complex. The annular ligament ⇨ is shown inferiorly. **(Right)** Coronal T2WI FS MR shows a high-grade tear of the common extensor tendon → with an intact RCL ↶.*

COMMON EXTENSOR TENDON INJURY, ELBOW

TERMINOLOGY

Synonyms

- Extensor tendinopathy, extensor tendinitis, lateral epicondylitis, tennis elbow

Definitions

- Acute or chronic injury to origin of common extensor tendon at its attachment to lateral humeral epicondyle
- Almost always related to chronic repetitive microtrauma leading to macroscopic tear
- "Lateral epicondylitis" is thus doubly inaccurate
  - Does not involve epicondyle except with reactive edema (pathology is primarily in tendon)
  - Not an "itis" because it is rarely inflammatory

IMAGING

General Features

- Best diagnostic clue
  - Abnormal signal or morphology in common extensor tendon on MR
- Location
  - Common extensor tendon origin at or just distal to lateral humeral epicondyle
- Size
  - May range from microscopic tears to complete disruption of tendon units

Radiographic Findings

- Usually none
- Chronic traction changes in cortex of lateral epicondyle
- May see calcific tendinopathy or heterotopic ossification in some chronic cases

CT Findings

- Tendon calcification, heterotopic ossification, or subcortical cyst formation better seen than on XR

MR Findings

- Increased size &/or signal in common extensor tendon origin
- Differentiation between normal and abnormal tendon may be difficult on T1WI MR
- Earliest signal changes best seen on PDWI ± fat suppression
- Focal fiber disruption appears as fluid signal within tendon substance on T2WI ± fat suppression
- May see reactive marrow edema in subchondral bone of lateral epicondyle
- Peritendinous edema
- Foci of intratendinous ossification show marrow signal on all sequences
- Lateral collateral ligament pathology often associated

Ultrasonographic Findings

- Grayscale ultrasound
  - Disruption of normal hypoechoic tendon architecture
  - Focal anechoic fluid with increased through-transmission in large tear
  - Tendon calcifications → increased echogenicity
- Color Doppler
  - Increased intratendinous and peritendinous flow

Imaging Recommendations

- Best imaging tool
  - MR
- Protocol advice
  - Combination of axial and coronal PDWI (for morphology) and T2WI FS MR (for focal tears and soft tissue edema)

DIFFERENTIAL DIAGNOSIS

Radial Collateral ± Lateral Ulnar Collateral Ligament Tear

- Ligaments deep to tendons
- Ligaments can be torn with tendons in severe injury

Lateral Epicondylar Fracture

- Traumatic injury due to direct blow
- Can result in avulsion of tendons and ligaments

Elbow Dislocation

- Extensor tendon may be damaged or torn in conjunction with ligament tears

Infection

- Due to direct or hematogenous spread
- May result in inflammation of extensor tendons, muscles, and surrounding tissues
- Abscess may mimic tendon tear
- Usually much more soft tissue edema than with tendinopathy

PATHOLOGY

General Features

- Involves proximal extent of forearm extensor units
- Extensor carpi radialis brevis (ECRB) most commonly involved

Staging, Grading, & Classification

- Nirschl staging of extensor tendinopathy (pathologic)
  - Stage 1: Inflammatory changes that are reversible
  - Stage 2: Irreversible pathologic changes to origin of extensor carpi radialis brevis (ECRB) muscle
  - Stage 3: Rupture of ECRB muscle origin
  - Stage 4: Secondary changes such as fibrosis or calcification
- Imaging classification
  - Grade I: Normal tendon surrounded by abnormal signal
  - Grade II: Abnormal signal ± morphology in tendon without complete disruption
  - Grade III: Complete disruption of tendon unit

Gross Pathologic & Surgical Features

- "Tendinitis" is misnomer; rarely has major inflammatory component
- Angiofibroblastic hyperplasia
  - Repetitive microtears and fibrous repair
  - Granulation tissue and neovascularization
  - Calcification
  - Necrosis
- Repaired tissue not as strong as normal tendon

COMMON EXTENSOR TENDON INJURY, ELBOW

CLINICAL ISSUES

Presentation

- Most common signs/symptoms
  - Lateral elbow pain = "tennis elbow"
    - More severe with elbow extension and supination
  - Focal tenderness 5-10 mm distal to lateral epicondyle
  - ± swelling
- Clinical profile
  - Classically described in tennis players
    - Associated with backhand swing
  - Seen with variety of activities

Demographics

- Age
  - Depends on activity
    - Usually in middle-aged or elderly
    - Peak: 35-55 years
    - Can be present in adolescents in sports
- Gender
  - M > F
- Ethnicity
  - No race predilection
- Epidemiology
  - Commonly associated with characteristic activities
    - Tennis backhand
      - Present in up to 50% of tennis players
    - Work-related repetitive motion
      - Holding tools > 1 kg
      - Handling loads > 20 kg 10x per day
      - Repetitive motions > 2 hours per day

Natural History & Prognosis

- Usually due to recurrent injury to extensor tendon
- Symptoms improve with rest, worsen with activity
- May progress to high-grade tear if not addressed

Treatment

- Usually responds well to activity cessation and conservative treatment
  - Rest
  - Ice
  - NSAIDs
  - ± wrist splinting
- Rehabilitation after pain subsides
- 95% respond to conservative therapy
- Corticosteroid injection
  - Provides short-term relief in most cases
- Autologous blood injection
  - Promotes healing and fibroblastic response
- More severe injuries may be surgically debrided
  - Remove areas of fibrous degeneration and necrosis, especially ECRB
  - Decortication of lateral epicondyle
  - Can be performed open, percutaneously, or arthroscopically
  - Allows earlier rehabilitation and resumption of activity
  - 85% report complete pain relief
  - Complications
    - If debridement taken too deep, may injure lateral collateral ligament complex → lateral elbow instability
    - Arthroscopic debridement → radial nerve injury
    - Persistent pain and limited activity

DIAGNOSTIC CHECKLIST

Consider

- History of repetitive activity or sports involvement
- Evaluate for associated ligamentous injury

Image Interpretation Pearls

- Focal tears best seen on axial T2WI MR

Reporting Tips

- Describe approximate percentage of tendon cross-section torn
  - Based on axial images

SELECTED REFERENCES

1. Cohen MS et al: Open and arthroscopic management of lateral epicondylitis in the athlete. Hand Clin. 25(3):331-8, 2009
2. De Zordo T et al: Real-time sonoelastography of lateral epicondylitis: comparison of findings between patients and healthy volunteers. AJR Am J Roentgenol. 193(1):180-5, 2009
3. Merrell G et al: Arthroscopic treatment of lateral epicondylitis. J Hand Surg Am. 34(6):1130-4, 2009
4. Scher DL et al: Lateral epicondylitis. Orthopedics. 32(4), 2009
5. Sampson S et al: Platelet rich plasma injection grafts for musculoskeletal injuries: a review. Curr Rev Musculoskelet Med. 1(3-4):165-74, 2008
6. Thornton SJ et al: Treatment of recalcitrant lateral epicondylitis with suture anchor repair. Am J Sports Med. 33(10):1558-64, 2005
7. Chung CB et al: MR imaging of tendon abnormalities of the elbow. Magn Reson Imaging Clin N Am. 12(2):233-45, vi, 2004
8. Miller TT et al: Comparison of sonography and MRI for diagnosing epicondylitis. J Clin Ultrasound. 30(4):193-202, 2002
9. Bredella MA et al: MR imaging findings of lateral ulnar collateral ligament abnormalities in patients with lateral epicondylitis. AJR Am J Roentgenol. 173(5):1379-82, 1999

COMMON EXTENSOR TENDON INJURY, ELBOW

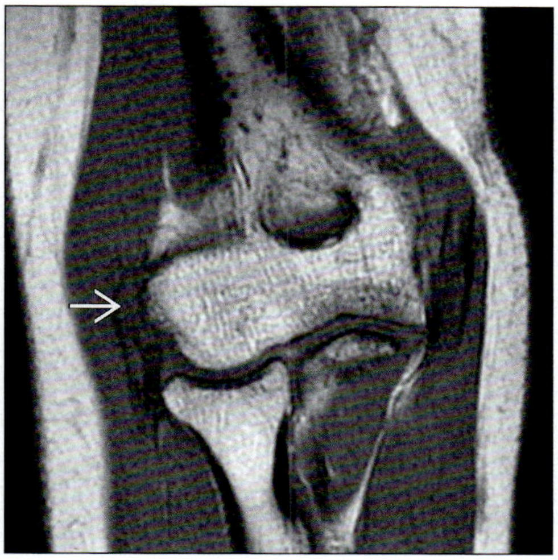

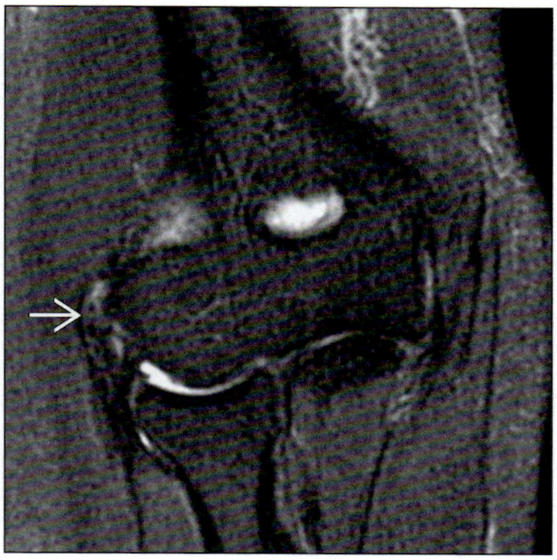

(Left) *Coronal T1WI MR shows mild thickening and heterogeneous signal within the common extensor tendon* ➔ *adjacent to the lateral humeral epicondyle. Such changes are unreliable on T1-weighted images due to possible magic angle effect, volume averaging, and low sensitivity for tendon pathology.* ***(Right)*** *Coronal T2WI FS MR from the same patient more clearly shows thickening and heterogeneity in the extensor tendon* ➔*, allowing the diagnosis of extensor tendinopathy to be made with confidence.*

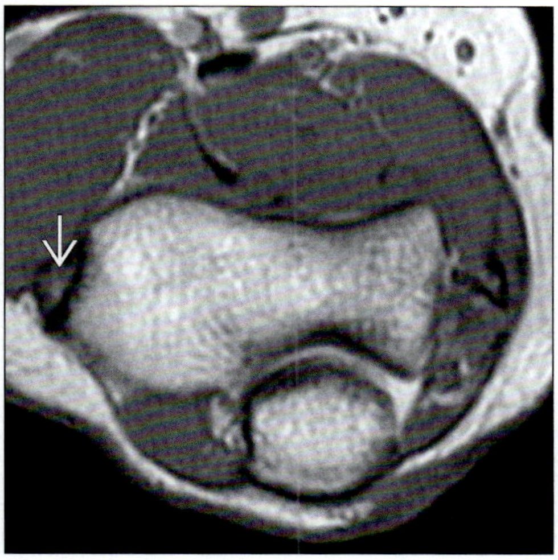

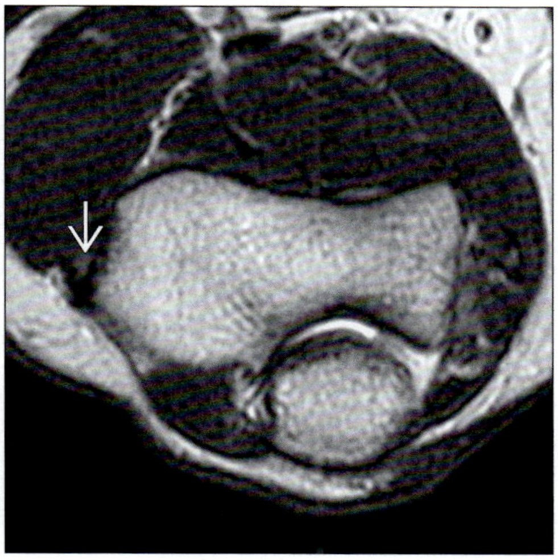

(Left) *Axial PDWI MR from the same patient shows mild elevation in signal in the common extensor tendon* ➔*. This may indicate pathology ranging from tendinopathy to a high-grade tear, and should be interpreted in conjunction with T2-weighted images.* ***(Right)*** *Axial T2WI MR in the same patient at the same level shows no fluid signal in the extensors* ➔*, excluding a high-grade tear and leading to the diagnosis of tendinopathy.*

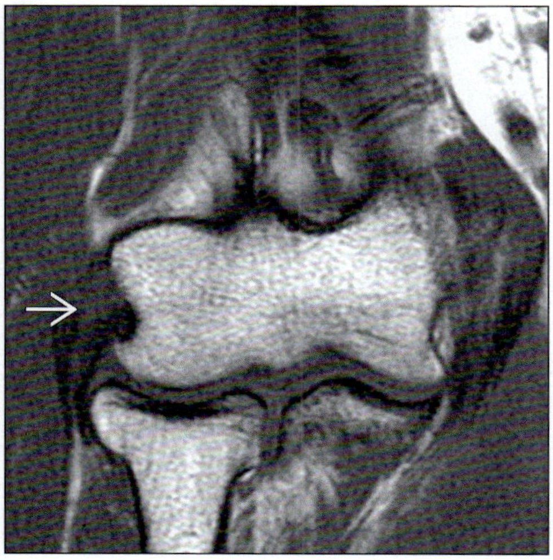

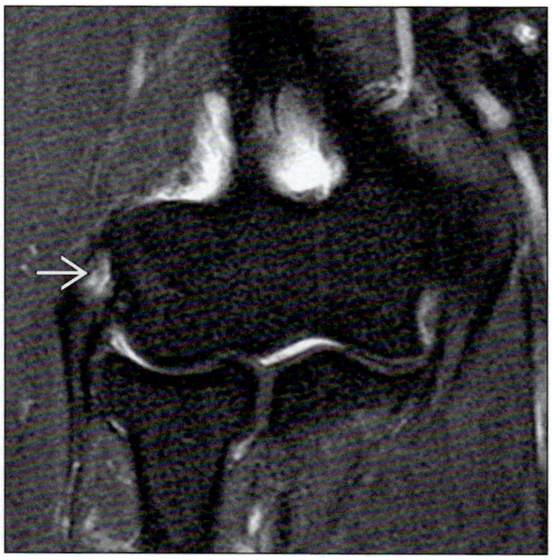

(Left) *Coronal T1WI MR in a different patient shows similar thickening and heterogeneity in the common extensor tendon* ➔*.* ***(Right)*** *Coronal T2WI FS MR in the patient at the same level shows more focal fluid signal* ➔ *in the proximal extensor unit, indicating focal fiber disruption at a macroscopic level. This represents extensor tendinopathy with a partial tear. The percentage of the tendon torn is best assessed on axial T2WI MR.*

COMMON EXTENSOR TENDON INJURY, ELBOW

(Left) *Axial PDWI MR in the same patient shows enlargement of and increased signal within the extensor tendon ➔. The distinction between tendinopathy and partial tear is difficult based on this short TE sequence.* ***(Right)*** *Axial T2WI MR in the same patient at the same level shows a small partial tear ➔ in the extensor tendon. Based on this axial image, the tear probably involves about 10 percent of the tendon substance in cross-section.*

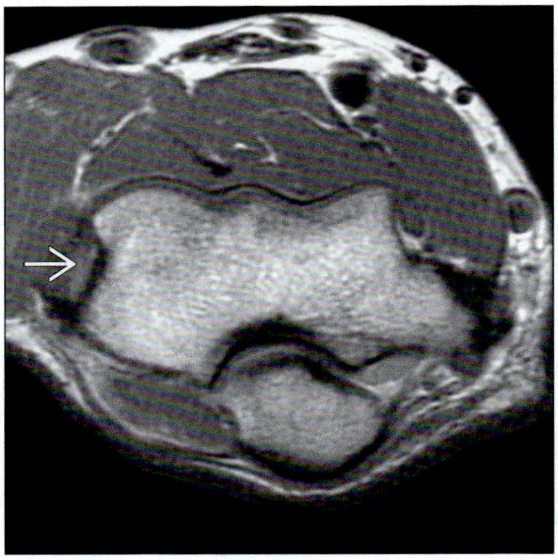

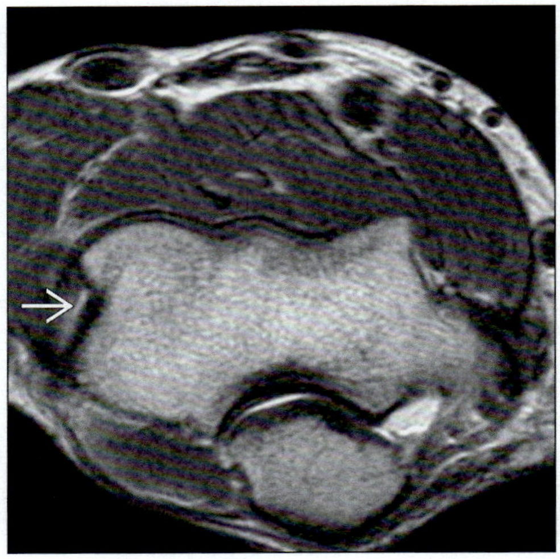

(Left) *Coronal T2WI FS MR demonstrates focal fiber disruption of the common extensor tendon ➔, primarily involving proximal and posterior portions of the combined unit. This is the extensor carpi radialis brevis (ECRB), the typical location for focal tears.* ***(Right)*** *Sagittal PDWI FS MR in the same patient confirms the posterior location of the extensor tear ➔ in the ECRB. Sagittal images are generally not as clinically useful as coronal and sagittal planes for this diagnosis.*

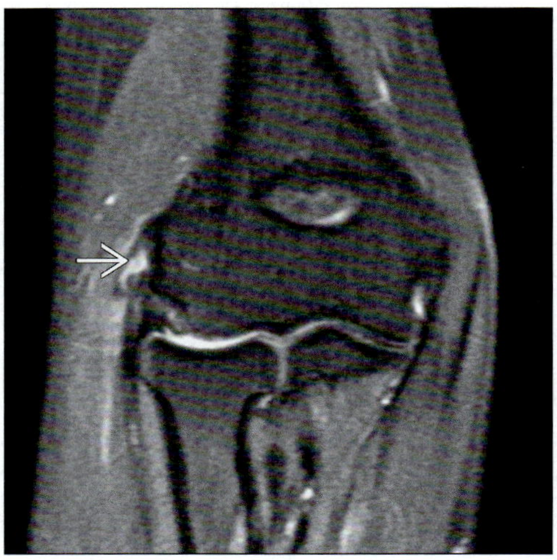

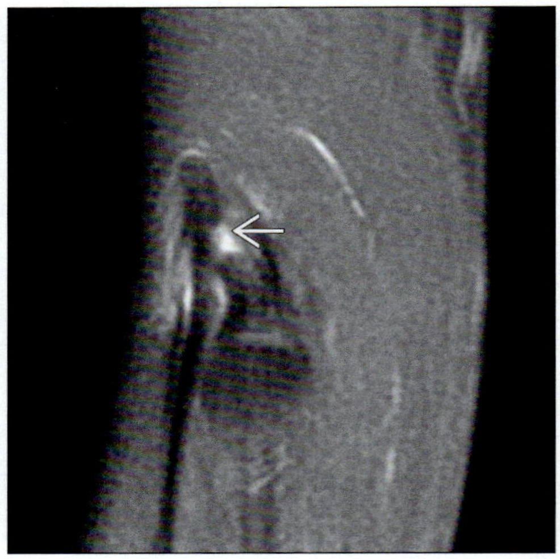

(Left) *Coronal MR arthrogram T1WI shows high-signal gadolinium infiltrating into a partial tear of the extensor tendon ➔, indicating a defect in the lateral elbow capsule. Note that the radial collateral ligament ➔ looks intact.* ***(Right)*** *Axial MR arthrogram T1WI FS in the same patient shows extravasated gadolinium injectate in the extensor tendon tear ➔. MR arthrography is not generally a useful tool in the evaluation of extraarticular structures such as the extensor tendons.*

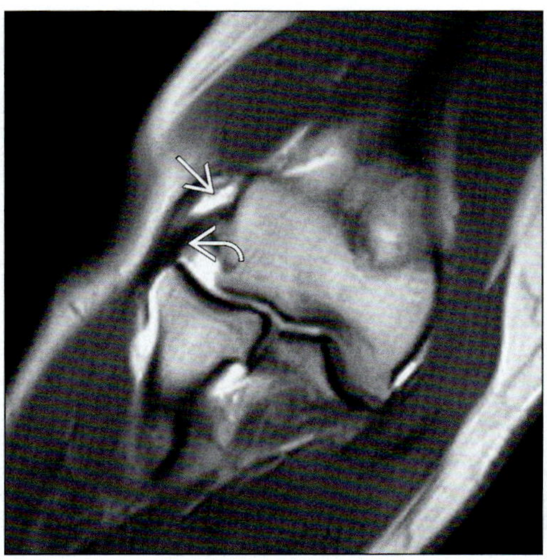

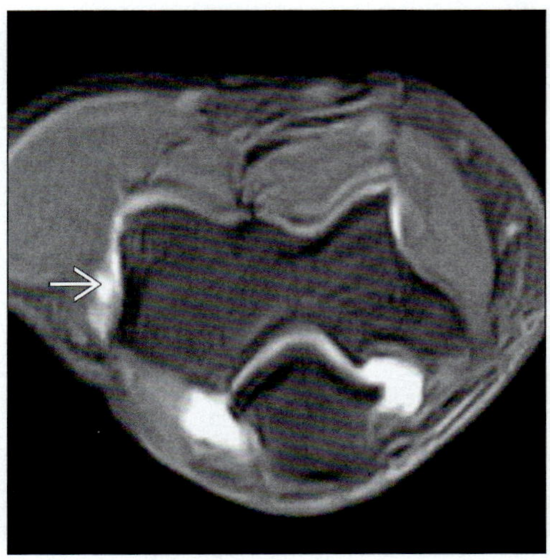

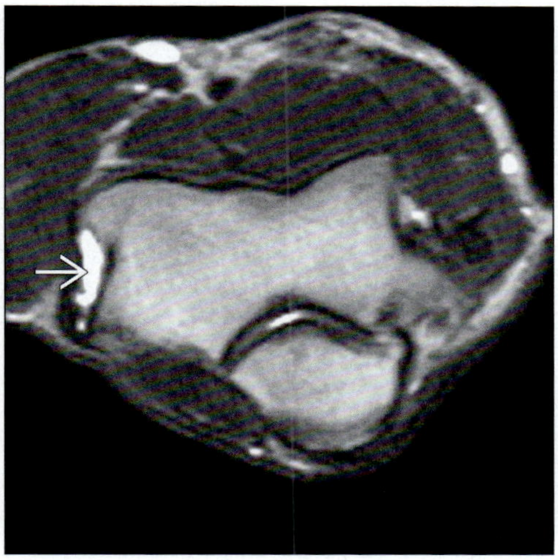

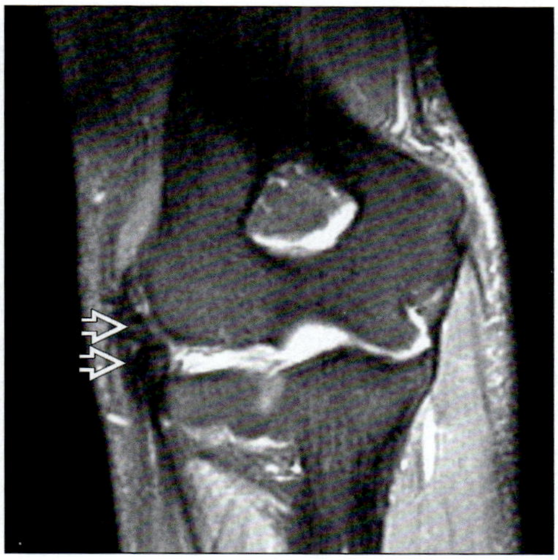

(Left) *Axial T2WI MR shows a high-grade partial tear ➡ of the common extensor tendon. About 80% of the tendon is torn and replaced by fluid.* ***(Right)*** *Coronal PDWI FS MR in a patient with a history of previous extensor tendon repair shows moderate susceptibility artifact ➡ due to hemosiderin deposition, suture, and fibrous scar tissue. Integrity of the tendon can be difficult to ascertain in this scenario. Fat-suppressed fast spin-echo sequences are best in this situation.*

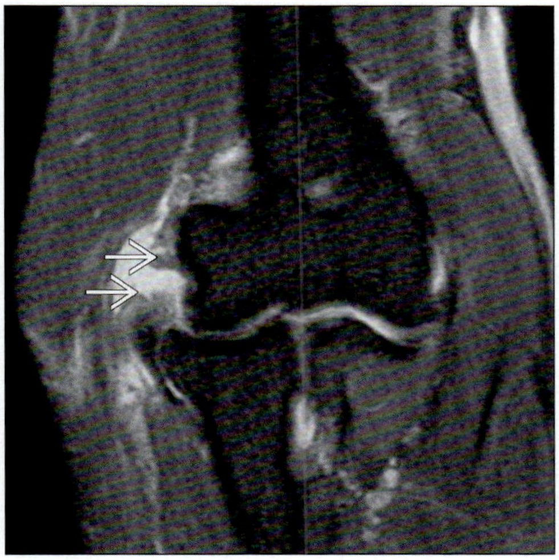

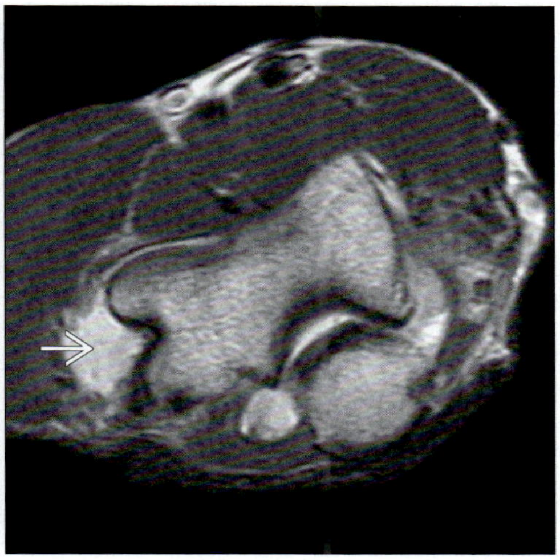

(Left) *Coronal T2WI FS MR shows a complete rupture of the extensor tendon complex ➡ with fluid filling the interfragmentary gap. The tendon ends are thickened and heterogeneous.* ***(Right)*** *Axial T2WI MR in the same patient shows fluid filling the normal location of the extensor tendon ➡ immediately lateral to the lateral humeral epicondyle. This is a complete tear of the tendon, with at least partial tear of the lateral collateral complex as well.*

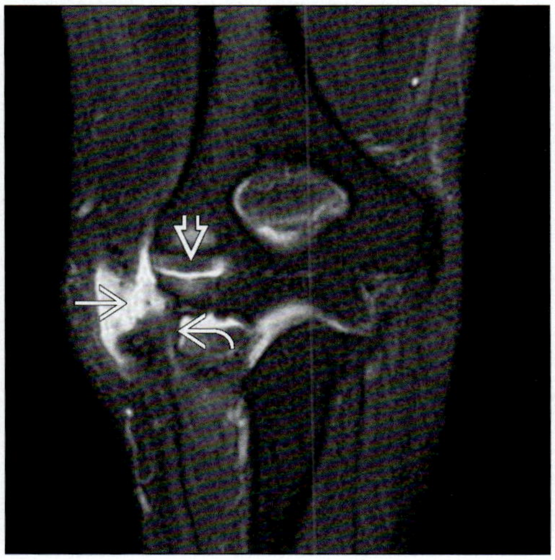

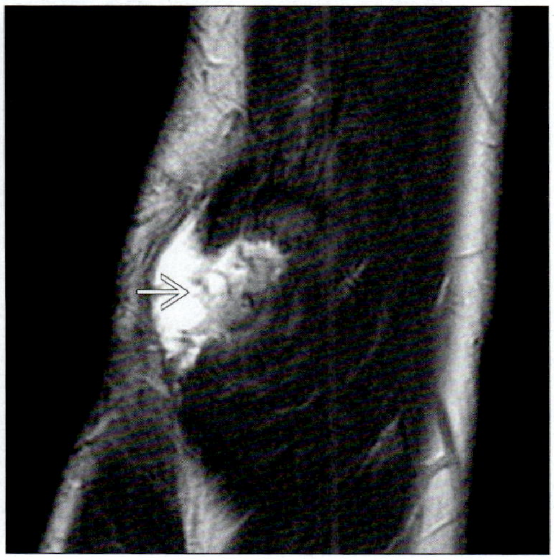

(Left) *Coronal T2WI FS MR of retear of a previously repaired common extensor tendon shows a fluid-filled defect ➡ at the site of disruption. Artifact ➡ from fixation hardware in the lateral humeral epicondyle prevents evaluation of the pericortical region. Radial collateral ligament ➡ remains intact.* ***(Right)*** *Sagittal T2WI MR in the same patient shows the large defect ➡ in the tendon. Because tears are usually due to underlying tendon degeneration, such repairs can fail.*

COMMON FLEXOR/PRONATOR TENDON INJURY, ELBOW

Key Facts

Terminology
- Flexor tendinopathy, flexor tendinitis, medial epicondylitis, golfer's elbow
- Acute or chronic injury to origin of common flexor tendon at its attachment to medial humeral epicondyle

Imaging
- Increased size &/or signal in common flexor tendon origin
- Focal fiber disruption appears as fluid signal within tendon substance on T2WI ± fat suppression
- May range from microscopic tears to complete disruption of tendon units
- Chronic traction changes in cortex of medial epicondyle

Pathology
- "Tendinitis" is misnomer; rarely has major inflammatory component
- Angiofibroblastic hyperplasia
- Repaired tissue not as strong as normal tendon

Clinical Issues
- Medial elbow pain
- More severe with elbow extension
- Symptoms improve with rest, worsen with activity
- May progress to high-grade tear if not addressed

Diagnostic Checklist
- Describe approximate percentage of tendon cross-section torn

__(Left)__ Graphic illustration of flexor tendinopathy of the elbow shows irregularity and thickening of the common flexor/pronator tendon ➔ just distal to its origin at the medial humeral epicondyle. Such tendinopathy is usually due to chronic repetitive microtrauma to the tendon. __(Right)__ Graphic illustration of a partial tear of the common flexor/pronator tendon complex ➔ at the medial humeral epicondyle shows focal fiber disruption without complete fiber disruption.

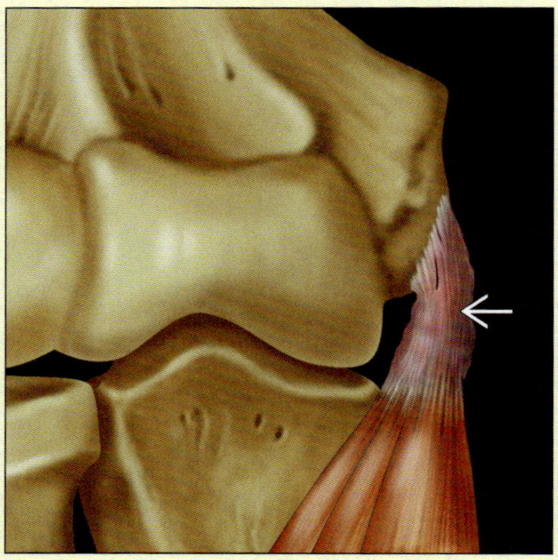

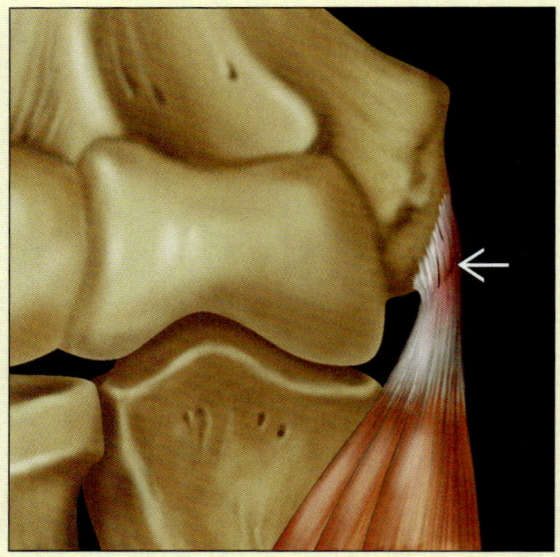

__(Left)__ Graphic illustration of complete avulsion of the common flexor/pronator tendon origin at the medial humeral epicondyle shows the broad site of attachment ➔ of this tendon complex. __(Right)__ Coronal T2WI FS MR gradient echo image of a partial tear of the flexor/pronator tendon unit shows focal high signal ➔ in the deep portion of the tendon origin. More superficial elements ➔ are intact. The degree of tendon disruption is best judged on axial images.

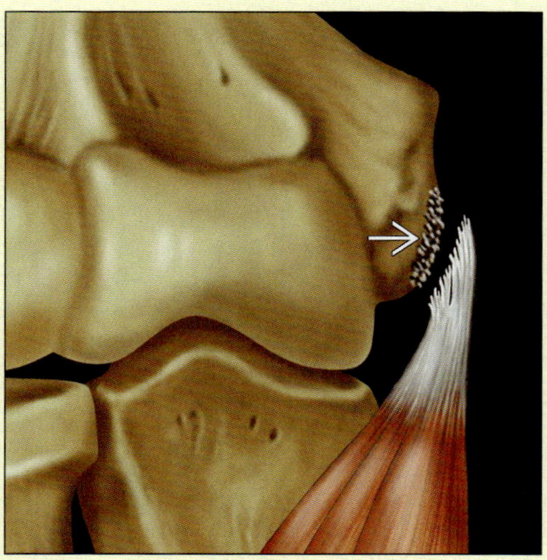

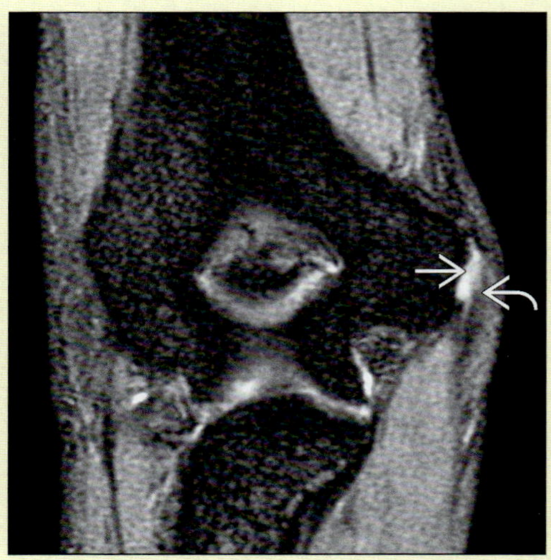

COMMON FLEXOR/PRONATOR TENDON INJURY, ELBOW

TERMINOLOGY

Synonyms

- Flexor tendinopathy, flexor tendinitis, medial epicondylitis, golfer's elbow

Definitions

- Acute or chronic injury to origin of common flexor tendon at its attachment to medial humeral epicondyle
- Pronator teres also originates here and can be injured

IMAGING

General Features

- Best diagnostic clue
  - Abnormal signal or morphology in common flexor/pronator tendon on MR
- Location
  - Common flexor/pronator tendon origin at or just distal to medial humeral epicondyle
- Size
  - May range from microscopic tears to complete disruption of tendon units

Radiographic Findings

- Usually none
- Chronic traction changes in cortex of medial epicondyle
- May see calcific tendinopathy or heterotopic ossification in some chronic cases

CT Findings

- Tendon calcification, heterotopic ossification, or subcortical cyst formation better seen than on XR

MR Findings

- Increased size &/or signal in common flexor/pronator tendon origin
- Earliest signal changes best seen on short TE sequences (PDWI, T1WI) without fat suppression
- Focal fiber disruption appears as fluid signal within tendon substance on T2WI ± fat suppression
- May see reactive marrow edema in subchondral bone of medial epicondyle
- Peritendinous edema
- Foci of intratendinous ossification show marrow signal on all sequences

Ultrasonographic Findings

- Grayscale ultrasound
  - Disruption of normal hypoechoic tendon architecture
  - Focal anechoic fluid with increased through-transmission in large tear
  - Tendon calcifications → increased echogenicity
- Color Doppler
  - Increased intratendinous and peritendinous flow

Imaging Recommendations

- Best imaging tool
  - MR
- Protocol advice
  - Combination of axial and coronal PDWI (for morphology) and T2WI FS MR (for focal tears and soft tissue edema)

DIFFERENTIAL DIAGNOSIS

Ulnar Collateral Ligament Tear

- Ligaments deep to tendons
- Ligaments can be torn with tendons in severe injury (dislocation)

Little Leaguer's Elbow

- Avulsion of medial epicondylar ossification center in skeletally immature individual
- Origins of ulnar collateral ligament and common flexor/pronator tendon attached to avulsed fragment
- In this sense, functionally equivalent to complete disruption of flexor tendon

Medial Epicondylar Fracture

- Direct blow may result in chip fracture of epicondyle
- If displaced, may mimic appearance of complete tear of flexor tendon

PATHOLOGY

General Features

- Deep and superficial flexor tendons and pronator teres tendon originate on broad shelf of medial humeral epicondyle
- Individual tendon units difficult to identify
- Most commonly involves pronator teres, flexor carpi radialis, palmaris longus

Staging, Grading, & Classification

- As with all tendon injuries
  - Grade 1: Normal tendon surrounded by abnormal signal
  - Grade 2: Abnormal signal ± morphology in tendon without complete disruption
  - Grade 3: Complete disruption of tendon unit

Gross Pathologic & Surgical Features

- "Tendinitis" is misnomer; rarely has major inflammatory component
- Condition is generally caused by repetitive microtrauma due to overuse with fibrous repair
- Angiofibroblastic hyperplasia
  - Repetitive microtears and fibrous repair
  - Granulation tissue and neovascularization
  - Calcification
  - Necrosis
- Repaired tissue not as strong as normal tendon

CLINICAL ISSUES

Presentation

- Most common signs/symptoms
  - Medial elbow pain
  - More severe with elbow flexion and forearm pronation
  - ± swelling
  - Focal tenderness over flexor tendons 20 mm distal to medial epicondyle
- Other signs/symptoms

COMMON FLEXOR/PRONATOR TENDON INJURY, ELBOW

- Ulnar neuropathy: 50%

Demographics

- Age
  - Depends on activity
    - Peak: 20-49 years
    - Can be present in adolescents in sports
- Gender
  - M:F = 2:1
- Ethnicity
  - No race predilection
- Epidemiology
  - Much less common than extensor tendinopathy
  - Commonly associated with characteristic activities
    - Improper golf swing
    - Overhand throwers
    - Tennis serve
    - Work-related repetitive motion
  - Often reported to begin with specific injury event
    - "Straw that breaks the camel's back" phenomenon of repetitive injury finally resulting in mechanical failure of tendon
  - May be related to high-force acute injury forcing elbow into valgus
    - Falls
    - Motor vehicle/motorcycle accidents
    - Work-related or industrial accidents

Natural History & Prognosis

- Usually due to recurrent injury to flexor/pronator tendon
- Symptoms improve with rest, worsen with activity
- May progress to high-grade tear if not addressed

Treatment

- Usually respond well to activity cessation and conservative treatment
  - Rest
  - Ice
  - NSAIDs
  - Bracing
- Rehabilitation
- Corticosteroid injection
  - Provides short-term symptomatic relief
- More severe injuries may be surgically debrided
  - Rarely necessary (< 10%)
  - Remove areas of fibrous degeneration or necrosis
  - Complications
    - If debridement taken too deep, may injure ulnar collateral ligament
    - Persistent or recurrent symptoms

DIAGNOSTIC CHECKLIST

Consider

- History of repetitive activity or sports involvement

Image Interpretation Pearls

- Focal tears best seen on axial T2WI MR

Reporting Tips

- Describe approximate percentage of tendon cross-section torn
  - Based on axial images

SELECTED REFERENCES

1. Rajeev A et al: Lateral compartment cartilage changes and lateral elbow pain. Acta Orthop Belg. 75(1):37-40, 2009
2. Rineer CA et al: Elbow tendinopathy and tendon ruptures: epicondylitis, biceps and triceps ruptures. J Hand Surg Am. 34(3):566-76, 2009
3. Park GY et al: Diagnostic value of ultrasonography for clinical medial epicondylitis. Arch Phys Med Rehabil. 89(4):738-42, 2008
4. Richard MJ et al: Traumatic valgus instability of the elbow: pathoanatomy and results of direct repair. J Bone Joint Surg Am. 90(11):2416-22, 2008
5. Shahabpour M et al: The effectiveness of diagnostic imaging methods for the assessment of soft tissue and articular disorders of the shoulder and elbow. Eur J Radiol. 65(2):194-200, 2008
6. Suresh SP et al: Medial epicondylitis: is ultrasound guided autologous blood injection an effective treatment? Br J Sports Med. 40(11):935-9; discussion 939, 2006
7. Tuite MJ et al: Sports-related injuries of the elbow: an approach to MRI interpretation. Clin Sports Med. 25(3):387-408, v, 2006
8. Banks KP et al: Overuse injuries of the upper extremity in the competitive athlete: magnetic resonance imaging findings associated with repetitive trauma. Curr Probl Diagn Radiol. 34(4):127-42, 2005
9. Kijowski R et al: Magnetic resonance imaging findings in patients with medial epicondylitis. Skeletal Radiol. 34(4):196-202, 2005
10. Kijowski R et al: Magnetic resonance imaging of the elbow. Part II: Abnormalities of the ligaments, tendons, and nerves. Skeletal Radiol. 34(1):1-18, 2005
11. Ciccotti MC et al: Diagnosis and treatment of medial epicondylitis of the elbow. Clin Sports Med. 23(4):693-705, xi, 2004
12. Descatha A et al: Medial epicondylitis in occupational settings: prevalence, incidence and associated risk factors. J Occup Environ Med. 45(9):993-1001, 2003
13. Martin CE et al: MR imaging of epicondylitis. Skeletal Radiol. 27(3):133-8, 1998
14. Fritz RC et al: Magnetic resonance imaging of the musculoskeletal system: Part 3. The elbow. Clin Orthop Relat Res. (324):321-39, 1996

COMMON FLEXOR/PRONATOR TENDON INJURY, ELBOW

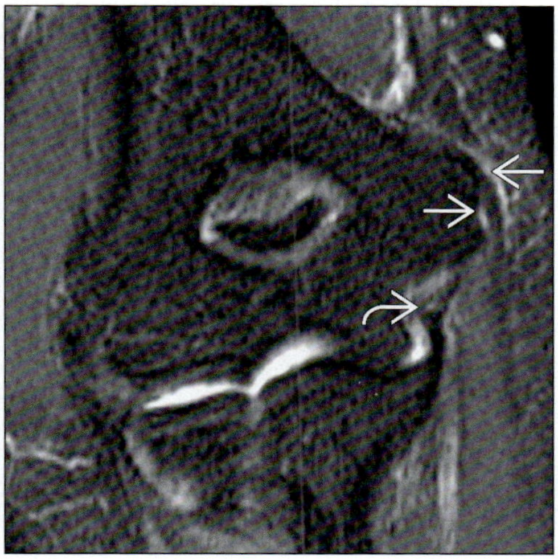

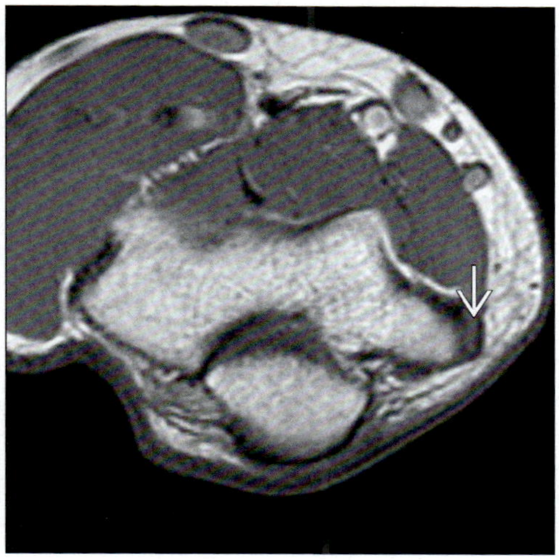

(Left) *Coronal T2WI FS MR demonstrates mild elevated signal ➔ around the common flexor tendon in a patient with medial elbow pain. The ulnar collateral ligament ➔ is intact.* ***(Right)*** *Axial PDWI MR in the same patient shows subtle elevated signal ➔ in the central fibers of the common flexor/pronator tendon adjacent to the medial humeral epicondyle. Axial fat-suppressed T2-weighted images (not shown) demonstrate no fluid signal within the tendon; thus, this is not a partial tear.*

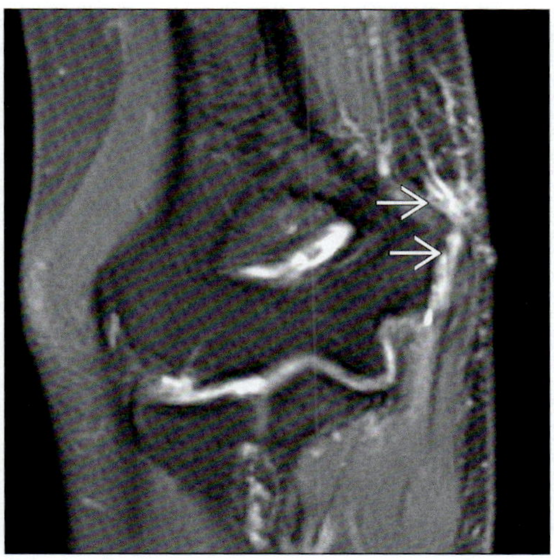

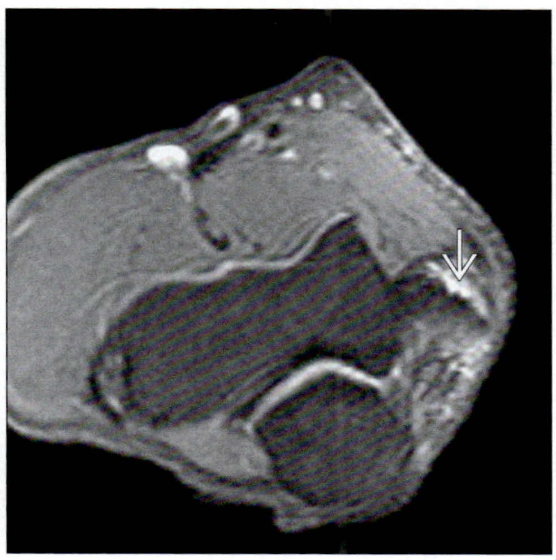

(Left) *Coronal T2WI FS MR demonstrates elevated signal within and around the origin of the common flexor/pronator tendon ➔ adjacent to the medial humeral epicondyle. Coronal images alone may make it difficult to determine the extent of a tendon tear in this location.* ***(Right)*** *Axial PDWI FS MR in the same patient better delineates the extent of injury to the flexor tendon; the elevated signal indicating a partial tear ➔ involves about 30% of the tendon cross-section.*

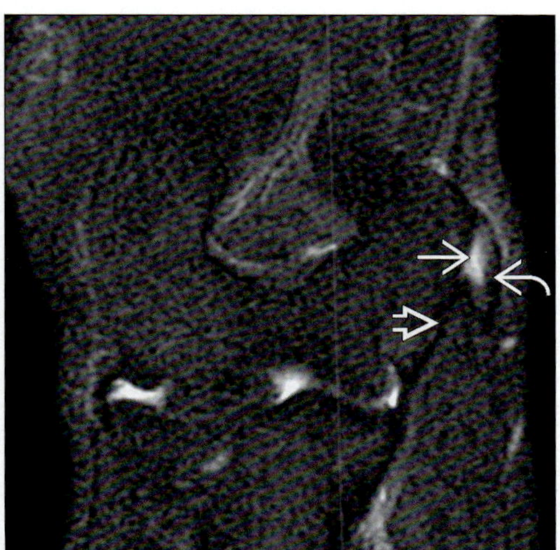

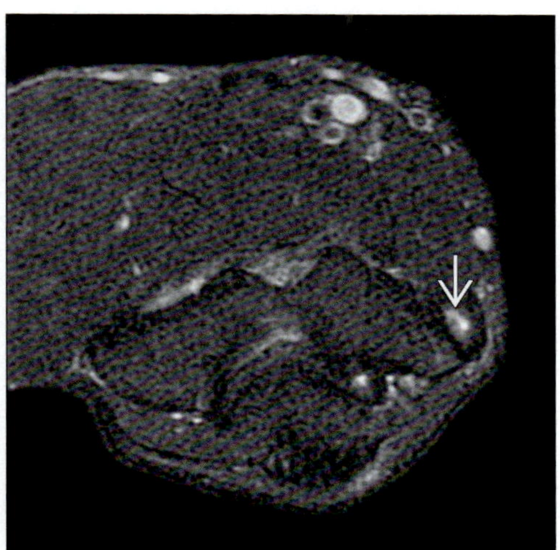

(Left) *Coronal T2WI FS MR demonstrates a well-defined partial tear of the common flexor/pronator tendon at its origin as a focus of fluid signal ➔ within the tendon substance. More peripheral tendon fibers ➔ are intact. The ulnar collateral ligament ➔ lies deep to the tendon complex, and is intact.* ***(Right)*** *Axial T2WI FS MR in the same patient confirms the intratendinous location of the high signal ➔, consistent with a partial tear of the tendon.*

COMMON FLEXOR/PRONATOR TENDON INJURY, ELBOW

(Left) *Coronal T2WI FS MR in a 17-year-old wrestler with medial elbow pain and valgus instability of the elbow shows a partial tear of the common flexor tendon → and diffuse edema in the proximal flexor muscles →. The ulnar collateral ligament → is only partially included on this image.* **(Right)** *Coronal T2WI FS MR more anteriorly in the same patient shows the striated edema → in the flexor muscles related to an acute strain injury. The UCL → is intact.*

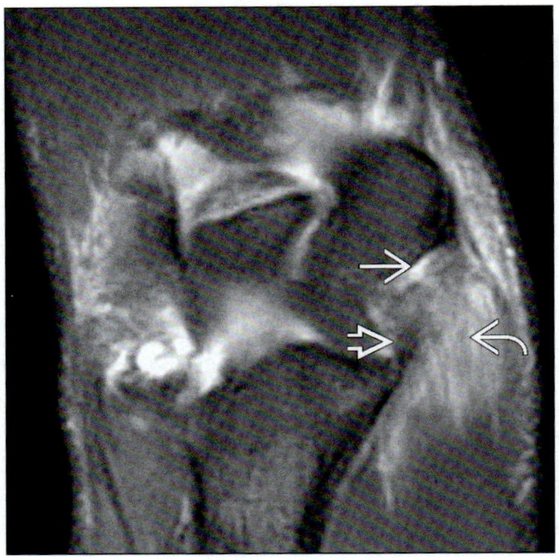

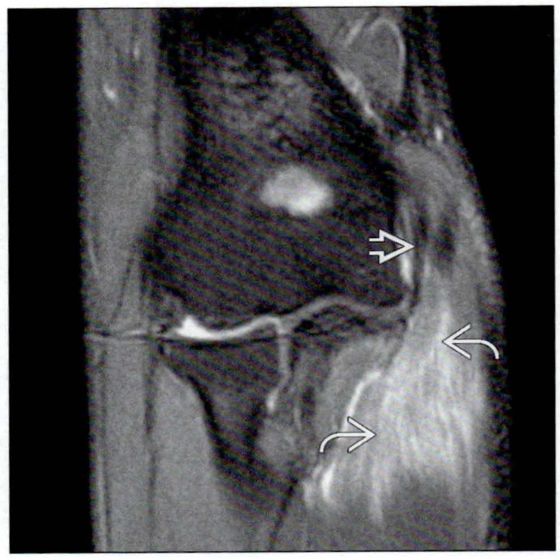

(Left) *Axial PDWI FS MR in the same patient shows partial separation of the common flexor tendon → from the medial humeral epicondyle with fluid → interposed between them, indicating a partial tear due to acute avulsion.* **(Right)** *Axial PDWI FS MR more distally in the same patient shows the striated intramuscular edema → in the common flexors indicating an acute strain injury. Such acute injuries are less common than repetitive overuse injuries in the flexor and extensor tendons.*

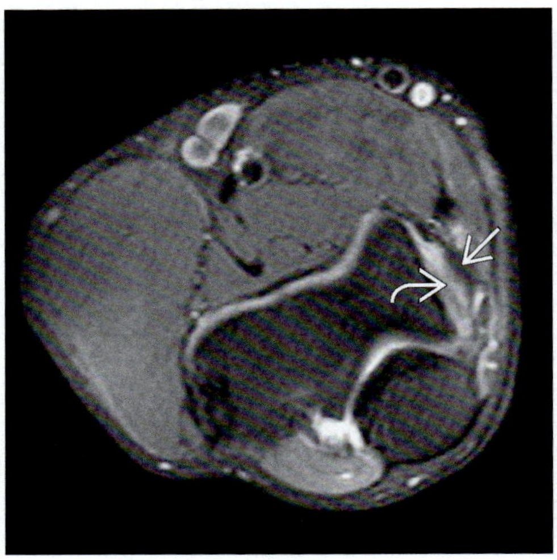

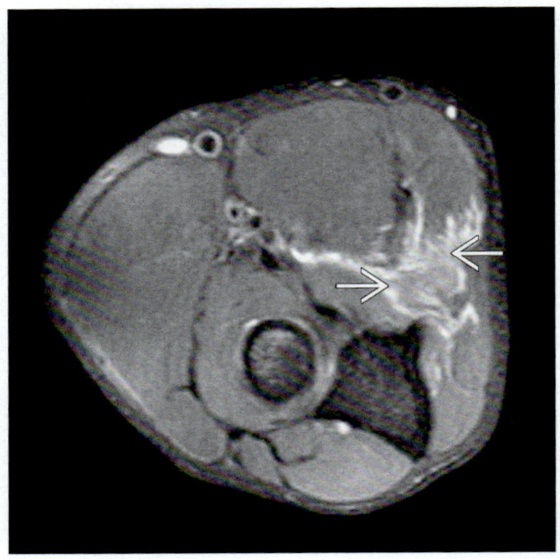

(Left) *Coronal T1WI C+ FS MR after severe elbow trauma shows complete disruption of the proximal flexor/pronator tendon → with associated ulnar collateral ligament disruption. This was likely due to a severe valgus force on the elbow.* **(Right)** *Axial T1WI C+ FS MR in the same patient shows the degree of tissue disruption in the medial elbow → to good advantage. The medial epicondylar origin of the common flexor tendon is unrecognizable as is the ulnar collateral ligament.*

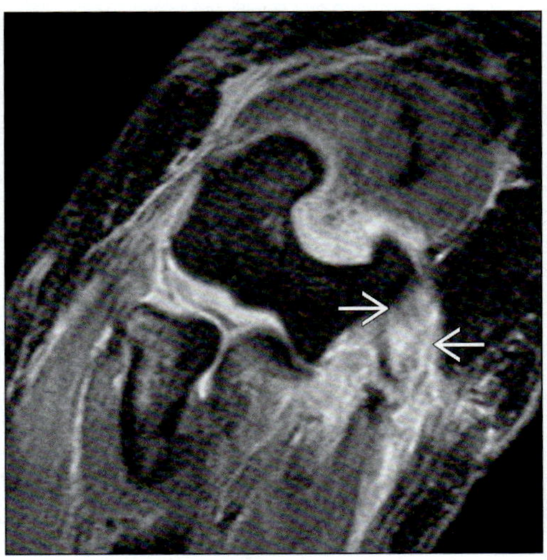

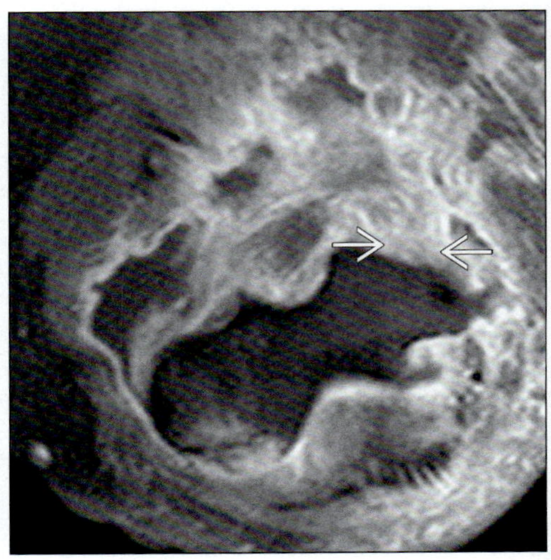

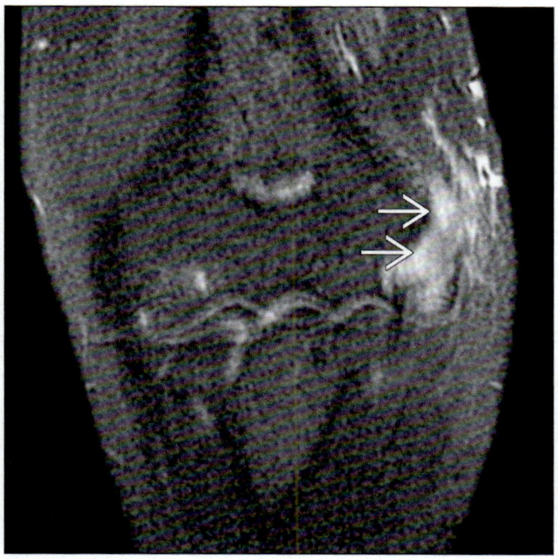

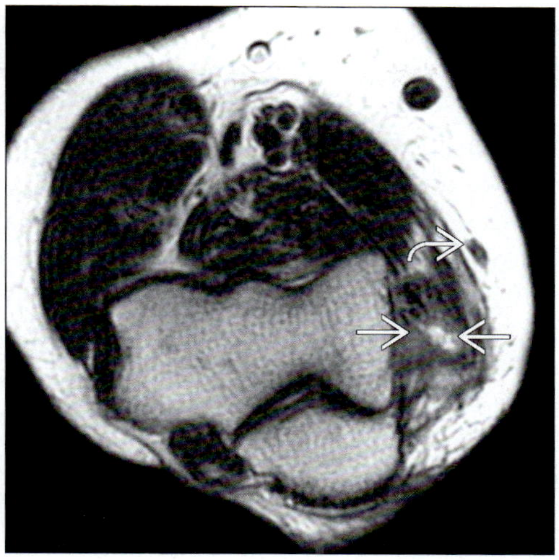

(Left) *Coronal T2WI FS MR status post ulnar nerve transposition surgery shows high-grade partial disruption of the common flexor tendon →.* ***(Right)*** *Axial T2WI MR in the same patient shows heterogeneous elevated signal → throughout the common flexor tendon due to partial tear after ulnar nerve transposition. The displaced nerve ↪ is seen in the subcutaneous fat of the elbow.*

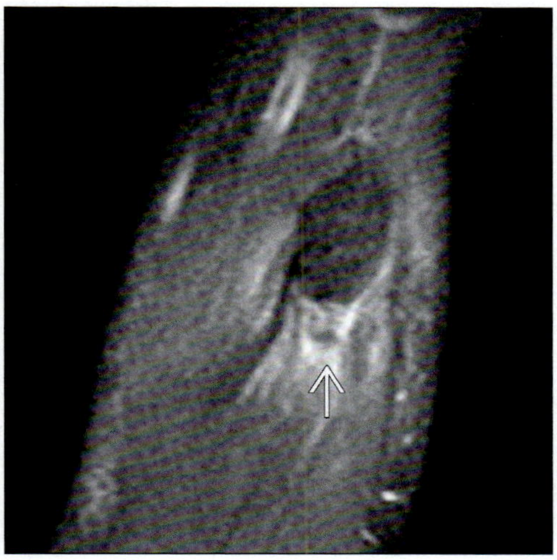

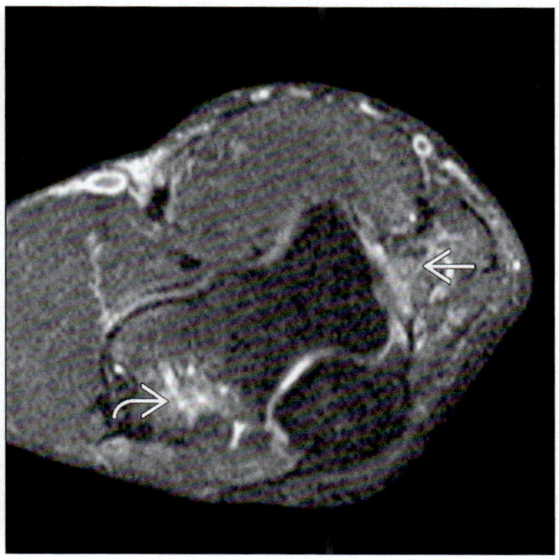

(Left) *Sagittal T2WI FS MR shows low-grade partial tear of the common flexor tendon → at its origin after a fall.* ***(Right)*** *Axial T2WI FS MR in the same patient shows the edema in the flexor muscles → due to valgus strain, and an impaction contusion in the posterior aspect of the capitellum ↪ due to a corresponding impaction force on the lateral aspect of the elbow. The posterior location of the capitellar contusion implies that the elbow was in extension during the injury.*

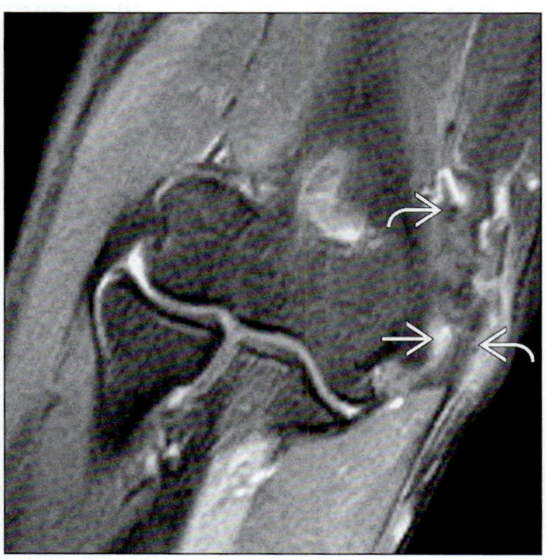

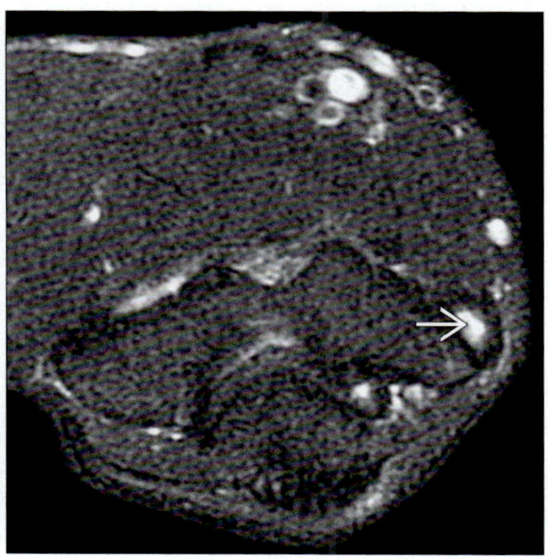

(Left) *Coronal PDWI FS MR in a patient status post prior debridement and repair of a tear of the common flexor/ pronator tendon unit shows elevated signal → at the medial epicondylar tendon attachment site, which is consistent with a recurrent partial tear. Postoperative irregularity ↪ is seen in the surrounding soft tissues.* ***(Right)*** *Axial T2WI FS MR in the same patient shows the well-defined focus of fiber disruption → in the central aspect of the tendon. This is a 50% partial tear.*

INTRAARTICULAR BODIES, ELBOW

Key Facts

Terminology

- Loose bodies, loose fragments, joint mice
- May be attached to synovium or free to move in joint

Imaging

- Most common in coronoid or olecranon fossae
- Pure chondral fragments not seen on radiographs
- Often associated with joint effusion
- Fragment appearance depends on composition
- Radiographs usually sufficient
- MR for small or nonossified fragments
- Often associated with joint effusion
- T2WI MR ± FS in multiple planes most sensitive
- Especially axial and sagittal
- T1WI MR useful to see larger fragments containing fatty marrow

Top Differential Diagnoses

- Os supratrochlear dorsale
- Patella cubiti
- Acute fracture
- Air bubbles (MR arthrogram)

Clinical Issues

- Acute osteochondral fragments may be restored to original location
- Larger more chronic bodies may need to be resected
- Associated osteochondral defects repaired

Diagnostic Checklist

- If well-defined osteochondral defect, donor site may be visible
- Measure large fragments, describe location

*(**Left**) Graphic illustration demonstrates multiple intraarticular bodies ➡ in the coronoid fossa of the elbow due to synovial chondromatosis. (**Right**) Lateral radiograph in a 14-year-old boy with a history of elbow locking demonstrates innumerable ossified bodies within the elbow joint. No significant degenerative changes are noted in the elbow or radiocapitellar joint. This is synovial osteochondromatosis, a benign metaplastic condition that results in marked intraarticular body formation.*

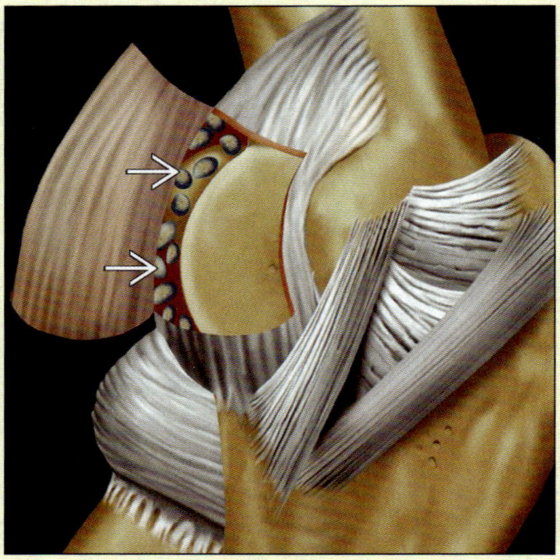

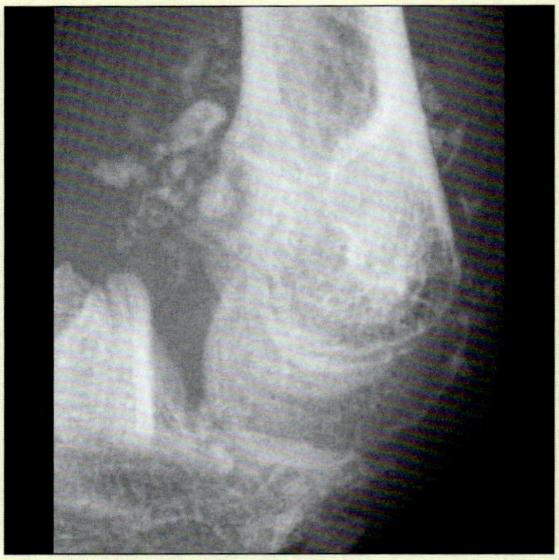

*(**Left**) Axial bone CT through the level of the humeral condyles shows several tiny osseous intraarticular bodies ➡ that were not visible on radiographs. A large joint effusion ➦ is present. CT is more sensitive than radiographs for small or poorly mineralized bodies. (**Right**) Sagittal T1WI MR shows advanced osteoarthritis in the capitellum and radial head. There are multiple corticated bodies ➡ posterior to the radiocapitellar joint. The lack of marrow signal in the bodies indicates sclerosis.*

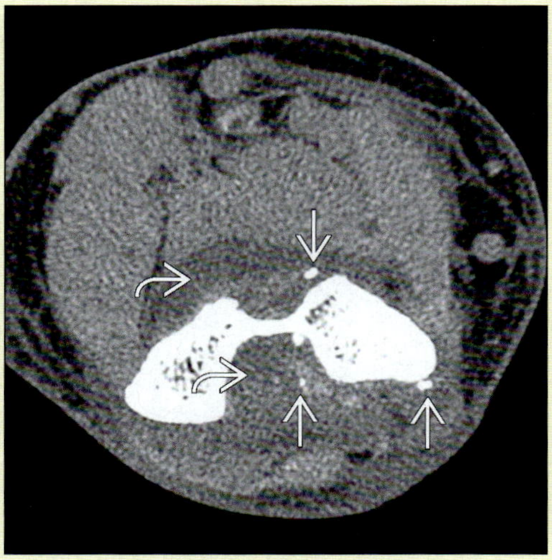

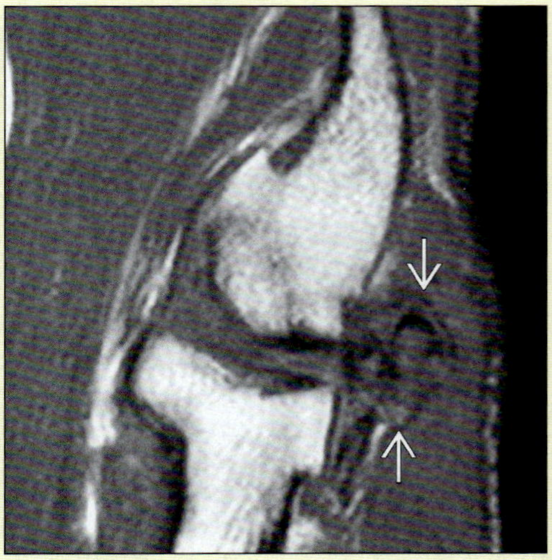

INTRAARTICULAR BODIES, ELBOW

TERMINOLOGY

Synonyms

- Loose bodies, loose fragments, joint mice

Definitions

- 1 or more chondral, osseous, or osteochondral fragments within joint space
- May be attached to synovium or free to move in joint; hence, "loose bodies" may not be accurate

IMAGING

General Features

- Best diagnostic clue
  - Ossified fragment(s) projecting in joint on radiographs
- Location
  - May be anywhere in elbow joint capsule
  - Most common in coronoid or olecranon fossae
- Size
  - May range from submillimeter to multiple centimeters in size
- Morphology
  - Acute: Sharply marginated; may contain bone, cartilage, or both
  - Chronic: Often smoother in contour; corticated; ± attached to synovium

Radiographic Findings

- Pure chondral fragments not seen on radiographs
- Ossific fragment(s) of variable size projecting within elbow joint capsule on all views
- Fragments may be smoothly corticated if chronic
- Often associated with joint effusion
- Donor site may be visible as subchondral lucency
  - Especially capitellum

CT Findings

- Similar to radiographs, but smaller fragments better seen
- Better ability to localize fragments in joint
  - CT arthrogram
    - Adds ability to detect cartilage defects
    - Better able to see unossified bodies

MR Findings

- T1WI
  - If containing marrow, high signal centrally, low signal cortical rim
  - If no marrow, usually low signal
  - Not as useful for identifying donor sites unless large osseous defect
- T2WI
  - If containing marrow, central signal follows normal fat marrow characteristics
  - If no marrow, low signal
- T2WI FS
  - Generally low signal on all sequences
  - Useful for identifying donor sites
- PD/intermediate
  - May discern chondral portion of osteochondral fragments
  - Useful for identifying donor sites
- T2* GRE
  - Intraarticular bodies may appear larger due to blooming artifact in cortical bone
  - Generally low signal unless significant fatty marrow is present
  - Useful for identifying donor sites
- T1WI C+ FS
  - No enhancement of intraarticular bodies
  - May see associated hyperenhancement of synovium
  - May help discern synovial attachment of fragments
- MR arthrography useful if
  - No significant joint effusion is present
  - Multiple small fragments are present
  - Integrity of donor site cartilage in question
- Fragment appearance depends on composition

Ultrasonographic Findings

- May see large bodies if either
  - Large effusion is present
  - Using saline arthrosonography

Imaging Recommendations

- Best imaging tool
  - Radiographs usually sufficient
  - MR for small or nonossified fragments
    - MR arthrography advocated by some; probably not necessary
- Protocol advice
  - T2WI MR ± FS in multiple planes is most sensitive
    - Especially axial and sagittal
  - T1WI MR is useful to see larger fragments containing fatty marrow

DIFFERENTIAL DIAGNOSIS

Os Supratrochleare Dorsale

- Congenital accessory ossification center at proximal olecranon
- Intraarticular
- May grow over time

Patella Cubiti

- Sesamoid bone in distal triceps

Acute Fracture

- Often associated with intraarticular bodies

Synovial Osteochondromatosis

- Intraarticular bodies may not be visible on radiograph
- Bodies may not follow marrow signal on MR

Synovial Hemangioma

- Tubular rather than rounded structures

Focal Nodular Synovitis

- Generally a single nodule
- Most commonly seen in knee, usually in infrapatellar fat

Focal Pigmented Villonodular Synovitis (PVNS)

- Contains hemosiderin, which may be confused with ossification on MR
- No mineralization on radiographs or CT
- Usually large effusion

INTRAARTICULAR BODIES, ELBOW

Lipoma Arborescens
- Prominent subsynovial fat with frond-like synovial projections

Synovial Plica
- Linear in shape
- Attach to synovium, usually in more than 1 location

Air Bubbles (MR Arthrogram)
- Signal void on all sequences
- Congregate in nondependent portion of joint

PATHOLOGY

Gross Pathologic & Surgical Features
- Acute fracture fragments
  - Normal hyaline cartilage ± underlying subchondral bone
- Chronic bodies
  - Generally ossified
  - Lamellated appearance due to deposition of mineral in layers (like a pearl in an oyster)
  - May become secondarily synovialized
    - Nonmobile
    - May see enhancing synovium adjacent to or surrounding body

CLINICAL ISSUES

Presentation
- Most common signs/symptoms
  - Pain
  - Limited motion
  - Crepitus
  - Effusion
  - Inflammation
- Other signs/symptoms
  - ± erosion of adjacent bone due to longstanding friction

Demographics
- Gender
  - M > F
- Epidemiology
  - Elbow is 2nd most common site of intraarticular bodies

Natural History & Prognosis
- May lead to accelerated osteoarthritis
- Progressive pain and limited motion
- Chronic joint effusion

Treatment
- Acute osteochondral fragments may be restored to original location
- Larger more chronic bodies may need to be resected
- Associated osteochondral defects repaired

DIAGNOSTIC CHECKLIST

Consider
- Presence of acute fracture or osteochondral defect

Image Interpretation Pearls
- If well-defined osteochondral defect, donor site may be visible
- Size of defect and fragment(s) should roughly match

Reporting Tips
- Measure large fragments, describe location
- Determine whether likely loose or attached to synovium

SELECTED REFERENCES

1. Dodson CC et al: Elbow arthroscopy. J Am Acad Orthop Surg. 16(10):574-85, 2008
2. Flury MP et al: Arthroscopic and open options for surgical treatment of chondromatosis of the elbow. Arthroscopy. 24(5):520-5, 2008
3. Winter M et al: Bilateral patella cubiti. A case report. J Bone Joint Surg Am. 88(2):415-7, 2006
4. Dubberley JH et al: The detection of loose bodies in the elbow: the value of MRI and CT arthrography. J Bone Joint Surg Br. 87(5):684-6, 2005
5. Osada D et al: Multiple loose bodies in the elbow showing a snow storm appearance at arthroscopy. J Orthop Sci. 10(4):422-4, 2005
6. Yorgancigil H et al: Multiple loose bodies in the joints: from snowstorm to hailstones. Arthroscopy. 20(8):e113-6, 2004
7. Bradley JP et al: Osteochondritis dissecans of the humeral capitellum. Diagnosis and treatment. Clin Sports Med. 20(3):565-90, 2001
8. Clasper JC et al: Arthroscopy of the elbow for loose bodies. Ann R Coll Surg Engl. 83(1):34-6, 2001
9. Miller JH et al: Detection of intraarticular bodies of the elbow with saline arthrosonography. Clin Radiol. 56(3):231-4, 2001
10. Mueller T et al: Primary synovial chondromatosis of the elbow. J Shoulder Elbow Surg. 9(4):319-22, 2000
11. Bianchi S et al: Detection of loose bodies in joints. Radiol Clin North Am. 37(4):679-90, 1999
12. Palmer WE: MR arthrography: is it worthwhile? Top Magn Reson Imaging. 8(1):24-43, 1996
13. Gudmundsen TE et al: Accessory ossicles in the elbow. Acta Orthop Scand. 58(2):130-2, 1987
14. Obermann WR et al: The os supratrochleare dorsale: a normal variant that may cause symptoms. AJR Am J Roentgenol. 141(1):123-7, 1983

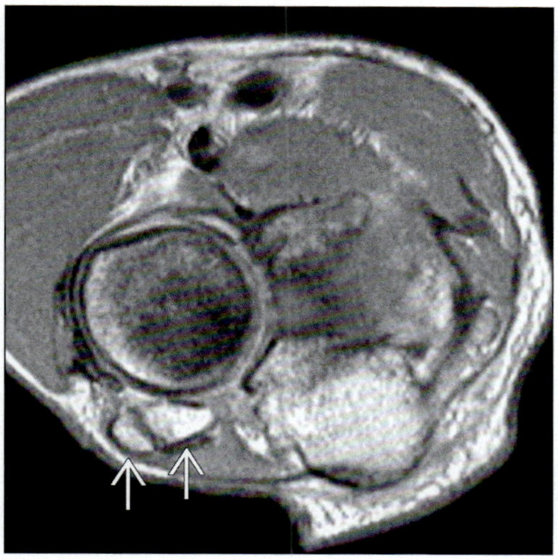

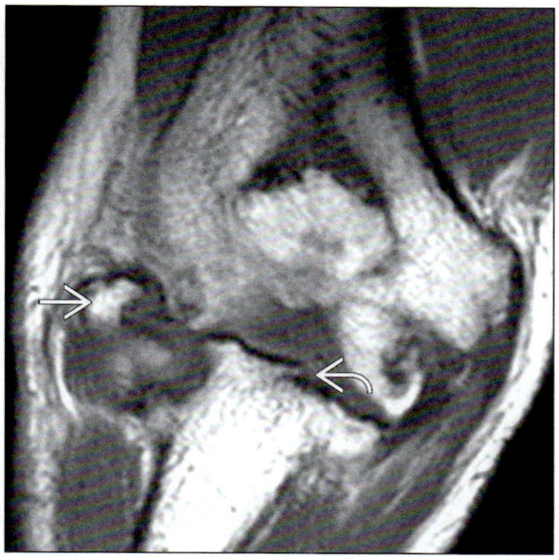

(Left) *Axial PDWI MR demonstrates corticated bodies posterior to the radial head. Fatty marrow in the bodies makes them easily seen on this short TE sequence.* ***(Right)*** *Coronal T1WI MR in the same patient shows high signal marrow in a posterolateral intraarticular body. Diffuse degenerative changes in the elbow are related to prior trauma and fractures, which is the likely etiology of the bodies.*

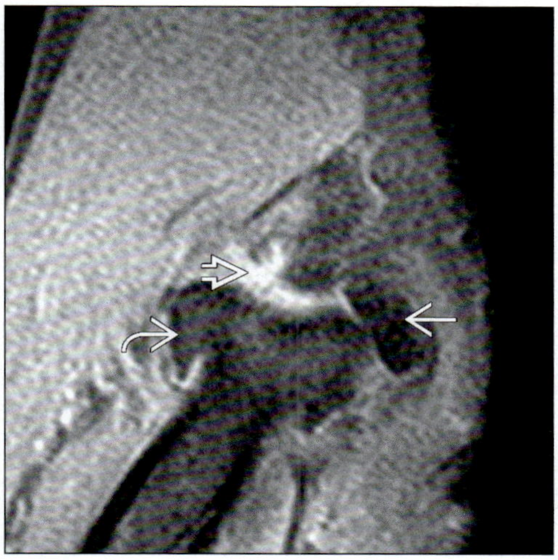

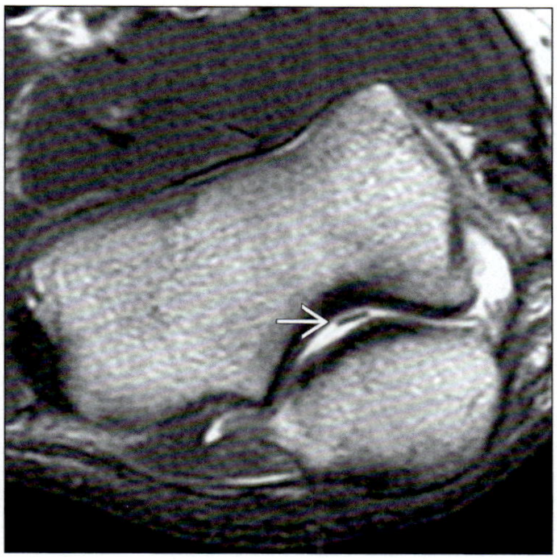

(Left) *Sagittal PDWI FS MR in the same patient shows post-traumatic deformity of the radial head and 1 of the large intraarticular bodies posterior to it. Fat-suppressed sequences highlight bodies against joint fluid, but make identification of marrow signal difficult.* ***(Right)*** *Axial T2WI MR shows a tiny intraarticular body in the elbow. A physiologic amount of joint fluid is sufficient to outline even small bodies.*

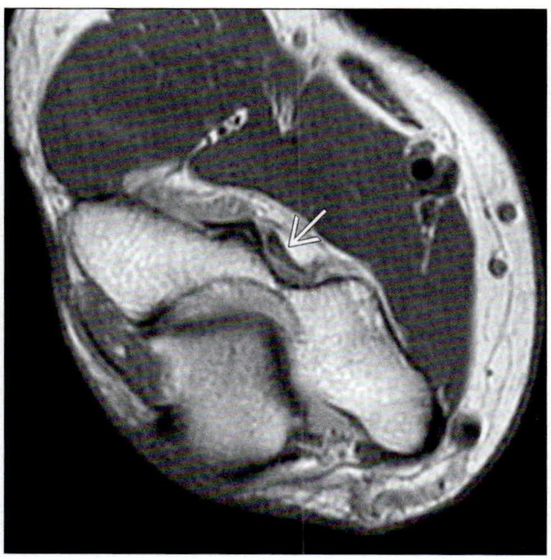

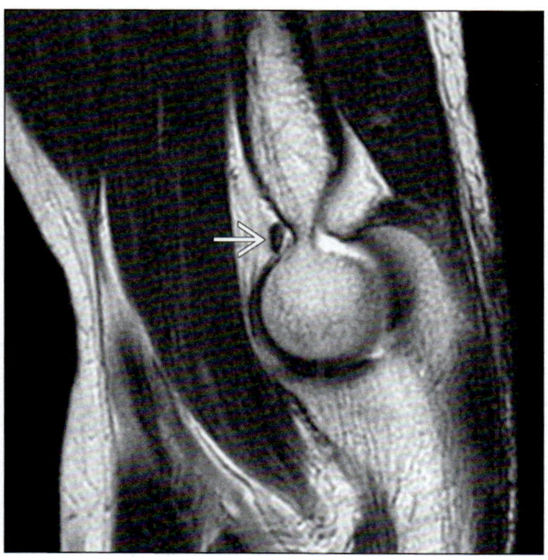

(Left) *Axial PDWI MR shows a sclerotic low signal intraarticular body in the coronoid fossa. No discrete donor site could be identified in this case.* ***(Right)*** *Sagittal T2WI MR in the same patient shows the small body very well despite the lack of significant joint fluid. MR arthrography is rarely required to identify intraarticular bodies in the elbow unless they are resynovialized, small, and in unusual locations.*

INTRAARTICULAR BODIES, ELBOW

*(**Left**) Sagittal T1WI MR in an adolescent baseball pitcher shows marked irregularity of the capitellar articular surface and a low signal body posterior to the radiocapitellar joint. (**Right**) Sagittal T2WI FS MR in the same patient at the same level better demonstrates the chondral defect in the capitellum, underlying marrow edema and cyst formation, and sclerotic intraarticular body. The body is well seen here due to the large effusion.*

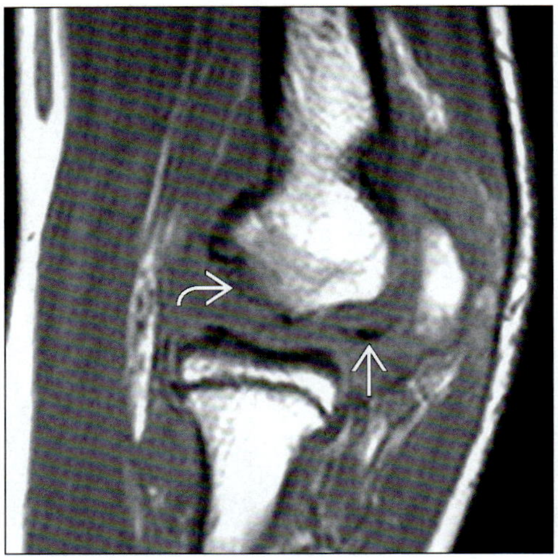

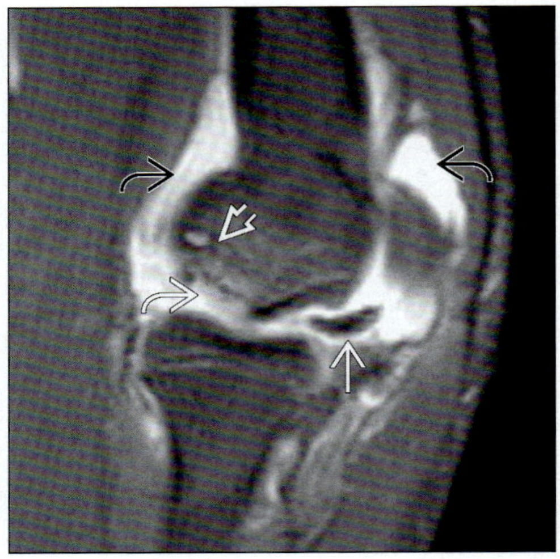

*(**Left**) Axial T2WI FS MR through the level of the humeral condyles demonstrates a curvilinear chondral fragment in the olecranon fossa with a large joint effusion. (**Right**) Axial PDWI MR shows the chondral nature of the fragment; if the body were osseous, more prominent cortex would be visible. An osteochondral fragment would have both bone and cartilage. This fragment is composed entirely of articular cartilage, and thus has medium intensity signal.*

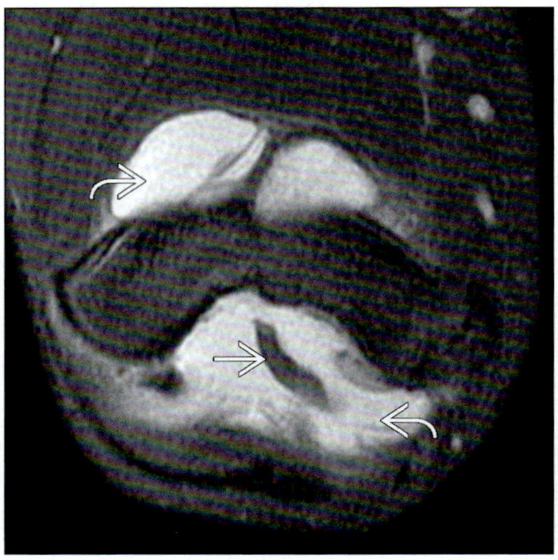

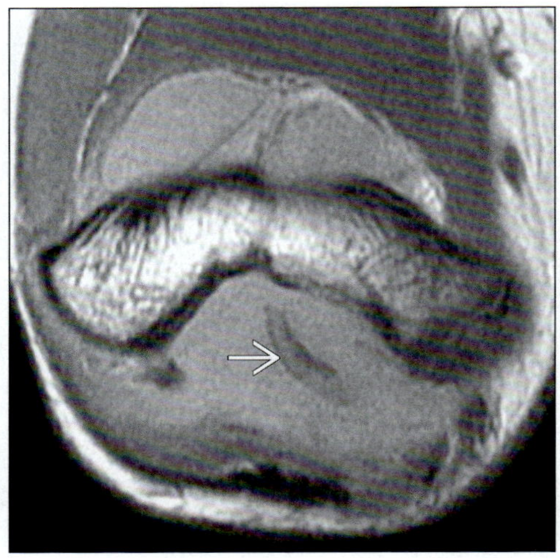

*(**Left**) Sagittal bone CT shows a comminuted fracture of the coronoid process and a small intraarticular body in the coronoid fossa. (**Right**) Sagittal bone CT from the same patient shown in soft tissue window more clearly shows a large joint effusion. Multiple small intraarticular bodies are present in the olecranon fossa that were not well seen on bone windows; small or less densely ossified bodies are often better seen this way.*

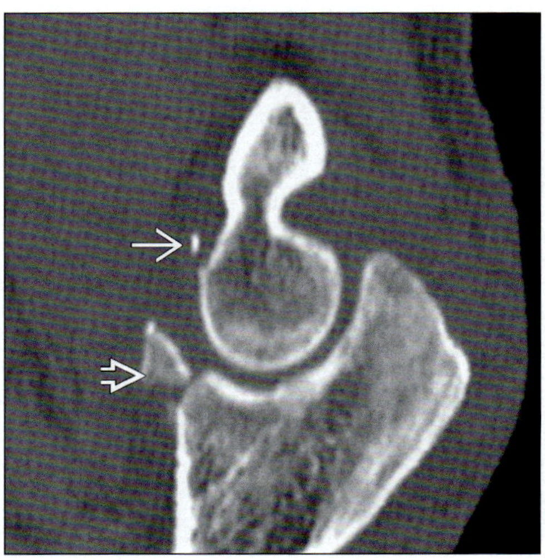

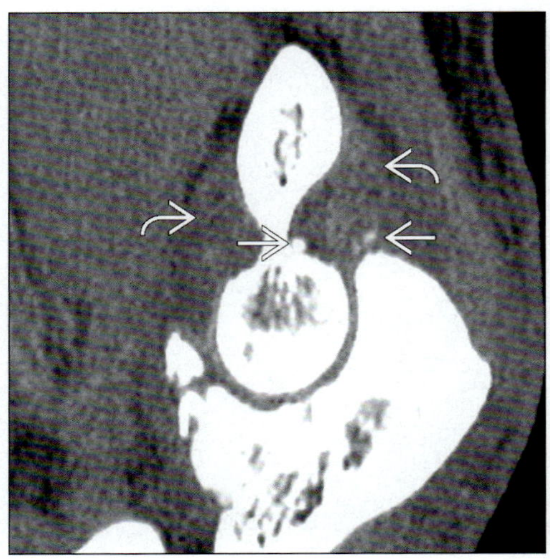

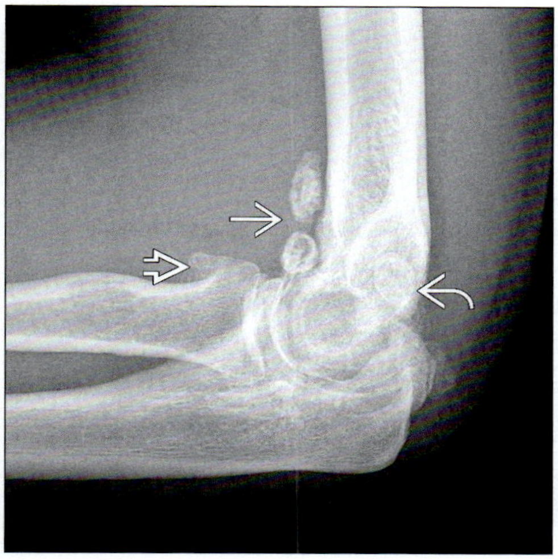

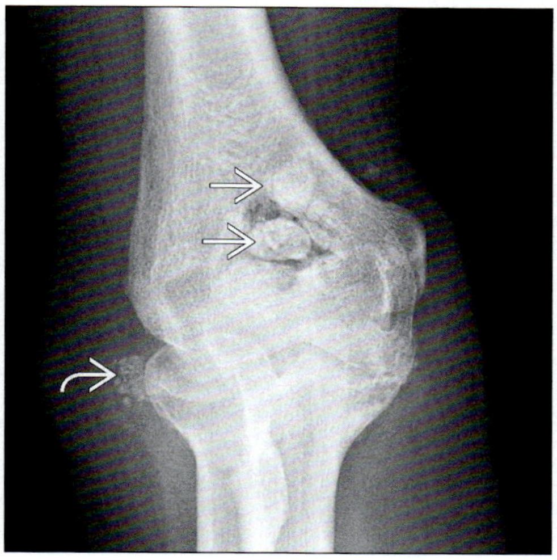

(Left) *Lateral radiograph shows multiple well-corticated bodies in the coronoid and olecranon fossae. There are degenerative changes in the radiocapitellar joint.* ***(Right)*** *AP radiograph in the same patient shows the mature cortication in several of the bodies indicating their chronic nature. In such cases, the original donor site is often indeterminate. The chronicity of smaller fragments lateral to the radial head is not as obvious.*

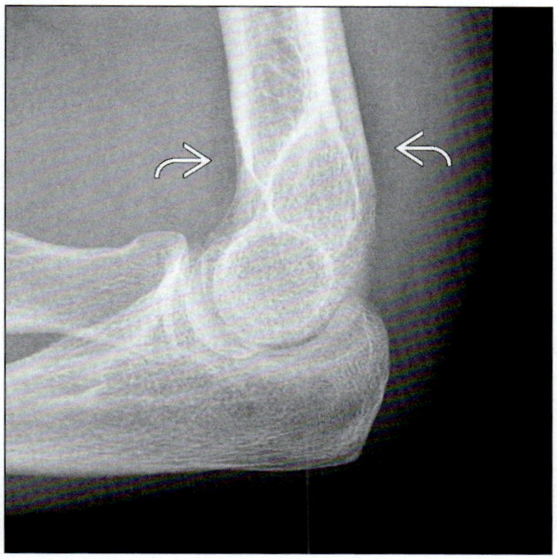

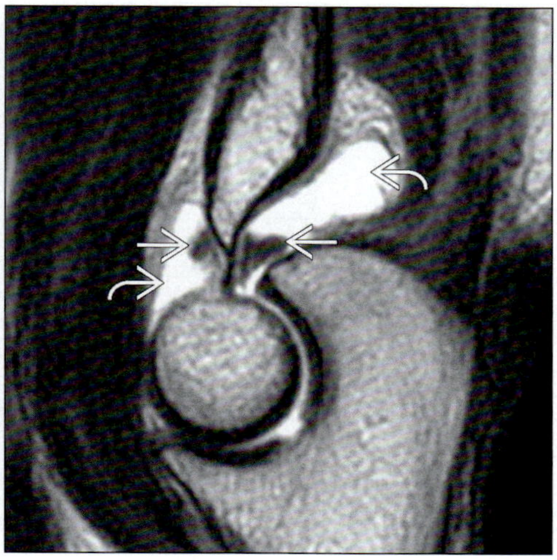

(Left) *Lateral radiograph in a 53-year-old woman with an acute injury to the elbow shows distended fat pads indicating an effusion. No fracture or intraarticular bodies are seen on this or other radiographs obtained.* ***(Right)*** *Sagittal T2WI MR in the same patient obtained the next day shows small intraarticular chondral fragments in the coronoid and olecranon fossae surrounded by joint fluid. Nonossified fragments will not be visible on radiographs.*

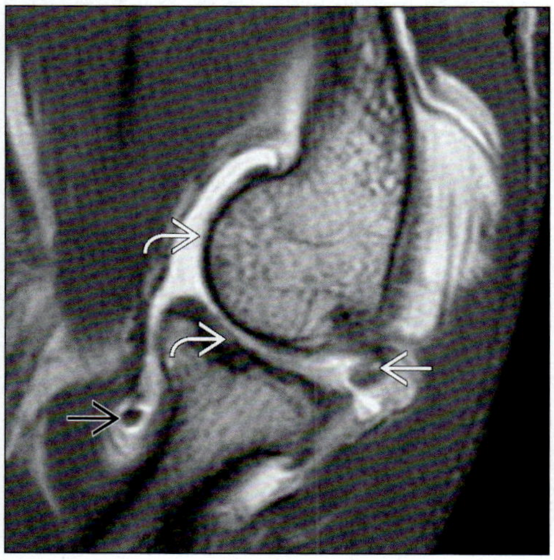

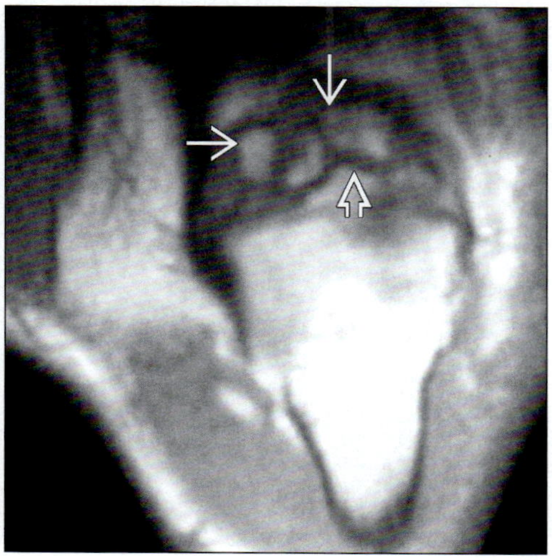

(Left) *Sagittal MR arthrogram T1WI shows extensive degenerative change in the radial head and capitellum from previous trauma. Intraarticular fragments are seen posterior to this joint and anterior to the radial neck. MR arthrography can make identification and characterization of donor sites easier.* ***(Right)*** *Coronal PDWI MR shows multiple ossicles in the olecranon fossa. The faceted way that some bodies articulate with the olecranon suggests os supratrochlear dorsale.*

OLECRANON BURSITIS

Key Facts

Terminology
- Student's elbow, dart thrower's elbow, miner's elbow

Imaging
- Focal fluid collection in subcutaneous tissues posterior to olecranon
- Typically signal consistent with simple fluid
- More complex fluid may be present with hemorrhage, infection, crystalline arthropathy
- Well-defined anechoic fluid collection with increased through-transmission posterior to olecranon

Top Differential Diagnoses
- Triceps tendinopathy/tear
    - Abnormal signal in triceps extending into adjacent soft tissues
- Neoplasm
- Hematoma (characteristic blood signal on MR)

Pathology
- Repetitive trauma
- Acute trauma
- Systemic inflammatory condition
- Infection
- Triceps tendinopathy

Clinical Issues
- General: RICE (rest, ice, compression, elevation)
- NSAIDs
- Corticosteroid injection if not septic
- Septic: Aspiration (for diagnosis and treatment)
- Bursectomy for recalcitrant cases

Diagnostic Checklist
- History of trauma
- Evidence of associated arthropathy

*(**Left**) Graphic illustration of the posterior elbow shows inflammation of the olecranon bursa ➔. The bursa is located between the point of the olecranon and the skin, just distal to the triceps tendon insertion ➔. (**Right**) Lateral radiograph shows soft tissue prominence ➔ over the olecranon ➔, consistent with olecranon bursitis. The findings are nonspecific, and could represent bursal inflammation due to trauma, infection, or systemic disease.*

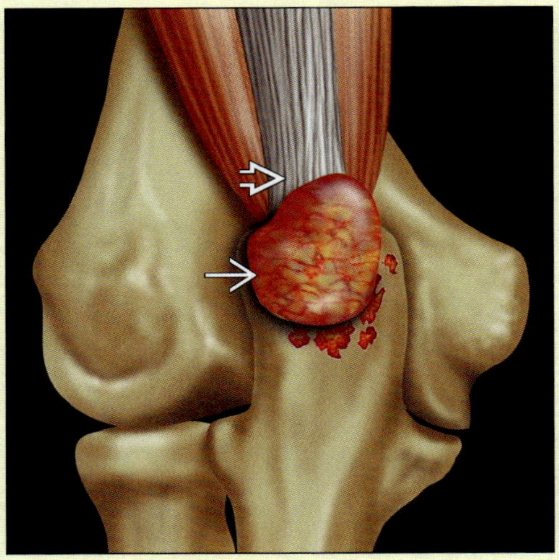

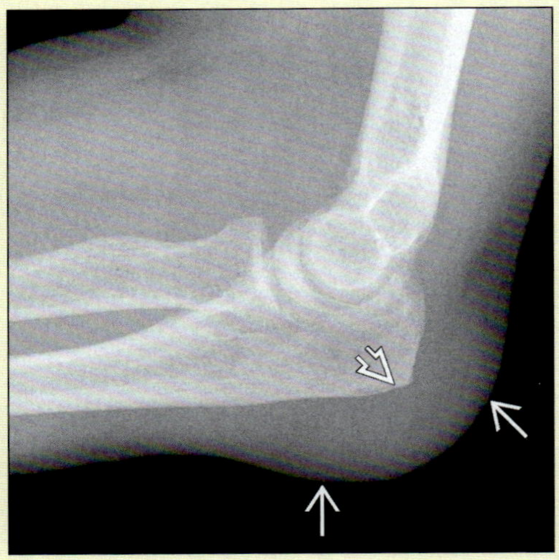

*(**Left**) Sagittal T2WI FS MR shows a large fluid collection ➔ in the posterior subcutaneous space of the elbow consistent with olecranon bursitis. Note also the mild inflammatory component ➔ in this case of infectious bursitis. (**Right**) Axial T2WI FS MR in the same patient shows the thin, well-defined border ➔ of the bursal collection. Subcutaneous inflammation ➔ defines the outer surface. Note that the bursa extends to the periosteal surface of the olecranon ➔.*

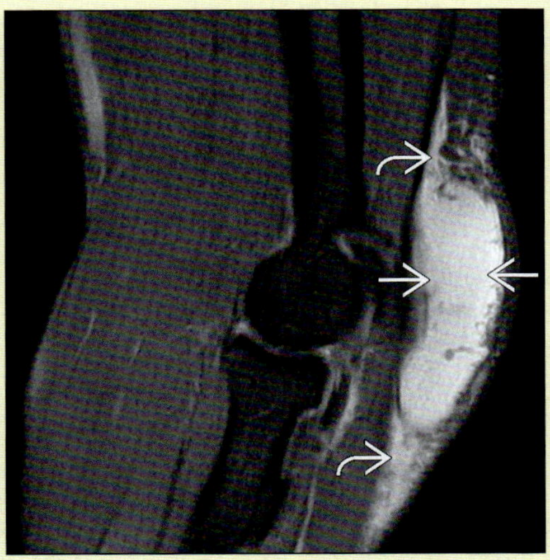

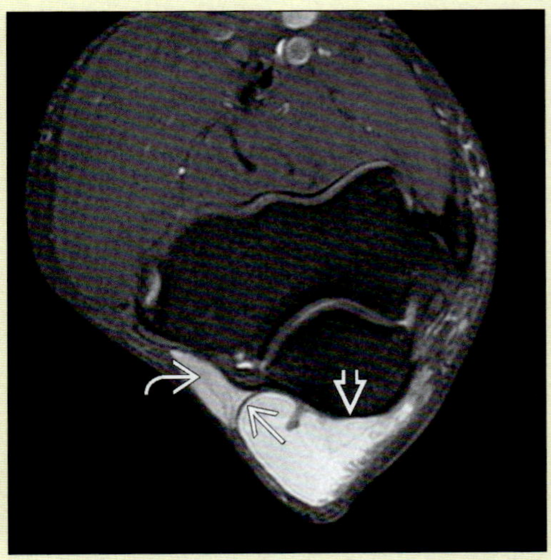

OLECRANON BURSITIS

TERMINOLOGY

Synonyms

- Student's elbow, dart thrower's elbow, miner's elbow

Definitions

- Swelling and enlargement of well-defined collection of fluid posterior to proximal olecranon

IMAGING

General Features

- Best diagnostic clue
  - Focal fluid collection in subcutaneous tissues posterior to olecranon
- Location
  - Subcutaneous tissues posterior to proximal olecranon between skin and bone

Radiographic Findings

- Focal soft tissue fullness posterior to olecranon on lateral view
- ± calcification, especially with gout or CPPD

MR Findings

- Typically signal consistent with simple fluid
  - Rim, if present, ↓ on all sequences
  - May be surrounded by soft tissue edema
  - ± reactive edema in subcortical bone in olecranon
  - Post-contrast MR: No enhancement of bursal contents
- More complex fluid may be present with hemorrhage, infection, crystalline arthropathy

Ultrasonographic Findings

- Well-defined anechoic fluid collection with increased through-transmission posterior to olecranon
- Complex fluid with low-level echoes in septic collections, gout, CPPD, hemorrhage

Imaging Recommendations

- Best imaging tool
  - MR

DIFFERENTIAL DIAGNOSIS

Triceps Tendinopathy/Tear

- Abnormal signal in triceps extending into adjacent soft tissues

Neoplasm

- Giant cell tumor of tendon sheath

Hematoma

- Characteristic blood signal on MR

PATHOLOGY

General Features

- Etiology
  - Repetitive trauma
  - Acute trauma
  - Systemic inflammatory condition (including rheumatoid arthritis, gout)
  - Infection
- Associated abnormalities
  - Triceps tendinopathy

Microscopic Features

- Aspiration of fluid to evaluate for
  - Gram stain for bacteria
  - Leukocyte count
  - Crystal identification

CLINICAL ISSUES

Presentation

- Most common signs/symptoms
  - Focal swelling posterior to olecranon
    - Often painless
    - Well demarcated
  - ± pain, exacerbated by pressure, at site

Demographics

- Epidemiology
  - Very common

Natural History & Prognosis

- Often persistent ± recurrent
- Most nonseptic cases resolve with conservative therapy ± 1-2 aspirations

Treatment

- General: RICE (rest, ice, compression, elevation)
- NSAIDs
- Corticosteroid injection if not septic
- Septic: Aspiration (for diagnosis and treatment)
  - Complications
    - Recurrent swelling and pain
    - Infection
    - Persistent drainage through aspiration tract
    - Ulnar nerve injury
- Bursectomy for recalcitrant cases

DIAGNOSTIC CHECKLIST

Consider

- History of trauma
- Associated arthropathy

Image Interpretation Pearls

- Search for other signs to identify disease process
  - Erosions in rheumatoid arthritis, gout
  - Hemarthrosis in clotting abnormalities
  - Marrow contusion or fracture in trauma

Reporting Tips

- Describe size of collection and complexity of fluid

SELECTED REFERENCES

1. Blankstein A et al: Ultrasonographic findings in patients with olecranon bursitis. Ultraschall Med. 27(6):568-71, 2006
2. Floemer F et al: MRI characteristics of olecranon bursitis. AJR Am J Roentgenol. 183(1):29-34, 2004
3. Stell IM: Management of acute bursitis: outcome study of a structured approach. J R Soc Med. 92(10):516-21, 1999
4. Stewart NJ et al: Surgical treatment of aseptic olecranon bursitis. J Shoulder Elbow Surg. 6(1):49-54, 1997

OLECRANON BURSITIS

(Left) *Axial T1WI MR demonstrates a mass ➔ in the posterior subcutaneous tissues posterior to the olecranon ⇨. High signal in the mass on T1-weighted imaging suggests hemorrhage; the patient had a history of trauma to this area. A marker capsule ↪ has been placed over the site of symptomatology.* ***(Right)*** *Axial T2WI MR in the same patient demonstrates uniform high signal within the hemorrhagic olecranon bursitis →. Note the relatively thin bursal rim ↪.*

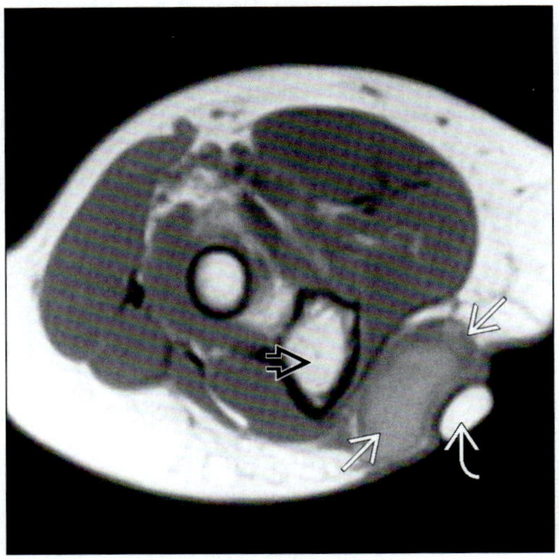

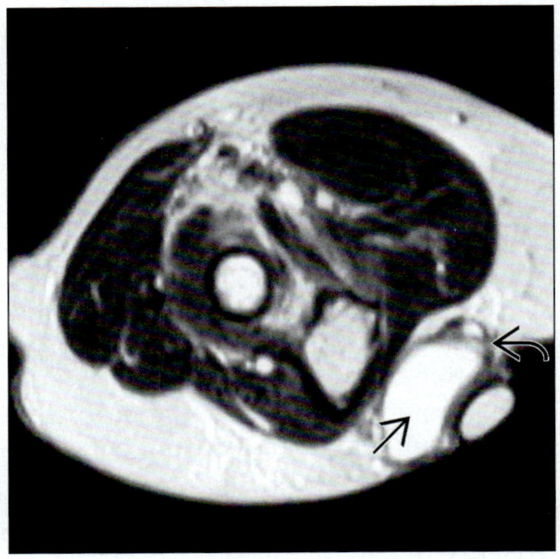

(Left) *Axial T2WI FS MR in the same patient demonstrates high signal in the bursa →, with mild soft tissue swelling ↪ surrounding the mass.* ***(Right)*** *Sagittal T2WI MR demonstrates a complex collection of varied signal intensity ➔ along the posterior soft tissues of the elbow and proximal forearm. This patient with hemorrhagic olecranon bursitis had suffered a direct blow injury to this area. The lower signal material inferiorly may represent more acute blood.*

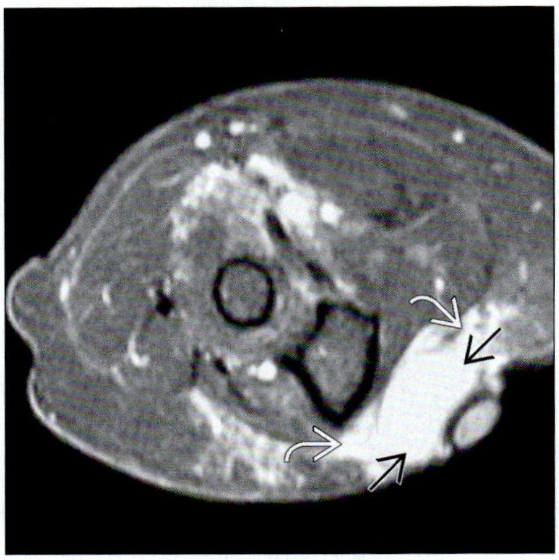

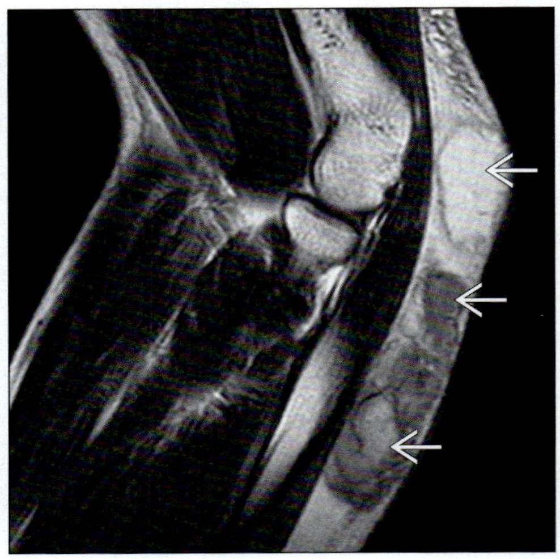

(Left) *Sagittal T1WI C+ FS MR in the same patient confirms that the central aspects of the collection ➔ do not enhance. The periphery of the bursa ↪ shows marked enhancement. This pattern of thin peripheral enhancement is typical of noninfectious bursitis.* ***(Right)*** *Lateral radiograph shows marked soft tissue fullness posterior to the olecranon ➔ with small central air collections ↪ within it. This patient with septic bursitis had lacerated this area several weeks earlier.*

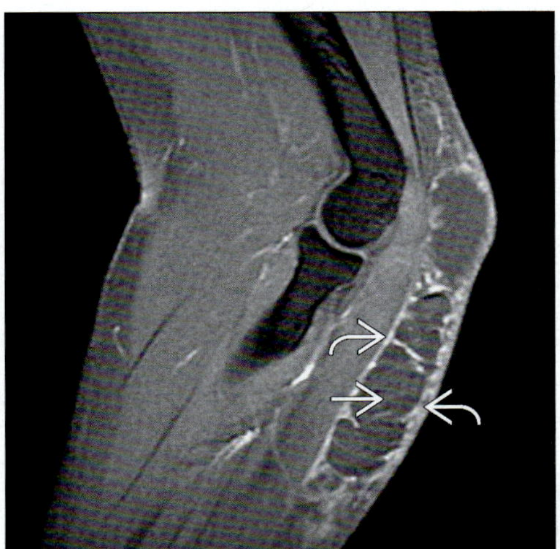

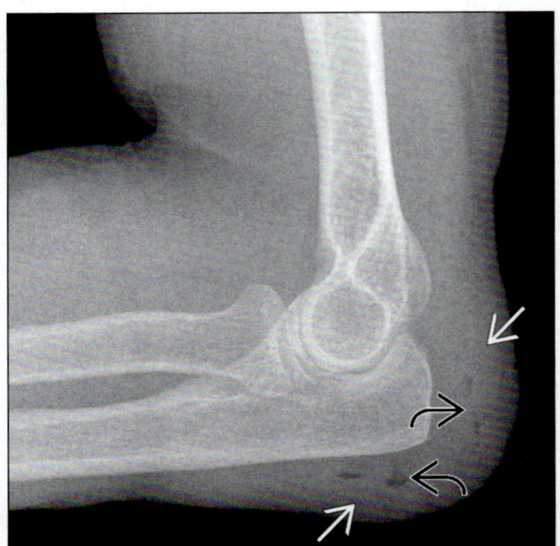

OLECRANON BURSITIS

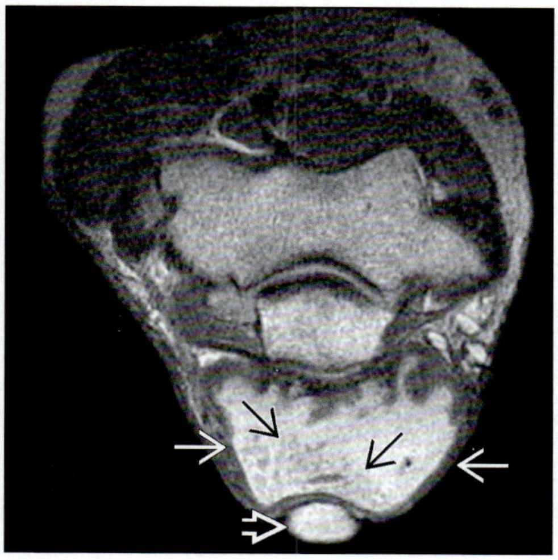

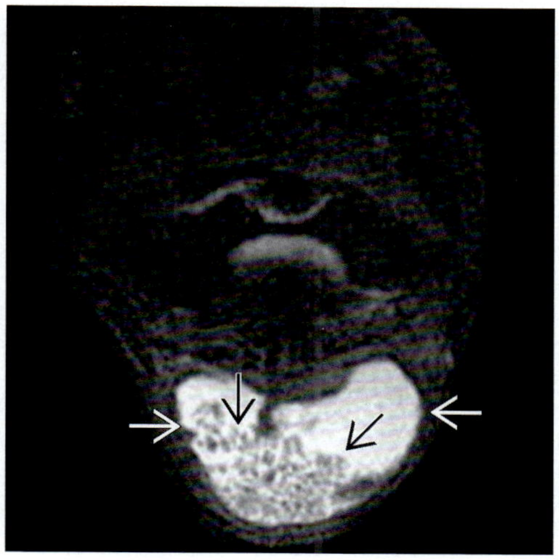

(Left) *Axial T2WI MR demonstrates a predominantly fluid signal mass* ➡ *posterior to the olecranon. A marker capsule is present on the skin* ⮕*. Multiple tiny bodies* ⇨ *are seen within the fluid in the inflamed olecranon bursa, likely debris from chronic synovitis &/or large urate crystals.* ***(Right)*** *Axial T2WI FS MR in the same patient better shows the multiple tiny bodies* ⇨ *within the large olecranon bursal fluid collection* ➡*.*

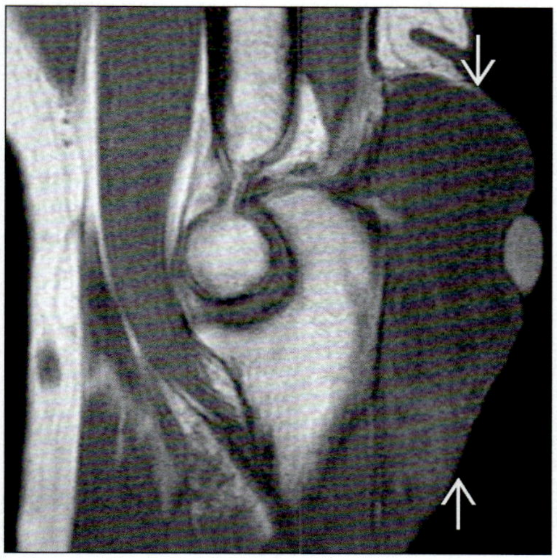

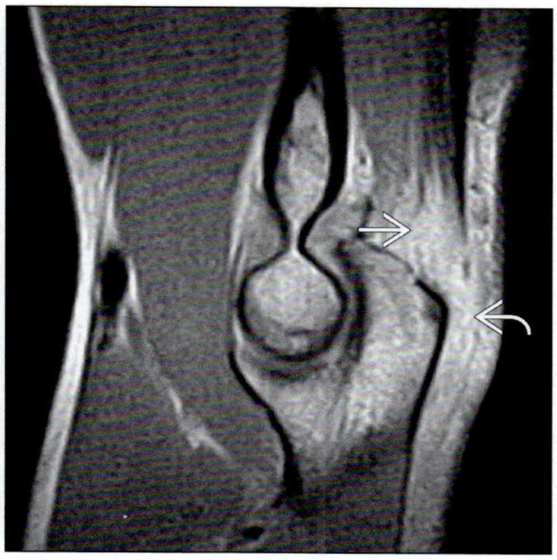

(Left) *Sagittal T1WI FS MR of the contralateral elbow in the same patient shows a similar mass, also consistent with olecranon bursitis* ➡*. When olecranon bursitis is present bilaterally, there is often an underlying systemic disease present, such as rheumatoid arthritis or gout (the latter in this case).* ***(Right)*** *Sagittal PDWI MR shows markedly elevated signal in the distal fibers of the triceps tendon* ➡ *extending into a poorly defined fluid collection* ↪ *posterior to the olecranon.*

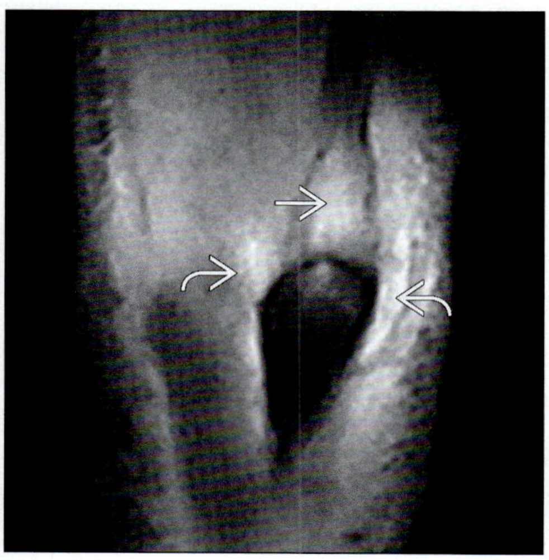

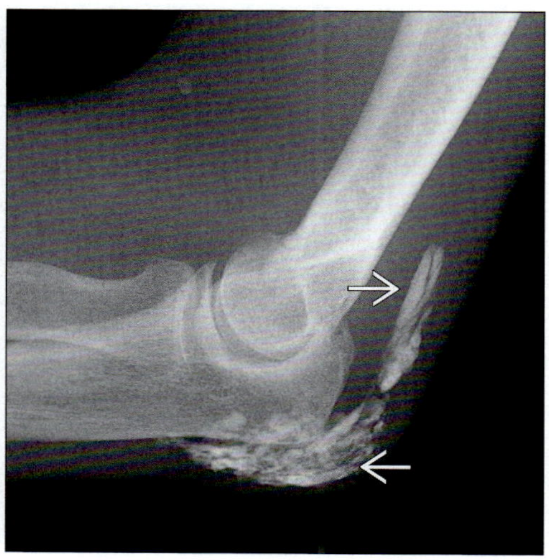

(Left) *Coronal T2WI FS MR in the same patient shows the disruption of the distal triceps fibers* ➡ *and fluid around the olecranon* ↪*. This patient with septic olecranon bursitis and tricipital tendinitis had suffered a bug bite in this location 2 weeks earlier.* ***(Right)*** *Lateral radiograph shows calcium hydroxyapatite dihydrate deposition (HADD) in the olecranon bursa* ➡*. This finding may follow trauma or may be spontaneous.*

SYNOVIAL FRINGE, ELBOW

Key Facts

Terminology
- Synovial fold, synovial plica, radiocapitellar meniscus
- Focal meniscal-like synovial thickening in lateral aspect of radiocapitellar articulation

Imaging
- May range from 1 to > 5 mm in thickness
- Usually triangular or band-like projection from posterolateral capsular margin of joint
- Low signal on all sequences due to fibrocartilaginous nature
- MR arthrogram may help in identifying and characterizing synovial fringe
- May be associated with chondral thinning at lateral capitellum or radial head

Top Differential Diagnoses
- Intraarticular body
- Normal redundant synovium
- Focal synovitis

Pathology
- Focal synovial hypertrophy due to repetitive injury

Clinical Issues
- Lateral elbow pain, especially with extension
- Snapping elbow with flexion/pronation
- Locking
- Adults, usually middle-aged
- Mild cases treated conservatively
- Surgical resection definitive

Diagnostic Checklist
- Describe thickness of plica and any associated chondromalacia or bone marrow edema

*(**Left**) Graphic illustration shows a synovial fringe → in the posterolateral corner of the elbow. The tapered portion of the triangular mass may extend into the radiocapitellar joint, leading to abrasion of the joint surfaces and ultimately to chondromalacia of the capitellum ⇩ and radius ↰. (**Right**) Coronal MR arthrogram gradient echo image in an 18-year-old baseball pitcher shows marked thickening of a synovial fringe → in the radiocapitellar joint.*

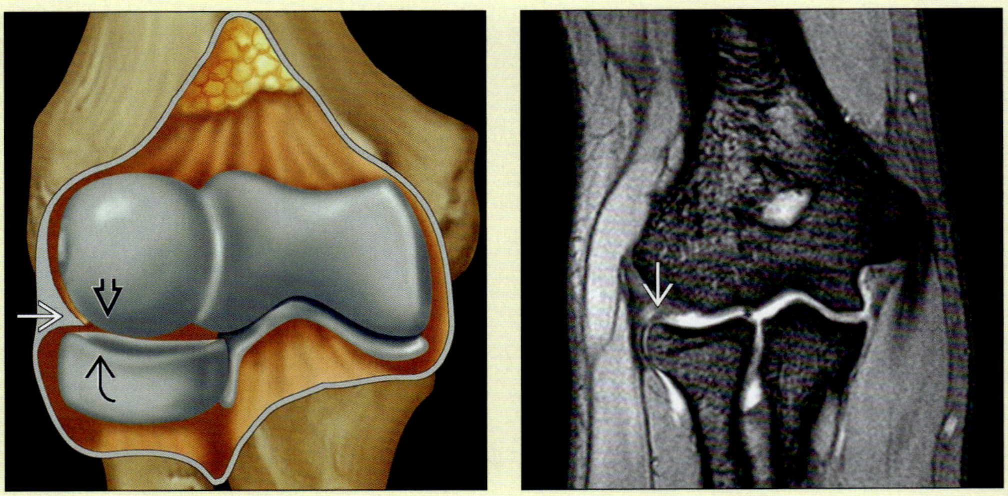

*(**Left**) Sagittal MR arthrogram T1WI FS in the same patient shows the triangular synovial fringe → posterior to the radiocapitellar joint. There may be a normal synovial excrescence here, but generally these are 1-2 mm or less in thickness. Note the capsular reflections ↰ above and below the fringe. (**Right**) Coronal T2WI FS MR shows a large synovial fringe → with associated degenerative chondromalacia in the capitellum ⇩. High signal in the fringe indicates irritation.*

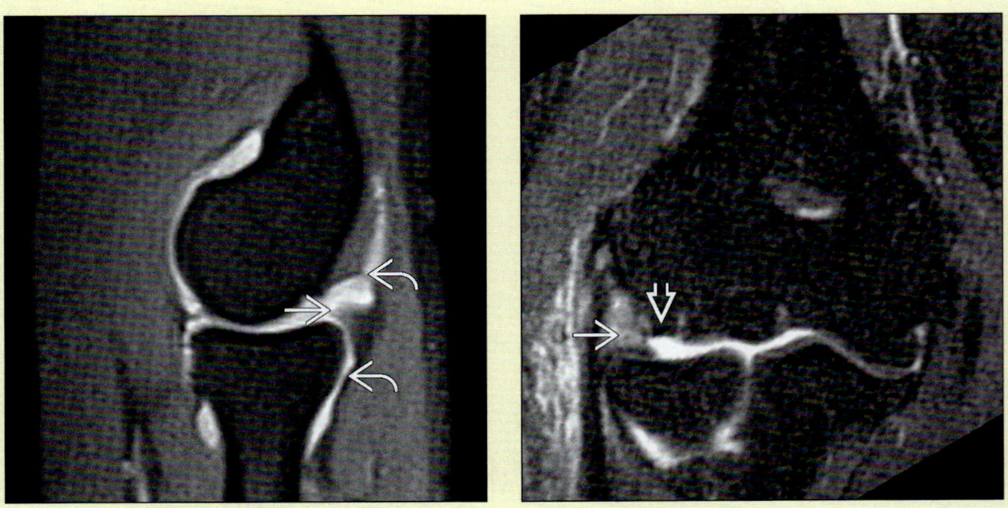

SYNOVIAL FRINGE, ELBOW

TERMINOLOGY

Synonyms

- Synovial fold, synovial plica, synovial shelf, radiocapitellar meniscus

Definitions

- Focal meniscal-like synovial thickening in lateral aspect of radiocapitellar articulation

IMAGING

General Features

- Best diagnostic clue
  - Triangular or rounded synovial projection extending into radiocapitellar articulation from posterolateral aspect of elbow joint capsule on MR
- Location
  - Posterolateral capsular margin of elbow joint at radiocapitellar joint
- Size
  - May range from 1 to > 5 mm in thickness
  - Diameter variable
- Morphology
  - Usually triangular or band-like projection from posterolateral capsular margin of joint
  - Similar in shape to meniscus of knee, with broad portion of triangle at capsular margin

MR Findings

- Low signal on all sequences due to fibrocartilaginous nature
  - May develop higher T2 signal due to chronic irritation and hyperemia
- Better seen when surrounded by joint fluid
- MR arthrogram may help in identifying and characterizing synovial fringe
- May be associated with chondral thinning at lateral capitellum or radial head

Imaging Recommendations

- Best imaging tool
  - MR
- Protocol advice
  - Coronal PDWI FS MR or T2WI FS MR best for identification
  - MR arthrography sometimes more definitive

DIFFERENTIAL DIAGNOSIS

Intraarticular Body

- More rounded

Normal Redundant Synovium

- Not usually seen without MR arthrography

Focal Synovitis

- May demonstrate ↑ T2 signal and post-contrast enhancement

PATHOLOGY

General Features

- Etiology
  - Focal synovial hypertrophy due to repetitive injury
- Associated abnormalities
  - Chondral thinning in capitellum ± radial head

Gross Pathologic & Surgical Features

- Triangular or shelf-like fibrosynovial thickening arising from inner margin of joint capsule
- Thick portion of triangle abuts capsule, thin portion projects into joint

CLINICAL ISSUES

Presentation

- Most common signs/symptoms
  - Lateral elbow pain, especially with extension
  - Snapping elbow with flexion/pronation
  - Locking

Demographics

- Age
  - Adults, usually middle-aged
- Gender
  - M = F

Natural History & Prognosis

- Usually good prognosis after resection
- Residual osteoarthritis in radiocapitellar joint if present before resection

Treatment

- Mild cases treated conservatively
  - Rest, NSAIDs
- Surgical resection definitive

DIAGNOSTIC CHECKLIST

Consider

- Intraarticular body

Reporting Tips

- Describe thickness of plica and any associated chondromalacia or bone marrow edema

SELECTED REFERENCES

1. Steinert AF et al: Snapping elbow caused by hypertrophic synovial plica in the radiohumeral joint: a report of three cases and review of literature. Arch Orthop Trauma Surg. Epub ahead of print, 2008
2. Fukase N et al: Usefulness of MRI for diagnosis of painful snapping elbow. Skeletal Radiol. 35(10):797-800, 2006
3. Kim DH et al: Arthroscopic treatment of posterolateral elbow impingement from lateral synovial plicae in throwing athletes and golfers. Am J Sports Med. 34(3):438-44, 2006
4. Ruch DS et al: The posterolateral plica: a cause of refractory lateral elbow pain. J Shoulder Elbow Surg. 15(3):367-70, 2006
5. Antuna SA et al: Snapping plicae associated with radiocapitellar chondromalacia. Arthroscopy. 17(5):491-5, 2001
6. Awaya H et al: Elbow synovial fold syndrome: MR imaging findings. AJR Am J Roentgenol. 177(6):1377-81, 2001
7. Clarke RP: Symptomatic, lateral synovial fringe (plica) of the elbow joint. Arthroscopy. 4(2):112-6, 1988

ANCONEUS EPITROCHLEARIS

Key Facts

Terminology

- Accessory muscle coursing from medial humeral epicondyle to medial olecranon, which may cause ulnar nerve compression

Imaging

- Roof of cubital tunnel
- Replacement of normal retinaculum of cubital tunnel with muscle on all axial images
- Oval-shaped muscle on axial and coronal images
- May vary from thin slip to large bulky muscle
- MR
  - Axial images key to making diagnosis
  - Follows normal muscle signal on all MR sequences
- No abnormal enhancement

Top Differential Diagnoses

- Prominent medial head of triceps
- Synovitis
- Neoplasm
- Hematoma

Clinical Issues

- Incidental finding
- Ulnar neuropathy
- All ages
- M = F
- Ulnar neuropathy, if present, may progress and require surgery
- Physical therapy, NSAIDs
- Surgical resection if nonresponsive to conservative treatment

Diagnostic Checklist

- Mass or synovitis
- Follows muscle signal on all MR sequences

*(**Left**) Graphic of anconeus epitrochlearis ➔, an accessory muscle that runs from the medial humeral epicondyle to the medial border of the olecranon and courses over a portion of the roof of the cubital tunnel. This normal variant can compress the ulnar nerve ➔ in the cubital tunnel, leading to ulnar neuropathy. (**Right**) Coronal T1WI MR demonstrates an accessory muscle, the anconeus epitrochlearis ➔, medial to the olecranon ⇨.*

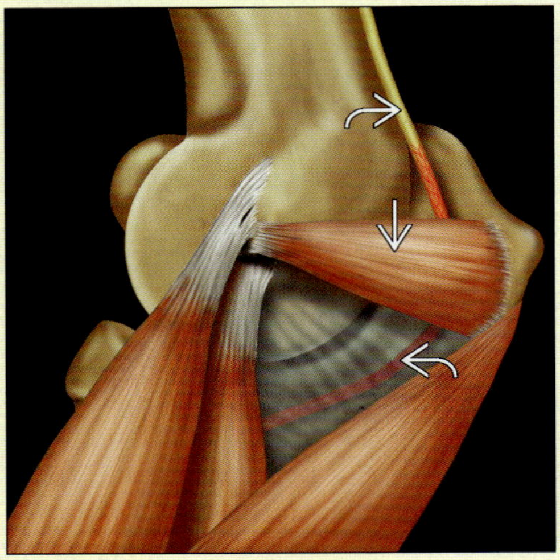

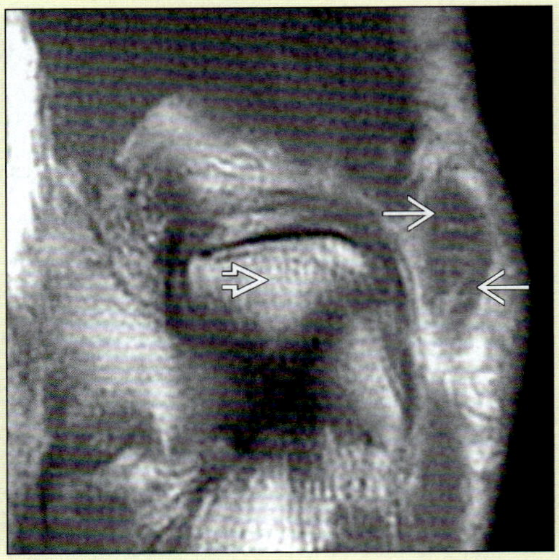

*(**Left**) Axial T1WI MR in the same patient shows the anconeus epitrochlearis ➔ forming the roof of the cubital tunnel posterior to the medial humeral condyle ⇨. The ulnar nerve ➔ resides within the cubital tunnel and compressive symptoms may result in ulnar neuropathy. (**Right**) Axial T2WI MR in the same patient at the same level confirms that the anconeus epitrochlearis ➔ follows normal muscle signal on all sequences, which confirms its identity.*

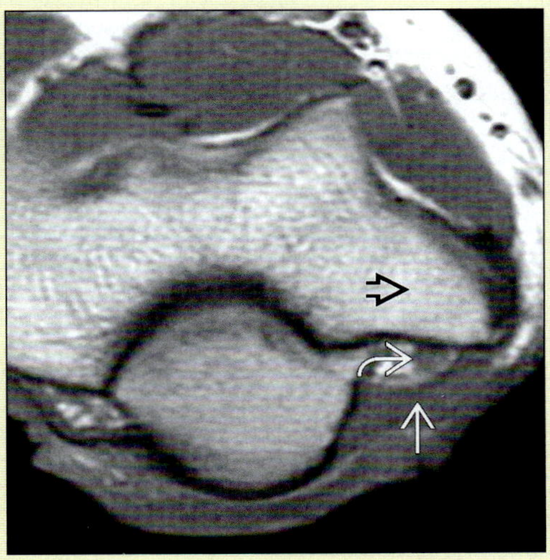

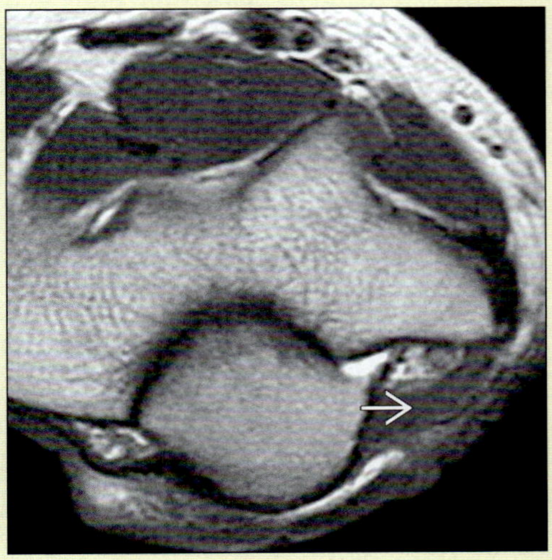

ANCONEUS EPITROCHLEARIS

TERMINOLOGY

Definitions

- Accessory muscle coursing from medial humeral epicondyle to medial olecranon, which may cause ulnar nerve compression

IMAGING

General Features

- Best diagnostic clue
  - Replacement of normal retinaculum of cubital tunnel with muscle on all axial images
- Location
  - Proximal roof of cubital tunnel
  - Extends from medial humeral epicondyle to medial rim of olecranon
- Size
  - May vary from thin slip to large bulky muscle
- Morphology
  - Oval-shaped muscle on axial and coronal images

Radiographic Findings

- None

CT Findings

- Ovoid muscle in normal location of roof of cubital tunnel

MR Findings

- Follows normal muscle signal on all MR sequences
- No abnormal enhancement

Ultrasonographic Findings

- Well-defined muscle tissue replacing some of the arcuate ligament in roof of cubital tunnel
- No hyperemia
- May see focal enlargement of ulnar nerve if secondary ulnar neuropathy present

Imaging Recommendations

- Best imaging tool
  - MR
- Protocol advice
  - Axial images key to making diagnosis

DIFFERENTIAL DIAGNOSIS

Prominent Medial Head of Triceps

- More vertical orientation than anconeus epitrochlearis
- Also follows muscle signal on all sequences
- Associated with "snapping triceps" syndrome

Synovitis

- Less well defined
- High signal on T2WI MR

Neoplasm

- Generally do not follow muscle signal on all sequences
- May enhance after contrast administration

Hematoma

- Will not follow muscle signal on all sequences
- Often ↑ T1 signal
- No central enhancement
- History of trauma

CLINICAL ISSUES

Presentation

- Most common signs/symptoms
  - Incidental finding
- Other signs/symptoms
  - Ulnar neuropathy

Demographics

- Age
  - All ages
- Gender
  - M = F

Natural History & Prognosis

- Ulnar neuropathy, if present, may progress and require surgery

Treatment

- Physical therapy, NSAIDs
- Surgical resection if nonresponsive to conservative treatment

DIAGNOSTIC CHECKLIST

Consider

- Mass or synovitis

Image Interpretation Pearls

- Follows muscle signal on all MR sequences

Reporting Tips

- Describe thickness and dimensions of accessory muscle
- Comment on any morphologic or signal changes in ulnar nerve

SELECTED REFERENCES

1. Husarik DB et al: Elbow nerves: MR findings in 60 asymptomatic subjects--normal anatomy, variants, and pitfalls. Radiology. 252(1):148-56, 2009
2. Jeon IH et al: MR imaging of edematous anconeus epitrochlearis: another cause of medial elbow pain? Skeletal Radiol. 34(2):103-7, 2005
3. Okamoto M et al: Diagnostic ultrasonography of the ulnar nerve in cubital tunnel syndrome. J Hand Surg [Br]. 25(5):499-502, 2000
4. Masear VR et al: Ulnar compression neuropathy secondary to the anconeus epitrochlearis muscle. J Hand Surg Am. 13(5):720-4, 1988
5. Dahners LE et al: Anconeus epitrochlearis, a rare cause of cubital tunnel syndrome: a case report. J Hand Surg Am. 9(4):579-80, 1984

ULNAR NERVE INJURY

Key Facts

Terminology

- Cubital tunnel syndrome
- Osborne (arcuate) ligament = retinaculum of cubital tunnel

Imaging

- Ultrasound or MR
- Enlarged and ↑ T2 signal intensity in nerve
- Fibrosis results in thick irregular retinaculum
- Edema or fatty atrophy of FDP, FCU, and intrinsic hand muscles

Pathology

- Overuse
- Subluxation
- Extrinsic compression
  - "Sleep palsy"
- Primary thickening of Osborne ligament
- Trauma
- Osseous abnormalities
  - Osteophytes
  - Fracture ± malunion
  - Intraarticular bodies
- Muscle abnormalities
  - Anconeus epitrochlearis
  - Snapping triceps
- Mass

Clinical Issues

- 2nd most common upper extremity entrapment neuropathy
- Ulnar nerve transposition
  - Subcutaneous placement: More prone to direct trauma
  - Submuscular placement: More prone to muscular impingement

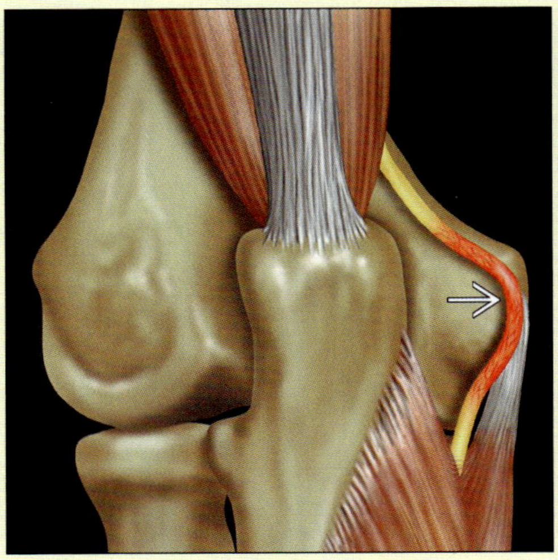
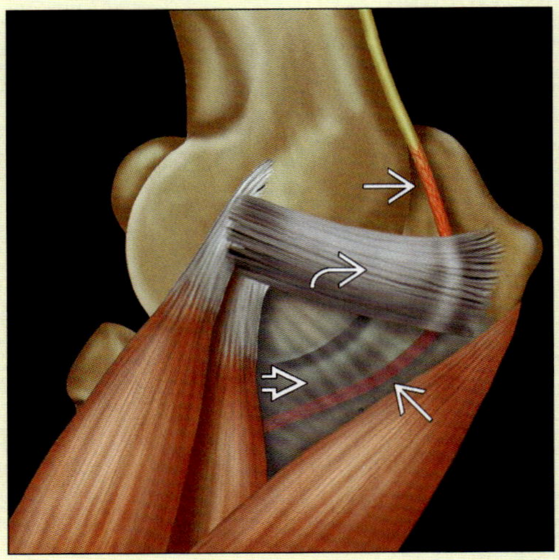

*(**Left**) Graphic from posterior view shows inflammation of the ulnar nerve ➔ in the cubital tunnel. The nerve stretches around the posterior aspect of the medial humeral epicondyle and is prone to a variety of extrinsic forces. (**Right**) Graphic seen from medial view shows the ulnar nerve ➔ passing through the cubital tunnel deep to Osborne ligament ➔ and the ulnar collateral ligament ➔.*

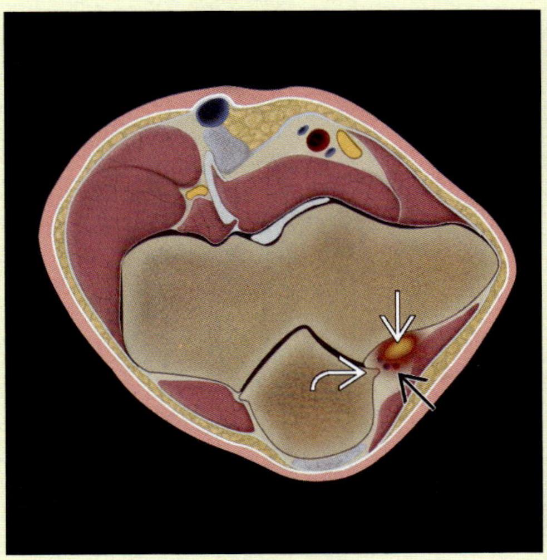
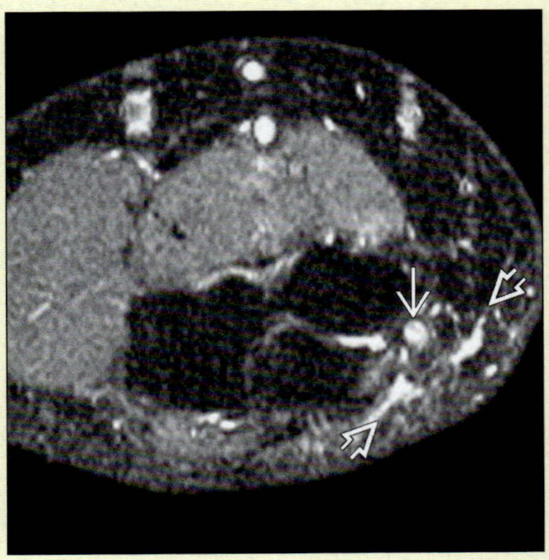

*(**Left**) Axial graphic of the cubital tunnel shows inflammation of the ulnar nerve ➔. Its superficial location makes the nerve vulnerable to trauma. Posterior recurrent ulnar vessels ➔ are located lateral to the nerve. Osteophytosis ➔ of the medial olecranon can impinge the nerve. (**Right**) Axial T2WI FS MR demonstrates mild enlargement and increased T2 signal in the ulnar nerve ➔ proximal to the cubital tunnel. There is mild surrounding subcutaneous edema ➔ posterior to the nerve.*

ULNAR NERVE INJURY

TERMINOLOGY

Synonyms

- Cubital tunnel syndrome
- Osborne (arcuate) ligament = retinaculum of cubital tunnel

Definitions

- Ulnar neuropathy due to trauma to nerve at level of elbow, within or around cubital tunnel
- Cubital tunnel: Anatomic space in posteromedial elbow through which ulnar nerve passes
- Osborne ligament: Fibrous sheath forming roof of cubital tunnel
  - a.k.a. cubital tunnel retinaculum or
  - Humeroulnar arcade (HUA)
- Arcade of Struthers: Thin fibrous band extending from medial head of triceps to medial intermuscular septum ~ 6-10 cm above medial humeral epicondyle

IMAGING

General Features

- Best diagnostic clue
  - Clinical ulnar neuropathy with associated morphologic changes in ulnar nerve on MR
  - Often with abnormal nerve conduction tests
- Location
  - Most commonly in cubital tunnel
- Morphology
  - Ulnar nerve (UN) usually rounded in cubital tunnel
  - With compression, nerve may be flattened
  - If swollen, nerve may be enlarged

Radiographic Findings

- Osteophyte formation in medial aspect of olecranon or posterior aspect of medial humeral epicondyle
- Valgus deformity of elbow

MR Findings

- Ulnar nerve
  - Enlarged and ↑ T2 signal intensity compared to normal
    - Highly variable and not very reliable finding
  - Hyperenhancement after gadolinium administration
    - Especially if neurogenic tumor present
  - Medial subluxation if imaged in flexion
- Cubital tunnel contents
  - Enlarged posterior recurrent ulnar artery or vein
  - Lymphadenopathy
- Transverse retinaculum
  - Normal Osborne retinaculum is thin and regular
  - Fibrosis results in thick irregular retinaculum
- Downstream muscle changes
  - Edema or fatty atrophy of flexor digitorum profundus (FDP), flexor carpi ulnaris (FCU), and intrinsic hand muscles

Ultrasonographic Findings

- Increased segmental diameter of ulnar nerve within cubital tunnel
  - Fusiform enlargement seen in long axis
- Ulnar nerve subluxation
  - Nerve pops out of cubital tunnel medially on elbow flexion
- Snapping triceps
  - Medial head of triceps slides into cubital tunnel on elbow flexion, compressing ulnar nerve

Imaging Recommendations

- Best imaging tool
  - MR or ultrasound
- Protocol advice
  - MR: Axial T1WI and T2WI FS MR best for gauging nerve size and extrinsic compression
    - Imaging in flexion may help identify UN subluxation
  - US: Long axis images best for identifying segmental nerve enlargement
    - Real-time evaluation in flexion and extension for nerve subluxation, snapping triceps

DIFFERENTIAL DIAGNOSIS

Normal Ulnar Nerve

- Size and signal of normal UN is highly variable

Posterior Recurrent Ulnar Artery and Vein

- Also present in cubital tunnel, may be mistaken for abnormal nerve

Lymphadenopathy

- Round or elliptical, not elongate like nerve

Other Masses

- Ganglion cyst
- Neoplasm

PATHOLOGY

General Features

- Ulnar nerve anatomy
  - Ulnar nerve derives from C8 and T1 roots and medial cord of brachial plexus
  - In proximal arm, follows axillary and brachial arteries through anterior compartment along medial intermuscular septum
  - In mid-arm, enters posterior compartment through medial intermuscular septum at arcade of Struthers
  - At elbow, enters cubital tunnel deep to Osborne ligament
  - In proximal forearm, passes between 2 heads of flexor carpi ulnaris muscle
  - Ulnar nerve supplies all intrinsic hand muscles except opponens pollicis, superficial head of flexor pollicis brevis (FPB), and lateral 2 lumbricals
  - At level of elbow, ulnar nerve gives off motor branches to FCU and ulnar 1/2 of FDP
- Cubital tunnel anatomy
  - Medial wall = medial humeral epicondyle
  - Lateral wall = olecranon
  - Floor = elbow joint capsule at posterior aspect of ulnar collateral ligament
  - Roof = Osborne ligament
- **Etiology**
  - Overuse
    - Repetitive valgus stress (e.g., throwing athletes)
      - Traction &/or friction injury to nerve

ULNAR NERVE INJURY

      - FCU hypertrophy
  - Subluxation
    - Nerve may suddenly "pop" out of medial epicondylar groove during elbow flexion
    - Likely related to insufficient Osborne ligament fibers or shallow epicondylar groove
  - Extrinsic compression
    - Intra- or perioperative
    - "Sleep palsy"
  - Primary thickening of Osborne ligament
  - Trauma
    - Operating jackhammer
  - Osseous abnormalities
    - Osteophytes
    - Fracture ± malunion
    - Intraarticular bodies
  - Muscle abnormalities
    - Anconeus epitrochlearis
    - Snapping triceps
  - Mass
    - Lymphadenopathy
    - Aneurysm of posterior recurrent ulnar artery
    - Ganglion
    - Osteochondroma
    - Neoplasm
  - Infection
    - Leprosy
- **Associated abnormalities**
  - Anconeus epitrochlearis
  - Snapping triceps muscle

CLINICAL ISSUES

Presentation

- Most common signs/symptoms
  - Compressive neuropathy
    - Medial elbow pain
      - Worse with elbow flexion
    - Numbness and paresthesias in medial hand and fingers
    - Weakness in affected muscles
      - Intrinsic hand muscles, sparing lateral 2 lumbricals, opponens pollicis, and superficial head of FPB
    - Tenderness to palpation at cubital tunnel
    - Claw-like deformity in hand
  - Ulnar nerve subluxation
    - Patient feels "snap" in medial elbow with elbow flexion
    - Symptoms of ulnar neuropathy same as above but worse with snap
    - UN may be felt undergoing medial subluxation with flexion on exam
- Other signs/symptoms
  - Decreased nerve conduction velocity at elbow

Demographics

- Age
  - Usually adults
  - May present in adolescent overhand throwing athletes
- Gender
  - M < F, except in throwing athletes
- Epidemiology
  - 2nd most common upper extremity entrapment neuropathy
  - Abnormalities in UN at cubital tunnel are present in 40% of asymptomatic individuals
  - Static compression of nerve by anconeus epitrochlearis in 10% of population

Natural History & Prognosis

- Slowly progressive symptoms unless related to specific episode of trauma
- Prognosis favorable after nerve decompression

Treatment

- Conservative
- Surgical
  - Resection of olecranon osteophytes
  - Resection of anconeus epitrochlearis
  - In situ decompression (release of Osborne ligament)
  - Ulnar nerve transposition
    - Nerve moved anterior and medial from cubital tunnel
    - May be placed deep to pronator teres muscle or in subcutaneous location
      - Subcutaneous placement: More prone to direct trauma
      - Submuscular placement: More prone to muscular impingement; adjacent to median nerve

DIAGNOSTIC CHECKLIST

Consider

- Previous ulnar nerve transposition surgery if nerve not seen in cubital tunnel

Image Interpretation Pearls

- Search for anconeus epitrochlearis as possible source of ulnar neuropathy

SELECTED REFERENCES

1. Cummins CA et al: Peripheral nerve injuries in baseball players. Neurol Clin. 26(1):195-215; x, 2008
2. Gellman H: Compression of the ulnar nerve at the elbow: cubital tunnel syndrome. Instr Course Lect. 57:187-97, 2008
3. Elhassan B et al: Entrapment neuropathy of the ulnar nerve. J Am Acad Orthop Surg. 15(11):672-81, 2007
4. Szabo RM et al: Natural history and conservative management of cubital tunnel syndrome. Hand Clin. 23(3):311-8, v-vi, 2007
5. Andreisek G et al: Peripheral neuropathies of the median, radial, and ulnar nerves: MR imaging features. Radiographics. 26(5):1267-87, 2006
6. Martinoli C et al: Sonography of entrapment neuropathies in the upper limb (wrist excluded). J Clin Ultrasound. 32(9):438-50, 2004
7. Grana W: Medial epicondylitis and cubital tunnel syndrome in the throwing athlete. Clin Sports Med. 20(3):541-8, 2001
8. Folberg CR et al: Cubital tunnel syndrome. Part I: Presentation and diagnosis. Orthop Rev. 23(2):136-44, 1994
9. Folberg CR et al: Cubital tunnel syndrome. Part II: Treatment. Orthop Rev. 23(3):233-41, 1994

ULNAR NERVE INJURY

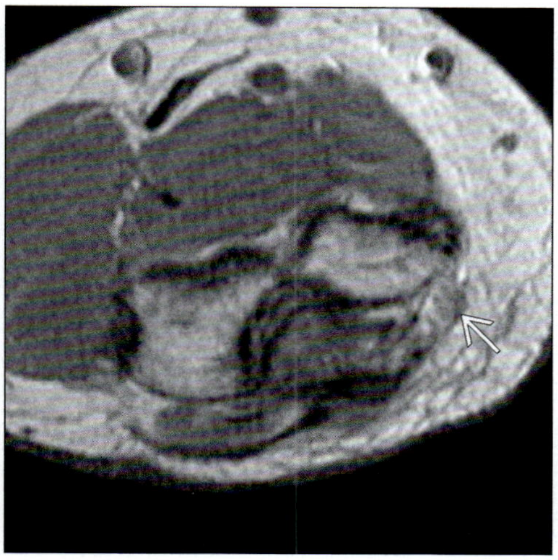

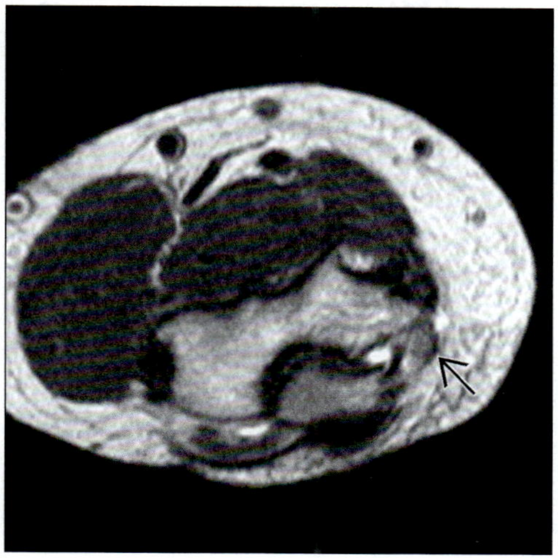

(Left) *Axial PDWI MR of the elbow in a patient with ulnar neuropathy after direct trauma to the elbow demonstrates mild enlargement of the ulnar nerve proximal to the cubital tunnel.* ***(Right)*** *Axial T2WI MR at the same level in the same patient shows enlargement and increased signal within the ulnar nerve. The main role for MR imaging in cases such as this is to exclude treatable causes of nerve compression, such as fracture or hematoma.*

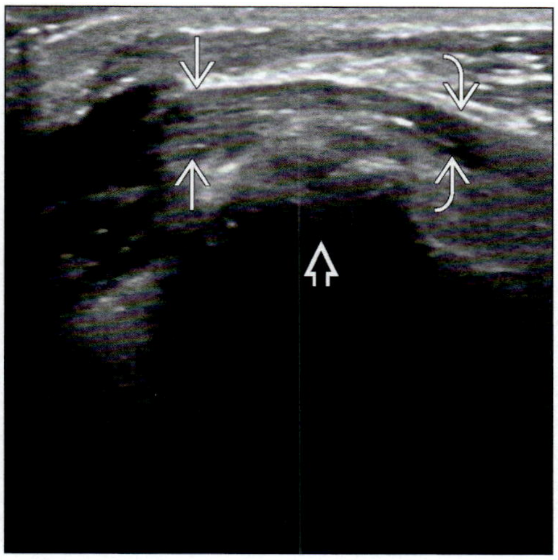

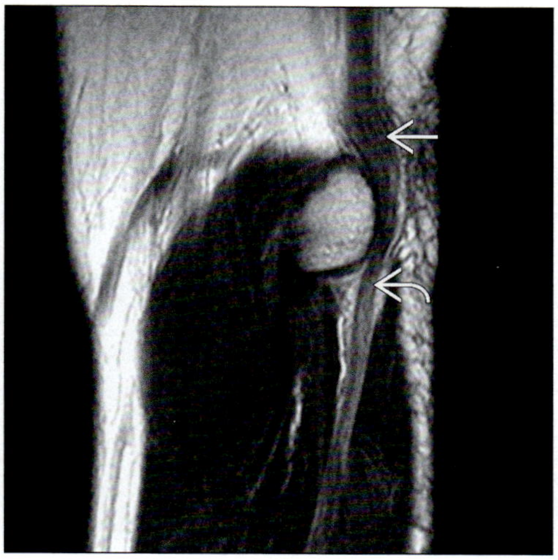

(Left) *Coronal ultrasound of the cubital tunnel in a patient with ulnar neuropathy demonstrates relative enlargement of the nerve proximal to the cubital tunnel. The nerve demonstrates normal diameter in the distal aspect of the cubital tunnel. Medial humeral epicondyle is seen.* ***(Right)*** *Sagittal T2WI MR shows enlargement of the ulnar nerve in a different patient with ulnar neuropathy. Diameter of the nerve proximal to the cubital tunnel is larger than it is distally.*

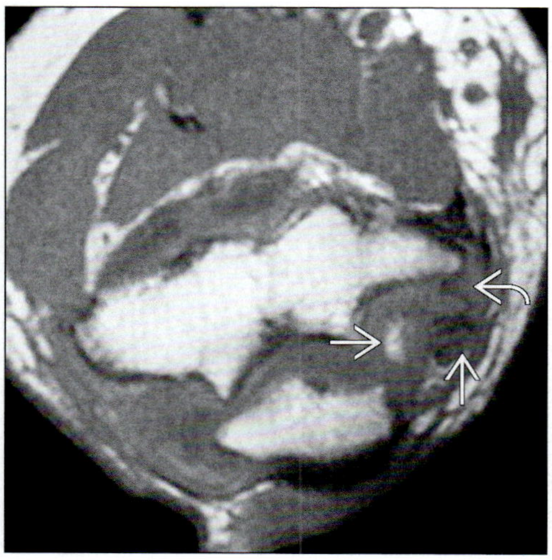

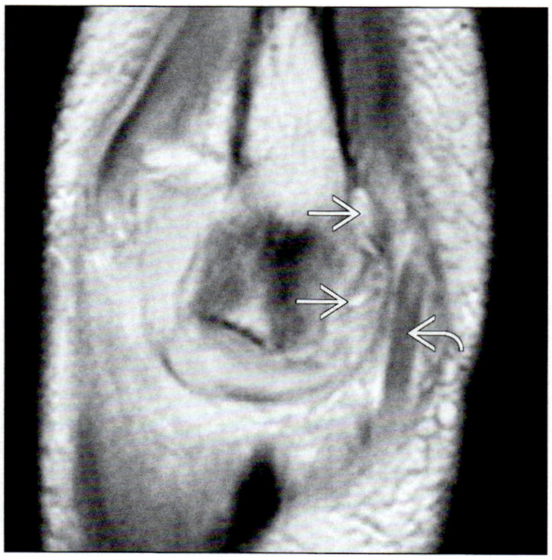

(Left) *Axial PDWI MR of the elbow in a patient with remote history of elbow fracture and ulnar neuropathy shows a collection of ossific fragments posteromedial to the elbow, distorting and displacing the ulnar nerve.* ***(Right)*** *Coronal PDWI MR through the posterior elbow in the same patient shows ossific fragments displacing the ulnar nerve medially. The nerve also appears swollen, but this is a somewhat subjective finding.*

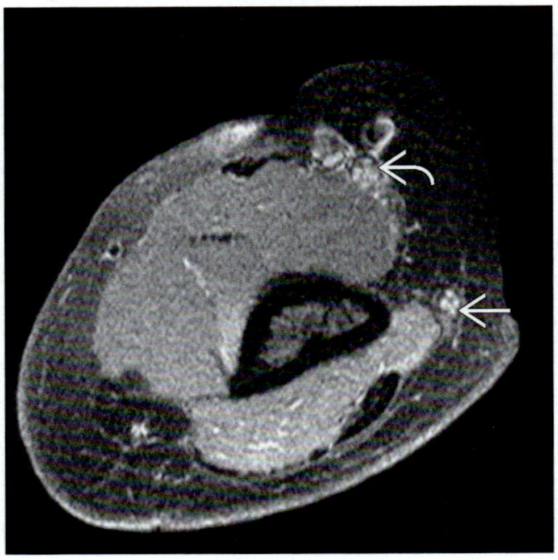

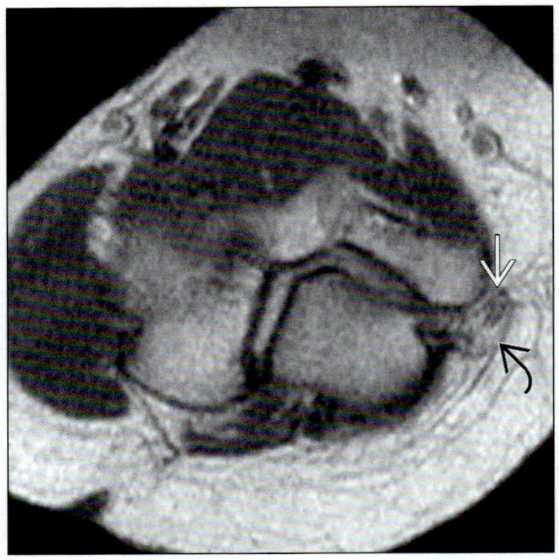

(Left) *Axial PDWI FS MR of the distal arm just above the elbow in a patient with median and ulnar neuropathy after trauma demonstrates elevated signal and enlargement in the median and ulnar nerves.* ***(Right)*** *Axial PDWI MR in a patient with snapping elbow symptoms demonstrates medial subluxation of the ulnar nerve within the cubital tunnel. The ligament of Osborne appears redundant, possibly due to chronic tension or old injury.*

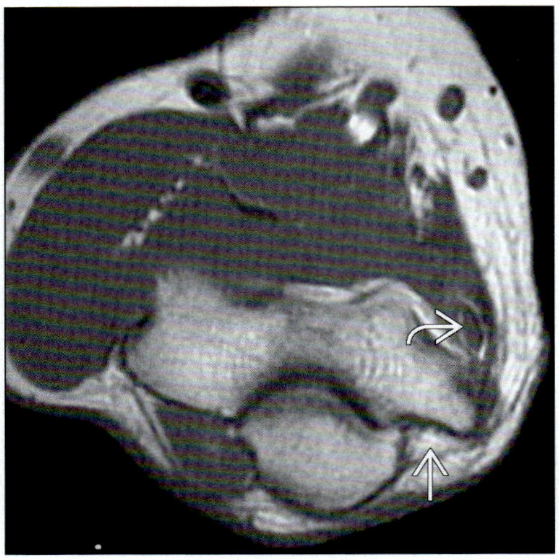

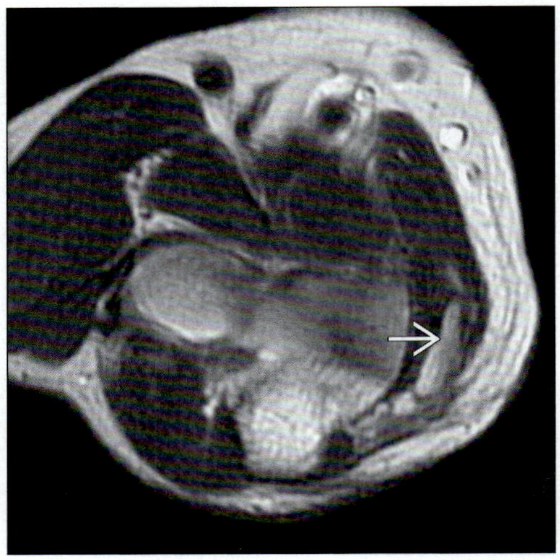

(Left) *Axial PDWI MR in a patient with recurrent ulnar neuropathy after ulnar nerve transposition surgery shows the intramuscular course of the nerve. The cubital tunnel, the normal location of the nerve, is empty.* ***(Right)*** *Axial T2WI MR in the same patient more distally demonstrates high signal within the swollen repositioned nerve. This is a known complication of intramuscular placement and may be due to compression by the flexor muscles.*

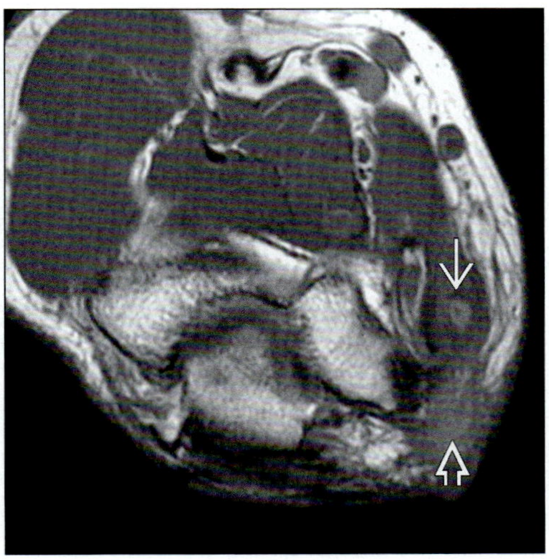

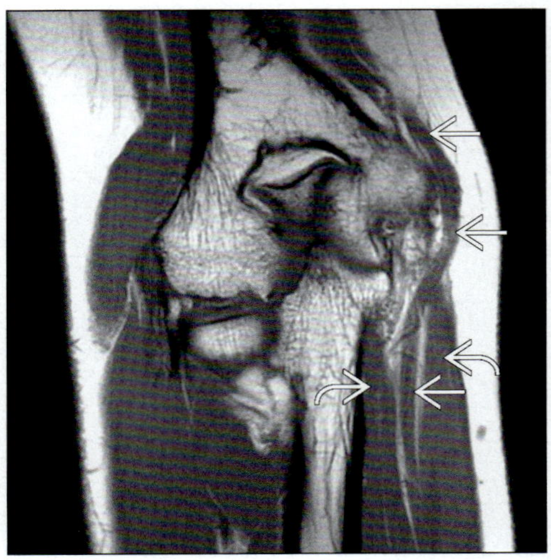

(Left) *Axial PDWI MR in a patient with ulnar neuropathy soon after ulnar nerve transposition surgery demonstrates a small postoperative hematoma at the site of cubital tunnel release. The swollen nerve is present within the pronator teres muscle.* ***(Right)*** *Coronal T1WI MR in the same patient demonstrates the course of the transposed nerve through the pronator teres muscle. Intramuscular placement of the nerve is useful in patients with little protective subcutaneous fat.*

ULNAR NERVE INJURY

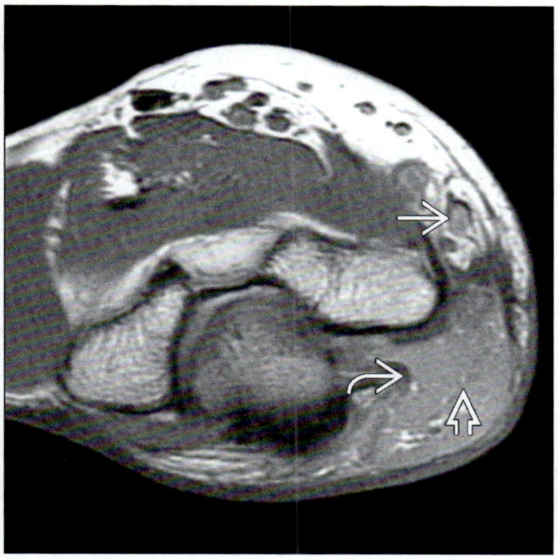

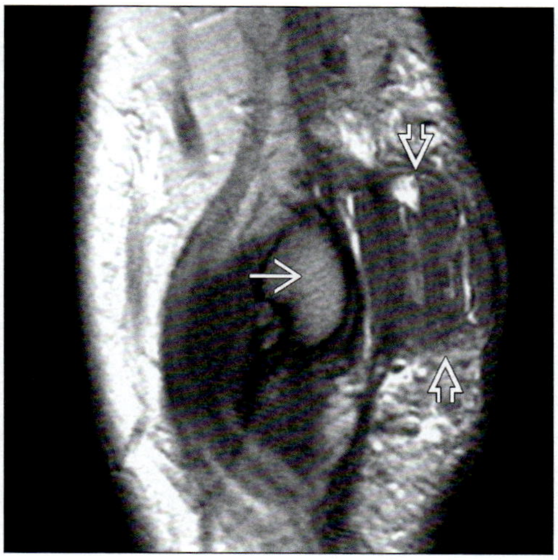

(Left) *Axial PDWI MR in a patient with ulnar neuropathy soon after ulnar nerve transposition surgery demonstrates a postoperative hematoma at the site of cubital tunnel release . The swollen nerve is present within the subcutaneous fat of the elbow. This represents the more common subcutaneous ulnar nerve transposition.* ***(Right)*** *Sagittal T2WI MR in the same patient shows the heterogeneous hematoma posterior to the medial epicondyle .*

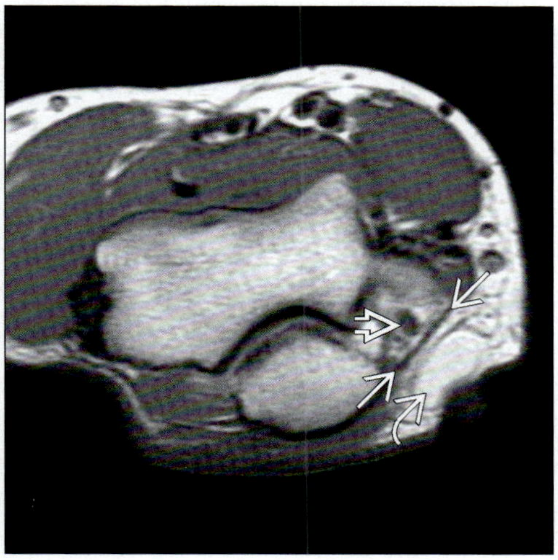

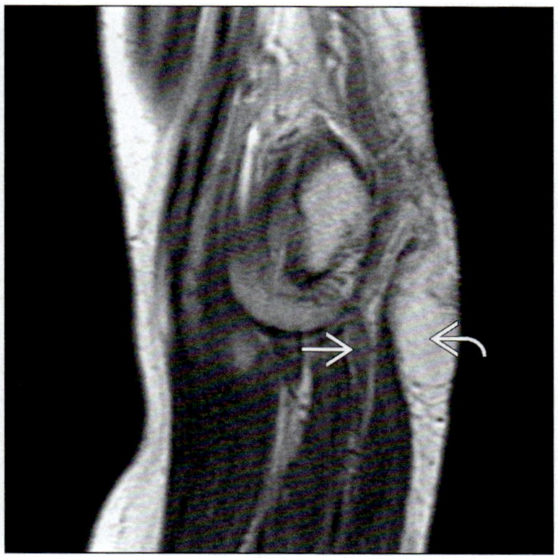

(Left) *Axial PDWI MR in a patient with ulnar neuropathy demonstrates a small subcutaneous lipoma superficial to the ligament of Osborne . The ulnar nerve demonstrates normal size and signal within the cubital tunnel. The patient's symptoms improved after resection of the lipoma.* ***(Right)*** *Sagittal T2WI MR in the same patient shows the lipoma possibly compressing the ulnar nerve within the cubital tunnel.*

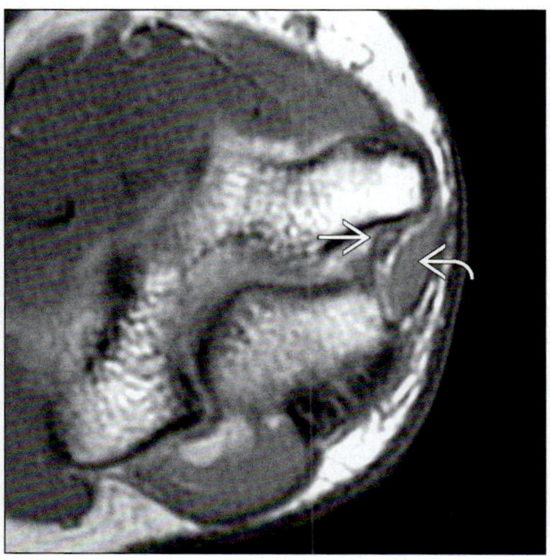

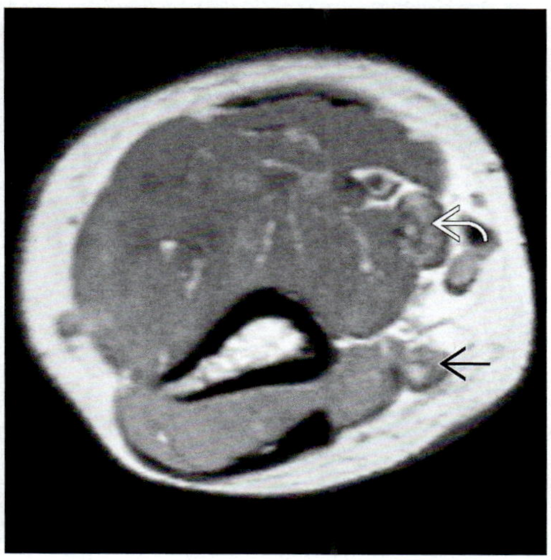

(Left) *Axial PDWI MR demonstrates a large anconeus epitrochlearis muscle superficial to the cubital tunnel. The ulnar nerve is normal, but the anomalous muscle can be a cause of extrinsic compression of the ulnar nerve in this location.* ***(Right)*** *Axial PDWI MR demonstrates neurofibromas of the median and ulnar nerves. This patient had neurofibromatosis and complained of median and ulnar neuropathy. Neoplasm is a relatively rare cause of ulnar neuropathy.*

MEDIAN NERVE INJURY

Key Facts

Terminology
- Anterior interosseous nerve (AIN) syndrome, Kiloh-Nevin syndrome, pronator syndrome, honeymooner's paralysis, supracondylar process syndrome

Imaging
- Site of impingement rarely identified on MR
- Rarely, mass or inflammation seen in course of median nerve or AIN to explain symptoms
- T2WI or PDWI FS MR in axial plane for detection of muscle edema ± atrophy

Pathology
- AIN syndrome
  - Tendinous origin of deep head of pronator teres (most common)
- Pronator syndrome
  - Entrapment between superficial and deep heads of pronator teres muscle
- May occasionally be due to
  - Extrinsic mass
  - Direct pressure ("honeymooner's paralysis")
- Supracondylar process syndrome
  - Median nerve compressed by ligament of Struthers

Clinical Issues
- Conservative therapy: ± effective
- Decompression surgery for severe or refractory cases

Diagnostic Checklist
- Median neuropathy in patients with supracondylar process
- Subtle muscle denervation in appropriate distribution suggests median neuropathy
- Exclude identifiable causes of nerve compression

(Left) Graphic illustration of inflammation of the median nerve at the proximal aspect of the pronator teres muscle, which is a common site of compression. This location is also near the lacertus fibrosus, or bicipital aponeurosis. Also seen are the brachial artery and the radial nerve. (Right) Coronal PDWI FS MR of the proximal forearm and antecubital fossa shows edema in the pronator teres due to compression of the median nerve.

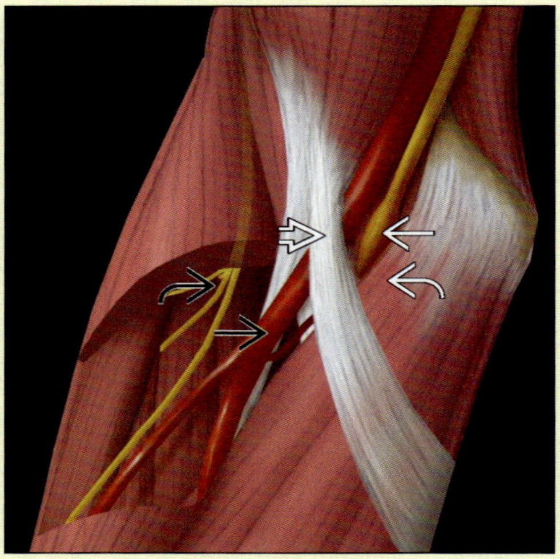

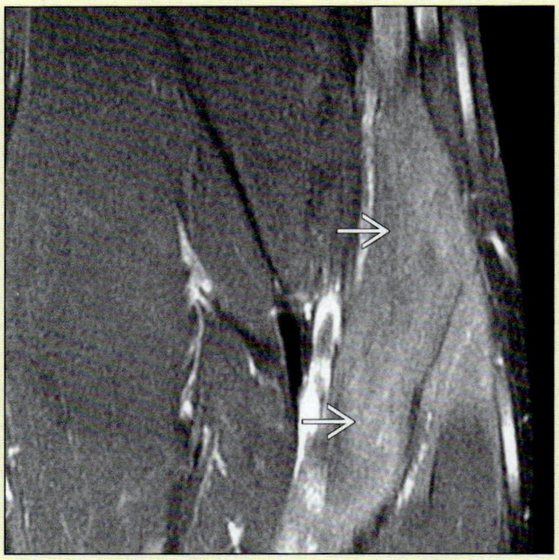

(Left) Graphic illustration of a cross-section of the proximal forearm shows an inflamed median nerve deep to the pronator teres muscle. (Right) Axial PDWI FS MR more distally in the same patient shows signal changes of denervation in the pronator teres, flexor carpi radialis, and a portion of the flexor digitorum superficialis muscles of the proximal forearm. Note that the flexor carpi ulnaris muscle, innervated by the ulnar nerve, is normal.

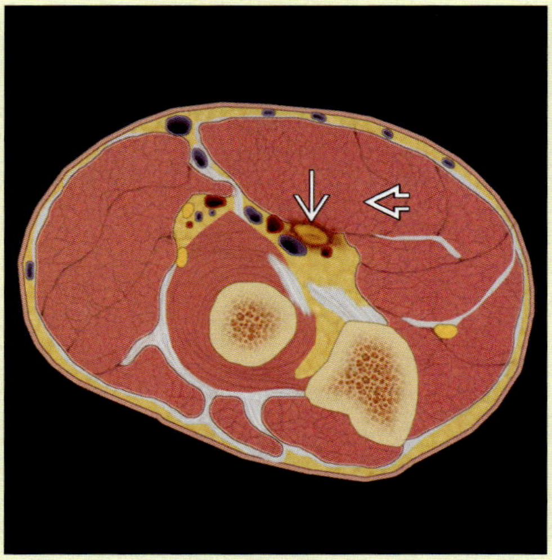

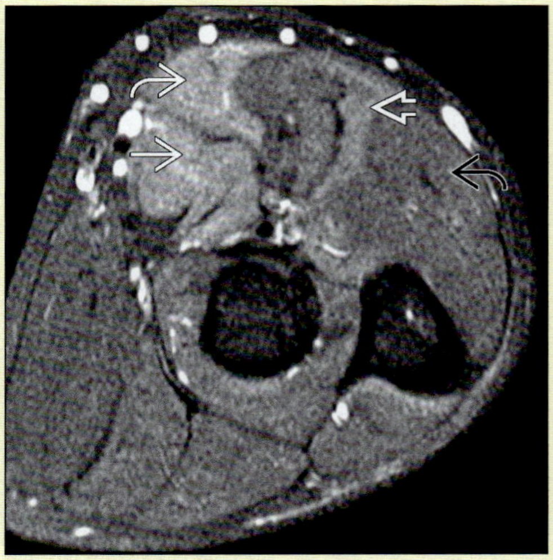

MEDIAN NERVE INJURY

TERMINOLOGY

Synonyms

- Anterior interosseous nerve (AIN) syndrome, Kiloh-Nevin syndrome, pronator syndrome, honeymooner's paralysis, supracondylar process syndrome

Definitions

- AIN syndrome
  - Denervation of flexor pollicis longus (FPL) and flexor digitorum profundus (FDP) muscles of 2nd and 3rd digits
  - ± pronator quadratus muscle involvement
  - Also called Kiloh-Nevin syndrome
- Pronator syndrome
  - Pain ± paresthesias in median nerve distribution
  - Motor nerve involvement is rare
- Supracondylar process syndrome
  - Rare median neuropathy caused by compression of nerve by ligament of Struthers arising from supracondylar process of humerus

IMAGING

General Features

- Best diagnostic clue
  - Denervation of affected muscles on MR
- Location
  - Injury or compression of median nerve at
    - Mid-arm
    - Distal humerus (supracondylar process/ligament of Struthers)
    - Pronator teres muscle

Radiographic Findings

- Midshaft humeral (Holstein-Lewis) fracture
- Supracondylar process of humerus (avian spur)

CT Findings

- Atrophy of affected muscles
- May see course of median nerve in relation to humeral fracture
- May be useful to evaluate subtle fracture in supracondylar process

MR Findings

- T1WI
  - Early: No T1WI MR changes
  - Late: Fat atrophy of muscles supplied by median or anterior interosseous nerves
- T2WI FS
  - Early: Swelling and ↑ signal in affected muscles
  - Late: Atrophy of affected muscles, usually with residual ↑ signal
- Site of impingement rarely identified on MR
  - Nerve size and signal usually unaltered
  - Pressure on nerve by normal structures not well delineated
  - Diagnosis made by indirect findings
- Rarely, mass or inflammation seen in course of median nerve or AIN to explain symptoms

Imaging Recommendations

- Best imaging tool
  - MR
- Protocol advice
  - T2WI or PDWI FS MR in axial plane for detection of muscle edema ± atrophy
    - May need to image down to level of wrist to include all affected muscles
  - T1WI in axial plane to detect fatty atrophy of muscle

DIFFERENTIAL DIAGNOSIS

Traumatic Muscle Edema

- Not in typical nerve distribution

Brachial Neuritis/Parsonage-Turner Syndrome

- May be difficult to differentiate

PATHOLOGY

General Features

- Associated abnormalities
  - Supracondylar process
    - Osseous excrescence arising anteriorly from distal humeral shaft
    - Rare in humans; common in reptiles, birds, and some mammals
    - Ligament of Struthers runs from supracondylar process to medial humeral epicondyle
- **Median nerve**
  - Arises from medial and lateral cords of brachial plexus
  - Includes contributions from C6-T1 roots
  - Follows course of axillary artery anterior to brachialis muscle
  - Enters antecubital fossa between pronator teres deep and superficial heads
  - Innervates intrinsic hand muscles
    - Abductor pollicis brevis
    - Opponens pollicis
    - Superficial head of flexor pollicis brevis
    - 1st and 2nd lumbricals
  - Innervates forearm muscles
    - Pronator teres
    - Flexor carpi radialis
    - Palmaris longus
    - Flexor digitorum superficialis
- **Anterior interosseous nerve (antebrachial nerve)**
  - Largest branch of median nerve
  - Course
    - Arises from median nerve just distal to pronator teres muscle
    - Passes between 2 heads of pronator teres muscle
    - Runs along volar surface of flexor digitorum profundus at interosseous membrane
    - Terminates in pronator quadratus muscle
  - Motor branch
    - Supplies FPL, pronator quadratus, and radial aspect of FDP muscles
  - Deep pain and proprioception fibers in radiocarpal, radioulnar, intercarpal, and carpometacarpal joints
  - No cutaneous sensory function
- **Martin-Gruber anastomosis**
  - Motor-only connection between median nerve or AIN and ulnar nerve

MEDIAN NERVE INJURY

  - Runs from median nerve or AIN in mid-forearm to ulnar nerve in distal forearm
  - Innervates
    - 1st dorsal interosseous muscle
    - Adductor pollicis
    - Abductor digiti minimi
- Present in 10-15% of population

Staging, Grading, & Classification

- **AIN syndrome**
  - Compression of AIN caused by
    - Tendinous origin of deep head of pronator teres (most common)
    - Enlarged bicipitoradial bursa
    - Aberrant or thrombosed radial artery in mid-forearm
    - Fascial band at origin of FDS
- **Pronator syndrome**
  - Entrapment of median nerve at 1 of 3 sites
    - Between superficial and deep heads of pronator teres muscle
    - At bicipital aponeurosis (lacertus fibrosus)
    - At origin of FDS
  - Usually due to either congenital variation in these structures, or acquired hypertrophy
  - May occasionally be due to
    - Extrinsic mass
    - Direct pressure ("honeymooner's paralysis")
    - Elbow fracture (Volkmann contracture)
- **Supracondylar process syndrome**
  - Median nerve compressed by ligament of Struthers
  - Occasionally due to fracture of supracondylar process

CLINICAL ISSUES

Presentation

- Most common signs/symptoms
  - **AIN syndrome**
    - Motor weakness in FPL, FDP of index and long fingers, ± pronator quadratus
      - Often acute onset
      - Circle sign: Patient cannot make "O" with thumb and index finger
      - Muscle atrophy compared to contralateral extremity
    - Pain along course of median nerve in volar forearm
    - No numbness
  - **Pronator syndrome**
    - Pain and numbness in volar forearm and hand
    - No weakness
    - May have pain on palpation of pronator teres
    - ± Tinel sign
    - EMG may be normal, or may show atrophy of pronator, flexor carpi radialis, and FDS
  - **Supracondylar process syndrome**
    - Very rare
    - Paresthesias and numbness in hand
      - Exacerbated by elbow extension
- Clinical profile
  - Prolonged pressure on arm or forearm
    - Heavy alcohol or drug use: Due to unconscious patient lying on arm for a prolonged period
    - Honeymooner's paralysis: Due to pressure of head of lover against patient arm while sleeping
    - Trauma
  - Overuse syndromes
    - Violinists
    - Machinists
    - Mechanics

Demographics

- Age
  - Children: Usually related to trauma
  - Adults: Repetitive motion injury
- Epidemiology
  - Unusual neuropathy

Natural History & Prognosis

- Symptoms may improve with cessation of offending activity

Treatment

- Conservative therapy: ± effective
- Decompression surgery for severe or refractory cases

DIAGNOSTIC CHECKLIST

Consider

- Median neuropathy in patients with supracondylar process

Image Interpretation Pearls

- Subtle muscle denervation in appropriate distribution suggests median neuropathy
- Exclude identifiable causes of nerve compression

SELECTED REFERENCES

1. Tsai P et al: Median and radial nerve compression about the elbow. Instr Course Lect. 57:177-85, 2008
2. Kim S et al: Role of magnetic resonance imaging in entrapment and compressive neuropathy--what, where, and how to see the peripheral nerves on the musculoskeletal magnetic resonance image: part 2. Upper extremity. Eur Radiol. 17(2):509-22, 2007
3. Andreisek G et al: Peripheral neuropathies of the median, radial, and ulnar nerves: MR imaging features. Radiographics. 26(5):1267-87, 2006
4. Lee MJ et al: Pronator syndrome and other nerve compressions that mimic carpal tunnel syndrome. J Orthop Sports Phys Ther. 34(10):601-9, 2004
5. Martinoli C et al: Sonography of entrapment neuropathies in the upper limb (wrist excluded). J Clin Ultrasound. 32(9):438-50, 2004
6. Lubahn JD et al: Uncommon nerve compression syndromes of the upper extremity. J Am Acad Orthop Surg. 6(6):378-86, 1998
7. Dawson DM: Entrapment neuropathies of the upper extremities. N Engl J Med. 329(27):2013-8, 1993
8. Eversmann WW: Proximal median nerve compression. Hand Clin. 8(2):307-15, 1992
9. Werner CO: The anterior interosseous nerve syndrome. Int Orthop. 13(3):193-7, 1989

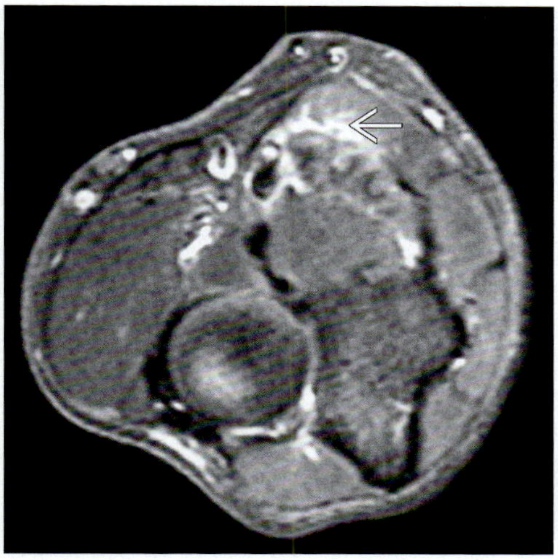

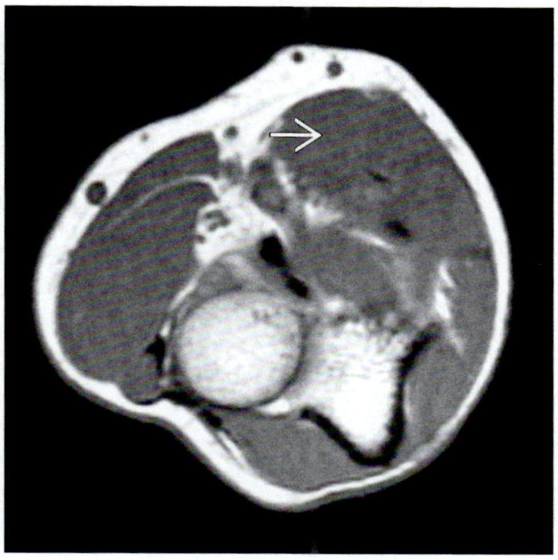

(Left) *Axial T2WI FS MR of the proximal forearm shows mild elevated signal indicating edema in the pronator teres muscle, a finding due to median nerve impingement. The patient had suffered a midshaft humeral fracture 2 weeks earlier.* ***(Right)*** *Axial T1WI MR of the same patient shows no atrophy of the pronator teres muscle due to the relatively short time course of the median nerve impingement. Over a period of time, fatty replacement of muscle fibers becomes evident.*

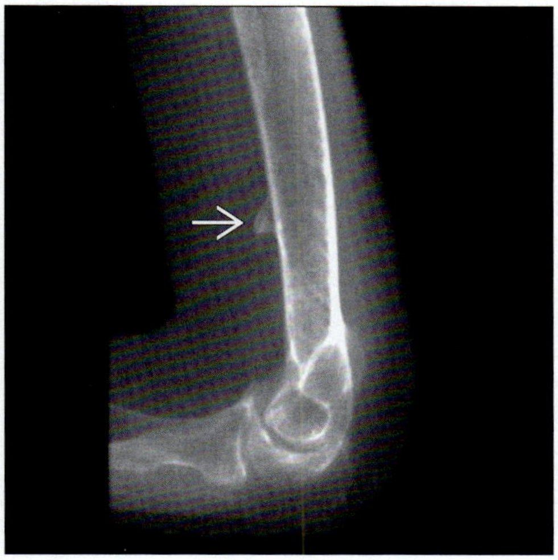

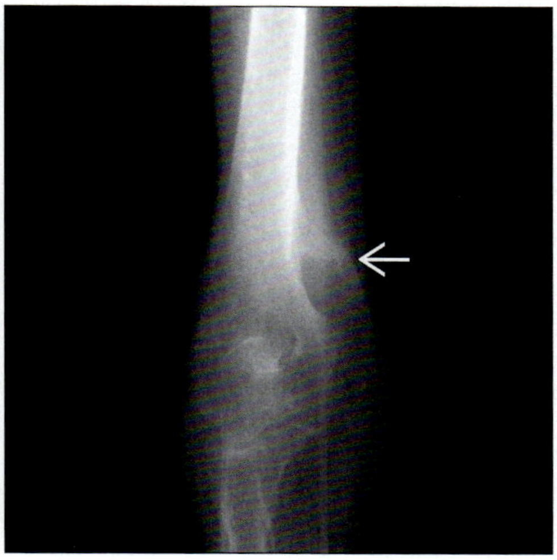

(Left) *Lateral radiograph demonstrates a well-defined osseous excrescence arising from the anterior humeral cortex. A fibrous band, the ligament of Struthers, may extend from the tip of this process to the medial humeral condyle. The median nerve passes under the arcade formed by this ligament, and contraction of the adjacent pronator teres can compress the nerve and result in median neuropathy.* ***(Right)*** *Anteroposterior radiograph in the same patient demonstrates the supracondylar process.*

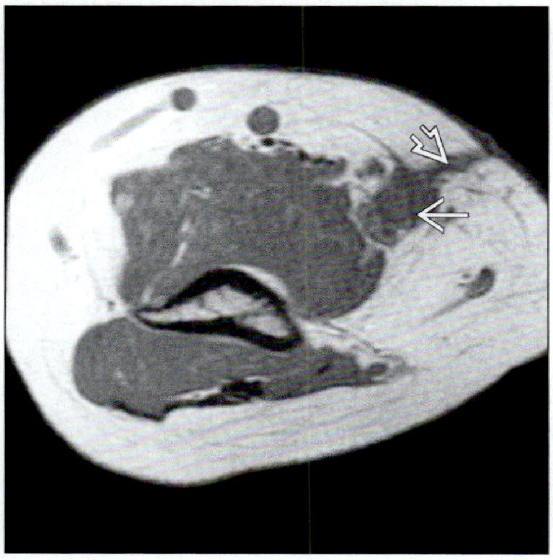

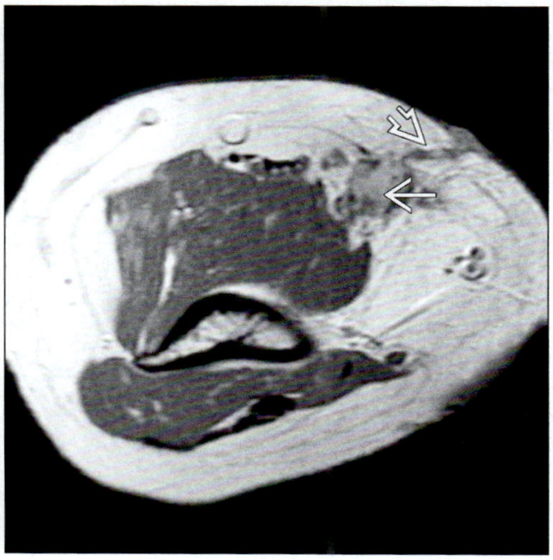

(Left) *Axial T1WI MR of the distal arm in a patient status post resection of a malignant schwannoma of the median nerve demonstrates a recurrent mass in the distribution of the median nerve. Subcutaneous scar from previous resection is present.* ***(Right)*** *Axial T1WI C+ MR in the same patient at the same level shows enhancement of the recurrent tumor mass and the surgical scar. Tumor is a rare cause of median neuropathy; most cases are due to compression by normal structures.*

RADIAL NERVE INJURY

Key Facts

Terminology

- Radial tunnel syndrome, posterior interosseous nerve (PIN) syndrome

Imaging

- Denervation changes in muscles supplied by radial nerve or posterior interosseous nerve
  - Early: Swelling and ↑ signal on T2WI in affected muscles
  - Later: Atrophy of affected muscles, usually with residual ↑ signal
- Site of nerve impingement rarely recognized on MR

Pathology

- Upper radial nerve compression: Humeral fracture
- Radial tunnel syndrome (RTS)/PIN syndrome
  - Anterolateral proximal forearm pain (more distal than "lateral epicondylitis")
  - No sensory deficit
  - Due to overuse injury
  - Various possible compression sites
- Wartenberg syndrome
  - Selective compression of radial sensory nerve
  - Distal radial forearm pain

Clinical Issues

- Conservative therapy usually ineffective for RTS, PIN syndrome

Diagnostic Checklist

- RTS with "lateral epicondylitis" but normal MR
- Actual nerve compression/abnormality rarely seen
- Downstream effects (muscle edema ± atrophy must be appreciated)
- Reporting tip: Try to narrow down likely site of impingement based on muscle groups affected

__(Left)__ Graphic shows radial nerve in the antecubital fossa. Above the radial tunnel, the nerve bifurcates into the radial sensory nerve, which runs more superficially, and the deep motor nerve, which becomes the posterior interosseous nerve. __(Right)__ Coronal T1WI MR shows the bifurcation of the radial nerve into the radial sensory nerve and the deep motor branch as it enters the radial tunnel. Portions of the leash of Henry are seen.

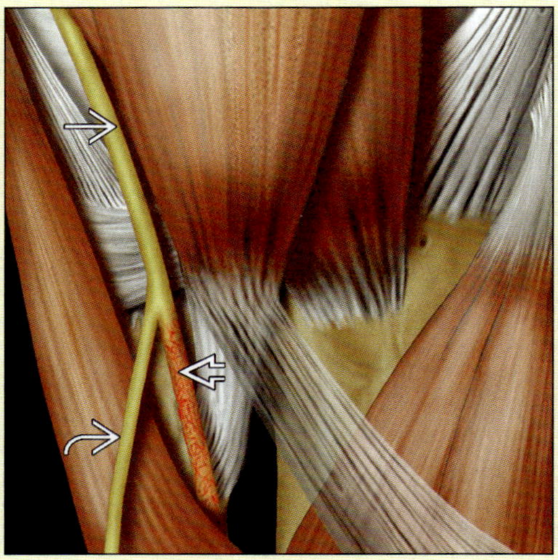

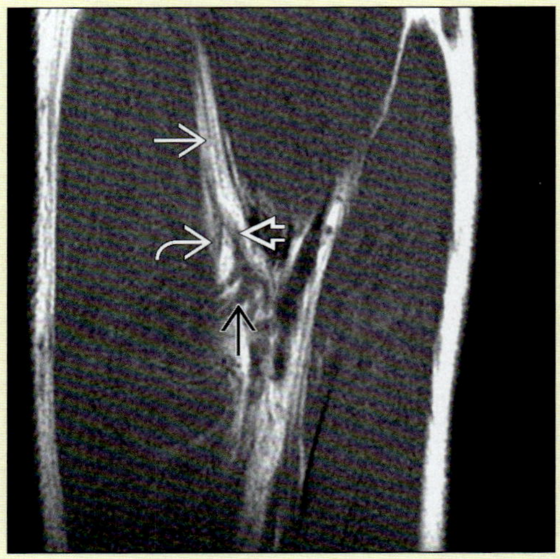

__(Left)__ Coronal T1WI MR slightly more posteriorly demonstrates the PIN surrounded by vessels of the leash of Henry as they emerge distal to the superficial head of the supinator. __(Right)__ Axial PDWI MR at the level of the radial neck shows the deep motor branch of the radial nerve passing through the space between the deep and superficial heads of the supinator muscle in the radial tunnel. Small vessels of the leash of Henry are visible adjacent to the nerve.

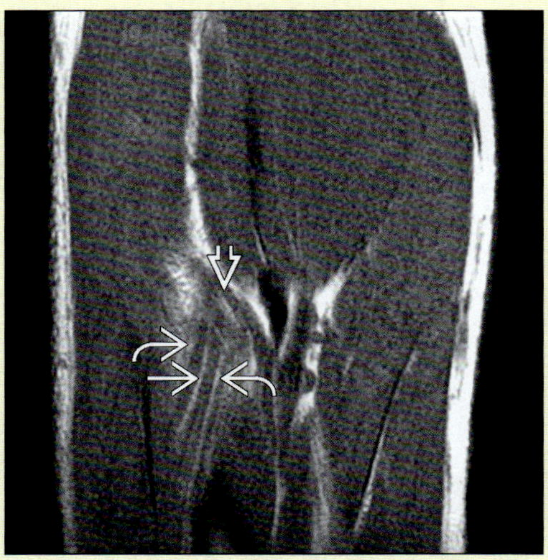

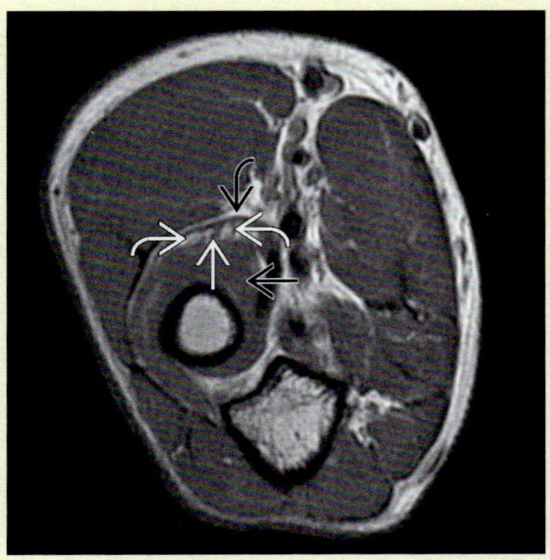

RADIAL NERVE INJURY

TERMINOLOGY

Abbreviations

- Radial tunnel syndrome (RTS)
- Posterior interosseous nerve (PIN)

Synonyms

- Radial tunnel syndrome, radial pronator syndrome, posterior interosseous nerve syndrome

Definitions

- Compression neuropathy of radial nerve or its branches in elbow region

IMAGING

General Features

- Best diagnostic clue
  - Denervation changes in muscles supplied by radial nerve or posterior interosseous nerve
- Location
  - Radial nerve (RN) can be compressed at several sites
    - High radial lesion
      - Above takeoff of posterior interosseous nerve
    - Radial tunnel (affects PIN)
    - Proximal forearm
      - Affecting medial or lateral branches of PIN, or
      - Radial sensory nerve (RSN)

Radiographic Findings

- Usually none

CT Findings

- Atrophy of affected muscles

MR Findings

- T1WI
  - Early: No T1WI MR changes
  - Later: Fat atrophy of muscles supplied by radial nerve or its branches (chiefly PIN)
- T2WI FS
  - Early: Swelling and ↑ signal on T2WI in affected muscles
  - Later: Atrophy of affected muscles, usually with residual ↑ signal
- Site of nerve impingement rarely recognized on MR
  - Nerve size and signal usually unaltered
  - Pressure on nerve by normal structures not well delineated
  - Diagnosis made by indirect findings
    - Atrophy or edema of affected muscles
- Rarely, mass seen in course of RN to explain symptoms

Imaging Recommendations

- Best imaging tool
  - MR
- Protocol advice
  - Axial T1WI and T2WI FS MR for evaluation of nerve &/or muscle edema and atrophy

DIFFERENTIAL DIAGNOSIS

Extensor Tendinopathy

- Pain located at level of lateral humeral epicondyle instead of distal to it
- Degenerative change or disruption in extensor tendon on MR

Focal Muscle Injury

- Does not usually follow nerve distribution
- Usually attributable to traumatic episode
- Symptoms time-limited

PATHOLOGY

General Features

- **Course of radial nerve**
  - RN arises from posterior cord of brachial plexus (C5-T1)
  - At cubital fossa, RN passes between brachialis and brachioradialis muscles
  - RN then divides into RSN and deep motor branch at level of supinator muscle
  - Motor branch enters radial tunnel (RT) proximal to radiocarpal joint
  - Passes deep to fascia, adjacent to radiocapitellar joint capsule
  - Passes through Leash of Henry (radial recurrent artery and venae comitantes)
  - Passes deep to fibrous edge of extensor carpi radialis brevis (ECRB)
  - Passes through arcade of Frohse (tendinous origin of supinator)
    - Present in 30-50% of population
  - Passes between superficial and deep heads of supinator muscle
  - On exiting RT under fascial bands of supinator, nerve referred to as PIN
  - PIN then divides into medial and lateral branches
- **Radial tunnel anatomy**
  - Floor: Radiocapitellar joint
  - Roof and lateral wall: Extensor carpi radialis longus (ECRL) and ECRB
  - Medial wall: Brachialis muscle proximally, biceps tendon distally

Staging, Grading, & Classification

- **Upper radial nerve compression**
  - Humeral fracture
    - Midshaft (Holstein-Lewis fracture)
    - Junction of middle and distal 1/3
  - Lateral intermuscular septum
  - Fibrous arch lateral head of triceps
  - Accessory subscapularis-teres-latissimus muscle
- **Radial tunnel syndrome (RTS) (or PIN syndrome)**
  - Due to overuse injury
  - 5 common compression sites
    - Tendon of ECRB
    - Fibrous bands at radiocapitellar joint
    - Leash of Henry
    - Arcade of Frohse (most common site of RN impingement)
    - Fibrous bands at and distal to distal supinator
- **Wartenberg syndrome**

RADIAL NERVE INJURY

- Selective compression of radial sensory nerve

CLINICAL ISSUES

Presentation

- Most common signs/symptoms
  - **High radial lesion**
    - Loss of functional forearm supination and flexion
    - Loss of wrist and finger extension
    - Sensory loss dorsal thumb, index and long fingers, and radial aspect of ring finger
  - **Radial tunnel/PIN syndrome**
    - Anterolateral proximal forearm pain (more distal than "lateral epicondylitis")
    - Symptoms worse with elbow extension and forearm pronation
    - Resisted active forearm supination and extension of long finger → pain
    - No sensory deficit
    - Loss of extension at MCP joints of thumb and fingers
    - Radial deviation of wrist with extension
    - Selective medial branch compression causes paralysis of
      - Extensor carpi ulnaris
      - Extensor digiti quinti
      - Extensor digitorum communis
    - Selective lateral branch compression causes paralysis of
      - Abductor pollicis longus
      - Extensor pollicis brevis
      - Extensor pollicis longus
      - Extensor indicis proprius
    - Compression above bifurcation causes combination of above
  - **Wartenberg syndrome**
    - Distal radial forearm pain
    - Dorsal radial hand paresthesias
    - Symptoms worse when tightly pinching thumb and index finger together
    - Tinel sign over radial sensory nerve
    - Hyperpronation of forearm → Tinel sign
- Other signs/symptoms
  - High percentage of patients with Wartenberg syndrome also have de Quervain tenosynovitis
- Clinical profile
  - Often overuse syndrome
    - Violinists
    - Laborers
  - Compression
    - Crutch use
    - Radial head fracture
  - Extrinsic mass
    - Ganglion
    - Lipoma
    - Arthritis (rheumatoid, septic)
    - Synovial osteochondromatosis

Demographics

- Age
  - Children and adults
- Gender
  - M > F
- Epidemiology
  - Least common of neuropathies of upper extremity

Natural History & Prognosis

- Usually progressive
- Usually requires surgical decompression (rarely spontaneously improves)
- Excellent prognosis if recovery of function occurs in less than 6 months

Treatment

- Conservative therapy usually ineffective for RTS, PIN syndrome
  - Often effective for Wartenberg syndrome
- Surgery
  - Release at site(s) of compression
    - Typically supinator muscle

DIAGNOSTIC CHECKLIST

Consider

- RTS in patient with "lateral epicondylitis" but normal MR

Image Interpretation Pearls

- Actual nerve compression or nerve abnormality rarely seen
- Downstream effects (muscle edema ± atrophy must be appreciated)

Reporting Tips

- Try to narrow down likely site of impingement based on muscle groups affected

SELECTED REFERENCES

1. Huisstede B et al: Interventions for treating the radial tunnel syndrome: a systematic review of observational studies. J Hand Surg Am. 33(1):72-8, 2008
2. Kim S et al: Role of magnetic resonance imaging in entrapment and compressive neuropathy--what, where, and how to see the peripheral nerves on the musculoskeletal magnetic resonance image: part 2. Upper extremity. Eur Radiol. 17(2):509-22, 2007
3. Andreisek G et al: Peripheral neuropathies of the median, radial, and ulnar nerves: MR imaging features. Radiographics. 26(5):1267-87, 2006
4. Stanley J: Radial tunnel syndrome: a surgeon's perspective. J Hand Ther. 19(2):180-4, 2006
5. Kijowski R et al: Magnetic resonance imaging of the elbow. Part II: Abnormalities of the ligaments, tendons, and nerves. Skeletal Radiol. 34(1):1-18, 2005
6. Bordalo-Rodrigues M et al: MR imaging of entrapment neuropathies at the elbow. Magn Reson Imaging Clin N Am. 12(2):247-63, vi, 2004
7. Martinoli C et al: Sonography of entrapment neuropathies in the upper limb (wrist excluded). J Clin Ultrasound. 32(9):438-50, 2004
8. Plate AM et al: Compressive radial neuropathies. Instr Course Lect. 49:295-304, 2000
9. Barnum M et al: Radial tunnel syndrome. Hand Clin. 12(4):679-89, 1996
10. Kleinert JM et al: Radial nerve entrapment. Orthop Clin North Am. 27(2):305-15, 1996

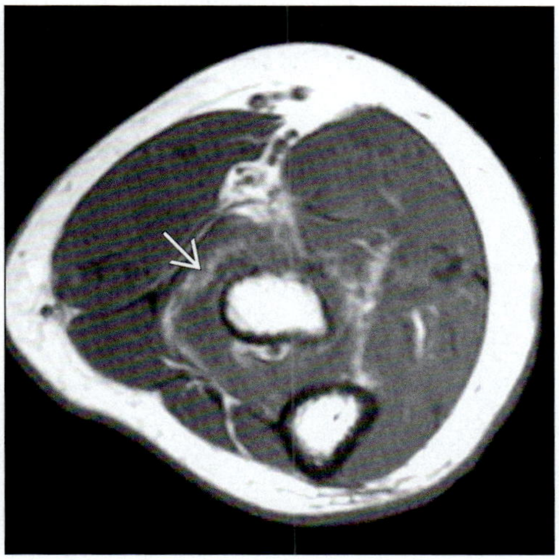

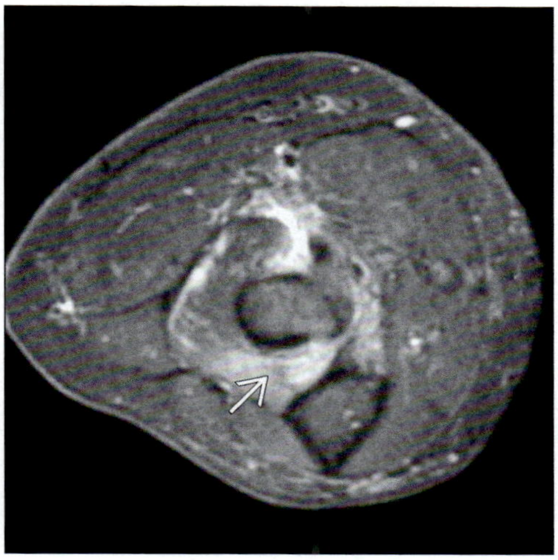

*(**Left**) Axial T1WI MR of the proximal forearm demonstrates mild fat atrophy in the supinator muscle ➡, a sign of denervation of the radial nerve. Other muscles demonstrate normal bulk and signal. Involvement of the supinator muscle suggests pathology in the radial nerve above the radial tunnel. (**Right**) Axial T2WI FS MR in the same patient at the same level demonstrates moderate elevated signal ➡, consistent with high radial neuropathy.*

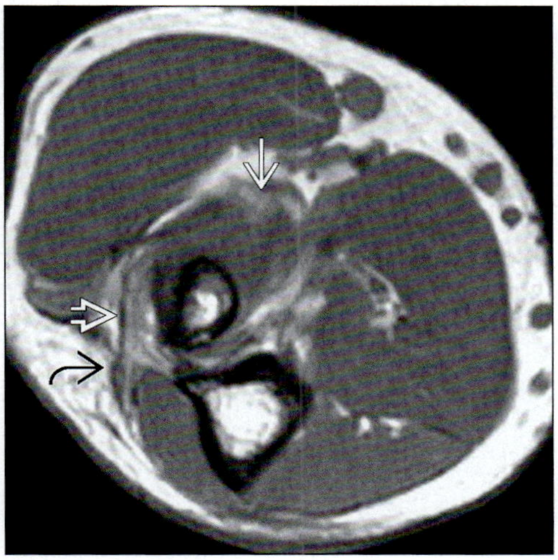

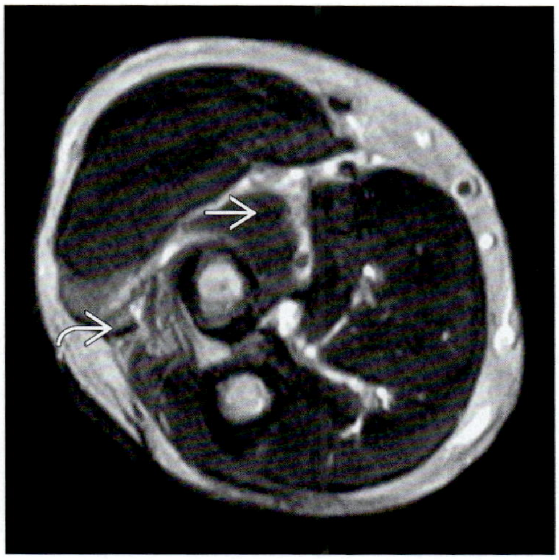

*(**Left**) Axial PDWI MR shows mild atrophy of the supinator ➡ and extensor digitorum ↗ muscles. Fibrosis ⇨ from biceps tendon repair 1 month earlier. The patient now presented with PIN syndrome symptoms. (**Right**) Axial T2WI MR in the same patient at the same level better shows the changes in the extensor digitorum muscle ↗ due to the elevated signal. Mild atrophy of the supinator ➡ is also present, but significant edema is not present in this muscle.*

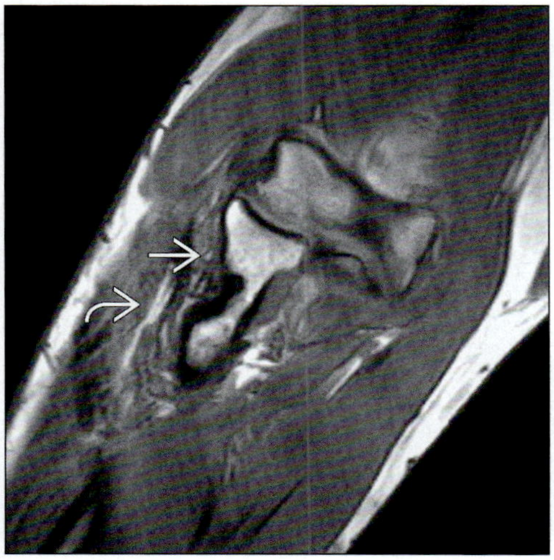

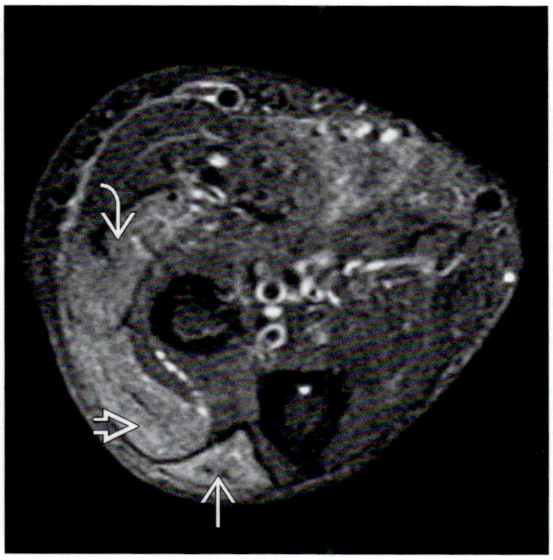

*(**Left**) Coronal T1WI MR in the same patient shows the atrophy of the extensor digitorum ↗ and supinator ➡ muscles after repair of the biceps tendon. The close proximity of the radial nerve to the biceps tendon may explain the appearance of neuropathy shortly after surgery. (**Right**) Axial STIR MR shows high signal involving the extensor carpi ulnaris ➡, extensor digitorum ⇨, and extensor carpi radialis longus and brevis ↗. Findings suggest PIN syndrome.*

RADIAL NERVE INJURY

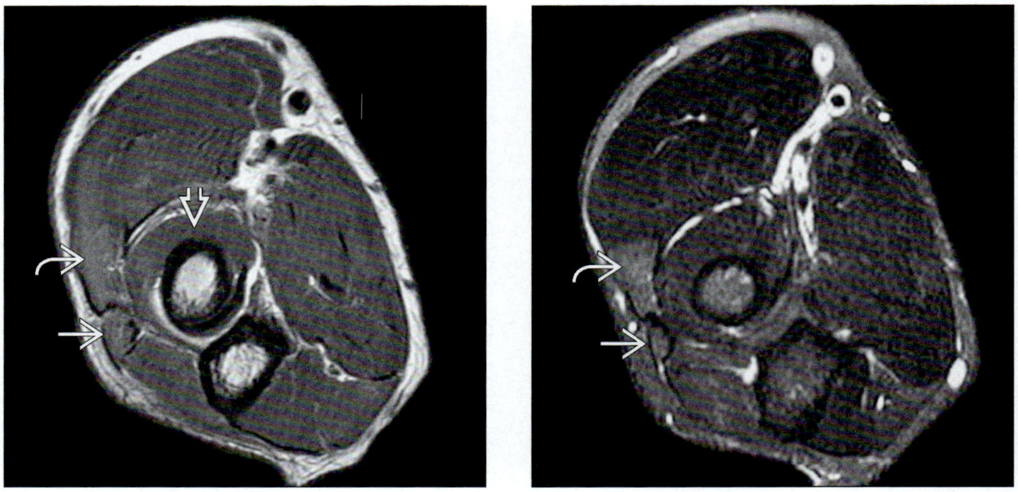

(Left) *Axial T1WI MR in a 42-year-old man with symptoms of posterior interosseous nerve syndrome shows mild atrophy and elevated signal in the extensor carpi ulnaris and extensor digitorum muscles. The supinator is normal.* ***(Right)*** *Axial T2WI FS MR in the same patient shows atrophy and signal elevation in the extensor carpi ulnaris and extensor digitorum muscles, consistent with the diagnosis of PIN syndrome.*

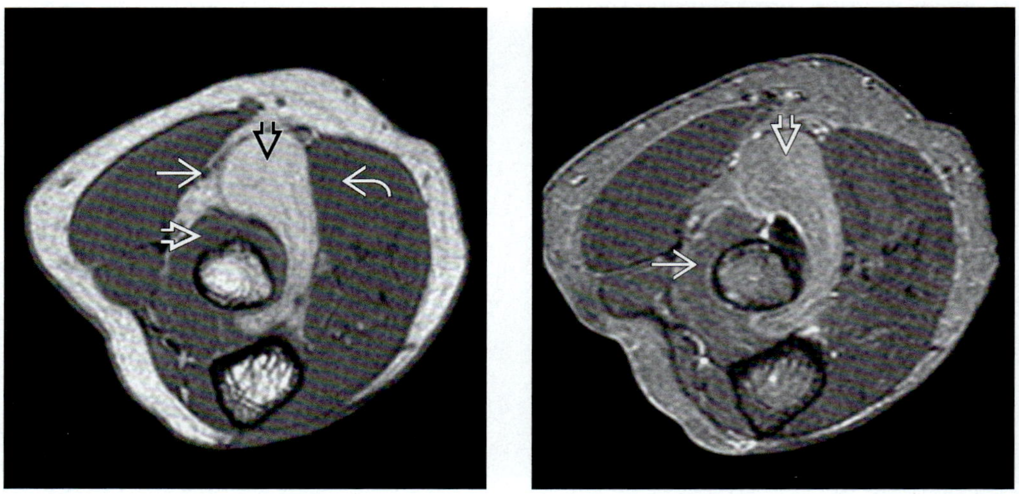

(Left) *Axial T1WI MR shows a large high signal mass consistent with a lipoma between the supinator and the pronator teres muscles, slightly displacing the radial nerve in its course proximal to the radial tunnel. The patient presented with mild radial neuropathy.* ***(Right)*** *Axial T2WI FS MR in the same patient confirms the lipomatous nature of the mass and shows mild elevated signal in the supinator muscle.*

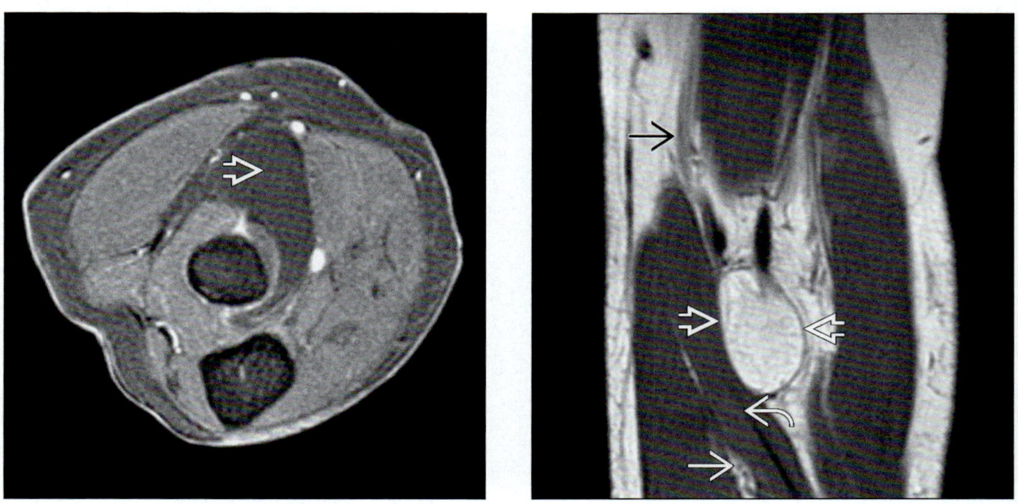

(Left) *Axial T1WI C+ FS MR in the same patient shows homogeneous fat signal in the mass with no internal enhancement, consistent with a simple lipoma.* ***(Right)*** *Coronal T1WI MR in the same patient shows the circumscribed nature of the mass and its relationship to the radial nerve (seen proximal to the radial tunnel) and the supinator muscle. The posterior interosseous nerve can be seen exiting the radial tunnel inferiorly.*

RADIAL NERVE INJURY

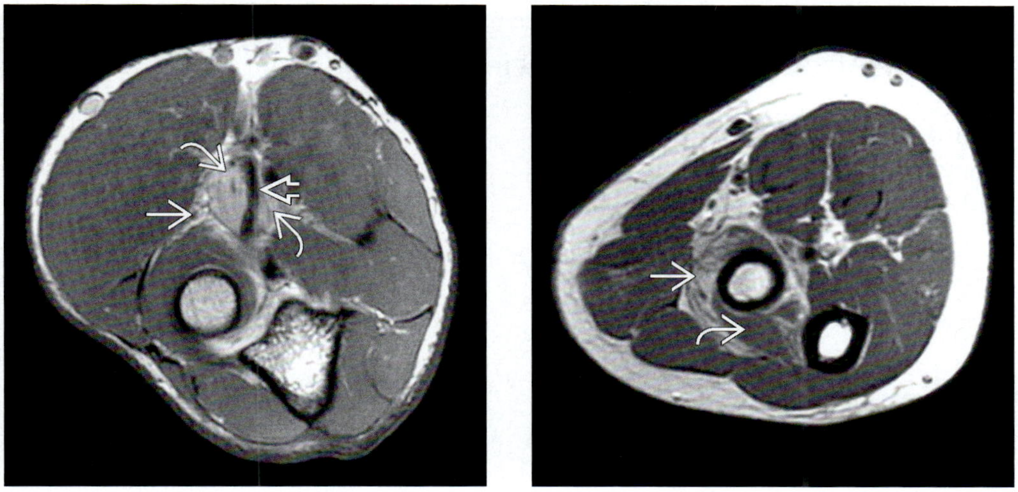

(Left) *Axial PDWI MR demonstrates bicipitoradial bursitis diffusely around the biceps tendon related to a low-grade tendon tear more distally (not shown). The fluid is adjacent to the radial nerve and in some cases can cause radial neuropathy.* ***(Right)*** *Axial T1WI MR shows significant atrophy of the superficial head of the supinator muscle, with preservation of the deep head. This patient underwent resection of the arcade of Frohse for radial tunnel syndrome 1 year earlier.*

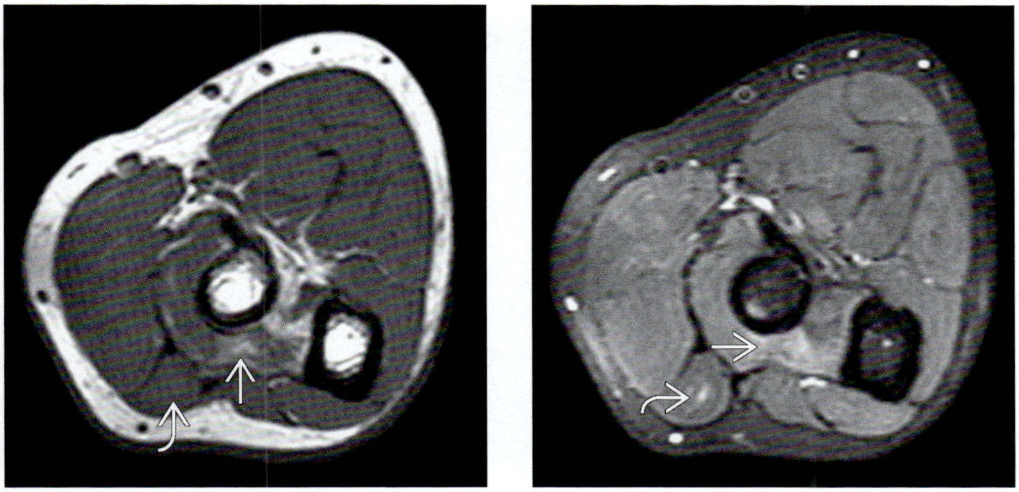

(Left) *Axial T1WI MR demonstrates atrophy of the supinator muscle with relative preservation of the extensor digitorum.* ***(Right)*** *Axial PDWI FS MR in the same patient at the same level shows atrophy and high signal in the supinator and minimal signal elevation in the extensor digitorum muscle without atrophy. This case shows the utility of fat-suppressed long TR sequences for detecting early changes in suspected neuropathy.*

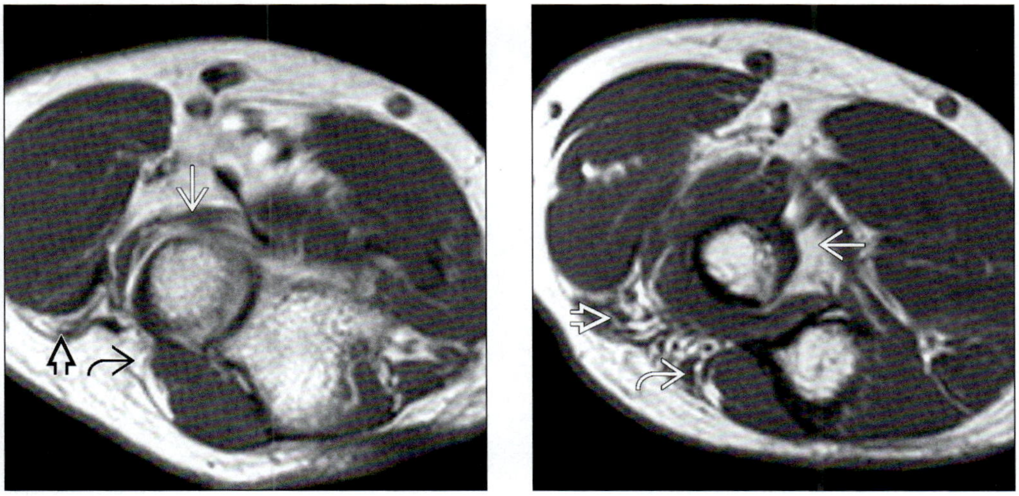

(Left) *Axial T2WI MR in a patient with clinical radial neuropathy several months after a humeral shaft fracture shows atrophy of the supinator, extensor carpi radialis longus, and extensor digitorum.* ***(Right)*** *Axial T2WI MR more inferiorly in the same patient shows the marbling of the extensor digitorum and extensor carpi radialis longus muscles, and fatty replacement of portions of the supinator. Involvement of the supinator points to a proximal nerve lesion.*

SECTION 4
Wrist and Hand

Introduction and Overview

Osseous Pathology

Distal Radius and Ulna

Wrist

Hand

Ligamentous Pathology

Wrist

Hand

Tendinous Pathology

Other Pathology

Neuropathies

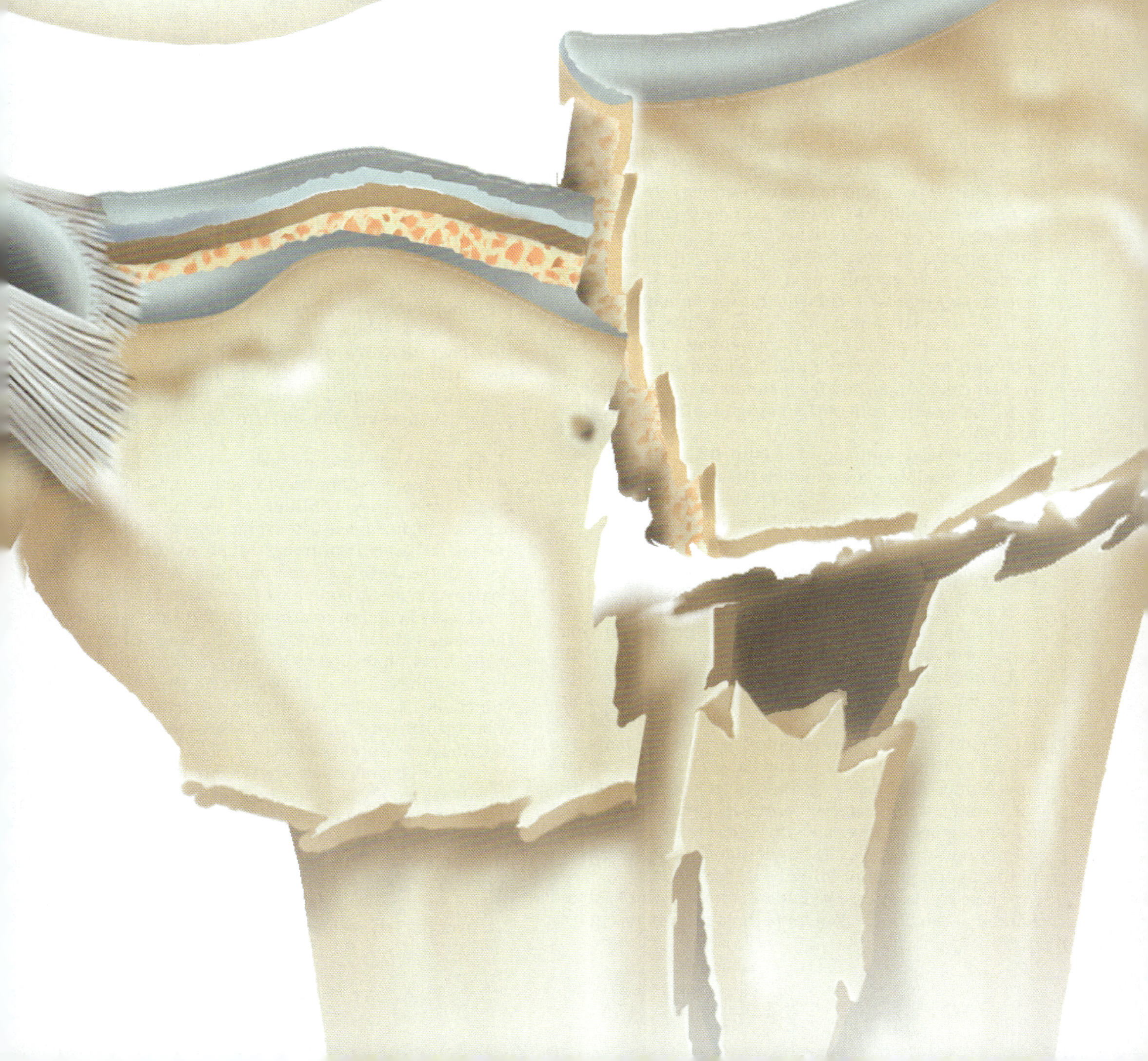

Overview

The distal upper extremity includes the distal forearm, wrist, hand, and fingers. While these are discretely separate entities anatomically, they function as and are often injured as a unit. This section reviews the common osseous and soft tissue injury patterns associated with each anatomic region with attention to the interplay between these regions. The imaging evaluation of these injury patterns is multimodality. Radiography and CT remain the mainstay of imaging of osseous injuries, while MR and ultrasound are key tools for delineation of soft tissue pathology.

Terminology and Conventions

Variable terms are used for describing locations in the wrist. In this text, the ulnar aspect of the wrist and hand are referred to as "medial" while the radial aspect is referred to as "lateral." The palmar aspect of the wrist and hand is described as "volar." For ease of viewing, the images are consistently positioned with the thumb to the left and the volar hand facing to the left. Transverse imaging is oriented with the palm down and the thumb to the left.

The tendons may be completely or partially torn in acute trauma. More gradual degenerative changes, commonly related to chronic repetitive microtrauma, are referred to as "tendinopathy" rather than "tendinosis" or "tendinitis" as in other sections of this book.

The literature is filled with eponyms and acronyms used in conjunction with wrist and hand trauma. One chapter is dedicated to a summary of the most commonly used terms. It is by no means comprehensive but may serve as a basic reference. Eponyms are often applied incorrectly. For instance, the term "Colles fracture" is frequently used to describe distal radial fractures regardless of actual fracture pattern, involvement of the articular surface, and age of the patient. An accurate description of the components of an injury, including location, angulation, displacement, apposition, and articular surface involvement are far more important to the care of the patient than assigning them the correct eponym.

Hardware terminology can also be confusing. Basic types of hardware are described with an emphasis on common complications of the plates, screws, and wires. The Herbert screw is referred to specifically because of its historical importance in treating scaphoid fractures.

Anatomic Considerations

The distal forearm refers to the distal 1/3 of the radius and ulna as they articulate with the proximal carpal row and to each other. The wrist refers to the distal radioulnar joint (DRUJ), proximal and distal carpal rows, and the carpometacarpal (CMC) joints. The hand variably refers to the CMC joints, metacarpophalangeal (MCP) joints, and the intervening metacarpal shafts; the thumb with MCP and interphalangeal (IP) joints and intervening phalanges, or the finger(s), including the MCP and interphalangeal joints and phalanges. There may be significant overlap with regard to imaging, and it is important to have a clear understanding of the portion of the anatomy that needs to be evaluated.

There are many ossicles and sesamoids in the wrist and hand that can be mistaken for pathology if not recognized as anatomic variants. A summary of the most common sesamoids and ossicles is provided.

The intrinsic and extrinsic ligaments of the wrist are essential for stability and mobility. Intrinsic ligaments are defined as those that connect a carpal bone to another carpal bone. While there are many intercarpal ligaments, the scapholunate ligament (SLL) (which resists volar flexion of the scaphoid and dorsal flexion of the lunate) and the lunotriquetral ligament (LTL) (which balances proximal carpal row motion in the opposite direction) are the most clinically significant. Extrinsic ligaments are defined as those that connect the carpal bones to adjacent structures, such as the distal radius, distal ulna, &/or metacarpals. The extrinsic ligaments provide gross stability to the wrist while the intrinsic ligaments "fine tune" wrist motion.

The triangular fibrocartilage complex (TFCC) is typically considered separately from the remainder of the wrist ligaments for ease of clinical evaluation and treatment. The TFCC is composed of the triangular fibrocartilage interposed between the distal ulna and the proximal carpal row, the dorsal and volar radioulnar ligaments, the (ulno)meniscal homologue, the ulnotriquetral and ulnolunate volar ligaments, and the ulnar collateral ligament. The latter is not a true ligament but is variably considered a component of the extensor carpi ulnaris subsheath &/or a medial portion of the ulnotriquetral ligament. The TFCC is integral to the stability of the DRUJ.

The flexor tendons course from the distal forearm into the wrist where the flexor digitorum profundus and superficialis tendons and flexor pollicis longus tendon pass through the carpal tunnel and into their respective digits. In each digit, the superficial tendon splits at the MCP joint, encircles the profundus tendon at the proximal interphalangeal (PIP) joint, and inserts on the volar medial phalangeal base. The profundus tendon continues distally to insert on the distal phalangeal base. The flexor apparatus is enclosed in a fibroosseous synovial-lined tunnel created by annular and cruciate pulleys, which provide mechanical advantage in finger motion.

The extensor tendons travel from the distal forearm into six discrete fibroosseous compartments along the dorsal wrist. They continue to the digits where each extensor digitorum tendon splits into proximal, central, and lateral bands, continuing distally to form the extensor hood, and inserting on the proximal, middle, and distal phalangeal bases to provide extension of the finger.

The key neural structures in the hand and wrist include the median, ulnar, and radial nerves. The median nerve is at risk for compression as it passes through the relatively rigid confines of the carpal tunnel. The ulnar nerve is vulnerable to injury as it enters Guyon canal in the volar medial wrist and bifurcates into deep (motor) and superficial (sensory) branches.

The radial and ulnar arteries branch into deep and superficial arches in the palmar of the hand. The ulnar artery is at particular risk for injury in a 2 cm segment as the superficial branch leaves Guyon canal and passes over the flexor retinaculum prior to entering the palmar aponeurosis. The radial artery is at risk as its superficial branch runs deep to the abductor pollicis brevis tendon just over the trapezoidal ridge.

Pathologic Considerations

The forearm, wrist, and hand are at high risk of injury, with fractures of the forearm accounting for 15-18% of all fractures in an emergency setting. These injuries may result from a high-energy injury, such as a motor vehicle accident, but are most commonly a result of low-energy injuries, such as injury during participation in athletics. The common causative mechanism across all injury patterns in the distal upper extremity is a fall on outstretched hand (FOOSH) injury.

The border digits (1st and 5th rays) are commonly injured because they are more exposed to catching and twisting trauma than are the middle three fingers.

The various osseous and soft tissue injury patterns are separated for ease of consideration and discussion. Fractures and injuries to the ligaments and tendons commonly occur in combination. It is important to remember this when evaluating a case and beware of "satisfaction of search" error, in which identifying one abnormality interferes with finding additional abnormalities.

Imaging Considerations

Radiologic evaluation is the keystone to evaluation of the traumatized distal upper extremity. The posteroanterior, oblique, and lateral views of the wrist, hand, and thumb are essential for adequate evaluation. Additional scaphoid, clenched fist, ulnar- and radial-deviated views, etc. may be added as indicated by clinical circumstance. True PA and lateral views of each injured digit are essential, particularly for the thumb.

- Neutral PA wrist view: Obtained with shoulder abducted 90°, elbow flexed 90°, forearm in neutral rotation, palm flat on table
- Neutral lateral wrist view: Obtained with elbow adducted to patient's side, wrist in neutral rotation, fingers extended. Volar border of pisiform should lie equidistant between volar borders of scaphoid tuberosity and capitate head
- True thumb AP (Roberts) view: Obtained with forearm flat on table, hand pronated 20° with dorsal thumb flat on cassette, and x-ray tube angled 10° distal to proximal

Fluoroscopy is useful for clinical evaluation of wrist motion and is particularly useful for evaluation of dynamic instability patterns, such as palmar midcarpal instability.

Conventional arthrography has been a mainstay of wrist ligament assessment. MR is frequently used for evaluation of the ligaments, but many studies indicate that MR imaging following the introduction of dilute gadolinium (direct MR arthrography) increases the sensitivity and specificity of MR imaging for ligament pathology in the wrist.

Multi-slice CT provides accurate and detailed 2- and 3-dimensional imaging of the wrist, hand, and fingers. CT may be obtained with a patient in a cast or splint. Positioning the patient with the extremity away from the body center will substantially decrease motion and splay artifact and will improve the quality of the images. Placing the patient prone with the affected extremity above the head ("superman" position) will reduce these artifacts. While imaging can be acquired with patient in an external fixation device, care should be taken to acquire thin-section images at an angle to the hardware to reduce metal artifact. Reformation software can provide additional detail to the imaging assessment. Obtaining direct long axis oblique images of the scaphoid is recommended for optimal visualization of scaphoid fracture.

Ultrasound provides real-time evaluation of the integrity and function of tendons and of the superficial portions of the intrinsic ligaments and TFC complex. Ultrasound of the wrist and hand is ideally acquired with a > 10 MHz small footprint linear probe with color Doppler. While this technique can provide valuable information, it is highly operator-dependent.

MR is a key imaging technique for evaluation of soft tissue pathology in the upper extremity. Proper positioning is essential for obtaining a high-quality study. Most manufacturers now offer dedicated wrist volume coils that allow patients to be scanned while lying prone with the extremity at the side. If such a coil is not available, the wrist should be positioned in magnetic isocenter with the arm overhead in the pronated "superman" position. Some patients cannot tolerate this position because of shoulder or elbow pain. When positioning the wrist and hand, care must be taken to keep the wrist alignment in a neutral position. Ulnar or radial deviation of the wrist may result in apparent carpal alignment that suggests an instability pattern when no such abnormality exists.

MR imaging evaluation of ligament integrity increases in sensitivity and specificity with the addition of direct gadolinium arthrography. Intravenous gadolinium injection (indirect arthrography) may also be useful in evaluation of ligament status and can also provide information regarding inflammation (such as tenosynovitis), hand masses and ganglia, and in the evaluation of viability of scaphoid fragments following fracture.

Imaging protocols for the wrist and hand MR vary widely. A standard set of sequences usually includes coronal T1- and fat-suppressed fluid-sensitive sequences, at least one sagittal sequence, and usually one or two axial sequences. Small imaging field-of-view and thin slice thickness are key factors in providing diagnostically satisfying images.

Selected References

1. Fayad LM et al: Pediatric skeletal trauma: use of multiplanar reformatted and three-dimensional 64-row multidetector CT in the emergency department. Radiographics. 29(1):135-50, 2009
2. Lisle DA et al: MR imaging of traumatic and overuse injuries of the wrist and hand in athletes. Magn Reson Imaging Clin N Am. 17(4):639-54, vi, 2009
3. Kaewlai R et al: Multidetector CT of carpal injuries: anatomy, fractures, and fracture-dislocations. Radiographics. 28(6):1771-84, 2008
4. Bencardino JT et al: Sports-related injuries of the wrist: an approach to MRI interpretation. Clin Sports Med. 25(3):409-32, vi, 2006
5. Cerezal L et al: MR imaging findings in ulnar-sided wrist impaction syndromes. Magn Reson Imaging Clin N Am. 12(2):281-99, vi, 2004
6. Rosner JL et al: Imaging of athletic wrist and hand injuries. Semin Musculoskelet Radiol. 8(1):57-79, 2004
7. Clavero JA et al: MR imaging of ligament and tendon injuries of the fingers. Radiographics. 22(2):237-56, 2002

(Left) *PA wrist radiograph obtained with the patient seated, shoulder abducted, and elbow flexed 90° shows the fingers are extended, and the palm is flat on the cassette. This standard image provides baseline information for diagnosis and treatment.* **(Right)** *Medially deviated PA radiograph in a patient with a suspected scapholunate ligament tear demonstrates the value of a special view, such as this stress view, added to routine imaging to accentuate widening of the scapholunate interval →.*

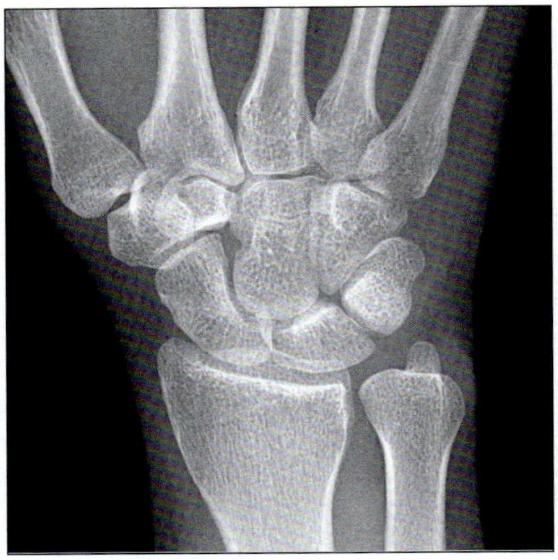

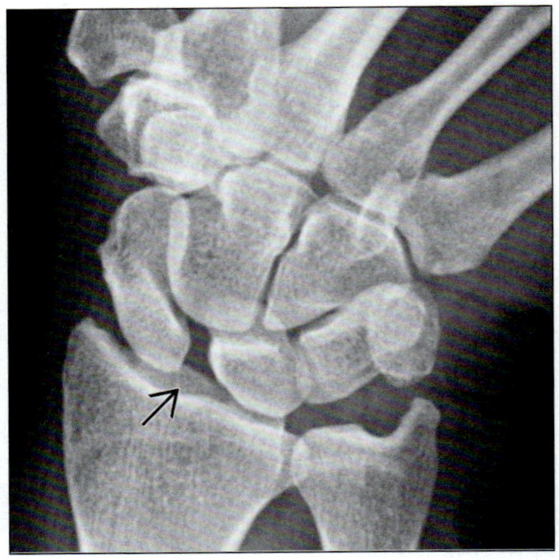

(Left) *Lateral wrist radiograph is properly positioned when the volar aspect of the pisiform ⇨ is centered equidistant between the volar borders of the scaphoid tuberosity → and capitate head ↩. This view is obtained with the arm adducted and the elbow flexed 90° with wrist in neutral position.* **(Right)** *Lateral wrist view allows evaluation of carpal stability by identifying capitate (dotted line), lunate (solid line), distal radius (dashed line), and scaphoid → to measure key carpal angles.*

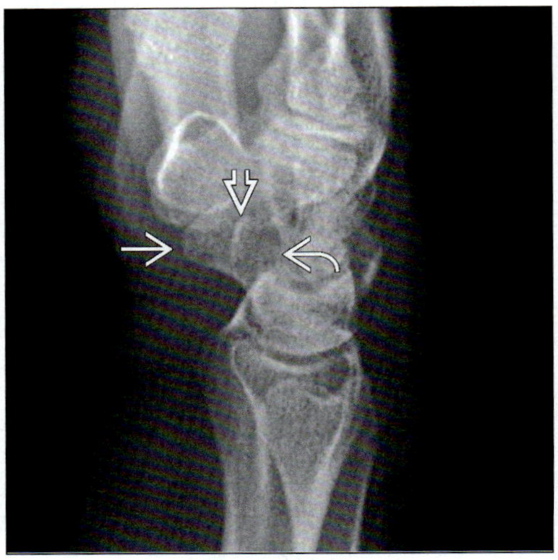

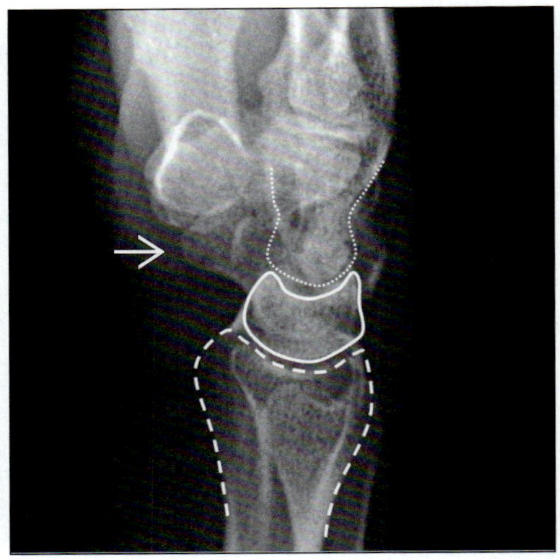

(Left) *PA radiograph shows a comminuted Colles fracture. Appropriate use of this eponym allows the clinician to make clinical decisions. The term "Colles fracture" implies a distal radial fracture with dorsal displacement and angulation of the distal fragment, with or without extension into radiocarpal → and distal radioulnar ⇨ joints, usually in the setting of osteopenia.* **(Right)** *Lateral radiograph from the same patient also shows significant loss of normal volar radial tilt →.*

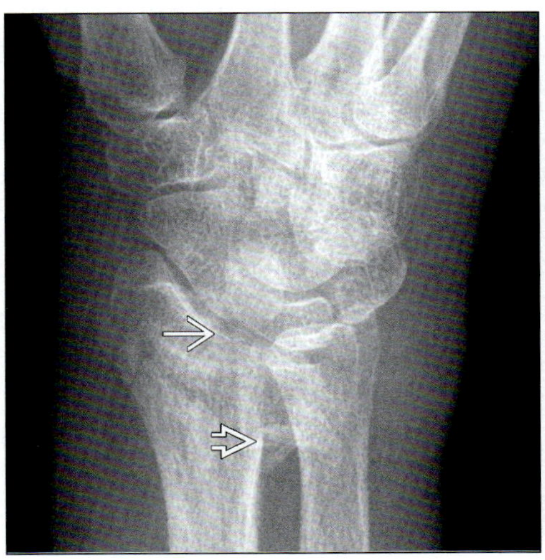

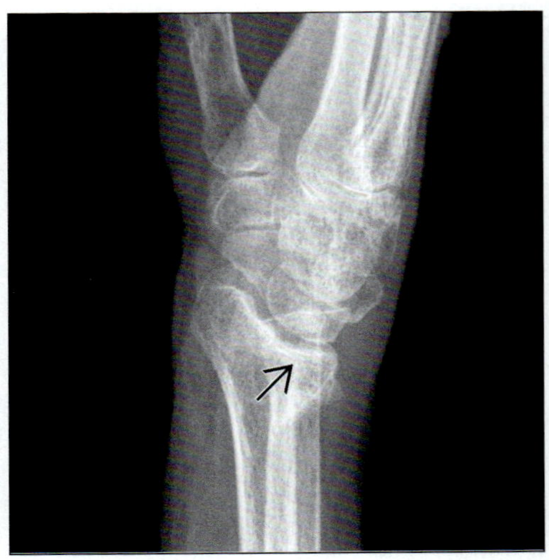

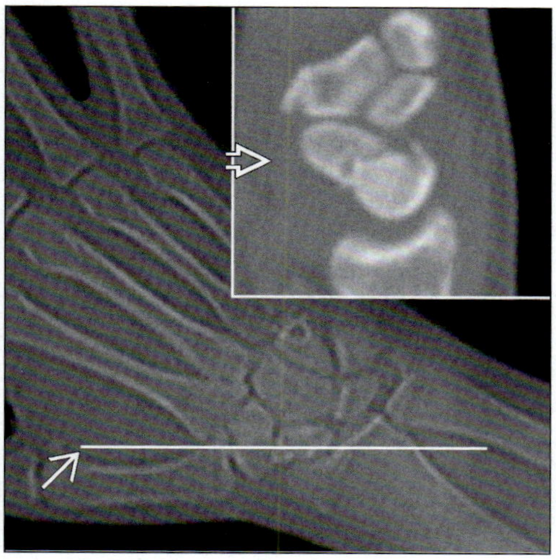

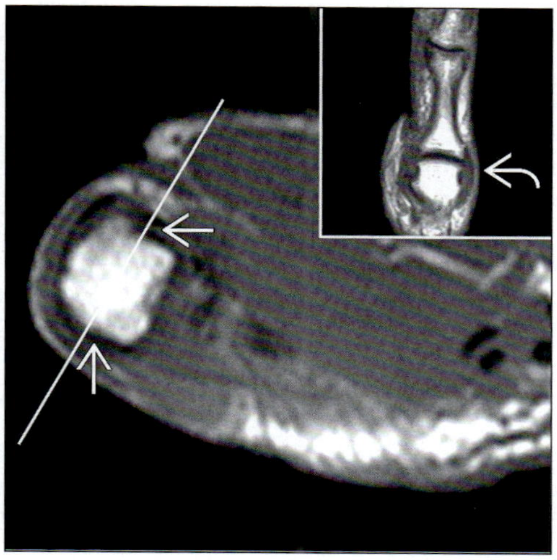

(Left) *Sagittal bone CT demonstrates a direct long axis sagittal image of the scaphoid, ideal for fracture evaluation. The patient is placed in "superman" position with scaphoid oriented parallel to scanner gantry.* ***(Right)*** *Axial T1WI MR demonstrates the correct method for prescribing true coronal images of the thumb. Axial localizer through the 1st MCP head is used. Coronal plane is oriented parallel to the interglenoid fossae to visualize of the collateral ligaments (inset).*

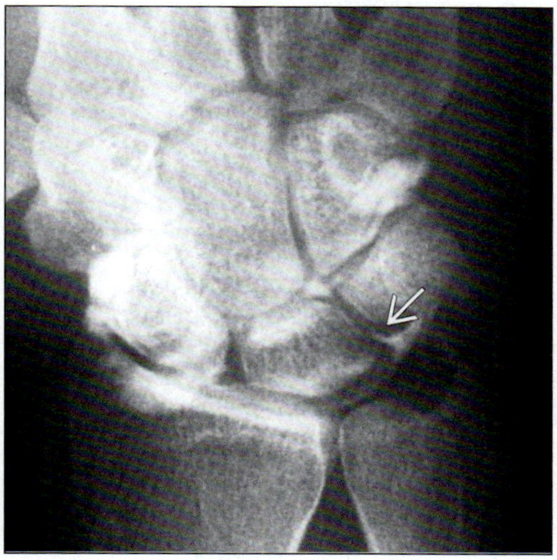

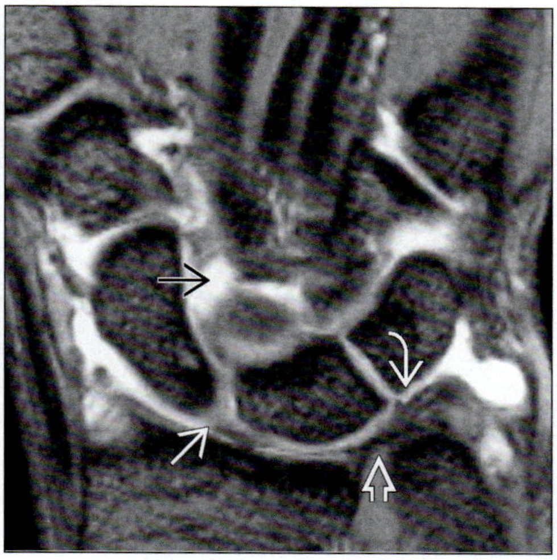

(Left) *PA fluoroscopy obtained after radiocarpal joint arthrogram shows contrast extending through the lunotriquetral interval into the midcarpal compartment indicating a tear of the lunotriquetral ligament.* ***(Right)*** *Coronal MR arthrogram T1WI FS shows gadolinium injected into the radiocarpal joint extending through the ulnar aspect of the lunotriquetral ligament and into the midcarpal compartment. The scapholunate ligament and triangular fibrocartilage are intact.*

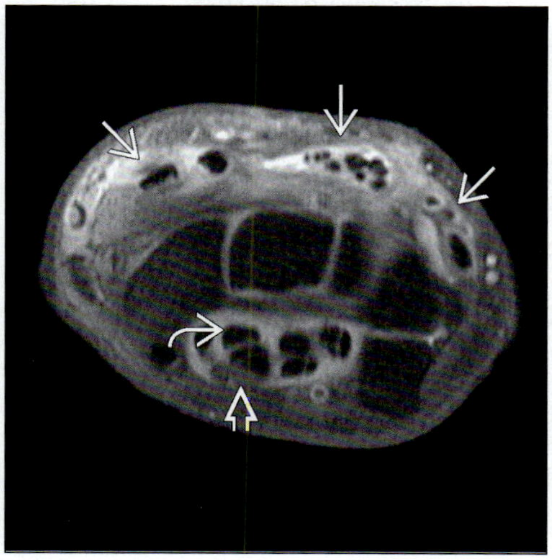

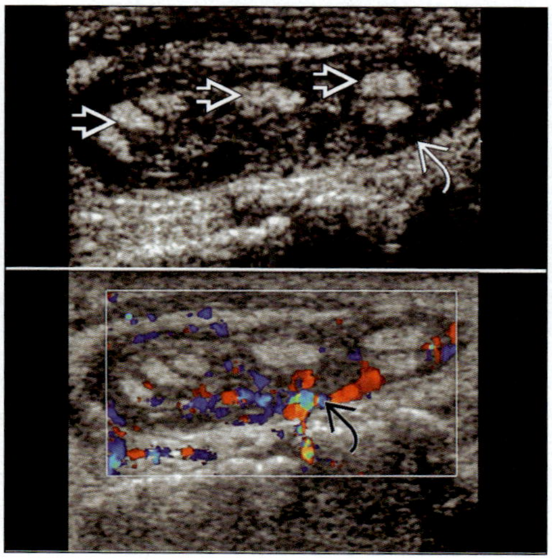

(Left) *Axial T1WI C+ FS MR demonstrates enhancement of the flexor and extensor tendon sheaths consistent with tenosynovitis. The tendons are normal. Note that the median nerve does not enhance.* ***(Right)*** *Axial grayscale ultrasound of the dorsal wrist (upper image) demonstrates normal extensor digitorum tendons surrounded by hypoechoic fluid of tenosynovitis. Color Doppler ultrasound (lower image) in a different patient shows hyperemia of chronic tenosynovitis.*

OSSICLES AND SESAMOIDS, HAND AND WRIST

Key Facts

Imaging

- Sesamoids
  - Volar only; near joints at areas of tendinous angulation
  - Embedded in capsule rather than tendon
  - 5 common locations
  - 2 at thumb MCP joint; present in 100%
  - 1 at thumb IP joint; present in ~ 75%
  - 1 at index MCP joint; present in ~ 35%
  - 1 at small MCP joint; present in ~ 80%
- Accessory ossicles
  - Numerous accessory bones
  - Not necessary to name each bone accurately but should recognize as normal variant
- Range in size from 1-10 mm
- Tend to be more radiodense when smaller

Top Differential Diagnoses

- Fracture
  - Acute: Sharply marginated edges; fragments "fit together" to complete adjacent bone geometry
  - Chronic: Margins corticated, smoothed but still "fit" to complete bone geometry
- Avascular necrosis
  - Radiodense; fragmented
- Heterotopic ossification
  - Appears "oversized" and irregular for location
  - May be difficult to distinguish if mature
- Soft tissue tumor calcification
  - Soft tissue mass present
  - Mineralization patchy, irregular

Diagnostic Checklist

- Fracture vs. normal variant
  - Look for smooth, corticated margins

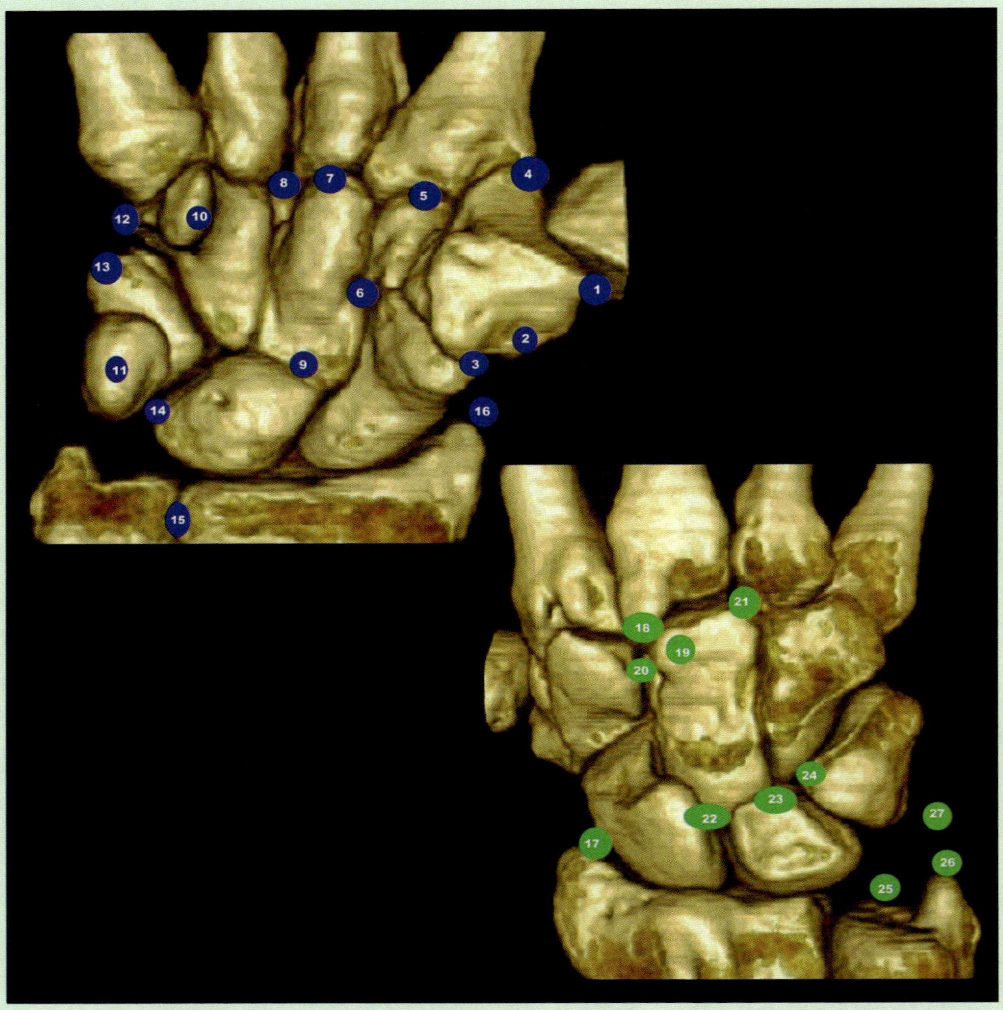

Graphic shows accessory ossicles. The volar (upper left) include 1) paratrapezium, 2) epitrapezium, 3) os radiale externum (APL sesamoid), 4) secondary trapezium, 5) secondary trapezoid, 6) os centrale carpi, 7) subcapitate, 8) os Gruber, 9) hypolunatum, 10) os hamuli proprium, 11) secondary pisiform, 12) os vesalianum, 13) os ulnare externum, 14) secondary triquetrum, 15) os radioulnare, and 16) os radiale styloideum. The dorsal (lower right) include 17) paranavicular, 18) os styloideum, 19) metastyloideum, 20) parastyloideum, 21) secondary capitate, 22) paranavicular, 23) epilunatum, 24) epitriquetrum (epipyramis), 25) os triangulare, 26) os ulnar styloideum, and 27) lunula.

OSSICLES AND SESAMOIDS, HAND AND WRIST

TERMINOLOGY

Abbreviations

- Abductor pollicis longus (APL)
- Carpometacarpal (CMC)
- Distal interphalangeal (DIP)
- Triangular fibrocartilage complex (TFCC)
- Interphalangeal (IP)
- Metacarpophalangeal (MCP)

Synonyms

- Carpal boss, carpe bossu

IMAGING

General Features

- Best diagnostic clue
  - Smooth, rounded, well-corticated margins
- Location
  - **Sesamoids**
    - Palmar (volar) location only; near joints at areas of tendinous angulation
    - Embedded in capsule rather than tendon
    - May be bilateral or unilateral
    - 5 common sesamoid locations
      - 2 at thumb MCP joint; present in 100%
      - 1 at thumb IP joint; present in ~ 75%
      - 1 at index MCP joint; present in ~ 35%
      - 1 at small MCP joint; present in ~ 80%
    - Less common locations
      - Middle and ring MCPs; index, middle, ring, and small DIPs
  - **Accessory ossicles**
    - May be dorsal, palmar, or nestled between carpal bones
    - Numerous accessory bones
      - Seen in ~ 1.5% of population
      - Not necessary to name each bone accurately but should recognize as normal variant
  - **Common ossicles**
    - Os styloideum
      - Dorsal to bases of 2nd and 3rd metacarpals; ± fused to metacarpals
      - Carpal boss: Dorsal protuberance at 3rd CMC joint; results from os styloideum or osteophytic ridging
    - Lunula
      - Centered in meniscal homologue
      - May fuse to ulnar styloid tip or be mistaken for styloid fracture
    - Os triangulare
      - Distal to ulnar fovea in region of TFCC
    - Os hamuli proprium
      - Palmar near hamate hook
    - Epilunatum
      - Dorsal to lunate; may appear to be within joint
    - Secondary trapezium
      - Nestled between 1st and 2nd metacarpal bases
    - Os radiale externum
      - Embedded in abductor pollicis longus
    - Os ulnare externum
      - Probable calcification in tendon or bursa rather than true ossicle
- Size
  - Sesamoids and accessory ossicles
    - Range from 1-10 mm
    - Tend to be more radiodense when smaller
- Morphology
  - Sesamoids
    - Oval, faceted with articular surface directed toward adjacent bone
  - Accessory ossicles
    - Smoothly corticated ossific body
    - Range from ovoid to triangular

Radiographic Findings

- Radiography
  - Sesamoids
    - Rounded corticated bone; articular surface toward adjacent bone
    - May be bipartite or multipartite
  - Accessory ossicles
    - Corticated bone, often interdigitating with adjacent normally contoured structures
    - No evidence of "subtraction" from adjacent bones or donor site

CT Findings

- Bone CT
  - Findings similar to radiography
  - Allows confirmation of ossicle corticated surfaces without adjacent fracture line
  - Look for evidence of degenerative disease

MR Findings

- Normal ossicles and sesamoids
  - Intact cortex without osseous or soft tissue edema

Imaging Recommendations

- Best imaging tool
  - Radiography, computed tomography
  - MR useful if symptoms correspond to ossicle
- Protocol advice
  - Radiography
    - Imaging in at least 2 planes allows precise location of ossicle or sesamoid
  - CT
    - Thin section; small FOV
  - MR
    - Use dedicated wrist coil
    - Correlate T1 (to identify ossicle of interest) and fluid-sensitive sequences (to identify pathology)

DIFFERENTIAL DIAGNOSIS

Fracture

- Acute: Sharply marginated edges; fragments "fit together" to complete adjacent bone geometry
- Chronic: Margins corticated, smoothed but still "fit" to complete bone geometry

Osteonecrosis

- Radiodense; fragmented
- Beware: Smaller ossicles and sesamoids tend to be radiodense normally

Heterotopic Ossification

- May be difficult to distinguish if mature
- Appears "oversized" and irregular for its location

OSSICLES AND SESAMOIDS, HAND AND WRIST

Lateral radiograph demonstrates a small rounded ossicle at the junction of the 2nd and 3rd metacarpal bases, just dorsal to the distal capitate, representing an os styloideum ➡. This ossicle may become symptomatic.

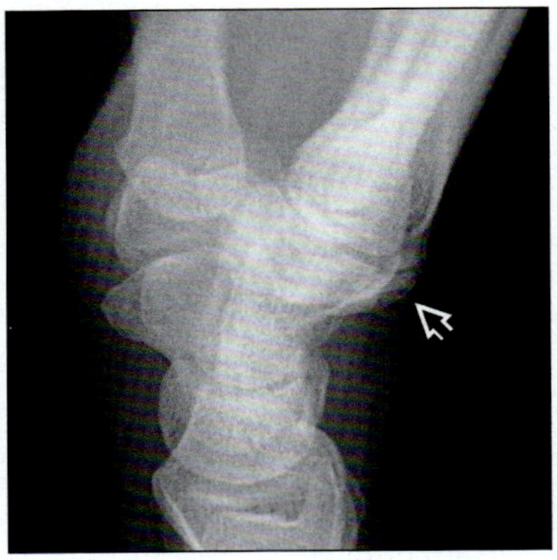

PA radiograph demonstrates the os styloideum ➡ at the base of the 3rd metacarpal. Its proximal cortex mimics a fracture ➡. The overlying tendons moving across the osseous protuberance may become irritated.

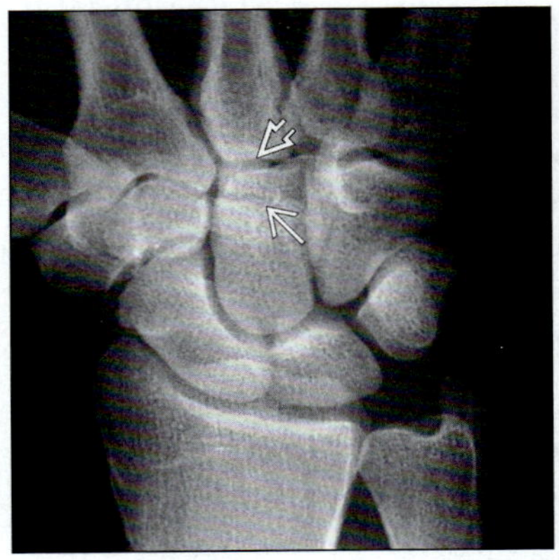

Sagittal reformation CT reveals a small os styloideum nestled behind the capitate/3rd metacarpal joint ➡. This was mistaken for a fracture following very minor trauma to the dorsal wrist.

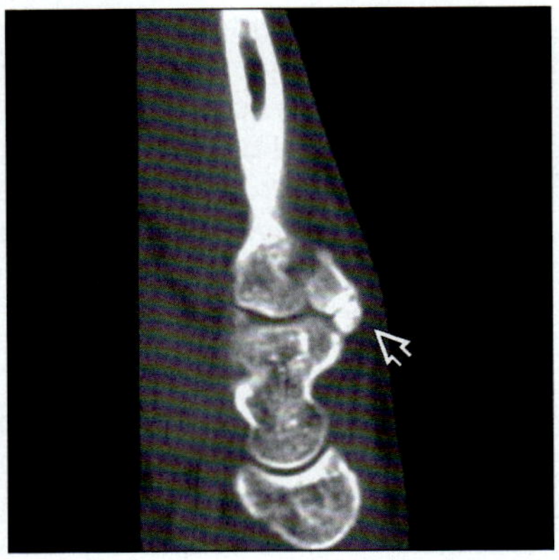

Axial STIR MR reveals symptomatic carpal bossing ➡ at the dorsal capitate/3rd metacarpal articulation with increased signal intensity of marrow edema. Note the impingement on the extensor tendon ➡.

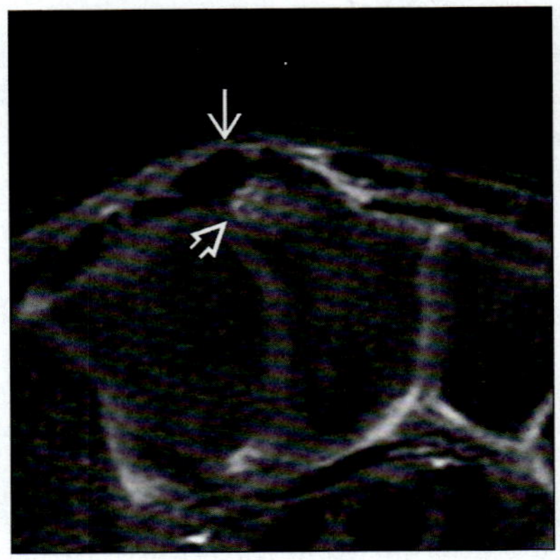

Os styloideum and the carpal boss.

Soft Tissue Tumor Calcification

- Soft tissue mass present
- Mineralization patchy, irregular
- Multiplanar imaging (CT or MR) provides best detail

CLINICAL ISSUES

Presentation

- Most common signs/symptoms
  - Sesamoids and ossicles
    - Typically asymptomatic, incidental finding
- Other signs/symptoms
  - Pain and focal tenderness
    - May be related to painful synchondroses or degenerative changes
    - May be related to extensor tendon impingement with bursitis, tenosynovitis

Natural History & Prognosis

- Sesamoids
  - Present in fetal development but ossified later in life
  - Age 13-18 years in males; 11-15 years in females
  - Order of appearance: Thumb MCP, thumb IP; ± index and small finger MCPs
- Accessory ossicles
  - May originate from
    - Normal ossification center
    - Secondary ossification center which failed to unite
    - Sequelae of repetitive or prior trauma (not a true normal variant, but may be indistinguishable)
    - Calcification in tendon or bursa
  - Appear more frequently in skeletal syndromes than in normal population

DIAGNOSTIC CHECKLIST

Consider

- Fracture vs. normal variant

Image Interpretation Pearls

- Look for smooth, corticated margins
- Consider MR to identify marrow edema as indication of pathology

Reporting Tips

- Typically not commented on in report unless symptomatic

SELECTED REFERENCES

1. Timins ME: Osseous anatomic variants of the wrist: findings on MR imaging. AJR Am J Roentgenol. 173(2):339-44, 1999
2. Lawson JP: International Skeletal Society Lecture in honor of Howard D. Dorfman. Clinically significant radiologic anatomic variants of the skeleton. AJR Am J Roentgenol. 163(2):249-55, 1994
3. Conway WF et al: The carpal boss: an overview of radiographic evaluation. Radiology. 156(1):29-31, 1985
4. Joseph J: The sesamoid bones of the hand and the time of fusion of the epiphyses of the thumb. J Anat. 85(3):230-41, 1951

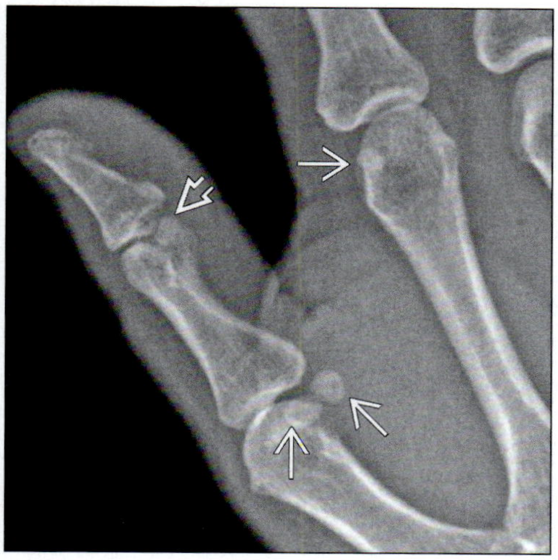

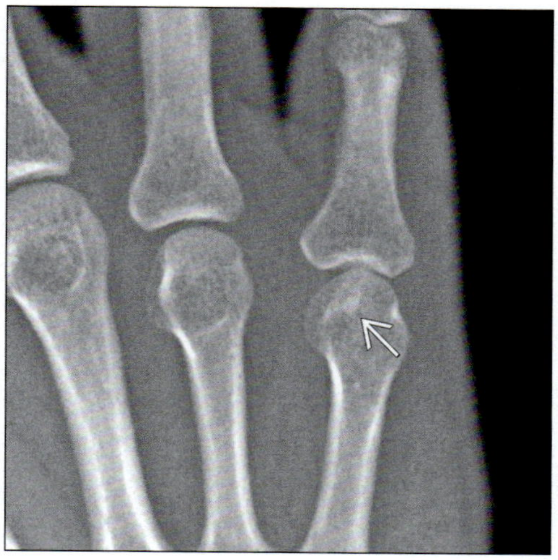

(Left) *AP radiograph shows normal sesamoids located at the 1st and 2nd MCPs and 1st IP. The size is variable. The sesamoids at the 1st MCP are the most consistently seen sesamoids in the hand.* **(Right)** *AP radiograph shows a normal sesamoid at the 5th MCP. This very small sesamoid may appear very dense because of its size. The larger sesamoids have better defined medullary spaces and cortical margins.*

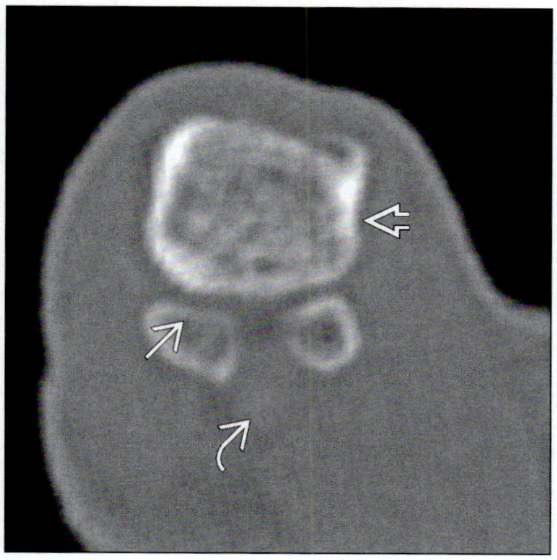

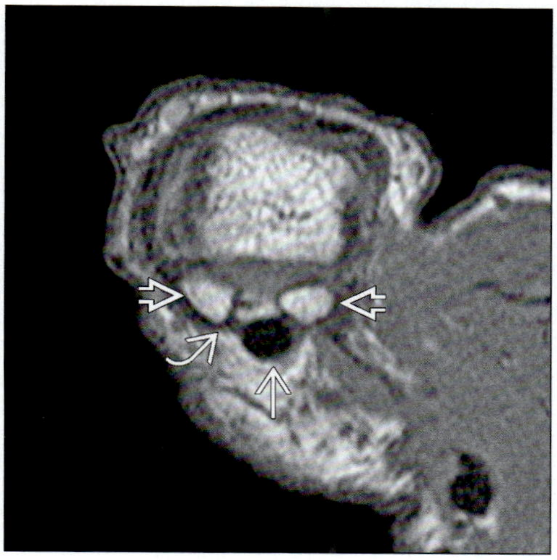

(Left) *Axial bone CT of 1st MCP shows smoothly corticated sesamoids with a faceted contour facing the metacarpal head. Note that the sesamoids arise within the MCP capsule rather than the adjacent flexor pollicis longus tendon.* **(Right)** *Axial T2WI MR in a different patient emphasizes the relationship of the flexor pollicis longus tendon to the 1st MCP joint and the adjacent discrete faceted sesamoids embedded in the 1st MCP capsule.*

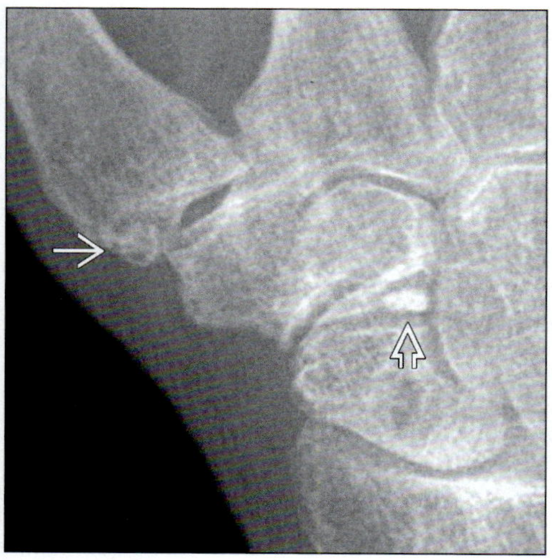

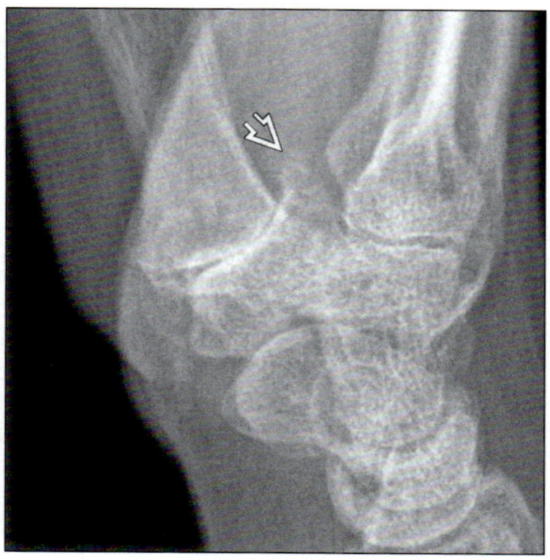

(Left) *AP radiograph demonstrates a prominent paratrapezium located immediately lateral to the 1st MCP joint. A smaller, slightly more dense os centrale carpi can be seen medially along the distal scaphoid articulation with the trapezoid.* **(Right)** *Lateral radiograph reveals a secondary trapezium immediately adjacent to the distal trapezium margin located between the 1st and 2nd metacarpal bases.*

OSSICLES AND SESAMOIDS, HAND AND WRIST

(Left) *Lateral radiograph reveals a large secondary trapezium just distal to the trapezium and medial 1st metacarpal base. A prominent epilunatum is also present. Its large size allows visualization of the cortical surfaces and medullary spaces.* ***(Right)*** *PA oblique radiograph reveals a secondary trapezium, defined as such due to its classic location between the 1st and 2nd metacarpals. There is 2nd degenerative arthropathy with prominent trapezium osteophytic ridging.*

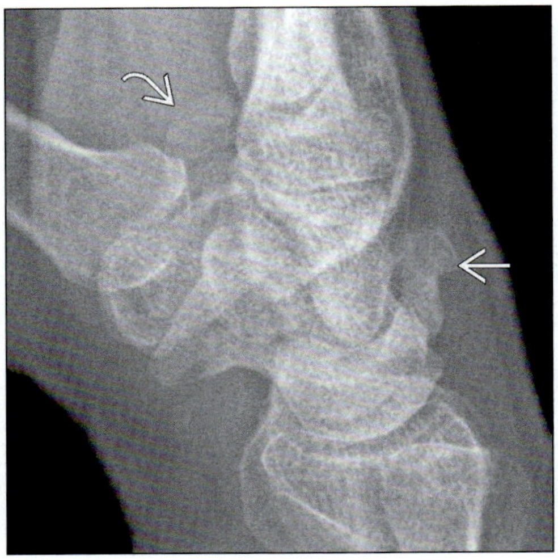

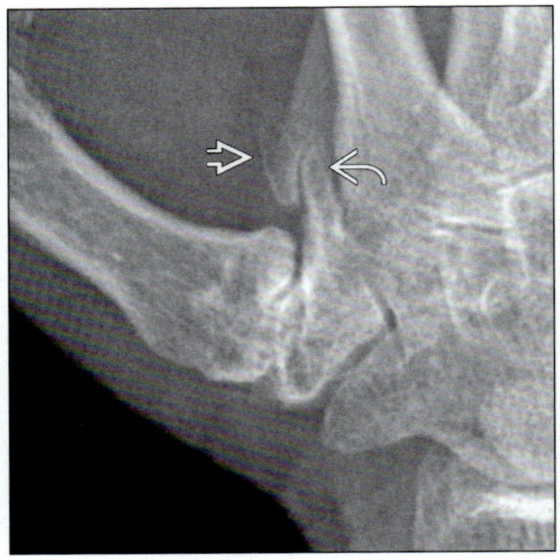

(Left) *PA radiograph demonstrates a small secondary trapezoid located distal and slightly medial to the distal trapezoid articular surface.* ***(Right)*** *PA oblique radiograph reveals paratrapezium located just lateral to the 1st MCP joint. The joint itself is arthritic with slight lateral subluxation of the 1st metacarpal, joint space narrowing, and small osteophytes. A large os triangulare is located just distal to the ulnar fovea.*

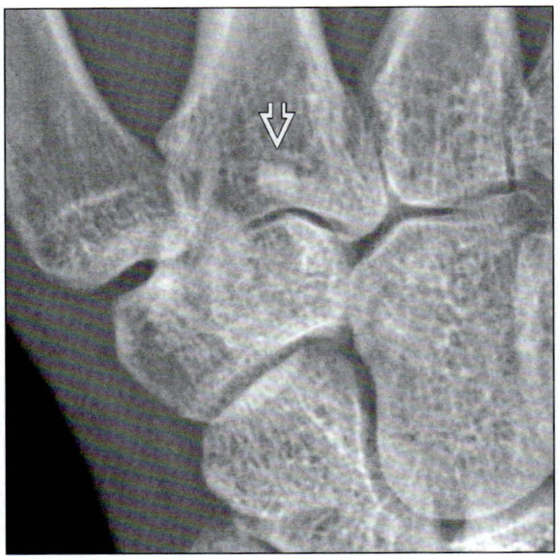

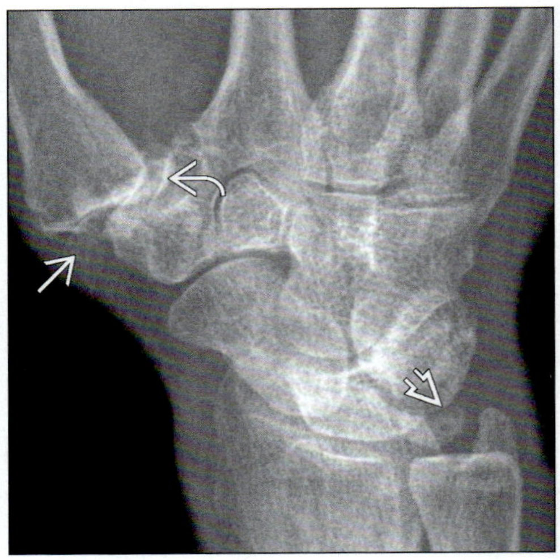

(Left) *Axial PDWI FS MR at the CMC joint reveals edema in a smoothly corticated ossicle medially at the capitate, representing a symptomatic secondary capitate.* ***(Right)*** *Sagittal T1WI MR in the same patient demonstrates the location of the secondary capitate located dorsally at the CMC joint between the capitate and hamate. Note the marker placed to indicate location of symptoms.*

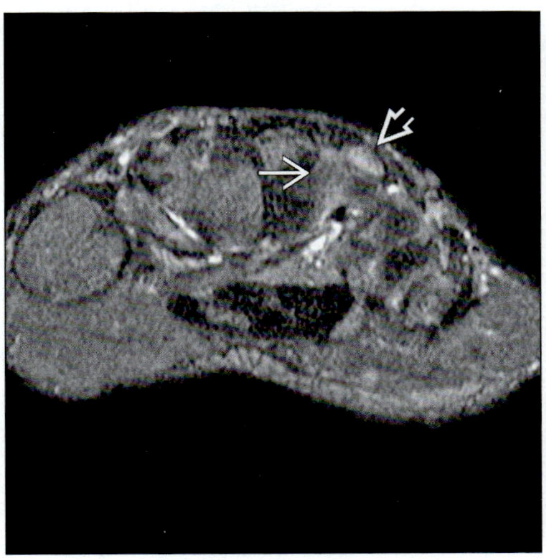

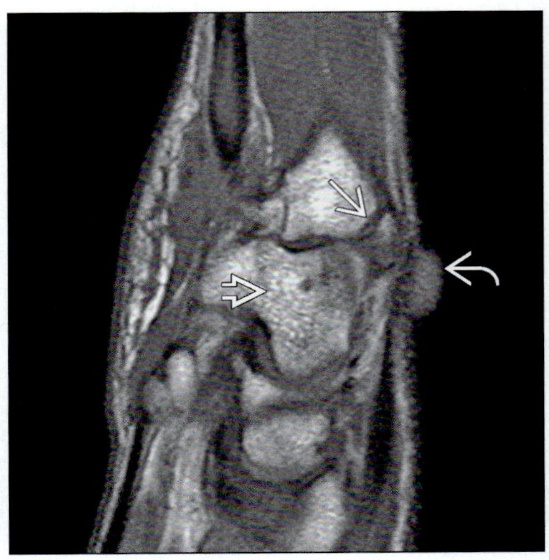

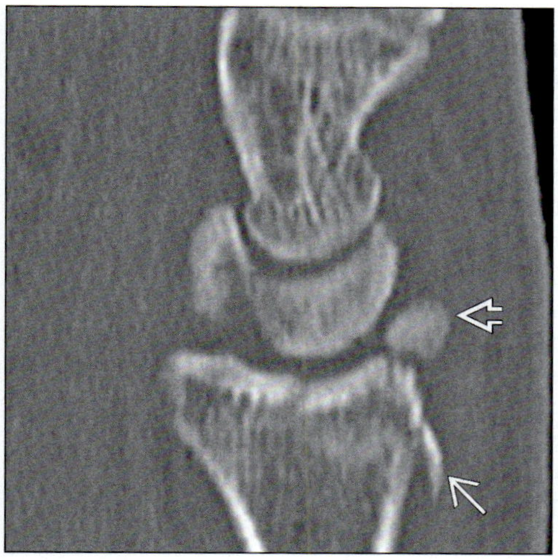

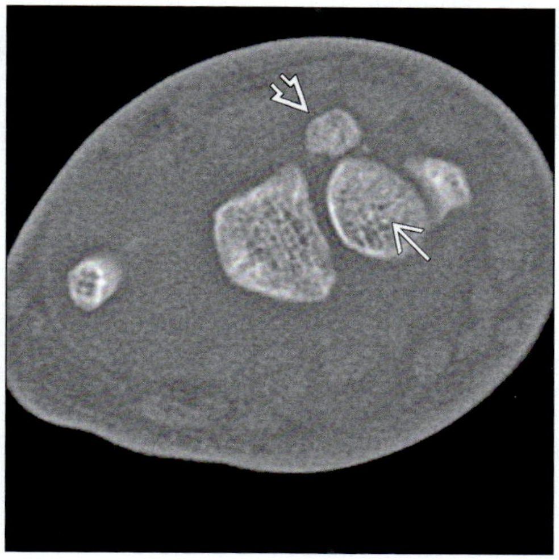

(Left) *Sagittal reformat bone CT demonstrates a comminuted fracture of the distal radius dorsal lip →. An epilunatum ⇨ is incidentally noted. It is clearly separate from the adjacent acute fracture. Its smoothly corticated surfaces and size distinguish it from the sharp margins of the acute fracture.* ***(Right)*** *Axial bone CT in the same patient demonstrates the epilunatum ⇨ relationship to the lunate →, "articulating" with the dorsal and radial cortex of the lunate.*

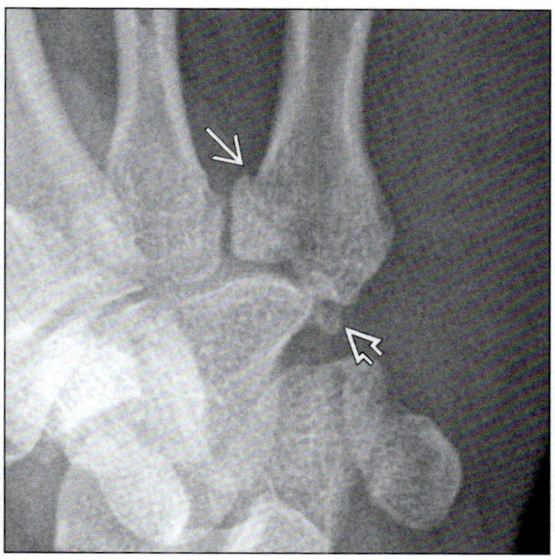

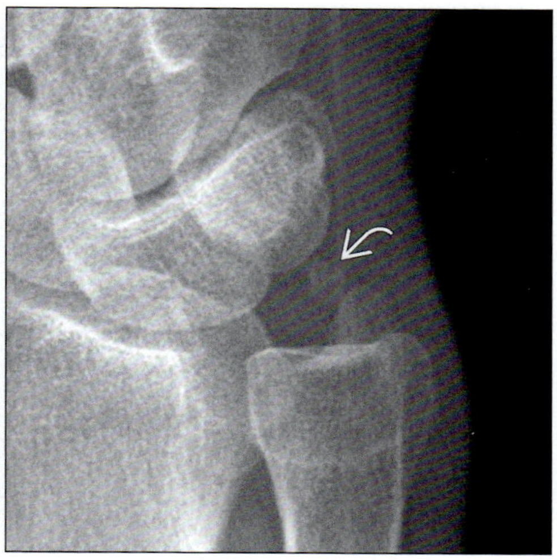

(Left) *Reverse oblique radiograph demonstrates an os hamuli proprium ⇨ located near the tip of the hook of the hamate. This was incidentally noted in this patient with an acute intraarticular 5th metacarpal base fracture →.* ***(Right)*** *PA radiograph demonstrates the lunula ↶, an accessory ossicle located distal to the ulnar styloid tip. This ossicle arises from an ossification center within the meniscal homologue. It may fuse with the styloid tip giving it an elongate appearance.*

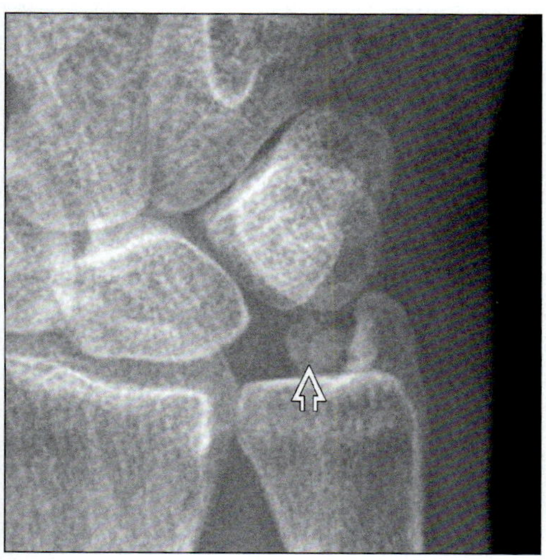

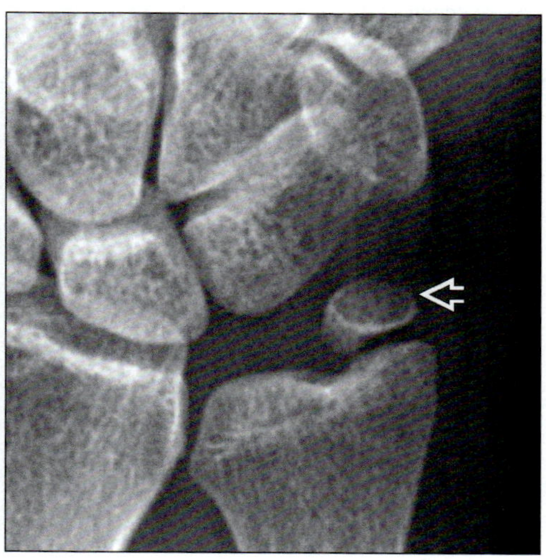

(Left) *PA radiograph demonstrates an os triangulare ⇨, just distal to the ulnar fovea and medial to the styloid process, thus distinguishing it from the lunula which is located distal to the styloid tip.* ***(Right)*** *PA radiograph demonstrates a persistent ulnar styloid ossification center ⇨. This corticated ossicle, seen in place of a normally contoured ulnar styloid, represents a failure of the ossification center to fuse with the distal ulnar during development.*

ACRONYMS AND EPONYMS, HAND AND WRIST

Key Facts

Terminology

- Summary of commonly used acronyms and eponyms related to trauma of hand and wrist
- Common acronyms
  - **DISI**: **D**orsal **i**ntercalated **s**egment **i**nstability
  - **SLAC wrist deformity**: **S**caphо**l**unate **a**dvanced **c**ollapse
  - **SNAC wrist**: **S**caphoid **n**onunion **a**dvanced **c**ollapse
  - **VISI**: **V**olar **i**ntercalated **s**egment **i**nstability
- Common eponyms
  - Baseball finger (mallet finger): Forced flexion of extended DIP results in unopposed DIP flexion
  - Boutonnière deformity: PIP flexion with DIP extension
  - Boxer fracture: 5th metacarpal neck fx
  - Coach's finger: PIP dislocation related to hyperextension injury when catching ball
  - Colles fracture: Distal radius metaphyseal fx with dorsal angulation
  - de Quervain tenosynovitis: 1st extensor compartment tenosynovitis
  - Gamekeeper's thumb (skier thumb): Thumb MCP UCL injury
  - Greenstick fracture: Incomplete long bone fx with plastic bowing on compression side and cortical break on tension side
  - Hitchhiker's thumb (duckbill deformity): Thumb IP hyperextension and MCP flexion
  - Jersey finger (rugby finger): FDP tendon avulsion results from sudden hyperextension of actively flexed finger
  - Nightstick fracture (parry fx): Isolated ulnar shaft fx; related to direct blow to ulna
  - Torus fracture: Juvenile incomplete cortical buckle fx; resembles torus contour of Grecian column

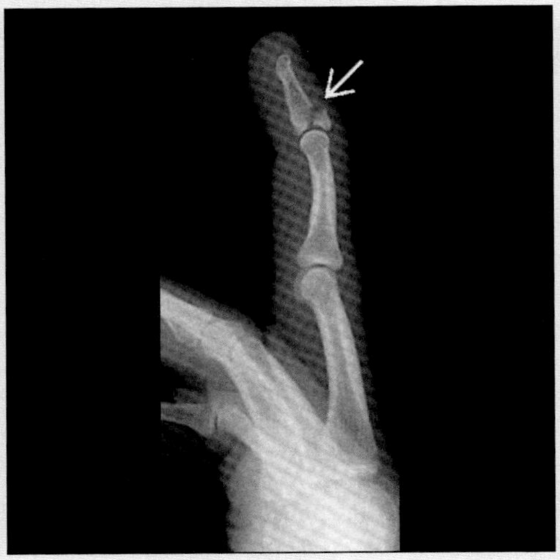

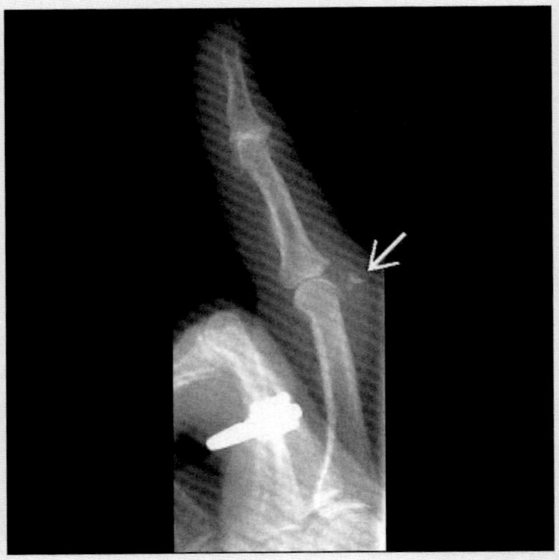

(Left) *Lateral radiograph reveals a baseball (mallet) finger. Forced flexion of an extended finger results in an avulsion ➔ where the extensor tendon inserts into the dorsal distal phalanx base.* ***(Right)*** *Lateral radiograph reveals an avulsion fracture ➔ of the extensor tendon central slip from the dorsal middle phalanx base. Note the developing boutonnière deformity with slight flexion of the PIP and extension of the DIP.*

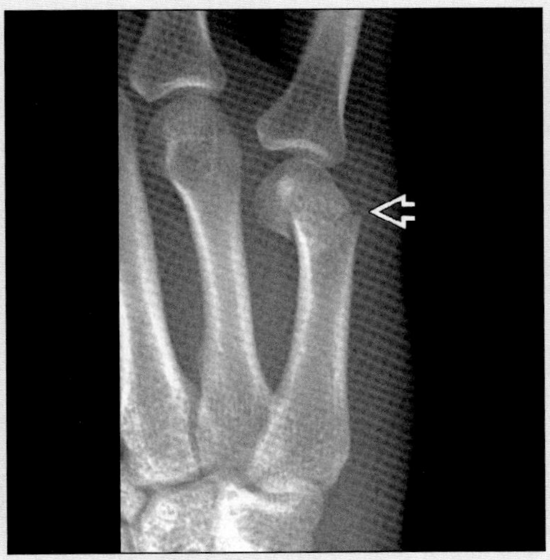

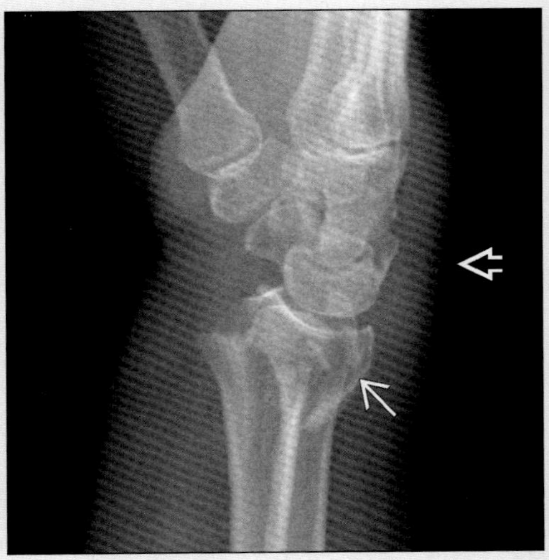

(Left) *Oblique radiograph demonstrates a boxer fracture in this 26-year-old man with poor impulse control and a poor roundhouse punch. This resulted in a 5th metacarpal neck fracture ➔ with volar angulation of the metacarpal head.* ***(Right)*** *Lateral radiograph demonstrate the "silver fork" deformity of an acute Colles fracture, with dorsal angulation and displacement ➔, as well as dorsal soft tissue swelling ➔ resulting in a contour similar to a dinner fork in profile.*

ACRONYMS AND EPONYMS, HAND AND WRIST

TERMINOLOGY

Acronyms

- **CLIP** = **C**apito**l**unate **i**nstability **p**attern: Related to radiocapitate ligament disruption
- **CRPS** (RSD) = **C**omplex **r**egional **p**ain **s**yndrome: Chronic, progressive, severe pain; swelling and skin changes may be seen after injury
- **DISI** = **D**orsal **i**ntercalated **s**egment **i**nstability: Lunate dorsiflexed > 15° (greater range may be allowed) compared to capitate; scapholunate ligament injury associated
- **SLAC wrist deformity** = **S**cholunate **a**dvanced **c**ollapse: Scapholunate interval widening with capitate invaginating between scaphoid and lunate
- **SNAC wrist** = **S**caphoid **n**onunion **a**dvanced **c**ollapse: Arthritis following scaphoid nonunion with carpal collapse
- **VISI** = **V**olar **i**ntercalated **s**egment **i**nstability: Lunate volarly flexed > 15° (greater range may be allowed) compared to capitate; lunotriquetral ligament injury associated

Eponyms

- **Barton fracture**: Intraarticular distal radius dorsal rim fx; carpals dislocate dorsally, remaining with fx fragment
- **Baseball finger** (mallet finger): Forced flexion of extended DIP with injury to extensor tendon terminal band, ± associated dorsal distal phalanx base avulsion fx, results in unopposed DIP flexion
- **Bennett fracture**: Oblique intraarticular fx of 1st metacarpal base resulting in small triangular volar lip fragment
- **Boutonnière deformity**: Extensor tendon central slip injury at or near insertion on middle phalanx, ± avulsion fx of dorsal middle phalanx base, results in PIP flexion with DIP extension
- **Boxer fracture**: 5th metacarpal neck fx, metacarpal head volar angulation
- **Boxer knuckle**: Extensor hood sagittal band tear at MCP with subluxation/dislocation of common extensor tendon, results from direct blow while joint flexed and finger ulnarly deviated
- **Chauffeur fracture** (Hutchinson fx): Oblique intraarticular radial styloid fx; injury sustained when early 20th century automobile backfired while chauffeur was handcranking engine and crankshaft recoiled, striking wrist
- **Coach's finger**: PIP dislocation, typically dorsal; ± volar plate injury; related to hyperextension injury from catching ball
- **Colles fracture** (Pouteau fx): Distal radius metaphyseal fx with dorsal angulation, ± dorsal displacement, ± ulnar styloid fx
- **de Quervain fracture**: Transscaphoid perilunate fracture-dislocation
- **de Quervain tenosynovitis** (washerwoman sprain, mother wrist): 1st extensor compartment (EPB and APL) tenosynovitis
- **Dupuytren contracture**: Hand/finger contracture due to palmar fascia/skin fibrosis
- **Essex Lopresti fracture**: Radial head fx with associated distal radioulnar joint dislocation
- **Galeazzi fracture** (reverse Monteggia fx): Distal radial shaft fx with DRUJ dislocation, ± ulnar styloid fx
- **Gamekeeper's thumb** (skier's thumb): Thumb MCP UCL injury, related to hyperabduction, ± avulsion fx
- **Greenstick fracture**: Incomplete long bone fx with plastic bowing on compression side and cortical break on tension side
- **Gymnast wrist**: Distal radial physeal injury with resultant osteolysis, related to repetitive trauma
- **Hitchhiker's thumb** (duckbill deformity): Deformity of thumb with IP hyperextension and MCP flexion; commonly seen in RA or after tendon injury
- **Hypothenar hammer syndrome**: Thrombosis of ulnar artery superficial palmar arch, related to repetitive trauma to heel of hand, ± associated ulnar artery aneurysm
- **Humpback deformity**: Dorsal angulation between proximal and distal fragments of scaphoid fx
- **Intersection syndrome**: Radial wrist extensor tenosynovitis at 1st (EPB/APL) and 2nd (ECRB/ECRL) extensor compartment intersection
- **Jersey finger** (rugby finger): FDP tendon avulsion; most commonly affects ring finger; results from sudden hyperextension of actively flexed finger (i.e., grabbing opponent's jersey)
- **Kienbock disease**: Lunate osteonecrosis
- **Lead-pipe fracture**: Incomplete long bone fx; cortical break on tension side and buckled cortex on compression side
- **Nightstick fracture** (parry fx): Isolated mid to distal ulnar shaft fx, related to direct blow to ulna
- **Preiser disease**: Scaphoid osteonecrosis
- **Reverse Barton fracture** (volar Barton fx): Intraarticular distal radius volar lip fx; carpals displace volarly with fx fragment
- **Rolando fracture**: Intraarticular comminutes 1st metacarpal base fx; results in volar & dorsal lip fragments, forms Y- or T-shaped fragment
- **Silver fork deformity**: Colles fx with dorsal displacement resulting in deformity like dinner fork
- **Skier's thumb** (gamekeeper thumb): Resulting from fall against ski pole
- **Smith fracture** (reverse Colles fx): Distal radial metaphyseal fx with volar angulation/displacement
- **Stener lesion**: Thumb UCL tear with retraction of torn ligament fibers proximal to adductor pollicis aponeurosis
- **Swan neck deformity**: PIP extension with DIP flexion; results from volar plate injury at PIP or extensor tendon injury at DIP
- **Terry Thomas sign**: Scapholunate ligament disruption with interspace widening; mimics space between front teeth (a.k.a. "David Letterman" sign)
- **Trigger finger**: Locking, catching, or popping of finger through range of motion due to tenosynovitis
- **Torus fracture**: Juvenile incomplete impaction fx resulting in cortical buckling; resembles torus contour of Grecian column
- **Volkmann contracture**: Hand/wrist flexion contractures due to forearm compartment syndrome

SELECTED REFERENCES

1. Green DP et al. Green's Operative Hand Surgery. 5th ed. Philadelphia: Elsevier Churchill Livingstone, 2005

(Left) *Lateral radiograph demonstrates DISI malalignment with lunate tilting dorsally, increasing lunate ➔ capitate ➔ angle as well as the scaphoid ➔ lunate angle. Scapholunate ligament tear is often associated.* ***(Right)*** *Lateral radiograph demonstrates a VISI malalignment with the lunate ➔ tilting volarly, decreasing the scaphoid ➔ lunate angle and increasing the capitate ➔ lunate angle. This is often associated with lunotriquetral ligament tear.*

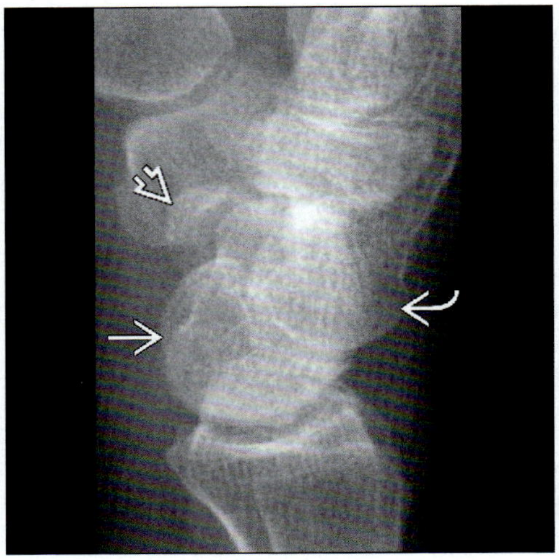

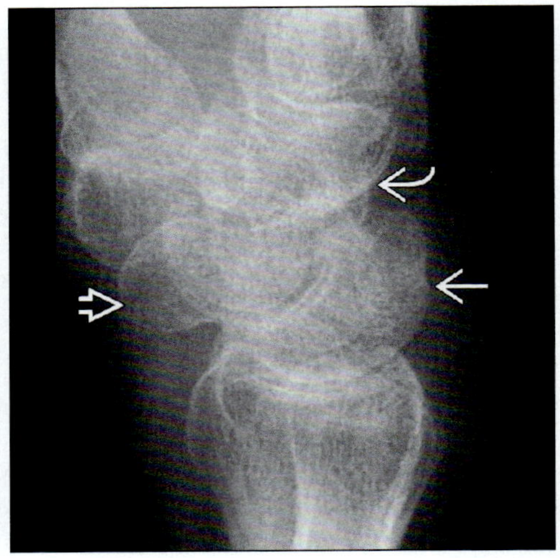

(Left) *Graphic depicts de Quervain syndrome or tenosynovitis. Inflammation and thickening of the APL ➔ and EPB ➔ tendon sheaths result in pain with range of motion, particularly in thumb abduction.* ***(Right)*** *Coronal GRE MR demonstrates a gamekeeper thumb with tear of the ulnar collateral ligament ➔ with fluid interposed between the proximal phalanx base ➔ and the detached ligament fibers. The aponeurosis is intact ➔.*

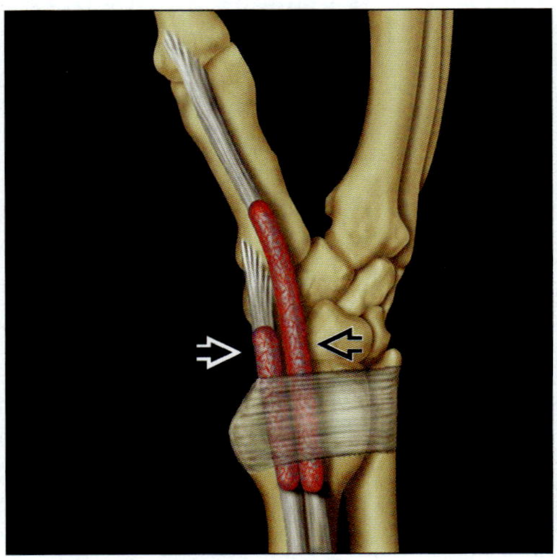

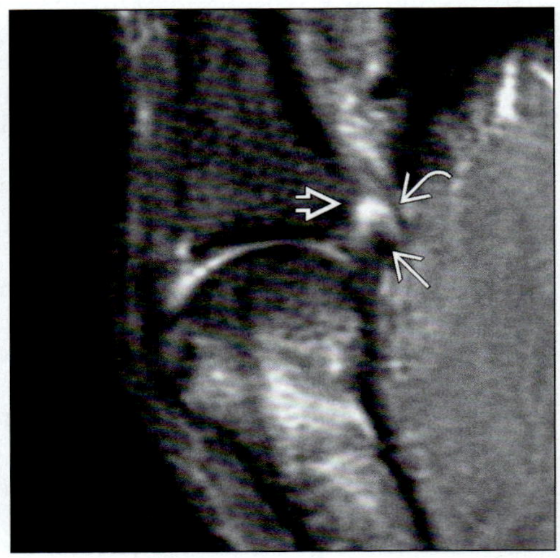

(Left) *Oblique radiograph in RA patient shows marked digit deformities related to volar plate failure with "swan neck" finger deformities ➔ (PIP extension with DIP flexion) & duckbill or hitchhiker deformity ➔ with IP hyperextension and MCP hyperflexion.* ***(Right)*** *Sagittal bone CT shows a humpback deformity ➔ of the scaphoid. This results from scaphoid fracture with acute dorsal angulation of the fracture fragments. While the fracture has been grafted and stabilized, it has healed with this persistent deformity.*

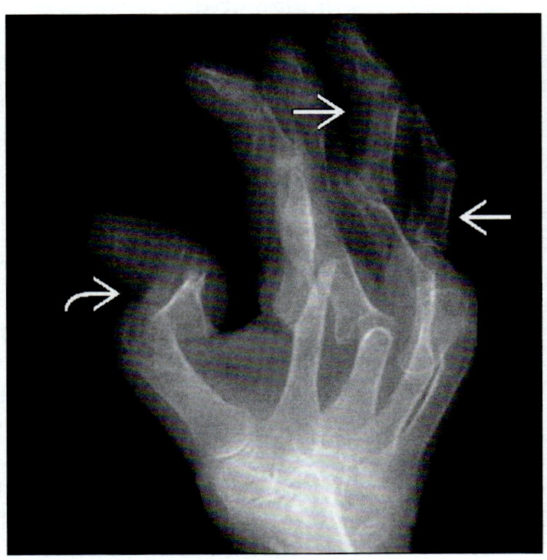

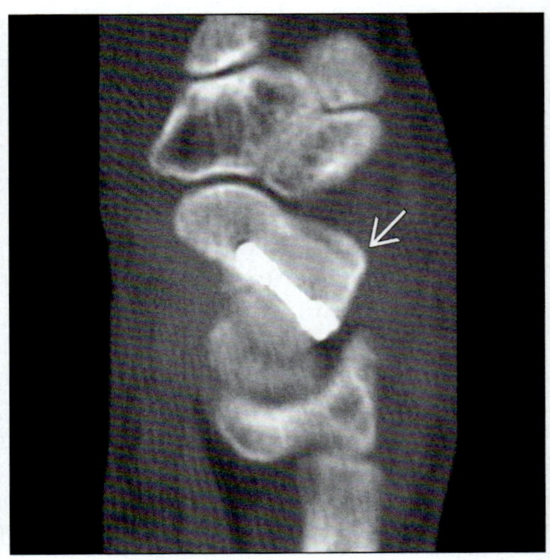

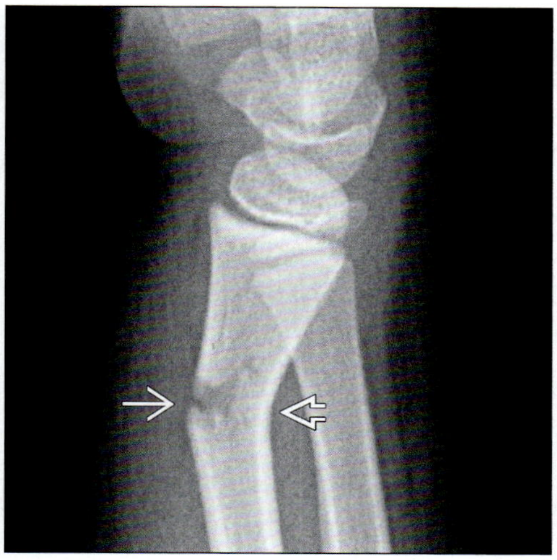

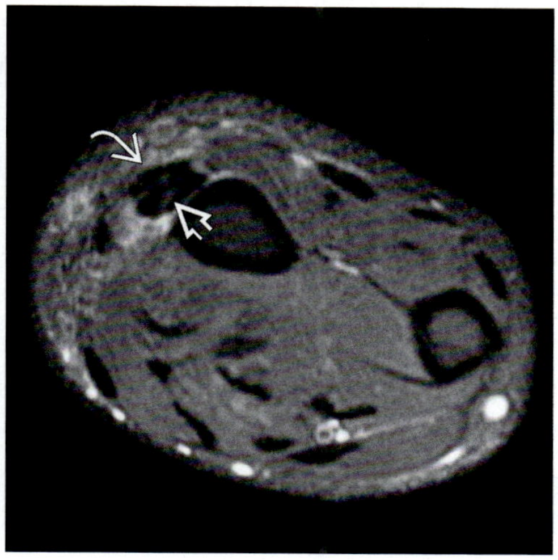

(Left) *Sagittal radiograph demonstrates a distal radius greenstick fracture, resulting from a fall on an outstretched hand. A cortical break occurs on the tension side of the fracture, while there is a plastic bowing deformity located on compression side.* ***(Right)*** *Axial PDWI FS MR demonstrates intersection syndrome in this 26-year-old competitive rower. Friction at the intersection between the 1st (EPB/APL) & 2nd (ECRB/ECRL) extensor tendon compartments results in tenosynovitis & soft tissue inflammation.*

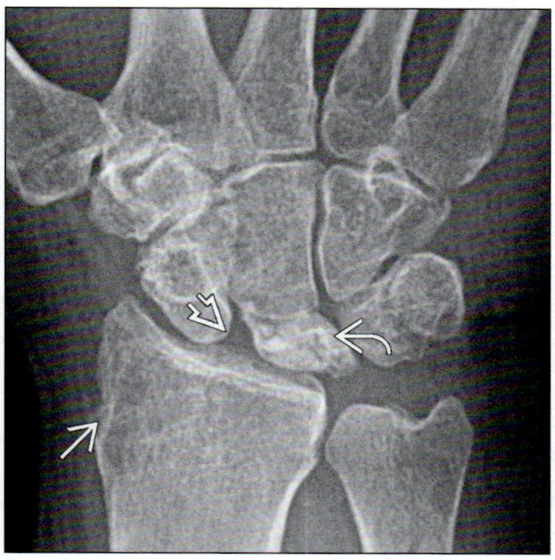

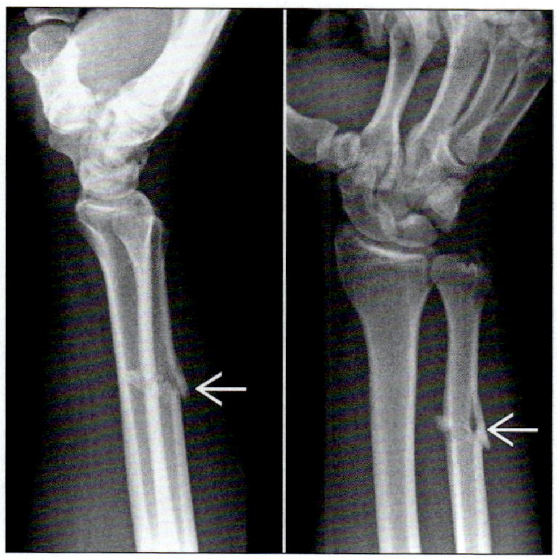

(Left) *PA radiograph demonstrates Kienbock disease with a radiodense, flattened, irregular lunate . Scapholunate interval widening suggests SL ligament tear. A subtle distal radial styloid fracture is also present.* ***(Right)*** *Lateral and oblique radiographs demonstrate an isolated ulnar shaft fracture resulting from a direct blow. This is known as a parry or nightstick fracture.*

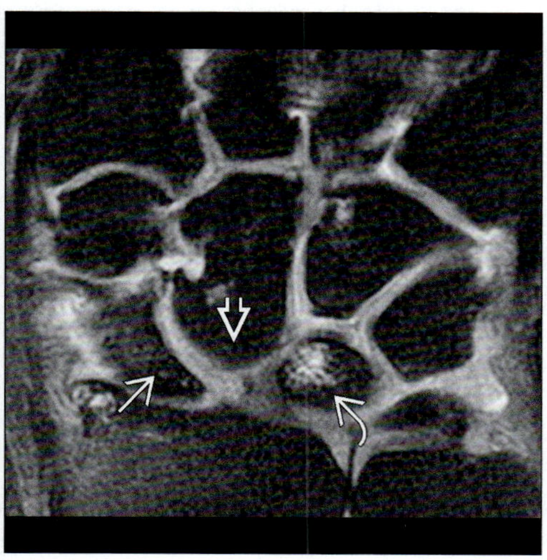

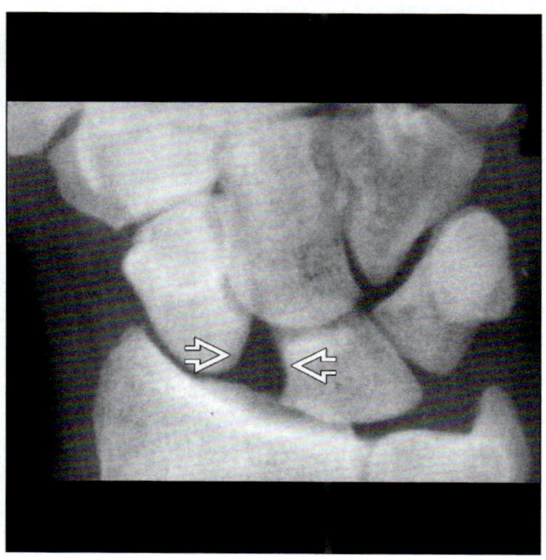

(Left) *Coronal STIR MR demonstrates SLAC wrist deformity resulting from chronic SL ligament tear with proximal migration of capitate interposed between the scaphoid and lunate .* ***(Right)*** *PA radiograph demonstrates widening of scapholunate interval due to SL ligament tear. The resultant gap is reminiscent of Terry Thomas (or David Letterman) tooth gap.*

JUVENILE DISTAL FOREARM FRACTURES

Key Facts

Imaging

- Torus/buckle fx: Cortical buckling, eccentric crimping
- Greenstick fracture: Transverse cortical break without fracture of opposite cortex
- Bowing fracture: Diaphyseal bowing; often seen in only 1 view
- Growth plate fracture: Salter-Harris type I and II most common
- Both-bone forearm fractures: May be complete fxs or combination of complete and incomplete fxs
- Galeazzi fracture/dislocation: Radial fracture with DRUJ disruption
- Fracture may involve 1 or both bones in various combinations

Clinical Issues

- Most common mechanism: FOOSH
- Healing potential
  - Torus/buckle fracture: Stable
  - Greenstick fracture: Unstable; may re-angulate; must re-image
  - Bowing fracture: Will remodel if bowing < 10°; young children
  - Growth plate fracture: Distal physes account for 75-80% of total forearm growth; excellent remodeling potential
  - Galeazzi fracture/dislocation: Heals with adequate shaft and DRUJ reduction

Diagnostic Checklist

- Is fracture on compression (torus) or tension (greenstick) side?
- If shaft fx(s) present, evaluate wrist & elbow integrity
- Confirm injury is consistent with history; exclude nonaccidental trauma

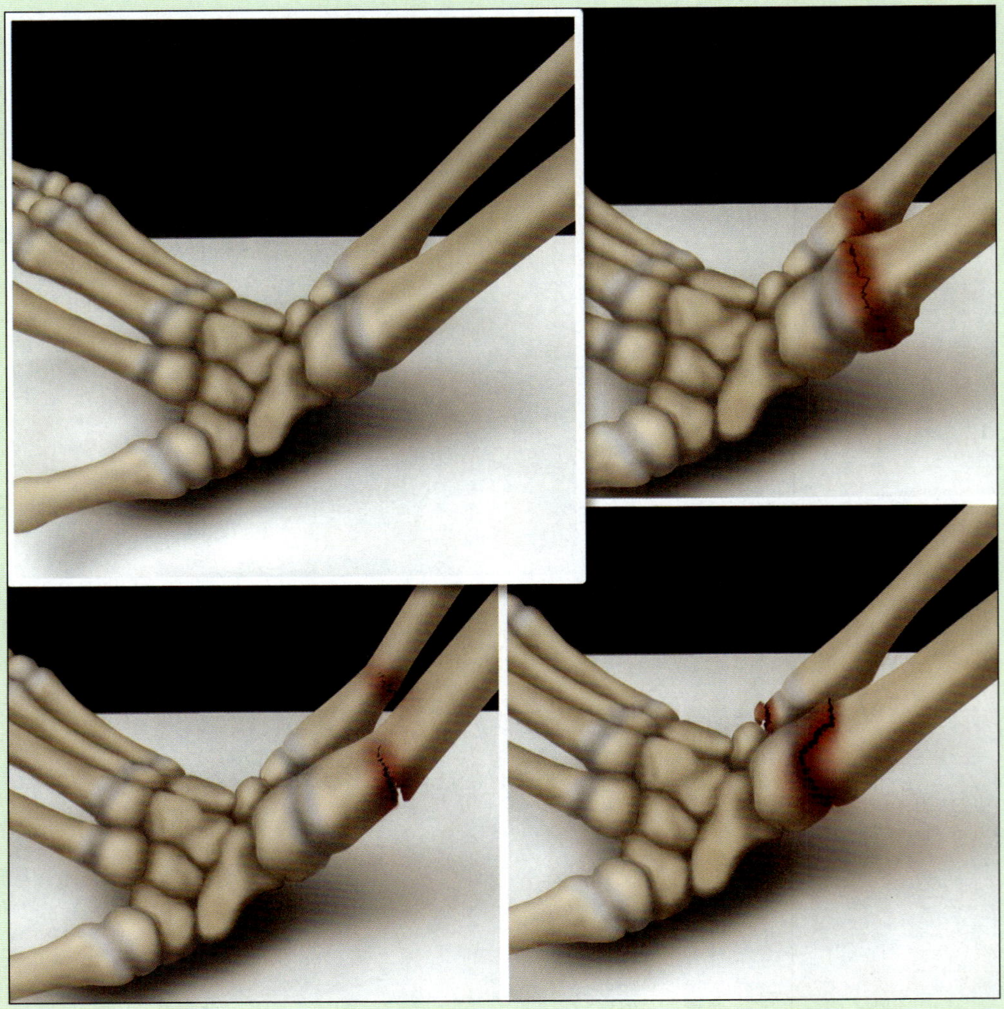

Falling on an outstretched hand (top left) may result in a forearm fracture. Resulting fractures in children are uniquely different than those occurring in adults due to the increased elasticity of the immature skeleton. The torus fracture (top right) results with low-energy injury, buckling the cortex. With increasing energy, the greenstick fracture (bottom left) may occur, with a cortical break along the tension side and plastic deformation on the compression side of this fracture. The growth plate is also at risk. A type II Salter-Harris fracture is the most common injury, with the fracture propagating across the growth plate and exiting through the metaphysis (bottom right).

JUVENILE DISTAL FOREARM FRACTURES

IMAGING

General Features

- Best diagnostic clue
  - Cortical buckling
  - Fracture of 1 or more cortices
  - Angulation or bowing
  - Pronator fat pad effacement
- Location
  - Torus: Distal metaphysis
  - Greenstick: Middle > distal > proximal
  - Growth plate: Distal physis
  - Both-bone: Distal = 75%, mid = 18%, proximal = 7%

Radiographic Findings

- Radiography
  - Fracture may involve 1 or both bones in various combinations
  - Torus/buckle fracture
    - Cortical buckling or eccentric crimping
    - Subtle sclerosis across metaphysis
    - Fracture occurs on compression side
  - Greenstick fracture
    - Transverse cortical break extends across bone; becomes longitudinal without fracture of opposite cortex
    - Fracture occurs on tension side
    - Lead pipe fx: Combination of greenstick and torus
  - Bowing fracture
    - Abnormal bowing of diaphysis
    - Usually seen in only 1 projection
    - May require comparison to unaffected forearm to evaluate
  - Growth plate fracture
    - Salter-Harris type I and II most common
    - Type I: Pronator quadratus fat pad displacement
    - Type II: Typically dorsally displaced; metaphyseal fragment may be very small
    - Type V: May not be identified until 6-12 months later when physeal fusion occurs
    - Radius fracture most common ± ulnar metaphyseal, distal epiphyseal, or styloid fracture
  - Both-bone forearm fractures
    - May be complete fracture or combination of complete and incomplete
  - Galeazzi fracture/dislocation
    - Radial shaft fracture
    - Distal radioulnar joint (DRUJ) subluxation/ dislocation

Imaging Recommendations

- Best imaging tool
  - Radiographs
    - Orthogonal views centered on fracture site
    - Radiographs not taken in plane of maximal deformity falsely minimize deformity
  - MR or ultrasound may be useful in assessing growth plate injury in children under 4 (epiphysis not yet ossified)
- Protocol advice
  - Imaging centered on forearm is inadequate to evaluate adjacent joint
  - Dedicated wrist &/or elbow imaging needed to evaluate joint integrity

PATHOLOGY

General Features

- Etiology
  - Deforming forces
    - Muscle attachments help determine fragment position
    - Biceps, supinator insert on proximal 1/3 radius: Supinate fragments
    - Pronator teres inserts mid 1/3 radius: Supinates proximal fragment; pronates distal fragment
    - Pronator quadratus inserts distal 1/3 radius: Neutral proximal fragment; pronates distal fragment when fracture distal to teres
    - Flexors pull distal fragments, producing dorsal bowing
  - Torus/buckle fracture
    - Cortex failure with compression
    - Occurs at transition between woven metaphyseal bone and lamellar diaphyseal bone
    - Tends to be stable fracture
  - Greenstick fracture
    - Fracture occurs on tension side (convexity of angulation) while plastic deformation occurs along concavity
    - Tends to be unstable
    - Dorsal fx sustained with forearm in pronation
    - Volar fracture sustained with forearm in supination
    - Lead pipe fracture: Combined compression (torus fracture) and angular (greenstick) forces
  - Bowing fracture
    - Plastic deformation results from forces great enough to bend bone beyond its natural resiliency without fracture
    - Microscopic fractures found along bowing concavity
    - Often seen in combination with complete greenstick fracture of other bone, bowing of other bone &/or dislocation
  - Growth plate fracture
    - Shearing forces of hyperextension and supination displace distal radial epiphysis dorsally
    - Distal radial fractures represent 30% of all physeal injuries
    - Often associated with ulnar greenstick fracture ± ulnar styloid fracture
  - Both-bone forearm fractures
    - Distal 1/3: May be combination of greenstick, torus, or physeal
      - Galeazzi fracture/dislocation: Displaced distal radial fracture with ulnar dislocation at DRUJ; 3x as common as Monteggia fracture
      - Axial loading on hyperpronated forearm
    - Mid 1/3: Displaced midshaft fracture with radial fracture proximal to ulnar fracture especially unstable; prone to re-displacement
      - Fractures unstable, difficult to reduce because of muscle mass
      - Rotational displacement: Look at difference in fracture margin cortical widths to identify rotation
    - Proximal 1/3: Proximal ulnar fracture with radial head dislocation (Monteggia fracture)

JUVENILE DISTAL FOREARM FRACTURES

- Proximal radial fracture with ulnar shaft fracture (Monteggia equivalent)

CLINICAL ISSUES

Presentation

- Most common signs/symptoms
  - Swelling, tenderness, deformity
  - Inability to rotate forearm
- Other signs/symptoms
  - If isolated radius fracture, examine DRUJ for dislocation (Galeazzi fracture/dislocation)

Demographics

- Age
  - Occur at all ages
    - Growth plate fx: Most common in 10-15 year olds
      - Average age: Boys = 12, girls = 11
    - Complete fxs: Most common in 10-14 year olds
- Gender
  - Under 4 years of age: Male = female
  - Over 4 years of age: Male > female
- Epidemiology
  - Most common mechanism
    - Fall on outstretched hand (FOOSH)
  - Sports
    - In-line skating, skateboarding, scooter riding, mountain biking, contact sports
    - Soccer: 79% from fall, 21% direct blow
    - Wheeled sports: 5% open fracture compared to 0.3% open fracture by other mechanisms
      - Skateboarding: Average age: 13 years (95% male)
      - Roller skating: Average age: 10.8 years (54% male)
      - Scooter riding: Average age: 9.7 years (64% male)

Natural History & Prognosis

- Torus/buckle fracture
  - Stable; typically heal without sequelae
- Greenstick fracture
  - May re-displace/angulate over time; must re-image at 1-2 weeks to evaluate
  - Recurrent angulation more likely with volar fracture or radial greenstick with intact ulna
  - Midshaft radial fracture proximal to ulnar fracture; > 50% will re-displace/angulate
- Bowing fracture
  - Remodeling occurs in younger child (< 5 years); bowing < 10°
  - Deformity may persist in older children or angulation > 15°
- Growth plate fracture
  - Distal physes account for 75-80% of total forearm growth; excellent remodeling potential
  - < 30% physeal displacement heals/remodels in 1.5 years
  - > 50% displacement may heal with deformity
  - Forced manipulation can damage growth plate
- Galeazzi fracture/dislocation
  - Heals with adequate shaft fx & DRUJ reduction
  - Disruption of DRUJ results in higher morbidity if overlooked
- Complications
  - Swelling or direct nerve/soft tissue entrapment
    - Volkmann ischemic contracture from untreated compartment syndrome
    - Nerve entrapment: Ulnar, median

Treatment

- Depends on fracture type and degree of deformity
- Nonsurgical: Cast or splint
  - Torus
    - Cast for comfort 3-4 weeks
  - Growth plate fracture
    - Manual reduction; long arm cast
    - Reassess periodically as growth proceeds (even after healing)
  - Greenstick fracture
    - May need to create complete fracture to get adequate reduction
    - Reduced by rotating forearm so palm directed toward fracture apex
    - Acceptable reduction: Infants up to 30° angulation; children up to 15°
  - Bowing fracture
    - Reduce under general anesthesia for bowing deformities > 10°
  - Galeazzi fracture/dislocation
    - Reduce radial fracture and DRUJ dislocation; cast in full supination
- Surgical treatment
  - Unstable fracture; displaced segments
  - Open fracture
  - Manual manipulation fails to realign fracture
  - Healing in malalignment

DIAGNOSTIC CHECKLIST

Consider

- Is fracture on compression (torus) or tension (greenstick) side?
- Confirm injury consistent with history given

Image Interpretation Pearls

- If shaft fx(s) present, evaluate wrist & elbow integrity

Reporting Tips

- Description to include
  - Location
  - Fracture completeness
  - Fragment displacement, angulation, rotation, bowing

SELECTED REFERENCES

1. Zalavras C et al: Pediatric fractures during skateboarding, roller skating, and scooter riding. Am J Sports Med. 33(4):568-73, 2005
2. Vorlat P et al: Bowing fractures of the forearm in children: a long-term followup. Clin Orthop Relat Res. (413):233-7, 2003
3. Vittas D et al: Angular remodeling of midshaft forearm fractures in children. Clin Orthop Relat Res. (265):261-4, 1991
4. Walsh HP et al: Galeazzi fractures in children. J Bone Joint Surg Br. 69(5):730-3, 1987
5. Light TR et al: The anatomy of metaphyseal torus fractures. Clin Orthop Relat Res. (188):103-11, 1984

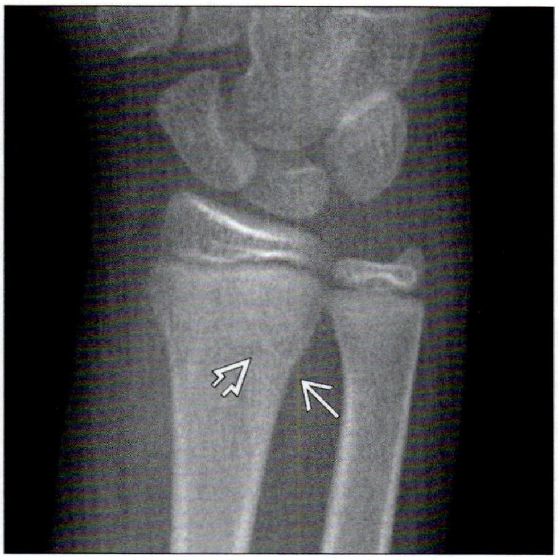

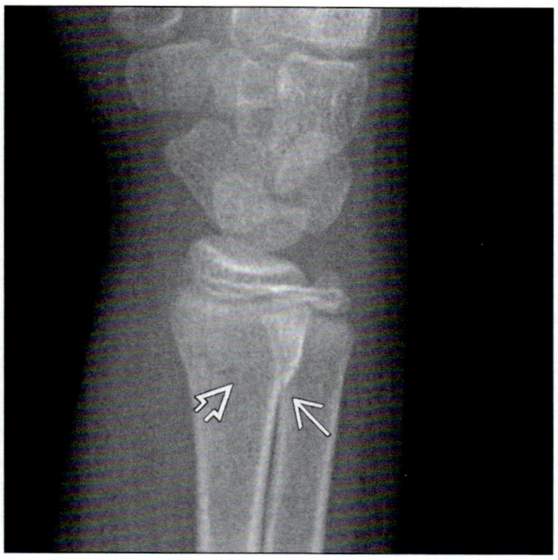

(Left) *AP radiograph of an 11 year old who fell on an outstretched hand while running on a soccer field demonstrates a very subtle radial cortical buckle → with a slight sclerotic line → across the medial metaphysis.* ***(Right)*** *Lateral oblique radiograph in the same patient clearly demonstrates posterolateral radius cortical crimping → and a metaphyseal fracture line →.*

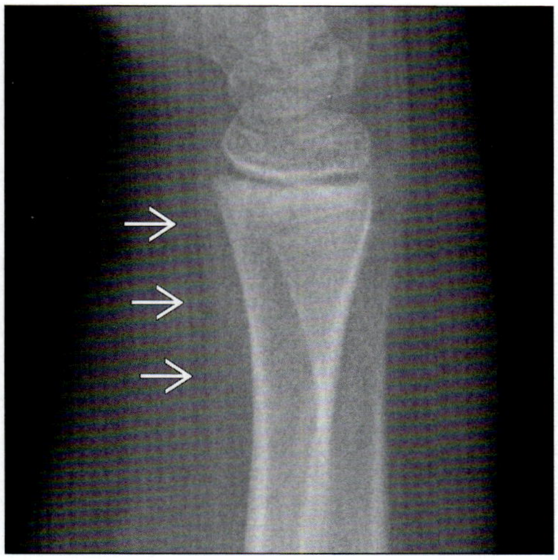

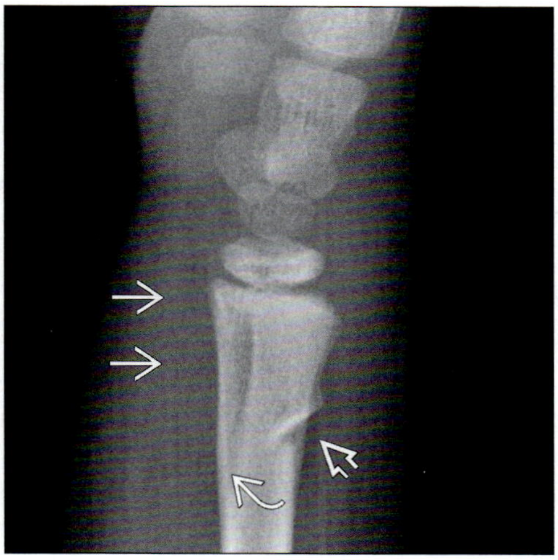

(Left) *Lateral radiograph demonstrates the typical appearance of the normal pronator quadratus fat pad →.* ***(Right)*** *Lateral radiograph shows pronator quadratus fat pad effacement → in a 7 year old with FOOSH injury. Cortical buckling → is prominent dorsally. This incomplete fracture hinges → at the volar surface.*

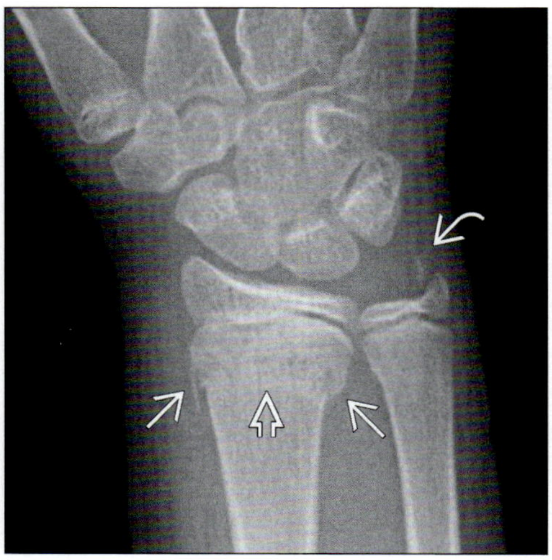

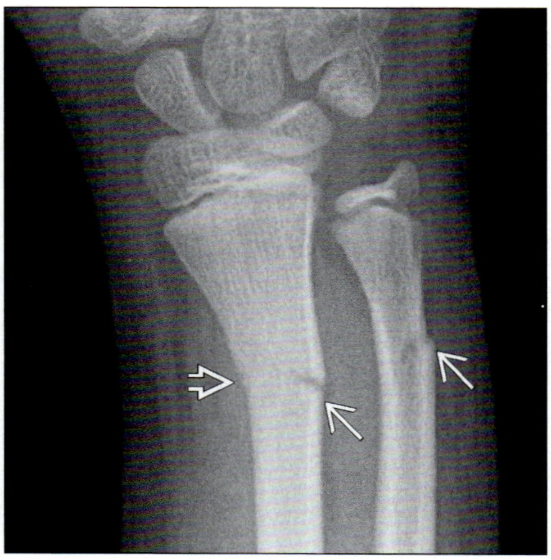

(Left) *AP radiograph in a 10 year old demonstrates bulging medial and lateral radial cortices → with band-like sclerosis → crossing the metaphysis in this torus fx. Note the associated ulnar styloid avulsion fx →.* ***(Right)*** *AP radiograph in a 12 year old shows incomplete greenstick fractures of distal radius & ulna with transverse fx lines → propagating laterally. Lateral cortices of both the radius → and ulna are intact. These unstable fxs may re-angulate & should be imaged 1-2 weeks after reduction & casting.*

JUVENILE DISTAL FOREARM FRACTURES

(Left) AP radiograph demonstrates a lead pipe fracture, a combination of greenstick ➡ and torus ⇨ resulting from angular and compressive forces. *(Right)* AP radiograph shows a distal radial metaphyseal fracture ➡ with slight displacement. Single bone fractures are not unusual, but careful inspection reveals a concomitant nondisplaced ulnar styloid fracture ↪.

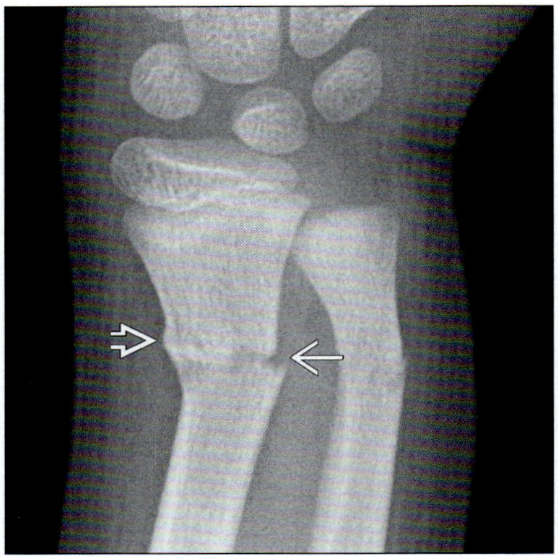

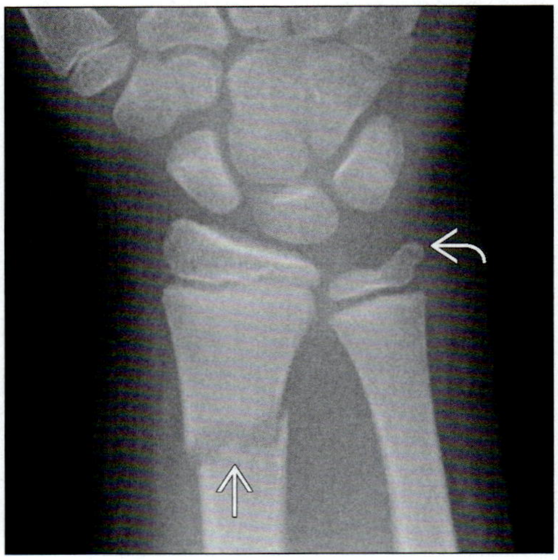

(Left) AP forearm radiograph in a 6 year old shows both-bone fractures. A greenstick distal ulnar diaphysis fx ➡ is seen with a concomitant radius bowing deformity (> 15°) ⇨. Ulnar fx may need to be completed to attain satisfactory alignment. *(Right)* Lateral oblique radiograph shows subtle posterior radius metaphysis fragment ➡ & slight physeal widening, typical of Salter-Harris II injury. FOOSH injury most often results in dorsal displacement ⇨ of the distal fx fragment, as in this case.

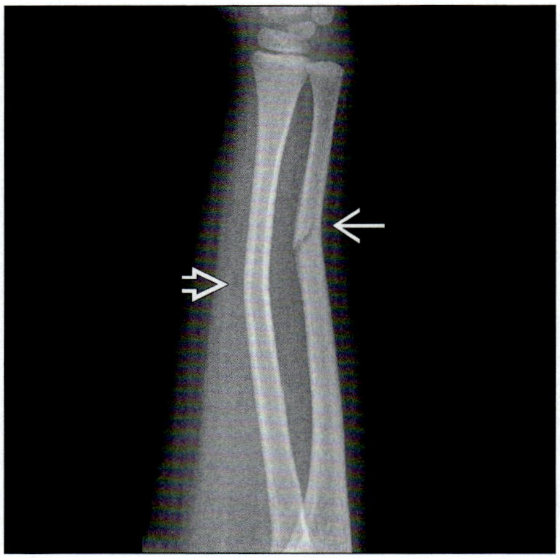

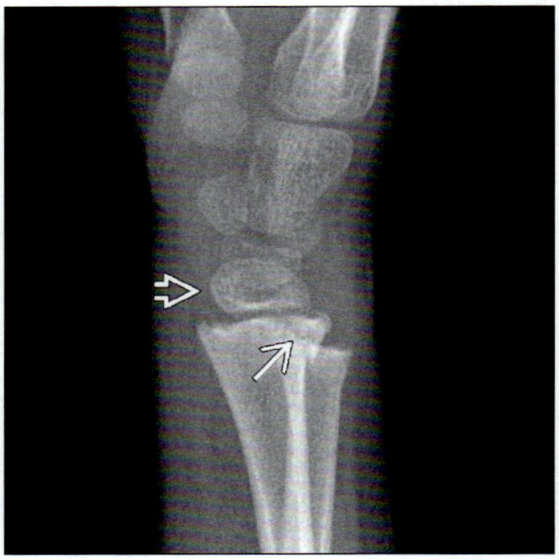

(Left) Lateral radiograph (in splint) shows dorsal physis displacement ⇨ and a small dorsal metaphyseal fragment ➡ in this acute Salter-Harris II fracture in a 12 year old. Splint obscured this fracture on all other views. *(Right)* Salter-Harris II fracture is demonstrated on this lateral oblique radiograph, with a volar metaphyseal fragment ➡ and volar displacement of radial epiphysis ⇨. This 8 year old sustained the fracture with fall on hand in supination.

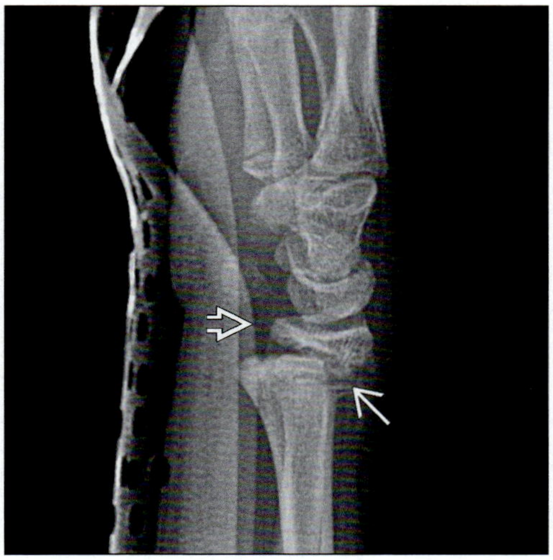

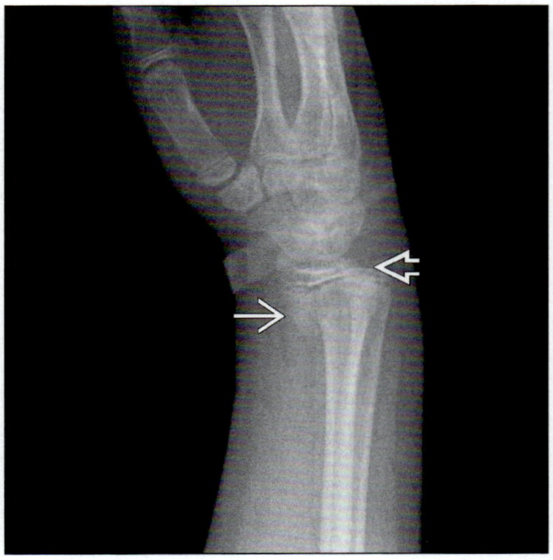

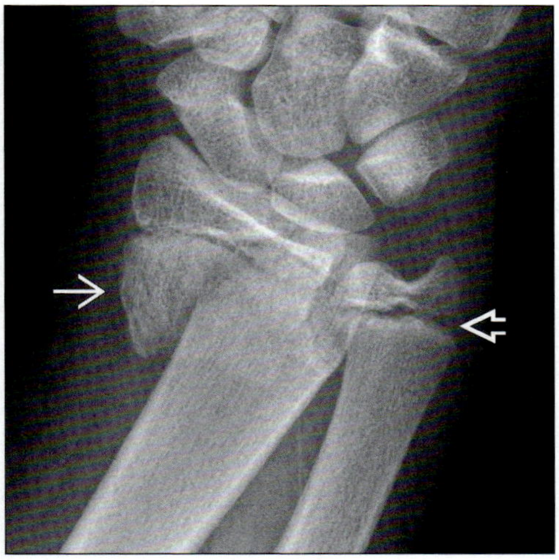

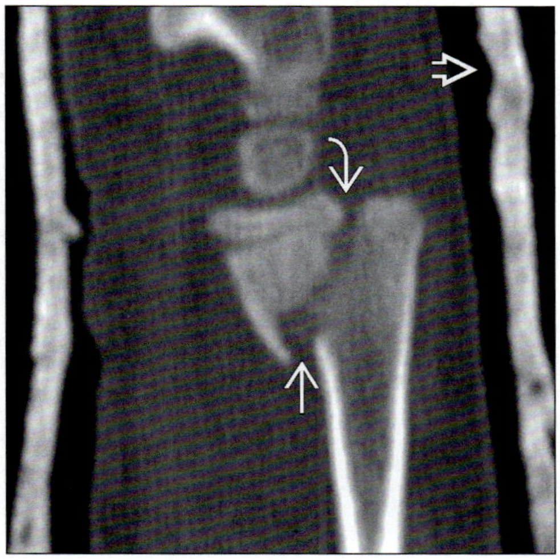

(Left) *AP oblique radiograph in a 14 year old shows a distal radius metaphyseal fracture ➡ and slight widening of ulnar physis ➡. The radiograph is limited due to fracture complexity.* ***(Right)*** *Sagittal reformation bone CT in the same patient confirms Salter-Harris IV fracture of radius, with fracture line running through metaphysis ➡, physis, and dorsal epiphysis ➡ with volar displacement of fracture fragments. CT was obtained with patient in cast ➡.*

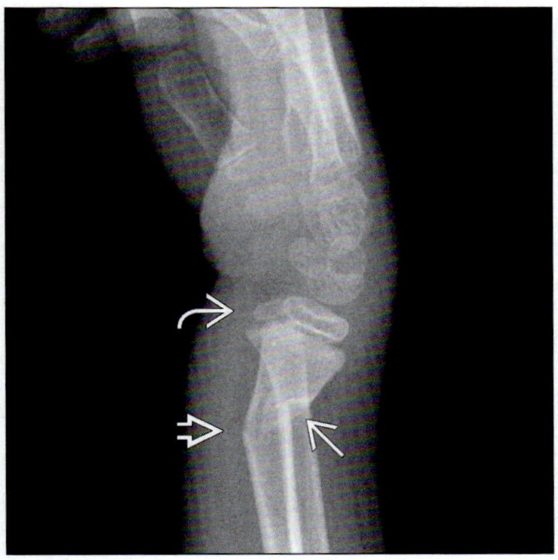

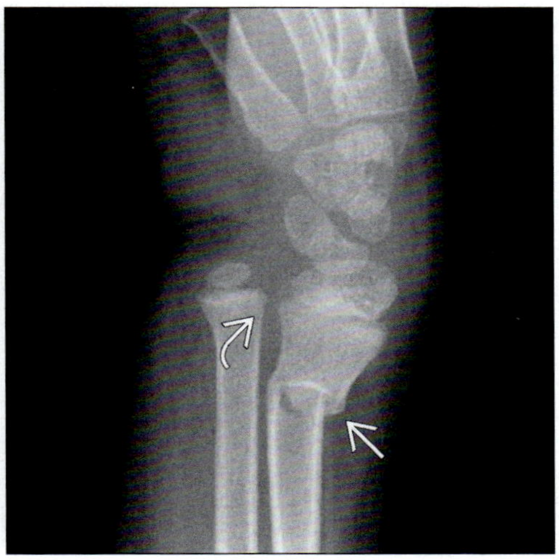

(Left) *Lateral radiograph in an 8 year old with FOOSH injury demonstrates a distal radius metaphyseal greenstick fracture, with dorsal cortical break ➡ while the volar cortex ➡ is bowed but remains intact. Note the volar position of the distal ulna ➡ compared to radius.* ***(Right)*** *AP oblique radiograph in the same patient confirms a Galeazzi fracture/dislocation with the distal radius greenstick fracture displaced dorsally & radially ➡ as well as DRUJ dislocation ➡ with an intact distal ulna.*

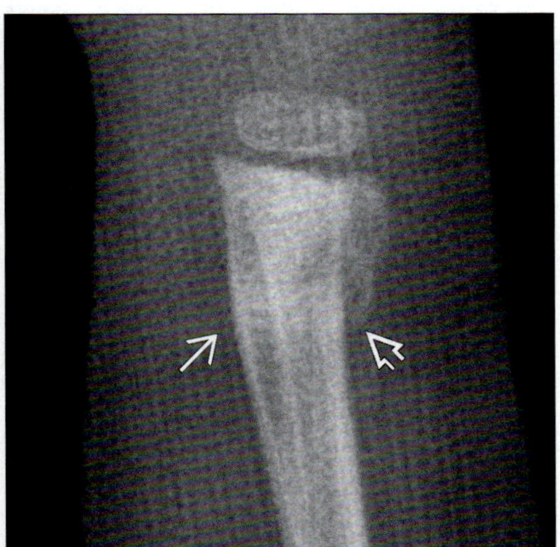

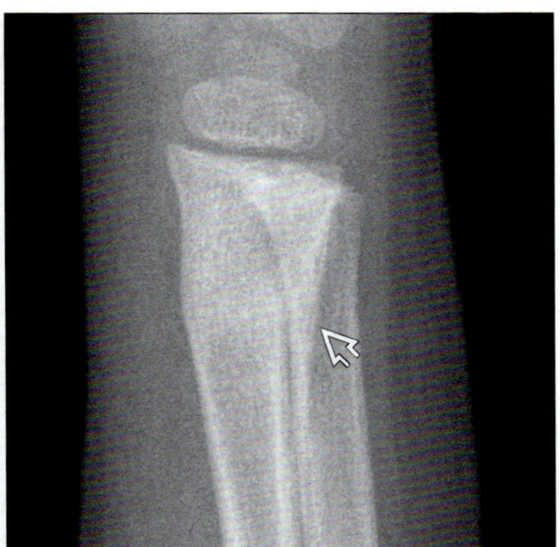

(Left) *Lateral radiograph shows the post-reduction imaging in a 6 year old who suffered a distal radial greenstick fracture following a fall on an outstretched hand. Note the dorsal cortical disruption ➡ while the volar cortex remains intact but bowed ➡. A fiberglass cast is present.* ***(Right)*** *Lateral radiograph in the same patient, taken after cast removal 4 weeks later, shows the fracture is healed & remodeling of the residual deformity is already occurring, particularly along the dorsal cortex ➡.*

DISTAL RADIUS FRACTURE

Key Facts

Terminology

- Colles: Transverse fracture through metaphysis with dorsal angulation and displacement
- Smith (reverse Colles): Transverse fracture through metaphysis with volar angulation and displacement
- Barton: Dorsal lip intraarticular fracture
- Reverse Barton: Volar lip intraarticular fracture
- Chauffeur (Hutchinson): Oblique intraarticular radial styloid fracture
- Die-punch: Lunate fossa impaction fracture
- Complex intraarticular fracture: Generally 3 intraarticular fragments, often with angulation
- Traumatic osteolysis: Irregular lysis of distal metaphysis adjacent to growth plate

Diagnostic Checklist

- Not all distal radial fracture are Colles, and should not be identified as such!
- Differences relate to fracture line placement and angular deformity, which must be described accurately
- Reduction technique and treatment varies for different fracture patterns
- All distal radial fractures should be specifically evaluated for
  - Loss of volar tilt
  - Loss of radial angle
  - Shortening of radius relative to ulna
  - Intraarticular extension
  - Diastasis/depression at intraarticular fracture site
  - Comminution: Angulation and displacement of major fragments
  - Ulnar styloid fracture
  - Disruption of carpal arcs to indicate more extensive injury
  - Associated injuries

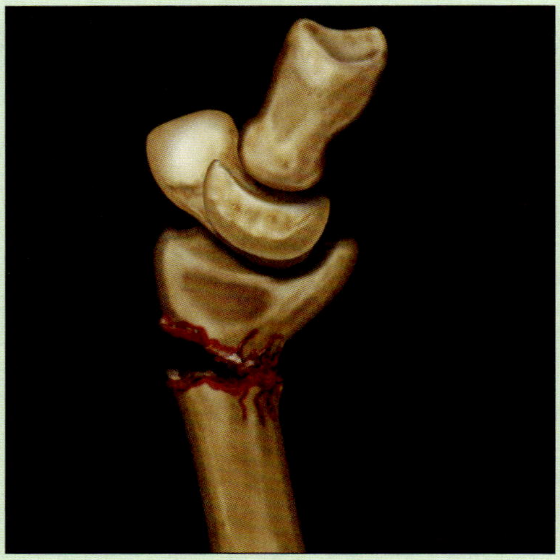

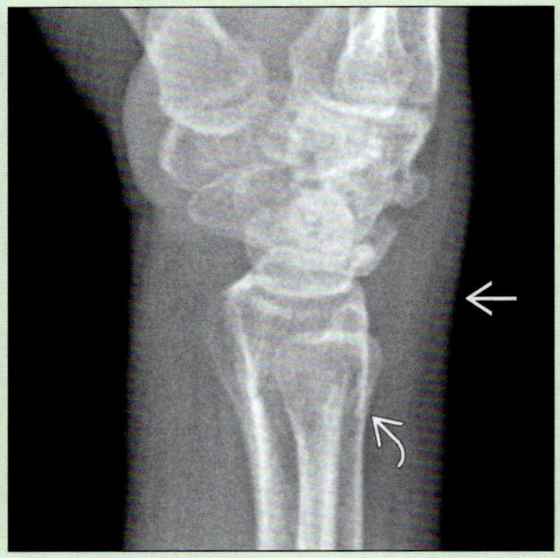

(Left) *Lateral graphic shows the configuration of a typical Colles fracture, resulting from a fall on an outstretched hand. This results in a dorsal angulation of the distal fragment and comminution at the dorsal cortex.* ***(Right)*** *Lateral radiograph demonstrates the dorsal cortical comminution ➔ of a very subtle extraarticular distal radial fracture, which results in loss of the normal volar tilt. The adjacent soft tissue swelling ➔ helps to focus the search for abnormality.*

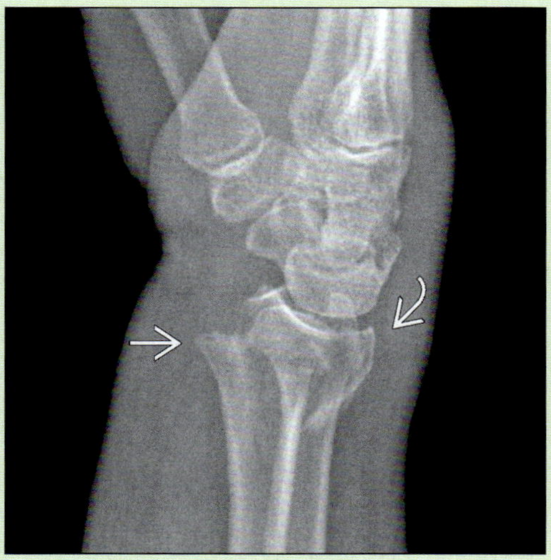

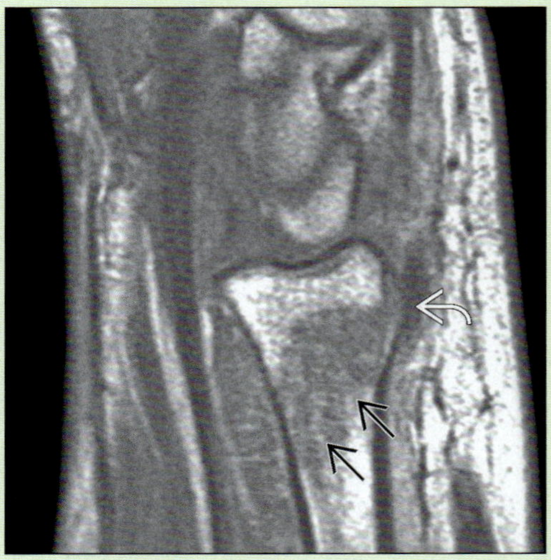

(Left) *Lateral radiograph shows a classic Colles fracture of the distal radius with the primary fracture line crossing the metaphysis 1 cm proximal to the radiocarpal joint ➔. The distal fracture fragment is dorsally angulated and displaced ➔.* ***(Right)*** *Sagittal T1WI MR reveals a band of low signal intensity indicating microtrabecular fracture ➔ crossing the distal radius in this 34-year-old woman who had an injury and ongoing pain with normal radiographs. Note the small dorsal cortical break ➔.*

DISTAL RADIUS FRACTURE

TERMINOLOGY

Synonyms

- Colles fx, Smith fx, Barton fx, reverse Barton fx, chauffeur fx, Hutchinson fx

Definitions

- Colles: Transverse fracture through metaphysis with dorsal angulation and displacement
  - ± intraarticular (most often)
  - ± associated ulnar styloid fracture
- Smith (reverse Colles): Transverse fracture through metaphysis with volar angulation and displacement
- Barton: Dorsal lip intraarticular fracture
- Reverse Barton: Volar lip intraarticular fracture
- Chauffeur (Hutchinson): Oblique intraarticular radial styloid fracture
- Complex intraarticular fracture: Generally 3 intraarticular fragments, often with angulation
- Die-punch: Lunate fossa impaction fracture
- Traumatic osteolysis: Irregular lysis of distal metaphysis adjacent to growth plate

IMAGING

Radiographic Findings

- Colles fracture
  - Associated with osteopenia
  - Transverse distal radial metaphyseal fracture
  - On lateral, major fracture fragments in dorsal angulation (reversal of normal distal radial volar tilt)
    - Angulation relates to severe comminution of dorsal aspect of radius
  - Pronator quadratus sign
    - Displacement/effacement of fat between pronator and flexor digitorum muscles
  - Intraarticular extension common
    - Tends to occur near ulnar aspect of distal radius
    - When subtle, intraarticular extension easiest to see on pronated oblique image
- Smith (reverse Colles) fracture
  - Bone density typically normal
  - Transverse distal radial metaphyseal fracture
  - On lateral, major fx fragments are in volar angulation (exaggeration of normal volar tilt)
  - May have intraarticular extension
- Barton fracture
  - Radial lip intraarticular fracture
  - May be dorsal or volar (reverse Barton)
  - Carpus follows diastatic lip fragment
    - On lateral, lunate and remainder of carpus are displaced, along with radial fracture fragment, relative to shaft of radius
  - Unstable fracture
- Chauffeur (Hutchinson) fracture
  - Oblique intraarticular radial styloid avulsion fracture
  - Fragment often associated with carpal displacement; check carpal arcs
  - Radiocarpal ligament avulsion; may be associated with ligament injury
- Die-punch fracture
  - Intraarticular impaction fracture centered in lunate fossa
  - Best seen on PA as disruption of carpal arc 1
- Complex intraarticular fracture
  - PA and pronated oblique show separate radial and ulnar styloid fragments
  - Pronated oblique may show dorsal and volar fragmentation of medial distal radius
  - Lateral shows dorsal or volar angulation
- Salter II fracture
  - PA often appears normal since epiphyseal plate is displaced but not fractured
  - Lateral view shows epiphyseal displacement as well as associated metaphyseal fragment
- Osteolysis of distal radial metaphysis
  - Widening and irregularity of epiphyseal plate; often bilateral

Imaging Recommendations

- Best imaging tool
  - 3 view radiograph of wrist: PA, lateral, and pronated oblique
  - CT if further definition of fracture displacement/angulation is required
  - MR if soft tissue evaluation is required

PATHOLOGY

General Features

- Etiology
  - Fall on outstretched hand (FOOSH) can result in many fracture and fracture-dislocation patterns
  - Most frequent injury pattern for FOOSH relates to patient age
    - 4-10 years: Distal radius and ulna transverse metaphyseal, often incomplete
    - 11-16 years: Distal radius, usually Salter II (dorsal displacement best seen on lateral view)
    - 17-40 years: Scaphoid, occasionally triquetrum or both
    - > 40 years: Colles
  - Distal radial fracture patterns represent spectrum of injuries based on
    - Direction of 3D loading
    - Magnitude and duration of force
    - Position of hand and wrist at injury
    - Biomechanical properties of affected ligaments and bones
  - Colles
    - Underlying osteoporosis makes bones vulnerable to fracture
    - Fall onto outstretched, dorsiflexed hand
    - Impact force aligned to long axis of radius
    - Tensile failure of cancellous metaphyseal bone on volar side and compressive failure on dorsal side
    - Ulnar styloid avulsed by TFCC in 60%
  - Complex distal radial fracture
    - High-energy axial compression forces
    - Force transmitted through lunate to medial radial articular surface
    - Articular surface often splits into 3 major fragments: Radial styloid and 2 medial fragments (dorsal and volar)
    - May have associated soft injuries and disrupted distal radioulnar joint (DRUJ)
  - Osteolysis of distal radial metaphysis

DISTAL RADIUS FRACTURE

Fracture Classifications

Classification	*Strengths*	*Weaknesses*
Frykman (1967): Intra- vs. extraarticular; ± ulnar styloid fracture	Considers fracture extension into radiocarpal & DRUJ	Fails to describe displacement, radial length, stability; fair interobserver agreement
Melone (1986): 5 fracture types based on 4 major radius components: Shaft, radial styloid, volar medial facet, dorsal medial facet, and degree/direction of displaced fragments	Emphasizes importance of lunate fossa, degree/direction of displaced fragments; may help direct treatment	Fair interobserver/intraobserver reproducibility
AO/ASIF (1986): 3 major groups: Extraarticular, partial articular, complete articular; increasing complexity results in 7 fracture pattern subtypes	Considers bone injury severity; may serve as basis for treatment; applicable to any joint	Complex; poor interobserver/intraobserver reproducibility
Mayo's universal (1990): 4 fracture types: Extraarticular and intraarticular + fracture extending into scaphoid fossa, lunate fossa or sigmoid notch; ± displacement, degree of stability, reducibility	Emphasizes reducibility by ligamentotaxis and fracture stability; may serve as treatment guide	Moderate interobserver agreement
Fernandez (1993): Description based on mechanism of injury: Bending, shearing, compression, avulsion, high energy with resultant combination injuries	Assists in treatment decisions and addresses prognosis	Fair interobserver agreement

- Salter I injury due to repetitive stress at wrist, generally in gymnast
- Associated abnormalities
  - Colles fracture: Because of osteoporosis, associated with humeral and femoral neck fractures

Staging, Grading, & Classification

- Many published classifications (see table)
  - Be familiar with preferences of surgeons
  - Recognize strengths and weaknesses of each system

CLINICAL ISSUES

Presentation

- Most common signs/symptoms
  - Pain, swelling, deformity following fall on outstretched hand

Demographics

- Age
  - Colles: > 40 years
  - Salter II and stress osteolysis (Salter I): 11-16 years
- Gender
  - Colles: M < F; others not gender specific
- Epidemiology
  - Distal radius is among most common fractures (15-18% of all fx seen in acute setting)
  - Colles fractures particularly common in elderly osteoporotic women
  - 44% of Salter Harris fractures occur in distal radius

Natural History & Prognosis

- Outcome depends on anatomic alignment
- Nonanatomic alignment may result in
  - Decreased grip strength
  - Decreased range of motion
  - Shortening of radius creates relative ulnar positive variance and associated carpal impaction
  - Secondary osteoarthritis
- May be complicated by complex regional pain syndrome (CRPS)

Treatment

- Stress osteolysis: Stop activity
- Others: Reduce; cast immobilization if anatomic
  - Reimage in 1-2 weeks to confirm reduction stability
- If reduction is nonanatomic, ORIF considered
  - Loss of radial length relative to ulna (> 3 mm)
  - Articular diastasis of fracture fragments (> 2 mm associated with osteoarthritis)
  - Loss of normal volar tilt or radial angle
- Normal anatomy should be restored; watch for
  - PA x-ray: Radius normally 0-2 mm longer than ulna
  - PA radiograph: Radial articular surface normally tilts 20-25° towards ulna
  - Lateral radiograph: Radial articular surface normally tilts 10-12° in volar direction
  - Lateral radiograph: Lunate should articulate concentrically with radial articular surface
  - Articular diastasis/depression ≥ 2 mm

DIAGNOSTIC CHECKLIST

Consider

- Not all distal radial fracture are Colles, and should not be identified as such!

Image Interpretation Pearls

- All distal radial fractures should be specifically evaluated for
  - Loss of volar tilt
  - Loss of radial angle
  - Shortening of radius relative to ulna
  - Intraarticular extension
  - Ulnar styloid fracture
  - Carpal arc disruption indicating more extensive injury

Reporting Tips

- Become familiar with and adhere to grading system used by referring clinician

SELECTED REFERENCES

1. Clayton RA et al: Association between decreased bone mineral density and severity of distal radial fractures. J Bone Joint Surg Am. 91(3):613-9, 2009

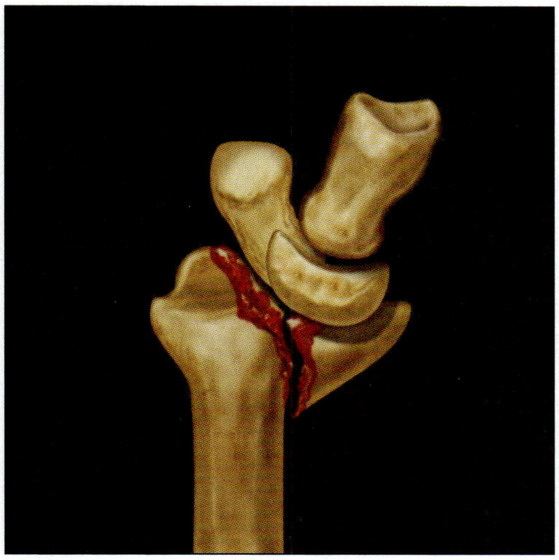

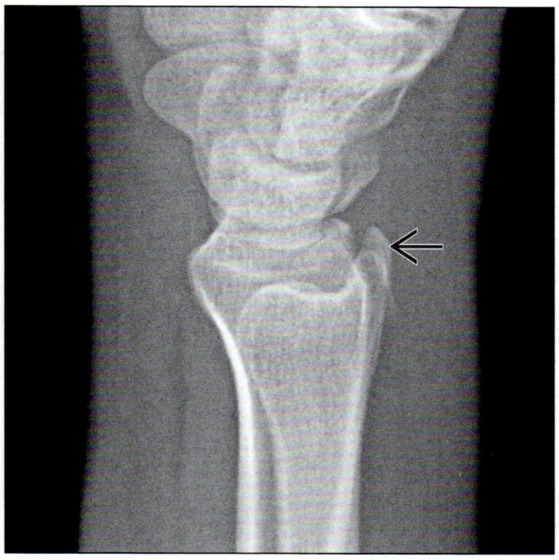

*(**Left**) Lateral graphic shows a classic Barton fracture with an intraarticular dorsal lip fracture of the distal radius. Note that the carpals move dorsally & proximally with the fracture fragment, & are no longer aligned with the radial shaft. (**Right**) Lateral radiograph reveals a small intraarticular fracture of the dorsal radial lip, representing a small Barton fracture resulting from a hyperextension injury. The fragment is minimally displaced and the carpal bones are aligned with the radial shaft.*

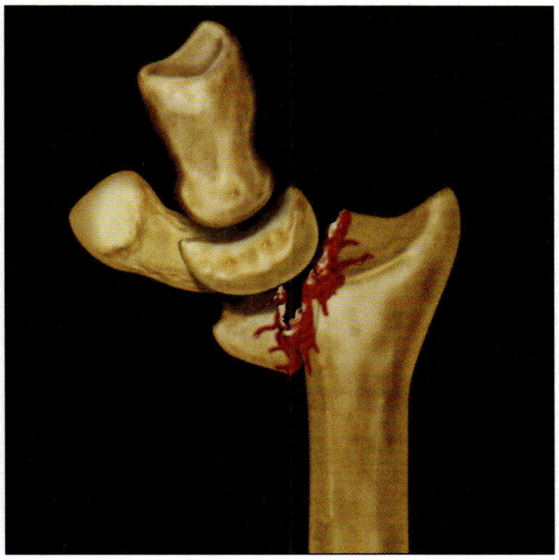

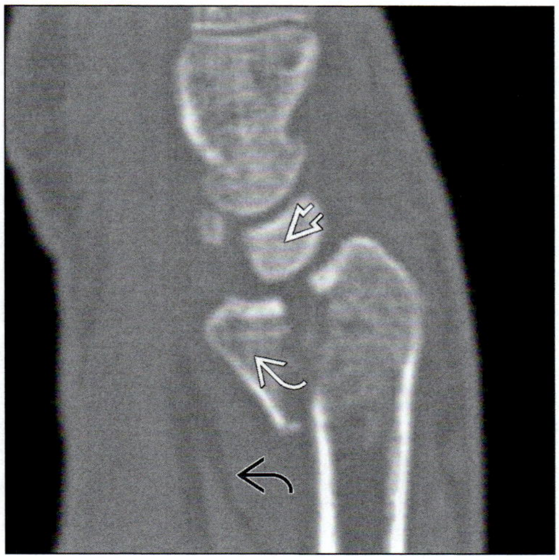

*(**Left**) Lateral graphic shows a reverse Barton (volar lip) fracture. The intraarticular fracture is located volarly, & the fragment is displaced volarly & proximally. As with the Barton fracture, the carpal bones move with the fragment. (**Right**) Sagittal reformat bone CT demonstrates a reverse Barton fracture with a volarly & proximally displaced fragment. The lunate is volarly subluxated as the carpus follows the fracture fragment. A swollen pronator quadratus displaces the adjacent fat pad.*

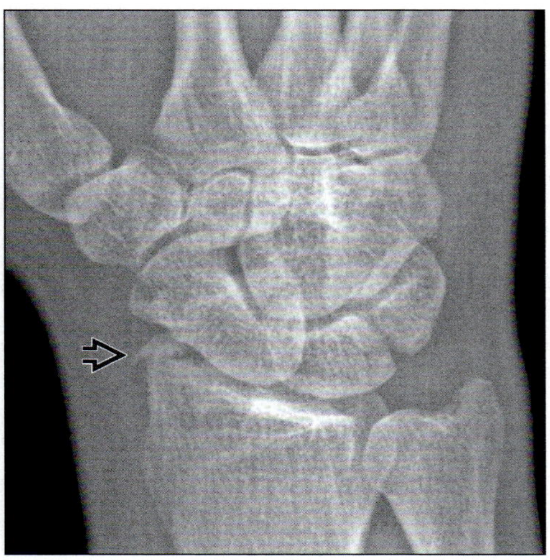

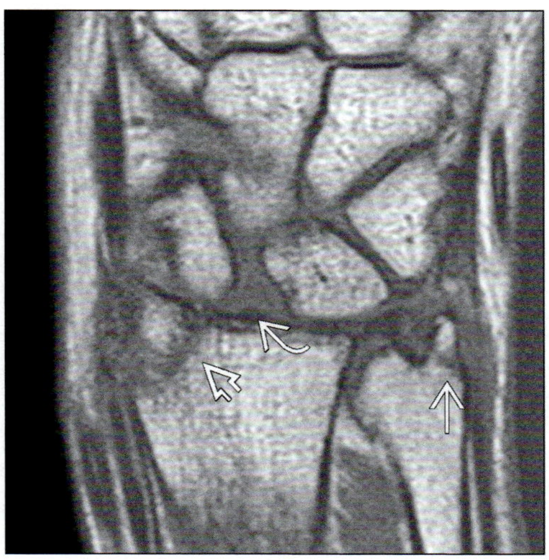

*(**Left**) Pronated oblique radiograph shows a small chauffeur (Hutchinson) fracture of the distal radial styloid. This is an avulsion fracture by the radiocarpal ligaments. (**Right**) Coronal T1WI MR shows a chauffeur fracture, an oblique intraarticular fracture crossing the radial styloid base. The fragment has accompanying scapholunate ligament injury, resulting in disruption of the proximal carpal arc & widening of the scapholunate interval. A small distal ulnar fracture is also present.*

DISTAL RADIUS FRACTURE

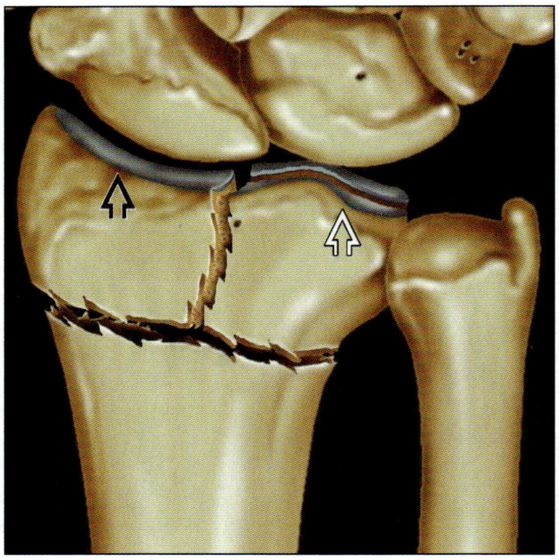

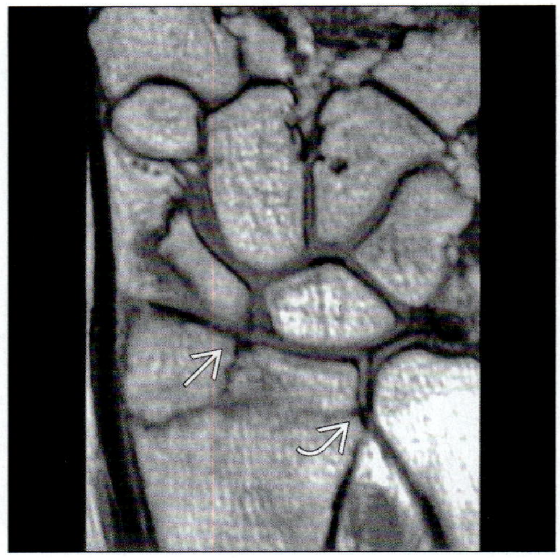

(Left) *Coronal graphic demonstrates intraarticular extension of a distal radial fracture into the radiocarpal joint separating the scaphoid fossa from the lunate fossa.* ***(Right)*** *Coronal T2WI MR shows a nondisplaced transverse distal radius metaphyseal fracture with intraarticular extension into both radiocarpal & distal radioulnar joints. There is no depression of the fragments & the articular surfaces of the scaphoid & lunate fossae are intact.*

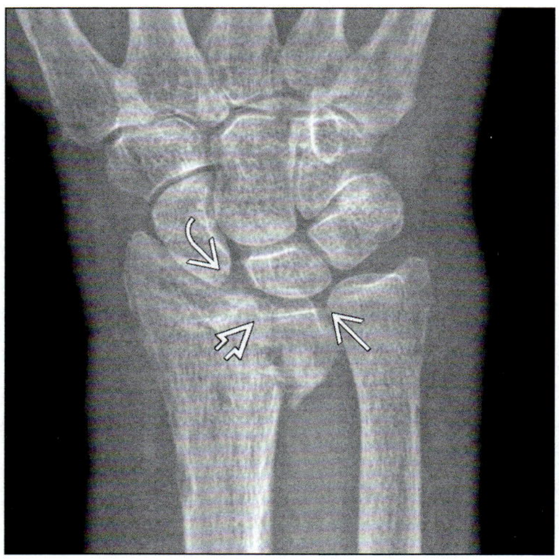

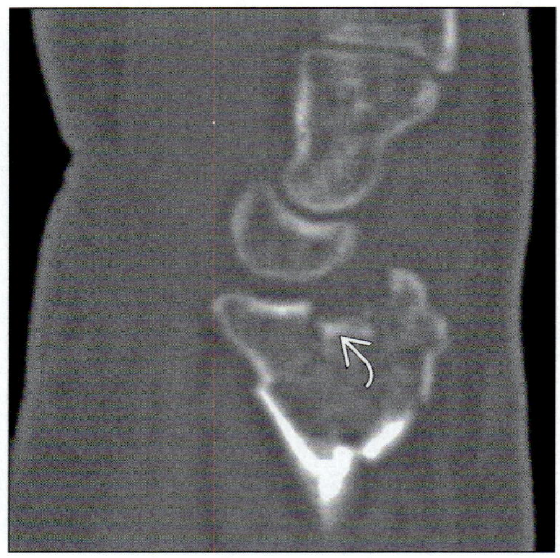

(Left) *PA radiograph reveals a subacute intraarticular distal radial fracture with marked depression of the lunate fossa. This die-punch fracture results from axial loading with the lunate "punching" into the distal radius at the lunate fossa. Radial shortening & probable scapholunate ligament tear are indicators of poor prognosis.* ***(Right)*** *Sagittal reformat bone CT shows marked comminution and 3 mm depression of the lunate fossa articular surface from lunate impaction (die-punch fracture).*

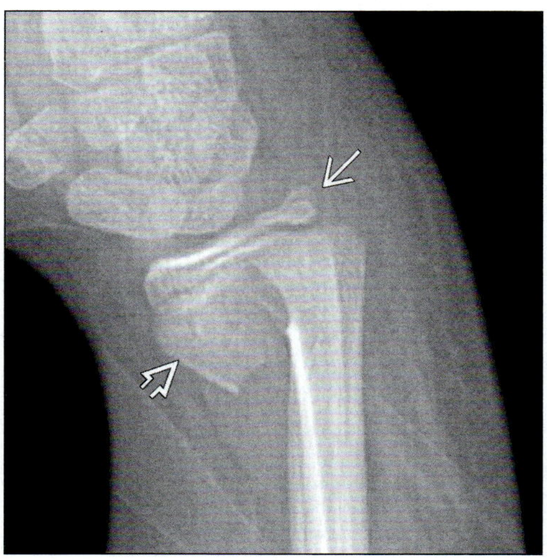

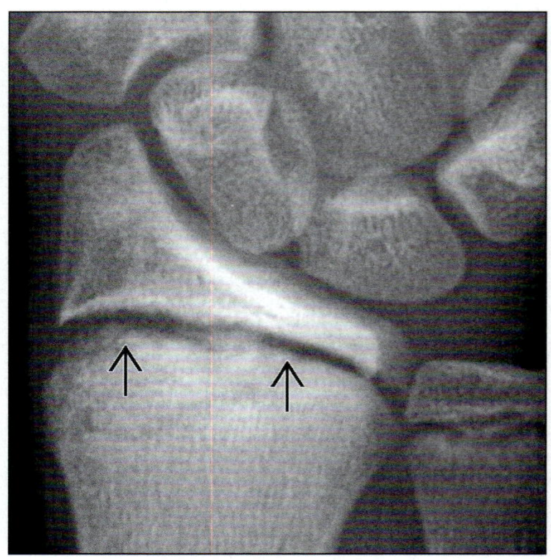

(Left) *Lateral radiograph shows a Salter-Harris type II fracture in a 9-year-old boy who fell from monkey bars. The fracture propagates across the physis, exiting through the metaphysis. This leaves an intact epiphysis attached to a large metaphyseal fragment.* ***(Right)*** *PA radiograph shows resorption of the distal radial metaphysis in a teenage gymnast. The appearance suggests rickets, but this pattern is also seen as a Salter I injury, related to significant repetitive stress.*

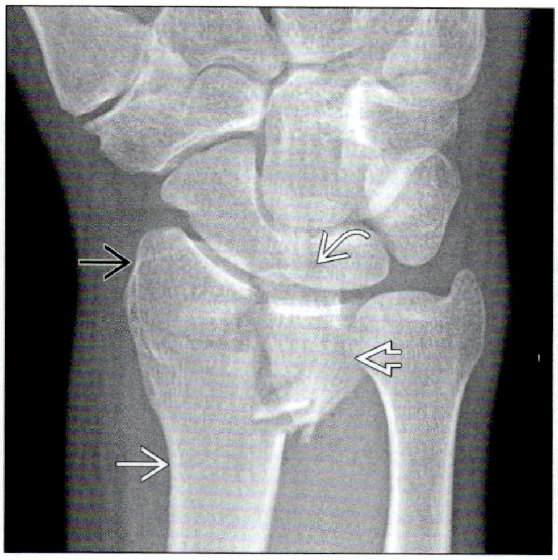
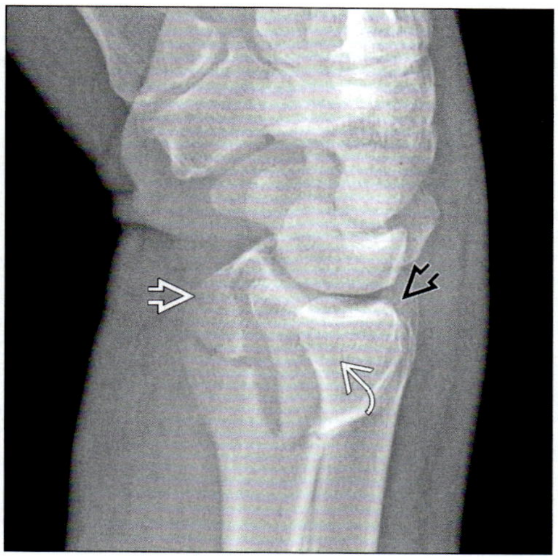

(Left) *PA radiograph in a 37-year-old man after a bike accident demonstrates comminution of distal radius resulting in 4 discrete fragments: Radial shaft →, radial styloid →, volar lunate fossa →, & dorsal lunate fossa →. This fracture extends into the radiocarpal joint but spares the distal radioulnar joint.* ***(Right)*** *Lateral radiograph in the same patient demonstrates loss of the normal volar tilt → & the slight splaying of the volar → & dorsal → portions of the lunate fossa fracture.*

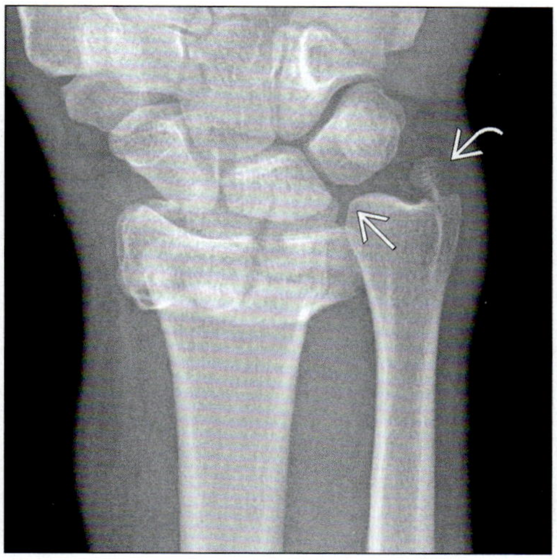
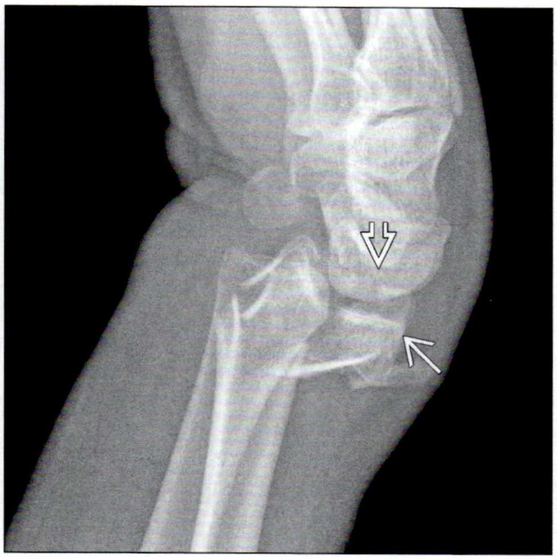

(Left) *PA radiograph in a 34-year-old man reveals a comminuted distal radial fracture with shortening of the distal radius & loss of normal radial angle, resulting in an ulnar plus variant →. The fracture extends into both radiocarpal & DRUJ joints with a fracture at the ulnar styloid base → as well.* ***(Right)*** *Lateral radiograph in the same patient reveals dorsal angulation and displacement → of the dorsal lunate fossa fragment. The carpals remain aligned → with the dorsal fragment, similar to Barton fracture.*

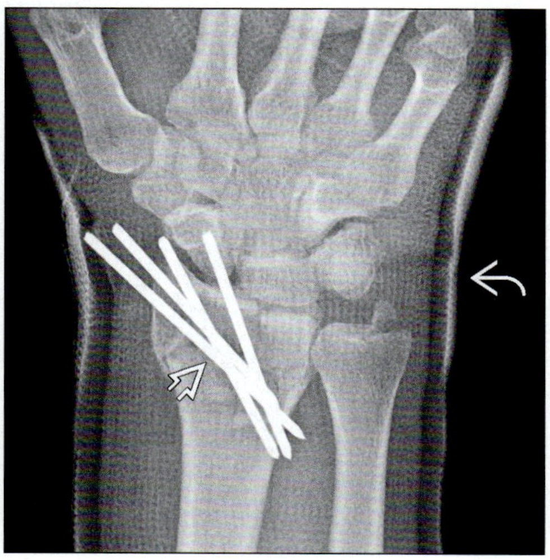
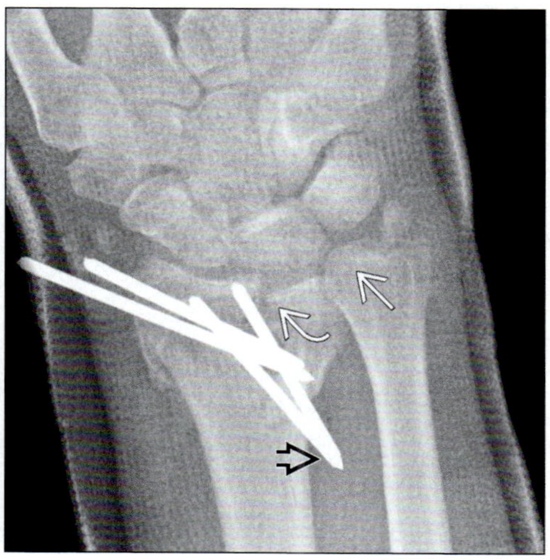

(Left) *PA radiograph in the same patient shows initial postoperative appearance following closed reduction with percutaneous pinning. Four pins cross the radial styloid & radial shaft →, restoring radial height & angulation. The fiberglass cast → is in place.* ***(Right)*** *PA radiograph in the same patient 6 weeks later demonstrates redisplacement of fracture fragments with migration of the pins →. There is a 4 mm step-off in the radial articular surface → & ulnar plus variance → has reappeared.*

DIE-PUNCH FRACTURE

Key Facts

Terminology
- Die-punch fracture: Impaction of carpal bone(s) into radius articular surface
- Principal mechanism of injury for intraarticular distal radius fractures

Imaging
- Best diagnostic clue
  - Depression of lunate (or scaphoid) fossa
- 4 major fragments: Shaft, radial styloid, dorsal lunate fossa, and volar lunate fossa
  - Dorsal and volar fragment displacement due to lunate impaction
  - ± volar metaphyseal "spike" fragment
- Bone CT with reformats best defines fx fragments

Pathology
- Fall on outstretched hand
  - Axial compression
  - Lunate driven proximally, impacting lunate fossa
  - Impaction results in comminuted intraarticular fracture
- Melone classification of intraarticular fractures: Most common are type II
  - IIa: Classic (most frequent) die punch; dorsal displacement and tilt of dorsal medial fragment
  - IIb: Volar medial component with greater amount of comminution and volar tilt

Clinical Issues
- Open reduction usually required for any fracture > type I (significant articular diastasis, shortening, and tilt)

Diagnostic Checklist
- Look for metaphyseal spike to differentiate types II and III fractures
- Look for associated ulnar styloid fracture

*(**Left**) Volar graphic shows a Melone type III die-punch fracture. The lunate fossa is impacted with a coronal fracture line ➡ separating the dorsal & volar portions & an additional volar fragment or "spike" ➡. There is also an avulsion of the ulnar styloid ➡. (**Right**) PA radiograph demonstrates a type IIa die-punch fx with an axial fracture across the metaphysis ➡, a sagittal fracture ⇨ separating the scaphoid & lunate fossae, & a coronal fracture splitting the lunate fossa into dorsal ⇨ & volar ➡ parts. There is an ulnar styloid fracture ➡.*

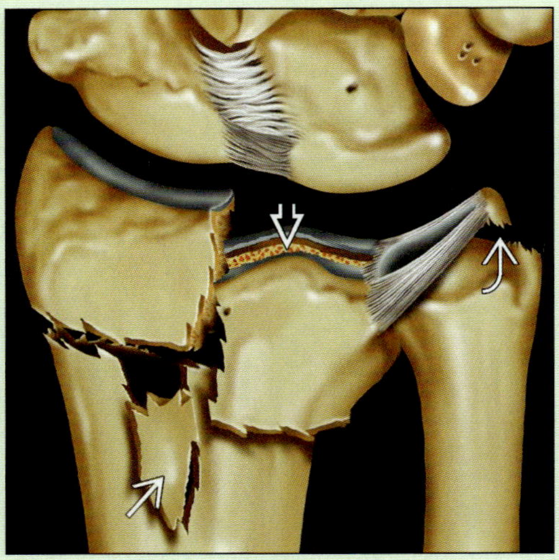

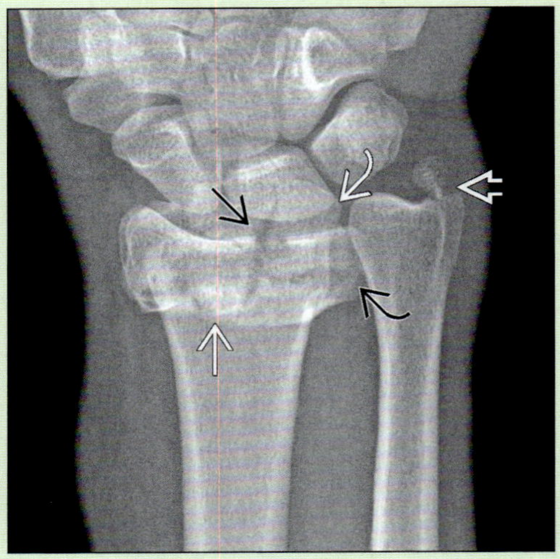

*(**Left**) Axial bone CT demonstrates the star-like appearance ⇨ of the intraarticular impaction fracture of the lunate fossa in this 71-year-old woman who fell on an outstretched hand. (**Right**) Sagittal T2WI FS MR shows the fractured cortex of the lunate fossa ⇨ with associated marrow edema. Note the diffuse edema of the lunate ⇨ which impacted ("punched") the distal radius causing the intraarticular fracture.*

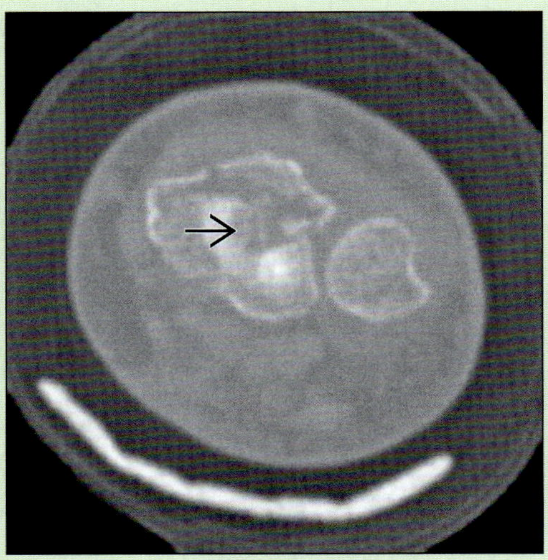

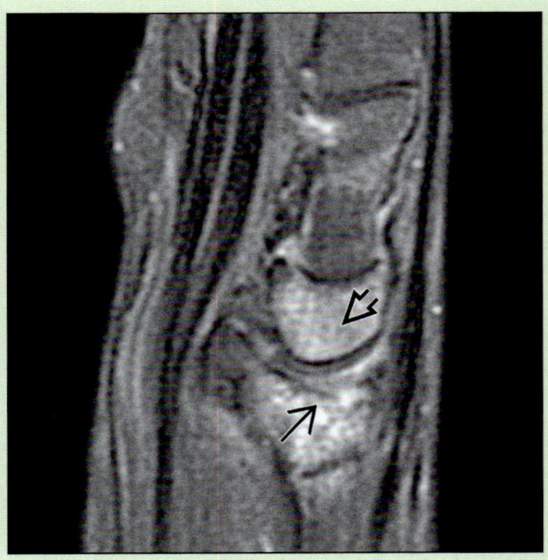

DIE-PUNCH FRACTURE

TERMINOLOGY

Synonyms
- Lunate fossa loading fracture, radiolunate fracture, scaphoid die-punch fracture

Definitions
- Die-punch fracture: Impaction of carpal bone(s) into radius articular surface
  - Splits articular surface: Coronal and sagittal planes
- Principal mechanism of injury for intraarticular distal radius fractures

IMAGING

General Features
- Best diagnostic clue
  - Depression of lunate (or scaphoid) fossa
- Location
  - Distal radial articular surface
    - Lunate fossa ± scaphoid fossa
- Morphology
  - Depressed, impacted, &/or comminuted
  - Dorsal &/or volar displacement of fragments

Radiographic Findings
- 4 major fragments: Shaft, radial styloid, dorsal lunate fossa, and volar lunate fossa
- Dorsal and volar fragment displacement due to lunate impaction
- ± volar metaphyseal "spike" fragment

CT Findings
- Bone CT with reformats best defines fx fragments
- Minimally displaced fx appears star-like in axial plane

MR Findings
- T1WI MR: Fx diastasis measured directly in 3 planes
- Fluid-sensitive sequences: Hyperintense edema usually seen in fracture fragments and in lunate
- Associated ligamentous or tendinous injury

Imaging Recommendations
- Best imaging tool
  - 3 view wrist radiographs: PA, lateral, oblique
    - Oblique may best delineate discrete fragments
- Protocol advice
  - Bone CT: Thin section imaging with reformations

DIFFERENTIAL DIAGNOSIS

Colles Fracture
- Metaphyseal fracture
- ± secondary intraarticular extension

PATHOLOGY

General Features
- Etiology
  - Fall on outstretched hand
    - Axial compression
    - Lunate driven proximally, impacting lunate fossa
    - Impaction results in comminuted intraarticular fracture of radius
  - Scaphoid may similarly impact scaphoid fossa
  - Double die-punch injury may occur with impaction of both scaphoid and lunate fossae

Staging, Grading, & Classification
- Melone classification of intraarticular distal radial fxs
  - Describes 4 components: Shaft, radial styloid, dorsal lunate fossa, volar lunate fossa
  - Type I: No significant displacement or angulation
  - Type II: Unstable fracture; most common type of die punch
    - IIa: Classic; dorsal displacement and tilt of dorsal medial fragment
    - IIb: Volar medial component with ↑ comminution and volar tilt
  - Type III: Similar to type II fracture with additional volar fragment or "spike"
  - Type IV: Wide diastasis of dorsal & volar lunate fxs
  - Type V: Severe comminution
- AO classification
  - Subtype B4: Impaction fracture of lunate or scaphoid fossa

CLINICAL ISSUES

Demographics
- Age
  - Low-energy injury: Older adult with osteopenia
  - High-energy injury: Younger adult

Natural History & Prognosis
- Low-energy injury: Articular incongruity generally has favorable outcome
- Potential complications
  - Articular incongruity leads to osteoarthritis
  - Ligamentous &/or tendinous injury
  - Arterial or nerve (ulnar, radial, median) injuries
  - Chronic regional pain syndrome

Treatment
- Conservative
  - Type I fracture treated with closed reduction
- Surgical
  - Open reduction usually required for any fracture > type I, relating to degree of articular diastasis, shortening, and tilt
  - May require combined internal & external fixation

DIAGNOSTIC CHECKLIST

Consider
- 4 discrete fracture fragments: Shaft, radial styloid, dorsal lunate fossa, volar lunate fossa
- Look for associated ulnar styloid fracture

Image Interpretation Pearls
- Look for volar spike to differentiate types II and III fractures

SELECTED REFERENCES

1. Trumble TE et al: Intra-articular fractures of the distal aspect of the radius. Instr Course Lect. 48:465-80, 1999
2. Melone CP Jr: Distal radius fractures: patterns of articular fragmentation. Orthop Clin North Am. 24(2):239-53, 1993

ULNAR STYLOID FRACTURE

Key Facts

Imaging

- Radiography
  - Fracture line crossing ulnar styloid, ± fragment displacement
  - Rarely isolated injury
- MR T2WI
  - Hypointense fracture line; may be obscured by surrounding edema (↑ signal intensity)
  - Associated injuries
- Best imaging tools
  - Radiography: Posteroanterior (PA) and oblique
  - MR: Nondisplaced fractures; associated TFCC tears

Pathology

- Fall on outstretched pronated hand
- Styloid tip fracture
  - Impaction: Styloid vs. carpals
  - Avulsion: Ulnar collateral ligament complex
  - Typically stable fracture
- Styloid base fracture
  - TFC ulnar attachment avulsion ± TFC tear
  - Distal radial fracture with DRUJ instability always has ulnar styloid base fracture
  - Potentially unstable if DRUJ or TFC injured

Clinical Issues

- Distal radius fxs account for 74% of forearm fxs
  - Associated ulnar styloid fracture in 61%
- Styloid fx common in adult women (60-80 years)
- TFC (ulnar) tears common in younger population

Diagnostic Checklist

- Describe ulnar styloid fracture location and degree of displacement
- If styloid base fracture present, evaluate for DRUJ and TFCC integrity

__(Left)__ Coronal graphic shows distal ulnar styloid fracture without displacement of the TFCC attachments. This tip fracture may result from impaction on the adjacent carpals or from avulsion of the ulnar collateral ligament complex. __(Right)__ Coronal T1WI MR obtained in a 39 year old with wrist swelling but otherwise normal radiographs demonstrates a nondisplaced comminuted distal radial → fracture. Note the nondisplaced ulnar styloid tip fracture →, likely from direct impact on the adjacent carpals.

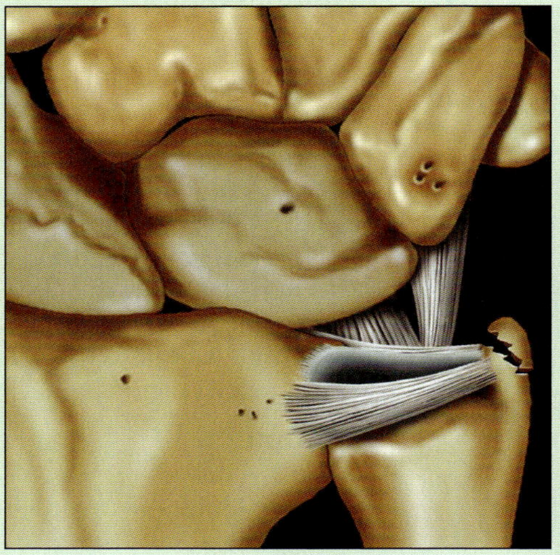

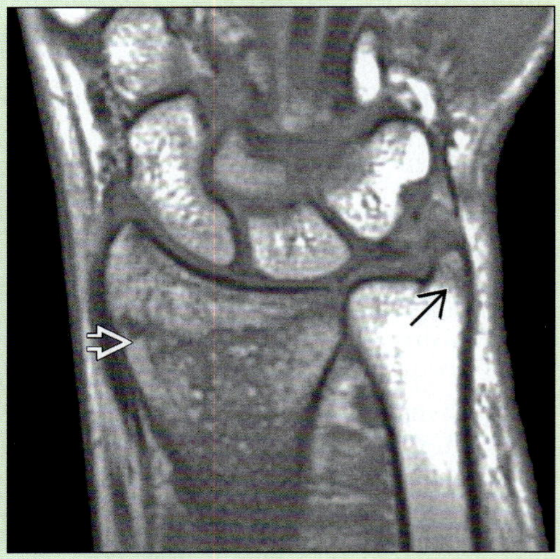

__(Left)__ PA radiograph shows a large ulnar styloid base fracture → with 2 mm displacement. There is an associated comminuted radial fracture →. Note the widening of the DRUJ → suggesting possible loss of joint stability. __(Right)__ Coronal T1WI MR demonstrates an ulnar styloid fracture nonunion →. The fracture fragment is well corticated as is the donor site →. The TFC ulnar attachment → remains intact along the lateral aspect of the fragment.

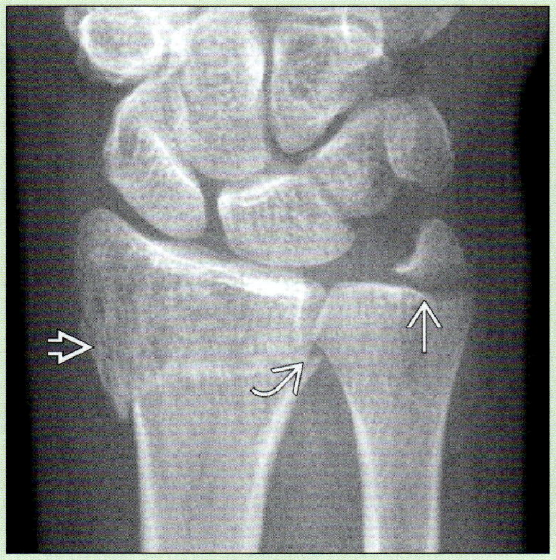

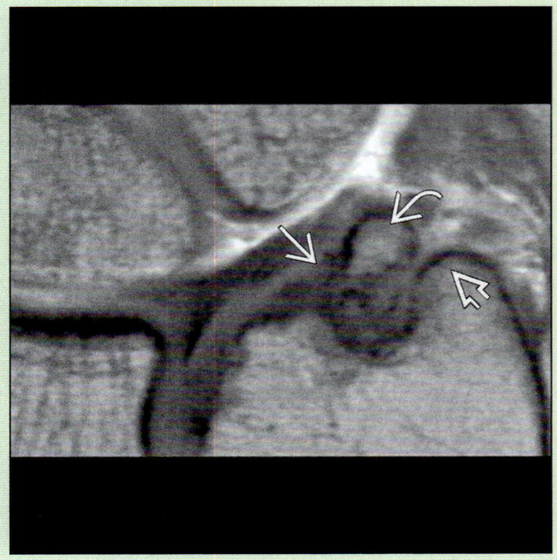

ULNAR STYLOID FRACTURE

IMAGING

General Features

- Best diagnostic clue
  - Fracture line crossing ulnar styloid, ± fragment displacement
  - Rarely isolated injury

Radiographic Findings

- Radiography
  - Fracture fragments vary greatly in size

MR Findings

- T1WI
  - Hypointense fracture line
  - ± associated injuries
- T2WI
  - Hypointense fracture line; may be obscured by surrounding edema (↑ signal intensity)
  - Associated injuries
    - Distal radius fracture
    - Soft tissue injury
      - TFC complex (TFCC) tears: ↑ signal intensity interposed between TFC and ulnar styloid base
      - Tendon injuries (e.g., extensor carpi ulnaris)
    - Distal radioulnar joint (DRUJ) instability
      - Ulnar subluxation with respect to sigmoid notch of radius
      - Interosseous membrane tear

Imaging Recommendations

- Best imaging tool
  - Radiography: Posteroanterior (PA) and oblique
  - MR: Nondisplaced fractures; associated TFCC tears
- Protocol advice
  - Imaging should be centered on radiocarpal joint

PATHOLOGY

General Features

- Etiology
  - Fall on outstretched pronated hand
  - May result in impaction or avulsion fracture

Staging, Grading, & Classification

- Commonly occurs with distal radius fracture
- Palmer classification for associated TFCC tears
  - IA: Central perforation
  - IB: Ulnar avulsion, ± distal ulnar fracture
  - IC: Distal avulsion
  - ID: Radial avulsion, ± sigmoid notch fracture
- Symptomatic ulnar styloid nonunion
  - Type 1: Fx with stable DRUJ
  - Type 2: Fx with DRUJ subluxation

Gross Pathologic & Surgical Features

- Styloid tip fracture
  - Impaction: Styloid vs. carpals
  - Avulsion: Ulnar collateral ligament complex
- Styloid base fracture
  - TFC ulnar attachment avulsion ± TFC tear
  - Distal radial fracture with DRUJ instability always has ulnar styloid base fracture

CLINICAL ISSUES

Presentation

- Most common signs/symptoms
  - Medial wrist pain

Demographics

- Age
  - Styloid fx common in adult women (60-80 years)
  - TFC (ulnar) tears common in younger population
- Gender
  - M < F
- Epidemiology
  - Distal radius fxs account for 74% of forearm fxs
    - Associated ulnar styloid fracture in 61%

Natural History & Prognosis

- Styloid tip fracture: Stable
- Styloid base fracture: Potentially unstable
  - Associated with DRUJ instability, TFCC tear, &/or tendon tear
  - ↑ nonunion risk
- Styloid fracture nonunion
  - Type 1 associated with stable DRUJ (TFC insertion intact)
  - Type 2 associated with unstable DRUJ
  - Rate of nonunion = 26%

Treatment

- Conservative
  - Isolated ulnar styloid tip fractures = closed reduction/immobilization
  - Nondisplaced fx base of styloid + stable DRUJ = immobilization
- Surgical
  - Ongoing debate regarding surgical intervention
    - Conventional approach
      - Open repair for fx > 2 mm displacement, severe ligament or TFC injury, unstable DRUJ
    - Recent research suggests 1° ulnar styloid base fx repair unnecessary if distal radius fx appropriately stabilized and soft tissue restraints intact
  - Symptomatic fracture nonunion
    - Type 1: Fragment excision relieves pain
    - Type 2: DRUJ stabilization, TFCC repair ± fragment excision

DIAGNOSTIC CHECKLIST

Reporting Tips

- Describe ulnar styloid fracture location and degree of displacement
- If styloid base fracture present, evaluate for DRUJ and TFCC integrity

SELECTED REFERENCES

1. Souer JS et al: Effect of an unrepaired fracture of the ulnar styloid base on outcome after plate-and-screw fixation of a distal radial fracture. J Bone Joint Surg Am. 91(4):830-8, 2009
2. May MM et al: Ulnar styloid fractures associated with distal radius fractures: incidence and implications for distal radioulnar joint instability. J Hand Surg Am. 27(6):965-71, 2002

DISTAL RADIOULNAR FRACTURE, POSTOPERATIVE

Key Facts

Imaging

- Fractures requiring surgical intervention
  - Highly comminuted, intraarticular distal radius ± ulna fracture; marked depression & angulation
  - Failure to achieve &/or maintain adequate reduction with closed manipulation
- Hardware complications
  - Loss of anatomic reduction
  - Change in position of K-wires, screws, plates
  - Hardware fracture
  - Lucency around pins or K-wires
- Radiography for baseline & serial evaluation
- CT &/or MR for problem solving
- Serial imaging is key

Clinical Issues

- **Treatment options** (used individually or in combination)
  - Casting/splinting
  - Kirschner ("K") wires
  - External fixation
  - Internal fixation
  - Bone grafting/biologics
- **Complications**
  - Loss of anatomic reduction
  - Loss of fixation
  - Hardware fracture (may be subtle)
  - Screws/plates impinging soft tissues
  - Infection
  - Cross-union fusion between radius and ulna

Diagnostic Checklist

- Initial reduction: Alignment, including articular surface congruity; hardware type and position
- Subsequent imaging: Alignment/hardware changes; progression of healing

*(**Left**) Graphic shows placement of 2 K-wires ➡ for reduction of a comminuted distal radius fracture. K-wires can be used to manipulate and align fragments prior to being placed through multiple fragments for fracture stabilization. These smooth wires can migrate and are often removed. (**Right**) PA radiograph shows 4 K-wires placed for anatomic reduction of a comminuted distal radius fracture. Wires pass through the radial styloid ➡ and are used to lever the medial fragments ➡ into position.*

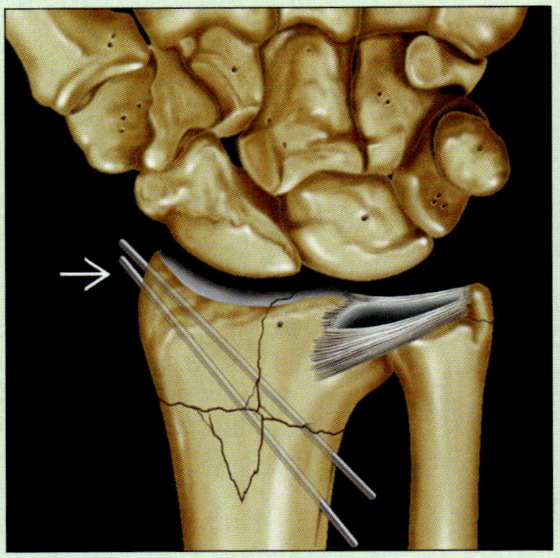

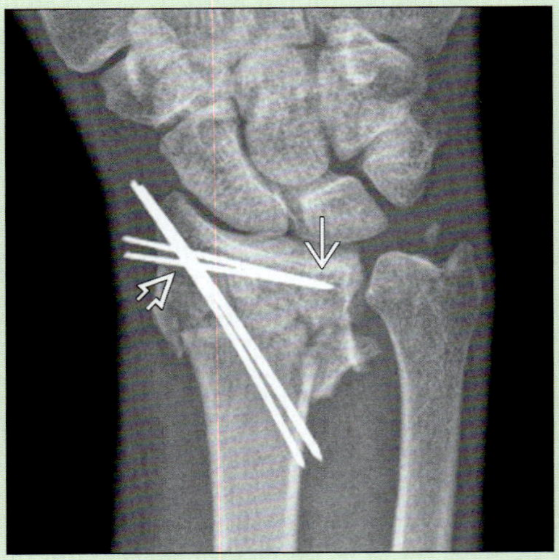

*(**Left**) PA radiograph shows an external fixator spanning a highly comminuted intraarticular distal radial fracture ➡. Pins in 2nd metacarpal and distal radial shaft secure the fixator. The 2nd radius pin was slightly bent ➡ during placement and may be a point of failure. (**Right**) PA radiograph shows K-wire, volar T-plate, and cast for fracture reduction and immobilization. There is persistent DRUJ instability evidenced by ulnar subluxation ➡ and an ulnar styloid base fracture ➡.*

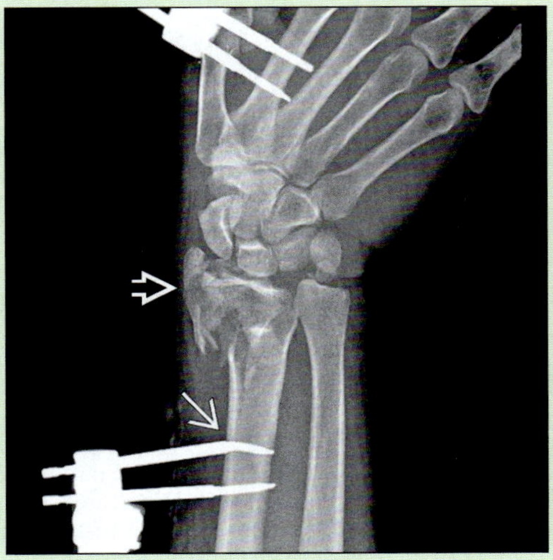

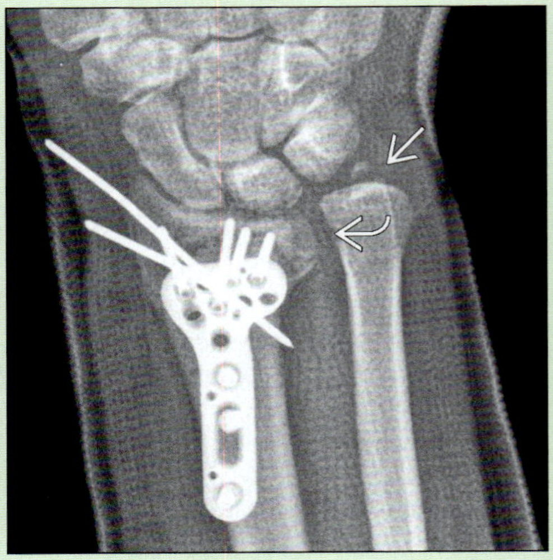

DISTAL RADIOULNAR FRACTURE, POSTOPERATIVE

IMAGING

General Features

- Best diagnostic clue
  - Fractures requiring surgical intervention
    - Highly comminuted ± intraarticular distal radial ± ulnar fractures with marked depression and angulation
    - Failure to achieve &/or maintain adequate reduction with closed manipulation
  - Hardware complications
    - Loss of anatomic reduction
    - Change in position of Kirschner wires (K-wires), screws, or plates
    - Hardware fracture
    - Lucency around pins or K-wires

Imaging Recommendations

- Best imaging tool
  - Radiography
    - Baseline at time of reduction
    - Evaluate maintenance of reduction & healing
    - Evaluate hardware integrity
- Protocol advice
  - Radiography
    - PA, lateral, & oblique views for serial imaging
  - CT
    - Thin section acquisition with reformation to allow 2D & 3D evaluation
    - Utilize metal artifact reduction techniques
  - MR
    - Utilize metal artifact reduction techniques
    - Implant imaging
      - Weakly magnetic implants rigidly fixed to body (screws, etc.) safe for immediate post-op MR
      - Torque (in off-center position) may result in vibrations felt by patient
    - External fixators
      - Composed of many diverse materials; not all are MR compatible
      - MR safety determined case by case

CLINICAL ISSUES

Natural History & Prognosis

- Treatment considerations
  - Stability of fracture depends on
    - Degree of comminution & fracture displacement
    - Articular surface congruity
  - Patient considerations
    - Age, medical condition, functional limits
    - Bone quality
  - Treatment goals
    - Restore articular congruity and osseous alignment
      - Radial length restoration
      - Normal radial volar tilt at radiocarpal joint
    - Maintain stability during healing

Treatment

- **Options**
  - Casting/splinting
    - Used when satisfactory alignment can be achieved & maintained
  - K-wires
    - Minimally invasive
    - Used to achieve alignment & stabilization
    - Often used in combination with casting, plating, or external fixation
  - External fixation
    - Minimal soft tissue disruption
    - May be combined with other techniques
  - Internal fixation
    - Stronger stabilization with rapid mobilization
    - Requires more extensive surgery
  - Bone grafting/biologics
    - Speed healing
    - Provides for lost bone matrix (in marked comminution)
- **Complications**
  - Initial hardware placement
    - Failure to achieve adequate reduction
    - Poor hardware purchase
    - Hardware in or limiting joint
  - Subsequent complications
    - Serial imaging is key
    - Loss of anatomic reduction
    - Loss of fixation
      - Hardware migration from original position
      - Lucency around screws, pins, or wires = motion
      - Plates not apposed to cortical surface
    - Hardware fracture (may be subtle)
      - Screws: At threads, head/shaft junction
      - Pins: At thread/shaft transitions
      - Plates: At screw holes
    - Screws/plates impinging soft tissues
    - Infection
      - Particularly problematic with external fixators or exposed K-wires
      - Lucency at hardware/bone interfaces
    - Cross-union fusion between radius & ulna
  - Clinical
    - Hardware irritation due to subcutaneous position
    - Post-fixation joint stiffness
    - Post-traumatic osteoarthritis
    - Chronic regional pain syndrome (CRPS)

DIAGNOSTIC CHECKLIST

Consider

- Factors that will determine need for surgery
- Complicating factors: Open wounds; tendon, ligament, or nerve injury

Reporting Tips

- Initial reduction: Alignment, including articular surface congruity; hardware type and position
- Subsequent imaging: Alignment/hardware changes; progression of healing

SELECTED REFERENCES

1. Chung KC et al: Trends in the United States in the treatment of distal radial fractures in the elderly. J Bone Joint Surg Am. 91(8):1868-73, 2009

DISTAL RADIOULNAR FRACTURE, POSTOPERATIVE

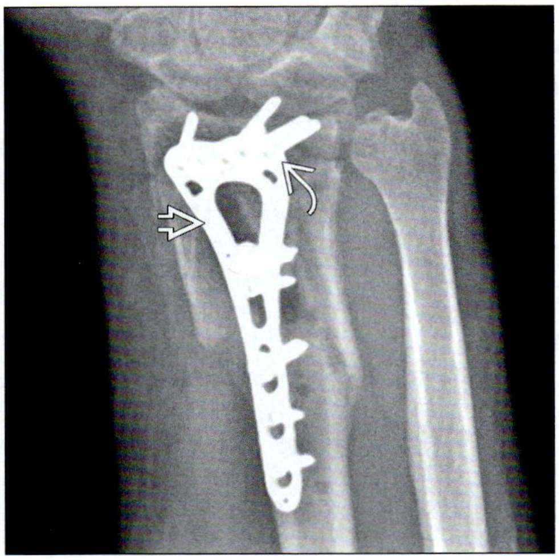

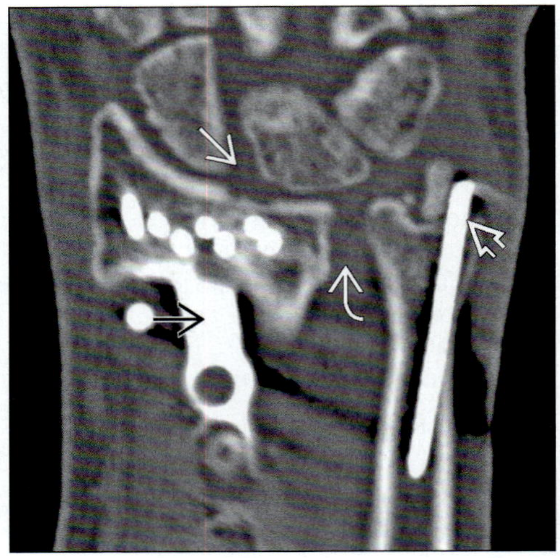

(Left) *Oblique radiograph shows a specialized volar plate & screw assembly placed for revision of a prior ORIF in which the hardware failed. This variable angle plate is secured with locking screws. It is designed to stabilize scaphoid and lunate fossae.* ***(Right)*** *Coronal reformat bone CT shows a volar T-plate stabilizing a healed distal radial fracture. A K-wire secures a nonunited ulnar styloid fx. Note widening of scapholunate interval and DRUJ, indicating ligamentous injury.*

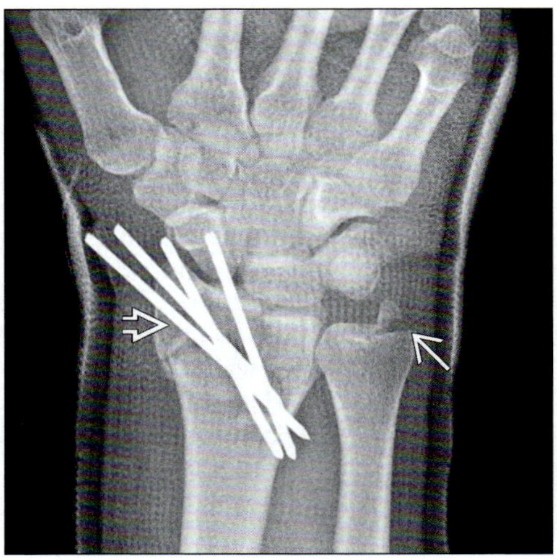

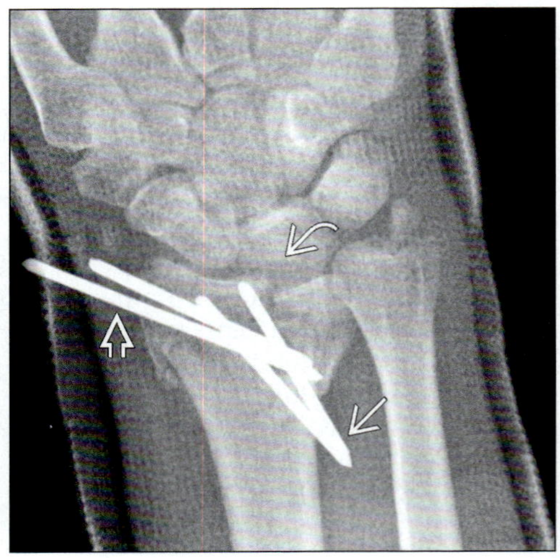

(Left) *PA radiograph taken immediately after fracture reduction shows 4 K-wires crossing the radial styloid with near anatomic reduction of fracture fragments. Circumferential cast is placed for additional immobilization. Note the ulnar styloid base fracture.* ***(Right)*** *PA radiograph from the same patient taken 2 weeks later shows migration of the K-wires and loss of fracture reduction. The 2 lateral wires have withdrawn, & the 2 medial wires have advanced proximally.*

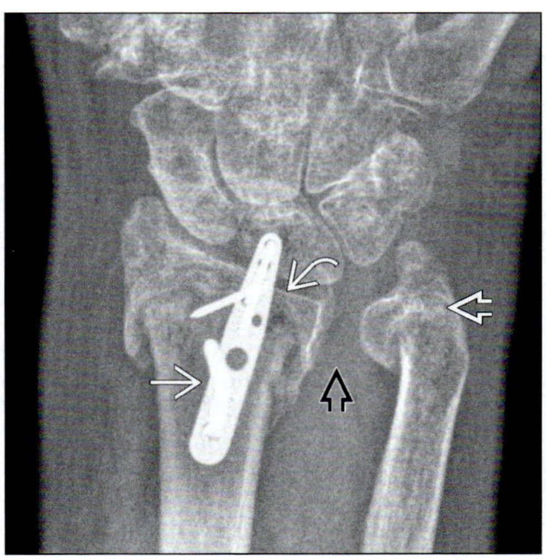

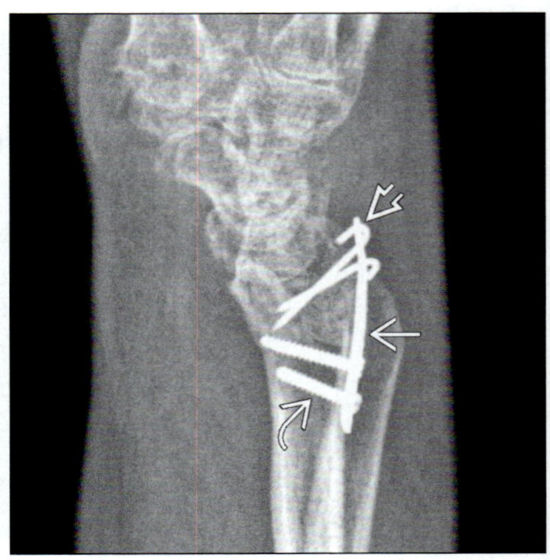

(Left) *PA radiograph demonstrates a dorsal plate placed to stabilize the lunate fossa fragment of a highly comminuted distal radial fracture. There is an associated distal ulnar fracture deformity. The fractures are healed with marked deformity and widening of the DRUJ.* ***(Right)*** *Lateral radiograph demonstrates a dorsal plate placed to stabilize a comminuted distal radial fracture. This plate is attached by 2 cancellous screws and 2 Kirschner wires.*

DISTAL RADIOULNAR FRACTURE, POSTOPERATIVE

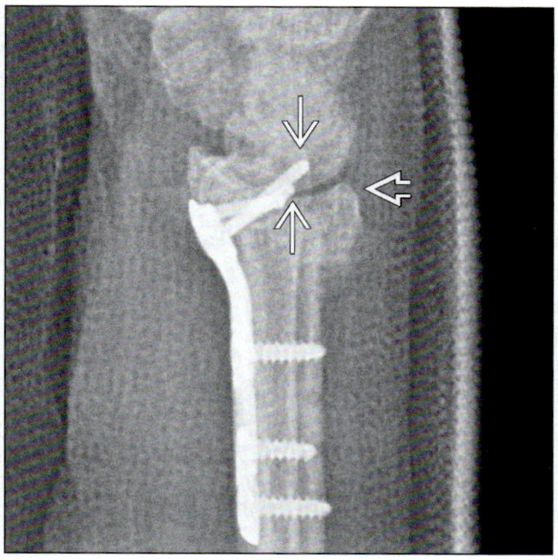

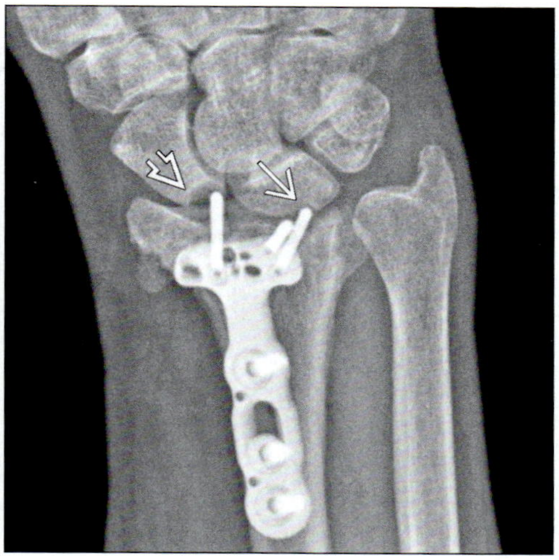

(Left) *Lateral radiograph obtained immediately following placement of a volar plate for radial fracture shows the 2 distal screws ➔ extending into the radiocarpal joint. The normal volar tilt has not been restored ➔.* ***(Right)*** *PA radiograph from the same patient obtained 6 months later shows the fracture is healed but misplaced screws impinge and remodel the proximal scaphoid ➔, likely disrupting the scapholunate ligament. There is impingement and erosion of the lunate as well ➔.*

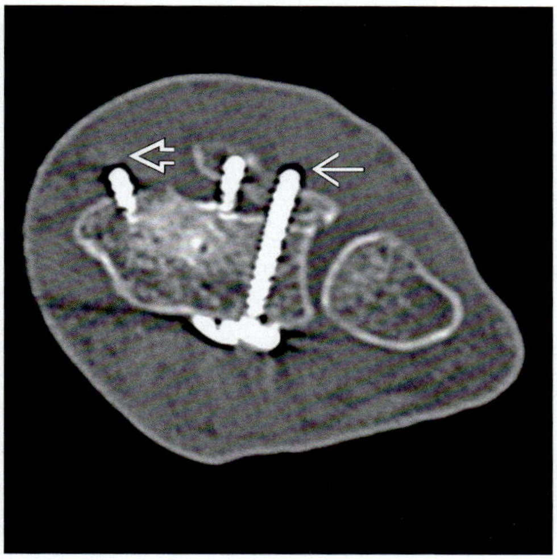

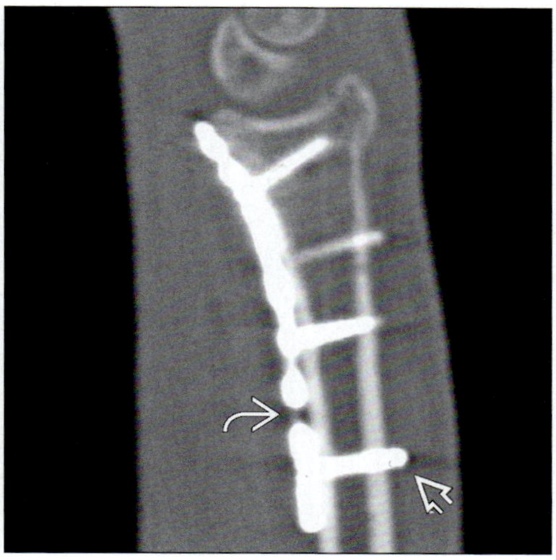

(Left) *Axial bone CT demonstrates screws securing a volar plate. The lateral screw directly impinges the extensor carpi radialis longus tendon ➔. The medial screw impinges extensor digitorum tendons in compartment 4 ➔.* ***(Right)*** *Sagittal reformat bone CT reveals a fully healed distal radius fracture successfully treated with ORIF. The patient had pain with wrist extension. The proximal bicortical screw directly impinges the extensor muscles ➔. Apparent interruption in plate ➔ is artifactual.*

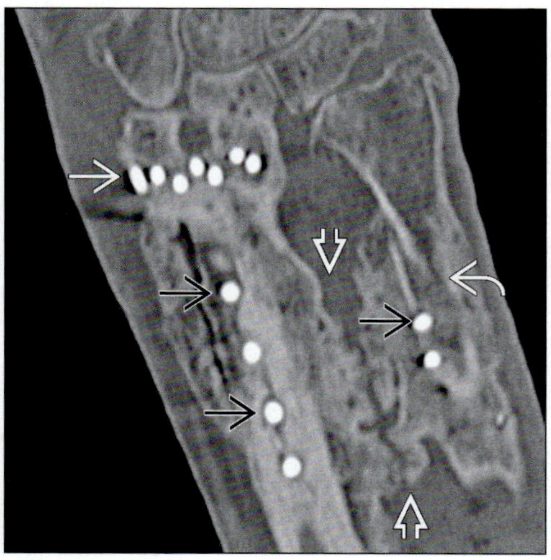

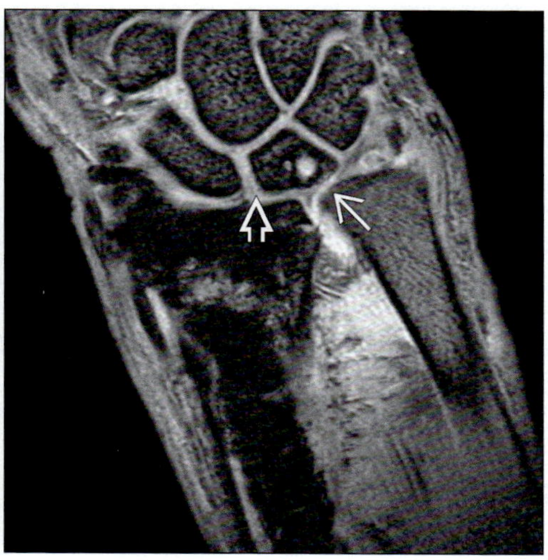

(Left) *Coronal reformat bone CT demonstrates cross-union ➔ with exuberant ossification along the interosseous membrane after ORIF for treatment of comminuted intraarticular distal radial ➔ and ulnar shaft ➔ fractures. Tips of screws ➔ are seen.* ***(Right)*** *Coronal gradient-echo MR shows tears of the TFCC ➔ and scapholunate ligament ➔ in this patient who has undergone ORIF of a distal radial fracture. Quality of the image is good despite the presence of metal artifact.*

TRAUMA-RELATED OSTEOLYSIS IN CHILDREN

Key Facts

Imaging

- Distal radius > > distal ulna
- Radiography
  - Physeal widening, irregularity, fragmentation
  - Metaphyseal cystic change
  - Ulnar plus variance
  - Metaphyseal band-like sclerosis paralleling growth plate
  - Delayed bone age
- MR
  - Allows evaluation of injury before radiographic findings present
  - Allows evaluation of adjacent osseous and ligamentous structures
- CT
  - Metaphyseal clefts (areas of focal growth arrest) may be more evident
  - Assists in quantifying bone bridging

Top Differential Diagnoses

- Carpal fracture
- Ligament tear
- Extensor carpi ulnaris tendinopathy

Pathology

- Female gymnasts at increased risk
  - Delayed bone age
  - High training intensity; poor nutrition, delayed menarche
  - Physes remain vulnerable longer
- Vulnerability accentuated
  - During periods of rapid growth
  - During puberty
- Growth plate strength
  - Weaker than articular cartilage
  - Less resistant to shear and tension forces than adjacent bone

*(**Left**) Graphic shows normal growth plate organization. The growth plate develops separately from epiphyseal ossification and is composed of 4 zones: Growth (chondroblasts divide & columnate), maturation (cells hypertrophy & calcified matrix penetrates between columns), transformation (calcification becomes more organized as metaphyseal vessels penetrate matrix), & remodeling. (**Right**) PA radiograph shows the radiographic appearance of normal epiphyseal, growth plate, & metaphyseal relationships.*

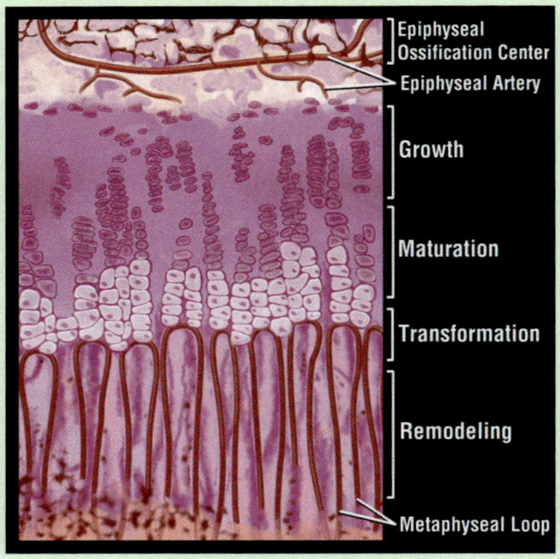

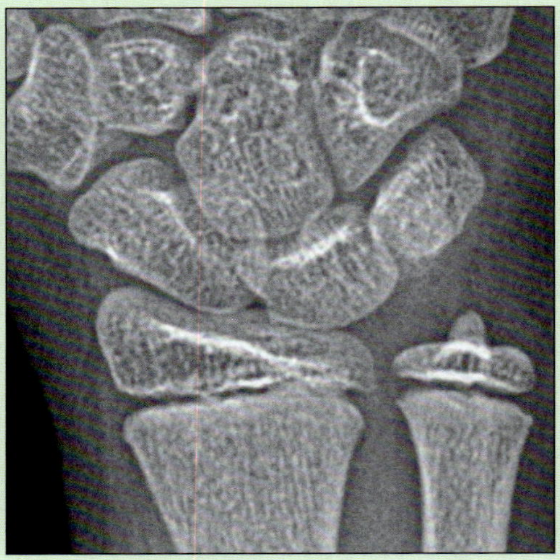

*(**Left**) Graphic shows growth plate changes with repetitive trauma. Poor metaphyseal vessel penetration into the transformation zone limits deposition of calcified matrix between hypertrophied chondroblasts, resulting in a widened, nonmineralized growth plate. This area is at particular risk of injury. (**Right**) AP radiograph shows radius growth plate irregularity and widening ➡ with band-like metaphyseal sclerosis ⇨ and small metaphyseal cystic changes ➡ in this young elite gymnast.*

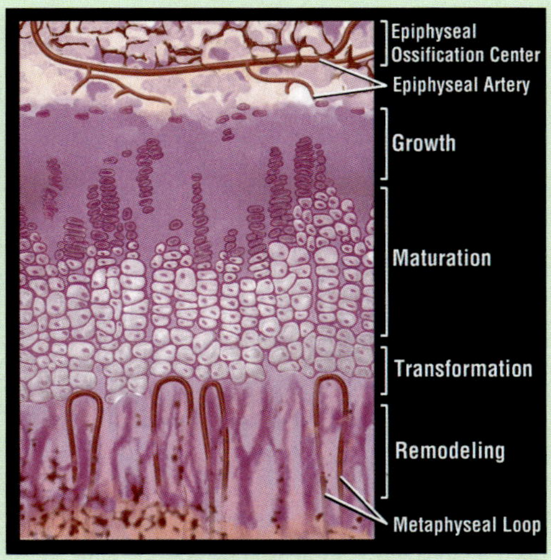

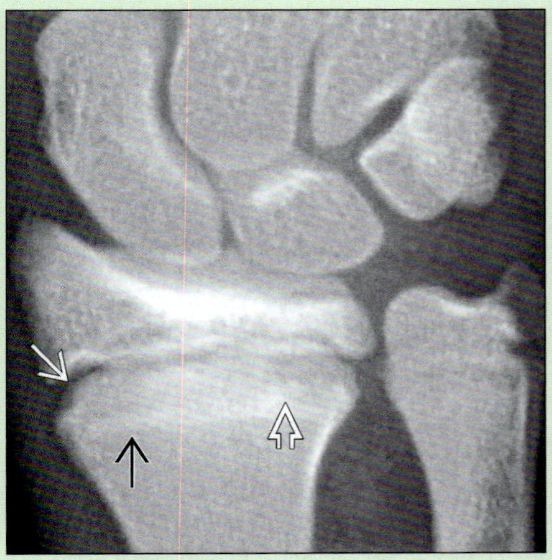

TRAUMA-RELATED OSTEOLYSIS IN CHILDREN

IMAGING

General Features

- Best diagnostic clue
  - Physeal widening, irregularity, fragmentation
  - Metaphyseal cystic change
  - Ulnar plus variance
- Location
  - Distal radius > > distal ulna

Radiographic Findings

- Radiography
  - Findings frequently bilateral
  - May be symmetric or asymmetric
  - Normal bone density
  - Physeal widening, irregularity, fragmentation
  - Metaphyseal band-like sclerosis paralleling growth plate
  - Metaphyseal margin flaring (beaking)
  - Delayed bone age
  - Ulnar plus variance

CT Findings

- Metaphyseal clefts (areas of focal growth arrest) may be more evident
- Assists in quantifying bone bridging

Bone Scan

- Best with pinhole collimation
- Increased uptake: May be difficult to distinguish normal from abnormal, as injury may be symmetric

MR Findings

- Allows evaluation of injury before radiographic findings present
- Allows evaluation of adjacent osseous and ligamentous structures
- Acute
  - Fluid-sensitive sequences: ↑ signal = marrow edema/contusion involving metaphysis and epiphysis
  - Increased signal also reflects ↑ volume of noncalcified cells, fibrous hyperplasia
- Subsequent
  - Physeal widening, irregularity, fragmentation
  - Physeal changes may be focal or diffuse
- Chronic
  - Ongoing marrow edema
  - Metaphyseal beaking
  - Ulnar plus variance

Imaging Recommendations

- Best imaging tool
  - MR to identify early bone injury
- Protocol advice
  - Radiography
    - Center radiographs at wrist to delineate osseous changes
    - Compare symptomatic side to asymptomatic side
  - MR
    - Gradient refocused imaging to maximize cartilage evaluation
    - Fluid-sensitive sequences for subtle marrow contusion

DIFFERENTIAL DIAGNOSIS

Radiographic

- Rickets
  - Typically affects multiple joints, particularly areas of rapid growth
  - Radiography
    - Diffuse osteopenia due to osteomalacia
    - Thin sclerotic metaphyseal bands (often multiple) representing growth arrest line
    - Marked growth plate widening and irregularity
    - Metaphyseal "cupping" and "beaking"
    - Marked bone age delay; growth retardation
    - Bowing deformity of long bones
- Infection
  - Typically asymmetric
  - Radiography
    - Focal metaphyseal widening is common
    - Periosteal reaction
    - Marked soft tissue swelling
    - Chronic changes
      - Cone-shaped epiphysis
      - Metaphyseal "cupping"
      - Involucrum development
      - Early growth plate closure
  - MR
    - Diffuse marrow edema
    - Focal fluid collection communicating with metaphyseal growth plate
    - Periosteal lifting, reaction

Clinical

- Carpal fracture
  - Radiography
    - Sclerosis; cortical break across scaphoid short axis
  - MR
    - Focal or diffuse marrow edema, ± cortical break
- Ligament tear
  - Radiography
    - Widening of scapholunate interval
  - MR
    - Discontinuity of scapholunate or lunotriquetral ligaments or triangular fibrocartilage
- Extensor carpi ulnaris tendinopathy
  - MR
    - Diffuse or focal tendon thickening, ± tenosynovitis

PATHOLOGY

Growth Plate Vulnerability

- Growth plate strength
  - Weaker than articular cartilage
  - Weaker than surrounding fibrous tissue
  - Less resistant to shear and tension forces than adjacent bone
- Vulnerability accentuated
  - During periods of rapid growth
  - During puberty
- Female gymnasts at increased risk
  - Delayed bone age
    - Due to high training intensity; poor nutrition, delayed menarche
  - Physes remain vulnerable longer

TRAUMA-RELATED OSTEOLYSIS IN CHILDREN

Growth Plate Injury Pattern

- Typical Salter 1 injury
  - Fracture occurs at maturation zone between uncalcified and calcified parts of hypertrophic zone
  - Leaves intact growth (germinal) layer connected to epiphysis
  - Leaves intact epiphyseal and metaphyseal circulation
  - Typically heals without sequelae
- Repetitive use Salter 1 injury
  - Maturation (hypertrophic) zone injury
    - Repetitive trauma delays metaphyseal vessel penetration into zone
    - Calcification delayed due to impaired circulation
    - Growth zone (feed by epiphyseal vessels) continues to add to hypertrophic zone
    - Noncalcified zone widens in somewhat irregular columns
    - Stress-induced fibrous hyperplasia also present
    - Results in radiographic widening of growth plate
  - Healing phase
    - Decreasing repetitive trauma allows metaphyseal vessel penetration
    - Hypertrophic zone calcifies in normal fashion
    - Growth plate (hypertrophic zone) restored to normal thickness
    - Premature closure
      - Small osseous bridges may transiently form without sequelae
      - More extensive bridging may result in growth deformity depending on remaining growth potential
    - Segmental overgrowth may result from "hypervascularity" of healing

CLINICAL ISSUES

Presentation

- Most common signs/symptoms
  - Radial wrist pain accentuated by active and passive dorsiflexion and axial loading
  - Symptoms worsen after weight-bearing exercise: Compressive, rotational forces
  - Most commonly seen in gymnasts but reported in other sports (e.g., badminton)

Demographics

- Age
  - Skeletally immature
  - Risk of injury increases around puberty
- Gender
  - M < F
- Epidemiology
  - Wrist pain
    - Ranges from 45-87% in gymnasts
    - May be present in all athletes at all levels of competition
    - Higher competitive level, more hours spent in practice = greater risk of injury
  - Physeal stress response
    - 51% demonstrate stress injury to distal radial growth plate
    - 7% demonstrate frank physeal widening
    - Radius more commonly affected than ulna

Natural History & Prognosis

- Risk factors
  - Growth delay may be seen (related to high athletic output, diet, etc.)
  - Female athletes may be amenorrheic
  - Delayed growth increases amount of time growth plate is at risk
  - Repetitive trauma during period of growth and development
- Mechanism of injury
  - Compressive, torsional, and tensile stresses in wrist in extremes of extension, pronation, and supination
  - Normal wrist extension: 60-75° (> 90° in gymnasts)
  - Compressive forces > 2x body weight for handspring
- Ulnar plus (+) variance
  - Seen in high percentage of gymnasts (both symptomatic and asymptomatic)
    - 50% with ulnar + variance are asymptomatic
  - May develop ulnar symptoms due to altered load bearing
  - May result in TFC tear, ulnocarpal impingement
- Madelung-like deformity
  - Unusual but reported
  - Results from asymmetric radius growth plate closure (ulnar portion prior to radial portion)

Treatment

- Majority will recover with rest and with no growth disturbance
  - Radiographic changes absent: 1 month off sport
  - Radiographic changes present: 3 months off sport
- Significant symptoms related to ulnar + variance may require ulnar shortening

DIAGNOSTIC CHECKLIST

Consider

- Clues to diagnosis
  - Delayed bone age
  - Ulnar plus variance
  - Ulnocarpal impingement

Image Interpretation Pearls

- Look for associated injury or abnormality (e.g., ulnar variance, carpal contusion, ligament or soft tissue injury)

SELECTED REFERENCES

1. DiFiori JP et al: Wrist pain, distal radial physeal injury, and ulnar variance in the young gymnast. Am J Sports Med. 34(5):840-9, 2006
2. Ogden JA: Skeletal Injury in the Child. 3rd ed. Berlin: Springer Verlag. 153-60, 2000
3. Caine D et al: Does repetitive physical loading inhibit radial growth in female gymnasts? Clin J Sport Med. 7(4):302-8, 1997
4. De Smet L et al: Gymnast wrist: an epidemiologic survey of ulnar variance and stress changes of the radial physis in elite female gymnasts. Am J Sports Med. 22(6):846-50, 1994

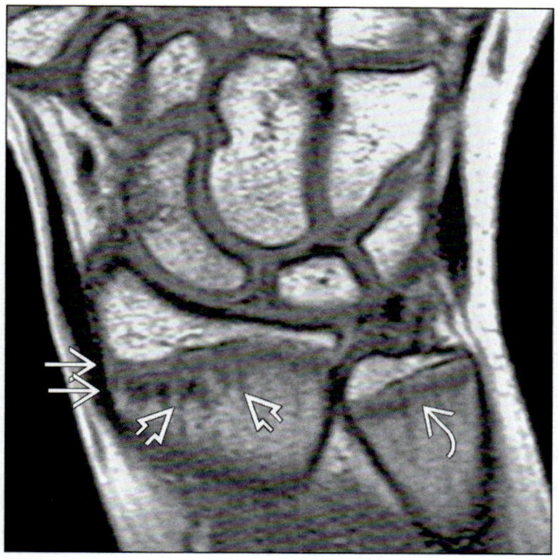
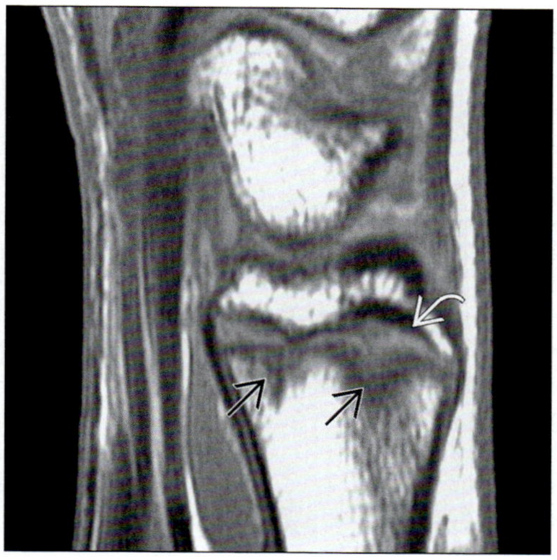

(Left) *Coronal T1WI MR shows radius growth plate widening → with irregular metaphyseal columns → related to repetitive trauma, which inhibits metaphyseal vessel ingrowth & delays calcification. Similar changes are noted in the ulnar growth plate →.* ***(Right)*** *Sagittal T1WI MR reveals growth plate widening →, particularly dorsally. This irregular metaphysis → is due to widening of the hypertrophic zone of the growth plate & may evolve to a more cystic appearance with continued trauma.*

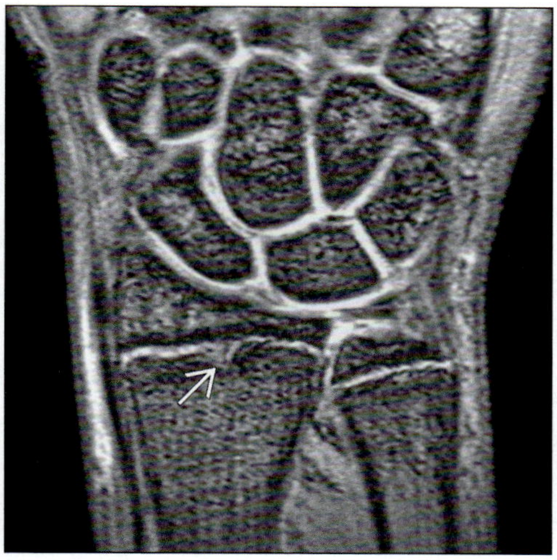
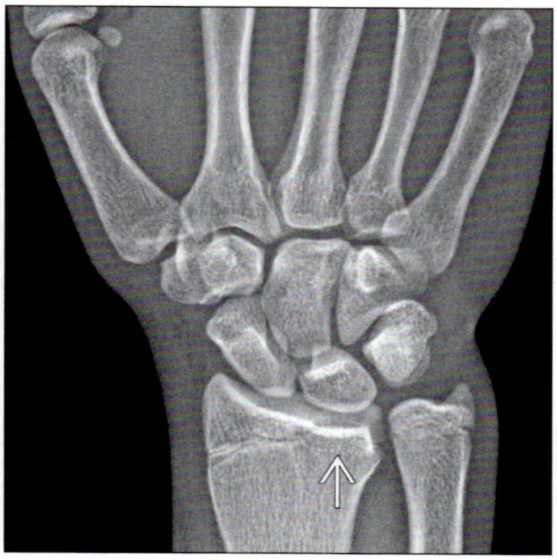

(Left) *Coronal gradient refocused MR image demonstrates focal physeal irregularity →, which may represent a bone bridge. These bridges may result in growth deformity if significant growth potential remains.* ***(Right)*** *PA radiograph in an elite female gymnast with delayed bone age (2 standard deviations below the mean) & amenorrhea. Medial radial growth plate is prematurely closed →, resulting in an ulnar plus variance. This early closure is related to the chronic repetitive stress of her sport.*

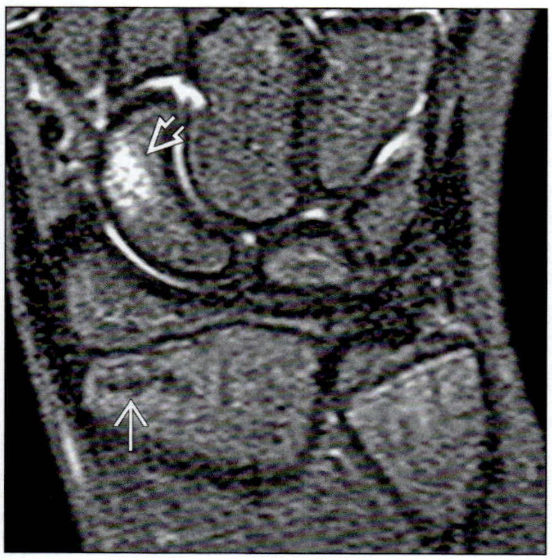
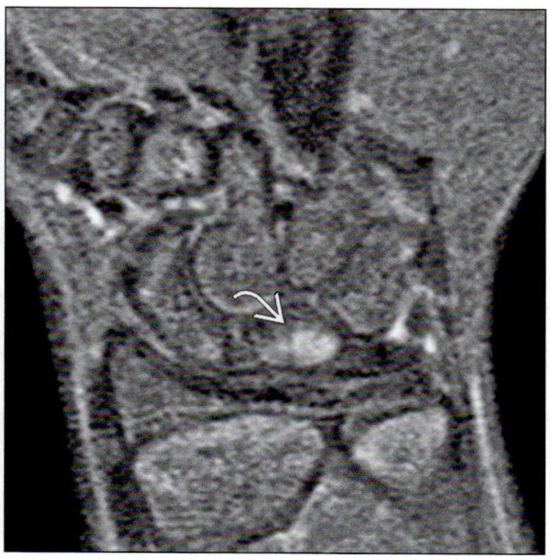

(Left) *Coronal STIR MR reveals ↑ signal of marrow contusion → in mid & distal scaphoid pole without a cortical break. Note the low signal band → paralleling the slightly irregular metaphysis in this athlete who was recovering from a prior growth plate injury.* ***(Right)*** *Coronal STIR MR reveals marrow contusion of lunate → in a 13-year-old gymnast. The subtle diffuse increased signal in the distal radius & ulna is normal. Do not forget that these athletes are at risk for other injuries that may clinically mimic osteolysis.*

DISTAL RADIOULNAR JOINT INSTABILITY

Key Facts

Terminology
- Distal radioulnar joint (DRUJ) instability evaluated as if ulna were mobile structure

Imaging
- Proper positioning essential for accurate diagnosis
- Direct sign
  - Ulnar head subluxation or dislocation
  - Dorsal: DRUJ widened on PA imaging
  - Volar: ↑ ulnar-radial overlap on PA imaging
- Indirect signs
  - Highly comminuted or malunited distal radial fracture
  - Dorsal radial angulation (> 20°)
  - Significant radial shortening (> 5 mm)
  - Displaced ulnar styloid base fracture
- Contralateral comparison may be useful
- CT: Best evaluation of DRUJ integrity

Clinical Issues
- Acute instability
  - Fracture: Deformity & pain
  - No fracture: Medial wrist pain & swelling
- Chronic instability
  - Painful, palpable clunk; grip weakness; ROM loss

Diagnostic Checklist
- Confirm appropriate radiographic position
- When in doubt, use multiplanar imaging (CT, MR) to visualize DRUJ
- Degree & direction of subluxation/dislocation
  - Associated osseous injury
  - Associated soft tissue injury
  - Distal radius fx alignment
  - Ulnar styloid fracture
- Adequacy of distal radius fracture reduction

*(**Left**) PA radiograph shows a comminuted distal radial fracture → with shortening of radius by > 5 mm, which puts tension on TFC complex. Ulnar base avulsion fracture → contributes to the DRUJ instability → as the TFC ulnar anchor is disrupted. (**Right**) Lateral radiograph from the same patient shows distal radius dorsal angulation > 20° →. Note the volar dislocation of the ulnar head →. Restoring distal radius alignment is important for long-term DRUJ integrity.*

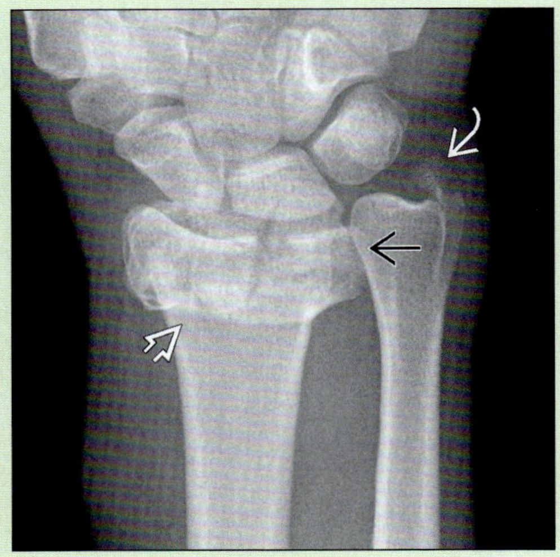

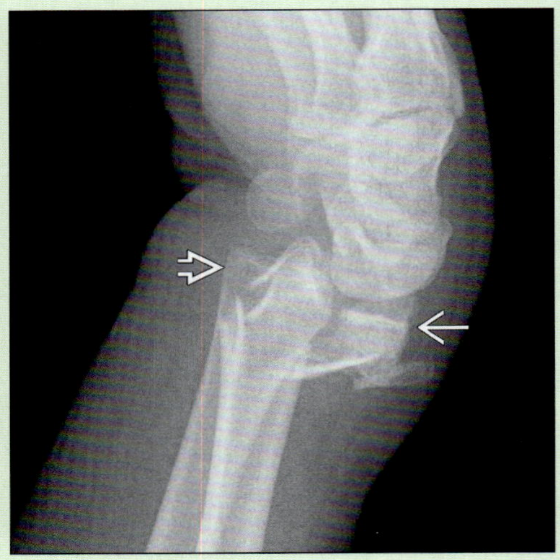

*(**Left**) AP graphic demonstrates the fractures that contribute to DRUJ instability with complex distal radius/ulna fractures. The radius fragments are depressed & angulated →. The TFC complex is disrupted when TFC avulses the ulnar styloid → and the ulnocarpal ligaments → are torn. (**Right**) PA radiograph reveals fracture stabilization with Kirschner wires through distal radius → and ulnar styloid → fractures. Further immobilization is provided with circumferential casting →.*

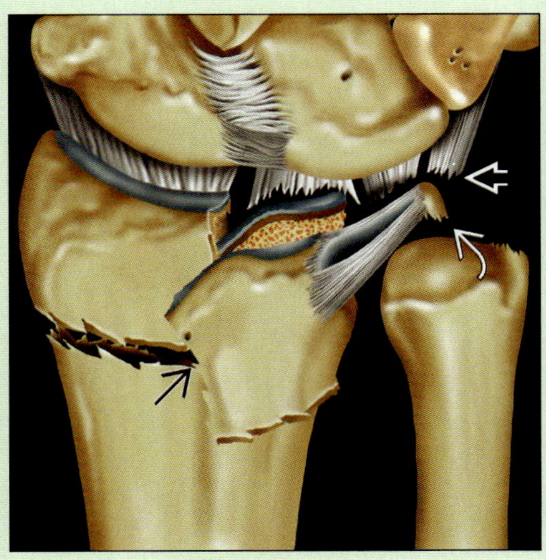

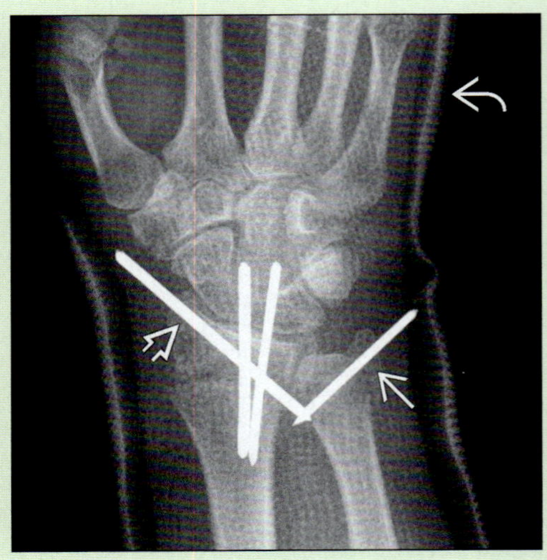

DISTAL RADIOULNAR JOINT INSTABILITY

TERMINOLOGY

Abbreviations

- Distal radioulnar joint (DRUJ)

Synonyms

- DRUJ instability

Definitions

- Disruption of normal relationship between distal radius and ulna

IMAGING

General Features

- Best diagnostic clue
  - Displacement of ulna with respect to sigmoid notch of radius
- Location
  - Dorsal displacement most common
- Morphology
  - By convention, DRUJ instability evaluated as if ulna were mobile structure
    - Anatomically, radius is mobile, rotating around static ulna
    - However, ulnar position is described relative to radius

Radiographic Findings

- Radiography
  - Proper positioning essential for accurate diagnosis
    - PA imaging
      - Shoulder abducted 90°; elbow flexed 90°; forearm & palm flat on cassette
      - Wrist in neutral position (no flexion, extension or medial-lateral deviation)
      - Correct position: Extensor carpi ulnaris (ECU) groove lies lateral to ulnar styloid long axis
    - Lateral imaging
      - Shoulder in neutral position; elbow flexed 90°; thumb toward ceiling
      - Wrist in neutral position (no flexion, extension, or medial-lateral deviation)
      - Correct position: Pisiform volar surface visualized midway between volar surfaces of capitate & scaphoid distal pole
      - Limited diagnostic value: 10° change in position results in DRUJ malalignment in normal wrist
    - Appropriate positioning may be difficult in setting of acute trauma
  - Normal DRUJ motion
    - Neutral
      - Dorsal & volar radioulnar ligaments are slightly lax
      - Imaging: Ulnar styloid points medially
    - Full pronation
      - Ulna shifts minimally dorsally
      - Mild ulnar minus variance
      - Dorsal radioulnar ligament key stabilizer
      - Imaging key: Ulnar styloid points volarly
    - Full supination
      - Ulna shifts minimally volarly
      - Mild ulnar plus variance
      - Volar radioulnar ligament key stabilizer
      - Imaging key: Ulnar styloid points dorsally
  - Direct sign of DRUJ disruption
    - Ulnar head subluxation or dislocation
      - Dorsal dislocation: Widened DRUJ on PA imaging
      - Volar dislocation: Increased ulnar-radial overlap on PA imaging
  - Indirect signs of DRUJ disruption
    - Highly comminuted distal radial fracture (intra- or extraarticular)
      - Dorsal radial angulation (> 20°)
    - Deformity of radial or ulnar shaft
      - Significant radial shortening (> 5 mm)
    - Displaced ulnar styloid base fracture
    - Displaced avulsion fracture from ulnar fovea
    - Malunited distal radius fracture
  - Contralateral side comparison may be useful

CT Findings

- Bone CT
  - Most accurate evaluation of DRUJ integrity
  - Imaging may be accomplished in cast or splint
- Measurement methods
  - Multiple methods
    - Radioulnar ratio (RUR): Most reproducible
      - Draw perpendicular line from volar (A) to dorsal (B) sigmoid notch rim
      - Draw line perpendicular to AB from ulnar head center; point of intersection (D)
      - RUR = AD/AB ± 2 standard deviations (nl pronation: 0.6 ± 0.05; neutral: 0.5 ± 0.04; nl supination: 0.37 ± 0.09)

MR Findings

- CT measurement methods applicable to axial MR
- Provides detailed soft tissue assessment
  - Triangular fibrocartilage (TFC) complex injury
    - Radioulnar ligaments
      - Volar radioulnar ligament tear → dorsal subluxation
      - Dorsal radioulnar ligament tear → volar subluxation
      - Ulnolunate & ulnotriquetral ligaments prevent dorsal displacement
    - Central disc tear
  - Extensor carpi ulnaris (ECU) tendon integrity
    - Entrapped tendon may limit reduction of acute DRUJ dislocation
      - ECU displaces medially around ulnar head or directly laterally into DRUJ
  - Interosseous membrane (IOM) integrity
    - Edema &/or tear of distal membrane
- Imaging challenges
  - Stabilizing wrist for imaging may reduce subtle subluxation
  - Wrist position (supination/pronation) affects study interpretation
    - Apparent ulnar variance
    - Apparent dorsal or volar subluxation

Imaging Recommendations

- Best imaging tool
  - Bone CT
- Protocol advice
  - Thin-section CT with reformations

DISTAL RADIOULNAR JOINT INSTABILITY

- Image patient in position of comfort

PATHOLOGY

General Features

- Etiology
  - Injury associated with fracture
    - Distal radius fracture most common
      - Intra- & extraarticular fracture with resultant marked dorsal angulation (> 20°)
      - Ulnar styloid base fracture
      - Ulnar head & sigmoid notch fractures ↑ complexity
      - Dorsal & ulnar radioulnar ligaments tear with > 5 mm radial shortening
      - TFC complex tears at peripheral (ulnar) attachments
      - Secondary stabilizers injured: IOM, ECU subsheath (ulnar collateral ligament), ulnocarpal & lunotriquetral ligaments
    - Galeazzi fracture
      - Distal radial shaft fx with DRUJ dislocation
      - DRUJ injury may be ligamentous &/or osseous
      - Unrecognized (untreated) DRUJ disruption results in higher morbidity
    - Essex-Lopresti fracture
      - Radial head fracture + DRUJ dislocation & IOM disruption
  - Isolated DRUJ injury without fracture rare
    - Dorsal dislocation most common
    - Hyperpronation + extension results in
      - Tightening of ECU & ulnar carpal ligaments
      - Ulnar head extension through dorsal capsule
      - TFC peripheral avulsion
      - Volar radioulnar ligament injury
    - Obstructions to reduction
      - ECU, extensor communis, extensor digiti minimi, &/or TFC
  - Chronic instability
    - Most common mechanism (without fracture)
      - Injury from fall on outstretched hand (hyperpronation) or forced rotation
    - May be due to distal radial fracture malunion
    - Residual dorsal angulation > 20° results in
      - Increased loading to distal ulna
      - Radioulnar incongruity
      - TFC complex tension/distortion

Staging, Grading, & Classification

- Fernandez classification of DRUJ injury
  - Type I: Stable DRUJ lesion
    - Ulnar tip avulsion or ulnar neck fracture
    - Primary stabilizers (TFCC & capsular ligaments) intact
    - Clinically stable; good prognosis
  - Type II: Unstable DRUJ lesion
    - Ulnar head dislocated or subluxated
    - Ulnar styloid base avulsion fracture
    - Unstable due to extensive TFCC tear
  - Type III: Potentially unstable DRUJ lesion
    - Disruption of joint surface at sigmoid notch or ulnar head

CLINICAL ISSUES

Presentation

- Most common signs/symptoms
  - Acute instability
    - Fracture: Deformity, pain
    - Without fracture: Medial wrist pain & swelling
  - Chronic instability
    - Painful, palpable clunk; grip weakness; range of motion (ROM) loss

Natural History & Prognosis

- Acute dislocation without fracture
  - If adequately reduced, intact secondary stabilizers will maintain stability during healing
  - If secondary stabilizers injured in original trauma, open surgery may be necessary
- DRUJ instability with fracture
  - Radial fracture reduction key to restoration of DRUJ
  - Residual dorsal angulation of distal radius results in limited DRUJ ROM
  - Malunion with dorsal angulation, shortening, & translation limits DRUJ function
- Ulnar styloid base fracture
  - Higher risk of DRUJ instability due to associated TFC complex injury/instability
- TFC complex injury may require surgical repair to avoid chronic long-term limitations

Treatment

- Determined by etiology
  - DRUJ instability/dislocations without fracture
    - Treated with reduction + immobilization
  - Dislocations with entrapped soft tissue
    - Open reduction ± TFCC repair
  - DRUJ dysfunction with fracture
    - Anatomic reduction of distal radius fracture ± ulnar styloid repair
    - Osteotomy to restore anatomy in fracture malunion

DIAGNOSTIC CHECKLIST

Consider

- Wrist position during imaging (pronation vs. supination)

Image Interpretation Pearls

- Confirm appropriate radiographic position
- When in doubt, use multiplanar imaging (CT, MR) to directly visualize DRUJ

Reporting Tips

- Degree & direction of subluxation/dislocation
- Associated osseous injury
- Associated soft tissue injury

SELECTED REFERENCES

1. Tsai PC et al: The distal radioulnar joint. Bull NYU Hosp Jt Dis. 67(1):90-6, 2009
2. Garrigues GE et al: Acute distal radioulnar joint instability. J Surg Orthop Adv. 17(4):262-6, 2008
3. Palmer AK: Triangular fibrocartilage complex lesions: a classification. J Hand Surg Am. 14(4):594-606, 1989

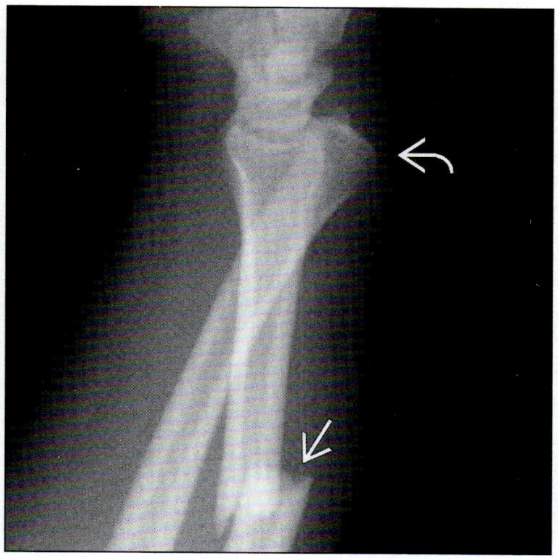

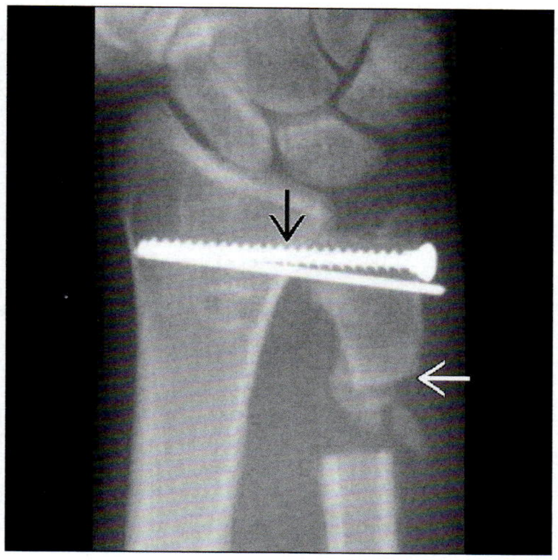

(Left) *Lateral radiograph shows a Galeazzi fracture-dislocation with distal radial shaft fracture ➡ and DRUJ dislocation ↷. Dorsal ulnar displacement is the most common dislocation pattern.* ***(Right)*** *PA radiograph shows DRUJ stabilization ⇨ following radial osteotomy ➡. Patient had Essex-Lopresti fracture-dislocation with a radial head fracture and disruption of the DRUJ. The DRUJ injury was not recognized at time of trauma, resulting in delayed treatment.*

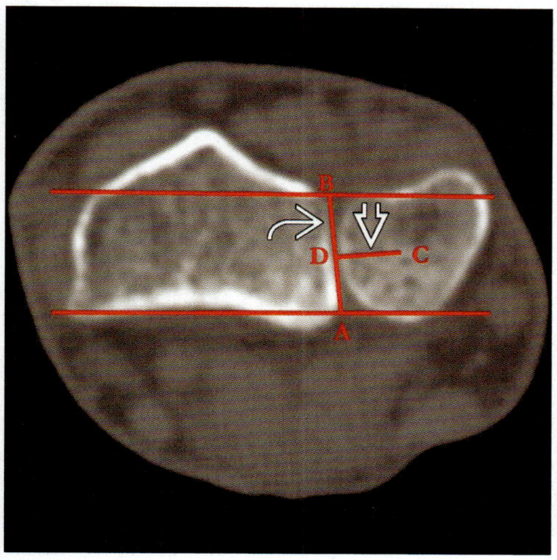

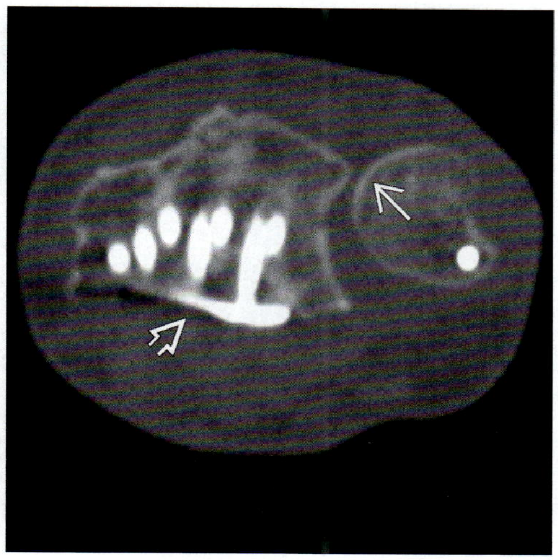

(Left) *Axial bone CT demonstrates radioulnar ratio method of evaluating DRUJ alignment. A line ↷ is drawn from the volar (A) to the dorsal (B) rim of the sigmoid notch. A perpendicular line ⇨ is then drawn from the ulnar head center (C) to the line AB. The ratio is AD/AB ± SD.* ***(Right)*** *Axial bone CT shows dorsal subluxation of the ulna ➡ relative to the radius. Despite reduction of the distal radius fracture ⇨, disruption of the volar radioulnar ligament allows ulnar subluxation.*

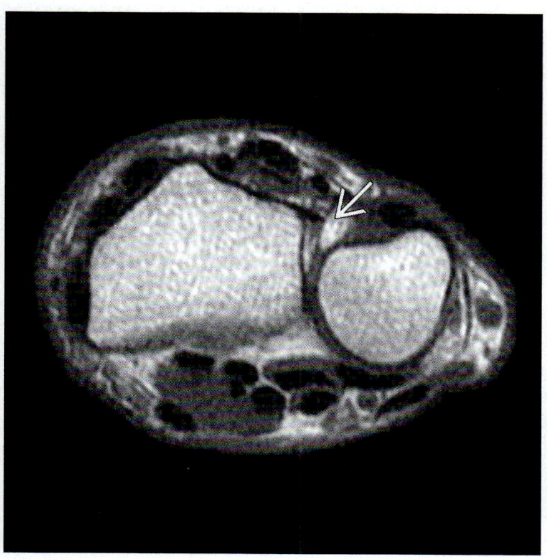

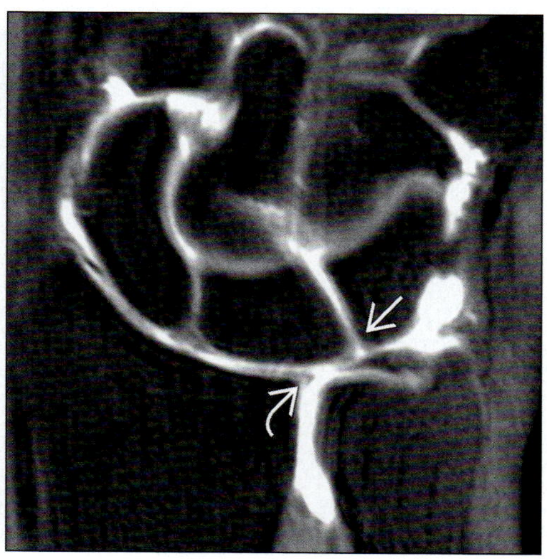

(Left) *Axial T1WI MR shows subtle volar translation of ulna due to tear of dorsal radioulnar ligament ➡. This patient fell on an outstretched hand but did not sustain a fracture. He has developed a painful wrist "clunk."* ***(Right)*** *Coronal MR arthrogram T1WI FS demonstrates a tear of the TFC at its radial attachment ↷. There is a lunotriquetral ligament tear ➡ as well. Patient had ulnar-sided pain with decreased grip strength due to DRUJ joint disruption.*

SCAPHOID FRACTURE

Key Facts

Imaging
- Radiographs
  - Use multiple views to evaluate
  - May be visible in only 1 view
- Thin-section CT in direct sagittal scaphoid long axis orientation
- MR for identifying radiographically occult fracture or evaluating vascularity of proximal pole fragment
- Imaging recommendations
  - Radiographs for initial evaluation
  - MR if initial radiographs are negative
  - CT for follow-up

Pathology
- Most common cause: Fall on outstretched hand
- Vascular supply
  - Proximal pole entirely dependent on intraosseous blood flow
  - Results in protracted healing or osteonecrosis

Clinical Issues
- Primary symptom: Anatomic snuffbox tenderness
- Healing occurs in ~ 90%
- Complications
  - "Humpback" deformity most common malunion
  - Delayed union: Incomplete union after 4 months
  - Nonunion: Failure to heal after 6 months; occurs in 5-10%
  - Osteonecrosis: Occurs 15-30% of all scaphoid fractures

Diagnostic Checklist
- Beware of radial deviated PA view; fracture line obscured
- Proximal pole ↑ density may be seen initially & resolves over time without ON

*(**Left**) Graphic demonstrates 3 typical patterns of scaphoid fracture. The mid-waist fracture ➡ is most common. The distal fracture ➡ is less common but typically heals without sequelae. The proximal pole fracture ➡ is least common but most problematic due to its tenuous blood supply. (**Right**) PA radiograph shows a complete transverse mid-waist fracture ➡. This fracture occurs as the scaphoid is trapped between capitate and radial styloid while being stabilized by volar ligaments.*

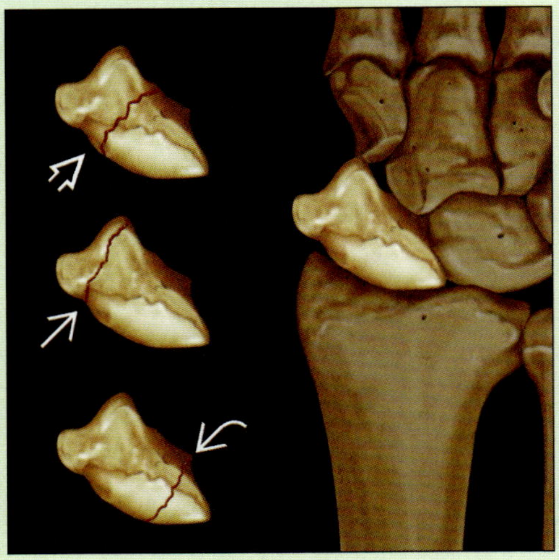

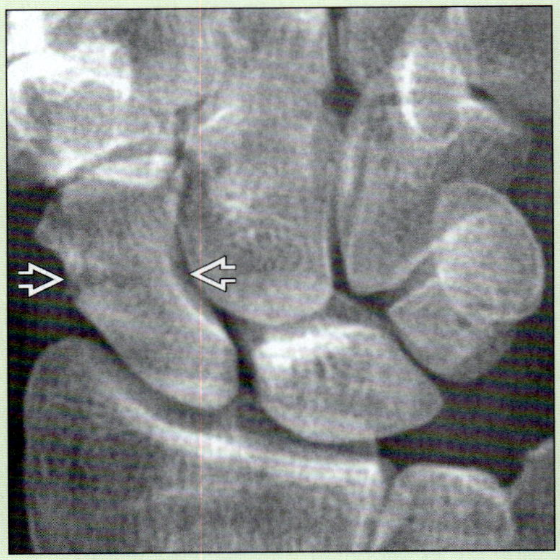

*(**Left**) PA radiograph in this 12 year old shows a complete transverse distal scaphoid fracture ➡. This patten is more common in children. This nondisplaced fracture will likely heal without complication due to the generous blood supply in the distal pole. (**Right**) PA radiograph demonstrates a complete transverse nondisplaced proximal pole fracture ➡. Fractures in this location tend to isolate the fragment from its key blood supply, putting it at risk for osteonecrosis.*

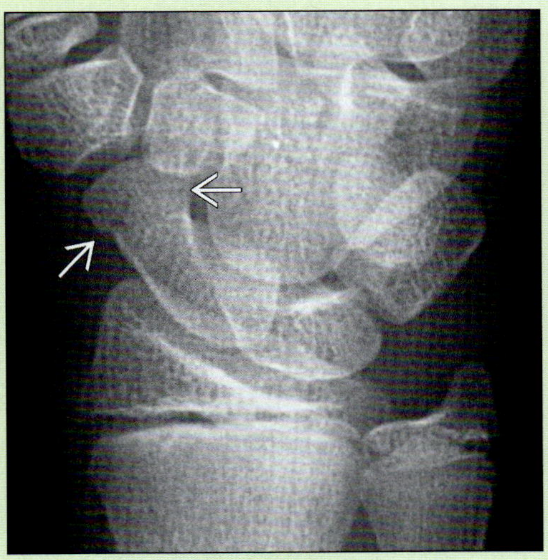

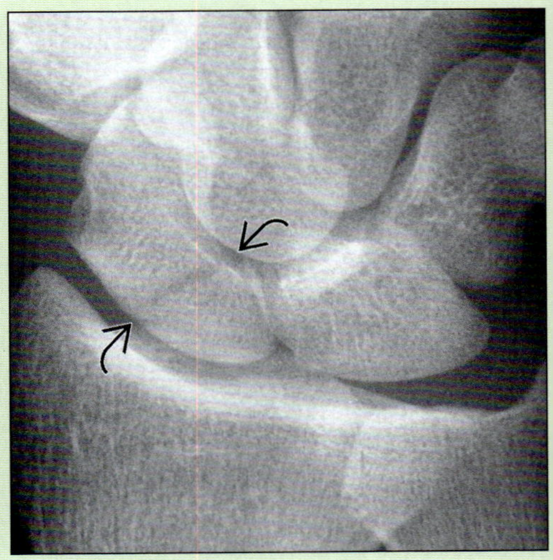

SCAPHOID FRACTURE

TERMINOLOGY

Synonyms
- Navicular fracture

IMAGING

General Features
- Best diagnostic clue
  - Cortical/trabecular disruption on radiographs
  - May be visible on only 1 view
- Location
  - Tubercle (distal volar prominence): 5%
  - Distal 1/3: 10%
  - Waist (mid 1/3): 70%
  - Proximal 1/3: 20%
- Morphology
  - Transverse or mildly oblique
  - Usually not comminuted or displaced

Radiographic Findings
- Use multiple views to evaluate
  - 75% visible on PA & pronated oblique views
  - Only 25% visible on all views
  - 2-5% not visible on initial imaging
- Standard views: PA, lateral, pronated oblique
- Navicular view: Ulnar-deviated PA → ↑ view of fx line
- Fisted PA: Accentuates fx line & associated injuries
- Lateral evaluates carpal & DRUJ alignment
- Radial-deviated PA obscures waist and fracture line
  - Patient tends to hold hand in radial deviation; must image in neutral or ulnar deviation
- Obliterated scaphoid fat stripe; soft tissue swelling suggestive but nonspecific

CT Findings
- Excellent fracture definition; evaluation of fx healing
- Delineates developing nonunion
  - Lack of osseous bridging, smoothly corticated margins
- Thin-section CT in direct sagittal scaphoid long axis orientation

MR Findings
- Acute fracture
  - T1: Fracture line ↓ signal intensity (SI)
  - Fluid-sensitive sequences: Fracture line may be ↓ or ↑ SI; surrounding diffuse marrow edema is ↑ SI
- Osseous vascularity (may be used to determine surgical status in painful subacute fracture)
  - Nonenhanced MR
    - Sensitivity = 36%, specificity = 78%, accuracy = 68%
    - Initial ↓ SI on fluid-sensitive sequences suggests ischemia; ± progression to osteonecrosis (ON)
    - Fragments not uniformly vascular or avascular
      - T1 & T2: Patchy signal may be present in both necrotic and viable bone
  - Contrast-enhanced MR
    - For determining viability of fragment: Sensitivity = 66%, specificity = 88%, accuracy = 83%
    - Post-gadolinium T1 imaging
      - ↑ SI compared to pre-contrast T1 suggests intact vascularity
      - ↓ SI suggests hypovascularity
      - Conspicuity increased with fat suppression

Bone Scintigraphy
- High sensitivity for fracture after 1st 48 hours
- Specificity limited by synovitis, prior arthritis, or injury

Imaging Recommendations
- Best imaging tool
  - Standard radiographs
  - MR if initial radiographs are negative
    - MR with contrast for fragment viability in painful subacute fracture
  - CT for follow-up assessment of bridging
- Protocol advice
  - Radiographs: Standard views; add additional views if routine radiographs negative
  - CT and MR: Direct sagittal imaging aligned parallel to scaphoid long axis

PATHOLOGY

General Features
- Etiology
  - Most common: Fall on outstretched hand
  - Forced dorsiflexion + radial deviation maximizes forces across scaphoid
- Relevant anatomy
  - Scaphoid is bridge between proximal & distal carpal rows
    - Injury to scaphoid may lead to carpal instability
  - Scaphoid surface largely cartilage-covered due to extensive articulations
  - Vascular supply
    - Dorsal scaphoid branch of radial artery
      - Enters at dorsal scaphoid ridge ligament attachments
      - Supplies 80% of total blood supply
    - Palmar vessels enter distal pole; supply 20% of vascularity
    - Proximal pole dependent entirely on intraosseous blood flow
      - Results in protracted healing or osteonecrosis
  - Ligament attachments
    - Scapholunate ligament
      - Intact: Proximal pole extends (with lunate) while distal pole stays flexed
      - Disrupted: May result in carpal instability
    - Volar ligaments act as fulcrum in mid-waist fracture; accentuate intrascaphoid angle

Staging, Grading, & Classification
- Multiple classification systems
  - Limited prognostic value
  - Useful to communicate findings
- Russe (1960)
  - Horizontal oblique (most compressive force)
  - Transverse (combined shear/compressive forces)
  - Vertical oblique (most shear force; most unstable)
- Herbert (1990)
  - A: Stable, acute
  - B: Unstable, acute
  - C: Delayed union
  - D: Established nonunion

SCAPHOID FRACTURE

CLINICAL ISSUES

Presentation

- Most common signs/symptoms
  - Anatomic snuffbox tenderness
- Other signs/symptoms
  - Pain with axial compression of thumb
  - Pain over distal scaphoid tubercle
  - Limited range of motion
  - Pain at end arc of motion, especially with flexion and radial deviation
  - Reduced grip strength

Demographics

- Age
  - 15-60 years
  - Most common in adolescents & young athletes
  - Distal 1/3 fracture more common in children than adults
- Gender
  - M > F
- Epidemiology
  - 70% of all carpal bone fractures
  - 5-12% associated with other fractures
    - Radial styloid, triquetrum, fracture-dislocation
    - Usually due to higher-impact trauma

Natural History & Prognosis

- **Healing**
  - 90% heal if treated promptly
  - Imaging: Progressive loss of fracture line visibility on radiograph &/or CT, often with smudgy sclerosis
  - Averaging healing time: 3-6 months (related to blood supply)
- **Malunion**
  - "Humpback" deformity: Most common malunion
    - Proximal pole "extends" with adjacent lunate while distal pole remains flexed
    - Results in increased intrascaphoid angle > 35°
      - Angle > 45° associated with poor outcome
- **Delayed union**
  - Incomplete osseous union after 4 months
  - Due to poor immobilization; failure to diagnose/treat promptly
  - Imaging: ↑ lucency along fracture line or rounded (cystic) focal lucencies suggesting motion
  - May still heal with ingrowth of granulation tissue and eventual bony bridging
- **Nonunion**
  - Failure to heal after 6 months; occurs in 5-10%
  - Factors: > 1 mm displacement, vertical oblique or proximal fracture, ON, smoking, associated ligament injury
  - Imaging: Persistent lucency along fracture line with smooth, rounded, sclerotic margins
  - T2WI MR: ↓ SI = callus or fibrous union, ↑ SI = fluid in fracture defect
  - Established nonunion may lead to scaphoid nonunion advanced collapse (SNAC)
    - Progressive carpal arthritis & collapse
      - Involves radioscaphoid, scaphocapitate, & capitolunate joints
      - Relative sparing of radiolunate joint
    - Proximal scaphoid fragment acts as extension of lunate
    - Often associated with dorsal intercalated segment instability (DISI)
- **Osteonecrosis**
  - 15-30% of all scaphoid fractures develop ON
  - Factors: Fx proximal to waist; ↑ proximal = ↑ risk
  - Imaging: Fragmentation & collapse; sclerosis around fracture may be seen in ON or with normal healing
  - MR imaging: ↓ SI on T1 & fluid-sensitive sequences suggests hypovascularity
  - Enhanced MR imaging: Lack of enhancement suggestive of ischemia
    - Not a reliable sign of ON or accurate predictor of impending nonunion
- **Complications**
  - Limited range of motion & pain; decreased grip strength; osteoarthritis
- **Associated injuries**
  - Scapholunate dissociation
  - Dorsal intercalated segment instability
  - Perilunate instability
  - Capitate head shear fracture
  - Triquetral fracture

Treatment

- Casting
  - Used in stable, nondisplaced fractures involving mid/distal scaphoid
- Surgical intervention
  - Used in unstable, displaced fracture or delayed union, symptomatic malunion/nonunion, osteonecrosis

DIAGNOSTIC CHECKLIST

Consider

- If clinical findings suspicious
  - May cast for 2 weeks; re-image to identify fracture line as resorption of fracture margins occurs in early healing
  - If waiting not acceptable, consider MR
- Direct sagittal scaphoid CT for evaluation of healing

Image Interpretation Pearls

- Fx may be evident only on single radiographic view
- Beware of radial deviated PA view; fracture line obscured
- Resorption/cystic change suggests impending nonunion but may be reversible
- Proximal pole ↑ density may be seen initially & resolves over time without ON
- MR (even with contrast) is not 100% accurate in determining fragment viability

SELECTED REFERENCES

1. Hansen TB et al: Cost-effectiveness of MRI in managing suspected scaphoid fractures. J Hand Surg Eur Vol. 34(5):627-30, 2009
2. Haisman JM et al: Acute fractures of the scaphoid. J Bone Joint Surg Am. 88(12):2750-8, 2006
3. Cerezal L et al: Usefulness of gadolinium-enhanced MR imaging in the evaluation of the vascularity of scaphoid nonunions. AJR Am J Roentgenol. 174(1):141-9, 2000

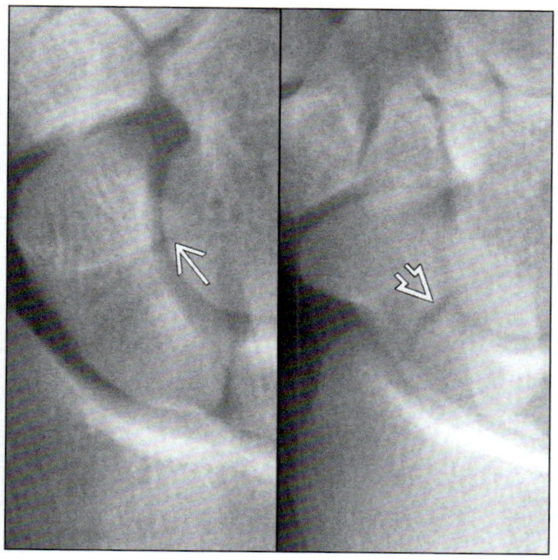

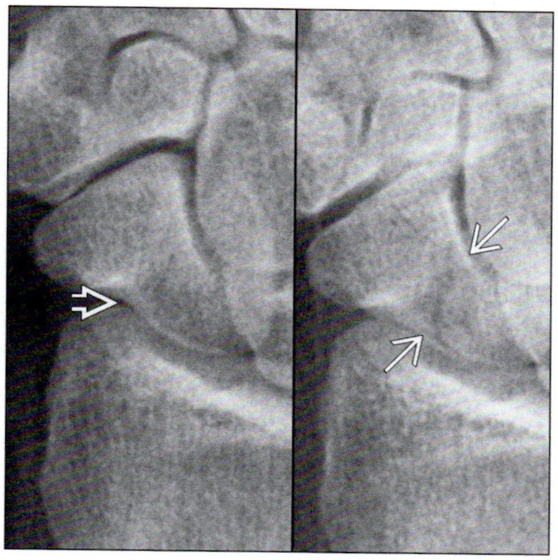

(Left) *PA and oblique radiographs show a mid-waist scaphoid fracture. The fracture is subtle on PA view* ➔ *but easily seen on oblique* ➔. *It is not uncommon to be able to visualize a scaphoid fracture on only a single view; navicular view is helpful.* **(Right)** *Oblique radiograph (and additional views) fail to show fracture* ➔ *immediately after injury. Three weeks later, radiographs clearly show a fracture line* ➔ *as the resorption phase of healing has increased the conspicuity of the fracture.*

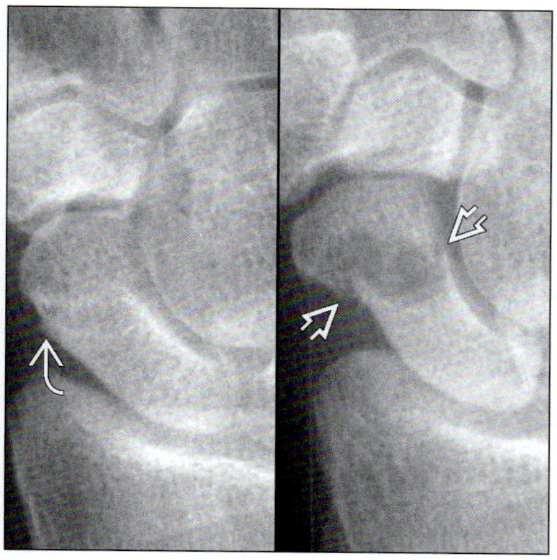

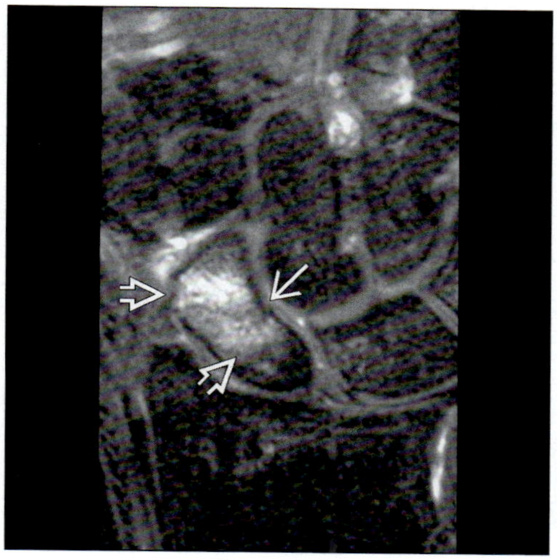

(Left) *PA x-ray immediately following trauma reveals subtle cortical irregularity* ➔ *along the scaphoid waist. PA x-ray obtained 4 months later shows delayed union of a slightly displaced distal pole fracture* ➔ *with cystic change. Granulation tissue may form with eventual bony bridging.* **(Right)** *Coronal PDWI FS MR shows linear low signal* ➔ *representing the fracture with hyperintense marrow edema* ➔ *surrounding the fracture and extending proximal and distal to the scaphoid fracture.*

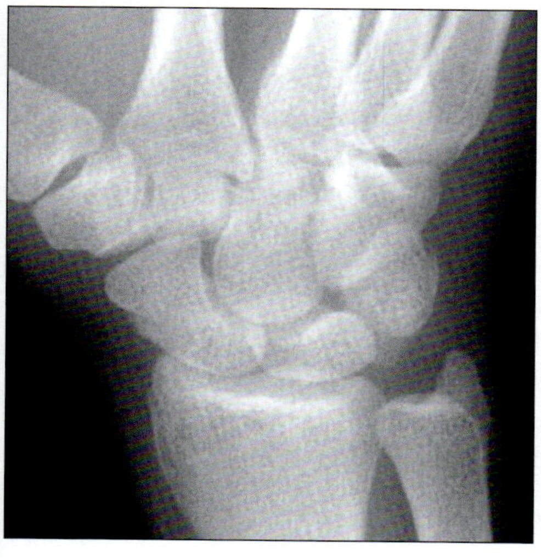

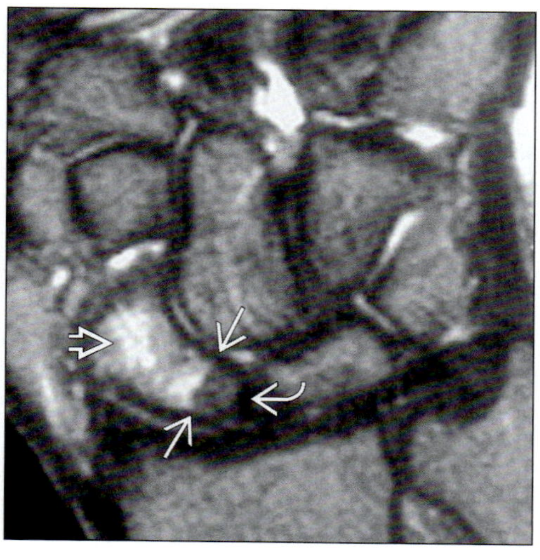

(Left) *Oblique radiograph (and multiple additional views) obtained following a FOOSH shows no evidence of a scaphoid fracture.* **(Right)** *Coronal T2WI FS MR from the same patient obtained 20 minutes later shows nondisplaced proximal scaphoid fracture* ➔ *with marrow edema in the distal fragment* ➔ *but not in the proximal fragment* ➔. *Absence of edema in the proximal pole, particularly in the setting of an acute injury, is highly suggestive of ischemia of that fragment & may lead to osteonecrosis.*

SCAPHOID FRACTURE

(Left) *CT scanogram obtained with the patient prone and the hand positioned with the scaphoid parallel to CT bore shows reference line for orientation of image acquisition parallel to the long axis of the scaphoid. This should also be done with MR.* ***(Right)*** *Direct sagittal bone CT from the same patient reveals an acute angulated displaced mid-waist scaphoid fracture. Thin-section CT allows full delineation of the fracture displacement (> 1 mm) and increased intrascaphoid angle.*

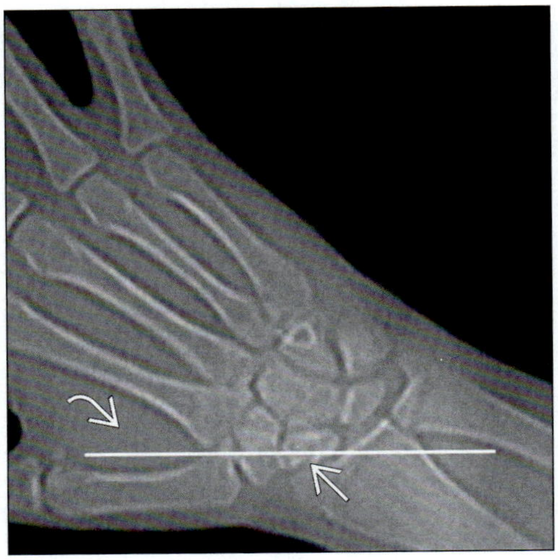

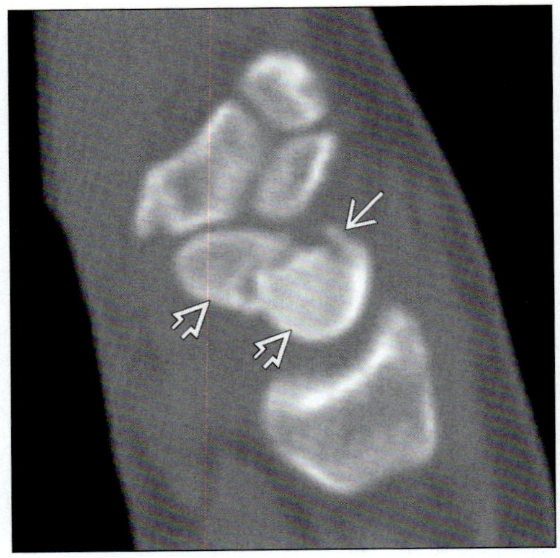

(Left) *Sagittal graphic shows scaphoid flexion with "humpback" deformity of the scaphoid. This results when the proximal fragment extends, maintaining its alignment with the lunate, while the distal fragment flexes.* ***(Right)*** *Sagittal T1WI MR demonstrates a low signal mid-waist scaphoid fracture line. There is an increased intrascaphoid angle, indicating minimal displacement of the fracture, which was not evident on radiographs.*

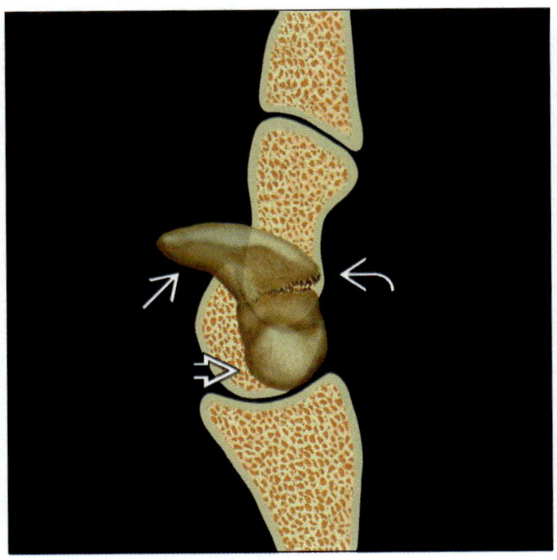

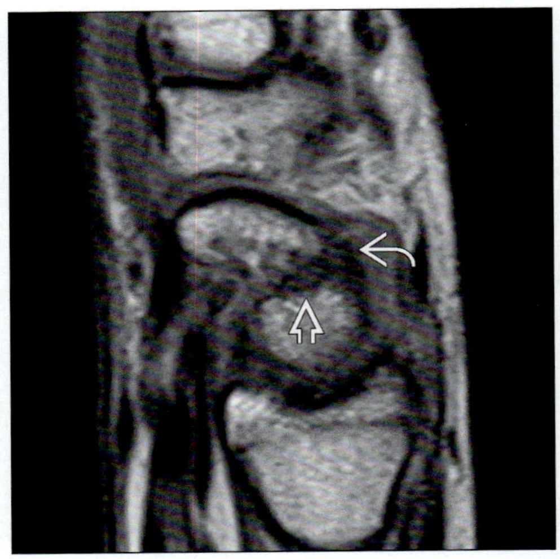

(Left) *Direct sagittal bone CT demonstrates a developing "humpback" deformity in this healing scaphoid fracture. This may not be symptomatic, but there is an increased risk of osteoarthritis due to the malunion.* ***(Right)*** *Coronal reformat bone CT demonstrates a healing proximal pole fracture. Note increased density of the proximal scaphoid. While sclerosis is typically present in osteonecrosis, it is not diagnostic for it. The vascularity may recover without sequelae.*

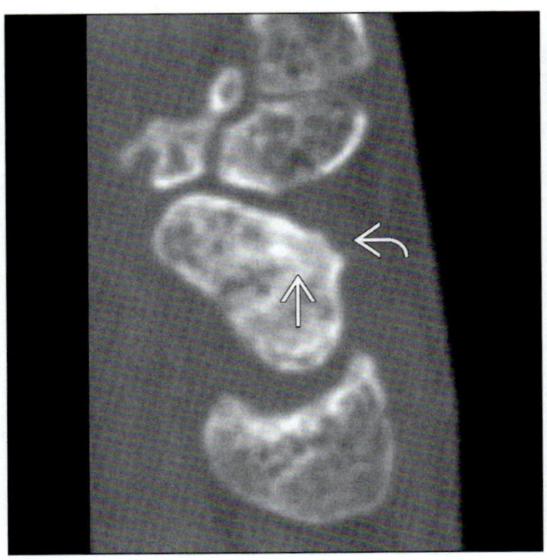

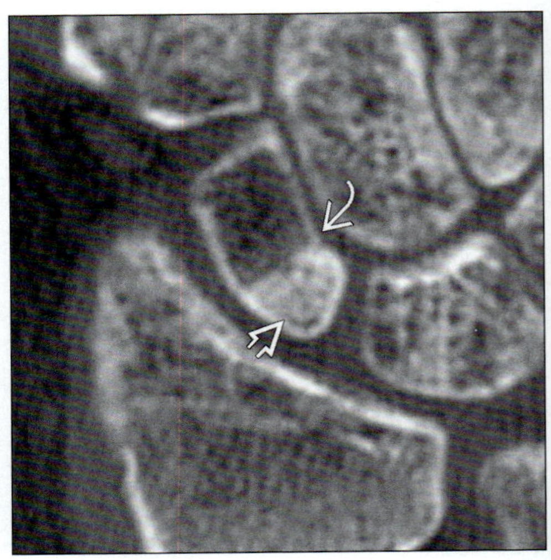

SCAPHOID FRACTURE

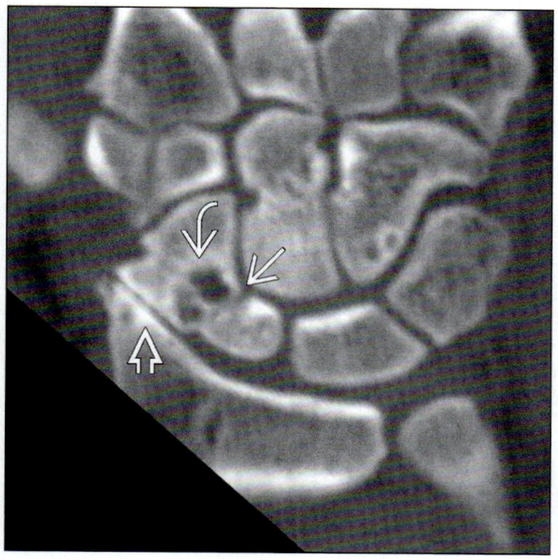

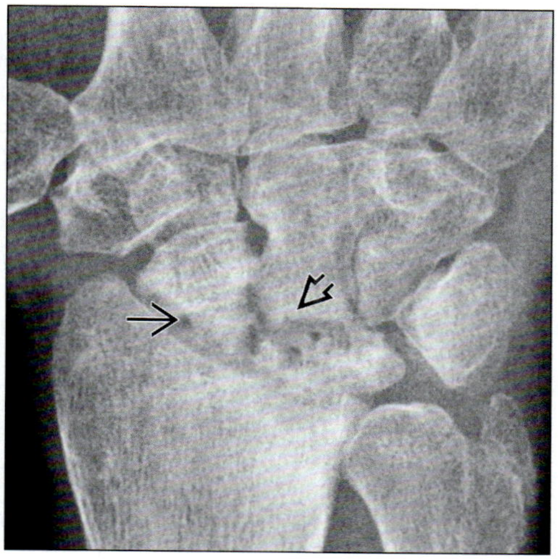

(Left) *Coronal reformat bone CT shows a proximal pole scaphoid fracture nonunion. The fracture margins are corticated →. Cystic changes → are seen along the fracture line. There is radioscaphoid joint space narrowing →, indicative of early stage SNAC wrist.* ***(Right)*** *PA radiograph demonstrates a proximal pole scaphoid fracture nonunion →. Advanced arthritis → involves the radiocarpal joint and proximal carpal row, typical of scaphoid nonunion advanced collapse (SNAC).*

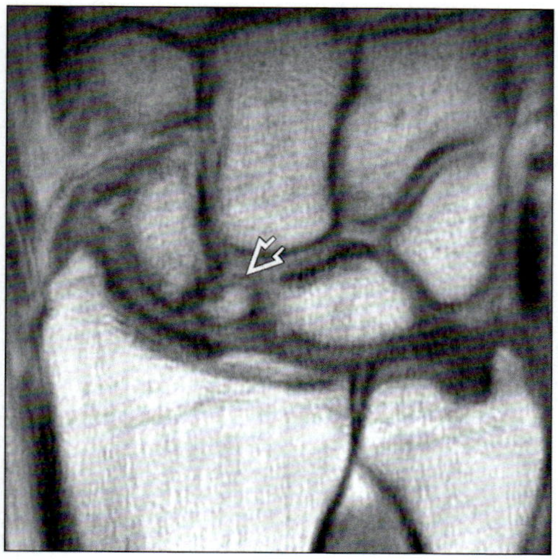

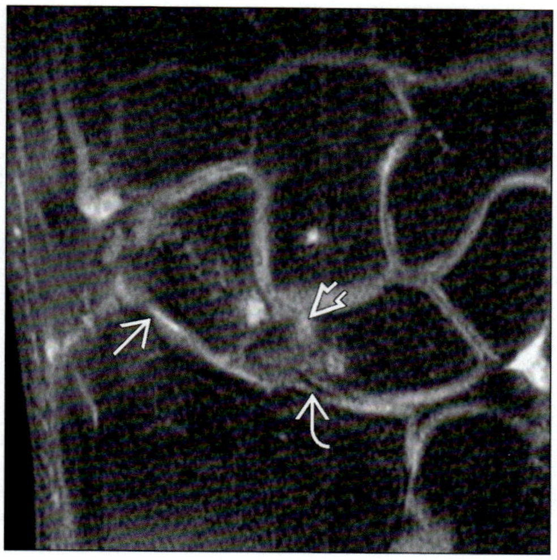

(Left) *Coronal T1WI MR demonstrates a very small proximal pole scaphoid fracture with a well-corticated fragment →. The marrow signal follows the high signal of normal fat marrow on T1 imaging.* ***(Right)*** *Coronal STIR MR from the same patient demonstrates mild patchy hyperintensity in the proximal pole fragment → suggesting that the fragment is viable. Note the cartilage thinning of the radial styloid cartilage →. The scapholunate ligament is intact →.*

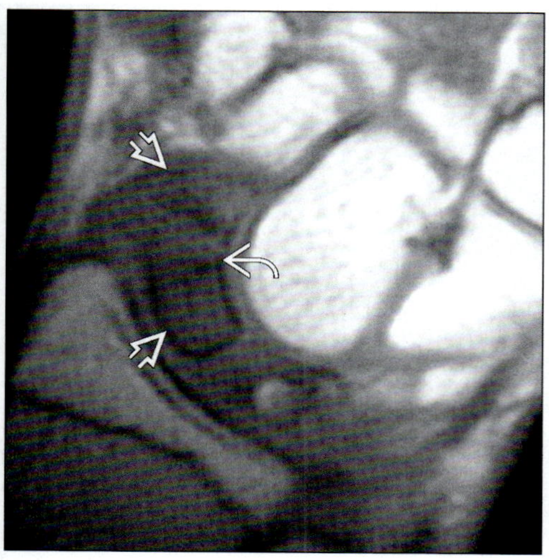

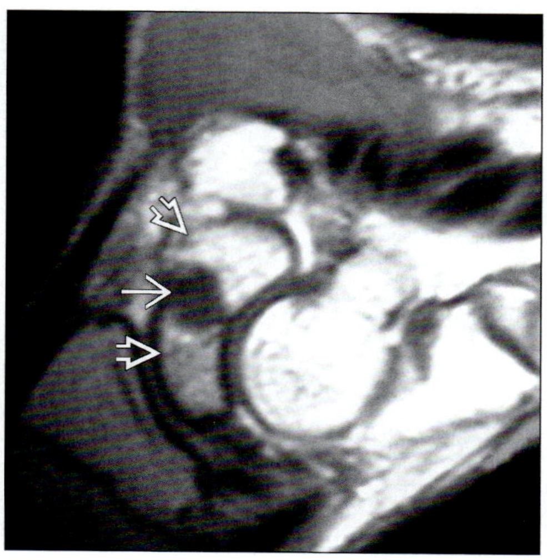

(Left) *Coronal T1WI MR shows the sclerotic margins → of a mid-waist scaphoid fracture nonunion. The proximal and distal pole fragments are hypointense →. This may result from edema or hypovascularity and fibrosis.* ***(Right)*** *Coronal T1WI C+ MR from the same patient shows avid enhancement in both fragments →, suggesting intact vascularity. Note the nonenhancing fibrous tissue → in the fracture gap. The use of fat suppression on pre- and post-contrast images may help see enhancement.*

SCAPHOID FRACTURE, POSTOPERATIVE

Key Facts

Imaging

- Healing: Anatomic alignment, osseous bridging
- Complications: No osseous bridging, carpal collapse, malalignment, poor hardware position
- Radiographs: Anatomic fracture apposition & alignment
  - Appropriate hardware positioning
  - Look for lucency at bone-hardware interface
- CT: Presence or absence of osseous bridging
- MR: Does not reliably determine ischemia or viability
  - Initial ↓ SI on fluid-sensitive sequences suggests ischemia
  - May or may not progress to osteonecrosis (ON)
  - Gadolinium-enhanced MR may improve assessment of vascular integrity

Clinical Issues

- Fracture location guides treatment
  - Distal pole fracture: Well vascularized, tends to heal well
  - Waist fracture: Nondisplaced vs. displaced
  - Nondisplaced: Treat closed; 90% heal
  - Displaced: High malunion or nonunion rate without surgery
  - Proximal pole fracture: Treat surgically unless contraindicated medically
- Choice of fixation related to fragment size
- Immobilization (closed treatment)
  - Advantages: Fracture healing without surgery
  - Disadvantages: Prolonged immobility & healing
- Surgery
  - Advantages: Rapid mobilization, ↑ fusion rates
  - Disadvantages: Surgical & anesthesia complications

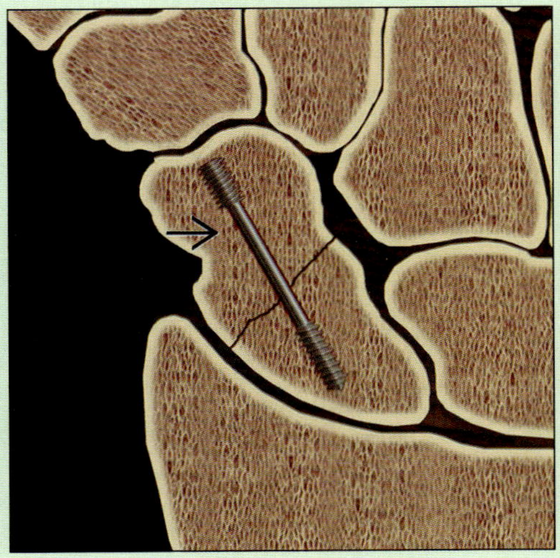

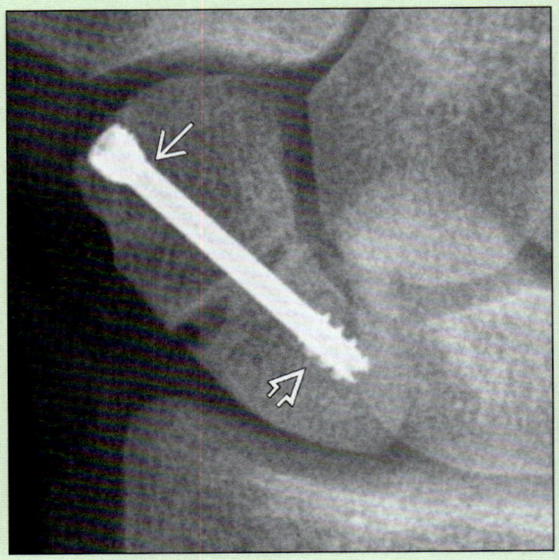

(Left) *Coronal graphic demonstrates the positioning of a Herbert screw →. This differentially threaded screw pulls proximal and distal fracture fragments into close apposition to facilitate healing.* ***(Right)*** *Oblique radiograph shows a cannulated differentially threaded screw spanning a subacute healing mid-waist fracture. Note the greater pitch → of the leading threads vs. the tighter pitch → of the trailing threads. This allows apposition and slight compression of fracture fragments.*

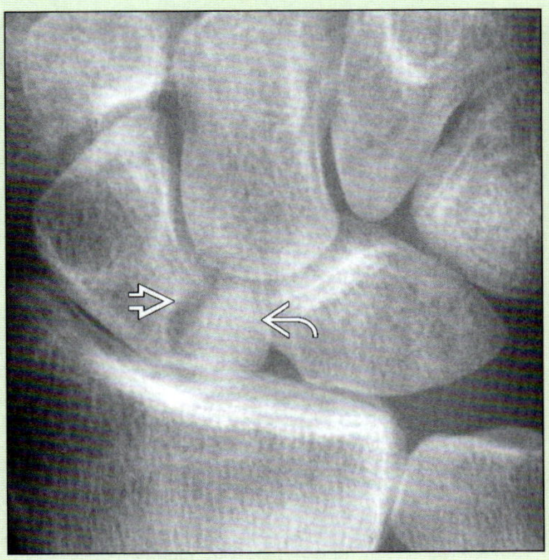

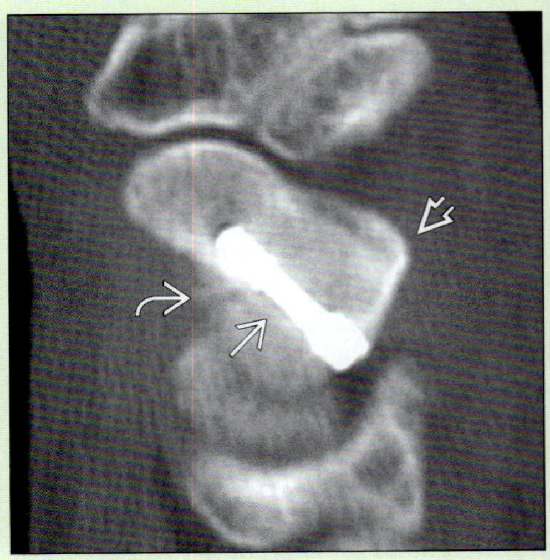

(Left) *PA radiograph shows a proximal pole fracture 4 months after immobilization with no significant bony bridging → and sclerosis of the proximal pole →, suggesting nonunion and developing osteonecrosis.* ***(Right)*** *Sagittal bone CT from the same patient obtained 4 months after grafting and internal fixation with a cannulated Herbert screw → shows the fracture → in mild apex dorsal angulation, constituting a "humpback" deformity →.*

SCAPHOID FRACTURE, POSTOPERATIVE

IMAGING

General Features

- Best diagnostic clue
  - Healing: Anatomic alignment, osseous bridging
  - Complications: No osseous bridging, carpal collapse, malalignment, poor hardware position

Radiographic Findings

- Anatomic fracture apposition & alignment
- Appropriate hardware positioning
- Progressive osseous bridging
- Hardware withdrawal or loosening

CT Findings

- Alignment
- Presence or absence of osseous bridging

MR Findings

- MR does not reliably determine ischemia or viability
  - Initial ↓ SI on fluid-sensitive sequences suggests ischemia
  - May or may not progress to osteonecrosis (ON)
- Initial unenhanced MR can be useful in predicting healing with bone grafting
  - If ↓ SI on T1 & T2WI MR: < 5% will heal
  - If ↓ SI on T1WI & slight ↓ SI on T2WI: 75% will heal
  - If slight ↓ SI on T1 & isointense on T2: 90% will heal
- Gadolinium-enhanced MR may improve assessment of vascular status
  - ↑ SI on post-contrast T1 suggests intact vascularity
  - ↓ SI on post-contrast T1 suggests hypovascularity
- Useful to track revascularization following vascularized graft; evaluate 3-4 months post-op

Imaging Recommendations

- Best imaging tool
  - CT to evaluate progress of healing
  - MR to evaluate soft tissues, ligaments, tendons
  - Radiography for serial evaluation
- Protocol advice
  - Obtain thin section CT with direct sagittal orientation paralleling scaphoid long axis
  - MR imaging obtained ± contrast
    - T1 fat suppression images in 2 planes
  - Hardware artifact reduced with metal artifact reduction techniques

PATHOLOGY

Gross Pathologic & Surgical Features

- Hardware
  - Kirschner wires
  - Cancellous compression screws
    - Herbert screws: Headless compression screw with proximal and distal threads of different pitch
    - Herbert-Whipple screws: Headless compression screw with cannulation
    - Bioabsorbable headless compression screws
  - Plate and screw fixation for wrist fusion
- Bone grafting
  - Cancellous graft
    - 94% success rate
    - If initial grafting fails, future surgery less likely to be successful
  - Vascularized graft
    - Improves blood supply & primary bone healing
    - Improves fusion rates; useful when prior fusion failure or proximal pole ON
    - Complications: Infection, graft extrusion, graft resorption, progressive degenerative changes
- Salvage procedures
  - Wrist fusion, scaphoidectomy, proximal row carpectomy (PRC)

CLINICAL ISSUES

Treatment

- Fracture location guides treatment
  - Distal pole fracture
    - Well vascularized, tends to heal well
    - Conservative treatment (4-8 weeks immobilization)
  - Waist fracture
    - Nondisplaced: Casted; 90% heal
      - x-ray at 12 weeks may be limited; consider CT
    - Displaced: High malunion or nonunion rate without surgery
  - Proximal pole fracture
    - Treat surgically unless contraindicated medically
    - Choice of fixation related to fragment size
- Indications for surgical intervention
  - All displaced fractures
    - Displacement > 1 mm, scapholunate angle > 60°, radiolunate angle > 15°, intrascaphoid angle > 35°
    - Some surgeons feel any visible fx line is displaced
  - Proximal pole fracture (regardless of displacement)
  - Open fractures
  - Fractures with associated perilunate injuries
  - Symptomatic malunion
  - Nonunion
  - Osteonecrosis
  - Lack of healing after 3-4 months of immobilization
  - Potential morbidity from prolonged immobilization
- Immobilization (closed treatment)
  - Advantages: Fracture healing without surgery
  - Disadvantages: Prolonged immobility & healing
- Surgery
  - Advantages: Rapid mobilization, ↑ fusion rates
  - Disadvantages: Surgical & anesthesia complications

DIAGNOSTIC CHECKLIST

Consider

- Indications for surgical intervention

Image Interpretation Pearls

- MR may be helpful for evaluation of vascular integrity

SELECTED REFERENCES

1. Waitayawinyu T et al: Management of scaphoid nonunions. Hand Clin. 26(1):105-17, 2010
2. Anderson SE et al: MR imaging of avascular scaphoid nonunion before and after vascularized bone grafting. Skeletal Radiol. 34(6):314-20, 2005

SCAPHOID FRACTURE, POSTOPERATIVE

(Left) *PA radiograph obtained with the wrist held in slight ulnar deviation shows a Herbert screw bridging a subacute midwaist scaphoid fracture. The distal screw tip aligns with the scaphoid cortex.* ***(Right)*** *PA radiograph with radial deviation from the same patient shows the screw appears "proud," with the distal end projecting over the scaphoid-trapezium articulation due to scaphoid foreshortening with radial deviation. Proper imaging positioning is crucial for hardware evaluation.*

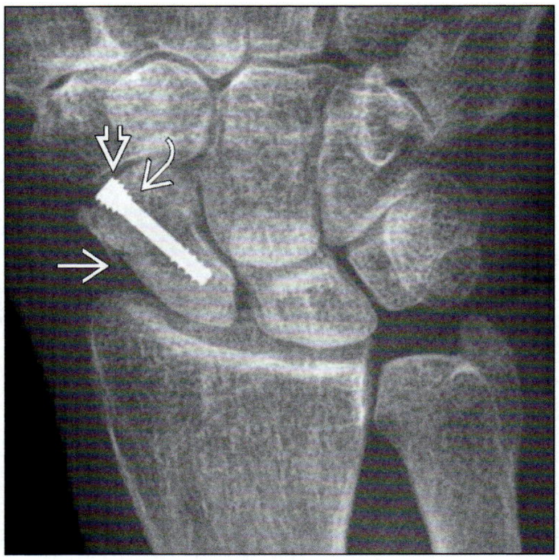

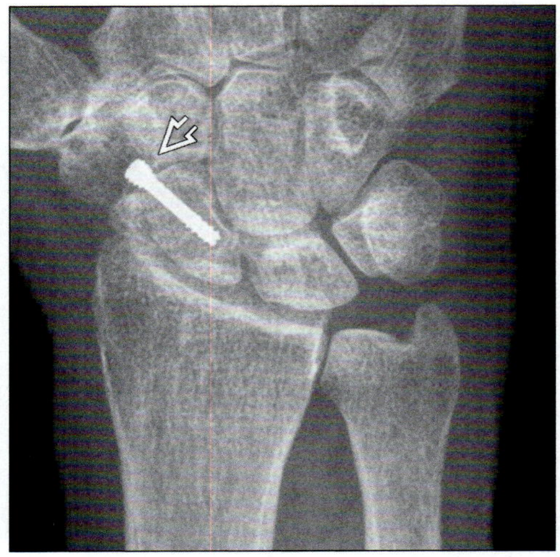

(Left) *Sagittal reformat bone CT reveals a proximal pole fracture stabilized with a Herbert screw. CT allows a detailed evaluation of the fracture line and hardware. No bone bridging is present.* ***(Right)*** *Coronal reformat bone CT demonstrates a healed scaphoid fracture with a retained compression screw. This fracture was treated with autologous cancellous bone graft harvested from the distal radius. The harvest site is evident.*

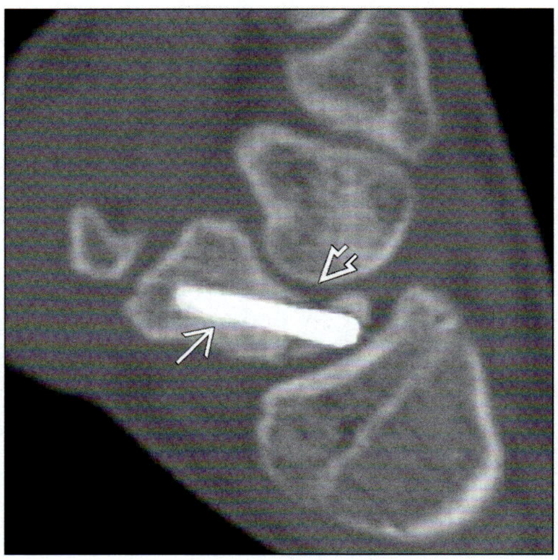

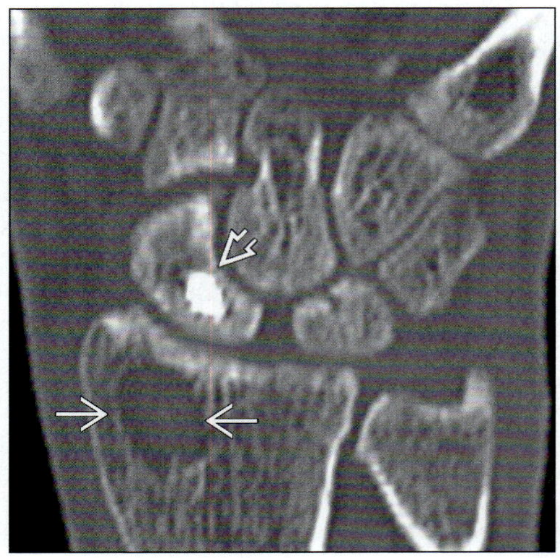

(Left) *Thin-section bone CT can be manipulated to maximize assessment of the fracture and associated hardware. This oblique axial image in a 15-year-old girl is reformatted along the K-wires to best evaluate the bone graft as well as the progress of healing of the underlying fracture.* ***(Right)*** *Coronal reformatted bone CT in the same patient demonstrates the healing harvest site for the vascularized bone graft used to treat this scaphoid fracture bridged by a screw.*

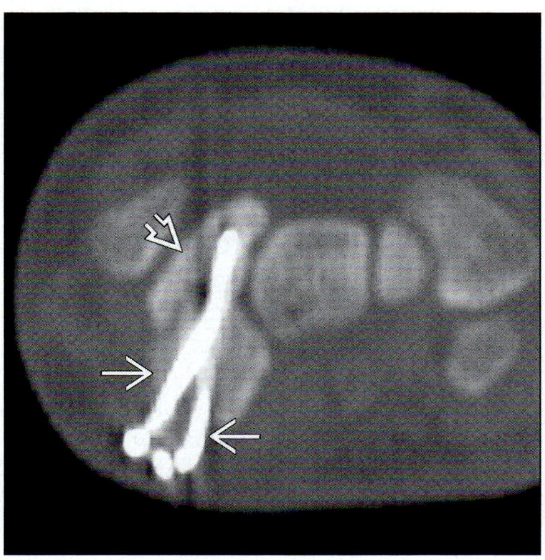

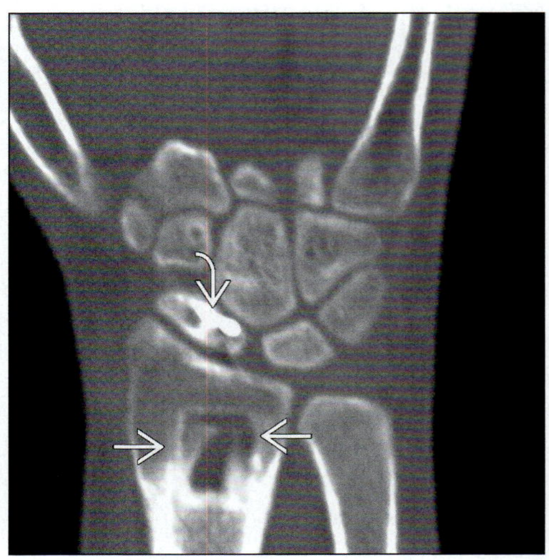

SCAPHOID FRACTURE, POSTOPERATIVE

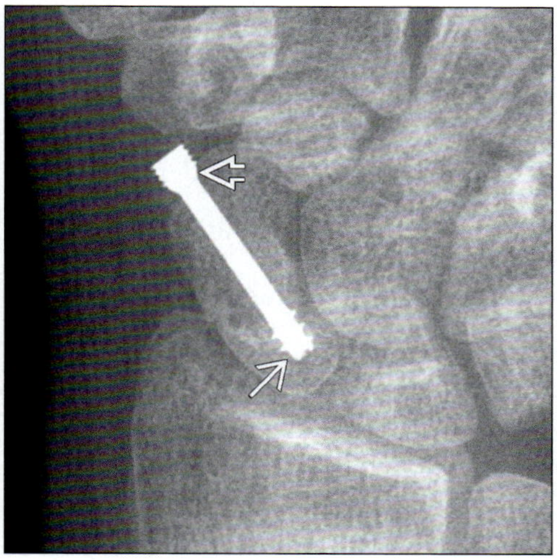

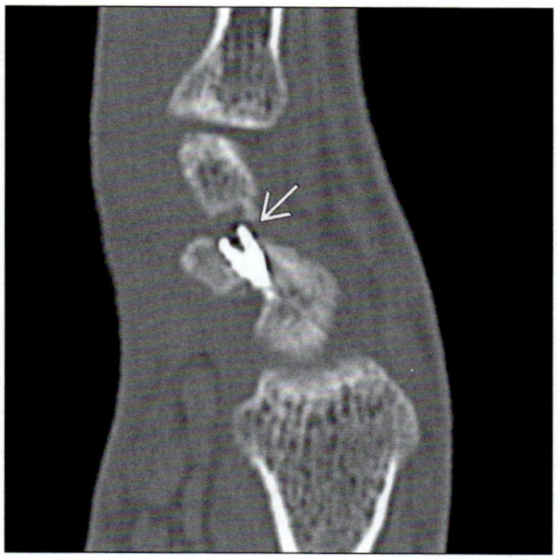

(Left) *PA radiograph shows a Herbert-Whipple screw placed to treat a proximal pole scaphoid fracture. The leading threads → barely engage the proximal pole fragment. The trailing threads → barely engage in the distal pole fragment. This positioning distracts rather than compresses the fragments.* ***(Right)*** *Sagittal reformat bone CT demonstrates the Herbert screw to be incompletely recessed. The mid-waist scaphoid fracture is healed, but the screw impinges → on the trapezoid.*

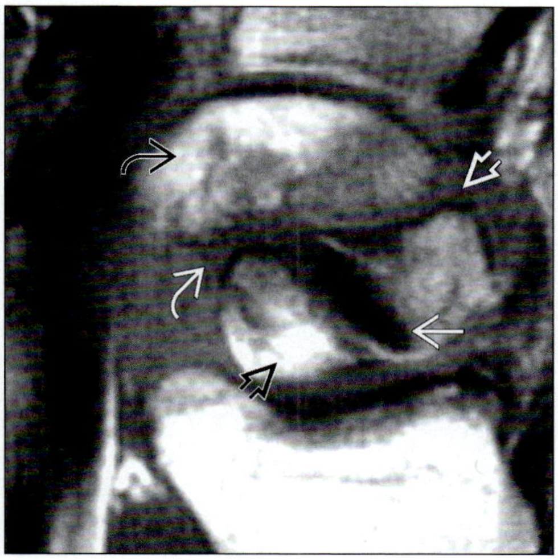

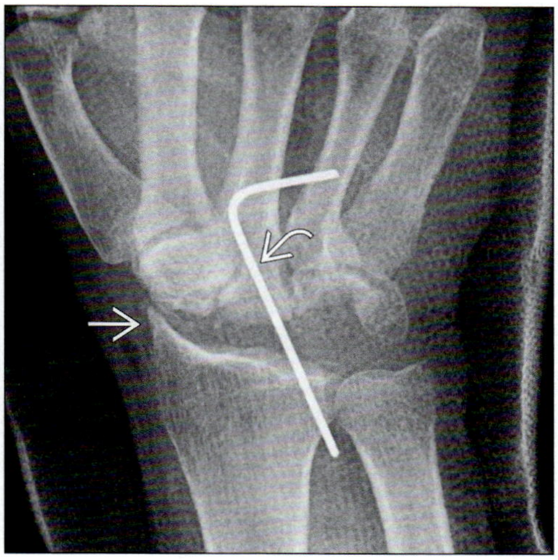

(Left) *Sagittal T1WI demonstrates a scaphoid nonunion with unsuccessful Herbert screw fixation →. The fracture is ununited → and malaligned with flexion of proximal → and distal → fragments, resulting in an increased intrascaphoid angle →.* ***(Right)*** *PA radiograph shows a proximal row carpectomy performed as a salvage procedure in a patient with scaphoid nonunion advanced collapse (SNAC) and radial osteophytosis →. Patient is casted with a single stabilizing K-wire →.*

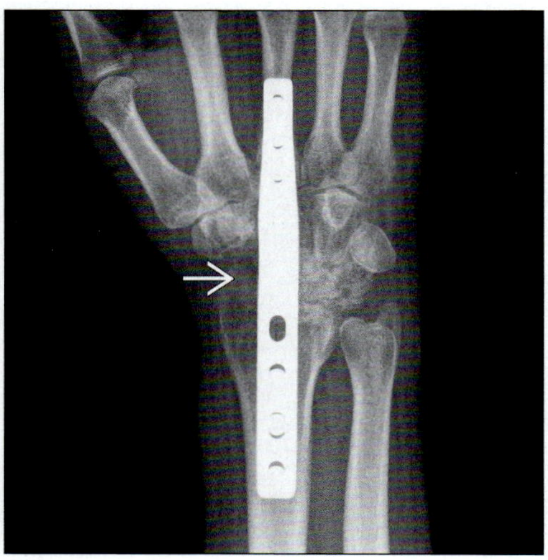

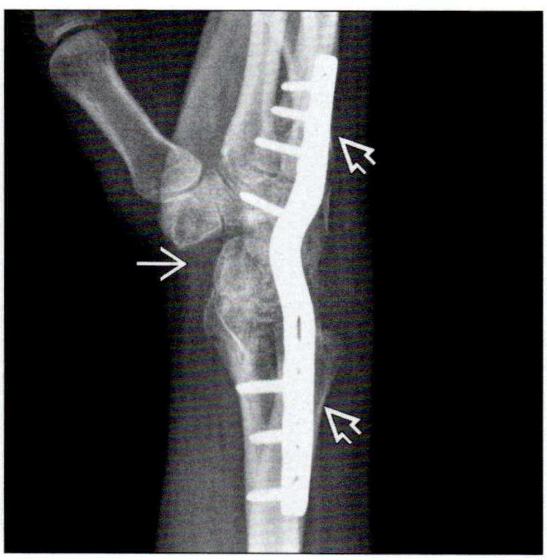

(Left) *PA radiograph demonstrates a scaphoidectomy → and wrist fusion in this 44 year old with a persistently symptomatic scaphoid nonunion. The fusion is stabilized with a dorsal plate secured in the radius and across the carpus to the base of the 3rd metacarpal. Note the diffuse disuse osteopenia.* ***(Right)*** *Lateral radiograph from the same patient shows the dorsal plate and screw assembly → spanning the carpus, securing the wrist fusion. The scaphoidectomy → is evident.*

CARPAL FRACTURE, OTHER THAN SCAPHOID

Key Facts

Terminology
- Isolated fractures rare; more often seen with perilunate fracture/dislocation injury

Imaging
- Cortical ± trabecular disruption
- Variable size and shape; rarely significantly displaced

Pathology
- Most common mechanism of injury: Axial loading on dorsiflexed wrist
  - Fall on outstretched hand; MVA or sports

Clinical Issues
- Carpal fractures = 14% of all hand fractures
- Presentation: Focal tenderness, soft tissue swelling, grip weakness
- Age: Young athletes; elderly osteoporotic patients
- Natural history
  - Generally heal satisfactorily with immobilization
  - Increased risk of osteonecrosis: Lunate, pisiform, capitate, hamate hook
  - Increased risk of nonunion: Capitate neck; hamate hook; transverse pisiform fracture
  - Intraarticular fractures into CMC joints at ↑ risk for developing early OA

Diagnostic Checklist
- Radiographs alone likely underestimate number/extent of carpal fractures
- CT identifies subtle & small fractures; conservative treatment may not be altered
- Beware of satisfaction of search
  - If carpal fracture identified, evaluate for other fractures or ligament injury

*(**Left**) Graphic of the medial wrist demonstrates the ligament attachments along the dorsal triquetrum, including the dorsal intercarpal ➔ and the radiotriquetral ➔ ligaments. These strong ligaments may avulse the dorsal cortex of the triquetrum. (**Right**) Lateral radiograph shows an acute triquetral avulsion fracture with minimal displacement of the curvilinear fragment ➔. Note the dorsal soft tissue swelling ➔. This fracture is typically only seen on the lateral view.*

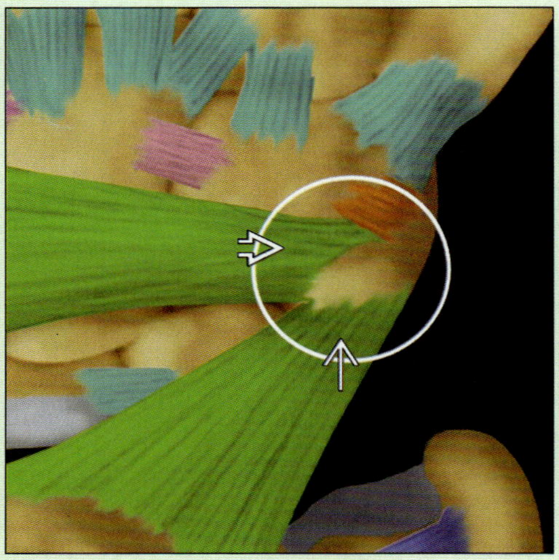

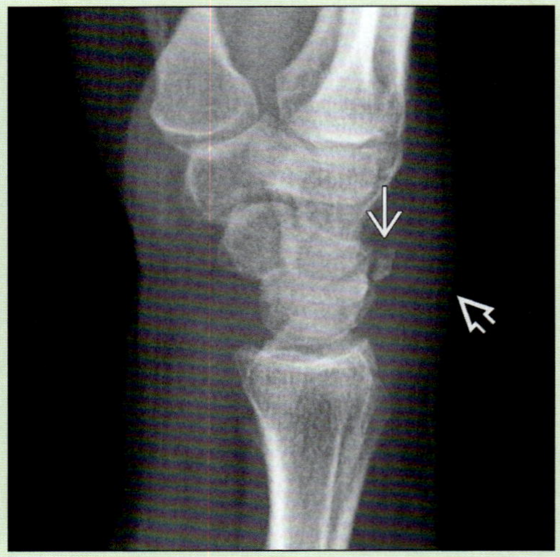

*(**Left**) Oblique radiograph demonstrates a transverse comminuted triquetral body fracture ➔. This fracture is typically seen with more extensive trauma and careful inspection of the surrounding osseous structures and carpal alignment is important to evaluate for a perilunate injury. (**Right**) Axial PDWI FS MR shows a coronal fracture through the dorsal aspect of the triquetrum ➔. Most triquetral fractures involve avulsion of a portion of the dorsal capsular attachment.*

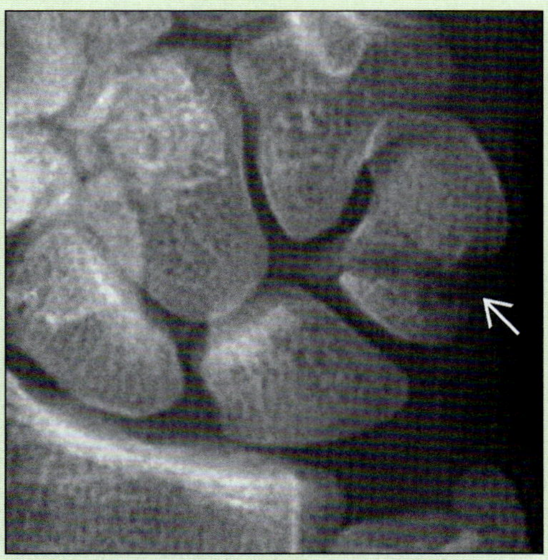

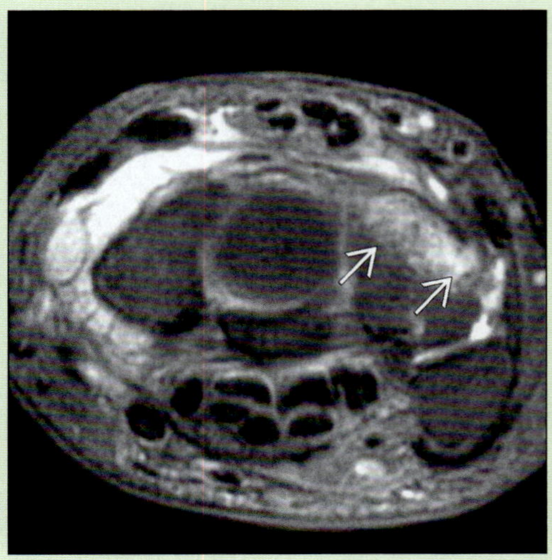

CARPAL FRACTURE, OTHER THAN SCAPHOID

TERMINOLOGY

Definitions

- Isolated fractures rare; more often seen with perilunate fracture/dislocation injury

IMAGING

General Features

- Location
  - Triquetrum
    - Most common: Dorsal cortex
    - Volar or body
  - Lunate
    - Most common: Volar pole
    - Dorsal pole, osteochondral (chip), sagittal or coronal body
  - Trapezium
    - Most common: Vertical transarticular
    - Volar medial (trapezial) ridge, transverse, dorsal radial tuberosity, comminuted
  - Capitate
    - Most common: Transverse body
    - Transverse proximal pole, sagittal or coronal
  - Hamate
    - Most common: Hook (tip, waist, or base)
    - Body: Medial tuberosity, sagittal, proximal pole or dorsal coronal
  - Pisiform
    - Most common: Transverse
    - Sagittal, pisotriquetral impaction or comminuted
  - Trapezoid (least common)
    - Dorsal rim or body
- Morphology
  - Variable, based on fracture type

Radiographic Findings

- Triquetrum
  - Dorsal cortical fracture best seen on lateral
  - Transverse or vertical fracture best seen on PA
  - Associated with perilunate injuries
- Lunate
  - Cortical break, offset; cortical double density
  - Subtle due to overlap of adjacent structures
  - Small avulsion fracture at ligament attachment sites
  - Associated with perilunate injuries; scapholunate (SL) or lunotriquetral (LT) tears; osteonecrosis (ON)
- Trapezium
  - Bett view shows trapezium without overlap of other carpals
    - Forearm and thumb flat on cassette with hand pronated 20°
  - Carpal tunnel view for volar avulsion
  - Associated with fractures of 1st metacarpal or scaphoid
- Capitate
  - Often nondisplaced
  - Proximal fragment may rotate 90-180°; proximal cortex appears rounded
  - Scaphocapitate syndrome
    - Capitate neck fracture with proximal fragment malrotation + scaphoid waist fracture
  - Associated with perilunate injuries; 3rd/4th CMC fracture-dislocation, proximal pole ON
- Hamate
  - Semisupinated oblique, carpal tunnel views for evaluation
  - Hook fracture: Absence of hook cortical "ring" on PA
  - Body fracture: Possible 4th/5th CMC disruption
  - Associated with perilunate injuries, 4th/5th metacarpal fractures, medial ridge of trapezium avulsion; proximal pole or hamate hook ON
- Pisiform
  - Supinated oblique, carpal tunnel views to evaluate
  - Osteochondral fragment in pisotriquetral joint suggests impaction injury
  - Last carpal bone to ossify; may have multiple ossification centers
  - Associated with other upper extremity injuries and may be overlooked; occasional ON
- Trapezoid
  - Isolated fracture is rare; obscured on routine radiographs
  - Associated with 2nd metacarpal fracture-dislocation

CT Findings

- Thin-section CT with reformats useful to demonstrate occult fracture, especially chip or avulsion
- CT highly sensitive and specific; should be undertaken if high index of clinical suspicion and normal radiographs

MR Findings

- T1WI MR: ↓ signal fracture line; may be subtle, obscured by surrounding edema
- Fluid-sensitive sequences: Cortical irregularity; discrete fracture line may be obscured by marrow edema
  - Marrow contusion: Diffuse ↑ signal without discrete fracture line with no cortical break
- Small avulsion/chip fracture: Fragment often not visualized; surrounding edema suggests injury
- May identify occult fractures, associated dislocation, ligament, tendon, or nerve injury
  - Evaluate ulnar nerve with pisiform or hamate fracture

Nuclear Medicine Findings

- Bone scan may show focal uptake

Imaging Recommendations

- Best imaging tool
  - Routine PA, lateral, oblique radiography
  - If high clinical suspicion, special radiographic views
- Protocol advice
  - CT should be obtained promptly if ↑ clinical suspicion and (-) radiographs
  - Consider MR or scintigraphy for occult injury

DIFFERENTIAL DIAGNOSIS

Carpal Fracture-Dislocation

- Evaluate carpal arcs on PA and coaxial alignment of radius/lunate/capitate on lateral views

Accessory Ossification Center

- Ossicle is rounded & well corticated

CARPAL FRACTURE, OTHER THAN SCAPHOID

PATHOLOGY

General Features

- Etiology
  - Most common: Axial loading on dorsiflexed wrist
    - Fall on outstretched hand; MVA or sports
  - Triquetrum
    - Dorsal cortical
      - Chip from direct impact from ulnar styloid or hamate from hyperextension
      - Avulsion of conjoined dorsal intercarpal &/or radiotriquetral ligaments
    - Volar fracture: Ulnotriquetral or lunotriquetral ligament avulsion
    - Body fracture: Seen with perilunate fracture-dislocation
  - Lunate
    - Volar pole: Lunotriquetral ligament avulsion with hyperextension injury
    - Dorsal pole: Shear injury or scapholunate ligament avulsion
    - Osteochondral (chip) fracture: Shear injury from lunate dislocation/subluxation
    - Sagittal body: Shear fracture from radiocarpal fracture dislocation
    - Coronal body: Short radiolunate ligament avulsion with hyperextension injury
  - Trapezium
    - Vertical transarticular: Axial loading through adducted 1st metacarpal
    - Volar medial (trapezial) ridge: Avulsion of transverse carpal ligament or direct blow
    - Transverse or dorsal radial tuberosity: Shear injury
    - Comminuted: High-energy injury
  - Capitate
    - Transverse body or proximal pole: Dorsiflexion
    - Sagittal or coronal: Axial load &/or hyperextension
  - Hamate: Seen in golf, tennis, squash
    - Hook fractures: Direct blow, repetitive trauma, avulsion FCU, &/or transverse carpal ligament
    - Body fractures: Direct ulnar blow, crush injury
    - Dorsal: 4th &/or 5th CMC fracture-dislocation (punching)
  - Pisiform
    - Transverse: Direct impaction or forceful contraction of flexor carpi ulnaris (FCU) as palm strikes ground
    - Sagittal or comminuted: Direct or pisotriquetral impaction
  - Trapezoid
    - Dorsal rim or body: Axial loading along 2nd metacarpal

CLINICAL ISSUES

Presentation

- Most common signs/symptoms
  - Focal tenderness
  - Soft tissue swelling
  - Grip weakness
- Other signs/symptoms
  - Hook of hamate or pisiform
    - Pain with resistance to flexion of 5th finger
    - Paresthesia of 4th & 5th fingers

Demographics

- Age
  - Young athletes; elderly osteoporotic patients
- Gender
  - Young males; elderly females
- Epidemiology
  - Carpal fracture = 14% of all hand fractures
  - Incidence of carpal fractures, isolated & multiple (n = 6390)
    - Scaphoid (79%), triquetrum (13%), lunate (3.9%), trapezium (2.5%), capitate (1.9%), hamate (1.7%), pisiform (1.3%), trapezoid (0.2%)

Natural History & Prognosis

- Generally heal satisfactorily with immobilization
- Increased risk of osteonecrosis: Lunate, pisiform, capitate, hamate hook
- Increased risk of nonunion: Capitate neck; hamate hook body or tip; transverse pisiform fracture
- Intraarticular fractures into CMC joints at ↑ risk for developing early OA

Treatment

- Immobilization (casting/splinting) usually sufficient
- Open reduction internal fixation (ORIF) for any fracture with associated carpal fracture-dislocation or open injury
- ORIF for lunate volar pole, trapezial intraarticular, large trapezoid, displaced capitate, or intraarticular hamate body fracture
- Resection of symptomatic fracture fragments: Triquetrum, pisiform, hamate hook

DIAGNOSTIC CHECKLIST

Consider

- Radiographs alone likely underestimate number/ extent of carpal fractures
- CT identifies subtle & small fractures; conservative treatment may not be altered

Image Interpretation Pearls

- If carpal fracture identified, evaluate for occult injuries
- Beware of satisfaction of search; isolated carpal fractures are rare
- CT sagittal & coronal reformats needed to confirm small fracture fragments
- MR may show edema, but fragments may be too small to visualize

SELECTED REFERENCES

1. Papp S: Carpal bone fractures. Hand Clin. 26(1):119-27, 2010
2. Kaewlai R et al: Multidetector CT of carpal injuries: anatomy, fractures, and fracture-dislocations. Radiographics. 28(6):1771-84, 2008
3. Rosner JL et al: Imaging of athletic wrist and hand injuries. Semin Musculoskelet Radiol. 8(1):57-79, 2004
4. Goldfarb CA et al: Wrist fractures: what the clinician wants to know. Radiology. 219(1):11-28, 2001

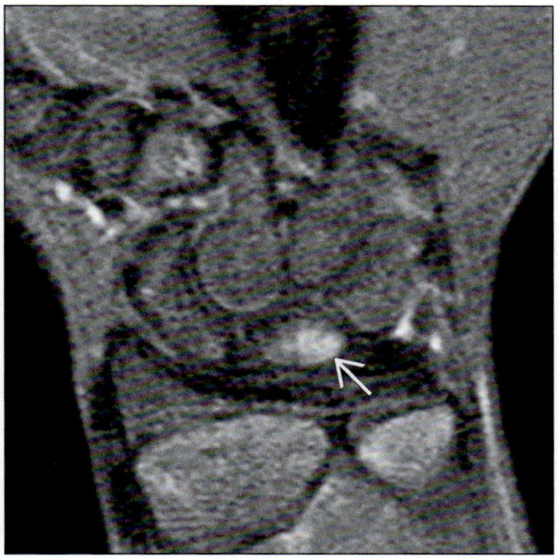
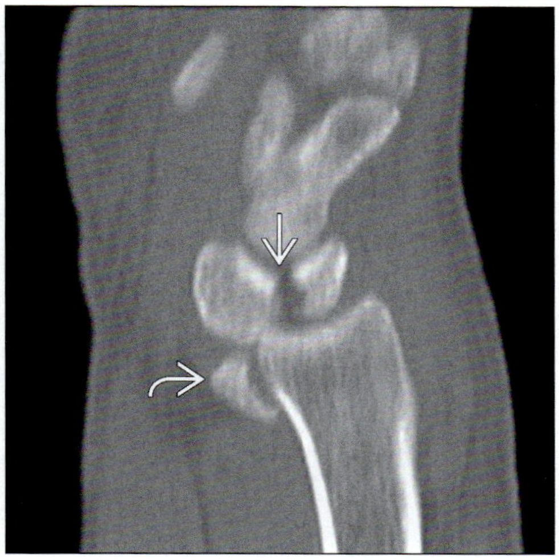

(Left) *Coronal STIR MR demonstrates diffuse high signal intensity in the proximal lunate. This focal contusion is the result of a direct blow to the palm in this 13 year old.* ***(Right)*** *Sagittal reformat bone CT demonstrates a vertical split fracture of the lunate due to an axial loading injury (fall on outstretched hand). Lunate fractures are usually very difficult to see on radiographs and are often not diagnosed until CT or MR are performed. A distal radial fracture fragment is present .*

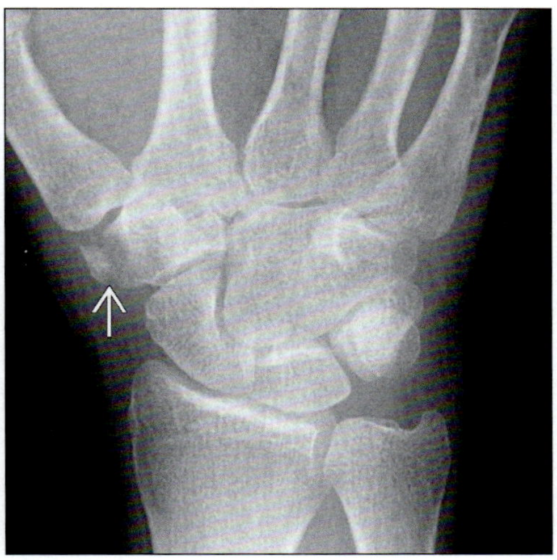
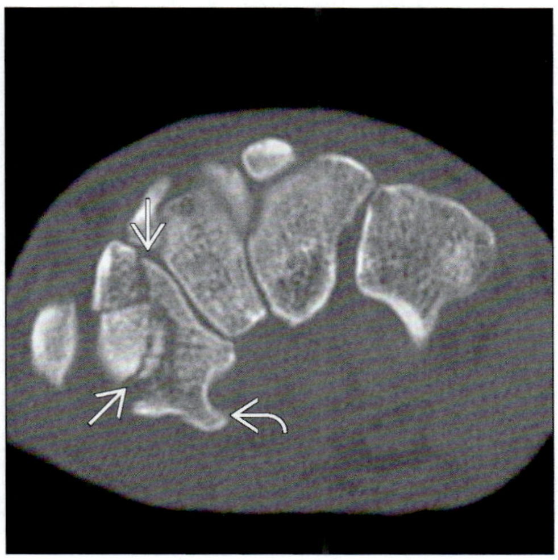

(Left) *PA radiograph shows a displaced intraarticular fracture of the trapezium lateral cortex, extending into the 1st carpometacarpal joint. This is the most common type of trapezial fracture.* ***(Right)*** *Axial bone CT obtained through a fiberglass cast demonstrates a comminuted fracture of the trapezium , which spares the tubercle . The complexity of carpal fractures is often difficult to appreciate on radiographs, and CT is helpful in assessing intraarticular extension of fractures.*

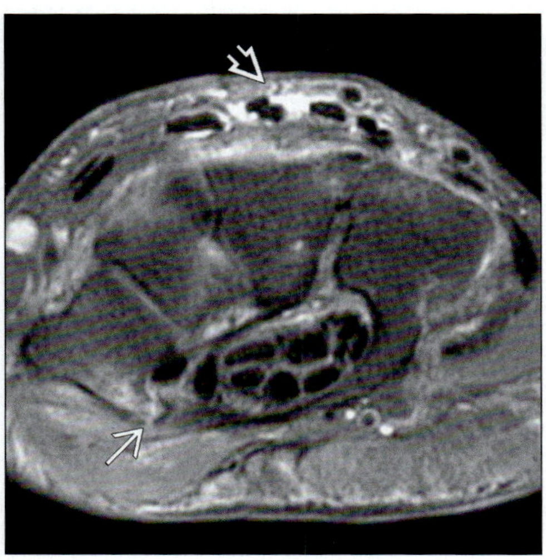
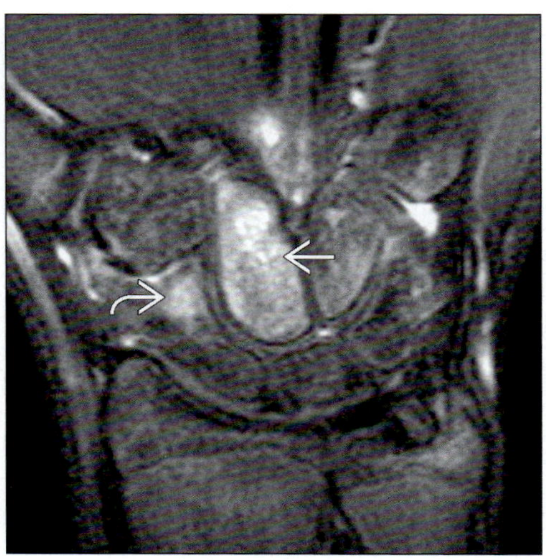

(Left) *Axial PDWI FS MR shows a fracture of the tip of the tubercle of the trapezium with surrounding marrow and soft tissue edema. Fractures in this location are rare and require CT or MR for diagnosis. Note the tenosynovitis of the extensor digitorum communis tendon sheath .* ***(Right)*** *Coronal T2WI FS MR shows diffuse marrow edema in the capitate and in the distal medial aspect of the scaphoid without a discrete fracture. This young patient fell on an outstretched hand.*

CARPAL FRACTURE, OTHER THAN SCAPHOID

(Left) *Sagittal T1WI MR demonstrates an irregular transverse fracture at the waist of the capitate. Patient had wrist pain for several weeks following an MVA. The fracture was not apparent on radiographs. A lunate fracture is also present.* ***(Right)*** *Axial T2WI MR demonstrates a coronal fracture of the capitate. This nondisplaced intraarticular fracture resulted from a fall on a slightly hyperextended hand. The patient had persistent pain, and radiographs were normal.*

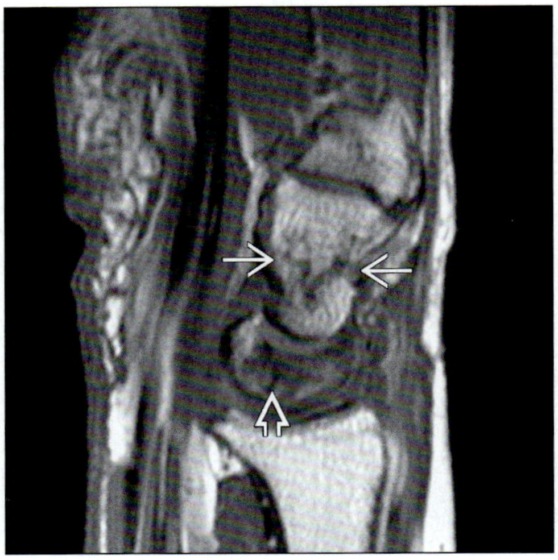

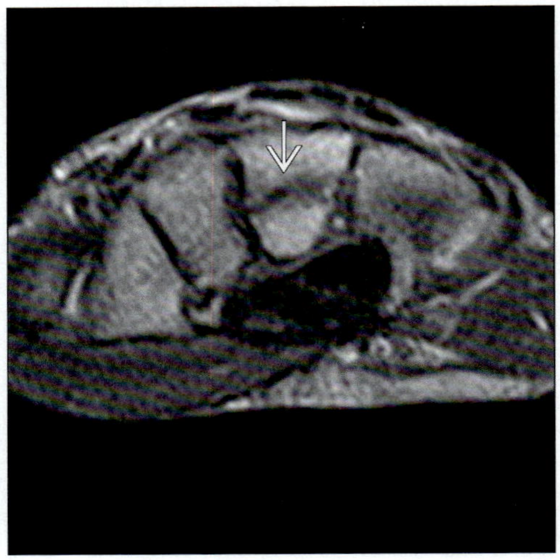

(Left) *Graphic shows a fractured capitate (left) with 90° proximal fragment rotation. Fragment rotation may be up to 180°. Capitate fractures are difficult to see on radiographs due to overlapping bones. Normal comparison is shown (right).* ***(Right)*** *Lateral radiograph shows a fracture at the waist of the capitate, a typical location for capitate fracture. The proximal fracture fragment has rotated 90°; the proximal cortex is seen directed volarly.*

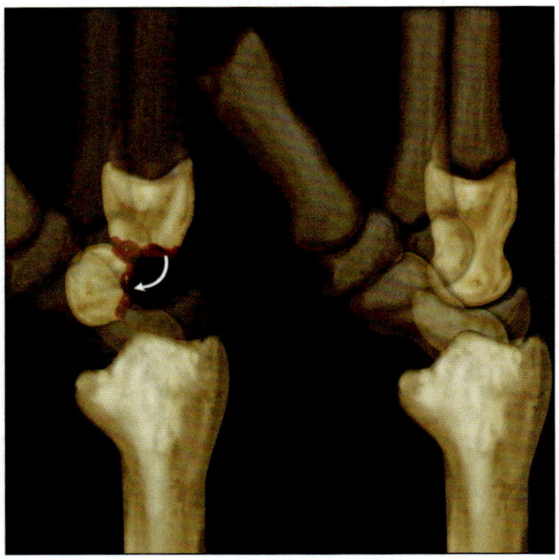

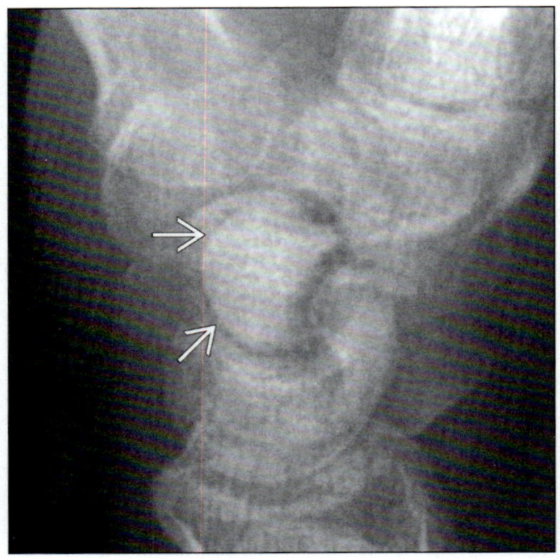

(Left) *Sagittal reformat bone CT demonstrates a comminuted intraarticular fracture of the dorsal hamate, resulting from multi-trauma in an MVA. The 4th metacarpal base impacted the distal hamate articular surface, and the carpometacarpal joint is disrupted.* ***(Right)*** *Axial PDWI FS MR shows marrow edema in the hook of the hamate. There is no discrete fracture line. This patient is a college baseball player with ulnar-sided wrist pain when swinging a bat. Pain resolved with rest.*

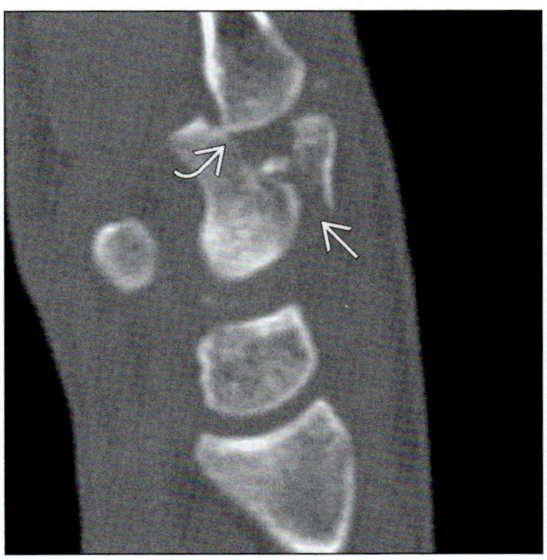

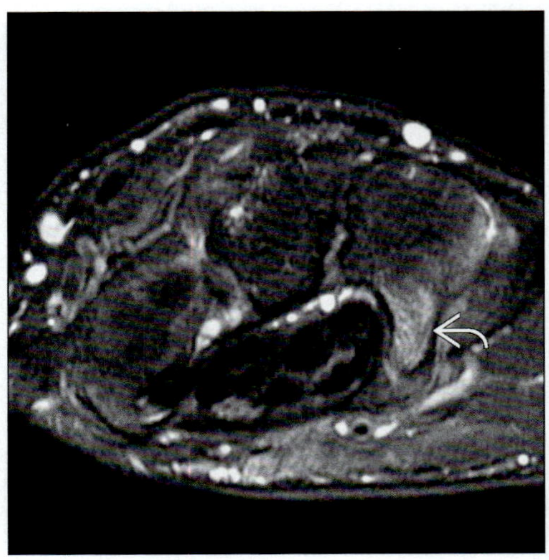

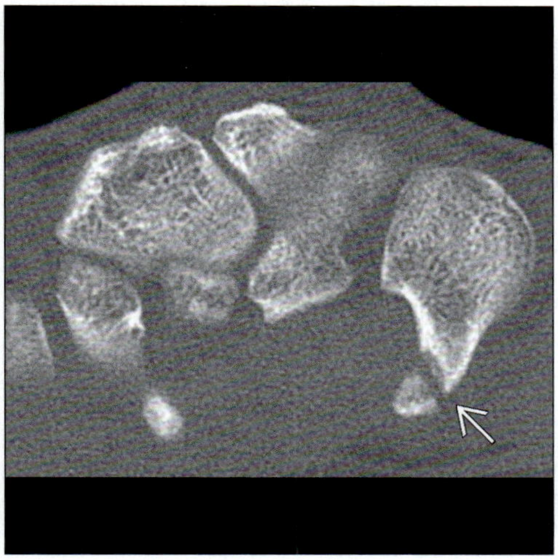

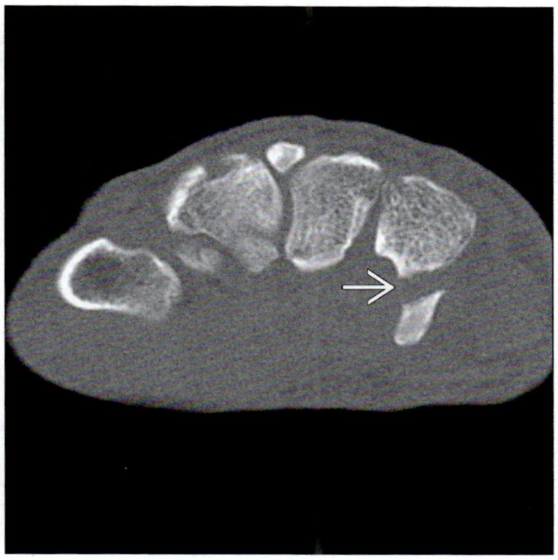

(Left) *Axial bone CT shows an acute fracture at the distal end of the hook of hamate ➡. This small distal fragment may be at risk for osteonecrosis as the vascular supply to the distal hamate hook is somewhat tenuous.* ***(Right)*** *Axial bone CT demonstrates a chronically ununited fracture of the hook of the hamate ➡. The fracture edges appear sclerotic. The resulting mobile fragment may impinge the ulnar nerve or vessels; this patient had ulnar neuropathy resolved by resection of the fragment.*

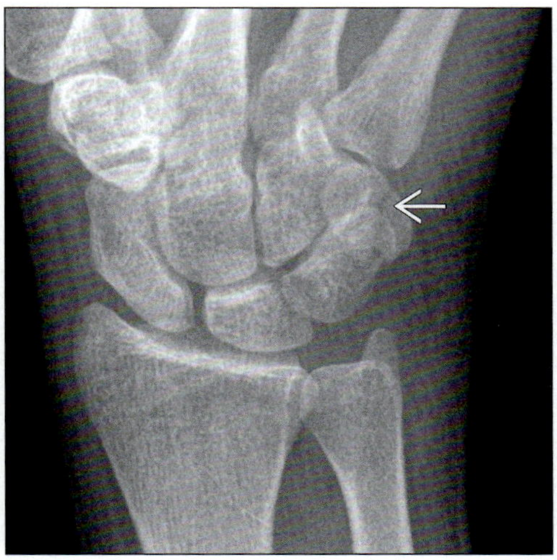

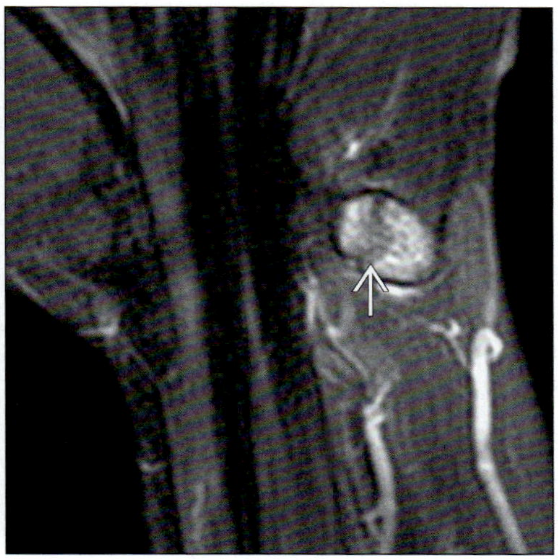

(Left) *Oblique radiograph shows a transverse fracture of the pisiform ➡. Pisiform fractures are often very subtle or invisible on radiographs, and careful scrutiny of this bone is necessary in a patient with ulnar-sided wrist pain.* ***(Right)*** *Coronal STIR MR shows a sagittally oriented fracture of the pisiform ➡. Marrow edema surrounds the fracture. Most fractures of the pisiform are oriented transversely; vertical fractures are usually due to a direct blow injury.*

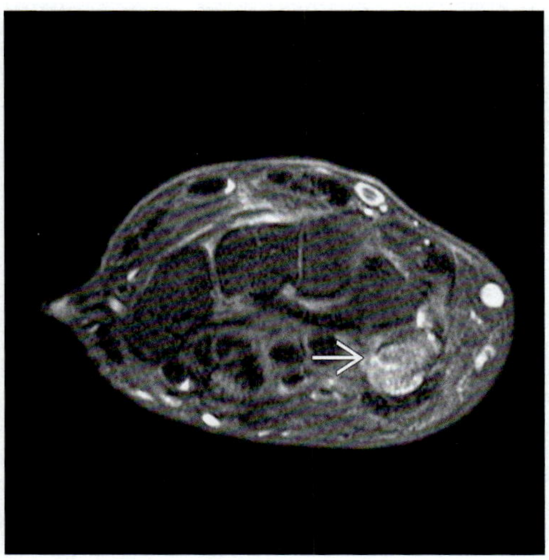

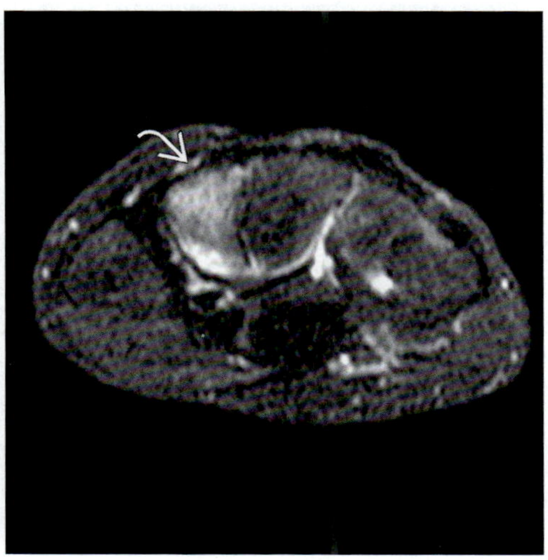

(Left) *Axial PDWI FS MR shows a transverse fracture of the pisiform ➡ with surrounding marrow edema. This fracture, like many pisiform fractures, was not seen on radiographs. This patient was a golfer and had ulnar wrist pain with his golf swing.* ***(Right)*** *Axial STIR MR reveals diffuse high signal intensity ➡ in the trapezoid. There is no discrete cortical fracture. This marrow contusion occurred after impaction of the 2nd finger into the trapezoid during a fall.*

CARPAL DISLOCATION

Key Facts

Terminology

- Dislocation: Complete dissociation of articular surfaces
- Subluxation: Partial dislocation of joint

Imaging

- Radiographs
  - PA: Loss of normal carpal arc relationships
  - Lateral: Disruption of coaxial radius-lunate-capitate relationships
  - Cortical disruption of 1 or more carpal bones
- Location
  - Lesser arc injury: Purely ligamentous involving intercarpal perilunate ligaments
  - Greater arc injury: Injury of perilunate carpal bones ± ligamentous injury
- Perilunate dislocation: Capitate dislocates dorsally; radiolunate articulation intact
- Midcarpal dislocation: Capitate dislocates dorsally; radiolunate subluxation
- Lunate dislocation: Capitate dislocates dorsally, moves proximally; lunate dislocates volarly & rotates 90°
- Axial fracture-dislocation: Wrist splits in 2 columns
- Scaphocapitate syndrome: Scaphoid and capitate neck fractures with malrotation of proximal capitate fragment
- Volar perilunate dislocation (rare)

Clinical Issues

- Focal tenderness and pain; wrist deformity
- Young adults; M > F

Diagnostic Checklist

- Carpal dislocations often go unrecognized in ER; must have high degree of suspicion

(Left) PA radiograph shows carpal (Gilula) arcs, the smoothly curving lines of a normally aligned wrist. Arcs 1 and 2 outline the proximal and distal articular surfaces of the proximal carpal row respectively. Arc 3 outlines capitate and hamate proximal articular surfaces. Disruption of these lines is a key to detection of carpal subluxation or dislocation. ***(Right)*** *Lateral radiograph shows the normal alignment of the radius, lunate, and capitate.*

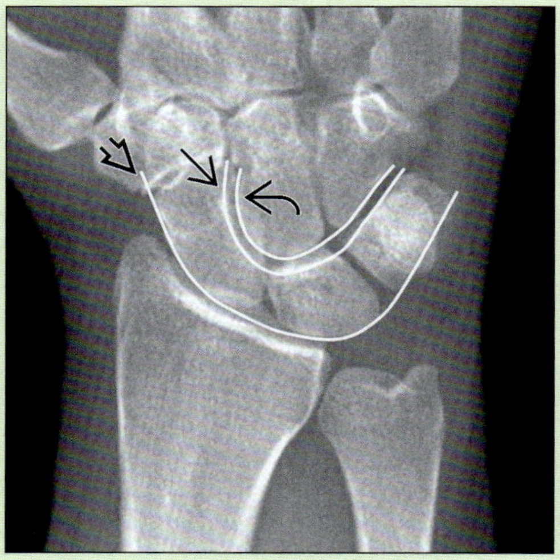

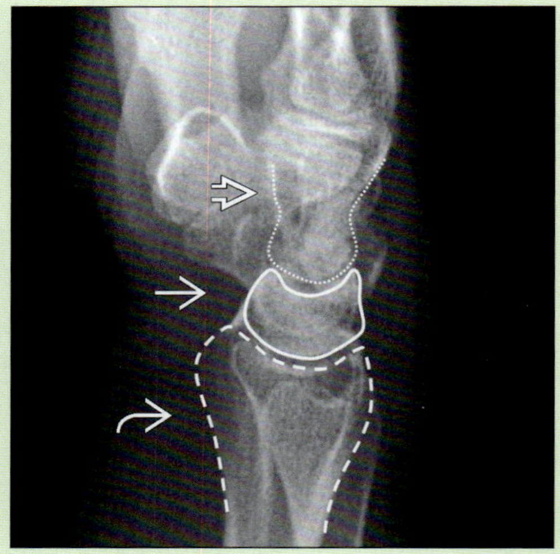

(Left) Lateral graphic shows a perilunate dislocation. The capitate dislocates from the lunate in a dorsal direction; the lunate maintains its normal relationship with the radius. The remaining carpals travel with the capitate. ***(Right)*** *Lateral radiograph shows a dorsal perilunate dislocation. There is dorsal displacement of the capitate but intact radial-lunate articulation. The remainder of the carpus travels with the capitate.*

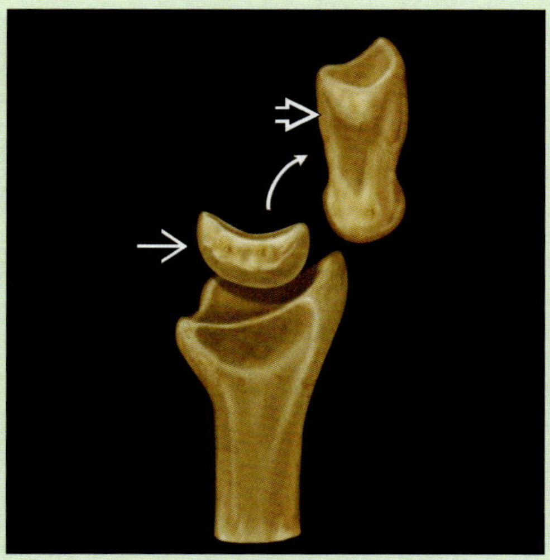

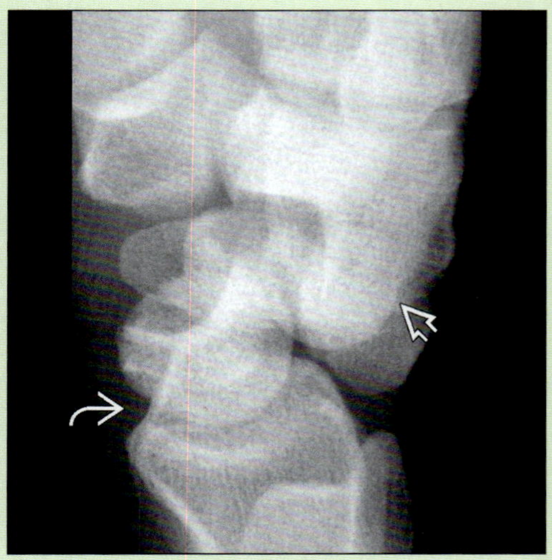

CARPAL DISLOCATION

TERMINOLOGY

Abbreviations
- Scapholunate ligament (SLL)
- Lunotriquetral ligament (LTL)

Definitions
- Dislocation: Complete dissociation of articular surfaces
- Subluxation: Partial dislocation of joint
- Complex fracture dislocation terminology
  - Name fractured carpals, inserting "trans" before each fractured bone
  - Name fractures in order (radial to ulnar)
  - Name direction of dislocation
  - Name type of dislocation
  - State "fracture-dislocation"
    - e.g., transradial, transscaphoid dorsal perilunate fracture-dislocation
- Carpal (Gilula) arcs: Smoothly curved lines paralleling
  - Arc 1: Proximal articular surfaces of proximal carpal row
  - Arc 2: Distal articular surfaces of proximal carpal row
  - Arc 3: Proximal surfaces of capitate and hamate
- Zone of vulnerability
  - Curved band of locations most at risk for injury with perilunate dislocation
    - Radial styloid, scaphoid waist, capitate waist, triquetrum, ulnar styloid

IMAGING

General Features
- Best diagnostic clue
  - PA: Loss of normal carpal arc relationships
  - Lateral: Disruption of coaxial radius-lunate-capitate relationships
  - Cortical disruption of 1 or more carpal bones
    - Important to evaluate for fractures before and after reduction of dislocations
- Location
  - Lesser arc injury: Purely ligamentous injury involving intercarpal perilunate ligaments
  - Greater arc injury: Fracture of 1 or more perilunate carpal bones ± ligamentous injury
- Size
  - Fracture fragments vary from small avulsions to large fracture-dislocations
- Morphology
  - Disorganization of carpal rows on PA and lateral radiographs

Radiographic Findings
- Posteroanterior
  - Disruption of smooth contours of carpal arcs
  - Lunate appears triangular ("candy corn" or "wedge of pie" appearance)
  - Dislocations easily missed
  - Cortical disruption of carpus around lunate
    - May be small chip or avulsion fractures or larger body fractures
- Lateral
  - Loss of normal radius-lunate-capitate alignment
    - Dorsal capitate displacement
    - Volar lunate rotation and subluxation or dislocation
    - Remainder of carpals follow capitate
  - Perilunate dislocation: Capitate dislocates dorsally; radiolunate articulation intact
  - Midcarpal dislocation: Capitate dislocates dorsally; radiolunate subluxation
  - Lunate dislocation: Capitate dislocates dorsally, moves proximally; lunate dislocates volarly, rotating 90° resulting in "spilled teacup" appearance
- Associated fracture patterns
  - Subtle cortical disruption
  - Avulsion fractures from intercarpal ligaments
  - Chip fractures from impacting carpal surfaces as bones dislocate (or relocate)
  - Transverse or longitudinal fracture lines
    - Most common: Scaphoid waist fracture
- Axial fracture-dislocation
  - Wrist splits in 2 columns
    - 1 column remains aligned with radius
    - Other column is unstable, displacing in radial or ulnar direction
    - Metacarpals follow displacement of their corresponding carpal bones
    - Neurovascular injury common
- Scaphocapitate syndrome
  - Scaphoid and capitate neck fractures with malrotation of proximal capitate fragment
- Volar perilunate dislocation (rare)
  - PA: "Crowded carpal" sign results from overlap of proximal & distal carpal rows
  - Lateral: Capitate subtly displaced volarly; lunate dorsally displaced ± lunate or scaphoid fracture

Imaging Recommendations
- Best imaging tool
  - True lateral radiograph essential to correct diagnosis
- Protocol advice
  - Image before & after reduction to identify fractures
  - Beware of satisfaction of search
    - If you see a carpal fracture, look carefully for malalignment or other fractures that would indicate more complex injury

CT Findings
- Thin-section bone CT with sagittal and coronal reformations
- Demonstrates radiographic features
- Assists in identification of small fracture fragments

MR Findings
- Confirm intrinsic ligamentous injuries
- Small fracture fragments may be inapparent; high signal marrow edema may be helpful in search for fractures
- Larger fractures: Low signal band with surrounding marrow edema

DIFFERENTIAL DIAGNOSIS

Carpal Instabilities
- PA radiograph may appear similar, with disrupted arcs 1 & 2 as well as lunate tilt (triangular appearance)

CARPAL DISLOCATION

- Diagnosis is secured on lateral; instability pattern shows abnormal angulation of carpals but no true dislocation

PATHOLOGY

General Features

- Etiology
  - High-energy injury: MVA, fall from height, sports-related injury, industrial accidents
    - Hand/wrist undergoes hyperextension, ulnar deviation, and intercarpal supination (rotatory motion between proximal and distal carpal rows)
  - Perilunate injury
    - Ligament/osseous injury around lunate
    - Structures fail in predictable pattern from radial to ulnar
    - Lesser arc injury
      - Pure ligamentous injury involving tears of SLL → lunocapitate ligament → LTL with subsequent lunate malposition
    - Greater arc injury
      - Arc passes through radial styloid, scaphoid waist, capitate waist, hamate body, triquetrum body, ulnar styloid
      - Osseous injury ± ligament disruption
      - Injury results in dorsal fracture-dislocation
      - Common pattern: Transscaphoid dorsal perilunate fracture-dislocation
    - Scaphocapitate syndrome
      - Results from direct blow to flexed dorsal wrist or fall on outstretched hand with wrist extended and in radial deviation
      - Results in scaphoid and capitate waist fractures
      - Proximal capitate fragment rotates 90-180°
  - Axial loading injury
    - Typically occur from anteroposterior compression forces; explosion or crush injury
    - Injury separates carpus into medial and lateral column, often on either side of capitate
  - Volar perilunate dislocation
    - Results from hyperflexion and supination
    - Often seen in association with lunate fracture

Staging, Grading, & Classification

- Lesser arc injury
  - Stage I: SLL tear with SL dissociation ± rotatory subluxation; opening of space of Poirier
  - Stage II: Capitate dorsal dislocation + dissociation of lunocapitate joint (perilunate dislocation)
  - Stage III: LTL fails; lunate separated from triquetrum (midcarpal dislocation)
  - Stage IV: Capitate migrates proximally; lunate is completely unrestrained & dislocates volarly
- Greater arc injury
  - Stage I: Transcaphoid fracture-dislocation
  - Stage II: Transcaphoid, transcapitate perilunate fracture-dislocation
  - Stage III: Transcaphoid, transcapitate, transtriquetral fracture-dislocation

CLINICAL ISSUES

Presentation

- Most common signs/symptoms
  - Focal tenderness and pain
  - Wrist deformity

Demographics

- Age
  - Young adults
- Gender
  - M > F
- Epidemiology
  - Greater arc injuries (fracture-dislocations) are 2x as frequent as lesser arc injuries (carpal dislocations)
  - In carpal dislocations, frequency of occurrence is inversely related to severity of ligamentous injury
    - Perilunate dislocations are most frequent & have least severe ligamentous damage
    - Lunate dislocations are least frequent & have most severe ligamentous damage

Natural History & Prognosis

- Significant impairment if failure to recognize severity of injury
- Even with adequate reduction and healing of fractures, long-term results usually show wrist compromise
  - Grip strength reduced
  - Instability
  - Associated osteoarthritis

Treatment

- Immediate reduction, often with distraction
- Closed reduction rarely satisfactory
- Internal fixation
  - Open reduction, arthroscopic reduction, & fluoroscopically aided percutaneous techniques can all be successful

DIAGNOSTIC CHECKLIST

Consider

- Carpal dislocations often go unrecognized in ER; must have high degree of suspicion

Image Interpretation Pearls

- Posteroanterior view: Evaluate carpal arcs for disruption
- Lateral view: Assess radius-lunate-capitate alignment

Reporting Tips

- Follow nomenclature rules (e.g., "transradial, transscaphoid dorsal perilunate dislocation")

SELECTED REFERENCES

1. Kaewlai R et al: Multidetector CT of carpal injuries: anatomy, fractures, and fracture-dislocations. Radiographics. 28(6):1771-84, 2008
2. Weil WM et al: Open and arthroscopic treatment of perilunate injuries. Clin Orthop Relat Res. 445:120-32, 2006
3. Gilula LA: Carpal injuries: analytic approach and case exercises. AJR Am J Roentgenol. 133(3):503-17, 1979

CARPAL DISLOCATION

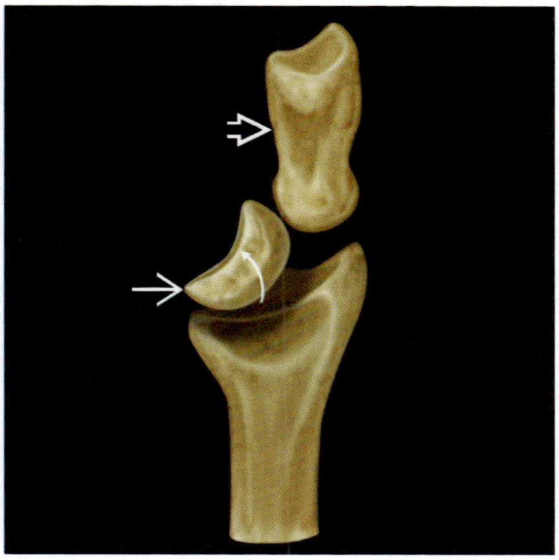

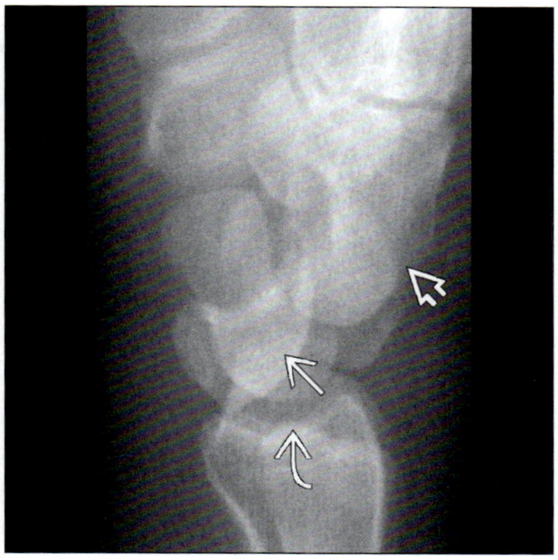

(Left) *Lateral graphic shows a midcarpal dislocation. This is the next step following perilunate dislocation. The dislocated capitate nudges the lunate volarly, but the lunate is not yet dislocated from the radius.* ***(Right)*** *Lateral radiograph shows a midcarpal dislocation, in which the posteriorly dislocated capitate migrates proximally, pushing the lunate slightly volarly. The lunate is not completely dislocated from the radius but is subluxated.*

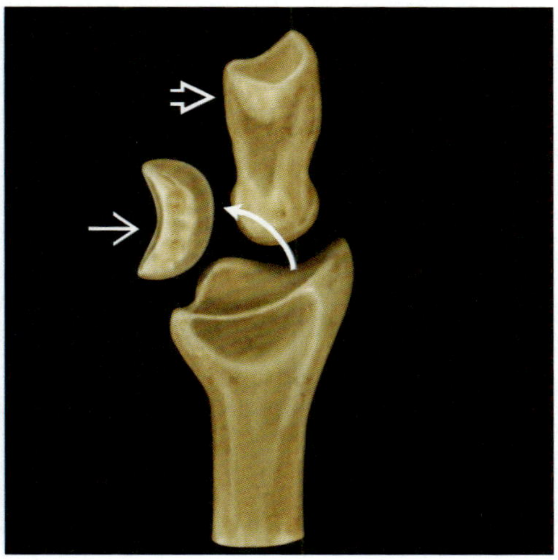

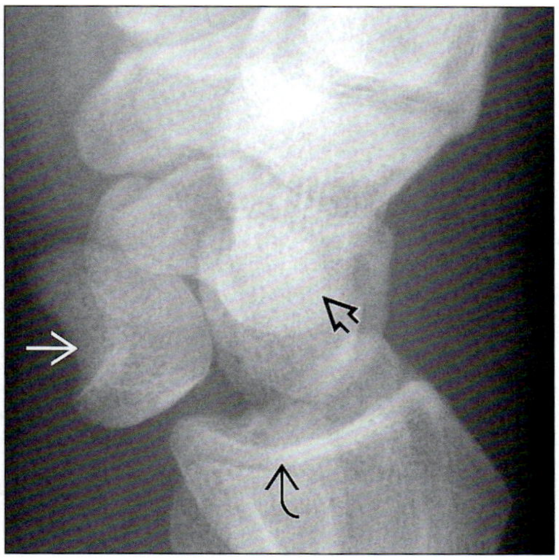

(Left) *Lateral graphic shows a lunate dislocation, the next stage of carpal dislocation. As the dislocated capitate migrates proximally toward the radius, it pushes the lunate volarly. As the lunate dislocates, it typically tilts 90° in the volar direction.* ***(Right)*** *Lateral radiograph shows a volar lunate dislocation, with dorsal dislocation of the capitate relative to the lunate, and the lunate dislocated volar to the radius. As the lunate dislocates, it rotates 90°.*

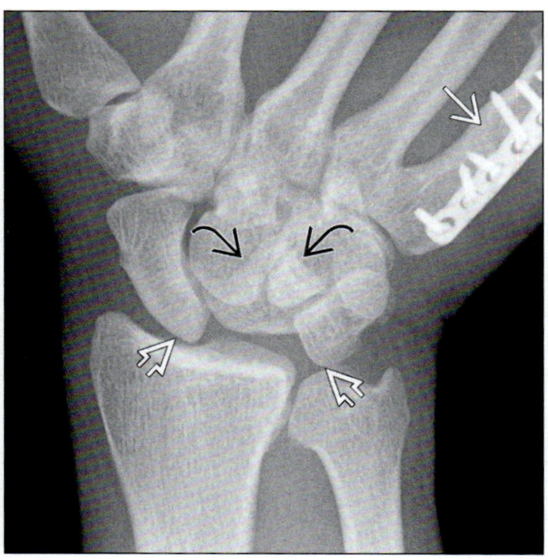

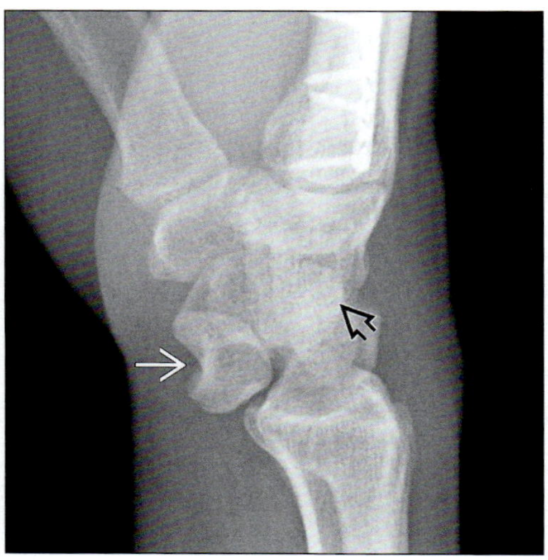

(Left) *PA radiograph was obtained as a follow-up evaluation of distal 5th metacarpal ORIF in a patient with continued pain. Note the disruption of the carpal arcs. The lunate is triangular in contour.* ***(Right)*** *Lateral radiograph from the same patient demonstrates a volar lunate dislocation with capitate nudging lunate volarly as it travels proximally. This purely ligamentous injury occurred at the time of the metacarpal fracture but was missed in earlier evaluation.*

CARPAL DISLOCATION

(Left) *PA radiograph demonstrates disruption of the carpal arcs with marked overlap of proximal and distal carpal rows, resulting in the "crowded carpal" sign.* ***(Right)*** *Lateral radiograph from the same patient shows a rare dorsal lunate dislocation. The capitate migrates proximally toward the radius, similar to the volar dislocation. This pattern indicates tremendous direct trauma on the volar lunate rather than the usual etiology of fall on outstretched hand.*

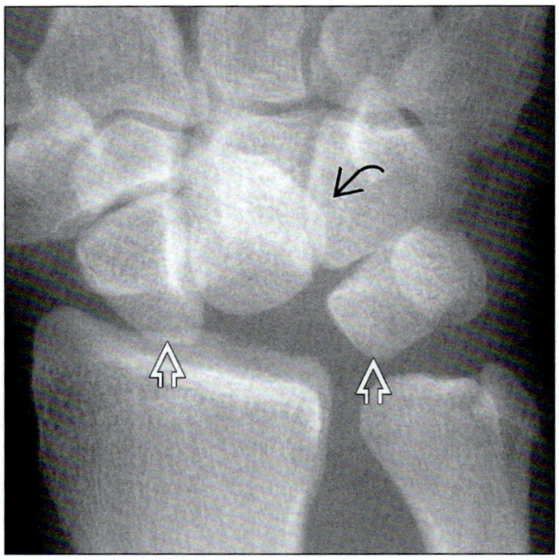

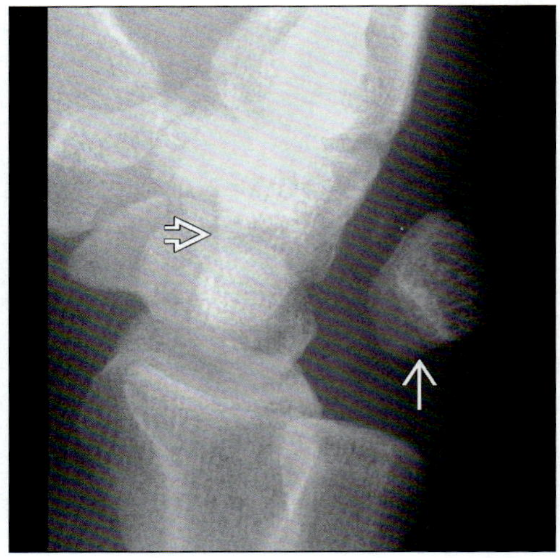

(Left) *PA radiograph demonstrates the pattern of lesser arc (ligamentous) and greater arc (osseous ± ligamentous) injuries. The shaded areas define the "zone of vulnerability," the region of the most common carpal injuries.* ***(Right)*** *Coronal reformat bone CT demonstrates a greater arc injury with a mid-waist scaphoid fracture and a highly comminuted proximal capitate fracture. There is slight widening of the scapholunate interval, suggesting associated ligament injury.*

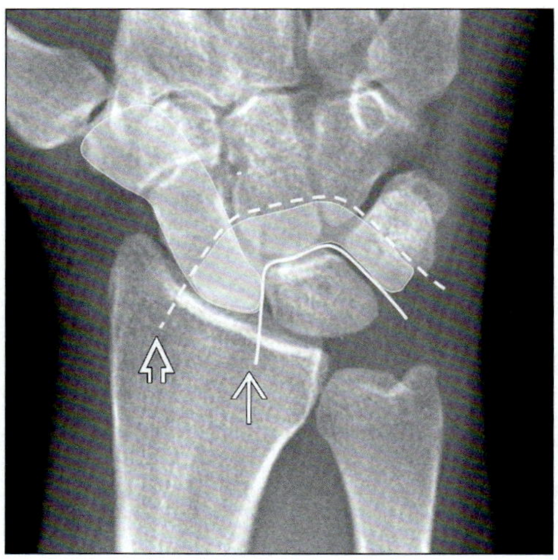

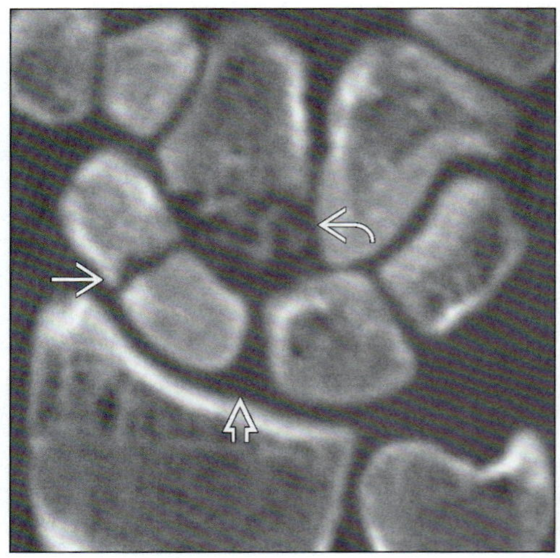

(Left) *PA oblique radiograph in a 38-year-old patient following a motor vehicle accident shows marked distortion of the carpals with a crescentic fracture fragment from the distal radius and a small avulsion fracture of the ulnar styloid.* ***(Right)*** *Lateral radiograph from the same patient demonstrates complete dorsal dislocation of the capitate while the lunate remains aligned with the distal radius. This represents a transradial transulnar dorsal perilunate fracture-dislocation.*

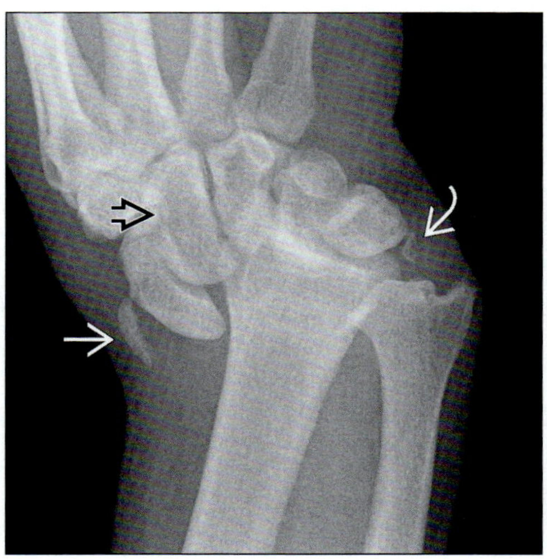

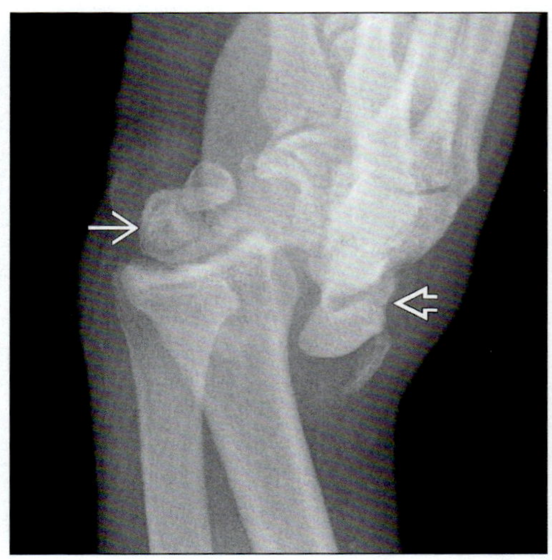

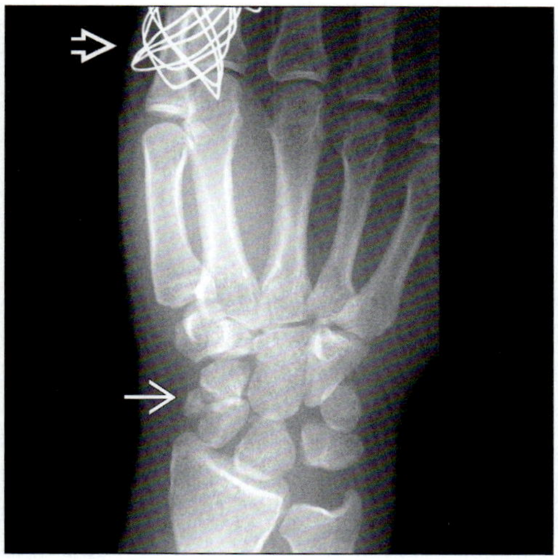

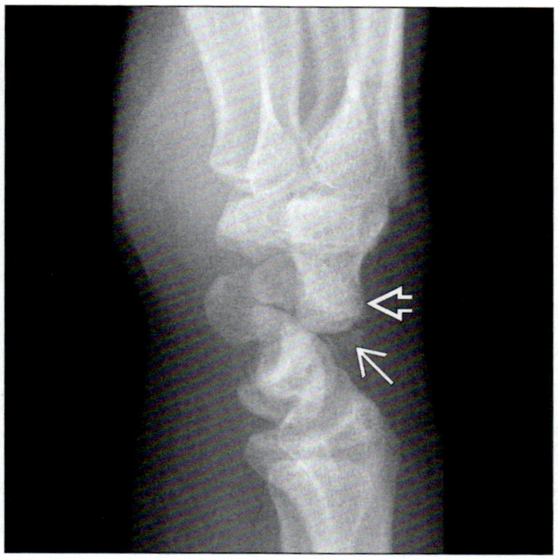

(Left) *PA radiograph shows a comminuted mid-waist scaphoid fracture ➔. The carpal arcs are disrupted, and the lunate is somewhat triangular. Note finger traps ➔ placed to distract the wrist and reduce malalignment.* ***(Right)*** *Lateral radiograph from the same patient confirms carpal dislocation with dorsal capitate displacement ➔, nudging the lunate volarly. Note the small capitate fracture fragment ➔. This injury is described as a transscaphoid transcapitate midcarpal fracture-dislocation.*

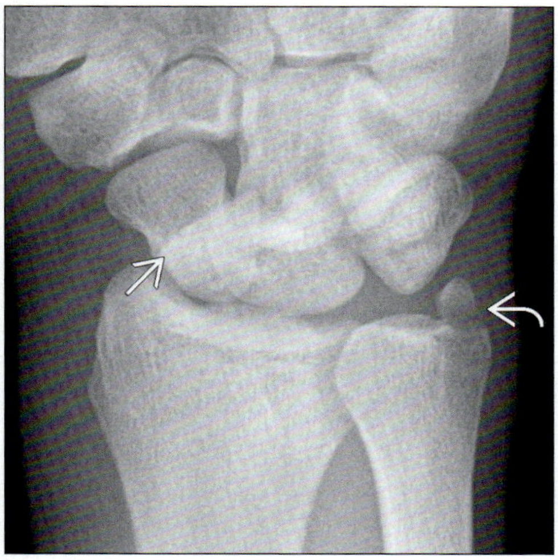

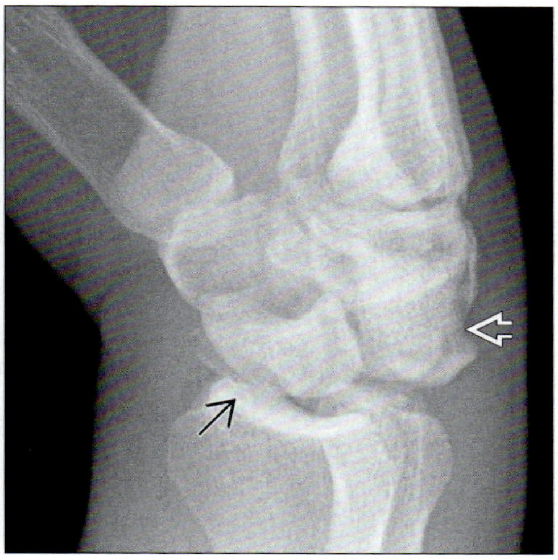

(Left) *PA radiograph demonstrates the disruption of the normal carpal arcs and slight overlap of proximal and distal carpal rows. Fractures of the scaphoid waist ➔ and ulnar styloid ➔ are also present.* ***(Right)*** *Lateral radiograph from the same patient confirms the dorsal dislocation ➔ of the capitate, while the radius-lunate relationship ➔ remains intact. This injury is classified as a transscaphoid transulnar dorsal perilunate fracture-dislocation.*

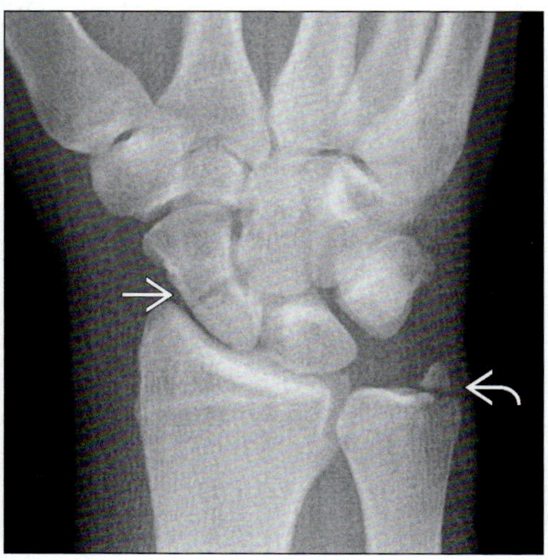

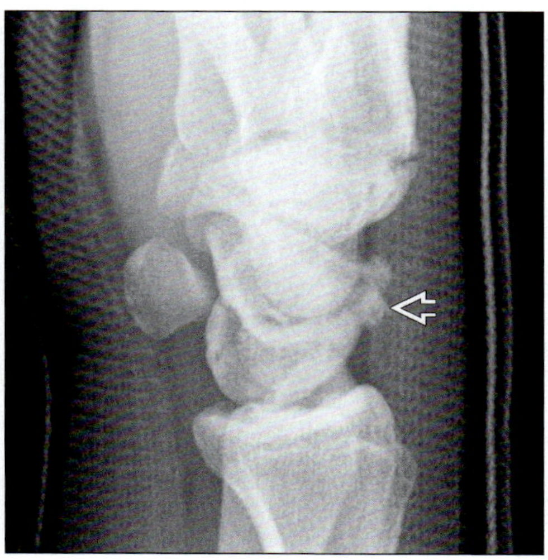

(Left) *PA radiograph from the same patient after reduction of carpal dislocation shows restoration of normal carpal arcs. Scaphoid ➔ and ulnar ➔ fractures are evident through cast material.* ***(Right)*** *Lateral radiograph from the same patient after reduction shows an additional fracture along the dorsal lip of the lunate ➔. This may have occurred at the initial injury or at reduction. This injury would be considered a transscaphoid translunate transulnar dorsal perilunate fracture-dislocation.*

CARPAL IMPACTION SYNDROMES

Key Facts

Imaging

- MR findings
  - T1WI: ↓ SI with edema, cysts, sclerosis
  - T2WI: ↑ SI with edema/hyperemia; ↓ SI with sclerosis
  - Signal is diffuse in marrow edema; focal in subchondral cysts
  - Cartilage loss
- Radiographic findings
  - Focal subchondral sclerosis and cyst formation
  - Joint space narrowing
- Location of imaging findings
  - Ulnocarpal abutment: Medial lunate, lateral triquetrum ± positive ulnar variance
  - Ulnar styloid impaction: Styloid tip or fragment against dorsal medial triquetral
  - Radial styloid impingement: Scaphoid tuberosity/ dorsal lateral radial rim
  - Dorsal impingement: Dorsal radial rim, scaphoid &/or lunate
  - Hamatolunate impaction: Proximal hamate, distal medial lunate

Top Differential Diagnoses

- Intraosseous ganglion
  - Randomly located in lunate
  - Sharp margins without perilesional edema
- Kienböck disease
  - Typically involves entire lunate
  - Occurs in isolation; adjacent joints intact
- Ulnar impingement
  - Significant ulnar shortening with impaction of distal ulna on distal radius

Clinical Issues

- Focal tenderness; restricted range of motion; grip strength loss

*(**Left**) Coronal graphic shows the injuries of ulnocarpal abutment. The TFC disc is thinned and torn. The articular cartilage of proximal lunate, triquetrum, and distal ulna is damaged, and the subchondral bone reacts with edema and sclerosis. The LTL may tear in more advanced disease. (**Right**) PA radiograph shows mild ulnar positive variance, with 2 small well-marginated subchondral cysts and adjacent sclerosis, typical of ulnocarpal abutment.*

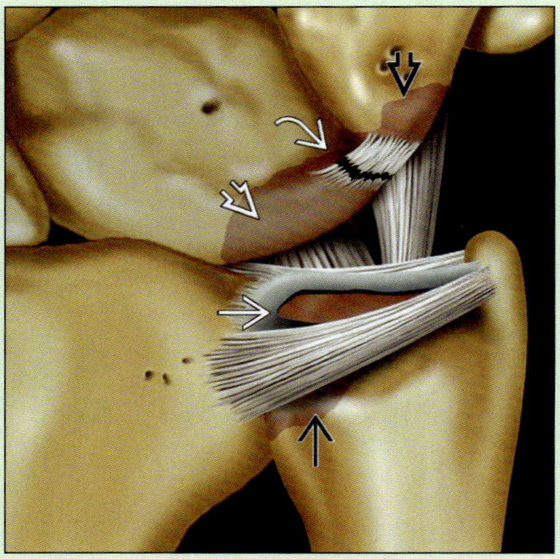

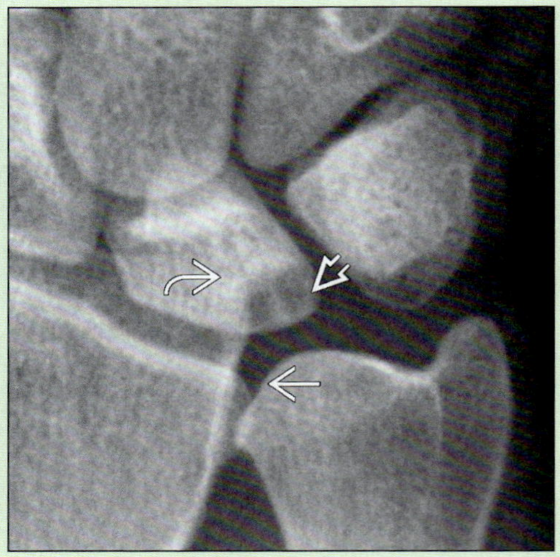

*(**Left**) Coronal PDWI MR shows mild ulnocarpal abutment with fraying of the central TFC. Subtle cartilage thinning is accompanied by ↓ SI along the proximal lunate representing marrow edema and a small developing subchondral cyst. (**Right**) Coronal GRE MR demonstrates significant positive ulnar variance with the TFC draped over the ulnar head and full thickness cartilage loss and cyst formation in the lunate. There is also disruption of the LTL.*

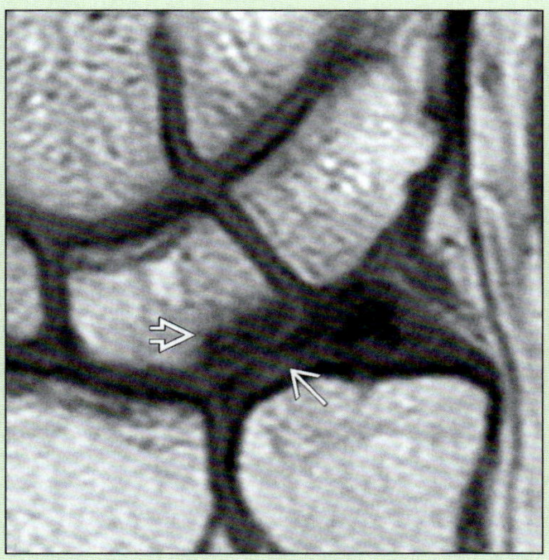

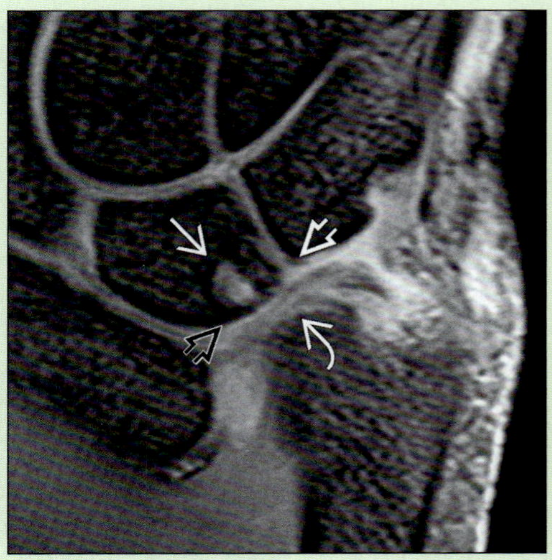

CARPAL IMPACTION SYNDROMES

TERMINOLOGY

Abbreviations

- Lunotriquetral ligament (LTL)
- Triangular fibrocartilage (TFC)

Synonyms

- Ulnocarpal abutment: Ulnocarpal loading, ulnar abutment, ulnolunate abutment, ulnar impaction syndrome

Definitions

- Ulnocarpal abutment: Medial wrist pain related to altered mechanics between ulna, TFC, lunate, and triquetrum
- Ulnar styloid impaction: Excessively long or fractured ulnar styloid impinges triquetrum
- Radial styloid impingement: Styloid impingement on dorsal scaphoid with forced radial deviation
- Dorsal impingement: Impingement of distal radius on scaphoid ± lunate
  - Scaphoid impaction syndrome: Impaction of dorsal scaphoid by dorsal radius
- Hamatolunate impaction: Medial wrist pain related to cartilage loss of proximal hamate

IMAGING

General Features

- Best diagnostic clue
  - Focal subchondral sclerosis and cyst formation
  - Joint space narrowing
  - Cartilage loss
  - Marrow edema
- Location
  - Ulnocarpal abutment: Medial lunate, lateral triquetrum ± positive ulnar variance
  - Ulnar styloid impaction: Styloid tip or fragment against dorsal medial triquetral
  - Radial styloid impingement: Scaphoid tuberosity/ dorsal lateral radial rim
  - Dorsal impingement: Dorsal radial rim, scaphoid ± lunate
  - Hamatolunate impaction: Proximal hamate, distal medial lunate
  - Rare: Triquetrohamate, lunotriquetral

Radiographic Findings

- Ulnocarpal abutment
  - Positive ulnar variance
    - Not necessary but most commonly seen
    - May have symptoms with neutral or negative ulnar variance as well
  - Subchondral cyst/sclerosis in proximal medial lunate, proximal lateral triquetrum, ulnar head
- Ulnar styloid impaction
  - Long styloid: Overall length > 6 mm
  - ± styloid contact with medial dorsal triquetrum
  - Separate entity from ulnocarpal abutment but may be seen in concert with it
  - Impaction secondary to ulnar styloid nonunion
    - Base or tip fracture nonunion
    - Medial triquetral sclerosis
- Radial styloid impingement
  - Rotatory scaphoid subluxation
  - May occur after triscaphe arthrodesis
- Dorsal impingement
  - Dorsal radial rim, scaphoid ± lunate osteophytes
  - Scaphoid impaction syndrome
    - Small ossicle or hypertrophic ridge contiguous with scaphoid rim (best seen on lateral view)
- Hamatolunate impaction
  - Medial (hamate) facet of lunate (type II lunate)
  - Subchondral sclerosis/cysts: Proximal hamate > distal medial lunate

CT Findings

- Similar to radiographs
- Thin section CT allows reformations for anatomic relationships

Nuclear Medicine Findings

- Increased uptake corresponds to symptomatic area
- Pinhole imaging → ↑ resolution

MR Findings

- General imaging findings
  - Signal intensity (SI) patterns
    - T1WI: ↓ SI with edema, cysts, sclerosis
    - T2WI: ↑ SI with edema/hyperemia; ↓ SI with sclerosis
  - Signal is diffuse in marrow edema, focal with subchondral cysts
- Ulnocarpal impaction
  - TFC thinning, or partial or complete tear
  - Cartilage loss along proximal medial lunate, proximal lateral triquetrum, ulnar head
  - Subchondral edema ± cysts
  - LTL tear
  - Osteophytic ridging
  - Positive ulnar variance
  - MR arthrography: Evaluation of cartilage and ligament integrity
- Ulnar styloid impaction
  - Long styloid process
  - Subchondral edema, cyst formation in ulnar tip ± proximal triquetrum
  - LTL tear
  - May be seen as separate entity or in conjunction with ulnocarpal abutment
  - Ulnar styloid impaction secondary to ulnar styloid nonunion
    - Tip fracture: TFC intact; fracture creates loose irritative body
    - Base fracture: TFC tear
- Dorsal impingement
  - Edema along dorsal radial rim ± impacted carpals
  - Dorsal capsular thickening, synovitis
- Hamatolunate impingement
  - Medial lunate facet (type II)
  - Chondromalacia, marrow edema in proximal hamate

Imaging Recommendations

- Best imaging tool
  - Radiographs to identify small ossicles, osteophytes, joint space narrowing, subchondral sclerosis, cyst formation
  - MR demonstrates detail of multi-stage process
    - Best to detail cartilage damage

CARPAL IMPACTION SYNDROMES

- Protocol advice
  - Consider MR arthrography for additional detail

DIFFERENTIAL DIAGNOSIS

Intraosseous Ganglion
- Randomly located in lunate
- Sharp margins without perilesional edema
- Often asymptomatic

Kienböck Disease
- Typically involves entire lunate
- Occurs in isolation; adjacent joints intact
- Associated with ulnar negative variance

Ulnar Impingement Syndrome
- Significant ulnar shortening with impaction of distal ulna on distal radius

PATHOLOGY

General Features
- Etiology
  - Significant loads transmitted to forearm through distal ulna via TFC
    - Neutral ulnar variance: 80% of force across radius; 20% across ulna
    - Load increases across ulna with
      - Pronation; power grip; change in ulnar variance; dorsal radial tilt
  - Ulnocarpal impaction predisposing factors
    - Positive ulnar variance
    - Malunion of distal radial fracture
    - Premature distal radial growth plate closure
    - Essex-Lopresti fracture
    - Previous surgical resection of radial head
    - May also occur in absence of structural abnormalities; most commonly from repetitive activities causing excessive ulnar carpal loading
  - Ulnar styloid impaction
    - Repeated triquetral impingement by long styloid or fracture nonunion
    - Forced ulnar deviation
  - Radial styloid impingement
    - Forced radial deviation; seen in tennis players & gymnasts
  - Dorsal impingement/scaphoid impaction syndrome
    - Occurs with forced dorsiflexion or repetitive hyperextension
    - Seen in gymnasts, weightlifters, excessive push-ups
  - Hamatolunate impaction
    - Altered mechanics with type II lunate → ↑ proximal hamate abrasion, particularly in ulnar deviation
    - Not seen in type 1 lunate (absent medial facet); see in only 5-8% of type II lunates

Staging, Grading, & Classification
- Palmer Classification of TFC tears
  - Type I: Traumatic
  - Type II: Degenerative
    - IIA: TFC wear
    - IIB: TFC wear & lunate ± ulnar chondromalacia
    - IIC: TFC perforation & lunate ± ulnar chondromalacia
    - IID: TFC perforation & lunate ± ulnar chondromalacia & LT ligament tear
    - IIE: TFC perforation & lunate ± ulnar chondromalacia & LT ligament tear & ulnocarpal arthritis
- Lunate configuration
  - Type I: No medial facet; ~ 35% prevalence
  - Type II: Medial (hamate) facet; size varies from 1-6 mm; ~ 65% prevalence

CLINICAL ISSUES

Presentation
- Most common signs/symptoms
  - Focal tenderness
  - Restricted range of motion
  - Grip strength loss
- Other signs/symptoms
  - Ulnocarpal abutment: Pain exacerbated by ulnar deviation, power grip, active pronation/supination
  - Radial styloid impingement: Pain exacerbated by radial deviation, pressure to scaphoid tuberosity
  - Dorsal impingement: Dorsal pain worsened with hyperextension
  - Scaphoid impaction: Point tenderness at dorsal scaphoid with flexion & ulnar deviation

Natural History & Prognosis
- Ulnocarpal abutment/ulnar styloid impaction
  - Advanced arthritis of medial wrist
  - ECU tendonitis/tenosynovitis

Treatment
- Symptomatic treatment
- Surgical intervention goals: Remove impinging osseous ridge; decrease loading

DIAGNOSTIC CHECKLIST

Consider
- Location of imaging findings correlates with clinical findings
- Ulnocarpal and ulnar styloid impaction can coexist

Image Interpretation Pearls
- Early pathologic changes may be inapparent on radiographs
- MR imaging key to identify early pathology, including TFC & LTL tears

SELECTED REFERENCES

1. Greenberg JA: Reconstruction of the distal ulna: instability, impaction, impingement, and arthrosis. J Hand Surg Am. 34(2):351-6, 2009
2. Bickel KD: Arthroscopic treatment of ulnar impaction syndrome. J Hand Surg Am. 33(8):1420-3, 2008
3. Henry M: Arthroscopic management of dorsal wrist impingement. J Hand Surg Am. 33(7):1201-4, 2008
4. Cerezal L et al: Imaging findings in ulnar-sided wrist impaction syndromes. Radiographics. 22(1):105-21, 2002
5. Viegas SF et al: Medial (hamate) facet of the lunate. J Hand Surg Am. 15(4):564-71, 1990

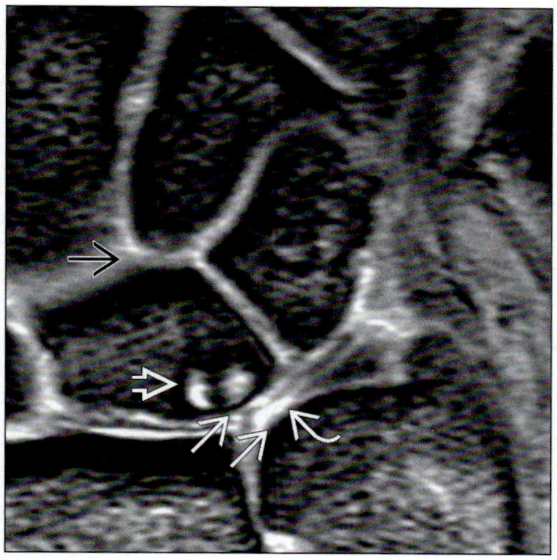
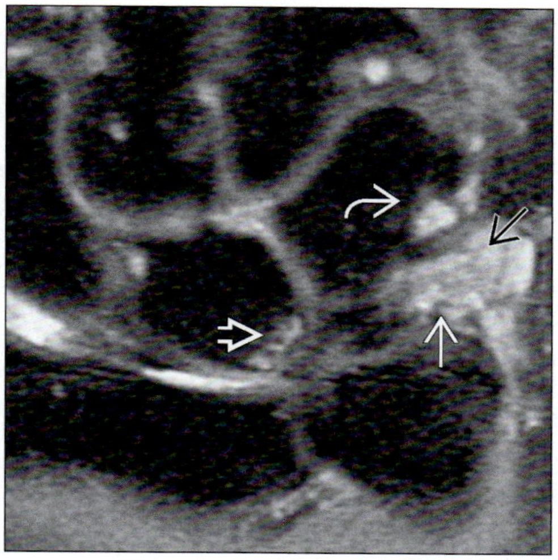

(Left) Coronal GRE MR shows a central TFC perforation and cartilage loss of the ulnar head and proximal lunate with cystic changes. The lunotriquetral ligament is intact. A type II lunate is present. *(Right)* Coronal STIR MR demonstrates the typical appearance of ulnar styloid impingement with edema in the ulnar styloid and lateral triquetrum and mild ulnocarpal abutment. Synovitis is present in the prestyloid soft tissues.

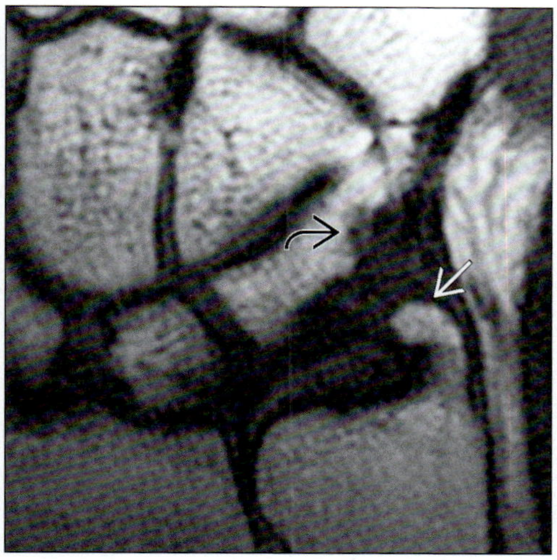
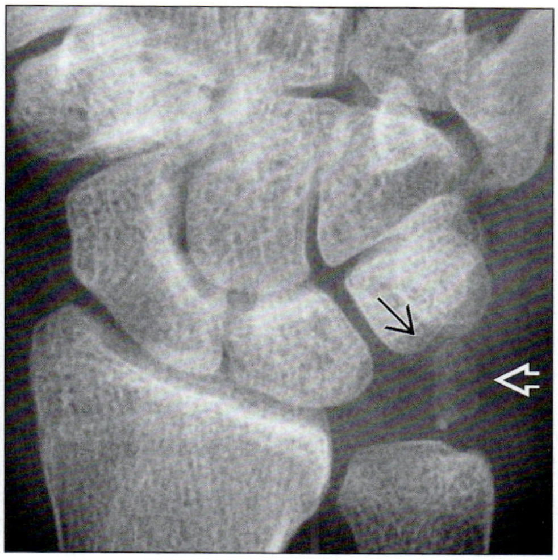

(Left) Coronal T1WI MR demonstrates sclerotic irregularity of the proximal triquetrum. The long and hook-shaped ulnar styloid impinges the dorsal medial triquetrum, damaging the cartilage and resulting in subchondral sclerosis. *(Right)* PA radiograph shows subtle impingement of the proximal triquetrum by a well-corticated rounded ulnar styloid base fracture nonunion. The patient had persistent pain, which may be due to direct osseous impaction or an accompanying TFC tear.

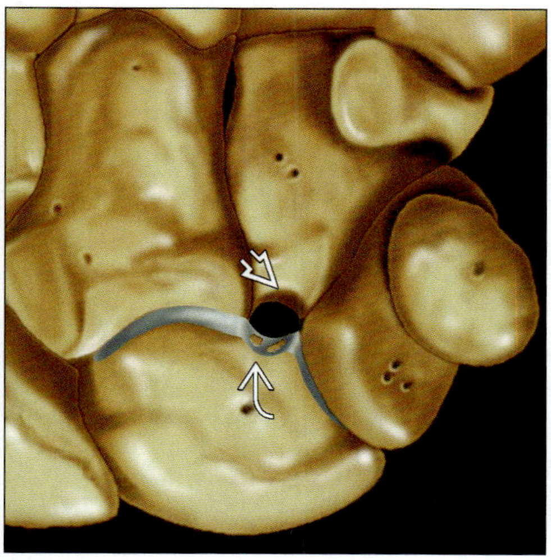
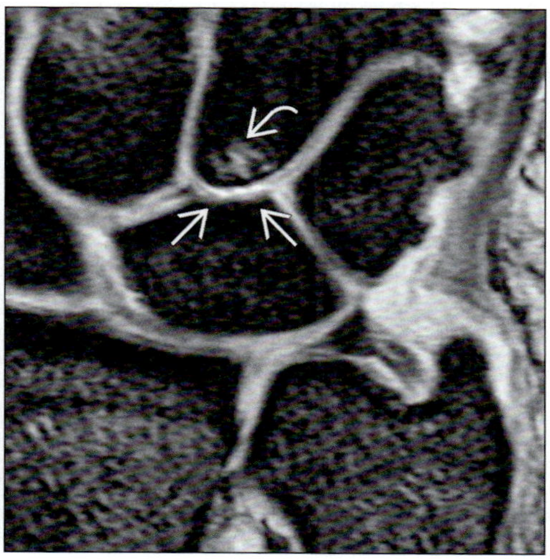

(Left) Coronal graphic shows the injuries of hamatolunate impaction syndrome. The lunate has a medial facet, which articulates directly with the hamate. There is full thickness cartilage loss of the proximal hamate and subchondral cyst formation. *(Right)* Coronal PDWI FS MR demonstrates a type II lunate with a medial facet measuring 5 mm. The articular cartilage of the distal lunate and proximal hamate is thinned with reactive subchondral marrow edema in the hamate.

ULNAR IMPINGEMENT SYNDROME

Key Facts

Terminology

- Ulnar impingement syndrome: Ulnar pain with shortened ulna impinging on distal radial shaft
    - **Not** synonymous with ulnocarpal abutment syndrome

Imaging

- General imaging features
    - Ulna shortened proximal to sigmoid notch
    - Distal radial shaft scalloping
    - Radioulnar convergence
- MR imaging
    - Thickened TFC (in congenital shortening)
    - Thinning and remodeling of radial shaft cortex
    - Cortical changes evident earlier on MR than on radiographs

Top Differential Diagnoses

- Ulnocarpal abutment
    - Positive ulnar variance
    - TFC tear
    - Subchondral sclerosis; cysts in lunate, triquetrum, and ulnar head

Pathology

- Ulnar shortening etiology
    - Idiopathic
    - Growth arrest of ulna as a child (e.g., fracture)
    - Excision of distal ulna (Darrach procedure)
- Previous forearm fracture ± surgery
- Radioulnar convergence results from
    - Loss of buttress effect of distal radioulnar joint
    - Interosseous membrane and pronator quadratus muscles pull radius/ulna together
    - Extensor pollicis brevis & abductor pollicis longus muscles pull ulna toward radius

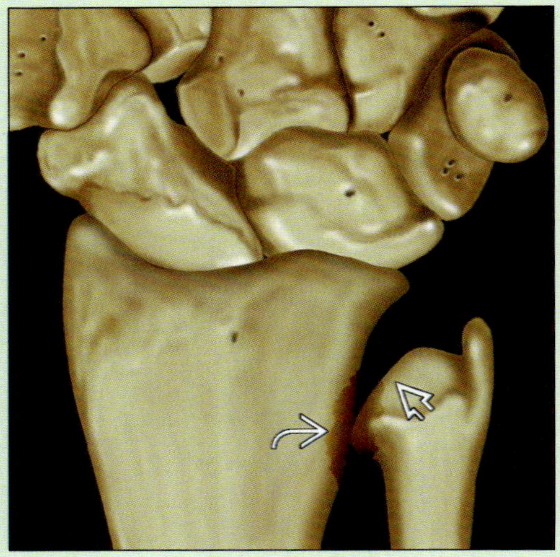
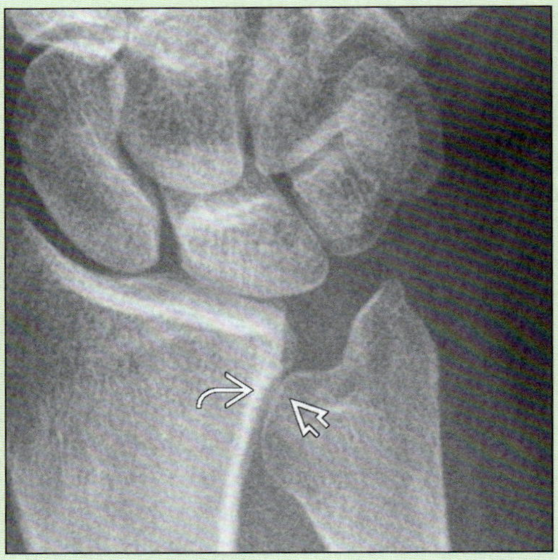

(Left) *Coronal graphic shows the classic appearance of ulnar impingement due to a congenitally short ulna. The ulnar head is slightly laterally tilted ⇨ and articulates with the radius, eroding the radial cortex ⮫.* ***(Right)*** *PA radiograph shows mild erosion of the distal radius just proximal to the sigmoid notch. The ulnar head is slightly tilted ⇨ and impinges the distal radius ⮫, creating this pseudoarthrosis. Ulnar shortening in this patient was related to a prior fracture deformity.*

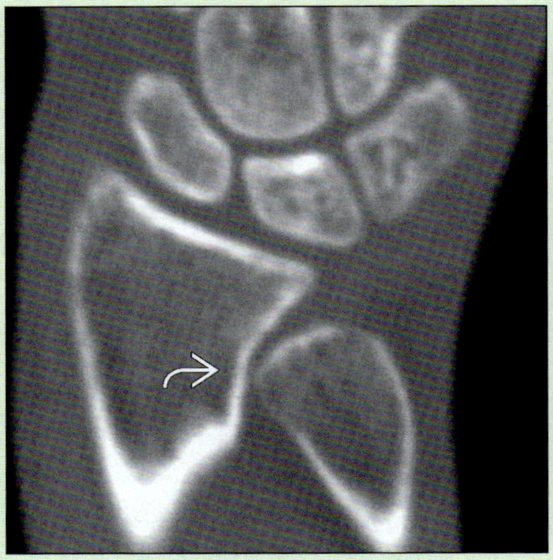
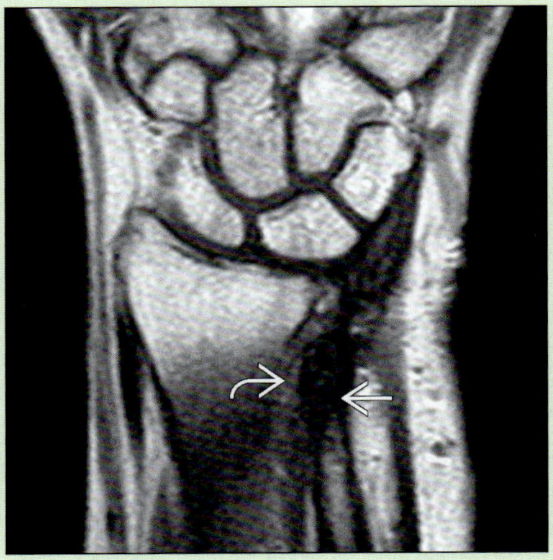

(Left) *Coronal reformat bone CT demonstrates radioulnar pseudoarthrosis with remodeling of the radial cortex ⮫. There is minimal subcortical sclerosis, which tends to be a late finding. The patient reported significant pain with repetitive supination.* ***(Right)*** *Coronal T2WI MR shows minimal mechanical erosion ⮫ of the distal radius in this patient who has undergone resection of the distal ulna ⇨ (Darrach procedure) for treatment of intractable pain from distal radioulnar instability.*

ULNAR IMPINGEMENT SYNDROME

TERMINOLOGY

Abbreviations
- Triangular fibrocartilage (TFC)

Definitions
- Ulnar impingement syndrome: Ulnar pain with shortened ulna impinging on distal radial shaft
  - **Not** synonymous with ulnocarpal abutment syndrome

IMAGING

General Features
- Best diagnostic clue
  - Ulna shortened proximal to sigmoid notch
  - Distal radial shaft scalloping
  - Radioulnar convergence

Radiographic Findings
- Markedly shortened ulna
  - Distal end of ulna proximal to radial sigmoid notch
- Scalloping and subcortical sclerosis of distal radius
  - Tends to be late finding
- Convergence of ulna with distal radius
  - Ulnar head appears to articulate with radius
  - Resected distal ulnar shaft appears to articulate with radius
- Osseous excrescence arising from distal ulna after resection
- Residua from prior trauma may be present

CT Findings
- Cortical scalloping and sclerosis

MR Findings
- Thickened TFC (in congenital shortening)
- Thinning and remodeling of radial shaft cortex
  - Occurs at level of distal ulna
- Cortical changes evident much earlier on MR than on radiographs

Imaging Recommendations
- Best imaging tool
  - Radiography
  - MR for early changes along radial cortex

DIFFERENTIAL DIAGNOSIS

Ulnocarpal Abutment
- Positive ulnar variance
- Subchondral sclerosis ± cysts
  - Proximal lunate
  - Distal ulna
  - Proximal triquetrum
- TFC tear

PATHOLOGY

General Features
- Etiology
  - Ulnar shortening
    - Idiopathic
    - Growth arrest of ulna as a child
      - Related to prior trama
      - Developmental abnormality, such as osteochondromatosis
    - Previous forearm fracture ± surgery
    - Excision of distal ulna (Darrach procedure)
      - Surgery used to treat symptomatic post-traumatic distal radioulnar joint
      - Correction of Madelung deformity
      - Rheumatoid arthritis

Gross Pathologic & Surgical Features
- Radioulnar convergence results from
  - Loss of buttress effect of distal radioulnar joint in shortened ulna
  - Interosseous membrane and pronator quadratus muscles pull radius and ulna together
  - Extensor pollicis brevis & abductor pollicis longus muscles pull ulna toward radius

CLINICAL ISSUES

Presentation
- Most common signs/symptoms
  - Disabling ulnar-sided pain
    - Symptoms are similar to ulnocarpal abutment but markedly more debilitating
  - Grip weakness
  - Clicking with pronation and supination
- Other signs/symptoms
  - Deformity at pseudoarthrosis
  - Forearm rotation causes compression of distal radioulnar joint
    - Increases symptoms
    - May produce grating sensation

Treatment
- If symptoms occur after ulnar shortening and resection
  - Management is difficult
  - Revision and stabilization of stump with soft tissue reconstruction
  - Resection of osteophytes
  - Ulnar head prosthesis
  - Ulnar lengthening &/or radial shortening

DIAGNOSTIC CHECKLIST

Consider
- Reason for original ulnar shortening
- Integrity of TFC complex

Image Interpretation Pearls
- Look for distal ulna directed at and articulating with radius proximal to sigmoid notch
- Look for subtle radial cortical erosion

SELECTED REFERENCES

1. De Witte PB et al: The Darrach procedure for post-traumatic reconstruction. Acta Orthop Belg. 75(3):316-22, 2009
2. Bell MJ et al: Ulnar impingement syndrome. J Bone Joint Surg Br. 67(1):126-9, 1985

METACARPAL FRACTURE AND DISLOCATION

Key Facts

Terminology

- Boxer fracture: 4th or 5th MC neck fracture; MC head volar angulation
- Bennett fracture: Oblique intraarticular fracture of 1st MC base, resulting in small triangular volar lip fragment
- Rolando fracture: Intraarticular comminuted 1st MC base fracture; often forms Y- or T-shaped fracture
- "Fight bite": Opponent's tooth breaks the skin over MCP

Imaging

- Metacarpal head: Comminuted intraarticular fracture
- Metacarpal neck: Dorsal angulation ± rotation of fracture
- Metacarpal shaft: Fragment override with distal fragment rotation
- CMC fracture-dislocation
  - Disruption of normal "parallel M" zigzag pattern of CMC joints
  - Overlap of 2nd-5th CMC joint margins
  - MC displacement (usually dorsal)
  - MC base ± adjacent carpal fractures

Clinical Issues

- 1st & 5th metacarpals most commonly injured
- Simple nonarticular MC fractures heal without complication when treated conservatively
- Majority of CMC dislocations involve multiple joints
- "Fight bite" may → septic arthritis

Diagnostic Checklist

- Significant MC shaft angulation: Look for CMC dislocation
- Dorsal fracture of capitate or hamate: Look for CMC joint subluxation/dislocation

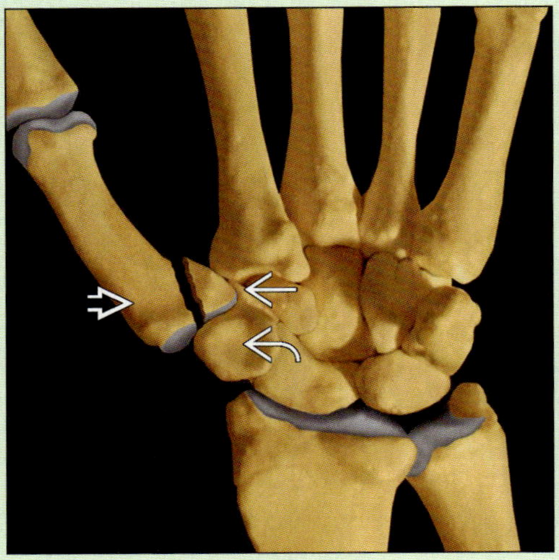
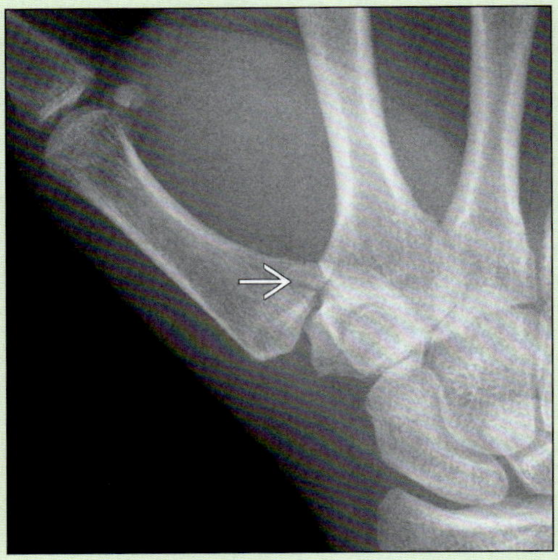

*(**Left**) AP graphic shows a typical Bennett fracture. The medial fracture fragment ➡ remains in anatomic position relative to the trapezium ➡. The remainder of the 1st MC ➡ subluxates laterally. (**Right**) PA radiograph demonstrates a subtle intraarticular fracture ➡ across the medial corner of the 1st metacarpal base. There is minimal displacement of the fragment. The lateral fragment is stabilized and held in place by the anterior oblique ligament.*

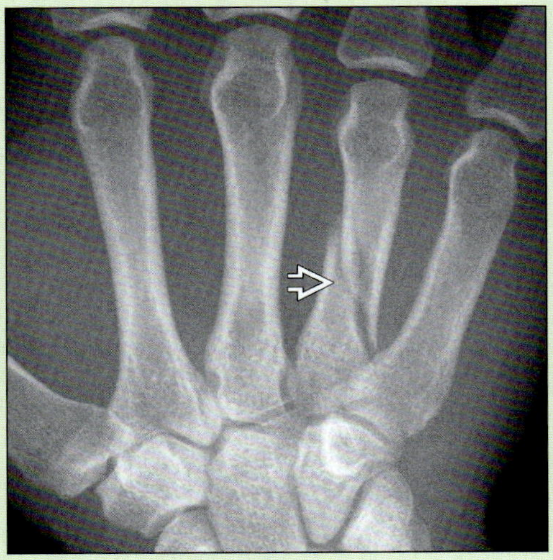
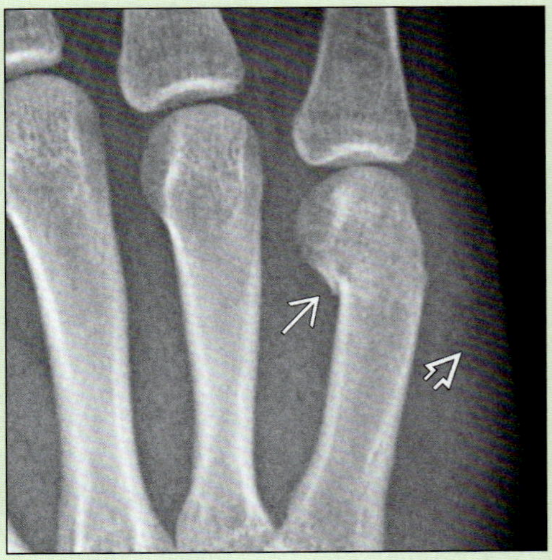

*(**Left**) PA radiograph shows a typical spiral fracture of the 4th MC shaft ➡ with slight medial displacement of the distal fragment. There is mild shortening of the shaft. This fracture results from torsional forces on the shaft. (**Right**) PA radiograph shows a typical boxer fracture of the 5th metacarpal neck ➡ with dorsal angulation at the fracture line. Note the indistinct fat planes ➡ in the adjacent soft tissues in this acute injury.*

METACARPAL FRACTURE AND DISLOCATION

TERMINOLOGY

Definitions

- Fracture &/or dislocation of metacarpal (MC) base, shaft, neck, or head
- Fracture/dislocation at carpometacarpal (CMC) joint (fracture can be at either carpal bone or MC base)
- Dislocation: Complete dissociation of articular surfaces
- Subluxation: Partial dissociation of articular surfaces
- Eponyms
  - Boxer fracture: 4th or 5th MC neck fracture; MC head volar angulation
  - Epibasal fracture: Stable 2-part fracture of 1st metacarpal shaft
  - Bennett fracture: Oblique intraarticular fracture of 1st MC base, resulting in small triangular volar lip fragment
  - Rolando fracture: Intraarticular comminuted 1st MC base fracture
    - Often forms Y- or T-shaped fracture
  - "Fight bite": Opponent's tooth breaks the skin over metacarpophalangeal (MCP) joint with injury to dorsal capsule, extensor tendon, ± MC head

IMAGING

General Features

- Best diagnostic clue
  - Discontinuity of metacarpal
- Location
  - Most common location by fracture type
    - MC head: 2nd MC
    - MC neck: 4th and 5th MC
    - MC shaft: 2nd MC
    - MC base: 1st and 5th MC
    - CMC fracture-dislocation: 4th & 5th MC
- Size
  - Variable; frequently small and very subtle
- Morphology
  - Transverse
    - 1st MC proximal shaft (epibasal)
    - MC neck with dorsal angulation
  - Oblique or spiral
    - 2nd-5th MC shafts; rare in 1st MC
  - Intraarticular
    - Most common in 1st and 5th MC
    - Bennett fracture: Pyramidal volar medial fragment stays with trapezium; remaining base subluxates dorsolaterally
    - Rolando fracture: Comminuted 1st MC base; fracture lines are Y- or T-shaped
  - Avulsion
    - 1st MC volar base
      - Abductor pollicis longus (APL)
  - Comminuted
    - Look for adjacent structure fractures
  - CMC dislocations
    - Dorsal (60%), volar (20%), ulnar (15%)
    - 5th CMC dislocation more commonly volar than others

Radiographic Findings

- Routine views
  - PA: Demonstrates CMC alignment, MC angulation, bone loss
  - Lateral: Identify degree of angulation, small avulsion fragments, dislocation
  - Oblique: Often key in identification of subtle fragments and dislocations
- Metacarpal head: Comminuted intraarticular fracture
- Metacarpal neck: Dorsal angulation ± rotation of fracture
- Metacarpal shaft: Fragment override with distal fragment rotation
  - Rotation difficult to assess radiographically; clinical evaluation more reliable
- Metacarpal base fracture
  - 1st MC: Medial palmar fragment; MC subluxation
  - 2nd-5th MC: Subtle cortical fragments
- CMC fracture-dislocation
  - Disruption of normal "parallel M" zigzag pattern of CMC joints
  - Overlap of 2nd-5th CMC joint margins
  - MC displacement (usually dorsal)
  - MC base ± adjacent carpal fractures
- "Fight bite"
  - Most common in 3rd MCP
  - Look for angular divot in dorsal metacarpal head
  - Can lead to septic arthritis

CT Findings

- Used in search for occult fracture
- Used to evaluate pre- ± post-reduction
  - Intraarticular fractures
  - Fracture-dislocation

MR Findings

- Useful to evaluate associated soft tissue injury
- Not typically used for fracture-dislocation evaluation

Imaging Recommendations

- Best imaging tool
  - Radiography
- Protocol advice
  - 2nd-5th MC injury: PA, lateral, oblique; center on hand
    - PA with palm flat to maximize CMC evaluation
    - Reverse oblique to visualize 4/5th CMC
  - 1st MC injury: PA, lateral, oblique; center on thumb
    - True lateral: Palm on table, hand/wrist pronated 15-35°, tube centered on 1st CMC joint
    - Routine views of hand all show the thumb in similar obliquity, & injury may be missed
  - Thin-section CT with reformations
    - Image area of interest obliquely; will increase cuts through area of interest and improve reformats

PATHOLOGY

Mechanisms of Injury

- 1st MC epibasal fracture
  - Occurs with longitudinal loading
  - Angulation due to tension from abductor pollicis brevis (APB), adductor pollicis (AP), & flexor pollicis brevis (FPB) on distal fragment
  - Abductor pollicis longus (APL) stabilizes proximal fragment

METACARPAL FRACTURE AND DISLOCATION

- Bennett/Rolando fracture
  - Occurs with axial loading on partially flexed thumb (as in a fist fight)
  - Volar medial fragment remains with trapezium via anterior oblique ligament
  - Remaining fragment often subluxates proximally, laterally, and dorsally
- Boxer fracture
  - Results from poor punching technique or hitting solid objects (such as a wall)
  - Dorsal angulation due to initial blow; intrinsic muscles flexes MC head
- MC transverse fracture
  - Axial loading or direct blow
  - Dorsal angulation due to interosseous muscles pulling distal fragment volarly
- MC oblique/spiral fracture
  - Results from torsional forces contributing to rotational malalignment
- MC comminuted fracture
  - Direct impact; often associated with soft tissue injury
- CMC fracture-dislocation
  - High-energy injury with torsional forces; associated with carpal fractures
- Isolated 5th MC fracture-dislocation
  - Dorsal: Most common
    - Analogous to Bennett fracture
    - Lateral base fragment stays with hamate; medial fragment pulled proximally by extensor carpi ulnaris
  - Volar dislocation
    - Volar radial type: All ligaments and tendons torn at attachment sites (more common)
    - Volar ulnar type: Pisometacarpal ligament and flexor carpi ulnaris tendon attachments remain intact

CLINICAL ISSUES

Presentation

- Most common signs/symptoms
  - Focal tenderness
  - Deformity following trauma
- Other signs/symptoms
  - Limited range of motion

Demographics

- Epidemiology
  - 1st & 5th MC most commonly injured
    - 5th MC fracture = 20% of hand fractures
  - CMC dislocations: Relatively rare
    - Majority are multiple (60%)
      - Involve 4th and 5th CMC in 30%
      - Involve 2nd through 5th CMC in 30%
    - Isolated CMC dislocations: 50% involve 5th CMC, 25% involve 2nd CMC

Natural History & Prognosis

- Simple nonarticular MC fractures heal without complication when treated conservatively
- MC neck fracture
  - The more distal the fracture, the greater the degree of dorsal angulation tolerated
  - 4th and 5th MC fractures: Dorsal angulation < 50° is functionally tolerated
  - 2nd and 3rd MC fractures: Dorsal angulation < 20° is functionally tolerated
- MC oblique/spiral shaft fracture
  - > 5 mm shortening → poor functional outcome
  - 1° rotation at MC shaft results in 5° rotation of fingertip
    - May result in difficulty making fist
- MC base/intraarticular fracture
  - Bennett or Rolando fracture: Inadequate reduction leads to malunion & secondary osteoarthritis
- CMC fracture-dislocation
  - 5th ray fracture: Inadequate reduction → loss of grip strength
- "Fight bite" may → septic arthritis

Treatment

- Extraarticular, minimally malaligned MC fractures treated with closed reduction
- Surgical intervention considered for
  - Intraarticular fractures
  - Irreducible ± malrotated fractures
  - Marked MC shortening (> 5 mm)
  - Open fractures
  - Fractures with extensive associated soft tissue injury
  - Multiple dislocations
  - Polytrauma
  - High performance athlete
- Bennett fracture: Involvement of > 20% articular surface → open reduction and internal fixation (ORIF)
- Rolando fracture: Highly comminuted fracture treated closed; less amenable to ORIF

DIAGNOSTIC CHECKLIST

Consider

- Significant MC shaft angulation: Look for CMC dislocation
- Dorsal fracture of capitate or hamate: Look for CMC joint subluxation/dislocation

Reporting Tips

- Important to quantify alignment
  - Angulation, rotation, and fragment override

SELECTED REFERENCES

1. Lisle DA et al: MR imaging of traumatic and overuse injuries of the wrist and hand in athletes. Magn Reson Imaging Clin N Am. 17(4):639-54, vi, 2009
2. Al-Qattan MM: The triad of multiple metacarpal fractures and/or dislocations of the fingers, severe hand swelling and clinical evidence of acute median nerve dysfunction. J Hand Surg Eur Vol. 33(3):298-304, 2008
3. Rettig AC: Athletic injuries of the wrist and hand: part II: overuse injuries of the wrist and traumatic injuries to the hand. Am J Sports Med. 32(1):262-73, 2004
4. Rosner JL et al: Imaging of athletic wrist and hand injuries. Semin Musculoskelet Radiol. 8(1):57-79, 2004
5. Lawlis JF 3rd et al: Carpometacarpal dislocations. Long-term follow-up. J Bone Joint Surg Am. 73(1):52-9, 1991

METACARPAL FRACTURE AND DISLOCATION

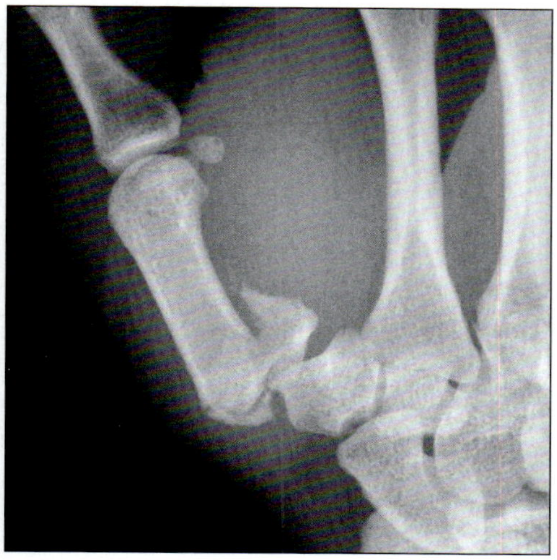

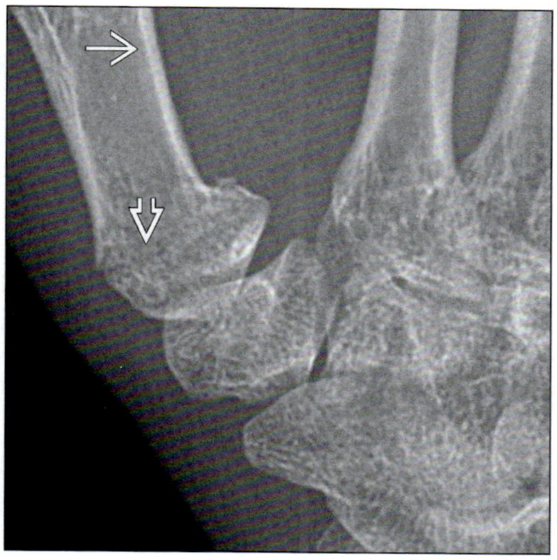

(Left) *PA radiograph demonstrates a comminuted intraarticular 1st metacarpal base fracture. This Rolando fracture results from axial loading with a partially flexed thumb. This 31 year old was injured during a fist fight.* **(Right)** *PA radiograph shows an extraarticular epibasal fracture ⇨ of the 1st metacarpal. There is apex lateral angulation as the distal fragment is pulled medially → by the AP, APB, and FPB tendons, while the proximal fragment is stabilized by the APL.*

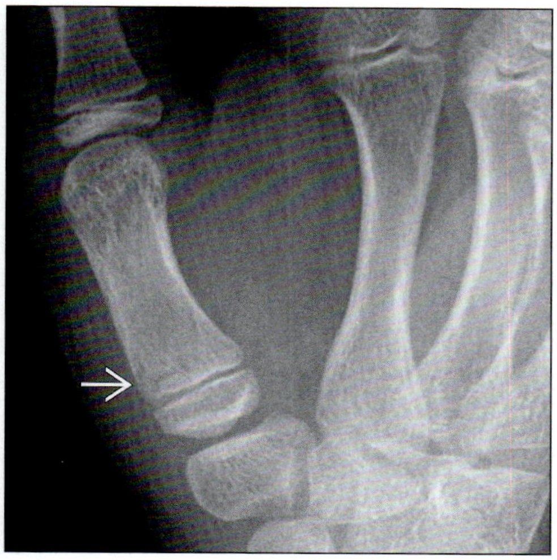

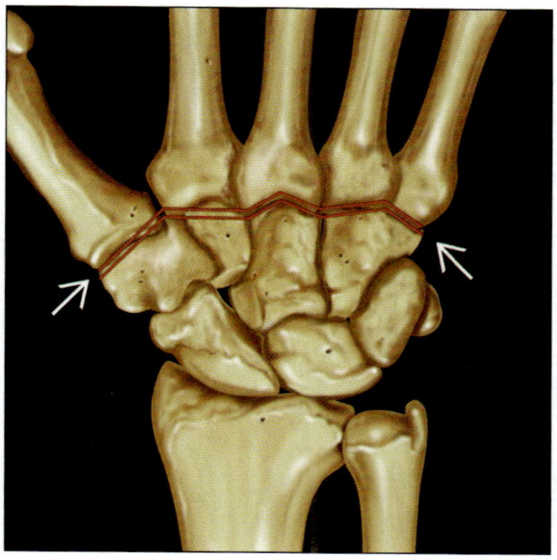

(Left) *AP radiograph shows a nondisplaced Salter-Harris type II fracture of the 1st MC growth plate →. This fracture resulted from a skiing injury with the fracture propagating across the physis from medial to lateral, exiting through the metaphysis.* **(Right)** *Posteroanterior graphic shows the "zigzag" or "M" pattern → outlining the carpometacarpal joints. Disruption of these articulations is usually subtle; checking this articulation carefully at each joint is crucial.*

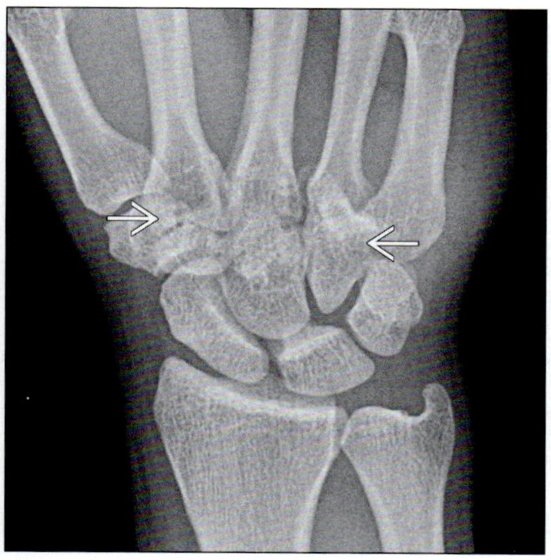

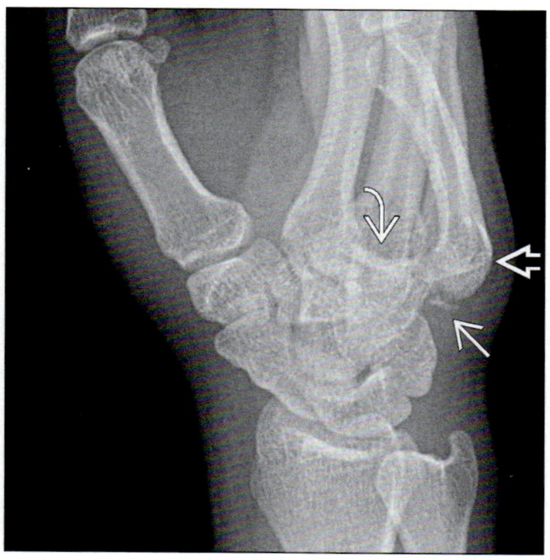

(Left) *PA radiograph reveals marked disruption of the normal CMC articular relationships → indicative of a significant trauma. Marked overlap of the articular surfaces represents fracture-dislocation of the 2nd-4th CMC joints. The fractures are obscured in this PA view.* **(Right)** *Suboptimal lateral radiograph from the same patient demonstrates the dorsally dislocated 3rd metacarpal ⇨ with a small fracture fragment →. Note the uncovered distal capitate ↷.*

METACARPAL FRACTURE AND DISLOCATION

*(**Left**) PA radiograph shows a complex medial hand injury with oblique fractures of the 3rd MC neck and 4th proximal phalanx. CMC alignment is also disrupted with comminuted intraarticular fractures of the 4th and 5th MC bases. (**Right**) PA radiograph reveals a nondisplaced oblique fracture of 5th MC shaft with a pisiform fracture. This unusual combination relates to the ECU and abductor digiti quinti tendon attachments on the pisiform and 5th MC base respectively.*

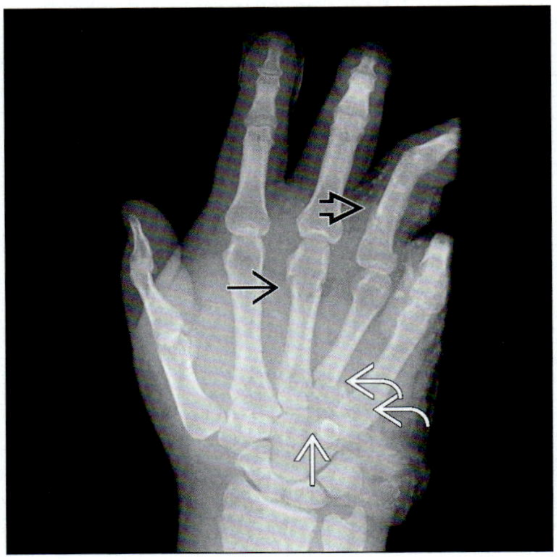

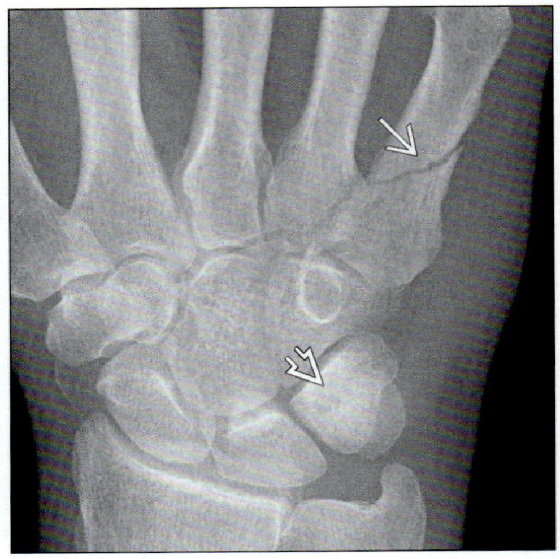

*(**Left**) PA radiograph shows normal alignment of the 1st through 3rd CMC joints. The 4th and 5th CMC joints are disrupted. The 5th MC base fracture is analogous to a Bennett fracture; the small lateral fragment is held in place by the intermetacarpal ligament. (**Right**) Coronal reformat bone CT demonstrates an intraarticular fracture of the 5th metacarpal base with the lateral fragment maintaining its relationship to the 4th metacarpal via the intermetacarpal ligament.*

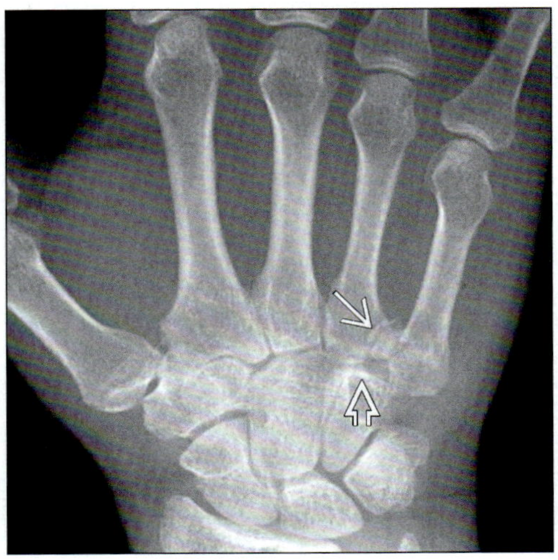

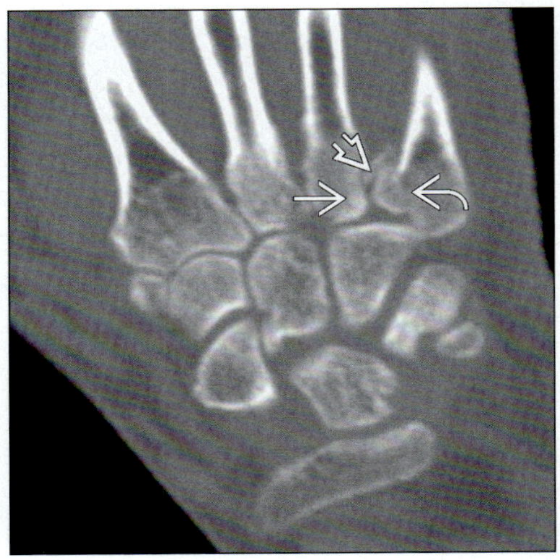

*(**Left**) Coronal STIR MR demonstrates diffuse marrow contusions in the 2nd, 3rd, and 4th MC bases as well as the trapezoid and distal medial hamate. This resulted from a high-energy injury related to gripping the steering wheel during an auto accident. (**Right**) Sagittal PDWI FS MR from the same patient shows a nondisplaced intraarticular fracture of the 3rd MC base with thin linear low signal representing discrete fracture lines surrounded by diffuse marrow edema.*

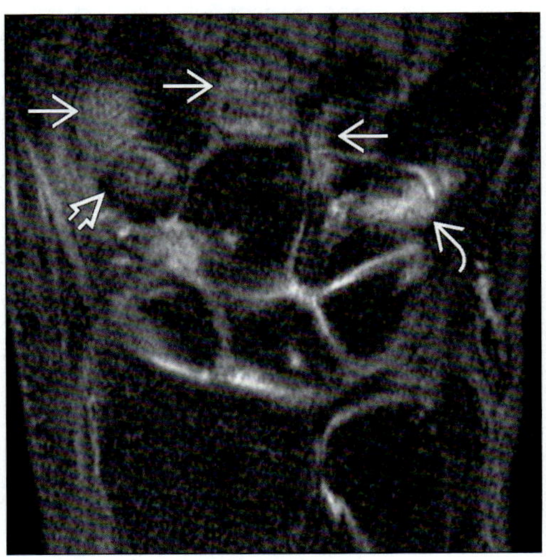

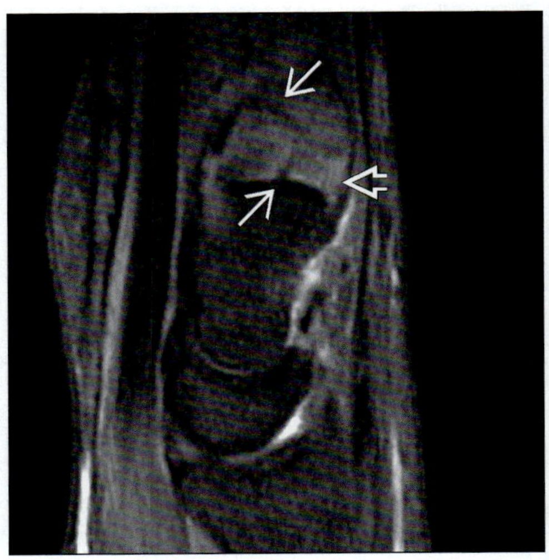

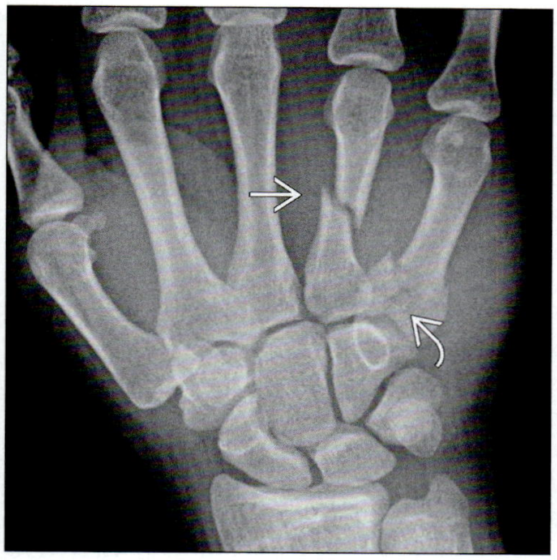

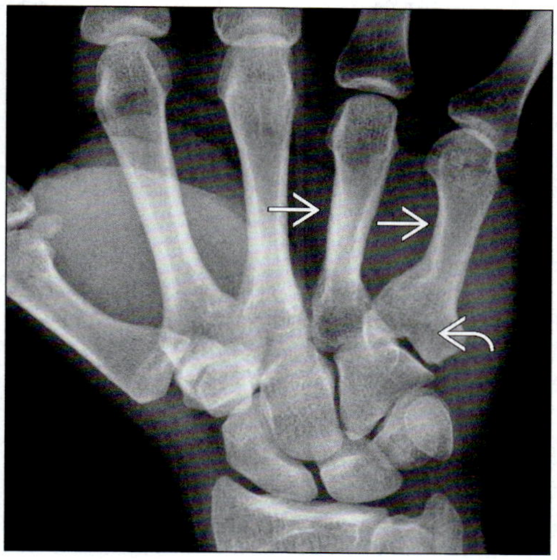

(Left) *PA radiograph shows an acute medial hand injury with an extraarticular 4th MC spiral shaft fracture and a comminuted intraarticular fracture of the proximal 5th MC. This injury occurred from a combination of torsional and axial impaction forces sustained during an auto accident.* ***(Right)*** *PA radiograph from the same patient 2 years later shows residual deformity with shortening of both 4th and 5th MC shafts and broadening and flattening of the proximal 5th metacarpal base.*

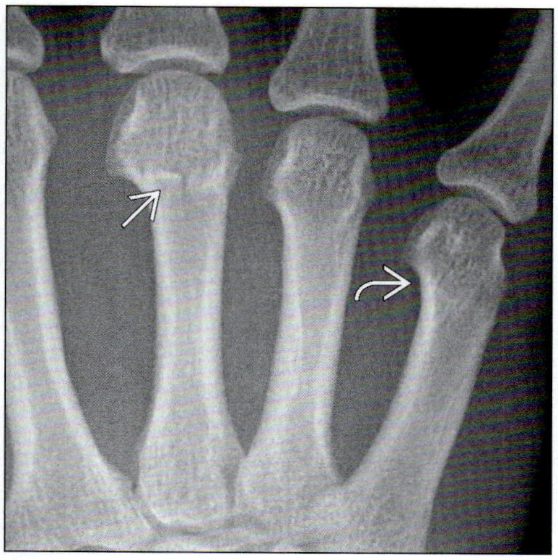

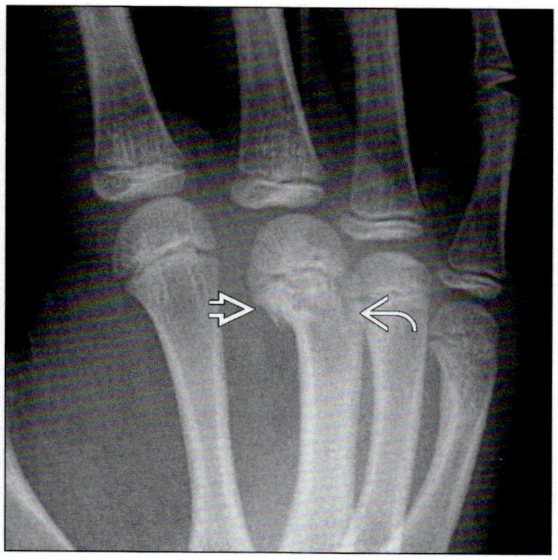

(Left) *PA radiograph shows a subtle acute nondisplaced 3rd metacarpal neck fracture in this young male who punched a wall. There is slight dorsal angulation of the 5th metacarpal neck, evidence of prior trauma from another punching incident.* ***(Right)*** *PA radiograph shows a Salter-Harris type II fracture of the 3rd metacarpal head resulting from a blow to a wall. This is a juvenile version of the boxer fracture. Note the dorsal angulation of the fracture.*

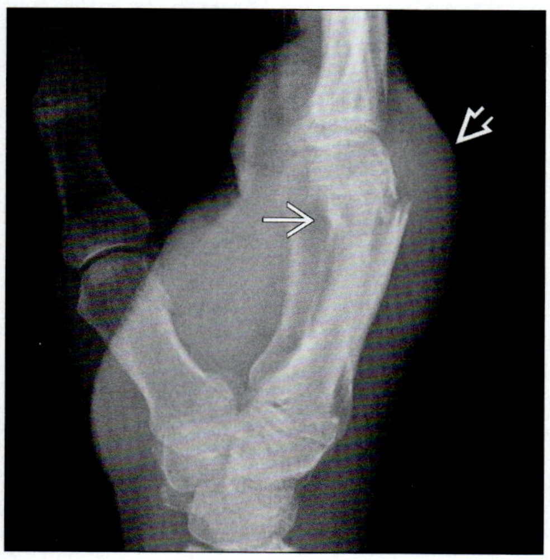

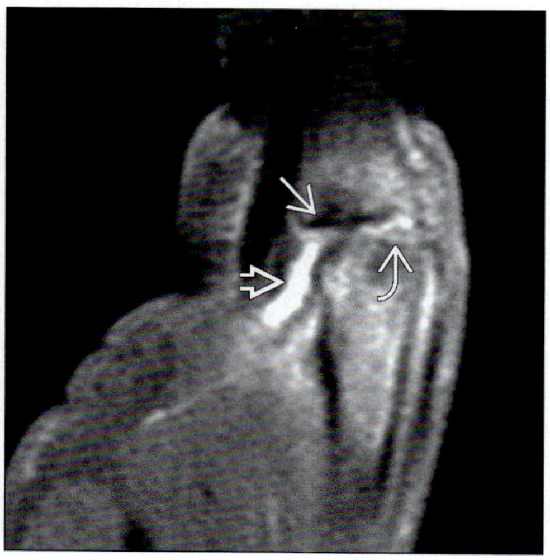

(Left) *Lateral radiograph shows transverse 2nd and 3rd MC neck fractures resulting from a fist fight. Marked soft tissue swelling suggests possible infection if the opponent's tooth broke the skin (a "fight bite").* ***(Right)*** *Sagittal PDWI FS MR shows sequelae of a "fight bite." There is a subtle impaction divot in the dorsal MC head from the opponent's tooth. There is septic arthritis from this unrecognized open fracture with articular cartilage loss and a joint effusion.*

FINGER FRACTURE AND DISLOCATION

Key Facts

Terminology

- Mallet (baseball, drop) finger: Forced flexion of extended distal interphalangeal (DIP) joint with injury to terminal extensor tendon ± avulsion fracture
- Boutonniere deformity: Extensor tendon central slip injury at or near insertion on to middle phalanx; results in proximal interphalangeal (PIP) joint flexion + DIP extension
- Coach's finger: PIP dislocation, typically dorsal; ± volar plate injury
- Gamekeeper's (skier's) thumb: Injury to ulnar collateral ligament (UCL) of 1st MCP joint
- Jersey (rugby) finger: Flexor digitorum profundus (FDP) tendon avulsion ± avulsion fracture

Imaging

- Distal phalanx injuries
  - Soft tissue distortion may suggest nailbed injury
  - Shaft: Longitudinal or transverse fracture
  - DIP joint dislocation dorsal or lateral
- Middle/proximal phalanx injuries
  - Volar plate injury
  - Collateral ligament injury: Asymmetric joint widening
  - PIP dislocation most commonly dorsal
- MCP injury of fingers
  - Proximal phalanx dorsal dislocation/subluxation; ± fracture
  - Look for sesamoid interposed in MCP
- Thumb MCP injury
  - Joint subluxation or widening suggests collateral ligament injury
  - UCL > > RCL injury

(Left) Sagittal graphic shows a mallet finger injury with avulsion of the terminal extensor tendon → from the dorsal base of the distal phalanx. This injury results from forced flexion of the DIP, often due to a blow to the finger from a baseball. ***(Right)*** *Lateral radiograph shows mallet finger injury resulting from a blow to the tip of the finger. The terminal extensor tendon avulses the fragment ↷, pulling it proximally. Note DIP joint flexion → and PIP joint hyperextension ⇨.*

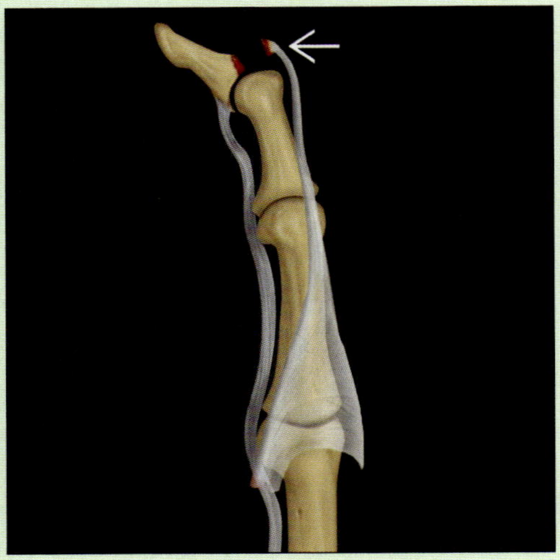

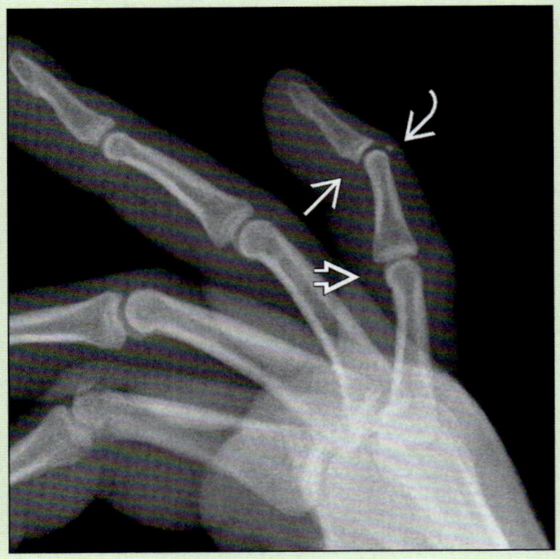

(Left) Lateral graphic shows a boutonniere deformity. There is a fracture → of the dorsal base of the middle phalanx resulting from an extensor central slip avulsion. The terminal extensor tendon ↶ remains intact. This injury results from a blow to the fingertip. ***(Right)*** *Sagittal PDWI FS MR shows an extensor central slip avulsion → with surrounding soft tissue swelling. There is a developing boutonniere deformity with the PIP joint flexion ⇨ and DIP extension ↷.*

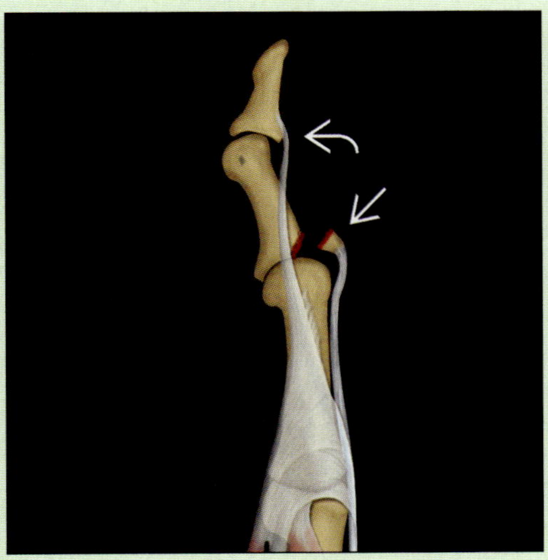

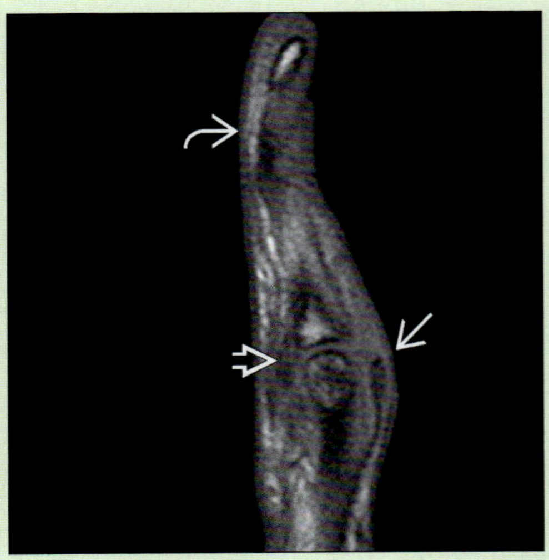

FINGER FRACTURE AND DISLOCATION

TERMINOLOGY

Definitions

- Fracture ± dislocation of phalanx base, shaft, neck, or head
  - Fracture-dislocation at metacarpophalangeal (MCP) joint involving metacarpal (MC) head or proximal phalanx base
- Eponyms
  - Mallet (baseball, drop) finger: Forced flexion of extended distal interphalangeal (DIP) joint with injury to terminal extensor tendon ± avulsion fracture
  - Boutonniere deformity: Extensor tendon central slip injury at or near insertion on middle phalanx; results in proximal interphalangeal (PIP) joint flexion + DIP extension
  - Coach's finger: PIP dislocation, typically dorsal; ± volar plate injury
  - Gamekeeper's (skier's) thumb: Injury to ulnar collateral ligament (UCL) of 1st MCP joint
    - Related to hyperabduction
    - ± avulsion fracture
  - Jersey (rugby) finger: Flexor digitorum profundus (FDP) tendon avulsion ± avulsion fracture
    - Most commonly affects 4th finger

IMAGING

General Features

- Best diagnostic clue
  - Fracture line through phalanx
- Morphology
  - Spiral/oblique: Shaft
  - Transverse: Shaft or neck
  - Comminuted
    - Fracture resulting from crush or high-energy injury
    - Terminal tuft
  - Triangular: Avulsion

Radiographic Findings

- Distal phalanx injuries
  - Tuft: Simple or comminuted fracture
    - Soft tissue distortion may suggest nailbed injury
  - Shaft: Longitudinal or transverse fracture
  - Intraarticular (DIP) injury
    - Jersey finger: Distal FDP injury; ± volar avulsion fragment; unable to flex DIP
    - Mallet finger: DIP flexion due to terminal extensor tendon injury; ± dorsal avulsion fragment
  - DIP joint dislocation: Typically dorsal or lateral
  - Epiphyseal injuries
    - Child: Salter-Harris type I or II
    - Adolescent: Salter-Harris type II or III
    - Displacement of fracture fragment may mimic dislocation
- Middle/proximal phalanx injuries
  - Intraarticular (PIP) injury
    - Condylar: Unicondylar, bicondylar, or comminuted fracture
      - May be missed clinically because finger still flexes
    - Base: Avulsion or impaction fracture
      - Boutonniere injury: PIP flexion with DIP extension due to extensor tendon central slip injury; ± dorsal avulsion middle phalanx base
      - Volar plate injury: Typically distal volar plate injured; ± volar avulsion of base of middle phalanx
      - Collateral ligament injury: Asymmetric joint widening; radial collateral ligament (RCL) > UCL; ± lateral avulsion fracture
      - Comminuted impaction (pilon) fracture: Central depression with splaying of articular margins
      - Lateral plateau fracture: Typically involves middle phalanx base
    - Dislocation: Dorsal PIP most common
      - Coach's finger: Dorsal dislocation of middle phalanx base on proximal phalanx head; ± volar plate injury
    - Physeal fractures: 85-90% are Salter-Harris II
      - Typically 5th finger with medial angulation
  - Extraarticular fractures
    - Subcapital: Common in children; frequently overlooked
      - Head fragment may displace & rotate 90° dorsally
    - Spiral/oblique fracture: Proximal phalanx most common
    - Transverse fracture: Middle phalanx most common
- MCP injury of fingers
  - Proximal phalanx dorsal dislocation/subluxation; ± avulsion or osteochondral shear fracture
    - ± joint space widening; medial or lateral angulation
    - Look for sesamoid interposed in MCP
      - Indicates volar plate injury
- Thumb MCP injury
  - Joint subluxation or widening suggests collateral ligament injury
    - Asymmetric widening of medial or lateral joint > 30° with flexion/extension
    - Difference between injured & uninjured joint > 15° suggests ligament injury
    - UCL injury 10x more frequent than RCL injury
    - Look for associated fracture fragment, typically from medial base of proximal phalanx
  - Mallet thumb: Rare; extensor pollicis longus tear; ± dorsal avulsion

CT Findings

- Delineate small fractures ± intraarticular pathology
- Evaluate healing ± hardware

MR Findings

- Not typically used to evaluate acute injury
- Used to evaluate ligament or tendon integrity
  - Evaluate irreducible fractures and dislocations
    - e.g., PIP dorsal dislocation with volar plate tear and entrapment
    - e.g., 1st MCP dislocation with MC head trapped between flexor pollicis brevis (FPB) heads

Ultrasonographic Findings

- Allows dynamic evaluation of tendons & ligaments

FINGER FRACTURE AND DISLOCATION

Imaging Recommendations

- Best imaging tool
  - Radiography
- Protocol advice
  - Radiograph standard views: True PA and lateral of digit of interest, oblique
    - Additional views as indicated
    - True AP and lateral of thumb
      - Standard positioning for hand x-rays results in thumb positioned similarly on all views
    - Repeat films should always be obtained after reduction of finger fracture or dislocation
      - Assess alignment
      - Identify subtle fractures not seen prior to reduction
  - MR: Small field of view; specialized coils
    - Proper plane selection key to evaluation
      - Coronal & sagittal views of hand do not provide anatomic planes of thumb
      - Acquisition technique for thumb: Use axial image for localization; coronal plane acquired from line drawn through 1st MC head lateral condylar margins

PATHOLOGY

Mechanisms of Injury

- Fractures
  - Tuft/comminuted fracture: Crush injury
  - Spiral/oblique fracture: Torsional forces
  - Transverse fracture: Direct blow
  - Pilon fracture: Axial loading
  - Avulsion fracture: Hyperflexion or hyperextension; ± dislocation
  - Physeal injury at DIP
    - Terminal extensor tendon attached to epiphysis; FDP is attached to metaphysis
    - Hyperflexion results in dorsal displacement of epiphyseal fragment while phalanx tip flexes
- Dislocation
  - Dorsal PIP dislocation: Hyperextension with longitudinal compression (ball against tip of finger)
  - Volar PIP dislocation: Rotatory longitudinal compression (catching finger in moving spin dryer)
  - Thumb MCP dislocation: Sudden forced radial deviation from fall on outstretched hand with thumb abducted (e.g., gripping ski pole)

CLINICAL ISSUES

Natural History & Prognosis

- Many finger fractures heal with conservative treatment
- Injuries may be viewed as trivial and neglected resulting in
  - Deformity and stiffness
    - Boutonniere deformity: PIP flexed with DIP extended; extensor central slip disruption
    - Swan neck deformity: PIP extended with DIP flexed following FDP disruption (mallet finger)
- Age influences outcome
  - Extraarticular fracture
    - Age 0-20 will regain 88% of normal mobility
    - Age 60-70 will regain < 60% of normal mobility
- Other issues affecting prognosis
  - Intraarticular fractures with fragment size > 2 mm
    - Some suggest stability & alignment more important than articular congruency for healing
  - Comminuted fractures
  - Unstable fractures with significant deformity
  - Soft tissue sleeve injury
  - Prolonged immobilization
- Distal tuft fractures often remain ununited but are asymptomatic

Treatment

- Closed reduction, splinting, and buddy taping (taping fractured finger to adjacent fingers) often sufficient
- Surgical intervention for
  - Irreducible or open fractures
  - Marked displacement or malrotation
  - Athlete's need to return to sport
- Potential complications
  - Nonunion
  - Malunion with shortening ± angulation
  - Loss of motion
  - PIP joint extensor lag
    - Adhesions of dorsal apparatus to proximal phalanx
    - Shortening of proximal phalanx
    - Apex volar deformity
  - Infection

DIAGNOSTIC CHECKLIST

Consider

- "Fight bite": Punching injury with tooth breaking skin of dorsal MCP joint capsule
  - May lead to septic arthritis
  - 3rd MCP most common

Image Interpretation Pearls

- Dislocation may be transient and partially reduced when imaging obtained
- Identify source of avulsed fragment

Reporting Tips

- Quantify alignment: Angulation, displacement, rotation, shaft length
- Intraarticular fractures: Define % of articular surface involved in injury
- Identify involvement of nail bed

SELECTED REFERENCES

1. Lisle DA et al: MR imaging of traumatic and overuse injuries of the wrist and hand in athletes. Magn Reson Imaging Clin N Am. 17(4):639-54, vi, 2009
2. Kijowski R et al: The role of ultrasound in the evaluation of sports medicine injuries of the upper extremity. Clin Sports Med. 25(3):569-90, viii, 2006
3. Rettig AC: Athletic injuries of the wrist and hand: part II: overuse injuries of the wrist and traumatic injuries to the hand. Am J Sports Med. 32(1):262-73, 2004
4. Rosner JL et al: Imaging of athletic wrist and hand injuries. Semin Musculoskelet Radiol. 8(1):57-79, 2004
5. Clavero JA et al: MR imaging of ligament and tendon injuries of the fingers. Radiographics. 22(2):237-56, 2002

FINGER FRACTURE AND DISLOCATION

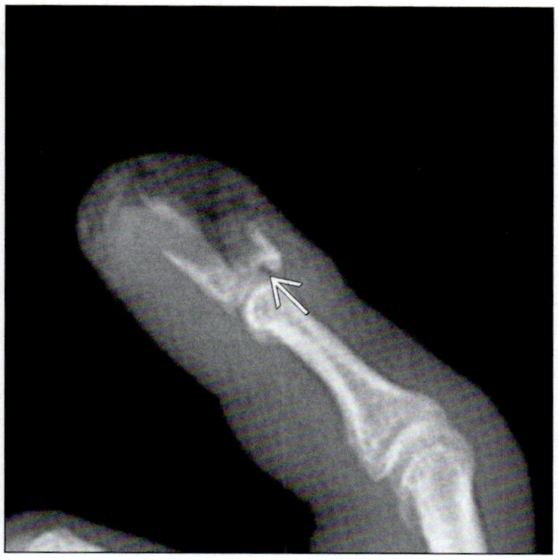

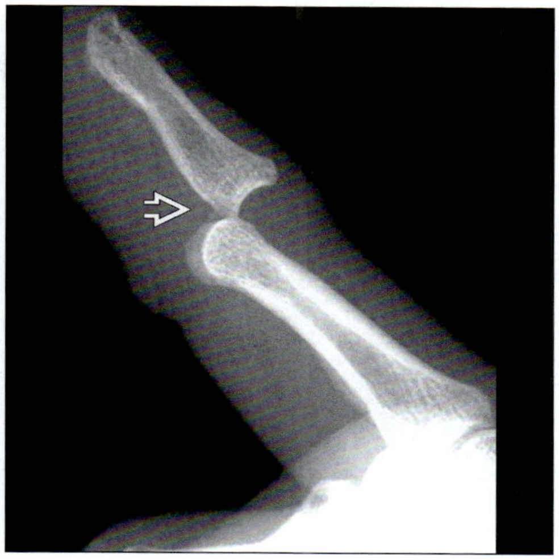

(Left) *Lateral radiograph shows a markedly comminuted open fracture of the distal phalanx due to a gunshot injury. Air tracks into the DIP joint indicating an open joint.* ***(Right)*** *Lateral radiograph demonstrates acute dorsal dislocation of the DIP joint. Dislocation of this joint is less common than dislocation of the PIP joint. Post-reduction radiographs should always be obtained to identify subtle fractures.*

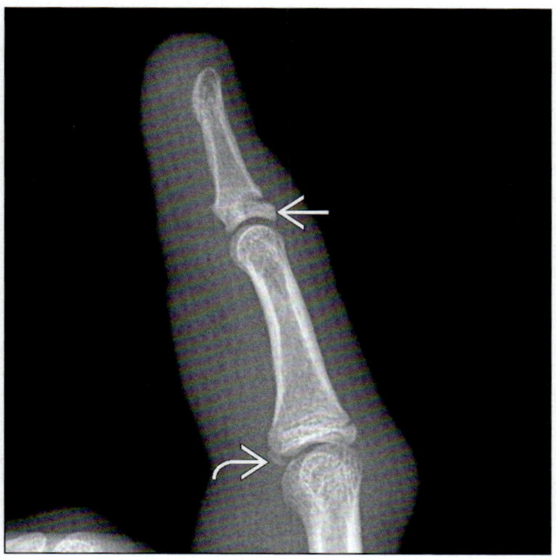

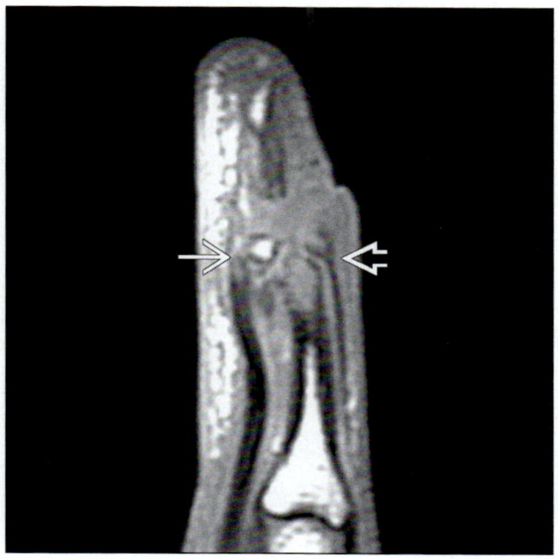

(Left) *Lateral radiograph demonstrates avulsion of the terminal extensor tendon at the dorsal base of the distal phalanx and a subtle crescentic fragment at the volar base of the middle phalanx representing a volar plate avulsion.* ***(Right)*** *Sagittal T1WI MR demonstrates a comminuted distal phalanx fracture resulting from a lawnmower injury. Note the FDP attached to the volar fracture fragment and the terminal extensor tendon attached to the volar fragment.*

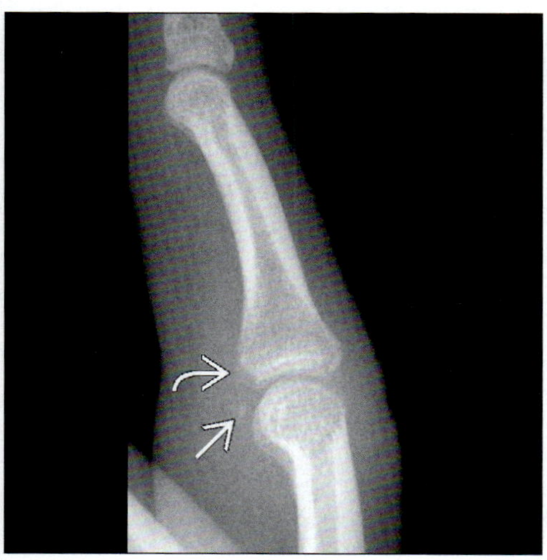

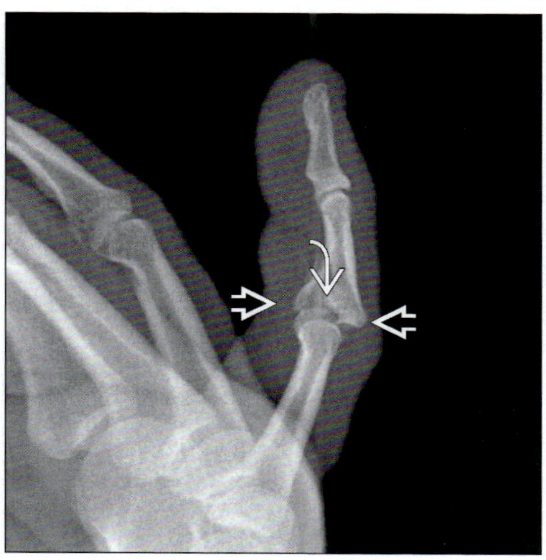

(Left) *Lateral radiograph following reduction of a PIP dislocation demonstrates a small triangular fragment volar to the joint resulting from avulsion of the distal insertion of the volar plate. This fracture was not apparent on initial imaging.* ***(Right)*** *Lateral radiograph shows central depression of the middle phalanx base articular surface and anterior and posterior splaying of the articular margins. This is a pilon fracture and results from axial loading injury.*

FINGER FRACTURE AND DISLOCATION

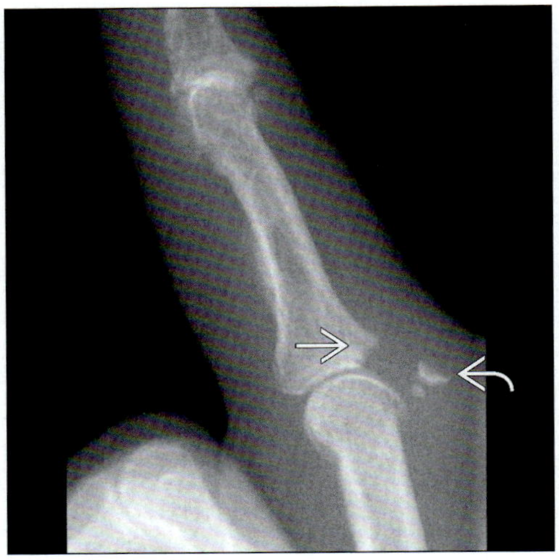

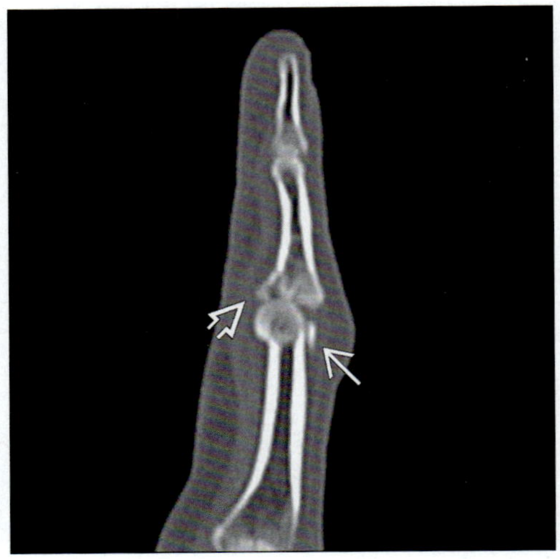

(Left) *Lateral radiograph shows a small fracture fragment originating from the dorsal base of the middle phalanx. The fragment is rotated > 90° and retracted slightly, as it remains attached to the extensor tendon central slip.* ***(Right)*** *Sagittal reformat bone CT demonstrates a pilon fracture of the base of the middle phalanx with splaying of the articular surface and a small fragment dorsal to the proximal phalanx head indicating an associated extensor central slip avulsion.*

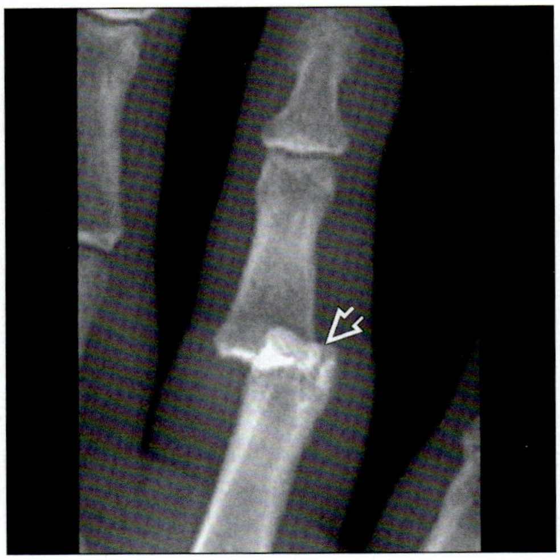

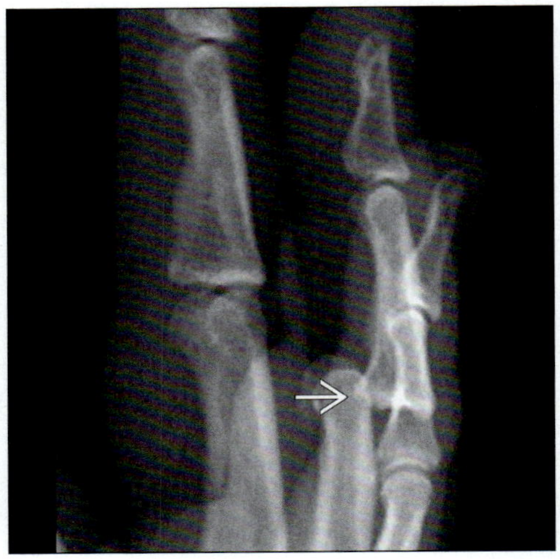

(Left) *PA radiograph demonstrates PIP joint dislocation with the middle phalanx overlapping the proximal phalanx. There is no evidence of fracture.* ***(Right)*** *Lateral radiograph from the same patient shows the 4th middle phalanx displaced dorsally and perched on the dorsal aspect of the proximal phalanx head. This dislocation is known as coach's finger and is commonly seen in ball-handling sports such as baseball, football, and basketball.*

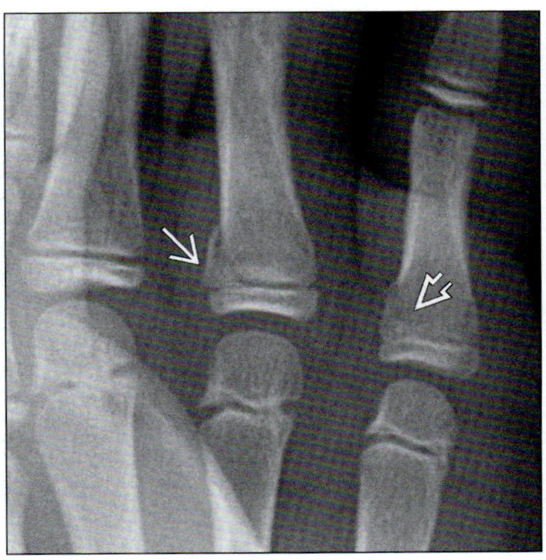

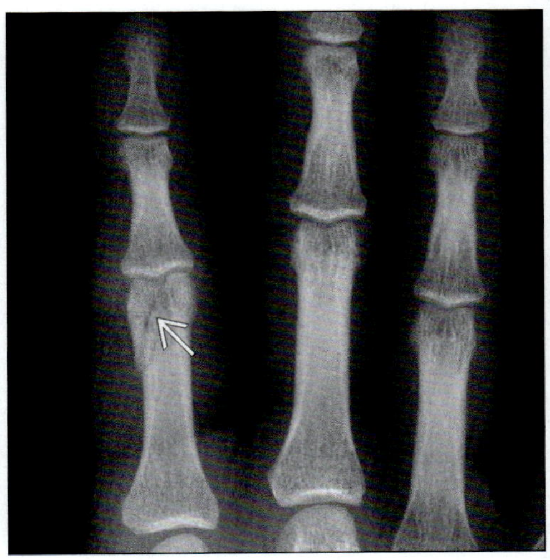

(Left) *PA radiograph shows Salter-Harris type II fractures of the 4th and 5th proximal phalanx fractures. Mild angulation typically accompanies growth plate injuries in the fingers.* ***(Right)*** *PA radiograph demonstrates an oblique unicondylar 2nd proximal phalanx head fracture. These fractures may be missed initially clinically because the patient may still be able to flex the injured finger.*

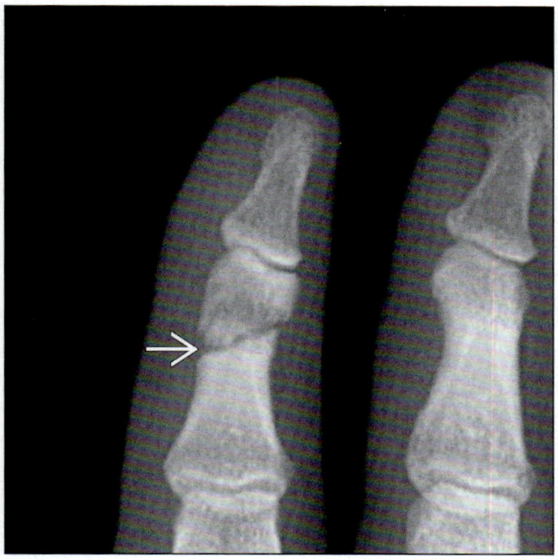
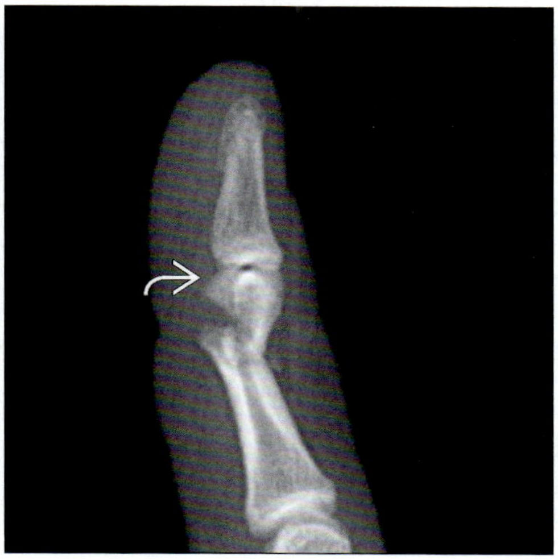

(Left) *AP radiograph of the 2nd finger after a dog bite shows an oblique fracture → of the distal shaft of the middle phalanx with mild displacement.* ***(Right)*** *Lateral radiograph from the same patient shows significant apex volar angulation of the fracture. The condyles → are intact, as is the DIP joint. The fracture is due to a focal crushing force. Bite injuries should always be treated as potential open wounds, with associated risk of sepsis.*

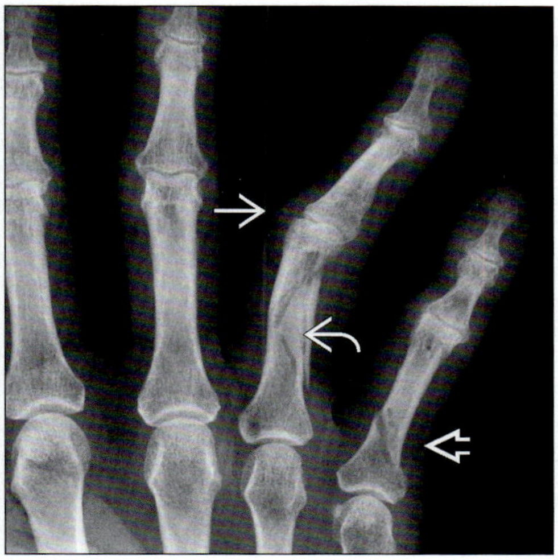
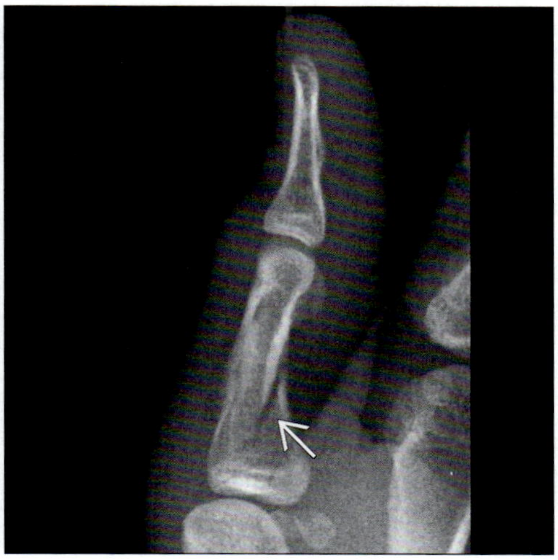

(Left) *PA radiograph demonstrates a comminuted fracture → of the 4th proximal phalanx. The 5th proximal phalanx demonstrates an oblique proximal shaft fracture →. There is a boutonniere deformity → of the 4th finger suggesting disruption of the central slip of the extensor tendon.* ***(Right)*** *Lateral radiograph demonstrates a 1st proximal phalanx fracture →. This oblique fracture results from torsional injuries and is unusual in the thumb because of the flexibility of this digit.*

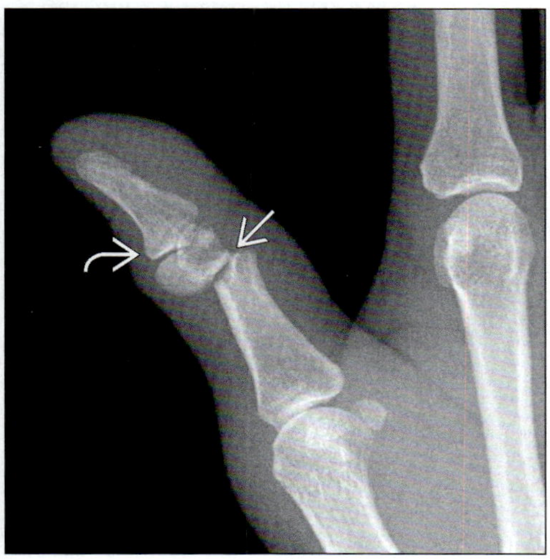
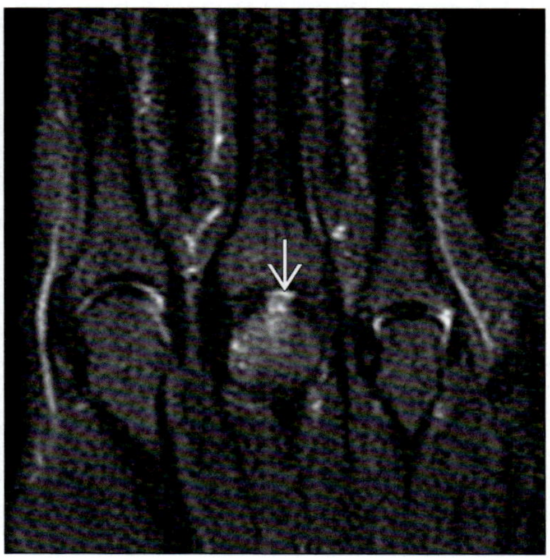

(Left) *Oblique radiograph of the thumb shows a subcondylar fracture → of the proximal phalanx with marked rotation of the condyles and distal phalanx →. Involvement of the articular surface in such cases is best gauged with CT. A CT was not obtained in this case, however.* ***(Right)*** *Coronal STIR MR shows an acute chondral fracture → of the central articular surface of the 3rd metacarpal head with underlying marrow edema in a patient who suffered trauma to the hand. Radiographs were normal.*

INTRINSIC LIGAMENT TEAR, WRIST

Key Facts

Terminology

- Intrinsic ligaments: Ligaments connecting 1 carpal bone to another
- Extrinsic ligaments: Capsular thickenings connecting carpals to metacarpals, radius, or ulna

Imaging

- Best diagnostic clues
  - Radiograph: Abnormal widening of interspace between carpals
  - Arthrography: Contrast flows between carpals from radiocarpal joint to midcarpal joint
  - MR: Complete ligament disruption
- PA radiographs
  - Scapholunate diastasis > 3 mm
  - Target or ring sign; scaphoid foreshortening if abnormally volarly flexed
- Lateral radiographs
  - Abnormal scaphoid volar flexion
- SLL: Typically tears at scaphoid attachment
  - DISI: Capitolunate angle > 30°; scapholunate angle > 60°
- LTL: Typically tears at triquetral attachment
  - VISI: Capitolunate angle > 30°; scapholunate angle < 30°

Clinical Issues

- SLL injury: Anatomic snuffbox tenderness
- LTL injury: Ulnar-sided wrist pain
- Grip weakness
- Treatment
  - Conservative: Cast or splint immobilization; nonsteroidal anti-inflammatory medication
  - Surgical: Closed reduction & internal fixation; open reduction repair/reconstruction; arthrodesis

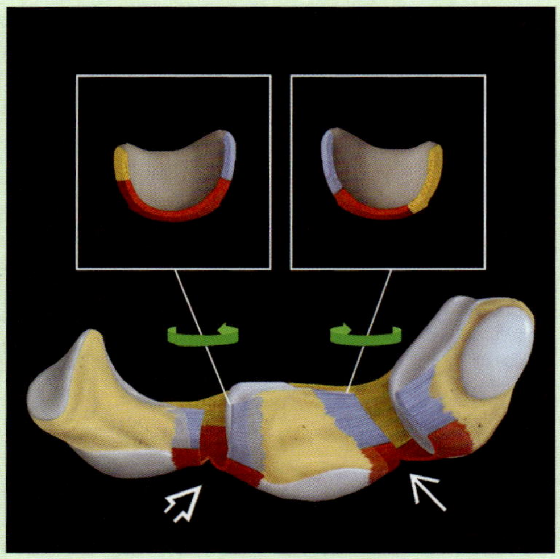
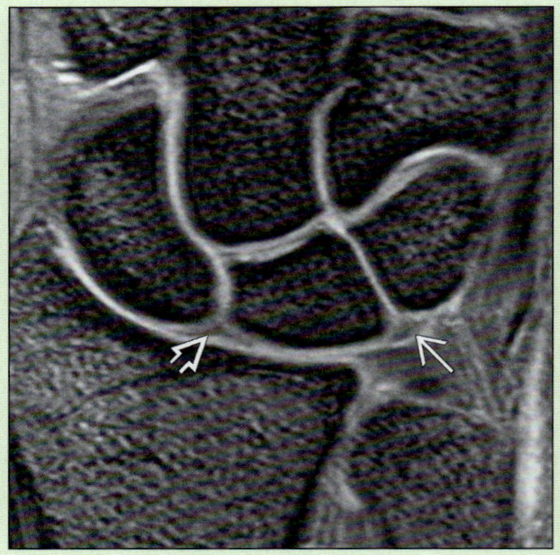

(Left) *Coronal graphic demonstrates normal anatomy of the scapholunate ligament ➔ with the dorsal (gold) portion thicker than volar (blue) portion. The lunotriquetral ligament ➔ has a volar (blue) portion, which is thicker than dorsal (gold) portion. The central membranous (red) portions of each ligament may perforate without symptoms.* ***(Right)*** *Coronal GRE MR delineates the normal triangular appearance of the intact membranous portions of the scapholunate ➔ and lunotriquetral ➔ ligaments.*

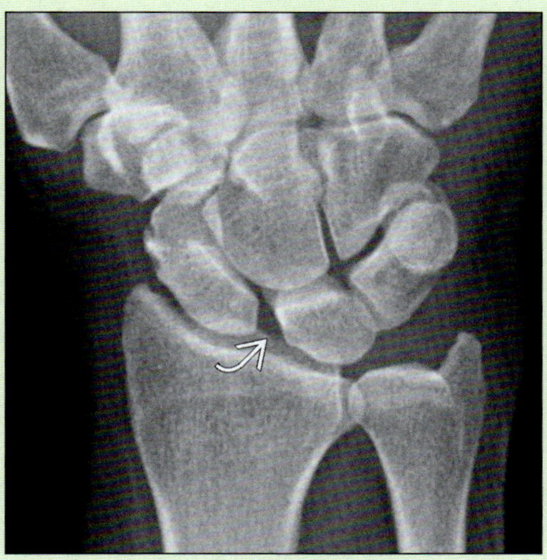
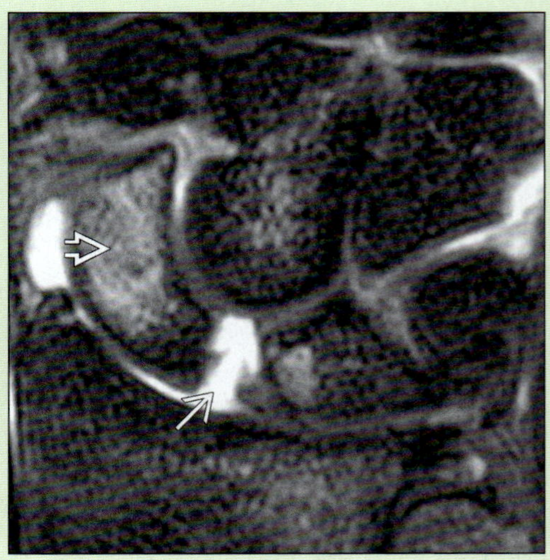

(Left) *PA radiograph reveals subtle widening ➔ of the scapholunate interval, often the result of complete disruption of the scapholunate ligament.* ***(Right)*** *Coronal T2WI FS MR shows marked widening of the scapholunate interval with a broad full-thickness tear ➔ of the midportion of the SLL. There is marrow edema ➔ in the scaphoid as a result of acute trauma.*

INTRINSIC LIGAMENT TEAR, WRIST

TERMINOLOGY

Synonyms

- Terry Thomas sign (David Letterman sign, Lauren Hutton sign): Scapholunate ligament disruption with interspace widening; mimics space between front teeth

Definitions

- Intrinsic ligaments: Ligaments connecting 1 carpal bone to another
- Extrinsic ligaments: Capsular thickenings connecting carpals to metacarpals, radius, or ulna
- Intercarpal ligament disruption
  - Most commonly relates to scapholunate ligament (SLL) and lunotriquetral ligament (LTL)

IMAGING

General Features

- Best diagnostic clue
  - Radiograph: Abnormal widening of interspace between carpals
  - Arthrography: Contrast flows between carpals from radiocarpal joint to midcarpal joint
  - MR: Complete ligament disruption
- Location
  - Scapholunate or lunotriquetral ligament

Radiographic Findings

- Posteroanterior view
  - Scapholunate diastasis > 3 mm
  - Target or ring sign; scaphoid foreshortening if abnormally volarly flexed
  - Lunate may have triangular appearance
  - Step-off at lunotriquetral proximal articular surface
- Lateral view
  - Abnormal scaphoid volar flexion
  - Radioscaphoid angle > 80°
  - Dorsal intercalated segmental instability (DISI): Scapholunate angle > 60°
  - Volar intercalated segmental instability (VISI): Scapholunate angle < 30°
- Clenched fist views: Accentuates joint space widening

Fluoroscopic Findings

- Allows dynamic evaluation of wrist motion
- Arthrography
  - Flow of contrast through scapholunate or lunotriquetral interspace
    - From radiocarpal to midcarpal compartment or vice versa

CT Findings

- Findings similar to radiography
- CT arthrography
  - Sensitivity: 80%, specificity: 100% for intrinsic ligament tear

MR Findings

- Ligament disruption
  - Ligament not visualized
  - Ligament fraying, thinning, or irregularity
  - Fluid traverses the ligament (communicating defect)
  - Widening of carpal interspace
  - Secondary cartilage loss
  - Synovitis
  - Soft tissue ganglion often seen in conjunction with SLL tear
  - Partial or complete tear seen on MR can be occult on fluoroscopic arthrogram
    - Delayed penetration of contrast into defect
    - Small tear not seen on fluoroscopy
- SLL: Typically tears at scaphoid attachment
  - Membranous portion may have degenerative perforation while dorsal & ulnar components intact
- LTL: Typically tears at triquetral attachment
  - Associated abnormalities: DRUJ instability; ulnocarpal abutment
  - Membranous portion may have degenerative perforation while dorsal & ulnar components intact
- Carpal instability
  - SL dissociation requires both intrinsic SLL and extrinsic radioscaphocapitate ligament tears
  - SLL tear may result in DISI
    - Capitolunate angle > 30°; scapholunate angle > 60°
  - LTL tear may result in VISI
    - Capitolunate angle > 30°; scapholunate angle < 30°
  - Beware of wrist positioning when evaluating alignment
    - Radiographs are more accurate than MR for assessment of alignment
- MR without contrast
  - Sensitivities and specificities range widely
  - SLL tear: Sensitivity ~ 92 %; specificity ~ 88%
  - LTL tear: Sensitivity ~ 84%; specificity ~ 75%
  - MR arthrography improves both sensitivity and specificity

Ultrasonographic Findings

- SLL: Dorsal ligament best seen
  - Look for ligament disruption; adjacent soft tissue ganglia
- LTL: Not well visualized by ultrasound

Imaging Recommendations

- Best imaging tool
  - MR arthrography
- Protocol advice
  - Radiography: PA, lateral, & oblique
    - Clenched fist views with radial and ulnar deviation
  - Arthrography
    - Volumes: Radiocarpal 2-3 cc; midcarpal 4-5 cc
    - Injectate for radiography: Iodinated contrast (180-300 mg I/ml)
    - Injectate for CT: Iodinated contrast < 300 mg I/ml to avoid beam hardening
    - Injectate for MR: 1:200 mix of gadolinium:normal saline
      - May use 50% iodinated contrast in saline instead of pure saline to allow for better fluoroscopic arthrogram
  - MR performed with small FOV (8-10 cm); dedicated wrist coil
    - Coronal & axial planes: T1, PDWI FS, ± GRE
    - Sagittal plane to evaluate alignment

- MR arthrography: T1WI FS in 3 planes
- Ultrasound: 12-14 MHz transducer; highly operator-dependent

PATHOLOGY

General Features

- Etiology
  - SLL injury
    - Excessive wrist extension and ulnar deviation
    - May result from fall on outstretched hand
  - LTL injury
    - Fall on dorsiflexed wrist with forearm forced into pronation
    - Sustained volar flexion or extreme dorsiflexion (e.g., weightlifters and gymnasts)
- Relevant anatomy
  - SLL: U-shaped ligament in cross section
    - Thicker dorsal transverse fibers; best seen on axial MR
    - Thinner volar transverse fibers; best seen on axial MR
    - Triangular, intermediate signal fibrocartilaginous membrane centrally; best seen on coronal MR
  - LTL: U-shaped ligament
    - Thicker volar transverse fibers
    - Thinner dorsal transverse fibers
    - Thin, intermediate signal fibrocartilaginous membrane centrally
  - Ligament function
    - Balance volar flexion tendency of scaphoid on lunate
    - Balance dorsiflexion tendency of triquetrum on lunate
    - Change in balance increases load on adjacent structures
  - Extrinsic ligaments provide gross stability
  - Intrinsic ligaments provide "fine tuning" of stability

Staging, Grading, & Classification

- SLL injury classification
  - Acute (< 6 weeks)
    - Stable: Partial tear
    - Unstable: Complete tear; instability may be dynamic or static
  - Chronic (> 6 weeks)
    - Stable, unstable, or fixed
    - Pancarpal arthropathy
    - Scapholunate advanced collapse (SLAC) wrist
- Perilunate instability
  - Stage I: SLL tear with SL dissociation ± rotary subluxation of scaphoid; opening of space of Poirier
  - Stage II: Capitate dorsal dislocation + dissociation of lunocapitate joint (perilunate dislocation)
  - Stage III: LTL fails; lunate separated from triquetrum (midcarpal dislocation)
  - Stage IV: Capitate migrates proximally; lunate is completely unrestrained & dislocates volarly

CLINICAL ISSUES

Presentation

- Most common signs/symptoms
  - Wrist pain ± swelling
  - SLL injury: Anatomic snuffbox tenderness
    - Scaphoid shift test (Watson test) results in pain on medial-to-lateral deviation
    - Audible lateral wrist click ± pain
  - LTL injury: Ulnar-sided wrist pain
    - Ulnar side compression test, lunotriquetral ballottement test
    - Painful medial wrist click
  - Grip weakness

Demographics

- Epidemiology
  - Associated with sports injuries (typically low-energy injury)
  - SLL tear: Most common intrinsic wrist ligament tear
  - LTL tear: Less common than SLL tear
    - Present in 13% of individuals > 40 years old

Natural History & Prognosis

- Scapholunate dissociation
- Carpal instability
  - Dorsal intercalated segment instability (DISI)
  - Volar intercalated segment instability (VISI)
  - Scapholunate advanced collapse (SLAC) wrist
- Degenerative wrist arthropathy

Treatment

- Conservative
  - Cast or splint immobilization
  - Nonsteroidal anti-inflammatories (NSAIDs)
- Surgical
  - Acute
    - Closed reduction with internal fixation
    - Open reduction; 1° repair/reconstruction
  - Chronic
    - Open reduction; reconstruction
    - Arthrodesis

DIAGNOSTIC CHECKLIST

Consider

- If intrinsic ligament tear found, examine extrinsic ligaments for accompanying abnormalities
- Attritional fenestrations may be delineated with arthrography ± multiplanar imaging
- MR may not be reliable for carpal angle measurement due to wrist positioning artifact

Image Interpretation Pearls

- Ligaments attach at cartilage rather than bone; transition may mimic a tear

SELECTED REFERENCES

1. De Filippo M et al: MDCT arthrography of the wrist: diagnostic accuracy and indications. Eur J Radiol. 74(1):221-5, 2010
2. Lisle DA et al: MR imaging of traumatic and overuse injuries of the wrist and hand in athletes. Magn Reson Imaging Clin N Am. 17(4):639-54, vi, 2009
3. Watanabe A et al: Ulnar-sided wrist pain. II. Clinical imaging and treatment. Skeletal Radiol. Epub ahead of print, 2009

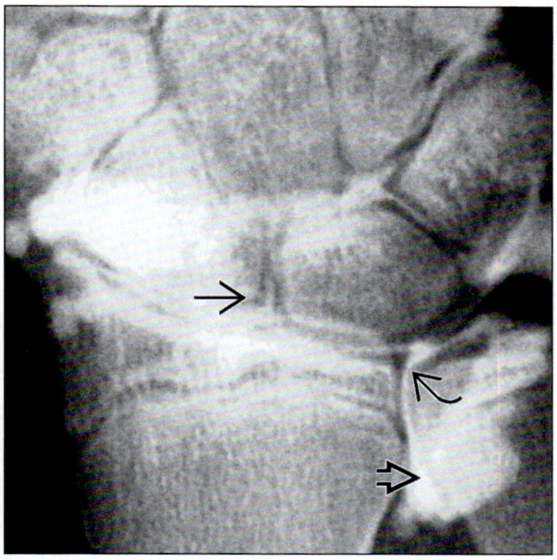

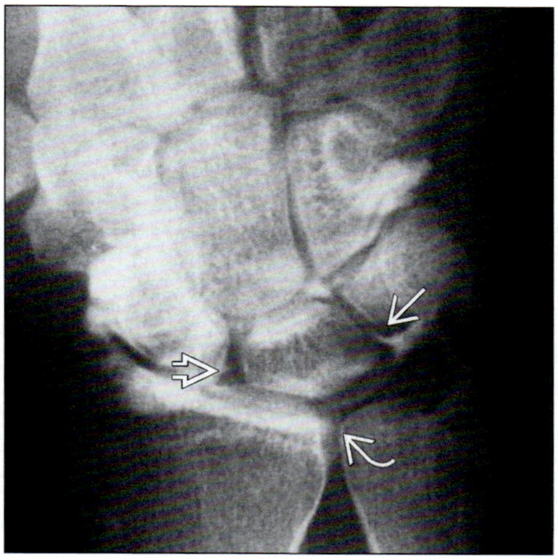

*(**Left**) PA arthrogram demonstrates a tear of the SLL after a fall on outstretched hand. A TFC tear is also present, with contrast in the distal radioulnar joint. (**Right**) PA arthrogram after radiocarpal injection demonstrates contrast extending through the lunotriquetral interval into the midcarpal compartment in this patient with ulnar-sided wrist pain. This confirms a communicating defect of the LTL. The scapholunate ligament and triangular fibrocartilage are intact.*

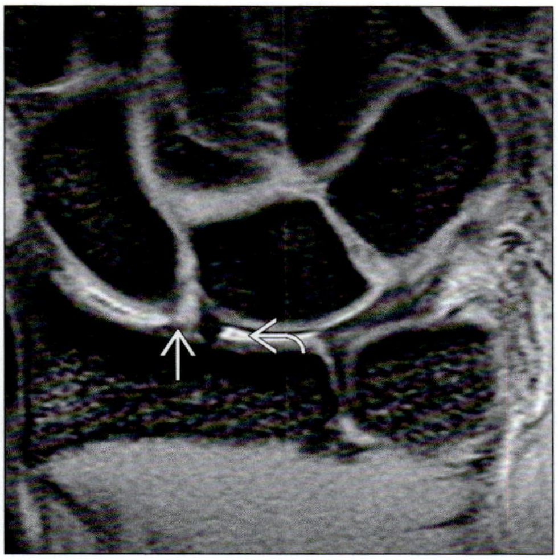

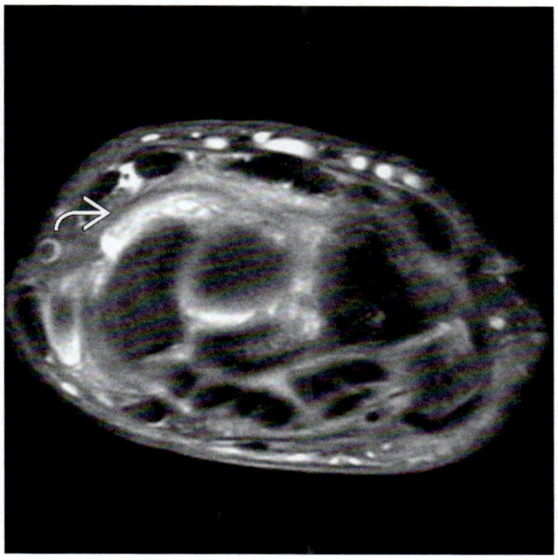

*(**Left**) Coronal GRE MR demonstrates tear of the scaphoid attachment of the SLL. Fluid from the radiocarpal joint flows through the resultant defect between the scaphoid and lunate. (**Right**) Axial T2WI FS MR from the same patient shows a slightly heterogeneous crescentic fluid collection along the dorsal lateral wrist, representing a soft tissue ganglion interdigitating with the dorsal extrinsic ligaments. Ganglia are often seen in association with intrinsic ligament tear.*

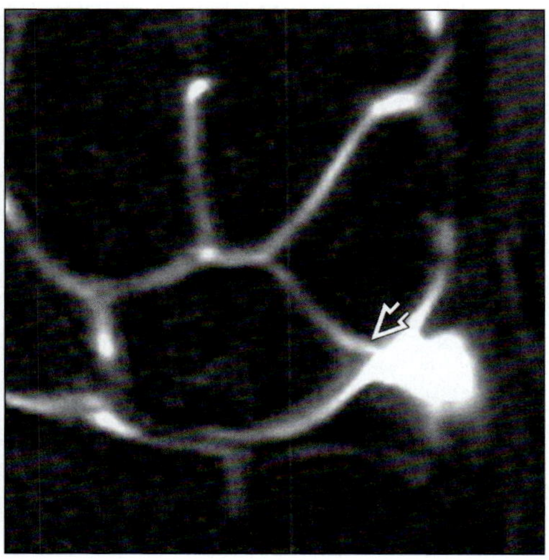

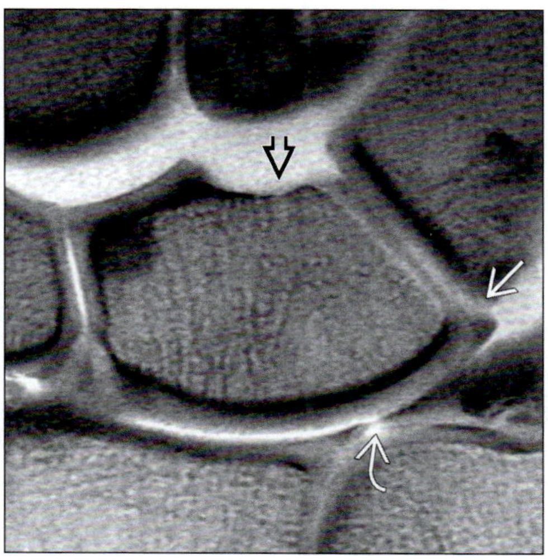

*(**Left**) Coronal MR arthrogram T1WI FS reveals a complete tear of the lunotriquetral ligament with flow of intraarticular contrast through the lunotriquetral interval. The tear is located at the triquetral portion of the ligament. (**Right**) Coronal MR arthrogram T1WI shows a tear of the LTL, with contrast crossing from the radiocarpal joint into the midcarpal joint. There is an ulnar positive variance and a TFC tear. A medial lunate facet with chondromalacia is also present.*

*(**Left**) Coronal MR arthrogram T1WI FS after radiocarpal joint arthrogram demonstrates a communicating defect ➡ in the lunotriquetral ligament. Acquisition using a knee coil with field-of-view of 12 cm results in poor signal-to-noise ratio. (**Right**) Coronal MR arthrogram T1WI using a field-of-view of 8 cm and a dedicated wrist coil shows a small ulnar-sided LTL tear ➡. Note significant improvement in image signal-to-noise ratio compared to the previous image.*

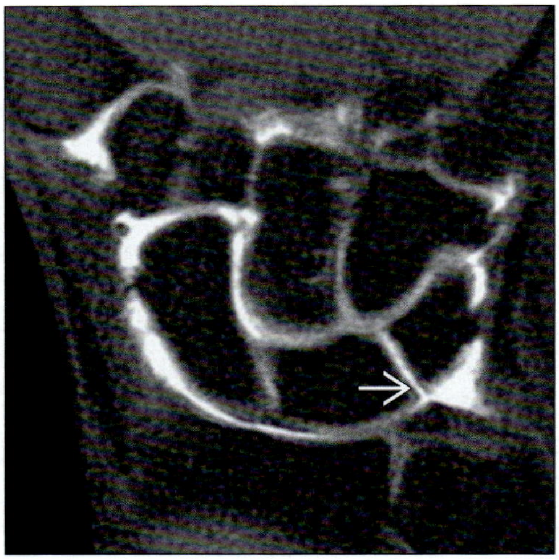

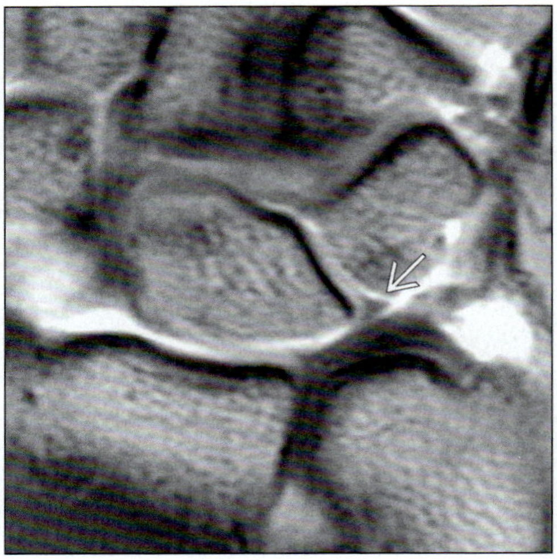

*(**Left**) PA arthrogram after radiocarpal joint injection shows contrast in the scapholunate interval ➡ extending into the midcarpal compartment consistent with a tear of the SLL. The size and location of the tear are not well determined, however. (**Right**) Sagittal MR arthrogram T1WI FS from the same patient shows that the SLL tear ➡ is at the scaphoid attachment of the ligament. The degree of involvement of the dorsal and ulnar bands and central substance is poorly delineated.*

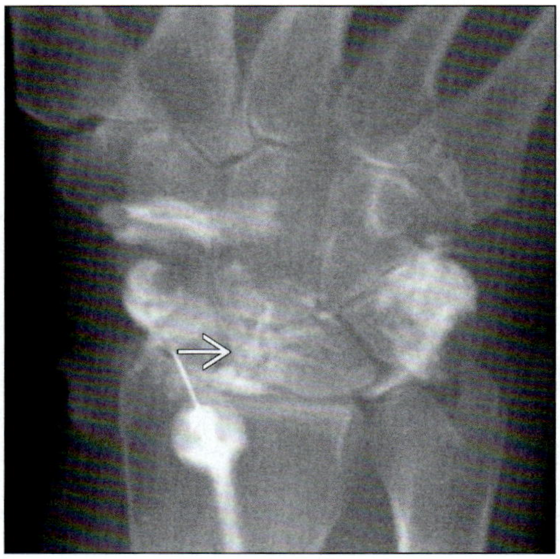

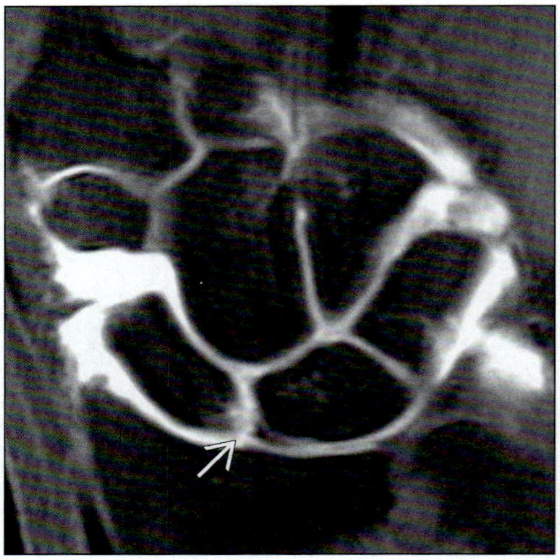

*(**Left**) Axial MR arthrogram T1WI FS from the same patient shows that the dorsal band of the scapholunate ligament ⇨ is the site of tear. The volar band ➡ is intact. Thin axial images can help delineate more precisely the nature of intrinsic ligament disruption. (**Right**) Coronal MR arthrogram T1WI FS shows a central perforation ➡ of the scapholunate ligament with intact scaphoid and lunate attachments. There is a central perforation ➡ in the triangular fibrocartilage as well.*

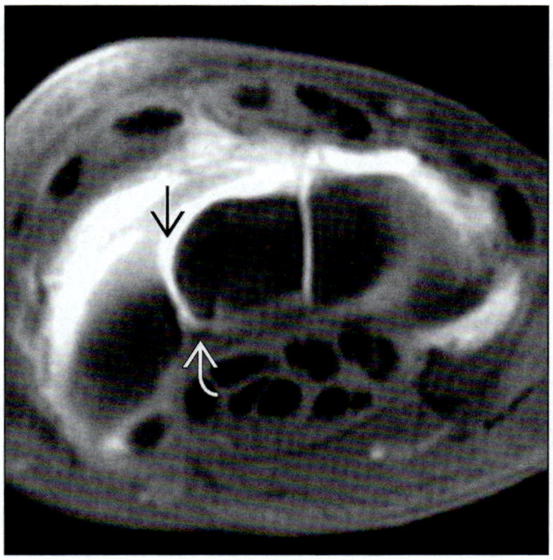

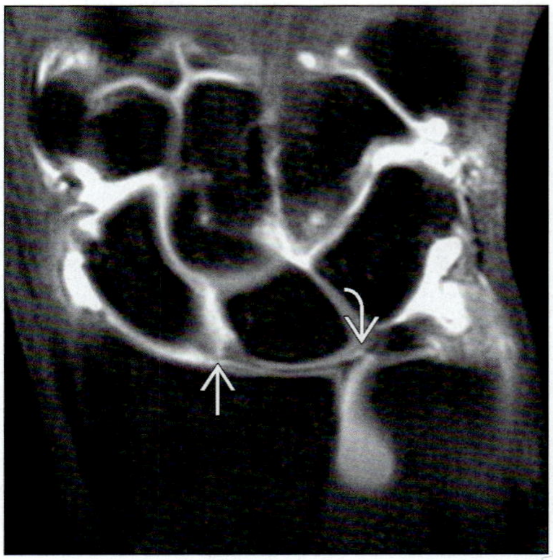

INTRINSIC LIGAMENT TEAR, WRIST

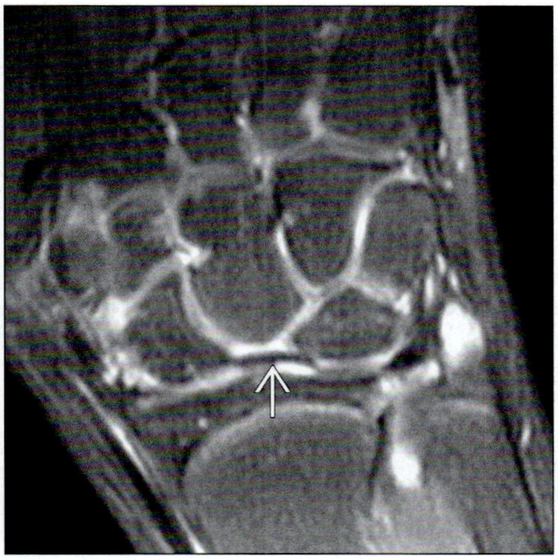

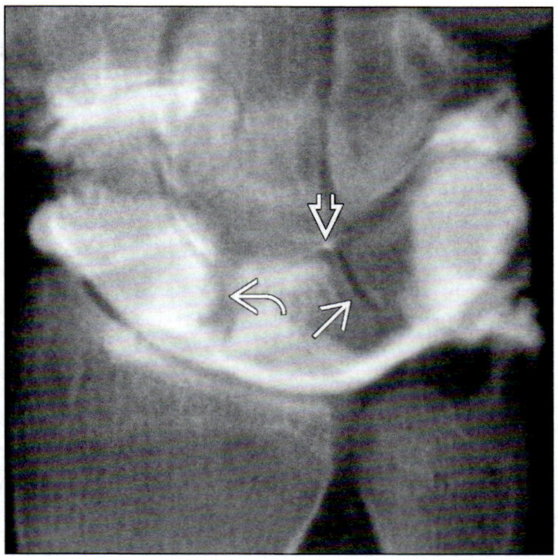

(Left) *Coronal T2WI FS MR from an 11 year old after a fall. Radiographs showed a widened SL interval. MR shows an intact but elongated SLL. Widening of an intercarpal space, particularly in children, does not always imply a ligament tear.* ***(Right)*** *PA arthrogram after injection of the radiocarpal joint demonstrates contrast in the midcarpal compartment; however, no contrast is seen in the scapholunate or lunotriquetral intervals to suggest the point of communication.*

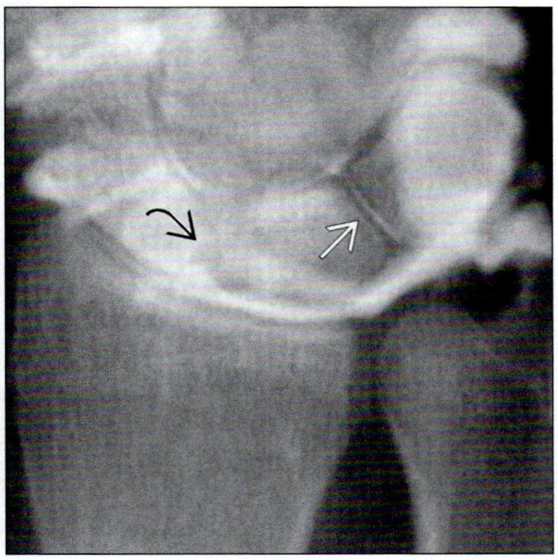

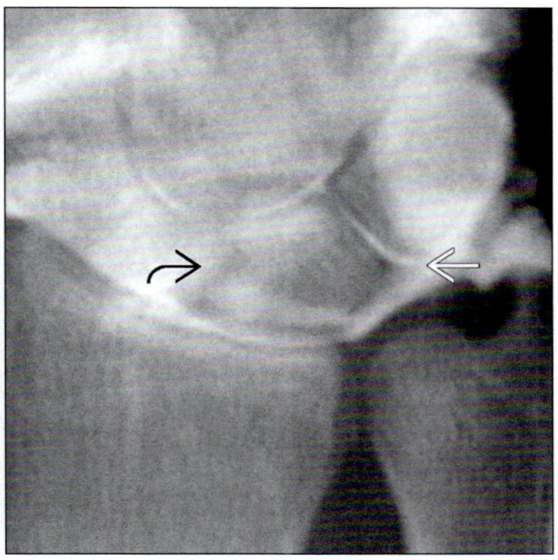

(Left) *PA arthrogram from the same patient after clenched-fist maneuver shows contrast filling the lunotriquetral interval but not the scapholunate interval, indicating a likely LTL tear.* ***(Right)*** *PA arthrogram from the same patient after radial and ulnar deviation shows the communicating LTL lesion. There is no contrast in the scapholunate interval, but as the midcarpal compartment fills, this space could fill retrograde without a SLL tear.*

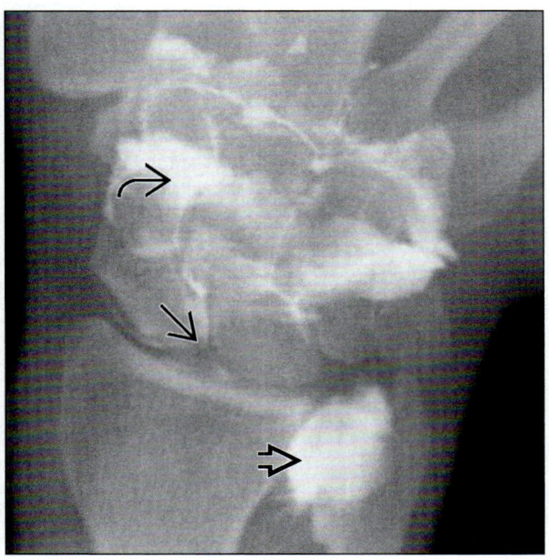

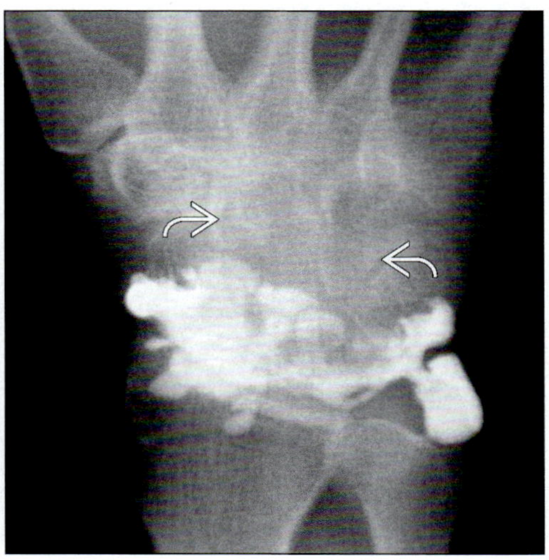

(Left) *PA arthrogram after distal radioulnar joint and midcarpal joint arthrography demonstrates a small amount of contrast entering the radiocarpal joint through a defect in the scapholunate ligament.* ***(Right)*** *PA arthrogram of the radiocarpal joint was obtained 3 hours later in the same patient. There is no contrast in the midcarpal compartment; a defect in the scapholunate ligament is not seen. Some ligament tears are unidirectional on arthrography.*

TRIANGULAR FIBROCARTILAGE COMPLEX INJURY

Key Facts

Imaging

- TFC tear
  - Partial: Most commonly located proximally
  - Complete: Full thickness disruption of TFC, typically along lateral margin
- Associated abnormalities
  - Cartilage thinning, loss along ulnar head, proximal lunate, proximal triquetrum
  - Degenerative subchondral reactive marrow edema, cysts
  - Lunotriquetral ligament complete disruption

Pathology

- Palmer classification
- Class 1: Traumatic
  - A: Central perforation
  - B: Ulnar avulsion
  - C: Distal avulsion (from carpus)
  - D: Radial avulsion
- Class 2: Degenerative (ulnocarpal abutment syndrome)
  - A: TFCC wear without perforation
  - B: TFCC wear + lunate &/or ulnar head chondromalacia
  - C: TFCC perforation + lunate &/or ulnar head chondromalacia
  - D: Class C + lunotriquetral ligament perforation
  - E: Class D + ulnocarpal arthropathy
- Palmar classification suggests lesions occur in isolation when, in fact, they probably occur in combination

Clinical Issues

- Conservative: Splinting, immobilization; anti-inflammatory medication
- Surgical: Repair, reattachment, ulnar shortening

__(Left)__ Coronal graphic demonstrates normal anatomy of the TFC complex. The triangular fibrocartilage ➔ arises from the radius and inserts along the ulnar styloid with dorsal ➔ and volar ➔ radioulnar ligaments. The ulnotriquetral ➔ and ulnocarpal ➔ ligaments are also components of the complex. The lunotriquetral ligament is shown as well ➔. __(Right)__ Coronal graphic demonstrates a tear of the central TFC ➔. This is a Palmer class 1A traumatic tear.

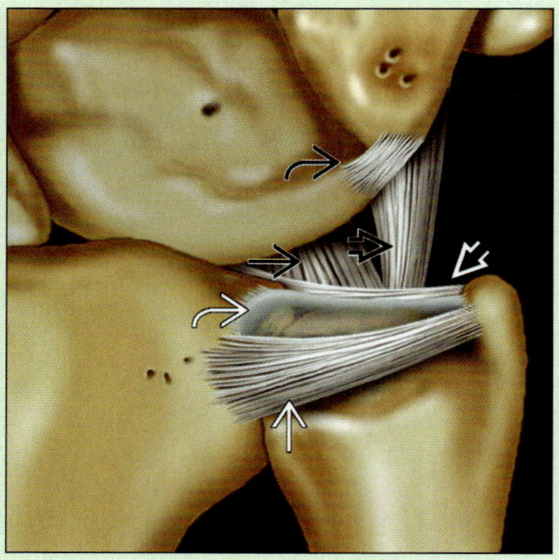

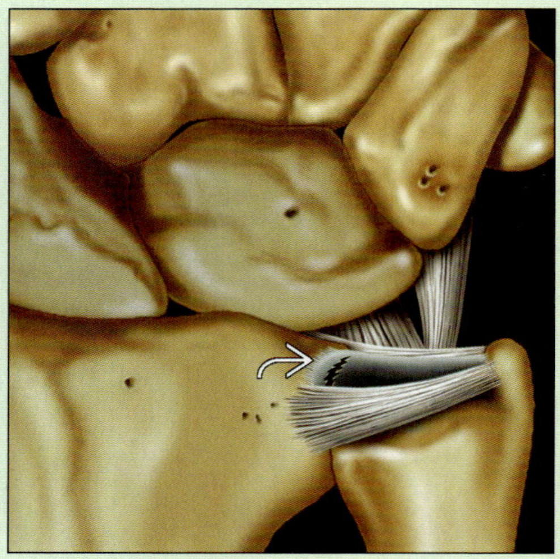

__(Left)__ AP radiograph after injection of contrast into the radiocarpal joint shows contrast flowing through a tear of the triangular fibrocartilage ➔ into the distal radioulnar joint ➔. This location is the most typical for a TFC complex tear. __(Right)__ Coronal MR arthrogram T1WI FS shows a large full thickness tear of triangular fibrocartilage ➔ with contrast injected into the radiocarpal joint diffusing into the distal radioulnar joint ➔. This is a large Palmer class 1A lesion.

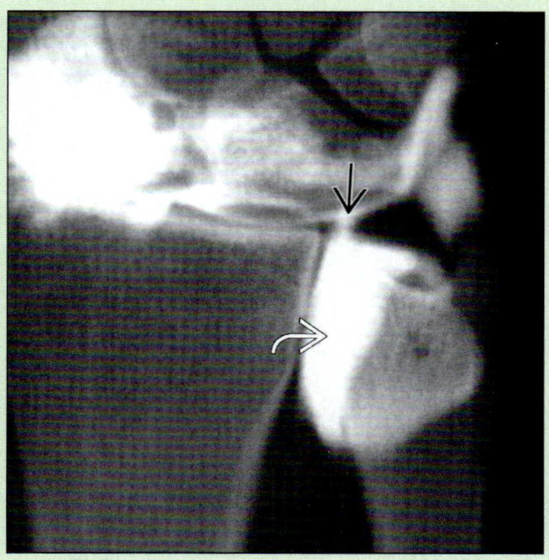

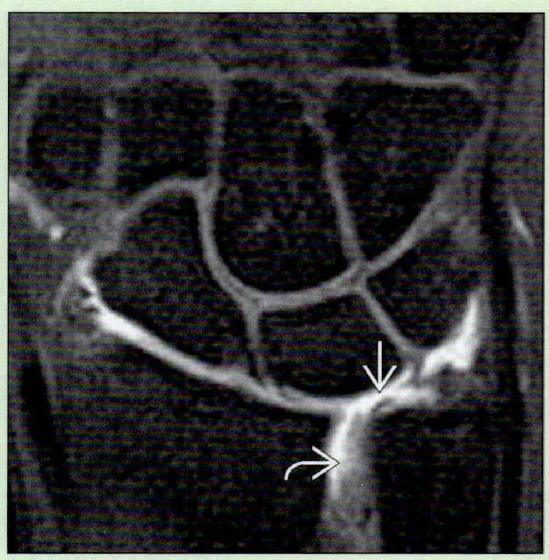

TRIANGULAR FIBROCARTILAGE COMPLEX INJURY

TERMINOLOGY

Abbreviations

- Triangular fibrocartilage (TFC)
- Triangular fibrocartilage complex (TFCC)

Definitions

- Complete TFC tear: Communicating defect between radiocarpal joint (RCJ) and distal radioulnar joint (DRUJ)
- Partial TFC tear: Noncommunicating defect

IMAGING

General Features

- Best diagnostic clue
  - Complete disruption of TFC complex
- Location
  - Tear may be located centrally or at radial or ulnar attachment
- Morphology
  - Tears may be slit-like, circular, irregular, or frayed

Radiographic Findings

- Ulnar variance: Positive more common than negative
- Step-off at lunotriquetral proximal articular surface
- Widening of DRUJ
- Can progress to ulnocarpal impaction
  - Ulnar positive variance
  - Subchondral sclerosis and cyst formation in distal ulna, proximal medial lunate, and proximal lateral triquetrum
- Ulnar styloid fracture

Fluoroscopic Findings

- Arthrography: Flow of contrast from radiocarpal joint to DRUJ or vice versa
  - Important to inject DRUJ separately as most partial tears occur along proximal surface of TFC

CT Findings

- DRUJ instability best evaluated on axial imaging in pronation, supination, and neutral positions
- CT arthrography
  - Sensitivity: 92%, specificity: 94% for TFC complex tear

MR Findings

- TFC tear
  - Partial
    - Most commonly located in proximal surface (DRUJ side)
    - Typically located along medial (ulnar) aspect of TFC
    - Irregular, frayed
  - Complete
    - Typically located along lateral (radial) aspect of TFC
    - Ulnar detachment
      - Need to distinguish from fibrovascular ligamentum subcruentum
  - Partial or complete tear seen on MR can be occult on fluoroscopic arthrogram
    - Delayed penetration of contrast into defect
    - Small tear not seen on fluoroscopy
- Associated abnormalities
  - Chondromalacia along ulnar head, proximal lunate, & proximal triquetrum
  - Degenerative subchondral reactive marrow edema & cysts
    - ↓ signal on T1WI in distal ulna, proximal lunate, & proximal triquetrum
    - Fluid-sensitive sequences: Intermediate to high signal corresponding to marrow edema, subchondral cysts
  - Lunotriquetral ligament (LTL) disruption
  - DRUJ instability
    - Volar subluxation in supination
    - Dorsal subluxation in pronation
- MR without contrast
  - Sensitivities and specificities range widely
  - MR arthrography improves sensitivity
    - Sensitivity: 96%, specificity: 100%

Ultrasonographic Findings

- Ulnar and dorsal radioulnar portions of TFC; meniscal homologue evident

Imaging Recommendations

- Best imaging tool
  - MR arthrography
- Protocol advice
  - Radiography: PA, lateral, & oblique views
  - Arthrography
    - Volumes: Radiocarpal 2-3 cc; DRUJ 1-2 cc
    - Injectate for radiography: Iodinated contrast (180-300 mg I/ml)
    - Injectate for CT: Iodinated contrast < 300 mg I/ml to avoid beam hardening
    - Injectate for MR: 1:200 mix of gadolinium:normal saline
      - May use 50% iodinated contrast in saline instead of pure saline to allow for better fluoroscopic arthrogram
  - MR performed with small FOV 8-10 mm; dedicated wrist coil
    - Coronal & axial T1, PDWI FS, ± GRE
    - MR arthrography: T1WI FS in 3 planes
  - Ultrasound: 12-14 MHz transducer; highly operator-dependent

PATHOLOGY

General Features

- Etiology
  - Traumatic injury
    - Axial loading with rotational stress
    - Axial loading with distraction of medial wrist structures
    - Fall on outstretched pronated hand
    - Injury may occur in sports, such as gymnastics, racquet sports, golf, boxing
    - Injury tends to be more common in ulnar neutral or positive variance
  - Degenerative injury
    - Repetitive loading of medial wrist

Staging, Grading, & Classification

- Palmer classification
  - Class 1: Traumatic

TRIANGULAR FIBROCARTILAGE COMPLEX INJURY

- A: Central perforation
- B: Ulnar avulsion
  - ± ulnar styloid base fracture
  - May be associated with DRUJ instability
- C: Distal avulsion (from carpus)
  - Ulnolunate or ulnotriquetral ligament avulsion
  - May be associated with carpal translocation
- D: Radial avulsion
  - ± sigmoid notch fracture

- Class 2: Degenerative (ulnocarpal abutment syndrome)
  - A: TFCC wear without perforation
  - B: TFCC wear
    - + lunate &/or ulnar head chondromalacia
  - C: TFCC perforation
    - + lunate &/or ulnar head chondromalacia
  - D: TFCC perforation
    - + lunate &/or ulnar head chondromalacia
    - + lunotriquetral ligament perforation
  - E: TFCC perforation
    - + lunate &/or ulnar head chondromalacia
    - + lunotriquetral ligament perforation
    - + ulnocarpal arthropathy
- Note: Palmar classification suggests lesions occur in isolation, but they probably occur in combination

Relevant Anatomy

- TFC complex is main stabilizer of DRUJ
- Triangular fibrocartilage complex
  - Triangular fibrocartilage disc
    - Triangular low signal structure
    - Arises from cartilage of medial distal radius
    - Relatively avascular
    - Variable thickness dependent on ulnar variance (shorter ulna → thicker TFC)
  - Volar radioulnar ligament: Volar thickening of TFC; inserts on ulnar styloid
  - Dorsal radioulnar ligament: Dorsal thickening of TFC; blends with volar radioulnar ligament to insert on ulnar styloid
    - Ligamentum subcruentum: Fibrovascular tissue interposed between radioulnar ligament and its ulnar attachment; intermediate signal
  - Ulnocarpal meniscal homologue
    - Low signal periphery with intermediate signal vascular center
    - Located medial to prestyloid recess
  - Subsheath of extensor carpi ulnaris (ECU)
  - Volar ligaments
    - Ulnolunate ligament
    - Ulnotriquetral ligament
  - Ulnar collateral ligament
    - Not a distinct ligament but a complex formed by medial ulnotriquetral ligament and ECU subsheath

CLINICAL ISSUES

Presentation

- Most common signs/symptoms
  - Ulnar-sided wrist pain
  - Focal swelling or deformity
  - Loss of DRUJ range of motion
  - Weakness or pain with supination lift test

Demographics

- Age
  - Traumatic tears more common < 40 years old
  - Degenerative tears common > 50 years old
- Epidemiology
  - Not all TFC tears are symptomatic
    - Controlled study of symptomatic wrist pain, contralateral (asymptomatic) wrist also studied
      - Complete tears: Present in 64% of symptomatic wrists, 46% of asymptomatic wrists; 69% bilateral
      - Partial tears: Present in 50% of symptomatic wrists, 27% of asymptomatic wrists; 39% bilateral

Natural History & Prognosis

- Progression of degenerative disease over time
- DRUJ instability
- Carpal translocation

Treatment

- Conservative
  - Splinting, immobilization
  - Anti-inflammatory medication
  - Rest
  - Job restriction
- Surgical
  - Repair
  - Reattachment
  - Ulnar shortening
  - Ulnar head excision
  - Limited wrist fusion
  - Complete wrist fusion

DIAGNOSTIC CHECKLIST

Consider

- Complete and partial tears may be present in asymptomatic patient population
- Asymptomatic TFC perforations common in adults > 50 years old

SELECTED REFERENCES

1. De Filippo M et al: MDCT arthrography of the wrist: diagnostic accuracy and indications. Eur J Radiol. 74(1):221-5, 2010
2. Buck FM et al: Ulnomeniscal homologue of the wrist: correlation of anatomic and MR imaging findings. Radiology. 253(3):771-9, 2009
3. Watanabe A et al: Ulnar-sided wrist pain. II. Clinical imaging and treatment. Skeletal Radiol. Epub ahead of print, 2009
4. Moser T et al: Multidetector CT arthrography of the wrist joint: how to do it. Radiographics. 28(3):787-800; quiz 911, 2008
5. Rüegger C et al: Peripheral tear of the triangular fibrocartilage: depiction with MR arthrography of the distal radioulnar joint. AJR Am J Roentgenol. 188(1):187-92, 2007
6. Zanetti M et al: Characteristics of triangular fibrocartilage defects in symptomatic and contralateral asymptomatic wrists. Radiology. 216(3):840-5, 2000
7. Palmer AK: Triangular fibrocartilage complex lesions: a classification. J Hand Surg Am. 14(4):594-606, 1989

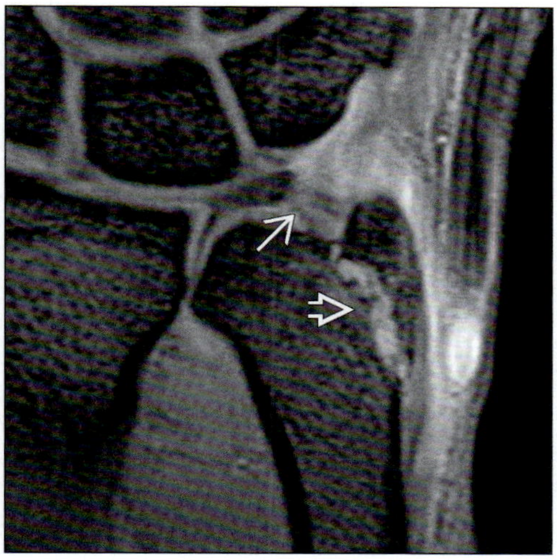
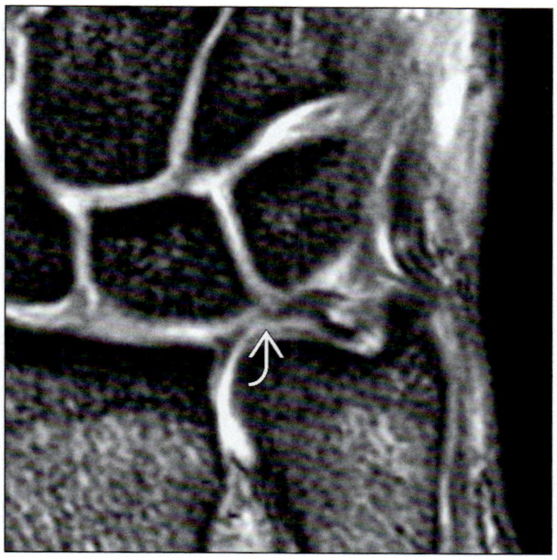

(Left) *Coronal GRE MR shows a traumatic peripheral partial TFC tear ➔. This is seen in association with an oblique ulnar styloid base fracture ➔. This is a subacute nondisplaced fracture and a Palmer class 1B ulnar avulsion tear.* ***(Right)*** *Coronal GRE MR reveals excoriation and thinning ➔ along the proximal surface of the triangular fibrocartilage without discrete perforation. This represents a degenerative Palmer class 2A lesion. Note positive ulnar variance.*

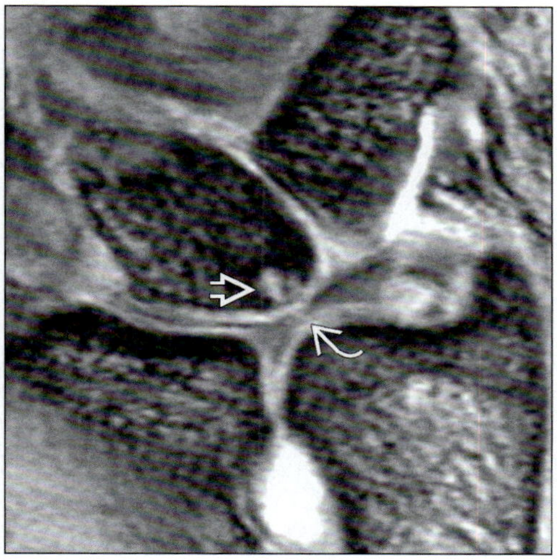
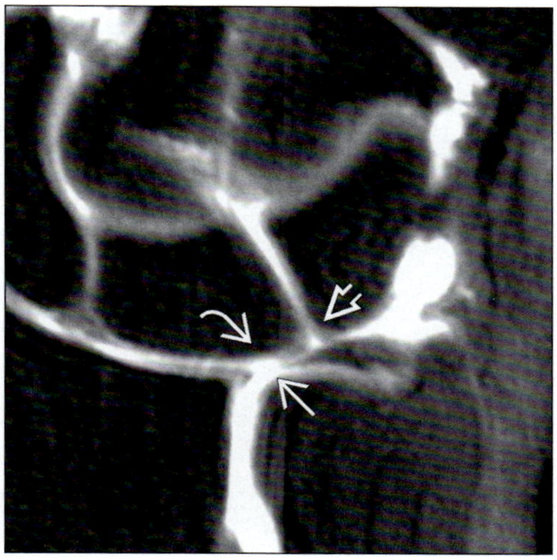

(Left) *Coronal GRE MR demonstrates thinning and irregularity of the proximal TFC ➔ with chondromalacia along the medial lunate and an associated subchondral cyst ➔. This represents a Palmer class 2B lesion.* ***(Right)*** *Coronal MR arthrogram T1WI FS reveals a full thickness tear ➔ of the central TFC. There is focal cartilage thinning along the ulnar head and adjacent lunate ➔. A lunotriquetral ligament tear ➔ is also present. This represents a Palmer class 2D injury.*

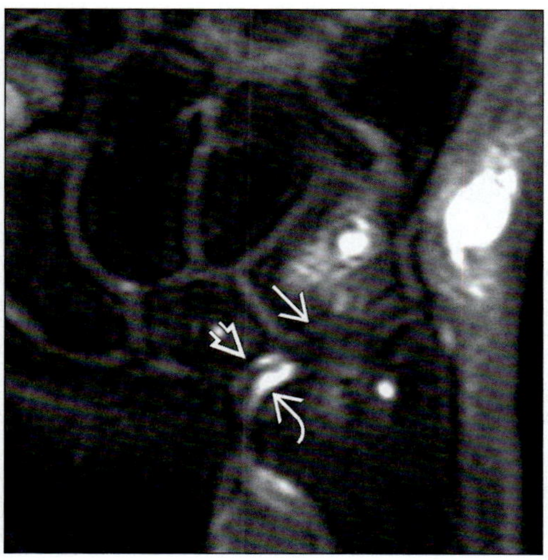
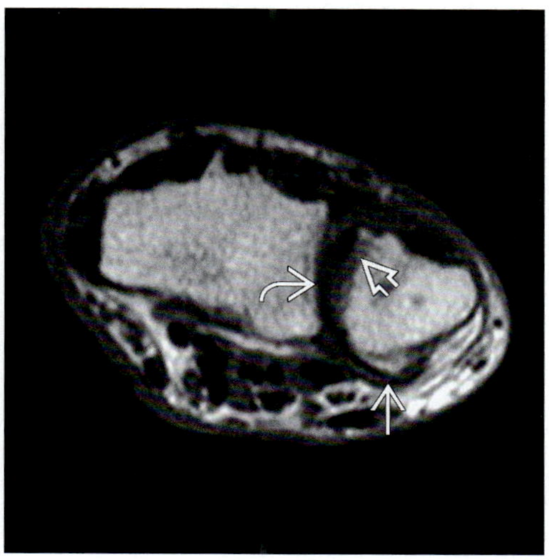

(Left) *Coronal STIR MR reveals a complete TFC tear ➔ with ulnar ➔ and triquetral ➔ cartilage loss. There was an associated lunotriquetral tear (not shown). This is a Palmer class 2E lesion.* ***(Right)*** *Axial T1WI MR from the same patient confirms this class 2E lesion with evidence of distal radioulnar joint space narrowing ➔, subchondral eburnation ➔, and osteophytic ridging. The ulna is slightly volarly subluxed ➔, consistent with volar ulnocarpal ligament insufficiency.*

TRIANGULAR FIBROCARTILAGE COMPLEX INJURY

(Left) *Coronal MR arthrogram T1WI demonstrates a partial tear → of the triangular fibrocartilage involving its distal articular surface medially. Most partial tears of the TFC involve the proximal articular surface and are therefore only well seen after DRUJ arthrography.* ***(Right)*** *Coronal MR arthrogram T1WI FS after radiocarpal joint injection shows an oblique linear full thickness tear → of the triangular fibrocartilage. The tear extends from the lateral proximal to the medial distal surface.*

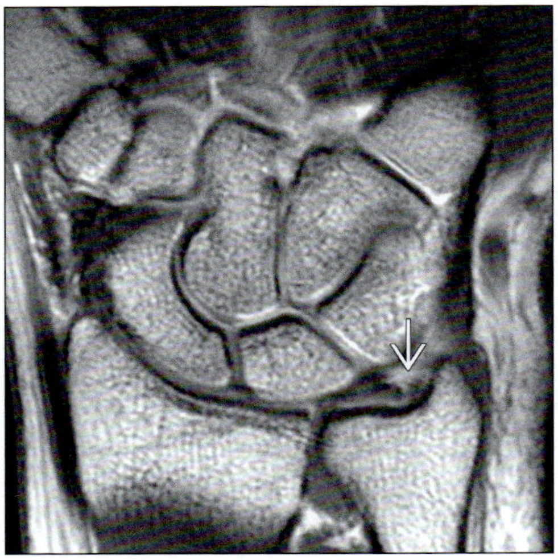

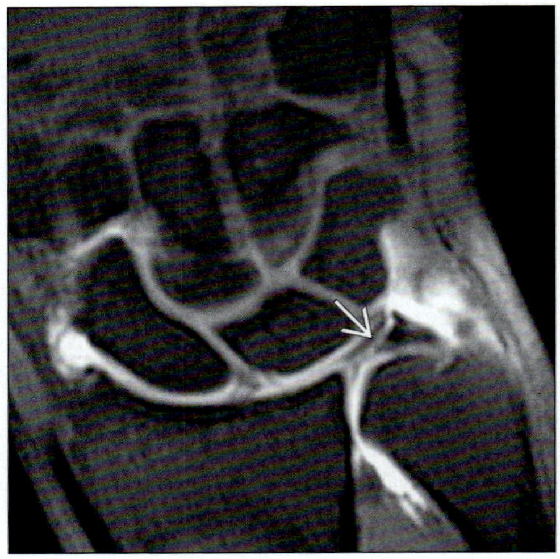

(Left) *Coronal MR arthrogram T1WI shows a communicating defect in the triangular fibrocartilage with contrast from the radiocarpal joint injection entering the distal radioulnar joint →. There is linear intermediate signal in the DRUJ → representing a displaced fragment of the TFC. There is also a small tear of the LTL →.* ***(Right)*** *Coronal MR arthrogram T2WI FS from the same patient shows the TFC fragment → with similar signal to the TFC →.*

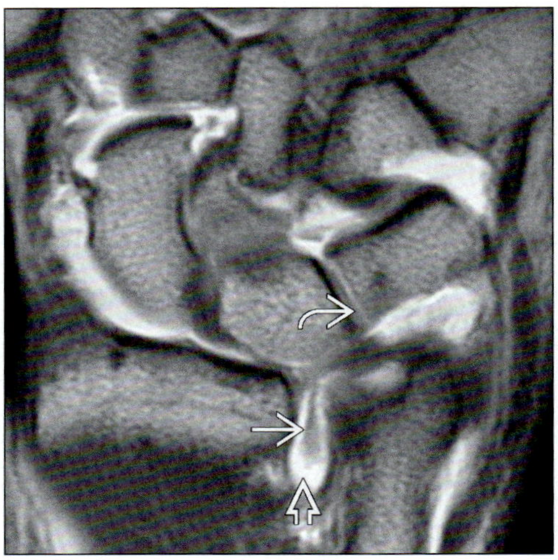

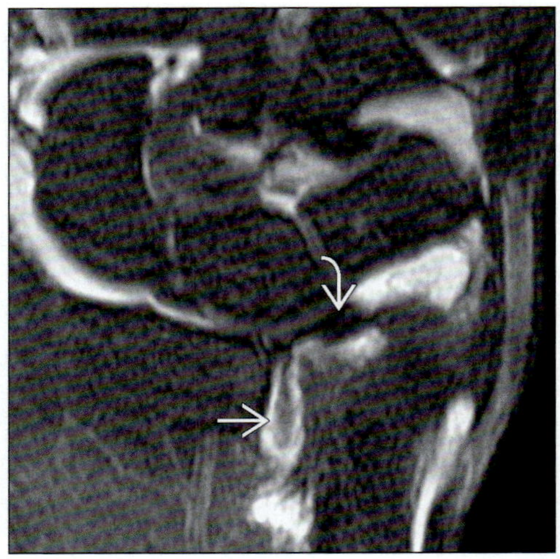

(Left) *Sagittal MR arthrogram T1WI from the same patient shows the origin of the flap fragment → from the triangular fibrocartilage →. Sagittal images can add value to the interpretation of MR arthrography.* ***(Right)*** *Axial MR arthrogram T1WI from the same patient shows the displaced fragment of fibrocartilage → filling part of the distal radioulnar joint. The patient had pain on pronation and supination of the wrist.*

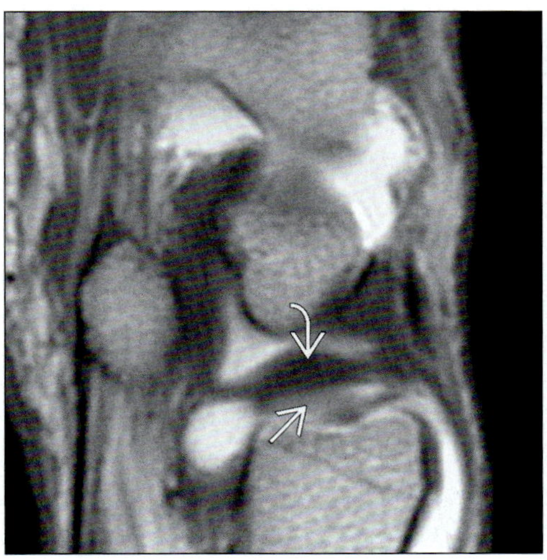

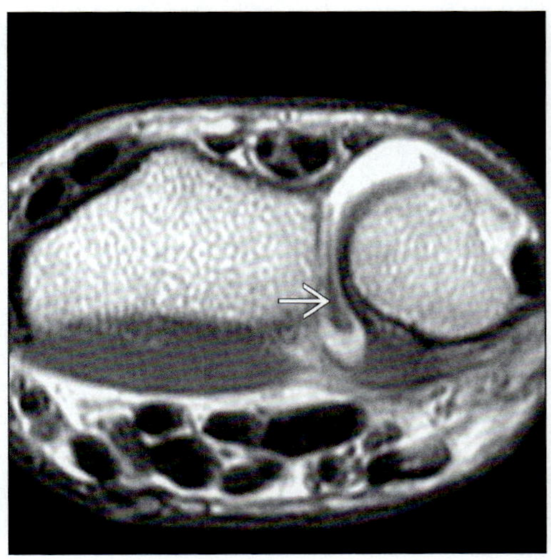

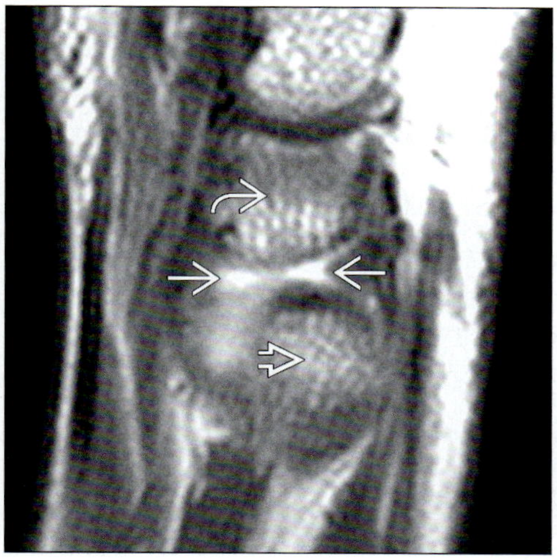

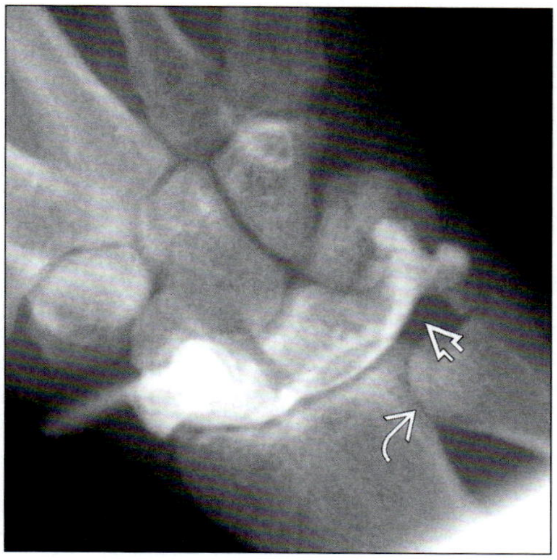

(Left) *Sagittal MR arthrogram T1WI shows a large defect in the midportion of the triangular fibrocartilage between the distal ulna and the lunate. Sagittal images help provide a complete understanding of the location and size of TFC tears.* ***(Right)*** *PA arthrogram during radiocarpal joint injection demonstrates no abnormal contrast entering the substance of the triangular fibrocartilage or in the distal radioulnar joint.*

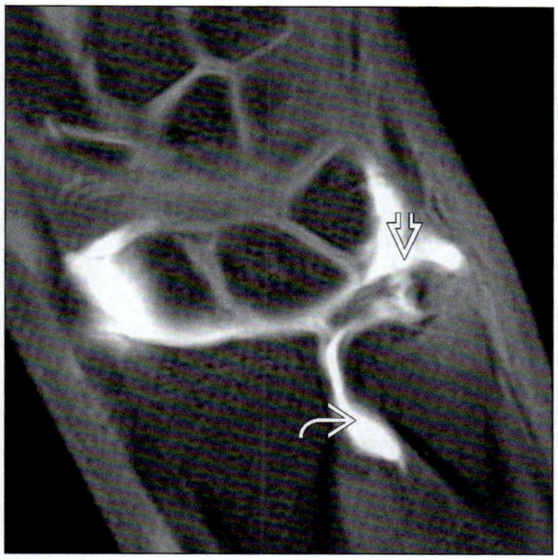

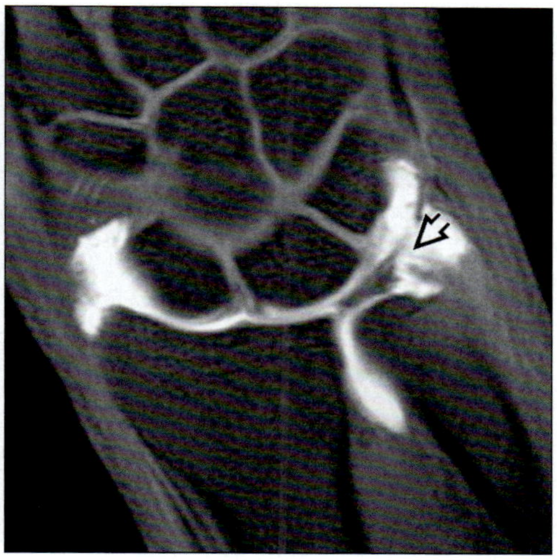

(Left) *Coronal MR arthrogram T1WI FS from the same patient shows contrast in the DRUJ and in the ulnar fibers of the TFC, but a communicating defect is not identified.* ***(Right)*** *Coronal MR arthrogram T1WI FS more dorsally from the same patient shows the triangular fibrocartilage tear at its most medial attachment. Delayed filling of the TFC and the DRUJ after arthrography can account for a "normal" fluoroscopic arthrogram and a ligament tear seen on MR arthrography.*

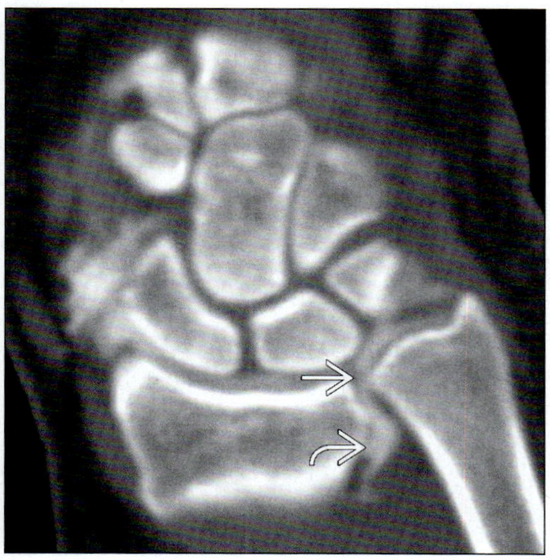

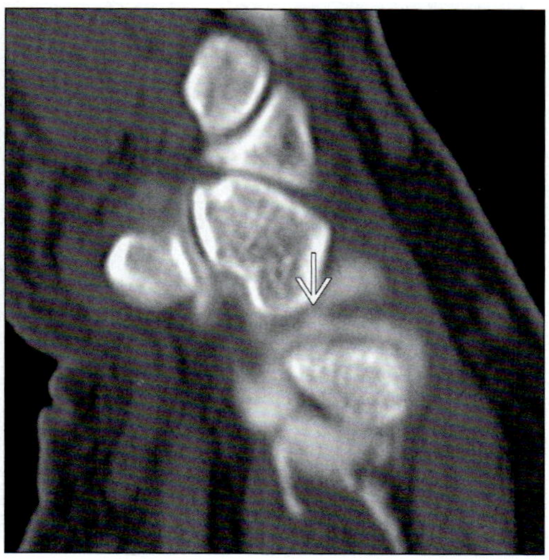

(Left) *Coronal reformat CT arthrogram after radiocarpal joint injection shows contrast extending through the radial aspect of the triangular fibrocartilage and filling the distal radioulnar joint. CT arthrography is highly accurate for TFC tear and can be used when patients are unable to undergo MR arthrography.* ***(Right)*** *Sagittal reformat CT arthrogram from the same patient shows the TFC tear in the central portion of its radial attachment.*

CARPAL INSTABILITY

Key Facts

Imaging

- Scapholunate (SLL) dissociation with rotary subluxation of scaphoid
  - Scapholunate widening > 3 mm; scapholunate angle on lateral > 60°
- DISI: Axes of radius, lunate, & capitate assume zigzag configuration
  - Results in abnormal angulations: Capitolunate angle > 30°, scapholunate angle > 60°
- VISI: Axes of radius, lunate, & capitate assume reverse zigzag configuration
  - Dynamic palmar midcarpal instability (PMCI) may progress to static instability (VISI)
  - Results in abnormal angulations: Scapholunate angle < 30° ± capitolunate angle > 30°
- CLIP: Dynamic dorsal midcarpal instability (MCI) results from radiocapitate laxity
- Carpal translocation: May occur in any direction, but medial (ulnar) most frequent
- SLAC wrist: Progressive carpal collapse
- SNAC wrist: Progressive carpal collapse with fracture nonunion

Top Differential Diagnoses

- Carpal dislocation
- Inflammatory arthritis

Clinical Issues

- Often overlooked initially as "wrist sprain" if no fracture seen at initial imaging at time of injury
- Scapholunate tear leads to most common types of carpal instabilities
- Ligament injuries resulting in instability may be either isolated or residua of prior perilunate injury
- Incidence: SLL dissociation > MCI > LTL instability

*(**Left**) Lateral graphic shows a DISI pattern. The lunate is tilted dorsally, while the capitate may become volarly angulated. The scaphoid often rotates volarly. This results in an increased capitolunate angle (> 30°) & an increased scapholunate angle (> 60°). (**Right**) Lateral radiograph shows a DISI pattern, with dorsal tilt of the lunate → & volar tilt of the scaphoid ⇨. The capitolunate angle measures > 30°, & the scapholunate angle measures 90°. Note that the bones are tilted but not subluxed or dislocated.*

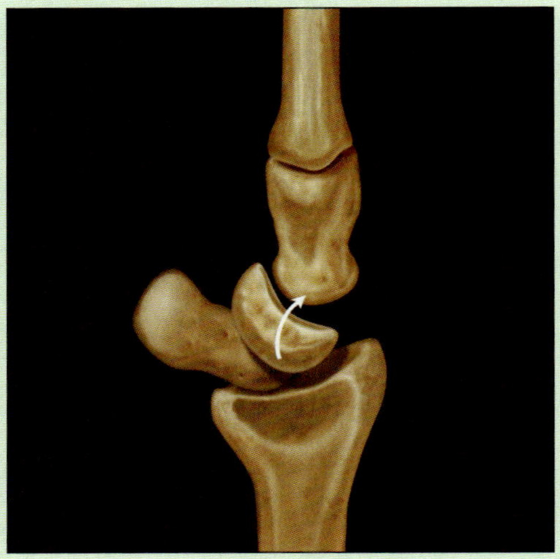

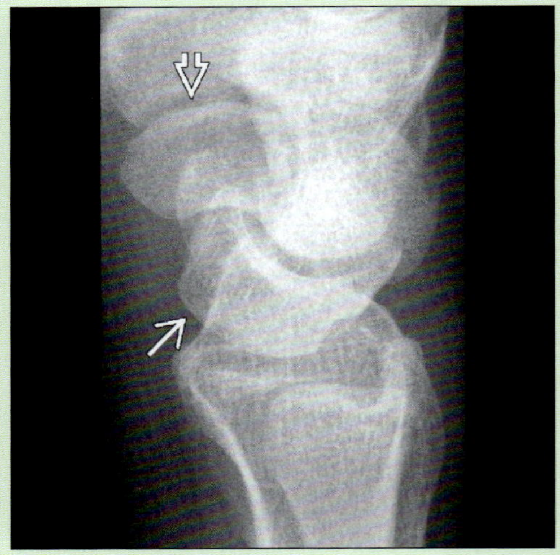

*(**Left**) Lateral graphic shows a VISI pattern. The lunate is volarly tilted (but continues to articulate with both the radius and the capitate). This results in a capitolunate angle > 30° and scapholunate angle < 30°. (**Right**) Lateral radiograph shows a VISI pattern. The lunate is tilted volarly →, resulting in a capitolunate angle that far exceeds 20°. The scapholunate angle is far reduced, approaching 0°. Note that this is an abnormality in angulation, not a dislocation.*

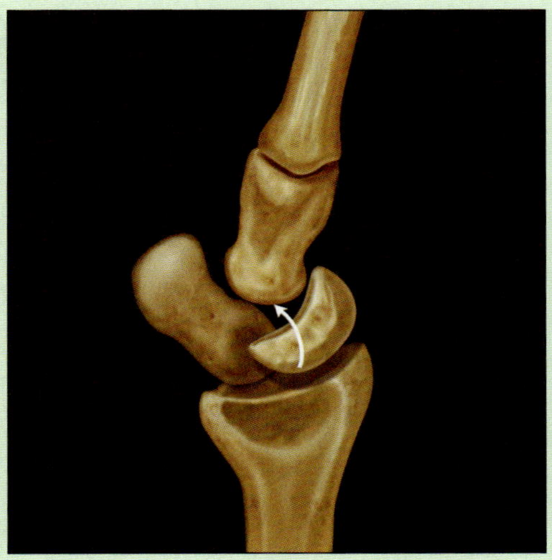

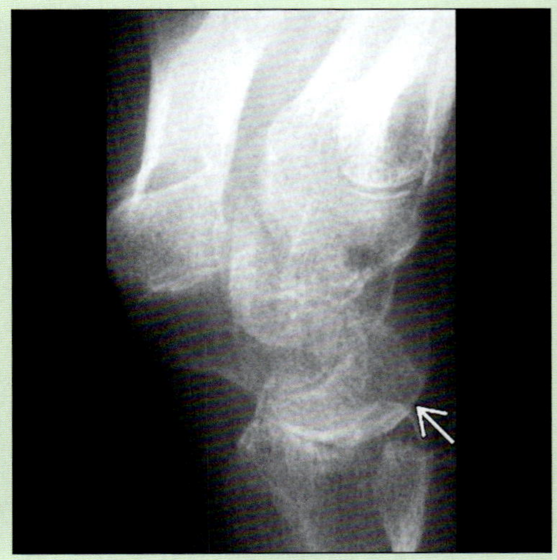

CARPAL INSTABILITY

TERMINOLOGY

Abbreviations

- Capitolunate instability pattern (CLIP)
- Dorsiflexion instability, dorsal intercalated segmental instability (DISI)
- Lunotriquetral ligament (LTL)
- Midcarpal instability (MCI)
- Palmar midcarpal instability (PMCI)
- Scapholunate advanced collapse (SLAC)
- Scapholunate ligament (SLL)
- Scaphoid nonunion advanced collapse (SNAC)
- Volar flexion instability, volar intercalated segmental instability (VISI)

Definitions

- Instability: Inability to withstand normal loading
- Static instability: Seen on routine images
- Dynamic instability: Requires motion (fluoroscopy) or force (clenched fist, deviation) to be detected
- Translocation: Shift of entire carpus from normal position relative to radius
- Dissociation: Abnormal motion between bones within same carpal row, often in association with interruption of corresponding interosseous ligament
- Intercalated segment instability: Instability between carpal rows (particular focus on alignment between lunate & capitate)

IMAGING

General Features

- Best diagnostic clue
  - Abnormal carpal alignment (static or dynamic) with associated intrinsic &/or extrinsic ligament injuries
- Morphology
  - Scapholunate dissociation with rotary subluxation of scaphoid
    - Scapholunate widening > 3 mm
    - Scapholunate angle on lateral > 60°
    - Disruption of scapholunate ligament with incompetent radioscaphocapitate ligament
  - DISI: Axes of radius, lunate, & capitate assume zigzag configuration
    - Dorsal angulation (dorsiflexion) of lunate relative to radius
    - Volar angulation of capitate relative to lunate
    - Rotary subluxation of scaphoid frequent
    - Lateral ligament abnormalities including SLL ± dorsal intercarpal ligament
    - Results in abnormal angulations: Capitolunate angle > 30°, scapholunate angle > 60°
  - VISI: Axes of radius, lunate, & capitate assume reverse zigzag configuration
    - Volar angulation (volar flexion) of lunate relative to radius
    - Dorsal angulation of capitate relative to lunate
    - Medial ligament abnormalities including LTL ± medial arcuate (triquetrohamatocapitate), dorsal radiocarpal, or volar radioscaphocapitate
    - Dynamic instability (PMCI) may progress to static instability (VISI)
    - Results in abnormal angulations: Scapholunate angle < 30° ± capitolunate angle > 30°
  - CLIP: Dynamic dorsal MCI resulting from radiocapitate laxity
    - May be seen with radial styloid fracture malunion
  - Triquetrohamate instability: Dynamic instability of medial wrist with pain & reproducible click
  - Carpal translocation: May occur in any direction, but medial (ulnar) most frequent
    - Radioscaphoid space widened; lunate translates > 50% medially from lunate fossa
    - Scapholunate, capitolunate angles normal
  - SLAC wrist: Progressive carpal collapse
    - Begins with radioscaphoid joint space narrowing
    - Progresses to arthrosis in triscaphe, capitolunate, and scaphocapitate joints
    - With traumatic and ligament laxity etiologies, radiolunate joint typically **not** involved
    - Seen in concert with scapholunate diastasis
  - SNAC wrist: Progressive carpal collapse with fracture nonunion
    - Scaphoid fracture nonunion
    - Begins with radial styloid beaking
    - Progresses to arthrosis in distal radioscaphoid joint, midcarpal joints
    - Minimal arthrosis of proximal scaphoid pole fragment-radius, radiolunate articulations

Radiographic Findings

- Posteroanterior view: Signs of instability
  - Loss of scapholunate articular parallelism; lunate may assume triangular appearance
  - Scapholunate diastasis > 3 mm for scapholunate dissociation
  - "Target" or "ring" sign; scaphoid foreshortening if abnormally volarly flexed
- Lateral view: Signs of instability
  - Dorsal or volar flexion of lunate with abnormal capitolunate angle (> 30°); scapholunate angle (< 30° or > 60°): VISI or DISI, respectively
  - Radioscaphoid angle > 80°: Abnormal scaphoid volar flexion (often seen with scapholunate dissociation & DISI)
- Clenched fist views may show dynamic instability
  - Obtain in PA, ulnar and radial deviation

Fluoroscopic Findings

- Manipulation under fluoroscopy may show dynamic instability or severity of static instability

MR Findings

- T1WI: Loss of osseous alignment; reactive change in bone
- Fluid-sensitive sequences
  - Intrinsic ligament tear or thinning
  - Synovitis or disruption of extrinsic ligament (especially volar radiocarpal)
- MR arthrogram (direct or indirect): Improves conspicuity of intrinsic/extrinsic ligament perforation/tears; cartilage thinning
- Wrist often placed in some degree of ulnar deviation during patient positioning
  - May give false impression of DISI
  - May obscure medial pathology, such as LTL tear

CARPAL INSTABILITY

Imaging Recommendations

- Best imaging tool
  - Radiographs for static instability
  - Fluoroscopy for dynamic instability
  - MR for evaluation of intrinsic & extrinsic ligaments
    - MR arthrography increases conspicuity
- Protocol advice
  - Manipulation maneuvers under fluoroscopy to evaluate for specific instabilities
    - Scapholunate dissociation: Widening of scapholunate interval, generally provoked with clenched fist or ulnar deviation
    - PMCI: Loss of normal physiologic VISI to DISI when moving from radial to ulnar deviation; reproducible clunk
    - CLIP: In lateral position, stabilize patient's forearm with wrist in slight ulnar deviation; grasp patient's hand at metacarpals & apply alternating dorsal-/volar-oriented force; observe for subluxation at capitolunate articulation
    - Triquetrohamate instability: During radial to ulnar deviation, proximal carpal row abruptly shifts from palmar flexion to dorsi flexion just before extreme ulnar flexion is achieved; hamate may slide against triquetrum

DIFFERENTIAL DIAGNOSIS

Carpal Dislocation

- Triangular appearance of lunate on PA radiograph & loss of articular parallelism may suggest dislocation
- Lateral view differentiates abnormal angulation from dislocation at the middle carpal column

Inflammatory Arthritis

- Pyrophosphate arthropathy may result in SLAC wrist; involves both radioscaphoid and radiolunate articulations
- Rheumatoid arthritis may result in any deformity, but particularly ulnar translocation & VISI

PATHOLOGY

General Features

- Etiology
  - Axial compression/hyperextension & intercarpal supination
  - FOOSH or backward fall on pronated hand
  - Often overlooked initially as "wrist sprain" if no fracture seen at initial imaging at time of injury
  - Inflammatory arthritis (particularly RA) can cause enough ligamentous damage to result in instability
    - Most common etiology of ulnar translocation
    - VISI more common than DISI
- Relevant anatomy
  - In neutral position, axes of radius, lunate, & capitate are colinear on lateral images
  - With flexion or extension, half the normal motion occurs between lunate & radius; half between capitate & lunate
  - Normal scapholunate angle on lateral: 30-60°
  - Normal capitolunate angle on lateral: ≤ 30°
  - Extrinsic carpal ligaments provide gross stability
  - Intrinsic carpal ligaments "fine tune" carpal stability

CLINICAL ISSUES

Presentation

- Most common signs/symptoms
  - Acute instability
    - Pain & swelling, related to site of injury
    - Painful clicking/clunk, reproducible
  - Chronic instability
    - History of "wrist sprain"
    - Pain & instability
    - Decreased grip strength
    - Painful clicks

Demographics

- Age
  - Adult population
- Gender
  - M = F in younger population
  - Older women with rheumatoid arthritis
- Epidemiology
  - Scapholunate tear leads to most common types of carpal instabilities
  - Ligament injuries resulting in instability may be either isolated or residua of prior perilunate injury
  - Incidence: SLL dissociation > MCI > LTL instability

Natural History & Prognosis

- Scapholunate dissociation may progress to SLAC wrist deformity
  - Medial scaphoid & lateral lunate collapse, with capitate interposed
- MCI (dynamic) can progress to VISI (static)

Treatment

- Conservative: Activity modification, NSAIDs, steroid injection, wrist immobilization
- Surgical if continued pain & instability
  - Acute: Closed reduction + internal fixation
  - Ligament reconstruction: Acute or chronic
  - Chronic: Capsular tightening
  - Chronic: Arthrodesis, such as 4 corner fusion involving lunate, triquetrum, capitate, hamate

DIAGNOSTIC CHECKLIST

Consider

- Clinical exam of characteristic sites of pain, clunk on exam, assessed in combination with imaging findings
- Clicking of carpus is common in asymptomatic wrists
- MR may not be reliable for carpal angle measurement due to wrist positioning

SELECTED REFERENCES

1. Magee T: Comparison of 3-T MRI and arthroscopy of intrinsic wrist ligament and TFCC tears. AJR Am J Roentgenol. 192(1):80-5, 2009
2. Bencardino JT et al: Sports-related injuries of the wrist: an approach to MRI interpretation. Clin Sports Med. 25(3):409-32, vi, 2006
3. Theumann NH et al: Association between extrinsic and intrinsic carpal ligament injuries at MR arthrography and carpal instability at radiography: initial observations. Radiology. 238(3):950-7, 2006
4. Cerezal L et al: Wrist MR arthrography: how, why, when. Radiol Clin North Am. 43(4):709-31, viii, 2005

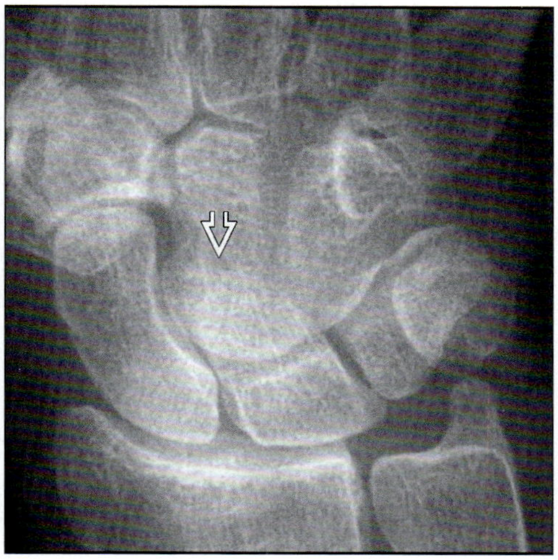

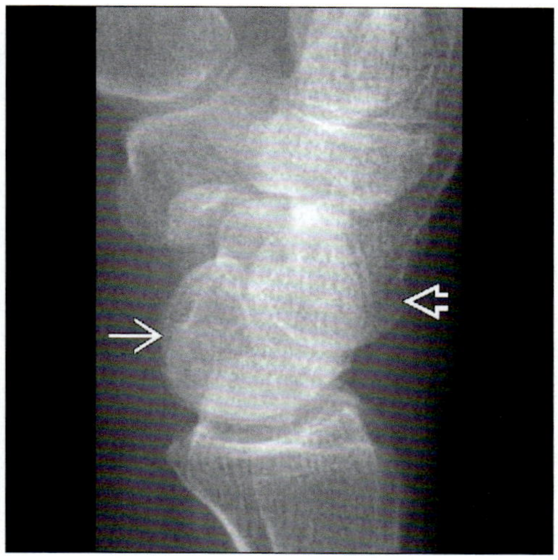

(Left) *PA radiograph in this case of DISI is relatively normal. The scapholunate and lunotriquetral relationships are maintained, but the lunate has a somewhat elongated contour ➡.* ***(Right)*** *Lateral radiograph in the same patient demonstrates lunate ➡ dorsal tilt without dislocation. The capitate ➡ is relatively volar tilted compared to the lunate with a capitolunate angle > 30° while the scapholunate angle measures 90°, defining the dorsal intercalated segment instability (DISI) pattern.*

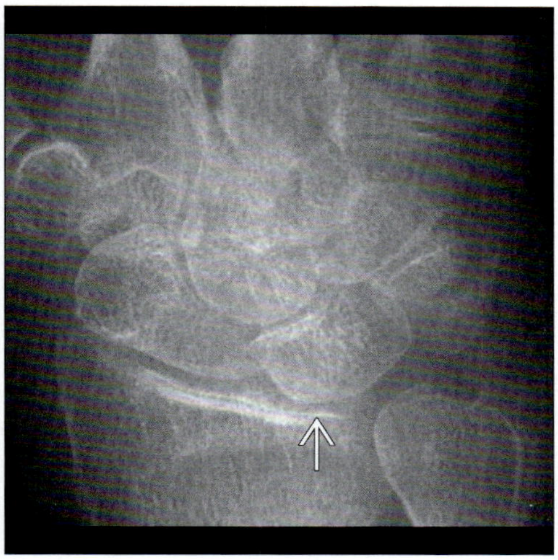

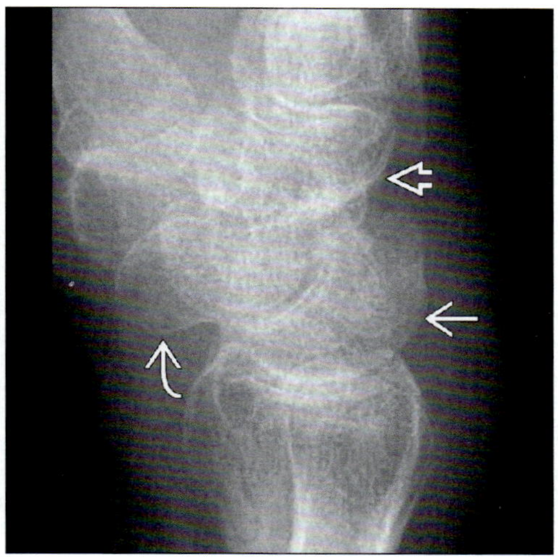

(Left) *PA radiograph demonstrates slight ulnar translocation ➡. The patient is osteopenic, and there is mild diffuse joint space narrowing. There is no evidence of subluxation or dislocation.* ***(Right)*** *Lateral radiograph in the same patient reveals volar tilt of the lunate ➡. The capitate ➡ is dorsally tilted, and this results in capitolunate angle of 90°. The scaphoid ➡ is tilted volarly, and the scapholunate angle measures 10°. This represents a volar intercalated segment instability.*

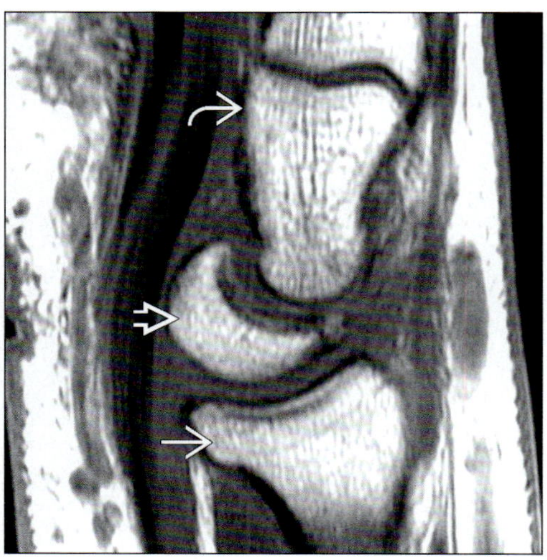

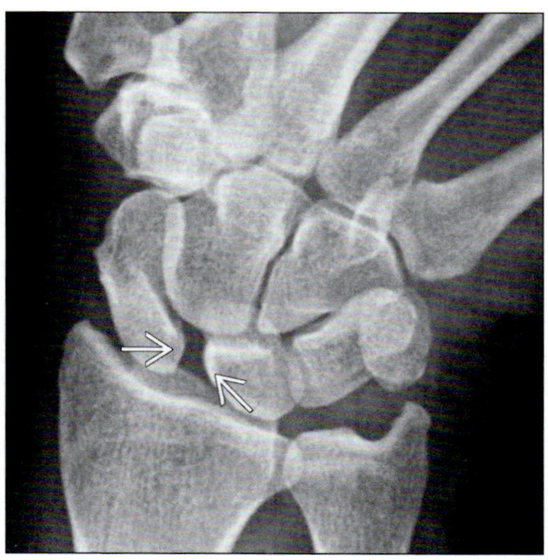

(Left) *Sagittal T1WI MR demonstrates dorsal angulation of the lunate ➡ relative to the radius ➡ and the capitate ➡, indicating DISI in this patient with a concomitant scapholunate ligament tear.* ***(Right)*** *PA radiograph (obtained in ulnar deviation) delineates and accentuates the scapholunate interval widening ➡, resulting from the disruption of the SLL. This widening suggests a developing scapholunate dissociation. The interval widening was not evident on PA neutral films.*

(Left) *Coronal PDWI FS MR demonstrates a complete tear of the scapholunate ligament from the scaphoid attachment with widening of the scapholunate interval, suggesting development of scapholunate dissociation.* **(Right)** *Coronal PDWI FS MR in the same patient, obtained just volar to the prior image, demonstrates the tear of the radioscaphocapitate ligament. The tear of this extrinsic ligament, in concert with the SLL tear, allows the development of the dissociation pattern.*

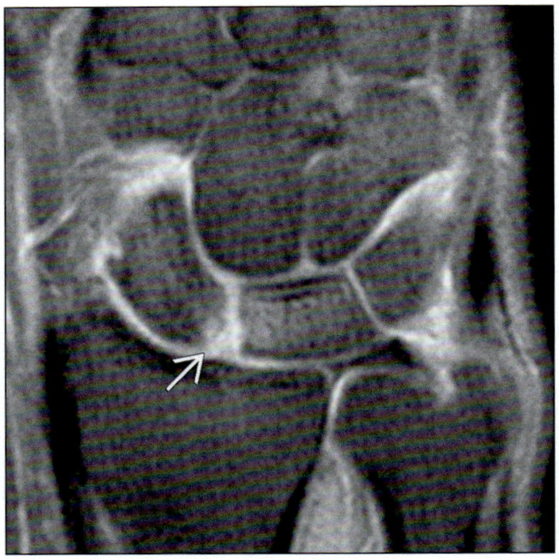

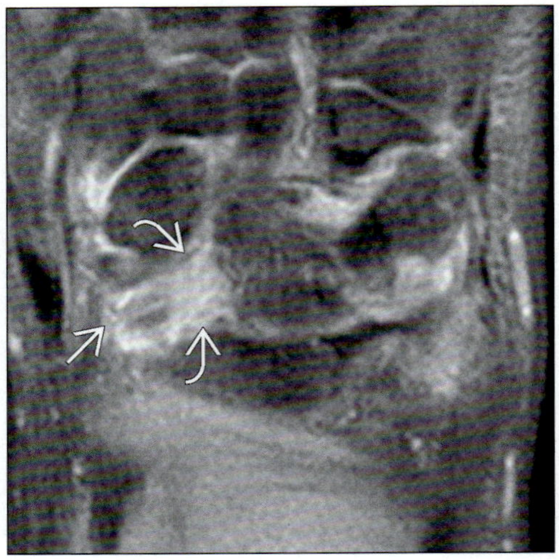

(Left) *Coronal PDWI FS MR shows a normal arcuate ligament. This is a stabilizing confluence of extrinsic ligaments, including radioscaphocapitate and triquetrohamatocapitate. Laxity of this ligament leads to MCI.* **(Right)** *Coronal MR arthrogram T1WI shows cartilage damage and subchondral sclerosis at distal lunate as it articulates with capitate. This pattern of arthrosis suggests dynamic midcarpal instability. Static radiographic imaging in early MCI typically demonstrates no malalignment.*

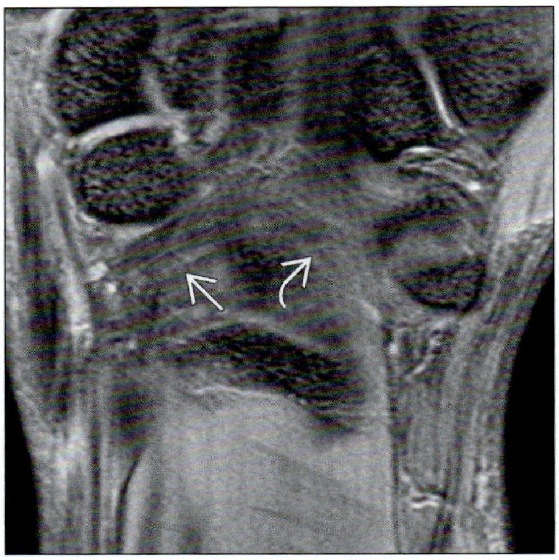

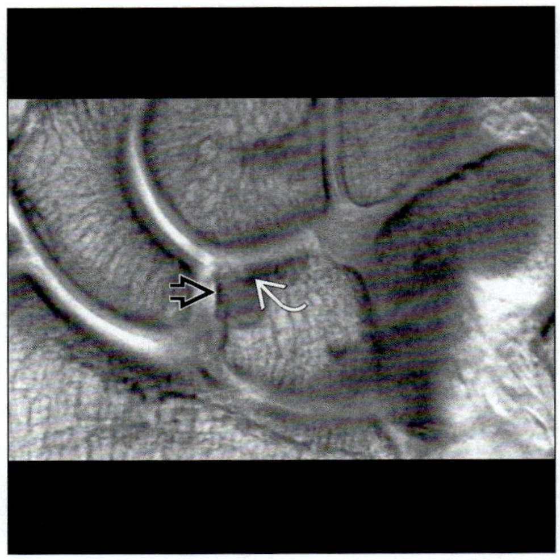

(Left) *Coronal STIR MR demonstrates complete disruption of the SLL with widening of the joint. Note the significant joint space narrowing of the capitolunate articulation as the capitate begins to migrate proximally in early SLAC wrist.* **(Right)** *Posteroanterior radiograph shows scapholunate dissociation. Abnormal volar subluxation of the scaphoid is inferred from the "target" or "ring" seen at the distal pole. There is associated arthrosis of the radiocarpal joint.*

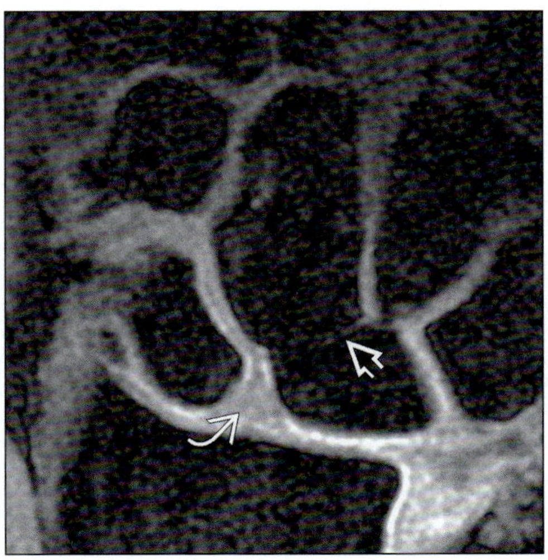

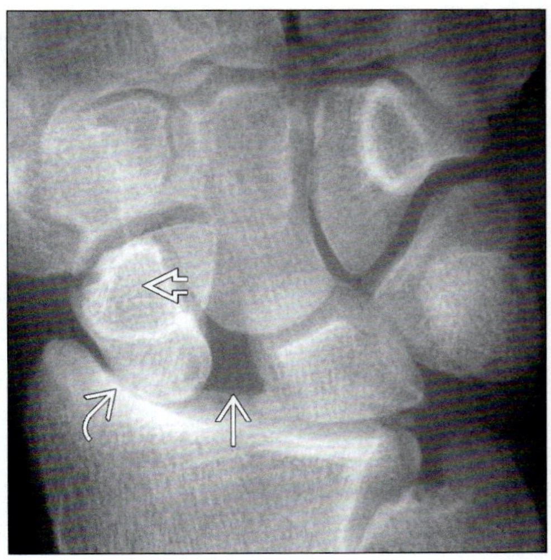

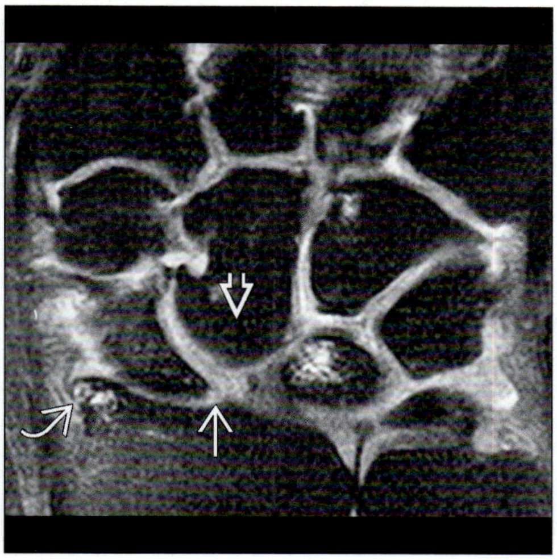

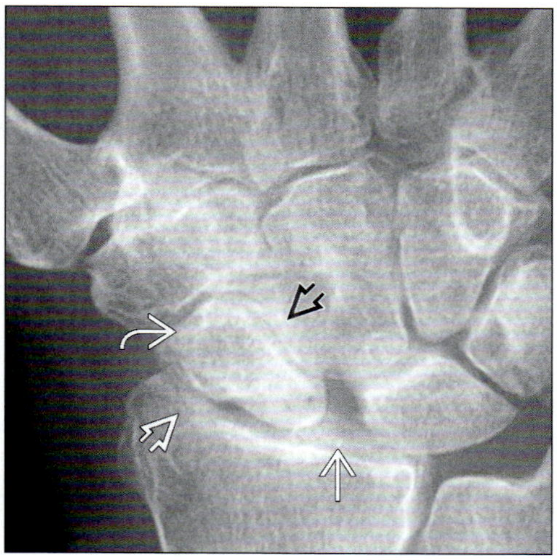

(Left) Coronal T1WI C+ FS MR shows scapholunate dissociation with chronic SLL tear ➔ and developing SLAC wrist. The capitate ➔ migrates proximally as midcarpal ligaments become lax. There is radioscaphoid osteophyte ➔ and cyst formation. (Right) PA radiograph shows a SLAC wrist deformity with widening of the scapholunate interval ➔ (SLL tear), rotary subluxation of the scaphoid ➔, and ↓ radioscaphoid ➔ and scaphocapitate ➔ joint space.

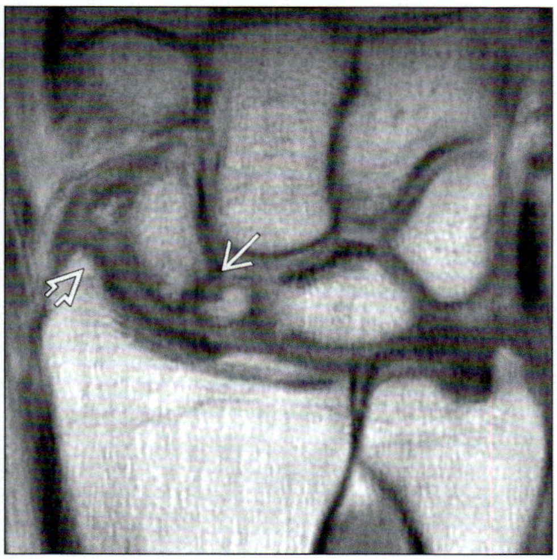

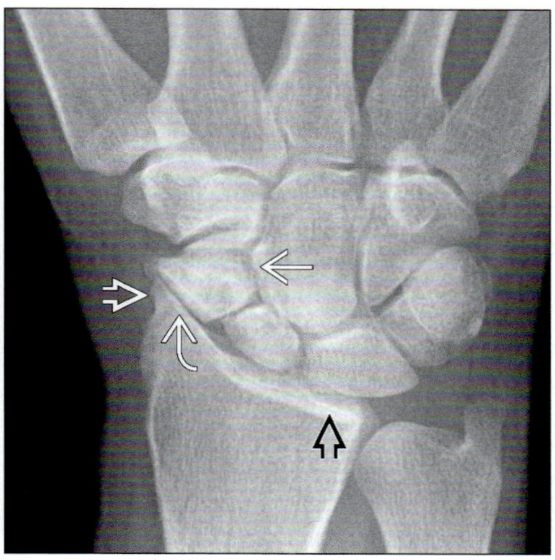

(Left) Coronal T1WI MR shows a scaphoid fracture nonunion ➔. The SLL is intact, but proximal scaphoid pole fragment moves with lunate. Note early beaking of the radial styloid ➔. This finding is typical of early scaphoid nonunion advanced collapse (SNAC). (Right) PA radiograph shows SNAC wrist with pronounced radial styloid beaking ➔ with radioscaphoid ➔ and scaphocapitate ➔ joint narrowing in this patient with midwaist scaphoid nonunion. The radiolunate joint ➔ is preserved.

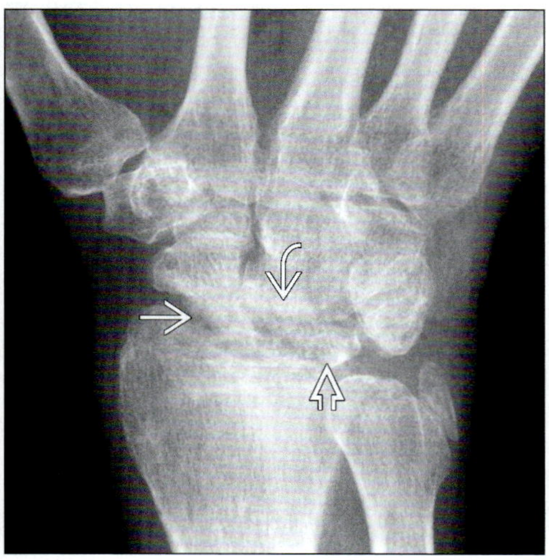

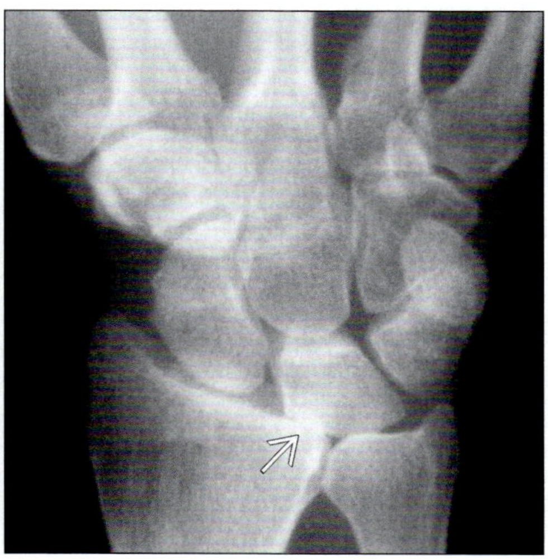

(Left) PA oblique radiograph shows the final stages of advanced collapse in a patient with scaphoid fracture nonunion. The capitate ➔ migrates proximally between the scaphoid ➔ and the lunate ➔. The lunate has collapsed as well. (Right) PA radiograph shows radiocarpal ➔ joint space narrowing. There is cartilage loss, and the carpus has translocated in a medial (ulnar) direction. The majority of the lunate overlies the ulna rather than the radius. This represents ulnar translocation of the carpus.

COLLATERAL LIGAMENT INJURY, FINGERS AND THUMB

Key Facts

Terminology

- Gamekeeper's (skier's) thumb: Thumb MCP UCL injury, related to hyperabduction, ± avulsion fracture
- Pseudo-boutonniere deformity: Results from PIP hyperextension + volar plate injury with intact extensor tendon; may cause PIP flexion contracture over time
- Stener lesion: Thumb UCL tear with retraction of torn ligament fibers proximal to adductor pollicis aponeurosis

Imaging

- MR: Ligament thickening, irregularity, discontinuity with joint gapping, subluxation, or dislocation
  - Collateral ligament tear: Ligament disruption ± avulsion fracture
  - Volar plate: Plate discontinuity ± avulsion fracture
- Radiographs
  - Radiographs often very benign in appearance
  - Medial or lateral joint angulation > 10° suggests collateral ligament injury
  - Gamekeeper's thumb: Valgus angulation > 30°

Top Differential Diagnoses

- Boutonniere deformity: Results from extensor central slip tear at PIP
- Pseudo-boutonniere deformity: Results from distal volar plate tear and PIP flexion contracture

Clinical Issues

- Displaced gamekeeper injury (with or without osseous avulsion) treated surgically
  - Includes Stener lesion

(Left) Sagittal graphic of a Stener lesion shows the thin adductor aponeurosis ➡ interposed between the torn, retracted UCL ➡ and its attachment site at the base of the proximal phalanx ↪. (Right) Coronal T1WI MR demonstrates the typical "yo-yo on a string" appearance of a Stener lesion. The torn, retracted UCL has a mass-like appearance (the yo-yo) ➡ as it retracts, and the adductor aponeurosis (the string) ➡ becomes interposed between the torn ligament and the joint line.

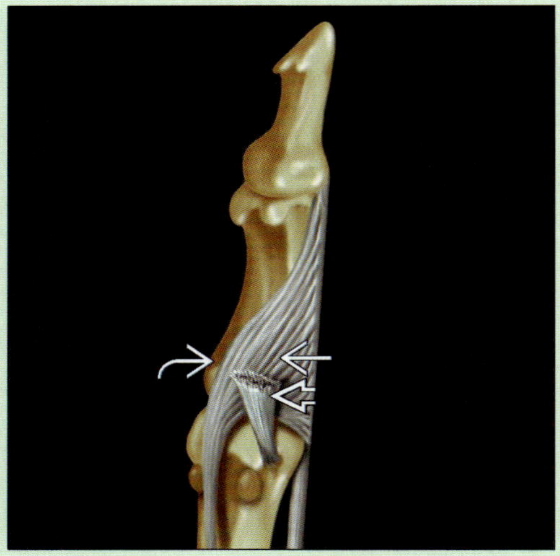

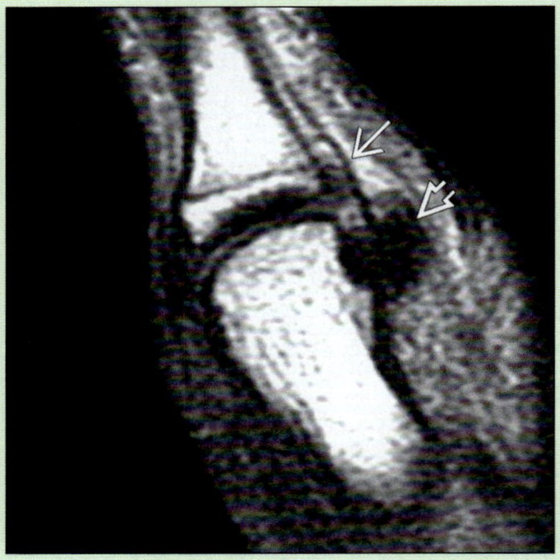

(Left) Sagittal graphic shows the normal volar plate ➡ located deep to the flexor tendon ➡ and the A3 pulley ↪ of the proximal interphalangeal joint. Avulsion of the volar plate usually occurs at the PIP and involves the distal fibers. (Right) Lateral radiograph of the finger from a 29-year-old woman with PIP joint pain after a hyperextension injury shows a subtle avulsion ➡ of the volar plate from the base of the middle phalanx. Volar plate avulsion fragments are often quite small.

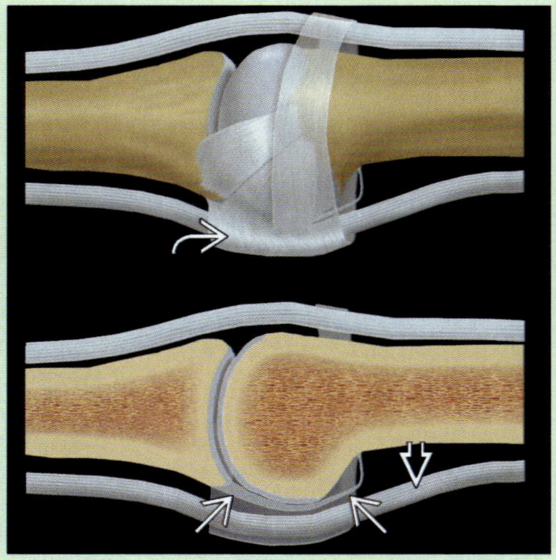

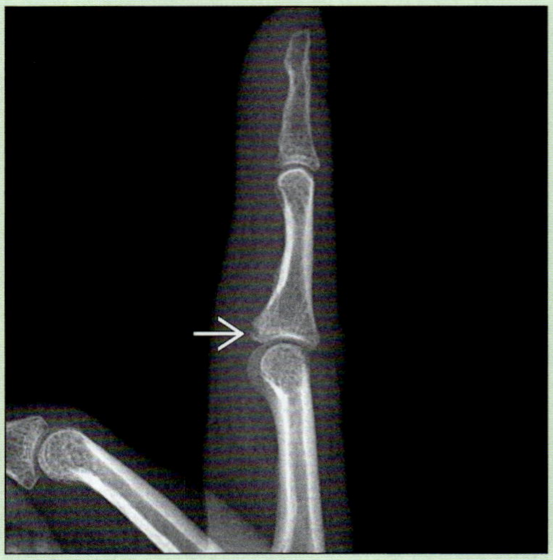

COLLATERAL LIGAMENT INJURY, FINGERS AND THUMB

TERMINOLOGY

Definitions

- Eponyms
  - Gamekeeper's (skier's) thumb: Sprain or tear of ulnar collateral ligament (UCL) of 1st metacarpophalangeal (MCP) joint related to hyperabduction injury, ± avulsion fracture
  - Boutonniere deformity: Extensor tendon central slip injury at or near insertion on middle phalanx, ± avulsion fracture of dorsal base of middle phalanx; results in proximal interphalangeal (PIP) joint flexion with distal interphalangeal (DIP) joint extension
  - Pseudo-boutonniere deformity: Results from PIP hyperextension with volar plate injury and intact extensor tendon; may cause PIP flexion contracture over time
  - Stener lesion: Gamekeeper's thumb with retraction of torn UCL fibers with interposition of adductor pollicis aponeurosis between UCL and 1st metacarpal (MC)
  - Swan neck deformity: PIP extension with DIP flexion; results from volar plate injury at PIP or extensor tendon injury at DIP

IMAGING

General Features

- Best diagnostic clue
  - Collateral ligament tear: Ligament disruption ± avulsion fracture
  - Volar plate disruption: Plate discontinuity ± avulsion fracture
- Location
  - May occur at any joint but most common in finger PIPs, thumb MCP
  - Finger PIP: Dorsal dislocation; distal volar plate avulsion ± collateral ligament injury
  - Finger MCP: Radial > ulnar collateral ligament
  - Thumb MCP: Ulnar > radial collateral ligament

Radiographic Findings

- Soft tissue swelling; obliteration of normal fat planes
- Inappropriate IP flexion or extension suggests ligament injury
- Medial or lateral joint angulation > 10° suggests collateral ligament injury
  - Gamekeeper's thumb: Valgus angulation > 30°
- Joint subluxation or dislocation
- Avulsion fracture: Triangular fragment
  - Displacement of UCL avulsion fragment at 1st MCP implies Stener lesion and is operative indication for many surgeons

MR Findings

- Collateral ligament injury
  - Obliteration of fat planes near injured ligament
  - Grade I injury (sprain): Periligamentous edema ± ligament thickening
  - Grade II injury (partial tear): Partial fiber disruption + surrounding edema
  - Grade III injury (complete tear): Ligament detachment or discontinuity
  - Fluid-sensitive sequences: ↑ signal intensity (SI) within ligament & surrounding soft tissues
  - IP injury
    - PIP injury > DIP injury
      - Ligament tear: Proximal portion in 85%
  - MCP injury
    - Gamekeeper's thumb: 1st MCP (thumb) UCL disruption
      - Ligament tears at distal (phalangeal) attachment
      - Osseous avulsion in 12%
      - MR: 90-100% sensitive, 67% specific
    - Stener lesion
      - UCL appears rounded & mass-like
      - Adductor aponeurosis interposed between retracted UCL and base of proximal phalanx ("yo-yo on a string")
      - MR: 90% sensitive, 94% specific
    - Finger MCP collateral ligament tears rare; RCL > UCL
- Volar plate injury
  - Thickening, irregularity, ± discontinuity of volar plate ligament
    - Common: Distal plate avulsion ± osseous fragment; ± subluxation
      - ~ 80% occur at distal attachment
  - Joint dislocation: PIP joint most common
    - Middle phalanx dislocated dorsally
- Look for interposition of torn collateral ligament or volar plate into joint space if joint irreducible
- Avulsion fracture
  - ± subtle marrow edema at donor site
  - Small fragments obscured by surrounding edema

Ultrasonographic Findings

- Collateral ligament injury
  - Partial tear: Ligament thickening
  - Complete tear: Detachment, discontinuity
    - Stener lesion: Retracted ligament fibers appear mass-like
      - Ultrasound: 88% sensitive, 83% specific
- Volar plate injury
  - Normal triangular contours distorted/disrupted

Imaging Recommendations

- Best imaging tool
  - MR or ultrasound
- Protocol advice
  - Standard radiographic views: PA, lateral, & oblique of hand
    - Thumb views: True AP & lateral
  - MR
    - Small imaging field of view
    - Dedicated wrist or finger coil
    - Include adjacent digits for comparison
    - Collateral ligaments best seen in axial & coronal planes
    - Volar ligaments best seen in axial & sagittal planes
    - Appropriate positioning for thumb evaluation
      - Axial localizer: Coronal plane acquired through 1st MC head parallel to collateral ligament

COLLATERAL LIGAMENT INJURY, FINGERS AND THUMB

PATHOLOGY

Staging, Grading, & Classification

- Ligament injury
  - Grade I: Microscopic ligament tear; stable joint
  - Grade II: Moderate fiber disruption; ± laxity
  - Grade III: Complete tear; unstable joint
- Volar plate injuries
  - Type I: Hyperextension; distal volar plate avulsion; stable
  - Type II: Dorsal dislocation; distal volar plate avulsion; complete split in collateral ligaments; may be unstable
  - Type III: Fracture-dislocation; volar plate disruption; collateral ligaments remain with volar plate; articular cartilage defect; unstable

Mechanisms of Injury

- Collateral ligament failure
  - Hyperabduction/adduction
  - Most common in PIP joint; rare at finger MCPs
    - Forced hyperextension (e.g., fall on outstretched hand)
    - Unstable injury if large avulsion fracture (> 40% articular surface)
      - Majority of collateral ligament(s) likely attached to fragment
  - Gamekeeper's (skier's) thumb/Stener lesion
    - Forced hyperabduction of 1st MCP joint while thumb flexed (e.g., fall while gripping ski pole)
    - 1st MCP joint: UCL > > RCL injury
- Volar plate failure
  - Hyperextension: Common in sports injury; low-energy injury
  - ± dorsal subluxation/dislocation
    - Most common in PIP due to long lever arm
    - Uncommon in DIP/thumb IP joint due to shorter lever arm
    - Finger MCP dislocations uncommon; distal volar plate injury ± unilateral collateral ligament tear (RCL > UCL)
- Complex forces
  - Rotational longitudinal compression of semiflexed PIP joint (e.g., finger caught in spinning dryer)
    - Volar subluxation with unilateral collateral ligament tear and partial avulsion of volar plate ± extensor mechanism injury

CLINICAL ISSUES

Demographics

- Epidemiology
  - Finger PIPs and thumb MCP are most commonly injured hand joints
  - Thumb UCL injuries: 7% of ski injuries
    - 29% become Stener lesions

Natural History & Prognosis

- With appropriate treatment and physical therapy
  - Complete recovery with full range of motion
  - May fully heal but with residual loss of strength & range of motion
- Delayed or undertreated injuries
  - Weakness
  - Loss of grip strength
  - Loss of range of motion
  - Osteoarthritis
- Untreated PIP distal volar plate injury may → PIP hyperextension with DIP flexion (swan neck deformity)
- Untreated PIP proximal volar plate injury may → PIP flexion (pseudo-boutonniere deformity)

Treatment

- Fingers
  - Conservative
    - Partial tears
    - Nondisplaced complete tears
  - Surgical indications
    - Failed conservative treatment
    - Open injury
    - Volar plate injury with large fracture
      - Typically stable if fragment < 40% of total articular surface
      - Unstable if > 40% of articular surface fracture
    - Irreducible subluxations/dislocation with release of entrapped ligaments, tendons, or osseous fragments
- Gamekeeper injury
  - Partial and nondisplaced UCL injury treated conservatively
  - Displaced UCL injury (with or without osseous avulsion) treated surgically
    - Includes Stener lesion
- Complications
  - Infection; particularly with open injuries
  - Loss of joint motion and grip strength
  - Osteoarthritis

DIAGNOSTIC CHECKLIST

Consider

- Radiographic appearance may be deceptively benign even with significant ligament injury
- Dislocations often reduced by the time imaging takes place

Image Interpretation Pearls

- Injuries often occur in combination

Reporting Tips

- Describe any displacement of gamekeeper avulsion fragment

SELECTED REFERENCES

1. Wessely K et al: Ultrasound findings of the locked metacarpophalangeal joint. Skeletal Radiol. 38(1):85-9, 2009
2. Ebrahim FS et al: US diagnosis of UCL tears of the thumb and Stener lesions: technique, pattern-based approach, and differential diagnosis. Radiographics. 26(4):1007-20, 2006
3. Clavero JA et al: MR imaging of ligament and tendon injuries of the fingers. Radiographics. 22(2):237-56, 2002
4. Hinke DH et al: Ulnar collateral ligament of the thumb: MR findings in cadavers, volunteers, and patients with ligamentous injury (gamekeeper's thumb). AJR Am J Roentgenol. 163(6):1431-4, 1994

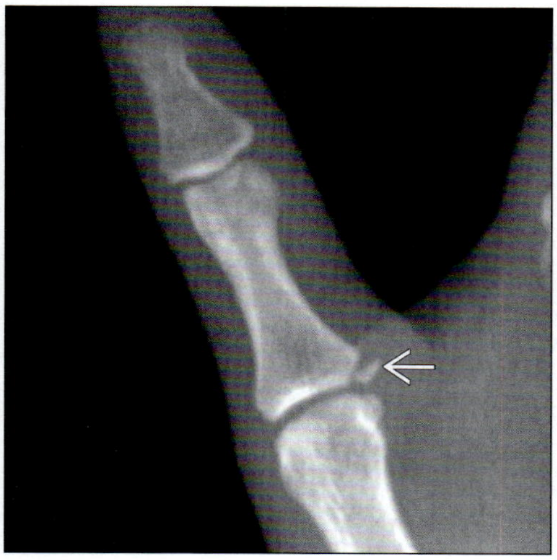

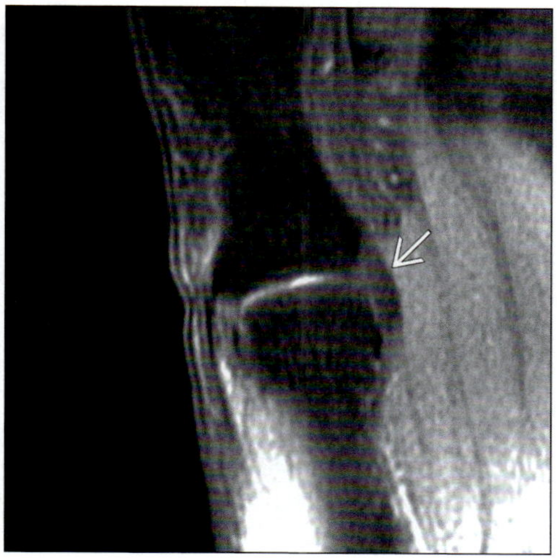

(Left) *PA radiograph of the thumb demonstrates an acute osseous avulsion → of the ulnar collateral ligament (UCL) from the base of the proximal phalanx. This is the functional equivalent of a gamekeeper's injury. Significant displacement of the fragment is taken to imply a Stener lesion and is typically an indication for surgery.* ***(Right)*** *Coronal STIR MR demonstrates a grade I sprain of the 1st MCP UCL →. There is diffuse thickening of the ligament without a discrete tear.*

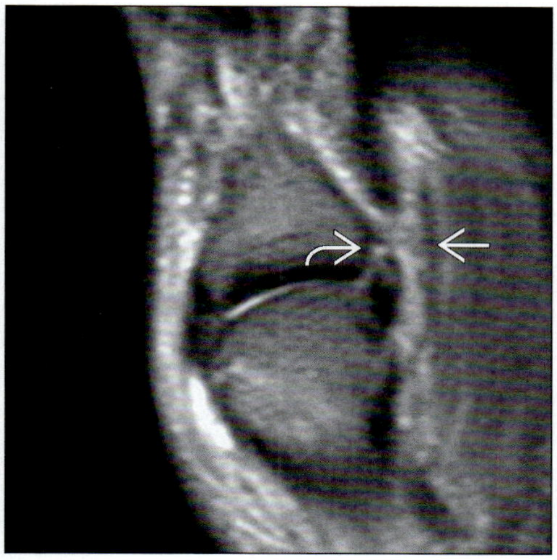

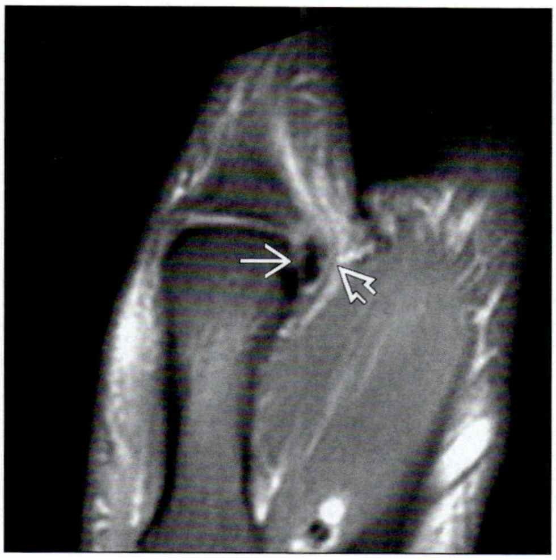

(Left) *Coronal PDWI FS MR demonstrates fluid at the distal bone-ligament interface ↷ of the 1st MCP UCL, consistent with complete disruption. The torn ligament is nondisplaced and remains deep to the adjacent adductor aponeurosis →.* ***(Right)*** *Coronal PDWI FS MR of the thumb demonstrates a torn UCL →. The intact aponeurosis ⇨ overlies the UCL. Because of its surgical ramifications, a Stener lesion must be excluded by MR or ultrasound when a UCL tear is suspected.*

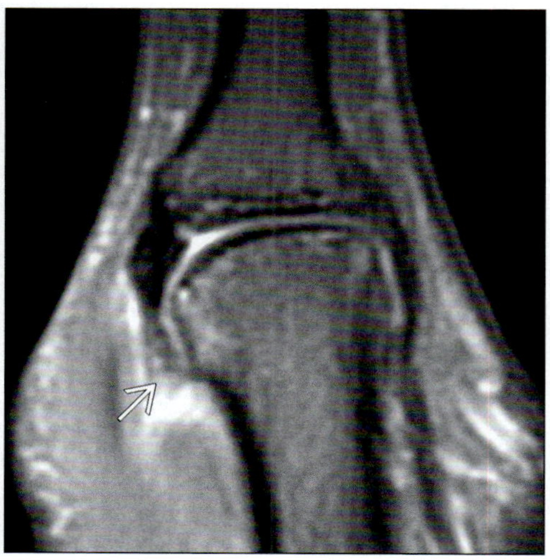

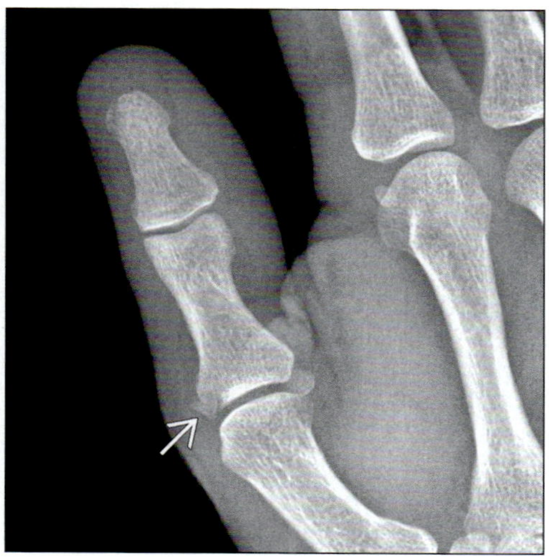

(Left) *Coronal T2WI FS MR demonstrates partial disruption → at the proximal insertion site of the radial collateral ligament (RCL) of the thumb MCP joint. Tears of the RCL occur due to forced hyperadduction and are far less common than UCL tears.* ***(Right)*** *AP radiograph of the thumb demonstrates a triangular avulsion fragment → at the lateral base of the proximal phalanx, indicating avulsion of the RCL of the 1st metacarpophalangeal joint. Patient had varus laxity on exam.*

COLLATERAL LIGAMENT INJURY, FINGERS AND THUMB

*(**Left**) AP radiograph from an 84-year-old woman with longstanding mixed connective tissue disorder shows lateral dislocations of the 2nd through 5th metacarpophalangeal joints and subluxation of the 1st MCP joint. There was no history of significant trauma; these subluxations are the result of severe chronic ligamentous laxity. (**Right**) PA radiograph shows a small avulsion fragment → at the attachment of the radial collateral ligament of the 3rd MCP joint. Such osseous avulsions are unusual.*

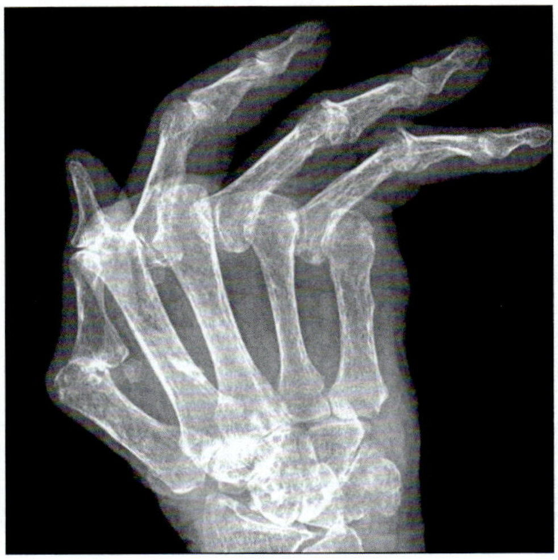

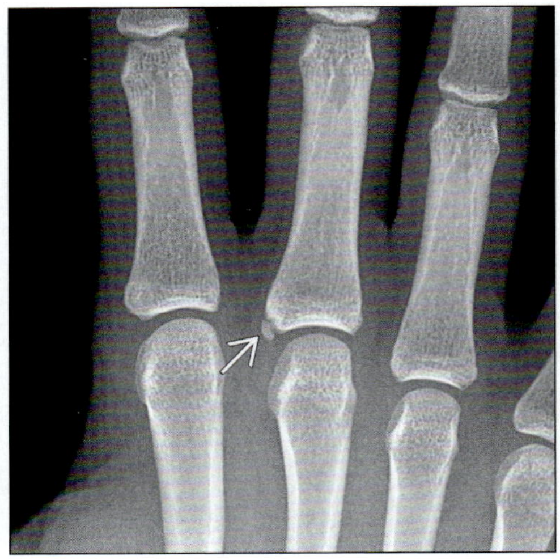

*(**Left**) Coronal T2WI FS MR shows thickening and heterogeneity of the radial collateral ligament → of the 2nd metacarpophalangeal joint in a 17-year-old athlete after trauma. There was mild valgus laxity of the joint on clinical exam. Triangular low signal in the metacarpal head ⇨ is artifactual. (**Right**) Coronal T2WI FS MR shows a more significant sprain (partial tear) → of the RCL of the 2nd MCP joint in a 26-year-old mechanic. Some ligament fibers are disrupted, but others remain intact.*

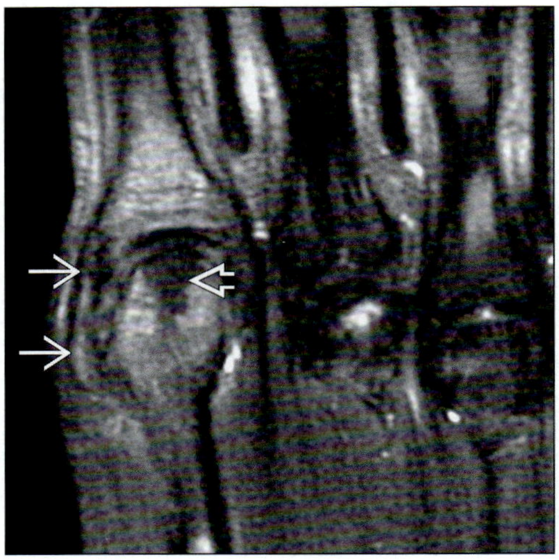

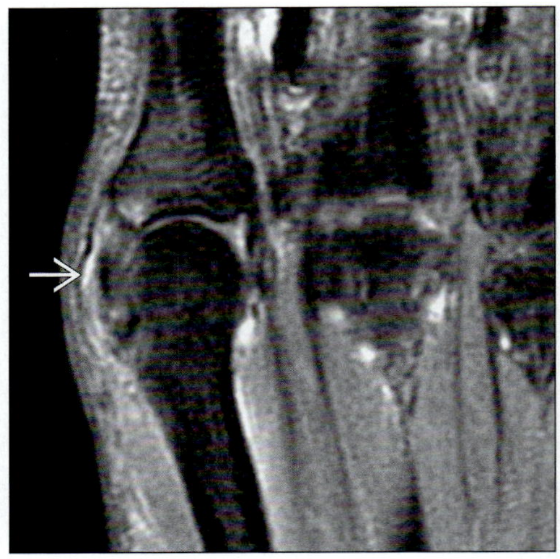

*(**Left**) Coronal T2* GRE MR shows complete disruption of the radial collateral ligament → of the 2nd metacarpophalangeal joint in a 33 year old who was in a motor vehicle crash. No normal ligamentous fibers are identified. There was gross valgus instability on exam. (**Right**) Coronal T2WI FS MR shows an osseous avulsion → of the ulnar collateral ligament of the 2nd MCP joint. Patient fell down stairs and had gross varus instability of the joint. Osseous collateral ligament avulsions are unusual.*

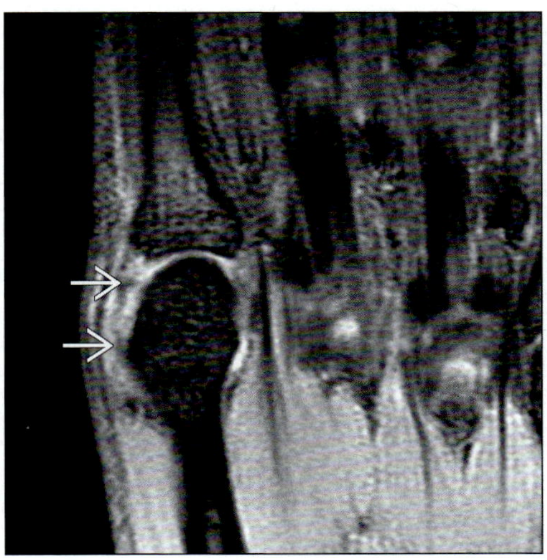

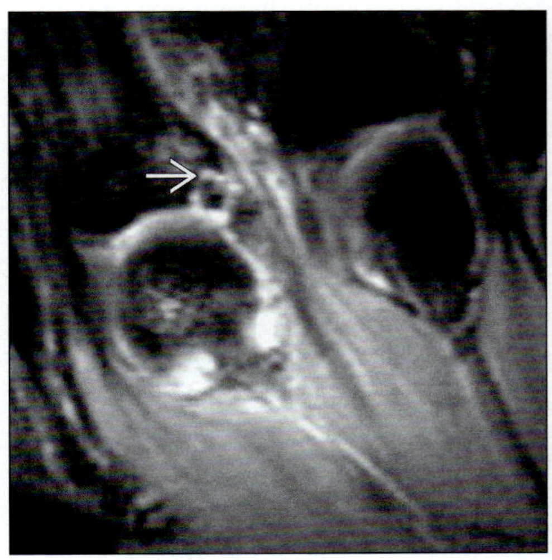

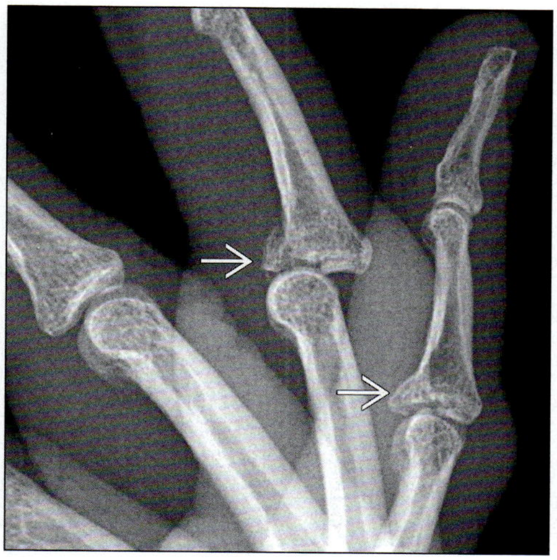

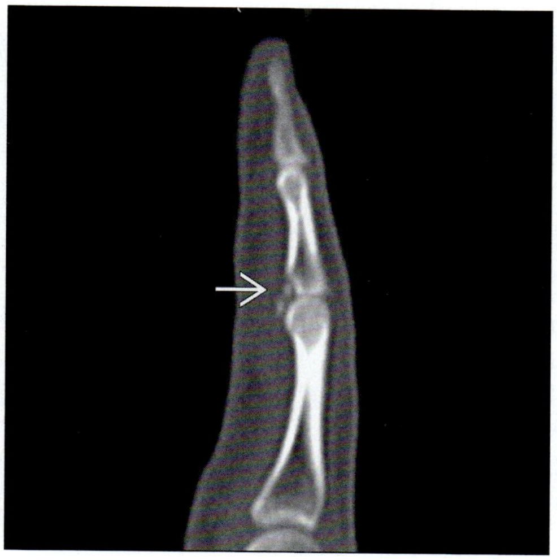

(Left) *Lateral radiograph shows large volar plate avulsion fractures at the PIP joints of the 4th and 5th fingers. There is dorsal subluxation of the 4th proximal interphalangeal joint.* ***(Right)*** *Sagittal reformat bone CT demonstrates small, subacute fragments at the base of the middle phalanx resulting from avulsion of the distal aspect of the volar plate. Left untreated, this may result in progressive flexion of the PIP and a pseudo-boutonniere deformity.*

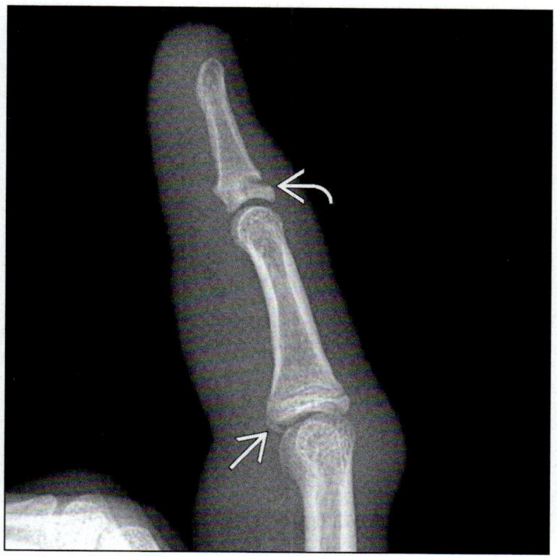

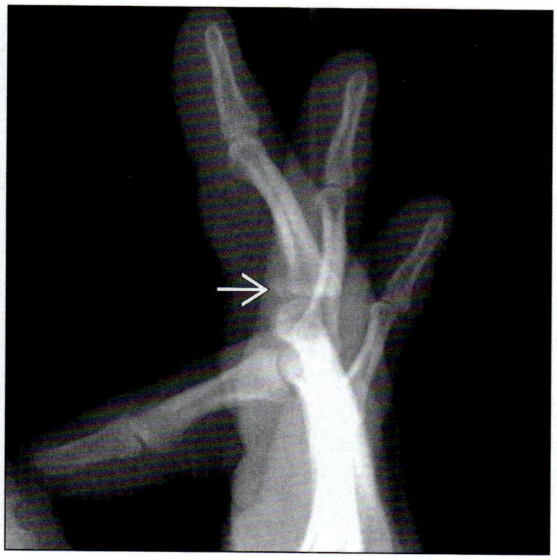

(Left) *Lateral radiograph demonstrates a small volar plate avulsion fragment at the PIP joint and a dorsal base avulsion fracture consistent with avulsion of the terminal extensor tendon at the DIP joint.* ***(Right)*** *Lateral radiograph demonstrates a volar plate avulsion of the 3rd PIP joint. The joint's middle phalanx is subluxed dorsally relative to the proximal phalanx, indicating severe laxity of the collateral ligaments as well.*

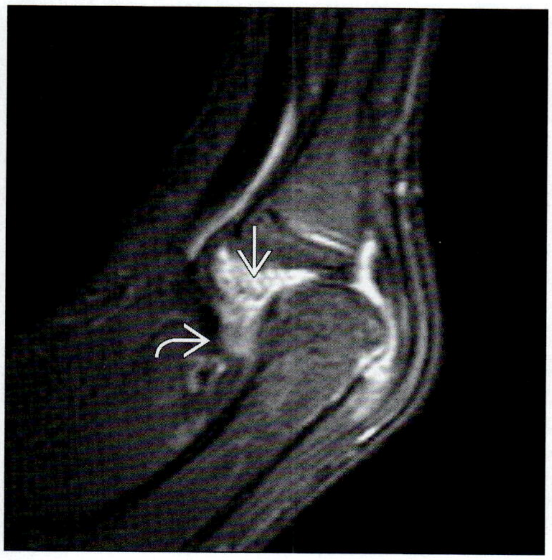

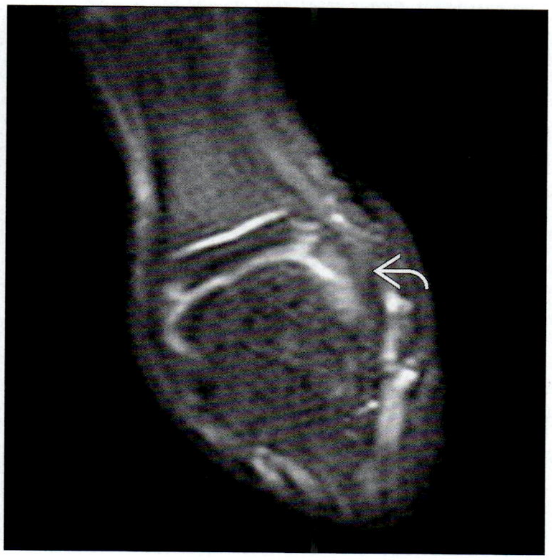

(Left) *Sagittal T2WI FS MR shows proximal disruption of the volar plate of the 1st metacarpophalangeal joint from a 14-year-old boy who suffered a hyperextension injury to the thumb while playing football. There is a large joint effusion.* ***(Right)*** *Coronal T2WI FS MR from the same patient shows edema around a thickened and irregular ulnar collateral ligament of the 1st MCP joint, consistent with a sprain.*

FLEXOR TENDON INJURY, WRIST AND FINGERS

Key Facts

Imaging

- Tendon laceration
  - Partial tear: Focal areas of ↑ signal intensity (SI) on T1- & T2WI; portion of tendon remains intact
  - Complete tear: Discontinuity of fibers ± retraction
- Pulley injury: A2 injury most common
  - Partial tear: Thickening of pulley without bowstringing
  - Complete tear: Tendon bowstrings away from phalanx
- Jersey finger: Disruption of FDP at volar base of distal phalanx ± avulsion fragment
- Tendinopathy/tenosynovitis
  - T1WI: Focal or diffuse tendon enlargement with ↑ intrasubstance signal
  - ↑ fluid in tendon sheath on fluid-sensitive sequences
  - Trigger finger: Focal tendon & sheath thickening ± A1 pulley thickening
- MR: Small field-of-view; dedicated wrist or finger coil
- Ultrasound: Tendon disruption; anechoic focus corresponds to tendon tear
  - Dynamic scanning to evaluate partial vs. complete tear

Pathology

- Laceration: Dependent on finger position at time of injury
  - In flexion: Tendon injury distal to skin laceration
  - In extension: Tendon and skin laceration will be adjacent to each other
- Pulley injuries: Extreme finger flexion with MCP extension, PIP flexion, and DIP extension
  - Results in progressive injury: A2 → A3 → A4
- Tendinopathy/tenosynovitis
  - May affect any tendon

*(**Left**) Coronal graphic of the volar 2nd finger shows fracture of the volar base of the distal phalanx resulting from avulsion of the FDP ➡. This injury is known as jersey finger because it occurs when a flexed finger is forcibly extended (e.g., a finger is caught in the opponent's jersey). (**Right**) Axial T1WI MR through the PIP joint of the 3rd finger shows rupture of the FDS ➡ due to a jersey injury. The profundus tendon ➡ remains intact. The tendons of the adjacent 2nd finger ➡ are normal.*

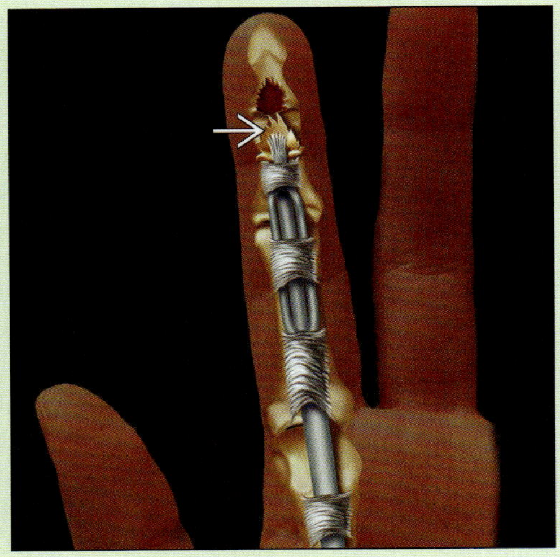

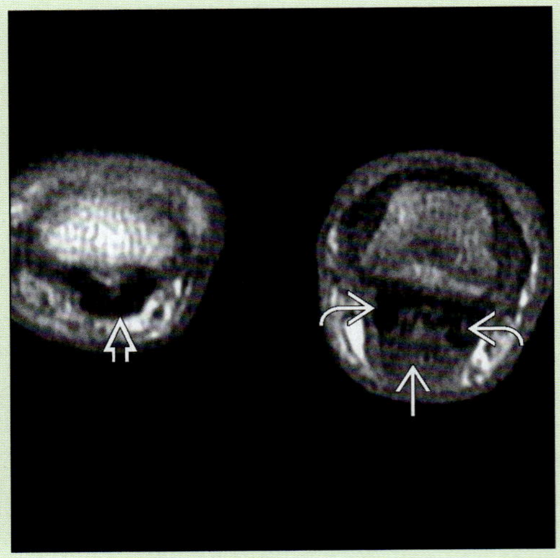

*(**Left**) Lateral graphic demonstrates failure of the A2 pulley ➡ resulting in a bowstring deformity. The flexor tendon ➡ pulls away from its normal relationship with the proximal phalanx. (**Right**) Sagittal T1WI MR demonstrates bowstringing of the taut FDP ➡ resulting from chronic disruption of the A2 ➡ and C1 ➡ pulleys. The torn pulleys are not visualized, but the region of tendon bowstringing indicates the areas of failure. The A3 pulley ➡ remains intact.*

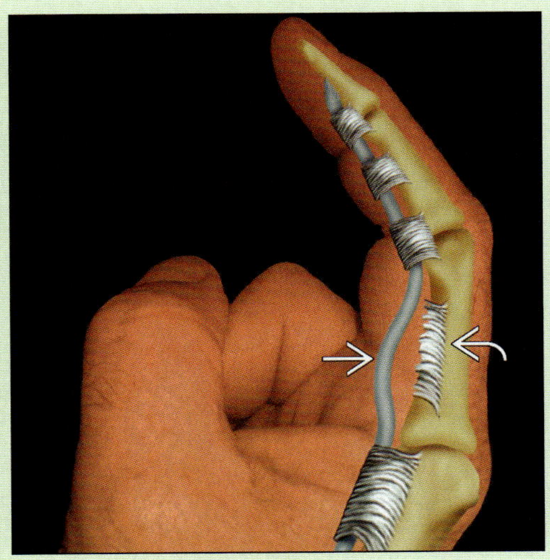

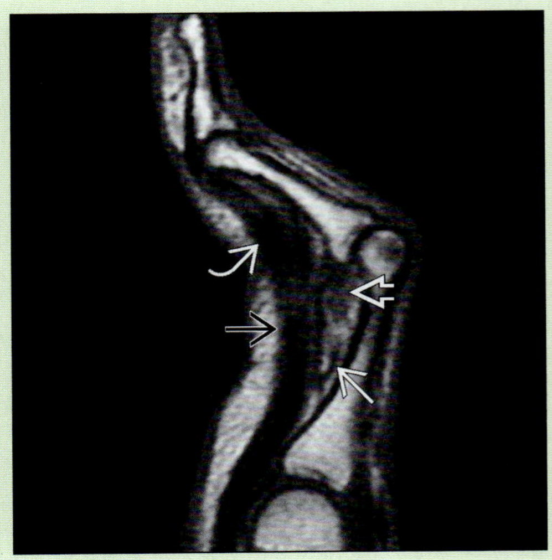

FLEXOR TENDON INJURY, WRIST AND FINGERS

IMAGING

Radiographic Findings
- Avulsion fracture

MR Findings
- Tendon tear or laceration
  - Partial tear: Focal areas of ↑ signal intensity (SI) on T1- & T2WI; portion of tendon remains intact
  - Complete tear: Discontinuity of fibers ± retraction
    - Flexor tendons retract more than extensor tendons when disrupted due to lack of tethering
  - Wound debris (micrometallic debris) may be present
- Pulley injury: A2 injury most common
  - Thickening of pulley (particularly A2) without bowstringing of flexor tendon indicates partial pulley tear
  - Tendon pulled away from proximal phalanx (bowstringing) indicates complete pulley tear
- Tendinopathy
  - T1WI: Focal or diffuse tendon enlargement with ↑ intrasubstance signal
  - ± tenosynovitis & edema in surrounding soft tissues
- Tenosynovitis
  - ↑ fluid in tendon sheath on fluid-sensitive sequences
    - When diameter of tendon sheath fluid > diameter of tendon → tenosynovitis
  - May see thickening or irregularity of sheath
  - T1WI FS MR C+: Enhancement of sheath
- Jersey finger: Disruption of flexor digitorum profundus (FDP) at volar base of distal phalanx ± avulsion fragment
- Trigger finger: Focal tendon ± tendon sheath thickening & A1 pulley thickening

Ultrasonographic Findings
- Tendon tear: Fiber disruption; anechoic focus corresponds to tendon tear
  - Dynamic scanning to evaluate partial vs. complete tear
- Pulley injury: Pulley thickening or disruption; bowstring sign
  - A2 & A4 pulleys most readily seen
- Tenosynovitis: Sheath distention; fluid may be anechoic or slightly complex
- Tendinopathy/partial intrasubstance tear: Hypoechoic regions in swollen tendon
- Dynamic scanning accentuates abnormalities

Imaging Recommendations
- Best imaging tool
  - MR
- Protocol advice
  - Radiography: PA, lateral, & oblique views
  - MR: Small field-of-view; dedicated wrist or finger coil
    - Sagittal & axial T1WI & PDWI FS
  - Ultrasound: Small footprint, high-frequency transducer (≥ 10 MHz) ± stand-off pad

PATHOLOGY

Staging, Grading, & Classification
- FDP avulsion classification
  - Type I: FDP retracts into palm
  - Type II: FDP ± avulsed fragment retracts to PIP
  - Type III: FDP attached to large avulsed fragment; retracts to A4 pulley
  - Type IV: Avulsed fragment caught at A4 pulley; FDP torn from fragment, retracted proximally

Relevant Anatomy
- Tendons
  - At carpal tunnel: FDP deep to flexor digitorum superficialis (FDS)
    - Thumb: Flexor pollicis longus (FPL) has its own fibroosseous canal
  - Distal to carpal tunnel: Directed to respective fingers
  - Proximal to metacarpophalangeal (MCP) joint: FDS superficial to FDP
    - Tendons enter fibroosseous tunnel just proximal to MCP
      - Tunnel extends from MCP to distal interphalangeal (DIP) joint
      - Floor created by phalanges & volar plates
      - Encircled by annular and cruciate pulleys
      - Lined by synovial sheath; tendons secured to sheath by vincula tendina (thin sheaths of connective tissue)
      - Tendon sheath and vincula are key to tendon nutrition & healing
  - At MCP: FDS splits, paralleling FDP
  - At proximal interphalangeal (PIP) joint: FDS encircles FDP, merging deep to FDP to insert on volar base of middle phalanx
  - At DIP: FDP inserts on volar base of distal phalanx
    - Thumb: FPL inserts on volar base of distal phalanx
- Pulley system
  - Supplies mechanical advantage by maintaining tendons close to axis of joint motion
  - Palmar aponeurosis pulley: Not as closely applied as finger pulleys
  - Annular pulleys: Broad, flat, transverse; keeps tendon close to bone for mechanical advantage
    - A1 at MCP; A2 at mid-proximal phalanx; A3 at PIP; A4 at mid-middle phalanx; A5 at DIP
    - A2 & A4: Largest and most consistent pulleys
  - Cruciate pulleys: Thin, fine, variable, crisscrossing; prevent tendon deformity during range of motion
    - C1 at head of proximal phalanx; C2 at base of middle phalanx; C3 at head of middle phalanx
  - Thumb pulleys: A1 at MCP; oblique pulley at mid-proximal phalanx; A2 at volar plate of interphalangeal (IP) joint

Mechanisms of Injury
- Laceration: Dependent on finger position at time of injury
  - In flexion: Tendon injury distal to skin laceration
  - In extension: Tendon and skin laceration adjacent to each other
  - Open injury typically located in midsubstance rather than at insertion of tendon
- Tendinopathy/tenosynovitis

FLEXOR TENDON INJURY, WRIST AND FINGERS

Flexor Tendon Anatomic Zones

Zone	*Finger*	*Thumb*
I	Distal to FDS insertion (includes C3, A5 pulleys)	Distal to oblique pulley
II	Tendon within fibroosseous tunnel (A1pulley to C3 pulley)	Proximal to A1 pulley to oblique pulley
III	Distal to transverse ligament (carpal tunnel) to A1 pulley	Distal to transverse ligament (carpal tunnel) to A1 pulley
IV	Carpal tunnel	Carpal tunnel
V	Myotendinous junction to proximal carpal tunnel	Myotendinous junction to proximal carpal tunnel

These zones define unique anatomic characteristics, which influence evaluation, treatment, and prognosis in flexor tendon injury.

- May affect any tendon related to repetitive trauma, direct blow, or inflammatory or infectious arthritides
- Flexor carpi radialis (FCR): Tendon travels through a tight fibroosseous tunnel at proximal trapezium; vulnerable to trauma and stenosing tenosynovitis
- Trigger finger: Thickened tendon & sheath catch on A1 pulley

- Pulley injury: Extreme finger flexion with MCP extension, PIP flexion, and DIP extension
  - Results in progressive injury: A2 → A3 → A4 (A1 injury is rare)
  - Common among rock climbers
- Jersey finger: Results from sudden hyperextension during active flexion; most common in 4th finger

CLINICAL ISSUES

Presentation

- Most common signs/symptoms
  - Tendon tear
    - Pain and swelling
    - Weakened grip
    - Inability to form a fully clenched fist
    - Incomplete flexion of affected digit
  - Tendinopathy/tenosynovitis: Focal tenderness; decreased grip strength; pain with ROM
  - Trigger finger: Painful snapping ± locking
- Other signs/symptoms
  - Tendon may be a palpable nodule if retracted

Demographics

- Epidemiology
  - Flexor tendon injury less common than extensor injury
  - Open injury more common than closed injury
  - Zone II injuries most common & have poorest prognosis

Natural History & Prognosis

- Evaluation, treatment, and prognosis determined by zone of injury
  - Zone I: Isolated FDP injury only; only DIP flexion lost
  - Zone II: Historically referred to as "no-man's land" because outcomes were poor
  - Zone II-V: Involve both FDP and FDS; loss of PIP and DIP flexion
- Prompt primary repair of tendon laceration facilitates good outcome
- Laceration may interrupt nutritional system of tendon, resulting in slow repair
  - Loss of synovial fluid or direct injury to vincula tendina or digital arteries
- Delay in treatment results in weakness & dysfunction
- Delay in treatment of pulley injuries results in fibrosis and flexion contractures
- Jersey finger may be missed in acute phase → poorer outcome

Treatment

- Conservative
  - Partial (nondisplaced) tear: Splinting
  - Tendinopathy/tenosynovitis: Splinting, NSAIDs, physical therapy
- Surgical intervention
  - All complete flexor tendon injuries should be surgically repaired
    - Some controversy regarding surgery for partial tear
    - Zone II injuries difficult to treat due to adhesions and fibrosis
- Surgical goals
  - Strong, undistorted tendon repair with uninjured gliding surfaces
  - Preservation of tendon nutrition
- Complications
  - Adhesions
  - Re-rupture of repaired tendon (most common on 7-10th day postop)

DIAGNOSTIC CHECKLIST

Consider

- Subtle malalignment may herald a more serious injury
  - Dislocations may be reduced by the time imaging takes place

Image Interpretation Pearls

- Beware of satisfaction of search
  - If flexor tendon injury present, look at volar plate and collateral ligament integrity as well

SELECTED REFERENCES

1. Lisle DA et al: MR imaging of traumatic and overuse injuries of the wrist and hand in athletes. Magn Reson Imaging Clin N Am. 17(4):639-54, vi, 2009
2. Schöffl VR et al: Finger pain in rock climbers: reaching the right differential diagnosis and therapy. J Sports Med Phys Fitness. 47(1):70-8, 2007

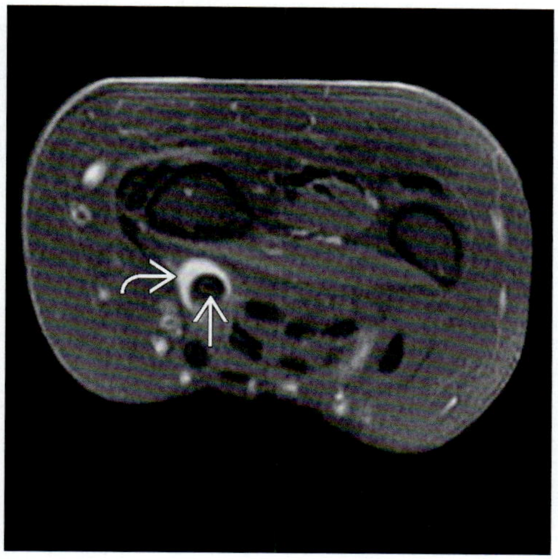
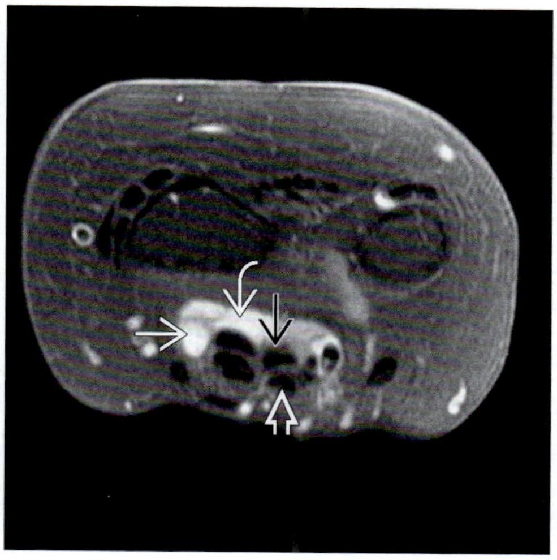

(Left) *Axial T2WI FS MR through the distal forearm demonstrates fluid [curved arrow] around a mildly thickened flexor pollicis longus (FPL) tendon [arrow]. Other flexor tendons are normal.* ***(Right)*** *Axial T2WI FS MR at the radiocarpal joint from the same patient shows absence of the flexor pollicis tendon [arrow] due to acute rupture. Fluid fills the defect in the tendon sheath and extends medially [curved arrow]. Flexor digitorum profundus [arrow] and superficialis [open arrow] tendons are normal.*

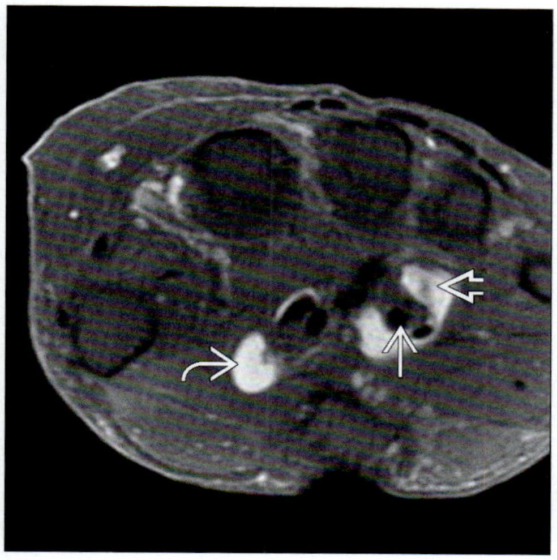
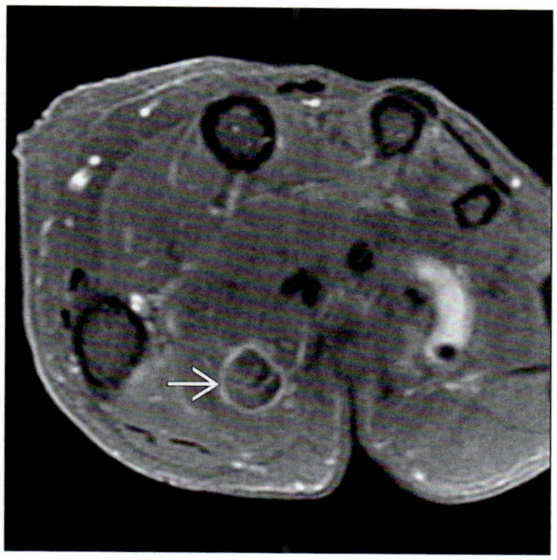

(Left) *Axial T2WI FS MR at the level of the proximal metacarpals from the same patient shows fluid [curved arrow] filling the empty sheath of the flexor pollicis longus tendon. There is fluid [open arrow] around the FDP and FDS [arrow] of the medial hand without tendon rupture.* ***(Right)*** *Axial T2WI FS MR at the level of the distal metacarpals from the same patient shows reconstitution of the flexor pollicis longus tendon [arrow]. The tendon is thickened and contains increased signal, consistent with acute edema from injury.*

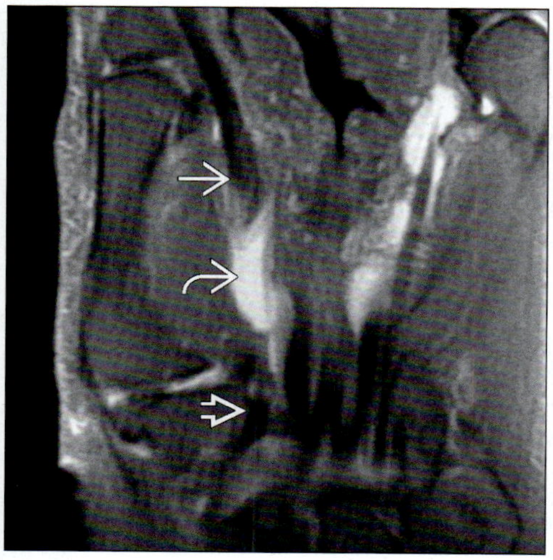
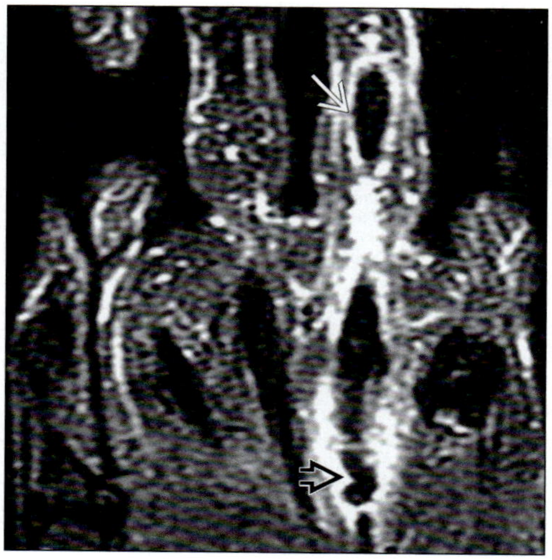

(Left) *Coronal T2WI FS MR from the same patient shows the distal portion of the ruptured flexor pollicis longus tendon [arrow] with fluid [curved arrow] filling the sheath at the site of rupture. The proximal portion of the tendon [open arrow] is at the thumb base. Axial images are more accurate than long-axis images for identifying site of tendon rupture.* ***(Right)*** *Coronal STIR MR shows an acute 4th FDP tear. The torn tendon is withdrawn to the PIP joint [arrow] (type II). The retracted tendon is redundant [open arrow].*

FLEXOR TENDON INJURY, WRIST AND FINGERS

*(**Left**) Sagittal T2WI FS MR shows rupture of the flexor digitorum profundus at the level of the proximal phalanx with retraction to the metacarpal. The flexor digitorum superficialis tendon is intact. (**Right**) Axial PDWI MR shows acute rupture of the palmaris longus tendon. The tendon is absent from its usual location in the volar lateral forearm, and there is amorphous soft tissue swelling in the gap. A skin marker indicates the site of pain.*

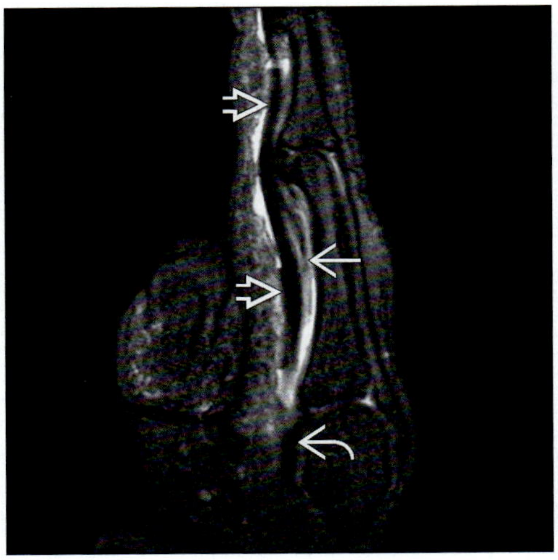

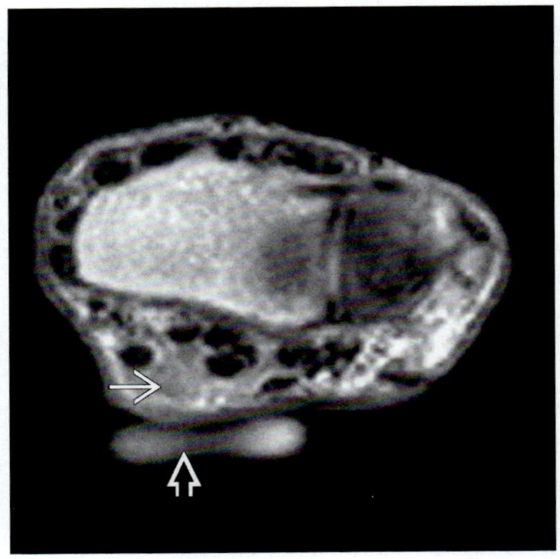

*(**Left**) Axial T2WI FS MR at the level of the metacarpal heads shows rupture of the 2nd ray flexor digitorum superficialis tendon with tenosynovitis. The profunda tendon is intact. (**Right**) Axial T1WI MR at the proximal metacarpal level shows partial tear of the flexor digitorum superficialis of the 5th ray. The tendon is thickened and irregular and demonstrates heterogeneous signal. The profunda tendon is intact.*

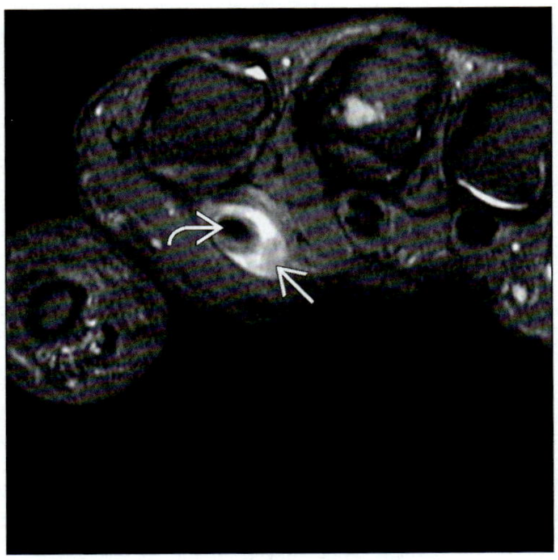

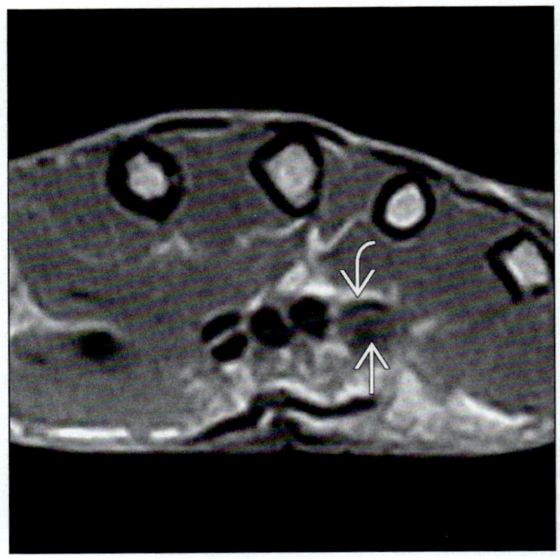

*(**Left**) Axial T2WI FS MR shows a longitudinal split tear of the flexor digitorum superficialis tendon of the 2nd finger. There is thinning of the profunda tendon representing a partial tear. Tendons of the 3rd finger are normal. (**Right**) Axial T1WI MR shows marked thickening and heterogeneity of the flexor digitorum profundus tendon of the 4th finger, consistent with a partial tear. The superficialis tendon is also somewhat thickened and irregular, likely a partial tear.*

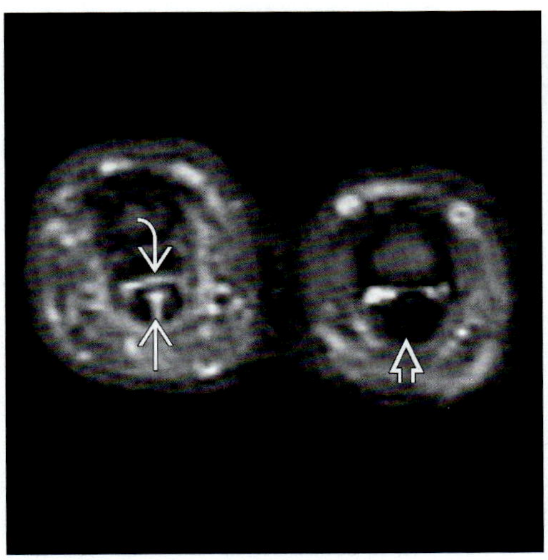

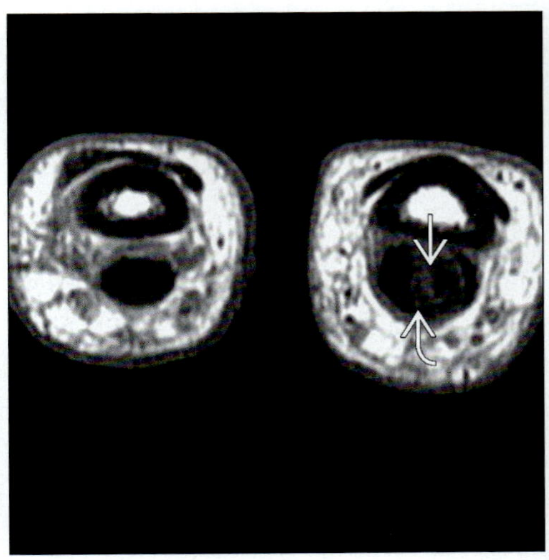

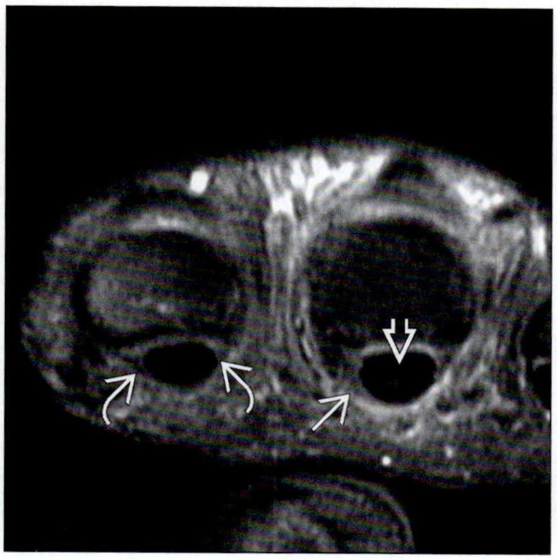

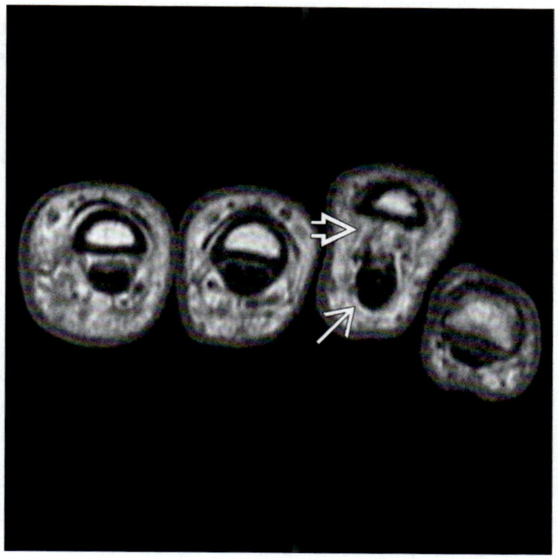

(Left) *Axial PDWI FS MR shows a tear of the A2 pulley ➔ of the 3rd finger with surrounding edema. The flexor tendons ➔ are normal. The A3 pulley of the 2nd finger ➔ is also normal.* ***(Right)*** *Axial T1WI MR shows volar displacement of the flexor tendons ➔ of the 4th finger relative to the phalangeal shaft, indicating an A2 pulley tear ➔. The absence of edema indicates that this bowstring deformity is a chronic process.*

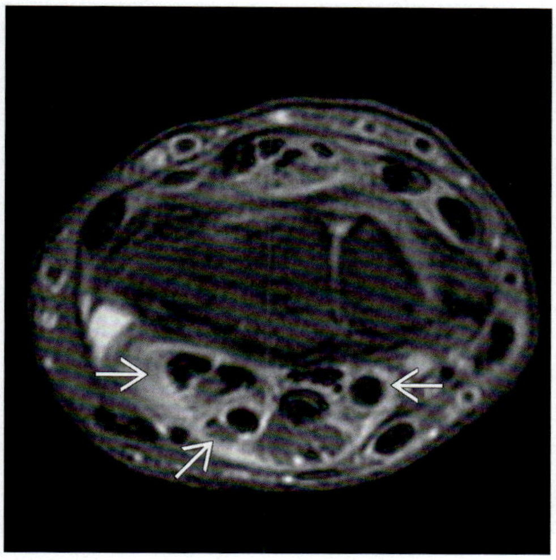

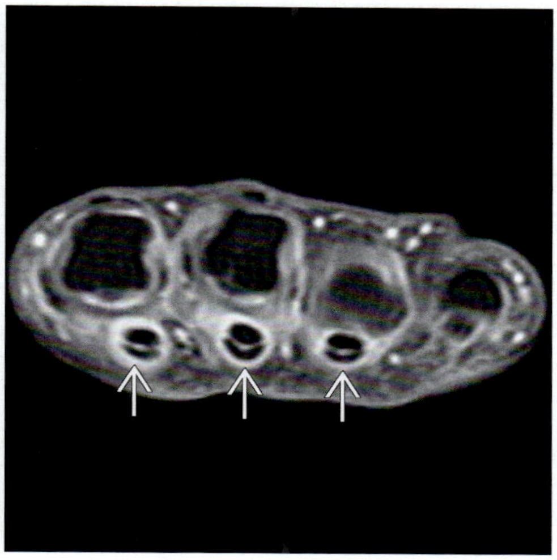

(Left) *Axial T2WI FS MR at the level of the radiocarpal joint shows diffuse edema ➔ surrounding the myotendinous junctions of the flexor units at the proximal aspect of the carpal tunnel. The tendons are intrinsically normal. This is an unusual manifestation of tenosynovitis.* ***(Right)*** *Axial T1WI C+ FS MR shows diffuse enhancement ➔ of the synovial sheaths of the flexor tendons, consistent with tenosynovitis. The tendons are normal.*

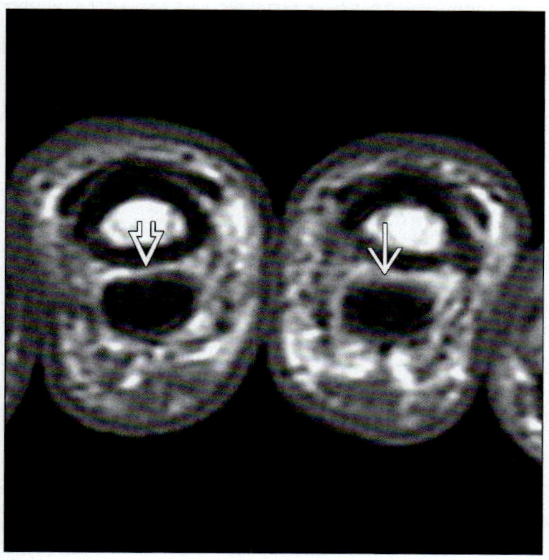

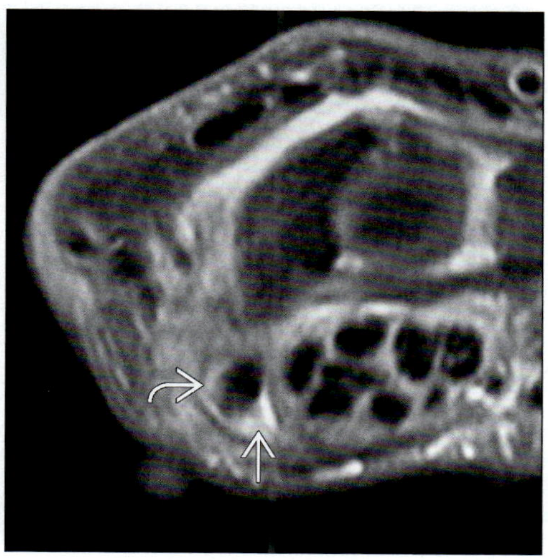

(Left) *Axial T1WI MR shows mild irregularity ➔ of the margins of the flexor tendons of the 4th finger without signal change. This is a subtle manifestation of tendinopathy. The adjacent 3rd finger flexors ➔ are normal.* ***(Right)*** *Axial PDWI FS MR demonstrates thickening and subtle increased signal ➔ in the flexor carpi radialis indicating a partial intrasubstance tear and mild tenosynovitis ➔. This injury results from repetitive gripping motion.*

EXTENSOR TENDON INJURY, WRIST AND FINGERS

Key Facts

Terminology

- Boxer knuckle: Extensor hood sagittal band tear at MCP with subluxation/dislocation of common extensor tendon, results from direct blow while joint flexed and finger ulnarly deviated
- Mallet finger (drop finger): Inability to extend DIP joint due to terminal extensor tendon avulsion
- Boutonniere deformity: PIP joint flexion with DIP joint extension due to extensor tendon central slip injury at or near insertion on middle phalanx
- de Quervain tenosynovitis (washerwoman's sprain, mother's wrist): 1st extensor compartment (EPB & APL) tenosynovitis
- Intersection syndrome: Radial wrist extensor tenosynovitis at 1st (EPB & APL) and 2nd (ECRB & ECRL) extensor compartment intersection

Imaging

- Tendon tear: Complete or partial tendon discontinuity
- Tendinopathy: Diffuse or focal tendon thickening ± ↑ intrasubstance signal
- Tenosynovitis: ↑ SI in tendon sheath
- Avulsion: Triangular bone fragment

Clinical Issues

- May maintain ROM with < 80% tendon lacerated
- Wrist extension maintained with isolated tendon loss
- Central slip injury: May be missed acutely as PIP joint still extends due to intact lateral bands
- Tendinopathy/tenosynovitis: Tenderness radiating proximally; decreased grip strength; pain with normal ROM

(Left) Lateral radiograph shows an intraarticular fracture of the dorsal base of the distal phalanx with a triangular fragment ➔. The fragment is avulsed by the extensor terminal tendon. This injury results from forced flexion against an extended finger and is known as a "mallet," "baseball," or "drop" finger. (Right) Sagittal T2WI FS MR shows avulsion of the extensor central slip ➔ from the dorsal base of the middle phalanx. There is no associated fracture.

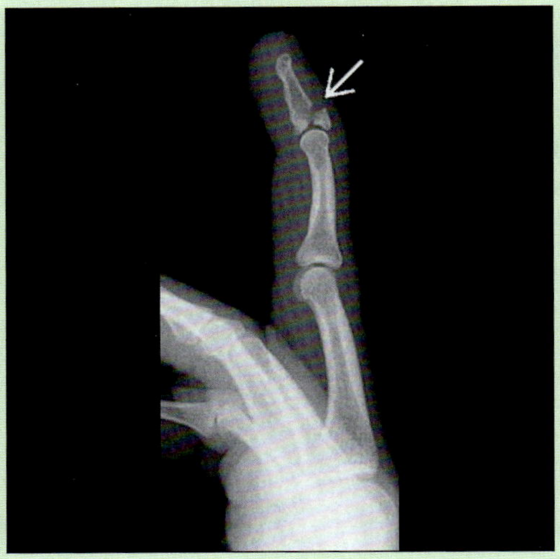

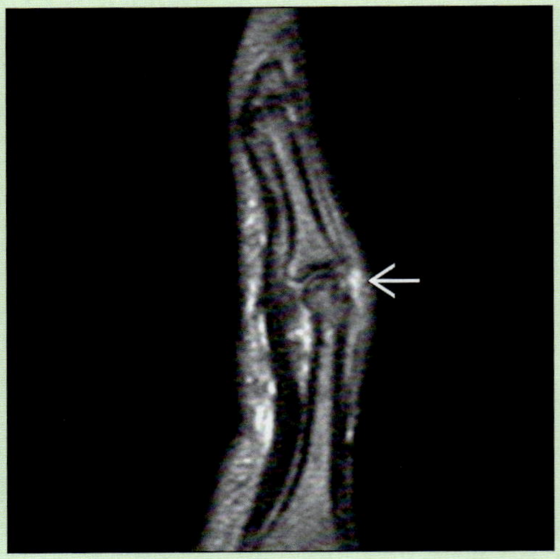

(Left) Axial graphic shows a cross section of the MCP joint. The medial sagittal band is torn ➔ allowing lateral subluxation of the extensor tendon ➔. This injury is known as boxer's knuckle. (Right) Axial STIR MR from a 23-year-old boxer with pain over the dorsal 3rd MCP shows disruption ➔ of the sagittal band allowing lateral subluxation of the extensor tendon ➔. Occasionally, such tears are not seen in neutral position and are better seen on MR obtained with the hand in a fisted position.

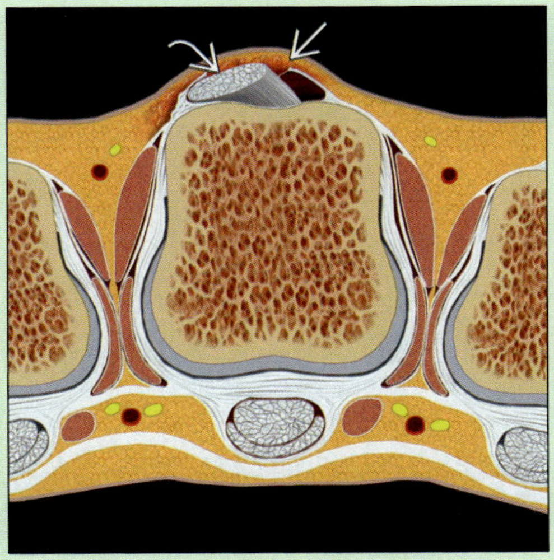

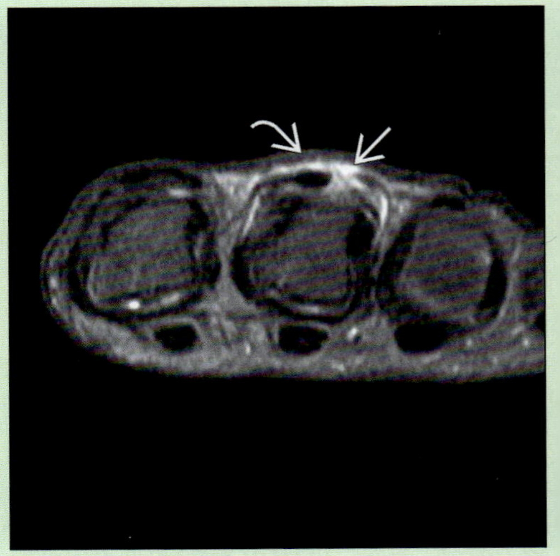

EXTENSOR TENDON INJURY, WRIST AND FINGERS

TERMINOLOGY

Definitions

- Boxer knuckle: Extensor hood sagittal band tear at metacarpophalangeal (MCP) joint with subluxation/ dislocation of common extensor tendon, results from direct blow while joint flexed and finger ulnarly deviated
- Mallet finger (drop finger): Inability to extend distal interphalangeal (DIP) joint due to terminal extensor tendon avulsion
- Boutonniere deformity: Proximal interphalangeal (PIP) joint flexion with DIP joint extension due to extensor tendon central slip injury at or near insertion on middle phalanx, ± avulsion fracture of dorsal base of middle phalanx
- de Quervain tenosynovitis (washerwoman's sprain, mother's wrist): 1st extensor compartment (extensor pollicis brevis [EPB] and abductor pollicis longus [APL]) tenosynovitis
- Fight bite: Opponent's tooth breaks skin over MCP joint with injury to dorsal capsule, extensor tendon ± metacarpal (MC) head
- Gamer thumb/finger (gamer grip, Nintendo thumb, Nintendo-itis, PlayStation thumb, iPod thumb, BlackBerry thumb, text-messaging syndrome): Repetitive injury resulting in pain + swelling & tendinopathy ± tenosynovitis
- Intersection syndrome: Radial wrist extensor tenosynovitis at 1st (APL & EPB) and 2nd (extensor carpi radialis brevis [ECRB] & extensor carpi radialis longus [ECRL]) extensor compartment intersection
- Swan-neck deformity: PIP joint extension with DIP joint flexion; results from volar plate injury at PIP or extensor tendon injury at DIP
- Duckbill deformity (hitchhiker thumb): Thumb interphalangeal (IP) joint hyperextension and MCP flexion; commonly seen in rheumatoid arthritis (RA), may also develop after tendon injury, results with stretching of volar plate at attachment of extensor tendon on distal phalanx (similar to swan-neck deformity)

IMAGING

General Features

- Best diagnostic clue
  - Tendon tear: Complete or partial tendon discontinuity
  - Tendinopathy: Diffuse or focal tendon thickening ± ↑ intrasubstance signal
  - Tenosynovitis: ↑ signal intensity (SI) in tendon sheath on T2WI
  - Avulsion: Triangular bone fragment

Radiographic Findings

- Soft tissue swelling
- Loss of fat planes
- Avulsion fracture

CT Findings

- Useful to evaluate associated injuries, such as fracture fragments

MR Findings

- Tendon injury
  - Complete tear: Discontinuity & distortion of tendon
    - T2WI: ↑ SI at site of tendon discontinuity
    - Surrounding edema & hemorrhage
  - Partial tear: Focal area of ↑ SI on T1- & T2WI; portion of tendon remains intact
- Tendinopathy
  - T1WI: Focal or diffuse tendon enlargement
  - T2/STIR: Intermediate SI in tendon
  - Most common: 6th compartment (extensor carpi ulnaris [ECU]) tendinopathy
    - Need to distinguish from ECU subluxation
    - Often early finding in rheumatoid arthritis
- Tenosynovitis
  - ↑ SI of fluid in sheath on T2WI
    - Diameter of tendon sheath fluid > diameter of tendon
      - Less fluid = physiologic
  - de Quervain tenosynovitis: 1st compartment (APL & EPB)
- Fight bite: MC capsule or tendon laceration typically proximal to skin laceration; high risk for infection
- Mallet deformity: ↑ SI in terminal tendon at dorsal base of distal phalanx ± fracture fragment
- Central slip injury: ↑ SI in tendon at dorsal base of middle phalanx ± fracture fragment
- Boxer knuckle: Sagittal band injury at MCP
  - Focal thickening, poor definition, or discontinuity of sagittal band with extensor tendon subluxation
- Intersection (crossover) syndrome
  - Proximal: 1st compartment (APL & EPB) crosses over 2nd (ECRB & ECRL) ~ 4 cm proximal to Lister tubercle
    - T2/STIR: ↑ SI surrounding myotendinous junctions of involved tendons & adjacent fat
  - Distal: 3rd compartment (EPL) crosses over 2nd (ECRB & ECRL) distal to retinaculum (less common)
    - T2/STIR: ↑ SI within tendon sheath & peritendinous soft tissues
- Extensor pollicis longus (EPL) rupture: Absence of tendon in 3rd extensor compartment
  - Potential complication of distal radial fracture or placement of fracture fixation devices

Ultrasonographic Findings

- Tendon tear: Fiber disruption; anechoic foci = tendon tear; identifiable retracted tendon stump
- Tenosynovitis: Sheath distention; fluid may be anechoic or slightly complex
- Tendinopathy: Hypoechoic regions in swollen tendon
- Dynamic scanning to accentuate abnormalities
  - Boxer knuckle: Tendon subluxation more evident when joint flexed

Imaging Recommendations

- Best imaging tool
  - MR

PATHOLOGY

Mechanisms of Injury

- Mallet finger: Tip of finger struck by or against object

EXTENSOR TENDON INJURY, WRIST AND FINGERS

Extensor Tendon Anatomic Zones

Zone	*Finger*	*Thumb*
I	DIP joint	IP joint
II	Middle phalanx	Proximal phalanx
III	PIP joint	MCP joint
IV	Proximal phalanx	Metacarpal
V	MCP joint	CMC joint/radial styloid
VI	Metacarpal	
VII	Dorsal retinaculum	
VIII	Distal forearm	

These zones define unique anatomic characteristics that influence evaluation, treatment, and prognosis in flexor tendon injury.

- Central slip injury: Blow to dorsum of middle phalanx, acute violent PIP joint flexion or volar dislocation of PIP joint
- Boxer knuckle: Direct blow to clenched fist with injury sustained at the MCP
- Fight bite: MCP flexed, contacts opponent's teeth; results in tendon laceration
- Rotational longitudinal compression of semi-flexed PIP joint
- Tendinopathy/tenosynovitis
  - Related to repetitive trauma, direct blow, inflammatory or infectious arthritides
  - May affect any tendon
    - Most common in areas where tendons are constricted, e.g., ECU as it passes over ulnar groove
  - de Quervain syndrome: 1st extensor compartment tendons travel over radial styloid in fibroosseous groove
    - Repetitive grasping motions (e.g., racket sports)
  - ECU: Tendon travels over ulnar styloid in fibroosseous tunnel
    - Repetitive ulnar deviation (e.g., golf, baseball)
  - Intersection syndrome: Inflammation resulting from friction as tendons from one extensor compartment cross over another

CLINICAL ISSUES

Presentation

- Most common signs/symptoms
  - May maintain ROM with < 80% tendon torn
  - Tendinopathy/tenosynovitis: Tenderness radiating proximally; decreased grip strength; pain with normal ROM
  - Intersection syndrome: Tenderness, erythema, edema; worse with flexion & extension

Demographics

- Epidemiology
  - Mallet finger: Most common closed sports injury
  - Injuries to extensor tendons more common than to flexors
  - de Quervain tenosynovitis: Most common stenosing tenosynovitis
  - EPL rupture: Distal radius fracture/repair potential complication

Natural History & Prognosis

- Evaluation, treatment, and prognosis considered by zone of injury
- Many injuries respond to conservative therapy
- Delay in treatment results in loss of function
  - Tendon tear: Limited range of motion from adhesions
  - Mallet finger: May progress to swan-neck deformity
  - Boutonniere deformity: May occur gradually after central slip injury

Treatment

- Conservative
  - Tendon injury/avulsion: Splinting
  - Mallet finger: Closed reduction if closed injury
  - Tendinopathy/tenosynovitis: Conservative therapy
- Surgical
  - Failed conservative therapy
  - Irreducible deformities
  - Large displaced intraarticular fracture
  - Open injuries
  - Tendon tear: Primary repair; particularly if > 40% of tendon torn
    - When considering tendon transfers: ECRB not used because resultant wrist extension will deviate radial or ulnar
- Complications
  - Contractures
  - Loss of range of motion
  - Wound infection

DIAGNOSTIC CHECKLIST

Consider

- Subtle malalignment may herald a more serious injury
- Dislocations may be reduced by the time imaging takes place

SELECTED REFERENCES

1. Lee RP et al: Extended MRI findings of intersection syndrome. Skeletal Radiol. 38(2):157-63, 2009
2. Shin AY et al: Ulnar-sided wrist pain: diagnosis and treatment. Instr Course Lect. 54:115-28, 2005
3. Lopez-Ben R et al: Boxer knuckle (injury of the extensor hood with extensor tendon subluxation): diagnosis with dynamic US--report of three cases. Radiology. 228(3):642-6, 2003
4. Clavero JA et al: MR imaging of ligament and tendon injuries of the fingers. Radiographics. 22(2):237-56, 2002

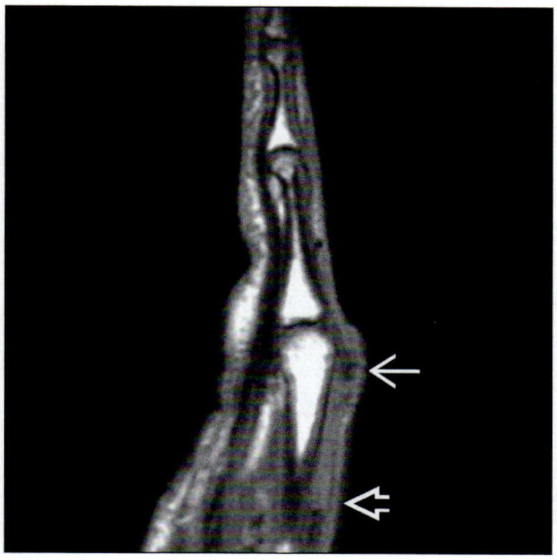

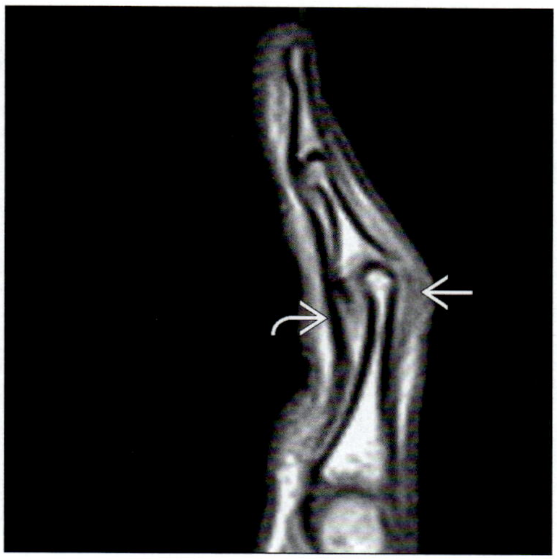

(Left) *Sagittal T1WI MR demonstrates a tear of the extensor tendon with retraction of the torn tendon fibers. The tendon was lacerated at the metacarpal head and retracted proximally.* ***(Right)*** *Sagittal T1WI MR shows complete disruption of the extensor tendon of the 5th finger after a hyperflexion injury. The retracted tendon is not identified, but there is soft tissue swelling at the level of the PIP joint. The flexor tendon is normal.*

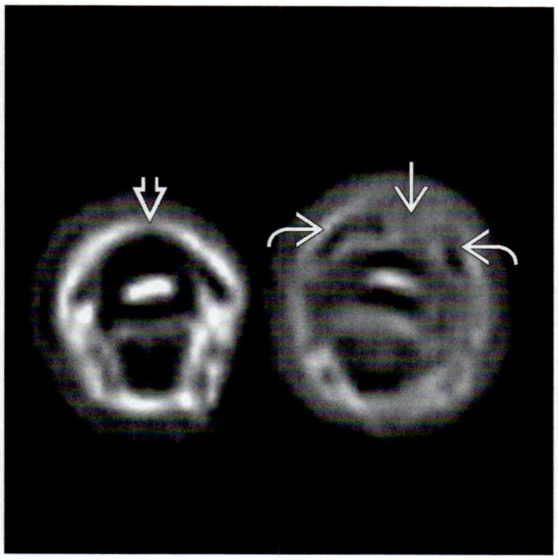

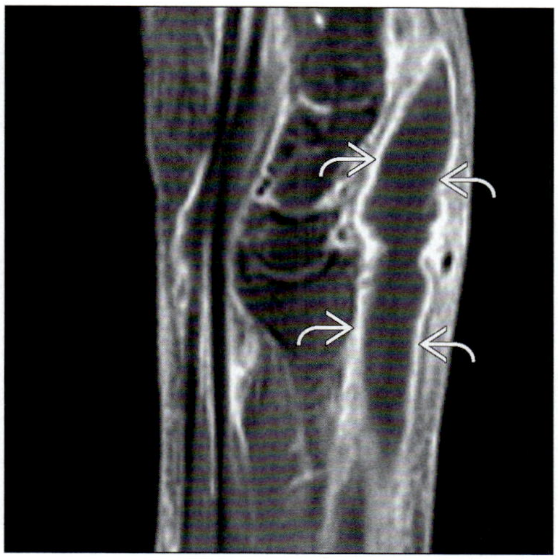

(Left) *Axial T1WI MR at the level of the PIP joints in the same patient shows the normal 4th finger extensor mechanism and the disrupted extensor tendon of the 5th finger. Note that the radial and ulnar conjoined tendons are intact at the site of central tendon rupture.* ***(Right)*** *Sagittal T1WI C+ FS MR shows massive septic tenosynovitis of the extensor tendons of the distal forearm, wrist, and hand. There is florid enhancement of the synovium surrounding the septic fluid.*

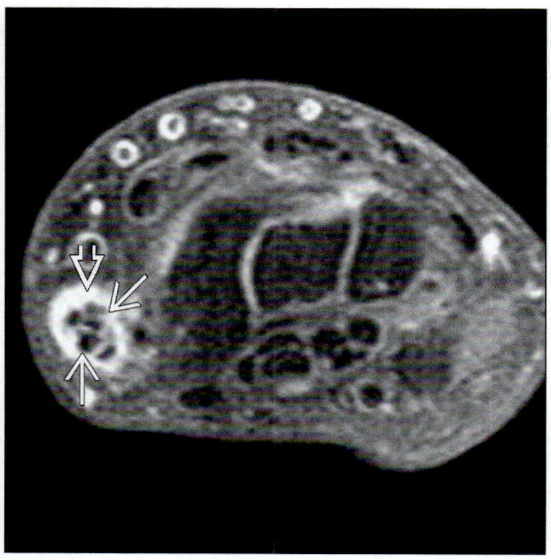

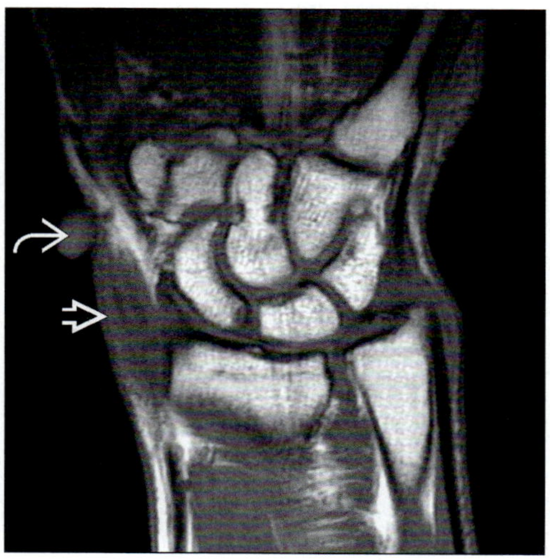

(Left) *Axial T2WI FS MR shows de Quervain tenosynovitis. Fluid distends the tendon sheaths surrounding the thickened and fragmented 1st extensor compartment tendons (abductor pollicis longus and extensor pollicis brevis).* ***(Right)*** *Coronal T1WI MR from a patient complaining of radial-sided wrist and thumb pain related to work activity shows a skin marker placed adjacent to thickened and hyperintense APL and EPB tendons consistent with de Quervain tenosynovitis.*

(Left) *Coronal STIR MR demonstrates thickening and increased intrasubstance signal of the extensor carpi ulnaris tendon → as it passes over the ulnar groove. There is associated tenosynovitis →.* ***(Right)*** *Axial T2WI MR demonstrates thickening of the extensor carpi ulnaris (ECU) tendon → with extensive surrounding tenosynovitis →. The ECU should be comma-shaped at the level of the distal ulna; a round cross-sectional appearance is abnormal and suggests tendinopathy.*

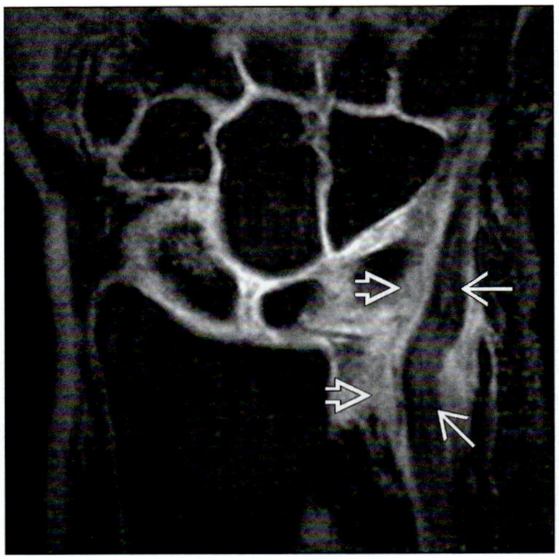

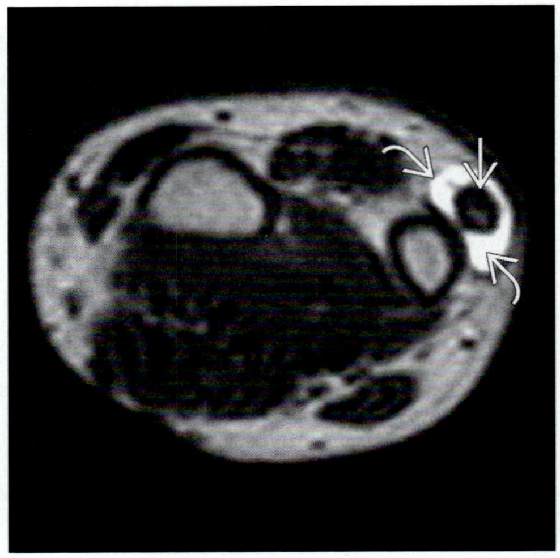

(Left) *Axial PDWI MR from a 47-year-old woman with painful snapping in the dorsal-ulnar aspect of the wrist when pronating and supinating the hand demonstrates ulnar (medial) subluxation of the extensor carpi ulnaris tendon → relative to the ulnar styloid →. This image was obtained with wrist in pronation; on supination, tendon returned to normal position.* ***(Right)*** *Coronal T1WI MR from the same patient shows medial displacement and diffuse thickening of the ECU →.*

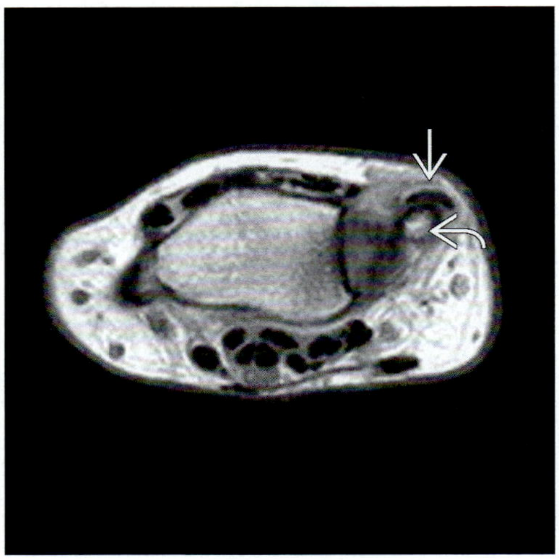

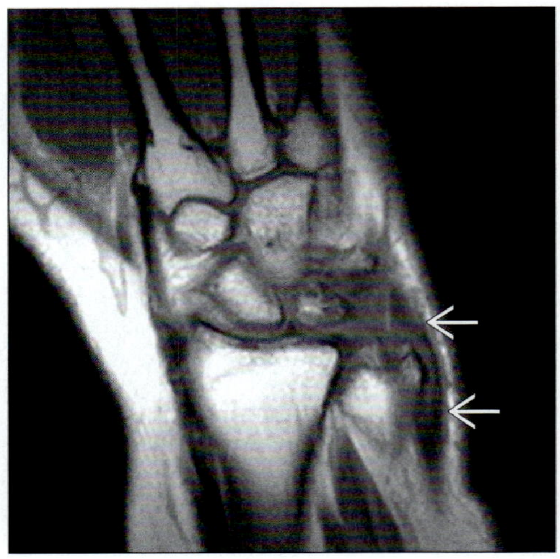

(Left) *Axial T2WI FS MR demonstrates disruption → of the lateral band on the radial aspect of the 4th PIP joint with fluid filling the gap. The extensor tendon → is intact. Patient was in a fist fight, and a laceration caused this disruption (a "fight bite").* ***(Right)*** *Sagittal T2WI FS MR shows extensive edema around the extensor tendon attachment to the base of the middle phalanx of the 2nd ray. This is a partial tear of the lateral slip →, which contributes to the conjoined tendon.*

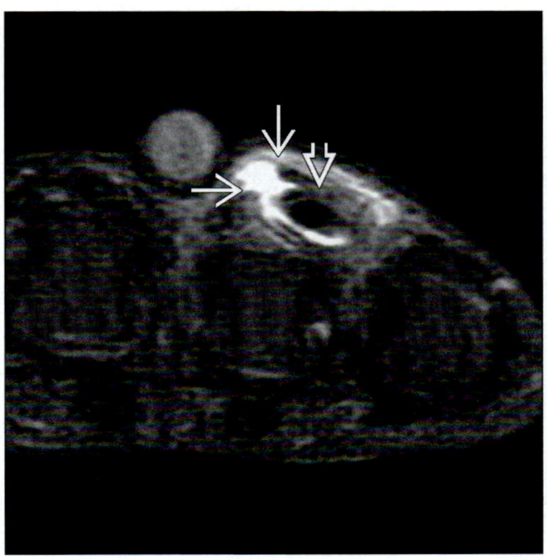

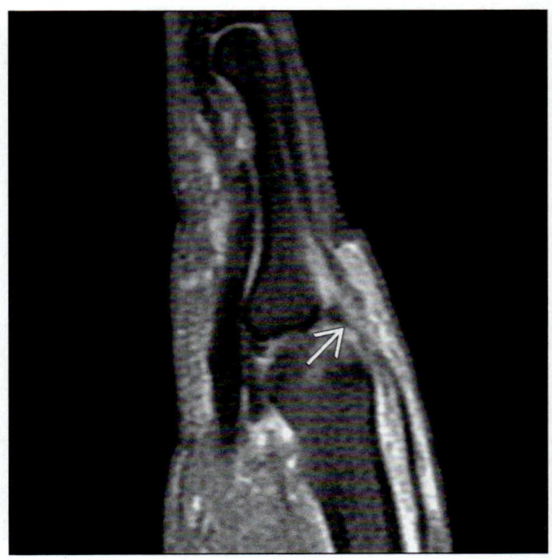

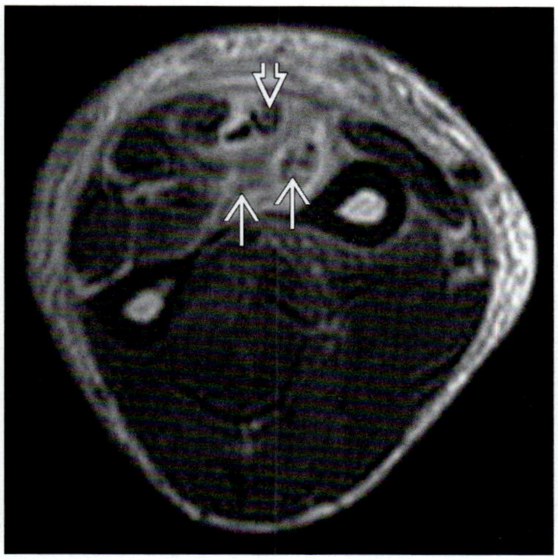

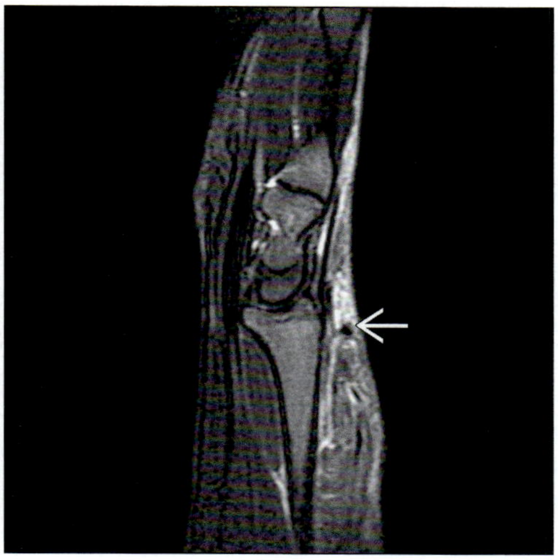

(Left) *Axial T2WI MR through the mid-forearm of a college football quarterback with sudden onset of severe dorsal forearm pain during practice shows marked edema and disorganization of the extensor pollicis longus & brevis and extensor digitorum muscles.* ***(Right)*** *Sagittal T2WI FS MR from the same patient shows complete disruption and retraction of the extensor pollicis brevis muscle to the level of the distal radius. The tendon was surgically repaired.*

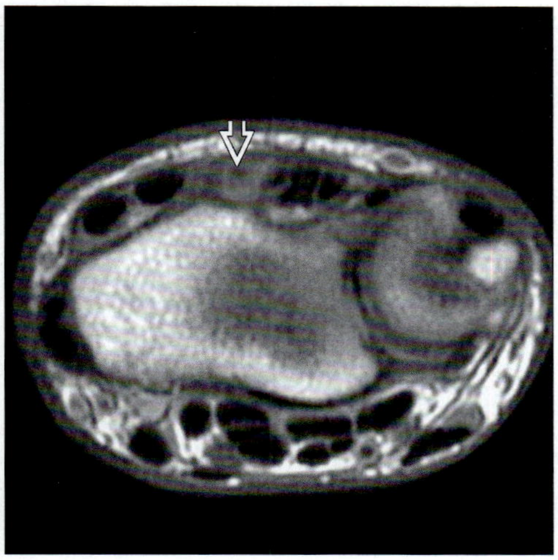

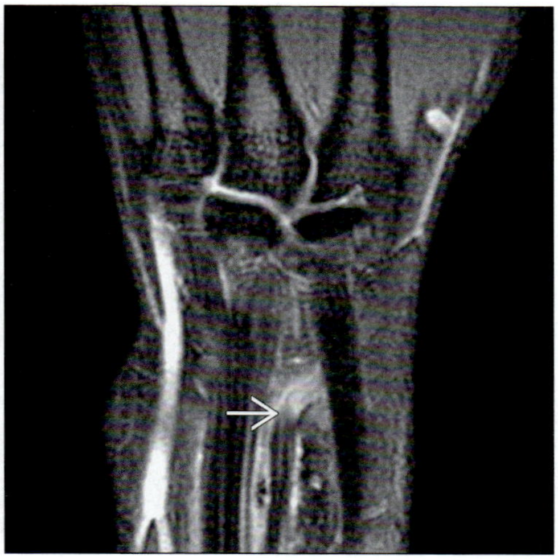

(Left) *Axial PDWI MR shows absence of the extensor pollicis longus tendon from its usual location in the 3rd extensor compartment. This patient had a distal radial fracture several months prior and now complained of thumb weakness.* ***(Right)*** *Coronal PDWI FS MR from the same patient shows complete disruption and retraction of the extensor pollicis longus in the distal forearm. This tendon in particular is vulnerable to trauma from distal forearm fracture and repair.*

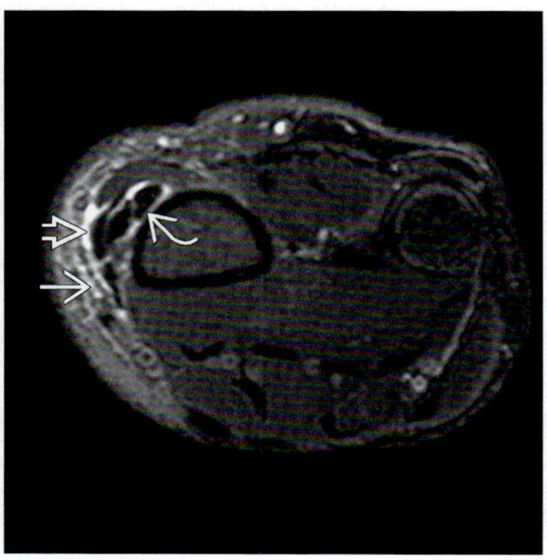

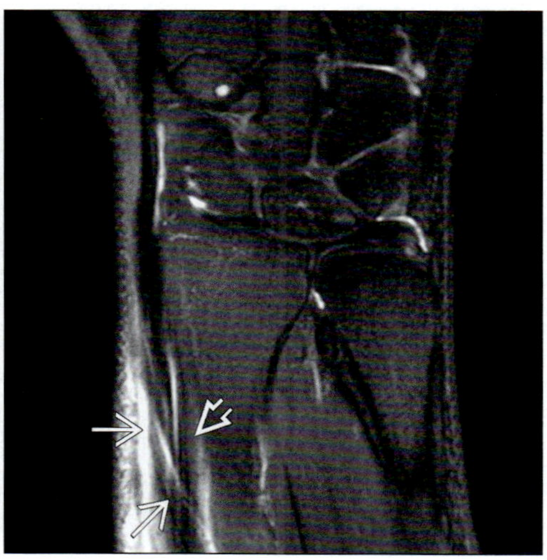

(Left) *Axial T2WI FS MR demonstrates proximal intersection syndrome with tenosynovitis of EPB and APL tendons as they cross over ECRB/ECRL tendons in the distal forearm proximal to the extensor retinaculum. Tendons may be normal, as in this case, or thickened.* ***(Right)*** *Coronal STIR MR from the same patient demonstrates soft tissue edema around the APL & EPB tendons as they pass over the ECRB & ECRL tendons proximal to the extensor retinaculum.*

GANGLION CYST, WRIST

Key Facts

Imaging
- Adjacent to joint or tendon sheath
- Dorsal or volar to wrist
- Smooth, well-defined margin; round, oblong, or teardrop-shaped
  - Often multilocular
  - Often see narrow neck extending to joint
- Homogeneous low signal intensity on T1WI
- High signal intensity on T2WI, STIR, ± signal heterogeneity
- Rim enhancement with gadolinium administration

Top Differential Diagnoses
- Synovitis
- Nerve sheath tumor
- Giant cell tumor tendon sheath

Pathology
- Often associated with ligament tears
- Narrow communication with joint at tear site
- Large cyst may compress median or ulnar nerve

Clinical Issues
- Reported in up to 50% of asymptomatic wrists based on MR evaluation
- Often resolve spontaneously
- Many ganglion cysts are asymptomatic
- Dorsal ganglion cysts are more likely to cause pain

Diagnostic Checklist
- Presence of ganglion cyst should raise suspicion for ligament tear
- Post-contrast images to differentiate cyst from solid mass

__(Left)__ Coronal graphic shows the typical appearance of a ganglion cyst ➡: A balloon-like fluid collection connected by a narrow neck ⮫ to the joint. In this case, the cyst arises from a perforation of the volar band of the scapholunate ligament ⇨. __(Right)__ Sagittal PDWI FS MR shows a ganglion cyst ➡ arising from a tear of the triangular fibrocartilage. The adjacent, distended pisiform-triquetral joint recess ➡ has a wide communication with the joint, unlike the ganglion cyst.

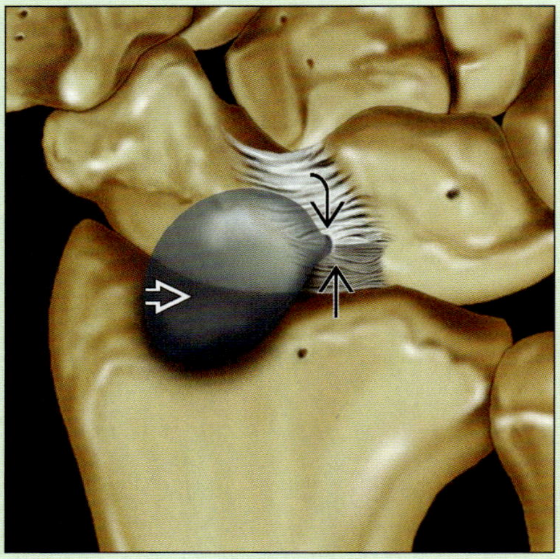

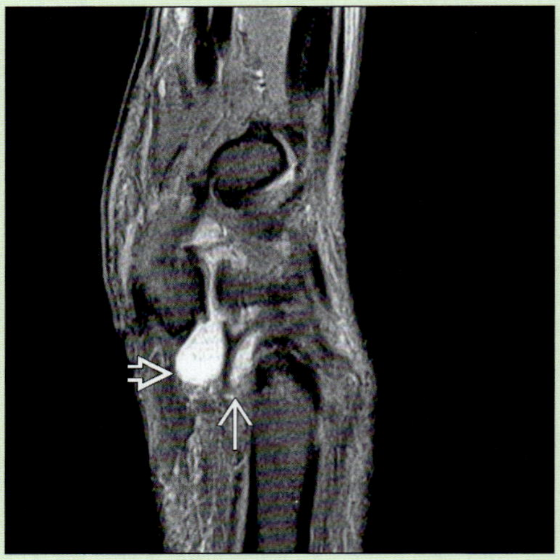

__(Left)__ Sagittal PDWI FS MR shows a multiloculated ganglion cyst ➡. The site of communication with the joint is not apparent in this plane. __(Right)__ Axial T2WI FS MR from the same patient shows a narrow neck ➡ connecting the ganglion cyst ⇨ with the joint. The communication is through a tear in the volar band of the scapholunate ligament. Cysts are a useful secondary sign of ligament tear.

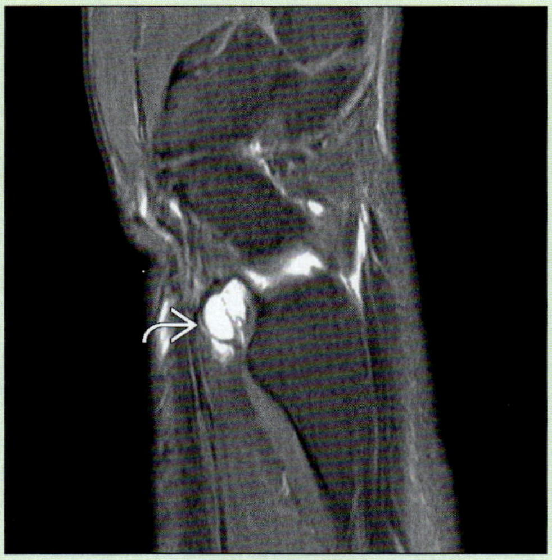

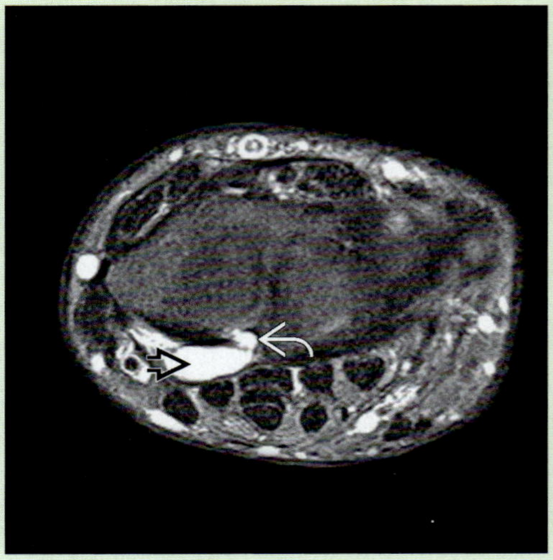

GANGLION CYST, WRIST

TERMINOLOGY

Synonyms

- Ganglion
- Synovial cyst

IMAGING

General Features

- Location
  - Adjacent to joint or tendon sheath
  - Dorsal or volar to wrist
- Morphology
  - Smooth, well-defined margin; round, oblong, or teardrop-shaped
    - May have lobulated shape conforming to adjacent structures
  - Often multilocular
  - Often see narrow neck extending to joint

MR Findings

- Unilocular or multilocular; each loculation shows
  - Homogeneous low signal intensity on T1WI
  - High signal intensity on T2WI, STIR, ± signal heterogeneity
  - Rim enhancement with gadolinium administration
- In many cases, ligament tear visible adjacent to cyst on MR

Ultrasonographic Findings

- Cysts are often multilocular and contain low-level echoes
  - Flow may be seen on Doppler evaluation

Imaging Recommendations

- Best imaging tool
  - MR

DIFFERENTIAL DIAGNOSIS

Synovitis

- Involves a larger area of joint
- Recesses often crescentic in shape
  - Side communication of joint recess with joint

Nerve Sheath Tumor

- Uniform enhancement with gadolinium
- Fusiform shape
- Parallels course of nerve

Giant Cell Tumor Tendon Sheath

- Heterogeneous signal intensity on T1 & T2WI MR
- ± areas of low signal intensity on all sequences due to hemosiderin deposition
- Intense heterogeneous enhancement with gadolinium

Rheumatoid Arthritis, Wrist and Hand

- Synovium irregular & thickened
- Erosions ± bone marrow edema usually visible on MR before visible on radiographs

PATHOLOGY

General Features

- Etiology
  - Fluid or mucin extravasates through joint capsule or tendon sheath and coalesces into cyst
- Associated abnormalities
  - Often associated with ligament tears
    - Narrow communication with joint at tear site
  - Large cyst may compress median or ulnar nerve

Gross Pathologic & Surgical Features

- Not truly a cyst (i.e., no epithelial lining)

CLINICAL ISSUES

Presentation

- Most common signs/symptoms
  - Many ganglion cysts are asymptomatic
  - Dorsal ganglion cysts are more likely to cause pain

Demographics

- Age
  - May occur at any age
  - Most common in young adults
- Gender
  - M < F
- Epidemiology
  - Reported in up to 50% of asymptomatic wrists based on MR evaluation

Natural History & Prognosis

- Often resolve spontaneously

Treatment

- Aspiration associated with 50% recurrence rate
  - Underlying ligament tear must be repaired

DIAGNOSTIC CHECKLIST

Consider

- Presence of synovial cyst should raise suspicion for ligament tear
- Post-contrast images to differentiate cyst from solid mass

SELECTED REFERENCES

1. Ergun T et al: Magnetic resonance imaging in the visualization of benign tumors and tumor-like lesions of hand and wrist. Curr Probl Diagn Radiol. 39(1):1-16, 2010
2. Goldsmith S et al: Magnetic resonance imaging in the diagnosis of occult dorsal wrist ganglions. J Hand Surg Eur Vol. 33(5):595-9, 2008
3. Teefey SA et al: Ganglia of the hand and wrist: a sonographic analysis. AJR Am J Roentgenol. 191(3):716-20, 2008
4. Anderson SE et al: MRI for differentiating ganglion and synovitis in the chronic painful wrist. AJR Am J Roentgenol. 186(3):812-8, 2006
5. Lowden CM et al: The prevalence of wrist ganglia in an asymptomatic population: magnetic resonance evaluation. J Hand Surg Br. 30(3):302-6, 2005
6. Thornburg LE: Ganglions of the hand and wrist. J Am Acad Orthop Surg. 7(4):231-8, 1999

GANGLION CYST, WRIST

(Left) Sagittal T2WI FS MR shows a multiseptate ganglion cyst → arising dorsal to the lunate and superficial to the extrinsic carpal ligaments ↪, the typical location for such cysts. Identifying the precise site of origin can be difficult. (Right) Sagittal T2WI FS MR shows a complex multilobular cyst ↪ arising from the pisiform-triquetral recess → of the wrist. Differentiation of large joint space excrescences from joint ganglia can be difficult and may be arbitrary.

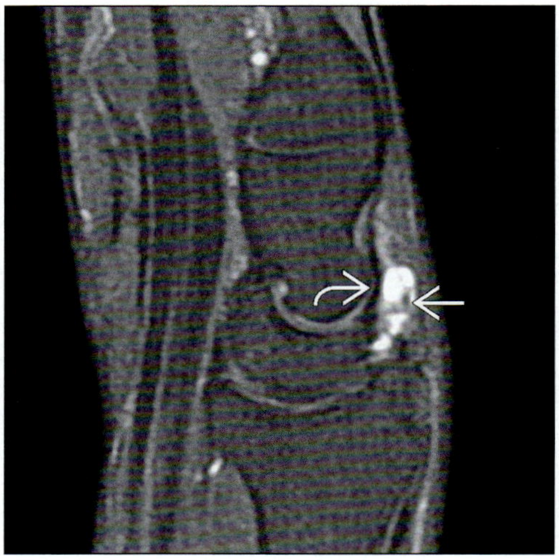

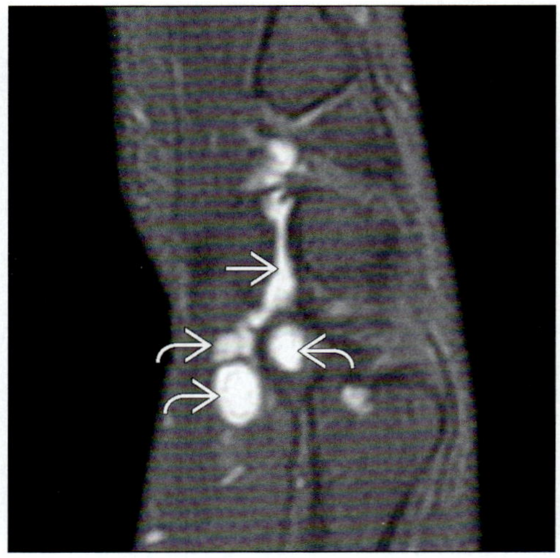

(Left) Axial T2WI FS MR shows a multicystic ganglion → in the volar radial aspect of the wrist. There is a tail of fluid ↪ tracking back to the radiocarpal joint, suggesting that the radiocarpal joint is the site of origin of the cyst. (Right) Coronal T2WI FS MR from the same patient shows the multicystic nature of the cyst →, helping distinguish it from a volar joint recess. This patient had pain localized to the volar radial wrist.

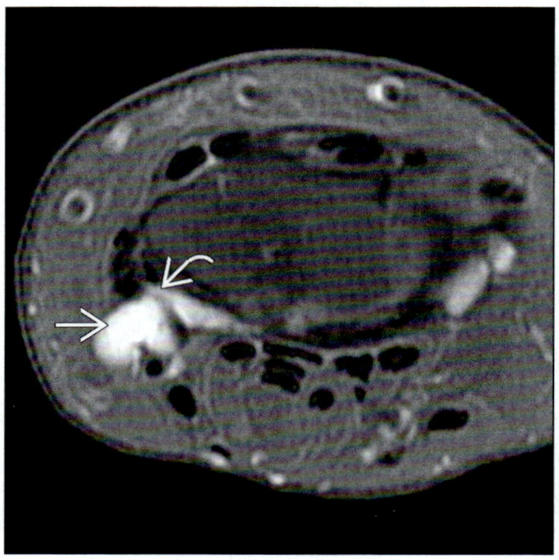

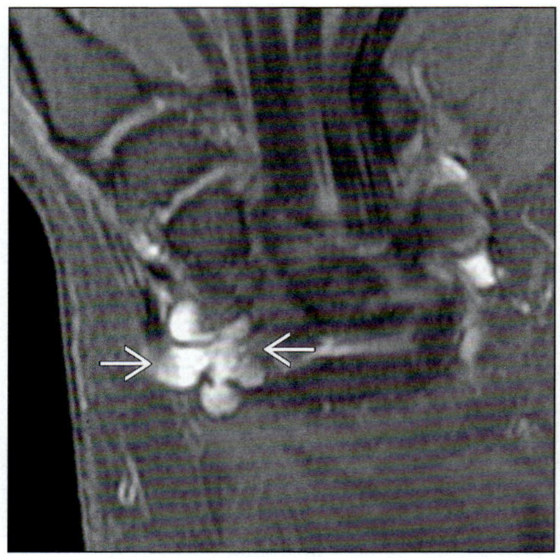

(Left) Axial PDWI MR shows a rounded high signal collection → volar to the triangular fibrocartilage (TFC) ↪. Ganglia in this location often arise as a result of tear of the TFC. (Right) Sagittal T2WI MR from the same patient shows the lobular fluid signal ganglion → volar to the TFC ↪. MR arthrography may be useful in such cases as a method to demonstrate a defect in the TFC and any potential communication of the wrist joint with the ganglion.

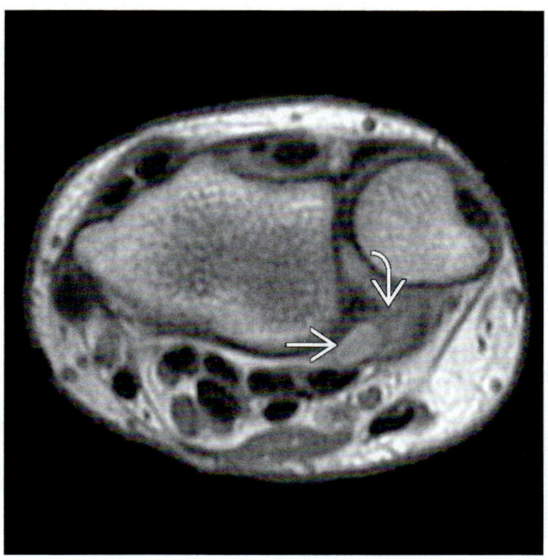

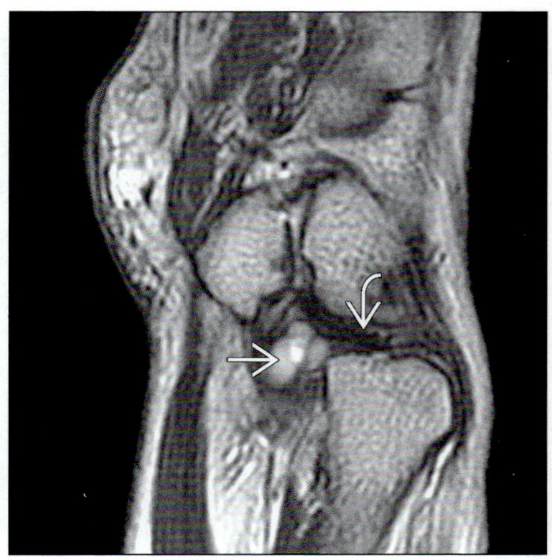

GANGLION CYST, WRIST

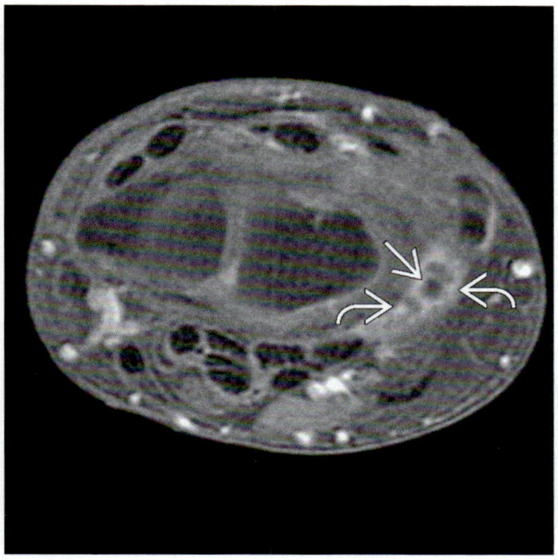
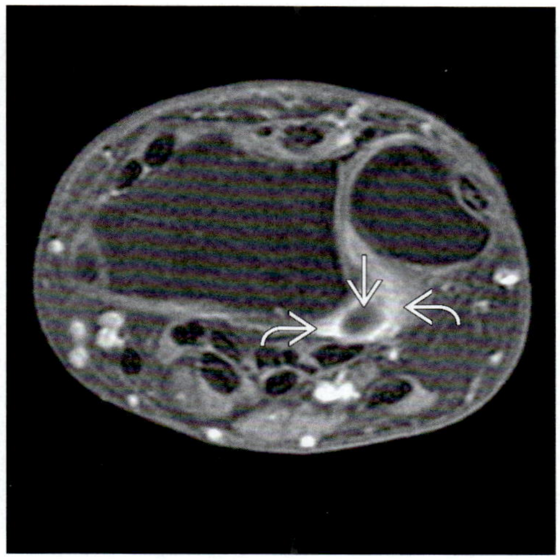

(Left) *Axial T1WI C+ FS MR from the same patient shows the neck of the ganglion* → *as a nonenhancing collection surrounded by enhancing synovial tissue* ↪ *extending into the TFC, the site of origin.* ***(Right)*** *Axial T1WI C+ FS MR more proximally from the same patient shows intense enhancement of synovial tissue* ↪ *surrounding the ganglion* →*. Post-contrast images can be a useful method to define the origin of the neck of a ganglion and therefore the likely underlying pathology.*

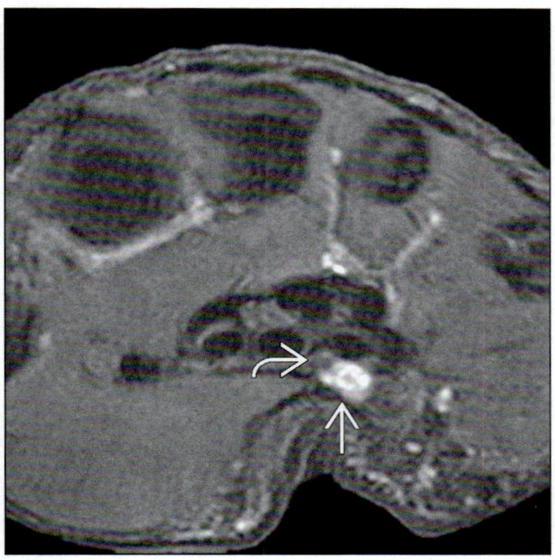
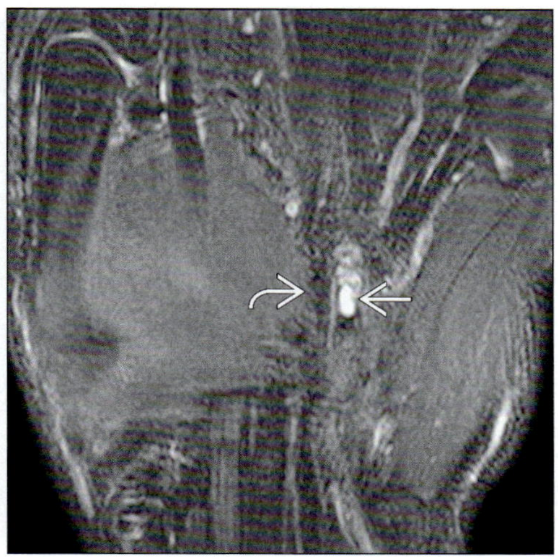

(Left) *Axial PDWI FS MR from a patient with ulnar neuropathy shows a small, well-defined, high signal mass* → *in the distal canal of Guyon adjacent to the ulnar nerve* ↪*.* ***(Right)*** *Coronal PDWI FS MR from the same patient shows the small ganglion* → *adjacent to the ulnar nerve* ↪*. Small ganglia can cause compression symptoms upon nerves or vessels when located in a confined space, such as the carpal tunnel (median nerve) or canal of Guyon (ulnar nerve, artery, and vein).*

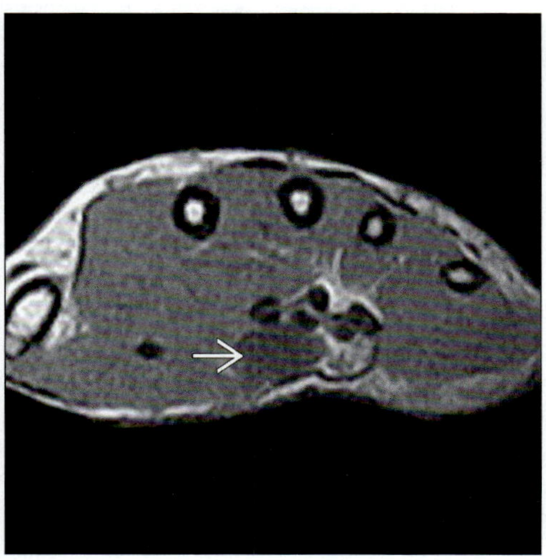
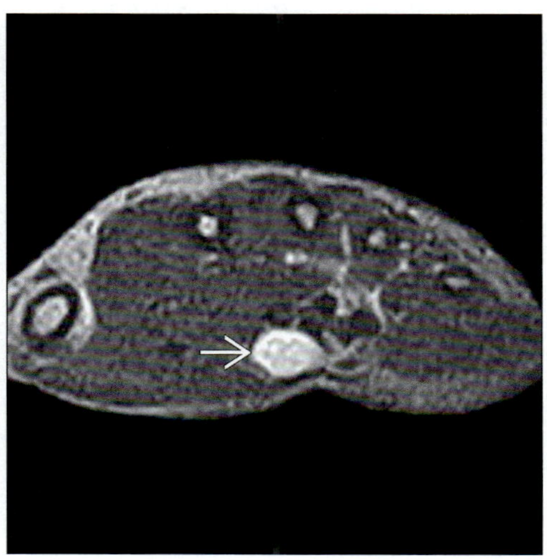

(Left) *Axial T1WI MR shows a rounded fluid signal ganglion* → *in the lateral aspect of the carpal tunnel. The cyst is adjacent to the median nerve, but no neuropathic symptoms were present in this case.* ***(Right)*** *Axial T2WI MR from the same patient shows fluid signal in the ganglion* → *without surrounding edema. Solid tumors may also occur in this location, and gadolinium enhanced images are needed for definitive diagnosis.*

VASCULAR INJURY, WRIST AND HAND

Key Facts

Terminology

- Hypothenar hammer syndrome: Ulnar artery superficial palmar arch injury; related to repetitive trauma to heel of hand, ± associated ulnar artery pseudoaneurysm
- Pseudoaneurysm: Saccular cystic dilatation of artery related to disruption of intima + media with hematoma; contained by adventitia

Imaging

- Hypothenar hammer syndrome
  - Complete ulnar artery occlusion on angiography
  - Ulnar artery pseudoaneurysm may occur
- Pseudoaneurysm
  - Saccular mass-like dilatation
  - Variable size with heterogeneous internal signal; ± flow void
- Angiography goals
  - Locate area of pathology: Proximal vs. distal
  - Define vessel abnormality: Spasm, thrombosis, &/ or aneurysm
  - Evaluate distal (digital) emboli

Top Differential Diagnoses

- Cystic adventitial disease
  - Eccentric cystic collections arising from outer vessel wall

Clinical Issues

- Hypothenar hammer syndrome
  - Pain in response to cold or repetitive movements; ± rest or night pain
  - Raynaud-like phenomenon
- Pseudoaneurysm
  - Palpable mass
  - Paresthesias may result from ulnar nerve compression

(Left) Coronal graphic demonstrates a pseudoaneurysm ➔ of the superficial branch of the ulnar artery located superficial to the flexor retinaculum ➲ and lateral to the hook of the hamate ➔. Ulnar nerve ➔ is adjacent to the artery. (Right) Coronal MR angiogram demonstrates a post-traumatic pseudoaneurysm ➔ of the ulnar artery in the volar wrist distal to Guyon canal where a 2 cm segment of the artery lies superficial to the flexor retinaculum.

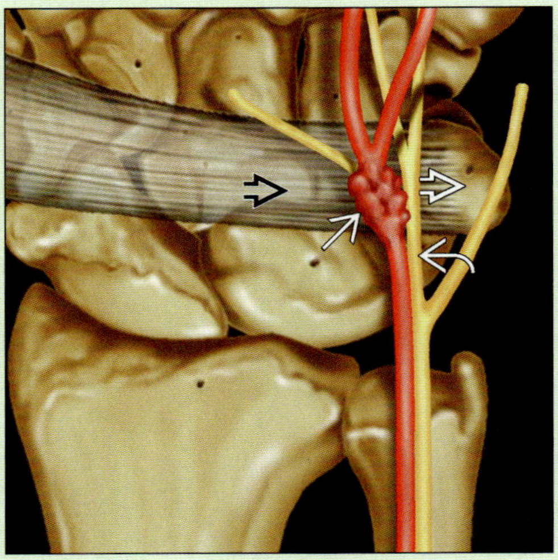

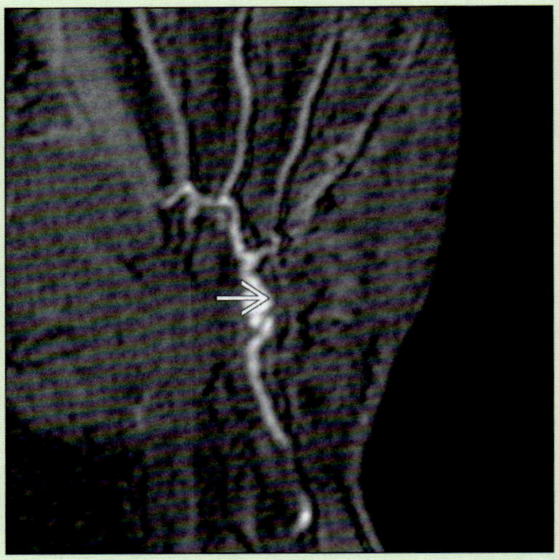

(Left) PA angiography shows decreased flow ➔ through the superficial branch of the ulnar artery, representing intravascular thrombus in this 50-year-old auto mechanic who reported pain in his index finger. A small saccular pseudoaneurysm ➔ is present distal to the vascular narrowing. (Right) PA angiography in the same patient demonstrates absence of flow ➲ in the 2nd ray digital arteries secondary to embolic occlusion in this patient with hypothenar hammer syndrome.

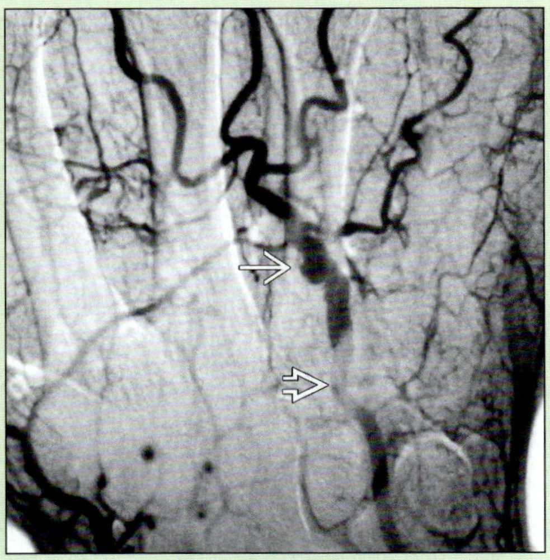

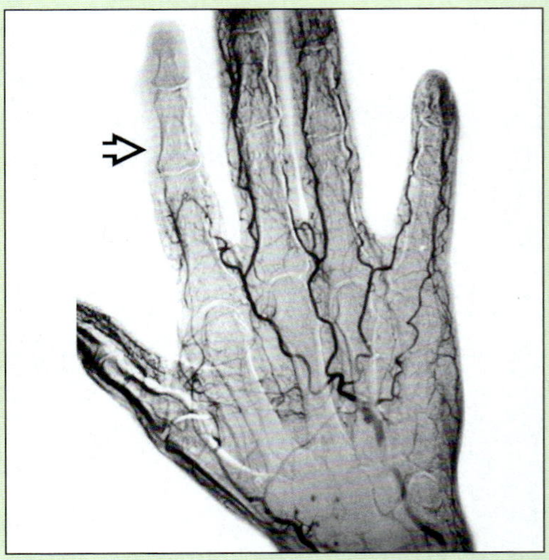

VASCULAR INJURY, WRIST AND HAND

TERMINOLOGY

Definitions

- Hypothenar hammer syndrome (HHS): Ulnar artery superficial palmar arch injury; related to repetitive trauma to heel of hand, ± associated ulnar artery pseudoaneurysm

IMAGING

General Features

- Best diagnostic clue
  - Hypothenar hammer syndrome: Complete occlusion of ulnar artery
  - Pseudoaneurysm: Saccular mass directly related to vessel
- Location
  - Ulnar artery as it forms superficial palmar arch

Angiographic Findings

- Hypothenar hammer syndrome
  - Aneurysm formation
  - Ulnar artery occlusion at hook of hamate
- May be accomplished with conventional, digital subtraction, CT angiography, or MR angiography

MR Findings

- Pseudoaneurysm
  - Eccentric saccular mass closely related to vessel
  - Thickened irregular wall
  - ± intralesional flow

Ultrasonographic Findings

- Hypothenar hammer syndrome
  - Diminished or absent blood flow
  - Thrombus formation
  - ± aneurysmal dilatation

Imaging Recommendations

- Best imaging tool
  - Angiography: Conventional, digital subtraction, MR or CT angiography

DIFFERENTIAL DIAGNOSIS

Cystic Adventitial Disease

- Eccentric cystic collections arising from outer vessel wall
- Generally does not impede flow

PATHOLOGY

General Features

- Etiology
  - Hypothenar hammer syndrome
    - Chronic blunt low-grade trauma
      - Work-related: Manual laborers who use heel of hand as makeshift hammer; jackhammers or vibrating tools
      - Athletics: Baseball (particularly catchers), tennis, golf, mountain biking, martial arts
  - Pseudoaneurysm
    - Single direct trauma such as stabbing, fall on outstretched hand ± fracture
    - Chronic blunt low-grade trauma as in hypothenar hammer syndrome

Relevant Anatomy

- Ulnar artery branches at Guyon canal
  - Deep palmar branch continues in Guyon canal
  - Superficial branch exits canal, crosses surface of hypothenar muscles (2 cm segment), enters palmar aponeurosis
    - Superficial segment vulnerable to trauma
    - Can be compressed against hamate hook
  - Superficial palmar arch provides main blood supply to digits
    - Variable connection to radial artery
    - Superficial palmar arch entirely supplied by ulnar artery in 37% of people
- Radial artery
  - At risk regions
    - Between 1st & 2nd metacarpals; artery covered only by flexor pollicis brevis & subcutaneous fat
    - Radial artery superficial branch covered by abductor pollicis brevis tendon & subcutaneous tissues; compressed against trapezoidal ridge

CLINICAL ISSUES

Presentation

- Most common signs/symptoms
  - Hypothenar hammer syndrome
    - Pain in response to cold or repetitive movements; ± rest or night pain
    - Raynaud-like phenomenon: Color changes including white (decreased blood flow) and blue (ischemia); red (erythema) **not** typical
    - Typically seen in dominant hand
  - Pseudoaneurysm
    - Palpable mass
    - Paresthesias may relate to compression of adjacent ulnar nerve

Natural History & Prognosis

- Often have accompanying neural injury
- May proceed to marked ischemia with ulceration, eschar formation, and gangrene
- May result in autoamputation

Treatment

- Medical
  - Smoking cessation
  - Altered activity to reduced continued trauma
  - Anticoagulation/antiplatelet therapy to reduce thrombosis & emboli
- Surgical
  - Techniques
    - Arterial ligation, resection, anastomosis, or reconstruction

SELECTED REFERENCES

1. Lisle DA et al: MR imaging of traumatic and overuse injuries of the wrist and hand in athletes. Magn Reson Imaging Clin N Am. 17(4):639-54, vi, 2009
2. Stepansky F et al: Dynamic MR angiography of upper extremity vascular disease: pictorial review. Radiographics. 28(1):e28, 2008

*(**Left**) Coronal PDWI FS MR demonstrates an irregular ectatic ulnar artery in a 32-year-old tennis player with hypothenar hammer syndrome. (**Right**) Transverse color Doppler ultrasound demonstrates a pseudoaneurysm with arterial flow through the intima and media of the vessel into a cavity contained by the adventitia. Complex color Doppler signal is due to turbulent flow in the pseudoaneurysm. The mass is partially thrombosed with no flow evident in the thickened wall.*

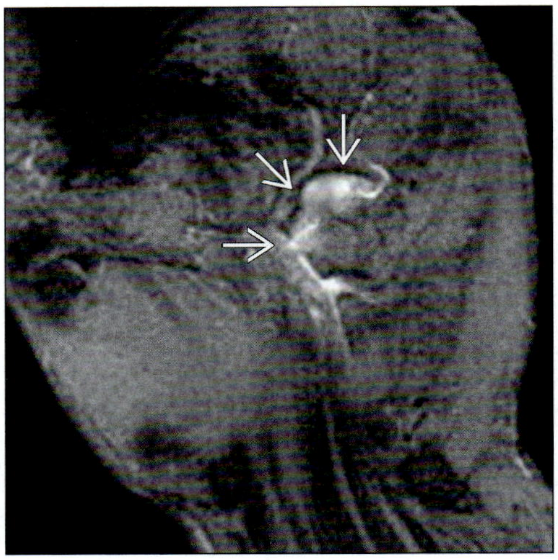

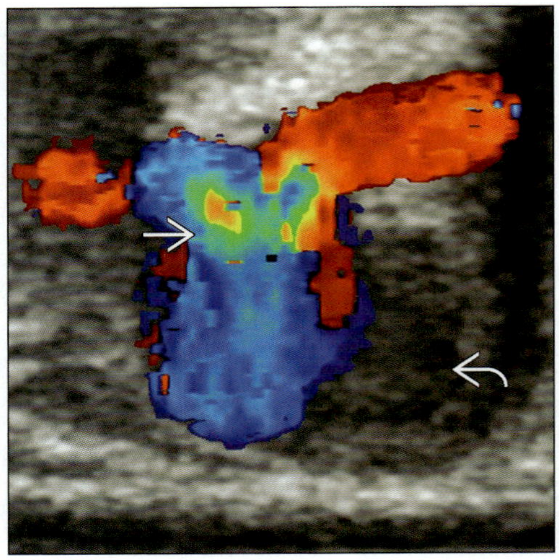

*(**Left**) Sagittal T1WI MR shows a uniformly low signal mass overlying the hook of the hamate and superficial to the flexor retinaculum. The mass represents an ulnar artery pseudoaneurysm due to hypothenar hammer syndrome. (**Right**) Axial T2WI MR from the same patient reveals the laminated appearance of the mass, representing heterogeneity associated with blood products of varying ages, including acute hemorrhage (high signal) and hemosiderin (low signal).*

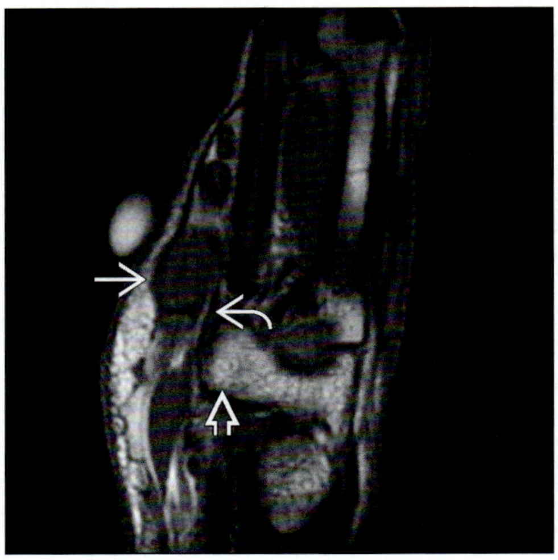

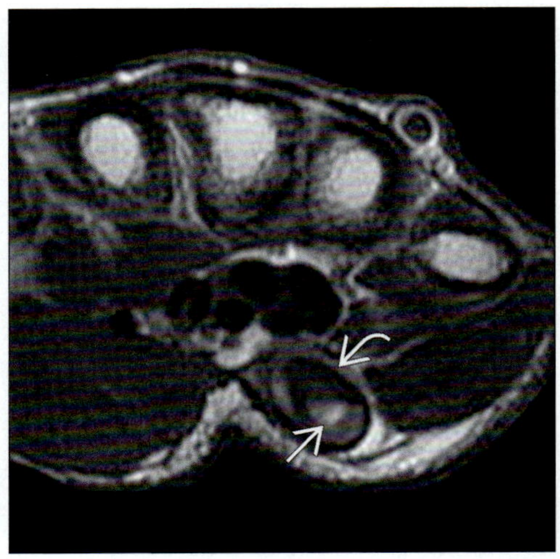

*(**Left**) Axial T2WI FS MR shows a thrombosed ulnar artery within the canal of Guyon directly volar to the hook of the hamate. Patient had hypothenar hammer syndrome with cold-induced numbness of the 2nd finger. Note location of ulnar nerve. (**Right**) Axial T1WI MR shows a thrombosed ulnar artery pseudoaneurysm containing hyperintense thrombus extending superficial to the canal of Guyon. The patient had a history of penetrating injury at the site.*

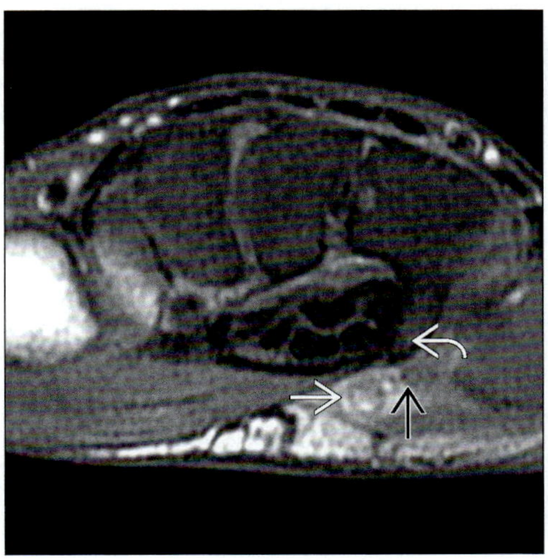

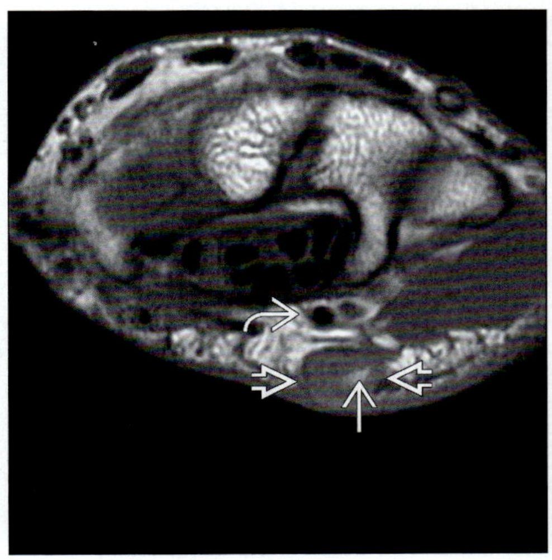

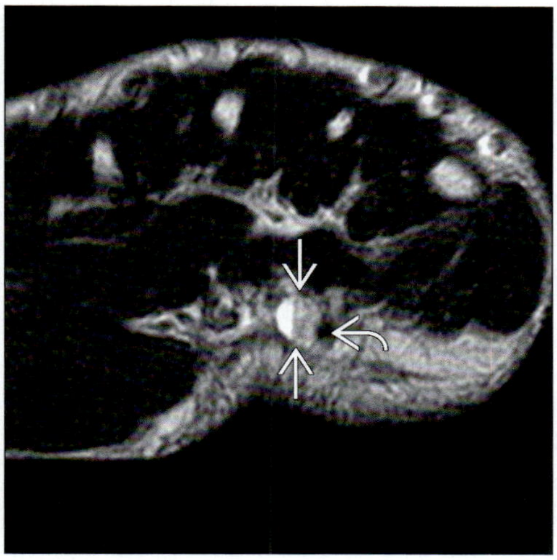

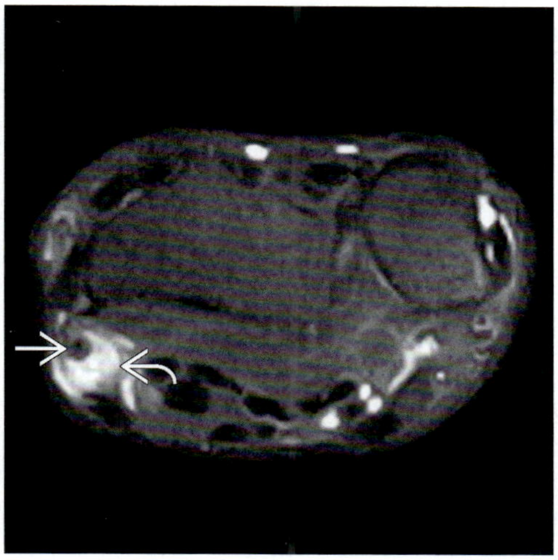

(Left) *Axial T2WI MR shows a pseudoaneurysm of the ulnar artery as it leaves the canal of Guyon to become the superficial palmar arch. There is thrombus within the pseudoaneurysm, which arises eccentrically from the ulnar artery.* ***(Right)*** *Axial T2WI FS MR shows cystic adventitial disease of the radial artery. The artery is intact and demonstrates normal flow void. Eccentric cystic collections arise from the adventitia of the artery but do not impede flow.*

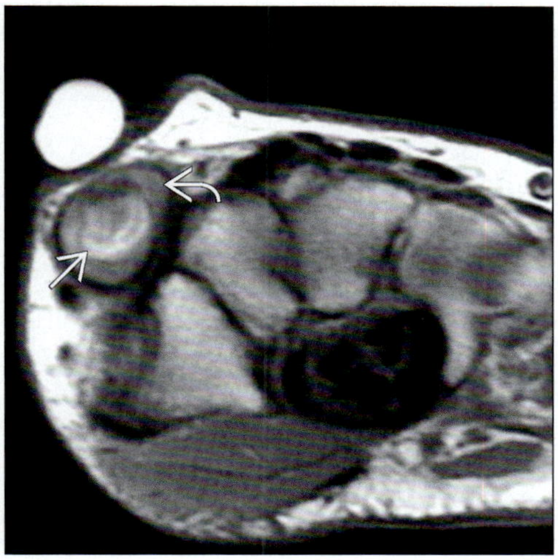

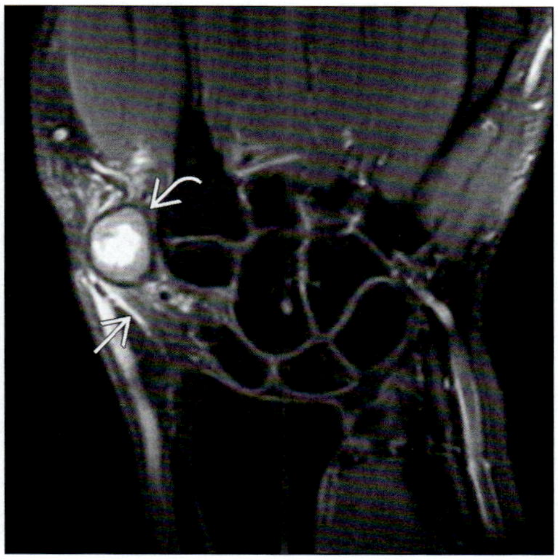

(Left) *Axial T1WI MR demonstrates a discrete mass along the dorsal lateral wrist. This heterogeneous mass has a laminated appearance with rings of intermediate and high signal related to blood products of varying ages.* ***(Right)*** *Coronal STIR MR from the same patient demonstrates the intimate relationship of this mass to the dorsal carpal branch of the radial artery. This represents a pseudoaneurysm resulting from a stabbing injury sustained at the 1st dorsal web space.*

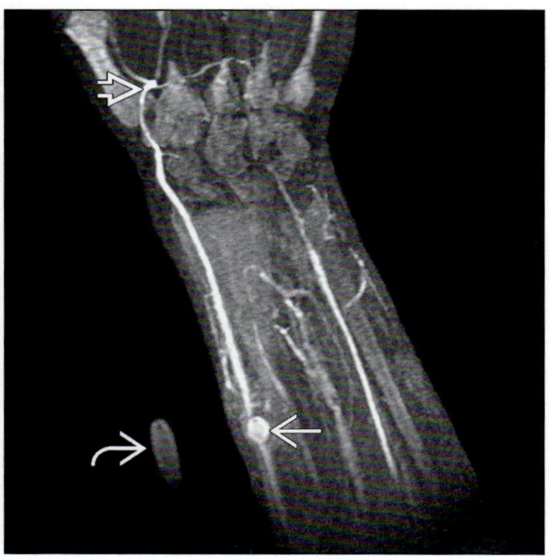

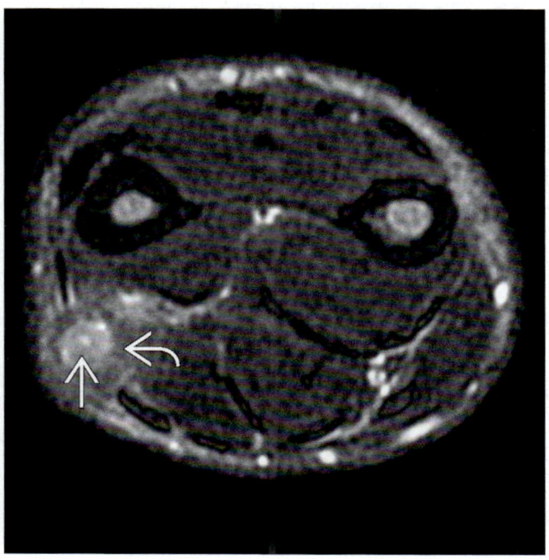

(Left) *Coronal MR arthrogram T1WI C+ FS in a patient who suffered a deep laceration from broken glass shows a pseudoaneurysm at the site indicated by the skin marker in the distal forearm. Note the deep palmar arch arising from the radial artery.* ***(Right)*** *Axial T2WI FS MR from the same patient shows the thrombosed radial artery pseudoaneurysm in the volar lateral aspect of the distal forearm surrounded by a thick fibrous rim.*

NERVE ENTRAPMENT SYNDROMES, WRIST

Key Facts

Terminology

- Carpal tunnel syndrome (CTS): Compression of median nerve at wrist
- Ulnar tunnel syndrome: Compression of ulnar nerve at wrist
- Wartenberg syndrome: Entrapment of superficial (sensory) branch of radial nerve

Imaging

- Carpal tunnel syndrome
  - Median nerve swelling proximal to carpal tunnel
  - Median nerve flattening in carpal tunnel
  - Flexor retinaculum thickening ± bowing
- Ulnar tunnel syndrome
  - Ulnar nerve focal or diffuse swelling ± hyperintense intraneural signal
- Wartenberg syndrome
  - Mass in area of radial nerve superficial branch

Pathology

- Median nerve located in volar lateral portion of carpal tunnel
- Ulnar nerve
  - Passes superficial to flexor retinaculum
  - Bifurcates at distal pisiform into deep and superficial branches
  - Deep (motor) branch dorsal, medial to hook of hamate

Clinical Issues

- Carpal tunnel syndrome: Paresthesias ± pain in 1st-3rd digits and radial aspect of 4th digit
- Ulnar tunnel syndrome: Paresthesias ± pain of 5th and medial aspect of 4th digit
- Wartenberg syndrome: Paresthesias ± pain in dorsal aspect of 1st web space

*(**Left**) Coronal graphic demonstrates the swollen median nerve ➡ deep to the flexor retinaculum ↪ and superficial to the flexor tendons and their sheaths ⇨. Because the carpal tunnel is a confined space, it is the most common site of median nerve compression. (**Right**) Axial STIR MR reveals intense edema in the median nerve ➡ as well as mild nerve enlargement. The flexor retinaculum is bowed volarly ↪ and mildly thickened. There are no obvious compressive lesions in the carpal tunnel.*

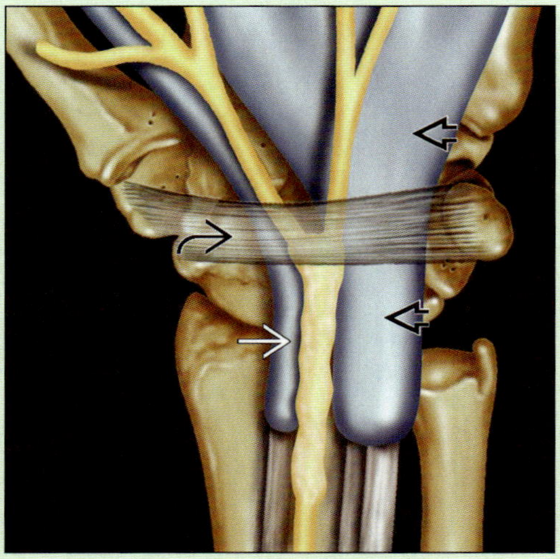

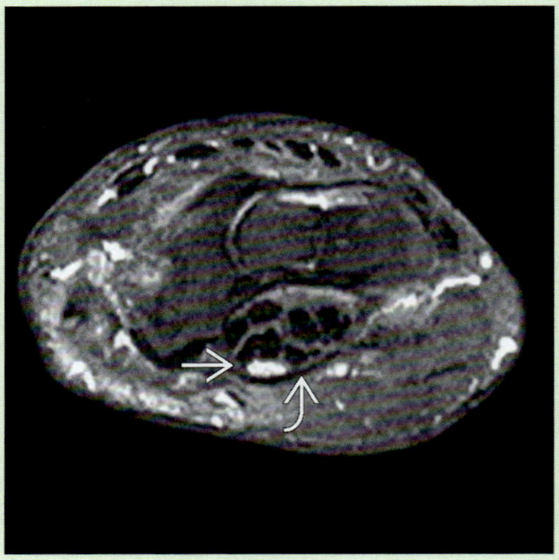

*(**Left**) Coronal graphic shows the course of the ulnar nerve. Proximal to the pisiform, the nerve contains both motor and sensory fibers ➡ (zone 1). Distal to the pisiform, the nerve bifurcates into deep (zone 2, motor) ➡ and superficial (zone 3, sensory) ↪ branches. (**Right**) Coronal PDWI FS MR shows a thin-walled, high signal ganglion cyst ⇨ compressing the ulnar nerve deep (motor) branch ➡ just beyond the bifurcation resulting in weakness without sensory changes.*

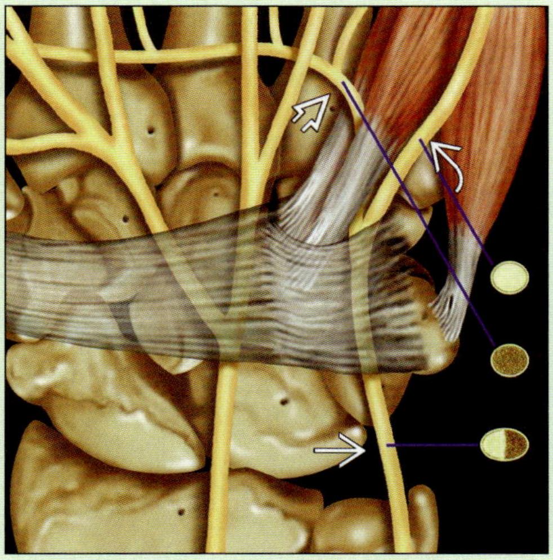

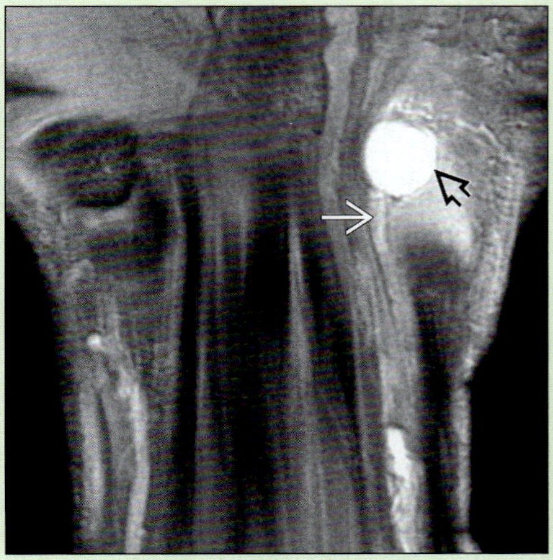

NERVE ENTRAPMENT SYNDROMES, WRIST

TERMINOLOGY

Synonyms

- Carpal tunnel syndrome: Median neuropathy
- Ulnar tunnel syndrome: Ulnar neuropathy, Guyon canal syndrome
- Wartenberg syndrome: Cheiralgia paresthetica
- Flexor retinaculum: Transverse carpal ligament

Definitions

- Carpal tunnel syndrome: Compression of median nerve at wrist
- Ulnar tunnel syndrome: Compression of ulnar nerve at wrist
- Wartenberg syndrome: Entrapment of superficial (sensory) branch of radial nerve

IMAGING

General Features

- Best diagnostic clue
  - Carpal tunnel syndrome (CTS)
    - Median nerve enlargement ± enhancement
    - Flexor retinaculum bowing
  - Ulnar tunnel syndrome
    - Ulnar nerve enlargement ± enhancement
    - Compressive extrinsic mass
  - Transection neuroma
    - Nodular and slightly heterogeneous; corresponds to region of prior trauma
- Location
  - CTS: Median nerve in carpal tunnel
  - Ulnar tunnel syndrome: Ulnar nerve in Guyon canal
  - Wartenberg syndrome: Radial nerve superficial branch at dorsolateral wrist
- Morphology
  - Carpal tunnel formed by transverse carpal ligament volar to carpal bones
    - Median nerve located in volar lateral portion of carpal tunnel
      - Supplies pronator teres, flexor carpi radialis, palmaris longus, flexor digitorum superficialis, & thenars (opponens pollicis, abductor pollicis brevis, & flexor pollicis brevis)
  - Ulnar nerve resides in Guyon canal (formed by pisiform and hamate)
    - Ulnar nerve passes superficial to flexor retinaculum (zone 1)
    - Bifurcates at distal pisiform into deep and superficial branches
    - Deep (motor) branch dorsal & medial to hook of hamate (zone 2)
      - Supplies hypothenar, interosseous, 3rd & 4th lumbricals, and adductor pollicis
    - Superficial (sensory) branch superficial to flexor retinaculum (zone 3)
      - Supplies skin of hypothenar eminence, 4th & 5th digits, & palmaris brevis
  - Radial nerve superficial branch
    - Sensory branch becomes subcutaneous 5-10 cm proximal to radial styloid in brachioradialis/ extensor carpi radialis longus interval
    - Bifurcates just prior to wrist
      - Supplies dorsal skin of wrist, hand, thumb, and fingers

Radiographic Findings

- Indirect findings, such as fracture or post-fracture deformity
  - CTS: Fractures of distal radius or capitate; perilunate dislocation
  - Ulnar tunnel syndrome: Hook of hamate fracture

MR Findings

- Carpal tunnel syndrome
  - Median nerve swelling
    - Nerve cross-sectional area (CSA) > 10 mm^2 (evaluated at pisiform level)
  - Median nerve flattening (evaluated at hamate level)
  - Increased intraneural signal on fluid-sensitive sequence
    - Signal change may be seen distal to carpal tunnel
    - Signal rarely abnormal proximal to DRUJ
  - ± nerve enhancement (no enhancement if nerve ischemic)
  - Flexor retinaculum thickening ± bowing (evaluated at hamate level)
  - Potential associated findings
    - Fracture of distal radius or capitate
    - Flexor tenosynovitis
    - Mass or ganglion
  - Thenar eminence muscle atrophy in late stages (downstream denervation)
- Postoperative CTS
  - Flexor retinaculum interruption along its medial margin near hamate hook
  - ± volar displacement of carpal tunnel contents
  - ± incomplete resection, scar formation, or persistent/recurrent flexor tenosynovitis
- Ulnar tunnel syndrome
  - Ulnar nerve focal or diffuse swelling
  - Hyperintense signal within ulnar nerve on fluid-sensitive sequence
  - Potential associated findings
    - Fracture of hook of hamate
    - Flexor tenosynovitis
    - Mass or ganglion
    - Vascular abnormality, such as pseudoaneurysm of ulnar artery
- Wartenberg syndrome
  - Mass in area of radial nerve superficial branch
    - Hematoma: Heterogeneous mass with blood products of varying ages
    - Mass or ganglion
  - Potential associated findings
    - de Quervain tenosynovitis with fluid in abductor pollicis longus (APL) & extensor pollicis brevis (EPB) tendon sheaths
    - Intersection syndrome (inflamed crossing point of extensor carpi radialis tendons over APL and EPB)

Ultrasonographic Findings

- General
  - Symptoms reproduced by transducer compression
  - Operator dependent
- CTS
  - Hypoechoic, swollen median nerve
  - Calculated nerve volume > 10 mm^2

NERVE ENTRAPMENT SYNDROMES, WRIST

- Change in CSA of 2 mm^2; measured proximal to and within carpal tunnel
- Reduced tendon sliding with finger motion
- Intraneural hypervascularity
- Ulnar tunnel syndrome
  - Hypoechoic ulnar nerve
- Transection neuroma
  - Heterogeneous but predominately hypoechoic mass in continuity with peripheral nerve
  - Symptoms reproduced with compression

Imaging Recommendations

- Best imaging tool
  - MR or ultrasound
- Protocol advice
  - MR: 8-12 cm field of view; specialized wrist coil
    - Axial T1 and fluid-sensitive sequences
    - Intravenous gadolinium for evaluation of inflammation ± mass
  - Ultrasound
    - Linear transducer 7-15 MHz; stand-off pad

PATHOLOGY

General Features

- Etiology
  - CTS: Any process that decreases area of fibroosseous tunnel or increases carpal tunnel content volume
    - Idiopathic (most common)
    - Systemic disorders, such as diabetes, thyroid disorders
    - Inflammatory processes, including tenosynovitis
      - Rheumatoid arthritis, gout, amyloid
      - May be due to repetitive motion injury
    - Masses, including neoplasm or ganglion
    - Anatomic anomalies, such as persistent medial artery or anomalous muscle
    - Trauma, including fracture
    - Pregnancy
    - Obesity
  - Ulnar tunnel syndrome
    - Compression from fracture fragment
    - Ulnar artery aneurysm or thrombosis
    - Anomalous muscle belly
      - Flexor digiti minimi brevis
    - Seen in golfers, baseball players, & cyclists
  - Wartenberg syndrome
    - Extrinsic compression due to tight wristwatch or handcuffs
    - Repetitive supination/pronation motions, e.g., twisting a screwdriver

CLINICAL ISSUES

Presentation

- Most common signs/symptoms
  - Carpal tunnel syndrome
    - Paresthesias ± pain in 1st-3rd digits and radial aspect of 4th digit
    - Tinel sign: Tapping median nerve → paresthesias
    - Phalen sign: Paresthesias in 1st-3rd digits with wrist flexion
  - Ulnar tunnel syndrome
    - Paresthesias ± pain of 5th and medial aspect of 4th digit
    - Tinel sign: Tapping ulnar nerve → paresthesias
    - Phalen sign: Paresthesias in 4th and 5th digits with wrist flexion
  - Wartenberg syndrome
    - Paresthesias ± pain along dorsal medial aspect of 1st digit and dorsal lateral aspect of 2nd digit (dorsal 1st web space)
    - Tinel sign: Tapping superficial radial nerve → paresthesias
    - Symptoms reproduced with supination/pronation of wrist
- Other signs/symptoms
  - CTS: Abductor pollicis brevis and opponens pollicis weakness (early) followed by atrophy (late)

Demographics

- Age
  - 50 years ± 15 years
- Gender
  - CTS: M:F = 1:6
- Epidemiology
  - CTS: Most common upper extremity neuropathy
    - Incidence: Up to 50% during adult life; 100% from repetitive motion activity
    - Bilateral in 10-40% of cases
    - Occurs in 1-10% of distal radial fractures
  - Ulnar tunnel syndrome less common than median nerve compression or ulnar entrapment at elbow

Natural History & Prognosis

- Symptoms may persistent for months to years
  - May gradually resolve over time
  - May progress; results in atrophy and weakness

Treatment

- Conservative
  - Wrist splint
  - Anti-inflammatory medication
  - Modification of activity
- Surgical
  - Removal of space-occupying lesion
  - Carpal tunnel or ulnar tunnel release

DIAGNOSTIC CHECKLIST

Consider

- CTS in patients with unexplained hand or wrist paresthesias, even if symptoms atypical for CTS
- "Double crush" phenomenon results from neural compression simultaneously at more than 1 site, e.g., cervical spine, axilla, or elbow in addition to wrist

SELECTED REFERENCES

1. Dang AC et al: Unusual compression neuropathies of the forearm, part I: radial nerve. J Hand Surg Am. 34(10):1906-14, 2009
2. Klauser AS et al: Carpal tunnel syndrome assessment with US: value of additional cross-sectional area measurements of the median nerve in patients versus healthy volunteers. Radiology. 250(1):171-7, 2009
3. Lanzetta M et al: Entrapment of the superficial branch of the radial nerve (Wartenberg's syndrome). A report of 52 cases. Int Orthop. 17(6):342-5, 1993

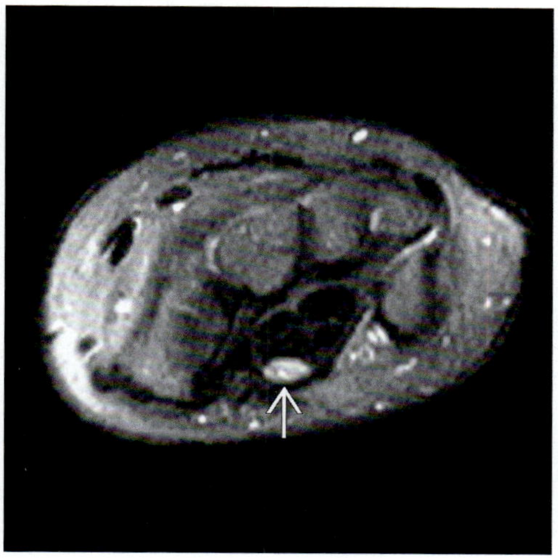
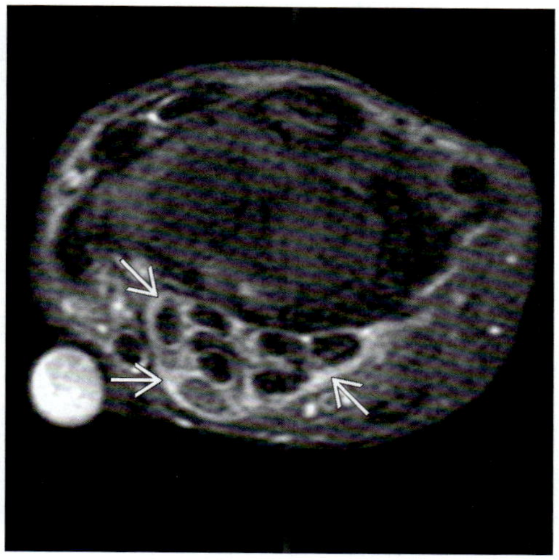

(Left) *Axial T2WI FS MR shows marked signal elevation in the median nerve ➔ in a patient with symptoms of carpal tunnel syndrome (CTS). Elevated T2 signal within and compression of the median nerve can be useful signs supporting the diagnosis of CTS but are neither sensitive nor specific for the diagnosis.* ***(Right)*** *Axial T2WI MR shows diffusely elevated signal ➔ around the flexor tendons in the carpal tunnel consistent with tenosynovitis in this patient with carpal tunnel syndrome.*

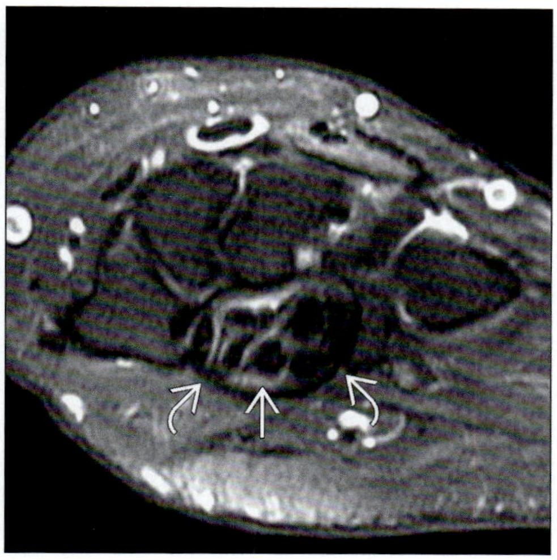
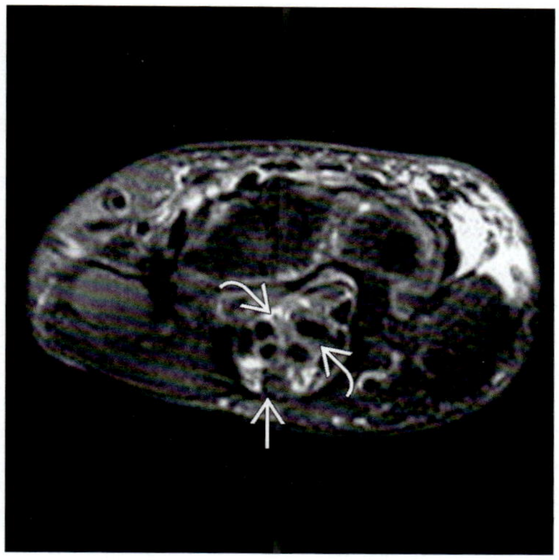

(Left) *Axial T2WI FS MR shows moderate volar bowing of the flexor retinaculum ➔ in this patient with flexor tenosynovitis and carpal tunnel syndrome. The median nerve ➔ is flattened but demonstrates normal signal.* ***(Right)*** *Axial T2WI FS MR from a patient with longstanding rheumatoid arthritis and carpal tunnel syndrome demonstrates diffuse flexor tenosynovitis ➔ within the carpal tunnel. The median nerve ➔ demonstrates normal size and signal.*

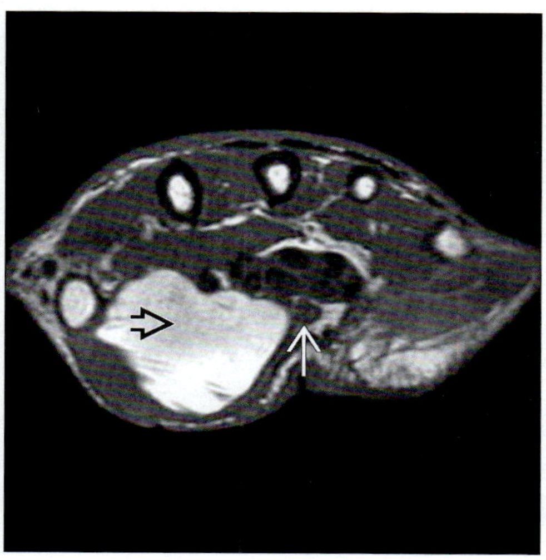
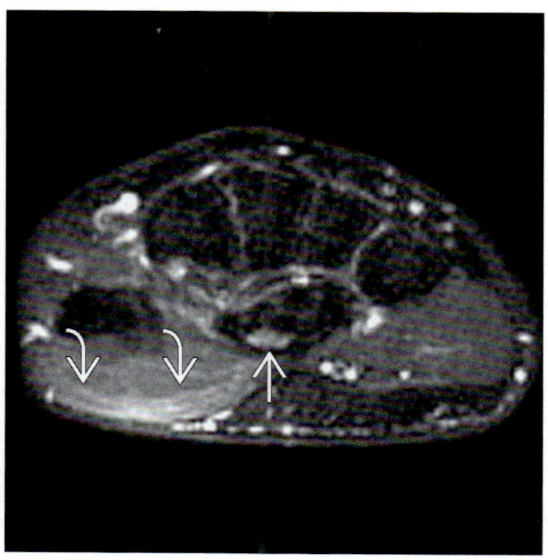

(Left) *Axial T1WI MR from a patient with clinical carpal tunnel syndrome shows a large, well-defined, fat signal mass ➔ in the thenar eminence, consistent with a lipoma displacing the median nerve ➔ medially at the distal aspect of the carpal tunnel.* ***(Right)*** *Axial T2WI FS MR from a patient with lateral hand numbness and thumb weakness shows compression of and increased signal in the median nerve ➔. There is edema and mild atrophy of the abductor pollicis brevis muscle ➔.*

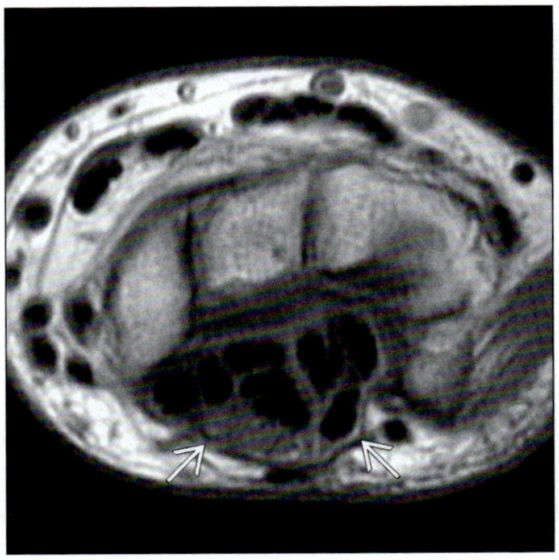

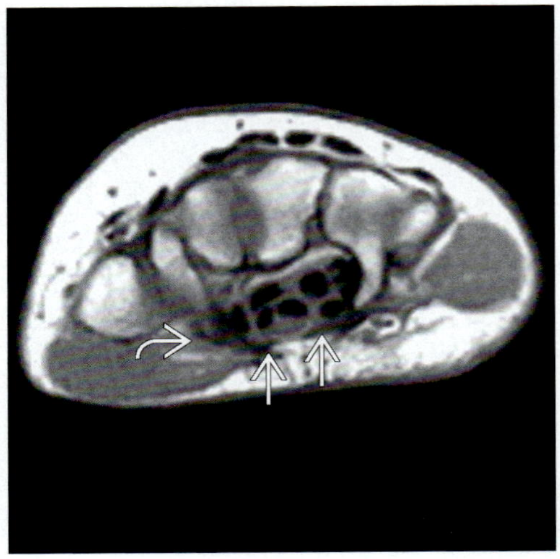

(Left) *Axial PDWI MR in a patient who had previously undergone carpal tunnel release (CTR) surgery for median neuropathy demonstrates volar eventration of the contents of the carpal tunnel through the patulous flexor retinaculum. This is a normal finding after CTR.* ***(Right)*** *Axial PDWI MR from a patient after carpal tunnel release shows a wide defect in the flexor retinaculum. Usually the defect is near the trapezial attachment.*

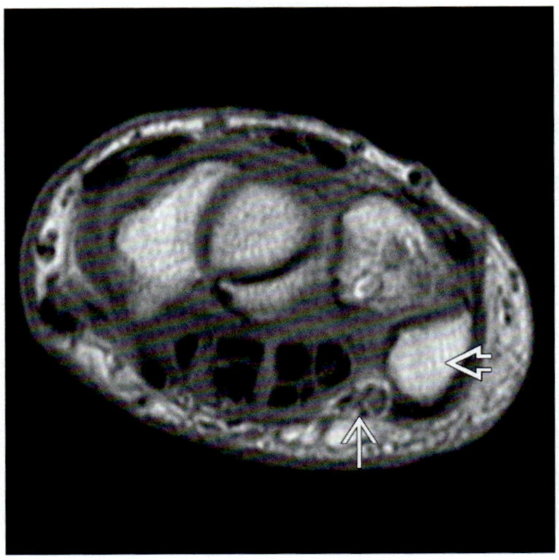

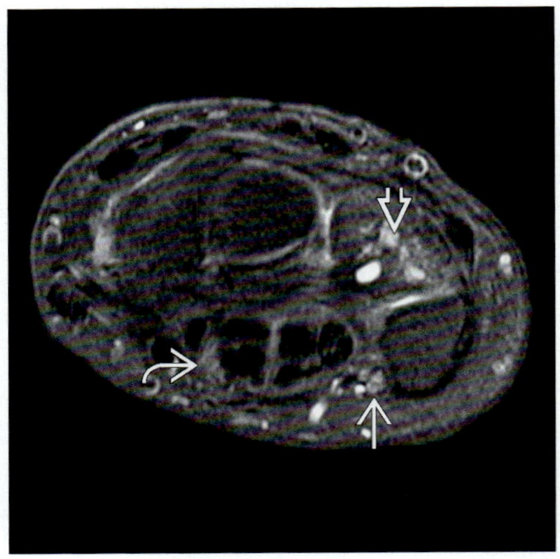

(Left) *Axial T1WI MR from a patient with ulnar neuropathy and positive Phalen test shows enlargement of the ulnar nerve at the level of the pisiform consistent with ulnar neuritis. As in many cases, there is no extrinsic compressive lesion.* ***(Right)*** *Axial T2WI FS MR from the same patient shows slightly elevated signal in the ulnar nerve indicating ulnar neuritis. Signal in the median nerve is normal. Degenerative cysts are present in the triquetrum.*

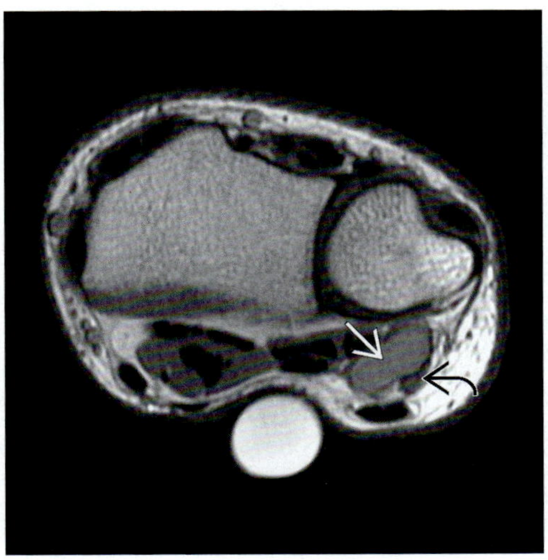

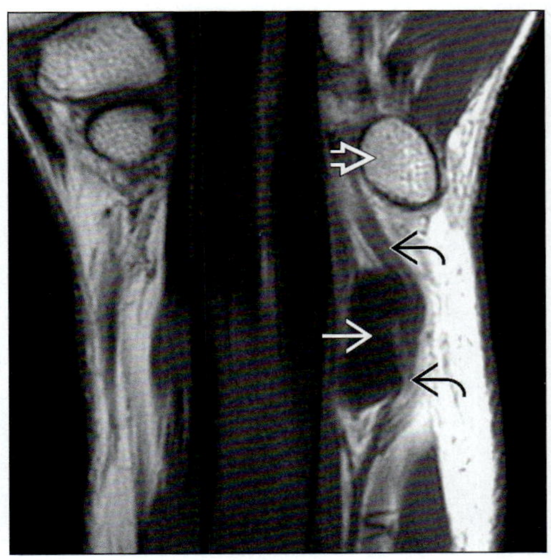

(Left) *Axial PDWI MR from a patient with ulnar neuropathy and hand weakness shows an ovoid, uniformly intermediate signal mass consistent with a ganglion volar to the ulna causing medial displacement of the ulnar nerve just proximal to the canal of Guyon.* ***(Right)*** *Coronal T1WI MR from the same patient shows the ganglion proximal to the level of the pisiform causing marked medial displacement of the ulnar nerve. Symptoms resolved with resection of the cyst.*

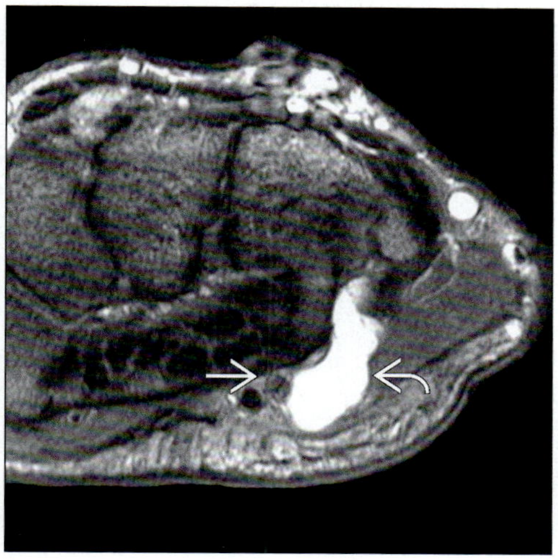

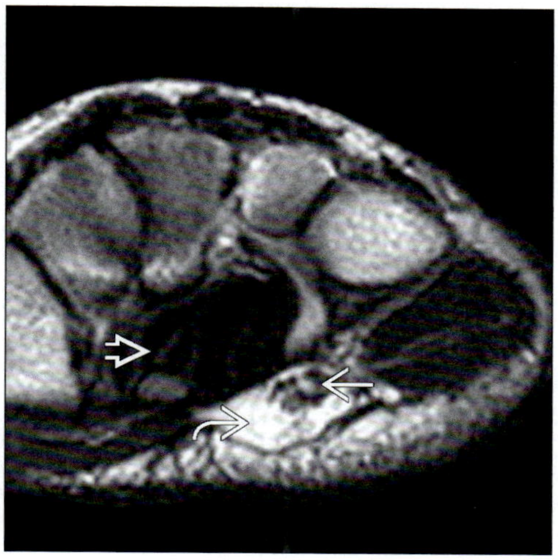

(Left) *Axial T2WI MR reveals a large, irregular, thin-walled ganglion in the canal of Guyon, compressing and laterally displacing the ulnar nerve.* ***(Right)*** *Axial T1WI MR demonstrates an ovoid fatty mass consistent with a lipoma in Guyon canal, encasing the neurovascular structures in the proximal canal. The patient presented with motor weakness and paresthesias in the 4th and 5th digits. The carpal tunnel is normal.*

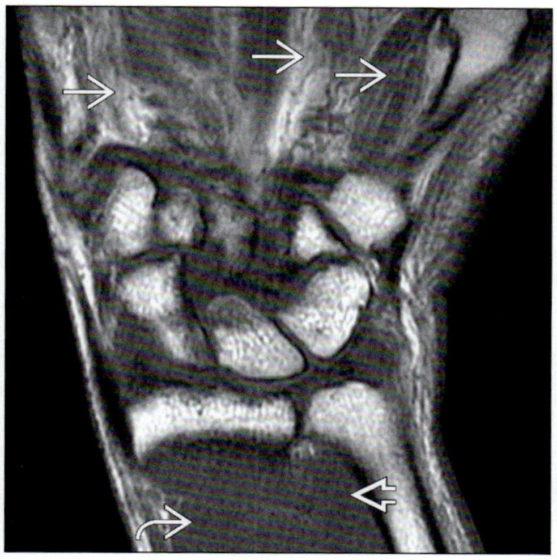

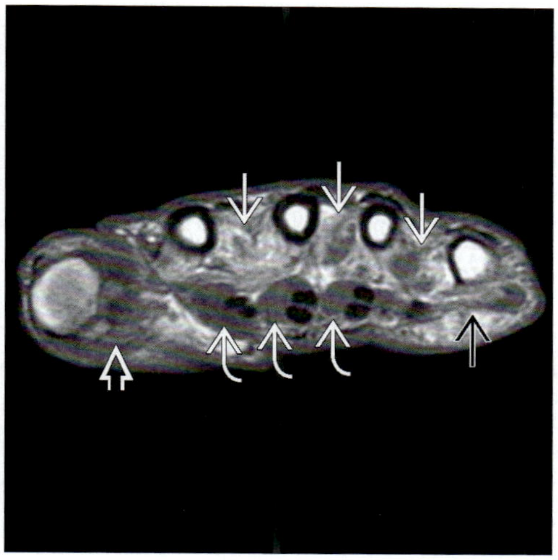

(Left) *Coronal T1WI MR in a patient with longstanding ulnar neuropathy shows marked diffuse atrophy of the intrinsic musculature of the hand. The pronator quadratus and flexor digitorum profundus are normal.* ***(Right)*** *Axial T1WI MR from the same patient shows marked atrophy of the dorsal and volar interosseous and the abductor digiti minimi muscles. The lumbricals and muscles of the thenar eminence are not affected.*

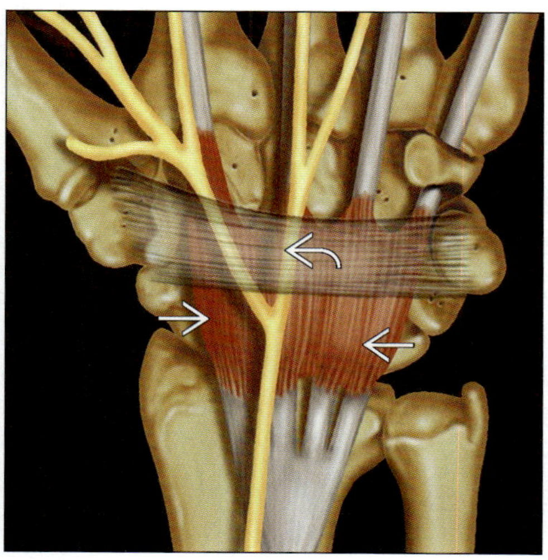

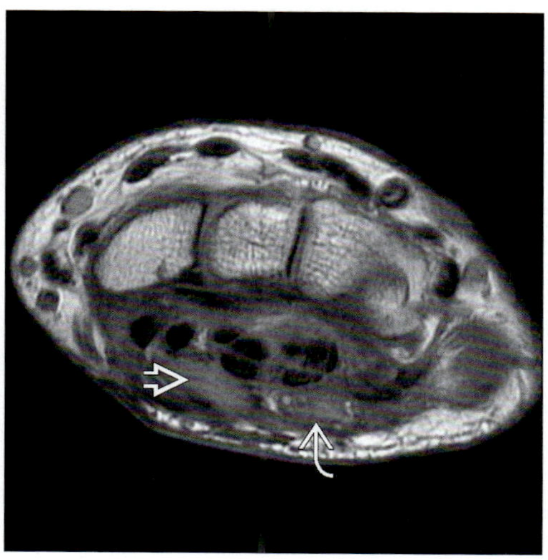

(Left) *Coronal graphic shows an anomalous muscle belly of the flexor digitorum superficialis. This muscle results in decreased carpal tunnel volume and can cause carpal tunnel syndrome due to compression of the medial nerve in this confined space.* ***(Right)*** *Axial PDWI MR in a patient with aggressive fibromatosis shows mildly elevated signal in the fibrous tissue entering the carpal tunnel and canal of Guyon. The patient had both median and ulnar neuropathy.*

SECTION 5
Hip and Pelvis

Introduction and Overview

Osseous Pathology

Hip and Proximal Femur

Pelvis

Postoperative Imaging

Labroligamentous Pathology

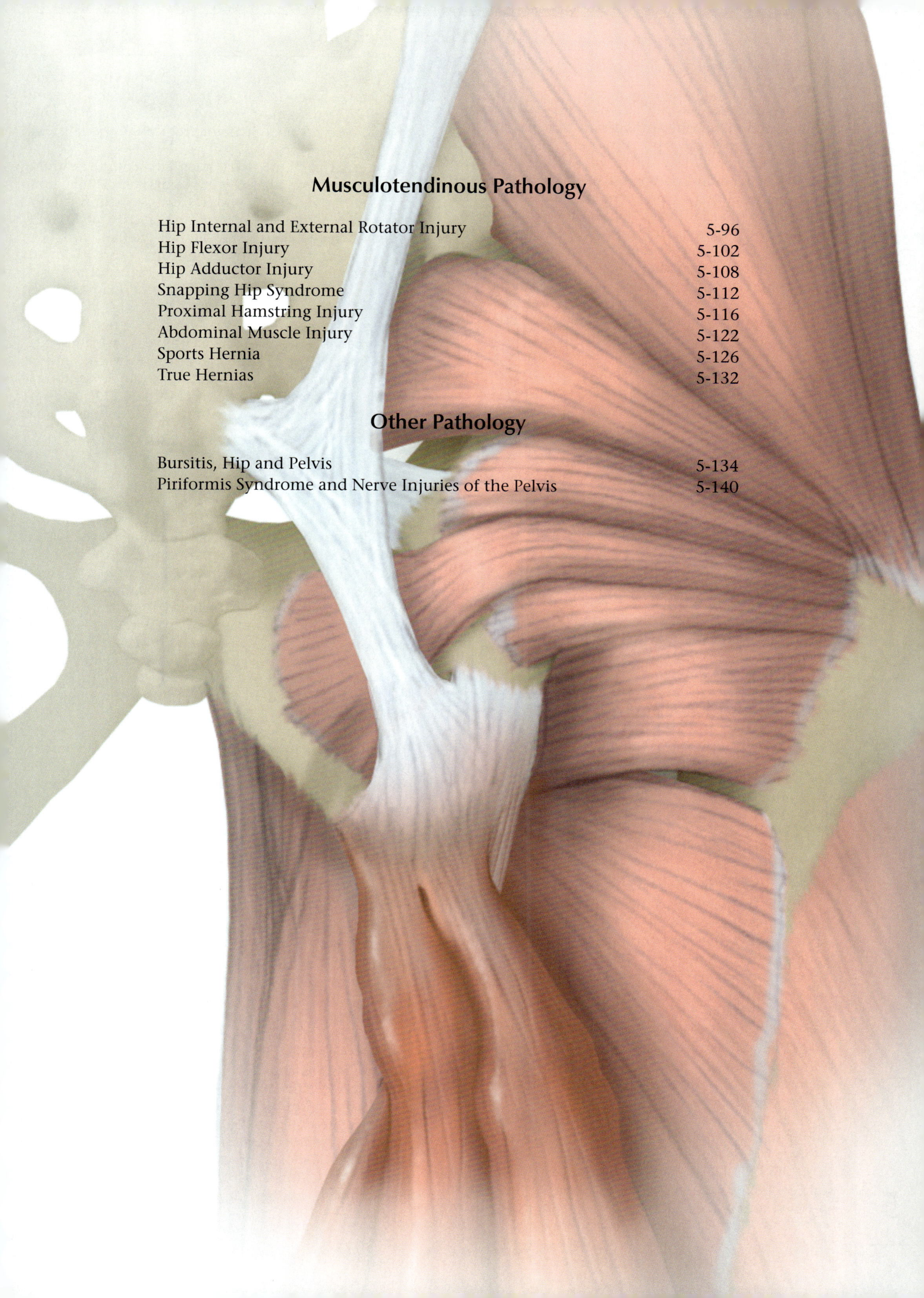

Musculotendinous Pathology

Other Pathology

Overview

No region of the musculoskeletal system has seen more therapeutic and imaging advances over the past ten years than the pelvic girdle and hips. Referrals for imaging assessment of previously unrecognized and untreatable conditions, including femoral acetabular impingement and athletic pubalgia, are increasing at a rapid pace, and it is essential for the radiologist to be familiar with morphologic variations and cutting edge therapies. At the same time, tried and true classification systems still play an important role in guiding therapeutic regimens in the setting of pelvic and hip trauma. Radiologists have more imaging options today than ever before, and selection of the most appropriate imaging modality and protocol is key to reaching an accurate diagnosis. This section reviews the imaging patterns that should be recognized in order for the radiologists to effectively play an integral role as a member of the healthcare team, ranging from fracture descriptors to dynamic biomechanical syndromes.

Terminology

A number of new terms have made their way into the imaging literature in recent years, and some of these have become standard in daily musculoskeletal practice. **Femoral acetabular impingement** is the general term referring to a group of articular disorders of the hip stemming from variant morphologies and subsequent abnormal biomechanical stresses. Many of these can now be diagnosed at a young age, aided by imaging, allowing for definitive treatment before the onset of secondary degenerative conditions. **Developmental dysplasia of the hip** is a term that has recently become increasingly relevant thanks to innovative new surgical techniques and an improved understanding of imaging patterns. Some patients have lesions with both impinging and dysplastic elements. Remote from the hip, sources of groin pain in athletes and young adults have been described with terms including the colloquial **sports hernia**, the anatomic **rectus abdominis/adductor aponeurosis**, and the clinical **athletic pubalgia**. In postoperative patients, terms including **pelvic floor repair** and **femoral head/ neck debridement** are relatively new in the surgical literature but are essential vocabulary for orthopaedic and sports medicine radiologists.

Anatomic Considerations

The bony pelvis is a rigid ring comprised of the sacrum and paired lateral components formed by the ilium, the ischium, and the pubis, which together are sometimes termed the innominate bone. Familiarity with this anatomy is necessary for accurate classification of pelvic fractures. The term "hip" generally refers to the femoral acetabular joint, but "hip pain" can also refer to a variety of extraarticular lesions. With chronic and overuse musculoskeletal injuries causing hip, groin, and pelvic pain, one of the great diagnostic challenges is anatomic identification of the pain source. Referred pain is very common with pelvic lesions and with internal derangement of the hip, often leading to confusion among both patients and examining clinicians regarding a target for imaging interrogation. With clinical groin pain, the radiologist may be best suited to identify the source of pain and direct the treatment plan toward the most appropriate subspecialist. The injured structure could be regional to the groin (such as a visceral injury that would be optimally treated by a general surgeon) or could be anatomically remote (such as the acetabular labrum, warranting treatment from a subspecialist hip orthopedic surgeon).

In recent years, there has been greater understanding of the complex anatomy of an interrelated group of osseous, tendinous, and visceral structures around the pubic symphysis called the rectus abdominis/ adductor aponeurosis. New knowledge has also been gained regarding how anatomic variations can impact biomechanical stresses during an otherwise normal range of motion. While an isolated synovial herniation pit can be seen in the femoral head as a normal variant, an osseous excrescence at the anterosuperior femoral head/ neck junction with an associated synovial herniation pit is now recognized as morphology associated with cam-type femoral acetabular impingement. As our understanding of complex anatomy and anatomical variation increases, we need to recognize these variations when they appear on imaging studies.

Pathologic Considerations

With high-velocity pelvic girdle trauma, isolated disruption at a single point in the biomechanical ring is unlikely (similar to breaking a hard pretzel). Thus, identification of a single pelvic fracture in the setting of trauma should lead to a thorough assessment of the pelvic ring in an attempt to identify additional fractures or joint disruptions. Once imaging is complete, the use of classification systems is important, both in helping to guide therapy and to assess the likelihood of associated visceral, soft tissue, and neurologic injuries. A primary goal of imaging the pelvis and hip should be to identify significant lesions and to then tailor the imaging study to allow for a thorough and complete description of these injuries and any associated lesions. Some observations, such as peritoneal fat extending into the inguinal canal, are unlikely to be the primary source of symptomatology. Other findings, including the presence of a **secondary cleft** or **osteitis pubis**, may be overlooked by an inexperienced reviewer but could be clinically significant. Radiologists must be comfortable enough with treatment options to recognize an expected postoperative appearance and differentiate it from a new injury or failed intervention.

Imaging Considerations

Radiographs of the pelvis are the first-line imaging modality in the settings of high-velocity trauma, hip fractures, and suspected developmental lesions. On an AP view of the pelvis, iliopubic, ilioischial, and iliopectineal (arcuate) lines should be assessed for incongruency or asymmetry. With pelvic fractures, Judet views or oblique views should be considered if a CT is not planned. It is wise to remember that a single radiographic view is a two-dimensional survey and is inadequate for complete evaluation of the musculoskeletal pelvis. Standing AP radiograph of the pelvis and frog leg lateral views of the hips are integral tools for evaluation of femoral acetabular impingement and developmental dysplasia of the hip. Radiographs can be normal in the setting of proximal femoral fracture, particularly for subcapital femoral neck fracture or pelvic insufficiency fracture.

When these entities are suspected and radiographs are equivocal, particularly in osteoporotic patients, MR is the imaging modality of choice.

CT has become the standard of care for imaging high-velocity pelvic trauma. With state of the art imaging systems, high resolution bone CT exams can be generated from standard soft tissue acquisitions obtained during a visceral trauma study. Post-processing for CT of the osseous pelvis includes coronal and sagittal reformats and a variety of 3D reconstruction techniques and can be useful for delineating fractures as well as morphologic variants, such as femoral acetabular impingement and acetabular dysplasia. In the setting of orthopedic hardware, metal CT protocols, with increased photon flux achieved through higher mAs, greater slice overlap, and tangential acquisition planes allow for high resolution imaging of the metal-bone interface. CT is the study of choice for suspected radiographically occult hardware failure, hardware loosening, and giant cell granulomatous response or particle disease.

Ultrasound has become an important modality for assessing myotendinous injuries around the pelvis. The ability to image while simultaneously performing dynamic maneuvers lends ultrasound unique capabilities in the musculoskeletal system. Musculoskeletal radiologists will increasingly need to possess ultrasound skills for the assessment of kinematic myotendinous lesions and for guidance of percutaneous therapeutic techniques.

MR is the workhorse modality for sports injuries around the pelvis, as well as arthropathies and degenerative conditions involving the proximal femur. Several different MR protocols are used in standard practice when imaging hip and pelvis lesions. With large body parts, homogeneous fat suppression is best achieved with inversion recovery (STIR) images. The built-in body coil of a high-field magnet can image large regions, but a phased array torso coil offers higher spatial resolution along with robust signal throughout the pelvic region. Musculoskeletal MR protocols generally need to identify lesions over a region spanning from the pubic symphysis to the greater trochanters and even the iliac crests. Thus, most MR protocols should include an axial T2WI FS sequence covering the entire pelvis, and a large field-of-view T1WI sequence is useful to exclude osseous lesions and infiltrating marrow processes.

A noncontrast bony pelvis MR protocol is quick and easy to perform, while also accurate for diagnosing a variety of osseous, muscular, and tendinous lesions. In the setting of painful lesions, including suspected radiographically occult hip fracture, stress or avulsion fracture, or osteonecrosis, a bony pelvis MR is often the most prudent imaging choice. A three or four sequence bony pelvis protocol can be implemented for emergency department patients or inpatients to confirm these diagnoses while minimizing impact on a busy MR schedule.

An athletic pubalgia MR protocol combines large field-of-view imaging of the entire pelvis with higher resolution imaging centered on the pubic symphysis. This noncontrast protocol adds a coronal oblique imaging plane prescribed along the plane of the anterior iliac wing and high resolution sagittal images through the pubic symphysis performed with a surface coil centered over the symphysis. This imaging strategy offers improved specificity for lesions involving the rectus abdominis/adductor aponeurosis and is the protocol of choice in young patients with groin pain or clinical athletic pubalgia.

A noncontrast hip MR protocol can be employed when a localized hip injury is suspected in a nonarthritic hip. Such a protocol should include large field-of-view imaging of the pelvis as well as smaller field-of-view sequences dedicated to the hip in question, including an oblique axial plane prescribed along the femoral neck on a coronal localizer. While arthrographic protocols have inherent advantages, high resolution noncontrast protocols can be sensitive and specific for labral tears and articular cartilage lesions.

Direct MR arthrography is the gold standard imaging modality for evaluation of internal derangement of the hip. MR arthrographic protocols offer high resolution imaging of the hip in three planes with fat-suppressed T1-weighted spin-echo sequences, as well as a more global view of the pelvis in the form of large field-of-view coronal and axial acquisitions. Oblique axial images are most suited to delineate anterosuperior femoral head/neck morphology. Intraarticular infusion of anesthetic during the arthrogram injection, with assessment of any change in symptoms during provocative hip maneuvers, can provide further diagnostic information.

Therapeutic Considerations

With advancements in therapy come new imaging observations. Similarly, new imaging observations can help tailor creative therapeutic regimens. Accurate and complete description of an acetabular labral tear can help direct the surgeon toward either a primary labral repair or labral and cartilage debridement and femoral head/neck cheilectomy. Imaging descriptions of "sports hernia" lesions have led to changes in surgical techniques for these lesions. New ultrasound-guided interventional techniques for myotendinous lesions around the hip and pelvis are constantly being developed. Ultrasound can, in one sitting, accurately identify diseased tendons, directly observe the impact these tendon injuries have on the biomechanical axis, and guide needle placement for therapeutic tenotomy, steroid injection, or platelet-rich plasma therapy. In this regard, radiologists can play a greater role in the treatment of tendinopathy and overuse injury affecting the hip and pelvis.

Selected References

1. Hodnett PA et al: MR imaging of overuse injuries of the hip. Magn Reson Imaging Clin N Am. 17(4):667-79, vi, 2009
2. Omar IM et al: Athletic pubalgia and "sports hernia": optimal MR imaging technique and findings. Radiographics. 28(5):1415-38, 2008
3. Pfirrmann CW et al: MR arthrography of acetabular cartilage delamination in femoroacetabular cam impingement. Radiology. 249(1):236-41, 2008
4. Parvizi J et al: Femoroacetabular impingement. J Am Acad Orthop Surg. 15(9):561-70, 2007

(Left) *Portable PA radiograph of the pelvis obtained in the trauma bay provides important but limited information regarding the patient's injury. There is a metallic structure overlying the sacrum ➔, but we have little insight as to its true position on this single, 2-dimensional exposure.* ***(Right)*** *Axial bone CT from the same patient shows the knife penetrating into the sacroiliac joint ➔. Tomographic acquisitions offer a 3-dimensional recapitulation of anatomy and pathology.*

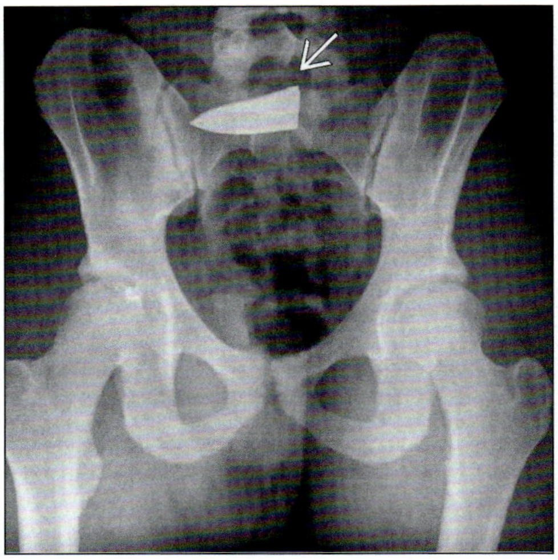

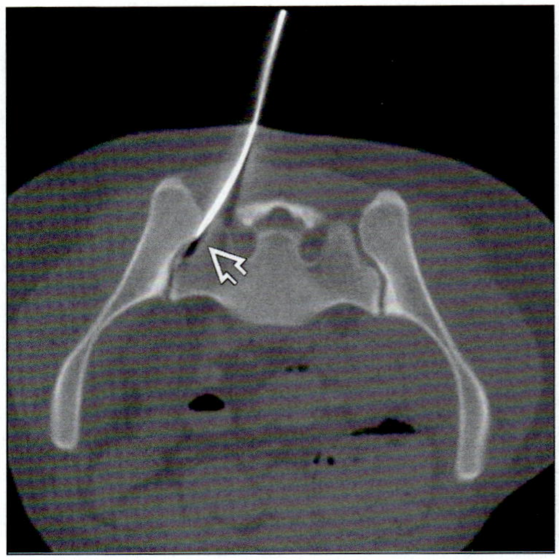

(Left) *Coronal radiograph of the hip in a 68-year-old woman after a low-velocity fall can initially appear normal, but with careful inspection shows a suspicious intertrochanteric lucency ➔.* ***(Right)*** *Coronal reformat bone CT from the same patient was performed for definitive evaluation, and a nondisplaced intertrochanteric fracture was confirmed ➔. The patient was taking alendronate for osteoporosis, which has been associated with proximal femoral insufficiency fracture.*

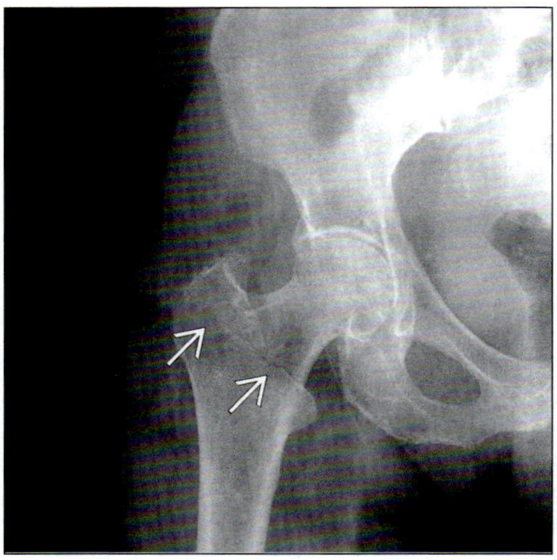

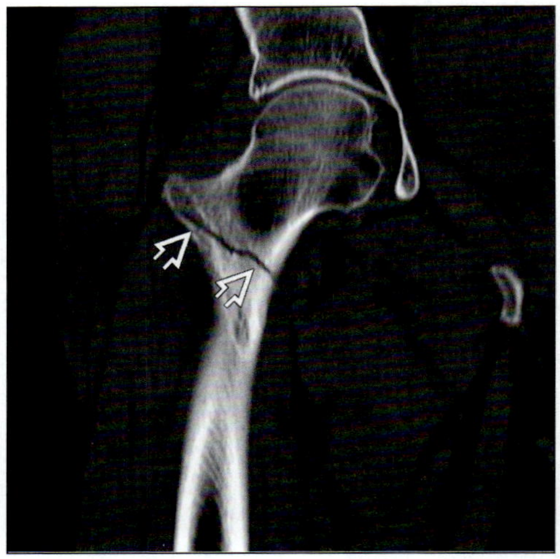

(Left) *Coronal reformat bone CT from a patient after an MVA shows a posterior acetabular fracture ➔ extending through the medial cortex of the ischium ➔.* ***(Right)*** *3D bone CT reformats were performed on a workstation to further elucidate fracture extent ➔ and ischial involvement ➔. While 3D reformats can be useful for describing gross morphology, they are incomplete representations of the acquired source data and should not be used in isolation for treatment planning.*

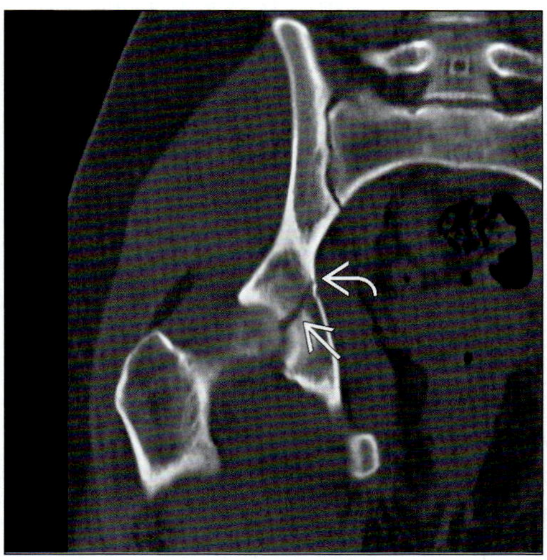

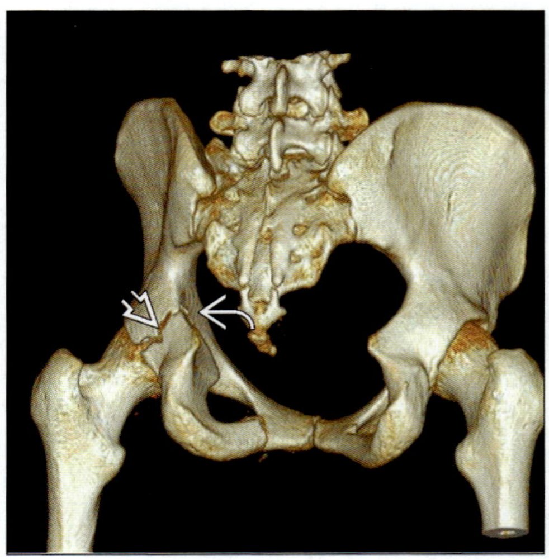

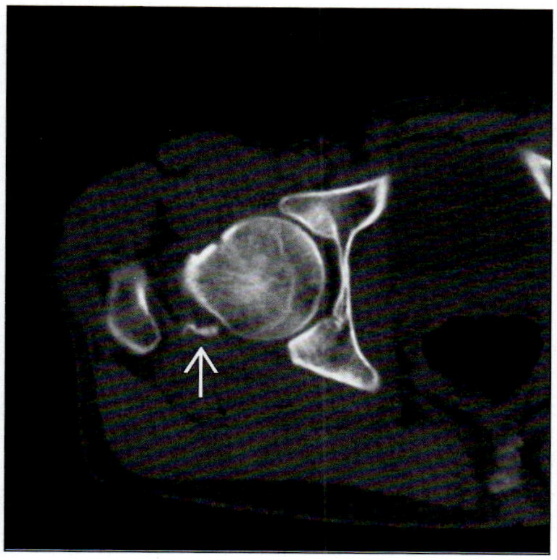

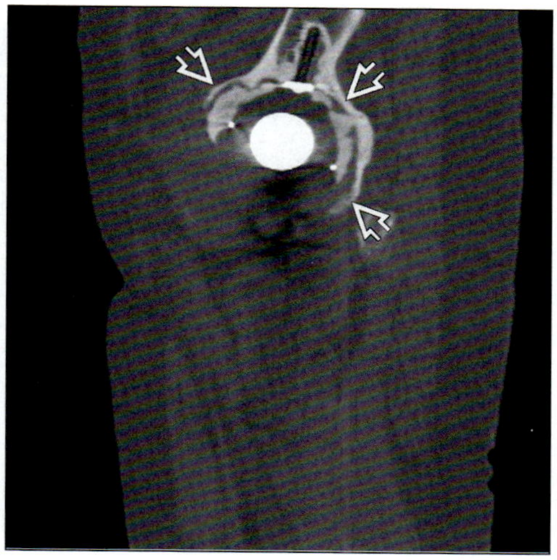

(Left) *Axial bone window CT from a trauma study post MVA shows a small osseous fragment within the hip joint without degenerative findings, indicating a transient posterior hip dislocation. Osseous pelvic structures should be assessed on each trauma CT study.* ***(Right)*** *Sagittal reformat bone CT using a metal protocol shows abnormal lucency between the cement and the acetabular fossa, indicating periprosthetic loosening in this elderly patient with worsening hip pain on ambulation.*

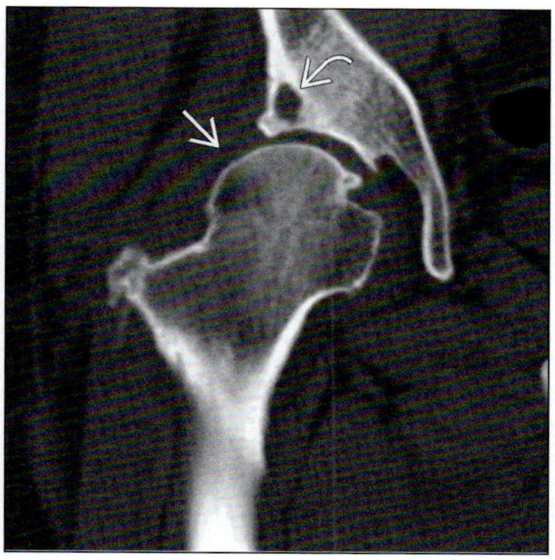

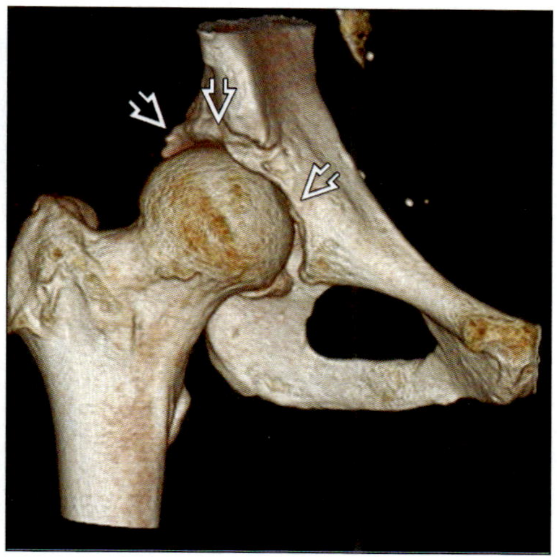

(Left) *Coronal reformat bone CT shows abnormal undercoverage of the femoral head indicating developmental dysplasia of the hip. There is an associated subchondral cyst formation and sclerosis in the acetabular roof indicating secondary osteoarthritis.* ***(Right)*** *Coronal oblique 3D reformat bone CT from the same study creates a virtual map of the deficient acetabular rim. Subtle morphologic variations at the hip can cause relative instability and subsequent premature osteoarthritis.*

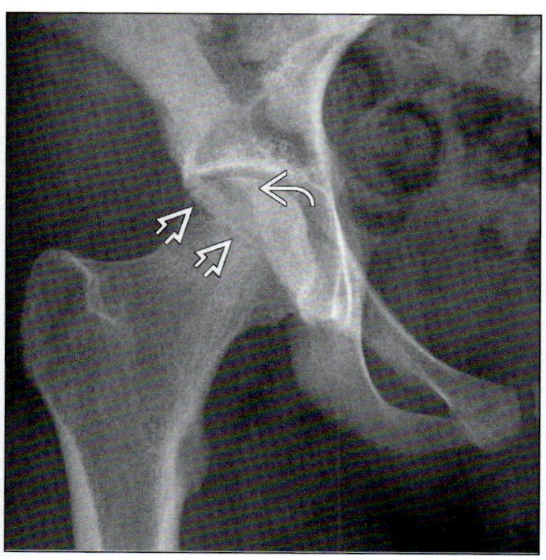

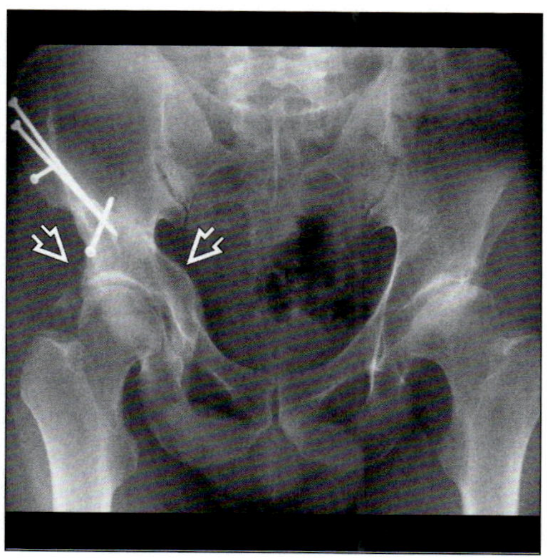

(Left) *AP standing radiograph of the hip in a 24 year old with symptoms of pincer-type impingement shows deficiency of the posterosuperior acetabulum. The posterior acetabular rim has a normal linear contour, but the superior region of the anterior acetabulum crosses lateral to the posterior rim, a morphology known as retroversion.* ***(Right)*** *Standing AP radiograph of the pelvis after posterior acetabular osteotomy (PAO) for acetabular retroversion shows a rebuilt acetabulum.*

*(**Left**) Sagittal MR arthrogram T1WI FS shows contrast undercutting the anterior acetabular labrum at its base →, indicating a detached anterior labrum tear. (**Right**) Axial oblique MR arthrogram T1WI FS shows an undersurface tear at the anterosuperior labrum →. Anterosuperior labral tears are often related to cam-type femoral acetabular impingement, whereas anterior labral tears are often detached at the labral base, related to a pincer mechanism.*

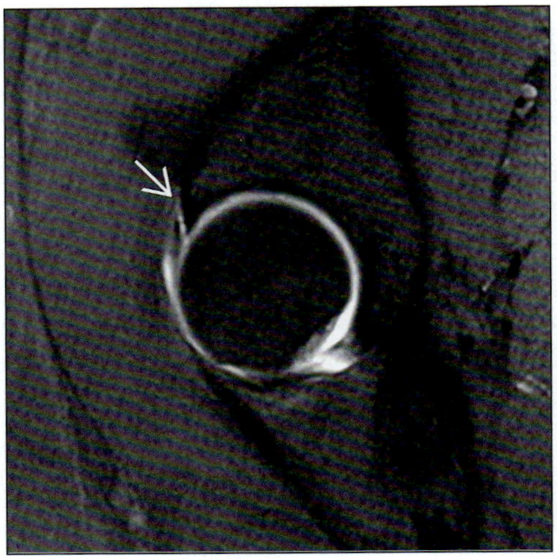

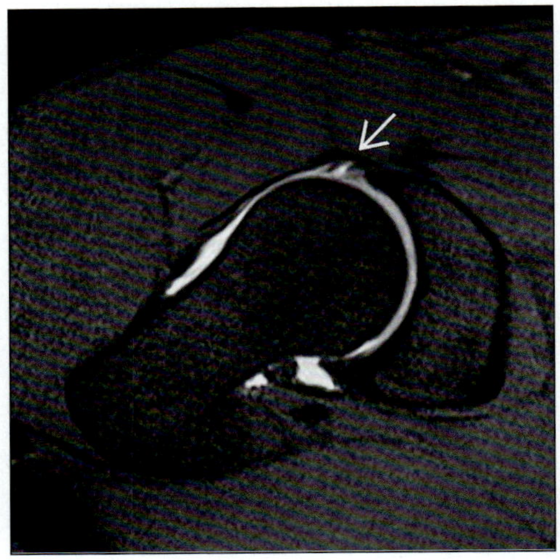

*(**Left**) Coronal MR arthrogram T1WI FS shows a complex anterosuperior labral tear →, cartilage loss on the adjacent acetabulum →, and osseous prominence at the femoral head/neck junction →, the triad of MR arthrographic findings associated with cam-type femoral acetabular impingement. (**Right**) Axial oblique MR arthrogram T1WI FS shows an abnormal femoral head/neck alpha angle as well as early cartilage loss at the anterosuperior acetabulum → in cam-type femoral acetabular impingement.*

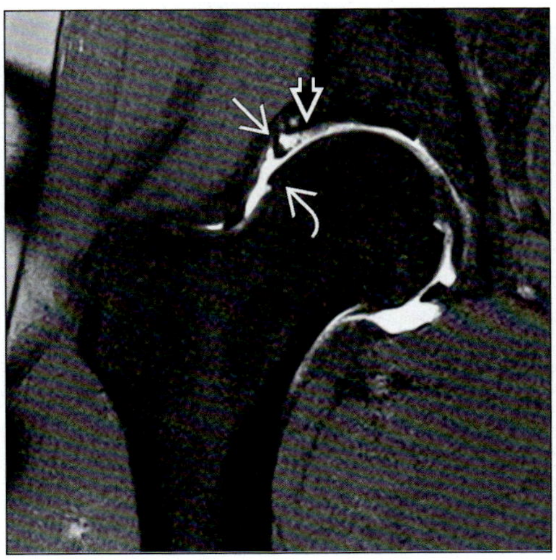

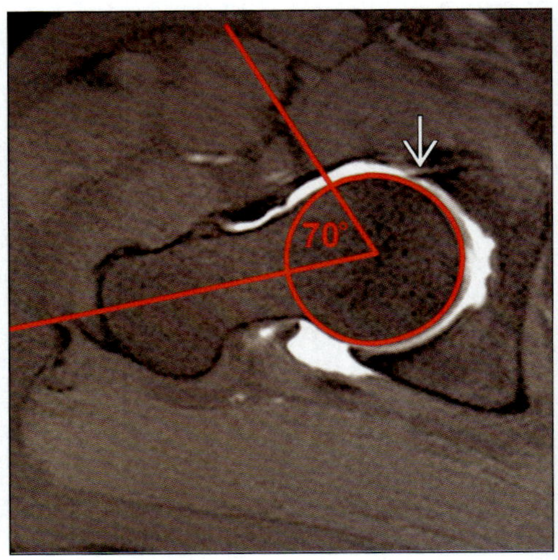

*(**Left**) AP radiograph from a 14-year-old soccer player with acute groin pain shows a displaced osseous fragment → superimposed over the femoral head, suggesting an apophyseal avulsion injury. (**Right**) Coronal STIR MR from the same patient shows avulsion of the rectus femoris apophysis →, with edema in the iliacus → and tensor fascia lata → origins, indicating a high-grade hip flexor injury. In young patients, immature apophyseal plates can be the weak point of the myo-teno-osseous axis.*

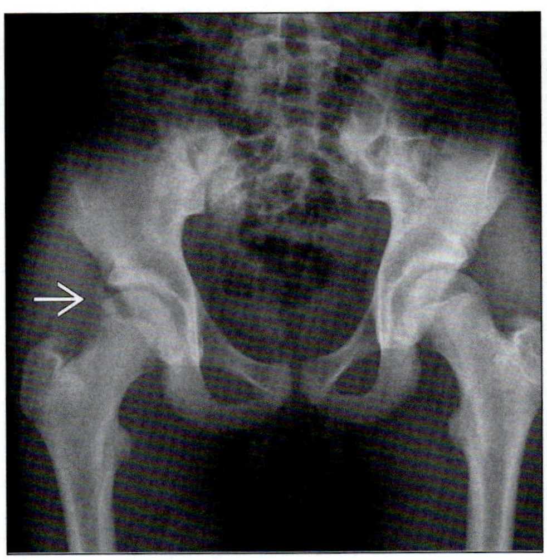

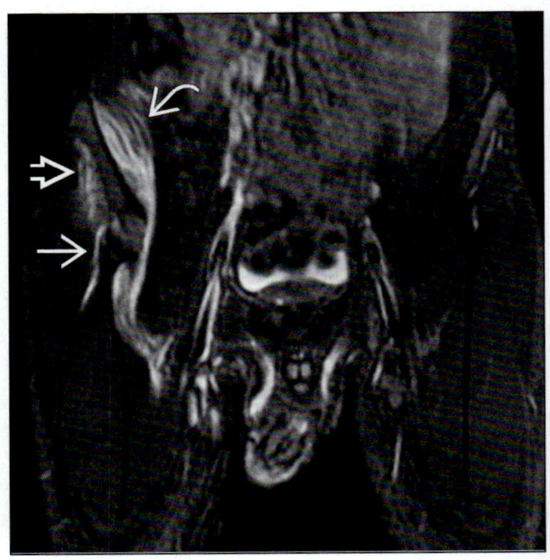

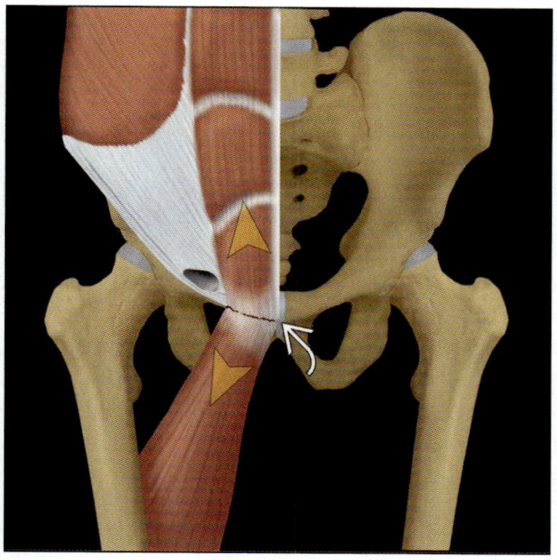

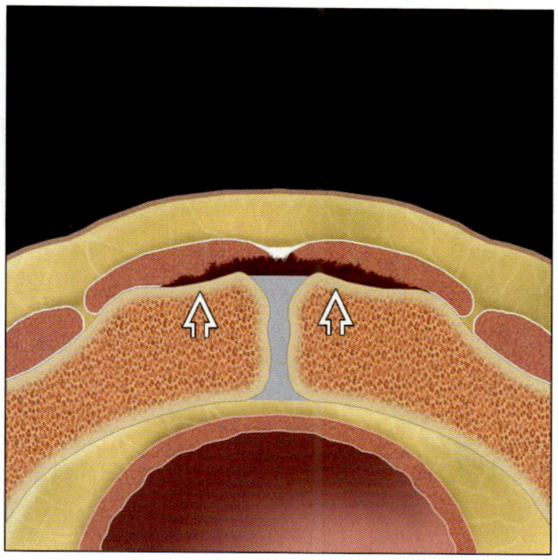

(Left) *AP graphic shows the rectus abdominis/adductor aponeurosis, the most common site for a "sports hernia" lesion ➢. The rectus abdominis attaches anteroinferiorly on the pubic tubercle, blending with the origin of the adductor longus and pectineus and creating a tangential biomechanical axis.* ***(Right)*** *Axial graphic shows a rectus abdominis/ adductor aponeurotic plate disruption ➢. In contrast to the unilateral aponeurosis lesion, this variant originates at midline and extends bilaterally.*

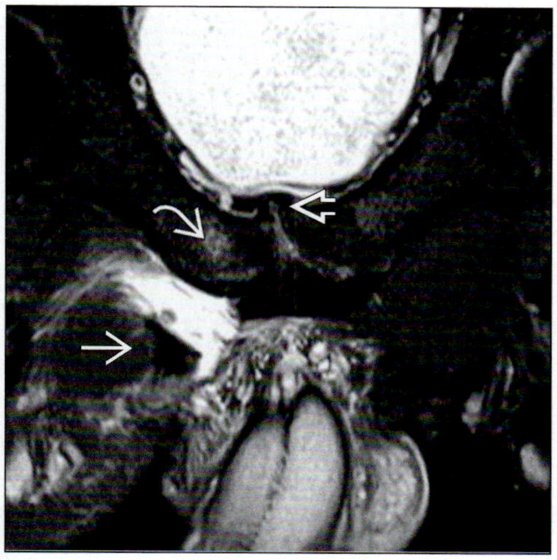

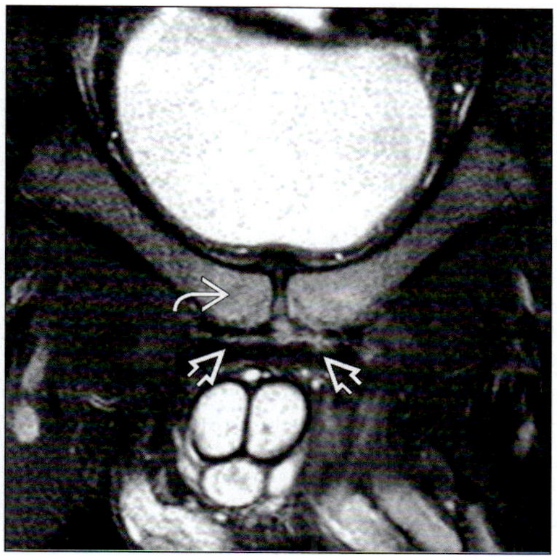

(Left) *Coronal oblique T2WI FS MR shows an acute rectus abdominis/ adductor aponeurosis tear with retraction of the adductor longus ➔, marrow edema at the site of rectus abdominis detachment ➢, and capsular hypertrophy, indicating chronic osteitis pubis ➢.* ***(Right)*** *Coronal oblique T2WI FS MR shows a rectus abdominis/adductor aponeurotic plate disruption. There is confluent bilateral aponeurotic detachment spanning the midline ➢ and subchondral marrow edema from acute osteitis pubis ➢.*

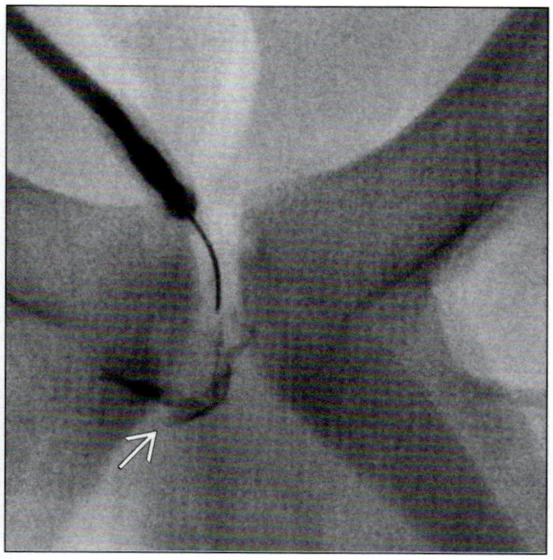

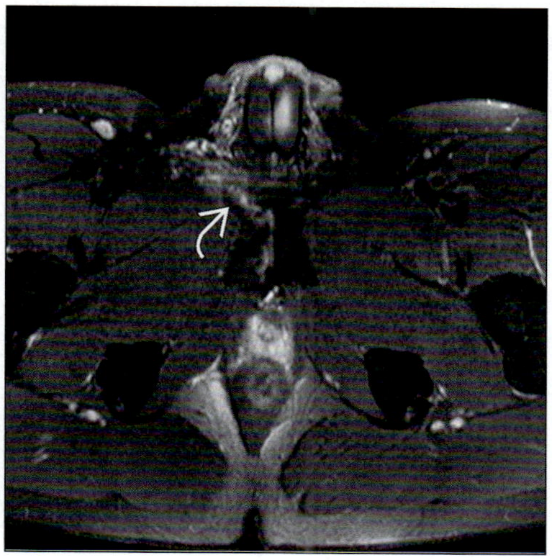

(Left) *AP spot radiograph from a pubic symphysis arthrogram and steroid injection in a baseball player with right-sided groin pain shows the classic secondary cleft ➔, an inferolateral extension of contrast from the primary symphyseal cleft on the side of pain.* ***(Right)*** *Axial T2WI FS MR from the same patient shows a tear at the rectus abdominis/ adductor aponeurosis ➢. Patient initially responded to the symphyseal injection, but symptoms returned and he ultimately underwent pelvic floor repair.*

FEMORAL HEAD FRACTURE

Key Facts

Terminology
- Fracture of femoral head ± articular cartilage disruption
- Related to
  - Posterior dislocation femoral head
  - Fatigue or insufficiency fracture

Imaging
- Commonly nondisplaced
- Can be occult on radiographs and CT
  - CT in setting of hip dislocation to identify fragments
- Atraumatic hip pain: Radiographs followed by MR
- T1WI: Low signal edema of subchondral fat marrow and hypointense fracture line
- T2WI: Fracture may manifest as marrow edema only
  - ± hypointense fracture line

Top Differential Diagnoses
- Femoral head osteonecrosis
- Rapidly progressive osteoarthritis

Pathology
- Pipkin classification

Diagnostic Checklist
- Associated with posterior hip dislocation and acetabular rim fracture
- Fatigue fracture of femoral head should be considered in young athlete
- Look for subtle anterior femoral head impaction fracture on CT for hip dislocation
- Insufficiency fracture in elderly patient with pain and normal radiograph
- Search for entrapped femoral head fragments within joint

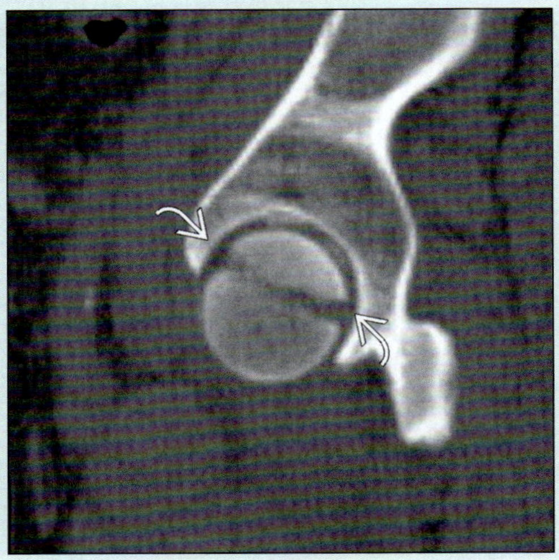
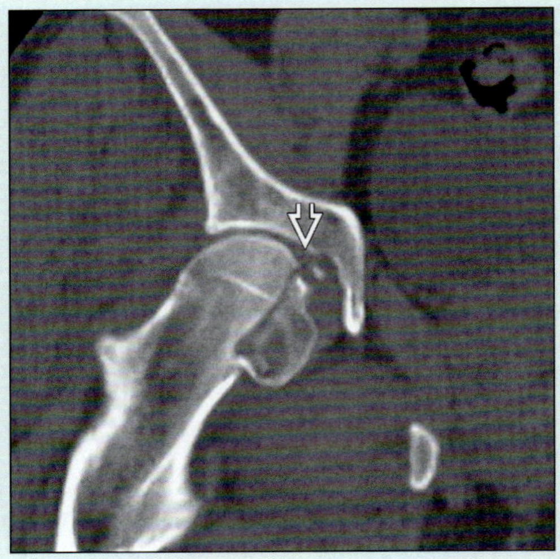

(Left) *Sagittal reformat bone CT after MVA shows a minimally displaced femoral head fracture extending to both anterior and posterior articular surfaces. It is important to assess for step-off at articular surfaces in 3 planes.* ***(Right)*** *Coronal reformat bone CT from the same patient reveals a step-off at the anterior femoral head articular surface with a small displaced fragment. The fracture extends below the ligamentum teres and is a Pipkin 1 femoral head fracture.*

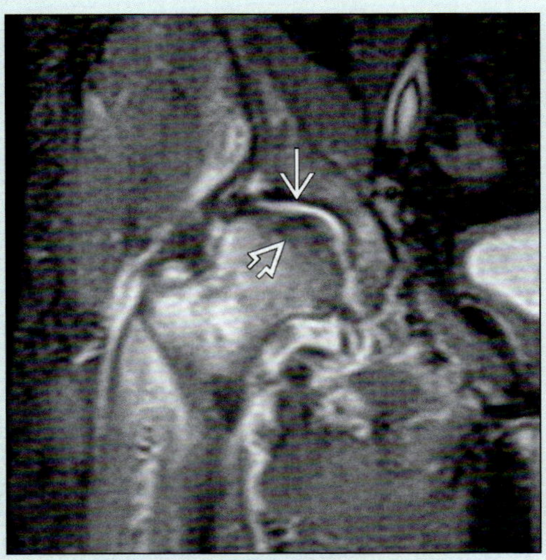
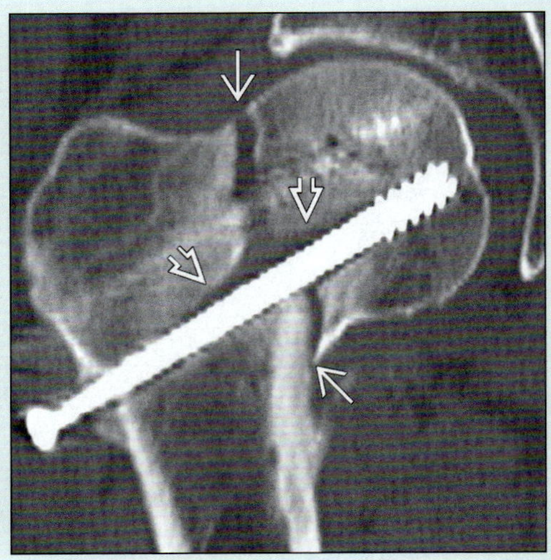

(Left) *Coronal STIR MR of a subchondral fracture of the femoral head shows focal collapse of the weight-bearing articular surface and a perpendicular, nondisplaced fracture line.* ***(Right)*** *Coronal reformat CT 12 months after open reduction and internal fixation of a Pipkin 3 femoral head fracture extending to the femoral neck shows nonunion of the fracture with no osseous bridging at the fracture site, and lucency around the cannulated screw suggesting hardware loosening.*

FEMORAL HEAD FRACTURE

TERMINOLOGY

Synonyms
- Subchondral fracture of femoral head
- Osteochondral fracture of hip
- Capital fracture

Definitions
- Fracture of femoral head
  - Usually subchondral

IMAGING

General Features
- Best diagnostic clue
  - Radiographs and CT
    - Subchondral impaction of femoral head
    - Linear fracture in femoral head (less common)
    - May be occult on radiographs and CT
  - MR
    - Intense marrow edema in femoral head out of proportion to osteoarthritis
    - ± hypointense fracture line, depending upon resolution
- Size
  - May range from tiny subchondral fracture to comminution of large portion of femoral head
- Morphology
  - Commonly nondisplaced
  - Subchondral fracture usually linear, may be curved
    - Parallels articular surface
  - Vertical or oblique linear fracture extending to articular surface

Radiographic Findings
- Traumatic fracture ranges from displaced fracture to subtle impaction at anterior femoral cortex
- Often occult on radiographs

CT Findings
- Hip dislocation
  - Associated femoral head fracture ranges from large displaced fracture to subtle impaction of anterior femoral head cortex
  - Associated posterior acetabular fracture
- Nondisplaced impaction fracture occult on CT

MR Findings
- T1WI
  - Low signal edema of subchondral fat marrow and hypointense fracture line
- STIR
  - Fracture may manifest as marrow edema only
  - ± hypointense fracture line, depending upon resolution

Imaging Recommendations
- Best imaging tool
  - Radiographs for initial evaluation
  - CT after hip dislocation to assess
    - Degree of posterior acetabular rim fragmentation
    - Presence of intraarticular fragments
  - For suspected nondisplaced fracture, radiographs followed by MR
    - Sensitivity of MR > CT for nondisplaced fracture
- Protocol advice
  - Coronal and axial CT and MR: Include both hips for assessment of symmetry

DIFFERENTIAL DIAGNOSIS

Chondral and Osteochondral Lesions, Hip
- Focal injury to articular cartilage and underlying subchondral bone

Osteonecrosis
- Crescentic hypointensity (not linear) meeting articular surface in 2 locations
  - Subtended fragment often has normal signal centrally
- Regional (not linear) hypointense subchondral marrow on T2/STIR
- Articular collapse and central femoral head sclerosis on radiograph

Rapidly Destructive Osteoarthritis of Hip
- Rapid (months) onset of severe hip pain and chondrolysis

Stress or Insufficiency Fracture
- Linear subchondral focus of low signal surrounded by marrow edema
- MR diagnosis

PATHOLOGY

General Features
- Etiology
  - Traumatic fracture
    - Oblique fracture from shearing force during hip dislocation
    - Anterior femoral head impaction from posterior hip dislocation
    - Generally high-velocity injuries, commonly due to motor vehicle crash
  - Fatigue (stress) fracture: Repetitive force on normal bone with overuse
- Associated abnormalities
  - Hip dislocation
    - Ipsilateral acetabular rim fracture
    - Concurrent head, abdominal, and pelvic trauma
  - Stress
    - Other foci of osseous stress response

Staging, Grading, & Classification
- Pipkin classification

CLINICAL ISSUES

Presentation
- Clinical profile
  - Traumatic
    - Posterior hip dislocation
  - Stress fracture
    - Elite athletes
    - Military recruits

Demographics
- Gender
  - M > F

FEMORAL HEAD FRACTURE

Pipkin 1: Fracture exits below ligamentum teres with no other associated osseous injury.

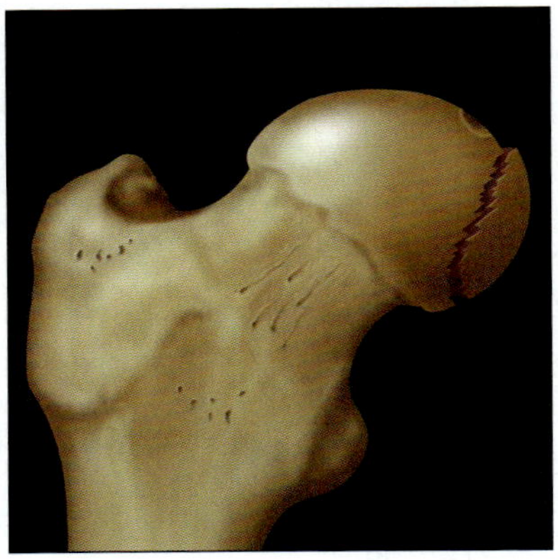

Pipkin 2: Fracture exits above ligamentum teres with no other associated osseous injury.

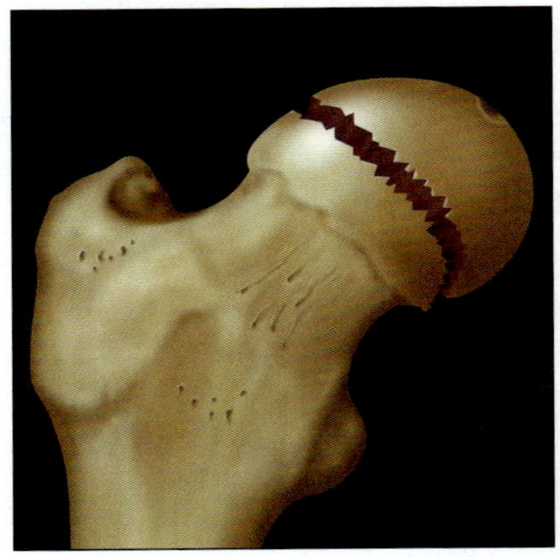

Pipkin 3: Fracture exits above or below ligamentum teres with an associated femoral neck fracture.

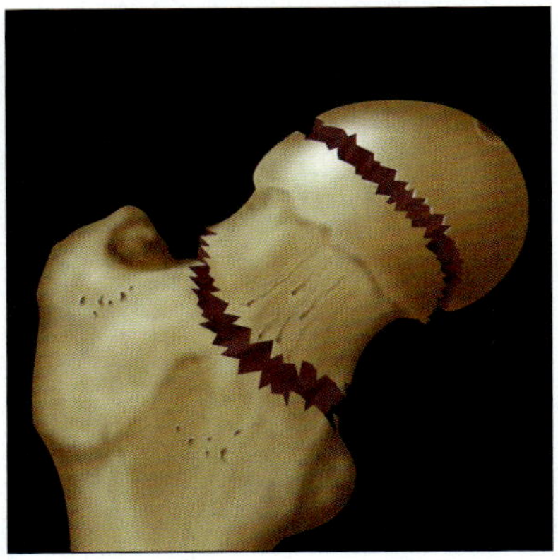

Pipkin 4: Axial plane cutaway view. Fracture exits above or below ligamentum teres with an associated acetabular fracture.

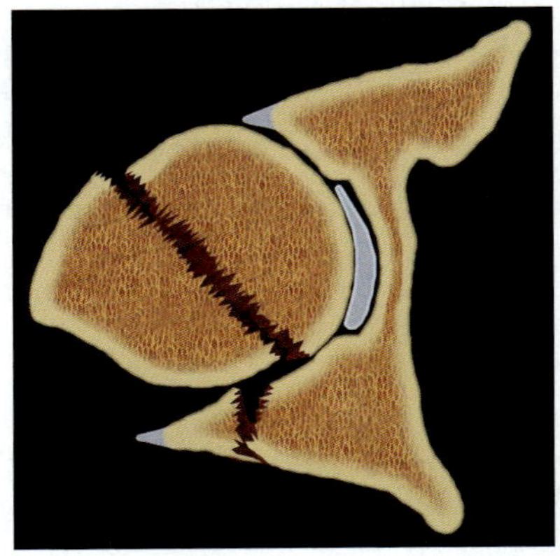

Pipkin 1 fracture is treated conservatively with closed reduction; displacement of an inferior fragment is acceptable. Pipkin 2 fracture may be treated conservatively or surgically, with closed reduction if fragments are near anatomic and congruent, or with ORIF if fragments displaced. Surgical treatment for Pipkin 3 fracture includes ORIF with rigid pin in young patients or hip arthroplasty in older patients. For Pipkin 4 fracture, surgical treatment is required: ORIF of acetabulum in young patients, total hip replacement and acetabuloplasty in older patients.

Natural History & Prognosis

- If nondisplaced, most heal with good outcome
- If displaced, open reduction and internal fixation (ORIF) and total hip replacement are options
  - Osteonecrosis (ON) of femoral head is primary concern
- Fatigue (stress): Good outcome if deformity prevented

Treatment

- Anatomic/near anatomic fracture reduction essential
  - Preserve function
  - Avoid ON and osteoarthritis
    - Incidence of ON increases significantly if reduction delayed > 12 hours
    - Maximum 2-3 attempts at closed reduction
- Traumatic and fatigue femoral head fractures
  - Pipkin 1: Conservative
    - Closed reduction preferred
    - Displaced inferior fragment acceptable if it does not involve weight-bearing portion of head
  - Pipkin 2: Conservative or surgical
    - Closed reduction only if fx fragments congruent
    - ORIF for displaced/unstable fracture fragments
  - Pipkin 3: Surgical
    - ORIF for younger patients (rigid pinning)
    - Hemiprosthesis for older patients
  - Pipkin 4: Surgical
    - ORIF for displaced/unstable acetabular and femoral head fractures in young patient
    - Total hip arthroplasty (THA) in older patient
  - Complications for fracture related to hip dislocation
    - ON, post-traumatic arthritis, recurrent dislocation, sciatic nerve injury
- Fatigue (stress) fracture
  - Cessation of activity until resolution
  - Malaligned fracture requires ORIF or THA

DIAGNOSTIC CHECKLIST

Consider

- Association of femoral head fractures with posterior dislocation and acetabular rim fractures
- Fatigue fracture in young patient with overexertion and pain

Image Interpretation Pearls

- Look for subtle anterior femoral head impaction fracture on CT for hip dislocation
- Include femoral head in search pattern on MR for femoral neck fractures

Reporting Tips

- Search for entrapped femoral head fragments within joint

SELECTED REFERENCES

1. Giannoudis PV et al: Management, complications and clinical results of femoral head fractures. Injury. 40(12):1245-51, 2009

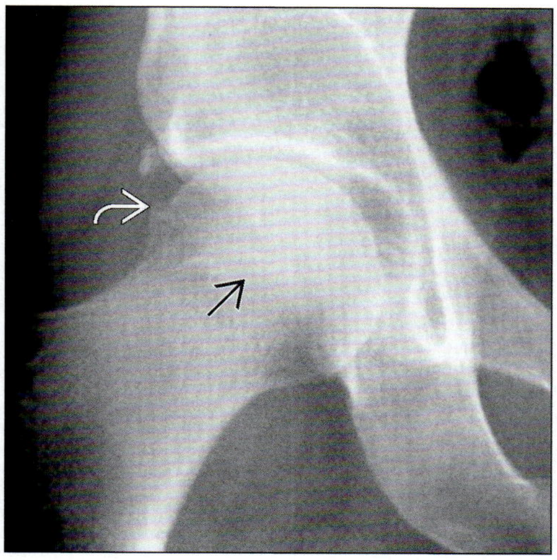

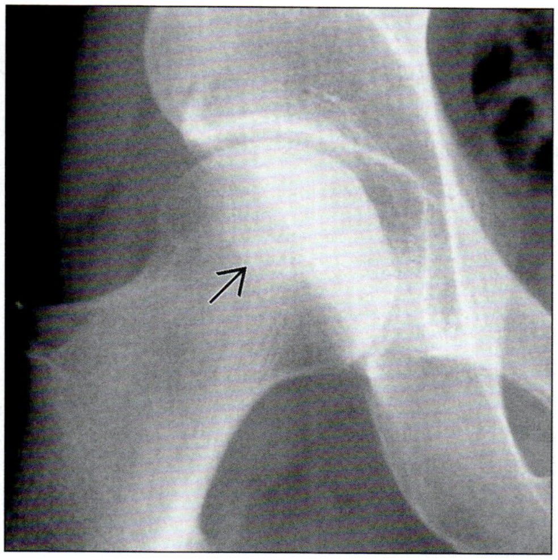

(Left) *AP radiograph shows an unusual fracture pattern seen in conjunction with early osteonecrosis of the hip. Note the central femoral head sclerosis ⇨ and a femoral head fracture with slight displacement ➡.* ***(Right)*** *AP radiograph of the hip in a young athlete after axial loading trauma shows focal sclerosis centrally in the femoral head ⇨ consistent with a nondisplaced fracture. Despite the subtle findings, fractures in this location are associated with femoral head osteonecrosis.*

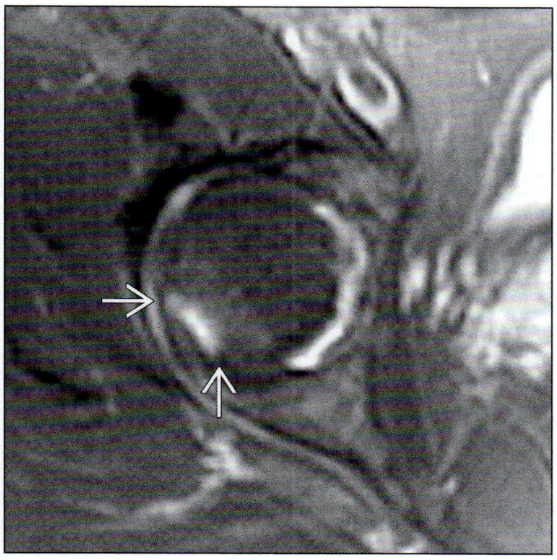

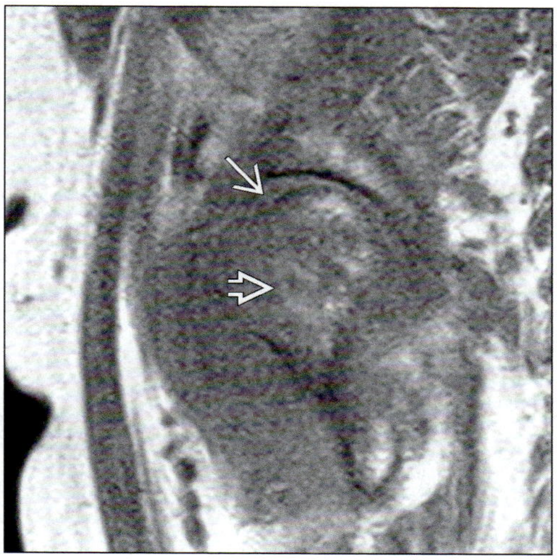

(Left) *Axial T2WI FS MR from a football player after an anterior hip subluxation shows a nondisplaced fracture of the posterior femoral head ➡. This subchondral fracture was radiographically occult, but the location raises concern for post-traumatic osteonecrosis.* ***(Right)*** *Sagittal T1WI MR shows a slightly displaced Pipkin 2 fracture of the femoral head ➡. Marrow signal is diffusely abnormal ➡, raising concern for rapidly destructive osteoarthritis of the hip.*

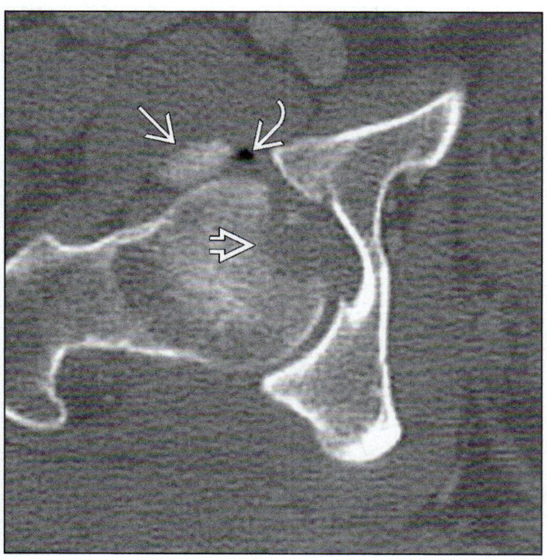

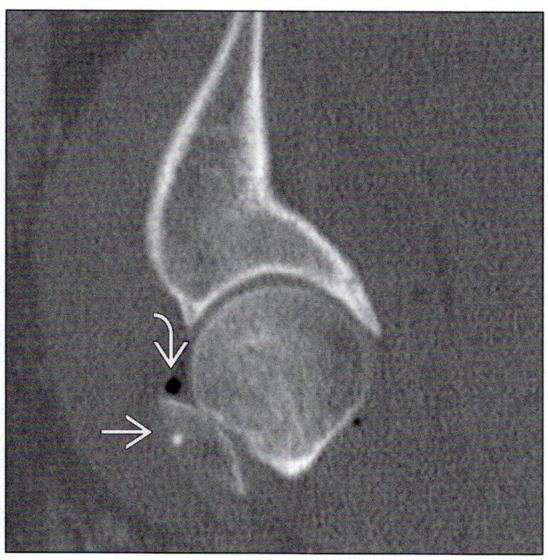

(Left) *Axial bone CT shows a large defect in the femoral head following hip dislocation ➡. The fracture is located below the level of the ligamentum teres; this represents a Pipkin 1 fracture. An osseous fragment is displaced anteriorly ➡. The air bubble located anteriorly in the joint ➡ is indicative of recent hip dislocation.* ***(Right)*** *Sagittal reformat bone CT in the same patient shows a large anterior osseous fragment ➡ adjacent to an air bubble ➡.*

FEMORAL NECK FRACTURE

Key Facts

Terminology
- Subcapital, midcervical or transcervical, basicervical fracture
- Intracapsular femoral neck fractures

Imaging
- Malalignment is diagnostic in most cases
- If nondisplaced, may be radiographically occult
- Sclerosis in typical fracture plane
- Asymmetry of lateral femoral head/neck contour
- Change of alignment of trabeculae
- Smudgy sclerosis from impaction
- If radiographs negative or equivocal and clinical suspicion high, noncontrast MR indicated
- Coronal STIR or T2 FS near 100% sensitivity by virtue of bone marrow edema

Pathology
- Usually osteoporotic insufficiency injury: Senile osteoporosis is most common
- Stress fracture in runners (cyclic loading + torsional force)
- Subcapital fracture: Garden classification
- Fall-related femoral neck fractures in elderly → high incidence of upper extremity fracture

Clinical Issues
- Pain in groin, anterior thigh, or knee
- Hip held in flexion, external rotation
- Subcapital fracture: Risk of displacement ± osteonecrosis
- Basicervical stress fracture: Risk of completion
- Garden IV fracture has > 40% nonunion rate & 30% rate of osteonecrosis

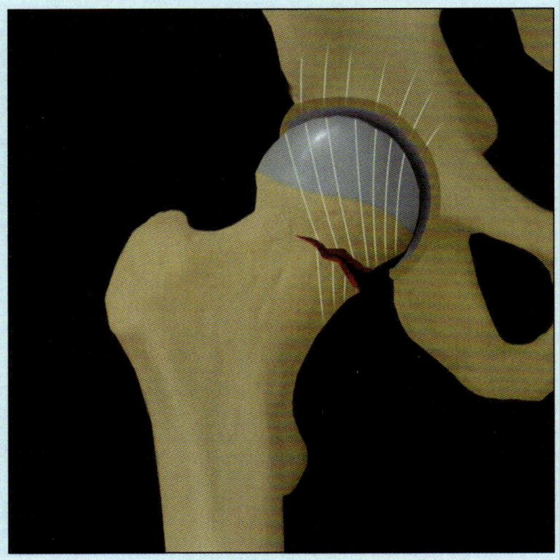
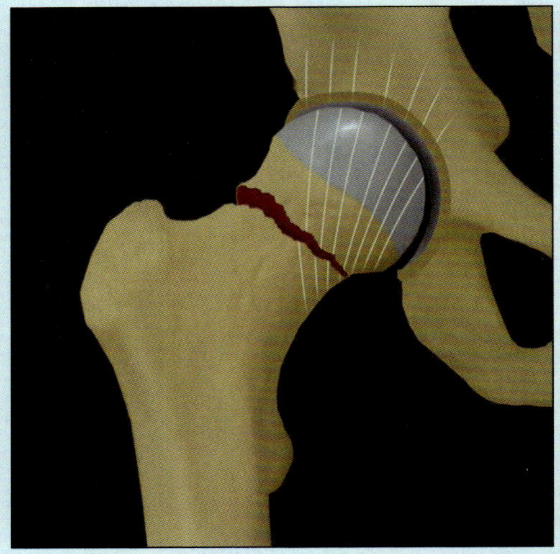

(Left) *Graphic of a Garden stage I fracture demonstrates an incomplete fracture involving either the lateral or medial cortex. There is malalignment of trabecula due to rotation of the femoral shaft relative to the head.* ***(Right)*** *Graphic of a Garden stage II fracture shows a complete fracture involving both the medial and lateral cortices without displacement. Femoral head/shaft alignment is preserved.*

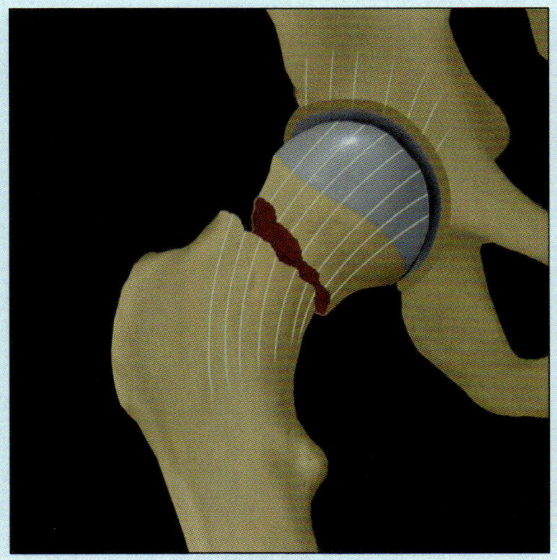
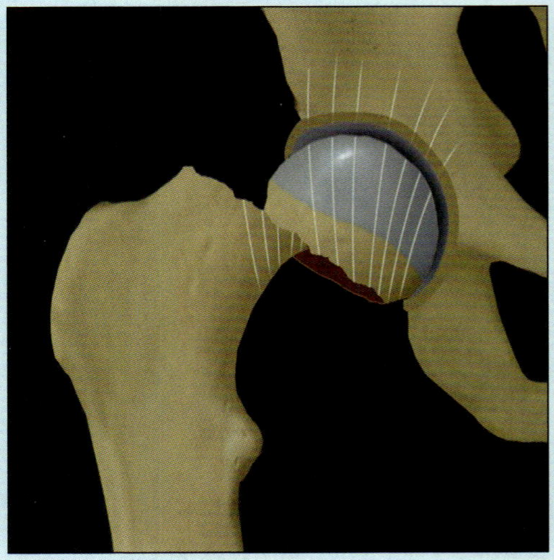

(Left) *Graphic of a Garden stage III fracture shows a complete fracture with mild displacement. This fracture most commonly demonstrates varus and posterior angulation and abduction, anterior rotation of the femoral head, and external rotation of the femoral shaft.* ***(Right)*** *Graphic of a Garden stage IV fracture shows complete displacement of the femoral shaft relative to the femoral head, often with external rotation and cephalad migration of the shaft.*

FEMORAL NECK FRACTURE

TERMINOLOGY

Definitions

- Intracapsular femoral neck fractures, due to either trauma, insufficiency, or stress
  - Subcapital: Femoral head/neck junction
  - Midcervical (transcervical): Midportion of femoral neck
  - Basicervical: Base of femoral neck

IMAGING

General Features

- Best diagnostic clue
  - Radiographic malalignment diagnostic
    - Asymmetry on AP pelvis helpful in setting of osteoarthritis
    - Varus angulation common
  - If high clinical suspicion of femoral neck fracture and equivocal radiographs → MR
- Location
  - Ranges from junction of femoral head & neck (subcapital) to just above lesser trochanter (basicervical)
- Size
  - Ranges from incomplete & nondisplaced to displaced & comminuted
- Morphology
  - Subcapital
    - Following plane of femoral head/neck junction, perpendicular to neck
    - May be nondisplaced or displaced
  - Transcervical (midcervical)
    - Mildly oblique, perpendicular to femoral neck
    - Commonly displaced in varus angulation
  - Basicervical
    - Transverse at medial cortex, extends obliquely to lateral cortex
    - Commonly displaced in varus angulation

Radiographic Findings

- Nondisplaced fractures may be subtle on x-ray
- Sclerosis in typical fracture plane
- Angulation of trabeculae
- Smudgy sclerosis from impaction
- Impacted fracture may be mistaken for ring osteophyte
- Asymmetry of lateral femoral head/neck contour or ring osteophyte
- Frank cortical offset
  - May only be seen in 1 plane
- Displaced subcapital fracture in varus may have permeative appearance, mimicking pathologic fracture

CT Findings

- CT is less sensitive than MR for nondisplaced fracture regardless of detector rows
  - Fracture plane often parallels axial imaging plane
    - Coronal and sagittal reformats necessary
  - Osteopenia and paucity of trabeculae obscure fx
  - CT can be helpful in setting of comminution, concomitant fractures, or for assessment of alignment

MR Findings

- T1WI
  - Hypointense fracture line and marrow replacement in coronal and sagittal planes
    - Thick and often incomplete hypointensity at medial cortex with stress fracture
- STIR
  - Coronal STIR or T2WI FS MR nearly 100% sensitive for fracture by virtue of bone marrow edema
  - Bone marrow edema extent distinguishes complete from incomplete fracture

Nuclear Medicine Findings

- Bone scan is insensitive in osteoporotic patients for 1st 72 hours

Imaging Recommendations

- Best imaging tool
  - Radiograph is 1st line of imaging
  - If radiographs negative or equivocal and clinical suspicion high, noncontrast MR indicated
    - 27-40% of cases with clinical suspicion and normal x-rays show hip or pelvic fracture on MR
    - 50-65% of cases with clinical suspicion and normal x-rays show soft tissue injury or fracture by MR
- Protocol advice
  - Image entire pelvis: Coronal STIR, T1 & axial T2 FS
    - Must include sacrum and obturator rami in search pattern
    - Can add sagittal and axial oblique T2 FS of hip in question for better resolution if desired

DIFFERENTIAL DIAGNOSIS

Femoral Head Fractures

- Involves subchondral region and often articular surface

Intertrochanteric Fracture

- Involving 1 or both femoral trochanters
- Can extend from lesser trochanter to femoral neck

Osteoarthritis

- Sclerotic ring osteophyte may mimic subcapital hip fracture
- MR diagnostic if radiographs equivocal

Osteonecrosis

- Crescentic density or hypointensity extending from subchondral femoral head

Transient Bone Marrow Edema Syndrome

- Marrow edema femoral head on MR without visible fracture line

Healed Fracture

- Can mimic acute fracture on radiographs but no marrow edema on MR

PATHOLOGY

General Features

- Etiology
  - Subcapital fracture
    - Fall or twisting injury

FEMORAL NECK FRACTURE

- Osteoporotic insufficiency most common
- Rarely, insufficiency from other etiologies
  - Anorexia or corticosteroids
- Midcervical/transcervical fracture
  - Rare fracture, requires significant trauma
- Basicervical fracture
  - Stress fracture in runners (cyclic loading + torsional force)
    - Coxa vara predisposes athlete to risk of fracture
  - Rarely, direct and significant trauma
- Associated abnormalities
  - Fall-related femoral neck fracture in elderly highly associated with upper extremity osteoporotic fracture
  - Femoral neck stress fracture associated with other osseous stress syndromes
  - Osteonecrosis of femoral head
    - Incidence increases with higher position of fracture on femoral neck

Staging, Grading, & Classification

- Subcapital fracture: Garden classification
  - Garden I: Incomplete with lateral impaction
    - Slight valgus orientation of femoral neck
  - Garden II: Complete but nondisplaced
  - Garden III: Complete with partial displacement
    - Slight varus orientation of fracture line
  - Garden IV: Complete and displaced
    - Femoral shaft externally rotated & telescoped
- Stress fracture: Blickenstaff/Morris classification
  - Type I: Endosteal or periosteal callus without fracture line
  - Type II: Discrete fracture line
  - Type III: Displaced fracture

CLINICAL ISSUES

Presentation

- Most common signs/symptoms
  - Pain in groin, anterior thigh, or knee increased with axial compression or greater trochanter percussion
  - Hip held in flexion, external rotation
- Other signs/symptoms
  - Inability to bear weight on affected extremity

Demographics

- Age
  - Subcapital: Elderly, osteoporotic
    - By age 80, 10% of Caucasian women & 5% of Caucasian men sustain a hip fracture
    - By age 90, these rates double
  - Midcervical and basicervical: Young adult
- Gender
  - Elderly osteoporotic fracture: F > M
  - Stress fracture in military population: M > F
  - Stress fracture in athletes: F > M (anorexia, amenorrhea, premature osteoporosis)
- Epidemiology
  - Insufficiency fractures (subcapital) related to osteoporosis
    - Extremely common in elderly; paralleling prevalence of osteoporosis
  - Stress fracture (basicervical) related to activity
    - 15% of runners develop stress fracture
    - 5-10% of stress fractures involve femoral neck

Natural History & Prognosis

- Subcapital fracture: Risk of displacement ± osteonecrosis
- Basicervical stress fracture: Risk of completion
- Basicervical complete fracture: Risk of delayed or nonunion
- Non-weight-bearing treatments put patient at risk for other morbidities (pneumonia, pulmonary embolus)
- Garden IV fracture has > 40% nonunion rate & 30% rate of osteonecrosis

Treatment

- Subcapital
  - Garden I or II: Percutaneous pinning
  - Garden III or IV: Hemiarthroplasty or total hip arthroplasty (THA)
  - Chronic fractures have higher risk of osteonecrosis so THA often indicated
- Basicervical
  - Traumatic or complete: Internal fixation
  - Stress reaction or incomplete fracture: Conservative; non-weight-bearing

DIAGNOSTIC CHECKLIST

Consider

- Femoral neck fractures common in athletes and in patients with senile osteoporosis
- Subcapital insufficiency & incomplete basicervical fatigue fractures may be missed on radiograph
- If hip fracture is clinically suspected and radiographs equivocal → MR

Image Interpretation Pearls

- Subcapital fracture often positioned in flexion & external rotation
  - Foreshortens femoral neck, making diagnosis difficult
- Ring osteophytes can mimic subcapital fracture
- Asymmetry of anterosuperior femoral head/neck angle on AP pelvis

Reporting Tips

- If radiographs equivocal, recommend MR

SELECTED REFERENCES

1. Davidovitch RI et al: Challenges in the treatment of femoral neck fractures in the nonelderly adult. J Trauma. 68(1):236-42, 2010
2. Kettunen JA et al: Hip fractures and femoral bone mineral density in male former elite athletes. Bone. 46(2):330-335, 2010
3. Aleem IS et al: Arthroplasty versus internal fixation of femoral neck fractures: a clinical decision analysis. Ortop Traumatol Rehabil. 11(3):233-41, 2009
4. Schemitsch E et al: Femoral neck fractures: controversies and evidence. J Orthop Trauma. 23(6):385, 2009
5. Schmidt AH et al: Optimal arthroplasty for femoral neck fractures: is total hip arthroplasty the answer? J Orthop Trauma. 23(6):428-33, 2009
6. Talbot JC et al: Femoral neck stress fractures in military personnel--a case series. J R Army Med Corps. 154(1):47-50, 2008

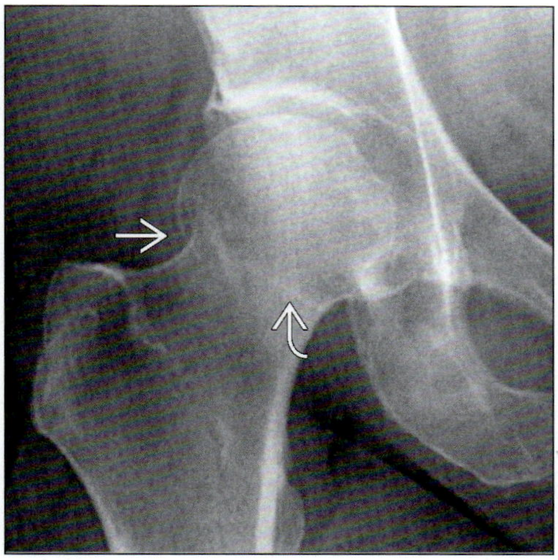

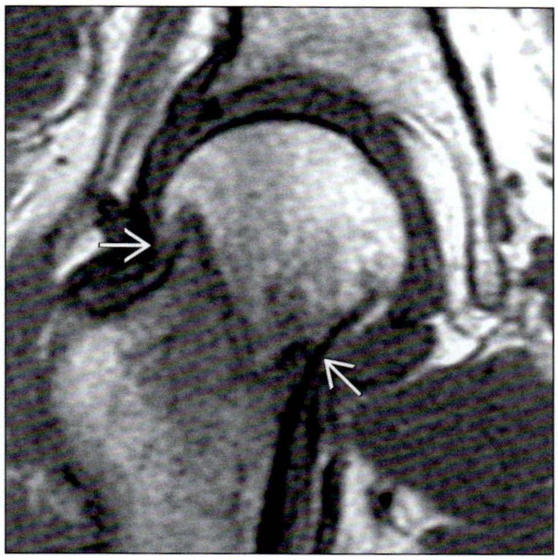

(Left) *AP radiograph from an elderly woman with groin pain after a fall shows a lateral cortical angulation and a linear sclerosis. These subtle abnormalities could be misinterpreted as a ring osteophyte related to osteoarthritis. There is mild valgus at the site of the fracture, which is otherwise nondisplaced.* ***(Right)*** *Coronal T1WI MR from the same patient confirms a jagged hypointense fracture line extending to both cortices in this Garden I subcapital fracture.*

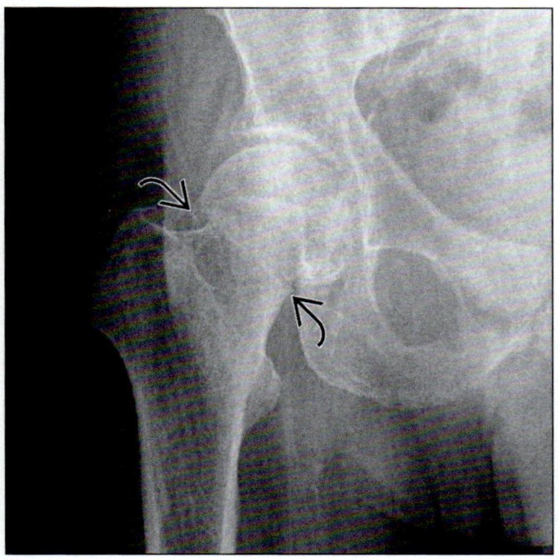

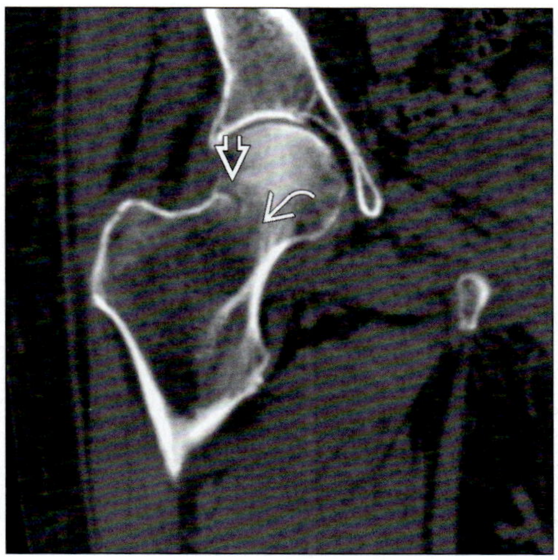

(Left) *AP radiograph shows a complete but nondisplaced Garden stage II femoral neck fx, with mild external rotation of the femoral shaft.* ***(Right)*** *Coronal reformat bone CT from an elderly male in the emergency department with hip pain & equivocal radiographs after a low-velocity fall shows a transcervical femoral neck fx with minimal displacement at its superolateral margin, consistent with a Garden II fracture. CT may be subtle in nondisplaced fracture; MR is more reliable.*

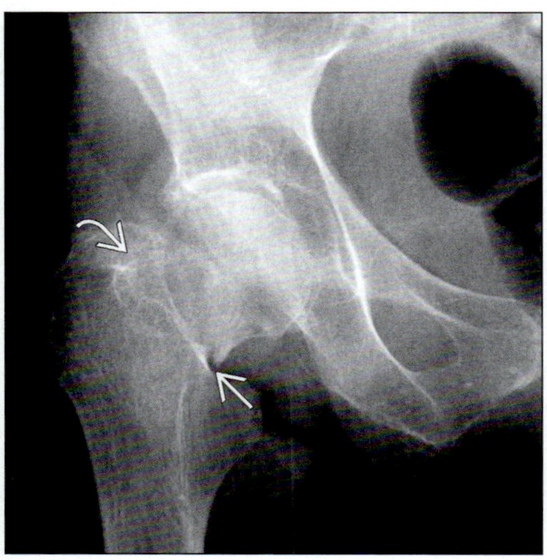

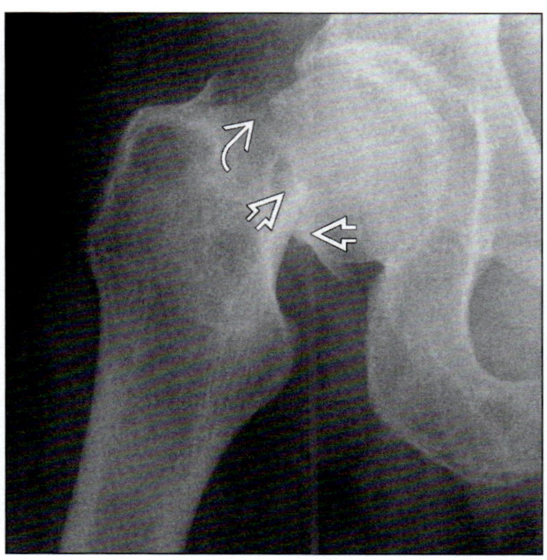

(Left) *AP radiograph shows a complete fracture of the subcapital femoral neck with mild displacement of the medial and lateral cortices. The fracture is in mild varus. This is a Garden III fracture.* ***(Right)*** *AP radiograph shows a complete subcapital fracture with varus angulation and anterior rotation of the femoral head relative to the shaft in a Garden III fracture. This fracture is somewhat more displaced and rotated than the previous case.*

FEMORAL NECK FRACTURE

*(**Left**) AP radiograph shows a complete subcapital femoral neck fracture with marked varus angulation and complete offset of the cortex. This injury could be characterized as a Garden III or IV injury depending on the degree of offset of the cortex; this is best judged on CT. (**Right**) Coronal reformat bone CT shows an impacted and displaced subcapital femoral neck fracture. There is near complete displacement of the femoral shaft making this a Garden IV fracture.*

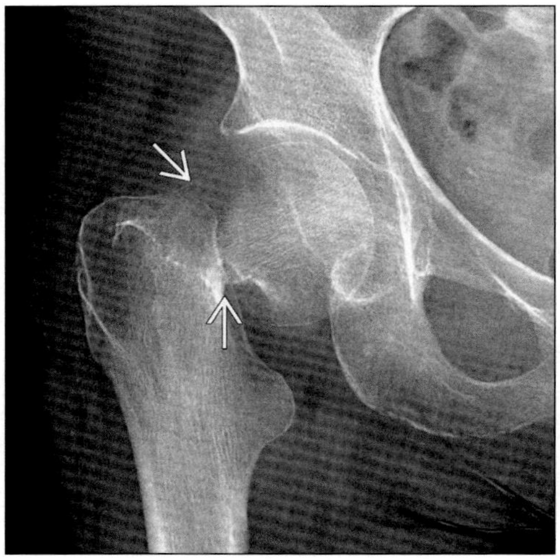

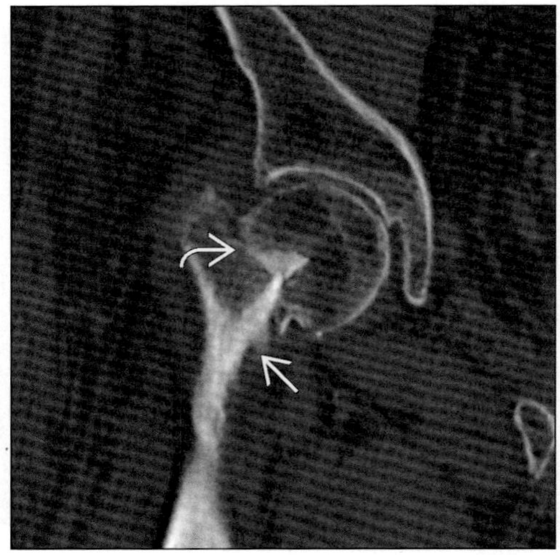

*(**Left**) AP radiograph in a 78 year old status post fall shows a buckle at the lateral cortex diagnostic of a Garden I subcapital fracture. This abnormality was missed, and the patient was sent home. (**Right**) AP radiograph from the same patient obtained 8 days later shows evolution to a Garden IV fracture, with varus angulation and displacement. A permeative appearance surrounding the fracture is seen frequently in osteoporotic patients and should not be mistaken for tumor.*

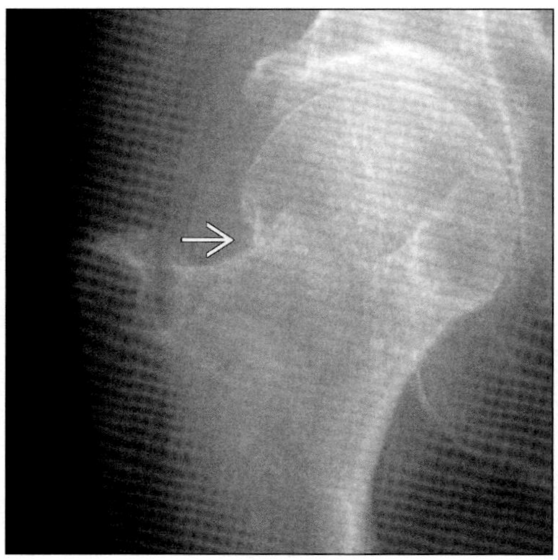

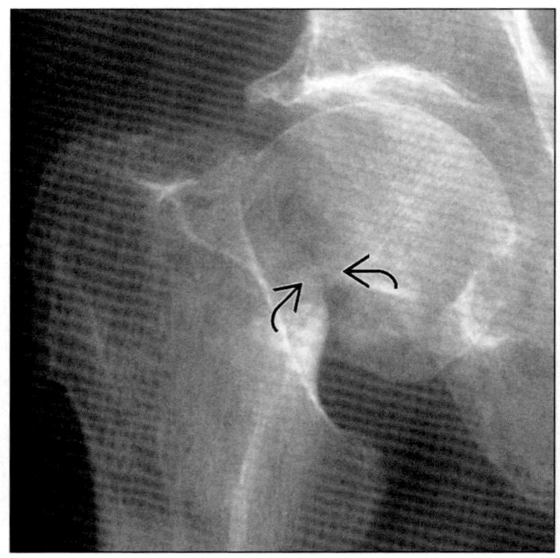

*(**Left**) AP radiograph shows very subtle subcortical sclerosis in the medial basicervical region of the femoral neck in this 63-year-old woman. This finding was missed on initial interpretation, and the patient was sent home. (**Right**) AP radiograph from the same patient 2 days later shows completion of the basicervical fracture, which has displaced and fallen into significant varus. A high degree of clinical suspicion warrants emergent MR to exclude a nondisplaced fracture.*

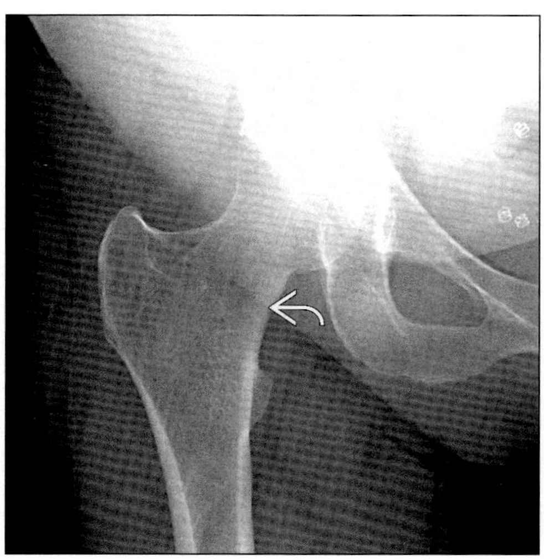

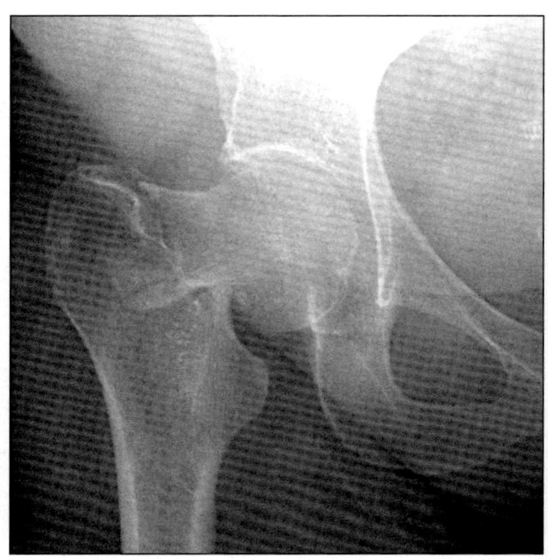

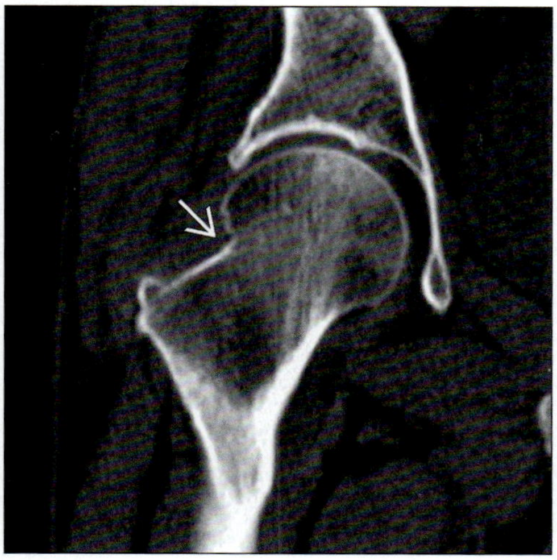

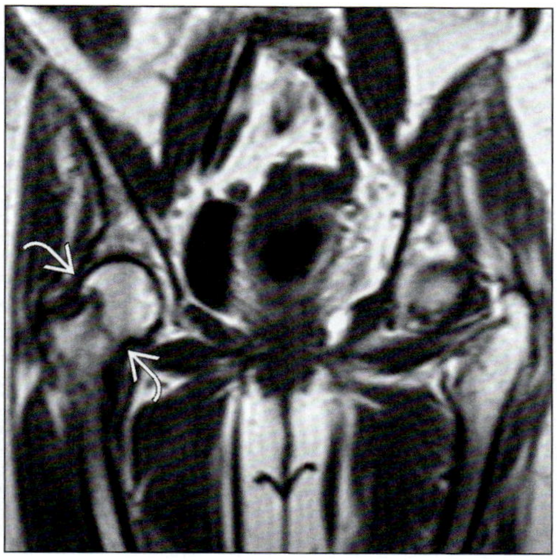

*(**Left**) Coronal reformat bone CT image from a 57-year-old woman with osteoporosis shows a subtle buckle in the lateral femoral head/neck junction ➡. With a high clinical suspicion, CT can be used for confirmation, but subtle Garden I fractures can be missed. (**Right**) Coronal T1WI MR from the same patient was acquired on an open system as part of a 3 sequence "fast" protocol due to claustrophobia but confirms the clinically suspected Garden I femoral neck fracture ↪.*

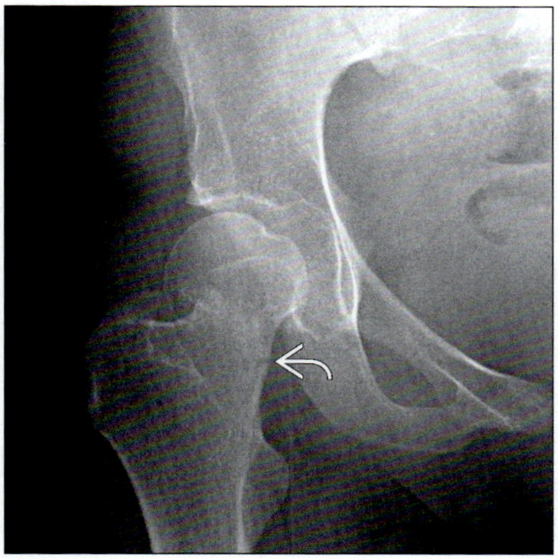

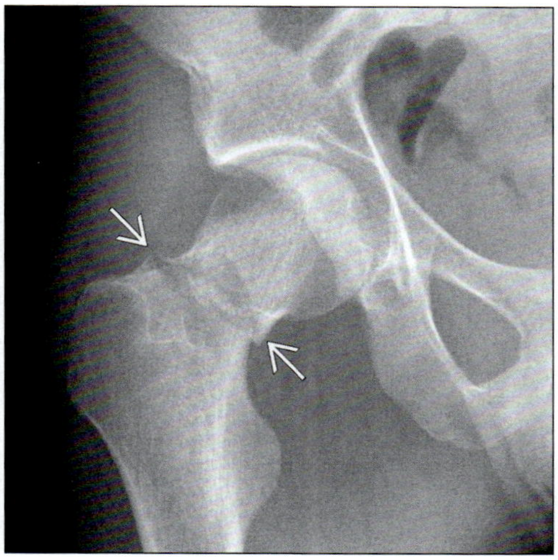

*(**Left**) AP radiograph from a 22-year-old man status post motorcycle crash shows a medial midcervical fracture line ↪ and mild valgus angulation of the femoral neck. Midcervical fractures often occur in young, active individuals with normal mineralization in the setting of high-energy trauma. Many are incomplete fractures. (**Right**) AP radiograph from an 18-year-old male after a motor vehicle crash shows a comminuted midcervical fracture ➡ of the femur. Overall alignment is preserved.*

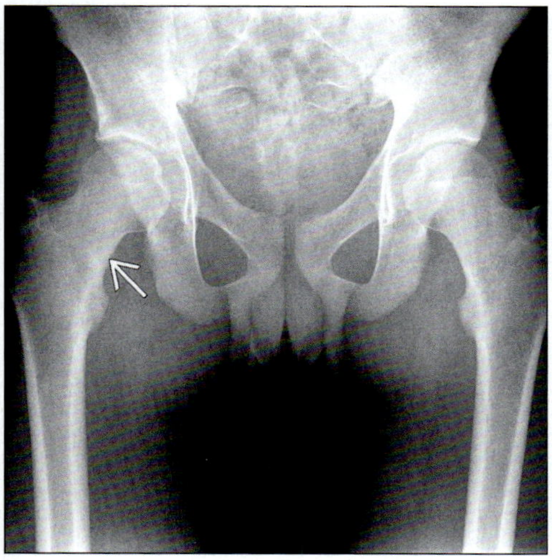

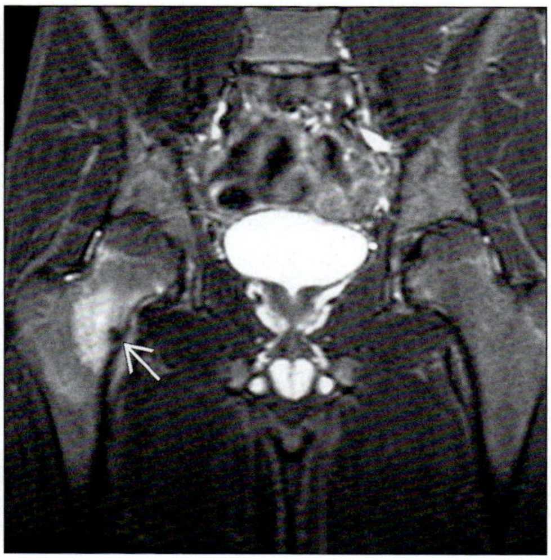

*(**Left**) AP radiograph from a 27-year-old runner with right hip pain was initially interpreted as normal. In retrospect, there is a subtle asymmetric subcortical transverse band of sclerosis ➡ in the basicervical portion of the right femoral neck. (**Right**) Coronal T2WI FS MR from the same patient 1 hour later confirms the presence of a medial basicervical stress fracture ➡ with surrounding marrow edema. This is a common location for stress fractures in runners.*

TROCHANTERIC AND INTERTROCHANTERIC FRACTURE

Key Facts

Terminology

- Lesser trochanteric fracture, greater trochanteric fracture, intertrochanteric fracture

Imaging

- If radiographs negative or equivocal and clinical suspicion high → MR
- If incomplete and nondisplaced, MR delineates extent
- Isolated greater trochanteric fracture can be gluteal avulsion
- Isolated lesser trochanteric fracture in adult raises concern for metastasis
- Radiographs 1st line of imaging
- Varus angulation
- Coronal and sagittal CT reformats for malalignment

Top Differential Diagnoses

- Pathologic fracture: Especially with primary involvement of lesser trochanter

Pathology

- Evans classification

Clinical Issues

- 90% > 65 years of age
- Incomplete intertrochanteric fracture: Risk of fracture completion
- Most treated surgically
- Elderly fall with hip pain
- Acute onset atraumatic pain in osteoporosis

Diagnostic Checklist

- Describe extent of intertrochanteric involvement with incomplete fractures

(Left) Coronal T1WI MR shows a nondisplaced oblique fracture line running from the greater to the lesser trochanter. This is an Evans type I fracture, a stable injury. (Right) AP radiograph from a dedicated hip series in an osteoporotic female shows displacement of both the lesser and greater trochanters consistent with an Evans type V fracture. This is an unstable injury. The major fragments are in varus angulation, which is typical.

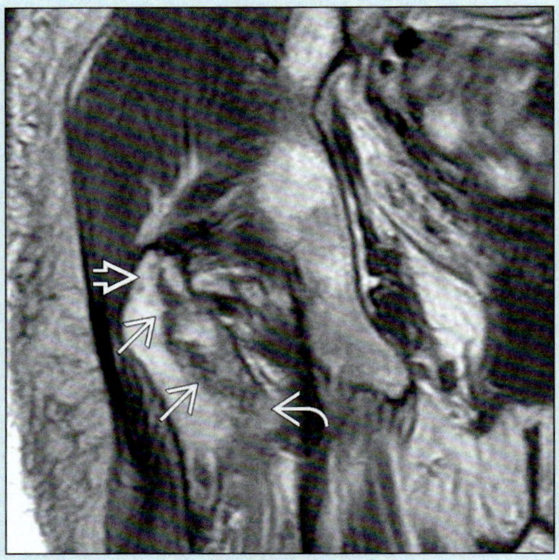

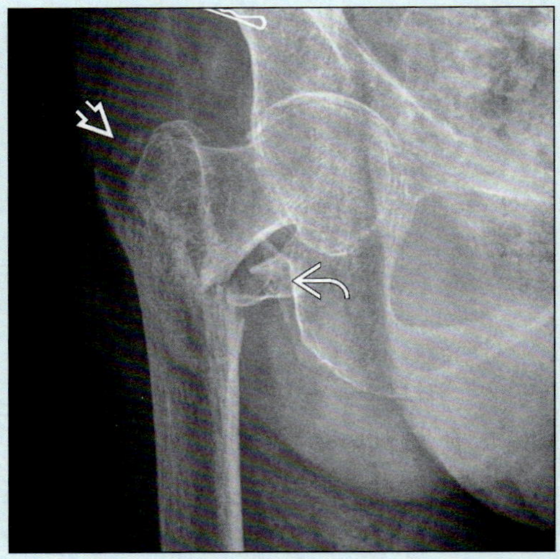

(Left) Coronal T2WI FS MR from a 17-year-old competitive swimmer shows a nondisplaced fracture of the greater femoral trochanter. Fracture is likely an avulsion injury due to gluteal tendon action. (Right) Axial T1WI MR in a patient with severe groin pain shows a destructive lesion in the lesser trochanter typical for an osseous metastasis. With hip or groin pain and normal radiographs, noncontrast MR is the modality of choice for identifying occult bone and soft tissue lesions.

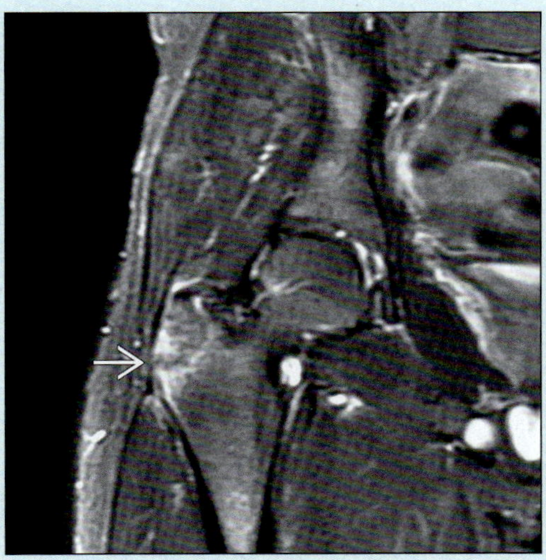

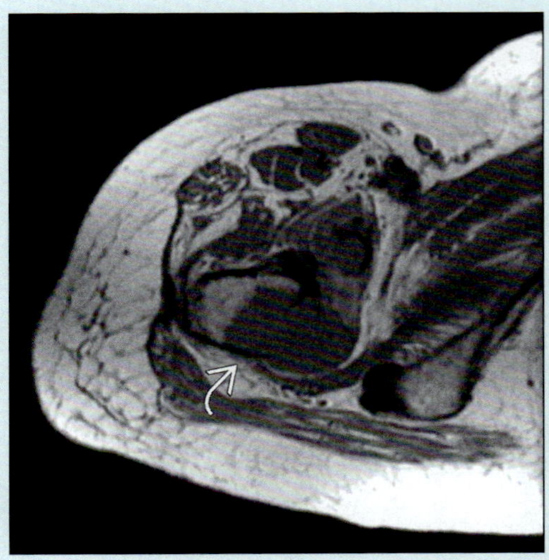

TROCHANTERIC AND INTERTROCHANTERIC FRACTURE

TERMINOLOGY

Synonyms

- Greater trochanteric fracture, lesser trochanteric fracture, trochanteric fracture, Evans fracture

Definitions

- Intertrochanteric fracture
  - Fracture extending from lesser trochanter (LT) to greater trochanter (GT) of hip
  - Can be complete or incomplete
- Isolated lesser trochanteric fracture
- Isolated greater trochanteric fracture

IMAGING

General Features

- Best diagnostic clue
  - Radiographic malalignment diagnostic in most cases
    - Varus angulation
    - Visible fracture line
    - Displacement of lesser trochanter
  - If incomplete and nondisplaced, MR diagnostic
- Location
  - Generally oblique plane between trochanters
    - Evans type R fracture: Reverse obliquity from lesser trochanter
  - Usually propagates superolateral from LT toward GT
- Morphology
  - Intertrochanteric fracture
    - Complete fracture extends from LT to GT
    - Almost invariably displaced
    - Incomplete fracture generally extends from LT vertically down femoral shaft
  - Greater trochanter fracture: Usually nondisplaced oblique
  - Lesser trochanter fracture: Usually displaced superomedially

Radiographic Findings

- Discontinuity of cortex at LT
- Visible oblique fracture line from LT toward GT
- Displacement of GT or LT
- Varus angulation
- Displaced GT less common
- Impaction less frequent than with femoral neck fx

CT Findings

- Discontinuity or displacement of cortex
- Asymmetry helpful: Image both hips
- Coronal/sagittal reformats should always be included

MR Findings

- T1WI
  - Hypointense fracture line
  - Displacement of lesser trochanter
- STIR
  - Bone marrow edema best delineates fracture extent

Imaging Recommendations

- Best imaging tool
  - Radiographs 1st line of imaging
  - If radiographs negative or equivocal and clinical suspicion high → MR
- Protocol advice
  - Image entire pelvis
    - Subset of patients will have sacral fracture or lower lumbar disease
  - Incomplete fx may need MR or CT to define extent

DIFFERENTIAL DIAGNOSIS

Basicervical Fracture

- Above intertrochanteric line

Pathologic Fracture

- Osteoporosis with all hip fractures
- Metastases with isolated LT fracture
- Demineralizing process, such as myeloma

PATHOLOGY

Staging, Grading, & Classification

- Evans classification

CLINICAL ISSUES

Presentation

- Most common signs/symptoms
  - Hip pain
  - Acute onset atraumatic pain in osteoporosis

Demographics

- Age
  - 90% > 65 years of age
- Gender
  - M:F = 1:3
- Epidemiology
  - Most common type of hip fracture

Natural History & Prognosis

- Incomplete intertrochanteric fracture: Risk of fracture completion
- All non-weight-bearing treatments put patient at risk for other morbidities (pneumonia, pulmonary embolus)

Treatment

- Most treated surgically
- Incomplete with < 50% width through shaft can be treated with non-weight-bearing
- Complete or > 50% width: Dynamic plate & screw fixation
- Evans III, IV, or V may require multiple plates

DIAGNOSTIC CHECKLIST

Consider

- Pathologic fracture from osteoporosis, metastatic disease, or myeloma
- Isolated LT fracture in adult should be considered pathologic until proven otherwise

Image Interpretation Pearls

- If incomplete fracture, consider MR to delineate extent
  - Apparent isolated greater trochanteric fracture on radiograph in elderly patient may in fact extend to intertrochanteric region

TROCHANTERIC AND INTERTROCHANTERIC FRACTURE

Evans I: Nondisplaced fracture with 2 fragments. This is a stable fracture and is generally treated with dynamic screw and plate fixation if extending more than 50% through the intertrochanteric region.

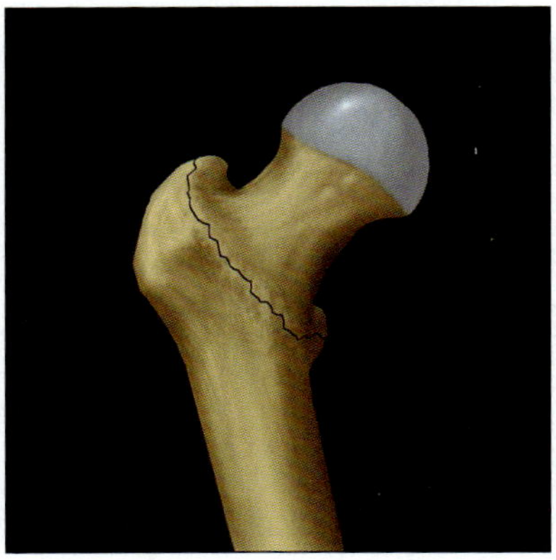

Evans II: Displaced fracture with 2 fragments. This can be a stable fracture if there is apposition of the fracture fragments at the medial cortical margin. Dynamic screw and plate fixation is standard.

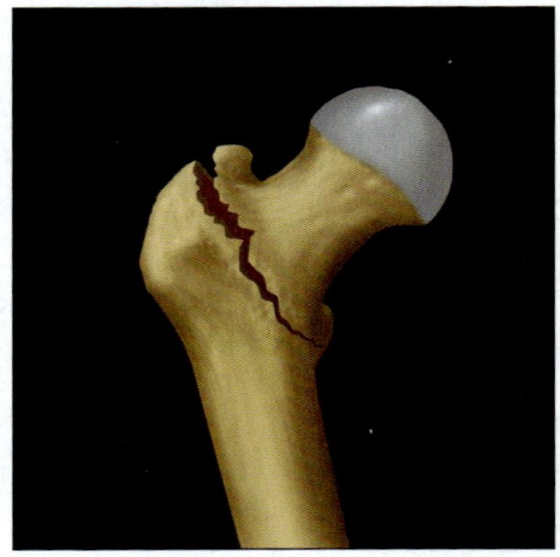

Evans III: Fracture with 3 or more fragments with comminution or displacement of the greater trochanteric fragment. This fracture is posterolaterally unstable and may be treated with a dynamic screw and 1 or more plates.

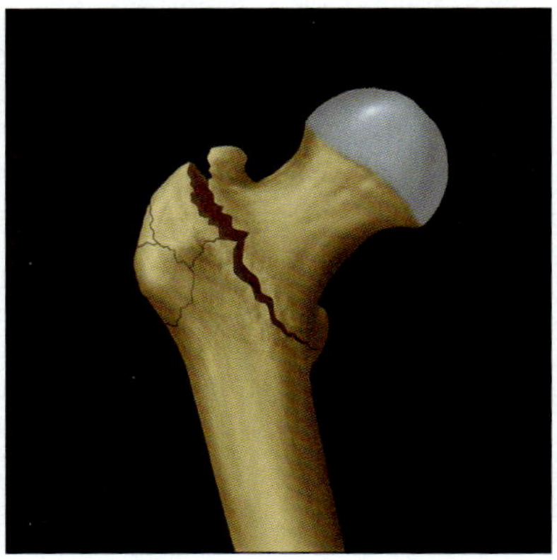

Evans IV: Fracture with 3 or more fragments with comminution or displacement of the lesser trochanteric fragment. This fracture is posteromedially unstable and may be treated with a dynamic screw and 1 or more plates.

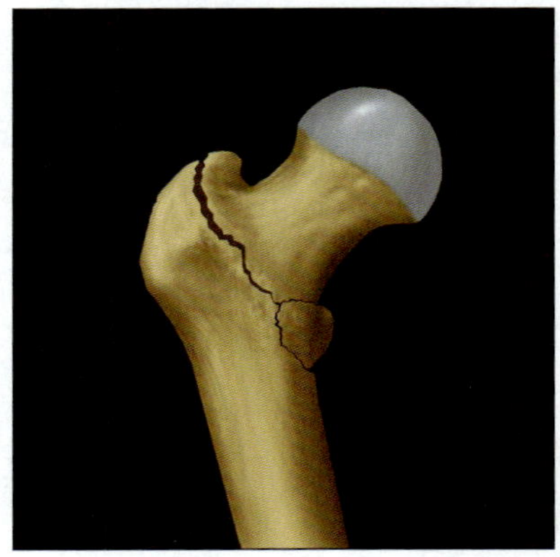

Evans V: Fracture with 4 or more fragments with comminution or displacement of both the greater and lesser trochanteric fragments. This fracture is unstable anteromedially and posterolaterally and fixation often requires 2 or more plates with a dynamic screw.

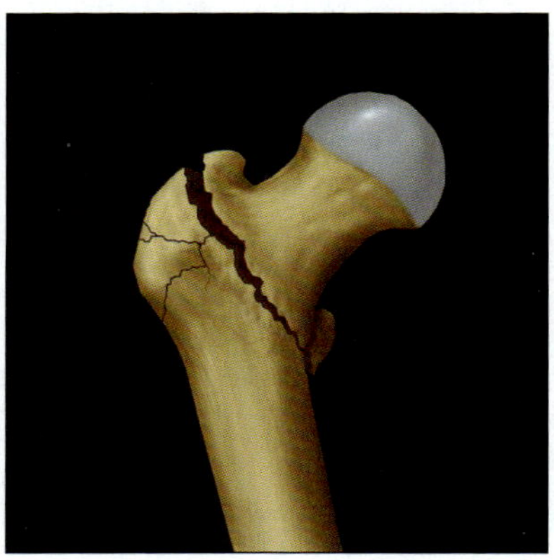

Evans R: Reversed obliquity fracture. This can be a stable fracture but is generally treated with open reduction and internal fixation using either a dynamic screw and plate device or an intramedullary pin.

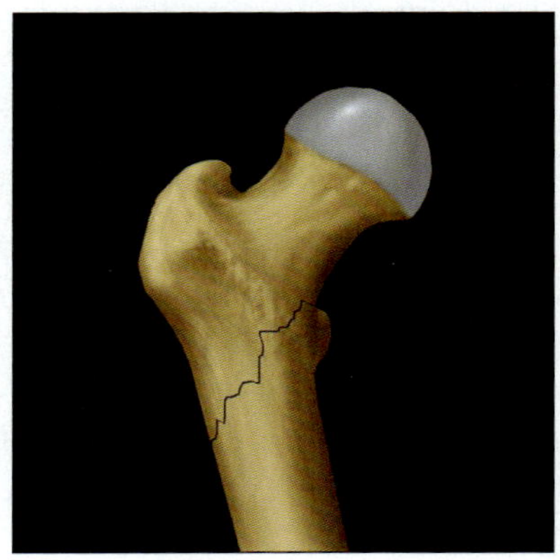

The Evans classification of intertrochanteric fractures provides guidance for appropriate therapy based on the number and displacement of fracture fragments.

  - If clinical concern for fracture and equivocal radiographs → MR pelvis

Reporting Tips

- Describe extent of intertrochanteric involvement with incomplete fractures
  - Helpful to describe extension from LT through femoral shaft

SELECTED REFERENCES

1. Kesmezacar H et al: Predictors of mortality in elderly patients with an intertrochanteric or a femoral neck fracture. J Trauma. 68(1):153-8, 2010
2. Ekström W et al: Quality of life after a stable trochanteric fracture--a prospective cohort study on 148 patients. J Orthop Trauma. 23(1):39-44, 2009
3. McGrory BJ et al: The use of locking plates for greater trochanteric fixation. Orthopedics. 32(12):917-20, 2009
4. Sankey RA et al: The use of MRI to detect occult fractures of the proximal femur: a study of 102 consecutive cases over a ten-year period. J Bone Joint Surg Br. 91(8):1064-8, 2009
5. Nguyen JT et al: Stress-related injuries around the lesser trochanter in long-distance runners. AJR Am J Roentgenol. 190(6):1616-20, 2008
6. James SL et al: Atraumatic avulsion of the lesser trochanter as an indicator of tumour infiltration. Eur Radiol. 16(2):512-4, 2006
7. Haidukewych GJ et al: Reverse obliquity fractures of the intertrochanteric region of the femur. J Bone Joint Surg Am. 83-A(5):643-50, 2001

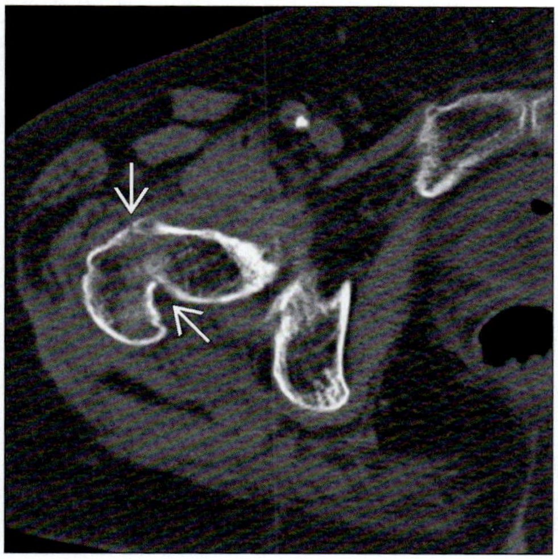

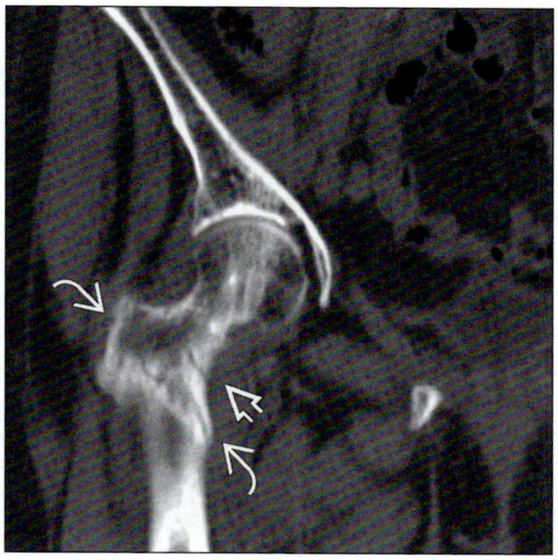

(Left) *Axial bone CT shows a subtle lucency through the intertrochanteric region of the femur that could easily be overlooked in this plane. This MVA victim had visceral pelvic injury and coronal and sagittal reformats were performed per trauma protocol.* ***(Right)*** *Coronal reformat bone CT from the same patient shows a nondisplaced complete Evans type I intertrochanteric fracture, with a 2nd incomplete basicervical femoral neck fracture occult on the axial acquisitions.*

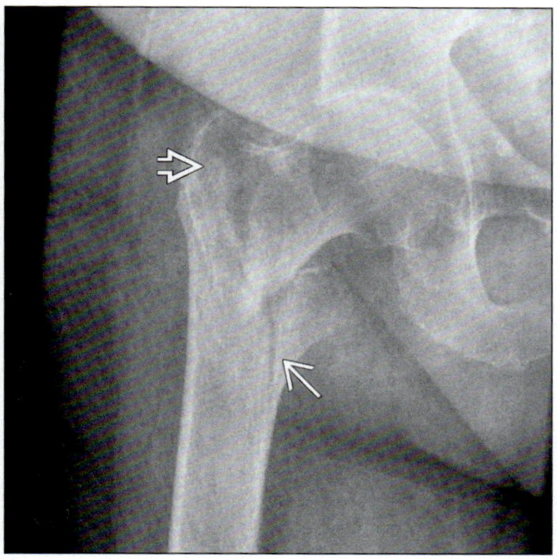

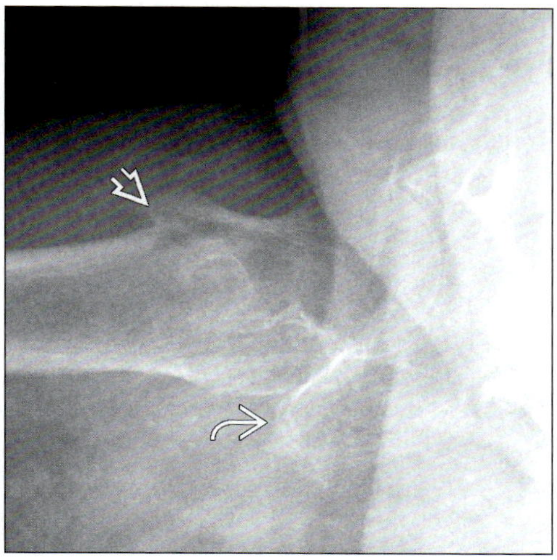

(Left) *AP radiograph of the hip demonstrates a fracture between the lesser trochanter and the greater trochanter with minimal angulation between the fracture fragments. This is an Evans type II fracture.* ***(Right)*** *Groin lateral radiograph from the same patient shows mild posterior displacement of the shaft relative to the femoral neck. There is a step-off at the anterior fracture margin. Orientation can be determined by locating the ischial tuberosity posterior to the hip.*

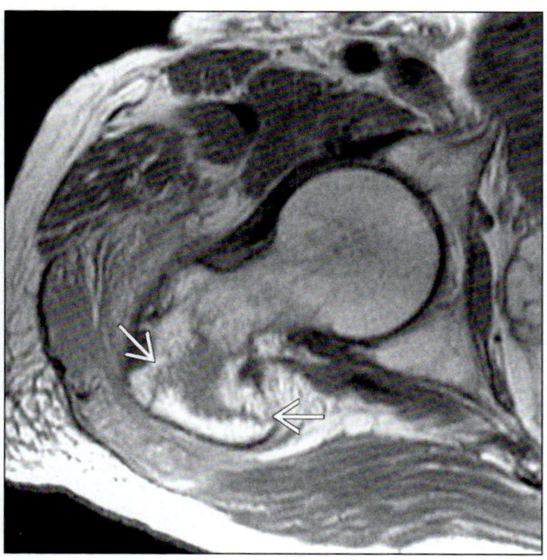

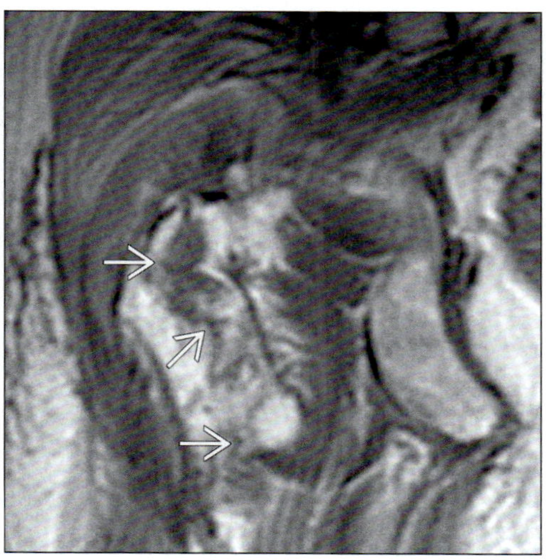

(Left) *Axial T1WI MR demonstrates a comminuted fracture of the greater femoral trochanter. The femoral neck is intact. The MR was performed because radiographs were read as normal and the patient could not bear weight.* ***(Right)*** *Coronal T1WI MR from the same patient shows extension of the fracture to the lesser trochanter, thus constituting an intertrochanteric fracture. Comminution of the greater trochanter makes this an Evans type III injury, which is posterolaterally unstable.*

TROCHANTERIC AND INTERTROCHANTERIC FRACTURE

*(**Left**) Coronal T1WI MR shows a comminuted fracture of the greater trochanter extending into a near complete intertrochanteric Evans type III fracture ➡. With continued weight-bearing, incomplete trochanteric fractures can progress to unstable fractures. (**Right**) AP radiograph shows an Evans type IV intertrochanteric fracture with displacement of the lesser tuberosity ➡ and varus angulation at the fracture line. The greater tuberosity ⇨ remains intact.*

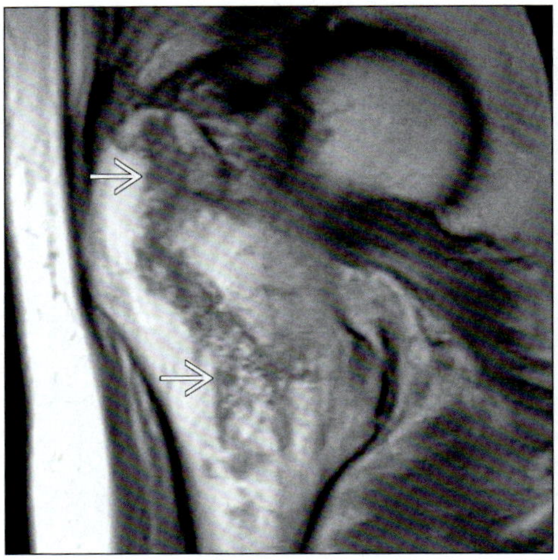

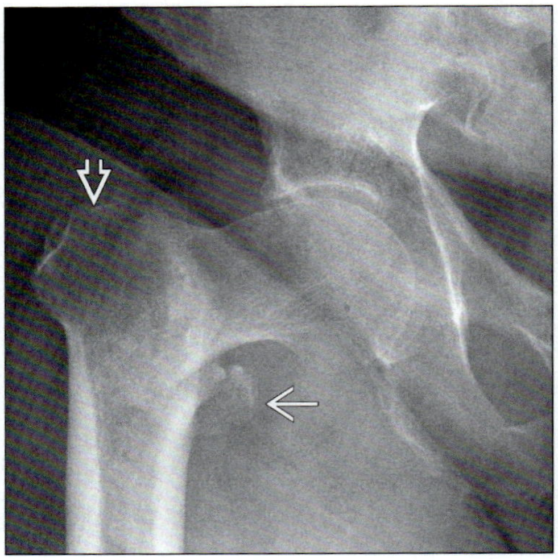

*(**Left**) AP radiograph from a 93-year-old woman with hip pain after a fall shows a subtle nondisplaced Evans type I intertrochanteric fracture ➡. The fracture was missed in the ER and the patient was sent home. (**Right**) AP radiograph from the same patient 3 days later shows conversion of the Evans type I fracture to an Evans type V with displacement of the greater ➡ and lesser ↪ trochanters. The stable type I fracture requires less extensive surgery than the unstable type V fracture.*

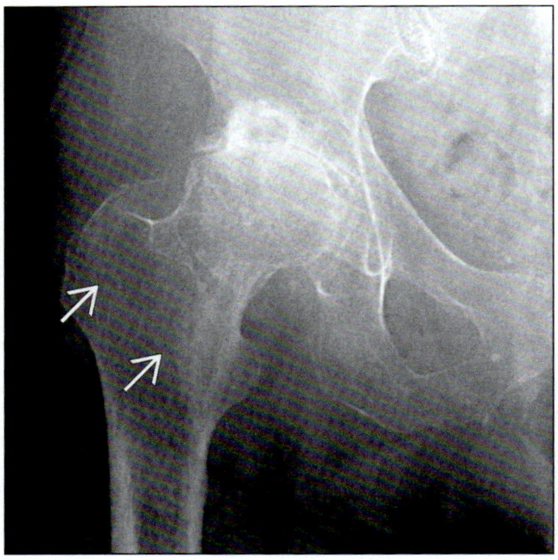

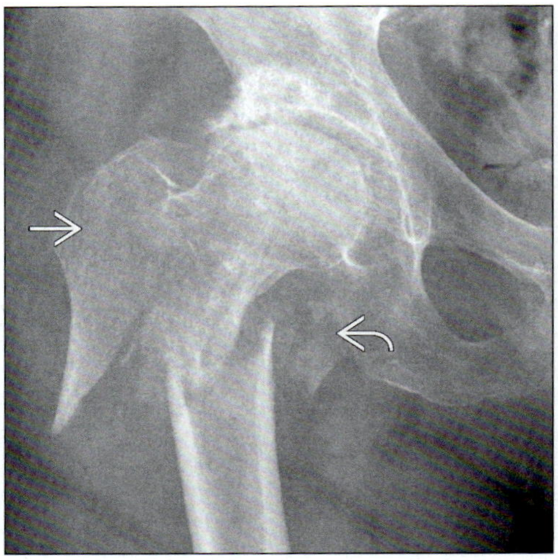

*(**Left**) AP radiograph demonstrates a reverse angle intertrochanteric fracture (Evans type R) in a 73-year-old man status post fall on the hip. Note that the fracture extends above the lesser trochanter ➡, differentiating it from a subtrochanteric fracture. There is varus angulation. (**Right**) AP radiograph shows a reverse angle intertrochanteric fracture (Evans type R) with separation of the lesser trochanter ↪. The fracture is in varus, which is typical for this type of injury.*

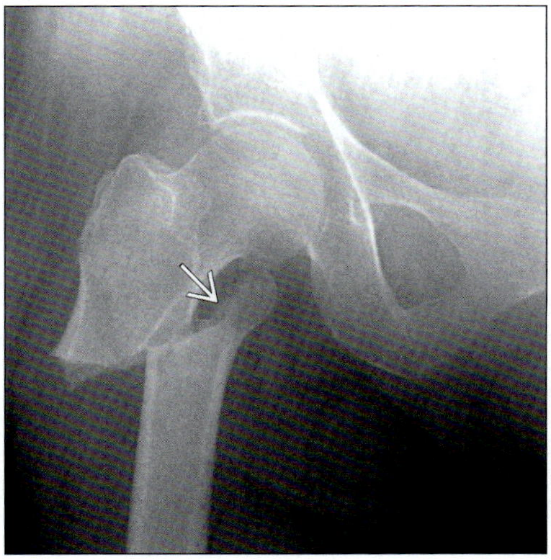

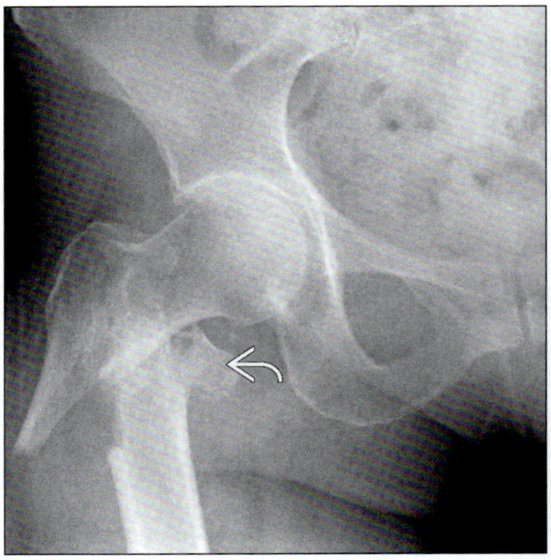

TROCHANTERIC AND INTERTROCHANTERIC FRACTURE

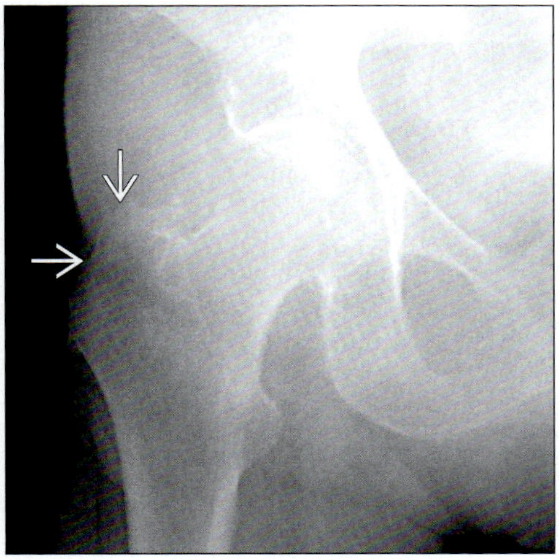

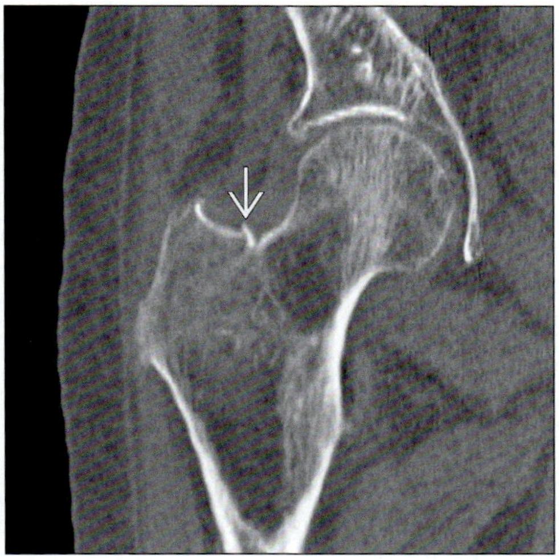

(Left) *AP radiograph from an elderly man with hip pain after a fall demonstrates subtle irregularity of the lateral cortex of the greater femoral trochanter ➡ that was not perceived on the original interpretation. The greater trochanter is often irregular due to chronic traction by the gluteal tendons.* ***(Right)*** *Coronal reformat bone CT from the same patient shows an acute cortical fracture of the superior greater trochanter ➡. Isolated greater trochanter fractures are stable injuries.*

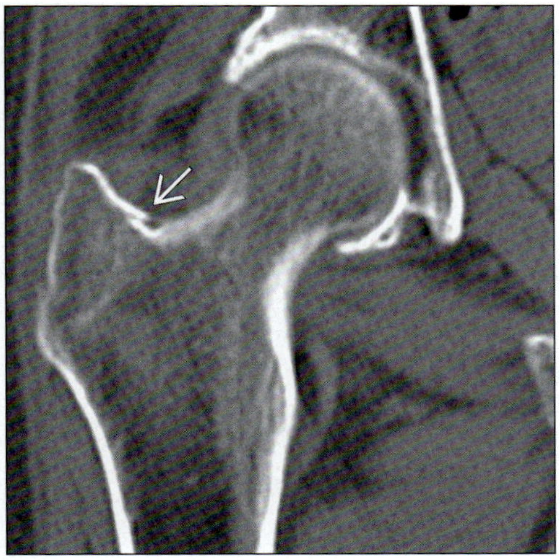

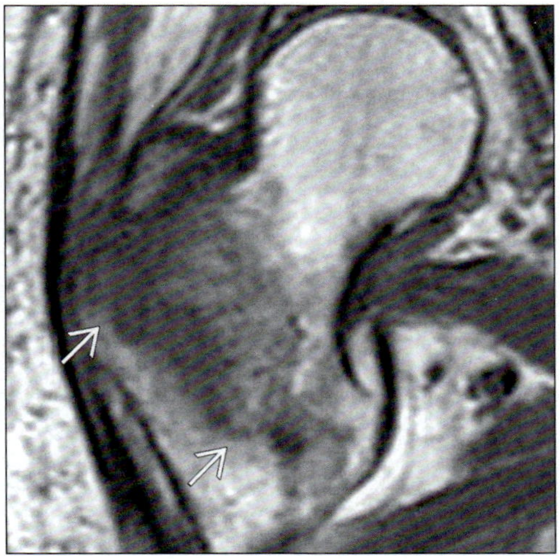

(Left) *Coronal reformat bone CT from an elderly patient with worsening hip pain on external rotation and refusal to ambulate shows a nondisplaced and possibly incomplete fracture ➡ of the greater trochanter.* ***(Right)*** *Coronal T1WI MR from the same patient 4 weeks later shows evolution of the incomplete fracture at the greater trochanter to a near complete intertrochanteric fracture ➡. With continued weight-bearing, incomplete trochanteric fractures can progress to unstable fractures.*

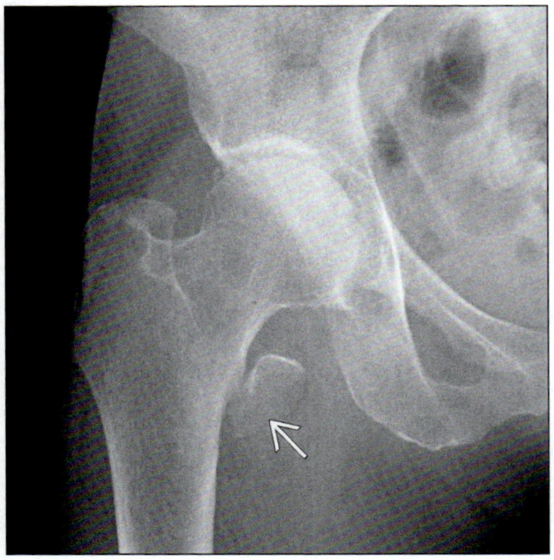

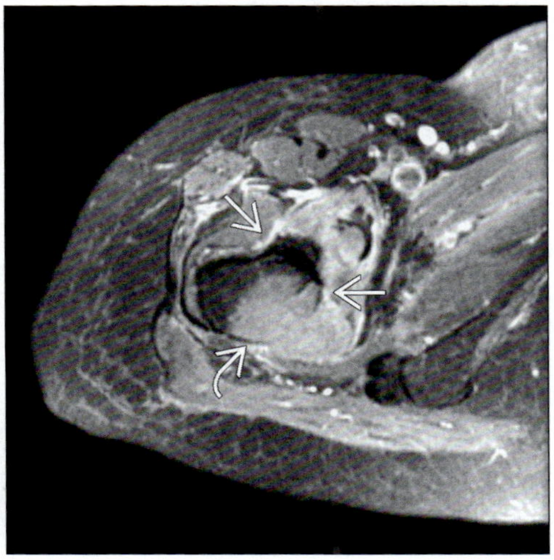

(Left) *AP radiograph in a patient with a history of breast cancer shows avulsion of the lesser trochanter ➡ with rotation and displacement of the avulsed fragment. The fracture edge is irregular and poorly defined; this is a pathologic fracture due to a breast carcinoma metastasis.* ***(Right)*** *Axial T1WI C+ FS MR acquired to assess tumor extent for possible ablation shows not only osseous metastasis in the lesser trochanter ➡ but a pathologic fracture ➡ as well.*

SUBTROCHANTERIC AND FEMORAL SHAFT FRACTURE

Key Facts

Terminology

- Subtrochanteric fracture: ≤ 5 cm below lesser trochanter
- Femoral shaft fracture: > 5 cm distal to lesser trochanter and proximal to distal metaphysis

Imaging

- Consider MR or CT if concern for pathologic fracture
- Lateral cortical thickening with stress/insufficiency
  - Generally at junction of proximal and distal thirds
- Can be nondisplaced or severely comminuted
- Often incomplete with stress or insufficiency fracture
- Radiographs generally diagnostic
- Often varus angulation
- Torsion displacement with spiral shaft fracture
- Femoral shaft fracture often accompanied by soft tissue injury
- Displacement dictated by tendinous attachments

Pathology

- Trauma, often high velocity in young patients
- Osteoporosis, metastases, or bisphosphonate therapy

Clinical Issues

- Consider metastatic work-up if no osteoporosis or trauma
- MVA most frequent setting in young patients
- ORIF generally indicated
- Bisphosphonates, particularly alendronate
- Minor trauma or fall with osteoporosis or metastasis
  - Elderly females most common

Diagnostic Checklist

- Look for focal cortical thickening to signal incomplete stress/insufficiency fracture
- Describe number of fragments and angulation

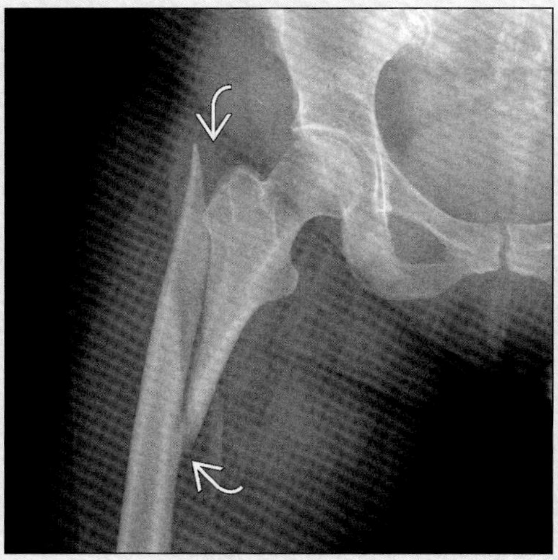

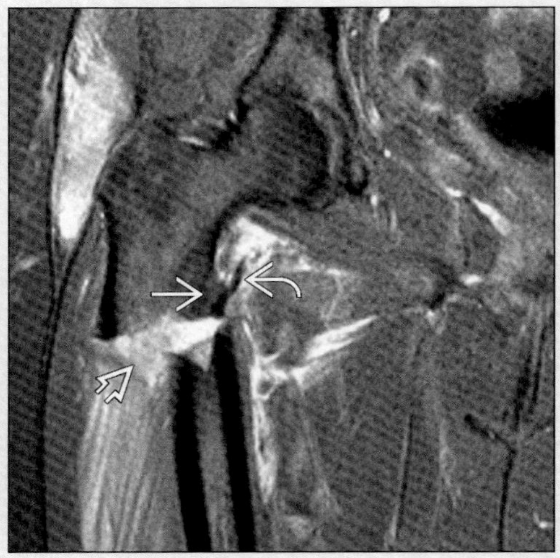

(Left) *AP radiograph of the hip from a perimenopausal woman with severe leg pain after an innocuous twisting trauma shows a comminuted Seinsheimer type V fracture ➔ involving the intertrochanteric and subtrochanteric regions.* ***(Right)*** *Coronal T2WI FS MR from a 74-year-old man demonstrates a displaced transverse subtrochanteric fracture ➔ with the iliopsoas tendon ➔ attached to the lesser trochanter ➔, which is attached to the proximal fragment. This is a Seinsheimer IIA fracture.*

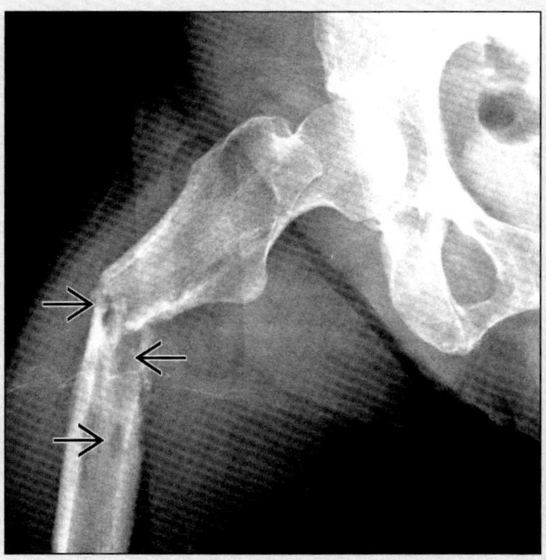

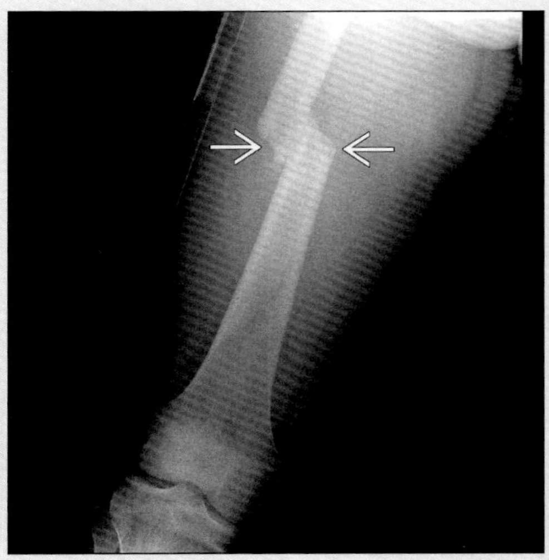

(Left) *AP radiograph demonstrates a subtrochanteric fracture with varus angulation. The fracture extends through a series of lucencies in the bone ➔; this is a pathologic fracture due to lung cancer metastases.* ***(Right)*** *AP radiograph of a 16-year-old boy after a motor vehicle crash shows a transverse oblique fracture of the midshaft of the femur ➔ with 1 shaft-width medial displacement of the distal major fragment. Femoral shaft fractures in healthy bone are due to high-energy trauma.*

SUBTROCHANTERIC AND FEMORAL SHAFT FRACTURE

TERMINOLOGY

Definitions

- Subtrochanteric fracture: ≤ 5 cm below lesser trochanter
- Femoral shaft fracture: > 5 cm distal to lesser trochanter and proximal to distal metaphysis

IMAGING

General Features

- Location
  - Subtrochanteric fracture: Proximal femoral shaft within 5 cm of lesser trochanter
  - Femoral shaft fracture: 3 locations
    - Proximal 1/3: 5 cm distal to lesser trochanter to diaphysis
    - Middle 1/3: Entirely diaphyseal
    - Distal 1/3: Distal diaphysis to supracondylar region
- Morphology
  - Often in varus angulation
  - Displacement often dictated by tendinous attachments
  - Can be incomplete with stress or insufficiency fracture
  - Femoral shaft fracture often has displaced butterfly fragment

Radiographic Findings

- Displacement and varus angulation
- Retraction with fragment overlap
- Torsion displacement with spiral shaft fracture
- Cortical thickening with stress/insufficiency fracture

CT Findings

- Lateral cortical thickening with incomplete fracture
- Frequently bilateral fractures when insufficiency
  - Usually at different stages and only 1 side may be symptomatic

Imaging Recommendations

- Best imaging tool
  - Radiographs generally diagnostic
  - Consider MR or CT if concern for pathologic fracture

DIFFERENTIAL DIAGNOSIS

Pathologic Fracture

- Atypical locations, permeative lesions, incomplete fractures

PATHOLOGY

General Features

- Etiology
  - Trauma, often high velocity in young patients
  - Can be innocuous trauma in osteoporotic patients
    - Increased suspicion with bisphosphonate therapy
  - Young children (< 5 years old): Suspect child abuse

Staging, Grading, & Classification

- Seinsheimer subtrochanteric femoral fracture classification
  - Based upon loss of medial cortex and subsequent stability
  - Dictates operative vs. nonoperative but limited in guiding surgery
- Russell-Taylor subtrochanteric femoral fracture classification
  - Useful in guiding operative therapy
    - Type 1: No involvement of piriformis fossa; often treated with intramedullary devices
    - Type 2: Involvement of piriformis fossa; often require ORIF with screw plate devices or fixed-angle implants
- Winquist-Hansen femoral shaft fracture classification
  - Based upon degree of comminution and percent of intact cortex
    - Type 0: No comminution, simple transverse or oblique
    - Type I: Cortex almost entirely intact with no comminution
    - Type II: Butterfly fragment with at least 50% of circumference of cortices of the 2 major fragments intact
    - Type III: Butterfly fragment with 50-100% of circumference of the 2 major fragments comminuted
    - Type IV: No intact cortex, extensive comminution

CLINICAL ISSUES

Presentation

- Other signs/symptoms
  - > 50% of femoral shaft fractures associated with other major trauma

Demographics

- Age
  - High-energy trauma: Adolescents and young adults
  - Low-energy trauma: Elderly osteoporotic adults in subtrochanteric region

Treatment

- ORIF generally indicated

DIAGNOSTIC CHECKLIST

Image Interpretation Pearls

- Watch for retraction and overlap of fragments
- Look for focal cortical thickening to suggest incomplete stress/insufficiency fracture
- Look for symmetric contralateral fracture

SELECTED REFERENCES

1. Ekström W et al: Quality of life after a subtrochanteric fracture: a prospective cohort study on 87 elderly patients. Injury. 40(4):371-6, 2009
2. Goh SK et al: Subtrochanteric insufficiency fractures in patients on alendronate therapy: a caution. J Bone Joint Surg Br. 89(3):349-53, 2007
3. Tornetta P 3rd: Subtrochanteric femur fracture. J Orthop Trauma. 16(4):280-3, 2002

SUBTROCHANTERIC AND FEMORAL SHAFT FRACTURE

Seinsheimer I: Nondisplaced transverse fracture with < 2 mm offset. This is a stable fracture and can be treated with immobilization and non-weight-bearing if incomplete.

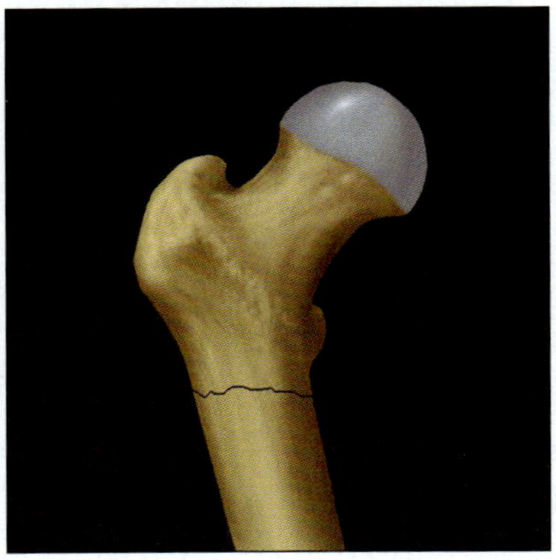

Seinsheimer IIA: 2-part transverse displaced fracture. This fracture has an intact lesser trochanter and can generally be treated with a dynamic screw and plate open reduction internal fixation (ORIF).

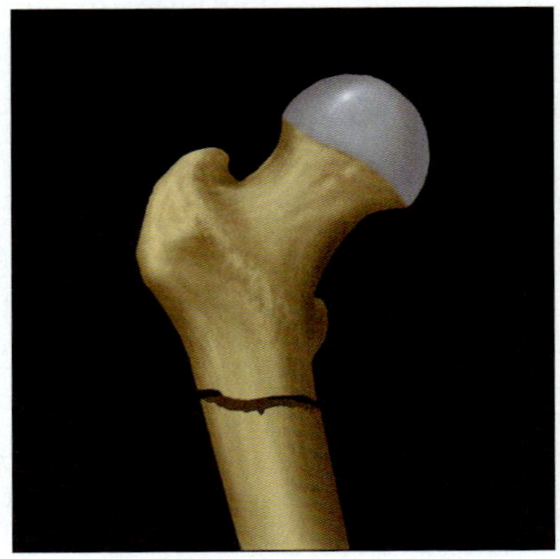

Seinsheimer IIB: 2-part spiral fracture with lesser trochanter attached to the proximal fragment. Similar to IIA but with more offset and often varus angulation. Dynamic screw and plate fixation is the most common treatment.

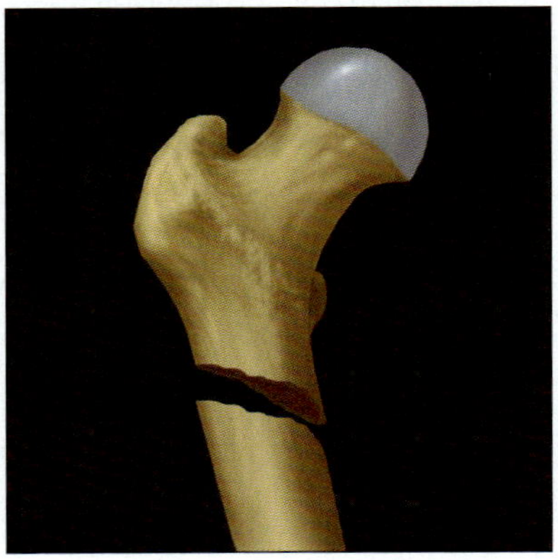

Seinsheimer IIB: 2-part spiral fracture angled such that the fracture extends to the lesser trochanter or lesser trochanter is attached to the distal fragment. An intramedullary nail can be considered with the lesser trochanter intact.

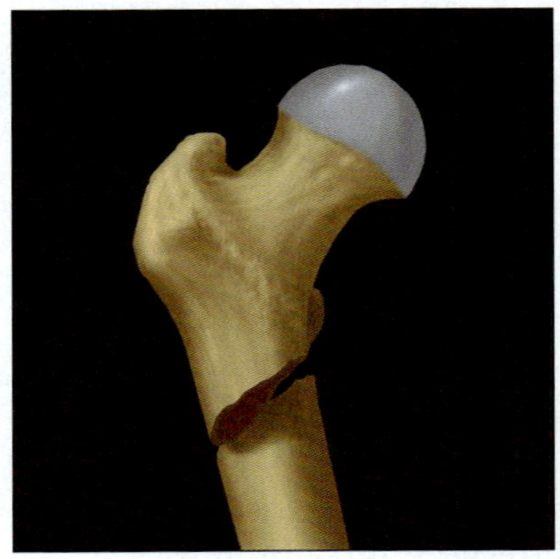

Seinsheimer IIIA: 3-part spiral fracture with lesser trochanter forming displaced 3rd fragment. The lesser trochanter and medial cortex are unstable, but implant failure and nonunion are uncommon.

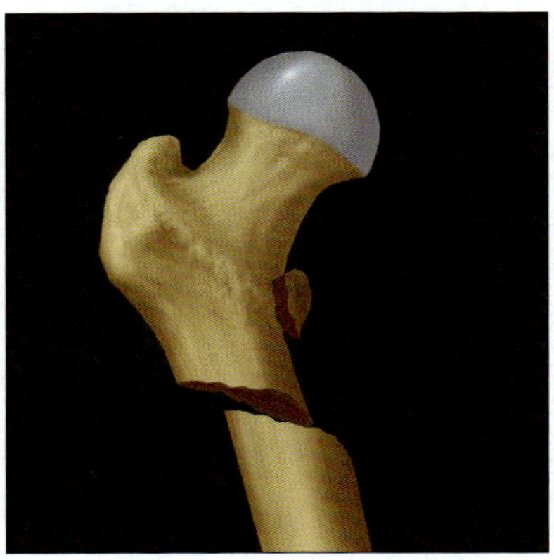

Seinsheimer IIIB: 3-part spiral fracture with 3rd part represented by a posterior butterfly fragment. Implant failure is more common in this group. Dynamic screw and blade plate fixation is standard.

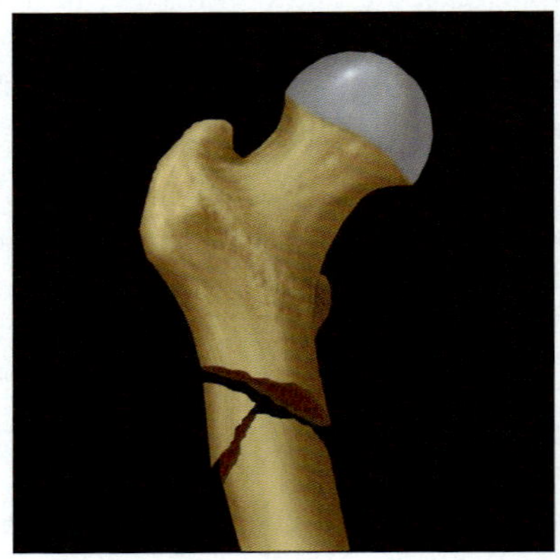

Seinsheimer IV: Comminuted fracture with 4 or more fragments. A 95° blade plate and condylar screw fixation should be considered in this unstable fracture.

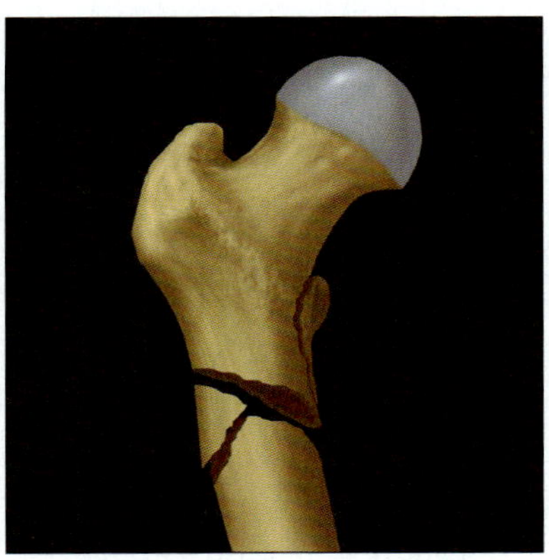

Seinsheimer V: Combination of subtrochanteric and intertrochanteric fractures. Two plates including a 95° blade plate and condylar screw may be necessary for adequate fixation of this complex fracture.

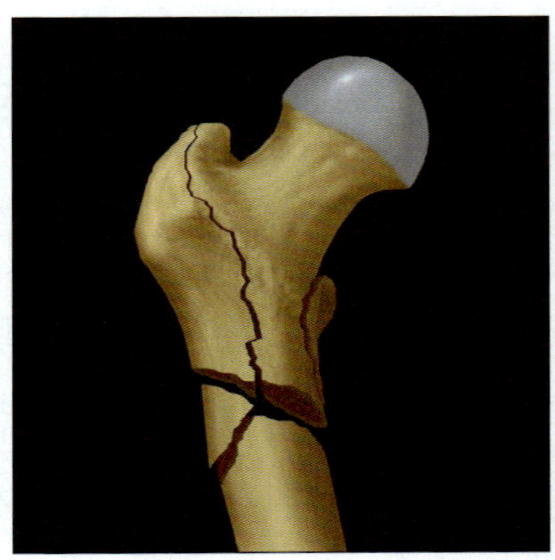

Seinsheimer's classification of subtrochanteric proximal femur fractures has important prognostic implications: Implant failure and nonunion are common in type IIIA and higher. Location of the lesser trochanter and number and location of fragments are important determinants in this grading system.

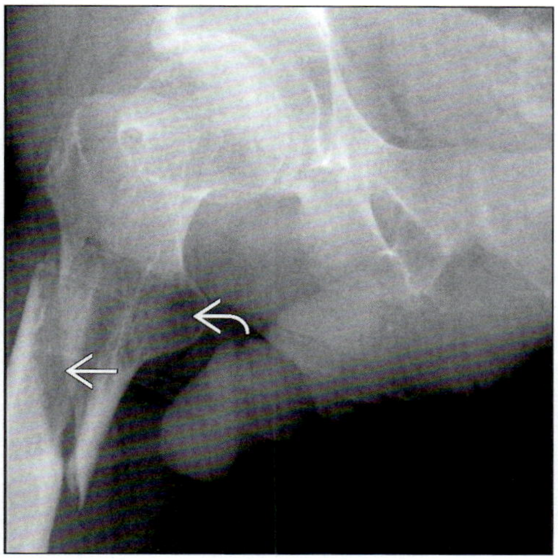

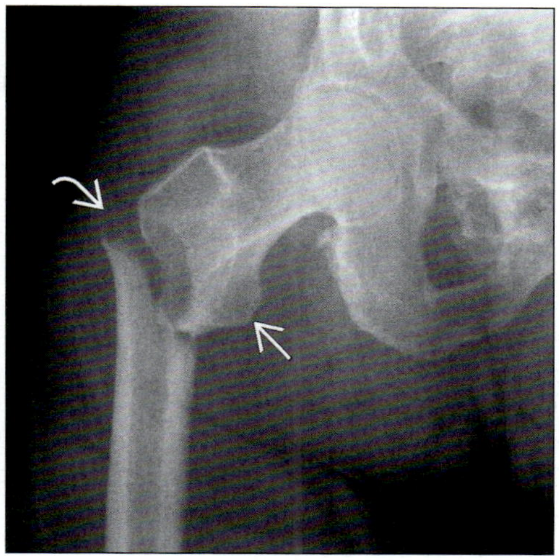

(Left) *AP radiograph shows a spiral subtrochanteric fracture ➔ extending obliquely below the lesser trochanter ➔. This is a Seinsheimer IIB injury.* ***(Right)*** *Portable AP radiograph of the hip obtained in the emergency department in a 73 year old shows a Seinsheimer IIB subtrochanteric fracture ➔ sparing the lesser trochanter ➔. This fracture is in varus angulation. Dynamic screw and plate fixation is the most common form of treatment for this injury.*

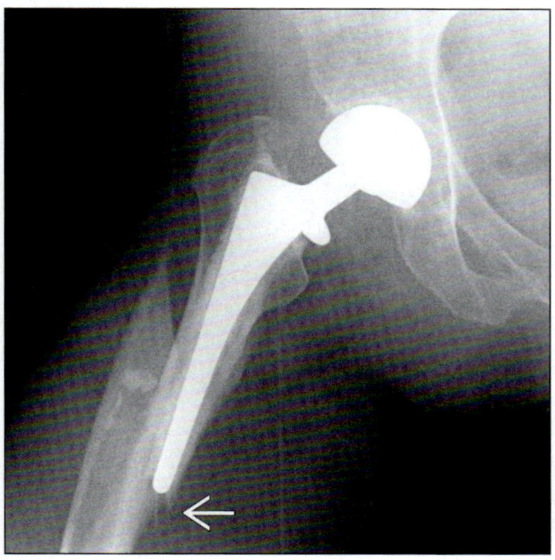

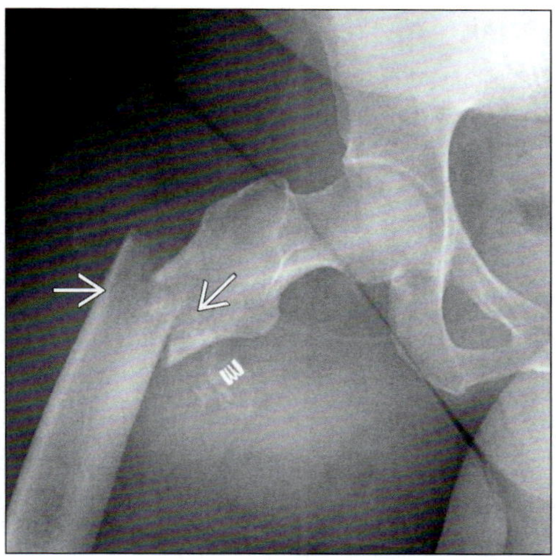

(Left) *AP radiograph shows a spiral subtrochanteric femoral fracture occurring around a cemented bipolar hemiprosthesis of the hip. These fractures usually extend inferiorly below the tip ➔ of the metal stem of the femoral component of the prosthesis.* ***(Right)*** *AP radiograph shows a subtrochanteric fracture of the femur with irregular lucent margins of the fragments ➔ suggesting a pathologic fracture. The fracture occurred with minimal trauma. This is a Seinsheimer IIIA fracture.*

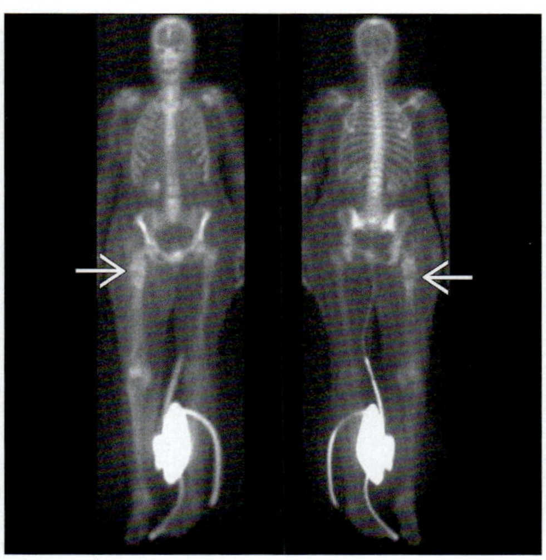

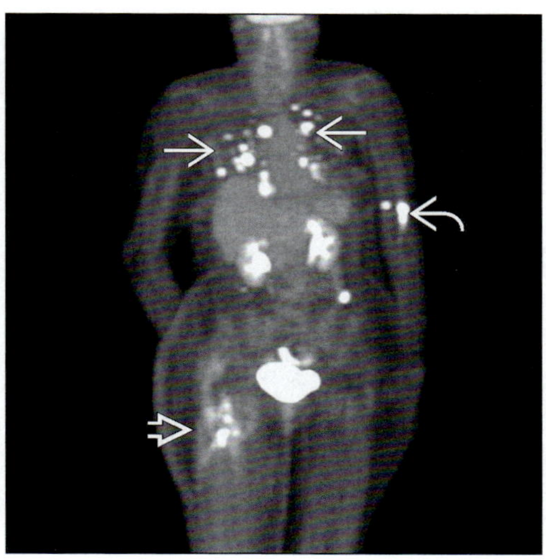

(Left) *AP bone scan from the same patient shows vague increased uptake in the region of the femoral fracture ➔ but no other osseous lesion. Subtrochanteric fracture without high-energy trauma raises suspicion for a pathologic lesion, particularly metastatic disease.* ***(Right)*** *AP projection from a PET scan of the same patient shows multiple lung masses ➔ and the femoral lesion ➔. Activity in the left antecubital fossa ➔ is the site of injection. This patient had metastatic breast carcinoma.*

SUBTROCHANTERIC AND FEMORAL SHAFT FRACTURE

(Left) *AP radiograph of the hip in an 81 year old on bisphosphonate therapy with groin pain was initially interpreted as normal. However, there is focal thickening of the lateral femoral cortex ➔, typical for an early femoral shaft insufficiency fracture.* ***(Right)*** *AP radiograph from the same patient obtained 8 days later upon return to the emergency department demonstrates completion of a transverse fracture through the proximal femoral shaft ➔, now with varus angulation and retraction.*

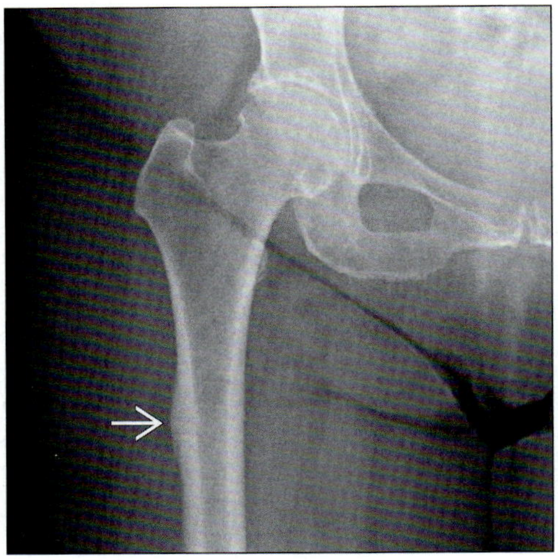

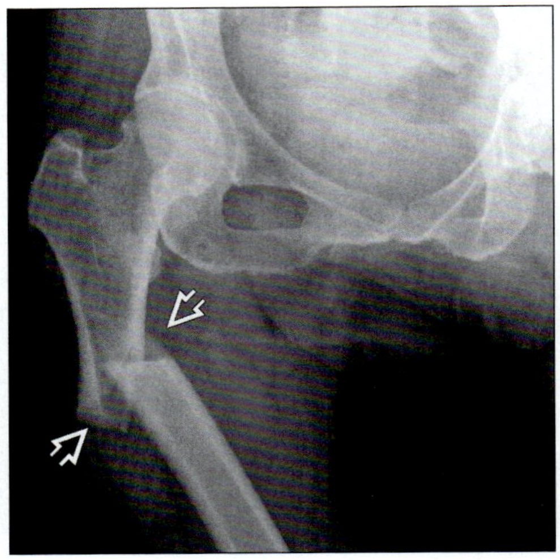

(Left) *Anteroposterior image from a technetium bone scan performed for osseous metastases in a patient with 2 years of bisphosphonate use demonstrates faint, symmetric uptake in the subtrochanteric region of the femora near the junction of the proximal and middle 1/3 ➔.* ***(Right)*** *Coronal reformat bone CT from the same patient shows bilateral lateral cortex thickening at the subtrochanteric region of the femora ➔, typical for insufficiency fractures seen with bisphosphonate use.*

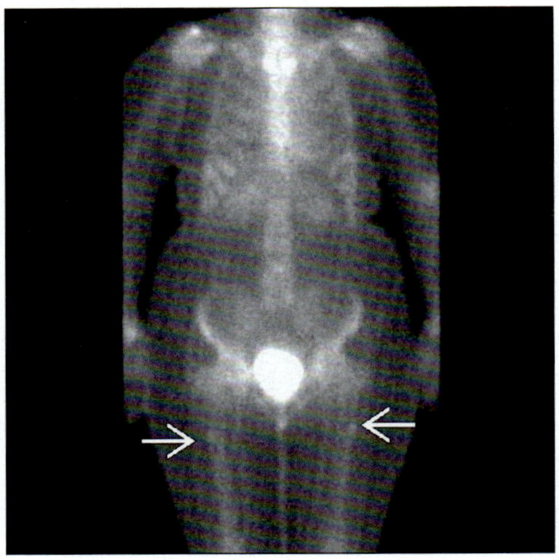

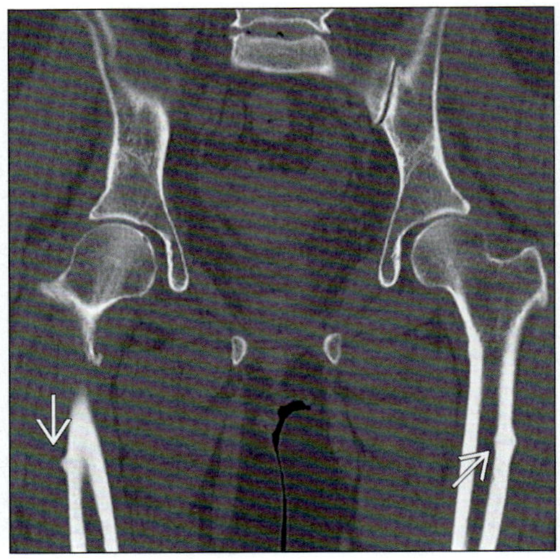

(Left) *AP external rotation radiograph shows an oblique fracture of the proximal femoral shaft with complete displacement of the fragments but little comminution. There is moderate fragment override. This is a Winquist-Hansen type 0 fracture, as there is no comminution.* ***(Right)*** *AP radiograph from the same patient after open reduction and internal fixation with an intramedullary rod and screw combination shows anatomic reduction of the fracture ➔.*

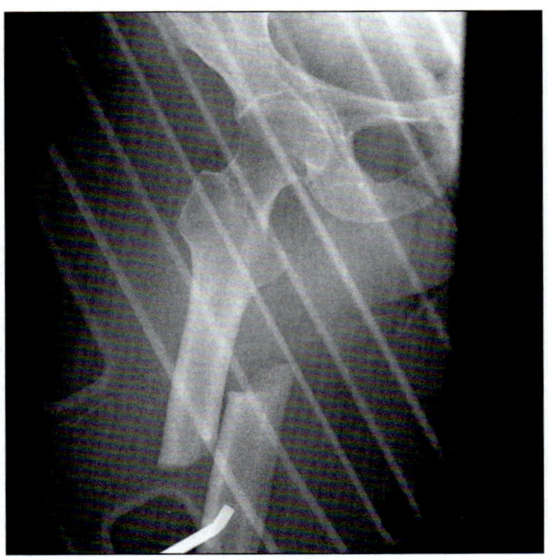

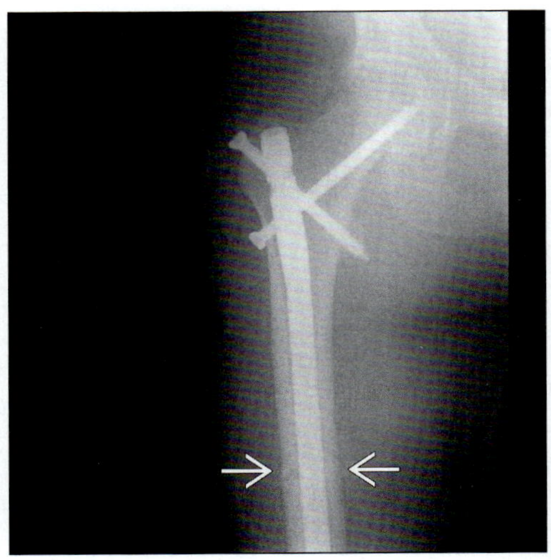

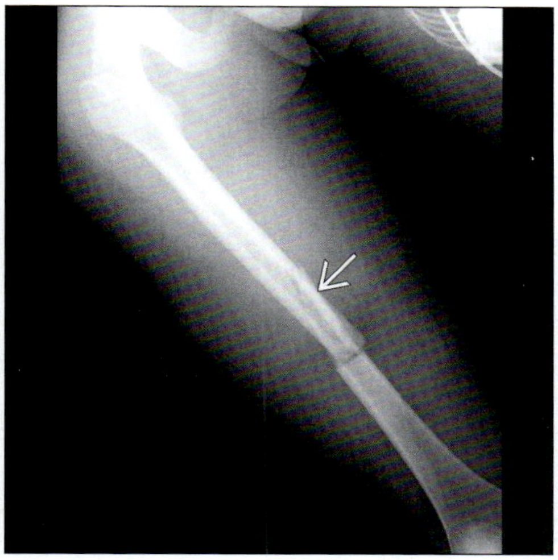
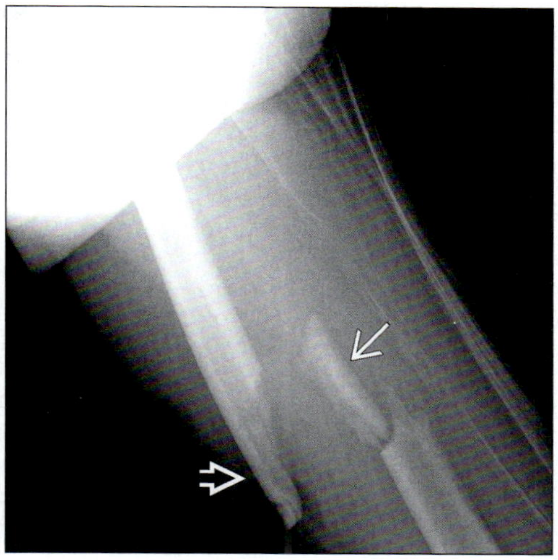

***(Left)** AP radiograph demonstrates a comminuted fracture of the shaft of the femur after a motor vehicle crash. There is a cortical butterfly fragment → projecting over the medial aspect of the fracture complex, but no displacement is seen on this image. **(Right)** Lateral radiograph from the same patient shows extensive displacement of the proximal fragment out of the skin ⇨. A large cortical butterfly fragment is seen posteriorly →. This is a Winquist-Hansen type IV fracture.*

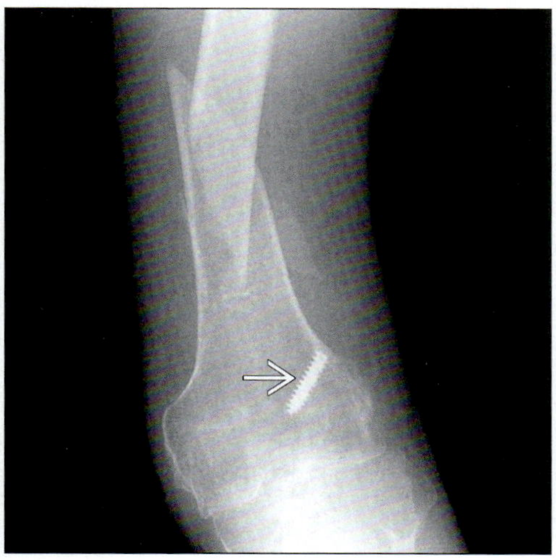
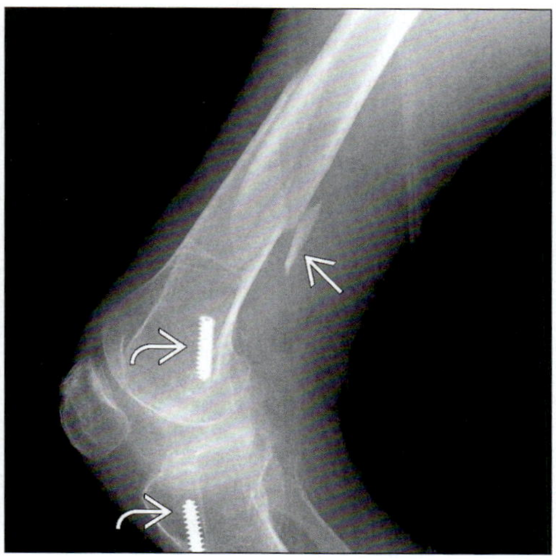

***(Left)** AP radiograph shows a comminuted spiral fracture of the distal femoral shaft in an osteoporotic patient. The fracture is in varus alignment. Interference screw → from previous ACL repair is present. **(Right)** Lateral radiograph from the same patient shows a posterior cortical butterfly fragment → at the fracture site. Interference screws ↷ are present in the knee from previous anterior cruciate ligament repair.*

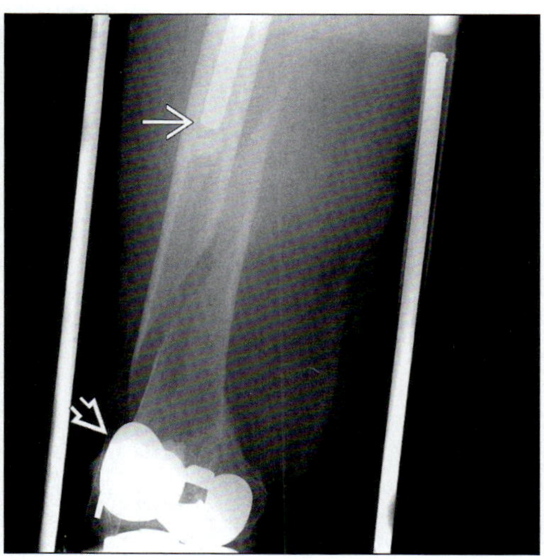
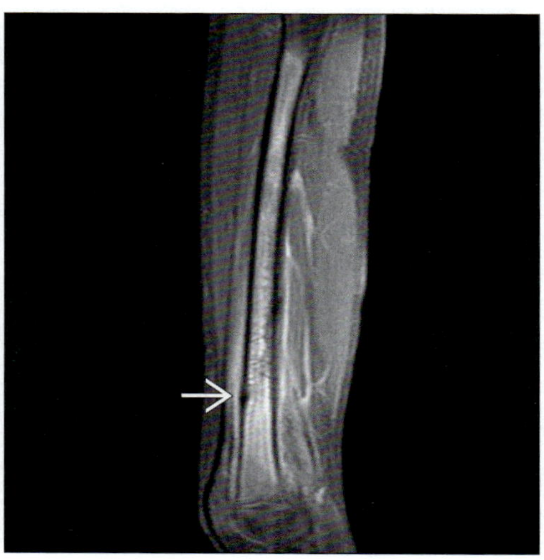

***(Left)** AP radiograph shows a spiral fx in an osteoporotic patient. The fracture is located between the stem of a hip prosthesis → and knee prosthesis ⇨. The components act as stress risers, transferring forces to the weaker bone around them. **(Right)** Sagittal T2WI FS MR from a 10-year-old boy with thigh pain shows a focal cortical thickening → with extensive surrounding marrow edema and soft tissue swelling. This is a stress fracture of the distal femoral shaft in this patient due to excessive skateboarding.*

HIP DISLOCATION

Key Facts

Imaging

- Posterior dislocation: 90% of hip dislocations
- Anterior dislocation: Obturator or iliac position
- Lack of congruence of femoral head within acetabulum
- Associated femoral head or acetabular rim fractures
- AP view is diagnostic, Judet (oblique) or lateral views confirmatory

Pathology

- Pipkin classification used for fracture of femoral head associated with hip dislocation

Clinical Issues

- Uncomplicated cases: Closed reduction successful (76-93%)
- Surgical treatment required for
  - Intraarticular bodies
  - Concomitant femoral head/neck fractures
- Often associated pelvic ring fractures & severe hemorrhage
- Osteonecrosis in up to 50% if not reduced within 24 hours
- Sciatic nerve injury

Diagnostic Checklist

- On follow-up radiographs, watch for early signs of osteonecrosis and consider performing MR
- With high-velocity motor vehicle crashes, check hip alignment on trauma CT abdomen/pelvis
- Direct posterior hip dislocation can be minimally displaced on single AP image
- Anterior hip dislocation iliac-type position can mimic posterior hip dislocation on AP image
- Dislocated head appears smaller than contralateral head in posterior dislocation, larger in anterior dislocation

*(**Left**) PA graphic shows posterior dislocation of the hip. The femoral head overlaps the posterior margin of the acetabulum ➔. The hip is in internal rotation. (**Right**) AP radiograph shows a traumatic posterior hip dislocation. Note that the right femoral head ➔ is not only abnormally cephalad in position but also appears smaller than the left, indicating that it is closer to the cassette. The hip is held in neutral rotation but is flexed and adducted.*

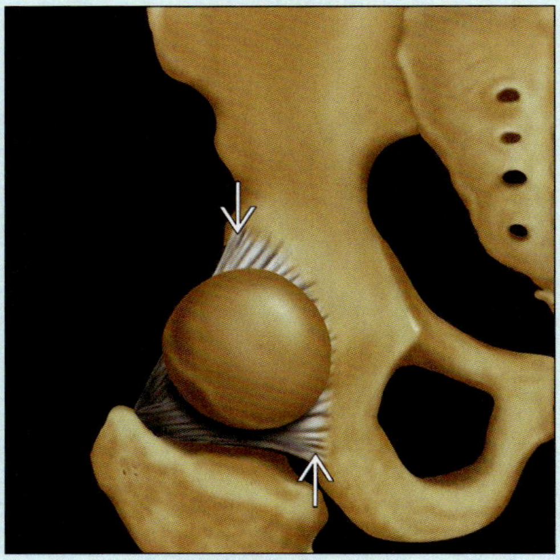

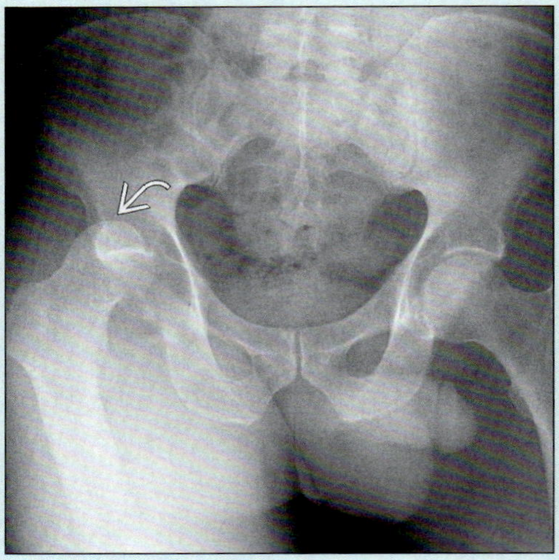

*(**Left**) AP graphic shows an obturator-type anterior hip dislocation with the femoral head anteromedial to the pubofemoral ligament ➔. The hip is in external rotation, and the femoral head is inferior to the acetabulum. (**Right**) AP radiograph shows the right femoral head abnormally positioned anteroinferior to the acetabulum with the femur externally rotated, flexed, and abducted. This is an obturator-type anterior hip dislocation. A trauma board is present under the patient.*

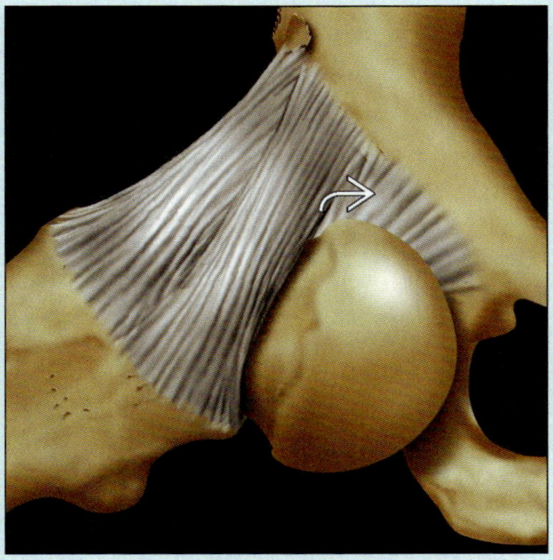

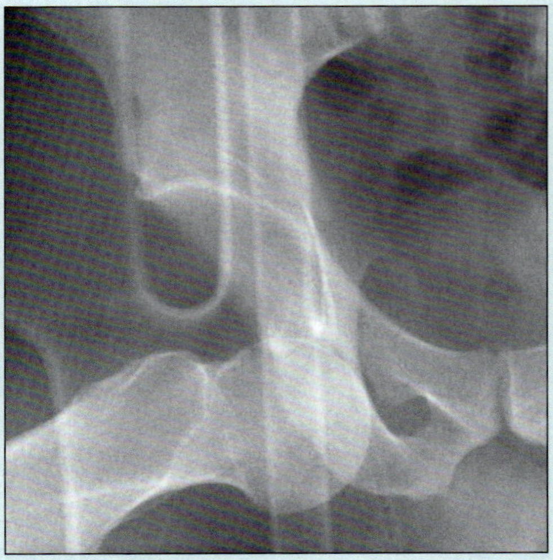

HIP DISLOCATION

TERMINOLOGY

Abbreviations

- Posterior hip dislocation (PHD)
- Anterior hip dislocation (AHD)

Synonyms

- Hip subluxation, femoroacetabular dislocation or subluxation

Definitions

- Disarticulation of femoral head relative to acetabulum
  - PHD: Femoral head posterior to acetabular rim
  - AHD: Femoral head anterior to acetabular rim
  - Transient dislocation: Femoroacetabular alignment spontaneously restored after dislocation

IMAGING

General Features

- Best diagnostic clue
  - Lack of congruence of femoral head within acetabulum
- Location
  - Posterior: Most common (90%)
  - Anterior: Uncommon (10%)
    - Obturator: Hip flexed, abducted & externally rotated, femoral head inferior
    - Iliac: Hip extended & externally rotated, femoral head superior
  - Medial: Misnomer; indicates impaction of femoral head into medial acetabular wall with fx

Radiographic Findings

- Posterior dislocation
  - Femoral head posterior, usually slightly superior to acetabulum
  - Internal rotation: Greater trochanter is profiled, lesser trochanter obscured
  - Dislocated femoral head appears smaller than contralateral head due to proximity to cassette
- Anterior dislocation: Obturator
  - Femoral head positioned medially & inferiorly, overlying obturator foramen
  - Femur in flexed, externally rotated, & abducted position
- Anterior dislocation: Iliac
  - Femoral head superior to acetabulum
  - Femur extended & externally rotated (lesser trochanter profiled, greater trochanter obscured)
- AP view is diagnostic, Judet (oblique) or lateral views confirmatory
- Associated femoral head or acetabular rim fractures

CT Findings

- Femoroacetabular malalignment on trauma CT
- Associated acetabular rim fragmentation
- Associated femoral head shear or impaction fx
- Intraarticular osseous fragments
- Intraarticular gas: Subtle collection may be only sign of reduced dislocation

MR Findings

- MR is generally not indicated for initial diagnosis
- Used later for identification of labral tear, chondral lesion, or osteonecrosis

Imaging Recommendations

- Best imaging tool
  - Radiograph (AP) makes initial diagnosis
  - CT delineates extent of fractures and fragmentation
- Protocol advice
  - CT: < 1 mm slice thickness, bone algorithm, coronal & sagittal reformats

DIFFERENTIAL DIAGNOSIS

Femoral Neck Fracture

- May appear clinically similar to PHD
- Femoroacetabular alignment preserved

Femoral Head Fracture

- Generally no associated acetabular fracture
- Anterior femoral head fx should raise concern for posterior dislocation

Femoral Head Osteonecrosis

- Curvilinear subchondral line
- Hypointense on all sequences

Femoral Head Insufficiency Fracture

- Femoral head marrow edema with no reciprocal acetabular findings

PATHOLOGY

General Features

- Etiology
  - High-velocity force required to dislocate hip
    - PHD frequently related to motor vehicle crashes
      - Dashboard injury drives flexed femoral head posteriorly
    - PHD in athlete falling directly onto flexed knee
    - AHD more common with alpine skiing injury
    - PHD with contralateral AHD reported in stepping from moving boat
- Associated abnormalities
  - Acetabular rim fracture
  - Femoral head fracture
    - Impaction (generally anterior cortex)
    - Shear fx
    - Avulsion of ligamentum teres at fovea
  - Intraarticular osseous fragments
  - Posterior acetabular labrum tear

Staging, Grading, & Classification

- Pipkin classification used for hip dislocation with fracture of femoral head
  - Pipkin I: Fracture of femoral head below central fossa
  - Pipkin II: Fracture of femoral head involving central fossa
  - Pipkin III: Fracture of femoral head extending to neck
  - Pipkin IV: Fractures of femoral head and posterosuperior acetabular rim

Gross Pathologic & Surgical Features

- Relevant anatomy

HIP DISLOCATION

- Iliofemoral, pubofemoral, ischiofemoral, transverse, femoral head ligaments maintain femoral head position
- Blood supply to femoral head via femoral neck from branches of circumflex arteries

CLINICAL ISSUES

Presentation

- Most common signs/symptoms
  - Hip pain & deformity following high-grade trauma
  - Restricted range of motion
- Clinical profile
  - Posterior dislocation
    - Hip & gluteal pain
    - Hip adducted, internally rotated
    - Lower extremity appears shortened
    - Lack of range of motion, inability to bear weight
    - Sciatic nerve injury
    - Hematoma, soft tissue swelling
  - Anterior dislocation: Obturator
    - Hip, gluteal, groin pain
    - Hip externally rotated, abducted, flexed
    - Lower extremity appears long
    - Lack of range of motion, inability to bear weight
    - Femoral nerve injury
    - Femoral artery injury: Pain, pallor, pulselessness of extremity
  - Anterior dislocation: Iliac
    - Hip, gluteal, groin pain
    - Hip extended & externally rotated
    - Lower extremity appears shortened
    - Lack of range of motion, inability to bear weight
    - Neurovascular injuries uncommon

Demographics

- Age
  - Traumatic injuries due to motor vehicle crashes < 40 years most common
  - Trauma secondary to falls > 65 years
- Gender
  - M > F

Natural History & Prognosis

- Orthopedic emergency
  - Reduce as soon as patient is otherwise stabilized
  - Osteonecrosis in up to 50% if not reduced within 24 hours
- High morbidity/mortality
  - Intraabdominal, intrapelvic, & head injuries associated with motor vehicle crash
  - Often associated pelvic ring fractures & severe hemorrhage
- Uncomplicated cases: Closed reduction successful (76-93%)
- Poorer prognosis when fractures present
- Recurrent dislocations if ligamentous injury, labral tear
- Complications
  - Osteoarthritis
  - Femoral head osteonecrosis
  - Sciatic nerve injuries (posterior) or femoral (anterior) nerve injury
  - Deep venous thrombosis

Treatment

- Conservative
  - Closed reduction
  - Maneuvers to recreate deforming force + apply longitudinal traction
    - Posterior dislocation: Flexion, adduction, internal rotation
      - Stimson maneuver (prone)
      - Allis maneuver (supine)
    - Anterior dislocation: Abduction, external rotation, extension
- Surgical treatment required for
  - Failed closed reduction
  - Intraarticular bodies
  - Interposed cartilage, labrum, ligament
  - Concomitant femoral head/neck fractures

DIAGNOSTIC CHECKLIST

Consider

- With high-velocity motor vehicle crash, check hip alignment on trauma CT abdomen/pelvis
- Direct posterior hip dislocation can be minimally displaced on single AP image
  - Assess congruence of femoral head and acetabulum
- Iliac-type AHD can mimic posterior hip dislocation on AP image
  - Differentiate direction of hip rotation
    - Posterior = internal rotation
    - Anterior = external rotation
  - Dislocated head appears smaller than contralateral head in posterior dislocation, larger in anterior dislocation

Image Interpretation Pearls

- After relocation, watch for
  - Fracture of head &/or acetabular rim
  - Increased distance between acetabulum & medial femoral head, indicating entrapped intraarticular fragments &/or soft tissue structures
- On follow-up radiographs, watch for early signs of osteonecrosis and consider performing MR

SELECTED REFERENCES

1. Herrera-Soto JA et al: Traumatic hip dislocations in children and adolescents: pitfalls and complications. J Am Acad Orthop Surg. 17(1):15-21, 2009
2. Pascarella R et al: Intra-articular fragments in acetabular fracture-dislocation. Orthopedics. 32(6):402, 2009
3. Philippon MJ et al: Arthroscopic findings following traumatic hip dislocation in 14 professional athletes. Arthroscopy. 25(2):169-74, 2009
4. Uzel AP et al: Hip subluxation associated with Pipkin II fracture: an injury to look for before reduction. Musculoskelet Surg. 93(1):33-5, 2009
5. Sah AP et al: Traumatic simultaneous asymmetric hip dislocations and motor vehicle accidents. Orthopedics. 31(6):613, 2008
6. Sahin V et al: Traumatic dislocation and fracture-dislocation of the hip: a long-term follow-up study. J Trauma. 54(3):520-9, 2003
7. Fairbairn KJ et al: Gas bubbles in the hip joint on CT: an indication of recent dislocation. AJR Am J Roentgenol. 164(4):931-4, 1995

HIP DISLOCATION

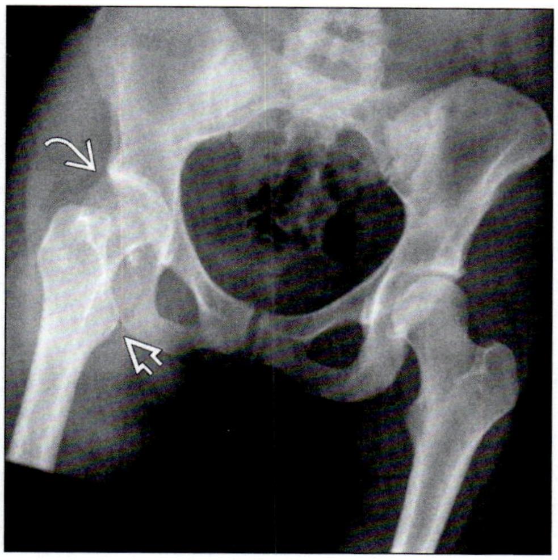

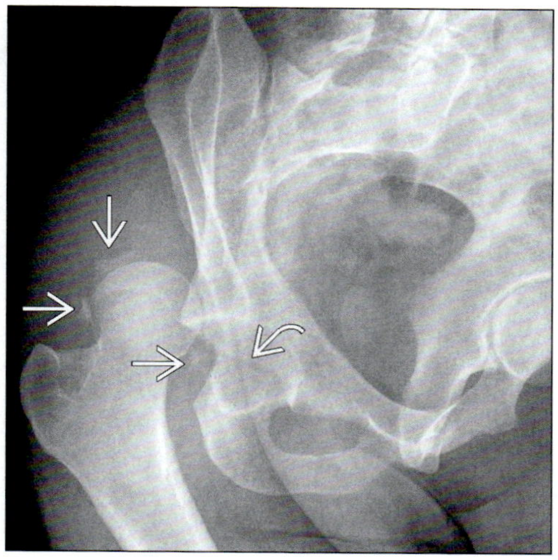

*(**Left**) AP radiograph shows cephalad displacement of the right femoral head but external rotation and extension of the femur, profiling the lesser trochanter, typical for an iliac-type anterior dislocation. (**Right**) Oblique (Judet view) radiograph shows a posterior hip dislocation. There are numerous fragments arising from the posterior wall of the acetabulum, and a large acetabular fracture line is visible. Judet views often demonstrate osseous fragments to better advantage.*

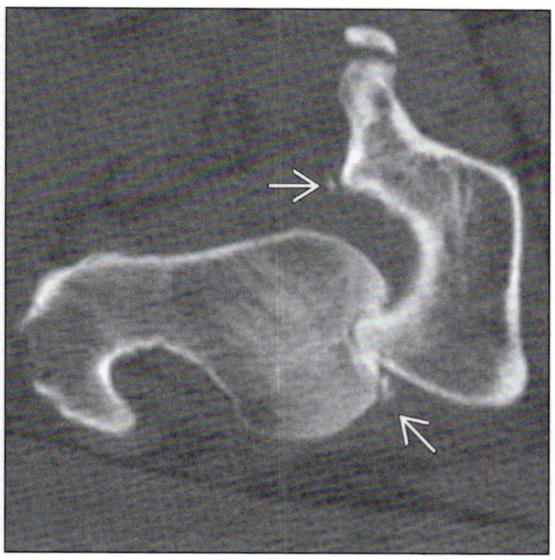

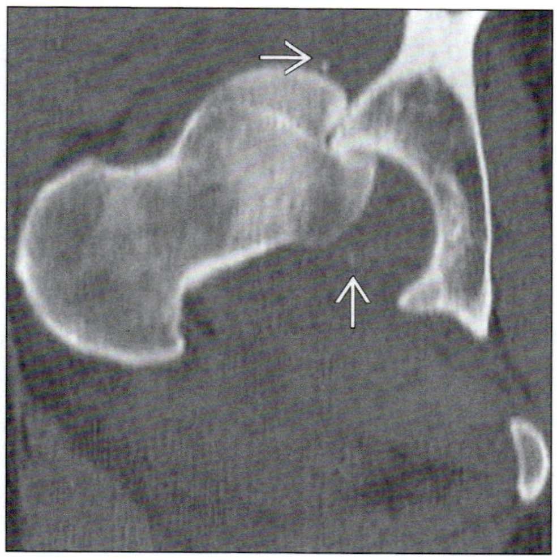

*(**Left**) Axial bone CT shows a posterior hip dislocation with impaction of the femoral head on the acetabular rim. Impacted fracture/dislocations such as this are harder to reduce without surgery. Small osseous fragments are present anteriorly and posteriorly. (**Right**) Coronal reformat bone CT from the same patient shows the impaction of the femoral head on the posterior superior aspect of the acetabulum. Tiny osseous fragments are present in the joint.*

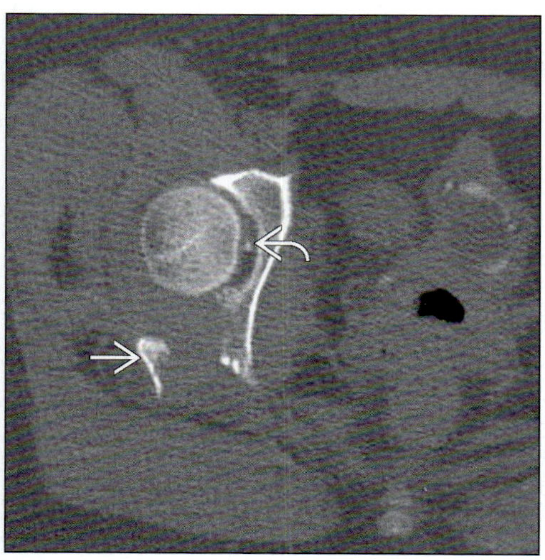

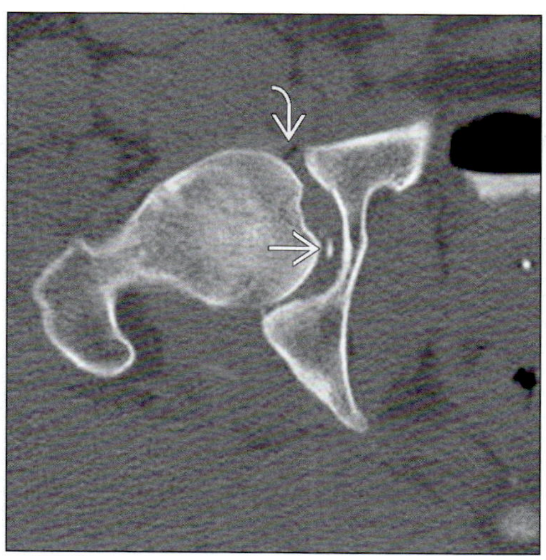

*(**Left**) Axial bone CT after posterior hip dislocation shows a large displaced fragment arising from the posterior acetabular wall, as well as a small intraarticular fragment in the joint. (**Right**) Axial bone CT after posterior hip dislocation shows incomplete reduction of the hip due to a small osseous fragment trapped in the joint. Such small fragments are usually not visible on radiographs. A tiny air bubble in the joint is a clue to recent dislocation.*

HIP DISLOCATION

(Left) AP radiograph shows bilateral posterior hip dislocations. There are also fractures of the left acetabulum ➡ and right pubic ramus ➡. Associated diastasis of the right SI joint ➡ makes this a Malgaigne (potentially unstable) fracture/dislocation. *(Right)* Axial bone CT shows bilateral femoral head fractures ➡ after bilateral hip dislocations. A small osseous fragment ➡ is seen posterior to the right acetabulum.

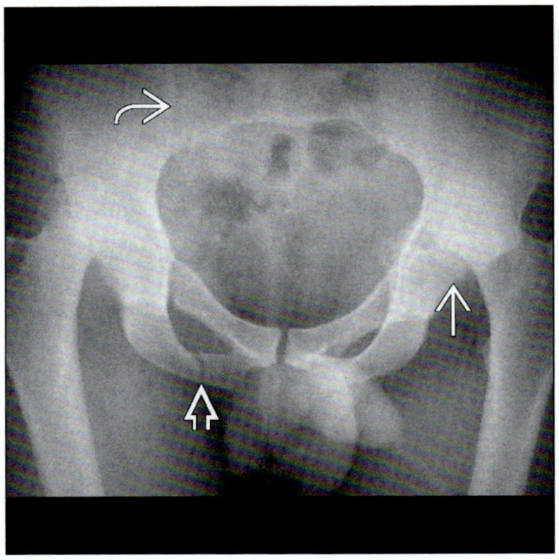

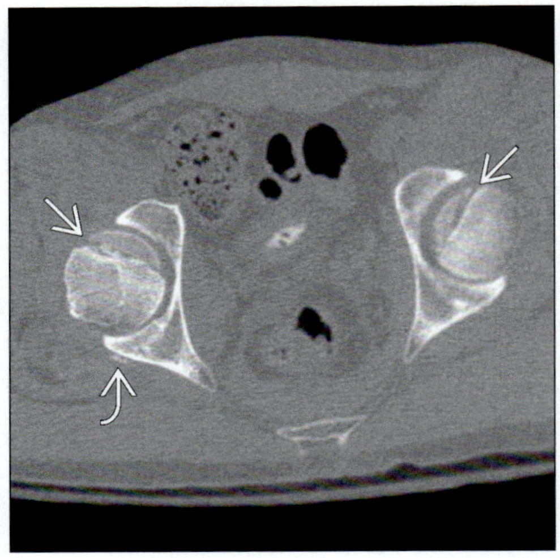

(Left) Portable AP radiograph demonstrates a posterior dislocation ➡ after total hip arthroplasty. The femur is internally rotated, which is typical for posterior dislocations of both native and prosthetic hips. Component dislocations are often reduced without surgery, but such dislocations are prone to recurrence. *(Right)* Groin lateral radiograph shows posterior dislocation of a total hip prosthesis ➡. A contralateral prosthesis ➡ is located normally.

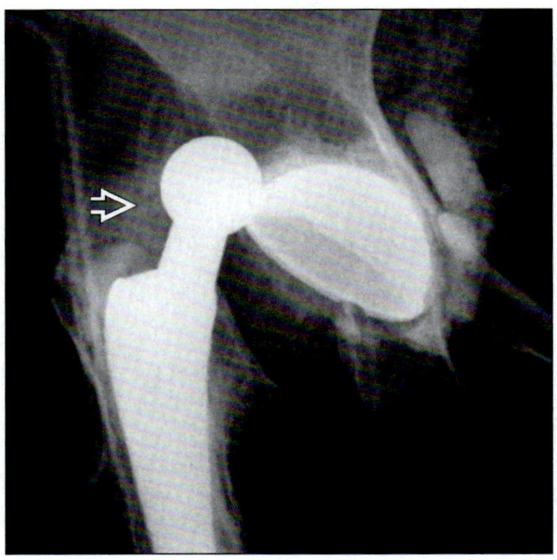

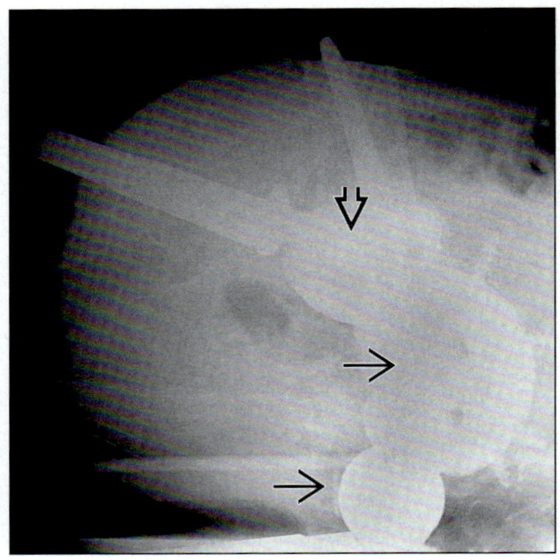

(Left) AP radiograph shows malalignment of a total hip prosthesis. The femoral head is superolateral to its expected position relative to the acetabular component. The hip is in external rotation with the lesser trochanter in profile ➡, a clue that this represents an iliac-type anterior dislocation and not a posterior dislocation. *(Right)* Groin lateral radiograph from the same patient shows that the dislocation is anterior.

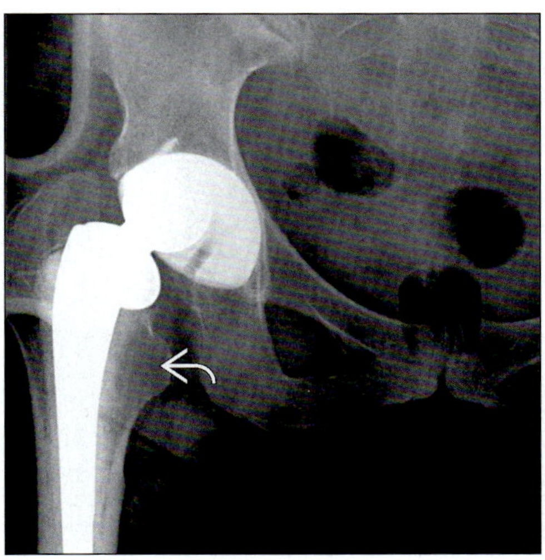

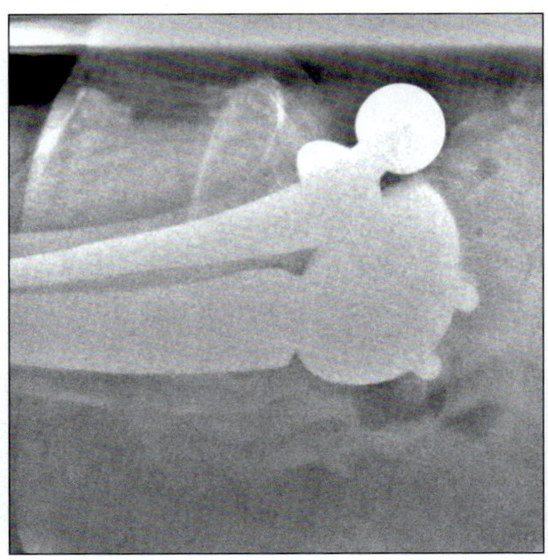

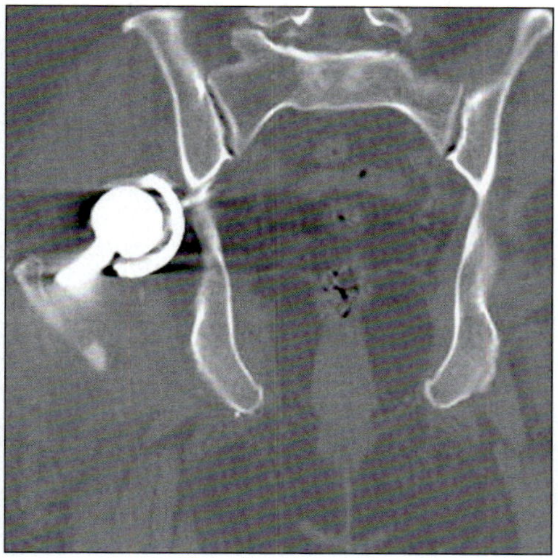

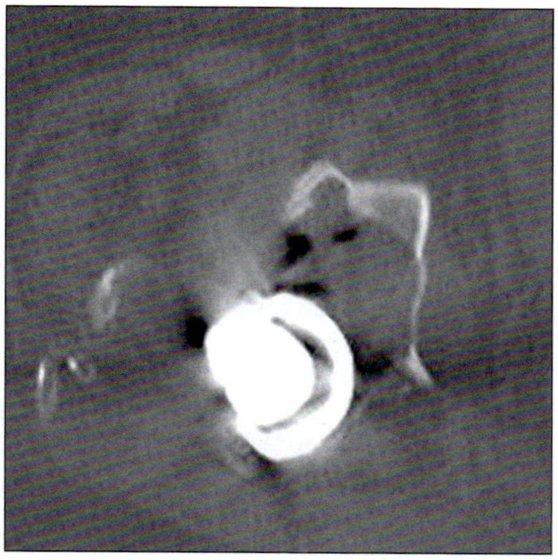

(Left) *Coronal reformat bone CT shows posterosuperior dislocation of both the femoral and acetabular components of a total hip prosthesis.* ***(Right)*** *Axial bone CT from the same patient shows posterosuperior dislocation of the entire prosthetic hip joint. Dislocation of the acetabular component with the femoral component is very unusual; most prosthesis dislocations involve dislocation of the femoral component relative to the acetabular component.*

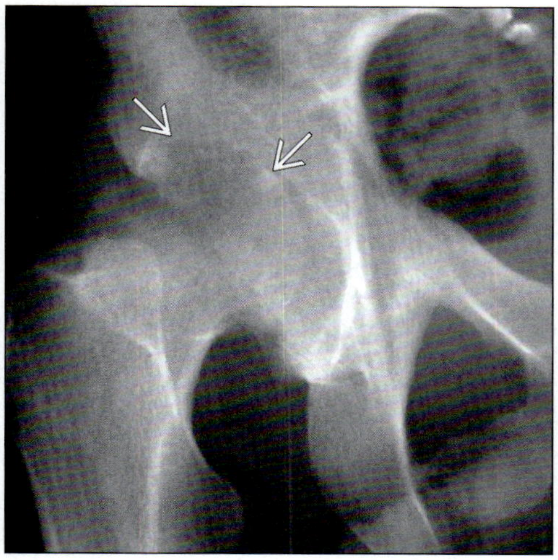

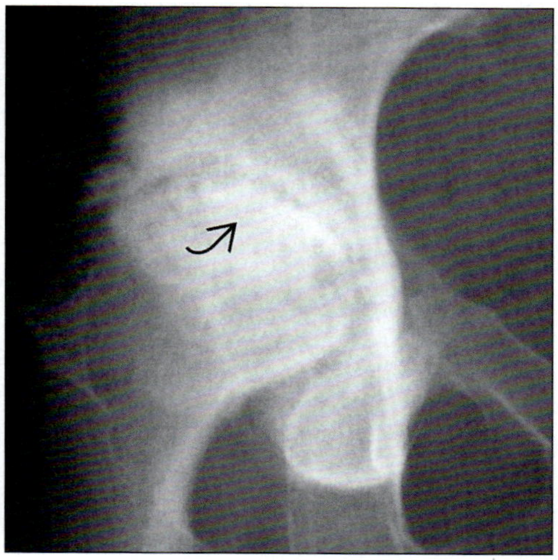

(Left) *A coned-in AP radiograph of the hip shows overlap of the femoral head and the superior acetabulum ➔ in a more subtle case of posterior hip dislocation. Trauma occurred more than 12 hours before this radiograph was obtained.* ***(Right)*** *AP radiograph from the same patient 1 year later shows flattening and sclerosis of the femoral head ➔, typical of osteonecrosis. Hips that remain dislocated longer than 12 hours are at significantly increased risk for developing femoral head osteonecrosis.*

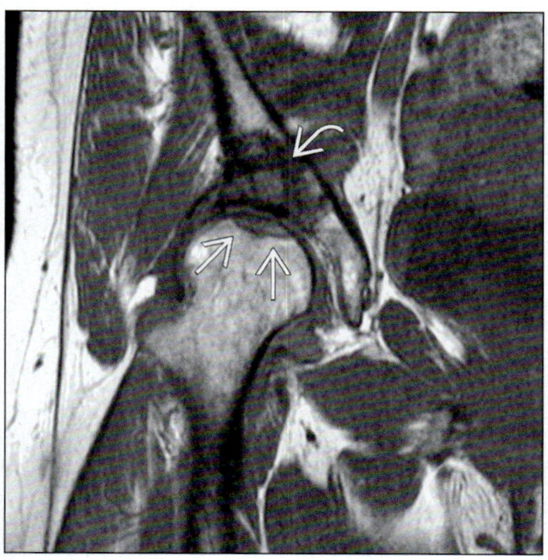

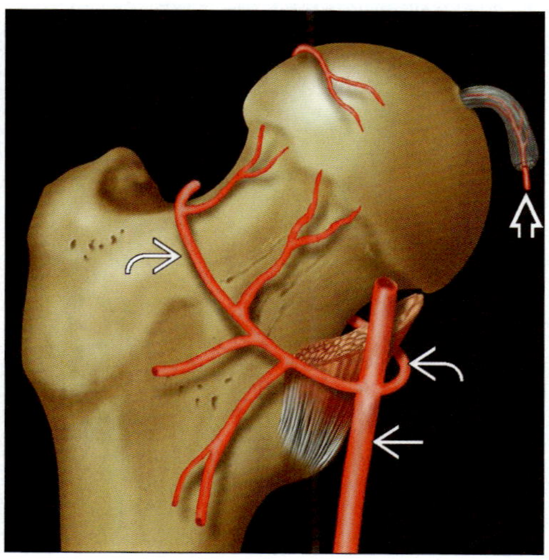

(Left) *Coronal T1WI MR more than 6 months after reduction of an anterior hip dislocation shows osteoarthritis of the anterior acetabulum ➔ and a subchondral focus of osteonecrosis ➔.* ***(Right)*** *Coronal graphic shows normal vascular supply of the proximal femur. The profunda femoral artery is shown medially ➔. The medial and lateral circumflex branches ➔ encircle the femoral neck. The artery of the ligamentum teres ➔ contributes to the blood supply of the femoral head.*

CHONDRAL AND OSTEOCHONDRAL INJURY, HIP

Key Facts

Terminology
- Breach in hyaline cartilage on articular surface ± involvement of underlying bone

Imaging
- Range from small cartilage fissure to large osteochondral lesion
- Look for underlying femoroacetabular impingement or congenital hip dysplasia
- Direct MR arthrography recommended for labral/cartilage injury without osteoarthritis
- Anterosuperior acetabulum most common location
- Anterior acetabulum and femoral head with posterior acetabular deficiency and anterior labral tear
- Superior acetabulum and medial femoral head with developmental dysplasia of hip (DDH)
- Posterior acetabulum and anterior femoral head with posterior dislocation/subluxation
- Weight-bearing femoral head with high-grade trauma

Pathology
- Outerbridge classification
- Noyes classification

Clinical Issues
- Can evolve to progressive chondral loss and osteoarthritis
- Arthroscopic debridement if symptomatic and no osteoarthritis

Diagnostic Checklist
- With arthrogram, use intraarticular anesthetic to confirm hip as source of pain

__(Left)__ Coronal T1 C+ FS MR arthrographic image shows intrasubstance contrast within the acetabular articular cartilage indicating a delaminating chondral lesion. __(Right)__ Coronal T1 C+ FS MR arthrographic image shows a normal supra-acetabular fossa at the superior acetabular chondral surface near its apex. This smoothly marginated chondral dimple and contour irregularity in the subchondral bone should not be mistaken for a cartilage defect. This appearance has also been termed a "stellate lesion."

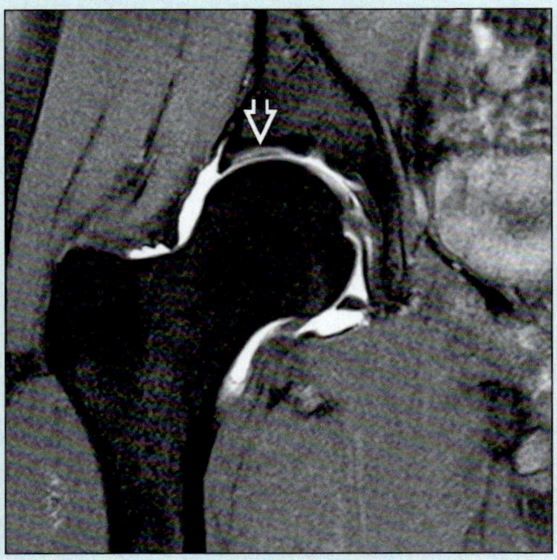

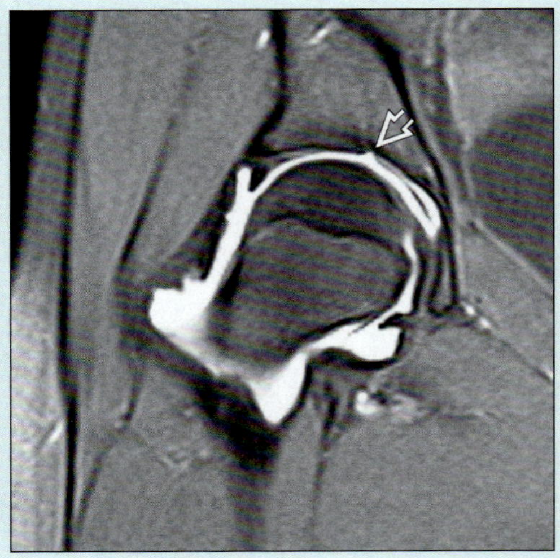

__(Left)__ Coronal T1WI MR from a noncontrast pelvis study in a runner with acute right hip and groin pain shows focal subchondral marrow hypointensity in the weight-bearing portion of the femoral head, which could reflect contusion or osteochondral lesion. __(Right)__ Axial oblique MR arthrogram T1WI FS from the same patient 6 weeks later shows a displaced osteochondral fragment with subjacent bone marrow edema, indicative of a traumatic osteochondral lesion of the femoral head.

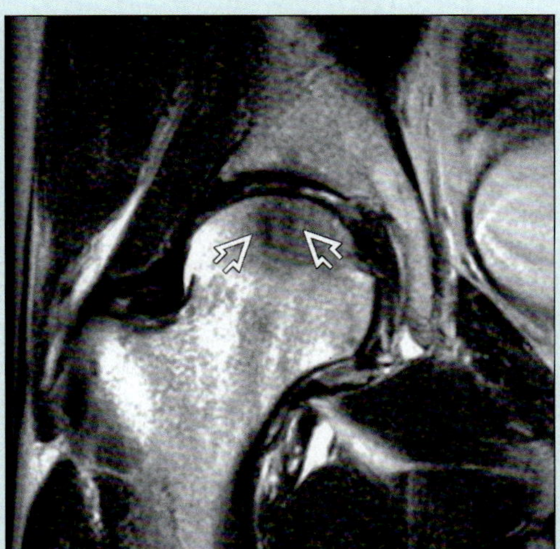

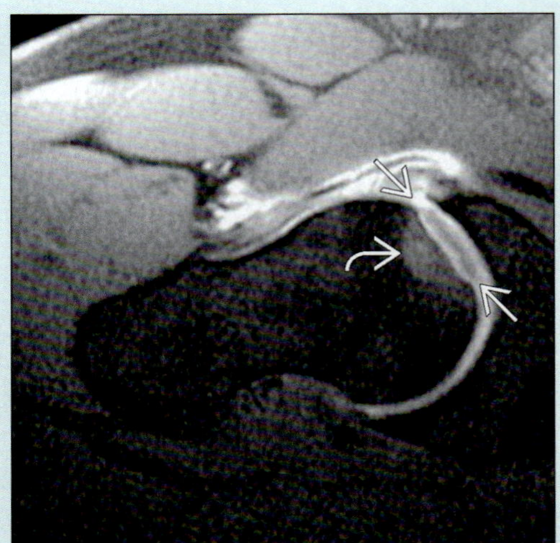

CHONDRAL AND OSTEOCHONDRAL INJURY, HIP

TERMINOLOGY

Abbreviations

- Osteochondritis dissecans (OCD)

Synonyms

- Chondrosis, osteochondral fracture, articular defect, chondromalacia

Definitions

- Breach in hyaline cartilage on articular surface ± involvement of underlying bone

IMAGING

General Features

- Best diagnostic clue
  - Abnormal subchondral density or signal on radiographs, CT, or MR
- Location
  - Can be on acetabulum, femoral head, or both
    - Anterosuperior acetabulum most common location
      - Especially with cam-type femoroacetabular impingement (FAI)
    - Anterior acetabulum and posterior femoral head with posterior acetabular deficiency and anterior labral tear
    - Superior acetabulum and medial femoral head with developmental dysplasia of hip (DDH)
    - Posterior acetabulum and anterior femoral head with posterior dislocation/subluxation
    - Weight-bearing femoral head with high-grade trauma
- Morphology
  - Partial thickness flap, delamination, diffuse thinning
  - Full thickness defects ± osseous involvement
    - Unstable osteochondral fragment less common than in ankle or knee

Radiographic Findings

- Radiography
  - Subchondral sclerosis indicates cartilage loss
  - Dense subchondral crescent can indicate traumatic osteochondral lesion
  - Morphology of underlying FAI or DDH

CT Findings

- Bone CT
  - CT arthrography
    - Similar to radiographic findings but more sensitive for sclerosis
    - 3D reformats can be helpful in demonstrating impingement morphologies
    - Can be useful in assessing for stability of osteochondral fragment
      - Contrast surrounding osteochondral fragment suggests instability

MR Findings

- T1WI
  - Subchondral hypointensity and heterogeneous or absent cartilage
- T2WI FS
  - Subchondral marrow edema
  - Chondral defect usually fluid signal
    - Chronic lesions fill with low signal fibrin
- PD/intermediate FS
  - Similar to T2WI FS
  - ↑ spatial resolution but less sensitivity than T2WI
- STIR
  - Sensitive for subchondral edema but poor specificity for grading
- MR arthrography
  - Sensitive for small chondral defects and associated labral tears
  - Must add T2WI FS to assess marrow edema

Imaging Recommendations

- Best imaging tool
  - MR with small field-of-view focused on hip
  - Direct MR arthrography standard for pre-arthritic cartilage lesions
- Protocol advice
  - T1 helps distinguish osteochondral lesion from geode
  - Fluid-sensitive fat-suppressed sequences in 3 planes
  - Include axial oblique down femoral next to assess for impingement morphology
  - Direct MR arthrography recommended for labral/ cartilage injury without osteoarthritis
    - T1WI FS in 3 planes

DIFFERENTIAL DIAGNOSIS

Osteochondral Contusion

- Subchondral marrow edema without cartilage defect

Femoral Head Fractures

- Discrete linear fracture in subchondral bone

Osteoarthritis

- Associated subchondral cysts

Supraacetabular Fossa or Stellate Crease

- Normal variant at apex of acetabulum (12 o'clock position)
  - Supraacetabular fossa is pit in articular surface
  - Associated cartilage deficiency is termed "stellate crease"

Sublabral Recess

- Generally anterior or anterosuperior
- Linear high signal without contrast imposition
- Prevalence = 18%

PATHOLOGY

General Features

- Acetabular cartilage lesion more common with degenerative loss; femoral head more common with trauma

Staging, Grading, & Classification

- Outerbridge classification
  - Adapted from arthroscopy
  - Based on cartilage signal on T2WI or PDWI FS MR
    - 0 = normal

CHONDRAL AND OSTEOCHONDRAL INJURY, HIP

AP graphic of cam-type femoroacetabular impingement (FAI) morphology shows the typical location of cartilage lesions on the anterosuperior acetabulum and at the anterosuperior femoral head.

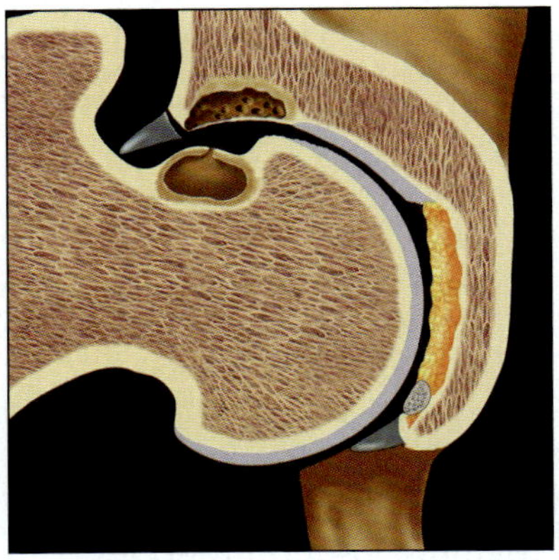

AP graphic of an adult dysplastic hip shows the associated chondral lesions. Undercoverage of the femoral head by the dysplastic acetabulum leads to instability and eventually to labral tears and articular cartilage lesions.

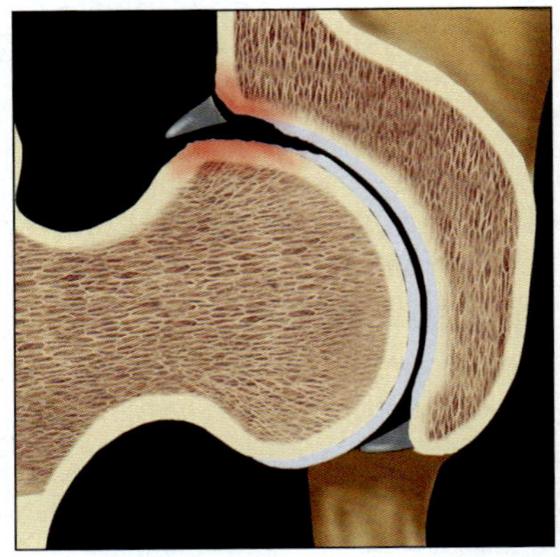

Sagittal graphic shows articular lesions associated with anterior labral tear. While anterior chondral lesions are common, posterior femoral and acetabular lesions can be seen with posterosuperior insufficiency and pincer-type FAI.

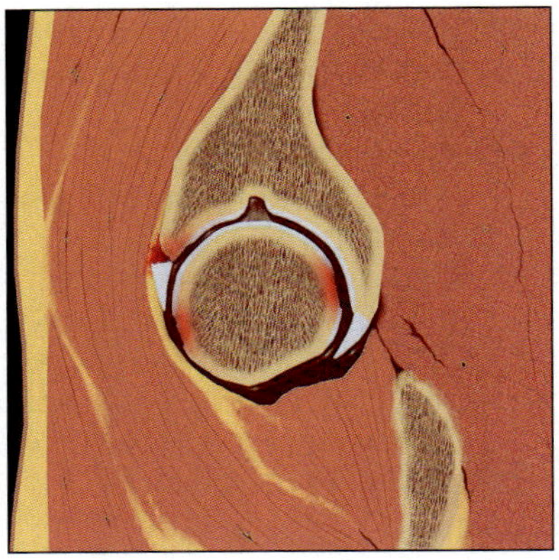

Oblique axial graphic shows typical cartilage injuries after posterior hip dislocation. Chondral injuries occur at the posterior acetabulum with a posterior labral tear as well as at the anterior femoral head in a coup-contrecoup pattern.

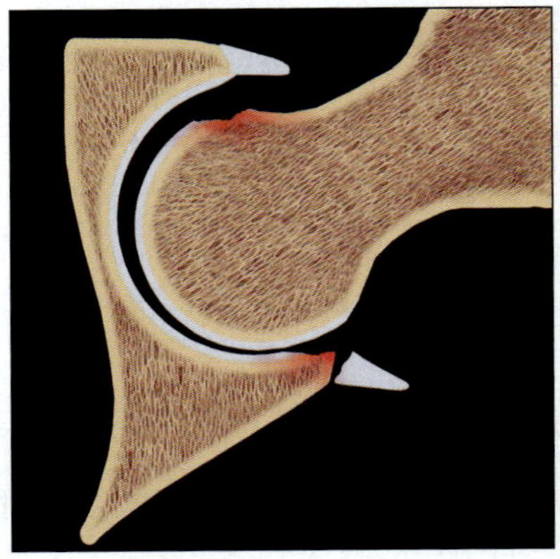

Typical locations for chondral injury in the hip.

  - 1 = cartilage softening/blistering; hyperintense signal without visible defect
  - 2 = shallow superficial ulceration, fibrillation, or fissuring < 50% cartilage depth
  - 3 = deep ulceration, fissuring, or chondral flap > 50% in depth without exposed bone
  - 4 = full thickness defect with exposed bone
- Noyes classification
  - Adapted from arthroscopy
  - Constructed to define degree of joint degeneration

CLINICAL ISSUES

Presentation

- Most common signs/symptoms
  - Pain with hip flexion or abduction/external rotation
  - ± pain with internal rotation
  - Ipsilateral groin is most common location of pain
  - Pain often radiates to ipsilateral knee
- Other signs/symptoms
  - Catching, clicking less common

Natural History & Prognosis

- Can evolve to progressive chondral loss and osteoarthritis

Treatment

- Arthroscopic debridement if symptomatic and no osteoarthritis
- Repair associated labral tear
- Treatment of underlying FAI or DDH
- Total hip arthroplasty if osteoarthritis and refractory pain

DIAGNOSTIC CHECKLIST

Consider

- Osteonecrosis of femoral head
- Look for FAI or DDH morphology

Reporting Tips

- With arthrogram, use intraarticular anesthetic to confirm hip as source of pain
  - If pain improves post injection, hip is source
  - If no change in pain post injection, look for pubic symphysis, tendon, or bursal lesion

SELECTED REFERENCES

1. Marshall N et al: Traumatic injuries of the hip. Magn Reson Imaging Clin N Am. 17(4):681-96, vi, 2009
2. Sekiya JK et al: Arthroscopic repair of delaminated acetabular articular cartilage in femoroacetabular impingement. Orthopedics. 32(9), 2009
3. Zaragoza E et al: Magnetic resonance imaging with gadolinium arthrography to assess acetabular cartilage delamination. Hip Int. 19(1):18-23, 2009
4. Studler U et al: MR arthrography of the hip: differentiation between an anterior sublabral recess as a normal variant and a labral tear. Radiology. 249(3):947-54, 2008
5. Tannast M et al: Hip damage occurs at the zone of femoroacetabular impingement. Clin Orthop Relat Res. 466(2):273-80, 2008

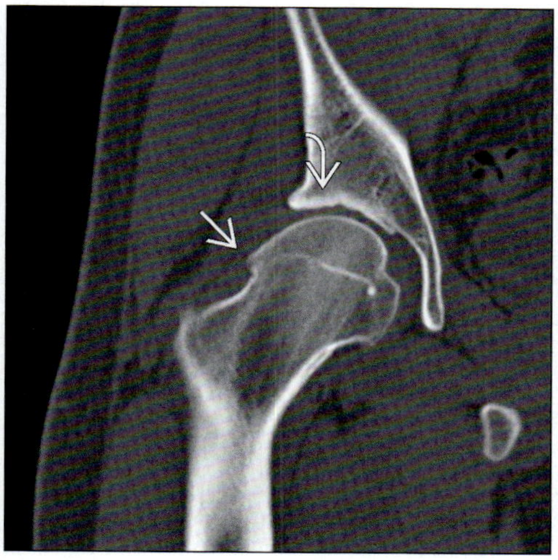

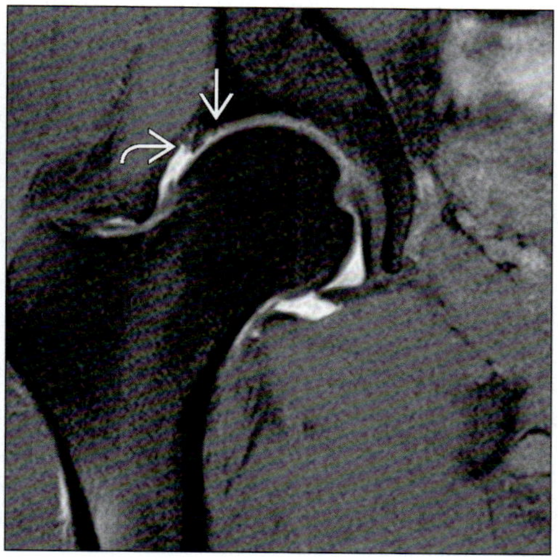

(Left) *Coronal reformat bone CT in a hockey goalkeeper shows the typical femoral head/neck morphology of cam-type femoroacetabular impingement (FAI) as well as subchondral sclerosis of the anterosuperior acetabulum, indicating overlying cartilage loss.* ***(Right)*** *Coronal MR arthrogram T1WI FS in same patient confirms scattered partial thickness cartilage defects in the anterosuperior acetabulum at the site of impingement as well as abnormal labral morphology.*

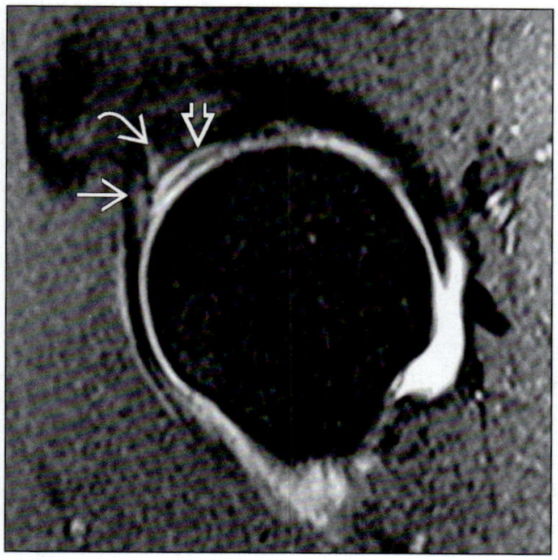

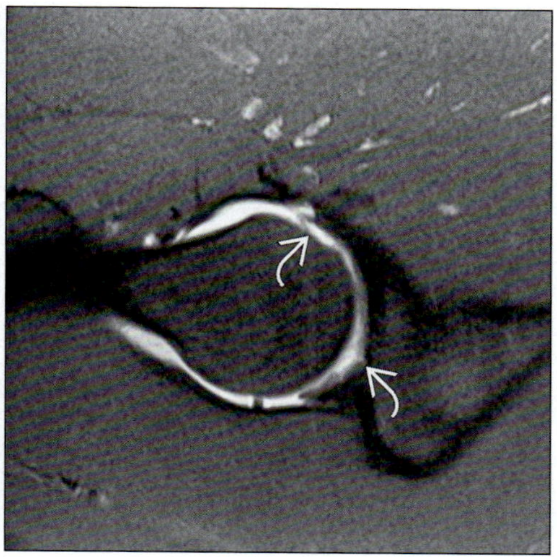

(Left) *Sagittal T1 C+ FS MR arthrographic image shows a delaminating chondral defect of the anterior acetabulum, subchondral marrow edema in the acetabulum, and a complex anterior acetabular labrum tear, the source of instability leading to the chondral loss.* ***(Right)*** *Axial oblique MR arthrogram T1WI FS from a 16-year-old ballerina with acetabular retroversion on radiographs and posterosuperior acetabular insufficiency shows chondral defects both anteriorly and posteriorly.*

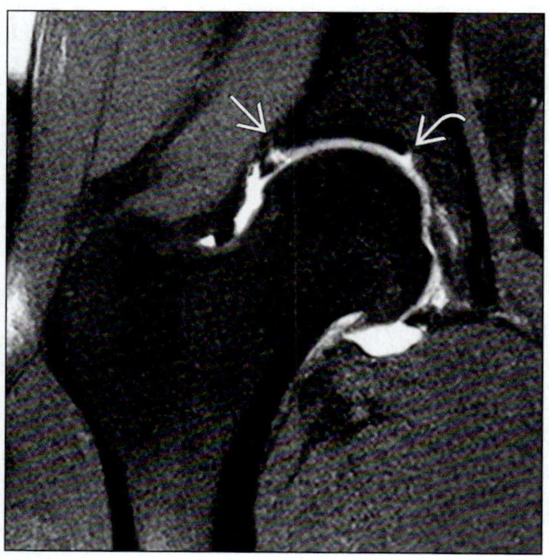

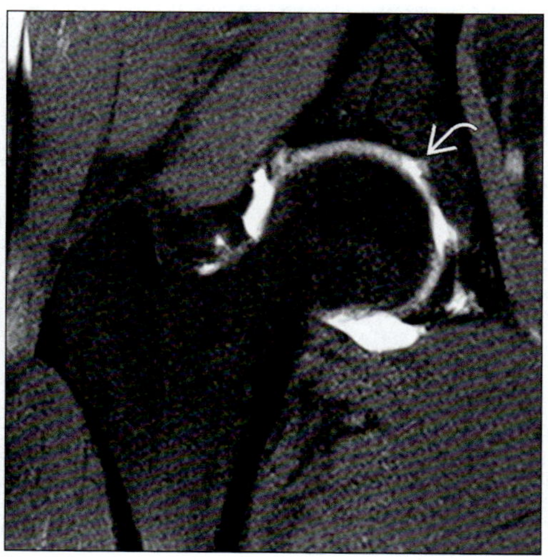

(Left) *Coronal MR arthrogram T1WI FS from a patient with a superior acetabular labral tear shows a supraacetabular fossa near the apex of the acetabulum or at the 12 o'clock position, a normal variant.* ***(Right)*** *Coronal MR arthrogram T1WI FS more posteriorly in the same patient shows a focal area of cartilage thinning subjacent to the supraacetabular fossa. This is considered a normal variant at arthroscopy, called the stellate crease.*

ACETABULAR FRACTURE

Key Facts

Terminology
- Fracture involving anterior, posterior, or both columns of acetabular fossa

Imaging
- Iliopectineal line disruption = anterior column fracture
- Ilioischial line disruption = posterior column fracture
- AP radiograph in emergency department or trauma bay
- Judet views if high suspicion in outpatients
- CT with reformats for trauma patients
  - Note sacroiliac, pubic symphysis, & pelvic ring injuries
- MR for suspected osteochondral injury

Pathology
- High-velocity trauma most common (MVA)
- Distraction and twisting hip injuries

Clinical Issues
- High-velocity trauma: Young adults
- Low-velocity trauma: Elderly
- 15-20% association with pelvic ring fracture
- Isolated posterior wall = 30-35% of acetabular fractures (most common)
- Both columns = 20-25% of acetabular fractures
- Mortality = 1-2% with surgery
- Prognosis & treatment based on classification
- Pelvic hematoma and soft tissue injury with MVA

Diagnostic Checklist
- Use Harris-Coupe and Judet-Letournel classifications
- Supplement with descriptive information
  - Biomechanical vector and likely soft tissue injuries
  - Articular step-off and displacement

*(**Left**) Graphic illustrations demonstrate normal acetabular column anatomy. On the left is a medial view, and on the right is a lateral view. Posterior column structures are in blue ➢, and anterior column structures are in red ➢. (**Right**) Lateral graphic representation shows a both-column fracture with fracture lines separating the columns and isolating the acetabulum from the sacrum. There is extension superiorly and inferiorly with little displacement. This is a Harris-Coupe type 2D injury.*

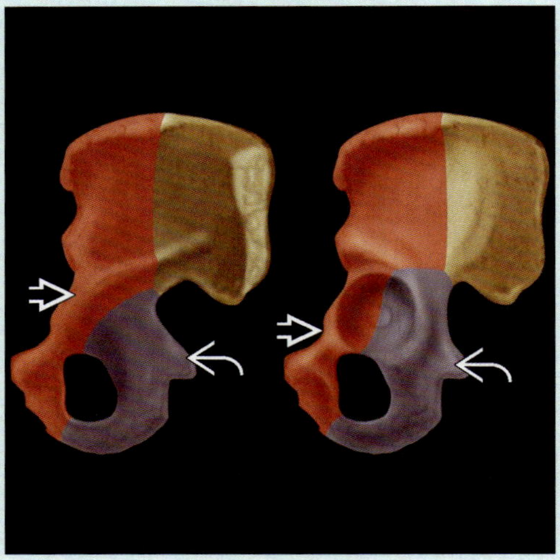

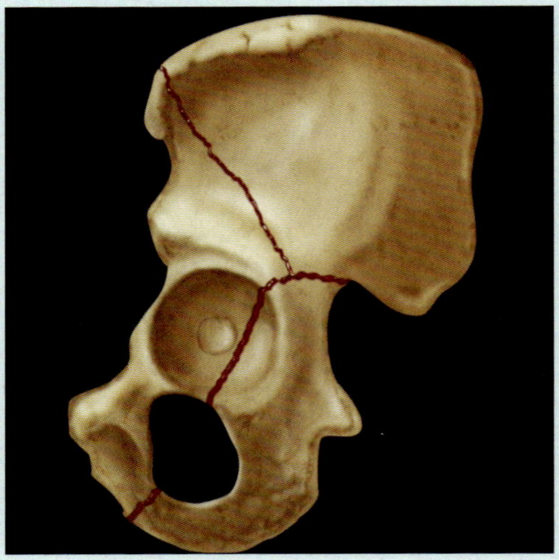

*(**Left**) Lateral graphic shows an anterior column acetabular fracture. While the fracture extends superiorly into the ileum ➢ and inferiorly into the pubic ramus →, there is no extension to the posterior column structures. This a Harris-Coupe type 1 injury or an elementary single column fracture in the Letournel system. (**Right**) Lateral graphic shows a posterior wall (rim) acetabular fracture. This is a low-grade Harris-Coupe type 0 or an elementary Letournel injury, and it carries a good prognosis.*

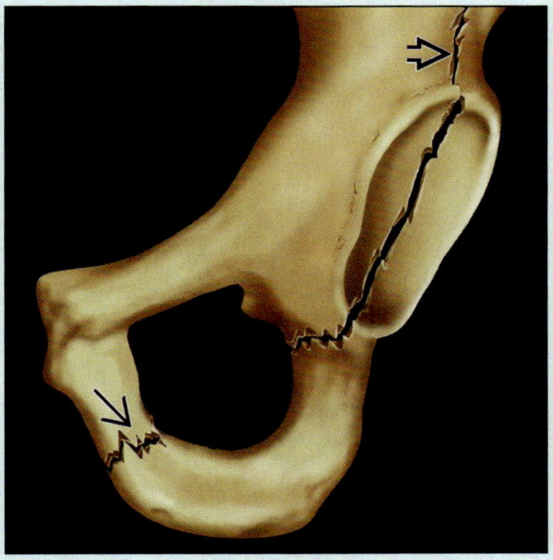

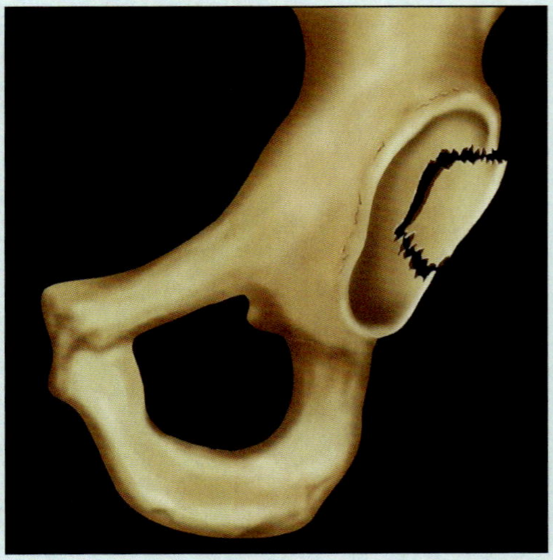

ACETABULAR FRACTURE

TERMINOLOGY

Definitions

- Fracture involving anterior, posterior, or both columns of acetabular fossa

IMAGING

General Features

- Best diagnostic clue
  - Iliopectineal line disruption = anterior column fracture
  - Ilioischial line disruption = posterior column fracture
- Location
  - Anterior column
    - Anterior half of acetabular cup
    - Superior pubic ramus
    - Anterior superior & inferior iliac spines
    - Anterior iliac crest
  - Posterior column
    - Posterior half of acetabular cup
    - Ischiopubic junction to greater sciatic notch
    - Obturator foramen
  - Sciatic buttress
    - Strut of bone extending from acetabulum toward sacrum
    - Links acetabulum to axial skeleton

Radiographic Findings

- AP radiograph "lines"
  - Iliopubic = anterior column
  - Ilioischial = posterior column
  - Anterior and posterior wall (rim)
  - Acetabular roof (dome)
  - Acetabular teardrop: Defines medial wall
- Judet view
  - Obturator oblique (internal oblique)
    - Assess anterior column and posterior rim
    - Obturator foramen en face
  - Iliac oblique (external oblique)
    - Assess posterior column and anterior rim
    - Iliac wing en face
- Obturator ring fracture
  - Involves anterior/posterior column
  - Both-column or T-shaped fracture
  - Not always acetabular fracture
- Iliac wing fracture
  - Anterior or both-column fracture
  - Not found in transverse/T-shaped fracture
  - Not always acetabular fracture
- Spur sign: Unique to both-column fracture
  - Inferolateral ilium still attached to sacrum, which projects lateral to acetabulum
- Iliopubic and ilioischial line disruption
  - Transverse or T-shaped fracture involving both anterior and posterior columns

CT Findings

- Bone CT
  - Multidetector CT with coronal, sagittal, and 3D reformats are standard of care
  - Harris-Coupe classification on axial images
  - Use reformats to describe articular step-off & displacement
  - Note sacroiliac, pubic symphysis, & pelvic ring injuries
  - Assess for potential joint instability

Imaging Recommendations

- Best imaging tool
  - AP radiograph initially
  - Judet views if high suspicion in outpatients
  - CT with reformats for trauma patients
  - MR for suspected osteochondral injury
- Protocol advice
  - CT: < 1 mm acquisitions with reformats
  - MR: Whole pelvis with dedicated hip sequences

DIFFERENTIAL DIAGNOSIS

Pelvic Ring Fracture

- Assess pubic rami, pubic symphysis, sacroiliac joints, & iliac wing
- Attention to sacral arcuate lines and sacral ala

Avulsion Fracture

- Anterior superior iliac spine: Sartorius and tensor fascia lata
- Anterior inferior iliac spine: Rectus femoris
- Ischial tuberosity: Hamstring complex

Proximal Femoral Fracture

- Femoral neck (subcapital) and intertrochanteric

Hip Dislocation

- 90% posterior
  - Associated with posterior rim fracture

Insufficiency Fracture

- Sacrum, pubic ramus, or femoral neck
- Often radiographically occult

Labral and Chondral Lesions

- May see marrow edema
- Direct MR arthrogram more specific

PATHOLOGY

General Features

- Etiology
  - High-velocity trauma most common (MVA)
  - Fall in elderly osteoporotic patient
  - Distraction and twisting hip injuries
  - Dashboard injury (flexed hip and knee) → posterior wall and column fracture
  - Direct blow to greater trochanter
    - Femur externally rotated → anterior fracture
    - Femur internally rotated → posterior fracture

Staging, Grading, & Classification

- Judet-Letournel classification
  - 5 simple fracture types & 5 associated (combination) fracture types
    - Anterior column
    - Posterior column
    - Anterior wall (rim)
    - Posterior wall (rim) isolated; most common

ACETABULAR FRACTURE

- Transverse
- Both columns
- Posterior column + posterior wall (rim)
- Transverse with posterior wall (rim) involvement
- T-shaped
- Anterior column with posterior hemi-transverse fracture
  - Most widely used historically
- Harris-Coupe classification
  - Based on axial CT
    - 0: Wall fractures only
    - 1: Single column fractures
    - 2: Both-column fractures
      - 2A: No extension
      - 2B: Superior extension
      - 2C: Inferior extension
      - 2D: Superior and inferior extension
    - 3: Floating acetabulum; separated from axial skeleton
  - Adds modern day practicality to Judet-Letournel classification
    - Includes biomechanical vector to guide surgery
- AO (fracture workgroup) classification
  - Type A: Wall (rim)/column fracture
  - Type B: T-shaped, transverse, hemi-transverse fracture
  - Type C: Both-column fracture
    - + dislocation/impaction/articular surface modifiers
  - More complex system than Judet-Letournel
    - Complexity limits practical application

CLINICAL ISSUES

Presentation

- Most common signs/symptoms
  - Hematoma anterior or lateral to hip
  - Posterior dislocation: Hip flexed & internally rotated
  - Anterior dislocation: Hip externally rotated & abducted
  - Dashboard knee injury: Patellar fracture, PCL injury
- Other signs/symptoms
  - Multisystem trauma
  - Sciatic nerve injury
    - 10-30% of acetabular fractures
    - More common with posterior dislocation
  - Superior gluteal artery injury: Rare

Demographics

- Epidemiology
  - 15-20% association with pelvic ring fracture
  - Isolated posterior wall = 30-35% of acetabular fractures (most common)
  - Both columns = 20-25% of acetabular fractures
  - Transverse and posterior wall = 20-25% of acetabular fractures

Natural History & Prognosis

- Good clinical outcome in up to 90% of minimally displaced fractures (< 1 mm offset)
- Arthritis: Incidence = 10-20% if < 1 mm fragment offset after surgery; 30-50% if 1-3 mm offset

Treatment

- Prognosis & treatment based on classification
  - Direction and amount of displacement
  - Degree of comminution
  - Chondral injury
  - Urgent surgery if
    - Irreducible dislocation
    - Increasing neurologic deficit
    - Associated vascular injury
    - Open fracture
- Surgical treatment: Nonurgent
  - Anatomic reduction and stable fixation
  - Instability or incongruity of femoral head or acetabulum
    - Fracture involving > 40% posterior wall
    - Wall fracture and instability
    - > 2 mm offset of weight-bearing portion of dome
    - Transverse fracture through weight-bearing portion of dome
    - Both-column fracture involving dome with incongruity
    - Bone or soft tissue fragments preventing congruity
- Nonoperative treatment
  - Minimally/nondisplaced congruous fracture
    - Includes secondarily congruous both-column fracture
  - Consider with < 2 mm articular surface offset
  - Poor medical/post-traumatic status
  - Severe osteoporosis/comminution
- Complications
  - Mortality = 1-2% with surgery
  - Pulmonary embolus ± symptomatic DVT = 2-6%
  - Pelvic hematoma and soft tissue injury with MVA
  - Heterotopic ossification very common
  - Osteoarthritis common
  - Sciatic nerve injury uncommon

DIAGNOSTIC CHECKLIST

Consider

- CT pelvis with high-velocity injury
  - Associated unstable pelvic fractures
- MR for dislocation/subluxation

Reporting Tips

- Use Harris-Coupe and Judet-Letournel classifications
  - Supplement with descriptive information
    - Biomechanical vector and likely soft tissue injuries
    - Articular step-off and displacement

SELECTED REFERENCES

1. Ferguson TA et al: Fractures of the acetabulum in patients aged 60 years and older: an epidemiological and radiological study. J Bone Joint Surg Br. 92(2):250-7, 2010
2. Giannoudis PV et al: Factors determining quality of life and level of sporting activity after internal fixation of an isolated acetabular fracture. J Bone Joint Surg Br. 91(10):1354-9, 2009
3. Durkee NJ et al: Classification of common acetabular fractures: radiographic and CT appearances. AJR Am J Roentgenol. 187(4):915-25, 2006
4. Harris JH Jr et al: Acetabular fractures revisited: part 1, redefinition of the Letournel anterior column. AJR Am J Roentgenol. 182(6):1363-6, 2004

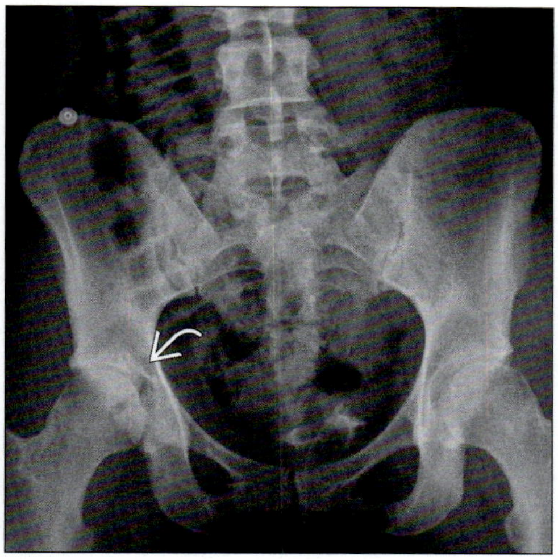

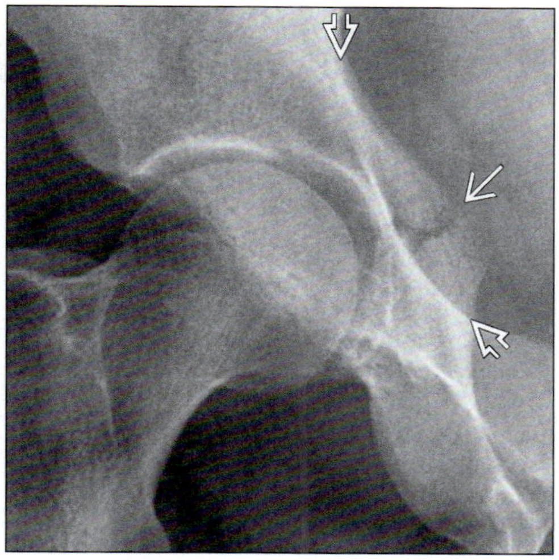

(Left) *AP radiograph of the pelvis was obtained after a "dashboard mechanism" trauma in a motor vehicle crash. A posterior column acetabular fracture can be definitively identified, but assessment of extent of the fracture is limited on this image.* ***(Right)*** *AP radiograph from a hip series in a 60-year-old osteoporotic woman after a fall shows a fracture extending through the ilioischial line but sparing the iliopectineal line, indicating a posterior column acetabular fracture.*

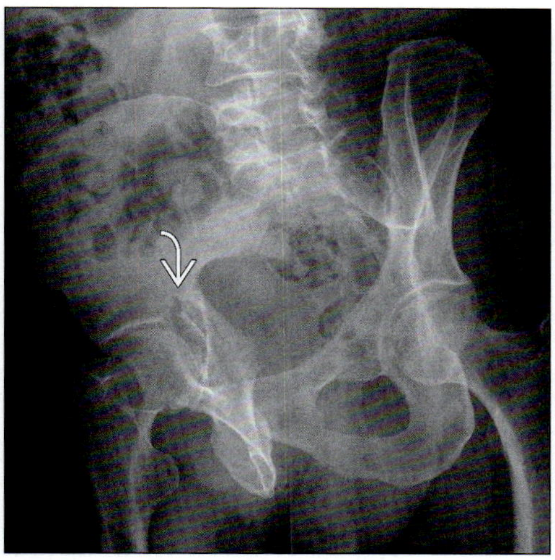

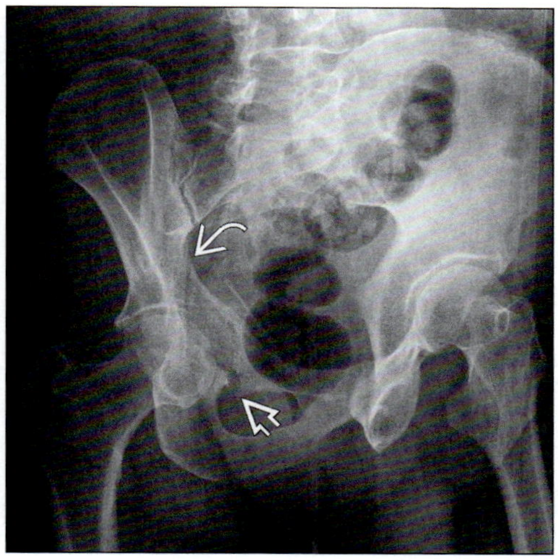

(Left) *AP internal oblique (iliac) radiograph shows a comminuted fracture through the posterior acetabular column extending superiorly into the ileum.* ***(Right)*** *AP external (obturator) oblique view from the same patient shows the iliac fracture and a displaced fracture extending through the obturator ring. The oblique sagittal orientation of the posterior column component with extension to the obturator ring make this a T-shaped or both-column fracture with inferior extension.*

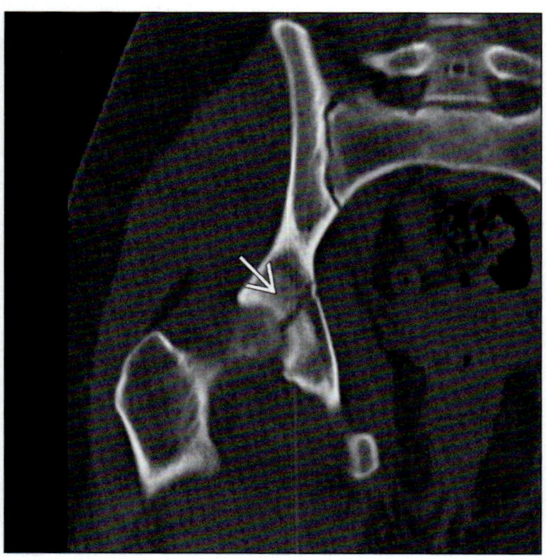

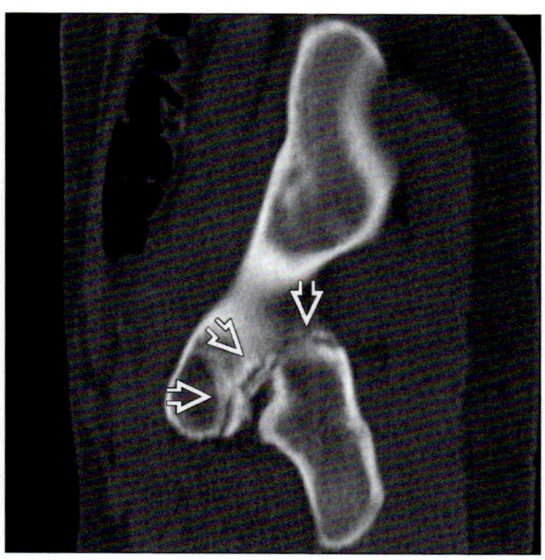

(Left) *Coronal reformat bone CT through the posterior column of the acetabulum in a motor vehicle crash victim shows a fracture extending to the weight-bearing portion of the acetabular cup. No significant step-off is seen.* ***(Right)*** *Sagittal reformat bone CT from the same patient shows extension of the isolated posterior column fracture through the posterior cortex of the ischium. Multiplanar CT reformats are very useful for delineating fracture displacement and step-off of articular surfaces.*

*(**Left**) Axial bone CT obtained after posterior hip dislocation shows an extensively comminuted posterior column fracture. Involvement of the mid-acetabular cup differentiates this single column fracture from an isolated posterior acetabular rim fracture. (**Right**) Sagittal reformat bone CT after a fall from a roof shows a nondisplaced posterior column fracture. Lack of articular surface displacement is associated with a better clinical prognosis.*

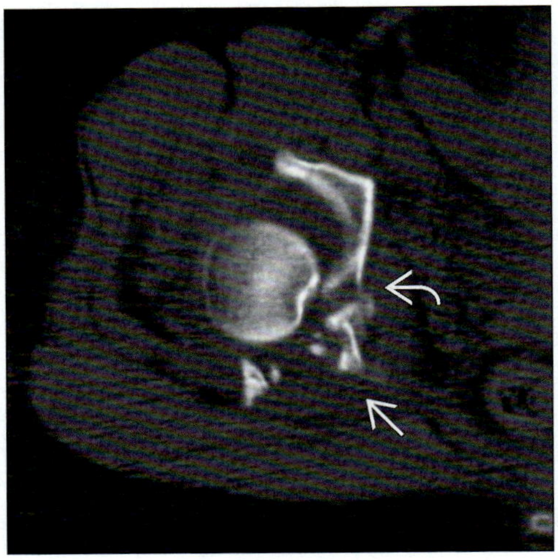

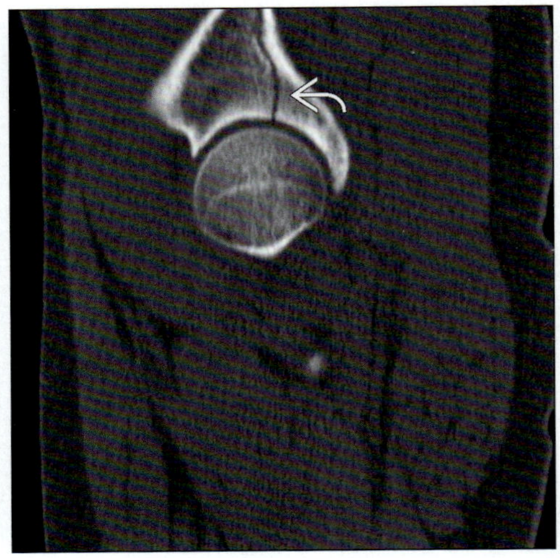

*(**Left**) Axial bone CT of a Harris-Coupe type 1 fracture isolated to the anterior column shows > 3 cm of articular step-off. Despite the low grade of this fracture, there is a strong likelihood of post-traumatic osteoarthritis due to the articular malalignment. (**Right**) Coronal bone CT reformat of an anterior column fracture shows involvement of the medial articular surface of the acetabulum. The presence of a small intraarticular fragment warrants surgical removal.*

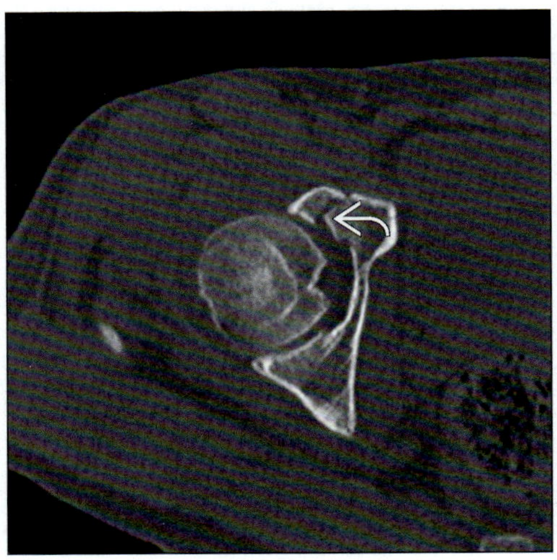

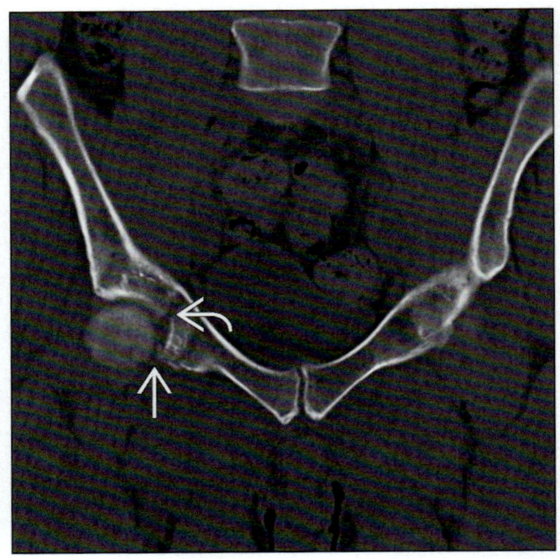

*(**Left**) Coronal T1WI MR for suspected femoral fracture shows instead a fracture that involves the posterior wall of the acetabulum with surrounding sclerosis, suggesting a subacute fracture. (**Right**) Sagittal T2WI FS MR from a football player with normal radiographs shows extensive subchondral marrow edema in the acetabulum. A nondisplaced fracture extending into the posterosuperior acetabular articular surface indicates an osteochondral impaction fracture.*

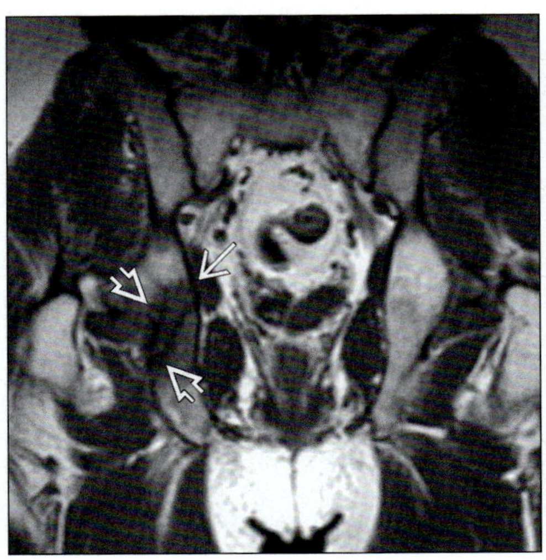

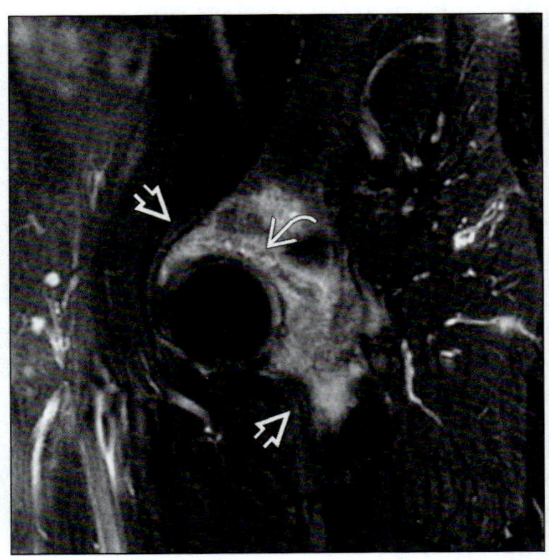

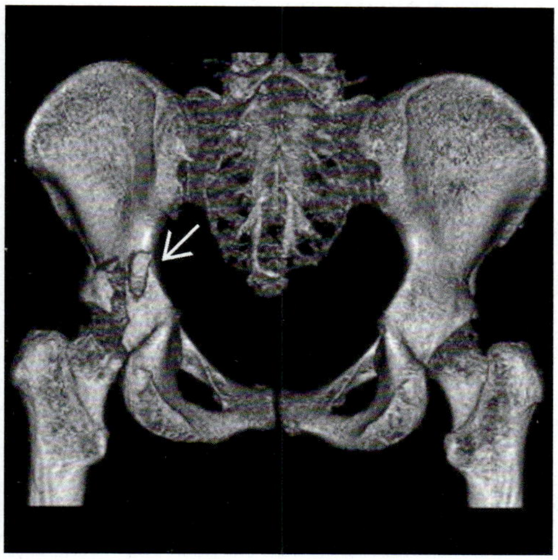
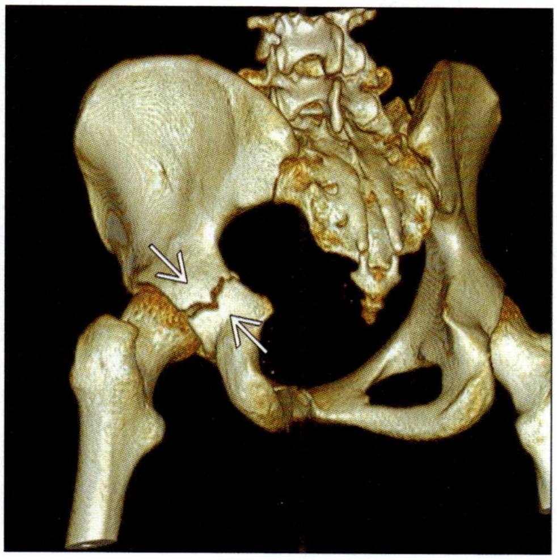

(Left) *Coronal reformat 3D bone CT from a patient status post dashboard injury shows extensive comminution of the posterior acetabular wall ➔. This fracture was isolated to the posterior column.* ***(Right)*** *Coronal oblique 3D bone CT reformat nicely delineates a posterior wall (rim) fracture with no significant displacement ➔. This is the most common type of acetabular fracture, is classified as Harris-Coupe grade 0, and has an excellent prognosis when there is no comminution or step-off.*

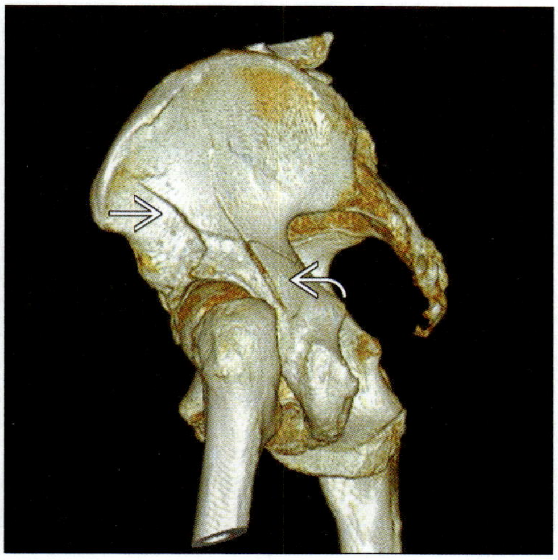
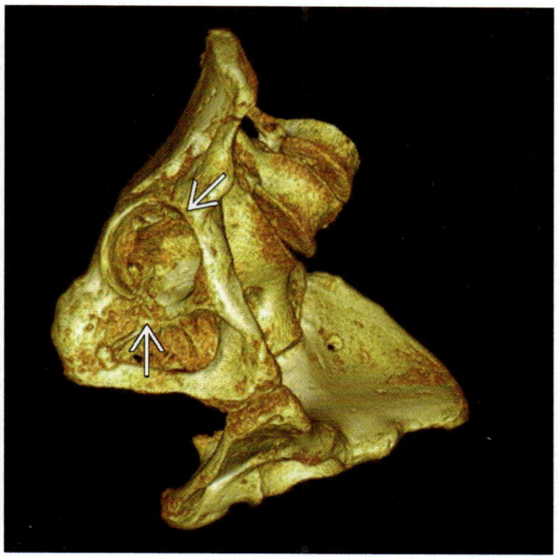

(Left) *Sagittal 3D bone CT reformat from an MVA patient shows a both-column acetabular fracture. The anterior column hemi-transverse component extends to the ileum ➔, and the posterior column component extends to the ischium ➔. There is no significant displacement in either column.* ***(Right)*** *Axial oblique 3D bone CT reformat with subtraction of the femur offers excellent detail of the acetabular articular surface in this both-column fracture ➔, creating a virtual map for surgical planning.*

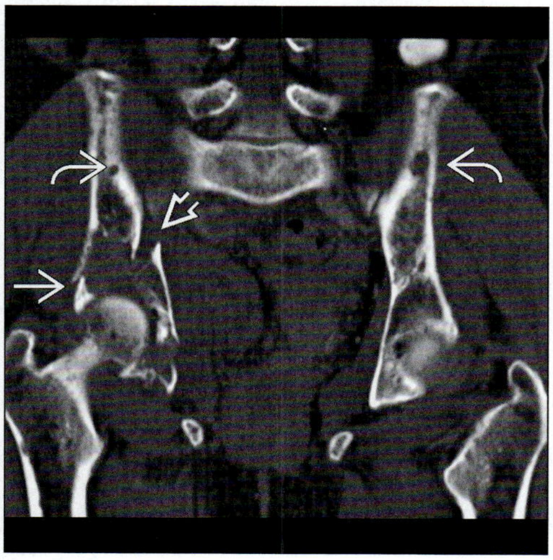
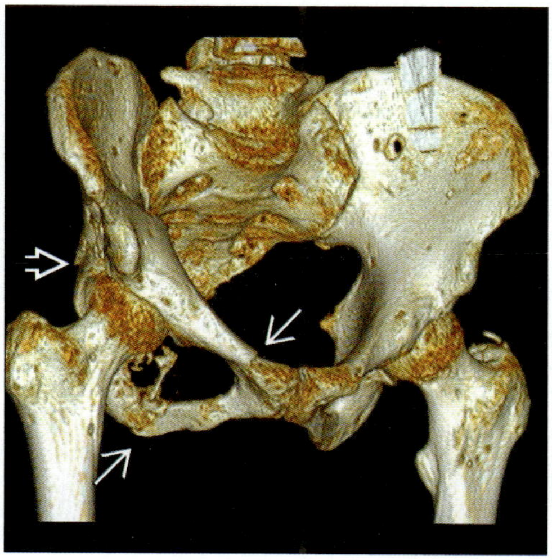

(Left) *Coronal bone CT reformat shows a both-column fracture with posterior rim ➔ and iliac ➔ extension. There are multiple lytic lesions ➔ in this patient with multiple myeloma and a pathologic fracture.* ***(Right)*** *Coronal oblique 3D reformat bone CT from the same patient shows pubic ramus fractures ➔ and propagation of the posterior wall fracture ➔ laterally through the ileum and ischium, segmenting acetabulum from the axial pelvis. This injury is termed a "floating acetabulum."*

PELVIC FRACTURE, STABLE

Key Facts

Terminology
- Fractures sparing posterior arch and pelvic ring
- Able to withstand normal physiological stress without displacement

Imaging
- Isolated fracture of innominate bone (ilium, ischium, or pubic)
- Iliac wing fracture not extending through pelvic ring
- Ipsilateral pubic ramus fractures
- Transverse fracture of sacrum or coccyx
- Radiographs usually adequate
  - AP of pelvis diagnostic in most cases
  - Inlet/outlet views if ramus fracture suspected
  - Judet views for iliac fracture
  - Nondisplaced posterior arch injuries may be stable if sacroiliac (SI) ligaments intact (includes some open book injuries)
- CT standard of care in high-velocity trauma

Top Differential Diagnoses
- Unstable pelvic fracture
- Myotendinous injury
- Sacral insufficiency fractures

Pathology
- Goal of classification is to determine stability and corrective forces needed for realignment of ring

Clinical Issues
- High-velocity trauma
- Elderly and osteoporotic patients with low-velocity trauma
- Sacral/coccyx pain with normal radiographs after fall
- Stable pelvic fractures with < 5 mm displacement typically recover in 6-8 weeks

__(Left)__ AP radiograph demonstrates a comminuted fracture of the right iliac wing in a man struck by a motor vehicle while walking. The ipsilateral sacroiliac joint appears intact, but more accurate evaluation is provided by CT. __(Right)__ Axial bone CT from the same patient shows comminution of the iliac wing fracture. The fracture does not enter the SI joint, and there is no diastasis of the joint. The sacrum and posterior arch are intact. This is an OTA A2 fracture.

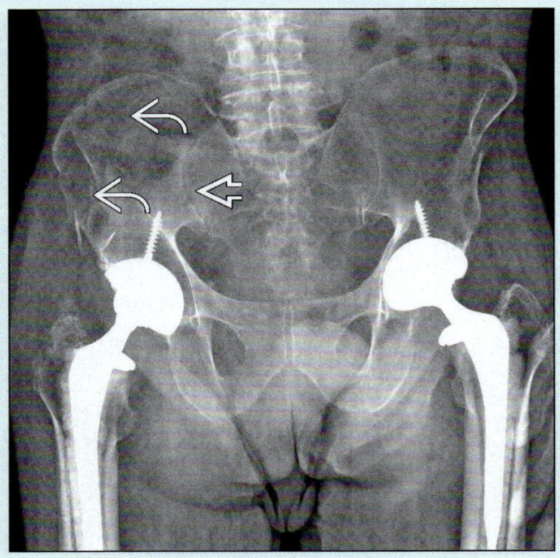

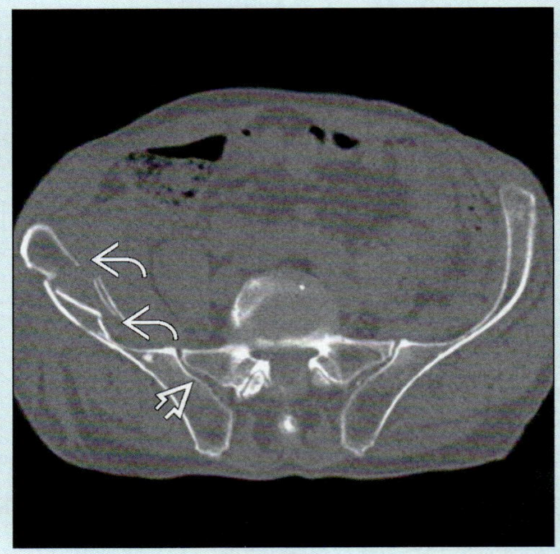

__(Left)__ AP radiograph shows ipsilateral superior and inferior pubic ramus fractures after a fall onto the right hip. The symphysis is intact. This is an OTA A2 fracture. __(Right)__ Axial bone CT shows an isolated fracture of the inferior pubic ramus. Fractures of the pubic rami without other bony injury in the pelvis should be considered stable pelvic fractures, but the sacrum and SI joints should be carefully evaluated for associated shear injuries that would render the fractures unstable.

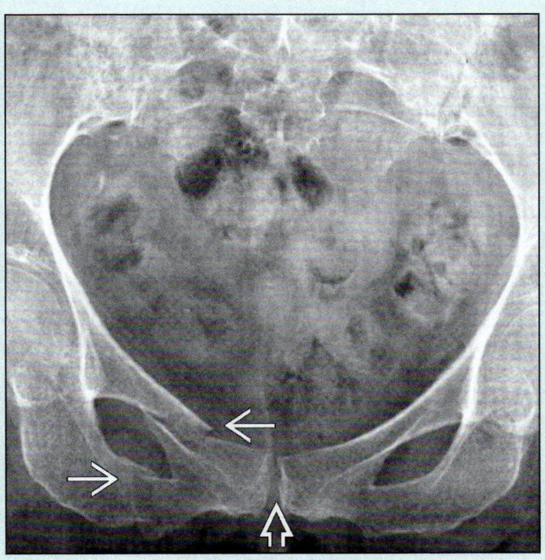

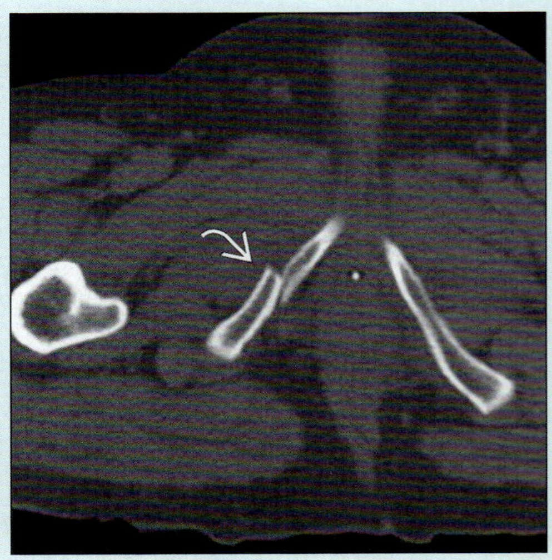

PELVIC FRACTURE, STABLE

TERMINOLOGY

Definitions

- Pelvic fracture not resulting in complete disruption of posterior arch
  - Able to withstand normal physiological stress without displacement
  - Includes
    - Avulsion fracture of innominate
    - Fracture of innominate bone involving rami or iliac wing (Duverney)
    - Pubic ramus fracture(s) (straddle fracture)
    - Transverse fracture of sacrum/coccyx
    - Incomplete disruption of posterior arch (open book) injuries may be included (partially stable; posterior sacroiliac ligaments intact)

IMAGING

General Features

- Best diagnostic clue
  - Posterior pelvic arch intact
- Location
  - Isolated fracture of innominate bone (ilium, ischium, or pubis)
  - Iliac wing fracture not extending through pelvic ring
  - Ipsilateral pubic ramus fractures
  - Transverse fracture of sacrum or coccyx
  - Nondisplaced posterior arch fracture may be stable if sacroiliac (SI) ligaments intact
- Morphology
  - Iliac wing fracture not extending through ring
    - Generally lateral compression force
  - Ramus fracture(s)
    - Vertical/longitudinal plane with vertical shear force
    - Horizontal/transverse plane with lateral compression force
  - Sacrum/coccyx fracture
    - Transverse plane with direct trauma from fall

Radiographic Findings

- AP view of pelvis diagnostic in most cases
  - Pelvic ring remains intact
- Inlet/outlet views if ramus fracture suspected
  - Radiograph tube angled 45° caudal (inlet) and 45° cranial (outlet) relative to pelvis
  - Documents AP (inlet) and craniocaudal (outlet) fracture displacement
- Judet views
  - Iliac oblique obtained at 45° to pelvis with side of interest down
    - Helpful with iliac wing fractures
  - Obturator oblique obtained at 45° to pelvis with side of interest up
    - Helpful with iliopubic fractures
- Sacral arcuate lines and SI joint preserved
- Can see minimal offset at pubic symphysis with pubic ramus fractures

CT Findings

- Accurately delineates osseous injury patterns
- Can identify lesions indicating unstable fractures
  - Occult posterior ring fractures
  - Diastasis at sacroiliac joints

MR Findings

- Not commonly indicated in setting of acute pelvic fracture
- Completion of pubic ramus stress fracture
- Ligamentous and tendinous injury

Imaging Recommendations

- Best imaging tool
  - Radiographs
    - Most fractures (90%) detected on AP view
    - Consider inlet/outlet and Judet views in trauma protocol if no CT
  - CT standard of care in high-velocity trauma
- Protocol advice
  - CT: Coronal, sagittal, and 3D reformats

DIFFERENTIAL DIAGNOSIS

Pelvic Fractures, Unstable

- Posterior arch partially or completely disrupted
- Contralateral/bilateral pubic ramus fractures

Sacral Stress Fracture

- Vertical fracture plane

Pubic Ramus Stress Fracture

- Often incomplete, sclerotic, without offset

Osteitis Pubis

- Subchondral productive changes around symphysis

PATHOLOGY

General Features

- Etiology
  - Iliac wing fracture (Duverney fracture)
    - Direct trauma from lateral compressive force, such as MVA or iatrogenic
  - Unilateral pubic ramus fracture
    - Transverse superior ramus fractures most common
    - Fall in elderly osteoporotic patients
  - Transverse sacrum and coccyx fracture
    - Direct trauma, often from fall in seated position
- Associated abnormalities
  - Sacroiliac shear injuries
  - Bilateral ramus fractures (straddle fracture, butterfly fracture)
  - Genitourinary (GU) and arterial injury
    - More common with unstable pelvic fractures
  - Rectal and sacral plexus injury
    - Transverse sacral fractures at S3 and above
- Relevant anatomy
  - Pelvic ring composed of 2 arches
    - Posterior arch
      - Located behind acetabular surface
      - Includes sacrum, SI joints & ligaments, and posterior ilium
    - Anterior arch
      - Anterior to acetabular surface
      - Includes pubic rami and symphysis

PELVIC FRACTURE, STABLE

Isolated iliac wing fractures (Duverney fractures) are considered stable pelvic fractures, but contralateral pelvic ring fractures are common and should be excluded with imaging.

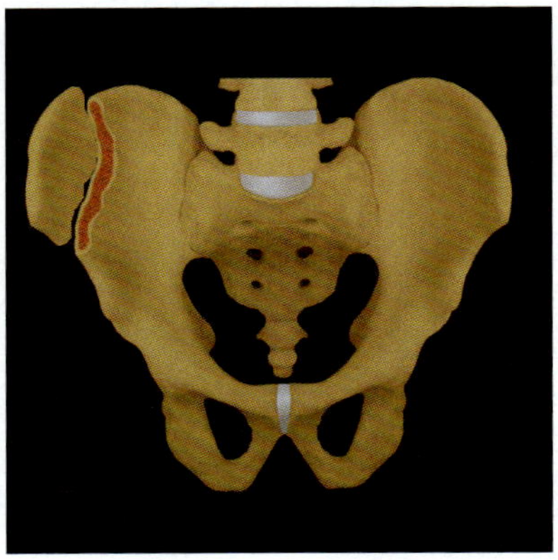

Isolated single pubic ramus fractures are stable but relatively uncommon in the setting of trauma. Completion of a pubic ramus stress fracture can lead to this deformity.

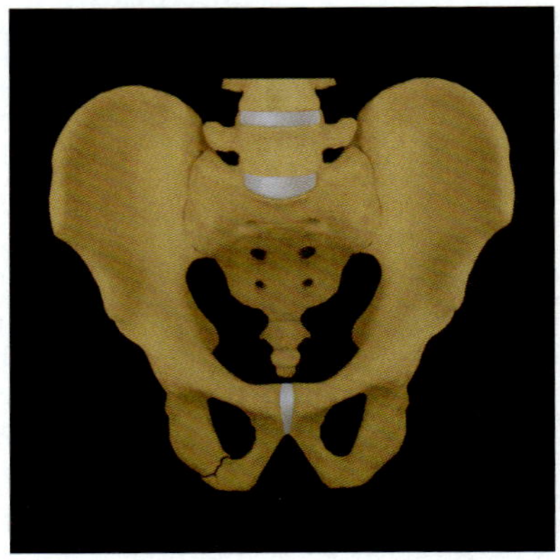

Transverse fractures through the sacrum or coccyx are considered stable, but associated vertical fracture components through the sacral ala, often seen in insufficiency fractures, must be excluded.

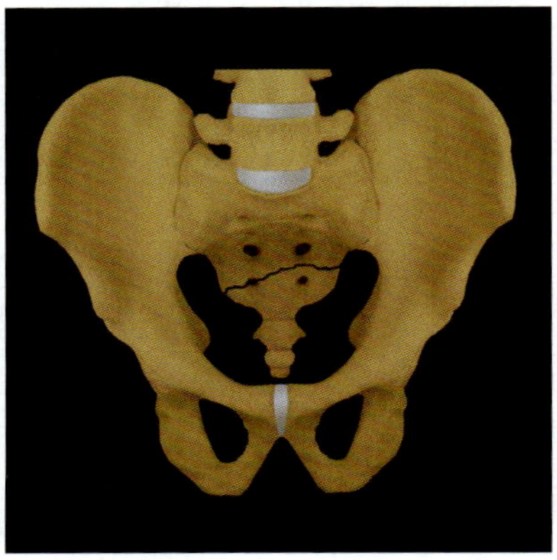

Fractures of ipsilateral superior and inferior pubic ramus fractures are stable fractures seen commonly with osteoporosis. Care should always be taken to exclude ipsilateral sacroiliac shear injuries that would change the classification to an unstable Malgaigne fracture.

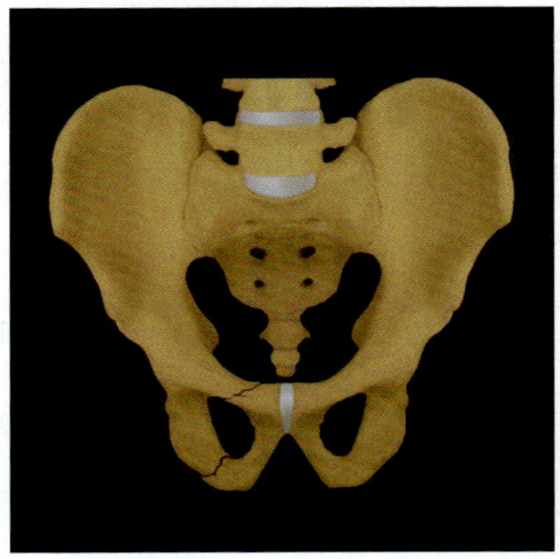

Common stable pelvic fractures include isolated iliac wing or pubic ramus fractures and transverse fractures of the sacrum and coccyx. Stable pelvic fractures can usually be treated conservatively. Unstable pelvic fractures involve the posterior pelvic ring and usually require surgical stabilization.

Staging, Grading, & Classification

- Goal of classification is to determine stability and corrective forces needed for realignment of ring
- Orthopedic Trauma Association (OTA) classification
  - Type A (stable): Lesion sparing (or with no displacement of) posterior arch
    - A1: Avulsion fracture of innominate bone
    - A2: Fracture of innominate bone involving rami or iliac wing (Duverney)
    - A3: Transverse fracture/dislocation of sacrum/coccyx
  - Type B (partially stable): Incomplete disruption of posterior arch (typically posterior SI ligaments intact)
  - Type C (unstable): Complete disruption of posterior arch (sacrum, SI ligaments, or posterior ilium)

CLINICAL ISSUES

Demographics

- Epidemiology
  - Pelvic fractures account for 3% of all fractures and 2% of orthopedic hospital admissions

Natural History & Prognosis

- Stable pelvic fractures with < 5 mm displacement typically recover in 6-8 weeks
  - Associated soft tissue injury may require longer healing time
- Displaced pubic ramus fractures have poorer prognosis due to associated posterior arch, vascular, and GU injuries
  - GU injuries (bladder rupture &/or urethral tear) occur with ~ 1/3 of bilateral ramus fractures
- Open fractures involving perineum, vagina, or bowel have up to 50% mortality
- Transverse sacral fracture above S3 → up to 30% associated sacral plexus injury

Treatment

- Conservative for uncomplicated stable pelvic fractures
- Unstable pelvic fractures commonly require ORIF

DIAGNOSTIC CHECKLIST

Consider

- CT to ensure no unstable pelvic ring fracture
- Watch for associated soft tissue injury with high-velocity trauma

Image Interpretation Pearls

- Consider inlet/outlet, Judet views or CT

SELECTED REFERENCES

1. Hak DJ et al: Sacral fractures: current strategies in diagnosis and management. Orthopedics. 32(10), 2009

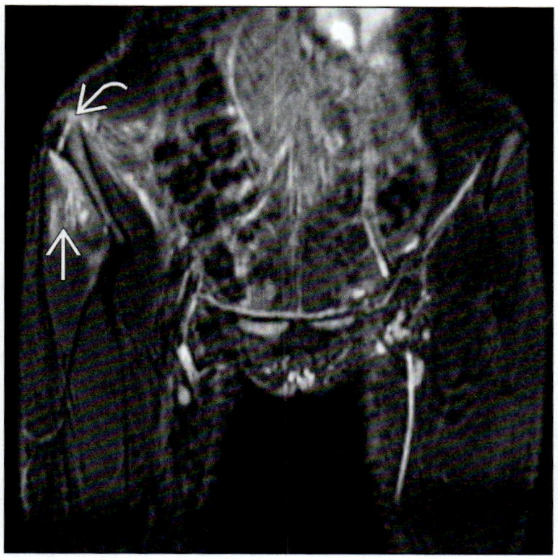

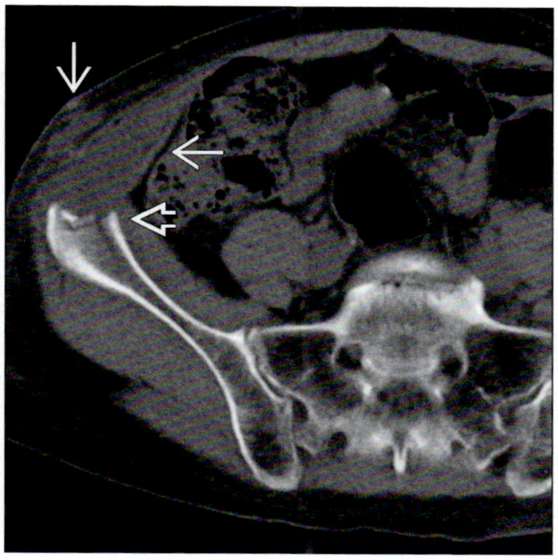

(Left) *Coronal STIR MR of the pelvis in a football player with direct anterolateral trauma shows an isolated fracture at the anterior superior iliac spine ➔, likely in part related to tendinous avulsion ➔, and therefore an OTA A1 fracture.* ***(Right)*** *Axial bone CT shows deformity of the anterior iliac wing consistent with a Duverney fracture ➔. Soft tissue injury ➔ and a penetrating trauma history make this a stable but open pelvic fracture requiring irrigation and debridement.*

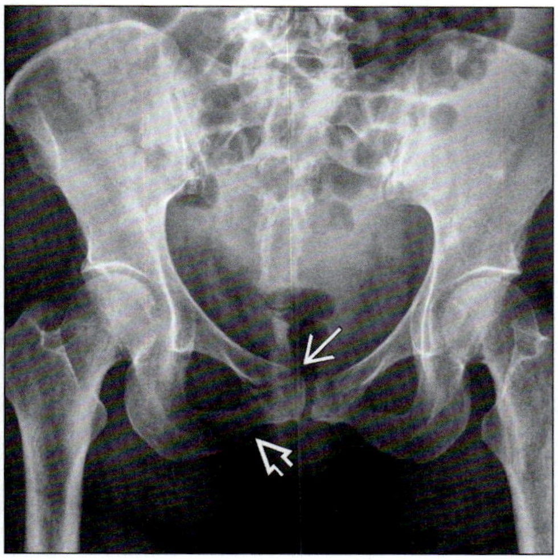

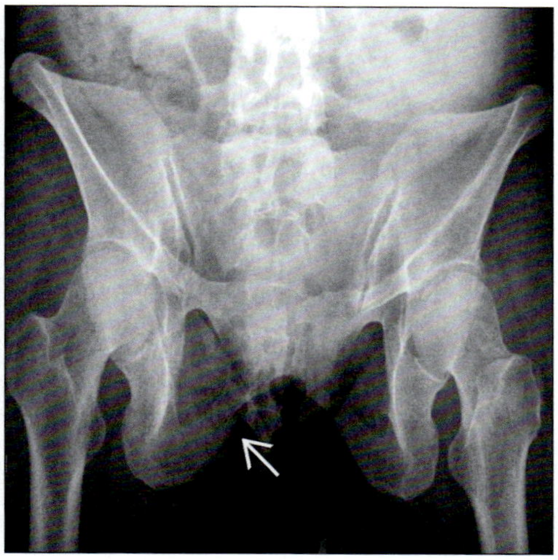

(Left) *Standard AP radiograph of the pelvis from an MVA patient shows a horizontal fracture of the superior pubic ramus ➔ and a vertical fracture of the inferior pubic ramus ➔.* ***(Right)*** *AP oblique radiograph of the pelvis with the tube angled 45° cephalad (outlet view) nicely demonstrates the sacrum, sacroiliac joints, and the inferior pubic rami, where there is a stable fracture ➔. Angled and oblique views of the pelvis can be helpful in delineating the extent of osseous injury.*

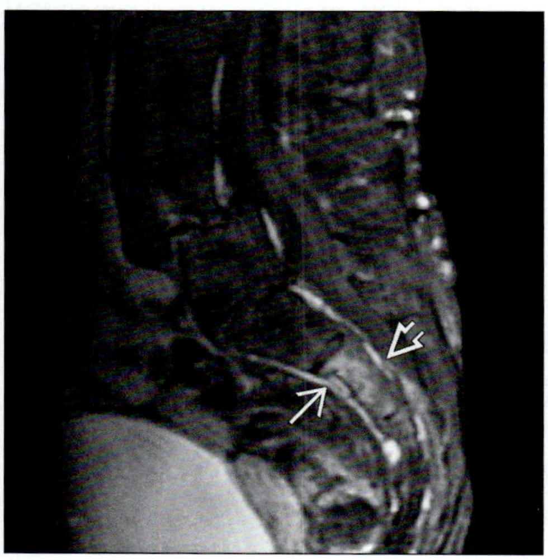

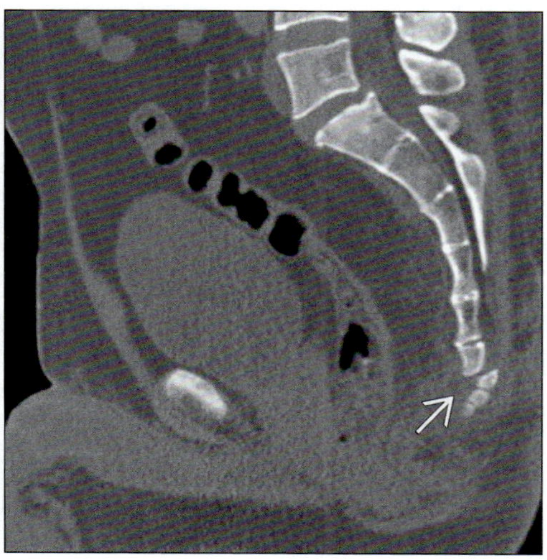

(Left) *Sagittal T2WI FS MR from an osteoporotic patient with severe low back and buttock pain after a low-velocity fall shows marrow edema throughout the S2 segment ➔ and buckling of the anterior cortex ➔, indicating a traumatic sacral fracture. This is an OTA A3 injury.* ***(Right)*** *Sagittal reformat bone CT from a young patient with coccydynia after a high-velocity fall while playing soccer demonstrates a displaced fracture between the 2nd and 3rd segments of the coccyx ➔.*

PELVIC FRACTURE, UNSTABLE

Key Facts

Terminology
- Partially stable fracture: Posterior integrity partially maintained; rotationally unstable, vertically stable
- Unstable fracture: Complete loss of posterior integrity; rotationally and vertically unstable

Imaging
- Posterior arch disruption (sacrum, posterior ilium, SI joint)
- Displaced (> 5 mm) pubic rami fractures or pubic symphysis diastasis
- Anterior-posterior compression
  - Pubic symphysis and SI joint
- Lateral compression
  - Transverse pubic ramus fractures and posterior fracture or dislocation
- Vertical shear
  - Vertical fractures of both pubic rami and posterior disruption
- CECT in trauma setting to evaluate osseous and visceral injury
- Coronal, sagittal, & 3D reformats of pelvis

Pathology
- Neurological, long bone, thoracic, visceral, vascular injury
- GU complications
- Young-Burgess classification
  - Focuses on degree of injury and direction of force
- Tile classification
  - Focuses on pelvic stability

Clinical Issues
- Classification guides external fixation and compression therapies

__(Left)__ AP graphic shows a vertical shear (Tile type C or Malgaigne) pelvis fracture. While this particular injury has bilateral pubic ramus fractures, it is the vertical orientation of these fractures and disruption of the right SI joint that indicates a vertical shear mechanism. __(Right)__ Axial graphic shows a completely unstable pelvic ring fracture with bilateral disruption of the SI joints and the sacrotuberous ligaments and symphyseal diastasis in this Tile type C3 injury.

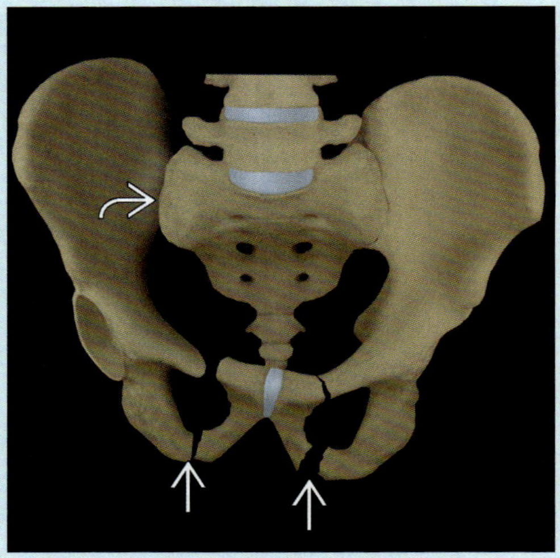

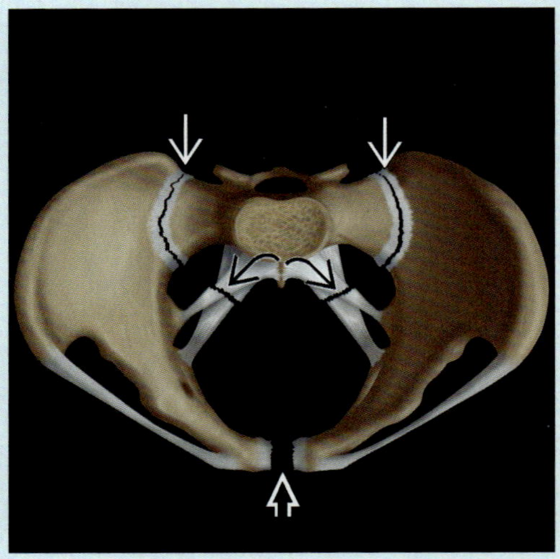

__(Left)__ Axial graphic of a unilateral, completely unstable Tile type C1.2 fracture shows malalignment at the pubic symphysis and complete disruption of ipsilateral sacroiliac and sacrotuberous ligaments. __(Right)__ Axial graphic of a partially unstable pelvic ring fracture shows fracture of the posterior ileum and disruption of the pubic symphysis, but the sacrotuberous ligaments remain intact in this Tile type B2.3 injury with rotational instability but vertical stability.

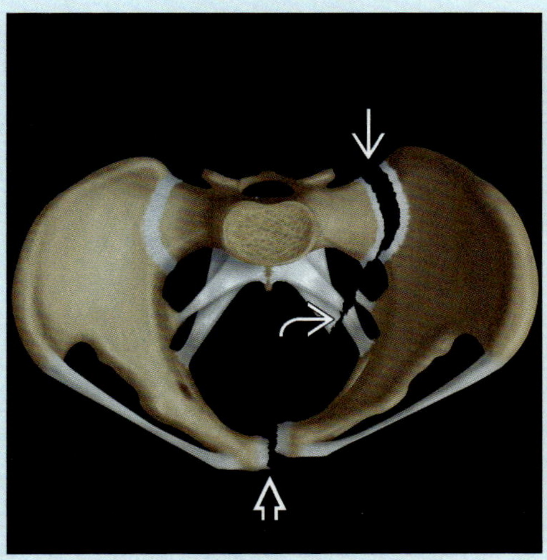

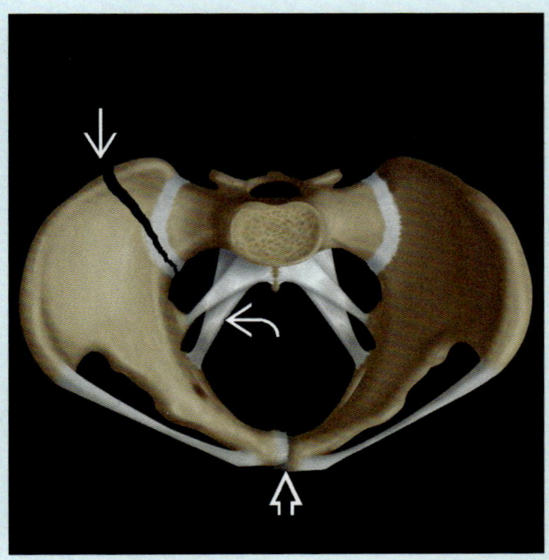

PELVIC FRACTURE, UNSTABLE

TERMINOLOGY

Synonyms

- Tile type B and C fractures

Definitions

- Partially stable fracture: Posterior pelvic integrity partially maintained; rotationally unstable, vertically stable
- Unstable fracture: Complete loss of posterior pelvic integrity; rotationally and vertically unstable

IMAGING

General Features

- Best diagnostic clue
  - Posterior arch disruption (sacrum, posterior ilium, sacroiliac [SI] joint)
  - Displaced (> 5 mm) pubic rami fractures or pubic symphysis diastasis
- Morphology
  - Anterior-posterior compression
    - Pubic symphysis and SI joint
    - Open book pelvis: Diastasis of SI joints and pubic symphysis
  - Lateral compression
    - Transverse pubic ramus fracture with posterior arch disruption
    - Most common pelvic fracture
    - Windswept pelvis
      - Bilateral posterior arch disruption, 1 side externally & 1 side internally rotated
  - Vertical shear
    - Vertical pubic ramus fx and posterior disruption
    - a.k.a. Malgaigne fracture
  - Bucket handle
    - Posterior arch disruption with contralateral pubic fracture
  - Crescent
    - Sacroiliac dislocation with fracture of iliac wing

Radiographic Findings

- Pubic symphysis widened > 1 cm after trauma
- Horizontal/transverse ramus fractures with medial rotation of hemipelvis in lateral compression
- Vertical/sagittal ramus fractures with lateral rotation of hemipelvis in AP compression
- Vertical/sagittal ramus fractures with vertical hemipelvis disruption & vertical shear

CT Findings

- CECT
  - Visceral or arterial injury → contrast blush
- Bone CT
  - Posterior arch & anterior ring fractures
- CTA
  - Major vascular injuries

Angiographic Findings

- Traumatic dissection, pseudoaneurysm, thrombosis, or vasospasm of injured artery

Other Modality Findings

- Retrograde urethrogram
  - Pelvic fracture: Posterior (membranous, prostatic) urethral injury
  - Straddle trauma: Anterior (bulbous) urethral injury

Imaging Recommendations

- Best imaging tool
  - CECT in trauma for osseous and visceral injury
  - In trauma bay: AP, inlet and outlet views
    - Inlet view: 40-45° caudal tilt shows anteroposterior and rotational displacement
    - Outlet view: 40-45° cephalad tilt shows vertical displacement and sacral fractures
- Protocol advice
  - Trauma protocol CT
    - Coronal, sagittal, & 3D reformats of pelvis

MR Findings

- Once stable, can be useful for visceral & vascular injury

DIFFERENTIAL DIAGNOSIS

Pelvic Fractures, Stable

- Tile type A

Stress and Avulsion Fractures

- No disruption of posterior ring

Acetabular Fractures

- Can be concomitant with pelvic ring fractures

PATHOLOGY

General Features

- Etiology
  - High-velocity trauma
    - MVA, motorcycle accident, fall from height
- Associated abnormalities
  - Neurological, long bone, thoracic, visceral, vascular
    - Closed head injury: 51%; splenic injury: 10%; long bone fracture: 48%
  - Genitourinary (GU) complications
    - With displaced ramus fractures, straddle fractures, pubic symphysis diastasis
  - Infectious complications
    - Open pelvic fracture involving perineum, vagina, or bowel: Up to 50% mortality
    - Open pelvic fracture not involving perineum, vagina, or bowel → < 5% mortality
  - Pelvic arterial structures at risk
    - Internal iliac artery and branches in sacroiliac region
    - Obturator and internal pudendal arteries in pubic ramus region

Staging, Grading, & Classification

- Young-Burgess classification
  - Focuses on degree of injury and direction of force
  - Lateral compression (LC): Transverse ramus fractures plus
    - I: Sacral impaction fracture on side of impact (stable)
    - II: Iliac wing fracture extending through ring (crescent fracture) on side of impact (partial to unstable)

- III: Type I or II injury on side of impact with contralateral open-book injury = "windswept" pelvis (partial to unstable)
- Anterior-posterior compression (APC): Symphyseal diastasis or vertical ramus fractures +
  - I: Symphyseal diastasis < 2.5 cm, ± anterior SI joint diastasis; felt to represent stretched but intact SI ligaments (stable)
  - II: Symphyseal diastasis > 2.5 cm, anterior SI joint diastasis; felt to represent torn anterior SI ligaments but intact posterior SI ligaments (partially stable)
  - III: Complete disruption of posterior arch including posterior SI ligaments (unstable)
- Vertical shear (VS): Symphyseal diastasis or vertical ramus fractures with complete disruption of posterior arch and vertical displacement (unstable)
- Combined mechanism injury (CMI): Any combination of above injuries

- Tile classification
  - Focuses on pelvic stability
  - A: Posterior arch intact (stable)
    - A1: Pelvic ring avulsion fracture
    - A2: Direct blow pelvic ring fracture
    - A3: Transverse sacral fracture
  - B: Partial posterior arch disruption (rotationally unstable, vertically stable)
    - B1: Anterior-posterior compression, unilateral
      - B1.1: Pubic and anterior SI diastasis
      - B1.2: Pubic diastasis and sacral fracture
    - B2: Lateral compression
      - B2.1: Anterior sacral buckle
      - B2.2: Partial SI fracture/subluxation
      - B2.3: Incomplete posterior iliac fracture
    - B3: Lateral compression with bilateral involvement
      - B3.1: Anterior-posterior compression, bilateral
      - B3.2: Anterior-posterior and lateral compression
      - B3.3: Bilateral lateral compression
  - C: Complete disruption of posterior arch (rotationally and vertically unstable)
    - C1: Vertical shear
      - C1.1: Displaced iliac fracture
      - C1.2: SI fracture/dislocation
    - C2: Vertical shear and anterior-posterior or lateral compression
    - C3: Bilateral vertical shear
- Denis classification of sacral fractures
  - Associates fracture with risk (%) of neurologic injury
    - Zone I: Alar fracture lateral to sacral foramina (6%)
    - Zone II: Transforaminal fracture (28%)
    - Zone III: Central fracture medial to sacral foramina (56%)

CLINICAL ISSUES

Presentation

- Most common signs/symptoms
  - Clinical assessment of stability: AP and lateral compression; leg length discrepancy
  - Indicators of severe visceral hemorrhage
  - Signs of GU injury
  - Neurologic complication
  - Internal open pelvic fracture

Demographics

- Epidemiology
  - Fracture patterns in order frequency
    - Lateral compression > anterior-posterior compression > combined mechanism injury > vertical shear

Natural History & Prognosis

- Mortality rate = 5-50% depending on severity of fracture, bleeding, and associated injuries

Treatment

- Classification guides external fixation and compression therapies
  - External fixation decreases mortality associated with hemorrhage
- Associated injuries play a role in guiding therapy
  - Stable pelvic ring (Tile 1)
    - Typically conservative treatment
  - Partially stable ring (Tile 2)
    - ± external fixation initially
    - APC II: Consider anterior arch fixation only
    - LC II, III: Variable, often with anterior and posterior arch fixation
  - Unstable ring (Tile 3)
    - ± external fixation initially
    - APC III, LC III, VS, CMI: Anterior and posterior arch fixation
- Hemorrhage
  - Pelvic bleeding generally approached with combination of transfusion and
    - External fixation to decrease pelvic volume
    - Retroperitoneal packing vs. catheter angiography and embolization

DIAGNOSTIC CHECKLIST

Consider

- Immediately identify associated life-threatening injuries
- Identify open fractures as mortality increases significantly

Reporting Tips

- Classify by injury extent and mechanism
- Log evidence for associated injuries

SELECTED REFERENCES

1. Papakostidis C et al: Pelvic ring injuries with haemodynamic instability: efficacy of pelvic packing, a systematic review. Injury. 40 Suppl 4:S53-61, 2009
2. Gardner MJ et al: Stabilization of unstable pelvic fractures with supraacetabular compression external fixation. J Orthop Trauma. 21(4):269-73, 2007
3. Suzuki T et al: Long-term functional outcome after unstable pelvic ring fracture. J Trauma. 63(4):884-8, 2007
4. Demetriades D et al: Pelvic fractures: epidemiology and predictors of associated abdominal injuries and outcomes. J Am Coll Surg. 195(1):1-10, 2002

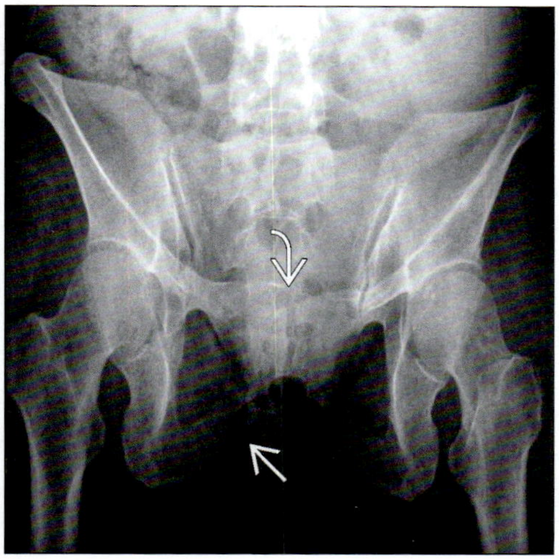

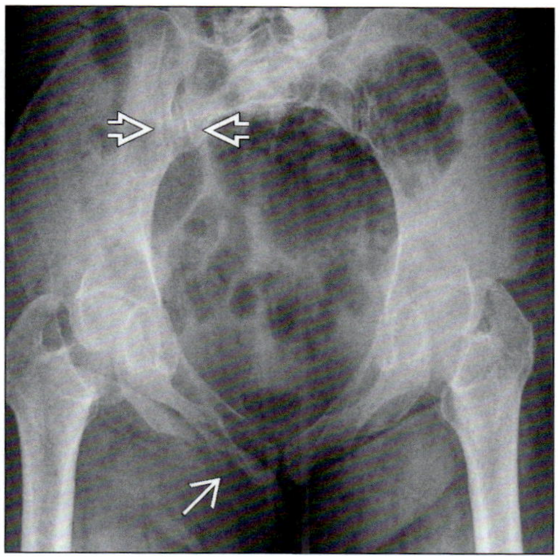

(Left) *AP outlet view radiograph shows malalignment at the pubic symphysis ➔ in this Tile type B partially unstable fracture. The outlet view is most useful for identifying vertical malalignment, but it also shows an inferior pubic ramus fracture in this case ➔.* ***(Right)*** *AP inlet view radiograph from the same patient shows fractures of both pubic rami with a transverse morphology ➔ as well as mild diastasis of the sacroiliac joint ➔. The inlet view is most useful for transverse malalignment.*

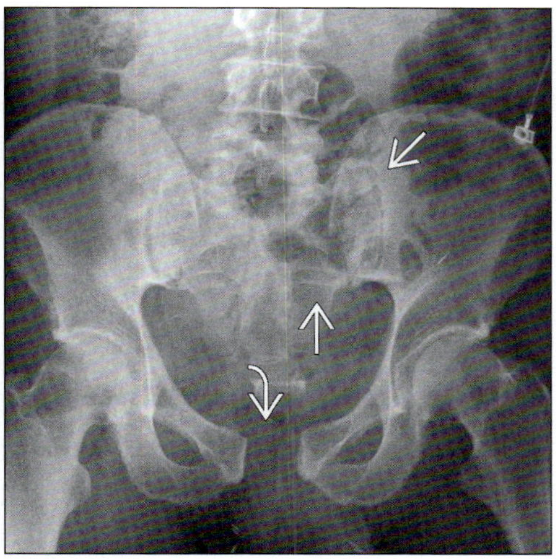

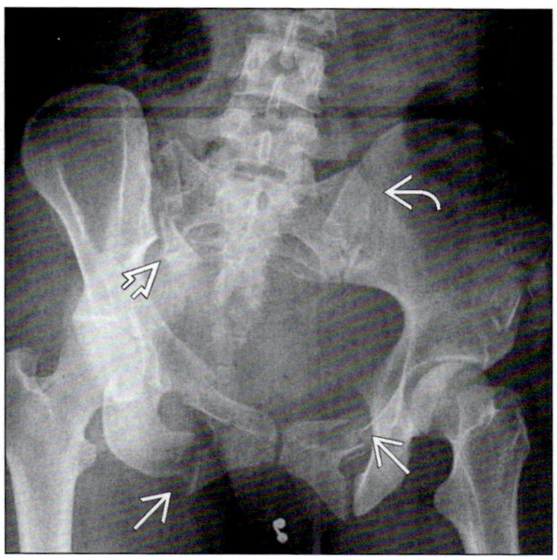

(Left) *AP radiograph of a partially stable AP compression pelvic fracture shows diastasis of the pubic symphysis ➔ and an oblique fracture of the sacral ala ➔. The posterior ring is completely disrupted in this Tile B1.2 fracture.* ***(Right)*** *AP radiograph of the pelvis obtained in the trauma bay shows bilateral pubic ramus fractures ➔, sacral fracture ➔ on the right, and sacroiliac joint disruption ➔ on the left, an unstable grade C2 vertical shear fracture.*

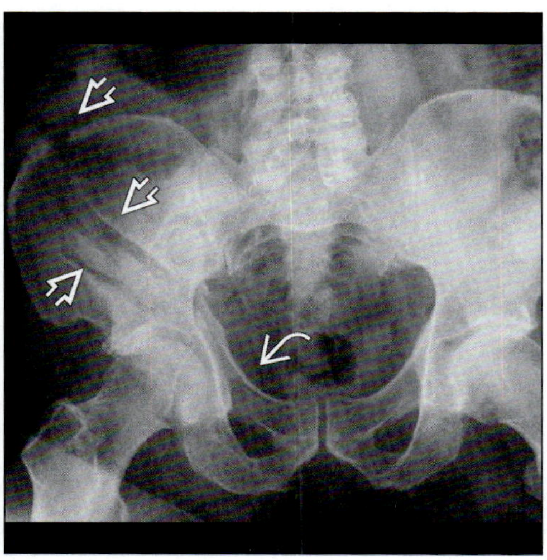

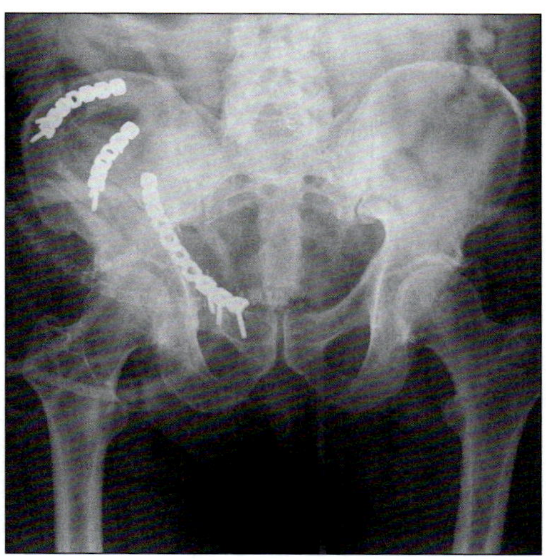

(Left) *AP radiograph from a motorcycle accident victim shows a lateral compression fracture extending through the iliac wing ➔ and a transverse fracture ➔ of the ipsilateral superior pubic ramus indicating a partially stable Tile B2.3 injury.* ***(Right)*** *AP radiograph from the same patient after surgery shows open reduction and internal fixation (ORIF) of the pubic ring fractures. Emergency surgery was performed to address active pelvic hemorrhage seen on contrast-enhanced pelvic CT.*

PELVIC FRACTURE, UNSTABLE

*(**Left**) AP radiograph of the pelvis from a fall victim after a retrograde urethrogram shows an abnormal pear-shaped morphology of the bladder, indicating pelvic hematoma. (**Right**) AP radiograph after a MVA shows bilateral posterior hip dislocations, an acetabular fracture on the left, and nondisplaced pubic ramus fractures on the right. Slight diastasis of the right sacroiliac joint makes this an unstable vertical shear Malgaigne fracture/dislocation.*

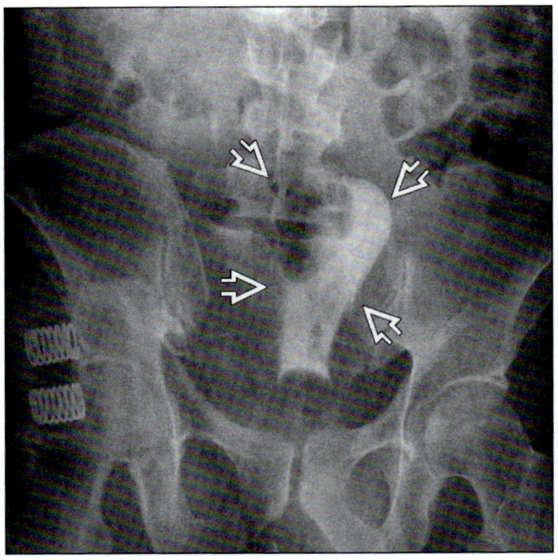

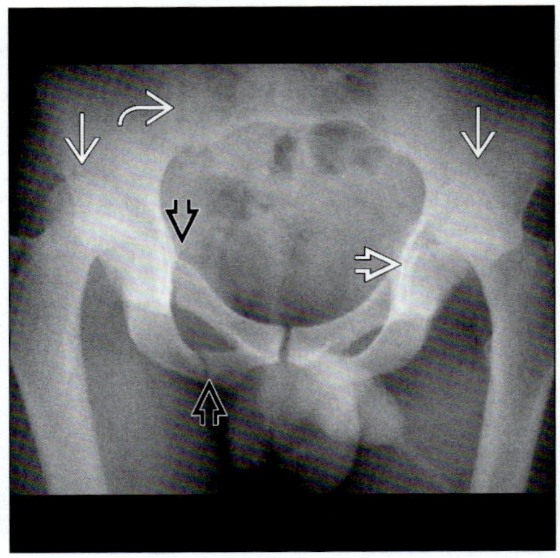

*(**Left**) AP radiograph shows complete left SI joint disruption, superior hemipelvis displacement, pubic symphysis diastasis, right pubic ramus fractures, and offset at the right SI joint, an APC Tile type C3 injury. (**Right**) Axial bone CT from the same patient better defines posterior arch injuries with internal rotation of left iliac bone (crescent fracture) and anterior right sacroiliac joint widening/external rotation. This configuration is termed a "windswept" pelvis.*

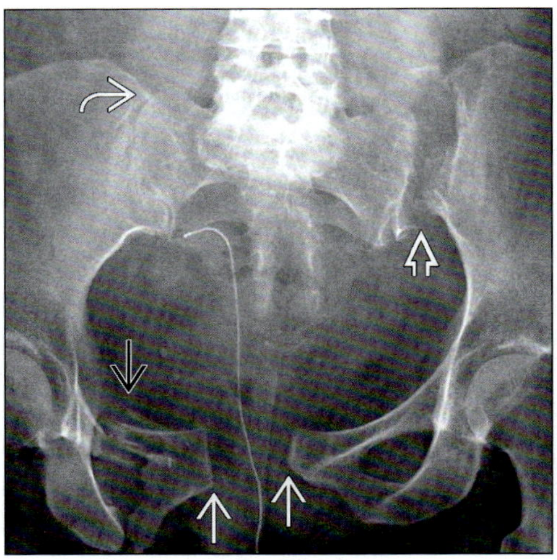

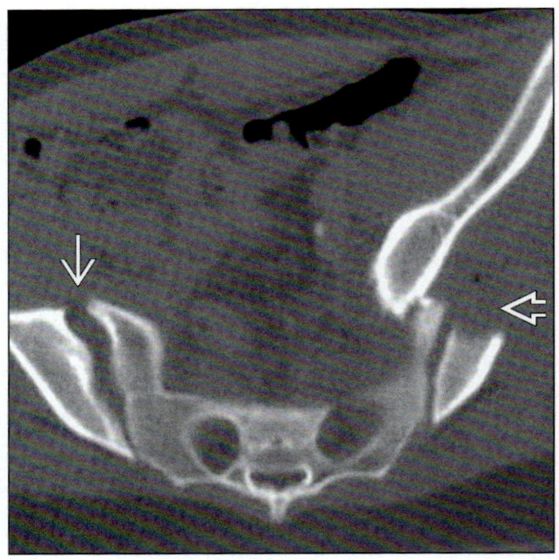

*(**Left**) AP radiograph shows an unstable Malgaigne pelvic fracture involving the anterior and posterior pelvic rings. There are also pubic ramus fractures on the right. Overlap of the femoral head and superior acetabulum was overlooked in this patient with a concomitant posterior hip dislocation. (**Right**) Coronal reformat bone CT from an MVA patient shows disruption of the sacroiliac joint. It is important to assess osseous structures of the pelvis on all trauma CTs.*

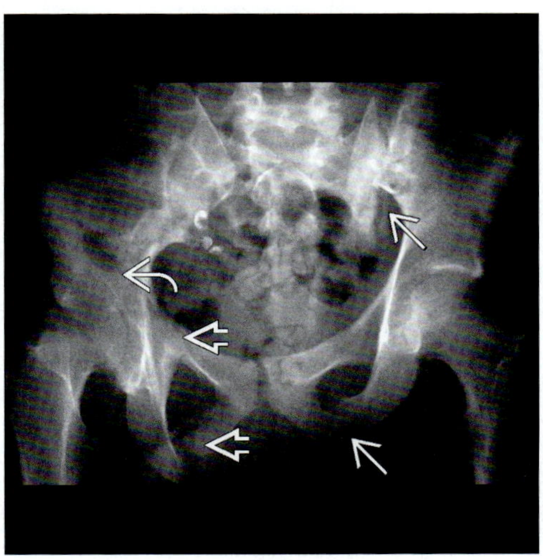

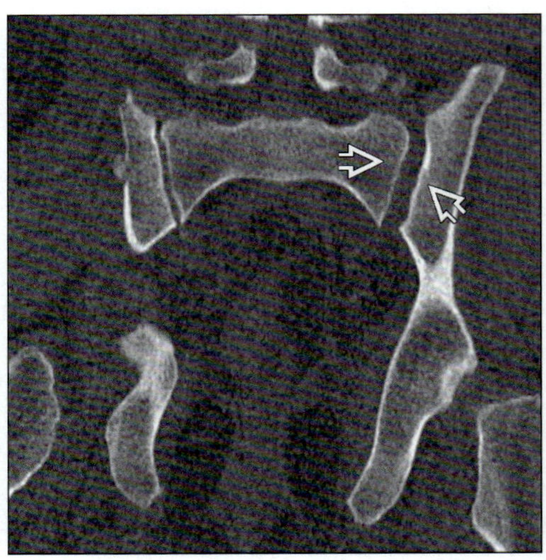

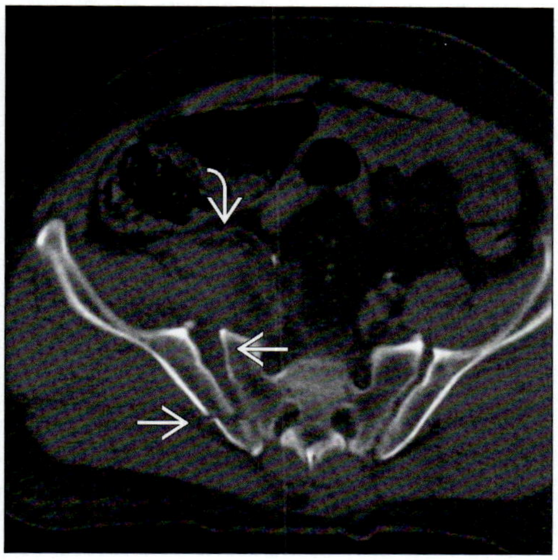

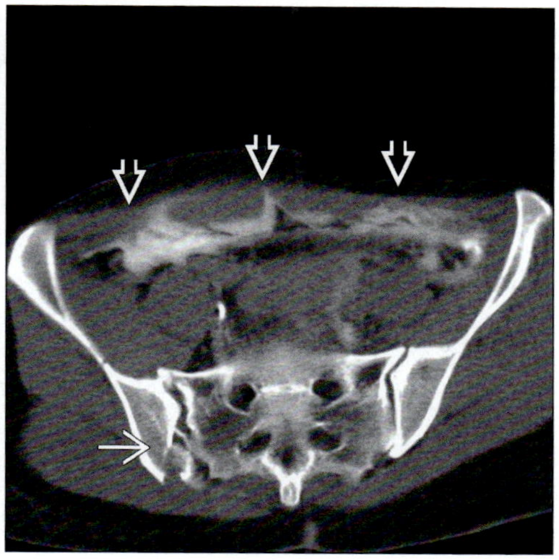

(Left) *Axial bone CT in a trauma patient shows a fracture-dislocation of the posterior arch. Asymmetry of soft tissue anterior to the right psoas muscle represents hematoma.* **(Right)** *Axial CECT from the same patient shows a blush of intraperitoneal contrast typical for active pelvic hemorrhage, along with the posterior arch fracture-dislocation. Visceral injuries, including pelvic hemorrhage and hematoma, are often associated with unstable pelvic fractures.*

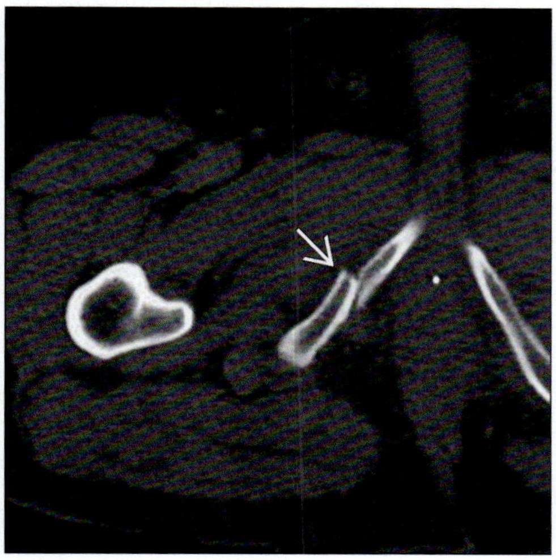

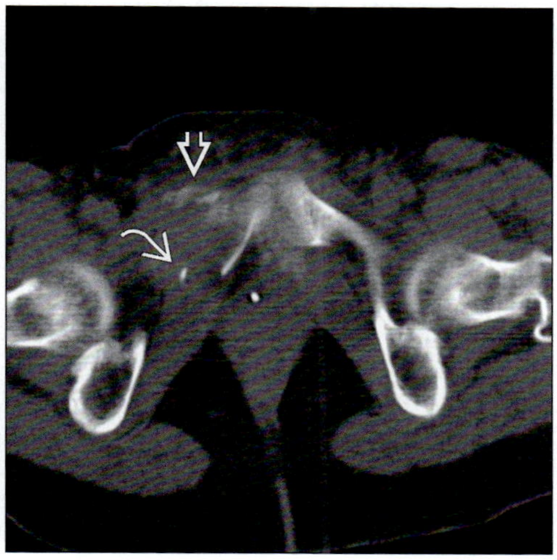

(Left) *Axial bone CT from a trauma patient shows a mildly displaced fracture of the inferior pubic ramus. There was posterior arch disruption and the patient's hemoglobin was dropping without a clear source of hemorrhage.* **(Right)** *Axial post-contrast CT from the same patient shows active contrast extravasation in the region of the internal pudendal artery and a blush of hemorrhage within the adductor muscle compartment. This patient underwent catheter embolization.*

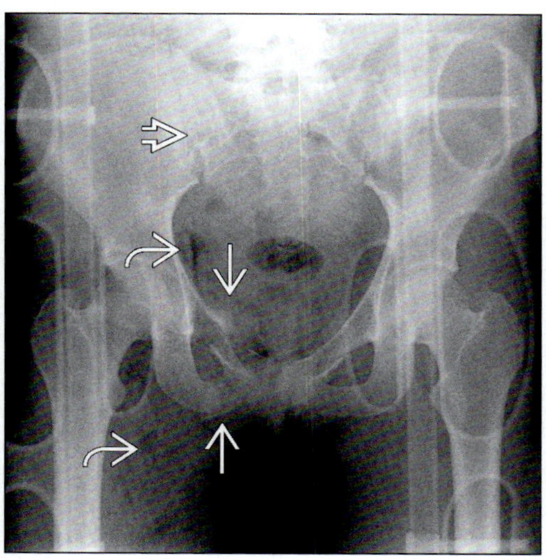

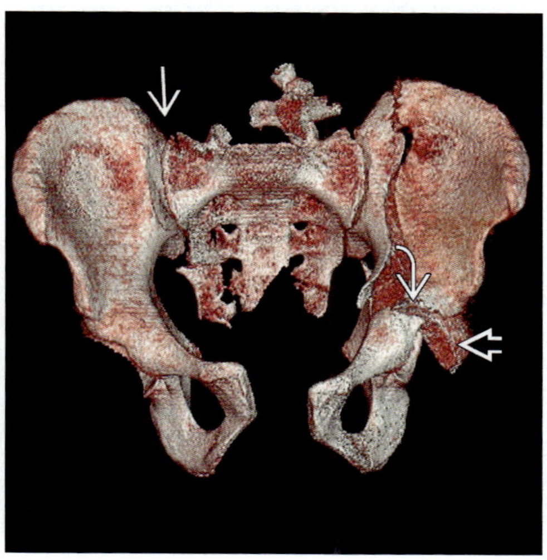

(Left) *AP radiograph obtained with patient on a trauma board demonstrates sacral and obturator fractures on the right. Gas collections seen in the deep pelvic fat and groin indicate open fractures.* **(Right)** *Anterior 3D reformat bone CT delineates the relationship between fracture fragments. While the right sacroiliac joint is disrupted, there is no vertical displacement of the right hemipelvis. The left iliac fracture extends to a displaced ischiopubic ramus fragment.*

PELVIC STRESS AND INSUFFICIENCY FRACTURE

Key Facts

Terminology
- Transient osteoporosis of hip (TOH)
- Transient bone marrow edema syndrome (TBME)
- Fatigue fracture: Trabecular and cortical disruption due to repetitive trauma
- Insufficiency fracture: Trabecular and cortical disruption from repetitive trauma and weakened bone

Imaging
- Sclerosis or bone marrow edema in typical locations
  - Sacral ala
  - Inferior > superior obturator rami
  - Superior acetabulum at lateral cortex
  - Subchondral femoral head
- Often occult on radiographs: MR yields diagnosis
- TOH (subchondral insufficiency fracture)
  - Intense marrow edema in femoral head ± extension into femoral neck
  - Small central subchondral focus of low signal on all sequences = fracture site
- Bone scan: Sensitive but not specific

Top Differential Diagnoses
- Metastatic disease
- Osteitis pubis
- Osteonecrosis of femoral head

Clinical Issues
- Most fractures heal with rest

Diagnostic Checklist
- Look for contralateral fracture or ipsilateral fracture in other ramus
- Intense marrow edema without fracture line treated as fracture

*(**Left**) AP graphic shows the typical configuration of sacral insufficiency fractures of elderly osteoporotic patients. Bilateral vertical alar fractures are often bridged by a horizontal component (most often around the S2 level), forming an H-shaped configuration. This has been termed the "Honda" sign. (**Right**) Coronal T2WI FS MR from a patient with sacral insufficiency fractures demonstrates extensive marrow edema around the vertical and horizontal components.*

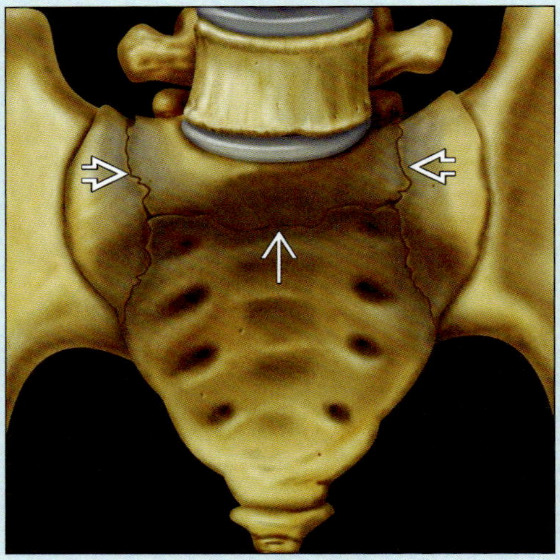

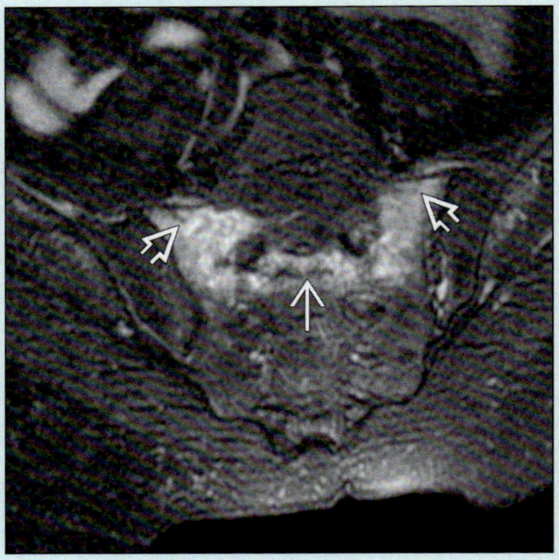

*(**Left**) AP graphic shows the typical pattern of insufficiency fractures of the superior and inferior obturator (pubic) rami. These fractures usually occur in conjunction with an ipsilateral or bilateral sacral fracture. (**Right**) AP radiograph of the pubic symphysis in an 81-year-old female with severe groin pain demonstrates superior and inferior pubic ramus insufficiency fractures with callus bone indicating early healing. There is offset at the pubic symphysis and osteitis pubis.*

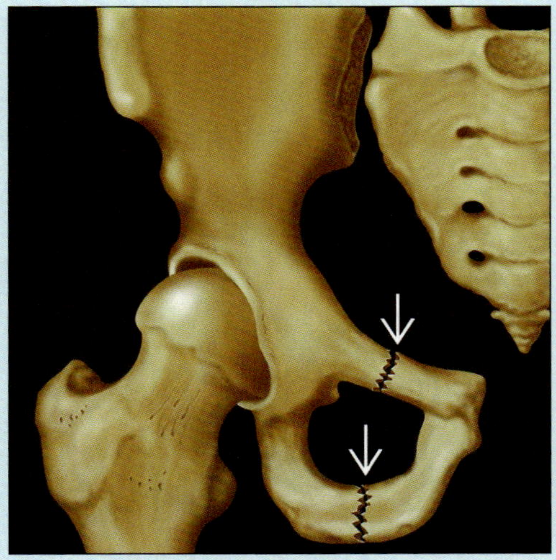

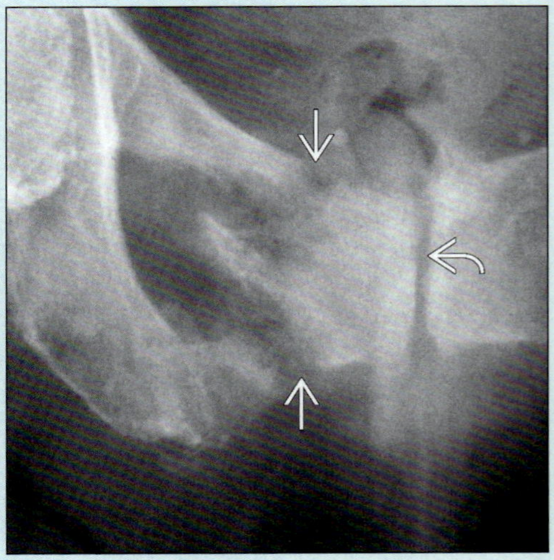

PELVIC STRESS AND INSUFFICIENCY FRACTURE

TERMINOLOGY

Abbreviations

- Transient osteoporosis of hip (TOH)
- Transient bone marrow edema syndrome (TBME)

Synonyms

- Pelvic fatigue fracture, overuse fracture, insufficiency fracture

Definitions

- Fatigue fracture: Trabecular and cortical disruption due to repetitive trauma
- Insufficiency fracture: Trabecular and cortical disruption from repetitive trauma and weakened bone
- Obturator rami = pubic rami
- TOH = TBME

IMAGING

General Features

- Best diagnostic clue
  - Sclerosis or bone marrow edema in typical locations
- Location
  - Sacral ala
  - Superior and inferior obturator rami
  - Superolateral acetabulum
  - Subchondral femoral head
  - Medial femoral neck
- Size
  - Sacral and ramus fractures may be small and nondisplaced or large and displaced
  - Supraacetabular and femoral head fractures usually small
    - Focal fracture in femoral head may be subtle
- Morphology
  - Parallel to biomechanical stress in obturator rami and sacral ala
  - Orthogonal to biomechanical stress in femoral head and neck and acetabulum

Radiographic Findings

- Radiography
  - Cortical thickening and sclerosis in typical locations
    - Sacral ala
      - May be unilateral or bilateral
      - Often occult on radiographs due to osteopenia and overlying bowel content
      - Often associated with ipsilateral obturator ramus fractures
    - Obturator rami
      - Inferior > superior, but often involves both
      - Often associated with ipsilateral sacral fracture
      - Bilateral parasymphysial in superior pubic rami
    - Superior acetabulum at lateral cortex
    - Femoral neck medial cortex
  - Femoral head
    - Diffuse osteopenia with focus of subchondral sclerosis
      - Usually in weight-bearing portion
      - ± subchondral collapse
    - TOH: Focal osteopenia of proximal femur with hip pain
  - Subacute radiographic findings
    - Progression to complete fracture with linear lucency and surrounding sclerosis
    - Can heal with exuberant callus and cortical enlargement

CT Findings

- Similar to radiographic findings
  - Can be helpful to assess extent of cortical disruption
- More sensitive than radiographs for sacral and femoral head fractures
  - May only see small cortical discontinuity in sacrum
    - Especially anterosuperiorly

MR Findings

- Fracture usually ↓ signal on all sequences with surrounding marrow edema
- May only see marrow edema without fracture line (stress reaction)
- Periostitis and surrounding soft tissue edema
- Most will have osseous stress response in another location in pelvis
  - Always look at sacrum in setting of obturator stress fracture
- Sacral insufficiency fracture
  - Vertical orientation
  - May involve 1 or both ala
  - Often with bridging transverse component in upper sacrum ("Honda" sign on coronal images)
  - May involve neural foramina
- TOH (subchondral insufficiency fracture)
  - Intense marrow edema in femoral head ± extension into femoral neck
  - Small central subchondral focus of low signal on all sequences = fracture site
  - Often associated joint effusion
  - Little to no marrow edema in acetabulum

Nuclear Medicine Findings

- Bone scan
  - Increased uptake on delayed-phase images
    - Sensitive but not specific
  - Typical locations and patterns of uptake similar to marrow edema on MR
    - H-shaped sacral fracture = "Honda" sign
  - Positive 2-3 weeks before radiographs
    - May be falsely negative in 1st 72 hours
  - Increased uptake persists for months in healing fractures

Imaging Recommendations

- Best imaging tool
  - Radiographs followed by MR if equivocal
- Protocol advice
  - Include whole pelvis for work-up of hip pain in elderly

DIFFERENTIAL DIAGNOSIS

Metastatic Disease

- Cortical and trabecular destruction
- Atypical locations for stress fracture

Osteitis Pubis

- Subchondral marrow edema
- Often osseous productive change

PELVIC STRESS AND INSUFFICIENCY FRACTURE

Osteonecrosis of Femoral Head
- Subchondral focus of low signal curves away from articular surface rather than parallel to it
- Normal marrow signal centrally surrounded by edema

Sacroiliitis
- Marrow edema on both sides of sacroiliac joint
- Subchondral osseous erosion

Osteomyelitis
- Overlying skin defect to bone
- Cortical destruction with medullary edema
- Different patient population

PATHOLOGY

General Features
- Etiology
  - Fatigue fracture: Repetitive loading of normal bone
    - Single event of trauma below threshold to cause fracture
  - Insufficiency fracture: Normal forces applied to abnormal, weakened bone
    - Osteoporosis most common
    - Bone can weaken from many metabolic conditions
      - Rheumatoid arthritis, radiation osteitis, renal osteodystrophy
- Vertebral compression fractures
  - Especially lumbar

Staging, Grading, & Classification
- Complete vs. incomplete fracture
  - Incomplete = extends to only 1 cortical surface
- Osseous stress response = focal bone marrow edema without fracture

CLINICAL ISSUES

Presentation
- Most common signs/symptoms
  - Pubic ramus fracture
    - Groin pain
    - Point tenderness at inferior pubic ramus
    - Standing sign: Unable to stand unsupported on affected leg
  - Acetabular stress fracture: Anterior hip pain
  - Sacral insufficiency fracture: Low back pain
  - TOH (TBME)
    - Relatively abrupt onset of hip pain without significant trauma
    - Classically described in middle-aged men & in women in 3rd trimester of pregnancy
    - Usually self-limited time course
      - Imaging findings may lag behind clinical changes
- Clinical profile
  - Runners
  - Elderly osteoporotic patients ± fall history
  - Transient bone marrow edema syndrome/transient osteoporosis of hip
    - Elderly patients, M > F
    - Pregnant women

Demographics
- Age
  - Stress fracture: Young adult
  - Insufficiency fracture: Older population (> 55 years)
- Gender
  - M < F
- Epidemiology
  - Insufficiency fracture incidence: 1% of females > 55 years old

Natural History & Prognosis
- Most fractures heal with rest
- Healing process slow: 4 weeks to 5 months
- Noncompliance can lead to nonunion
- Complications (due to immobility)
  - Deep venous thrombosis
  - Decubitus ulcer formation

Treatment
- Conservative
  - Rest
  - Cessation of training regimen
  - Nonunion: Most heal with rest
- Percutaneous sacroplasty
  - Injection of polymethyl methacrylate with CT or fluoroscopic guidance
- Percutaneous transiliosacral screw stabilization
- Surgical
  - Rare except nonunion or displaced fracture

DIAGNOSTIC CHECKLIST

Consider
- Complete vs. incomplete fracture
- Look for contralateral stress fracture
- Look for ipsilateral fractures in sacrum &/or obturator rami

Image Interpretation Pearls
- Intense marrow edema without fracture line treated as fracture

Reporting Tips
- Report evidence of healing response

SELECTED REFERENCES

1. Campbell SE et al: Imaging of stress injuries of the pelvis. Semin Musculoskelet Radiol. 12(1):62-71, 2008
2. Zajick DC et al: Spectrum of MRI findings in clinical athletic pubalgia. Semin Musculoskelet Radiol. 12(1):3-12, 2008
3. Bencardino JT et al: Imaging of hip disorders in athletes. Radiol Clin North Am. 40(2):267-87, vi-vii, 2002
4. Spitz DJ et al: Imaging of stress fractures in the athlete. Radiol Clin North Am. 40(2):313-31, 2002
5. Otte MT et al: MR imaging of supra-acetabular insufficiency fractures. Skeletal Radiol. 26(5):279-83, 1997
6. Hill PF et al: Stress fracture of the pubic ramus in female recruits. J Bone Joint Surg Br. 78(3):383-6, 1996
7. Peh WC et al: Imaging of pelvic insufficiency fractures. Radiographics. 16(2):335-48, 1996
8. Hayes CW et al: MR imaging of bone marrow edema pattern: transient osteoporosis, transient bone marrow edema syndrome, or osteonecrosis. Radiographics. 13(5):1001-11; discussion 1012, 1993

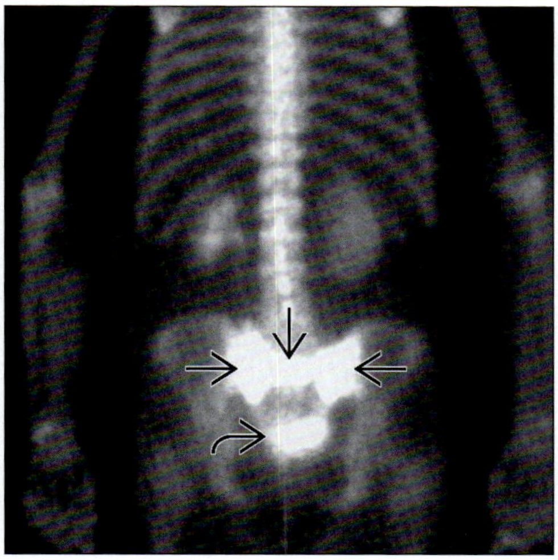

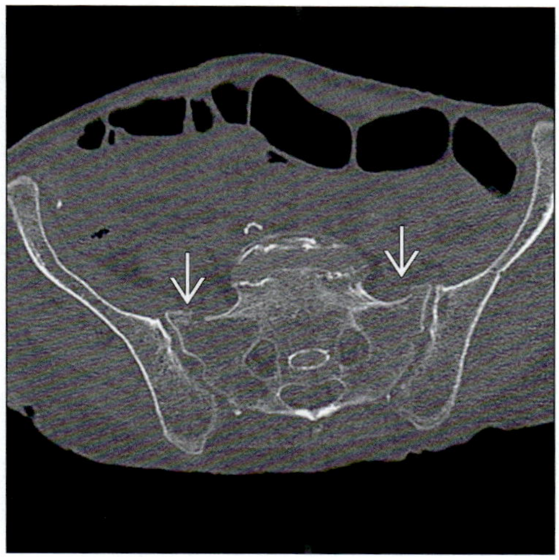

(Left) *PA projection from a bone scan shows the typical "H" configuration → of a sacral insufficiency fracture, the "Honda" sign. Focal activity more inferiorly → is in the bladder. Bone scan is sensitive for insufficiency fracture, but may be falsely negative in the 1st 72 hours after injury.* ***(Right)*** *Axial bone CT from an osteoporotic woman with back pain shows bilateral sacral insufficiency fractures →. These fractures usually begin superiorly and anteriorly in the sacrum.*

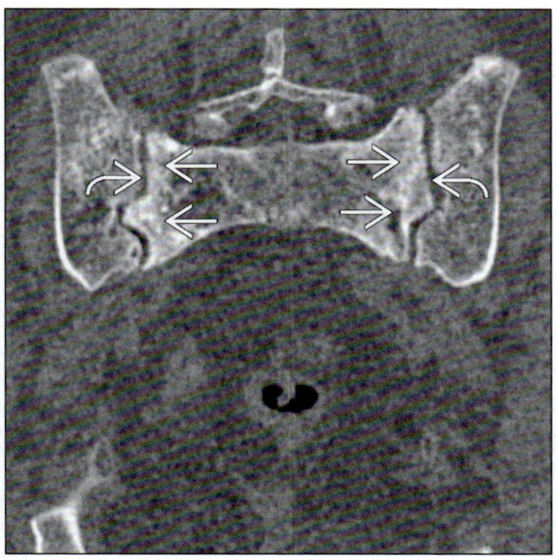

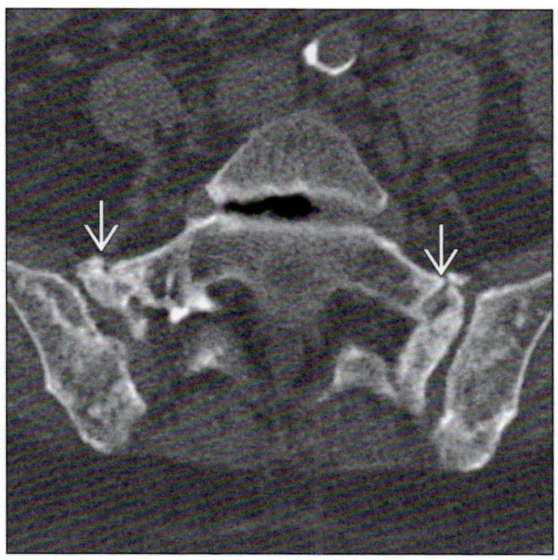

(Left) *Coronal reformat bone CT shows typical vertical bilateral sacral alar insufficiency fractures → due to osteopenia. These fractures tend to parallel the sacroiliac joints →.* ***(Right)*** *Axial bone CT from the same patient shows the bilateral sacral alar insufficiency fractures →. Sclerosis around the fractures indicates healing; patients often present with subacute insufficiency fractures as there is often not a discrete episode of trauma, but a gradual onset of pain.*

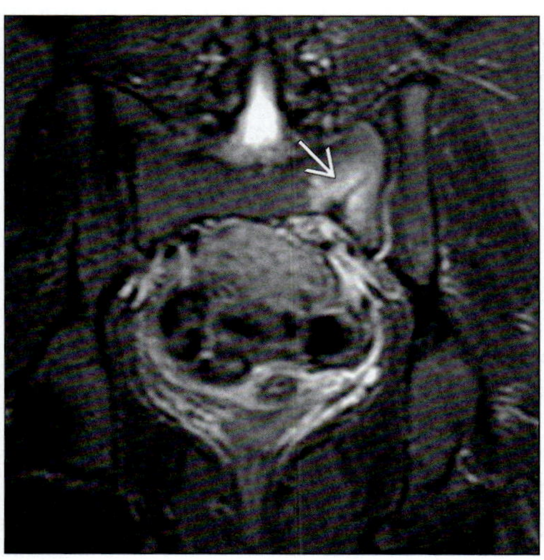

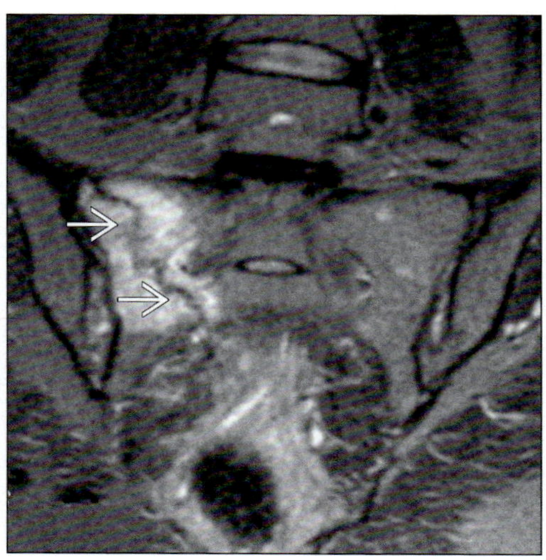

(Left) *Coronal T2WI FS MR from an 8-year-old girl with hip and low back pain related to extensive horseback riding shows an oblique stress fracture → of the sacrum. The patient improved with conservative therapy.* ***(Right)*** *Coronal T2WI FS MR shows a more extensive unilateral stress fracture → of the sacrum. Sacral stress (fatigue) fractures in young patients are often well delineated and focal, while insufficiency fractures are often bilateral and harder to delineate because of osteopenia.*

*(**Left**) Axial bone CT from an 81-year-old man with hip pain shows an acute comminuted fracture → of the base of the superior pubic ramus on the right entering the hip joint. Underlying osteopenia is present and is the most likely etiology of this insufficiency fracture. (**Right**) Axial bone CT more inferiorly from the same patient demonstrates an ipsilateral inferior ramus fracture →. This pattern of fracture should be actively sought in elderly osteoporotic patients with hip pain.*

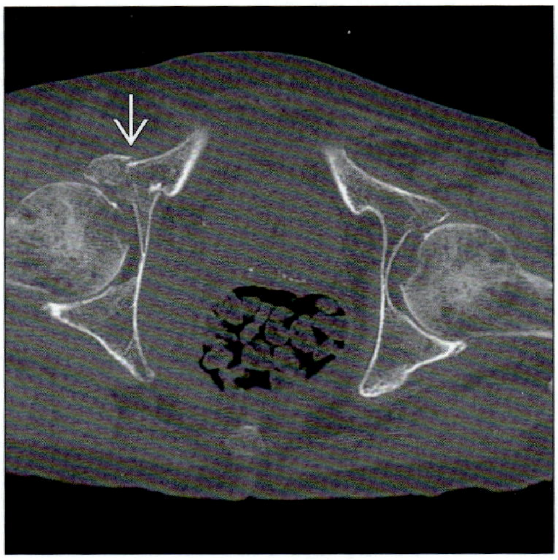

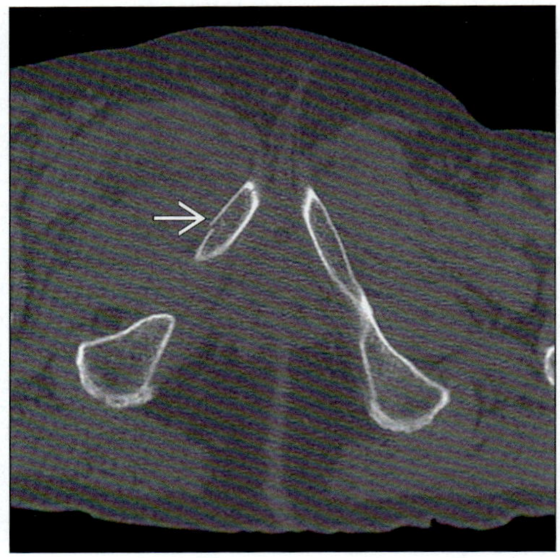

*(**Left**) Axial T2WI FS MR from a 77-year-old woman after a mild fall 1 week earlier shows marrow and soft tissue edema around an acute insufficiency fracture of the base of the pubic ramus →. Radiographs were negative. (**Right**) Axial T2WI FS MR more inferiorly in the same patient shows a nondisplaced ipsilateral inferior pubic ramus fracture →. This pattern of fracture is often associated with sacral fracture; the whole pelvis should be imaged in elderly patients with hip pain.*

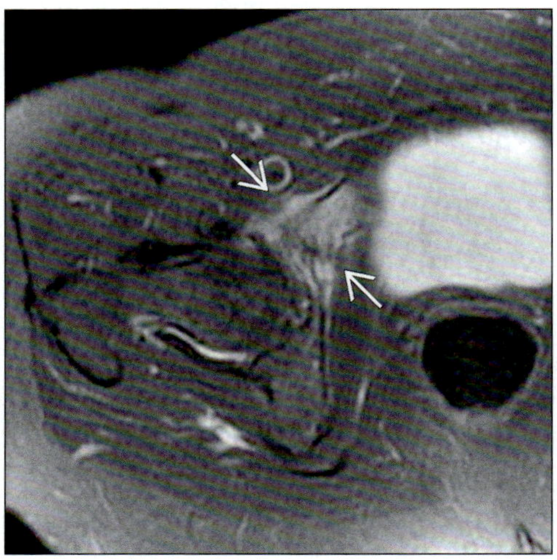

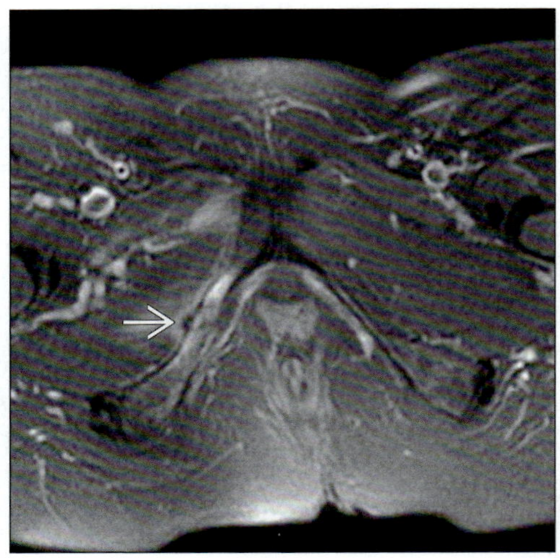

*(**Left**) Coronal T2WI FS MR shows stress edema → without a discrete fracture line in the pubic bones of a 19-year-old cross country runner with pubic pain. There is mild soft tissue swelling surrounding the pubic bones. The patient improved with rest. (**Right**) Axial T2WI FS MR from a 15-year-old soccer player with debilitating right-sided groin pain shows bone marrow edema around an incompletely ossified physis ↷ at the medial acetabulum, a Salter-Harris type I stress fracture.*

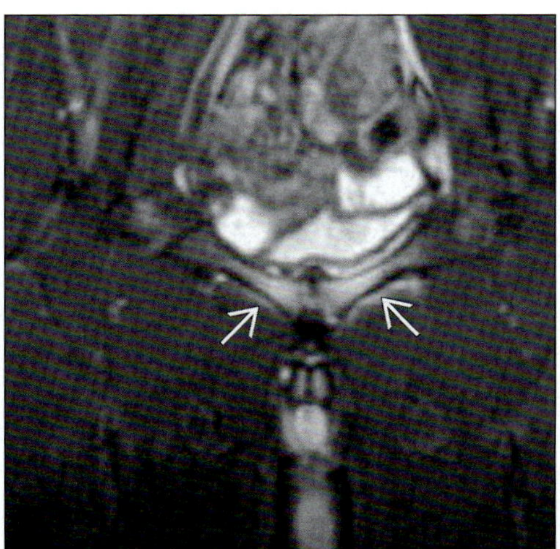

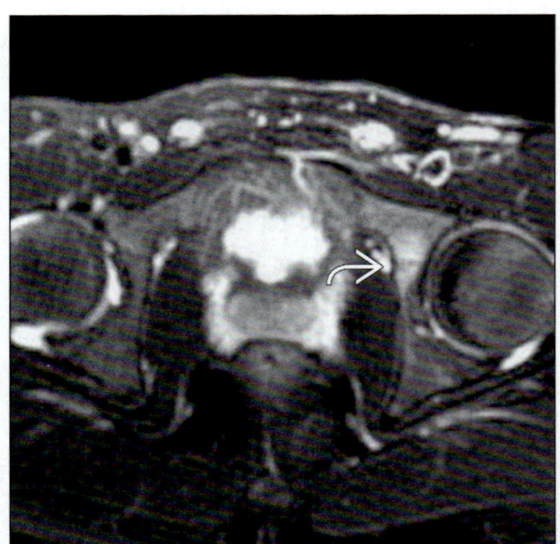

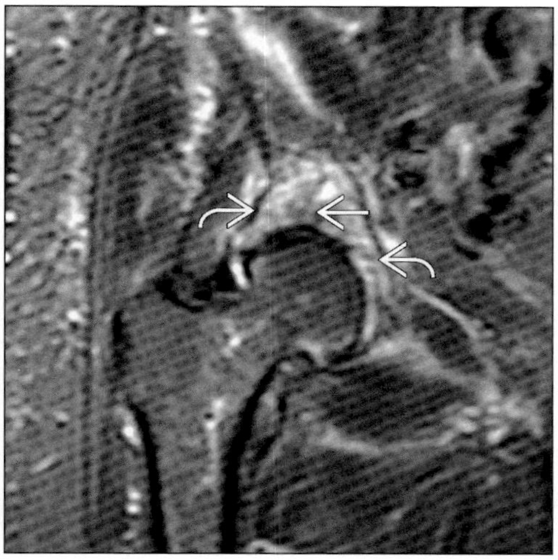

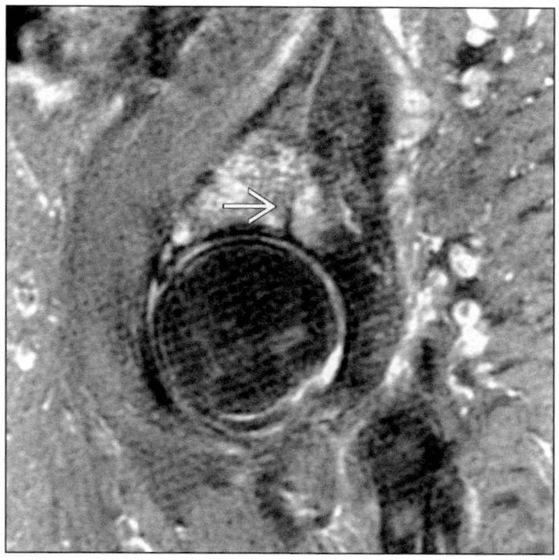

(Left) *Coronal T2WI FS MR from a 78-year-old man with sudden onset of hip pain after stepping off a curb demonstrates extensive marrow edema in the acetabulum surrounding a subtle vertical insufficiency fracture. The femoral head is normal.* ***(Right)*** *Sagittal PDWI FS MR from the same patient better illustrates the vertical acetabular insufficiency fracture, which extends into the joint. Most supraacetabular insufficiency fractures parallel the articular surface.*

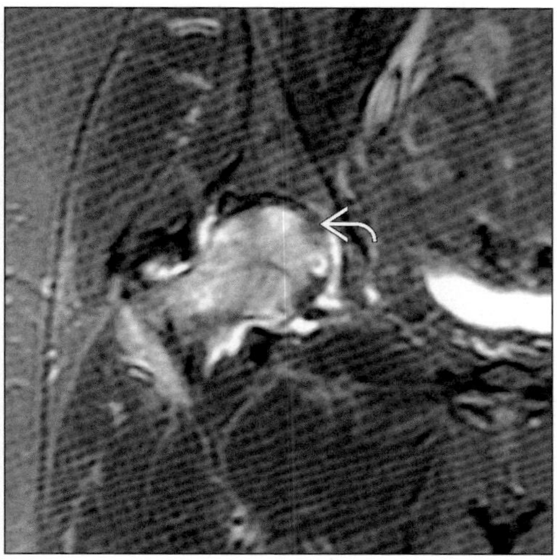

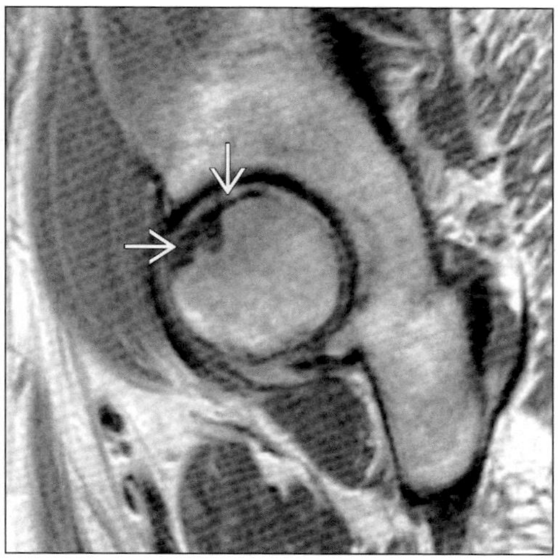

(Left) *Coronal T2WI FS MR from a 55-year-old woman shows diffuse elevated marrow signal throughout the femoral head and neck surrounding a small subchondral focus of low signal, indicating an insufficiency fracture in this elderly woman. There is a small hip effusion. Acetabular marrow signal is normal.* ***(Right)*** *Sagittal PDWI MR from the same patient shows the linear low signal subchondral insufficiency fracture paralleling the femoral articular surface.*

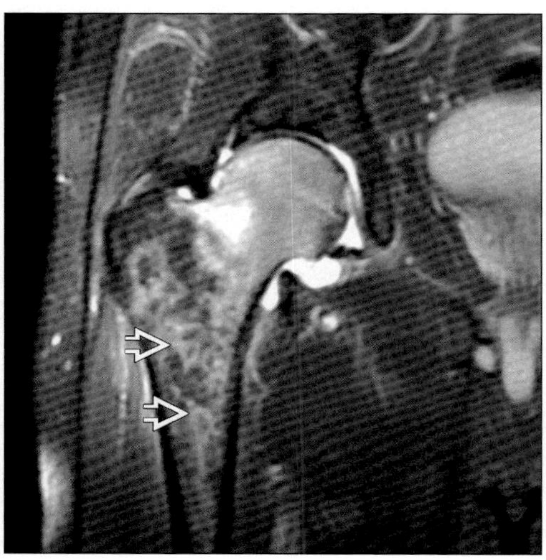

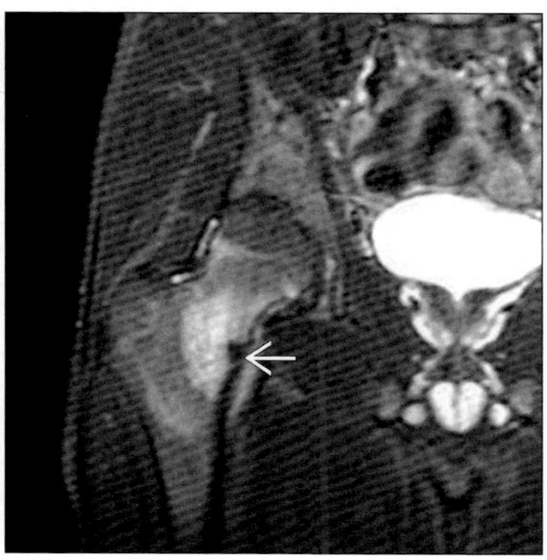

(Left) *Coronal MR arthrogram T2WI FS from a 67-year-old man shows intense marrow edema throughout the femoral head and neck with patchy edema in the proximal femoral shaft. A discrete subchondral focus of sclerosis is not identified on this image; however, the presence of such intense marrow edema should prompt a careful search for an insufficiency fracture. This is an example of TBME.* ***(Right)*** *Coronal T2WI FS MR shows a typical medial basicervical femoral neck stress fracture in a runner.*

OSTEITIS PUBIS, POST-TRAUMATIC

Key Facts

Terminology
- Subchondral osseous changes at pubic symphysis
- Acute changes of bone marrow edema
- Chronic changes of osseous resorption or osseous production

Imaging
- Often occult to radiographs
- Subchondral sclerosis in advanced cases
- Bone marrow edema on both sides of pubic symphysis on MR
- Secondary cleft
- Bone scan sensitive but not specific or descriptive

Top Differential Diagnoses
- Subenthesial/subtendinous reactive marrow edema
- Pubic ramus stress fracture
- Osteomyelitis

Pathology
- Initially, inflammatory osteitis, often from repetitive microtrauma
- Later, osteoarthritis-like picture
- Fragmentation of pubic tubercles or apophyses

Clinical Issues
- Young athletic males or females
- Postpartum females of any age
- Can progress to debilitating osseous injury

Diagnostic Checklist
- Look for rectus abdominis/adductor aponeurosis pathology
- Describe acute vs. chronic findings
- Describe extent and asymmetry
- Note any osseous resorption or stress fracture

*(**Left**) Coronal graphic illustrates the findings of osteitis pubis. Osseous edema ➡ is usually bilateral but can be asymmetric. Edema is most intense in a subchondral location, decreasing as it extends away from the symphysis. (**Right**) Coronal STIR MR from a 26-year-old female professional sprinter with severe groin pain, 18 months post partum, demonstrates intense subchondral marrow edema on both sides of the pubic symphysis ➡ without osseous erosion or osseous productive change.*

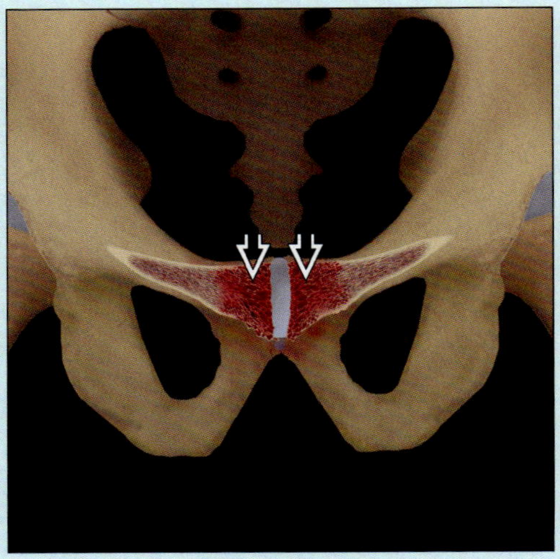

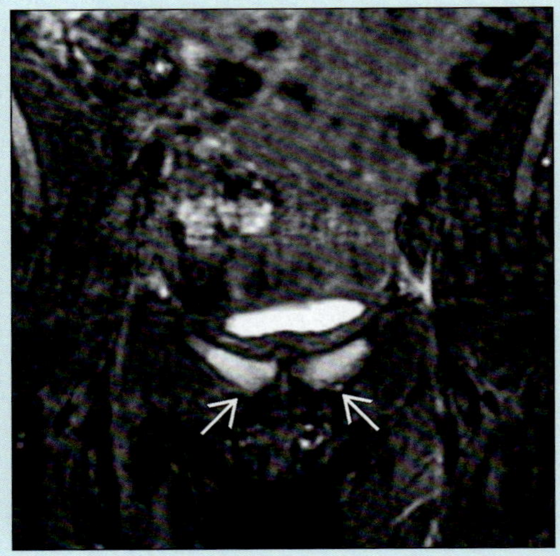

*(**Left**) Axial graphic of acute osteitis pubis shows bone marrow edema ➡ spanning the pubic symphysis joint anterior to posterior, in contrast to other sources of bone marrow edema, such as subenthesial reaction in sports hernia or linear stress fracture. (**Right**) Axial T2WI FS MR shows symmetric marrow edema spanning the subchondral symphysis pubis anterior to posterior ➡ in a pattern typical for acute osteitis pubis. Osseous edema in osteitis pubis is most prevalent in a subchondral location.*

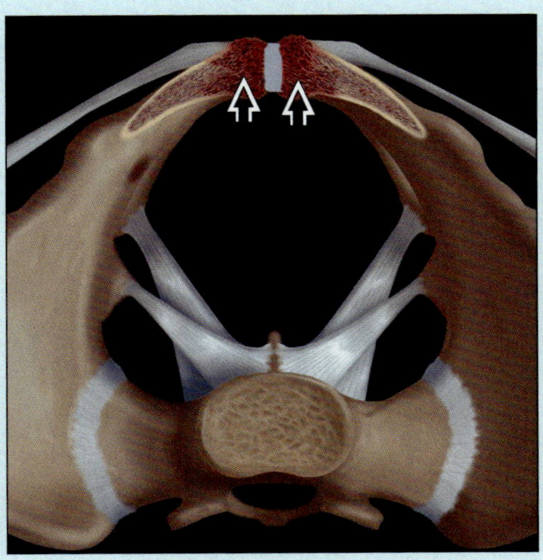

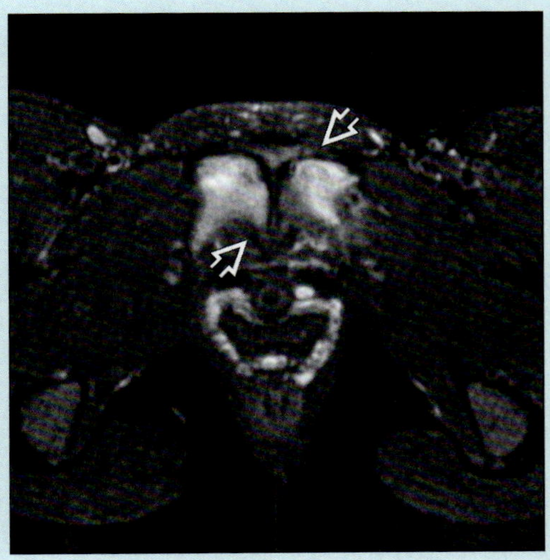

OSTEITIS PUBIS, POST-TRAUMATIC

TERMINOLOGY

Definitions

- Subchondral osseous changes at pubic symphysis
- Post-traumatic OP may result from
  - Trauma with destabilization of pubic symphysis
  - Pregnancy (destabilization 2° to ligamentous laxity)
  - Sports hernia (tear of inferior rectus abdominus, extending into origin of adductor longus)

IMAGING

General Features

- Best diagnostic clue
  - Bone marrow edema on both sides of pubic symphysis on MR

Radiographic Findings

- Acute edema occult on radiographs
- Subacute: Subchondral resorption, often unilateral
  - Associated with rectus abdominis/adductor aponeurosis lesion
- Chronic: Subchondral sclerosis, osteophytes in advanced cases

MR Findings

- T1WI
  - Bone marrow edema ± sclerosis, osteophytes, resorption
- T2WI FS
  - Bone marrow edema on both sides of symphysis pubis ± osseous resorption
  - Secondary cleft
    - T2 hyperintense curvilinear signal extending inferolaterally from primary cleft of symphysis
      - Signal ≤ fluid intensity on T2WI
    - Unilateral or bilateral

Imaging Recommendations

- Protocol advice
  - Small field-of-view images of pubic symphysis; consider angling coronal to pubic symphysis

DIFFERENTIAL DIAGNOSIS

Avulsion Injury at Pubis

- Adductor tendons originate at apophysis along inferior pubic symphysis
- Apophysis ossified and unfused from ~ age 15-25 and at risk for avulsion
- Crescentic fragment difficult to see; leaves irregular inferior pubic symphysis surface, mimicking resorption

Sports Hernia

- May be exacerbating factor in developing osteitis pubis; may also mimic osteitis pubis with marrow edema but no other osseous injury

Stress Fracture/Reaction

- ± hypointense fracture line
- Marrow edema extends a variable distance lateral to symphysis
- May be unilateral
- Can be concomitant with osteitis pubis

Osteomyelitis

- Uncommon
- Post partum, post radiation therapy, post prostatectomy, IV drug users

PATHOLOGY

General Features

- Initially inflammatory osteitis, often from repetitive microtrauma
  - Destabilization of symphysis from rectus abdominis/adductor aponeurosis injury
  - Ligamentous laxity from childbirth or pregnancy
  - Excessive athletic activity with running and twisting
- Later, develop osteoarthritis-like picture

CLINICAL ISSUES

Presentation

- Most common signs/symptoms
  - Unilateral or bilateral groin pain
    - Exacerbated by activity
      - Running and twisting mechanisms common
    - Aching "bone" pain
  - Often have history of destabilizing lesion
    - Pregnancy, childbirth
    - Rectus abdominis/adductor (sports hernia) lesions

Demographics

- Age
  - Young athletic adults
  - Postpartum females of any age
  - Elderly osteoporotic adults with asymmetric gait

Natural History & Prognosis

- Can progress to debilitating osseous injury
  - Stress fracture; subchondral osseous resorption

Treatment

- Cessation of activity
- Stabilization of underlying lesion

DIAGNOSTIC CHECKLIST

Consider

- History of athletic activity
- Recent pregnancy
- "Sports hernia" history

Image Interpretation Pearls

- Secondary cleft on MR
- Look for rectus abdominis/adductor aponeurosis pathology

Reporting Tips

- Describe acute vs. chronic findings
- Describe extent and asymmetry
- Note any osseous resorption or stress fracture

SELECTED REFERENCES

1. Omar IM et al: Athletic pubalgia and "sports hernia": optimal MR imaging technique and findings. Radiographics. 28(5):1415-38, 2008

OSTEITIS PUBIS, POST-TRAUMATIC

(Left) *AP radiograph from an American football player with persistent bilateral groin pain shows only minimal subchondral sclerosis and early resorption at the pubic symphysis that could easily be overlooked.* ***(Right)*** *AP image from a Tc-99 MDP bone scan from the same patient shows intense radiotracer uptake around the pubic symphysis in a pattern similar to the subchondral bone marrow edema seen on MR in acute osteitis pubis. Osteitis pubis is often occult on radiographs.*

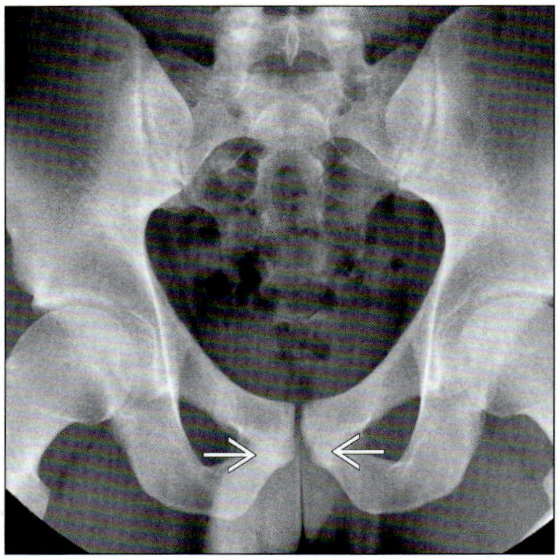

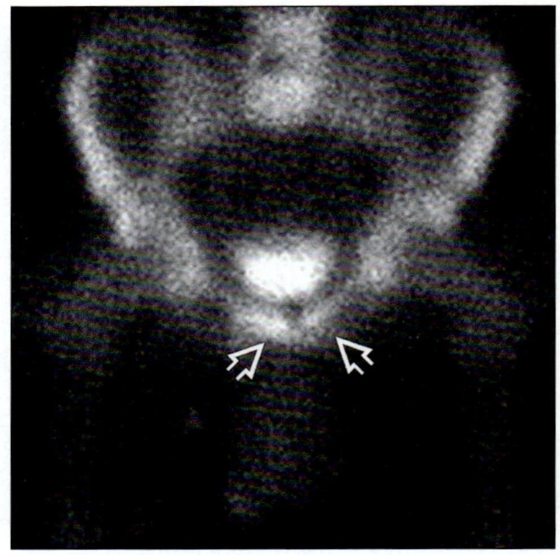

(Left) *Coronal T1WI MR shows that the subchondral bone is eroded and sclerotic in this 18-year-old soccer player who had been noncompliant to prescribed therapy for 1 year after being diagnosed with osteitis pubis.* ***(Right)*** *Axial T2WI FS MR from the same patient shows subchondral erosion as well as bilateral and asymmetric bone marrow edema and osseous productive change in a pattern mimicking an aggressive arthritis.*

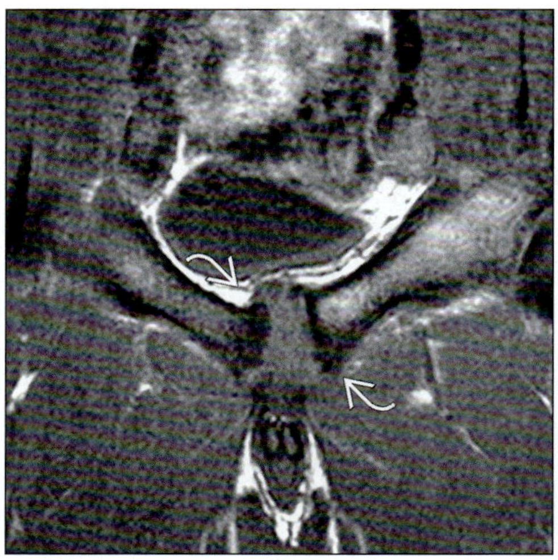

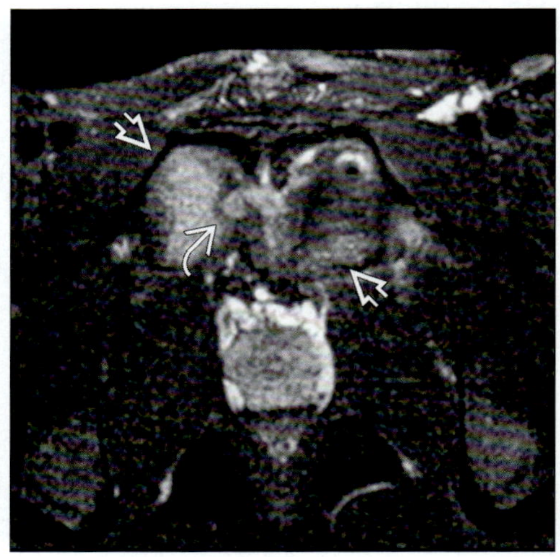

(Left) *Coronal T2WI FS MR from an athlete with left-greater than right-sided groin pain shows subchondral bone marrow edema, also greater on the left. A curvilinear fluid signal band lateral to the symphysis indicates a secondary cleft. A unilateral secondary cleft is a reliable indicator of the side of symptomatology.* ***(Right)*** *AP radiograph shows asymmetric left-sided resorption of the pubic symphysis and rami in this athlete with left groin pain.*

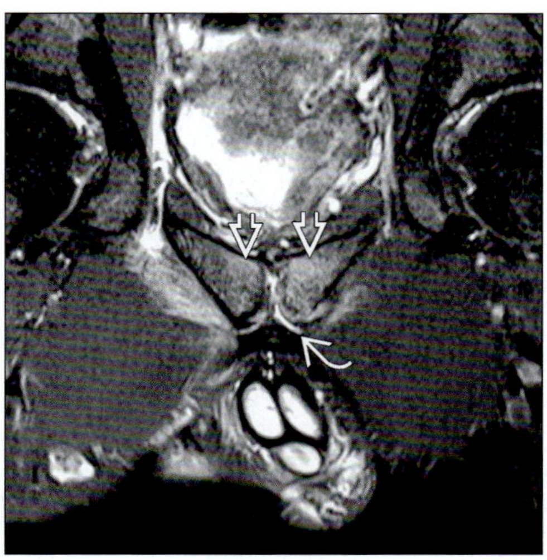

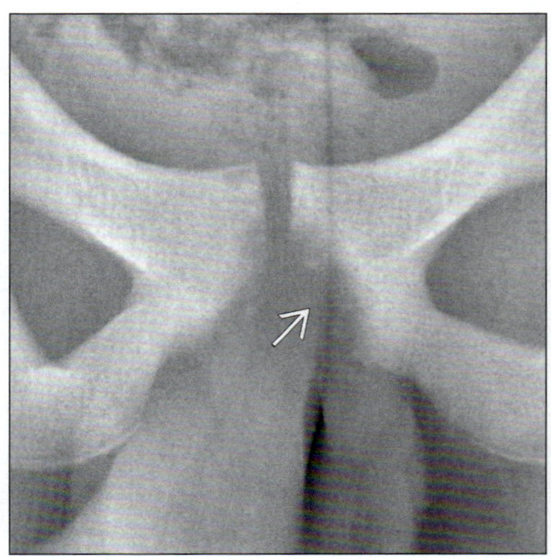

OSTEITIS PUBIS, POST-TRAUMATIC

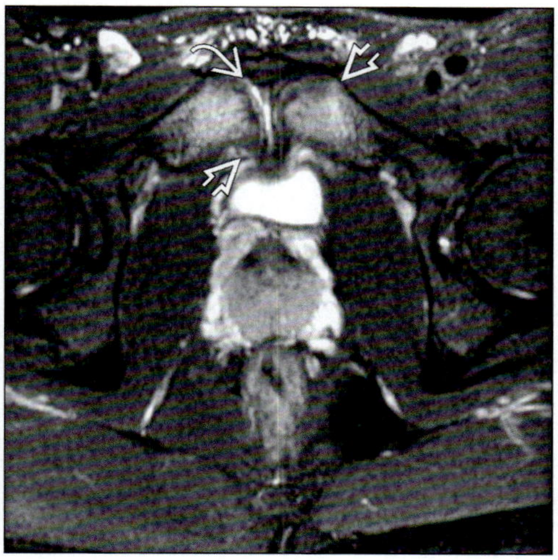

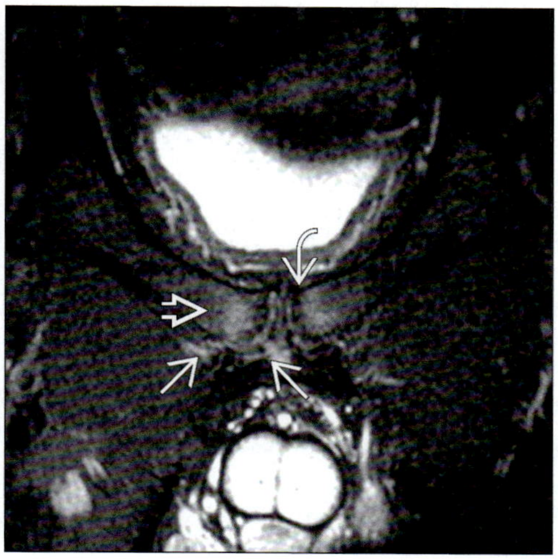

***(Left)** Axial T2WI FS MR from a hockey player with right-sided groin pain shows the osteitis pubis findings of subchondral bone marrow edema spanning the pubic symphysis joint from anterior to posterior. There is a right-sided secondary cleft. **(Right)** Coronal oblique T2WI FS MR from the same patient shows marrow edema, sclerosis, and visible detachment of the right rectus abdominis/ adductor aponeurosis, sometimes referred to as a "sports hernia."*

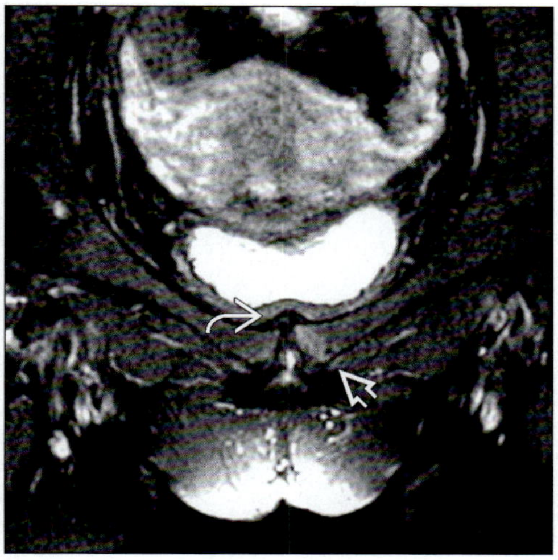

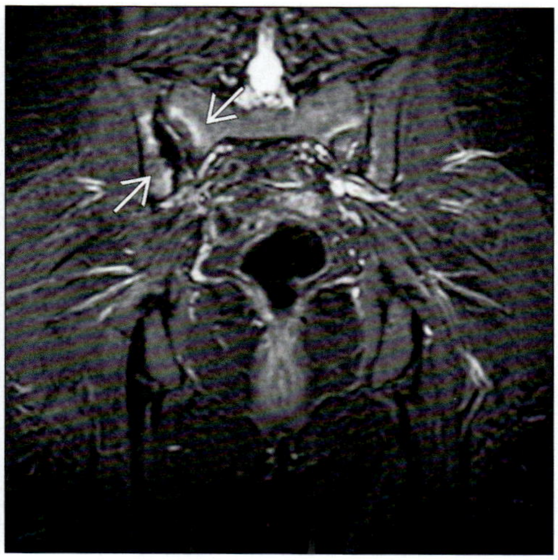

***(Left)** Coronal oblique T2WI FS MR from a 32-year-old woman with left groin & right hip pain shows asymmetric left subchondral bone marrow edema, as well as pubic symphysis capsular hypertrophy with early productive change, suggesting osteitis pubis related to a destabilizing trauma. **(Right)** Coronal STIR MR, same patient, demonstrates right sacroiliitis. Osteitis pubis and sacroiliitis are common in postpartum patients, such as this one, imaged 7 months after delivery.*

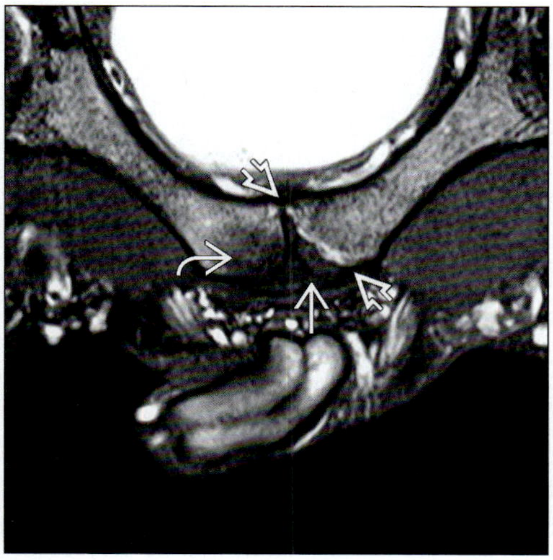

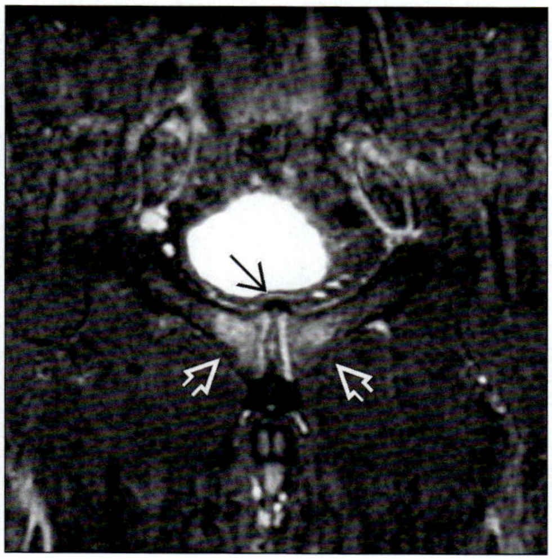

***(Left)** Coronal oblique T1WI MR from a 19-year-old lacrosse player with pubalgia shows irregularity of the left symphysis pubis margin with an ununited, abnormally hypointense left pubic tubercle apophysis and reciprocal bone marrow edema on the right. Osteitis pubis can manifest as pubic tubercle apophysitis in a young athlete. **(Right)** Coronal STIR MR shows a more chronic and mature pattern of osteitis pubis with marrow edema and osseous productive change.*

PELVIC AVULSION FRACTURE/APOPHYSITIS

Key Facts

Terminology

- Avulsion: Injury to tendon or bone at tendinous attachment due to single or repetitive trauma from myotendinous contraction

Imaging

- Thin crescentic calcific fragment at tendinous attachment on radiographs
  - Anteroposterior generally most useful radiographic view for pelvic avulsions
  - If necessary, contralateral views distinguish open apophysis from avulsion fracture
  - No radiographic abnormality seen if apophysis not yet ossified (< age 15)
- Healing response is exuberant callus
- MR: Soft tissue and osseous edema in typical location
- Location
  - Anterior superior iliac spine (ASIS): Sartorius
  - Anterior inferior iliac spine (AIIS): Rectus femoris long head
  - Inferior pubic symphysis: Adductor tendon group
  - Ischial tuberosity: Hamstring group
  - Iliac crest: Oblique abdominal muscles and tensor fascia lata

Top Differential Diagnoses

- Osteosarcoma
- Stress fracture
- Pathologic fracture: Lesser trochanter avulsion in adults is pathologic until proven otherwise

Clinical Issues

- Avulsion fractures = 13.4% of pelvic fractures in children
- Most injuries heal with conservative treatment
- Poor healing related to continued activity results in chronic pain

__(Left)__ AP graphic of the pelvis shows osseous avulsive injury of the sartorius tendon at the anterior superior iliac spine. This injury is common in kickers. __(Right)__ Coronal STIR MR from a 21-year-old lacrosse player with a sartorius origin avulsion shows asymmetric bone marrow edema at the right anterior superior iliac spine (ASIS) without fragmentation. Displacement of the fragment, if any, is best illustrated on sagittal MR images.

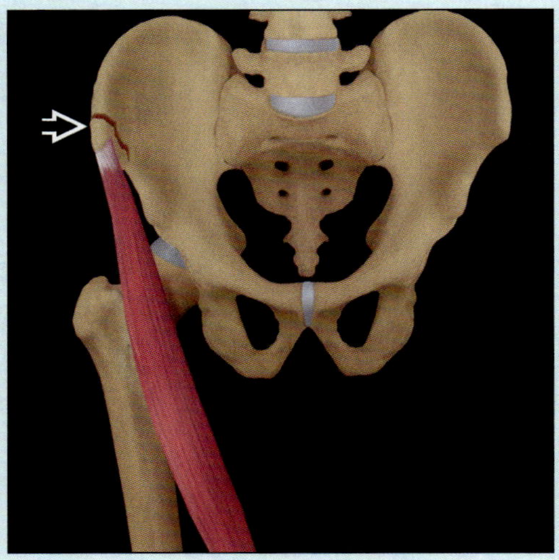

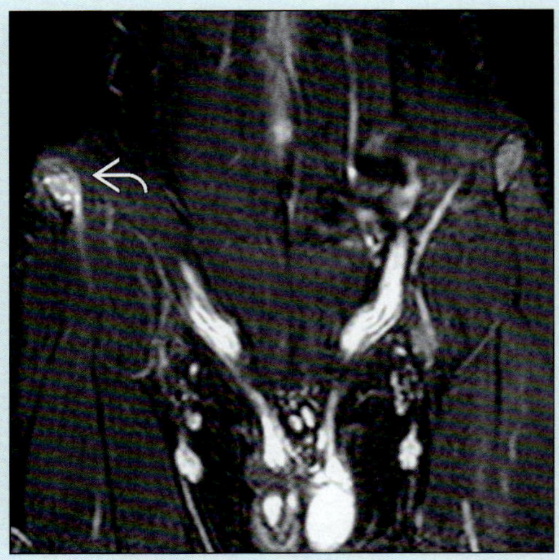

__(Left)__ PA graphic illustrates avulsion of the ischial tuberosity apophysis, the origin site of the hamstring tendons. This injury is common in hurdlers, jumpers, and dancers. __(Right)__ Coronal STIR MR from a 16-year-old female soccer player with chronic proximal right hamstring origin pain shows intense subenthesial bone marrow edema in the right ischial tuberosity as well as cortical fragmentation typical for chronic avulsion.

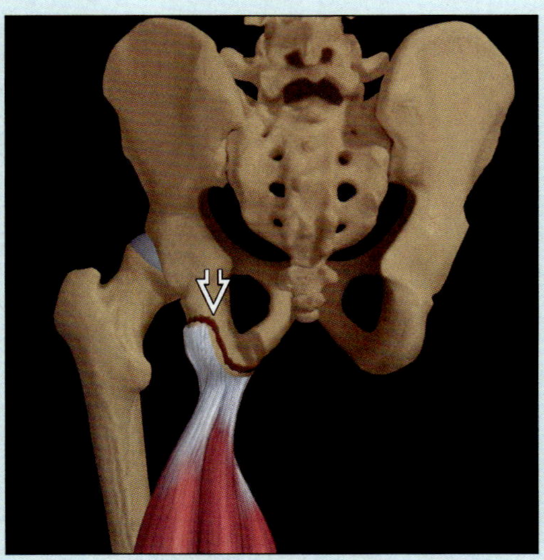

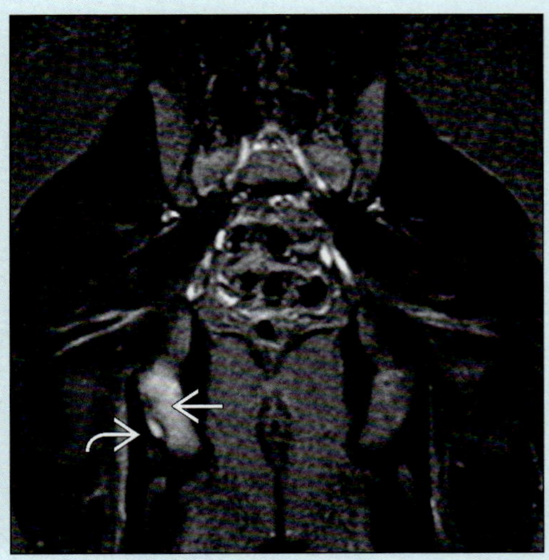

PELVIC AVULSION FRACTURE/APOPHYSITIS

TERMINOLOGY

Synonyms

- Tug injury, tug lesion, pelvic avulsion

Definitions

- Avulsion: Injury to tendon or bone at tendinous attachment due to single or repetitive trauma from myotendinous contraction
- Apophysis: Growth center at tendinous attachment
  - Does not contribute to length of bone
- Apophysitis: "Inflammatory" changes at apophysis 2° to trauma
  - Injury occurs prior to ossification of apophysis

IMAGING

General Features

- Best diagnostic clue
  - Thin crescentic calcific fragment at tendinous attachment on radiographs
  - Soft tissue and osseous edema in typical location on MR
- Location
  - Anterior superior iliac spine (ASIS): Sartorius
  - Anterior inferior iliac spine (AIIS): Rectus femoris long head
  - Inferior pubic symphysis: Adductor tendon group
  - Ischial tuberosity: Hamstring group
  - Less common avulsion sites
    - Iliac crest: Oblique abdominal muscles and tensor fascia lata
    - Greater trochanter: Gluteus medius more frequently involved than gluteus maximus and minimus
    - Lesser trochanter: Iliopsoas
- Size
  - Variable; relates to age, traumatic force, & osseous development of apophysis
- Morphology
  - Can be purely tendinous, osseous with fragment attached to tendon, or apophyseal in skeletally immature

Radiographic Findings

- Apophysitis not visualized on radiograph
  - No radiographic abnormality seen if apophysis not yet ossified (< age 15)
- Avulsions: Anteroposterior generally most useful radiographic view
  - If necessary, contralateral views distinguish open apophysis from avulsion fracture
- Osseous fragment identified adjacent to donor site
  - Can be crescentic (in profile) or discoid (en face) in shape
  - Most fragments (ASIS, ischial tuberosity, pubic ramus) are displaced < 1 cm from donor site
  - AIIS may be displaced > 1 cm from donor site (distally)
  - Lesser trochanter often displaced several cm
  - Pubic ramus avulsion may require oblique (Judet) view to visualize
- Healing response is exuberant callus
  - 4-6 weeks after injury: Callus is immature osteoid & may give appearance of aggressive tumor
    - Do not mistake eccentric avulsive callus for malignant periostitis or tumor bone formation
  - > 6 weeks, osteoid begins to mature, resulting in bony excrescence at site
    - Bony excrescence can be chronic or can resorb over time

CT Findings

- Similar to radiographic findings; could provide additional 3D detail
- Immature callus may simulate aggressive lesion on CT and PET/CT

MR Findings

- T1WI
  - Avulsions
    - Avulsed osseous fragment can show normal hyperintense marrow or hypointense cortex
    - Adjacent soft tissue and hematoma signal may obscure small apophyseal fragment
    - Hypointense osseous and soft tissue edema
    - Mixed intensity hematoma adjacent to traumatic avulsion
  - Apophysitis
    - Asymmetric tendinous or apophyseal enlargement 2° to low SI edema
- T2WI FS
  - Avulsions &/or apophysitis
    - Hyperintense fluid, soft tissue edema, & hematoma at site of injury
    - Strain pattern of hyperintensity at closest musculotendinous unit
  - Avulsions
    - Hyperintense marrow edema at donor site
    - Fluid signal at injured growth plate
    - Avulsed osseous fragment may be obscured; correlate with radiograph
- MR provides soft tissue detail and is diagnostic in most cases
  - Shows pure tendon avulsions and apophysitis prior to apophyseal ossification (radiographs are normal)

Imaging Recommendations

- Best imaging tool
  - Radiographs when apophysis is at least partially ossified
  - MR for unossified apophysis or nonosseous avulsion
- Protocol advice
  - Coronal & axial of entire pelvis for symmetry
  - Sagittal plane helpful for rectus femoris injuries (AIIS) & thigh adductor origins

DIFFERENTIAL DIAGNOSIS

Osteosarcoma

- Amorphous osteoid formation
- Cortical destruction
- Large soft tissue mass

Tendinopathy

- Thickening of tendon origin & insertions
- Intermediate signal on T1WI & T2WI MR
- Within spectrum of avulsive injury

PELVIC AVULSION FRACTURE/APOPHYSITIS

Stress Fracture
- Distinct locations from avulsive injuries
- Superior & inferior obturator rami as opposed to ischial tuberosity or pubic tubercle

Pathologic Fracture
- Lesser trochanter avulsion in older adults is a special case: Must be considered pathologic until proven otherwise

PATHOLOGY

General Features
- Etiology
  - Adolescents: Traction on unfused apophysis
  - Fracture secondary to single forceful concentric or eccentric muscle contraction
  - Fracture related to chronic or repetitive microtrauma (AIIS in kicking athletes)
  - Fracture secondary to extreme passive stretch (dancers, gymnasts, yoga enthusiasts)
- Relevant anatomy: Explains the common age range for pelvic avulsion fractures
  - Pelvic apophyses do not ossify until middle teen years (~ age 14)
  - Pelvic apophyses do not fuse until young adult (up to age 25)

Gross Pathologic & Surgical Features
- Displacement of apophyses limited by adjacent structures
  - ASIS limited by other tendinous attachments and lateral inguinal ligament
  - AIIS limited by dual origin of rectus femoris (reflected head)
  - Ischial apophysis limited by sacrotuberous ligament
  - Pubic tubercle limited by rectus abdominis

Microscopic Features
- Hematoma
- Revascularization & resorption of devascularized bone fragment
- Callus and fibrous scar tissue
- Biopsy should be avoided
  - May show mitoses and osteoid in healing bone
  - Simultaneous osteoclastic and osteoblastic activity can mimic aggressive process

CLINICAL ISSUES

Presentation
- Most common signs/symptoms
  - Local pain at apophysis or tendon attachment related to activity
  - Swelling and pain after activity
  - Palpable mass or discoloration secondary to hematoma
  - Limitation of activity
  - Point tenderness
    - Iliac crest avulsion can mimic appendicitis
  - Pain & weakness of involved muscle when placed under stress
  - Altered gait or loss of foot speed
- Clinical profile
  - ASIS: Kicking sports, jumping, or running
  - AIIS: Kicking sports
  - Ischial tuberosity: Dancing, gymnastics, sprinting, hurdling, & football

Demographics
- Age
  - 14-25 years
- Gender
  - M > F, but younger age in females due to earlier skeletal maturation
- Epidemiology
  - Avulsion fractures = 13.4% of children's pelvic fx

Natural History & Prognosis
- Most injuries heal with conservative treatment
- Recovery slow: Average 4-8 weeks, up to 4 months
- Poor healing related to continued activity results in chronic pain
  - Can result in asymmetric growth plate fusion

Treatment
- Conservative
  - Initial bedrest → non-weight-bearing + crutches
  - Ice, anti-inflammatories
  - Physical therapy: Gradual strengthening + progressive weight-bearing
  - Re-injury related to premature resumption of activity
- Surgical
  - Reattachment of severely displaced avulsed fragment
  - Chronic pain: Resection of weak bony union & surgical reattachment
  - Evacuation of hematoma and tendon debridement

DIAGNOSTIC CHECKLIST

Consider
- Avulsions particularly in 15-25 year age range
- Atraumatic avulsion in adult: Pathologic or insufficiency fracture (especially lesser trochanter)

Image Interpretation Pearls
- Asymmetry can be helpful: Image contralateral side when necessary
- Location is key; scan soft tissues in region of ASIS, AIIS, pubis, & ischial tuberosity

Reporting Tips
- Describe status of physeal maturation/fusion on MR
- Look for injury to related regional tendinous attachments
  - Rectus abdominis with thigh adductors
  - Tensor fascia lata with sartorius

SELECTED REFERENCES

1. Hébert KJ et al: MRI appearance of chronic stress injury of the iliac crest apophysis in adolescent athletes. AJR Am J Roentgenol. 190(6):1487-91, 2008
2. Sanders TG et al: Avulsion injuries of the pelvis. Semin Musculoskelet Radiol. 12(1):42-53, 2008
3. Zajick DC et al: Spectrum of MRI findings in clinical athletic pubalgia. Semin Musculoskelet Radiol. 12(1):3-12, 2008

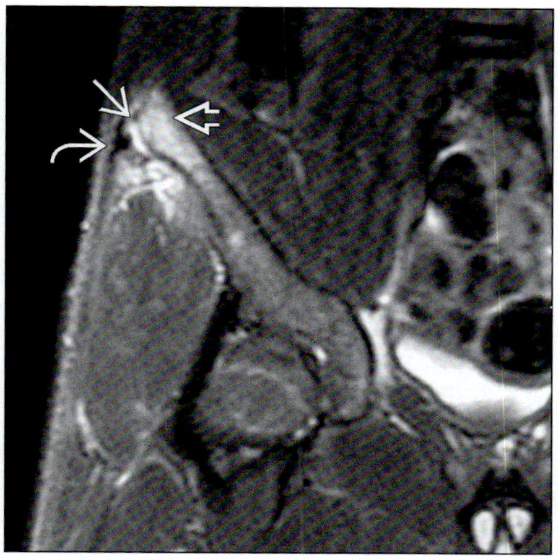

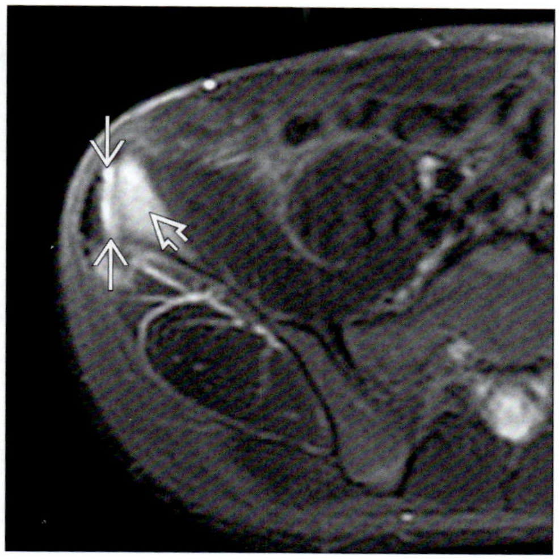

(Left) *Coronal T2WI FS MR shows partial avulsion of the origin of the sartorius tendon ➔ from the anterior superior iliac spine (ASIS) ➔ in a 16-year-old athlete. There is high signal fluid ➔ in the cleft, indicating an acute injury. There is little displacement; it is not surprising that the radiographs were negative.* ***(Right)*** *Axial T2WI FS MR from the same patient shows marrow edema ➔ in the iliac crest at the site of injury. Fluid in the cleft ➔ delineates the site of avulsion.*

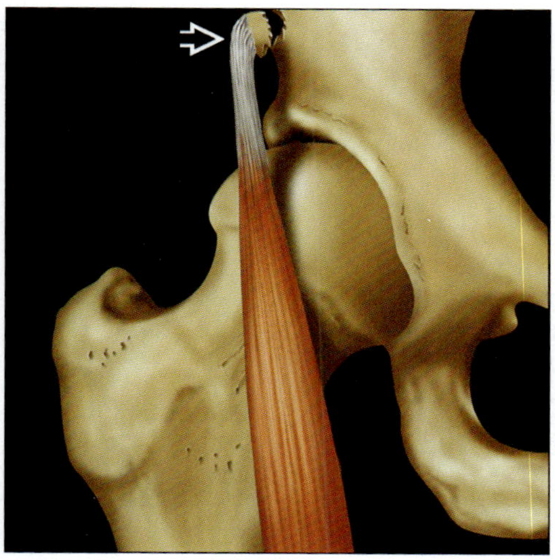

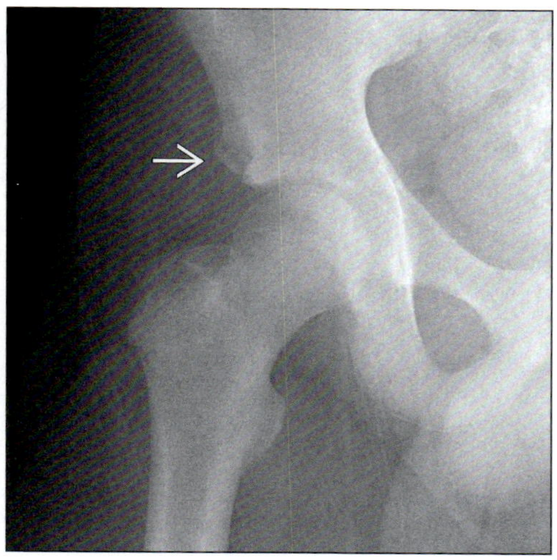

(Left) *Anterior graphic shows avulsion of the anterior inferior iliac spine ➔, the origin of the long head of the rectus femoris. As with most pelvic avulsion injuries, this is most commonly seen in older children and adolescents whose apophyseal attachments are still unfused.* ***(Right)*** *AP radiograph shows a mildly displaced avulsion of the anterior inferior iliac spine (AIIS) ➔, the site of origin of the long head tendon of the rectus femoris.*

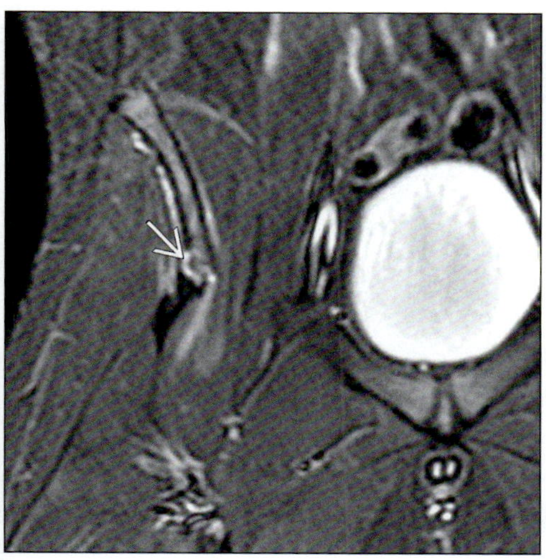

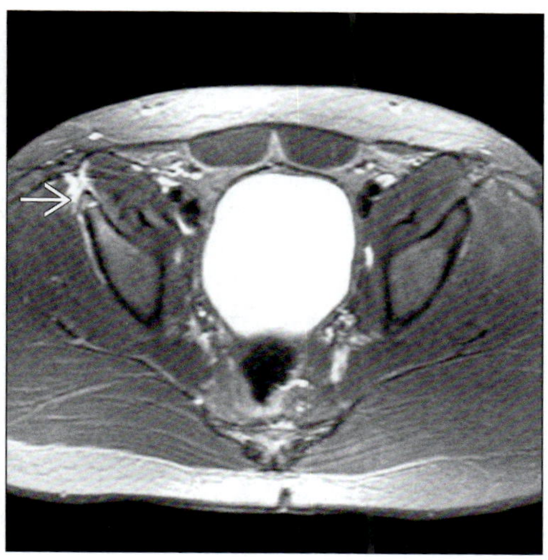

(Left) *Coronal T2WI FS MR shows avulsion of the AIIS in a 15-year-old football player. The patient complained of hip pain, and radiographs were read as normal. Soft tissue edema surrounds the site of avulsion, with a small cleft of high signal fluid ➔ interposed between the bone and the tendon.* ***(Right)*** *Axial T2WI FS MR from the same patient shows a small triangular avulsion fragment ➔ arising from the AIIS. Pelvic avulsions can be subtle, and comparison to the opposite side is helpful.*

PELVIC AVULSION FRACTURE/APOPHYSITIS

(Left) *Frog leg lateral radiograph from a 15-year-old male soccer player with pain on hip flexion exacerbated by kicking shows an amorphous ossific structure superolateral to the acetabulum, suggesting avulsion of the partially ossified AIIS apophysis, the site of attachment of the long head of rectus femoris.* ***(Right)*** *AP radiograph shows a triangular bone fragment arising from the superior acetabulum, consistent with avulsion of the reflected head tendon of the rectus femoris.*

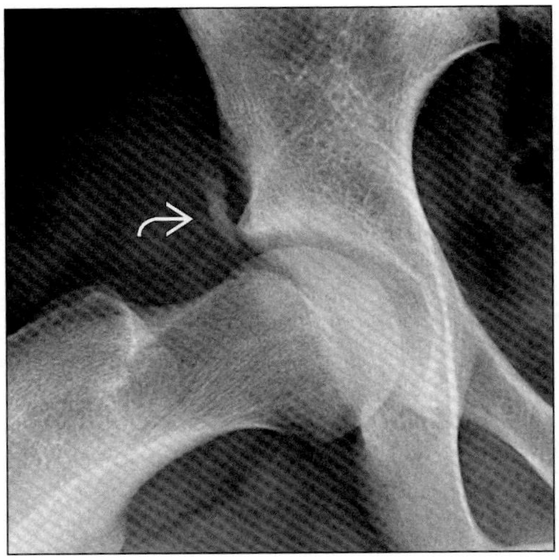

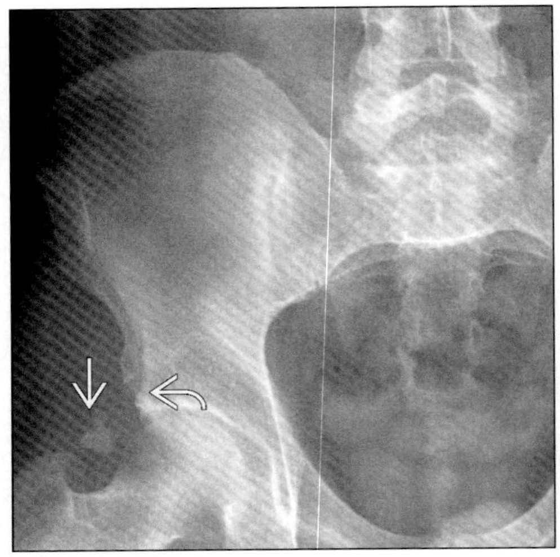

(Left) *Axial T2WI FS MR after a water-skiing injury in this 28 year old shows a partially torn biceps/semitendinosus tendon origin with an absent semimembranosus tendon at the level of the ischial tuberosity indicating an acute hamstring avulsion.* ***(Right)*** *AP radiograph of the ischial tuberosity in a 45-year-old runner with chronic pain at the right hamstring origin shows exuberant periostitis and osseous reparative response, typical for a chronic avulsive injury related to overuse.*

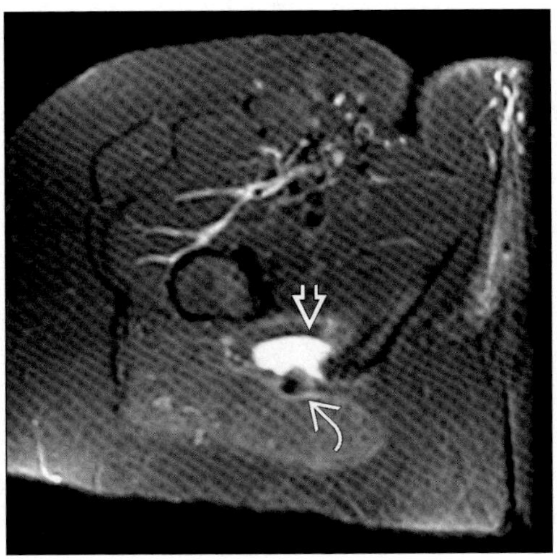

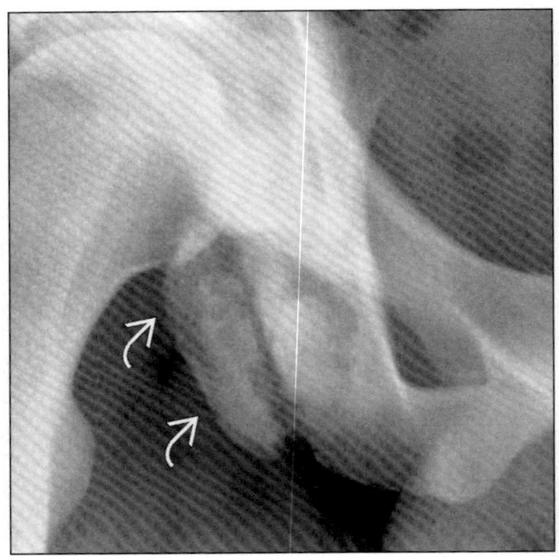

(Left) *Coronal oblique T2WI FS MR from a professional baseball pitcher shows an osseous avulsion of the adductor longus and the pectineus from the inferior ramus.* ***(Right)*** *Sagittal T2WI FS MR 2 cm off midline from a basketball player with acute injury shows soft tissue edema around the rectus abdominis/adductor aponeurosis and adductor longus avulsion from the pubic tubercle. Adductor avulsions are associated with rectus abdominis/adductor aponeurosis tears ("sports hernia").*

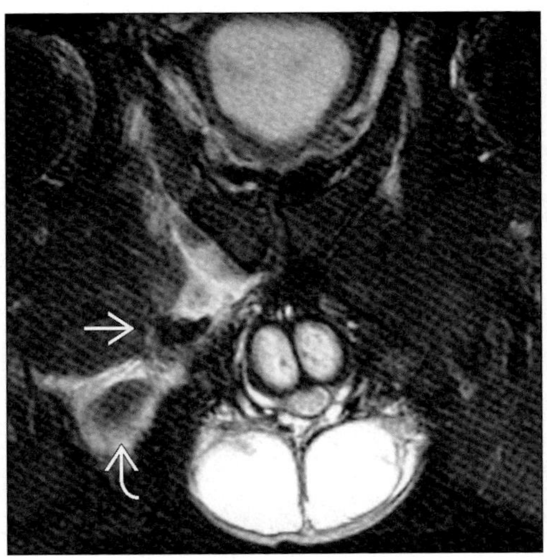

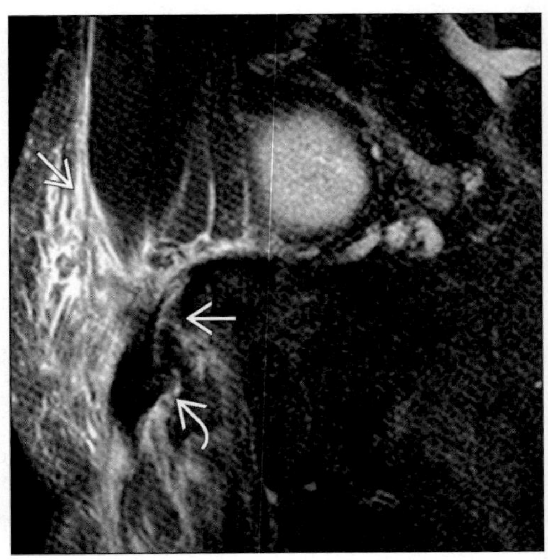

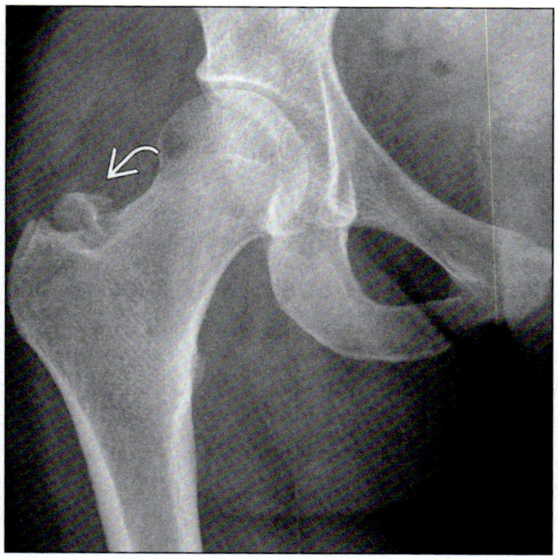

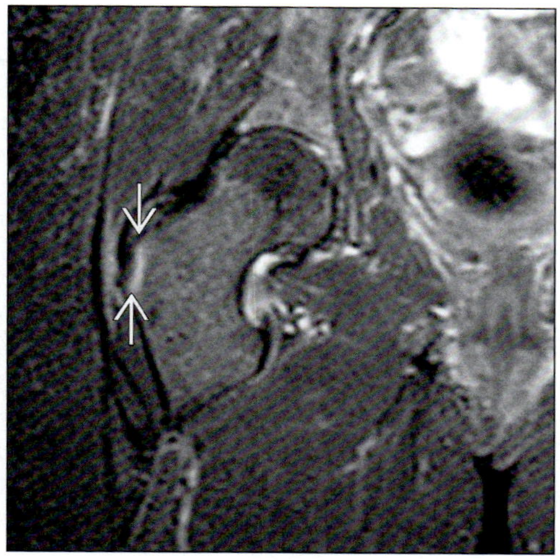

(Left) *AP radiograph of the hip from a 25 year old after a forceful external rotator contraction against internal rotation stress shows a traumatic osseous avulsion of the greater trochanter at the gluteus medius insertion .* ***(Right)*** *Coronal T2WI FS MR from a 17-year-old competitive swimmer shows a nondisplaced avulsion fracture of the greater trochanter at the site of attachment of the gluteus maximus, probably due to repetitive hip extension. Lack of significant adjacent edema suggests chronicity.*

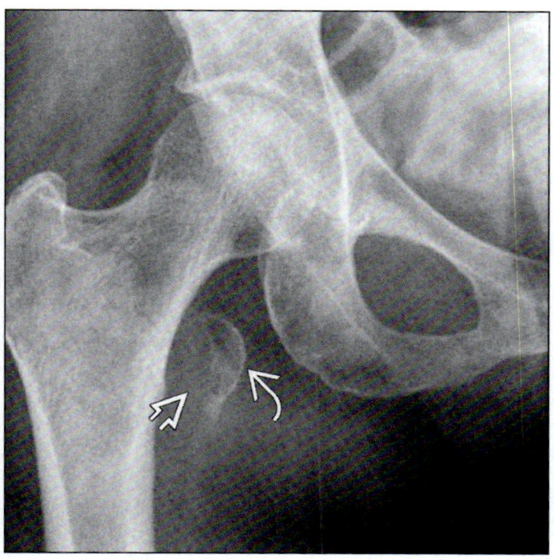

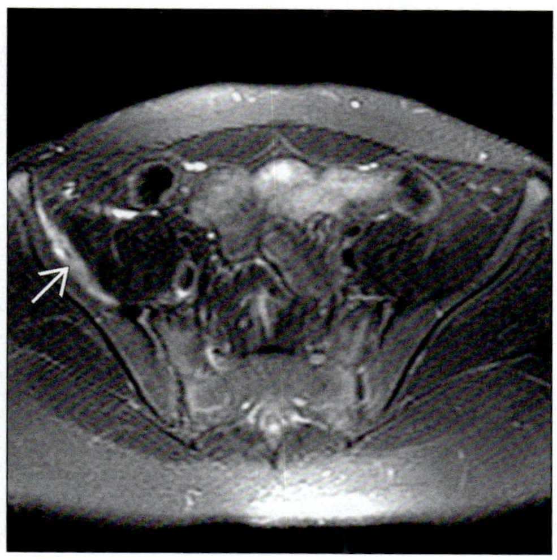

(Left) *AP radiograph of the pelvis in a 48 year old with lung cancer shows an avulsed and displaced lesser trochanter . Note the ill-defined and permeative appearance of the lateral margin of the avulsed fragment , confirming the diagnosis of a pathologic fracture.* ***(Right)*** *Axial T2WI FS MR demonstrates acute avulsion of a portion of the right iliacus muscle from the iliac crest. Heterogeneous signal represents blood and fluid at the site of injury. Compare to the normal contralateral side.*

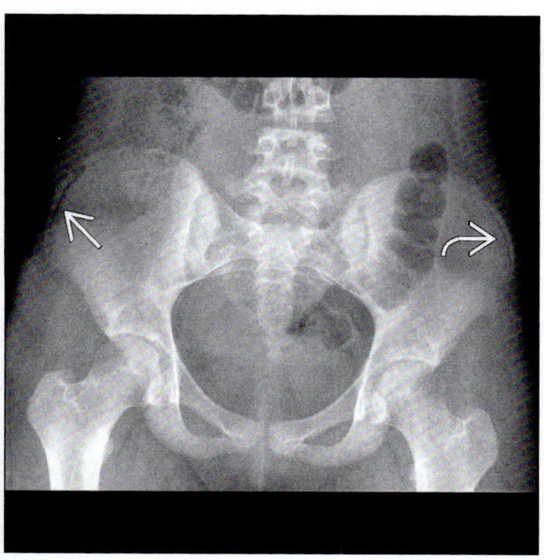

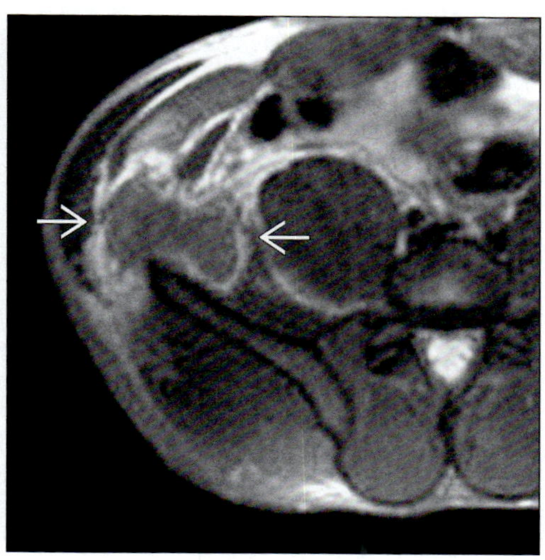

(Left) *AP radiograph demonstrates subtle asymmetric separation of right iliac crest apophysis compared to the normal left side in this 17 year old. This is the site of attachment of the oblique abdominal muscles and tensor fascia lata.* ***(Right)*** *Axial PDWI MR from a football quarterback with flank pain 5 days after a twisting & hyperextension injury shows a hypointense hematoma at the site of external oblique muscle avulsion. Large hematomas can delay healing with pelvic avulsions.*

PELVIS/HIP/FEMUR TRAUMA, POSTOPERATIVE APPEARANCE

Key Facts

Imaging

- Plates + screws
- Interfragmentary screws
- Dynamic hip screw
- Intramedullary nail/rod
- Fixed angle blade plate
- External fixator
- Joint replacement
- Radiographs used for routine surveillance
- CT more accurate in evaluation of fracture healing
- MR often diagnostic of abnormalities occurring adjacent to fixation
- Ultrasound useful for fluid collections around hardware

Clinical Issues

- Radiographic signs of fracture healing are less conspicuous after rigid internal fixation compared to casting
  - Peripheral callus often absent
  - 1st see resorption at fracture site
  - Followed by gradual blurring of fracture line
  - Fracture line may still be faintly visible when patient is clinically healed
- Complications
  - Malreduction, incorrect hardware placement
  - Hardware fracture, displacement, or loosening
  - Infection
  - Nonunion
  - Osteonecrosis
  - Rapidly progressive osteoarthritis
  - Heterotopic ossification

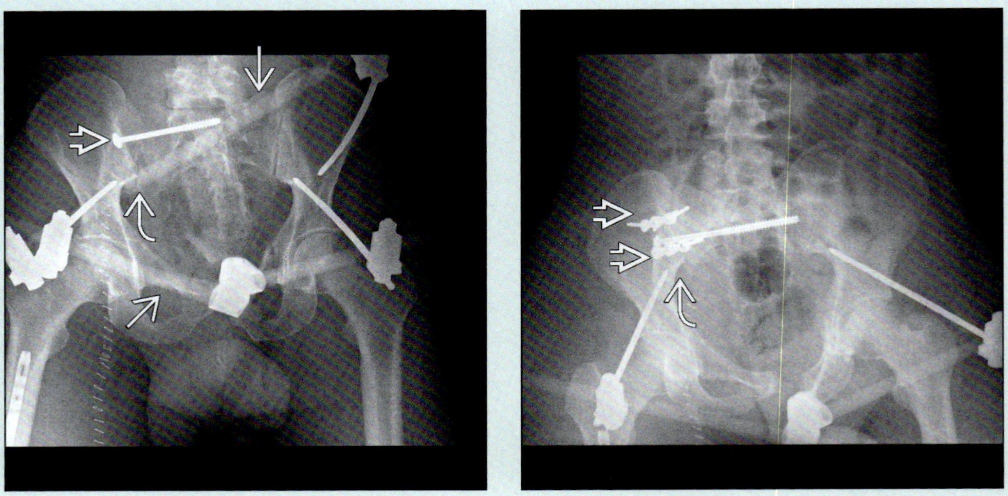

__(Left)__ AP radiograph shows initial fixation of pelvis ring disruption with a single right sacroiliac screw ➔. Right sacroiliac joint remains widened ↪. A doubled anterior fixator ➔ has been used for anterior stabilization. Left hemipelvis remains displaced superiorly. __(Right)__ AP radiograph after revision of fixation shows replacement of screw and placement of plates ➔ with reduction of sacroiliac joint ↪. Pubic symphysis diastasis was subsequently reduced with plate and screws.

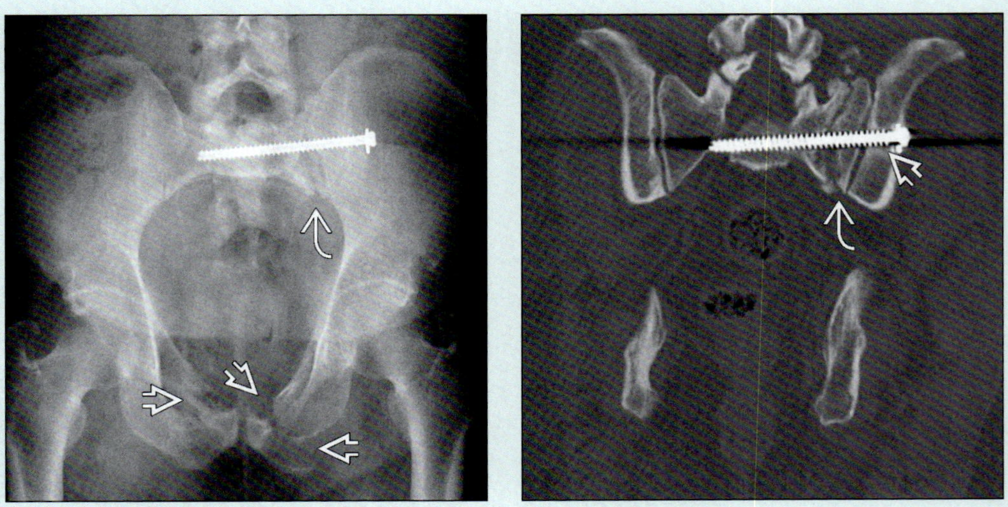

__(Left)__ AP radiograph shows apparent sacral nonunion ↪ without evidence of screw loosening. Pubic fractures ➔ are nonunited as well. __(Right)__ Coronal bone CT confirms sacral fracture nonunion ↪. The fracture margins are corticated, and no bridging bone is present. Increased lucency ➔ is seen around the fixation screw. If more than 1 mm of lucency is seen around the fixation screw, motion is usually present.

PELVIS/HIP/FEMUR TRAUMA, POSTOPERATIVE APPEARANCE

IMAGING

Hardware Types

- Plates + screws
  - Contoured to underlying bone
  - Screws may be "locked" into plate to improve rigidity
- Interfragmentary screws
  - Large screws are often used as primary fixation across sacroiliac joints
    - May extend across 1 joint or both joints
  - 3-4 screws often used as primary fixation of femoral neck fracture
    - Often cannulated
    - Placed percutaneously from lateral approach over a guide wire
- Dynamic hip screw
  - Screw placed from lateral margin of trochanter to femoral head
  - Screw can telescope within outer sleeve
    - Allows for impaction at fracture site
  - Used for intertrochanteric and subtrochanteric femoral fractures
- Intramedullary nail/rod
  - Medullary canal must be reamed prior to placement
  - Femoral nails slightly curved to match anterior curve of femur
  - Interlocking screws provide rotational stability
    - Can slow healing because of reduction of load-bearing across fracture site
  - Cephalomedullary nail has interlocking screws in femoral neck and shaft
  - Femoral shaft fractures may be fixed with antegrade or retrograde nail
    - Retrograde nail may cause knee pain
- Fixed angle blade plate
  - Older method of fixation; highly rigid
  - Often used as salvage after failure of primary fixation
- External fixator
  - Often used anteriorly for pelvic ring disruption
- Joint replacement
  - Femoral neck fractures in elderly patients may do better with hemiarthroplasty than with ORIF

Imaging Recommendations

- Best imaging tool
  - Radiographs used for routine surveillance
  - CT more accurate in evaluation of fracture healing
  - MR often diagnostic of abnormalities occurring adjacent to fixation
  - Ultrasound useful for soft tissue abnormalities

CLINICAL ISSUES

Natural History & Prognosis

- Radiographic signs of fracture healing are less conspicuous after rigid internal fixation compared to casting
  - Peripheral callus often absent
  - 1st see resorption at fracture site
  - Followed by gradual blurring of fracture line
  - Fracture line may still be faintly visible when patient is clinically healed
    - i.e., no pain at fracture site with stress on bone

Hardware Complications

- Malreduction
  - Search for bone fragments preventing reduction
- Hardware fracture or displacement
- Hardware loosening
  - Thin rim of lucency surrounding screw is normal on digital imaging
    - > 1 mm suggests loosening
  - Lucency usually becomes wider at 1 end of screw
    - "Windshield wiper" appearance due to toggling motion
- Infection
  - Lucency around screw
    - Rounded or linear
    - May appear similar to hardware loosening
  - Soft tissue infection
- Incorrect hardware placement
  - Sacroiliac screws may traverse spinal canal
  - Fracture fragments may not be engaged by hardware
- Loss of reduction
  - Weight-bearing films useful after reduction to evaluate for stability
  - Fragments may migrate in immediate postoperative period

Fracture Nonunion

- Be cautious in diagnosis of nonunion
- Fracture healing may be slow because of patient age, nutrition, medications, or general poor health

Osteonecrosis

- Complication of femoral neck fracture

Rapidly Progressive Osteoarthritis

- May be due to cartilage injury or malreduction

Heterotopic Ossification

- May bridge joint and limit motion

SELECTED REFERENCES

1. Leonardsson O et al: Long-term follow-up of replacement compared with internal fixation for displaced femoral neck fractures: results at ten years in a randomised study of 450 patients. J Bone Joint Surg Br. 92(3):406-12, 2010
2. Fayad LM et al: Value of 3D CT in defining skeletal complications of orthopedic hardware in the postoperative patient. AJR Am J Roentgenol. 193(4):1155-63, 2009
3. Buckwalter KA et al: Multichannel CT Imaging of Orthopedic Hardware and Implants. Semin Musculoskelet Radiol. 10(1):86-97, 2006
4. Sofka CM et al: Musculoskeletal imaging update: current applications of advanced imaging techniques to evaluate the early and long-term complications of patients with orthopedic implants. HSS J. 2(1):73-7, 2006
5. Jacobson JA et al: Musculoskeletal sonography of the postoperative orthopedic patient. Semin Musculoskelet Radiol. 6(1):67-77, 2002

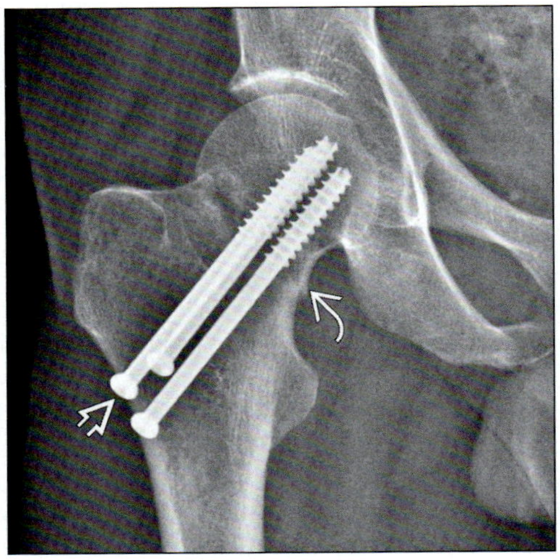

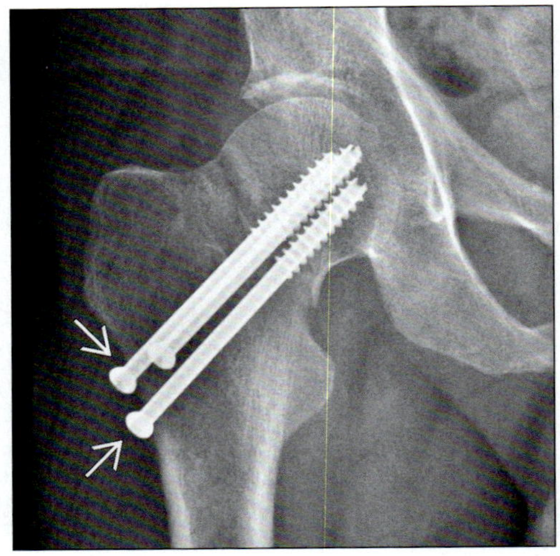

*(**Left**) AP radiograph shows fixation of femoral neck fracture in a 59-year-old man with 3 cannulated screws. Cannulation creates the "dimple" in the screw head. There is mild valgus alignment across the fracture (normal neck/shaft angle is 120-135°). (**Right**) AP radiograph from the same patient 2 months later shows scant callus and slight displacement of the fracture. Two of the screws are now slightly proud (i.e., protrude beyond the cortex) due to displacement of the fracture.*

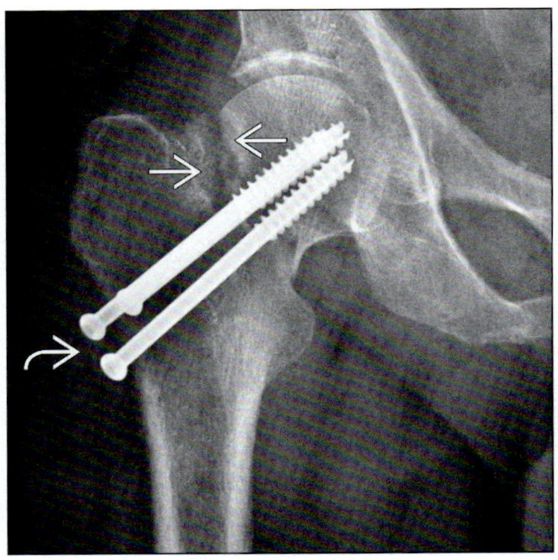

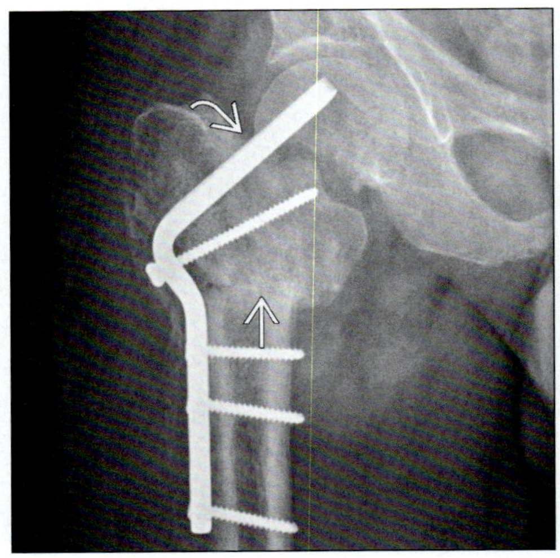

*(**Left**) AP radiograph from the same patient 4 months after initial injury shows further displacement of the fracture and increasing prominence of screw heads. Cortication of fracture margins is worrisome for impending nonunion. (**Right**) AP radiograph from the same patient was acquired 2 months after subtrochanteric osteotomy to reduce varus deformity and hardware revision with blade plate fixator. Osteotomy shows early healing, but femoral neck fracture has displaced around the blade.*

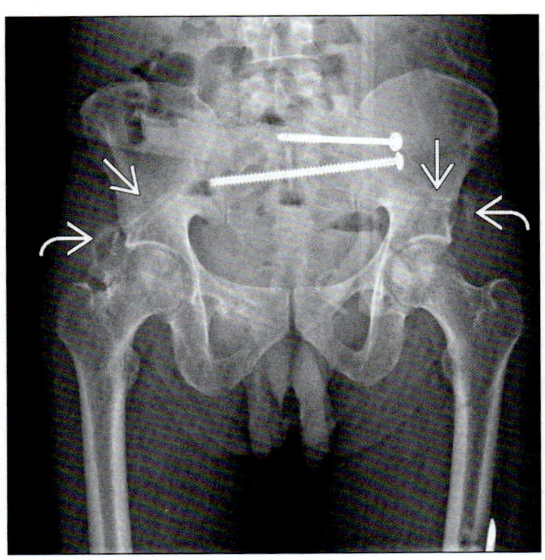

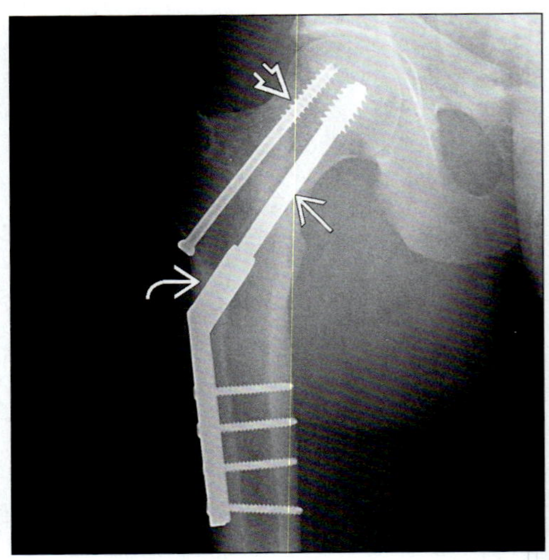

*(**Left**) AP radiograph shows heterotopic ossification adjacent to tracks from prior anterior fixators. (**Right**) AP radiograph shows dynamic hip screw (DHS) used for intertrochanteric and basicervical fractures. The screw can telescope within the sheath and may protrude lateral to the femur when there is significant telescoping. The derotational screw is employed for basicervical fractures to provide rotational stability.*

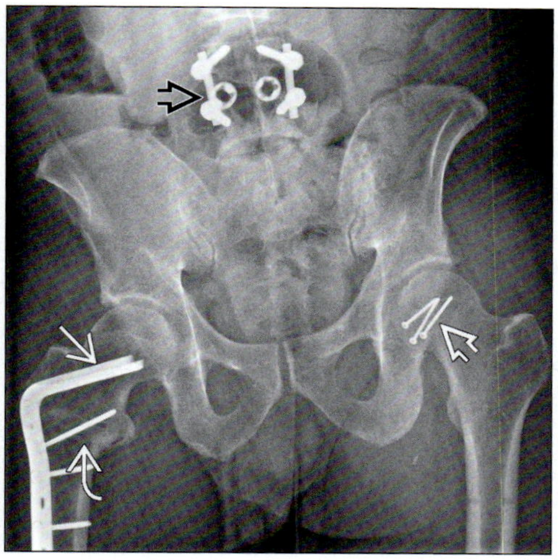

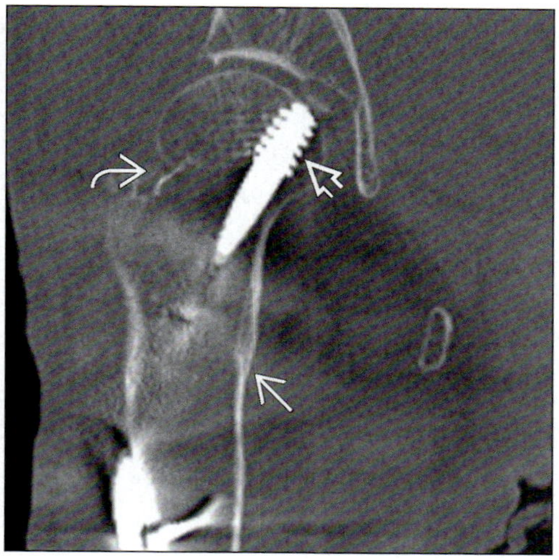

(Left) *AP radiograph shows blade plate fixator for fixation of a high subtrochanteric fracture . Interfragmentary screws in the left femoral head fix a Pipkin II femoral head fracture. There has been L4-5 fusion .* ***(Right)*** *Coronal bone CT shows dynamic hip screw across a healed intertrochanteric fracture . Note eccentric position of femoral neck screw , normally placed in center of head. An acute femoral neck fracture has developed adjacent to the screw.*

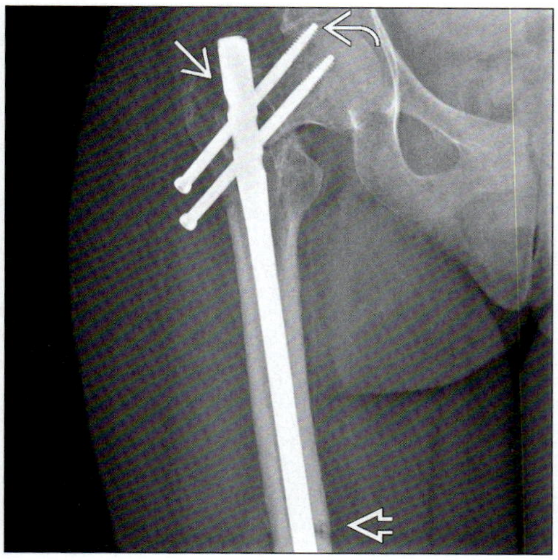

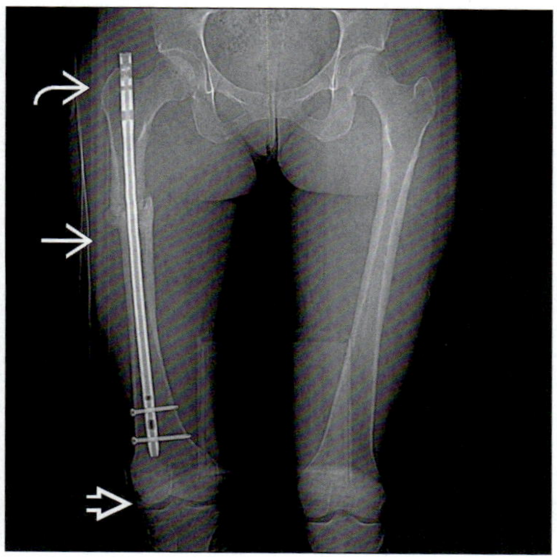

(Left) *AP radiograph shows a cephalomedullary nail across a femoral shaft fracture . The patient presented with new hip pain, and a basicervical femoral neck fracture was seen with varus alignment and cutting out of a fixation screw through the femoral cortex.* ***(Right)*** *AP radiograph shows antegrade femoral nail across a femoral shaft fracture . The femur is shortened. Malrotation can be recognized by the combination of internal hip rotation and external knee rotation .*

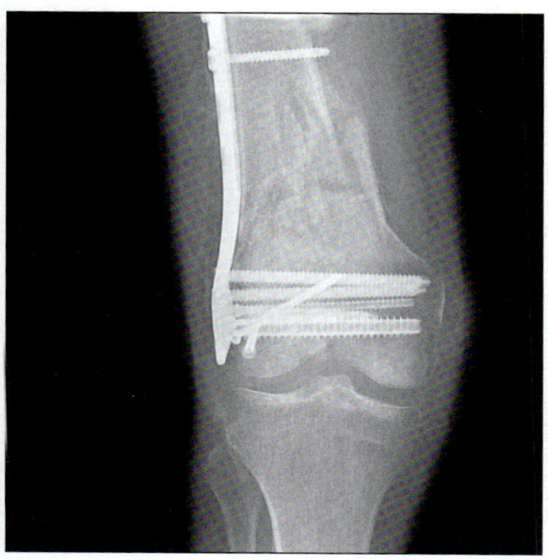

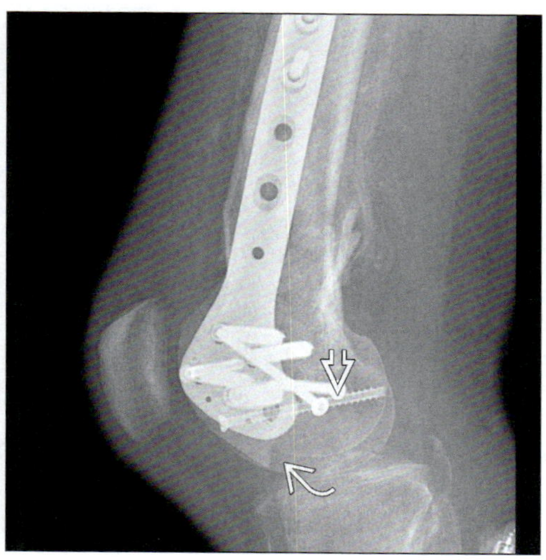

(Left) *AP radiograph shows ORIF of distal femur fracture with lateral fixation plate and screws. Articular surface appears well aligned on this view.* ***(Right)*** *Lateral radiograph from the same patient shows stepoff of the lateral femoral condylar articular surface across a coronally oriented fracture (Hoffa fragment). Fracture has been bridged with a separate interfragmentary screw , but it became displaced postoperatively.*

FAI AND ADULT DDH, POSTOPERATIVE APPEARANCE

Key Facts

Imaging

- Periacetabular osteotomy (PAO)
  - Bone contour broadened at site of healed osteotomy
  - May result in coxa profunda or retroversion
- Femoral osteochondroplasty (cheilectomy)
  - Abrupt change in contour of anterior femoral head and neck
  - ± greater trochanteric osteotomy
- Subtrochanteric femoral osteotomy
  - Hardware tracks, deformity of proximal femur usually visible
- Postoperative labrum
  - Blunted, small labrum or retained limbus labrum
  - Irregular regions of increased signal intensity in labrum
  - Recurrent tear difficult to distinguish from postoperative changes
- Postoperative hyaline cartilage
  - Acute trauma or ongoing microtrauma may cause new cartilage defect
- Chiari osteotomy
  - Acetabular roof often flat & irregular
- CT arthrography preferred for internal derangement if large amount of metal in situ
- MR arthrography has higher sensitivity if only a small amount of metal present

Top Differential Diagnoses

- Fracture

Clinical Issues

- Complications: Osteotomy nonunion, iatrogenic coxa profunda, recurrent labral tear, intraarticular adhesions, stress fracture, infection, progressive osteoarthritis, sciatic nerve palsy, osteonecrosis

*(**Left**) Lateral intraoperative photograph shows characteristic abrupt step-off at the site of resection of cam lesion of femoral head. Range of motion is checked intraoperatively to ensure that enough bone is removed. (**Right**) AP radiograph in the same patient shows femoral osteochondroplasty (cheilectomy) fixed with 2 fully threaded screws. Bone defect is present at the lateral and anterior aspect of the head/neck junction.*

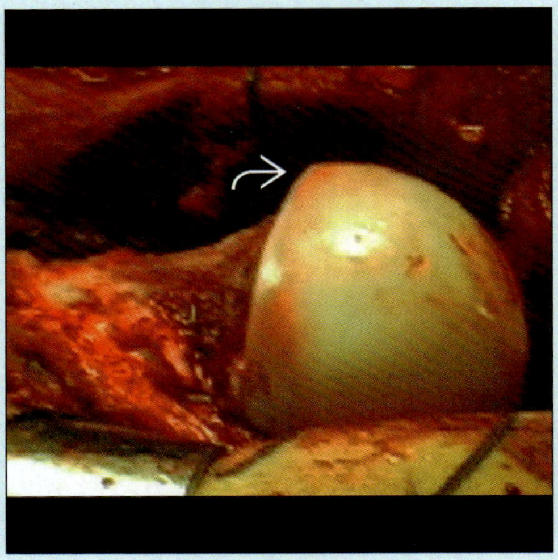

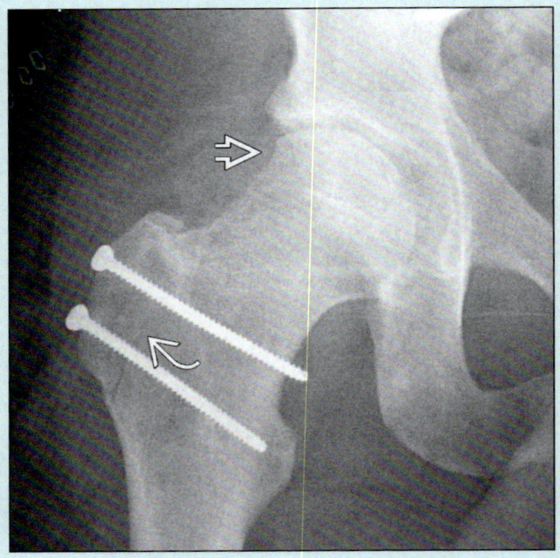

*(**Left**) Axial oblique MR arthrogram T1WI FS shows a region of resected bone. Labral tear is seen adjacent to prior labral repair. The anterior joint capsule is typically thickened and difficult to puncture with a small gauge arthrography needle after FAI surgery. (**Right**) Sagittal MR arthrogram PDWI FS shows femoral head resection site and recurrent complex labral tear. Low signal region in acetabulum is at the site of bioabsorbable suture from labral repair.*

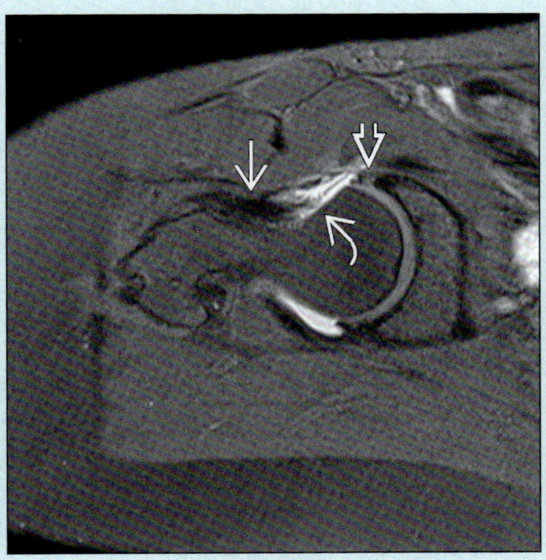

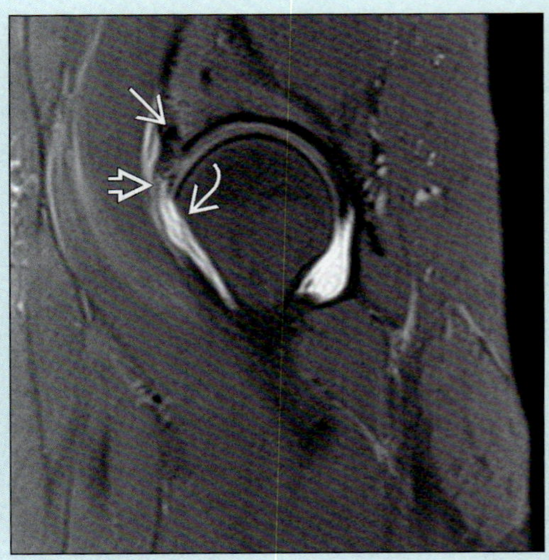

FAI AND ADULT DDH, POSTOPERATIVE APPEARANCE

TERMINOLOGY

Definitions

- Periacetabular osteotomy (PAO)
  - Performed for pincer-type femoral acetabular impingement (FAI) or developmental dysplasia of hip (DDH)
  - Usually a triple innominate osteotomy
    - Variants: Ganz, Bernese, Steele, Pemberton
    - Osteotomies of ilium, ischium, and pubis isolate acetabulum and allow realignment
  - Increases anterior and lateral femoral head coverage
- Salter osteotomy
  - Supraacetabular iliac osteotomy performed at age < 6 years
  - Acetabulum, pubic bone, and ischium rotated on pubic symphysis
- Chiari (shelf) osteotomy
  - Oblique supraacetabular osteotomy
  - Nonarticular lateral margin of ilium becomes lateral acetabular roof
  - Acetabulum medialized
- Femoral osteochondroplasty (cheilectomy): Surgical recontouring of head/neck junction
  - Improves head/neck offset in cam-type FAI
  - Open or arthroscopic
- Subtrochanteric femoral osteotomy
  - Performed in childhood to redirect femoral head into acetabulum

IMAGING

Radiographic Findings

- Periacetabular osteotomy (PAO)
  - Bone contour broadened at site of healed osteotomy
  - Hardware may be retained or removed after healing
  - May result in coxa profunda or retroversion
- Chiari osteotomy
  - Acetabular roof often flat & irregular
- Femoral osteochondroplasty (cheilectomy)
  - Abrupt change in contour of anterior femoral head and neck
- Subtrochanteric femoral osteotomy
  - Mild deformity of proximal femur seen in adulthood

CT Findings

- Best method for evaluation of osseous nonunion

MR Findings

- Repaired labrum
  - Blunted, small labrum
    - Irregular regions of increased signal intensity in labrum
      - Recurrent tear difficult to distinguish from postoperative change
- Postoperative hyaline cartilage
  - Acute trauma or ongoing microtrauma may cause new cartilage defect
- Prior Chiari osteotomy
  - Fibrous tissue fills site of acetabular augmentation
  - Often has irregular, fibrillated appearance on MR arthrogram
- Femoral osteochondroplasty site
  - Minimal marrow edema
  - Angular contour defect at surgical site
  - ± greater trochanteric osteotomy

Imaging Recommendations

- Best imaging tool
  - CT arthrography preferred for internal derangement if large amount of metal in situ
    - Decreased metal artifact compared to MR
  - MR arthrography has higher sensitivity if only a small amount of metal present

DIFFERENTIAL DIAGNOSIS

Fracture

- Line more jagged/irregular than osteotomy line

CLINICAL ISSUES

Complications of Surgery

- Osteotomy nonunion
- Iatrogenic coxa profunda
  - PAO for DDH may result in lateral or anterior overcoverage of femoral head
- Recurrent labral tear
  - Difficult to distinguish from labral repair
  - Labral tears tend to recur if underlying abnormal morphology not adequately addressed
- Intraarticular adhesions
  - At sites of femoral osteochondroplasty or ligamentum teres resection
  - May cause soft tissue impingement
- Stress fracture
  - Femoral neck after osteochondroplasty
- Infection
- Progressive osteoarthritis
- Osteonecrosis
- Sciatic nerve palsy

SELECTED REFERENCES

1. Clohisy JC et al: Surgical treatment of femoroacetabular impingement: a systematic review of the literature. Clin Orthop Relat Res. 468(2):555-64, 2010
2. Beck M: Groin pain after open FAI surgery: the role of intraarticular adhesions. Clin Orthop Relat Res. 467(3):769-74, 2009
3. Matheney T et al: Intermediate to long-term results following the Bernese periacetabular osteotomy and predictors of clinical outcome. J Bone Joint Surg Am. 91(9):2113-23, 2009
4. Parvizi J et al: Arthroscopy for labral tears in patients with developmental dysplasia of the hip: a cautionary note. J Arthroplasty. 24(6 Suppl):110-3, 2009
5. Vukasinovic Z et al: Triple pelvic osteotomy for the treatment of residual hip dysplasia. Analysis of complications. Hip Int. 19(4):315-22, 2009
6. Murray KA et al: Radiographic imaging for treatment and follow-up of developmental dysplasia of the hip. Semin Ultrasound CT MR. 22(4):306-40, 2001

FAI AND ADULT DDH, POSTOPERATIVE APPEARANCE

*(**Left**) AP arthrogram shows protrusio acetabuli after periacetabular osteotomy performed for developmental dysplasia of the hip (DDH). (**Right**) Coronal reformat CT arthrogram from the same patient shows limbus labrum . This is an enlarged labrum seen in DDH. It is susceptible to mucoid degeneration and tear. Subchondral cysts in the acetabulum are attributable to pincer-type femoral acetabular impingement (FAI).*

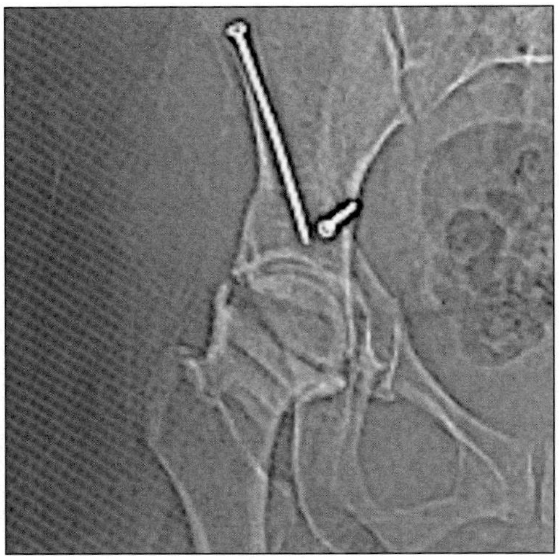

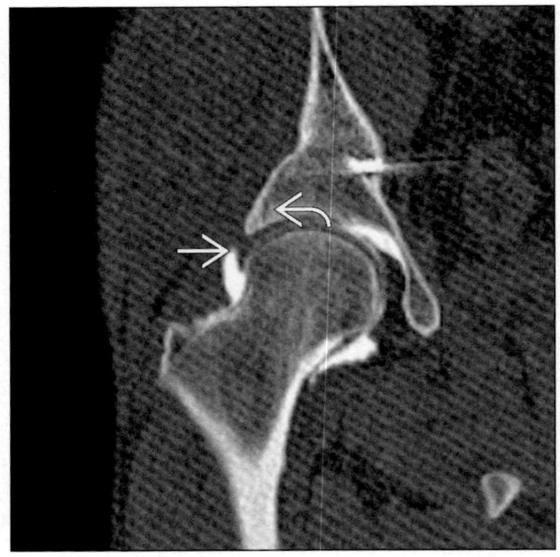

*(**Left**) AP radiograph in a 20-year-old patient status post femoral and periacetabular osteotomies shows characteristic mild deformities at healed iliac , pubic , and ischial osteotomies. Despite early intervention, patient has severe premature osteoarthritis . (**Right**) Coronal reformat CT arthrogram of the left hip shows healed supraacetabular osteotomies performed for DDH. Coxa profunda is present bilaterally and on the left is increased by limbus labrum .*

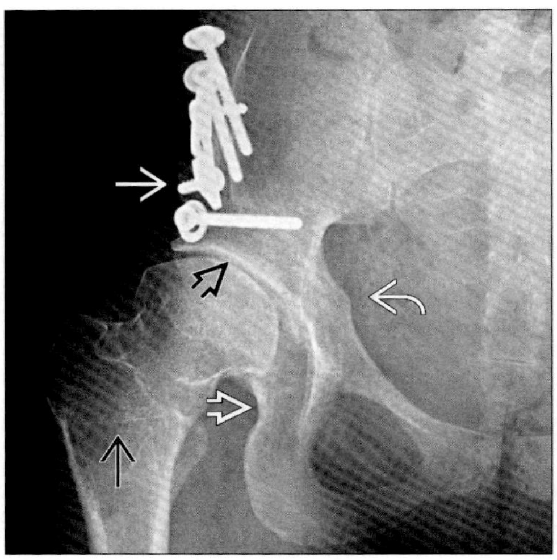

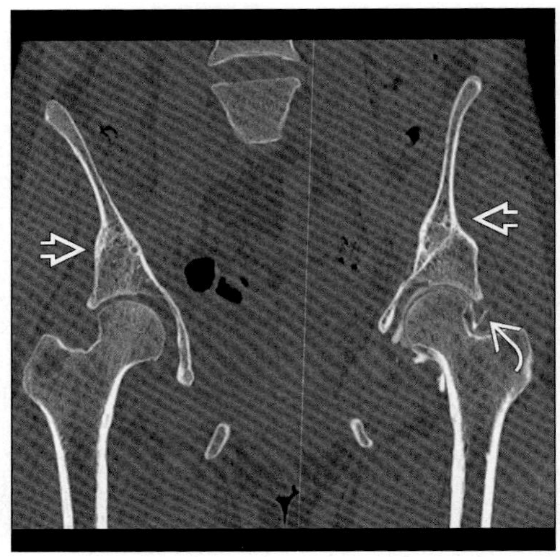

*(**Left**) AP arthrogram shows osseous irregularity at site of Chiari osteotomy. This deepens the acetabulum using lateral ilium, which is normally not articular. The augmented acetabular roof tends to be flat and irregular. Note secondary growth disturbance of the femoral head . (**Right**) Coronal reformat CT arthrogram from the same patient shows healed osteotomy . Fibrous tissue fills the region below the surgical augmentation. Clefts in the fibrous tissue fill with contrast .*

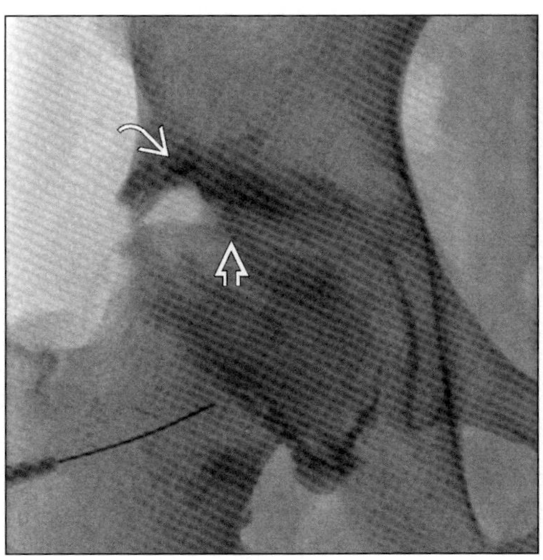

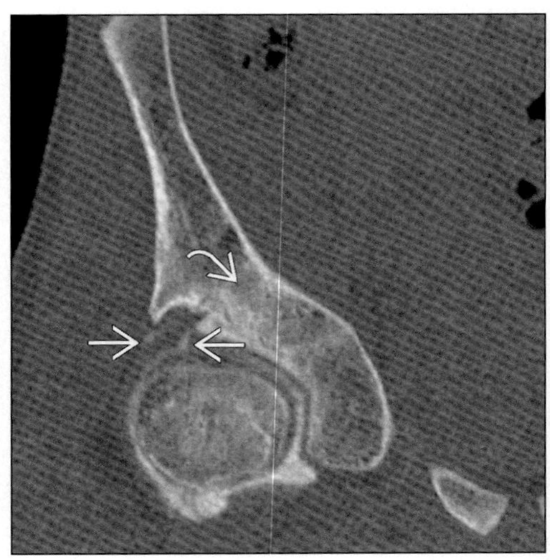

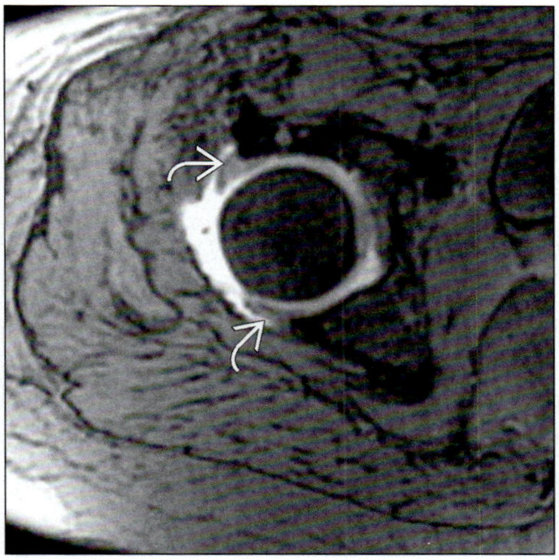

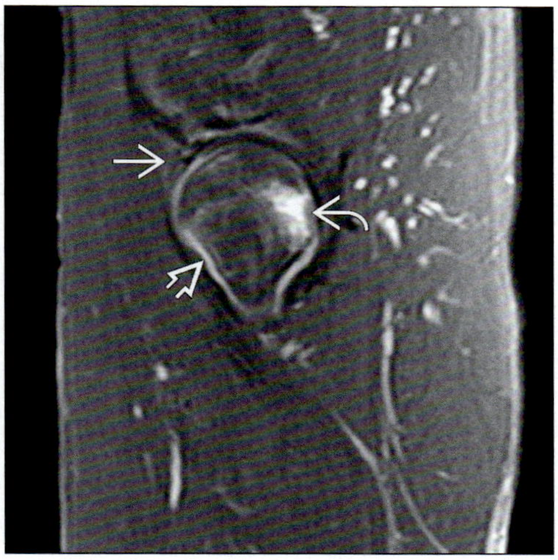

(Left) *Axial MR arthrogram gradient echo T2WI shows acetabular retroversion after DDH surgery. Note the large limbus labrum . It contains multiple areas of abnormal signal intensity and, at the time of revision surgery, showed extensive degeneration.* ***(Right)*** *Sagittal PDWI FS MR shows a stress fracture occurring 6 weeks after femoral osteochondroplasty . The patient was playing soccer against medical advice. Repaired labrum appears irregular.*

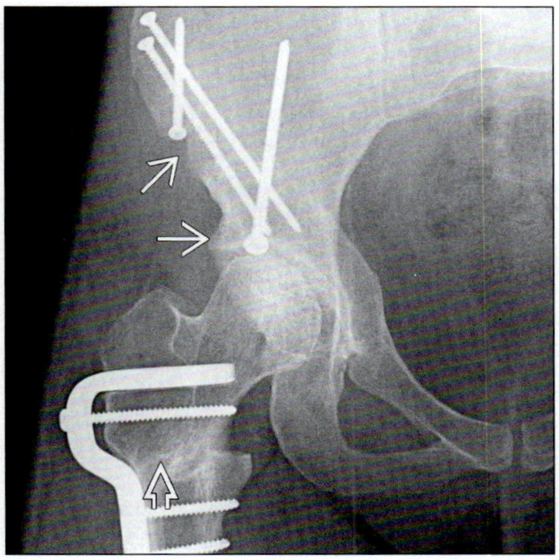

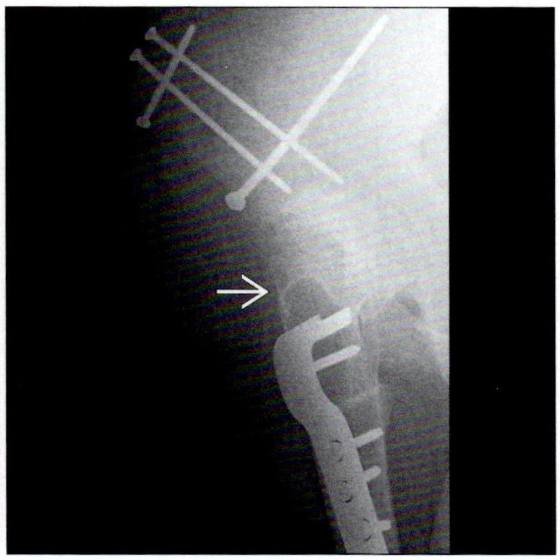

(Left) *AP radiograph shows healed subtrochanteric and periacetabular osteotomies for DDH. Alignment appears normal, and hardware is intact.* ***(Right)*** *Lateral radiograph from the same patient shows loss of femoral neck anteversion. Anterior head-neck offset is markedly decreased . Patient presented with recurrent hip pain and was found to have recurrent femoral acetabular impingement. She was treated with osteochondroplasty.*

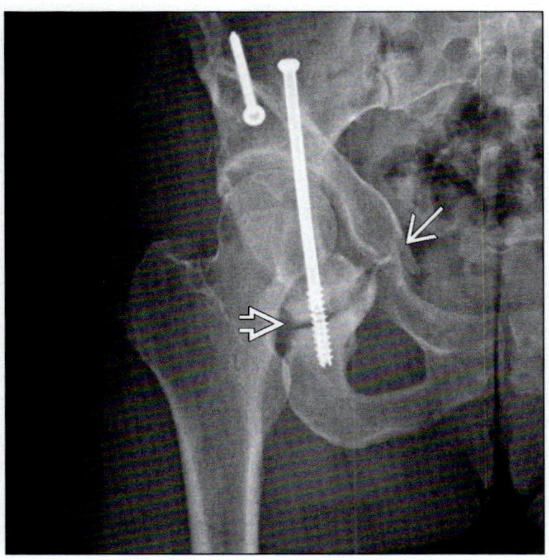

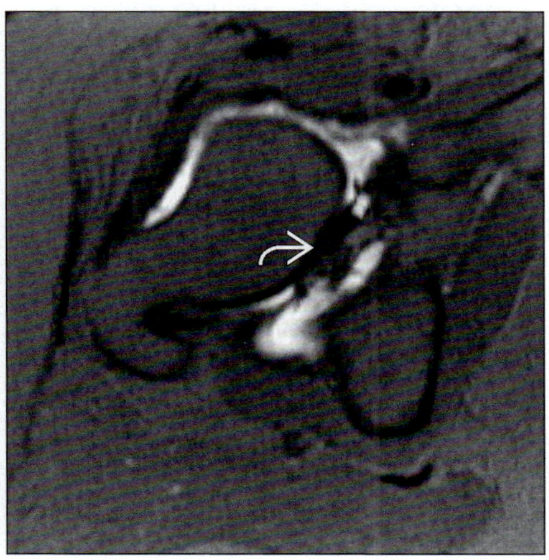

(Left) *AP radiograph shows periacetabular osteotomy with ischial nonunion . Marked sclerosis at the nonunion site is due to motion at the site. Pubic osteotomy is healed.* ***(Right)*** *Axial MR arthrogram T1WI FS shows inferomedial scar tissue at the site of ligamentum teres resection, performed as part of DDH surgery in early childhood. This can result in internal impingement of the hip and can cause intraarticular snapping hip syndrome.*

ACETABULAR LABRAL TEAR

Key Facts

Terminology

- Tear, avulsion, or detachment of acetabular labrum

Imaging

- Direct MR arthrography is standard of care for evaluation of labral tear in absence of OA
- Hyperintensity within normally hypointense labral substance on T2WI
- Contrast extending into labral substance on direct MR arthrogram
- Anterosuperior undersurface most common
- Anterior undersurface or detachment at anterior base
- Posterior detachment with hip dislocation or subluxation
- Look for FAI and DDH morphologies
- Acquire large FOV imaging of entire pelvis region
- Acquire sequences dedicated to hip in question

Pathology

- Degenerative: Associated with cam- and pincer-type FAI, DDH, and OA
- Traumatic: Ranges from twisting injury to hip dislocation
- Czerny classification based upon direct MR arthrography
- Blankenbaker clockface system for extent of tear

Clinical Issues

- Progression of labral tear is typical

Diagnostic Checklist

- Assess subjacent acetabular articular cartilage
- Localize labral tear on acetabulum
- Delineate extent of labral tear
- Identify any labral detachment from acetabular rim

__(Left)__ Coronal graphic shows a superior acetabular labral tear ➡. These tears are almost always associated with articular cartilage lesions of the subjacent acetabulum. With cam-type FAI, tears generally involve the anterosuperior region and extend from the labral undersurface. __(Right)__ Sagittal graphic shows a detached anterior labral tear ➡ at the base of the labrum. This type of tear is more common in women and presents with pain on hip flexion. This diagnosis is hard to make without direct MR arthrography.

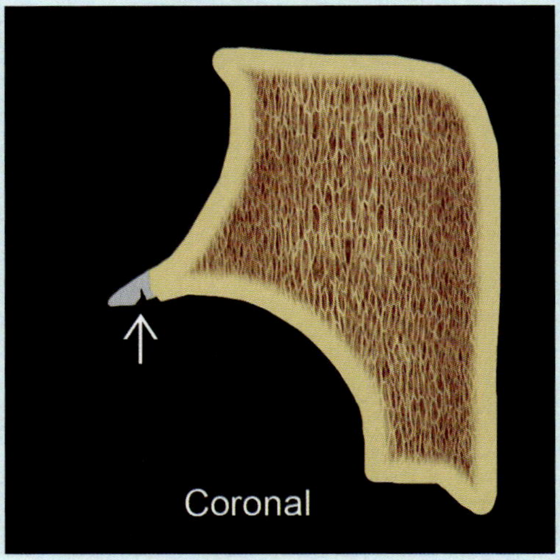

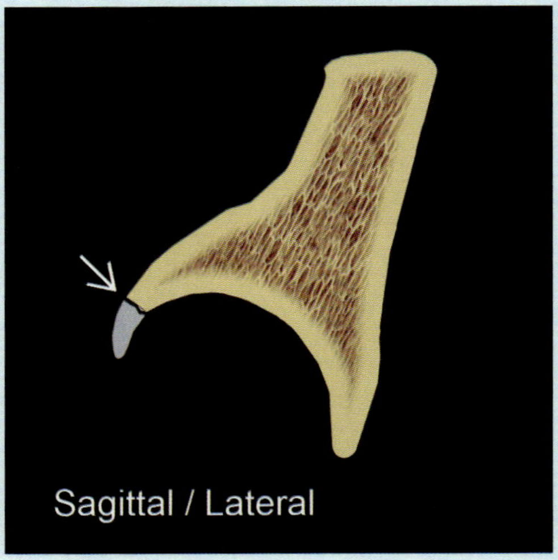

__(Left)__ Axial graphic shows a posterior acetabular labrum tear ➡. These are generally traumatic and most commonly related to hip subluxation or dislocation. There can be labral detachment, displacement, or periosteal stripping. __(Right)__ Sagittal graphic shows an undersurface tear ➡ of the anterior labrum. These tears are prone to instability of labral fragments. There is often articular cartilage loss of the anterior acetabulum, and there is an association with pincer-type FAI.

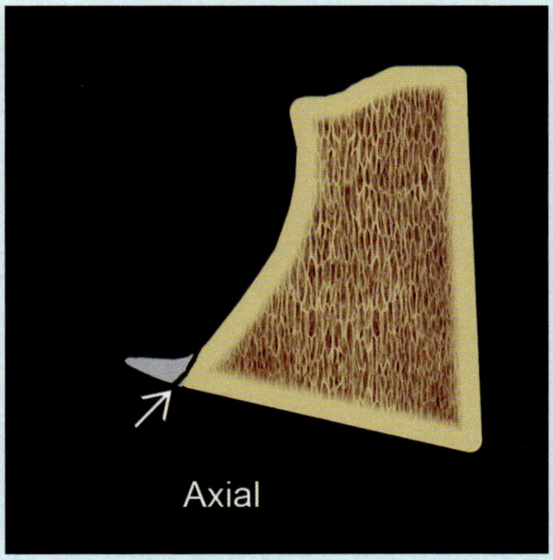

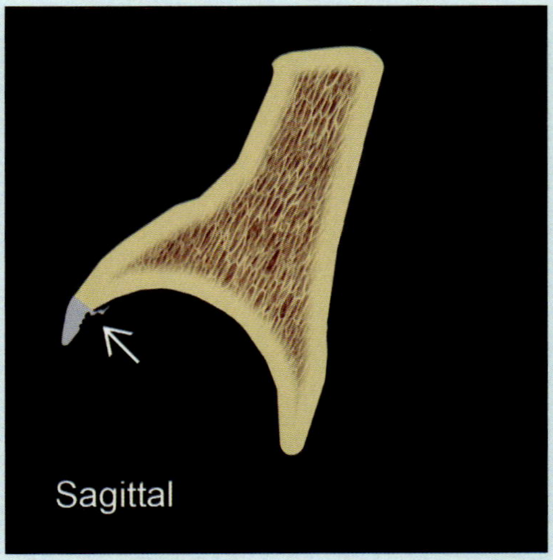

ACETABULAR LABRAL TEAR

TERMINOLOGY

Definitions

- Tear, avulsion, or detachment of acetabular labrum
- Degeneration, fissuring, or wearing of labrum

IMAGING

General Features

- Best diagnostic clue
  - Hyperintensity in labral substance on T2WI
  - Contrast extending into labrum on direct MR arthrogram
- Location
  - Anterosuperior undersurface most common
  - Anterior undersurface or detachment at anterior base
  - Superior undersurface with osteoarthritis (OA)
  - Posterior seen with hip dislocation or subluxation
- Morphology
  - Oblique or linear, extending from undersurface
  - Diffuse degeneration &/or delamination
  - Detachment at labral base
  - Through-and-through tear
  - Bucket-handle or displaced fragment

Radiographic Findings

- Radiography
  - Insensitive for labral tear

Fluoroscopic Findings

- Arthrography: Contrast extends into labral substance
- Symptom relief after intraarticular anesthetic indicates hip as pain source

CT Findings

- Direct CT arthrography
  - Appropriate modality when patient is not MR compatible

MR Findings

- T1WI
  - Normal labrum
    - Hypointense
    - Covers articular cartilage at acetabular margin
  - Diffuse intermediate signal not reaching undersurface = degeneration
  - Fluid or contrast in labrum from undersurface = tear
  - Diastasis between acetabular rim and labrum = tear
  - Loss of normal triangular morphology suggests tear
- T2WI FS
  - Hyperintensity within labrum generally = tear
    - Hyperintense signal to undersurface = tear
  - Loss of triangular morphology = tear
  - Joint fluid dissecting through labrum = detached tear
  - With anterosuperior tears, look for adjacent cartilage loss
- T1WI C+ FS
  - MR arthrography
    - Direct MR arthrography is standard of care for labral tear
      - Articular contrast extends into labral defect
      - Joint distension and traction improve sensitivity
    - Abnormal labrum morphology = tear
      - Also excellent specificity for partial thickness articular cartilage lesions
    - Look for morphologies suggesting developmental dysplasia of hip (DDH) or femoral acetabular impingement (FAI)
  - Indirect MR arthrography may help as alternative but lacks joint distension

Imaging Recommendations

- Best imaging tool
  - Direct MR arthrogram for young patients
    - Sensitivity as low as 20% on noncontrast, over 90% with MR arthrography
  - Noncontrast MR for older patients, osteoarthritis
- Protocol advice
  - Phased array torso coil over pelvis
  - Large field-of-view (FOV) imaging of entire pelvis region
    - Coronal STIR, coronal T1, axial T2 FS
    - Sources of referred pain and extraarticular pathology
      - Tendon injury, osteitis pubis, sports hernia, sacroiliitis, fracture
  - Sequences dedicated to hip in question
    - Coronal, sagittal, and axial oblique prescribed down femoral neck on coronal localizer
    - FOV = 14-18 with arthrogram, 16-24 with noncontrast
  - Direct MR arthrogram
    - Fat-suppressed T1-weighted sequences in 3 planes, including axial oblique
    - Can acquire radial images by adapting MRCP protocol
  - Noncontrast protocol
    - T2 or PD FS in 3 planes, including axial oblique

DIFFERENTIAL DIAGNOSIS

Degenerative Labrum

- Intrasubstance intermediate signal
- No breach in undersurface margin
- Considered tear by some criteria

Normal Labral Attachment

- Normal intermediate cartilage signal at labral base
- No contrast or fluid extension

Sublabral Recess

- Incidence = 18%
- Located anteriorly
  - Sharp margins paralleling acetabular rim
  - Intermediate signal without interposed contrast
- Reported posterosuperiorly (rare)

Perilabral Recess

- Fluid between superior labral margin and capsule (normal)

Osteoarthritis, Hip

- Secondary to labral tear or cause of labral tear
- Labral tear may be symptomatic issue

ACETABULAR LABRAL TEAR

PATHOLOGY

General Features

- Etiology
  - Degenerative with cam- and pincer-type FAI, DDH, and osteoarthritis
  - Traumatic: From twisting injury or hip dislocation
  - Acute, traumatic: Extreme hip rotation and flexion
    - Soccer, hockey, golf, gymnastics, dancing, kickboxing
  - Developmental: Detachment of labrum with os acetabulare in skeletally immature patient
- Associated abnormalities
  - Os acetabulare
  - Paralabral cysts
  - Articular cartilage defects

Staging, Grading, & Classification

- Czerny classification based upon direct MR arthrography
  - Stage IA: Hyperintense signal with no communication to articular surface; perilabral sulcus
  - Stage IB: IA with no perilabral sulcus
  - Stage IIA: Contrast extension into articular surface and perilabral sulcus
  - Stage IIB: IIA but no perilabral sulcus
  - Stage IIIA: Labral detachment and visualized perilabral sulcus
  - Stage IIIB: IIIA with no perilabral sulcus
- Blankenbaker clockface system for extent of tear
  - 3 o'clock = anterior
  - 6 o'clock = transverse ligament
- Lage arthroscopic classification system
  - Focus on etiology and morphology
  - Poor correlation with MR

CLINICAL ISSUES

Presentation

- Most common signs/symptoms
  - Ipsilateral groin pain
- Clinical profile
  - Pain exacerbated by hip flexion or internal rotation suggests anterior tear
  - Pain exacerbated by abduction/external rotation suggests anterosuperior tear
  - Clicking and locking less common than with glenoid labrum tear
  - Internally snapping hip generally related to labral tear

Demographics

- Age
  - 20-50 years: Commonly primary source of pain
    - Beyond 50, labral tears are generally sequelae of degenerative joint disease
  - Age spectrum: 18-75 years
  - Young athletes more likely to sustain acute, traumatic tears
  - Middle-aged patients more likely to have degenerative tears
  - Older patients likely to have tears with osteoarthritis
  - Femoroacetabular impingement and dysplasia cause tears at any age
- Gender
  - M < F for anterior tears
  - M > F for anterosuperior tears and cam-type FAI
- Epidemiology
  - 28% of asymptomatic individuals have labral abnormalities (includes labral degeneration)

Natural History & Prognosis

- Progression of labral tear is typical
- Development of osteoarthritis especially as part of femoroacetabular impingement
  - Vicious cycle: Labral tear causes osteoarthritis; osteoarthritis causes progression of tear

Treatment

- Conservative
  - Activity modification (limit flexion + internal rotation)
  - Anti-inflammatory medication
  - Intraarticular steroid injection
- Surgical
  - Arthroscopic or open
  - Debridement: Trim irregular edges and leave blunted morphology
  - Labral repair: Re-attach labrum to acetabular rim or to more proximal labrum

DIAGNOSTIC CHECKLIST

Consider

- Labral tear as part of FAI or DDH
- Extraarticular & referred sources of hip/groin pain
- Direct MR arthrography with intraarticular anesthetic

Reporting Tips

- Localize labral tear on acetabulum
- Delineate extent of labral tear
- Identify any labral detachment from acetabular rim
- Identify associated paralabral cysts
- Describe lesions of subjacent articular cartilage

SELECTED REFERENCES

1. Potter HG et al: High resolution noncontrast MRI of the hip. J Magn Reson Imaging. 31(2):268-78, 2010
2. Zlatkin MB et al: Acetabular labral tears and cartilage lesions of the hip: indirect MR arthrographic correlation with arthroscopy--a preliminary study. AJR Am J Roentgenol. 194(3):709-14, 2010
3. Kang C et al: Acetabular labral tears in patients with sports injury. Clin Orthop Surg. 1(4):230-5, 2009
4. Studler U et al: MR arthrography of the hip: differentiation between an anterior sublabral recess as a normal variant and a labral tear. Radiology. 249(3):947-54, 2008
5. Blankenbaker DG et al: Classification and localization of acetabular labral tears. Skeletal Radiol. 36(5):391-7, 2007
6. Toomayan GA et al: Sensitivity of MR arthrography in the evaluation of acetabular labral tears. AJR Am J Roentgenol. 186(2):449-53, 2006
7. Kassarjian A et al: Triad of MR arthrographic findings in patients with cam-type femoroacetabular impingement. Radiology. 236(2):588-92, 2005
8. Petersilge C: Imaging of the acetabular labrum. Magn Reson Imaging Clin N Am. 13(4):641-52, vi, 2005
9. Zoga AC et al: Technical considerations in MR imaging of the hip. Magn Reson Imaging Clin N Am. 13(4):617-34, v, 2005

ACETABULAR LABRAL TEAR

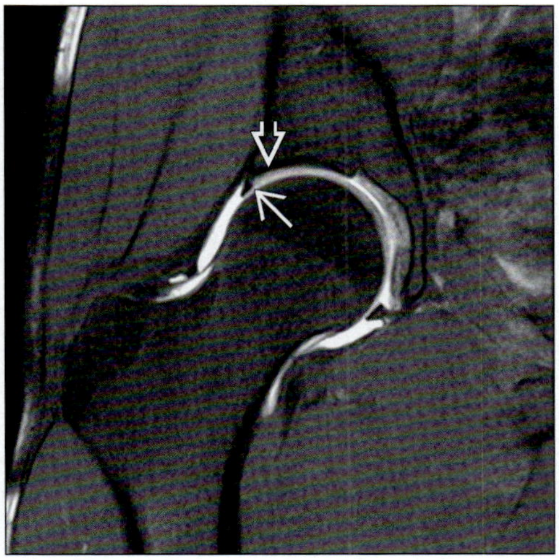

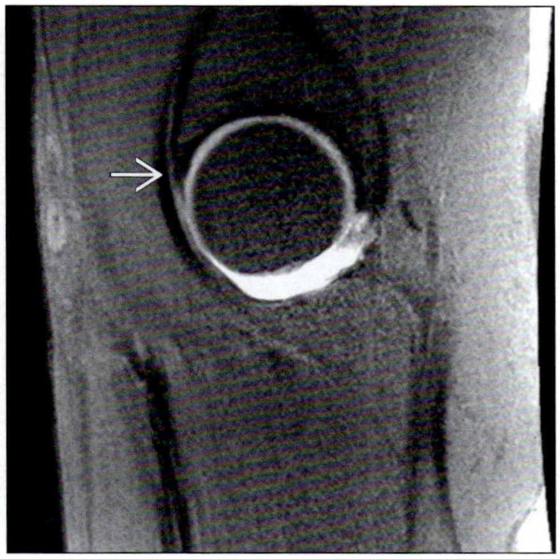

(Left) *Coronal MR arthrogram T1WI FS shows the normal superior acetabular labrum. The labrum itself should be dark and triangular in all planes* ➔*. At the apex of the acetabular roof, the labrum blends with the more hypointense articular cartilage* ➔*.* ***(Right)*** *Sagittal MR arthrogram T1WI FS shows the normal anterior labrum. The undersurface of the dark, triangular labrum* ➔ *should be continuous with the acetabular rim, with no contrast entering the labral substance.*

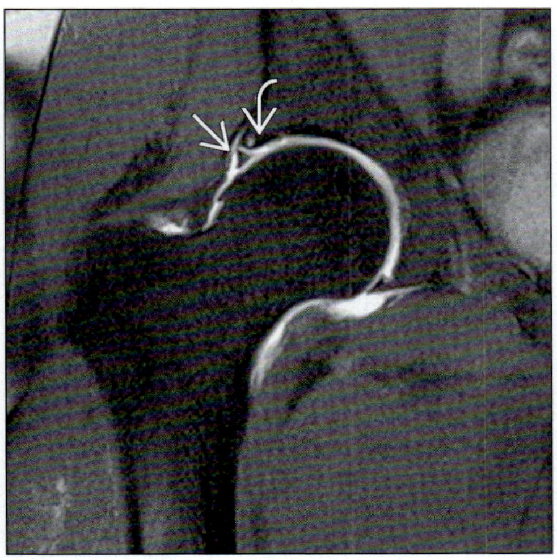

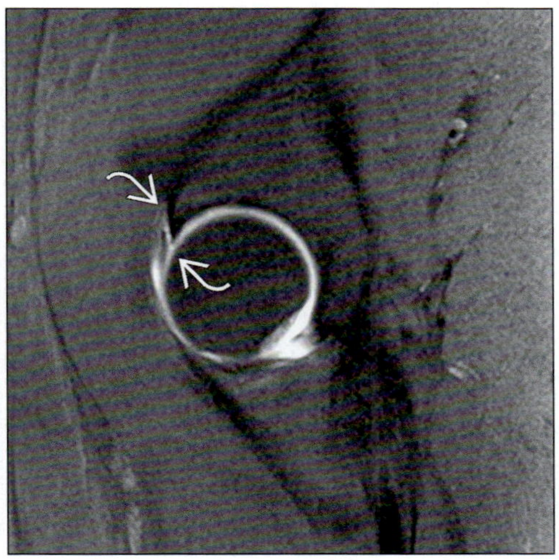

(Left) *Coronal MR arthrogram T1WI FS shows an obvious superior acetabular labrum tear. Contrast undermines the detached labrum* ➔*, and there is cartilage loss and an evolving geode on the acetabular rim* ➔ *in this patient with cam-type FAI.* ***(Right)*** *Sagittal MR arthrogram T1WI FS demonstrates an obvious anterior labral tear with detachment. Contrast passes all the way through the base of the labrum* ➔ *in this professional baseball player with pain on hip flexion.*

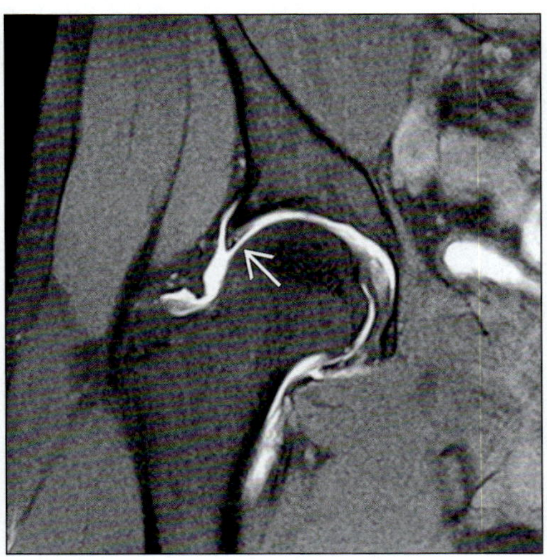

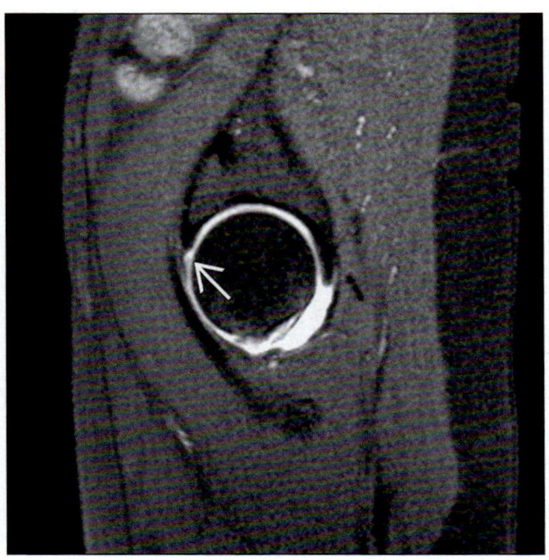

(Left) *Coronal MR arthrogram T1WI FS demonstrates a subtle superior labral tear. A small defect at the undersurface of the labrum* ➔ *was found to be a labral tear at arthroscopy in this 18-year-old hockey player with pain on external hip rotation.* ***(Right)*** *Sagittal MR arthrogram T1WI FS shows an abnormal contour at the labral undersurface* ➔ *without detachment. At arthroscopy, the defect extended to an intrasubstance tear in this female collegiate lacrosse player.*

ACETABULAR LABRAL TEAR

*(**Left**) Sagittal T2WI FS MR from a noncontrast hip MR shows a paralabral cyst extending from the region of the anterior labrum, indicating the presence of a labral tear. (**Right**) Axial T2WI FS MR from the same patient localizes the paralabral cyst to an extension from the anterior labral base , but the tear is still not visualized. Noncontrast MR protocols often do not provide the signal necessary to identify acetabular labral tears in the absence of paralabral cysts or osteoarthritis.*

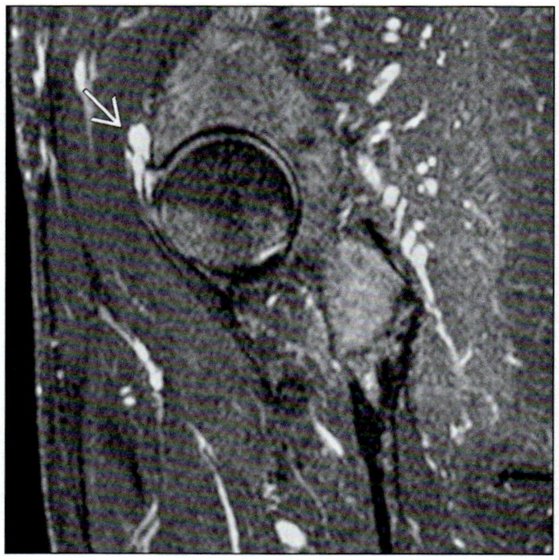

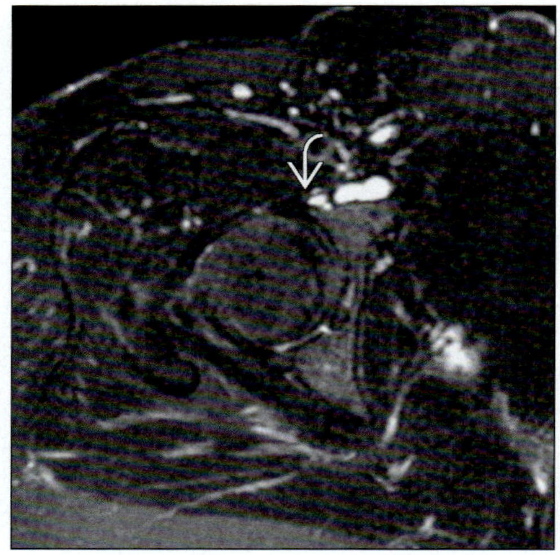

*(**Left**) Sagittal T2WI FS MR from a 25-year-old woman with pain on thigh flexion shows a focus of abnormal hyperintense signal in the anterior labrum indicating a tear , with extension of the paralabral cyst (proved contiguous on other images) distal along the hip flexor tendons . (**Right**) Coronal STIR MR from a 47 year old with anterior hip pain shows fluid along the medial iliac wing . The fluid could be tracked distal to the anterosuperior acetabular labrum, representing a large paralabral cyst.*

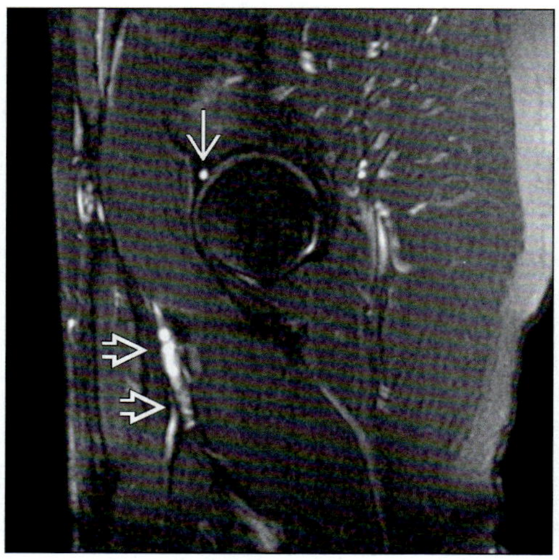

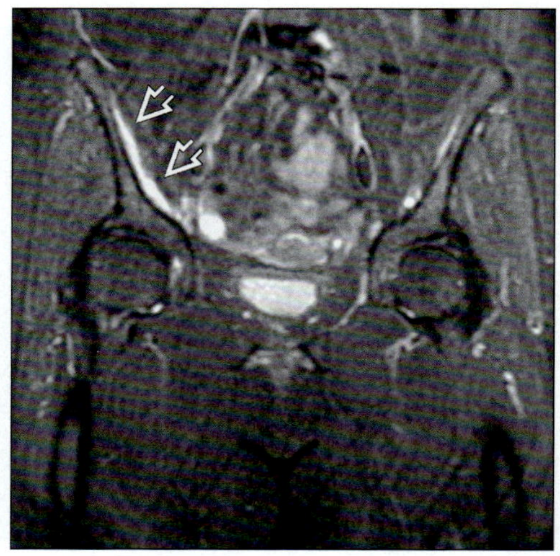

*(**Left**) Sagittal indirect MR arthrogram T1WI FS generates relatively low signal to noise ratio but shows maceration of the anterior labrum , a large enhancing paralabral cyst extending into the iliopsoas bursa , and a degenerative cyst in the acetabulum. (**Right**) Axial PDWI MR acquired at 3T in a professional football player with buttock pain shows a torn posterior labrum with capsular detachment from the posterior acetabulum after a posterior hip subluxation.*

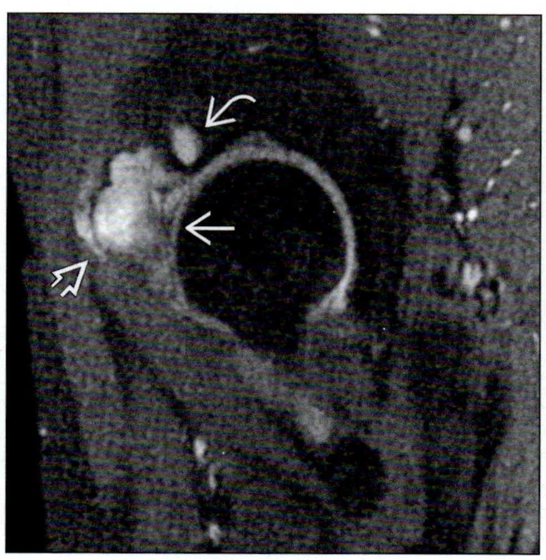

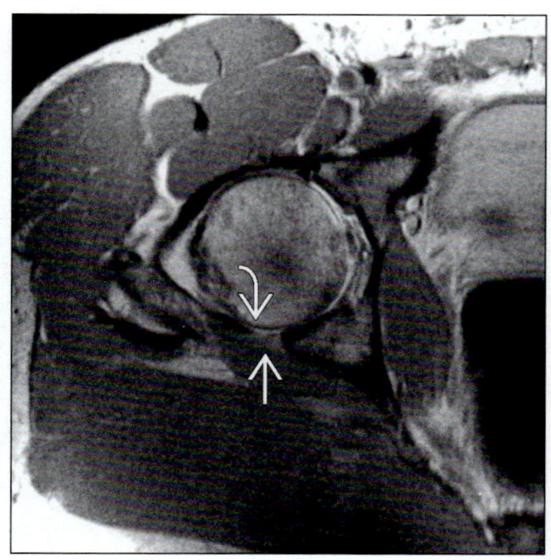

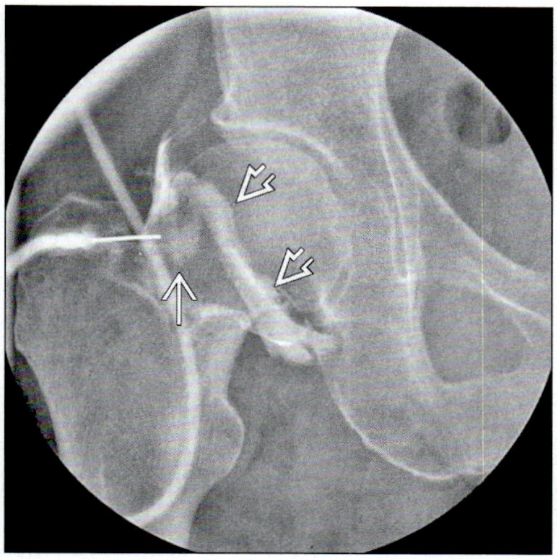

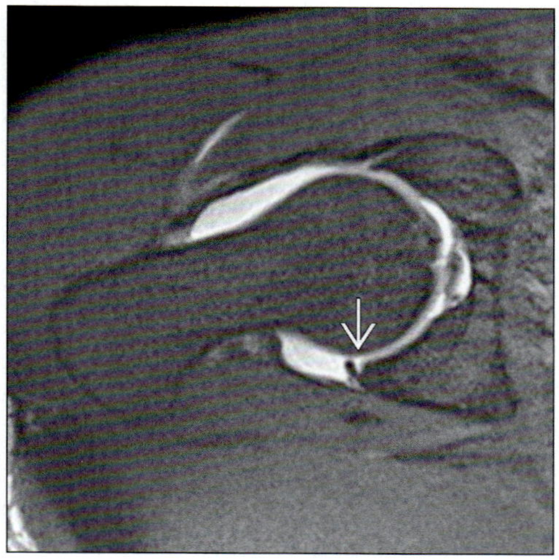

***(Left)** AP fluoroscopic spot radiograph from a pre-MR arthrogram shows a collection of contrast with a vertical configuration related to an initial iliopsoas bursal injection → but more contrast forming a reassuring "turtleneck" ⇒ around the femoral neck indicating correct intraarticular needle position. **(Right)** Axial oblique MR arthrogram T1WI FS from a wrestler with hip pain on external rotation shows a focal partial detachment → of the posterior labrum from the acetabular rim.*

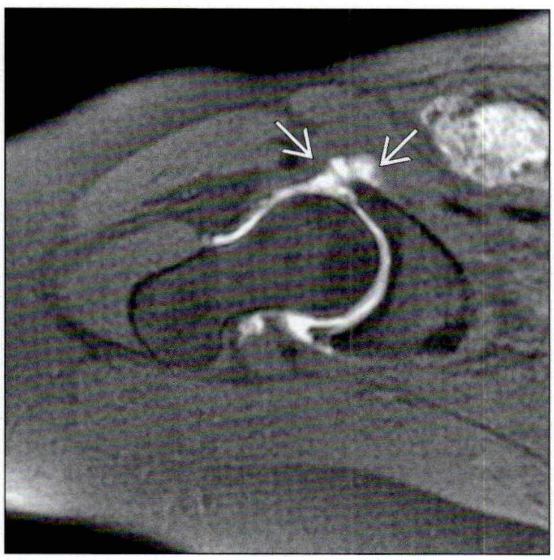

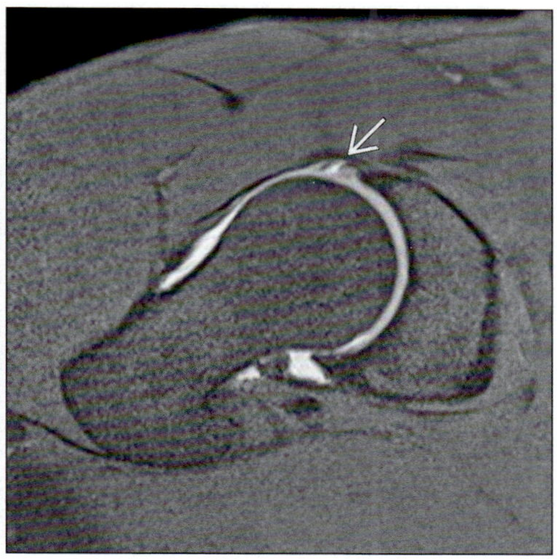

***(Left)** Axial oblique MR arthrogram T1WI FS shows contrast pooling in a multilobulated structure at the anterosuperior acetabulum →. This is the typical appearance of a paralabral cyst with a labral tear at direct MR arthrography. **(Right)** Axial oblique MR arthrogram T1WI FS shows a detached anterosuperior acetabular labral tear → in a high school soccer player. The acetabular labrum should be homogeneously hypointense and triangular in all arthrographic imaging planes.*

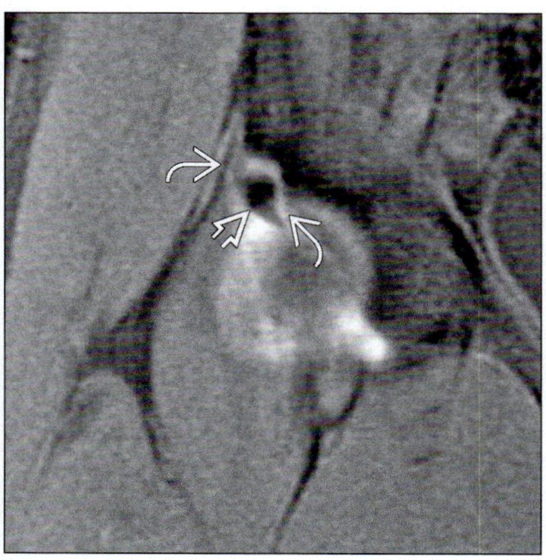

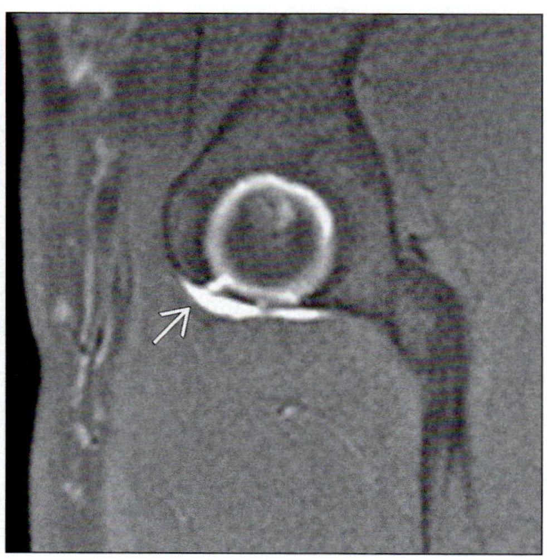

***(Left)** Coronal MR arthrogram T1WI FS at the anterior hip in a professional hockey player with coxa saltans and 2 previous negative noncontrast MR studies shows contrast ↷ lifting the anterosuperior labrum and an acetabular ossicle ⇒ from the acetabular rim at the site of the symptoms. **(Right)** Sagittal MR arthrogram T1WI FS from a competitive swimmer with weakness on hip flexion shows focal disruption of the transverse ligament from the anteromedial aspect of the acetabular labrum →.*

FEMORAL ACETABULAR IMPINGEMENT AND DYSPLASIA

Key Facts

Terminology

- Cam-type FAI: Osseous prominence at femoral head/neck junction leads to impingement on anterosuperior acetabulum
- Pincer-type FAI: Osseous prominence of acetabulum leads to impingement on femoral neck
- DDH: Spectrum of developmental/modeling failure of acetabulum leading to abnormal acetabular morphology and undercoverage of femoral head
- Some have elements of both cam- and pincer-type FAI

Imaging

- Variant femoral head/neck morphology
  - Osseous excrescence or prominence in cam-type FAI
- Variant acetabular morphology
  - Acetabular retroversion with pincer-type FAI and DDH
  - Coxa profunda: Acetabulum is too deep, leading to pincer-type FAI
  - Deficiency of acetabulum in DDH
- MR arthrographic findings of acetabular labral tear and cartilage loss
  - Triad of MR arthrographic findings with cam-type FAI: Labral tear, cartilage defect, abnormal alpha angle
  - Overgrown labrum with DDH

Pathology

- Trauma to labrum, cartilage, and bone dictated by morphology

Clinical Issues

- Most "idiopathic" hip osteoarthritis attributable to FAI or DDH

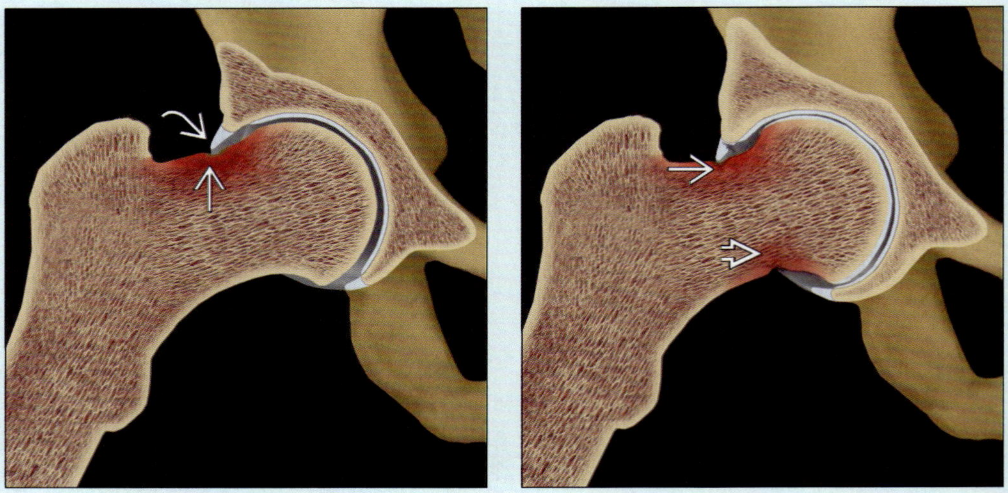

(Left) *Coronal graphic of cam-type femoral acetabular impingement (FAI) shows the point of impact between an abnormal femoral head/neck junction ➔ and the anterosuperior acetabular rim ➔. Repeated impaction in this location will lead to focal labral tearing and articular cartilage loss.* ***(Right)*** *Coronal graphic of pincer-type FAI shows coup-contrecoup points of impaction on both the superior ➔ and inferior ⇨ aspects of the femoral neck due to overcoverage by an abnormally deep acetabulum.*

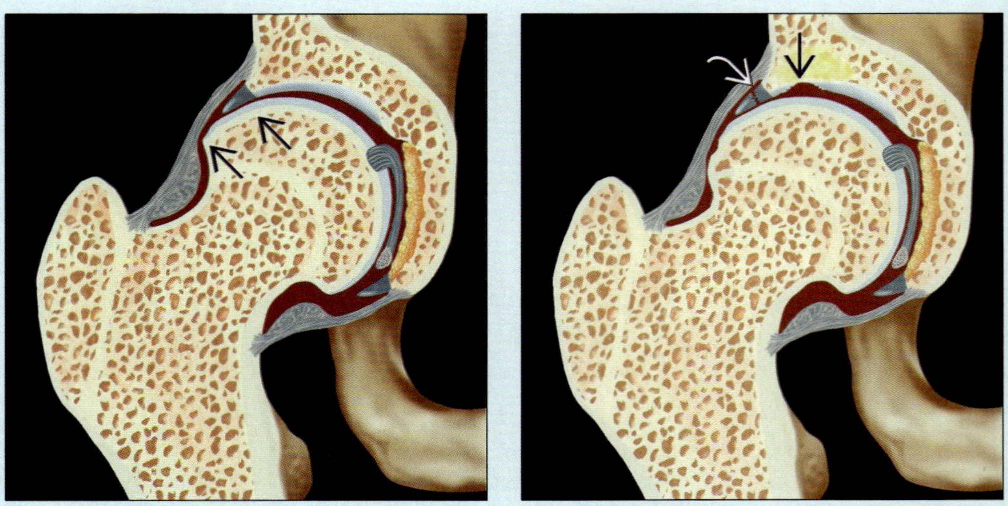

(Left) *Coronal graphic of a juvenile hip shows relative undercoverage of the femoral head ⇨ by the acetabulum, indicating developmental dysplasia of the hip (DDH).* ***(Right)*** *Coronal graphic of the same hip years into adulthood shows stable deficiency at the dysplastic acetabular rim, but evolution of articular cartilage loss in the superior acetabulum ⇨, decreased superior hip joint space, and a degenerative acetabular labrum tear ➔.*

FEMORAL ACETABULAR IMPINGEMENT AND DYSPLASIA

TERMINOLOGY

Abbreviations

- Femoral acetabular impingement (FAI)
- Developmental dysplasia of hip (DDH)

Definitions

- FAI: Morphologic abnormality at femoral head/neck junction or acetabulum resulting in osseous &/or labral and cartilaginous impingement
  - Cam-type: Prominence at femoral head/neck junction impinges anterosuperior acetabulum
    - Bump at anterosuperior femoral head/neck junction: May result from eccentric skeletal maturation
    - Flat anterosuperior femoral head/neck junction: May reflect subclinical SCFE
  - Pincer-type: Prominence of anterior acetabulum leads to impingement on femoral neck
    - Coxa profunda: Acetabulum too deep with excessive surface area
    - Posterior acetabular deficiency: Femoral head shifts posteriorly with hip flexion increasing coverage by acetabulum anteriorly
  - May have both cam- and pincer-type FAI
    - Combination with cam FAI common with acetabular retroversion
- DDH: Spectrum of modeling failure of acetabulum leading to undercoverage of femoral head

IMAGING

General Features

- Best diagnostic clue
  - Abnormal morphology at proximal femur and acetabular rim with associated labral tear and cartilage lesions
- Location
  - Cam-type FAI: Femoral head/neck junction
  - Pincer-type FAI: Acetabular rim
  - DDH: Acetabular rim

Radiographic Findings

- Variant femoral head/neck morphology
  - Osseous excrescence or prominence in cam-type FAI
    - Accentuated with frog leg lateral view (flexion + external rotation)
    - Coxa magna: Short, broad femoral neck with mushroom-shaped femoral head
  - Flattening along femoral head/neck junction in cam-type FAI
    - "Pistol grip" deformity
  - Synovial herniation pit in cam- or pincer-type FAI
- Variant acetabular morphology
  - Os acetabulare with cam- or pincer-type FAI
    - Accessory ossicle at lateral acetabular margin
  - Acetabular retroversion: Pincer-type FAI and DDH
    - Crossover of anterior and posterior acetabular rims on standing AP radiograph
    - Posterosuperior acetabulum is deficient
    - → anterior pincer impingement with flexion
  - Coxa profunda: Deep acetabulum in pincer-type FAI
    - Overgrowth of acetabulum
  - Protrusio acetabulum
  - Deficiency of acetabulum in DDH
    - Flat or upturned acetabular roof
    - Acetabular sourcil (lateral margin of subchondral cortical line) is upturned
    - Inadequate coverage of femoral head
      - Femoral head extrusion index (> 25%) or abnormal center-edge angle of Wiberg
- Osteoarthritis (OA)
  - Ultimately can occur in setting of any of these morphologies
  - Goal should be to treat variant morphology and prevent osteoarthritis

MR Findings

- T1WI
  - MR adaptation of radiographic findings
  - FAI: Low signal subchondral degenerative change
  - DDH: Overgrowth of acetabular labrum
- T2WI
  - Articular cartilage loss, subchondral marrow edema, and cystic change in typical locations
    - Anterosuperior acetabulum with cam-type FAI
    - Superior and inferomedial femoral neck with pincer-type FAI
    - Anterosuperior femoral head and superior acetabulum with DDH
- T1WI C+
  - MR arthrographic findings of acetabular labral tear and cartilage loss
    - Anterosuperior femoral head/neck junction with cam-type FAI
      - Triad of MR arthrographic findings with cam-type FAI: Labral tear, cartilage defect, abnormal alpha angle
    - Superior and anterior with DDH
    - Overgrown labrum with DDH
    - Anterior with retroversion and posterior acetabular insufficiency
    - Extensive labral tears with pincer-type FAI
- Coronal sequences
  - Labral tear, cartilage lesions, & herniation pits with cam-type FAI
  - Undercoverage, labral & cartilage lesions with DDH
- Sagittal sequences
  - Labral tear & cartilage loss with pincer-type FAI and retroversion
- Axial oblique sequences
  - Abnormal alpha angle, labral tear, & cartilage loss with cam-type FAI
  - Alpha angle drawn from center of femoral head
    - 1 ray along femoral neck
    - 1 ray to location of head/neck bump
      - Point where cortex extends beyond circle placed on femoral head

Imaging Recommendations

- Best imaging tool
  - Well-positioned pelvic radiograph for morphology
    - Frog leg lateral view, Dunn view, false profile are additive
  - MR (direct MR arthrography) for sequelae of morphologic abnormalities
- Protocol advice
  - Axial oblique prescribed along femoral neck

FEMORAL ACETABULAR IMPINGEMENT AND DYSPLASIA

DIFFERENTIAL DIAGNOSIS

Slipped Capital Femoral Epiphysis
- Can look like cam-type impingement
- Skeletally immature

Legg-Calvé-Perthes
- When mature, can manifest as cam-type FAI or DDH
- Flattening of articular surface of femur

Osteonecrosis, Hip
- Subchondral crescent with double line sign on MR
- Surface morphology normal in early stages, prior to osteophyte production

Isolated Acetabular Labral Tears
- Unilateral or asymmetric
- Other imaging findings to indicate trauma
- Normal femur and acetabulum morphology

Osteoarthritis, Hip
- Can arise 2° to any of previously listed processes
- Subchondral cysts, sclerosis, osteophytes, joint narrowing

PATHOLOGY

General Features
- Etiology
  - Trauma to labrum, cartilage, and bone dictated by morphology
    - Repetitive microtrauma from eccentric pressures and impaction with FAI
    - Repetitive microtrauma from instability in DDH
    - Leads to labral tears and articular cartilage degeneration
      - Exacerbates degenerative process by adding further instability
- Associated abnormalities
  - Paralabral cysts
  - Concomitant pubic symphysis injury in athletes with cam-type FAI

Staging, Grading, & Classification
- Chondral wear can be graded with Outerbridge or Noyes systems
- Labral tear can be graded with Czerny or Blankenbaker systems

Gross Pathologic & Surgical Features
- Absent or thinned articular cartilage in locations of impingement
- Synovial herniation pits filled with loose fibromyxoid tissue

CLINICAL ISSUES

Presentation
- Most common signs/symptoms
  - Cam-type FAI: Abduction, external rotation pain
  - Pincer-type FAI: Flexion, internal rotation pain
  - DDH: Pain with variable maneuvers
    - Giving way or internal snapping hip
- Other signs/symptoms
  - Groin pain with activity common with labral tear
- Clinical profile
  - Cam-type FAI: Tall males
    - Possible association with activities during skeletal immaturity: Ice hockey
  - Pincer-type FAI: Young patients with retroversion, or middle-aged females
  - DDH: Any age; M < F
  - Most "idiopathic" hip OA attributable to FAI or DDH
    - Incidence of OA ↑ with age and activity

Natural History & Prognosis
- If unrecognized can lead to OA and hip replacement
- Activity modification may stabilize symptoms
- Cartilage lesions and labral tears accelerate OA

Treatment
- Conservative
  - Non-weight-bearing (short term)
  - Anti-inflammatory medication
  - Core muscle strengthening (decrease hip stress)
  - Non-loading activities (bicycle, elliptical)
- Surgical
  - Labral and chondral debridement
  - Labral repair
  - Femoral cheilectomy: Removal of osseous excrescence at femoral head/neck junction
  - Periacetabular osteotomy for retroversion
  - THR for painful osteoarthritis

DIAGNOSTIC CHECKLIST

Consider
- Standing AP radiograph of pelvis for femoral and acetabular morphology
- Direct MR arthrography for labral and cartilage lesions
  - Axial oblique sequence for alpha angle

Image Interpretation Pearls
- Look for bilateral symmetric morphologic variations

Reporting Tips
- Consider morphologic source of soft tissue trauma

SELECTED REFERENCES

1. Jessel RH et al: Radiographic and patient factors associated with pre-radiographic osteoarthritis in hip dysplasia. J Bone Joint Surg Am. 91(5):1120-9, 2009
2. Kang C et al: Acetabular labral tears in patients with sports injury. Clin Orthop Surg. 1(4):230-5, 2009
3. Leunig M et al: The concept of femoroacetabular impingement: current status and future perspectives. Clin Orthop Relat Res. 467(3):616-22, 2009
4. Pfirrmann CW et al: MR arthrography of acetabular cartilage delamination in femoroacetabular cam impingement. Radiology. 249(1):236-41, 2008
5. Blankenbaker DG et al: The painful hip: new concepts. Skeletal Radiol. 35(6):352-70, 2006
6. Bredella MA et al: MR imaging of femoroacetabular impingement. Magn Reson Imaging Clin N Am. 13(4):653-64, 2005
7. Kassarjian A et al: Triad of MR arthrographic findings in patients with cam-type femoroacetabular impingement. Radiology. 236(2):588-92, 2005
8. Abe I et al: Acetabular labrum: abnormal findings at MR imaging in asymptomatic hips. Radiology. 216(2):576-81, 2000

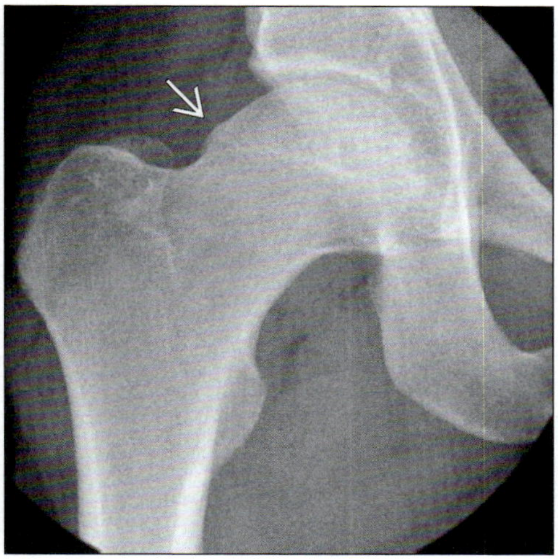

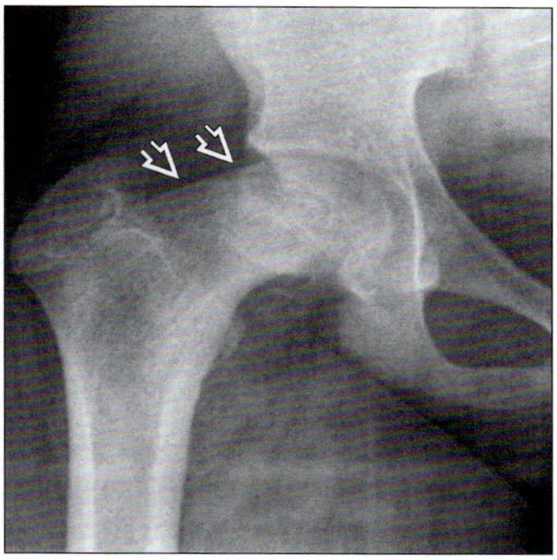

(Left) *AP hip radiograph in a 19-year-old male athlete with impingement symptoms shows an abnormal osseous prominence at the anterosuperior femoral head/neck junction or coxa magna morphology. This is 1 of the morphologies associated with cam-type FAI.* ***(Right)*** *AP radiograph of the hip in a 17 year old with impingement symptoms demonstrates a developmentally abnormal flat anterosuperior femoral head/neck junction, or the "pistol grip" morphology associated with cam-type FAI.*

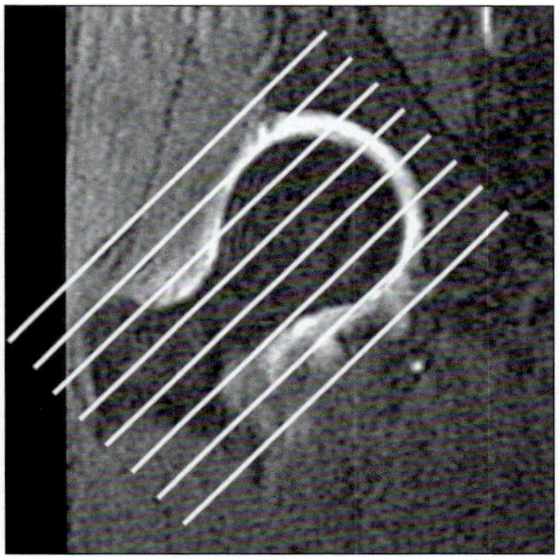

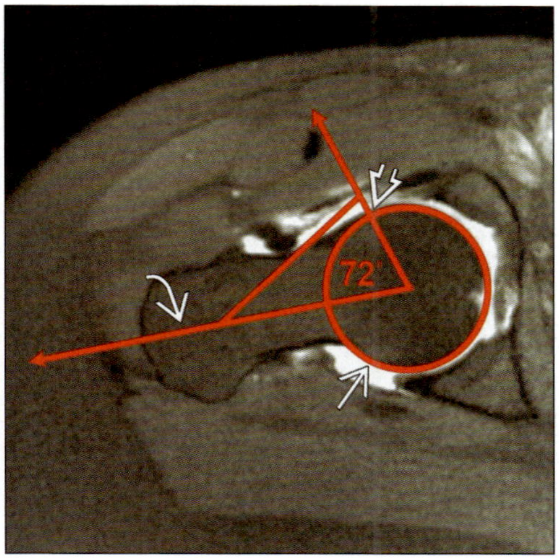

(Left) *Coronal MR arthrogram localizer image used to prescribe the axial oblique MR plane helpful in the evaluation of cam-type FAI.* ***(Right)*** *Axial oblique T1WI C+ FS MR in cam-type FAI shows an abnormal alpha angle generated with rays from the center of a circle placed on the femoral head cortex along the femoral neck, and to the point where the anterosuperior cortex erupts from the circle. Reproducibility of the MR alpha angle has been questioned, but 70° and above is clearly abnormal.*

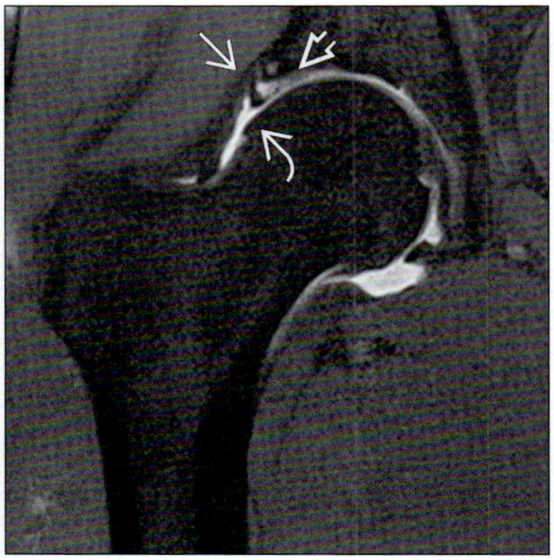

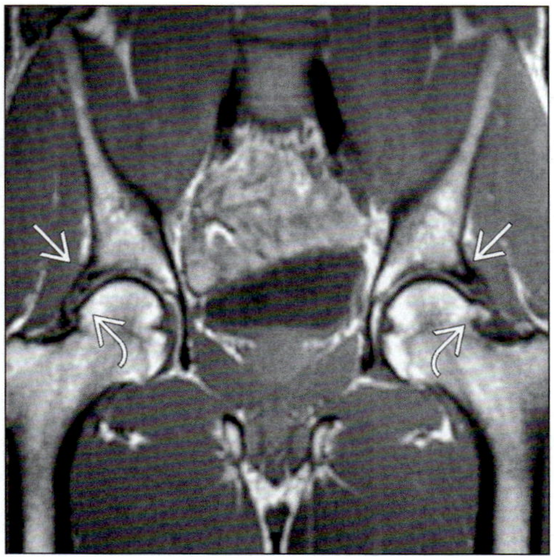

(Left) *Coronal MR arthrogram T1WI FS shows the triad of MR findings associated with cam-type femoral acetabular impingement: A complex labral tear, cartilage loss on the anterosuperior acetabulum, and an osseous bump at the femoral head/neck junction.* ***(Right)*** *Coronal T1WI MR of the pelvis from an NBA player with chronic groin pain shows huge osseous excrescences at the femoral head/neck junctions and complex labral tears, suggesting chronic cam-type FAI.*

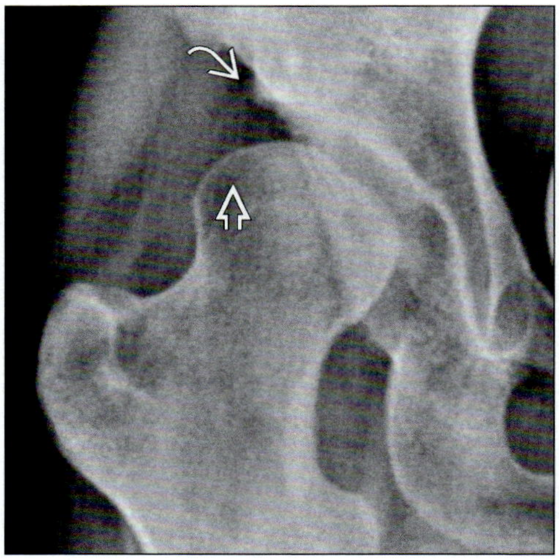

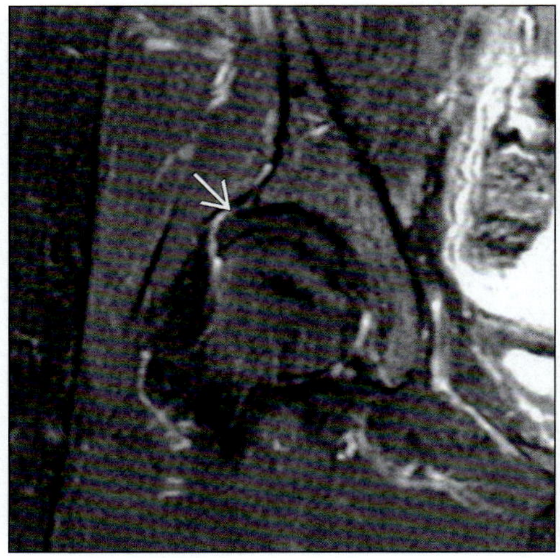

(Left) *AP radiograph from an adult patient with developmental dysplasia of the hip (DDH) shows an abnormal upturned sourcil as well as an uncovered femoral head. Severe cases of DDH can be diagnosed easily on radiographs.* ***(Right)*** *Coronal T2WI FS MR from the same patient shows an extremely large, overgrown acetabular labrum. With developmental dysplasia of the hip, the hyperplastic labrum takes on the responsibility of replacing the deficient osseous acetabulum.*

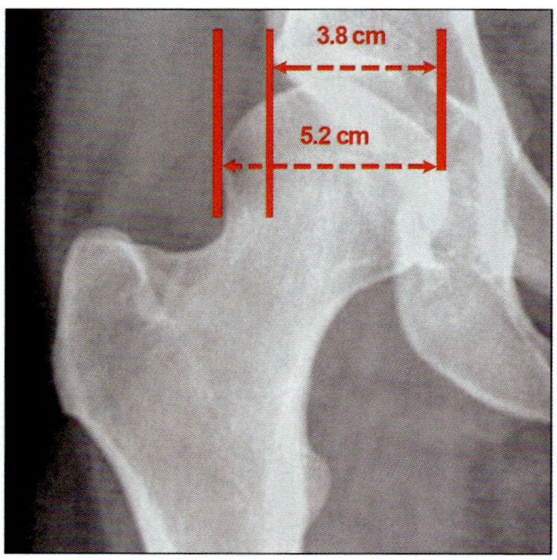

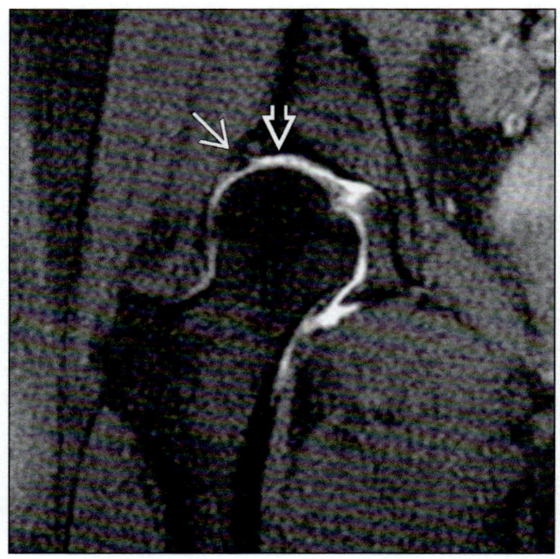

(Left) *Standing AP radiograph of the hip shows an abnormal femoral head extrusion index (percentage of uncovered femoral head) in a more subtle case of DDH. In this case, 3.8/(5.2-3.8) x 10 = an index of 27; index values above 25 are considered abnormal.* ***(Right)*** *Coronal T2WI FS MR from the same patient, a 31 year old with chronic hip pain, shows cartilage loss in the superior acetabulum and a labral tear, likely reflecting microinstability related to DDH.*

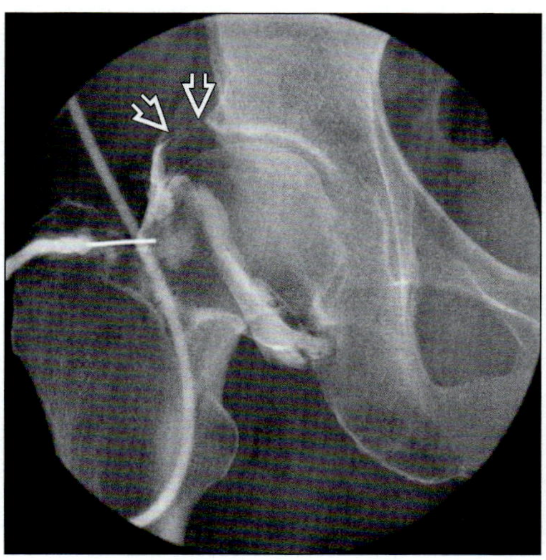

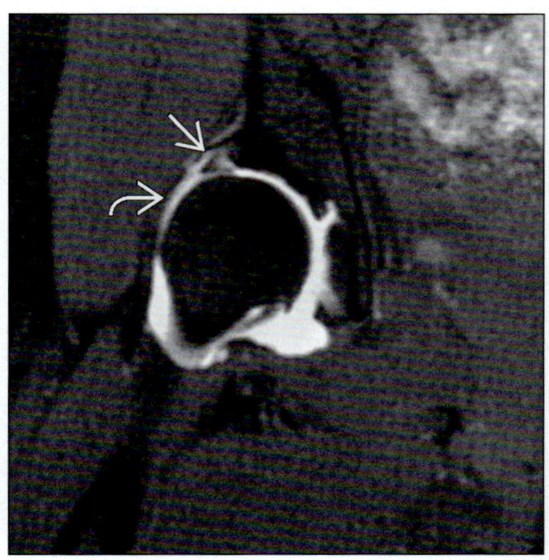

(Left) *Coronal fluoroscopic spot image from a direct MR arthrogram shows contrast outlining a very large superior acetabular labrum in a patient with femoral head undercoverage and developmental dysplasia of the hip.* ***(Right)*** *Coronal MR arthrogram T1WI FS acquired with the hip in abduction and external rotation shows a torn and laterally displaced labrum as well as an uncovered femoral head, supporting the concept of femoral acetabular instability in this patient with DDH.*

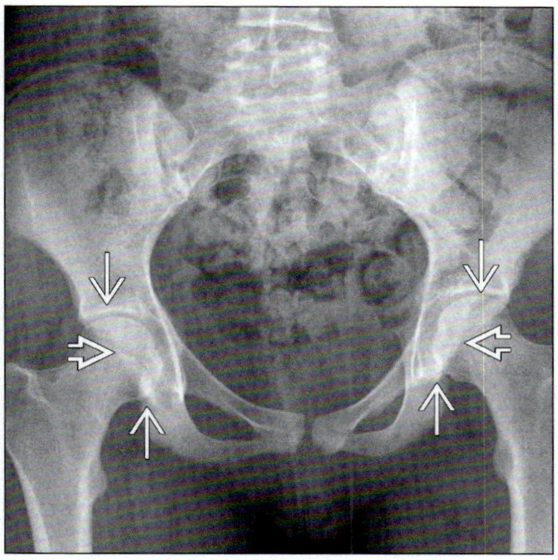

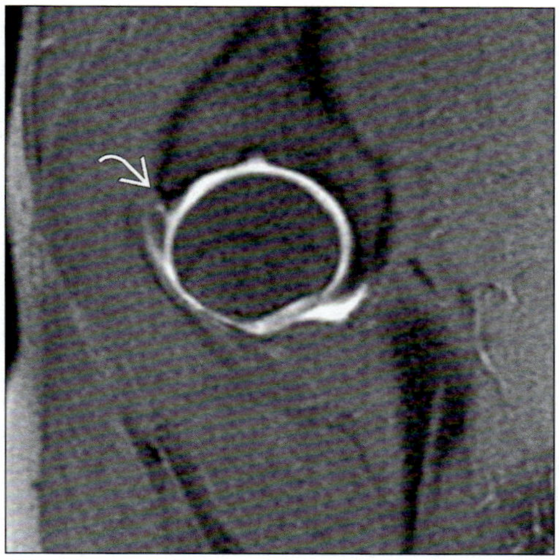

(Left) *Standing AP radiograph from a 14-year-old girl with groin pain shows bilateral acetabular retroversion: The posterior acetabulum ➔ crosses medial to the anterior acetabulum ⇨ at the superior joint margin. This posterosuperior insufficiency leads to pincer-type FAI anteriorly.* ***(Right)*** *Sagittal MR arthrogram T1WI FS from the same patient shows disruption and detachment of the anterior labrum ↷ at its base, a common lesion in the setting of acetabular retroversion.*

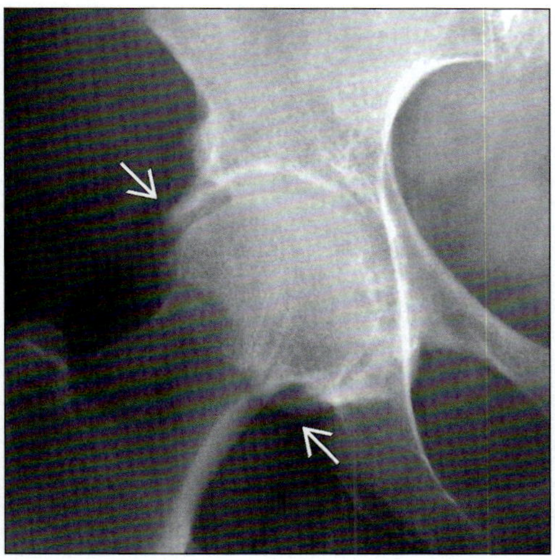

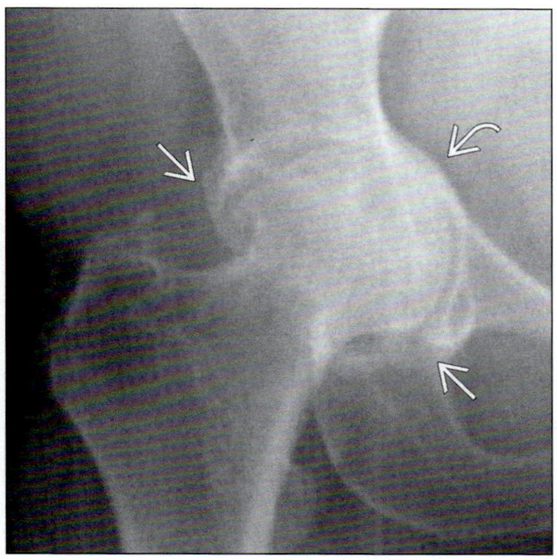

(Left) *AP radiograph of the hip in a 45 year old with chronic groin pain shows overcoverage of the femoral head by an enlarged, deep acetabular cup ➔. This morphology leads to the coup-contrecoup sites of impingement typical in pincer-type FAI.* ***(Right)*** *AP radiograph in a 38 year old with rheumatoid arthritis also shows overcoverage of the femoral head by an enlarged acetabulum ➔. In this case, acetabular protrusio ↷ has led to deepening of the cup and secondary pincer-type FAI.*

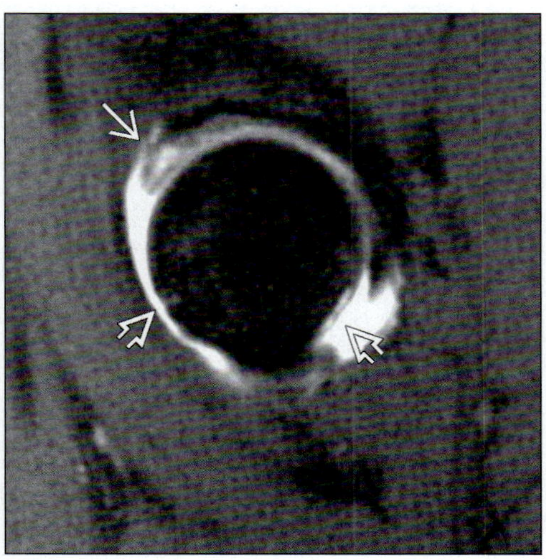

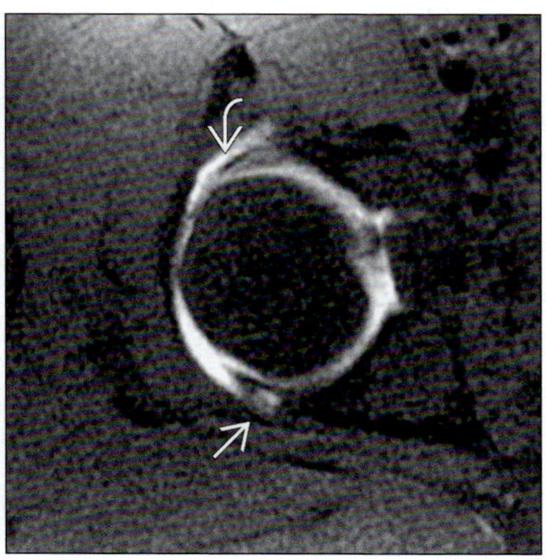

(Left) *Sagittal MR arthrogram T1WI FS shows a complex and extensive labral tear ➔ with cortical indentations ⇨ both anteriorly and posteriorly on the femoral neck. Coup-contrecoup sites of impingement on the femoral neck and complex labral tears are common with pincer-type FAI.* ***(Right)*** *Axial MR arthrogram T1WI FS from the same patient shows the labral tear extending posteriorly ➔ as well as relative overcoverage of the femoral head by the acetabulum ↷, indicating pincer-type FAI.*

Key Facts

Imaging

- Direct MR arthrography for re-tear, postoperative anatomy
- Intravenous contrast MR for infection, synovitis
- Look for cheilectomy (osteoplasty)
- Assess osseous contours
- Labrum can have heterogeneous signal but look for
  - Jagged contours in labrum to suggest re-tear
  - Frank detachment from acetabulum
- Hip capsular hypertrophy expected
- Look for remote sources of pain
- Comparison with preoperative study using similar protocol when possible

Top Differential Diagnoses

- Try to distinguish post-debridement granulation tissue in labrum from re-tear

Pathology

- Arthroscopic debridement of labral tears
- Debridement to vascular region of labrum
- Acetabular cartilage often debrided in conjunction
- Labral repair
- Suture or suture anchors attach labrum to proximal labrum or acetabulum
- Cheilectomy (osteoplasty)
  - Remodeling of anterosuperior femoral head/neck junction
  - Used with underlying cam-type FAI
- Ligamentum teres often debrided at arthroscopy

Clinical Issues

- Persistent or recurrent groin or hip pain post labral debridement or repair
- Occasionally coxa saltans (snapping hip)

*(**Left**) Sagittal MR arthrogram T1WI FS in a 26-year-old woman after labral repair for a detached tear at anterior labrum base shows susceptibility artifact anterior to the acetabular rim representing suture ➡, but there is no detachment or re-tear allowing contrast to traverse the repair. (**Right**) Sagittal MR arthrogram T1WI FS from a 23 year old after anterior labrum debridement shows heterogeneous intrasubstance signal ➡, typical for granulation tissue. There is no re-tear.*

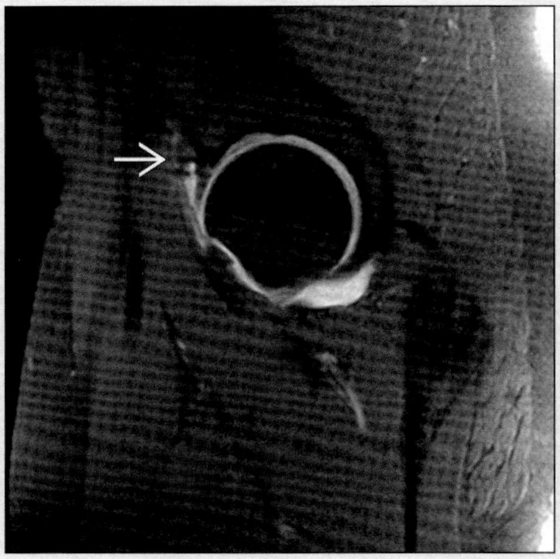

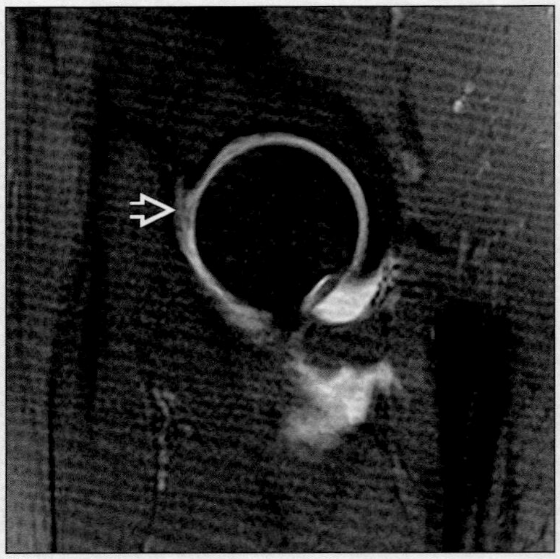

*(**Left**) Coronal STIR MR from a hip arthrogram in a patient with prior labral debridement shows asymmetric capsular thickening anterolaterally on the right ➡, an expected finding with prior hip arthroscopy. Note normal contralateral capsule ➡. (**Right**) Axial oblique MR arthrogram T1WI FS after labral debridement and cheilectomy shows surgical remodeling of the anterosuperior femoral head/neck junction ➡ and heterogeneous signal in the labrum ➡.*

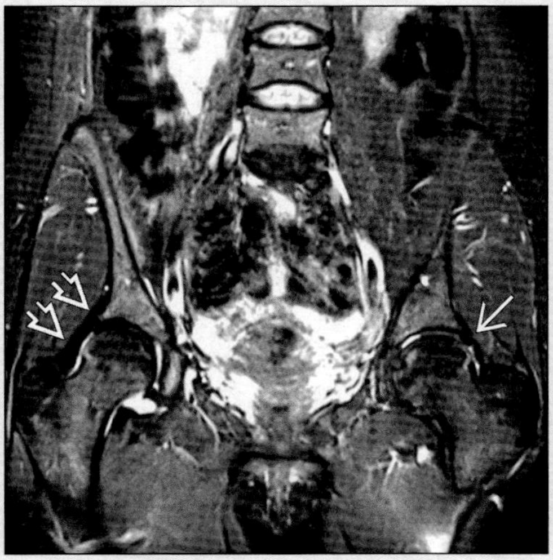

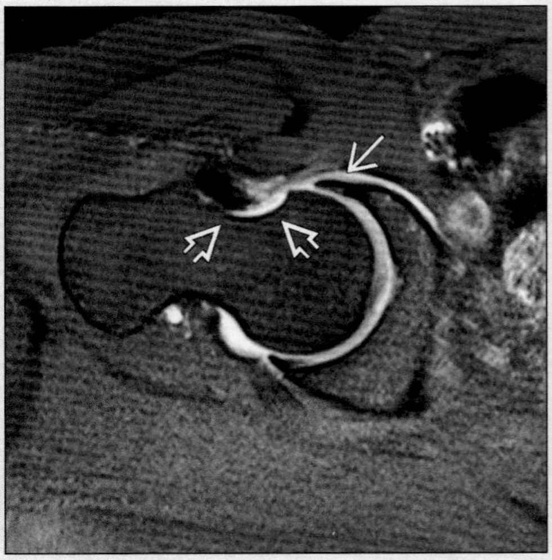

HIP LABRAL TRAUMA, POSTOPERATIVE

IMAGING

General Features

- Best diagnostic clue
  - Thickened hip capsule
  - Heterogeneous labral signal
  - Surgical contour at anterosuperior femoral head/neck junction
- Location
  - Anterior, anterosuperior, or posterior aspect of hip

Imaging Recommendations

- Best imaging tool
  - Direct MR arthrography for re-tear, postoperative anatomy
  - Intravenous contrast MR for infection, synovitis
- Protocol advice
  - Include axial oblique sequence on MR
    - Prescribed along femoral neck on coronal localizer

Radiographic Findings

- Look for cheilectomy (femoral neck osteoplasty)
  - Extrinsic remodeling at anterosuperior femoral head/neck junction
- Sutures, suture anchors (rare)

MR Findings

- Assess osseous contours (femoral head/neck & acetabulum)
  - Cheilectomy
  - Productive change indicating osteoarthritis
- Labrum can have heterogeneous signal but look for signs of re-tear
  - Jagged contours in labrum suggest re-tear
    - Contrast entry into labral substance on MR arthrography
  - Cartilage flaps and fissures
  - Frank detachment from acetabulum
- Hip capsular hypertrophy expected
  - Look for scar or fibrosis in surrounding soft tissues
- Ligamentum teres
  - Often debrided at arthroscopy
  - Compare with contralateral hip if possible
  - Describe morphology
- Comparison with preoperative study using similar imaging protocol when possible

DIFFERENTIAL DIAGNOSIS

Acetabular Labral Cartilaginous Lesions

- No history of previous surgery

Loose Bodies, Hip

- Look for chondral donor site

PATHOLOGY

General Features

- Etiology
  - Arthroscopic debridement of labral tears
    - Jagged contours debrided
    - Debridement to vascular region of labrum
      - Facilitate healing of labral defect
    - Acetabular cartilage often debrided in conjunction
  - Labral repair
    - Suture or suture anchors attach labrum to proximal labrum or acetabulum
    - Common at base of anterior labrum
  - Cheilectomy (femoral neck osteoplasty)
    - Remodeling of anterosuperior femoral head/neck junction
    - Used with underlying cam-type femoroacetabular impingement (FAI)
    - Cheilectomy techniques widely variable
    - Generally accompanies labral debridement
  - Ligamentum teres often debrided or resected at arthroscopy

CLINICAL ISSUES

Presentation

- Most common signs/symptoms
  - Persistent or recurrent groin or hip pain post labral debridement or repair
  - Occasionally coxa saltans (snapping hip)

Demographics

- Epidemiology
  - > 90% of patients report improvement in symptoms after hip arthroscopy

Treatment

- Intraarticular steroid injection
- Repeat arthroscopy

DIAGNOSTIC CHECKLIST

Consider

- Osseous stress injury
- Pubic symphysis injury
- Hip flexor tendinopathy
- Bursitis

Image Interpretation Pearls

- Expect thick capsule, heterogeneous labrum

SELECTED REFERENCES

1. Larson CM et al: Arthroscopic debridement versus refixation of the acetabular labrum associated with femoroacetabular impingement. Arthroscopy. 25(4):369-76, 2009
2. Ziegert AJ et al: Comparison of standard hip MR arthrographic imaging planes and sequences for detection of arthroscopically proven labral tear. AJR Am J Roentgenol. 192(5):1397-400, 2009
3. Zoga AC et al: Technical considerations in MR imaging of the hip. Magn Reson Imaging Clin N Am. 13(4):617-34, v, 2005

LIGAMENTUM TERES INJURY

Key Facts

Terminology
- Ligamentum teres: Flattened structure from fovea of femur to borders of acetabular notch

Imaging
- Normal ligament is hypointense and homogeneous at femoral fovea
- Generally redundant on neutral images
- Broadens, flattens, and divides into 2 or 3 bundles at acetabulum
- Serves as conduit for artery in childhood
- Direct MR arthrogram provides excellent visualization
- Enlarged, elongated with DDH

Pathology
- Complete tears
- High-velocity trauma
- Partial tears
- Associated with labral tears
- Degeneration
- Osteoarthritis

Clinical Issues
- Catching, locking, clicking
- Giving way
- Arthroscopic debridement of injured portion of ligament

Diagnostic Checklist
- Abnormally large ligament at foveal attachment = degeneration
- Partial tear if 2 bundles at fovea
- If absent, consider prior arthroscopy

*(**Left**) Coronal graphic shows a normal ligamentum teres. The ligamentum teres ➔ is generally broad with several bundles at its acetabular attachment ⇨ and tapering over its course to the femur. (**Right**) Coronal MR arthrogram T1WI FS shows a normal ligamentum teres. The foveal attachment ⇨ should be hypointense and homogeneous without intrasubstance fluid or contrast. The ligament ➔ is generally redundant and flaccid on all MR sequences.*

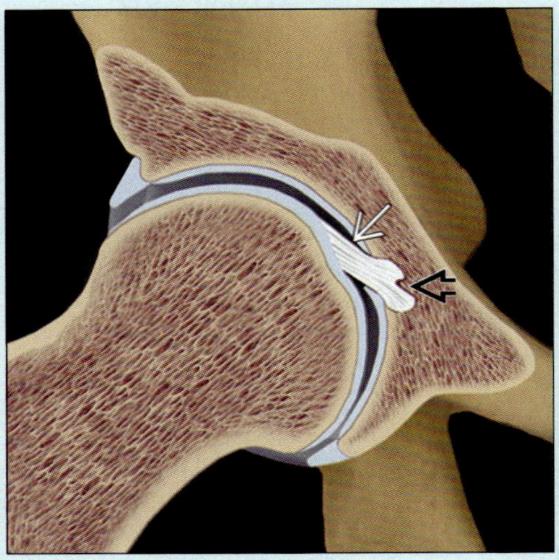

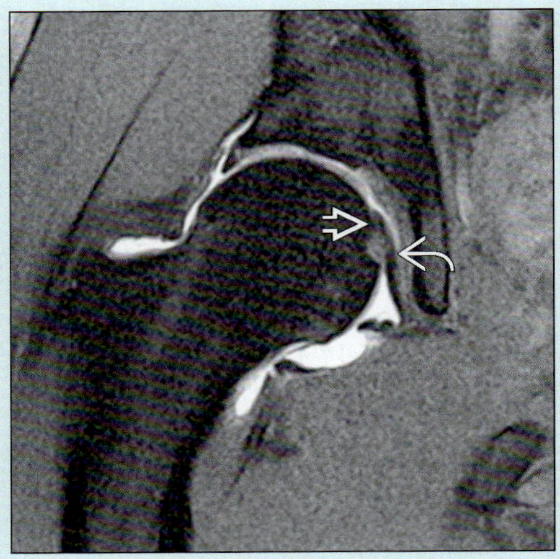

*(**Left**) Coronal MR arthrogram T1WI FS shows an arthroscopically proven ligamentum teres tear. There is a near complete intrasubstance tear ➔ at the foveal attachment. The patient had a history of hip clicking and giving way after a fall, and symptoms were relieved after debridement. (**Right**) Axial oblique MR arthrogram T1WI FS shows a complete ligamentum teres rupture. The ligament ⇨ is displaced from its insertion at the femoral fovea, where there is a visible disruption ➔.*

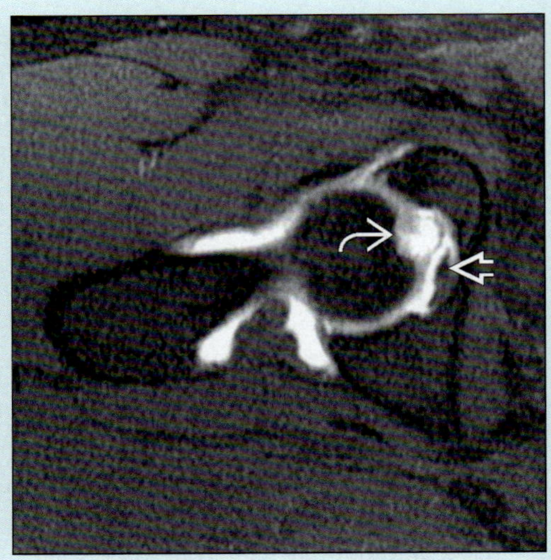

LIGAMENTUM TERES INJURY

TERMINOLOGY

Synonyms

- Ligamentum teres femoris

Definitions

- Flattened structure from fovea of femur to borders of acetabular notch

IMAGING

General Features

- Best diagnostic clue
  - Complete or partial disruption of ligamentum teres on MR arthrography
  - Enlargement of ligament at its foveal attachment
- Morphology
  - Serves as conduit for artery in childhood
    - Foveolar artery helps supply femoral head epiphysis from age 9 through osseous maturity
      - Injury could contribute to osteonecrosis in child
    - Occludes in adulthood
  - Enlarged and elongated with developmental dysplasia of hip (DDH)

MR Findings

- Normal ligament is hypointense and homogeneous at femoral fovea
- Generally redundant on neutral position images
  - More taut with abduction and external rotation of hip
  - Only fully taut with dislocation of hip
- Broadens, flattens, and divides into 2 or 3 bundles at acetabulum

Imaging Recommendations

- Best imaging tool
  - Direct MR arthrogram
- Protocol advice
  - Coronal, sagittal, and axial oblique sequences

DIFFERENTIAL DIAGNOSIS

Acetabular Labral Tears

- Often concomitant with ligamentum teres injury

Loose Bodies, Hip

- No origin or insertion
- Look for articular cartilage donor site

Externally Snapping Hip

- Iliopsoas or sartorius tendon pathology

Primary Synovial Process

- Diffuse throughout joint
  - Pigmented villonodular synovitis (PVNS)
  - Synovial chondromatosis

PATHOLOGY

General Features

- Etiology
  - Complete tear
    - High-velocity trauma
    - Forceful thigh abduction injury
  - Partial tear
    - Longstanding hip pain
    - Associated with labral tears
  - Degeneration
    - Osteoarthritis
- Associated abnormalities
  - Labral tears
  - Osteoarthritis

Staging, Grading, & Classification

- Gray and Villar classification
  - Complete tear
  - Partial tear
  - Degeneration

Gross Pathologic & Surgical Features

- Degeneration at foveal attachment

CLINICAL ISSUES

Presentation

- Most common signs/symptoms
  - Catching, locking, clicking
  - Giving way
  - Chronic hip pain
  - Pain with thigh abduction
- Other signs/symptoms
  - Approximately 15% of ligamentum teres abnormal at arthroscopy
  - With DDH surgery, ligamentum teres can obstruct reduction
    - Excision or transfer of ligamentum teres

Treatment

- Arthroscopic debridement of injured portion of ligament

DIAGNOSTIC CHECKLIST

Consider

- Synovitis, intraarticular bodies
- If ligamentum teres absent, consider prior arthroscopy

Image Interpretation Pearls

- Abnormally large ligament at foveal attachment = degeneration
- Partial tear if 2 bundles at fovea

Reporting Tips

- Lower threshold for diagnosing ligamentum teres tear if history of clicking and giving way

SELECTED REFERENCES

1. Bardakos NV et al: The ligamentum teres of the adult hip. J Bone Joint Surg Br. 91(1):8-15, 2009
2. Philippon MJ et al: Arthroscopic findings following traumatic hip dislocation in 14 professional athletes. Arthroscopy. 25(2):169-74, 2009
3. Sampatchalit S et al: Degenerative changes in the ligamentum teres of the hip: cadaveric study with magnetic resonance arthrography, anatomical inspection, and histologic examination. J Comput Assist Tomogr. 33(6):927-33, 2009
4. Byrd JW et al: Traumatic rupture of the ligamentum teres as a source of hip pain. Arthroscopy. 20(4):385-91, 2004

HIP INTERNAL AND EXTERNAL ROTATOR INJURY

Key Facts

Terminology

- Rotator cuff of hip; hip rotators

Imaging

- Gluteus maximus, medius, and minimus
- Deep external rotators: Piriformis, superior and inferior gemelli, obturator internus and externus, quadratus femoris
- Tensor fascia lata
- Apophysitis, tendinopathy at origin
- Enthesopathy at the greater tuberosity
- Calcific tendinopathy common in gluteus medius
- Greater trochanter avulsion fracture (rare)
- Insertional strain: Peritendinous hyperintensity at greater trochanter
  - Analogous to insertional rotator cuff strain
- Myotendinous junction strain: Pennate, feathery muscle hyperintensity
- Muscle belly strain: More proximal hyperintensity, often extending from broad origin
- Complete tear: Hyperintense gap
- Chronic tendinopathy: Enlarged, hypointense tendon, often with peritendinous hyperintensity
- Atrophy
  - Gluteus medius > deep external rotators

Pathology

- Gluteus medius atrophy associated with hip arthroplasty

Clinical Issues

- Gluteus medius avulsion associated with hip arthroplasty using lateral approach

Diagnostic Checklist

- Bilateral lesions common

(Left) PA graphic shows the gluteus medius ⇨ and minimus → (gluteus maximus removed), which provide either internal or external rotation depending on hip position. Piriformis →, gemelli →, quadratus femoris ↷, and obturator internus → are the deep external rotators. (Right) Coronal T2WI FS MR shows edema involving the lateral border of the gluteus maximus → and edema in the obturator externus ⇨, indicating acute rotator myotendinous strain injury.

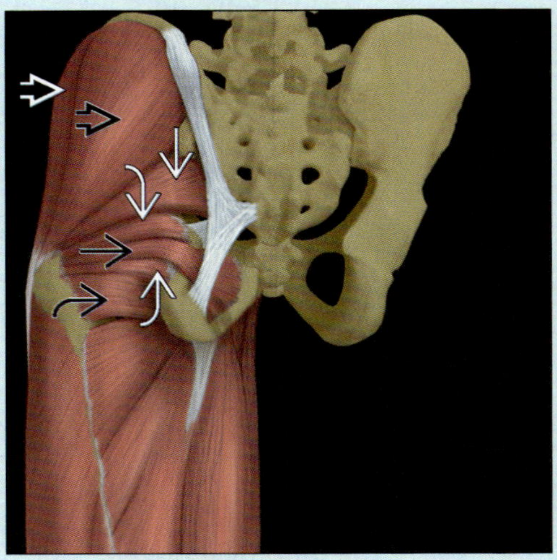

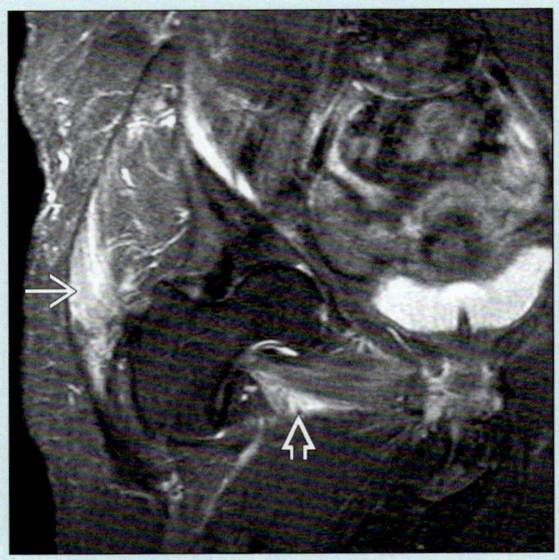

(Left) Coronal PDWI FS MR shows significant edema in and around the tendon and myotendinous junction of the gluteus medius ↷ due to athletic activity. (Right) Sagittal PDWI FS MR from the same patient shows a focal partial tear → of the gluteus medius tendon insertion with high signal fluid filling the defect. There is surrounding tendinopathy. The gluteus medius is the most commonly injured external rotator of the hip.

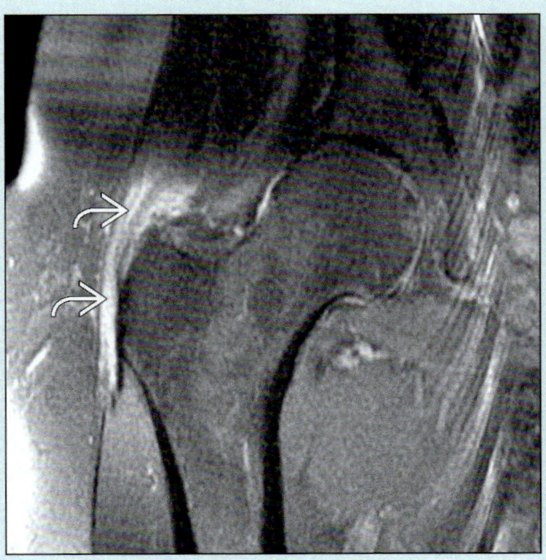

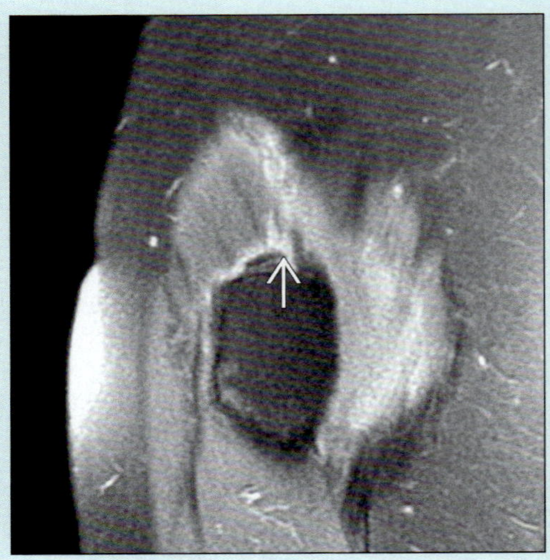

HIP INTERNAL AND EXTERNAL ROTATOR INJURY

TERMINOLOGY

Synonyms

- Rotator cuff of hip; hip rotators

Definitions

- Injury anywhere along myo-teno-osseous axis of hip rotator(s)
  - Gluteus maximus, medius, and minimus
  - Deep external rotators
  - Tensor fascia lata

IMAGING

General Features

- Best diagnostic clue
  - Fluid signal at myotendinous junction of hip rotator
- Location
  - Generally within 10 cm of greater trochanter
  - Gluteus maximus
    - Origin: Posterior gluteal line of upper inner ilium
      - Additional origins from sacrum, coccyx, erector spinae aponeurosis, sacrotuberous ligament, & gluteal aponeurosis
    - Insertion: Tensor fascia lata and gluteal tuberosity of subtrochanteric femur
    - Actions: Thigh adduction and external rotation, hip extension, standing pelvis stabilization
  - Gluteus medius
    - Origin: Outer surface of ileum between anterior and posterior gluteal line and gluteal aponeurosis
    - Insertion: Lateral surface of greater trochanter
    - Actions: Leg flexed = external rotation, leg neutral = abduction, leg extended = internal rotation
    - Most analogous to supraspinatus muscle of shoulder
  - Gluteus minimus
    - Origin: Outer surface of ileum between anterior and inferior gluteal line and margin of greater sciatic notch
    - Insertion: Anterior border of greater trochanter and hip capsule
    - Actions: Leg flexed = external rotation, leg neutral = abduction, leg extended = internal rotation
  - Deep external rotators
    - Piriformis
    - Superior and inferior gemelli
    - Obturator internus and externus
    - Quadratus femoris
    - Origins: Sacrum, sacrotuberous ligament, ischium (spine and tuberosity), pubic inner surface, obturator membrane
    - Insertion: Greater trochanter of femur
  - Tensor fascia lata
    - Origin: Anterior superior iliac spine (ASIS)
    - Insertion: Lateral tibial condyle
    - Action: Lower extremity external rotation
- Morphology
  - Myotendinous junction (MTJ) strain (partial tear)
    - Most common acute injury
    - Gluteus medius > gluteus maximus > deep external rotators
  - Insertional strain
    - Analogous to insertional rotator cuff strain
    - High-grade avulsion with significant trauma
      - Tendinous > osseous in young patients
      - Osseous avulsion more common with elderly (osteoporosis or bisphosphonate therapy)
  - Muscle belly strain
    - Diffuse injury to proximal-mid myotendinous axis
  - Insertional tendinopathy
    - Analogous to insertional rotator cuff tendinopathy of shoulder
  - Atrophy
    - Gluteus medius > deep external rotators
  - Apophysitis, tendinopathy at origin
    - Tensor fascia lata at ASIS

Radiographic Findings

- Enthesopathy at the greater tuberosity
- Calcific tendinopathy common in gluteus medius
  - Hydroxyapatite deposition disease (HADD)
- Calcification of hematoma; myositis ossificans
- Greater trochanter avulsion fracture (rare)

CT Findings

- Similar to radiographic findings
- Often bilateral enthesopathy/calcific tendinopathy

MR Findings

- T1WI
  - Chronic tendinopathy: Enlarged, hypointense tendon
  - Fatty atrophy: Diffuse hyperintensity of muscle belly
  - Nodular hypointensity seen with enthesopathy, HADD
  - Heterogeneous hematoma with high-grade injury
- T2WI FS
  - Insertional strain: Peritendinous hyperintensity at site of attachment
  - Myotendinous junction strain: Pennate, feathery muscle hyperintensity
  - Muscle belly strain: More proximal hyperintensity, often extending from broad origin
  - Complete tear: Hyperintense gap
  - Insertional tendinopathy: Large, hypointense tendon, often with peritendinous hyperintensity
  - Atrophy: Diffusely hyperintense muscle
  - Nodular hypointensity with enthesopathy, HADD
  - Subenthesial marrow edema at greater trochanter
  - Heterogeneous hematoma with high-grade strain
- STIR
  - Similar to T2WI FS findings
  - Homogeneous fat suppression helps differentiate edema from artifact at periphery of field
  - Hyperintense &/or fragmented ASIS apophysis
- T1WI C+ FS
  - Insertional strain/tendinopathy: Peritendinous enhancement
  - Myotendinous or muscle belly injury: Enhancement follows edema pattern
  - HADD: Intense enhancement around tendon

Ultrasonographic Findings

- Sensitive and specific for tendinous and myotendinous injury
  - Chronic tendinopathy: Enlarged hypoechoic tendon

HIP INTERNAL AND EXTERNAL ROTATOR INJURY

- Hypoechogenicity along myotendinous axis
- HADD: Hypoechoic shadowing calcium
- Hypoechoic tendon gap with high-grade injuries

- Insensitive for marrow edema at greater trochanter
- Modality of choice for guiding percutaneous therapy

Imaging Recommendations

- Best imaging tool
  - MR
- Protocol advice
  - MR bony pelvis protocol
    - Coronal T1 and STIR of entire pelvis
    - Axial T2WI FS of entire pelvis
    - Sagittal T2WI FS of region in question
  - CT: Image entire pelvis for comparison
    - Coronal and sagittal reformats

Nuclear Medicine Findings

- HADD hypermetabolic on FDG PET
- HADD and myositis ossificans can show increased Tc-99 MDP uptake

DIFFERENTIAL DIAGNOSIS

Trochanteric Bursitis

- Fluid distending 1 of 6 bursae at greater trochanter
- Associated with chronic external rotator tendinopathy
- Calcific bursitis: HADD dissects into bursa

Hip Flexor Injury

- Overlap with internal rotator injury

Internal Derangement of Hip

- Labral tear, cartilage loss, osteoarthritis

Femoral Head Osteonecrosis

- Signal abnormality subchondral in head, not trochanter

Transient Osteoporosis of Hip

- Femoral head insufficiency fracture

Contusion/Hematoma

- Direct trauma
- Overlying subcutaneous edema with rotator edema

Metastases, Bone Marrow

- Lesser trochanter > greater trochanter
- Destructive osseous lesion

PATHOLOGY

Staging, Grading, & Classification

- Proximal vs. distal strain
- Myotendinous injury grading

Gross Pathologic & Surgical Features

- Rotator muscle atrophy
  - Gluteus medius atrophy associated with hip arthroplasty

Muscle Group Function

- Gluteus minimus & medius contribute to ability to stand on 1 leg
  - Anterior fibers of gluteus medius and minimus are responsible for internal rotation
  - Posterior fibers of gluteus medius and minimus are responsible for external rotation

CLINICAL ISSUES

Presentation

- Most common signs/symptoms
  - External rotators: Pain in extension phase of gait cycle
    - Exacerbated with active femoral abduction & external rotation
  - Internal rotator: Pain in flexion phase of gait cycle
    - Exacerbated with active internal rotation of extremity
- Other signs/symptoms
  - Exquisite tenderness with tendinopathy or HADD
  - Gluteus medius avulsion associated with hip arthroplasty using lateral approach

Demographics

- Avulsion injuries in young with high-velocity trauma
- Myotendinous strain in "weekend warriors"
  - Often bilateral; delayed onset muscle soreness (DOMS)

Natural History & Prognosis

- Generally responds to conservative therapy
- Can be nagging, chronic injury

Treatment

- Conservative treatment most common
- Percutaneous therapies gaining popularity
  - US-guided steroid injection
  - US-guided tenotomy/debridement
  - Platelet-rich plasma (PRP) injection (controversial)
- Primary surgical repair for complete avulsion
  - May be related to hip arthroplasty
  - Hardware fixation for osseous avulsion

DIAGNOSTIC CHECKLIST

Consider

- Calcific tendinopathy or bursitis with chronic pain
- Bilateral gluteus medius strain in "weekend warrior"
- Ultrasound-guided injection for chronic lesions

Reporting Tips

- Identify calcifications & enthesophytes
- Note bone marrow edema at insertion site

SELECTED REFERENCES

1. Labrosse JM et al: Effectiveness of ultrasound-guided corticosteroid injection for the treatment of gluteus medius tendinopathy. AJR Am J Roentgenol. 194(1):202-6, 2010
2. Housner JA et al: Sonographically guided percutaneous needle tenotomy for the treatment of chronic tendinosis. J Ultrasound Med. 28(9):1187-92, 2009
3. Rasch A et al: Persisting muscle atrophy two years after replacement of the hip. J Bone Joint Surg Br. 91(5):583-8, 2009
4. Kong A et al: MRI and US of gluteal tendinopathy in greater trochanteric pain syndrome. Eur Radiol. 17(7):1772-83, 2007

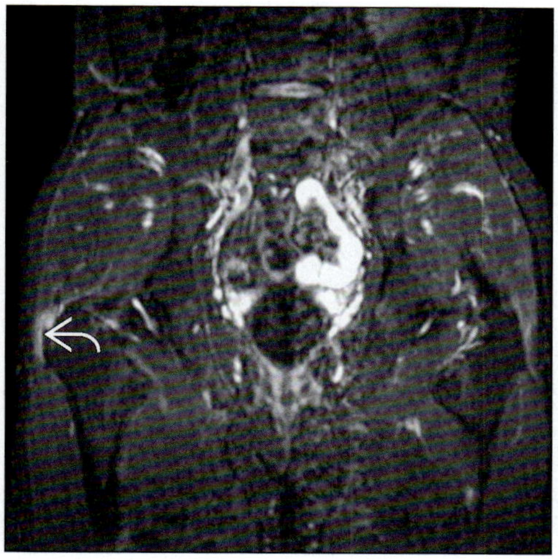

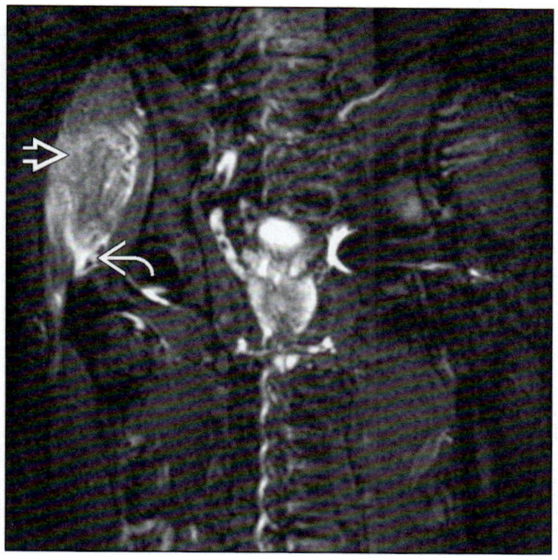

(Left) *Coronal STIR MR from a patient with lateral hip pain and tenderness at the greater trochanter shows focal soft tissue edema around the distal gluteus medius tendon ➔, typical of overuse tendinopathy.* ***(Right)*** *Coronal STIR MR from an ultimate Frisbee player shows a high-grade acute myotendinous strain of the gluteus medius. There is feathery edema within the gluteus medius muscle belly ➔ and a nearly complete disruption of the distal gluteus medius tendon ➔.*

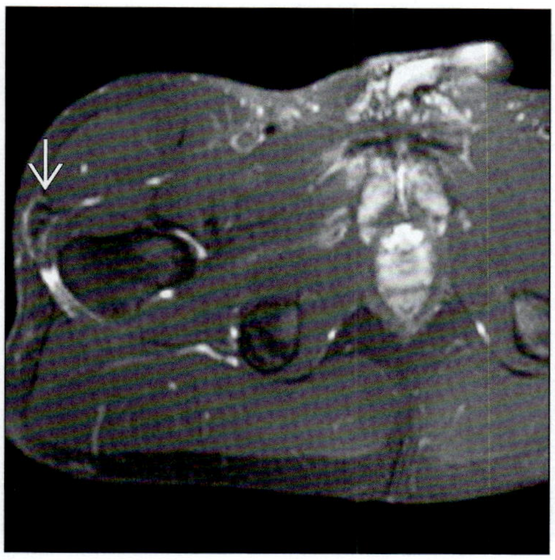

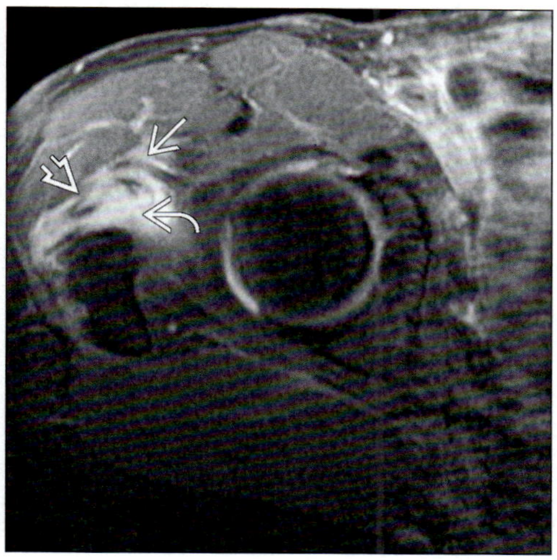

(Left) *Axial T2WI FS MR from a professional basketball player with chronic lateral hip pain shows an enlarged hypointense distal gluteus medius tendon ➔, indicating chronic tendinopathy.* ***(Right)*** *Axial T2WI FS MR shows severe acute-on-chronic tendinopathy at the gluteus minimus ➔ and gluteus medius ➔ insertions. There is peritendinous edema at the insertion sites of these internal rotators with mass effect on regional muscles, consistent with reactive bursitis ➔.*

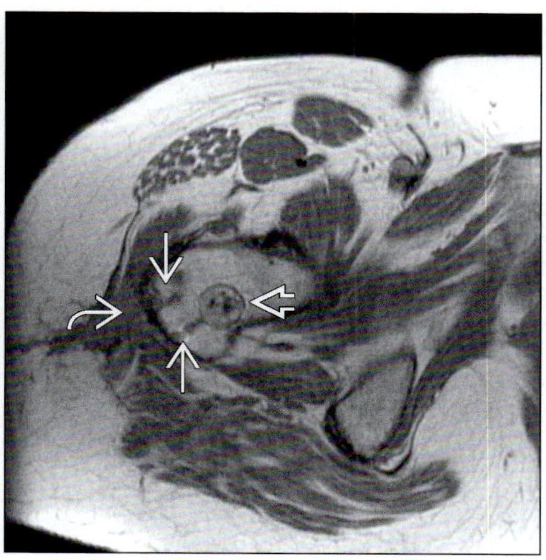

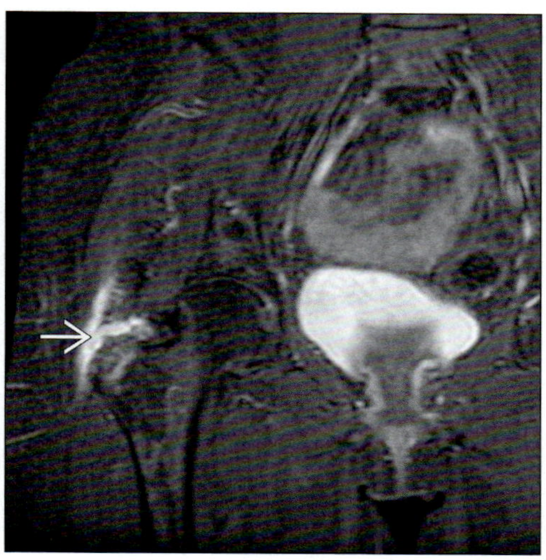

(Left) *Axial PDWI MR from a 64-year-old man with hip pain after a fall following recent gluteal tendon repair surgery shows a defect ➔ at the insertion of the gluteus maximus. Pin tracks ➔ from tendon reattachment are seen at the site of avulsion. An intertrochanteric rod has been previously removed from the femur ➔.* ***(Right)*** *Coronal T2WI FS MR from the same patient shows the tendon gap filled with fluid ➔. Gluteal tears can occur after hip replacement or fracture fixation.*

*(**Left**) Coronal STIR MR of the pelvis from a 40 year old with acute buttock pain after running shows focal edema and partial muscle fiber disruption within the most distal region of the gluteus maximus →, consistent with an acute grade 2 strain. (**Right**) Axial T2WI FS MR from the same patient shows feathery edema within the gluteus maximus muscle → and periosteal avulsion from the gluteal tuberosity of the subtrochanteric femur → in this acute distal external rotator strain injury.*

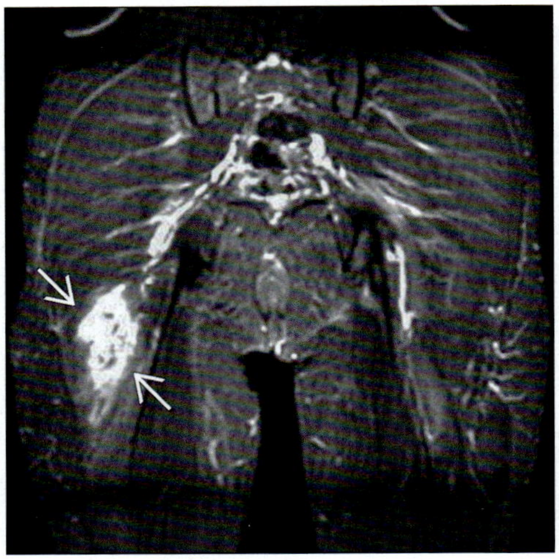

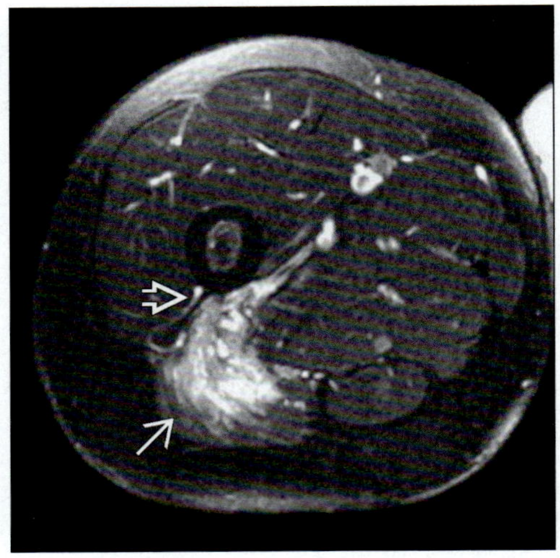

*(**Left**) Axial T2WI FS MR shows an acute focal partial tear → of the gluteus maximus at its ilial origin, with feathery edema extending throughout the muscle belly →. This is a relatively unusual location for gluteal injury. (**Right**) Axial T2WI FS MR shows focal disruption → of the gluteus minimus near its myotendinous junction. Most injuries to the gluteal units are located in or near the distal tendinous insertions, and are seen in the setting of chronic underlying tendinopathy.*

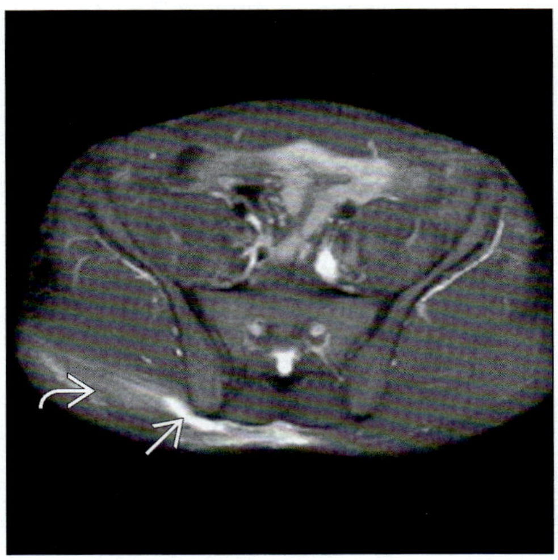

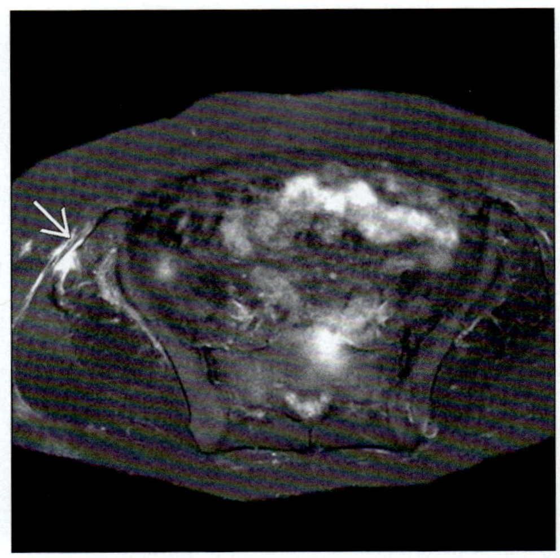

*(**Left**) Axial T1WI MR shows marked atrophy of the gluteus medius → and minimus → myotendinous units after an old injury. The gluteus maximus → is moderately atrophic as well. There is a chronic hamstring tear →. (**Right**) AP radiograph shows amorphous calcification → in the region of the gluteus medius insertion, typical for calcific bursitis. Hydroxyapatite deposits on the tendon and ultimately dissects into the adjacent trochanteric bursa.*

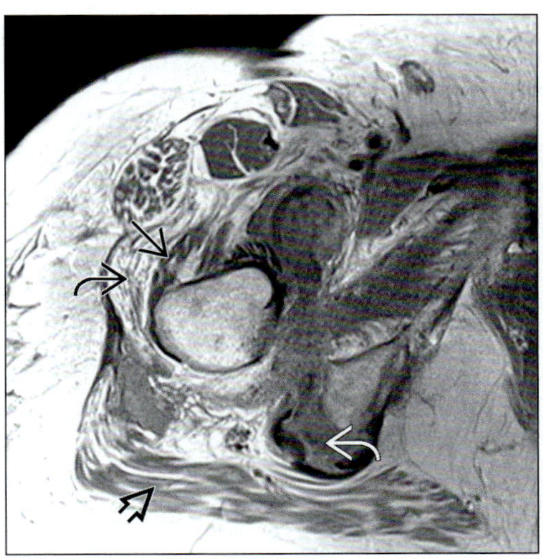

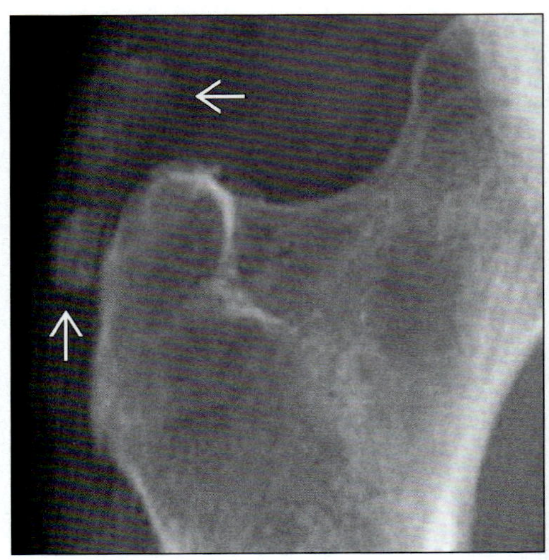

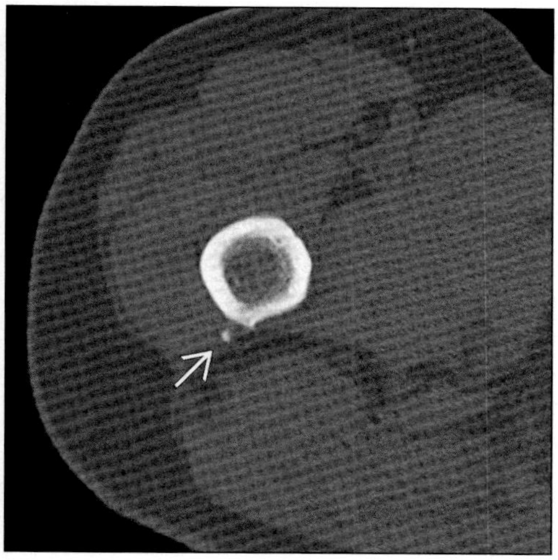

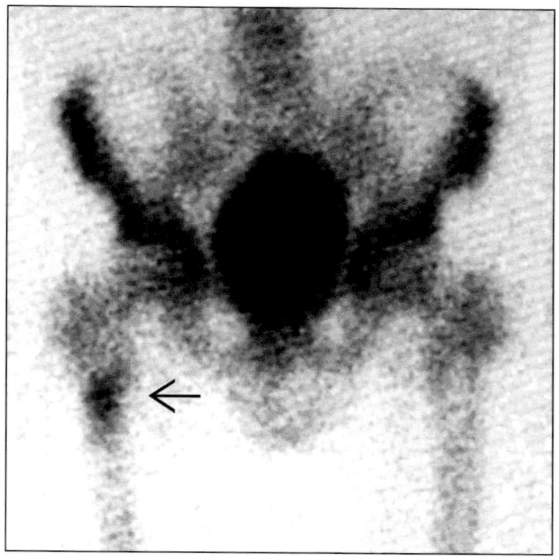

***(Left)** Axial bone CT image from a prostate cancer patient with hip pain and concern for a pathologic fracture shows no osseous abnormality. The focus of calcium at the distal gluteus maximus tendon → was initially overlooked. **(Right)** AP image from a Tc-99 MDP bone scan in the same patient shows increased radiotracer uptake → in the subtrochanteric femur. Hydroxyapatite deposition disease can exhibit increased uptake on bone scan and hypermetabolic activity on FDG PET studies.*

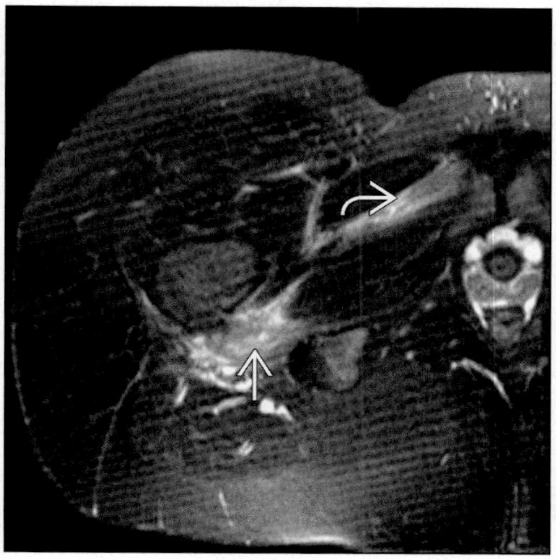

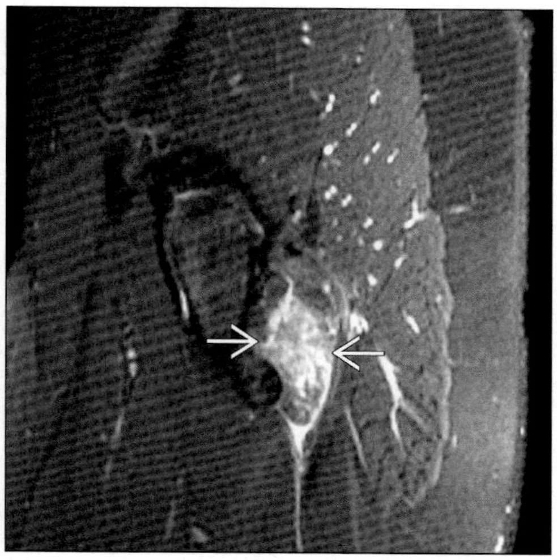

***(Left)** Axial T2WI FS MR from a 48-year-old man after a slip and fall shows a tear → of the quadratus femoris myotendinous junction with hematoma formation in the gap. There is mild edema → in the intermuscular fat of the adductor compartment. **(Right)** Sagittal T2WI FS MR from the same patient shows high-grade injury with muscle fiber disruption → in the quadratus femoris muscle. Isolated injury to this muscle is relatively uncommon.*

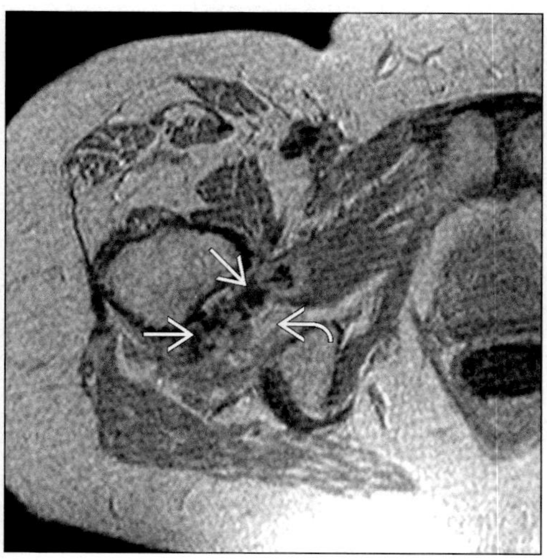

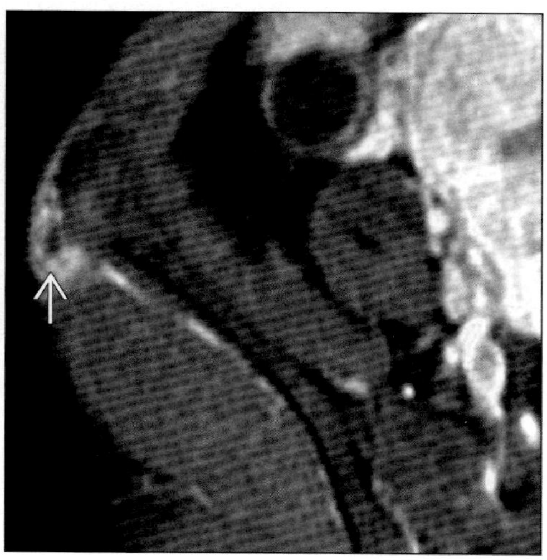

***(Left)** Axial T1WI MR shows atrophy of the quadratus femoris muscle → with fat signal indicating chronic muscle loss. Nodular foci of low signal → at its tendinous insertion site are due to calcium hydroxyapatite deposition. **(Right)** Axial T2WI FS MR from a runner with chronic iliac crest pain on external rotation shows a large tensor fascia lata origin → with peritendinous edema, consistent with apophyseal tendinopathy.*

HIP FLEXOR INJURY

Key Facts

Terminology
- Myotendinous, teno-osseous, or muscle belly lesion with pain on thigh flexion

Imaging
- Most commonly occult on radiographs
- Avulsive/teno-osseous: Tendon attachment on bone
- Myotendinous: Proximal or distal muscle-tendon junction (MTJ)
- Muscle belly: Midportion of myotendinous unit (MTU)
- T2WI FS is workhorse sequence (with STIR) for muscle and tendon injury

Pathology
- Injury often at weakest point of MTU

Clinical Issues
- Once injured, recurrent injury at same site common
- Activities with repetitive flexion predispose to flexor injuries
  - Psoas: Hockey, soccer, rugby, swimming
  - Iliacus: Gymnastics, hockey
  - Sartorius: Kickboxing, yoga
  - Rectus femoris: Sprinting, football, baseball, basketball, any kicking activity
- Activity modification to prevent recurrent/persistent injury

Diagnostic Checklist
- Image-guided minimally invasive intervention
- Grade myotendinous and muscle belly strain
- Describe location of center of injury (myotendinous, tendon origin, muscle belly)

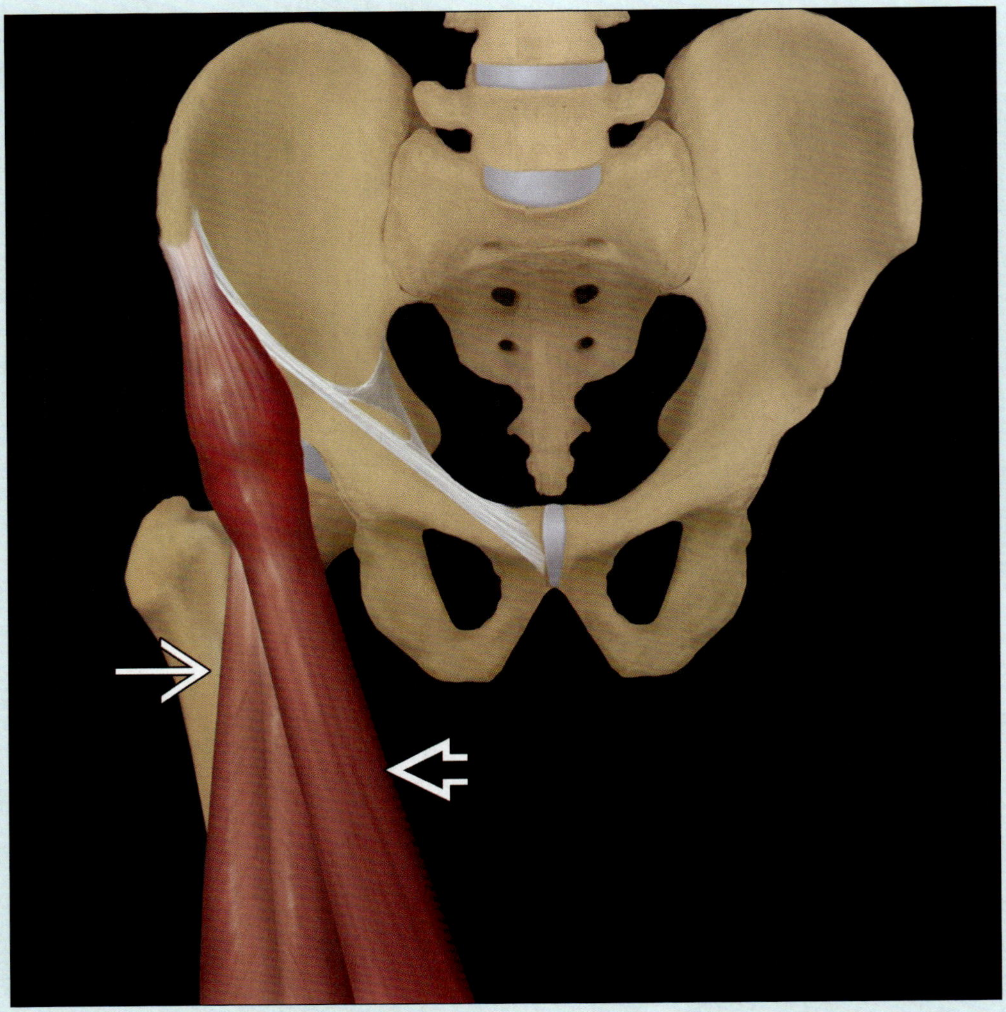

AP graphic shows commonly injured hip flexors. The rectus femoris ➡ can be injured anywhere from its origin at the anterior inferior iliac spine to its proximal and distal myotendinous junctions. The sartorius ⇨ is often symptomatic with repetitive hip flexion and internal rotation activities, such as kicking above the level of the waist. Hip flexor injuries range from acute myotendinous or teno-osseous tears to chronic overuse syndromes. Identification of the cause of these injuries often plays an important role in generating an effective treatment plan. Imaging can also be used to direct minimally invasive treatments for chronic hip flexor injury.

HIP FLEXOR INJURY

TERMINOLOGY

Synonyms

- Hip flexor, flexor strain, anterior hip strain

Definitions

- Myotendinous, teno-osseous, or muscle belly lesion causing pain on active thigh flexion

IMAGING

General Features

- Best diagnostic clue
  - Soft tissue edema in distribution of 1 or more thigh flexors on MR
    - Edema follows distribution of a myotendinous unit (MTU)
  - Enthesopathy or avulsion at thigh flexor attachments on radiographs
- Location
  - Iliacus
    - Iliac fossa at medial iliac crest to lesser trochanter
  - Psoas
    - T12 and lumbar vertebral bodies/transverse processes to lesser trochanter
  - Sartorius
    - Anterior superior iliac spine (ASIS) to anteromedial tibia (pes anserinus)
  - Rectus femoris
    - Anterior inferior iliac spine (AIIS) to patella
- Morphology
  - Avulsive/teno-osseous: Tendon attachment on bone
  - Myotendinous: Proximal or distal muscle-tendon junction (MTJ)
  - Muscle belly: Midportion of MTU

Radiographic Findings

- Most commonly occult on radiographs
- Enthesopathy with chronic tendinopathy
- Calcific tendinopathy (hydroxyapatite deposition) with chronic tendinopathy
- Myositis ossificans with remote high-grade injury
- Osseous avulsion
  - AIIS, lesser trochanter, ASIS
    - Consider pathologic avulsion at lesser trochanter

CT Findings

- Similar to radiographic findings
- Hematoma at site of injury
- Axial plane shows MTU in cross section

MR Findings

- T1WI
  - Hypointense fluid with acute injury
  - Hypointense calcium or hypoxic tendon with chronic injury
  - Heterogeneous hematoma
- T2WI FS
  - Workhorse sequence (with STIR) for muscle and tendon injury
  - Avulsion: Edema in proximal tendon and bone
  - Myotendinous: Edema centered at junction
  - Muscle belly: Edema within muscle sheath with pennate (feathery) morphology
  - Heterogeneous hematoma near center of injury
- STIR
  - Findings same as T2WI FS
  - Can provide homogeneous fat suppression over larger region
    - Excellent for assessing symmetry
- T1WI C+
  - Injury will enhance with intravenous contrast
  - Hematomas can enhance peripherally: Can be mistaken for or obscure soft tissue mass
- Soft tissue edema along MTU
- Associated soft tissue injury in other muscle compartments
- Concomitant pubic symphysis/sports hernia lesion

Ultrasonographic Findings

- Fluid around MTJ indicates strain
- Enlarged tendon in chronic injury
- Frank gap in high-grade injuries
- Useful for guidance of therapeutic injections

Imaging Recommendations

- Best imaging tool
  - Noncontrast MR of pelvis ± thighs
  - Focused musculoskeletal ultrasound
- Protocol advice
  - Fluid-sensitive sequence in 3 planes
  - Coronal and axial large field-of-view (FOV) sequences for symmetry
  - Sagittal and axial small FOV sequences for resolution
    - Include only side of injury
  - Consider a plane paralleling the injured structure for grading
  - Athletic pubalgia protocol generally useful
  - Include fluid-sensitive and large FOV sequences on hip direct MR arthrogram protocol

DIFFERENTIAL DIAGNOSIS

Avulsion Fracture

- Radiographs/CT for displaced osseous fragment

Apophysitis

- Skeletally immature
- Edema in apophysis and associated tendon

Rectus Abdominis/Adductor Aponeurosis (Sports Hernia) Lesion

- Injury/edema centered at pubic tubercle
- Associated with hip flexor injuries

Adductor Tendon Injury

- Centered near midline/medial thigh

Acetabular Labral Tears

- Anterior tear at labral base mimics flexor injury

Bursitis

- Iliopsoas bursitis mimics flexor tendon injury

PATHOLOGY

Staging, Grading, & Classification

- Avulsive (teno-osseous) injury

HIP FLEXOR INJURY

- Complete vs. incomplete
- Tendinous vs. osseous
- Myotendinous injury
  - Grade 1: Fluid surrounding MTJ without disruption
  - Grade 2: Partial disruption (some fibers) and surrounding fluid
  - Grade 3: Complete disruption of MTU
    - Generally with measurable gap
    - Tear often irregular, oblique to all imaging planes
  - Measure hematoma
- Muscle belly injury
  - Grade 1-3 as with myotendinous injury
  - Measure hematoma in muscle sheath
- Acute hip flexor strain
  - Soft tissue edema and tear without underlying tendinopathy
- Chronic hip flexor
  - Hypointense large hypoxic tendon
  - Peritendinous ± osseous edema
  - Repetitive flexion activity (e.g., ice skating)
  - Tendinopathy, fibrosis, and contracture in cerebral palsy

Gross Pathologic & Surgical Features

- Injury at weakest point of MTU
  - Apophyseal growth plate in skeletally immature
  - MTJ in all groups

CLINICAL ISSUES

Presentation

- Most common signs/symptoms
  - Pain on active thigh/hip flexion
  - Asymmetric weakness with active flexion
  - Point tenderness at site of injury
    - Exquisite bone tenderness with avulsive lesion
    - Thickened, cord-like tendon/MTJ with chronic lesion
- Other signs/symptoms
  - Ecchymosis from hematoma
  - Palpable mass from hematoma/edema
  - Once injured, recurrent injury at same site common

Demographics

- Age
  - Avulsions/apophysitis in young
  - Muscle belly in high-level athletes (3rd-4th decade)
  - Tendon avulsions in older, "weekend warrior" types
  - Chronic tendinopathy/calcific tendinopathy in middle-aged patients

Natural History & Prognosis

- Activities with repetitive flexion predispose to flexor injury
  - Mechanism often predicts structure involved
    - Psoas: Hockey, soccer, rugby, swimming
    - Iliacus: Gymnastics, hockey
    - Sartorius: Kickboxing, yoga
    - Rectus femoris: Sprinting, football, baseball, basketball, any kicking activity
- Calcific tendinopathy often multifocal
- Hematoma can be painful & delay healing
- Bilateral symmetric grade 1 strain: Delayed onset muscle soreness (DOMS)
- Partial lesions can → complete with continued activity

Treatment

- Teno-osseous (avulsive) lesion
  - Conservative if incomplete
  - Consider primary surgical repair if complete, retracted, or skeletally mature
  - Stop activity with apophysitis, tendinopathy
  - Ultrasound-guided tenotomy (microperforation/debridement)
  - Directed tendon sheath steroid injection for pain relief
  - Platelet-rich plasma (PRP) injection for chronic lesion
- Myotendinous lesions
  - Rest, physical therapy
  - Stretching regimens to prevent re-injury
  - Injury grade helps predict recovery time
  - Consider image-guided evacuation of large hematoma
- Muscle belly lesion
  - Rest, physical therapy, stretching
- Activity modification to prevent recurrent/persistent injury
- Physical therapy to accelerate healing: Massage, whirlpool, ultrasound therapy

DIAGNOSTIC CHECKLIST

Consider

- Activity or avocation as source of hip flexor injury
- Image-guided minimally invasive intervention for chronic lesion
- Concomitant adductor and sports hernia lesion

Image Interpretation Pearls

- Grade myotendinous and muscle belly strain
- With acute injury, identify indicators of underlying chronic tendinopathy

Reporting Tips

- Describe center of injury (myotendinous, tendon origin, muscle belly)
- Measure any fluid-filled gap in long and short axis

SELECTED REFERENCES

1. Bui KL et al: Iliopsoas injury: an MRI study of patterns and prevalence correlated with clinical findings. Skeletal Radiol. 37(3):245-9, 2008
2. Sanders TG et al: Avulsion injuries of the pelvis. Semin Musculoskelet Radiol. 12(1):42-53, 2008
3. Zajick DC et al: Spectrum of MRI findings in clinical athletic pubalgia. Semin Musculoskelet Radiol. 12(1):3-12, 2008
4. Niemuth PE et al: Hip muscle weakness and overuse injuries in recreational runners. Clin J Sport Med. 15(1):14-21, 2005
5. Tyler TF et al: The association of hip strength and flexibility with the incidence of adductor muscle strains in professional ice hockey players. Am J Sports Med. 29(2):124-8, 2001

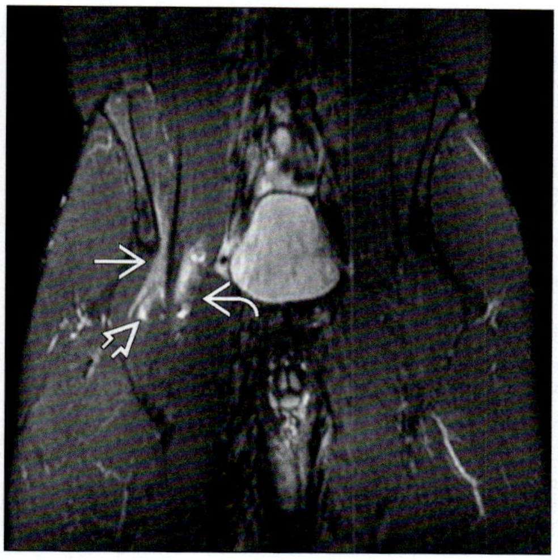

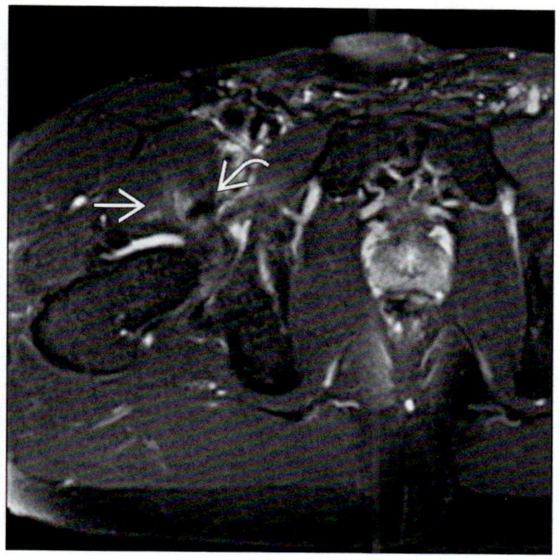

(Left) *Coronal STIR MR shows an acute hip flexor strain in a football player. There is muscle edema at the distal myotendinous junction (MTJ) of the iliacus → with a focus of fluid, suggesting fiber disruption →. There is also a grade 1 strain of the psoas at the same level without fiber disruption →.* ***(Right)*** *Axial T2WI FS MR from a hockey player shows a chronic hip flexor injury. There is enlargement of the distal iliopsoas tendon → with peritendinous edema →.*

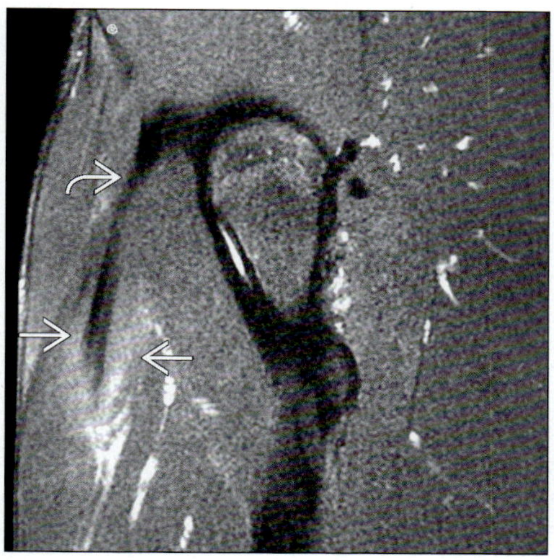

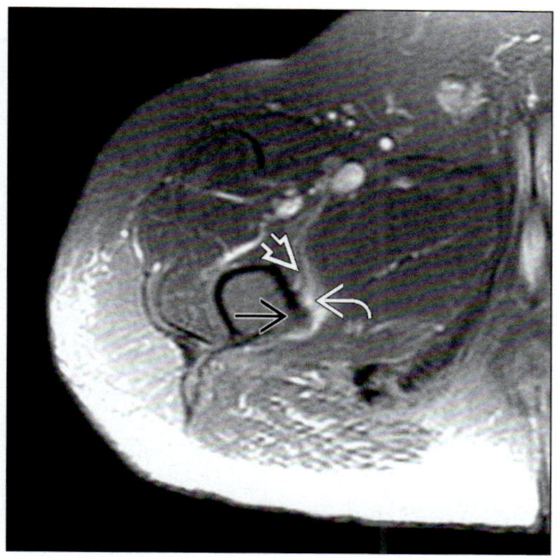

(Left) *Sagittal T2WI FS MR from a 16-year-old high school football player demonstrates edema → around the proximal myotendinous junction (MTJ) of the iliopsoas unit consistent with a grade 1 strain injury. The tendon → is intact.* ***(Right)*** *Axial T2WI MR demonstrates a high-grade partial avulsion → of the insertion of the iliopsoas tendon on the lesser femoral trochanter →. A few wispy fibers → remain intact, but hip flexor function was almost absent.*

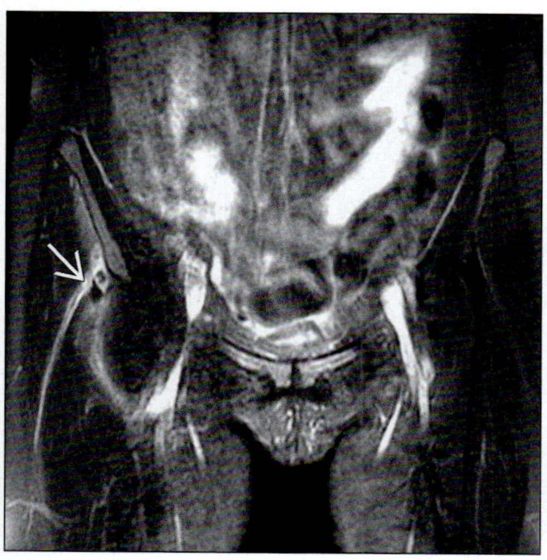

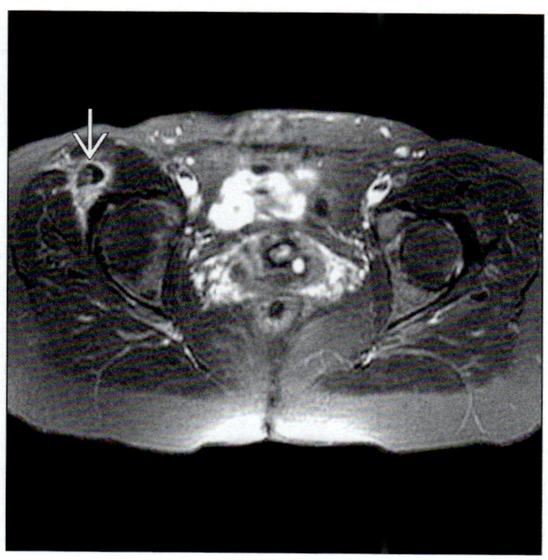

(Left) *Coronal T2WI FS MR demonstrates avulsion of the rectus femoris long head tendon from the anterior inferior iliac spine (AIIS). A small osseous fragment → is slightly separated from the native bone. There is surrounding edema.* ***(Right)*** *Axial T2WI MR from the same patient shows the displaced and rotated avulsion fragment → with surrounding soft tissue edema. Because the reflected head tendon (which arises above the acetabulum) remains intact, the rectus retains some function.*

HIP FLEXOR INJURY

(Left) *Axial T2WI FS MR from a middle-aged woman with worsening hip flexion pain while speed-walking shows lobulated hypointensity* ➔ *at the anterior inferior iliac spine or proximal rectus femoris, consistent with calcific tendinopathy (hydroxyapatite deposition disease).* **(Right)** *Axial bone CT from a 47 year old with chronic hip pain demonstrates a fusiform calcification* → *in the distal iliopsoas tendon, consistent with calcium hydroxyapatite dihydrate deposition (CPPD).*

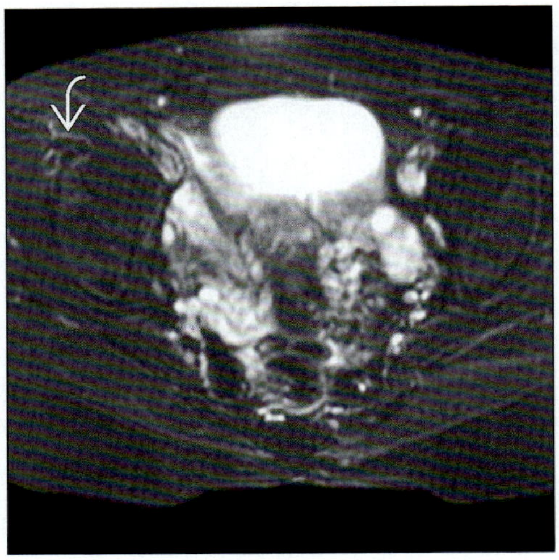

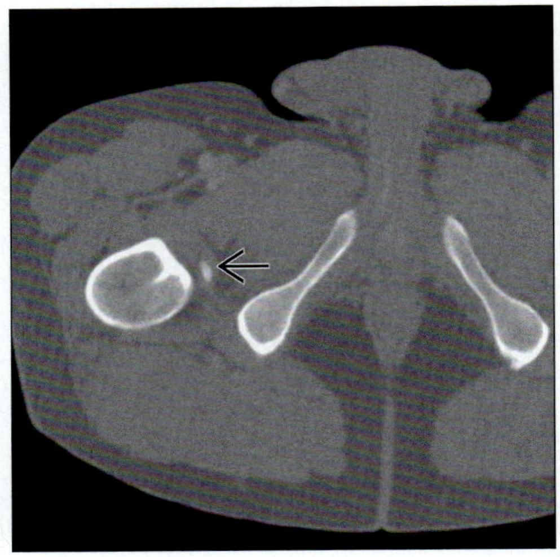

(Left) *Coronal STIR MR from a gymnast with severe pain on active hip flexion shows fluid signal* ➔ *outlining the iliacus muscle, compatible with an acute grade 2 avulsive-type strain of the iliacus.* **(Right)** *Axial T2WI FS MR from the same patient shows fluid* ➔ *surrounding the proximal iliacus muscle and avulsive soft tissue edema* ➔ *at the sartorius and tensor fascia lata origin. Acute flexor injuries are commonly associated with tendon injuries in adjacent compartments.*

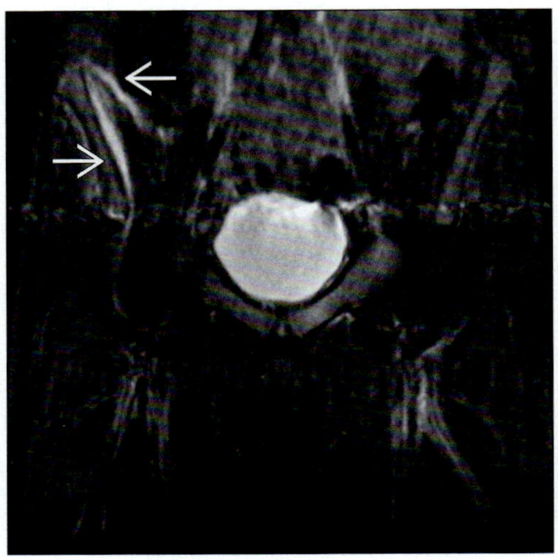

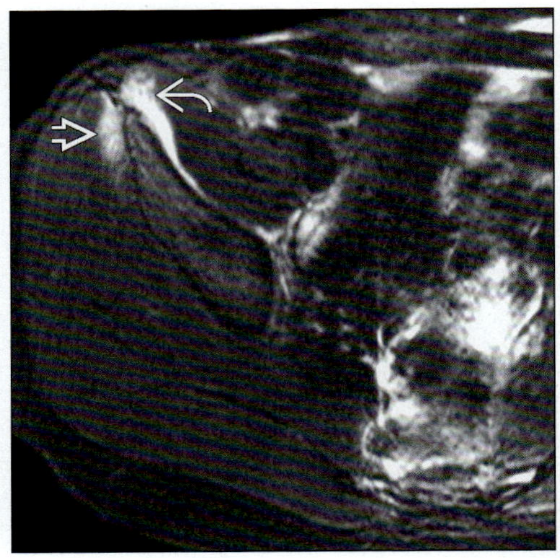

(Left) *Coronal STIR MR from a patient with pain on hip flexion after bilateral femoral head resurfacing shows lobulated hypointensity involving the proximal rectus femoris tendon* ➔*, typical of post-traumatic calcific tendinopathy.* **(Right)** *Sagittal T1 C+ FS MR from a runner with hip flexor weakness shows a hypointense mass* ➔ *within the rectus femoris muscle belly with peripheral enhancement* ➔*. Calcification on radiographs was diagnostic for myositis ossificans.*

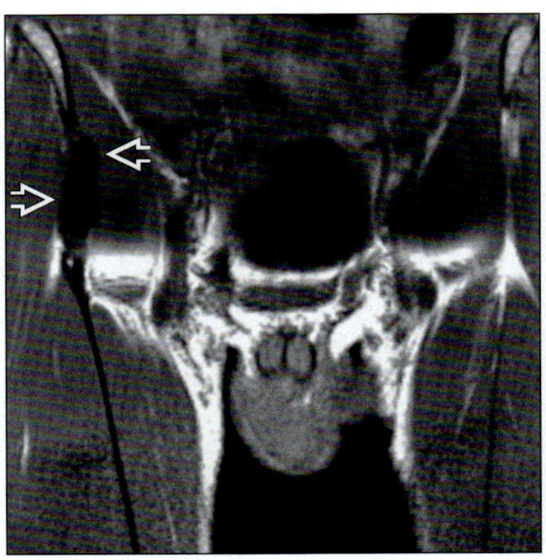

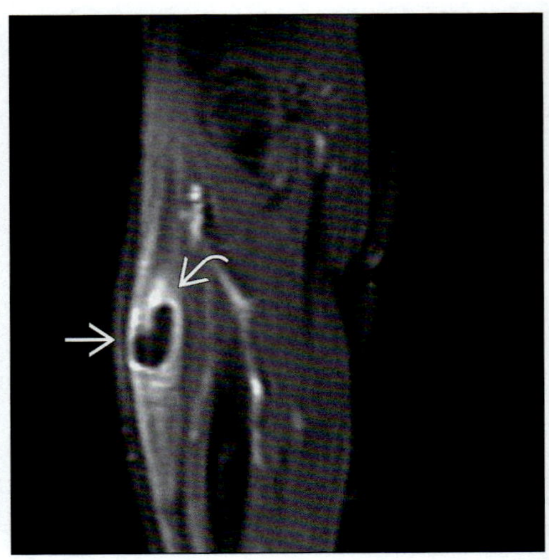

HIP FLEXOR INJURY

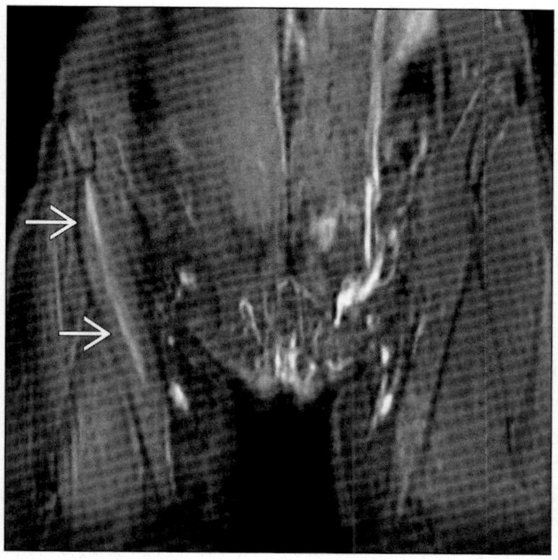

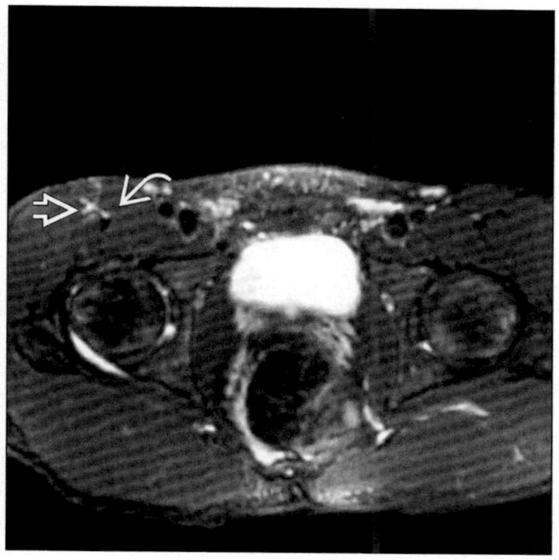

(Left) *Coronal STIR MR shows an acute sartorius strain in a female kickboxer with streaky edema → in the proximal muscle and tendon. The sartorius serves as an accessory hip flexor and a hip adductor.* ***(Right)*** *Axial T2WI FS MR from an NFL player with decreased foot speed and a clinical concern for "sports hernia" shows asymmetric enlargement of the rectus femoris at its proximal myotendinous junction ↶ with surrounding soft tissue edema ⇨, typical for an acute-on-chronic hip flexor strain.*

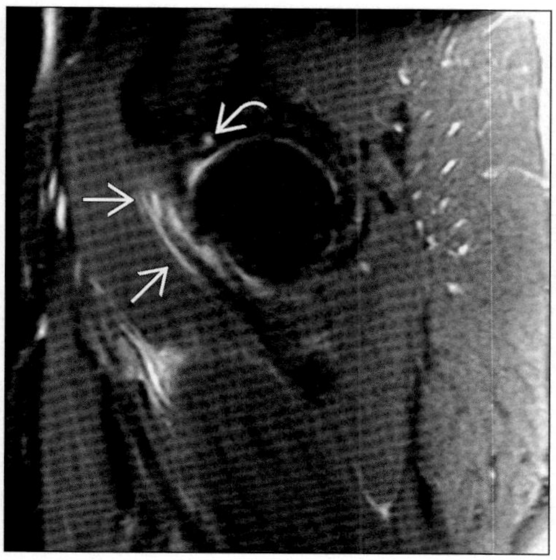

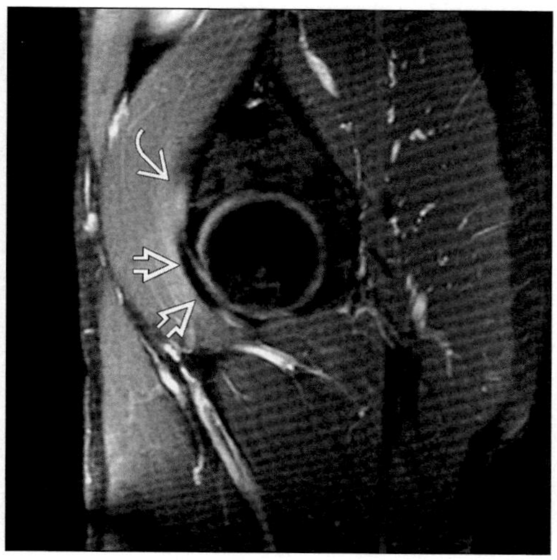

(Left) *Sagittal T2WI FS MR in a runner with an acute flexor strain shows peritendinous edema at the distal iliopsoas myotendinous junction and tendon →. A subchondral cyst in the anterosuperior acetabulum indicates concomitant degenerative joint disease ↶.* ***(Right)*** *Sagittal T2WI FS MR of a chronic hip flexor injury shows a mildly enlarged but very hypointense distal iliopsoas tendon ⇨, suggesting hypoxic tendinopathy. There is mild edema at the more proximal myotendinous junction ↶.*

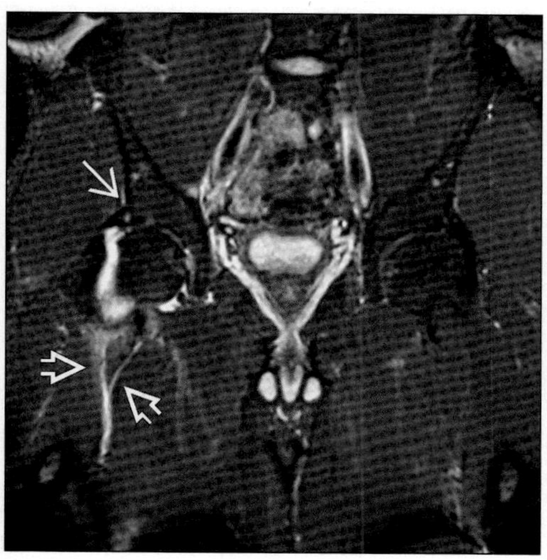

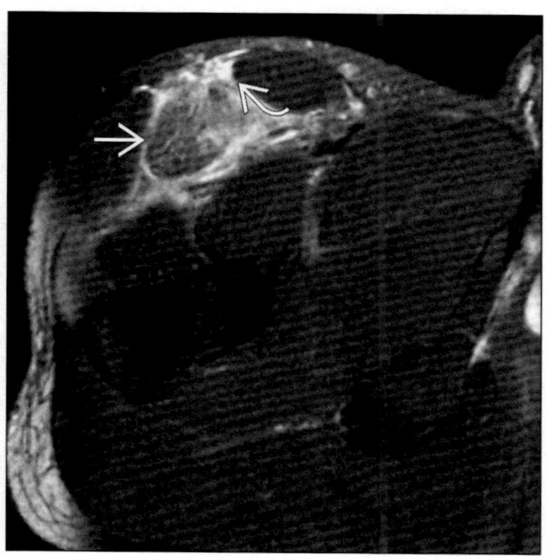

(Left) *Coronal STIR MR from a hockey player with an acute flexor strain shows edema ⇨ tracking along the distal iliopsoas myotendinous junction and a right hip effusion and labral tear →.* ***(Right)*** *Axial T2WI FS MR from a baseball player after an acute thigh flexor injury while running shows feathery edema → throughout the muscle belly of the rectus femoris, as well as focal fluid ↶ between the rectus femoris and the sartorius. These findings indicate a grade 2 strain and evolving hematoma.*

HIP ADDUCTOR INJURY

Key Facts

Terminology
- Injury of adductors or internal rotators sparing rectus abdominis

Imaging
- Soft tissue edema in adductor compartment on MR
- Avulsed fragment at pubic tubercle (rare)
- Acute injury: Identify center of soft tissue edema
  - Proximal at tendon origin
  - More distal at myotendinous junction
- Direction of ↑ T2 signal indicates tendon/muscle primarily involved
- Grade strains on T2 or STIR
- Large, hypointense on all sequences = hypoxic tendinopathy
- Bony pelvis or athletic pubalgia protocol
- Cover both thighs in axial plane for symmetry

Pathology
- Adductor avulsions
- Adductor myotendinous strains
- Chronic adductor origin tendinopathy
- Adductor compartment hematoma

Clinical Issues
- Unilateral proximal, medial thigh pain
- Nagging, aching pain with degenerative tendinopathy
- Acute: Young athletes
- Chronic: Middle-aged, weekend warrior-type athletes
- Difficult to distinguish sports hernia from isolated adductor lesions

Diagnostic Checklist
- Use available clinical history as guide to description
- Measure adductor tendon retraction

(Left) Coronal STIR MR from a hockey player with an acute groin injury shows a grade 2 partial tear at the adductor longus origin ➡ and a grade 1 strain of the pectineus ➡. The rectus abdominis/adductor aponeurosis ➡ appears intact, and there is no osteitis pubis. (Right) Sagittal T2WI FS MR from an athletic pubalgia protocol shows an intact rectus abdominis/adductor aponeurosis ➡, but there is a muscle strain more posteriorly involving the adductor brevis near its origin ➡.

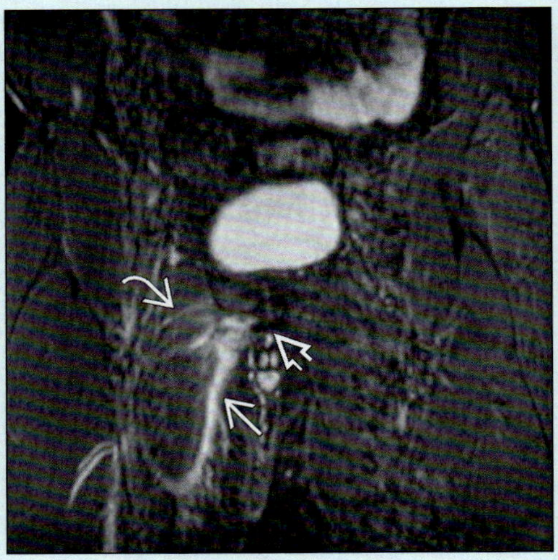

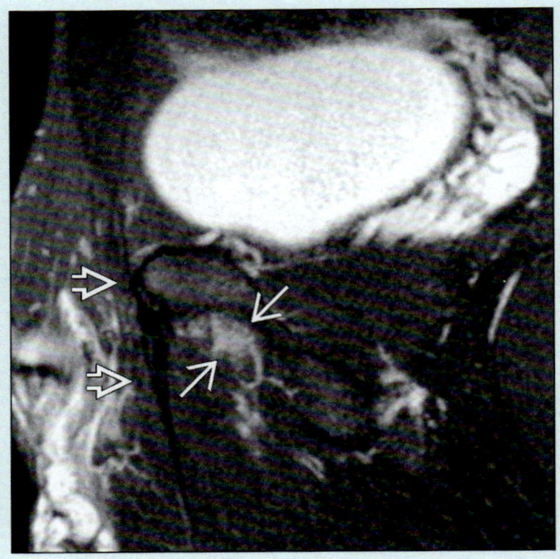

(Left) Coronal oblique T1WI MR from a middle-aged kickboxer with nagging pain shows marked enlargement of the right adductor longus tendon at its origin ➡ with intratendinous black signal ➡, typical for hydroxyapatite deposition and calcific tendinopathy. (Right) Axial T2WI FS MR from a skier who caught his ski tip while jumping from a helicopter shows a large hematoma ➡ within the adductor compartment creating mass effect. The avulsed and displaced adductor longus ➡ is hypointense.

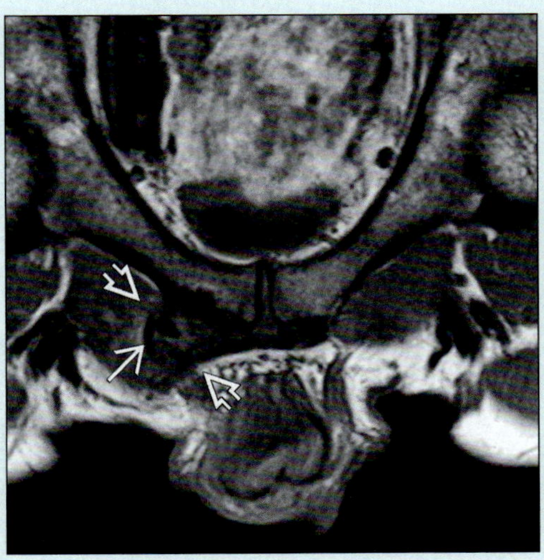

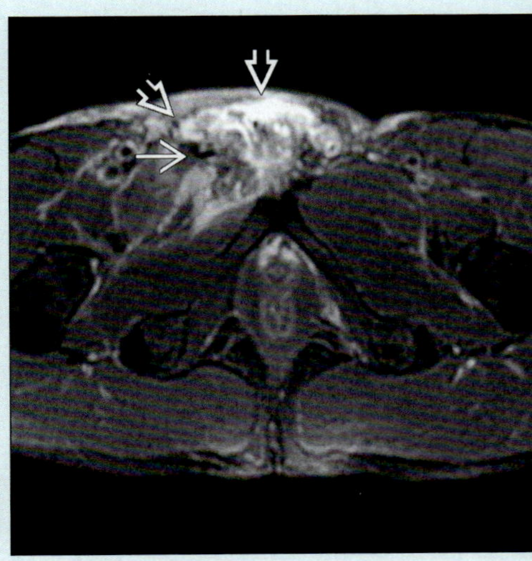

HIP ADDUCTOR INJURY

TERMINOLOGY

Synonyms

- Adductor syndrome

Definitions

- Injury of adductors or internal rotators of hip sparing rectus abdominis

IMAGING

General Features

- Best diagnostic clue
  - Soft tissue edema in adductor compartment on MR
    - Can be mild or extensive
    - Unilateral or bilateral
- Location
  - Adductor origin > myotendinous junction and muscle belly
- Size
  - Low-grade strain to complete avulsion

Radiographic Findings

- Avulsed fragment at pubic tubercle (rare)
- Calcific tendinopathy of adductors
- Associated osteitis pubis

CT Findings

- Confirmation of radiographic findings
- Asymmetric thigh compartment hematoma

MR Findings

- T1WI
  - Avulsed osseous fragment, osteitis pubis
  - Enthesopathy at pubic tubercle
  - Hyperintense blood products with acute/subacute hematoma
- T2WI FS
  - Acute injury: Identify center of soft tissue edema
    - Proximal at tendon origin
    - More distal at myotendinous junction
    - Mid-muscle belly or throughout compartment
    - Direction of ↑ T2 signal indicates tendon/muscle primarily involved
    - Grade strains on T2 or STIR
      - Grade 1: Peritendinous edema without visible tear
      - Grade 2: Partial tear without complete disruption
      - Grade 3: Complete tendon or muscle tear, often with retraction
    - Secondary cleft indicates adductor involvement
      - Inferolateral extension of hyperintense signal from symphysis
      - May indicate aponeurosis injury
  - Chronic injury: Identify muscle or tendon asymmetry
    - Enlargement, atrophy, nodularity
    - Peritendinous T2 hyperintensity = peritendinitis or strain
    - Large, hypointense on all sequences = hypoxic tendinopathy
    - Focal black signal can reflect calcific tendinopathy
- STIR
  - Multiple bilateral edematous muscles = delayed onset muscle soreness (DOMS)
- Standard protocol for adductor/internal rotator injury
- Bony pelvis or athletic pubalgia protocol

Ultrasonographic Findings

- Sensitive for adductor tendon injury
- Insensitive for associated pubic symphysis pathology

Imaging Recommendations

- Best imaging tool
  - MR
- Protocol advice
  - Athletic pubalgia protocol for proximal injury
  - MR thighs for more distal injury
    - Always cover both thighs in axial plane for symmetry
    - Vitamin E marker at site of pain
    - Gracilis strain tends to include mid to distal thigh

DIFFERENTIAL DIAGNOSIS

Sports Hernia

- More common than isolated adductor injury
- Rectus abdominis involvement can be subject to surgical interpretation
  - Lesions called adductor avulsion by some, aponeurosis injury by others
  - Watch carefully for subtle rectus abdominis involvement
- Association with osteitis pubis and secondary cleft

Osseous Injury

- Inferior pubic ramus stress fracture or fall from height
- Ischiopubic growth plate stress in young
- Osteitis pubis: Associated with adductor or aponeurosis injury

Hip Flexor Injuries

- Resistance to flexion more than adduction
- Accessory hip flexor injury can be clinically confusing
  - Sartorius, iliacus
- Flexor origin apophysitis in young

Soft Tissue Tumors

- Rare in femoral neurovascular region of thigh
- Consider contrast study if no trauma or overuse

Internal Derangement of Hip

- Consider anesthetic arthrogram

Hernia

- Femoral hernia in females can cause adductor symptoms

Adductor Myositis and Myonecrosis

- Status post pelvic radiotherapy
- Chronic sacral decubitus ulcers

PATHOLOGY

General Features

- Associated abnormalities
  - Adductor compartment hematoma
    - With high-grade aponeurosis tears

HIP ADDUCTOR INJURY

- Can be source of delayed healing
- If large, can contribute to compartment syndrome
- Can calcify, leaving chronic, painful mass

- Many adductor lesion variants in sports medicine literature
  - Adductor avulsions
    - Adductor longus from pubic tubercle most common
    - Forceful thigh adduction with extension at waist
      - Jumping in basketball, track and field
  - Adductor myotendinous strains
    - Generally > 1 muscle involved
      - Adductor longus > pectineus > adductor brevis > gracilis
    - Can be chronic issue with overuse syndromes
      - Hockey goalies
      - Equestrians
    - Suspect ipsilateral aponeurosis lesion
    - Most common cause of pain 2-6 months post pelvic floor repair
      - Postop PT regimens designed to maximize adductor flexibility
  - Chronic adductor origin tendinopathy
    - Overuse injury
    - Pectineus & adductor brevis less frequently involved
  - "Baseball pitcher/hockey goalie syndrome"
    - Myotendinous junction strain with epimysial (muscle sheath) injury preventing healing
    - Common with groin pain after pelvic floor repair
    - Requires operative debridement
  - Calcific adductor variant
    - Hydroxyapatite deposition in degenerated tendon
    - Kickboxing, hurdling
- Tendon rupture related to long-term ciprofloxacin therapy (rare)

Gross Pathologic & Surgical Features

- Muscle location from anterior to posterior: Pectineus, adductor longus, adductor brevis, gracilis

CLINICAL ISSUES

Presentation

- Most common signs/symptoms
  - Frequency of injury: Adductor longus > pectineus > adductor brevis > gracilis
  - Unilateral proximal & medial thigh pain
  - Resistance to adduction
  - Nagging, aching pain with degenerative tendinopathy
  - Difficult to distinguish sports hernia from isolated adductor lesion
    - Adductor tendinopathy can evolve to aponeurosis lesion
    - Patients initially treated conservatively often ultimately have pelvic floor repair
    - Secondary cleft associated with both lesions
- Acute: Young athletes
- Acute: High-grade trauma
- Chronic: Middle-aged, weekend warrior-type athletes

Treatment

- Regimens range from conservative to operative
  - Primary adductor tendon repair uncommon
    - Re-attachment of avulsed osseous fragment
  - Evacuation of adductor hematoma and tendon debridement
    - Often in conjunction with rectus abdominis stabilization
    - Percutaneous catheter drainage
  - Directed steroid injection for chronic tendinopathy
  - Ultrasound-guided tenotomy; needle debridement
  - Strengthening and flexibility programs
- More likely to respond to conservative therapy than aponeurosis lesion

DIAGNOSTIC CHECKLIST

Consider

- More extensive rectus abdominis/adductor aponeurosis lesion
  - Look for confluent lesion extending above and below symphysis
- With no injury or overuse, consider soft tissue mass
- Hypoxic tendinopathy with large, hypointense tendon

Image Interpretation Pearls

- Use available clinical history as guide to description

Reporting Tips

- Measure adductor tendon retraction
- Identify hypoxic/calcific tendinopathy

SELECTED REFERENCES

1. Schlegel TF et al: Success of nonoperative management of adductor longus tendon ruptures in National Football League athletes. Am J Sports Med. 37(7):1394-9, 2009
2. Meyers WC et al: Experience with "sports hernia" spanning two decades. Ann Surg. 248(4):656-65, 2008
3. Omar IM et al: Athletic pubalgia and "sports hernia": optimal MR imaging technique and findings. Radiographics. 28(5):1415-38, 2008
4. Topol GA et al: Regenerative injection of elite athletes with career-altering chronic groin pain who fail conservative treatment: a consecutive case series. Am J Phys Med Rehabil. 87(11):890-902, 2008
5. Zajick DC et al: Spectrum of MRI findings in clinical athletic pubalgia. Semin Musculoskelet Radiol. 12(1):3-12, 2008
6. Zoga AC et al: Athletic pubalgia and the "sports hernia": MR imaging findings. Radiology. 247(3):797-807, 2008
7. Strauss EJ et al: Analysis of the cross-sectional area of the adductor longus tendon: a descriptive anatomic study. Am J Sports Med. 35(6):996-9, 2007
8. Rizio L 3rd et al: Adductor longus rupture in professional football players: acute repair with suture anchors: a report of two cases. Am J Sports Med. 32(1):243-5, 2004
9. Robinson P et al: Adductor-related groin pain in athletes: correlation of MR imaging with clinical findings. Skeletal Radiol. 33(8):451-7, 2004
10. Meyers WC et al: Surgical management of chronic lower abdominal and groin pain in high-performance athletes. Curr Sports Med Rep. 1(5):301-5, 2002

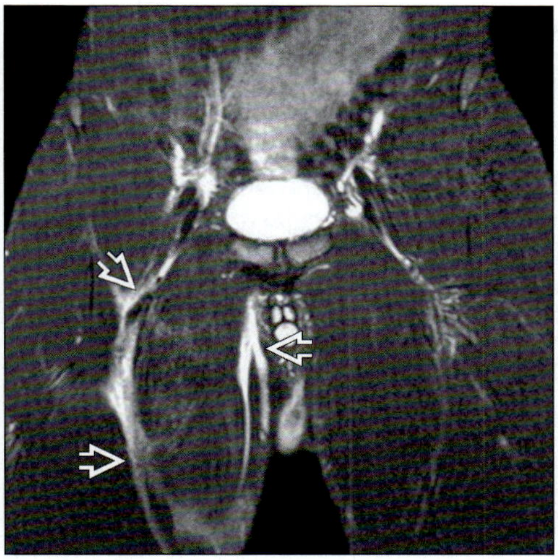

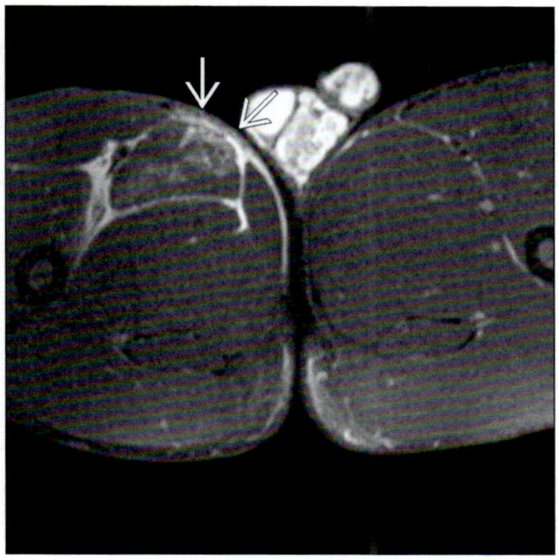

(Left) *Coronal STIR MR from a professional hockey player with nagging groin pain demonstrates perimuscular edema throughout the adductor compartment, indicating a chronic strain.* ***(Right)*** *Axial T2WI FS MR from the same patient shows indistinctness of the anteromedial aspect of the adductor longus muscle sheath. This patient went to surgery where a portion of the adductor longus had herniated through its epimysium, an injury termed "baseball pitcher/hockey goalie syndrome."*

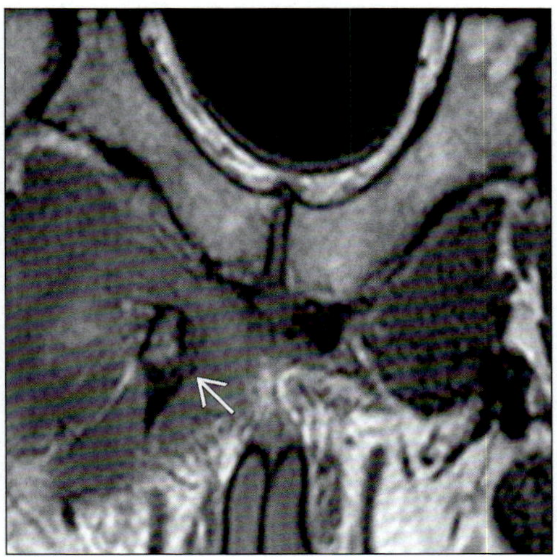

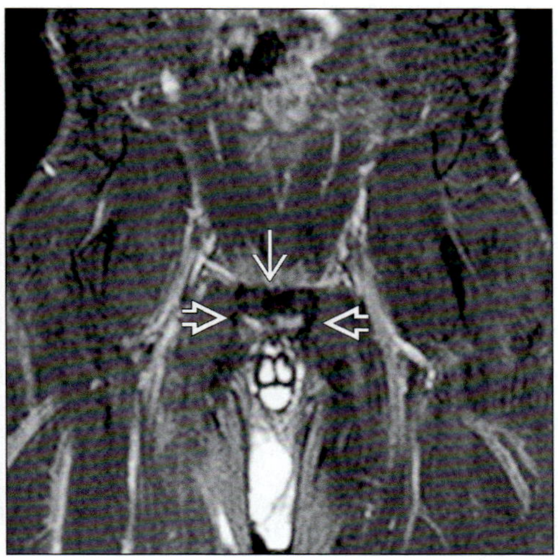

(Left) *Coronal oblique T1WI MR from a professional basketball player after a thigh abduction injury while rebounding shows an osseous fragment avulsed from the pubic tubercle. Osseous avulsions of the adductor longus are rare.* ***(Right)*** *Coronal STIR MR from a professional soccer player with a long history of bilateral groin pain and previous adductor tendon steroid injections shows partial tears at the origins of both adductor longus tendons, but the pubic symphysis is normal.*

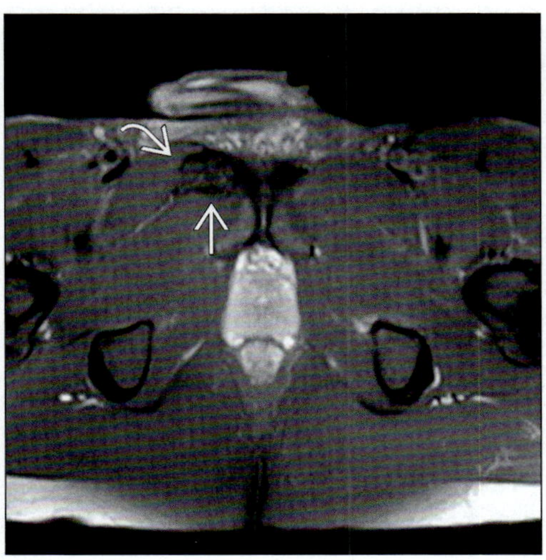

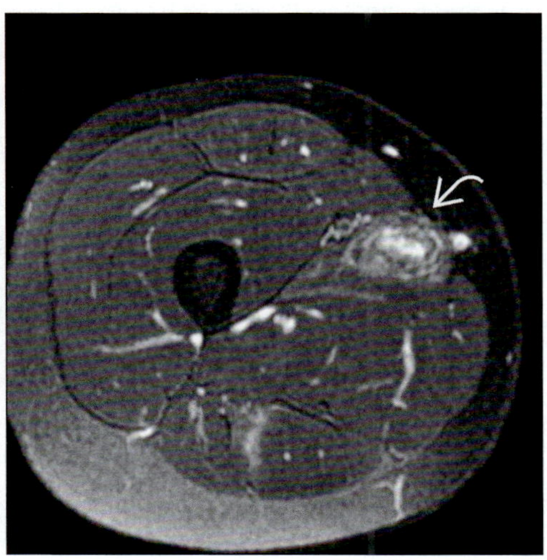

(Left) *Axial T2WI FS MR from a hurdler with nagging right-sided groin pain shows marked enlargement at the origins of the right adductor longus and pectineus. Hypoxic tendinopathy can be a debilitating degenerative syndrome.* ***(Right)*** *Axial T2WI FS MR from a runner with worsening thigh pain after a grade 2 strain at the proximal adductor longus myotendinous junction diagnosed by MR 4 weeks prior shows an organizing hematoma creating mass effect within the adductor longus muscle.*

SNAPPING HIP SYNDROME

Key Facts

Terminology

- Snapping hip, coxa saltans
- Painful snapping, popping, or clicking sensation with hip motion

Imaging

- MR of external snapping: Peritendinous hyperintensity in region of greater trochanter
- MR of internal snapping: Peritendinous hyperintensity around distal iliopsoas tendon
  - Usually in region of iliopectineal eminence
- Intraarticular: Labral tear, intraarticular bodies, ligamentum teres injury
- Dynamic ultrasound diagnostic in most cases
  - Subluxation of involved tendon (snapping) during range of motion

Pathology

- External: Proximal iliotibial band or distal gluteus maximus pops over greater trochanter
  - Occurs as hip extends from flexed position
- Internal: Iliopsoas tendon pops over iliopectineal eminence and anterior femoral head
  - Occurs with internal rotation > external rotation
- Intraarticular: Any internal derangement causing abnormal motion of intraarticular structures
- Commonly combination of variant anatomy and overuse syndromes

Clinical Issues

- Young adults: 15-40 years old
- Ultrasound-guided steroid injection for diagnosis and treatment
- Tendon debridement and bursectomy

*(**Left**) Coronal graphic shows iliopsoas impingement on the iliopectineal eminence during extension of the leg. The tendon passes under the inguinal ligament. (**Right**) Sagittal T2WI FS MR shows soft tissue edema around the iliopsoas tendon as it passes over the anterior acetabulum near the iliopectineal eminence. Edema in this location is suggestive of internal snapping hip syndrome in the appropriate clinical setting.*

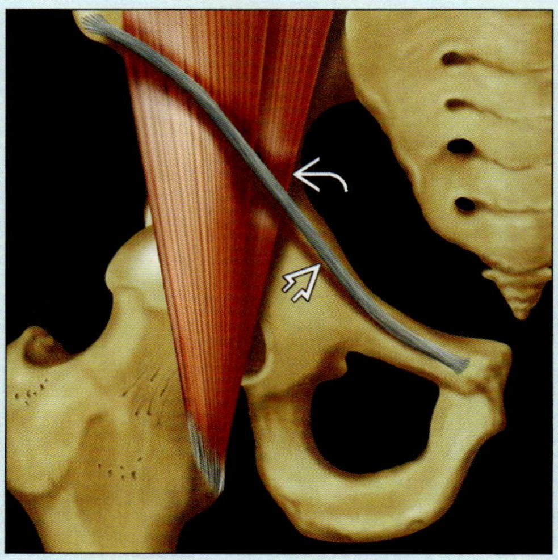

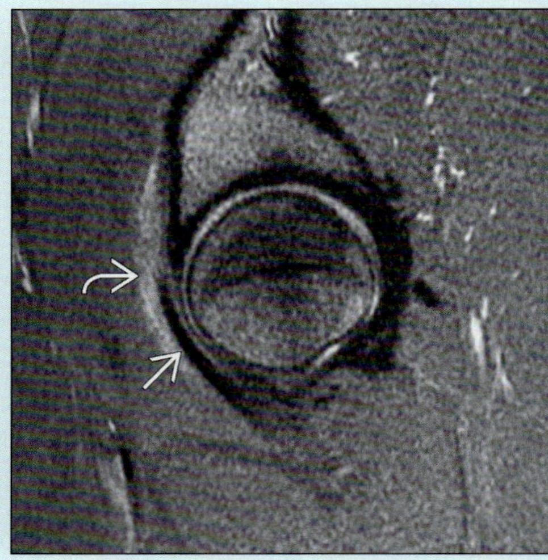

*(**Left**) Coronal graphic shows contact of the iliotibial band with the anterior margin of the greater femoral trochanter; this anatomy can result in external snapping hip syndrome with repetitive motion. (**Right**) Axial T2WI FS MR shows myotendinous edema involving the tensor fascia lata with thickening of the iliotibial band just below the level of the greater trochanter. This patient's externally snapping hip symptoms resolved after debridement of the iliotibial tract.*

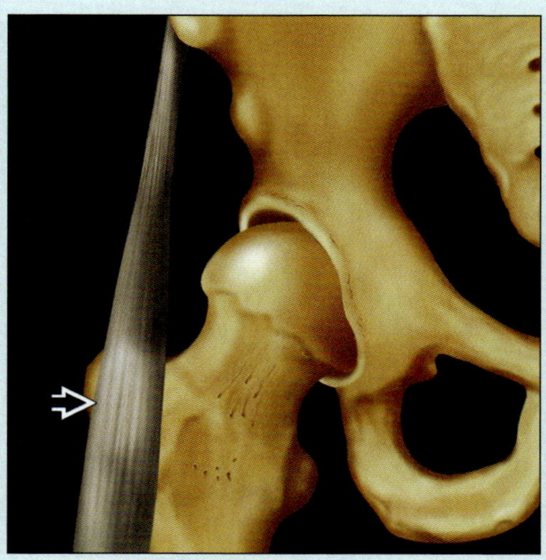

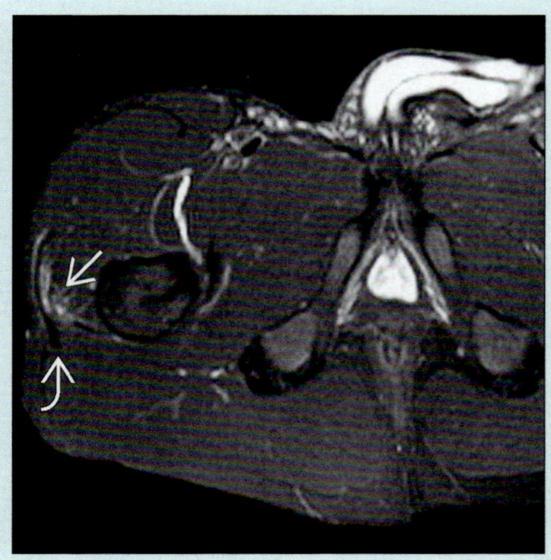

SNAPPING HIP SYNDROME

TERMINOLOGY

Synonyms
- Snapping hip, coxa saltans

Definitions
- Painful snapping, popping, or clicking sensation with hip motion

IMAGING

General Features
- Best diagnostic clue
  - Soft tissue edema around iliopsoas tendon anterior to hip or around iliotibial band at greater femoral trochanter
- Location
  - External: Iliotibial band or gluteus maximus
  - Internal: Iliopsoas tendon
  - Intraarticular: Labral tear, intraarticular bodies, ligamentum teres injury

Radiographic Findings
- Radiography
  - Generally normal with external and internal sources
  - Calcific bodies and osteoarthritis with intraarticular snapping

MR Findings
- T1WI
  - External: Hypointense thickening of proximal iliotibial band
  - Internal: Hypointense thickening of distal iliopsoas tendon
- T2WI
  - External
    - Peritendinous hyperintensity in region of greater trochanter
    - Hyperintense greater trochanteric bursa
  - Internal
    - Peritendinous hyperintensity around distal iliopsoas tendon (especially anterior to hip joint at iliopectineal eminence)
    - Hyperintense iliopsoas bursa
- T1WI C+
  - Peritendinous enhancement of involved structure

Ultrasonographic Findings
- Dynamic ultrasound diagnostic in most cases
  - Subluxation of involved tendon (snapping) during range of motion
  - Painful: Patients often allow maneuver only once

Imaging Recommendations
- Best imaging tool
  - Noncontrast MR: Secondary findings
  - Contrast-enhanced MR: Peritendinous enhancement
  - Dynamic ultrasound: Diagnostic but painful

DIFFERENTIAL DIAGNOSIS

Iliopsoas Bursitis
- Associated with internal snapping hip

Greater Trochanteric Bursitis
- Associated with external snapping hip

PATHOLOGY

General Features
- Etiology
  - Commonly combination of variant anatomy and overuse syndromes
  - External: Proximal iliotibial band or distal gluteus maximus pops over greater trochanter
    - Occurs as hip extends from flexed position
  - Internal: Iliopsoas tendon pops over iliopectineal eminence and anterior femoral head
    - Occurs with internal rotation > external rotation
  - Intraarticular: Any internal derangement causing abnormal motion of intraarticular structures
- Intraarticular snapping
  - Detached or displaced labral tear
  - Intraarticular bodies
  - Ligamentum teres rupture
  - Os acetabulare
  - Osteoarthritis

CLINICAL ISSUES

Presentation
- Most common signs/symptoms
  - Palpable snap during hip flexion or rotation

Demographics
- Age
  - Young adults: 15-40 years old

Treatment
- Conservative
- Minimally invasive
  - Ultrasound-guided steroid injection
- Surgical
  - Tendon debridement and bursectomy
  - Tendon lengthening procedures
  - Arthroscopy for intraarticular lesions

DIAGNOSTIC CHECKLIST

Consider
- Internal vs. external snapping by clinical exam
- MR arthrography for evaluation of intraarticular snapping

Image Interpretation Pearls
- Dynamic ultrasound can confirm diagnosis

SELECTED REFERENCES

1. Hodnett PA et al: MR imaging of overuse injuries of the hip. Magn Reson Imaging Clin N Am. 17(4):667-79, vi, 2009
2. Yoon TR et al: Clinical results of multiple fibrous band release for the external snapping hip. J Orthop Sci. 14(4):405-9, 2009
3. Deslandes M et al: The snapping iliopsoas tendon: new mechanisms using dynamic sonography. AJR Am J Roentgenol. 190(3):576-81, 2008

*(**Left**) Axial PDWI FS MR after failed gadolinium arthrogram attempt shows a lobular fluid collection anterior to the acetabular margin and lateral to the iliopsoas tendon. The patient complained of anterior snapping in the hip with rotational motion. Injected contrast is present in the anterior pericapsular tissues. (**Right**) Coronal T2WI FS MR from the same patient shows the lobular bursal fluid collection lateral to the iliopsoas myotendinous junction.*

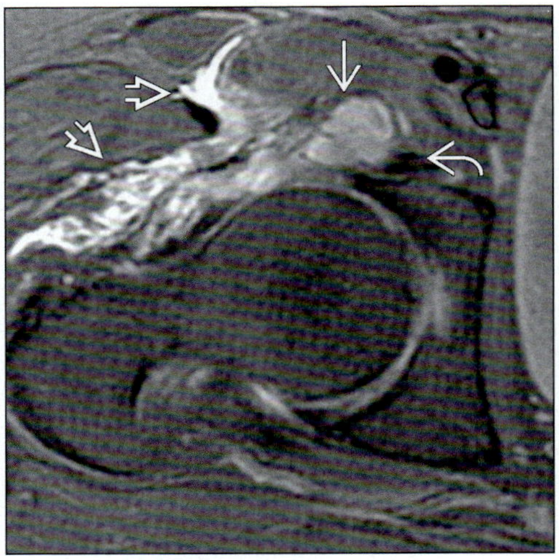

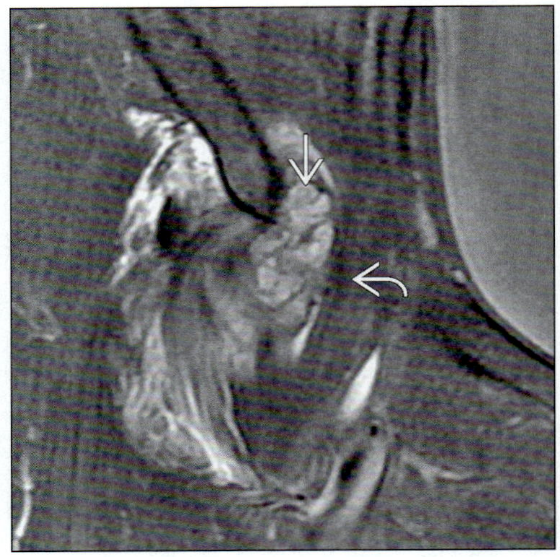

*(**Left**) Sagittal T2WI FS MR from the same patient shows the iliopsoas tendon passing anterior to the hip joint. The large iliopsoas bursal fluid collection lies anterior and lateral to the tendon. Inflamed bursal collections are often associated with internal snapping hip syndrome. (**Right**) Coronal T2WI FS MR from a young woman with painful popping at the anterior hip with internal and external rotation shows mild soft tissue edema around an enlarged iliopsoas tendon.*

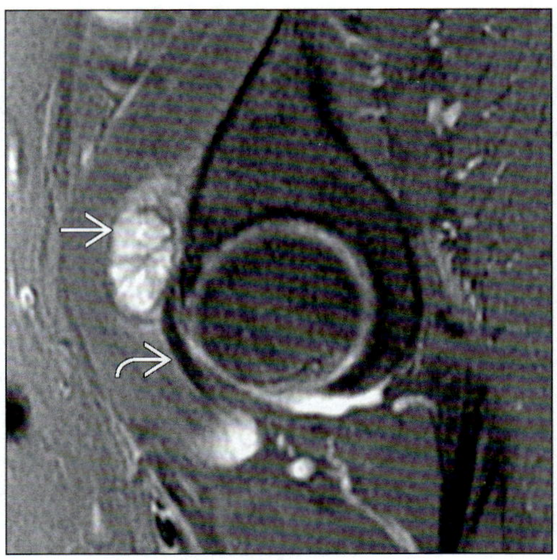

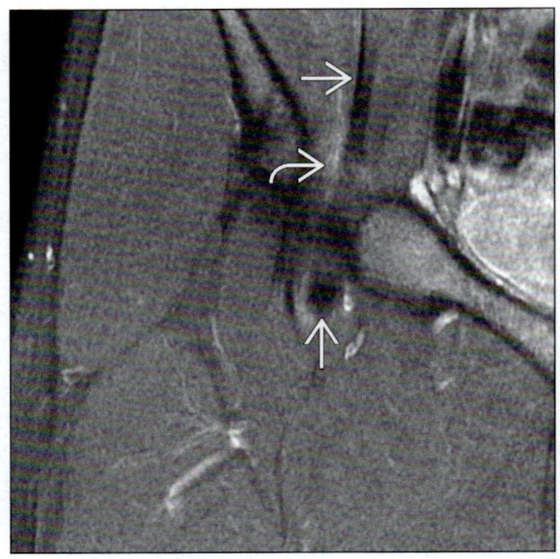

*(**Left**) Axial T2WI FS MR from the same patient shows the static MR findings most commonly seen with a symptomatic internal snapping hip. The distal iliopsoas tendon is enlarged and exhibits abnormal ovoid morphology. (**Right**) Axial T2WI FS MR more inferiorly in the same patient shows the iliopsoas tendon with surrounding edema. The iliopsoas is snapping over a thickened anterior hip capsule in this case.*

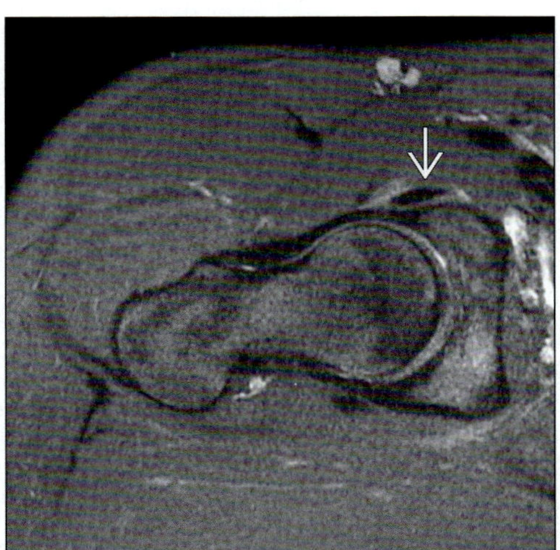

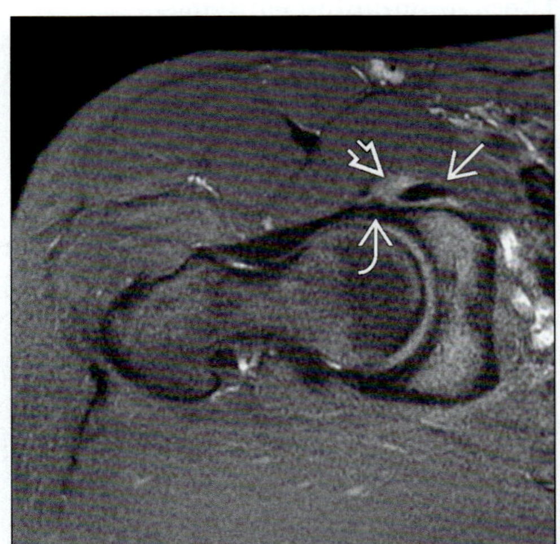

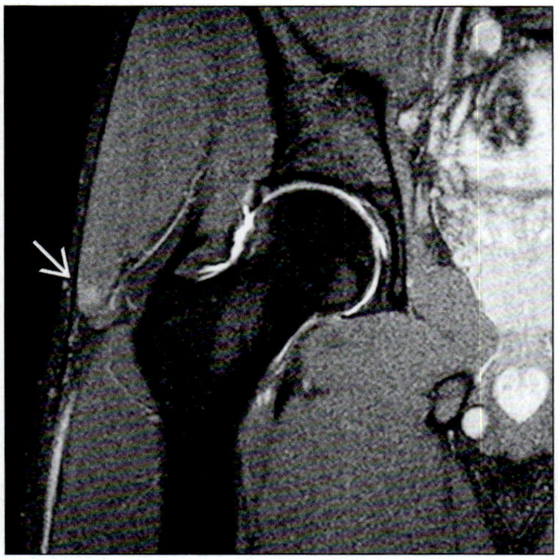

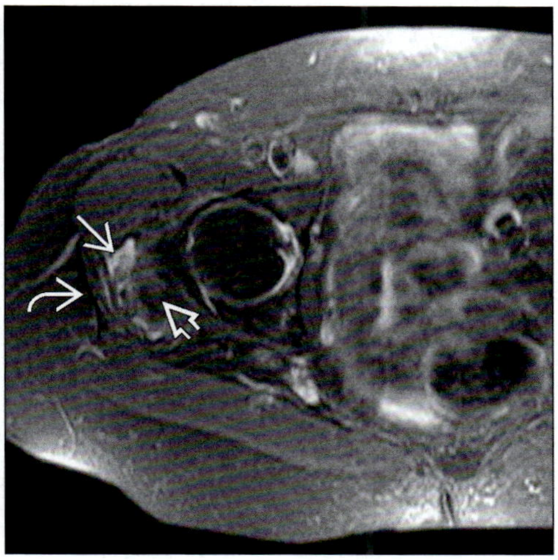

(Left) *Coronal T2WI FS MR from a professional basketball player with recurrent painful snapping at the greater trochanter shows focal soft tissue edema at the proximal iliotibial band.* ***(Right)*** *Axial T2WI FS MR from a 64-year-old woman with painful snapping sensation in the lateral hip shows significant edema between the iliotibial band and the greater femoral trochanter. This is the typical constellation of findings in external snapping hip.*

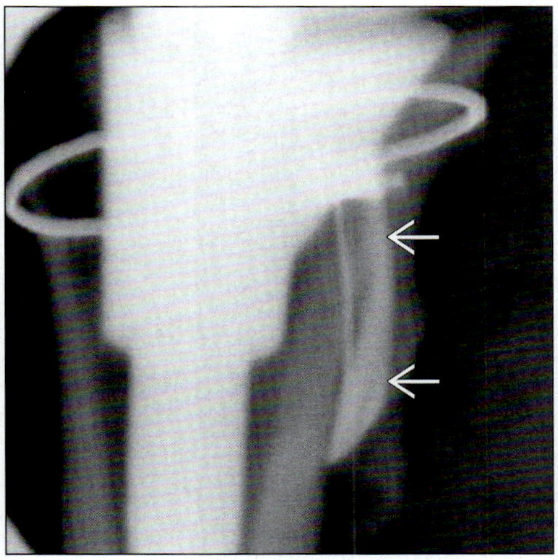

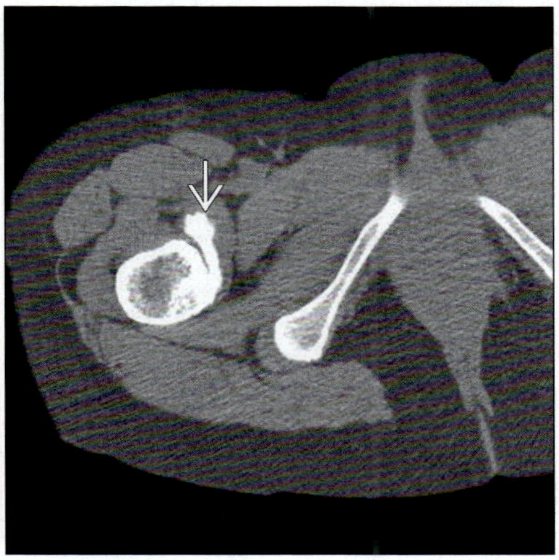

(Left) *AP radiograph demonstrates contrast in the iliopsoas tendon sheath after steroid/anesthetic + contrast injection for snapping hip. The snapping tendon can be observed under fluoroscopy during hip flexion; anesthetic should alleviate the symptoms. This can also be done under ultrasound guidance.* ***(Right)*** *Axial CT following iliopsoas tendon injection shows contrast in the tendon sheath at its insertion. Injection under CT does not allow for observation of tendon motion.*

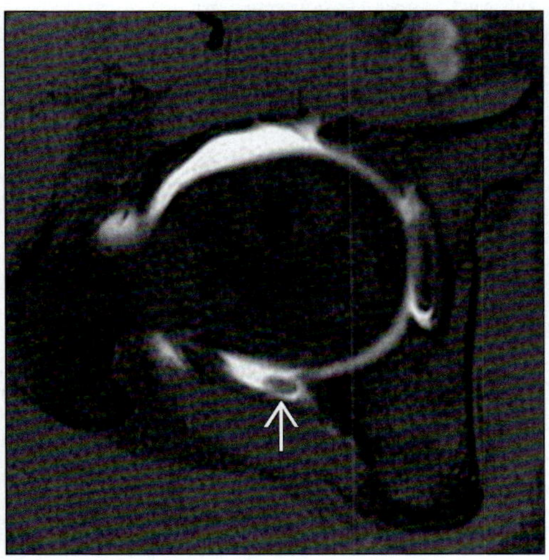

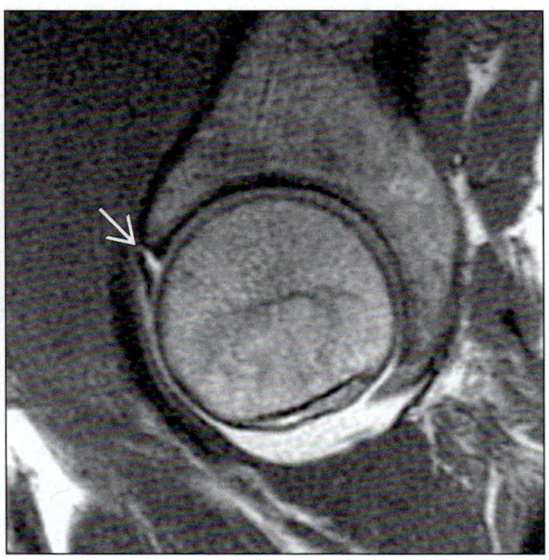

(Left) *Axial MR arthrogram T1WI FS shows a chondral fragment in the posterior hip joint in a 17-year-old high school football player complaining of a painful popping hip. Intraarticular causes of snapping hip are best diagnosed using MR arthrography. This patient's pain resolved with intraarticular anesthetic injection.* ***(Right)*** *Sagittal MR arthrogram T1WI shows a full thickness tear of the anterosuperior acetabular labrum in a young patient with a popping sensation in the hip.*

PROXIMAL HAMSTRING INJURY

Key Facts

Terminology
- Pulled hamstring, hamstring strain

Imaging
- Conjoined tendon origin
  - Biceps femoris and semitendinosus
- Semimembranosus origin
  - Anterior to conjoined tendon
- MR: Diffuse edema throughout muscle belly
  - Perimyotendinous edema with acute strain
  - Missing tendon at ischial tuberosity
  - Fluid gap at tendon origin with tendon retraction
  - Bone marrow edema at tendon origin
  - Muscle sheath or intermuscular hematoma
- Tendon avulsion of 1 unit combined with myotendinous junction strain of another muscle is common

Pathology
- Grade 1: Fluid around myotendinous junction but no fiber disruption
- Grade 2: Partial tear
- Grade 3: Complete disruption

Clinical Issues
- Muscle belly and myotendinous strains heal with conservative therapy
- Consider primary reattachment of high-grade proximal avulsion

Diagnostic Checklist
- Skin marker over site of maximal tenderness
- Measure length of T2 hyperintensity in muscle to predict recovery time
- Look for calcium in tendon

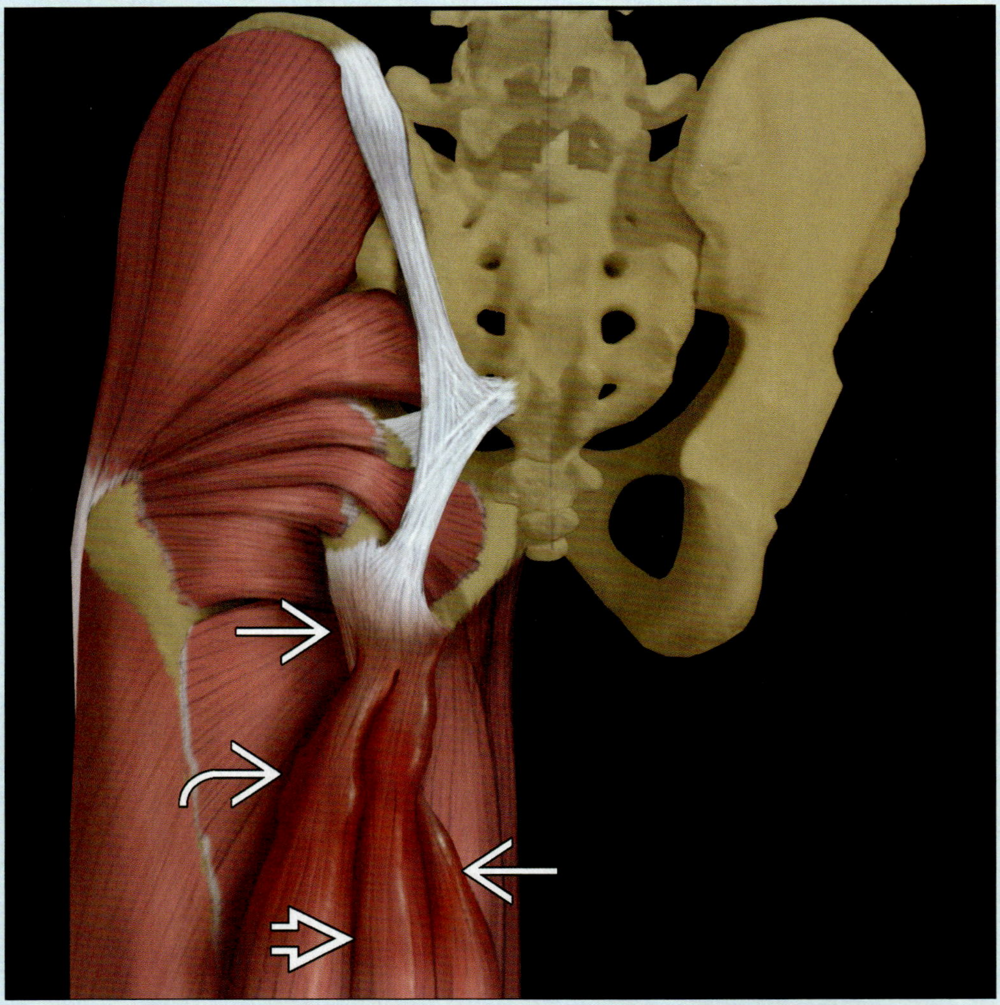

Posterior graphic shows the origin of the proximal hamstring complex. The semimembranosus ➡ originates slightly anteriorly; the tendon crosses medially to form the medial portion of the hamstring muscle complex. The semitendinosus ➡ and biceps femoris ➡ have a conjoined origin posterolateral to that of the semimembranosus. The semitendinosus tracks between the biceps femoris and the semimembranosus in the thigh. The short head of the biceps femoris originates more distally from the mid-femur and linea aspera. The biceps femoris inserts on the fibula, while the semimembranosus and semitendinosus insert on the tibia. Injury can occur anywhere along this anatomical route.

PROXIMAL HAMSTRING INJURY

TERMINOLOGY

Synonyms

- Pulled hamstring, hamstring strain

Definitions

- Myo-teno-osseous injury of biceps femoris (BF), semitendinosus (ST), or semimembranosus (SM) with at least some fiber disruption
- Degenerative tendinopathy or enthesopathy involving biceps femoris, semitendinosus, or semimembranosus

IMAGING

General Features

- Best diagnostic clue
  - Edema around hamstring complex myotendinous junction (MTJ) or tendon origin
  - Fluid gap within hamstring complex after injury
- Location
  - Conjoined tendon origin
    - Biceps femoris and semitendinosus
    - Posterior to semimembranosus tendon
    - Injury may occur in tendon, bone, or apophysis
  - Semimembranosus origin
    - Anterior to conjoined tendon
    - Injury may occur in tendon, bone, or apophysis
  - Myotendinous junction strain
    - Proximal or distal
    - Biceps femoris, semitendinosus, or semimembranosus
  - Muscle belly
    - Biceps femoris, semitendinosus, or semimembranosus
  - Distal strain
    - Biceps femoris long or short head, semimembranosus, or semitendinosus
  - Involvement of multiple muscles or tendons is common
- Morphology
  - Interstitial tear of tendon or muscle
  - Frank tendinous avulsion
    - Conjoined tendon avulsion: "2 tendon" injury (BF and ST)
    - Conjoined tendon + SM: "3 tendon" injury
  - Osseous avulsion (rare)
  - Tendon avulsion of 1 unit combined with myotendinous junction strain of another muscle is common

Radiographic Findings

- Radiography
  - Most often radiographically occult
  - Acute osseous avulsion
  - Calcific tendinopathy with chronic lesion
  - Calcification of hematoma/myositis ossificans

MR Findings

- T1WI
  - Hypointense peritendinous fluid
  - Osseous avulsion
  - Asymmetric fluid around muscle/tendon
  - Abnormally hypointense tendon: Chronic injury
  - Heterogeneous hematoma
- T2WI
  - Diffuse edema throughout muscle belly
  - Peri-myotendinous edema with acute strain
    - Proximal > distal
    - Pennate (feathery) pattern of T2 hyperintensity
  - Missing tendon at ischial tuberosity
  - Fluid gap at tendon origin with tendon retraction
  - Bone marrow edema at tendon origin
  - Muscle sheath or intermuscular hematoma
- Calcific tendinopathy: Generally ↓ signal on all sequences, may have surrounding edema

Imaging Recommendations

- Best imaging tool
  - MR
- Protocol advice
  - Essential for patient to localize injury with marker
  - Center on marker
  - Coronal STIR, axial T2WI FS of entire pelvis or both thighs
    - Depends on site of pain
  - Higher resolution, smaller field-of-view (FOV) axial T2WI FS or PDWI FS of injured side
  - Sagittal T2WI FS centered on injury

Ultrasonographic Findings

- Sensitive and specific for location and grade
- Insensitive for osseous edema

DIFFERENTIAL DIAGNOSIS

Apophysitis

- Repetitive avulsion at skeletally immature bone
- Apophyseal junction weaker than tendon or muscle

Hip Internal Derangement

- Stress fracture, osteonecrosis, labrum/cartilage lesions

Pelvis Stress Fracture

- Inferior pubic ramus/ischium mimic hamstring injury

Sciatic Nerve Compression

- Piriformis syndrome
- Asymmetry on axial & coronal sequences

PATHOLOGY

General Features

- Etiology
  - Hip extension against resistance
  - Hip flexion with extended knee
  - Repetitive microtrauma with running & jumping
  - Inflexibility, especially with sprinting
  - Muscle/tendon fatigue from repetitive trauma
  - Functional imbalance between different hamstring units or between hamstrings & quadriceps
  - Delayed onset muscle soreness (DOMS) predisposes to strain
- Associated abnormalities
  - Thigh adductor tendon injury
  - Gastrocnemius origin strain with distal hamstring injury
- Relevant anatomy
  - Semimembranosus
    - Origin: Anteromedial ischial tuberosity
    - Insertion: Posteromedial proximal tibia

PROXIMAL HAMSTRING INJURY

- Semitendinosus
  - Origin: Posterolateral ischial tuberosity with biceps
  - Insertion: Proximal medial tibia (with gracilis & sartorius as part of pes anserinus)
- Biceps femoris
  - Long head origin: Posterolateral ischial tuberosity (with semitendinosus)
  - Short head origin: Linea aspera and posterior femur
  - Insertion: Proximal lateral fibular head (as conjoined tendon with fibular collateral ligament) and lateral tibia
- Innervation: Tibial branch of sciatic nerve (except short head of BF: Peroneal branch)

Staging, Grading, & Classification

- Grade myotendinous junction strains on fluid-sensitive MR
  - Grade 1: Fluid around myotendinous junction but no fiber disruption
  - Grade 2: Partial tear
  - Grade 3: Complete disruption
- Proximal vs. distal

Gross Pathologic & Surgical Features

- Acute injury
  - Traumatic disruption of fibers along myo-teno-osseous axis
- Chronic tendinopathy
  - Degenerative process with hypoxic cell death
  - Calcific tendinopathy and intratendinous mucinous degeneration
  - Myofibroblasts proliferate; no inflammatory cells
- Acute injury superimposed on chronic tendinopathy common

CLINICAL ISSUES

Presentation

- Most common signs/symptoms
  - Proximal: Gluteal pain and point tenderness
  - Distal: Pain and tenderness proximal to popliteal region
  - Acute injury: Sharp, stabbing pain
  - Chronic injury: Dull, nagging ache
  - Painful when sitting
- Other signs/symptoms
  - Pain & tightness with forward kicking motion
  - Painful static stretch of hamstring group
  - Feeling of impending muscle pull
- Clinical profile
  - Activities requiring bursts of speed and repetitive hip stretch
    - Football, soccer, basketball, sprinting, rowing, kickboxing

Demographics

- Age
  - Young athletes
  - Weekend warrior and middle-aged athletes
  - Older patients with innocuous trauma

Natural History & Prognosis

- Muscle belly and myotendinous strains heal with conservative therapy
  - Recovery process lengthy
    - Up to 6 months
  - Elite sprinting athletes heal more slowly than large, plodding athletes
  - Length of T2 hyperintense signal within muscle helps predict return to play
- Large hematoma delays healing
- Avulsive injuries respond better to interventional therapies

Treatment

- Conservative
  - Massage, whirlpool, ultrasound therapy
  - Gradually progressive muscle strengthening and stretching
- Percutaneous
  - Aspiration of large hematomas
  - Steroid injections may help symptoms but may weaken tendon & cause tear
  - Ultrasound-guided tenotomy for chronic lesions
  - Platelet-rich plasma (PRP) injections controversial
- Surgical
  - Consider primary reattachment of high-grade proximal avulsion
    - 2 or 3 tendon tears with retraction
  - Limited role in tendinopathy: Debridement

DIAGNOSTIC CHECKLIST

Consider

- Skin marker over site of maximal tenderness

Image Interpretation Pearls

- Localize and grade lesion

Reporting Tips

- Specify muscles/tendons involved
- Measure retraction or gap with high-grade injury
- Measure length of T2 hyperintensity in muscle to predict recovery time
- Look for calcium in tendon

SELECTED REFERENCES

1. Carmichael J et al: Avulsion of the proximal hamstring origin. Surgical technique. J Bone Joint Surg Am. 91 Suppl 2:249-56, 2009
2. Lempainen L et al: Proximal hamstring tendinopathy: results of surgical management and histopathologic findings. Am J Sports Med. 37(4):727-34, 2009
3. Davis KW: Imaging of the hamstrings. Semin Musculoskelet Radiol. 12(1):28-41, 2008
4. Silder A et al: MR observations of long-term musculotendon remodeling following a hamstring strain injury. Skeletal Radiol. 37(12):1101-9, 2008
5. Wood DG et al: Avulsion of the proximal hamstring origin. J Bone Joint Surg Am. 90(11):2365-74, 2008
6. Koulouris G et al: Hamstring muscle complex: an imaging review. Radiographics. 2005 May-Jun;25(3):571-86. Review. Erratum in: Radiographics. 25(5):1436, 2005

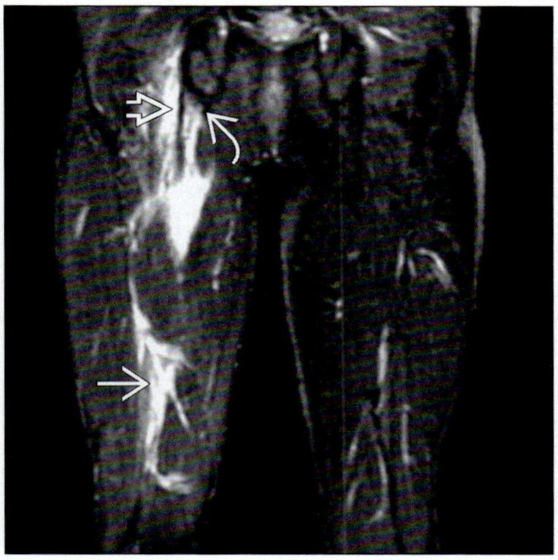

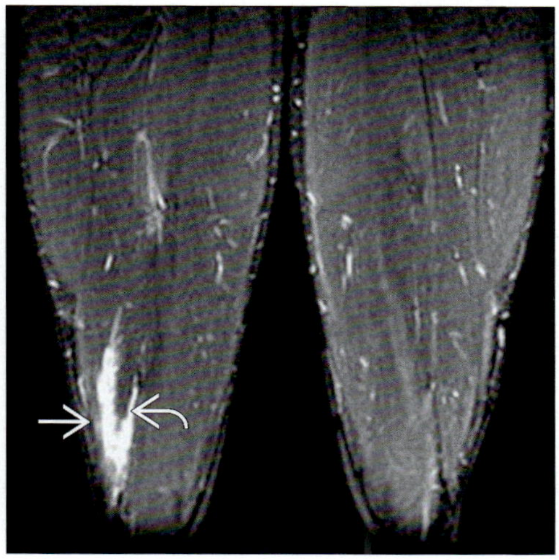

(Left) *Coronal STIR MR shows a high-grade proximal hamstring injury. The semimembranosus is avulsed from the ischial tuberosity. The conjoined tendon origin remains partially intact. There is extensive associated muscle injury, with edema extending to the level of the distal myotendinous junction.* ***(Right)*** *Coronal STIR MR showing a distal biceps femoris myotendinous strain in a professional football player. Partial fiber disruption makes this a grade 2 injury.*

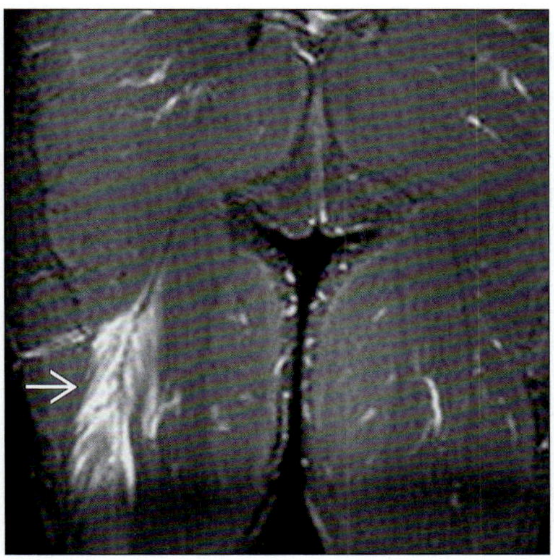

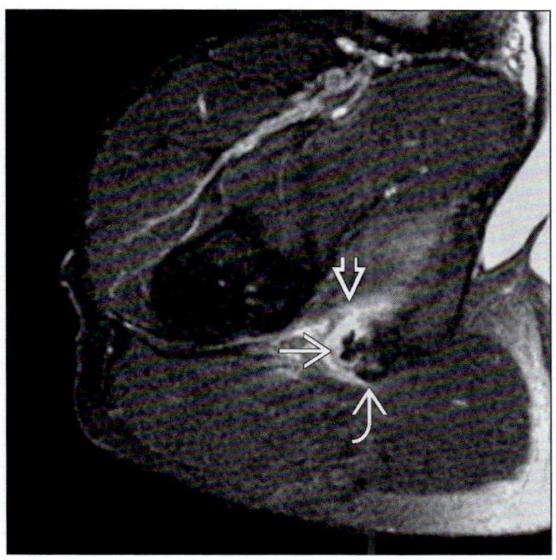

(Left) *Coronal STIR MR of a grade 1 distal myotendinous junction biceps femoris strain shows a pennate, feathery pattern of edema outlining muscle and tendon fibers.* ***(Right)*** *Axial T2WI FS MR shows an acute-on-chronic partial proximal hamstring injury. The semimembranosus is partially torn and enlarged, suggesting underlying hypoxic tendinopathy. The conjoined tendon is intact but also somewhat enlarged. Peritendinous fluid confirms an element of acute injury.*

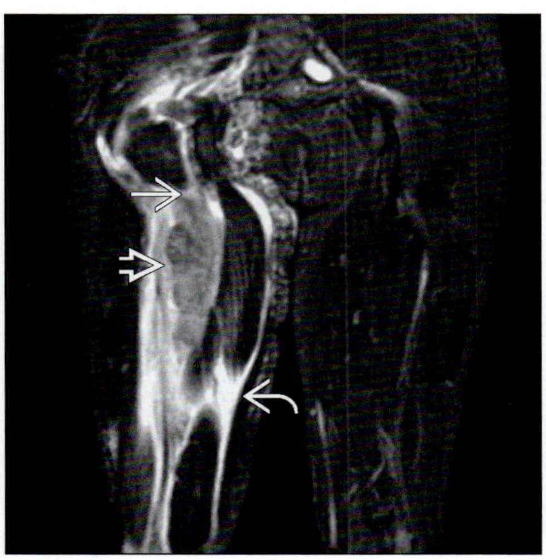

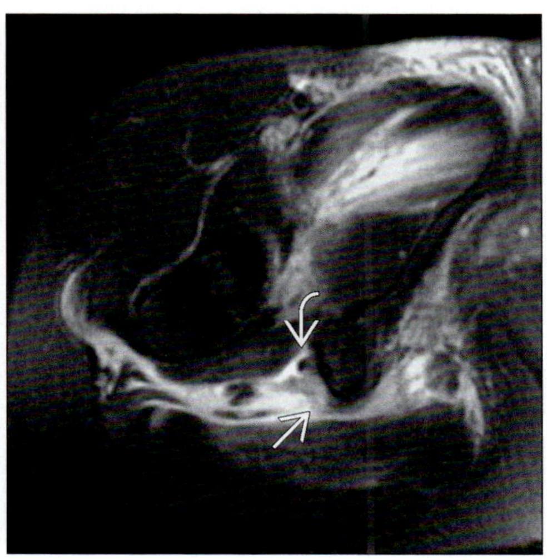

(Left) *Coronal STIR MR from a 44-year-old in-line skater with acute posterior thigh pain shows avulsion of the conjoined tendon from the ischium with a large hematoma and a more distal muscle belly strain of the semimembranosus, a common pattern of high-grade hamstring strain.* ***(Right)*** *Axial T2WI FS MR shows a complete 3 tendon proximal hamstring avulsion. A portion of the torn semimembranosus is visible, but there is no sign of the conjoined tendon, indicating retraction.*

PROXIMAL HAMSTRING INJURY

*(**Left**) Axial T2WI FS MR through the mid-thigh from a 57 year old who was injured water-skiing shows a high-grade distal myotendinous junction strain. The biceps tendon is torn and retracted, and there is fluid around the strained semimembranosus and edema throughout the semitendinosus, also indicating strain. (**Right**) Axial T1WI MR from the same patient shows several hyperintense regions within the injury bed, indicating a large hematoma.*

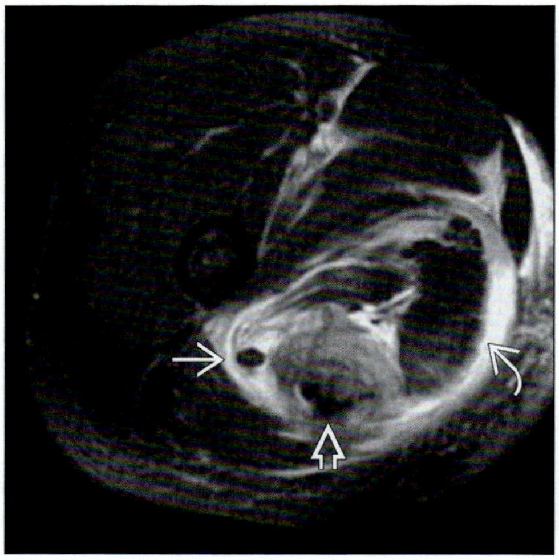

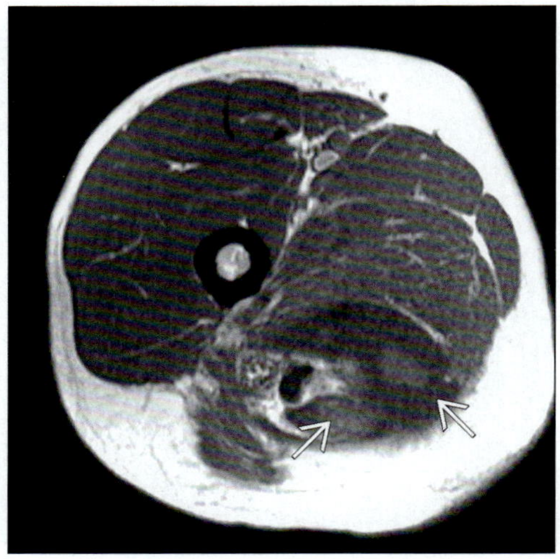

*(**Left**) Axial T1WI MR of the mid-thigh from a 38 year old with chronic posterior thigh pain and a normal lumbar spine MR shows atrophy of the biceps femoris muscle and nodularity of the tendon, indicating chronic tendinopathy from a remote hamstring strain. (**Right**) Axial T2WI FS MR from a 48-year-old recreational athlete with pain when sitting but no acute injury shows enlarged, hypointense hamstring tendon origins with peritendinous edema, compatible with chronic tendinopathy.*

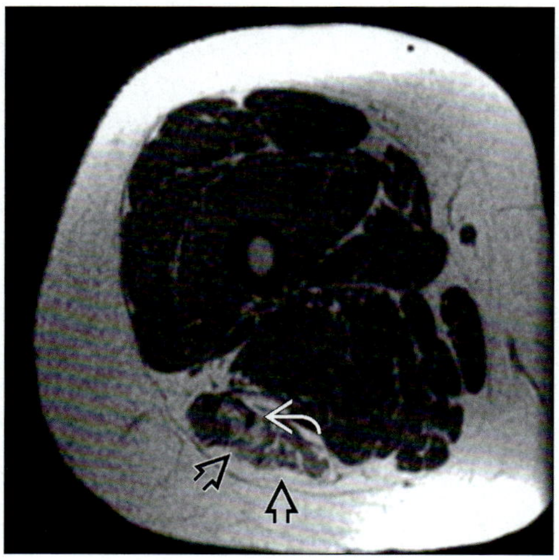

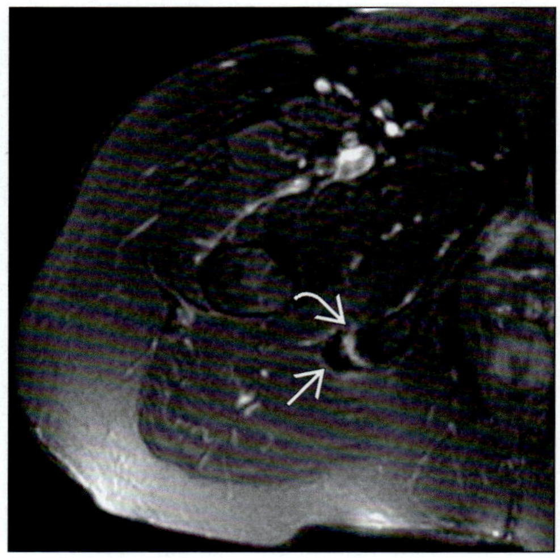

*(**Left**) Sagittal T2WI FS MR from a soccer player with an acute proximal myotendinous hamstring strain 3 weeks earlier shows a very large hematoma with mass effect and nodular hypointensity indicating evolving myositis ossificans. (**Right**) Coronal T2WI FS MR shows a large osseous avulsion of the entire hamstring origin from the ischial tuberosity in a 16-year-old track athlete with acute-on-chronic buttock pain when running. Hematoma fills the resulting gap.*

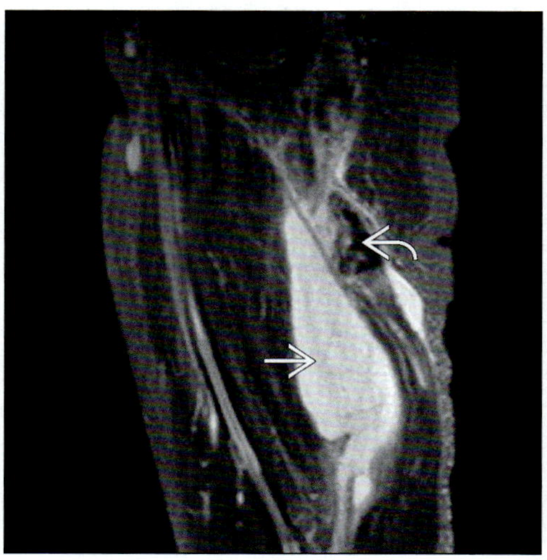

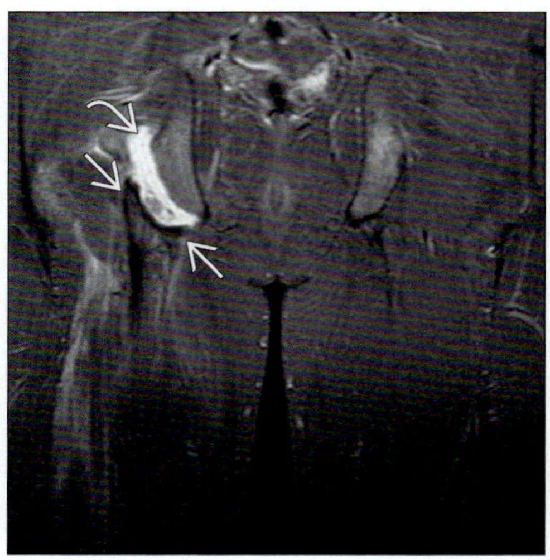

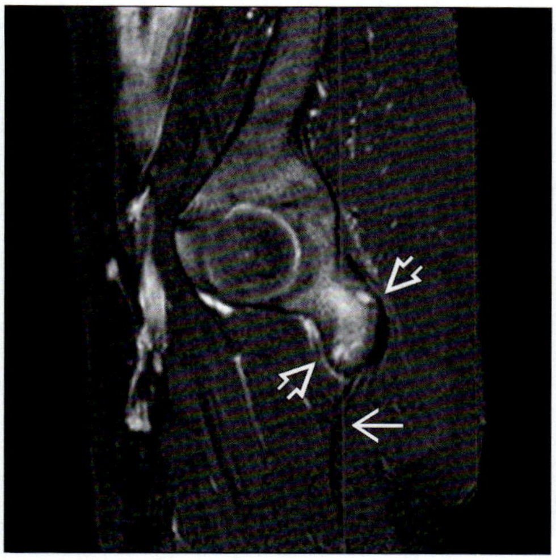

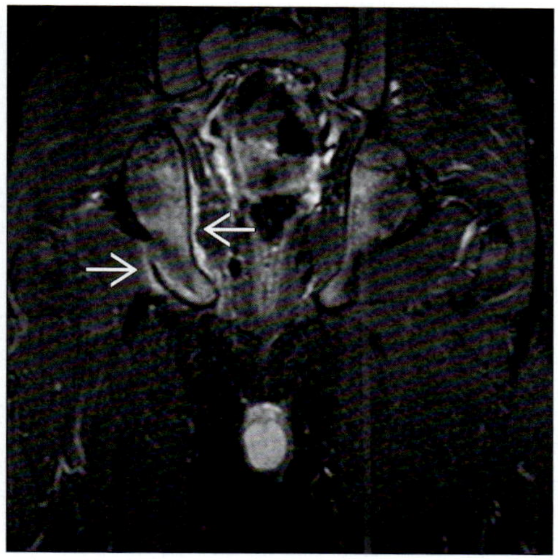

***(Left)** Sagittal T2WI FS MR from a 15-year-old soccer player with proximal hamstring pain shows a normal biceps femoris tendon →. Marrow edema spans the ischial tuberosity apophysis ⇨, typical of overuse apophysitis. **(Right)** Coronal STIR MR from the same patient shows periosteal edema → along the medial and lateral aspects of the ileum and ischium indicating an element of periosteal sleeve avulsion. Immature osseous structures are at risk for avulsive injuries.*

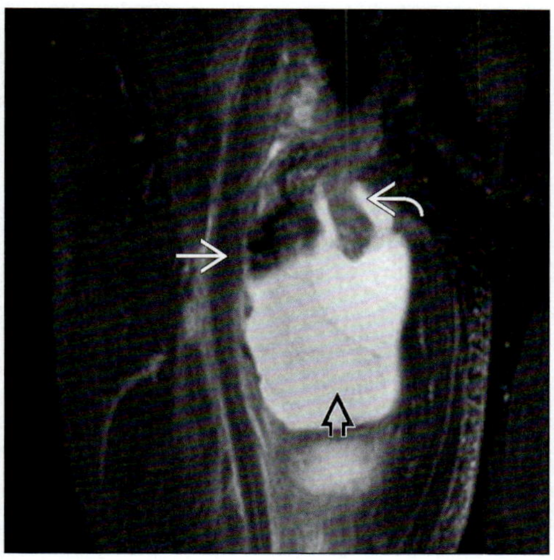

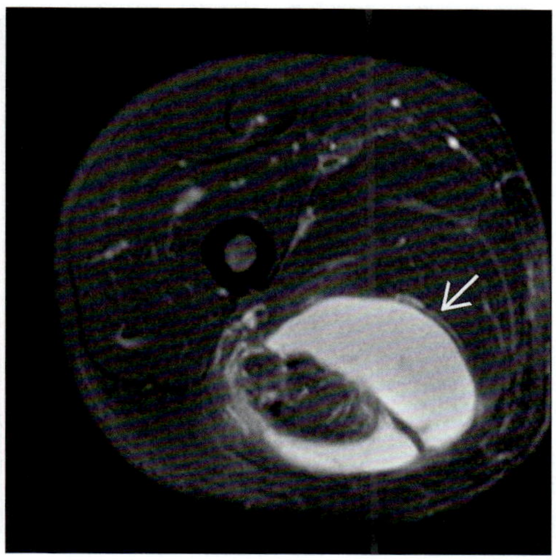

***(Left)** Coronal T2WI FS MR from a professional football player with a severe proximal hamstring strain shows complete disruption of the biceps femoris →, semitendinosus, and semimembranosus tendons ↶ and a very large hematoma ⇨. **(Right)** Axial T2WI FS MR from the same patient shows the liquified hematoma → centered in the semimembranosus muscle sheath. Large hematomas related to hamstring injury can delay healing. This patient had a surgical evacuation of the hematoma.*

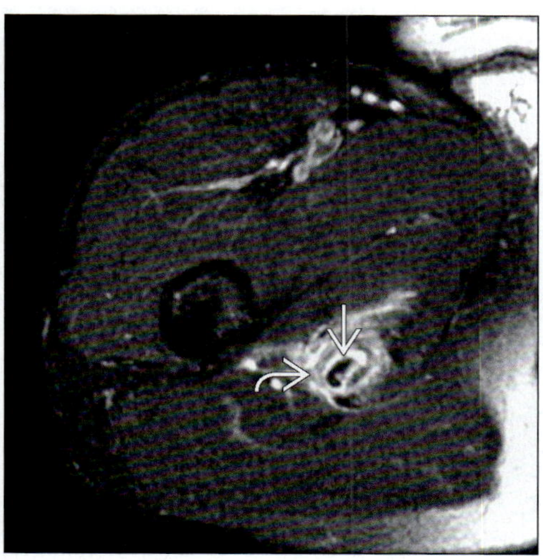

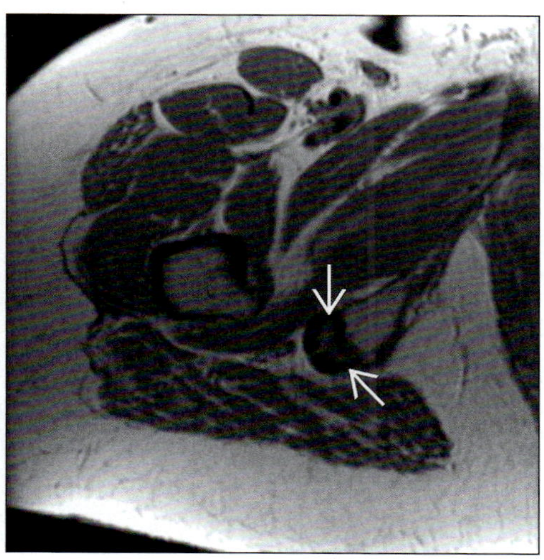

***(Left)** Axial T2WI FS MR from a college football player with an acute proximal hamstring strain shows fluid ↷ surrounding an enlarged retracted semimembranosus tendon →. Fluid seen surrounding an injured tendon on short axis images is a reliable indicator of a strain injury and is called the "central tendon complex." **(Right)** Axial T1WI MR from a 36 year old with ischial pain shows intermediate signal → at the origin of both hamstring tendons, consistent with chronic hypoxic tendinopathy.*

ABDOMINAL MUSCLE INJURY

Key Facts

Terminology
- Muscle, ligament, or avulsive osseous injury at abdominal wall
- Oblique strain, flank strain

Imaging
- Serratus anterior attachment on flank
- External oblique anterolaterally
- Rectus abdominis
  - Lower costal margin to aponeurosis
- Can avulse bone or cartilage at lower costal margin
- Strain generally follows fiber direction of myotendinous unit
- Noncontrast MR with torso coil
- Coronal and axial fat-suppressed sequences of entire trunk
- Pennate pattern of T2 hyperintensity = muscle strain

Top Differential Diagnoses
- Rectus abdominis/adductor aponeurosis lesion
- Visceral abdominal pathology
- Abdominal wall hernia

Pathology
- Grade myotendinous strain
- Extent of injury delineated by T2 hyperintensity
- Tendinous or osseous avulsions

Clinical Issues
- Acute onset oblique or flank pain after trauma
- Nagging pain with repetitive activity
- Throwing, swinging, shooting
- Hyperextension/twisting at trunk
- Ecchymosis often mirrors distribution of strain
- Can have protracted course if activity is continued

*(**Left**) AP external rotation graphic shows a typical site of abdominal oblique muscle strain, in this case centered at the serratus anterior ➡. Upper oblique strains are often associated with forceful arm abduction, as in swinging a baseball bat, or with a rowing mechanism, as in crew. (**Right**) AP external rotation graphic shows normal abdominal wall musculature but with edema within the latissimus dorsi ➡. There is overlap between thoracic cage myotendinous strains and abdominal wall muscle injury.*

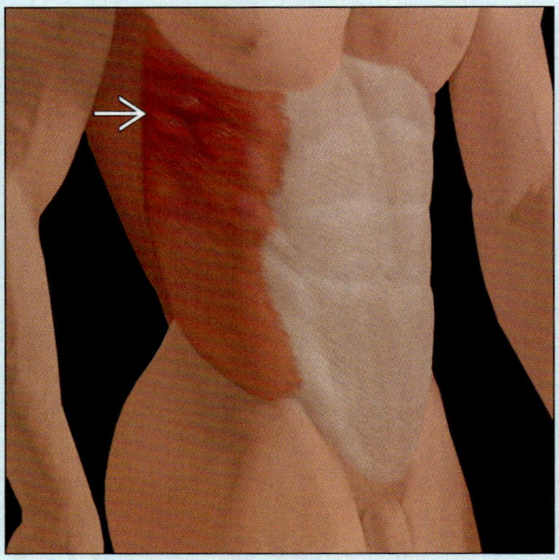

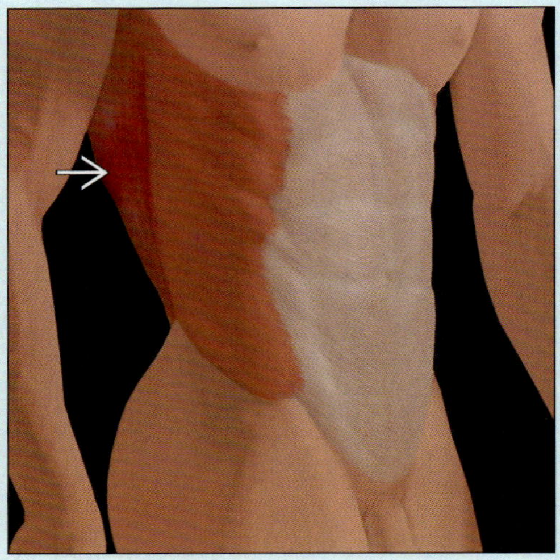

*(**Left**) AP external rotation graphic shows injury at the external oblique muscle and its aponeurosis ➡, a common location for acute and chronic abdominal muscle strains. The injuries in this location often occur with twisting mechanisms at the waist, as in football, or with forceful thigh flexion against resistance, often in hockey players. (**Right**) AP external rotation graphic shows a lower abdominal strain primarily involving the rectus abdominis ➡, which can be isolated or related to a sports hernia lesion.*

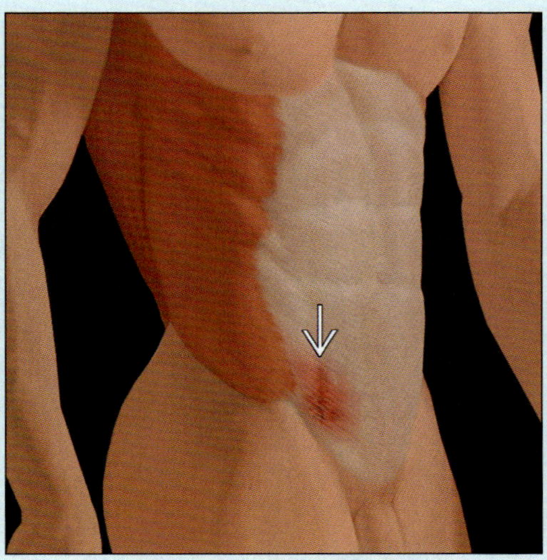

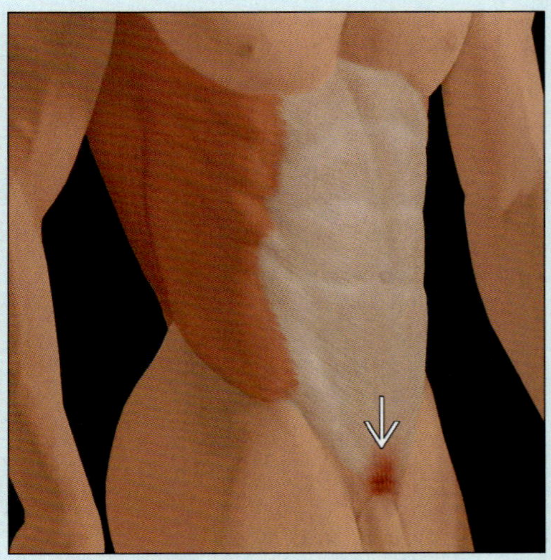

ABDOMINAL MUSCLE INJURY

TERMINOLOGY

Synonyms

- Oblique strain, flank strain

Definitions

- Muscle, ligament, or avulsive osseous injury at abdominal wall

IMAGING

General Features

- Best diagnostic clue
  - T2 hyperintense signal along myotendinous axis
- Location
  - Anywhere in abdominal wall
    - Serratus anterior attachment on flank
      - Throwing, swinging bat or stick
      - Associated with intercostal sprain
    - External oblique anterolaterally
      - Hyperextension/twisting at trunk
      - At external oblique aponeurosis: Hockey groin
      - With costochondral sprain in rowers
    - Rectus abdominis
      - Lower costal margin to aponeurosis
      - Can avulse bone or cartilage at costal margin
    - Internal oblique less common
- Morphology
  - Strain generally follows fiber direction of a myotendinous unit

Radiographic Findings

- Osseous or periosteal avulsions (rare)

MR Findings

- T1WI
  - Insensitive for muscle strain
  - Useful to assess for fracture or avulsion
  - Muscle atrophy in chronic strain
  - Occasional secondary hernia
- T2WI FS
  - Pennate hyperintense pattern = muscle strain
  - T2 hyperintensity mirrors extent of injury
  - Hematoma with high-grade injuries
  - Avulsive marrow edema at tendinous attachment
  - Intercostal and costochondral cartilage edema
  - Displaced rib fracture in axial plane (rare)

Ultrasonographic Findings

- Sensitive and specific for flank strains
- Useful to guide therapeutic injections

Imaging Recommendations

- Best imaging tool
  - Noncontrast MR with torso coil
- Protocol advice
  - Coronal and axial sequences of entire trunk
    - Identify focal injury
    - Assess symmetry
  - Small field-of-view sagittal images of injured region
  - Vitamin E markers at site of pain

DIFFERENTIAL DIAGNOSIS

Rectus Abdominis/Adductor Aponeurosis Lesion

- Lower: Centered at pubic symphysis or tubercle

Costochondral/Intercostal Fracture

- Direct trauma as opposed to avulsive

Visceral Abdominal Pathology

- Blunt trauma or idiopathic

Abdominal Wall Hernia

- Visceral contents protrude through defect

Thoracic Cage Muscle Injury

- Latissimus dorsi or teres major edema

PATHOLOGY

Staging, Grading, & Classification

- Grading of myotendinous strains
  - Grade 1: Peritendinous fluid
  - Grade 2: Partial tear
  - Grade 3: Complete tear
- Extent of injury delineated by T2 hyperintensity
- Tendinous or osseous avulsions

CLINICAL ISSUES

Presentation

- Most common signs/symptoms
  - Oblique or flank pain after acute strain
  - Nagging pain with repetitive activity
    - Throwing, swinging, shooting
- Other signs/symptoms
  - Ecchymosis often mirrors injury distribution

Natural History & Prognosis

- Can have protracted course if activity is continued
- Avulsion injuries tend to heal without intervention

Treatment

- Heat, ice, massage, whirlpool
- NSAIDs for hematoma, ecchymosis
- US-guided steroid injections or tenotomy
- Platelet-rich plasma (PRP) injections controversial

DIAGNOSTIC CHECKLIST

Consider

- MR of torso for acute or chronic flank pain
- Intercostal muscle sprain or rib fracture

Reporting Tips

- Identify primary muscle or tendon involved
- Grade myotendinous strain

SELECTED REFERENCES

1. Obaid H et al: Sonographic appearance of side strain injury. AJR Am J Roentgenol. 191(6):W264-7, 2008
2. Johnson R: Abdominal wall injuries: rectus abdominis strains, oblique strains, rectus sheath hematoma. Curr Sports Med Rep. 5(2):99-103, 2006

ABDOMINAL MUSCLE INJURY

(Left) Coronal STIR MR from a professional baseball player with acute flank pain after swinging and missing shows a pennate pattern of soft tissue edema following a serratus distribution ➡ as well as intercostal muscle edema ➡. (Right) Axial T2WI FS MR from the same patient shows edema tracking along the serratus anterior ➡ and focal disruption at the anterolateral costochondral junction ➡, or an oblique muscle strain with an avulsive costochondral lesion.

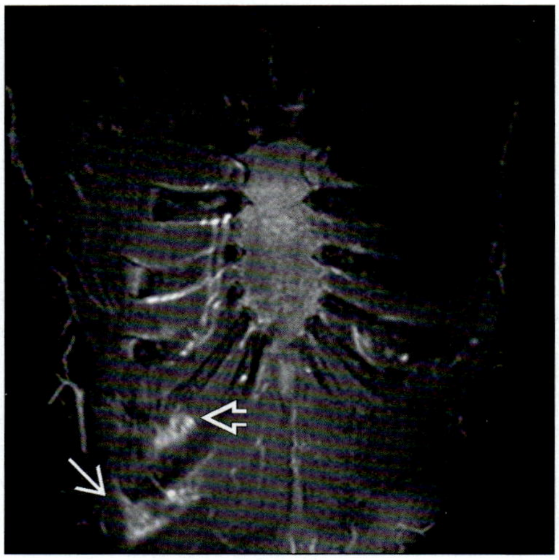

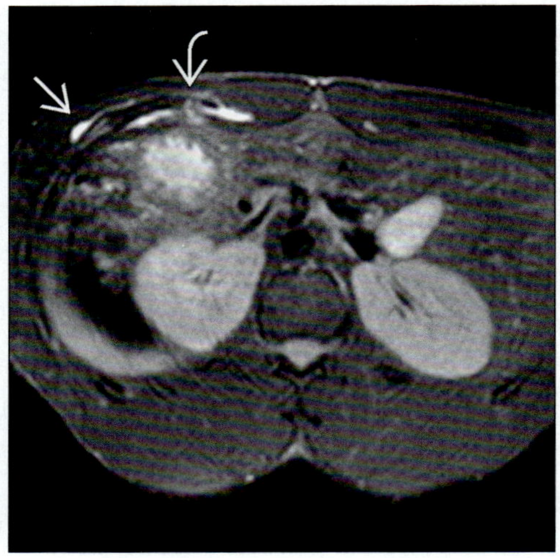

(Left) Coronal STIR MR from a professional football pass rusher after a high-velocity twisting injury at the torso shows extensive soft tissue edema involving the lower serratus and the anterolateral external oblique ➡. (Right) Axial T2WI FS MR from the same patient shows a severe oblique muscle strain ➡ with an incomplete fracture at the lower anterior rib cage ➡. Costal stress fractures are associated with repetitive oblique strains from activities, including rowing crew.

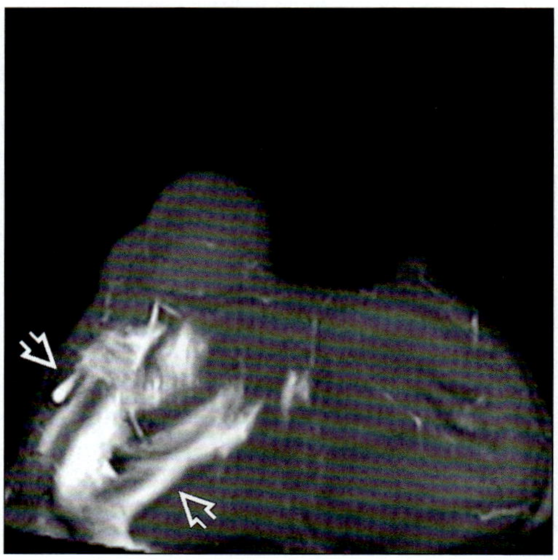

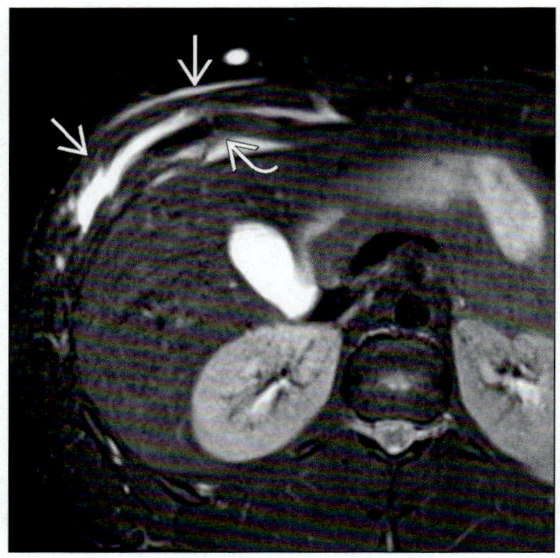

(Left) Coronal STIR MR from a football cornerback with oblique pain shows feathery soft tissue edema throughout the lateral portion of the external oblique muscle consistent with a grade 1 strain ➡. There is also a low-grade strain of the contralateral rectus abdominis ➡. (Right) Coronal T2WI FS MR more anteriorly in the same patient confirms a strain of the lower anterior external oblique ➡ and mild edema in the deeper internal oblique muscle ➡.

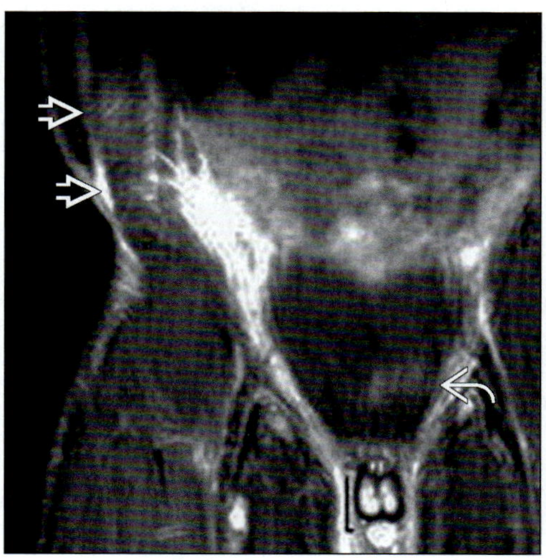

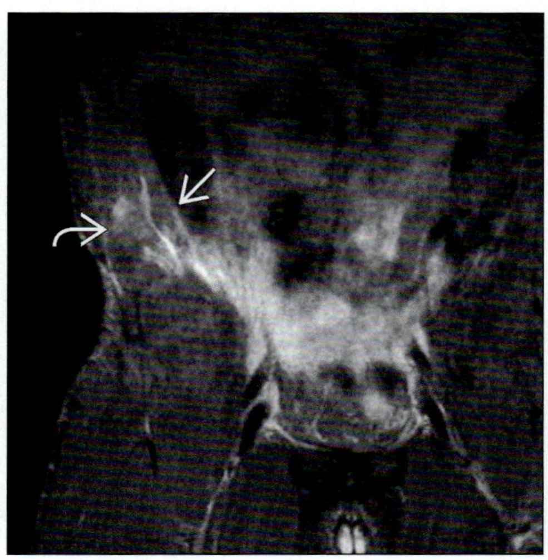

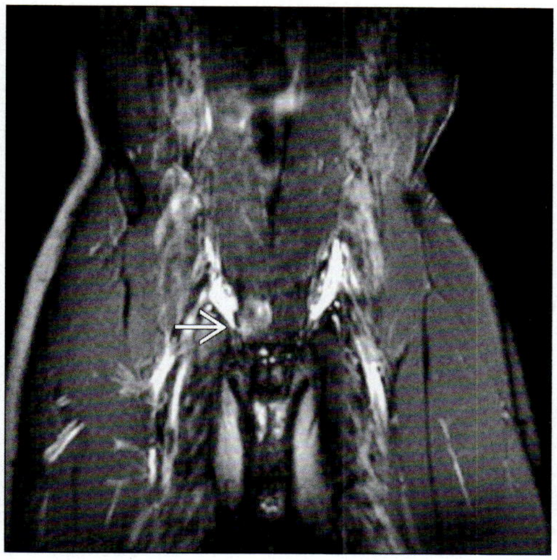

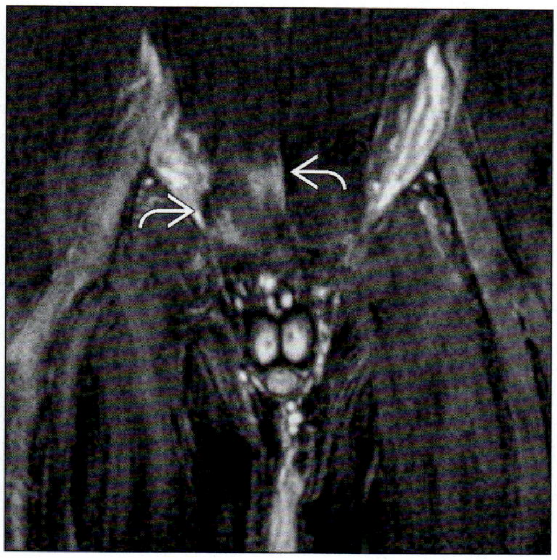

(Left) *Coronal STIR MR from a football quarterback after a hyperextension and twisting injury at the waist shows a pennate pattern of edema in the rectus abdominis muscle.* ***(Right)*** *Coronal T2WI FS MR from the same patient 6 weeks later shows persistent soft tissue edema throughout the lower aspect of the rectus abdominis without a focal high-grade tear or avulsion. Lower rectus abdominis strains have been associated with the evolution of rectus abdominis/adductor aponeurosis injuries.*

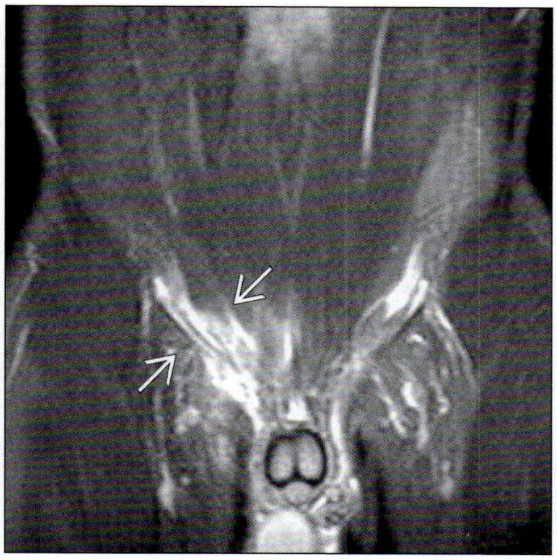

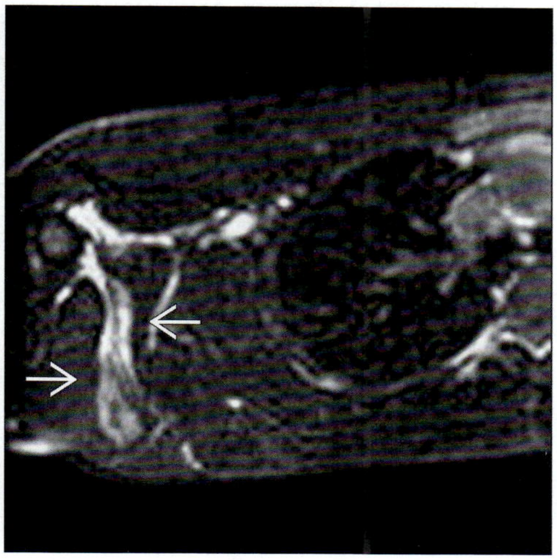

(Left) *Coronal STIR MR from a professional hockey player shows edema tracking along the external oblique aponeurosis, where the external oblique approximates the region of the deep inguinal ring. A recurrent, nagging abdominal muscle strain in this location has been termed hockey groin.* ***(Right)*** *Axial T2WI FS MR from a baseball pitcher with flank pain on throwing. Abdominal oblique strain was suspected; MR shows teres major edema. Thoracic strains can mimic abdominal muscle injury.*

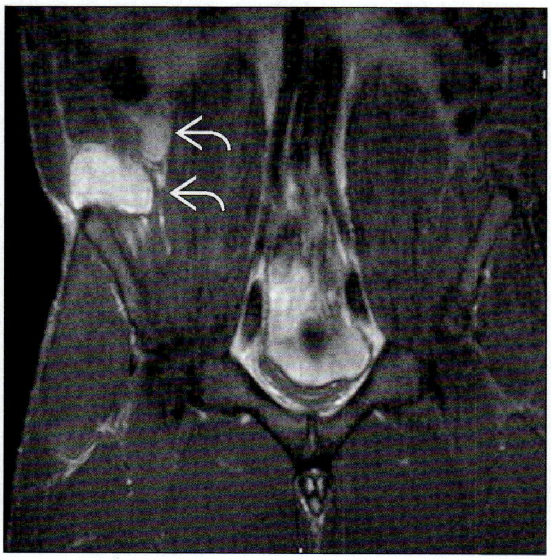

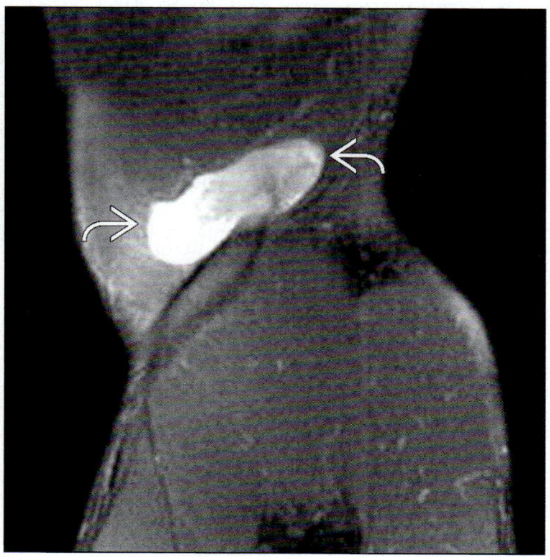

(Left) *Coronal STIR MR from a college football quarterback after an extreme twisting injury shows a complex hematoma in the region of the lower external and internal oblique muscles.* ***(Right)*** *Sagittal T2WI FS MR in the same patient 1 week later shows the evolution of a large abdominal muscle hematoma. With severe abdominal muscle injuries, resultant hematomas can contribute to delayed healing.*

SPORTS HERNIA

Key Facts

Terminology

- Tear or detachment of inferior rectus abdominis extending into adductor longus origin

Imaging

- MR using athletic pubalgia protocol; findings include
  - Subenthesial marrow edema at anteroinferior pubic tubercle
  - Secondary cleft on coronal oblique
  - Tenoperiosteal disruption of aponeurosis on sagittal
  - Sometimes frank disruption at aponeurosis
- Location of injury
  - Lateral edge rectus abdominis/adductor aponeurosis lesion
  - Midline rectus abdominis/adductor aponeurotic plate lesion
  - Rectus abdominis/adductor aponeurosis tear
- Look for source of referred groin pain

Pathology

- Most commonly related to athletic activity
  - Trauma with high-grade extension at waist
- Pregnancy and childbirth may contribute
- Osteitis pubis common: ~ 50%
- Concomitant internal derangement of ipsilateral hip in up to 25%

Clinical Issues

- Usually young athletic patients (18-40 years old)
- Most lesions respond to conservative regimens
- Invasive techniques have evolved over past decade
  - Operative pelvic floor repair

Diagnostic Checklist

- Distinguish midline from lateral edge aponeurosis lesion; describe extent of detachment

*(**Left**) AP graphic shows the aponeurosis between the inferior rectus abdominis ➔ (waist flexor) and the adductor tendon group origin ➔ (thigh adduction and trunk flexion). An injury to 1 of the structures along this biomechanical axis (yellow arrowheads) often leads to destabilization of the other. (**Right**) Sagittal graphic shows disruption of the rectus abdominis/ adductor aponeurosis from the anteroinferior pubis near midline ➔, a "sports hernia" lesion.*

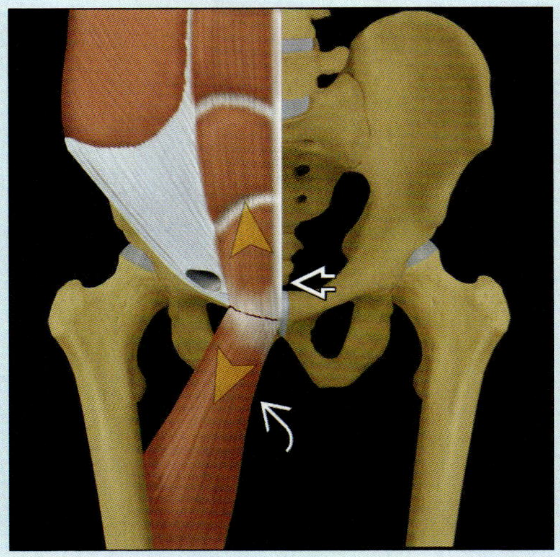

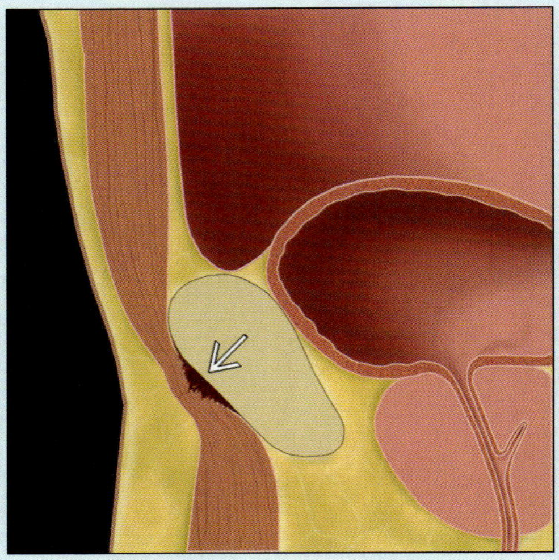

*(**Left**) Axial graphic shows detachment ➔ at the lateral edge of the rectus abdominis/ adductor aponeurosis. This focal teno-osseous injury causes groin pain with activity and tenderness in the region of the superficial ring. (**Right**) Axial graphic shows a midline disruption ➔ of the rectus abdominis/adductor aponeurotic plate. This "sports hernia" lesion spans the pubic symphysis and presents with bilateral activity-induced groin pain. Associated osteitis pubis is frequent with this lesion.*

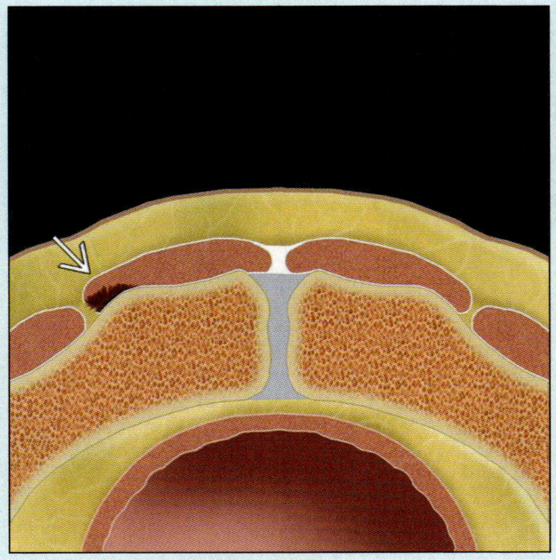

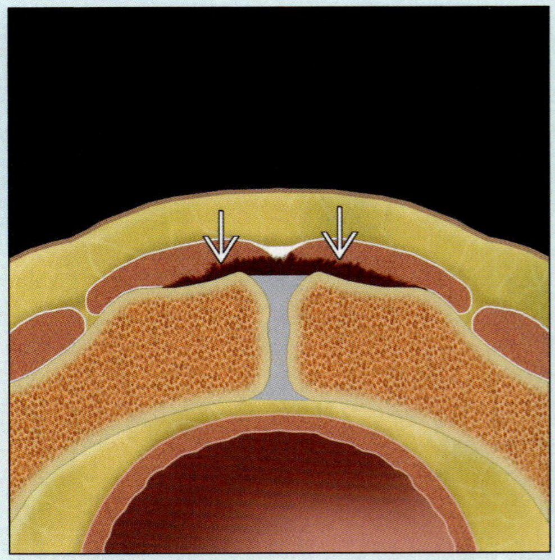

SPORTS HERNIA

TERMINOLOGY

Synonyms

- Sportsman's hernia, rectus abdominis/adductor aponeurosis injury, aponeurosis lesion
- Athletic pubalgia: Clinical synonym but not correctly used as imaging term

Definitions

- Tear or detachment of inferior rectus abdominis extending into adductor longus origin
  - Aponeurosis (lateral portion) or aponeurotic plate (midline) can be detached from pubis, strained, frankly torn, or any combination
    - Detachments often occur with strain; frank tears less common but still prevalent
    - Given this spectrum, injury is termed "lesion" within the text, but injuries are more specifically described on images

IMAGING

General Features

- Best diagnostic clue
  - Hyperintense T2 signal at anteroinferior aspect of pubic symphysis
- Location
  - Midline or at lateral edge of rectus abdominis attachment
- Size
  - Ranges from subtle strain to large tear with retraction
- Morphology
  - Defect can widen superficial inguinal ring

Imaging Recommendations

- Best imaging tool
  - MR with athletic pubalgia protocol
- Protocol advice
  - Surface coil over anterior pelvis centered on symphysis
  - 3 large field-of-view (FOV) sequences of pelvis
    - Cor STIR, cor T1WI, axial T2WI FS
      - For referred sources of pubalgia
  - 3 small FOV sequences centered on symphysis
    - Coronal oblique T2WI FS
      - Prescribed along arcuate line of pelvis from sagittal localizer
    - Coronal oblique PDWI
    - Sagittal T2WI FS

Radiographic Findings

- Insensitive for acute lesions
- Erosions or osseous productive change (osteitis pubis) if longstanding
- Look for hip lesions (femoroacetabular impingement [FAI], developmental dysplasia of hip [DDH]) that frequently appear as concomitant lesions

MR Findings

- T1WI
  - ± osseous productive change at symphysis
  - ± asymmetric rectus abdominis muscle
    - Common if there is aponeurosis lesion after herniorrhaphy
  - Evidence for prior surgery
- T2WI FS
  - Tenoperiosteal disruption of aponeurosis on sagittal images
    - Unilateral: Involves lateral edge of aponeurosis
    - Midline: Rectus abdominis/adductor aponeurotic plate lesion
      - Can be symmetric or asymmetric
  - Sometimes frank tear at aponeurosis
  - Retraction of adductor longus from pubic tubercle
  - Secondary cleft on coronal oblique images
    - Inferolateral extension of symphyseal cleft
    - Not necessarily fluid signal: Hyperintense signal equal to primary cleft
    - Can be subaponeurotic or subperiosteal
    - Correlates with site of pain
  - Subenthesial marrow edema at pubic tubercle
    - If unilateral, correlates with site of pain
  - Enlargement of superficial inguinal ring on axial images
- STIR
  - Look for source of referred groin pain
    - Hip, adductors, flexors, apophysitis, hernia
  - Osteitis pubis

Ultrasonographic Findings

- Sensitive for rectus abdominis and adductor lesions in experienced hands
- Can use Valsalva imaging for incipient hernia

DIFFERENTIAL DIAGNOSIS

Internal Derangement of Hip

- Look for morphologic signs (FAI, DDH, osteoarthritis)

Hip Adductor Injury

- Rectus abdominis normal
- Ranges from degenerative tendinopathy to myotendinous strain

Pelvic Stress and Insufficiency Fractures

- Pubic ramus stress fracture
- Ischiopubic stress response with running
- Osteitis pubis: Associated with aponeurosis lesions

Hip Flexor Injury

- Iliopsoas & rectus femoris most common
- Sartorius & iliacus less common

Visceral Pelvic Pathology

- Adnexal lesions
- Endometriosis
- Inflammatory bowel disease

True Hernias

- Uncommon in setting of athletic pubalgia

Piriformis Syndrome & Pelvic Nerve Injuries

- Pudendal, ilioinguinal

PATHOLOGY

General Features

- Etiology
  - Most commonly related to athletic activity

- Repeated extending and twisting at waist
  - Soccer, football, hockey, lacrosse, rodeo
- Repetitive stress across pubic symphysis
  - Distance running
- Pregnancy and childbirth
- Trauma with high-grade extension at waist
- Associated abnormalities
  - Osteitis pubis common: ~ 50%
    - Immature pubic apophyses often injured in young patients (16-25 years)
  - Osseous stress lesions
    - Asymmetric osteitis pubis on MR
  - Concomitant ipsilateral hip lesion (FAI or DDH) in up to 25%
  - Adductor compartment myotendinous injury
  - Secondary widening of superficial inguinal ring
    - True hernias very rare: < 2%

Staging, Grading, & Classification

- Lateral edge rectus abdominis/adductor aponeurosis lesion
  - Abnormal morphology (fraying) or detachment of rectus abdominis near inguinal ring
  - Confluent interstitial tear of adductor longus origin
  - Can be bilateral
- Rectus abdominis/adductor aponeurotic plate lesion
  - Both rectus abdominis muscles lifted from pubic symphysis at midline
    - Extension toward lateral edges variable
  - Can be symmetric or asymmetric
  - Often unilateral or bilateral adductor longus origin degeneration
  - Frequently associated with osteitis pubis
    - Symphyseal destabilizing lesion
- Rectus abdominis/adductor aponeurosis tear
  - Disruption of rectus abdominis from adductor longus and pubic tubercle
  - Also known as adductor avulsion
  - Retraction variable
    - Best measured on coronal oblique
  - Single episode of high-grade trauma

Gross Pathologic & Surgical Features

- Rectus abdominis destabilized from superior pubic ramus
  - Rectus abdominis atrophies with chronic lesions
- Adductor tendinopathy ranges from degeneration to interstitial tear and complete tear

CLINICAL ISSUES

Presentation

- Most common signs/symptoms
  - Unilateral or bilateral groin pain exacerbated by activity
    - Chronic or acute
  - Tenderness at pubic tubercle
  - Resistance to active thigh adduction
  - No visible or palpable hernia
  - ± deep perineal pain with associated osseous stress response
- Other signs/symptoms
  - Pain when running while rotated at waist
- Clinical profile
  - High-level athletes most common
    - Prevalent injury in professional and collegiate football players
    - Adductor lesions: Professional soccer players

Demographics

- Age
  - Young athletic patients most common (18-40 years old)
- Gender
  - M > F
  - Frequency of rectus abdominis/adductor aponeurotic plate lesion: M < F

Natural History & Prognosis

- Historically, sports hernia → premature end of athletic career
  - Patients can often function in day-to-day life without treatment
- Lesions without osteitis pubis have better prognosis

Treatment

- Most lesions respond to conservative regimens
  - A subset of patients will have refractory symptoms
- Invasive techniques have evolved over past decade
  - Operative pelvic floor repair
    - Primary restabilization of inferior rectus abdominis
    - Adductor longus origin often debrided
  - Ilioinguinal nerve release or neurectomy
  - Mesh abdominal wall repair losing favor
  - Platelet-rich plasma (PRP) injection at aponeurosis being explored

DIAGNOSTIC CHECKLIST

Consider

- Use dedicated athletic pubalgia protocol

Image Interpretation Pearls

- Marrow edema at pubic tubercle or presence of secondary clefts increase sensitivity
- Look for concomitant hip lesion

Reporting Tips

- Distinguish midline from lateral edge aponeurosis injury; describe extent of detachment

SELECTED REFERENCES

1. Meyers WC et al: Experience with "sports hernia" spanning two decades. Ann Surg. 248(4):656-65, 2008
2. Omar IM et al: Athletic pubalgia and "sports hernia": optimal MR imaging technique and findings. Radiographics. 28(5):1415-38, 2008
3. Shortt CP et al: Anatomy, pathology, and MRI findings in the sports hernia. Semin Musculoskelet Radiol. 12(1):54-61, 2008
4. Zajick DC et al: Spectrum of MRI findings in clinical athletic pubalgia. Semin Musculoskelet Radiol. 12(1):3-12, 2008
5. Zoga AC et al: Athletic pubalgia and the "sports hernia": MR imaging findings. Radiology. 247(3):797-807, 2008
6. Cunningham PM et al: Patterns of bone and soft-tissue injury at the symphysis pubis in soccer players: observations at MRI. AJR Am J Roentgenol. 188(3):W291-6, 2007

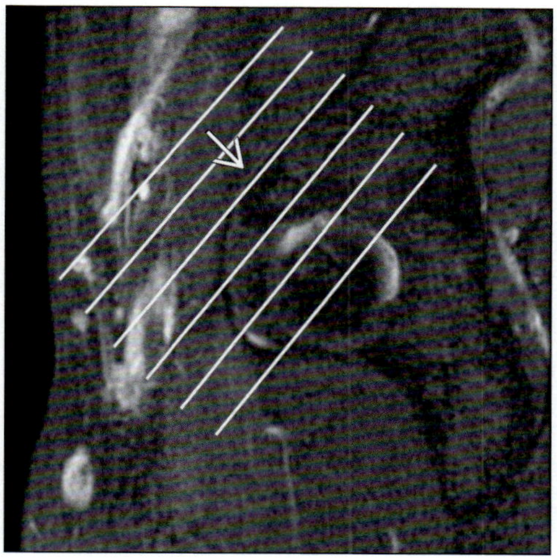

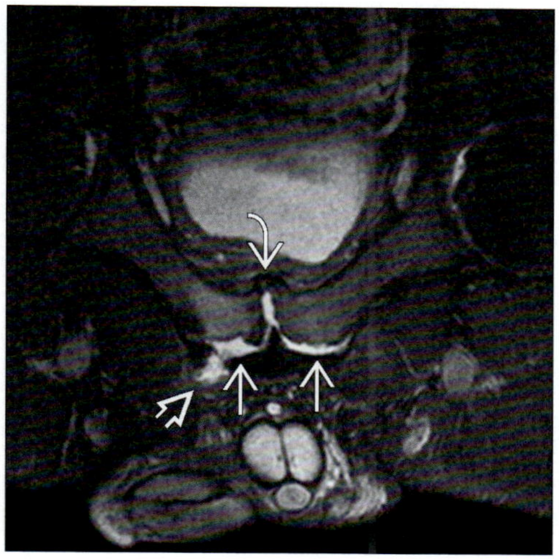

(Left) *Sagittal T2WI FS MR shows reference lines used to prescribe the coronal oblique MR sequences for an athletic pubalgia MR protocol. The arcuate line of the pelvis parallels the anterior border of the ilium ➔.* ***(Right)*** *Coronal oblique T2WI FS MR from a professional basketball player with chronic right- > left-sided groin pain shows bilateral secondary clefts ➔ confluent with the symphysis, osteitis pubis ➔, and extension of the tear into the adductor longus tendon on the right ➔.*

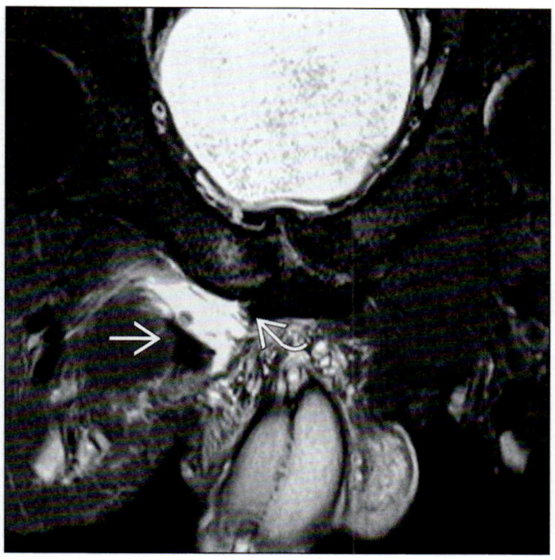

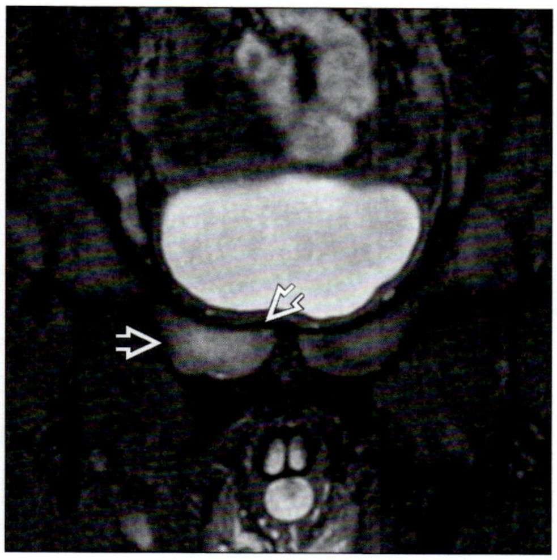

(Left) *Coronal oblique T2WI FS MR from a professional football player after an acute injury while running and rotating at the waist shows an amputated contour at the lateral edge of the rectus abdominis ➔ with retraction of the torn adductor longus ➔; the rectus is almost always involved when the adductors tear.* ***(Right)*** *Coronal oblique T2WI FS MR from a football defensive back with chronic groin pain and diagnosis of "sports hernia" shows subenthesial marrow edema ➔ indicating early rectus detachment.*

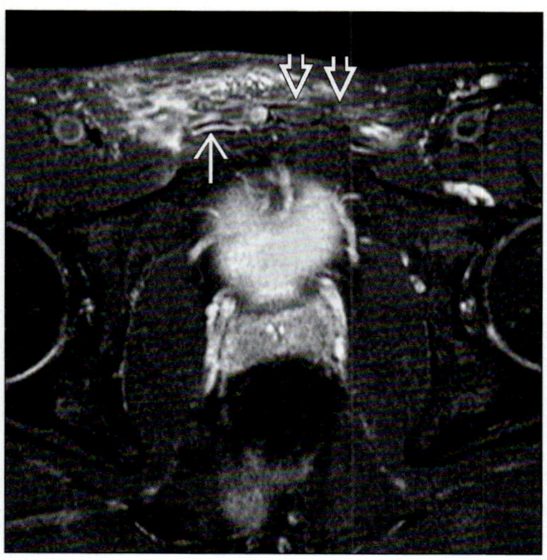

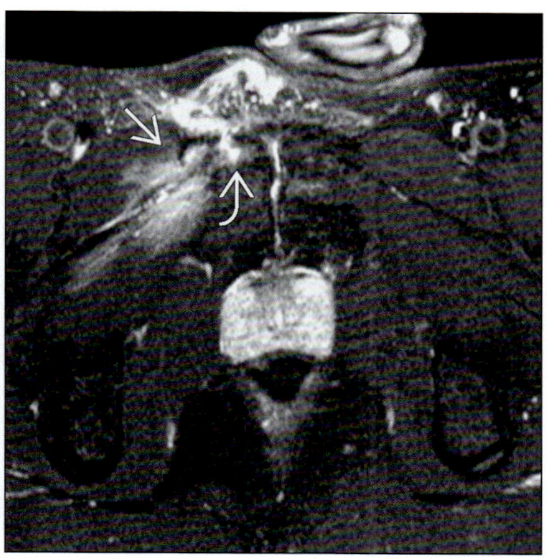

(Left) *Axial T2WI FS MR from a college lacrosse player with an acute injury shows edema ➔ within the right rectus abdominis (RA) at the level of the symphysis indicating a destabilizing aponeurosis injury. Note the normal medial and lateral RA heads on the left ➔.* ***(Right)*** *Axial T2WI FS MR acquisition from the same study 10 mm caudal to the previous shows a frank tear at the right rectus abdominis/ adductor aponeurosis ➔, just deep to the superficial inguinal ring with adductor edema ➔.*

SPORTS HERNIA

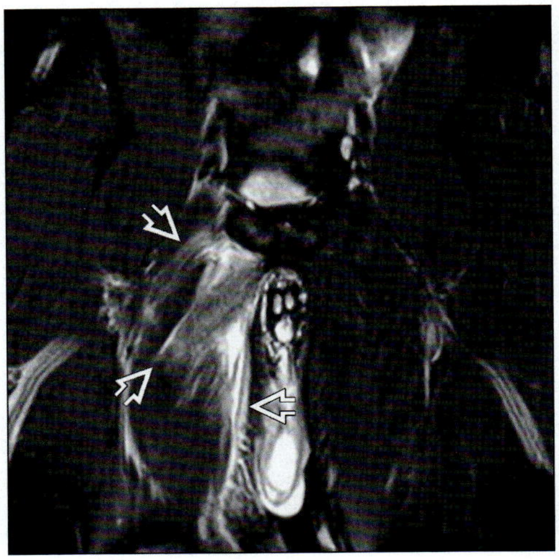
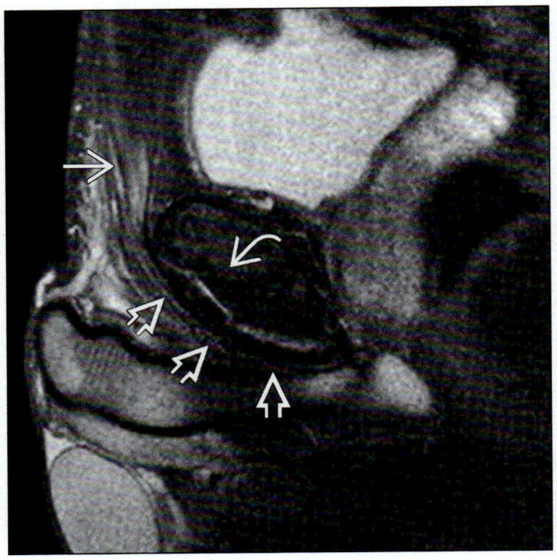

(Left) *Coronal STIR MR from a professional baseball catcher with an acute injury shows extensive soft tissue edema ➡ throughout the adductor compartment that might initially be classified as an isolated adductor injury.* ***(Right)*** *Sagittal T2WI FS MR from the same patient centered on the pubic region just off midline reveals an elongated region of tenoperiosteal detachment ➡ at the rectus abdominis/ adductor aponeurosis from the pubis ➡, as well as a strain of the superior rectus abdominis muscle ➡.*

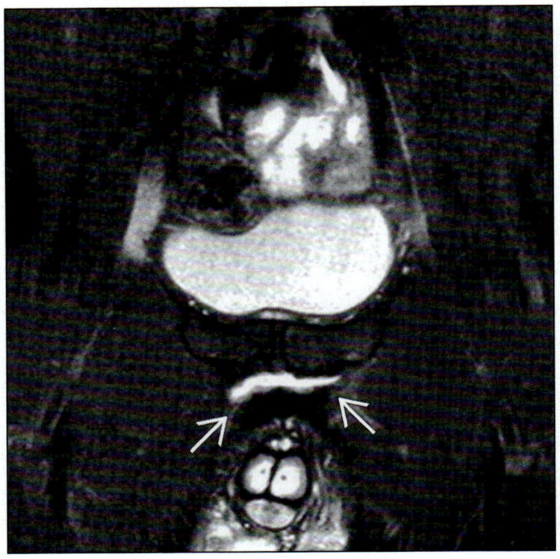
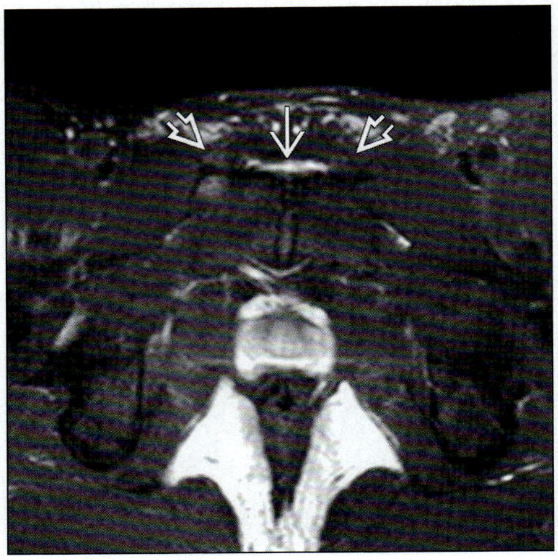

(Left) *Coronal oblique T2WI FS MR from a college football player with bilateral groin pain after 6 months of limited activity for osteitis pubis shows complete detachment ➡ of the rectus abdominis/ adductor aponeurotic plate.* ***(Right)*** *Axial T2WI FS MR from a soccer player shows a rectus abdominis/ adductor aponeurotic plate detachment ➡. The lateral edges of the aponeuroses remain intact ➡ in this less severe injury, but the patient still has groin pain with running and twisting at the waist.*

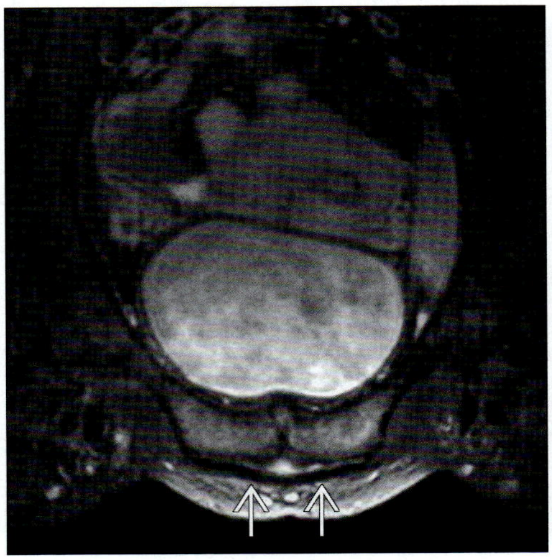
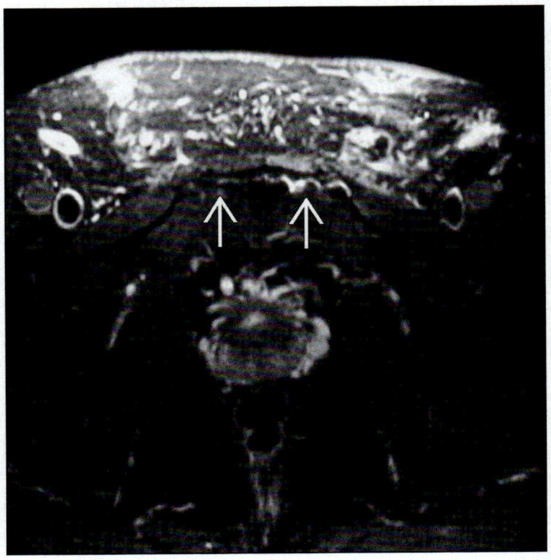

(Left) *Coronal oblique T2WI FS MR from a professional sprinter with midline groin pain shows focal detachment of the midline rectus abdominis/adductor aponeurotic plate ➡ without adductor tendon involvement.* ***(Right)*** *Axial T2WI FS MR from the same patient 6 months later shows susceptibility artifact at the aponeurotic plate compatible with a bilateral pelvic floor repair ➡. There was no pain at the site of initial injury or repair but had a myotendinous strain in the adductor compartment.*

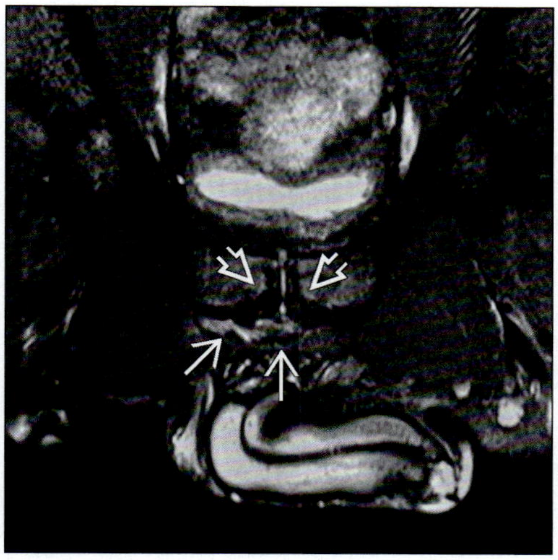

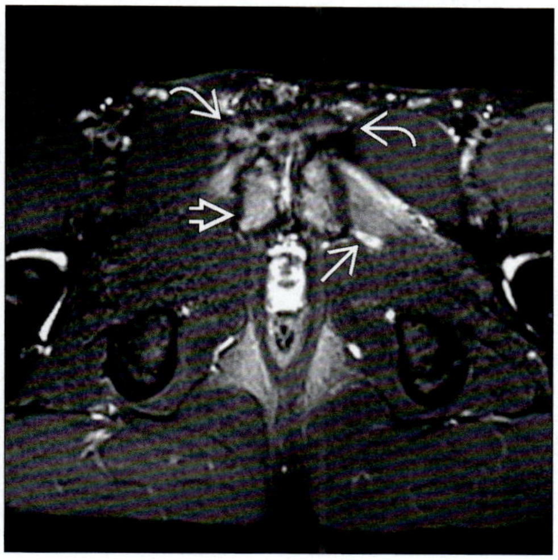

(Left) *Coronal oblique T2WI FS MR from a soccer player with acute-on-chronic left > right groin pain shows disruption of the aponeurosis on the right with a secondary cleft ➔ extending across midline and sclerosis ➔ indicating chronic osteitis pubis.* ***(Right)*** *Axial T2WI FS MR from the same patient shows a right-sided aponeurosis tear ➔, left-sided adductor tendon injury ➔, and pubic marrow edema ➔. Bilateral pelvic floor repair was performed.*

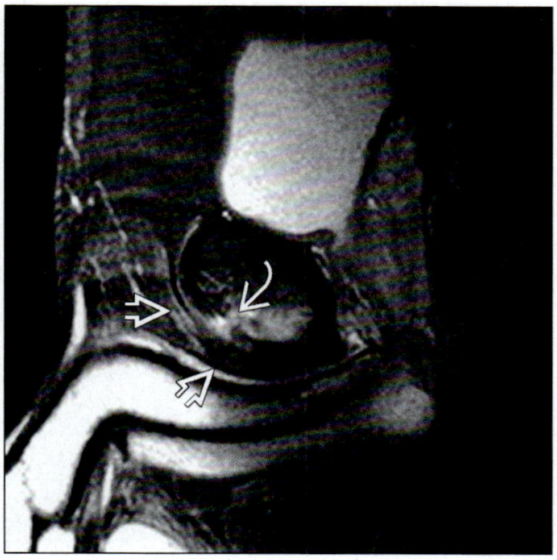

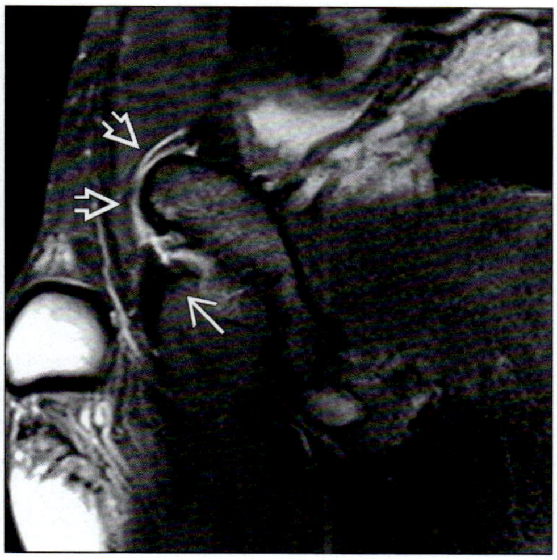

(Left) *Sagittal T2WI FS MR just off midline in a rodeo cowboy shows detachment of the aponeurotic plate ➔ with a focal disruption of the periosteum at the anteroinferior pubic ramus ➔ typical of an aponeurosis lesion.* ***(Right)*** *Sagittal T2WI FS MR 15 mm lateral to midline from a hockey player with a unilateral aponeurosis lesion shows detachment ➔ of the inferior rectus abdominis at its lateral edge from the pubic tubercle and detachment of the adductor longus tendon origin ➔.*

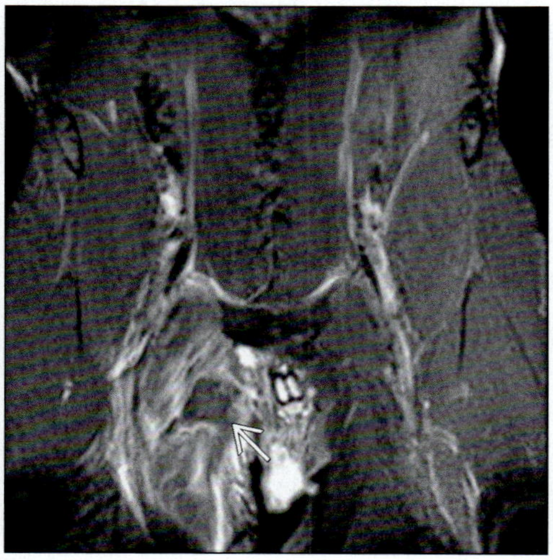

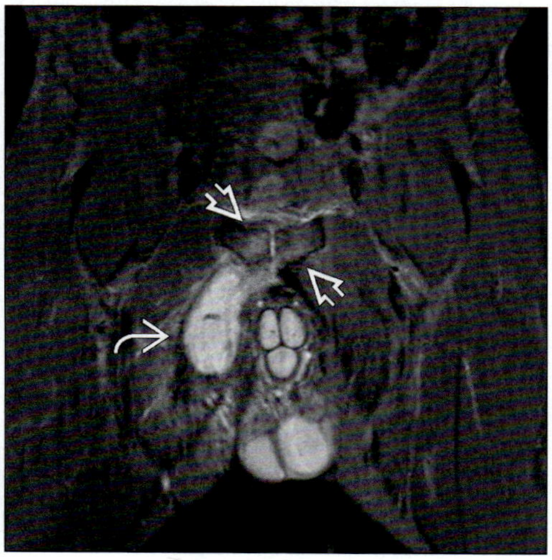

(Left) *Coronal STIR MR from a baseball pitcher after an acute injury shows frank avulsion ➔ of the adductor longus tendon origin with strain edema throughout the adductor compartment. There is no osteitis pubis.* ***(Right)*** *Coronal STIR MR from the same patient 6 weeks later shows the formation of a large adductor compartment hematoma ➔ and marrow edema around the pubic symphysis ➔. Disruption of the pubic symphysis tendon attachments can lead to symphysis instability and osteitis pubis.*

TRUE HERNIAS

Key Facts

Terminology
- Abdominal wall hernia, groin hernia

Imaging
- CT: Fat, bowel, or omentum protruding from peritoneum in typical location
  - Consider contrast if concern for strangulation
- MR: Hyperintense peritoneal fat through defect
- Ultrasound: Best tool for incipient or sliding hernias
- Hernia descriptors
  - Fatty: Contain only peritoneal fat
  - Enteric: Contain fat and bowel
  - Omental: Contain fat and omentum
  - Incarcerated: Herniated structures entrapped, irreducible without surgery
  - Strangulated: Herniated structures lose vascular supply from torsion or pressure

Pathology
- Direct inguinal: Through weakening in transversalis fascia
- Indirect inguinal: Nonclosure of internal inguinal ring after testicular descent
- Femoral: Inferior to inguinal ligament through femoral canal
- Spigelian: Between lateral edge of rectus abdominis and semilunar line
- Pantaloon: Simultaneous direct and indirect inguinal hernias

Clinical Issues
- Can enlarge, leading to bowel herniation and strangulation
- Most repaired surgically

*(**Left**) Axial T1WI MR shows an indirect inguinal hernia containing peritoneal fat ➡. The inguinal canal closely approximates the lateral edge of the rectus abdominis inferiorly ↪, contributing to the confusion surrounding the term "sports hernia," which is actually a lesion involving the rectus abdominis. (**Right**) Sagittal T2WI FS MR of an entrapped indirect inguinal hernia shows a knuckle of small bowel ➡ protruding through the inguinal canal and inflamed with surrounding soft tissue edema ↪.*

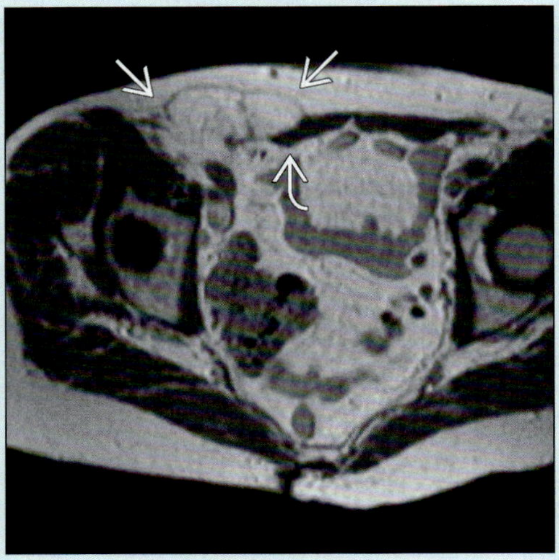

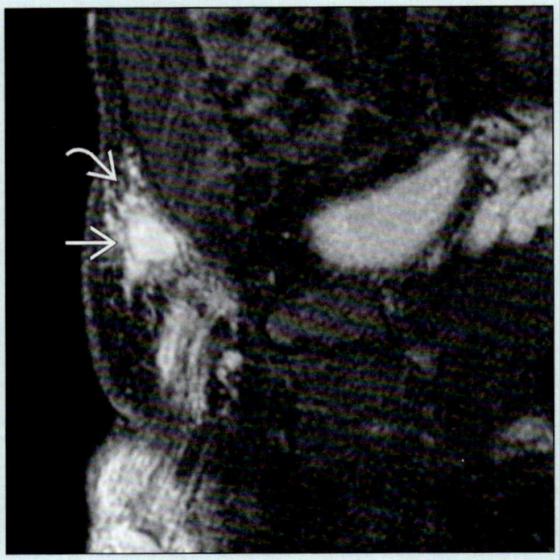

*(**Left**) Coronal T1WI MR shows a small fat-containing direct inguinal hernia ➡. Susceptibility artifact ↪ near the deep inguinal rings bilaterally reflects prior mesh herniorrhaphy. (**Right**) Axial noncontrast CT shows a spigelian hernia. Small bowel ➡ extends through a gap in the abdominal wall just lateral to an atrophic rectus abdominis muscle ⇨. Spigelian hernias most commonly occur caudal to the linea arcuata, below which the rectus abdominis has no posterior muscle sheath.*

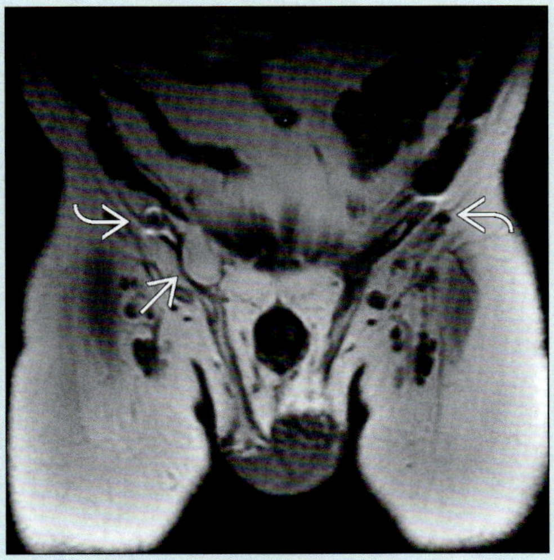

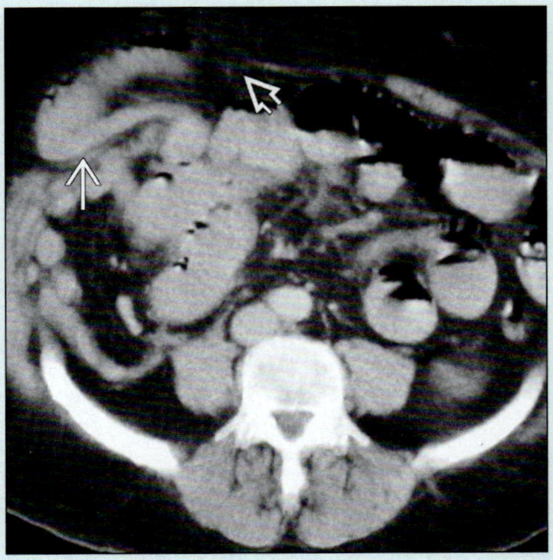

TRUE HERNIAS

TERMINOLOGY

Definitions

- Protrusion of intraperitoneal structures through congenital or acquired abdominal or pelvic wall defect

IMAGING

General Features

- Best diagnostic clue
    - Fat, bowel, or omentum protruding from peritoneum in typical location
- Morphology
    - Fatty: Contain only peritoneal fat
    - Enteric: Contain fat and bowel
        - Richter: Only a portion of bowel wall
        - Sliding: Bowel herniates and reduces with motion or on physical exam
    - Omental: Contain fat and omentum
    - Incarcerated: Herniated structures entrapped, irreducible without surgery
    - Strangulated: Herniated structures lose vascular supply from torsion or pressure

CT Findings

- Fat &/or bowel in hernia sac in typical location

MR Findings

- T1WI
    - Hyperintense peritoneal fat through defect
- T2WI FS
    - Hyperintense bowel and enteric fluid extending through defect
    - Edema around defect can indicate inflammation/ strangulation

Ultrasonographic Findings

- Best tool for incipient or sliding hernia
    - Dynamic imaging with and without Valsalva
- Power Doppler for vascularity, vitalization with suspected strangulation

Imaging Recommendations

- Protocol advice
    - Consider intravenous contrast if concern for strangulation
    - Consider prone and Valsalva ultrasound imaging with sliding hernias

DIFFERENTIAL DIAGNOSIS

Rectus Abdominis/Adductor Aponeurosis Lesion

- Sometimes referred to as sports hernia
- No true hernia of visceral contents
- Rectus abdominis deficiency enlarges superficial ring

Abdominal Muscle Strain

- Rarely related to true hernia
- Trauma history

Hydrocele

- Excess fluid in scrotum via inguinal canal

Spermatic Cord Lipoma

- Imaging overlap with fatty indirect inguinal hernia
- Fat adherent to spermatic cord within inguinal canal

PATHOLOGY

Staging, Grading, & Classification

- Direct inguinal: Through weakening in transversalis fascia
    - Medial to inferior epigastric artery
    - Posterior inguinal wall in Hesselbach triangle
- Indirect inguinal: Nonclosure of internal inguinal ring after testicular descent
    - Lateral to inferior epigastric artery
    - Most common groin hernia
- Femoral: Inferior to inguinal ligament through femoral canal
    - M << F
- Spigelian: Between lateral edge of rectus abdominis and semilunar line
    - Region lacking rectus abdominis posterior sheath
- Pantaloon: Simultaneous direct and indirect inguinal hernias

CLINICAL ISSUES

Presentation

- Most common signs/symptoms
    - Indirect inguinal: Asymptomatic groin bulge on exam in childhood
    - Direct inguinal: Tender groin bulge in adulthood
    - Femoral: Painful medial thigh bulge in females

Natural History & Prognosis

- Can enlarge, leading to bowel herniation and strangulation

Treatment

- Most repaired surgically
    - Primary closure
    - Mesh patch or plug repair

DIAGNOSTIC CHECKLIST

Consider

- Rectus abdominis/adductor aponeurosis lesion with history of sports hernia
- Strangulation with painful hernia

Image Interpretation Pearls

- Valsalva maneuver increases sensitivity

SELECTED REFERENCES

1. Cherian PT et al: The diagnosis and classification of inguinal and femoral hernia on multisection spiral CT. Clin Radiol. 63(2):184-92, 2008
2. Djuric-Stefanovic A et al: The accuracy of ultrasonography in classification of groin hernias according to the criteria of the unified classification system. Hernia. 12(4):395-400, 2008
3. Omar IM et al: Athletic pubalgia and "sports hernia": optimal MR imaging technique and findings. Radiographics. 28(5):1415-38, 2008

Key Facts

Imaging
- Asymmetric fluid collection in typical bursal locations
  - Iliopsoas
  - Obturator externus
  - Trochanteric
  - Ischial
- Radiographs generally normal with hip/pelvis bursae
- MR findings
  - Lobulated fluid signal in typical location
  - Hyperintense blood products
  - Capsular thickening with friction, trauma
  - Infection: Occasional neck extending from infectious source, rim enhancement after contrast

Pathology
- Septic bursitis
  - Hip and lumbar spine sources most common
- Hemorrhagic bursitis
  - Generally post-traumatic
- Calcific bursitis
  - Hypointense focus within distended bursa
- Consider PVNS if single location

Clinical Issues
- Iliopsoas: Pain on hip flexion
- Trochanteric: Point tenderness at greater trochanter
- Chronic, multifocal bursitis common with rheumatoid arthritis
- Percutaneous aspiration/steroid injection

Diagnostic Checklist
- Look for mass effect on surrounding structures
- Describe complexity of fluid, thick rind/capsule, connection to joint or other structure

*(**Left**) Coronal graphic shows an inflamed iliopsoas bursa ➔ arising anterior to the hip joint capsule. The iliopsoas muscle ➔ has been cut away. (**Right**) Coronal STIR MR from a hockey player with sharp hip pain on flexion shows a sharply marginated fluid collection ➔ surrounding the distal iliopsoas tendon ➔ in a location typical for iliopsoas bursitis. The iliopsoas bursa can become inflamed with repeated hip flexion activities. The rectus femoris tendon ➔ is uninvolved in this case.*

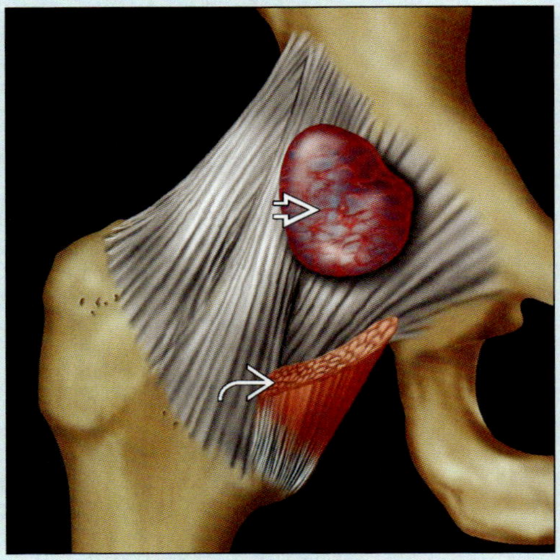

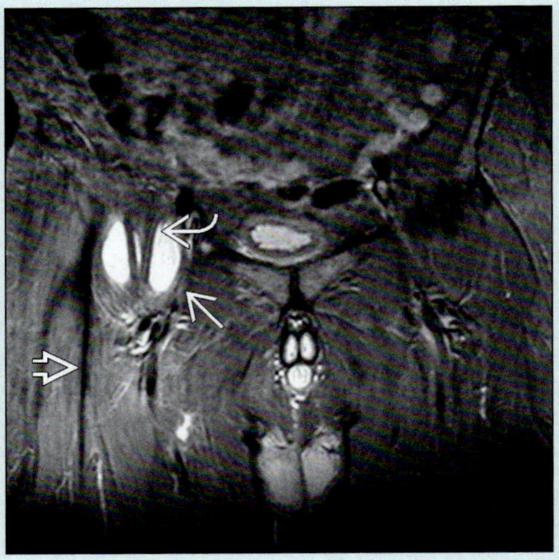

*(**Left**) Axial graphic at the level of the mid hip shows a small iliopsoas bursa ➔ anterior to the hip joint. The femoral nerve ➔, artery ➔, and vein ➔ lie medial to the bursa. The hip connects to the iliopsoas bursa in 10-15% of people. (**Right**) Axial MR arthrogram T1WI FS from a patient with osteoarthritis shows contrast entering the iliopsoas bursa ➔ through a thin neck ➔ connecting it to the anterior hip joint. Such connection is not always pathologic.*

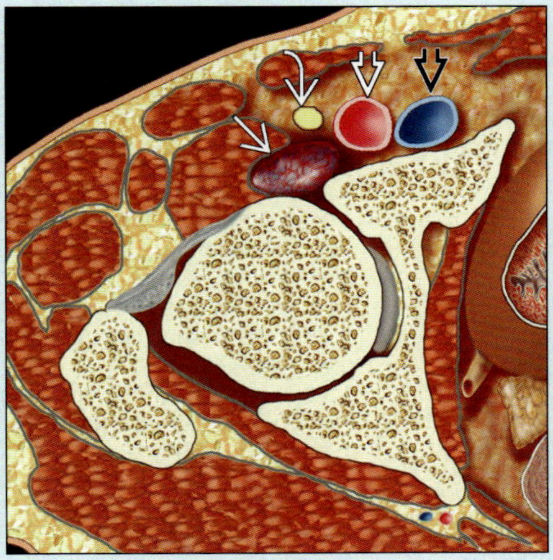

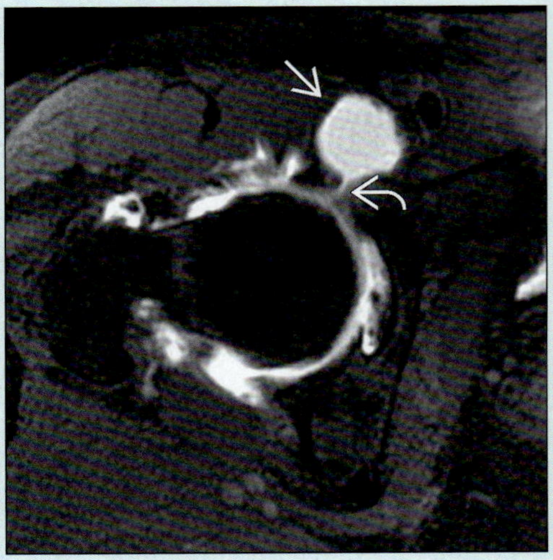

BURSITIS, HIP AND PELVIS

TERMINOLOGY

Synonyms

- Hip bursitis, groin bursitis

Definitions

- Abnormal distension ± inflammation of anatomic synovium-lined sac in musculoskeletal soft tissues
  - Anatomic bursae located where friction occurs between muscle/tendon and bone

IMAGING

General Features

- Best diagnostic clue
  - Asymmetric fluid collection in typical bursal locations
- Location
  - Iliopsoas
    - Between anterior hip capsule and distal iliopsoas tendon
    - Continuous with hip joint in 10-15% of all people
    - Also called subpsoas bursa
  - Obturator externus
    - Between obturator externus muscle/tendon and ischium
    - Continuous with hip joint in 5-10% of all people
  - Trochanteric
    - Between greater trochanter and internal or external rotator tendon insertions
    - 2 anatomic bursae at each gluteal tendon insertion
  - Ischial
    - Between hamstring tendon origins and adjacent ischium
  - Iliopectineal, iliofemoral, obturator internus are less common sites of bursal abnormalities
  - Bursae exist at almost every tendon insertion around pelvis and hip
- Morphology
  - Simple
    - Synovial fluid with low viscosity
  - Complex
    - Synovitis, synechiae, calcium products
    - Can indicate septic bursitis
  - Hemorrhagic
    - Complex with blood products of various ages
  - Calcific
    - Hydroxyapatite deposition disease (HADD) within synovial fluid

Radiographic Findings

- Radiographs generally normal with hip/pelvis bursae
- Focal soft tissue prominence in region of bursa
- Globular, cloud-like calcification with HADD

CT Findings

- Similar to radiographic findings
- Asymmetric soft tissue structure in typical location
  - Fluid density
  - Thin, smooth rim of peripheral enhancement with intravenous contrast
- Axial plane useful with iliopsoas bursitis

MR Findings

- T1WI
  - Lobulated fluid signal ± complexity
  - Hyperintense blood products
- T2WI FS
  - Lobulated fluid signal in typical location
  - Internal hypointensity indicates complexity
  - Capsular thickening with friction, trauma
  - Bilateral bursitis common
  - Thin neck may connect bursa to source
    - Iliopsoas: Often from hip joint
    - Infection: Occasional neck extending from infectious source
      - Lumbar discitis/osteomyelitis
      - Septic hip
- STIR
  - Homogeneous fat suppression allows for assessment of lesions at edge of field of view
    - e.g., bilateral trochanteric bursitis
- T1WI C+ FS
  - Peripheral rim enhancement
  - Center of distended bursa should not enhance
  - Enhancement extending into surrounding soft tissues suggests infection

Ultrasonographic Findings

- Sensitive for bursitis; specific for location of pain
- Optimal modality for therapeutic injection ± aspiration

Imaging Recommendations

- Best imaging tool
  - MR
- Protocol advice
  - Coronal and axial views of entire pelvis
    - Always check for bilateral bursitis
  - Sagittal through region in question
  - MR with contrast if concern for soft tissue tumor

DIFFERENTIAL DIAGNOSIS

Soft Tissue Tumor

- Synovial sarcoma, malignant fibrous histiocytoma, myxoid liposarcoma
- Location usually atypical for bursitis
- Tumors show internal enhancement
- With complex bursitis, consider contrast MR

Hip Internal and External Rotator Injury

- Trochanteric bursitis associated with gluteus medius & minimus tendinopathy

Hip Flexor Injury

- Iliopsoas bursitis associated with hip flexor tendinopathy and strain

Paralabral Cyst

- Can extend into iliopsoas bursa
- Generally more microlobulated than bursitis
- Labral tear associated with paralabral cyst

Idiopathic Contrast Extravasation

- Direct MR arthrography
- Contrast into iliopsoas and obturator externus bursae normal in 10-15% of individuals

BURSITIS, HIP AND PELVIS

Infection/Septic Joint
- Septic bursitis generally seen with hip effusion
- Enhancement of thick rind of reactive tissue
- Bursitis can become secondarily infected

Hematoma
- Superficial soft tissue edema
- Complex collection in atypical location for bursitis

Internal Derangement of Hip
- Clinical overlap, but distinct imaging findings
- Iliopsoas and obturator externus can be continuous with hip on direct arthrogram

PATHOLOGY

General Features
- Normal bursa
  - Fluid sac lined with synovial cells
  - Located at regions of tendon friction
- Bursitis
  - Abnormal distension of bursal sac
  - Source of pain ± tenderness
    - Mass effect
    - Inflammation of regional tendons & muscles

Staging, Grading, & Classification
- Bland bursitis
  - Simple synovial fluid
  - Homogeneous fluid signal
  - Most commonly from repetitive friction and overuse
- Synovitic bursitis
  - Complex aseptic collection
  - Consider pigmented villonodular synovitis (PVNS) if single location
  - Consider rheumatoid arthritis if multi-joint process
  - Bursitis related to amyloid arthropathy
- Hemorrhagic bursitis
  - Generally post-traumatic
  - Complex blood products
- Calcific bursitis
  - HADD of tendon erupts into bursa
  - Hypointense focus on MR within distended bursa
- Septic bursitis
  - Imaging features similar to complex bursitis
  - Fluid signal or enhancing neck to source of infection
    - Hip and lumbar spine sources most common
  - Primary septic bursitis rare
    - If iliopsoas bursa infected, hip considered infected

Gross Pathologic & Surgical Features
- Most commonly normal synovial fluid at aspiration
- Attempt to biopsy will yield fluid

CLINICAL ISSUES

Presentation
- Most common signs/symptoms
  - Iliopsoas: Pain on hip flexion
  - Trochanteric: Point tenderness at greater trochanter
  - Obturator externus: Perineal, buttock, or groin pain
  - Ischial: Buttock pain with sitting
  - Iliopsoas and trochanteric bursitis are in differential for post hip arthroplasty pain
- Other signs/symptoms
  - Fever & leukocytosis raise concern for septic bursitis

Natural History & Prognosis
- Waxing and waning symptoms common
- Can become secondarily infected when chronic
- Progressive pain with HADD
- Chronic, multifocal bursitis common with rheumatoid arthritis

Treatment
- Conservative regimens
  - Ice, massage, NSAIDs
- Percutaneous aspiration/steroid injection
  - Ultrasound guidance
  - Steroid injection advocated by some
    - Exclude infection 1st
  - Fluoroscopy-guided bursal injections possible
  - Aspirate as much fluid as possible prior to injection
- Surgical bursectomy
  - Refractory bursitis
  - Alleviates mass effect

DIAGNOSTIC CHECKLIST

Consider
- Paralabral cyst extending into iliopsoas bursa
- Septic bursitis secondary to septic hip
- Use of intravenous contrast for complex bursitis
  - Location of lesion (at a bursal site) and enhancement pattern should differentiate complex bursitis from solid mass lesion
- Ultrasound-guided percutaneous therapy

Image Interpretation Pearls
- Look for mass effect on surrounding structures, especially neurovascular

Reporting Tips
- Describe complexity of fluid, thick rind/capsule, connection to joint or other structure

SELECTED REFERENCES

1. Farmer KW et al: Trochanteric bursitis after total hip arthroplasty: incidence and evaluation of response to treatment. J Arthroplasty. 25(2):208-12, 2010
2. Murphy CL et al: Giant iliopsoas bursitis: a complication of chronic arthritis. J Clin Rheumatol. 16(2):83-5, 2010
3. Hodnett PA et al: MR imaging of overuse injuries of the hip. Magn Reson Imaging Clin N Am. 17(4):667-79, vi, 2009
4. Parziale JR et al: Iliopsoas bursitis. Am J Phys Med Rehabil. 88(8):690-1, 2009
5. Rowand M et al: Clinical inquiries. How should you treat trochanteric bursitis? J Fam Pract. 58(9):494-500, 2009
6. Williams BS et al: Greater trochanteric pain syndrome: a review of anatomy, diagnosis and treatment. Anesth Analg. 108(5):1662-70, 2009
7. Murphey MD et al: Pigmented villonodular synovitis: radiologic-pathologic correlation. Radiographics. 28(5):1493-518, 2008
8. Tibor LM et al: Differential diagnosis of pain around the hip joint. Arthroscopy. 24(12):1407-21, 2008

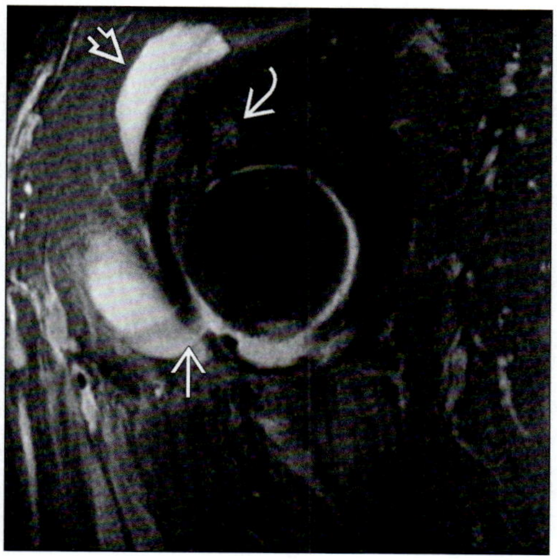

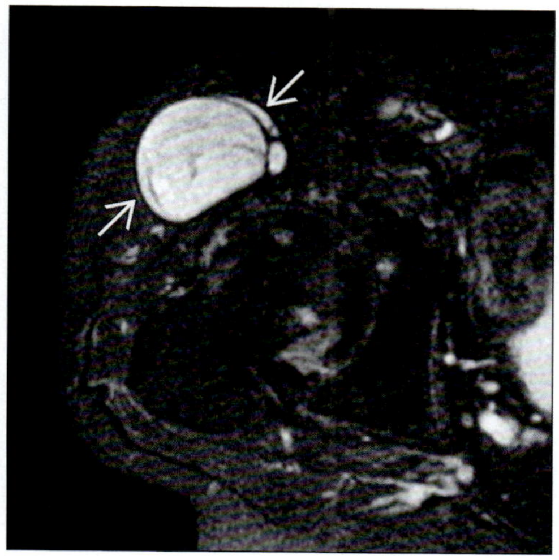

(Left) *Sagittal T2WI FS MR shows iliopsoas bursitis secondary to a hip effusion. Continuity between the hip and the bursa is demonstrated here, with more loculated bursitis proximal to the joint. There is subchondral marrow edema, suggesting cartilage loss as a source of the effusion.* ***(Right)*** *Axial T2WI FS MR shows a tense, distended iliopsoas bursa in a patient with subacute anterior hip trauma and severe pain with hip flexion. Mass effect from distended bursae is often painful.*

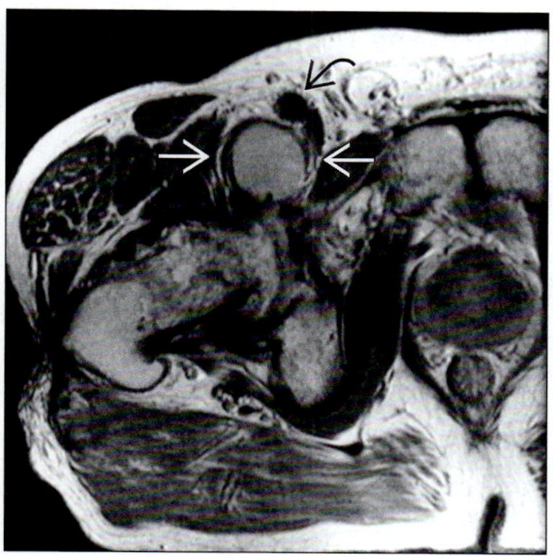

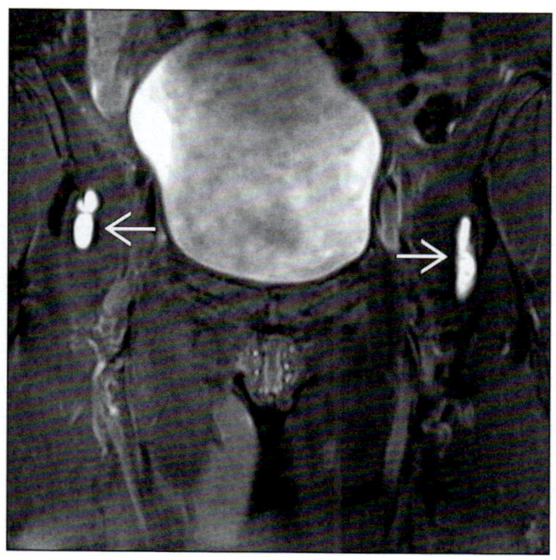

(Left) *Axial T1WI MR from a noncontrast hip protocol shows a smoothly marginated, lobulated structure extending anteriorly from the region of the iliopsoas bursa. Hemorrhagic bursitis can exhibit moderate T1 hyperintensity, as it does in this case. Note the hypointense vessels as an indicator of fluid signal.* ***(Right)*** *Coronal T2WI FS MR shows bilateral iliopsoas bursal fluid in a patient with posterior hip and back pain. Such small bursal fluid collections may be incidental.*

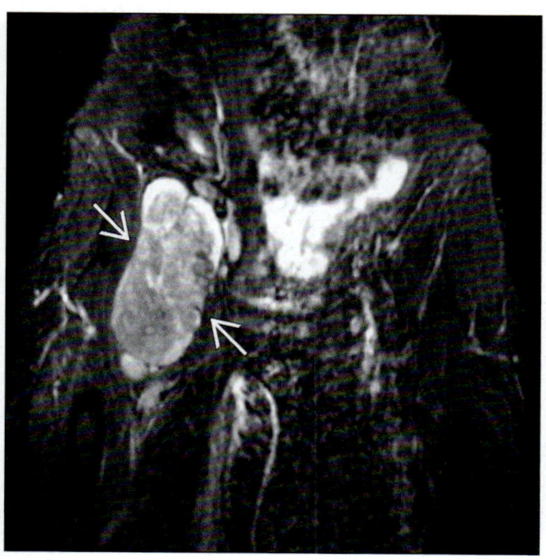

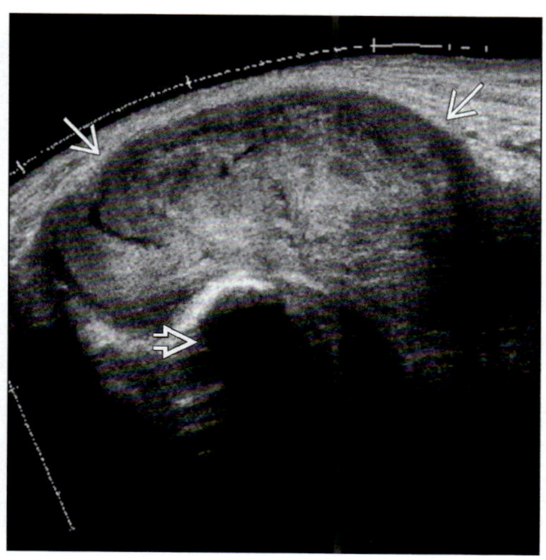

(Left) *Coronal STIR MR of the pelvis from a 30 year old with hip flexion pain shows a large mass distending the iliopsoas bursa. Synovitis within a bursa can be post-traumatic or may be related to a primary synovial process or inflammatory arthropathy.* ***(Right)*** *Transverse ultrasound from the same patient confirms a hyperechoic mass filling and distending the bursa superficial to the shadowing femoral head. This was a case of pigmented villonodular synovitis (PVNS) within the iliopsoas bursa.*

*(**Left**) Coronal STIR MR from a woman with gluteal pain while sitting shows a circumscribed fluid collection just posterior to the subtrochanteric region of the femur, typical for gluteus maximus bursitis. (**Right**) Axial T1WI C+ FS MR from a 54-year-old woman with bilateral lateral hip pain shows linear enhancement at both greater femoral trochanters. The enhancement on the left surrounds a small crescentic fluid collection. This is bilateral trochanteric bursitis.*

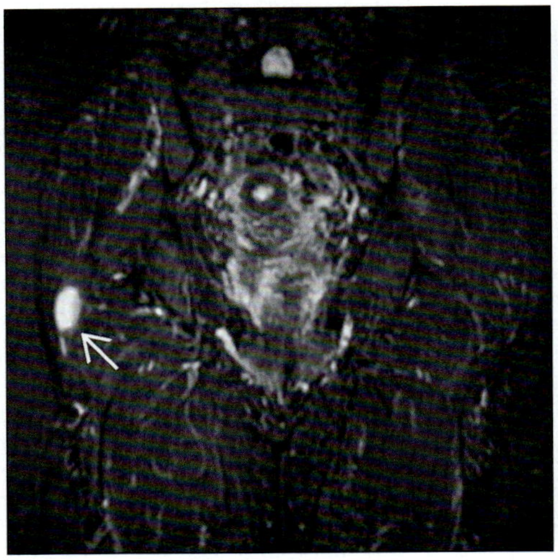

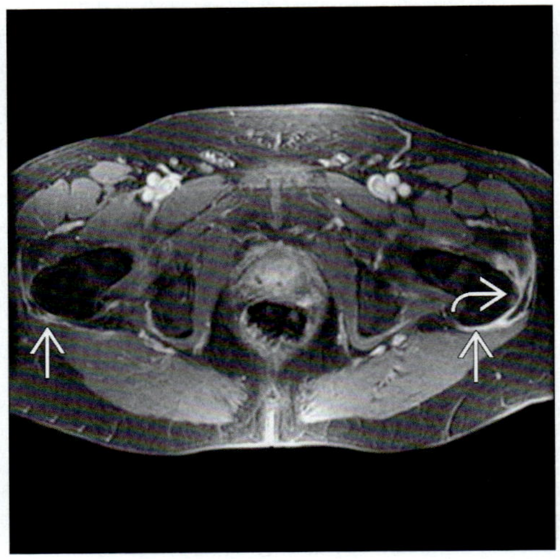

*(**Left**) Coronal STIR MR from a patient who was found down & incapacitated shows trochanteric bursitis but also diffuse edema throughout the gluteal musculature suggesting extrinsic pressure or trauma as the etiology. (**Right**) Sagittal T2WI FS MR from the same patient displays the trochanteric bursitis but also shows iliopsoas bursitis. There is amorphous hypointensity within the greater trochanteric tendon attachments, indicating HADD as a contributor to the bursitis.*

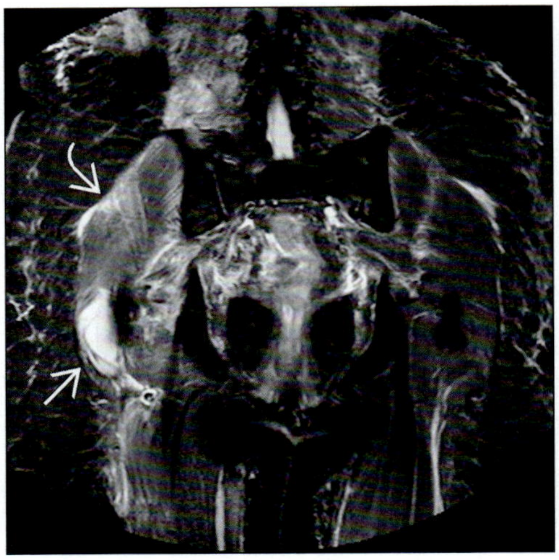

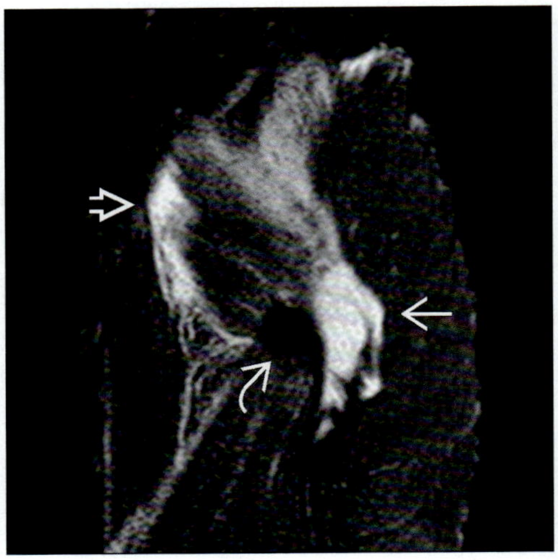

*(**Left**) Axial T2WI FS MR from a young athlete with pain on hip rotation shows fluid tracking along the obturator muscle group in a pattern typical for primary obturator externus bursitis. (**Right**) Coronal T1WI C+ FS MR from a direct arthrogram protocol shows injected contrast material extending from the hip joint along the obturator externus muscle, pooling in the obturator externus bursa. The obturator externus bursa is continuous with the hip joint in ~ 5% of all individuals.*

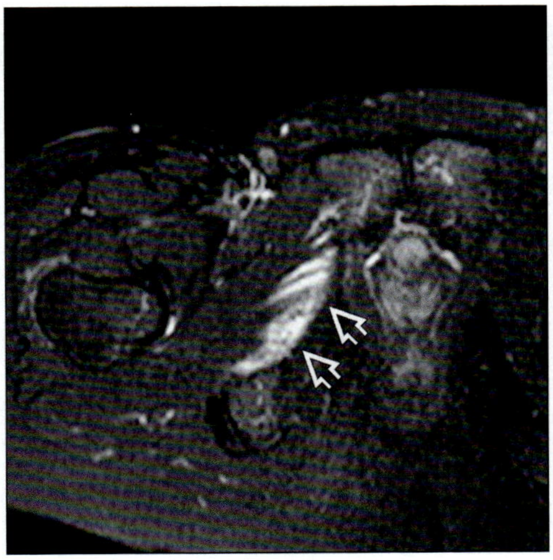

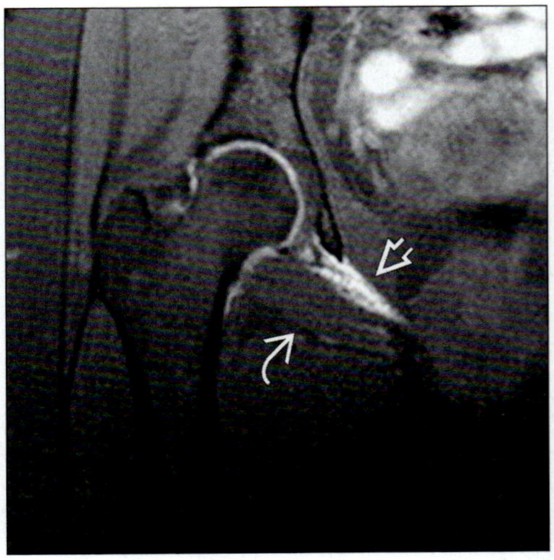

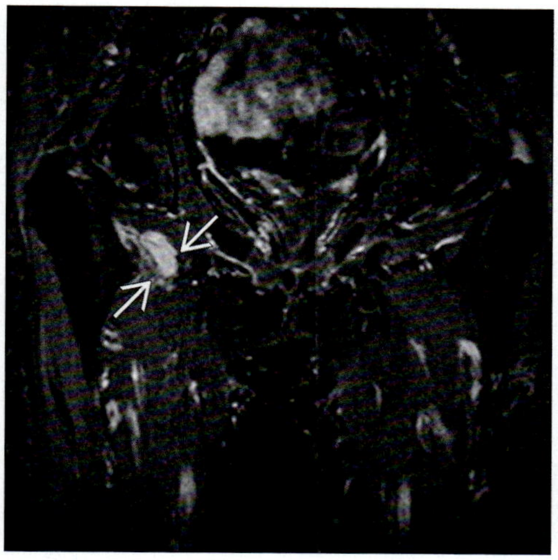

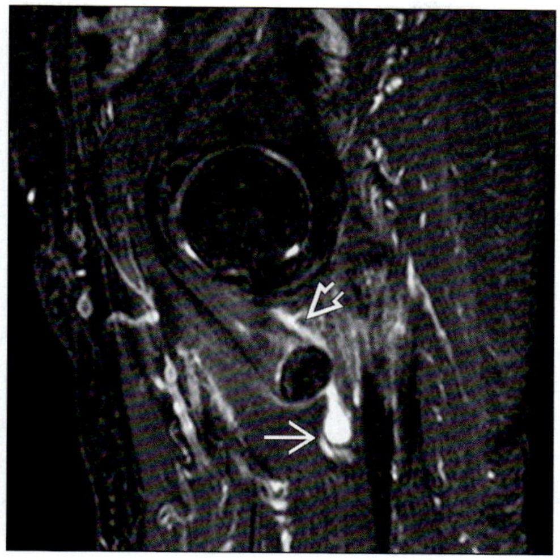

(Left) *Coronal STIR MR from a 51-year-old woman with pain on external thigh rotation shows a lobulated fluid collection ➔ interposed between the ischial tuberosity and the lesser trochanter of the femur.* ***(Right)*** *Sagittal T2WI FS MR from the same patient demonstrates a morphologic neck ⇨ from the obturator externus bursa in the upper thigh to the ischial bursa where there is more mass effect ➔. Anatomic bursae are often continuous with each other.*

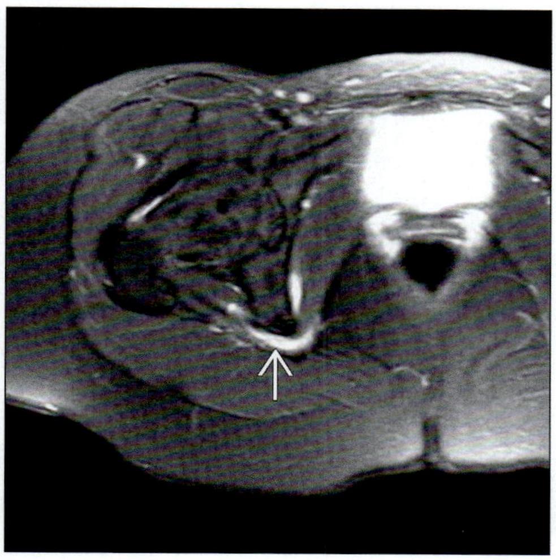

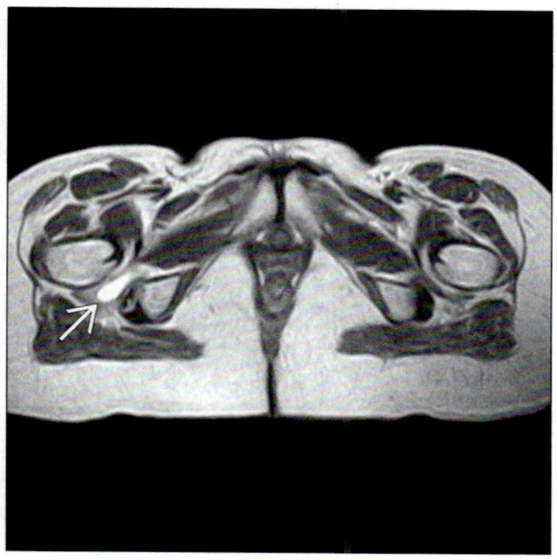

(Left) *Axial T2WI FS MR shows a crescentic fluid collection ➔ posterior to the ischial tuberosity and obturator internus muscle in a patient with ischial bursitis. The patient complained of buttock pain when riding a horse.* ***(Right)*** *Axial T2WI MR shows a well-defined fluid collection ➔ interposed between the lesser trochanter of the femur and the ischial tuberosity. This is an example of obturator external bursitis. The patient had hip pain with rotation.*

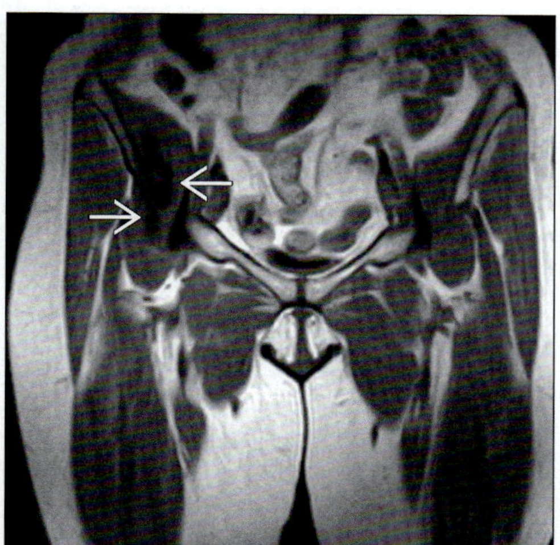

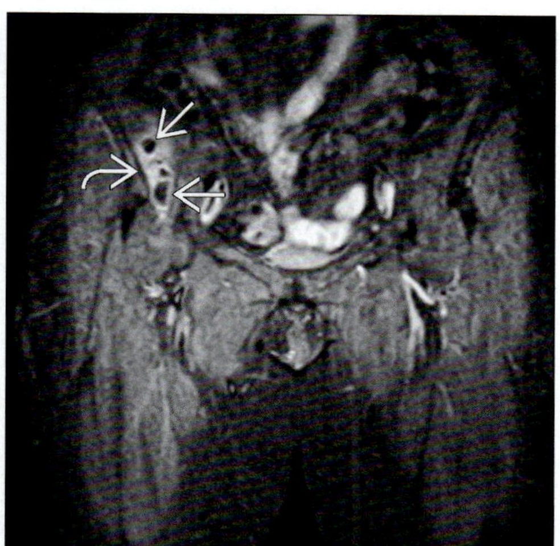

(Left) *Coronal T1WI MR from a patient with hip flexor pain shows asymmetric nodular hypointensity in the region of the iliopsoas tendon ➔. This degree of hypointensity suggests the presence of calcium. Fluid is similar in intensity to muscle on this image, and a collection is not visible.* ***(Right)*** *Coronal STIR MR from the same study shows several hypointense foci ➔ within a distended, hyperintense iliopsoas bursa ↪, typical for hydroxyapatite deposition disease or calcific bursitis.*

PIRIFORMIS SYNDROME AND NERVE INJURIES OF THE PELVIS

Key Facts

Terminology
- Piriformis syndrome: Neuritis of proximal sciatic nerve due to compression by piriformis muscle

Imaging
- Sciatic: Greater sciatic notch
- Iliohypogastric: Lateral margin of rectus abdominis near deep ring
- Ilioinguinal: External oblique aponeurosis
- Genitofemoral: Posterior abdominal/pelvic wall
- Pudendal: Between sacrotuberous and sacrospinalis ligaments
- Obturator: Obturator canal
- Lateral femoral cutaneous: At inguinal ligament
- Femoral: Femoral triangle below inguinal ligament
- Saphenous: Adductor canal
- Muscle/ligament hypertrophy causes compression
- Muscle/tendon atrophy downstream
- MR neurography
  - Large field of view coronal & axial T1 & T2WI FS
- MR C+: Asymmetric perineural enhancement
- Ultrasound: Visible entrapment/impingement with maneuvers

Pathology
- Trauma to gluteal, inguinal, or low abdominal region
- Variant muscle or nerve anatomy

Clinical Issues
- Image-guided injections
- Most cases respond to conservative treatment
- M:F = 1:6 for piriformis syndrome

Diagnostic Checklist
- Look for asymmetric muscle morphology or signal
- Sciatica type pain without neurologic defect: Piriformis syndrome

__(Left)__ Coronal graphic from anterior perspective shows entrapment and inflammation of sciatic nerve secondary to piriformis muscle hypertrophy. In some patients, the sciatic nerve may pass through part of the piriformis, which can lead to nerve compression. __(Right)__ Coronal T1WI MR shows a split or cleaved piriformis morphology that has been associated with piriformis syndrome when a division of the sciatic nerve passes between the piriformis bundles.

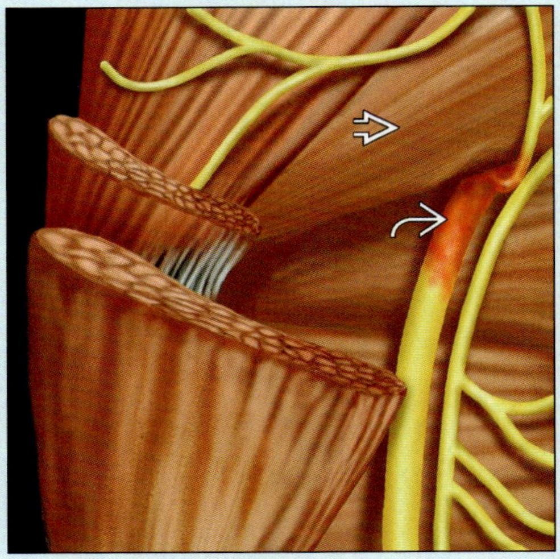

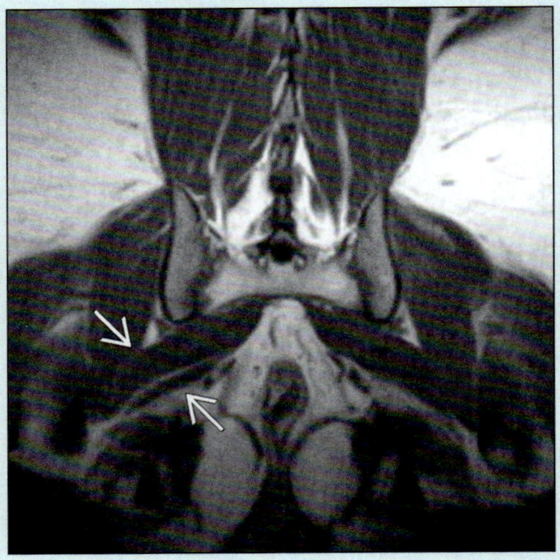

__(Left)__ Coronal T1WI MR from a neurography protocol shows a normal appearance of the lower sciatic nerves but mild asymmetry of the piriformis muscles with fatty atrophy on the right. __(Right)__ Coronal STIR MR from the same study shows perineural soft tissue edema around the sciatic nerve below the level of the piriformis muscle and the greater sciatic notch. Nerve entrapment syndromes in the pelvis can manifest as asymmetric muscle enlargement or asymmetric muscle atrophy on MR.

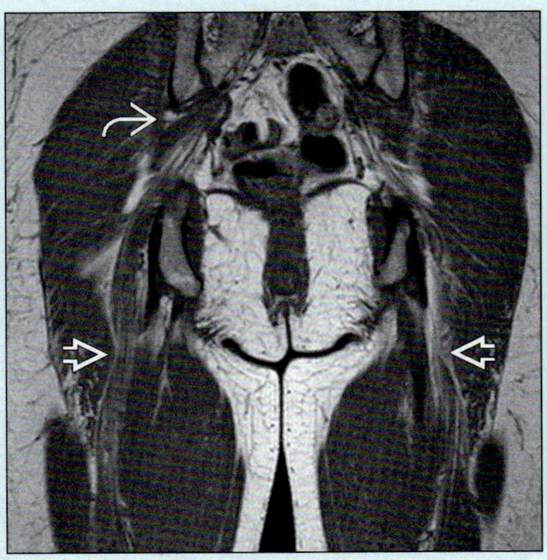

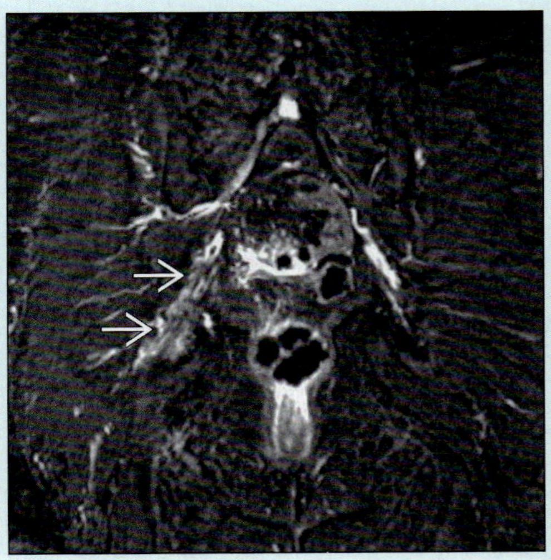

PIRIFORMIS SYNDROME AND NERVE INJURIES OF THE PELVIS

TERMINOLOGY

Synonyms
- Pseudosciatica, wallet sciatica, deep gluteal syndrome

Definitions
- Entrapment or compression of, or trauma to, lumbosacral plexus or pelvic nerve causing symptoms in innervated tissues
- Piriformis syndrome: Neuritis of proximal sciatic nerve due to compression by piriformis muscle

IMAGING

General Features
- Best diagnostic clue
  - Muscle atrophy within a neurotome
  - Piriformis or sciatic asymmetry
- Location
  - Sciatic: Greater sciatic notch
    - Between inferior piriformis and superior gemellus
    - At site where nerve pierces piriformis
  - Iliohypogastric: Lateral margin of rectus abdominis near deep ring
  - Ilioinguinal: External oblique aponeurosis
  - Genitofemoral: Posterior abdominal/pelvic wall
  - Pudendal: Between sacrotuberous and sacrospinal ligaments
  - Obturator: Obturator canal
  - Lateral femoral cutaneous: At inguinal ligament
  - Femoral: Femoral triangle below inguinal ligament
  - Saphenous: Adductor canal
- Morphology
  - Muscle/ligament hypertrophy
    - Source of entrapment/impingement
    - Piriformis = enlarged, pyramid-shaped
      - Generally asymmetric
  - Nerve enlargement
  - Muscle/tendon atrophy
    - Product of nerve entrapment/impingement
    - Atrophy often spans several dermatomes
- Many entrapment syndromes show no imaging findings

Radiographic Findings
- Radiography
  - Generally noncontributory
  - Calcified mass as source of entrapment

CT Findings
- NECT
  - Asymmetry of pelvic muscle
    - Hypertrophic if impinging
    - Atrophic if innervated by impinged nerve
  - Asymmetric muscle density
  - Asymmetrically enlarged nerve

MR Findings
- T1WI
  - Muscle hypertrophy
  - Effacement of fat in greater sciatic notch
  - Hyperintense atrophy in downstream muscles
  - Asymmetric sciatic enlargement ± hypointensity
- T2WI FS
  - Perineural hyperintensity
  - Muscle hyperintensity with early atrophy
  - Muscle hypointensity with late atrophy
  - Extrinsic mass lesion
- T1WI C+ FS
  - Asymmetric perineural enhancement

Imaging Recommendations
- Best imaging tool
  - MR of pelvis or focused ultrasound
- Protocol advice
  - MR neurography
    - Large field of view coronal & axial T1 & T2WI FS
      - Center on midline to allow side-to-side comparison
    - Angle along plexus involved
      - Along sacrum on sagittal localizer for lumbosacral plexus
    - Intravenous contrast when findings equivocal

Ultrasonographic Findings
- Enlarged, hypoechoic nerve
- Visible entrapment/impingement with evocative maneuvers

DIFFERENTIAL DIAGNOSIS

Hamstring Muscle Injury
- Hamstring muscle/tendon edema

Lumbar Radiculopathy
- Physical exam indicating lumbar root source
- Disc, facet disease on MR
- Recent paraspinal muscle strain

Sacroiliitis
- SI joint-centered imaging findings
- Radiographs ± MR

Hip Internal and External Rotator Injury
- Pain exacerbated by range of motion
- MR findings at greater tuberosity

Bursitis
- Fluid collection on MR

Stress Fracture
- Bone marrow edema on MR

Intrinsic Hip Pathology
- Relief with intraarticular anesthetic

Sports Hernia
- Associated with ilioinguinal nerve injury
- Secondary cleft, visible defect

PATHOLOGY

General Features
- Etiology
  - Trauma to gluteal, inguinal, or low abdominal region
    - Piriformis strain: Sciatic
    - Perineal trauma (bicycle seat): Pudendal
    - External oblique strain: Ilioinguinal
    - Transverse abdominis/internal oblique strain: Iliohypogastric

PIRIFORMIS SYNDROME AND NERVE INJURIES OF THE PELVIS

- Retroperitoneal hematoma: Genitofemoral
- Seatbelt trauma at anterior iliac crest: Lateral femoral cutaneous
- Variant muscle or nerve anatomy
  - Enlarged slip of piriformis where sciatic pierces it
  - Divisions of sciatic splitting piriformis
- Overuse and muscle/ligament hypertrophy
- Surgery
  - Cesarian section, abdominoplasty, herniorrhaphy: Ilioinguinal > genitofemoral > iliohypogastric
  - Hip arthroplasty with polymethylmethacrylate: Femoral
  - Hip and thigh surgery: Piriformis syndrome
- Pregnancy
  - Iliohypogastric, ilioinguinal, genitofemoral, lateral femoral cutaneous, femoral, pudendal
- Tumor compression: Any nerve can be affected
  - Neurofibromatosis common
  - Perineural cyst
- Prolonged sitting: Sciatic
  - Bilateral piriformis syndrome
- Diabetic amyotrophy: Femoral
- Herpetic infection: Pudendal
- Associated abnormalities
  - High-velocity pelvic trauma

Staging, Grading, & Classification

- Piriformis syndrome
  - Pain at greater sciatic notch in knee extension + hip flexion
    - Lasègue sign
  - Pain worse with contraction, improved with traction
  - Large, sausage-shaped piriformis
  - Split or cleaved piriformis morphology
  - Gluteal muscle atrophy

Gross Pathologic & Surgical Features

- Enlarged, edematous sciatic nerve
- Impinging mass lesion
- Piriformis split by sciatic nerve

Microscopic Features

- Inflammation
- Interstitial myofibrositis

CLINICAL ISSUES

Presentation

- Most common signs/symptoms
  - Chronic gluteal, groin, perineal, or thigh pain
  - Asymmetric weakness with muscle atrophy
- Clinical profile
  - Pain with hip maneuvers in piriformis syndrome
    - Passive hip flexion/internal rotation (Freiberg sign)
    - Resisted abduction/external rotation (Pace sign)
  - Tenderness in region of impingement and innervation
  - Pain with Valsalva maneuver
  - Dyspareunia, pain radiating to genitals
  - Prior hip surgery/arthroplasty: Internal/external rotators
  - Prior herniorrhaphy: Rectus abdominis, thigh adductors
  - Pregnant, postpartum females

Demographics

- Age
  - Piriformis syndrome: 18-55 years
  - Other nerve injuries in any age
- Gender
  - M:F = 1:6 for piriformis syndrome
  - Pregnancy associated with many lesions
- Epidemiology
  - Initially nagging pain
  - Ultimately muscle atrophy and weakness

Natural History & Prognosis

- Most cases respond to conservative treatment
- Functional biomechanical deficits
  - Tight piriformis muscle
  - Tightness, weakness of rotators, adductors
  - Sacroiliac joint hypomobility

Treatment

- Conservative
  - Physical therapy/stretching
  - NSAIDs
- Image-guided injections
  - Target suspected impingement/entrapment site
  - Anesthetic ± steroid
  - Botulinum toxin (Botox)
- Transcutaneous electrical nerve stimulation (TENS)
- Acupuncture
- Surgical
  - Release of piriformis from greater tuberosity insertion
  - Decompression of fibrosis surrounding nerve
  - Neurectomy for chronic groin pain

DIAGNOSTIC CHECKLIST

Consider

- Sciatica-type pain without neurologic defect: Piriformis syndrome

Image Interpretation Pearls

- Look for asymmetric muscle morphology or signal

SELECTED REFERENCES

1. Chhabra A et al: MR neurography of neuromas related to nerve injury and entrapment with surgical correlation. AJNR Am J Neuroradiol. Epub ahead of print, 2010
2. Jawish RM et al: Anatomical, clinical and electrical observations in piriformis syndrome. J Orthop Surg Res. 5(1):3, 2010
3. Lee FC et al: High-resolution ultrasonography in the diagnosis and intraoperative management of peripheral nerve lesions. J Neurosurg. Epub ahead of print, 2010
4. Halpin RJ et al: Piriformis syndrome: a real pain in the buttock? Neurosurgery. 65(4 Suppl):A197-202, 2009
5. Fishman LM et al: Piriformis syndrome: diagnosis, treatment, and outcome--a 10-year study. Arch Phys Med Rehabil. 83(3):295-301, 2002

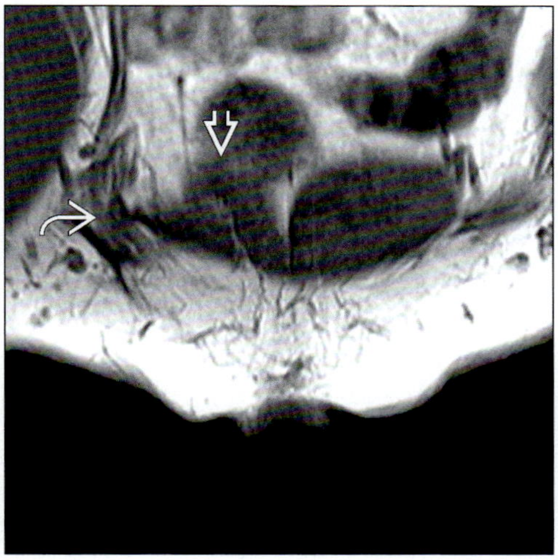

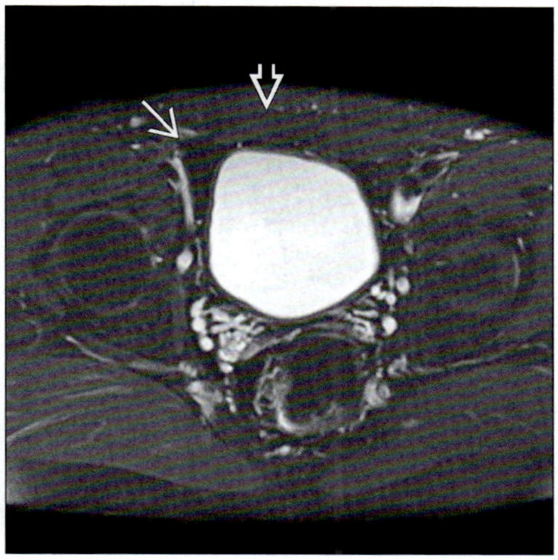

(Left) *Coronal oblique T1WI MR from an athletic pubalgia protocol shows asymmetric atrophy of the inferior right rectus abdominis tendon* ⇨ *and scarring at the inguinal ring* ↷ *compatible with prior herniorrhaphy.* ***(Right)*** *Axial T2WI FS MR from the same patient shows right rectus abdominis atrophy* ⇨ *causing secondary enlargement of the superficial inguinal ring* →*. Findings confirm nerve injury in the ilioinguinal, femoral, or genitofemoral distribution, not specific for a single lesion.*

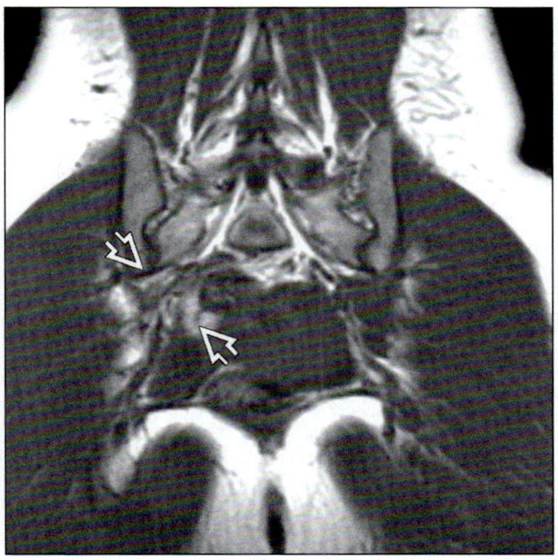

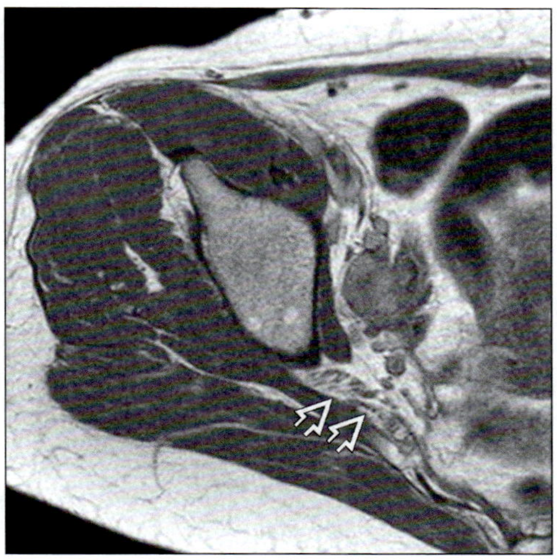

(Left) *Coronal T1WI MR shows near-complete atrophy of the right piriformis* ⇨ *in a patient with longstanding pain with sitting and suspected sciatic nerve impingement.* ***(Right)*** *Axial T1WI MR from the same study shows a large, fasciculated, fat-infiltrated sciatic nerve* ⇨ *in the bed of the atrophic piriformis muscle. With chronic nerve entrapment syndromes, neural structures can become devitalized and ultimately atrophic, leaving few treatment options and little hope for recovery.*

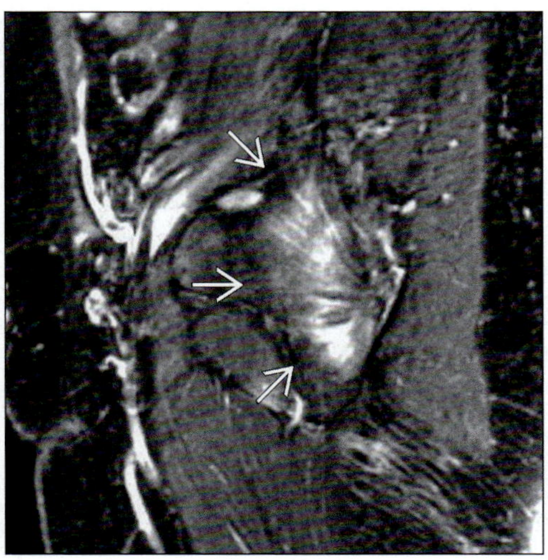

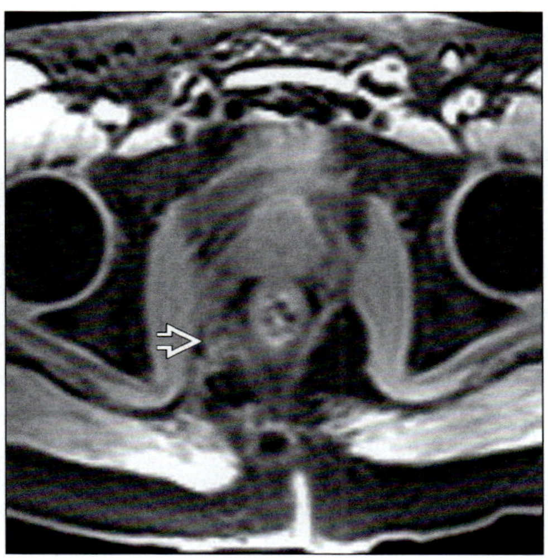

(Left) *Sagittal T2WI FS MR shows a feathery pattern of muscle edema* → *throughout the quadratus femoris and inferior gemellus muscles in a patient with pain on external rotation and no reported trauma, suggesting a femoral nerve entrapment syndrome.* ***(Right)*** *Axial T1WI C+ FS MR from a patient with severe perineal and scrotal pain shows intense asymmetric enhancement* ⇨ *of the pudendal nerve adjacent to the rectum. An inflammatory neurofibrosis attributed to a herpes infection was found at surgery.*

SECTION 6
Knee

Introduction and Overview

Osseous and Chondral Pathology

Ligamentous Pathology

Meniscal Pathology

Tendon Pathology

Other Pathology

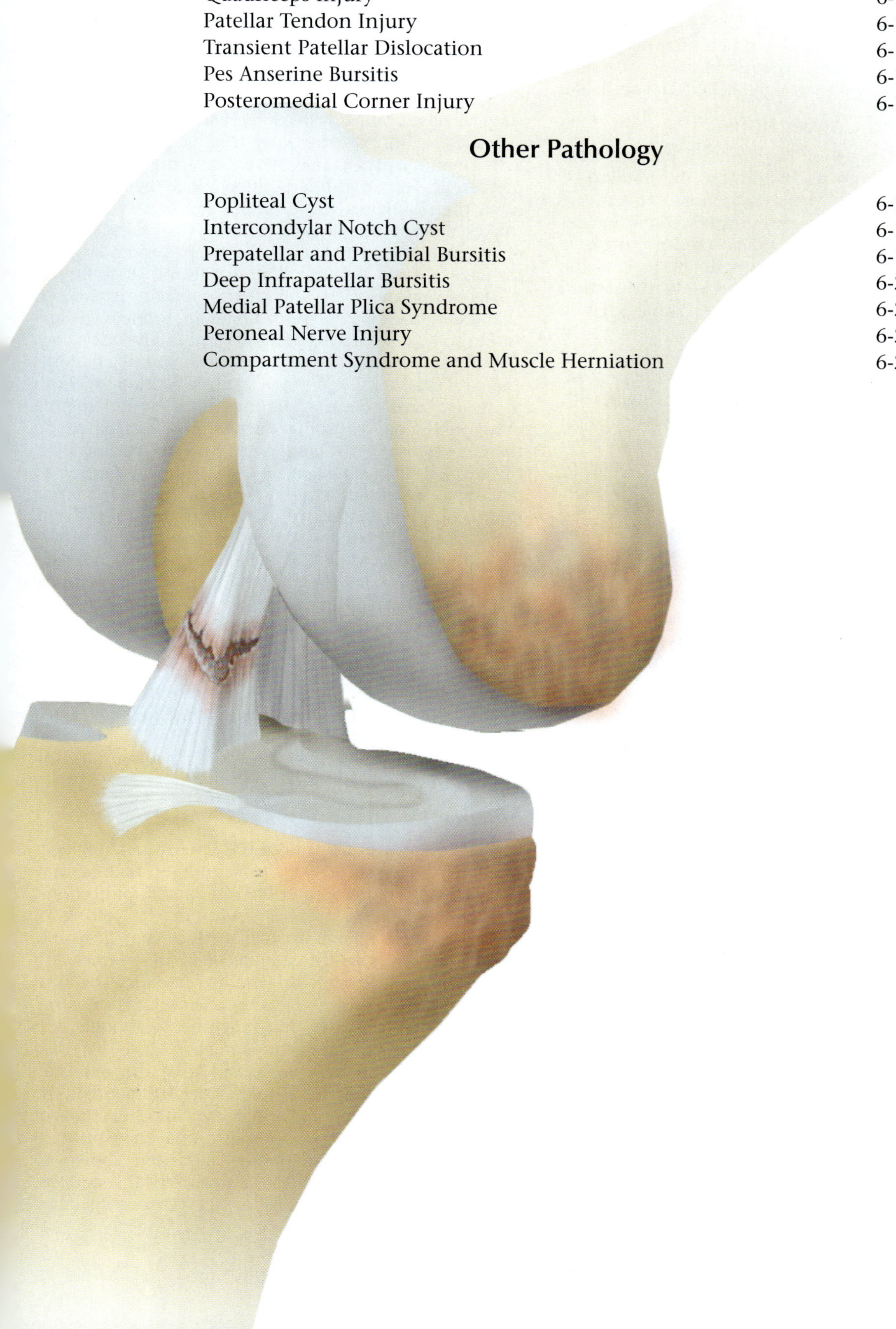

Overview

The knee is the most commonly imaged joint in most practices, particularly in the realm of magnetic resonance (MR) imaging. Achieving a functional comfort level with the relevant anatomy, pathology, and imaging techniques is imperative in order for the radiologist to add value to the diagnostic work-up of the patient with knee pain. Because the literature regarding knee imaging is bountiful, there is at once both an opportunity and an obligation for the radiologist to stay current with the latest techniques and reading strategies, lest s/he fall behind. This section explores the full range of knee trauma pathology seen in a modern practice, using the latest published data available.

Terminology and Conventions

Degenerative changes in a tendon are referred to as "tendinopathy" and not as "tendinitis" or "tendinosis" in an effort to stay true to the appropriate etymologic meanings of these terms. As elsewhere in this book, images have all been presented as right-sided in order to preserve uniformity. The term "spontaneous osteonecrosis of the knee" is not used but is instead covered under the section on osteochondral injury, as this imaging finding is felt by most authors to be due to shear trauma.

Anatomic Considerations

The femorotibial (knee) joint is a simple hinge, with very little rotational motion occurring at the articulation in normal physiologic motion. A few degrees of external tibial rotation occur in terminal extension, serving to "lock" the knee and reduce the need of constant muscular action to help hold the knee in this position during standing (this is sometimes referred to as the "screw-home" mechanism). The popliteus muscle serves to rotate the femur externally during initiation of flexion in order to "unlock" the knee. The patella is a large sesamoid bone in the quadriceps tendon complex and articulates with the trochlear groove of the femur in order to increase fulcrum length of the quadriceps units and reduce friction between the tendons and the femur.

The anterior and posterior cruciate ligaments serve to balance each other in the full range of flexion and extension of the knee. The anterior cruciate (ACL) prevents anterior translation of the tibia relative to the femur and is taut predominantly in knee extension. The posterior cruciate (PCL) prevents posterior tibial translation; it is taut primarily in knee flexion. The 2 ligaments also resist rotational knee motion and complement each other in that function. The medial collateral ligament resists valgus forces, and the lateral collateral ligament complex resists varus forces. The posterolateral corner ligament complex is a series of mostly capsular thickenings that serve to stabilize this important part of the joint.

The menisci are fibrocartilage wedges conforming to the shapes of the tibial articular condyles and serve to cushion the impact of the femur on the tibia during weight-bearing. The medial meniscus is larger and has a larger radius of curvature than its lateral counterpart. The medial meniscus is also more firmly attached to the bones than is the lateral meniscus, allowing more lateral meniscal motion during flexion and extension. The menisci derive their blood supply from a vascular pedicle that enters at the capsular margin of the meniscus. Vascularity within the meniscus becomes progressively more sparse toward the free edge and diminishes in general in older patients.

Because of its superficial location, the peroneal nerve is the only commonly injured nerve in the knee region. It lies close to the skin as it curves around the proximal fibula and can be compressed either by internal processes (ganglia, osteophytes, etc.) or by extrinsic trauma.

Pathologic Considerations

Injury to the knee is a common occurrence across the age spectrum and thus results in a high frequency of imaging studies in a typical practice. Radiographs and CT imaging are useful for characterizing osseous trauma, and surgical classification systems have been provided where appropriate. Injury to the knee is often related to sports activity and, as such, may be either acute or the result of chronic repetitive microtrauma. In either case, it is often the ligaments, tendons, and cartilage of the knee that bear the brunt of such injuries. In particular, tears of the menisci of the knee and focal or diffuse articular cartilage defects account for significant disability in modern Western society, and accurate imaging evaluation of these injuries may serve to target appropriate therapy and in some cases may obviate surgical intervention.

An understanding of the typical injury patterns encountered in the knee will aid the radiologist, both in recognizing common injuries and in anticipating more subtle but clinically relevant findings based on their association with these injury patterns. For instance, the pivot-shift mechanism of knee injury (often due to making a sudden cut while running or due to a clipping injury) not only frequently causes disruption of the ACL but is often associated with vertical longitudinal tears of the posterior horns of the medial and lateral menisci, avulsion of the lateral capsular ligament, and injury to the posterolateral corner ligaments. These associated injuries may be subtle on MR imaging, and anticipating their presence based on pattern of injury will serve the radiologist well. On the other hand, the presence of a small avulsion from the lateral tibial rim on radiography (Segond injury) may appear to be inconsequential at first glance, but its very high association with ACL tear should lead to further evaluation with more advanced imaging (MR) almost routinely.

Imaging Considerations

Radiographic evaluation of the knee usually includes 3 standard views, but in the setting of trauma may be limited to anteroposterior and lateral projections. A cross-table lateral view is useful in the setting of acute trauma, as a large lipohemarthrosis may be visible as fat-fluid levels in the suprapatellar joint recess and serve as an indicator for intraarticular fracture. The axial patellofemoral (sunrise) view is useful for evaluating patellofemoral arthritis and alignment, though less valuable in the setting of acute trauma.

Arthrography of the knee was once the mainstay of radiologic diagnosis of internal derangement of the knee but has been supplanted by MR. Injection of dilute gadolinium compounds into the knee prior to MR imaging (MR arthrography) has been shown to provide substantial benefit in certain clinical situations, particularly in the evaluation of the postoperative meniscus. Indirect MR arthrography (intravenous gadolinium injection with delayed MR imaging) has been shown by some authors to be useful in evaluation of postoperative cartilage and meniscal lesions.

CT of the knee is commonly used to characterize complex intraarticular fractures, particularly those of the tibial plateaus. Because they may be anatomically complex, these fractures are difficult to fully evaluate with routine radiography. Linear or complex polyspiral tomography used to be the mainstay in the evaluation of such fractures, but the machines used to create these images have largely been replaced by CT. MR imaging of intraarticular fractures can be useful in depicting associated soft tissue injury, but MR often lacks the spatial resolution to depict small osseous fragments.

Ultrasound of the knee is generally limited to the study of superficial tendons (typically the extensor mechanism) or of cystic collections (popliteal or Baker cysts). Some reports of evaluation of meniscal pathology using high-definition ultrasound have been published, but this is certainly not yet in the armamentarium of the typical radiology practice.

MR imaging serves as the mainstay of imaging for internal derangement of the knee. A thorough familiarity with the appearance of normal anatomy, common variants, and pathologic processes in the knee is incumbent on the radiologist attempting to interpret these studies. MR imaging of the knee is performed with a rigid volume coil with the knee near isocenter of the magnetic field, serving to maximize both signal-to-noise ratio and field homogeneity. The use of a transmit-receive knee coil further improves on these features. The recent advent of multichannel extremity coils has led to even greater improvements in image quality, and they can also be used to improve time efficiency of imaging.

Imaging of the menisci: Special mention of MR imaging strategies as they regard the menisci is warranted. Because they are composed of relatively homogeneous fibrocartilage, the normal menisci should demonstrate very low signal on all imaging sequences. Degeneration or tear of a meniscus is diagnosed based on abnormal signal within the meniscal substance. In the absence of prior meniscal surgery, any elevated signal that is seen to clearly extend to the superior or inferior articular surface or free edge of the meniscus fulfills the MR criteria for a tear. Some authors have argued that such contact with the articular surface of the meniscus should be confirmed on at least 2 MR images. Morphologic changes in the menisci may also herald a tear; blunting of the free edge, separation or displacement of a portion of the meniscus, or absence of meniscal substance in its expected location are all consistent with a tear in the absence of a history of surgery. In the postoperative meniscus, signal can extend to the surface of the meniscus due to resection of meniscal tissue down to an area of central degeneration, and the presence of fluid or arthrographically injected contrast into the meniscus may be necessary to diagnose a tear in this situation.

Imaging Protocols

Because of anatomic constraints, no special positioning maneuvers are commonly used. Imaging planes are straightforward, with coronal and sagittal images based on the position of the knee in the coil. A useful landmark for the prescription of coronal images is a line drawn through the furthest posterior edges of the femoral condyles on the appropriate axial image. Sagittal images are then prescribed perpendicular to the coronals. The use of 3 Tesla magnets has increased and with them have come the promise of higher spatial resolution, chemical composition information, and faster imaging times.

While the choices of pulse sequences available to perform MR imaging of the knee is vast and growing constantly, certain basic principles are worth bearing in mind. Meniscal imaging is best accomplished using short TE (time of echo) sequences, such as T1- or proton density-weighted imaging, in order to maximize signal in a meniscal tear and minimize echo blurring. The use of fast spin-echo sequences to evaluate the menisci is now considered mainstream, as the earlier concerns about echo blur and spatial resolution limitations have been solved by equipment manufacturers. Fluid-sensitive sequences, usually employing spectral fat suppression, are useful for the evaluation of marrow processes, cystic collections, and soft tissue injury. Experienced radiologists tend to find their favorite combinations of planes and sequences, and a typical diagnostic knee MR examination is comprised of 4-6 such sequences using at least 1 sequence in each major imaging plane (sagittal, coronal, and axial). General rules of thumb include using thin slice profiles for the evaluation of the menisci (no greater than 3-4 mm thickness, with interslice gap minimized to no greater than 1 mm) and the use of short TE sequences in both sagittal and coronal planes. Some authors advocate the addition of spectral fat suppression to these meniscal-sensitive sequences, claiming superior sensitivity for meniscal tears. Broader experience suggests that it may be that the use of fat suppression in this setting is best left simply to subjective preference.

Optimal visualization of articular cartilage is best accomplished using sequences that emphasize the difference of appearance between fluid and cartilage; many such combinations have been described and are part of the basic armamentarium of imaging sequences employed for routine knee imaging. On the other hand, the evaluation of meniscal and articular cartilage pathology has led to an explosion of specialized MR pulse sequences designed to provide increased sensitivity and specificity for subtle injuries to these important structures. Many have entered the mainstream lexicon of practicing radiologists, while others remain largely experimental. A brief overview of some of these techniques is presented in the chapters on articular cartilage imaging and postoperative cartilage imaging. Unfortunately, each magnet manufacturer identifies their particular versions of these types of sequences with proprietary names or acronyms, so that one is best served by becoming familiar with the capabilities of the magnets used in his/her practice setting.

Selected References

1. Miller TT: MR imaging of the knee. Sports Med Arthrosc. 17(1):56-67, 2009
2. Griffin N et al: High resolution imaging of the knee on 3-Tesla MRI: a pictorial review. Clin Anat. 21(5):374-82, 2008
3. Lee MJ et al: Ultrasound of the knee. Semin Musculoskelet Radiol. 11(2):137-48, 2007
4. Chung CB et al: MR arthrography of the knee: how, why, when. Radiol Clin North Am. 43(4):733-46, viii-ix, 2005

*(**Left**) Lateral radiograph shows the normal knee. The femorotibial joint is a simple hinge, with very little rotational motion. The patella ➔ is a large sesamoid bone that reduces friction between the extensor tendons and the femur. (**Right**) PA graphic shows the popliteal muscle ➔ coursing from the proximal posteromedial tibia to its insertion on the lateral aspect of the femur. The muscle serves to "unlock" the knee from terminal extension during knee flexion.*

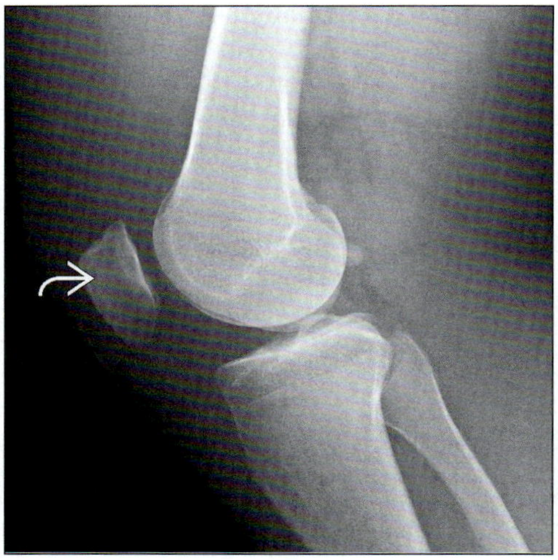

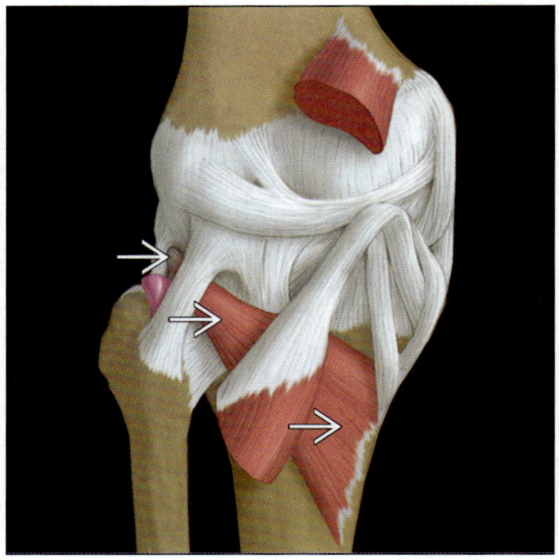

*(**Left**) Lateral graphic shows the extensor mechanism of the knee, composed of the quadriceps muscles and tendons ➔, the patella ➔, the patellar tendon ➔, and the medial and lateral retinacula ➔. (**Right**) Sagittal PDWI MR shows the anterior ➔ and posterior ➔ cruciate ligaments. The ligaments have complimentary morphology and function; the ACL is taut in extension and resists anterior tibial motion; the PCL is taut in flexion and resists posterior tibial motion.*

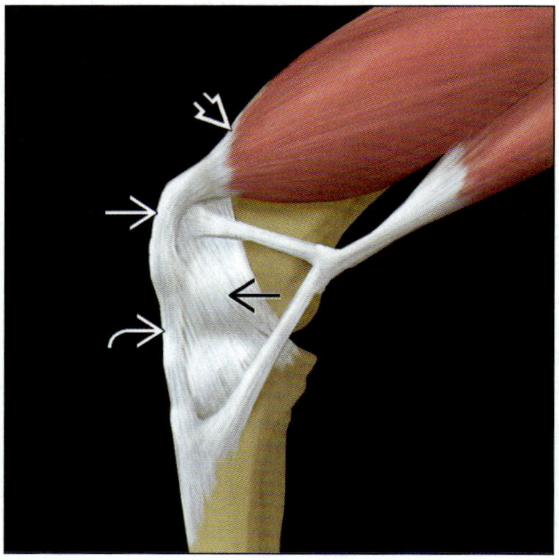

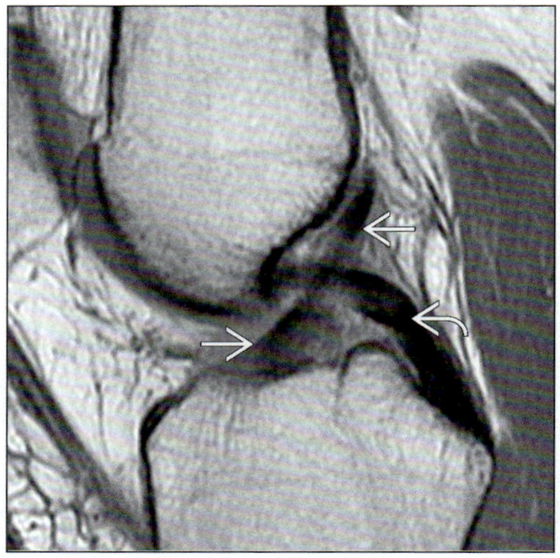

*(**Left**) Coronal T1WI MR shows the normal fibular collateral ligament (FCL) ➔ extending from the lateral femoral condyle to the styloid of the fibular head. The ligament is the primary restraint against varus force in the knee. (**Right**) Coronal T2WI FS MR shows the normal iliotibial band (ITB) ➔, which extends from the hip to Gerdy tubercle on the lateral tibia. The medial collateral ligament (MCL) ➔ runs from the medial femoral condyle to the medial tibia, inserting 6-7 cm below the knee.*

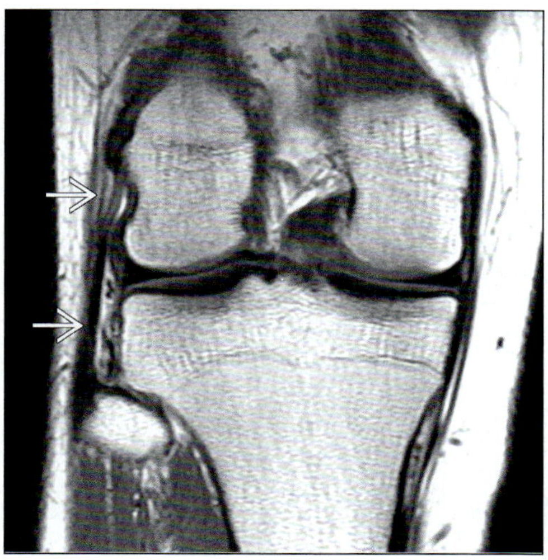

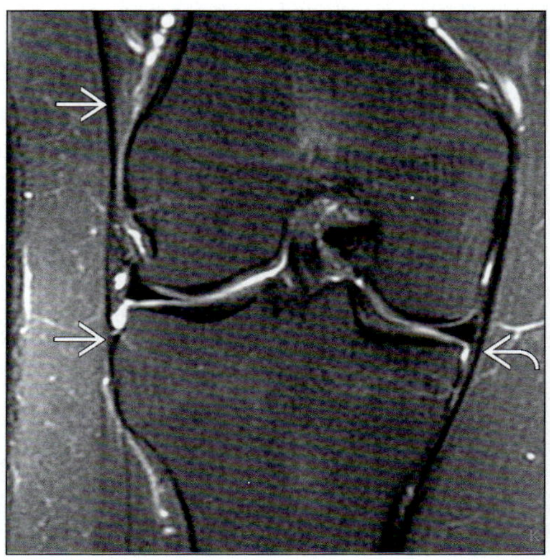

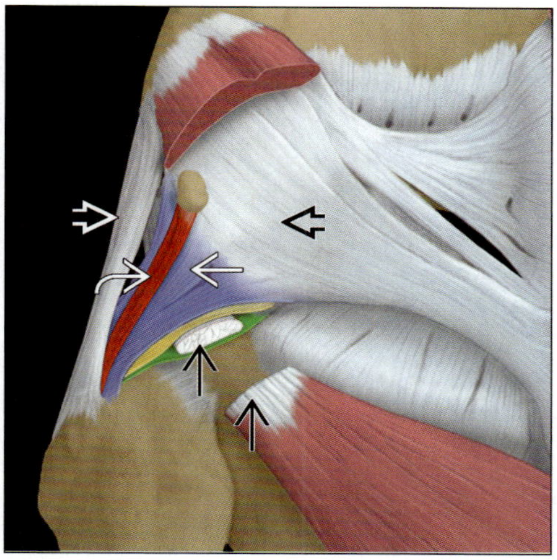
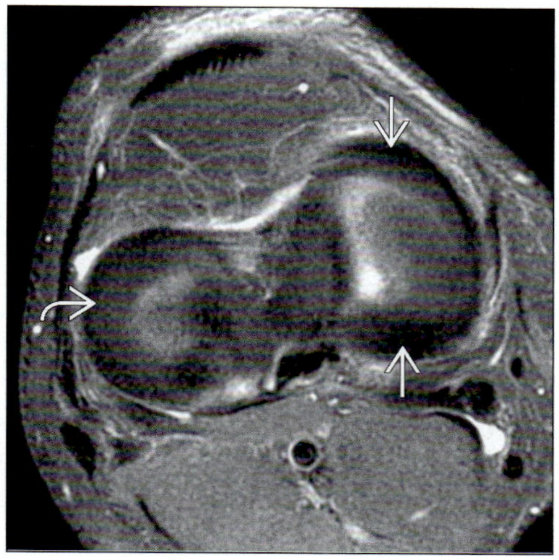

(Left) *PA graphic of the posterolateral corner of the knee demonstrates the popliteus tendon, oblique popliteal ligament, fibular collateral ligament, arcuate ligament, and fabellofibular ligament.* ***(Right)*** *Axial PDWI FS MR shows the different configurations of the medial and lateral menisci. The medial meniscus has a longer anteroposterior axis and is more C-shaped; the lateral meniscus is more symmetric and O-shaped.*

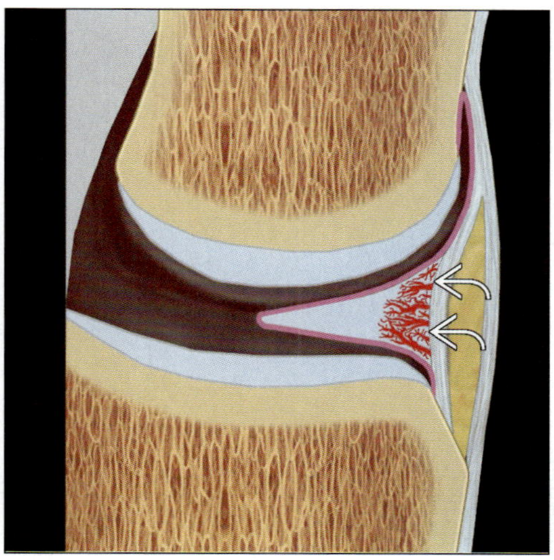
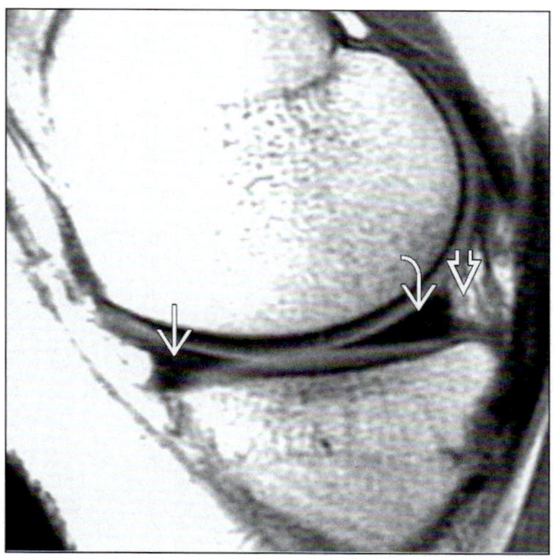

(Left) *Coronal graphic shows the vascularity of the medial meniscus. The vessels feeding the meniscus arise from the capsular surface and are generally limited to the peripheral 1/3 of the meniscus (the "red zone").* ***(Right)*** *Sagittal PDWI MR shows a normal medial meniscus. The anterior and posterior horns are triangular with no intrinsic signal. The posterior meniscocapsular junction contains fat but should not contain fluid.*

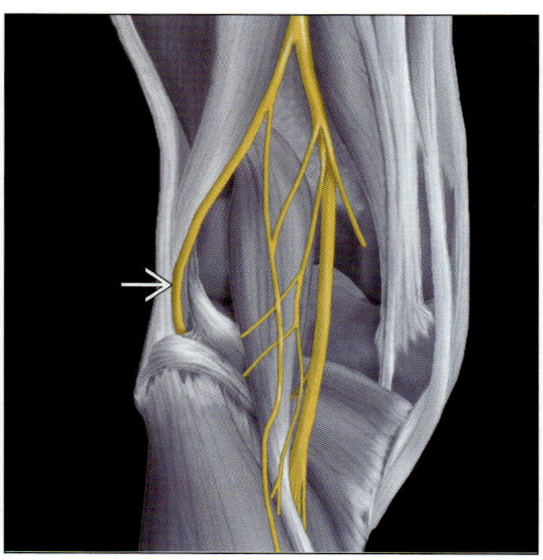
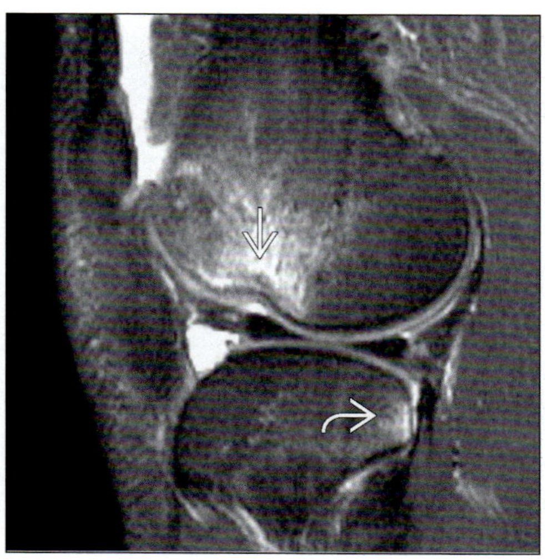

(Left) *PA graphic of the nerves around the knee shows the superficial course of the common peroneal nerve as it extends laterally around the fibular head. The nerve is prone to trauma due its vulnerable position.* ***(Right)*** *Sagittal PDWI FS MR shows subchondral marrow contusions in the posterior lateral tibia and at the terminal sulcus of the lateral femoral condyle. This is the typical "footprint" of a pivot-shift injury and indicates a high likelihood of ACL tear.*

(Left) *Cross-table lateral radiograph after acute knee trauma shows a fat-fluid level in the suprapatellar recess of the joint, indicating an intraarticular fracture. Because fat floats on water, the interface is seen with a horizontal beam only. The lateral tibial plateau fracture can be seen as a double cortical density.* ***(Right)*** *AP radiograph from the same patient shows the double cortical density of the impacted lateral tibial fracture.*

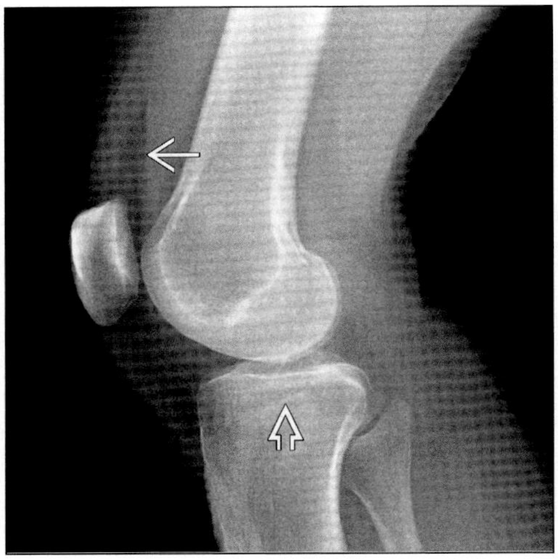

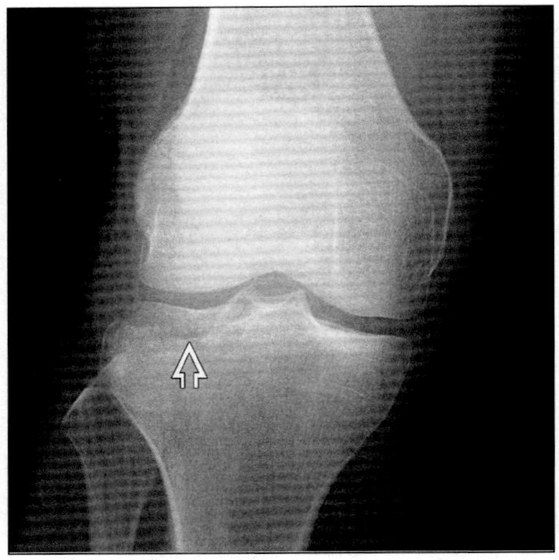

(Left) *AP radiograph after severe knee trauma shows a complex intraarticular fracture of the proximal tibia, but anatomic detail is limited.* ***(Right)*** *Coronal reformat bone CT from the same patient shows the extent of the articular surface defect in the tibia and the location and position of fracture fragments. Reformatted images created on multislice CT scanners in use today can provide exquisite detail of osseous anatomy in any plane desired.*

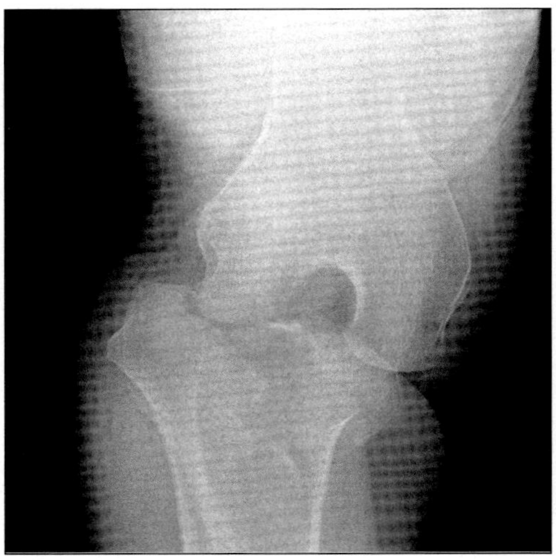

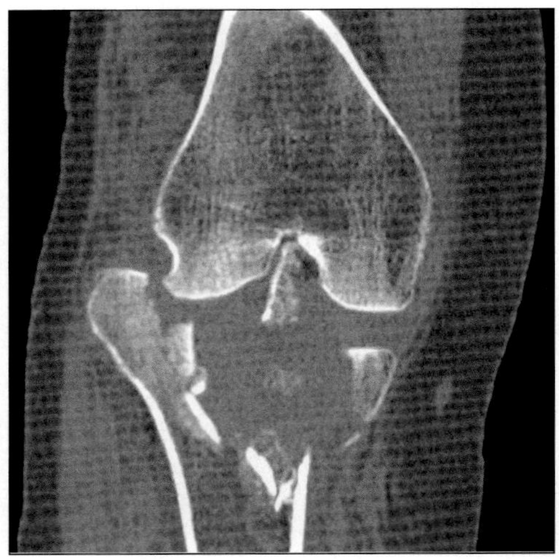

(Left) *Sagittal PDWI FS MR shows an impacted comminuted fracture of the lateral tibial plateau, with surrounding marrow edema. While MR imaging is not the optimal imaging modality to evaluate the morphology of the fracture, the integrity of adjacent soft tissues is well seen. The intact lateral meniscus is seen on this image.* ***(Right)*** *Coronal PDWI MR shows a sagittally oriented fracture of the lateral tibial plateau containing an entrapped lateral meniscus fragment.*

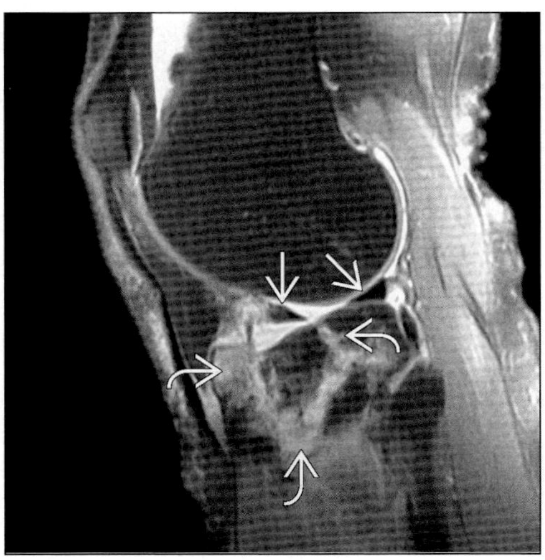

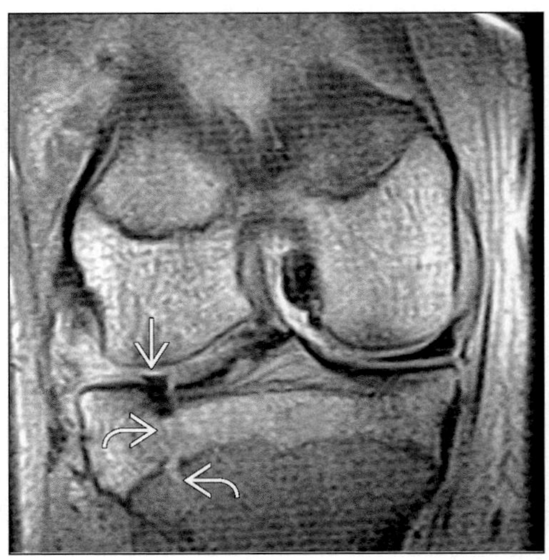

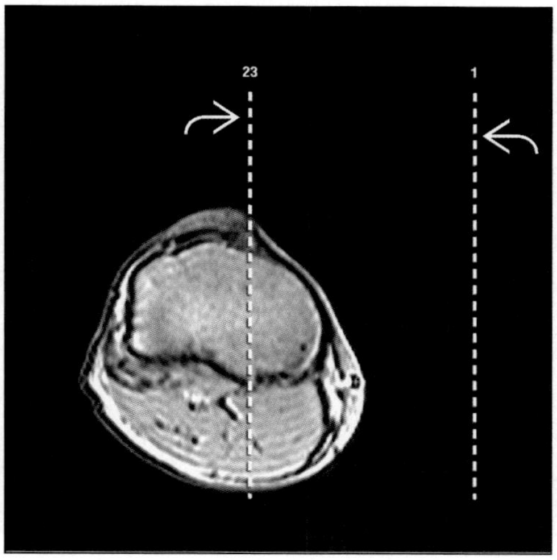

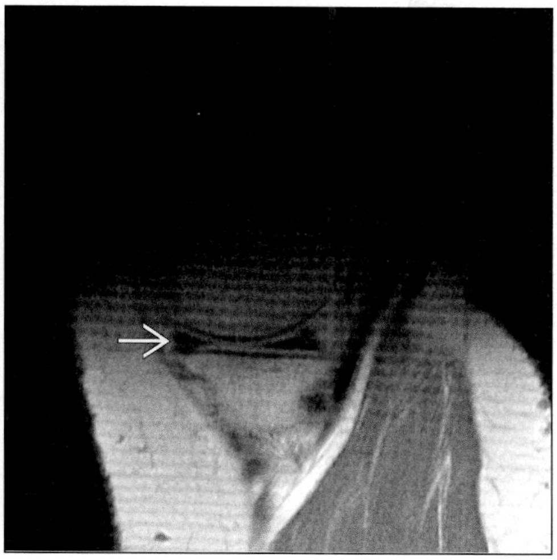

(Left) *Axial gradient echo scout image of the knee shows lines ➔ indicating the prescription of sagittal images. Some scanners have a "default to midline" function, which was not deactivated prior to the prescription and acquisition of the subsequent images.* ***(Right)*** *Sagittal PDWI MR shows the result of poor positioning of the knee coil; there is no signal in the distal thigh, and evaluation of the medial meniscus ➔ is severely limited by very low signal-to-noise ratio.*

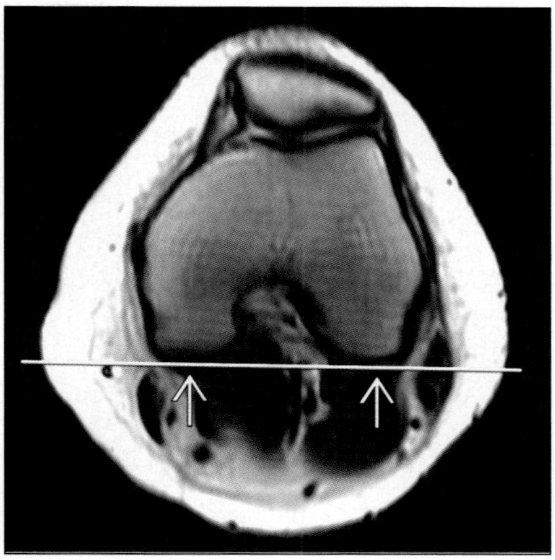

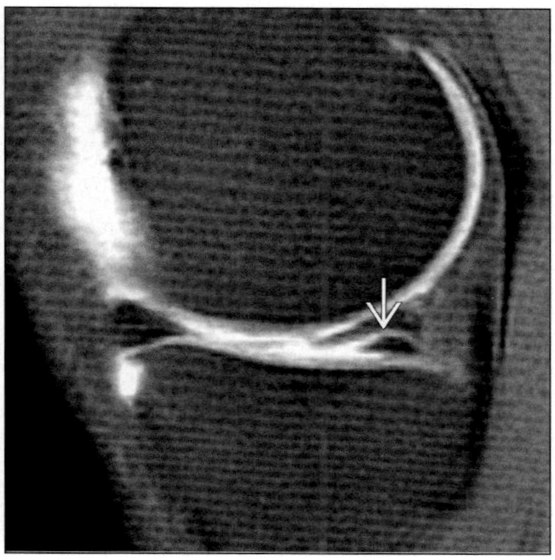

(Left) *Axial gradient echo scout image shows the prescription line ➔ for the creation of coronal images. The scan plane is drawn through the most posterior extent of the medial and lateral femoral condyles. Sagittal images are then prescribed perpendicular to this plane.* ***(Right)*** *Sagittal MR arthrogram T1WI FS in a patient status post medial meniscal tear resection shows contrast entering a linear defect ➔ in the meniscus, confirming a retear.*

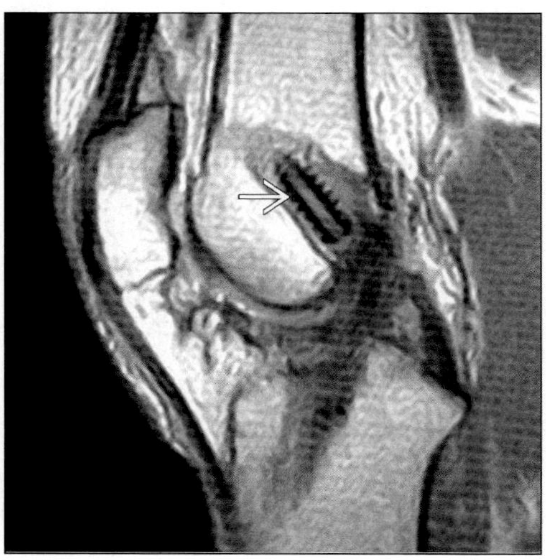

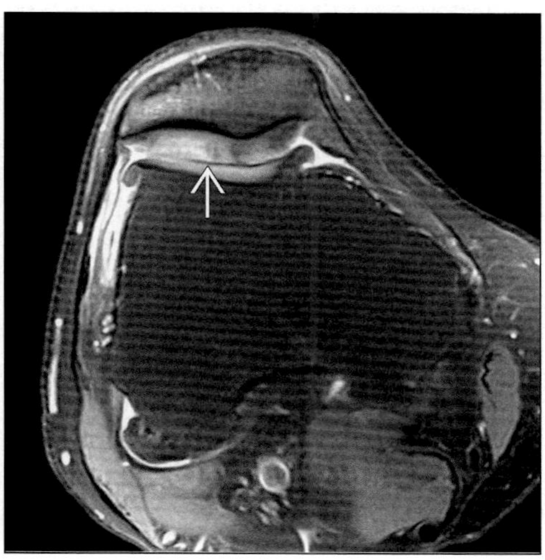

(Left) *Sagittal PDWI MR obtained on a low-field magnet (0.2T) after anterior cruciate ligament reconstruction shows the bioabsorbable interference screw ➔ in the femoral graft tunnel. While low-field strength provides less anatomic detail and lower signal-to-noise ratio than 1.5T systems, they also cause less susceptibility artifact.* ***(Right)*** *Axial PDWI FS MR performed on a 3T magnet shows the anatomic detail possible with high-field imaging. Note the fibrillation of the patellar cartilage ➔.*

DISTAL FEMORAL FRACTURE

Key Facts

Imaging
- Supracondylar and intercondylar fracture usually well seen on AP and lateral radiographs as linear lucency ± comminution
- Condylar fracture may be radiographically subtle, particularly if in coronal plane
- Radiographs
- CT useful for grading prior to surgery

Top Differential Diagnoses
- Floating knee
- Knee dislocation

Pathology
- Supracondylar
  - Usually comminuted
- Intercondylar
  - "T" or "Y" configuration
  - Resulting incongruity of knee/patellofemoral joint
- Condylar
  - Sagittal or coronal fracture line in a single condyle
  - Often require CT to identify and characterize

Clinical Issues
- Bimodal distribution
  - Young adults: High-energy trauma (e.g., MVA)
  - Elderly: Low-energy falls
  - 1% incidence of supracondylar fracture after knee arthroplasty
- Represent 4-7% of all femoral fracture
- Almost always treated surgically

Diagnostic Checklist
- Intercondylar extension of supracondylar fracture
- Underlying pathologic lesion
- Report and measure articular surface stepoff

*(**Left**) Graphic illustration of distal femoral fracture types in frontal projection shows a T-type supracondylar fracture ➔ with intercondylar extension ➔, a sagittal fracture of the medial condyle ➔, and a coronal fracture of the lateral condyle ➔. (**Right**) Coronal bone CT shows a T-shaped supracondylar fracture ➔ with a vertical component ➔ extending into the intercondylar notch. Note the disruption of the femoral articular surface ➔.*

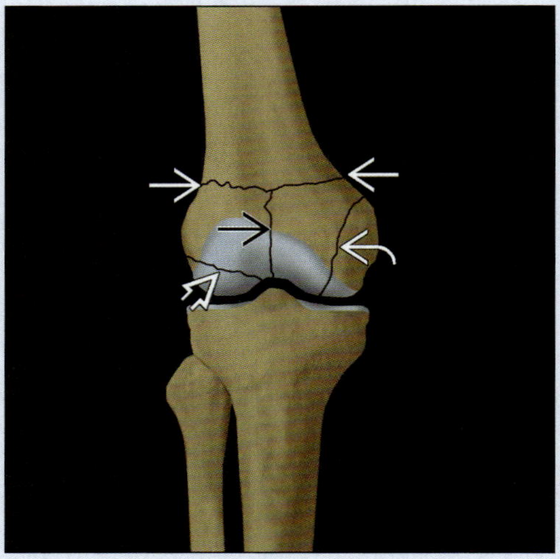

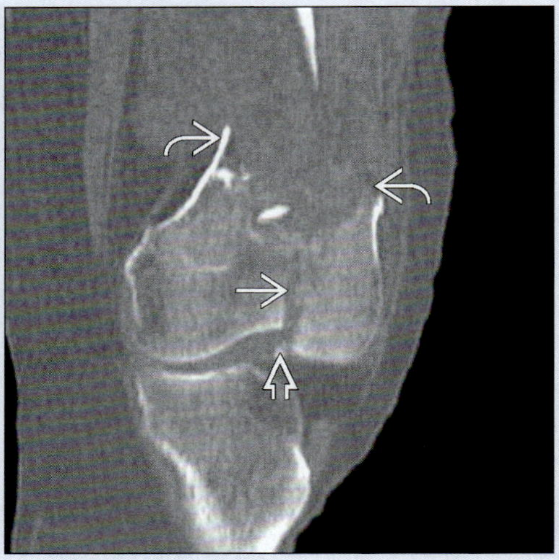

*(**Left**) Graphic illustration of distal femoral fracture types in lateral projection shows a supracondylar fracture ➔ and a coronal fracture of the lateral condyle ➔. Note that sagittally oriented fractures cannot be seen in this projection. (**Right**) Sagittal bone CT in the same patient as the previous CT shows the transverse component ➔ of the fracture offset. Note that the vertical portion of the fracture line ➔ runs partially coronally, and cannot be seen in the coronal plane.*

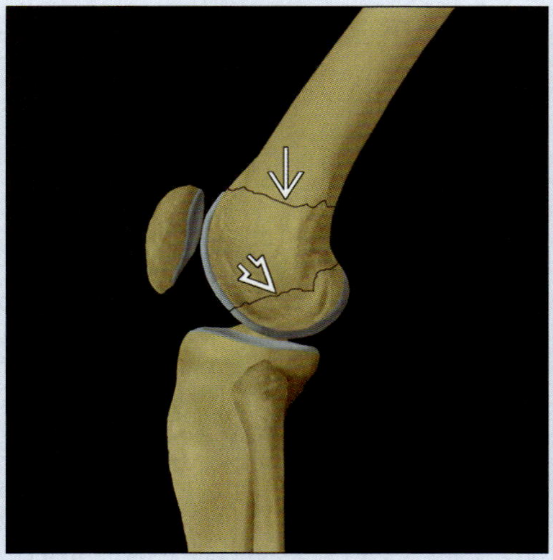

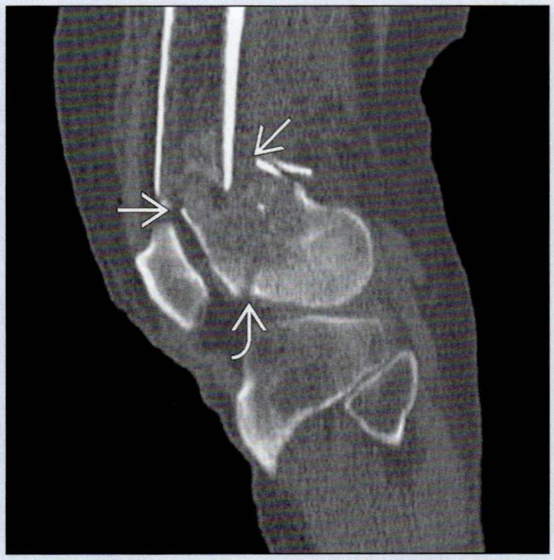

DISTAL FEMORAL FRACTURE

TERMINOLOGY

Synonyms

- Supracondylar femoral fracture

Definitions

- Fracture of femoral condyles &/or distal femoral metaphysis

IMAGING

General Features

- Best diagnostic clue
  - Linear fracture line extending into or through distal femur on radiographs
- Location
  - Distal femoral shaft, metaphysis, physis, &/or condyles
    - Fracture characterized by involvement of these areas
- Size
  - May range from nondisplaced stress fracture to large complex multipart fracture involving shaft, metaphysis, &/or condyles
- Morphology
  - May range from small Salter-Harris fractures to complex comminuted fractures with intraarticular extension

Radiographic Findings

- Supracondylar and intercondylar fracture usually well seen on AP and lateral radiographs as linear lucency ± comminution
  - Often angulated and displaced
- Condylar fracture may be radiographically subtle, particularly if in coronal plane
  - Oblique views may be required to see some condylar fractures
- Offset of femoral articular surface (intercondylar and condylar fracture)
- Hemarthrosis
  - Always with acute intercondylar or condylar fracture
  - Sometimes with acute supracondylar fracture

CT Findings

- Better for characterizing fracture by AO classification
- Condylar fractures well seen as linear sagittal or coronal fracture line in a single condyle
- In skeletally immature patient, may detect radiographically occult Salter-Harris injury

MR Findings

- Not usually indicated except to assess associated soft tissue injury
  - Ligament tear (cruciates)
  - Meniscal tear
  - Neurovascular injury (especially popliteal artery)
- In skeletally immature patients, may be best method to detect presence and extent of Salter-Harris injury
- Best method to detect nondisplaced stress injury

Imaging Recommendations

- Best imaging tool
  - Radiographs generally sufficient
  - MR for suspected Salter-Harris or stress injury
- Protocol advice
  - Radiographs: Oblique views may help with nondisplaced condylar fractures
  - CT: Useful for grading prior to surgery
    - Must have coronal and sagittal reformatted images to assess displacement
  - MR: Coronal and sagittal T2WI FS key to identifying growth plate involvement in skeletally immature patient

DIFFERENTIAL DIAGNOSIS

Floating Knee

- Associated proximal tibial fracture, isolating knee joint

Knee Dislocation

- Severely displaced supracondylar femoral fracture may mimic dislocation, especially on lateral radiographs

PATHOLOGY

Staging, Grading, & Classification

- Supracondylar
  - Transverse or slightly oblique orientation
  - Usually comminuted
  - May be accompanied by vertical intercondylar fracture line
  - Often extend into knee joint
  - Popliteal artery injury may occur
- Intercondylar
  - "T" or "Y" configuration
  - Sometimes simple vertical fracture line
  - Results in incongruity of knee or patellofemoral joint
  - If condyles widely separated, patella may be trapped between them
- Condylar
  - Sagittal or coronal fracture line in a single condyle
  - Often require CT to identify and characterize
- AO classification
  - **Extraarticular fracture**
    - A1: Simple
    - A2: Metaphyseal wedge
    - A3: Complex metaphyseal
  - **Partial intraarticular fracture**
    - B1: Sagittal fracture lateral condyle
    - B2: Sagittal fracture medial condyle
    - B3: Coronal fracture
  - **Complete intraarticular fracture**
    - C1: Simple articular and metaphyseal fragments
    - C2: Simple articular fracture with comminuted metaphyseal component
    - C3: Comminuted articular and metaphyseal components

CLINICAL ISSUES

Presentation

- Most common signs/symptoms
  - Knee pain and swelling
  - Inability to bear weight
- Other signs/symptoms

- Decreased posterior tibial artery pulse if popliteal artery injury
  - Present in up to 33% of supracondylar femoral fractures in 1 study
- Peroneal neuropathy
  - Unusual
  - May be related to surgery or casting

Demographics

- Age
  - Bimodal distribution for severe fractures
    - Young adults: High-energy trauma (e.g., MVA)
    - Elderly: Low-energy falls
  - Stress fractures
    - Children and adolescents
      - Typically Salter-Harris injuries
    - High-level athletes
- Gender
  - Young patients: M > F
  - Elderly: F > M
- Ethnicity
  - No ethnic predisposition
- Epidemiology
  - Represent 4-7% of all femoral fracture
  - 1% incidence of supracondylar fracture after knee arthroplasty

Natural History & Prognosis

- Stress fractures heal with conservative therapy
  - Rest
    - Modify or cease offending activity if identifiable
  - Ice
  - NSAIDs
- Post-surgical outcomes for severe fractures
  - 9% nonunion
  - 4% fixation failure
  - 3% infection
  - 13% revision surgery

Treatment

- Nondisplaced stress fracture or Salter-Harris fracture may be treated with non-weight-bearing
- Otherwise, essentially always treated surgically
  - Extraarticular fractures (A1, A2, A3) or fragmented condyles (C1, C2) → dynamic condylar screw or condylar blade plate
  - Condylar fractures (B1, B2, B3) → T buttress plate
  - Very comminuted fractures (C3) → may be difficult to secure all fragments
    - Often involves combination of external and internal fixation
- Fracture above knee arthroplasty → revision arthroplasty

DIAGNOSTIC CHECKLIST

Consider

- Intercondylar extension of supracondylar fracture
- Underlying pathologic lesion
- Neurovascular assessment

Reporting Tips

- Report and measure articular surface stepoff

SELECTED REFERENCES

1. Chettiar K et al: Supracondylar periprosthetic femoral fractures following total knee arthroplasty: treatment with a retrograde intramedullary nail. Int Orthop. 33(4):981-5, 2009
2. Kolb K et al: The condylar plate for treatment of distal femoral fractures: a long-term follow-up study. Injury. 40(4):440-8, 2009
3. Smith EL et al: Supracondylar femur fracture after knee manipulation: a report of 3 cases. Orthopedics. 32(1):18, 2009
4. Herrera DA et al: Treatment of acute distal femur fractures above a total knee arthroplasty: systematic review of 415 cases (1981-2006). Acta Orthop. 79(1):22-7, 2008
5. Hutchinson PH et al: Complete and incomplete femoral stress fractures in the adolescent athlete. Orthopedics. 31(6):604, 2008
6. Ross K et al: Bilateral femoral supracondylar stress fractures in a cross country runner. Orthopedics. 31(8):803, 2008
7. Wu CC: Femoral supracondylar malunions with varus medial condyle and shortening. Clin Orthop Relat Res. 456:226-32, 2007
8. Gocke TV: Case of the month. Salter-Harris type I fracture to the supracondylar femur physis. JAAPA. 19(10):72, 2006
9. Su ET et al: A proposed classification of supracondylar femur fractures above total knee arthroplasties. J Arthroplasty. 21(3):405-8, 2006
10. Abendschein W: Periprosthetic femur fractures--a growing epidemic. Am J Orthop. 32(9 Suppl):34-6, 2003
11. Smith NC et al: Supracondylar fractures of the femur in children. J Pediatr Orthop. 21(5):600-3, 2001
12. Stover M: Distal femoral fractures: current treatment, results and problems. Injury. 32 Suppl 3:SC3-13, 2001
13. Seligson D: Treatment of supracondylar fractures of the femur. J Trauma. 49(2):360, 2000
14. Muralikuttan KP et al: Supracondylar stress fracture of the femur. Injury. 30(1):66-7, 1999
15. Schatzker J: Fractures of the distal femur revisited. Clin Orthop Relat Res. (347):43-56, 1998
16. Albert MJ: Supracondylar Fractures of the Femur. J Am Acad Orthop Surg. 5(3):163-171, 1997
17. Marsh JL et al: Supracondylar fractures of the femur treated by external fixation. J Orthop Trauma. 11(6):405-10; discussion 411, 1997
18. Karpman RR et al: Supracondylar femoral fractures in the frail elderly. Fractures in need of treatment. Clin Orthop Relat Res. (316):21-4, 1995
19. Newman JH: Supracondylar fractures of the femur. Injury. 21(5):280-2, 1990

DISTAL FEMORAL FRACTURE

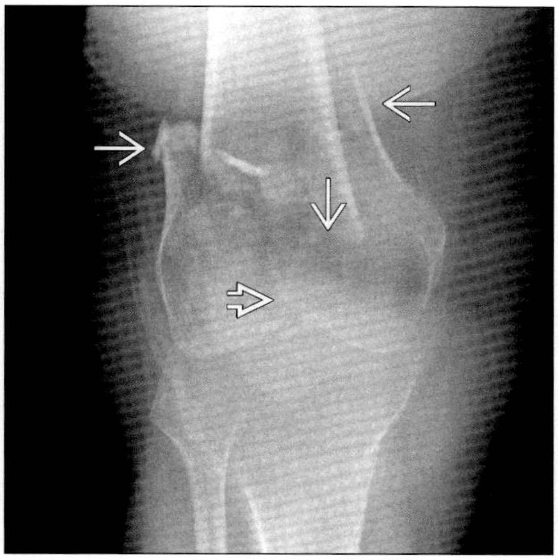

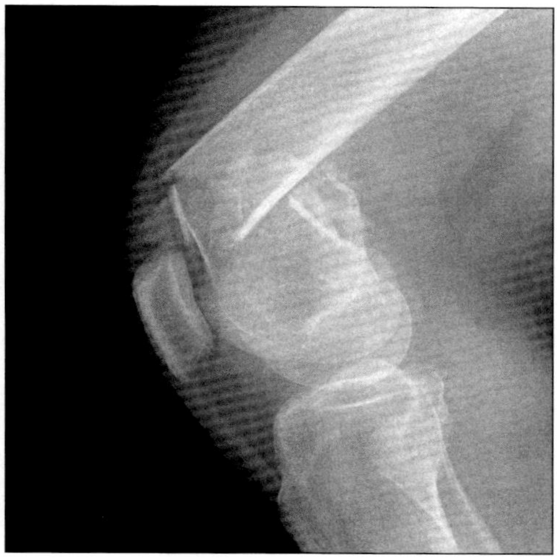

(Left) *AP radiograph of the knee in a 35-year-old man after a motorcycle accident demonstrates marked comminution of the distal femur. There is a subtle vertical component to the fracture complex that enters the articular surface.* ***(Right)*** *Lateral radiograph in the same patient shows the extent of comminution of the distal femur, as well as significant angular deformity. Note that the vertical portion of the fracture cannot be seen in the lateral projection.*

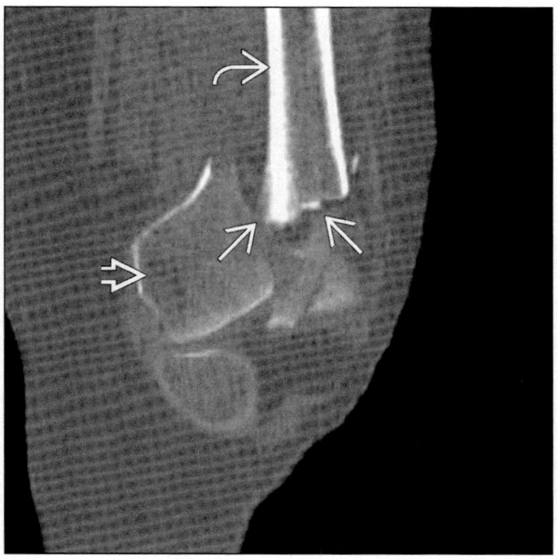

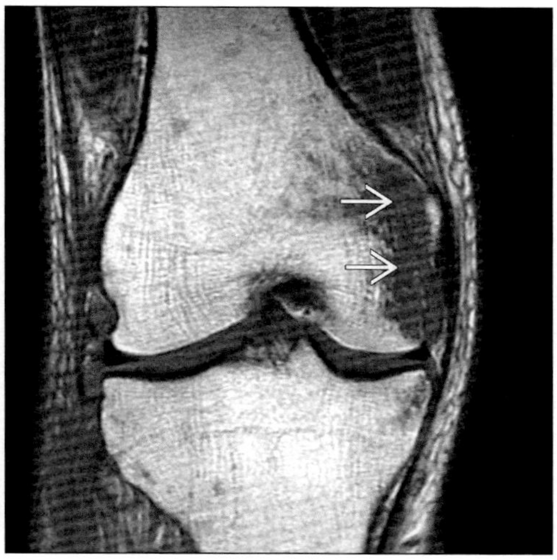

(Left) *Coronal bone CT in the same patient shows the comminution at the supracondylar portion of the fracture. There is lateral displacement of the lateral femoral condyle relative to the distal femora shaft.* ***(Right)*** *Coronal T1WI MR shows a vertical nondisplaced fracture of the medial femoral condyle. There is no definite involvement of the articular surface.*

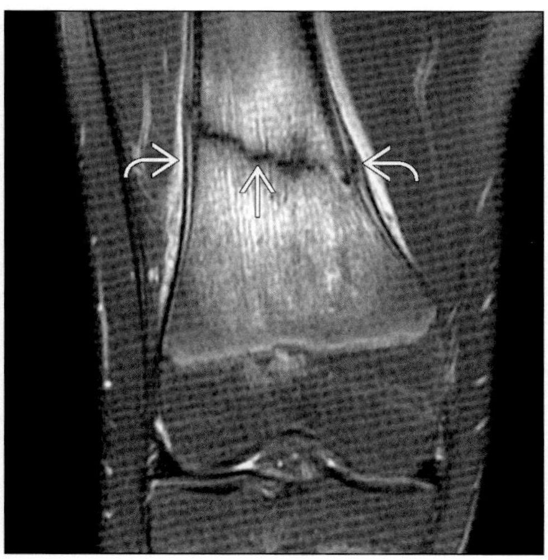

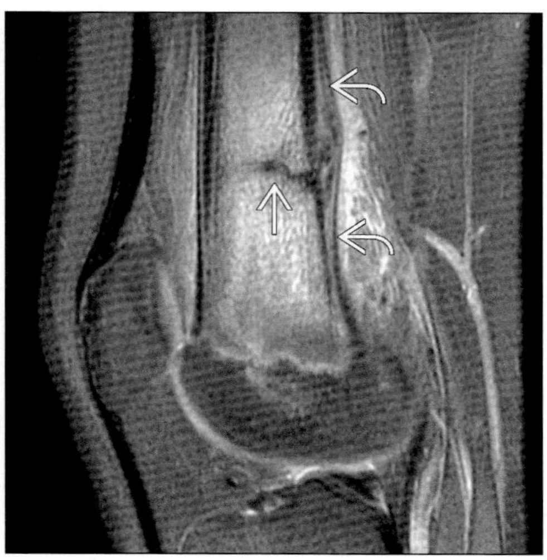

(Left) *Coronal T2WI FS MR in a 13-year-old runner with chronic thigh pain demonstrates a transverse supracondylar stress fracture surrounded by marrow edema. Note the periosteal new bone formation, suggesting a subacute time course for this injury.* ***(Right)*** *Sagittal PDWI FS MR in the same patient shows the incomplete nature of the femoral shaft fracture, which does not extend to the anterior cortex. Thick periosteal new bone is seen posteriorly.*

DISTAL FEMORAL FRACTURE

*(**Left**) AP radiograph in an 84-year-old woman after a fall shows marked comminution of the distal femur. Osteopenia likely played a significant causative role in this fracture pattern. There is subtle stepoff of the lateral condylar articular surface . (**Right**) Lateral radiograph in the same patient shows the degree of comminution of the supracondylar aspect of the distal femur, as well as posterior displacement of the femoral condyles relative to the shaft.*

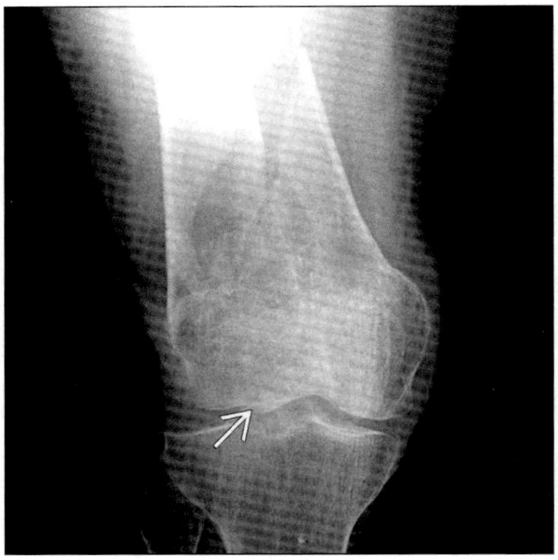

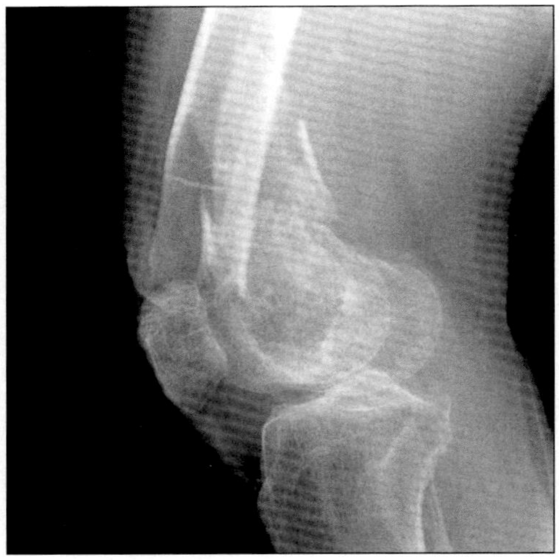

*(**Left**) Coronal bone CT in the same patient illustrates the comminution of the distal femur and the subtle vertical fracture extending into and disrupting the lateral femoral condylar articular surface . Hemorrhage in the marrow space should not be mistaken for tumor. (**Right**) Sagittal bone CT in the same patient shows the collection of blood products in the distal femoral marrow space due to the acute fracture. The sharp edge of the hemorrhage can be mistaken for a mass.*

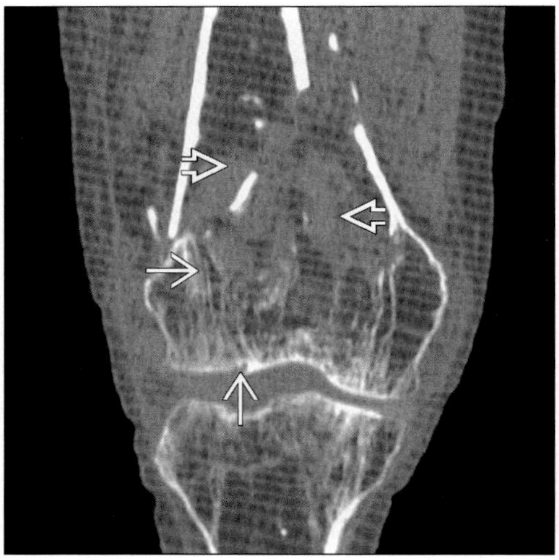

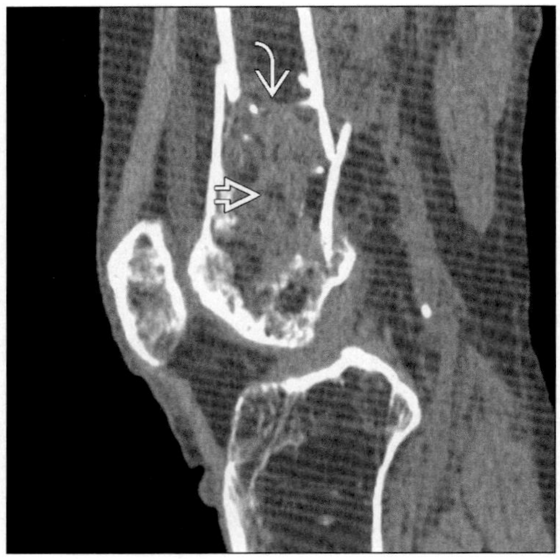

*(**Left**) Coronal T1WI MR in a 14 year old demonstrates low signal in the metaphyseal and epiphyseal marrow on either side of the medial femoral condylar growth plate. This is a Salter-Harris I stress fracture. Radiographs were normal. (**Right**) Coronal T2WI FS MR in a 13 year old shows elevated signal in the distal physis, consistent with a Salter-Harris injury. A small metaphyseal fragment makes this a type II injury. There is periosteal elevation above the fragment.*

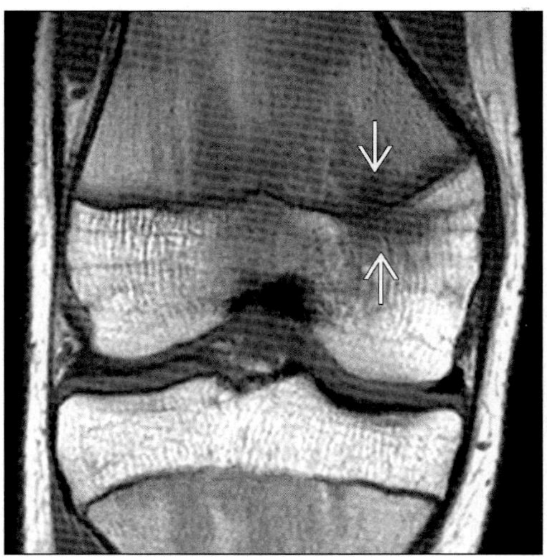

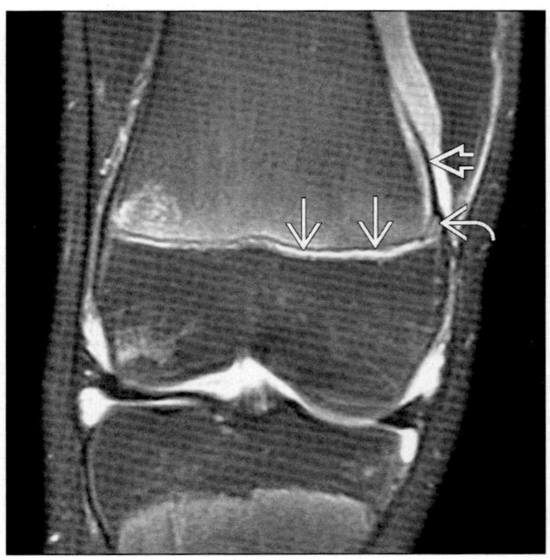

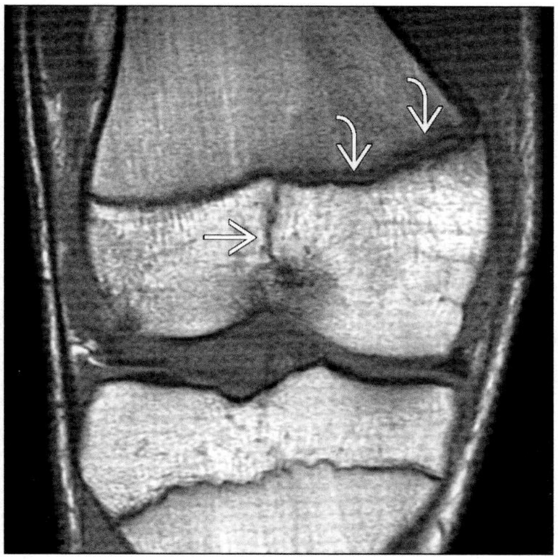

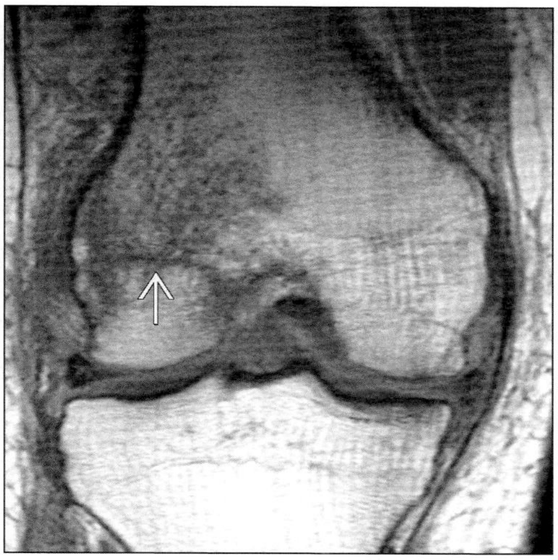

(Left) *Coronal T1WI MR shows a vertical fracture line extending through the intercondylar notch in the distal femoral metaphysis. There is some widening of the medial aspect of the growth plate, but the metaphysis is normal. This is a Salter-Harris III fracture.* ***(Right)*** *Coronal PDWI MR shows an isolated fracture in the lateral femoral condyle with surrounding marrow edema. These fractures can be very hard to see on conventional radiographs.*

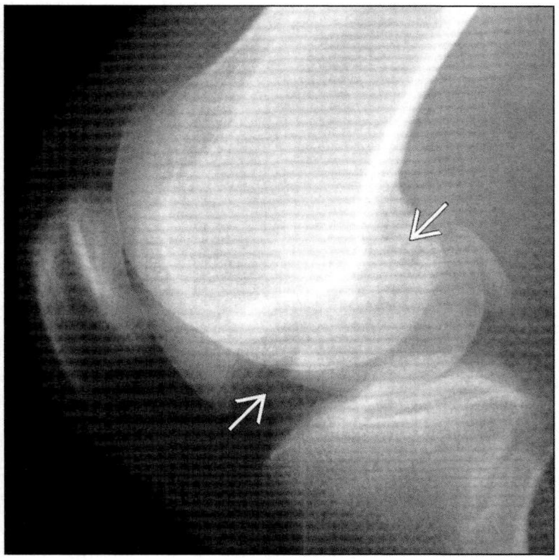

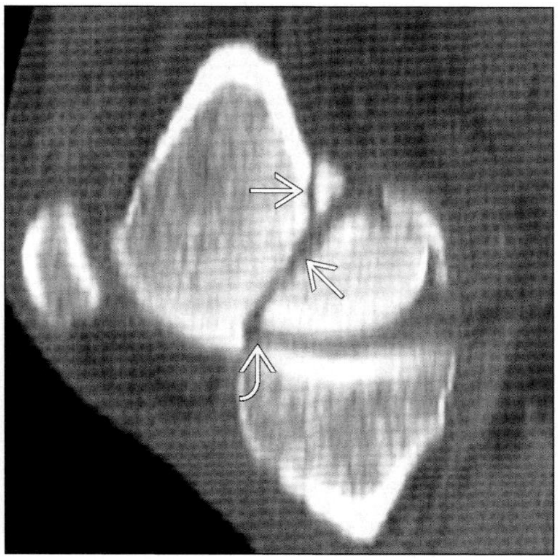

(Left) *Lateral radiograph shows a coronally oriented fracture of the medial femoral condyle, with moderate displacement of the articular surface. The fracture was poorly seen on AP radiograph, which is typical for this type of fracture.* ***(Right)*** *Sagittal bone CT in the same patient demonstrates the comminuted oblique coronal fracture of the medial femoral condyle. Note the offset of the anterior articular surface.*

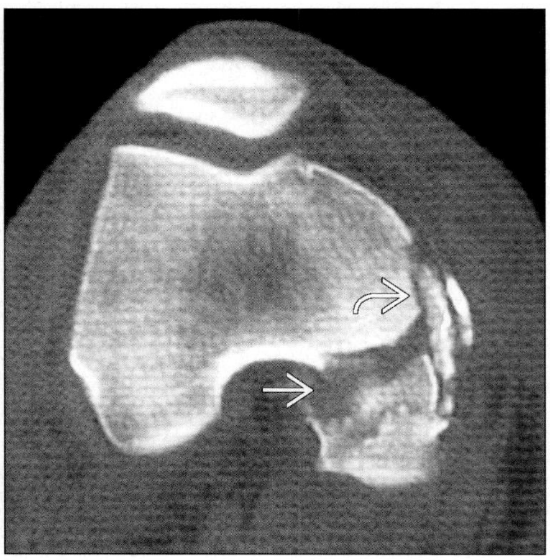

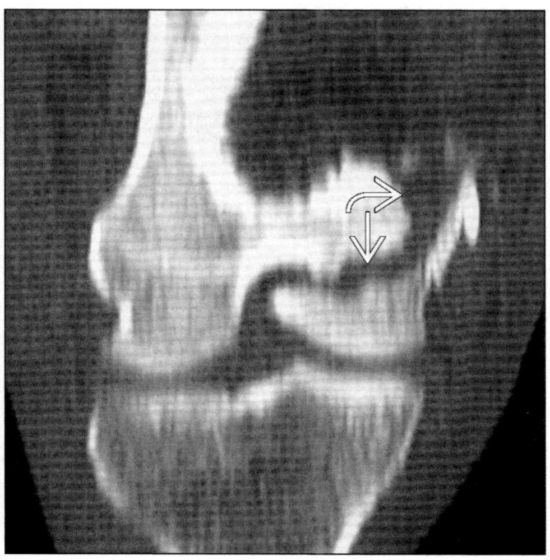

(Left) *Axial bone CT in the same patient shows the primary fracture line in the coronal plane, with moderate cortical comminution. A sagittally oriented component is also present along the medial edge of the medial condyle.* ***(Right)*** *Coronal bone CT in the same patient shows the coronal and sagittal components of the fracture. CT is very useful for surgical planning in distal femoral fractures.*

TIBIAL PLATEAU FRACTURE

Key Facts

Terminology

- Lateral tibial plateau fx (a.k.a. "bumper" or "fender" fracture)

Imaging

- Radiographs performed initially
  - Internal oblique may be added to routine series if plateau fracture is suspected
  - Cross-table lateral should be requested to look for lipohemarthrosis
- Morphology varies by patient age
  - Younger patients: Vertical split fracture
  - Older osteoporotic patients: Depressed fractures with variable extension
- CT with reformats detects radiographically occult fx
- CT with reformats or MR should be performed in all plateau fractures; radiographs underestimate both degree of depression and grade of fracture
- Soft tissue injuries often underestimated without MR

Pathology

- Trauma with axial load ± bending force
- Fall, with twisting or valgus force
- Motor vehicle accident
- Auto vs. pedestrian, "bumper" or "fender" fracture (etiology in 25%)
- 75-80% of tibial plateau fractures are lateral
- Schatzker classification; generally I-III are low energy, IV-VI are high energy

Clinical Issues

- Goal of treatment is anatomic reduction and early mobilization
- Depends on comorbidities; may opt for nonsurgical treatment if
  - Diastasis < 3-4 mm
  - Depression < 4-5 mm

(Left) Coronal bone CT shows a Schatzker I tibial plateau fracture. The split lateral plateau without impaction at the articular surface is a high-energy fracture when it occurs in young patients. Note the impacted cortical bone fragment → wedged within the split fracture, which must be removed in order to obtain an anatomic reduction. (Right) Sagittal PD FSE MR in a different patient shows a depressed lateral tibial plateau fracture. Other images (not shown) confirmed the lateral split with depression, typical of Schatzker II.

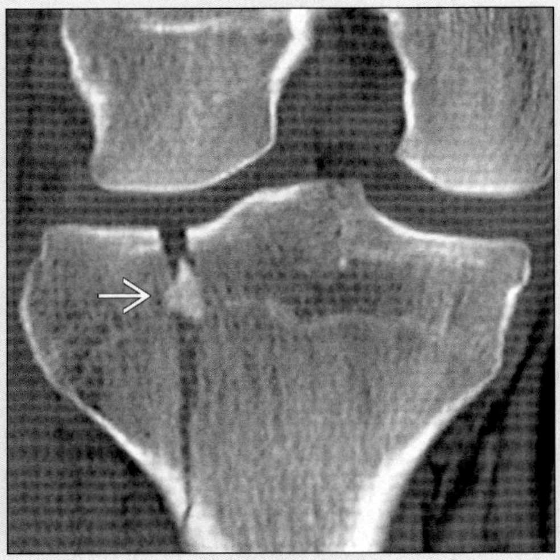

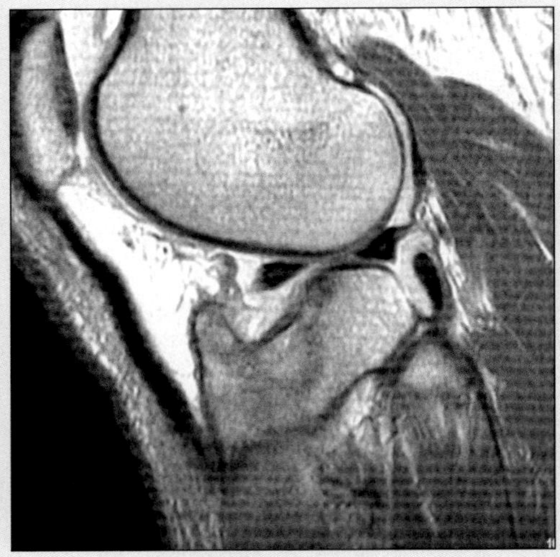

(Left) Coronal T2WI MR shows a depressed fragment in the lateral plateau →. It is not uncommon for valgus injury to have an associated medial collateral injury, seen here as complete disruption of the superficial fibers ⇨. (Right) Sagittal T2WI MR of the same patient demonstrates the split portion of the fracture →, rupture of the popliteus, and arcuate ligament disruption ⇨. This lateral split fracture with depression is a Schatzker II; associated soft tissue injury requires MR for diagnosis.

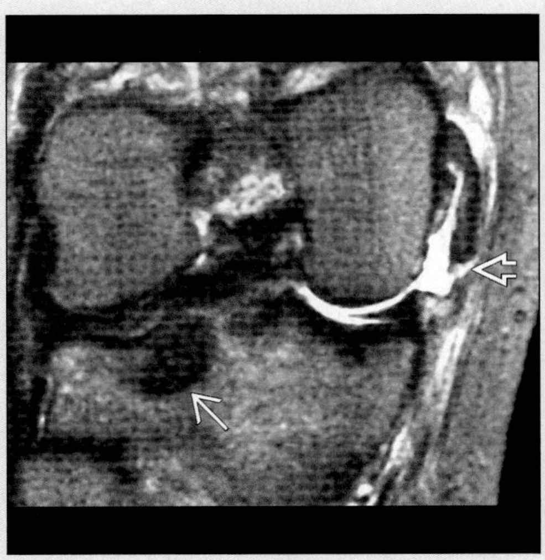

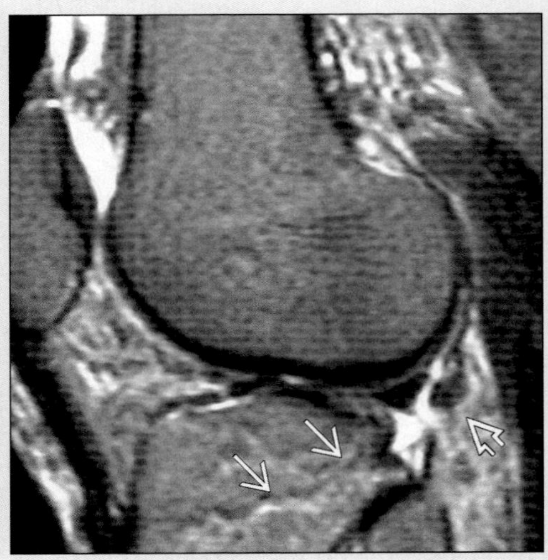

TIBIAL PLATEAU FRACTURE

TERMINOLOGY

Synonyms

- Lateral tibial plateau fx (a.k.a. "bumper" or "fender" fx)

IMAGING

General Features

- Best diagnostic clue
  - Fat-fluid level in suprapatellar bursa
  - Oblique or vertical linear fx line on radiograph
  - Sclerotic horizontal line below level of cortex of either lateral or medial tibial plateau
- Morphology
  - Younger patients: Vertical split fracture
  - Older osteoporotic patients: Depressed fractures with variable extension

Radiographic Findings

- Most plateau fractures can be diagnosed by conventional radiography, especially if depressed fragment present
- Lipohemarthrosis seen on cross-table lateral or axial view of patella
  - Fat-fluid line
  - Lipohemarthrosis will not be seen on conventional lateral radiograph
- Oblique radiograph may enhance detection
- Depressed fragment may be seen as horizontal sclerotic line below level of cortex, without lucent fx line

CT Findings

- Bone CT
  - Assists in diagnosis of radiographically occult fx
  - Confirms anatomic relationship of fx fragments in complex cases
    - Complete description of fracture line extension required: Inferior, vertical, horizontal, medial, and lateral propagation
    - Accurate measurement of size and extent of plateau fragment depression
  - Surgical planning for either elevation of depressed fragments or for Schatzker type IV-VI fx

MR Findings

- T1WI
  - Excellent for evaluation of fracture line propagation
    - Decreased signal intensity fracture line
    - Decreased signal intensity of surrounding edema
  - Low signal intensity cortical bone depressed into central cancellous bone
  - High-signal fat in lipohemarthrosis
- PD/intermediate FS
  - Evaluate for associated internal derangement (menisci, ligament, cartilage damage)
- Fluid-sensitive sequences (T2WI ± FS, STIR)
  - Decreased signal intensity fracture line with surrounding high signal edema
  - Fracture line itself may be obliterated by edema
  - Contralateral collateral ligament tear
  - Effusion with fluid-fluid level

Imaging Recommendations

- Best imaging tool
  - Radiographs performed initially
    - Internal oblique may be added to routine series if plateau fracture is suspected
    - Cross-table lateral → lipohemarthrosis
  - CT with reformats or MR should be performed in all plateau fractures; radiographs underestimate both degree of depression and grade of fracture
    - Soft tissue injuries underestimated without MR
- Protocol advice
  - Multidetector CT
    - High mAs (milliamperage-second)
    - Small axial slice thickness (≤ 1 mm) with coronal and sagittal reformats

PATHOLOGY

General Features

- Etiology
  - Trauma with axial load ± bending force
    - Fall + twisting or valgus force
    - Motor vehicle accident
    - Auto vs. pedestrian, "bumper" or "fender" fracture (etiology in 25%)
- Associated abnormalities
  - Lateral plateau fx: Valgus force → disruption of anterior cruciate and medial collateral ligaments
  - Medial plateau fracture: High energy, so may also result in disruption of
    - Lateral plateau
    - Posterior cruciate ligament
    - Posterolateral corner structures
    - Lateral collateral ligament
    - Popliteal artery (rare)

Gross Pathologic & Surgical Features

- 75-80% of tibial plateau fractures are lateral
  - Medial plateau is stronger than lateral
  - Varus stresses (leading to medial plateau fx) less common than valgus stresses (leading to lateral plateau fx) due to carrying angle of knee and protection by other extremity
- Lateral plateau fractures have different morphology in young and elderly patients
  - Elderly: Usually low energy → depressed fragments
  - Young: High energy → split or wedge fx pattern

CLINICAL ISSUES

Presentation

- Most common signs/symptoms
  - Knee effusion, pain, inability to bear weight following trauma

Demographics

- Gender
  - Older patients: M < F; related to osteoporosis
  - Younger patients: M > F; related to high-energy sports injuries

Natural History & Prognosis

- If uncorrected, depressed or nonreduced fragments and ligamentous instability result in osteoarthritis

Treatment

- Goal: Anatomic reduction and early mobilization

TIBIAL PLATEAU FRACTURE

Schatzker I: Split fracture with no depression (usually younger patients)

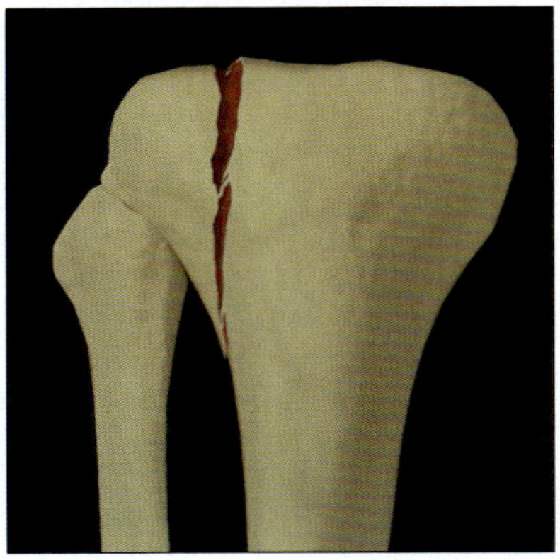

Schatzker II: Lateral split/wedge fracture with depression of weight-bearing portion (usually older patients with osteoporosis)

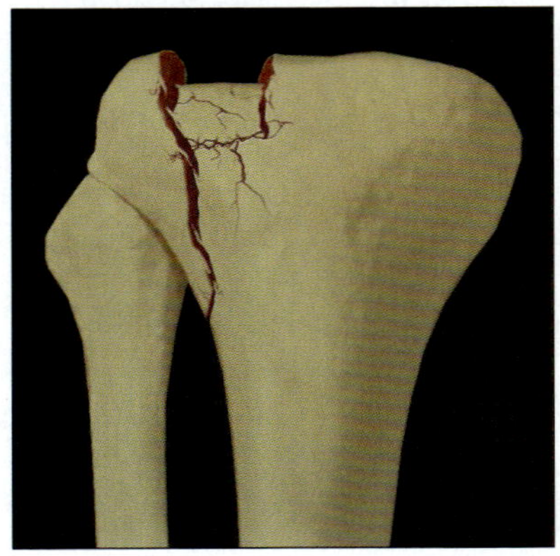

Schatzker III: Focal depression of articular surface, no associated split (elderly, osteoporotic patients)

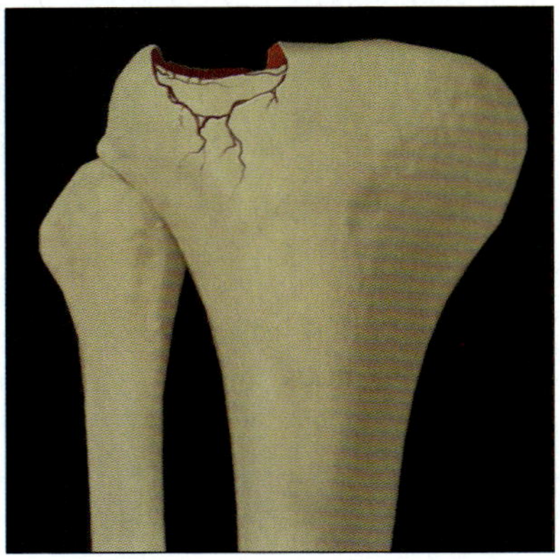

Schatzker IV: Medial plateau split, with or without depression; may involve tibial spines; associated soft tissue injuries & poor prognosis

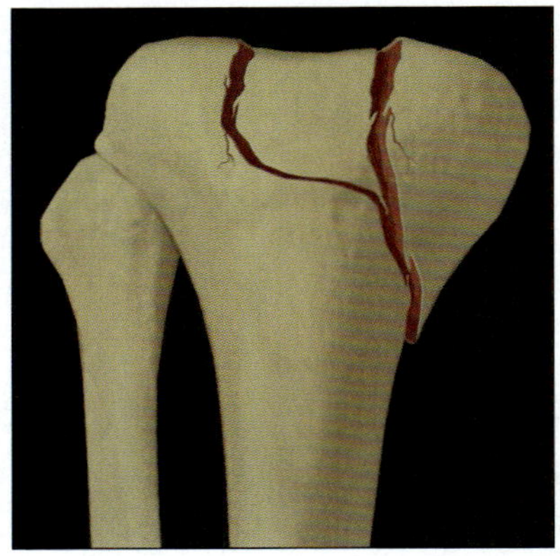

Schatzker V: Split fracture of both medial and lateral plateau (bicondylar)

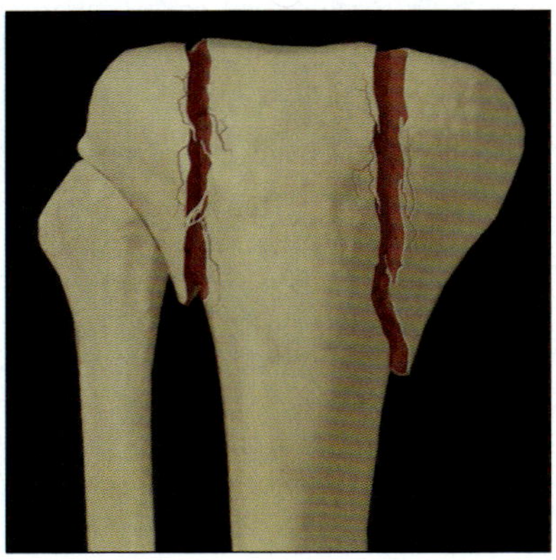

Schatzker VI: Bicondylar split fracture with dissociation of metaphysis from diaphysis

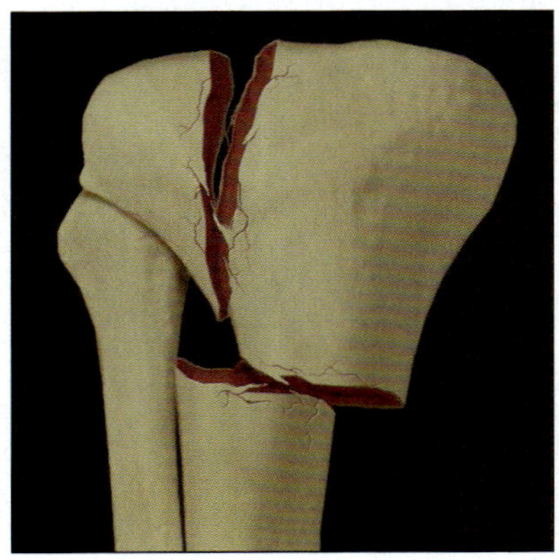

- Depends on comorbidities; may opt for nonsurgical Rx if diastasis < 3-4 mm and depression < 4-5 mm

DIAGNOSTIC CHECKLIST

Image Interpretation Pearls

- Watch for lipohemarthrosis as hint of occult intraarticular fracture
- Watch for sclerotic line within cancellous subchondral bone, which indicates occult depressed fragment
- Evaluate and measure degree of depression with CT; injuries are underestimated with radiographs

SELECTED REFERENCES

1. Markhardt BK et al: Schatzker classification of tibial plateau fractures: use of CT and MR imaging improves assessment. Radiographics. 29(2):585-97, 2009
2. Musahl V et al: New trends and techniques in open reduction and internal fixation of fractures of the tibial plateau. J Bone Joint Surg Br. 91(4):426-33, 2009
3. W Mui L et al: Comparison of CT and MRI in patients with tibial plateau fracture: can CT findings predict ligament tear or meniscal injury? Skeletal Radiol. 2006
4. Dirschl DR et al: Injury severity assessment in tibial plateau fractures. Clin Orthop Relat Res. (423):85-92, 2004
5. Shepherd L et al: The prevalence of soft tissue injuries in nonoperative tibial plateau fractures as determined by magnetic resonance imaging. J Orthop Trauma. 16(9):628-31, 2002
6. Yacoubian SV et al: Impact of MRI on treatment plan and fracture classification of tibial plateau fractures. J Orthop Trauma. 16(9):632-7, 2002

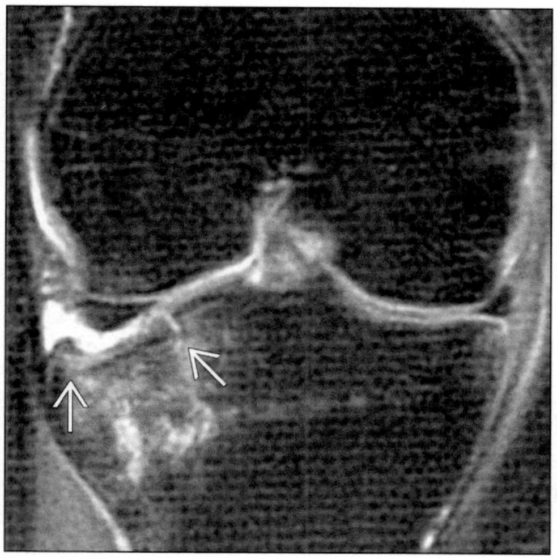
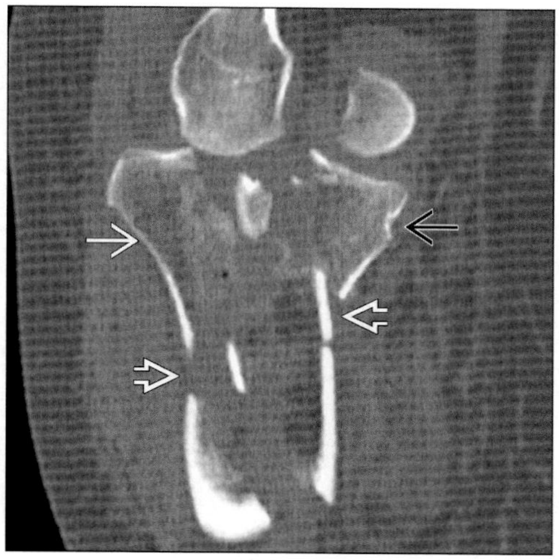

(Left) *Coronal T2WI FS MR shows an isolated lateral tibial impaction fracture. There was no associated vertical split. This lateral impaction fracture is classified as a Schatzker III injury.* ***(Right)*** *Coronal bone CT shows vertical split fractures of the medial and lateral tibial plateaus, impacted fragments, and a more distal transverse tibial metaphyseal fracture. This combination is a Schatzker VI injury. Note the displacement of the plateau relative to the femur; soft tissue injuries were severe.*

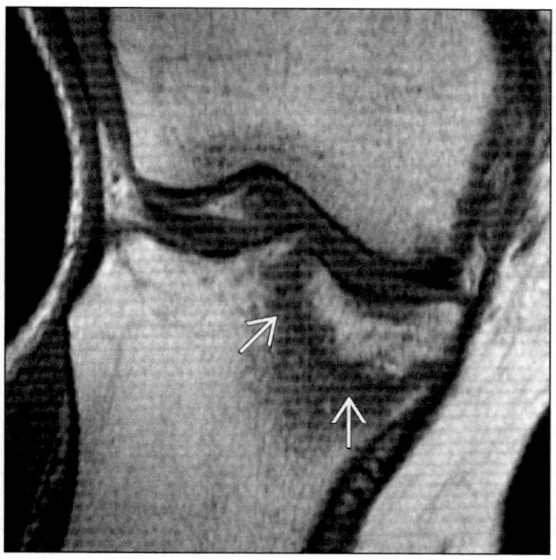
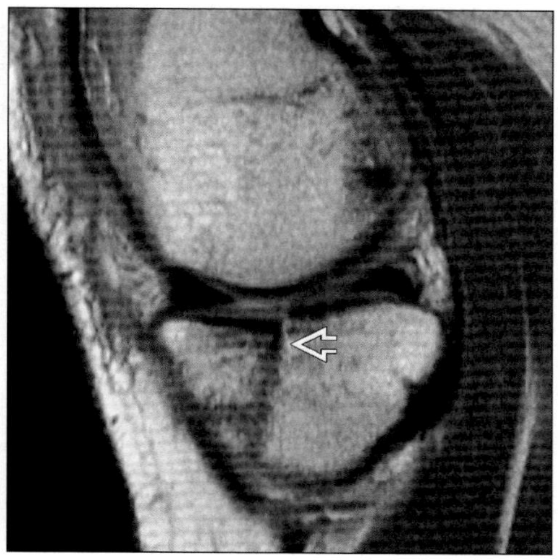

(Left) *Coronal T1WI MR is distorted by metal in external fixator, but the U-shaped fracture line is well seen. The articular surface of the medial plateau appears preserved. The medial location, without lateral involvement, makes this a Schatzker IV fracture.* ***(Right)*** *Sagittal PD FSE MR in the same patient shows coronal split extending to articular surface and slight articular depression of the anterior portion of the medial plateau. This medial split, with or without depression, is a Schatzker IV fracture.*

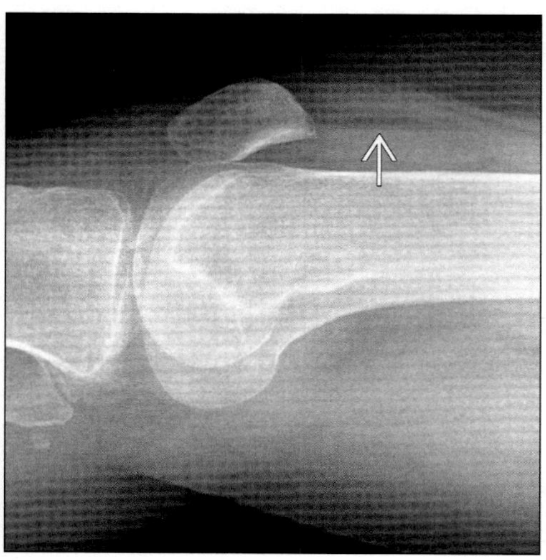
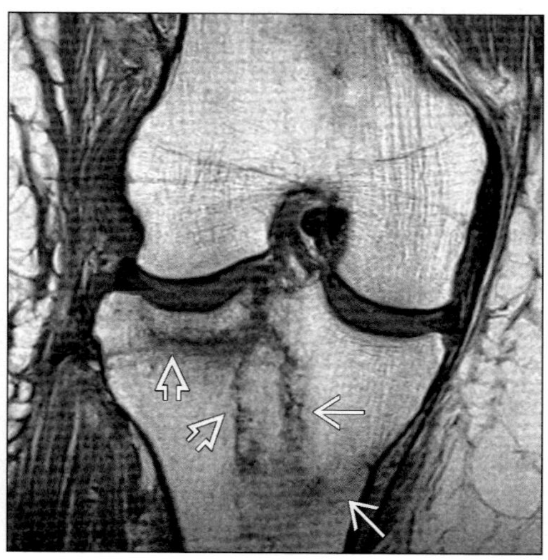

(Left) *Cross table lateral radiograph shows a clear fat-blood level within the suprapatellar bursa, indicating an intraarticular fracture. The remainder of the radiographs showed only subtle articular cortex irregularity, without definite fracture.* ***(Right)*** *Coronal T1WI MR shows a medial tibial split fracture, as well as a lateral plateau split and depressed fracture. This combination of bicondylar fractures, without more distal transverse metaphyseal fracture, is classified as a Schatzker V fracture.*

TIBIOFEMORAL DISLOCATION

Key Facts

Terminology

- Knee dislocation

Imaging

- Rotary dislocation may be subtle
- Associated fractures
- Marrow contusion
- Ligamentous disruption
- Meniscal tear: Relatively common
- Posterior capsule disruption: Common, especially with posterior dislocation
- Evaluation of popliteal artery required if diminished pulses following reduction
  - CT or MR angiography may be used, but question accuracy for small intimal tears
  - Controversy about vascular imaging with intact pulses after dislocation reduction

Pathology

- Usually due to high-force trauma
  - Motor vehicle accident
  - Fall from great height
- Occasionally due to lower force trauma
- Open joint injury: 20-30%
- Popliteal artery injury in 30%
- Peroneal nerve injury in 15-40%
- Major ligamentous injury
  - ACL (100%), PCL (85%)
  - MCL (35%), FCL (71%)

Diagnostic Checklist

- Popliteal artery and vein injury may be present even with intact pulses
- Search for subtle avulsion fractures that herald ligament disruption

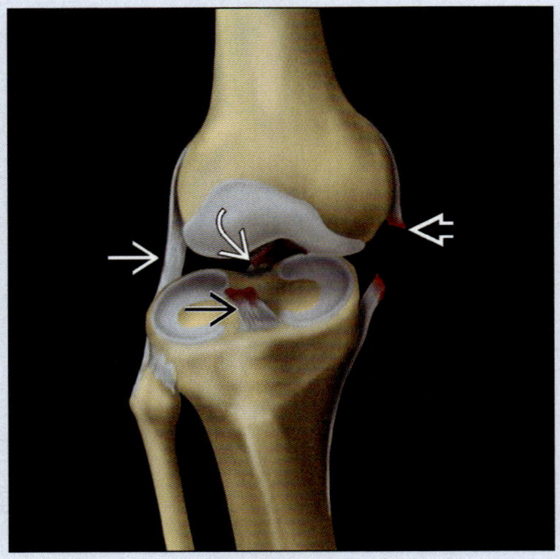
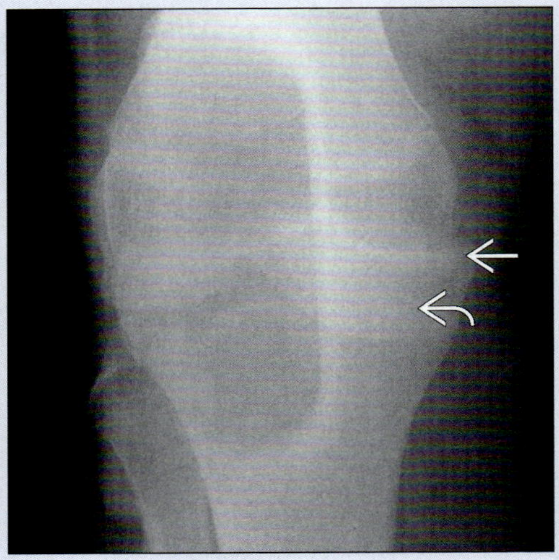

(Left) *Graphic of anterolateral knee dislocation shows disruption of the anterior ⇒ and posterior ⇒ cruciate ligaments, and the medial collateral ligament ⇒. The fibular collateral ligament ⇒ is intact.* ***(Right)*** *AP radiograph of the knee with the patient on a backboard shows overlap of the medial femoral condyle ⇒ and medial tibial plateau ⇒, indicating a knee dislocation. There is no significant medial or lateral tibial shift in this case.*

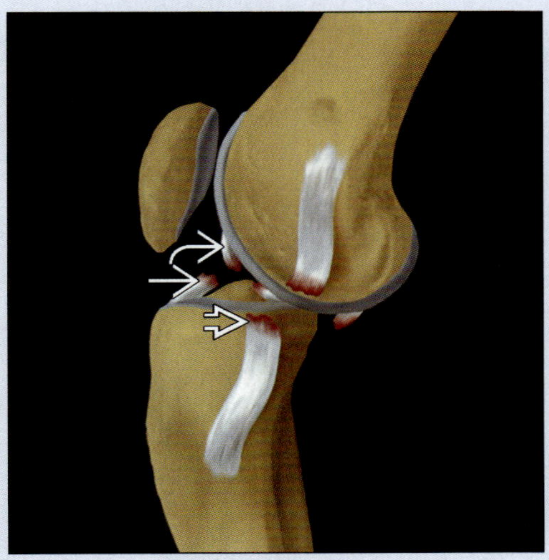
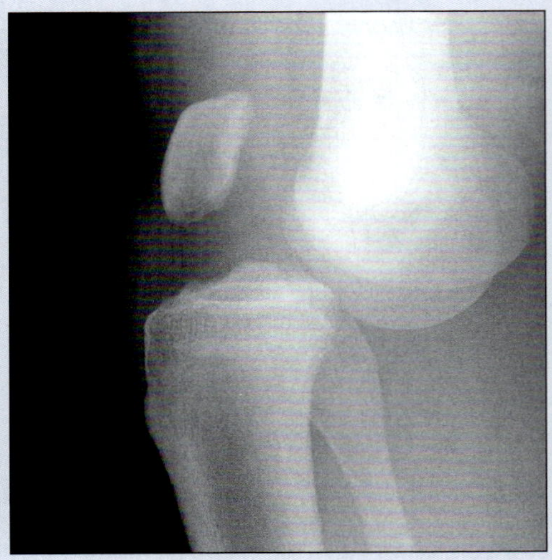

(Left) *Graphic of anterolateral knee dislocation seen from medial side shows disruption of the anterior ⇒ and posterior ⇒ cruciate ligaments and the medial collateral ligament ⇒.* ***(Right)*** *Lateral radiograph in the same patient as previous radiograph shows an anterior tibiofemoral dislocation. No fracture is visible. Uncomplicated knee dislocations are often reduced by EMTs in the field, and prereduction radiographs are therefore not available.*

TIBIOFEMORAL DISLOCATION

TERMINOLOGY

Abbreviations

- Anterior cruciate ligament (ACL)
- Posterior cruciate ligament (PCL)
- Medial collateral ligament (MCL)
- Fibular collateral ligament (FCL)

Synonyms

- Knee dislocation

Definitions

- Disruption of tibiofemoral joint

IMAGING

General Features

- Best diagnostic clue
  - Displacement of proximal tibial articular surface relative to femoral condyles
- Location
  - Knee joint

Radiographic Findings

- Usually obvious displacement of tibia relative to femur
- Rotary dislocation may be subtle
  - Lateral displacement of tibia < 1/4 width of femoral condyle
  - Medial femoral condyle projects medial to tibia on AP view
  - Medial joint space widened
  - Tibia rotated relative to femur on lateral view
- Associated fractures
  - Ligament avulsion
    - ACL: Usually anterior tibial spine
    - PCL: Posterior tibial plateau
    - MCL: Medial femoral condyle
    - FCL: Usually fibular styloid, may be lateral femoral condyle
  - Tibial plateau
  - Fibular head
  - Femoral condyle fracture uncommon
- Joint effusion less common due to rupture of joint capsule

CT Findings

- May be helpful to identify subtle fractures, particularly after relocation

MR Findings

- Usually knee reduced prior to MR imaging
- Ligamentous disruption
  - ACL tear: May be intrasubstance or avulsion (latter usually tibial)
  - PCL tear: Usually intrasubstance
  - MCL tear: Usually intrasubstance
  - Lateral collateral ligament complex and posterolateral corner
    - Variable involvement
    - Usually see either medial or lateral collateral ligament involvement, rarely both
  - Meniscal tear: Relatively common
    - Medial: 53%
    - Lateral: 70%
  - Fractures and marrow contusions
    - Locations depend on direction of dislocation and associated ligamentous injury
  - Posterior capsule disruption: Common, especially with posterior dislocation

Imaging Recommendations

- Best imaging tool
  - Radiographs
  - MR useful to characterize associated ligamentous injury
- Protocol advice
  - Evaluation of popliteal artery required with diminished pulses after reduction
    - Conventional angiography
    - CT or MR angiography may be used, but question accuracy for small intimal tears
  - Controversy about vascular imaging with intact pulses after dislocation reduction

DIFFERENTIAL DIAGNOSIS

Floating Knee

- Fractures of distal femur and proximal tibia, isolating knee joint

Multiligament Knee Injury Without Dislocation

- Probably part of a spectrum of injuries

PATHOLOGY

General Features

- Etiology
  - Usually due to high-force trauma
    - Motor vehicle accident
    - Fall from great height
  - Occasionally due to lower force trauma
    - Step in hole and twist leg
    - Blow to knee
    - Sports injury
- Associated abnormalities
  - Fracture
    - Uncommon
    - May see tibial avulsion of ACL attachment
    - Severe fracture-dislocations rare
  - Open joint injury: 20-30%
  - Major ligamentous injury
    - ACL disruption: 100%
    - PCL disruption: 85%
    - MCL disruption: 35%
    - FCL disruption: 71%
    - Popliteal tendon disruption: 47%
    - At least 3 major ligament disruptions likely required for dislocation
  - Vascular injury
    - Popliteal artery injury in 30%
    - Higher incidence if concomitant peroneal nerve injury
  - Peroneal nerve injury in 15-40%

Staging, Grading, & Classification

- Classified according to direction of displacement of tibia relative to femur
  - Anterior
    - Most common (31%)

- Most commonly associated with popliteal artery injury
- Due to hyperextension
  - Cadaver study: > 30° hyperextension required
- Posterior
  - Less common than anterior (25%)
  - May be associated with popliteal artery injury
  - Often associated with extensor mechanism injury
  - Due to crush injury/direct anterior blow to tibia
    - Dashboard injury
- Medial and lateral
  - Unusual
    - Lateral = 13%
    - Medial = 3%
  - Infrequently associated with arterial injury
- Rotary (or posterolateral)
  - Unusual (4%)
  - Medial femoral condyle forced through defect in medial knee joint capsule
  - Arterial injury unusual

CLINICAL ISSUES

Presentation

- Most common signs/symptoms
  - Deformity of knee
    - May reduce spontaneously
    - Often reduced in field by EMTs, etc.
  - Gross instability of knee
- Other signs/symptoms
  - Decreased or absent pulse in leg and foot due to popliteal artery injury
    - Intact pulses and capillary refill does not exclude significant vascular injury

Demographics

- Gender
  - M > F

Natural History & Prognosis

- Peroneal nerve injury: Half are permanent
- Delayed vascular injury
  - Thrombosis of popliteal artery or vein

Treatment

- Can usually be reduced by closed manipulation
  - Reduction may be limited by trapped tissues
- Rotary dislocation usually requires open reduction due to rent in medial capsule trapping medial femoral condyle
- Contemporaneous or delayed repair of disrupted ligaments

DIAGNOSTIC CHECKLIST

Consider

- Popliteal artery and vein injury even with intact pulses

Image Interpretation Pearls

- Search for subtle avulsion fractures that herald ligament disruption

SELECTED REFERENCES

1. Harb A et al: The MR dimple sign in irreducible posterolateral knee dislocations. Skeletal Radiol. 38(11):1111-4, 2009
2. Levy BA et al: Controversies in the treatment of knee dislocations and multiligament reconstruction. J Am Acad Orthop Surg. 17(4):197-206, 2009
3. Levy BA et al: Decision making in the multiligament-injured knee: an evidence-based systematic review. Arthroscopy. 25(4):430-8, 2009
4. Nicandri GT et al: Practical management of knee dislocations: a selective angiography protocol to detect limb-threatening vascular injuries. Clin J Sport Med. 19(2):125-9, 2009
5. Robertson A et al: Dislocation of the knee. J Bone Joint Surg Br. 88(6):706-11, 2006
6. Helgeson MD et al: Initial evaluation of the acute and chronic multiple ligament injured knee. J Knee Surg. 18(3):213-9, 2005
7. Giannoulias CS et al: Knee dislocations: management of the multiligament-injured knee. Am J Orthop. 33(11):553-9, 2004
8. Klineberg EO et al: The role of arteriography in assessing popliteal artery injury in knee dislocations. J Trauma. 56(4):786-90, 2004
9. Stannard JP et al: Vascular injuries in knee dislocations: the role of physical examination in determining the need for arteriography. J Bone Joint Surg Am. 86-A(5):910-5, 2004
10. Barnes CJ et al: Does the pulse examination in patients with traumatic knee dislocation predict a surgical arterial injury? A meta-analysis. J Trauma. 53(6):1109-14, 2002
11. Perron AD et al: Orthopedic pitfalls in the ED: vascular injury associated with knee dislocation. Am J Emerg Med. 19(7):583-8, 2001
12. Brautigan B et al: The epidemiology of knee dislocations. Clin Sports Med. 19(3):387-97, 2000
13. Hayes CW et al: Mechanism-based pattern approach to classification of complex injuries of the knee depicted at MR imaging. Radiographics. 20 Spec No:S121-34, 2000
14. Hegyes MS et al: Knee dislocation. Complications of nonoperative and operative management. Clin Sports Med. 19(3):519-43, 2000
15. Twaddle BC et al: MRI in acute knee dislocation. A prospective study of clinical, MRI, and surgical findings. J Bone Joint Surg Br. 78(4):573-9, 1996
16. Yu JS et al: Complete dislocation of the knee: spectrum of associated soft-tissue injuries depicted by MR imaging. AJR Am J Roentgenol. 164(1):135-9, 1995
17. Kremchek TE et al: Traumatic dislocation of the knee. Orthop Rev. 18(10):1051-7, 1989
18. Walls RM et al: Traumatic dislocation of the knee. J Emerg Med. 1(6):527-31, 1984

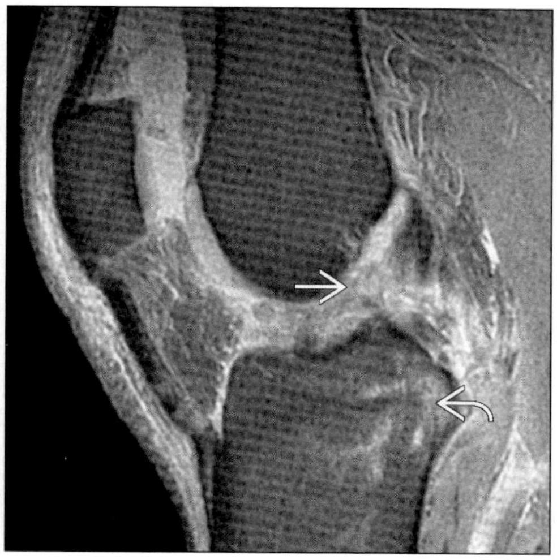

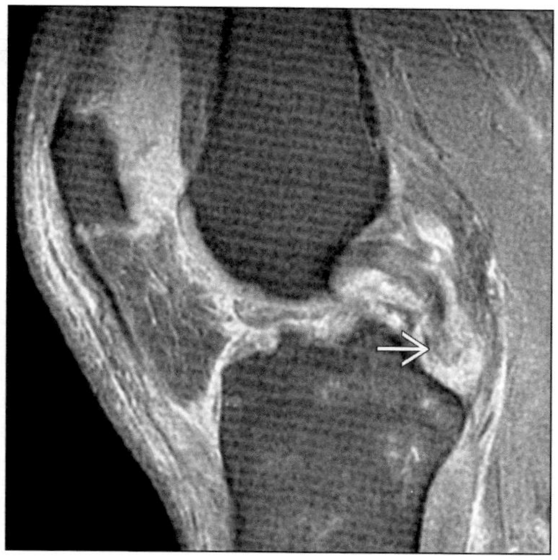

(Left) *Sagittal PDWI FS MR after anterolateral knee dislocation and closed reduction shows disruption of the anterior cruciate ligament (ACL) →. Mild marrow edema is seen in the posterior tibia →.* ***(Right)*** *Sagittal PDWI FS MR slightly more medial in the same patient shows disruption of the posterior cruciate ligament (PCL) →. Both cruciates are disrupted in the vast majority of knee dislocations.*

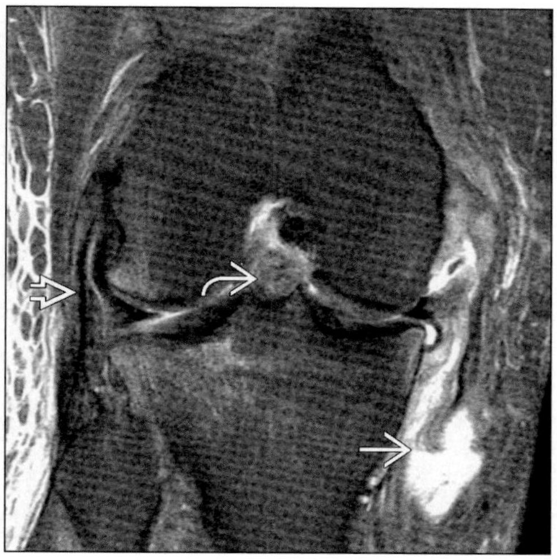

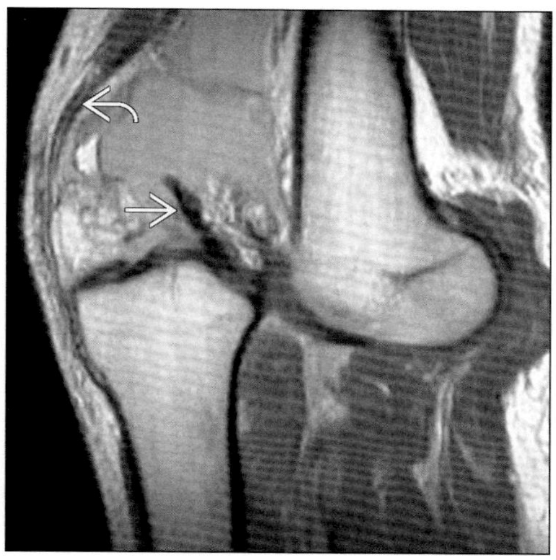

(Left) *Coronal T2WI FS MR in the same patient shows distal avulsion of the medial collateral ligament (MCL) → with fluid surrounding the avulsed ligament. The torn ACL → is seen as amorphous gray signal in the intercondylar notch. The fibular collateral ligament (FCL) is intact →.* ***(Right)*** *Sagittal PDWI MR shows complete anterior displacement of the tibia relative to the femur. The disrupted PCL → is well seen. The lateral aspect of the extensor mechanism → is intact.*

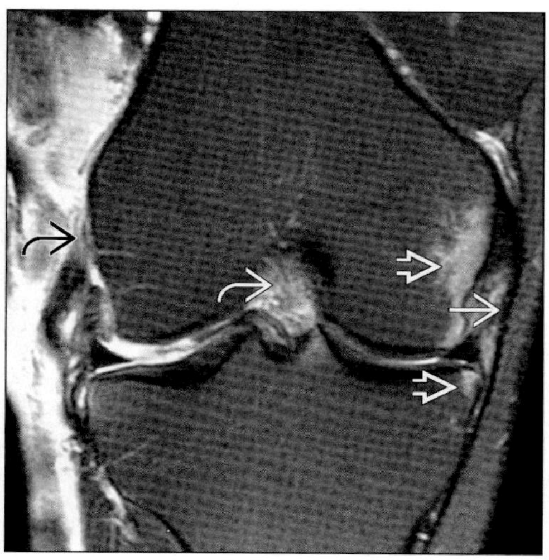

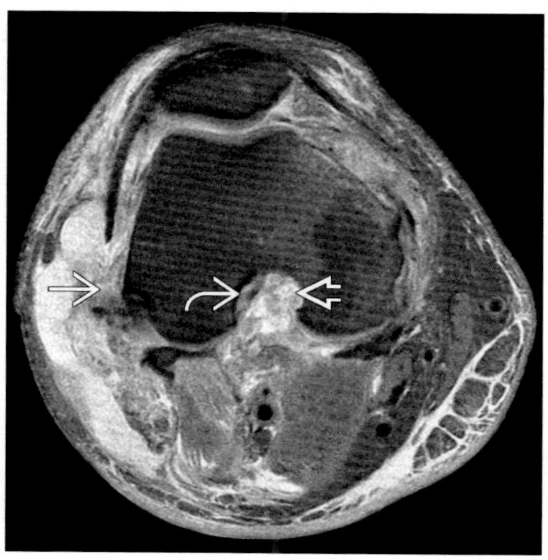

(Left) *Coronal T2WI FS MR after anterior knee dislocation shows disruption of the FCL → with hematoma formation in the surrounding soft tissues. The intercondylar notch is filled with amorphous gray signal → due to disruption of both the ACL and PCL. Impaction contusions are present in the medial femoral condyle and medial tibia →. The MCL is intact →.* ***(Right)*** *Axial PDWI FS MR in the same patient shows disruption of the FCL →, ACL →, and PCL →.*

TIBIOFEMORAL DISLOCATION

(Left) *AP radiograph of the knee demonstrates fractures of the lateral tibial plateau and proximal fibula. There is lateral shift of the tibia relative to the femur. Loss of congruency of the tibiofemoral joint indicates knee dislocation; the direction of dislocation (anterior vs. posterior) must be determined on the lateral radiograph.* ***(Right)*** *Lateral radiograph in the same patient confirms posterior dislocation of the knee. Tibial fracture is again seen.*

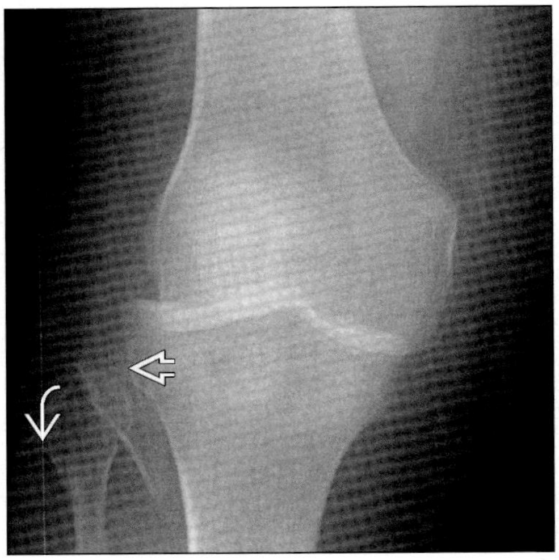

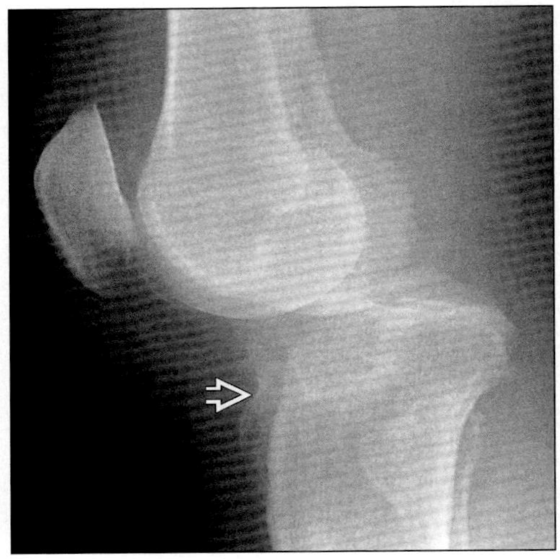

(Left) *AP radiograph shows lateral dislocation of the tibia and patella relative to the femur. No fracture is identified.* ***(Right)*** *Lateral radiograph in the same patient shows that the tibia is displaced posteriorly relative to the femur, with a slight rotational component. This injury falls in the spectrum between the unusual lateral dislocation and the rare rotary or posterolateral dislocation. Vascular injury is less common in this setting than with anterior dislocation.*

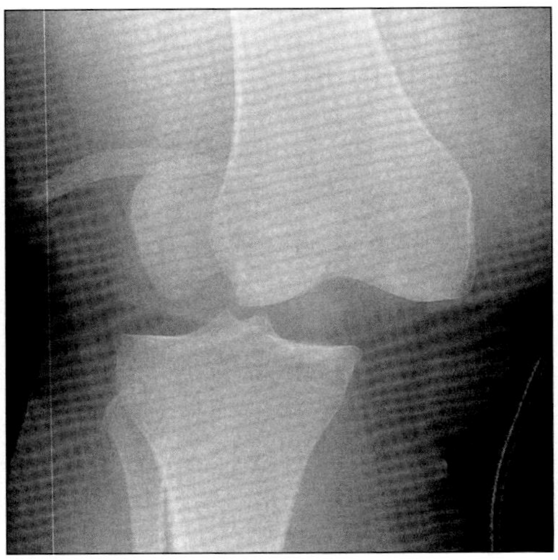

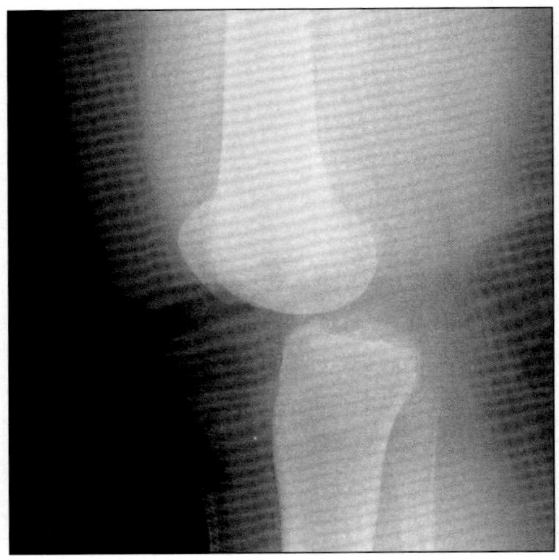

(Left) *AP radiograph shows a lateral tibiofemoral dislocation. The patella travels with the tibia. There is significant valgus deformity of the knee. No fracture is visible.* ***(Right)*** *Coronal T1WI MR shows lateral dislocation of the tibiofemoral joint. The cruciate and medial collateral ligaments are disrupted. The iliotibial band is intact.*

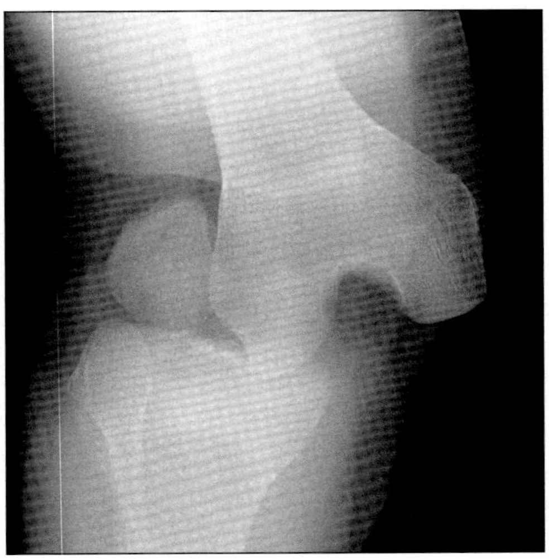

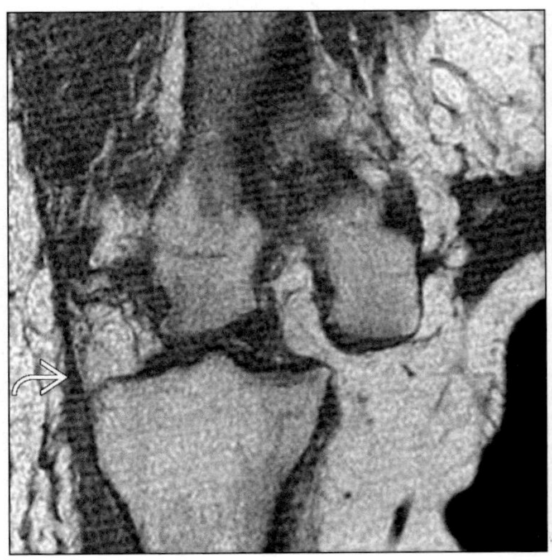

TIBIOFEMORAL DISLOCATION

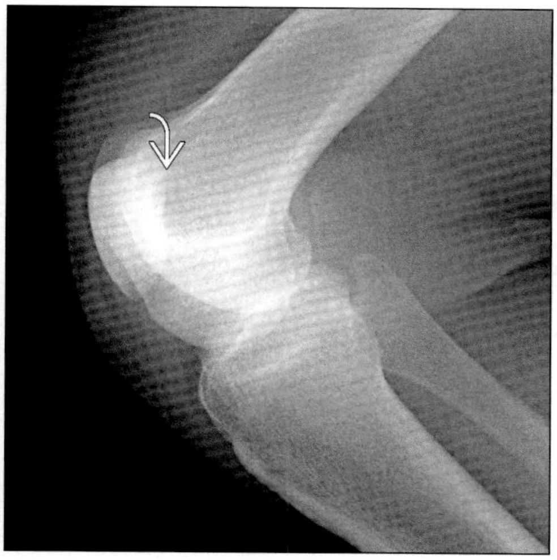

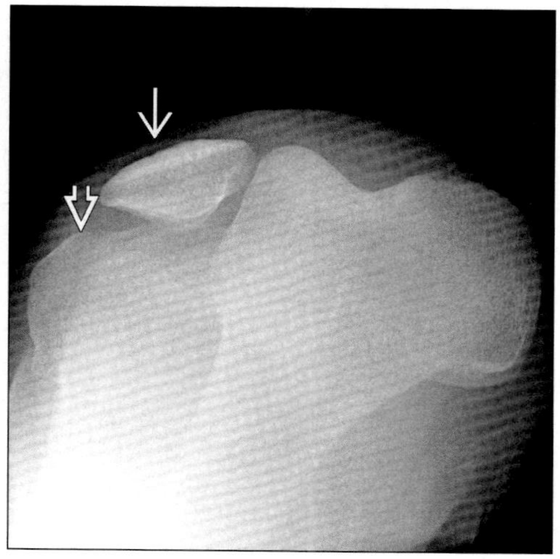

(Left) *Lateral radiograph demonstrates no fracture. There is subtle incongruity of the patellofemoral joint →, with the patella projecting posterior to the trochlear groove. The tibiofemoral joint is also incongruous, indicating knee dislocation.* ***(Right)*** *Axial sunrise radiograph in the same patient shows lateral dislocation of the tibia → and patella → relative to the femur. No fracture is observed.*

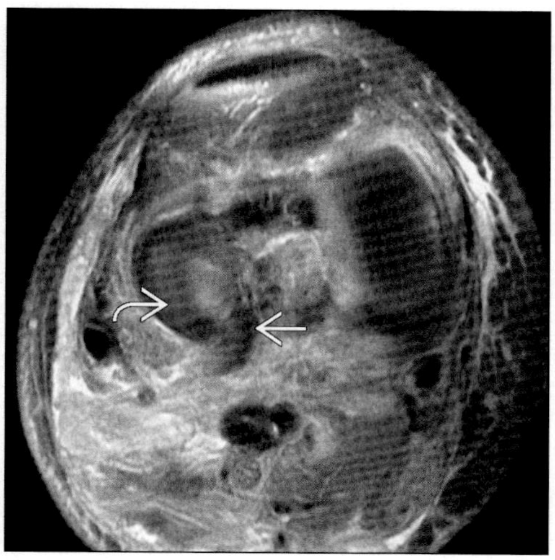

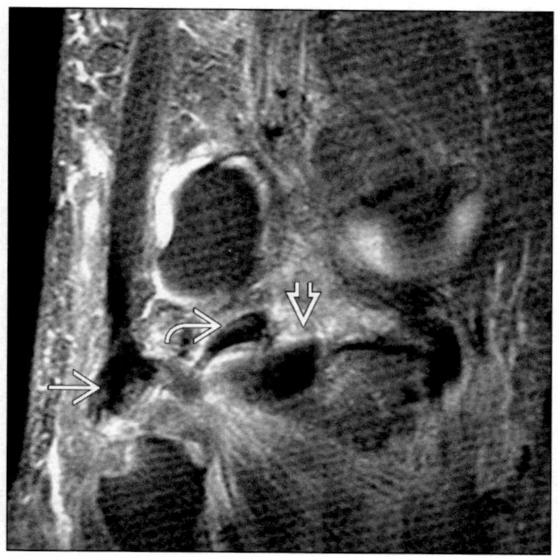

(Left) *Axial PDWI FS MR after knee dislocation shows interposition of the popliteal tendon → into the lateral compartment of the knee. This prevented adequate closed reduction of the dislocation. The lateral meniscus → is seen adjacent to the tendon.* ***(Right)*** *Coronal T2WI FS MR from the same patient shows avulsion of the conjoined tendon attachment to the fibular head →. The PCL is torn →. The popliteal tendon → enters the posterior joint compartment and is entrapped here.*

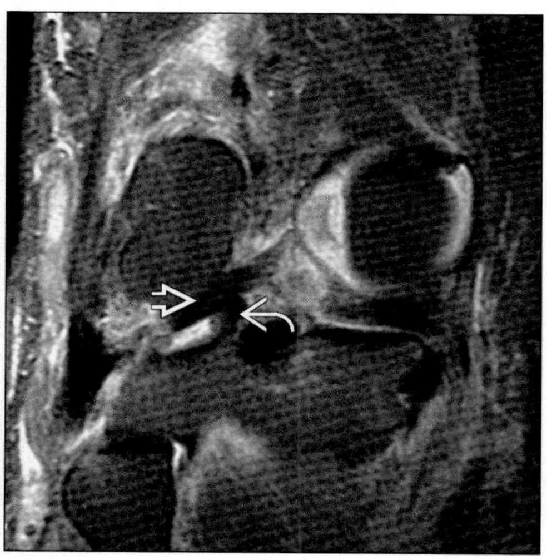

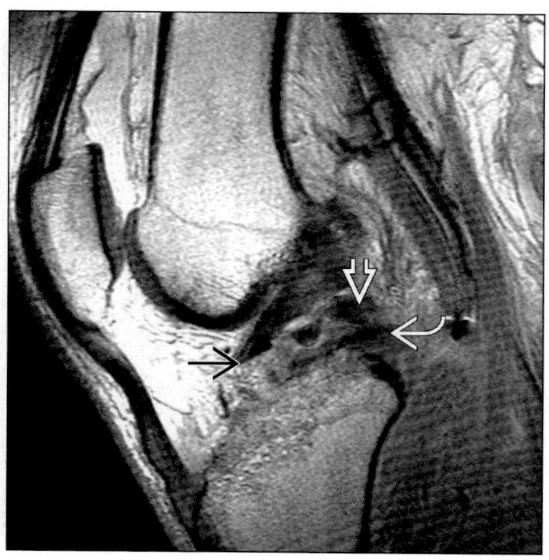

(Left) *Coronal T2WI FS MR more anteriorly in the same patient shows the popliteal tendon → interposed between the tibia and the lateral meniscus →. This was confirmed at surgery.* ***(Right)*** *Sagittal PDWI MR shows avulsion of the tibial attachment of the ACL →. The popliteal tendon → is seen interposed between the tibia and the lateral meniscus →.*

PROXIMAL TIBIOFIBULAR JOINT AND PROXIMAL FIBULAR FRACTURE

Key Facts

Imaging

- Radiographs
- Oblique views may be helpful
- Fibular neck/proximal shaft fracture → ankle radiographs to exclude Maisonneuve injury

Pathology

- PTFJ dislocation classified by direction of displacement
  - Anterolateral: Most common
  - Posteromedial
  - Superior
  - Inferior
- Fibular head fracture due to 3 mechanisms
  - Direct blow
  - Valgus force to knee (associated tibial plateau fracture)
  - Varus force to knee (disruption of fibular collateral ligament ± biceps femoris tendon; associated peroneal nerve injury)
- Fibular neck/proximal shaft fracture
- Maisonneuve injury (associated disruption of ankle syndesmosis)

Clinical Issues

- Anterolateral PTFJ dislocation: Parachutists, other jumpers
- Posteromedial PTFJ dislocation: Direct blow due to MVA, horseback riding
- Recurrent subluxation of PTFJ
  - Idiopathic
  - Older children or adolescents

Diagnostic Checklist

- Image interpretation pearl: Compare to opposite side for subtle displacement

__(Left)__ Graphic illustration of anterolateral dislocation of the proximal tibiofibular joint shows rupture of the anterior syndesmotic ligament ➡ and anterolateral displacement of the fibular head relative to the tibia. __(Right)__ AP radiograph illustrating anterolateral dislocation of the fibular head relative to the proximal tibia, indicating disruption of the proximal tibiofibular joint. Normally, most of the fibular head overlies the proximal tibia in this plane.

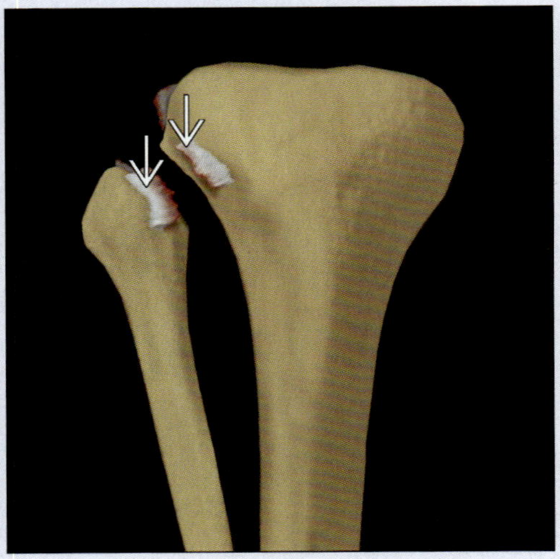

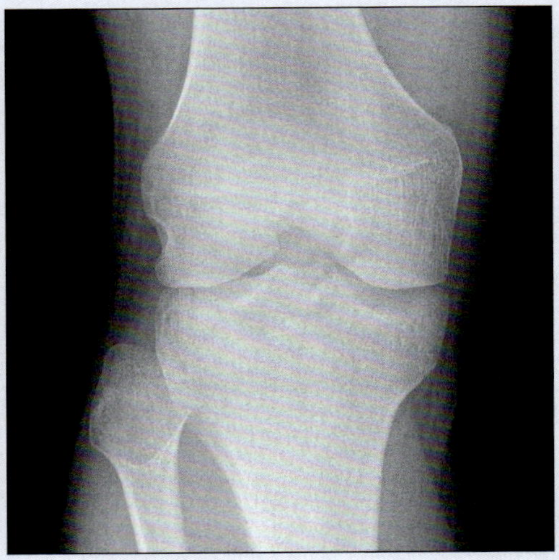

__(Left)__ Graphic illustration of anterolateral dislocation of the proximal tibiofibular joint demonstrates the anterior migration of the fibular head relative to its normal position, and rupture of the anterior ➡ and posterior ➡ syndesmotic ligaments. __(Right)__ Lateral radiograph in the same patient above shows anterolateral dislocation of the proximal tibiofibular joint. Normally, only a portion of the fibular head should overlie the tibia in the lateral projection.

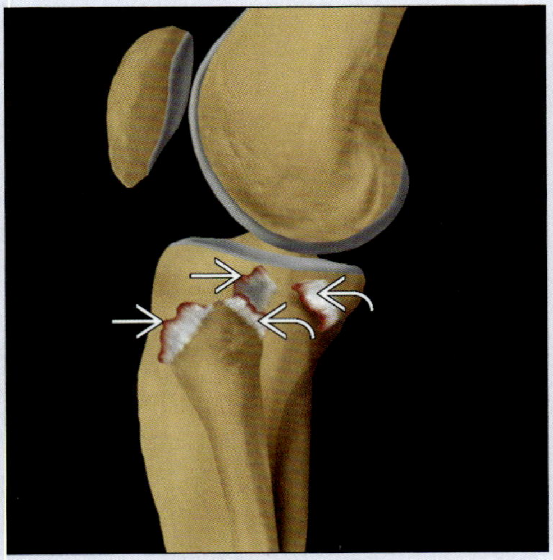

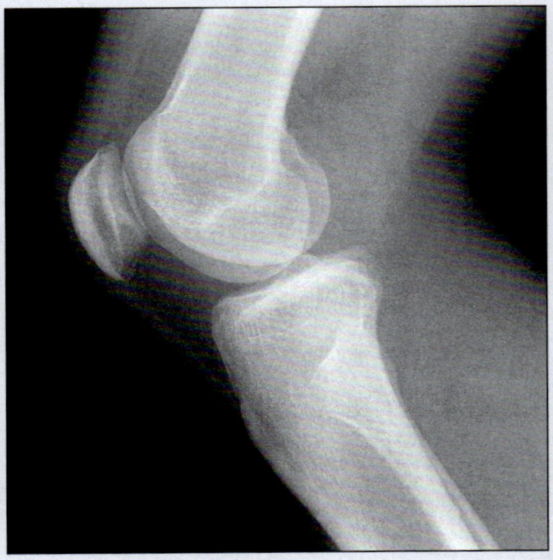

PROXIMAL TIBIOFIBULAR JOINT AND PROXIMAL FIBULAR FRACTURE

TERMINOLOGY

Abbreviations

- Proximal tibiofibular joint (PTFJ)

Definitions

- Injury to proximal tibiofibular articulation including dislocation, intraarticular fracture, or ligament disruption

IMAGING

General Features

- Best diagnostic clue
  - Displacement of proximal fibula relative to tibia
  - Lucent fracture line in proximal fibular shaft

Radiographic Findings

- Fracture
  - Lucent fracture line in fibular head or proximal shaft
    - Simple or comminuted
    - ± associated lateral tibial plateau fracture
    - ± Maisonneuve ankle injury
- Dislocation
  - Normal anatomy
    - AP view: Fibular head overlaps lateral tibial cortex
    - Lateral view: Only portion of fibular head projects over posterior tibial cortex
  - Anterolateral dislocation
    - AP view: Fibular head seen with minimal or no overlap on tibia
    - Lateral view: Fibular head displaced anteriorly, completely overlying tibia
  - Posteromedial dislocation
    - AP view: Fibular head lies behind tibia
    - Lateral view: Fibular head projects posterior to tibia
  - Superior dislocation
    - Superior displacement of fibular head so that styloid process is at level of knee joint

CT Findings

- More sensitive for subtle PTFJ subluxation or fracture

MR Findings

- Sprain or disruption of syndesmotic ligaments
- Displacement of fibular head relative to proximal tibia
- Fracture of tibia or fibula extending into syndesmotic joint

Imaging Recommendations

- Best imaging tool
  - Radiographs
- Protocol advice
  - Oblique views may be helpful
  - Fibular neck/proximal shaft fracture → ankle radiographs to exclude Maisonneuve injury

PATHOLOGY

Staging, Grading, & Classification

- **PTFJ dislocation** classified by direction of displacement
  - Anterolateral
    - Most common
    - Fall onto leg while in sitting position
  - Posteromedial
    - Much less frequent
    - Direct blow
    - Peroneal nerve injury
  - Superior
    - Invariably associated with fracture-dislocations of ankle or distal tibial fracture
- **Fibular head fracture** due to 3 mechanisms
  - Direct blow
    - Simple or comminuted fracture
  - Valgus force to knee
    - Associated tibial plateau fracture
    - Medial collateral ligament injury
  - Varus force to knee
    - Avulsion of styloid process (arcuate sign)
    - Disruption of fibular collateral ligament ± biceps femoris tendon
    - Associated peroneal nerve injury
- **Fibular neck/proximal shaft fracture**
  - Usually result of external rotation injury at ankle (Maisonneuve injury)

CLINICAL ISSUES

Presentation

- Clinical profile
  - Anterolateral PTFJ dislocation: Parachutists, other jumpers
  - Posteromedial PTFJ dislocation: Direct blow due to MVA, horseback riding

Natural History & Prognosis

- Recurrent subluxation of PTFJ
  - Idiopathic
    - Older children or adolescents
    - Usually girls
    - Responds to conservative therapy

Treatment

- Usually related to associated injury

DIAGNOSTIC CHECKLIST

Consider

- Maisonneuve injury in fibular neck/proximal shaft fracture
- Lateral tibial plateau fracture in fibular head fracture
- Fibular collateral ligament ± biceps tendon injury in styloid avulsion

Image Interpretation Pearls

- Compare to opposite side for subtle displacement

SELECTED REFERENCES

1. Park S et al: Compartment syndrome in tibial fractures. J Orthop Trauma. 23(7):514-8, 2009
2. Sekiya JK et al: Instability of the proximal tibiofibular joint. J Am Acad Orthop Surg. 11(2):120-8, 2003
3. Veth RP et al: The abnormal proximal tibiofibular joint. Arch Orthop Trauma Surg. 102(3):167-71, 1984

__(Left)__ Coronal T2WI FS MR shows disruption of the proximal tibiofibular joint (PTFJ) with edema within and around the syndesmotic ligaments ➔. There are marrow contusions ➔ in the medial fibular head and posterior medial tibia. __(Right)__ Axial PDWI FS MR in the same patient demonstrates disruption of the anterior ➔ and posterior ➔ syndesmotic ligaments of the PTFJ. The fibula is displaced laterally relative to its normal position.

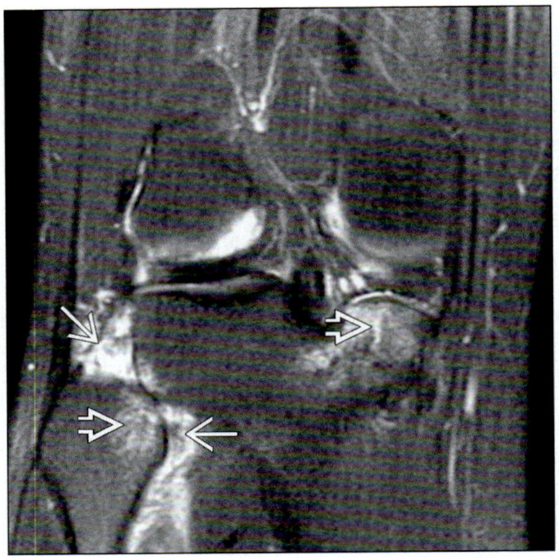

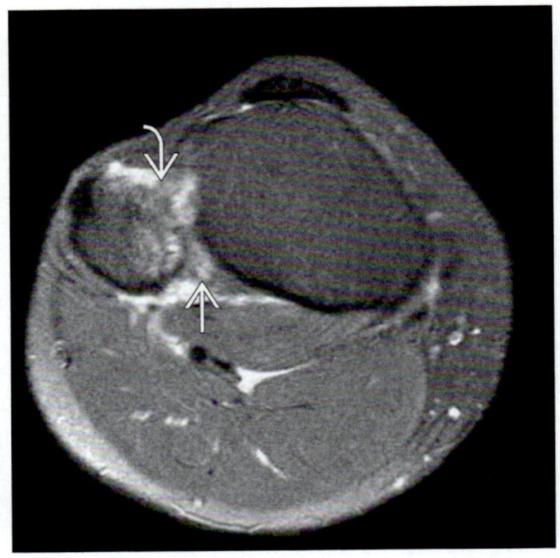

__(Left)__ AP radiograph shows a nondisplaced transverse fracture of the fibular head ➔. No other radiographic abnormality is evident. The finding suggests varus injury, and should raise suspicion for incompetence of the fibular collateral ligament. __(Right)__ Coronal T2WI FS MR shows a small stress fracture ➔ in the proximal medial fibula extending into the proximal tibiofibular joint. There is surrounding marrow edema. This is an unusual location for isolated stress injury.

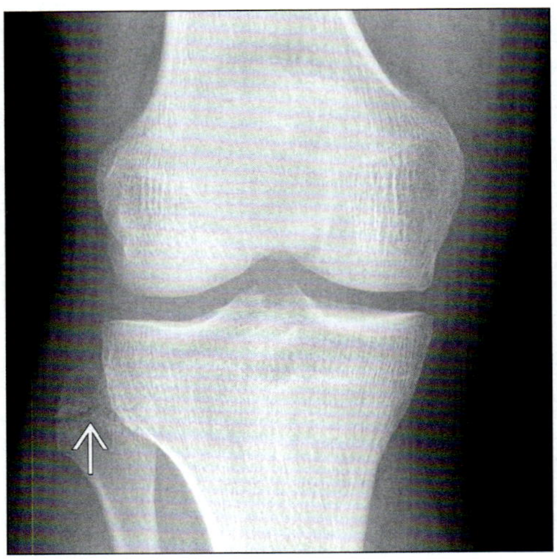

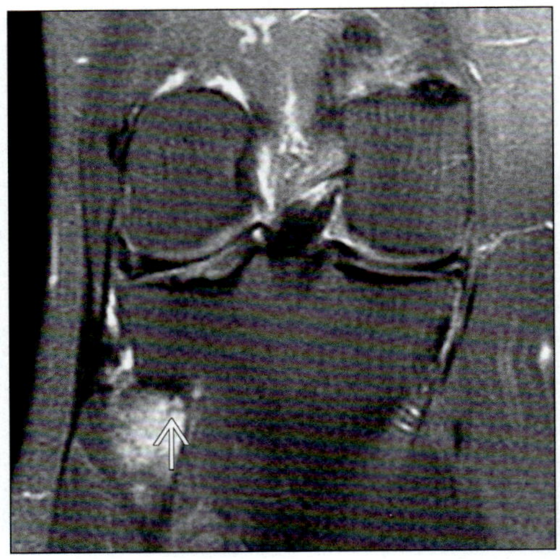

__(Left)__ Coronal T1WI MR demonstrates a fracture ➔ in the proximal fibula extending into the PTFJ. The longitudinal orientation of the fracture suggests a compression mechanism due to valgus force. __(Right)__ Coronal T2WI FS MR in the same patient shows an impaction contusion ➔ in the subchondral bone of the lateral tibial plateau, and a femoral avulsion ➔ of the medial collateral ligament. A valgus force at the time of injury explains this distribution of findings.

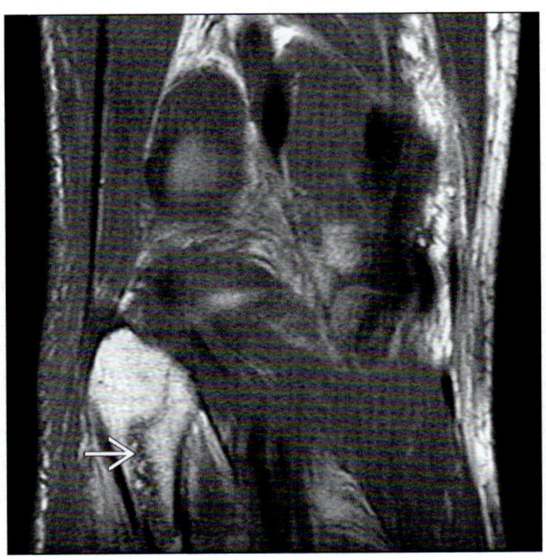

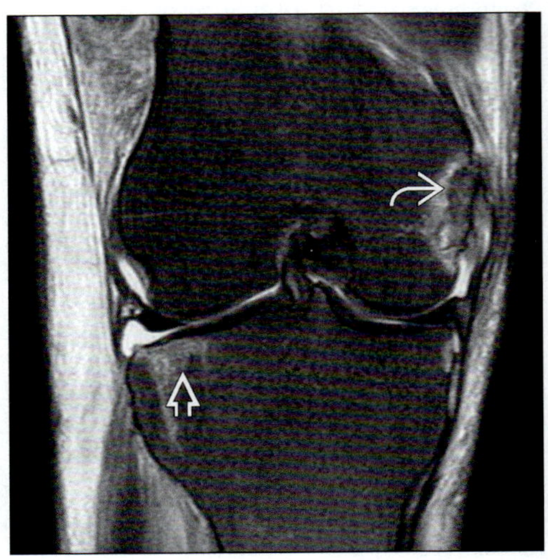

PROXIMAL TIBIOFIBULAR JOINT AND PROXIMAL FIBULAR FRACTURE

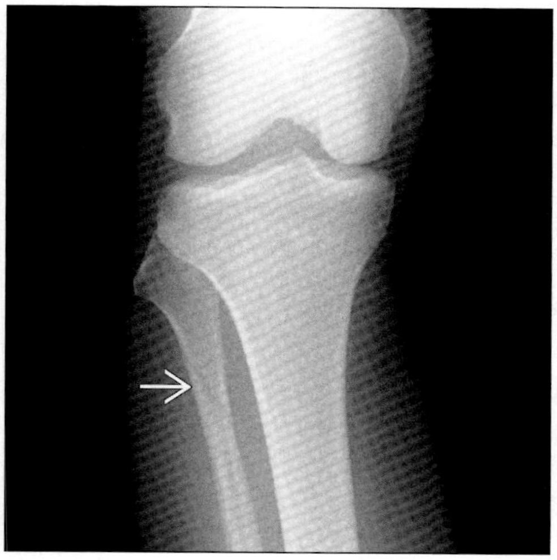

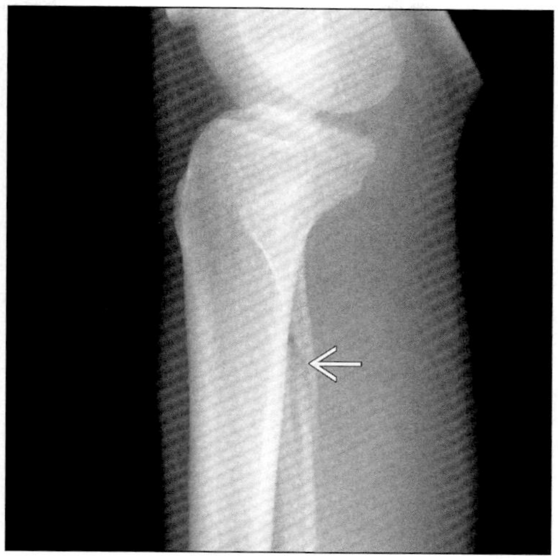

(Left) *AP radiograph demonstrates a spiral fracture of the proximal tibial shaft →. The knee is otherwise normal.* ***(Right)*** *Lateral radiograph in the same patient demonstrates the oblique spiral fracture → of the proximal tibia. A proximal tibial shaft fracture, particularly with spiral orientation, should prompt a search for associated ankle injury, as such fractures are often part of a Maisonneuve complex.*

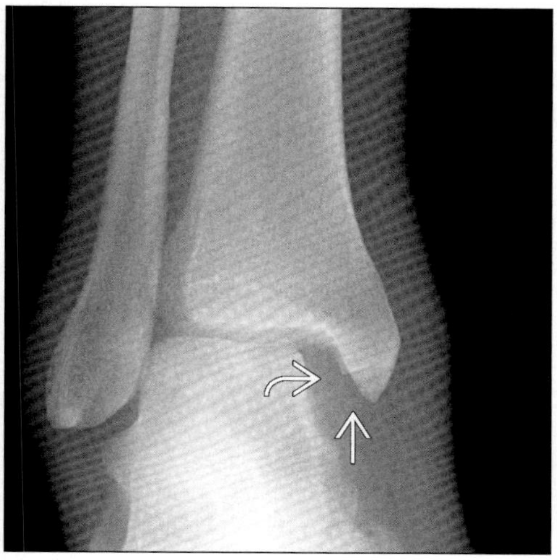

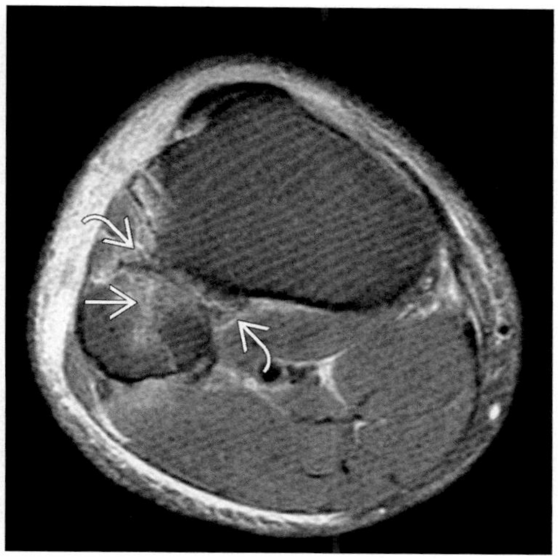

(Left) *Oblique radiograph of the ankle in the same patient shows widening of the medial ankle mortise ↷ with a small avulsion fragment → adjacent to the medial malleolar tip, confirming the presence of a Maisonneuve injury. Disruption of the interosseous ligament of the leg is present between the 2 injuries.* ***(Right)*** *Axial T2WI FS MR shows a fracture of the proximal fibula → entering the proximal tibiofibular joint. There is edema in the syndesmotic ligaments ↷.*

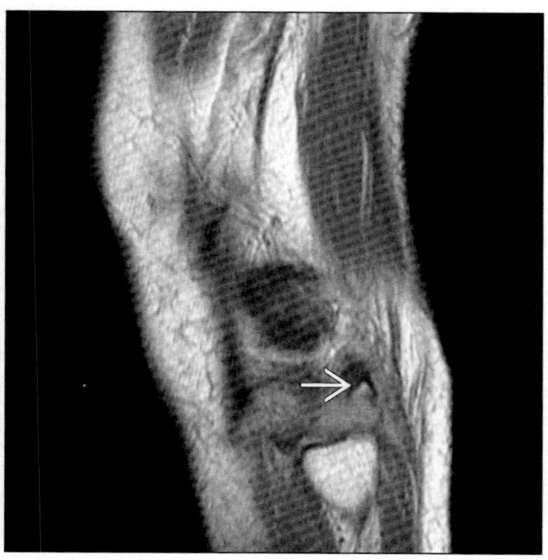

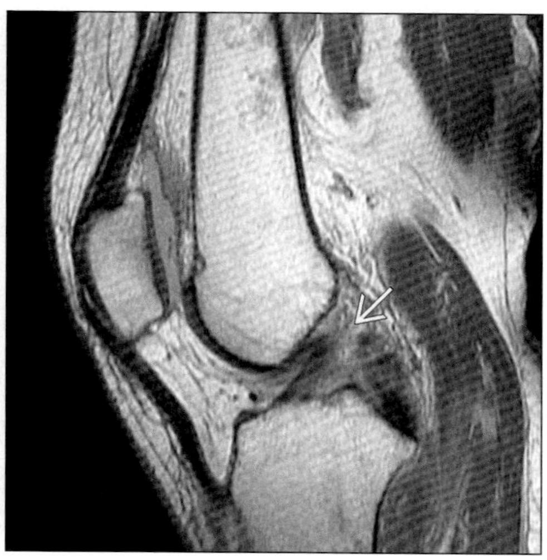

(Left) *Sagittal PDWI MR shows avulsion of the tip of the fibular styloid process →, indicating avulsion of the conjoined tendon (attachment of the fibular collateral ligament and biceps femoris tendon).* ***(Right)*** *Sagittal PDWI MR more medially in the same patient shows disruption of the posterior cruciate ligament (PCL) →. Injury to the posterolateral corner of the knee worsens the prognosis of PCL injury, and should be sought when evaluating images of PCL tear.*

PATELLAR FRACTURE

Key Facts

Imaging

- Transverse (50-60%)
  - Transverse fracture usually midportion of patella
- Stellate (comminuted) fracture (30-35%)
- Vertical (12-17%)
- Marginal avulsion
  - Avulsion fractures may occur at any patellar border
- Osteochondral injury
- Periosteal sleeve injury (children)
- Lateral radiographs do not detect vertical or osteochondral fracture
- MR to evaluate
  - Osteochondral injury
  - Periosteal sleeve avulsion
- CT for highly comminuted fracture
- Sunrise view to visualize vertical fracture

Top Differential Diagnoses

- Bipartite/multipartite patella
- Failure of fat suppression
- Dorsal defect of patella
- Chondromalacia patella
- Sinding-Larsen-Johansson syndrome

Pathology

- Direct blow to patella
- Indirect injury
- Tensile forces generated by quadriceps mechanism
- Secondary to ACL reconstruction

Clinical Issues

- Surgical indications
  - > 2 mm incongruity at articular surface
  - > 3 mm separation of fracture fragments
  - Disruption of extensor mechanism

*(**Left**) Lateral graphic shows a transverse fracture with separation caused by the opposing forces of the quadriceps → and patellar ↪ tendons. The patella is often the "weakest link" in the extensor chain. (**Right**) Lateral radiograph demonstrates a widely distracted transverse fracture of the midportion of the patella. Note the redundant quadriceps → and patellar ↪ tendons due to a loss of tension on the extensor mechanism. Such fractures are usually due to tensile overload.*

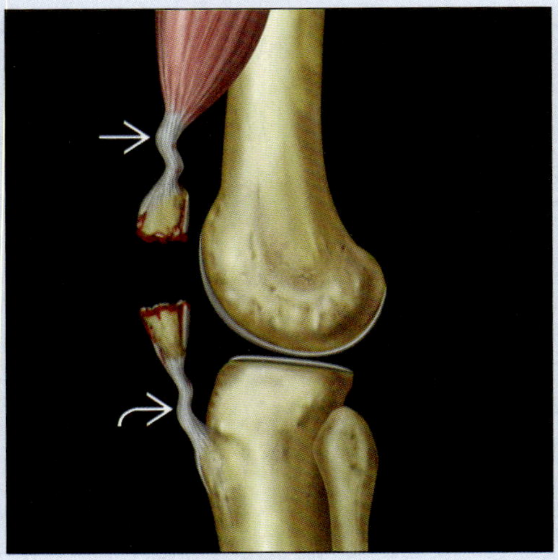

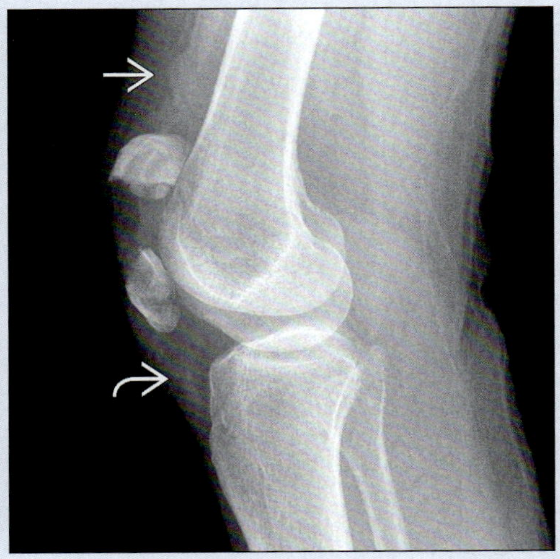

*(**Left**) Coronal graphic shows a transverse fracture through the inferior 1/2 of the patella →. Transverse fractures of the patella usually occur between the middle of the bone and the inferior quarter. Distraction of the fragments varies depending on the integrity of the medial and lateral patellar retinacula. (**Right**) AP radiograph in the same patient as previous radiograph demonstrates the sharp inferior edge → of the superior fracture fragment. The inferior fragment ↪ is difficult to see.*

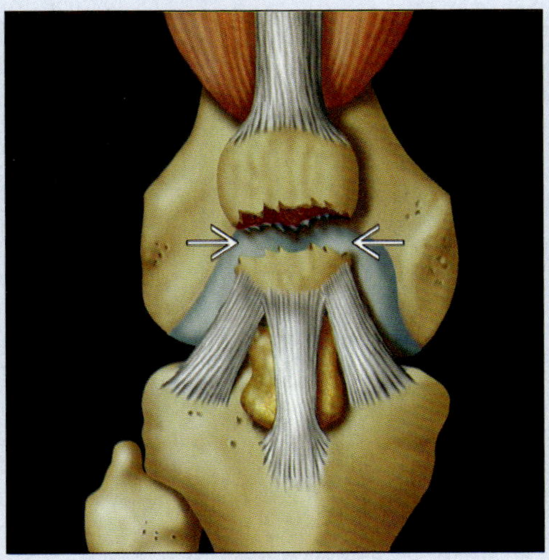

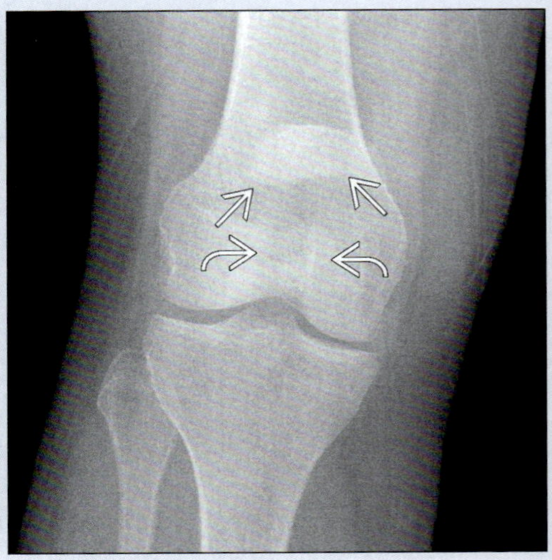

PATELLAR FRACTURE

IMAGING

General Features

- Best diagnostic clue
  - Lucent fracture line ± fragment displacement on radiographs
  - Patella alta
    - Look for small inferior pole avulsion
  - Patella baja
    - Look for small superior pole avulsion
- Location
  - Transverse fracture usually at midportion of patella
  - Avulsion fractures may occur at any patellar border
- Size
  - Ranges from tiny avulsion fragment to large displaced fragments
- Morphology
  - Transverse (50-60%)
  - Stellate (comminuted) fracture (30-35%)
  - Vertical (12-17%)
  - Marginal avulsion (quadriceps tendon, patellar tendon, retinacular insertions)
  - Osteochondral injury
  - Periosteal sleeve injury (children)
    - Superior or inferior pole avulsion attached to unossified cartilage and articular surface component
    - Most common patellar fracture < 16 years

Radiographic Findings

- Lateral radiograph
  - Transverse fracture line ± diastasis
  - Fracture displacement and comminution
  - Articular surface incongruity
  - Small avulsed fragment
    - Polar: Superior or inferior
    - ± visible ossified avulsion in sleeve fracture
  - Patella alta
    - High patella (patellar height:patellar tendon height < 0.8)
    - Suggests patellar tendon rupture
    - Look for small inferior pole fracture
    - Also seen with sleeve fracture
  - Patella baja
    - Low patella (patellar height:patellar tendon length > 0.8)
    - Suggests quadriceps tendon rupture
    - Look for small superior pole fracture
    - May be seen in sleeve fracture
  - Soft tissue mass of retracted tendons
  - Joint effusion
- AP and Merchant (sunrise) radiographs
  - Fracture displacement and comminution
  - Vertical fracture line

CT Findings

- Similar to radiographs, with better sensitivity for small avulsion fragments
- Better for detection of articular surface offset
- Patellofemoral alignment assessment
- Intraarticular bone fragments
- Associated femur/tibia fracture

MR Findings

- Fracture line: Low signal on T1 and T2WI
- Surrounding marrow edema
- Articular cartilage disruption
- Joint effusion
- ± tear of medial/lateral retinaculum
- ± ↑ T2 signal/laxity/discontinuity at patella tendon & quadriceps
- Osteochondral fracture
  - Evidence of transient lateral patellar dislocation
    - Bone bruise (↑ T2 FS signal) of medial patella &/or lateral femoral condyle
    - ± tear/injury (↑ T2 signal/discontinuity) of medial retinaculum, medial patellofemoral ligament (MPFL), & vastus medialis obliquus (VMO) muscle
  - ↓ signal fragment in joint space or adjacent to patella on all sequences
  - Concave deformity of inferomedial patella
- Sleeve fracture
  - Separation through ossifying cartilage joining patella and tendon
  - Posterior articular cartilage (sleeve) of patella follows polar fragment & separates from patella
  - ↓ signal cartilage ± bone fragment in joint space or adjacent to patella on all sequences

Imaging Recommendations

- Best imaging tool
  - Lateral radiographs most sensitive but do not detect vertical or osteochondral fracture
  - MR to evaluate
    - Osteochondral injury
    - Periosteal sleeve avulsion
    - Associated tendon and retinacular disruption
  - CT for highly comminuted fracture
- Protocol advice
  - Radiographs
    - Sunrise view to visualize vertical fracture

DIFFERENTIAL DIAGNOSIS

Bipartite/Multipartite Patella

- Secondary ossification center(s) in superolateral patella
- Smooth margins on radiographs
- Osseous fragments not contiguous, but overlying cartilage is smooth
- 2-8% of population (but only 2% of these are symptomatic)
- M:F = 9:1
- 40-45% bilateral: Consider contralateral knee radiograph
- Synchondrosis of 2° ossification centers can be injured

Dorsal Defect of Patella

- Osseous defect in posterior patella covered by articular cartilage
- Developmental variant

Chondromalacia Patella

- No fracture line

Sinding-Larsen-Johansson Disease
- Variable signal change ± fragmentation of patellar edge

Failure of Fat Suppression
- Seen on axial > sagittal PDWI or T2WI FS sequences
- Artifactual ↑ signal mimics marrow edema
- No edema seen on other sequences

PATHOLOGY

General Features
- Etiology
  - Direct blow to patella
    - Common cause of stellate and vertical fractures
    - More comminuted
    - Less displaced
    - More articular cartilage damage
  - Indirect injury
    - Tensile forces generated by quadriceps mechanism
    - Forced extension on flexed knee
    - Transverse fracture
    - Periosteal sleeve injury in children
    - Patellar retinaculum tear → fracture diastasis
  - Secondary to ACL reconstruction
    - Autologous tibia-patellar tendon-patella bone graft harvesting weakens patella
    - Usually transverse fracture 8-10 weeks post-op
    - Associated with direct or indirect trauma
    - Occurs in 1% of patients
  - After total knee prosthesis
    - Occurs in about 1%
    - May be any configuration, but transverse most common
  - Pathologic fracture
- Associated abnormalities
  - Direct blow
    - Chondral/osteochondral fracture of trochlear groove of femur
    - Intraarticular bodies
  - Tensile injury
    - Quadriceps or patellar tendon rupture
  - Transient patellar dislocation
    - Medial retinacular sprain or tear
    - Joint effusion
    - Typical pattern of marrow contusions
      - Lateral subcortical aspect of lateral femoral condyle &/or inferomedial patella

CLINICAL ISSUES

Presentation
- Most common signs/symptoms
  - Pain, swelling, & ↓ extension strength
  - Large joint effusion or hemarthrosis
  - Inability to perform straight-leg raise
  - Point tenderness over patella
  - Defect in smooth patellar cortex on palpation

Demographics
- Age
  - Periosteal sleeve avulsion most common in children
- Epidemiology
  - 1% of all skeletal fractures
  - Majority are direct fractures
  - Osteochondral fracture with 5% of patellofemoral dislocations

Natural History & Prognosis
- Persistent pain and arthritis in approximately 50%
- Good/excellent surgical results reported in 50-80%
- Patellectomy → knee instability & ↓ extension strength

Treatment
- Surgery
  - Indications
    - > 3 mm separation of fracture fragments
    - > 2 mm incongruity at articular surface
    - Disruption of extensor mechanism
    - Open fracture
  - Goals: Restore extensor mechanism & reduce complications (arthritis)
  - Techniques: Tension band, cerclage wire, screw fixation, partial/total patellectomy
    - Modified tension band = most stable fixation of transverse fracture
- Nonoperative: Splint/cast in extension for 4-6 weeks

DIAGNOSTIC CHECKLIST

Consider
- Disruption of extensor mechanism with patella alta or baja
- Transient patellar dislocation with medial patellar avulsion

Image Interpretation Pearls
- AP radiograph often falsely negative
- Tiny avulsion fragment may underlie major soft tissue injury

Reporting Tips
- Identify fragment displacement > 3 mm

SELECTED REFERENCES

1. Dupuis CS et al: Injuries and conditions of the extensor mechanism of the pediatric knee. Radiographics. 29(3):877-86, 2009
2. Sheth NP et al: Periprosthetic patellar fractures. J Bone Joint Surg Am. 89(10):2285-96, 2007
3. Drabicki RR et al: Stress fractures around the knee. Clin Sports Med. 25(1):105-15, ix, 2006
4. Gupta RR et al: Patellar sleeve fractures in children: a case report and review of the literature. Am J Orthop. 35(7):336-8, 2006
5. Tay GH et al: Indirect patella fractures following ACL reconstruction: a review. Acta Orthop. 77(3):494-500, 2006
6. Chun KA, et al: Patellar fractures after total knee replacement. AJR Am J Roentgenol. 185(3):655-60, 2005
7. Hunt DM et al: A review of sleeve fractures of the patella in children. Knee. 12(1):3-7, 2005
8. Carpenter JE et al: Fractures of the patella. Instr Course Lect. 43:97-108, 1994
9. Sonin AH: Magnetic resonance imaging of the extensor mechanism. Magn Reson Imaging Clin N Am. 2(3):401-11, 1994
10. Yu JS et al: MR imaging of injuries of the extensor mechanism of the knee. Radiographics. 14(3):541-51, 1994

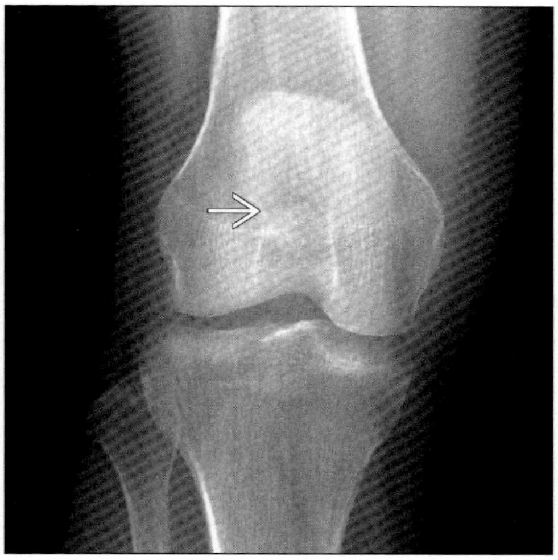

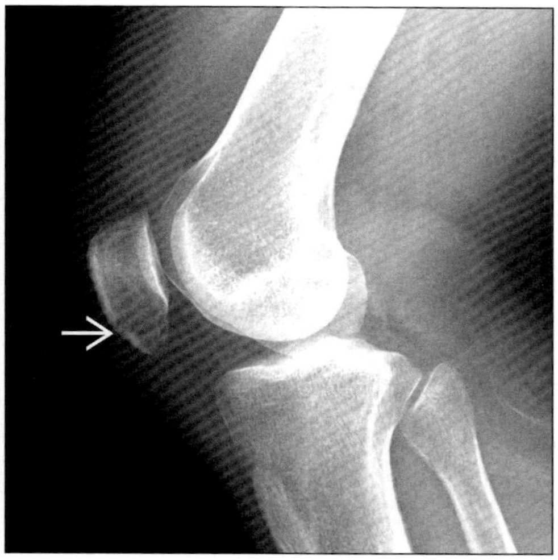

*(**Left**) AP radiograph of the knee demonstrates a nondisplaced fracture ➡ of the inferior portion of the patella. Nondisplaced fractures are often easier to see in the anteroposterior projection than are displaced fractures. (**Right**) Lateral radiograph in the same patient shows the transverse nondisplaced fracture ➡ in the inferior patella. Transverse fractures are often due to forced extension against a contracted quadriceps.*

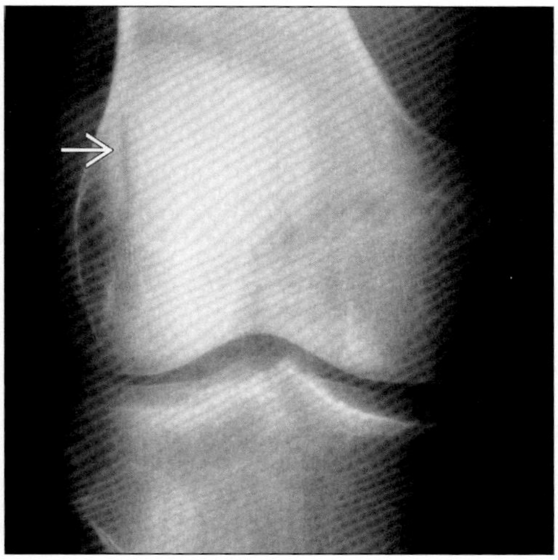

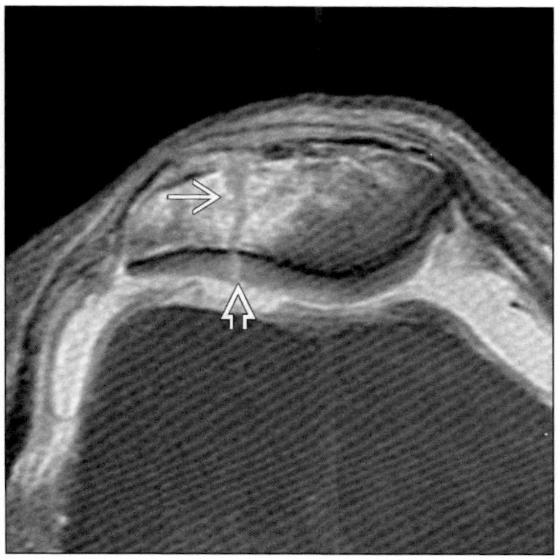

*(**Left**) AP radiograph demonstrates a vertical fracture of the patella ➡ due to a direct blow injury. The sharp margins of the fracture differentiate this from a bipartite patella. (**Right**) Axial PDWI FS MR after a fall onto the knee demonstrates a vertical fracture ➡ through the patella. The edges of the fracture fragments are sharp and without sclerosis. There is extensive marrow edema and disruption of the articular cartilage ⇨. Compare to bipartite patella below.*

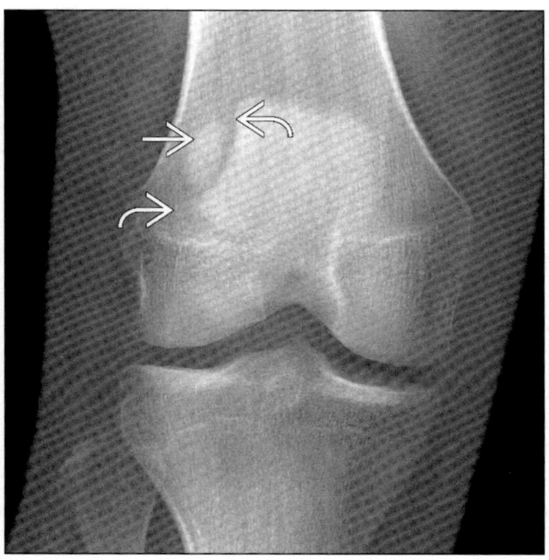

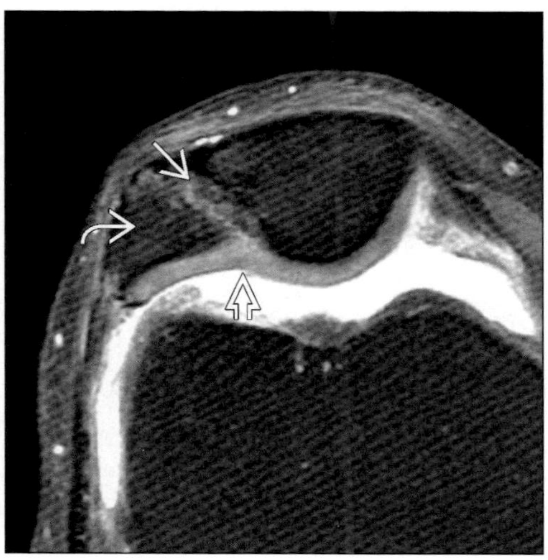

*(**Left**) AP radiograph shows a bipartite patella. The edges of the accessory ossification center ➡ are smooth and regular. The defect in the patella ↪ is often larger than the secondary center, unlike a fracture that would fit together. (**Right**) Axial PDWI FS MR shows bipartite patella. Note the irregular synchondrosis ➡ between the accessory ossification center ↪ and the native patella. There is no edema around the site. The overlying articular cartilage is intact ⇨.*

(Left) *Axial PDWI FS MR demonstrates marrow edema around the synchondrosis of a bipartite patella. This finding can be symptomatic, and is usually due to motion at the site, or to direct trauma. The overlying cartilage is intact.* ***(Right)*** *Axial bone CT shows an unusual coronally oriented patellar fracture. There was some comminution of this fracture, which occurred as a result of a direct blow injury. Note the offset of the articular surface.*

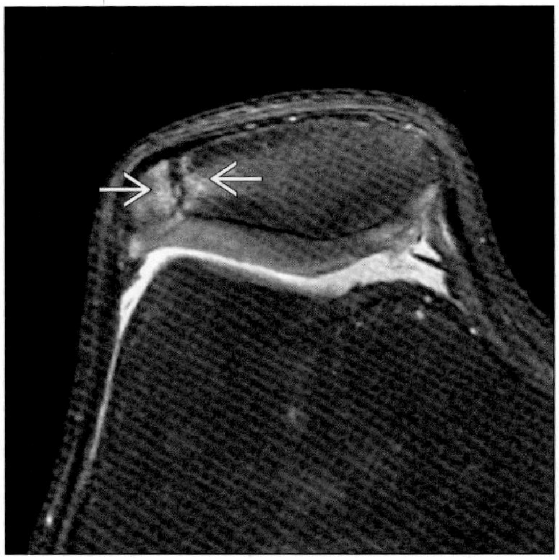

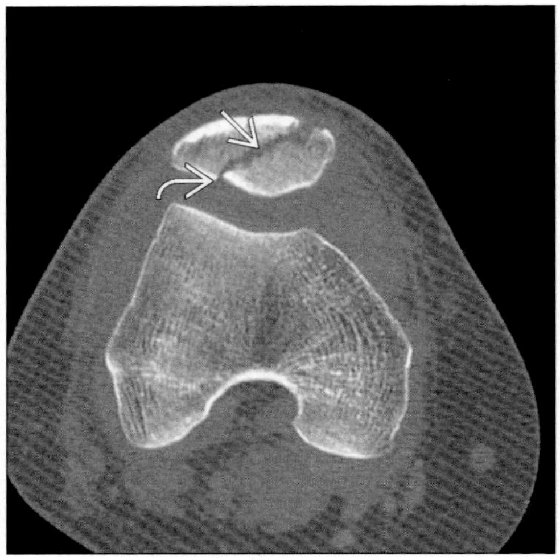

(Left) *Sagittal reformat bone CT shows a fracture of the patella around the patellar component of a total knee prosthesis. Extensive beam-hardening artifact is present from the femoral component of the prosthesis.* ***(Right)*** *Sagittal reformat bone CT from a 15-year-old boy after knee trauma demonstrates a defect in the articular surface of the patella and a small intraarticular body. These findings were not visible on radiographs.*

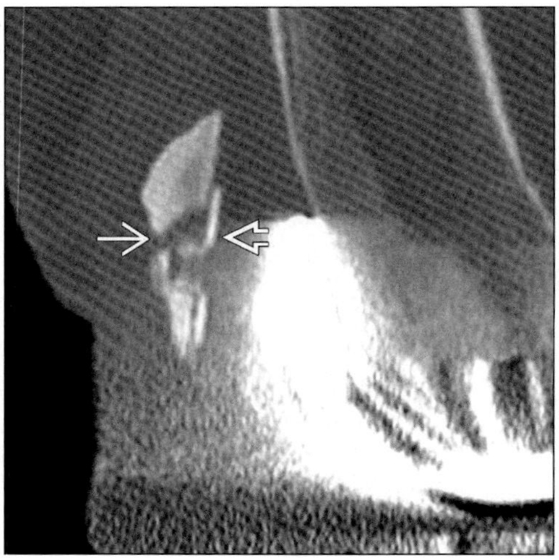

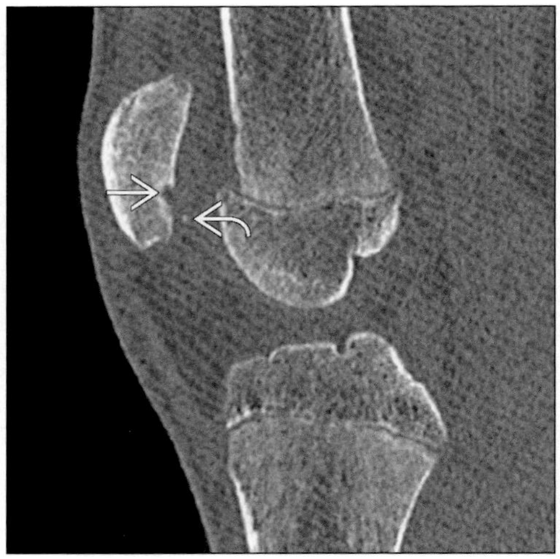

(Left) *Sagittal PDWI MR of the knee demonstrates a periosteal sleeve avulsion of the quadriceps tendon from the superior pole of the patella. Note the fluid interposed between the avulsed periosteum and the patellar cortex.* ***(Right)*** *Sagittal PDWI FS MR of the knee after a direct blow to the anterior knee demonstrates an impacted osteochondral fracture of the trochlea of the femur and marrow edema in the patella due to a fracture.*

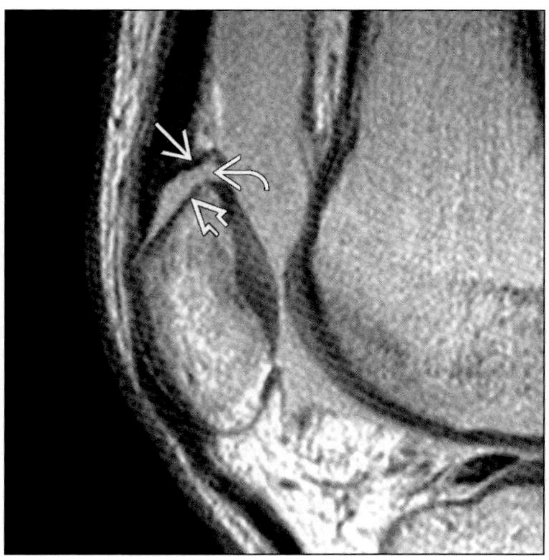

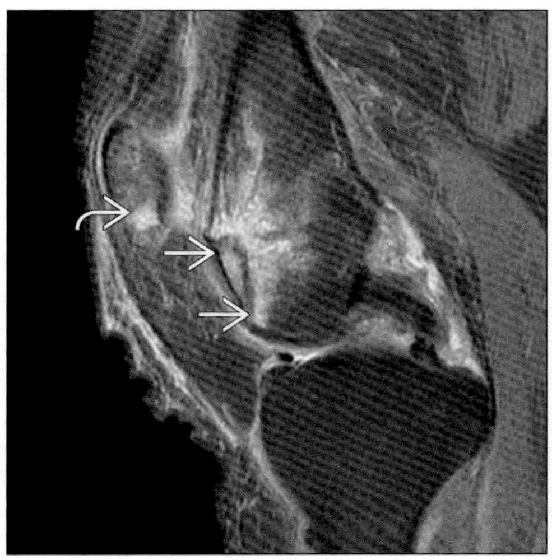

PATELLAR FRACTURE

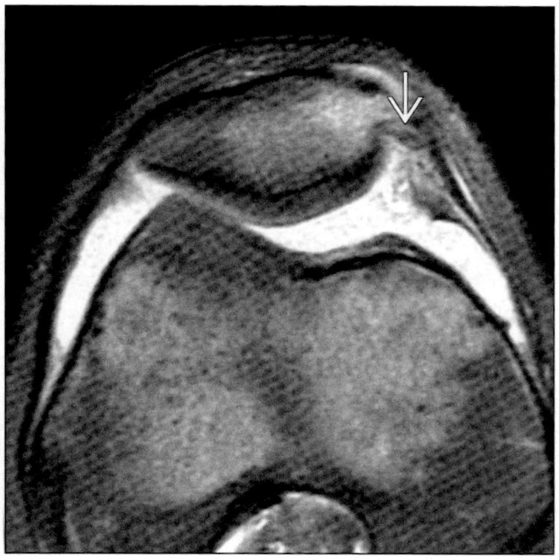

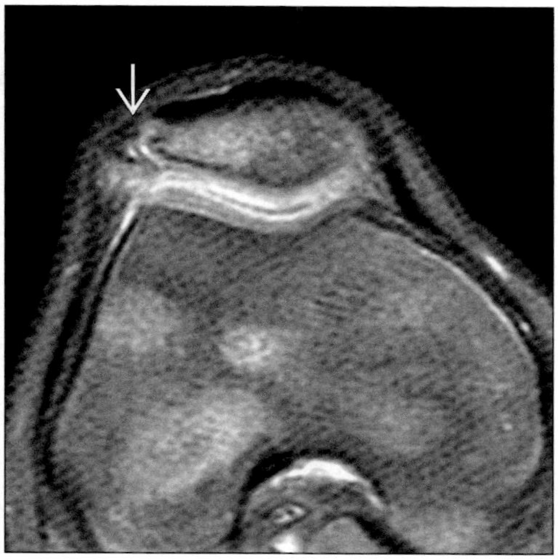

(Left) *Axial PDWI FS MR after transient patellar dislocation in a 15 year old demonstrates avulsion of the medial patellar cortex → at the site of attachment of the medial retinaculum. There is surrounding marrow edema and a joint effusion.* ***(Right)*** *Axial PDWI FS MR demonstrates avulsion of the lateral pole of the patella → due to traction stress on the vastus lateralis in a field-and-track athlete. Such avulsions are often radiographically occult.*

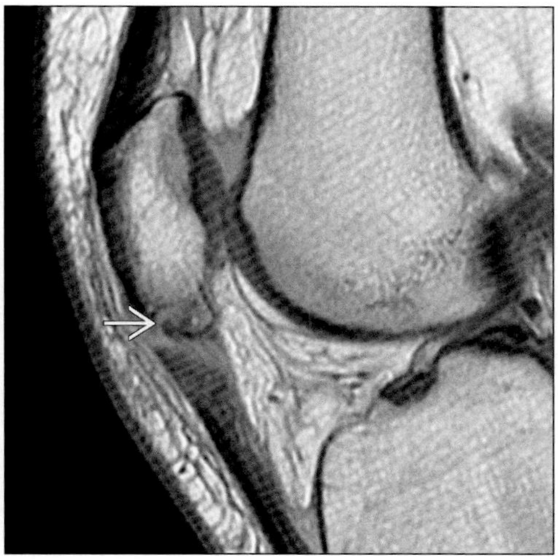

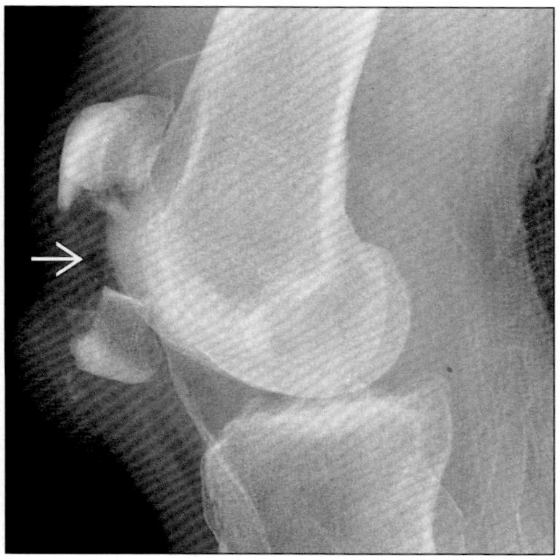

(Left) *Sagittal PDWI MR demonstrates osseous avulsion of the inferior pole of the patella → associated with patellar tendon injury. These injuries are usually the result of traction mechanism, but may be seen after a direct blow to the anterior knee.* ***(Right)*** *Lateral radiograph demonstrates a comminuted and widely distracted patellar fracture. Air in the knee joint → indicates that this is an open injury; immediate surgical repair is mandated.*

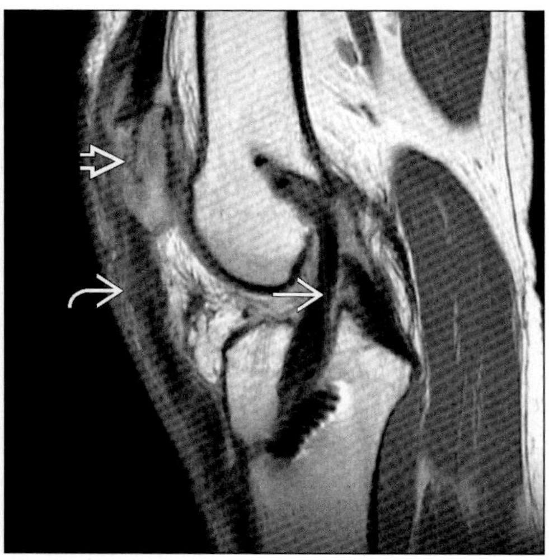

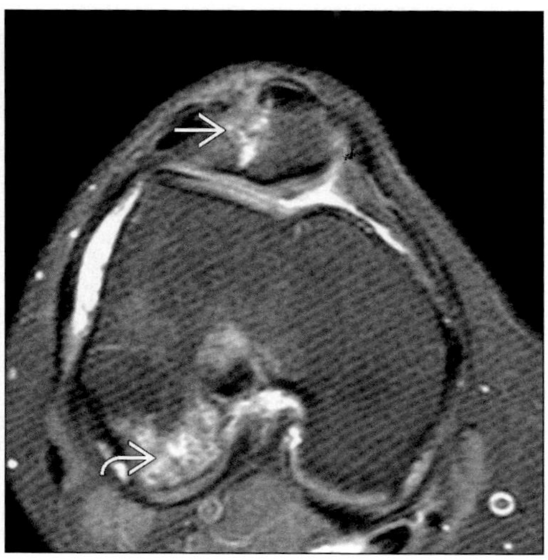

(Left) *Sagittal PDWI MR in a patient status post recent anterior cruciate ligament repair → with patellar tendon autograft who was in a motor vehicle accident. Note the postoperative change in the patellar tendon →. There is a vertical fracture in the patella →.* ***(Right)*** *Axial T2WI FS MR in the same patient shows a vertical nondisplaced fracture in the patella → due to a direct blow. Marrow edema in the posterior lateral femoral condyle → is also present.*

AVULSIVE INJURY, KNEE

Key Facts

Imaging

- Common sites
  - Distal ACL: Anterior medial tibial spine
  - Distal PCL: Posterior aspect of tibial plateau
  - Proximal MCL: Medial femoral condyle distal to adductor tubercle
  - Deep fibers distal MCL (coronary ligament): Proximal medial tibia at joint line (reverse Segond)
  - Superficial fibers distal MCL: Medial tibia 5 cm below joint
  - Distal FCL (conjoined tendon with biceps femoris): Fibular head
  - LCL: Lateral tibial rim (Segond injury)
  - ITB: Gerdy tubercle
  - Arcuate ligament complex: Tip of fibular head
  - Proximal patellar tendon: Inferior pole of patella (Sinding-Larsen-Johansson syndrome)
  - Distal patellar tendon: Tibial tubercle (Osgood-Schlatter disease)
  - MPFL: Medial pole of patella
- Small cortical avulsion fragment may be hard to see
- Presence of fragment distinguishes from pure ligamentous/tendinous injury
- Radiographs can identify most avulsion injuries
- MR to more accurately identify associated injuries

Top Differential Diagnoses

- Direct blow fracture
- Accessory ossicle
- Osteophyte

Diagnostic Checklist

- Associated injuries (e.g., ACL tear with Segond avulsion, cruciate and other posterolateral structures with arcuate avulsion)

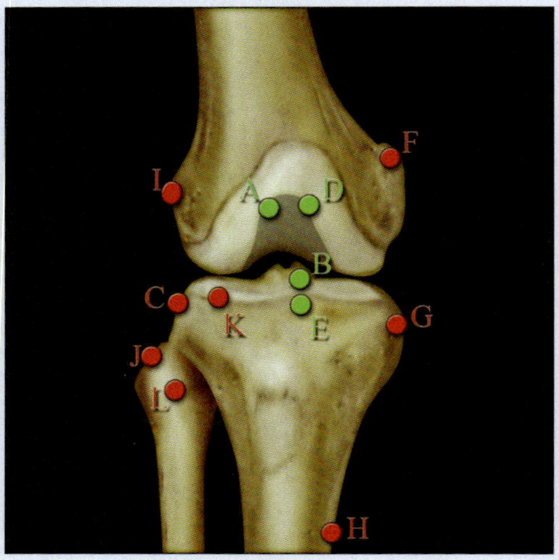

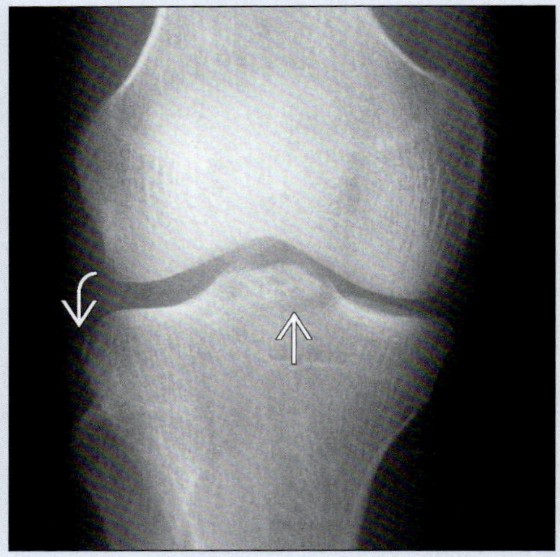

(Left) *Graphic shows common sites of osseous avulsion in the knee: A = ACL origin, B = ACL insertion, C = Segond, D = PCL origin, E = PCL insertion, F = MCL origin, G = coronary ligament, H = MCL insertion, I = FCL origin, J = arcuate ligament, K = ITB insertion, and L = conjoined tendon.* ***(Right)*** *AP radiograph demonstrates avulsion of the tibial attachment of the ACL → and a Segond avulsion → of the lateral capsular ligament. Segond injuries have a very high association with ACL tears.*

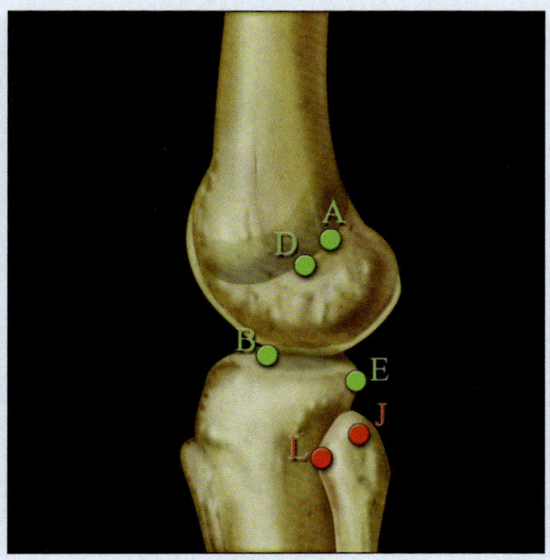

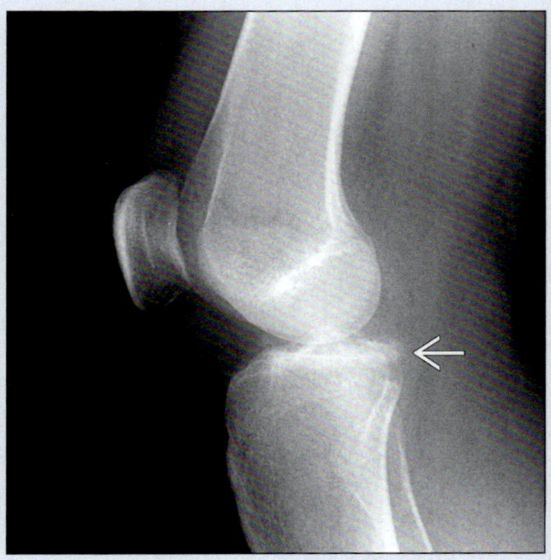

(Left) *Graphic shows common sites of avulsion in the knee from a lateral view: A = ACL origin, B = ACL insertion, D = PCL origin, E = PCL insertion, J = arcuate ligament, and L = conjoined tendon.* ***(Right)*** *Lateral radiograph demonstrates osseous avulsion of the tibial attachment of the posterior cruciate ligament (PCL) →. Usually related to hyperextension, this fracture accounts for less than 10% of PCL injuries. Avulsions are easier to repair than primary ligament tears.*

AVULSIVE INJURY, KNEE

TERMINOLOGY

Abbreviations

- Anterior/posterior cruciate ligament (ACL/PCL)
- Medial collateral ligament (MCL)
- Lateral capsular ligament (LCL)
- Fibular collateral ligament (FCL)
- Iliotibial band (ITB)
- Medial patellofemoral ligament (MPFL)

Definitions

- Arcuate sign: Avulsion of tip of fibular styloid at attachment of arcuate ligament complex
- Osgood-Schlatter disease: Chronic repetitive avulsive injury at patellar tendon insertion on tibial tubercle
- Segond fracture: Avulsion of lateral tibial rim cortex due to avulsion of lateral capsular ligament
  - Sometimes involves ITB more anteriorly
  - Strong association with ACL tear
- Reverse Segond fracture: Avulsion of deep capsular component of MCL on medial tibial rim
  - Associated with tear or tibial avulsion of PCL
- Sinding-Larsen-Johansson syndrome: Spectrum of overuse injuries affecting origin of patellar tendon at inferior pole of patella
  - Includes osseous or periosteal sleeve avulsion of inferior patella

IMAGING

General Features

- Best diagnostic clue
  - Small displaced osseous fragment at site of attachment of ligament or tendon
- Location
  - Cortical insertion of ligaments or tendons at various sites around knee
  - ACL
    - Distal (common): Anterior medial tibial spine
    - Proximal (uncommon): Lateral posterior superior intercondylar notch
  - PCL
    - Distal (common): Posterior aspect of tibial plateau
    - Proximal (uncommon): Medial anterior superior intercondylar notch
  - MCL
    - Proximal: Medial femoral condyle distal to adductor tubercle
    - Distal
      - Deep fibers: Proximal medial tibia at joint line (reverse Segond)
      - Superficial fibers: Medial tibia 5 cm below joint
  - FCL
    - Proximal: Lateral femoral condyle
    - Distal (conjoined tendon with biceps femoris): Fibular head
  - LCL: Lateral tibial rim (Segond injury)
  - ITB: Gerdy tubercle (anterolateral tibia at joint line)
  - Arcuate ligament complex: Tip of fibular head
  - Semimembranosus tendon: Posteromedial tibial rim
  - Patellar tendon
    - Proximal: Inferior pole of patella (Sinding-Larsen-Johansson syndrome)
    - Distal: Tibial tubercle (Osgood-Schlatter disease)
  - MPFL: Medial pole of patella
- Size
  - Fragments vary from tiny to > 1 cm

Radiographic Findings

- Flake of bone at attachment site of ligament or tendon

CT Findings

- Similar to radiographs; ↑ sensitivity for small fragments

MR Findings

- Small cortical avulsion fragment may be hard to see
- Presence of fragment distinguishes from pure ligamentous/tendinous injury
- Occasionally see marrow edema in donor bone

Imaging Recommendations

- Best imaging tool
  - Radiographs can identify most avulsion injuries
  - MR to more accurately identify associated injuries
- Protocol advice
  - Include sunrise view if transient patellar dislocation suspected

DIFFERENTIAL DIAGNOSIS

Ligament Tear

- May tear near osseous attachment

Direct Blow Fracture

- Cortical fracture without ligament detachment

Heterotopic Ossification

- May form along ligament or tendon attachments

Intraarticular Body

- Within joint

Osteophyte

- Continuous with cortical outline of bone

CLINICAL ISSUES

Presentation

- Most common signs/symptoms
  - Pain at avulsion site
  - Symptoms and exam findings in keeping with disruption of avulsed structure

DIAGNOSTIC CHECKLIST

Consider

- Associated injuries (e.g., ACL tear with Segond avulsion, cruciate and other posterolateral structures with arcuate avulsion)

Image Interpretation Pearls

- Consider avulsion to be "tip of iceberg"; small fracture fragment may connote more important ligament or tendon injury

SELECTED REFERENCES

1. Gottsegen CJ et al: Avulsion fractures of the knee: imaging findings and clinical significance. Radiographics. 28(6):1755-70, 2008

*(**Left**) Lateral radiograph in a 9 year old shows an acute avulsion of the tibial attachment of the ACL ➡. Note the elongate shape of the fragment in the anteroposterior axis of the knee, which is typical. A joint effusion is present ➡. (**Right**) AP radiograph in the same patient shows the avulsion fragment ➡ superimposed on the tibial spines. The size and displacement of the fragment are important arbiters for surgical decision-making and should be discussed in the radiologic report.*

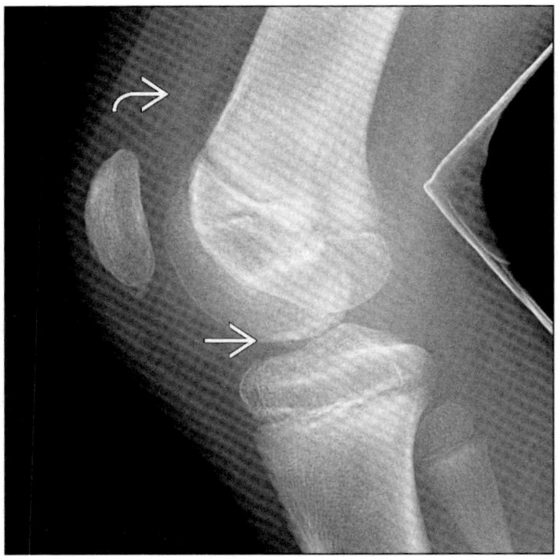

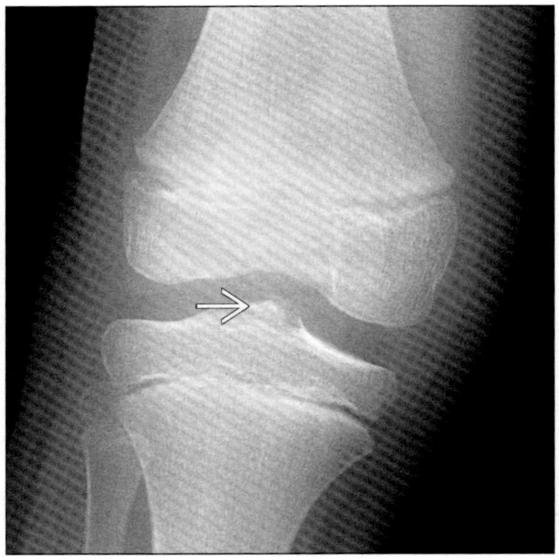

*(**Left**) Sagittal PDWI MR from the same patient illustrates the avulsed bone fragment ➡ and its relationship to the native tibia ➡. Some ACL fibers are included in this image ➡. The PCL is also seen ➡. (**Right**) AP radiograph demonstrates the subtle findings of a tibial avulsion of the posterior cruciate ligament (PCL). The curvilinear lucency ➡ projecting over the tibial spines is often the only finding on AP radiographs.*

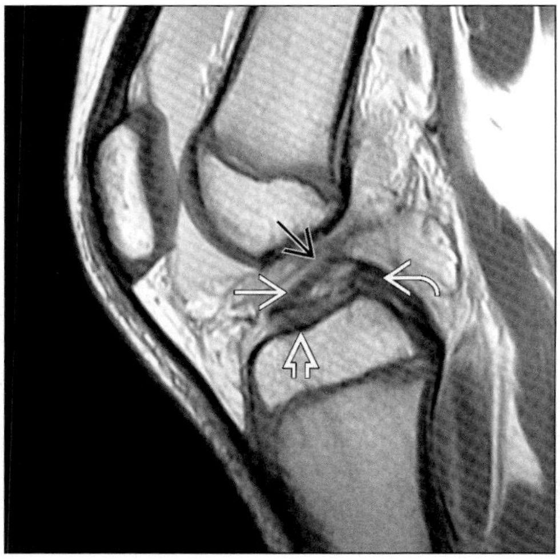

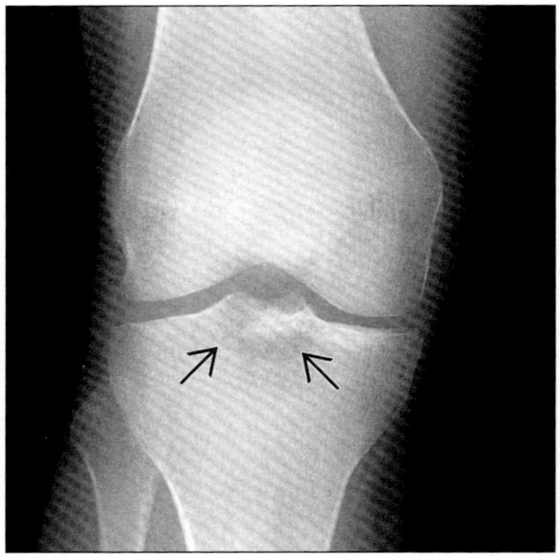

*(**Left**) Sagittal PDWI MR shows a comminuted avulsion fracture ➡ of the distal PCL attachment. This injury is often the result of a hyperextension mechanism. (**Right**) Sagittal bone CT shows osseous avulsion of the femoral attachment of the PCL ➡, a very uncommon injury. Fragments were not well seen on radiographs. Osseous avulsion of the PCL usually occurs at its tibial attachment but comprises less than 10% of PCL injuries in adults.*

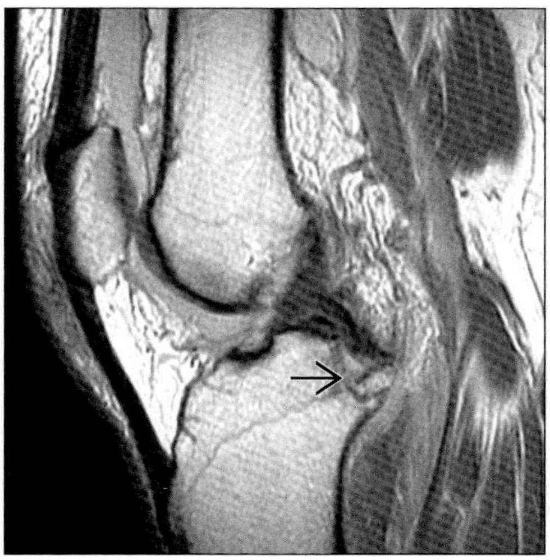

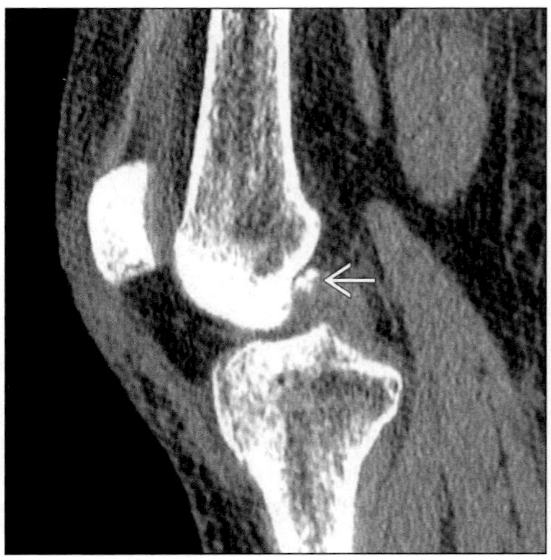

AVULSIVE INJURY, KNEE

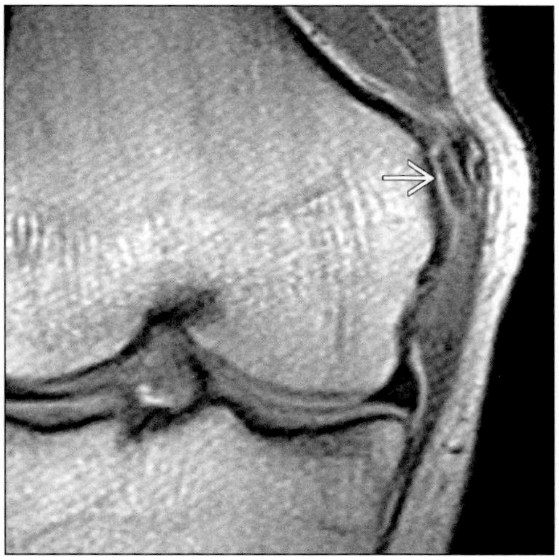

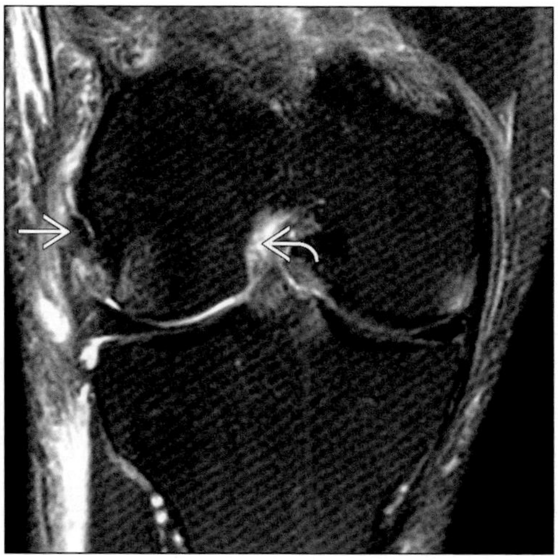

*(**Left**) Coronal T1WI MR shows a cortical avulsion of the femoral attachment of the medial collateral ligament ➔. Most MCL tears are midsubstance. (**Right**) Coronal T2WI FS MR in a patient with an anterior cruciate ligament tear ➔ shows subtle cortical avulsion of the femoral origin of the fibular collateral ligament ➔ as well.*

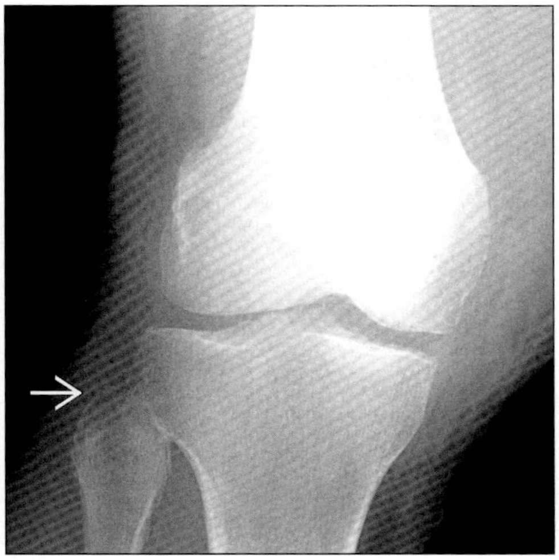

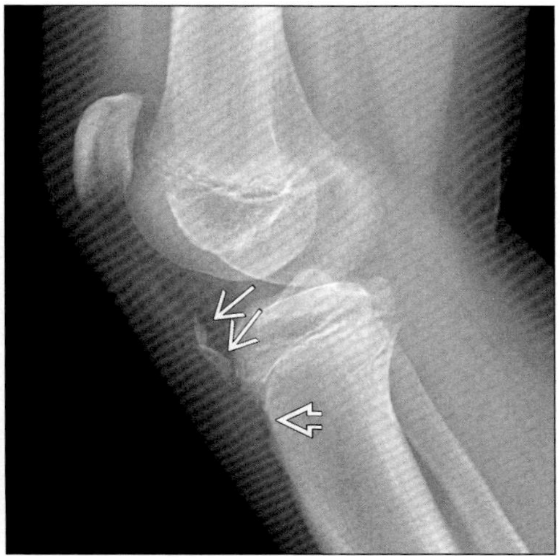

*(**Left**) Oblique radiograph shows the arcuate sign, a curvilinear bone fragment ➔ arising from the tip of the fibular styloid. This finding indicates avulsion of the arcuate ligament complex and injury to the posterolateral corner structures and has a high association with cruciate ligament injury. (**Right**) Lateral radiograph shows fragmentation and proximal displacement of tibial tubercle ➔ and defect at the fragment donor site ➔.*

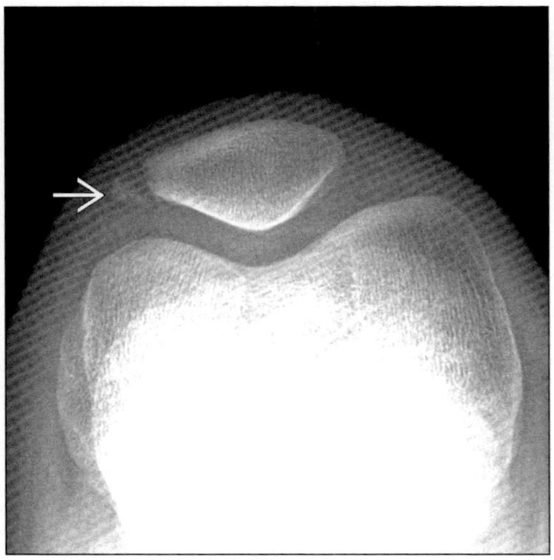

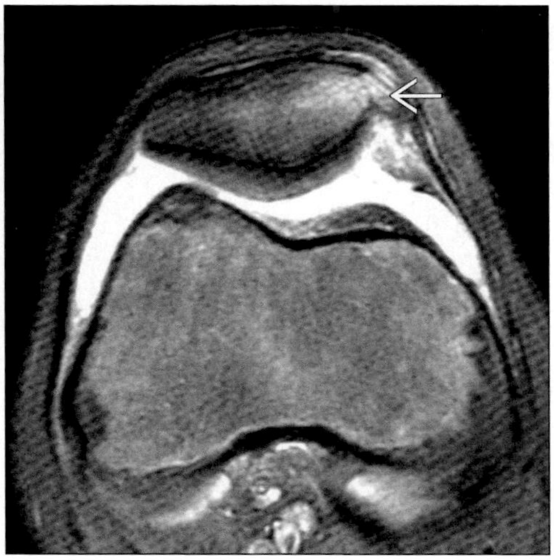

*(**Left**) Sunrise radiograph of the knee demonstrates an avulsion fragment ➔ arising lateral to the patella, signifying disruption of the lateral patellar retinaculum. When located medially, this indicates medial patellofemoral ligament avulsion and implies transient patellar dislocation. (**Right**) Axial PDWI FS MR in an 11 year old after knee injury shows avulsion of the medial patellofemoral ligament (MPFL) from the patella ➔. This is usually due to transient patellar dislocation.*

STRESS INJURY, LEG

Key Facts

Terminology

- Stress fracture, shin splints, medial tibial stress syndrome
- Fatigue fracture: Fracture due to abnormal stress on normal bone
- Insufficiency fracture: Fracture due to normal stress on abnormal bone
- Shin splints: Refers to a clinical condition of tibial pain in legs usually associated with running
  - Also known as medial tibial stress syndrome
  - Represents a range of injuries from stress edema to periostitis to frank cortical fracture of distal 2/3 of tibia

Imaging

- Tibia: Posterior proximal 1/3 > distal 1/3 > middle 1/3
- Fibula: Distal 1/3 > proximal 1/3
- Linear sclerosis
- Periosteal new bone formation
- Low signal fracture line in medullary bone surrounded by marrow edema
- Periosteal new bone: Low signal on all sequences
- Radiograph 1st line of diagnosis
- MR if radiographs negative and stress injury suspected

Clinical Issues

- Insidious onset of pain with activity
- Progression to linear fracture if activity persists

Diagnostic Checklist

- Look for underlying osteopenia in elderly patients with stress injury

__(Left)__ Coronal T2WI FS MR in a 14-year-old runner with knee and leg pain demonstrates a transverse band of low signal sclerosis in the proximal tibia surrounded by marrow edema. __(Right)__ Sagittal PDWI MR from the same patient shows that the transverse fracture is confined to the posterior subcortical aspect of the proximal tibia. There is marked cortical thickening as well, much of which likely represents periosteal new bone formation.

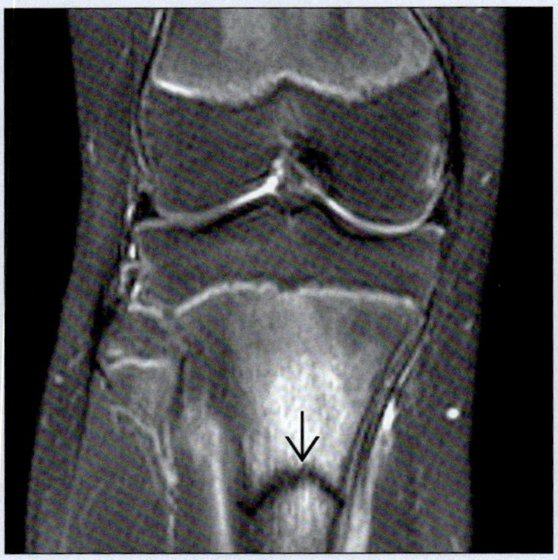

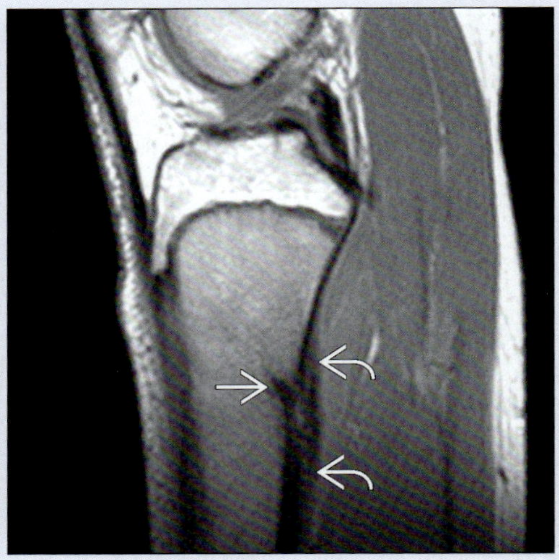

__(Left)__ Sagittal T2WI FS MR from the same patient shows the extent of marrow edema surrounding the fracture. There is also significant soft tissue edema surrounding the periosteal bone reaction. __(Right)__ Axial PDWI FS MR from the same patient better shows the degree of periosteal new bone formation involving about 1/2 of the circumference of the tibia. Soft tissue reaction mirrors the extent of the periosteal new bone formation. The medullary fracture is posterior.

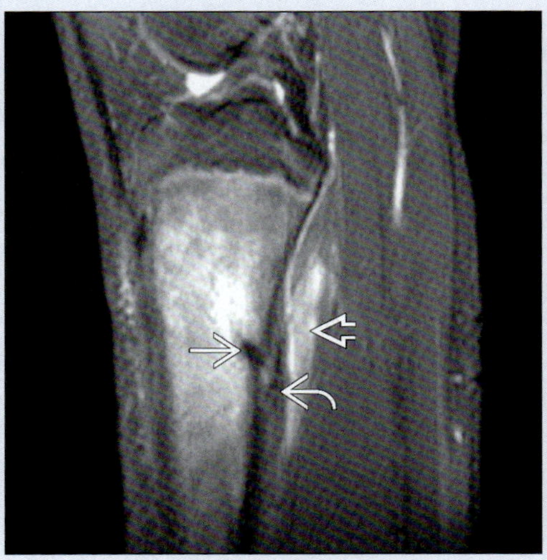

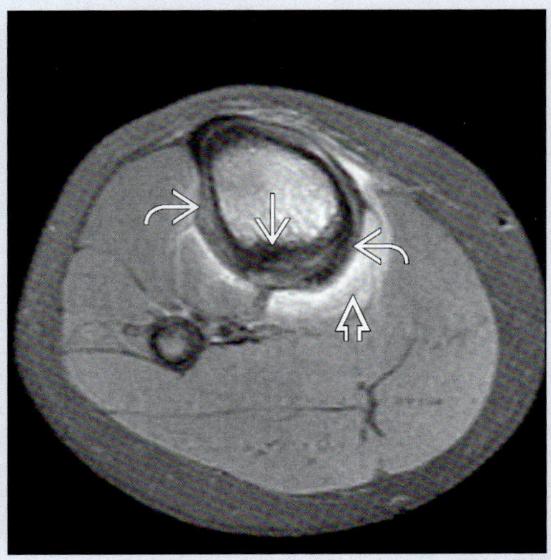

STRESS INJURY, LEG

TERMINOLOGY

Synonyms

- Stress fracture, shin splints, medial tibial stress syndrome

Definitions

- Fatigue fracture: Fracture due to abnormal stress on normal bone
- Insufficiency fracture: Fracture due to normal stress on abnormal bone
- Shin splints: Refers to a clinical condition of tibial pain in legs usually associated with running
  - Also known as medial tibial stress syndrome
  - Represents a range of injuries from stress edema to periostitis to frank cortical fracture of distal 2/3 of tibia

IMAGING

General Features

- Best diagnostic clue
  - Band-like fracture line in cortical &/or medullary bone of tibia or fibula
- Location
  - May occur anywhere in leg
  - Tibia: Posterior proximal 1/3 > distal 1/3 > middle 1/3
    - Often emanates from cortical channel for large medullary artery in posterior distal tibia
  - Fibula: Distal 1/3 > proximal 1/3
- Size
  - May range from tiny subtle cortical infraction to elongate fracture line several cm in length
- Morphology
  - Often linear, but can be curvilinear

Radiographic Findings

- Linear sclerosis
- Periosteal new bone formation
- Cortical lucency
- Soft tissue swelling

CT Findings

- Similar to radiographs
- More sensitive for fracture lines and periosteal new bone

MR Findings

- Spectrum of findings
  - Regional marrow edema without fracture
  - Periosteal &/or endosteal marrow edema without fracture
  - Periosteal &/or endosteal new bone
  - Cortical or medullary fracture line
  - May progress to frank linear fracture ± displacement
  - Healing phase: Cortical thickening
  - Cortical thickening may persist indefinitely after fracture heals
- Low signal fracture line in medullary bone surrounded by marrow edema
  - Often transverse
  - Can be longitudinal
- Cortical fracture may be hard to appreciate
  - Low signal fracture within low signal cortex
  - Sometimes fracture demonstrates ↑ signal on T2WI MR
- Periosteal new bone: Low signal on all sequences
  - Variable reactive soft tissue edema surrounding bone

Imaging Recommendations

- Best imaging tool
  - Radiograph 1st line of diagnosis
  - MR if radiographs negative and stress injury suspected
- Protocol advice
  - Combination of axial and long axis (coronal and sagittal) T1WI and T2WI FS MR necessary to see cortical fx, periosteal new bone, and linear medullary fx

Nuclear Medicine Findings

- Bone scan
  - Focal increased radiotracer uptake at fracture site on 3 phases
  - May be longitudinal near cortex
  - Less specific than MR

DIFFERENTIAL DIAGNOSIS

Normal Cortical Thickening

- Anterior tibial cortex often thicker than other areas
- Fibular cortex also asymmetric

Normal Tibial Nutrient Artery Channel

- Large penetrating osseous artery enters posterior tibial cortex in distal 1/3
- Travels superiorly a variable distance
- Channel may serve as stress riser and thus a starting point for stress injury

Compartment Syndrome

- Condition of increased intracompartmental pressure in leg with associated edema in muscles and fascia
- Bones generally normal

Osteomyelitis

- Often associated with marrow edema, cortical thickening or lucency, and soft tissue swelling
- Differentiation must be made by history and lab values

Osteoid Osteoma

- Intracortical lucency surrounded by thickening and edema
- Calcified nidus in cortical lucency
- No linear fracture
- Pain worse at night and relieved by aspirin

Surface Osteosarcoma

- Periosteal osteosarcoma: Lamellated periosteal new bone
  - ± marrow involvement
- Parosteal osteosarcoma: Thick ossified mass of bone arising from periosteal surface
  - Early tumor has amorphous bone formation and periosteal reaction which may mimic stress reaction/fracture
  - Some marrow involvement usually present

STRESS INJURY, LEG

PATHOLOGY

General Features

- Etiology
  - Bone responds to repetitive stress by remodeling
    - Cortical bone remodels by resorption followed by building of new trabeculae
    - Stress injury occurs during period when resorption has diminished trabeculae
  - Stress fractures evolve over time
    - Initially stress distributed over long segment (stress reaction)
    - Focal trabecular failure in this zone may occur (stress fracture)
    - This focal failure causes deformation that increases load on adjacent trabeculae
      - Process spreads along lines of stress
    - Muscle fatigue may play a role
      - Fatigued muscles provide less stress shielding of bone

Staging, Grading, & Classification

- Imaging staging
  - Stress edema without fracture
    - May be confined to pericortical zones in early phase
  - Focal cortical, subcortical, or medullary fatigue or insufficiency fracture
  - Displaced fracture

CLINICAL ISSUES

Presentation

- Most common signs/symptoms
  - Insidious onset of pain with activity
  - Onset often associated with change in duration or type of activity
- Other signs/symptoms
  - Swelling and focal tenderness

Demographics

- Age
  - Fatigue fracture: Young, active adults or adolescents
  - Insufficiency fracture: Middle-aged or elderly adults
- Gender
  - Fatigue fracture: F > M
  - Insufficiency fracture: F > > M
- Ethnicity
  - Insufficiency fractures: White, Asian > African-American
  - Fatigue fractures: No ethnic predilection
- Epidemiology
  - Fatigue fracture
    - Most common in runners
    - Also seen in soccer players, dancers, gymnasts
  - Insufficiency fracture
    - Typically elderly females with low BMI

Natural History & Prognosis

- Progression to linear fracture if activity persists
- Resolution with cessation of activity

Treatment

- Generally conservative
  - Cessation of offending activity
  - Limited weight-bearing
  - Ice
  - Physical therapy
- Must treat underlying osteopenia in insufficiency fracture

DIAGNOSTIC CHECKLIST

Consider

- Look for underlying osteopenia in elderly patients with stress injury

Image Interpretation Pearls

- Biopsy should be avoided, as healing bone may suggest low-grade osteosarcoma on microscopy

Reporting Tips

- Describe extent of cortical and medullary involvement

SELECTED REFERENCES

1. Sourlas I et al: Proximal tibial stress fractures associated with primary degenerative knee osteoarthritis. Am J Orthop. 38(3):120-4, 2009
2. Gaeta M et al: Diagnostic imaging in athletes with chronic lower leg pain. AJR Am J Roentgenol. 191(5):1412-9, 2008
3. Heyworth BE et al: Lower extremity stress fractures in pediatric and adolescent athletes. Curr Opin Pediatr. 20(1):58-61, 2008
4. Jowett AJ et al: Medial malleolar stress fracture secondary to chronic ankle impingement. Foot Ankle Int. 29(7):716-21, 2008
5. Woods M et al: Magnetic resonance imaging findings in patients with fibular stress injuries. Skeletal Radiol. 37(9):835-41, 2008
6. Drabicki RR et al: Stress fractures around the knee. Clin Sports Med. 25(1):105-15, ix, 2006
7. Fredericson M et al: Stress fractures in athletes. Top Magn Reson Imaging. 17(5):309-25, 2006
8. Ruohola JP et al: Fatigue bone injuries causing anterior lower leg pain. Clin Orthop Relat Res. 444:216-23, 2006
9. Wall J et al: Imaging of stress fractures in runners. Clin Sports Med. 25(4):781-802, 2006
10. Young AJ et al: Evaluation and treatment of tibial stress fractures. Clin Sports Med. 25(1):117-28, x, 2006
11. Kaeding CC et al: Management and return to play of stress fractures. Clin J Sport Med. 15(6):442-7, 2005
12. Aoki Y et al: Magnetic resonance imaging in stress fractures and shin splints. Clin Orthop Relat Res. (421):260-7, 2004
13. Tuan K et al: Stress fractures in athletes: risk factors, diagnosis, and management. Orthopedics. 27(6):583-91; quiz 592-3, 2004
14. Wilder RP et al: Overuse injuries: tendinopathies, stress fractures, compartment syndrome, and shin splints. Clin Sports Med. 23(1):55-81, vi, 2004
15. Spitz DJ et al: Imaging of stress fractures in the athlete. Radiol Clin North Am. 40(2):313-31, 2002
16. Gibbon WW: Shin splints. Radiology. 207(3):826-7, 1998
17. Shearman CM et al: Longitudinal tibial stress fractures: a report of eight cases and review of the literature. J Comput Assist Tomogr. 22(2):265-9, 1998
18. Anderson MW et al: Shin splints: MR appearance in a preliminary study. Radiology. 204(1):177-80, 1997
19. Batt ME: Shin splints--a review of terminology. Clin J Sport Med. 5(1):53-7, 1995

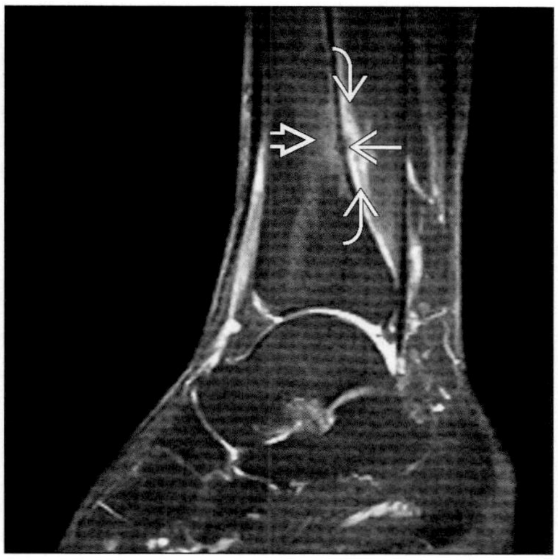

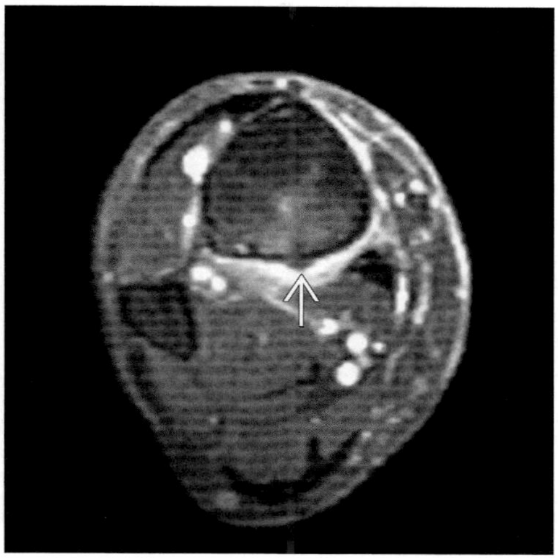

(Left) *Sagittal T2WI MR in a runner with distal leg pain shows a posterior cortical stress fracture with surrounding periosteal edema and mild endosteal marrow edema.* ***(Right)*** *Axial T2WI FS MR from the same patient shows the focal fracture and callus formation in the posterior tibial cortex. Stress fractures of the leg may occur in almost any part of the cortex or may not involve the cortex at all.*

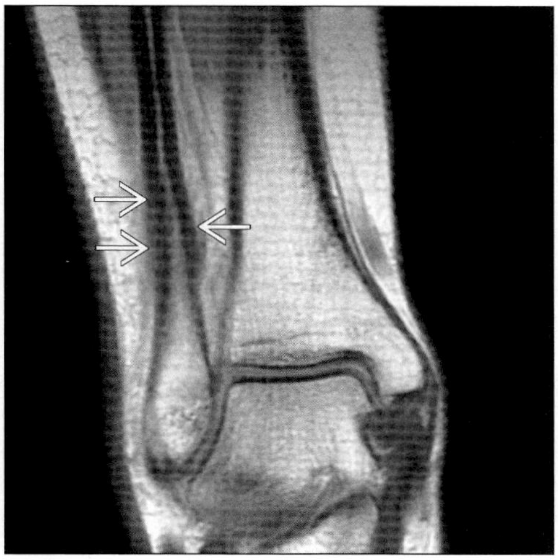

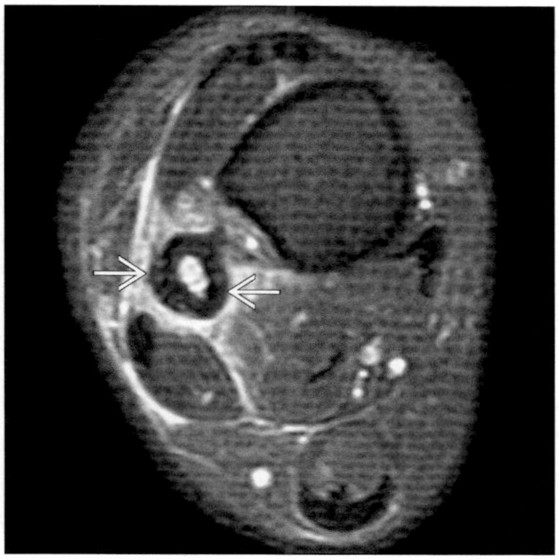

(Left) *Coronal PDWI MR in a 45-year-old woman who recently began exercising and complains of leg pain shows diffuse cortical thickening of the distal tibial shaft. A discrete fracture line is not observed.* ***(Right)*** *Axial T2WI FS MR from the same patient shows diffuse cortical thickening of the fibula. There is surrounding soft tissue swelling and edema of the marrow within the medullary space at this level. The tibia is normal.*

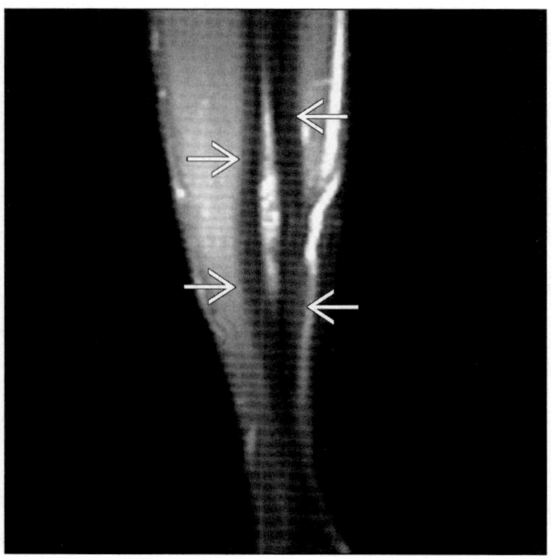

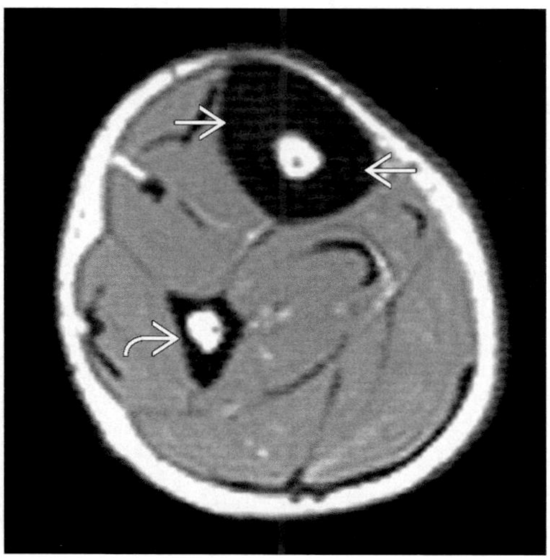

(Left) *Sagittal PDWI FS MR of the distal leg in a 68-year-old woman with chronic leg pain shows marked diffuse cortical thickening in the tibia, with mild associated marrow edema.* ***(Right)*** *Axial PDWI MR from the same patient shows the extensive cortical thickening of the tibia. There is no fracture line. The fibular cortex is normal. Healing or healed stress fractures in long bones often result in cortical thickening; it is more often focal than circumferential.*

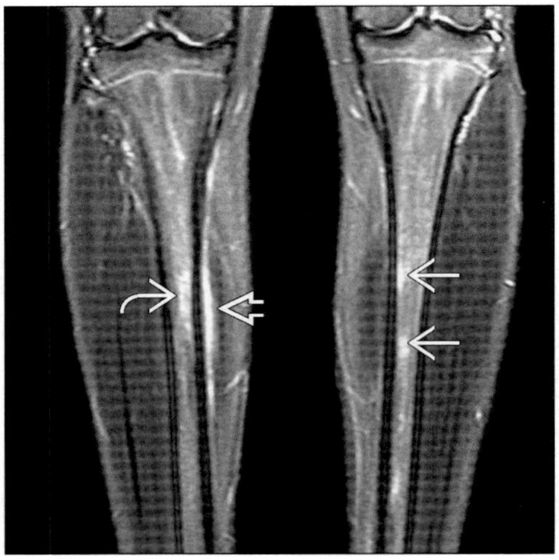

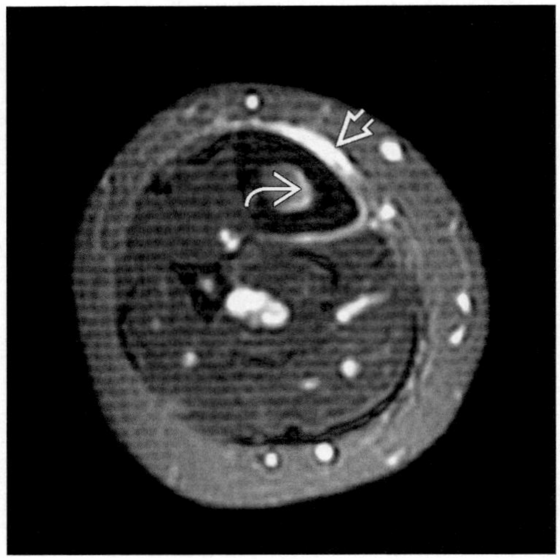

(Left) *Coronal STIR MR of both legs in a 14-year-old marathoner with right leg pain shows diffuse endosteal and periosteal edema centered around the mid-medial right tibial cortex. There are subtle areas of patchy marrow edema in the left leg as well.* ***(Right)*** *Axial STIR MR of the right leg from the same patient shows the periosteal and endosteal edema. The cortex is normal, and there is no visible fracture line. This pattern of early stress reaction is common in runners.*

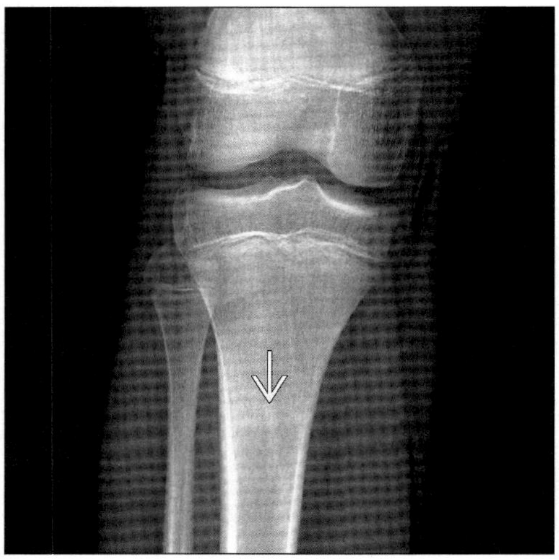

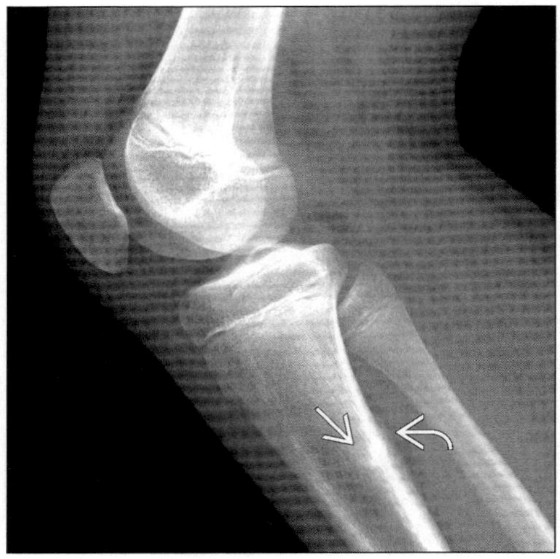

(Left) *AP radiograph in a 15-year-old runner with leg pain shows a band of sclerosis in the proximal tibia, indicating a healing stress fracture.* ***(Right)*** *Lateral radiograph from the same patient shows the band of sclerosis, as well as a focal area of cortical thickening posteriorly. Stress fractures in runners may or may not extend to the cortex. Any cortical surface may be involved, so multiple views are required to fully evaluate the injury.*

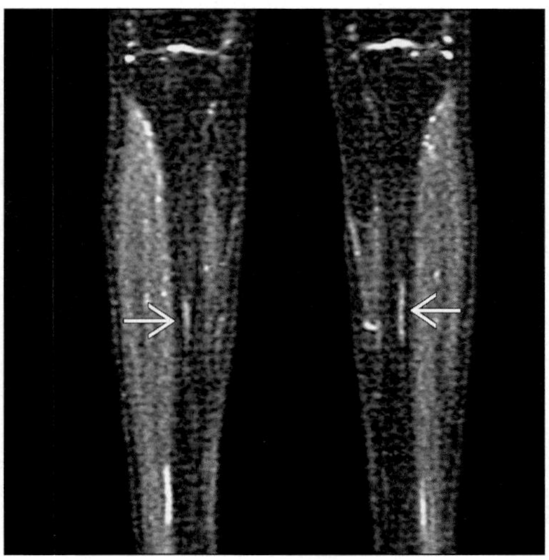

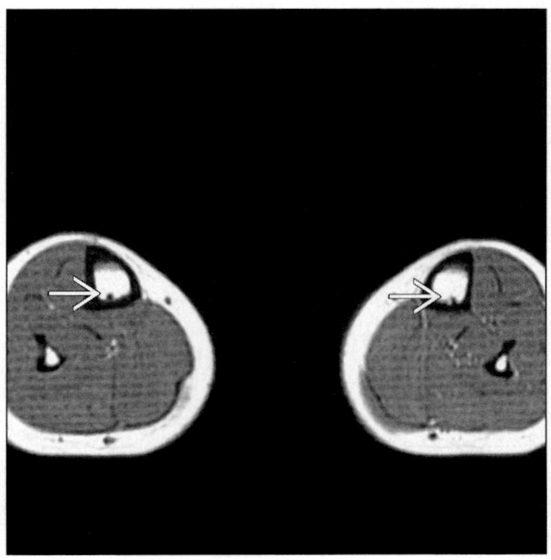

(Left) *Coronal STIR MR in a runner with bilateral leg pain shows linear foci of increased signal intensity in the midportion of both tibiae. This finding was initially interpreted as bilateral longitudinal stress fx.* ***(Right)*** *Axial T1WI MR from the same patient shows subcortical foci of low signal corresponding to the finding on the coronal image. These are medullary arteries and enter the tibia through characteristic foramen in the posterior cortex; they should not be mistaken for fractures.*

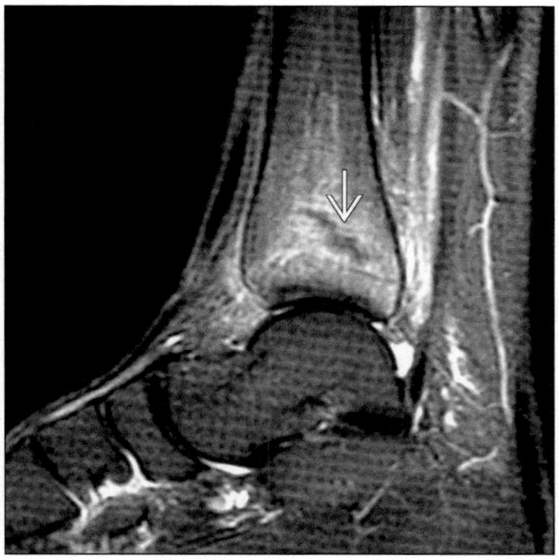

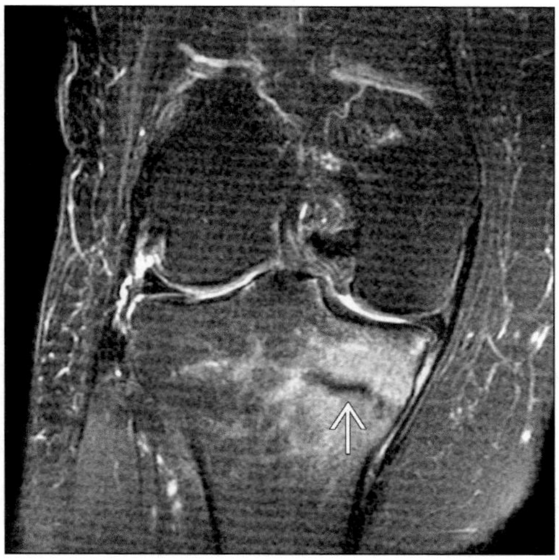

(Left) *Sagittal T2WI C+ FS MR in a 17-year-old runner shows a linear fatigue fracture in the medullary space of the distal tibia surrounded by marrow edema. There is diffuse soft tissue swelling as well.* ***(Right)*** *Coronal T2WI FS MR in a 57-year-old woman with knee pain after beginning a new exercise regimen shows a medullary insufficiency fracture in the medial tibia with surrounding marrow edema. Stress fractures do not always involve cortical bone.*

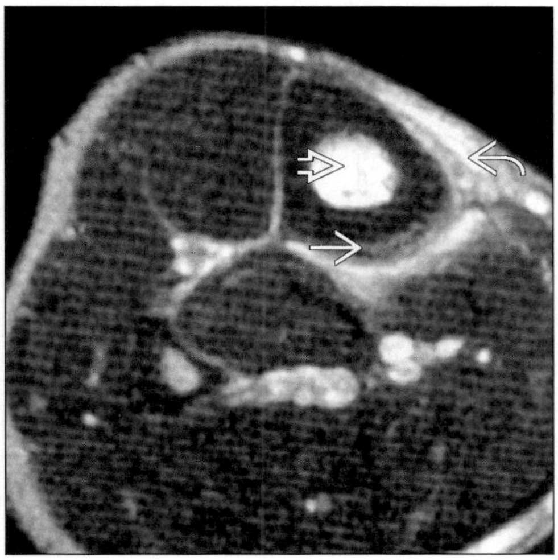

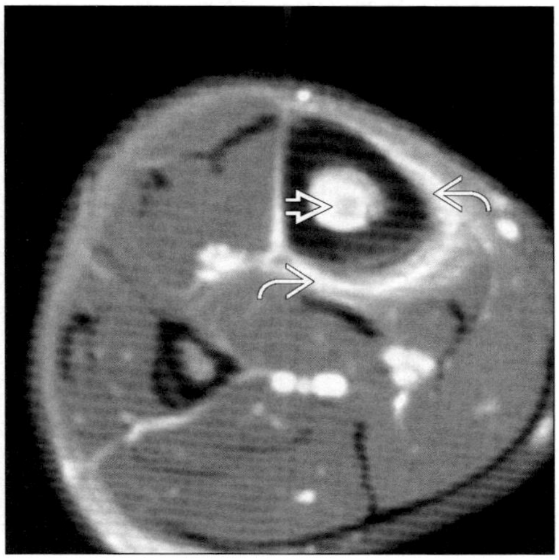

(Left) *Axial T2WI MR shows lamellated periosteal new bone in the posterior tibial cortex. There is surrounding soft tissue edema and diffuse marrow edema.* ***(Right)*** *Axial T1WI C+ FS MR from the same patient shows the robust enhancement in the marrow and periosteum and surrounding soft tissues typical of acute fracture. Care must be taken not to interpret enhancement as infection or tumor; trauma also results in significant enhancement in the acute and healing phases.*

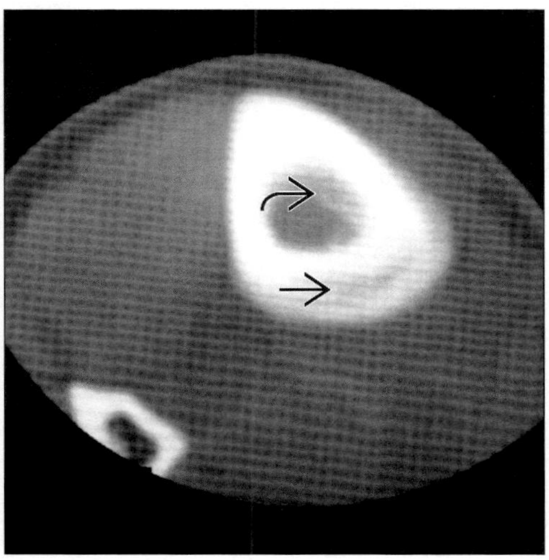

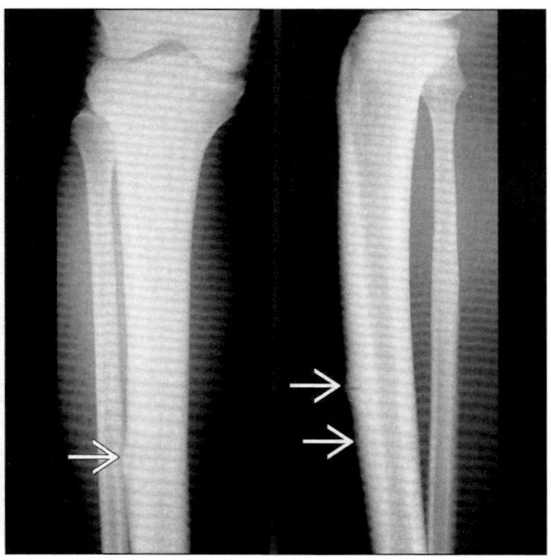

(Left) *Axial bone CT from the same patient shows rarefaction of the posterior cortex at the site of fracture. Endosteal new bone formation is seen anteromedially and was not appreciated on the MR images.* ***(Right)*** *Anteroposterior and lateral radiographs in a 26-year-old basketball player with 1 month of leg pain shows transverse linear stress fractures of the anterior tibial cortex. This location for stress fracture is somewhat unusual.*

TODDLER FRACTURE

Key Facts

Terminology
- Lower extremity stress fracture in young children, usually as they begin walking

Imaging
- Femur: Metaphyseal or diaphyseal fracture
- Tibia
  - Type I: Nondisplaced spiral fracture of midshaft, often better seen on obliques
  - Type II: Buckling of proximal anterior cortex, transverse fracture of posterior cortex
- Fibula: Buckle or bowing deformity of distal shaft
- Talus: Sclerotic band along short axis of neck
- Calcaneus
  - Vertical: Sclerotic fracture line parallel to calcaneal apophysis
  - Horizontal: Sclerotic band parallel to plantar cortex
- Cuboid: Sclerotic band parallel to calcaneocuboid joint
- Metatarsals: Buckle fracture of base(s), especially 1st ray

Top Differential Diagnoses
- Child abuse
- Osteomyelitis
- Osteoid osteoma
- Leukemia or neuroblastoma

Clinical Issues
- Toddler refusing to walk/bear weight on affected extremity
- Peak: 1-3 years

Diagnostic Checklist
- Child abuse: Any lower extremity fracture in a child not yet walking

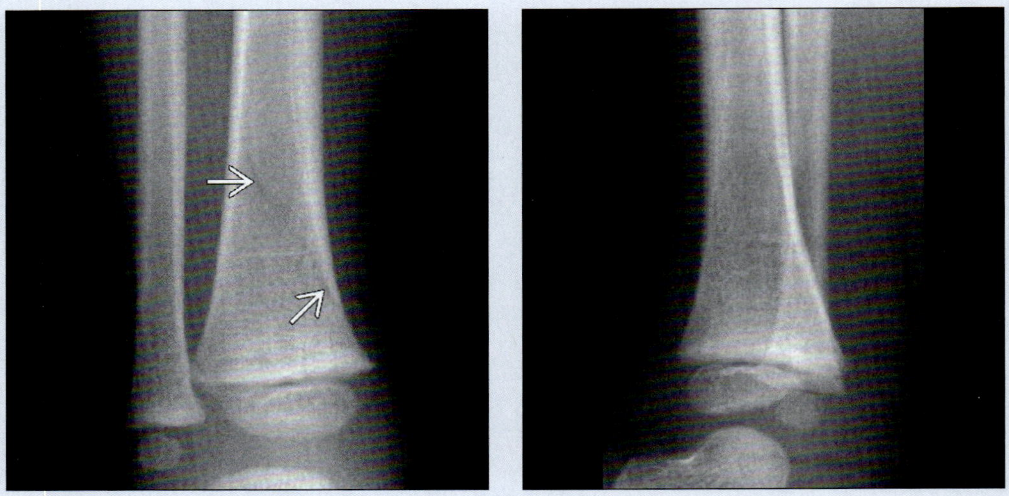

***(Left)** Anteroposterior radiograph shows a typical nondisplaced spiral toddler fracture of the distal tibial shaft ➔ in a young child just beginning to walk. The child suddenly refused to bear weight on the extremity, and was focally tender over the distal leg. **(Right)** Lateral radiograph in the same patient fails to demonstrate the fracture. It is not unusual for nondisplaced spiral fractures such as this to only be visible on 1 radiographic projection; oblique views should also be obtained.*

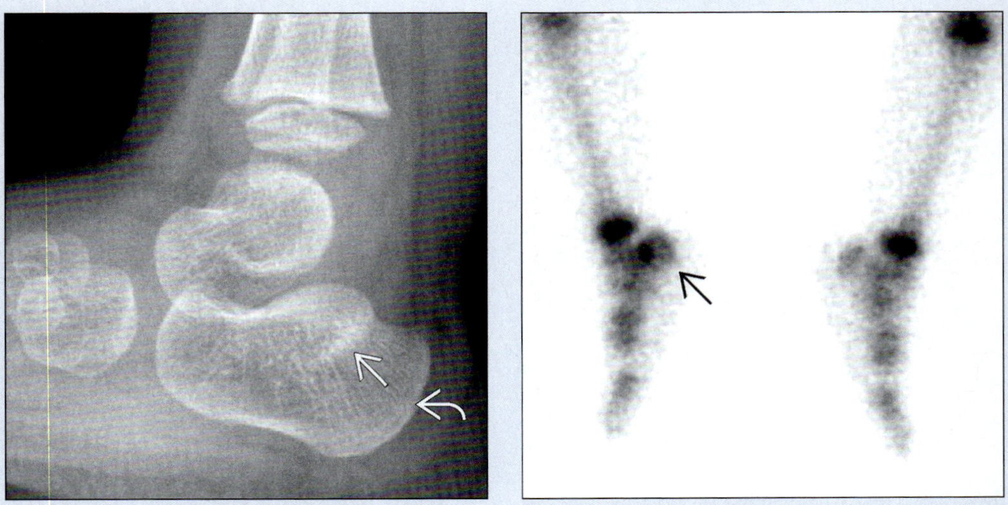

***(Left)** Lateral radiograph of the hindfoot in a child who began complaining of foot pain 14 days earlier shows a band of sclerosis ➔ paralleling the plane of the posterior apophysis ➔. This is a typical vertical calcaneal toddler fracture. **(Right)** Lateral bone scan in the same patient shows focal asymmetric radiotracer uptake ➔ in the posterior calcaneus, corresponding to the site of fracture. Bone scan is often positive a week or more before a fracture is evident on radiographs.*

TODDLER FRACTURE

TERMINOLOGY

Definitions

- Lower extremity stress fracture in young children, usually as they begin walking

IMAGING

General Features

- Best diagnostic clue
  - Spiral fracture of tibial midshaft in child of appropriate age
  - Sclerotic band in hindfoot or metatarsals
- Location
  - Femur: Diaphysis or metaphysis
  - Tibia: Midshaft > proximal shaft
  - Fibula: Distal diametaphysis
  - Talus: Neck and body
  - Calcaneus
  - Cuboid: Near calcaneocuboid joint
  - Metatarsals

Radiographic Findings

- Radiography
  - May be normal for up to 7-10 days after onset of symptoms
  - Bowing deformity without discrete fracture line may occur in tibia or fibula
  - Specific sites
    - Femur: Occurs at earlier age than tibia and hindfoot
    - Tibia
      - Type I: Nondisplaced spiral fracture of midshaft, often better seen on obliques
      - Type II: Buckling of proximal anterior cortex, transverse fracture of posterior cortex
    - Fibula: Buckle or bowing deformity of distal shaft
    - Talus: Sclerotic band along short axis of neck
    - Calcaneus
      - Vertical: Sclerotic fracture line parallel to calcaneal apophysis
      - Horizontal: Sclerotic band parallel to plantar cortex
    - Cuboid: Sclerotic band parallel to calcaneocuboid joint
    - Metatarsals: Buckle fx of base(s), especially 1st ray

Imaging Recommendations

- Best imaging tool
  - Radiographs for initial evaluation and follow-up
  - Bone scan or MR if radiographs equivocal or negative with strong clinical suspicion
- Protocol advice
  - Include oblique views in radiographic series

DIFFERENTIAL DIAGNOSIS

Child Abuse

- Findings may mimic toddler fracture
- High degree of suspicion appropriate in children not yet walking

Osteomyelitis

- Most common < 5 years of age

Osteoid Osteoma

- Focal cortical thickening with rounded lucent central nidus

Leukemia or Neuroblastoma Metastasis

- Horizontal lucent line(s) in metaphysis
- Typically wider and less well defined than stress fx

Septic Arthritis

- Age peak during infancy

PATHOLOGY

General Features

- Etiology
  - As children learn to walk, lower extremity is exposed to new axial loading stress
  - Mechanics of walking unrefined
    - Falls frequent
    - Torsional component
  - Jumping from height

CLINICAL ISSUES

Presentation

- Most common signs/symptoms
  - Toddler refusing to walk/bear weight on affected extremity
  - Painful to direct palpation

Demographics

- Age
  - 9 months to 5 years
  - Peak: 1-3 years

Natural History & Prognosis

- Rapid healing without deformity

Treatment

- Generally conservative

DIAGNOSTIC CHECKLIST

Consider

- Child abuse: Any lower extremity fracture in a child not yet walking
- Comparison to unaffected extremity for subtle bowing deformity

Reporting Tips

- If child abuse considered, **must** alert treating physician immediately

SELECTED REFERENCES

1. Sanchez TR et al: MR imaging of pediatric trauma. Magn Reson Imaging Clin N Am. 17(3):439-50, v, 2009
2. Jadhav SP et al: Commonly missed subtle skeletal injuries in children: a pictorial review. Emerg Radiol. 15(6):391-8, 2008
3. Swischuk LE et al: Frequently missed fractures in children (value of comparative views). Emerg Radiol. 11(1):22-8, 2004
4. John SD et al: Expanding the concept of the toddler's fracture. Radiographics. 17(2):367-76, 1997

OSTEOCHONDRAL INJURY, KNEE

Key Facts

Terminology
- Osteochondritis dissecans (OCD), osteochondral fracture (OCF)
- Injury resulting in fracture line extending through articular cartilage and some portion of underlying bone

Imaging
- Usually curved fracture line extending though cartilage and variable amount of subchondral bone
- MR ± direct arthrography is best imaging tool

Top Differential Diagnoses
- Chondral injury
- Subchondral fracture
- Osteoarthritis
- Osteonecrosis

Pathology
- In situ injuries generally divided into stable vs. unstable injuries
  - Stable: Nondisplaced OCF not likely to dislodge
  - Unstable: Non- or mildly displaced OCF with findings suggesting likelihood of future displacement
  - Different indicators for juvenile vs. adult form

Clinical Issues
- Stable OCF may become unstable if further traumatized
- Unstable OCF may dislodge and form intraarticular bodies

Diagnostic Checklist
- Describe location of lesion
- Describe size and depth of lesion
- Identify any associated intraarticular bodies

__(Left)__ Graphic illustration of an osteochondral fracture ➡ in the medial femoral condyle shows extension through the articular cartilage ➡ at both ends of the U-shaped lesion. __(Right)__ Coronal T1WI MR in a 15-year-old boy demonstrates a stable osteochondral fracture ➡ in the lateral weight-bearing surface of the medial femoral condyle. This is the classic appearance of "osteochondritis dissecans," an osteochondral fracture occurring in this location in adolescence.

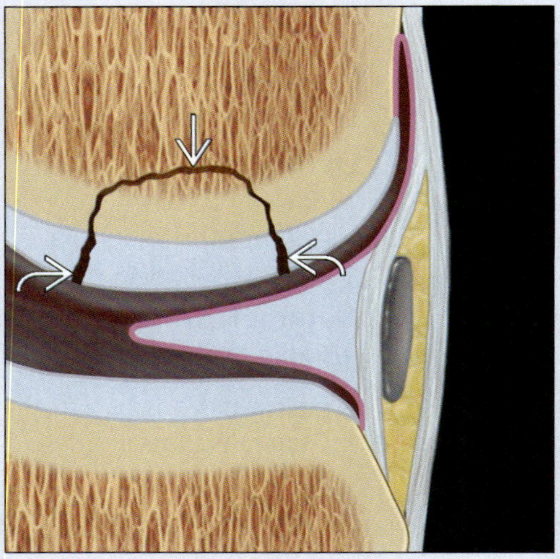

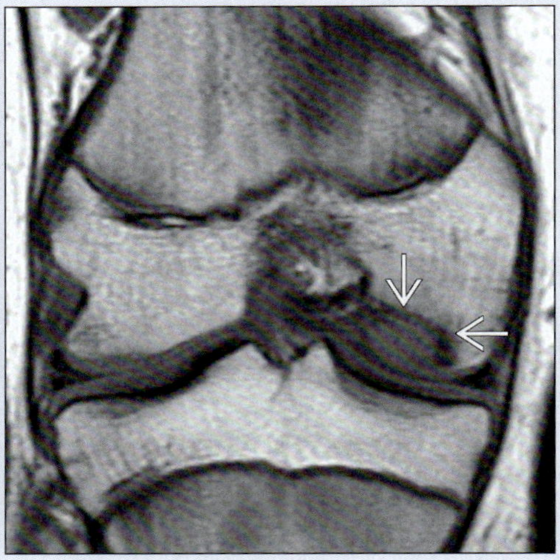

__(Left)__ Coronal T2WI FS MR in the same patient shows mild marrow edema ➡ surrounding the osteochondral fracture, indicating its acute or subacute nature. The lack of fluid in the interspace ➡ between the fragment and the native condyle indicates that this is a stable injury. __(Right)__ Sagittal PDWI MR shows the preserved cartilage ➡ over the lesion, indicating its stable nature. The transchondral extension of such lesions may be difficult to see after the acute phase.

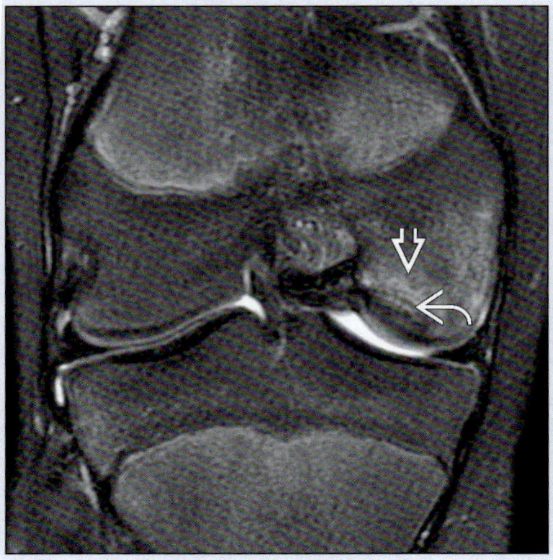

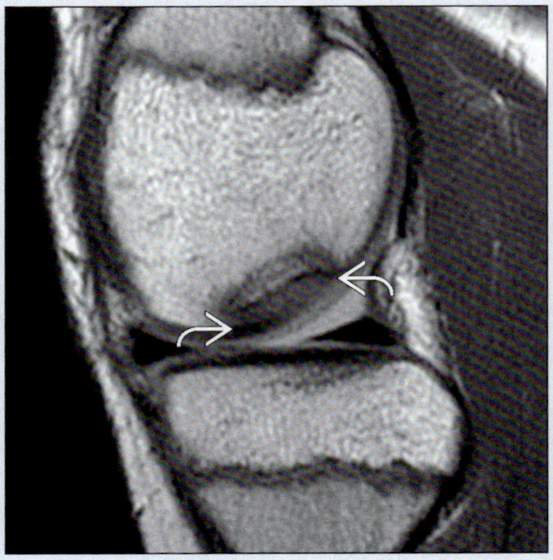

OSTEOCHONDRAL INJURY, KNEE

TERMINOLOGY

Abbreviations
- Osteochondral fracture (OCF), osteochondritis dissecans (OCD)

Synonyms
- Osteochondritis dissecans, osteochondral fracture

Definitions
- Injury resulting in fracture line extending through articular cartilage and some portion of underlying bone
- May or may not result in displaced fragment
- Osteochondritis dissecans
  - Term historically used to describe osteochondral fracture of knee in adolescents
  - Some believe this is also due to a component of osteonecrosis
  - Medial femoral condyle in 85%
    - Lateral weight-bearing aspect in 70% of these

IMAGING

General Features
- Best diagnostic clue
  - Fracture line extending through articular cartilage and into subchondral bone
- Location
  - Load-bearing articular surfaces
  - Lateral weight-bearing aspect of medial femoral condyle common in adolescents
    - "Osteochondritis dissecans"
- Size
  - May range from a few millimeters to several centimeters in diameter
- Morphology
  - Usually curved fracture line extending though cartilage and variable amount of subchondral bone

Radiographic Findings
- Acute
  - Nondisplaced: Subtle lucent curved fracture line in articular surface of bone
  - Displaced: Focal well-defined lucency in articular bone plate ± visible intraarticular body
- Chronic
  - Focal lucency in articular bone plate
  - May be ill-defined depending on healing response
  - Often associated intraarticular bodies

CT Findings
- Similar to radiographs, with more accurate depiction of size and depth of articular defect
- More sensitive for intraarticular bodies

MR Findings
- Acute
  - Nondisplaced
    - Curvilinear fracture line extending through subarticular bone plate
    - Both ends of fracture usually contact bone-cartilage junction
    - May or may not identify propagation of fracture through articular cartilage
    - Often with surrounding marrow edema
    - Stable lesion
      - No fluid extending completely around subtended osteochondral fragment
    - Unstable lesion
      - Fluid extends completely around fragment
      - Fragment may be slightly displaced
  - Displaced
    - Joint fluid fills defect in articular cartilage and part of subchondral bone
      - Well-defined, sharp-edged borders ("shouldered")
    - Often with surrounding marrow edema
    - Intraarticular fragments may contain bone, cartilage, or both
    - Other associated marrow &/or cartilage injuries often present
- Chronic
  - Signs of instability depend on patient age
    - Skeletally mature (adult) patients
      - Fluid-signal rim surrounding osteochondral fragment
      - Associated cyst formation
    - Skeletally immature (juvenile) patients
      - ↑ T2 signal surrounding osteochondral fragment with same intensity as joint fluid
      - ↓ signal rim surrounds ↑ signal rim on T2WI (fibrous tissue or sclerosis)
      - Multiple breaks in subchondral bone plate
      - Multiple or large (> 5 mm) subchondral cysts
  - Empty defect
    - Fluid signal fills variable-sized osteochondral defect
    - Edges may be smooth and sclerotic (low signal on all sequences)
    - Defect may fill partially or completely with low signal fibrous tissue
  - Intraarticular bodies

Imaging Recommendations
- Best imaging tool
  - MR
  - Some advocate MR arthrography for better delineation of stable vs. unstable lesions
- Protocol advice
  - T2WI FS or PDWI FS useful for delineation of subtle defects in cartilage and for fluid deep to fragment indicating instability
  - Other sequences also advocated (3D T1WI FS, gradient echo, etc.)
  - Specific sequences subject to personal preference

DIFFERENTIAL DIAGNOSIS

Chondral Injury
- Involves only cartilage, not underlying bone

Subchondral Fracture
- Fracture line in subchondral bone without disruption of overlying cartilage

Osteoarthritis
- Chronic lesions may be indistinguishable

OSTEOCHONDRAL INJURY, KNEE

Osteonecrosis
- Subchondral curvilinear sclerosis &/or edema surrounding normal marrow
- Overlying cartilage intact

Normal Variant
- Child may have cortical irregularity involving posterior weight-bearing portions of femoral condyle
- Overlying cartilage is normal
- Will develop normal appearance with skeletal maturation

PATHOLOGY

Staging, Grading, & Classification
- MR grading
  - Stable: Nondisplaced osteochondral fracture (OCF) not likely to dislodge
  - Unstable: Nondisplaced or mildly displaced OCF with findings suggesting likelihood of future displacement
    - Different indicators for juvenile vs. adult form
  - Grossly displaced injury: Empty defect with osteochondral fragment(s) elsewhere in joint
- Surgical classification
  - Category 1 (girls < 11, boys < 13 years of age): Conservative treatment
  - Category 2 (girls 11-15, boys 13-17 years of age): Treatment based on looseness of lesion
  - Category 3 (skeletally mature): Based on grade
    - Grade 1: Abnormal MR, intact articular surface at arthroscopy
    - Grade 2: Disrupted articular surface with no fragment looseness
    - Grade 3: Loose but in situ fragment
    - Grade 4: Detached fragment

Gross Pathologic & Surgical Features
- Fracture line extends through full thickness of articular cartilage, subchondral bone plate, and variable amount of medullary bone
- Fracture usually curvilinear, entering and exiting articular surface at ends of arc of fracture

Microscopic Features
- Osteonecrosis has been described on microscopic examination of these lesions, but this is true for many fractures

CLINICAL ISSUES

Presentation
- Most common signs/symptoms
  - Pain and swelling
  - Locking/restricted motion due to intraarticular fragments
- Other signs/symptoms
  - Joint effusion

Demographics
- Age
  - Children, adolescents, and young active adults
- Gender
  - M:F = 2:1
- Epidemiology
  - Femoral condyles most frequent site of osteochondral fracture
- May be genetic predisposition to developing osteochondral fractures

Natural History & Prognosis
- Stable OCF may become unstable if further traumatized
- Unstable OCF may dislodge and form intraarticular bodies
- More chronically
  - Empty osteochondral defect develops smoother edges
  - Defect may partially or completely fill with fibrin scar
  - Irregular articular surface results in osteoarthritis
- Prognosis much better in children than in skeletally mature patients

Treatment
- Conservative
  - Non-weight-bearing
  - Restriction of motion
- Drilling (encourages revascularization)
- Pinning (stabilize in situ lesions)
  - Steel vs. bioabsorbable pins
- Excision of fragment and loose bodies
- Screw fixation
- Osteochondral grafting (autograft or allograft)

DIAGNOSTIC CHECKLIST

Consider
- Stable vs. unstable lesion based on appropriate criteria for skeletally mature vs. immature patient

Reporting Tips
- Describe location of lesion
- Describe size and depth of lesion
- Identify any associated intraarticular bodies

SELECTED REFERENCES

1. Uozumi H et al: Histologic findings and possible causes of osteochondritis dissecans of the knee. Am J Sports Med. 37(10):2003-8, 2009
2. Kijowski R et al: Juvenile versus adult osteochondritis dissecans of the knee: appropriate MR imaging criteria for instability. Radiology. 248(2):571-8, 2008
3. Potter HG et al: Magnetic resonance imaging of cartilage repair. Sports Med Arthrosc. 16(4):236-45, 2008
4. Walsh SJ et al: Large osteochondral fractures of the lateral femoral condyle in the adolescent: outcome of bioabsorbable pin fixation. J Bone Joint Surg Am. 90(7):1473-8, 2008
5. Ramnath RR et al: MR appearance of SONK-like subchondral abnormalities in the adult knee: SONK redefined. Skeletal Radiol. 33(10):575-81, 2004
6. Alleyne KR et al: Management of osteochondral injuries of the knee. Clin Sports Med. 20(2):343-64, 2001
7. Farmer JM et al: Chondral and osteochondral injuries. Diagnosis and management. Clin Sports Med. 20(2):299-320, 2001
8. Loredo R et al: Imaging of osteochondral injuries. Clin Sports Med. 20(2):249-78, 2001

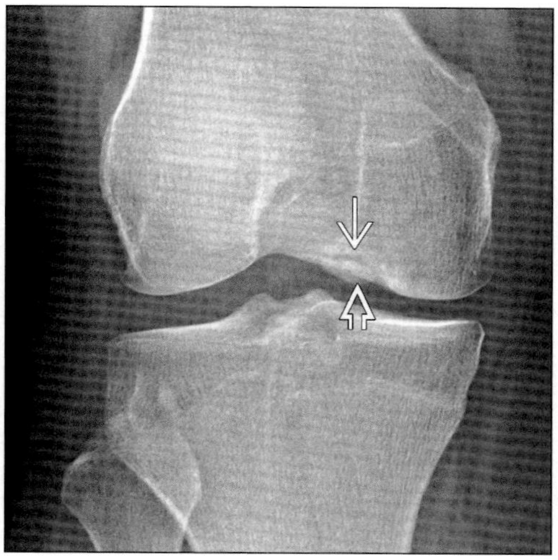

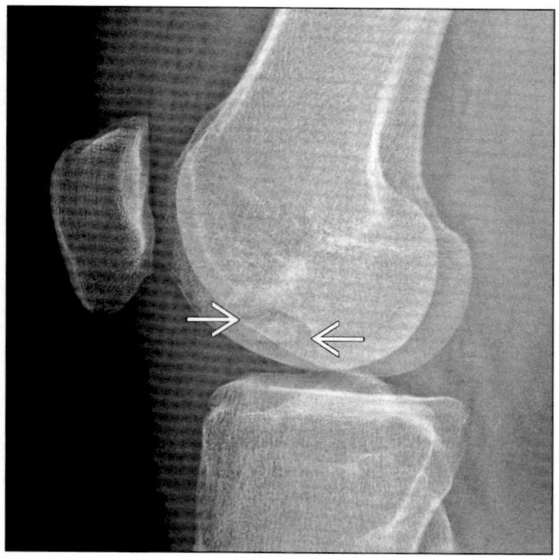

(Left) *AP radiograph of a 24-year-old man with chronic knee pain shows a crescentic lucency in the subchondral bone of the medial femoral condyle surrounding a sclerotic focus. This is the classic appearance of chronic osteochondral fracture. The lucency represents unossified cartilage.* ***(Right)*** *Lateral radiograph in the same patient shows the curvilinear osteochondral fracture. If such a defect is empty, intraarticular bodies should be sought.*

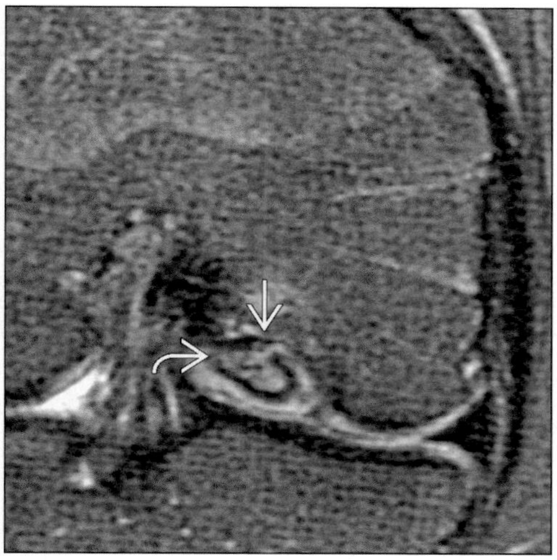

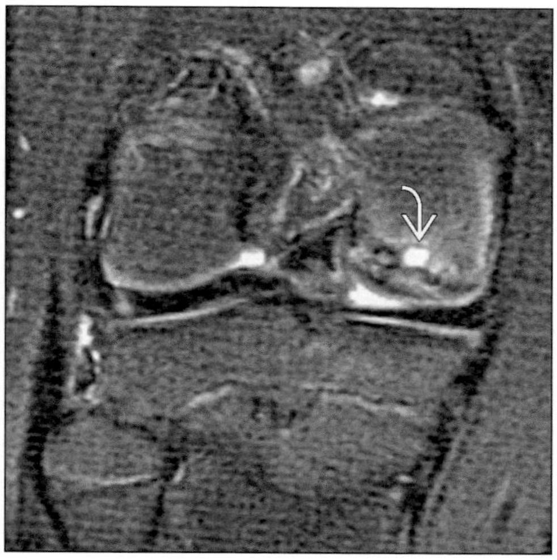

(Left) *Coronal T2WI FS MR in the same patient shows the corticated fragment with no fluid interposed between it and the native condyle. This is a chronic stable lesion.* ***(Right)*** *Coronal T2WI FS MR in the same patient shows underlying single cyst formation in the subchondral bone measuring 5 mm. The presence of a large cyst suggests instability, but the lack of fluid undermining the fragment suggests stability. The patient was treated successfully with conservative management.*

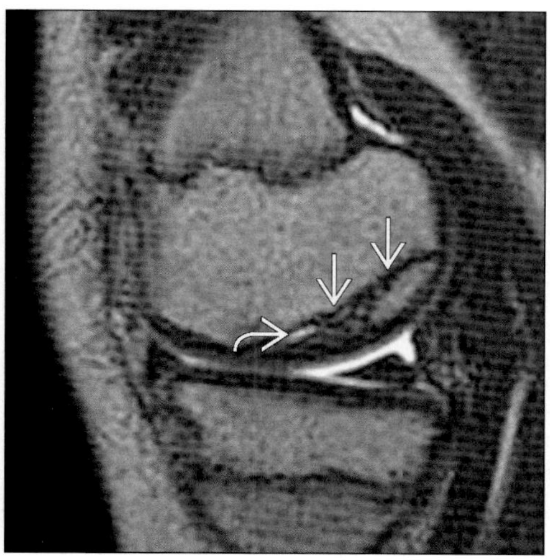

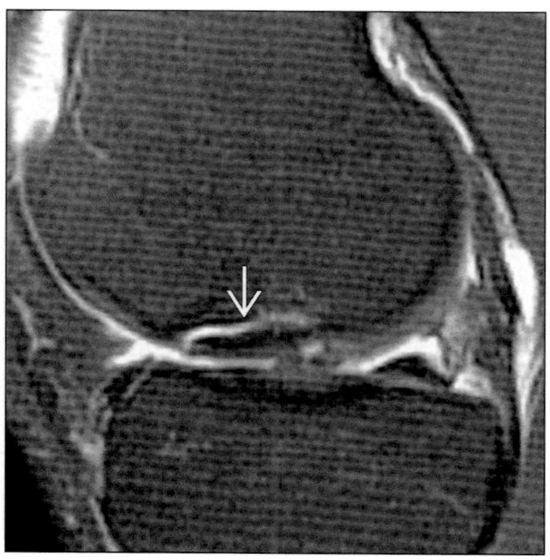

(Left) *Sagittal T2WI MR in a 14 year old shows a large chronic osteochondral fracture, with no surrounding edema. A small amount of fluid is present in the cleft between the fragment and the condyle, but most of this space is dry. This is a stable lesion.* ***(Right)*** *Sagittal MR arthrogram T1WI FS shows extension of contrast into the majority of the cleft in an osteochondral fracture, consistent with an unstable lesion.*

OSTEOCHONDRAL INJURY, KNEE

(Left) *Sagittal PDWI MR shows an acute osteochondral fracture of the lateral tibial articular surface with minimal impaction of the fracture. There is moderate surrounding marrow edema and a large lipohemarthrosis.* ***(Right)*** *Coronal T2WI FS MR in the same patient shows the slightly displace osteochondral fracture. Extension through the articular cartilage is better shown on this T2-weighted sequence.*

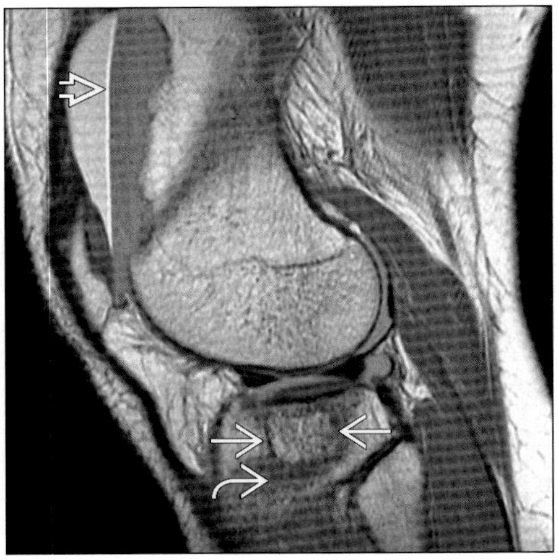

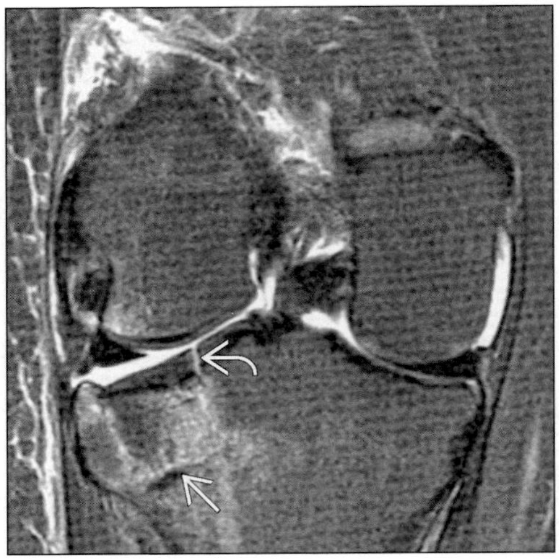

(Left) *Cross-table lateral radiograph after acute trauma to the knee demonstrates a fat-fluid level in the joint, consistent with an intraarticular fracture. There is a subtle defect in the medial trochlear articular surface, consistent with a displaced osteochondral fracture.* ***(Right)*** *Axial T2WI FS MR in the same patient confirms the osteochondral fracture in the medial trochlear ridge. There is surrounding marrow edema and a joint effusion is present.*

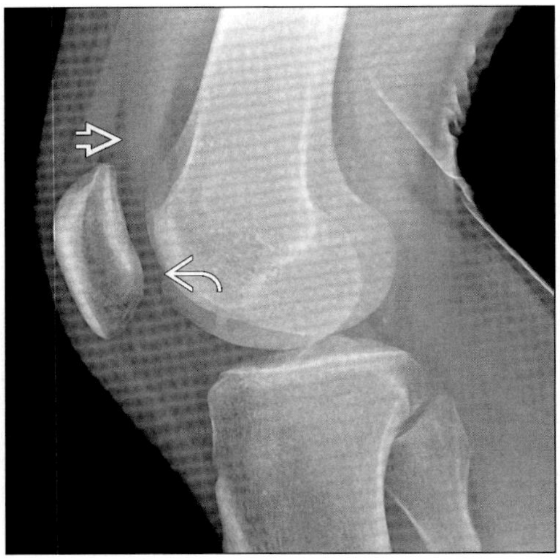

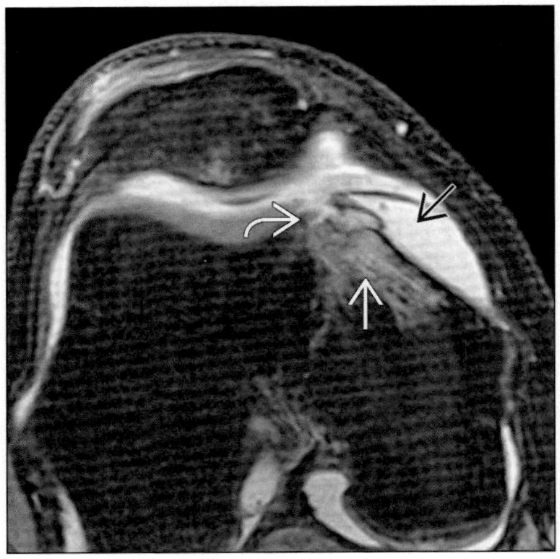

(Left) *Sagittal T2WI MR shows a slightly displaced osteochondral fracture of the medial femoral condyle. There is sclerosis of the edge of the fragment and in the native femur. Fluid surrounding the fragment has signal intensity equal to that of joint fluid. These features all suggest instability of the lesion.* ***(Right)*** *Coronal T2WI FS MR from the same patient confirms the fluid completely surrounding the fragment. Lack of marrow edema suggests chronicity.*

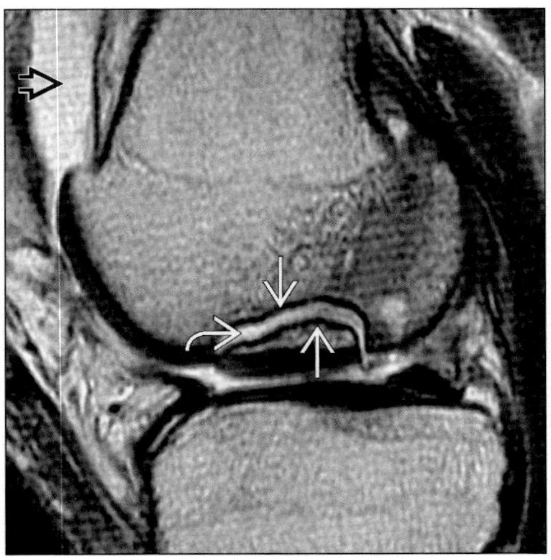

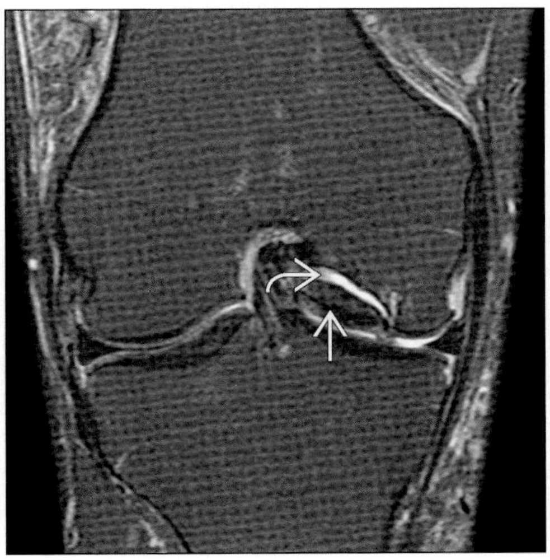

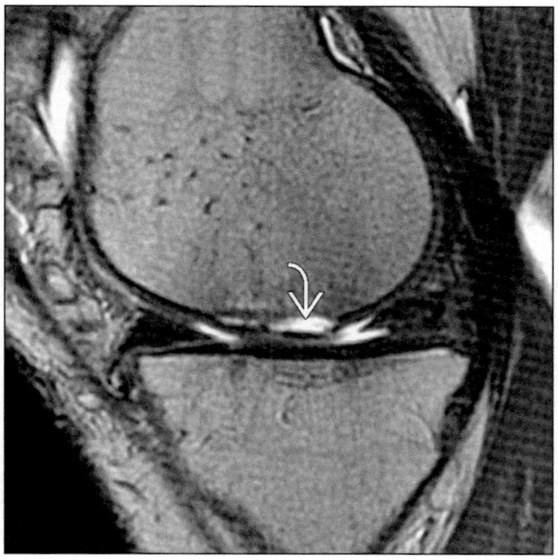

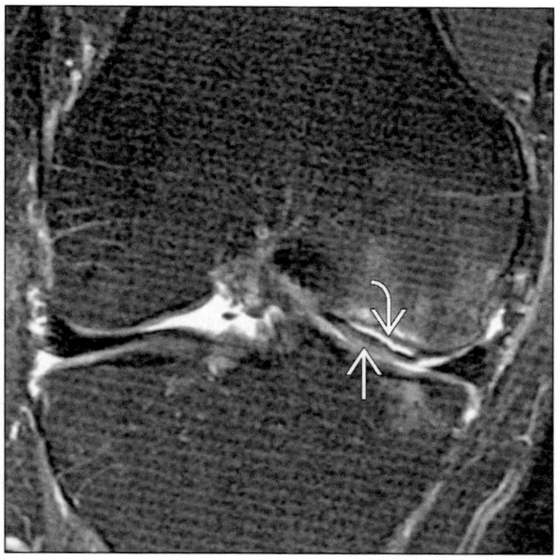

(Left) *Sagittal T2WI MR demonstrates an unstable osteochondral fracture of the medial femoral condyle, with a large amount of fluid between the native condyle and the osteochondral fragment.* ***(Right)*** *Coronal T2WI FS MR in the same patient shows mild displacement of the osteochondral fragment, with fluid filling the vacated space. There is mild marrow edema. This fragment is likely to dislodge completely at some point and become loose if not surgically stabilized.*

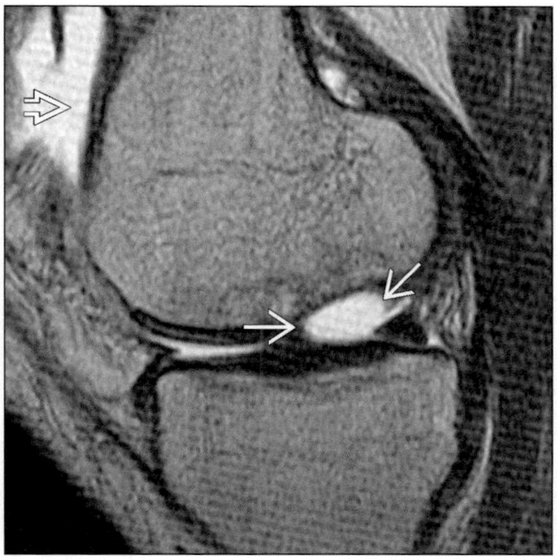

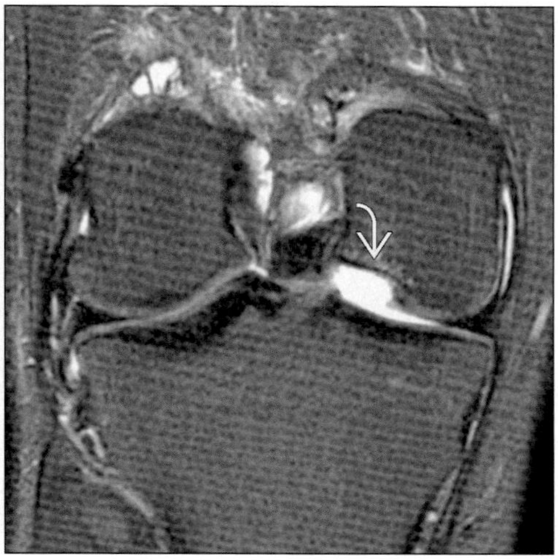

(Left) *Sagittal T2WI MR demonstrates an osteochondral defect in the posterior surface of the medial femoral condyle, indicating a previous osteochondral fracture with displacement of the fragment. A joint effusion is present.* ***(Right)*** *Coronal T2WI FS MR in the same patient confirms the subacute nature of the osteochondral defect , with mild sclerosis around the rim of the crater and associated reactive marrow edema.*

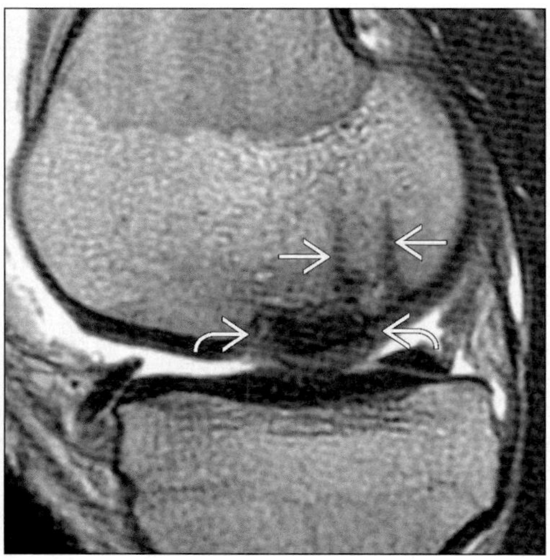

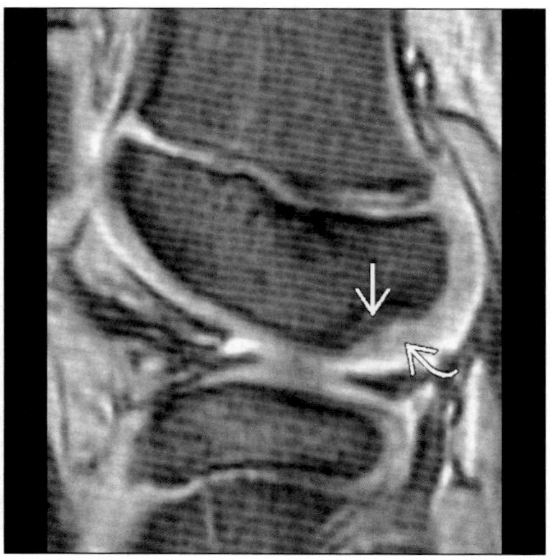

(Left) *Sagittal MR arthrogram T1WI after in situ fixation of an osteochondral fracture of the medial femoral condyle with bioabsorbable pins shows no contrast undermining the fragment to suggest instability. The cartilage surrounding the lesion is therefore intact.* ***(Right)*** *Sagittal T2* GRE MR in a child shows a "defect" in the subchondral bone of the posterior lateral femoral condyle, with intact overlying cartilage. This represents a normal variant of growth, and will remodel.*

CHONDRAL INJURY, KNEE

Key Facts

Terminology

- Cartilage fracture, cartilage tear, chondral flap, chondral fissure

Imaging

- Usually in weight- or stress-bearing surfaces
- Acute injury results in lesion with abrupt borders
- More chronic lesions may be more smoothly marginated due to healing response
- Defect usually filled with fluid
- Displaced fragments may be in several locations
- Joint effusion
- Some advocate MR arthrography for better sensitivity for small defects and intraarticular bodies
- Wide variety of pulse sequences have been advocated for cartilage lesion detection
- Specific pulse sequence selection is often matter of personal preference

Pathology

- Grade 0: Normal
- Grade I: Softening
- Grade II: Partial thickness defect < 50% of cartilage depth
- Grade III: Partial thickness defect > 50% of cartilage depth
- Grade IV: Full thickness defect ± underlying osseous abnormality

Clinical Issues

- Lesions may heal or enlarge ± deepen

Diagnostic Checklist

- Look for associated intraarticular bodies
- Lesion location
- Lesions size
- Lesion grade

__(Left)__ Graphic illustration shows a full thickness (grade IV) chondral defect → in the posterior weight-bearing portion of the medial femoral condyle. __(Right)__ Sagittal PDWI MR shows a full thickness (grade IV) chondral defect → in the posterior weight-bearing portion of the medial femoral condyle. There is a cartilage fragment → still located within the defect. Irregularity of the posterior horn of the medial meniscus → represents a tear.

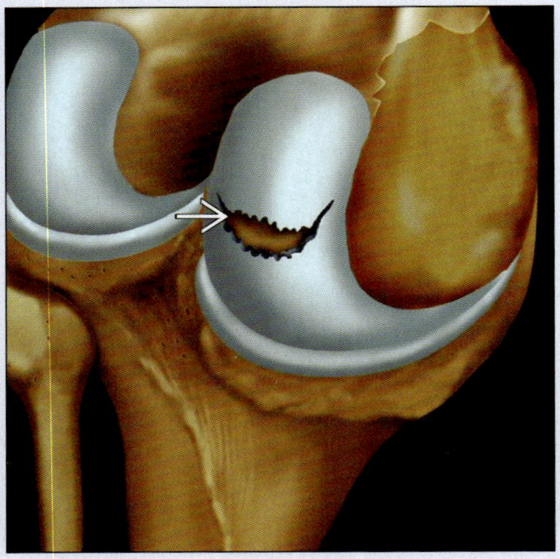

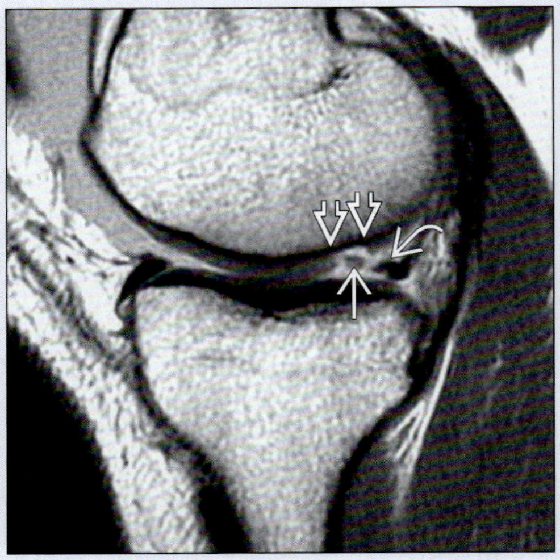

__(Left)__ Graphic illustration shows a partial thickness chondral defect → in the medial femoral condyle. __(Right)__ Coronal MR arthrogram PDWI FS shows a partial thickness chondral defect → in the lateral weight-bearing aspect of medial femoral condyle. This lesion involves slightly more than 50% of the total cartilage depth, making it grade III. Arthrography is generally not necessary to accurately grade articular cartilage lesions.

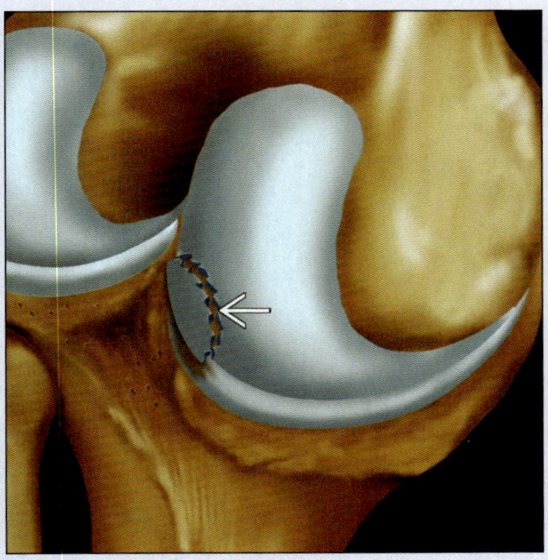

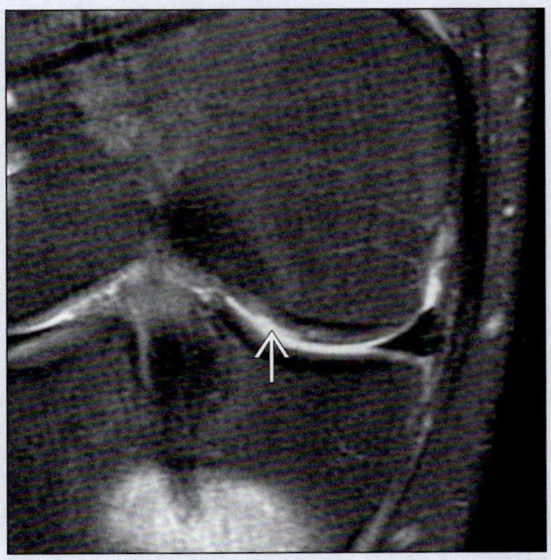

CHONDRAL INJURY, KNEE

TERMINOLOGY

Synonyms

- Cartilage fracture, cartilage tear, chondral flap, chondral fissure

Definitions

- Injury to hyaline cartilage surfaces in knee joint sparing underlying bone

IMAGING

General Features

- Best diagnostic clue
  - Well-demarcated chondral defect with or without underlying reactive marrow change on MR
- Location
  - Usually in weight- or stress-bearing surfaces
- Size
  - May range from 1 mm to several centimeters in size
- Morphology
  - Acute injury results in lesion with abrupt borders ("shouldered" lesion)
  - More chronic lesions may be more smoothly marginated due to healing response
  - Different types of lesions
    - Fibrillation: "Crab meat" appearance of cartilage on MR and arthroscopy
    - Fissuring: Shallow or deep linear defects in cartilage
    - Flap tear: Peel-back lesion of cartilage, may be partial or complete thickness
    - Chondral fracture: Well-demarcated lesion involving partial or complete thickness of cartilage

Radiographic Findings

- Usually none
- Indirect findings include effusion ± soft tissue swelling

CT Findings

- Cannot see nonmineralized cartilage or intraarticular bodies without arthrogram
- CT arthrogram highly accurate for detection and characterization of small high-grade chondral defects and intraarticular bodies

MR Findings

- Acute injury
  - Well-defined defect in articular (hyaline) cartilage surface
  - Defect usually filled with fluid
  - ± associated subchondral marrow edema
    - Especially if full thickness defect
  - Displaced fragments may be in several locations
    - Remain in situ (fluid surrounds detached fragment in defect)
    - Within joint adjacent to defect
      - Likely due to small point of attachment
    - Remotely displaced within joint
  - Joint effusion
    - Helpful for identifying small defects and associated small intraarticular bodies
- Chronic injury
  - Over time, defect is filled by scar
    - Scar is type I collagen instead of type II as found in normal hyaline cartilage
    - While scar cartilage looks same as normal articular cartilage, its compressive properties and resistance to shear stress are inferior
  - Eventually may be undetectable on MR imaging
  - Sharp edges become smoother
  - Underlying marrow edema resolves
  - Chronic subchondral sclerosis may be seen
  - Intraarticular bodies may
    - Resorb
    - Become synovialized
    - Remain loose within joint

Imaging Recommendations

- Best imaging tool
  - MR
  - Some advocate MR arthrography for better sensitivity for small defects and intraarticular bodies
- Protocol advice
  - Wide variety of pulse sequences have been advocated for cartilage lesion detection
  - Some are highly specific for cartilage imaging
  - Specific pulse sequence selection is often matter of personal preference

DIFFERENTIAL DIAGNOSIS

Osteochondral Injury

- Involves underlying bone in addition to articular cartilage

Degenerative Joint Disease (Osteoarthritis)

- Usually not as sharply marginated
- Secondary signs of degeneration present
- Subchondral cysts
- Subchondral sclerosis
- Osteophytosis

Magic Angle Artifact

- ↑ signal on short TE sequences (T1, PDWI) in areas where cartilage surface is ≈ 55° to main magnetic field (B°)
- Diminishes or disappears on longer TE sequences (T2WI)

PATHOLOGY

Staging, Grading, & Classification

- Several surgical and imaging grading systems used
- Become familiar with system used by your surgeons
- Generally based on modified Outerbridge grading system
  - Grade 0: Normal
  - Grade I: Softening
  - Grade II: Partial thickness defect < 50% of cartilage depth
  - Grade III: Partial thickness defect > 50% of cartilage depth
  - Grade IV: Full thickness defect ± underlying osseous abnormality

Gross Pathologic & Surgical Features

- Corresponds to arthroscopic and imaging grading system

Microscopic Features

- Physical defects in ground substance
- Disruption of chondrocytes
- Healing replaces type II collagen with type I collagen, which does not have same tensile properties

CLINICAL ISSUES

Presentation

- Most common signs/symptoms
  - Pain, particularly with motion
  - Swelling
  - Crepitus/catching/locking
- Other signs/symptoms
  - Joint effusion

Demographics

- Age
  - Younger, active patients
- Gender
  - M > F

Natural History & Prognosis

- Lesions may heal spontaneously
  - Healed areas do not have same mechanical properties of native cartilage
  - Increased stress on underlying bone
- Lesions may enlarge ± deepen
  - Leads to osteoarthritis

Treatment

- Conservative
  - Rest
  - Ice
  - NSAIDs
  - Physical therapy
- Glucosamine/chondroitin injections
  - Little scientific evidence of efficacy
- Microabrasion
  - Roughening of underlying bone to induce bleeding response
  - Fibrin clot transforms to type I collagen scar
- Chondroplasty ("pick" procedure)
  - Similar to microabrasion
  - Trephination of underlying bone to induce bleeding
  - Microfracture or drilling of bone often performed with this procedure
- Autologous osteochondral transplant (AOT)
  - Harvesting of osteochondral graft from non-weight-bearing surface
  - Reimplantation into affected site
  - Also referred to as OATS procedure
- Autologous chondrocyte implantation (ACI)
  - Chondrocytes harvested from patient
  - Cultured ex vivo, grown on synthetic scaffold
  - Reimplanted under periosteal membrane into defect
- Stem cell/platelet-rich plasma/whole blood injection
  - Early literature promising
  - Large-scale studies not published

DIAGNOSTIC CHECKLIST

Image Interpretation Pearls

- Look for associated intraarticular bodies
- Underlying marrow edema may indicate more acute lesion

Reporting Tips

- Lesion location
- Lesions size
- Lesion grade

SELECTED REFERENCES

1. Van Assche D et al: Autologous chondrocyte implantation versus microfracture for knee cartilage injury: a prospective randomized trial, with 2-year follow-up. Knee Surg Sports Traumatol Arthrosc. 18(4):486-95, 2010
2. Black BR et al: Cartilage imaging in sports medicine. Sports Med Arthrosc. 17(1):68-80, 2009
3. Gobbi A et al: Biological approaches for cartilage repair. J Knee Surg. 22(1):36-44, 2009
4. Gold GE et al: Recent advances in MRI of articular cartilage. AJR Am J Roentgenol. 193(3):628-38, 2009
5. Kon E et al: Arthroscopic second-generation autologous chondrocyte implantation compared with microfracture for chondral lesions of the knee: prospective nonrandomized study at 5 years. Am J Sports Med. 37(1):33-41, 2009
6. Levine DW et al: Characterized chondrocyte implantation results in better structural repair when treating symptomatic cartilage defects of the knee in a randomized controlled trial versus microfracture. Am J Sports Med. 37(3):e3; author reply e4, 2009
7. Mathieu L et al: Knee MR-arthrography in assessment of meniscal and chondral lesions. Orthop Traumatol Surg Res. 95(1):40-7, 2009
8. Mithoefer K et al: Clinical efficacy of the microfracture technique for articular cartilage repair in the knee: an evidence-based systematic analysis. Am J Sports Med. 37(10):2053-63, 2009
9. Tohyama H et al: Atelocollagen-associated autologous chondrocyte implantation for the repair of chondral defects of the knee: a prospective multicenter clinical trial in Japan. J Orthop Sci. 14(5):579-88, 2009
10. Trattnig S et al: MR imaging of cartilage and its repair in the knee--a review. Eur Radiol. 19(7):1582-94, 2009
11. Gaissmaier C et al: Growth and differentiation factors for cartilage healing and repair. Injury. 39 Suppl 1:S88-96, 2008
12. Hettrich CM et al: Cartilage repair: third-generation cell-based technologies--basic science, surgical techniques, clinical outcomes. Sports Med Arthrosc. 16(4):230-5, 2008
13. Kon E et al: Second generation issues in cartilage repair. Sports Med Arthrosc. 16(4):221-9, 2008
14. McNickle AG et al: Overview of existing cartilage repair technology. Sports Med Arthrosc. 16(4):196-201, 2008
15. Pelttari K et al: The use of mesenchymal stem cells for chondrogenesis. Injury. 39 Suppl 1:S58-65, 2008
16. Williams Iii RJ et al: Cartilage repair procedures: clinical approach and decision making. Instr Course Lect. 57:553-61, 2008
17. Hennig A et al: Osteochondral allografts in the treatment of articular cartilage injuries of the knee. Sports Med Arthrosc. 15(3):126-32, 2007
18. Noyes FR et al: A system for grading articular cartilage lesions at arthroscopy. Am J Sports Med. 17(4):505-13, 1989
19. Outerbridge RE et al: The problem of chondromalacia patellae. Clin Orthop Relat Res. (110):177-96, 1975

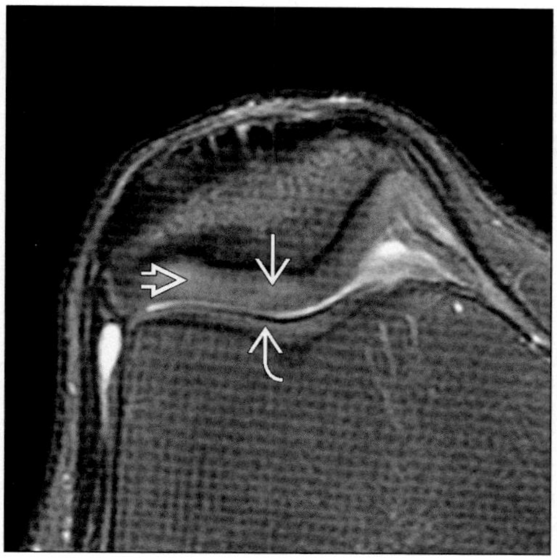

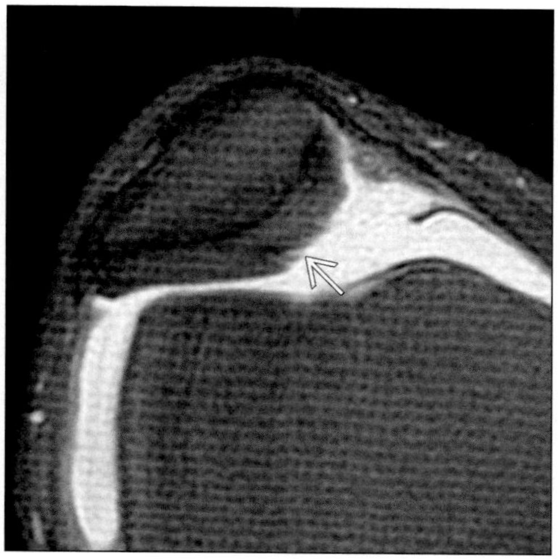

(Left) *Axial PDWI FS MR shows normal patellar and trochlear cartilage. Patellar cartilage is the thickest in the body, averaging 6 mm in depth. Areas of mild signal elevation within the patellar cartilage may represent grade I chondral changes.* ***(Right)*** *Axial T2WI FS MR shows grade II chondral fibrillation in the patellar apex. Grade II changes indicate cartilage defects extending less than 50% of the total depth of the cartilage.*

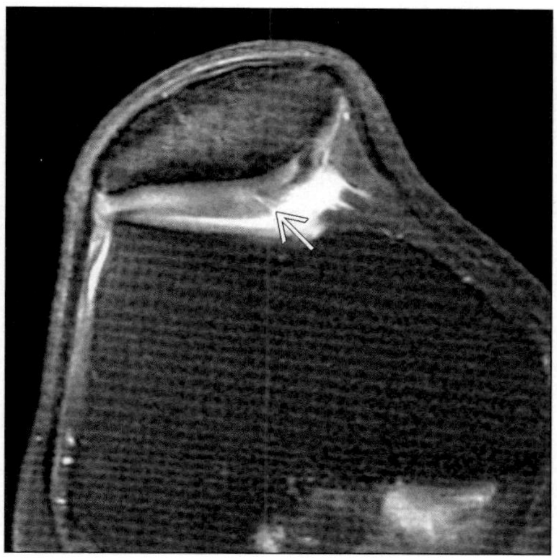

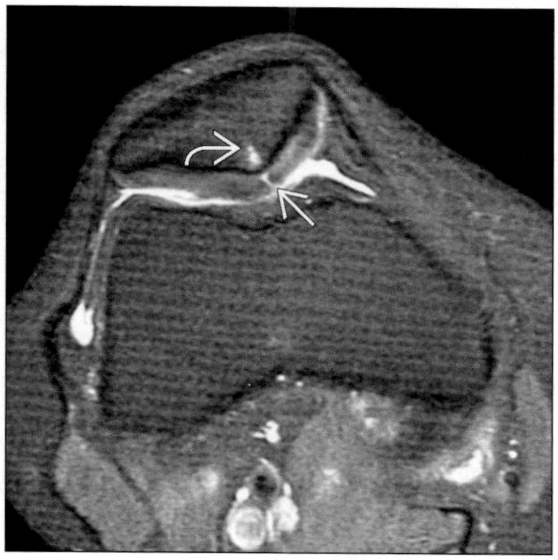

(Left) *Axial PDWI FS MR shows a grade III chondral defect in the patellar apex. Grade III defects are characterized by depth greater than 50% but less than 100% of total cartilage thickness.* ***(Right)*** *Axial PDWI FS MR shows a focal linear grade IV chondral defect in the patellar apex. Fluid tracks into the defect. Grade IV chondral defects extend through the full thickness of the articular cartilage. There is mild reactive marrow edema in the subchondral bone.*

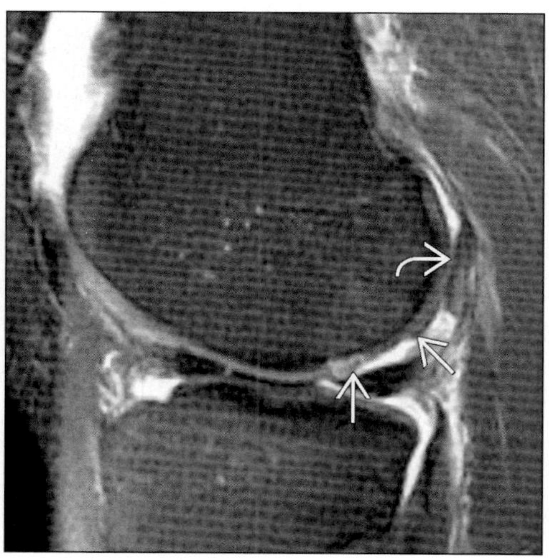

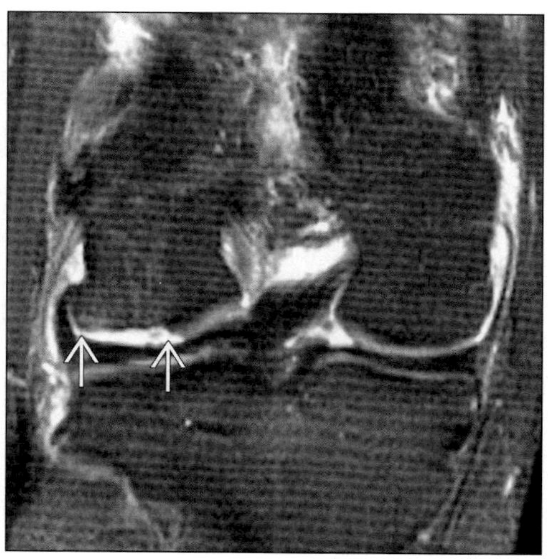

(Left) *Sagittal PDWI FS MR shows an elongate well-defined grade IV chondral defect in the posterior weight-bearing portion of the lateral femoral condyle with little reactive marrow reaction. A large chondral fragment is seen in the joint posterior to the condyle.* ***(Right)*** *Coronal T2WI FS MR in the same patient shows the well-marginated chondral defect. Acute chondral fractures are often sharply marginated like this one; in such cases, intraarticular fragments should be sought.*

(Left) *Coronal T2WI FS MR demonstrates a grade IV chondral defect in the central weight-bearing portion of the medial tibia ➔ with extensive underlying marrow edema ➔. There is likely a small subchondral insufficiency fracture ➔ as well. This patient was a 71-year-old woman with osteopenia.* ***(Right)*** *Coronal T2WI FS MR shows an acute grade IV chondral fracture of the lateral tibial plateau ➔. Marrow edema ➔ and sharp margins of the defect suggest acuity.*

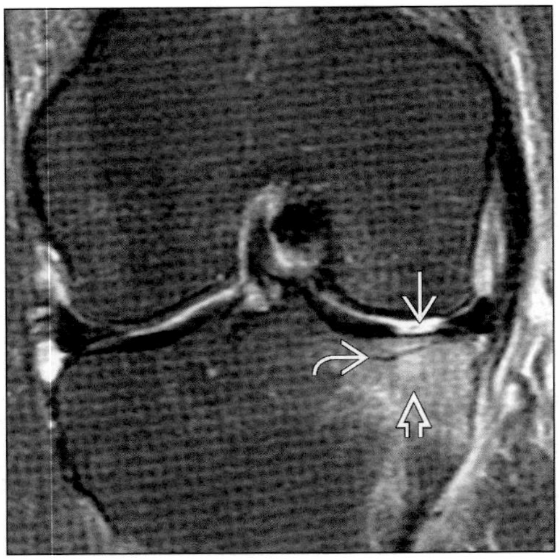

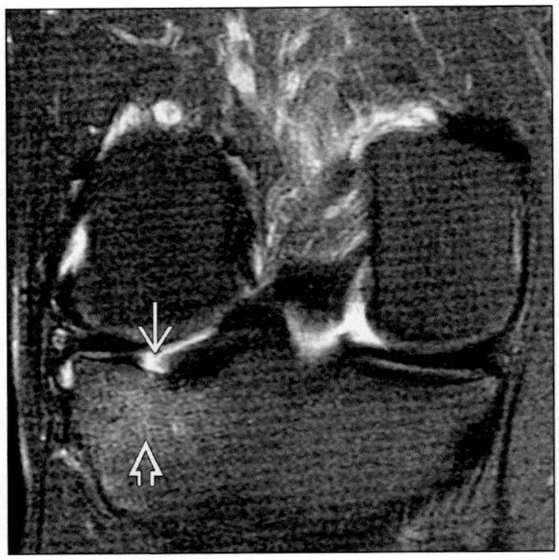

(Left) *Axial PDWI FS MR shows a large full thickness flap tear ➔ of the apical cartilage of the patella. The displaced fragment may eventually separate and become a free fragment.* ***(Right)*** *Sagittal T2WI MR shows a full thickness displaced flap tear ➔ of the lateral femoral condylar cartilage. The U-shaped fragment is peeled back from the underlying grade IV defect ➔ and is pointed anteriorly. Flap tears are inherently unstable and should be specifically noted.*

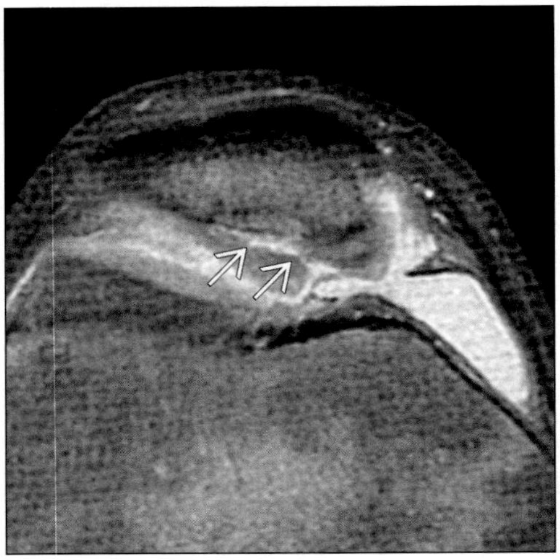

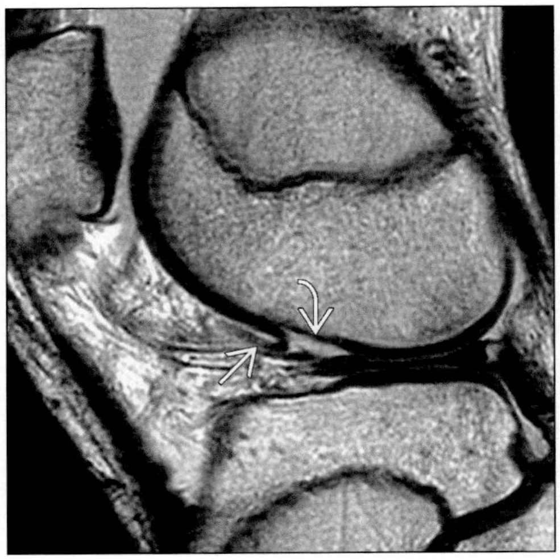

(Left) *Axial PDWI FS MR in a college football player shows an acute-on-chronic injury pattern in the patella with an acute grade IV chondral defect ➔ seen adjacent to a more chronic lesion that has healed with scar tissue ➔. Underlying subchondral cysts ➔ indicate the chronic nature of the lesion.* ***(Right)*** *Coronal MR arthrogram PDWI shows a partially healed chondral defect in the lateral tibial plateau ➔. The scar is thinner than the native cartilage adjacent to it ➔.*

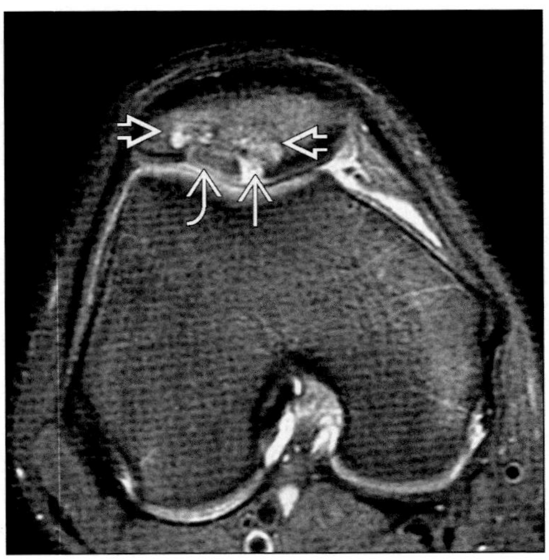

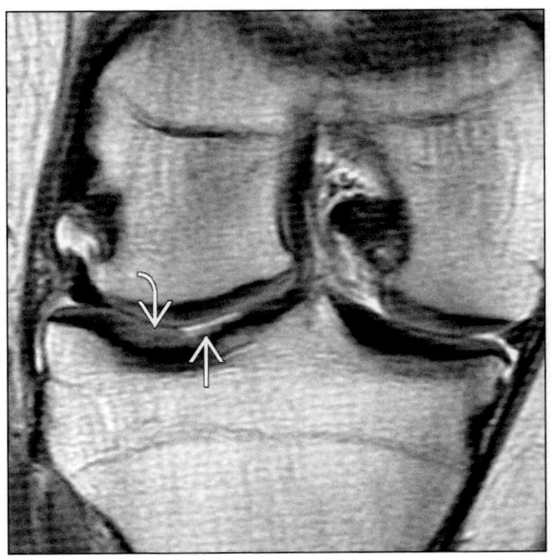

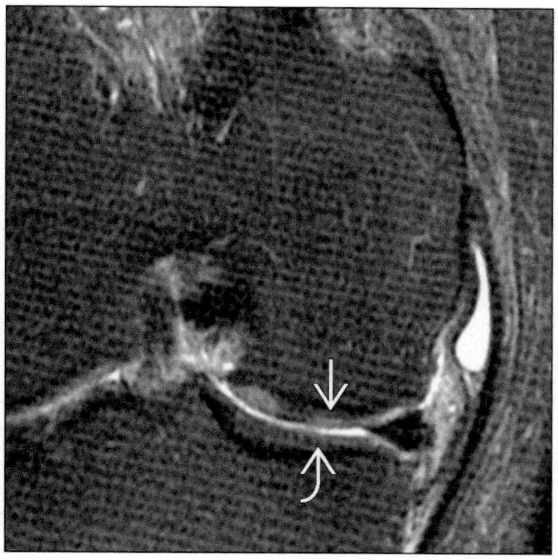

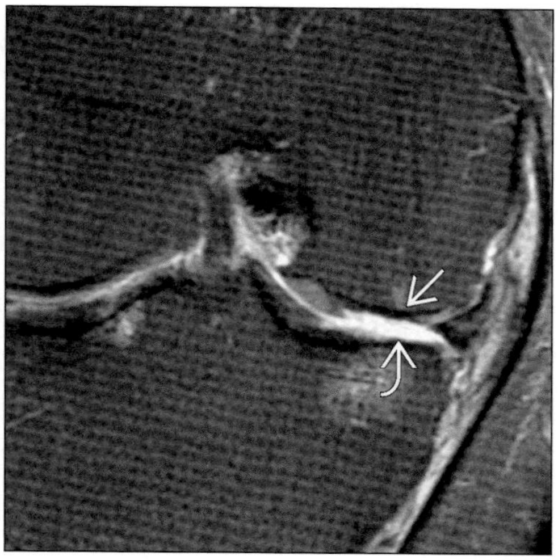

(Left) *Coronal T2WI FS MR in a 50-year-old woman prior to arthroscopy for medial meniscal tear shows normal medial femoral* ➔ *and tibial* ➔ *articular cartilage surfaces.* ***(Right)*** *Coronal T2WI FS MR in the same patient 7 months later and 6 months after interval arthroscopy shows full thickness cartilage defects in the medial femoral condyle* ➔ *and medial tibia* ➔*, with reactive marrow edema in both areas. Post-arthroscopic chondral fracture may be due to altered weight-bearing mechanics.*

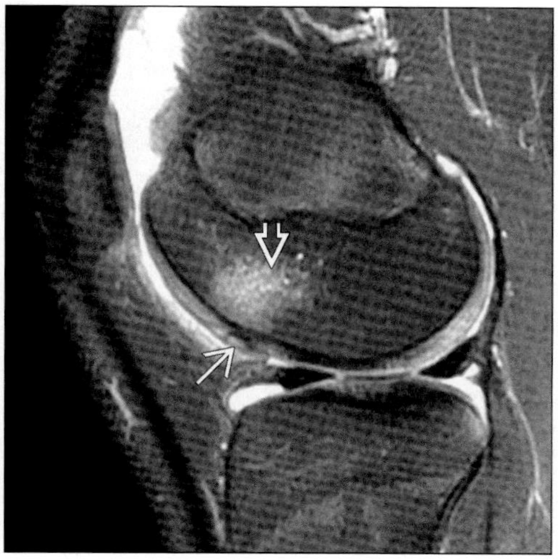

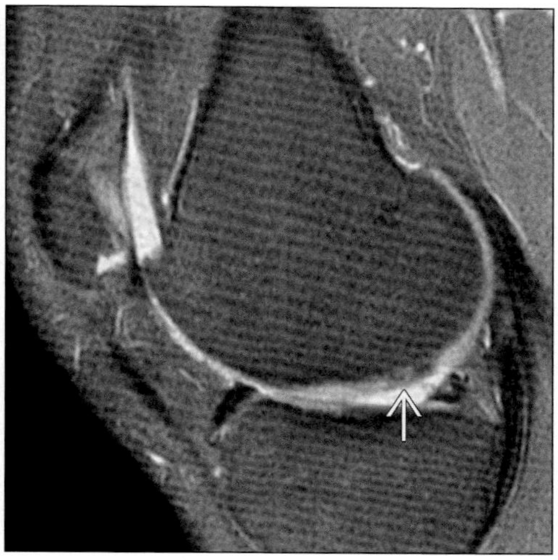

(Left) *Sagittal PDWI FS MR shows a subacute chondral fracture of the lateral femoral condyle. Underlying marrow edema* ➔ *and clinical history suggest a recent injury, but some filling in of the defect* ➔ *is commensurate with a subacute time course. Over time, chondral fractures heal by deposition of scar tissue.* ***(Right)*** *Sagittal PDWI FS MR shows a healed grade IV chondral fracture* ➔ *of the medial femoral condyle. Mild heterogeneity of the cartilage is present.*

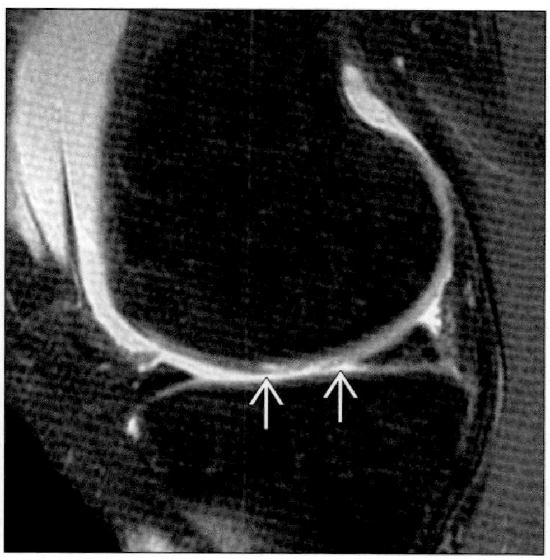

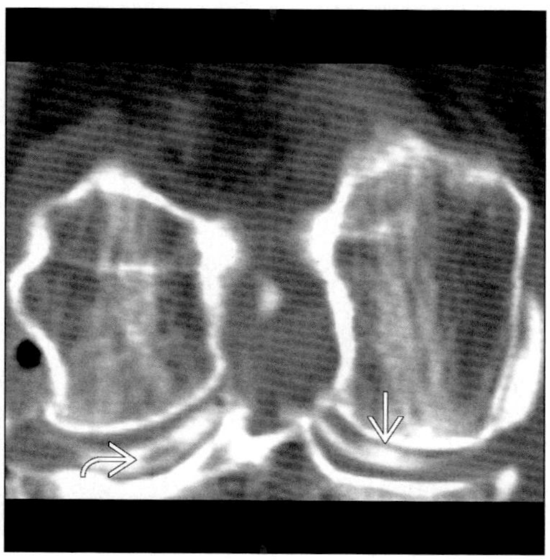

(Left) *Sagittal MR arthrogram T1WI FS shows an irregular grade III chondral fissure/flap tear* ➔ *in the medial femoral condyle. While not necessary to diagnose chondral defects, MR arthrography can help differentiate persistent defects from healed lesions in cases of follow-up after initial injury.* ***(Right)*** *Coronal bone CT arthrogram shows a well-defined full thickness chondral defect* ➔ *in the medial femoral condyle, with a chondral fragment seen in the lateral joint compartment* ➔*.*

SUBCHONDRAL FRACTURE, KNEE

Key Facts

Terminology

- Spontaneous osteonecrosis of knee (SONK), insufficiency fracture, stress fracture of knee

Imaging

- Weight-bearing articular surfaces of femoral condyles &/or tibial plateau
- Medial femoral condyle most common
- Linear or curvilinear fracture line in medullary bone of femoral condyles or tibial plateau near articular surface
- Surrounding marrow edema in acute or subacute cases
- Overlying articular cartilage often intact

Top Differential Diagnoses

- Osteochondral fracture
- Osteonecrosis
- Marrow signal subtended by sclerotic line is normal (follows fat signal on all sequences)
- Transient bone marrow edema syndrome
- Osteoarthritis

Pathology

- Insufficiency fracture: Normal stress on abnormally weak (osteopenic) bone
- Fatigue fracture: Abnormal stress on normal bone
- Previously held theory of osteonecrosis due to vascular insufficiency in subchondral bone falling out of favor

Clinical Issues

- Middle-aged and elderly adults
- M < F
- Conservative treatment

__(Left)__ Graphic shows a small subchondral insufficiency fracture → in the weight-bearing surface of the medial femoral condyle. The overlying cartilage → is intact. __(Right)__ Sagittal T2WI MR in a 71-year-old osteoporotic woman with sudden onset of knee pain shows a curvilinear focus of subchondral sclerosis → indicating a subchondral insufficiency fracture. Overlying cartilage → is intact. There is a small joint effusion →.

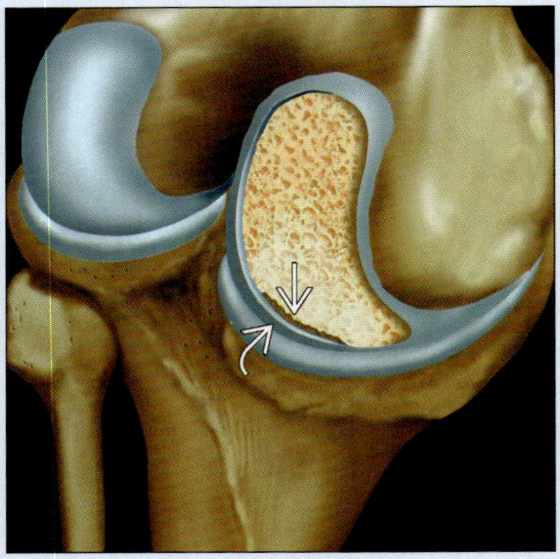

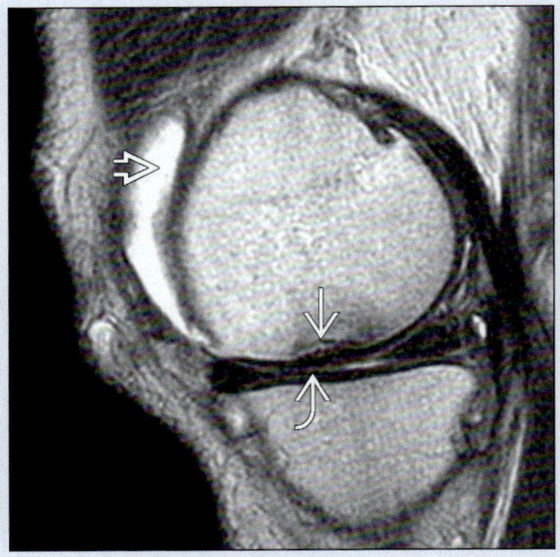

__(Left)__ Coronal T1WI MR in the same patient shows the sclerotic fracture line → in the subchondral bone of the medial femoral condyle. There is surrounding marrow edema →. A small horizontal cleavage tear of the medial meniscus → is present. __(Right)__ Coronal T2WI FS MR at the same slice position in the same patient shows the extensive marrow edema surrounding the slightly impacted subchondral fracture line →. There is mild edema in the tibial rim → that is degenerative.

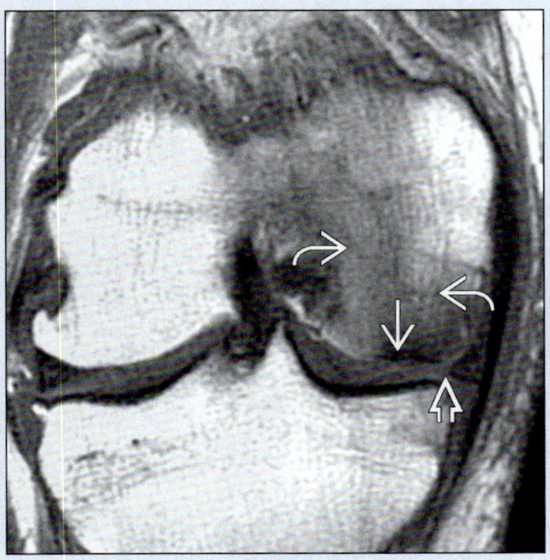

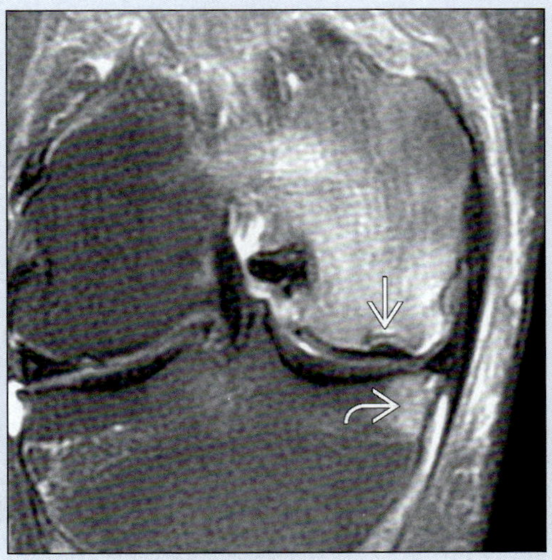

SUBCHONDRAL FRACTURE, KNEE

TERMINOLOGY

Synonyms

- Spontaneous osteonecrosis of knee (SONK), insufficiency fracture, stress fracture of knee, Ahlbäck disease

Definitions

- Medullary fracture of subchondral bone of distal femur or proximal tibia

IMAGING

General Features

- Best diagnostic clue
  - Linear or curvilinear fracture line with surrounding marrow edema in subchondral bone of distal femur or proximal tibia
- Location
  - Weight-bearing articular surfaces of femoral condyles &/or tibial plateau
  - Medial femoral condyle most common
- Size
  - May range from small focus of sclerosis to large linear fracture line
- Morphology
  - May be rounded, curvilinear, or linear

Radiographic Findings

- Often normal
- Subtle radiolucency in subchondral bone
- May see subchondral sclerosis due to trabecular impaction or during healing phase
- Diffuse osteopenia often underlying condition
- Chronic lesion may show cortical deformity

CT Findings

- Similar to radiographs but more sensitive for subtle sclerosis and trabecular disruption

MR Findings

- Linear or curvilinear fracture line in medullary bone of femoral condyles or tibial plateau near articular surface
  - Fracture does not extend through cortex, but remains subchondral
  - May be better seen on T1WI or T2WI FS MR
    - Must examine both sequences for fracture line
  - Small subchondral fractures may only appear as focus of sclerosis
    - Low signal on all sequences
    - Located centrally in region of edema
    - Small lesions almost always immediately subchondral
  - Occasional impaction deformity of cortex
- Surrounding marrow edema in acute or subacute cases
- Overlying articular cartilage often intact
  - No direct extension of fracture through adjacent cartilage
  - Fracture may occur in area of preexisting chondral degeneration, so cartilage may appear abnormally thin and irregular
- Association with medial meniscal tear
- Joint effusion in acute cases

Imaging Recommendations

- Best imaging tool
  - MR
- Protocol advice
  - T1WI and T2WI FS in multiple planes to maximize detection of marrow edema and fracture line

Nuclear Medicine Findings

- Bone scan
  - Focal increased radiotracer uptake in subchondral bone
  - Often delayed 48-72 hours after onset of pain

DIFFERENTIAL DIAGNOSIS

Osteochondral Fracture

- Fracture line extends beyond bone into cortex and articular cartilage
- Overlying cartilage is disrupted
- Usually younger patient age group than subchondral fracture
- Prognosis poorer than for subchondral fracture

Osteonecrosis

- Curvilinear subchondral line of sclerosis ± surrounding marrow edema
- Central marrow signal subtended by sclerotic line is normal, following fat signal on all sequences

Marrow Contusion

- Cloud-like marrow edema without discrete fracture line
- Usually near point of physical impact

Transient Bone Marrow Edema Syndrome

- Diffuse cloud of subchondral marrow edema
- No fracture line
- No history of trauma
- Classically described in distinct populations
  - Young women in 3rd trimester of pregnancy
  - Elderly men
- Diagnosis now questioned
  - May be precursor to subchondral fracture

Osteoarthritis

- Subchondral foci of sclerosis
- Associated with severe cartilage loss
- May have associated marrow edema
  - Usually over smaller area

PATHOLOGY

General Features

- Etiology
  - Insufficiency fracture: Normal stress on abnormally weak bone
    - Osteopenia
      - Senile
      - Metabolic disorders
      - Disuse
    - Osteomalacia
  - Fatigue fracture: Abnormal stress on normal bone
    - Military recruits
    - Athletes

SUBCHONDRAL FRACTURE, KNEE

- Previously held theory of osteonecrosis due to vascular insufficiency in subchondral bone falling out of favor
  - Fractures often demonstrate areas of "osteonecrosis" on microscopic examination
  - Morphologic features more in keeping with insufficiency fracture
- Questionable association with recent arthroscopic surgery
  - Often referred to as "post-arthroscopy osteonecrosis"
  - Proposed mechanisms
    - Direct trauma
    - Laser energy
    - High intraarticular fluid pressure
    - Ischemia due to tourniquet application
- Associated abnormalities
  - Often associated with medial meniscal tear

Gross Pathologic & Surgical Features

- Slightly impacted fracture line in subchondral medullary bone
- Hemorrhage and edema

CLINICAL ISSUES

Presentation

- Most common signs/symptoms
  - Knee pain and swelling
  - Often acute onset
- Other signs/symptoms
  - Joint effusion

Demographics

- Age
  - Middle-aged and elderly adults
- Gender
  - M < F
- Ethnicity
  - Caucasian and Asian > African-American

Natural History & Prognosis

- Generally heals without sequelae after conservative treatment
- May leave residual sclerotic focus or small subchondral impaction defect
- Can result in secondary osteoarthritis
  - Predominance of medial femoral condyle involvement leads to medial-predominant osteoarthritis

Treatment

- Conservative
  - Rest
  - NSAIDs for pain and swelling
- May progress if not rested
- Treatment of underlying osteopenia if present

DIAGNOSTIC CHECKLIST

Consider

- Underlying osteopenia

Image Interpretation Pearls

- Carefully examine subchondral bone for small fracture site in cases of subchondral marrow edema
- Correlate with radiographs to confirm osteopenia

Reporting Tips

- Describe deformity of cortex if present

SELECTED REFERENCES

1. Cetik O et al: Risk of osteonecrosis of the femoral condyle after arthroscopic chondroplasty using radiofrequency: a prospective clinical series. Knee Surg Sports Traumatol Arthrosc. 17(1):24-9, 2009
2. Fotiadou A et al: Acute nontraumatic adult knee pain: the role of MR imaging. Radiol Med. 114(3):437-47, 2009
3. Roemer FW et al: MRI-detected subchondral bone marrow signal alterations of the knee joint: terminology, imaging appearance, relevance and radiological differential diagnosis. Osteoarthritis Cartilage. 17(9):1115-31, 2009
4. Kattapuram TM et al: Spontaneous osteonecrosis of the knee. Eur J Radiol. 67(1):42-8, 2008
5. MacDessi SJ et al: Subchondral fracture following arthroscopic knee surgery. A series of eight cases. J Bone Joint Surg Am. 90(5):1007-12, 2008
6. Takeda M et al: Spontaneous osteonecrosis of the knee: histopathological differences between early and progressive cases. J Bone Joint Surg Br. 90(3):324-9, 2008
7. Pape D et al: Postarthroscopic osteonecrosis of the knee. Arthroscopy. 23(4):428-38, 2007
8. Yates PJ et al: Early MRI diagnosis and non-surgical management of spontaneous osteonecrosis of the knee. Knee. 14(2):112-6, 2007
9. Muscolo DL et al: Medial meniscal tears and spontaneous osteonecrosis of the knee. Arthroscopy. 22(4):457-60, 2006
10. Kidwai AS et al: Radiologic case study. Spontaneous osteonecrosis of the knee reclassified as insufficiency fracture. Orthopedics. 28(3):236, 333-6, 2005
11. Lecouvet FE et al: MR imaging of epiphyseal lesions of the knee: current concepts, challenges, and controversies. Radiol Clin North Am. 43(4):655-72, vii-viii, 2005
12. Ramnath RR et al: MR appearance of SONK-like subchondral abnormalities in the adult knee: SONK redefined. Skeletal Radiol. 33(10):575-81, 2004
13. Yamamoto T et al: Spontaneous osteonecrosis of the knee: the result of subchondral insufficiency fracture. J Bone Joint Surg Am. 82(6):858-66, 2000
14. Yamamoto T et al: Subchondral insufficiency fracture of the femoral head and medial femoral condyle. Skeletal Radiol. 29(1):40-4, 2000
15. Ecker ML et al: Spontaneous osteonecrosis of the knee. J Am Acad Orthop Surg. 2(3):173-178, 1994

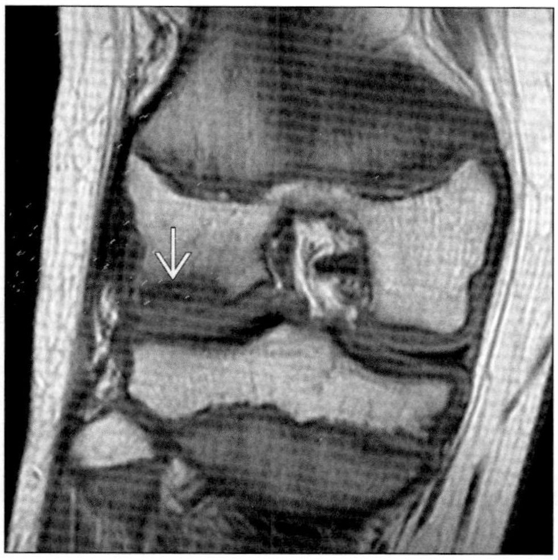

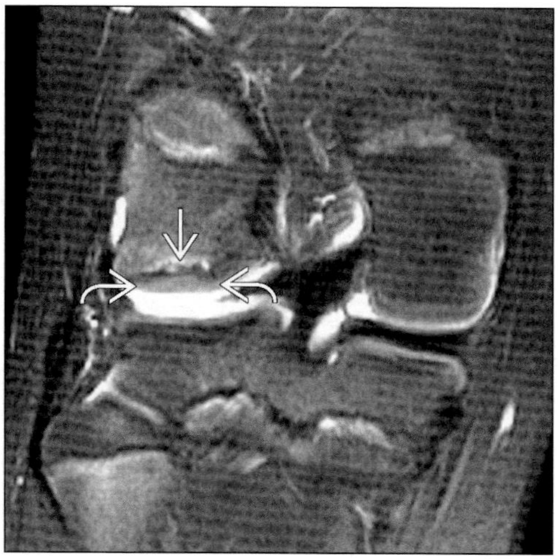

(Left) *Coronal T1WI MR in a 14-year-old boy demonstrates a broad, well-defined subchondral focus of low signal suggesting a fracture* ➔. ***(Right)*** *Coronal T2WI FS MR in the same patient shows high signal in the fracture plane* ➔ *suggesting a subacute lesion. There is minimal surrounding marrow edema. The overlying cartilage* ➔ *is intact; thus, this is not an osteochondral fracture but is a purely subchondral fracture, and therefore carries a better prognosis.*

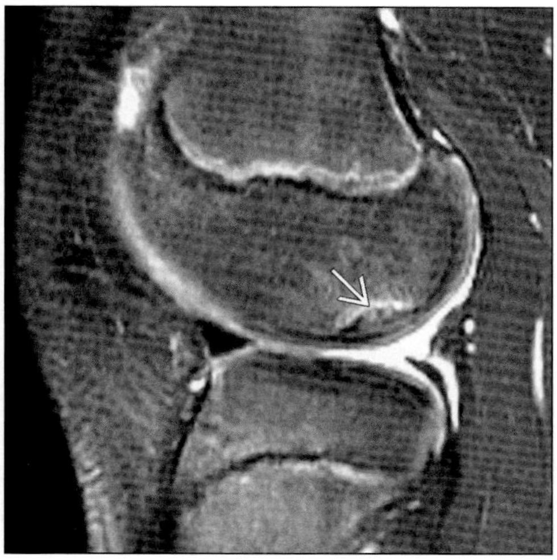

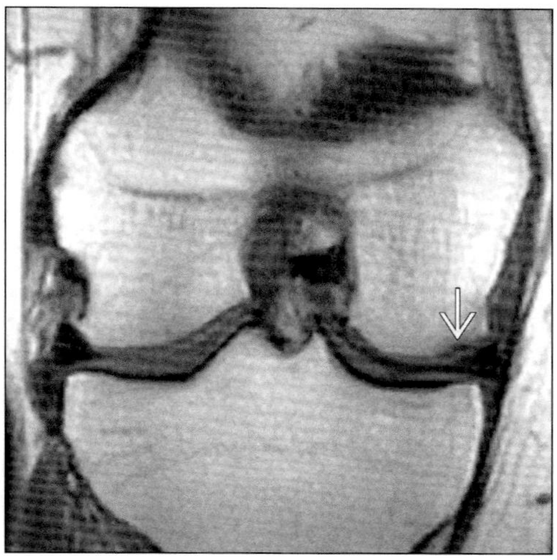

(Left) *Sagittal PDWI FS MR in the same patient confirms the subchondral nature of the fracture* ➔. *Most lesions in this age group involve the overlying cartilage as well; careful evaluation of cartilage is necessary to differentiate between these 2 entities.* ***(Right)*** *Coronal T1WI MR from a 60-year-old man shows a linear subchondral focus of sclerosis* ➔ *in the medial femoral condyle not associated with marrow edema. This is the remnant of an old subchondral fracture and is a permanent scar.*

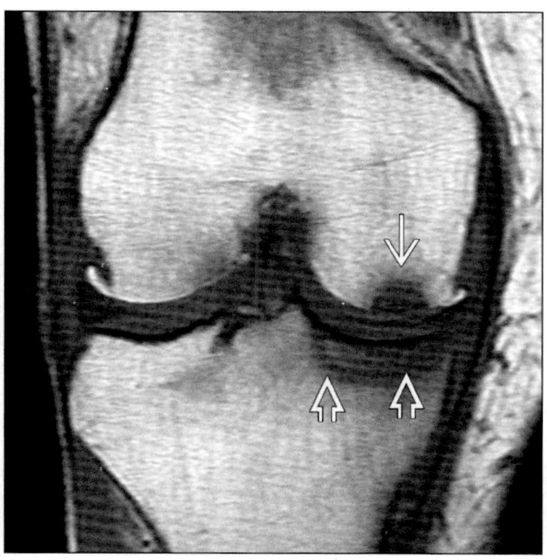

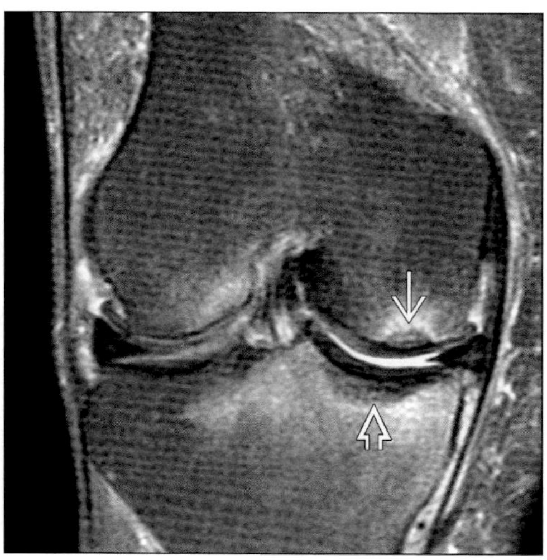

(Left) *Coronal T1WI MR in the same patient shows the crescentic shape of the femoral lesion* ➔, *suggesting an insufficiency fracture. The linear fracture line in the tibia* ➔ *is better appreciated.* ***(Right)*** *Coronal T2WI FS MR in the same patient better shows the subchondral fracture line in the medial femoral condyle* ➔, *but the tibial fracture line* ➔ *is less well seen than on the T1-weighted image.*

SUBCHONDRAL FRACTURE, KNEE

(Left) *Sagittal PDWI MR shows a small curvilinear focus of low signal* ➔ *in the posterior non-weight-bearing portion of the lateral femoral condyle. This is an unusual location for fatigue or insufficiency fracture.* ***(Right)*** *Axial PDWI FS MR in the same patient confirms the subchondral fracture line* ➔ *and the extensive surrounding marrow edema. The patient was a gymnast, and the injury may have occurred during knee flexion. Although this is a subchondral fx, this patient does not fit the usual patient profile.*

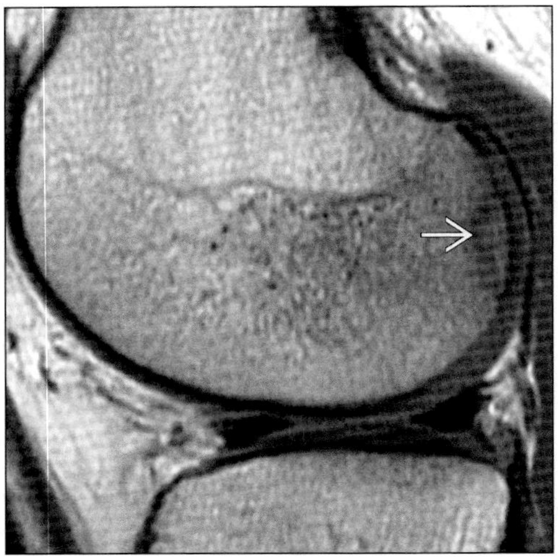

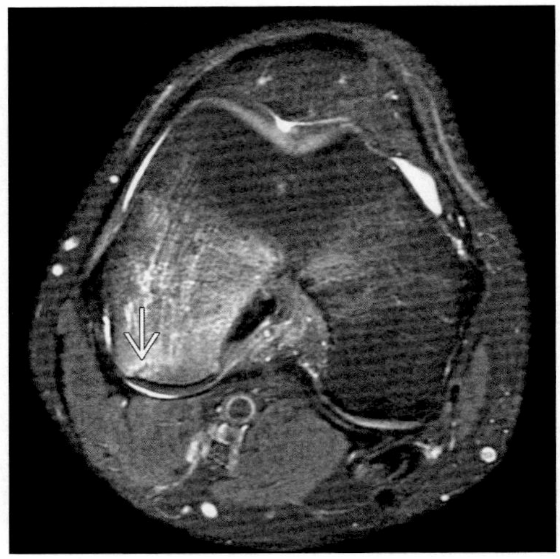

(Left) *Coronal T1WI MR in a 55-year-old man who is an avid runner shows a linear subchondral fracture* ➔ *in the lateral femoral condyle. The patient felt sudden pain in the knee while running.* ***(Right)*** *Coronal T2WI FS MR in the same patient shows the fracture line* ➔ *surrounded by marrow edema. The medial femoral condyle is normal. Note the soft tissue edema* ➦ *adjacent to the lateral femoral condyle; this is common after acute fracture.*

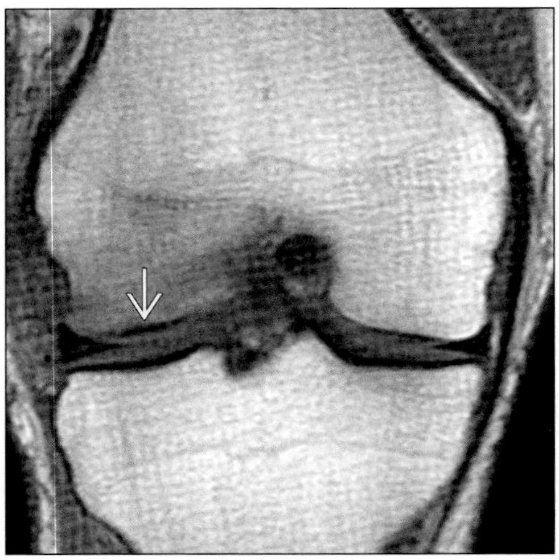

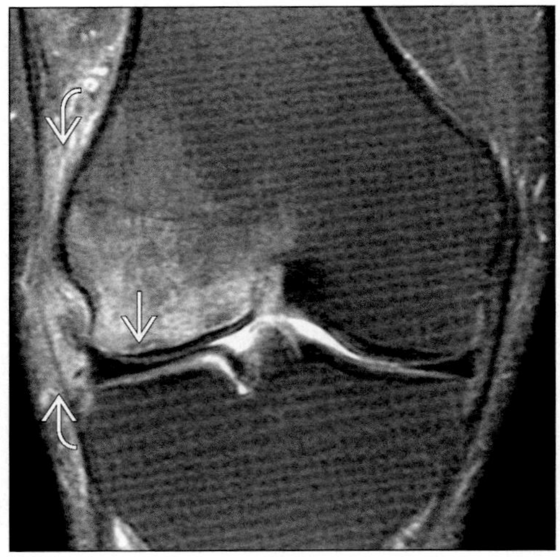

(Left) *Coronal T1WI MR in the same patient 3 months later shows subtotal resolution of the lateral femoral condylar fracture* ➔*, with a small residual sclerotic scar. There is a new subchondral fracture of the medial femoral condyle* ➦*.* ***(Right)*** *Coronal T2WI FS MR in the same patient at the same time as the previous image confirms the reduction in marrow edema of the lateral condyle* ➔*, with marked edema around the new fracture* ➦ *indicating its acute nature.*

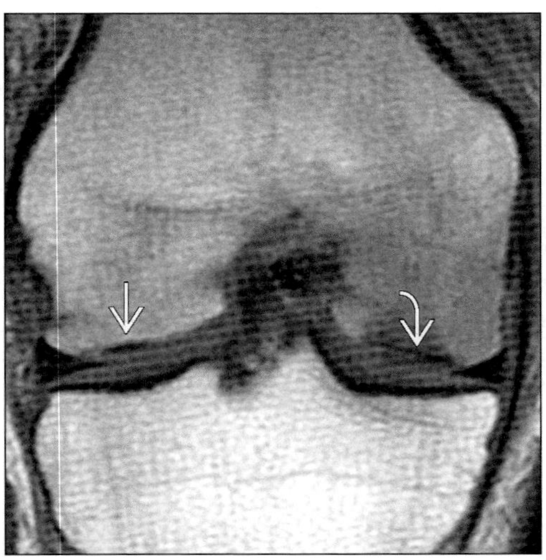

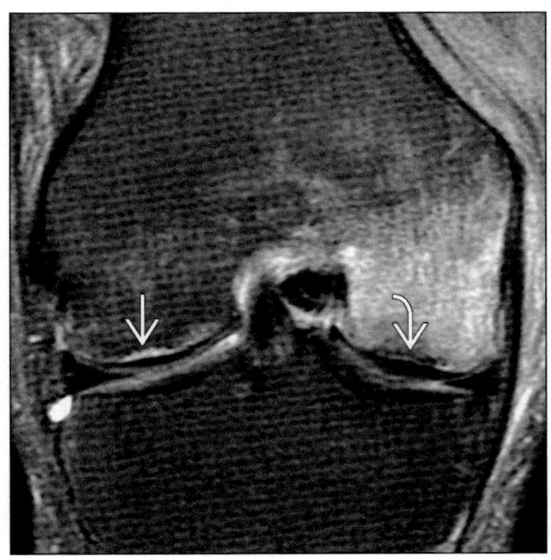

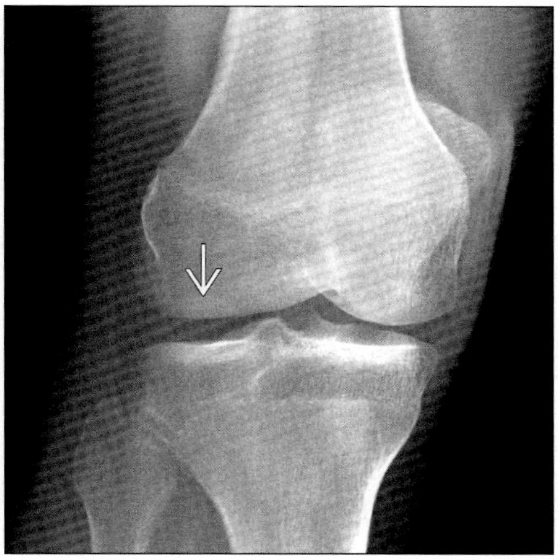

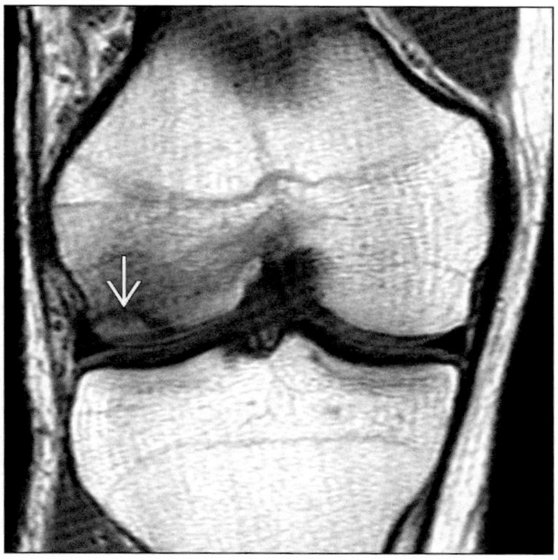

(Left) *Oblique radiograph in a 35-year-old runner with knee pain shows a subtle crescentic zone of sclerosis → in the subchondral bone of the lateral femoral condyle.* ***(Right)*** *Coronal T1WI MR in the same patient on the same day shows a well-demarcated curvilinear subchondral fracture line → in the lateral femoral condyle corresponding to the sclerotic focus on the radiograph. Surrounding marrow edema implies a relatively acute or subacute process.*

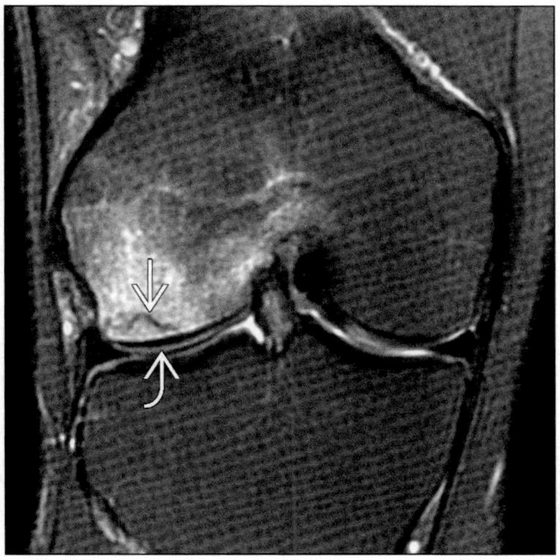

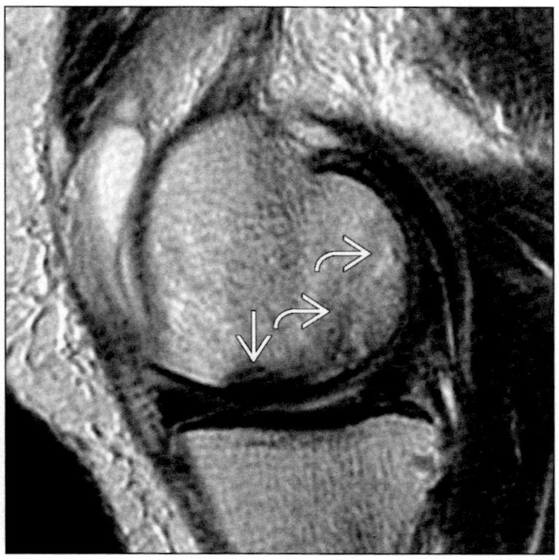

(Left) *Coronal T2WI FS MR in the same patient shows the subchondral fracture → and surrounding marrow edema. The overlying cartilage ↪ is intact.* ***(Right)*** *Sagittal T2WI MR in a 79-year-old woman with acute onset of knee pain and a history of steroid use shows a sclerotic focus → in the subchondral weight-bearing portion of the medial femoral condyle. More posteriorly, there is another more subtly demarcated focus of abnormal marrow signal ↪.*

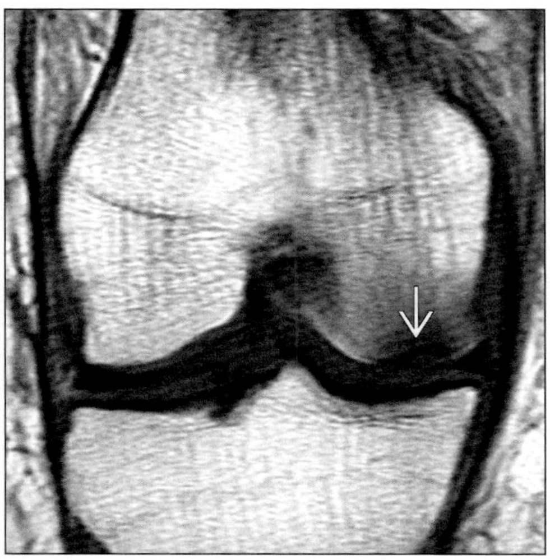

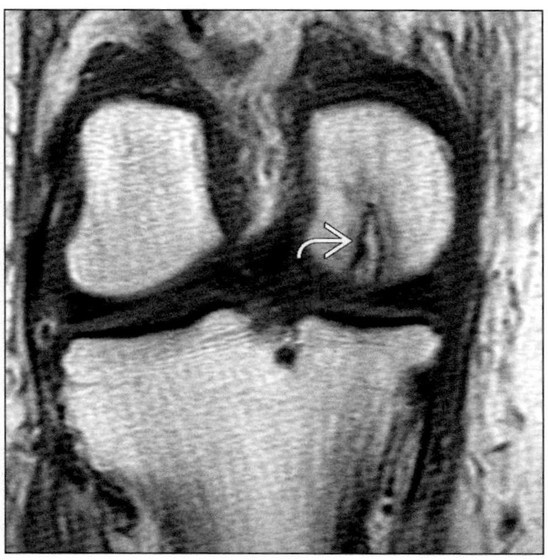

(Left) *Coronal T1WI MR in the same patient shows the more anterior lesion to be a subchondral insufficiency fracture →. Some still refer to this lesion as spontaneous osteonecrosis of the knee (SONK), but most such cases are fractures.* ***(Right)*** *Coronal T1WI MR from the same patient more posteriorly shows the 2nd lesion ↪. There is a sclerotic border with normal central marrow edema; this is chronic osteonecrosis.*

Key Facts

Imaging

- **Marrow stimulation techniques**
  - Abrasion arthroplasty and microfracture
  - Acute: Chondral defect fills with fluid
  - Progressive filling in of defect with type I collagen scar (fibrocartilage rather than hyaline cartilage)
- **Autologous osteochondral transplant (AOT)**
  - Fatty marrow in graft and surrounding bone from 6 weeks to 9 months after transplant
  - By 1-2 years, healing complete
- **Autologous chondrocyte implantation (ACI)**
  - Chondrocytes demonstrate fluid signal
  - Marrow edema resolves by 1 year
- Preferred imaging technique: MR ± arthrography
- Direct MR arthrography best for evaluation of articular cartilage congruity after AOT
- Indirect MR arthrography has been recommended to assess healing of ACI graft

Pathology

- **Marrow stimulation techniques**
  - Puncture of subchondral bone releases multipotential stem cells that form fibrocartilaginous scar
  - Resulting scar does not have biomechanical properties of normal cartilage
- **Biodegradable pin fixation**
  - Pins resorb in 6-24 months
- **Osteochondral transplantation** (usually AOT)
  - Several fragments may be placed to fill 1 large defect (mosaicplasty)
- **Autologous chondrocyte implantation** (ACI)
  - Patient's own cartilage harvested from non-weight-bearing surface
  - Periosteal cover sewn or glued over defect

__(Left)__ AP graphic shows sites of osteochondral graft harvest ➡ in the non-weight-bearing portion of the lateral femoral condyle for an AOT procedure. Placement of graft plugs ➡ is depicted in a medial femoral condylar defect. __(Right)__ Sagittal MR arthrogram T1WI FS after AOT demonstrates the rectangular osseous portion of the graft ➡ incorporated into the native bone. The overlying articular cartilage ➡ is flush with the native hyaline cartilage surface, creating a smooth, congruent articular surface.

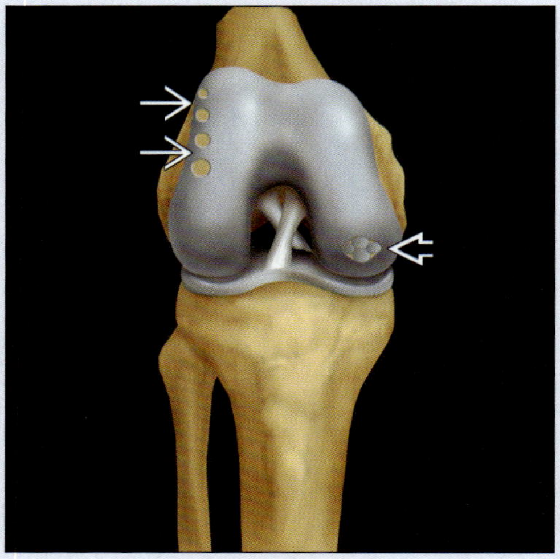

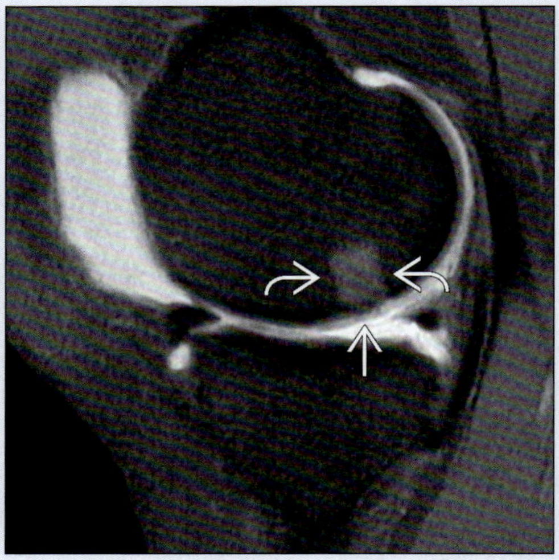

__(Left)__ AP graphic shows steps in an ACI procedure. Chondral defect in the lateral femoral condyle ➡ is covered with a periosteal flap ➡ derived from the proximal tibia ➡. Previously harvested and cultured chondrocytes are then injected ➡ into the defect under the flap. __(Right)__ Sagittal PDWI MR after ACI in the medial femoral condylar articular surface shows periosteal flap hypertrophy, causing thickening of the width of the graft ➡.

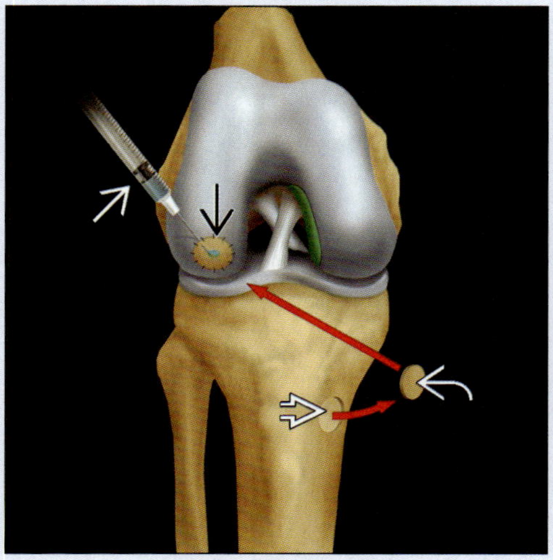

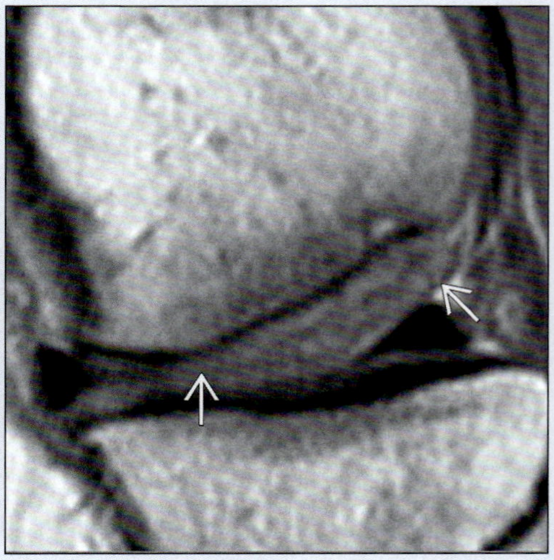

ARTICULAR CARTILAGE: POSTOPERATIVE IMAGING

TERMINOLOGY

Abbreviations

- Osteochondral autograft transplant surgery (OATS)
- Autologous osteochondral transplant (AOT)
- Autologous chondrocyte implantation (ACI)

Synonyms

- Mosaicplasty: Trade name for 1 method of performing osteochondral transplant of multiple small plugs

Definitions

- Any of a variety of surgical techniques to repair or replace damaged or missing articular cartilage surfaces in knee
- **Marrow stimulation techniques**
  - Appropriate for lesions < 4 cm^2
  - Puncture of subchondral bone releases multipotential stem cells that form fibrocartilaginous scar
  - Resulting scar does not have biomechanical properties of normal cartilage
  - Long-term preservation of joint function not documented
  - Types of procedures; may be used in combination
    - Abrasion arthroplasty
      - Underlying subchondral bone is roughened using arthroscopic burr to promote bleeding and subsequent scar formation in chondral defect
    - Microfracture ("pick procedure")
      - Multiple small holes are created in subchondral bone with a pick or drill to promote bleeding → fibrocartilaginous scar
      - Most commonly performed marrow stimulation procedure
    - Subchondral drilling
      - Drill used to trephine subchondral bone and promote bleeding → fibrocartilaginous scar
- **Biodegradable pin fixation**
  - Biodegradable pins used to stabilize lesion
    - In situ unstable osteochondral lesions
    - Chondral flap tears
  - Pins resorb in 6-24 months
- **Osteochondral transplantation**
  - Appropriate for lesions 1-4 cm^2
  - Osteochondral plugs inserted into lesion
  - Autologous osteochondral transplant (AOT)
    - Harvested from non-weight-bearing surface, usually lateral aspect of anterior lateral femoral
    - Depends on presence of useable donor cartilage
    - Possibility of donor site morbidity
  - Allogenic graft
    - Cadaveric graft material
    - Higher incidence of immune response and graft failure
    - Possibility of disease transmission from donor to host
  - Several fragments may be placed to fill 1 large defect (mosaicplasty)
  - Spaces between plugs filled with fibrocartilaginous grout
- **Autologous chondrocyte implantation** (ACI)
  - Appropriate for lesions 2-12 cm^2
  - Patient's own cartilage harvested from non-weight-bearing surface
  - Cells grown in culture for 3-5 weeks to produce chondrocytes
  - Periosteal cover sewn or glued over defect
    - Newer techniques involve using bilayer collagen membrane
  - Cultured chondrocytes injected under periosteal cover
  - Resulting cartilage thought to have biomechanical properties like native cartilage
- **Matrix-associated autologous chondrocyte transplantation**
  - 3D biologic scaffold optimized for culture of seeded chondrocytes implanted in defect by mini-arthrotomy
  - Trimmed to fit defect and glued in place

IMAGING

General Features

- Location
  - Generally, procedures performed on load-bearing cartilage surfaces
    - Medial and lateral femoral condyles most common
- Size
  - Lesions range in size from a few millimeters to > 12 cm^2
- Morphology
  - Margins may be ill defined after abrasion chondroplasty
  - Margins usually well defined after AOT or ACI surgery
    - Over time, margins of repair often fade and may become unrecognizable

CT Findings

- Border of AOT donor and recipient site usually well defined 1-2 years or more after surgery
- CT arthrography can be used to assess articular surface if patient not able to undergo MR

MR Findings

- **Marrow stimulation techniques**
  - Abrasion arthroplasty
    - Acute
      - Chondral defect fills with fluid
      - May have subchondral marrow edema
    - Progressive filling in of defect with type I collagen scar
    - Often eventually completely fills in, leaving little trace of original injury
    - By 1-2 years, healing complete
  - Microfracture
    - Small foci of linear low signal on T1 in subchondral bone (from trephination)
    - Overlying chondral defect filled with blood/fluid
    - Progressive filling in of defect with type I collagen scar
    - Often eventually completely fills in, leaving little trace of original injury
  - Scar often appears identical to normal hyaline cartilage on all sequences, making distinction of repair site difficult

  - Overgrowth of subchondral bone deep to chondral defect in 25-49%
    - Does not inherently result in bad outcome
    - Overlying cartilage surface may still be congruent with native bone
- **Autologous osteochondral transplant** (AOT)
  - Fatty marrow in graft and surrounding bone from 6 weeks to 9 months after transplant
    - Edema progressively decreases during this period
    - Marrow edema can persist for up to 2 years
  - Cyst formation in surrounding marrow may indicate poor incorporation
  - Cartilage surface assessment
    - Congruence, subsidence, protrusion (termed "proud"), rotation
  - Donor site evaluation
    - Marrow edema in early phase
    - Reverts to normal marrow signal within 6-9 months
- **Autologous chondrocyte implantation** (ACI)
  - Proliferative phase (0-6 weeks after surgery)
    - Chondrocytes demonstrate fluid signal
      - Bright enhancement of graft with intravenous contrast
    - Intense subchondral marrow edema
    - Irregular subchondral bone plate
  - Transition phase (7-26 weeks after surgery)
    - Decreasing T2 signal in graft
    - Decreasing marrow edema
    - Persistent fluid signal at margins of graft
  - Remodeling phase (27-52 weeks after surgery)
    - Chondrocyte signal stabilizes
      - May remain heterogeneous
    - Marrow edema resolves by 1 year
    - Margins become low signal on all sequences
  - Complications
    - Fibrous overgrowth (hypertrophy) of periosteal flap
    - Delamination (separation of periosteal flap from underlying chondrocytes)
      - Graft material may completely disengage from graft site and become free in joint
    - Poor integration of graft

Imaging Recommendations

- Best imaging tool
  - MR ± arthrography
    - Direct MR arthrography best for evaluation of articular cartilage congruity after AOT
    - Indirect MR arthrography has been recommended to assess healing of ACI graft
  - Advanced MR techniques (usually performed at 3T)
    - T2 mapping
    - T1rho
    - Sodium imaging
- Protocol advice
  - High-resolution technique
    - Small field-of-view
    - Transmit-receive volume coil
  - Individual MR imaging sequences subject to personal preference of radiologist
  - MR arthrography may have advantages in detection of persistent marginal defect after AOT

PATHOLOGY

Microscopic Features

- Despite appearing similar or identical to normal hyaline cartilage, scar formation after marrow stimulation results in formation of fibrocartilage
  - Type I collagen (fibrocartilage) instead of type II (normal hyaline cartilage)
  - Inferior mechanical properties for load-bearing and shear stress resistance
  - Higher propensity to reinjury than normal cartilage

DIAGNOSTIC CHECKLIST

Consider

- Time course after surgery in assessing significance of findings on MR

Reporting Tips

- Evaluate donor site for healing
- Evaluate graft or surgical site
  - Incorporation versus instability
  - Cartilage thickness
  - Cartilage delamination
- Evaluate for loose bodies or arthrofibrosis

SELECTED REFERENCES

1. Van Assche D et al: Autologous chondrocyte implantation versus microfracture for knee cartilage injury: a prospective randomized trial, with 2-year follow-up. Knee Surg Sports Traumatol Arthrosc. 18(4):486-95, 2010
2. Gold GE et al: Recent advances in MRI of articular cartilage. AJR Am J Roentgenol. 193(3):628-38, 2009
3. Kon E et al: Matrix-assisted autologous chondrocyte transplantation for the repair of cartilage defects of the knee: systematic clinical data review and study quality analysis. Am J Sports Med. 37 Suppl 1:156S-66S, 2009
4. Potter HG et al: New techniques in articular cartilage imaging. Clin Sports Med. 28(1):77-94, 2009
5. Trattnig S et al: MR imaging of cartilage and its repair in the knee--a review. Eur Radiol. 19(7):1582-94, 2009
6. Van Assche D et al: Physical activity levels after characterized chondrocyte implantation versus microfracture in the knee and the relationship to objective functional outcome with 2-year follow-up. Am J Sports Med. 37 Suppl 1:42S-49S, 2009
7. Welsch GH et al: T2 and T2* mapping in patients after matrix-associated autologous chondrocyte transplantation: initial results on clinical use with 3.0-Tesla MRI. Eur Radiol. Epub ahead of print, 2009
8. Choi YS et al: MR imaging of cartilage repair in the knee and ankle. Radiographics. 28(4):1043-59, 2008
9. McNickle AG et al: Overview of existing cartilage repair technology. Sports Med Arthrosc. 16(4):196-201, 2008
10. Potter HG et al: Magnetic resonance imaging of cartilage repair. Sports Med Arthrosc. 16(4):236-45, 2008
11. Welsch GH et al: Cartilage T2 assessment at 3-T MR imaging: in vivo differentiation of normal hyaline cartilage from reparative tissue after two cartilage repair procedures--initial experience. Radiology. 247(1):154-61, 2008
12. Tins BJ et al: Autologous chondrocyte implantation in knee joint: MR imaging and histologic features at 1-year follow-up. Radiology. 234(2):501-8, 2005
13. Recht M et al: MR imaging of cartilage repair procedures. Skeletal Radiol. 32(4):185-200, 2003

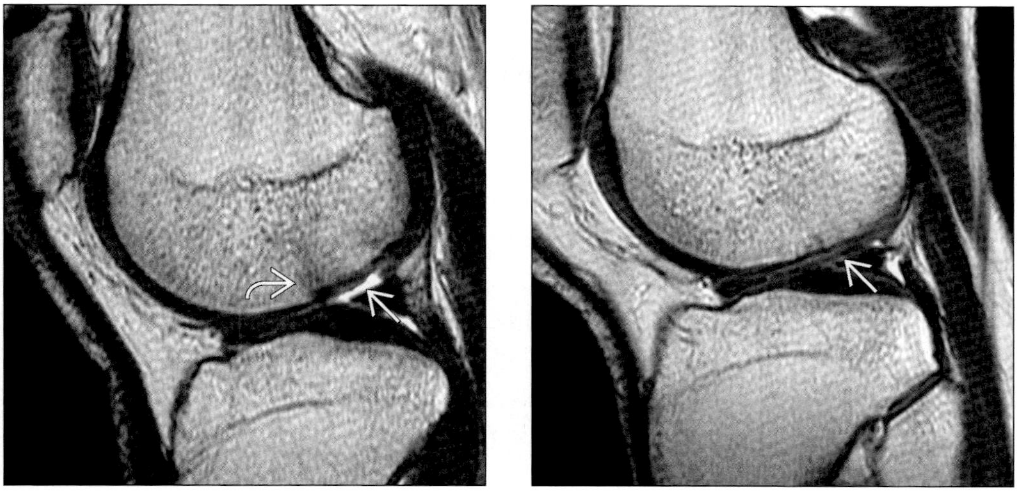

(Left) *Sagittal T2WI MR 1 month after pick arthroplasty (microfracture) shows a grade III chondral defect in the lateral femoral condyle with sharp margins. Linear pick tracks are present.* ***(Right)*** *Sagittal T2WI MR from the same patient 11 months later shows filling in of the defect with scar cartilage (fibrocartilage). Although the scar tissue looks very similar to (and is often indistinguishable from) native hyaline cartilage, its mechanical properties are inferior.*

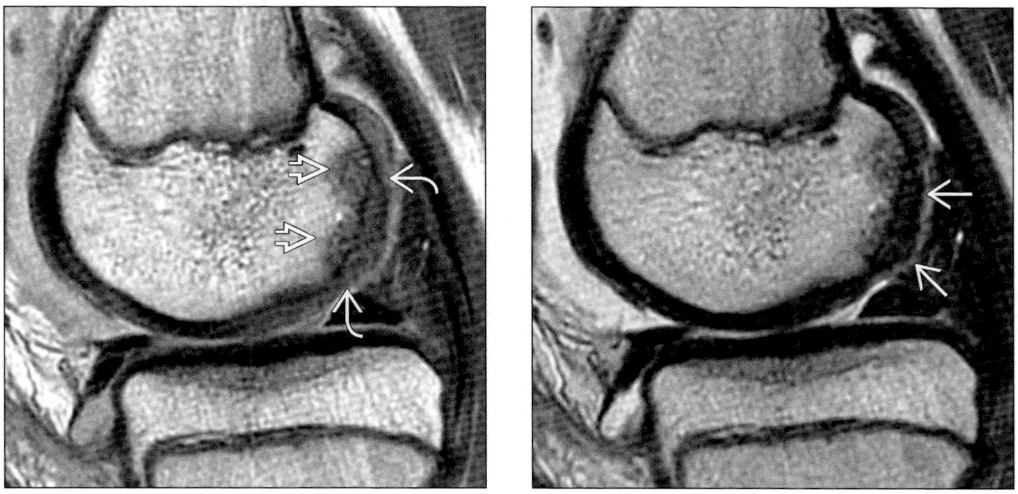

(Left) *Sagittal PDWI MR after abrasion chondroplasty using a burr in the posterior portion of the medial femoral condyle shows thick, irregular repair cartilage at the site of the former defect. This represents fibrocartilage and not hyaline cartilage. There is subchondral sclerosis at the site.* ***(Right)*** *Sagittal T2WI MR from the same patient shows the irregularity of the repair fibrocartilage at the site of repair. Note that repair cartilage looks similar to native cartilage.*

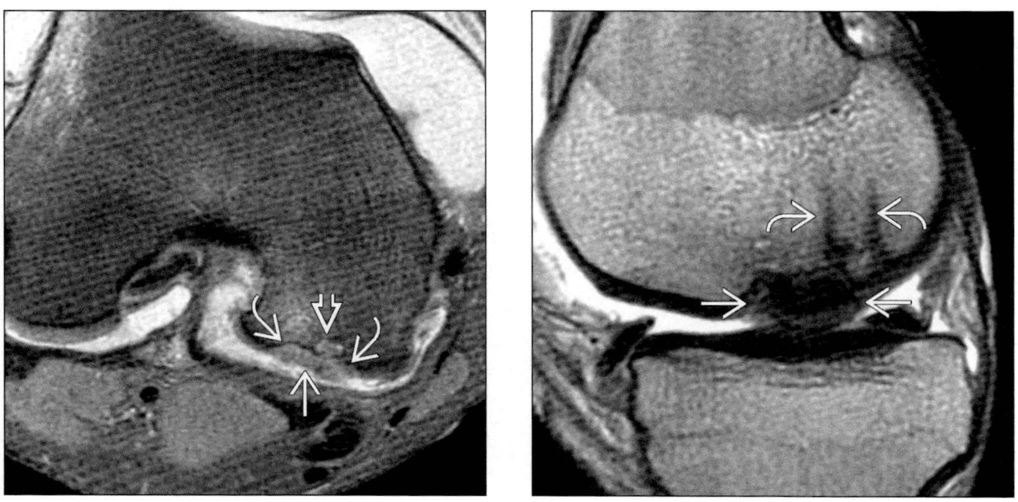

(Left) *Axial PDWI FS MR from the same patient shows the site of previous repair. There is an area of indentation in the subchondral bone related to surgical resection down to bleeding bone. The scar cartilage is somewhat irregular, and there are underlying subchondral cysts.* ***(Right)*** *Sagittal T2WI MR after repair of a displaced osteochondral fracture of the medial femoral condyle (MFC) using bioabsorbable pins shows no fluid tracking into the interface around the fragment. Pin tracks are present.*

__(Left)__ Coronal T2WI FS MR after pin fixation of an osteochondral fracture shows good incorporation of the fragment with no fluid entering the interface between the fragment and the native condyle. There is mild diffuse marrow edema throughout the MFC. __(Right)__ Coronal T2WI FS MR after repair of an osteochondral fracture of the MFC shows fluid tracking into part of the junction between the fragment and the native condyle, indicating incomplete healing or an unstable fragment.

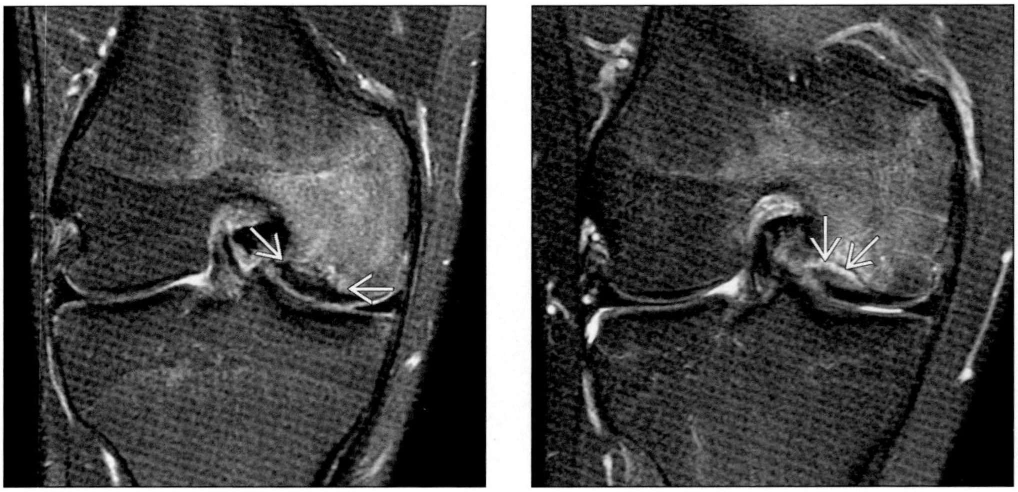

__(Left)__ Sagittal T2WI MR 6 months after AOT in the posterior aspect of the MFC shows mild residual high signal in the graft. Overlying cartilage has subsided slightly compared to the adjacent native articular surface, leaving a potential for locking or delamination. __(Right)__ Coronal T1WI MR from the same patient shows the healing MFC AOT graft. A lateral femoral condylar graft performed 18 months earlier shows more mature incorporation of the osseous plug.

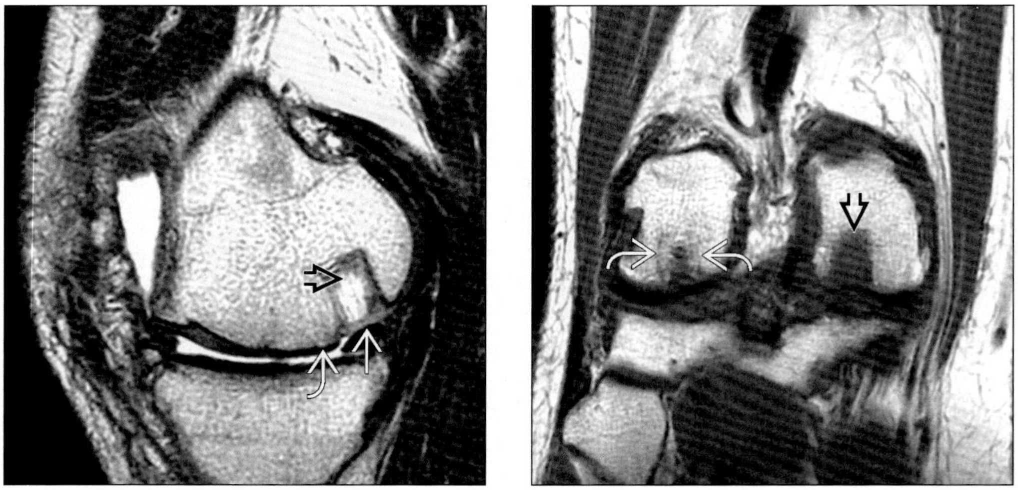

__(Left)__ Sagittal MR arthrogram T1WI after AOT in the MFC shows good congruency of the transplant articular cartilage with the native chondral surface. __(Right)__ Coronal MR arthrogram PDWI from the same patient shows seamless transition from the native articular cartilage of the MFC to the osteochondral allograft. Direct MR arthrography is the most accurate imaging technique for evaluating the articular surface after AOT.

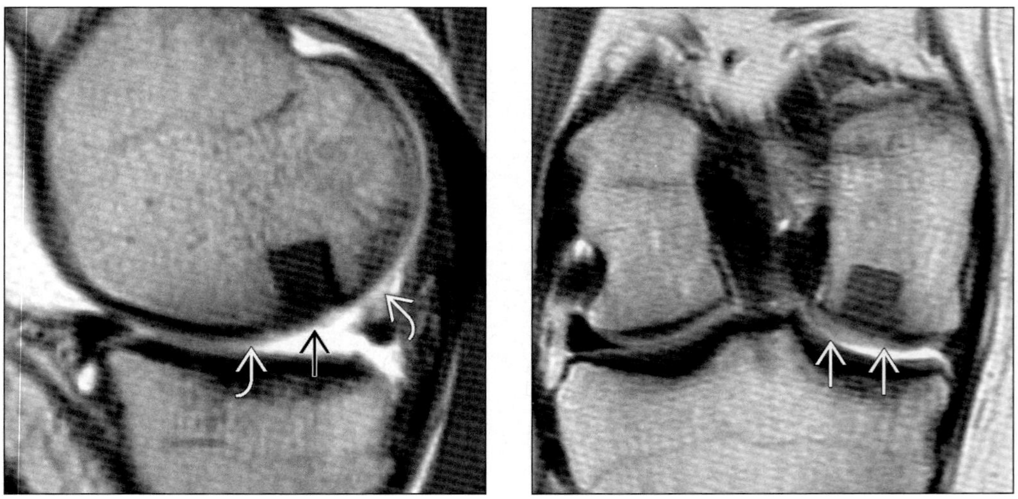

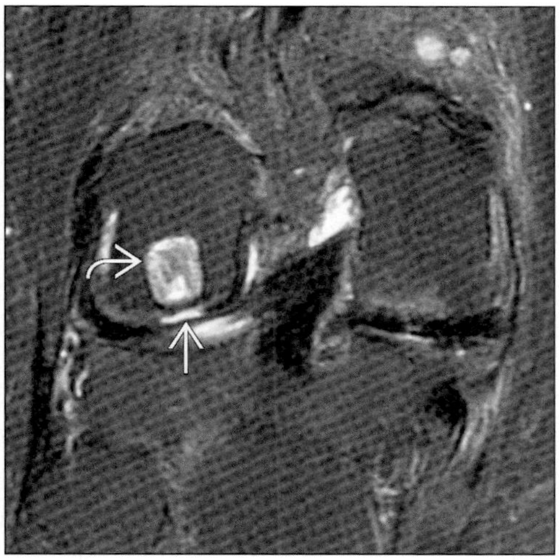

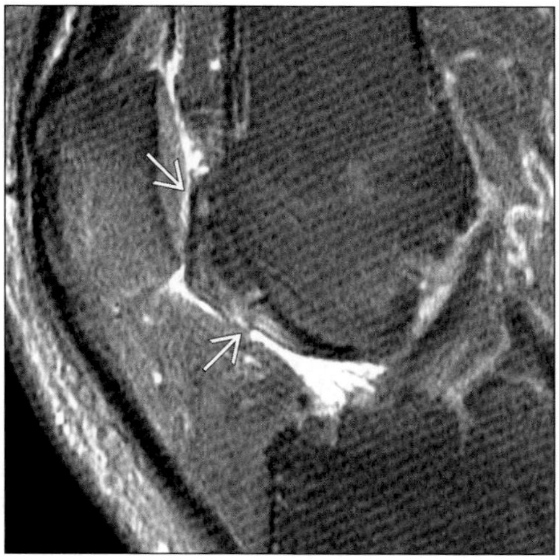

(Left) *Coronal T2WI FS MR 8 months after AOT surgery in the lateral femoral condyle shows residual marrow edema in the graft plug ↪. Edema in and around the graft can be seen up to 2 years after surgery. There is fluid ➔ undermining the articular surface of the graft, confirmed at surgery for graft failure.* ***(Right)*** *Sagittal T2WI FS MR after autologous chondrocyte implantation surgery shows hypertrophy of the periosteal graft covering ➔. (Courtesy C. Winalski, MD, Cleveland Clinic).*

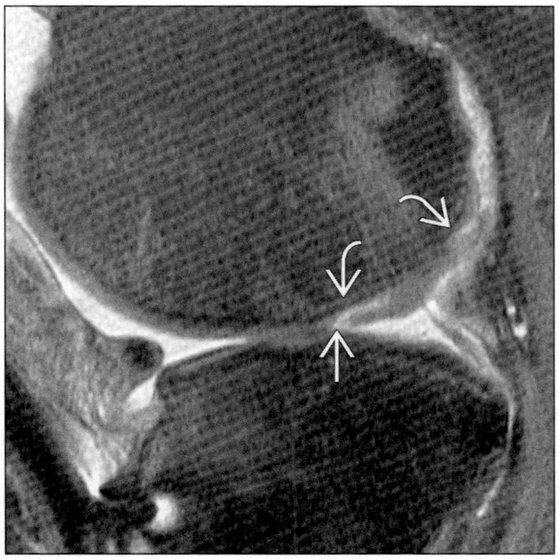

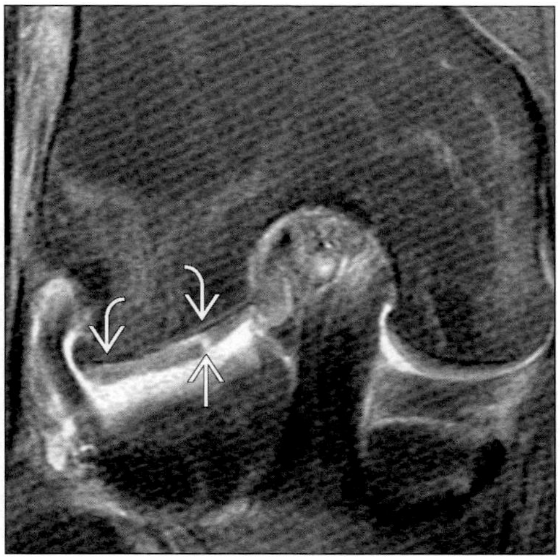

(Left) *Sagittal MR indirect arthrogram PDWI FS after ACI surgery on the posterior lateral femoral condyle shows a deep fissure ➔ extending through the anterior border of the graft ↪, indicating partial delamination of the graft.* ***(Right)*** *Coronal MR indirect arthrogram T1WI FS from the same patient shows the site of partial delamination ➔, with contrast entering the fissure. The remainder of the graft ↪ remains in good position. (Courtesy C. Winalski, MD, Cleveland Clinic).*

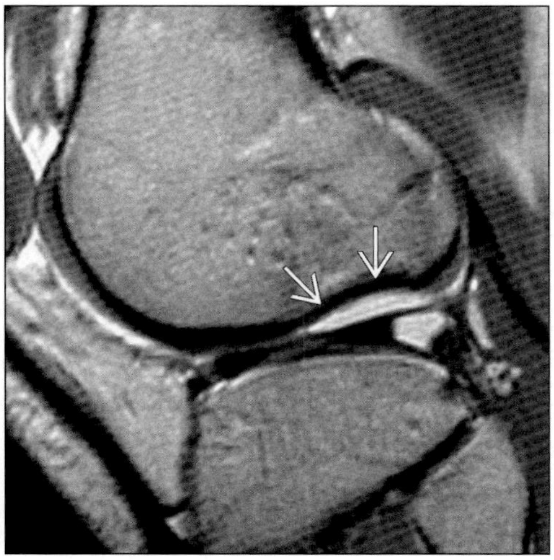

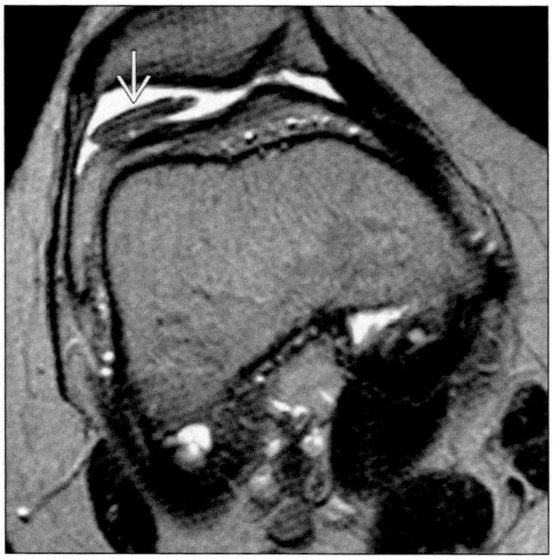

(Left) *Sagittal T2WI MR from a patient with new knee pain and locking 6 months after autologous chondrocyte implantation as treatment for a large osteochondral defect in the lateral femoral condyle shows a defect at the ACI site ➔. The implanted chondrocytes and overlying periosteal flap are not seen.* ***(Right)*** *Axial T2WI MR from the same patient shows the delaminated graft ➔ in the suprapatellar recess of the knee joint. Graft displacement is the end result of ACI delamination.*

ANTERIOR CRUCIATE LIGAMENT INJURY

Key Facts

Imaging

- Most tears occur in proximal or mid portion of ligament
- ACL consists of 2 discrete fiber bundles
- ACL serves as primary restraint against anterior tibial displacement
- Deep sulcus sign: Deepening of terminal sulcus of lateral femoral condyle indicates impaction due to pivot shift injury
- Segond fracture: Avulsion of lateral capsular ligament from lateral tibial rim
- Complete tear: MR sensitivity 90-98%, specificity 90-100%, accuracy 90-95%
- Chronically disrupted ligament: "Empty notch" sign of fat in expected location of ACL
- Torn or partially torn ligament often rounded or amorphous in shape
- Loss of normal alignment of ligament
- Osseous avulsion of ligament at tibia
- Associated marrow contusions
  - Posterolateral tibial plateau
  - Lateral femoral condyle at terminal sulcus
- Some advocate additional oblique sagittal T2WI FS MR images along long axis of ACL

Pathology

- Meniscal tear (50-80%)
- Lateral capsular ligament disruption (Segond injury)
- Posterolateral corner injury

Clinical Issues

- Conservative treatment for low-grade partial ACL tear
- Tears involving < 25% cross section of ligament have favorable prognosis
- Surgical repair success rate: 75-95%

*(**Left**) AP internal rotation graphic shows disruption of the mid portion of the anterior cruciate ligament (ACL) ➔ with marrow contusions in the posterior lateral tibia ⇨ and lateral femoral condyle ⇨ related to a pivot-shift injury mechanism. (**Right**) Sagittal PDWI FS MR demonstrates an acute complete tear of the anterior cruciate ligament ➔. There are no discernible fibers in the proximal 1/2 of the ligament, and there is extensive edema in the proximal intercondylar notch.*

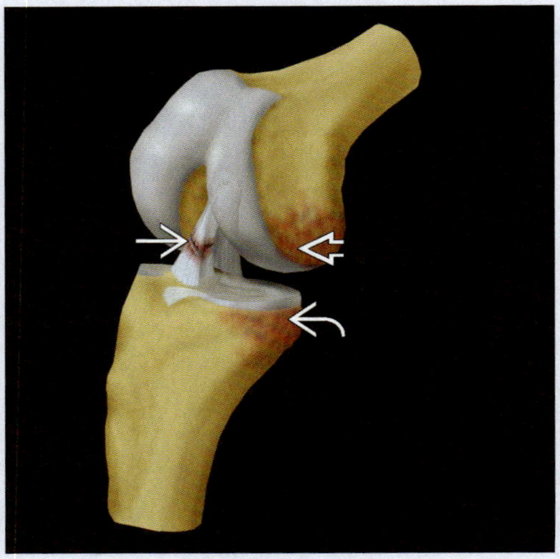

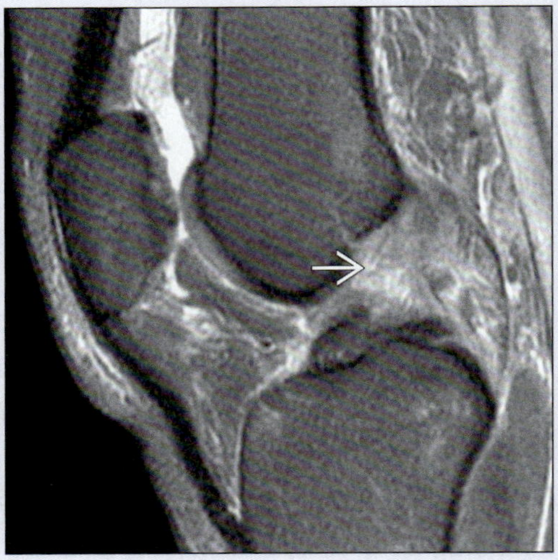

*(**Left**) Coronal T2WI FS MR from the same patient shows disruption of the ACL ➔. Vague gray density is seen in the proximal aspect of the notch, but no intact ligament fibers are seen. There is marrow edema in the tibia, particularly posterolaterally ⇨. (**Right**) Axial T2WI FS MR from the same patient shows complete disruption of the ACL ➔ in the lateral aspect of the intercondylar notch. A marrow contusion is seen at the terminal sulcus of the lateral femoral condyle ⇨.*

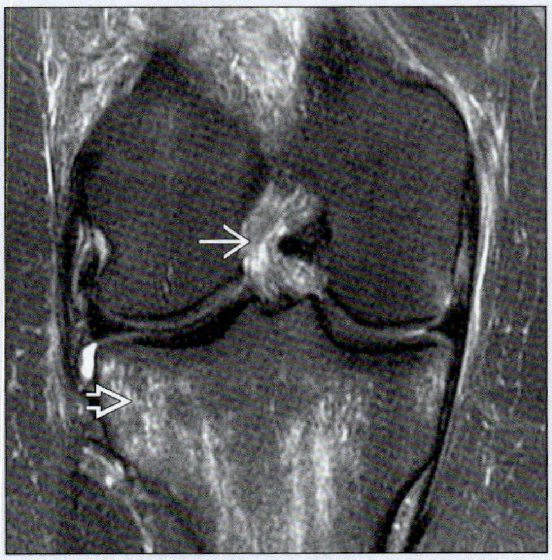

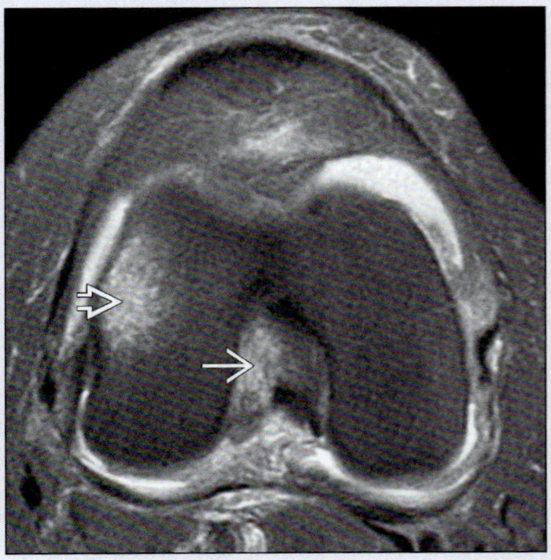

ANTERIOR CRUCIATE LIGAMENT INJURY

TERMINOLOGY

Definitions

- Partial or complete functional disruption of anterior cruciate ligament (ACL) mechanism

IMAGING

General Features

- Best diagnostic clue
  - Disruption of anterior cruciate ligament fibers on MR
  - Amorphous high signal on T1WI and T2WI in all or part of ACL
- Location
  - Most tears occur in proximal or mid portion of ligament
  - Osseous avulsion
    - Tibial attachment more commonly avulsed than femoral
- Morphology
  - ACL is intracapsular and extrasynovial
  - Proximal attachment: Posterior lateral wall of intercondylar notch
  - Distal attachment: Tibial articular surface just lateral to anterior tibial spine
  - ACL consists of 2 discrete fiber bundles
    - Posterolateral bundle
      - Larger than anteromedial
      - Taut in extension: Primary restraint of anterior tibial motion with knee in extension
    - Anteromedial bundle
      - Smaller than posterolateral
      - Taut in flexion: Primary restraint of anterior tibial motion with knee in flexion (anterior drawer test)
  - ACL serves as primary restraint against anterior tibial displacement
    - Provides greatest restraint with knee in extension
    - Provides secondary restraint against internal rotation and varus/valgus stress

Radiographic Findings

- Radiography
  - Deep sulcus sign: Deepening of terminal sulcus of lateral femoral condyle indicates impaction due to pivot-shift injury
  - Osteochondral fracture of posterolateral aspect of tibial plateau
  - Segond fracture: Avulsion of lateral capsular ligament from lateral tibial rim
    - Close to 100% association with ACL disruption
  - Anterior displacement of tibia relative to femur (radiographic "drawer" sign)
    - Implies injury to secondary restraints as well as ACL

MR Findings

- MR accuracy for ACL injury
  - Complete tear: Sensitivity 90-98%, specificity 90-100%, accuracy 90-95%
  - Partial tear: Sensitivity 40-75%, specificity 62-89%
- Direct signs of tear: Disruption of some or all of ACL fibers
  - Normal ligament bundles demonstrate uniform low signal on all sequences with fibrofatty tissue between bundles
  - Complete tear
    - Acutely disrupted fibers demonstrate elevated signal, particularly on T1WI or PDWI MR
    - Chronically disrupted ligament: "Empty notch" sign of fat in expected location of ACL
  - Partial tear
    - Best judged in axial plane
    - Partial tears can be divided into "stable" and "unstable" tears based on appearance
    - Stable tears do well with conservative therapy
      - ACL retains elliptical shape on axial MR images
      - Attenuated but intact ligament
      - Increased intrasubstance signal without disruption
    - Unstable tears likely to need surgical repair; more likely to lead to chronic instability
      - Absent ligament on axial MR images (on spectrum with complete ACL tear)
      - Absence of 1 bundle
      - Cloud-like mass of high signal in location of ACL
- Indirect sign of tear: Morphologic change in ACL
  - Normal morphology
    - Normal ACL is straight (taut) with knee in passive extension
    - Normal ACL is ovoid in cross section, with 2 distinct bundles, on axial MR images
      - Distal 1/3 of ligament has diverging fibers with fat between them
  - Torn or partially torn ligament often rounded or amorphous in shape
  - Chronic tear: Attenuated or absent ACL, no edema
    - Chronically torn ACL may fibrose to PCL, mimicking intact ligament
- Indirect sign of tear: Loss of normal alignment of ligament
  - ACL normally runs parallel to Blumensaat line (intercondylar notch roof angle)
  - Torn ACL may "sag" and drape onto posterior cruciate ligament (PCL)
- Indirect sign of tear: Associated marrow contusions
  - Pivot shift mechanism of ACL injury results in typical contusion pattern
    - Posterolateral tibial plateau, with variable medial extension
    - Lateral femoral condyle terminal sulcus region ± impaction of cortex
    - 50-95% sensitive
      - Tibial contusion more reliable than femoral contusion
    - Occasionally seen without complete ACL tear in adolescents
  - Due to hyperextension mechanism of injury
    - Anterior tibial plateau and anterior femoral condyles
- Osseous avulsion of ligament
  - Much more common in children
  - Almost always at tibial attachment
    - Fragment may be displaced &/or rotated
- Osseous avulsion of ligament
  - Much more common in children

ANTERIOR CRUCIATE LIGAMENT INJURY

- Almost always at tibial attachment
  - Fragment may be displaced &/or rotated
- Proximal avulsion: Lateral wall, intercondylar notch
  - Osseous fragment usually small, difficult to see

DIFFERENTIAL DIAGNOSIS

Complete vs. Partial ACL Tear

- Clinically relevant in determining need for surgery

ACL Mucinous Degeneration ± Ganglion Formation

- High intrasubstance signal on fluid-sensitive sequences
- Often well marginated if ganglion present
- Difficult to distinguish from partial tear without appropriate history

PATHOLOGY

General Features

- Associated abnormalities
  - Meniscal tear (50-80%)
    - Vertical peripheral tears common
    - Lateral > medial
  - Lateral capsular ligament disruption (Segond injury)
  - Medial collateral ligament (MCL) injury
  - Fibular collateral ligament (FCL)
  - Posterolateral corner injury
    - Usually requires earlier repair (< 3 weeks) than isolated ACL tear
    - Increased rate of ACL repair failure
  - PCL injury (often due to extreme hyperextension or knee dislocation)
  - Chondral fracture
    - Especially in posterolateral tibia and lateral femoral condyle
- Functional considerations

Staging, Grading, & Classification

- Complete tear: Ligament functionally incompetent
  - Some fibers may remain morphologically intact
- Partial tear
  - High grade (unstable): Abnormal Lachman but not completely disrupted
    - Usually ≥ 50% of ligamentous cross section disrupted
    - Tears involving 50-75% of ligament → high likelihood of progression to complete tear
  - Low grade (stable): Some laxity on exam but defined endpoint on anterior drawer test
    - Usually < 25% of ligamentous cross section disrupted
- Mucinous degeneration: Fiber degeneration ± intraligamentous cyst formation
  - May result from previous partial tear
  - Appearance may remain stable for many years or may worsen with time

CLINICAL ISSUES

Presentation

- Most common signs/symptoms
  - Feeling of instability in knee, particularly when pivoting on fixed foot
  - Anterior drawer test: Forward translation of tibia in 90° flexed knee
  - Lachman test: Drawer test performed with knee in 15-30° of flexion
  - Pivot-shift test: Valgus force applied to knee during flexion/extension
  - KT-1000 arthrometry: Quantitates degree of anterior tibial laxity
- Clinical profile
  - Athletic injury due to torsional strain on knee
    - Football, basketball, soccer, & alpine skiing injuries
  - "Pop" in knee during sports activity with acute onset of pain, instability, and swelling in knee

Demographics

- Gender
  - Prevalence: F:M = 2.4-9.7:1
- Epidemiology
  - Most commonly disrupted knee ligament: 95,000 ACL disruptions/year in USA
  - Partial tears represent 10-43% of ACL injuries

Natural History & Prognosis

- Low-grade (stable) partial tears do well with no significant instability or progression
- High-grade partial tears often → complete rupture
- Uncorrected ACL tear
  - Increased risk of meniscal tear
  - Chronic osteoarthritis

Treatment

- Conservative
  - Low-grade (stable) partial ACL tear
    - Tears involving < 25% cross section of ligament have favorable prognosis
  - Elderly or inactive patients
- Surgical repair
  - Success rate overall: 75-95%
  - Primary: Usually reserved for avulsion
  - Reconstruction
    - Autograft
      - Patellar tendon
      - Hamstring
    - Allograft
    - Synthetic graft

DIAGNOSTIC CHECKLIST

Consider

- Associated injuries (ligaments, menisci, chondral fx)

Image Interpretation Pearls

- Use axial MR images to determine partial vs. complete

Reporting Tips

- Discuss location of tear (proximal, middle, distal)

SELECTED REFERENCES

1. Aglietti P et al: Comparison between single-and double-bundle anterior cruciate ligament reconstruction: a prospective, randomized, single-blinded clinical trial. Am J Sports Med. 38(1):25-34, 2010

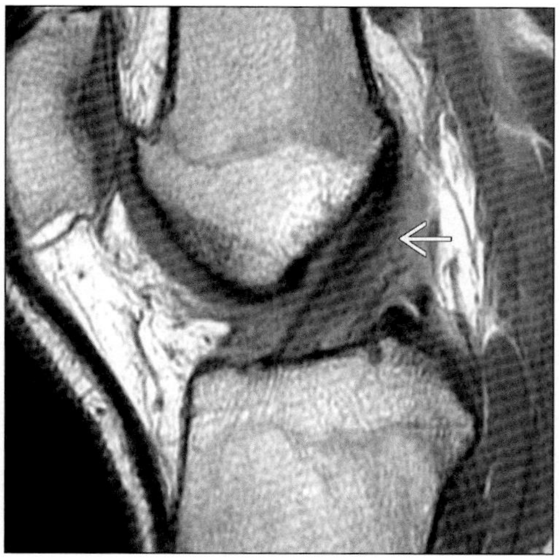

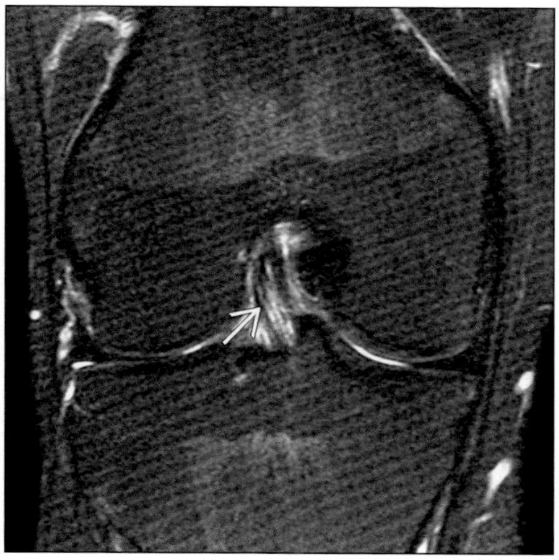

*(**Left**) Sagittal PDWI MR shows a low-grade interstitial tear (grade I sprain) of the ACL in a professional basketball player. While the ligament maintains its normal course through the intercondylar notch, there is some heterogeneity and fiber delamination ➔ within the substance of the ligament. (**Right**) Coronal T2WI FS MR from the same patient shows mild fiber delamination and edema ➔ in the mid and distal portions of the ACL. The normal unwinding of the bundles in the distal ACL should not be edematous.*

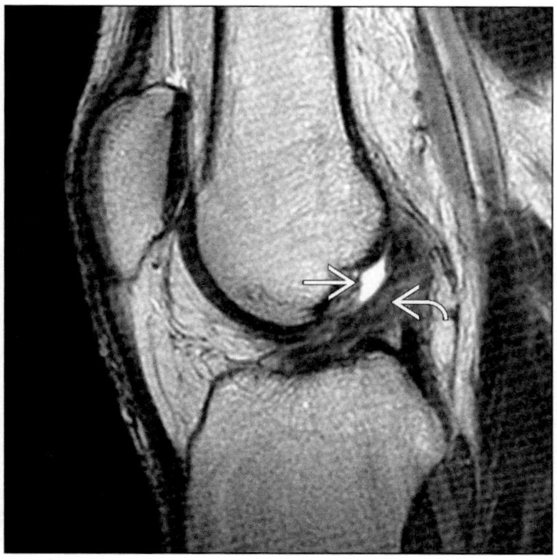

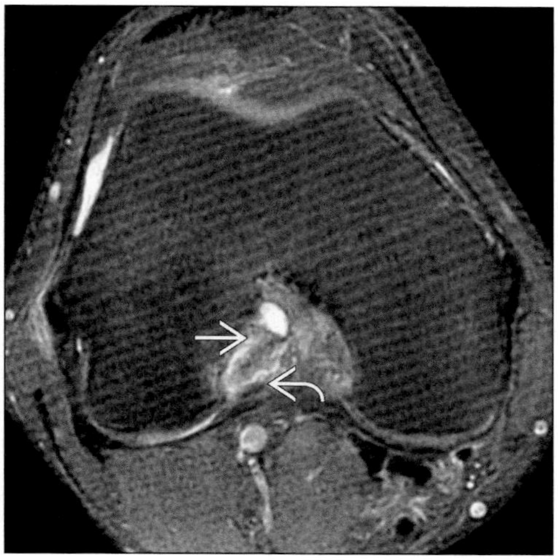

*(**Left**) Sagittal T2WI MR shows high-grade partial tear of the anterior cruciate ligament ➔ with intact posterior fibers ↪ remaining. There were mild marrow contusions in the posterolateral tibia and lateral femoral condyle (not shown) indicating significant pivot-shift injury. (**Right**) Axial PDWI FS MR from the same patient shows partial disruption of the ACL ➔. A small portion of the posterolateral bundle ↪ remains intact. Axial images provide the best evaluation of individual bundles.*

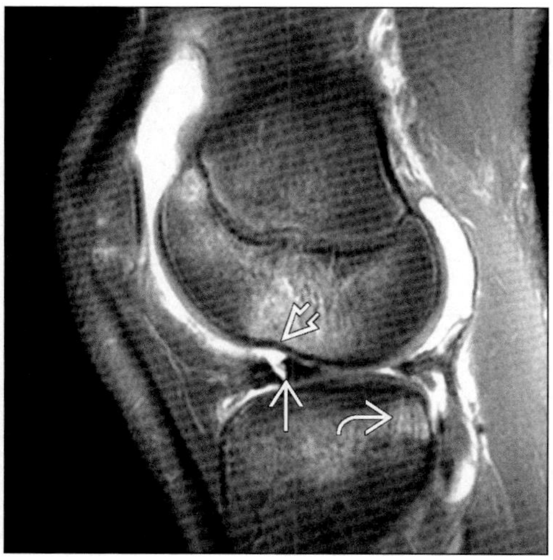

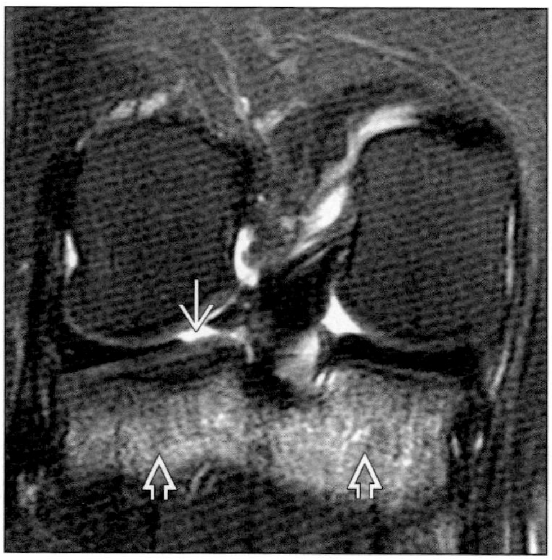

*(**Left**) Sagittal PDWI FS MR after acute ACL tear (not shown) demonstrates the classic marrow contusion pattern in the posterior lateral tibial plateau ↪ and in the region of the terminal sulcus of the lateral femoral condyle ⇨. There is also a displaced lateral meniscal tear with a flipped fragment ➔. (**Right**) Coronal T2WI FS MR after acute ACL tear shows a posterior tibial contusion ⇨; edema may extend beyond the expected posterolateral tibia. There is also a displaced tear of the posterior horn medial meniscus ➔.*

ANTERIOR CRUCIATE LIGAMENT INJURY

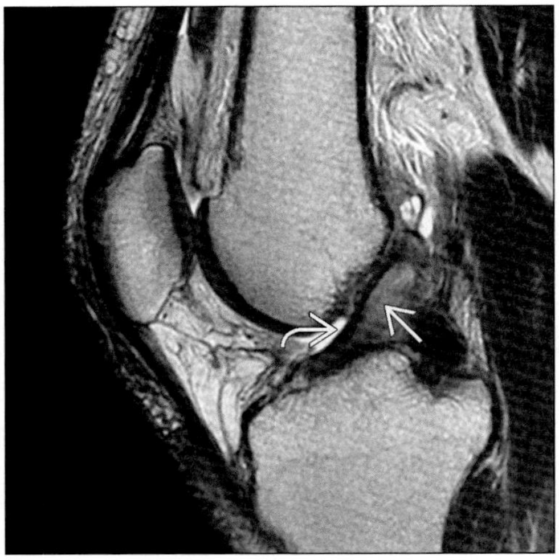

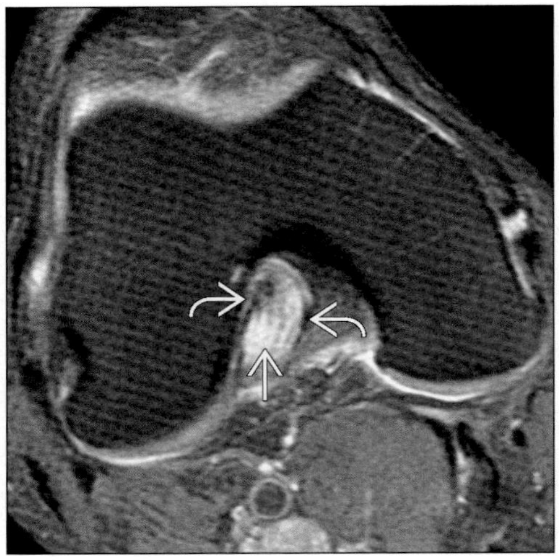

(Left) *Sagittal T2WI MR shows mucinous degeneration and ganglion formation in the proximal aspect of the ACL. The anterior fibers remain intact. The patient had no symptoms or signs of ACL insufficiency.* ***(Right)*** *Axial PDWI FS MR from the same patient shows the expanded contour of the ACL due to the presence of a mucinous cyst within the ligament. Intact ACL fibers are seen medially and laterally. Mucinous degeneration is usually seen in the setting of previous ACL partial tear.*

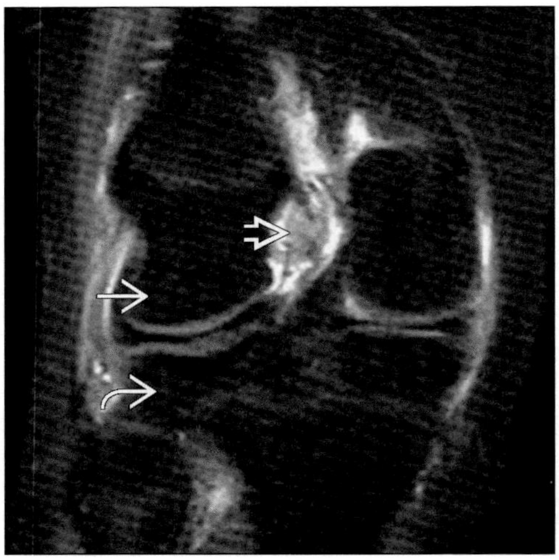

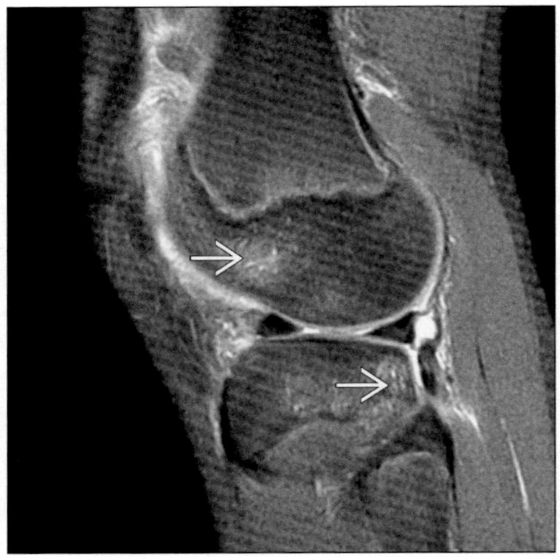

(Left) *Coronal T2WI FS MR shows acute disruption of the ACL. There is no associated marrow edema in the posterior lateral tibia or lateral femoral condyle. Absence of such contusions is very unusual in the setting of acute ACL tear.* ***(Right)*** *Sagittal PDWI FS MR from a 14-year-old girl shows typical pivot-shift mechanism marrow contusions. The ACL was intact on MR and by clinical exam. This can be seen in adolescents, probably due to increased elasticity of the ligament.*

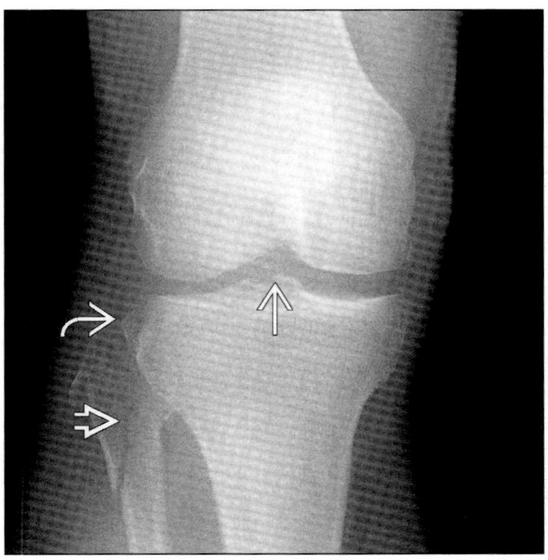

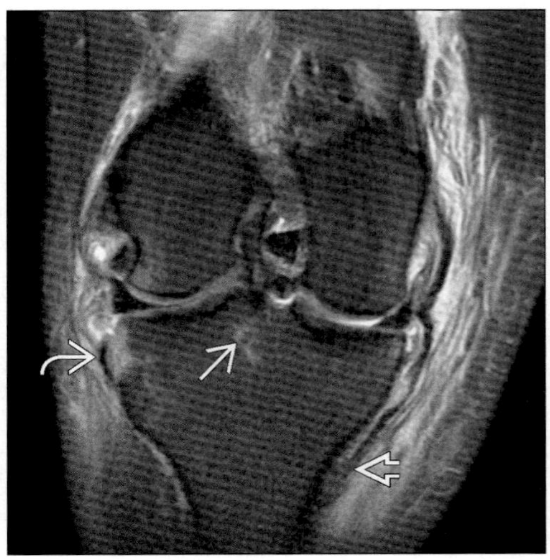

(Left) *AP radiograph shows avulsion of the tibial attachment of the ACL, as well as a comminuted fracture of the proximal fibula. There is osseous avulsion of the lateral rim of the tibia, a Segond injury. Segond injury has a very high association with ACL tear.* ***(Right)*** *Coronal T2WI FS MR from the same patient shows the Segond avulsion, as well as disruption of the distal medial collateral ligament. The ACL avulsion is partially seen, located more anteriorly.*

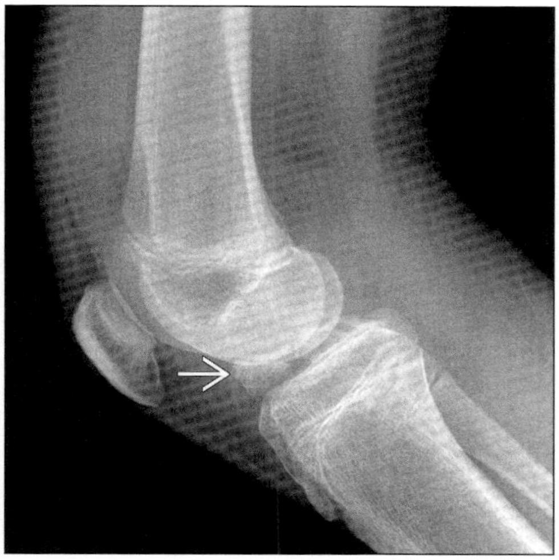

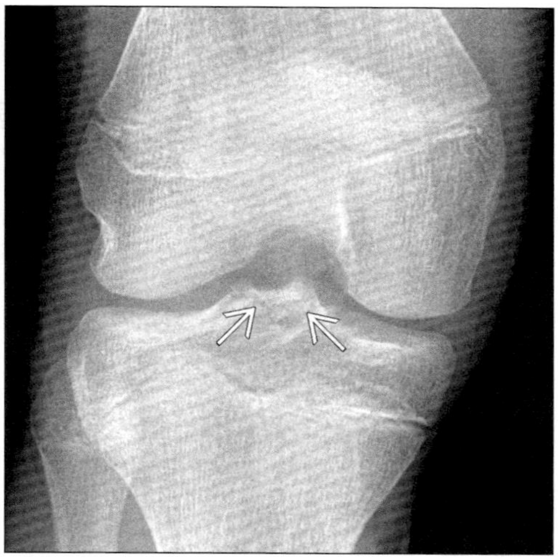

(Left) *Lateral radiograph shows osseous avulsion of the tibial attachment of the anterior cruciate ligament → in a 14-year-old boy. The fragment is displaced and rotated.* ***(Right)*** *AP radiograph from the same patient shows the displaced avulsion fragment → in the intercondylar notch. Careful scrutiny of radiographs is necessary to identify this injury when the fragment is less displaced. ACL avulsion is more common in children than in adults.*

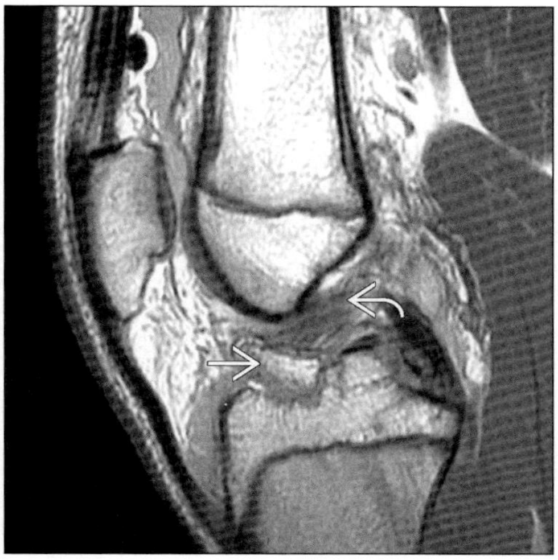

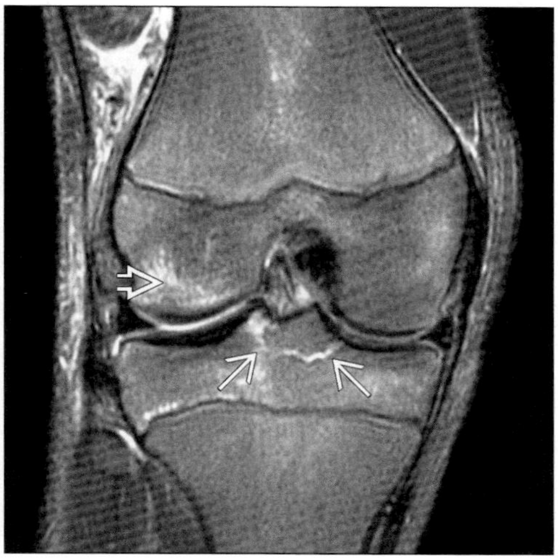

(Left) *Sagittal PDWI MR from the same patient shows the rotated avulsion fragment → of the tibial attachment of the ACL. There is some intrinsic heterogeneity and thickening of the ACL as well →.* ***(Right)*** *Coronal T2WI FS MR from the same patient shows the avulsion injury → of the ACL attachment on the anterior tibial spines. There is an associated impaction contusion → in the terminal sulcus region of the lateral femoral condyle.*

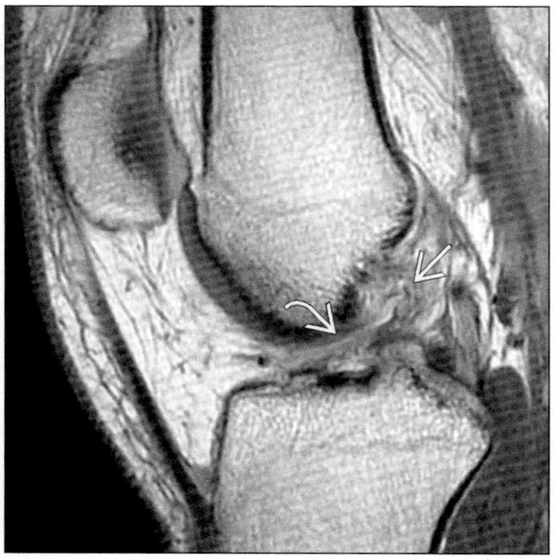

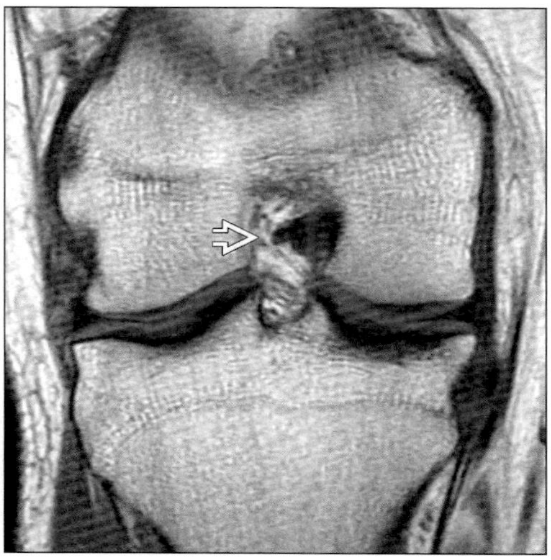

(Left) *Sagittal PDWI MR shows chronic disruption of the anterior cruciate ligament. Fibers of the ACL are redundant and atrophic →. The distal portion of the ligament → is more horizontal in orientation than normal. There is no edema in the intercondylar notch to suggest acute injury.* ***(Right)*** *Coronal T1WI MR from the same patient shows the "empty notch" sign of a chronically torn ACL. Fibers are scarred down onto adjacent structures, and fat fills the lateral aspect of the notch →.*

ANTERIOR CRUCIATE LIGAMENT: POSTOPERATIVE IMAGING

Key Facts

Terminology

- "Cyclops" lesion: Focal synovial hypertrophy anterior to reconstructed ACL

Imaging

- Graft
  - Relatively uniform low signal in graft for 1st few weeks after surgery
  - Graft begins "ligamentization" about 6 weeks after surgery
  - Areas of increased signal on T2WI MR are normal > 4 years after surgery
- Donor site patellar tendon autograft seen on MR
  - Patellar tendon central 1/3 defect (site of graft harvest)
- Graft rupture
  - Associated marrow contusions
- Improper graft placement
  - Proper graft function depends on isometry, which relies on recreation of original anatomy
- Internal (notch) impingement
  - Graft contacts roof of intercondylar notch
- "Cyclops" lesion
  - Localized arthrofibrosis anterior to ACL graft
  - Limits terminal knee extension
- Entrapped ACL stump
  - Distal remains of ACL interposed between lateral femoral condyle and lateral tibia
- Tibial tunnel syndrome
  - Fluid or cysts in graft tunnel ± pain
- Patellar fracture at graft harvest site
- Chemical fat suppression may fail because of field distortion due to metal hardware
  - Substitute STIR or non-fat-suppressed images
- Oblique sagittal images along plane of graft preferred by some

*(**Left**) Sagittal graphic of anterior cruciate ligament reconstruction shows anterior intercondylar notch roof impingement. The anterior aspect of the notch roof ⇨ rubs against the graft during normal knee motion, resulting in ligamentous fraying ➡. (**Right**) Sagittal PDWI MR shows intercondylar notch roof impingement of an ACL graft. There is posterior bowing of the graft and relatively focal increased signal intensity ➡ at the point of contact with the notch roof ⇨.*

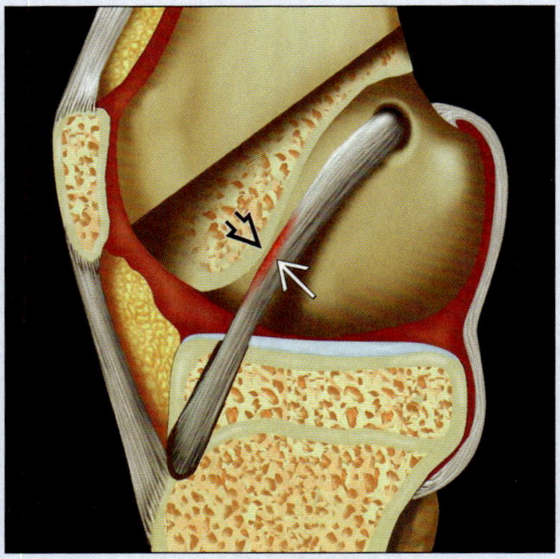

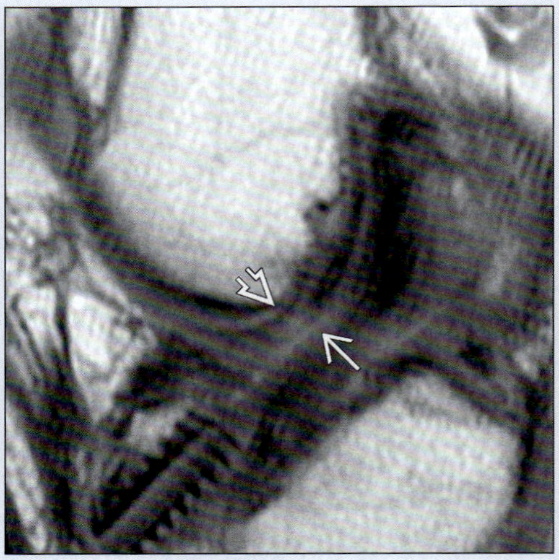

*(**Left**) Sagittal graphic shows focal anterior arthrofibrosis ("cyclops" lesion) ➡ after anterior cruciate ligament reconstruction. When large and firm, such a collection can lead to limited terminal extension of the knee joint. (**Right**) Sagittal PDWI MR shows focal anterior arthrofibrosis ("cyclops" lesion) ➡ in a patient 1 year after ACL repair. The graft ↪ is separate from the lesion and is normal. The patient complained of pain and limited knee extension.*

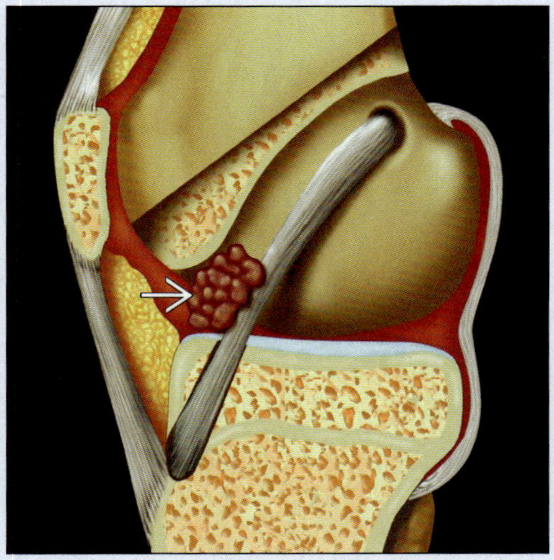

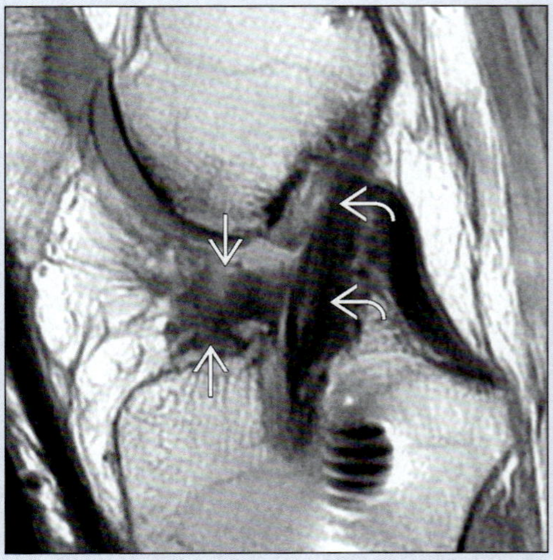

ANTERIOR CRUCIATE LIGAMENT: POSTOPERATIVE IMAGING

TERMINOLOGY

Abbreviations

- Anterior cruciate ligament (ACL)

Definitions

- Graft isometry
  - Refers to maintenance of constant length and tension of graft through normal range of motion
- Blumensaat line: Slope of intercondylar notch roof in sagittal plane
- Slope angle: Sagittal graft angle, measured relative to Blumensaat line
- Cyclops lesion
  - Focal synovial hypertrophy and arthrofibrosis anterior to reconstructed ACL
  - "Hard" nodule: Restricts terminal knee extension
  - "Soft" cyclopoid scar: Less likely to cause motion restriction

IMAGING

General Features

- Morphology
  - Patellar tendon graft
    - Bone-tendon-bone graft harvested from middle 1/3 of tendon
    - Bone plugs at either end of graft allow bone-to-bone graft incorporation
    - Secured with interference screws
  - Hamstring graft
    - Harvested from semitendinosus ± gracilis
    - 4 strand graft braided together
    - Secured with endobuttons, staples, interference screws, or screw and washer
  - Synthetic graft
    - 75% polyglycolic acid filaments
    - Wrapped in synovium
  - Double bundle graft
    - Technique seeks to recreate native morphology of ACL with separate posterolateral and anteromedial bundles

MR Findings

- **Normal postoperative appearance**
  - Graft
    - Relatively uniform low signal in graft for 1st 3 months after surgery
    - Graft begins "ligamentization" 2-3 months after surgery
      - Increased signal on short TE MR sequences
      - May become slightly thickened
    - Areas of increased signal on T2WI MR are normal > 4 years after surgery
      - Such areas do not correlate with graft instability or impingement
      - Seen in majority of patients with good tunnel placement
  - Graft tunnels
    - Marrow edema around tunnels persists for several weeks/months
    - Fluid within tunnels usually resolves by 18 months
  - Positioning
    - Proper graft function depends on isometry, which relies on recreation of original anatomy
    - Femoral tunnel in posterior notch within 2 mm of posterior cortex
    - Tibial tunnel just posterior to junction of Blumensaat line with tibia
    - Graft should be slightly oblique to vertical plane on coronal MR images, roughly parallel to Blumensaat line on sagittal
  - Donor site (patellar tendon autograft)
    - Patellar tendon central 1/3 defect (site of graft harvest)
    - Tendon thickened and irregular for 12-18 months post surgery
    - Mild marrow edema in patella and tibial tubercle resolves after several weeks
- **Complications**
  - Graft rupture
    - Focal fiber disruption
      - May be partial or complete
    - Horizontal orientation of graft fibers in sagittal plane
    - Laxity or graft fibers
    - Surrounding soft tissue swelling
    - Associated marrow contusions
      - Posterolateral tibial plateau
      - Lateral femoral condyle near terminal sulcus
    - Joint effusion
  - Mucinous degeneration of graft
    - Thickening of graft
    - Elevated signal in graft on T1WI and T2WI MR
    - No evidence of high-grade fiber disruption
    - Found in 60% of grafts in cadaveric study
    - Not indicative of graft failure
  - Improper graft placement (best judged on sagittal MR images)
    - Femoral tunnel too far posterior
      - "Over the top" positioning
      - Graft length and tension increase in extension
    - Femoral tunnel placed too far anterior
      - Graft length and tension increase in flexion
    - Tibial tunnel placed too far posterior
      - Graft appears vertical
      - Decreased anterior tibial stabilization
    - Tibial tunnel placed too far anterior
      - Intercondylar notch roof impingement of graft
  - Internal (notch) impingement
    - Graft contacts roof of intercondylar notch
    - Graft may be posteriorly bowed at point of contact
    - ± increased signal in graft
    - Seen best on sagittal MR images
  - "Cyclops" lesion
    - Localized arthrofibrosis anterior to ACL graft
    - Low to intermediate signal on T1WI and T2WI MR
    - May see attachment to graft
    - Often surrounded by fluid
    - Prevents complete extension of knee
  - Entrapped ACL stump
    - Distal remains of ACL interposed between lateral femoral condyle and lateral tibia
    - Limits terminal knee extension
    - May appear similar to cyclops lesion

- More linear than cyclops
- Demonstrable attachment to ACL insertion

- Tibial tunnel syndrome
  - Cyst formation in tibial graft tunnel
    - May extrude anterior to tibia and present as mass
  - Widening of tunnel due to "windshield wiper" effect of graft
    - Back-and-forth motion of graft during flexion-extension widens tunnel
    - More common when intraosseous graft fixation not performed
- Iliotibial band friction syndrome
  - Due to contact with dislodged or fragmented cross pins with hamstring graft
- Hardware complications
  - Loosening or displacement of screws or pins
  - Fragmentation of bioabsorbable pins or screws
- Donor site complications
  - Bone-tendon-bone patellar graft
    - Patellar fracture
    - Patellar tendon rupture (rare)
  - Hamstring harvest associated with little morbidity
- Infection
  - Cumulative incidence = 0.1-0.9%
  - MR identifies drainable fluid collections, synovitis, erosions, and marrow edema

Imaging Recommendations

- Best imaging tool
  - MR
- Protocol advice
  - Chemical fat suppression may fail because of field distortion due to metal hardware
    - Substitute STIR or non-fat-suppressed images
  - Oblique sagittal images along plane of graft preferred by some

PATHOLOGY

Gross Pathologic & Surgical Features

- Cyclops lesion
  - Focal fibrosynovial tissue in anterior intercondylar notch
  - At arthroscopy, appears as "head" with central area of discoloration; hence "cyclops" lesion

CLINICAL ISSUES

Presentation

- Most common signs/symptoms
  - Graft disruption
    - Anterior tibial shift and instability
      - Similar to primary ACL disruption
    - Early failure (< 6 months post surgery) due to poor positioning or incorporation failure
    - Late failure due to repeat trauma
  - Cyclops lesion or ACL stump entrapment
    - Limited terminal extension of knee
    - Anterior knee pain
  - Notch impingement
    - Limitation of terminal knee extension
  - Arthrofibrosis
    - Limited, painful range of motion

Demographics

- Epidemiology
  - Symptomatic osteoarthritis after ACL reconstruction = 13.6-21.5%
    - More prevalent when associated with meniscectomy
  - Recurrent ACL instability = 1-8%

Natural History & Prognosis

- Internal graft impingement may result in eventual graft rupture

Treatment

- Cyclops
  - Resection is curative
- Notch impingement
  - "Notchplasty" to open anterior superior aspect of notch and relieve impingement

DIAGNOSTIC CHECKLIST

Consider

- Elevated T2WI MR signal seen in majority of grafts > 4 years after surgery not associated with functional failure

Image Interpretation Pearls

- Differentiate cyclops from ACL stump entrapment
  - Stump demonstrates clear attachment to ACL tibial insertion

Reporting Tips

- Evaluate slope angle of graft in sagittal plane relative to Blumensaat line

SELECTED REFERENCES

1. Bencardino JT et al: MR imaging of complications of anterior cruciate ligament graft reconstruction. Radiographics. 29(7):2115-26, 2009
2. Casagranda BU et al: Normal appearance and complications of double-bundle and selective-bundle anterior cruciate ligament reconstructions using optimal MRI techniques. AJR Am J Roentgenol. 2009 May;192(5):1407-15. Review. Erratum in: AJR Am J Roentgenol. 192(6):1454, 2009
3. Pritchett JW: Assisted reproduction of the anterior cruciate ligament. J Knee Surg. 22(4):325-30, 2009
4. Lee GH et al: The incidence of acute patellar tendon harvest complications for anterior cruciate ligament reconstruction. Arthroscopy. 24(2):162-6, 2008
5. Pombo MW et al: Anatomic double-bundle anterior cruciate ligament reconstruction: where are we today? Arthroscopy. 24(10):1168-77, 2008
6. McCauley TR: MR imaging evaluation of the postoperative knee. Radiology. 234(1):53-61, 2005
7. Papakonstantinou O et al: Complications of anterior cruciate ligament reconstruction: MR imaging. Eur Radiol. 13(5):1106-17, 2003
8. Huang GS et al: Acute anterior cruciate ligament stump entrapment in anterior cruciate ligament tears: MR imaging appearance. Radiology. 225(2):537-40, 2002
9. Hodler J et al: The cruciate ligaments of the knee: correlation between MR appearance and gross and histologic findings in cadaveric specimens. AJR Am J Roentgenol. 159(2):357-60, 1992

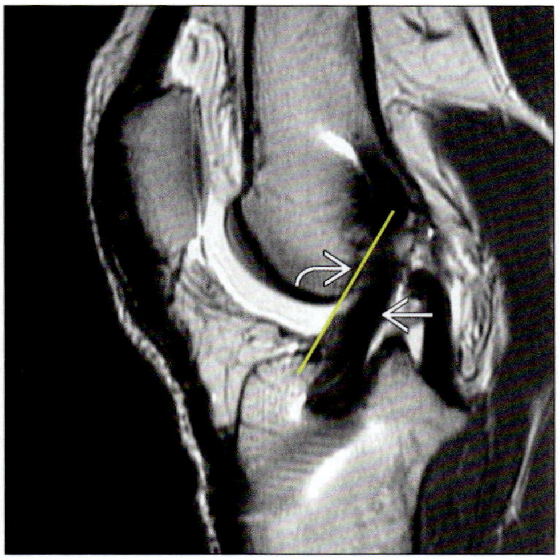
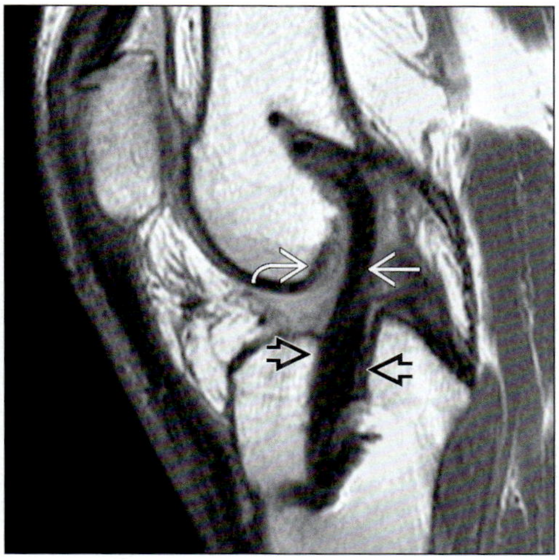

(Left) *Sagittal MR arthrogram T1WI shows correct placement of ACL tunnels resulting in ACL graft slope angle* ➔ *just behind Blumensaat line* ➔*, the projection created by the angle of the roof of the intercondylar notch toward the tibia.* ***(Right)*** *Sagittal PDWI MR shows posterior placement of tibial graft tunnel* ➔ *resulting in vertical orientation of ACL graft* ➔*. Note that the slope of the intercondylar notch roof* ➔ *projects a shallower angle.*

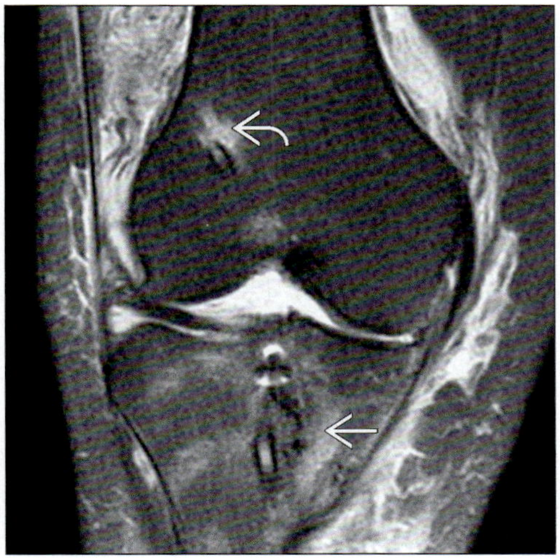
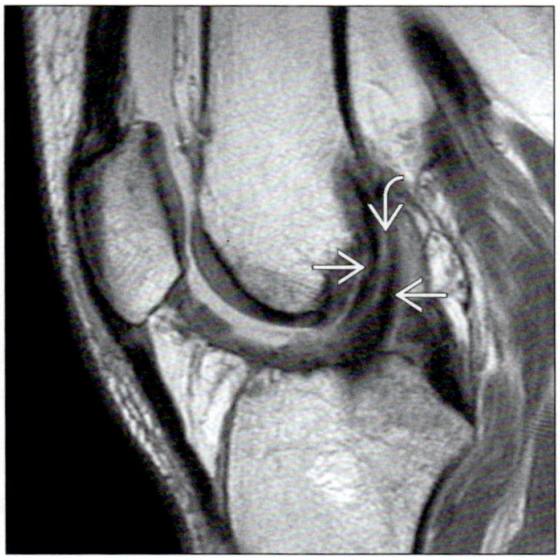

(Left) *Coronal T2WI FS MR obtained 1 day after ACL repair (due to a new injury) shows marrow edema around the femoral* ➔ *and tibial* ➔ *graft tunnels. Such edema may remain for months after graft placement.* ***(Right)*** *Sagittal PDWI MR illustrates the normal appearance of a multibundle hamstring graft for ACL repair. Note the separate bundles* ➔ *separated by a thin band of higher signal* ➔*. Several slips of semimembranosus and gracilis tendon are used to create such grafts.*

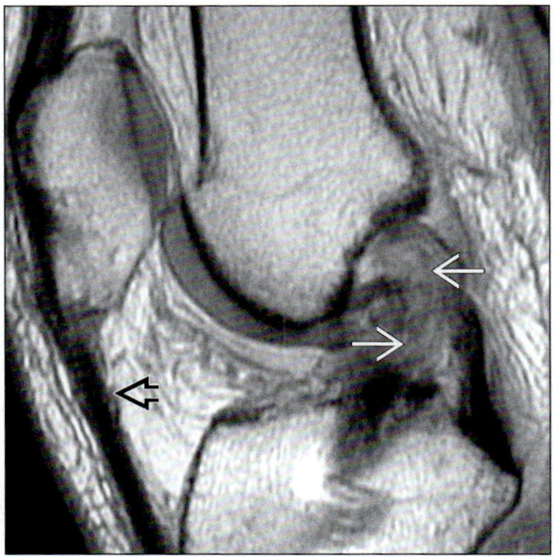
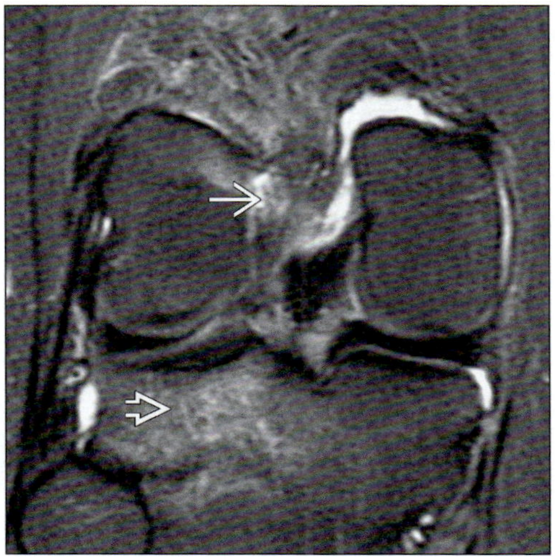

(Left) *Sagittal PDWI MR shows acute complete rupture of a bone-tendon-bone graft ACL repair. The disrupted graft* ➔ *shows thickening, amorphous elevated signal, loss of fiber structure, laxity, and posterior bowing. Mild irregularity of the patellar tendon* ➔ *is due to prior graft harvest.* ***(Right)*** *Coronal T2WI FS MR from the same patient shows elevated signal in the proximal graft substance* ➔ *and an impaction contusion in the posterolateral tibial plateau* ➔*, confirming an acute injury.*

(Left) *Coronal T2WI FS MR shows acute rupture of an ACL graft with fluid surrounding the graft fragments ➡ in the lateral aspect of the intercondylar notch.* ***(Right)*** *Sagittal PDWI MR shows a partial tear of a 4-bundle hamstring ACL graft. Two of the strands ➡ remain intact and follow their normal course to the femoral tunnel ➡. The 2 anterior strands ➡ are torn and have flopped into the anterior notch, creating a limitation to terminal extension.*

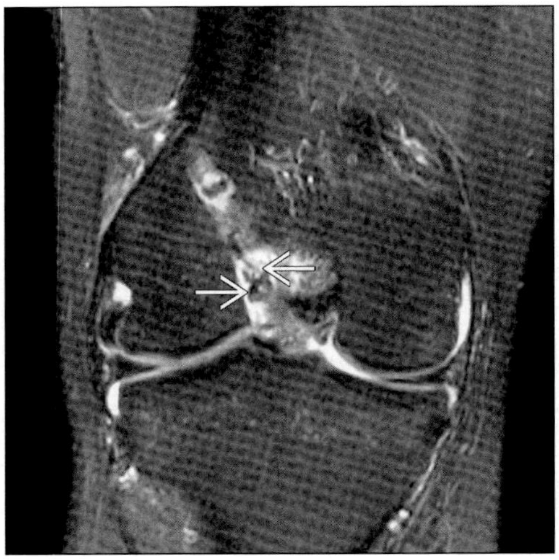

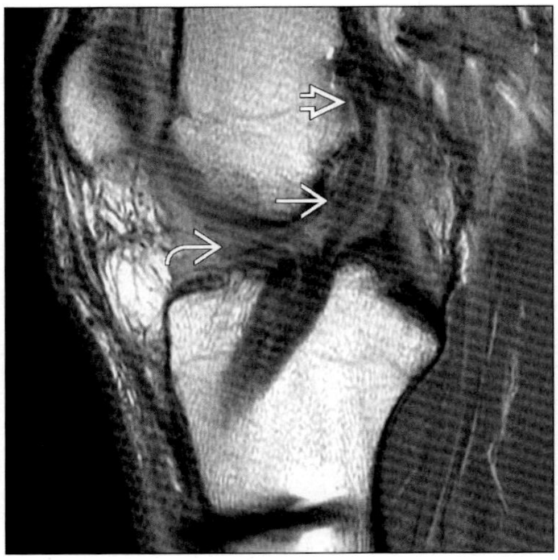

(Left) *Sagittal PDWI MR shows complete rupture of a hamstring-origin ACL graft ➡. A bioabsorbable interference screw ➡ secures the distal graft in the tibial tunnel, and 2 bioabsorbable pins ➡ secure the proximal end. Note the lack of metal artifact around the hardware.* ***(Right)*** *Sagittal PDWI MR shows chronic rupture of an ACL graft. There is anterior tibial translation, giving the appearance of notch impingement ➡.*

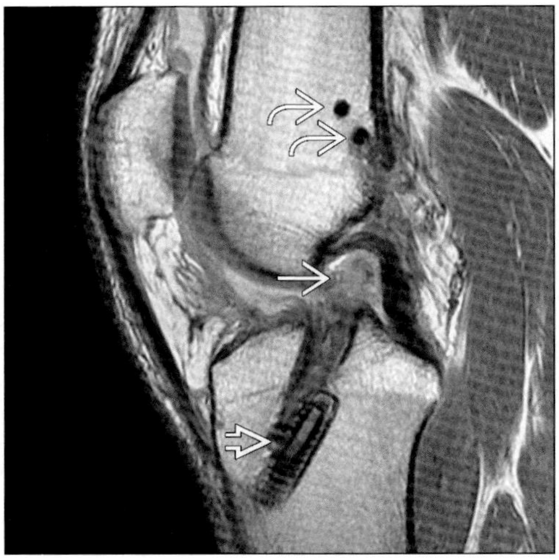

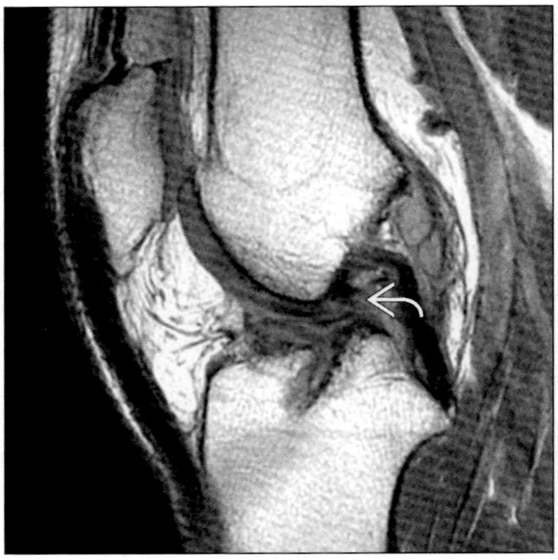

(Left) *Coronal T1WI MR shows a large osseous ridge ➡ in the lateral wall of the intercondylar notch distorting and displacing the ACL graft ➡. Lateral wall impingement is treated by notchplasty, where such bony prominences are excised, usually at the time of graft placement.* ***(Right)*** *Sagittal PDWI MR shows both intercondylar notch roof impingement of an ACL graft ➡ and formation of a "cyclops" lesion ➡. Note the intermediate signal intensity in the cyclops collection.*

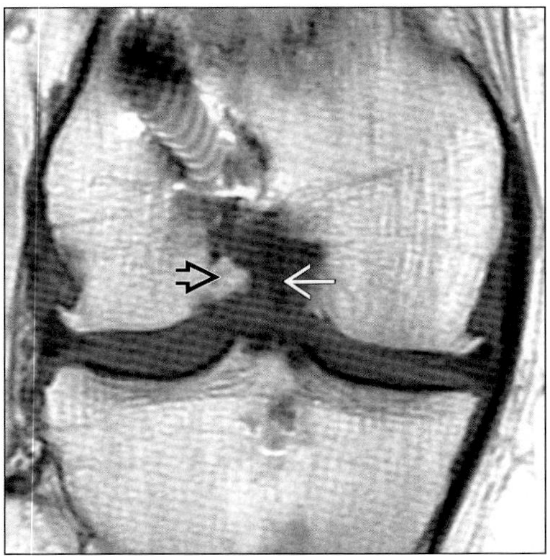

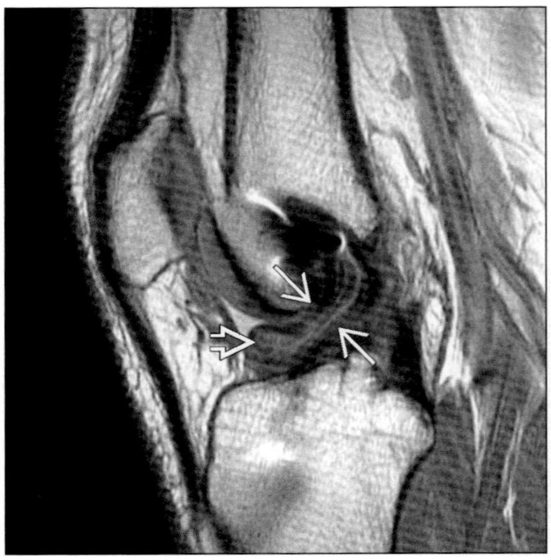

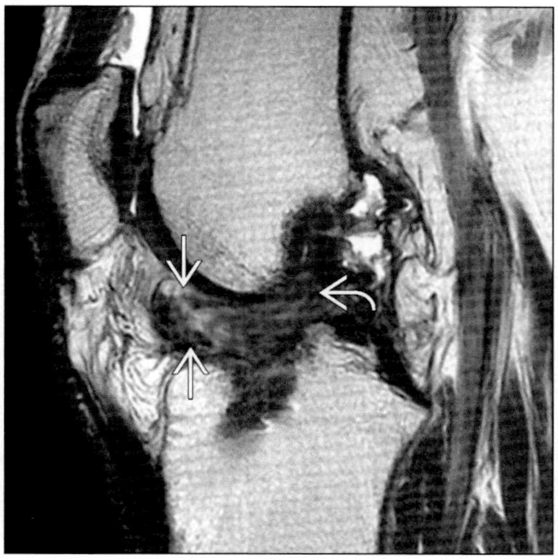

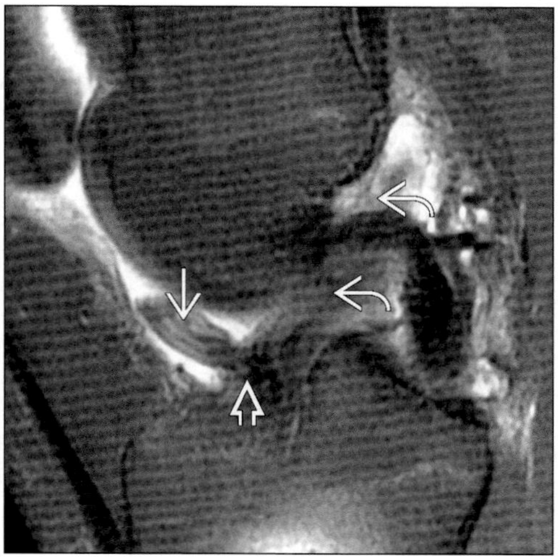

(Left) *Sagittal T2WI MR shows a large "cyclops" lesion → anterior to the ACL graft →. Typically measuring about 1 cm in size, this lesion measured about 2 cm and caused significant mechanical obstruction to motion.* ***(Right)*** *Sagittal PDWI FS MR shows an intact ACL graft → derived from hamstring elements. A well-defined linear focus of low signal → is seen anterior to the graft with attachment to the anterior tibial spine →. This is the entrapped stump of the native ACL.*

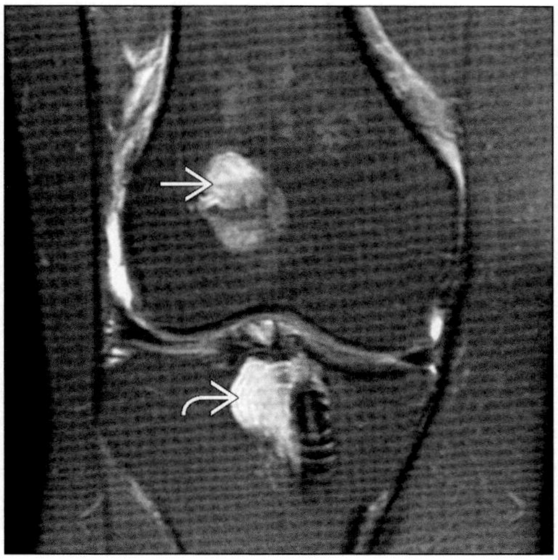

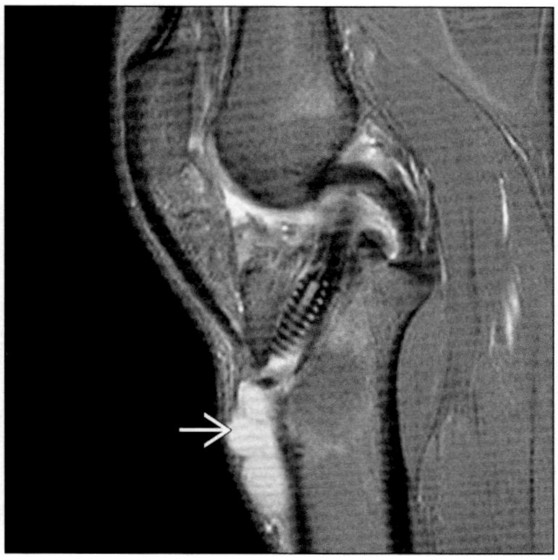

(Left) *Coronal T2WI FS MR shows fluid signal collections in the femoral → and tibial → graft tunnels after ACL repair with hamstring graft. Such ganglion cysts may enlarge over time and result in pain and abnormal graft motion.* ***(Right)*** *Sagittal PDWI FS MR shows a large tibial graft tunnel cyst → extruding through the anterior tibial cortex into the subcutaneous tissues and presenting as a painful mass. Such cysts must be resected, as they tend to continue to expand.*

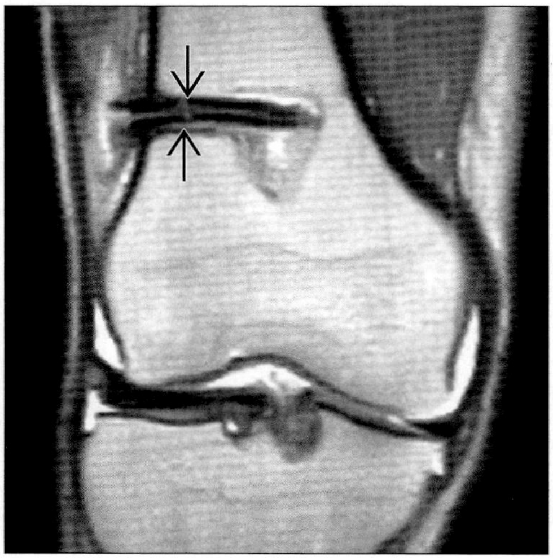

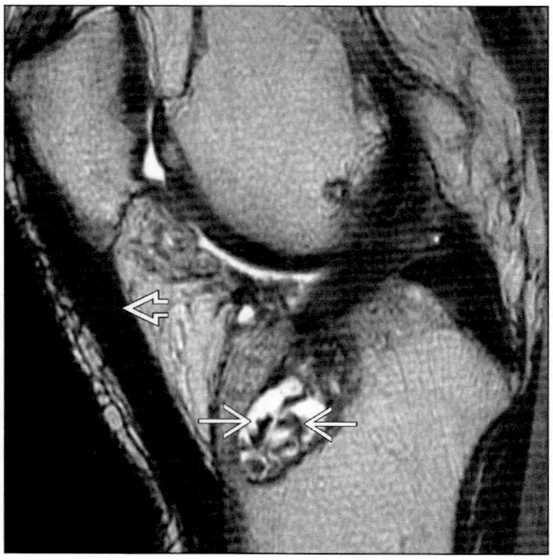

(Left) *Coronal MR arthrogram T1WI after ACL reconstruction shows a fracture → in a bioabsorbable femoral screw.* ***(Right)*** *Sagittal T2WI MR shows fragments of a bioabsorbable interference screw → within a cyst in the tibial tunnel of an ACL graft. Designed to reabsorb after graft incorporation, bioabsorbable fixation devices can loosen and fragment before graft incorporation, thus rendering the graft at risk for failure. Patellar tendon is thick after graft harvest →.*

POSTERIOR CRUCIATE LIGAMENT INJURY

Key Facts

Terminology

- Posterior cruciate ligament (PCL)

Imaging

- PCL serves as primary restraint to posterior displacement of tibia in 30-90° of knee flexion
- Intrasubstance tear: Usually around genu of ligament
- Avulsion: Posterior tibial attachment
- MR accuracy for PCL tear: 96-100%
- Disruption of some or all of PCL fibers
- Osseous avulsion of ligament
- Associated marrow contusions
  - Anterior tibial plateau + anterior femoral condyles

Pathology

- Most commonly due to posterior displacement of tibia relative to femur with knee in flexion
  - Dashboard injury: Flexed knee strikes dashboard during sudden deceleration
  - Fall on flexed knee
- Hyperextension
- 40% are isolated injuries
- Posterolateral corner injury
  - Significantly poorer outcome if both not fixed

Clinical Issues

- Posterior drawer test: Posterior tibial translation with patient supine and knee flexed 90°
- Reverse pivot shift test: Valgus force applied to tibia with foot externally rotated and knee ranged into extension
- Left untreated, ↑ incidence of osteoarthritis
- Surgical repair required if associated with posterolateral corner injury or multiligamentous injury

(Left) Sagittal graphic shows disruption of the genu (bend) of the posterior cruciate ligament ➡. There is posterior translation of the tibia relative to the femur due to a direct blow ➡ to the tibia in a flexed knee (dashboard injury). *(Right)* Coronal oblique graphic from the posterior medial aspect of the knee shows disruption ➡ of the vertical portion of the posterior cruciate ligament. Most tears occur in the genu of the ligament.

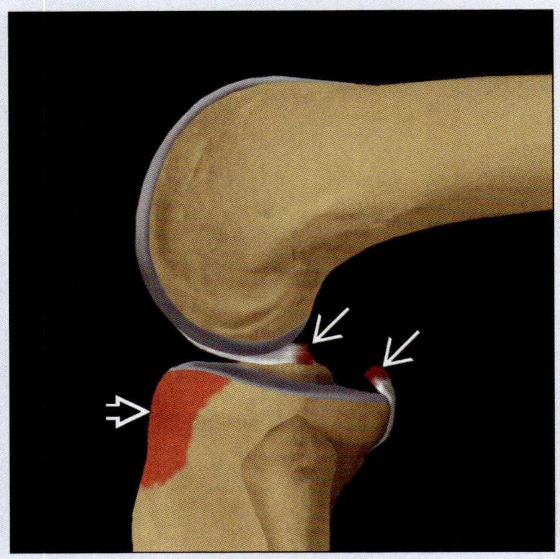

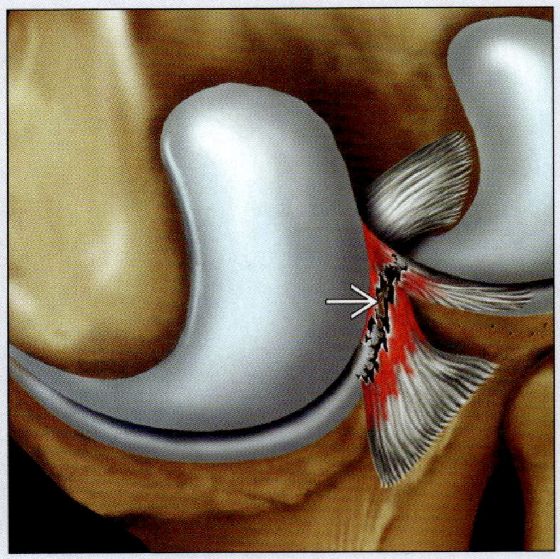

(Left) Sagittal PDWI MR shows disruption of the vertical portion of the posterior cruciate ligament (PCL) ➡. There are no intact fibers in the area of disruption, and the ligament here is thickened and demonstrates elevated signal. The vertical portion ➡ remains intact. *(Right)* Axial PDWI FS MR from the same patient shows disrupted fibers in the vertical portion of the PCL ➡. Fluid in the popliteal tendon sheath ➡ should not be mistaken for injury to the posterolateral corner stabilizing structures.

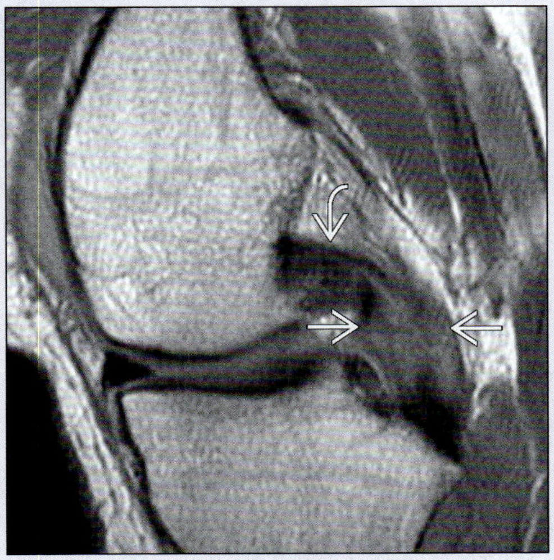

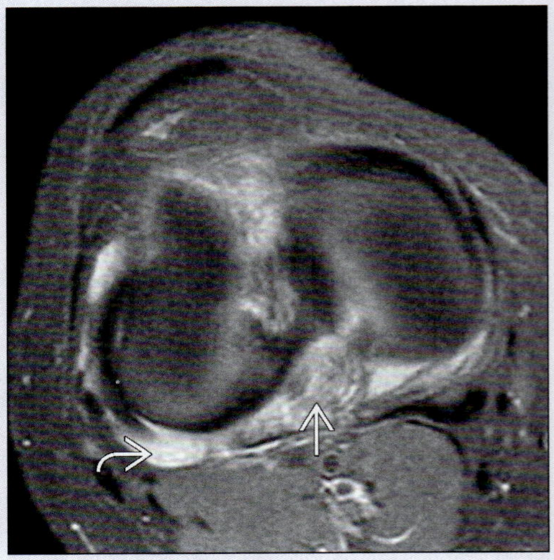

POSTERIOR CRUCIATE LIGAMENT INJURY

TERMINOLOGY

Abbreviations

- Posterior cruciate ligament (PCL)

Synonyms

- PCL tear

Definitions

- Partial or complete disruption of posterior cruciate ligament (PCL)

IMAGING

General Features

- Best diagnostic clue
  - Discontinuous PCL fibers on sagittal MR images
- Location
  - Intrasubstance tear: Usually around genu (bend) of ligament
    - Smaller percentage occur in vertical portion
      - Usually near tibial attachment
    - May occur in horizontal portion
      - Usually midway between genu and femoral attachment
  - Avulsion: Posterior tibial attachment
- Size
  - Tears range from microscopic fiber disruption to complete discontinuity of ligament
- Morphology
  - PCL is intracapsular and extrasynovial
  - PCL 2x as thick and strong as anterior cruciate ligament (ACL)
  - Proximal attachment: Anteromedial wall of lateral femoral condyle in intercondylar notch
  - Distal attachment: Central fovea of posterior proximal tibia
    - Located 1 cm distal to joint line
  - PCL consists of 2 discrete fiber bundles
    - Posteromedial bundle: Taut in extension
      - Stronger than anterolateral bundle
    - Anterolateral bundle: Taut in flexion
  - PCL serves as primary restraint to posterior displacement of tibia in 30-90° of knee flexion

Radiographic Findings

- Avulsion of tibial attachment
- Anterior impaction fractures
- Posterior sag of tibia relative to femur
- "Reverse Segond" fracture
  - Avulsion of deep fibers of medial collateral ligament (MCL) from medial tibial rim
  - Rare finding

MR Findings

- MR accuracy for PCL tear: 96-100%
- Disruption of some or all of PCL fibers
  - Use multiple imaging planes
  - Complete tear: Focal lack of PCL fibers on T1WI or PDWI MR with thickening of ligament
    - Usually around genu of PCL
  - Partial tear: Partial disruption of fibers with increased signal on T1WI and PDWI MR
  - Chronic PCL tear: Lack of normal PCL fibers in expected location, no edema
- Morphologic changes in PCL
  - Normal PCL has curved shape on sagittal MR images with knee in passive extension
  - Thickening of ligament with amorphous borders
  - Focal disruption in smooth curve of ligamentous border
- Osseous avulsion of ligament
  - May be displaced or nondisplaced
  - Often associated with marrow edema
  - Size of fragment may be small or quite large
- Periligamentous edema: Not as profound as in ACL tear
- Associated marrow contusions
  - Posterior tibial displacement mechanism
    - Anterior tibial plateau
  - Hyperextension mechanism
    - Anterior tibial plateau + anterior femoral condyles
    - May also see rupture of posterior joint capsule
- Joint effusion

Imaging Recommendations

- Best imaging tool
  - MR
- Protocol advice
  - Sagittal PDWI or T2WI MR best for overall integrity of PCL
  - Coronal images useful for evaluation of horizontal portion
  - Axial images useful for evaluation of vertical portion

DIFFERENTIAL DIAGNOSIS

Mucinous Degeneration of PCL without Tear

- Abnormally high signal on fluid-sensitive sequences and enlargement of PCL ± cyst formation
- May be due to remote injury

Normal Meniscofemoral Ligaments (Humphrey and Wrisberg)

- Arise from posterior horn of lateral meniscus and attach to medial wall of intercondylar notch near PCL

PATHOLOGY

General Features

- Etiology
  - Most commonly due to posterior displacement of tibia relative to femur with knee in flexion
    - Dashboard injury: Flexed knee strikes dashboard during sudden deceleration
    - Fall on flexed knee
  - Hyperextension
    - Often associated with ACL tear
  - Knee dislocation: 80% have PCL tear
  - Valgus and external rotation of knee on planted but unloaded foot
- Associated abnormalities
  - 40% are isolated injuries
  - ACL tear
    - Hyperextension injury
    - Knee dislocation
  - Posterolateral corner injury

POSTERIOR CRUCIATE LIGAMENT INJURY

        - Significant increase in poor outcome without surgery to correct both
    - Meniscal tear: Less frequent than with ACL tear
    - Posterior joint capsule rupture
        - Hyperextension injury

Staging, Grading, & Classification

- Partial intrasubstance tear
    - Some fibers remain intact
    - Divided into clinically stable or unstable types
    - MR only about 67% accurate in delineating partial from complete PCL tears
- Complete intrasubstance tear
- Osseous avulsion: 10% of cases
    - Tibial attachment most common
    - Femoral "peel-back" lesion (cortical avulsion of the femoral attachment of PCL)
- Mucinous degeneration
    - May be result of previous partial tear
    - Often seen in elderly patients
    - May be associated with intraligamentous cysts

CLINICAL ISSUES

Presentation

- Most common signs/symptoms
    - Vague feeling of knee instability, especially when walking down stairs
    - Posterior drawer test: Posterior tibial translation with patient supine and knee flexed 90°
        - Grade 1: Tibia lax compared to contralateral side, but it remains anterior to femoral condyles
        - Grade 2: Tibia translates to level of anterior margin of femoral condyles
        - Grade 3: Tibia translates posterior to anterior margin of femoral condyles
    - Posterior sag (Godfrey) test: Tibia passively sags posteriorly in position for posterior drawer test
    - Reverse pivot shift test: Valgus force applied to tibia with foot externally rotated and knee ranged into extension
- Clinical profile
    - Dashboard injury in motor vehicle crash
    - Sports injury: Fall on flexed knee (football, soccer)

Demographics

- Age
    - Usually adolescents and adults
- Gender
    - M > F
- Epidemiology
    - Accounts for 5-20% of knee ligament injuries

Natural History & Prognosis

- Left untreated, ↑ incidence of osteoarthritis
    - Especially medial and patellofemoral compartments
- Some literature suggests long-term increased incidence of meniscal tears

Treatment

- Conservative: Often applied to isolated PCL tear
- Surgical repair
    - Required if associated with posterolateral corner injury or multiligamentous injury
    - Uses bone-tendon-bone allograft from patellar tendon or Achilles tendon allograft
    - Double-bundle repairs being investigated

DIAGNOSTIC CHECKLIST

Consider

- Evaluate integrity of posterolateral corner in setting of acute PCL tear

Image Interpretation Pearls

- Use all available imaging planes to determine extent of PCL tear

Reporting Tips

- Describe pertinent positive and negative ancillary findings related to menisci, cartilage lesions, posterolateral corner, and ACL

SELECTED REFERENCES

1. Chen B et al: Double-bundle posterior cruciate ligament reconstruction using a non-hardware suspension fixation technique and 8 strands of autogenous hamstring tendons. Arthroscopy. 25(7):777-82, 2009
2. Hermans S et al: Long-term results of isolated anterolateral bundle reconstructions of the posterior cruciate ligament: a 6- to 12-year follow-up study. Am J Sports Med. 37(8):1499-507, 2009
3. McAllister DR et al: Posterior cruciate ligament biomechanics and options for surgical treatment. Instr Course Lect. 58:377-88, 2009
4. Sorrentino F et al: Role of high-resolution ultrasonography without and with real-time spatial compound imaging in evaluating the injured posterior cruciate ligament: preliminary study. Radiol Med. 114(2):312-20, 2009
5. Jung YB et al: Characterization of spontaneous healing of chronic posterior cruciate ligament injury: Analysis of instability and magnetic resonance imaging. J Magn Reson Imaging. 27(6):1336-40, 2008
6. Rodriguez W Jr et al: MRI appearance of posterior cruciate ligament tears. AJR Am J Roentgenol. 191(4):1031, 2008
7. Bellelli A et al: Magnetic resonance imaging of posterior cruciate ligament injuries: a new classification of traumatic tears. Radiol Med. 111(6):828-35, 2006
8. Cosgarea AJ et al: Posterior cruciate ligament injuries: evaluation and management. J Am Acad Orthop Surg. 9(5):297-307, 2001
9. Sherman PM et al: MR imaging of the posterior cruciate ligament graft: initial experience in 15 patients with clinical correlation. Radiology. 221(1):191-8, 2001
10. Shelbourne KD et al: Magnetic resonance imaging of posterior cruciate ligament injuries: assessment of healing. Am J Knee Surg. 12(4):209-13, 1999
11. Sonin AH et al: MR imaging of the posterior cruciate ligament: normal, abnormal, and associated injury patterns. Radiographics. 15(3):551-61, 1995
12. Fanelli GC et al: The posterior cruciate ligament arthroscopic evaluation and treatment. Arthroscopy. 10(6):673-88, 1994
13. Patten RM et al: Complete vs partial-thickness tears of the posterior cruciate ligament: MR findings. J Comput Assist Tomogr. 18(5):793-9, 1994
14. Sonin AH et al: Posterior cruciate ligament injury: MR imaging diagnosis and patterns of injury. Radiology. 190(2):455-8, 1994
15. Miller MD et al: Posterior cruciate ligament injuries. Orthop Rev. 22(11):1201-10, 1993
16. Barrett GR et al: Operative management of acute PCL injuries with associated pathology: long-term results. Orthopedics. 14(6):687-92, 1991

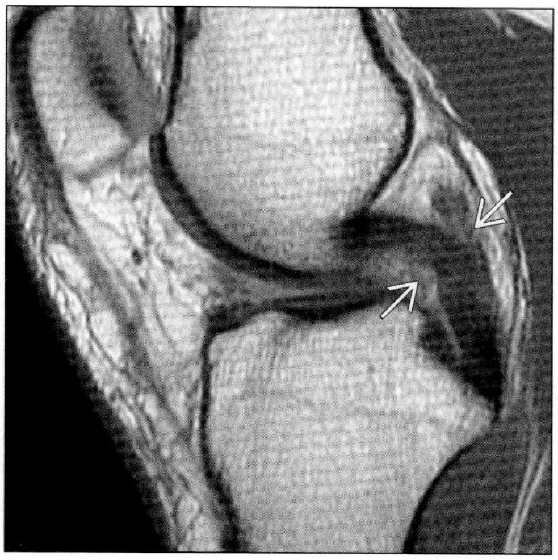

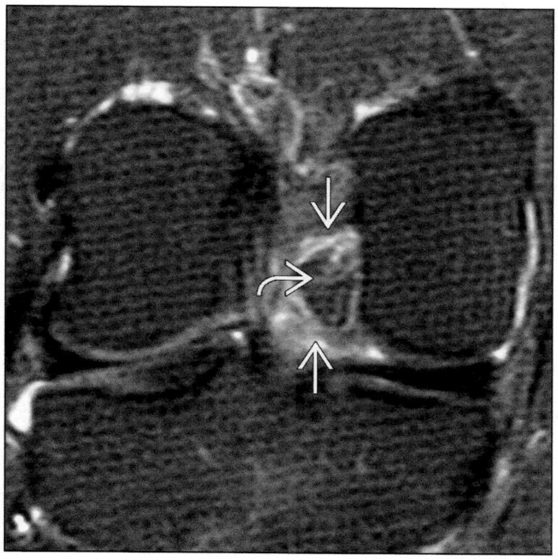

(Left) *Sagittal PDWI MR shows low-grade irregularity and signal heterogeneity at the genu of the posterior cruciate ligament (PCL). There is only a low-grade fiber disruption; this is a low-grade sprain.* ***(Right)*** *Coronal T2WI FS MR from the same patient shows mild periligamentous edema around the slightly swollen PCL. Magic angle artifact can cause spurious elevated signal in the PCL genu on short-TE sagittal images; confirmation of pathology with T2-weighted sequences is wise.*

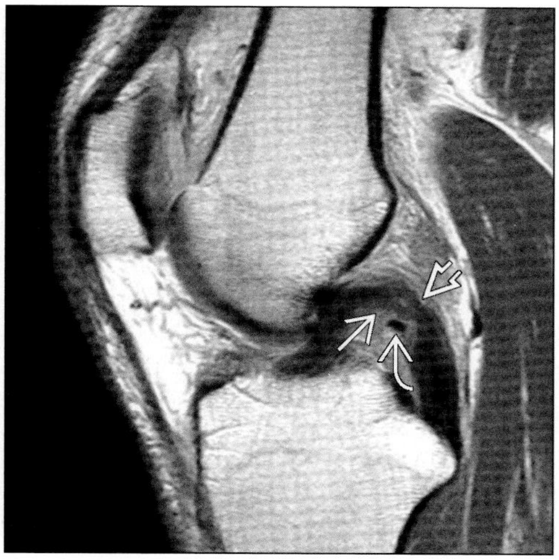

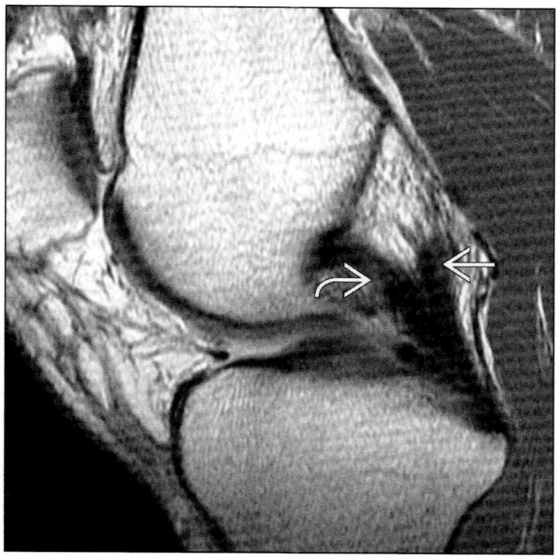

(Left) *Sagittal PDWI MR shows partial disruption of the genu of the PCL. The posterior portion of the ligament is relatively preserved, but the anterior fibers are disrupted. The anterior meniscofemoral ligament of Humphrey is also seen and should not be mistaken for a portion of the PCL.* ***(Right)*** *Sagittal PDWI MR shows an old partial tear of the PCL, with anterior fibers remaining intact and posterior fibers disrupted and posteriorly displaced.*

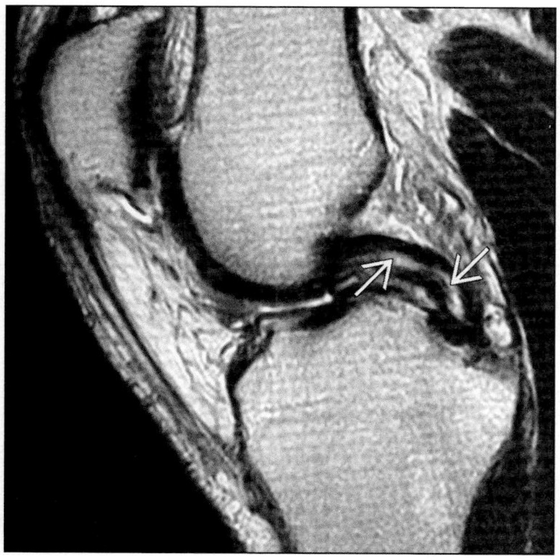

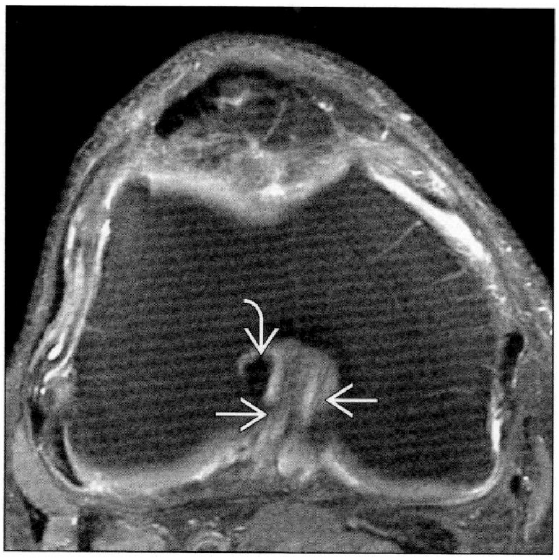

(Left) *Sagittal T2WI MR shows elevated signal in the central portion of the PCL, consistent with an old partial tear. Anterior and posterior fibers remain intact. This appearance can also suggest intraligamentous mucinous degeneration.* ***(Right)*** *Axial PDWI FS MR from the same patient shows a widened heterogeneous horizontal segment of the PCL indicating fiber disruption and mucinous change. The anterior cruciate ligament is seen in the lateral aspect of the notch.*

POSTERIOR CRUCIATE LIGAMENT INJURY

(Left) *Sagittal T2WI MR shows a focal tear of the PCL with fluid in the gap between the ligament fragments . The lack of ligament thickening and surrounding soft tissue swelling are in keeping with a chronic time course.* ***(Right)*** *Sagittal PDWI MR demonstrates the typical appearance of an acute PCL rupture. The genu and vertical portion are thickened, heterogeneous, and disorganized. The tibia is shifted posteriorly relative to the femur.*

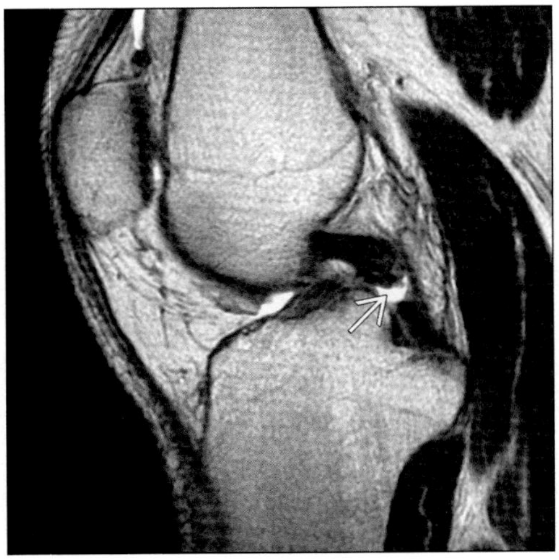

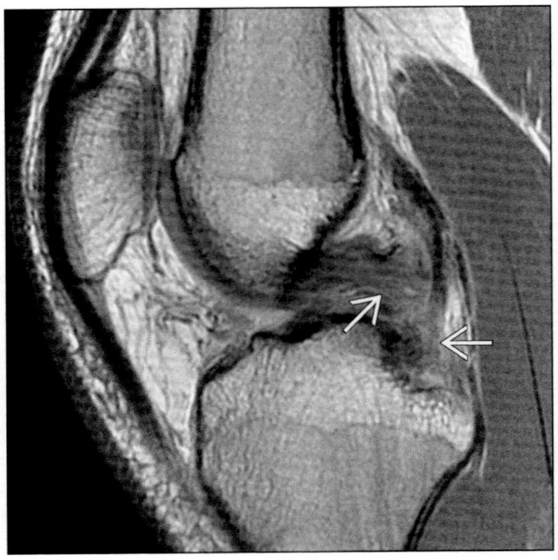

(Left) *Sagittal PDWI MR illustrates a tibial avulsion of the PCL attachment with surrounding marrow and soft tissue edema. The PCL itself is otherwise intact. Edema in the popliteus muscle belly indicates injury to the posterolateral corner, which should be further evaluated using T2W FS MR images through the area.* ***(Right)*** *AP radiograph demonstrates PCL avulsion from the tibia with mild fragment rotation. Nondisplaced avulsions like this can be subtle on radiographs.*

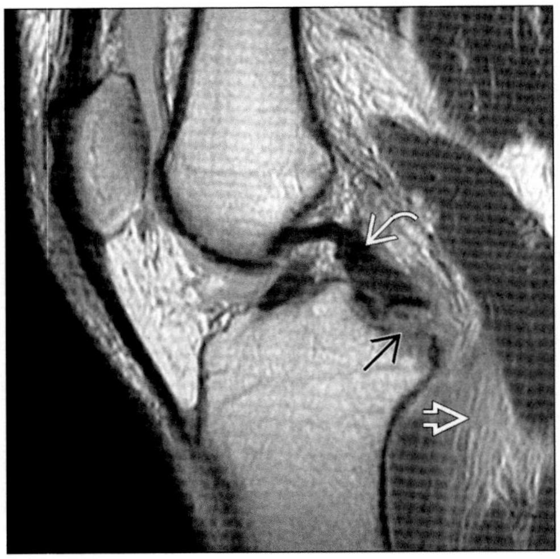

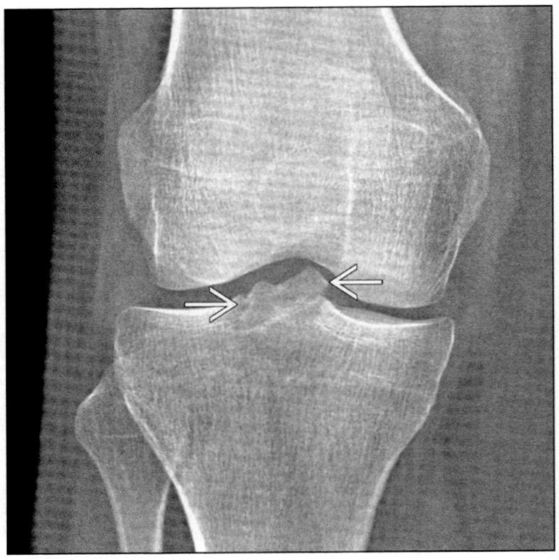

(Left) *Coronal bone CT shows PCL avulsion from the tibia with mild fragment displacement. Most PCL avulsions result in little displacement of the subtended osseous fragment.* ***(Right)*** *Coronal T2WI FS MR shows an acute tibial avulsion of the PCL with mild surrounding marrow edema. The PCL itself is intact. Nondisplaced osseous avulsions of the PCL have a favorable prognosis, and are generally treated conservatively.*

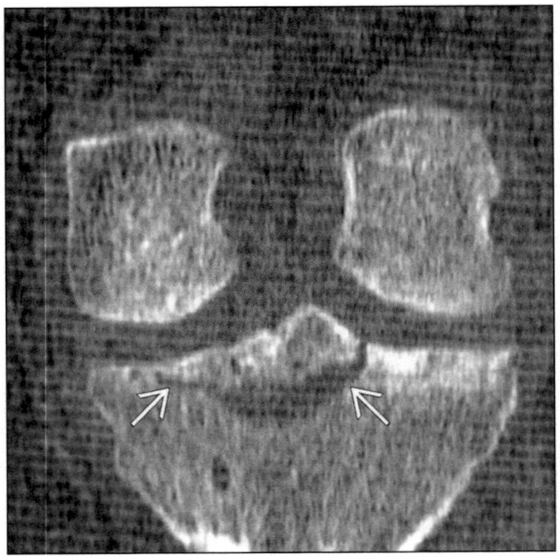

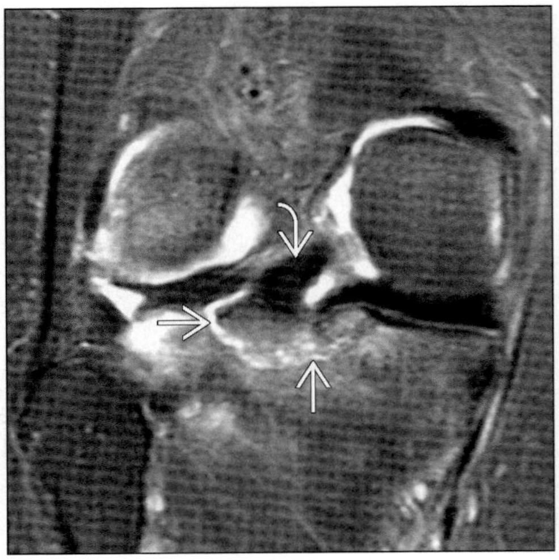

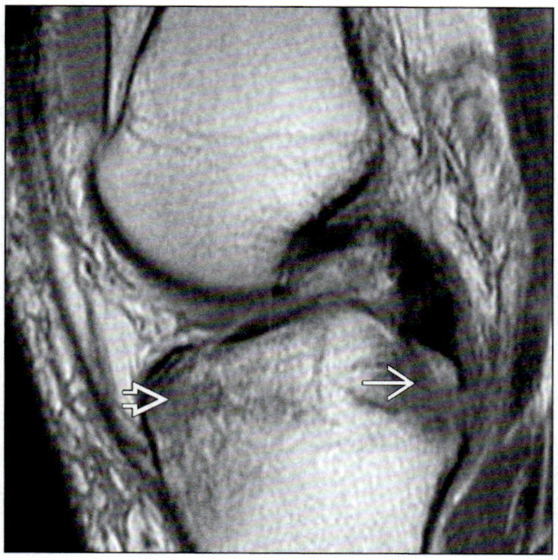

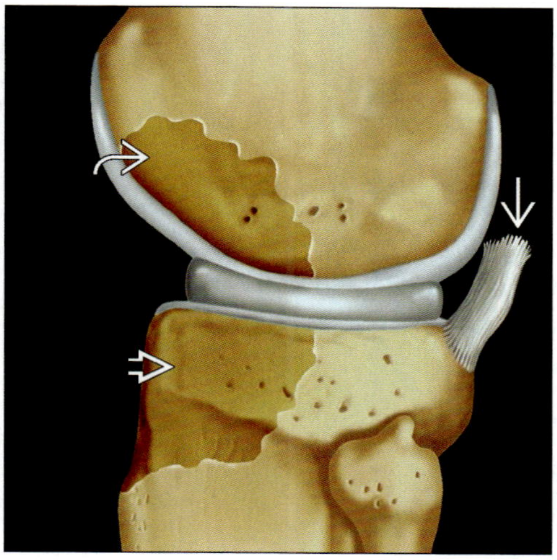

(Left) *Sagittal PDWI MR after acute PCL avulsion shows the avulsion site ➔ with mild displacement and an impaction contusion ➔ in the anterior lateral tibia due to a direct blow "dashboard" impaction injury.* ***(Right)*** *Sagittal graphic shows the typical distribution of marrow contusions in the anterior proximal tibia ➔ and anterior lateral femoral condyle ➔ due to a hyperextension injury resulting in disruption of the posterior cruciate ligament ➔.*

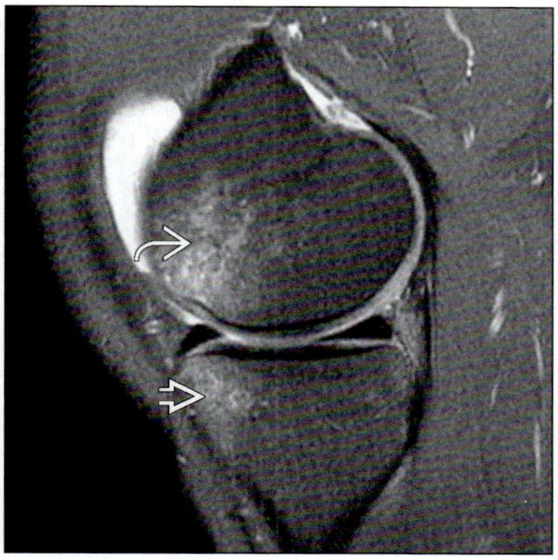

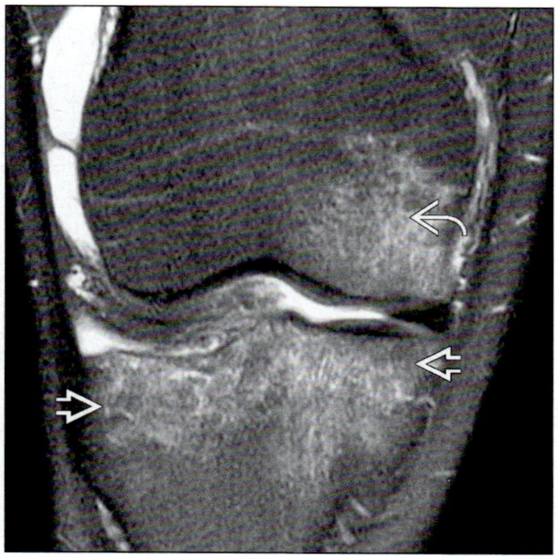

(Left) *Sagittal PDWI FS MR in a patient with posterior cruciate ligament tear due to hyperextension injury shows the typical marrow contusion pattern of edema in the anterior aspects of the tibial condyles ➔ and femoral condyles ➔.* ***(Right)*** *Coronal T2WI FS MR from the same patient shows the diffuse nature of the hyperextension contusion in the proximal tibia ➔. The femoral contusion is isolated to the medial femoral condyle ➔ in this case, but such contusion may involve either or both condyles.*

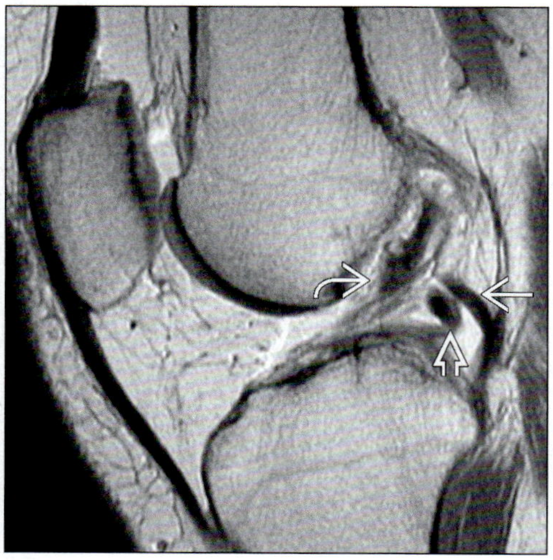

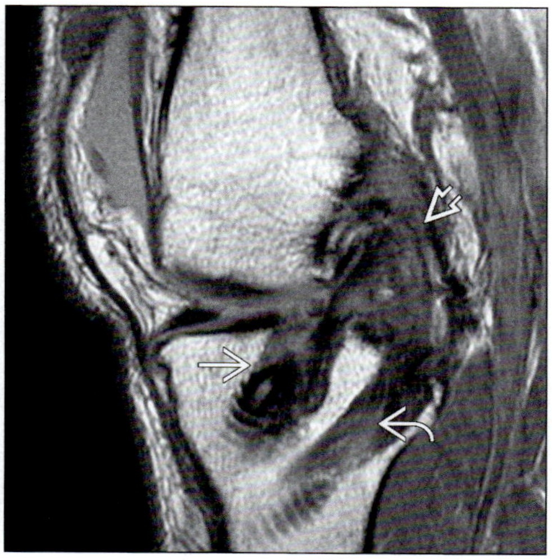

(Left) *Sagittal PDWI MR shows chronic atrophy and disruption of the anterior ➔ and posterior ➔ cruciate ligaments after previous knee dislocation. A large displaced medial meniscal fragment ➔ is also present.* ***(Right)*** *Sagittal PDWI MR after previous ACL ➔ and PCL ➔ reconstruction shows re-rupture of the PCL ➔ after an acute injury. When the PCL is torn as part of a multiligament injury, the importance of primary repair is significantly greater.*

MEDIAL COLLATERAL LIGAMENT, KNEE

Key Facts

Terminology
- Medial collateral ligament (MCL)
- Tibial collateral ligament tear

Imaging
- Focal disruption more likely in mid- or distal portion of MCL
- Deep layer: Meniscofemoral and meniscotibial ligaments
- Superficial layer
  - Proximal attachment: Posterior aspect of medial femoral condyle about 5 cm above joint line
  - Distal attachment: Medial tibial shaft, about 6-7 cm below joint line
- "Reverse Segond" avulsion of tibial insertion of deep fibers of MCL (rare)
  - Associated with posterior cruciate ligament (PCL) injury
- Pellegrini-Stieda disease
  - Linear calcification or ossification of MCL
- MCL bursitis
  - Focal fluid collection between layers of MCL
  - Often seen with intact MCL
- Chronic MCL injury (fibrosis)
  - Thickened ligament, especially proximal fibers
- T2WI FS MR in coronal and axial planes best define ligamentous integrity and status of adjacent structures

Pathology
- "Clipping injury" in sports

Clinical Issues
- Usually heals spontaneously
- Surgical repair considered in setting of multiligamentous injury

(Left) Coronal oblique graphic shows mild thickening of the proximal fibers of the superficial component of the medial collateral ligament ➔ with surrounding swelling ↪. (Right) Coronal oblique graphic shows laxity in the distal fibers of the medial collateral ligament ➔ consistent with a partial tear or distal avulsion. There is proximal periligamentous edema ↪. Inclusion of the distal insertion of the ligament on the tibia is crucial in accurately grading MCL injury in MR images.

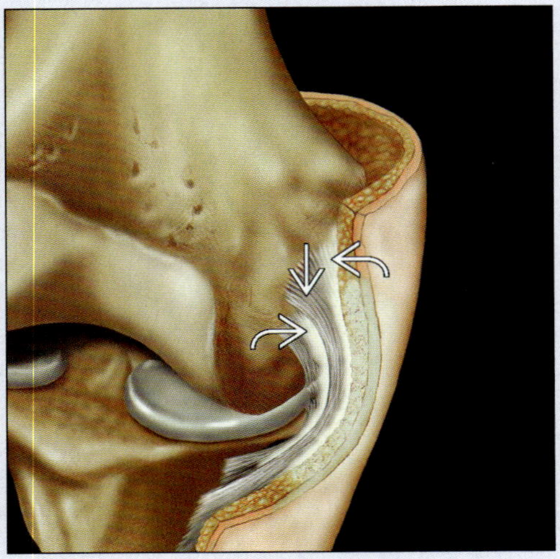

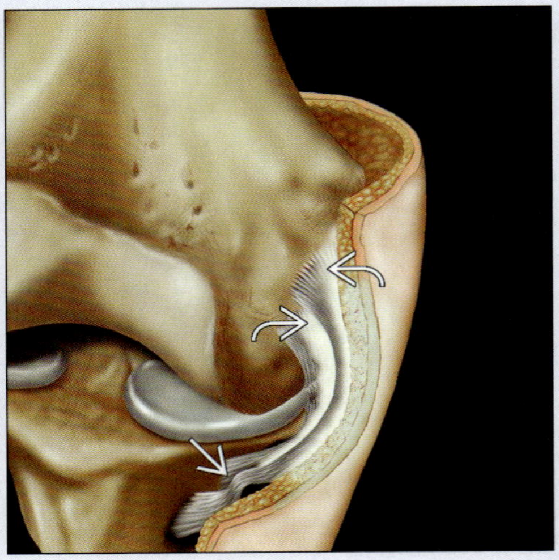

(Left) Coronal oblique graphic shows complete disruption of the superficial ➔ and deep ↪ components of the proximal MCL. Tears of the superficial component may be located in the proximal, middle, or distal portions of the ligament. Tears of the deep component are often located adjacent to the meniscus. (Right) Coronal oblique graphic demonstrates cortical avulsion ➔ of the proximal fibers of the medial collateral ligament, an unusual injury.

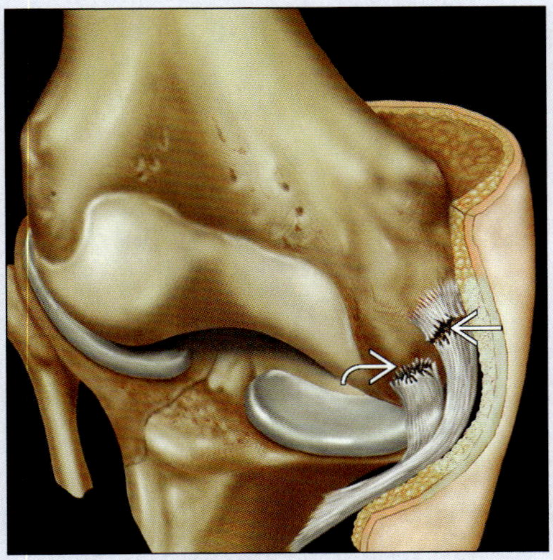

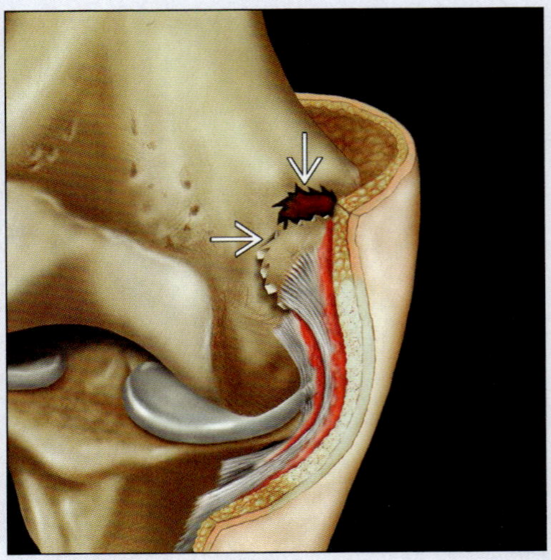

MEDIAL COLLATERAL LIGAMENT, KNEE

TERMINOLOGY

Abbreviations
- Medial collateral ligament (MCL)

Synonyms
- Medial collateral ligament tear
- Medial collateral ligament sprain
- Tibial collateral ligament tear

Definitions
- Partial or complete fiber disruption of medial collateral ligament of knee

IMAGING

General Features
- Best diagnostic clue
  - Abnormal signal around ± within MCL on coronal MR
- Location
  - Sprain may manifest as diffuse periligamentous edema
  - Focal disruption more likely in mid- or distal portion of MCL
- Morphology
  - Deep layer
    - Comprised of meniscofemoral and meniscotibial (coronary) ligaments
    - Anchor medial meniscus to adjacent bones
    - Attach directly to juxtaarticular surfaces
  - Superficial layer
    - Proximal attachment: Posterior aspect of medial femoral condyle about 5 cm above joint line
    - Distal attachment: Medial tibial shaft, about 6-7 cm below joint line
  - Posterior oblique ligament (POL)
    - Oblique fibers of posterior superficial layer
    - Suggested by some to be discrete ligament
    - Arises from adductor tubercle of femur (posterior and superior to medial epicondyle)
    - Distal end attaches to posterior horn of medial meniscus and blends into posterior joint capsule and semimembranosus sheath

Radiographic Findings
- Radiography
  - "Reverse Segond" avulsion of tibial insertion of deep fibers of MCL (rare)
    - Associated with posterior cruciate ligament (PCL) injury
    - Associated with peripheral medial meniscal tear
  - Pellegrini-Stieda disease
    - Linear vertical calcification or ossification of portion of MCL
    - Usually affects proximal fibers near medial femoral condyle
    - May or may not attach to femur
    - Seen best on AP projection

MR Findings
- Normal ligament
  - Well-defined linear signal void on all sequences seen best on coronal MR images
  - No direct attachment to midportion of medial meniscus
- Grade I injury (low-grade sprain)
  - Soft tissue swelling around MCL (superficial > deep tissues)
  - Best seen on T2WI FS MR coronal and axial sequences
  - Ligament intrinsically normal
- Grade II injury (partial tear or high-grade sprain)
  - Elevated signal in all or part of MCL
    - Best seen on T1WI or PDWI MR coronal sequences
  - Some fibers remain intact
  - Associated periligamentous soft tissue swelling
- Grade III injury (complete tear)
  - Focal disruption of MCL
  - Surrounding edema in acute setting
  - May be associated with osseous avulsion at attachment sites (uncommon)
  - Best seen on T2WI FS MR coronal sequences
- Associated findings
  - Marrow contusions
    - "Clipping injury": Direct blow to lateral knee causes valgus stress
      - Contusions in lateral femoral condyle ± lateral tibia
      - MCL injury
  - Other ligament injuries
    - Most commonly anterior cruciate ligament (ACL)
    - In case of "reverse Segond" avulsion, PCL
- Chronic MCL injury (fibrosis)
  - Thickened ligament, especially proximal fibers
  - Low signal on all sequences
  - No surrounding marrow edema
  - May be calcified or ossified (Pellegrini-Stieda disease)
- MCL bursitis
  - Focal fluid collection between superficial and deep layers of MCL
  - Usually well-defined borders
  - Often seen with intact MCL
  - Seen best on axial and coronal T2WI FS MR

Ultrasonographic Findings
- Loss of anisotropy of ligament
- Focal ligamentous anechoic defect in complete tear

Imaging Recommendations
- Best imaging tool
  - MR
- Protocol advice
  - T2WI FS MR in coronal and axial planes best define ligamentous integrity and status of adjacent structures

DIFFERENTIAL DIAGNOSIS

Osteoarthritis
- Chronic valgus deformity
- Leads to traction stress on MCL and ligamentous edema

MEDIAL COLLATERAL LIGAMENT, KNEE

Direct Blow Injury

- Associated superficial soft tissue swelling ± medial marrow contusions

Medial Meniscal Tear

- Peripheral tear may mimic disruption of meniscofemoral or meniscotibial ligaments

PATHOLOGY

General Features

- Etiology
  - Acute injury related to valgus force applied to knee
    - May occur in isolation, or as part of more complex injury mechanism
    - Frequently seen as part of pivot-shift mechanism of ACL injury
  - Often due to direct lateral blow to knee
    - "Clipping injury" in sports
    - ACL injury often associated
  - Repetitive trauma can lead to stretching or disruption of MCL
    - Frogleg kick used by breast stroke swimmers
- Associated abnormalities
  - Other ligamentous injuries
    - Especially ACL
    - "Unhappy triad": Combination of MCL tear, ACL tear, and medial meniscal tear
      - Less common than popularly thought
    - PCL and medial meniscal tear in "reverse Segond" injury
  - Meniscal tear
    - Medial meniscus: Due to traction forces
    - Lateral meniscus: Due to compression forces

Staging, Grading, & Classification

- MR grading
  - Grade 0: Normal
  - Grade I: Edema around ligament, but ligament intrinsically normal
  - Grade II: Abnormal signal within ligament, but no complete disruption
  - Grade III: Complete disruption of ligament
- Clinical grading
  - American Medical Association Committee on Medical Aspects of Sports
    - Grade I: 0-5 mm medial joint opening
    - Grade II: 5-10 mm medial joint opening
    - Grade III: > 10 mm medial joint opening
  - O'Donoghue grading
    - Grade I: Few fibers torn, clinically stable
    - Grade II: Incomplete tear, no pathologic laxity
    - Grade III: Complete tear, pathologic laxity

Gross Pathologic & Surgical Features

- Pellegrini-Stieda disease
  - Dystrophic calcification or ossification of superficial layer of MCL ± adductor magnus tendon
  - Related to previous injury

CLINICAL ISSUES

Presentation

- Most common signs/symptoms
  - Medial knee pain
  - Instability
    - Subjective instability on movement of knee
    - Abnormal laxity on exam with widening of medial joint space

Natural History & Prognosis

- Usually heals spontaneously
- May take up to 1 year to regain full ligament strength in cases of complete tear

Treatment

- Most treated conservatively
- Surgical repair considered in setting of multiligamentous injury

DIAGNOSTIC CHECKLIST

Consider

- Associated meniscal tear (medial or lateral)
- Avulsion of meniscotibial ligament ("reverse Segond")

Image Interpretation Pearls

- Make sure tibial attachment is included in coronal field of view (6-7 cm below joint line)

SELECTED REFERENCES

1. Chen L et al: Medial collateral ligament injuries of the knee: current treatment concepts. Curr Rev Musculoskelet Med. 1(2):108-13, 2008
2. Wen DY et al: MRI description of knee medial collateral ligament abnormalities in the absence of trauma: edema related to osteoarthritis and medial meniscal tears. Magn Reson Imaging. 25(2):209-14, 2007
3. Azar FM: Evaluation and treatment of chronic medial collateral ligament injuries of the knee. Sports Med Arthrosc. 14(2):84-90, 2006
4. Mendes LF et al: Pellegrini-Stieda disease: a heterogeneous disorder not synonymous with ossification/calcification of the tibial collateral ligament-anatomic and imaging investigation. Skeletal Radiol. 35(12):916-22, 2006
5. Pressman A et al: A review of ski injuries resulting in combined injury to the anterior cruciate ligament and medial collateral ligaments. Arthroscopy. 19(2):194-202, 2003
6. Escobedo EM et al: The "reverse Segond" fracture: association with a tear of the posterior cruciate ligament and medial meniscus. AJR Am J Roentgenol. 178(4):979-83, 2002
7. Loredo R et al: Posteromedial corner of the knee: MR imaging with gross anatomic correlation. Skeletal Radiol. 28(6):305-11, 1999
8. De Maeseneer M et al: Normal and abnormal medial meniscocapsular structures: MR imaging and sonography in cadavers. AJR Am J Roentgenol. 171(4):969-76, 1998
9. Reider B: Medial collateral ligament injuries in athletes. Sports Med. 21(2):147-56, 1996
10. Cameron JC et al: Management of medial collateral ligament laxity. Orthop Clin North Am. 25(3):527-32, 1994

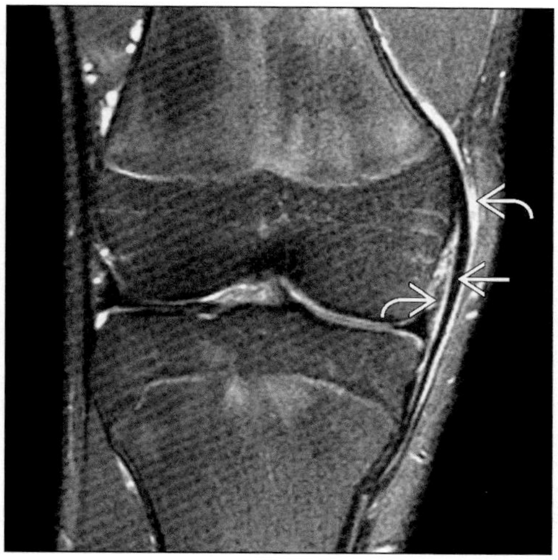

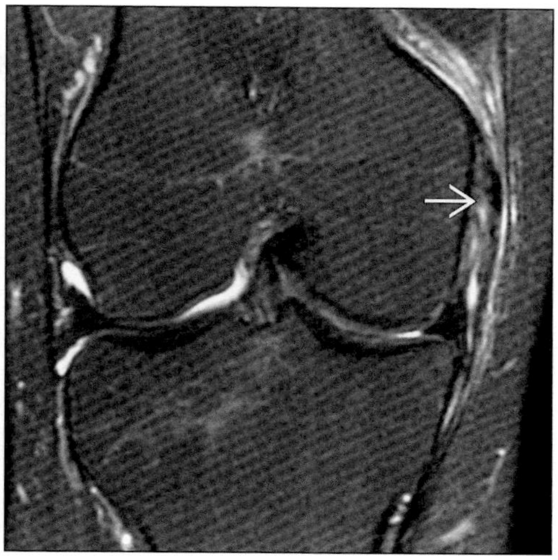

(Left) *Coronal T2WI FS MR demonstrates a grade I sprain of the medial collateral ligament. The ligament itself → demonstrates normal morphology and signal but is surrounded by mild soft tissue edema →. Clinically this corresponds to a stable ligament.* ***(Right)*** *Coronal T2WI FS MR shows a grade II sprain of the proximal MCL. There is increased size and signal in the proximal aspect of the ligament → with mild surrounding edema, but there is no complete disruption.*

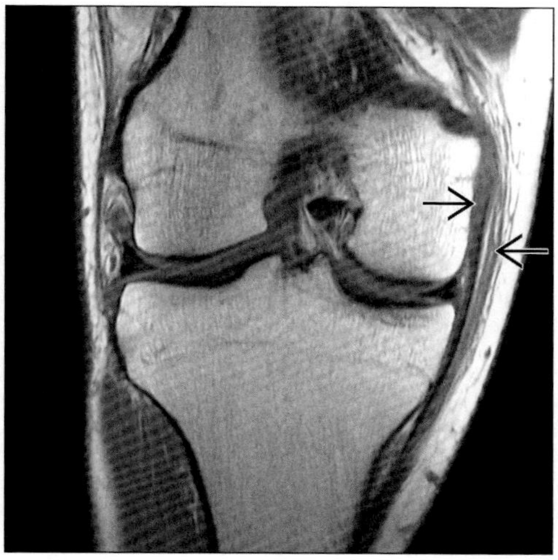

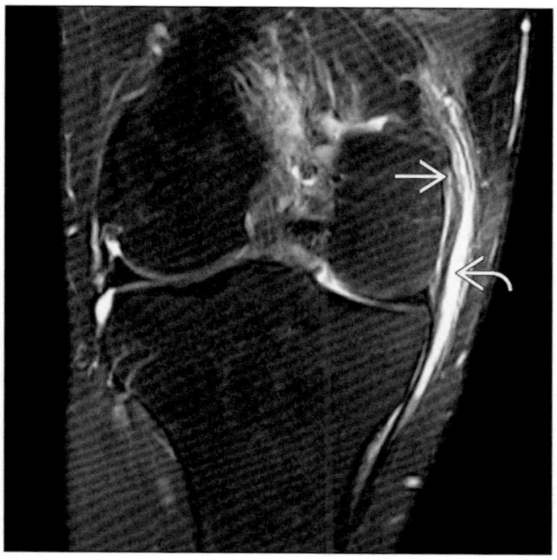

(Left) *Coronal T1WI MR shows mildly elevated signal → around the MCL consistent with injury, but the extent of injury is difficult to determine.* ***(Right)*** *Coronal T2WI FS MR from the same patient shows extensive swelling → around the MCL, with partial disruption of the proximal portion of the ligament →. Grade II injuries may manifest as attenuation of the ligament or as increased intrasubstance signal, or both. T2-weighted images are usually more accurate than T1 images for grading injury.*

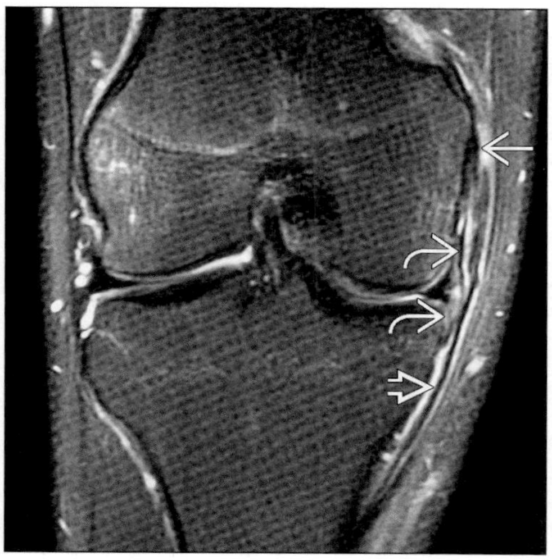

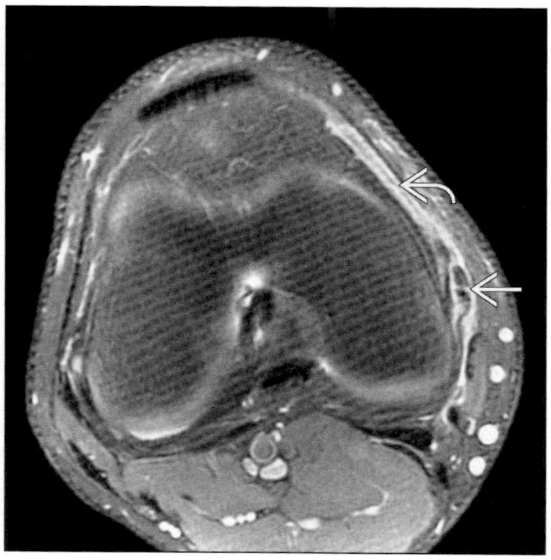

(Left) *Coronal T2WI FS MR shows partial disruption of the superficial fibers of the MCL proximally → and edema around the distal fibers →. The deep fibers (meniscofemoral and meniscotibial ligaments) are intact →.* ***(Right)*** *Axial PDWI FS MR from the same patient shows delamination of the MCL → with diffuse soft tissue swelling present along the course of the ligament and extending anteriorly along the medial patellar retinaculum →.*

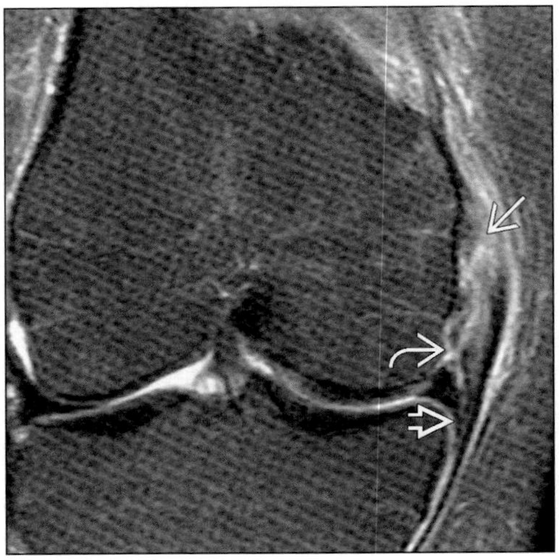

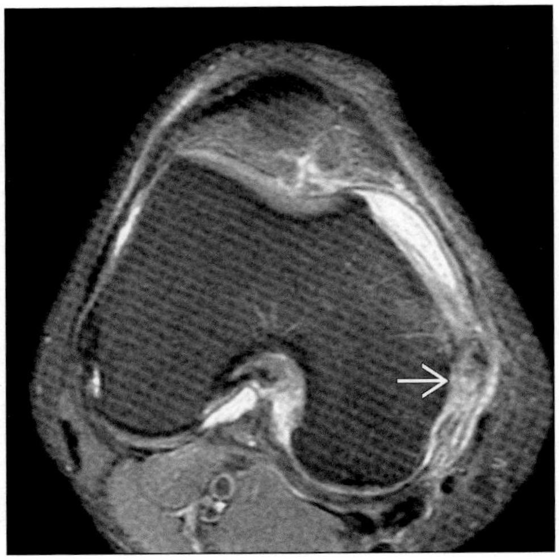

(Left) *Coronal T2WI FS MR shows complete proximal disruption of the superficial fibers of the medial collateral ligament ➔ and of the meniscofemoral ligament ↪. The meniscotibial ligament ➔ is intact.* ***(Right)*** *Axial PDWI FS MR from the same patient shows the extent of delamination and fiber disruption in the proximal MCL ➔. Axial MR images may help grade MCL injuries when coronal images are indeterminate.*

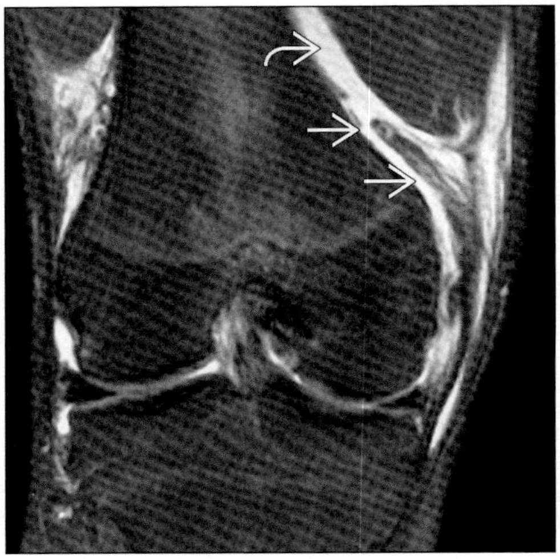

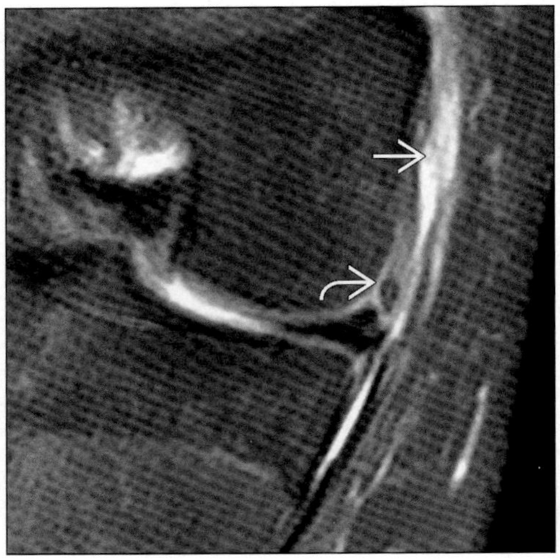

(Left) *Coronal T2WI FS MR shows a more extensive proximal avulsion of the MCL from the femur ➔ with fluid dissecting between the MCL and the medial femoral cortex, and extending superiorly along the medial border of the vastus medialis muscle ↪.* ***(Right)*** *Coronal T2WI FS MR shows a complete tear of the proximal MCL ➔ with surrounding edema. Tear of the meniscofemoral attachment has resulted in avulsion of a small portion of the medial meniscus ↪.*

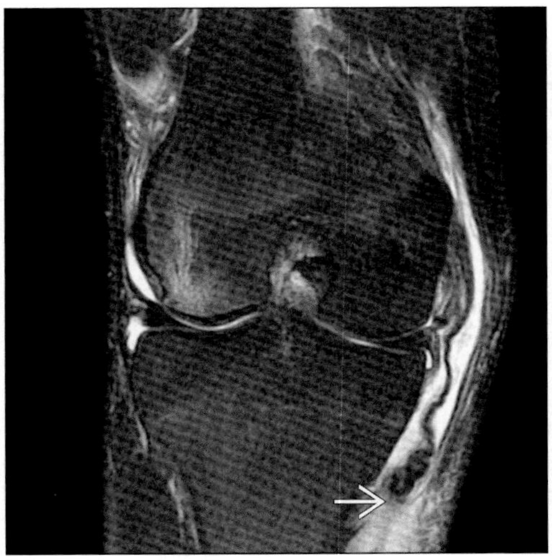

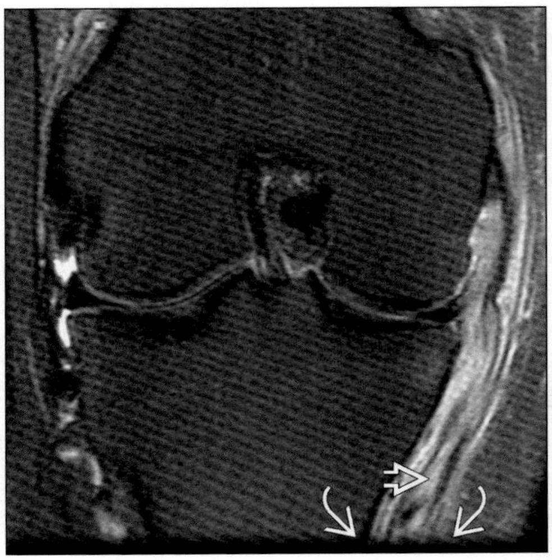

(Left) *Coronal T2WI FS MR shows complete distal disruption of the MCL with retraction and redundancy of the distal fibers ➔. Moderate edema is seen along the entire course of the ligament.* ***(Right)*** *Coronal T2WI FS MR shows a high-grade tear of the distal MCL ➔, but improper coil placement has resulted in cut-off of signal ↪ above the tibial insertion of the ligament. The MCL inserts about 5-7 cm below the knee joint, and this area must be included in the scan volume.*

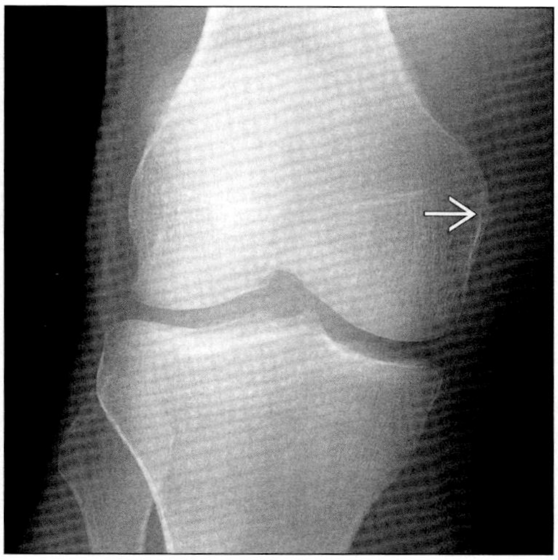

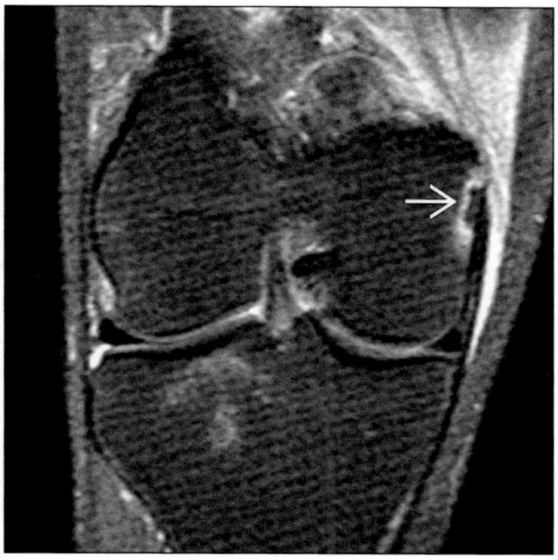

(Left) *AP radiograph in a patient with medial knee pain after injury demonstrates an area of subtle cortical irregularity in the medial femoral condyle.* ***(Right)*** *Coronal T2WI FS MR from the same patient shows cortical avulsion of the MCL from the medial femoral condyle. This is a relatively unusual injury and is often difficult to detect on radiographs.*

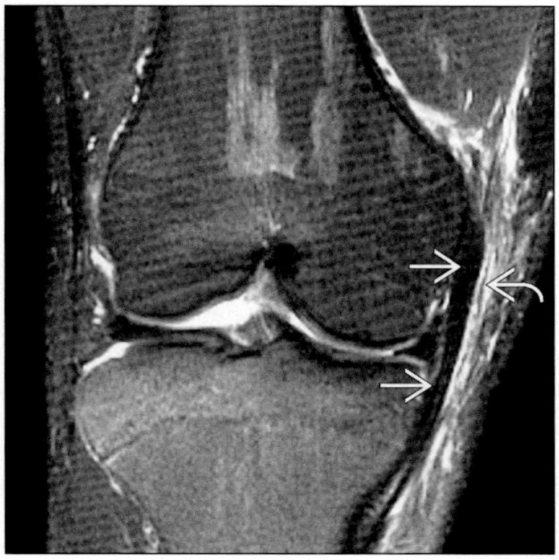

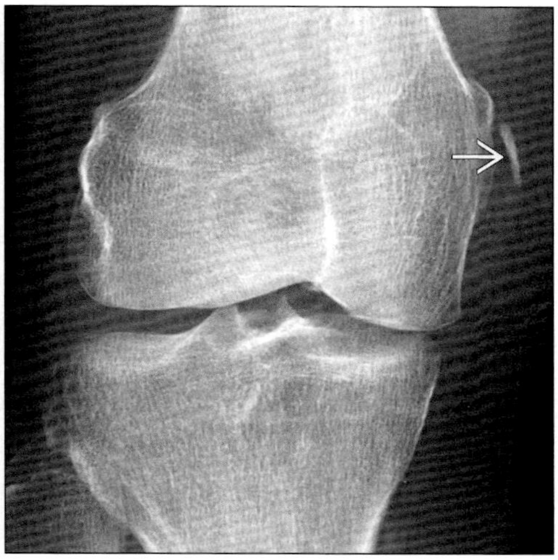

(Left) *Coronal T2WI FS MR shows marked thickening of the MCL consistent with fibrosis from previous injury. Soft tissue swelling superficial to the MCL indicates a new low-grade injury.* ***(Right)*** *Coronal oblique radiograph in a patient with chronic medial knee pain demonstrates a linear focus of osseous density medial to the medial femoral condyle. This is a Pellegrini-Stieda lesion or chronic dystrophic ossification of the MCL from previous injury.*

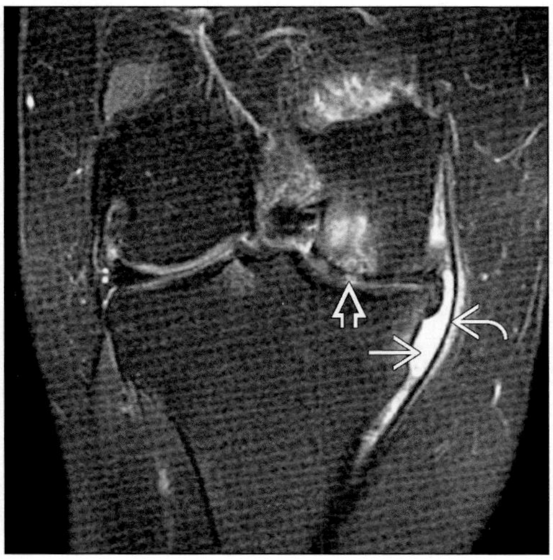

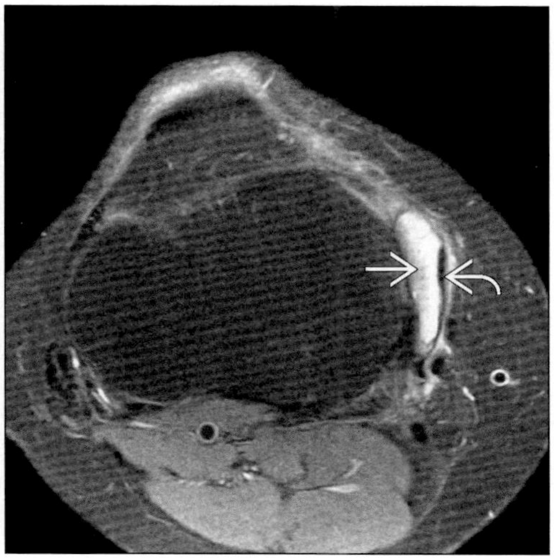

(Left) *Coronal T2WI FS MR in a patient with chronic medial knee pain shows a well-defined fluid collection located between the superficial fibers of the MCL and the tibia and medial meniscus. There are severe degenerative changes in the medial femoral condyle, and there is medial extrusion of the meniscus.* ***(Right)*** *Axial PDWI FS MR from the same patient shows the well-defined bursal fluid collection deep to the MCL, consistent with MCL bursitis.*

LATERAL COLLATERAL LIGAMENT COMPLEX, KNEE

Key Facts

Imaging

- Coronal and axial T2WI FS MR show lateral ligamentous pathology best
- Lateral collateral ligament complex composed of 2 discrete structures
  - Iliotibial band (ITB)
  - Fibular collateral ligament (FCL)
  - Anterior oblique band (AOB): Offshoot of FCL extending to lateral rim of tibia
  - Ligaments may be injured separately or in combination
- Avulsion fragments from fibula
  - Small curved cortical fragment from tip of fibular styloid: Avulsion by arcuate ligament (arcuate sign)
  - Larger fragment from fibular head or neck: More likely avulsion by conjoined tendon
  - May be associated with PCL tear
- Segond injury: Avulsion of insertion of AOB or ITB
  - Strong association with ACL tear

Pathology

- Varus stress ± external rotation injury to knee
- FCL rarely torn in isolation

Clinical Issues

- Varus instability
- Anterolateral knee instability
  - ITB and ACL tears
- Posterolateral knee instability
  - PCL, posterolateral corner, and FCL injuries
- Treatment
  - Isolated FCL tears treated conservatively
  - Combination injury with posterolateral corner structures may require open repair
  - Associated cruciate ligament injury usually requires open repair of cruciate ± elements of LCLC

__(Left)__ Sagittal graphic shows avulsion of the distal aspect of the fibular collateral ligament (FCL) from the fibular head. Rarely an isolated injury, FCL tear is often seen in association with injury to the cruciate ligaments &/or posterolateral corner structures. __(Right)__ Sagittal PDWI FS MR of the lateral aspect of the knee demonstrates complete avulsion of the FCL from the fibular head with surrounding soft tissue swelling.

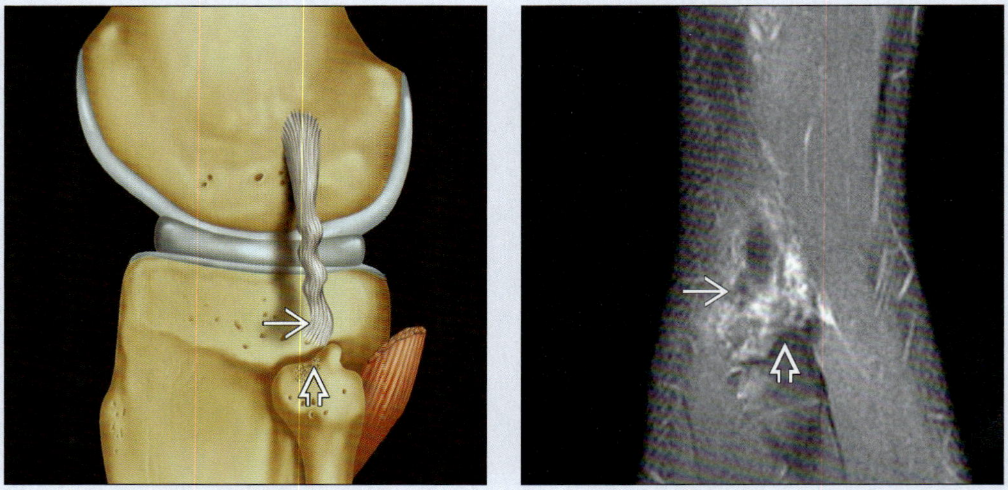

__(Left)__ Coronal T2WI FS MR from the same patient shows elevated signal at the ruptured attachment site of the fibular collateral ligament to the fibular head. The proximal portion of the FCL attaches to the lateral femoral condyle and is intact. __(Right)__ Axial PDWI FS MR from the same patient demonstrates disruption of the fibular collateral ligament just proximal to its insertion on the fibular head. The lateral joint capsule is intact.

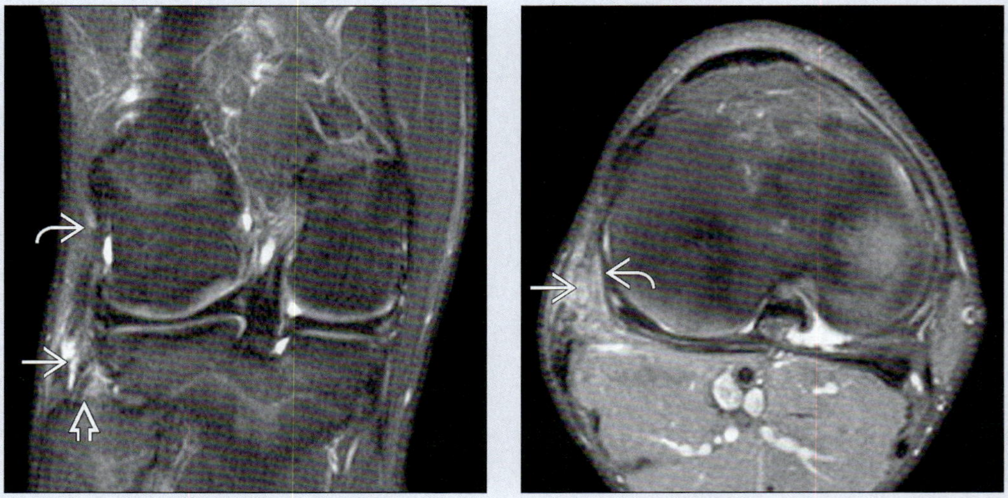

LATERAL COLLATERAL LIGAMENT COMPLEX, KNEE

TERMINOLOGY

Abbreviations

- Lateral collateral ligament complex (LCLC)
- Fibular collateral ligament (FCL)
- Anterior oblique band (AOB)
- Iliotibial band (ITB)

Synonyms

- ITB: Iliotibial tract (ITT)

Definitions

- Injury to lateral ligament complex of knee, including
  - Iliotibial band
  - Fibular collateral ligament (sometimes referred to as "lateral collateral ligament")
    - Anterior oblique band: Fibers of FCL extending anteriorly and inferiorly to lateral rim of tibia
      - Common site of Segond avulsion

IMAGING

General Features

- Best diagnostic clue
  - Abnormal signal within ± around elements of lateral collateral ligament complex
- Morphology
  - Lateral collateral ligament complex composed of 2 discrete structures (from anterior to posterior)
    - Iliotibial band
      - Tendinous extension of tensor fascia lata muscle
      - Some fibers continuous with vastus lateralis, contributes to lateral patellar retinaculum
      - Inserts on Gerdy tubercle on anterolateral aspect of tibial plateau
    - Fibular collateral ligament
      - Robust ligament extending from lateral femoral condyle to form part of conjoined tendon (with biceps femoris tendon)
      - Attaches to lateral aspect of fibular head
      - Primary restraint against varus force and against internal rotation of knee
      - Anterior oblique band: Offshoot of FCL extending to lateral rim of tibia

Radiographic Findings

- Avulsion of proximal fibular head
  - Usually transverse fracture with large proximal fragment
    - Smaller curved cortical fragment (arcuate sign): More likely avulsion of arcuate ligament complex
  - May be associated with posterior cruciate ligament (PCL) tear
- Segond injury: Avulsion of tibial insertion of AOB or ITB
  - Very high association with anterior cruciate ligament (ACL) tear and medial meniscal tear

CT Findings

- Similar to radiographs but more sensitive for small avulsion fragments

MR Findings

- Ligaments may be injured separately or in combination
- Iliotibial band
  - Thickening: Usually in distal aspect of tendon
  - Increased intrinsic signal: Indicates fiber disruption or degeneration
  - Peritendinous edema: Best seen on T2WI FS MR in coronal and axial planes
  - Osseous avulsion: Segond variant in which Gerdy tubercle attachment is pulled off
    - Best seen on coronal T1WI ± T2WI FS MR
    - Often associated with ACL tear
- Anterior oblique band
  - Normally very thin, may only be seen on single coronal or sagittal image
  - Ligament tear: Absence of identifiable ligament fibers on coronal T2WI FS MR
  - Segond avulsion: Best seen on coronal T1WI ± T2WI FS MR
- Fibular collateral ligament
  - Normally well-defined cord-like ligament in all imaging planes
  - Edema within ± around ligament → sprain
  - Focal disruption of fibers with no identifiable ligament present → complete tear
  - Occasional disruption at fibular head attachment site
    - Forms conjoined tendon with distal biceps femoris tendon
    - Avulsion implies disruption of biceps as well
    - Often associated with posterolateral corner injury
  - Associated marrow contusions in medial knee due to varus stress

Imaging Recommendations

- Best imaging tool
  - MR
- Protocol advice
  - Coronal and axial T2WI FS MR show lateral ligamentous pathology best
  - Sagittal T2WI FS MR occasionally helps define pathology at conjoined tendon insertion
  - Coronal and sagittal T1WI MR may better demonstrate osseous avulsion injury

DIFFERENTIAL DIAGNOSIS

Iliotibial Band Syndrome

- Edema between ITB and lateral femoral condyle
- ITB usually intrinsically normal
- Seen in long-distance runners or cyclists

Anterior Cruciate Ligament (ACL) Tear

- Segond avulsion of AOB or ITB may be associated with ACL injury complex

Posterolateral Corner Injury

- Arcuate avulsion of fibular styloid implies injury to arcuate ligament complex

Direct Blow Contusion

- Associated with predominant superficial soft tissue swelling
- ± marrow contusions in lateral compartment

LATERAL COLLATERAL LIGAMENT COMPLEX, KNEE

PATHOLOGY

General Features

- Etiology
  - Varus stress ± external rotation injury to knee
  - FCL rarely torn in isolation
    - Combined with posterolateral corner injury
    - Combined with cruciate ligament injury (ACL > PCL)
      - Varus stress and hyperextension → PCL and FCL injury
- Associated abnormalities
  - Segond avulsion
    - ACL tear: 75-100%
    - Meniscal tear: 66-75%

Staging, Grading, & Classification

- MR grading system
  - Grade 0: Normal
  - Grade I: Edema around intrinsically normal ligament (mild sprain)
  - Grade II: Abnormal signal within and around ligament without complete disruption (severe sprain or partial tear)
  - Grade III: Complete disruption of ligament

CLINICAL ISSUES

Presentation

- Most common signs/symptoms
  - Posterolateral or lateral knee pain
  - Varus instability
  - Anterolateral knee instability
    - ITB and ACL tears
  - Posterolateral knee instability
    - PCL, posterolateral corner, and FCL injuries
- Other signs/symptoms
  - Positive anterior drawer or pivot-shift tests if associated with ACL tear
  - Positive posterior drawer test if associated with PCL tear

Demographics

- Epidemiology
  - Isolated FCL tear represents 2% of knee ligament injuries
  - FCL injury combined with other ligament injury = 4% of knee ligament injuries

Natural History & Prognosis

- Isolated injury to LCLC usually heals by fibrosis
- Varus instability may persist, particularly if associated with cruciate ligament injury

Treatment

- Isolated FCL tears treated conservatively
- Combination injury with posterolateral corner structures may require open repair
- Associated cruciate ligament injury usually requires open repair of cruciate ± elements of LCLC

DIAGNOSTIC CHECKLIST

Consider

- Associated cruciate ligament injury
  - Especially ACL tear with Segond avulsion

Image Interpretation Pearls

- Evaluate posterolateral corner structures if LCLC injured

SELECTED REFERENCES

1. Clarke JV et al: The reproducibility and repeatability of varus stress radiographs in the assessment of isolated fibular collateral ligament and grade-III posterolateral knee injuries. J Bone Joint Surg Am. 91(2):485; author reply 485-6, 2009
2. Valkering KP et al: Segond fracture. J Am Coll Surg. 208(4):646, 2009
3. Gottsegen CJ et al: Avulsion fractures of the knee: imaging findings and clinical significance. Radiographics. 28(6):1755-70, 2008
4. Beall DP et al: Magnetic resonance imaging of the collateral ligaments and the anatomic quadrants of the knee. Radiol Clin North Am. 45(6):983-1002, vi, 2007
5. Bahk MS et al: Physical examination and imaging of the lateral collateral ligament and posterolateral corner of the knee. Sports Med Arthrosc. 14(1):12-9, 2006
6. Murphy KP et al: Surgical treatment of acute lateral collateral ligament and posterolateral corner injuries. Sports Med Arthrosc. 14(1):23-7, 2006
7. Lee J et al: Arcuate sign of posterolateral knee injuries: anatomic, radiographic, and MR imaging data related to patterns of injury. Skeletal Radiol. 32(11):619-27, 2003
8. Juhng SK et al: MR evaluation of the "arcuate" sign of posterolateral knee instability. AJR Am J Roentgenol. 178(3):583-8, 2002
9. Campos JC et al: Pathogenesis of the Segond fracture: anatomic and MR imaging evidence of an iliotibial tract or anterior oblique band avulsion. Radiology. 219(2):381-6, 2001
10. Recondo JA et al: Lateral stabilizing structures of the knee: functional anatomy and injuries assessed with MR imaging. Radiographics. 20 Spec No:S91-S102, 2000
11. Weber WN et al: Lateral tibial rim (Segond) fractures: MR imaging characteristics. Radiology. 180(3):731-4, 1991
12. Dietz GW et al: Segond tibial condyle fracture: lateral capsular ligament avulsion. Radiology. 159(2):467-9, 1986

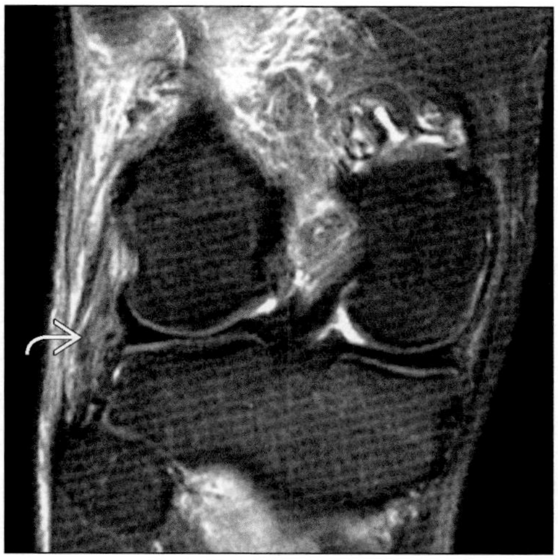

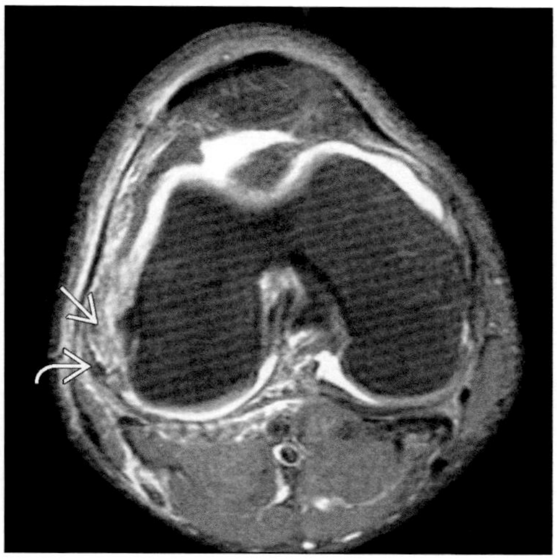

(Left) *Coronal T2WI FS MR shows a high-grade partial tear of the fibular collateral ligament with some fibers remaining intact.* ***(Right)*** *Axial PDWI FS MR from the same patient shows high-grade partial disruption of the midportion of the FCL with a few intact fibers posterolaterally. The degree of disruption of a ligament is best assessed in cross-section (in this case, the axial plane).*

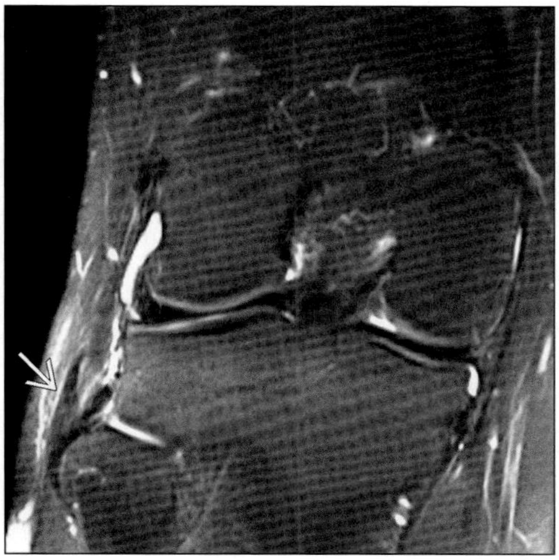

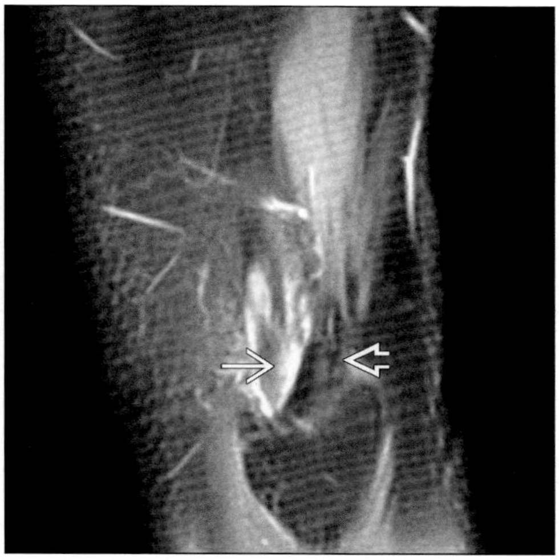

(Left) *Coronal T2WI FS MR shows mild thickening of the distal FCL without high-grade rupture. Coronal images may under- or overestimate ligament injury because of the oblique course of the ligament.* ***(Right)*** *Sagittal T2WI FS MR from the same patient better shows disruption of anterior fibers of the FCL and preservation of posterior fibers of the conjoined tendon, consistent with a partial tear. The biceps femoris tendon also makes up part of the conjoined tendon.*

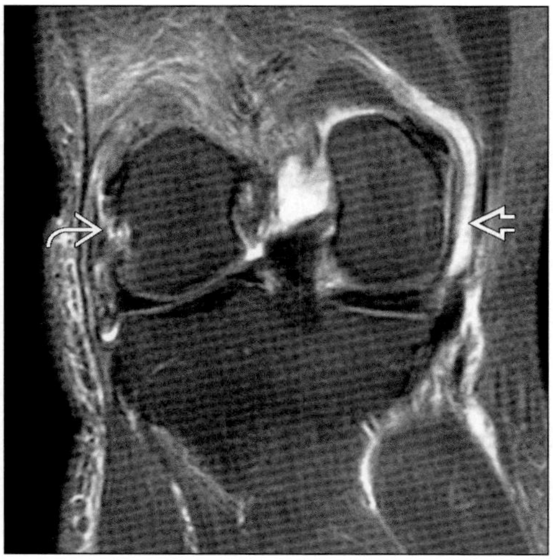

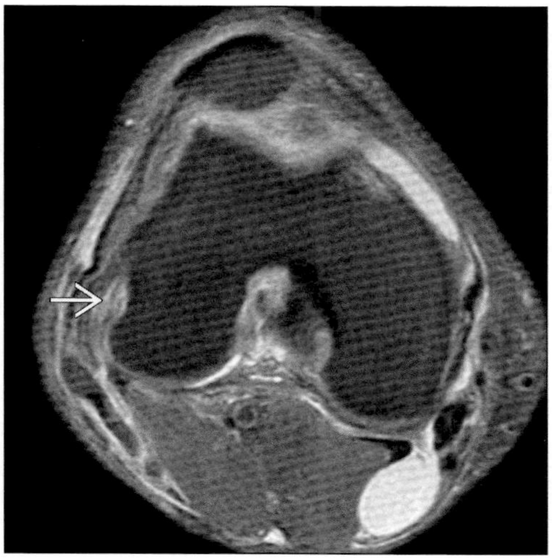

(Left) *Coronal T2WI FS MR shows a partial tear of the proximal end of the FCL. Edema is also present in the medial aspect of the knee related to a direct blow injury causing varus stress.* ***(Right)*** *Axial PDWI FS MR from the same patient shows partial disruption of the proximal aspect of the FCL. When grading ligament tears, an approximate percentage of fiber disruption can be a useful measurement for the treating surgeon. This case is about 75% disrupted.*

*(**Left**) Coronal T2WI FS MR shows a retracted and redundant fibular collateral ligament ➔ due to a previous distal tear. The relative lack of edema around the ligament indicates a less acute time course. There are severe degenerative changes in the lateral compartment of the knee ➔. (**Right**) Coronal T2WI FS MR shows an acute osseous avulsion of the femoral attachment of the FCL ➔, a relatively unusual injury. Edema surrounds the ligament.*

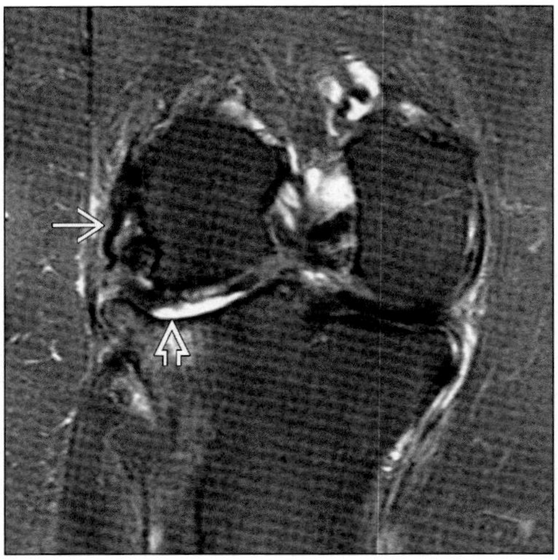

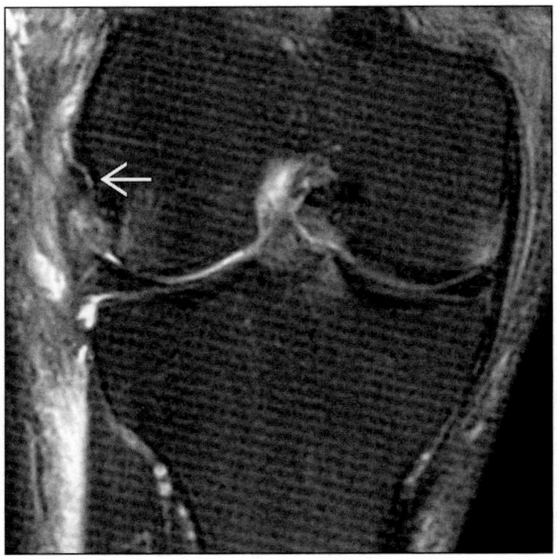

*(**Left**) Lateral radiograph shows a sizable avulsion of the fibular styloid/head ➔ commensurate with disruption by the conjoined tendon (FCL and biceps tendon). This avulsed fragment is larger than the expected thin crescentic appearance of arcuate fragment seen with avulsion by the arcuate ligament. (**Right**) Sagittal PDWI MR shows an osseous avulsion of the proximal fibular head ➔, which results in disruption of the conjoined tendon ➔. The patient also had disruption of the anterior cruciate ligament.*

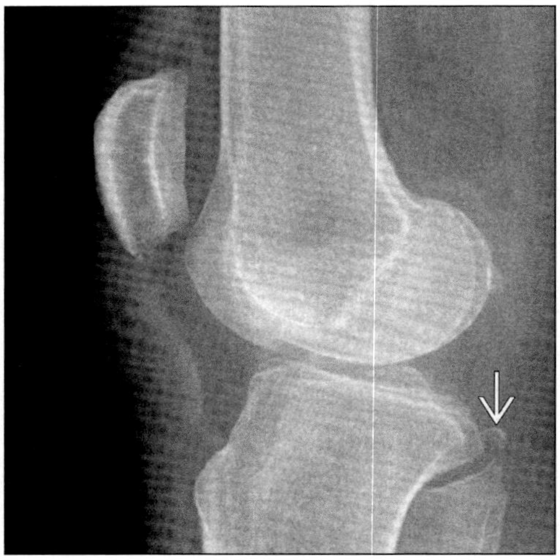

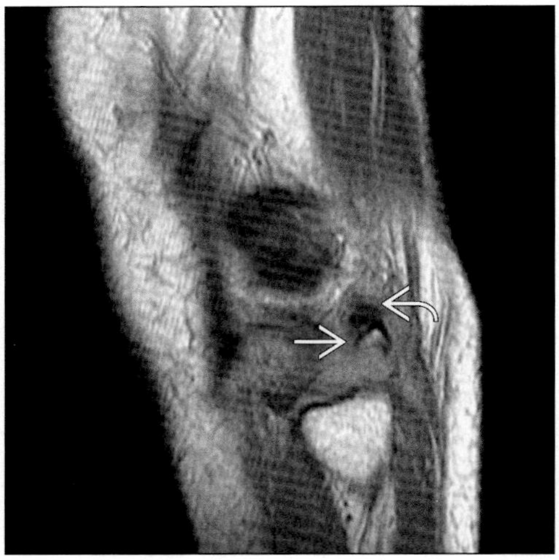

*(**Left**) Coronal T2WI FS MR shows disruption of the anterior oblique band from its tibial attachment (Segond injury) ➔, disruption of the ACL ➔, and partial tear of the medial collateral ligament ➔ after knee dislocation. Dislocation often involves rupture of 3 major stabilizing ligaments of the knee. (**Right**) Coronal T1WI MR from the same patient shows a transverse avulsion of the fibular styloid/head ➔ indicating conjoined tendon/ FCL disruption. Part of the biceps femoris ➔ is seen more proximally.*

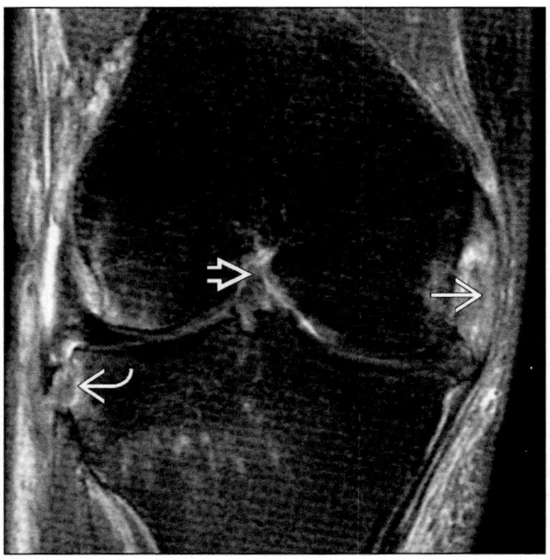

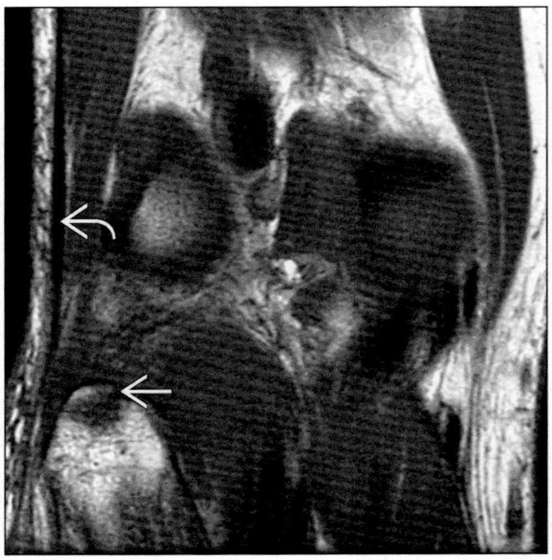

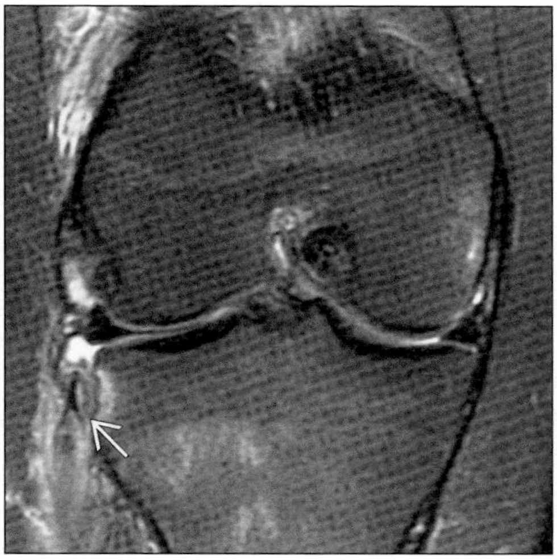

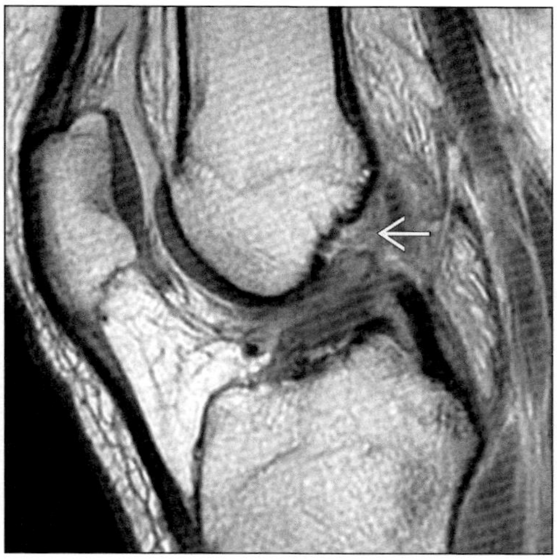

(Left) *Coronal T2WI FS MR shows acute osseous avulsion of the tibial attachment of the anterior oblique band (AOB) of the FCL → (Segond injury) after a twisting injury. Mild marrow edema is present at the site of avulsion, which is not always present in such injuries.* ***(Right)*** *Sagittal PDWI MR from the same patient shows disruption of the anterior cruciate ligament (ACL) → at its proximal extent. The presence of a Segond injury has a very strong association with ACL tear.*

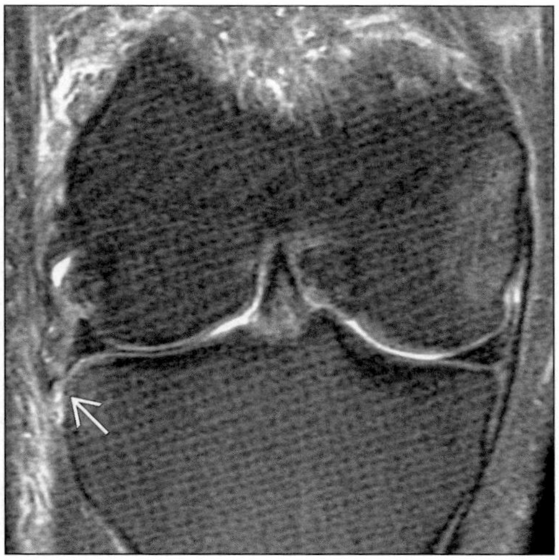

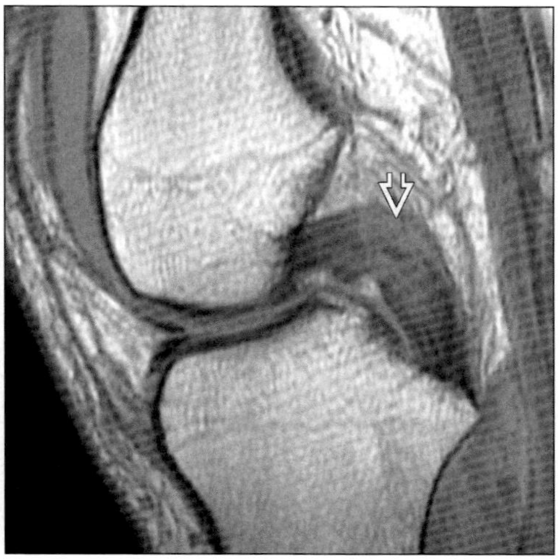

(Left) *Coronal T2WI FS MR shows osseous avulsion of the anterior oblique band of the FCL → from its tibial attachment; this is a Segond injury and is more often associated with ACL than PCL injury.* ***(Right)*** *Sagittal PDWI MR from the same patient demonstrates rupture of the posterior cruciate ligament (PCL) →. In the setting of acute PCL rupture, it is important to assess for other ligament injuries, as management of PCL injury often hinges on the degree of associated injury.*

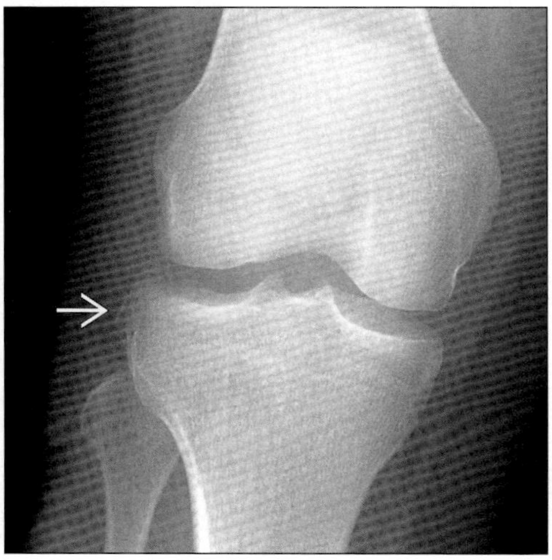

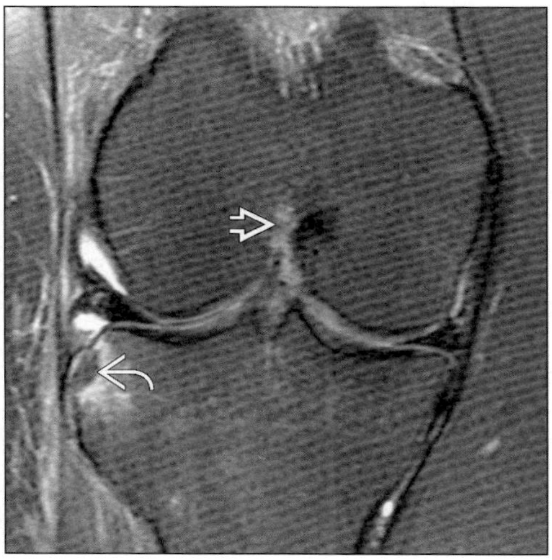

(Left) *AP radiograph shows a Segond avulsion of the lateral tibial rim → indicating disruption of the anterior oblique band of the FCL. The major significance of this finding is that it is very highly associated with ACL disruption; subsequent MR of this patient (not shown) proved an ACL tear.* ***(Right)*** *Coronal T2WI FS MR shows osseous avulsion of the iliotibial band (ITB) from Gerdy tubercle on the tibia →. This is a Segond variant and also associated with ACL tear →.*

POSTEROLATERAL CORNER INJURY

Key Facts

Terminology

- Posterolateral corner (PLC)
- Arcuate complex: Group of ligaments with common attachment to fibular styloid

Imaging

- Discrete structures not always identifiable
- Some are anatomically inconstant across population
- Injury may manifest as only vague edema in region
- Arcuate sign: Thin, curved cortical rim of bone avulsed from tip of fibular styloid
- Popliteal tendon and muscle injury
  - Usually involves muscle belly or myotendinous junction
- Biceps femoris/conjoined tendon injury
  - Best seen on coronal and axial T2WI FS MR
- Oblique coronal imaging suggested by some if PLC integrity in question

Pathology

- Anteromedial and varus force to proximal tibia
- Noncontact external rotation and hyperextension injury
- Seen with posterior > anterior cruciate ligament injury
- Impaction contusions in anterior aspects of medial femoral condyle and medial tibia

Clinical Issues

- High-grade PLC injury may cause ongoing instability
- Grade II PLC injury or combined with cruciate tear → primary repair of posterolateral corner
- Failure to address high-grade PLC injury at time of cruciate repair → increased incidence of cruciate repair failure

(Left) Coronal graphic of the posterolateral corner shows a tear of the popliteus muscle and of the lateral limb of the arcuate ligament. Popliteal hiatus is more proximal. (Right) Coronal T2WI FS MR from a patient with an acute anterior cruciate ligament tear shows rupture of the popliteus tendon and lateral band of the arcuate ligament. Individual ligamentous structures of the posterolateral corner may not be readily identifiable with high-grade injury.

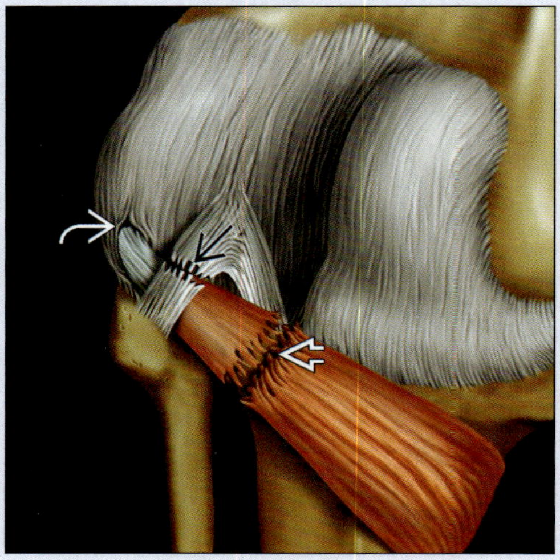

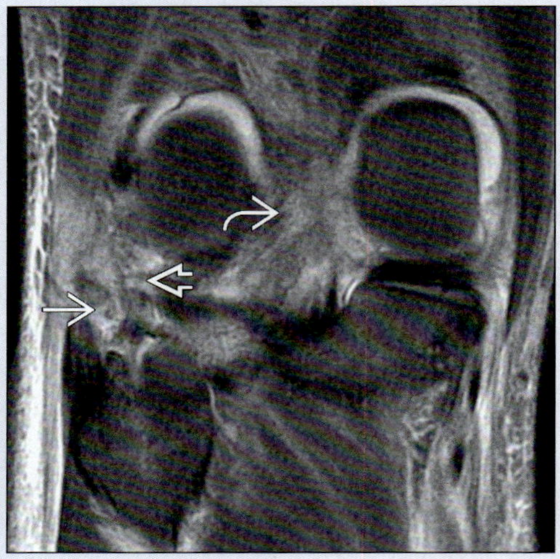

(Left) Axial graphic of the posterolateral corner shows the popliteal tendon passing through the popliteal hiatus. The arcuate ligament, fabellofibular ligament, biceps femoris tendon, and fibular collateral ligament are also shown. (Right) Axial PDWI FS MR from a patient with low-grade posterolateral corner injury shows popliteal tendon, arcuate ligament lateral band, fabellofibular ligament, biceps femoris, and fibular collateral ligament.

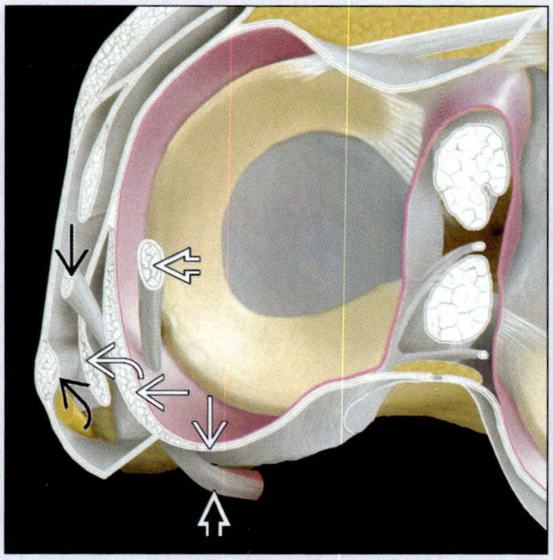

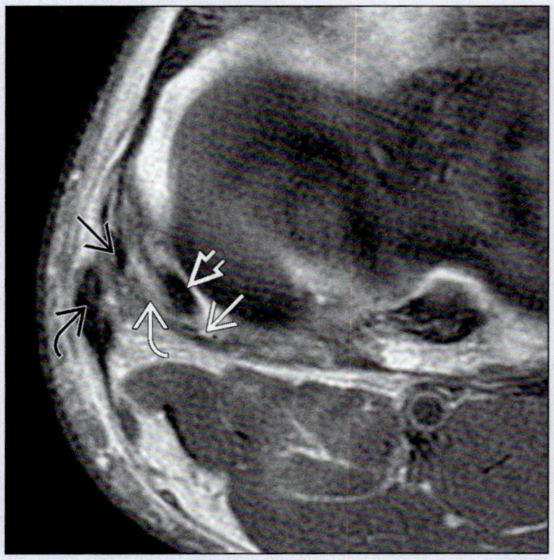

POSTEROLATERAL CORNER INJURY

TERMINOLOGY

Synonyms

- Posterolateral instability
- Arcuate complex injury

Definitions

- Injury to group of structures stabilizing the posterolateral corner (PLC) of the knee
  - Posterolateral joint capsule (most ligaments below are focal capsular thickenings)
    - Arcuate ligament
    - Popliteofibular ligament
    - Fabellofibular ligament
    - Oblique popliteal ligament
  - Popliteal (popliteus) tendon
  - Lateral gastrocnemius and biceps femoris muscles and tendons
  - Fibular collateral ligament (FCL) sometimes included
- Arcuate complex: Group of ligaments with common attachment to fibular styloid
  - Arcuate
  - Popliteofibular
  - Fabellofibular

IMAGING

General Features

- Best diagnostic clue
  - Edema in posterolateral corner of knee on T2WI FS MR
  - Overt disruption of major ligamentous or tendinous structures in PLC
- Morphology
  - Discrete structures not always identifiable
    - Some are anatomically inconstant across population
  - Injury may manifest as only vague edema in region

Radiographic Findings

- Arcuate sign: Thin, curved cortical rim of bone avulsed from tip of fibular styloid
  - Insertion of arcuate complex (arcuate, fabellofibular, and popliteofibular ligaments)
- Larger fragment originating from fibular head often indicates conjoined tendon avulsion
  - Attachments of FCL and biceps femoris tendon

MR Findings

- Arcuate complex injury
  - Injury implied by ↑ signal intensity on T2WI MR ± FS adjacent to space between posterior horn of lateral meniscus, popliteal tendon, and fibular styloid
  - Individual fiber bundles sometimes difficult to identify even when normal
  - Arcuate avulsion best seen on coronal or sagittal T2SI FS MR
- Popliteal tendon and muscle injury
  - Usually involves muscle belly or myotendinous junction
    - Best seen on axial T2WI FS MR
    - Increased size and signal in muscle ± focal disruption with surrounding edema
  - Occasional avulsion of femoral attachment
    - Best seen on coronal T1WI and T2WI FS MR
- Lateral gastrocnemius injury
  - Relatively rare injury
  - Edema and increased size of gastrocnemius muscle origin with surrounding edema
  - Complete avulsion of femoral origin usually associated with knee dislocation
  - Associated with fabellofibular ligament tear
- Biceps femoris/conjoined tendon injury
  - Best seen on coronal and axial T2WI FS MR
  - Low-grade injury: Increased signal around tendon ± tendon thickening
  - High-grade injury: Partial or complete tear or osseous avulsion from fibula
    - Avulsed fragment much larger than that associated with arcuate complex

Imaging Recommendations

- Best imaging tool
  - MR
- Protocol advice
  - T2WI FS MR in multiple planes with thin sections
  - Oblique coronal imaging suggested by some if PLC integrity in question
    - Prescribed off routine sagittal MR images
    - Scan plane parallel to long axis of popliteal tendon as seen on sagittal images

DIFFERENTIAL DIAGNOSIS

Lateral Meniscal Tear

- Posterior horn tears can be associated with isolated rupture of 1 or both popliteomeniscal fascicles

Normal Popliteal Hiatus

- Entry point for popliteal tendon into joint posterior to lateral meniscus
- May be mistaken for traumatic defect in capsule or meniscus

Fluid in Popliteal Tendon Sheath

- Small amount of fluid not uncommon

PATHOLOGY

General Features

- Etiology
  - Anteromedial and varus force to proximal tibia
    - Results in hyperextension of knee
  - Noncontact external rotation and hyperextension injury
  - Knee dislocation
- Associated abnormalities
  - Cruciate ligament injury
    - Posterior > anterior cruciate ligament
  - Fibular collateral ligament injury
    - Increases likelihood of varus instability
  - Knee dislocation
  - Impaction contusions in anterior aspects of medial femoral condyle and medial tibia

Staging, Grading, & Classification

- Clinical grading system

POSTEROLATERAL CORNER INJURY

- Grade I: Increased external tibial rotation without varus instability
- Grade II: Increased external tibial rotation + moderate varus laxity
- Grade III: Increased external tibial rotation + complete varus laxity

- MR grading system
  - Roughly divided into low- and high-grade injuries
    - Low-grade injury: Edema without complete disruption of popliteus tendon or popliteofibular ligament
    - High-grade injury: Disruption of popliteus, popliteofibular ligament, ± FCL

Relevant Anatomy

- Posterolateral corner of knee stabilized by passive and active stabilizing elements
  - **Passive stabilizers**
    - Arcuate ligament
      - Y-shaped or triangular thickening of knee joint capsule
      - Arises from tip of fibular styloid
      - Lateral limb ascends to lateral head of gastrocnemius
      - Medial limb joins fibers of oblique popliteal ligament of Winslow and popliteal muscle, forms roof of popliteal hiatus with superior popliteomeniscal fascicle
    - Popliteofibular ligament
      - Fibular head origin of popliteal muscle
      - Splits to become superior and inferior popliteomeniscal fascicles, which attach lateral meniscus to popliteal tendon sheath
      - Most important restraining structure in posterolateral corner
    - Fabellofibular ligament
      - Runs from fibular styloid tip to fabella (if present)
    - Oblique popliteal ligament of Winslow
      - Arises from tibial insertion of semimembranosus
      - Courses obliquely superiorly and laterally to origin of lateral gastrocnemius
    - Coronary ligament
      - Capsular attachment of lateral meniscus to tibia
      - Contains popliteal hiatus through which popliteal tendon passes
    - Fibular collateral ligament
      - Arises from lateral femoral condyle
      - Attaches to fibular head (with biceps femoris tendon) as part of conjoined tendon
  - **Active stabilizers**
    - Lateral head of gastrocnemius
    - Popliteal muscle and tendon
      - Arises in lateral femoral condyle anterior and inferior to FCL origin
      - Also takes origin on lateral meniscus, withdrawing meniscus during knee flexion and rotation
      - Inserts on posteromedial aspect of proximal tibia
    - Biceps femoris
      - Short and long heads combine above knee into common tendon and insert with FCL as conjoined tendon on fibular head

CLINICAL ISSUES

Presentation

- Most common signs/symptoms
  - Posterolateral knee pain
  - Backward buckling of knee when walking or standing

Natural History & Prognosis

- Isolated PLC injuries usually do well without surgical intervention
- High-grade PLC injury may cause ongoing instability
- Combined with cruciate ligament injury, likely to complain of knee instability
  - More prone to accelerated osteoarthritis
  - Failure to address high-grade PLC injury at time of cruciate repair → increased incidence of cruciate repair failure

Treatment

- Low-grade isolated injury treated conservatively
- Grade II PLC injury or combined with cruciate tear → primary repair of posterolateral corner
  - Cruciate repair in same setting if torn

DIAGNOSTIC CHECKLIST

Consider

- Look for posterolateral corner injury with cruciate ligament tear, especially PCL

Image Interpretation Pearls

- Oblique coronal MR imaging may provide more detailed evaluation of specific structures but rarely provides additional clinically relevant information

Reporting Tips

- Categorize injuries as low- or high-grade

SELECTED REFERENCES

1. Kim JG et al: Correlation between the rotational degree of the dial test and arthroscopic and physical findings in posterolateral rotatory instability. Knee Surg Sports Traumatol Arthrosc. 18(1):123-9, 2010
2. Peduto AJ et al: Popliteomeniscal fascicles: anatomic considerations using MR arthrography in cadavers. AJR Am J Roentgenol. 190(2):442-8, 2008
3. Vinson EN et al: The posterolateral corner of the knee. AJR Am J Roentgenol. 190(2):449-58, 2008
4. Stannard JP et al: Reconstruction of the posterolateral corner of the knee. Arthroscopy. 21(9):1051-9, 2005
5. Huang GS et al: Avulsion fracture of the head of the fibula (the "arcuate" sign): MR imaging findings predictive of injuries to the posterolateral ligaments and posterior cruciate ligament. AJR Am J Roentgenol. 180(2):381-7, 2003
6. Munshi M et al: MR imaging, MR arthrography, and specimen correlation of the posterolateral corner of the knee: an anatomic study. AJR Am J Roentgenol. 180(4):1095-101, 2003
7. Yu JS et al: Posterolateral aspect of the knee: improved MR imaging with a coronal oblique technique. Radiology. 198(1):199-204, 1996

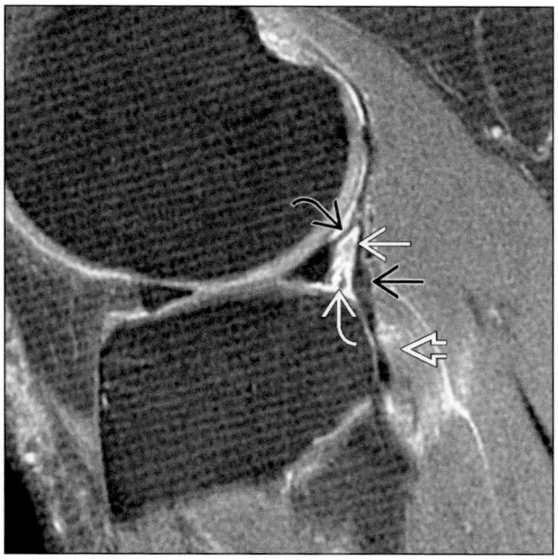

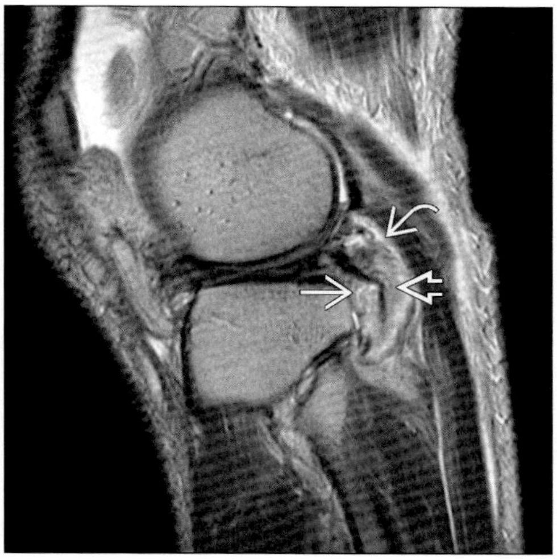

(Left) *Sagittal PDWI FS MR shows a low-grade injury to the myotendinous junction of the popliteus. The popliteofibular and arcuate ligaments remain intact. There is a tear of the inferior popliteomeniscal fascicle, but the superior fascicle is intact.* ***(Right)*** *Sagittal T2WI MR shows a high-grade partial tear of the popliteus myotendinous junction. Popliteofibular and arcuate ligaments are torn.*

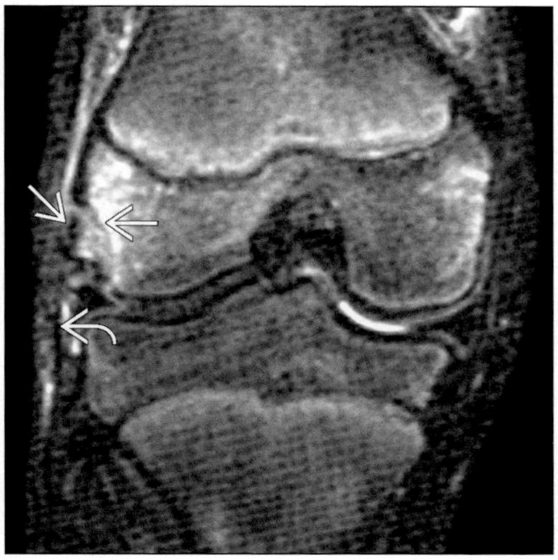

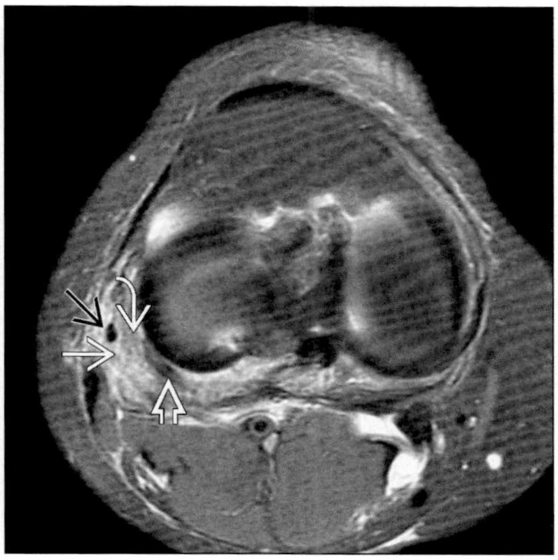

(Left) *Coronal T2WI FS MR shows avulsion of the femoral attachment of the popliteus tendon with marrow edema at the donor site of the lateral femoral condyle. Popliteofibular ligament is intact.* ***(Right)*** *Axial PDWI FS MR shows rupture of the arcuate and popliteofibular ligaments in a mid-grade injury to the posterolateral corner. The popliteus tendon is intact. Fibular collateral ligament is normal.*

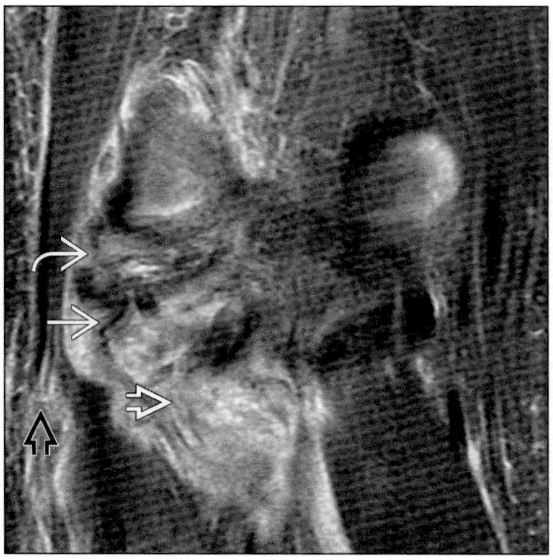

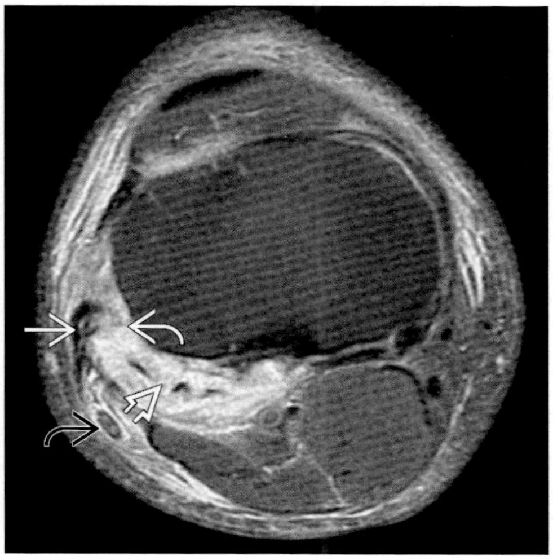

(Left) *Coronal T2WI FS MR shows a high-grade posterolateral corner injury with disruption of the popliteus myotendinous junction, fabellofibular ligament, and arcuate ligament complex. The biceps femoris is partially torn distally.* ***(Right)*** *Axial PDWI FS MR from the same patient shows marked swelling and edema in the popliteus muscle and arcuate ligament complex. Conjoined tendon remains intact. Peroneal nerve is surrounded by edema.*

ILIOTIBIAL BAND FRICTION SYNDROME

Key Facts

Terminology
- ITB friction syndrome, ITB syndrome
- Chronic inflammatory response to friction between iliotibial band and lateral femoral condyle (LFC)
- Lateral synovial recess (LSR): Adventitial bursal extension of knee joint superolaterally

Imaging
- Amorphous area of edema may be small or several cm in size
- Ill-defined area of elevated signal on T2WI or T2WI FS MR between LFC and ITB at their closest approximation
  - Best seen on coronal and axial sequences
  - More difficult to discern on T1WI MR
- May be associated with thickening of, or abnormal signal in, ITB
- Occasional reactive marrow edema in adjacent LFC

Top Differential Diagnoses
- Fluid in lateral knee joint recess
- Lateral collateral ligament complex injury
- Direct blow contusion
- Lateral femoral condyle fracture
- Lateral meniscal cyst
- Excessive lateral pressure syndrome

Clinical Issues
- Anterolateral knee pain with activity
- Long-distance runners, cyclists
- Responds well to cessation of offending activity and anti-inflammatory therapy

Diagnostic Checklist
- ITB syndrome leads to ill-defined soft tissue swelling
- Well-defined fluid superolateral to knee may be joint recess

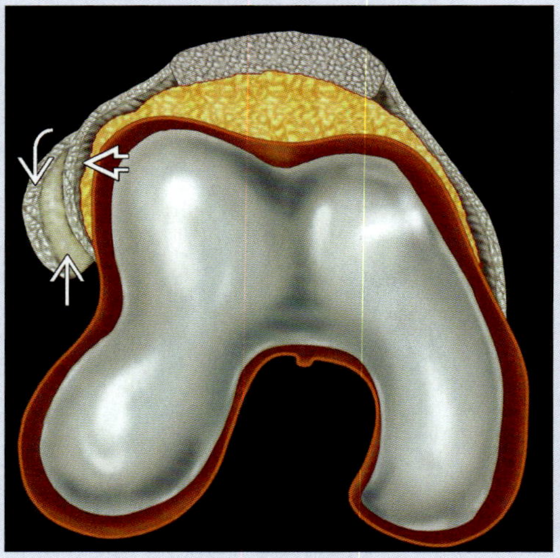
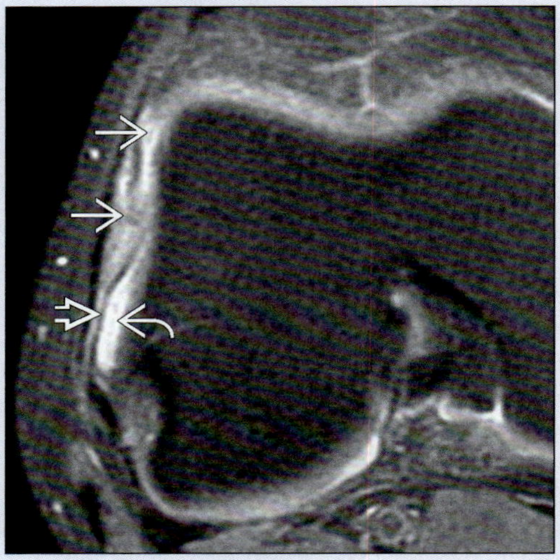

*(**Left**) Axial graphic shows inflammation of the lateral synovial recess ➡, a space between the iliotibial band ➡ and the lateral reflection of the knee joint synovium ➡. Inflammation of the recess is caused by repetitive friction between the ITB and the condyle during long-distance running or cycling. (**Right**) Axial PDWI FS MR shows findings of ITB friction syndrome. Ill-defined edema in the lateral synovial recess ➡ is separated from joint fluid ➡ by the synovial reflection ➡.*

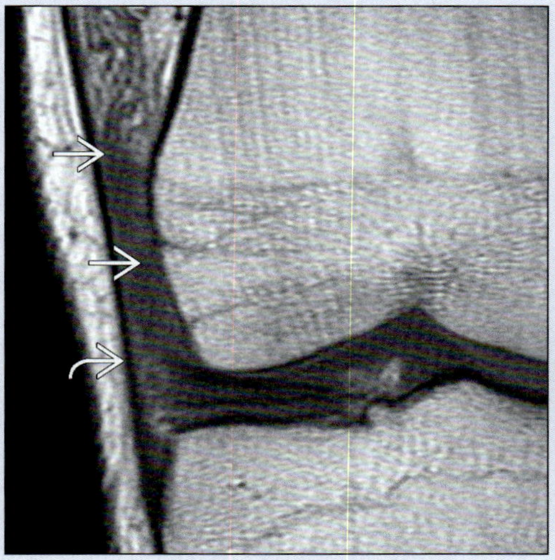
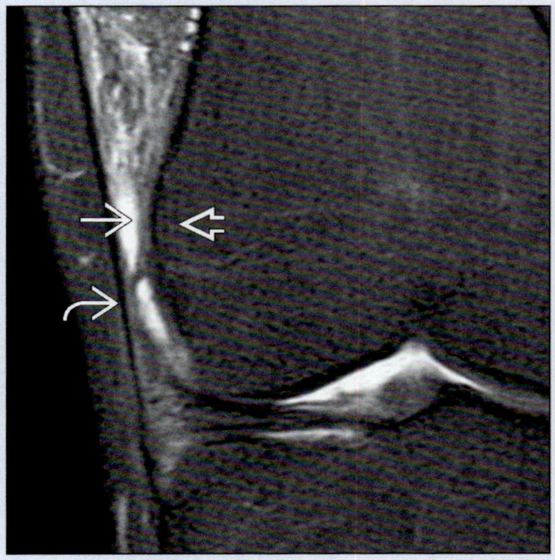

*(**Left**) Coronal T1WI MR from the same patient shows infiltration of the fat in the lateral synovial recess ➡. There is no thickening of the iliotibial band ➡ in this case; some patients may manifest thickening of, or increased signal in, the ITB along with ITB friction syndrome. (**Right**) Coronal T2WI FS MR from the same patient shows intense edema ➡ between the ITB ➡ and the lateral femoral condyle ➡ indicative of ITB friction syndrome.*

ILIOTIBIAL BAND FRICTION SYNDROME

TERMINOLOGY

Abbreviations

- Iliotibial band (ITB)

Synonyms

- ITB friction syndrome, ITB syndrome

Definitions

- Chronic inflammatory response to friction between iliotibial band and lateral femoral condyle (LFC)
  - Occurs as response to repetitive overuse injury
- Lateral synovial recess (LSR): Adventitial bursal extension of knee joint superolaterally
  - Lies between ITB and LFC

IMAGING

General Features

- Best diagnostic clue
  - Increased signal intensity on T2WI MR ± FS in fat space between ITB and LFC
- Location
  - Superolateral to knee joint
  - Lateral to anterolateral prominence of LFC
  - Deep to distal fibers of iliotibial band
  - Usually superior and superficial to lateral synovial joint recess
- Size
  - Amorphous area of edema may be small or several cm in size

MR Findings

- Ill-defined area of elevated signal on T2WI or T2WI FS MR between LFC and ITB at their closest approximation
  - Best seen on coronal and axial sequences
  - More difficult to discern on T1WI MR
- May be associated with thickening of, or abnormal signal in, the ITB
- Occasional reactive marrow edema in adjacent LFC
- Late in process, more defined bursal fluid collection may form at same site

Imaging Recommendations

- Best imaging tool
  - MR
- Protocol advice
  - Coronal and axial T2WI FS MR sequences most sensitive

DIFFERENTIAL DIAGNOSIS

Fluid in Lateral Knee Joint Recess

- Well-defined borders
- Connection to knee joint identifiable

Lateral Collateral Ligament Complex Injury

- Abnormal signal centered around ± within lateral ligaments

Direct Blow Contusion

- Associated superficial soft tissue swelling
- History of trauma

Lateral Femoral Condyle Fracture

- Marrow edema more prominent feature

Lateral Parameniscal Cyst

- Unusual to extend this high
- Associated lateral meniscal tear

Excessive Lateral Pressure Syndrome

- Edema in fat pad inferolateral to patella
- Due to patellofemoral maltracking with lateral patellar subluxation

PATHOLOGY

General Features

- Etiology
  - Repetitive motion
  - Contact between ITB and LFC greatest at 20-30° of flexion

Gross Pathologic & Surgical Features

- Adventitial bursal extension from knee joint extends into fat deep to ITB (lateral synovial recess)
  - Tissue becomes inflamed due to repetitive trauma as ITB slides past lateral femoral condyle

CLINICAL ISSUES

Presentation

- Most common signs/symptoms
  - Anterolateral knee pain with activity
- Clinical profile
  - Long-distance runners, cyclists

Natural History & Prognosis

- Responds well to cessation of offending activity and anti-inflammatory therapy

Treatment

- Conservative

DIAGNOSTIC CHECKLIST

Consider

- ITB syndrome in long-distance runner with lateral knee pain

Image Interpretation Pearls

- ITB syndrome leads to ill-defined soft tissue swelling
- Well-defined fluid superolateral to knee may be joint recess

SELECTED REFERENCES

1. Hariri S et al: Treatment of recalcitrant iliotibial band friction syndrome with open iliotibial band bursectomy: indications, technique, and clinical outcomes. Am J Sports Med. 37(7):1417-24, 2009
2. Vasilevska V et al: Magnetic resonance imaging signs of iliotibial band friction in patients with isolated medial compartment osteoarthritis of the knee. Skeletal Radiol. 38(9):871-5, 2009
3. Murphy BJ et al: Iliotibial band friction syndrome: MR imaging findings. Radiology. 185(2):569-71, 1992

ILIOTIBIAL BAND FRICTION SYNDROME

__(Left)__ Coronal T2WI FS MR shows an elongate focus of elevated signal → lateral to the lateral femoral condyle → and deep to the iliotibial band →. The collection is not well marginated and has signal less intense than joint fluid →. __(Right)__ Axial PDWI FS MR from the same patient shows mild edema → in the lateral synovial recess between the iliotibial band → and the lateral femoral condyle →. This is mild ITB friction syndrome.

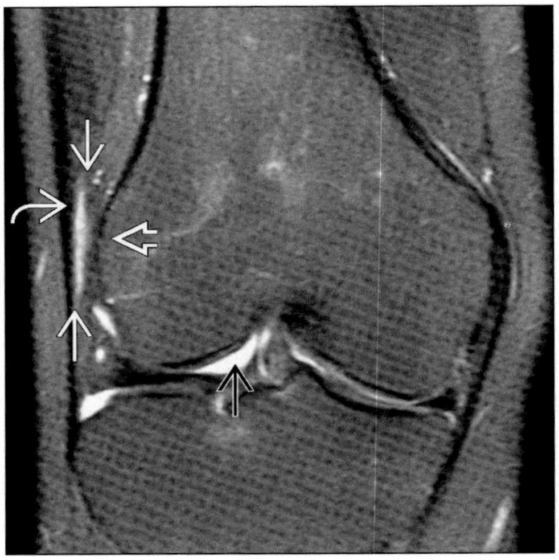

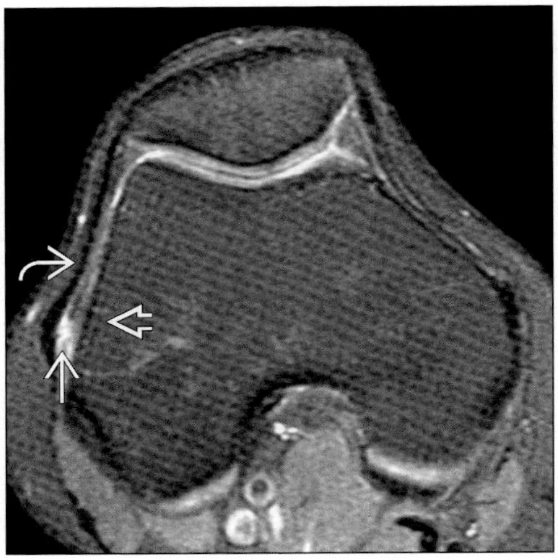

__(Left)__ Coronal T2WI FS MR shows more diffuse edema → in the lateral synovial recess of the knee. This patient is a long-distance runner and complains of lateral knee pain when running. Note the poorly marginated appearance of the edema; this should not be mistaken for a joint recess. __(Right)__ Axial PDWI FS MR from the same patient confirms the wispy edema in the lateral synovial recess →, consistent with ITB friction syndrome. Note that joint fluid → is much higher in signal.

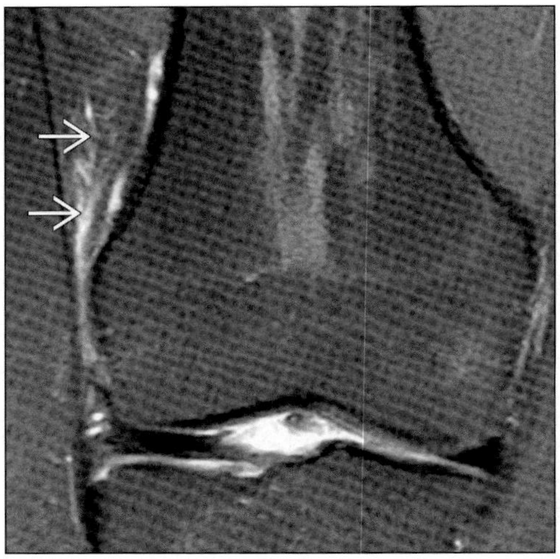

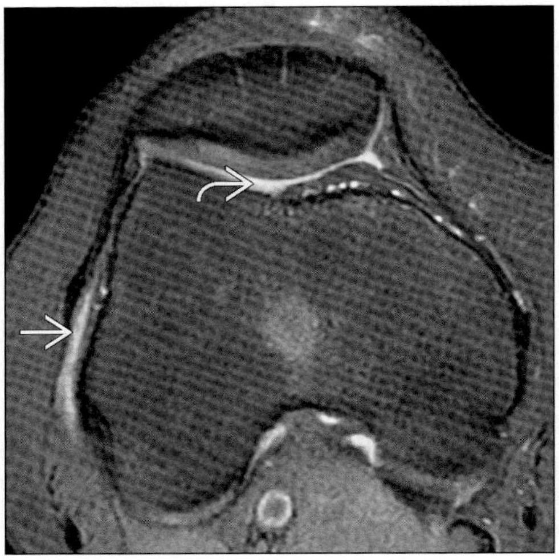

__(Left)__ Coronal T2WI FS MR shows a more circumscribed collection of fluid → in the lateral synovial recess in this cyclist with anterolateral knee pain while riding. Over time, a more bursa-like fluid collection can form here due to repeated episodes of inflammation. __(Right)__ Sagittal T2WI MR from the same patient shows the lobular outpouchings of bursal fluid → between the iliotibial band → and the lateral femoral condyle →.

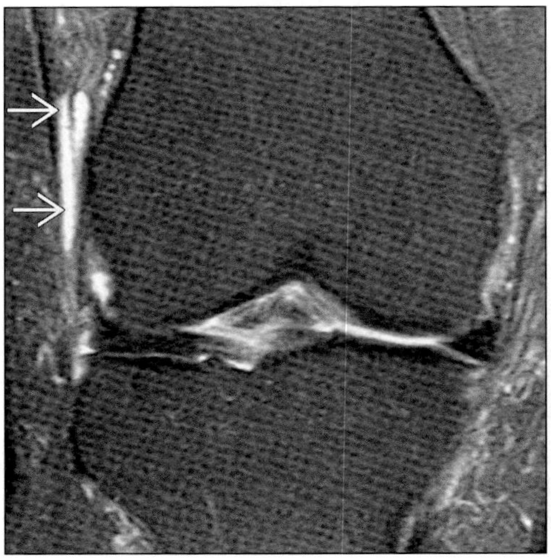

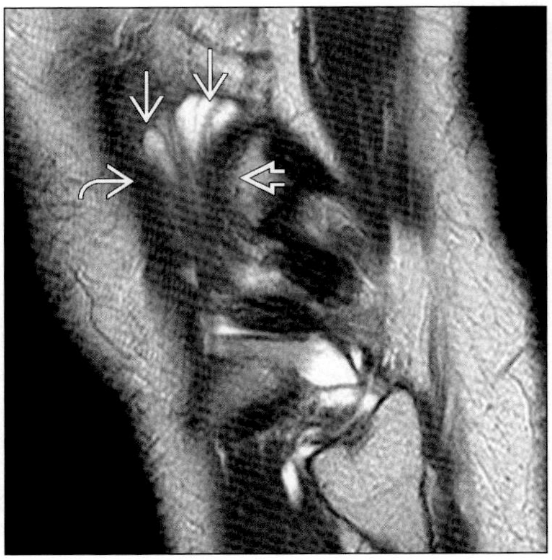

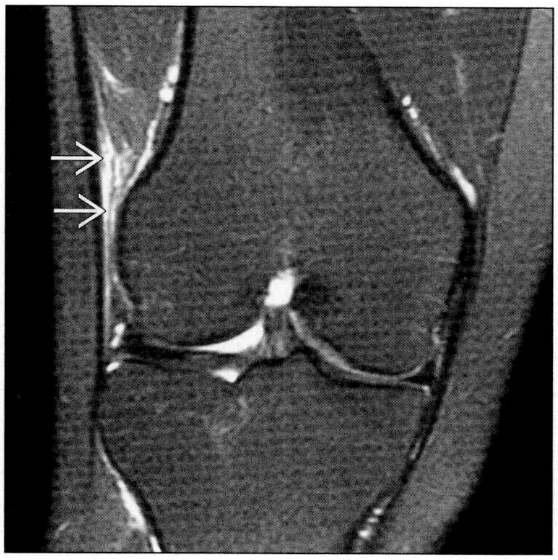

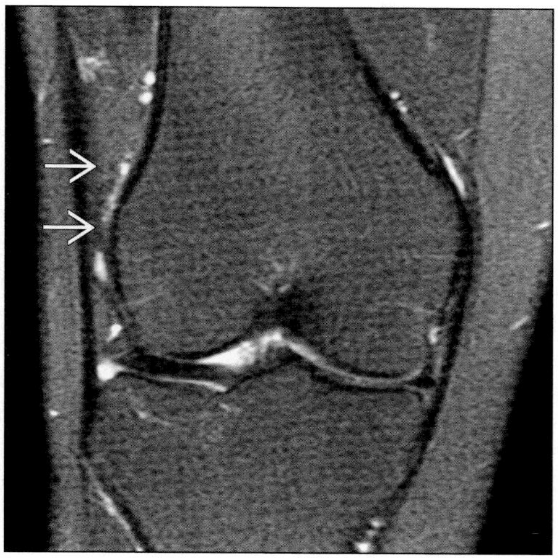

(Left) *Coronal T2WI FS MR in a 25-year-old man who has taken up cross-country running and presents with anterolateral knee pain shows amorphous edema → in the lateral synovial recess, typical of iliotibial band friction syndrome.* ***(Right)*** *Coronal T2WI FS MR from the same patient 1 year earlier shows no edema in the recess →; this study was performed for other reasons and before the patient started running.*

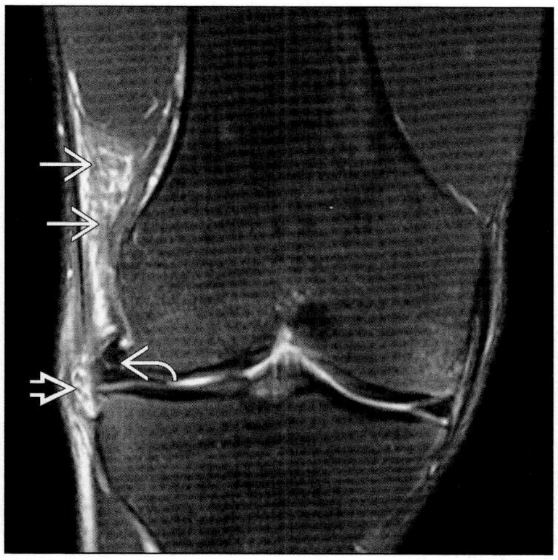

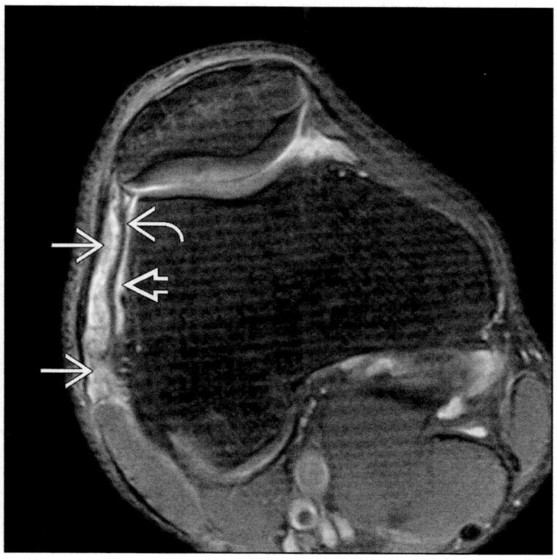

(Left) *Coronal T2WI FS MR shows a more severe case of ITB friction syndrome →. There is also a horizontal cleavage tear of the lateral meniscus ↷, which communicates with a parameniscal cyst ⇨.* ***(Right)*** *Axial PDWI FS MR from the same patient shows the extensive elevated T2 signal → in the lateral synovial recess, indicative of ITB friction syndrome. The lateral synovial joint reflection ↷ is outlined by the edema and a small amount of fluid in the lateral joint recess ⇨.*

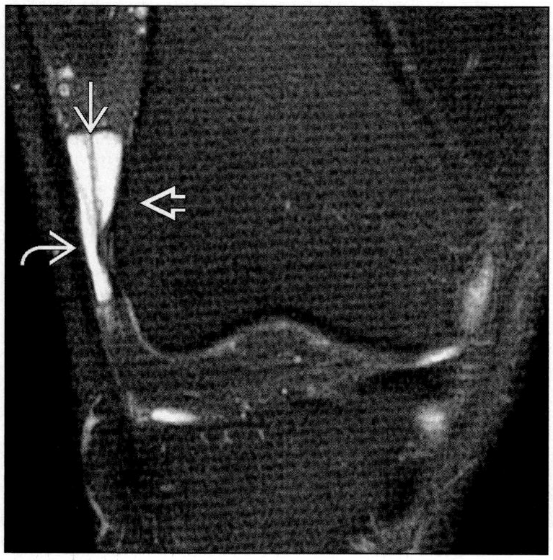

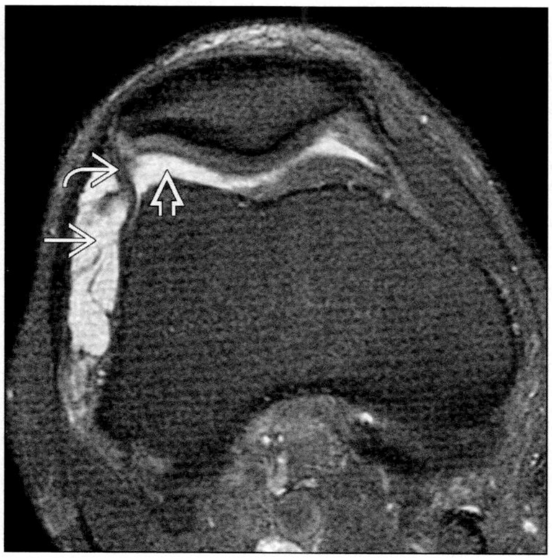

(Left) *Coronal T2WI FS MR shows a well-defined fluid collection → between the ITB ↷ and the lateral femoral condyle ⇨. Patient is a marathoner and complains of anterolateral knee pain while running. This is an unusual case of iliotibial band bursitis; it is likely the chronic sequela of ITB friction syndrome.* ***(Right)*** *Axial PDWI FS MR from the same patient shows the septated ITB bursa → separated from the knee joint ⇨ by the synovial reflection ↷ of the lateral joint recess.*

MENISCAL PITFALLS AND VARIANTS

Key Facts

Imaging

- Popliteal hiatus
  - Passage of popliteal tendon through joint capsule adjacent to LM
  - Border may be mistaken for vertical meniscal tear on coronal or sagittal images
- Meniscofemoral ligament (Humphrey ± Wrisberg)
  - Extends from posterior horn of LM to medial wall of intercondylar notch
  - Ligament of Humphrey: Extension of ligament anterior to posterior cruciate ligament (PCL)
  - Ligament of Wrisberg: Extension of ligament posterior to PCL
- Intermeniscal ligament of Winslow (transverse lig.)
  - Runs from anterior horn of lateral meniscus to anterior horn of MM
  - Points of attachment may be mistaken for vertical tear
- Meniscomeniscal ligament
  - Variable ligament extending from posterior horn of LM to anterior horn of MM or occasionally from anterior LM to posterior MM
  - Runs between anterior and posterior cruciate ligaments
- Normal meniscal vascular pedicle
  - Peripheral (red zone) vascular pedicle of menisci more prominent in children and adolescents
  - Confined to peripheral 1/3 of meniscus, does not contact articular surfaces
- Anterior medial meniscal root variants
  - Insertion of anterior root of MM may extend more inferiorly than remainder of meniscus
- Meniscal flounce
  - Wavy contour of free edge of meniscus seen on sagittal MR images
  - Usually affects medial meniscus

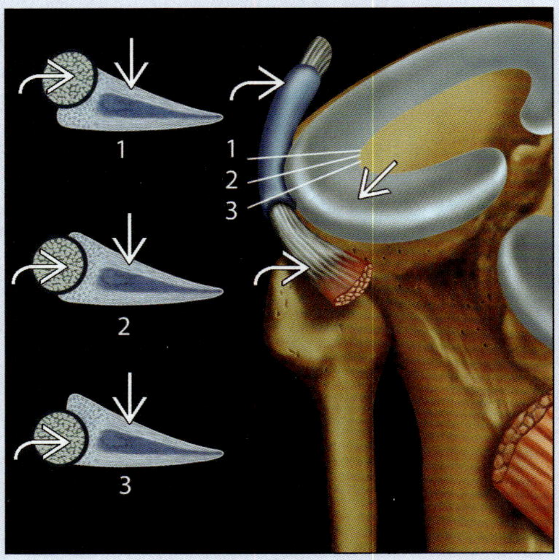

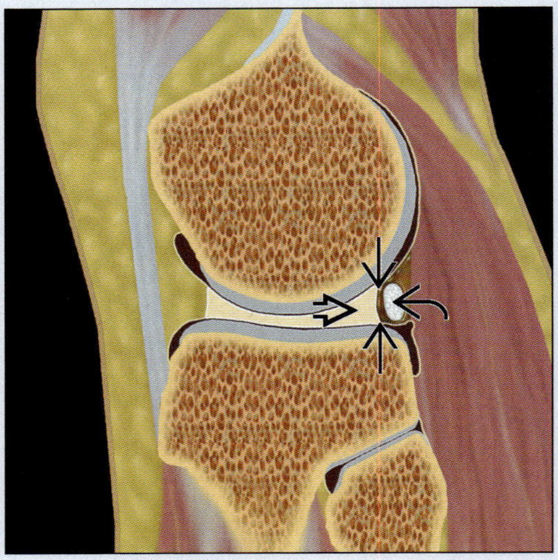

(Left) *Oblique graphic shows the popliteal hiatus with reference to the lateral meniscus. As the popliteal tendon ➔ courses from anterosuperior to posteroinferior, it appears in cross section to interrupt the capsular attachment of the lateral meniscus ➔. The popliteomeniscal fascicles are formed at the margins of this interruption.* ***(Right)*** *Sagittal graphic shows the popliteal hiatus ➔ as a gap between the popliteal tendon ➔ and the posterior horn of the lateral meniscus ➔.*

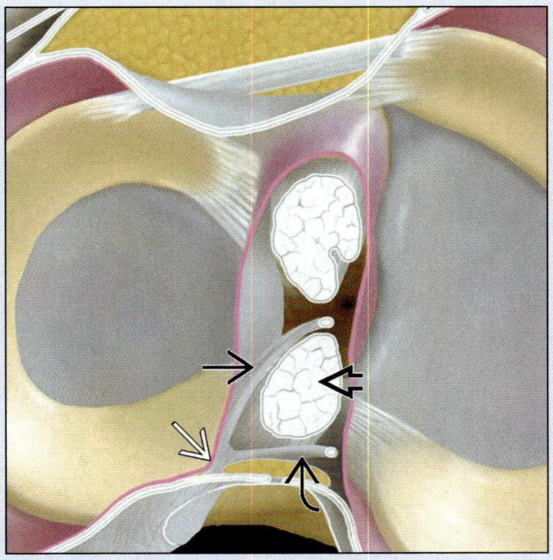

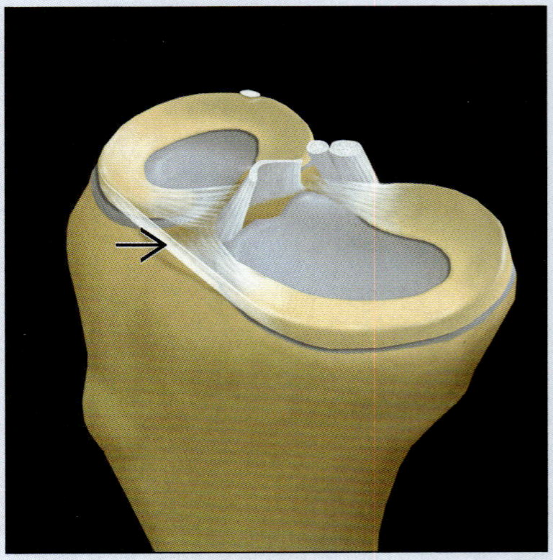

(Left) *Axial graphic shows the components of the meniscofemoral ligament as they course from its attachment to the posterior horn of the lateral meniscus ➔ to the medial wall of the intercondylar notch. The ligament of Humphrey ➔ extends anterior to the posterior cruciate ligament ➔, and the ligament of Wrisberg ➔ extends posterior to it.* ***(Right)*** *Oblique graphic shows the transverse intermeniscal ligament ➔ running between the anterior horns of the medial and lateral menisci.*

MENISCAL PITFALLS AND VARIANTS

TERMINOLOGY

Abbreviations

- Medial meniscus (MM)
- Lateral meniscus (LM)

Definitions

- Variety of anatomic conditions and variations that may be mistaken for pathology on MR

IMAGING

MR Findings

- **Popliteal hiatus**
  - Passage of popliteal tendon through joint capsule adjacent to LM
  - Border may be mistaken for vertical meniscal tear on coronal or sagittal images
- **Meniscofemoral ligament (Humphrey ± Wrisberg)**
  - Extends from posterior horn of LM to medial wall of intercondylar notch
    - Ligament of Humphrey: Extension of ligament anterior to posterior cruciate ligament (PCL)
    - Ligament of Wrisberg: Extension of ligament posterior to PCL
    - Attachment to LM may mimic vertical tear
- **Intermeniscal ligament of Winslow** (transverse lig.)
  - Runs from anterior horn of lateral meniscus to anterior horn of MM
  - Points of attachment may be mistaken for vertical tear
- **Meniscomeniscal ligament**
  - Variable ligament extending from posterior horn of LM to anterior horn of MM or occasionally from anterior LM to posterior MM
  - Runs between anterior and posterior cruciate ligaments
  - Function unknown
- **Normal meniscal vascular pedicle**
  - Peripheral (red zone) vascular pedicle of menisci more prominent in children and adolescents
    - Especially in posterior horns
  - MM > LM
  - Confined to peripheral 1/3 of meniscus, does not contact articular surfaces
  - May be mistaken for globular tear
- **Anterior medial meniscal root variants**
  - Insertion of anterior root of MM may extend more inferiorly than remainder of meniscus
- **Meniscal flounce**
  - Wavy contour of free edge of meniscus seen on sagittal MR images
  - Due to relaxed tension on meniscus during knee extension
  - Usually affects medial meniscus

Imaging Recommendations

- Best imaging tool
  - MR

DIFFERENTIAL DIAGNOSIS

Postoperative Meniscus Change

- May mimic vascular pedicle signal or meniscal flounce

Meniscal Degeneration

- May mimic vascular pedicle
- Typically seen in middle-aged or elderly patients

Popliteomeniscal Fascicles

- May mimic posterior horn LM tear or popliteal hiatus may be confused with tear

Extrusion of Meniscus

- Blunting of free edge and distortion of normal morphology due to loss of longitudinal fiber strength of meniscus

CLINICAL ISSUES

Demographics

- Epidemiology
  - Meniscofemoral ligament
    - Overall incidence (70-100%)
      - Incidence of posterior (Wrisberg) variety (35-76%)
      - Incidence of anterior (Humphrey) variety (26-50%)
      - Incidence of both present (6-17%)
  - Transverse (intermeniscal) ligament
    - Incidence (58%)
  - Meniscomeniscal ligament (1-4%)

DIAGNOSTIC CHECKLIST

Consider

- Be alert to anatomic variants in areas typically affected
  - Especially attachments of menisci to ligaments and bone
- Normal vascular pedicle in children with prominent signal in posterior MM that does not extend to articular surface

SELECTED REFERENCES

1. Kim HK et al: Oblique meniscomeniscal ligament: a normal variant. Pediatr Radiol. 39(6):634, 2009
2. de Abreu MR et al: Meniscofemoral ligaments: patterns of tears and pseudotears of the menisci using cadaveric and clinical material. Skeletal Radiol. 36(8):729-35, 2007
3. Park LS et al: Posterior horn lateral meniscal tears simulating meniscofemoral ligament attachment in the setting of ACL tear: MRI findings. Skeletal Radiol. 36(5):399-403, 2007
4. Kim YG et al: An arthroscopic analysis of lateral meniscal variants and a comparison with MRI findings. Knee Surg Sports Traumatol Arthrosc. 14(1):20-6, 2006
5. Sproule JA et al: Altered signal intensity in the posterior horn of the medial meniscus: an MR finding of questionable significance. Arch Orthop Trauma Surg. 125(4):267-71, 2005
6. Erbagci H et al: An MRI study of the meniscofemoral and transverse ligaments of the knee. Surg Radiol Anat. 24(2):120-4, 2002
7. Cothran RL Jr et al: MR imaging of meniscal contusion in the knee. AJR Am J Roentgenol. 177(5):1189-92, 2001
8. Mesgarzadeh M et al: MR imaging of the knee: expanded classification and pitfalls to interpretation of meniscal tears. Radiographics. 13(3):489-500, 1993
9. Kaplan PA et al: MR of the knee: the significance of high signal in the meniscus that does not clearly extend to the surface. AJR Am J Roentgenol. 156(2):333-6, 1991

*(**Left**) Axial PDWI FS MR shows the popliteal hiatus, the opening in the knee joint capsule through which the popliteal tendon passes. The hiatus is immediately adjacent to the posterior horn of the lateral meniscus and can be a source of diagnostic error. (**Right**) Coronal T2WI FS MR shows the margin of the popliteal hiatus as a high signal line extending vertically between the popliteal tendon and the lateral meniscus.*

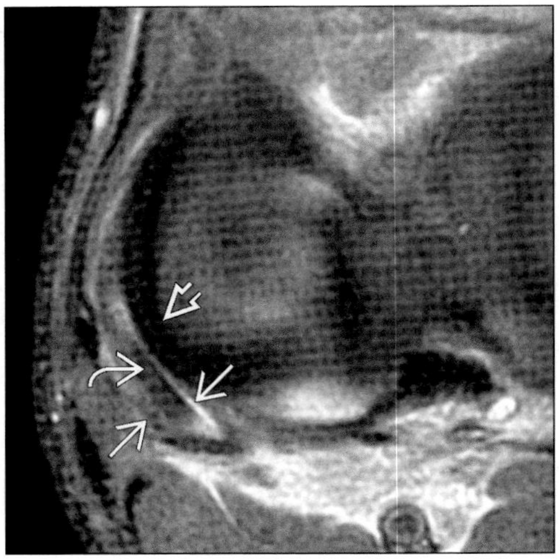

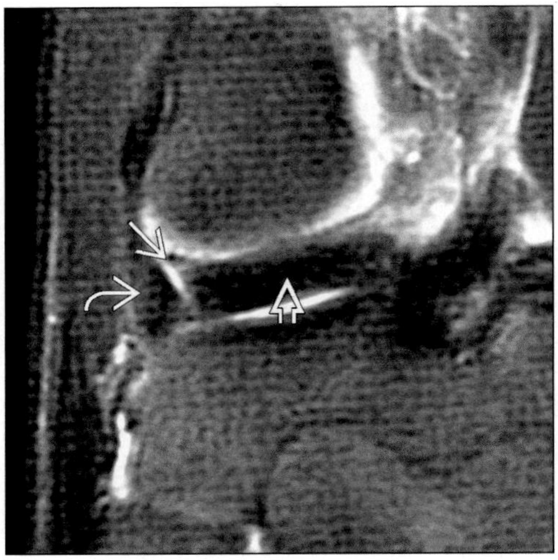

*(**Left**) Sagittal PDWI MR shows the margin of the popliteal hiatus at the point where the popliteal tendon passes through it to enter the knee joint capsule. This normal anatomic feature should not be mistaken for a vertical tear of the lateral meniscus. (**Right**) Sagittal PDWI FS MR shows slight irregular contour of the superior articular surface of the posterior horn of the lateral meniscus, without a linear focus of signal abnormality.*

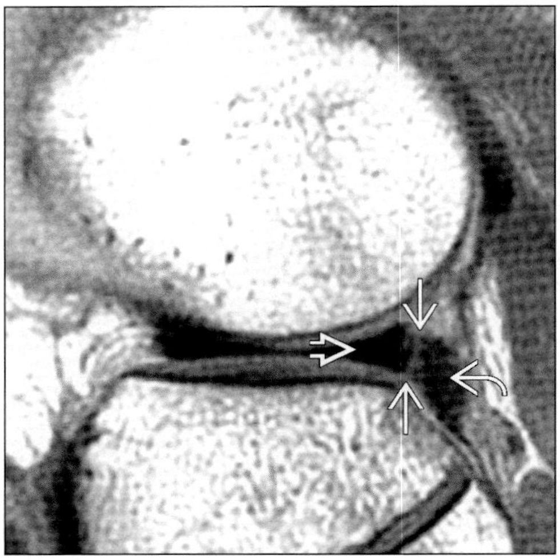

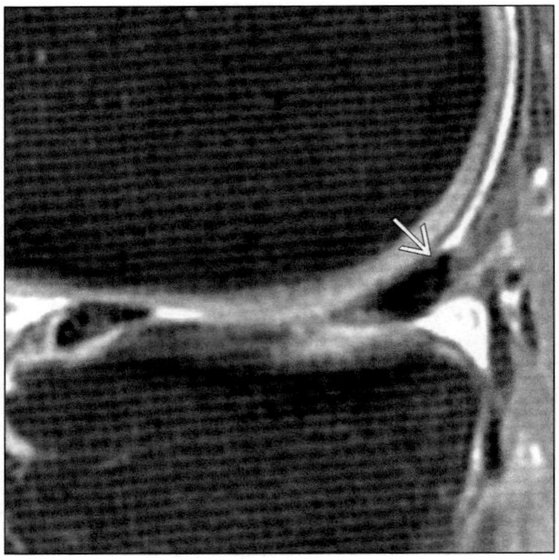

*(**Left**) Sagittal PDWI FS MR more medially in the same patient shows a linear focus of elevated signal in the same portion of the posterior horn. This is the junction point of the meniscus with the meniscofemoral ligament. (**Right**) Sagittal PDWI FS MR more medially in the same patient more clearly shows the separation of the posterior horn of the lateral meniscus from the meniscofemoral ligament. Fluid is interposed between them.*

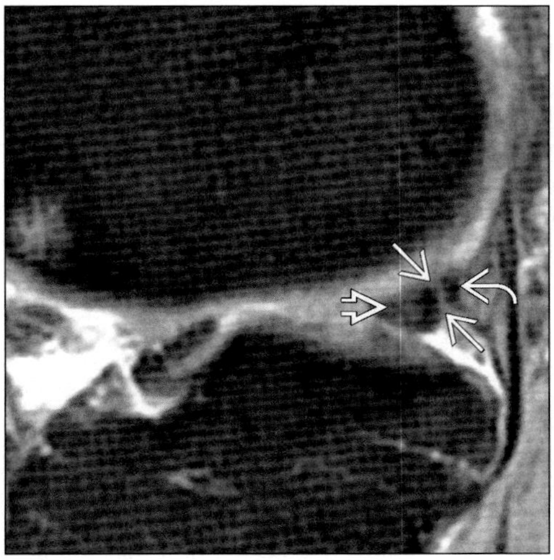

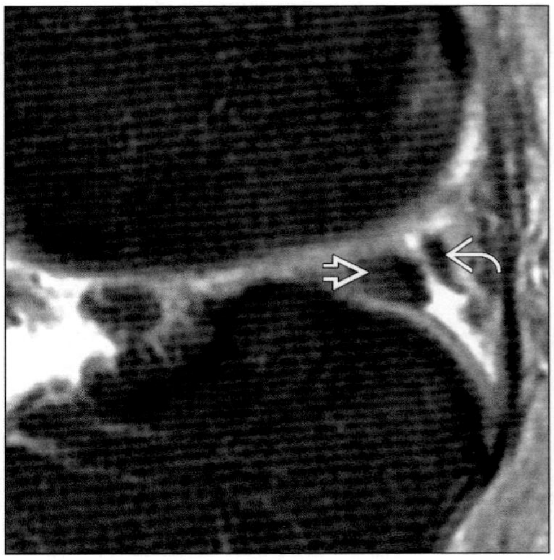

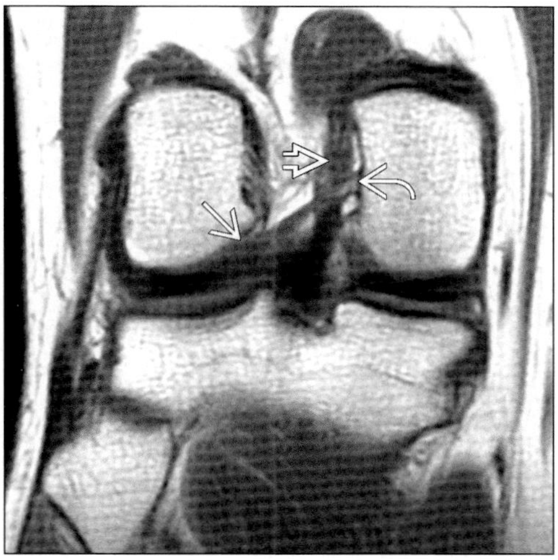

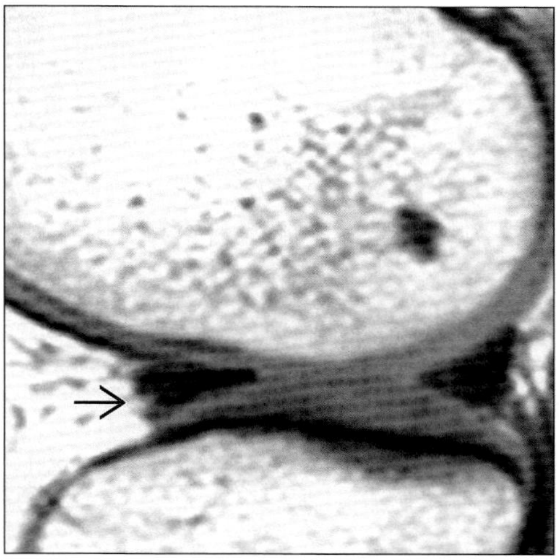

(Left) *Coronal T1WI MR shows the meniscofemoral ligament extending from the posterior horn of the lateral meniscus to the medial wall of the intercondylar notch. The ligament may pass anterior or posterior to the posterior cruciate ligament (PCL), or both. If it passes anterior to the PCL, it is termed a ligament of Humphrey; if posterior, it is a ligament of Wrisberg.* ***(Right)*** *Sagittal PDWI MR shows mild capsular irregularity in the anterior horn of the lateral meniscus (LM).*

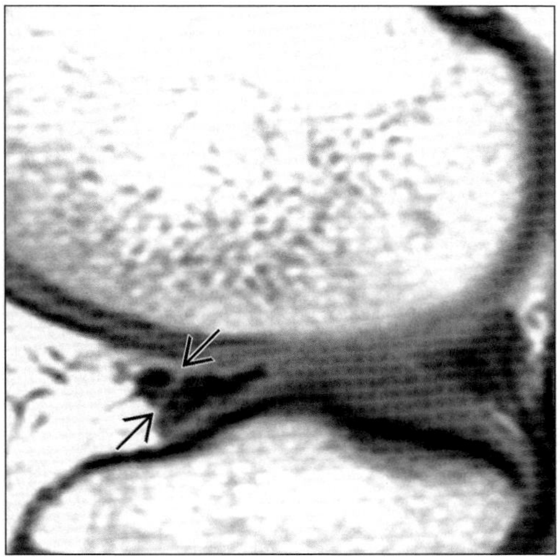

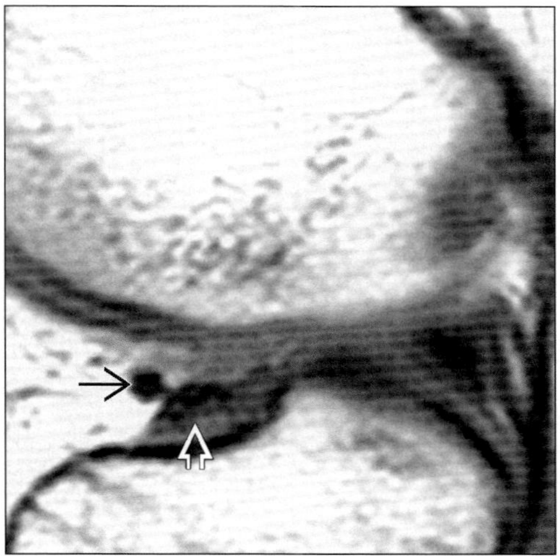

(Left) *Sagittal PDWI MR more medially in the same patient shows an oblique linear focus of signal extending across the anterior horn of the LM. This is the typical orientation of the point of attachment of the transverse intermeniscal ligament; tears are usually oriented at a different angle.* ***(Right)*** *Sagittal PDWI MR more medially in the same patient shows that the linear signal represents the junction of the anterior horn with the transverse intermeniscal ligament.*

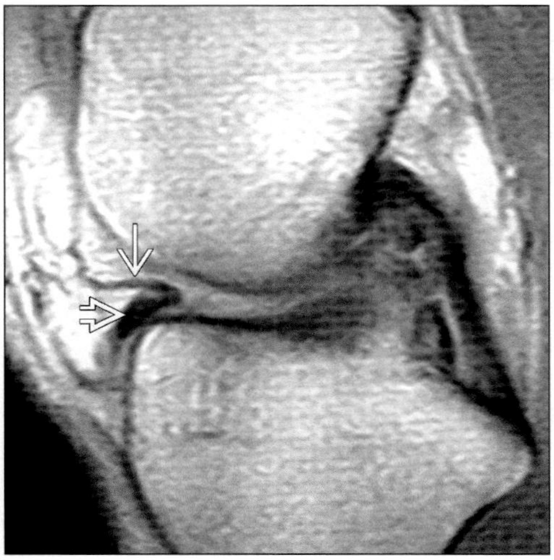

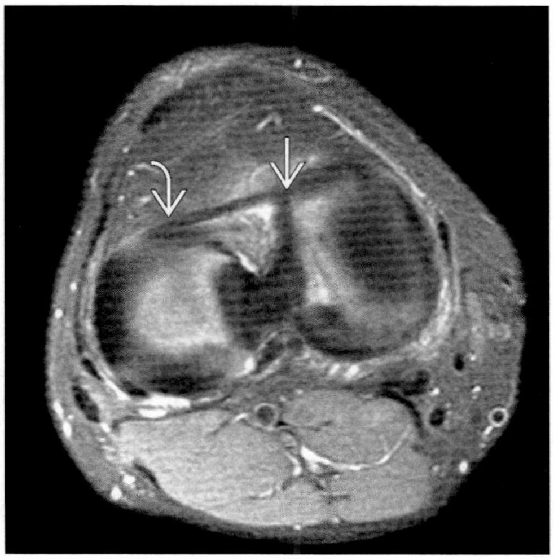

(Left) *Sagittal PDWI MR shows the junction of the transverse intermeniscal ligament with the anterior horn of the medial meniscus. The junction of the ligament with the lateral meniscus tends to be confused with a vertical tear more often than this medial junction.* ***(Right)*** *Axial PDWI FS MR shows the transverse intermeniscal ligament of Winslow running from the anterior horn of the medial meniscus to the anterior horn of the lateral meniscus. The ligament is present in about 58% of people.*

*(**Left**) Axial PDWI MR shows a meniscomeniscal ligament (MML) running from the posterior horn of the lateral meniscus to the anterior horn of the medial meniscus. This ligament courses between the ACL and PCL. Occasionally the MML may run from posteromedial to anterolateral. (**Right**) Sagittal PDWI MR shows the posterior and anterior horns of the medial meniscus (MM). No discrete tear is seen in either component. Image was obtained on a low-field magnet.*

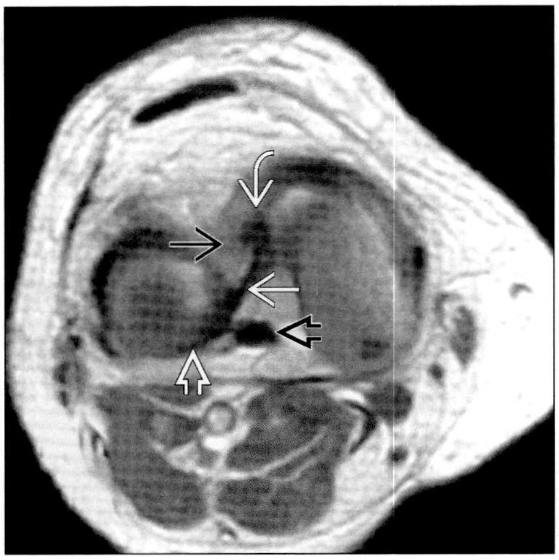

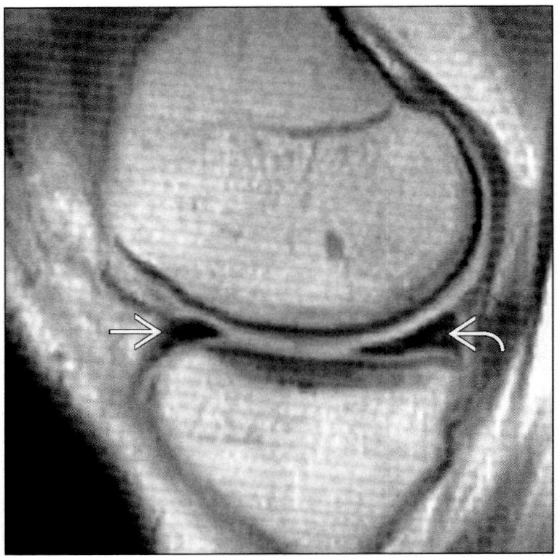

*(**Left**) Sagittal PDWI MR more laterally in the same patient shows extension of the anterior horn of the MM along the anterior margin of the tibia as the meniscus approaches its anterior root attachment. (**Right**) Sagittal PDWI MR more laterally in the same patient shows the anterior root attachment of the MM extending inferior to the joint line. This is a normal variant and should not be mistaken for a displaced tear. The transverse intermeniscal ligament is present at the joint line.*

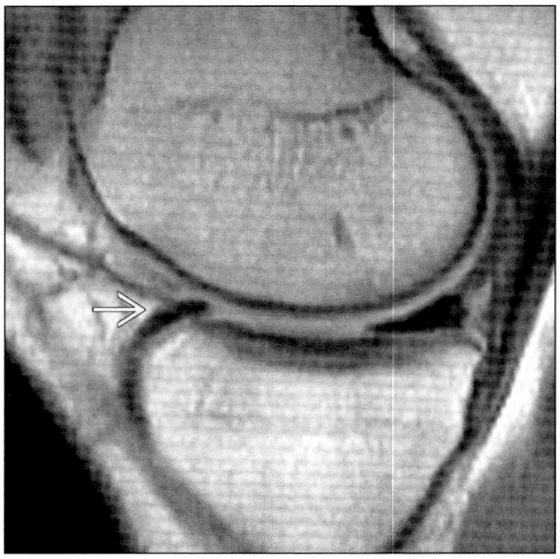

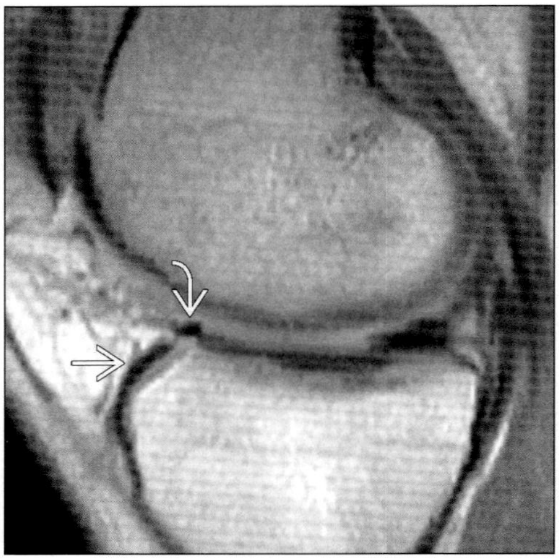

*(**Left**) Sagittal PDWI FS MR from a 15-year-old boy shows the normal linear signal related to the vascular pedicle of the meniscus in the posterior horn of the MM. Typically, this signal is confined to the peripheral 1/3 of the meniscus, does not contact the articular surfaces, and often contacts the peripheral (capsular) surface of the meniscus. (**Right**) Coronal T1WI MR from the same patient shows the typical meniscal vascular pedicle, which should not be mistaken for a tear.*

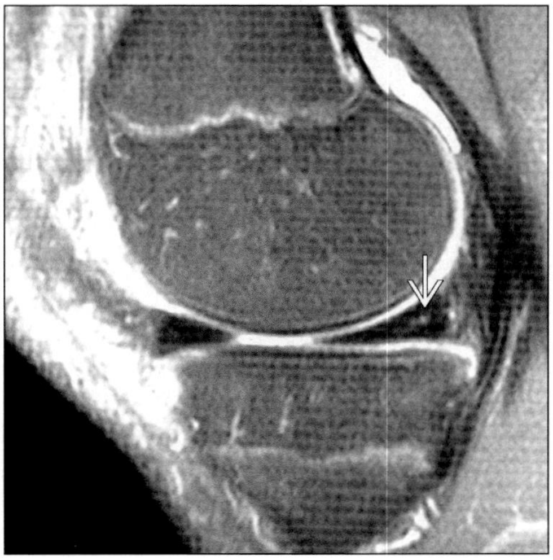

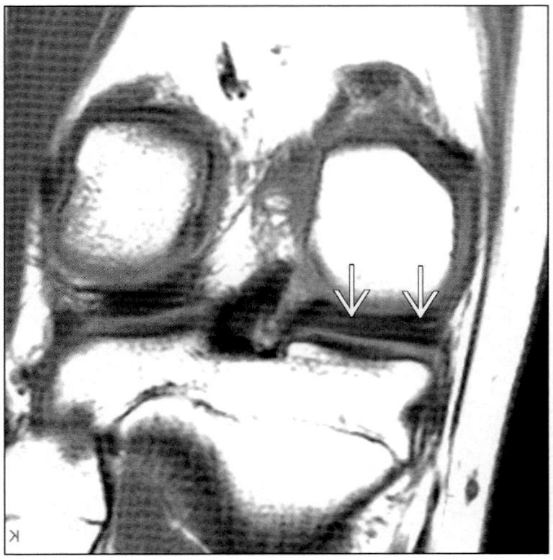

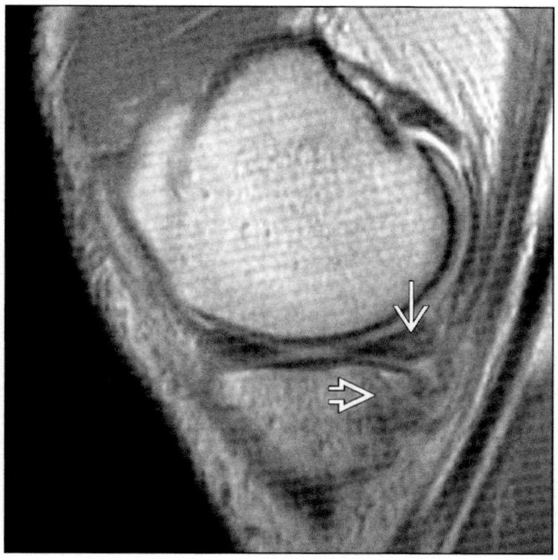

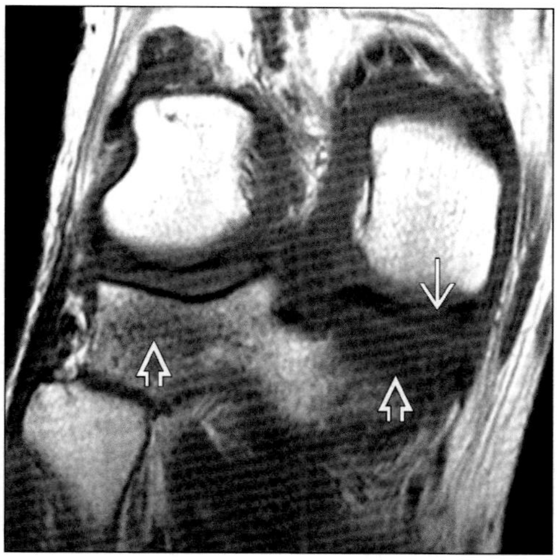

(Left) *Sagittal PDWI MR from a 47-year-old man 2 days after a fall demonstrates marrow edema in the posterior medial tibia* ⇨ *and elevated signal* → *in the posterior horn of the medial meniscus. The meniscal signal does not contact an articular surface.* ***(Right)*** *Coronal T1WI MR from the same patient demonstrates posterior tibial contusions* ⇨ *and globular intrameniscal signal in the medial meniscus posterior horn* →*. The setting of trauma and lack of extension of signal to meniscal surface suggests meniscal contusion.*

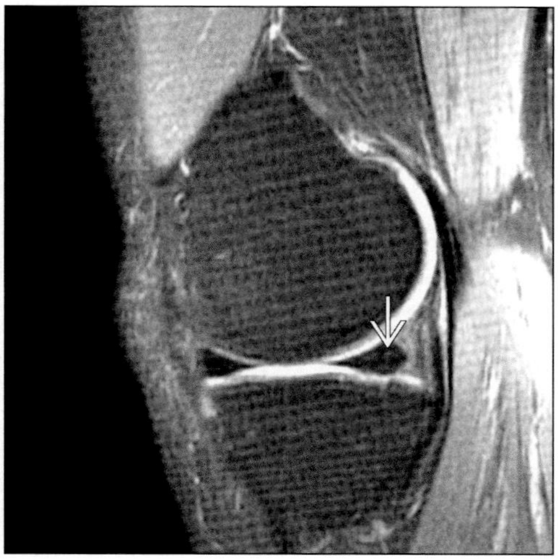

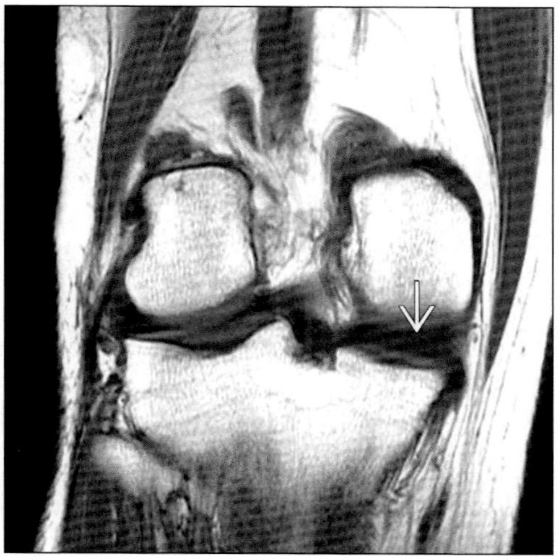

(Left) *Sagittal PDWI FS MR from the same patient 14 months later obtained for different reasons shows resolution of the intrameniscal signal in the posterior horn of the medial meniscus* →*.* ***(Right)*** *Coronal T1WI MR from the same patient 14 months after injury shows resolution of the posterior horn medial meniscal signal* →*. Transient intrameniscal signal that does not extend to an articular surface and is related to acute trauma is consistent with a meniscal contusion.*

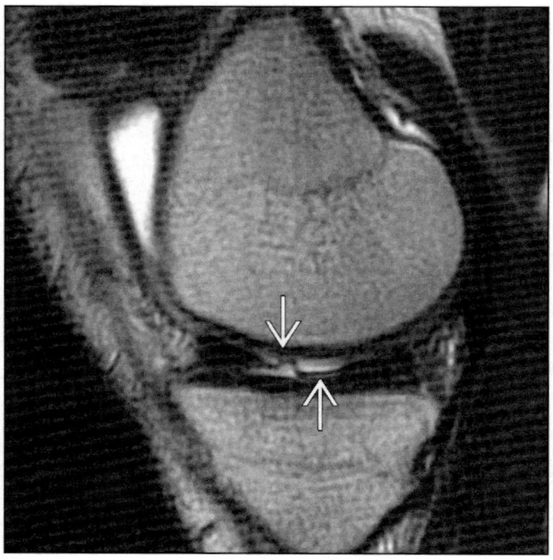

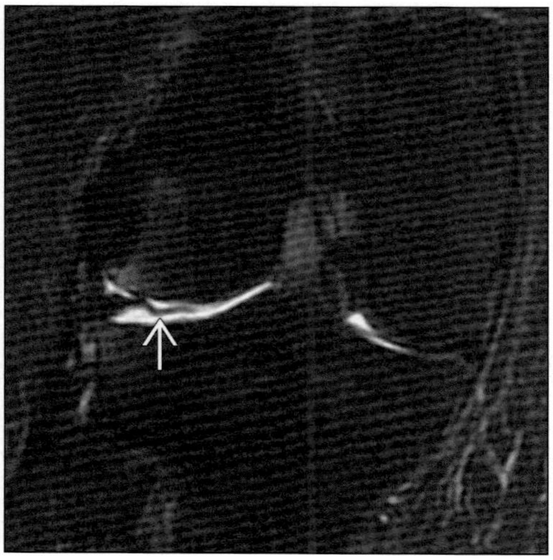

(Left) *Sagittal T2WI MR shows an undulating shape to the free edge of the medial meniscus* →*. At arthrography, no tear was identified. This appearance is typical of meniscal flounce, a transient morphologic state of the meniscus likely related to laxity of the free edge when the knee is in passive extension.* ***(Right)*** *Coronal T2WI FS MR demonstrates pronounced flounce of the lateral meniscus* →*. This finding is more common medially than laterally and more often observed on sagittal images.*

DISCOID MENISCUS

Key Facts

Terminology

- Congenitally large meniscus due to failure of resorption of its central portion during development
- Incomplete (partial) discoid meniscus: Incomplete resorption of central portion

Imaging

- Lateral > > medial
- > 14 mm meniscal width (medial to lateral) on coronal mid-meniscal image
- Coronal images: Loss of normal meniscal tapering wedge shape on central slices
- Sagittal images: "Too many bowties" sign of continuous meniscal substance from anterior to posterior (depends on slice and skip width)
- Coronal images most accurate in assessing for discoid meniscus

Top Differential Diagnoses

- Meniscal bucket-handle tear
- Flipped meniscus
- Other displaced meniscal tears

Pathology

- Failure of resorption of central meniscal substance during embryogenesis results in discoid morphology

Clinical Issues

- Pain, clicking, snapping of meniscus
- Commonly presents in adolescence
- 5% of population
- Bilateral in 20%
- Discoid meniscus prone to tear
- Arthroscopic saucerization of meniscus to resect discoid central portion if symptomatic in adult

*(**Left**) Axial graphic shows a discoid lateral meniscus → with a small central area of meniscal resorption ↩. Most discoid menisci are not complete discs; there is variable central embryologic resorption. (**Right**) Coronal T1WI MR shows a discoid lateral meniscus, with meniscal substance extending to the intercondylar notch →. On this mid-condylar image, the menisci should not extend more than 14 mm in medial-to-lateral dimension. The medial meniscus is partially discoid ↩.*

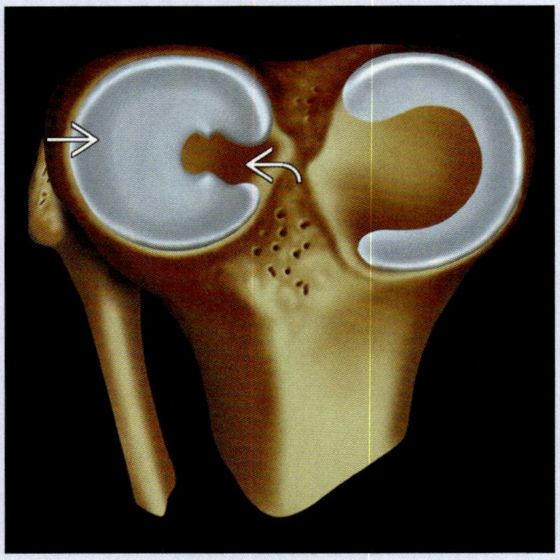

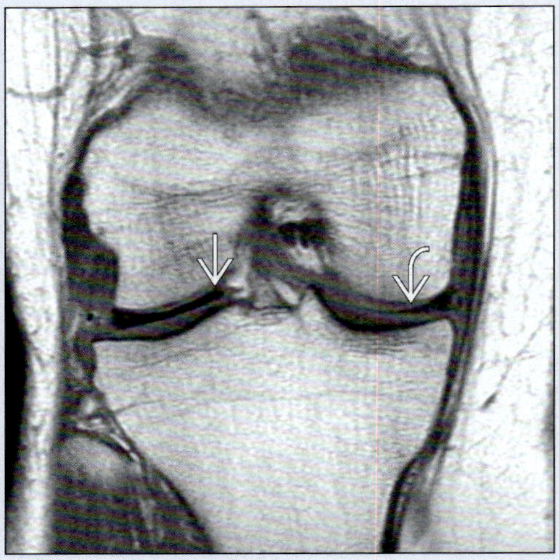

*(**Left**) Axial graphic shows a partial discoid lateral meniscus. There is moderate resorption of the central substance ↩, but the meniscus is still much larger than the normal medial meniscus ⇨. (**Right**) Coronal T2WI FS MR shows a partial discoid lateral meniscus. The mid-body of the meniscus is substantially larger than normal, but it comes to a tapering point → before reaching the intercondylar notch. The medial meniscus ⇨ demonstrates a normal configuration.*

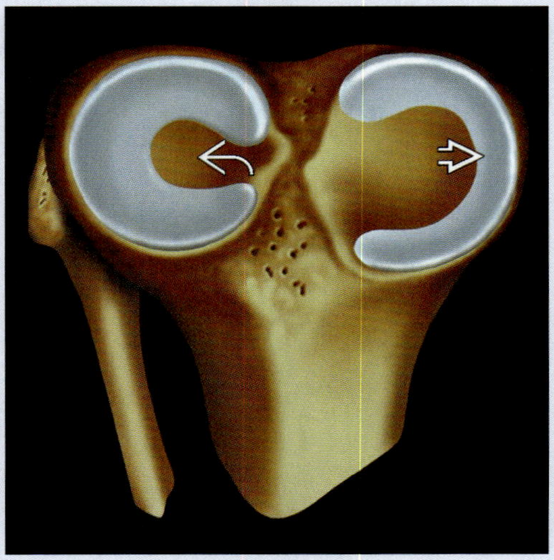

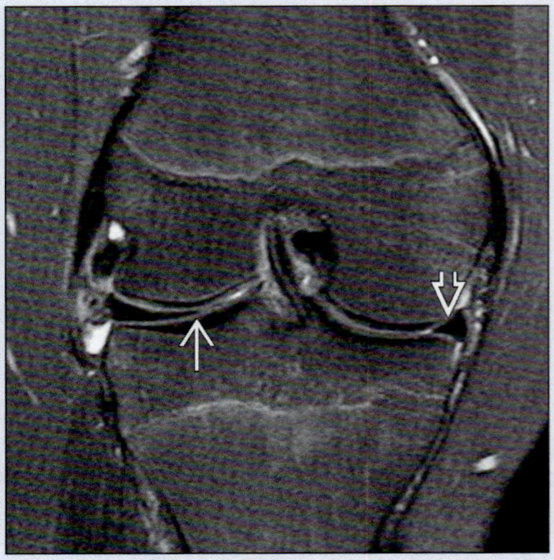

DISCOID MENISCUS

TERMINOLOGY

Definitions

- Congenitally large meniscus due to failure of resorption of its central portion during development
- Incomplete (partial) discoid meniscus: Incomplete resorption of central portion

IMAGING

General Features

- Best diagnostic clue
  - Abnormally large meniscus with loss of normal wedged shape in cross section
  - Central portion of meniscus covers > 50% of tibial condylar articular surface
- Location
  - Lateral > > medial
- Size
  - > 14 mm meniscal width (medial to lateral) on coronal mid-meniscal image
- Morphology
  - Disc-shaped meniscus without normal central tapering
  - Partial discoid meniscus usually tapers centrally but does so closer than normal to intercondylar notch

MR Findings

- Coronal images: Loss of normal meniscal tapering wedge shape on central slices
  - Partial discoid meniscus will taper but with medial-lateral width larger than normal (> 14 mm)
- Sagittal images: "Too many bowties" sign of continuous meniscal substance from anterior to posterior

Imaging Recommendations

- Best imaging tool
  - MR
- Protocol advice
  - Coronal images most accurate in assessing for discoid meniscus

DIFFERENTIAL DIAGNOSIS

Meniscal Bucket-Handle Tear

- Gap between peripheral meniscal remnant and displaced central fragment on coronal images

Flipped Meniscus

- Absent or severely truncated posterior horn

Other Displaced Meniscal Tears

- Parrot-beak/flap tear may result in central displacement of meniscal tissue

PATHOLOGY

General Features

- Etiology
  - Failure of resorption of central meniscal substance during embryogenesis results in discoid morphology

Staging, Grading, & Classification

- Watanabe classification
  - Complete: No resorption of central substance; meniscus extends into intercondylar notch
  - Incomplete: Partial resorption of central substance; no extension into intercondylar notch
  - Wrisberg type: No posterolateral meniscal attachment to tibia
    - Only attachment of posterior horn is to meniscofemoral ligament of Wrisberg
    - Unstable meniscus

CLINICAL ISSUES

Presentation

- Most common signs/symptoms
  - Pain, clicking, snapping of meniscus
  - Locking: More common in children

Demographics

- Age
  - Often asymptomatic during childhood
  - Commonly presents in adolescence
- Gender
  - M = F
- Epidemiology
  - 5% of population
  - Bilateral in 20%

Natural History & Prognosis

- Development of intrameniscal mucinous degeneration or meniscal cyst
- Discoid meniscus prone to tear

Treatment

- Conservative in children with no tear
- Arthroscopic saucerization of meniscus to resect discoid central portion if symptomatic in adult
- Repair/resection of tear

DIAGNOSTIC CHECKLIST

Consider

- Displaced meniscal tear with fragment filling central cavity of meniscus

Image Interpretation Pearls

- Scrutinize discoid meniscus for tear

Reporting Tips

- Categorize as complete or incomplete discoid type
- Suggest evaluation of contralateral knee

SELECTED REFERENCES

1. Ahn JH et al: A novel magnetic resonance imaging classification of discoid lateral meniscus based on peripheral attachment. Am J Sports Med. 37(8):1564-9, 2009
2. Good CR et al: Arthroscopic treatment of symptomatic discoid meniscus in children: classification, technique, and results. Arthroscopy. 23(2):157-63, 2007
3. Singh K et al: MRI appearance of Wrisberg variant of discoid lateral meniscus. AJR Am J Roentgenol. 187(2):384-7, 2006
4. Tachibana Y et al: Discoid medial meniscus. Arthroscopy. 19(7):E12-8, 2003
5. Ryu KN et al: MR imaging of tears of discoid lateral menisci. AJR Am J Roentgenol. 171(4):963-7, 1998

*(**Left**) Coronal T2WI FS MR demonstrates displacement of the central portion of a discoid lateral meniscus with the fragment → extending into the intercondylar notch. (**Right**) Coronal T2WI FS MR more anteriorly in the same patient shows the displaced central lateral meniscal fragment rejoining the anterior horn →; the intervening images demonstrated separation of the peripheral portion of the meniscus from the central fragment (not shown). This is a bucket-handle tear of a discoid meniscus.*

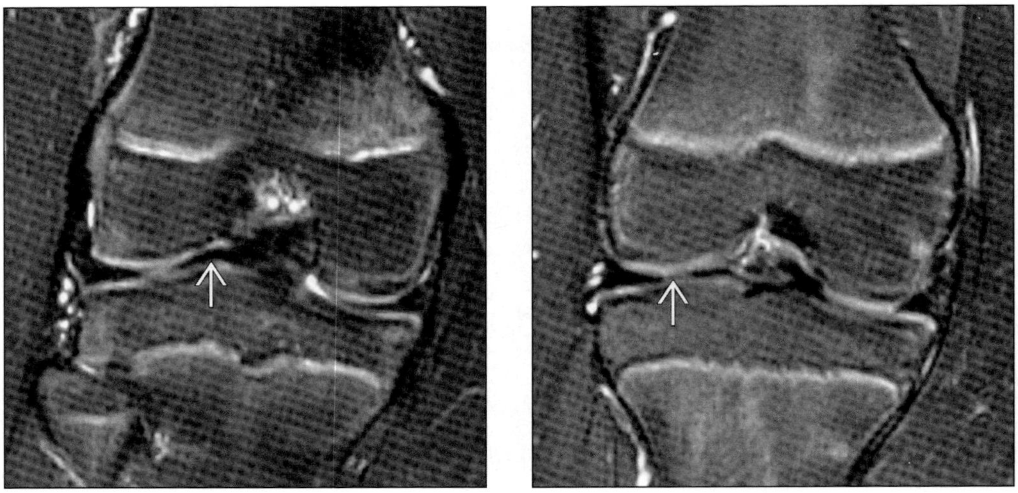

*(**Left**) Axial PDWI FS MR from the same patient shows the large displaced central fragment → extending into the notch. The residual peripheral zone of the lateral meniscus → is slightly larger than normal, but combined with the displaced fragment, the meniscus is much larger than normal, confirming its discoid nature. (**Right**) Coronal T2WI FS MR shows abrupt truncation of the free edge of the lateral meniscus →, consistent with a tear. The medial meniscus is partially discoid →.*

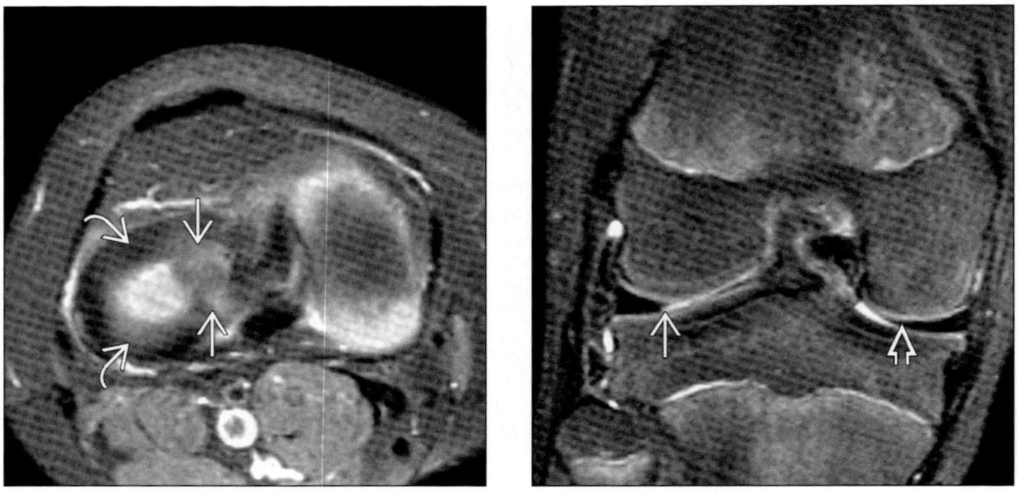

*(**Left**) Axial PDWI FS MR from the same patient shows what appears to be relatively normal size of the lateral meniscus →. The medial meniscus posterior horn → is much larger than expected, consistent with its partial discoid nature. (**Right**) Sagittal T2WI MR from the same patient demonstrates a large C-shaped meniscal fragment → in the lateral suprapatellar joint recess. At surgery, this was found to arise from a torn discoid lateral meniscus.*

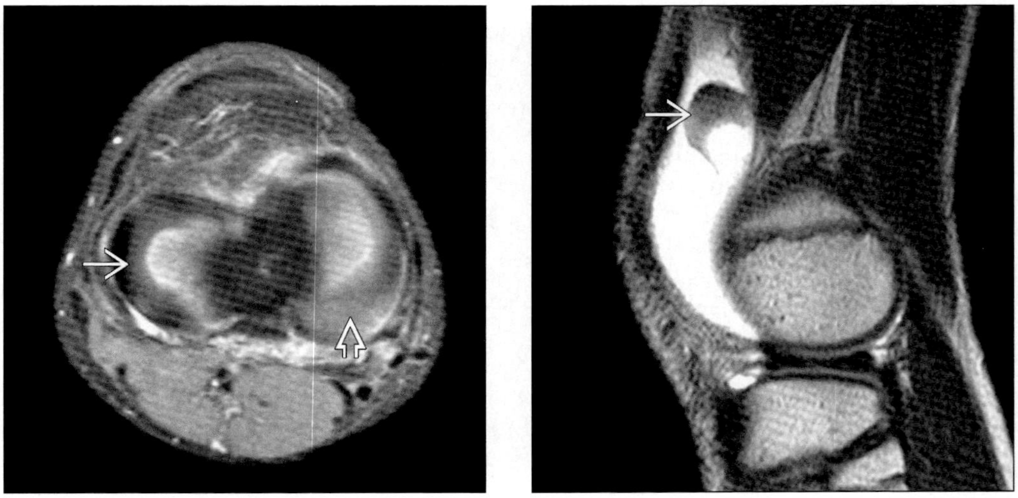

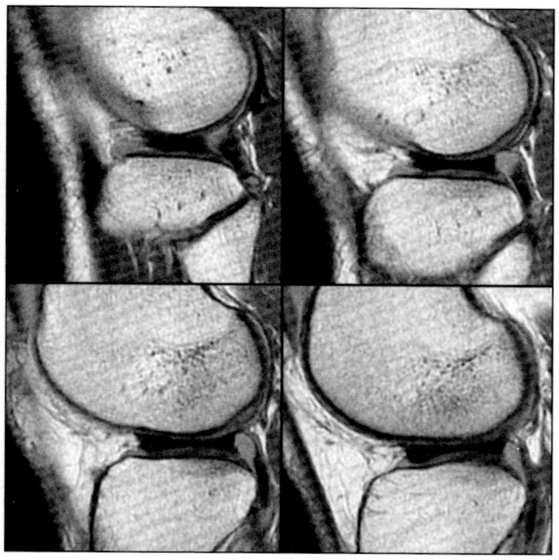

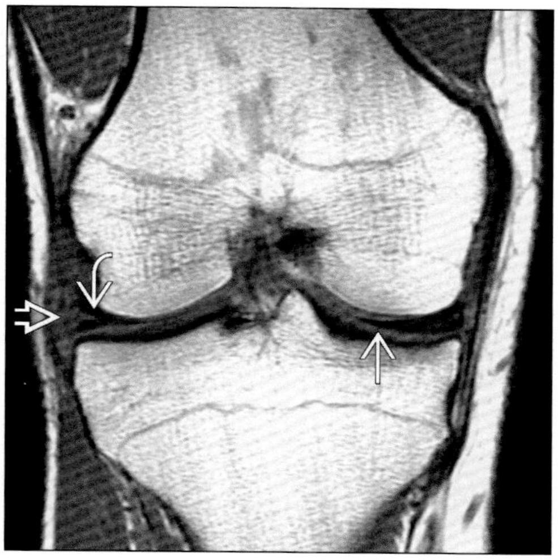

(Left) *Consecutive sagittal PDWI MRs show the "too many bow ties" sign of a discoid lateral meniscus. Using 3-4 mm thick sagittal images, there usually should not be more than 3 images of contiguous meniscal substance before the anterior and posterior horns are separate; in this patient, there were 5 (the most medial image was not included).* ***(Right)*** *Coronal T1WI MR shows a partial discoid medial meniscus →. There is also a horizontal tear of the lateral meniscus → with parameniscal cyst →.*

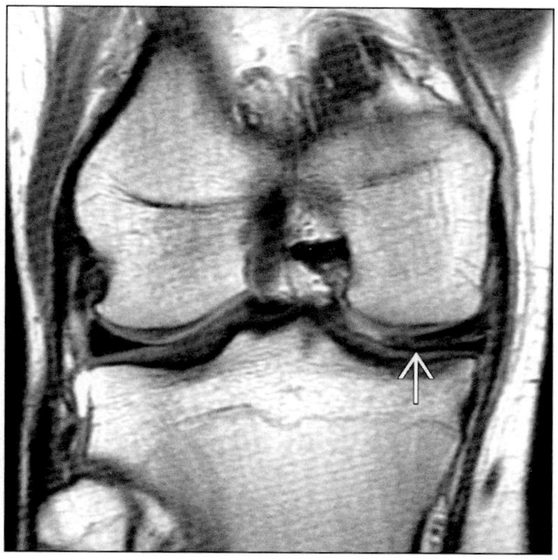

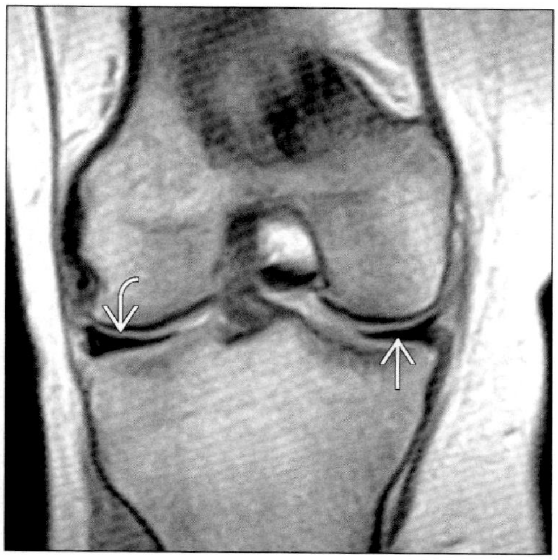

(Left) *Coronal PDWI MR shows a discoid medial meniscus with a horizontal cleavage tear →. Discoid menisci are more prone to tear than normal menisci due to increased axial and shear forces transmitted to the unresorbed central portions of the fibrocartilage.* ***(Right)*** *Coronal T1WI MR shows partial discoid medial → and lateral → menisci. The existence of both in the same knee is unusual. Lateral menisci are more commonly discoid than are medial menisci.*

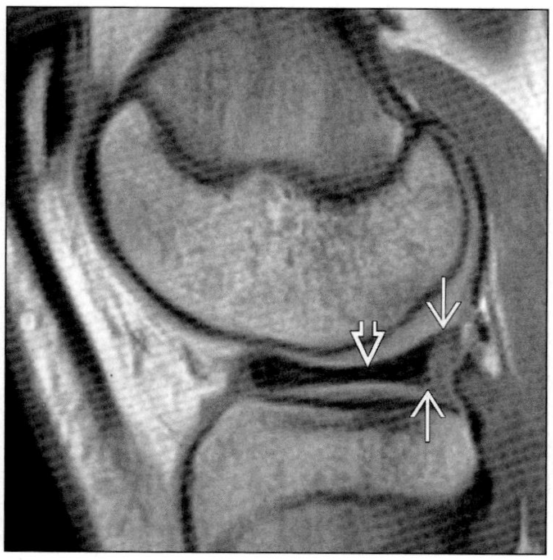

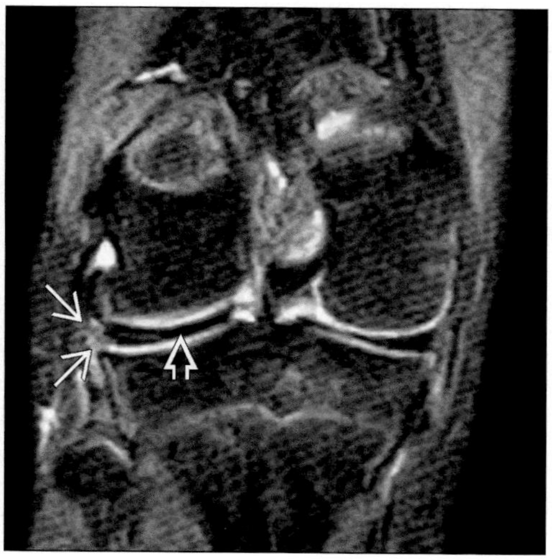

(Left) *Sagittal PDWI MR shows a discoid lateral meniscus → with complete absence of meniscotibial and meniscofemoral ligaments, including absent posterosuperior and anteroinferior popliteomeniscal fascicles →. This is a Wrisberg-type discoid lateral meniscus, the most unusual variant.* ***(Right)*** *Coronal T2WI FS MR from the same patient confirms the absence of the lateral coronary ligaments →, consistent with a Wrisberg-type discoid meniscus →.*

MENISCAL DEGENERATION

Key Facts

Terminology
- Intrasubstance degeneration, mucoid degeneration, mucinous degeneration, fraying
- Internal disruption of normal meniscal architecture ± mucinous degeneration without extension to articular surface
- Irregularity of meniscal free edge without linear tear

Imaging
- Increased signal on short TE MR sequences within meniscal substance with no extension to articular surface
- Most common in posterior horn of medial meniscus
- May be linear, globular, or complex in shape
- Fraying: Irregularity of free edge
- Direct MR arthrography
  - No extension of contrast into meniscus
- Indirect MR arthrography
  - Area of intrameniscal signal may enhance due to hypervascularity

Top Differential Diagnoses
- Meniscal tear
- Normal vascular pedicle of meniscus
- Postoperative change
- Meniscal contusion

Pathology
- Grading of meniscal signal not currently used; old literature described grades 1-3 depending on morphology and location of signal

Clinical Issues
- Usually asymptomatic
- Almost universal in elderly
- May progress to tear
  - Especially horizontal cleavage tear

(Left) Sagittal graphic shows central mucinous degeneration of the meniscus ➔ due to collagen breakdown and loss of fiber integrity. There is no linear tear. (Right) Sagittal PDWI FS MR from a 74-year-old man demonstrates diffuse hazy increased signal ➔ throughout the posterior horn of the medial meniscus. Careful scrutiny of sagittal and coronal images, including short TE sequences (T1WI or PDWI), is necessary to exclude a tear that extends to the articular surface of the meniscus.

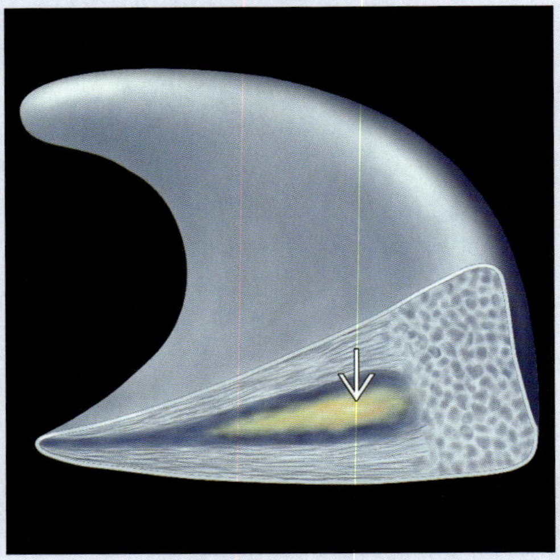

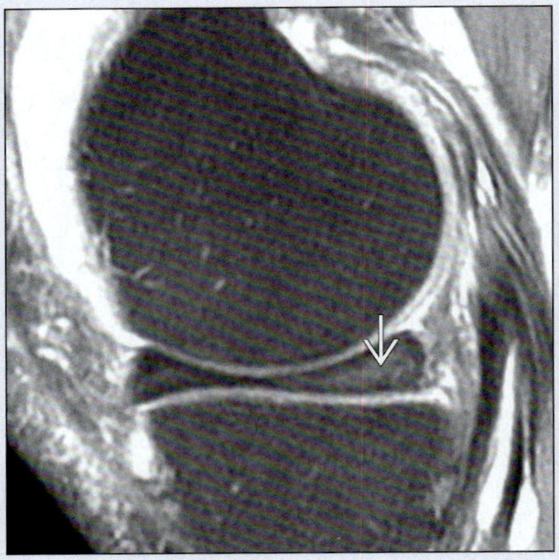

(Left) Coronal graphic shows fraying of the free edge of the meniscus ➔ consisting of fibrillation and multiple tiny irregularities in the normally smooth, tapered surface. This finding is often associated with osteoarthritis and may be a precursor to the formation of a discrete horizontal or radial tear. (Right) Coronal T1WI MR shows irregular blunting of the free edge of the medial meniscus ➔ consistent with fraying. Note the mild osteophytosis ➔ indicating early osteoarthritis.

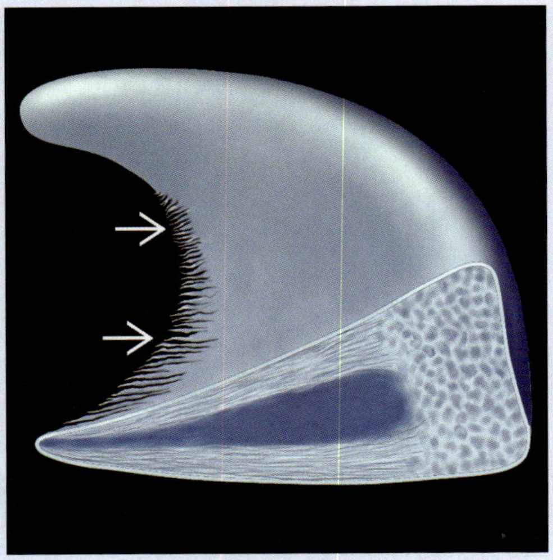

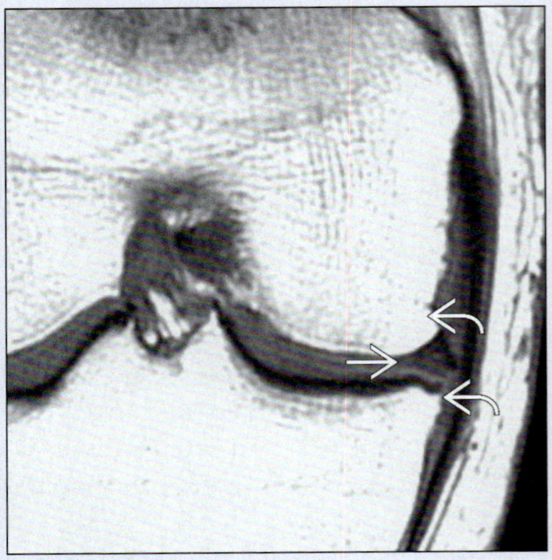

MENISCAL DEGENERATION

TERMINOLOGY

Synonyms

- Intrasubstance degeneration, mucoid degeneration, mucinous degeneration; fraying

Definitions

- Internal disruption of normal meniscal architecture ± mucinous degeneration without extension to articular surface
- Irregularity of meniscal free edge without linear tear

IMAGING

General Features

- Best diagnostic clue
  - Increased signal on short TE MR sequences within meniscal substance with no extension to articular surface
  - Fraying: Irregularity of free edge
- Location
  - Most common in posterior horn or body of medial meniscus
- Morphology
  - May be linear, globular, or complex in shape

MR Findings

- T1WI
  - Globular or linear signal located within substance of meniscus
  - No extension of this signal to articular surface of meniscus
  - Free edge irregularity
- T2WI FS
  - Intrameniscal signal usually decreases on T2WI
  - May disappear altogether with FS
  - Focal intrameniscal high signal if associated with intrameniscal cyst
  - Fraying: Free edge irregularity may be subtle
- Direct MR arthrography
  - No extension of contrast into meniscus
- Indirect MR arthrography
  - Area of intrameniscal signal may enhance due to hypervascularity

Imaging Recommendations

- Best imaging tool
  - MR
- Protocol advice
  - Best seen on short TE images (T1WI, PDWI) without FS

DIFFERENTIAL DIAGNOSIS

Meniscal Tear

- Abnormal signal clearly extends to articular surface of meniscus

Normal Vascular Pedicle of Meniscus

- Posterior horn of medial meniscus
- Limited to peripheral 1/3 of meniscus
- More prominent in children and adolescents

Postoperative Change

- Often associated with morphologic meniscal defect
- History of prior meniscal surgery

Meniscal Contusion

- Related to acute trauma
- Transient in nature

Magic Angle Artifact

- Occurs where meniscal fibers are at 55° to main magnetic field
- Disappears on long TE sequences

Meniscal Ossicle

- Follows fat signal on all sequences

PATHOLOGY

Staging, Grading, & Classification

- Grading of meniscal signal
  - Not currently used; old literature described grades 1-3 signal depending on morphology and location

CLINICAL ISSUES

Presentation

- Most common signs/symptoms
  - Usually asymptomatic

Demographics

- Age
  - Older adults

Natural History & Prognosis

- May progress to tear
  - Especially horizontal cleavage tear

Treatment

- Conservative

DIAGNOSTIC CHECKLIST

Consider

- Vascular pedicle in children and adolescents
- Contusion following acute meniscal trauma

Image Interpretation Pearls

- Confirm or exclude prior meniscal surgical history

SELECTED REFERENCES

1. Galea A et al: The accuracy of magnetic resonance imaging scanning and its influence on management decisions in knee surgery. Arthroscopy. 25(5):473-80, 2009
2. Low AK et al: Clinical significance of intrasubstance meniscal lesions on MRI. J Med Imaging Radiat Oncol. 52(3):227-30, 2008
3. Sproule JA et al: Altered signal intensity in the posterior horn of the medial meniscus: an MR finding of questionable significance. Arch Orthop Trauma Surg. 125(4):267-71, 2005
4. Raunest J et al: Magnetic resonance imaging (MRI) and arthroscopy in the detection of meniscal degenerations: correlation of arthroscopy and MRI with histology findings. Arthroscopy. 10(6):634-40, 1994

MENISCAL HORIZONTAL TEAR

Key Facts

Terminology
- Horizontal cleavage tear (HCT)
- Cleavage tear, horizontal cleavage tear; meniscal fraying

Imaging
- Horizontally oriented linear elevated intrameniscal signal clearly extending to articular surface
- Usually starts at free edge and propagates peripherally
- Ranges from fraying of free edge to full thickness bifurcation of meniscus
- Often more visible on short TE sequences
- May be associated with parameniscal cyst or occasionally intrameniscal cyst
- If full thickness, inferior fragment may be displaced
- Subluxation of meniscus
- Adjacent osteoarthritic changes

Top Differential Diagnoses
- Meniscal degeneration
- Normal meniscal vascular pedicle
- Postoperative change

Pathology
- Related to degeneration of meniscus
- Starts as free edge defect

Clinical Issues
- Not always symptomatic
- Often seen in older adults
- Debridement or saucerization to resect away torn portion of meniscus down to stable rim

Diagnostic Checklist
- Look for associated parameniscal cyst
- Describe meniscal zones involved in tear

*(**Left**) Sagittal graphic shows a horizontal cleavage tear of a meniscus. The tear propagates from the free edge of the meniscus → peripherally and may reach the capsular margin →. Horizontal tears often occur in menisci with underlying mucinous degeneration →. (**Right**) Sagittal PDWI FS MR shows a horizontal cleavage tear of the posterior horn of the medial meniscus. Linear signal within the substance of the meniscus clearly contacts the inferior articular surface →.*

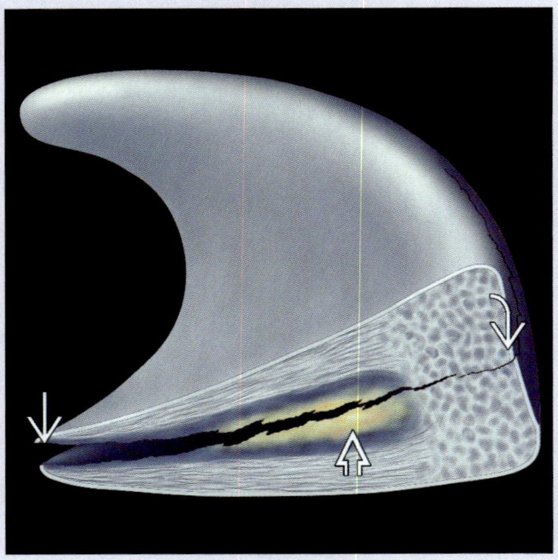

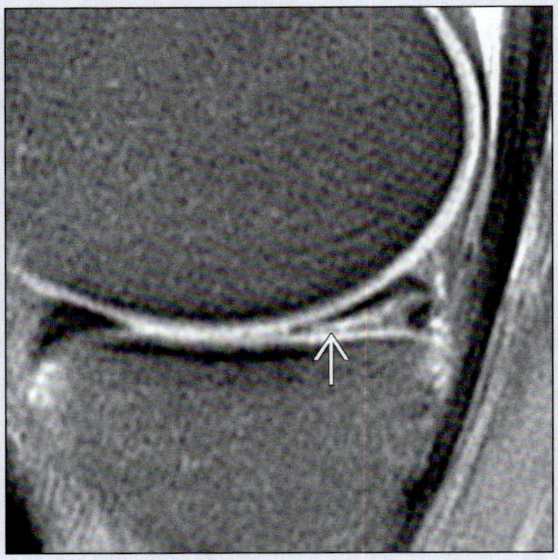

*(**Left**) Coronal T1WI MR from the same patient shows a broad band of increased signal → within the body and posterior horn of the medial meniscus, contacting the free edge and extending peripherally. (**Right**) Coronal T2WI FS MR from the same patient better shows the horizontal morphology of the tear →. A small fragment arising from the undersurface of the meniscus is displaced into the inferior parameniscal recess →.*

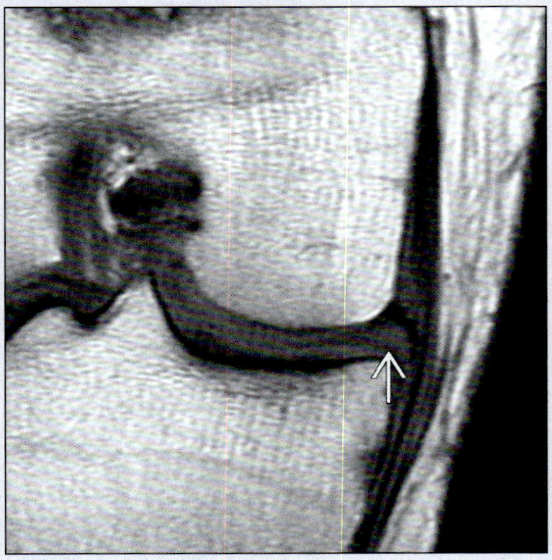

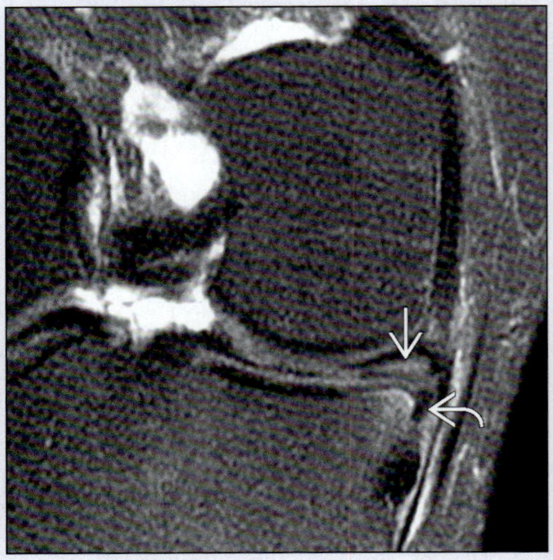

MENISCAL HORIZONTAL TEAR

TERMINOLOGY

Abbreviations

- Horizontal cleavage tear (HCT)

Synonyms

- Cleavage tear, horizontal cleavage tear; meniscal fraying

Definitions

- Meniscal tear occurring in the axial (horizontal) plane, transecting through the long axis of meniscus
- Meniscal red zone: Outer 1/3 of meniscus, vascular portion of meniscus
- Meniscal red-white junction: Middle 1/3 of meniscus between peripheral zone and free edge
- Meniscal white zone: Inner 1/3 of meniscus near free edge, nonvascular

IMAGING

General Features

- Best diagnostic clue
  - Horizontally oriented linear elevated intrameniscal signal clearly extending to articular surface on at least 1 image
- Location
  - Most common in body and posterior horn of medial or lateral meniscus
  - Usually starts at free edge and propagates peripherally
- Size
  - Ranges from fraying of free edge to full thickness bifurcation of meniscus

MR Findings

- Horizontal or horizontal oblique band of elevated signal within meniscal substance contacting an articular surface
- Often more visible on short TE sequences (PDWI, T1WI)
- Does not disappear on T2WI MR
- May be associated with parameniscal cyst or occasionally intrameniscal cyst
- Because of plane of tear, may not see tear extending to articular surface on peripheral sagittal images
- If full thickness, inferior fragment may be displaced
  - Usually peripherally
- Subluxation of meniscus
  - Loss of function of circumferential fibers → peripheral subluxation of meniscal substance out of joint compartment
- Adjacent osteoarthritic changes
  - Osteophytes, cartilage loss, marrow edema

Imaging Recommendations

- Protocol advice
  - Coronal short TE MR sequences most accurate

DIFFERENTIAL DIAGNOSIS

Meniscal Degeneration

- Does not contact articular surface

Normal Meniscal Vascular Pedicle

- Confined to red zone (peripheral 1/3)

Postoperative Meniscus Change

- Must solicit history of prior meniscal surgery

PATHOLOGY

General Features

- Etiology
  - Related to degeneration of meniscus
- Associated abnormalities
  - Often associated with osteoarthritis
  - May be associated with parameniscal cyst

CLINICAL ISSUES

Presentation

- Most common signs/symptoms
  - Not always symptomatic

Demographics

- Age
  - Often seen in older adults
- Epidemiology
  - Most common type of meniscal tear

Natural History & Prognosis

- Free edge fraying may progress to linear HCT
- Nondisplaced HCT may progress to displaced tear

Treatment

- Conservative
  - May be appropriate for free edge fraying or small nondisplaced asymptomatic tears
- Surgical
  - Debridement or saucerization to resect away torn portion of meniscus down to stable rim
  - May result in residual intrameniscal signal contacting postoperative meniscal surface

DIAGNOSTIC CHECKLIST

Consider

- HCT may be an incidental finding, particularly in elderly patient with osteoarthritis

Image Interpretation Pearls

- Look for associated parameniscal cyst

Reporting Tips

- Describe meniscal zones involved in tear

SELECTED REFERENCES

1. Englund M et al: The role of the meniscus in knee osteoarthritis: a cause or consequence? Radiol Clin North Am. 47(4):703-12, 2009
2. Galea A et al: The accuracy of magnetic resonance imaging scanning and its influence on management decisions in knee surgery. Arthroscopy. 25(5):473-80, 2009
3. De Smet AA et al: Clinical and MRI findings associated with false-positive knee MR diagnoses of medial meniscal tears. AJR Am J Roentgenol. 191(1):93-9, 2008

MENISCAL HORIZONTAL TEAR

(Left) *Sagittal T2WI MR shows an oblique linear focus of increased signal* ➔ *in the posterior horn of the medial meniscus and a 2nd questionable focus* ➔ *of signal abnormality more anteriorly.* ***(Right)*** *Sagittal PDWI FS MR at the same slice location in the same patient clearly shows the linear signal contacting the articular surface of the meniscus* ➔*, thus meeting the MR criteria for a horizontal tear. PDWI sequences are generally more accurate than T2WI for evaluation of meniscal tears.*

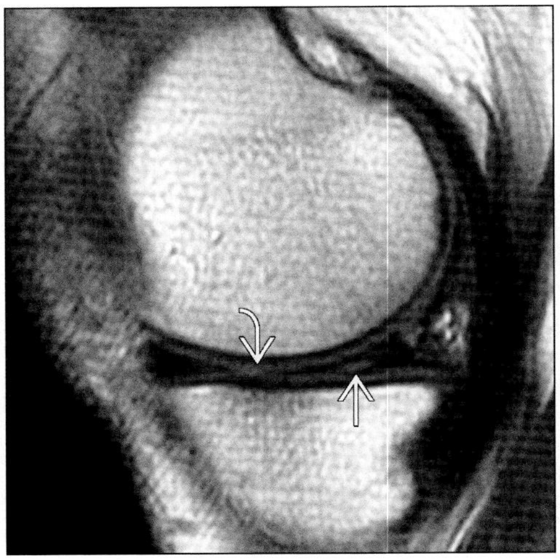

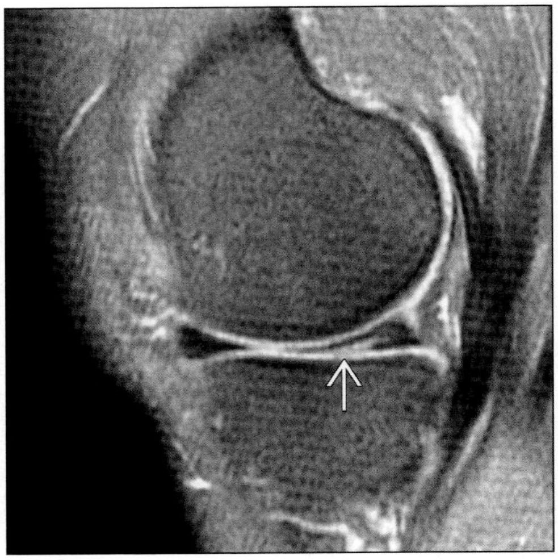

(Left) *Coronal T2WI FS MR from the same patient confirms the horizontal cleavage tear of the medial meniscus* ➔ *and demonstrates an adjacent multiseptated parameniscal cyst* ➔*. Horizontal tears are the most common tears to be associated with such cysts.* ***(Right)*** *Sagittal PDWI FS MR shows a horizontal cleavage tear of the lateral meniscus* ➔*. Because of the plane of section, sagittal images through the periphery of the meniscus may not show contact with the meniscal surface.*

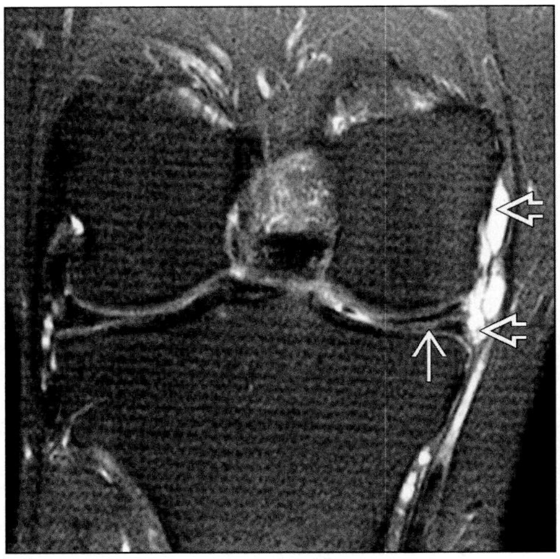

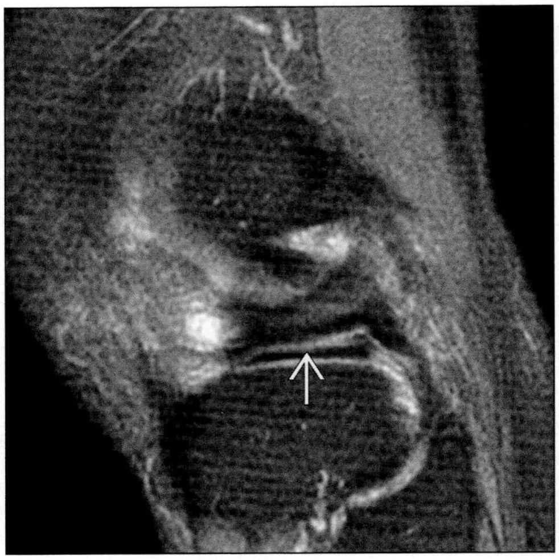

(Left) *Sagittal PDWI MR shows a subtle linear signal focus in the posterior horn of the lateral meniscus contacting the inferior articular surface* ➔*.* ***(Right)*** *Coronal PDWI MR from the same patient confirms the presence of a large horizontal cleavage tear* ➔*. Meniscal tears should always be evaluated in sagittal and coronal planes in order to best estimate their morphology and extent. Demonstration of contact of meniscal signal with an articular surface may only be appreciated on 1 or 2 images.*

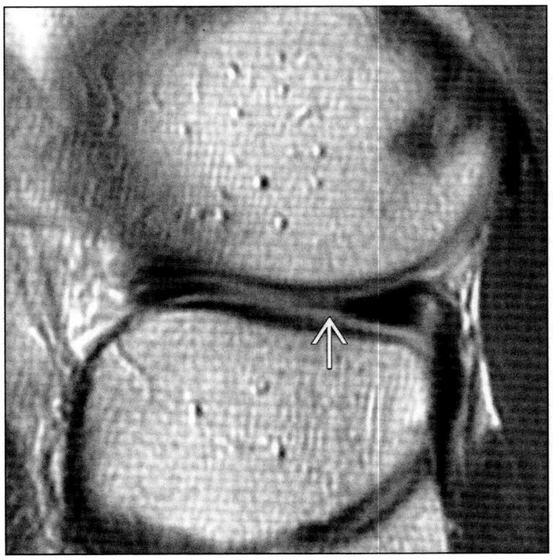

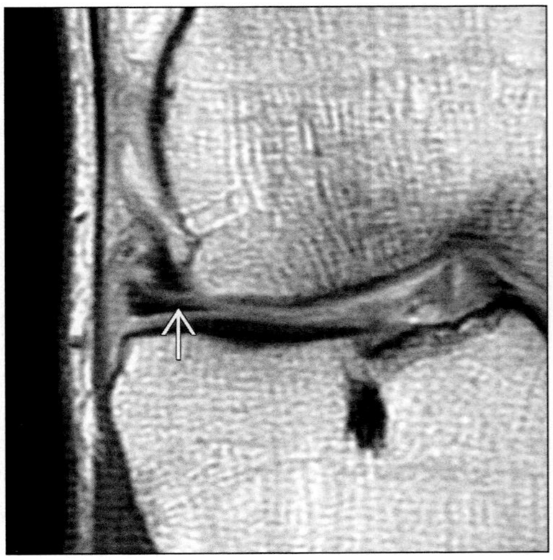

MENISCAL HORIZONTAL TEAR

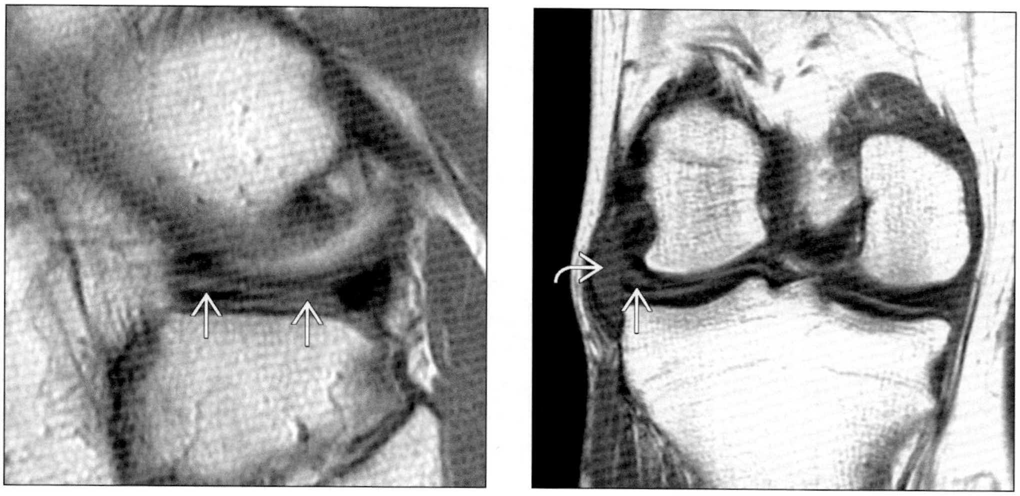

(Left) *Sagittal PDWI MR through the peripheral aspect of the lateral meniscus demonstrates a horizontal band of abnormal signal consistent with a horizontal cleavage tear. Volume averaging artifact can be mistaken for a horizontal cleavage tear but is rarely this long and well defined.* ***(Right)*** *Coronal T1WI MR from the same patient confirms the horizontal tear of the lateral meniscus and demonstrates a large parameniscal cyst with which the tear communicates.*

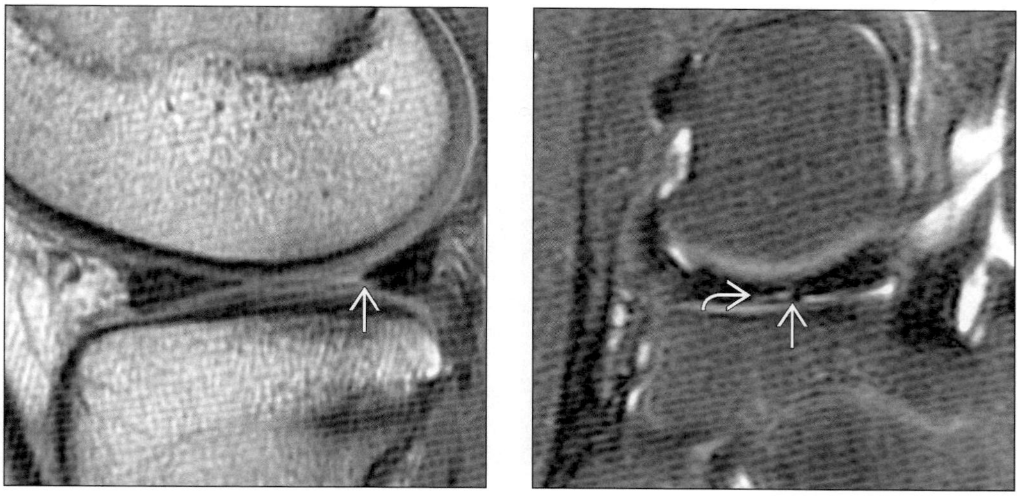

(Left) *Sagittal PDWI MR demonstrates a subtle linear defect in the substance of the posterior horn of the lateral meniscus, contacting the inferior articular surface. This horizontal tear is smaller and less obvious than a typical cleavage tear.* ***(Right)*** *Coronal T2WI FS MR from the same patient confirms the presence of a horizontal tear in the undersurface of the posterior horn of the lateral meniscus; a small displaced fragment is present inferiorly.*

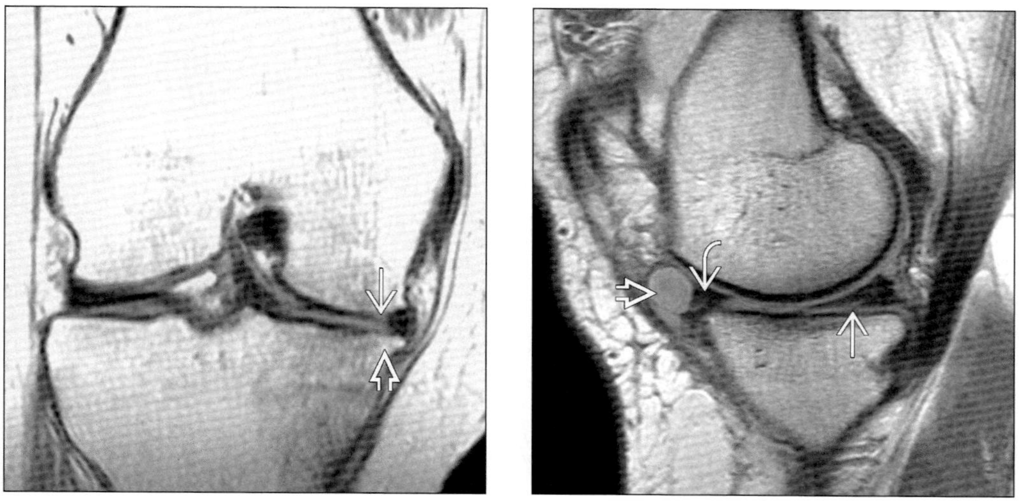

(Left) *Coronal T1WI MR shows irregular truncation of the free edge of the medial meniscal body. This finding is often due to a shallow horizontal cleavage tear, as it was in this case. Osteoarthritic changes are present, including an osteophyte.* ***(Right)*** *Sagittal PDWI MR shows a horizontal cleavage tear in the medial meniscus with components of the tear visible in the posterior and anterior horns. There is an associated parameniscal cyst at the anterior aspect of the tear.*

MENISCAL RADIAL TEAR

Key Facts

Terminology
- Radial tear (RT)
- Vertical meniscus tear oriented perpendicular to long axis of meniscus
- Includes posterior root tear

Imaging
- Focal truncation of free edge of meniscus
  - Seen best on coronal images
- Truncation of posterior root
  - Especially medial meniscus (MM)
- Best seen on coronal images as vertical defect through most posterior aspect of meniscus
- "Gap" sign on sagittal images
- "Ghost meniscus" sign: Sagittal image through plane of RT shows no meniscal tissue
- T2WI more specific for morphologic meniscal change
- If large RT, may be associated with marrow edema due to loss of meniscal cushioning

Top Differential Diagnoses
- Postoperative change
- Subluxation of meniscus
- Meniscal degeneration without tear

Pathology
- Lateral meniscus (LM) associated with anterior cruciate ligament (ACL) tear

Clinical Issues
- Posterior root tear of MM may → accelerated osteoarthritis

Diagnostic Checklist
- Consider posterior medial meniscal root tear in elderly patient with sudden onset of pain

*(**Left**) Axial graphic shows a partial thickness radial tear ➔ of the lateral meniscus at the junction of the body and posterior horn. Radial tears begin at the free edge of the meniscus and propagate a variable distance toward the peripheral (capsular) margin. (**Right**) Axial PDWI FS MR shows a full thickness radial tear ➔ of the junction of the anterior horn and body of the lateral meniscus. Thin section (3 mm) axial fluid-sensitive images often demonstrate radial tears well.*

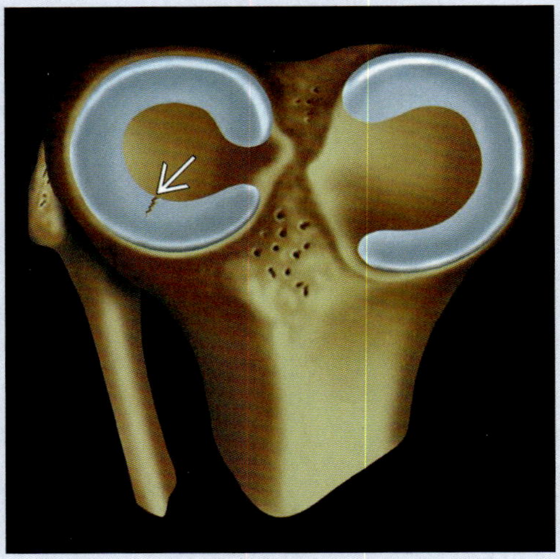

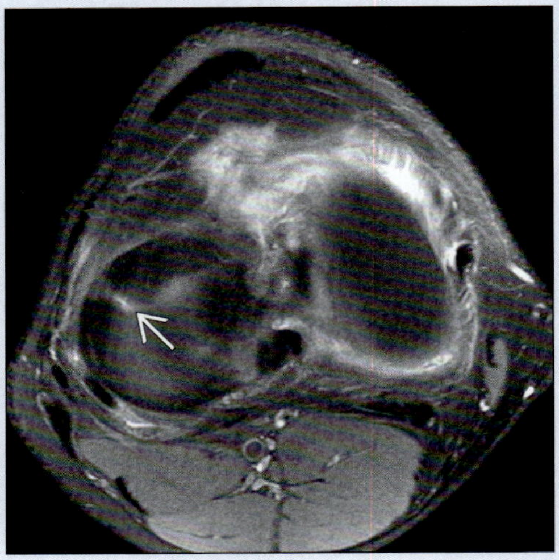

*(**Left**) Sagittal PDWI MR from the same patient shows a small rectangular "gap" ➔ in the free edge of the lateral meniscus, consistent with a radial tear. This finding may be subtle with small radial tears, but it is fairly specific when identified. (**Right**) Coronal T2WI FS MR from the same patient shows truncation of the free edge of the lateral meniscus ➔, indicating a radial tear. This finding may be seen postoperatively or with degenerative meniscal subluxation.*

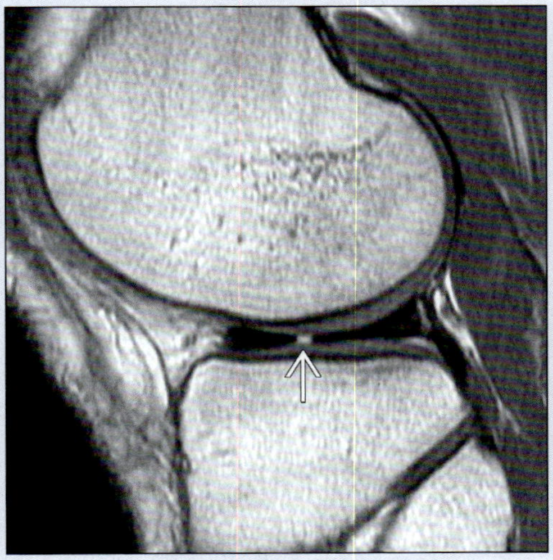

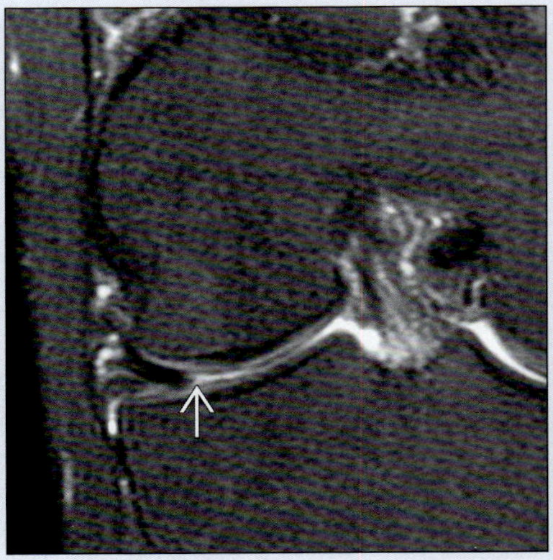

MENISCAL RADIAL TEAR

TERMINOLOGY

Abbreviations

- Radial tear (RT)

Definitions

- Vertical meniscus tear oriented perpendicular to long axis of meniscus
- Includes posterior root tear

IMAGING

General Features

- Best diagnostic clue
  - Abrupt truncation of free edge of meniscus
- Location
  - Free edge of meniscus
  - Posterior root of meniscus
- Size
  - May range from subtle truncation of meniscal free edge to large, full thickness meniscal discontinuity seen on several images
- Morphology
  - Linear or triangular defect oriented vertically through short axis of meniscus

CT Findings

- CT arthrography
  - Contrast entry into focal free edge meniscal defect

MR Findings

- Focal truncation of free edge of meniscus
  - Most typically in mid-body of meniscus
  - Seen best on coronal images
    - Short TE sequences most sensitive
    - T2WI more specific for morphologic meniscal change
  - "Gap" sign on sagittal images
    - Linear or rectangular defect in meniscal free edge
- Truncation of posterior root
  - Especially medial meniscus (MM)
  - Best seen on coronal images as vertical defect through most posterior aspect of meniscus
  - "Ghost meniscus" sign: Sagittal image through plane of radial tear shows no meniscal tissue
- Vertical tear seen on axial image
  - Linear or V-shaped radially oriented defect in free edge with variable peripheral extension
  - If full thickness (through peripheral margin) → may see displacement of fragments
- If large radial tear, may be associated with marrow edema due to loss of meniscal cushioning

Imaging Recommendations

- Best imaging tool
  - MR
- Protocol advice
  - Small tears of meniscal body seen best on coronal short TE sequences
  - Posterior root tears seen best on coronal T2WI ± FS
  - Both can be confirmed on thin section axial images

DIFFERENTIAL DIAGNOSIS

Postoperative Change

- Must obtain history of previous meniscal surgery

Subluxation of Meniscus

- May result in apparent blunting of free edge

Meniscal Degeneration without Tear

- Free edge irregularity common

PATHOLOGY

General Features

- Etiology
  - RT of meniscal body often due to acute trauma with axial loading
  - Posterior root tears
    - Lateral meniscus (LM) associated with anterior cruciate ligament (ACL) tear
    - Medial meniscus seen as degenerative finding
- Associated abnormalities
  - Often associated with osteoarthritis

CLINICAL ISSUES

Presentation

- Most common signs/symptoms
  - Joint line pain and tenderness
  - Small free edge tears may be asymptomatic

Natural History & Prognosis

- May enlarge and extend to peripheral zone
- May progress to flap tear ± displacement
- Posterior root tear of MM may → accelerated osteoarthritis

Treatment

- Conservative: If small RT and asymptomatic
- Surgical: Saucerization (wide debridement of free edge tear to stable meniscal tissue)

DIAGNOSTIC CHECKLIST

Consider

- Posterior medial meniscal root tear in elderly patient with sudden onset of pain

Image Interpretation Pearls

- Blunting of free edge of meniscus best confirmed using T2WI

SELECTED REFERENCES

1. Jones C et al: Repair of the posterior root of the medial meniscus. Knee. 17(1):77-80, 2010
2. De Smet AA et al: MR diagnosis of posterior root tears of the lateral meniscus using arthroscopy as the reference standard. AJR Am J Roentgenol. 192(2):480-6, 2009
3. Lee YG et al: Magnetic resonance imaging findings of surgically proven medial meniscus root tear: tear configuration and associated knee abnormalities. J Comput Assist Tomogr. 32(3):452-7, 2008
4. Harper KW et al: Radial meniscal tears: significance, incidence, and MR appearance. AJR Am J Roentgenol. 185(6):1429-34, 2005

MENISCAL RADIAL TEAR

(Left) Sagittal PDWI FS MR shows a tiny linear focus of abnormal signal in the free edge of the lateral meniscus, consistent with a small partial thickness radial tear. Adjacent images were normal. *(Right)* Axial PDWI FS MR from the same patient shows the tiny partial thickness radial tear of the lateral meniscus. Small tears may enlarge over time and are saucerized at surgery to a stable rim of meniscal tissue to prevent this.

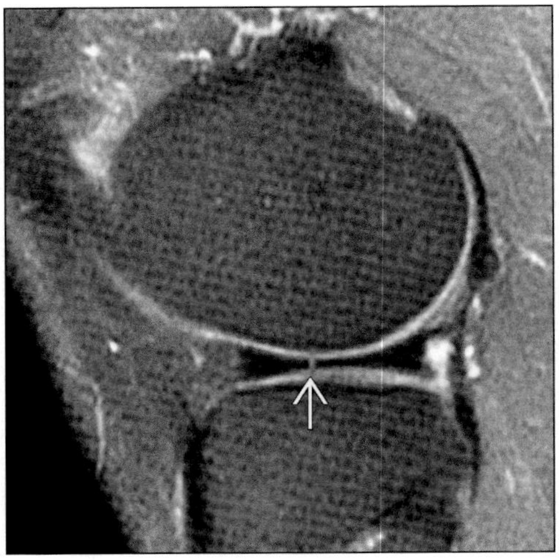

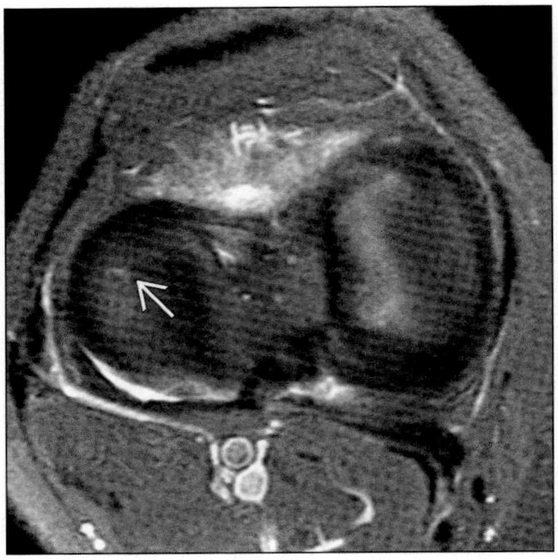

(Left) Sagittal PDWI FS MR shows a square focus of fluid signal causing an abrupt cut-off of the lateral meniscal contour, consistent with a radial tear with displacement. Marrow contusion in the posterolateral tibial plateau is due to concurrent tear of the anterior cruciate ligament. *(Right)* Sagittal PDWI FS MR more medially in the same patient shows the displaced radial tear of the lateral meniscus.

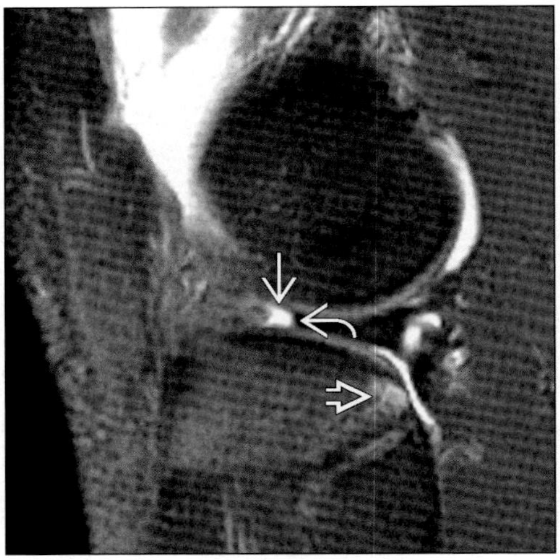

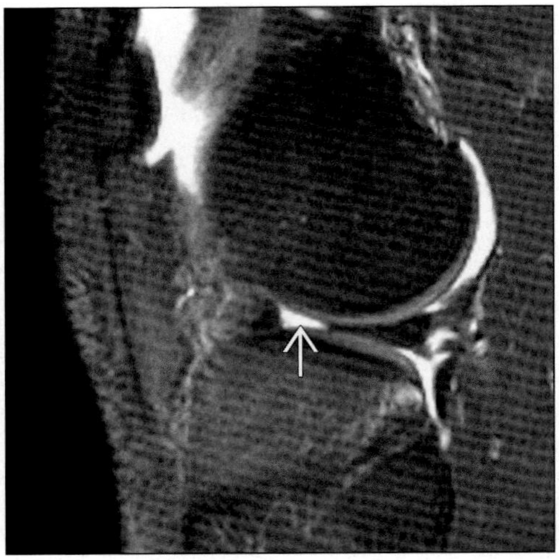

(Left) Axial PDWI FS MR from the same patient shows the displaced radial tear of the lateral meniscus. Full thickness radial tears have a propensity to displace, resulting in loss of longitudinal meniscal fiber integrity. *(Right)* Coronal T2WI FS MR shows a displaced radial tear of the posterior root of the lateral meniscus. An impaction contusion and focal chondral fracture of the underlying lateral tibial plateau suggest an axial load force as the cause of the meniscal tear.

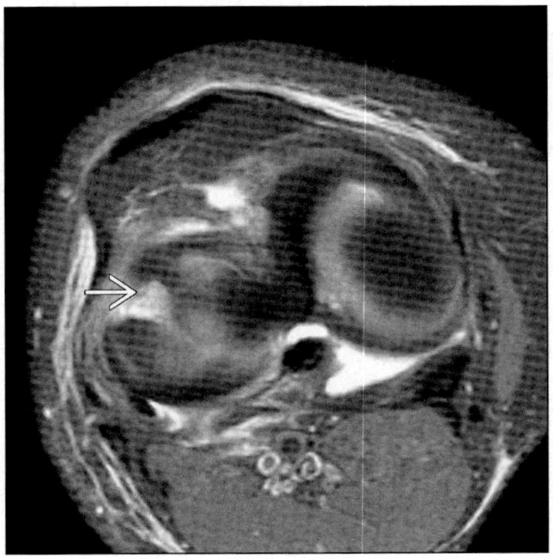

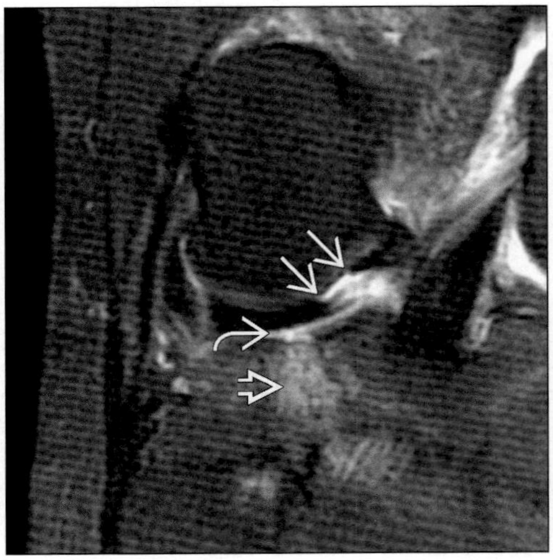

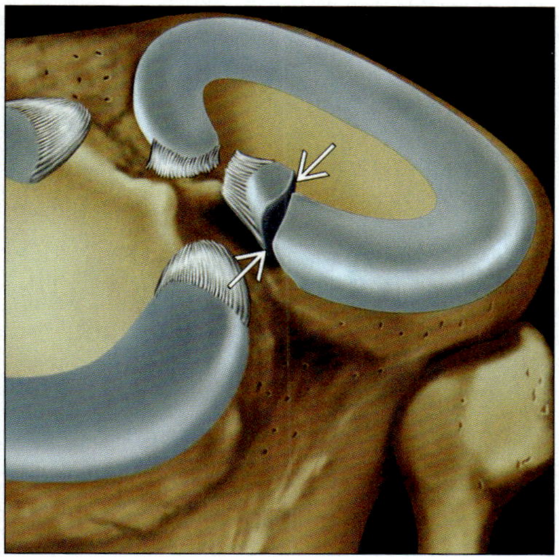

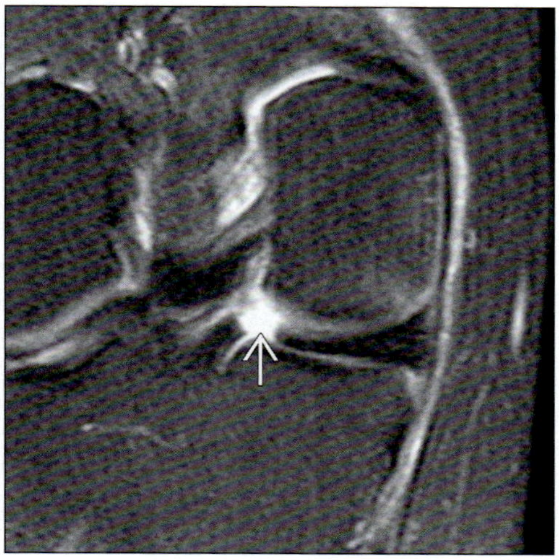

(Left) *Oblique axial graphic shows a radial tear of the posterior root of the medial meniscus ➔. This tear is commonly seen in the setting of osteoarthritis and probably occurs due to increased axial loading of the meniscus.* ***(Right)*** *Coronal T2WI FS MR shows a displaced radial tear ➔ in the posterior root of the medial meniscus. This is a relatively common finding in middle-aged or elderly patients with osteoarthritis and sudden onset of medial knee pain.*

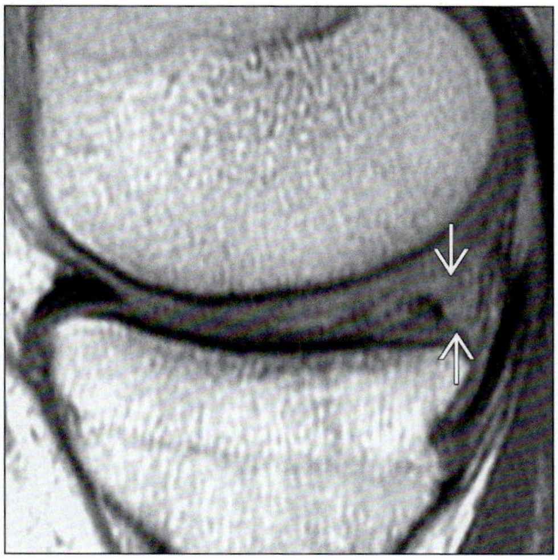

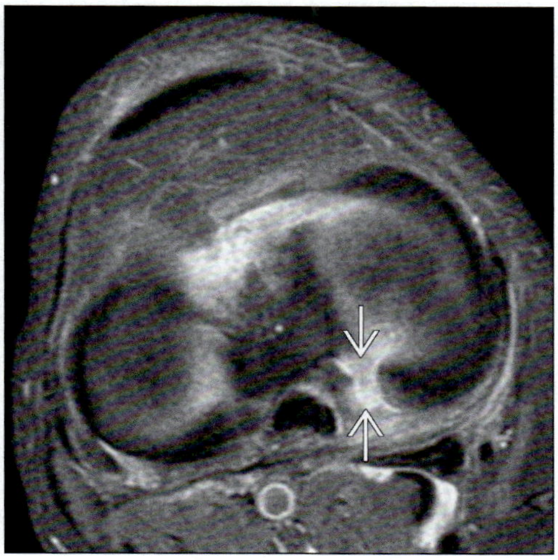

(Left) *Sagittal PDWI MR from the same patient obtained through the plane of the posterior horn radial tear shows the "ghost meniscus" sign of absent meniscal tissue ➔. Slices immediately medial and lateral to this location showed normal posterior horn tissue.* ***(Right)*** *Axial PDWI FS MR from the same patient shows the displaced posterior root tear of the medial meniscus ➔. These tears are not repairable and may lead to accelerated osteoarthritis.*

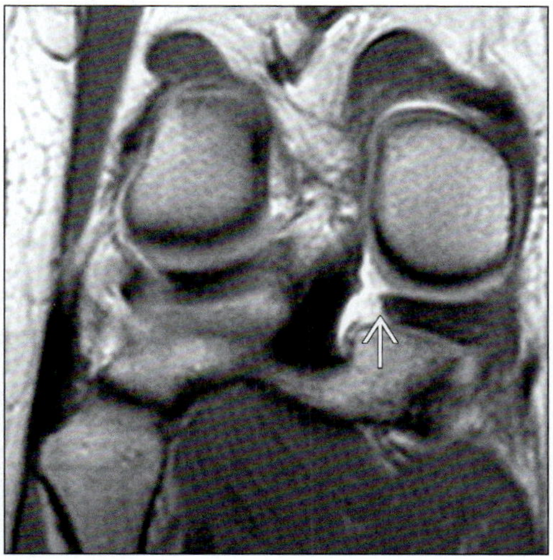

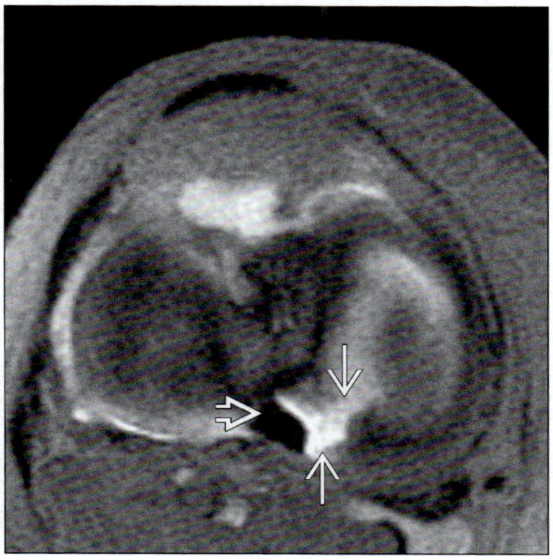

(Left) *Coronal MR arthrogram T1WI shows abrupt truncation of the posterior root of the medial meniscus ➔, consistent with a radial tear. The posterior root fragment is not seen on this image.* ***(Right)*** *Axial MR arthrogram T1WI FS from the same patient shows the displaced radial tear of the posterior root ➔. The posterior cruciate ligament ➔ is seen more laterally; the posterior root fragment is not identified and is probably loose in the joint.*

MENISCAL VERTICAL LONGITUDINAL TEAR

Key Facts

Terminology

- Vertical longitudinal tear (VLT)
- Longitudinal tear

Imaging

- Typically in outer 1/3 of meniscus (red zone)
- Uncommon in white zone (near free edge)
- Isolated involvement of inferior > superior articular surface
- Correlate short TE coronal and sagittal sequences to identify full extent of tear

Top Differential Diagnoses

- Popliteal hiatus
- Meniscocapsular separation
- Attachment of meniscofemoral ligament
- Attachment of transverse intermeniscal ligament
- Postoperative meniscus change

Pathology

- Often related to a discrete traumatic event
- Usually originates in posterior horn, advances anteriorly
- High incidence with ACL tear

Clinical Issues

- Adolescents and young, active adults
- Small, nondisplaced red zone tears often heal spontaneously
- Mesial fragment may displace → bucket-handle tear
- Large meniscal resection → accelerated osteoarthritis
- Meniscal repair: Red zone tears that are stable to arthroscopic probing
- Meniscal resection: Unstable red zone tears, red-white or white zone tears

__(Left)__ Sagittal graphic shows a vertical longitudinal tear ➡ in the red-white junction of a meniscus. Vertical tears usually arise in the posterior horn and may propagate a variable distance anteriorly. They are commonly seen in the setting of anterior cruciate ligament tears and may occur in either the medial or lateral meniscus. __(Right)__ Sagittal PDWI MR shows a vertical longitudinal tear in the red-white zone of the medial meniscus ➡. Vertical tears are more common in the red zone of the meniscus.

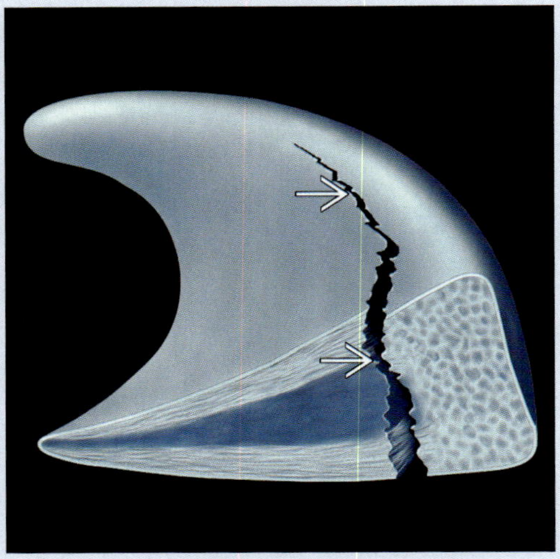

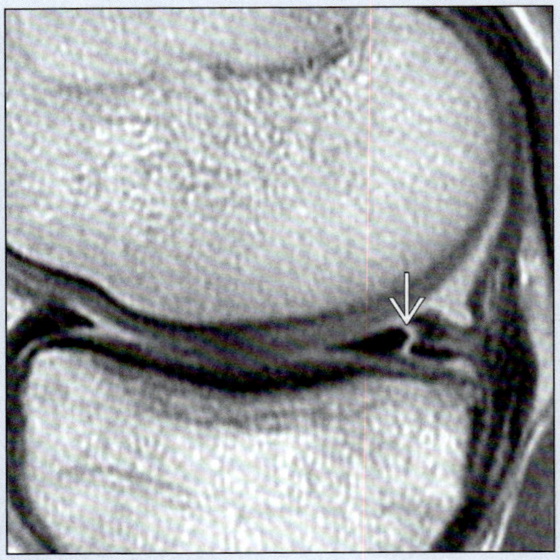

__(Left)__ Axial PDWI FS MR from the same patient shows the vertical longitudinal tear in the posterior aspect of the medial meniscus ➡. When vertical tears extend along a larger portion of the meniscus, they may displace and become bucket-handle tears. A radial tear is present in the posterior root of the lateral meniscus ➡. __(Right)__ Sagittal PDWI MR shows a red zone vertical longitudinal tear in the posterior horn of the lateral meniscus ➡.

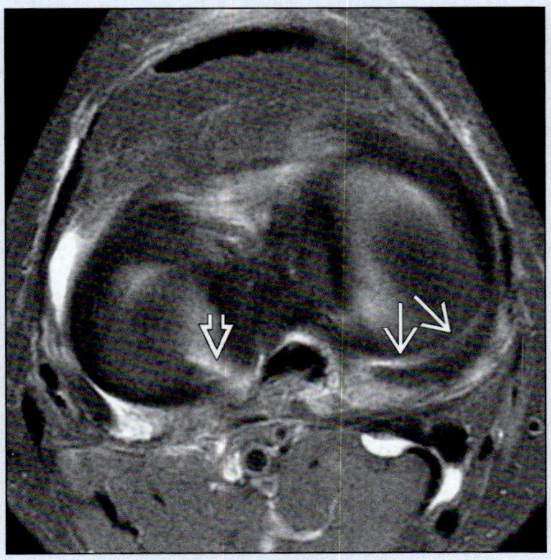

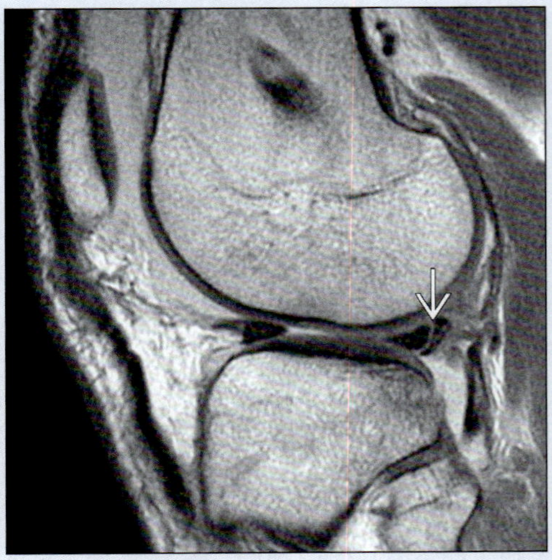

MENISCAL VERTICAL LONGITUDINAL TEAR

TERMINOLOGY

Abbreviations

- Vertical longitudinal tear (VLT)

Synonyms

- Longitudinal tear

Definitions

- Tear along vertical long axis of meniscus
- Meniscal red zone: Peripheral 1/3 of meniscus; vascular zone
- Meniscal red-white junction: Middle 1/3 of meniscus
- Meniscal white zone: Central 1/3 of meniscus, near free edge

IMAGING

General Features

- Best diagnostic clue
  - Linear elevated intrameniscal signal
    - Contacts 1 or both articular surfaces of meniscus
    - Extends through variable portion of meniscus from anterior to posterior
- Location
  - Typically in outer 1/3 of meniscus (red zone)
  - Usually starts in posterior horn
    - ± propagation into body and anterior horn
- Size
  - May range from 1-2 mm in length to involvement of entire meniscus
  - May range from 1-2 mm in depth to full height of meniscus
- Morphology
  - Vertically oriented tear of meniscal substance extending to superior &/or inferior articular surface
  - Tears extending through long segment of meniscus may displace
    - Bucket-handle tear with displacement of mesial fragment into intercondylar notch
      - Medial > lateral meniscus
    - Flipped meniscus with posterior horn displaced anteriorly
      - Lateral > medial meniscus

CT Findings

- CT arthrography
  - Contrast extends into vertical linear defect in superior &/or inferior articular surface
    - Best seen on coronal and sagittal reformatted images

MR Findings

- Linear increased intrameniscal signal oriented along vertical axis of meniscus contacting superior &/or inferior articular surface
- Best seen on short TE sequences (PDWI)
  - Posterior horn tears best seen on sagittal images
  - Body tears best seen on coronal images
  - Thin section axial images may also show VLT
- Often stays bright on T2WI, but less sensitive than PDWI
- Most commonly located in peripheral 1/3 (red zone) of meniscus
- Uncommon in white zone (near free edge)
- Usually starts in posterior horn
  - Variable propagation anteriorly into body and anterior horn
- Isolated involvement of inferior > superior articular surface
- Involvement of both superior and inferior articular surfaces may be focal

Imaging Recommendations

- Best imaging tool
  - MR
- Protocol advice
  - Correlate short TE coronal and sagittal sequences to identify full extent of tear

DIFFERENTIAL DIAGNOSIS

Popliteal Hiatus

- Posterior lateral meniscus

Meniscocapsular Separation

- Injury occurs at peripheral border of meniscus
- Usually in posterior horn of medial meniscus

Attachment of Meniscofemoral Ligament

- Posterior lateral meniscus
- More mesial than popliteal hiatus

Attachment of Transverse Intermeniscal Ligament

- Especially at anterior horn of lateral meniscus

Postoperative Meniscus Change

- Must solicit history of prior meniscal surgery
- May demonstrate elevated signal extending to meniscal surface on short TE sequences
- Should not stay bright on T2WI

PATHOLOGY

General Features

- Etiology
  - Often related to discrete traumatic event
  - Usually originates in posterior horn, advances anteriorly
  - Due to increased axial load combined with shear force
  - May be caused by pivot-shift mechanism of injury
    - Also causes anterior cruciate ligament (ACL) tear
- Associated abnormalities
  - High incidence with ACL tear
    - Medial and lateral menisci

CLINICAL ISSUES

Presentation

- Most common signs/symptoms
  - Pain and tenderness along joint line
- Other signs/symptoms
  - Locking or giving way if displaced fragment
- Clinical profile
  - Often seen in athletes after acute injury

MENISCAL VERTICAL LONGITUDINAL TEAR

Demographics

- Age
  - Adolescents and young, active adults
- Gender
  - M > F
- Epidemiology
  - 2nd most common tear pattern
  - Posterior horn VLT often seen in setting of
    - Acute ACL tear
    - Chronically ACL deficient knee

Natural History & Prognosis

- Small, nondisplaced red zone tears often heal spontaneously
- May propagate longitudinally (posterior → anterior)
- Mesial fragment may displace → bucket-handle tear
  - Usually requires extension from inferior to superior articular surface (i.e., full thickness tear)
- Large meniscal resection → accelerated osteoarthritis

Treatment

- Conservative: Small, nondisplaced red zone tears can heal spontaneously
- Surgical
  - Meniscal repair: Red zone tears that are stable to arthroscopic probing
  - Meniscal resection: Unstable red zone tears, red-white or white zone tears

DIAGNOSTIC CHECKLIST

Consider

- VLT in setting of acute ACL tear or ACL-deficient knee

Image Interpretation Pearls

- Careful scrutiny of posterior horns near meniscocapsular junction

Reporting Tips

- Describe location, depth, and length of tear
- Measure any displacement of meniscal fragments

SELECTED REFERENCES

1. Alatakis S et al: MR imaging of meniscal and cartilage injuries of the knee. Magn Reson Imaging Clin N Am. 17(4):741-56, vii, 2009
2. De Smet AA et al: MR diagnosis of posterior root tears of the lateral meniscus using arthroscopy as the reference standard. AJR Am J Roentgenol. 192(2):480-6, 2009
3. DeHaan A et al: Evaluation of success of a meniscus repair device for vertical unstable medial meniscus tears in ACL-reconstructed knees. Orthopedics. 32(4), 2009
4. Fetzer GB et al: Potential market for new meniscus repair strategies: evaluation of the MOON cohort. J Knee Surg. 22(3):180-6, 2009
5. Grossman JW et al: Comparison of the accuracy rates of 3-T and 1.5-T MRI of the knee in the diagnosis of meniscal tear. AJR Am J Roentgenol. 193(2):509-14, 2009
6. Laundre BJ et al: MRI accuracy for tears of the posterior horn of the lateral meniscus in patients with acute anterior cruciate ligament injury and the clinical relevance of missed tears. AJR Am J Roentgenol. 193(2):515-23, 2009
7. Stärke C et al: Meniscal repair. Arthroscopy. 25(9):1033-44, 2009
8. De Smet AA et al: Clinical and MRI findings associated with false-positive knee MR diagnoses of medial meniscal tears. AJR Am J Roentgenol. 191(1):93-9, 2008
9. De Smet AA et al: Clinical, MRI, and arthroscopic findings associated with failure to diagnose a lateral meniscal tear on knee MRI. AJR Am J Roentgenol. 190(1):22-6, 2008
10. Mustonen AO et al: MRI of acute meniscal injury associated with tibial plateau fractures: prevalence, type, and location. AJR Am J Roentgenol. 191(4):1002-9, 2008
11. Pujol N et al: Meniscal healing after meniscal repair: a CT arthrography assessment. Am J Sports Med. 36(8):1489-95, 2008
12. Vinson EN et al: Association of peripheral vertical meniscal tears with anterior cruciate ligament tears. Skeletal Radiol. 37(7):645-51, 2008
13. de Abreu MR et al: Meniscofemoral ligaments: patterns of tears and pseudotears of the menisci using cadaveric and clinical material. Skeletal Radiol. 36(8):729-35, 2007
14. Boxheimer L et al: Characteristics of displaceable and nondisplaceable meniscal tears at kinematic MR imaging of the knee. Radiology. 238(1):221-31, 2006
15. Majewski M et al: Midterm and long-term results after arthroscopic suture repair of isolated, longitudinal, vertical meniscal tears in stable knees. Am J Sports Med. 34(7):1072-6, 2006
16. Terzidis IP et al: Meniscal tear characteristics in young athletes with a stable knee: arthroscopic evaluation. Am J Sports Med. 34(7):1170-5, 2006
17. Tandogan RN et al: Analysis of meniscal and chondral lesions accompanying anterior cruciate ligament tears: relationship with age, time from injury, and level of sport. Knee Surg Sports Traumatol Arthrosc. 12(4):262-70, 2004
18. Jee WH et al: Meniscal tear configurations: categorization with MR imaging. AJR Am J Roentgenol. 180(1):93-7, 2003
19. Papachristou G et al: Isolated meniscal repair in the avascular area. Acta Orthop Belg. 69(4):341-5, 2003
20. Binfield PM et al: Patterns of meniscal tears associated with anterior cruciate ligament lesions in athletes. Injury. 24(8):557-61, 1993
21. Roeddecker K et al: Arthroscopic repair of traumatic longitudinal meniscal tears. A 3 to 5-year follow-up. Surg Endosc. 7(1):46-51, 1993
22. Paletta GA Jr et al: Patterns of meniscal injury associated with acute anterior cruciate ligament injury in skiers. Am J Sports Med. 20(5):542-7, 1992
23. Hanks GA et al: Repair of peripheral meniscal tears: open versus arthroscopic technique. Arthroscopy. 7(1):72-7, 1991

MENISCAL VERTICAL LONGITUDINAL TEAR

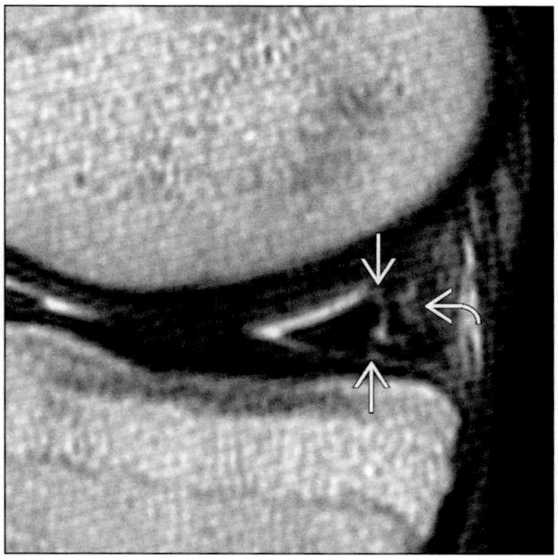

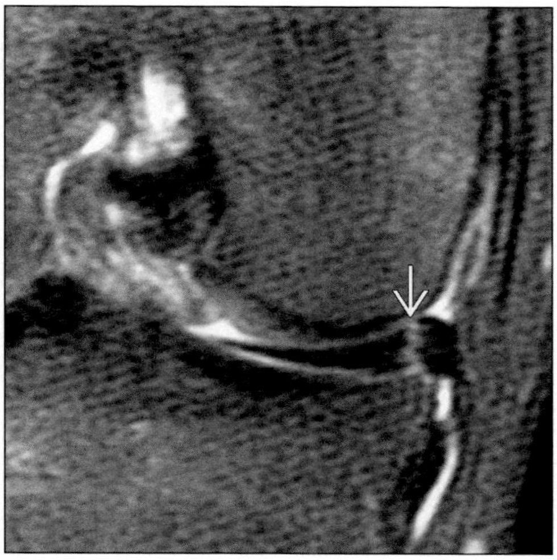

(Left) *Sagittal T2WI MR shows a vertical longitudinal tear → in the posterior horn of the medial meniscus. There is irregularity of the peripheral fragment →, indicating some complexity in the tear.* ***(Right)*** *Coronal T2WI FS MR from the same patient shows the tear → propagating through the red zone of the posterior horn. Since longitudinal tears are more commonly located in the posterior horn of the medial meniscus, they are less commonly seen on coronal images than on sagittal images.*

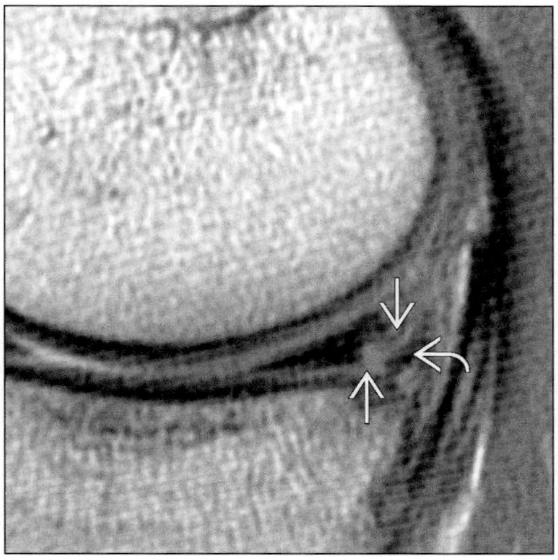

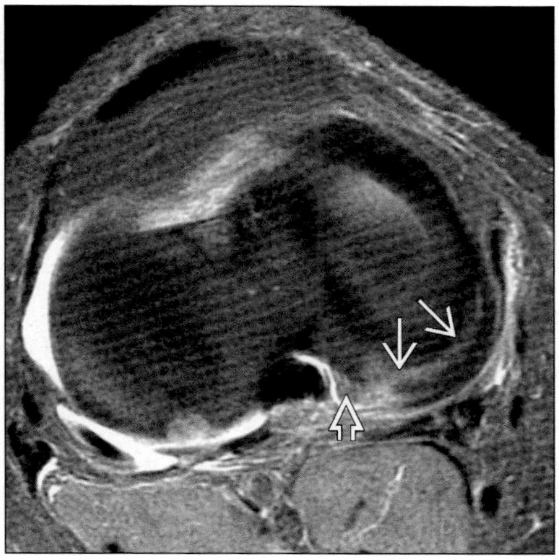

(Left) *Sagittal PDWI MR shows a vertical longitudinal tear of the posterior medial meniscus →. This tear is slightly displaced. The small peripheral meniscal fragment → distinguishes this tear from a meniscocapsular separation.* ***(Right)*** *Axial PDWI FS MR from the same patient shows the longitudinal nature of the tear → extending through the posterior horn of the medial meniscus. The tear extends into the posterior root of the meniscus →.*

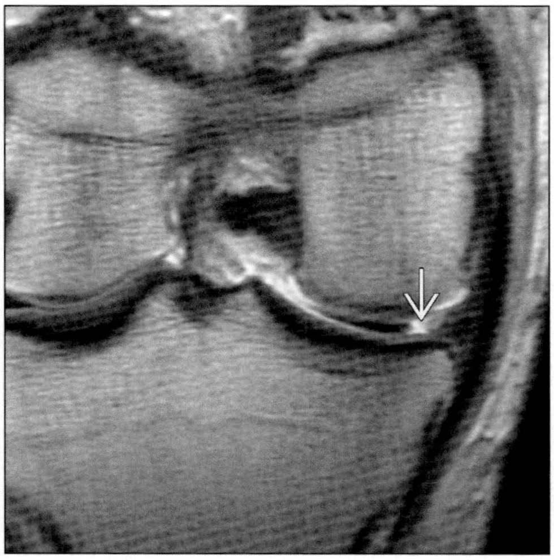

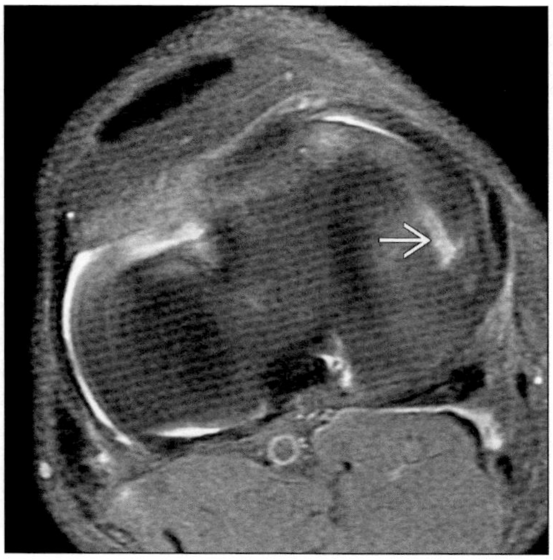

(Left) *Coronal MR arthrogram T1WI shows a slightly displaced vertical longitudinal tear of the medial meniscus →. This tear is in the red-white zone of the meniscus.* ***(Right)*** *Axial PDWI FS MR from the same patient shows the longitudinal tear → in the mid-body of the medial meniscus. This tear is unusual in that it is located in the red-white zone instead of the more typical location of the peripheral (red) zone. Tears are more likely to heal spontaneously in the vascular red zone.*

MENISCAL VERTICAL LONGITUDINAL TEAR

*(**Left**) Sagittal PDWI FS MR shows a globular vertical tear in the red zone of the posterior medial meniscus ➔. The precise morphology of the tear is difficult to discern from this image. (**Right**) Coronal T2WI FS MR from the same patient shows that the peripheral signal is a vertical longitudinal tear in the posterior horn of the meniscus ➔. This is not a meniscocapsular separation, as a thin rim of red zone meniscal tissue ↪ remains attached to the capsule.*

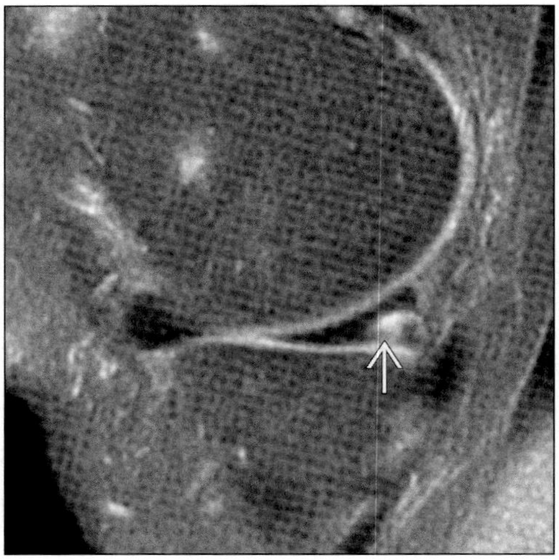

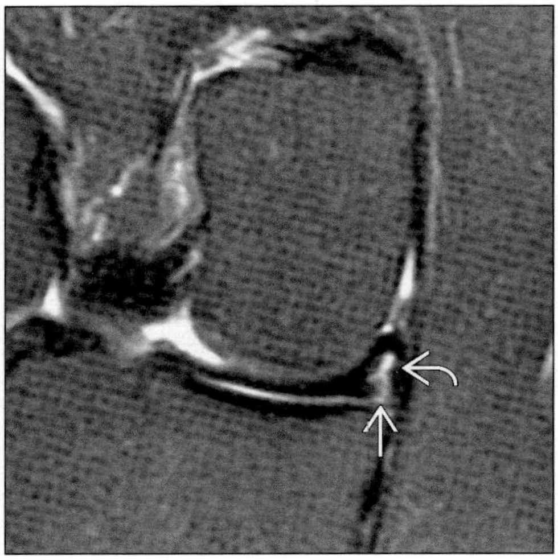

*(**Left**) Axial PDWI FS MR from the same patient shows the slightly displaced vertical longitudinal tear ➔ extending from the posterior horn to the mid-body of the medial meniscus. Axial images can be helpful in determining the extent of longitudinal tears. (**Right**) Sagittal PDWI MR shows an acute disruption of the anterior cruciate ligament (ACL) ⇨. Peripheral longitudinal meniscal tears are often seen with ACL tears, and particular attention should be paid to the posterior horns in this regard.*

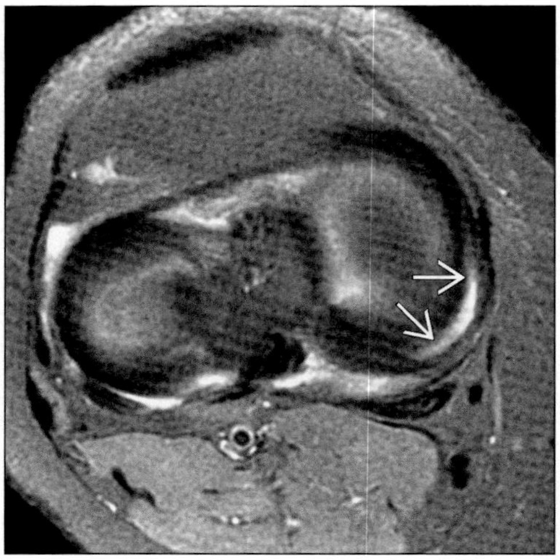

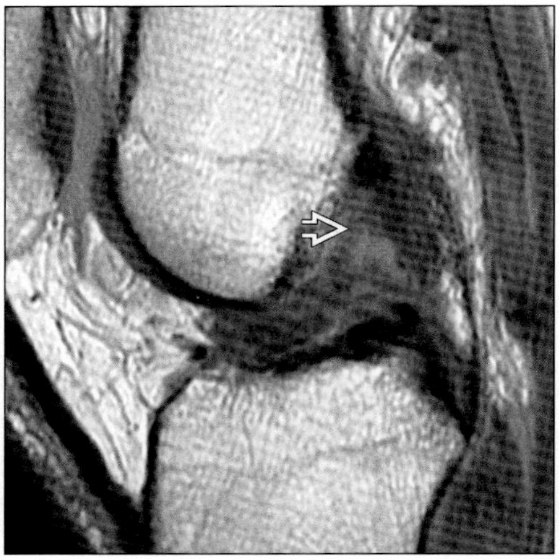

*(**Left**) Sagittal PDWI MR from the same patient shows a nondisplaced vertical longitudinal tear in the red zone of the medial meniscus posterior horn ➔. A thin rim of peripheral meniscal tissue ↪ differentiates this from a meniscocapsular separation. (**Right**) Sagittal PDWI MR from the same patient shows a complex vertical longitudinal tear in the lateral meniscus posterior horn ➔. Marrow contusions related to the ACL tear are present in the tibia and lateral femoral condyle ⇨.*

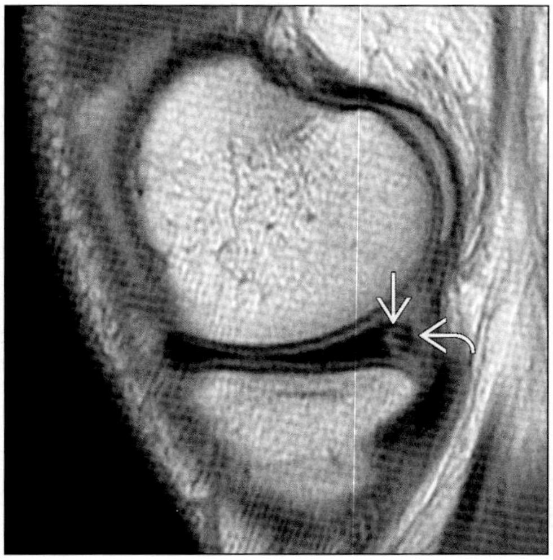

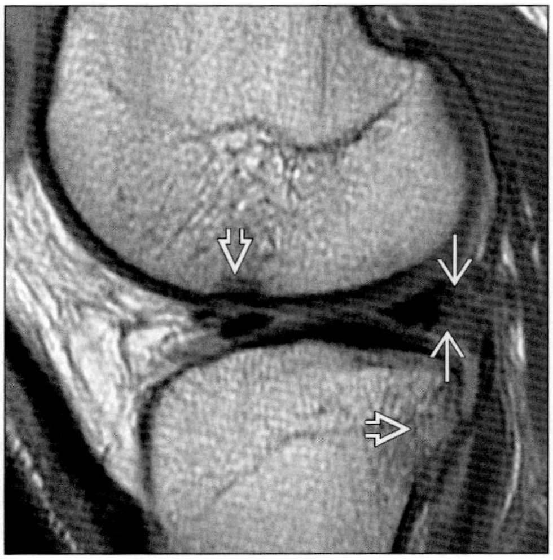

MENISCAL VERTICAL LONGITUDINAL TEAR

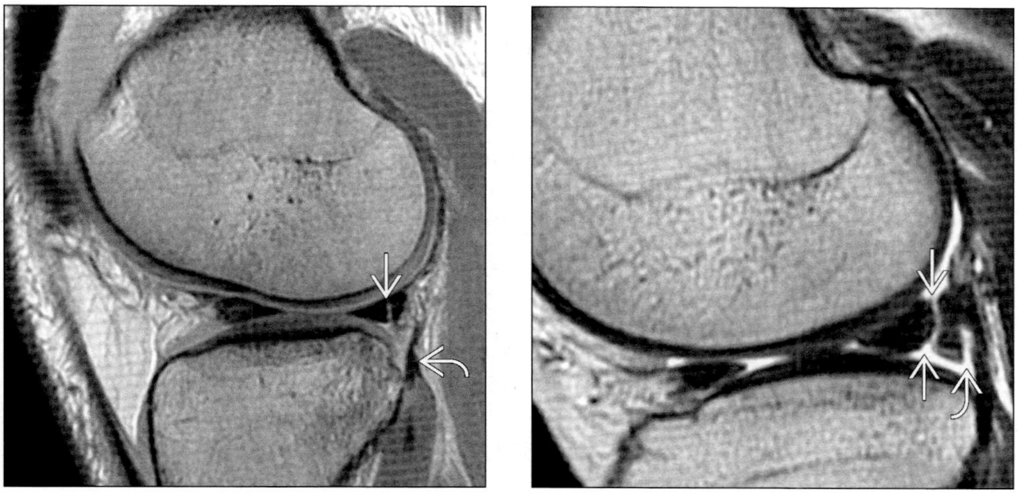

(Left) *Sagittal PDWI MR shows a linear full thickness vertical longitudinal tear of the posterior horn of the lateral meniscus. The tear is medial to the normal popliteal hiatus. The popliteal tendon can be seen more posteriorly.* ***(Right)*** *Sagittal T2WI MR shows a slightly displaced vertical longitudinal tear in the posterior horn of the lateral meniscus, with fluid tracking into the tear. A 2nd cleft of fluid is seen between the meniscus and the posterior joint capsule, indicating partial meniscocapsular tear.*

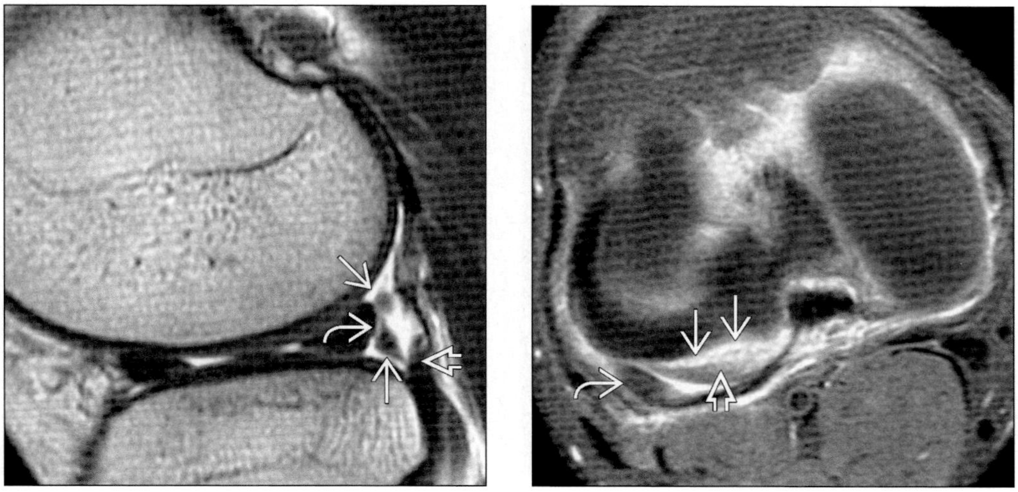

(Left) *Sagittal T2WI MR more laterally from the same patient shows mild fragmentation of the peripheral fragment of the vertical longitudinal tear of the posterior horn of the lateral meniscus. The tear is approaching the popliteal hiatus; the popliteal tendon can be seen entering the joint capsule more posteriorly.* ***(Right)*** *Axial PDWI FS MR from the same patient shows the displaced vertical longitudinal tear and displaced fragment extending toward the popliteal hiatus.*

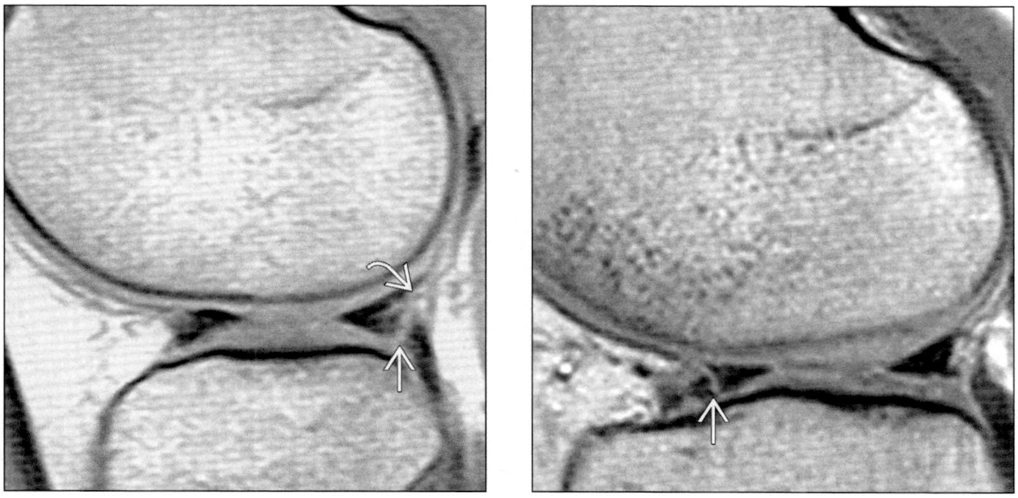

(Left) *Sagittal PDWI MR shows a vertical longitudinal tear in the extreme peripheral portion of the posterior lateral meniscus. The superior popliteomeniscal fascicle is also disrupted.* ***(Right)*** *Sagittal PDWI MR shows a vertical longitudinal tear of the anterior horn of the lateral meniscus. The normal junction of the anterior horn and the transverse intermeniscal ligament assumes a more oblique course, angled anteroinferior to posterosuperior.*

MENISCAL BUCKET-HANDLE TEAR

Key Facts

Terminology

- Bucket-handle tear (BHT)
- Vertical longitudinal meniscal tear with mesial displacement of free edge fragment toward or into intercondylar notch
- Displaced fragment maintains anterior and posterior attachments to peripheral remnant

Imaging

- Medial > lateral meniscus
- Truncation and small size of body of meniscus on coronal images
- Meniscal fragment in mesial aspect of compartment or displaced into intercondylar notch
- Fragment can be traced to peripheral attachments anteriorly and posteriorly
- Small size of anterior and posterior horns on sagittal images
- "Double PCL" sign not particularly sensitive for BHT
- Vertical longitudinal tear in peripheral zone of posterior horn as extension of BHT
- Coronal short TE images most accurate

Top Differential Diagnoses

- Displaced flap tear of meniscus
- Intraarticular body
- Meniscofemoral ligament
- Chronic anterior cruciate ligament tear

Clinical Issues

- Often sudden onset of symptoms after trauma
- Urgent surgery

Diagnostic Checklist

- Identify affected zone of meniscus (vascular vs. nonvascular) if possible

__(Left)__ Oblique coronal graphic with the knee hyperflexed shows a displaced bucket-handle tear of the medial meniscus. A vertical longitudinal tear of the meniscus separates and a displaced mesial fragment remains attached anteriorly and posteriorly. __(Right)__ Coronal T2WI FS MR shows a displaced BHT of the medial meniscus. The fragment is in the intercondylar notch. The peripheral fragment is small and irregular. This is a red zone tear and was repaired.

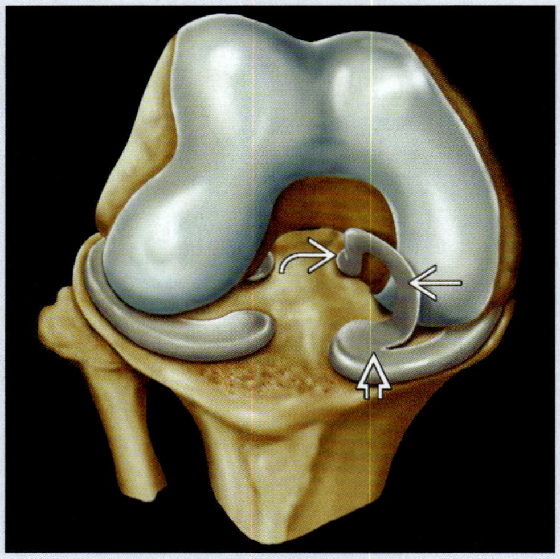

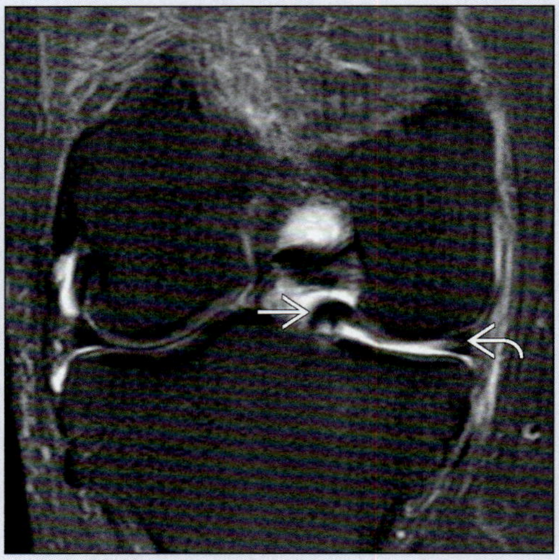

__(Left)__ Sagittal PDWI MR from the same patient shows an irregular contour of a small posterior horn of the medial meniscus. This is the "tip of the iceberg" of a more complex tear, and the remnant posterior horn may have a variety of appearances in this setting. __(Right)__ Sagittal PDWI MR more laterally in the same patient shows the displaced meniscal fragment creating the "double PCL" sign. This sign is not sensitive for BHT. Posterior cruciate ligament is posterior to fragment.

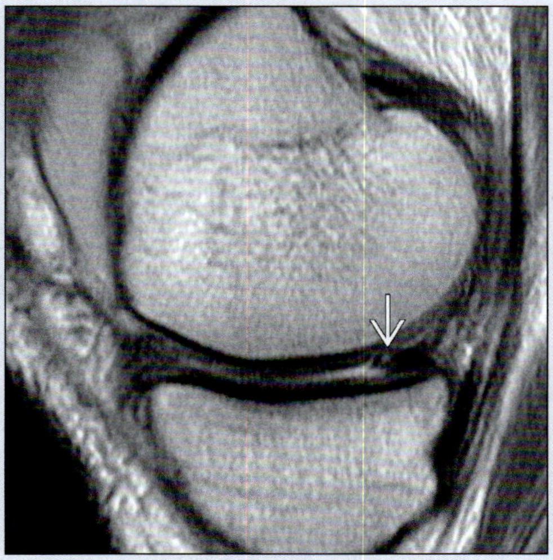

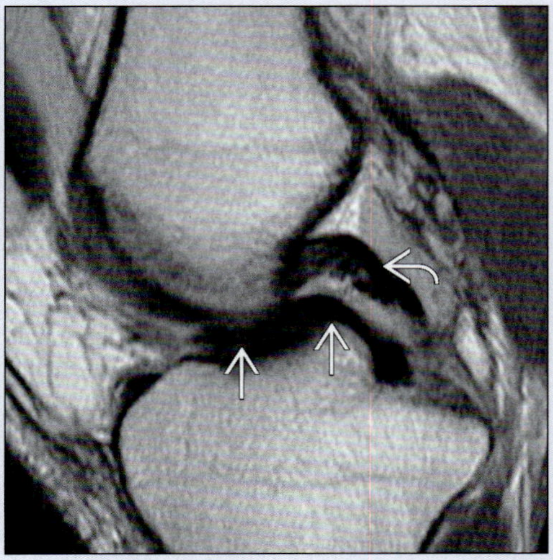

MENISCAL BUCKET-HANDLE TEAR

TERMINOLOGY

Abbreviations

- Bucket-handle tear (BHT)

Definitions

- Vertical longitudinal meniscal tear with mesial displacement of free edge fragment toward or into intercondylar notch
- Displaced fragment maintains anterior and posterior attachments to peripheral remnant
- Broken bucket-handle tear: Anterior or posterior attachment of displaced fragment disrupted

IMAGING

General Features

- Best diagnostic clue
  - Displacement of central meniscal substance mesially toward intercondylar notch on coronal MR images
- Location
  - Medial > lateral meniscus
- Morphology
  - Peripheral portion of meniscus remains in normal location (attached to joint capsule)
  - Central substance of meniscus remains attached to peripheral portion anteriorly and posteriorly
    - Broken bucket-handle tear: 1 attachment torn
  - Central fragment displaced mesially away from peripheral portion and toward intercondylar notch

MR Findings

- Coronal images
  - Truncation and small size of body of meniscus
  - Meniscal fragment in mesial aspect of compartment or displaced into intercondylar notch
  - Fragment can be traced to peripheral attachments anteriorly and posteriorly
    - Broken bucket-handle tear: 1 attachment torn
- Sagittal images
  - Small size of anterior and posterior horns
  - "Double PCL" sign
    - Displaced meniscal fragment in intercondylar notch roughly parallels posterior cruciate ligament (PCL)
      - Disc fragment anterior and inferior to PCL
      - Not particularly sensitive for BHT
  - Vertical longitudinal tear in peripheral zone of posterior horn as extension of BHT

Imaging Recommendations

- Best imaging tool
  - MR
- Protocol advice
  - Coronal short TE images most accurate

DIFFERENTIAL DIAGNOSIS

Displaced Flap Tear of Meniscus

- Fragment not attached both anteriorly and posteriorly

Intraarticular Body

- Chondral or osteochondral fragment may migrate into intercondylar notch
- No defect in meniscus

Meniscofemoral Ligament

- Extends from posterior horn of lateral meniscus to superior medial wall of intercondylar notch

Chronic Anterior Cruciate Ligament Tear

- Distal stump may lie horizontally in intercondylar notch

CLINICAL ISSUES

Presentation

- Most common signs/symptoms
  - Pain along joint line
  - Locking/limited terminal extension
  - Often sudden onset of symptoms after trauma
- Clinical profile
  - Young active patient; often related to sports activity

Demographics

- Age
  - Adolescents and young active adults
- Gender
  - M > F

Natural History & Prognosis

- Displaced fragment usually does not spontaneously reduce

Treatment

- Emergent surgery
  - Repair if tear in peripheral "red zone" of vascular meniscal tissue
  - Excision of displaced fragment if tear is in nonvascular free edge

DIAGNOSTIC CHECKLIST

Consider

- Other displaced meniscal tear types

Image Interpretation Pearls

- Confirm anterior and posterior attachments of displaced fragment to peripheral meniscal remnant before diagnosing BHT

Reporting Tips

- Identify affected zone of meniscus (vascular vs. nonvascular) if possible

SELECTED REFERENCES

1. Venkatanarasimha N et al: Potential pitfalls of a double PCL sign. Skeletal Radiol. 38(8):735-9, 2009
2. Makdissi M et al: MRI-negative bucket-handle tears of the lateral meniscus in athletes: a case series. Knee Surg Sports Traumatol Arthrosc. 14(10):1012-6, 2006
3. Dorsay TA et al: Bucket-handle meniscal tears of the knee: sensitivity and specificity of MRI signs. Skeletal Radiol. 32(5):266-72, 2003
4. Wright DH et al: Bucket-handle tears of the medial and lateral menisci of the knee: value of MR imaging in detecting displaced fragments. AJR Am J Roentgenol. 165(3):621-5, 1995
5. Singson RD et al: MR imaging of displaced bucket-handle tear of the medial meniscus. AJR Am J Roentgenol. 156(1):121-4, 1991

MENISCAL BUCKET-HANDLE TEAR

(Left) *Series of coronal T2WI FS MRs from posterior (1) to anterior (4) shows the appearance of a displaced BHT of the medial meniscus. The bucket-handle fragment remains attached at the posterior horn ➡ as well as the nondisplaced peripheral portion of the meniscus anteriorly ➡. The displaced "bucket-handle" meniscal fragment is located within the intercondylar notch ➡.* ***(Right)*** *Axial PDWI FS MR from the same patient shows the anterior ➡ and posterior ➡ attachment points and the fragment ➡.*

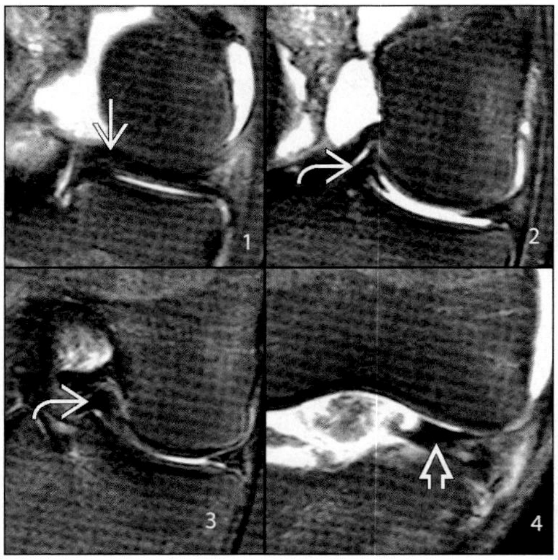

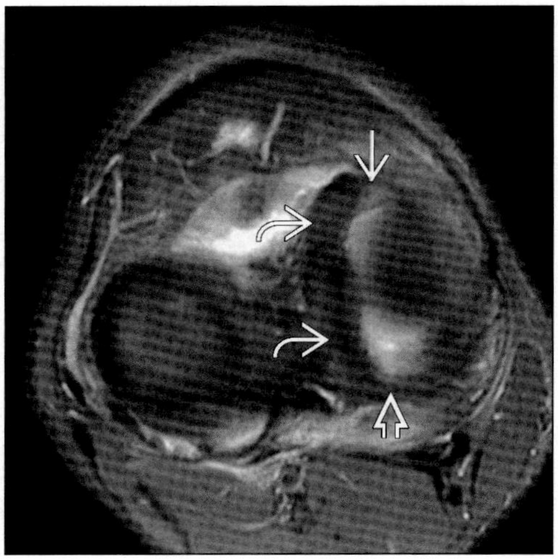

(Left) *Coronal T2WI FS MR demonstrates a BHT of the medial meniscus with a very small displaced fragment ➡. Without careful scrutiny of successive coronal images, this tear might be overlooked, as the peripheral fragment ➡ appears normal in shape and size. Fragment was resected because the tear was in the meniscal white zone.* ***(Right)*** *Sagittal PDWI FS MR from the same patient shows a very thin displaced meniscal fragment ➡ in the intercondylar notch, a variant of the "double PCL" sign.*

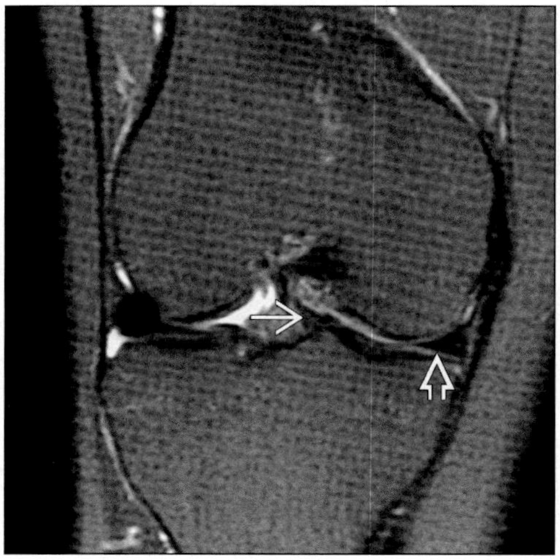

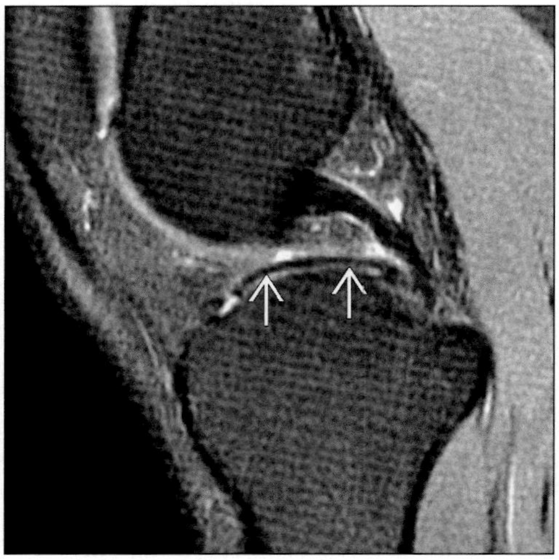

(Left) *Coronal MR arthrogram T1WI in a patient who underwent previous medial meniscal repair for BHT, now presenting with recurrent pain and locking in the knee, shows a large displaced medial meniscal fragment ➡ in the notch. The peripheral fragment ➡ is small. At surgery the meniscus separated along the previous repair line.* ***(Right)*** *Coronal PDWI MR shows a large displaced BHT of the medial meniscus ➡, with no residual peripheral meniscal substance present ➡. ACL graft is present ➡.*

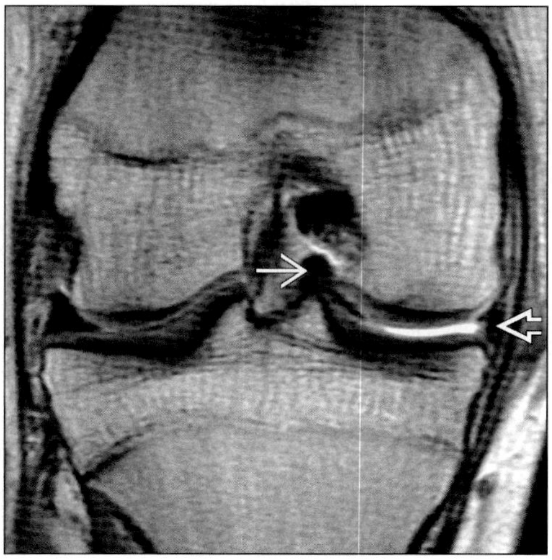

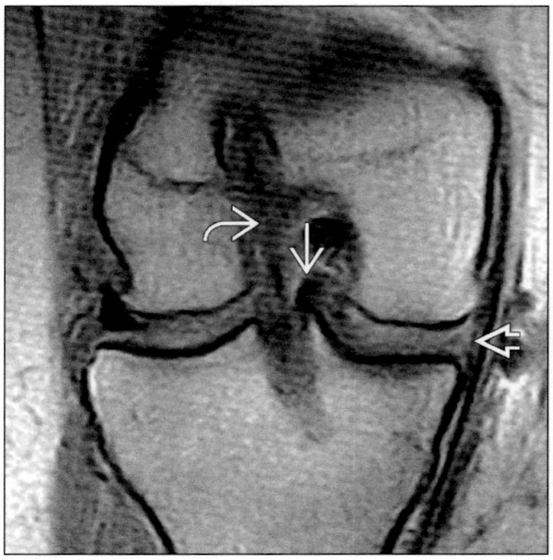

MENISCAL BUCKET-HANDLE TEAR

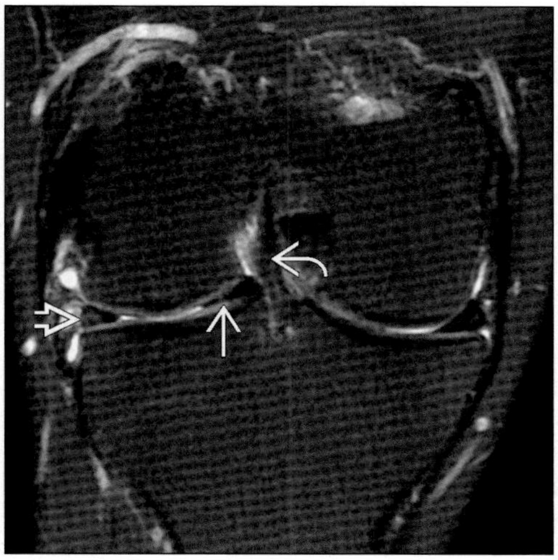

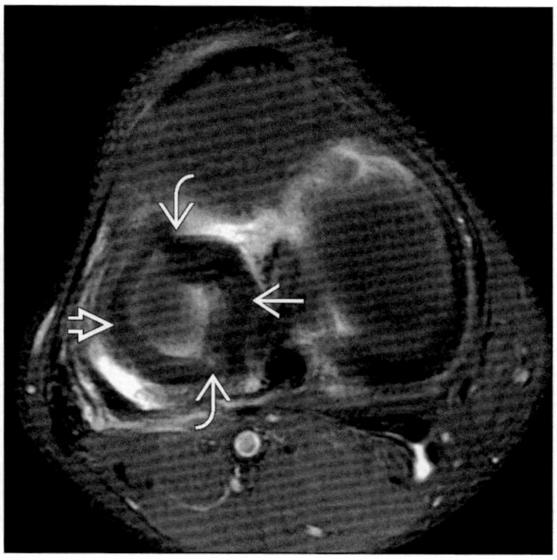

(Left) *Coronal T2WI FS MR demonstrates a displaced bucket-handle tear of the lateral meniscus. The displaced fragment extends into the intercondylar notch and abuts the anterior cruciate ligament (ACL) . The peripheral fragment is truncated.* ***(Right)*** *Axial PDWI FS MR from the same patient shows the large displaced fragment of the bucket-handle tear, with attachment points to the peripheral fragment well seen.*

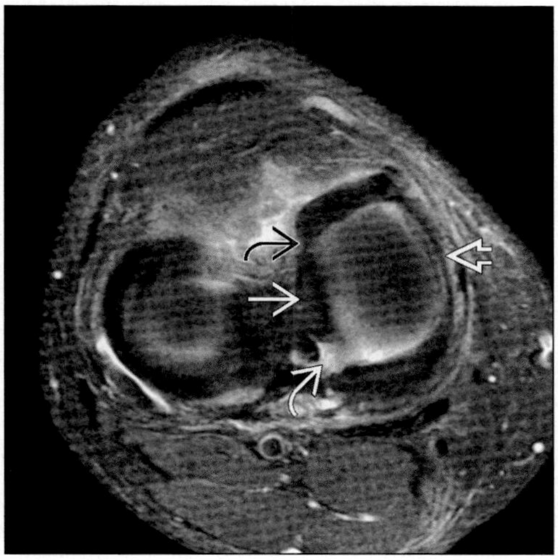

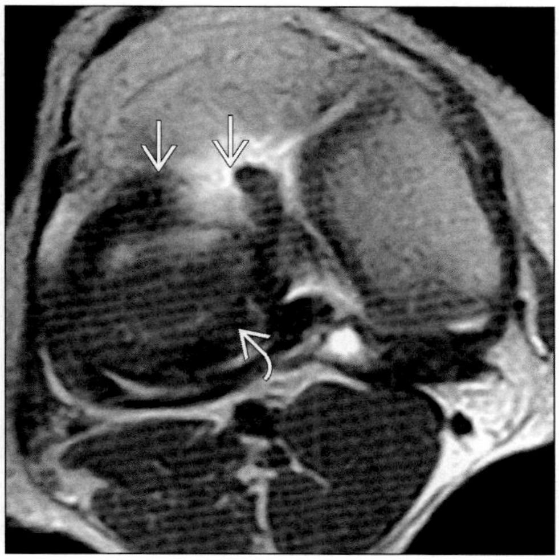

(Left) *Axial PDWI FS MR shows a broken bucket-handle tear of the medial meniscus. The thin peripheral fragment and larger displaced fragment are identified; the anterior connection is intact, but the posterior connection is disrupted.* ***(Right)*** *Axial T2WI MR demonstrates a broken bucket-handle tear of the lateral meniscus. The posterior attachment of the displaced fragment to the peripheral fragment is intact, but the anterior attachment is disrupted.*

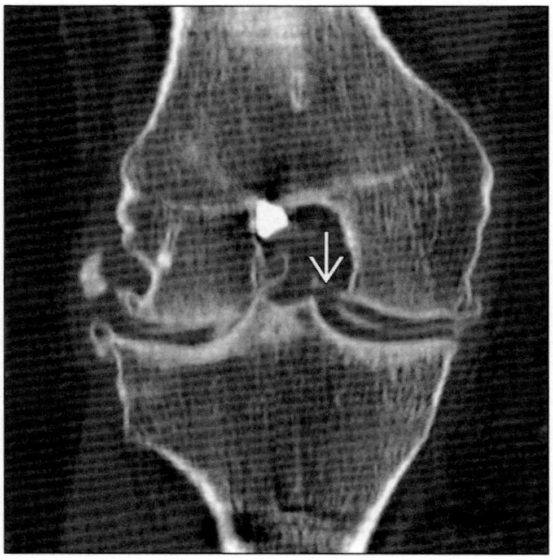

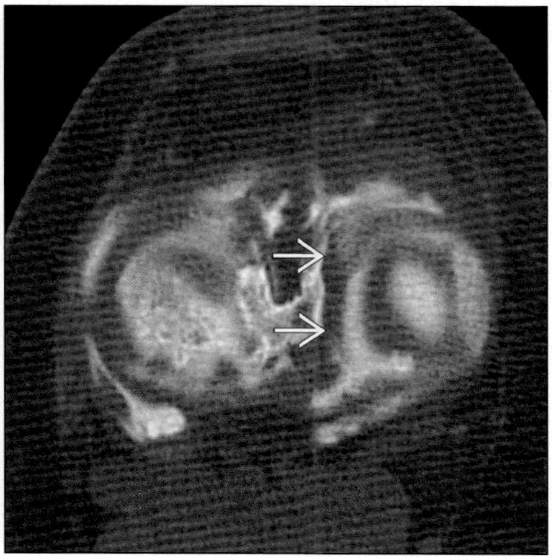

(Left) *Coronal reformat bone CT arthrogram shows a displaced BHT of the medial meniscus. The displaced fragment is seen adjacent to the medial tibial spine. It may be difficult to distinguish between displaced meniscal fragments and other types of noncalcified intraarticular bodies on CT arthrography because their densities may be similar.* ***(Right)*** *Axial bone CT arthrogram from the same patient shows the bucket-handle fragment displaced toward the intercondylar notch.*

OTHER DISPLACED MENISCAL TEARS

Key Facts

Terminology

- Flap tear (parrot-beak), flipped meniscus, complex displaced tear
- Bucket-handle tear discussed separately

Imaging

- Flap (parrot-beak) tear: Most common in posterior medial meniscus
  - Oblique radial tear, often with longitudinal component
  - V-shaped defect best seen on axial images
- Flipped meniscus: Most commonly involves displacement of posterior horn of lateral meniscus anteriorly
  - "Double meniscus" sign anteriorly: Anterior fragment is normal anterior horn, and posterior fragment is displaced posterior horn
  - Severely truncated or absent posterior horn
  - Best seen on sagittal images
- "Too much meniscus" sign: Overabundance of meniscal tissue in part of knee suggests displaced fragment arising from another part of meniscus
- "Absent meniscus" sign: No meniscal tissue in expected location suggests displaced fragment

Clinical Issues

- May lead to limited motion and increased pain if not resected
- May lead to early osteoarthritis
- Arthroscopic debridement

Diagnostic Checklist

- Account for whole volume of meniscus (i.e., sum of fragments should equal expected size of normal whole meniscus)
  - Displaced meniscal tear if portion of meniscus is absent and no history of prior meniscal surgery

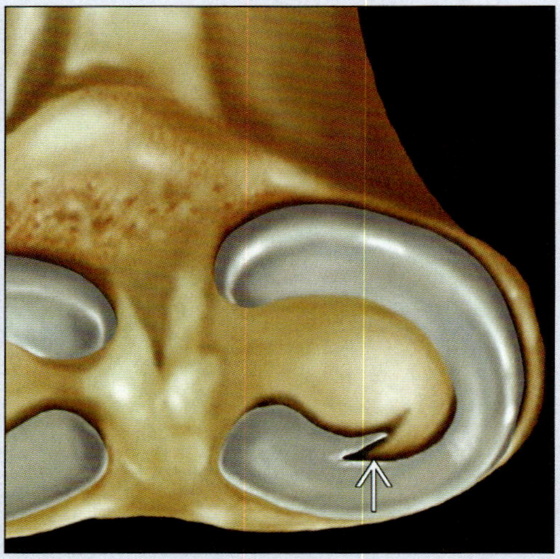

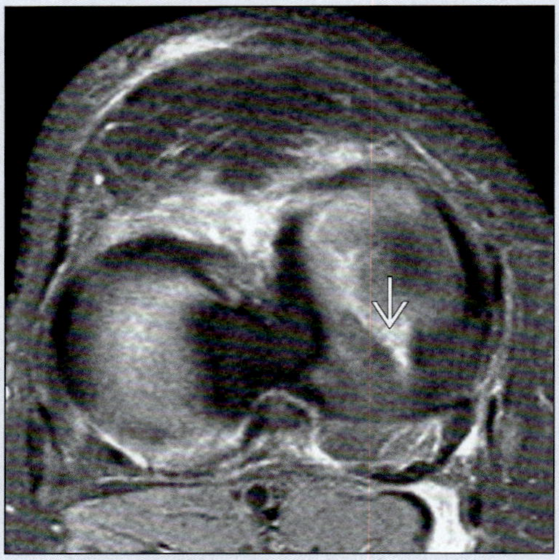

(Left) *Axial graphic shows a flap or parrot-beak tear of the posterior horn of the medial meniscus. This is an oblique radial or longitudinal tear entering the free edge of the meniscus with displacement of the edge, resulting in a defect shaped like a parrot's beak or the letter "V" ➔.* ***(Right)*** *Axial PDWI FS MR shows a parrot-beak tear of the posterior horn of the medial meniscus ➔. The V-shaped morphology of the tear is best seen on axial images.*

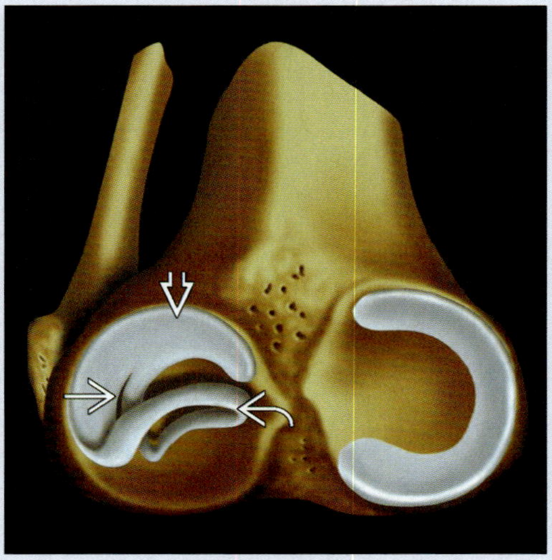

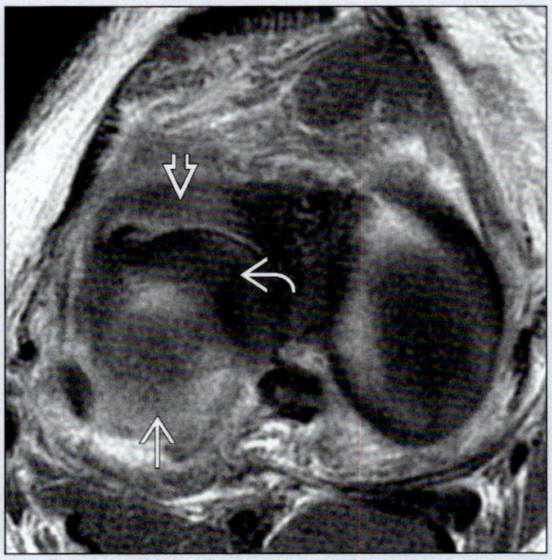

(Left) *Axial graphic demonstrates a lateral flipped meniscus. In this tear, a large portion of the posterior horn ➔ is displaced anteriorly and lies adjacent to the anterior horn ➔. This usually involves a longitudinal tear of the meniscus ➔ and disruption of its posterior capsular attachments.* ***(Right)*** *Axial PDWI FS MR shows a flipped lateral meniscus. The displaced posterior horn fragment ➔ is seen posterior to the anterior horn ➔. The posterior compartment is empty ➔.*

OTHER DISPLACED MENISCAL TEARS

TERMINOLOGY

Synonyms

- Flap tear (parrot-beak), flipped meniscus, complex displaced tear

Definitions

- Tear of meniscus in which 1 or more meniscal fragments are displaced from their normal location
- Bucket-handle tear discussed separately

IMAGING

General Features

- Best diagnostic clue
  - Meniscal tear with displacement of 1 or more meniscal fragments
- Location
  - Flap tear: Most common in posterior medial meniscus
  - Flipped meniscus: Most commonly involves displacement of posterior horn of lateral meniscus anteriorly
- Size
  - Usually larger tears
- Morphology
  - Flap tear (a.k.a. parrot-beak tear)
    - Oblique short-axis tear with displacement, resulting in V-shaped defect of free edge
  - Flipped meniscus
    - A portion of meniscus flipped into abnormal position that is not a typical bucket-handle
    - Usually posterior horn of meniscus displaced anteriorly, with fragment lying adjacent to anterior horn
      - Lateral > > medial meniscus
  - Complex displaced tear
    - Any of a variety of other morphologies of displaced meniscal substance

MR Findings

- Flap (parrot-beak) tear
  - Oblique radial with longitudinal component
  - Signal extends to free edge of meniscus with separation of borders of tear
  - V-shaped defect best seen on axial images (resembles parrot's beak)
- Flipped meniscus
  - "Double meniscus" sign anteriorly
    - 2 triangular meniscal fragments adjacent to one another
    - Anterior fragment is normal anterior horn
    - Posterior fragment is displaced posterior horn
  - Severely truncated or absent posterior horn
  - Best seen on sagittal images
- "Too much meniscus" sign: Overabundance of meniscal tissue in part of knee suggests displaced fragment arising from another part of meniscus
  - Donor site must be identified
- "Absent meniscus" sign: No meniscal tissue in expected location
  - If no history of surgery, assume displaced tear and search for displaced fragment(s)

Imaging Recommendations

- Best imaging tool
  - MR
- Protocol advice
  - Short TE (T1WI and PDWI ± FS) in multiple planes should be examined to assess nature and extent of complex tear and displacement of meniscal fragments

DIFFERENTIAL DIAGNOSIS

Meniscal Bucket-Handle Tear

- Vertical longitudinal tear with mesial displacement of fragment that remains attached to peripheral portion of meniscus anteriorly and posteriorly

Postoperative Meniscus Change

- Saucerization or partial resection of meniscus may result in irregular shape of meniscus
- Should not see separate fragment

Intraarticular Body

- Chondral or other intraarticular bodies may lie adjacent to meniscus and mimic displaced meniscal fragment

CLINICAL ISSUES

Presentation

- Most common signs/symptoms
  - Pain and swelling; joint line tenderness
  - May experience locking or extension lag due to entrapped meniscal fragment

Natural History & Prognosis

- May lead to limited motion and increased pain if not resected
- If fragment separates, may become loose in joint
  - May ossify and become intraarticular body

Treatment

- Arthroscopic debridement

DIAGNOSTIC CHECKLIST

Consider

- Displaced meniscal tear if portion of meniscus is absent and no history of prior surgery
- Account for whole volume of meniscus (i.e., sum of fragments should equal expected size of normal whole meniscus)

SELECTED REFERENCES

1. McKnight A et al: Meniscal tears with displaced fragments: common patterns on magnetic resonance imaging. Skeletal Radiol. 39(3):279-83, 2010
2. Boxheimer L et al: Characteristics of displaceable and nondisplaceable meniscal tears at kinematic MR imaging of the knee. Radiology. 238(1):221-31, 2006
3. Vande Berg BC et al: Meniscal tears with fragments displaced in notch and recesses of knee: MR imaging with arthroscopic comparison. Radiology. 234(3):842-50, 2005
4. Lecas LK et al: Inferiorly displaced flap tears of the medial meniscus: MR appearance and clinical significance. AJR Am J Roentgenol. 174(1):161-4, 2000

OTHER DISPLACED MENISCAL TEARS

(Left) *Sagittal PDWI MR shows a complex tear of the posterior horn of the medial meniscus with a very small remnant of normal meniscal tissue →. A "ghost meniscus" appearance of irregular elevated signal → is seen anterior to the posterior horn.* ***(Right)*** *Sagittal PDWI MR more medially in the same patient shows an irregular elongate meniscal fragment → extending anteriorly from the posterior horn →. The anterior horn → is normal.*

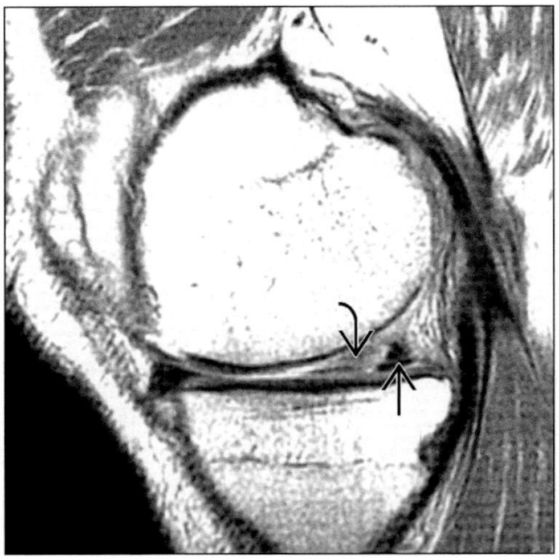

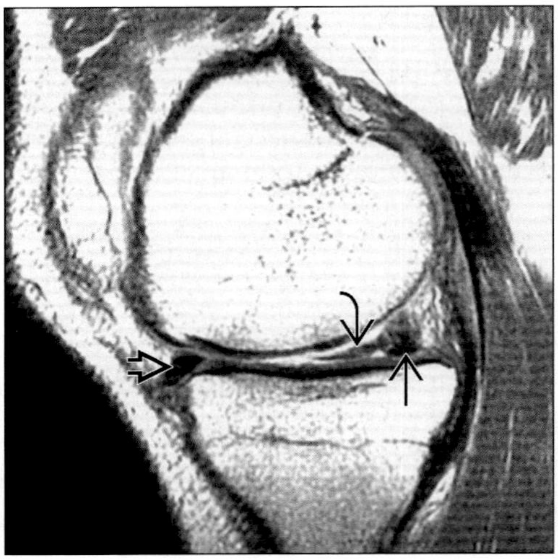

(Left) *Coronal T2WI FS MR from the same patient shows an oblique vertical tear → extending through the posterior horn of the medial meniscus. There is subchondral marrow edema → in the adjacent tibia. A linear chondral fissure → is present in the lateral femoral condyle.* ***(Right)*** *Axial PDWI FS MR from the same patient shows a displaced parrot-beak tear → of the posterior medial meniscus. There is lateral displacement of the posterior root fragment →.*

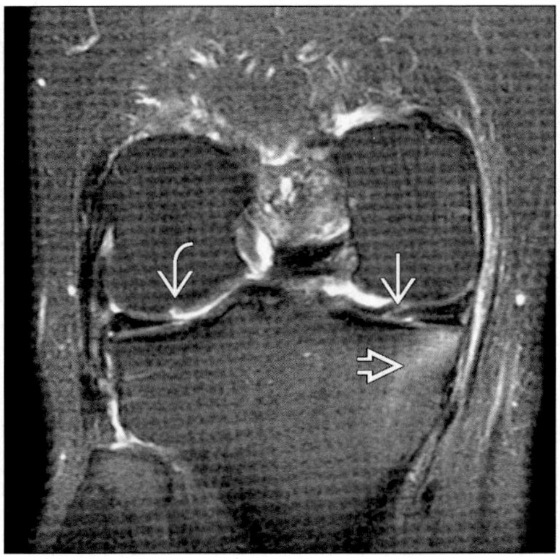

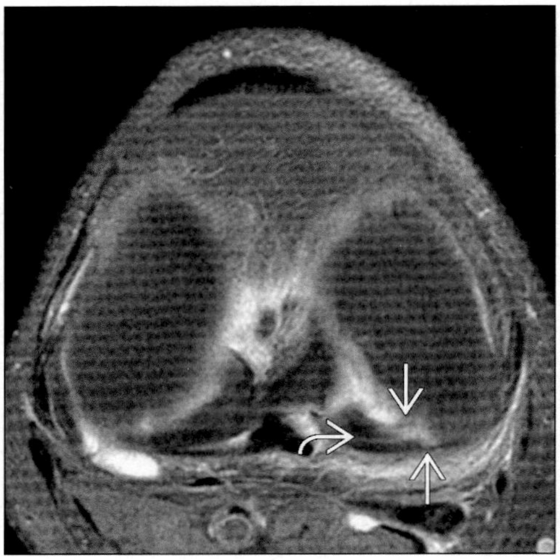

(Left) *Coronal T2WI FS MR shows the "absent meniscus" sign in the posterior aspect of the lateral compartment → in a patient with an acute tear of the anterior cruciate ligament →. There is an impaction contusion in the posterior lateral tibial plateau →.* ***(Right)*** *Sagittal PDWI MR from the same patient shows the "double meniscus" sign of a flipped meniscus. The posterior horn fragment → is displaced out of its normal location → and lies posterior to the anterior horn →.*

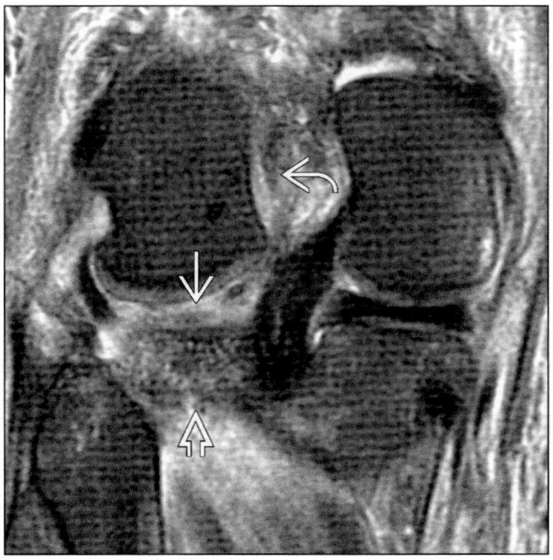

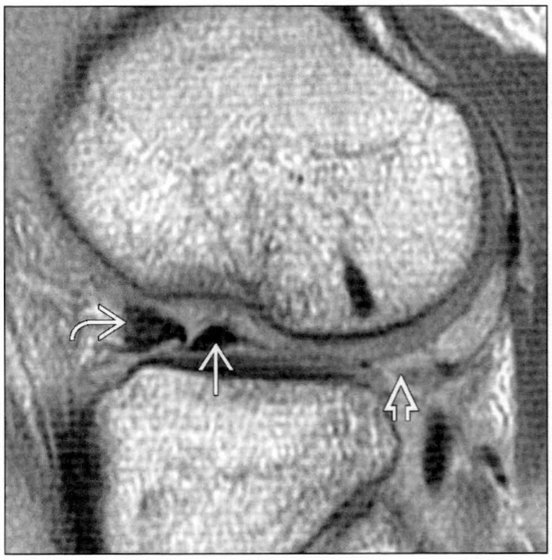

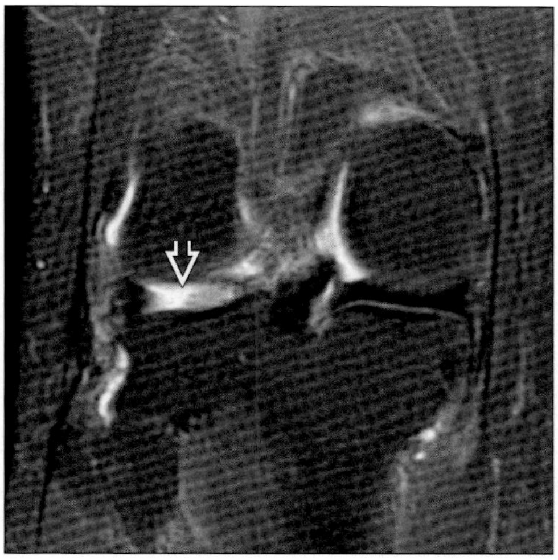

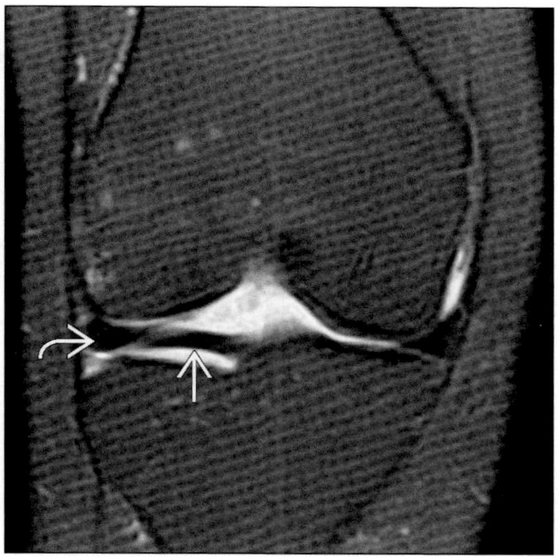

(Left) *Coronal T2WI FS MR shows an empty lateral compartment consistent with the "absent meniscus" sign in the posterior aspect of the knee, indicating a displaced tear of the lateral meniscus. In the absence of any prior surgical history, this finding should prompt a thorough search for the displaced fragment.* ***(Right)*** *Coronal T2WI FS MR more anteriorly in the same patient shows the "too much meniscus" sign with the displaced posterior horn fragment medial to the anterior horn.*

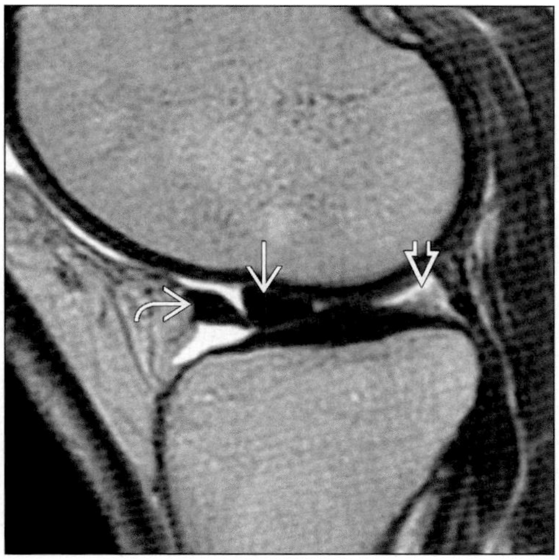

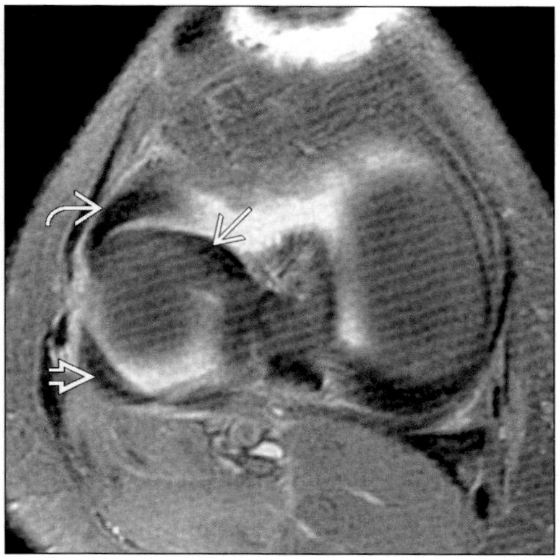

(Left) *Sagittal T2WI MR from the same patient shows the "double meniscus" sign. The displaced posterior horn fragment is present immediately posterior to the normal anterior horn, leaving the posterior aspect of the compartment empty, the "absent meniscus" sign.* ***(Right)*** *Axial PDWI FS MR from the same patient shows the displaced posterior horn fragment behind the normal anterior horn. The popliteus tendon remains in its normal position.*

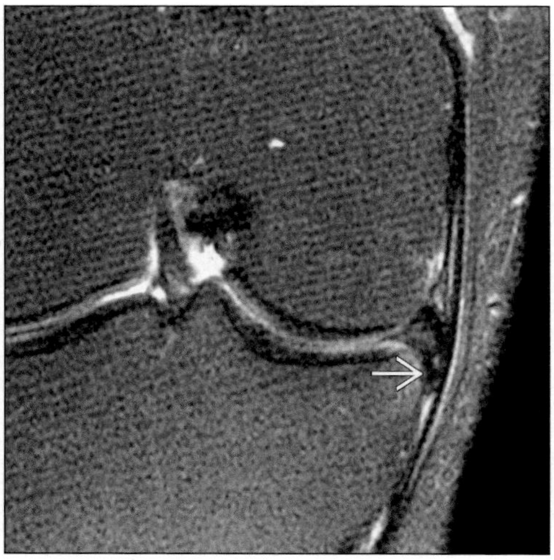

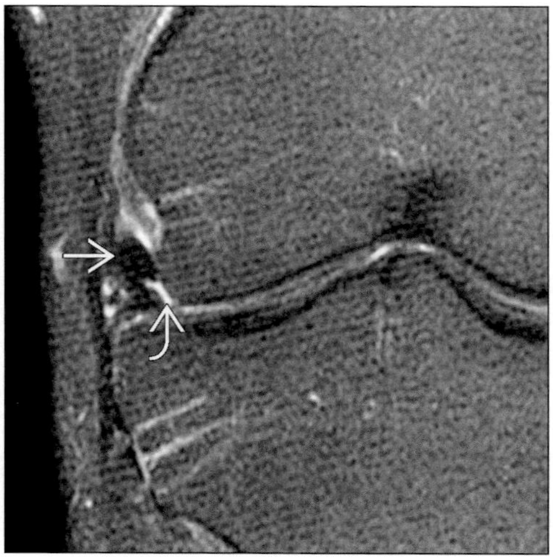

(Left) *Coronal T2WI FS MR demonstrates a horizontal cleavage tear of the medial meniscus with displacement of a portion of the undersurface of the meniscus into the inferior parameniscal recess. These tears can be subtle and should be actively sought.* ***(Right)*** *Coronal T2WI FS MR demonstrates a horizontal tear of the lateral meniscus with a displaced fragment in the suprameniscal recess. There is truncation of the free edge of the meniscus, a clue to a displaced fragment.*

OTHER DISPLACED MENISCAL TEARS

*(**Left**) Sagittal T2WI MR shows a complex displaced tear of the medial meniscus. There is a "double meniscus" sign in the anterior compartment, with a displaced fragment lying posterior and superior to the normal anterior horn. The posterior horn appears relatively normal. (**Right**) Sagittal T2WI MR more laterally in the same patient shows an attachment point between the displaced fragment and the anterior horn.*

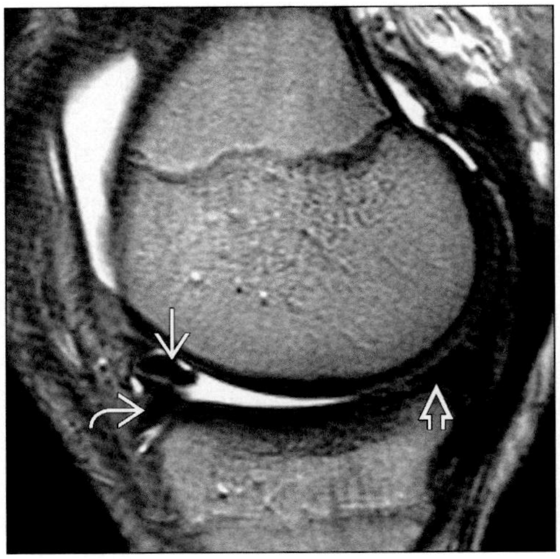

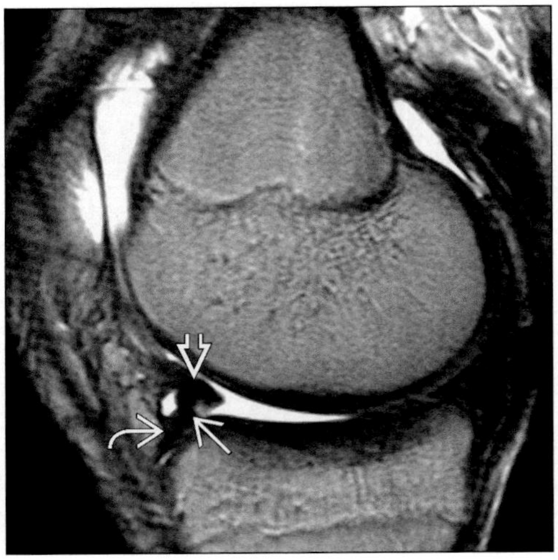

*(**Left**) Coronal T2WI FS MR from the same patient shows the displaced medial meniscal fragment superolateral to the anterior horn. (**Right**) Axial PDWI FS MR from the same patient shows truncation and irregularity of the free edge of the body of the medial meniscus indicating the site of tear. The displaced fragment is seen anterolaterally. In this tear, the displaced fragment arises not from the posterior horn but from the body of the meniscus.*

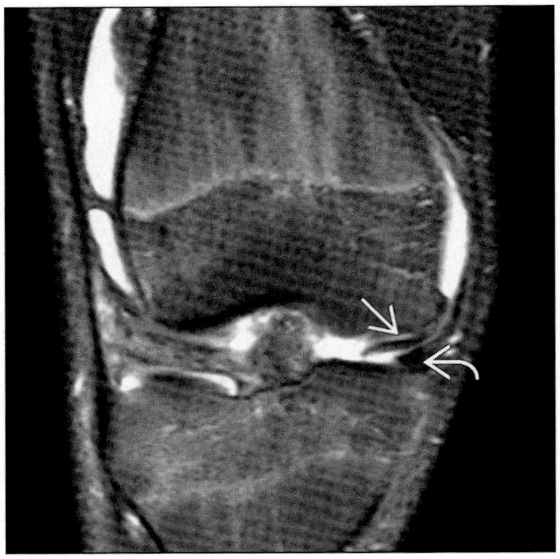

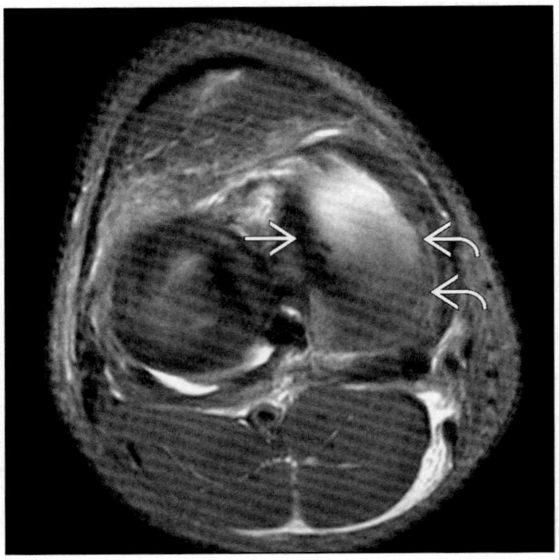

*(**Left**) Sagittal PDWI MR shows a radial defect in the free edge of the medial meniscal body. A subtle indentation in the superior articular margin of the anterior horn is also present. (**Right**) Axial PDWI FS MR from the same patient shows the tear as a displaced flap complex; there are separate parrot-beak deformities anteriorly and in the mid-body of the meniscus. The posterior horn is normal.*

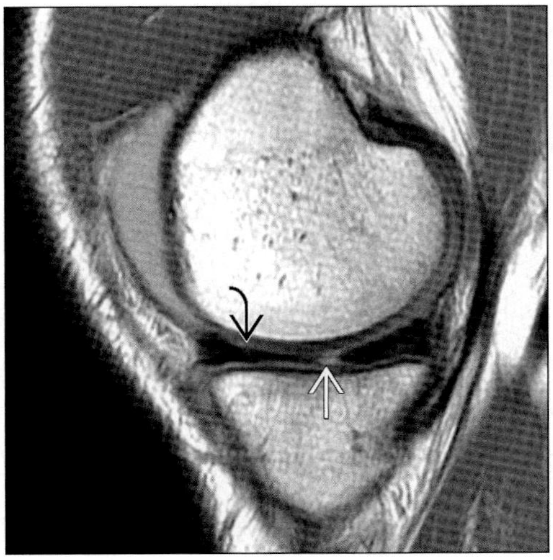

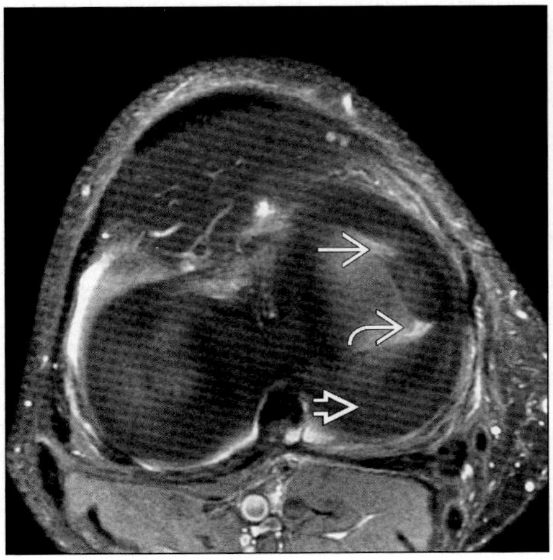

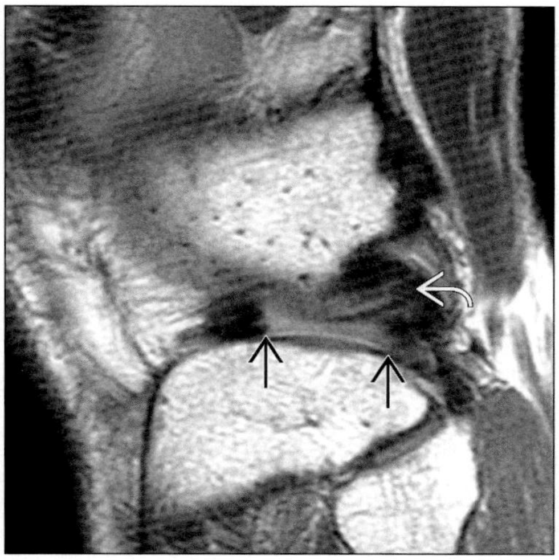

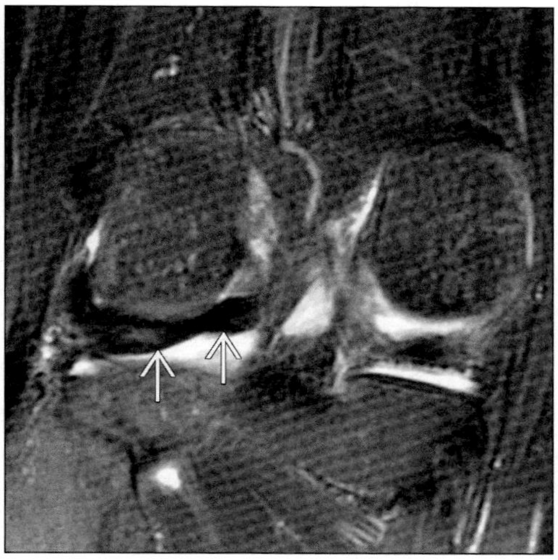

(Left) *Sagittal PDWI MR shows a large defect in the lateral meniscal body → with truncation of the anterior and posterior horn remnants, the "absent meniscus" sign. There is a large displaced fragment of meniscus → superiorly in the posterior compartment.* ***(Right)*** *Coronal T2WI FS MR through the posterior knee from the same patient shows "too much meniscus" in the lateral compartment →. This finding indicates a displaced fragment arising elsewhere and should prompt a search for a donor site.*

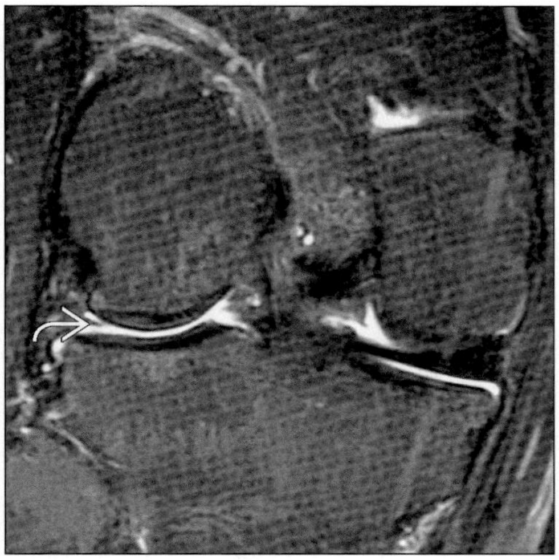

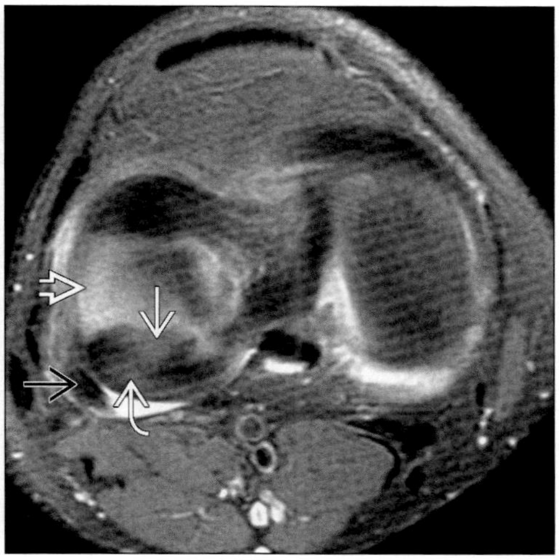

(Left) *Coronal T2WI FS MR more anteriorly in the same patient shows absence of meniscal tissue → in the lateral compartment, the "absent meniscus" sign indicating a displaced tear.* ***(Right)*** *Axial PDWI FS MR from the same patient shows displacement of a large lateral meniscal fragment → arising from the meniscal body → and lying in the posterior aspect of the knee adjacent to the posterior horn →. The normal popliteus tendon → should not be mistaken for a meniscal fragment.*

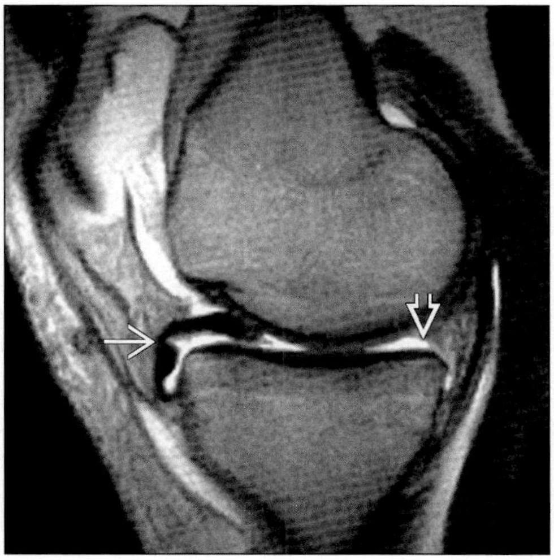

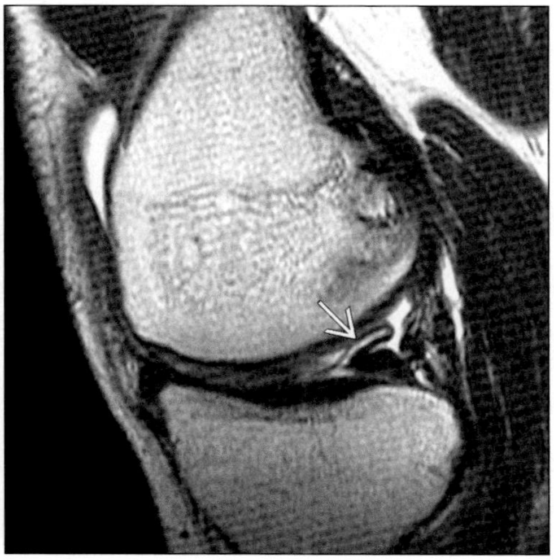

(Left) *Sagittal T2WI MR shows a complex displaced medial meniscal tear. There is absence of meniscal tissue in the posterior aspect of the compartment → with a large, irregular fragment → in the anterior aspect of the knee.* ***(Right)*** *Sagittal T2WI MR demonstrates a serpentine configuration → of a displaced posterior horn tear of the medial meniscus. Displaced tears can have a wide variety of appearances, and bizarrely shaped fragments may result.*

COMPLEX MENISCAL TEAR

Key Facts

Terminology
- Stellate meniscal tear

Imaging
- Meniscal tear with tear propagating in more than 1 plane
- Medial > lateral meniscus
- Posterior horn and body > anterior horn
- Most frequently horizontal + radial tear
  - Horizontal component best seen on coronal images
  - Radial component may be seen on coronal or sagittal images
- Vertical + radial tear
  - Vertical component best seen on sagittal images
  - Radial component best seen on sagittal or coronal images
- Stellate tear
  - Characterized by multiple extensions of tear, often emanating from central point, extending in multiple directions
- Short TE sequences (T1WI, PDWI ± FS) best for identifying and characterizing meniscal tears

Top Differential Diagnoses
- Meniscal horizontal tear
- Meniscal radial tear
- Meniscal longitudinal tear
- Meniscal degeneration
- Postoperative change

Pathology
- Often associated with osteoarthritis

Clinical Issues
- Arthroscopic debridement

(Left) *Sagittal graphic of a complex meniscal tear shows vertical ⇨ and horizontal ➡ components in the tear. Complex tears can manifest in a variety of shapes, and a portion of the meniscus may displace. Complex tears are often seen in the setting of osteoarthritis.* ***(Right)*** *Sagittal MR arthrogram T1WI FS shows a stellate tear ➡ of the posterior horn of the medial meniscus. The tear has horizontal and vertical components and contacts the superior and inferior articular meniscal surfaces.*

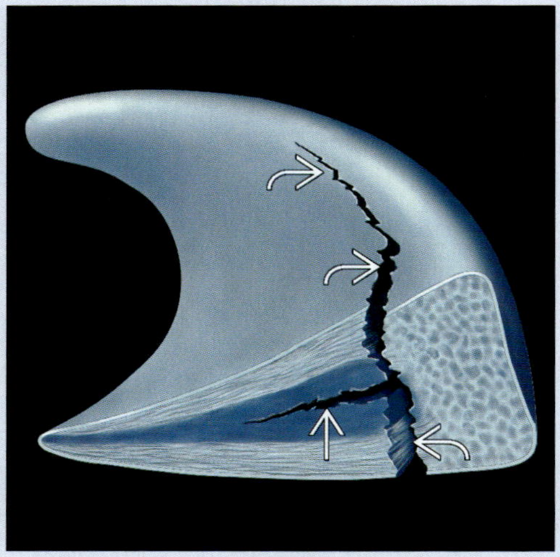

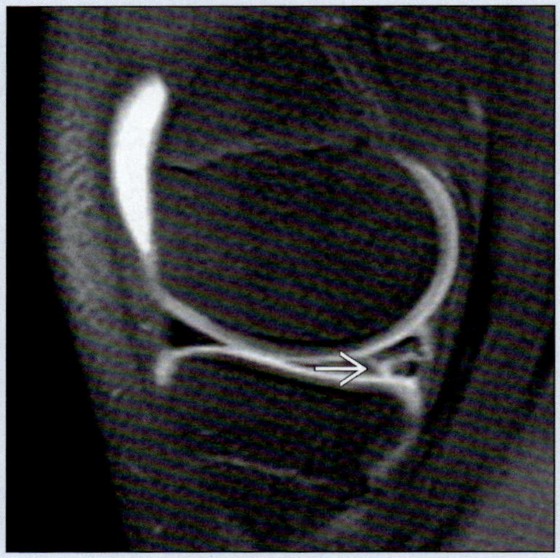

(Left) *Coronal MR arthrogram T2WI FS from the same patient shows a broad horizontal component ➡ and a short radial component ⇨. Stellate tears often propagate in multiple planes, and evaluation of all image sets is required to fully characterize the tear.* ***(Right)*** *Sagittal PDWI MR shows a complex tear of the posterior horn of the medial meniscus ➡, which extends in a circular pattern from the inferior articular surface of the red zone, isolating a central peripheral fragment ⇨.*

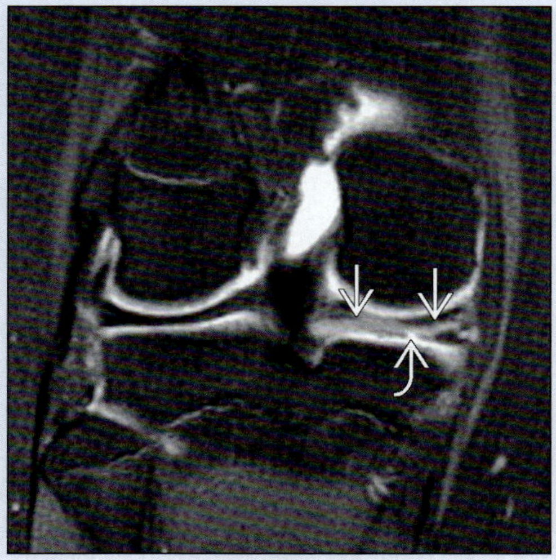

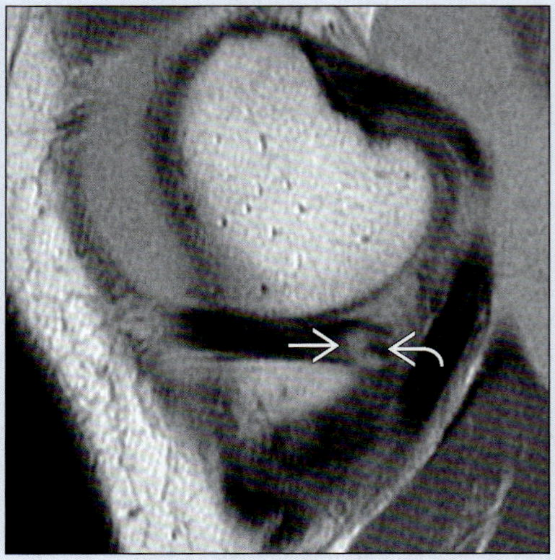

COMPLEX MENISCAL TEAR

TERMINOLOGY

Synonyms

- Stellate meniscal tear

Definitions

- Tears of menisci characterized by more than 1 type of morphology

IMAGING

General Features

- Best diagnostic clue
  - Meniscal tear with tear propagating in more than 1 plane
- Location
  - Medial > lateral meniscus
  - Posterior horn and body > anterior horn
- Size
  - Usually larger tears
- Morphology
  - Tears composed of a combination of 2 or more simple morphologies
    - Horizontal cleavage + radial, longitudinal + radial, etc.

CT Findings

- CT arthrography demonstrates entry of contrast into meniscus in multiple planes

MR Findings

- Abnormal signal in meniscal substance extending to 1 or more articular surface of meniscus, with tear propagating in more than 1 plane
  - Most frequently horizontal + radial tear
    - Horizontal component best seen on coronal images
    - Radial component may be seen on coronal or sagittal images
  - Vertical + radial tear
    - Vertical component best seen on sagittal images
    - Radial component best seen on sagittal or coronal images
  - Stellate tear
    - Characterized by multiple extensions of tear, often emanating from central point, extending in multiple directions

Imaging Recommendations

- Best imaging tool
  - MR
- Protocol advice
  - Short TE sequences (T1WI, PDWI ± FS) best for identifying and characterizing meniscal tears

DIFFERENTIAL DIAGNOSIS

Meniscal Horizontal Tear

- Tear limited to horizontal plane of meniscus

Meniscal Radial Tear

- Tear plane limited to short axis of meniscus

Meniscal Longitudinal Tear

- Tear limited to longitudinal vertical axis of meniscus

Meniscal Degeneration

- Globular or linear signal within meniscus not contacting articular surface

Postoperative Change

- May demonstrate complex signal ± extension to meniscal surface
- Must correlate with prior surgical history

PATHOLOGY

General Features

- Associated abnormalities
  - Often associated with osteoarthritis

Staging, Grading, & Classification

- If possible, characterized by individual planes of tear observed
  - Horizontal + radial, longitudinal + radial, etc.

CLINICAL ISSUES

Presentation

- Most common signs/symptoms
  - Knee pain and joint line tenderness

Natural History & Prognosis

- Complex tears less likely to heal spontaneously than other nondisplaced tears

Treatment

- Arthroscopic debridement

DIAGNOSTIC CHECKLIST

Consider

- Complex tear in setting of osteoarthritis

Image Interpretation Pearls

- Evaluate morphology of meniscal tear in 3 planes to identify multiple planes of propagation

Reporting Tips

- Describe tears in terms of planes of propagation

SELECTED REFERENCES

1. Lee YG et al: Magnetic resonance imaging findings of surgically proven medial meniscus root tear: tear configuration and associated knee abnormalities. J Comput Assist Tomogr. 32(3):452-7, 2008
2. Mustonen AO et al: MRI of acute meniscal injury associated with tibial plateau fractures: prevalence, type, and location. AJR Am J Roentgenol. 191(4):1002-9, 2008
3. Mohan BR et al: Reliability of clinical diagnosis in meniscal tears. Int Orthop. 31(1):57-60, 2007
4. Coulier B: Direct 3D imaging of the knee menisci during 16-row multislice CT arthrography. JBR-BTR. 89(6):291-7, 2006
5. Costa CR et al: Medial meniscus extrusion on knee MRI: is extent associated with severity of degeneration or type of tear? AJR Am J Roentgenol. 183(1):17-23, 2004
6. Jee WH et al: Meniscal tear configurations: categorization with MR imaging. AJR Am J Roentgenol. 180(1):93-7, 2003
7. Rubman MH et al: Technical considerations in the management of complex meniscus tears. Clin Sports Med. 15(3):511-30, 1996

MENISCOCAPSULAR SEPARATION

Key Facts

Terminology

- Meniscocapsular separation (MCS)
- Disruption of capsular attachment of meniscus

Imaging

- Usually posterior horn of medial meniscus
- Linear disruption at meniscocapsular junction extending circumferentially a variable distance around peripheral aspect of meniscus
  - Ranges from a few mm to > 1 cm circumference
- Abnormal meniscocapsular junction
  - Best seen on sagittal T2WI MR
  - Widening of meniscocapsular interval ≥ 5 mm
  - Irregularity of capsular margin of meniscus
  - ↑ signal on T2WI MR
  - Penetration of injected contrast into meniscocapsular interval on MR arthrography
- Associated abnormalities
  - Tear of adjacent meniscal tissue
  - Tear of meniscofemoral or meniscotibial ligaments
  - Medial collateral ligament (MCL) superficial fiber disruption
  - Fracture or osseous contusion

Top Differential Diagnoses

- Normal parameniscal recesses of joint
- Vertical longitudinal tear of meniscus
- MCL bursitis
- MCL tear

Clinical Issues

- Small MCS (< 1 cm circumference): Conservative
- Large MCS (> 1 cm in circumference): Surgical repair

Diagnostic Checklist

- If even a tiny portion of peripheral meniscus is attached to capsule, call it a vertical meniscal tear

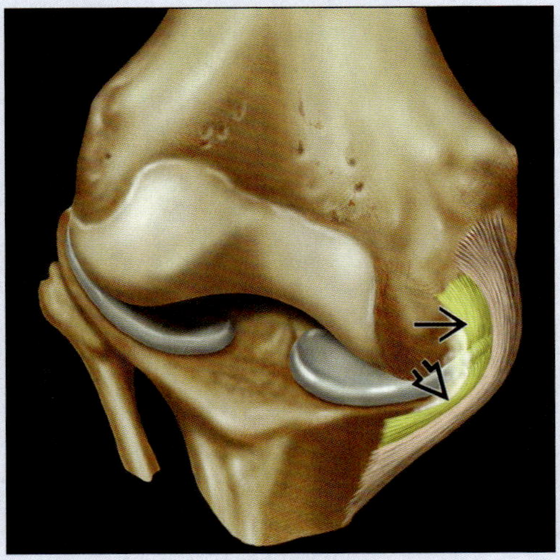

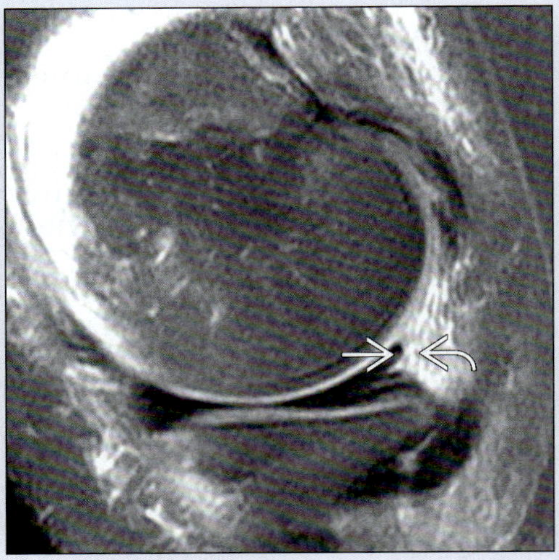

(Left) *Coronal oblique graphic shows separation of the capsule ⇨ from the meniscus at the meniscocapsular junction and surrounding edema ⇨ in the posteromedial aspect of the medial meniscus.* ***(Right)*** *Sagittal PDWI FS MR shows acute meniscocapsular separation (MCS) ➡ in the posteromedial corner of the knee, with diffuse, irregular soft tissue swelling and loss of the normal peripheral margin of the posterior horn of the medial meniscus ➡. This is the most common location for MCS.*

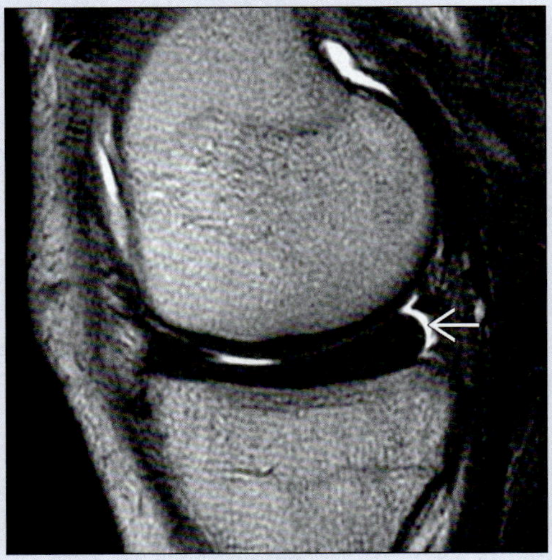

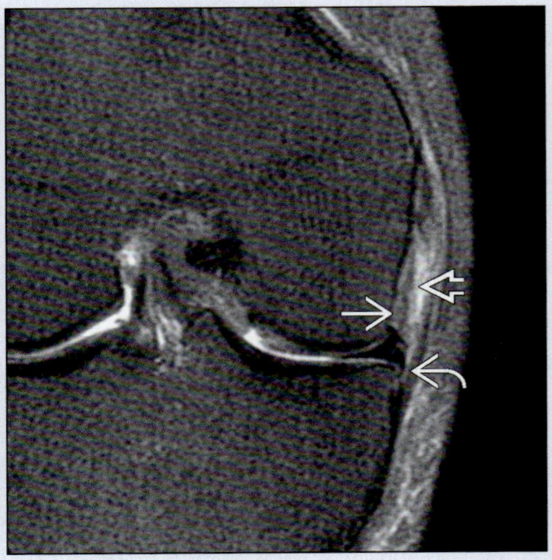

(Left) *Sagittal T2WI MR shows chronic MCS ➡ of the posterior horn of the medial meniscus. The margins of the meniscus and capsule have become smooth due to healing. This space should be filled with fibrovascular tissue at the meniscocapsular junction.* ***(Right)*** *Coronal T2WI FS MR shows disruption of the meniscofemoral ligament ➡ consistent with a focal MCS. The meniscotibial ligament ➡ is intact. Edema is present ⇨ between the layers of the medial collateral ligament.*

MENISCOCAPSULAR SEPARATION

TERMINOLOGY

Abbreviations

- Meniscocapsular separation (MCS)

Definitions

- Disruption of capsular attachment of meniscus

IMAGING

General Features

- Best diagnostic clue
  - Abnormal widening ± ↑ T2 signal between peripheral aspect of meniscus and joint capsule
- Location
  - Usually posterior horn of medial meniscus
- Size
  - Ranges from a few mm to > 1 cm
- Morphology
  - Linear disruption at meniscocapsular junction extending circumferentially a variable distance around peripheral aspect of meniscus

MR Findings

- Abnormal meniscocapsular junction
  - Best seen on sagittal T2WI MR
  - Widening of meniscocapsular interval ≥ 5 mm
  - Irregularity of capsular margin of meniscus
  - ↑ signal on T2WI MR
- Tear of meniscofemoral or meniscotibial ligaments
  - Affects body of meniscus
- Tear of adjacent meniscal tissue
- Medial collateral ligament (MCL) superficial fiber disruption
- Fracture or osseous contusion
- **Direct MR arthrography**
  - Penetration of injected contrast into meniscocapsular interval
  - Widened meniscocapsular interval ≥ 5 mm

Imaging Recommendations

- Best imaging tool
  - MR
- Protocol advice
  - Sagittal T2WI FS MR best imaging sequence

DIFFERENTIAL DIAGNOSIS

Normal Parameniscal Recesses of Joint

- Fluid within normal superior and inferior parameniscal recesses may mimic MCS

Vertical Longitudinal Tear of Meniscus

- Very peripheral tear leaves only small rim of meniscal tissue attached to capsule
- May be very difficult to distinguish from MCS

MCL Bursitis

- Fluid between superficial and deep portions of MCL

MCL Tear

- Disruption of superficial fibers may also disrupt deep layer (meniscofemoral and meniscotibial ligaments)

PATHOLOGY

General Features

- Etiology
  - Usually due to trauma
    - Shear or impaction force
    - Valgus stress may rupture meniscofemoral &/or meniscotibial ligaments
- Associated abnormalities
  - Often associated with disruption of anterior cruciate ligament (ACL) &/or MCL

Gross Pathologic & Surgical Features

- Disruption of fibrous attachments of meniscus to joint capsule
- Hemorrhage and edema

CLINICAL ISSUES

Presentation

- Most common signs/symptoms
  - Pain and tenderness at site of tear
  - Similar to meniscal tear
- Other signs/symptoms
  - Findings related to associated tear of ACL or MCL

Demographics

- Age
  - Young adults
- Gender
  - M > F

Natural History & Prognosis

- Small tears (< 1 cm) usually heal with time due to rich blood supply to peripheral zone of meniscus
- Large tear may → unstable meniscus

Treatment

- Small MCS (< 1 cm in circumference): Conservative
- Large MCS (> 1 cm in circumference): Surgical repair

DIAGNOSTIC CHECKLIST

Consider

- MCS vs. vertical longitudinal tear of peripheral meniscus

Image Interpretation Pearls

- If even a tiny portion of peripheral meniscus is attached to capsule, diagnose as a vertical meniscal tear

SELECTED REFERENCES

1. De Maeseneer M et al: Medial meniscocapsular separation: MR imaging criteria and diagnostic pitfalls. Eur J Radiol. 41(3):242-52, 2002
2. George J et al: Radiological classification of meniscocapsular tears of the anterolateral portion of the lateral meniscus of the knee. Australas Radiol. 44(1):19-22, 2000
3. De Maeseneer M et al: Normal and abnormal medial meniscocapsular structures: MR imaging and sonography in cadavers. AJR Am J Roentgenol. 171(4):969-76, 1998
4. Rubin DA et al: Are MR imaging signs of meniscocapsular separation valid? Radiology. 201(3):829-36, 1996

POPLITEOMENISCAL FASCICLES

Key Facts

Terminology
- Lateral meniscal struts, hypermobile lateral meniscus
- Thin bands of connective tissue running between posterior horn of lateral meniscus and lateral joint capsule
- Posterosuperior and anteroinferior fascicles form roof and floor, respectively, of popliteal hiatus
- Functionally important stabilizers of lateral meniscus
- Fascicles prevent excessive lateral movement of meniscus and possible entrapment

Imaging
- Thin, low signal bands of tissue on all sequences
- Best seen on sagittal images through popliteal hiatus
- Disruption seen as
  - Focal discontinuity in 1 or both popliteomeniscal fascicles
  - Complete absence of ligamentous tissue in expected location of 1 or both fascicles
- Vertical longitudinal tear of posterior horn of lateral meniscus may be associated
  - Occurs at attachment point of 1 or both popliteomeniscal fascicles

Top Differential Diagnoses
- Lateral meniscal tear
- Popliteus tendon injury
- Popliteofibular ligament injury

Pathology
- Lateral meniscus posterior horn tear
- Anterior cruciate ligament tear

Clinical Issues
- Popliteomeniscal fascicle tear has 96% specificity for posterior horn lateral meniscal tear

__(Left)__ Sagittal graphic of the popliteal hiatus in the posterolateral corner of the knee demonstrates the superior and inferior popliteomeniscal fascicles extending from the lateral meniscus to the joint capsule. The lateral portion of the inferior fascicle merges with the popliteofibular ligament. Popliteus tendon passes between the fascicles. __(Right)__ Sagittal T2WI MR shows disruption of the inferior popliteomeniscal fascicle. Superior fascicle is normal.

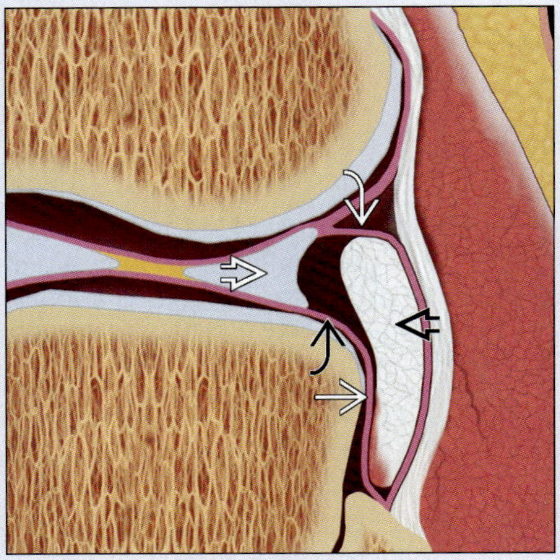

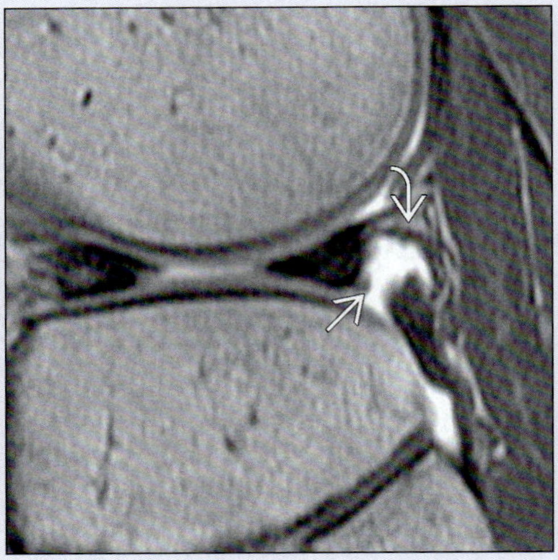

__(Left)__ Sagittal PDWI MR in a patient with a torn anterior cruciate ligament (ACL) (not shown) shows disruption of the superior and inferior popliteomeniscal fascicles. Disruption of a fascicle has a 25% positive predictive value (PPV) for ACL tear. __(Right)__ Sagittal MR arthrogram T1WI shows a peripheral (red zone) tear of the posterior horn of the lateral meniscus. Both meniscal fascicles are also torn. Torn fascicles have a 96% specificity for lateral meniscal tear.

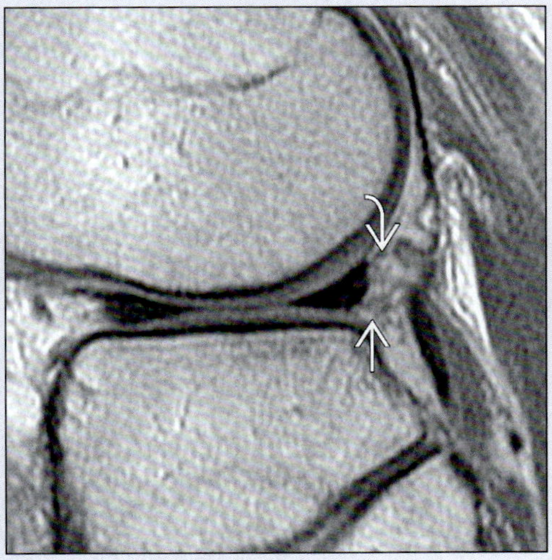

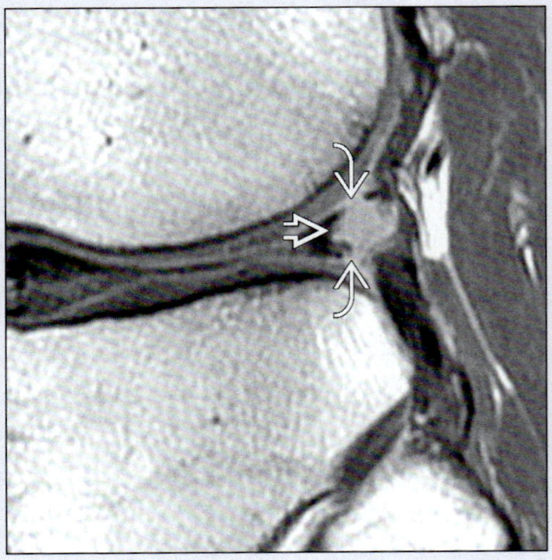

POPLITEOMENISCAL FASCICLES

TERMINOLOGY

Synonyms

- Lateral meniscal struts, hypermobile lateral meniscus

Definitions

- Thin bands of connective tissue running between posterior horn of lateral meniscus and lateral joint capsule

IMAGING

General Features

- Location
  - Posterosuperior popliteomeniscal fascicle
    - Originates at superior margin of posterior horn of lateral meniscus
    - Courses posteriorly, forming roof of popliteal hiatus
    - Attaches to posterior joint capsule
  - Anteroinferior popliteomeniscal fascicle
    - Originates at inferior margin of posterior horn of lateral meniscus
    - Courses inferolaterally, forming floor of popliteal hiatus
    - Attaches to myotendinous junction of popliteal muscle
- Morphology
  - Functionally important stabilizers of lateral meniscus
    - Fascicles prevent excessive lateral movement of meniscus and possible entrapment
    - Variable thickness

MR Findings

- Thin, low signal bands of tissue on all sequences
- Best seen on sagittal images through popliteal hiatus
- Disruption seen as
  - Focal discontinuity in 1 or both popliteomeniscal fascicles
  - Complete absence of ligamentous tissue in expected location of 1 or both fascicles
- ± diffuse soft tissue swelling in posterolateral corner
- Vertical longitudinal tear of posterior horn of lateral meniscus
  - Occurs at attachment point of 1 or both popliteomeniscal fascicles
  - Tear may be quite small

Imaging Recommendations

- Best imaging tool
  - MR
- Protocol advice
  - Thin section sagittal T2WI MR

DIFFERENTIAL DIAGNOSIS

Lateral Meniscal Tear

- Vertical longitudinal tear in posterior horn may spare fascicular attachment

Popliteus Tendon Injury

- Disruption of popliteus tendon without disruption of adjacent popliteomeniscal fascicles

Popliteofibular Ligament Injury

- Located more inferiorly, extends to fibular styloid

PATHOLOGY

General Features

- Etiology
  - Combination of compression and shear (twisting) forces most likely mechanism
- Associated abnormalities
  - Lateral meniscus posterior horn tear
  - Popliteomeniscal fascicle tear has up to 25% positive predictive value for anterior cruciate ligament tear

Staging, Grading, & Classification

- Isolated popliteomeniscal fascicle tear (relatively unusual)
- Fascicle tear associated with lateral meniscal tear

CLINICAL ISSUES

Presentation

- Most common signs/symptoms
  - Isolated tear of popliteomeniscal fascicles
    - Posterolateral knee pain
    - Locking of knee
    - "Hypermobile lateral meniscus"

Demographics

- Epidemiology
  - Popliteomeniscal fascicle tear has
    - 31% sensitivity for lateral meniscal tear
    - 96% specificity for posterior horn lateral meniscal tear

Treatment

- Arthroscopic repair of associated lateral meniscal tear
- Open repair of popliteomeniscal fascicles has been described

DIAGNOSTIC CHECKLIST

Consider

- Associated tears of lateral meniscus &/or anterior cruciate ligament

Image Interpretation Pearls

- Meniscal tear may be very subtle

SELECTED REFERENCES

1. Laundre BJ et al: MRI accuracy for tears of the posterior horn of the lateral meniscus in patients with acute anterior cruciate ligament injury and the clinical relevance of missed tears. AJR Am J Roentgenol. 193(2):515-23, 2009
2. Peduto AJ et al: Popliteomeniscal fascicles: anatomic considerations using MR arthrography in cadavers. AJR Am J Roentgenol. 190(2):442-8, 2008
3. LaPrade RF et al: Popliteomeniscal fascicle tears causing symptomatic lateral compartment knee pain: diagnosis by the figure-4 test and treatment by open repair. Am J Sports Med. 33(8):1231-6, 2005
4. Blankenbaker DG et al: Usefulness of two indirect MR imaging signs to diagnose lateral meniscal tears. AJR Am J Roentgenol. 178(3):579-82, 2002

PARAMENISCAL CYST

Key Facts

Terminology

- Fluid-filled cyst extending beyond confines of meniscus, arising from meniscal tear
- Meniscal cyst

Imaging

- Almost always arises along peripheral border of meniscus
- Most common segments
  - Posterior horn of medial meniscus
  - Body ± anterior horn of lateral meniscus
- Combination of short and long TE sequences to best distinguish meniscal tear and associated parameniscal cyst, respectively

Pathology

- Always associated with meniscal tear
  - Most commonly horizontal cleavage type

Clinical Issues

- Pisani sign: Palpable joint line mass that disappears with knee flexion
- Arthroscopic resection of cyst and repair or resection of meniscal tear
- Likely to recur if underlying meniscal tear not repaired or resected

Diagnostic Checklist

- Presence of cyst in typical location (posterior medial compartment or anterolateral aspect of lateral compartment) should strongly raise suspicion for associated meniscal tear
- If typical cyst encountered but meniscal tear not definitely identified, should suggest likely presence of nonvisualized tear

*(**Left**) Coronal graphic shows a parameniscal cyst arising at the peripheral margin of the meniscus related to a horizontal cleavage tear. Joint fluid tracks through the tear and fills the cyst. (**Right**) Coronal T1WI MR shows a broad horizontal cleavage tear in the posterior horn of the medial meniscus leading to a large soft tissue collection medial to the meniscus. While the appearance is suggestive of parameniscal cyst, T2-weighted images are necessary to exclude other masses.*

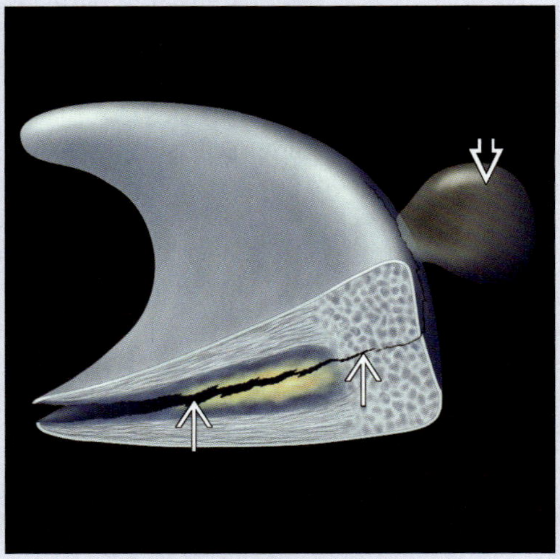

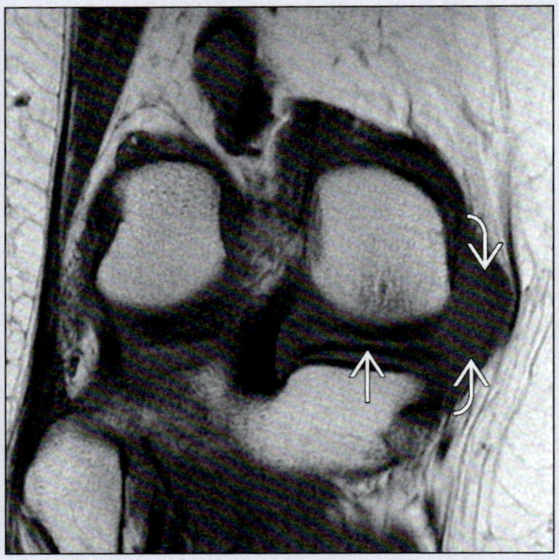

*(**Left**) Coronal T2WI FS MR from the same patient shows the cystic nature of the mass and mild septation within it. (**Right**) Axial PDWI FS MR from the same patient shows the extent of the cyst; there is a small component lateral to the main portion, posterior to the joint line. Axial images are very useful in gauging the extent of these cysts, particularly as they extend around the curved junctions of the meniscal body with the anterior or posterior horns.*

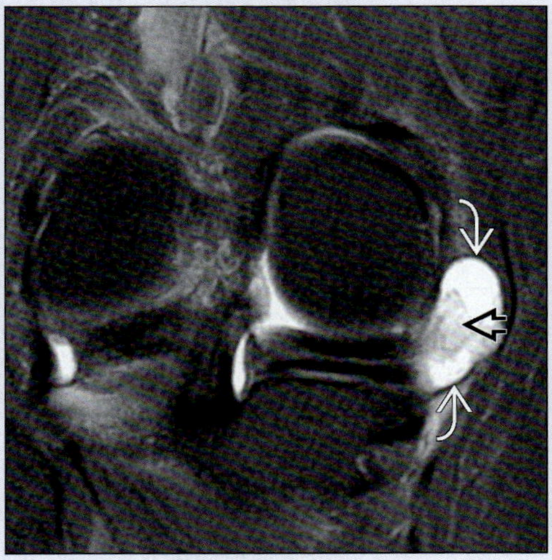

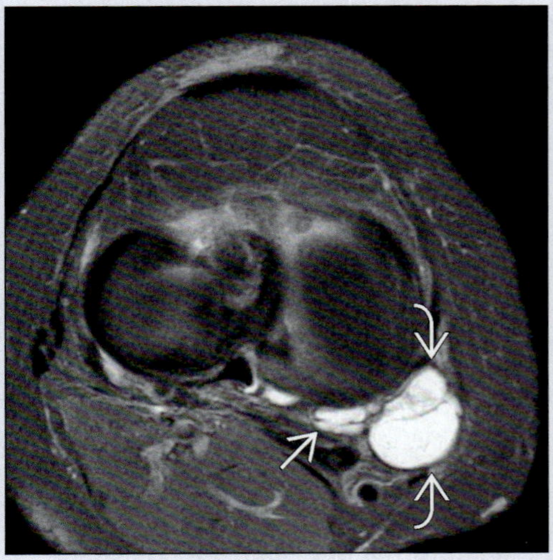

PARAMENISCAL CYST

TERMINOLOGY

Synonyms

- Meniscal cyst

Definitions

- Fluid-filled cyst extending beyond confines of meniscus, arising from meniscal tear

IMAGING

General Features

- Best diagnostic clue
  - Cystic mass arising from meniscal tear
- Location
  - Almost always arises along peripheral border of meniscus
  - Most common segments
    - Posterior horn of medial meniscus
    - Body ± anterior horn of lateral meniscus
- Size
  - Ranges from a few mm to several cm
- Morphology
  - Lobular in contour
  - May contain septations

MR Findings

- Lobular fluid signal mass with well-defined borders arising along peripheral margin of medial or lateral meniscus
- Usually traceable to meniscal tear
  - Most commonly horizontal cleavage tear

Ultrasonographic Findings

- If large cyst present, can be identified as cystic collection arising from joint line

Imaging Recommendations

- Best imaging tool
  - MR
- Protocol advice
  - Combination of short and long TE sequences to best distinguish meniscal tear and associated parameniscal cyst, respectively

DIFFERENTIAL DIAGNOSIS

Joint Ganglion

- Arises from joint capsule
- Not traceable to meniscal tear

Normal Joint Recess

- Parameniscal recesses may appear rounded

Bursitis

- Medial collateral ligament (MCL) bursa
  - Within leaves of MCL
- Semimembranosus-tibial collateral ligament bursitis
  - U-shaped collection around insertion of semimembranosus tendon

Cystic Mass

- Hemangioma
- Synovial sarcoma
- Hematoma

PATHOLOGY

General Features

- Etiology
  - Results from transmission of joint fluid through horizontal cleavage or radial meniscal tear into parameniscal tissues
  - Likely related to "ball-valve" mechanism where fluid enters cyst but cannot escape
- Associated abnormalities
  - Always associated with meniscal tear
    - Most commonly horizontal cleavage type

CLINICAL ISSUES

Presentation

- Most common signs/symptoms
  - Pain at joint line
  - ± palpable mass
    - Pisani sign: Palpable joint line mass that disappears with knee flexion

Demographics

- Age
  - Older adults
- Epidemiology
  - 7% prevalence in surgical meniscectomy

Natural History & Prognosis

- Grows slowly over time
- Likely to recur if underlying meniscal tear not repaired or resected

Treatment

- Arthroscopic resection of cyst and repair or resection of meniscal tear

DIAGNOSTIC CHECKLIST

Consider

- Meniscal tear if cystic collection seen adjacent to meniscus

Image Interpretation Pearls

- Presence of cyst in typical location (posterior medial compartment or anterolateral aspect of lateral compartment) should strongly raise suspicion for associated meniscal tear
  - Careful scrutiny of short TE sequences in these areas necessary for exclusion of associated tear

Reporting Tips

- If typical cyst encountered but meniscal tear not definitely identified, should suggest likely presence of nonvisualized tear

SELECTED REFERENCES

1. Chang A: Imaging-guided treatment of meniscal cysts. HSS J. 5(1):58-60, 2009
2. Hulet C et al: Arthroscopic treatment of 105 lateral meniscal cysts with 5-year average follow-up. Arthroscopy. 20(8):831-6, 2004
3. Campbell SE et al: MR imaging of meniscal cysts: incidence, location, and clinical significance. AJR Am J Roentgenol. 177(2):409-13, 2001

PARAMENISCAL CYST

(Left) *Sagittal T2WI MR shows a small parameniscal cyst with small septations arising from a horizontal tear in the anterior horn of the lateral meniscus. This is a common location for the formation of parameniscal cysts.* **(Right)** *Sagittal T2WI MR more laterally in the same patient shows the well-marginated nature of the cyst and the debris and septations within it. The meniscus was partially discoid and the tear somewhat complex.*

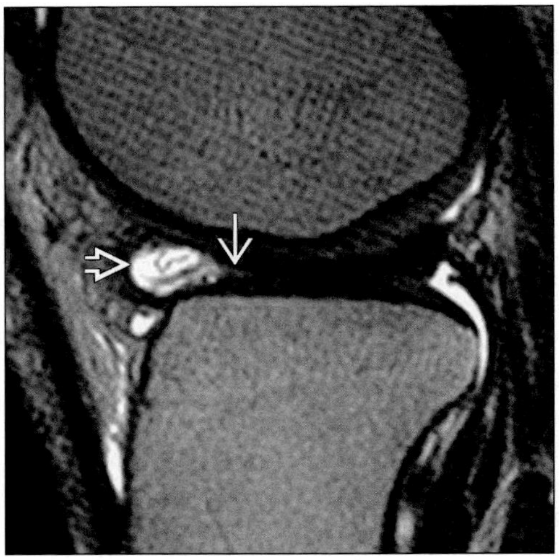

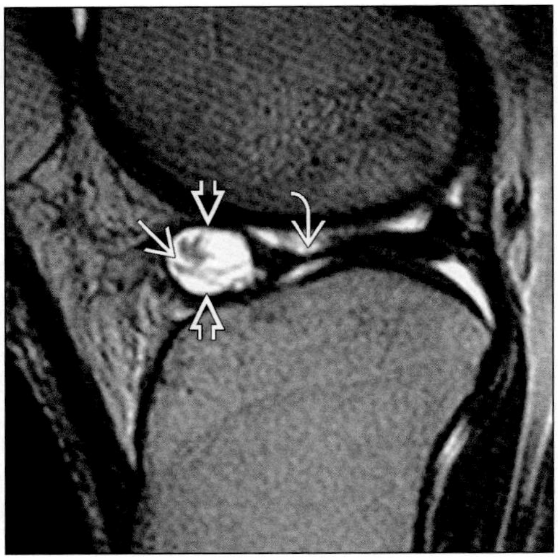

(Left) *Axial PDWI FS MR from the same patient shows the extent of the parameniscal cyst, which lies between the midline of the joint and the junction of the anterior horn and body of the lateral meniscus. Axial images are very useful in demonstrating the size and extent of these cysts.* **(Right)** *Sagittal PDWI MR shows a small parameniscal cyst arising from a small horizontal cleavage tear in the anterior horn of the medial meniscus. This is a relatively uncommon location for cysts.*

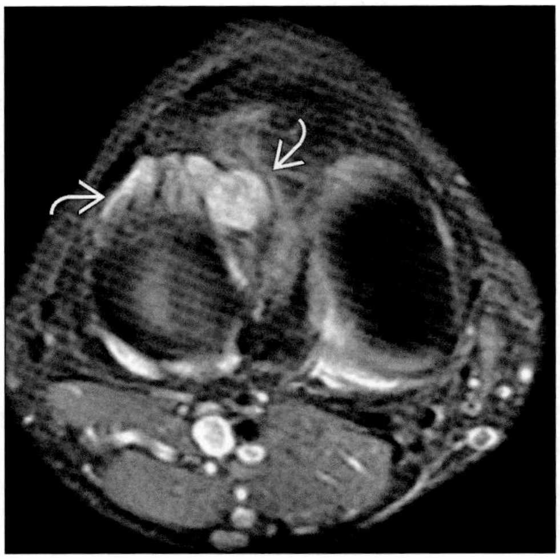

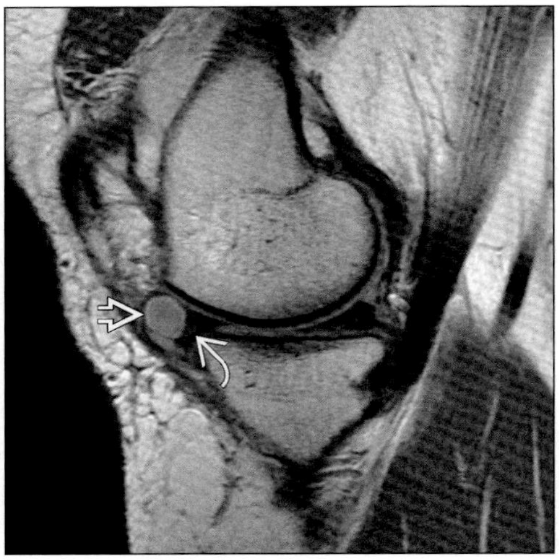

(Left) *Sagittal PDWI MR shows a small parameniscal cyst arising from an oblique horizontal tear in the posterior horn of the medial meniscus.* **(Right)** *Sagittal T2WI MR from the same patient shows fluid signal in the cyst and in the tear. It is less common to see the tear well on T2-weighted images. If no tear can be identified on MR, the suggestion should still be made that such a cyst in a characteristic location could be due to a nonvisualized tear.*

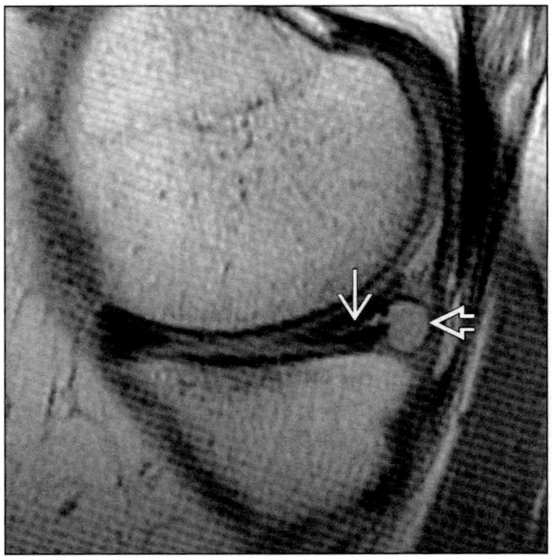

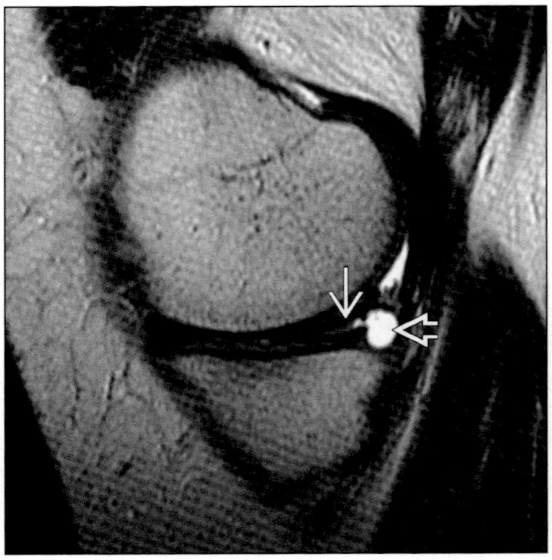

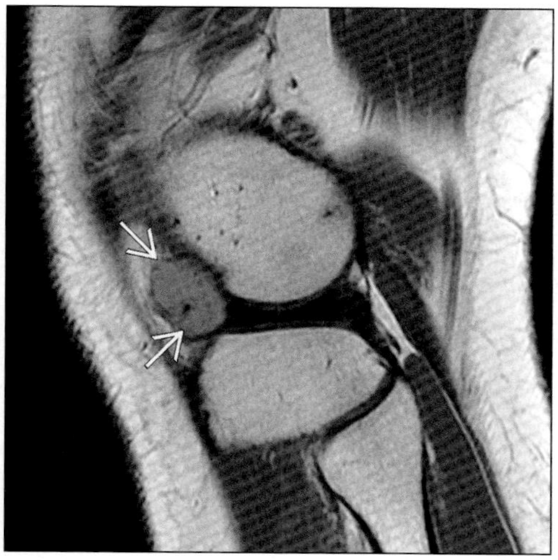

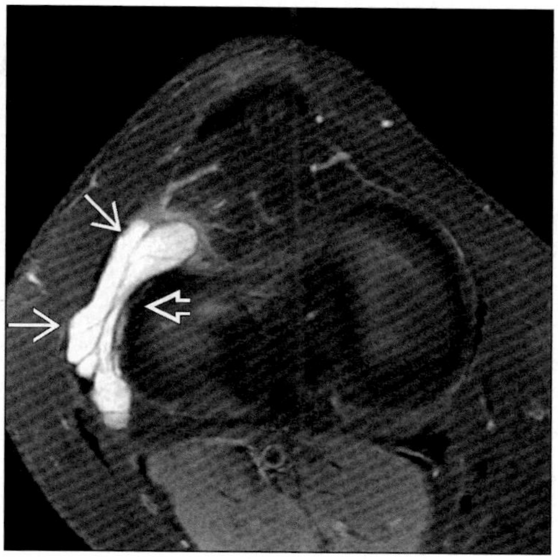

(Left) *Sagittal PDWI MR from a 19-year-old woman with anterolateral knee pain demonstrates a rounded hyperintense mass anterolateral to the lateral meniscus. No definite meniscal tear could be identified at MR imaging.* ***(Right)*** *Axial PDWI FS MR from the same patient shows an elongate septated cyst conforming to the peripheral surface of the lateral meniscus . At arthroscopy, the patient had a small horizontal tear of the lateral meniscus leading to the large cyst.*

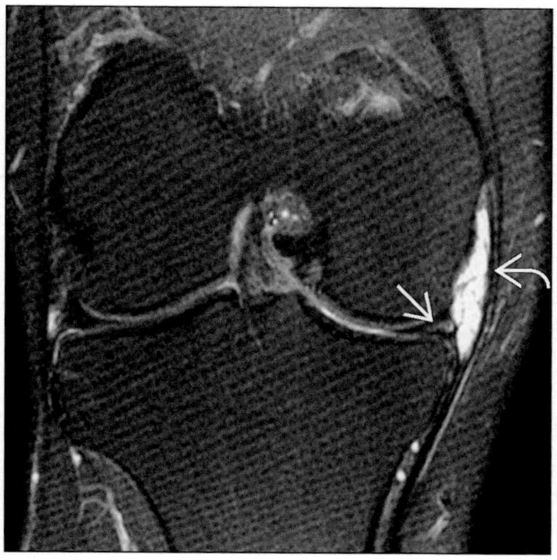

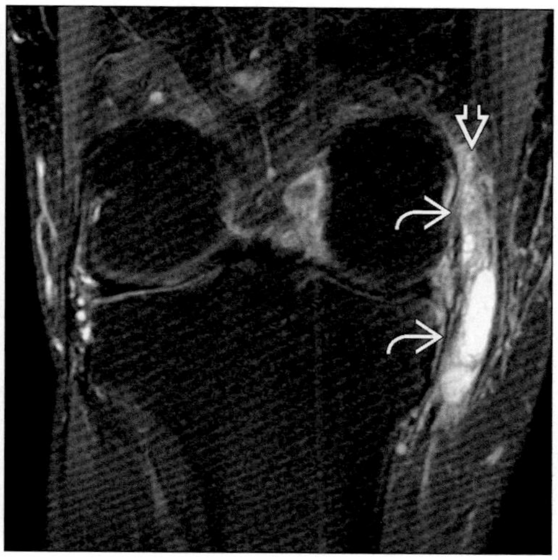

(Left) *Coronal T2WI FS MR shows a horizontal cleavage tear of the body of the medial meniscus leading to a large parameniscal cyst. The cyst dissects superior and inferior to the joint line but does not extend superficial to the medial collateral ligament (MCL) superficial fibers .* ***(Right)*** *Coronal T2WI FS MR shows a large irregular parameniscal cyst that has dissected through and superficial to the MCL . The cyst is partially ruptured superiorly .*

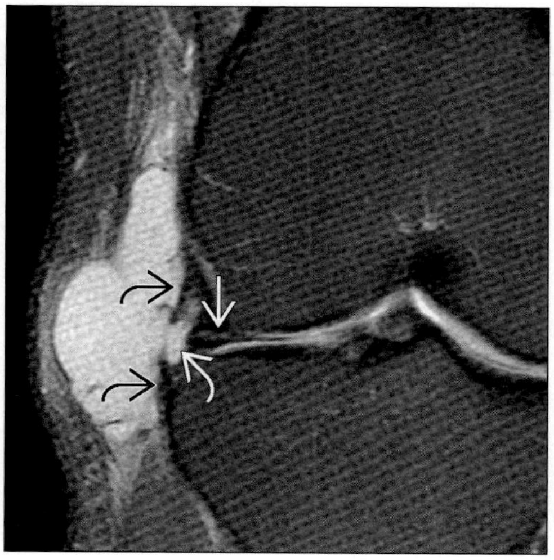

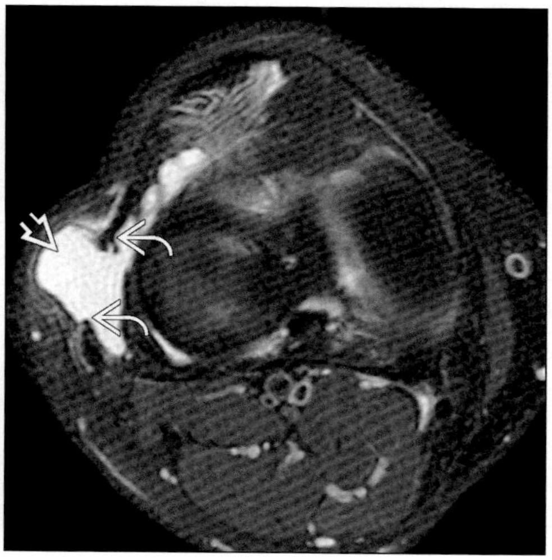

(Left) *Coronal T2WI FS MR shows a large parameniscal cyst arising from the mid-body of the lateral meniscus due to a horizontal tear . A small portion of the cyst remains deep to the lateral capsular ligament , but most of the cyst has extended through the ligament.* ***(Right)*** *Axial PDWI FS MR from the same patient shows the superficial extension of the lateral parameniscal cyst . There is a large defect in the lateral collateral ligament complex .*

INTRAMENISCAL CYST

Key Facts

Terminology
- Focal mass-like fluid collection within substance of meniscus

Imaging
- May be in medial or lateral meniscus
  - More common in posterior horn or body of meniscus
- Rounded or ovoid in shape
- Follows fluid on all imaging sequences
- Fully contained within substance of meniscus
- Usually adjacent to a linear meniscal tear
  - Most often horizontal cleavage tear

Top Differential Diagnoses
- Meniscal horizontal tear
- Meniscal degeneration
- Meniscal ossicle
- Parameniscal cyst
- Magic angle artifact

Pathology
- Probably cystic expansion of degenerative meniscal tear due to fluid build-up within meniscal defect
- Can be connected to a parameniscal cyst

Clinical Issues
- Joint line pain similar to meniscal tear
- May slowly enlarge over time
- Resection of cyst and resection or repair of underlying meniscal tear

Diagnostic Checklist
- Meniscal tear as underlying cause of intrameniscal cyst
- Measure and report size of cyst

*(**Left**) Coronal T1WI MR shows globular intermediate signal in the body of the medial meniscus →. Differential diagnosis includes mucinous degeneration, tear, magic angle artifact, and cyst. (**Right**) Coronal T2WI FS MR from the same patient shows a well-defined fluid signal collection → within the confines of the meniscus consistent with an intrameniscal cyst. The posterior horn and body of the medial meniscus is the most common site for this finding.*

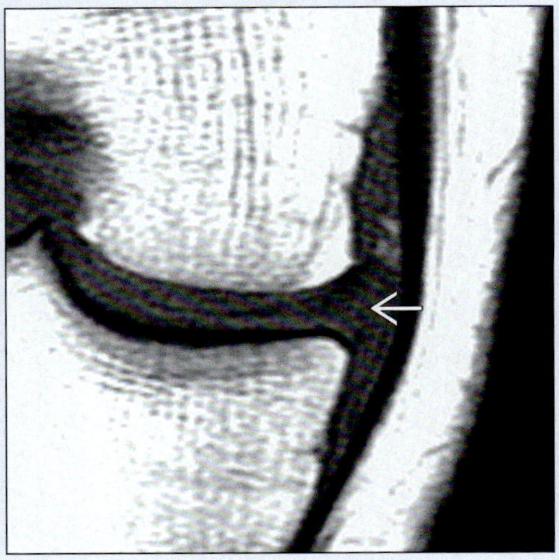

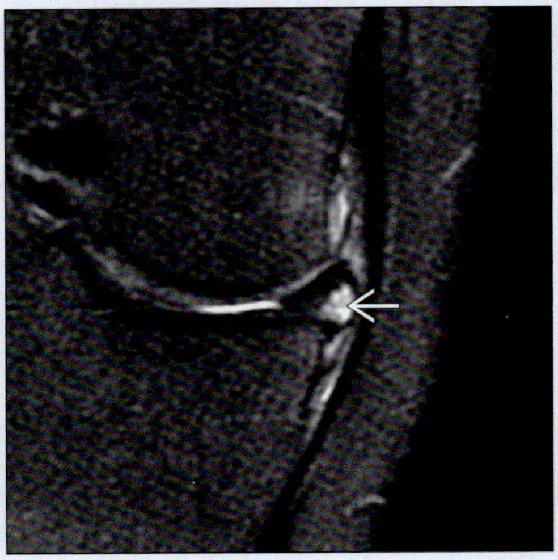

*(**Left**) Sagittal T2WI MR shows a well-defined fluid-signal collection → adjacent to a more linear lower signal collection → in the anterior horn of the medial meniscus. This is an intrameniscal cyst adjacent to (and probably arising from) a horizontal cleavage tear. A small posterior horn parameniscal cyst → is present posteriorly. (**Right**) Sagittal T2WI MR shows an intrameniscal cyst → in the anterior horn of the lateral meniscus. This was due to a horizontal cleavage tear.*

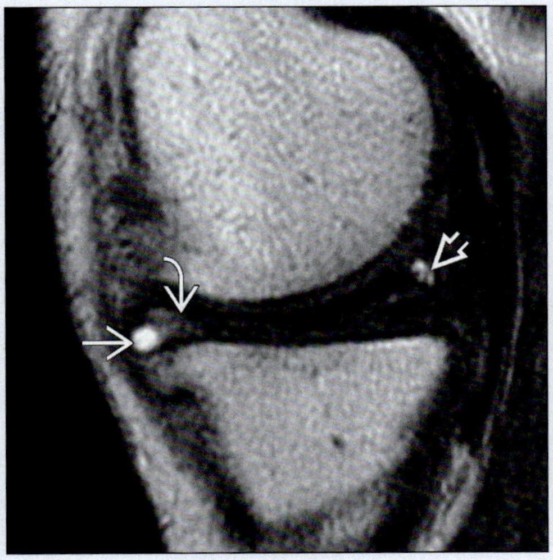

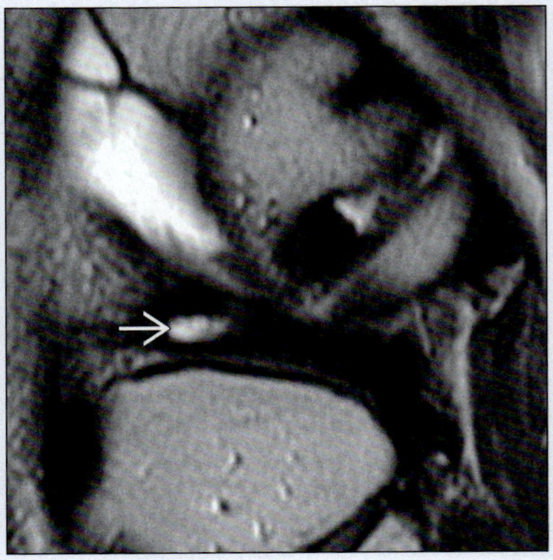

MENISCAL OSSICLE

Key Facts

Terminology
- Ossification within meniscus of knee

Imaging
- Focus of bone within meniscal substance
- Posterior horn, medial > lateral meniscus
- May range from tiny to > 1 cm
- Ovoid or triangular
- Radiographs
  - Usually corticated
  - Usually ovoid or triangular
- MR
  - Generally follows normal marrow signal on all sequences
  - Surrounded by meniscal tissue
  - Meniscus may be normal or torn
- Radiographs confirm osseous nature of lesion
- Oblique radiographs may be necessary

Top Differential Diagnoses
- Osteophyte
- Intraarticular body
- Meniscal calcification: Generally CPPD

Pathology
- Probably related to previous meniscal trauma

Clinical Issues
- Local pain and swelling similar to meniscal tear
- May be asymptomatic
- Arthroscopic resection if symptomatic

Diagnostic Checklist
- Meniscal ossicle if well-defined noncystic focus present in substance of meniscus on MR
- Always correlate suspected meniscal ossicle seen on MR with radiographs

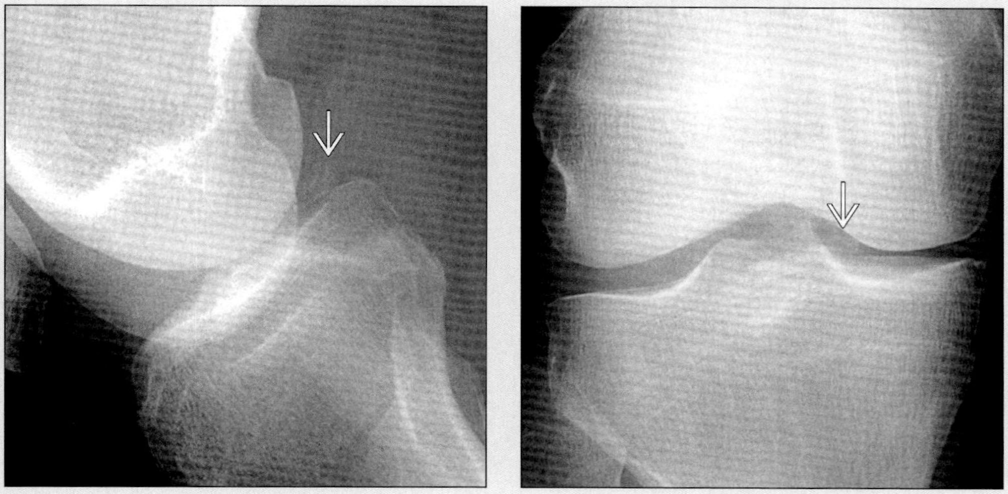

(Left) *Lateral radiograph demonstrates a well-marginated ossicle ➡ projecting posteriorly at the joint line. The object is ovoid and faintly calcified.* ***(Right)*** *AP radiograph from the same patient shows that the ossicle ➡ is within the medial compartment of the knee, corresponding to the expected location of the posterior root of the medial meniscus. One theory of meniscal ossicle formation is that they represent old avulsions of the tibial attachment of the posterior horn.*

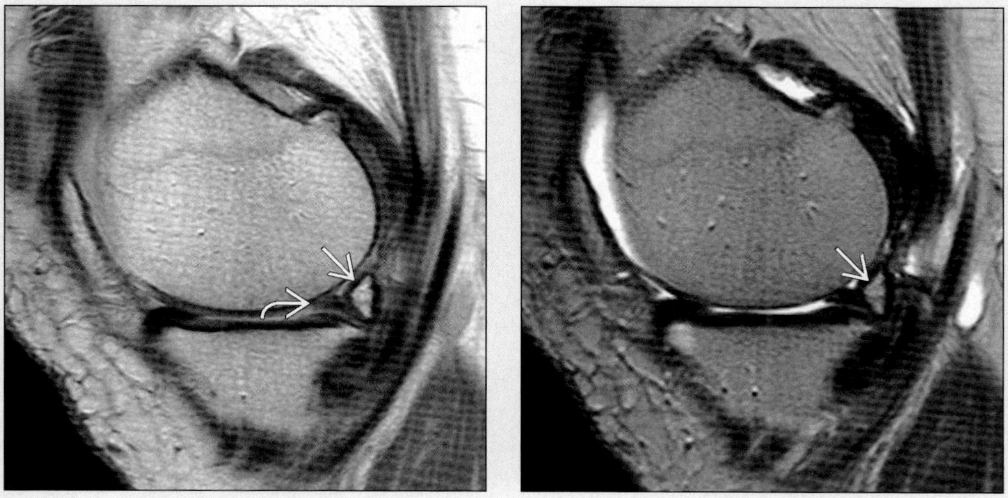

(Left) *Sagittal PDWI MR shows a triangular corticated ossicle ➡ within the posterior horn of the medial meniscus. There is a mild degenerative horizontal cleavage tear ➡ of the adjacent meniscus.* ***(Right)*** *Sagittal T2WI MR from the same patient shows the internal signal of the ossicle ➡ follows that of marrow fat in the larger bones. Patient complained of knee pain and clicking; resection of the meniscal ossicles resulted in resolution of the symptoms.*

POSTOPERATIVE IMAGING OF MENISCI

Key Facts

Terminology

- Meniscal allograft transplantation (MAT)

Imaging

- No meniscal abnormality
  - Minimal resection
  - Primary repair of meniscal tear with no resection
- Morphologic change
  - Meniscal volume diminished
  - Blunting/truncation of free edge
- Signal abnormalities
  - Residual globular or linear signal within meniscus
  - May extend to surgical border even in absence of residual or recurrent tear
  - Visible on short TE (T1WI, PDWI) sequences
  - Significantly diminishes on long TE (T2WI)
  - Indirect MR arthrogram may → false-positive findings due to enhancement of repair tissue
- Meniscal allograft
  - Small linear surgical defects in tibia at sites of meniscal attachment
  - Transplanted meniscus often subluxes out of joint more than native meniscus
- Recurrent or residual meniscal tear
  - Recurrent tears may occur at location of prior meniscal resection or at new location
  - Linear or globular signal extending to articular surface of meniscus visible on T2WI
  - Contrast enters tear on direct MR arthrography
- Micrometallic artifact
- Complications
  - Subchondral insufficiency fracture
  - Chondral fracture
  - Synovitis
  - Arthrofibrosis

*(**Left**) Sagittal graphic demonstrates the origin of signal in a postoperative meniscus. The figure on the left shows a complex tear ➔ and the line of resection ➔ beyond the tear margin. The figure on the right shows postoperative meniscus with central degenerative signal now extending to new articular surface ➔. (**Right**) Sagittal MR arthrogram T1WI FS after previous medial meniscal surgery shows linear signal ➔ extending to inferior surface. No contrast enters the meniscus.*

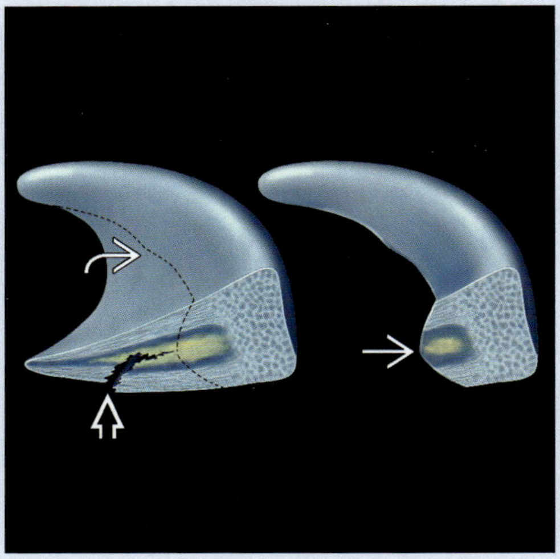

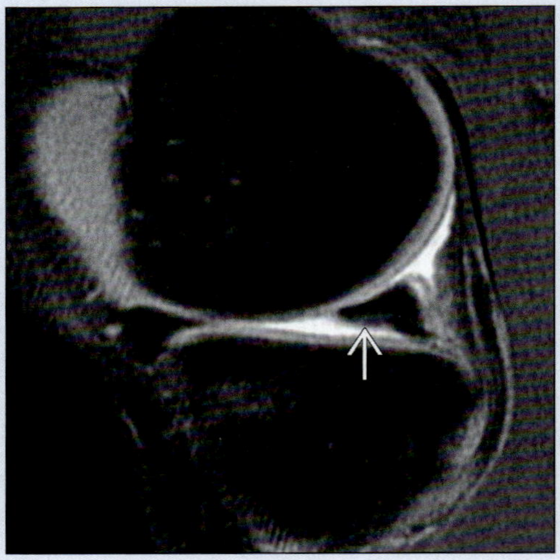

*(**Left**) Coronal MR arthrogram T1WI from the same patient shows elevated intrameniscal signal ➔ in the postoperative meniscus extending to the inferior articular surface. No contrast enters the meniscus. Findings are typical of previous meniscal resection without a recurrent tear. (**Right**) Coronal MR arthrogram T2WI FS from the same patient confirms the lack of fluid entering the meniscus. The postoperative margin of the meniscus is smooth and regular.*

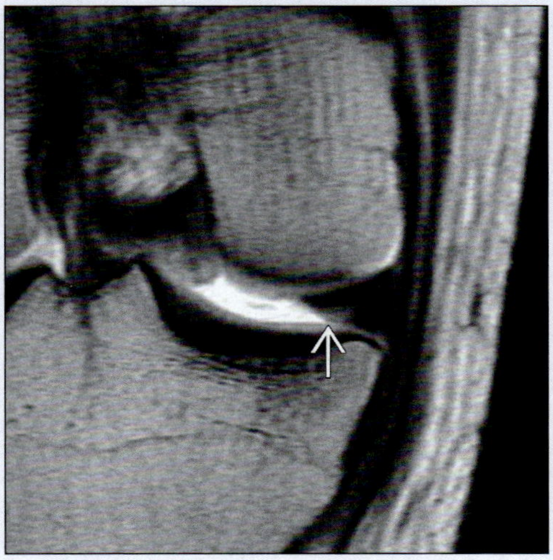

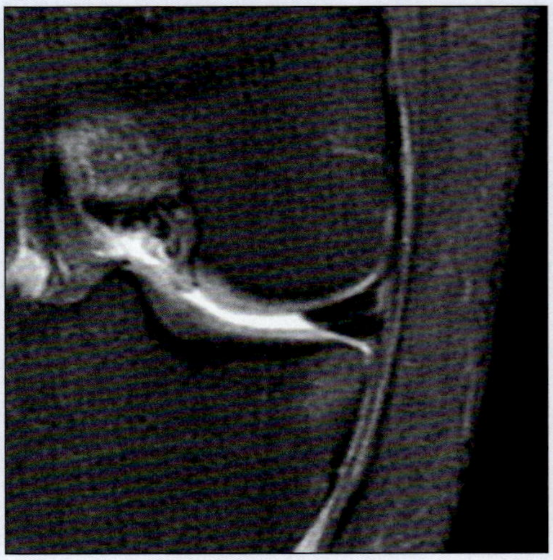

POSTOPERATIVE IMAGING OF MENISCI

TERMINOLOGY

Abbreviations

- Meniscal allograft transplantation (MAT)

Synonyms

- Status post (s/p) meniscectomy

Definitions

- Previous meniscal repair or previous partial or complete meniscal resection (meniscectomy)
- Meniscal allograft: Transplantation of cadaveric meniscal tissue to replace badly torn native meniscus

IMAGING

General Features

- Best diagnostic clue
  - Truncation or decreased size of meniscus
  - Micrometallic artifact in knee joint from previous surgery
- Morphology
  - Postoperative meniscus may have variety of appearances
    - Normal
      - Primary repair of meniscal tear
      - Resection of small amount of meniscal tissue
    - Blunted free edge
    - Diminished volume of meniscus
    - Absent meniscus

CT Findings

- CT arthrography
  - Small ± blunted meniscus
  - Smooth postsurgical edge
  - Infrapatellar scar with arthroscopic portals

MR Findings

- **No meniscal abnormality**
  - Postoperative meniscus may appear normal if
    - Resection was minimal
    - Primary repair of meniscal tear with no resection
      - Usually for vertical longitudinal tear in red zone
- **Morphologic change**
  - Meniscal volume diminished
  - Blunting/truncation of free edge
    - Especially after resection of horizontal cleavage tear
- **Signal abnormalities**
  - Residual globular or linear signal within meniscus
  - May extend to surgical border even in absence of residual or recurrent tear
    - Surgery debrides meniscus beyond margin of macroscopic tear
    - Surgical margin extends to underlying meniscal degeneration, which contains abnormal signal
    - Fibrovascular healing response at site of resection may also have elevated signal extending to surface
    - High signal extending to surface at site of surgery may persist for 1 year following meniscal repair
    - Visible on short TE (T1WI, PDWI) sequences
      - Specificity for recurrent or residual tear (50%)
      - Sensitivity (70%)
    - Signal significantly diminishes on long TE (T2WI)
    - Indirect MR arthrogram may result in false-positive due to enhancement of repair tissue
- **Meniscal allograft**
  - Small linear surgical defects in tibia at sites of meniscal attachment
    - Capsular margin
    - Anterior and posterior root entheses
      - ± bone plugs
  - Susceptibility artifact
  - Transplanted meniscus often subluxes out of joint more than native meniscus
- **Recurrent or residual meniscal tear**
  - Recurrent tears may occur at location of prior meniscal resection or at new location
  - Linear or globular high signal extending to articular surface of meniscus visible on T2WI
    - Specificity for recurrent or residual tear (88-92%)
    - Sensitivity (41-69%)
  - Contrast enters tear on direct MR arthrography
    - Specificity for recurrent or residual tear (90%)
    - Sensitivity (88-95%)
- **Micrometallic artifact**
  - Related to drilling or burring with microscopic metallic particles causing susceptibility artifact
  - May be located at site of meniscal surgery ± elsewhere in knee
- **Linear fibrosis in infrapatellar fat**
  - Focal scar at entry points of arthroscopic instruments medial and lateral to patellar tendon
- **Complications**
  - Subchondral insufficiency fracture
    - Linear low signal on all sequences surrounded by marrow edema
    - Usually in femoral condyle adjacent to postoperative meniscus
  - Chondral fracture
    - Focal well-marginated chondral defect with subjacent marrow edema
    - Usually in femoral condyle adjacent to postoperative meniscus
  - Synovitis
    - Thick, irregular synovial tissue in margins of joint, especially near surgical site
    - Enhances after IV gadolinium administration
  - Arthrofibrosis
    - Nodular or diffuse low PDWI signal synovial thickening ± linear or mass-like fibrosis in infrapatellar fat
    - Limited joint distension on direct MR arthrography

Imaging Recommendations

- Best imaging tool
  - Direct MR arthrography (88% accuracy for retear vs. 66% for conventional MR)

DIFFERENTIAL DIAGNOSIS

Meniscal Degeneration

- Mild blunting of free edge may be degenerative finding in absence of prior surgery

Displaced Meniscal Tear

- Small or blunted meniscal remnant

- Displaced meniscal fragment may be attached to peripheral remnant or loose in joint

PATHOLOGY

General Features

- Meniscal allograft
  - Harvested and trimmed ± associated bone plugs at anterior and posterior root attachment sites
  - Meniscus sutured in place along capsular margin
  - Holes drilled in recipient tibia at sites of root attachments
  - Bone plugs from graft sutured into drilled holes

CLINICAL ISSUES

Presentation

- Most common signs/symptoms
  - Recurrent meniscal tear
    - Recurrent joint line pain
    - Crepitus
    - Positive McMurray test
    - If displaced fragment → locking
    - Effusion
- Other signs/symptoms
  - Knee pain related to secondary chondral fracture

Natural History & Prognosis

- Most patients do well after meniscal surgery
  - Often deteriorate clinically over time due to underlying meniscal degeneration
- Complications
  - If large meniscal resection → increased bone-on-bone contact → accelerated osteoarthritis
  - Postoperative subchondral insufficiency fracture
    - Previously termed "post-meniscectomy osteonecrosis"
  - Occasional acute chondral fracture
  - Postoperative synovitis
  - Postoperative arthrofibrosis
  - Meniscal allograft
    - Severe graft subluxation
    - Detachment of graft
    - Tear of graft

DIAGNOSTIC CHECKLIST

Consider

- Blunted or small meniscus may represent postoperative change rather than new tear
  - Other signs of arthroscopy (fibrosis in Hoffa fat pad) may confirm
- ↑ signal extending to surface persists post-arthroscopically on short TE sequences; decreases accuracy of diagnosing new tear

Image Interpretation Pearls

- Never diagnose recurrent meniscal tear on basis of short TE sequences alone
  - Must compare with T2WI for presence of fluid in tear

SELECTED REFERENCES

1. Cardello P et al: Retears of postoperative knee meniscus: findings on magnetic resonance imaging (MRI) and magnetic resonance arthrography (MRA) by using low and high field magnets. Skeletal Radiol. 38(2):149-56, 2009
2. Lee DH et al: Midterm outcomes after meniscal allograft transplantation: comparison of cases with extrusion versus without extrusion. Am J Sports Med. Epub ahead of print, 2009
3. Miao Y et al: MRI signal changes in completely healed meniscus confirmed by second-look arthroscopy after meniscal repair with bioabsorbable arrows. Knee Surg Sports Traumatol Arthrosc. 17(6):622-30, 2009
4. Vance K et al: Magnetic resonance imaging of the postoperative meniscus. Arthroscopy. 25(5):522-30, 2009
5. Ciliz D et al: Evaluation of postoperative menisci with MR arthrography and routine conventional MRI. Clin Imaging. 32(3):212-9, 2008
6. Lee DH et al: Evaluation of meniscus allograft transplantation with serial magnetic resonance imaging during the first postoperative year: focus on graft extrusion. Arthroscopy. 24(10):1115-21, 2008
7. MacDessi SJ et al: Subchondral fracture following arthroscopic knee surgery. A series of eight cases. J Bone Joint Surg Am. 90(5):1007-12, 2008
8. Pujol N et al: Meniscal healing after meniscal repair: a CT arthrography assessment. Am J Sports Med. 36(8):1489-95, 2008
9. Genovese E et al: Follow-up of collagen meniscus implants by MRI. Radiol Med. 112(7):1036-48, 2007
10. Gopez AG et al: MR imaging of the postoperative meniscus: repair, resection, and replacement. Semin Musculoskelet Radiol. 10(3):229-40, 2006
11. Muscolo DL et al: Medial meniscal tears and spontaneous osteonecrosis of the knee. Arthroscopy. 22(4):457-60, 2006
12. De Smet AA: MR imaging and MR arthrography for diagnosis of recurrent tears in the postoperative meniscus. Semin Musculoskelet Radiol. 9(2):116-24, 2005
13. Hantes ME et al: Evaluation of meniscal repair with serial magnetic resonance imaging: a comparative study between conventional MRI and indirect MR arthrography. Eur J Radiol. 2004 Jun;50(3):231-7. Erratum in: Eur J Radiol. 53(2):321, 2005
14. McCauley TR: MR imaging evaluation of the postoperative knee. Radiology. 234(1):53-61, 2005
15. Toms AP et al: Imaging the post-operative meniscus. Eur J Radiol. 54(2):189-98, 2005
16. Encalada I et al: Osteonecrosis after arthroscopic meniscectomy using radiofrequency. Arthroscopy. 20(6):632-6, 2004
17. DeFalco RA et al: Osteonecrosis of the knee after arthroscopic meniscectomy and chondroplasty: a case report and literature review. Am J Sports Med. 31(6):1013-6, 2003
18. Magee T et al: MR arthrography of postoperative knee: for which patients is it useful? Radiology. 229(1):159-63, 2003
19. Peters G et al: The current state of meniscal allograft transplantation and replacement. Knee. 10(1):19-31, 2003
20. Davis KW et al: MR imaging of the postoperative meniscus of the knee. Semin Musculoskelet Radiol. 6(1):35-45, 2002
21. Nakamura N et al: Subchondral microfracture of the knee without osteonecrosis after arthroscopic medial meniscectomy. Arthroscopy. 18(5):538-41, 2002
22. Recht MP et al: MR imaging of the postoperative knee: a pictorial essay. Radiographics. 22(4):765-74, 2002
23. Johnson TC et al: Osteonecrosis of the knee after arthroscopic surgery for meniscal tears and chondral lesions. Arthroscopy. 16(3):254-61, 2000
24. Totty WG et al: Imaging the postoperative meniscus. Magn Reson Imaging Clin N Am. 8(2):271-83, 2000
25. Lim PS et al: Repeat tear of postoperative meniscus: potential MR imaging signs. Radiology. 210(1):183-8, 1999

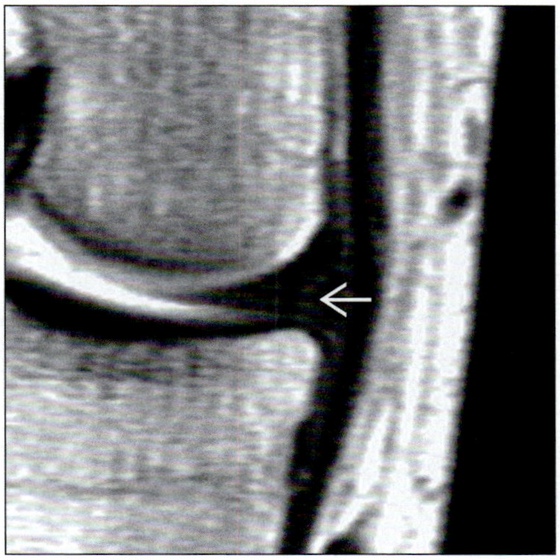

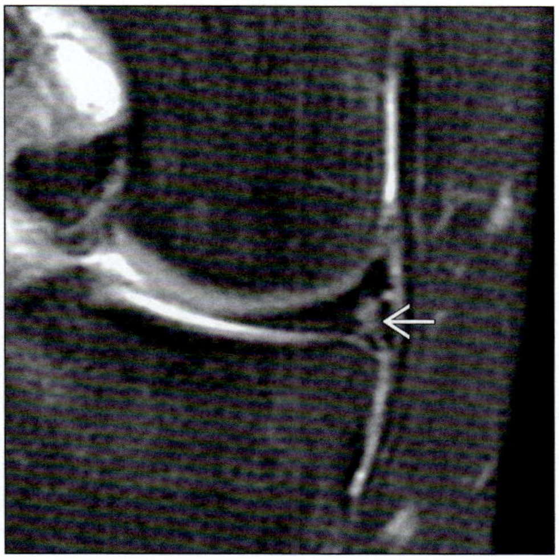

(Left) *Coronal MR arthrogram T1WI after surgical repair of a vertical longitudinal tear of the medial meniscal red zone shows faint linear intrameniscal signal* ➔*, but no contrast enters the meniscus.* ***(Right)*** *Coronal MR arthrogram T2WI FS from the same patient better shows the residual signal* ➔ *that extends to the inferior articular surface in the meniscus at the site of repair. No contrast enters the meniscus. Signal may remain at the site of previous meniscal repair for up to a year.*

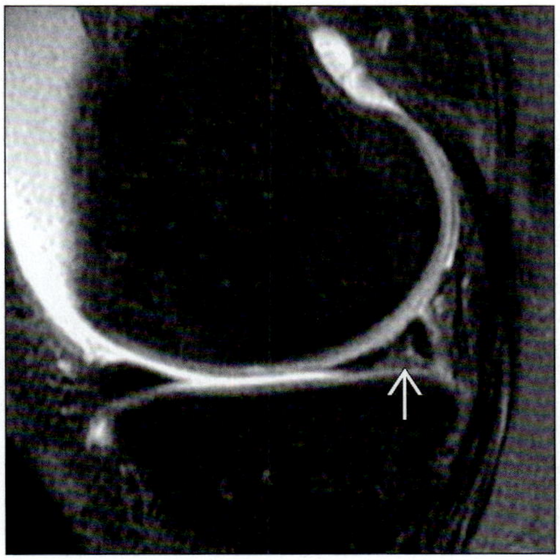

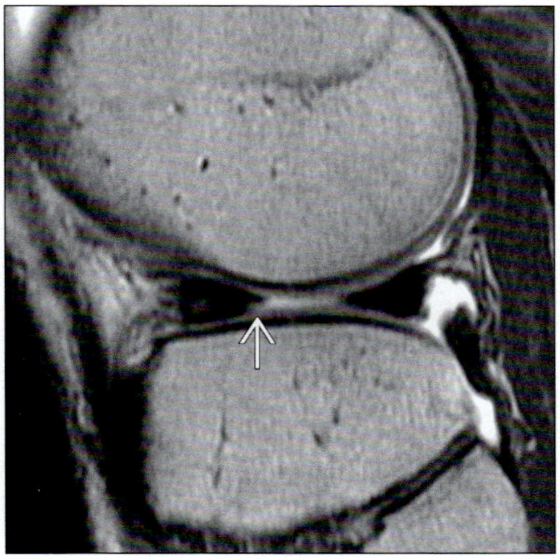

(Left) *Sagittal MR arthrogram PDWI FS from the same patient shows the residual intrameniscal signal* ➔ *extending to the inferior articular surface in the posterior horn. The presence of signal in the meniscus after surgery without entry of contrast into a defect on direct MR arthrography is a common finding and does not indicate a retear.* ***(Right)*** *Sagittal MR arthrogram T1WI after partial resection of the lateral meniscus for a tear shows mild irregularity of the free edge* ➔ *without a linear tear.*

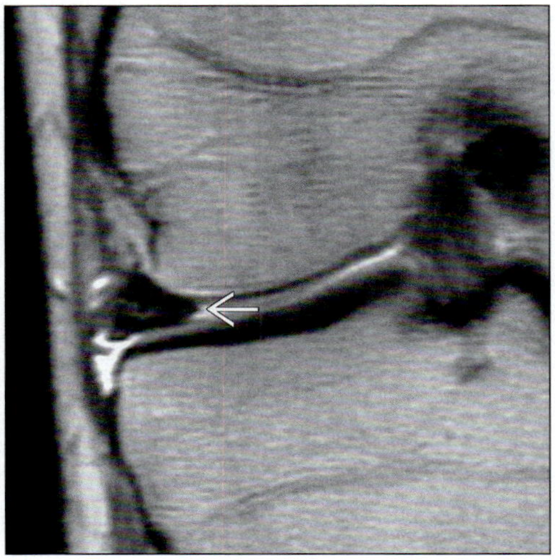

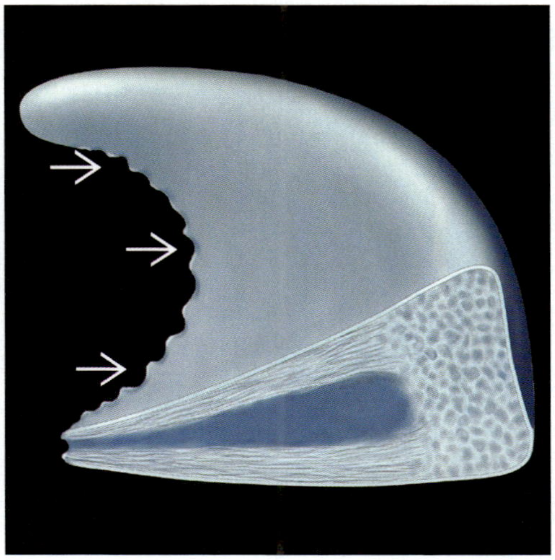

(Left) *Coronal MR arthrogram PDWI from the same patient shows mild irregularity of the free edge at the site of resection* ➔ *without a linear defect to suggest a tear. The postoperative meniscal border is usually smooth and regular; this finding suggests either incomplete smoothing at surgery or a tiny retear.* ***(Right)*** *Coronal graphic shows irregularity of the free edge of a meniscus* ➔ *after surgical trimming. Most meniscal surgery achieves a smooth, regular postoperative edge.*

(Left) *Sagittal MR arthrogram T2WI after large partial medial meniscal resection for a tear shows a very small residual posterior horn remnant ➔. The margins of the remnant are smooth, and there is no entry of contrast into the meniscus to suggest a recurrent tear.* ***(Right)*** *Coronal MR arthrogram PDWI from the same patient shows the very small meniscal remnant ➔. The triangular shape and smooth margins of the meniscus have been preserved. No contrast enters the meniscal substance.*

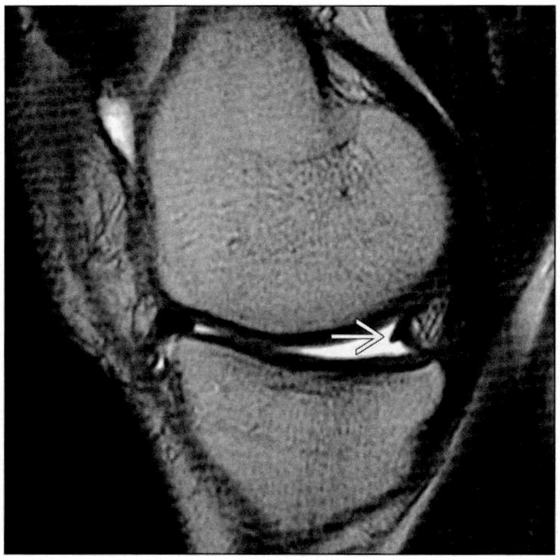

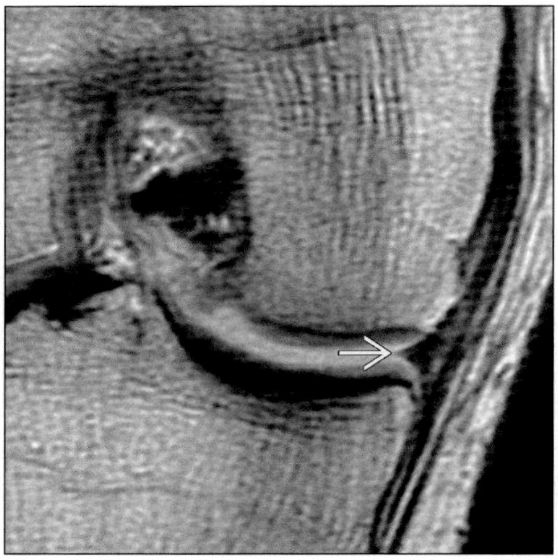

(Left) *Sagittal PDWI MR after previous partial medial meniscectomy shows linear signal in the posterior horn extending to the articular surface ➔. The presence of surfacing signal on short TE sequences in a postoperative meniscus is not sufficient to diagnose a recurrent tear.* ***(Right)*** *Sagittal T2WI MR from the same patient shows joint fluid entering the meniscus ➔. This finding on a long TE sequence confirms the presence of a recurrent meniscal tear, which was identified at surgery.*

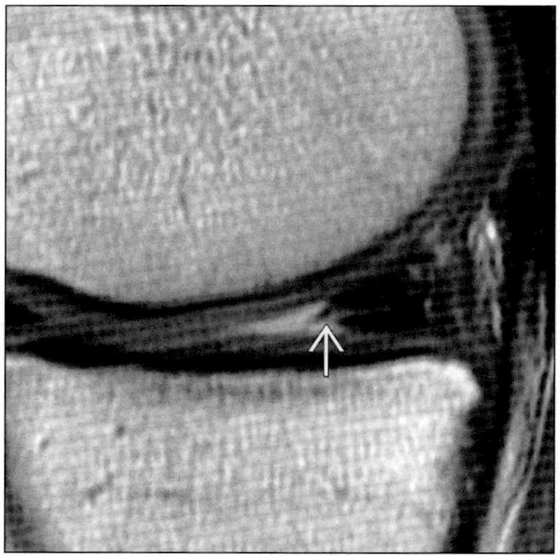

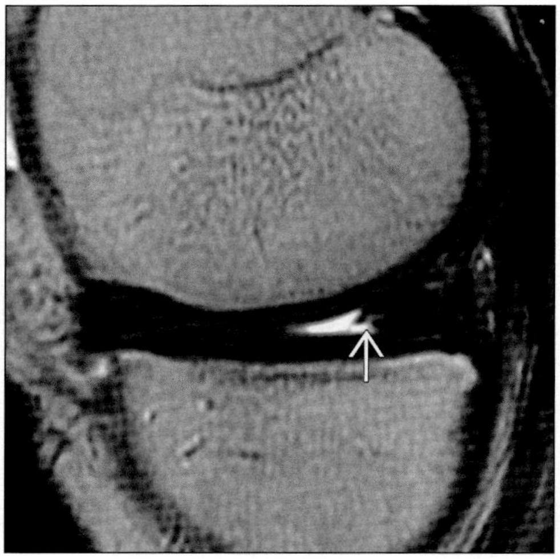

(Left) *Coronal MR arthrogram PDWI after previous high-grade partial resection of the medial meniscus shows irregularity of the meniscal margin ➔ with a small amount of contrast entering the substance of the meniscus, consistent with a shallow retear.* ***(Right)*** *Sagittal MR arthrogram T1WI FS from the same patient shows undersurface irregularity in the meniscal remnant with a small amount of contrast entering the defect ➔. A shallow tear was confirmed at surgery.*

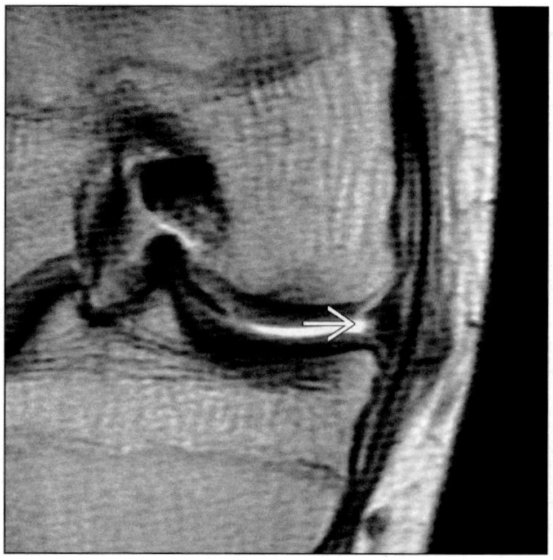

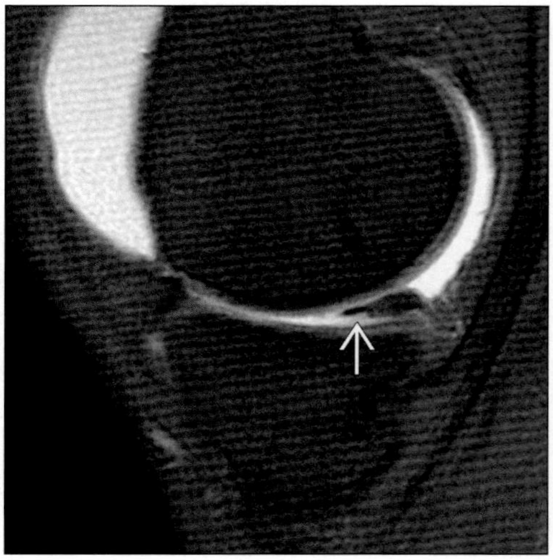

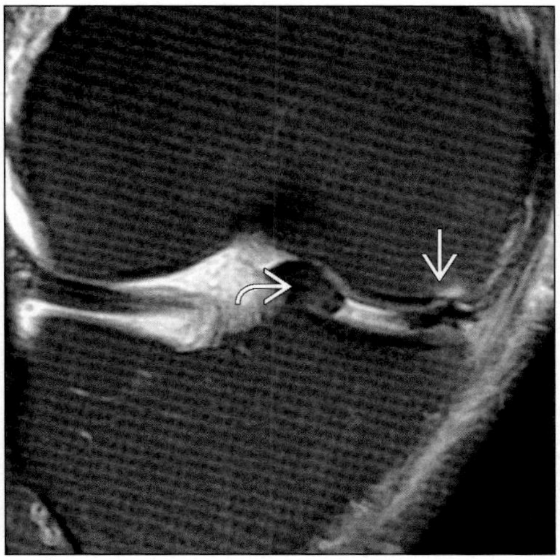

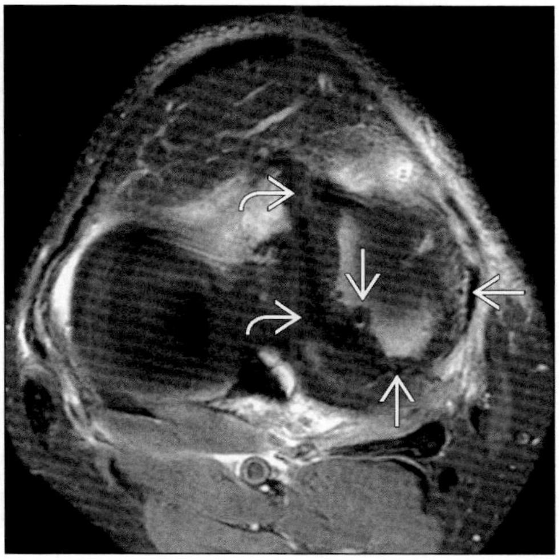

(Left) *Coronal T2WI FS MR after previous repair of a displaced bucket-handle tear of the medial meniscus shows a recurrent displaced tear. A large fragment extends into the notch. Postoperative micrometallic artifact is present at the site of previous repair.* ***(Right)*** *Axial PDWI FS MR from the same patient shows the large displaced recurrent tear and metallic artifact. At surgery, the meniscus separated at the site of previous repair in the red zone.*

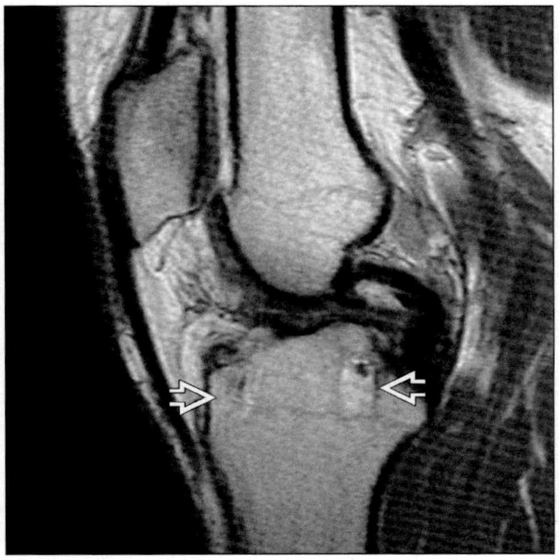

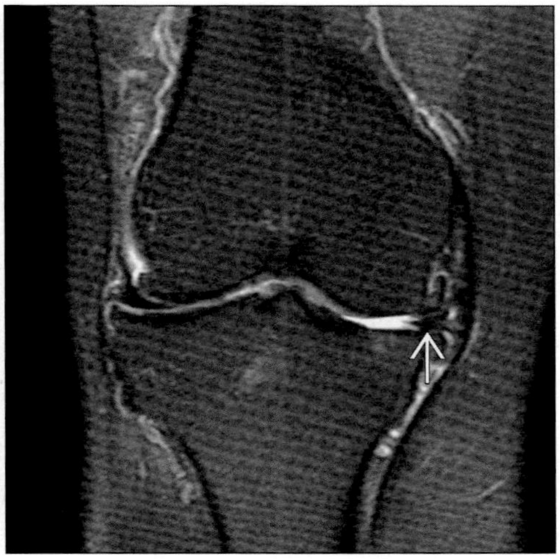

(Left) *Sagittal PDWI MR after medial meniscal allograft surgery shows that the bone plugs at the anterior and posterior root insertion sites are well incorporated.* ***(Right)*** *Coronal T2WI FS MR from the same patient shows moderate medial extrusion of the transplanted meniscus out of the compartment. There is associated osteoarthritis. Meniscal extrusion and accelerated osteoarthritis are common complications after meniscal allograft transplantation surgery.*

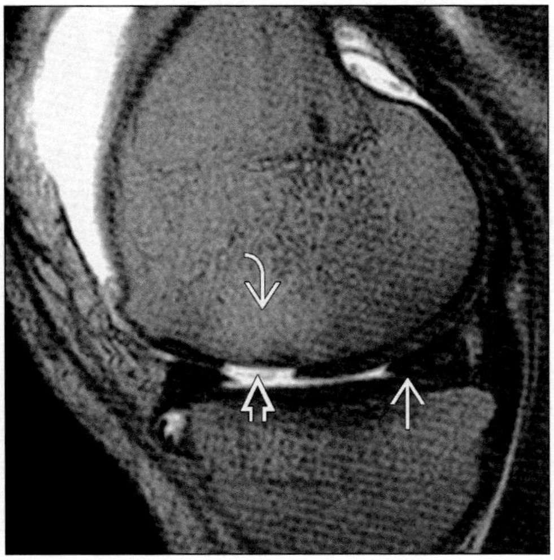

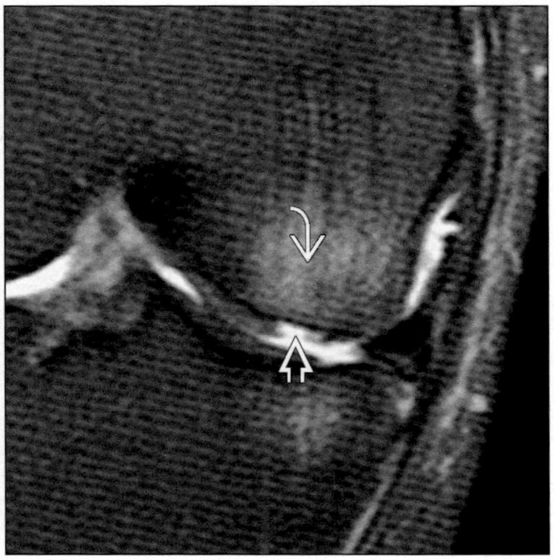

(Left) *Sagittal T2WI MR 3 months after medial meniscal surgery shows a large, well-marginated, full thickness chondral fracture of the medial femoral condyle with underlying marrow edema. The postoperative meniscus is intact.* ***(Right)*** *Coronal T2WI FS MR from the same patient shows the chondral fracture and underlying marrow edema. This likely occurred due to the lack of intervening meniscal tissue after the resection and is a fairly common complication after meniscectomy.*

QUADRICEPS INJURY

Key Facts

Terminology

- Quadriceps rupture, quadriceps tear, quad tear

Imaging

- May occur anywhere in distal anterior thigh, most commonly involving tendons of vastus intermedius
- Tears begin centrally and progress peripherally
- Patella baja (low patella) with high-grade tendon rupture
- Normal quadriceps tendon has laminar appearance
- **Tendinopathy**: Thickening and heterogeneity of 1 or more tendon units of the quadriceps mechanism
- **Partial tendon tear**: Focal partial discontinuity of 1 or more tendon units of the quadriceps mechanism
- **Complete tendon tear**: Focal discontinuity of 1 or more tendon units of quadriceps mechanism
- **Intramuscular injury**
    - Often occurs at myotendinous junction of 1 or more quadriceps tendon units
    - Larger field of view than usual for knee MR when evaluating quadriceps injury

Top Differential Diagnoses

- Patellar tendon injury
- Patellar fracture
- Transient patellar dislocation

Pathology

- Underlying chronic tendon-weakening condition often present
    - e.g., chronic renal disease, rheumatoid arthritis, gout, steroid use

Clinical Issues

- Early surgical repair indicated for acute complete rupture

__(Left)__ AP graphic of the knee shows complete disruption of the quadriceps mechanism just above the patella. The patellar tendon and retinacula remain intact. __(Right)__ Coronal T2WI FS MR shows complete rupture of the quadriceps mechanism just above the patella. There is surrounding hematoma, indicating an acute injury. Most large ruptures of the quadriceps occur in patients with underlying predisposing conditions; this patient had renal failure.

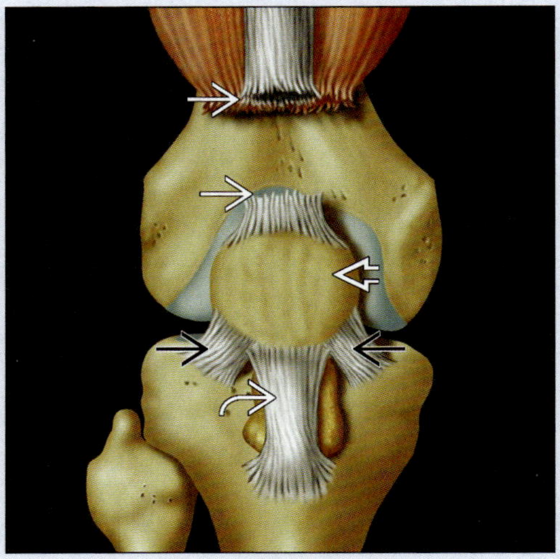

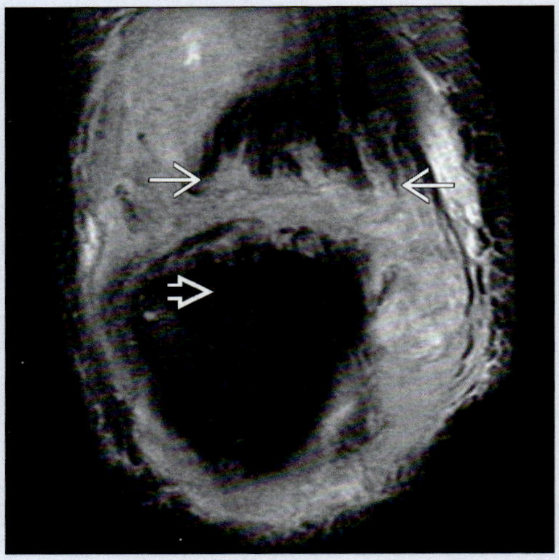

__(Left)__ Sagittal T2WI MR in the same patient shows complete rupture of the quadriceps just above the patella. There is mild retraction of the tendon due to muscle contraction, and hematoma is interposed between the tendon and the patella. Note mild redundancy of the patellar tendon due to laxity. __(Right)__ Axial PDWI FS MR in the same patient shows the large hematoma filling most of the anterior thigh; no tendinous tissue is identified in the region of the quadriceps.

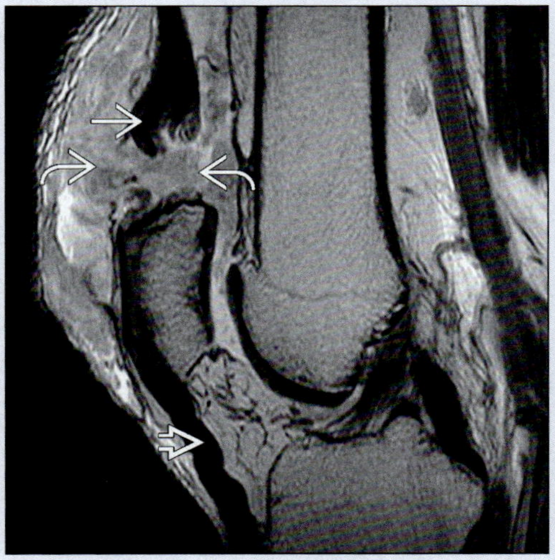

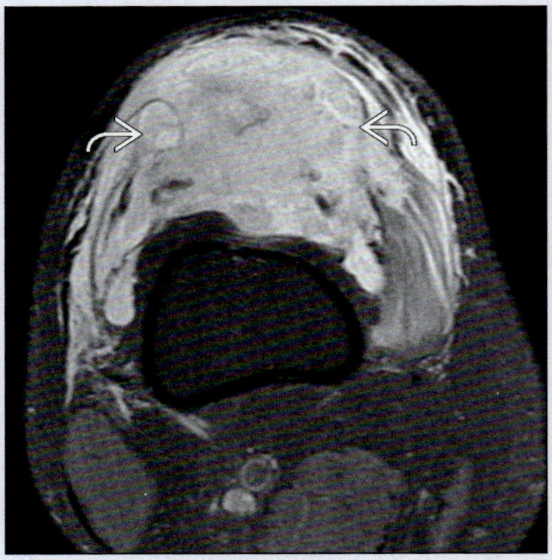

QUADRICEPS INJURY

TERMINOLOGY

Synonyms

- Quadriceps rupture, quadriceps tear, quad tear

Definitions

- Partial or complete disruption of distal myotendinous units of quadriceps mechanism of thigh

IMAGING

General Features

- Best diagnostic clue
  - Focal discontinuity of a portion of quadriceps tendon
- Location
  - May occur anywhere in distal anterior thigh
    - Most commonly involves tendon of vastus intermedius
  - Tears begin centrally and progress peripherally
- Size
  - Small partial tears may be a few mm in diameter
  - Large tears may be several cm in size
- Morphology
  - Quadriceps mechanism made up of
    - Rectus femoris (RF): Anterior and midline
    - Vastus intermedius (VI): Posterior and midline
    - Vastus lateralis (VL): Lateral to midline
    - Vastus medialis (VM): Medial to midline
      - Vastus medialis obliquus (VMO): Distal portion of VM with more horizontally oriented fibers

Radiographic Findings

- Patella baja (low patella) with high-grade tendon rupture
- Avulsion fragment from superior patella
- Suprapatellar soft tissue swelling
- Calcification in quadriceps tendons

CT Findings

- Similar to radiographs, but more sensitive for small avulsions and calcifications

MR Findings

- Normal quadriceps tendon has laminar appearance
  - May be bi-, tri-, or quadrilaminar on sagittal MR images
  - Fat intervenes between layers of tendon
- **Tendinopathy**
  - Thickening and heterogeneity of 1 or more tendon units of quadriceps mechanism
  - ± mildly elevated intratendinous signal
  - Best seen on axial and sagittal images
- **Partial tendon tear**
  - Focal partial discontinuity of 1 or more tendon units of quadriceps mechanism
  - Thinning of quadriceps
  - Focal fluid signal within quadriceps mechanism
  - Surrounding soft tissue swelling if acute
  - ± associated intramuscular edema
  - Best seen on axial fluid-sensitive sequences
- **Complete tendon tear**
  - Focal discontinuity of 1 or more tendon units of quadriceps
  - ± associated intramuscular edema
  - Inferior retraction of patella with large tear
  - Redundancy of patellar tendon
  - ± associated marrow edema in superior patella
  - Findings best seen on T2WI FS MR
- **Intramuscular injury**
  - Focal ↑ signal on T2WI MR
    - May be ovoid or linear
    - Usually along long axis of muscle
  - Often occurs at myotendinous junction of 1 or more quadriceps tendon units
  - Occasionally occurs more proximally
    - May need to increase imaging field of view to include more proximal thigh
  - Isolated vastus medialis obliquus injury may indicate transient patellar dislocation

Ultrasonographic Findings

- Accurate for partial or complete tears
- Hypo- or anechoic focus in expected location of quadriceps muscle or tendon

Imaging Recommendations

- Best imaging tool
  - MR
- Protocol advice
  - Include sagittal and axial T2WI FS MR
  - Larger field of view (FOV) than usual for knee MR when evaluating quadriceps injury
    - May need to use torso coil instead of extremity coil

DIFFERENTIAL DIAGNOSIS

Patellar Tendon Injury

- Patella retracted superiorly
- Abnormal signal in patellar tendon

Patellar Fracture

- Abnormal signal and morphology in patella
- Focal fracture line

Postsurgical Fibrosis

- Can result in thickened tendons with irregular borders
- No inflammation
- Appropriate history of surgery

Transient Patellar Dislocation

- Edema or disruption of medial patellar retinaculum
- Edema in vastus medialis obliquus
- Marrow contusions
  - Lateral subcortical marrow of lateral femoral condyle
  - Inferomedial patella
- Chondral fracture of patellar apex or lateral trochlear ridge
- Joint effusion ± intraarticular bodies

PATHOLOGY

General Features

- Etiology
  - Acute eccentric contraction (muscle contraction against fixed force with muscle lengthening)

QUADRICEPS INJURY

  - Foot planted and knee partially flexed
  - Standing from sitting position
- Associated abnormalities
  - Underlying chronic tendon-weakening condition often present
    - Chronic renal disease
    - Rheumatoid arthritis
    - Gout
    - Exogenous steroid use
    - Connective tissue disease
      - Marfan disease
      - Ehlers-Danlos
  - Association with prior knee surgery
    - Total knee arthroplasty
    - Lateral retinacular release

Staging, Grading, & Classification

- Grade I: Swelling of tendon ± surrounding swelling, no focal macroscopic disruption
- Grade II: Focal fiber disruption in 1 or more tendons without complete discontinuity
  - May present as focal tendon thinning
- Grade III: Focal complete disruption of 1 or more tendon units

Gross Pathologic & Surgical Features

- Most tears are acute injuries superimposed on chronic degeneration of tendon
  - Mucinous
  - Fibrous
  - Lipomatous

CLINICAL ISSUES

Presentation

- Most common signs/symptoms
  - Anterior thigh and knee pain and swelling
  - Freely mobile patella
  - Palpable defect in distal quadriceps tendons
  - Extensor lag on exam
- Other signs/symptoms
  - Inferiorly displaced patella

Demographics

- Age
  - 5th and 6th decade
- Gender
  - M = F
- Epidemiology
  - Unilateral rupture relatively rare but more common than patellar tendon rupture
  - Bilateral ruptures very rare, almost always due to underlying systemic condition

Natural History & Prognosis

- Low-grade partial tears usually heal without residua
- High-grade partial tears may progress to full thickness tears if not protected from additional stress
- Full thickness tears progress to atrophy and immobility with permanent loss of function if not repaired

Treatment

- Partial tears often treated conservatively
- Early surgical repair indicated for acute complete rupture
- Chronic complete tears often warrant surgical correction if < 1 year after injury

DIAGNOSTIC CHECKLIST

Consider

- Midsubstance muscle injury if distal quadriceps intact (requires large field of view imaging)

Image Interpretation Pearls

- Axial and sagittal MR images both contribute to identifying affected units

Reporting Tips

- Identify individual tendon units affected if possible

SELECTED REFERENCES

1. Basic-Jukic N et al: Spontaneous tendon ruptures in patients with end-stage renal disease. Kidney Blood Press Res. 32(1):32-6, 2009
2. Brooks P: Extensor mechanism ruptures. Orthopedics. 32(9), 2009
3. Yun HH et al: Calcific tendinitis of the rectus femoris. Orthopedics. 32(7):490, 2009
4. Bimmel R et al: Quadriceps tendon rupture and periosteal sleeve avulsion fracture of the proximal patella pole in an active adolescent. Orthopedics. 31(7):716, 2008
5. LaRocco BG et al: Ultrasound diagnosis of quadriceps tendon rupture. J Emerg Med. 35(3):293-5, 2008
6. McKinney B et al: Intra-articular knee injuries in patients with knee extensor mechanism ruptures. Knee Surg Sports Traumatol Arthrosc. 16(7):633-8, 2008
7. Pfirrmann CW et al: Quadriceps tendinosis and patellar tendinosis in professional beach volleyball players: sonographic findings in correlation with clinical symptoms. Eur Radiol. 18(8):1703-9, 2008
8. Sonin AH: Magnetic resonance imaging of the extensor mechanism. Magn Reson Imaging Clin N Am. 2(3):401-11, 1994
9. Yu JS et al: MR imaging of injuries of the extensor mechanism of the knee. Radiographics. 14(3):541-51, 1994

QUADRICEPS INJURY

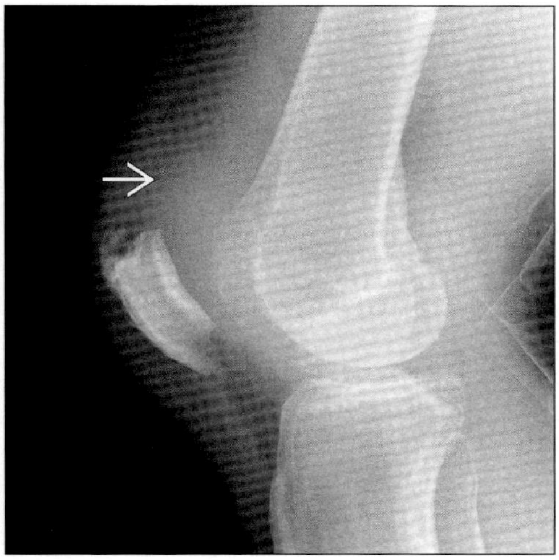

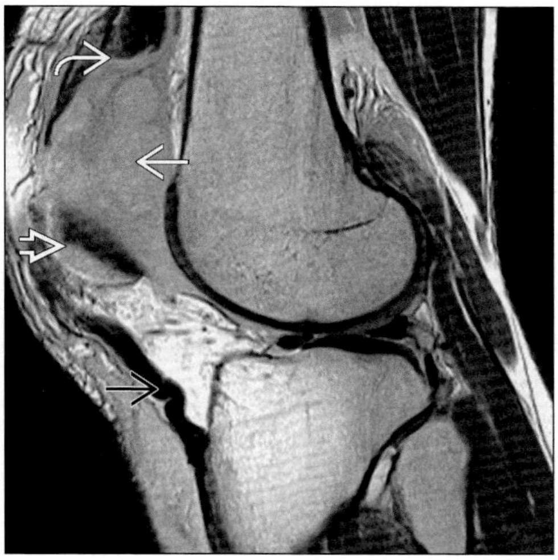

(Left) *Lateral radiograph in an alpine skier after a fall demonstrates a large soft tissue mass in the anterior distal thigh with distal displacement of the patella (patella baja) and patellar rotation. These findings suggest quadriceps rupture in the appropriate clinical setting.* ***(Right)*** *Sagittal PDWI MR in the same patient shows a large hematoma at the site of quadriceps tendon rupture. Patella baja and tilt are present, as is patellar tendon redundancy.*

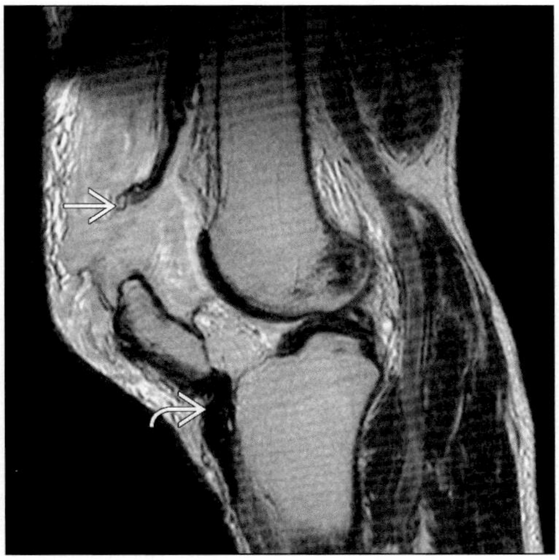

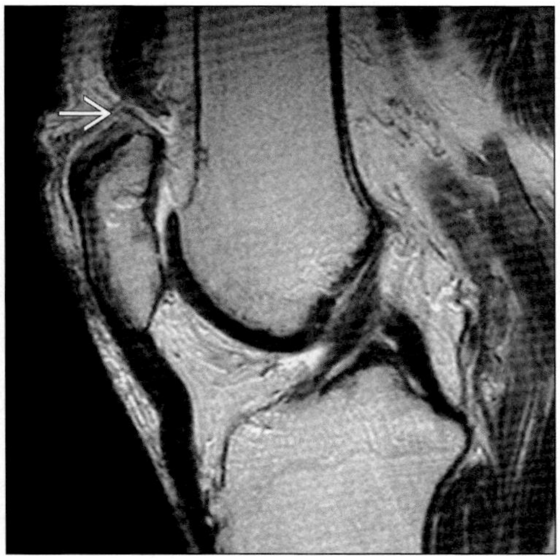

(Left) *Sagittal T2WI MR shows complete quadriceps rupture in the right knee of a 68-year-old man with chronic renal disease and on dialysis. There is patella baja and rotation, with foreshortening of the patellar tendon.* ***(Right)*** *Sagittal T2WI MR of the left (opposite) knee in the same patient at the same time demonstrates a nondisplaced tear of the quadriceps mechanism. Concurrent bilateral quadriceps ruptures are very unusual and usually occur due to underlying disease.*

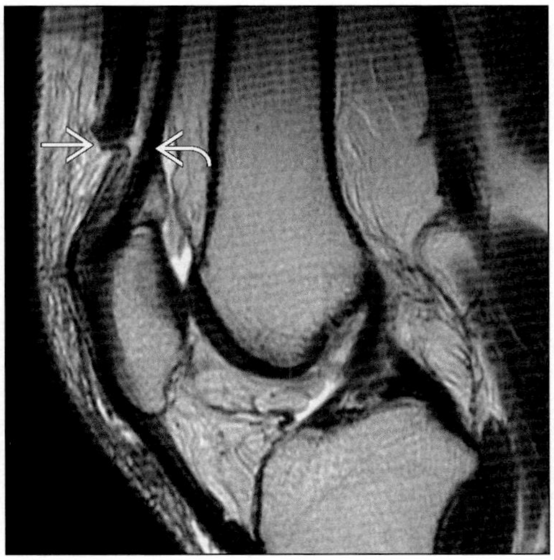

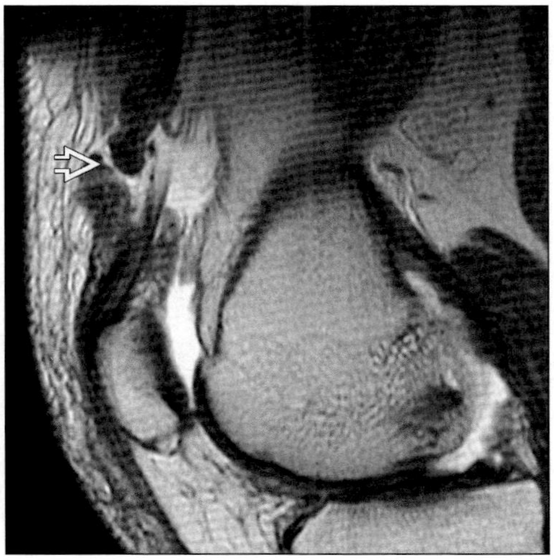

(Left) *Sagittal T2WI MR shows partial tear of the quadriceps. The rectus femoris tendon is torn and retracted, but the vastus intermedius unit remains intact.* ***(Right)*** *Sagittal T2WI MR more medially in the same patient shows disruption of the vastus medialis portion of the quadriceps mechanism. The full extent of a quadriceps injury may be under- or overestimated on sagittal images unless all images in the series are carefully scrutinized.*

(Left) *Coronal T2WI FS MR in the same patient shows the tear in the vastus medialis with sparing of the vastus lateralis and intermedius. The rectus femoris is located more anteriorly and out of the plane of imaging.* ***(Right)*** *Axial PDWI FS MR in the same patient shows disruption of the vastus medialis and rectus femoris tendons, with preservation of the vastus lateralis and vastus intermedius units.*

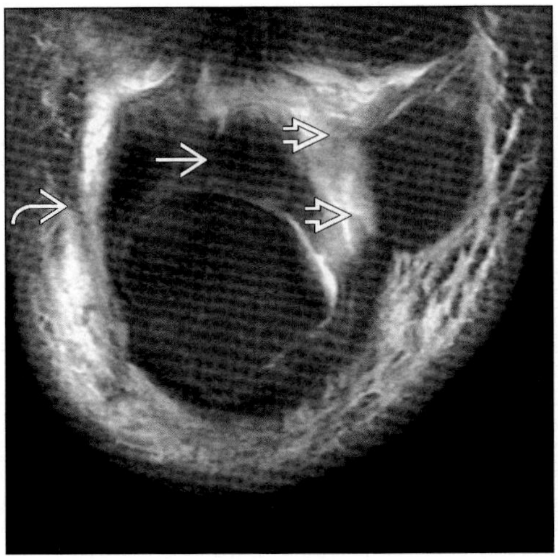

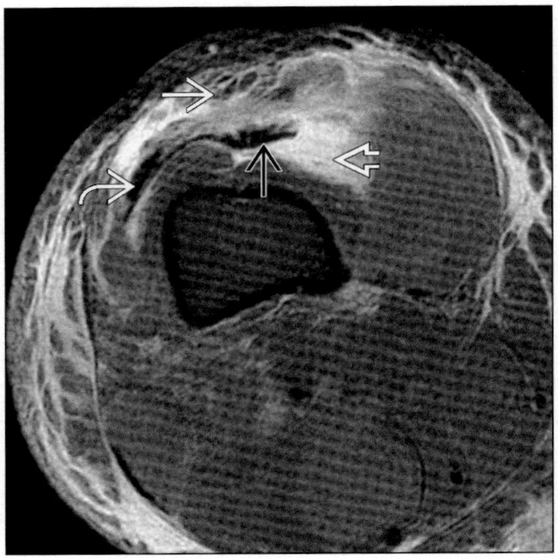

(Left) *Sagittal T2WI MR with a large field of view shows a high-grade partial tear of the quadriceps that spares only the vastus intermedius tendon. The rectus femoris is torn and retracted.* ***(Right)*** *Axial PDWI FS MR in the same patient shows disruption of the vastus lateralis and medialis and the rectus femoris, with surrounding hematoma. Only the vastus intermedius remains intact. This is unusual in that the vastus intermedius is often the 1st unit torn.*

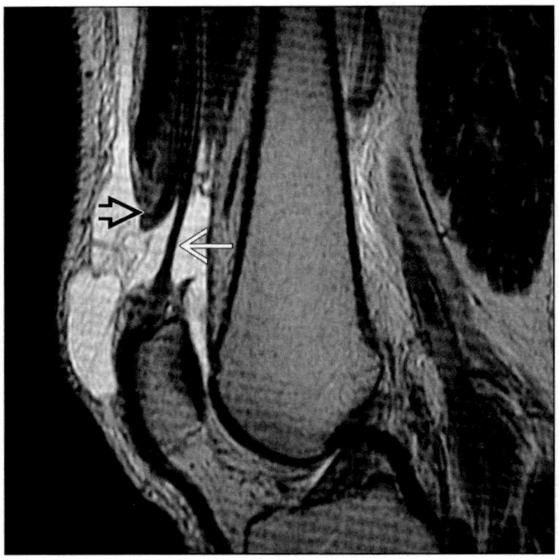

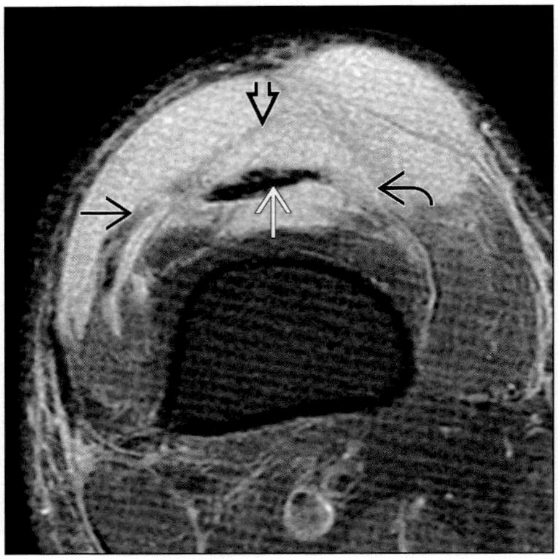

(Left) *Sagittal T2WI MR shows a partial quadriceps tear involving the vastus intermedius with a low-grade partial tear of the distal rectus femoris. This is a more typical pattern of partial tear than seen in the previous case.* ***(Right)*** *Axial PDWI FS MR in the same patient shows disruption of the vastus intermedius, with sparing of the vastus lateralis and medialis and partial tear of the rectus femoris.*

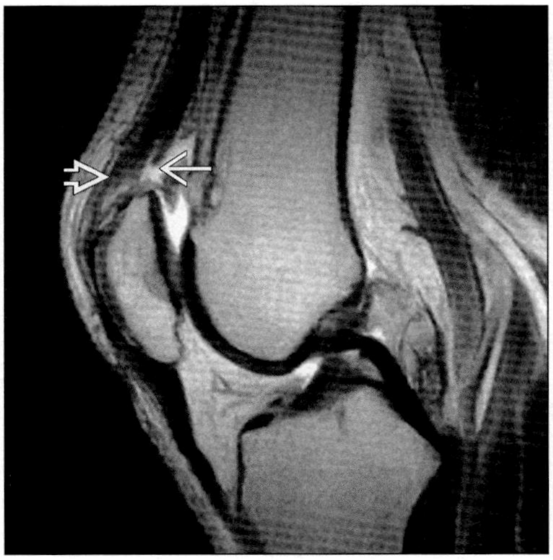

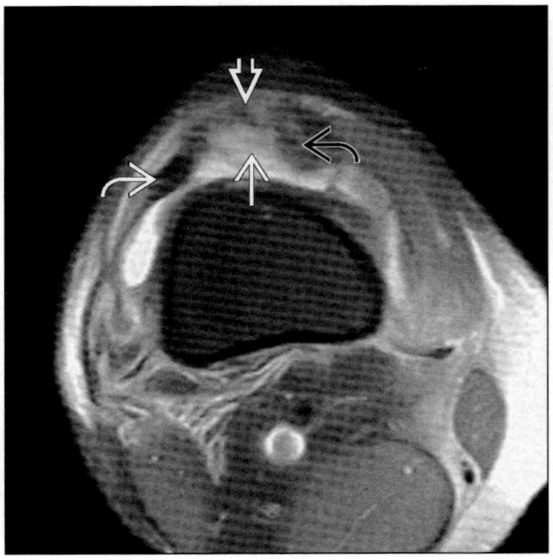

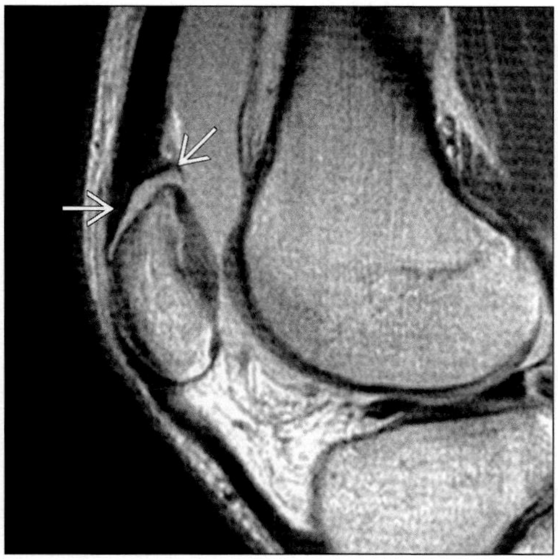

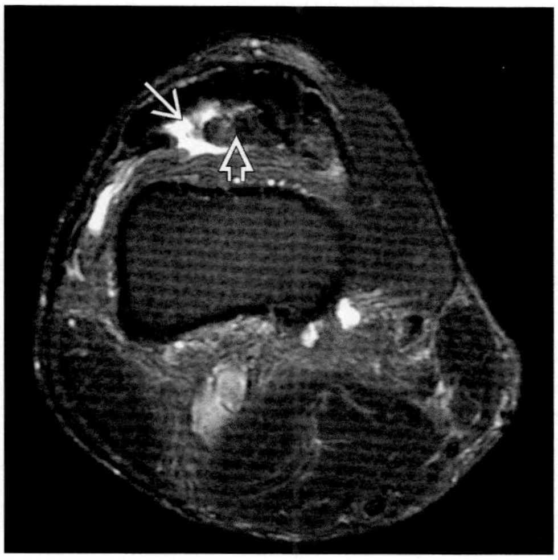

(Left) *Sagittal PDWI MR shows periosteal sleeve avulsion of the distal insertion of the quadriceps mechanism ➔. A thin rim of low signal periosteum remains attached to the distal end of the tendon complex, which cannot be seen on radiographs.* ***(Right)*** *Axial T2WI FS MR shows a partial tear of the vastus lateralis ➔ just above its insertion on the patella. The defect is filled with fluid. There is a low-grade partial tear of the vastus intermedius ⇨.*

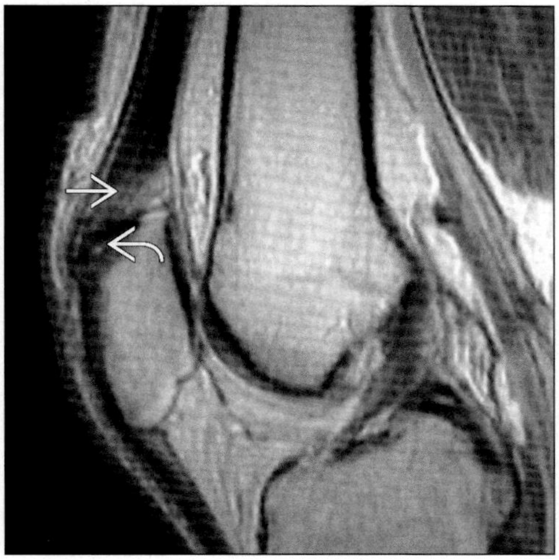

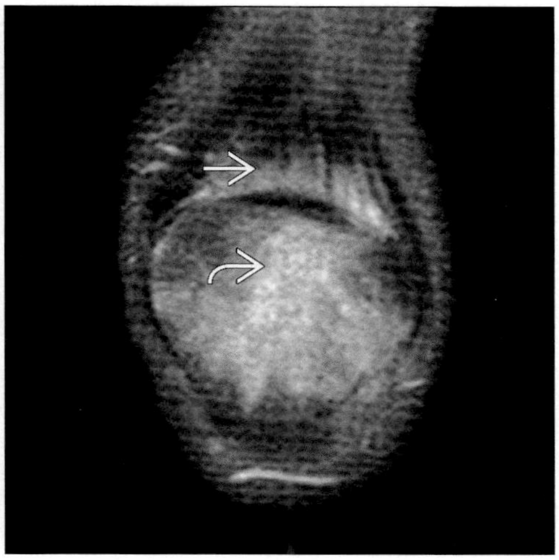

(Left) *Sagittal PDWI MR shows mildly elevated signal in the distal quadriceps tendon ➔ without high-grade fiber disruption. The overall architecture of the tendon remains intact. There is mild irregularity of the superior pole of the patella ↪.* ***(Right)*** *Coronal T2WI FS MR in the same patient shows elevated signal in the attaching fibers of the quadriceps ➔ and diffuse marrow edema in the patella ↪. These findings represent acute low-grade strain injury.*

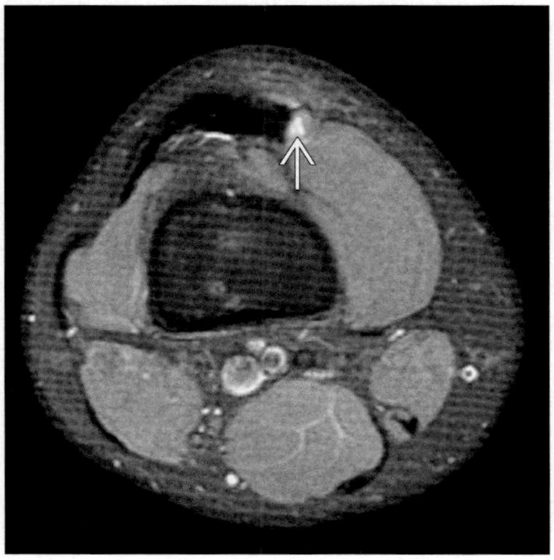

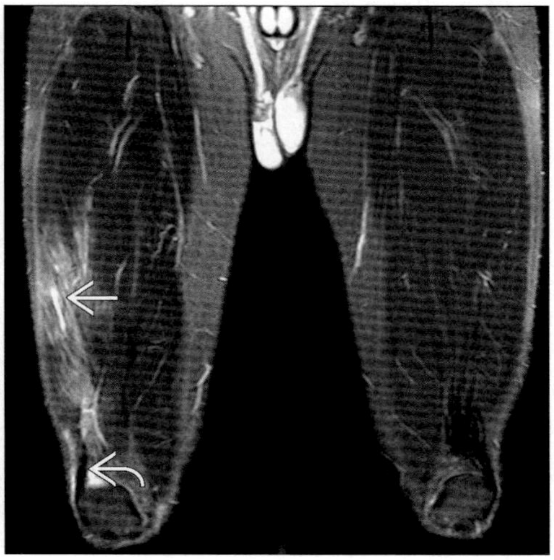

(Left) *Axial PDWI FS MR shows a small focal partial tear in the myotendinous junction of the vastus medialis obliquus ➔ with superficial soft tissue swelling. The remainder of the distal quadriceps mechanism remains intact. Findings are most likely due to an episode of transient patellar dislocation.* ***(Right)*** *Coronal T2WI FS MR obtained using a body coil and large field of view shows intramuscular injury to the vastus lateralis ➔. The tendon is intact ↪.*

PATELLAR TENDON INJURY

Key Facts

Terminology
- Patellar tendinopathy, patellar tendinosis, patellar tendon tear, jumper's knee, Osgood-Schlatter syndrome, Sinding-Larsen-Johansson syndrome

Imaging
- Location
  - Proximal 1/3 of patellar tendon most common with sports injury
  - Midportion of tendon torn with underlying medical conditions (diabetes, etc.)
- **Patellar tendinopathy**
  - Tendon thickening and signal heterogeneity
- **Partial tear of patellar tendon**
  - Focal incomplete defect in patellar tendon
- **Complete tear of patellar tendon**
  - Usually in proximal portion of tendon
- **Osgood-Schlatter syndrome**
  - Thickening of distal patellar tendon, usually with increased signal
  - Enlarged and irregular tibial tubercle
- **Sinding-Larsen-Johansson syndrome**
  - Thickening of proximal patellar tendon, usually with increased signal
  - Irregularity of inferior pole of patella
- Ultrasound useful to gauge extent of tendon retraction

Pathology
- Chronic overuse, usually due to sports activity (especially jumping): Jumper's knee
- Chronic medical conditions resulting in weakened tendons

Clinical Issues
- Complete tear requires surgical repair and is often considered orthopedic emergency

*(**Left**) Coronal graphic shows a tear of the proximal patellar tendon. The medial and lateral patellar retinacula are intact, as is the quadriceps tendon. (**Right**) Coronal T2WI FS MR through the patella shows a complete tear of the patellar tendon in its proximal extent just below the patella. There is diffuse surrounding soft tissue swelling and hematoma, as this was an acute injury.*

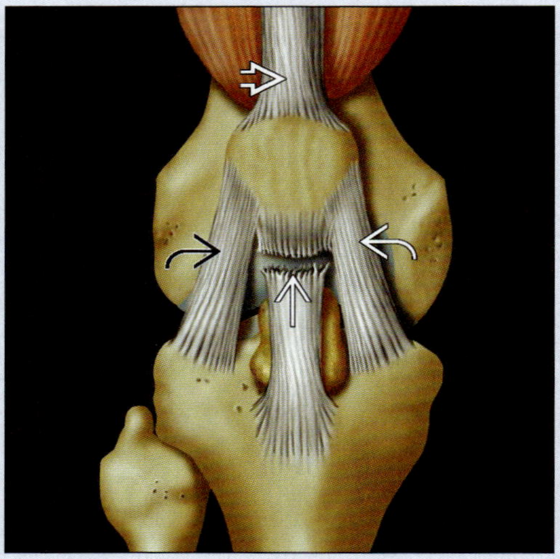

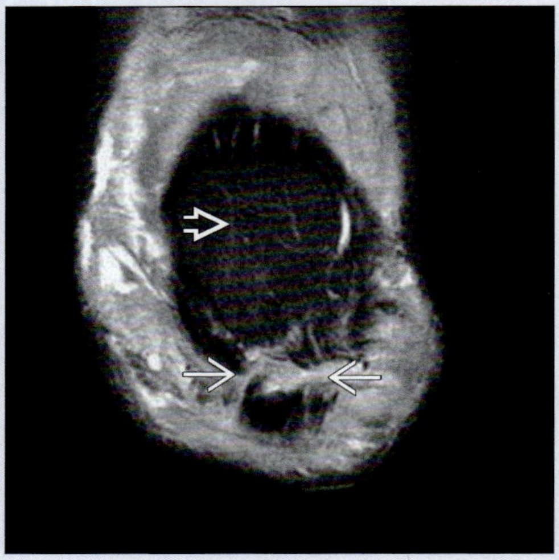

*(**Left**) Sagittal T2WI MR from the same patient shows complete disruption of the proximal patellar tendon with hematoma in the interval between the ruptured tendon and the patella. The quadriceps tendon is slightly redundant, indicating laxity. (**Right**) Axial T2WI FS MR from the same patient shows complete rupture of the patellar tendon with hematoma filling the anterior soft tissues of the knee. Mild patella alta results in an empty trochlear groove.*

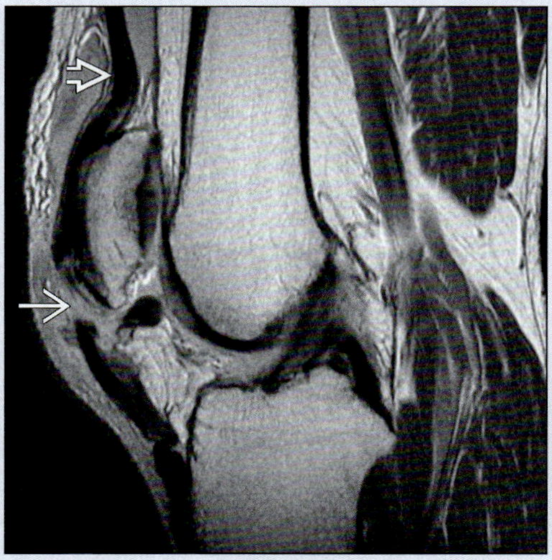

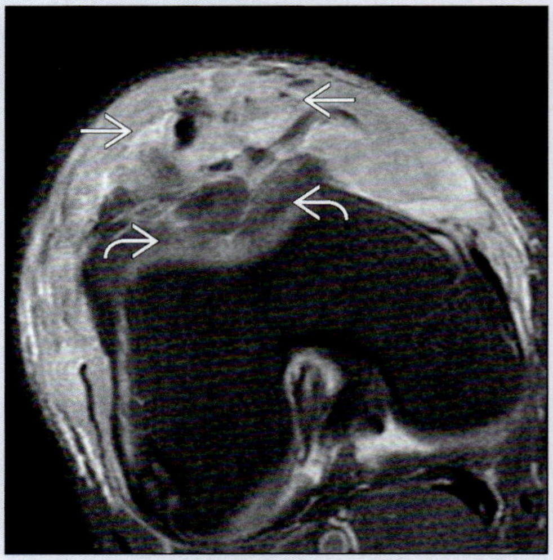

PATELLAR TENDON INJURY

TERMINOLOGY

Synonyms

- Patellar tendinopathy, patellar tendinosis, patellar tendon tear, jumper's knee, Osgood-Schlatter syndrome, Sinding-Larsen-Johansson syndrome

IMAGING

General Features

- Best diagnostic clue
  - Focal signal disruption in substance of patellar tendon
- Location
  - Proximal 1/3 of patellar tendon most common with sports injury
    - Posterior fibers of proximal tendon
  - Midportion of tendon torn with underlying medical conditions (diabetes, etc.)
  - Distal patellar tendon ± tibial tubercle: Children > > adults
  - Inferior pole of patella: Uncommon
- Size
  - Ranges from microscopic fiber degeneration to full thickness tears with multi-centimeter gaps

Radiographic Findings

- Radiography
  - Soft tissue swelling in region of patellar tendon
  - Patella alta (proximal positioning) with complete tendon tear
  - Avulsion of inferior patellar pole
  - Irregularity and overgrowth of tibial tubercle in Osgood-Schlatter syndrome (OS)
  - Irregularity of inferior patella ± elongation of inferior pole in Sinding-Larsen-Johansson syndrome (SLJ)

MR Findings

- **Patellar tendinopathy**
  - Tendon thickening and signal heterogeneity
    - Not as bright as fluid on T2WI MR ± FS
  - No focal tendon defect
- **Partial tear of patellar tendon**
  - Focal incomplete defect in patellar tendon
  - May be incomplete in anterior to posterior or medial to lateral direction
  - Often associated with underlying tendinopathy
- **Complete tear of patellar tendon**
  - Focal complete rupture of patellar tendon
  - Usually in proximal portion of tendon
  - May include osseous avulsion fragment
    - Usually from inferior pole of patella
- **Osgood-Schlatter syndrome**
  - Thickening of distal patellar tendon, usually with increased signal
  - Enlarged and irregular tibial tubercle ± fragmentation
    - ↑ marrow signal on T2WI MR
    - Cortical irregularity or erosion
  - Surrounding soft tissue swelling
  - Fluid in deep infrapatellar bursa
- **Sinding-Larsen-Johansson syndrome**
  - Thickening of proximal patellar tendon, usually with increased signal
  - Irregularity of inferior pole of patella
    - ↑ marrow signal on T2WI MR
    - Cortical irregularity or erosion
  - Surrounding soft tissue swelling

Ultrasonographic Findings

- Thickened hypoechoic tendon
- Extent of tendon retraction can be evaluated

Imaging Recommendations

- Best imaging tool
  - MR
  - Ultrasound useful to gauge extent of tendon retraction
- Protocol advice
  - Sagittal and axial T2WI FS MR best for
    - Assessment of location and extent of tear
    - Differentiation of tendinopathy from tear

DIFFERENTIAL DIAGNOSIS

Postsurgical Change

- Anterior cruciate ligament (ACL) graft harvest
- Arthroscopic portals
- Patellar tendon transposition (Fulkerson osteotomy)
- Previous patellar tendon repair

Patellar Fracture

- Marrow edema in patella ± fx line
- Mild changes in adjacent patellar tendon

Quadriceps Tendon Tear

- Inferior retraction of patella
- Thickening and redundancy of patellar tendon

Magic Angle Artifact

- Occurs in portion of tendon at 55° to main magnetic field
- High signal on short TE sequences, resolves on long TE
- No associated tendon thickening or pain

PATHOLOGY

General Features

- Etiology
  - **Tendinopathy**
    - Chronic overuse, usually due to sports activity (especially jumping): Jumper's knee
    - Chronic medical conditions resulting in weakened tendons, which tear with lesser force
      - Diabetes
      - Chronic renal failure (especially on hemodialysis)
      - Rheumatoid arthritis
      - Gout
      - Collagen-vascular disease (Ehlers-Danlos, Marfan, etc.)
  - **Enthesopathy (apophysitis)**
    - Overuse injury at attachment points of tendon, usually in childhood or adolescence

PATELLAR TENDON INJURY

- Osgood-Schlatter syndrome: Tibial tubercle attachment
- Sinding-Larsen-Johansson syndrome: Inferior patellar attachment

- **Acute tear**
  - High-grade force in athletics
    - Fall on flexed knee with foot planted → eccentric contraction
  - Associated with posterior knee dislocation
  - Often superimposed on underlying chronic tendinopathy
  - May be seen in setting of previous surgery
    - Total knee arthroplasty
    - ACL reconstruction (with middle 1/3 patellar tendon graft harvest)
- Steroid injections increase risk for tendon rupture

- Associated abnormalities
  - Patellofemoral maltracking
    - Chronic overload on patellar tendon because of strain from quadriceps muscle group
      - Results in microtears & focal degeneration
    - May have role in weakening tendon
      - Degeneration & deterioration are predisposing factor to tendon rupture

Gross Pathologic & Surgical Features

- Thickened, indurated tendon edges with discontinuity if complete tear

Microscopic Features

- Collagen degeneration without significant influx of inflammatory cells
  - "Tendinopathy" is more appropriate term than "tendinitis"

CLINICAL ISSUES

Presentation

- Most common signs/symptoms
  - Complete tear: Pain, palpable defect in patellar tendon, extension lag
  - Tendinopathy: Pain & tenderness over patellar tendon, especially with resisted extension
- Other signs/symptoms
  - Patella alta
- Clinical profile
  - Jumper's knee: Basketball and tennis players
  - Acute tendon rupture: Hemodialysis patients, long-term steroid use

Demographics

- Age
  - Acute tendon rupture: Physically active patients < 40 years
    - Older patients with underlying disease predisposing to tendon weakness
  - Osgood-Schlatter syndrome
    - Boys: 10-15 years of age
    - Girls: 8-13 years of age
- Gender
  - M > F

Natural History & Prognosis

- Tendinopathy, OS, and SLJ will heal/improve with cessation of jumping sports
  - OS and SLJ may → tendon ossification
- Complete tear requires surgical repair and is often considered orthopedic emergency

Treatment

- Conservative
  - Tendinopathy, OS, SLJ
  - Rest, ice, cessation of inciting activity, physical therapy
- Surgical repair
  - Complete tear: Repair using end-to-end sutures
  - Persistent pain with tendinopathy: Maquet procedure
    - Vertical osteotomy tibial apophysis, interposed bone graft to elevate anterior tibial tubercle
    - Procedure has moderate risk of nonunion

DIAGNOSTIC CHECKLIST

Consider

- Previous patellar tendon surgery if tendon thickened or thinned

Image Interpretation Pearls

- Axial images more accurate for discerning degree of injury

Reporting Tips

- Describe location and degree of tear

SELECTED REFERENCES

1. Brooks P: Extensor mechanism ruptures. Orthopedics. 32(9), 2009
2. Peace KA et al: Imaging the infrapatellar tendon in the elite athlete. Clin Radiol. 61(7):570-8, 2006
3. McGrory JE: Disruption of the extensor mechanism of the knee. J Emerg Med. 24(2):163-8, 2003
4. McKoy BE et al: Acute tibial tubercle avulsion fractures. Orthop Clin North Am. 34(3):397-403, 2003
5. Pagnano MW: Patellar tendon and quadriceps tendon tears after total knee arthroplasty. J Knee Surg. 16(4):242-7, 2003
6. Panni AS et al: Overuse injuries of the extensor mechanism in athletes. Clin Sports Med. 21(3):483-98, ix, 2002
7. Casey MT Jr et al: Neglected ruptures of the patellar tendon. A case series of four patients. Am J Sports Med. 29(4):457-60, 2001
8. Kasten P et al: Rupture of the patellar tendon: a review of 68 cases and a retrospective study of 29 ruptures comparing two methods of augmentation. Arch Orthop Trauma Surg. 121(10):578-82, 2001
9. Sonin AH et al: MR imaging appearance of the extensor mechanism of the knee: functional anatomy and injury patterns. Radiographics. 15(2):367-82, 1995
10. Sonin AH: Magnetic resonance imaging of the extensor mechanism. Magn Reson Imaging Clin N Am. 2(3):401-11, 1994
11. Yu JS et al: MR imaging of injuries of the extensor mechanism of the knee. Radiographics. 14(3):541-51, 1994
12. Schweitzer ME et al: The patellar tendon: thickening, internal signal buckling, and other MR variants. Skeletal Radiol. 22(6):411-6, 1993
13. Bodne D et al: Magnetic resonance images of chronic patellar tendinitis. Skeletal Radiol. 17(1):24-8, 1988
14. Kricun R et al: Patellar tendon rupture with underlying systemic disease. AJR Am J Roentgenol. 135(4):803-7, 1980

PATELLAR TENDON INJURY

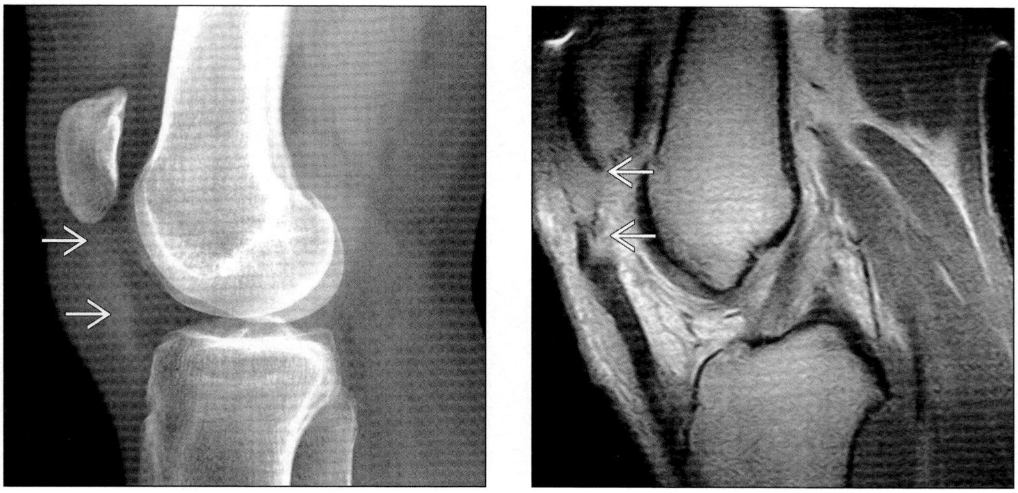

(Left) *Lateral radiograph in a patient with acute patellar tendon rupture shows patella alta (elevated patella) and soft tissue swelling around a poorly defined patellar tendon* ➔. *The normal patellar tendon should have a smooth posterior margin.* ***(Right)*** *Sagittal PDWI MR shows an acute proximal patellar tendon rupture* ➔ *in a 25-year-old basketball player. There is superior retraction of the patella (indicating the complete nature of the tendon tear) and intervening hematoma.*

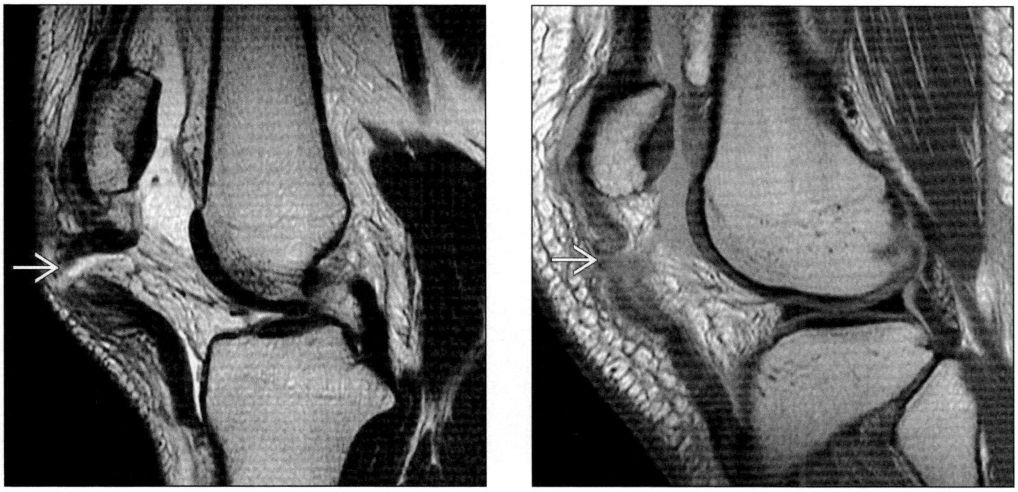

(Left) *Sagittal T2WI MR shows a midsubstance tear of the patellar tendon* ➔. *The underlying tendon is somewhat thickened, suggesting chronic tendinopathy.* ***(Right)*** *Sagittal PDWI MR shows midsubstance delamination and high-grade partial rupture of the patellar tendon* ➔. *The whole tendon is thick and somewhat heterogeneous, indicating underlying tendinopathy. This patient had rheumatoid arthritis.*

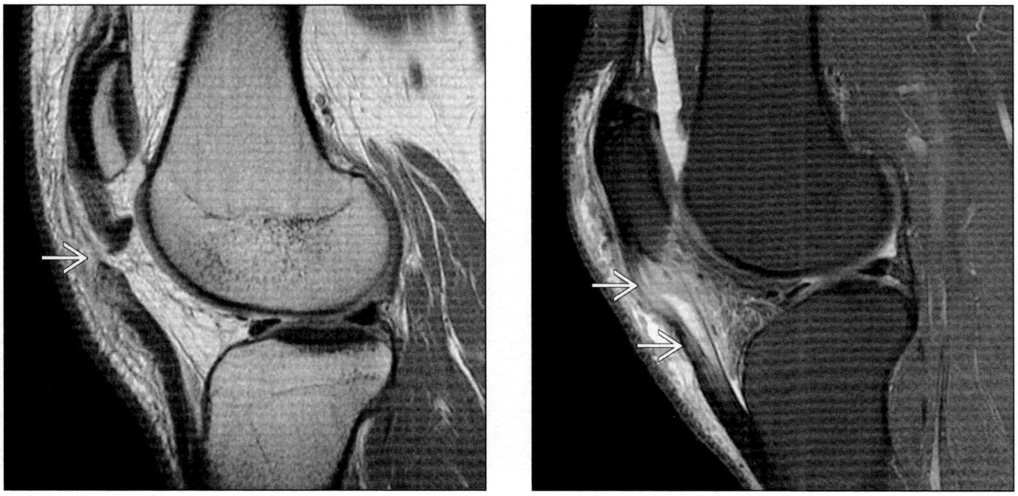

(Left) *Sagittal PDWI MR shows a subacute midsubstance rupture of the patellar tendon* ➔. *The underlying tendon is otherwise normal, and there is little surrounding edema. The patient was involved in a motor vehicle accident 3 weeks earlier.* ***(Right)*** *Sagittal PDWI FS MR shows high-grade delamination of the proximal and midportion of the patellar tendon* ➔ *with surrounding soft tissue edema. This was due to acute trauma a day earlier.*

PATELLAR TENDON INJURY

*(**Left**) Sagittal T2WI MR shows marked lengthening and irregularity of the patellar tendon with diffuse thinning of the tendon as a result of diffuse tendon delamination and stretching due to repetitive injuries. There is marked patella alta as a result. (**Right**) Axial PDWI FS MR from the same patient shows the marked thinning of the patellar tendon in its midsubstance. The central portion of the tendon should be substantially thicker than the adjacent patellar retinacula .*

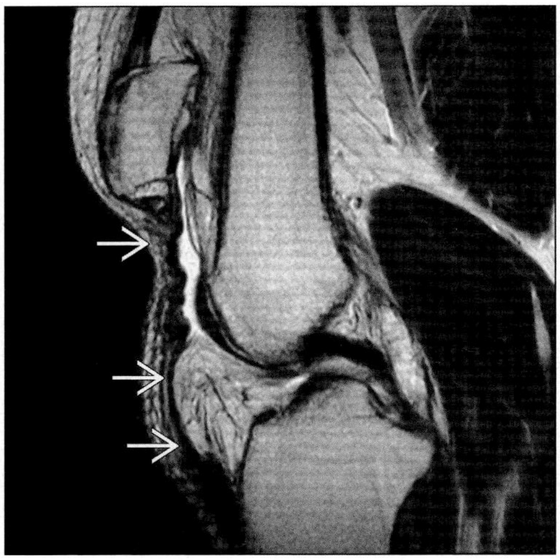

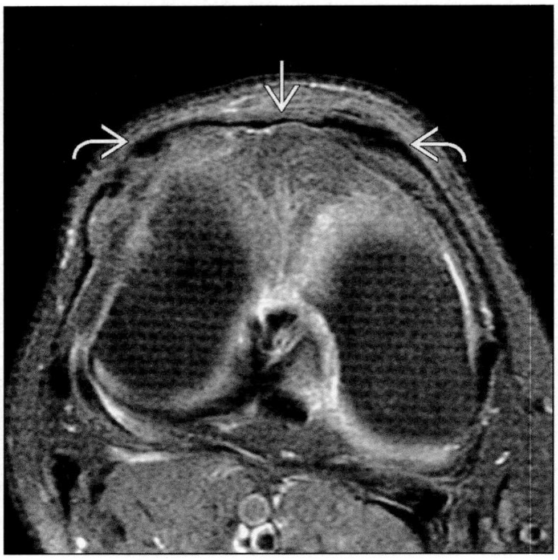

*(**Left**) Sagittal T2WI MR shows proximal retraction and redundancy of the patellar tendon with patella alta. Findings are due to a tendon tear more distally. (**Right**) Sagittal T2WI MR more laterally in the same patient demonstrates the location of the tear at its tibial insertion . This is a relatively unusual injury in an adult and is the functional equivalent of Osgood-Schlatter syndrome of adolescence.*

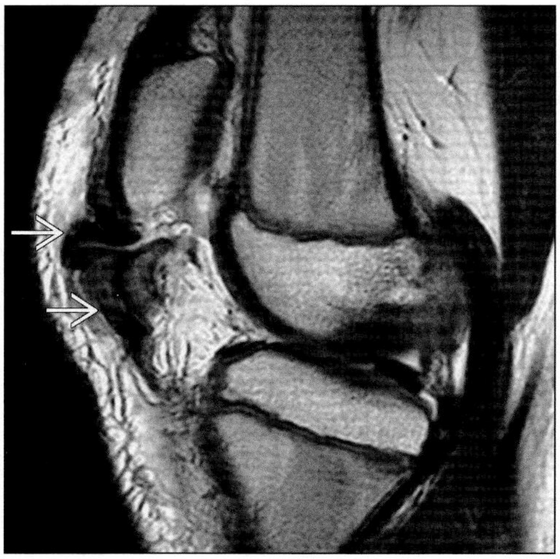

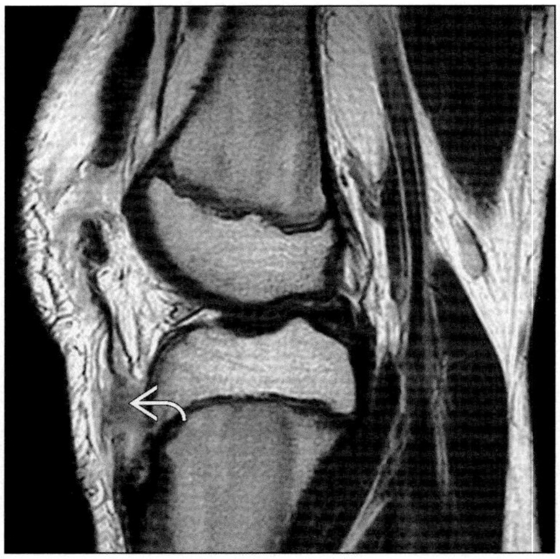

*(**Left**) Sagittal T2WI MR from a 13-year-old basketball player shows thickening and irregularity of the proximal patellar tendon and irregularity and elongation of the inferior pole of the patella , consistent with Sinding-Larsen-Johansson syndrome. (**Right**) Coronal T2WI FS MR from the same patient shows elongation of the inferior pole of the patella with marrow edema and a small partial tear of the proximal patellar tendon . This is the functional equivalent of jumper's knee.*

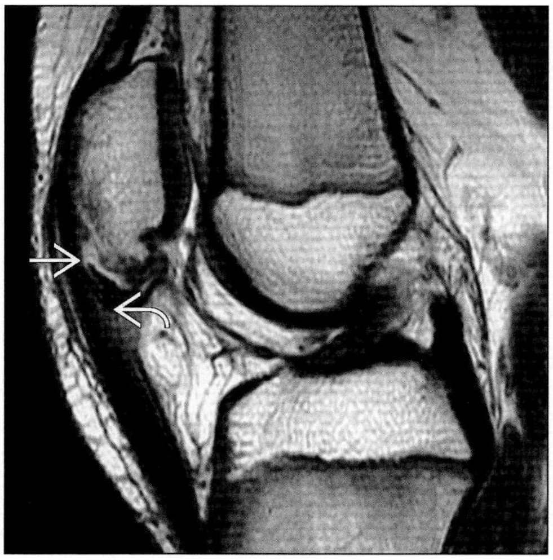

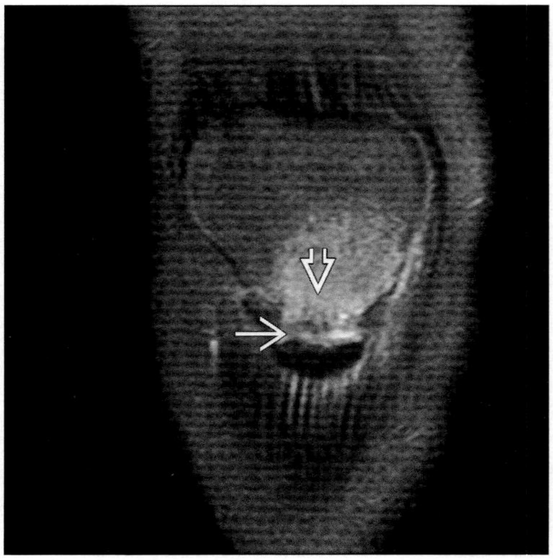

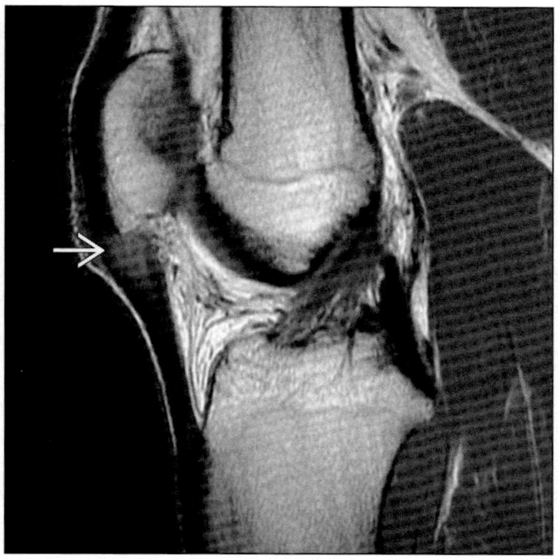

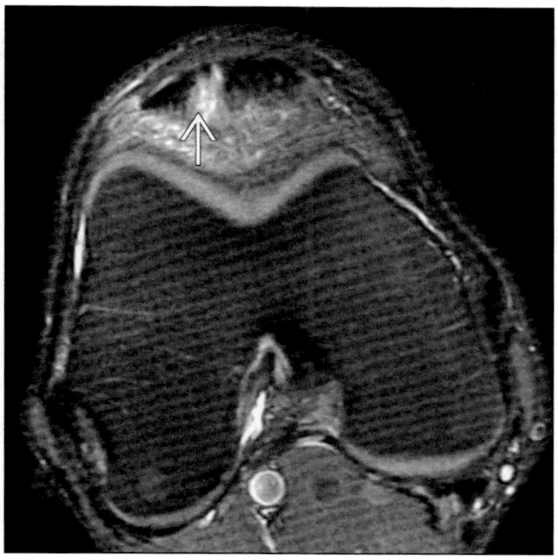

(Left) *Sagittal PDWI MR from a 22-year-old professional basketball player shows diffuse thickening and elevated signal in the proximal patellar tendon just below its origin on the patella* ➔. ***(Right)*** *Axial PDWI FS MR from the same patient shows focal high signal in the proximal aspect of the patellar tendon* ➔ *consistent with severe jumper's knee with partial tear. Milder cases of this entity do not demonstrate focal fluid signal within the tendon on T2WI.*

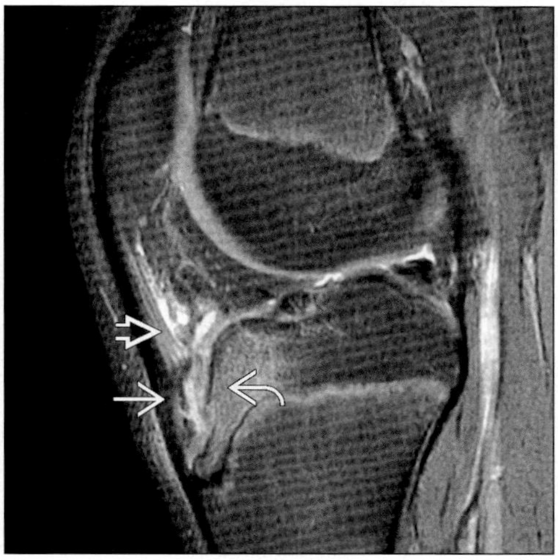

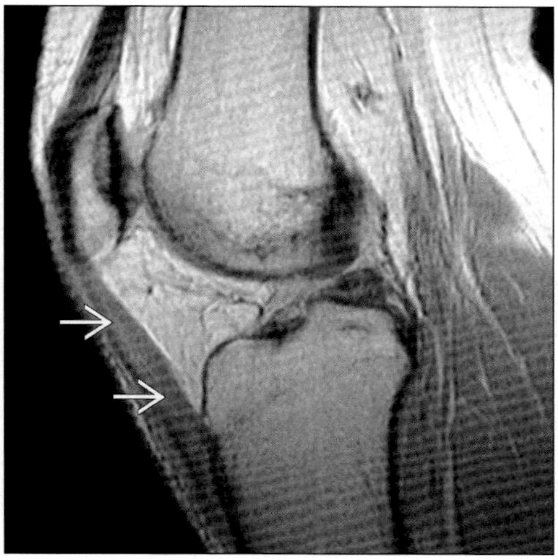

(Left) *Sagittal PDWI FS MR from a 12-year-old boy demonstrates the classic changes of Osgood-Schlatter syndrome. There is marrow edema and cortical irregularity of the tibial tubercle* ➔ *with a cortical fragment avulsed* ➔. *The distal patellar tendon* ➔ *is mildly thickened and demonstrates elevated signal.* ***(Right)*** *Sagittal PDWI MR shows diffuse thickening and elevated signal throughout the patellar tendon* ➔ *consistent with tendinopathy.*

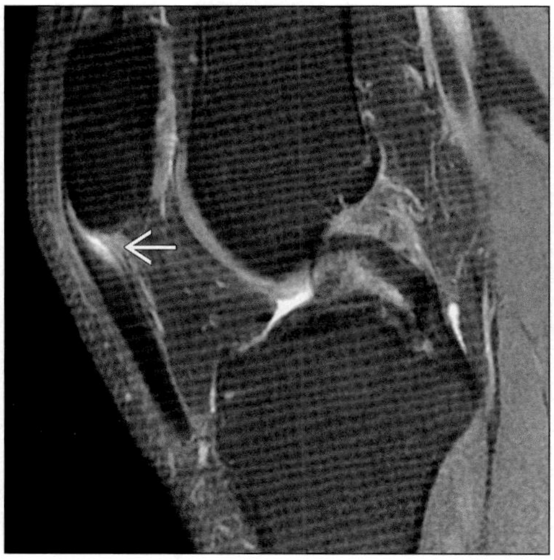

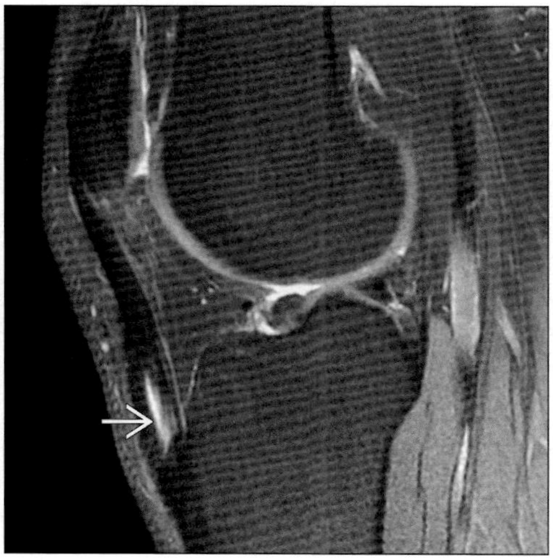

(Left) *Sagittal PDWI FS MR demonstrates partial fiber disruption in the deep portion of the proximal patellar tendon* ➔ *consistent with a partial tear.* ***(Right)*** *Sagittal PDWI FS MR from the same patient slightly more laterally shows a partial intrasubstance tear of the distal aspect of the tendon* ➔ *as well. The entire tendon is somewhat thickened and irregular, representing chronic underlying patellar tendinopathy, and is thus predisposed to tear.*

TRANSIENT PATELLAR DISLOCATION

Key Facts

Terminology
- Transient patellar dislocation (TPD)
- Lateral dislocation of patella out of femoral trochlear groove due to twisting injury of knee

Imaging
- Marrow contusions in classic distribution
  - Inferomedial patella
  - Lateral aspect of lateral femoral condyle (LFC)
- Edema or disruption of medial patellofemoral ligament (MPFL)
- Joint effusion
- Intraarticular bodies (if chondral fracture present)

Top Differential Diagnoses
- Patellar fracture
- Direct blow injury to lateral femoral condyle
- Chondromalacia patella (osteoarthritis)
- Quadriceps tendon tear

Pathology
- Acute event: May be single episode
- Chronic repetitive patellofemoral dislocations
  - Subset of patients anatomically predisposed to patellofemoral instability

Clinical Issues
- Medial knee pain after injury
- Lateral patellar dislocation usually reduces spontaneously with knee extension following injury

Diagnostic Checklist
- TPD in setting of typical marrow contusion pattern
- Look carefully for chondral injury ± intraarticular bodies

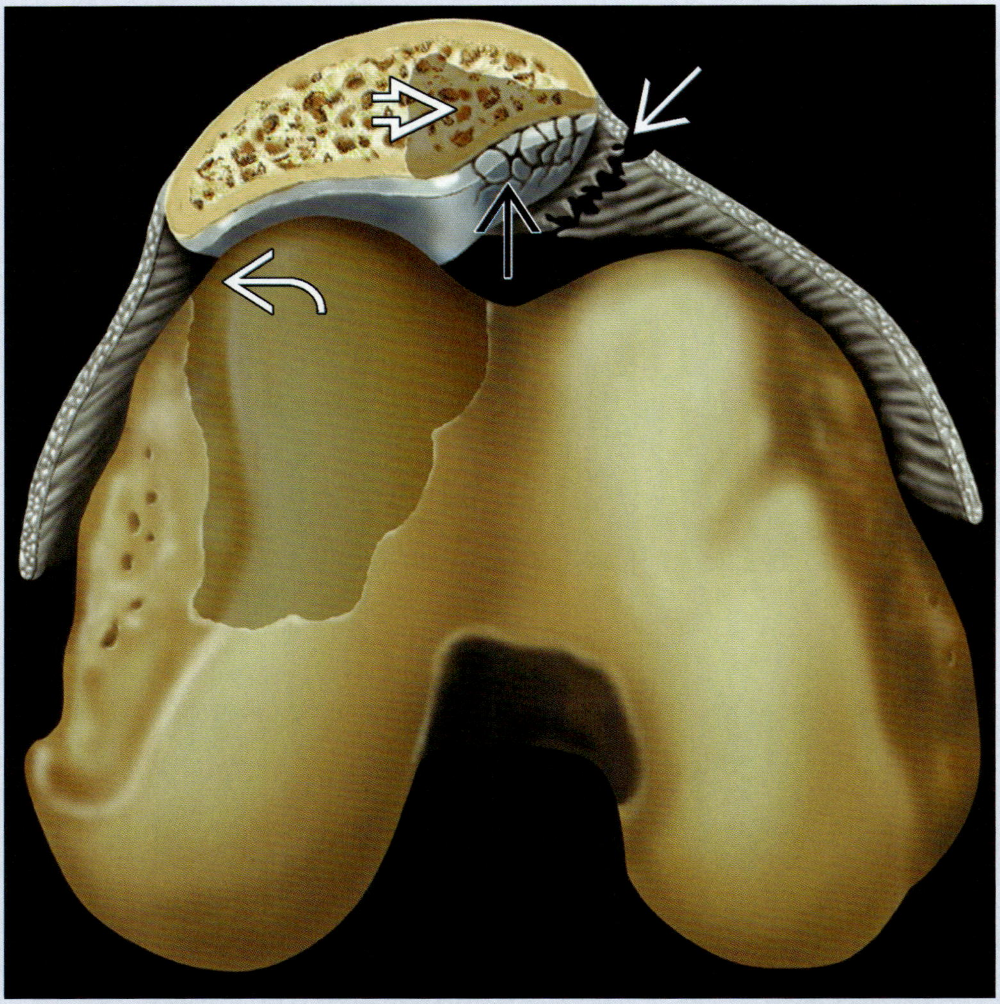

Axial graphic shows the classic constellation of findings after transient patellar dislocation (TPD): The medial patellofemoral ligament (MPFL) is torn ➡, there are subchondral marrow contusions in the medial patella ➡ and lateral femoral condyle ➡, and chondral injury to the patella ➡. Acute events usually demonstrate a large effusion ± intraarticular bodies. Most injuries are transient and reduce spontaneously upon knee extension. Predisposing factors include a congenitally shallow trochlear groove, a Wilberg type III patella (very prominent lateral facet), patella alta, genu valgum, and ligamentous laxity.

TRANSIENT PATELLAR DISLOCATION

TERMINOLOGY

Abbreviations

- Transient patellar dislocation (TPD)

Synonyms

- Patellar dislocation, patellar subluxation

Definitions

- Lateral dislocation of patella out of femoral trochlear groove due to twisting injury of knee
- Usually transient in nature: Spontaneous reduction in most cases
- Pertinent anatomy: Medial patellofemoral ligament (MPFL) = variable focal thickening in medial patellar retinaculum
  - Primary stabilizer against lateral patellar dislocation

IMAGING

General Features

- Best diagnostic clue
  - Marrow contusions in classic distribution
    - Inferomedial patella
    - Lateral aspect of lateral femoral condyle (LFC)
    - Both sites of contusion need not be present
- Location
  - Patella almost invariably dislocates laterally
- Morphology
  - Stabilizers of patellofemoral alignment
    - Passive: Medial and lateral patellar retinacula
    - Active: Muscles of quadriceps, especially vastus medialis obliquus (VMO)

Radiographic Findings

- May rarely see persistent lateral patellar dislocation
- Subtle flake of avulsed bone at medial pole of patella due to MPFL disruption
  - Seen best on axial sunrise view
- Joint effusion
- Osteochondral fragment(s) in joint (rare)

CT Findings

- Similar to radiographs, more sensitive for small avulsion fragments or intraarticular bodies

MR Findings

- Edema or disruption of medial patellar retinaculum/ MPFL
  - May occur anteriorly (patellar attachment), midsubstance, or posteriorly (medial femoral condylar attachment)
- Marrow contusions
  - 1 or (more commonly) both
    - Inferomedial patella
    - Anterolateral aspect of lateral femoral condyle
- Chondral fracture: At sites of marrow contusions
- Joint effusion
- Intraarticular bodies (if chondral fracture present)
- Excessive lateral pressure syndrome
  - Lateral patellar tilt
  - Fat edema inferolateral to patella
  - Focal chondral loss in LFC or lateral patella

Ultrasonographic Findings

- Disruption of MPFL diagnosed as focal gap in fibrous tissue

Imaging Recommendations

- Best imaging tool
  - MR
- Protocol advice
  - Axial PDWI or T2WI FS MR for retinacular injury, marrow contusions, chondral defects in patella or lateral trochlear ridge
  - Sagittal PDWI or T2WI FS MR for chondral defect in LFC
    - T2WI FS MR in 3 planes for associated intraarticular bodies

DIFFERENTIAL DIAGNOSIS

Patellar Fracture

- Marrow edema ± fx line most prominent feature

Direct Blow Injury to Lateral Femoral Condyle

- Marrow edema and overlying soft tissue swelling
- MPFL intact

Chondromalacia Patella (Osteoarthritis)

- Rounded margins of chondral defects
- No soft tissue swelling
- MPFL normal

Quadriceps Tendon Tear

- Especially VMO
- May be present in conjunction with TPD

PATHOLOGY

General Features

- Etiology
  - Internal rotation of knee on planted foot with knee flexed (indirect trauma)
    - Basketball, football, soccer
  - Lateral blow to flexed knee (direct trauma)
  - Excessive lateral pressure syndrome
    - Seen in adolescents and adults
    - Persistent lateral patellar tilt causing wear on lateral trochlea and lateral patellar facet
    - Results in premature cartilage breakdown
    - May be due to previous MPFL or VMO injury, anatomic variation
  - Predisposing factors
    - Shallow trochlear groove
    - Large Q angle (relative valgus alignment of extensor mechanism)
    - Genu valgum
    - Patella alta
    - Ligamentous laxity
    - Muscle weakness (inactivity, chronic neuromuscular disease)
- Associated abnormalities
  - Often associated with acute chondral fracture of patella &/or LFC
  - Strain or tear of vastus medialis obliquus (VMO) muscle
- **Mechanics of injury**

TRANSIENT PATELLAR DISLOCATION

- Flexed knee forced into valgus with foot planted → increased tension on lateral patellar retinaculum → lateral dislocation of patella over lateral trochlear ridge → disruption of MPFL
- Active tension of VMO contraction pulls patella back medially, where inferomedial portion of patella strikes lateral cortex of LFC → contusions in these bones
- Extension of knee due to pain and restricted motion → spontaneous reduction of patella back into trochlear groove
- Chondral injury can occur during initial dislocation event or at relocation

Staging, Grading, & Classification

- Acute event: May be single episode
- Chronic repetitive patellofemoral dislocations
  - Subset of patients anatomically predisposed to patellofemoral instability
  - More likely to experience recurrent dislocations
  - More likely to benefit from surgical therapy
- Wilberg classification of patellar shape
  - Type I: Concave facets, symmetrical and equal in size (10%)
  - Type II: Lateral facet size > medial facet; lateral facet concave (65%)
  - Type III: Marked lateral facet predominance (25%)

CLINICAL ISSUES

Presentation

- Most common signs/symptoms
  - Medial knee pain after injury
    - Due to retinacular injury; often clinically mistaken for medial collateral ligament or medial meniscal injury
  - Anterior and medial swelling
  - Persistent lateral patellar dislocation (rare)
    - Indicates trapped fragment or tissue preventing reduction
- Clinical profile
  - Usually an isolated traumatic event
    - Basketball, football, soccer players
  - Can be recurrent in patients anatomically predisposed
    - Shallow trochlear groove
    - Patellar morphology

Demographics

- Age
  - Children > adults
- Gender
  - F slightly > M

Natural History & Prognosis

- Lateral patellar dislocation usually reduces spontaneously with knee extension following injury
- Torn MPFL → ↑ incidence of recurrent dislocations

Treatment

- Conservative therapy usually applied for solitary event
  - Immobilization and NSAIDs followed by rehab
- Surgery
  - Lateral retinacular release: Reduces lateral tension on patella
  - Medial retinacular plication: Increases medial tension on patella
  - Fulkerson osteotomy: Anteromedial tibial tubercle transfer
  - Arthroscopy to repair chondral injury and retrieve intraarticular bodies

DIAGNOSTIC CHECKLIST

Consider

- TPD in setting of typical marrow contusion pattern

Image Interpretation Pearls

- Look carefully for chondral injury ± intraarticular bodies

Reporting Tips

- Describe morphology of trochlear groove and patella
- Describe location of MPFL tear (if possible)

SELECTED REFERENCES

1. Camanho GL et al: Conservative versus surgical treatment for repair of the medial patellofemoral ligament in acute dislocations of the patella. Arthroscopy. 25(6):620-5, 2009
2. Kapur S et al: Acute knee dislocation: review of an elusive entity. Curr Probl Diagn Radiol. 38(6):237-50, 2009
3. Kitamura K et al: Avulsion-tear type medial patellofemoral ligament injury with a small bony fragment in lateral patellar dislocation. J Trauma. 66(1):271-3, 2009
4. Le Corroller T et al: Transient medial patellar dislocation: injury patterns at US and MR imaging. Skeletal Radiol. 38(5):519-23, 2009
5. Nietosvaara Y et al: Acute patellar dislocation in children and adolescents. Surgical technique. J Bone Joint Surg Am. 91 Suppl 2 Pt 1:139-45, 2009
6. Andrish J: The management of recurrent patellar dislocation. Orthop Clin North Am. 39(3):313-27, vi, 2008
7. Christiansen SE et al: Isolated repair of the medial patellofemoral ligament in primary dislocation of the patella: a prospective randomized study. Arthroscopy. 24(8):881-7, 2008
8. Colvin AC et al: Patellar instability. J Bone Joint Surg Am. 90(12):2751-62, 2008
9. Panagopoulos A et al: MPFL reconstruction for recurrent patella dislocation: a new surgical technique and review of the literature. Int J Sports Med. 29(5):359-65, 2008
10. Robinson RJ et al: Wii knee. Emerg Radiol. 15(4):255-7, 2008
11. Feibel RJ et al: Irreducible lateral patellar dislocation: the importance of impaction fracture recognition. J Emerg Med. 33(1):11-5, 2007
12. Sanders TG et al: MRI of osteochondral defects of the lateral femoral condyle: incidence and pattern of injury after transient lateral dislocation of the patella. AJR Am J Roentgenol. 187(5):1332-7, 2006
13. Elias DA et al: Imaging of patellofemoral disorders. Clin Radiol. 59(7):543-57, 2004
14. O'Reilly MA et al: Sonographic appearances of medial retinacular complex injury in transient patellar dislocation. Clin Radiol. 58(8):636-41, 2003
15. Sanders TG et al: Medial patellofemoral ligament injury following acute transient dislocation of the patella: MR findings with surgical correlation in 14 patients. J Comput Assist Tomogr. 25(6):957-62, 2001
16. Sonin AH et al: MR imaging appearance of the extensor mechanism of the knee: functional anatomy and injury patterns. Radiographics. 15(2):367-82, 1995

TRANSIENT PATELLAR DISLOCATION

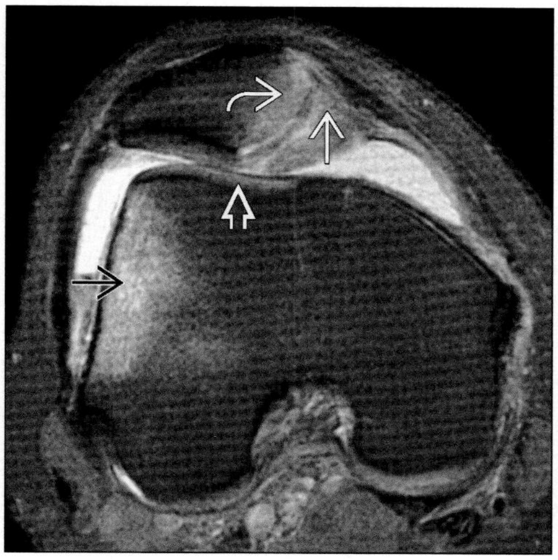

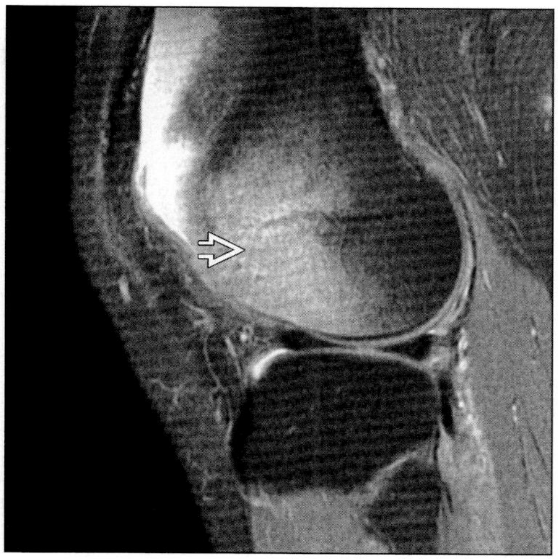

(Left) *Axial PDWI FS MR shows the classic MR imaging findings of transient patellar dislocation (TPD), with marrow edema in the subcortical aspects of the lateral femoral condyle (LFC) and the medial patella. There is a tear of the anterior aspect of the medial patellofemoral ligament (MPFL), and a joint effusion is present. The trochlear groove is very shallow.* ***(Right)*** *Sagittal PDWI FS MR from the same patient shows the broad LFC contusion pattern of TPD.*

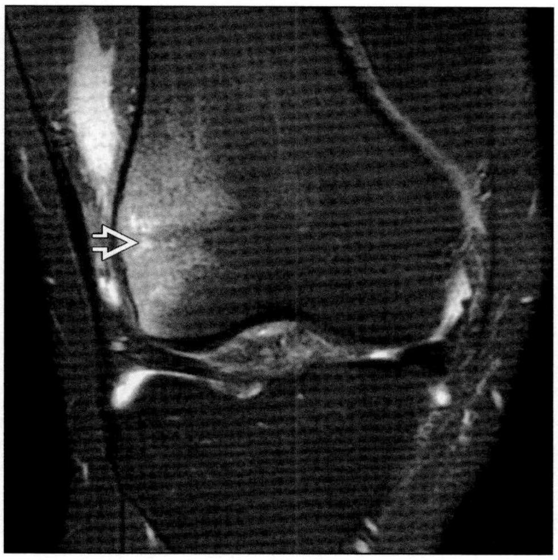

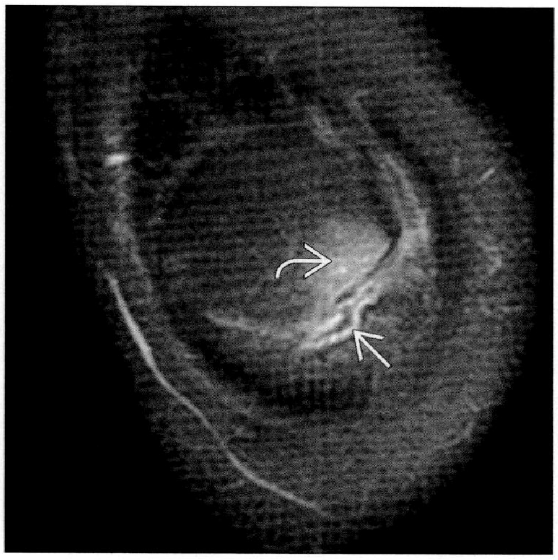

(Left) *Coronal T2WI FS MR from the same patient shows the typical marrow contusion in the lateral femoral condyle, located near the impaction point along the lateral cortex.* ***(Right)*** *Coronal T2WI FS MR more anteriorly in the same patient shows the inferomedial location of the patellar contusion associated with TPD. A focal tear of the MPFL is also seen and may be associated with a small cortical avulsion of the patellar cortex.*

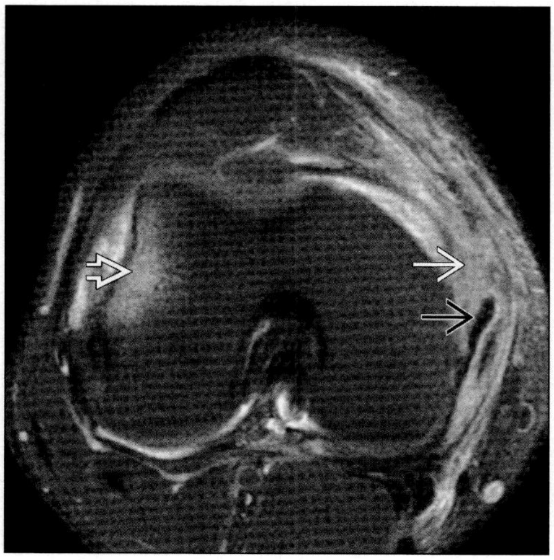

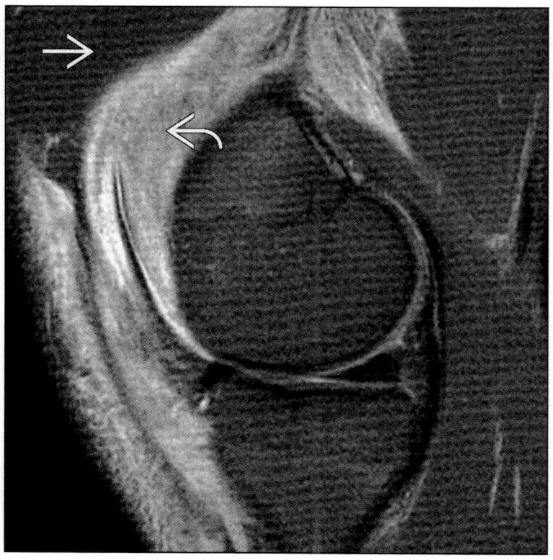

(Left) *Axial PDWI FS MR after transient patellar dislocation shows the typical subcortical LFC contusion and a tear of the MPFL. In this case, the MPFL tear is located more posteriorly near its junction with the medial collateral ligament (MCL).* ***(Right)*** *Sagittal PDWI FS MR from the same patient shows hematoma from MPFL disruption displacing the vastus medialis obliquus muscle away from the medial femoral condyle.*

*(**Left**) Axial PDWI FS MR after transient patellar dislocation shows MPFL tear anteriorly with partial avulsion of the medial patellar cortex. Note the very shallow trochlear groove, which predisposes to TPD. (**Right**) Axial PDWI FS MR after TPD shows a large, slightly displaced chondral fragment arising from the medial patella. There is a partial tear of the anterior MPFL, and a large joint effusion is present. The trochlear groove is very shallow.*

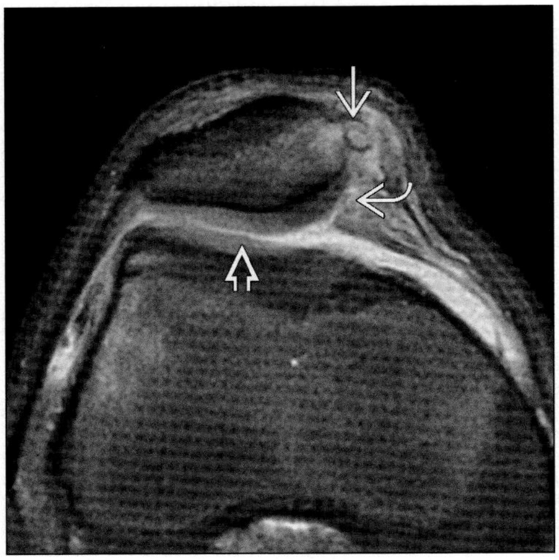

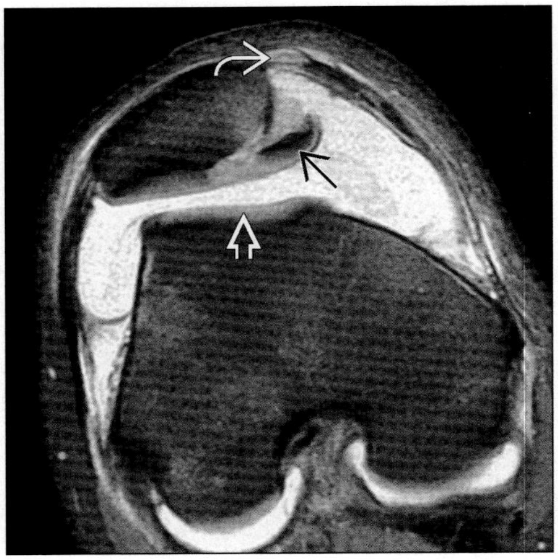

*(**Left**) Axial PDWI FS MR after TPD shows a large irregular grade IV chondral defect in the patellar apex and a large effusion. The fragment arising from the defect is not seen in this image and must be sought. Partial tear of the anterior MPFL is present. (**Right**) Axial PDWI FS MR from the same patient more inferiorly shows the displaced chondral fragment in the medial joint recess. Typical LFC contusion is present.*

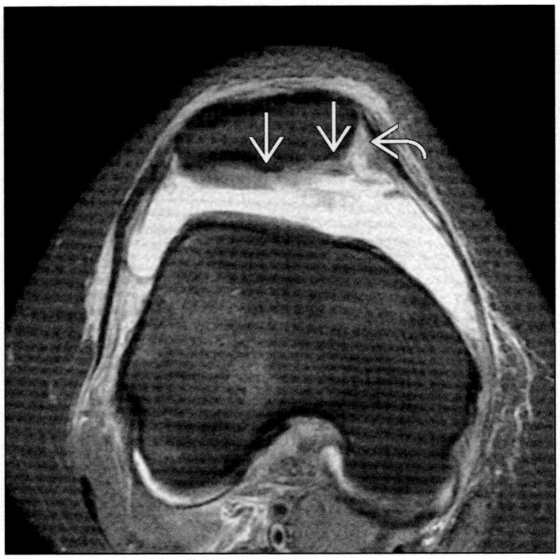

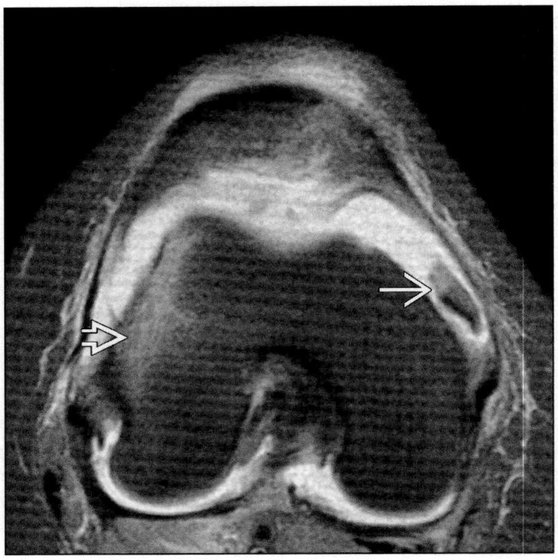

*(**Left**) Axial PDWI FS MR shows a focal high-grade chondral fracture in the patellar apex after TPD. Also noted are a LFC contusion and a joint effusion. (**Right**) Coronal T2WI FS MR from the same patient demonstrates a chondral fracture in the lateral femoral condyle as well. While chondral fractures are not uncommon in the setting of TPD, seeing both patellar and lateral femoral condylar chondral fractures in the same patient is unusual.*

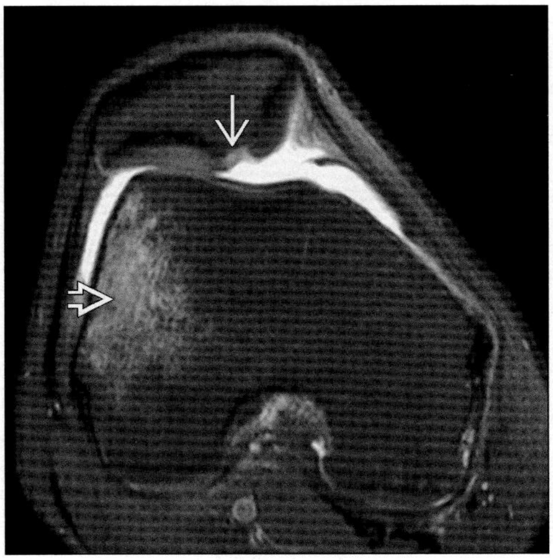

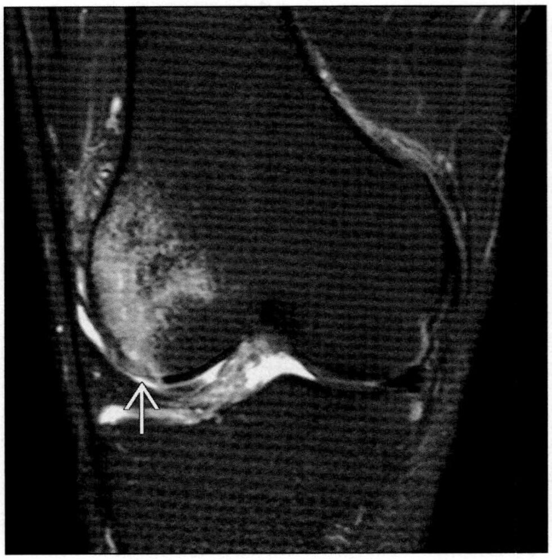

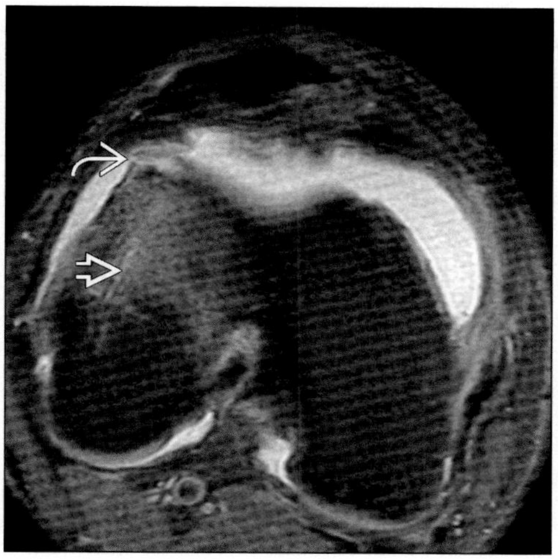

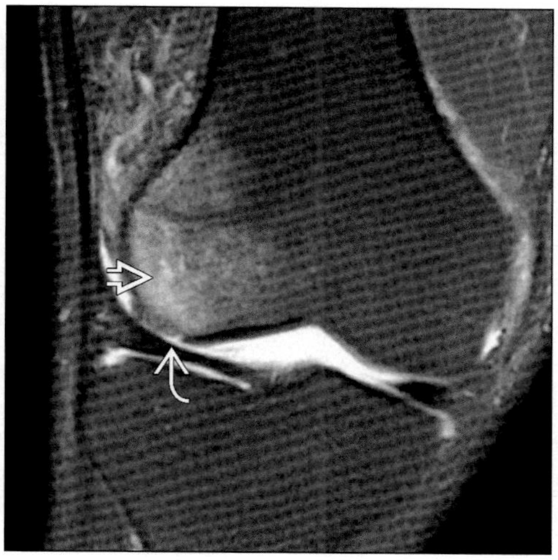

(Left) *Axial PDWI FS MR immediately after an episode of transient patellar dislocation shows diffuse marrow edema in the lateral aspect of the LFC with a focal grade IV chondral fracture in the lateral trochlear ridge.* ***(Right)*** *Coronal T2WI FS MR from the same patient shows the focal chondral defect and the more diffuse underlying marrow edema. This area should be scrutinized in all imaging planes, as chondral defects here can be subtle.*

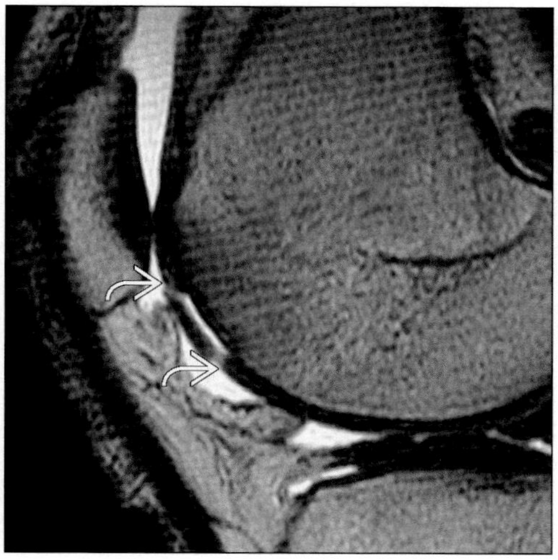

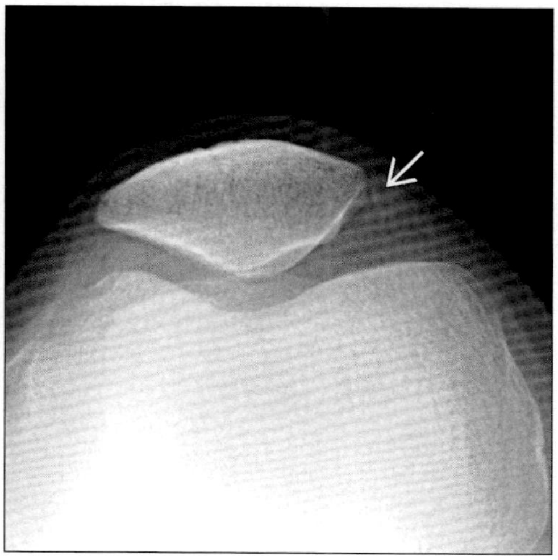

(Left) *Sagittal T2WI MR from the same patient shows an in situ unstable chondral fracture of the lateral trochlear ridge. Chondral fractures in the setting of transient patellar dislocation may occur in the patella or in the lateral femoral condyle.* ***(Right)*** *Axial sunrise view radiograph shows a small cortical avulsion fragment arising from the medial pole of the patella, consistent with a tear of the MPFL. This finding is consistent with recent TPD.*

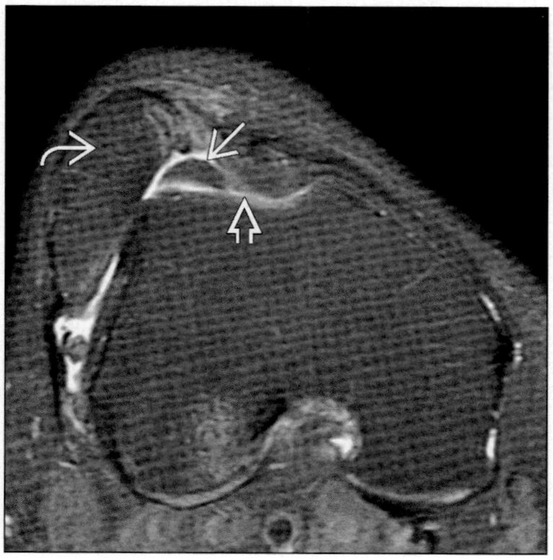

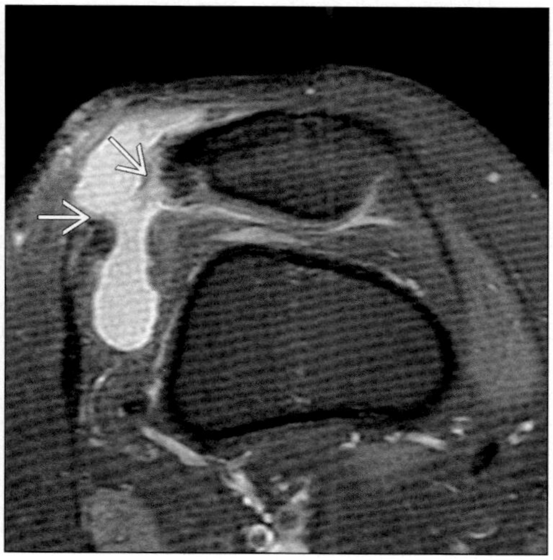

(Left) *Axial PDWI FS MR shows persistent lateral displacement of the patella out of the trochlear groove after traumatic dislocation. Spontaneous relocation did not occur, and therapeutic reduction in the ED failed due to a trapped fragment in the joint.* ***(Right)*** *Axial PDWI FS MR in a patient with a history of recurrent lateral patellar dislocation and lateral retinacular release shows a large defect in the lateral retinaculum with joint fluid extending to the subcutaneum.*

PES ANSERINE BURSITIS

Key Facts

Terminology
- Pes anserinus: Tendons of sartorius, gracilis, and semitendinosus as they pass around posteromedial aspect of proximal tibia

Imaging
- Well-defined fluid collection between pes anserinus tendons and proximal medial tibia
- Usually centered 2 cm below medial knee joint line
- Best recognized on axial and coronal fluid-sensitive MR sequences

Top Differential Diagnoses
- Semimembranosus-tibial collateral ligament bursitis
- Popliteal cyst
- Tenosynovitis of pes anserinus tendons
- Normal medial parameniscal joint recesses
- Parameniscal cyst
- Knee joint ganglion cyst

Clinical Issues
- Athletes
- Due to overuse and repetitive activity
- Overweight middle-aged women
- Likely due to valgus knee deformity and chronic stress
- Improves with cessation of activity in athletes
- Treatment
  - Conservative in most cases
  - Refractory or infected cases → surgical resection

Diagnostic Checklist
- Consider pes anserine bursitis in setting of medial knee pain below joint line
- Exclude connection to medial meniscus or knee joint

(Left) *Sagittal graphic of the medial aspect of the knee shows inflammation of the pes anserine bursa ➔ located below the knee joint, between the tibia and the tendons of the pes anserinus (sartorius ➔, gracilis ➔, and semitendinosus ➔). The tendon of the semimembranosus ➔ is located posterior and lateral to the bursa.* ***(Right)*** *Sagittal T2WI MR shows a large lobular fluid collection ➔ deep to the tendons of the pes anserinus ➔.*

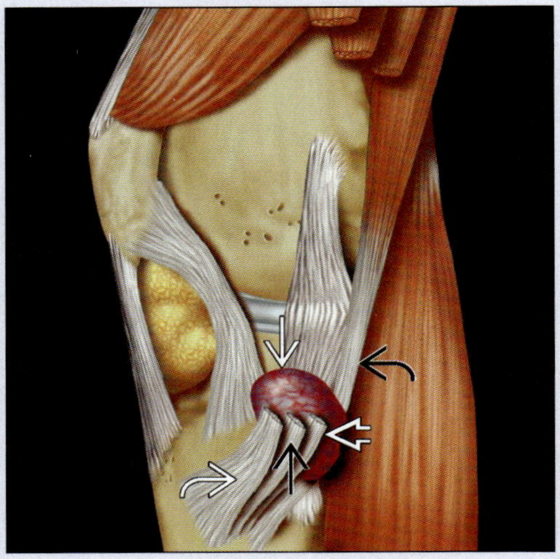

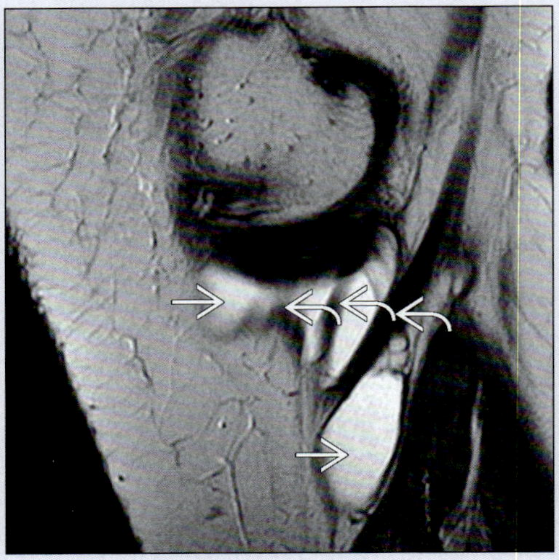

(Left) *Coronal T2WI FS MR from the same patient shows a very large pes anserine bursal fluid collection ➔ adjacent to the medial tibia and deep to the tendons of the pes anserinus ➔. Pes anserine bursitis typically begins slightly below the knee joint line and may extend several centimeters distally.* ***(Right)*** *Axial PDWI FS MR from the same patient shows the lobular nature of the bursal fluid collection; note how the pes anserine tendons are partially enveloped by the bursa ➔.*

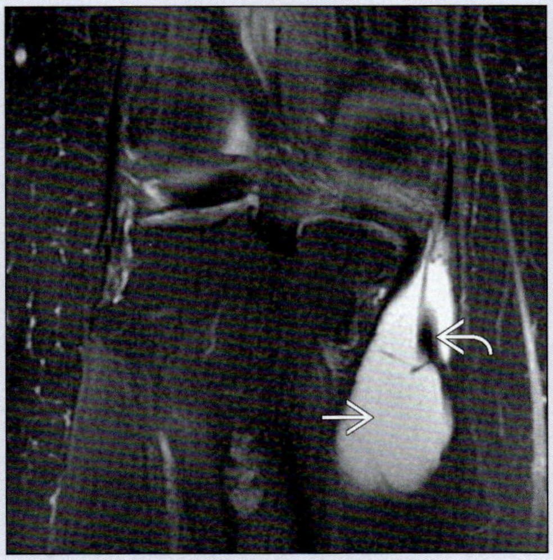

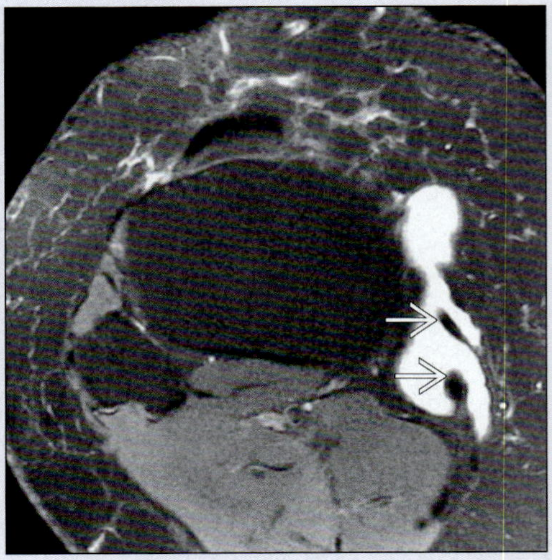

PES ANSERINE BURSITIS

TERMINOLOGY

Definitions

- Pes anserinus: Tendons of sartorius, gracilis, and semitendinosus as they pass around posteromedial aspect of proximal tibia
- Pes anserine bursitis: Focal bursal collection between pes anserinus tendons and tibia

IMAGING

General Features

- Best diagnostic clue
  - Well-defined fluid collection between pes anserinus tendons and proximal medial tibia
- Location
  - Usually centered 2 cm below medial knee joint line
- Size
  - May range from a few millimeters to several centimeters in diameter
- Morphology
  - Round or lobular outline

CT Findings

- Fluid-density lobular mass between tendons of pes anserinus and proximal tibia

MR Findings

- Lobular fluid signal mass with well-defined thinly marginated borders between tendons of pes anserinus and proximal medial tibia
- May be associated with tenosynovitis of tendons of pes

Ultrasonographic Findings

- Well-defined anechoic fluid collection with increased through-transmission deep to pes anserinus tendons below medial knee joint

Imaging Recommendations

- Best imaging tool
  - MR
- Protocol advice
  - Best recognized on axial and coronal fluid-sensitive MR sequences

DIFFERENTIAL DIAGNOSIS

Semimembranosus-Tibial Collateral Ligament Bursitis

- Located slightly more superior
- Forms "U" around distal semimembranosus tendon

Popliteal Cyst

- Forms more superiorly and medially between medial gastroc and semimembranosus

Tenosynovitis of Pes Anserinus Tendons

- Fluid centered around tendon(s) and not deep to them

Normal Medial Parameniscal Joint Recesses

- Small amounts of fluid collect near meniscofemoral and meniscotibial ligaments

Parameniscal Cyst

- Usually demonstrates connection to medial meniscal tear

Knee Joint Ganglion Cyst

- May arise superior or inferior to medial meniscus and extend along soft tissue planes

PATHOLOGY

General Features

- Etiology
  - Repetitive motion causes friction between medial tibial condyle and pes anserine tendons → inflammation of pes anserine bursa
  - Inflammation of pes anserine bursa can be seen in other settings, such as rheumatoid arthritis or infection
- Associated abnormalities
  - Can be associated with tenosynovitis of tendons of pes anserinus

CLINICAL ISSUES

Presentation

- Most common signs/symptoms
  - Medial knee pain below joint line
- Clinical profile
  - Athletes
    - Due to overuse and repetitive activity
  - Overweight middle-aged women
    - Likely due to valgus knee deformity and chronic stress

Natural History & Prognosis

- Improves with cessation of activity in athletes
- Slowly progressive if activity not modified

Treatment

- Conservative in most cases
- Refractory or infected cases → surgical resection

DIAGNOSTIC CHECKLIST

Consider

- Pes anserine bursitis in setting of medial knee pain below joint line

Image Interpretation Pearls

- Exclude connection to medial meniscus or knee joint

SELECTED REFERENCES

1. Clapp A et al: Radiologic case study. Pes anserine bursitis. Orthopedics. 31(4):306, 407-8, 2008
2. Marra MD et al: MRI features of cystic lesions around the knee. Knee. 15(6):423-38, 2008
3. Rennie WJ et al: Pes anserine bursitis: incidence in symptomatic knees and clinical presentation. Skeletal Radiol. 34(7):395-8, 2005
4. Rothstein CP et al: Semimembranosus-tibial collateral ligament bursitis: MR imaging findings. AJR Am J Roentgenol. 166(4):875-7, 1996
5. Forbes JR et al: Acute pes anserine bursitis: MR imaging. Radiology. 194(2):525-7, 1995

POSTEROMEDIAL CORNER INJURY

Key Facts

Terminology

- Posteromedial corner extends from posterior aspect of medial collateral ligament to medial edge of posterior cruciate ligament
- Major components
  - Semimembranosus muscle and tendon and its aponeurotic expansions
  - Posterior oblique ligament
  - Posteromedial aspect of medial meniscus

Imaging

- Best seen on coronal and axial T2WI FS or PDWI FS MR

Pathology

- Valgus stress and internal rotation of knee ± hyperextension
- Isolated PMC injury (12%)
- PMC + ACL tear (78%)
- MCL injured in 33% of PMC injuries
- Posterior oblique ligament
  - Discrete fiber bundle running posterior to MCL
  - Originates at adductor tubercle on medial femoral condyle (posterior and superior to MCL origin)
  - Attaches to medial meniscus and medial tibia

Clinical Issues

- Anteromedial rotatory instability (AMRI)
  - Increased laxity to abduction stress testing in 30° of knee flexion + anterior rotatory subluxation of medial tibial plateau on femur
- Primary surgical repair suggested if PMC injury associated with cruciate tear

Diagnostic Checklist

- PMC injury with ACL tear, peripheral medial meniscus tear, or MCL tear

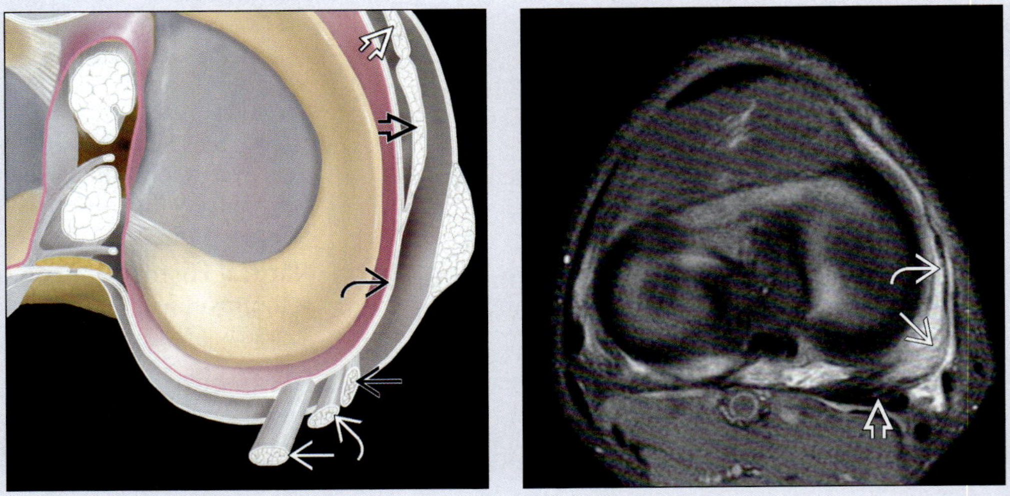

(Left) *Axial graphic of the posteromedial corner of the knee demonstrates the tendons of the semimembranosus (SM) ➡, semitendinosus ➡, and gracilis ➡. The vertical fibers of the medial collateral ligament (MCL) ➡ are seen as separate from the oblique portion ➡. The posterior oblique ligament (POL) ➡ runs between the oblique portion of the MCL and the posterior capsule.* ***(Right)*** *Axial PDWI FS MR shows edema in the POL ➡. The MCL oblique fibers ➡ and SM ➡ are intact.*

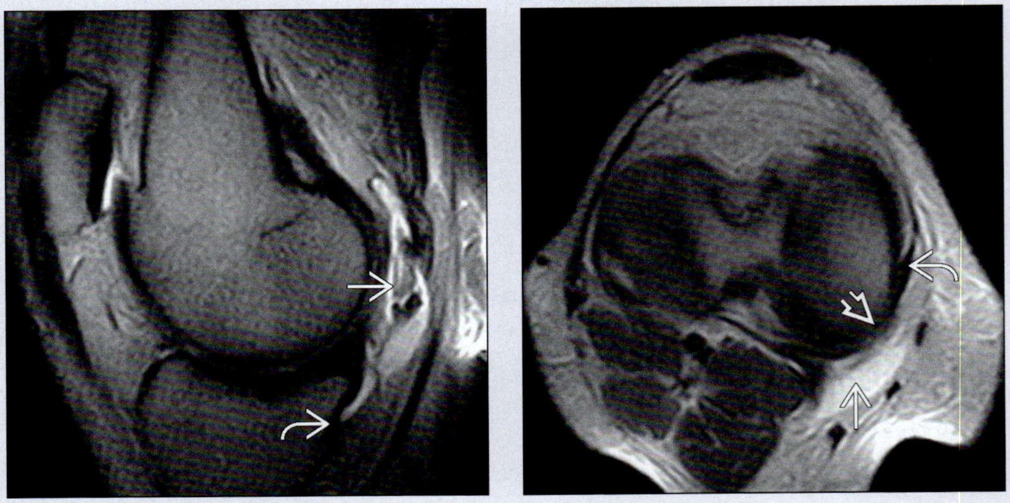

(Left) *Sagittal T2WI MR shows complete rupture of the semimembranosus tendon ➡ several centimeters above its tibial insertion ➡, with tendon retraction and surrounding hemorrhage and edema.* ***(Right)*** *Axial T2WI MR from the same patient at the level of the joint shows fluid in the expected location of the semimembranosus tendon ➡; the tendon is retracted more superiorly. MCL oblique fibers ➡ and POL ➡ are intact.*

POSTEROMEDIAL CORNER INJURY

TERMINOLOGY

Abbreviations
- Posteromedial corner (PMC)

Definitions
- Posteromedial corner extends from posterior aspect of medial collateral ligament to medial edge of posterior cruciate ligament
- Major components
  - Semimembranosus muscle and tendon and its aponeurotic expansions
  - Posterior oblique ligament
  - Posteromedial aspect of medial meniscus

IMAGING

General Features
- Best diagnostic clue
  - Partial or complete disruption of tibial attachment of semimembranosus
  - Soft tissue edema in posteromedial corner of knee around joint line

MR Findings
- Best seen on coronal and axial T2WI FS or PDWI FS MR
- Semimembranosus (SM)
  - Edema within ± around distal SM at or near its attachment to tibia
  - Disruption of 1 or more aponeurotic slips of distal SM
- Posterior oblique ligament (POL)
  - Edema within ± around POL
  - Disruption of POL
- Disruption of posteromedial capsular attachment of medial meniscus
  - Often seen best on sagittal PDWI or T2WI ± FS MR

Imaging Recommendations
- Best imaging tool
  - MR
- Protocol advice
  - Coronal and axial T2WI FS MR best for delineating major structures of PMC

DIFFERENTIAL DIAGNOSIS

Medial Meniscal Tear
- Peripheral tear may mimic peripheral meniscal detachment as part of PMC injury
- Thin rim of peripheral meniscus seen attached to capsule

Meniscocapsular Separation
- Isolated disruption of peripheral medial meniscal attachment, usually posteriorly
- No associated injury to posterior oblique ligament

PATHOLOGY

General Features
- Etiology
  - Valgus stress and internal rotation of knee ± hyperextension
- Associated abnormalities
  - Isolated PMC injury (12%)
  - PMC + ACL tear (78%)
  - PMC + PCL (2%)
  - MCL injured in 33% of PMC injuries

Gross Pathologic & Surgical Features
- Posteromedial corner composed of 3 main elements
  - Semimembranosus unit and its aponeurotic expansions
  - Posterior oblique ligament
    - Discrete fiber bundle running posterior to MCL
    - Originates at adductor tubercle on medial femoral condyle (posterior and superior to MCL origin)
    - Attaches to medial meniscus and medial tibia
- Medial head of gastrocnemius lends support to PMC

CLINICAL ISSUES

Presentation
- Most common signs/symptoms
  - Posteromedial knee pain
  - Anteromedial rotatory instability (AMRI)
    - Increased laxity to abduction stress testing in 30° of knee flexion + anterior rotatory subluxation of medial tibial plateau on femur

Natural History & Prognosis
- When PMC injury combined with ACL tear, failure to repair PMC at same time as ACL reconstruction may lead to increased incidence of
  - ACL graft failure
  - Medial meniscus tear

Treatment
- Primary surgical repair suggested if PMC injury associated with cruciate tear

DIAGNOSTIC CHECKLIST

Consider
- PMC injury with ACL tear, peripheral medial meniscus tear, or MCL tear

SELECTED REFERENCES

1. Lind M et al: Anatomical reconstruction of the medial collateral ligament and posteromedial corner of the knee in patients with chronic medial collateral ligament instability. Am J Sports Med. 37(6):1116-22, 2009
2. House CV et al: Posteromedial corner injuries of the knee. Clin Radiol. 62(6):539-46, 2007
3. Fanelli GC et al: Surgical treatment of acute medial collateral ligament and posteromedial corner injuries of the knee. Sports Med Arthrosc. 14(2):78-83, 2006
4. Loredo R et al: Posteromedial corner of the knee: MR imaging with gross anatomic correlation. Skeletal Radiol. 28(6):305-11, 1999

*(**Left**) Sagittal PDWI FS MR shows diffuse edema in the posterior oblique ligament and disruption of the posterior aspect of the meniscotibial ligament. There is a marrow contusion in the adjacent medial tibial condyle. Semimembranosus tendon is intact. (**Right**) Coronal T2WI FS MR from the same patient shows diffuse edema in the posterior oblique ligament in the posteromedial corner. Posterior tibial contusion is related to concurrent ACL tear (not shown).*

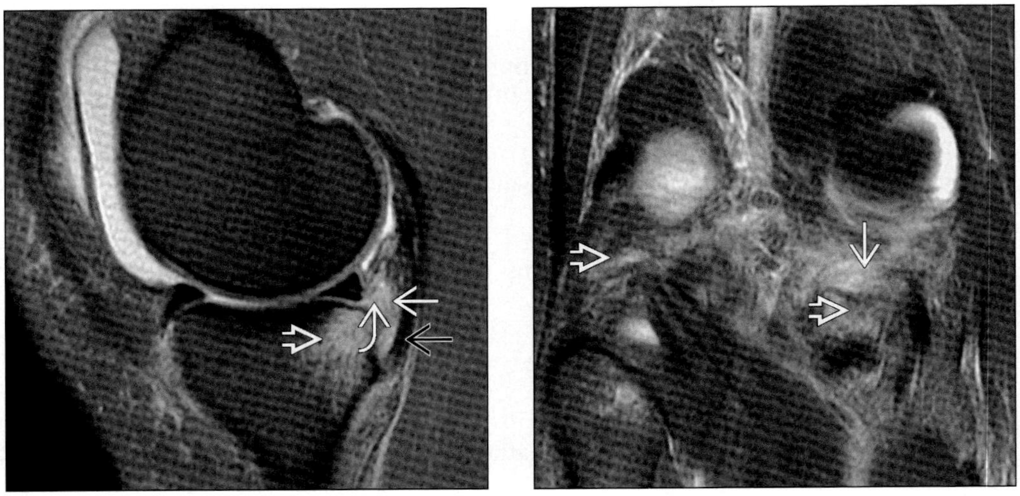

*(**Left**) Axial PDWI FS MR from the same patient shows a low-grade partial tear of the distal fibers of the semimembranosus tendon not well seen on sagittal or coronal images due to the oblique course of the tendon at its insertion. Tibial contusion from ACL tear extends across the posterior tibia. (**Right**) Axial PDWI FS MR shows edema around the oblique fibers of the MCL and partial tear of the posterior oblique ligament (POL). Vertical portion of MCL is partially torn as well.*

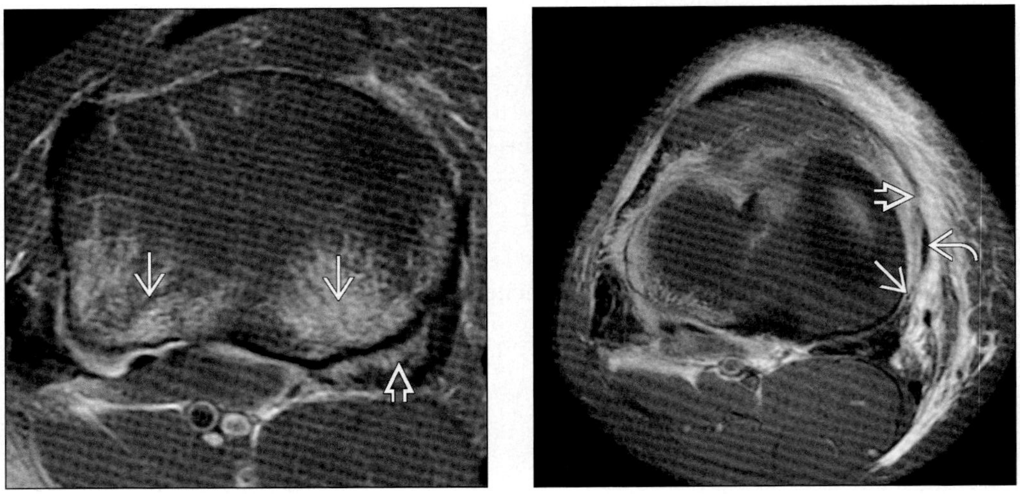

*(**Left**) Axial PDWI FS MR shows a partial tear of the distal semimembranosus tendon, complete tear of the vertical MCL fibers, and intact oblique MCL fibers. Posterior oblique ligament fibers demonstrate mild edema. (**Right**) Sagittal PDWI FS MR from the same patient demonstrates partial disruption and delamination of the distal semimembranosus tendon and edema in the tibial condyle. Low-grade edema in the posterior oblique ligament is present.*

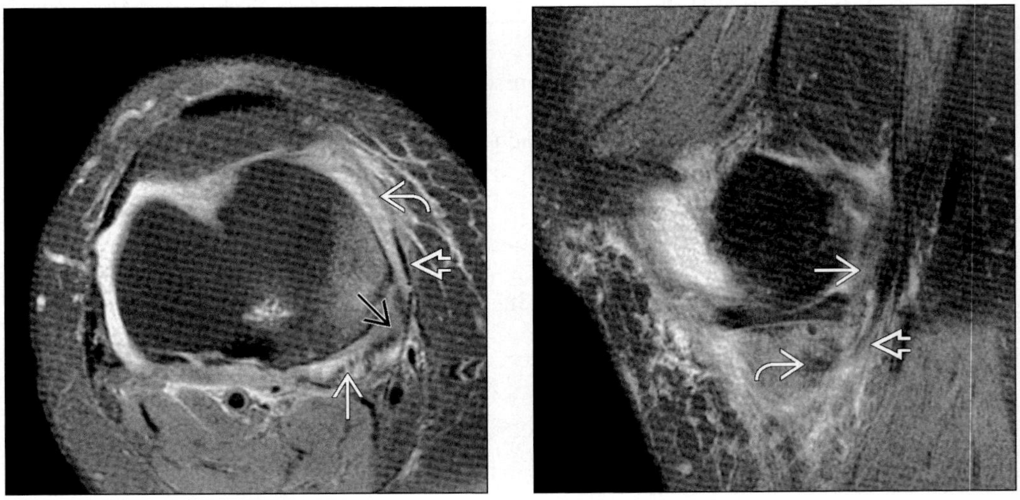

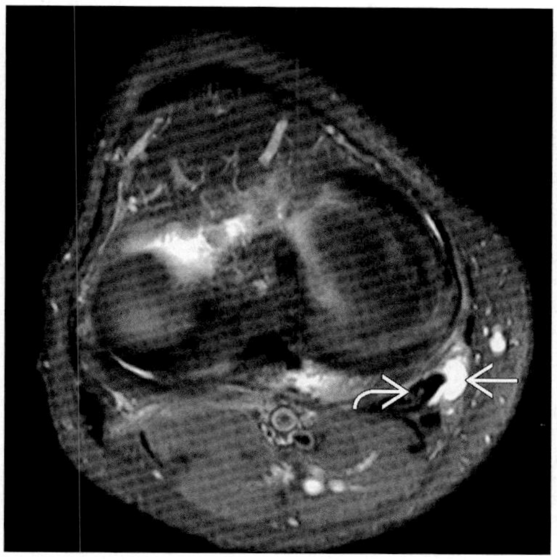

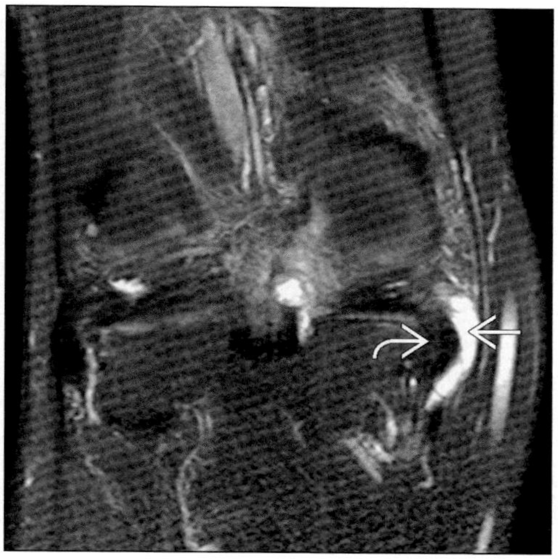

(Left) *Axial PDWI FS MR shows a small fluid collection in the semimembranosus-tibial collateral ligament (SM-TCL) bursa medial to the semimembranosus tendon at the joint line.* ***(Right)*** *Coronal T2WI FS MR from the same patient shows the typical C- or U-shaped SM-TCL bursa appearance adjacent to the semimembranosus. This collection must be distinguished from a popliteal cyst, normal joint recess, or parameniscal cyst. Its constant relationship to the SM tendon is the best clue.*

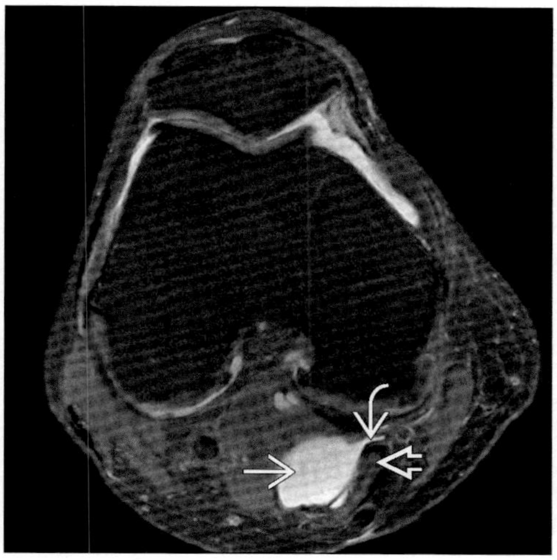

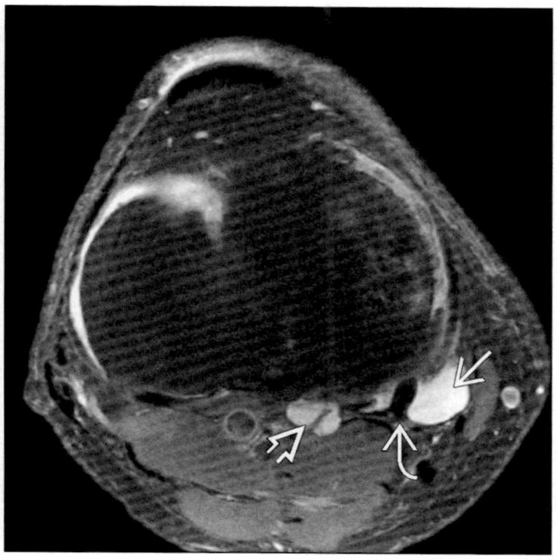

(Left) *Axial PDWI FS MR shows a popliteal cyst posteromedial to the knee just above the joint line. As is typical, the cyst arises from the gastrocnemius-semimembranosus hiatus and thus lies lateral to the semimembranosus tendon.* ***(Right)*** *Axial PDWI FS MR more inferiorly (at the joint line) in the same patient shows a different fluid collection located medial to the semimembranosus tendon. There is a small joint ganglion further laterally.*

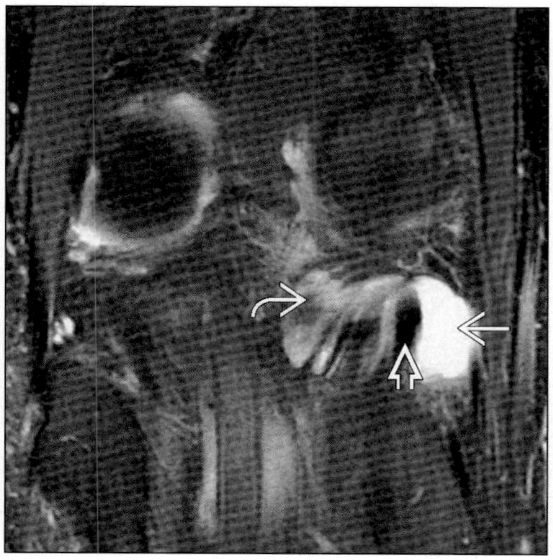

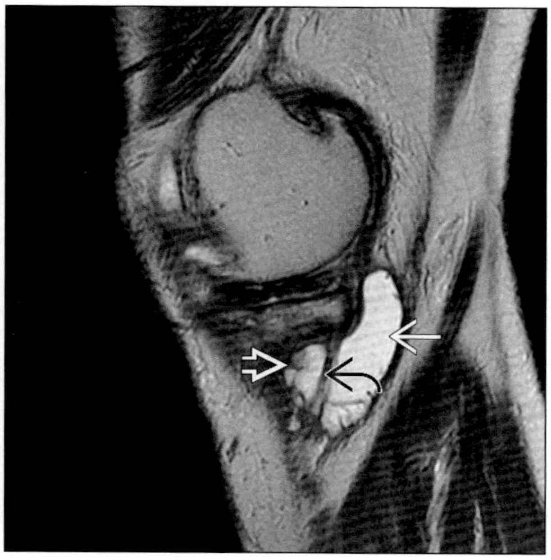

(Left) *Coronal T2WI FS MR from the same patient shows the semimembranosus-tibial collateral ligament (SM-TCL) bursa distended with fluid. The SM-TCL bursa is medial to the distal semimembranosus tendon. A joint ganglion is located lateral to the tendon.* ***(Right)*** *Sagittal T2WI MR from the same patient shows the superficial and deep components of the semimembranosus-tibial collateral ligament bursa, with the semimembranosus tendon dividing the 2 portions.*

POPLITEAL CYST

Key Facts

Terminology

- Synonyms: Baker cyst, semimembranosus-gastrocnemius bursa

Imaging

- Definition: Fluid-signal mass or masses in posteromedial popliteal fossa traceable to interval between semimembranosus (SM) and medial head of gastrocnemius (GN)
  - May dissect into adjacent muscle planes
- MR arthrography: Communication with joint
- US: Well-marginated fluid collection with increased through-transmission traceable to SM-GN hiatus in posteromedial popliteal fossa
  - Internal echoes if contains debris, hemorrhage, infection
- Complications
  - May be complicated by hemorrhage or infection
  - If ruptured or leaking → surrounding ill-defined edema

Top Differential Diagnoses

- Semimembranosus-tibial collateral ligament bursitis
- Ganglion
- Popliteal artery aneurysm
- Cystic adventitial disease of popliteal artery
- Deep venous thrombosis

Pathology

- Presence of fluid in SM-GN hiatus is very common but not necessarily pathologic
- Definition of popliteal cyst as pathologic finding based on size is somewhat arbitrary

Clinical Issues

- MR: 5-50% prevalence

*(**Left**) Axial graphic shows the typical location of a popliteal (Baker) cyst →, extending from near the posterior joint capsule through a hiatus between the medial head of the gastrocnemius → and the semimembranosus →. (**Right**) Axial PDWI FS MR shows the typical morphology of a popliteal cyst extending through the semimembranosus-gastrocnemius (SM-GN) hiatus → located between the semimembranosus → and medial gastrocnemius → tendons.*

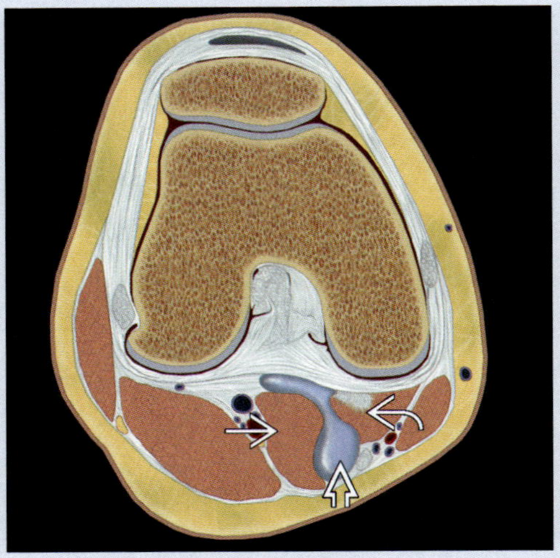

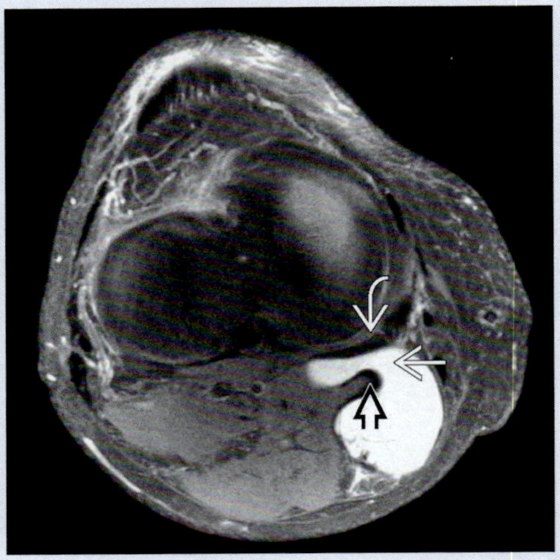

*(**Left**) Coronal T2WI FS MR from the same patient shows a popliteal cyst → medial to the medial margin of the gastrocnemius muscle →. A small amount of edema inferior to the cyst → indicates mild leakage of fluid through the lining of the cyst. (**Right**) Sagittal T2WI MR from the same patient shows the cyst extending through the SM-GN hiatus →. Usually more slit-like, the hiatus is widely patent in this patient, which affords a better appreciation of its location.*

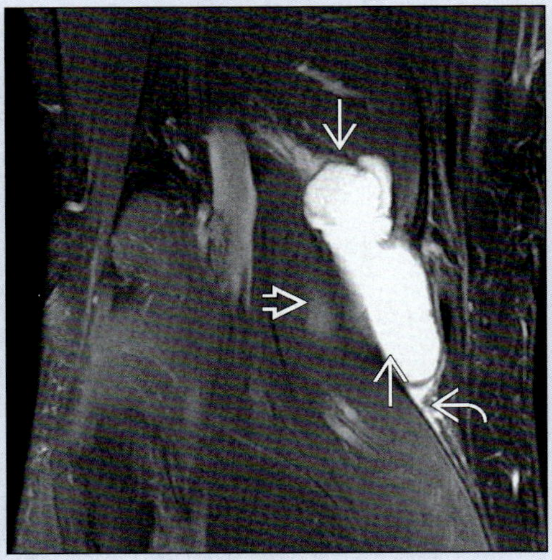

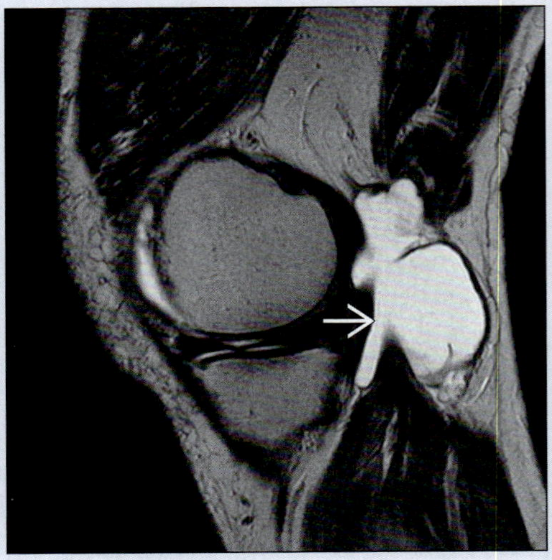

POPLITEAL CYST

TERMINOLOGY

Synonyms

- Baker cyst, semimembranosus-gastrocnemius bursa

Definitions

- Fluid distention of knee joint recess in posteromedial popliteal fossa

IMAGING

General Features

- Best diagnostic clue
  - Fluid-filled mass or masses in posteromedial popliteal fossa traceable to interval between semimembranosus (SM) and medial head of gastrocnemius (GN)
- Location
  - Arises between tendons of medial head of SM and GN posterior to medial femoral condyle just above joint line
  - Usually extends along medial margin of medial GN muscle
  - Portion of cyst may lie anterior to gastrocnemius
  - May extend variable distance above &/or below joint
- Size
  - May range from 1 to > 40 cm^3
- Morphology
  - Usually ovoid, with neck extending to point of origin between SM and GN tendons (SM-GN hiatus)
  - SM-GN hiatus about 15-20 mm long
    - May be covered by thin membrane in 70% of patients

Radiographic Findings

- If large, can see soft tissue mass in medial popliteal fossa
- Arthrography: Communicates with joint

CT Findings

- Fluid-density mass in medial popliteal fossa with point of origin between SM and GN tendons

MR Findings

- Lobular uni- or multicompartmental fluid signal mass traceable to hiatus between SM and GN tendons posterior to medial femoral condyle
  - Low signal on T1WI MR
  - High signal on T2WI MR
- May extend variable distance superior or inferior to site of origin
- May dissect into adjacent muscle planes
- If ruptured or leaking → surrounding ill-defined edema
  - Usually limited to intermuscular planes
  - May dissect into muscles
- May be complicated
  - Hemorrhage
    - Variable signal on T1WI and T2WI MR depending on stage of blood products
    - May demonstrate fluid-fluid levels
  - Debris or intraarticular bodies
    - Discrete foci of (usually) low signal on T1WI and T2WI MR within fluid
    - If ossified bodies contain marrow → high signal on T1WI MR
- MR arthrography
  - Communication with joint
    - Occasionally difficult to document if SM-GN hiatus obstructed

Ultrasonographic Findings

- Well-marginated fluid collection with increased through-transmission traceable to SM-GN hiatus in posteromedial popliteal fossa
- Internal echoes if contains debris, hemorrhage, infection
- Surrounding soft tissue edema if ruptured
- Often identified as incidental finding on venous duplex Doppler study

Imaging Recommendations

- Best imaging tool
  - MR or US
- Protocol advice
  - May need enlarged field-of-view if cyst is large
    - Extension into distal thigh or proximal leg

DIFFERENTIAL DIAGNOSIS

Semimembranosus-Tibial Collateral Ligament Bursitis

- U-shaped fluid collection around distal semimembranosus tendon
- No connection to knee joint

Ganglion

- May arise anywhere
- No connection to GN-SM hiatus

Popliteal Artery Aneurysm

- Saccular or fusiform enlargement of popliteal artery
- Complex flow signal

Cystic Adventitial Disease of Popliteal Artery

- 1 or more cystic collections in wall of popliteal artery

Deep Venous Thrombosis

- Enlargement of popliteal vein(s) containing complex signal
- Surrounding soft tissue swelling

PATHOLOGY

General Features

- Etiology
  - Popliteal recess may form as result of herniation of synovium through SM-GN hiatus
  - Increased intraarticular pressure in knee → passage of fluid through SM-GN hiatus into popliteal recess
  - Fluid trapped in cyst due to closure of slit-like SM-GN hiatus when knee extended
  - Over time, fluid accumulates in cyst and cyst enlarges
  - Enlargement of cyst may also contribute to closure of SM-GN hiatus due to pressure effect
- Associated abnormalities
  - Often seen with intraarticular knee pathology
    - Osteoarthritis, meniscal tear

POPLITEAL CYST

- Common with inflammatory conditions
  - Rheumatoid arthritis, gout, infection

Staging, Grading, & Classification

- Presence of fluid in SM-GN hiatus is very common but not necessarily pathologic
- Definition of popliteal cyst as a pathologic finding based on size is somewhat arbitrary

Gross Pathologic & Surgical Features

- Almost always communicates with knee joint, therefore really a joint recess
- Point of communication between SM and GN tendons may act as 1-way valve
  - Neck of recess between tendons becomes taut with knee extension, preventing return of fluid to joint
  - Fibrin or debris in joint may act as ball valve at neck
- As some fluid in cyst is reabsorbed or passes back into knee joint, remaining cyst contents are concentrated
  - May become very thick and gelatinous

Microscopic Features

- Cyst lined by synovium, as it is an extension of knee joint
- Secondary inflammation with hyperemia, inflammatory cells, fibrous change

CLINICAL ISSUES

Presentation

- Most common signs/symptoms
  - Posteromedial aching knee pain
    - Worse with activity
  - Palpable/visible mass in medial popliteal fossa
    - May enlarge with activity or prolonged standing
- Other signs/symptoms
  - Intraarticular symptoms related to underlying conditions
    - Effusion, meniscal tear, osteoarthritis

Demographics

- Epidemiology
  - MR: 5-50% prevalence
    - Prevalence higher with underlying knee pathology
    - Up to 12% prevalence in pediatric population
  - Cadaveric dissection: 30% prevalence

Natural History & Prognosis

- May regress or enlarge over time
- Usually flares in concert with acute knee inflammation

Treatment

- Conservative
  - Knee wrap, NSAIDs, ice
- Percutaneous aspiration ± corticosteroid injection
  - Variable success, as SM-GN hiatus remains patent and cyst usually reforms
- Surgical
  - Excision with oversewing of SM-GN hiatus

DIAGNOSTIC CHECKLIST

Consider

- Underlying intraarticular knee pathology as etiology for recurrent popliteal cyst

Image Interpretation Pearls

- Document connection to SM-GN hiatus to differentiate from other cystic masses
- Be certain to consider and differentiate from other cystic-appearing lesions in similar location (contrast may be useful)
  - Popliteal aneurysm
  - Mass extending into SM-GN bursa from knee joint
    - Pigmented villonodular synovitis
    - Synovial chondromatosis
  - Soft tissue tumor arising in popliteal fossa

Reporting Tips

- Describe any atypical features
  - Hemorrhage within cyst
  - Rupture of cyst
  - Dissection of cyst away from normal position within SM-GN bursa
  - Intramuscular dissection of cyst

SELECTED REFERENCES

1. Chatzopoulos D et al: Baker's cysts in knees with chronic osteoarthritic pain: a clinical, ultrasonographic, radiographic and scintigraphic evaluation. Rheumatol Int. 29(2):141-6, 2008
2. Marra MD et al: MRI features of cystic lesions around the knee. Knee. 15(6):423-38, 2008
3. Ahn JH et al: Arthroscopic cystectomy for popliteal cysts through the posteromedial cystic portal. Arthroscopy. 23(5):559, 2007
4. Acebes JC et al: Ultrasonographic assessment of Baker's cysts after intra-articular corticosteroid injection in knee osteoarthritis. J Clin Ultrasound. 34(3):113-7, 2006
5. Fritschy D et al: The popliteal cyst. Knee Surg Sports Traumatol Arthrosc. 14(7):623-8, 2006
6. Fang CS et al: Intramuscular dissection of Baker's cysts: report on three cases. Skeletal Radiol. 33(6):367-71, 2004
7. McCarthy CL et al: The MRI appearance of cystic lesions around the knee. Skeletal Radiol. 33(4):187-209, 2004
8. Rupp S et al: Popliteal cysts in adults. Prevalence, associated intraarticular lesions, and results after arthroscopic treatment. Am J Sports Med. 30(1):112-5, 2002
9. Torreggiani WC et al: The imaging spectrum of Baker's (Popliteal) cysts. Clin Radiol. 57(8):681-91, 2002
10. Friedman L et al: Ultrasound of the knee. Skeletal Radiol. 30(7):361-77, 2001
11. Handy JR: Popliteal cysts in adults: a review. Semin Arthritis Rheum. 31(2):108-18, 2001
12. Sansone V et al: Popliteal cysts and associated disorders of the knee. Critical review with MR imaging. Int Orthop. 19(5):275-9, 1995
13. Pathria MN et al: Ultrasonography of the popliteal fossa and lower extremities. Radiol Clin North Am. 26(1):77-85, 1988

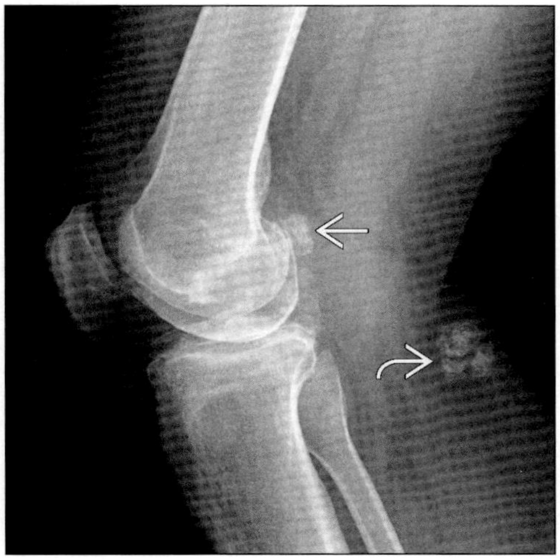
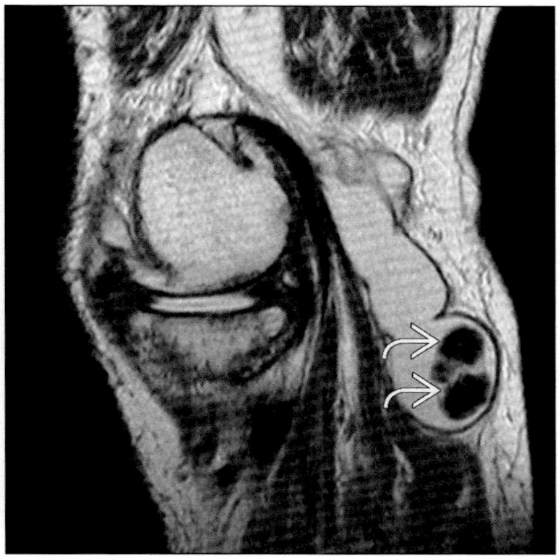

(Left) *Lateral radiograph shows changes of osteoarthritis with multiple intraarticular bodies present in the posterior aspect of the joint. Several calcified bodies are also present far posterior to the knee joint ; these were located medially on the AP view (not shown).* ***(Right)*** *Sagittal T2WI MR from the same patient shows the posterior calcified bodies within a large popliteal cyst, which extends posterior and inferior to the knee joint line.*

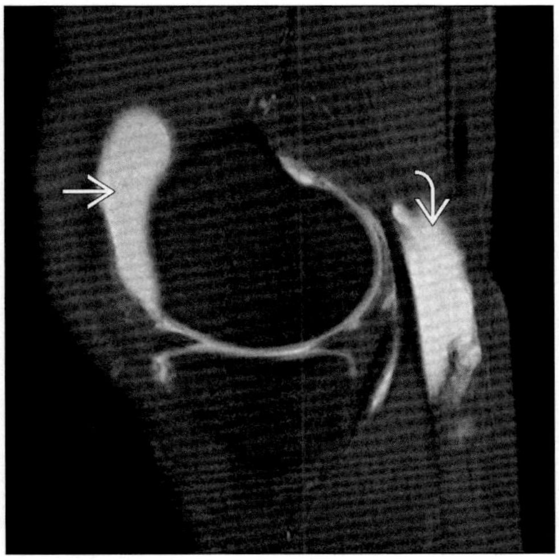
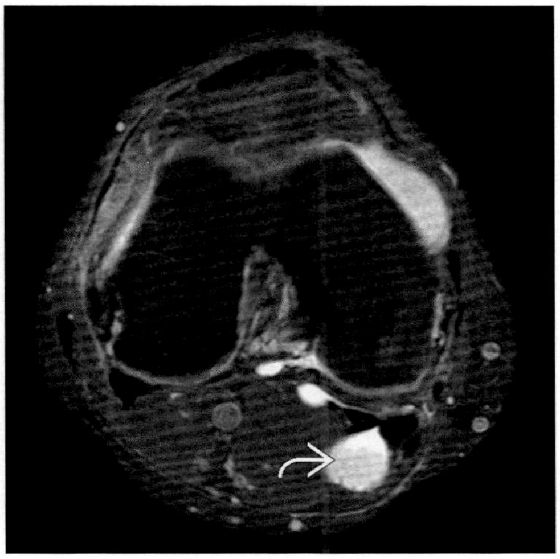

(Left) *Sagittal MR arthrogram T1WI FS shows contrast in the knee joint and passage of injected contrast in a popliteal cyst , confirming a patent connection between the joint and the cyst. While essentially all Baker cysts communicate with the knee in their formation, the communication may be blocked due to a variety of causes.* ***(Right)*** *Axial PDWI FS MR from the same patient shows the Baker cyst . Contrast within the cyst must be confirmed with a T1-weighted sequence.*

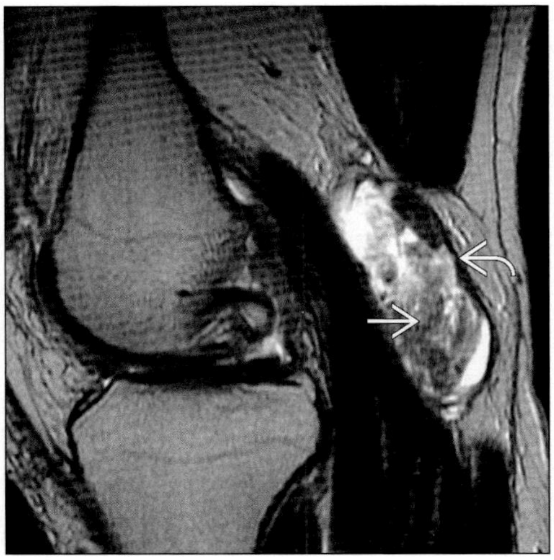
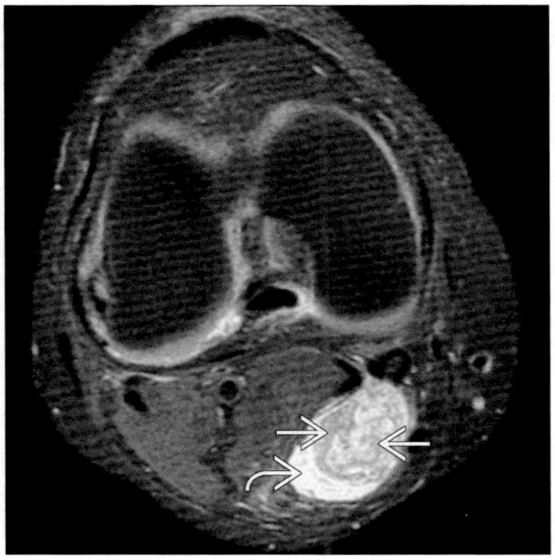

(Left) *Sagittal T2WI MR demonstrates irregular debris within a popliteal cyst. The lining of the cyst is fairly thin , suggesting that the debris is free-floating and not of synovial origin.* ***(Right)*** *Axial PDWI FS MR from the same patient shows fibrinous debris in the popliteal cyst, manifested by irregular intermediate signal. Simple fluid is present more posteriorly in the cyst . There is no knee joint effusion in this case.*

POPLITEAL CYST

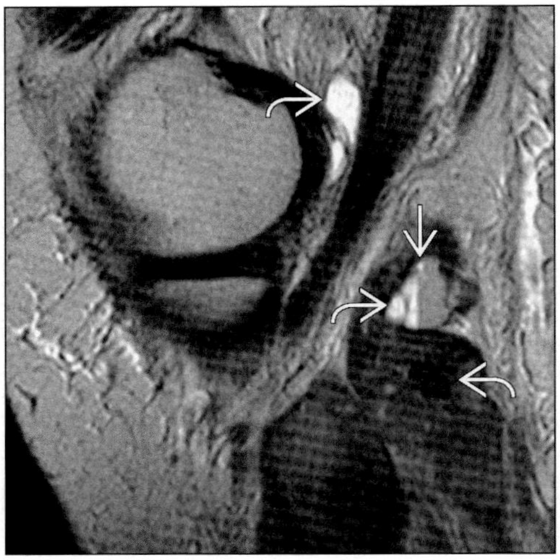

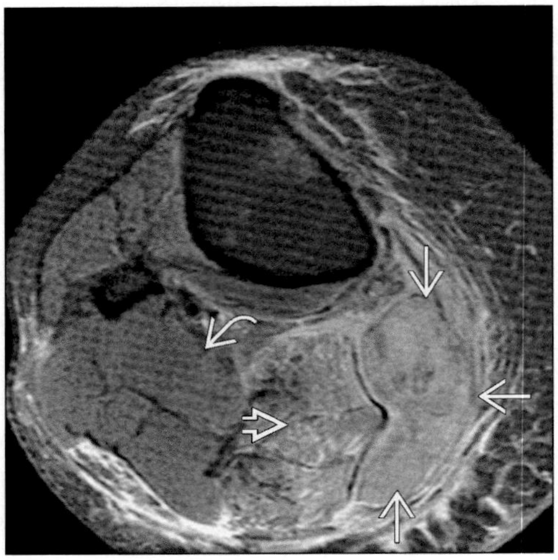

(Left) *Sagittal T2WI MR shows a complex Baker cyst with multiple components that demonstrate variable signal characteristics. A fluid-fluid level is present in 1 of the collections, strongly suggesting a hemorrhagic component of the cyst.* ***(Right)*** *Axial PDWI FS MR of the proximal leg from the same patient shows a large hematoma which dissects distally, medial to the medial gastrocnemius muscle. Petechial hemorrhage is seen within the muscle. The lateral gastrocnemius is normal.*

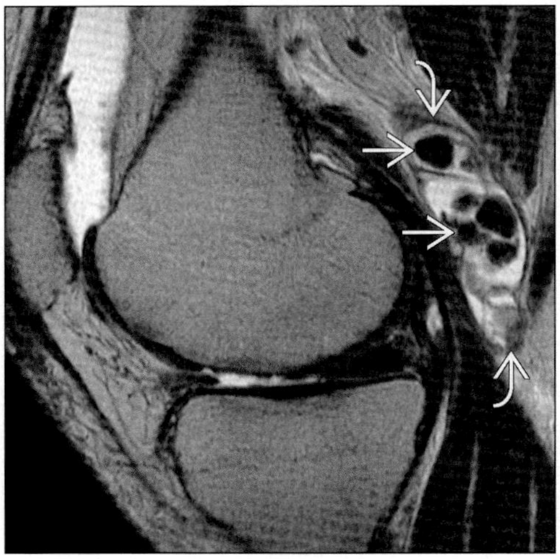

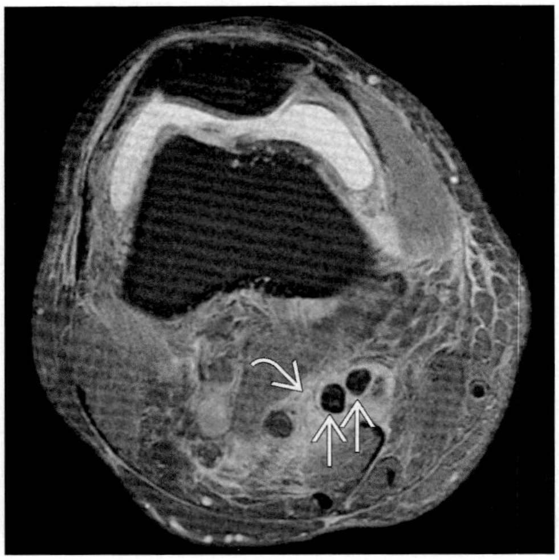

(Left) *Sagittal T2WI MR shows multiple low signal bodies within a popliteal cyst. In this case, the bodies are not all dependent in the fluid and may be adherent to the synovial lining of the cyst.* ***(Right)*** *Axial PDWI FS MR from the same patient shows the bodies within a portion of the cyst above the knee joint. The intermediate signal in this portion of the cyst suggests inflamed synovium instead of simple fluid.*

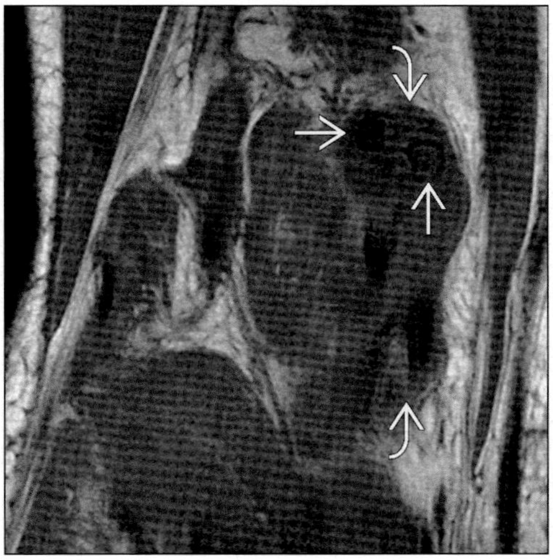

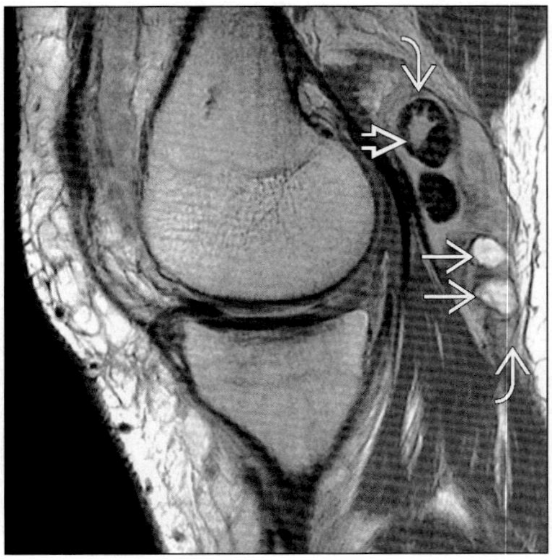

(Left) *Coronal T1WI MR from the same patient shows the low signal bodies within the popliteal cyst. It can be difficult to identify intracystic bodies on T1-weighted MR images unless they contain fat marrow signal.* ***(Right)*** *Sagittal PDWI MR shows another patient with multiple osseous bodies in a popliteal cyst. In this case, some of the bodies are sclerotic and demonstrate low signal; others contain fat marrow and demonstrate high signal.*

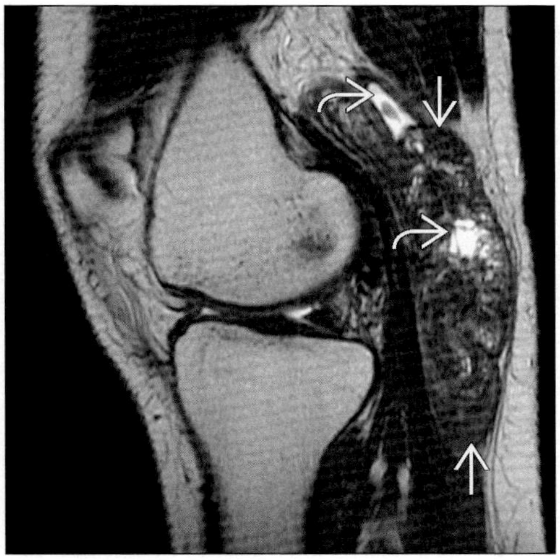

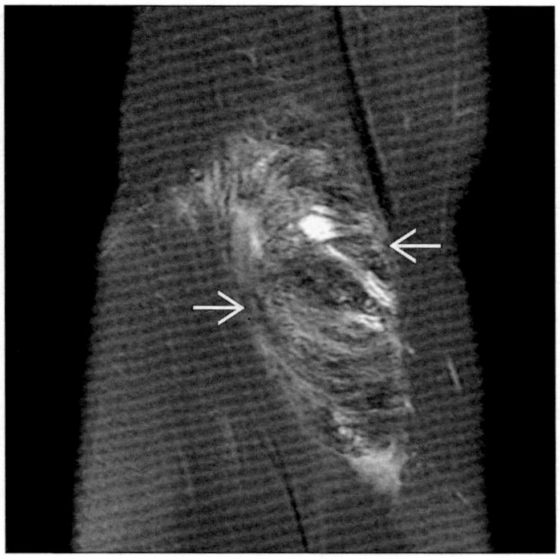

(Left) *Sagittal T2WI MR demonstrates marked synovial overgrowth in a large popliteal cyst. The material demonstrates low signal, indicating fibrous or hemorrhagic material. There are a few small pockets of fluid. There is no synovial hypertrophy or effusion in the knee joint.* ***(Right)*** *Coronal T2WI FS MR from the same patient shows the thick, low signal synovial hypertrophy present in the popliteal cyst. Surgical resection yielded pigmented villonodular synovitis (PVNS).*

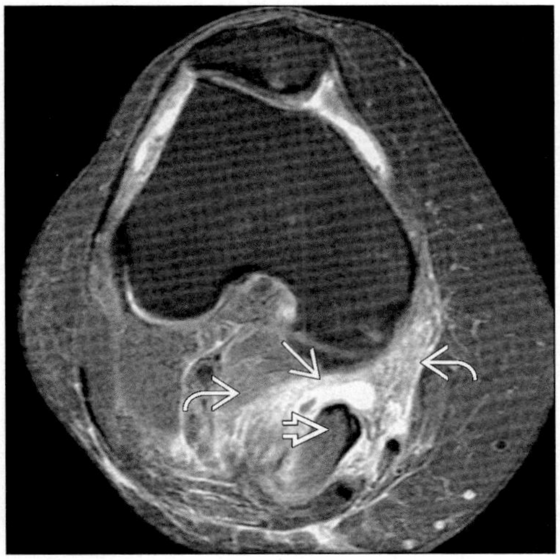

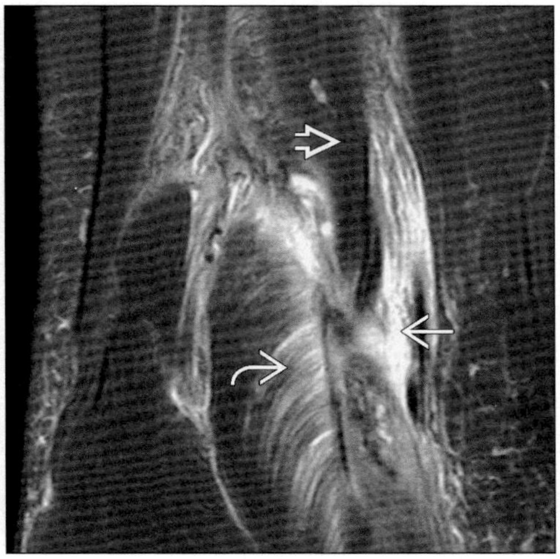

(Left) *Axial PDWI FS MR demonstrates rupture of a popliteal cyst with edema tracking into the medial gastrocnemius muscle and surrounding intermuscular planes. The semimembranosus is normal.* ***(Right)*** *Coronal T2WI FS MR from the same patient shows edema from the ruptured popliteal cyst in the intermuscular septae of the popliteal fossa. A portion of the cyst is present. There is feathery intramuscular edema in the medial gastroc. The semimembranosus is normal.*

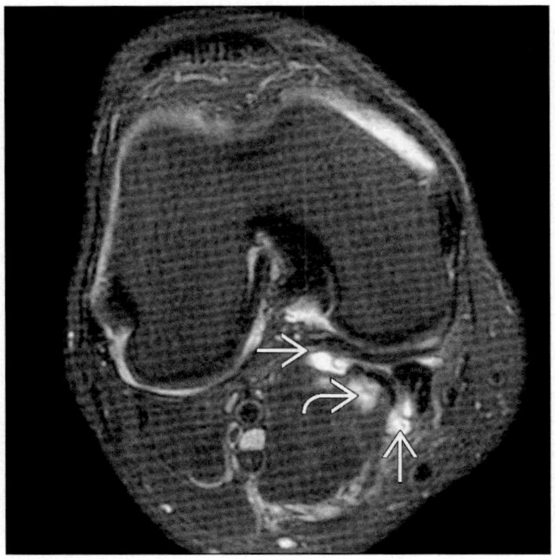

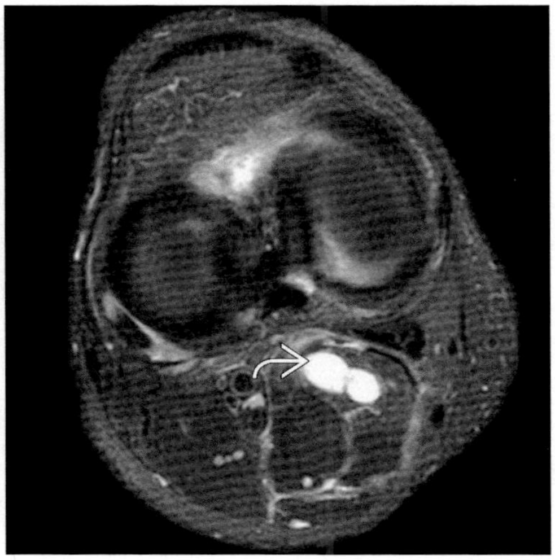

(Left) *Axial PDWI FS MR shows a small Baker cyst in a dumbell shape within the SM-GN bursa, with intramuscular extension/ dissection into the medial gastrocnemius muscle belly.* ***(Right)*** *Axial PDWI FS MR from the same patient more inferiorly shows a well-defined bilobular cystic collection within the medial gastrocnemius muscle as a result of dissection of the popliteal cyst into the muscle. In this case, there was no rupture of the cyst.*

INTERCONDYLAR NOTCH CYST

Key Facts

Terminology
- Synonyms: Intercruciate cyst

Imaging
- MR
  - Low signal on T1WI MR
  - Fluid-sensitive sequences best identify lesion as cystic
  - Homogeneous high signal on T2WI MR
  - May have a few septations
  - May be 1 large cyst or several small cysts
  - Marrow edema in walls of notch if cyst inflamed or due to pressure
  - T1WI + C MR important to confirm fluid content vs. solid mass
- Location: Lobular fluid signal mass located in intercondylar notch
  - Typically between anterior and posterior cruciate ligaments

Top Differential Diagnoses
- Mucinous degeneration of cruciate ligaments
  - Confined to borders of enlarged ligament
- Normal joint fluid
  - Joint fluid can extend between cruciates if synovial lining is breached

Clinical Issues
- Arthroscopic excision of symptomatic cysts

Diagnostic Checklist
- Try to identify site of origin of cyst
  - Partial tear of cruciate ligament may be underlying source
- Check for other sequela of previous knee trauma

__(Left)__ Sagittal graphic shows an intercondylar notch cyst located posterior to the proximal fibers of the anterior cruciate ligament (ACL). __(Right)__ Sagittal T2WI MR shows a well-defined fluid signal ganglion in the posterior aspect of the intercondylar notch between the ACL and posterior cruciate ligament (PCL). A septation is seen within the cyst. Such intralesional septations are common.

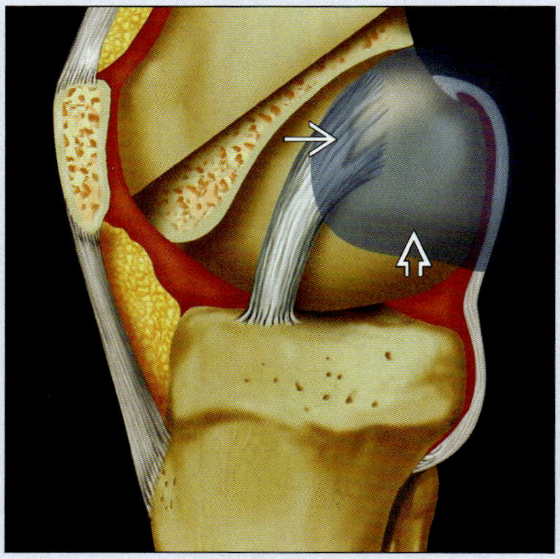

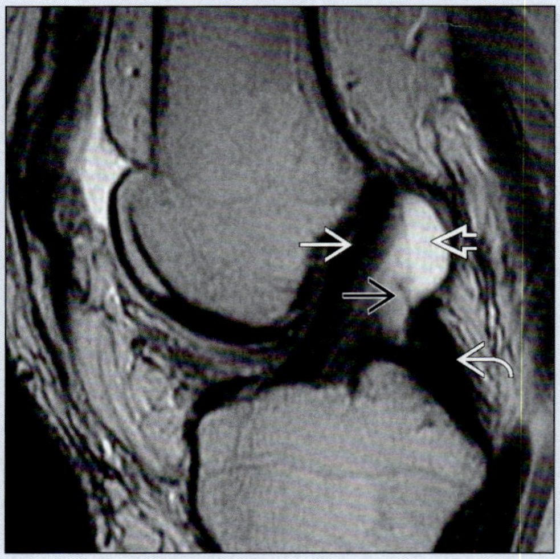

__(Left)__ Coronal T1WI MR from the same patient shows the fluid signal cyst just medial to the ACL and above and lateral to the PCL. T1WI MR is useful in characterizing the fluid nature of the collection and to define the perilesional margins. __(Right)__ Axial PDWI FS MR from the same patient shows a thin margin of synovial tissue around the edge of the cyst. The margins of the cyst are sharp, and the internal signal is very homogeneous. The lesion lies adjacent to the ACL.

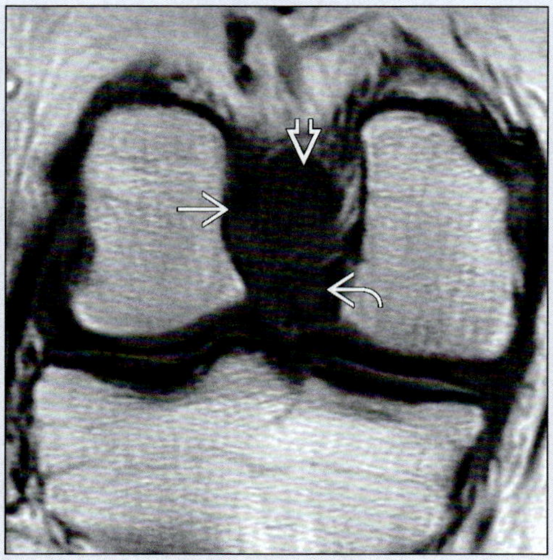

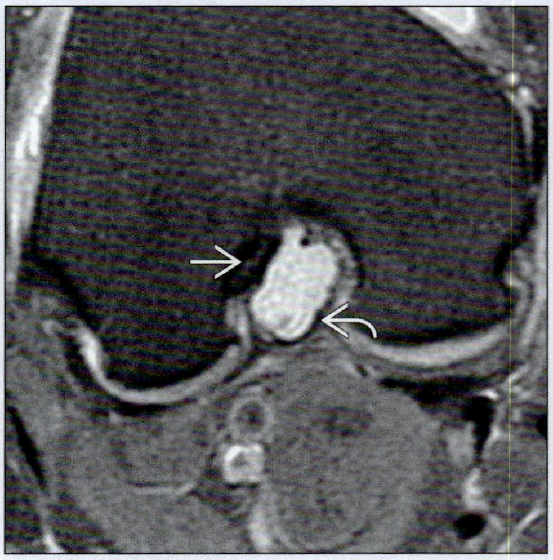

INTERCONDYLAR NOTCH CYST

TERMINOLOGY

Synonyms

- Intercruciate cyst

Definitions

- Focal fluid collection located between cruciate ligaments in intercondylar notch

IMAGING

General Features

- Best diagnostic clue
  - Lobular fluid signal mass located in intercondylar notch
- Location
  - Typically between anterior and posterior cruciate ligaments
- Size
  - May range from very small to quite large
- Morphology
  - Lobular fluid collection with well-defined borders
  - Thin or imperceptible synovial lining
  - Extends around ligaments usually without distorting them

CT Findings

- Fluid attenuation mass in intercondylar notch
- No enhancement

MR Findings

- Well-defined fluid signal mass located in intercondylar notch
  - Homogeneous low signal on T1WI MR
  - Homogeneous high signal on T2WI MR
    - May have a few septations
  - Often wraps around cruciate ligaments
  - Little if any enhancement (lining may enhance if inflamed)
  - May be 1 large cyst or several small cysts
- Marrow edema in walls of notch if cyst inflamed or due to pressure
- Partial tear of cruciate ligament may be underlying source
- Other sequela of previous knee trauma may be present

Imaging Recommendations

- Best imaging tool
  - MR
- Protocol advice
  - Fluid-sensitive sequences best identify cyst
  - T1WI + C MR important to confirm fluid content vs. solid mass

DIFFERENTIAL DIAGNOSIS

Mucinous Degeneration of Cruciate Ligaments

- Confined to borders of enlarged ligament

Normal Joint Fluid

- Joint fluid can extend between cruciates if synovial lining is breached

PATHOLOGY

Gross Pathologic & Surgical Features

- Fluid collection lined by synovium
- Typically arise from alar folds covering cruciate ligaments or infrapatellar fat pad

CLINICAL ISSUES

Presentation

- Most common signs/symptoms
  - Knee pain
    - Worse with activity
  - Limited knee motion if cyst is large
  - Palpable mass (unusual)

Demographics

- Epidemiology
  - Present in 0.2-1% of knee MR exams, 0.6% of knee arthroscopies

Natural History & Prognosis

- May slowly enlarge over time

Treatment

- Arthroscopic excision of symptomatic cysts
  - 95% good-excellent results
  - No recurrences reported
- CT-guided aspiration

DIAGNOSTIC CHECKLIST

Consider

- Partial tear of ACL or PCL with intercondylar ganglion
- Be certain the lesion is not a solid mass
  - May require T1WI + C if not classic in appearance or location

Reporting Tips

- Try to identify site of origin of cyst
- Watch for other sequela of previous knee trauma

SELECTED REFERENCES

1. Fernandes JL et al: Mucoid degeneration of the anterior cruciate ligament: magnetic resonance imaging findings of an underdiagnosed entity. Acta Radiol. 49(1):75-9, 2008
2. Marra MD et al: MRI features of cystic lesions around the knee. Knee. 15(6):423-38, 2008
3. Beaman FD et al: MR imaging of cysts, ganglia, and bursae about the knee. Radiol Clin North Am. 45(6):969-82, vi, 2007
4. Bergin D et al: Anterior cruciate ligament ganglia and mucoid degeneration: coexistence and clinical correlation. AJR Am J Roentgenol. 182(5):1283-7, 2004
5. Krudwig WK et al: Intra-articular ganglion cysts of the knee joint: a report of 85 cases and review of the literature. Knee Surg Sports Traumatol Arthrosc. 12(2):123-9, 2004
6. McCarthy CL et al: The MRI appearance of cystic lesions around the knee. Skeletal Radiol. 33(4):187-209, 2004
7. Zantop T et al: Intra-articular ganglion cysts of the cruciate ligaments: case report and review of the literature. Arch Orthop Trauma Surg. 123(4):195-8, 2003
8. Kang CN et al: Intra-articular ganglion cysts of the knee. Arthroscopy. 15(4):373-8, 1999

INTERCONDYLAR NOTCH CYST

(Left) *Axial T2WI FS MR shows a small intercruciate ganglion located between the ACL and PCL. The cyst is sharply marginated with a thin synovial lining and contains small septations.* ***(Right)*** *Sagittal T2WI MR from the same patient shows the ganglion partially wrapped around the PCL. It may be difficult to determine the location of such cysts on sagittal images.*

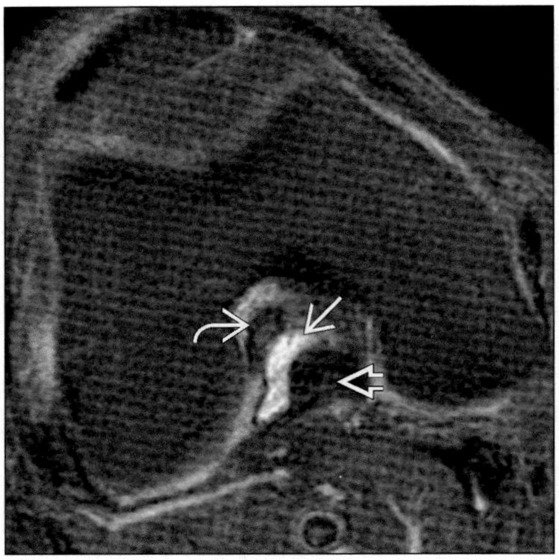

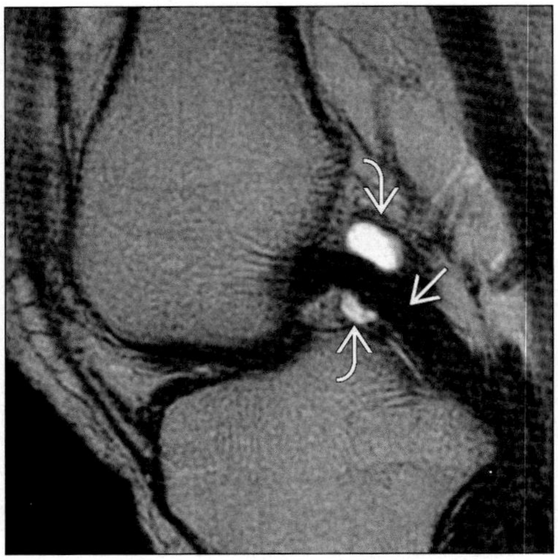

(Left) *Coronal T2WI FS MR from the same patient shows the intercruciate location of the cyst to good advantage. The ACL is present in the lateral notch, and the PCL is seen medially.* ***(Right)*** *Axial T2WI FS MR shows a small intercruciate ganglion. Despite its small size, the lesion contains several septations and is comprised of multiple lobulations. This cyst may have originated from the distal fibers of the ACL. The PCL is posterior to the cyst.*

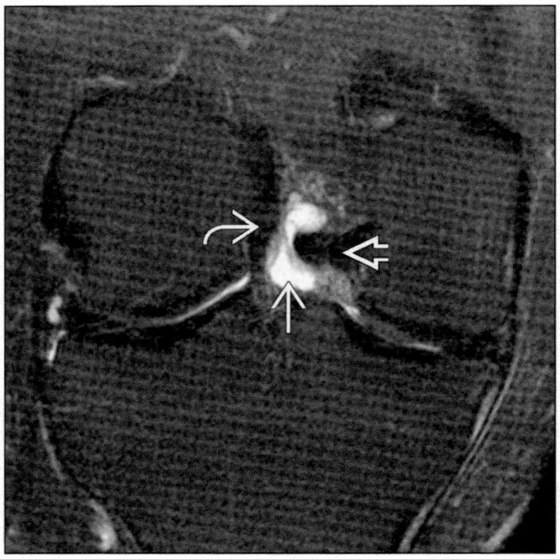

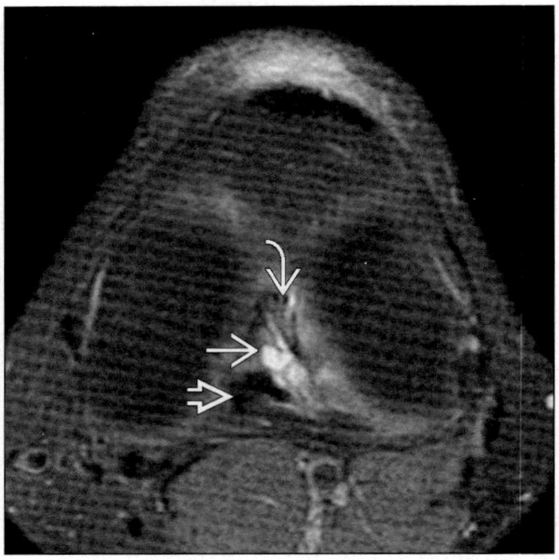

(Left) *Coronal T2WI FS MR shows a large intercruciate cyst with lobular contours and mild surrounding edema. When such cysts get large they may become inflamed or irritate adjacent structures. In this example, there is mass effect on the ACL, causing lateral bowing of the ligament.* ***(Right)*** *Axial PDWI FS MR from the same patient shows the relatively poorly defined borders of the cyst due to surrounding inflammation.*

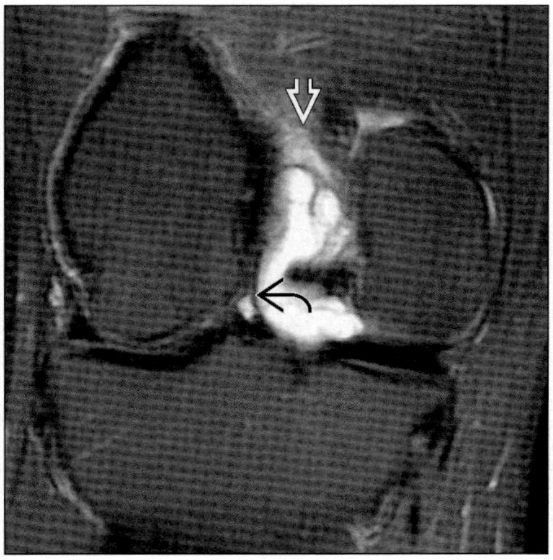

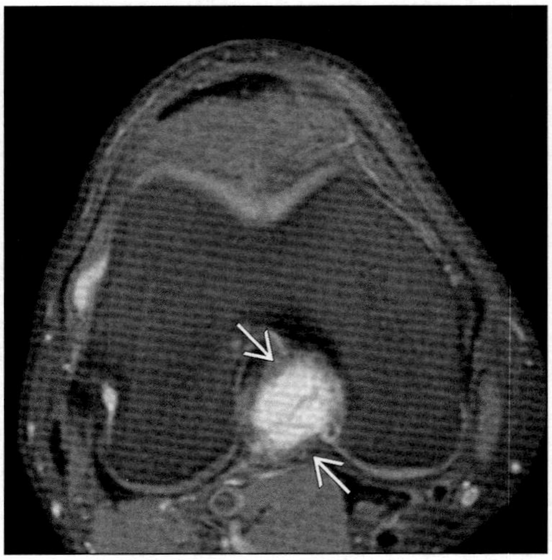

INTERCONDYLAR NOTCH CYST

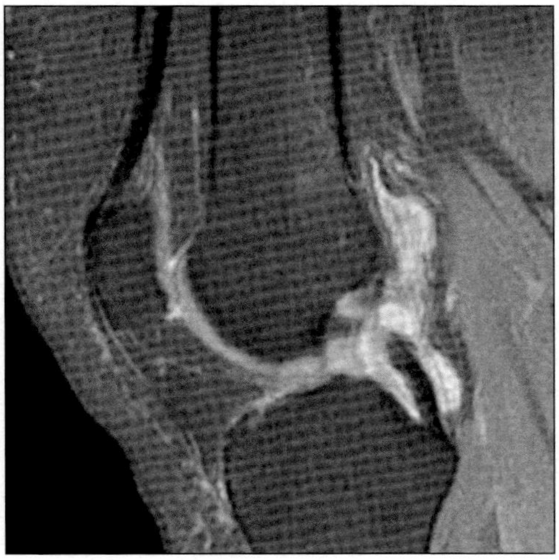

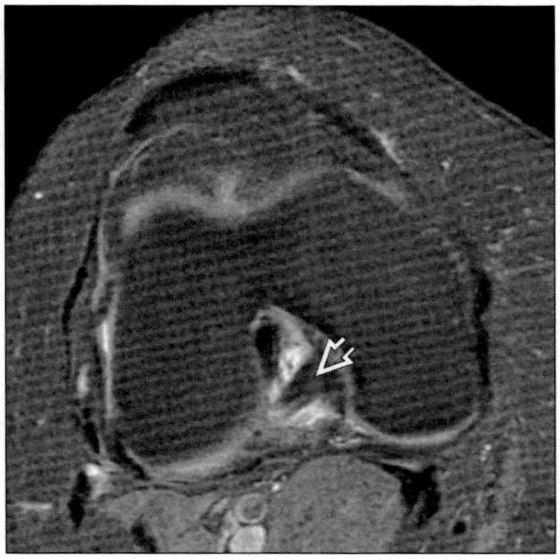

(Left) *Sagittal PDWI FS MR shows an irregular fluid signal collection arising from the intercondylar notch and extending superiorly into the distal thigh. The margins are somewhat ill defined.* ***(Right)*** *Axial PDWI FS MR from the same patient shows the origins of the cyst around the PCL ➔. Some delaminated fibers in the PCL may indicate a prior partial tear as the etiology of this large collection. In many cases, a precise point of origin cannot be identified.*

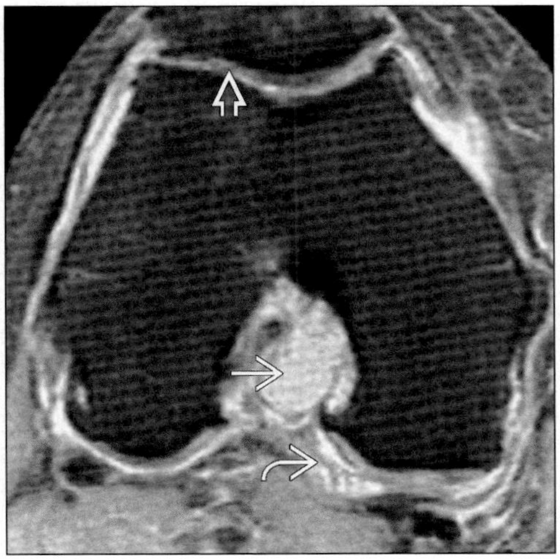

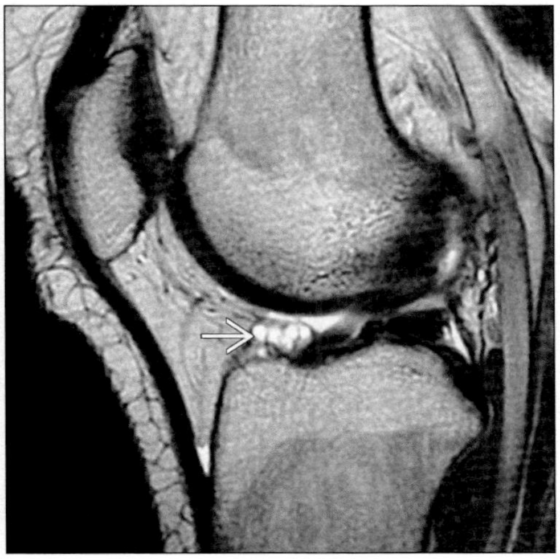

(Left) *Axial PDWI FS MR shows a partially marginated intercruciate cyst ➔ with surrounding edema. There are multiple small cystic collections ➔ in the posterior joint, which may be connected to the larger cyst. High-grade degenerative changes are present in the patellofemoral articulation ➔. Intercruciate cysts may be more common in patients with osteoarthritis.* ***(Right)*** *Sagittal T2WI MR shows a multilobular cyst ➔ in the anterior aspect of the notch.*

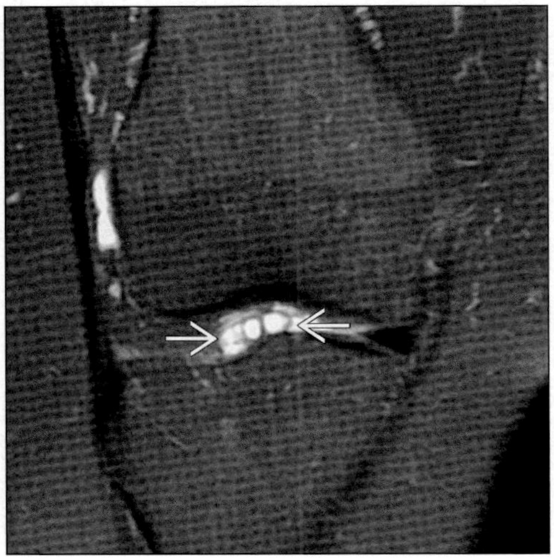

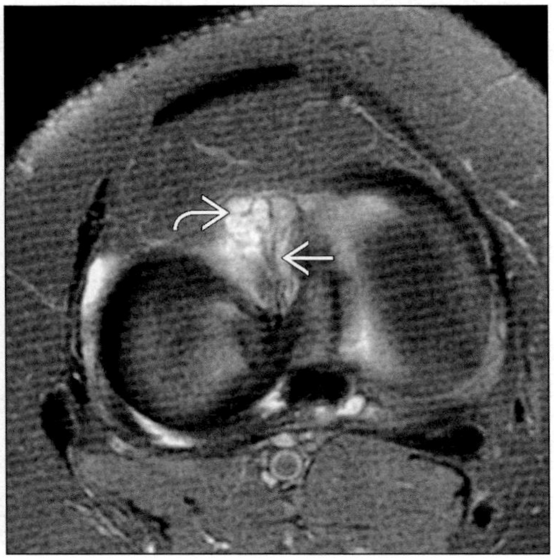

(Left) *Coronal T2WI FS MR from the same patient shows the multiseptated nature of the anterior notch cyst ➔.* ***(Right)*** *Axial PDWI FS MR from the same patient shows the cyst ➔ in the anterior aspect of the intercondylar notch. Delamination of the distal ACL ➔ may serve as a site of origin for the cyst. Anterior notch cysts are often seen adjacent to the tibial attachment of the ACL, suggesting a relationship to the ligament.*

PREPATELLAR AND PRETIBIAL BURSITIS

Key Facts

Terminology
- Washerwoman's knee, housemaid's knee, preacher's knee

Imaging
- Prepatellar bursitis
  - Well-defined mass in subcutaneous space anterior to patella
  - Signal usually follows fluid but may be complicated if hemorrhagic or infected
- Pretibial bursitis
  - Well-defined fluid signal mass anterior to distal patellar tendon &/or tibial tubercle in subcutaneous space
  - ± surrounding soft tissue edema, tibial tubercle marrow edema, and patellar tendinopathy
- Best imaging tool: MR or ultrasound

Pathology
- Chronic repetitive trauma due to friction or prolonged pressure
- Acute trauma (direct blow or penetrating injury)
- Postsurgical
- Related to underlying disease (gout, rheumatoid arthritis)

Clinical Issues
- Domestic cleaners, clergy, carpet layers
- Treatment options
  - Conservative: Rest, NSAIDs, ice
  - Steroid injection
  - Aspiration and antibiotic therapy for infection
  - Surgical: Resection for recalcitrant cases, recurrent infection

*(**Left**) Sagittal graphic shows inflammation of the prepatellar bursa ➡. The location of a normal pretibial bursa is shown ➦ in this image. (**Right**) Sagittal T2WI MR shows a small well-defined fluid collection ➦ anterior to the patella consistent with prepatellar bursitis. Such collections usually result from chronic friction or compression and are frequently associated with job activities performed by house cleaners, carpet layers, and others who work on their knees.*

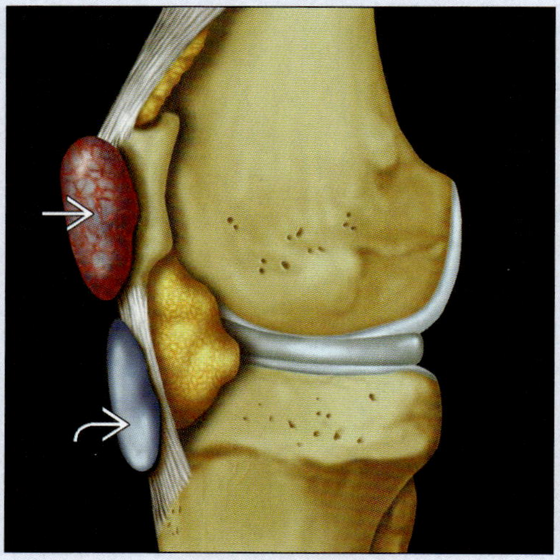

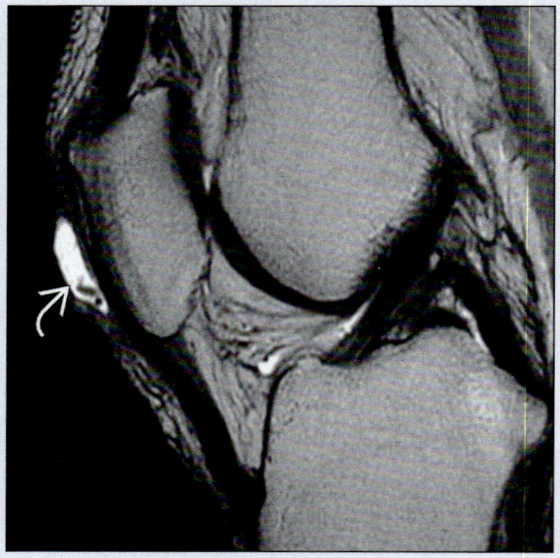

*(**Left**) Sagittal graphic shows inflammation of the pretibial bursa ➡. The prepatellar bursa ➦ is normal in this image. (**Right**) Sagittal PDWI FS MR shows a well-defined fluid collection ➡ anterior to the distal patellar tendon and just superior to the tibial tubercle ⇨ consistent with pretibial bursitis. Like prepatellar bursitis, pretibial bursitis is often seen in the setting of repetitive trauma due to kneeling and is sometimes referred to as "preacher's knee."*

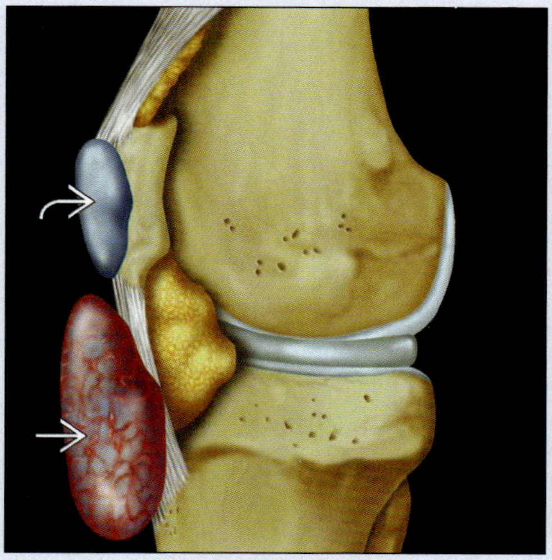

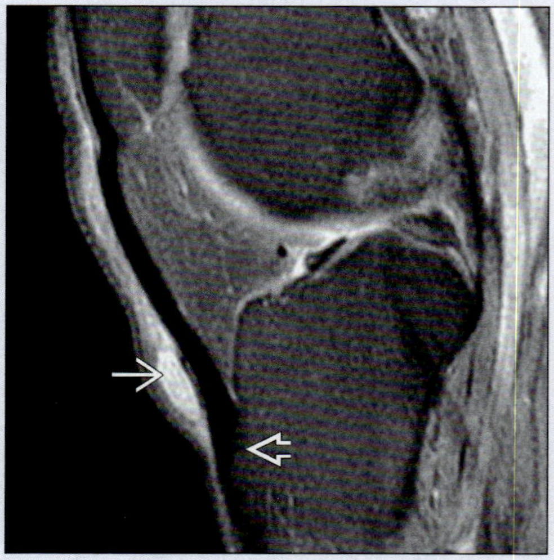

PREPATELLAR AND PRETIBIAL BURSITIS

TERMINOLOGY

Synonyms

- Washerwoman's knee, housemaid's knee, preacher's knee

Definitions

- Inflammation of bursae anterior to patella (prepatellar) or tibial tubercle (pretibial)

IMAGING

General Features

- Best diagnostic clue
  - Focal fluid signal collection anterior to patella (prepatellar) or tibial tubercle (pretibial)
- Location
  - Prepatellar bursitis: Anterior to patella in subcutaneous space
  - Pretibial bursitis: Anterior to distal patellar tendon &/or tibial tubercle

Radiographic Findings

- Soft tissue mass ± swelling anterior to patella or proximal tibia
- Associated calcifications, especially if due to gout

CT Findings

- Similar to radiographs but more sensitive for small fluid collections and calcifications

MR Findings

- Prepatellar bursitis
  - Well-defined mass in subcutaneous space anterior to patella
  - Thin low signal border (bursal synovium), all sequences
  - Signal usually follows fluid but may be complicated if hemorrhagic or infected
    - Simple fluid: ↓ T1 signal, ↑ T2 signal
    - Hemorrhage: May have ↑ T1 signal, T2 signal variable
    - May contain "rice bodies" (small fibrinous bodies with intermediate signal usually seen in rheumatoid arthritis)
  - ± surrounding soft tissue swelling
  - ± marrow edema in patella
- Pretibial bursitis
  - Well-defined fluid signal mass anterior to distal patellar tendon &/or tibial tubercle in subcutaneous space
    - May be complicated by hemorrhage, infection, rice bodies
  - Thin low signal border (bursal synovium)
  - ± surrounding soft tissue edema, tibial tubercle marrow edema, and patellar tendinopathy
- Infectious bursitis
  - May appear more irregular and inhomogeneous
  - Contrast imaging may define drainable portions

Ultrasonographic Findings

- Well-defined fluid collection in subcutaneous space anterior to patella or proximal tibia

Imaging Recommendations

- Best imaging tool
  - MR or ultrasound
- Protocol advice
  - MR: T1WI and T2WI necessary to characterize nature of fluid (simple vs. complex)
    - Sagittal and axial planes most useful

DIFFERENTIAL DIAGNOSIS

Knee Joint Ganglion

- Should see connection to joint

Soft Tissue Mass

- Hemangioma: Enhances after contrast
- Pigmented villonodular synovitis (PVNS): Lobular outline, low signal on all sequences

PATHOLOGY

General Features

- Etiology
  - Chronic repetitive trauma due to friction or prolonged pressure
  - Acute trauma (direct blow or penetrating injury)
  - Postsurgical
  - Related to underlying disease (gout, rheumatoid arthritis)

CLINICAL ISSUES

Presentation

- Most common signs/symptoms
  - Anterior knee pain and swelling
- Other signs/symptoms
  - May have other signs of underlying disease (rheumatoid arthritis, gout)
- Clinical profile
  - Professional activities associated with prolonged kneeling
    - Domestic cleaners, clergy, carpet layers

Natural History & Prognosis

- Insidious onset
- Often resolves with cessation of inciting activity

Treatment

- Conservative: Rest, NSAIDs, ice
  - Steroid injection
  - Aspiration and antibiotic therapy for infection
  - Treatment of underlying arthropathy if appropriate
- Surgical: Resection for recalcitrant cases, recurrent infection

DIAGNOSTIC CHECKLIST

Consider

- Related professional activities as underlying cause

SELECTED REFERENCES

1. Ike RW: Chemical ablation as an alternative to surgery for treatment of persistent prepatellar bursitis. J Rheumatol. 36(7):1560, 2009
2. Marra MD et al: MRI features of cystic lesions around the knee. Knee. 15(6):423-38, 2008

*(**Left**) Lateral radiograph in a patient who struck his knee against the dashboard in a motor vehicle crash shows an irregular soft tissue mass ➡ throughout the anterior knee. (**Right**) Sagittal T2WI MR from the same patient shows a large subcutaneous hematoma ➡ throughout the anterior subcutaneous space. Relatively low signal and heterogeneous architecture suggest the presence of blood products in this large bursal collection.*

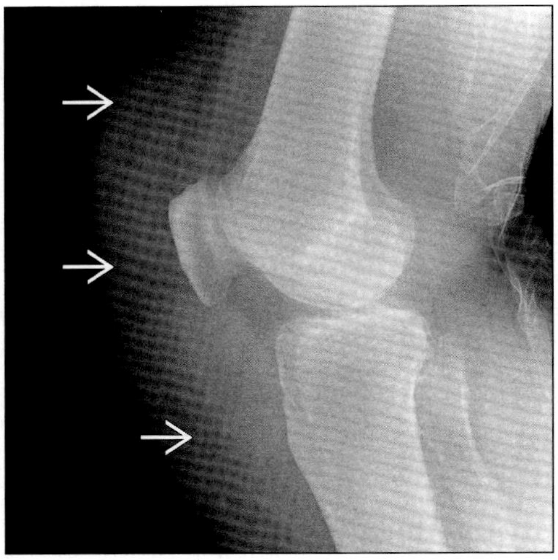

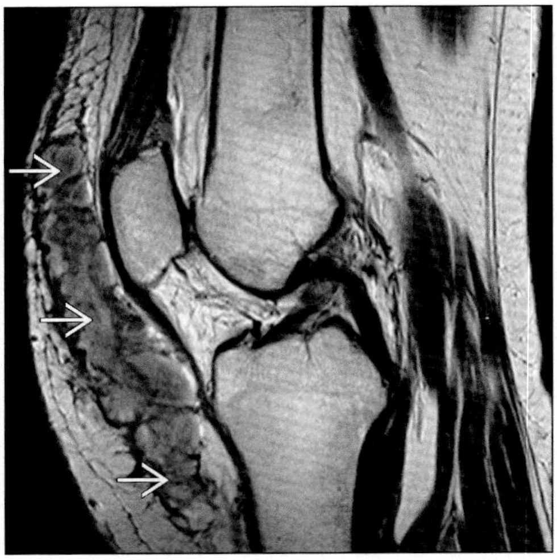

*(**Left**) Axial PDWI FS MR from the same patient confirms the subcutaneous prepatellar location of the hematoma ➡. A simple bursal fluid collection would be expected to have more homogeneous signal; this appearance is more typical of hematoma. (**Right**) Sagittal PDWI FS MR demonstrates a prepatellar hematoma ➡ in a high school lacrosse player. Note the surrounding soft tissue edema, indicating the hematoma's acute traumatic etiology. Such intense soft tissue swelling is uncommon in chronic bursitis.*

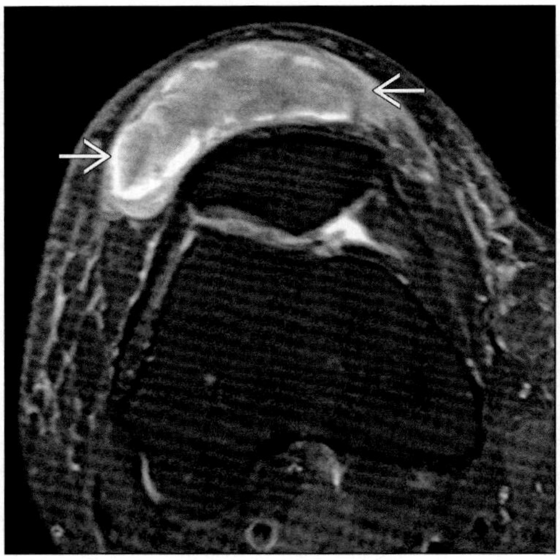

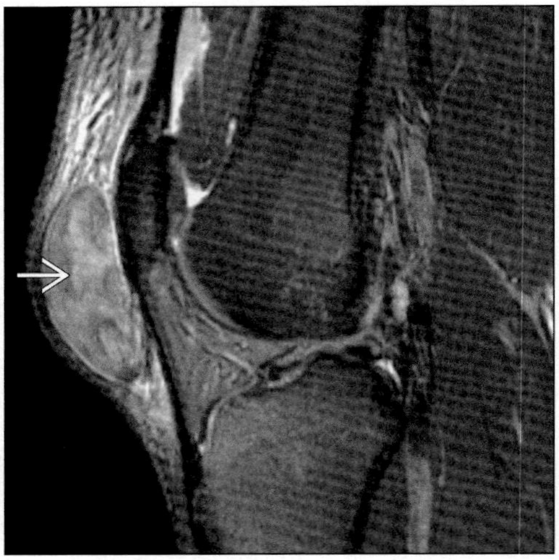

*(**Left**) Sagittal PDWI MR in a patient with rheumatoid arthritis (RA) and a large anterior knee mass demonstrates a complex prepatellar fluid collection containing rice bodies ➡. (**Right**) Sagittal T2WI MR from the same patient better shows the rice bodies ➡ within the prepatellar bursal collection. These small fibrinous bodies can occur with a variety of inflammatory conditions but are commonly seen in RA. Note the chronic fibrous rim ⮫ of the bursal collection.*

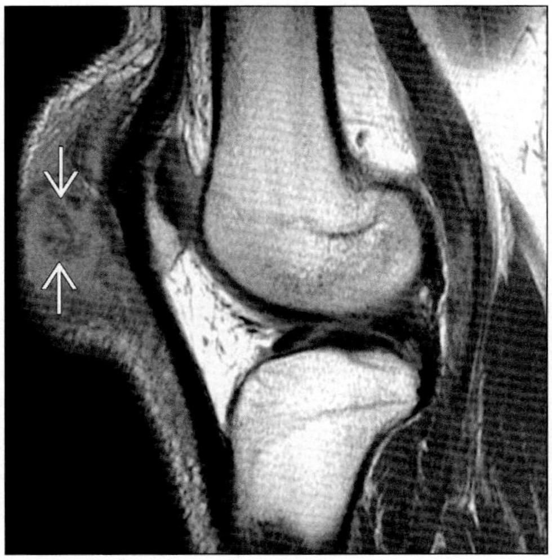

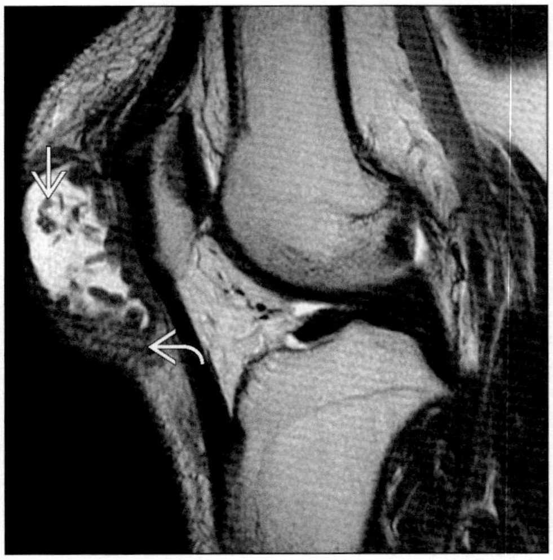

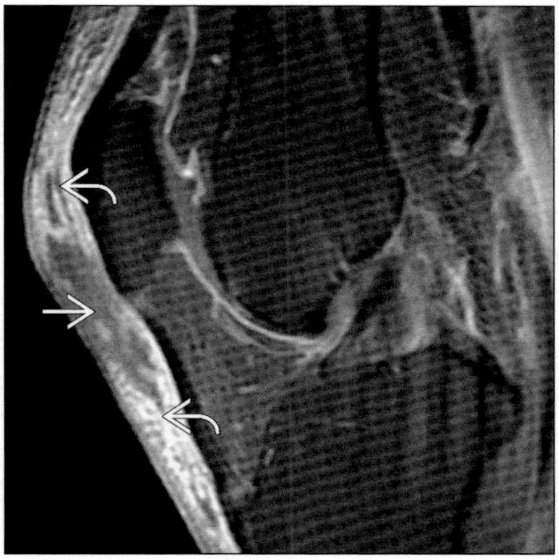

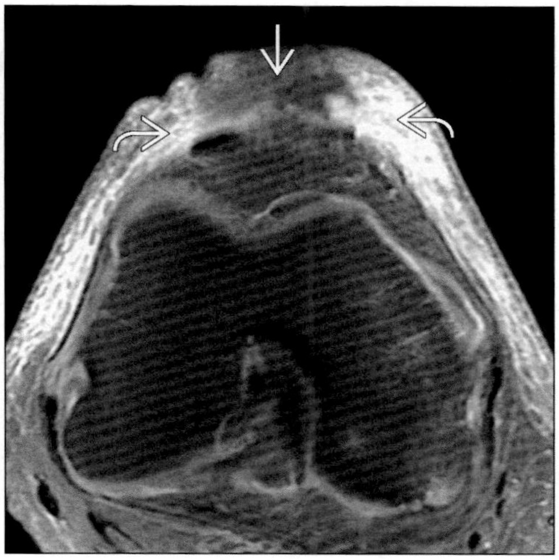

*(**Left**) Sagittal T1WI C+ FS MR shows an irregular fluid collection in the prepatellar space ➔ surrounded by intensely enhancing tissue ➦. This patient had developed progressive swelling and redness here 2 weeks after penetrating trauma. This was a prepatellar abscess. (**Right**) Axial T1WI C+ FS MR from the same patient shows the irregular nonenhancing abscess ➔ surrounded by enhancing inflammatory tissue ➦. Contrast can be useful in defining drainable portions of abscesses.*

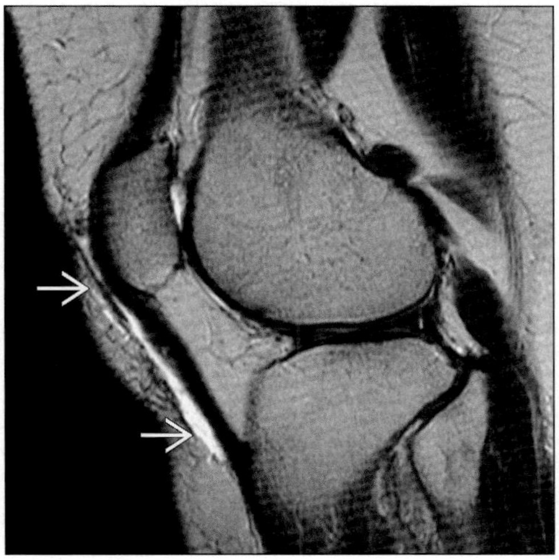

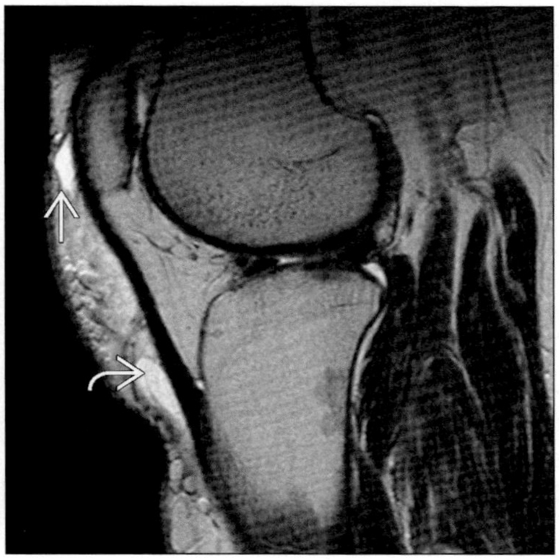

*(**Left**) Sagittal T2WI MR shows a superficial bursal fluid collection ➔ extending from the prepatellar region to the pretibial region. This illustrates the continuity of the subcutaneous space anterior to the knee; inflammation or infection in the prepatellar space can spread to the pretibial space or vice versa. (**Right**) Sagittal T2WI MR shows separate prepatellar ➔ and pretibial ➦ bursal fluid collections. Bursitis in these locations may be confluent, or separate as in this case.*

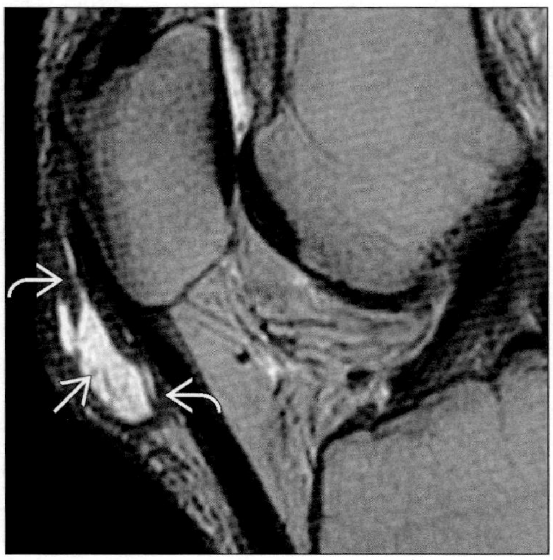

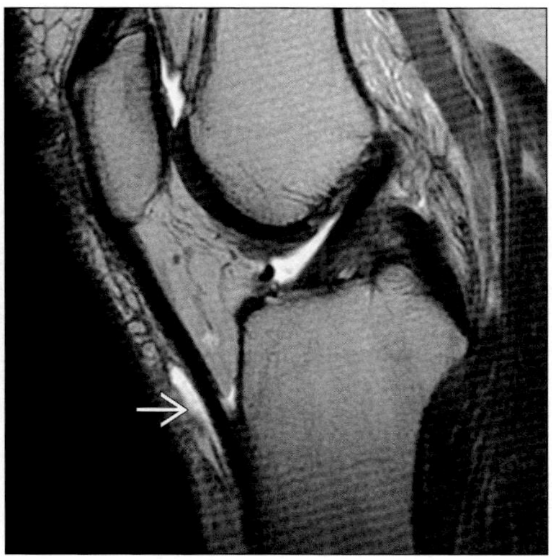

*(**Left**) Sagittal T2WI MR demonstrates a bursal fluid collection anterior to the inferior pole of the patella ➔. Most cases of prepatellar bursitis are located more superiorly. A thick fibrous rim ➦ is present around the collection, suggesting chronicity. (**Right**) Sagittal T2WI MR shows a mild case of pretibial bursitis. An elongate fluid collection ➔ is present anterior to the distal patellar tendon in the subcutaneous fat. The tendon itself is normal.*

DEEP INFRAPATELLAR BURSITIS

Key Facts

Terminology

- Retropatellar bursitis
- Inflamed, fluid-filled bursa located between distal patellar tendon and anterior cortex of proximal tibia

Imaging

- Best imaging tool: MR
- Small fluid collections in this location can be physiologic and therefore normal
- Well-defined fluid-signal collection located deep to distal patellar tendon
- ± surrounding peribursal swelling
- Sagittal T2WI MR ± fat suppression best characterizes size and location of bursal fluid

Pathology

- Probably related to repetitive motion injury in many cases
- Secondary to inflammatory conditions
- Direct blow trauma
- Can be seen status post patellectomy
- May become secondarily infected
- Bursa present in all knees
- No connection to knee joint
- Seen in 71% of patients with Osgood-Schlatter syndrome

Clinical Issues

- Anterior knee pain just above level of tibial tubercle
- Most cases respond well to conservative therapy
- Percutaneous injection of steroids and anesthetic may be beneficial

*(**Left**) Sagittal T2WI MR in an 11-year-old soccer player with anterior knee pain demonstrates a triangular fluid-signal collection ➔ deep to the distal patellar tendon ➔. This is the typical appearance of deep infrapatellar bursitis. (**Right**) Axial PDWI FS MR from the same patient shows the bursal fluid collection ➔ between the patellar tendon ➔ and the tibia ➔. Note that the collection projects further to the lateral side of the tendon than it does medially.*

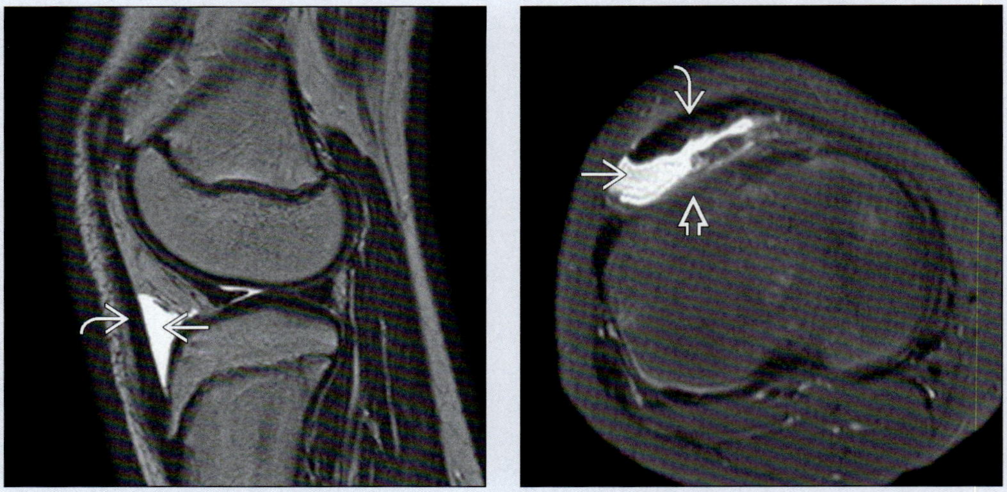

*(**Left**) Sagittal T2WI MR shows mild inflammation of the deep infrapatellar bursa ➔ in a patient with Osgood-Schlatter syndrome (edema in the tibial tubercle ➔ due to enthesopathy of the distal patellar tendon). Most patients with Osgood-Schlatter syndrome have bursitis here. (**Right**) Sagittal T2WI MR in a 13 year old shows mild inflammation of the deep infrapatellar bursa ➔. Collections smaller than this are often present as an incidental finding and should not be considered pathologic.*

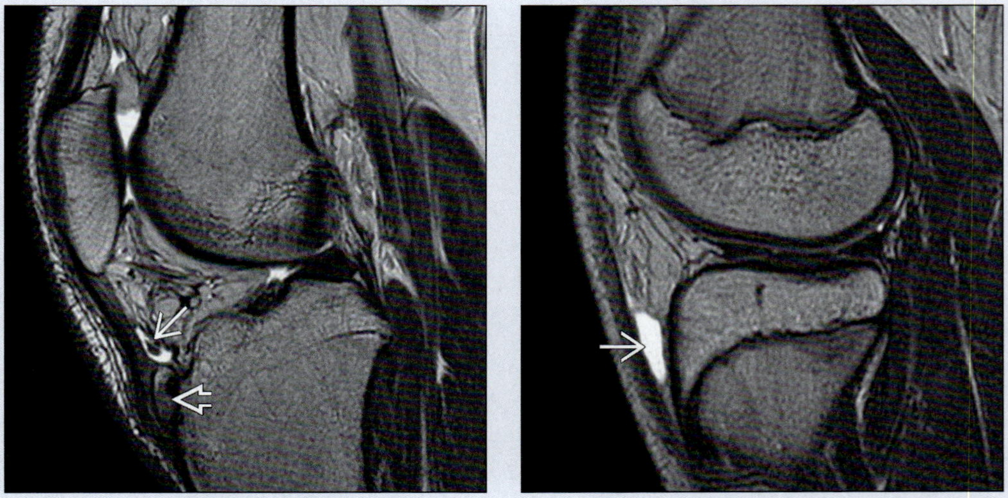

DEEP INFRAPATELLAR BURSITIS

TERMINOLOGY

Synonyms

- Retropatellar bursitis

Definitions

- Inflamed, fluid-filled bursa located between distal patellar tendon and anterior cortex of proximal tibia

IMAGING

General Features

- Best diagnostic clue
  - Fluid collection deep to distal portion of patellar tendon
- Location
  - Between anterior tibial shaft and distal aspect of patellar tendon just above tibial tubercle
- Size
  - May range from a few mm to several cm in diameter
  - Small fluid collections in this location can be physiologic and therefore normal
- Morphology
  - Often triangular in shape due to inherent confines of retropatellar space

Radiographic Findings

- Large collections may be visible on lateral radiograph as soft tissue mass

CT Findings

- Fluid-density collection deep to patellar tendon just above tibial tubercle

MR Findings

- Well-defined fluid signal collection located deep to distal patellar tendon
  - Low signal on T1WI MR and homogeneous high signal on T2WI MR
- ± surrounding peribursal swelling
- May be associated with Osgood-Schlatter syndrome
  - Marrow edema in tibial tubercle and adjacent soft tissue swelling
  - Thickening and increased signal in distal patellar tendon
- Adjacent tibial marrow edema
  - Especially when associated with inflammatory arthritis

Ultrasonographic Findings

- Hypo- or anechoic fluid collection deep to distal portion of patellar tendon

Imaging Recommendations

- Best imaging tool
  - MR
- Protocol advice
  - Sagittal T2WI MR ± fat suppression best characterizes size and location of bursal fluid

PATHOLOGY

General Features

- Etiology
  - Probably related to repetitive motion injury in many cases
  - Secondary to inflammatory conditions
  - Direct blow trauma
  - Can be seen status post patellectomy
- Associated abnormalities
  - Seen in 71% of patients with Osgood-Schlatter syndrome
  - May be present with gout, rheumatoid arthritis
  - May become secondarily infected
  - Hemorrhagic bursitis associated with direct blow trauma

Gross Pathologic & Surgical Features

- Bursa present in all knees
- Synovial-lined compartment slightly wider than adjacent patellar tendon
- No connection to knee joint
- Divided into anterior and posterior compartments by partial septum of infrapatellar fat

CLINICAL ISSUES

Presentation

- Most common signs/symptoms
  - Anterior knee pain just above level of tibial tubercle
  - May be palpable lateral to distal patellar tendon

Natural History & Prognosis

- Usually improves with conservative therapy

Treatment

- Most cases respond well to conservative therapy
- Percutaneous injection of steroids and anesthetic may be beneficial
- Surgery for refractory cases

DIAGNOSTIC CHECKLIST

Consider

- Deep infrapatellar bursitis in patients with Osgood-Schlatter syndrome

SELECTED REFERENCES

1. Marra MD et al: MRI features of cystic lesions around the knee. Knee. 15(6):423-38, 2008
2. Aydingoz U et al: The deep infrapatellar bursa: prevalence and morphology on routine magnetic resonance imaging of the knee. J Comput Assist Tomogr. 28(4):557-61, 2004
3. LaPrade RF: The anatomy of the deep infrapatellar bursa of the knee. Am J Sports Med. 26(1):129-32, 1998
4. Klein W: Endoscopy of the deep infrapatellar bursa. Arthroscopy. 12(1):127-31, 1996
5. Taylor PW: Inflammation of the deep infrapatellar bursa of the knee. Arthritis Rheum. 32(10):1312-4, 1989
6. Lagier R et al: Bilateral deep infrapatellar bursitis associated with tibial tuberosity enthesopathy in a case of juvenile ankylosing spondylitis. Rheumatol Int. 5(4):187-90, 1985

MEDIAL PATELLAR PLICA SYNDROME

Key Facts

Terminology
- Medial shelf, Aoki edge, Iino's band

Imaging
- Well-defined band of tissue extending from medial joint capsule to infrapatellar fat pad
  - Low signal on all sequences
- Joint effusion
- Secondary changes in medial femoral condylar trochlear ridge or medial patellar facet (rare)
- Medial parapatellar plica best seen on axial and sagittal fluid-sensitive sequences
- 4 types of plicae
  - Suprapatellar
  - Medial
  - Lateral
  - Infrapatellar (ligamentum mucosum)
- Ultrasound diagnostic accuracy: 88%

Top Differential Diagnoses
- Normal joint fold
- Normal (uninflamed) medial patellar plica
- Synovial mass

Pathology
- Recurrent trauma to plica causes hypertrophy and synovitis → inflammatory symptoms and pain

Clinical Issues
- Intermittent anteromedial knee pain
- Snapping or clicking on motion
- "Pseudolocking" or catching
- Inflamed plica may be palpable
- Conservative: Stretching and strengthening, bracing, NSAIDs
- Surgical: Arthroscopic resection

*(**Left**) Axial graphic shows a thickened and elongated medial parapatellar plica interposed between the patella and trochlea. The plica undergoes repetitive trauma in this location. (**Right**) Axial PDWI FS MR shows an edematous, elongated and thickened medial parapatellar plica extending into the patellofemoral joint. Medial plicae can be associated with medial patellofemoral degeneration, not seen here. In this case, there are unrelated degenerative changes in the lateral patella.*

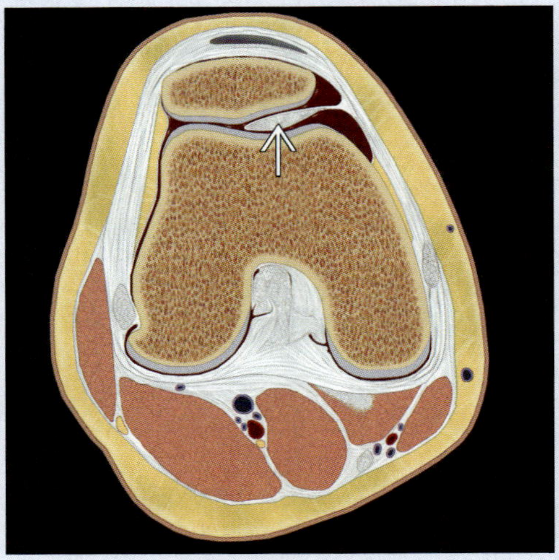

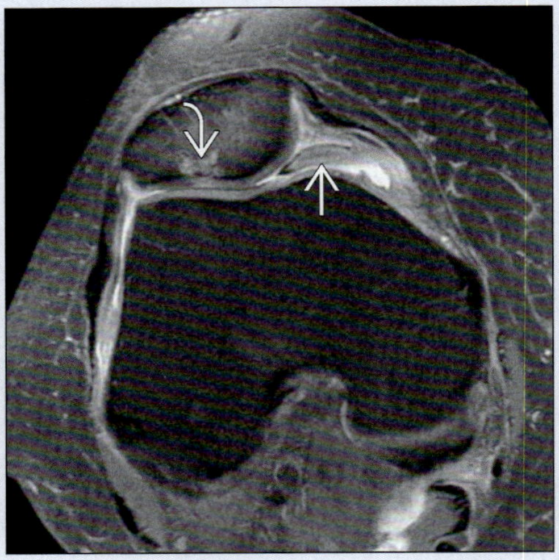

*(**Left**) Sagittal graphic demonstrates a suprapatellar plica extending from the anterior femoral metaphysis to the fat posterior to the quadriceps tendon. An infrapatellar plica (or ligamentum mucosum) is also demonstrated. (**Right**) Sagittal T2WI MR shows a thickened and irregular suprapatellar plica which continues as a medial parapatellar plica. Even when inflamed, suprapatellar plica rarely cause significant symptoms.*

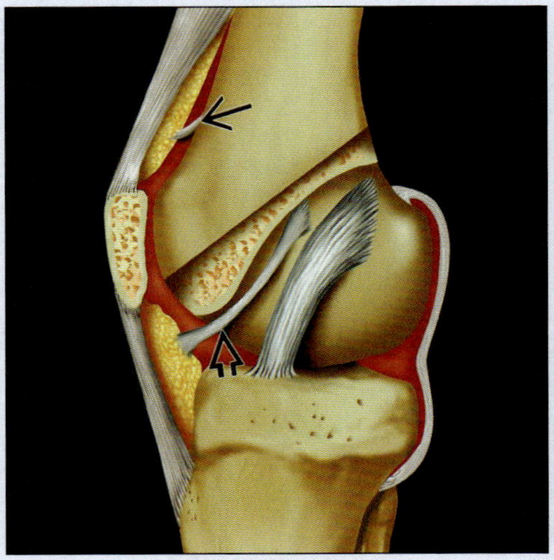

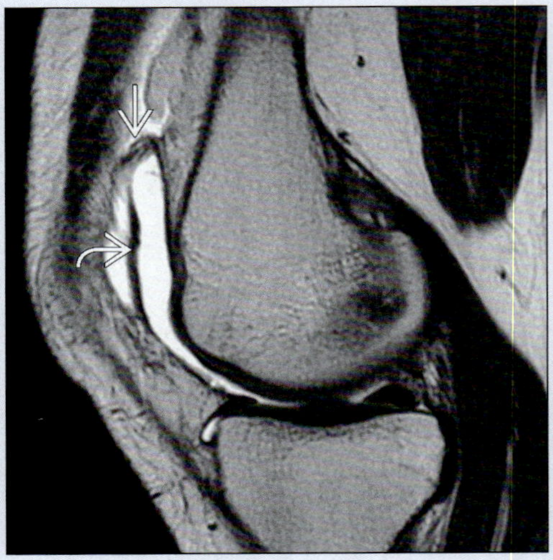

MEDIAL PATELLAR PLICA SYNDROME

TERMINOLOGY

Synonyms

- Medial shelf, Aoki edge, Iino's band

Definitions

- Inflammation of medial parapatellar plica causing knee pain and crepitus

IMAGING

General Features

- Best diagnostic clue
  - Thickening of medial parapatellar plica with concordant pain in medial aspect of knee
- Location
  - 4 types of plicae
    - Suprapatellar
      - 11-90% reported incidence
      - Runs from anterior femoral metaphysis to medial joint capsule
    - Medial
      - 18-60% reported incidence
      - Runs from medial joint capsule to infrapatellar fat pad
    - Lateral
      - Rarest synovial plica (< 1% incidence)
    - Infrapatellar (ligamentum mucosum)
      - Most common plica; 65% incidence
      - Extends from intercondylar notch to infrapatellar fat
- Morphology
  - Coronally oriented embryonic remnant of synovial tissue running from medial aspect of knee joint capsule above patella and inserting on infrapatellar fat

MR Findings

- Well-defined band of tissue extending from medial joint capsule to infrapatellar fat pad
  - Low signal on all sequences
  - Incidental finding in many patients (thickness < 2-3 mm)
- Joint effusion
- Secondary changes in medial femoral condylar trochlear ridge or medial patellar facet (rare)
  - Reactive marrow edema
  - Chondral thinning

Ultrasonographic Findings

- Diagnostic accuracy: 88%

Imaging Recommendations

- Best imaging tool
  - MR or ultrasound
- Protocol advice
  - Medial parapatellar plica best seen on axial and sagittal fluid-sensitive sequences

DIFFERENTIAL DIAGNOSIS

Normal Joint Fold

- Normal folds in synovial lining of joint may simulate thickened medial plica

Normal (Uninflamed) Medial Patellar Plica

- Presence of medial patellar plica not pathologic in itself
- Measures 1-2 mm in thickness

Synovial Mass

- Hemangioma
- Pigmented villonodular synovitis (PVNS)
- Synovial chondromatosis

PATHOLOGY

General Features

- Etiology
  - Recurrent trauma to plica causes hypertrophy and synovitis → inflammatory symptoms and pain
- Associated abnormalities
  - Meniscal tear, patellar tendinopathy, Osgood-Schlatter syndrome, Sinding-Larsen-Johansson syndrome

Staging, Grading, & Classification

- Munzinger classification of medial parapatellar plica
  - A: Cord-like
  - B: Shelf-like, does not cover medial femoral condyle
  - C: Shelf-like, covers medial femoral condyle
  - D: Double insertion

CLINICAL ISSUES

Presentation

- Most common signs/symptoms
  - Intermittent anteromedial knee pain
  - Snapping or clicking on motion
  - "Pseudolocking" or catching
  - Inflamed plica may be palpable

Treatment

- Conservative: Stretching and strengthening, bracing, NSAIDs
- Surgical: Arthroscopic resection

DIAGNOSTIC CHECKLIST

Consider

- Inflamed medial parapatellar plica in patients with medial knee pain

Image Interpretation Pearls

- Evaluate for erosion or chondromalacia in medial trochlear ridge

SELECTED REFERENCES

1. Paczesny L et al: Medial plica syndrome of the knee: diagnosis with dynamic sonography. Radiology. 251(2):439-46, 2009
2. Shetty VD et al: Clinical diagnosis of medial plica syndrome of the knee: a prospective study. J Knee Surg. 20(4):277-80, 2007
3. Jee WH et al: The plica syndrome: diagnostic value of MRI with arthroscopic correlation. J Comput Assist Tomogr. 22(5):814-8, 1998
4. Dupont JY: Synovial plicae of the knee. Controversies and review. Clin Sports Med. 16(1):87-122, 1997

MEDIAL PATELLAR PLICA SYNDROME

__(Left)__ Axial PDWI FS MR shows a thick, irregular, and undulating medial parapatellar plica extending deep into the trochlear notch, and interposed between the medial trochlea and patella. Such plicae can become inflamed and thickened due to repetitive trauma during knee motion, setting up a vicious cycle of trauma and inflammation. __(Right)__ Sagittal T2WI MR from the same patient shows the well-defined plica arising from the medial joint and attaching to a suprapatellar plica.

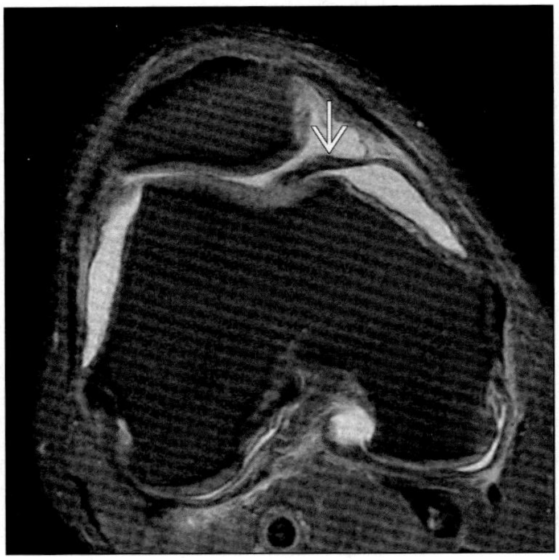

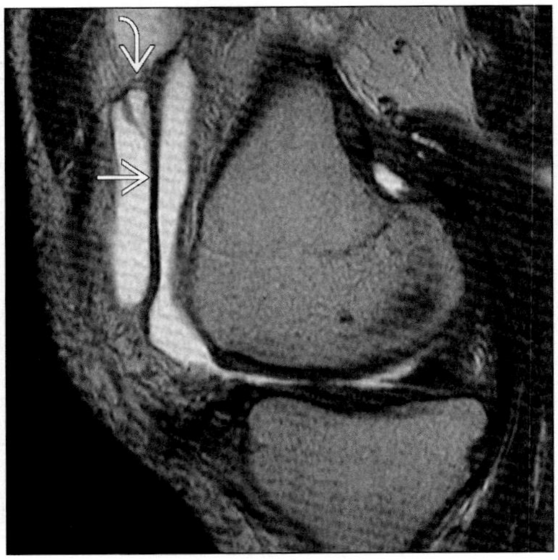

__(Left)__ Axial PDWI FS MR after transient patellar dislocation shows the typical lateral femoral condylar and inferomedial patellar marrow contusions related to that injury. There is a mildly thickened medial parapatellar plica, which may be an incidental finding in this patient. __(Right)__ Coronal T2WI FS MR shows a portion of a medial parapatellar plica. It is less common to see such plica in the coronal plane, as the plica is oriented coronally.

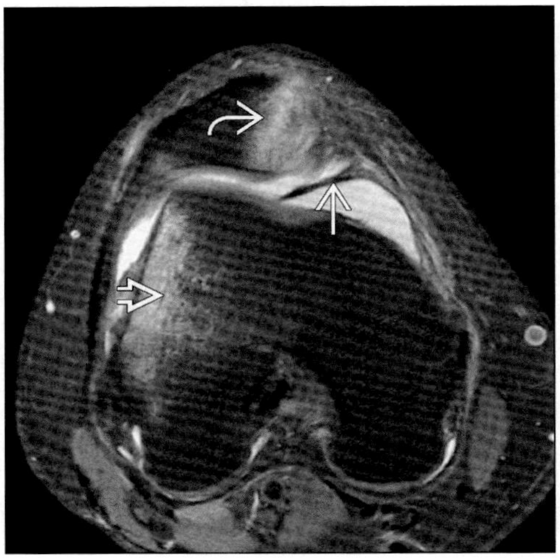

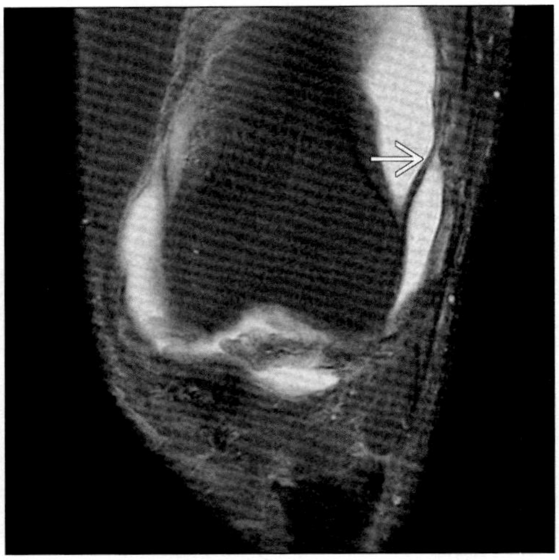

__(Left)__ Axial PDWI FS MR shows a slight subsynovial prominence in the medial aspect of this knee, but a discrete synovial plica is not present. __(Right)__ Axial PDWI FS MR from the same patient 1 year later after interval posterior cruciate ligament repair now shows a thick and slightly hyperintense medial parapatellar plica in the medial joint recess. It is possible that surgical trauma caused inflammation of the plica, which caused it to become more visible.

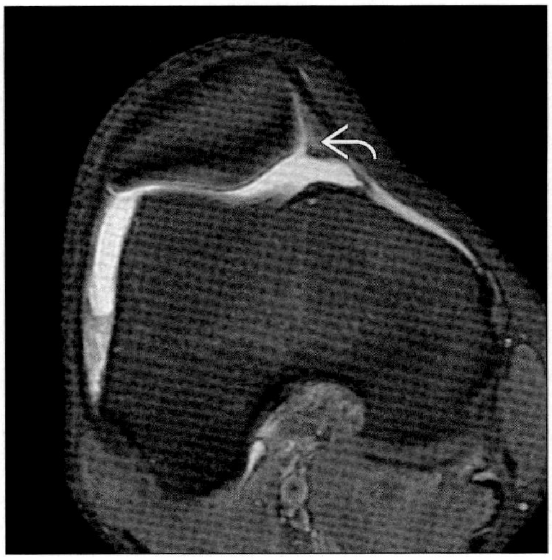

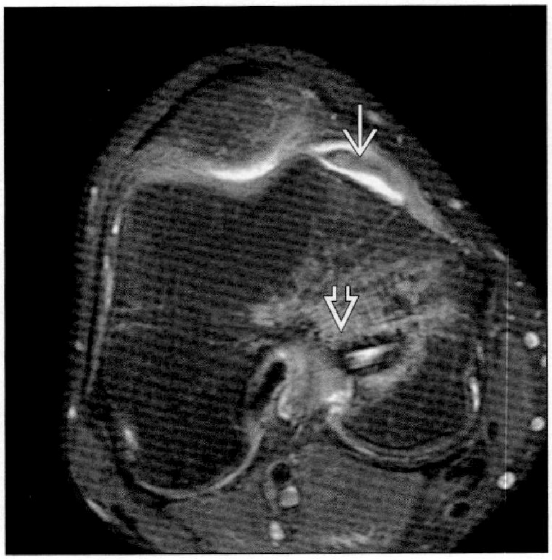

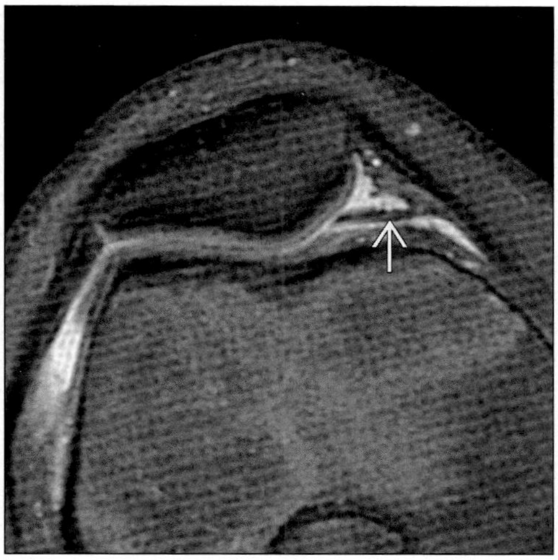

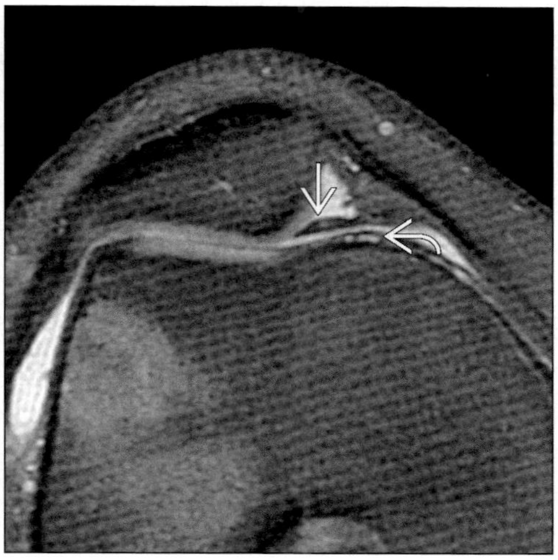

(Left) *Axial PDWI FS MR shows a mildly thickened medial parapatellar plica arising from the medial joint capsule.* ***(Right)*** *Axial PDWI FS MR more inferiorly from the same patient shows extension of the plica across the medial trochlear ridge and toward the patellar apex. Note that there is only slightly more than a physiologic amount of fluid in the joint; a large effusion need not be present to identify and characterize synovial plicae in the knee.*

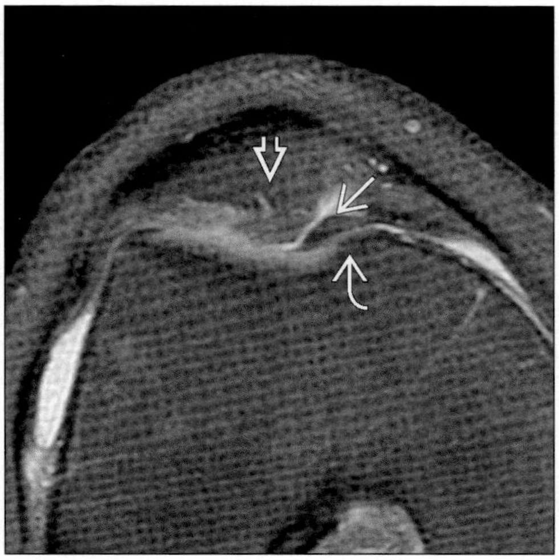

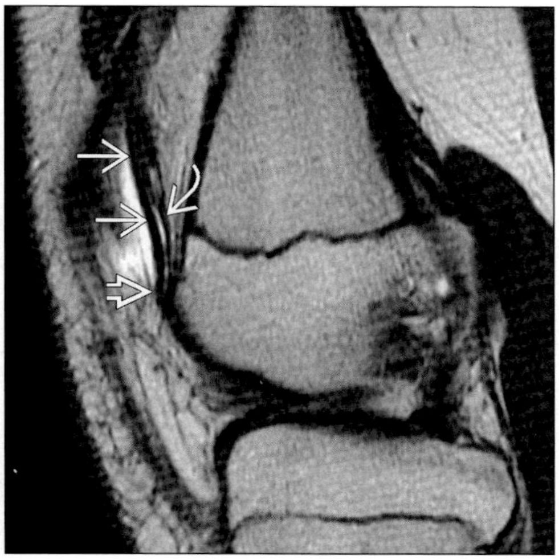

(Left) *Axial PDWI FS MR more inferiorly in the same patient shows the thickened medial parapatellar plica interposed between the medial femoral condylar articular surface and the inferior patella.* ***(Right)*** *Sagittal T2WI MR from the same patient shows the thickened medial parapatellar plica as a distinct structure just anterior to the posterior synovial lining of the joint. Point of contact with the medial femoral condyle is seen inferiorly.*

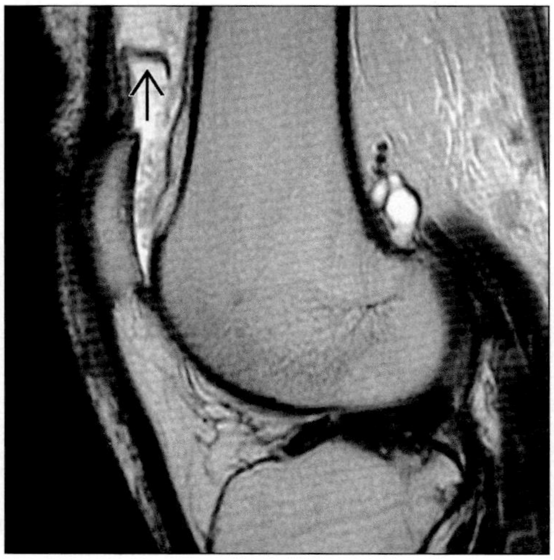

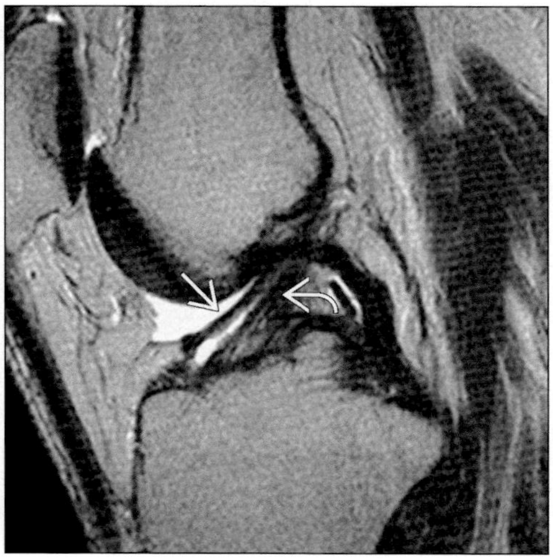

(Left) *Sagittal T2WI MR shows a thickened suprapatellar plica. In this case, the plica is incomplete, as it does not contact the posterior wall of the superior joint recess on this or any other image. These plicae are rarely symptomatic.* ***(Right)*** *Sagittal T2WI MR shows a prominent infrapatellar plica (ligamentum mucosum) paralleling but anterior to the anterior cruciate ligament. This plica rarely becomes acutely symptomatic.*

PERONEAL NERVE INJURY

Key Facts

Imaging
- Most common site of injury: Lateral to fibular head
- T2WI FS and T1WI MR to assess nerve size and signal and muscle atrophy and edema
- Intrinsic nerve abnormality
- Intrinsic neoplasm of nerve
- Extrinsic compression of nerve
- Downstream muscle changes
  - Atrophy ± edema
  - Affects: PL, PB, EHL, EDL, AT, peroneus tertius

Top Differential Diagnoses
- Direct blow injury
- Polyneuropathy

Pathology
- Direct nerve trauma
- Chronic compression
- Secondary to osseous trauma
- Iatrogenic
- Intrinsic neoplasm of nerve
- Compression by extrinsic mass
- Excessive weight loss ("slimmer's disease")

Clinical Issues
- Weakness in ankle dorsiflexors (foot drop)
- Numbness or paresthesias in dorsum of foot
- Pain variable, usually related to underlying injury
- Most common compressive neuropathy of lower extremity

Diagnostic Checklist
- Must exclude causes of extrinsic compression of nerve

(Left) Axial graphic of the posterolateral corner demonstrates an inflamed peroneal nerve ➔ as it extends around the fibular head ➔. The nerve is superficial to the fabellofibular ligament ➔. (Right) Axial PDWI FS MR shows a lobular ganglion ➔ extending anteriorly and posteriorly from the proximal tibiofibular joint, its site of origin. The normal course of common peroneal nerve at this level places it between the fibula ➔ and the posterior portion of the cyst.

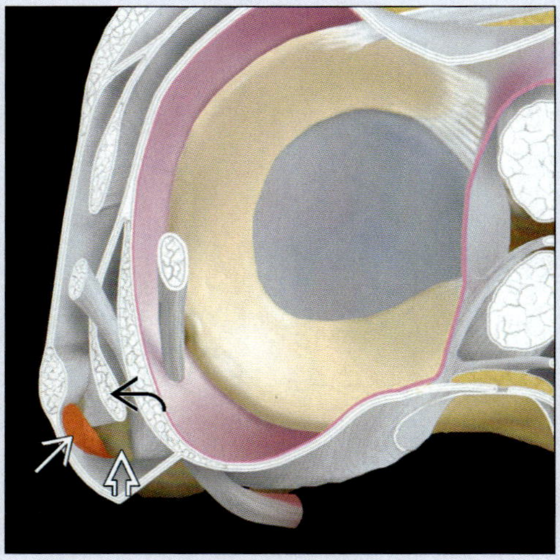

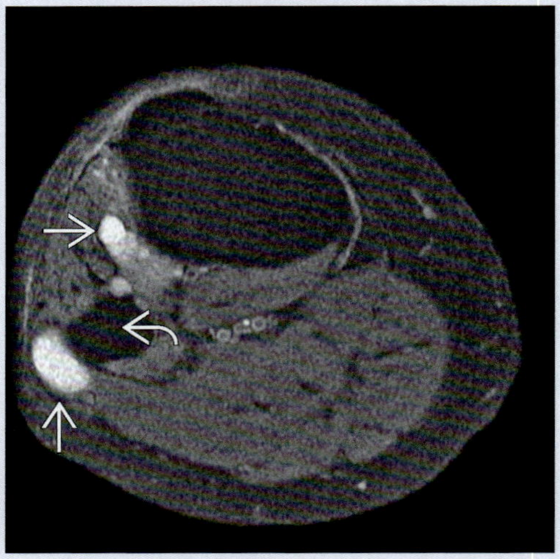

(Left) Axial PDWI FS MR from the same patient more inferiorly shows the extensive nature of the cyst ➔. Given its size and location, compression of the common peroneal nerve is highly likely. (Right) Axial PDWI FS MR in the same patient in the mid-leg shows edema and mild atrophy in the anterior tibialis ➔ and extensor digitorum longus ➔ muscles. There is sparing of the peroneus longus and brevis units ➔. Compression of the nerve may affect some or all of the downstream muscles.

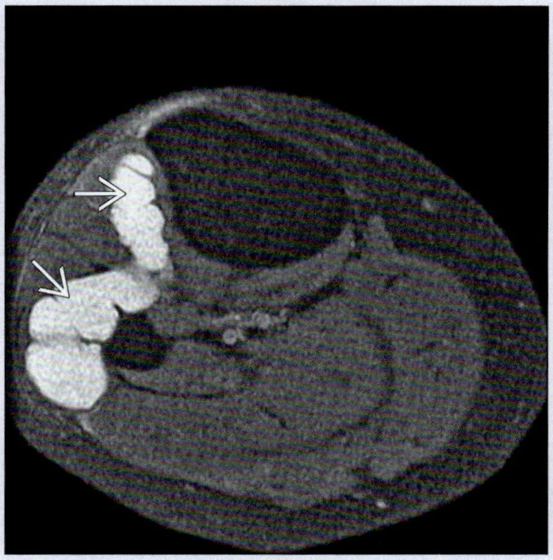

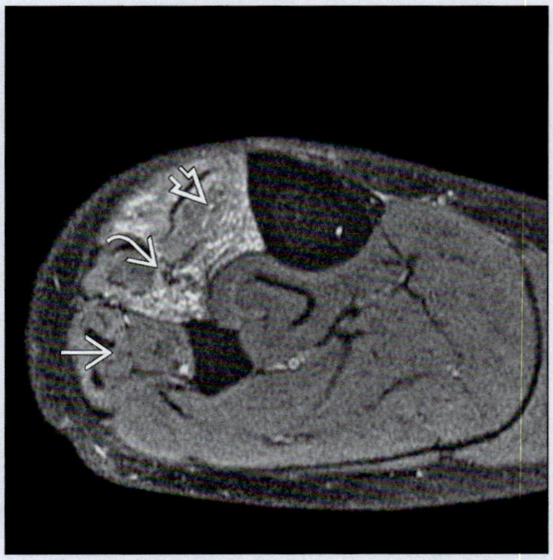

PERONEAL NERVE INJURY

TERMINOLOGY

Synonyms

- Strawberry picker's palsy, slimmer's disease

Definitions

- Injury to or entrapment of common peroneal nerve in its course through knee region

IMAGING

General Features

- Best diagnostic clue
  - Downstream atrophy in muscles supplied by peroneal nerve
- Location
  - Most common site of injury: Lateral to fibular head
- Morphology
  - Common peroneal nerve arises from sciatic nerve in mid- to distal thigh
  - Travels to popliteal fossa laterally, deep to biceps femoris muscle
    - Gives off sural communicating branch and lateral cutaneous nerve of calf
  - Courses around fibular neck and passes through a fibro-osseous opening in superficial head of peroneus longus muscle
    - Nerve can be bent at acute angle at this point
    - Fibrous anchoring tethers nerve to fibula here
  - Divides into superficial and deep peroneal nerves in proximal leg
    - Superficial peroneal nerve
      - Innervates peroneus longus (PL) and brevis (PB) muscles
      - Splits into medial and lateral terminal sensory branches, innervating most of dorsum of foot except 1st dorsal interspace
    - Deep peroneal nerve
      - Descends between and innervates anterior tibialis (AT) and extensor hallucis longus (EHL) muscles
      - Also innervates extensor digitorum longus (EDL) and peroneus tertius (PT) muscles

Imaging Recommendations

- Best imaging tool
  - MR
- Protocol advice
  - T2WI FS, T1WI, post-gadolinium T1WI FS MR to assess nerve size and signal and muscle atrophy and edema
  - Specific MR neurogram sequence (fat-suppressed heavily T2WI MR) may be helpful for intrinsic abnormality of nerve

Radiographic Findings

- Fracture of distal femur, proximal fibula, or lateral tibial plateau
- Knee dislocation
- Osseous mass

MR Findings

- Intrinsic nerve abnormality
  - Enlargement or ↑ T2 signal in peroneal nerve due to impingement or stretch injury
  - Intrinsic neoplasm of nerve
    - Neurofibroma or schwannoma
    - Pre- and post-gadolinium T1WI FS MR may be useful to distinguish borders of tumor
- Extrinsic compression of nerve
  - Osseous trauma
  - Mass or neoplasm
- Downstream muscle changes
  - Atrophy ± edema
  - Affects: PL, PB, EHL, EDL, AT, PT
    - If nerve injured in thigh, may affect short head of biceps femoris

Ultrasonographic Findings

- Enlarged peroneal nerve
  - ± nerve hyperemia
- Presence of intrinsic or extrinsic mass

DIFFERENTIAL DIAGNOSIS

Direct Blow Injury

- Muscle edema without atrophy

Polyneuropathy

- May see neuropathic changes in other muscle groups
- Etiologies
  - Diabetes
  - Amyotrophic lateral sclerosis
  - Amyloid disease

PATHOLOGY

General Features

- Etiology
  - **Direct nerve trauma**
    - Motor vehicle accident, laceration, gunshot wound
  - **Chronic compression**
    - Habitual leg-crossing (strawberry picker's palsy)
    - Compression in debilitated or immobilized patient
      - Bed railing, hard mattress, anesthetized patients
    - Casting for fracture
    - In athletes due to muscle hypertrophy
    - During childbirth due to prolonged squatting
  - **Secondary to osseous trauma**
    - Fracture of lateral tibial plateau, proximal fibula
    - Knee dislocation
  - **Iatrogenic**
    - After anterior cruciate ligament reconstruction
      - Related to femoral tunnel placement
    - Tibial osteotomy
    - Total knee arthroplasty
    - Arthroscopy
  - **Intrinsic neoplasm of nerve**
    - Benign or malignant peripheral nerve sheath tumor
      - Neurofibroma
      - Schwannoma
    - Nerve ganglion
  - **Compression by extrinsic mass**
    - Hematoma
    - Neoplasm
    - Ganglion

- Most commonly from proximal tibiofibular joint
- Lateral parameniscal cyst
    - Popliteal cyst
    - Vascular abnormality
  - **Excessive weight loss (slimmer's disease)**
    - Due to rapid loss of fat pad over fibular head
- Associated abnormalities
  - Osseous trauma to distal femur, proximal leg, or knee

CLINICAL ISSUES

Presentation

- Most common signs/symptoms
  - Weakness in ankle dorsiflexors (foot drop)
  - Numbness or paresthesias in dorsum of foot
  - Tinel sign over fibular head
- Other signs/symptoms
  - Pain variable, usually related to underlying injury
  - Abnormal nerve conduction velocity

Demographics

- Age
  - Adults > children
- Gender
  - M = F
- Epidemiology
  - Most common compressive neuropathy of lower extremity

Natural History & Prognosis

- May spontaneously resolve or may progress
  - Depends on underlying etiology

Treatment

- Conservative
  - Identify and eliminate underlying cause if possible
  - Bracing, splinting
  - NSAIDs or corticosteroids
- Surgery
  - Excision of compressing neoplasm or ganglion
  - Exploration of nerve for severe compressive neuropathy
    - Recommended if fail 3-4 months of conservative therapy

DIAGNOSTIC CHECKLIST

Consider

- Peroneal neuropathy in setting of unexplained foot drop or atrophy of muscles of anterior leg

Image Interpretation Pearls

- Must exclude causes of extrinsic compression of nerve

Reporting Tips

- Look for early signs of peroneal neuropathy in setting of lateral knee fracture or knee dislocation

SELECTED REFERENCES

1. Anderson AW et al: Common peroneal nerve neuropraxia after arthroscopic inside-out lateral meniscus repair. J Knee Surg. 22(1):27-9, 2009
2. Hall MP et al: Risk of iatrogenic injury to the peroneal nerve during posterolateral femoral tunnel placement in double-bundle anterior cruciate ligament reconstruction. Am J Sports Med. 37(1):109-13, 2009
3. Kapur S et al: Acute knee dislocation: review of an elusive entity. Curr Probl Diagn Radiol. 38(6):237-50, 2009
4. Papoutsidakis A et al: Peroneal nerve damage by bicortical tibial screw in ACL reconstruction. Knee Surg Sports Traumatol Arthrosc. Epub ahead of print, 2009
5. Sanger JR et al: Peroneal nerve compression by lateral gastrocnemius flap. J Plast Reconstr Aesthet Surg. 62(8):e280-2, 2009
6. Baima J et al: Evaluation and treatment of peroneal neuropathy. Curr Rev Musculoskelet Med. 1(2):147-53, 2008
7. Bruzzone M et al: The Risk of Direct Peroneal Nerve Injury Using the Ranawat "Inside-Out" Lateral Release Technique in Valgus Total Knee Arthroplasty. J Arthroplasty. Epub ahead of print, 2008
8. Frank DA et al: Peroneal nerve palsy secondary to ganglion cyst of the proximal tibiofibular joint in a pediatric patient. Orthopedics. 31(1):86, 2008
9. Johnson ME et al: Neurologic and vascular injuries associated with knee ligament injuries. Am J Sports Med. 36(12):2448-62, 2008
10. Ersozlu S et al: Common peroneal nerve palsy from an untreated popliteal pseudoaneurysm after penetrating injury. J Vasc Surg. 45(2):408-10, 2007
11. Phisitkul P et al: Complications of locking plate fixation in complex proximal tibia injuries. J Orthop Trauma. 21(2):83-91, 2007
12. Horan J et al: Proximal tibiofibular dislocation. Emerg Med J. 23(5):e33, 2006
13. Bottomley N et al: Displacement of the common peroneal nerve in posterolateral corner injuries of the knee. J Bone Joint Surg Br. 87(9):1225-6, 2005
14. Dellon AL: Postarthroplasty "palsy" and systemic neuropathy: a peripheral-nerve management algorithm. Ann Plast Surg. 55(6):638-42, 2005
15. Iverson DJ: MRI detection of cysts of the knee causing common peroneal neuropathy. Neurology. 65(11):1829-31, 2005
16. Niall DM et al: Palsy of the common peroneal nerve after traumatic dislocation of the knee. J Bone Joint Surg Br. 87(5):664-7, 2005
17. Krivić A et al: Lesion of the common peroneal nerve during arthroscopy. Arthroscopy. 19(9):1015-8, 2003
18. Trappeniers L et al: Peroneal nerve injury in three patients with knee trauma: MR imaging and correlation with anatomic findings in volunteers and anatomic specimens. Eur Radiol. 13(7):1722-7, 2003
19. Johnson DS et al: Common peroneal nerve palsy following knee arthroscopy. Arthroscopy. 15(7):773-4, 1999
20. Mihalko WM et al: Transient peroneal nerve palsies from injuries placed in traction splints. Am J Emerg Med. 17(2):160-2, 1999

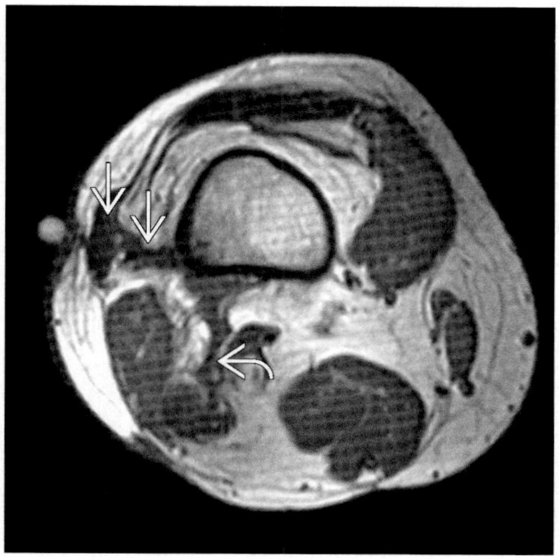

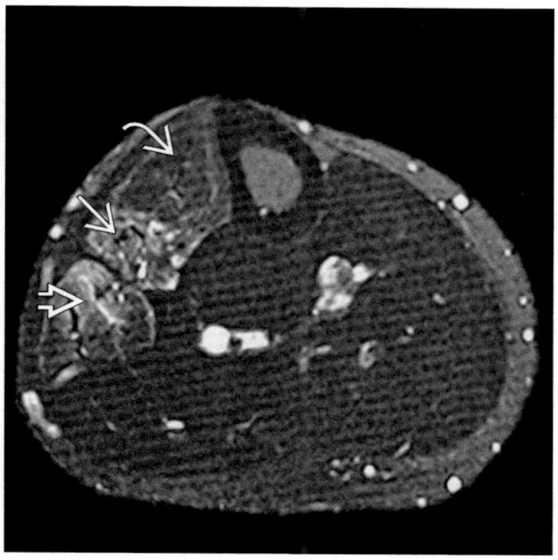

***(Left)** Axial T1WI MR just above the knee in a patient status post gunshot wound to the distal thigh and peroneal neuropathy shows the track of the bullet extending to the femur and scar tissue encasing the common peroneal nerve. **(Right)** Axial T2WI FS MR of the mid-leg in the same patient shows edema and mild atrophy in the extensor digitorum longus and peroneus muscles , and mild edema in the anterior tibialis muscle consistent with peroneal neuropathy.*

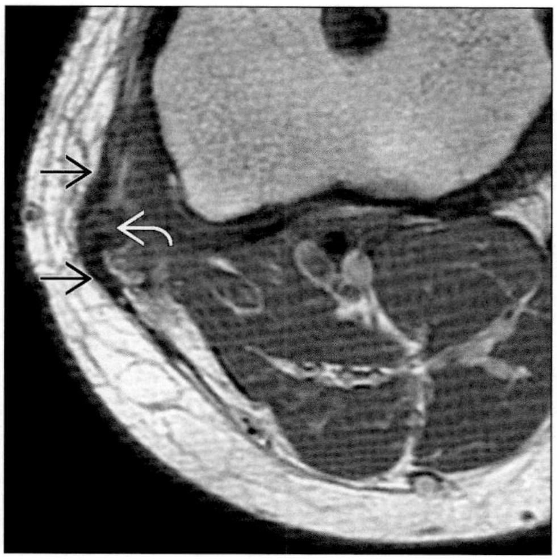

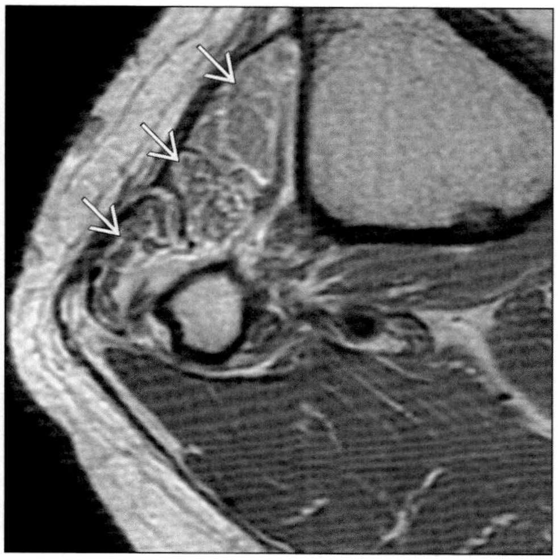

***(Left)** Axial T1WI MR in a patient who presented with symptoms of peroneal neuropathy after anterior cruciate ligament reconstruction shows extensive scar formation along the lateral soft tissues related to placement of the femoral graft tunnel. The common peroneal nerve is quite enlarged and demonstrates mildly elevated signal. **(Right)** Axial T1WI MR more inferiorly in the same patient shows significant fat atrophy of the muscles in the peroneal nerve distribution .*

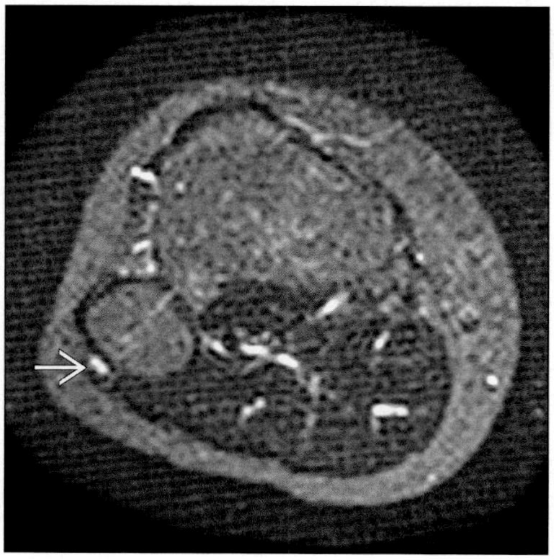

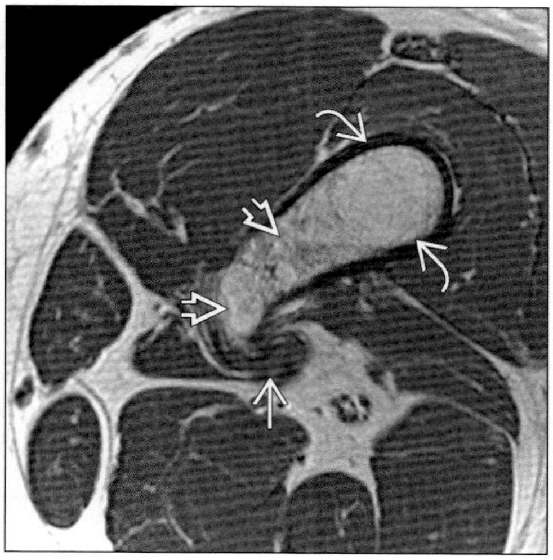

***(Left)** Axial T2WI FS MR in a patient with clinical peroneal neuropathy shows elevated signal in the common peroneal nerve at the point where it rounds the fibular neck. This is the most common site of trauma to the nerve. **(Right)** Axial T1WI MR through the distal thigh shows a large pedunculated osteochondroma arising from the distal femoral shaft . The tip of the osteochondroma abuts and distorts the common peroneal nerve .*

COMPARTMENT SYNDROME AND MUSCLE HERNIATION

Key Facts

Terminology
- Condition in which perfusion pressure falls below tissue pressure in closed anatomic space

Imaging
- Muscle edema within affected compartment on MR
- Axial T1WI and T2WI FS MR to assess muscle morphology and edema
- T1WI FS + Gad MR to assess for myonecrosis
- Most common: Anterior, lateral, superficial, deep posterior compartments of leg

Top Differential Diagnoses
- Muscle strain
- Myositis
- Deep vein thrombosis
- Diabetic muscle infarct
- Denervation hypertrophy

Pathology
- Acute compartment syndrome etiologies
  - Fracture (especially tibia, forearm, and calcaneus)
  - Muscle injury (trauma, weightlifting, sports activity)
  - Crush injuries
  - Splints or casts
  - Surgical positioning (especially lithotomy position)
  - Lying motionless on hard surface (drug or alcoholic stupor)
- Chronic compartment syndrome
  - Exercise-induced

Clinical Issues
- Crescendo pain out of proportion to injury
- Nerve/muscle ischemia > 12 hr → permanent injury
- Surgical

*(**Left**) Frontal graphic depicts edema and swelling in tibialis anterior →, as is seen in anterior compartment syndrome. (**Right**) Axial T2WI FS MR several days after MVA shows marked swelling and elevated signal in the muscles of the anterior compartment of the leg (tibialis anterior, extensor hallucis longus, and extensor digitorum longus) →. There is mildly elevated signal ↪ in the deep flexors as well. The patient complained of severe anterior leg pain.*

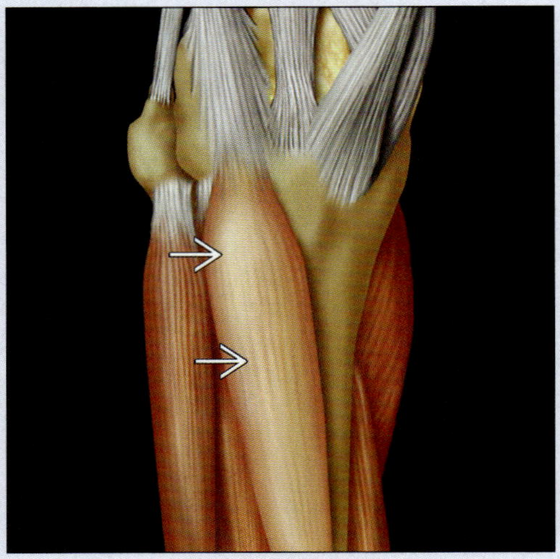

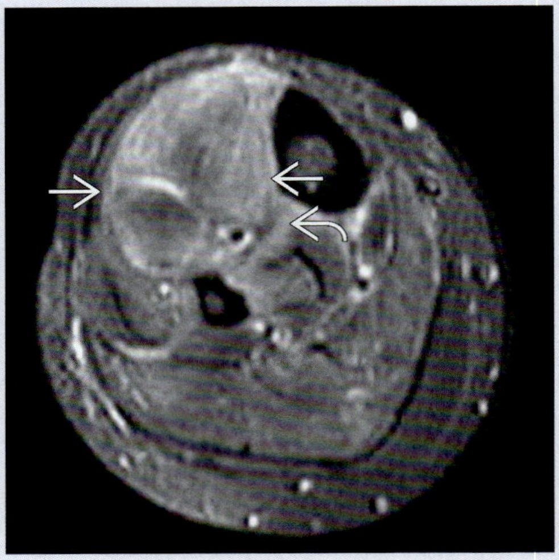

*(**Left**) Axial T1WI FS MR in the same patient shows enlargement of the anterior compartment muscles →. A rim of mildly increased signal in the superficial fascia ↪ may indicate petechial hemorrhage. (**Right**) Axial T1WI C+ FS MR in the same patient shows extensive enhancement in the periphery of the muscle compartment ↪ but no enhancement in the muscles themselves →. This indicates severe muscle ischemia and requires emergent fasciotomy.*

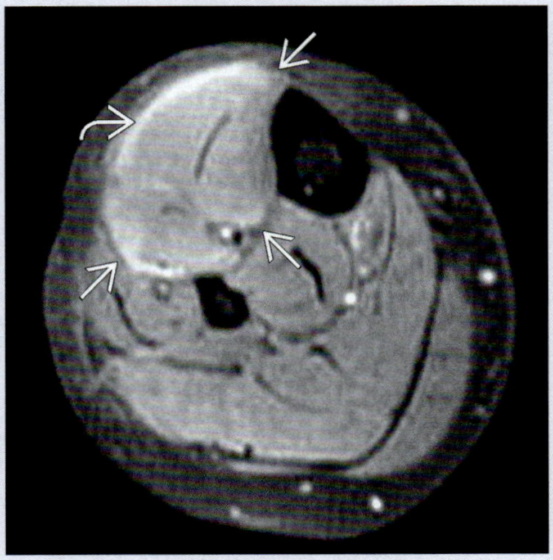

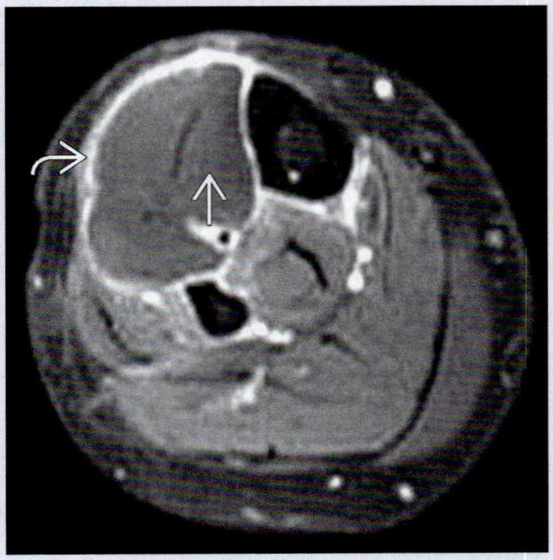

COMPARTMENT SYNDROME AND MUSCLE HERNIATION

TERMINOLOGY

Synonyms

- Acute compartment syndrome
- Chronic (exertional) compartment syndrome (exertional CS, recurrent CS, subacute CS)

Definitions

- Condition in which perfusion pressure falls below tissue pressure in closed anatomic space
- Damage to soft tissues within anatomically confined space resulting from elevated pressure within that space

IMAGING

General Features

- Best diagnostic clue
  - Muscle edema within affected compartment on MR imaging
- Location
  - Most common: Anterior, lateral, superficial, deep posterior compartments of leg
  - Medial, central, lateral, interosseous, calcaneal compartments of foot
  - Flexor, extensor compartments of forearm
  - Thigh (exertional compartment syndrome)
  - Gluteal, deltoid: Less frequent, often related to prolonged surgical positioning
- Morphology
  - MR signal abnormality is detected along full length of muscle or muscles in compartment
  - Combined leg and foot compartment syndromes may occur

MR Findings

- T1WI
  - Decreased signal intensity within edematous muscle
  - Loss of normal muscle striations
  - Enlargement of affected muscle group
  - ± increased signal in subacute hemorrhage
  - Hyperintense fat in chronic muscle atrophy
- Fluid-sensitive sequences
  - Hyperintense diffusely within cross-sectional anatomic boundaries of affected muscle
    - Affects variable proximal/distal extent of muscle
  - Hyperintense fluid/hemorrhage/edema also between muscles in fascial planes
  - ± subcutaneous tissue hyperintense edema
- Post-contrast imaging shows low signal regions of myonecrosis
- Transfacial muscle herniation
  - Herniation of muscle through enlarged vascular fenestrations of superficial fascia
  - Anterior tibialis most commonly affected
  - Usually due to direct trauma but can be related to muscle hypertrophy

Imaging Recommendations

- Best imaging tool
  - MR highly suggestive but not diagnostic
- Protocol advice
  - Axial T1WI and T2WI FS MR to assess muscle morphology and edema
  - T1WI FS + Gad MR to assess for myonecrosis
  - Focal muscle herniation may be best seen with active muscle contraction during MR

Ultrasonographic Findings

- Pulsed Doppler
  - May be useful in characterizing blood flow within affected compartment

DIFFERENTIAL DIAGNOSIS

Muscle Strain

- Partial tear of muscle/musculotendinous unit
- Feathery/diffuse edema on fluid-sensitive sequences

Cellulitis

- Subcutaneous edema without muscle compartment involvement or fascial extension

Myositis

- May see associated abscess formation

Delayed Onset Muscle Soreness (DOMS)

- Increased signal in 1 or more muscles occurring 1-3 days after exertion
- No significant muscle enlargement
- Resolves spontaneously

Deep Vein Thrombosis

- Neurovascular bundle perivascular edema

Myositis Ossificans/Rhabdomyolysis

- Earliest stages of these may appear as muscle edema
- Myositis progresses to form calcifications by 4-8 weeks

Diabetic Muscle Infarct

- Muscle and surrounding fascial edema
- May see decreased vascularity or areas of tissue necrosis on T1WI FS MR + Gad
- Associated with end-stage diabetes

Denervation Hypertrophy

- Enlarged and edematous muscle
- Fatty atrophy within muscle, late finding
- Onset more insidious; painless
- MR abnormality seen 2-4 weeks post denervation

Longitudinal Stress Fracture

- May have edema and swelling around fracture site

PATHOLOGY

General Features

- Etiology
  - Acute compartment syndrome etiologies
    - Increased pressure within compartment
      - Fracture (especially tibia, forearm, and calcaneus)
      - Muscle injury (trauma, weightlifting, sports activity)
      - Crush injuries
      - Hemorrhage (trauma, anticoagulation, hemophilia)
      - Androgen abuse (muscle hypertrophy)
    - Increased pressure from external source
      - Splints or casts

COMPARTMENT SYNDROME AND MUSCLE HERNIATION

- Surgical positioning (especially lithotomy position)
- Lying motionless on hard surface (drug or alcoholic stupor)
- Burns
- Ski boots

- 2 theories of acute compartment syndrome relating to compromised blood flow
  - Arteriovenous gradient theory
    - Compromised blood flow → hypotension, increased vascular resistance → increased tissue pressure (extravasated blood from initial trauma)
  - Ischemia-reperfusion theory
    - Compromised blood flow → ischemia → impaired cellular metabolism → interstitial edema → increased compartmental pressure
- Chronic compartment syndrome
  - Exercise-induced
  - Pain during exercise that subsides with rest

Staging, Grading, & Classification

- Acute: Resting intracompartmental pressures > 30 mmHg
- Chronic
  - Resting compartment pressure ≥ 15 mmHg
  - 1 minute post-exercise compartment pressure ≥ 30 mmHg
  - 5 minute post-exercise compartment pressure ≥ 20 mmHg

CLINICAL ISSUES

Presentation

- Most common signs/symptoms
  - Crescendo pain out of proportion to injury
    - Deep, aching
    - Worsened by passive stretching of involved muscles
  - Sensation of tightness
  - Weakness of involved muscles
- Other signs/symptoms
  - Paresthesias and numbness are unreliable early symptoms
  - Sensory nerves affected before motor nerves
  - Distal pulses usually intact
  - Pain, paresthesia, pallor, pulselessness, and poikilothermia ("5 P's") are uncommon until late-stage compartment syndrome
  - ↑ serum creatine kinase (CK) and myoglobinuria
- Clinical profile
  - Acute
    - Most frequently related to fractures and casting
    - Uncommon in athletic population
    - May have history of recent surgery, especially if in prolonged lithotomy position
    - Prolonged lying on hard surface (drug/alcoholic stupor)
  - Chronic
    - Exercise-related: Long distance runners, skiers, basketball and soccer players
    - Often bilateral
    - Resolves when exercise is stopped

Demographics

- Epidemiology
  - Open tibial fractures: 20% complicated by compartment syndrome
  - Tibial fractures account for 45% of cases of compartment syndrome
  - Acute leg compartment syndrome most commonly related to tibial fracture
  - Muscle hernias in 40% of chronic compartment syndrome
  - 5% of calcaneal fractures develop compartment syndrome of foot

Natural History & Prognosis

- Nerve/muscle ischemia > 12 hr → permanent injury
- Complications
  - Reperfusion injury (cellular activity restored → worsening of preexisting cellular damage)
  - Rhabdomyolysis → renal failure
  - Volkmann contracture (due to forearm flexor compartment syndrome)

Treatment

- Surgical
  - Acute: Fasciotomy performed based on clinical scenario
  - Chronic: Surgical decompression/fasciotomy

DIAGNOSTIC CHECKLIST

Consider

- Compartment syndrome should be considered in case of casted tibial fracture with disproportionate pain

Image Interpretation Pearls

- Look for associated muscle herniation

Reporting Tips

- Report muscle changes in terms of compartmental anatomy

SELECTED REFERENCES

1. Andreisek G et al: T2*-weighted and arterial spin labeling MRI of calf muscles in healthy volunteers and patients with chronic exertional compartment syndrome: preliminary experience. AJR Am J Roentgenol. 193(4):W327-33, 2009
2. Baker JF et al: Foot drop--an unusual presentation of exertional compartment syndrome. Clin J Sport Med. 19(3):236-7, 2009
3. McCaffrey DD et al: Acute compartment syndrome of the anterior thigh in the absence of fracture secondary to sporting trauma. J Trauma. 66(4):1238-42, 2009
4. Drey IA et al: Acute compartment syndrome of the calf presenting after prolonged decubitus position. Orthopedics. 31(2):184, 2008
5. Shadgan B et al: Diagnostic techniques in acute compartment syndrome of the leg. J Orthop Trauma. 22(8):581-7, 2008
6. Konstantakos EK et al: Diagnosis and management of extremity compartment syndromes: an orthopaedic perspective. Am Surg. 73(12):1199-209, 2007
7. Bong MR et al: Chronic exertional compartment syndrome: diagnosis and management. Bull Hosp Jt Dis. 62(3-4):77-84, 2005

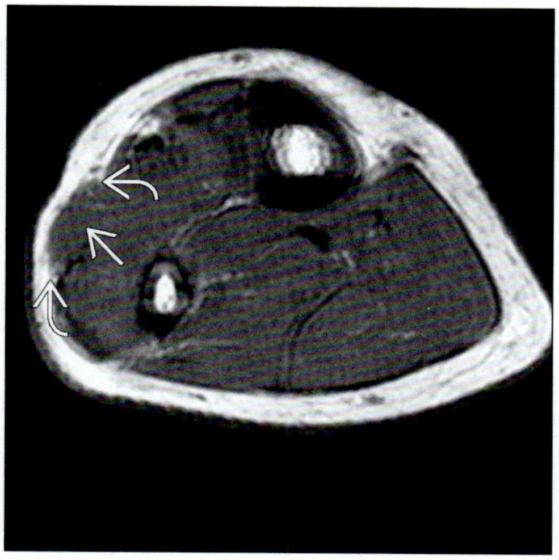

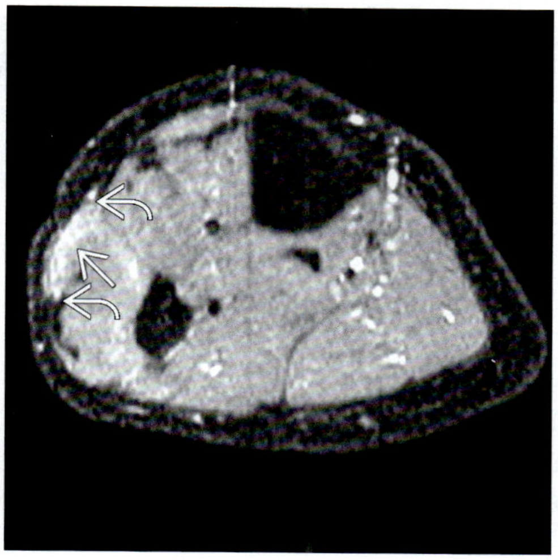

(Left) *Axial T1WI MR in a 15-year-old lacrosse player with anterolateral leg pain after running demonstrates a bulge of tissue isointense to muscle ➡ beyond the superficial fascial margins ↪, where the marker capsule has been placed. Motion artifact is due to the patient holding foot in dorsiflexion during scan.* ***(Right)*** *Axial PDWI FS MR in the same patient confirms the extrusion of extensor digitorum muscle tissue ➡ through the superficial fascia ↪ of the anterior compartment.*

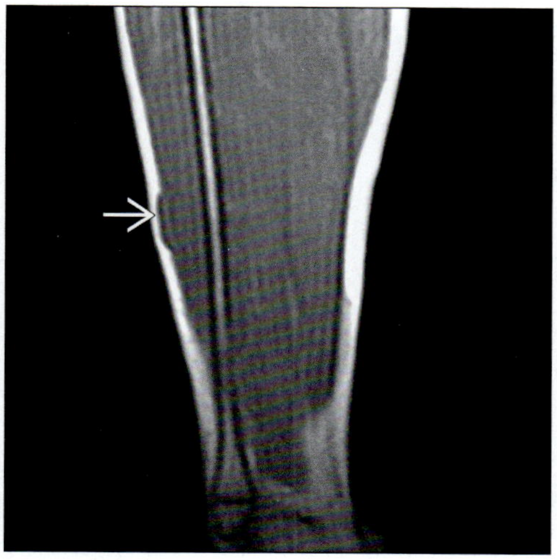

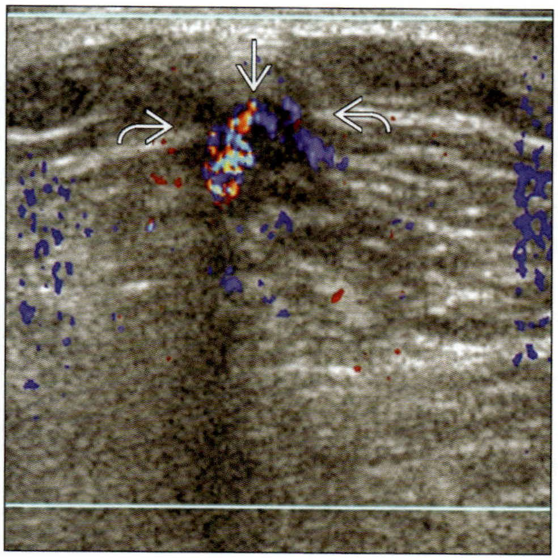

(Left) *Coronal T1WI MR in the same patient confirms the focal transfascial muscle herniation ➡. Such herniations are often transient and may not be visible on static imaging without activation of the affected muscles (in this case, foot dorsiflexion was necessary).* ***(Right)*** *Sagittal ultrasound in a patient with recurrent anterior leg pain when running shows a hypervascular portion of tibialis anterior ➡ herniating through the superficial fascia ↪ with active dorsiflexion.*

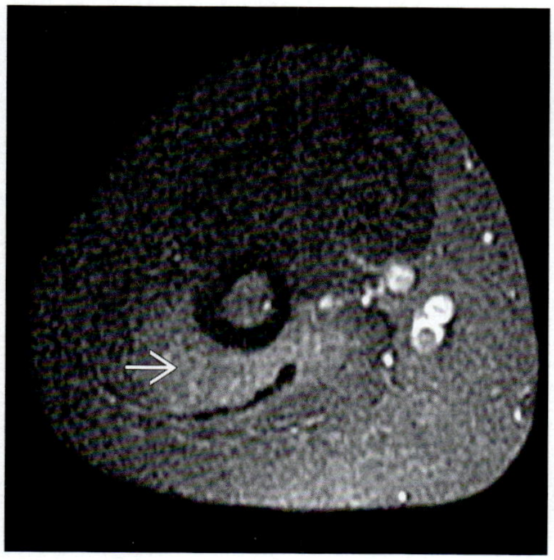

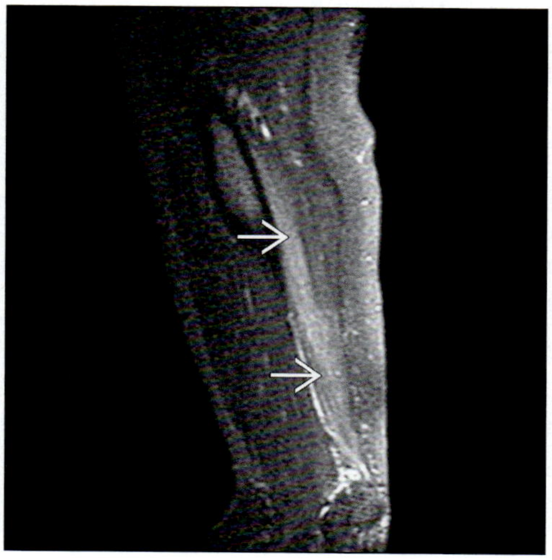

(Left) *Axial T2WI FS MR of the upper arm in a weightlifter with posterior arm pain occurring about 48 hours after exertion shows elevated signal ➡ in the triceps muscle. This appearance following exercise has been termed "delayed onset muscle soreness" (or DOMS), and should not be mistaken for compartment syndrome.* ***(Right)*** *Sagittal T2WI FS MR in the same patient shows the well-defined elongate nature of the edema in the triceps muscle ➡.*

SECTION 7
Ankle and Foot

Introduction and Overview

Osseous Injuries

Ligament Injuries

Musculotendinous Injuries

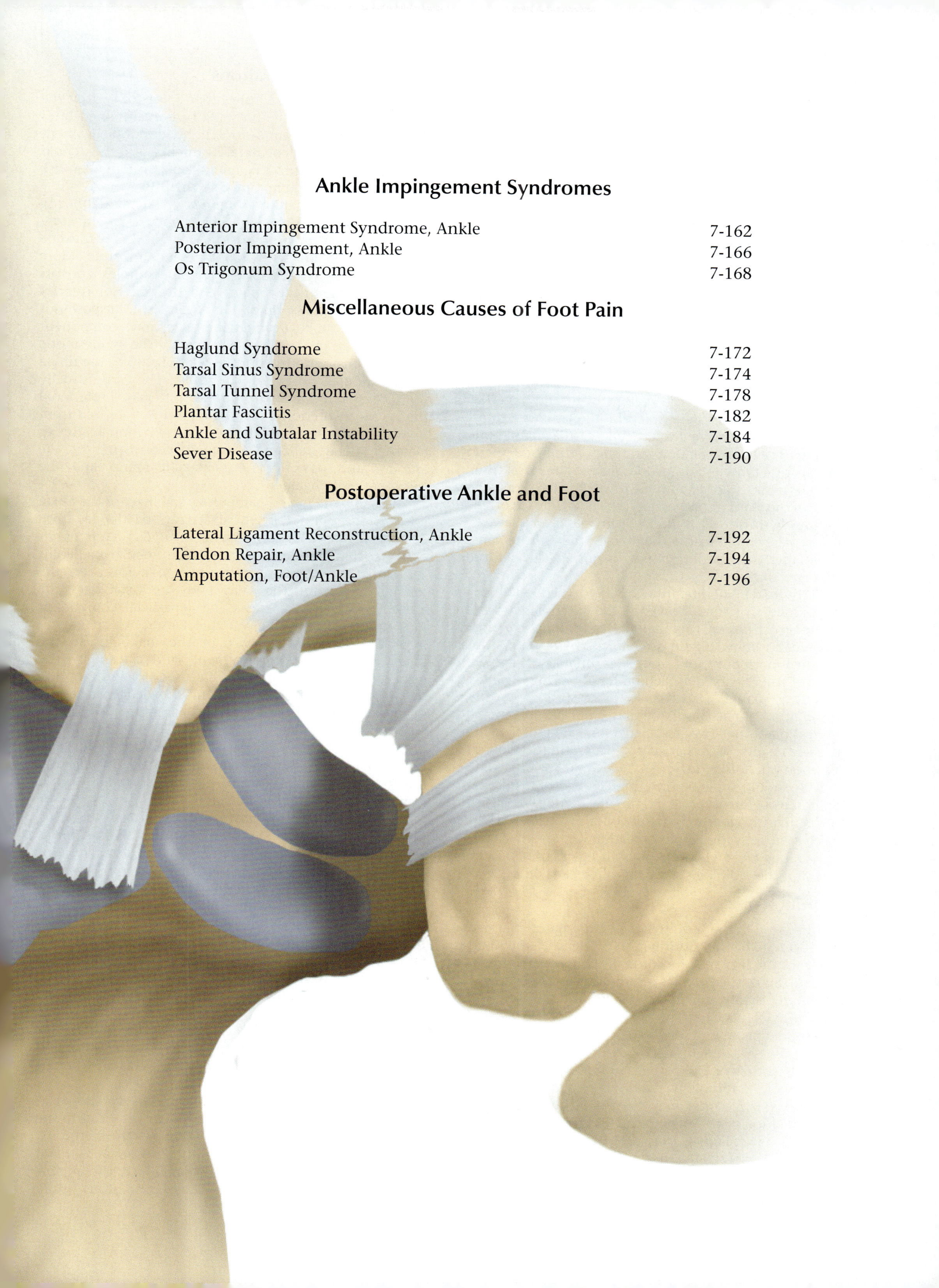

Ankle Impingement Syndromes

Miscellaneous Causes of Foot Pain

Postoperative Ankle and Foot

Overview

Injuries to the ankle and foot are very common and often subtle. The radiologist needs an intimate knowledge of the normal anatomical relationships to make the correct diagnosis; for that reason, a discussion of pertinent anatomy is included for each diagnosis in this section.

Injuries to the ankle and foot are often multiple. Satisfaction of search is therefore a real danger in evaluating images. When a finding is made, it is sometimes helpful to try to ascertain the mechanism of injury, so a search can be made for other injuries caused by that same mechanism. However, preconceptions about mechanism of injury may bias observations of the images, and the best rule is simply to continue scrutinizing an imaging study using a standard search pattern so that nothing will be missed.

For example, the search pattern on an ankle MR might start with the ankle ligaments, examining the lateral collateral, then the deltoid deep and superficial components, then the distal tibiofibular syndesmosis. Attention is then turned to articular cartilage and bones, then to the extrinsic tendons and their retinacula, then to the intrinsic muscles and plantar fascia. The tarsal tunnel and tarsal sinus are examined next. Finally, a search is made of the "corners of the film." The best time to correlate findings with history is after the images are reviewed. If history is reviewed before the images, there is the danger of glossing over findings that do not fit the clinically suspected diagnosis.

Early ankle MR literature often focused on injuries to a single structure. However, the interrelated nature of the structures in this region has become ever more apparent as knowledge has advanced. A posterior tibial tendon abnormality, for example, is often associated with tears of the flexor retinaculum and spring ligament, as well as plantar fasciitis and preexisting pes planus. Information about associated abnormalities is critical to surgical planning. The "associated abnormalities" section in each chapter is designed to focus attention on this common phenomenon.

MR should not be interpreted without pertinent radiographs. Small bone fragments are often difficult to distinguish from soft tissue injuries on MR but are easily seen on radiographs. In addition, weight-bearing radiographs often yield clues about ligamentous injuries that alert the radiologist to areas of concern on MR.

Terminology

In the ankle and foot, the axial and sagittal planes are considered to be the continuation of the axial and sagittal planes of the leg, and the coronal plane is perpendicular to them. Since the foot has a normal plantar declination, planes are usually angled slightly from those used in the leg. The axial plane is along the long axis of the talus, and the coronal plane is perpendicular to the talus.

Because MR imaging is often obtained in positions other than that of the patient supine with dorsiflexion of the foot, terminology that does not depend on the standard axial/coronal may be used. The short axis of the foot displays the metatarsals in cross section, while the long axis plane is similar in orientation to a PA radiographic view of the foot.

Anatomic Considerations

The ankle joint is a pure hinge joint formed by the tibia, fibula, and talus. Its stability is maintained by 3 major ligamentous complexes: The syndesmotic ligament complex, the lateral collateral ligaments, and the deltoid ligament. Injury to these ligaments puts the patient at risk for instability and premature osteoarthritis.

There are 2 separate groupings of lateral stabilizing ankle ligaments. Although the ligaments are in close proximity, their functions are different, and they must be approached on MR with that difference in mind. More superiorly, the syndesmotic ligament complex consists of the syndesmotic ligament and the anterior inferior and posterior inferior tibiofibular ligaments. The posterior inferior tibiofibular ligament has a separate component inferiorly, the intermalleolar ligament. The syndesmotic ligaments maintain the normal relationship of the fibula in the fibular notch of the tibia. They are usually injured with an eversion (pronation) stress or a rotational injury. More inferiorly, the ankle joint is stabilized from inversion, anterior translation, and rotational stresses by the lateral collateral ligament, consisting of the anterior and posterior talofibular ligaments and the calcaneofibular ligament.

Medially, the main ligamentous stabilizer of the ankle joint is the deltoid ligament. It can be divided into deep and superficial components. The deep component extends from the deep surface of the medial malleolus (primarily the posterior colliculus) to a fovea on the medial surface of the talar body and stabilizes the ankle from eversion stress. The superficial component has multiple, somewhat variable bands, which extend from the superficial surface of the malleolus to the talus, navicular, spring ligament, and calcaneus. The term deltoid ligament is often used to indicate the deep component only, but as our knowledge of deltoid ligament injuries improves, this oversimplification should be avoided.

The foot is divided into the hindfoot (talus and calcaneus), the midfoot (remaining tarsal bones), and the forefoot (metatarsals and phalanges). It has both a transverse and a longitudinal arch.

The subtalar joint consists of 3 facets: Posterior, middle, and anterior. The posterior subtalar facet forms a separate joint cavity, continuous with the ankle joint in about 15% of the population. Fifty percent of weight-bearing is through the posterior subtalar joint. Inversion, eversion, and a gliding motion occur at the posterior subtalar joint. The middle and anterior subtalar facets have a common joint cavity and communicate with the talonavicular joint, and this complex is often called the talocalcaneonavicular joint. The subtalar joint is stabilized by interosseous ligaments, as well as the deltoid and the calcaneofibular ligament.

The talonavicular and calcaneocuboid joints have separate joint cavities but act together in supination and pronation, and they are collectively known as the Chopart joint.

The tarsometatarsal joints are collectively known as the Lisfranc joint. The Lisfranc ligament is crucial to the stability of the Lisfranc joint. It extends from the 1st cuneiform to the medial base of the 2nd metatarsal and can be injured with relatively minor trauma due to forcible plantarflexion of the midfoot. This ligamentous injury can often be detected by lateral subluxation of the 2nd tarsometatarsal joint on weight-bearing radiographs, but the diagnosis is frequently missed.

Pathologic Considerations

The Ottawa rules were designed to avoid unnecessary imaging of the ankle and foot. Radiographs of the ankle are recommended only if the patient is unable to bear weight for 4 steps both immediately after the injury and in the emergency department, along with tenderness to palpation at the distal 6 cm of the posterior tibia, medial malleolus, or lateral malleolus. Radiographs of the foot are recommended if the patient is unable to bear weight for 4 steps both immediately after the injury and in the emergency department, and has tenderness to palpation along the 5th MT or the navicular. Unfortunately, the foot rules do not include criteria to aid in recognition of the important and often missed isolated Lisfranc ligament rupture (which presents with tenderness and soft tissue swelling at the medial margin of the 2nd MT base).

Stress fractures are common in the foot. In the metatarsals, they cause a sharp, focal pain, and diagnosis can be made clinically even when radiographs are normal. However, in the hind- and midfoot they often present with vague, ill-defined pain. MR is very useful in identifying the fracture and excluding soft tissue and articular causes of pain.

Radiographic Considerations

When possible, radiographs should be obtained with the patient standing, so that alignment can be assessed. Careful positioning of the patient is also important, and injuries can easily be missed if the radiographs were suboptimally positioned. A technologist's manual, such as *Merrill's Atlas of Radiographic Positioning and Procedures*, should be in every center where orthopedic imaging is performed. The standard ankle series consists of AP, mortise (15° internal oblique), and lateral views. The standard foot series consists of AP, oblique, and lateral views. Additional views are sometimes useful, such as the axial (Harris) view for calcaneal fractures and the hindfoot alignment view.

Ligament injuries can often be inferred on radiographs based on osseous injury &/or malalignment. For instance, an isolated fracture of the medial or posterior malleolus implies injury to the distal tibiofibular syndesmosis, often accompanied by a Maisonneuve fracture of the proximal fibula.

CT Considerations

CT scan is performed using thin (0.6-1 mm) multidetector scanning of the hindfoot or forefoot with sagittal and coronal reformations. A soft tissue algorithm should be obtained in addition to bone algorithm to permit diagnosis of tendon and ligament injuries.

3D CT has limited utility but can be helpful in understanding complex fracture patterns. The author's preference is for a volume rather than a surface rendering, because bone detail is better seen with volume rendering.

Bone Scan Considerations

Tc-99m bone scanning has a very limited role in evaluating trauma of the foot and ankle. It will show areas of hyperemia and areas of increased osteoblast activity, but these are nonspecific parameters. In cases of cryptogenic foot pain, MR is a more effective tool for identifying the site and etiology of the patient's pain.

MR Considerations

MR with FOV covering the entire foot is almost never indicated. Much higher resolution will be achieved with smaller FOV (10-14 cm), centered on the area of concern. At our institution we have standard protocols for the hindfoot, midfoot, and forefoot. Each protocol requires about 25 minutes and includes sagittal T1 and STIR images, axial T1 and PD FS, and coronal PD and T2 FS. Slice thickness varies from 2.5-4 mm.

The hindfoot protocol is performed using a multichannel knee coil with the patient supine and toes pointing up. The FOV is chosen to include an area from above the ankle joint to the cuneiforms, in order to include the entire distal course of the posterior tibial and peroneal tendons.

The midfoot protocol also utilizes the knee coil, but the patient is usually prone. The area of coverage is from the talar neck to the proximal metatarsal shafts.

The metatarsal protocol includes the region from the Lisfranc joint to the proximal phalanges, and a flexible surface coil is often used. The patient is prone.

The toe protocol utilizes a wrist coil or a flexible surface coil with a small FOV centered on the area of concern. Thinner slices are used, often covering only the toe of concern and the 2 adjacent toes. The patient is supine.

MR is usually performed for the evaluation of chronic ankle and foot pain and disability. Radiographs and CT are the primary means for diagnosis of acute injuries. However, MR has a role to play in acute injuries to tendons or the Lisfranc ligament.

Cartilage in the foot is thin and often tightly curved, and osteochondral injuries are commonly missed on MR. MR arthrography improves sensitivity in the diagnosis of hyaline cartilage injury, and may also improve visualization of ligament injuries.

Therapeutic Injections

Therapeutic joint or tendon sheath injections performed with fluoroscopic or ultrasound guidance are useful in the diagnosis and treatment of ankle pain. Injection can assist the surgeon in verifying pain generators; if performed for this purpose, care should be taken to evaluate for communications between joints. For instance, if the ankle and subtalar joints communicate, pain relief after injection is a less specific finding.

Selected References

1. Cerezal L et al: MR arthrography of the ankle: indications and technique. Radiol Clin North Am. 46(6):973-94, v, 2008
2. Collins MS: Imaging evaluation of chronic ankle and hindfoot pain in athletes. Magn Reson Imaging Clin N Am. 16(1):39-58, v-vi, 2008
3. Frank ED et al: Merrill's Atlas of Radiographic Positioning and Procedures. Philadelphia: Elsevier, 2007
4. Choplin RH et al: CT with 3D rendering of the tendons of the foot and ankle: technique, normal anatomy, and disease. Radiographics. 24(2):343-56, 2004
5. Newman JS: Diagnostic and therapeutic injections of the foot and ankle. Semin Roentgenol. 39(1):85-94, 2004
6. Stiell IG et al: Decision rules for the use of radiography in acute ankle injuries. Refinement and prospective validation. JAMA. 269(9):1127-32, 1993

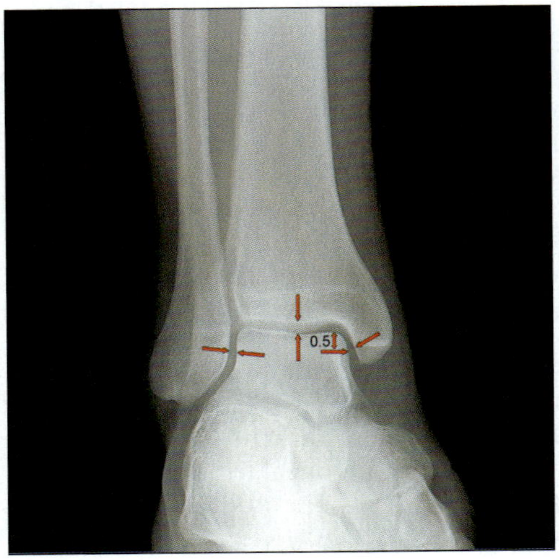

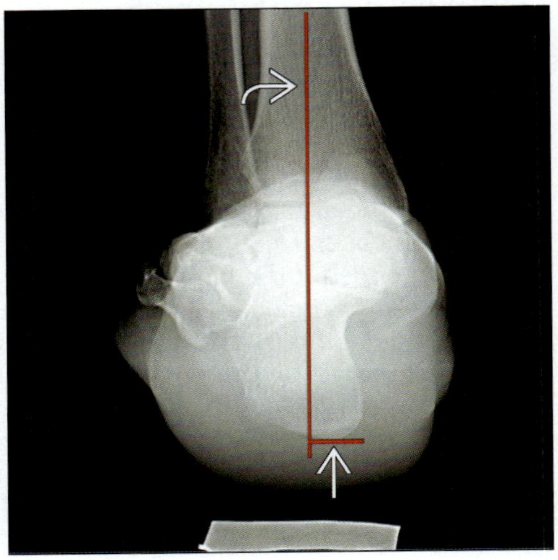

(Left) *Mortise radiograph shows equal joint spaces medially, laterally, and superiorly. The joint margins should be congruent and measure less than 4 mm. 4-6 mm is equivocal, and > 6 mm indicates ligament tear.* ***(Right)*** *Hindfoot alignment view, patient standing. Toes are against the radiographic cassette, which is angled 20° from vertical, and x-ray beam is angled 20° caudad from heel to toes. The distance from the lowest point of the heel to the tibial axis is < 8 mm in normal, asymptomatic patients.*

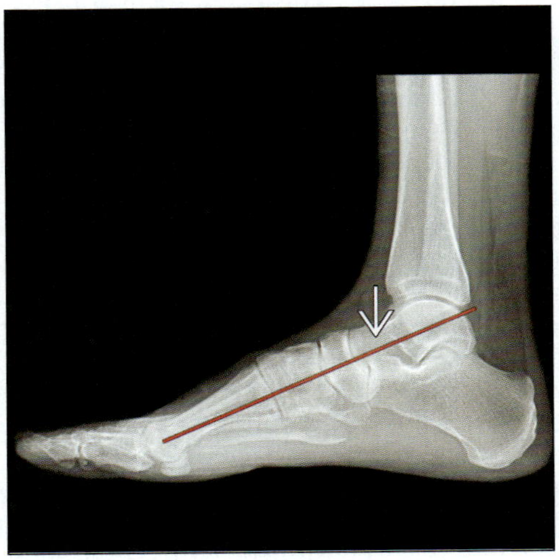

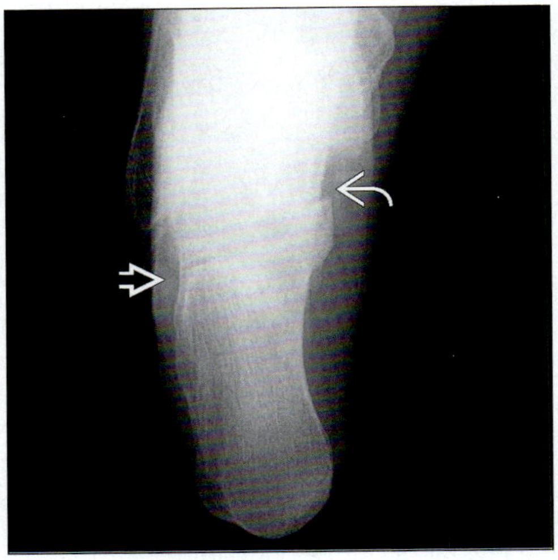

(Left) *Lateral radiograph shows the talar axis. In the normal foot, the talar axis and the 1st metatarsal (MT) axis are aligned. If the talar axis extends below the 1st MT axis, pes planus is present. If the talar axis is above the 1st MT axis, pes cavus is present.* ***(Right)*** *The os calcis (Harris) view is obtained with the x-ray beam directed at a 45° caudal angle posteriorly, with the foot dorsiflexed. This view depicts the middle and posterior subtalar joints.*

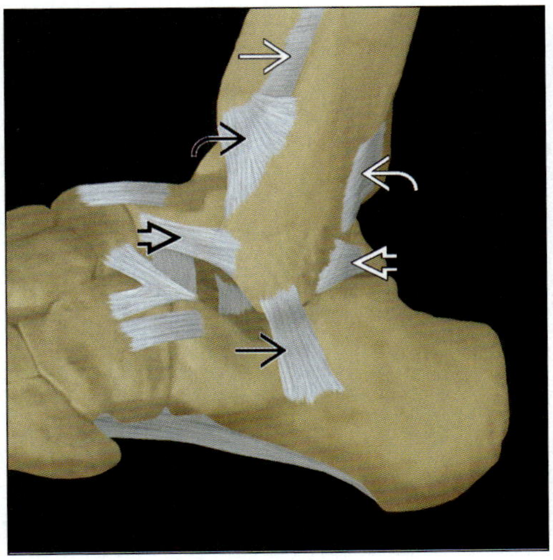

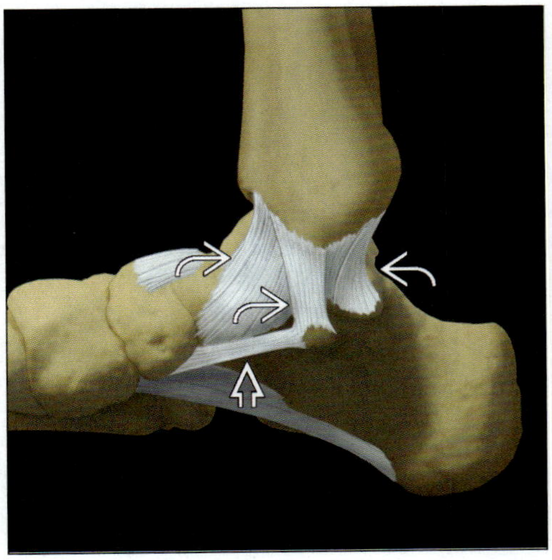

(Left) *Graphic of the lateral ankle shows the syndesmotic ligament complex: Anterior and posterior inferior tibiofibular ligaments and interosseous ligament. The lateral collateral ligament complex is located more inferiorly; it includes the anterior and posterior talofibular ligaments and the calcaneofibular ligament.* ***(Right)*** *Graphic shows superficial portions of deltoid ligament, with attachments to the calcaneus, talus, navicular, and spring ligament.*

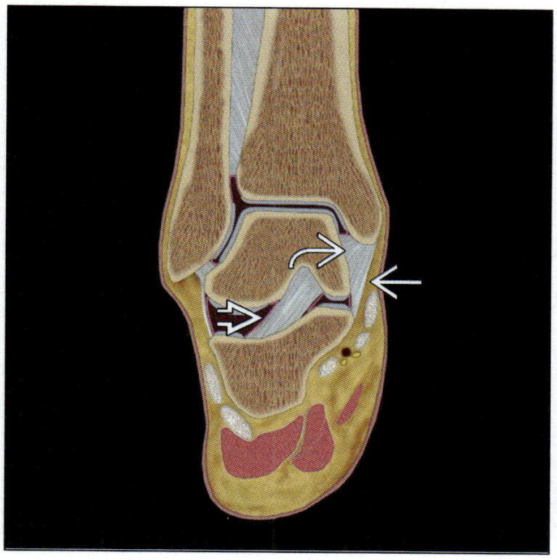

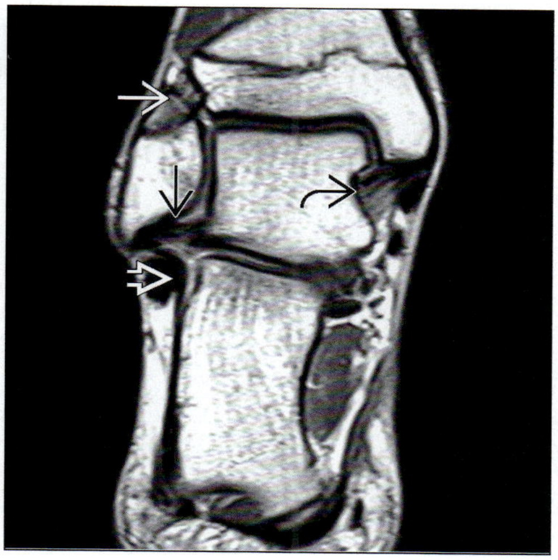

(Left) *Coronal graphic shows deep deltoid ligament extending from medial malleolus to medial surface of talar body. A portion of the superficial deltoid, attaching to the sustentaculum tali, is also seen. Interosseous ligament is located in the sinus tarsi.* ***(Right)*** *Coronal T1WI MR in a 14-year-old boy shows normal deep deltoid, anterior inferior tibiofibular, calcaneofibular, and posterior talofibular ligaments.*

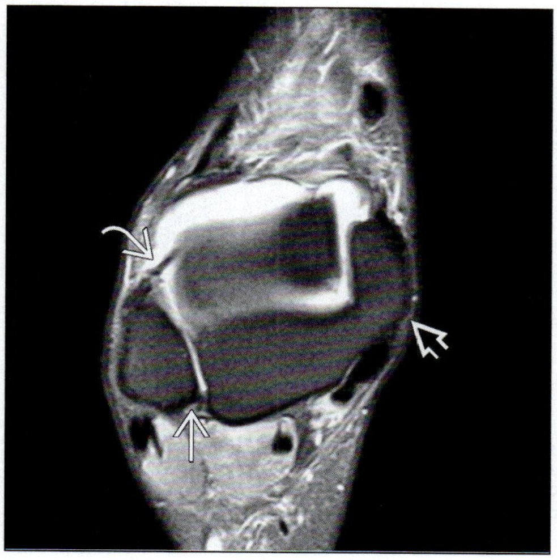

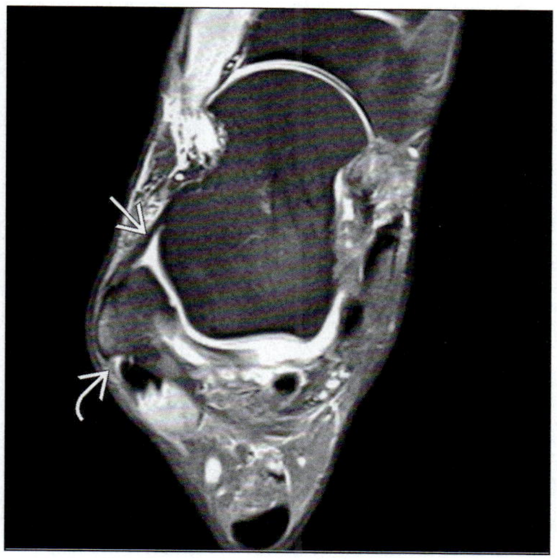

(Left) *Axial MR arthrogram T1WI FS shows the normal appearance of the anterior and posterior tibiofibular ligaments. Note normal, rounded anterior joint contour. Flexor retinaculum is seen attaching to the posteromedial corner of the tibia.* ***(Right)*** *Axial MR arthrogram T1WI FS shows the normal anterior talofibular ligament, visible on the most superior cut where the talar neck is seen. Peroneal retinaculum inserts on the posterolateral corner of the fibula.*

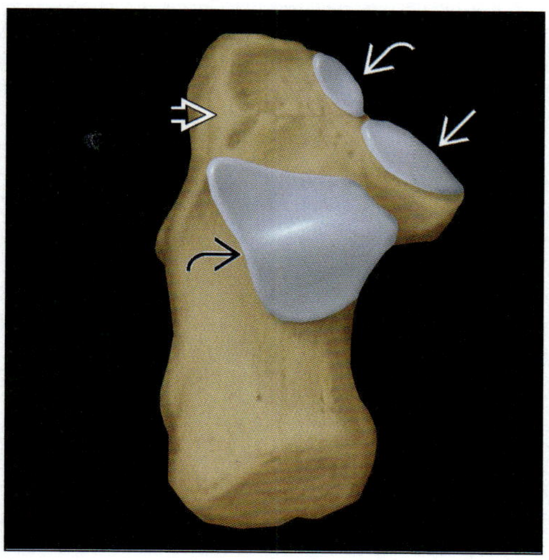

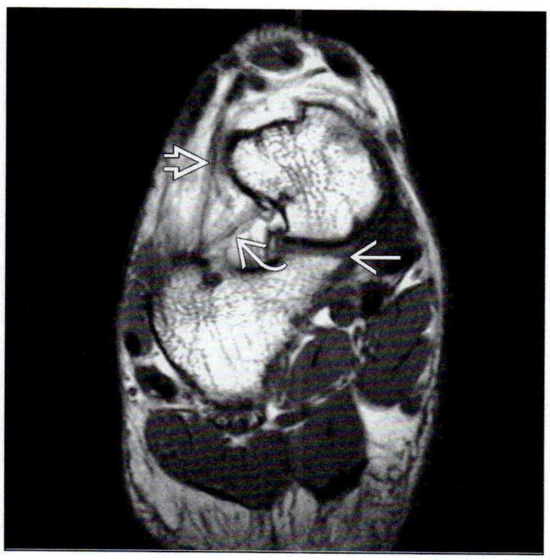

(Left) *Graphic of the dorsal surface of the calcaneus shows the floor of the tarsal sinus, a fat-filled space between the calcaneus and the talus. It lies anterior to the posterior subtalar joint and lateral to the middle and anterior subtalar joints.* ***(Right)*** *Coronal T1WI MR shows the cervical ligament in the tarsal sinus. Extensor retinaculum has 3 limbs inserting in the tarsal sinus near the cervical ligament. Middle subtalar facet is seen medially.*

*(**Left**) Medial graphic shows the posterior tibial, flexor digitorum longus, and flexor hallucis longus tendons, which course beneath the flexor retinaculum within the tarsal tunnel. (**Right**) Lateral graphic shows the peroneus brevis and longus tendons, which act to evert the foot. Achilles tendon flexes the ankle. Peroneus tertius and extensor digitorum longus, extensor hallucis longus, and tibialis anterior tendons are extrinsic foot extensors.*

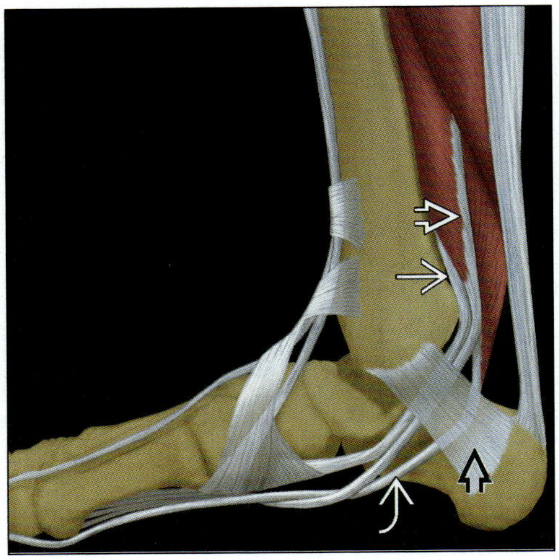

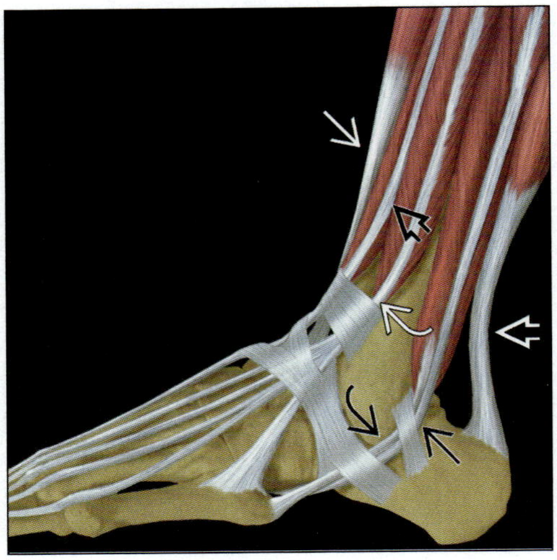

*(**Left**) Graphic shows the superficial layer of plantar foot muscles: Abductor digiti minimi, flexor digitorum brevis, and abductor hallucis. (**Right**) Graphic shows the 2nd layer of plantar muscles. The arrangement of interactions between flexors is well seen: Quadratus plantae inserts onto the flexor digitorum longus tendon. The lumbrical muscles originate from the individual slips of the extensor digitorum brevis tendon.*

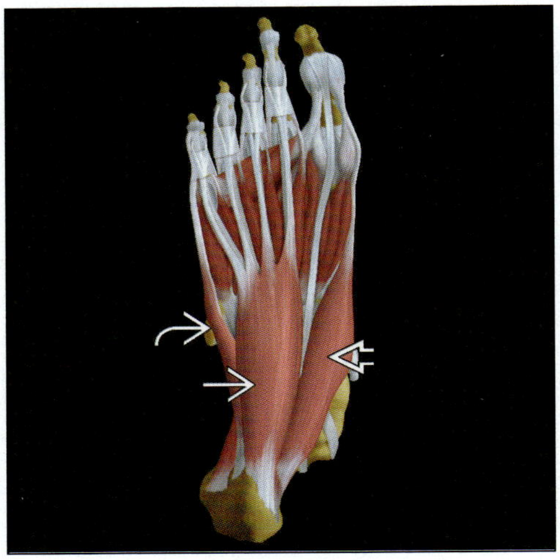

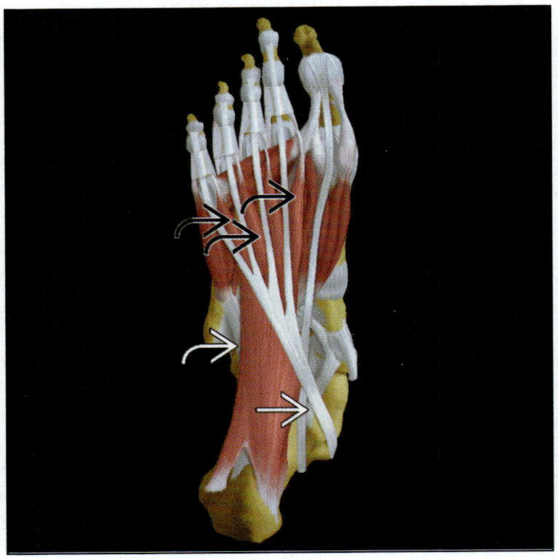

*(**Left**) Graphic shows the 3rd layer of plantar muscles. The adductor hallucis has oblique and transverse heads. The flexor hallucis brevis has medial and lateral heads. The flexor hallucis longus tendon passes between the heads of the flexor hallucis brevis. (**Right**) Coronal graphic shows the osseous architecture of the transverse arch of the foot and the triangular shape of the cuneiforms. The integrity of the transverse arch is in large part dependent on the Lisfranc ligament.*

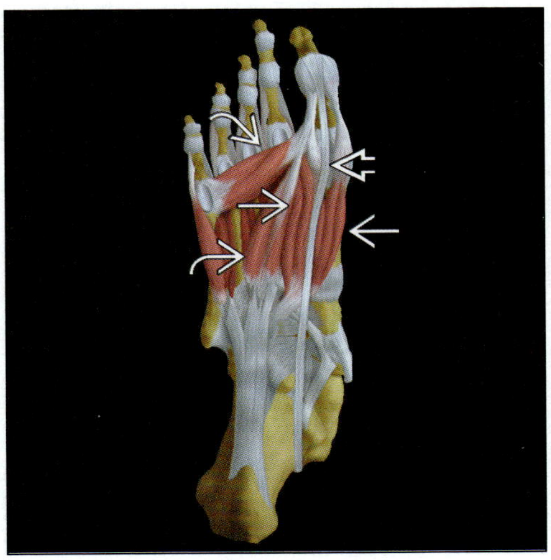

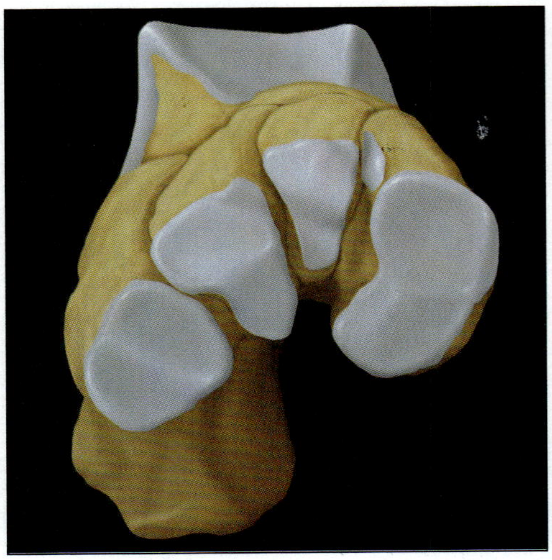

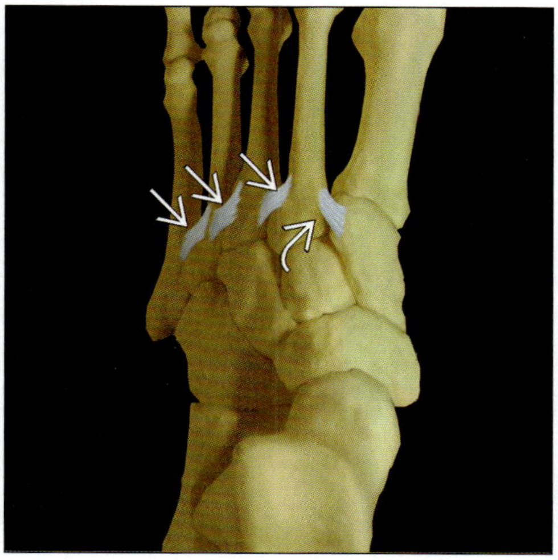

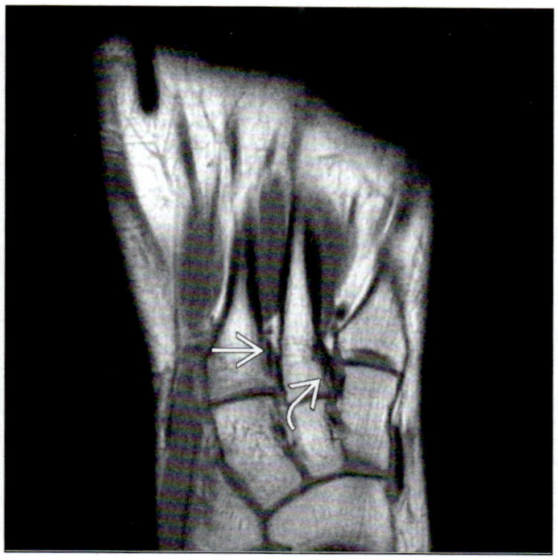

(Left) *Graphic of the dorsum of the foot shows the dorsal component of the Lisfranc ligament. This ligament has dorsal, interosseous, and planar components and is a crucial stabilizer of the Lisfranc joint. Intermetatarsal ligaments are also seen.* ***(Right)*** *Axial T1WI MR shows the interosseous component of the Lisfranc ligament and 1 intermetatarsal ligament. The 3 components of the Lisfranc ligament are reliably visualized on MR when the imaging plane is along the axis of the foot.*

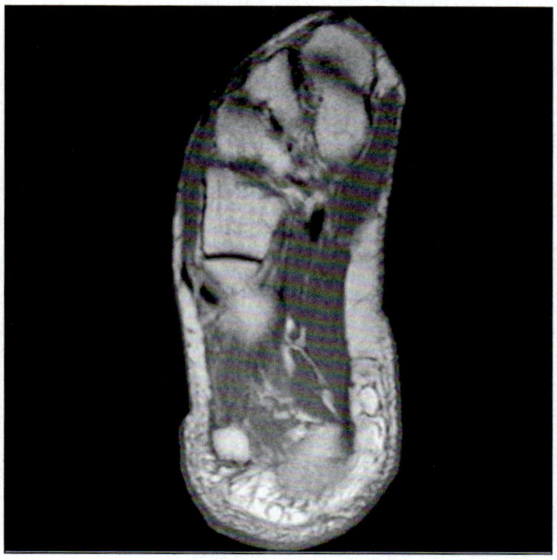

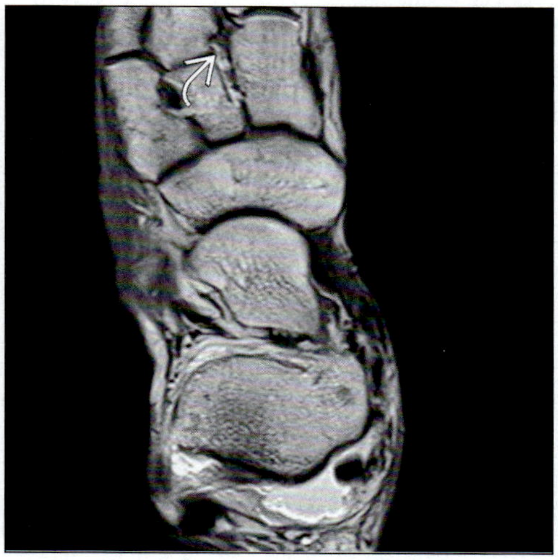

(Left) *Axial T1WI MR through the cuneiforms was obtained without angling the slices. Since the foot is plantarflexed, the image is oblique to the axis of the foot, and the midfoot articulations were not well seen on any images in this sequence. MR and CT should be angled with the talar axis.* ***(Right)*** *Axial T1WI MR also through the cuneiforms, but angled along the talar axis. Using this anatomic imaging plane, the midfoot articulations and the dorsal band of the Lisfranc ligament are well seen.*

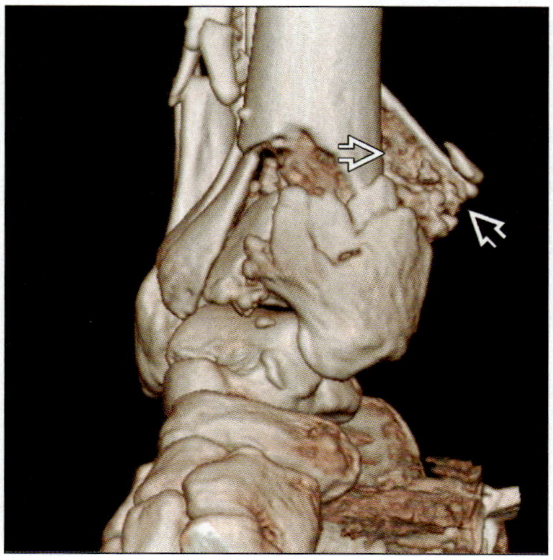

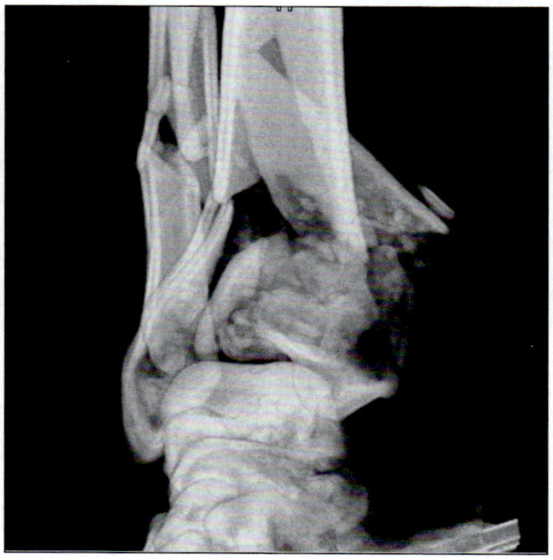

(Left) *Bone CT 3D surface rendering shows a pilon fracture. It is difficult on a surface rendering to understand the relationships between fracture fragments, since the fracture is partly obscured by overlying bone. In addition, the irregularity of the bone fragments tends to cause a confusingly granular appearance of the trabecular bone.* ***(Right)*** *Bone CT 3D volume rendering in the same patient more clearly shows the fracture extent and the relationship between fragments.*

ACCESSORY OSSICLES, ANKLE AND FOOT

Key Facts

Terminology

- Accessory center of ossification
  - Variably present ossific center closely associated with bone
  - Sometimes merges with parent bone at skeletal maturity
  - May be injured in sports or with chronic stress
- Sesamoid
  - Ossicle arising within tendon, improving its mechanical advantage
- Synchondrosis
  - Usually found between accessory center of ossification and parent bone
  - Cartilage bridge between 2 bones

Imaging

- Accessory center of ossification
  - Closely apposed to parent bone
  - Joined to parent bone by synchondrosis
- Sesamoid
  - Usually ovoid
  - May form part of synovial joint and have 1 or more articular facets
- MR: Ossicles usually show signal intensity of fat
- Bone marrow edema is important sign of injury to ossicle
  - Fracture through ossicle or through synchondrosis
  - Impingement of ossicle
- Other signs of accessory ossicle injury
  - Widened, fluid-filled synchondrosis
  - Abnormality of tendon inserting on ossicle

Top Differential Diagnoses

- Fracture
- Pseudoepiphysis

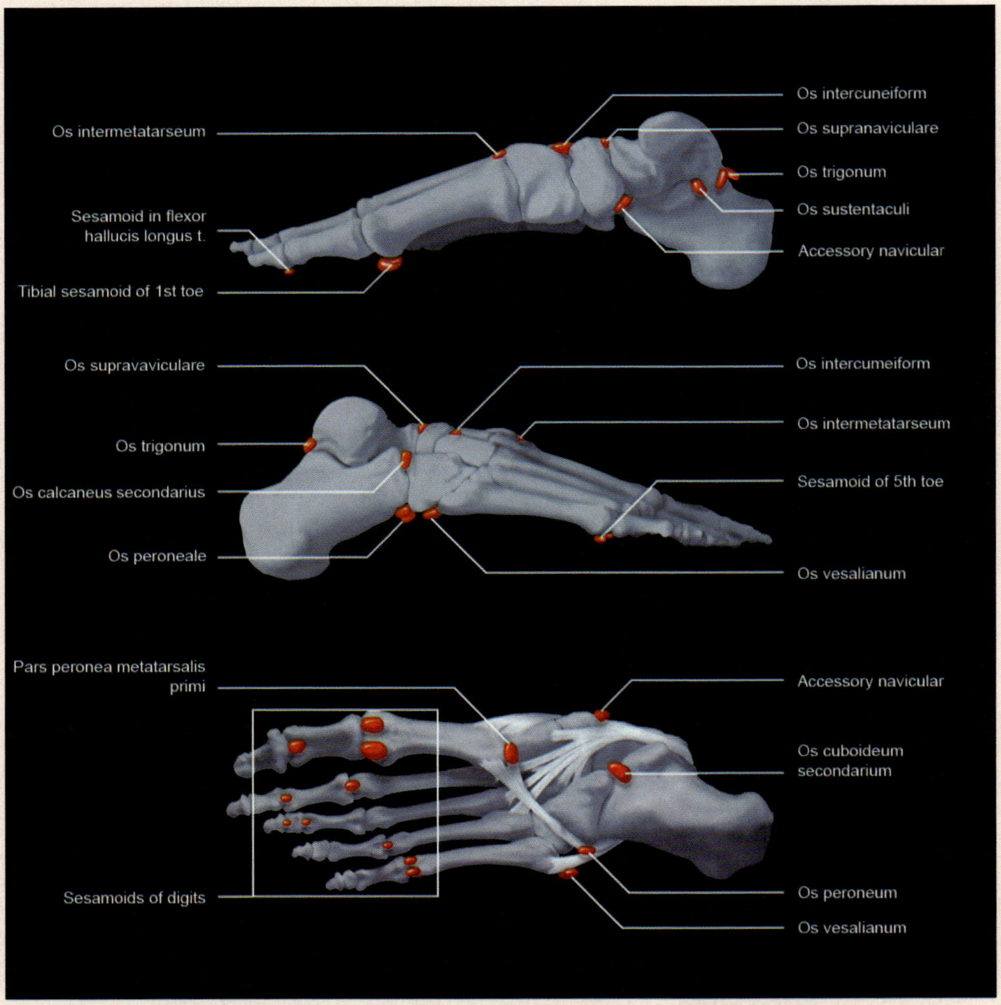

Graphic shows the most common accessory centers of ossification and sesamoids of the foot. The ossicles are constant in position, which aids in distinguishing an ossicle from a fracture. A chronic, nonunited fracture fragment is corticated on all surfaces, mimicking an ossicle. Even when chronic, fractures tend to have jagged margins, which are useful in distinguishing them from ossicles that are smoothly marginated. Some accessory ossicles in the foot can be symptomatic (most commonly the os trigonum, the accessory navicular, and the medial malleolar accessory ossification center).

ACCESSORY OSSICLES, ANKLE AND FOOT

TERMINOLOGY

Definitions

- **Accessory center of ossification**: Variably present ossific center associated with bone
- **Synchondrosis**: Cartilage bridge between 2 bones
  - Commonly found between accessory center of ossification and parent bone
- **Sesamoid**: Ossicle arising within tendon, improving its mechanical advantage

IMAGING

General Features

- Location
  - **Medial malleolus accessory center**
    - At anterior colliculus of medial malleolus
    - Usually unites with tibia at skeletal maturity
  - **Os subfibulare**
    - At anteroinferior margin of fibula
    - Extends inferior to normal position of fibular tip
  - **Os trigonum**
    - At posterior process of talus
    - "Teardrop" shape, apex anterior
  - **Os sustentaculi**
    - At posterior margin of sustentaculum tali
    - Rare
  - **Os calcis secondarius**
    - Adjacent to anterior process of calcaneus
    - On dorsal aspect of foot
    - Triangular or trapezoidal shape
    - Most supposed ossicles in this location are probably nonunited chip fractures
  - **Os supranaviculare**
    - At dorsal, proximal margin of navicular
    - Most supposed ossicles in this location are probably nonunited chip fractures
  - **Type I accessory navicular**
    - Sesamoid in posterior tibial tendon, near attachment to navicular
  - **Type II accessory navicular (a.k.a. os tibiale externum)**
    - At plantar margin of median eminence of navicular
    - Forms flat facet with parent navicular
    - Posterior tibial t. inserts primarily on ossicle
  - **Type III accessory navicular (a.k.a. cornuate navicular**)
    - At median eminence of navicular
    - Assimilated accessory of ossification
    - Enlarged median eminence (cornuate navicular)
  - **Cuboides secondarium**
    - Proximal, medial margin of cuboid
    - Between cuboid and navicular
    - On plantar aspect of foot
    - Trapezoidal shape
  - **Os peroneum (os peroneal)**
    - Sesamoid in peroneus longus tendon, occurs near lateral margin of cuboid
    - Often bipartite or multipartite
    - Periostitis often seen
      - Due to DISH, psoriatic or reactive arthritis, or tendinopathy
  - **Os intercuneiform**
    - Between 1st and 2nd cuneiforms
    - On dorsal aspect of foot
    - Thin "teardrop" shape
  - **Pars peronea metatarsalis primi**
    - Between base of 1st metatarsal and 1st cuneiform
    - On plantar aspect of foot
  - **Os vesalianum**
    - Accessory ossification center for styloid process at base of 5th metatarsal
    - Ovoid, larger than normal styloid process
    - Synchondrosis is horizontally oriented
      - Normal apophysis in this region is vertically oriented
  - **Sesamoids of 1st metatarsophalangeal joint**
    - Beneath 1st metatarsal head
    - Each sesamoid has facet on its dorsal margin
      - Sesamoid facets articulate with facets at plantar aspect of 1st metatarsal head
      - Part of 1st metatarsophalangeal joint
    - Medial sesamoid in tendons of flexor hallucis brevis (medial head) and abductor hallucis
    - Lateral sesamoid in tendons of flexor hallucis brevis (lateral head) and adductor hallucis
    - Medial sesamoid rarely congenitally absent
    - Sesamoid sclerosis may indicate chronic stress
  - **Other sesamoids of toes**
    - Variably present at metatarsophalangeal and interphalangeal joints
  - **Os intermetatarseum**
    - Between 1st and 2nd metatarsal bases, at dorsal margin
    - Thin "teardrop" shape
- Morphology
  - Accessory center of ossification
    - Closely apposed to parent bone
    - Usually has flat facet where it forms synchondrosis with parent bone
  - Sesamoid
    - Usually ovoid or round
    - May form part of synovial joint; may have 1 or more articular facets
    - If bipartite, both parts usually ovoid and do not fit closely together

Radiographic Findings

- Characteristic location and shape for each ossicle
- Entire ossicle has smooth, corticated margins
- Ossicle + parent bone usually larger than if accessory center not present
- If synchondrosis injured, accessory center may be displaced relative to parent bone

CT Findings

- Shows cortication of margins better than radiographs
  - Aids in differentiation between ossicle and fracture

MR Findings

- Ossicle usually shows fatty bone marrow signal intensity
  - May be difficult to see because marrow looks like adjacent soft tissue fat
  - Always correlate with radiographs
- Bone marrow edema is important sign of injury to ossicle
  - Fracture through ossicle

- Injury to synchondrosis between ossicle and parent bone
  - Often seen in type II accessory navicular or accessory center for medial malleolus
- Impingement of ossicle
  - Often seen in os trigonum
- Sclerosis
  - Chronic stress response
  - Low signal intensity on all sequences
- Normal synchondrosis shows cartilage signal intensity
  - Low signal intensity on T1WI, intermediate on T2WI FS
- Injured synchondrosis likely to fluid between ossicle and parent bone
  - No data found to indicate whether fluid may normally be found between os and parent bone
    - Author believes that a separate os need not have a synchondrosis, suggesting that fluid between them can be normal
    - If there is a question of normal separate os vs. synchondrosis disruption, rely on presence or absence of edema on fluid-sensitive sequence to differentiate

Imaging Recommendations

- Best imaging tool
  - Common incidental finding on radiographs
  - MR useful to identify signs of injury, impingement

Nuclear Medicine Findings

- Bone scan
  - Increased radionuclide activity at accessory ossicle is sign of ossicle injury
  - Superseded by MR, which gives more information about ossicle injury and associated soft tissue injuries

DIFFERENTIAL DIAGNOSIS

Fracture

- Located along lines of force
- Jagged margins, acute angles at corners
- If acute, margins are not corticated
- Fracture fragments fit together like pieces of jigsaw puzzle

Pseudoepiphysis (Accessory Epiphysis)

- Complete or incomplete accessory epiphysis of metatarsal or phalanx
  - Incomplete accessory epiphysis looks like cleft in bone
- Occurs at opposite end of bone from epiphysis
  - e.g., distal end of 1st metatarsal, proximal end of 2nd to 5th metatarsals
- Fuses at skeletal maturity

CLINICAL ISSUES

Presentation

- Most common signs/symptoms
  - Usually asymptomatic but several causes of symptoms
    - Impingement
    - Synchondrosis injury
    - Stress or traumatic fracture
  - Os trigonum
    - Pain, limited motion with extreme plantar flexion
    - Flexor hallucis longus impingement
  - Type II accessory navicular
    - Synchondrosis injury
    - Posterior tibial tendinosis
    - Posterior tibial tendon tear
  - Type III accessory navicular
    - Impingement with certain footwear (e.g., ski boots)
  - Accessory center, medial malleolus
    - Usually painful at ages 10-12
  - Os vesalianum
    - Impingement against shoes if large
    - Pain due to pull of peroneus brevis chronically injuring synchondrosis
  - Os peroneum
    - May impinge against cuboidal sulcus
    - Proximal displacement indicates peroneus longus rupture
  - Sesamoids of 1st toe
    - Commonly show arthritis when patient has osteoarthritis or rheumatoid arthritis
    - Proximal displacement indicates rupture of sesamophalangeal apparatus
  - Sesamoids may be painful 2° to stress or acute fx

DIAGNOSTIC CHECKLIST

Consider

- Type I accessory navicular not cause of tendon abnormalities
- Type II accessory navicular associated with posterior tibial tendon dysfunction

Image Interpretation Pearls

- Many ossicles are bipartite or multipartite
- Smooth, rounded contour is classic morphology
  - Some ossicles have flat articular facets
  - Some ossicles have "teardrop" shape

Reporting Tips

- Prudent to mention accessory ossicles in report to prevent their being misinterpreted as fractures

SELECTED REFERENCES

1. Cohen BE: Hallux sesamoid disorders. Foot Ankle Clin. 14(1):91-104, 2009
2. Nakasa T et al: Painful os intermetatarseum in athletes: report of four cases and review of the literature. Arch Orthop Trauma Surg. 127(4):261-4, 2007
3. Chao W: Os trigonum. Foot Ankle Clin. 9(4):787-96, vii, 2004
4. Ugolini PA et al: The accessory navicular. Foot Ankle Clin. 9(1):165-80, 2004
5. Kono T et al: Symptomatic os subfibulare caused by accessory ossification: a case report. Clin Orthop Relat Res. (399):197-200, 2002
6. Hodge JC: Anterior process fracture or calcaneus secundarius: a case report. J Emerg Med. 17(2):305-9, 1999
7. Maffulli N et al: Traumatic lesions of some accessory bones of the foot in sports activity. J Am Podiatr Med Assoc. 80(2):86-90, 1990
8. Ogden JA et al: Accessory ossification patterns and injuries of the malleoli. J Pediatr Orthop. 10(3):306-16, 1990

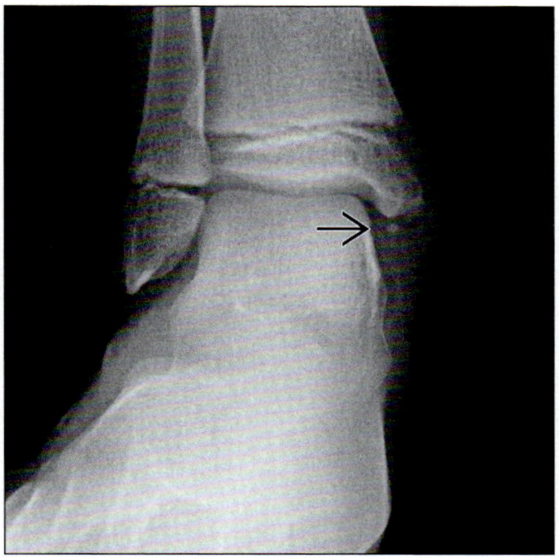

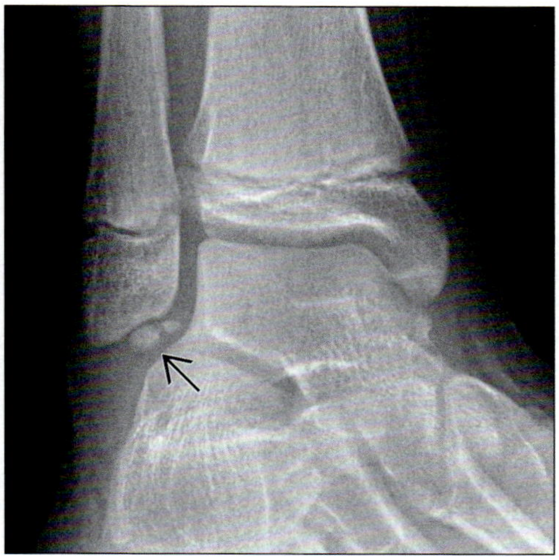

(Left) *Anteroposterior radiograph shows a small accessory center of ossification for the medial malleolus ⇨ involving the anterior colliculus. It may be a source of medial ankle pain in adolescents.* ***(Right)*** *Mortise view radiograph shows a bipartite os subfibulare ⇨ at the anterior margin of the fibular tip. The adjacent margin of the fibula is concave. The os subfibulare may grow to be quite large, resulting in lateral impingement of the ankle.*

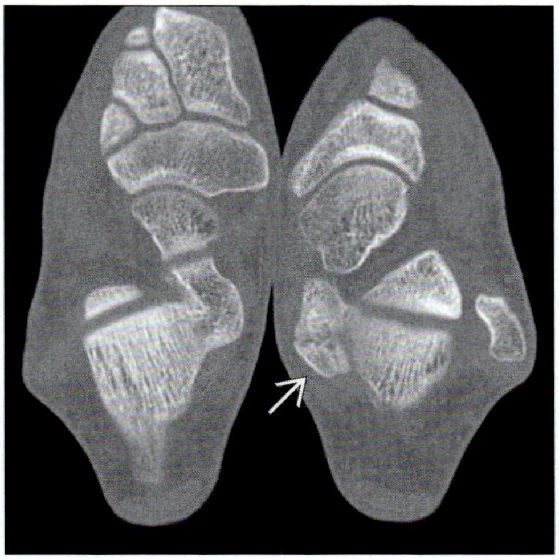

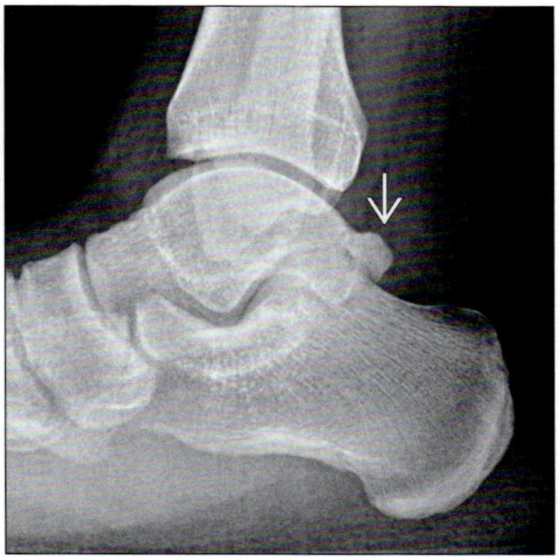

(Left) *Axial bone CT shows an os sustentaculi at the posterior margin of the sustentaculum tali ➡. This ossicle is rare and may result in tarsal tunnel syndrome.* ***(Right)*** *Lateral radiograph shows an os trigonum ➡, a separate ossification center for posterior process of talus. A small synchondrosis unites it with the parent navicular. A fracture of the posterior process is easily mistaken for an os trigonum, but the fracture line should be more distinct than is seen here.*

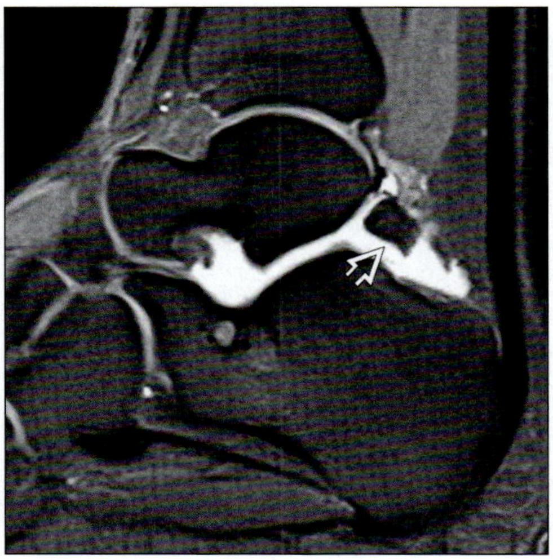

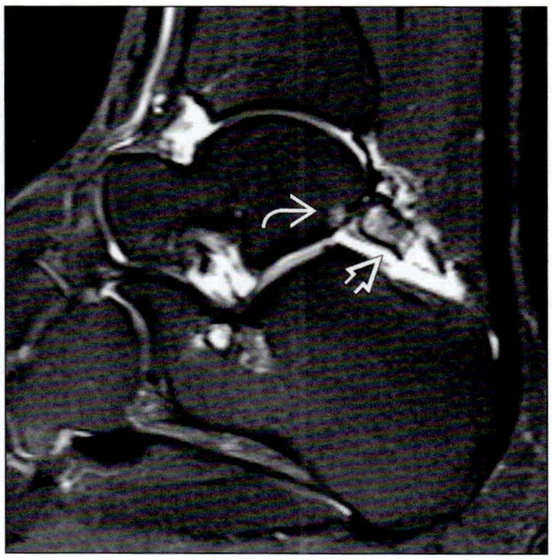

(Left) *Sagittal T1WI FS MR arthrogram shows an os trigonum ➡. Contrast has been injected into the posterior subtalar joint. Contrast extending between ossicle and parent talus may indicate disruption of the synchondrosis. No data is available defining whether fluid may normally extend between the talus and a separate os.* ***(Right)*** *Sagittal T2WI FS MR arthrogram, same patient, shows edema in the os ➡ and adjacent talus ➡, reflecting injury. This patient presented with posterior ankle impingement.*

(Left) *Oblique radiograph shows a rounded os calcis secundarium at the anterior process of the calcaneus. The bifurcate ligament (between calcaneus, navicular, and cuboid) attaches to this ossicle.* ***(Right)*** *Lateral radiograph shows a triangular os supranaviculare at the proximal margin of the navicular. Os supranaviculare is less common than fracture of the navicular rim.*

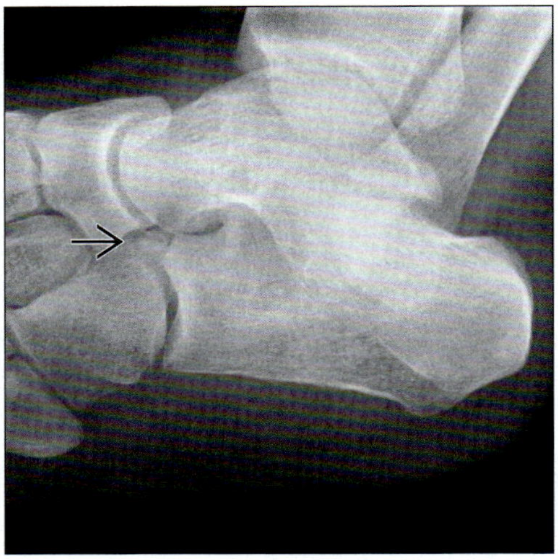

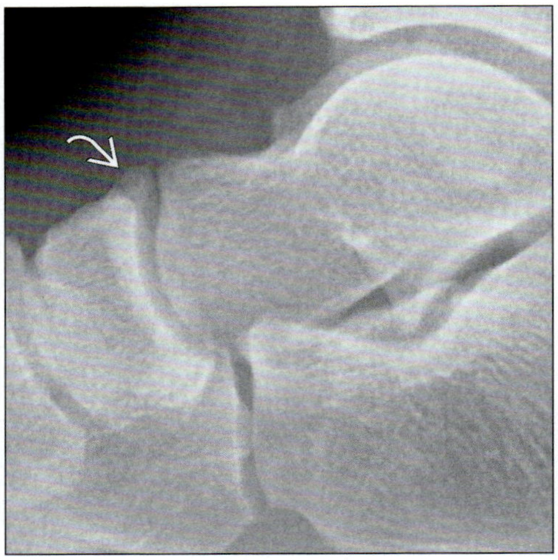

(Left) *Coronal PD FSE MR shows the rare os cuboideum. Like the os calcis secundarium, it is adjacent to the calcaneocuboid joint, but unlike the os calcis secundarium, it is at the plantar aspect of the foot. It has an angular, somewhat trapezoidal shape.* ***(Right)*** *Axial PDWI FS MR shows the trapezoidal os cuboideum lying between the calcaneus and the navicular.*

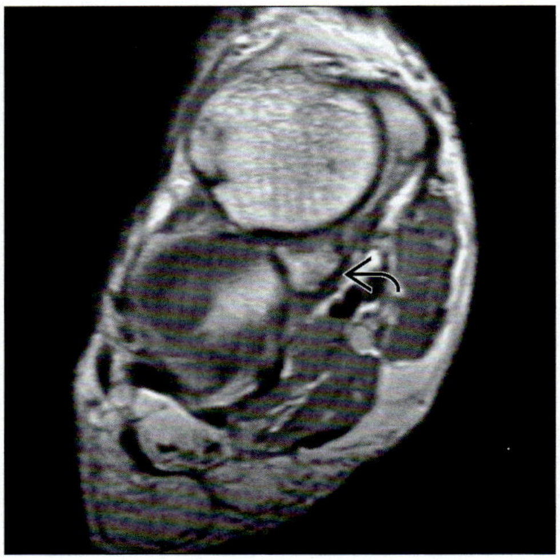

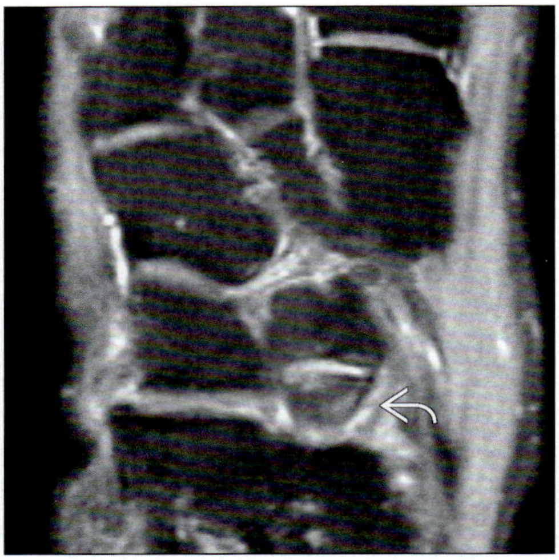

(Left) *Lateral radiograph shows the plantar position of type II accessory navicular. The narrow, lucent cleft between the ossicle and parent navicular represents the synchondrosis.* ***(Right)*** *Sagittal T2WI MR shows a large type II accessory navicular adjacent to the insertion of the tibialis posterior tendon on the navicular. Fluid in the synchondrosis between the navicular and the native navicular is usually associated with chronic repetitive stress injury at this site.*

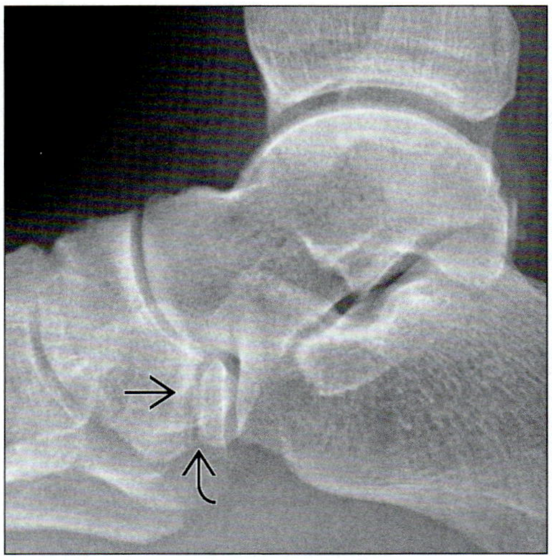

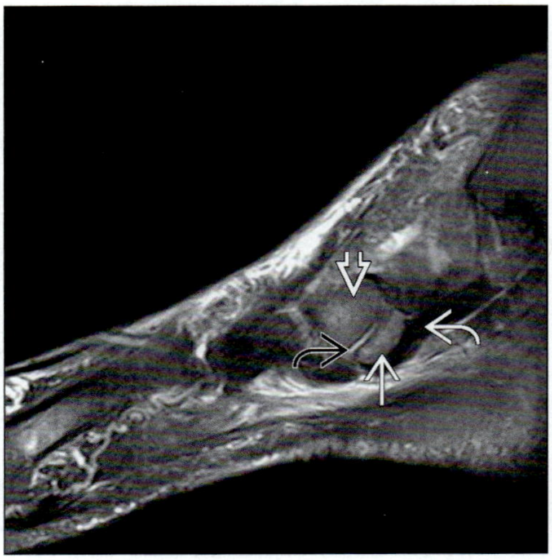

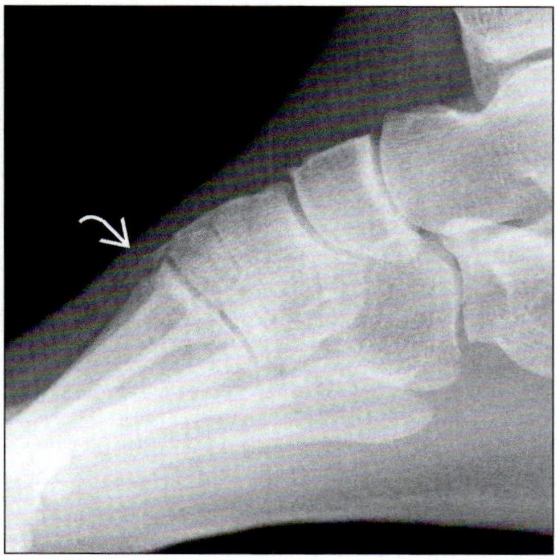

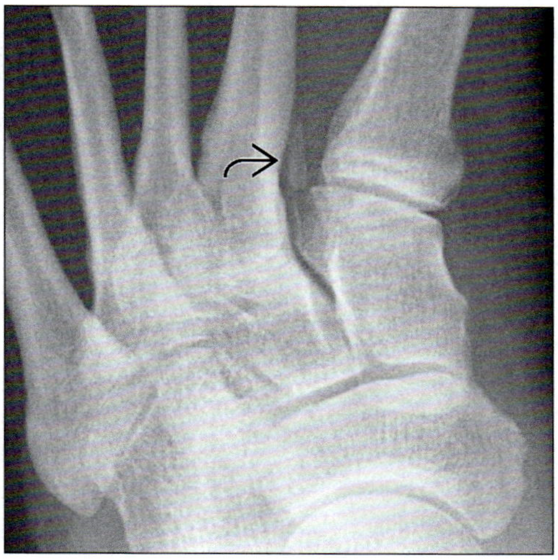

(Left) *Lateral radiograph shows a characteristic thin, teardrop-shaped os intermetatarseum. The alignment of the 1st and 2nd tarsometatarsal joints is normal.* ***(Right)*** *Anteroposterior radiograph shows an os intermetatarseum between the bases of the 1st and 2nd metatarsals. This ossicle is not reported to cause symptoms and is sometimes mistaken for an avulsion fracture due to Lisfranc ligament avulsion.*

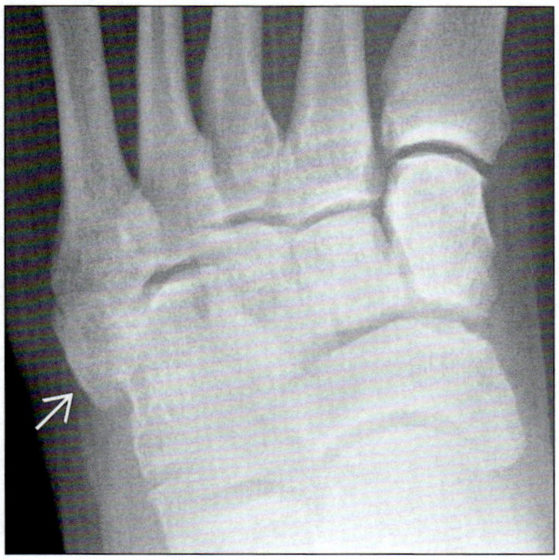

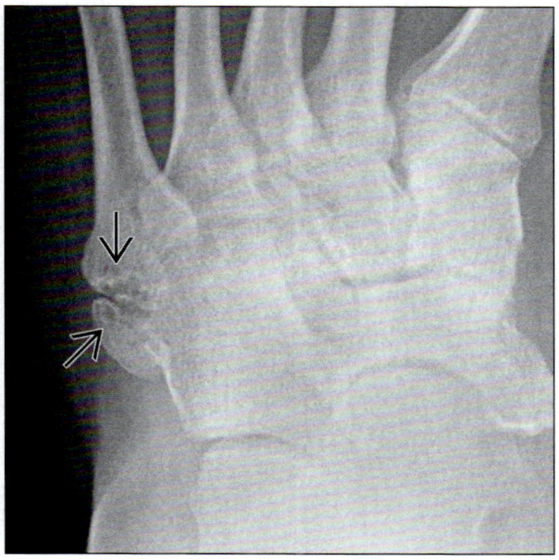

(Left) *Anteroposterior radiograph shows a large os vesalianum, markedly larger than the normal styloid process.* ***(Right)*** *Anteroposterior radiograph shows cysts centered at the synchondrosis between an os vesalianum and 5th metatarsal. These are a common sign of motion at the synchondrosis and often are associated with a painful synchondrosis.*

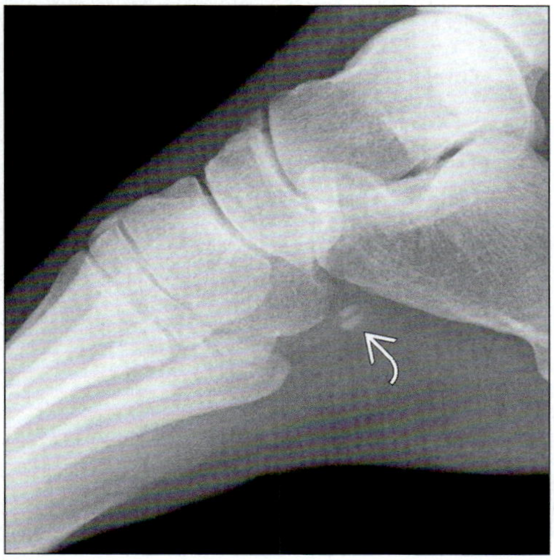

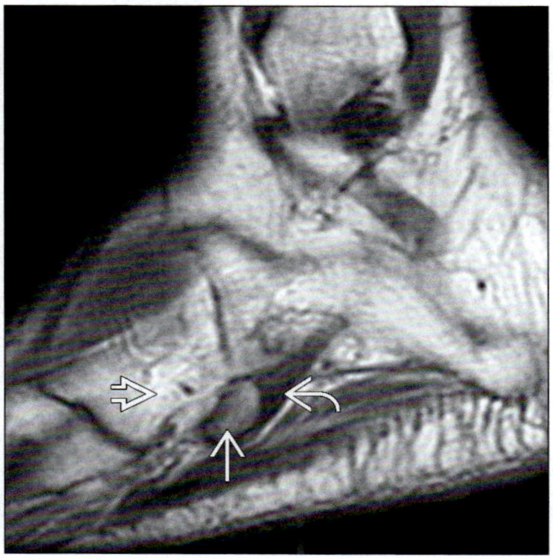

(Left) *Lateral radiograph shows a bipartite os peroneum (a.k.a. os peroneal), a sesamoid in the peroneus longus tendon. This ossicle may become displaced with peroneus longus rupture and is often involved with calcific tendinopathy and DISH.* ***(Right)*** *Sagittal T1WI MR demonstrates a corticated ossicle within the peroneus longus tendon, plantar and lateral to the cuboid, consistent with an os peroneum.*

PILON FRACTURE

Key Facts

Terminology

- Tibial plafond: Horizontally oriented, weight-bearing articular surface of distal tibia
- Pilon fracture: Axial load injury of tibial plafond

Imaging

- Radiographs usually underestimate fracture severity
  - Oblique fracture involving plafond
  - Fracture may be visible on only 1 view if nondisplaced
- CT scan shows true extent of injury

Pathology

- Axial load, ± twisting, valgus, varus
- 75-85% have fracture of fibula
- 30-50% have other fractures due to axial load

Clinical Issues

- Reported nonunion rates vary from 4-36%
- 50% of patients have arthritis 2 years after injury

Diagnostic Checklist

- CT report should include
  - Number & location of major fragments at level of joint
  - Amount, location of articular depression
  - Associated tendon injury or entrapment (use soft tissue algorithm and window)
  - Disruption of syndesmosis, bone fragments in syndesmosis
- Be careful to look for other axial load injuries in ankle and foot
- Flake-like fractures of medial or lateral malleoli suggest associated retinacular avulsion

*(**Left**) Frontal graphic shows longitudinal fracture ➔ originating in the lateral diaphysis and developing extensive comminution at the joint surface. The typical associated fracture of the distal fibular shaft ➔ is also present. (**Right**) Axial bone CT shows common pilon pattern of 3 major articular fragments: Anterolateral ➔, anteromedial ➔, and posterior ➔. As is typically seen, small, depressed, comminuted fragments are located centrally.*

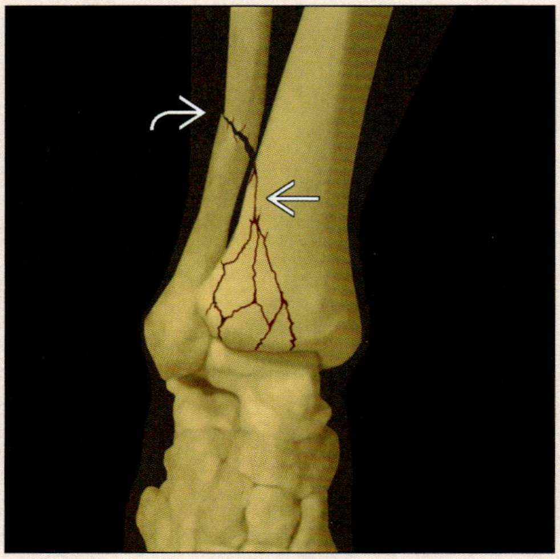

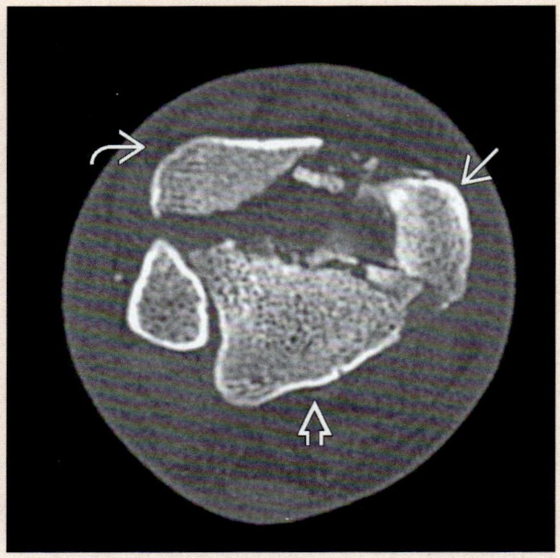

*(**Left**) AP bone CT 3D reformation shows a high-energy pilon fracture with central depression ➔ and diaphyseal-metaphyseal dissociation ➔. Fibular fracture is also comminuted. (**Right**) Sagittal soft tissue algorithm CT in the same patient shows the posterior tibial tendon (PTT) ➔ trapped between bone fragments. This information is important in surgical planning so that the PTT can be extricated and protected.*

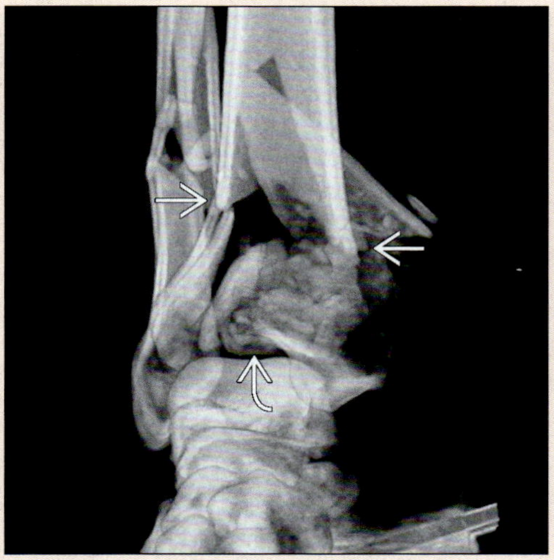

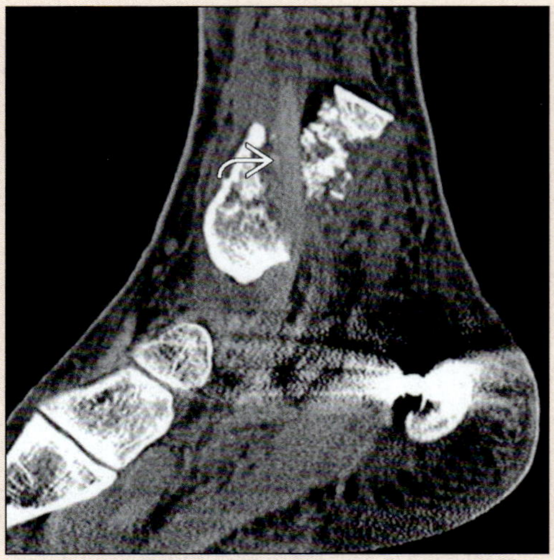

PILON FRACTURE

TERMINOLOGY

Definitions

- Tibial plafond: Horizontally oriented, weight-bearing articular surface of distal tibia
- Pilon fracture: Axial load injury involving tibial plafond

IMAGING

General Features

- Best diagnostic clue
  - Longitudinally oriented fracture of distal tibia on radiographs
    - Always involves articular surface

Radiographic Findings

- Longitudinally oriented fracture extends to articular surface
  - Fracture may be visible on only 1 view if nondisplaced
  - If fractures extend through both medial and lateral cortices of tibia, create metaphyseal-diaphyseal dissociation
- Usually comminuted
  - Radiographs underestimate degree of articular involvement, comminution
- ± articular depression
  - Tibia may be shortened relative to fibula
  - Estimate degree of depression of articular surface on all 3 ankle views
  - Talar dome may be interposed between tibial plafond fragments
- Syndesmosis and fibula are often injured
  - 75-85% have associated distal fibular fracture
  - ± widened tibiofibular syndesmosis
- Radiographs may show injury to retinacula at ankle
  - Flake of bone at lateral margin of distal fibula: Peroneal retinaculum avulsion
  - Flake of bone at medial margin of distal tibia: Flexor retinaculum or superficial deltoid avulsion

CT Findings

- Articular surface most commonly divided into 2-3 major fragments ± bony debris
  - e.g., anteromedial, anterolateral, posterior
  - Location of these fragments important in planning surgical fixation
- Extent of comminution, depression often greater than expected on radiographs
- "Die punch" fragment of centrally depressed articular surface often present
  - Often rotated about 90°, displaced into medullary cavity
- Injury of tibiofibular syndesmosis may be more significant than expected from radiographs
  - Fibula should fit snugly in fibular notch of tibia; separation implies syndesmotic rupture
  - Syndesmotic ligaments may be difficult to see due to surrounding hematoma
- Soft tissue injuries seen that may be clinically unsuspected
  - Entrapment of posterior tibial or flexor digitorum tendons between fracture fragments
  - Avulsion of flexor or peroneal retinaculum

MR Findings

- MR not indicated in acute management
- Slightly superior visualization of soft tissue injuries compared to CT
- Shows cartilage injuries not apparent on CT scan
  - Cartilage injuries usually evaluated at time of surgery
- May show associated talar dome injury (chondral or osteochondral fracture)

Imaging Recommendations

- Best imaging tool
  - CT scan with multiplanar reformations

DIFFERENTIAL DIAGNOSIS

Fractures of Posterior or Medial Malleoli

- Due to twisting injury rather than axial load
- May extend into tibial plafond
  - Spare majority of weight-bearing surface
- Look for associated injuries to indicate twisting vs. axial load mechanism

Tibial Shaft Fracture

- Intraarticular extension of distal tibial shaft fracture may be inapparent on radiographs
  - Changes surgical management
- If extension to joint is suspected, CT scan is recommended

Ankle Dislocation

- May occur due to any injury mechanism, including axial load
- Pre-reduction films show joint incongruity
- If only post-reduction films available, need to know history

Triplane Fracture

- Twisting injury occurring with partially fused growth plate, Salter-Harris type 4
- Sagittally oriented fracture line epiphysis, horizontal fracture through nonfused physis, oblique coronal fracture through metaphysis

Tillaux Fracture

- a.k.a. Tillaux-Chaput fracture
- Avulsion of anterolateral corner of tibial plafond by anterior tibiofibular ligament
- Usually seen in children as Salter-Harris 3 fracture (juvenile Tillaux fracture)

Earle Fracture

- a.k.a. Volkmann fracture
- Avulsion posterolateral corner of tibial plafond by posterior tibiofibular ligament
- Usually considered as type of posterior malleolar fracture

PATHOLOGY

General Features

- Etiology
  - Axial load, ± twisting, valgus, varus
    - Fall from height

PILON FRACTURE

- Motor vehicle accident with sudden stop (↑ incidence with airbags)
- Fall forward with trapped foot
- Associated abnormalities
  - Fibular fracture (75-85% of pilon fractures)
  - Ankle dislocation
  - Other fractures due to axial load (30-50% of pilon fractures)
    - Talus, calcaneus, midfoot, tibial shaft or plateau, distal femur, pelvis, spine, skull
    - Fractures of contralateral limb
  - Tibial shaft fracture may have unsuspected extension to articular surface
    - Intraarticular extension alters management
    - Most important reason to perform CT scan for distal tibial shaft fractures
  - Cartilage injury to plafond and talar dome
    - Usually evaluated at time of open reduction internal fixation (ORIF)
    - MR not indicated in acute setting but may be used for assessment if clinical outcome is poor
  - Disruption of tibiofibular syndesmosis
  - Entrapment of extrinsic flexor tendons of foot
  - Avulsion of flexor or superior peroneal retinaculum
  - Degloving injury to soft tissues
  - Compartment syndrome
    - Signs: Pain > expected, sensory changes in 1st webspace, weak toe dorsiflexion

Staging, Grading, & Classification

- Ruedi-Allgöwer classification
  - Type I: Plafond fracture without significant displacement (low energy)
  - Type II: Plafond fracture with minimal articular incongruity
  - Type III: Plafond fracture with significant comminution and articular depression (high energy)
- Other important considerations
  - Metaphyseal-diaphyseal dissociation
  - Spiral extension into tibial shaft
  - Ligamentous injuries, entrapment of tendons

CLINICAL ISSUES

Presentation

- Clinical profile
  - Most common: Male adult in 3rd to 5th decade

Demographics

- Epidemiology
  - 3-10% of tibial fractures
  - < 1% of lower extremity fractures

Natural History & Prognosis

- 50% of patients have arthritis 2 years after injury
  - 5% ultimately receive arthrodesis
- Ankle stiffness common
- Entrapment of posterior tibial tendon into fracture may not be seen at time of ORIF
  - Current surgical techniques use small incisions, carefully placed for optimal reduction and minimal soft tissue compromise
  - Posteromedial corner of tibia often not exposed
- Unrecognized tendon entrapment leads to tendon dysfunction, incomplete healing of fracture
- CT scan may be performed after ORIF if anatomic alignment not achieved
  - Malreduction may be due to fragment separating syndesmosis
  - Articular surface may be separated by rotated fragment in medullary canal
  - Hardware may be malpositioned
  - Tendon injury or subluxation may be visible

Treatment

- Options, risks, complications
  - Internal or external fixation usually performed
  - 5-10% develop skin necrosis
    - Skin over anteromedial tibia has poor vascularity
  - Postoperative infection (8-20%)
  - Nonunion rates 4-36% in different series

DIAGNOSTIC CHECKLIST

Consider

- Always evaluate for concomitant soft tissue injury

Image Interpretation Pearls

- Look for other axial load injuries in ankle and foot
- Evaluate contralateral ankle and foot for unsuspected injuries
- Flake-like fractures of medial or lateral malleoli suggest associated retinacular avulsion

Reporting Tips

- On CT scan, report
  - Number of major fragments at level of joint
  - Amount, location of articular depression
    - > 2-3 mm is significant
  - Associated tendon injury or entrapment (use soft tissue algorithm and window)
  - Disruption of syndesmosis
- On follow-up images, report
  - Shortening of tibia &/or fibula
  - Congruity of joint surface
  - Fracture healing
  - Hardware complications
  - Osteoarthritis

SELECTED REFERENCES

1. Tarkin IS et al: An update on the management of high-energy pilon fractures. Injury. 39(2):142-54, 2008
2. Chen SH et al: Long-term results of pilon fractures. Arch Orthop Trauma Surg. 127(1):55-60, 2007
3. Topliss CJ et al: Anatomy of pilon fractures of the distal tibia. J Bone Joint Surg Br. 87(5):692-7, 2005
4. Tornetta P 3rd et al: Axial computed tomography of pilon fractures. Clin Orthop Relat Res. (323):273-6, 1996
5. Ruwe PA et al: Pilon fractures of the distal tibia. Orthop Rev. 22(9):987-96, 1993

PILON FRACTURE

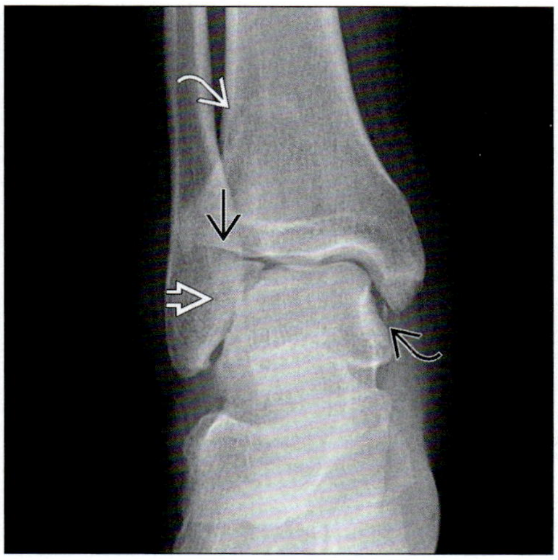

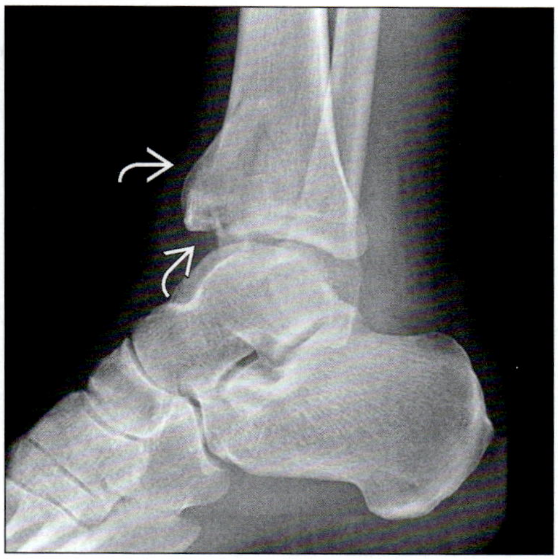

(Left) *Anteroposterior radiograph shows slight articular depression. Extension of fracture to the lateral metaphysis is evident. There is an associated transchondral fracture of the talar dome and an avulsion of the inner surface of the medial malleolus, both indicating valgus stress in addition to axial load.* ***(Right)*** *Lateral radiograph in the same patient shows a pilon fracture primarily involving the anterior tibia.*

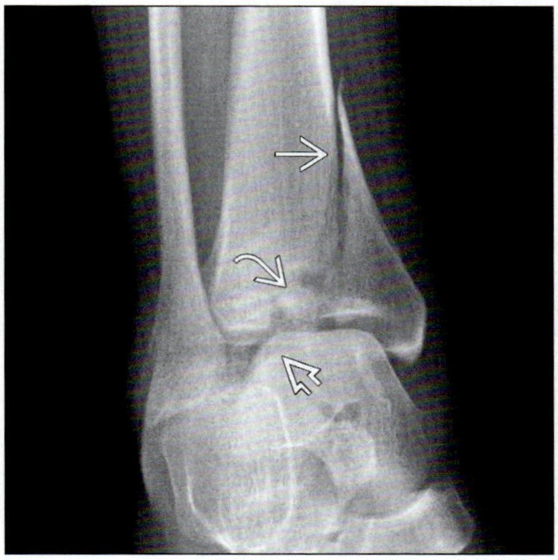

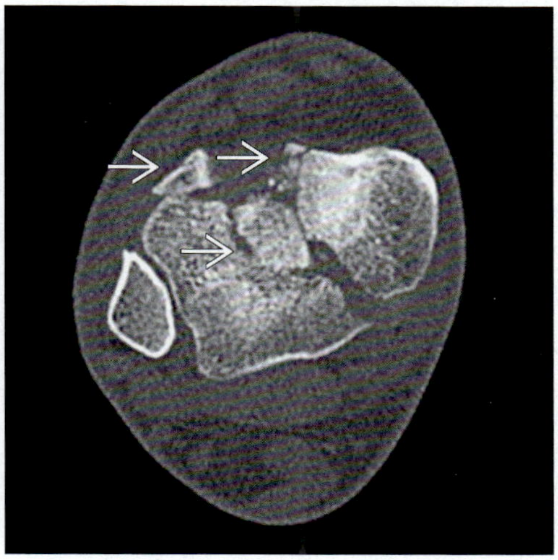

(Left) *Mortise view radiograph shows fracture originating at medial tibial diaphysis, resulting in a single large medial fragment, central comminution that includes a depressed "die punch" fragment, and a large lateral fragment. Osteochondral fracture of the lateral talar dome is difficult to see.* ***(Right)*** *Axial bone CT in the same patient shows 2 main fragments at the level of the articular surface of the plafond, as well as smaller comminuted fragments centrally.*

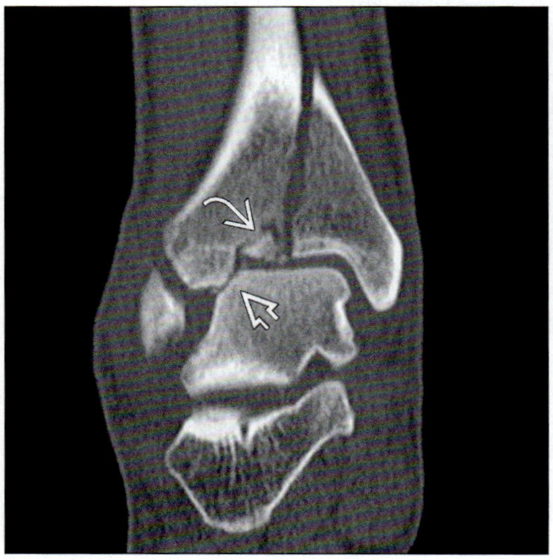

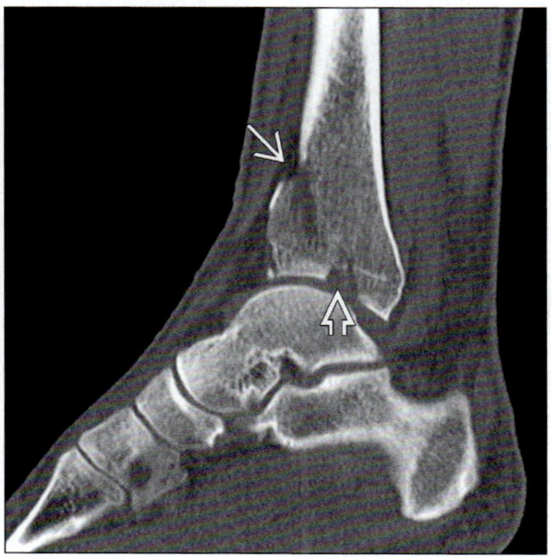

(Left) *Coronal bone CT from the same patient shows the size of the centrally located articular fragment and the degree of depression of the medial and central tibial plafond relative to the lateral fragment. The talar dome is impacted on the lateral tibial fragment.* ***(Right)*** *Sagittal bone CT in the same patient shows the size of the central articular defect and the orientation of the fracture through the anteromedial tibia.*

MALLEOLAR FRACTURE

Key Facts

Terminology

- Fractures of distal tibia and fibula are divided into
  - Pilon: Due to axial load
  - Malleolar: Due to shear or twisting forces

Imaging

- Mortise view shows fractures better than AP view
- If isolated fracture of medial or posterior malleolus, suspect Maisonneuve fracture of proximal fibula
- Lateral view underutilized
  - Best shows posterior malleolus fracture
  - Often shows Weber B & C fibular fractures well
  - Some LM and PM fractures only seen on lateral

Top Differential Diagnoses

- Pilon fracture
- Ankle dislocation
- Ankle sprain
- 5th metatarsal styloid process fracture
- Lateral process talus fracture
- Osteochondral injuries, ankle
- Accessory ossicles, ankle and foot

Pathology

- Weber classification
  - Type A fracture: Fibular fracture below level of tibial plafond
  - Type B fracture: Fibular fracture at level of tibial plafond
  - Type C fracture: Fibular fracture above level of tibial plafond

Diagnostic Checklist

- Isolated fracture of MM or PM indicates syndesmosis injury
  - Ankle is unstable
  - ± Maisonneuve fracture

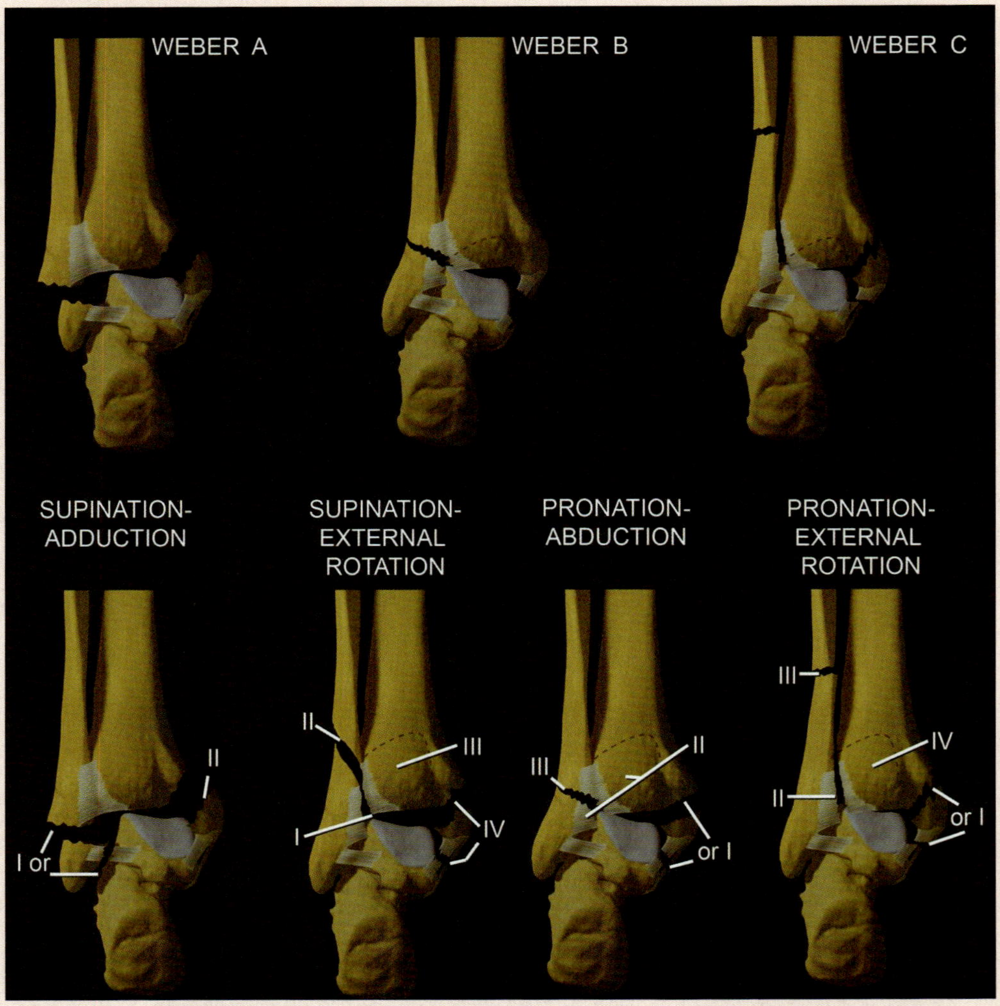

Graphics show the correspondence between Weber classification (top row) and Lauge-Hansen classification (bottom row). Note that a Weber type B fracture may be due to either supination-external rotation or pronation-abduction force. The numerals on the Lauge-Hansen graphics refer to the stage of the injury. Medial malleolar fractures are shown for Weber classification, but need not be present as classification is based on the level of the lateral malleolar fracture. Maisonneuve fractures are a type of Weber C or pronation-external rotation fracture.

MALLEOLAR FRACTURE

TERMINOLOGY

Definitions

- Fractures of distal tibia and fibula are divided into
  - Pilon fractures: Due to axial load
    - May secondarily involve malleoli
  - Malleolar fractures: Due to shear or twisting forces
- Maisonneuve fracture: Fracture of proximal fibula due to pronation injury of ankle
  - Reflects injury to tibiofibular syndesmosis and interosseous membrane
- Pott or Dupuytren fracture (obsolete): Fracture of distal fibula above syndesmosis
- Earle (Volkmann) fracture: Fracture of lateral margin of posterior malleolus (PM)
  - Due to avulsion of posterior inferior tibiofibular ligament (PITFL)
- Chaput-Tillaux fracture: Fracture of lateral margin of anterior tibia
  - Due to avulsion of anterior inferior tibiofibular ligament (AITFL)

IMAGING

General Features

- Morphology
  - Depends on mechanism of injury
    - Distraction force results in transverse fracture line or avulsion of ligamentous attachment(s)
    - Compression force results in oblique or comminuted fracture

Radiographic Findings

- Fractures often subtle, nondisplaced
- 3 view ankle series (with mortise) has higher accuracy than 2 view
  - Weber B fracture better seen on mortise view
- Lateral view underutilized
  - Best shows posterior malleolus fracture
  - Often shows Weber B & C fibular fractures well
  - Some lateral malleolar (LM) and PM fractures only seen on lateral
- Radiographic signs of deltoid ligament rupture
  - Best seen on stress or weight-bearing views
  - Medial gutter (a.k.a. clear space) wider than 4 mm
  - Lateral shift of talar dome
  - Talar tilt (i.e., loss of parallelism of articular surfaces)
- Signs of rupture of tibiofibular syndesmosis
  - Overlap between tibia and fibula < 1 mm on mortise view, < 1 cm on AP view
  - Displacement of fibula out of fibular notch of tibia
- Maisonneuve injury
  - Medial malleolus (MM) or PM fracture but no distal fibular fracture
  - Widened mortise without fracture
  - Proximal fibular fracture

CT Findings

- Occasionally useful to
  - Define extent of injury
  - Confirm fractures not definitively seen on radiograph

Imaging Recommendations

- Best imaging tool
  - Radiographs

DIFFERENTIAL DIAGNOSIS

Pilon Fracture

- Axial load injury
- Fracture of tibial metaphysis extending into tibial plafond
  - Usually see impaction, comminution of articular surface
- Posterior malleolar fractures also involve tibial plafond
  - But short, oblique fracture line limited to posterior plafond
- MM fractures may extend into medial margin of plafond

Ankle Dislocation

- Displacement of talus relative to tibia
- Need not see complete joint incongruity to diagnose dislocation
- Usually occurs in association with malleolar fractures

Ankle Sprain

- Injury to lateral collateral (ATFL, CFL, PTFL) ligaments
- Often have small chip fracture from avulsion of CFL

Malleolar Stress Fracture

- More common in LM, rare in MM
- Isolated fracture, incomplete

Salter-Harris Fractures

- Fracture involving growth plate in skeletally immature

Talus Fracture, Lateral Process

- Present with lateral ankle pain due to inversion injury

5th Metatarsal Styloid Process Fracture

- Area should be routinely included in ankle series

Osteochondral Injuries, Ankle

- Often subtle on radiographs

Accessory Ossicles, Ankle and Foot

- Os subfibulare
- Accessory MM ossification center

PATHOLOGY

General Features

- Associated abnormalities
  - Ankle ligament injuries
  - Avulsion of peroneal or flexor retinaculum

Staging, Grading, & Classification

- 2 major classification systems: Lauge-Hansen and Weber
- Lauge-Hansen classification
  - Based on mechanism of injury
  - 2 word descriptor for each type of fracture
    - 1st descriptor relates to force acting on talus
    - 2nd descriptor relates to position of foot at time of injury
  - 4 main types of fracture
    - Supination-adduction (SA)

MALLEOLAR FRACTURE

- Stage I: Transverse fracture of fibula below plafond (Weber type A) or rupture of lateral collateral ligaments
- Stage II: Stage I + vertical/oblique fracture of tibia
- May also have impaction fracture of medial tibial plafond
- Supination-external rotation (SER) = most common type of injury
  - Stage I: Rupture of AITFL
  - Stage II: Stage I + spiral fracture of fibula at level of plafond (Weber type B) or slightly above plafond (Weber type C)
  - Stage III: Stages I-II + rupture of PITFL or fracture of PM
  - Stage IV: Stages I-III + fracture of MM or rupture of deep deltoid ligament
- Pronation-abduction (PA)
  - Stage I: Fracture of MM or rupture of deltoid ligament
  - Stage II: Stage I + rupture of AITFL and PITFL or avulsion of tibial attachments (syndesmosis above plafond still intact)
  - Stage III: Stages I-II + fracture of distal fibula above plafond (Weber type B)
- Pronation-external rotation (PER)
  - Stage I: Fracture of MM or rupture of deltoid ligament
  - Stage II: Stage I + rupture of AITFL
  - Stage III: Stages I-II + oblique fracture of distal fibula above level of plafond
  - Stage IV: Stages I-II + rupture of PITFL or avulsion of PM
- Weber (properly known as AO-Danis-Weber) classification
  - Based on location of medial cortical involvement of fibular fracture
  - Type A: Fibular fracture below level of tibial plafond
    - Horizontal orientation
    - Corresponds to Lauge-Hansen SA injury
    - Unstable only if accompanied by MM fracture
  - Type B: Fibular fracture at level of tibial plafond (usually extending superiorly)
    - Most common type
    - Spiral orientation
    - Corresponds to Lauge-Hansen SER or PA
  - Type C: Fibular fracture above level of tibial plafond
    - Often but not always associated with injury of syndesmosis
    - Fracture here may also occur due to direct blow or as stress fracture
- Weaknesses of Weber classification
  - Does not address fractures of MM or PM
  - Type C not necessarily associated with unstable syndesmosis

Relevant Anatomy

- Ankle mortise: Pure hinge joint (mortise and tenon)
- Medial malleolus (MM)
  - Medial margin of distal tibia
- Lateral malleolus (LM)
  - Distal fibula from level of tibial plafond to inferior fibular tip
  - Anterior ligamentous attachments
    - Anterior inferior tibiofibular ligament (AITFL)
    - Anterior talofibular ligament (ATFL)
    - Calcaneofibular ligament (CFL)
  - Posterior ligamentous attachments
    - Posterior inferior tibiofibular ligament (PITFL)
    - Posterior talofibular ligament (PTFL)
    - Intermalleolar ligament (IML)
- Posterior malleolus (PM)
  - Less well demarcated than lateral and medial malleoli
  - Posterior lip of distal tibia
  - PITFL attaches here

CLINICAL ISSUES

Presentation

- Most common signs/symptoms
  - Unable to bear weight, focal bone tenderness over affected malleolus
- **Ottawa rules** help predict risk of fracture with ankle injury
  - Radiographs needed if patient has 1 of following
    - Focal bone tenderness at posterior margin of LM or MM, < 6 cm above tip
    - Focal tenderness at navicular or base of 5th MT
    - Inability to bear weight immediately and in ED

Treatment

- Closed reduction vs. ORIF depending on deformity & instability
- Restoration of fibular length & alignment of lateral malleolus key to ORIF
- Syndesmotic screws always loosen and are removed to prevent screw fracture
- Posterior malleolar fractures fixed if "large" (> 20-25% of articular surface)

DIAGNOSTIC CHECKLIST

Consider

- Ligament disruption is mechanical equivalent of malleolar avulsion

Image Interpretation Pearls

- Isolated fracture of MM or PM indicates disrupted syndesmosis and unstable ankle
  - ± Maisonneuve fracture

Reporting Tips

- Weber classification is standard

SELECTED REFERENCES

1. Gougoulias N et al: Supination-external rotation ankle fractures: stability a key issue. Clin Orthop Relat Res. 468(1):243-51, 2010
2. Jowett AJ et al: Medial malleolar stress fracture secondary to chronic ankle impingement. Foot Ankle Int. 29(7):716-21, 2008
3. Shariff SS et al: Lauge-Hansen classification--a literature review. Injury. 37(9):888-90, 2006
4. Kennedy JG et al: An evaluation of the Weber classification of ankle fractures. Injury. 29(8):577-80, 1998

MALLEOLAR FRACTURE

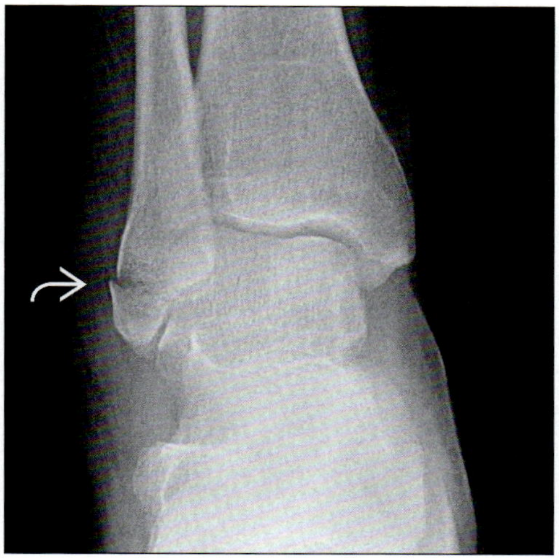

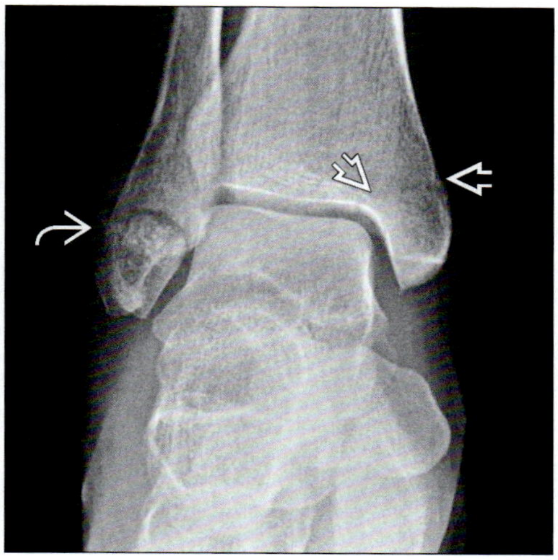

(Left) *AP radiograph shows a horizontal fracture of the lateral malleolus (LM) with no injury medially. This is a Weber type A fracture, or Lauge-Hansen supination-adduction I injury. Severe lateral soft tissue swelling is centered at the level of the fracture.* ***(Right)*** *Anteroposterior radiograph shows a Weber A fracture, below the level of the tibial plafond, and an oblique fracture of the medial malleolus (MM). This is a Lauge-Hansen II supination-adduction injury.*

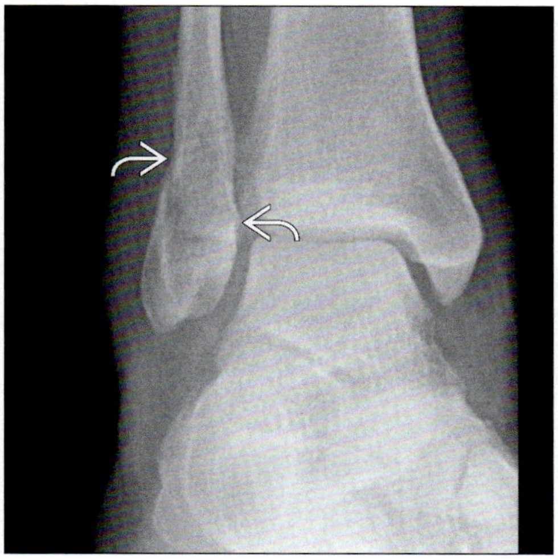

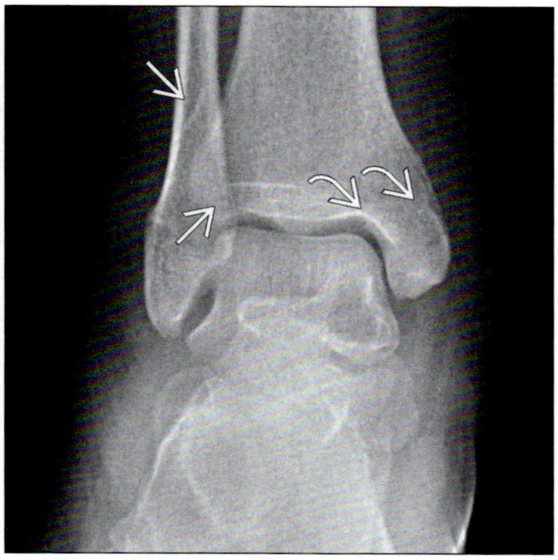

(Left) *Mortise view radiograph shows minimally displaced Weber B fracture. The fracture is classified as type B based on its level at the medial cortex of the fibula, as this is where the syndesmotic ligaments attach to the fibula. Laterally, the fracture always extends proximal to the tibial plafond.* ***(Right)*** *Anteroposterior radiograph shows Weber type B fracture, where medial and lateral malleolar fractures appear nondisplaced and are difficult to visualize.*

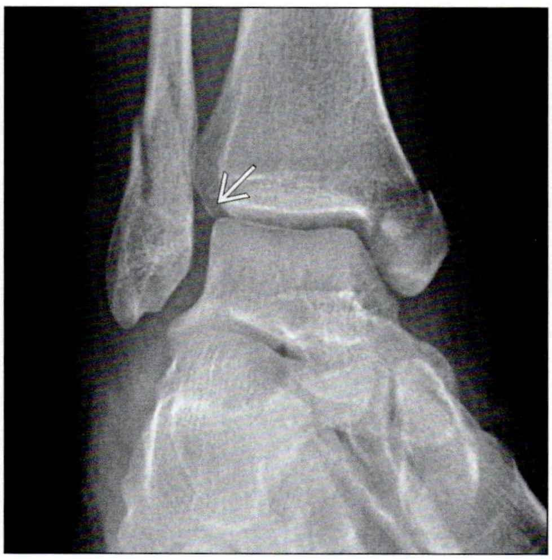

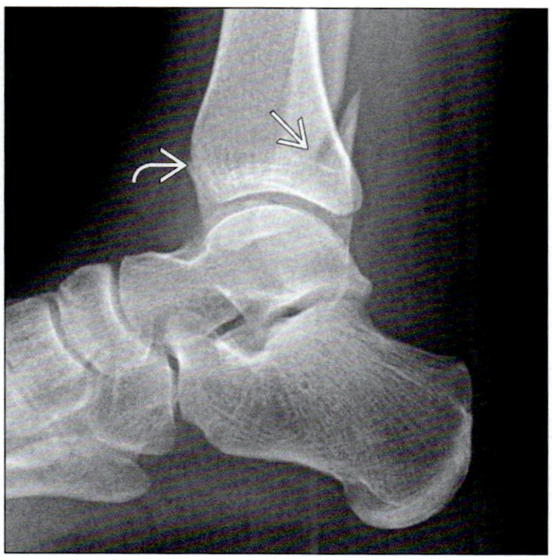

(Left) *Mortise view radiograph of the same patient better shows the bimalleolar fractures. In addition, a small Earle fragment is present. This is Lauge-Hansen SER III.* ***(Right)*** *Lateral radiograph of the same patient illustrates the Weber B fracture well. The lateral radiograph shows the Weber B fracture plane in profile, and is best for assessing healing on follow-up radiographs. The medial malleolar fracture is typically difficult to see on lateral radiographs.*

MALLEOLAR FRACTURE

(Left) *AP radiograph with eversion stress shows Weber B fibular fracture with widening of medial ankle gutter indicating deltoid ligament tear. Normal gutter width is < 4 mm; 4-6 mm is borderline, and > 6 mm is definitive for deltoid tear. Deltoid ligament tear may occur with Weber B or C fractures or Maisonneuve fracture (a type of Weber C fx).* **(Right)** *AP eversion stress radiograph of a Weber C (LH PER IV) fracture shows a widened medial gutter due to deltoid ligament injury.*

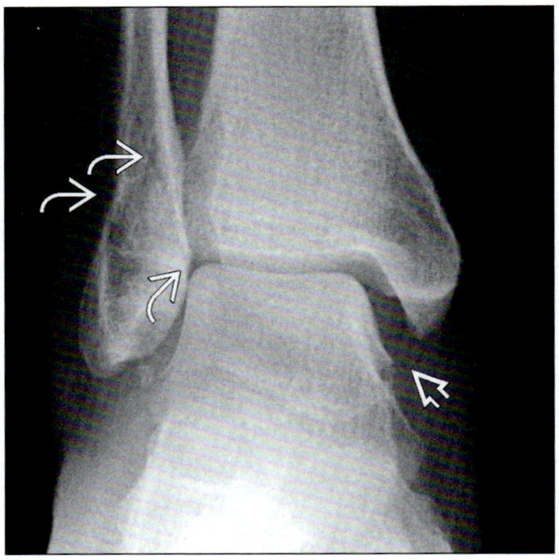

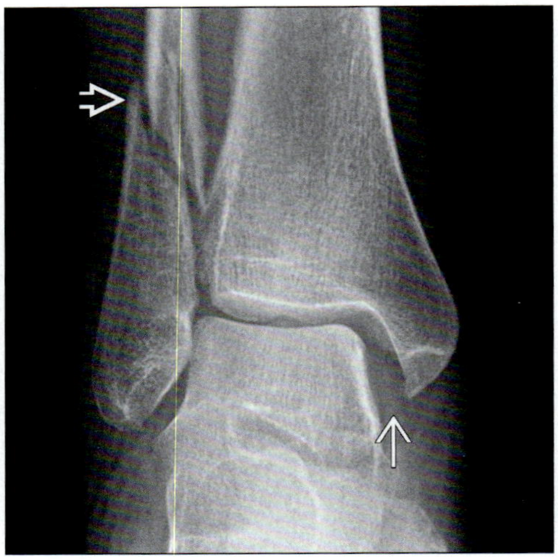

(Left) *Axial soft tissue algorithm CT shows Weber B and PM fractures. CT was obtained because closed reduction of the fx was unsuccessful, and shows the fragment which prevented reduction. There is osseous avulsion of the flexor retinaculum .* **(Right)** *Sagittal bone CT in the same patient best shows the entrapped fragment . The large PM fragment can be distinguished from pilon fx by the associated injuries (not visible on this slice) indicating a twisting injury mechanism.*

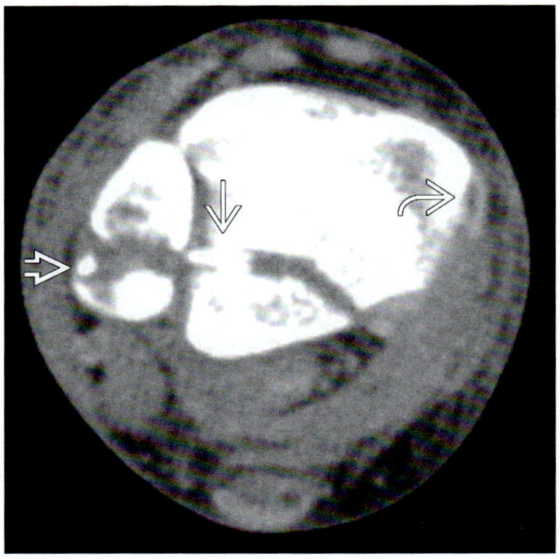

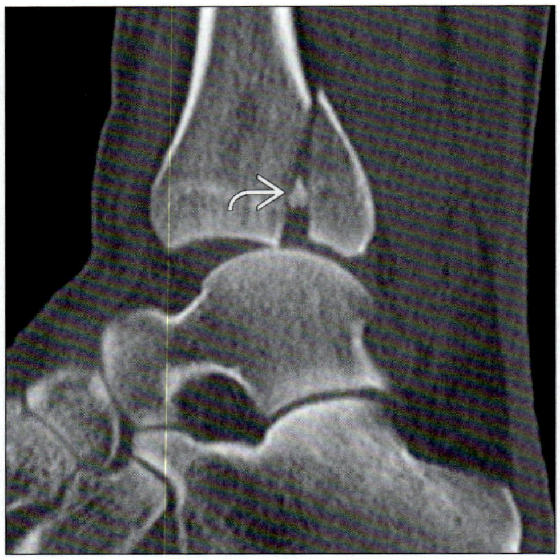

(Left) *Coronal bone CT in the same patient shows Weber B fracture and avulsion fracture of the flexor retinaculum . Flexor retinacular injuries are important because they destabilize the posterior tibial tendon.* **(Right)** *Lateral radiograph shows Weber type C fibular fracture with PM fx due to avulsion of the posterior inferior tibiofibular ligament. Small posterior malleolar fractures (< 20-25% of joint surface on lateral view) are usually not repaired surgically.*

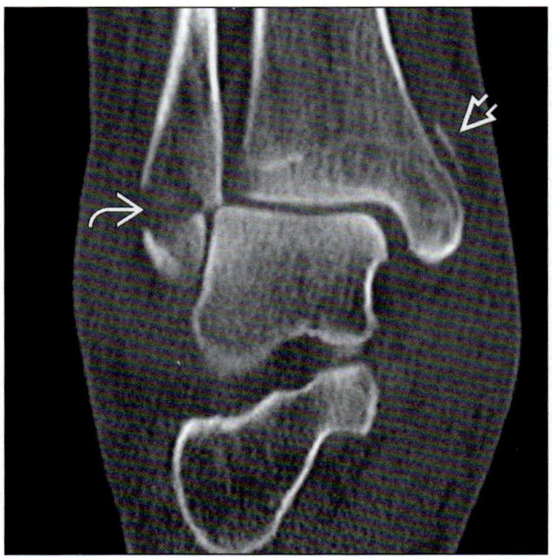

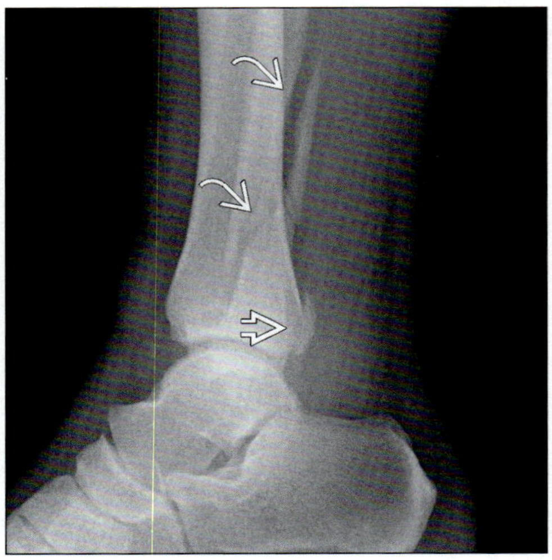

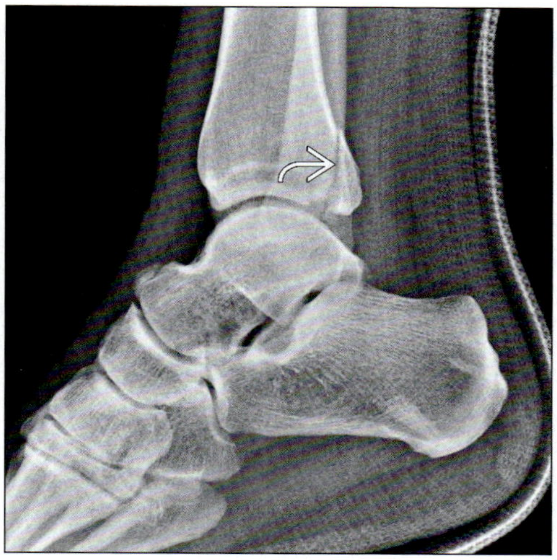

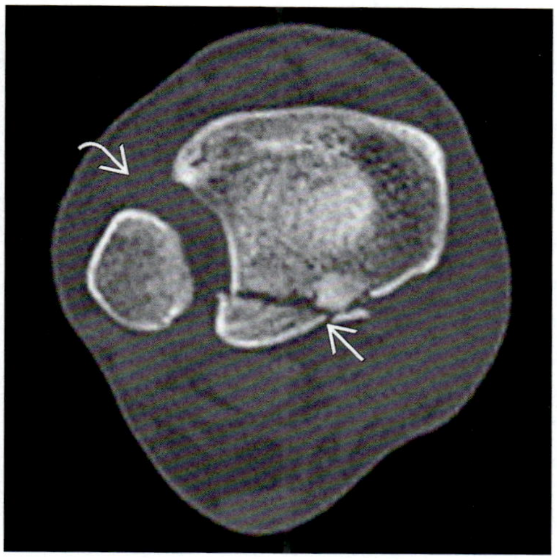

(Left) *Lateral radiograph shows isolated PM fracture . AP and mortise views showed intact medial and lateral malleoli. Isolated PM fx indicates tibiofibular syndesmosis injury. Maisonneuve fracture is sometimes present as well, even in the absence of proximal leg pain.* ***(Right)*** *Axial bone CT in the same patient shows PM fracture and widened tibiofibular syndesmosis . CT is not usually performed for this injury, but this patient had a complex contralateral foot injury.*

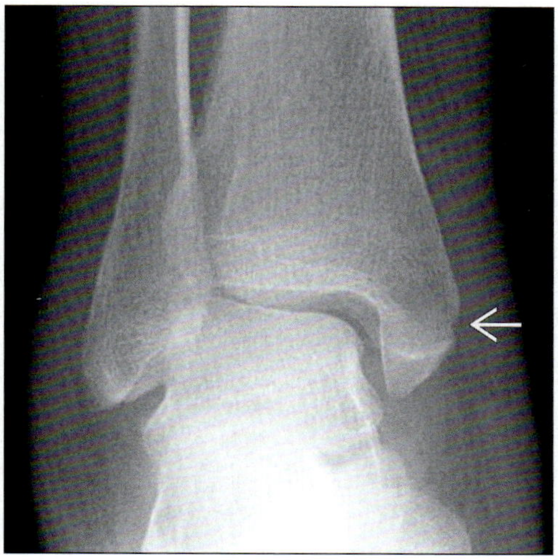

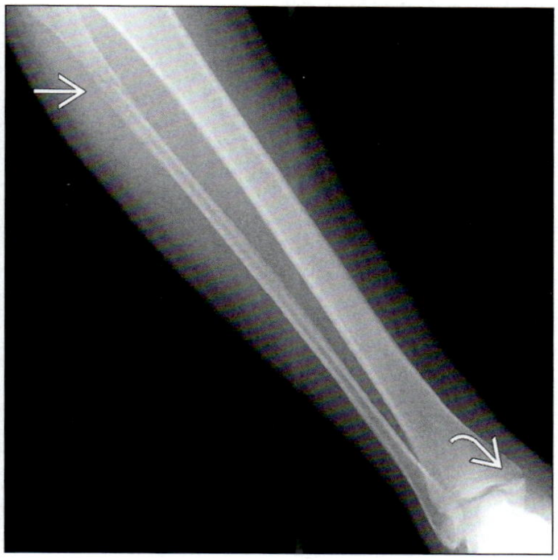

(Left) *Anteroposterior radiograph demonstrates a transverse fracture of the medial malleolus and talar tilt. There is lateral soft tissue swelling without visible fracture. Findings suggest deltoid ligament injury and raise suspicion for Maisonneuve injury.* ***(Right)*** *AP radiograph of the leg from the same patient shows a Maisonneuve and medial malleolar fractures. There is an associated syndesmotic tear to the level of the fibular fracture. The ankle mortise is unstable.*

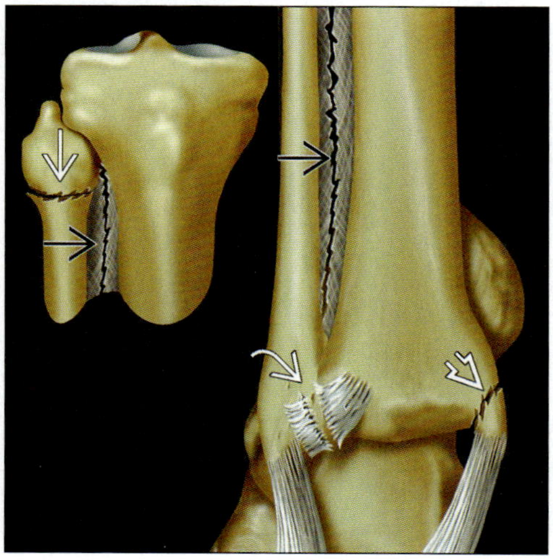

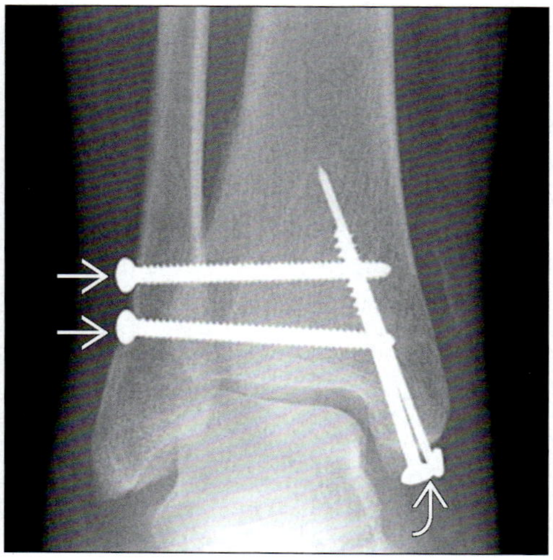

(Left) *Graphic shows how Maisonneuve fracture occurs. Pronation force at the ankle causes medial malleolus fx (or deltoid rupture) & syndesmosis tear . The force propagates superiorly, causing tear of interosseous membrane & proximal fibular fx.* ***(Right)*** *AP radiograph after ORIF of Maisonneuve fx shows syndesmotic screws and reduction of the medial malleolar fragment with pin and screw . Syndesmosis screws always loosen and are usually removed to prevent screw fracture.*

ANKLE DISLOCATION

Key Facts

Terminology
- Term fracture-dislocation is used loosely to mean any ankle fracture with significant displacement of tibiotalar articulation
- Bosworth fracture-dislocation (rare)
  - Proximal margin of fibular fracture fragment locked behind tibia

Imaging
- May dislocate in any direction
  - Posterior most common
- Fracture fragments may be widely displaced
  - Most commonly originate at malleoli or talar dome
- Presenting film may be post-reduction; look for subtle malalignment suggesting prior dislocation

Top Differential Diagnoses
- Subtalar dislocation
- Pilon fracture
- Malleolar fracture without dislocation

Pathology
- Associated injuries
  - Malleolar fractures (almost always present)
  - Injuries to collateral and syndesmotic ligaments
  - Impaction injuries to tibial plafond or talar dome
  - Injury to posterior tibial vessels or nerve
  - Avulsion of flexor or peroneal retinacula
  - Foot fractures resulting in "floating ankle"

Diagnostic Checklist
- Use CT to look for injuries to tendons, ligaments, retinacula

(Left) *Graphic shows lateral fracture-dislocation of the ankle with avulsion of the medial malleolus ⇨ and shear fracture of the lateral malleolus ⇨. Ankle dislocations occur without fracture or may be associated with malleolar or pilon fractures, subtalar dislocation, &/or talar fractures. These may occur in any direction, but are most commonly posterior.* ***(Right)*** *AP radiograph shows lateral fracture-dislocation with oblique distal tibial fracture ⇨ and fibular shaft fracture ⇨.*

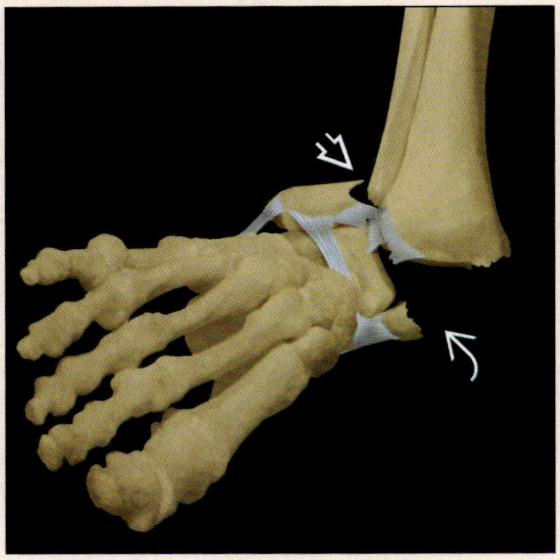

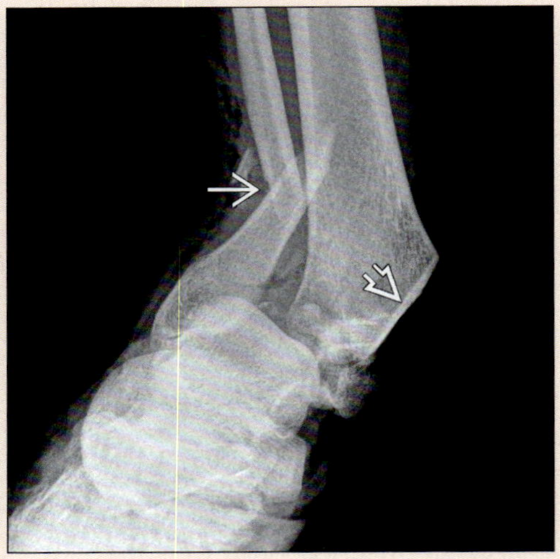

(Left) *Lateral radiograph in the same patient shows tibial plafond ⇨ displacement relative to talar dome ⇨. Calcaneal fracture ⇨ is also present.* ***(Right)*** *AP bone CT 3D reformation after reduction shows extent of bone defect in tibial plafond ⇨. CT is used after reduction of ankle dislocation to better define the complex bony injuries. Traction pins ⇨ cause mild image degradation but do not preclude adequate fracture evaluation.*

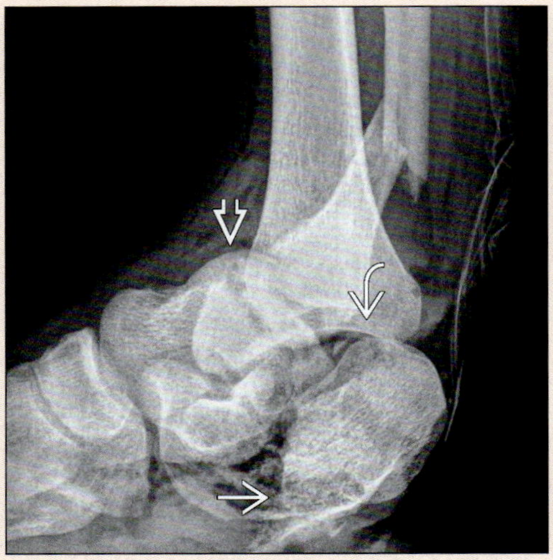

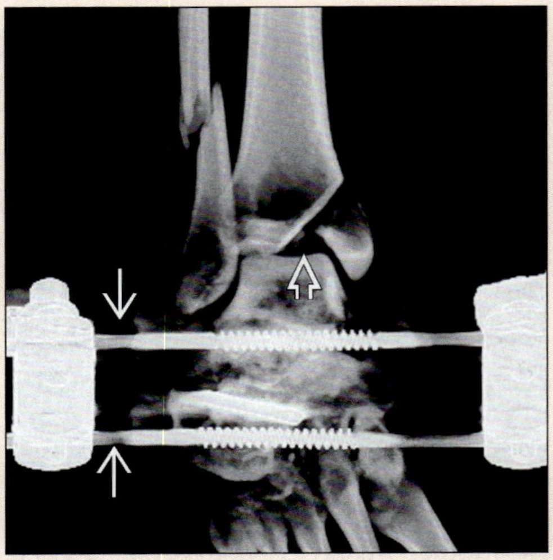

ANKLE DISLOCATION

TERMINOLOGY

Definitions

- Term fracture-dislocation is used loosely for any ankle fracture with significant joint displacement
  - i.e., complete dislocation need not be present
- Bosworth fracture-dislocation (rare)
  - Proximal margin of fibular fracture fragment locked behind tibia
- Earle fragment (a.k.a. Volkmann fragment): Fracture of posterolateral corner of tibia
- Chaput-Tillaux fragment: Fracture of anterolateral corner of tibia

IMAGING

Radiographic Findings

- Radiography
  - Fracture fragments may be widely displaced
    - Usually originate at malleoli, tibial plafond, or talar dome
  - Presenting film may be post-reduction; look for subtle malalignment suggesting prior dislocation
    - Widened medial clear space
    - Widened tibiofibular syndesmosis
    - Displacement of talus relative to tibia
      - Lateral margin of body of talus should align with fibular notch of tibia
    - Loss of superior clear space of ankle joint on AP radiograph
      - Indicates anterior or posterior subluxation of talus
    - Displacement of fibula relative to tibia on lateral radiograph

CT Findings

- Used to fully characterize fx, evaluate for soft tissue injury
- Performed after reduction

Imaging Recommendations

- Best imaging tool
  - Radiographs in acute setting
  - CT for preoperative planning

DIFFERENTIAL DIAGNOSIS

Subtalar Dislocation

- Foot deformity may clinically appear similar

Malleolar Fracture

- May occur with or without ankle dislocation

Pilon Fracture

- Impaction of tibial plafond due to axial load

PATHOLOGY

General Features

- Associated abnormalities
  - Fractures: Malleoli, pilon, talus, & other foot fractures
  - Ligamentous injuries: Deltoid, tibiofibular syndesmosis, lateral collateral ligaments
  - Injury to contralateral ankle or foot
  - Posterior tibial vessel injury
  - Chondral or osteochondral fracture of talus
  - Impaction injury of tibial plafond
  - Flexor retinaculum injury
  - Peroneal retinaculum injury
  - Tendon injury
  - Compartment syndrome

Staging, Grading, & Classification

- Posterior dislocation
  - Most common, due to posterior force on foot
- Anterior dislocation
  - Posterior force on tibia, foot in fixed position
  - ± fracture of body of talus, anterior lip of tibia
- Lateral dislocation
  - Fall landing on everted foot
  - Vertically oriented fracture of lateral malleolus
  - Horizontal fracture of medial malleolus or rupture of deltoid ligament
- Medial dislocation
  - Fall landing on inverted foot
  - Vertically oriented fracture of medial malleolus
  - Horizontal fracture of lateral malleolus or rupture of lateral ligaments
- Superior dislocation
  - Occurs with severe pilon injury of distal tibia
  - Talus displaced superiorly into fractured tibia

CLINICAL ISSUES

Natural History & Prognosis

- High incidence of premature osteoarthritis

Treatment

- Treated with open reduction and internal fixation

DIAGNOSTIC CHECKLIST

Consider

- Use CT to look for injuries to tendons, ligaments, retinacula
- Evaluate for concomitant injuries in foot, which might result in "floating ankle"

SELECTED REFERENCES

1. Gogi N et al: Anterior dislocation of the tibio-talar joint without diastasis or fracture--a case report. Foot Ankle Surg. 14(1):47-9, 2008
2. Mulligan ME: Ankle and foot trauma. Semin Musculoskelet Radiol. 4(2):241-53, 2000
3. Bos M et al: Acute disruption of the posterior tibial tendon associated with open-fracture dislocation of the ankle. Foot Ankle Int. 18(12):823-6, 1997
4. Wroble RR et al: Ankle dislocation without fracture. Foot Ankle. 9(2):64-74, 1988

ANKLE DISLOCATION

(Left) AP radiograph shows lateral ankle fracture dislocation, with Weber type B fracture of the fibula →, avulsion of posterolateral tibia (Volkmann or Earle fragment) ⇒, and avulsion of deltoid ligament →. This is a pronation-abduction type injury. **(Right)** Lateral radiograph of the same patient shows osteochondral fracture due to impaction of talar dome → on posterior margin of tibia, as well as fibular fracture → and posterior malleolar fracture ⇒.

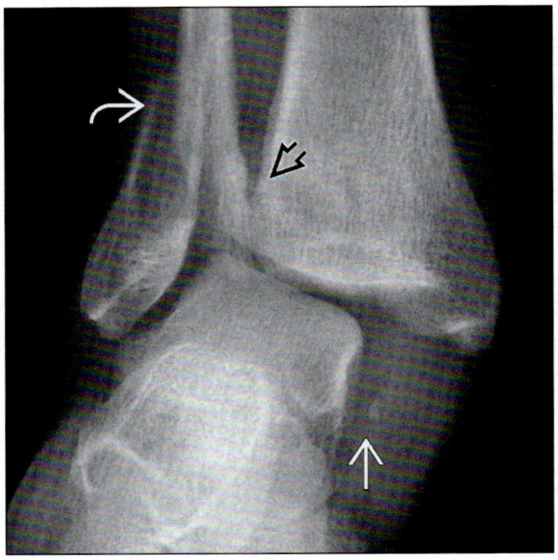

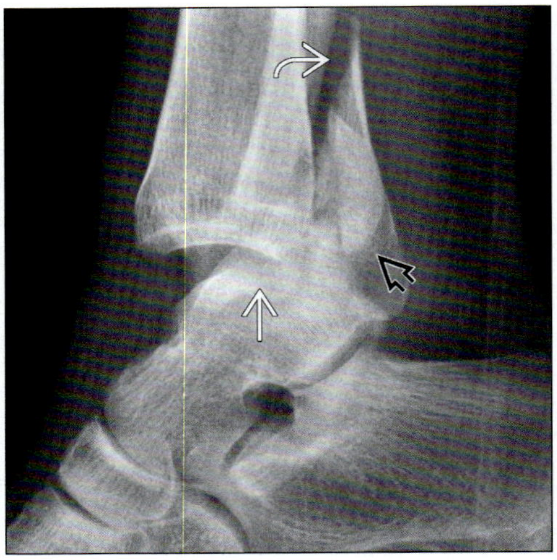

(Left) Anteroposterior radiograph shows loss of superior joint space and abnormal overlap of tibia with talus indicating subluxation of the joint, but the extent of the injury is not apparent. **(Right)** Lateral radiograph in the same patient shows complete posterior displacement of talus relative to tibia. There is a fracture of the posterior malleolus of the tibia →. Injury occurred falling down stairs. Only sequela of the injury 3 years later was slight morning stiffness.

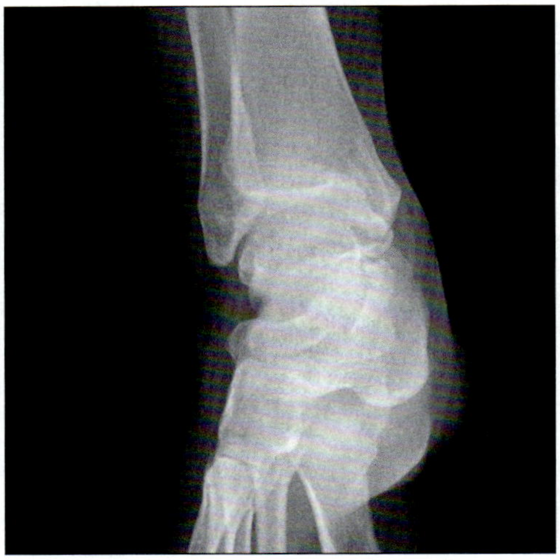

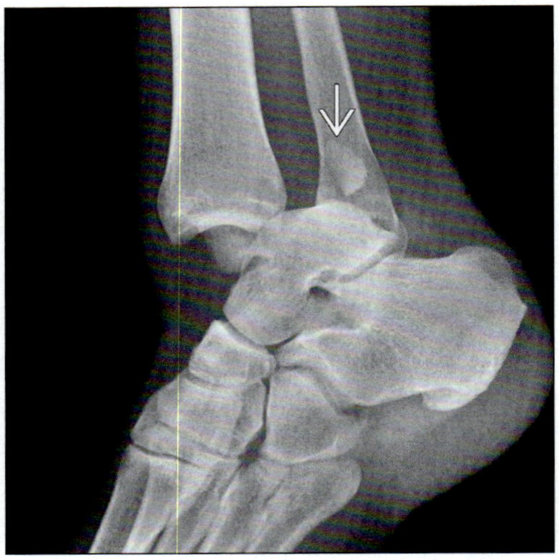

(Left) Anteroposterior radiograph shows medial fracture dislocation due to pronation-adduction injury. Talus is displaced medially. Medial malleolar fracture extends into the plafond →, and there is impaction of the plafond. Osteochondral fragment in lateral gutter ⇒ is seen from donor site in the lateral talar dome →. Fibula is intact. **(Right)** Lateral radiograph of the same patient shows fracture of posterior malleolus → and impaction of anteromedial tibial plafond →.

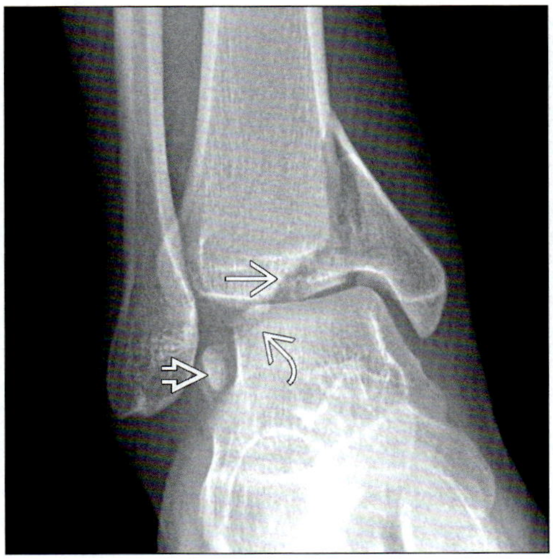

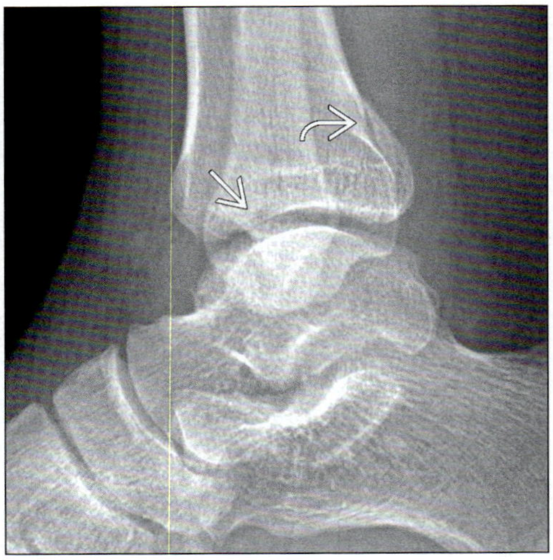

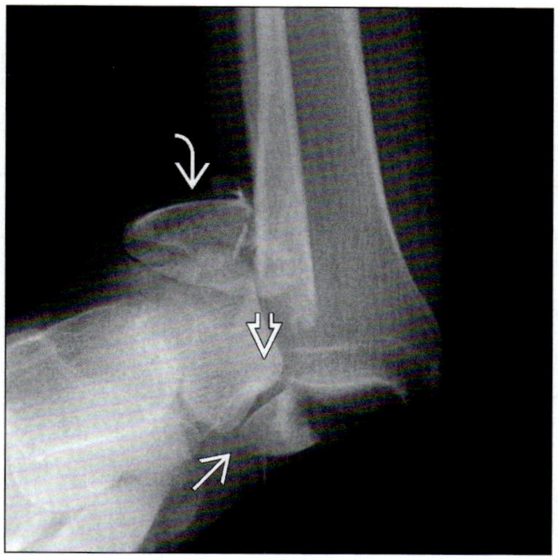

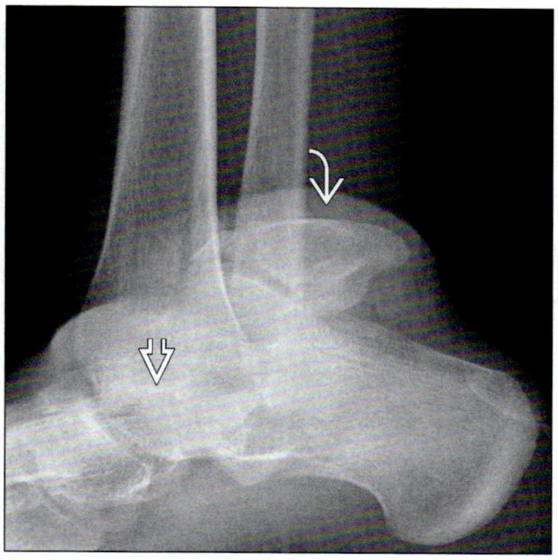

(Left) *Anteroposterior radiograph shows ankle dislocation due to pronation-abduction force. The medial malleolus and lateral malleolus are markedly displaced laterally but retain relatively normal position relative to talus, indicating intact collateral ligaments. Fracture of tibial plafond is poorly seen.* ***(Right)*** *Lateral radiograph in same patient shows displacement of lateral malleolus. Tibial plafond overlies the calcaneus, and the talus cannot be assessed.*

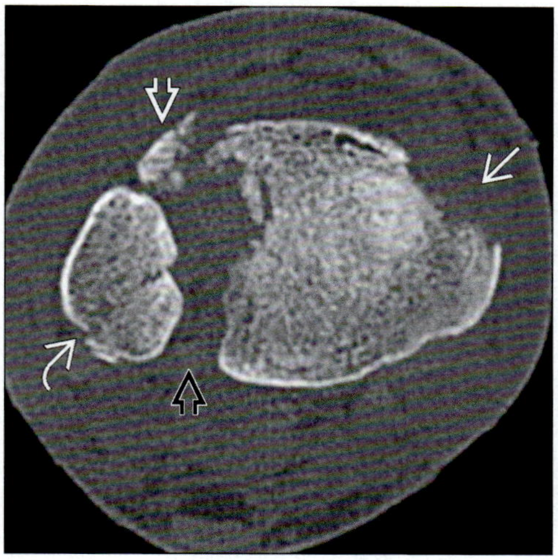

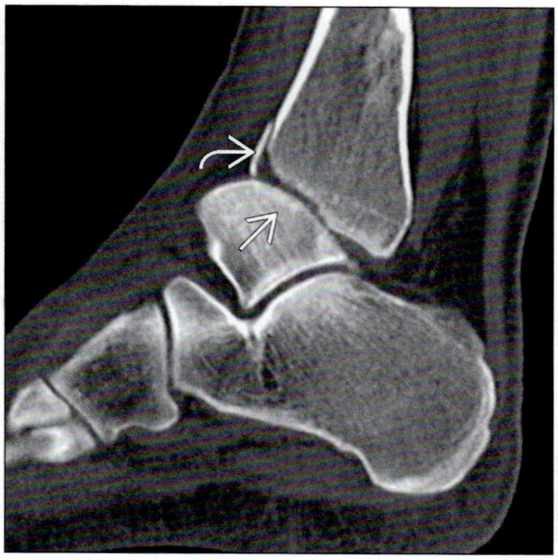

(Left) *Axial bone CT in the same patient shows medial malleolus defect and fracture of anterolateral tibial plafond with avulsion of anterior tibiofibular ligament. Syndesmosis widening indicates posterior tibiofibular ligament tear. Oblique fibular fracture has been reduced.* ***(Right)*** *Sagittal bone CT (different patient) shows displacement of talus into tibial plafond defect following closed reduction of anterior dislocation. Osteochondral injury of talar dome is present.*

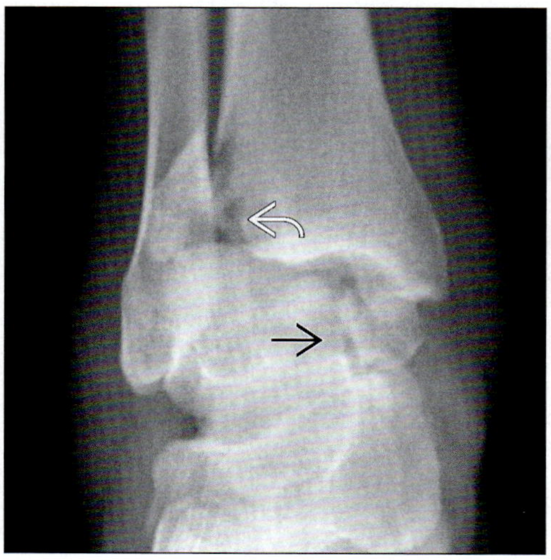

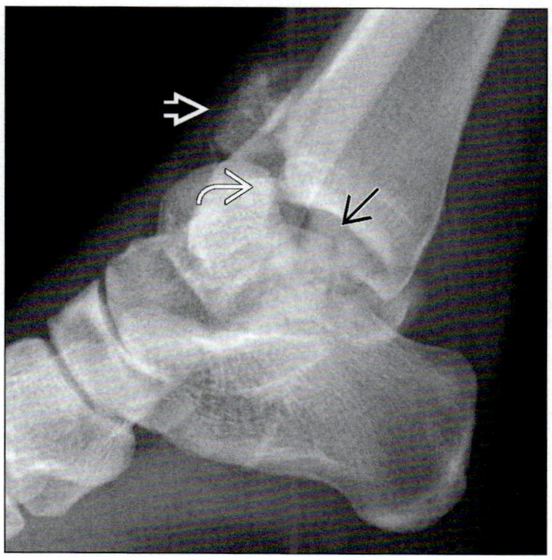

(Left) *Anteroposterior radiograph shows anterior fracture-dislocation, which includes a fracture of the body of the talus, and a fracture of the anterolateral tibial corner (Chaput-Tillaux fragment).* ***(Right)*** *Lateral radiograph of the same patient shows the talus impacted on the anterior lip of the tibia, a large fragment arising from the anterior tibial plafond, and a comminuted fracture of the body of the talus.*

OSTEOCHONDRAL INJURY, ANKLE

Key Facts

Terminology

- Osteochondral lesion (OCL): Umbrella term for focal injury of articular cartilage and underlying bone
- Osteochondral injury (or fracture): Usually reserved for acute osteochondral lesion
- Osteochondritis (or osteochondrosis) dissecans (OCD): Chronic osteochondral lesion

Imaging

- Most common at posteromedial or anterolateral talar dome
- Variable appearances on radiographs, CT
  - Crescentic, sclerotic fracture line
  - Focal defect or concavity of subchondral bone plate
  - Subarticular teacup-shaped lucency
- Variable appearance on MR depending on chronicity, stage of injury
  - Crescentic fracture line well seen
  - Unstable lesions: Fluid between fragment and underlying bone
  - Acute lesion: Focal cartilage lesion, bone marrow edema
  - Chronic lesion: Cartilage and marrow may normalize but fracture is still visible
- Arthrography increases sensitivity of CT and MR

Top Differential Diagnoses

- Arthritis with subchondral cyst formation
- Osteonecrosis, talus

Diagnostic Checklist

- Primary concerns of surgeons are location, size, and stability of lesion, rather than staging systems

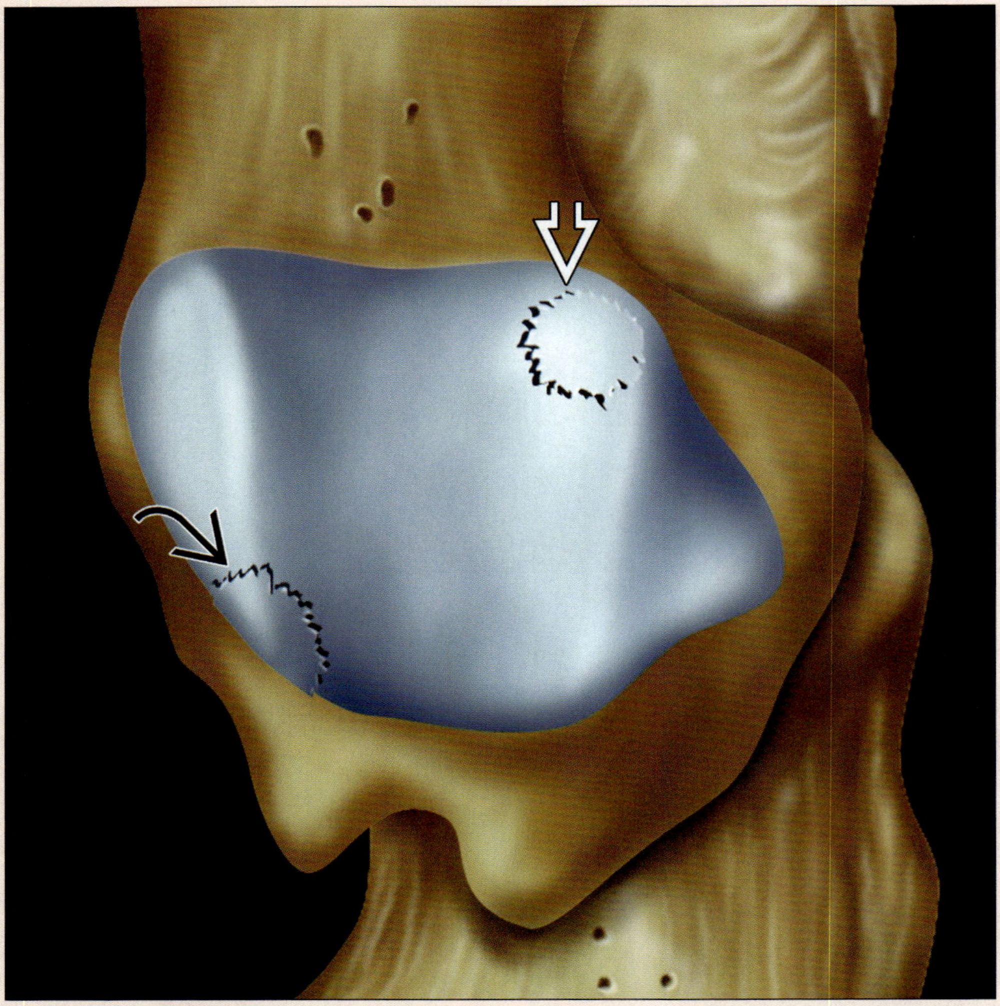

Graphic shows talar dome from above, with the tibia and fibula removed. The 2 most common sites of osteochondral lesion of the talus are posteromedial ➚ and anterolateral ➡. Both occur with an inversion injury. The anterolateral lesion occurs due to shearing of the talar dome against the tibial plafond and tends to be a thin flake of cartilage ± bone. The posteromedial lesion occurs due to impaction of the dome against the plafond and tends to have a deep, bowl-shaped configuration involving cartilage and subchondral bone.

OSTEOCHONDRAL INJURY, ANKLE

TERMINOLOGY

Definitions

- Osteochondral lesion (OCL): Umbrella term for focal injury of articular cartilage and underlying bone
  - Osteochondral lesion of talus (OLT)
- Osteochondral injury (or fracture): Usually reserved for acute osteochondral lesion
- Osteochondritis (or osteochondrosis) dissecans (OCD): Chronic osteochondral lesion

IMAGING

General Features

- Location
  - Most commonly involve posteromedial or anterolateral talar dome
  - May also involve tibial plafond
    - Often in same relative location as talar dome lesion
  - May primarily involve vertically oriented medial or lateral margins of talar body
- Size
  - Usually < 15 mm
- Morphology
  - Lateral talar OCL: Shallow flake of avulsed cartilage/bone
  - Medial talar OCL: Deep, rounded lesion due to impaction

Radiographic Findings

- Abnormal subchondral bone plate may have several different appearances
  - Crescentic, sclerotic subchondral fracture line
    - Exits through subchondral bone plate at both margins
  - Focal defect in subchondral bone plate
  - Focal concavity of articular surface
- Subarticular teacup-shaped lucency
  - Sharply demarcated, sclerotic rim
- Intraarticular bony fragment ± visible intraarticular bone donor site
- Radiographs will not detect Berndt and Harty stage 1 lesions

Arthrography Findings

- Arthrography generally performed with either CT or MR for better visualization
  - MR more sensitive to marrow abnormalities
- Cartilage defect
  - Contrast outlines cartilage defect as focal concavity
    - Sharp margins seen in acute injuries
    - Rounded margins seen in chronic injuries
  - Caveat: Cartilage contour may normalize in chronic injuries
    - Defect fills with fibrocartilage
    - Scar has abnormal, soft consistency at time of surgery/arthroscopy
- Unstable cartilage or osteocartilaginous fragment
  - Contrast extends beneath fragment
- Cystic degeneration of OCL
  - Fibrocartilage may form overlying "cyst"
  - So-called cyst is often filled with fibrous material, not fluid
  - Therefore, variable extension of arthrographic contrast into "cyst"

CT Findings

- Same findings as radiographs but more clearly evident
- Arthrography must be added to see stage 1 lesions
- CT arthrogram shows cartilage defect, contrast extending into bony defect
  - Cartilage may appear intact, but cyst or sclerotic fracture line visible

MR Findings

- Correspond fairly well to surgical Berndt and Harty staging system stages 1-5
  - Some variability found, attributable to age of lesion
- Bone marrow edema may be present at any stage
  - Radiates away from articular surface
  - Low signal intensity on T1WI, high signal intensity on T2WI, STIR
  - Edema diminishes and may resolve in chronic lesions
- Cartilage defects may heal in chronic lesions
  - Cartilage still abnormal to arthroscopic probe
  - Can make MR diagnosis based on bone abnormalities, even if cartilage appears intact
- Stage 1
  - Decreased T2 signal intensity in articular cartilage &/or cartilage defect
- Stage 2
  - Crescentic fracture line entering and exiting through subchondral bone plate
  - ± decreased T2 signal intensity in articular cartilage
  - ± focal cartilage defect
  - ± bone marrow edema
- Stage 3
  - Same signs as for stage 2 + native or arthrographic fluid extends beneath loose osteochondral fragment
- Stage 4
  - Concave defect in articular surface
  - ± loose body in joint
- Stage 5
  - Well-demarcated "cyst"
    - Low signal intensity on T1WI, high signal on T2WI
    - Often enhances after IV contrast because filled with fibrous tissue
    - Variably surrounded by bone marrow edema

Imaging Recommendations

- Best imaging tool
  - MR arthrography offers highest accuracy
- Protocol advice
  - Resolution needs to be maximized because of thinness of cartilage
    - 3-3.5 mm slices, high matrix, small FOV
  - Abnormality may be visible in only 1 imaging plane
  - PD FS tends to give best cartilage definition on nonarthrographic study
  - Arthrograms should include T1WI in all 3 planes and at least 1 fluid-sensitive sequence

OSTEOCHONDRAL INJURY, ANKLE

DIFFERENTIAL DIAGNOSIS

Subchondral Cyst Formation due to Arthritis

- Tends to have spherical rather than teacup shape
- May occur in osteoarthritis or inflammatory arthritis
- Osteoarthritis: Osteophytes are present, cartilage loss usually more extensive but nonuniform
- Inflammatory arthritis: Erosions, uniform cartilage loss are present
- Bone marrow edema not useful distinguishing sign

Erosion due to Inflammatory Arthritis

- Usually less sharply marginated
- 1st erosions occur in bare area (margins) of joint
- Diffuse, uniform cartilage loss usually present

Osteonecrosis, Talus

- Subarticular sclerotic, serpentine line
- Cartilage normal until late, when subchondral collapse occurs

Talus Body Fracture

- Fracture extends from superior to inferior surface of talus

Osteochondral Lesion at Other Sites

- May have similar clinical presentation with OCL of subtalar, talonavicular, or calcaneocuboid joint
- Less common than ankle joint

PATHOLOGY

General Features

- Etiology
  - Inversion injury
    - Tear of anterior talofibular ligament always present
    - Thin flake of cartilage &/or bone sheared off lateral margin of talar dome
    - Rounded area of compression occurs at medial margin of talar dome
    - Small areas of injury may also occur in ankle gutters by this mechanism
  - Axial load injury
    - Impaction of talar dome &/or tibial plafond
- Associated abnormalities
  - Lateral collateral ligament injury usually present
  - May occur with any ankle fracture

Staging, Grading, & Classification

- **Berndt and Harty staging system**
  - Stage 1: Injury to articular cartilage with subchondral bone bruise
  - Stage 2: Crescentic transchondral fracture line, stable or partly detached
  - Stage 3: Detached osteochondral fragment in situ
  - Stage 4: Osteochondral fragment displaced from donor site
  - Stage 5 (later addition to classification; a.k.a. stage 2A): Cystic degeneration of osteochondral lesion

Gross Pathologic & Surgical Features

- Early or partially healed injury: Cartilage appears intact but is soft to arthroscopic probe
- Stable injury: Defect visible but adjacent cartilage, bone not displaceable with probe
- Unstable injury: Cartilage &/or bone fragment is unstable to probe

CLINICAL ISSUES

Presentation

- Most common signs/symptoms
  - Acute episode of ankle sprain with persistent pain
  - Chronic, dull, aching pain
  - Clicking, locking in ankle joint
    - Reflects loose fragment or associated ligament tear

Natural History & Prognosis

- May resolve spontaneously
  - MR shows resolving bone marrow edema
- May develop chronic pain, loose bodies, osteoarthritis
- Many loose bodies in ankle resorb

Treatment

- Drilling chondroplasty, debridement
- Osteochondral allograft
- Mosaicplasty

DIAGNOSTIC CHECKLIST

Image Interpretation Pearls

- Cartilage may appear to normalize in chronic OCL
- Chronic lesions often seen as cysts at corner of talar dome
- Be alert for lesions in ankle gutters, which are often missed

Reporting Tips

- Primary concerns of surgeons are location, size, and stability of lesion, rather than staging systems

SELECTED REFERENCES

1. Elias I et al: Osteochondral lesions of the talus: change in MRI findings over time in talar lesions without operative intervention and implications for staging systems. Foot Ankle Int. 27(3):157-66, 2006
2. Berndt AL et al: Transchondral fractures (osteochondritis dissecans) of the talus. J Bone Joint Surg Am. 86-A(6):1336, 2004
3. Mintz DN et al: Osteochondral lesions of the talus: a new magnetic resonance grading system with arthroscopic correlation. Arthroscopy. 19(4):353-9, 2003
4. Higashiyama I et al: Follow-up study of MRI for osteochondral lesion of the talus. Foot Ankle Int. 21(2):127-33, 2000
5. De Smet AA et al: Value of MR imaging in staging osteochondral lesions of the talus (osteochondritis dissecans): results in 14 patients. AJR Am J Roentgenol. 154(3):555-8, 1990

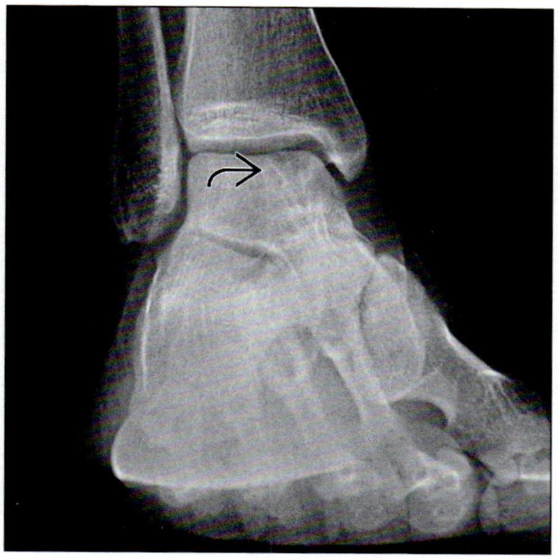

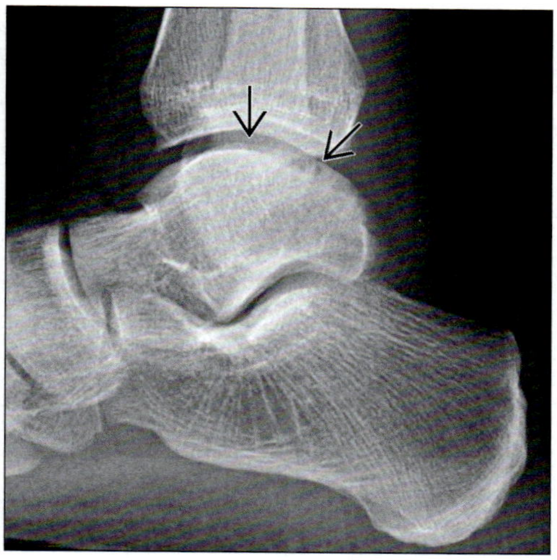

(Left) *Mortise view radiograph shows bowl-shaped lucency and sclerotic border typical of medial talar dome OCL. Subchondral bone plate appears intact on this view.* ***(Right)*** *Lateral radiograph in the same patient shows that the lesion is posterior to the apex of the talar dome. This is the most common location for medial lesions. There is slight flattening of subchondral bone plate overlying the lucent lesion. Cystic change was found in the lesion at time of arthroscopy.*

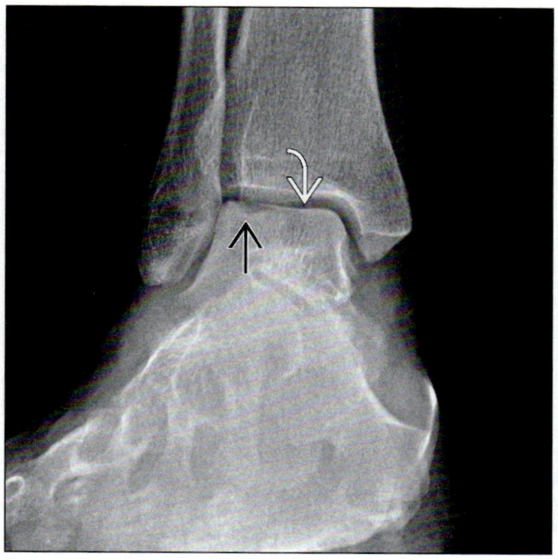

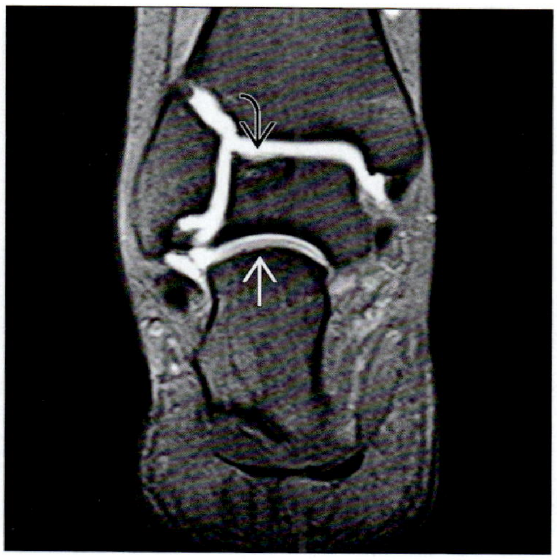

(Left) *AP radiograph shows typical thin crescent of lateral OCL, surrounded by a sclerotic rim of reactive bone. The subchondral bone plate is well seen medial to the lesion but disappears overlying the lesion, indicating cortical disruption.* ***(Right)*** *Coronal MR arthrogram T1WI FS in the same patient shows cortical depression and irregular cartilage contour. Note the extension of contrast into the subtalar joint, which is a normal variant occurring in 15% of patients.*

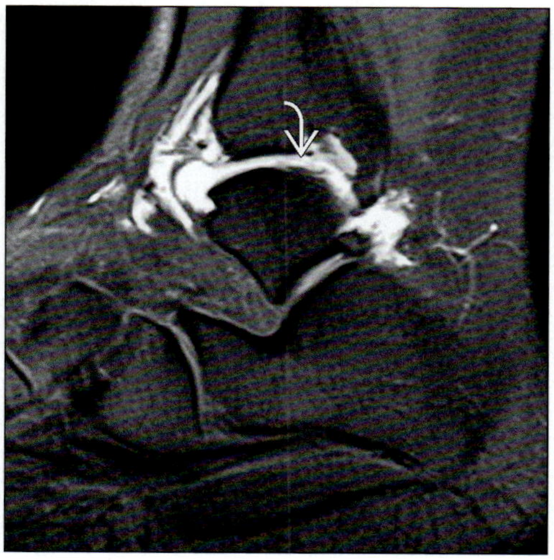

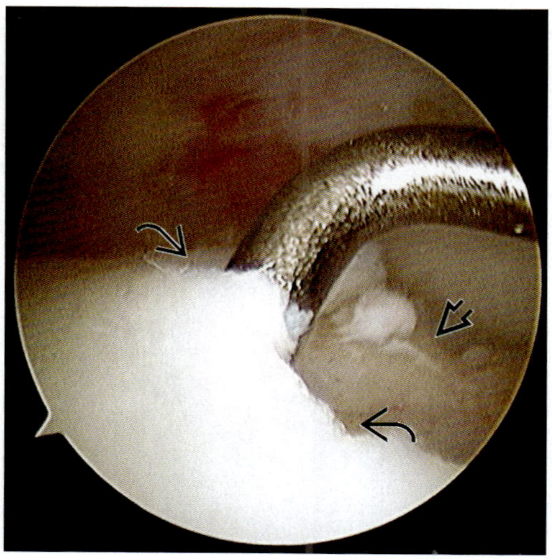

(Left) *Sagittal MR arthrogram T1WI FS in the same patient shows contrast under cartilage fragment, indicating that the fragment is unstable. Sagittal images are often best for identifying cartilage flaps, especially if they are on a portion of the talar dome that is curving posteriorly.* ***(Right)*** *Arthroscopic photograph in the same patient shows a probe lifting the unstable cartilage. More posteriorly, unstable cartilage has already been removed, exposing the underlying bone.*

OSTEOCHONDRAL INJURY, ANKLE

*(**Left**) Graphic shows cystic degeneration of an OCL, which forms a sharply demarcated cavity in the medial dome. An OCL with cystic degeneration is unlikely to heal without debridement. (**Right**) Sagittal bone CT arthrogram shows cystic degeneration of OCL. The bone cavity is sharply demarcated. The subchondral bone plate is depressed, but in this chronic lesion the overlying cartilage appears relatively normal. Arthroscopy showed soft, ballotable cartilage and a large cavity.*

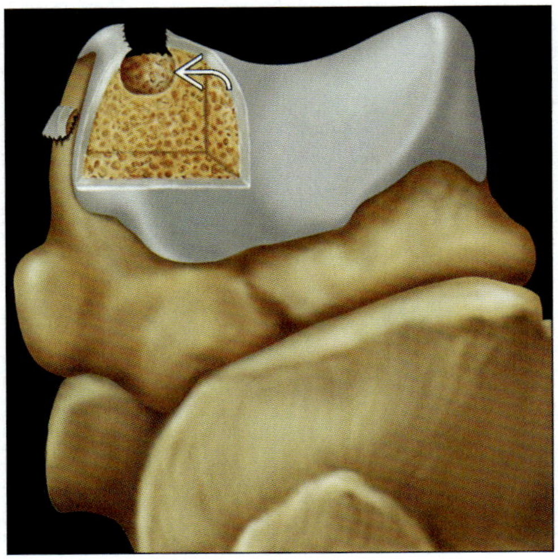

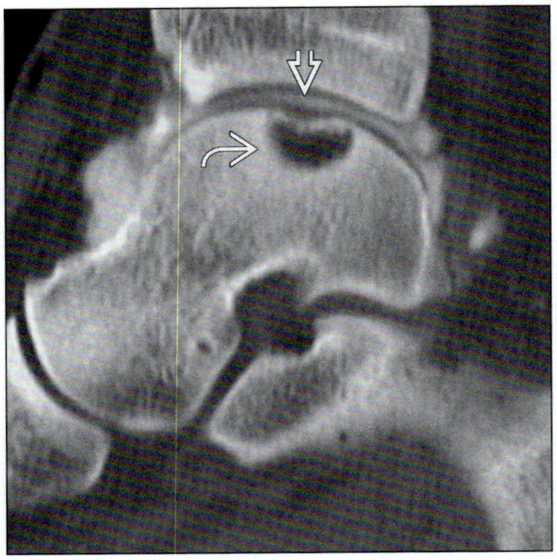

*(**Left**) Coronal T1 C+ FS MR shows diffuse enhancement of "cystic" OCL of the talar dome. OCL cavities often fill with loose reparative tissue, which enhances after contrast administration. (**Right**) Sagittal STIR MR in the same patient shows that cystic OCL can be multiloculated. Note the depression of subchondral bone plate but relatively normal-appearing cartilage.*

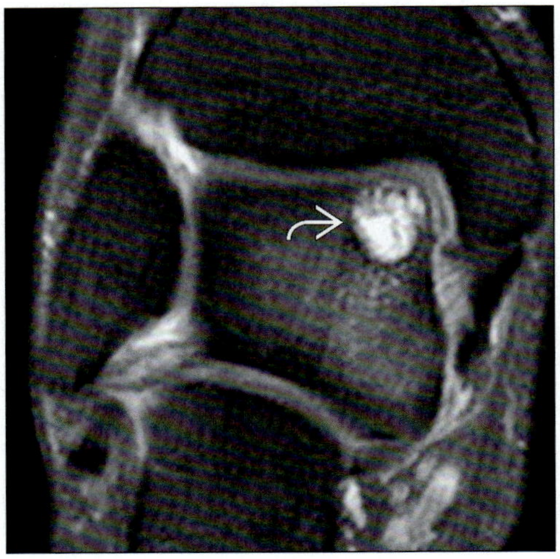

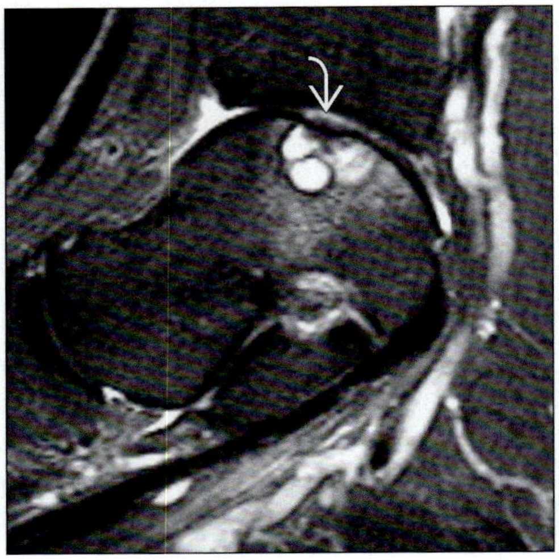

*(**Left**) Coronal T2WI FS MR shows bone marrow edema underlying a slightly depressed lateral corner of the talar dome. The articular surface appears smooth, but chondromalacia was found at the time of arthroscopy. (**Right**) Coronal T2WI FS MR shows a partly displaced OCL. Arthroscopy confirmed unstable OCL in situ, with underlying cystic change. Bone marrow edema at the posterior subtalar joint and medial malleolus may reflect injury or abnormal hindfoot biomechanics.*

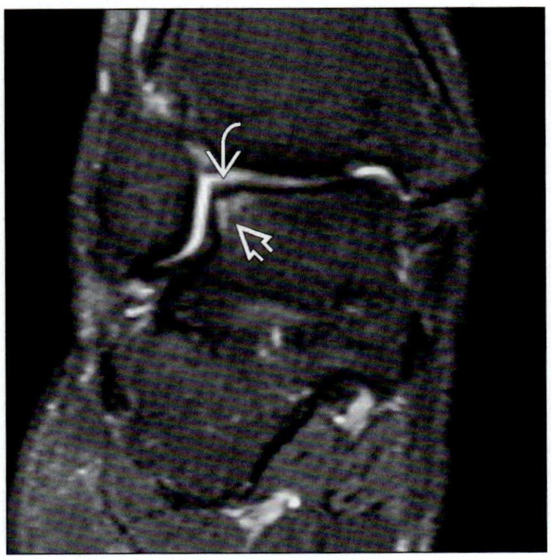

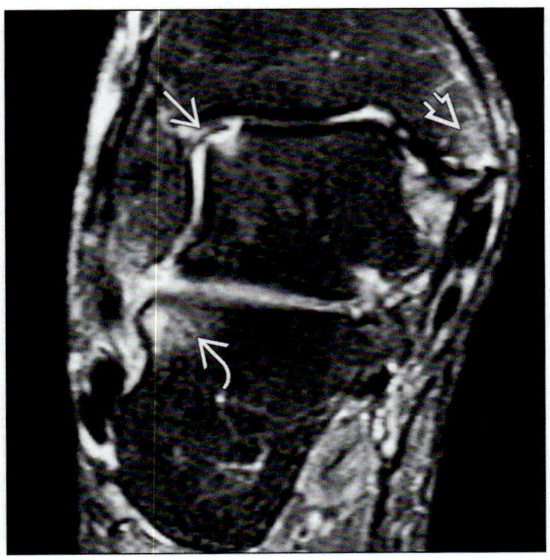

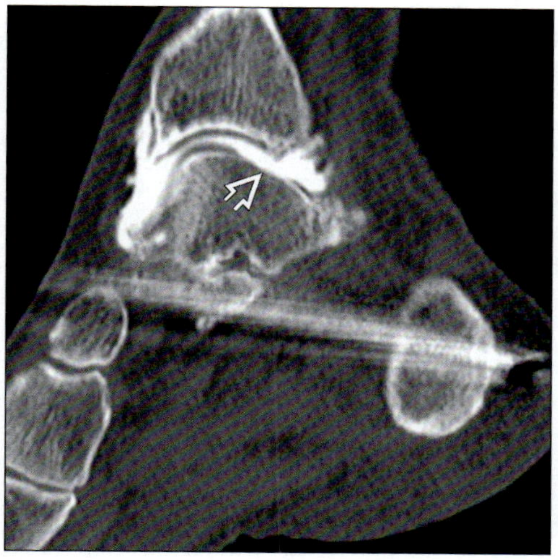

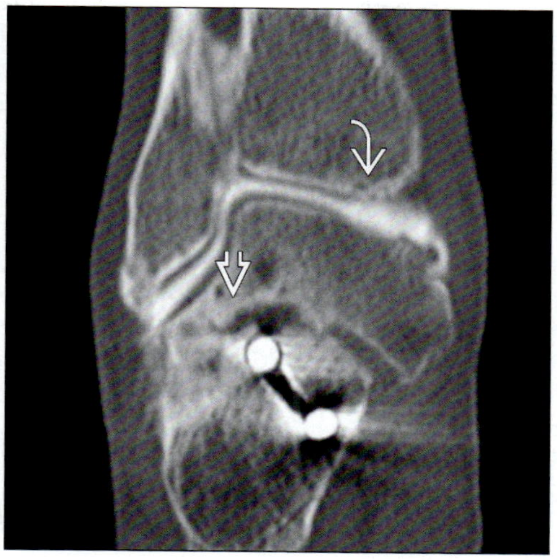

(Left) *Sagittal bone CT arthrogram shows cartilage defect with smooth, curved margins typical of a chronic cartilage injury. Acute injuries tend to be more sharply marginated.* ***(Right)*** *Coronal CT arthrogram shows a talar dome lesion and a kissing lesion of the talar dome. This patient has a failed subtalar fusion.*

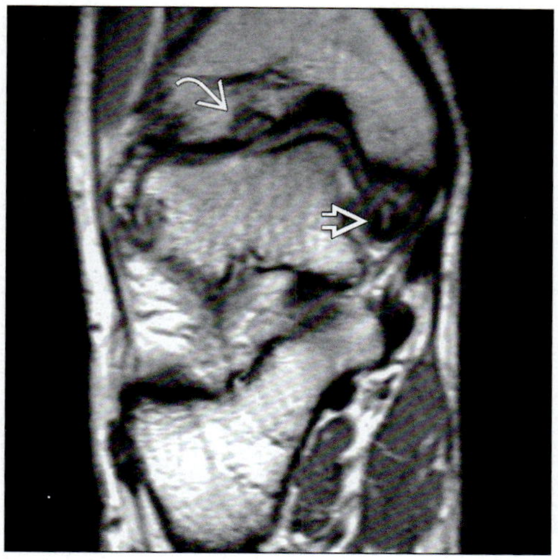

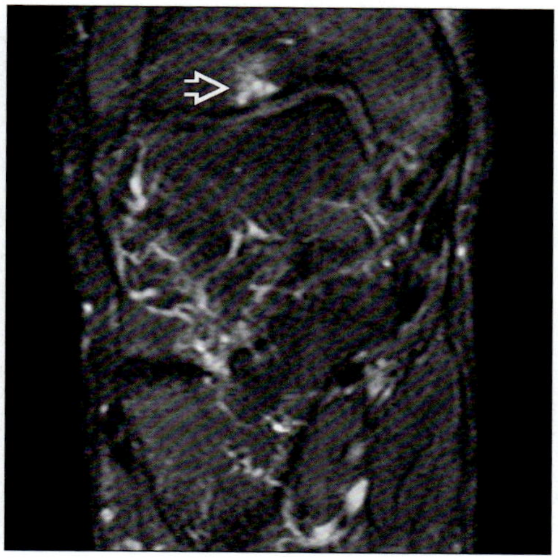

(Left) *Coronal PDWI MR shows a rounded OCL of the tibial plafond in a patient with chronic pain after bimalleolar fracture. Note the nonunited fracture of the medial malleolus.* ***(Right)*** *Coronal T2WI FS MR in the same patient shows bone marrow edema in OCL. Cartilage of tibial plafond is difficult to see on MR without arthrography, so cartilage defects can be underestimated. At arthroscopy, overlying cartilage was completely denuded.*

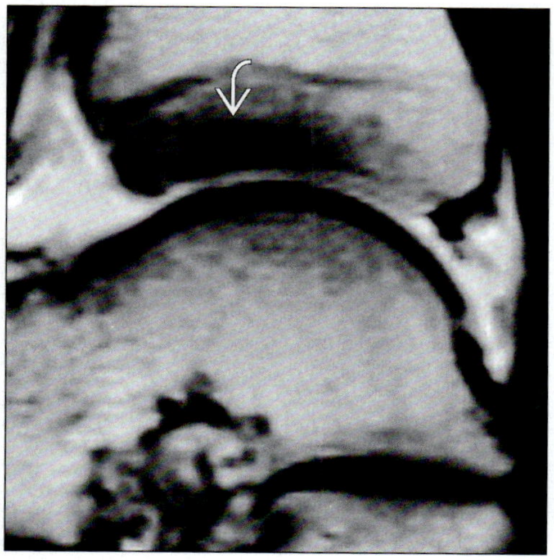

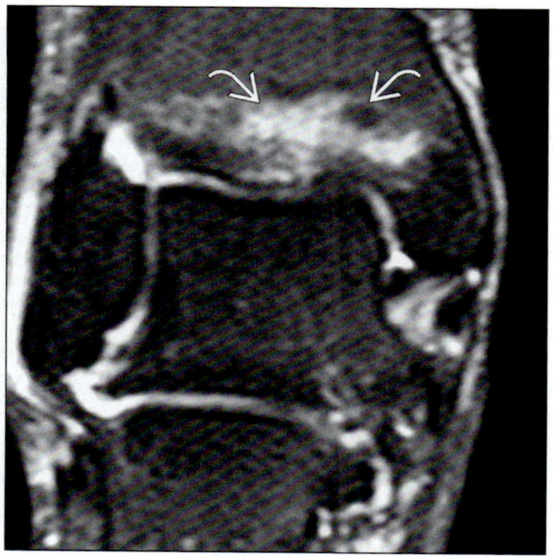

(Left) *Sagittal MR arthrogram T1WI shows subchondral sclerosis and loss of tibial articular cartilage, consistent with OCL after an axial load injury. Arthroscopy confirmed focal cartilage loss in this region, and no evidence of osteoarthritis was found elsewhere in the joint.* ***(Right)*** *Coronal T2WI FS MR in the same patient shows a horizontal band of abnormal signal reflecting the axial load mechanism with trabecular impaction.*

FRACTURE OF TALAR BODY AND PROCESS

Key Facts

Imaging

- Talar body fracture
  - Vertically oriented line in sagittal or coronal plane &/or
  - Crush injury along subtalar joint (best seen on CT)
- Lateral process fracture
  - Notoriously difficult to see on radiographs; often visible only on single view
  - Transversely, vertically, or obliquely oriented
  - Tip of process may rotate anteriorly
  - Lateral process may seem to be absent if fracture comminuted
  - Lateral soft tissue swelling is centered beneath tip of fibula, mimicking ankle sprain
- Posterior process fracture
  - Vertically oriented in coronal plane
  - Jagged margin helps distinguish from os trigonum
  - Posterior process fracture should raise concern for associated lateral process fracture

Top Differential Diagnoses

- Ankle sprain
- Talus neck fracture
- Os trigonum

Clinical Issues

- Lateral process fracture = 15% of ankle injuries in snowboarders
  - Also occurs with inversion injuries of other causes
  - Often misdiagnosed clinically as ankle sprain
- High incidence of post-traumatic osteoarthritis for body, lateral process fractures

Diagnostic Checklist

- Biggest cause of missed lateral process fracture: Poor positioning on lateral radiograph of ankle

*(**Left**) Graphic shows fracture of the lateral process of the talus, viewed from above, with the tibia and fibula removed. The dorsal surface of the lateral process of the talus articulates with the fibula, and the inferior surface articulates with the calcaneus. (**Right**) Lateral radiograph shows fracture of the lateral talar process. On lateral radiographs, the lateral process forms a triangle pointing inferiorly, and the fracture is typically across the triangle's base.*

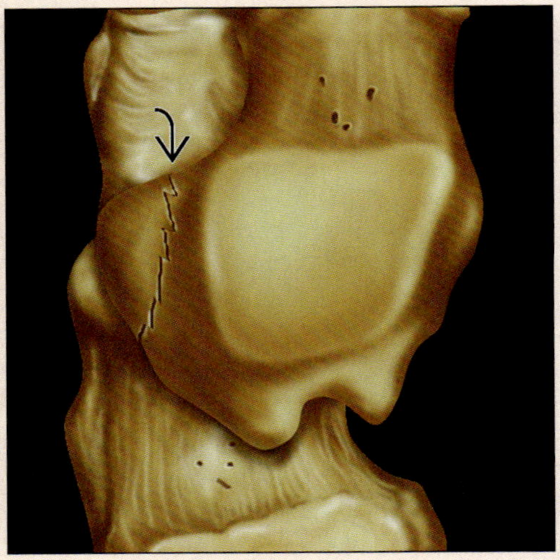

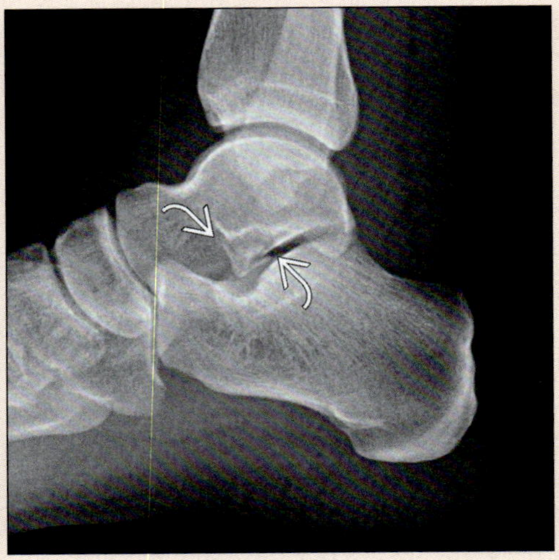

*(**Left**) Sagittal bone CT in the same patient more easily shows the fracture line and the involvement of the subtalar joint. (**Right**) Coronal bone CT in same patient shows that the fracture involves a greater portion of articular surface than might be suspected from the radiographs. In this case, the lateral process is displaced into the tarsal sinus.*

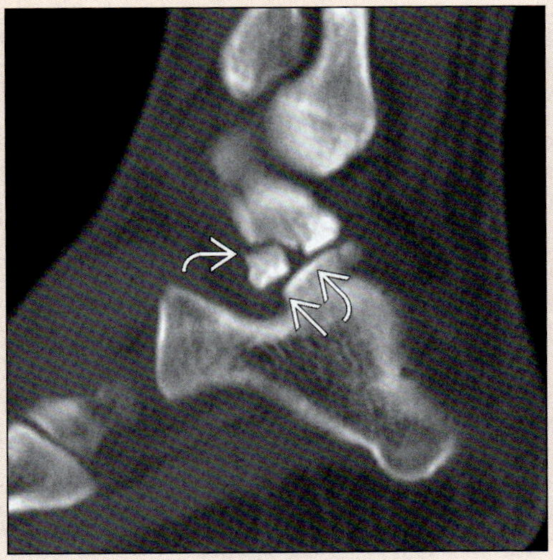

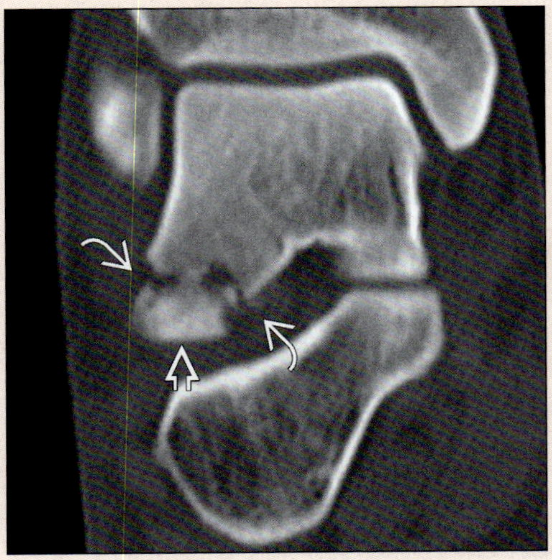

FRACTURE OF TALAR BODY AND PROCESS

IMAGING

General Features

- Location
  - Usually involves articular surface at ankle or subtalar joint
- Morphology
  - Body fracture
    - Vertical split of talus in coronal or sagittal plane
    - Crush injury centered on posterior subtalar joint
  - Lateral process fracture
    - Chip fracture (nonarticular): Vertically oriented
    - Intraarticular fracture: Obliquely oriented
      - Occasionally comminuted
      - Involves posterior subtalar joint and lateral gutter of ankle
  - Posterior process fracture: Coronally oriented
    - Mimics os trigonum

Radiographic Findings

- Talar body fracture
  - Vertical split-type fracture on AP or lateral view, usually displaced
  - Crush injury along subtalar joint poorly seen on radiographs
- Lateral process fractures
  - Notoriously difficult to see on radiographs
  - Lateral view
    - Lateral process should form "V" shape closely fitting into angle of Gissane
    - When lateral process is fractured, "V" may be rotated anteriorly
    - Lateral process may seem to be absent if fracture comminuted
    - Will not be visible if lateral radiograph is malrotated
      - Posterior margin of fibula should be at posterior margin of tibia
  - AP or mortise view
    - Lateral process extends beneath tip of fibula
    - May see lucency through process
    - May see avulsion flake lateral to process
    - Lateral soft tissue swelling is centered beneath tip of fibula, mimicking ankle sprain
- Posterior process fractures
  - Jagged margin helps distinguish from os trigonum
  - Usually minimally displaced
  - Seen on lateral view only

CT Findings

- Body fracture
  - Usually complex
- Lateral process fracture
  - Often involves more of subtalar joint than is evident on radiographs

MR Findings

- Missed fractures may present to MR for chronic pain

Imaging Recommendations

- Best imaging tool
  - CT preferable in acute setting
- Protocol advice
  - Coronal, sagittal reformations very helpful
    - Coronal images perpendicular to subtalar joint

DIFFERENTIAL DIAGNOSIS

Ankle Sprain

- Also shows lateral soft tissue swelling, joint effusion
- Lateral process fx often misdiagnosed as ankle sprain

Talus Neck Fracture

- Fracture line distal to talar body fracture

Os Trigonum

- Ossicle is well-corticated, unlike acute fracture

PATHOLOGY

General Features

- Etiology
  - Talus body fracture: Fall from height
  - Lateral process fracture: Inversion injury
    - 15% of ankle injuries in snowboarders
  - Posterior process fracture: Plantar flexion injury
- Associated abnormalities
  - Talus body fracture associated with
    - Ankle or subtalar dislocation
    - Other lower extremity fractures
    - Burst or compression fractures of spine
  - Lateral process fracture associated with
    - Posterior process fracture
    - Peroneal retinaculum avulsion
  - Posterior process fracture associated with
    - Lateral process fracture
    - Flexor hallucis longus tenosynovitis

Staging, Grading, & Classification

- Lateral process fracture
  - Type I: Nonarticular chip fracture
  - Type II: Intraarticular, single fracture line
  - Type III: Intraarticular, comminuted

CLINICAL ISSUES

Natural History & Prognosis

- High risk of post-traumatic osteoarthritis with fractures of body, lateral process

DIAGNOSTIC CHECKLIST

Image Interpretation Pearls

- Biggest cause of missed lateral process fracture: Poor positioning on lateral radiograph of ankle

SELECTED REFERENCES

1. Ebraheim NA et al: Clinical outcome of fractures of the talar body. Int Orthop. 32(6):773-7, 2008
2. Langer P et al: Incidence and pattern types of fractures of the lateral process of the talus. Am J Orthop. 37(5):257-8, 2008
3. von Knoch F et al: Fracture of the lateral process of the talus in snowboarders. J Bone Joint Surg Br. 89(6):772-7, 2007
4. Giuffrida AY et al: Pseudo os trigonum sign: missed posteromedial talar facet fracture. Foot Ankle Int. 24(8):642-9, 2003
5. Ebraheim NA et al: Posteromedial process fractures of the talus. Foot Ankle Int. 16(11):734-9, 1995

FRACTURE OF TALAR BODY AND PROCESS

(Left) *Coronal bone CT shows the typical appearance of a talus body fracture, with a sagittally oriented split fracture and marked articular depression. These are high-energy fractures, and patients usually have additional axial load injuries, such as the calcaneus fracture seen here.* ***(Right)*** *Axial bone CT in the same patient shows the sagittal fracture with comminution of the medial portion of the talar body. Calcaneal fracture involves the lateral margin of the subtalar joint.*

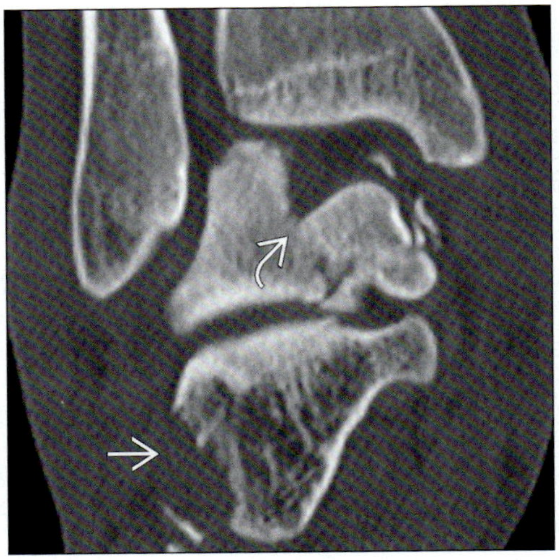

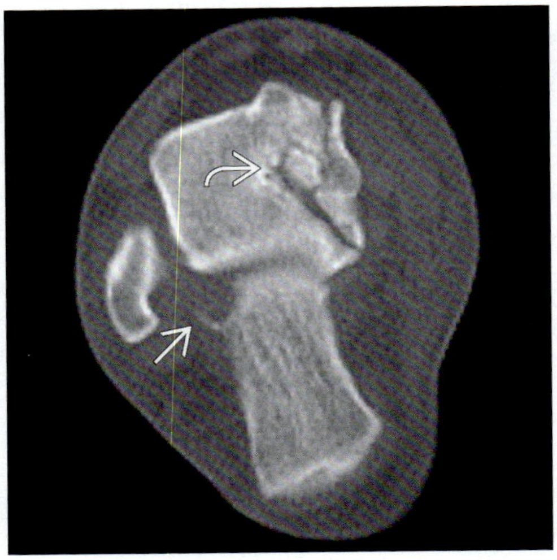

(Left) *Coronal bone CT shows a sagittal fracture through the lateral margin of the talus. Unlike an osteochondral fracture, which has a semilunar shape centered on the subchondral bone plate, it is linear in morphology and traverses the entire height of the talus from superior to inferior.* ***(Right)*** *Coronal bone CT shows a crush injury to the inferior portion of the talus body, with extensive disruption to the posterior subtalar joint. This unusual injury occurred due to a fall from a height.*

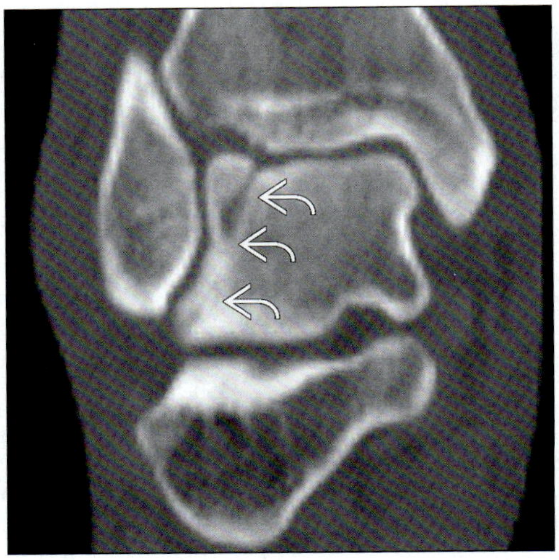

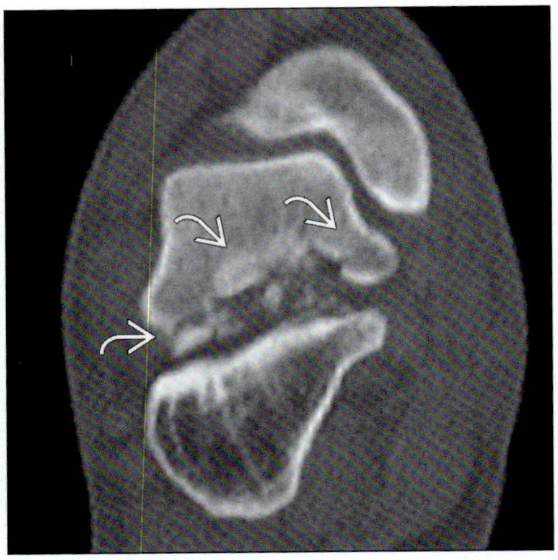

(Left) *Lateral radiograph shows a transversely oriented fracture near the tip of the lateral process of the talus and a fracture of the posterior process of the talus. These 2 fractures often occur together.* ***(Right)*** *Mortise view radiograph in the same patient shows fracture extending into the subtalar joint and the lateral ankle gutter. In many patients, the fracture is visible on only a single view.*

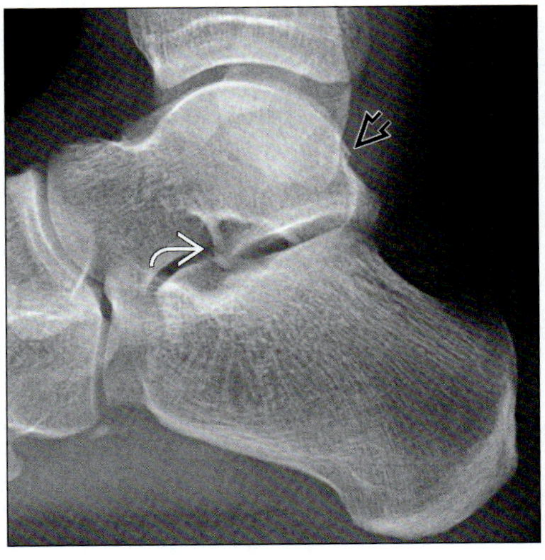

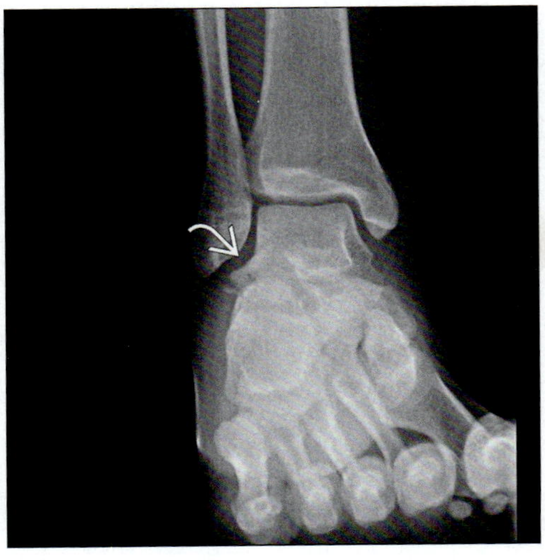

FRACTURE OF TALAR BODY AND PROCESS

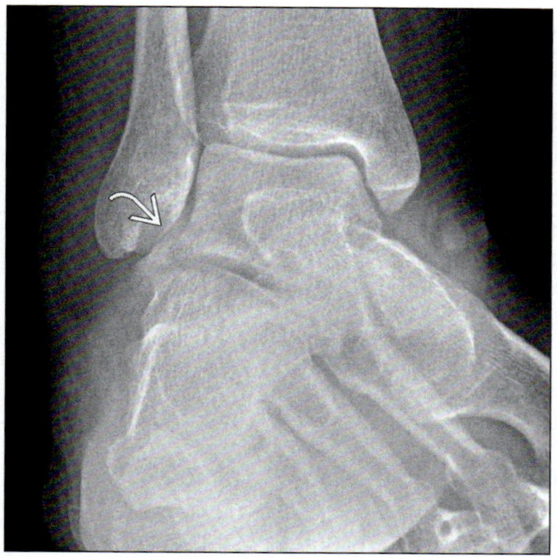

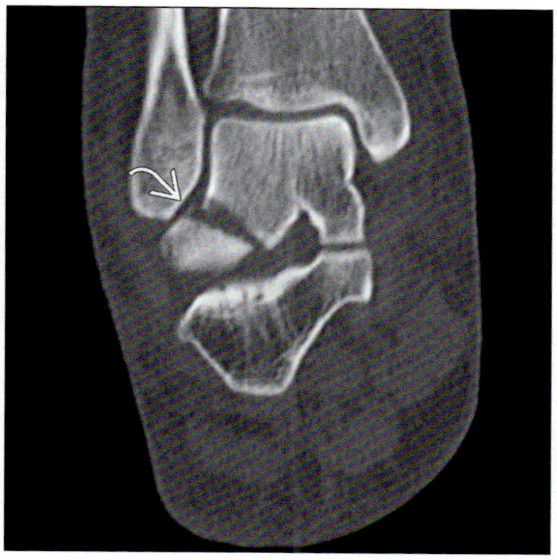

(Left) *Mortise view radiograph shows a lateral process talus fracture that appears to involve only a small portion of the posterior subtalar joint. Due to the complex anatomy of the joint, however, only a small portion of the joint surface is visible. CT scan is indicated to determine the extent of articular involvement.* ***(Right)*** *Coronal bone CT in the same patient shows that the lateral process fragment is large and involves a large portion of the joint.*

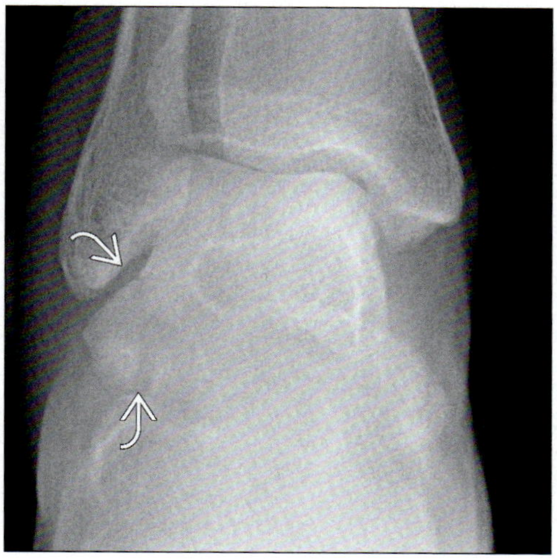

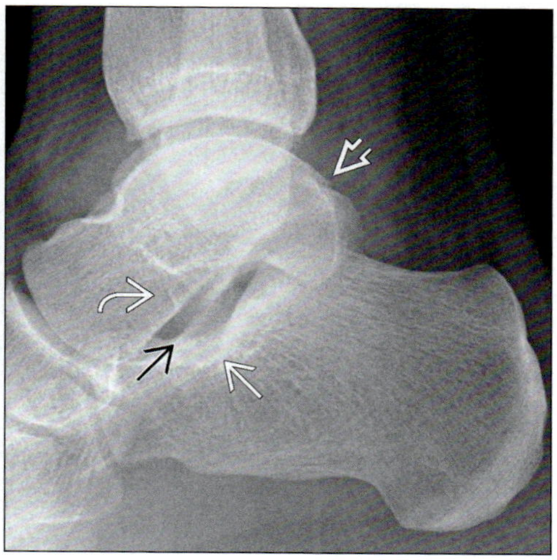

(Left) *AP radiograph shows a large lateral process fragment, which is rotated outwardly.* ***(Right)*** *Lateral radiograph in the same patient shows the lateral process displaced anteriorly, and its tip is no longer congruent with the angle of Gissane of the calcaneus. The posterior process of the talus is also fractured.*

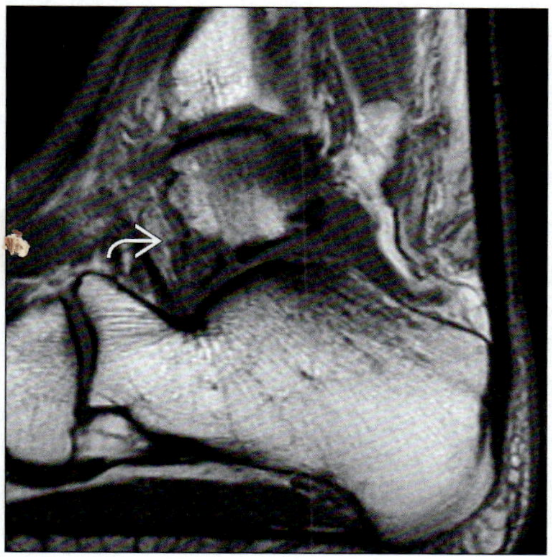

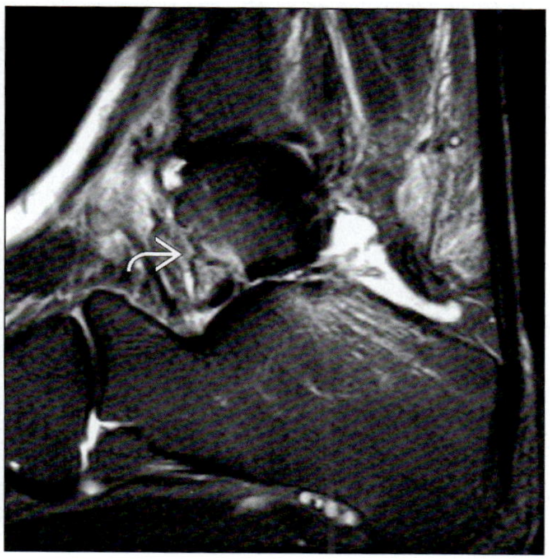

(Left) *Sagittal T1WI MR shows a chronic, nonunited fracture of the lateral process. Fracture was missed on initial radiographs. The patient returned 6 months later with persistent pain, and MR was performed at that time.* ***(Right)*** *Sagittal STIR MR in the same patient shows bone marrow edema surrounding the fracture. No callus is traversing the fracture line. The patient was treated with excision of the fragment.*

TALUS NECK AND HEAD FRACTURE

Key Facts

Imaging

- Talar neck fracture
  - Nondisplaced fractures difficult to see
  - When fracture displaced, subtalar joint usually subluxated or dislocated
  - Body of talus often rotated relative to ankle &/or subtalar joint
  - Talar neck may appear shortened
- Talar head fracture
  - Fracture line usually best seen on oblique view
- "Hawkins" sign
  - Develops 4-9 weeks after fracture
  - Lucent band paralleling subchondral bone plate
  - If present, reliably indicates that osteonecrosis (ON) will not develop
  - However, if absent, bone may still be vascularized
  - May involve only part of talar dome ("partial Hawkins" sign)

Top Differential Diagnoses

- Talar body fracture
- Talus dislocation
- Chip fracture of dorsal talar neck

Pathology

- Hawkins classification of neck fractures type I-IV
  - ON, infection risk increase with ↑ grade
  - Type I: Nondisplaced
  - Type II: Displaced + disruption of subtalar joint
  - Type III: Displaced + disruption of subtalar and ankle joints
  - Type IV: Displaced + disruption of subtalar, ankle, and talonavicular joints

Clinical Issues

- Radiographic osteonecrosis does not necessarily progress to articular collapse

*(**Left**) Lateral radiograph shows a nondisplaced (Hawkins type I) fracture of the neck of the talus ➡. These fractures are uncommon and difficult to see on radiographs. Often only the dorsal portion of the fracture line is visible. (**Right**) Lateral radiograph shows a nondisplaced fracture ➡ (Hawkins type I) extending vertically through the talar neck. Slight separation of the fracture dorsally makes this case easier to detect than many fractures of this type.*

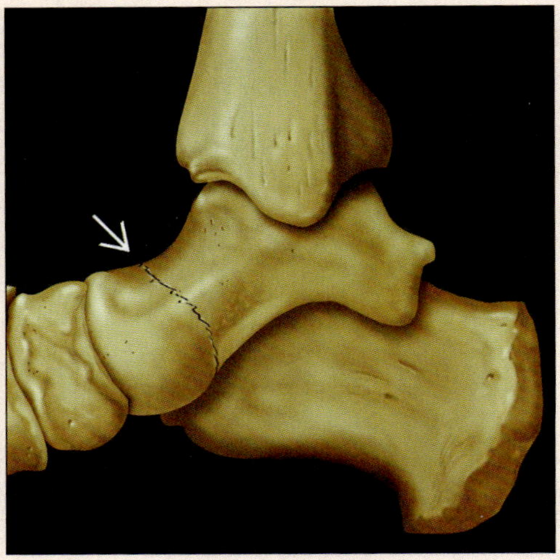

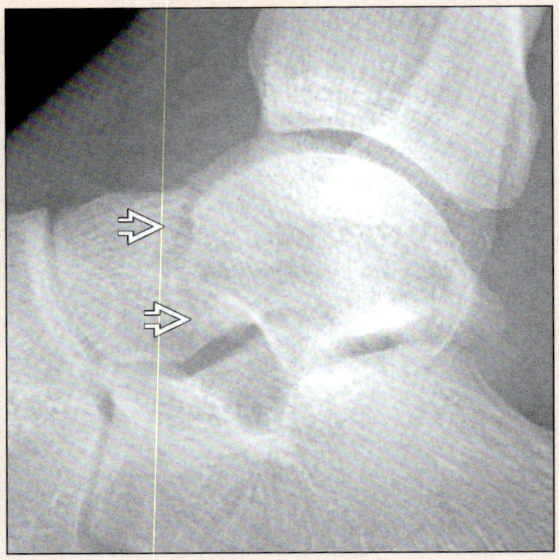

*(**Left**) Axial bone CT shows a nondisplaced (Hawkins type I) fracture ➡. CT may show unsuspected subluxation of subtalar joint, changing classification from Hawkins I to Hawkins II. (**Right**) Mortise view radiograph shows talar neck fracture ➡ and incongruity of the posterior subtalar joint ➡, indicating a Hawkins type II fracture. There are also small bony fragments at the lateral process of the talus ➡.*

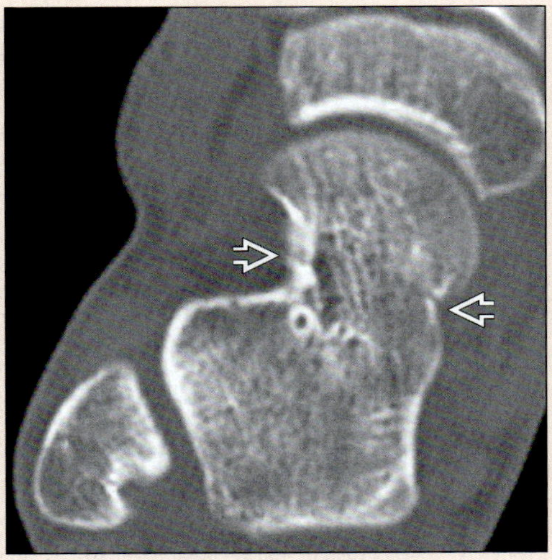

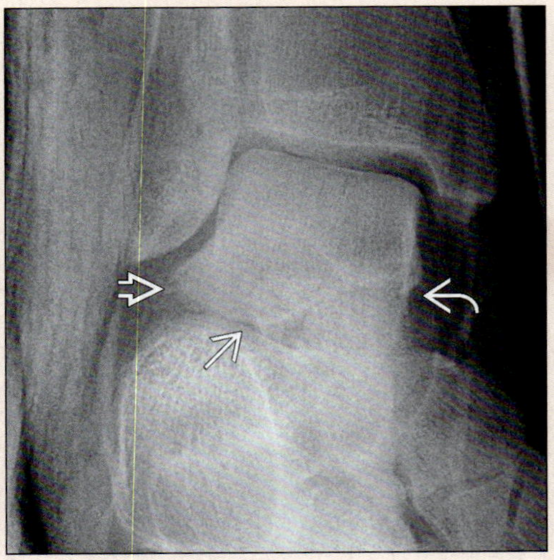

TALUS NECK AND HEAD FRACTURE

IMAGING

General Features

- Best diagnostic clue
  - Lucent fracture line or dorsal cortical offset of talar neck on lateral radiographs
- Morphology
  - Talar neck fractures vertically oriented
    - Usually midportion of neck
    - May extend into body of talus
  - Talar head fractures obliquely oriented, intraarticular

Radiographic Findings

- Talar neck fracture
  - Nondisplaced fractures difficult to see
  - If fracture displaced, subtalar joint usually disrupted
    - Look for loss of parallelism of joint surfaces
  - Body of talus often rotated relative to ankle &/or subtalar joint
  - Talar neck may appear shortened
  - In type III fractures, talar body may be extruded posterior to ankle
- Talar head fracture
  - Fracture line usually best seen on oblique radiograph of foot
  - Frequently see subluxation of talonavicular joint
  - May have associated cuboid fracture
- "Hawkins" sign
  - Develops 4-9 weeks after fracture
  - Lucent band paralleling subchondral bone plate of talar dome due to hyperemia
  - Form of disuse osteopenia, indicates that bone is vascularized
  - If present, reliably indicates that osteonecrosis (ON) will not develop
  - However, if absent, bone may still be vascularized
    - Abnormal Hawkins sign does not indicate inevitable ON or impending collapse
  - May involve only part of talar dome ("partial Hawkins" sign)
    - Area without Hawkins sign (generally less well-vascularized lateral dome of talus) is at risk for ON

CT Findings

- Frequently shows comminution of talar body not visible on radiographs
- Valuable in showing adduction deformity at neck
- Often results in change in Hawkins classification

Imaging Recommendations

- Best imaging tool
  - CT demonstrates occult fractures

DIFFERENTIAL DIAGNOSIS

Talus Dislocation

- May occur ± talus fracture

Dorsal Capsular Avulsion at Talar Neck

- Occurs at talar attachment of ankle joint capsule
- Horizontally oriented

PATHOLOGY

General Features

- Etiology
  - Talar neck fracture
    - Upward force on dorsiflexed foot
  - Talar head fracture
    - Compressive force on dorsiflexed foot
    - Direct blow
- Associated abnormalities
  - Other foot fractures are commonly present

Staging, Grading, & Classification

- Hawkins classification
  - Low interobserver variability, useful in predicting outcomes and directing treatment
  - Hawkins type I: Nondisplaced fracture of talar neck
    - 10% risk of osteonecrosis
  - Hawkins type II: Displaced fracture of talar neck + subluxation or dislocation of posterior subtalar joint
    - 30% risk of osteonecrosis
  - Hawkins type III: Displaced fracture of talar neck + subluxation or dislocation of posterior subtalar joint + subluxation or dislocation of ankle joint
    - 90% risk of osteonecrosis
  - Hawkins type IV: Displaced fracture of talar neck + subluxation or dislocation of posterior subtalar joint + subluxation or dislocation of ankle joint + dislocation of talonavicular joint
    - 90% risk of osteonecrosis

CLINICAL ISSUES

Natural History & Prognosis

- Risk of ON and infection increase with Hawkins grade
- Radiographic ON does not necessarily progress to articular collapse
- Talar head fracture has high risk of talonavicular osteoarthritis

DIAGNOSTIC CHECKLIST

Image Interpretation Pearls

- Displaced talar neck fracture implies subtalar joint disruption
- Adduction deformity at neck may be missed clinically and on radiographs
  - Neck appears short on radiographs
- Talar head fracture often missed on radiographs because it is rare

Reporting Tips

- Hawkins classification should always be used
- On follow-up, be alert for signs of ON or infection

SELECTED REFERENCES

1. Tezval M et al: Prognostic reliability of the Hawkins sign in fractures of the talus. J Orthop Trauma. 21(8):538-43, 2007
2. Adelaar RS et al: Avascular necrosis of the talus. Orthop Clin North Am. 35(3):383-95, xi, 2004
3. Vallier HA et al: Talar neck fractures: results and outcomes. J Bone Joint Surg Am. 86-A(8):1616-24, 2004

TALUS NECK AND HEAD FRACTURE

*(**Left**) AP radiograph shows a nondisplaced fracture of the talar neck. Talar neck fractures are especially difficult to see on AP radiographs, but can usually be detected with careful windowing on PACS. (**Right**) Lateral radiograph shows a Hawkins type II talar neck fracture extending into the talar body. There is subluxation of the posterior subtalar joint. Talar body and process fractures are commonly present with displaced talar neck fractures.*

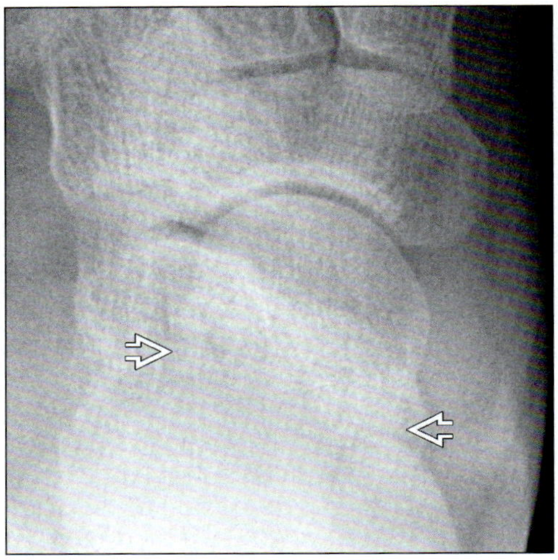

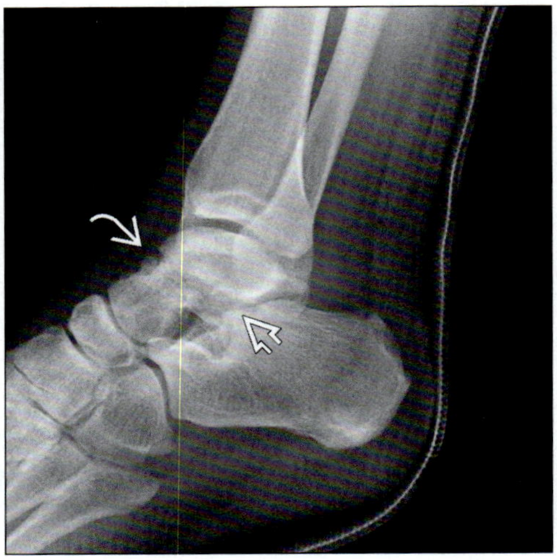

*(**Left**) Lateral radiograph shows a Hawkins type II fracture. The displaced fracture of the talar neck results in apparent shortening of the neck. There is subluxation of the subtalar joint, a fracture of the lateral process, and a nondisplaced intraarticular tongue-type fracture of the calcaneus. (**Right**) AP radiograph in the same patient shows the talar head completely displaced medial to the talar neck. This patient suffered the injury in a fall from a roof.*

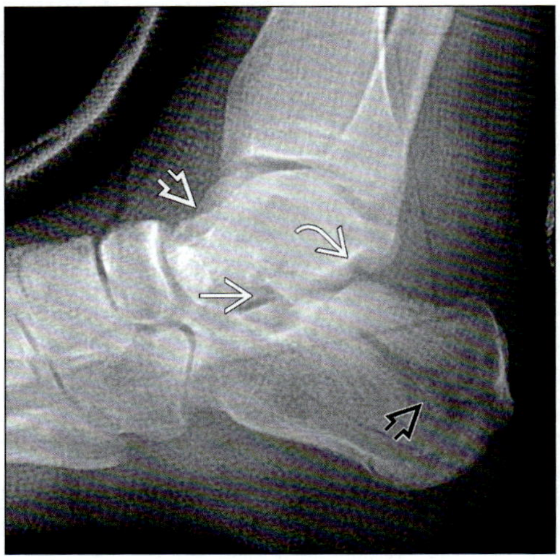

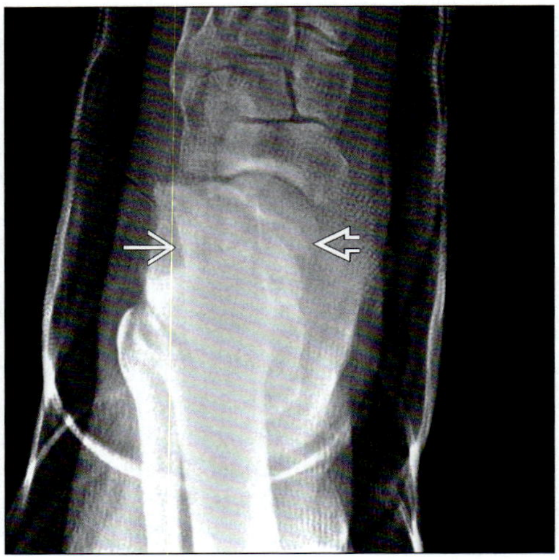

*(**Left**) Axial bone CT in the same patient shows the displacement, impaction, and rotation of the fracture, and the subluxation of the posterior subtalar joint. (**Right**) Coronal bone CT in the same patient shows the talar head displaced beneath the medial malleolus. A bone fragment is displaced into the tarsal sinus.*

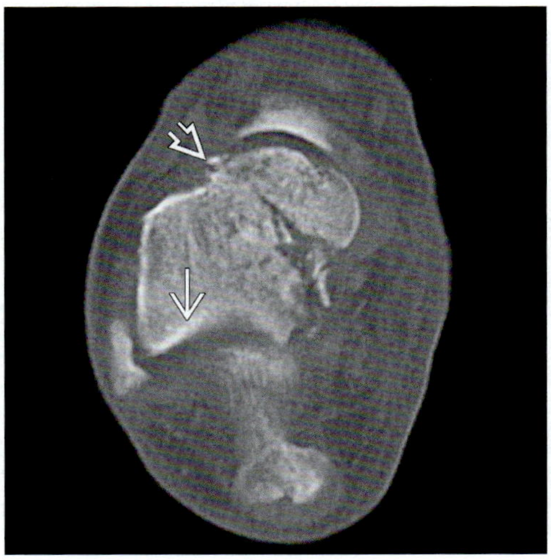

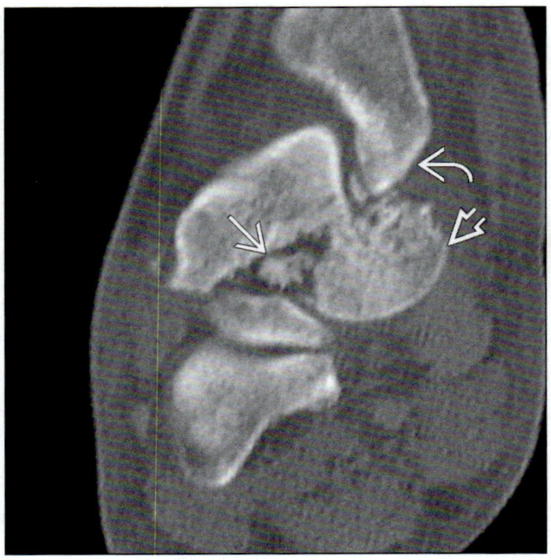

TALUS NECK AND HEAD FRACTURE

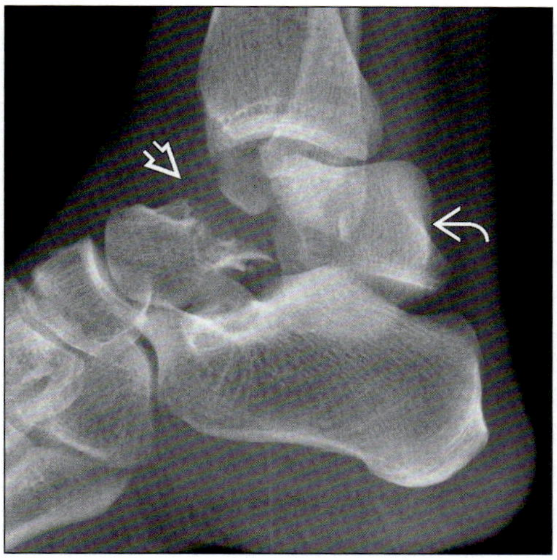

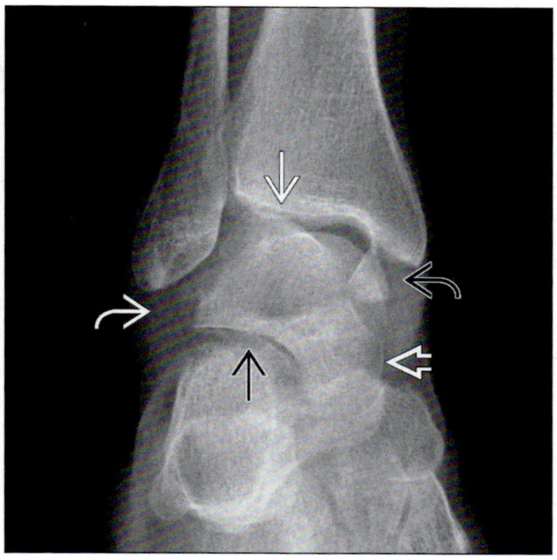

(Left) *Lateral radiograph shows a Hawkins type III fracture. The body of the talus is rotated 90° relative to the tibia and calcaneus, indicating complete dislocation of both joints. The talonavicular joint remains intact.* ***(Right)*** *AP radiograph in the same patient shows the posterior process of the talus rotated laterally and the talar neck projecting medially. The ankle and subtalar joints are completely incongruent. The medial malleolus is fractured.*

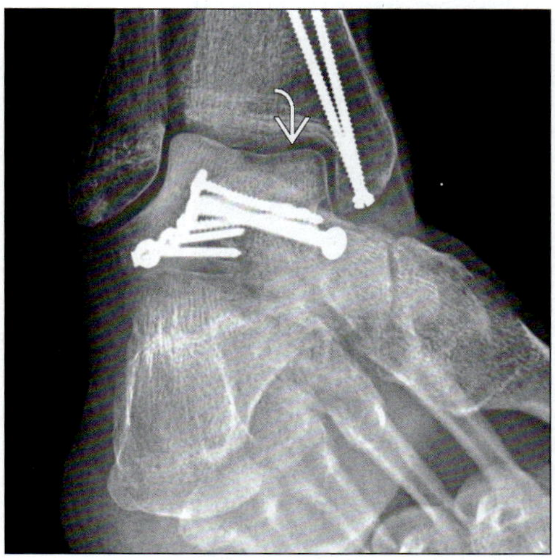

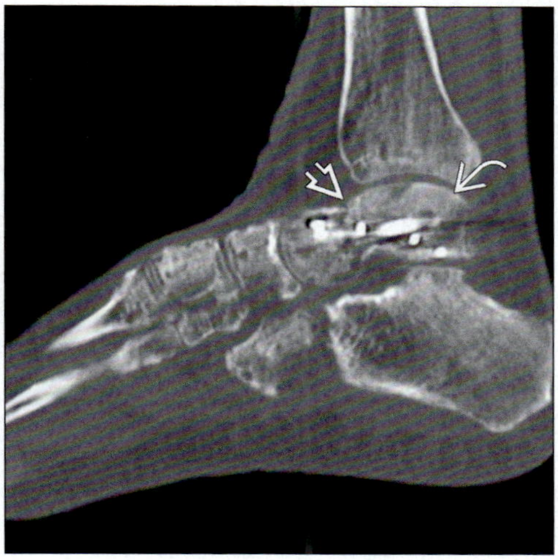

(Left) *Mortise view radiograph shows fixation of a talar neck fracture in a patient with a partial Hawkins sign. The medial portion of the talar body is better vascularized than the lateral, where the Hawkins sign is absent.* ***(Right)*** *Sagittal bone CT shows disuse osteopenia of the foot and sclerosis of the talar dome 3 months after fixation of a Hawkins type III fracture. Talar dome sclerosis indicates poor vascularity, but does not necessarily progress to articular surface collapse.*

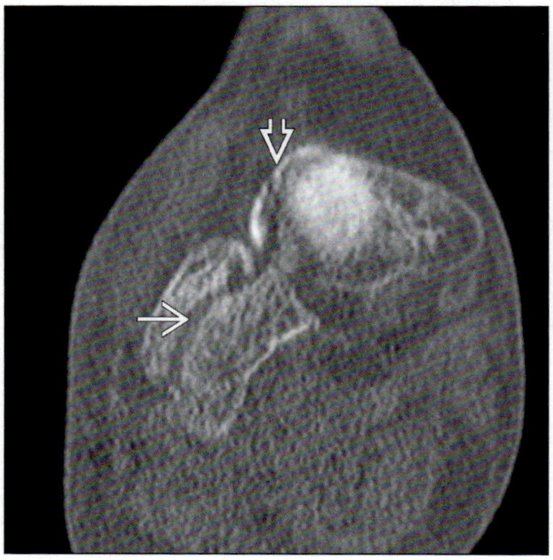

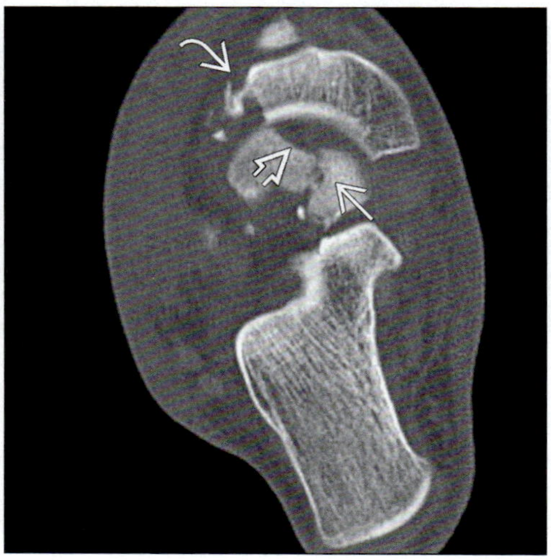

(Left) *Coronal bone CT shows a vertical fracture of the talar head and a cuboid fracture along the same plane. Talar head fractures are often associated with cuboid fracture &/or talonavicular joint injury.* ***(Right)*** *Axial bone CT shows comminuted fx of the talar head with extensive depression of the articular surface and subluxation of the talonavicular joint. There is an associated fracture of the navicular.*

TALUS DISLOCATION

Key Facts

Imaging

- Subtalar joint dislocation is most common type
  - Lateral dislocation is more common than medial
  - Abnormal superimposition of talus and calcaneus on lateral radiograph
  - Naked facet sign: Subtalar articular surface separated from calcaneus
  - Rotation of talus relative to calcaneus
- Talonavicular joint dislocation
  - Rare in isolation, usually see other fractures &/or dislocations
  - Sometimes involves entire Chopart joint (talonavicular + calcaneocuboid)
- Total (pantalar) dislocation
  - Talus extruded from all articulations

Top Differential Diagnoses

- Ankle dislocation
- Talar neck fracture
- Chopart dislocation

Pathology

- Injury to tibial nerve, posterior tibial artery and veins
  - Early reduction critically important to protect neurovascular structures
- Fractures of foot, ankle are commonly present

Diagnostic Checklist

- Always call report to requesting physician
- Pre-reduction radiographs often poor due to difficulty positioning patient
  - Repeat radiographs after reduction to fully characterize injury
- Do not delay reduction to obtain CT

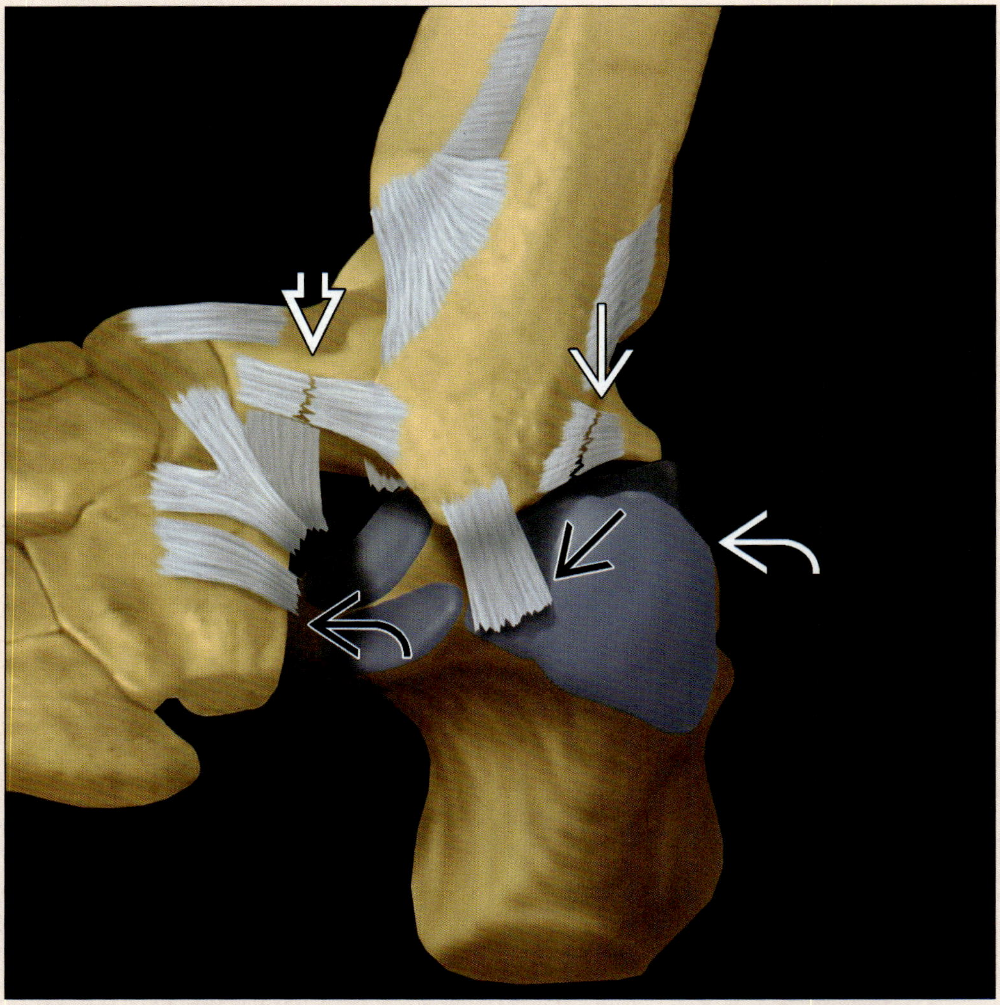

Graphic shows subtalar dislocation with medial rotation of the calcaneus, so that the naked articular surface ➡ of the calcaneus faces laterally. This injury usually occurs when the patient falls and lands on an inverted foot. The anterior talofibular ➡, calcaneofibular ➡, and posterior talofibular ➡ ligaments are torn. The talus remains normally aligned with the ankle and talonavicular joints. The calcaneocuboid joint ➡ is dislocated.

TALUS DISLOCATION

TERMINOLOGY

Definitions

- Chopart joint: Talonavicular and calcaneocuboid articulations

IMAGING

General Features

- Location
  - Most commonly at subtalar joint
    - ± involvement of talonavicular joint
  - May have isolated talonavicular dislocation
- Morphology
  - Subtalar joint dislocation
    - Medial dislocation is more common than lateral dislocation
    - Anterior or posterior dislocation is very rare
  - Talonavicular joint dislocation
    - Sometimes involves entire Chopart joint
  - Total (pantalar) dislocation
    - Talus extruded from all articulations
  - Subtalar subluxation
    - May occur in ballet dancers due to chronic repetitive stress

Radiographic Findings

- Partial superimposition of talus and calcaneus on lateral radiograph
- Rotation of talus relative to ankle, calcaneus
- Loss of joint congruity
- Naked facet sign
  - Subtalar articular facet visible but not articulating with calcaneus
- Rarely, extrusion of talus from all articulations

CT Findings

- Commonly shows associated fractures not visible on radiographs
- Performed after reduction of dislocation
- May show persistent subluxation after attempted reduction

Imaging Recommendations

- Best imaging tool
  - Radiographs
- Protocol advice
  - Do not expect good radiographs; deformity often precludes them
  - Rapid reduction is more important than good radiographs
  - Repeat radiographs should be obtained post reduction
    - Evaluate reduction
    - Evaluate fractures (often unsuspected on prereduction images)

DIFFERENTIAL DIAGNOSIS

Ankle Dislocation

- Subtalar joint may be obscured as in subtalar dislocation
  - Due to inward or outward rotation of hindfoot

Talar Neck Fracture

- Hawkins type II fracture associated with subtalar subluxation/dislocation
- Hawkins type III fracture associated with disruption of both subtalar and ankle joints
- Hawkins type IV fracture associated with pantalar dislocation

Chopart Dislocation

- Dislocation of talonavicular + calcaneocuboid joints

PATHOLOGY

General Features

- Etiology
  - Due to fall, landing on inverted or everted foot
- Associated abnormalities
  - Injury to tibial nerve, posterior tibial artery and veins
  - Fractures of foot, ankle are commonly present
    - Due to impaction or avulsion

Relevant Anatomy

- Talus articulates with tibia and fibula, navicular, calcaneus
- Subtalar joint has 3 facets: Posterior, middle, and anterior
  - Posterior facet largest; 50% of weight-bearing across this joint
- Posterior tibial neurovascular bundle passes behind medial malleolus, then through tarsal tunnel

CLINICAL ISSUES

Presentation

- Most common signs/symptoms
  - Swollen, painful foot; inability to walk
  - Foot deformity
    - Calcaneus displaced relative to leg

DIAGNOSTIC CHECKLIST

Reporting Tips

- Always call report to requesting physician
  - Early reduction is critically important to protect neurovascular structures

SELECTED REFERENCES

1. Van Opstal N et al: Traumatic talus extrusion: case reports and literature review. Acta Orthop Belg. 75(5):699-704, 2009
2. Wagner R et al: Talar dislocations. Injury. 35 Suppl 2:SB36-45, 2004
3. Bibbo C et al: Missed and associated injuries after subtalar dislocation: the role of CT. Foot Ankle Int. 22(4):324-8, 2001
4. Kanda T et al: Anterior dislocation of the subtalar joint: a case report. Foot Ankle Int. 22(7):609-11, 2001
5. Tucker DJ et al: Lateral subtalar dislocation: review of the literature and case presentation. J Foot Ankle Surg. 37(3):239-47; discussion 262, 1998
6. DeLee JC et al: Subtalar dislocation of the foot. J Bone Joint Surg Am. 64(3):433-7, 1982

TALUS DISLOCATION

(Left) *Lateral radiograph shows complete talar dislocation from ankle, subtalar, and talonavicular articulations. The talus has rotated almost 90°. This patient was injured during a motorcycle accident.* ***(Right)*** *AP radiograph in the same patient shows the head of the talus facing laterally. A fracture of the medial malleolus is also present.*

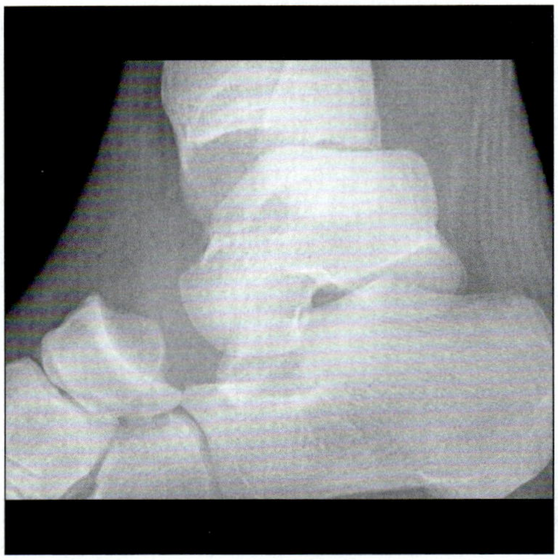

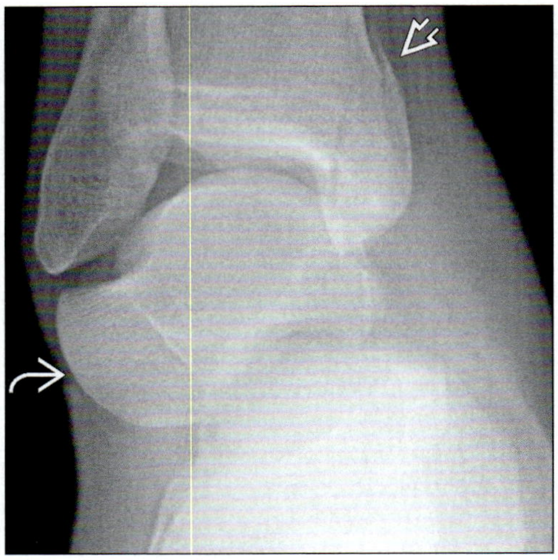

(Left) *Lateral radiograph shows anterior dislocation of the talus. The head of the talus overlies the navicular. The talar articular surface at the posterior subtalar joint is displaced anterior to the corresponding calcaneal articular surface.* ***(Right)*** *AP radiograph in the same patient shows the calcaneus to be medial to the talus and tibia, and the talar head is lateral to the navicular.*

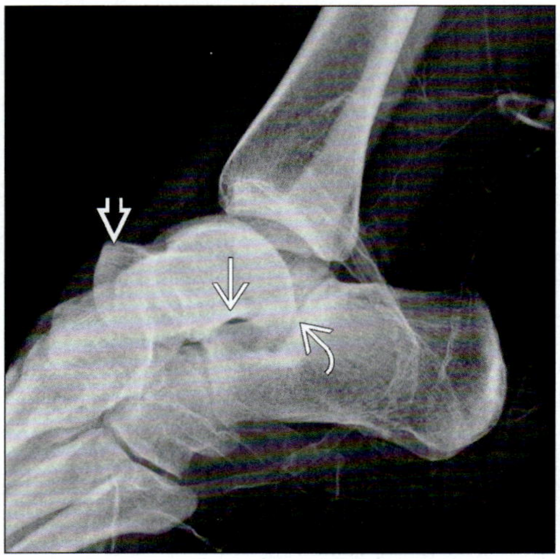

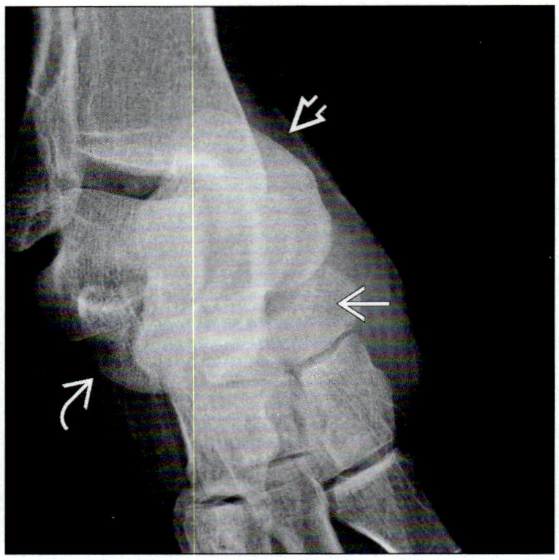

(Left) *Lateral radiograph shows the posterior talar articular facet overlying the calcaneus, inferior to the calcaneal articular surface. The middle facet does not articulate with the sustentaculum tali. There is a fracture of the posterior malleolus and a fracture of the posterior talar body.* ***(Right)*** *AP radiograph in the same patient shows the calcaneus dislocated medially and the naked articular facet of the talus at the posterior subtalar joint.*

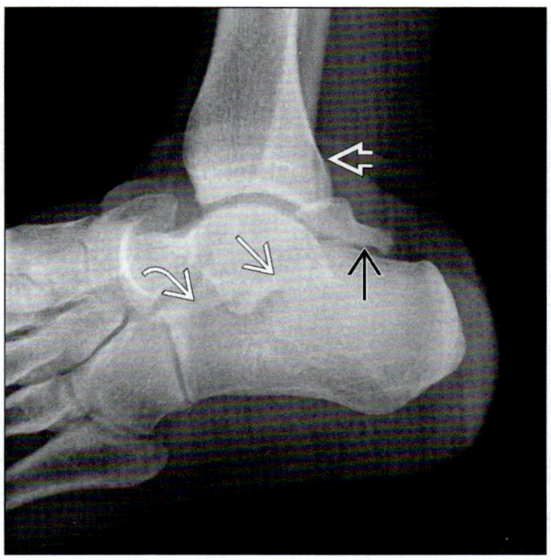

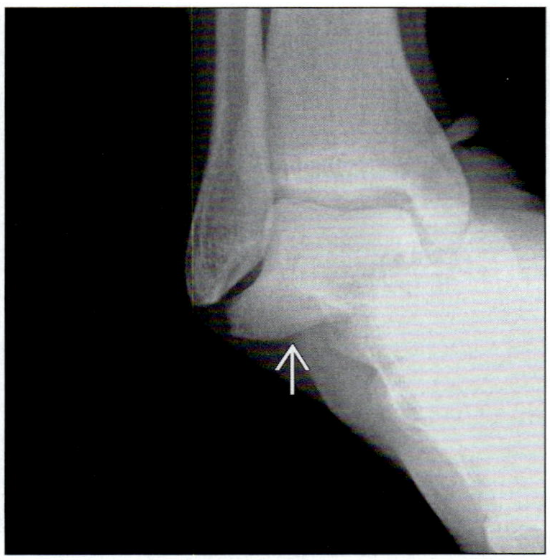

TALUS DISLOCATION

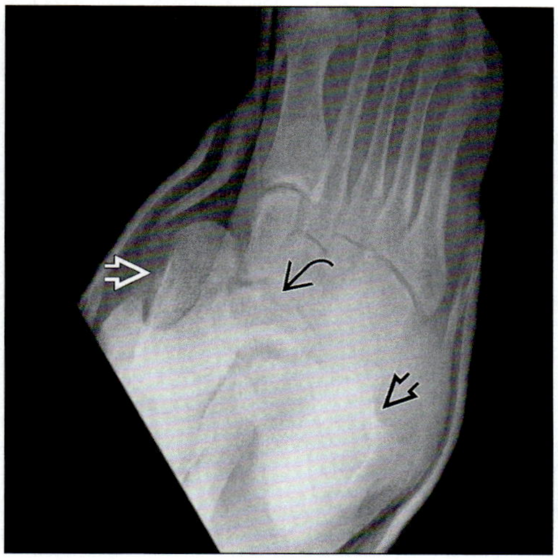

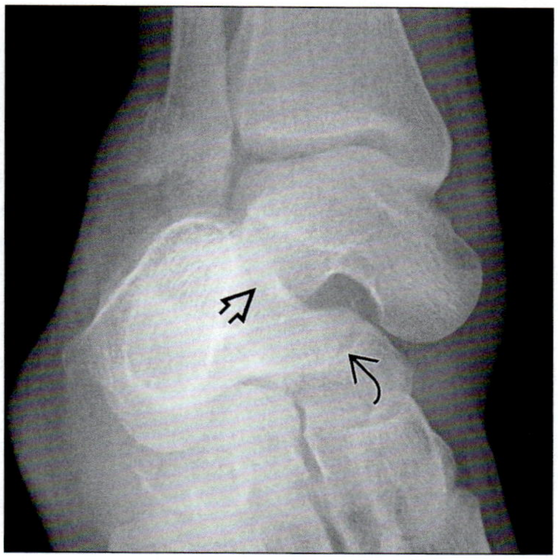

(Left) *AP radiograph of the foot shows the talus medially and anteriorly positioned, while the navicular and calcaneus, together with the remainder of the foot, are displaced laterally.* ***(Right)*** *Mortise view radiograph in the same patient shows the posterior subtalar articular surface of the calcaneus and the articular surface of the navicular displaced laterally. Lateral subtalar dislocation is less common than medial dislocation.*

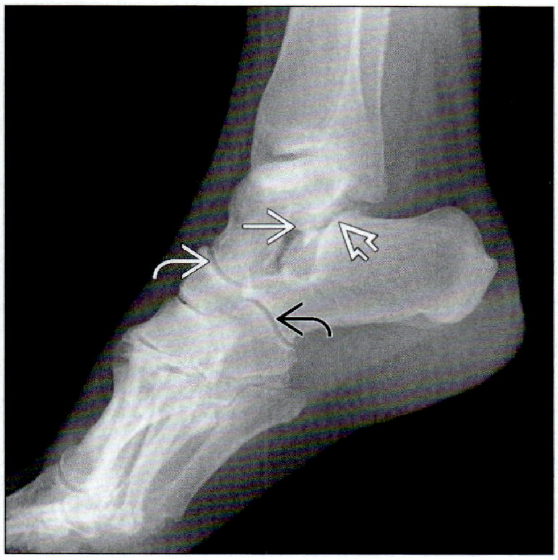

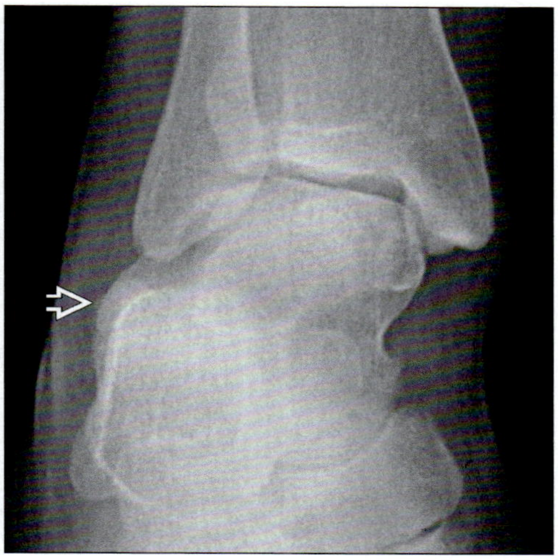

(Left) *Lateral radiograph obtained after attempted reduction of subtalar dislocation shows a fracture of the lateral process and an incompletely reduced joint. The talonavicular and calcaneocuboid joints are intact; these 2 articulations comprise the joint of Chopart.* ***(Right)*** *AP radiograph in the same patient shows the lateral position of the calcaneus relative to the talus. CT was recommended to determine abnormality preventing reduction.*

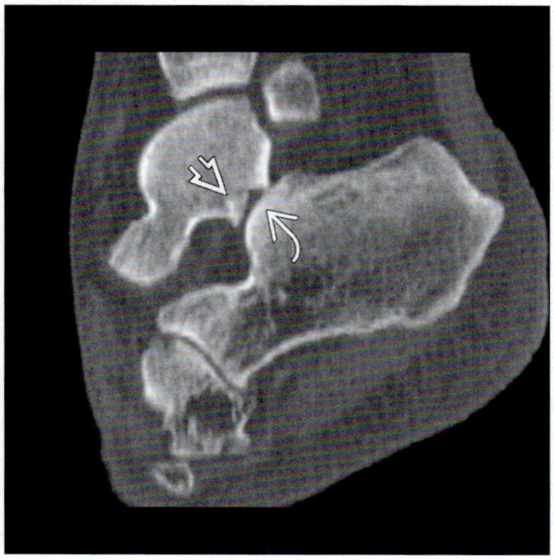

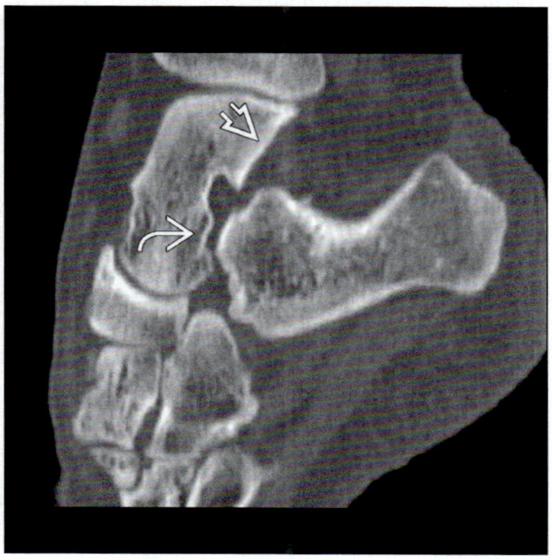

(Left) *Sagittal bone CT in the same patient shows the fracture of talus impacted against the posterior subtalar facet of the calcaneus and preventing reduction.* ***(Right)*** *Sagittal bone CT in the same patient shows the articular surfaces of the talus at the posterior and middle facets to be displaced relative to their expected calcaneal articulation. Bone CT is usually not performed in cases of subtalar dislocation because of the urgent need for rapid reduction in order to prevent vascular compromise.*

CALCANEUS FRACTURE, INTRAARTICULAR

Key Facts

Terminology

- Böhler angle: On lateral radiograph, angle between line drawn from posterior margin of calcaneal tuberosity and posterior margin of subtalar joint, and line from anterior margin of calcaneus to posterior margin of subtalar joint
  - Normal: 20-40°

Imaging

- Main (shear) fracture line in sagittal plane through posterior subtalar joint
- Comminuted (compression) fracture lines often radiate away from central fracture line
- Medial fragment of variable size; includes sustentaculum tali (constant fragment)
- Posterolateral fragment includes calcaneal tuberosity
- Böhler angle usually decreased
- Axial view (Harris) used to assess articular involvement
- CT findings
  - Comminution usually more extensive than evident on radiographs
  - Evaluate number of sagittal fracture lines entering posterior subtalar joint
  - Evaluate degree of depression of articular fragments
  - Soft tissue windows useful for associated tendon injuries

Pathology

- Associated injuries
  - Other axial load fractures of lower extremities, pelvis, spine
  - Tear of peroneal retinaculum
  - Entrapment of peroneal or flexor hallucis longus tendons into fracture

__(Left)__ Graphic shows a lateral view of an axial load injury of the calcaneus, involving the posterior subtalar joint. The posterior subtalar articular surface is depressed by a coronally oriented fracture, which depresses the Böhler angle. __(Right)__ Sagittal bone CT through the lateral portion of an intraarticular calcaneal fracture shows the articular surface is not only depressed, but rotated anteriorly > 45°. Multiple secondary fracture lines radiate out from the point of impaction.

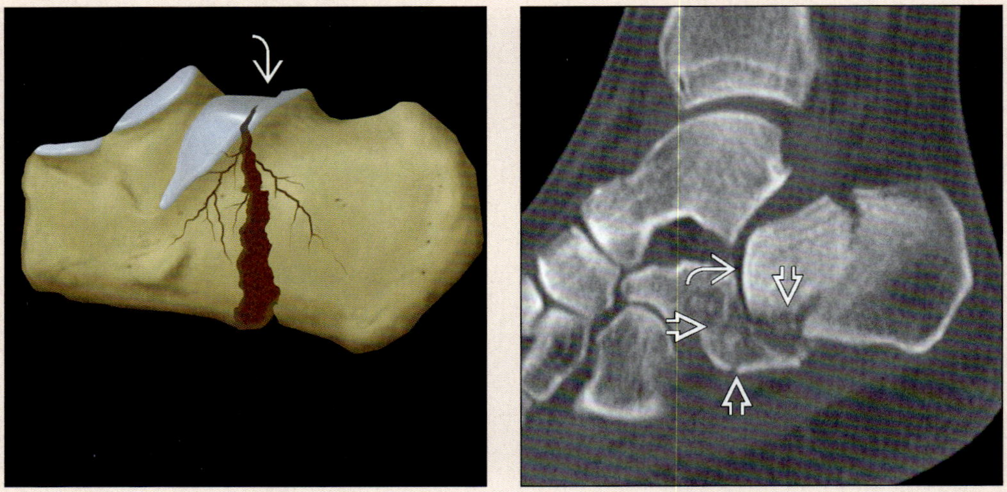

__(Left)__ Sagittal bone CT further medially in the same patient shows stellate appearance of the fracture, with multiple lines radiating out from the point of impact. Note talar osteochondral injury. __(Right)__ Coronal bone CT in the same patient shows the depressed, rotated fragment of articular surface, which splays the mediolateral dimension of the calcaneus. Comminution of the medial wall is another important finding for surgical planning.

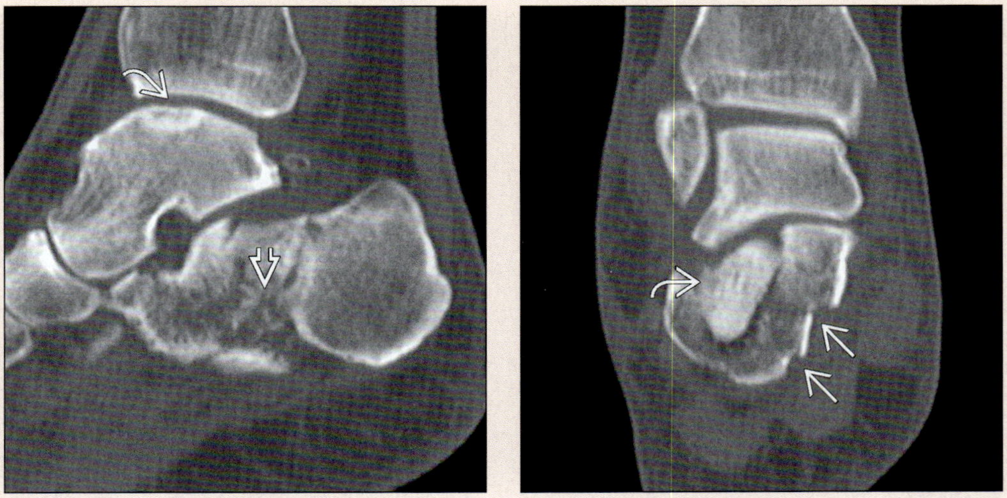

CALCANEUS FRACTURE, INTRAARTICULAR

TERMINOLOGY

Definitions

- Böhler angle: On lateral radiograph, angle between line drawn from posterior margin of calcaneal tuberosity and posterior margin of subtalar joint, and line from anterior margin of calcaneus to posterior margin of subtalar joint
  - Normal: 20-40°
- Gissane angle: On lateral radiograph, concavity of calcaneus at anterior margin of posterior subtalar joint
  - Measurement of this angle does not add information; not commonly performed

IMAGING

General Features

- Morphology
  - Main (shear) fracture line in sagittal plane through posterior subtalar joint
  - Comminuted (compression) fracture lines often radiate away from central fracture line
  - Medial fragment of variable size; includes sustentaculum tali (constant fragment)
  - Posterolateral fragment includes calcaneal tuberosity
  - Depressed portion of subtalar joint called thalamic fragment
  - Fracture may extend into calcaneocuboid joint &/or to posterior process of calcaneus

Radiographic Findings

- Böhler angle usually decreased
- Calcaneal height lost in central portion
- Articular surface of calcaneus at subtalar joint rotated into more vertical orientation
  - Forms sclerotic crescent beneath articular surface
- Lucent fracture line may be visible
- Axial (Harris or Harris-Beath) view: Shows articular depression, splaying of calcaneus
- Broden view (40° internal oblique): Shows depression of articular surface

CT Findings

- Comminution usually more extensive than evident on radiographs
- Evaluate degree of comminution of medial and lateral calcaneal wall
  - Useful in surgical planning fixation
- Look for fragments entrapped in medullary cavity, which may impede reduction
- Coronal images key to Sanders classification
  - Evaluate number of sagittal fracture lines entering posterior subtalar joint
  - Evaluate degree of depression of articular fragments
- Use sagittal reformations to measure Böhler angle
- Soft tissue windows useful for associated tendon injuries

Imaging Recommendations

- Best imaging tool
  - CT scan with sagittal, coronal reformations
- Protocol advice
  - Broden view: Foot internally rotated 40°
    - Exposures obtained with differing cephalad beam angulation show different parts of joint
    - 40° angulation shows anterior portion of joint
    - 10° angulation shows posterior portion of joint
  - Axial view
    - Foot dorsiflexed
    - x-ray beam centered just above posterior process of calcaneus
    - x-ray beam angled 35-45° caudad
  - CT
    - Coronal reformations should be perpendicular to neck of talus
    - Always obtain soft tissue algorithm in addition to bone algorithm

DIFFERENTIAL DIAGNOSIS

Insufficiency Fracture

- Usually spares subtalar articular surface
- May affect posterior process or central portion of calcaneus

Pathologic Fracture

- Commonly involves articular surface
- Most commonly due to simple cyst or lipoma located centrally in calcaneus

Subtalar Dislocation

- Incongruity of subtalar joint
- Fracture of talus commonly associated

Sustentaculum Tali Fracture

- Occasionally seen as isolated fracture, more commonly associated with intraarticular fracture

PATHOLOGY

General Features

- Associated abnormalities
  - Other ipsilateral and contralateral axial load injuries
  - Tear of superior peroneal retinaculum
    - Often unrecognized at time of injury
    - Leads to tendon subluxation, lateral ankle pain
    - May see flake of bone at retinacular attachment on radiographs or CT
    - Retinaculum discontinuous on CT, usually at fibular attachment
  - Entrapment of peroneal tendons into fracture
  - Peroneal tendon stenosing tenosynovitis due to hematoma, scar
  - Entrapment of flexor hallucis longus into fracture (uncommon)

Staging, Grading, & Classification

- Essex-Lopresti classification
  - Tongue type
    - Horizontal fracture extends posteriorly to tuberosity of calcaneus
  - Joint depression type
    - Fracture lines centrally located
- Sanders classification
  - Classified by number of fracture lines extending through articular surface of posterior subtalar joint
    - I: Nondisplaced
    - II: Single displaced fracture line

CALCANEUS FRACTURE, INTRAARTICULAR

The number of fracture fragments is designated by a Roman numeral; location is categorized by a letter as shown.

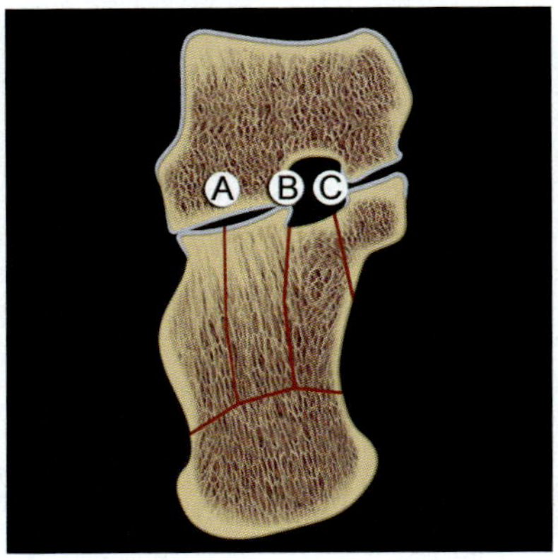

A type IIA fracture has a single displaced fracture line extending into the lateral 1/3 of the joint.

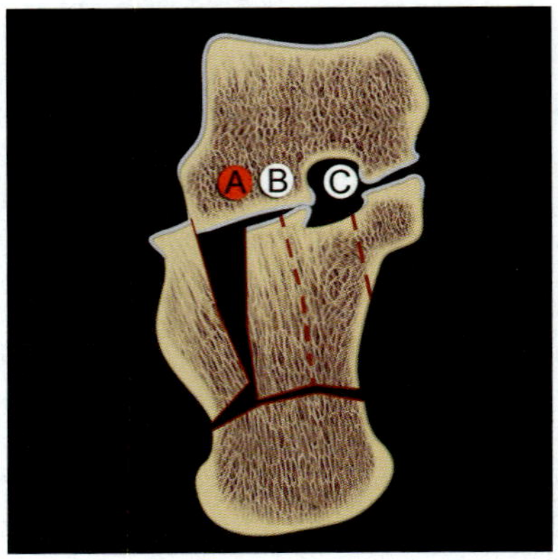

A type IIIBC fracture has displaced fracture lines extending into the middle and medial 1/3 of the joint.

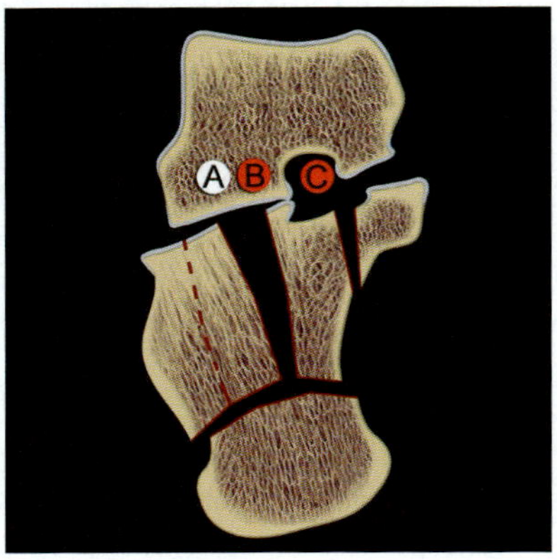

A type IVABC fracture has displaced fracture lines extending into the lateral, middle, and medial 1/3 of the joint.

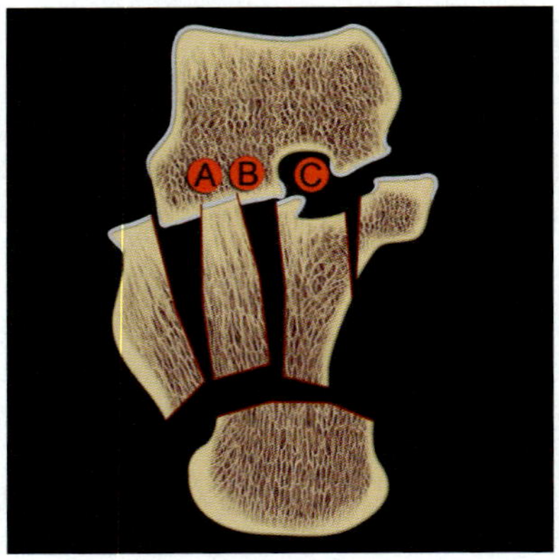

Sanders classification is based on the number and location of displaced intraarticular fracture lines. Type I fractures are nondisplaced, type II have 1 displaced intraarticular fracture line, type III have 2 displaced intraarticular fracture lines, and type IV have 3 displaced intraarticular fracture lines.

        - III: 2 displaced fracture lines
        - IV: Highly comminuted
    - Further classified by position of main fracture line through posterior subtalar joint
        - A: Fracture through lateral 1/3 of joint
        - B: Fracture through middle 1/3 of joint
        - C: Fracture through medial 1/3 of joint

CLINICAL ISSUES

Presentation

- Clinical profile
    - Due to fall from elevated height, landing on heel
    - Commonly bilateral
    - 75% of calcaneal fractures

Natural History & Prognosis

- Most develop subtalar osteoarthritis
- May undergo autofusion of subtalar joint

Treatment

- Usually treated with ORIF unless nondisplaced or comorbidity present

DIAGNOSTIC CHECKLIST

Image Interpretation Pearls

- Fracture may be present despite normal Böhler angle
- High index of suspicion for other foot fractures

Reporting Tips

- Important findings to report on radiographs
    - Depression of Böhler angle
    - Extension to calcaneocuboid joint, posterior process
    - Degree of comminution of lateral, medial walls (use axial view)
- Important findings to report on CT scans
    - Sanders classification
    - Extension to calcaneocuboid joint, posterior process
    - Degree of comminution of medial and lateral walls
    - Comminuted intraarticular fragments, which may impede reduction
    - Peroneal retinaculum injury
    - Tendon entrapment or laceration

SELECTED REFERENCES

1. Schepers T et al: Calcaneal fracture classification: a comparative study. J Foot Ankle Surg. 48(2):156-62, 2009
2. Daftary A et al: Fractures of the calcaneus: a review with emphasis on CT. Radiographics. 25(5):1215-26, 2005
3. Sanders R: Displaced intra-articular fractures of the calcaneus. J Bone Joint Surg Am. 82(2):225-50, 2000
4. Rosenberg ZS et al: Peroneal tendon injury associated with calcaneal fractures: CT findings. AJR Am J Roentgenol. 149(1):125-9, 1987

CALCANEUS FRACTURE, INTRAARTICULAR

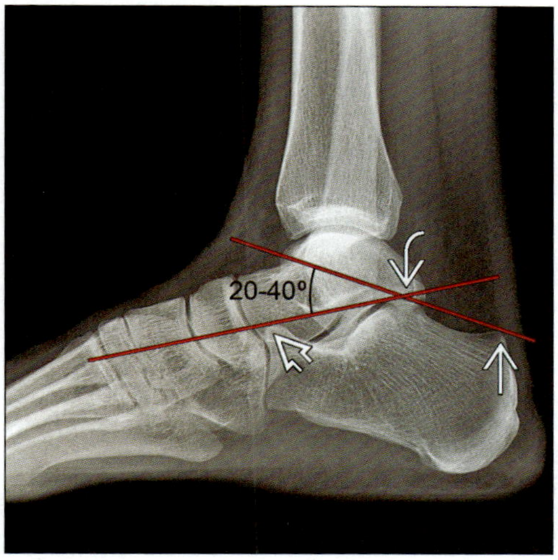

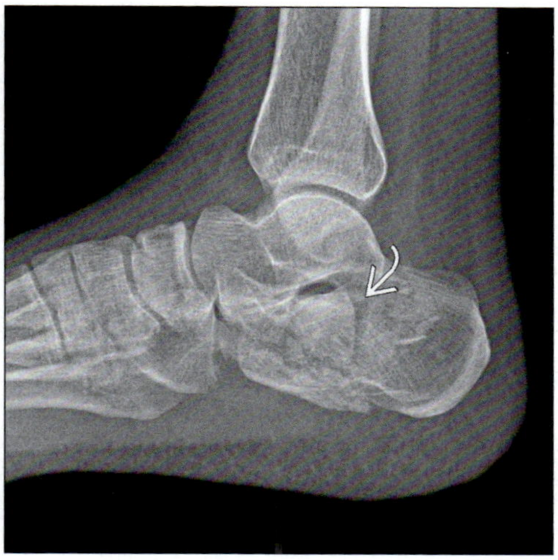

(Left) *Lateral radiograph shows a normal Böhler angle, between lines drawn from the posterior process to the posterior margin of the subtalar joint, and from the anterior process of the calcaneus to the posterior margin of the subtalar joint.* ***(Right)*** *Lateral radiograph shows loss of Böhler angle in a severely comminuted and depressed calcaneal fracture. Note the large, depressed fragment of the subchondral bone plate of the posterior subtalar joint.*

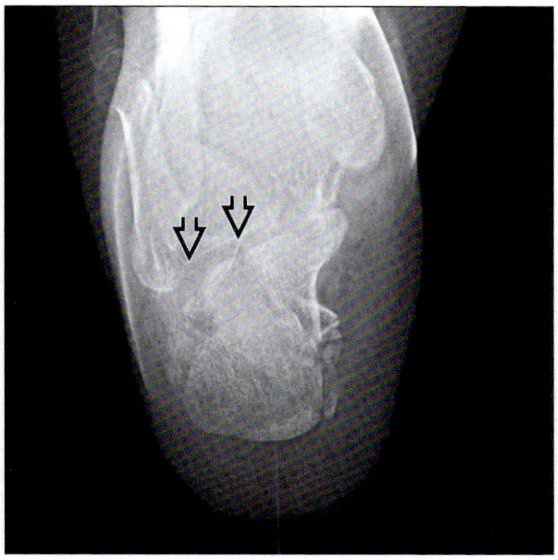

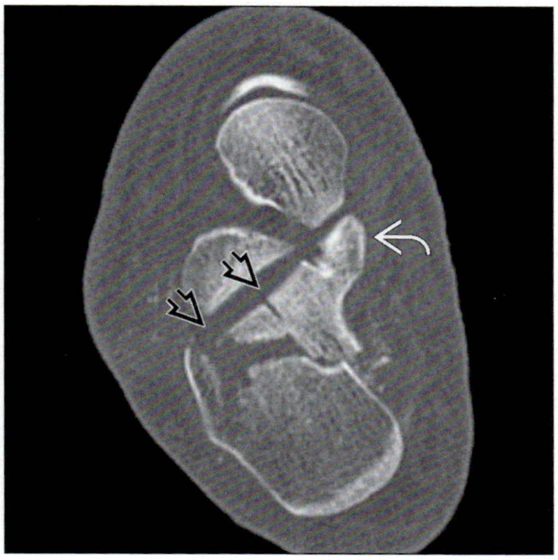

(Left) *Axial (Harris) radiograph in the same patient shows 2 depressed fracture lines entering the posterior subtalar joint. This is type IIIAB in the Sanders classification.* ***(Right)*** *Axial bone CT in the same patient confirms a type IIIAB fracture. Note the lateral displacement of the posterior process fragment relative to constant fragment (sustentaculum tali).*

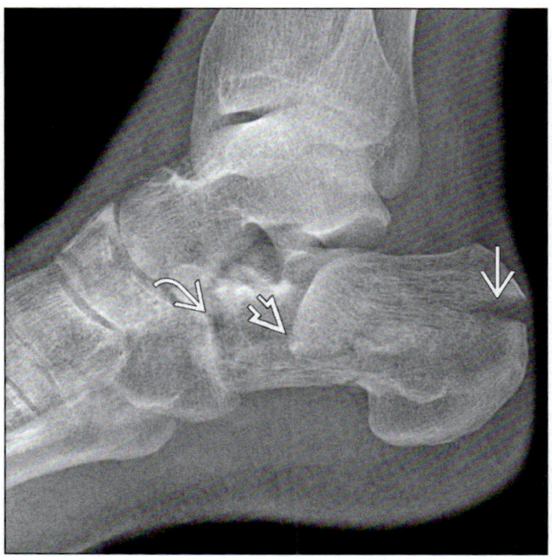

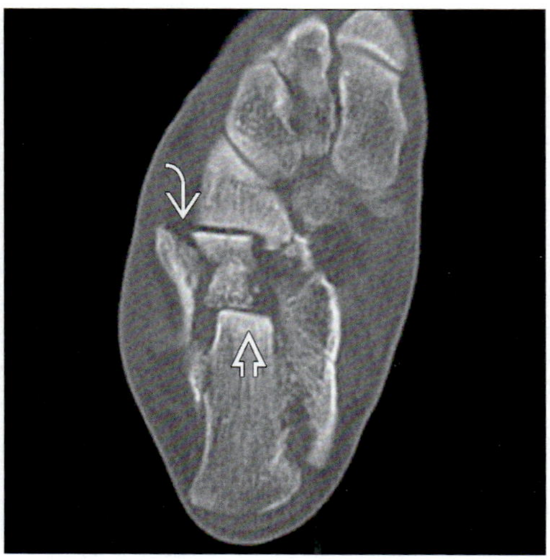

(Left) *Lateral radiograph shows an intraarticular fracture extending to the posterior process, proving to be a Essex-Lopresti tongue-type fracture. A large fragment of subchondral bone plate is depressed and rotated anteriorly. The fracture also involves the calcaneocuboid joint.* ***(Right)*** *Axial bone CT in the same patient shows the rotated subchondral bone plate, located centrally in the calcaneus, and involvement of the calcaneocuboid joint.*

CALCANEUS FRACTURE, INTRAARTICULAR

*(**Left**) Lateral radiograph shows an intraarticular fracture with a preserved Böhler angle. Posteriorly, the fracture extends to the posterior process of calcaneus →. A more vertically oriented fracture line → is difficult to see. (**Right**) Sagittal bone CT in the same patient better shows the fracture line extending through the angle of Gissane →, in addition to the posterior process fracture → evident on radiographs.*

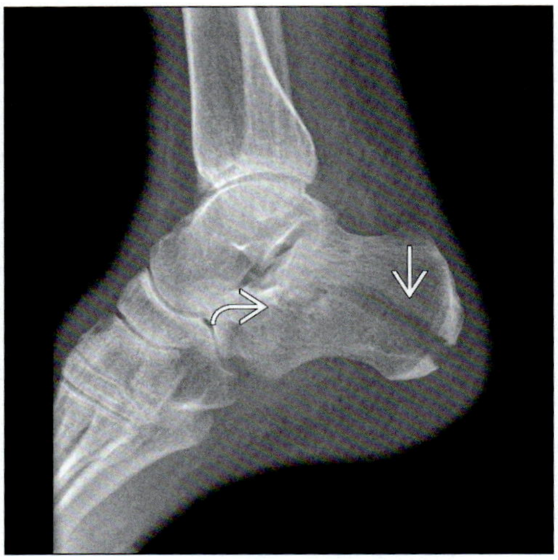

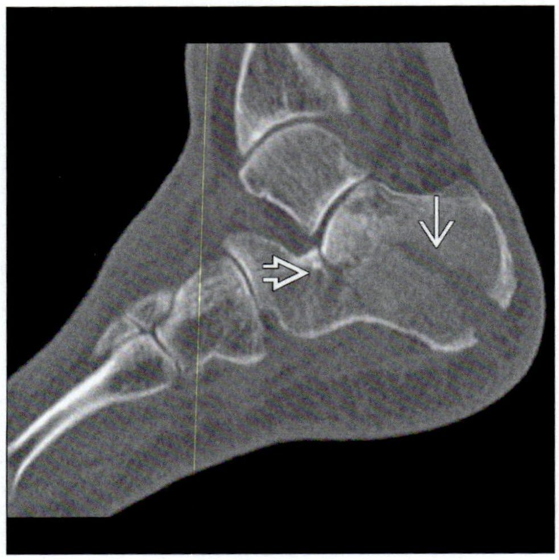

*(**Left**) Axial radiograph in the same patient shows the difficulty of assessing the subtalar joint on an axial view when the patient's mobility is limited. Only 1 horizontally oriented fracture line → is visible. (**Right**) Coronal bone CT in the same patient shows 2 nondisplaced fracture lines → through the posterior subtalar joint. This is a type I fracture in the Sanders classification. Note the horizontal fracture → through the calcaneus below the joint level.*

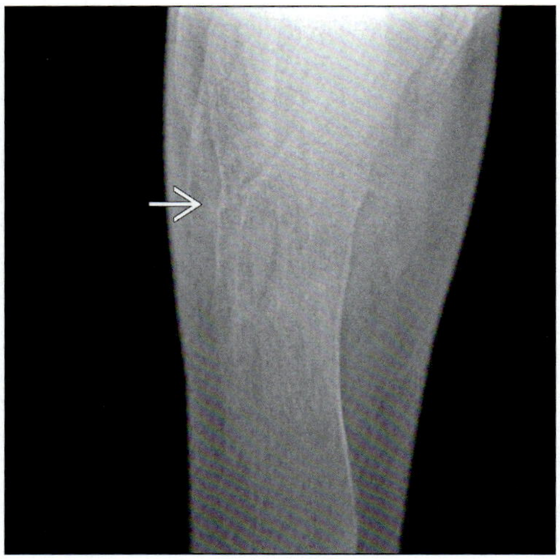

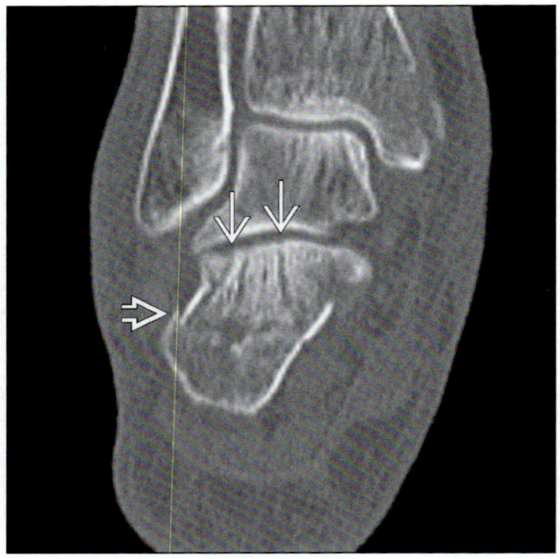

*(**Left**) Sagittal bone CT shows more severe depression at the angle of Gissane, with anterior rotation of the articular surface and comminution of the midportion of the body of the calcaneus. The fracture extends to the calcaneocuboid joint and posterior process. (**Right**) Coronal bone CT in the same patient shows the fracture to be a Sanders IIA. Diastasis of the sagittally oriented fracture line → results in splaying of the calcaneus. A horizontal fracture line → is also present.*

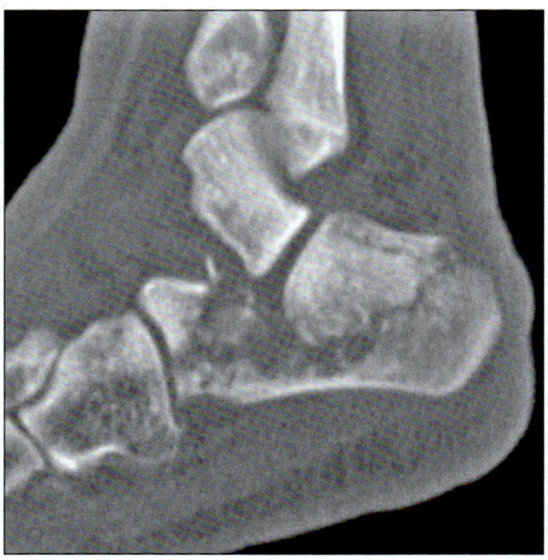

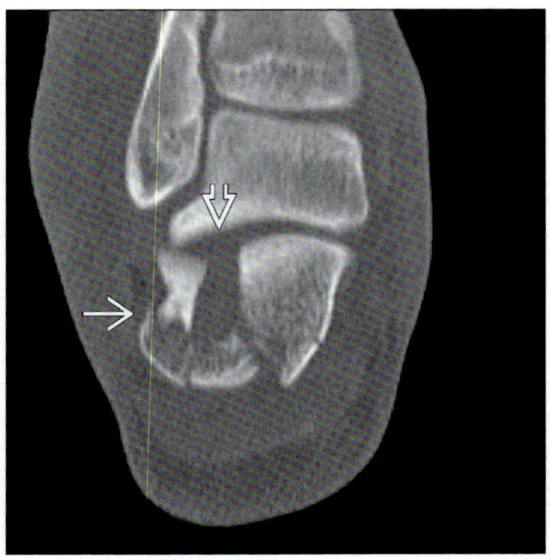

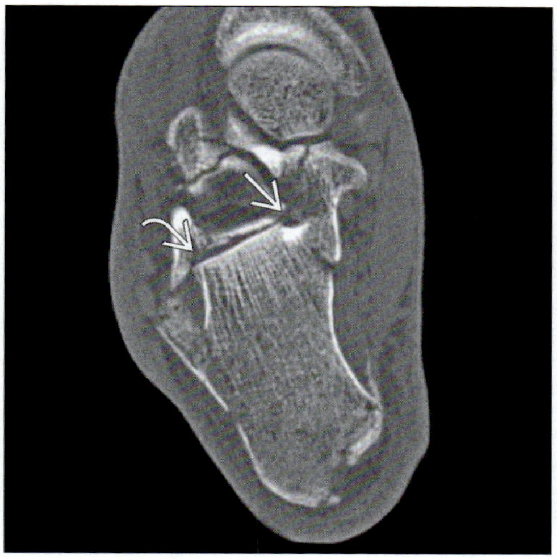

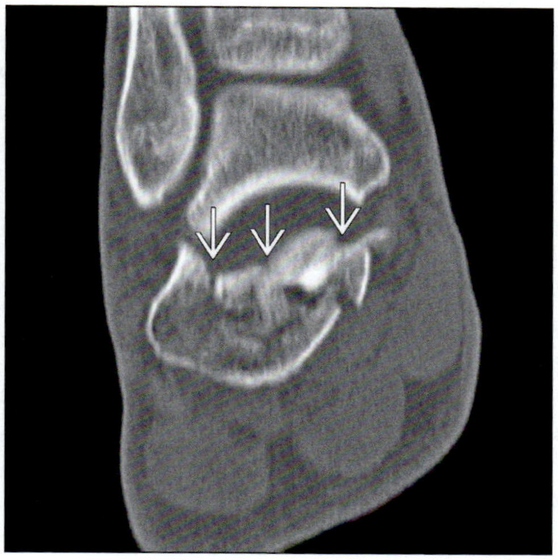

(Left) *Axial bone CT shows a calcaneal fracture extending into the posterior subtalar joint laterally ➔ and medially ➔. Comminution of the subtalar joint can be underestimated on axial slices.* ***(Right)*** *Coronal bone CT in the same patient shows 3 sites ➔ where fractures extend intraarticularly and marked depression of the central portion of the calcaneus at the subtalar joint. This is a Sanders type IVABC.*

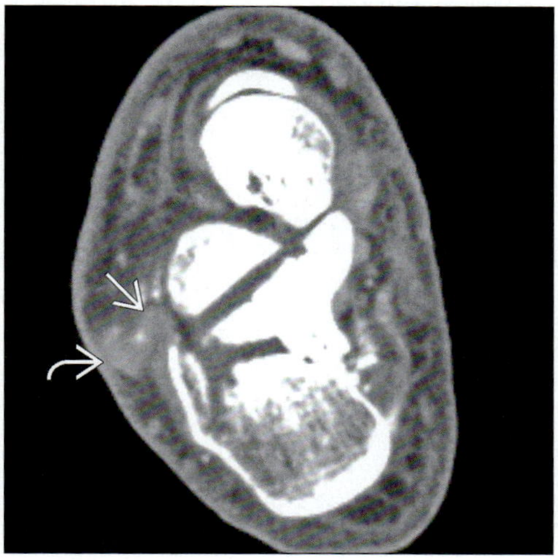

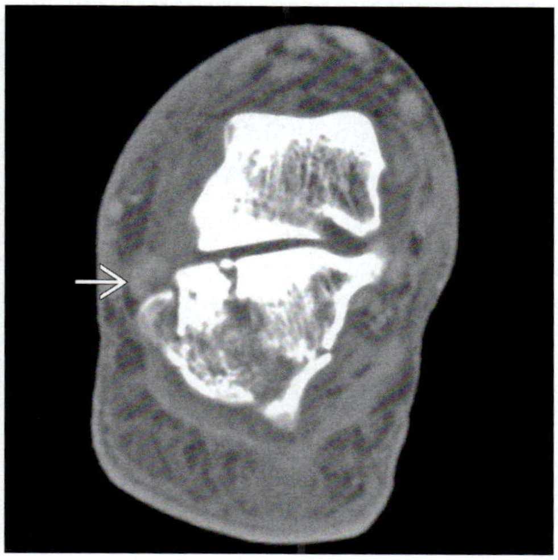

(Left) *Axial soft tissue CT shows peroneus brevis ➔ displaced between a bone fragment at the lateral process of the talus. The peroneus longus ➔ is normally located.* ***(Right)*** *Coronal soft tissue CT in a different patient shows both peroneal tendons ➔ above the fractured peroneal tubercle. CT algorithm should always be employed to evaluate for tendon injuries, which are often seen in calcaneal fractures. Patients may develop stenosing peroneal tenosynovitis as a result of tendon injury.*

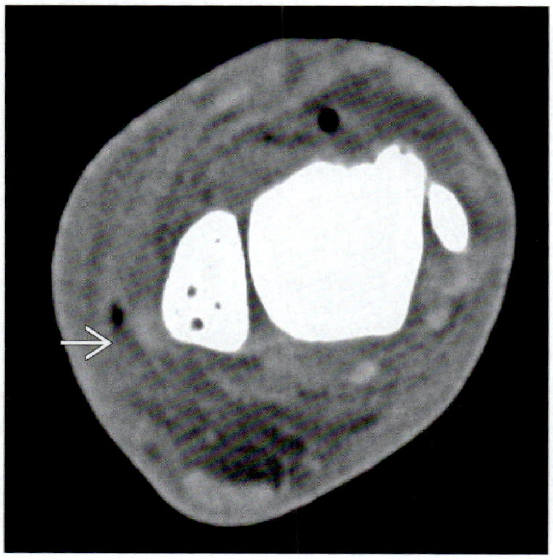

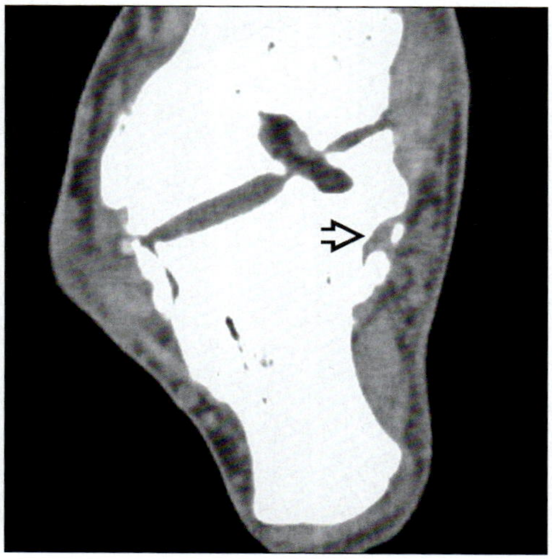

(Left) *Axial soft tissue CT in a patient with a calcaneal fracture shows dislocation of peroneal tendons ➔ at the level of the fibula due to avulsion of the peroneal retinaculum at the time of the calcaneal fracture.* ***(Right)*** *Axial soft tissue CT in another calcaneal fracture shows flexor hallucis longus ⇨ tendon entrapped by bone fragments as it courses through the tarsal tunnel. Tarsal tunnel syndrome (tibial neuropathy due to nerve entrapment) may also occur from bone fragments in this region.*

CALCANEUS FRACTURE, NONARTICULAR

Key Facts

Imaging

- Anterior process fracture
  - Fracture line perpendicular to process
  - Best seen on oblique view of foot
- Achilles tendon avulsion
  - Fracture line obliquely oriented across posterior process
  - Since Achilles tendon attaches on inferior 1/2 of process, a large fragment is avulsed and displaced superiorly with Achilles
- Impaction fracture of posterior process
  - Sclerotic line seen at inferior border of calcaneus
  - Often best seen on axial (Harris) views
- Stress fracture of calcaneus
  - Sclerotic line perpendicular to primary trabeculae of posterior process
- Avulsion of extensor digitorum brevis origin
  - Thin flake of bone seen from lateral, anterior margin of body of calcaneus
  - May be visible on AP view of ankle &/or AP view of foot
- Insufficiency fracture due to lipoma or simple cyst
  - Sclerotic or lucent, vertically oriented line through nonaggressive lesion in central calcaneus

Top Differential Diagnoses

- Plantar fasciitis
- Achilles tendon tear
- Os calcis secundarium
- Ankle sprain

Pathology

- Associated conditions
  - Simple cyst or lipoma of central calcaneus
  - Diabetes mellitus (posterior process fracture)

*(**Left**) Graphic of the lateral aspect of calcaneus shows the most common nonarticular calcaneal fractures: Anterior process ➔, extensor digitorum brevis avulsion ➔, Achilles tendon avulsion ➔, and stress fracture ➔. (**Right**) Oblique radiograph shows anterior process fracture ➔. The angular margins and lack of cortication distinguish the fracture from an accessory center of ossification (os calcis secundarium), which may also occur in this location.*

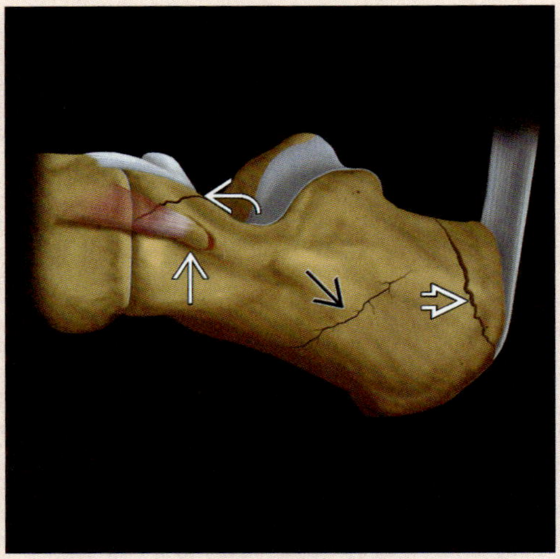

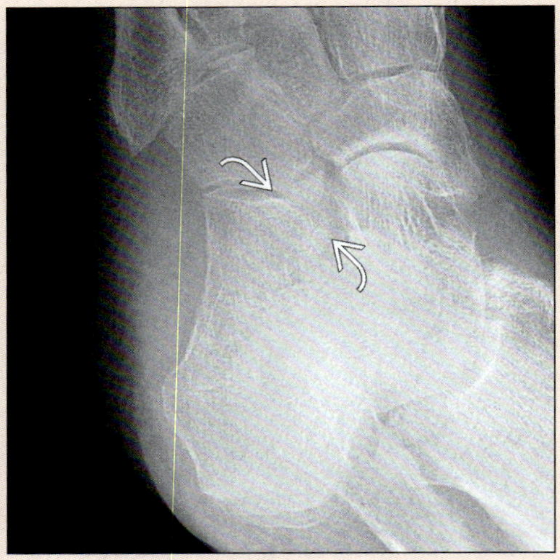

*(**Left**) Lateral radiograph shows a fracture of the inferior calcaneus ➔ due to a fall, landing on the heel. These fractures are uncommon, and intraarticular fractures are more often seen with this mechanism. (**Right**) Sagittal T1WI MR shows insufficiency fractures of the calcaneus ➔, talar neck ➔, and lateral process ➔ in a patient with osteoporosis. Insufficiency fractures are often multiple.*

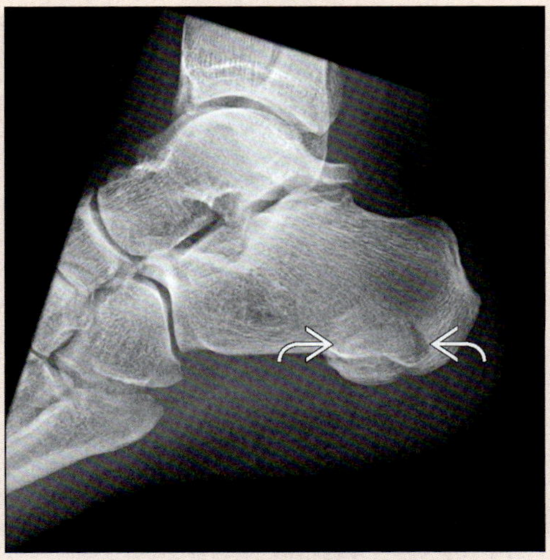

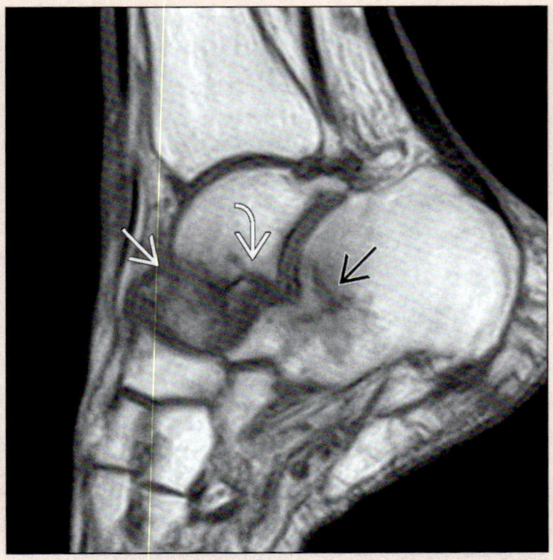

CALCANEUS FRACTURE, NONARTICULAR

TERMINOLOGY

Definitions

- Anterior process: Nonarticular wedge-shaped projection adjacent to calcaneocuboid joint
- Posterior process: Large, squared projection of calcaneus posterior to subtalar joint

IMAGING

General Features

- Location
  - Usually seen in 1 of 3 locations
    - Anterior process
    - Posterior process
    - Origin of extensor digitorum brevis

Radiographic Findings

- **Anterior process fracture**
  - Fracture line perpendicular to process
  - Best seen on oblique view of foot
  - Sometimes visible on lateral view
- **Avulsion of extensor digitorum brevis origin**
  - Thin flake of bone seen from lateral, anterior margin of body of calcaneus
  - May be visible on AP view of ankle &/or AP view of foot
- **Achilles tendon avulsion**
  - Fracture line obliquely oriented
  - Since Achilles tendon attaches on inferior 1/2 of process, a large fragment is avulsed and displaced superiorly with Achilles
- **Impaction fracture of posterior process**
  - Sclerotic line seen at inferior border of calcaneus
  - Often best seen on axial (Harris) views
- **Stress fracture of calcaneus**
  - Sclerotic line perpendicular to primary trabeculae of posterior process (most frequent location)
  - Occasionally lies immediately below and parallel to posterior subtalar joint articular surface
    - More difficult to visualize on radiograph since calcaneus is more sclerotic in region of sustentaculum and posterior facet
- **Insufficiency fracture due to lipoma or simple cyst**
  - Sclerotic or lucent, vertically oriented line through nonaggressive lesion in central calcaneus

CT Findings

- CT not usually performed, but findings as per radiographs

MR Findings

- Low signal intensity fracture line
- Surrounded by edema, high signal intensity on T2WI, STIR; low signal intensity on T1WI

Imaging Recommendations

- Best imaging tool
  - Radiographs
- Protocol advice
  - Impaction fractures of posterior process often best seen on axial (Harris) views

DIFFERENTIAL DIAGNOSIS

Plantar Fasciitis

- Pain at plantar margin of calcaneus due to stress fracture may mimic fasciitis

Achilles Tendon Tear

- Much more common than bone avulsion

Os Calcis Secundarium

- Small ossicle at anterior process of calcaneus

Ankle Sprain

- Avulsion of anterior process or extensor digitorum brevis origin presents like ankle sprain

PATHOLOGY

General Features

- Etiology
  - Anterior process fracture
    - Inversion injury with avulsion of bifurcate ligament
    - Eversion injury with crush of process between navicular and calcaneus
  - Extensor digitorum brevis fracture
    - Inversion injury
  - Achilles tendon avulsion
    - Hyperdorsiflexion injury
  - Impaction of plantar calcaneus
    - Direct blow
- Associated abnormalities
  - Simple cyst or lipoma of central calcaneus
    - Predispose patients to stress fracture
  - Diabetes mellitus
    - Predisposes patients to avulsion of posterior process of calcaneus

CLINICAL ISSUES

Presentation

- Most common signs/symptoms
  - Stress fracture: Aching pain, increased by exercise
  - Anterior process or extensor digitorum brevis avulsion fracture: Focal pain after inversion injury

Demographics

- Posterior calcaneal tuberosity avulsion usually occurs in diabetic patients

Treatment

- Conservative management
- Occasionally, excision of symptomatic nonunion of anterior process fracture

SELECTED REFERENCES

1. Beavis RC et al: Avulsion fracture of the calcaneal tuberosity: a case report and literature review. Foot Ankle Int. 29(8):863-6, 2008
2. Daftary A et al: Fractures of the calcaneus: a review with emphasis on CT. Radiographics. 25(5):1215-26, 2005
3. Squires B et al: Fractures of the tuberosity of the calcaneus. J Bone Joint Surg Br. 83(1):55-61, 2001
4. Trnka HJ et al: Fracture of the anterior superior process of the calcaneus: an often misdiagnosed fracture. Arch Orthop Trauma Surg. 117(4-5):300-2, 1998

(Left) *Lateral radiograph shows a fracture ➔ traversing the base of the anterior process. In this case, the fracture extends into the calcaneocuboid joint.* ***(Right)*** *Oblique radiograph shows a typical small chip fracture ➔ of the anterior process. The fracture involves the origin of the bifurcate ligament, between the anterior process, the navicular, and the cuboid. The fracture is difficult to tell from the os calcis secundarium, which also occurs in this area.*

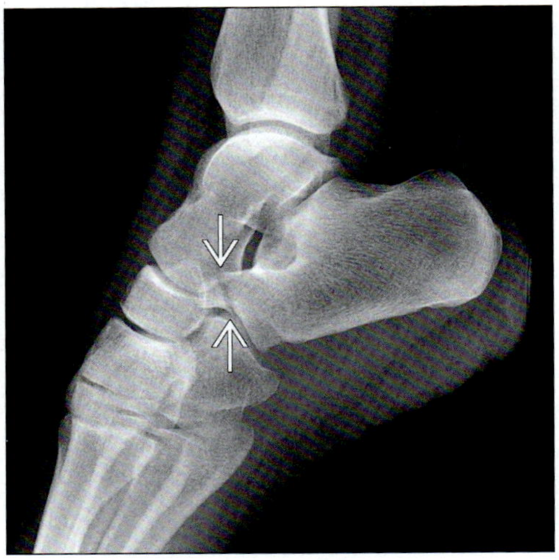

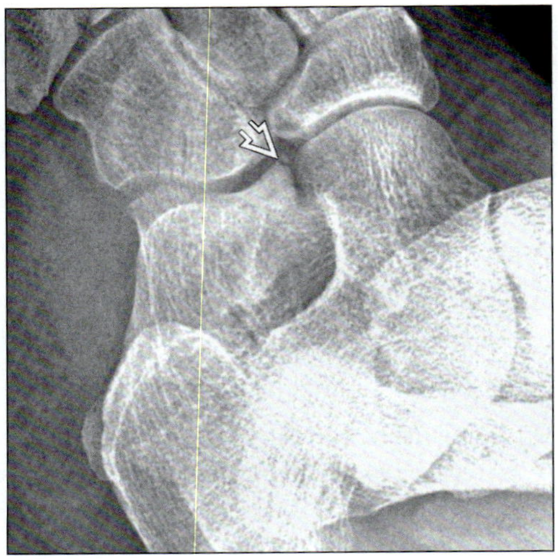

(Left) *Sagittal T2WI FS MR shows an acute fracture of the anterior process of the calcaneus ➔ with surrounding marrow edema. Note the mild edema in the dorsal fibers of the bifurcate ligament ➔.* ***(Right)*** *Sagittal bone CT shows an anterior process calcaneus fracture ➔. Small fragments in the tarsal sinus ➔ are due to associated extensor digitorum brevis avulsion. Both fractures are due to inversion injury and are not infrequently associated.*

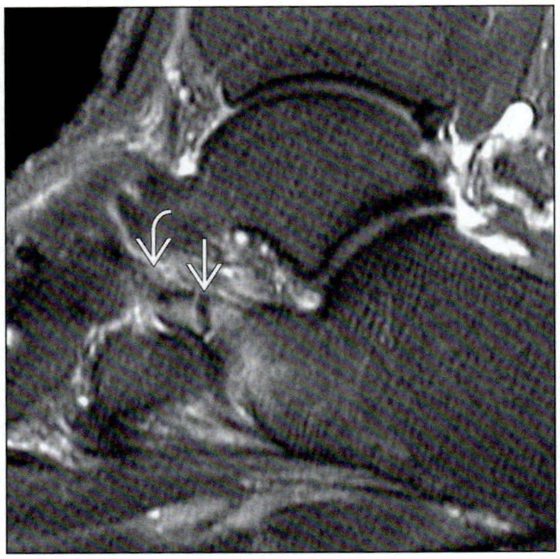

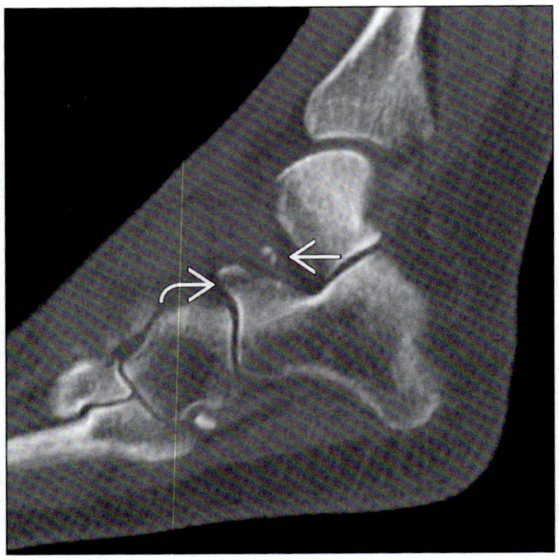

(Left) *AP radiograph shows an avulsed fragment of the extensor digitorum brevis origin ➔. Severe soft tissue swelling is present laterally ➔ but is centered further inferiorly than soft tissue swelling due to an ankle sprain.* ***(Right)*** *AP radiograph of the foot shows a small chip fragment ➔ due to extensor digitorum brevis avulsion. These fragments are often visible on the AP view of the foot or the AP view of the ankle, but not both.*

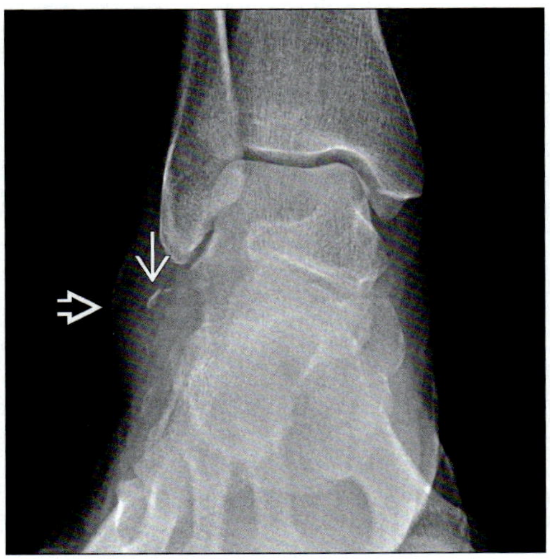

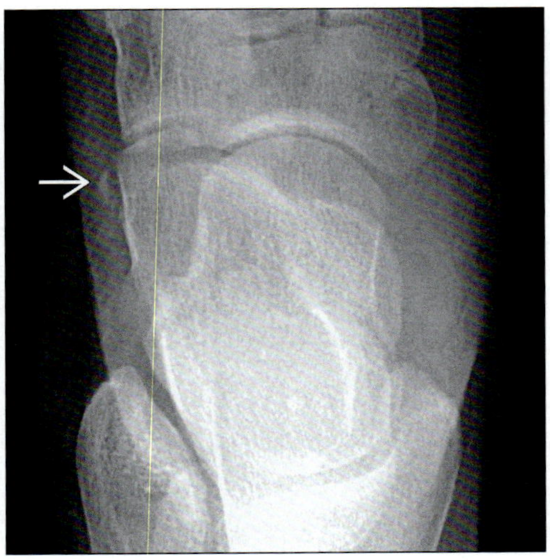

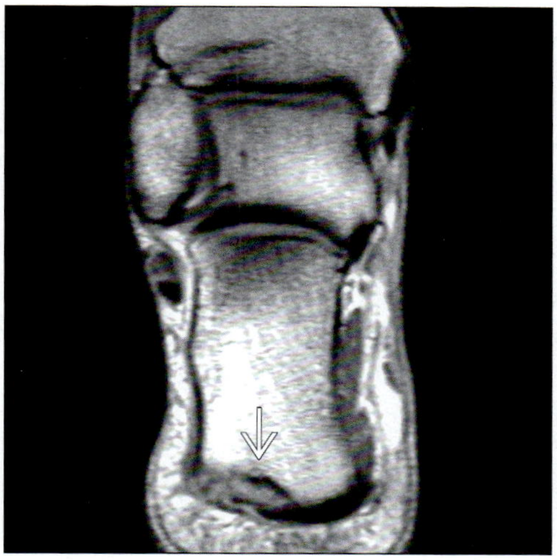

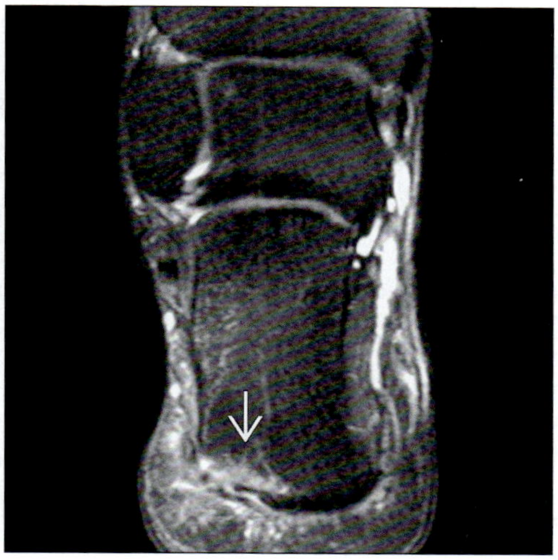

(Left) *Coronal PDWI MR shows an impaction fracture* ➔ *of the inferior margin of the calcaneus, which presented with heel pain initially diagnosed as plantar fasciitis.* ***(Right)*** *Coronal T2WI FS MR in the same patient shows bone marrow edema* ➔*, but the fracture line is obscured. Without the corresponding PD image, the erroneous diagnosis of reactive arthropathy or severe plantar fasciitis might be made.*

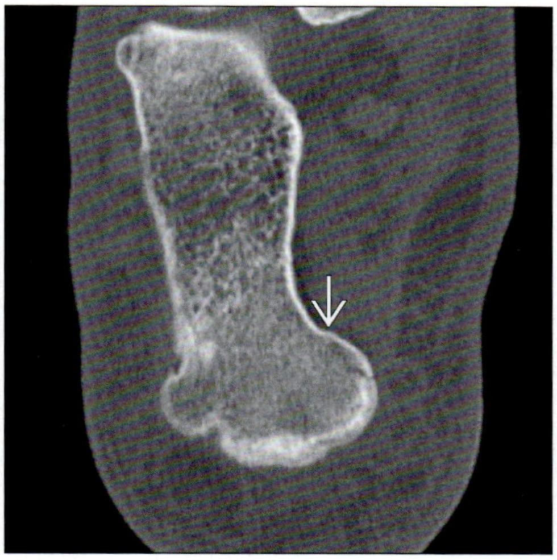

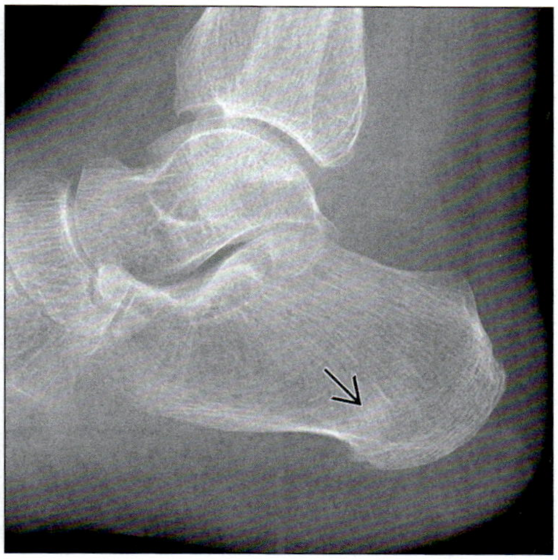

(Left) *Axial bone CT shows a small cortical fx* ➔ *due to a direct blow to the plantar aspect of the heel. This patient slipped on the stairs and landed on her heel.* ***(Right)*** *Lateral radiograph shows a subtle sclerotic line* ➔ *in the posterior process of the calcaneus of an osteopenic patient. The key to recognizing the fracture is to recognize that stress fractures are perpendicular to the normal primary trabeculae of the posterior process.*

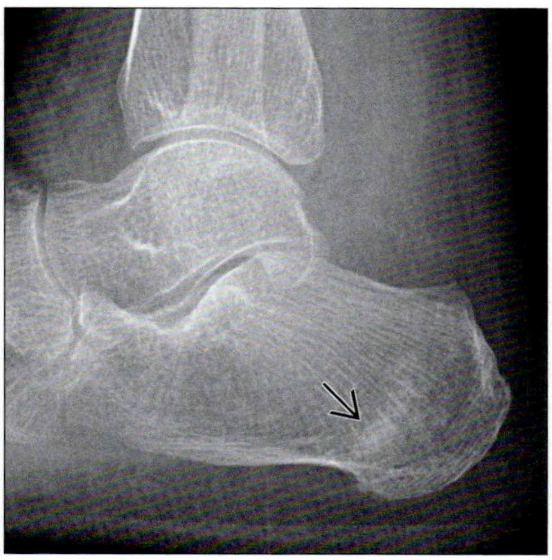

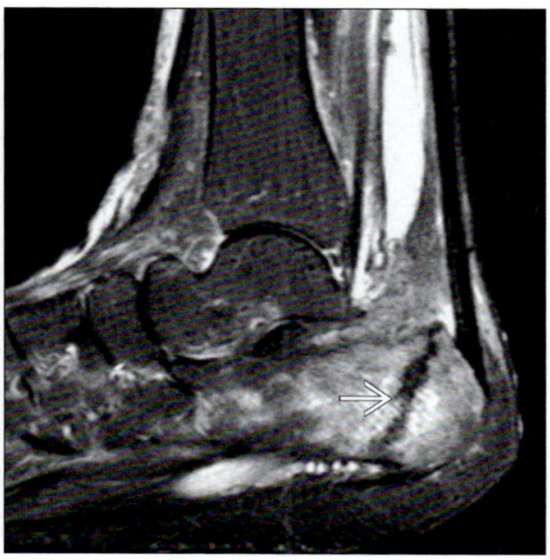

(Left) *Lateral radiograph in the same patient 2 weeks later shows that the sclerotic line* ➔ *is now thicker and longer. Despite prompt recognition, the stress fracture continued to progress over the next several months.* ***(Right)*** *Sagittal STIR MR in the same patient 2 months later shows marrow edema surrounding the completed stress fracture* ➔*. The normal trabecular architecture is preserved, helping to distinguish this case from a pathologic fracture due to a bone marrow infiltrative process.*

NAVICULAR FRACTURE AND DISLOCATION

Key Facts

Imaging

- Median eminence fracture: Best seen on AP + medial oblique views
  - Due to avulsion of median eminence by posterior tibial tendon
  - Often associated with crush-type injury of calcaneocuboid joint
- Body fracture: Usually in both sagittal and horizontal planes
  - Due to major trauma
  - Associated with other midfoot injuries
  - Radiographs underestimate comminution
  - Navicular compression may be only sign on radiographs
- Dorsal cortical avulsion fracture: Chip fracture avulsed by talonavicular joint capsule
  - Due to ankle inversion injury
  - Distinguish from accessory center of ossification
- Stress fracture: Sagittal fracture of mid navicular, incomplete or complete
  - Due to repetitive stress, seen in high-level athletes
  - Radiographs are 33% sensitive
  - Most common radiographic finding is ill-defined sclerosis of navicular
  - MR recommended to detect in athletes with exercise-related midfoot soreness
- Dislocation: Joint disruption obvious, associated fracture may be difficult to see

Top Differential Diagnoses

- Os supranaviculare
- Accessory navicular (os tibiale externum)
- Köhler disease
- Müller-Weiss disease

*(**Left**) Graphic shows a longitudinal fracture → of the navicular due to an anteroposterior force on the foot. There are 4 common types of navicular fracture: Longitudinal (often associated with dislocation), median eminence avulsion, chip fracture, and stress fracture. (**Right**) AP radiograph shows a comminuted navicular fracture →, with talar head impacted into the navicular. There is an associated Chopart injury, with medial displacement of the calcaneocuboid → and talonavicular joints.*

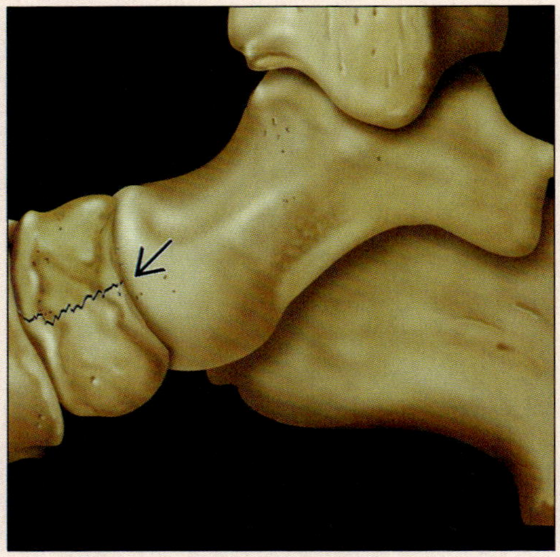

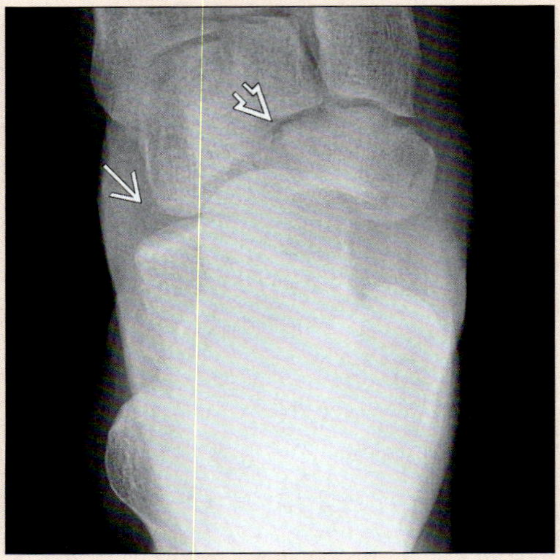

*(**Left**) Oblique radiograph shows a median eminence navicular fracture →. This fx may be associated with crush injury of the cuboid or the anterior margin of the calcaneus → ("nutcracker injury"). (**Right**) Axial bone CT shows a complete sagittal stress fracture → of the navicular. Navicular stress fractures are often occult on radiographs. MR is very sensitive for reactive marrow change due to stress injury, while the discrete fracture line is often better seen with CT.*

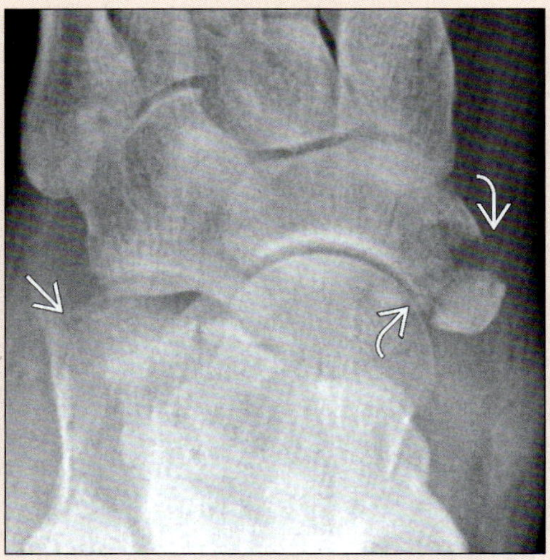

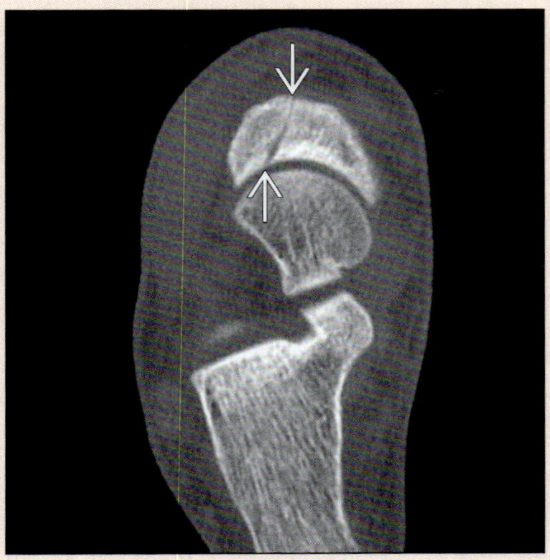

NAVICULAR FRACTURE AND DISLOCATION

IMAGING

General Features

- Best diagnostic clue
  - Fractures
    - Navicular tuberosity: Median eminence avulsed, coronal fracture line
    - Dorsal cortical avulsion: Small triangular fragment at talonavicular joint
    - Body: Flattening, irregularity of body
    - Stress: Ill-defined sclerosis on radiographs, ± sagittal fracture line mid navicular
  - Dislocation: Loss of joint congruity
- Morphology
  - 4 types of fractures
    - Navicular tuberosity avulsion: Posterior tibial tendon avulsion
    - Dorsal cortical avulsion: Talonavicular capsular avulsion
    - Body: Undisplaced or displaced
    - Stress: Partial or complete fracture of body in sagittal plane
  - 4 types of dislocations
    - Talonavicular joint
    - Complete dislocation
      - Navicular extruded dorsally
    - Chopart joint dislocation
      - Talonavicular + calcaneocuboid joint
    - Complex midfoot dislocation
      - Often see dislocations and fractures of multiple midfoot bones

Radiographic Findings

- Often missed due to relative rarity, physician unfamiliarity, and multiple concurrent injuries
- Median eminence avulsion: Best seen on AP + medial oblique views
  - Median eminence usually displaced proximally by pull of posterior tibial tendon
- Body
  - Navicular compression may be only sign on radiographs
  - Radiographs underestimate comminution
  - Horizontal fracture component through body seen on lateral view
- Stress: Often normal, radiographs 33% sensitive
  - Most common finding is ill-defined sclerosis of navicular
  - Fracture line may be visible on anteroposterior radiograph
- Dislocation: Usually best seen on lateral view

CT Findings

- Bone CT
  - Detect radiographically occult fractures
    - Navicular, cuneiform, metatarsal base fractures often associated with dislocations
  - Delineate extent/pattern of fracture prior to surgery
  - Navicular body fractures usually show extensive comminution
    - Inferior portion of body displaced from superior portion
    - Best seen on coronal reformations
  - Stress fractures visible, but CT is less sensitive than MR
    - Fracture always in sagittal plane, through midportion of navicular
    - See cleft in navicular, usually incomplete
      - Originates at dorsal margin
      - Has sclerotic margins

MR Findings

- T1WI
  - Low signal intensity fracture line and edema
- T2WI
  - High signal intensity fracture line and edema
- High sensitivity and specificity for stress fracture

Imaging Recommendations

- Best imaging tool
  - Radiographs initial standard
  - CT best for traumatic fracture
  - MR best for suspected stress fracture
- Protocol advice
  - Radiographs: AP, lateral, and oblique views
  - CT: Thin slices (≤ 1 mm) angled through plane of talonavicular joint; reformats in other planes
  - MR: Coronal, axial T1WI and T2WI perpendicular to navicular

DIFFERENTIAL DIAGNOSIS

Os Supranaviculare

- Small, rounded bone at dorsal and proximal margin of navicular

Accessory Navicular (Os Tibiale Externum)

- Type I: Rounded sesamoid in posterior tibial tendon
- Type II: Corticated accessory center of ossification, forms flat facet against median eminence
  - May be avulsed by posterior tibial tendon

Köhler Disease

- Self-limiting osteonecrosis of navicular in child < 6 years
- Collapsed navicular or normal navicular with increased density + fragmentation

Müller-Weiss Disease

- Spontaneous osteonecrosis of navicular in adult
- Sclerotic navicular with flattening of lateral portion
- May be due to unrecognized stress fracture

PATHOLOGY

General Features

- Etiology
  - Navicular tuberosity: Posterior tibial tendon pull from acute trauma consisting of pronation, external rotation, or dorsiflexion
  - Dorsal cortical avulsion: Caused by twisting force (usually eversion), capsular avulsion of talonavicular joint
  - Body: Caused by indirect axial load
    - Fall from height
    - Motor vehicle accidents
  - Stress: Caused by repetitive impact load
    - Push-off activities predispose: Jumping + sprinting
    - Central 1/3 of navicular is avascular (↑ risk of stress fracture + nonunion)

NAVICULAR FRACTURE AND DISLOCATION

  - Junction of medial 2/3 + lateral 1/3 of navicular experiences most shear stress
- Dislocation: Caused by indirect axial load or twisting injury
  - Major trauma
- Associated abnormalities
  - Median eminence (tuberosity) avulsion associated with
    - Crush injury of calcaneus or cuboid at calcaneocuboid joint ("nutcracker injury")
    - Spring ligament tear
  - Navicular body fracture associated with
    - Talonavicular dislocation
    - Other midfoot fractures
  - Talonavicular dislocation associated with
    - Dislocation of calcaneocuboid joint
  - Naviculocuneiform dislocation associated with
    - Other midfoot fracture and dislocations

Staging, Grading, & Classification

- Stress fractures: CT-based classification
  - Type I: Dorsal cortical break
  - Type II: Propagation into navicular body
  - Type III: Propagation into another cortex
- Sangeorzan classification of navicular body fractures
  - Type I: Coronal fracture with no dislocation
  - Type II: Dorsolateral or plantar-medial fracture with medial forefoot displacement
  - Type III: Comminuted fracture with lateral forefoot displacement

CLINICAL ISSUES

Presentation

- Clinical profile
  - Navicular median eminence fracture
    - Acute pain + swelling along medial arch of foot
  - Dorsal cortical avulsion: Pain + swelling at fracture
  - Body fracture or dislocation
    - Extensive midfoot trauma
    - High-velocity injury
  - Stress fracture
    - Insidious onset of poorly defined midfoot pain
    - Resolves with rest, returns when activity resumed
    - Minimal swelling

Demographics

- Epidemiology
  - Dorsal avulsion fracture is most common: 47-67% of navicular fractures
  - Body fracture or dislocation is commonly associated with other midfoot injuries
  - Navicular stress fracture
    - 14-35% of all stress fractures in 1 series
    - 1st described in military recruits
    - 59% are track and field athletes
      - Especially jumpers and sprinters
    - Leg length discrepancy
    - Uncommon in general population

Natural History & Prognosis

- Navicular median eminence
  - Heals with fracture reduction and immobilization
- Dorsal cortical avulsion: Full recovery at 4-6 months after injury with conservative treatment
- Body: Osteoarthritis is common complication
  - All navicular body fractures with ≥ 1 mm displacement require open reduction and internal fixation
- Stress
  - Delayed diagnosis is common
  - 86% of nondisplaced stress fractures heal
  - Up to 14% treated correctly do not return to previous activities
  - Time to return to sport
    - Nonoperative management: 3-4 months
    - Operative management: 5-6 months
      - Screw fixation
  - Complications of stress fracture
    - Progression to complete fracture
    - Delayed union and nonunion
    - Talonavicular osteoarthritis
    - Osteonecrosis

Treatment

- Median eminence: Surgery if > 1 cm proximal displacement, immobilization if nondisplaced
- Dorsal cortical avulsion: Immobilization + restricted weight-bearing for 6-12 weeks
- Body
  - Undisplaced: Immobilization + restricted weight-bearing for 8-12 weeks
  - Displaced: Surgery for ≥ 1 mm displacement
- Stress
  - Immobilization + restricted weight-bearing for 6 weeks, then functional rehabilitation
  - Surgery: Complete/comminuted fractures, nonhealing incomplete fractures

DIAGNOSTIC CHECKLIST

Consider

- Stress fracture in athlete with persistent midfoot pain
- Complications of untreated or undertreated navicular fractures can be career-ending for elite athletes

Image Interpretation Pearls

- Normal radiograph does not rule out navicular fx
- Often seen on only 1 radiographic view

SELECTED REFERENCES

1. Coris EE et al: Tarsal navicular stress fractures. Am Fam Physician. 67(1):85-90, 2003
2. Jameson BH et al: Bilateral navicular body fractures. J Trauma. 54(6):1231-4, 2003
3. Pinney SJ et al: Fractures of the tarsal bones. Orthop Clin North Am. 32(1):21-33, 2001
4. Prokuski LJ et al: Challenging fractures of the foot and ankle. Radiol Clin North Am. 35(3):655-70, 1997
5. Davis CA et al: Midtarsal fracture-subluxation. Case report and review of the literature. Clin Orthop Relat Res. (292):264-8, 1993
6. Nyska M et al: Fractures of the body of the tarsal navicular bone: case reports and literature review. J Trauma. 29(10):1448-51, 1989

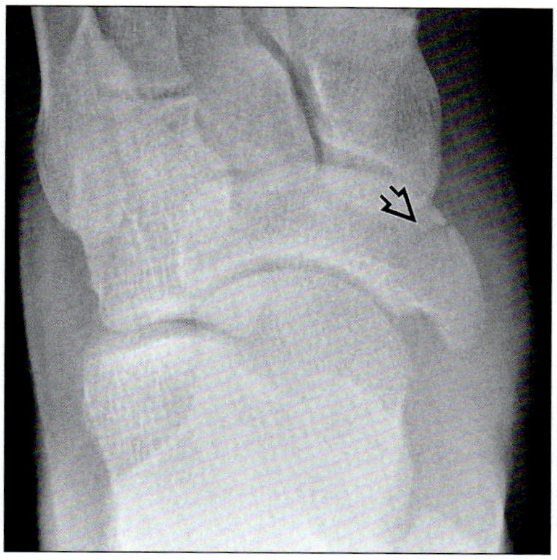

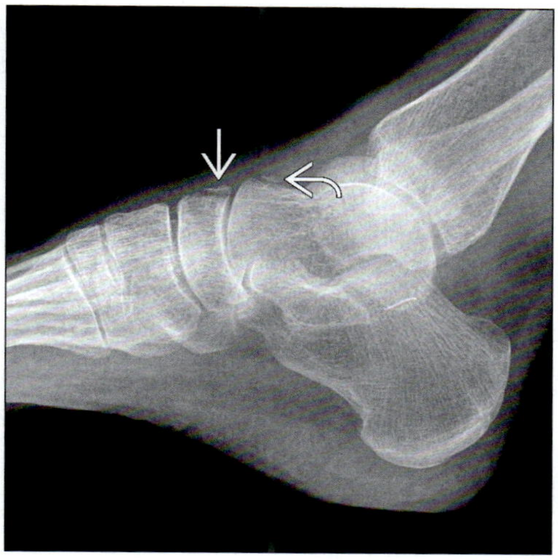

*(**Left**) AP radiograph shows a nondisplaced avulsion fracture of median eminence of the navicular (tuberosity fracture). The fracture can be distinguished from an accessory navicular by its jagged, nonsclerotic margins. (**Right**) Lateral radiograph shows chip fractures due to avulsion of the dorsal talar and navicular attachments of the talonavicular joint capsule. Typically 1 or the other site is involved; this case is unusual in that it involves both.*

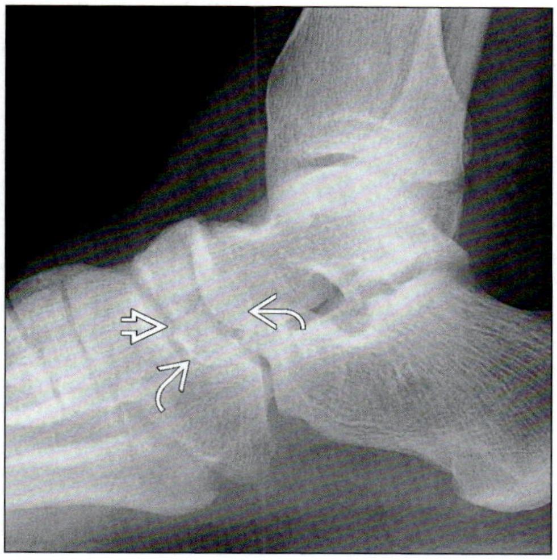

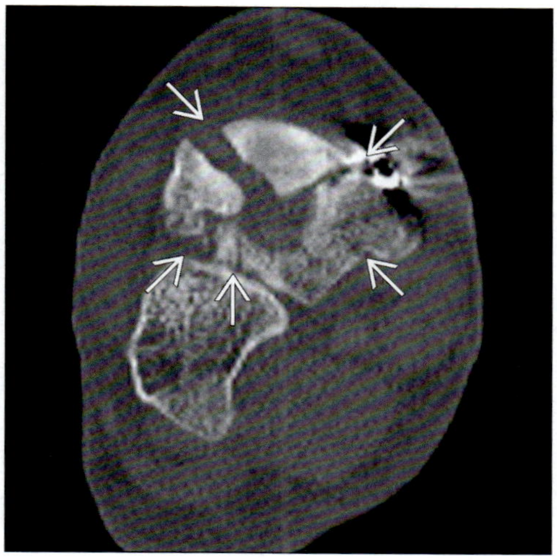

*(**Left**) Lateral radiograph shows a horizontally oriented navicular fracture, with dorsal subluxation of the navicular. There are multiple small, comminuted fragments of the inferior navicular, which are difficult to see on radiographs. (**Right**) Coronal bone CT in the same patient shows stellate pattern of fracture, with 2 large pieces dorsally and many small comminuted fragments through the plantar portion of the bone. Fracture is due to an anteroposterior force on the foot.*

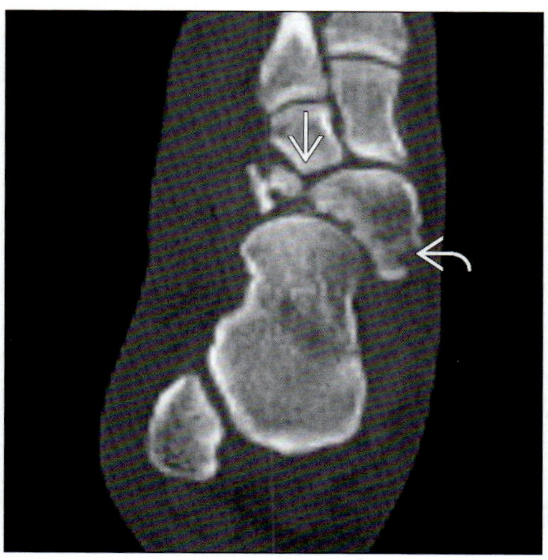

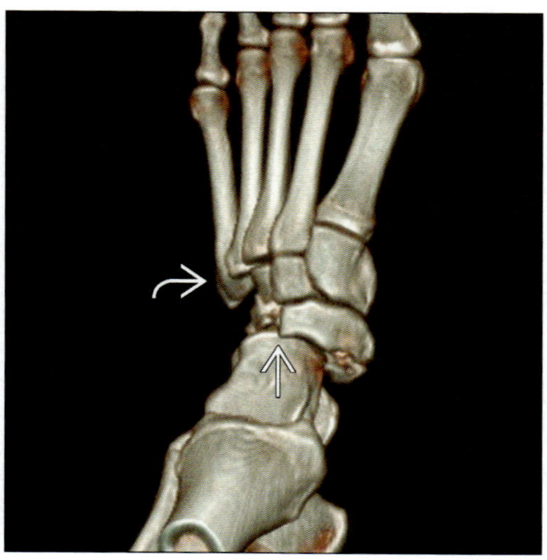

*(**Left**) Axial bone CT shows sagittally oriented main fracture line through navicular and avulsion of median eminence. Talus is impacted into the center of the navicular. (**Right**) AP bone CT 3D reformation of the same patient shows displacement of the talus into the shattered navicular. Note dislocation of 4th and 5th tarsometatarsal joints. Navicular fractures are commonly associated with other midfoot injuries.*

NAVICULAR FRACTURE AND DISLOCATION

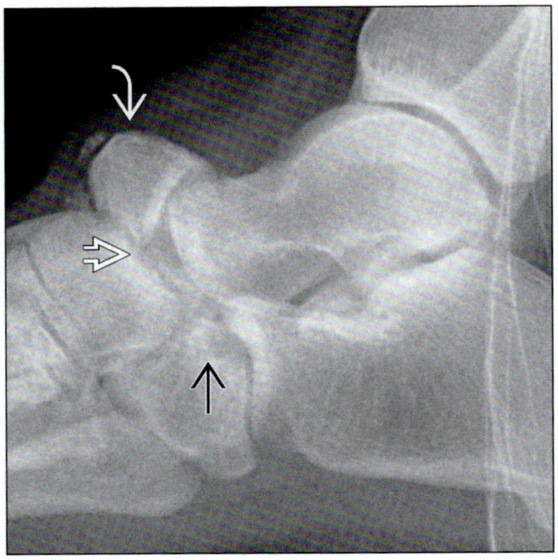

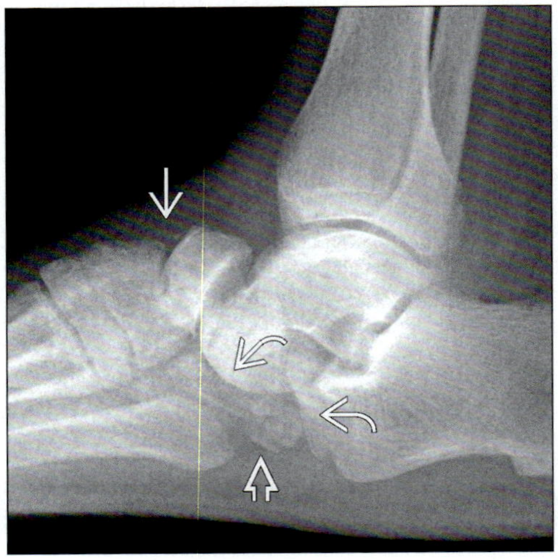

(Left) *Lateral radiograph shows dorsal extrusion of the navicular → due to a compressive force directed posteriorly from the toes. A horizontally oriented fracture line → is present, and the inferior portion of the navicular is difficult to see because of the extensive comminution. There is also a cuboid fracture →.* ***(Right)*** *Lateral radiograph shows Chopart joint dislocation →, a comminuted fracture of the cuboid →, and dislocation of the naviculo-cuneiform joint →.*

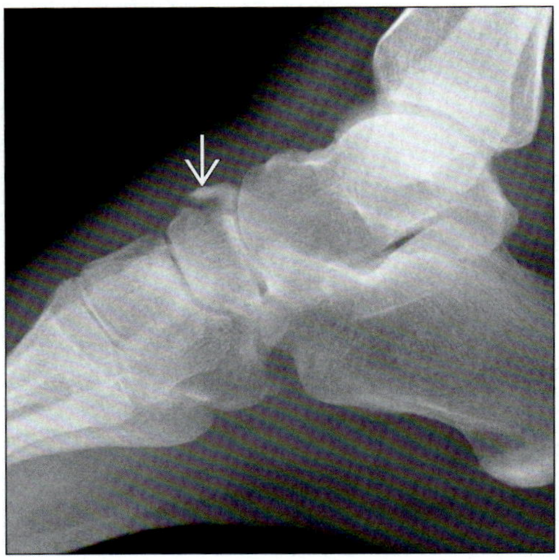

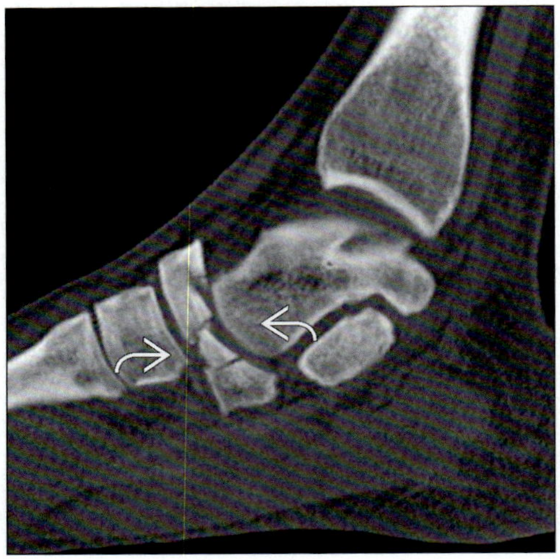

(Left) *Lateral radiograph demonstrates a large fragment → arising from the dorsal aspect of the navicular. This probably reflects a compressive injury. Most avulsion fragments in this region are significantly smaller.* ***(Right)*** *Sagittal bone CT shows compression of the navicular → and marked bony comminution. This fracture pattern has a poor prognosis, with nearly universal development of osteoarthritis.*

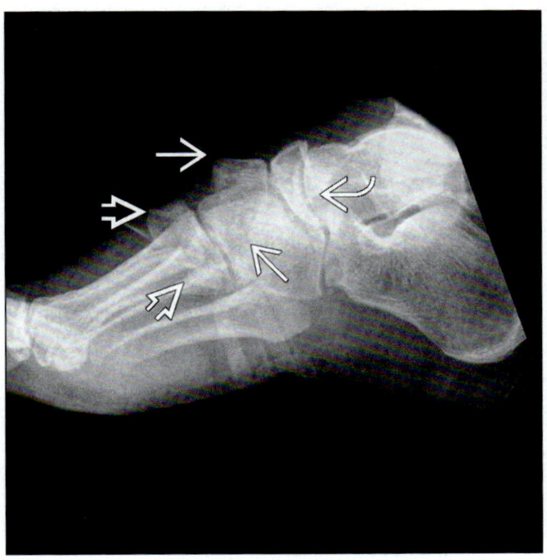

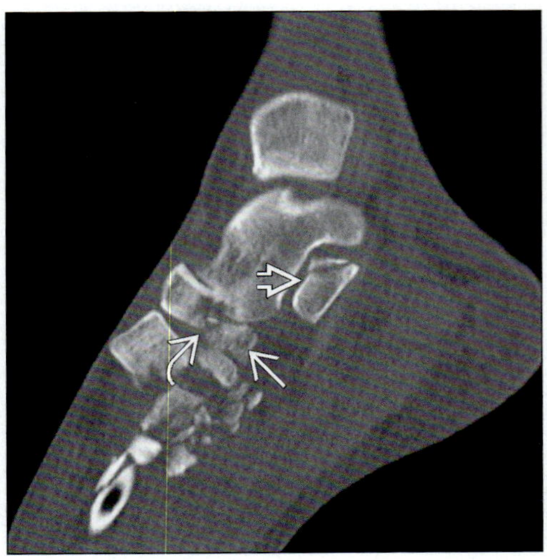

(Left) *Lateral radiograph shows compression of the navicular → consistent with fracture, but the fracture lines are not visible. There are also fractures of the cuneiforms → and metatarsals →.* ***(Right)*** *Sagittal bone CT in the same patient shows the horizontal component of the navicular fracture → and comminution of the inferior portion of the bone →. Fracture of the sustentaculum tali → is present in addition to the fractures visible on previous radiograph.*

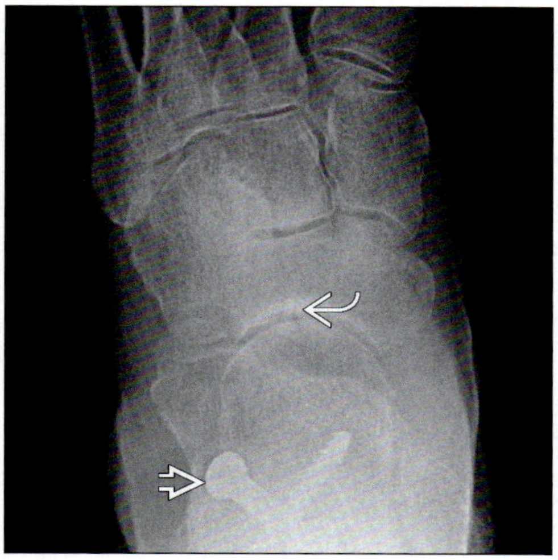

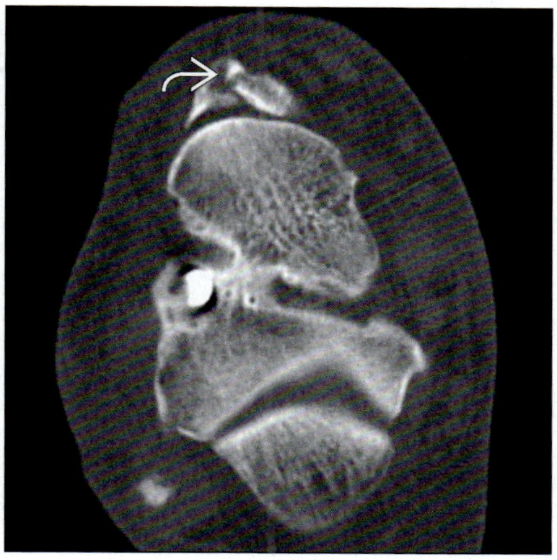

(Left) *AP radiograph shows a subtle stress fracture, with a faint lucent line in the navicular, involving the proximal but not the distal cortex. This patient was status post subtalar fusion and developed midfoot pain.* ***(Right)*** *Axial bone CT in the same patient shows a stress fracture at the dorsal margin of the navicular. The stress fracture only involved the dorsal margin of the navicular, where stress fractures typically begin. This area is easily overlooked on CT or MR scan.*

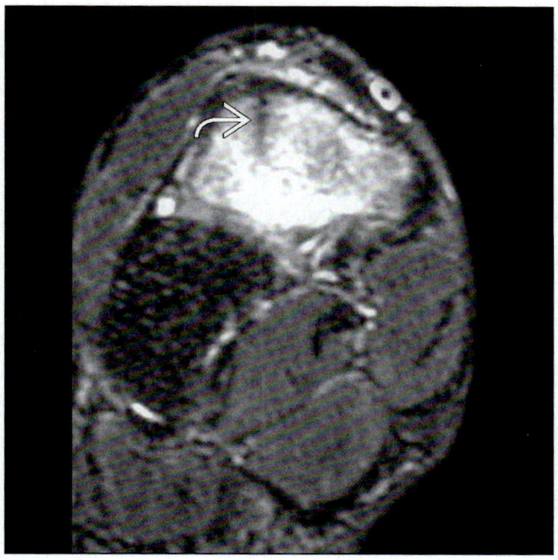

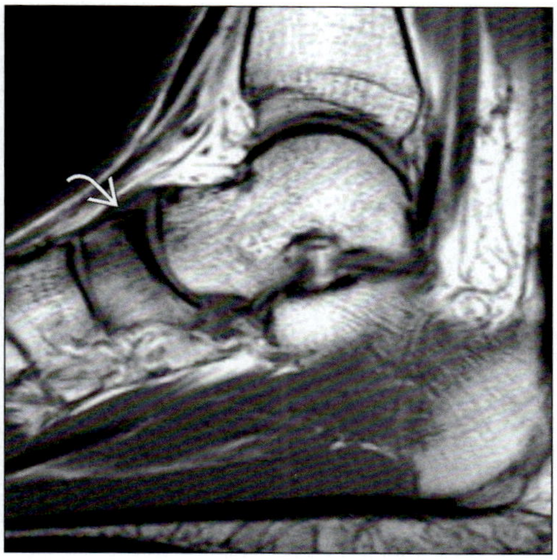

(Left) *Coronal T2WI FS MR shows stress fracture of the dorsal navicular, with bone marrow edema occupying the entire bone. If images are obtained with an oblique plane rather than perpendicular to the talus and navicular, the fracture line may be obscured.* ***(Right)*** *Sagittal T1WI MR shows dorsal sclerosis of the navicular and mild decrease in signal intensity in the remainder of the bone. These findings are nonspecific and must be correlated to coronal images to diagnose the stress fracture.*

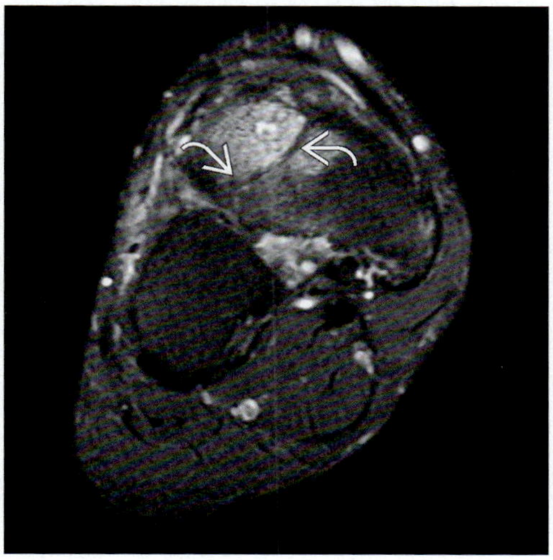

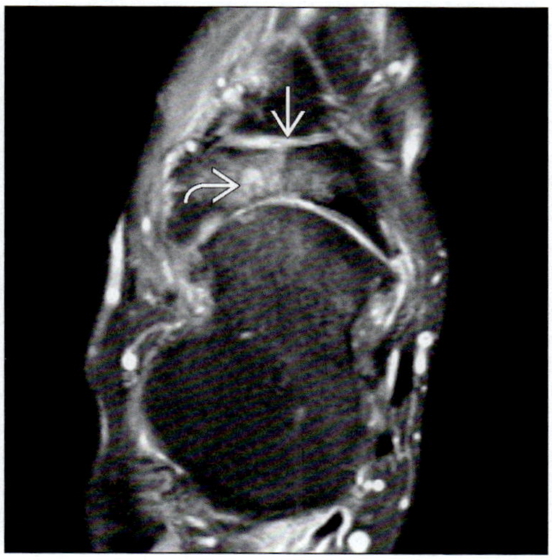

(Left) *Coronal T2WI FS MR shows a an advanced stress fracture, where the fracture line nearly completely traverses the navicular. Edema is less intense than in the preceding case, reflecting the chronicity of the findings.* ***(Right)*** *Axial PDWI FS MR in same patient shows fracture line, with surrounding bone marrow edema and bone sclerosis. Osteoarthritis has developed in this chronic stress fracture, with osteophytes and subchondral cysts.*

CUBOID FRACTURE

Key Facts

Imaging
- Marginal fracture: Sclerotic line parallels articular surface
  - Occurs at proximal or distal margin
- Central fracture: Impaction of lateral cortex, deformity of cuboid sulcus
- Stress fracture: Usually occult on radiographs, seen on MR
  - MR often shows stellate configuration
- Location
  - Fracture: Proximal or distal rim or mid body
  - Dislocation: Chopart or Lisfranc joint or complete extrusion of cuboid

Top Differential Diagnoses
- Lisfranc dislocation
- Isolated talonavicular dislocation

Pathology
- Associated injuries
  - Other midfoot fractures
  - Navicular median eminence fracture (nutcracker injury)
  - Dislocation of Chopart &/or Lisfranc joint associated with cuboid fracture

Clinical Issues
- Cuboid fracture often overshadowed clinically by associated injuries
- Stress fracture: Vague midfoot pain, limp

Diagnostic Checklist
- Fractures difficult to see on radiographs, look for deformity or sclerosis of cuboid
- Associated articular injuries of greater prognostic importance than cuboid fracture

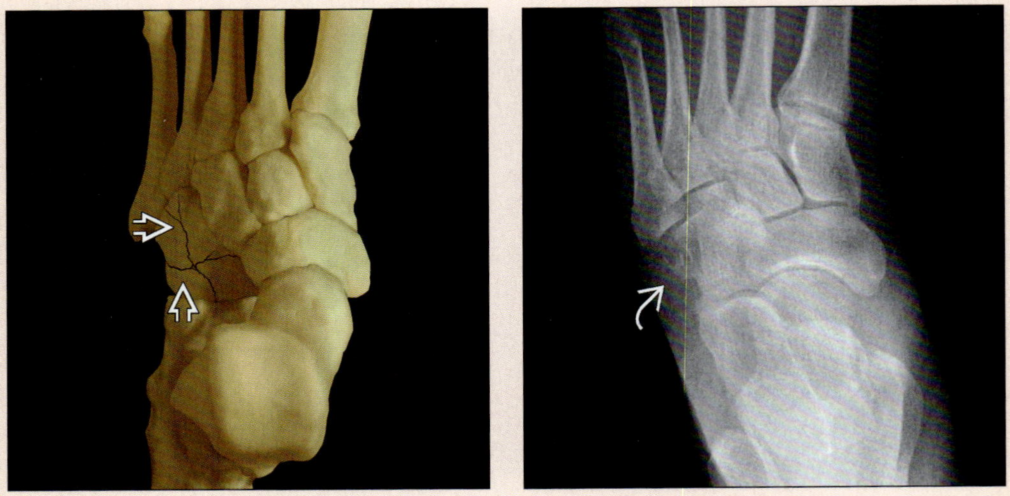

(Left) *Graphic of the dorsum of the foot shows the commonly seen stellate pattern of a cuboid fracture due to anteroposterior force on the cuboid, for instance when the forefoot impacts against a fixed object. Cuboid fractures rarely occur in isolation, and a search should always be made for other midfoot fractures or dislocations.* ***(Right)*** *AP radiograph shows a comminuted cuboid fracture. 5th metatarsal is displaced proximally because of cuboid impaction.*

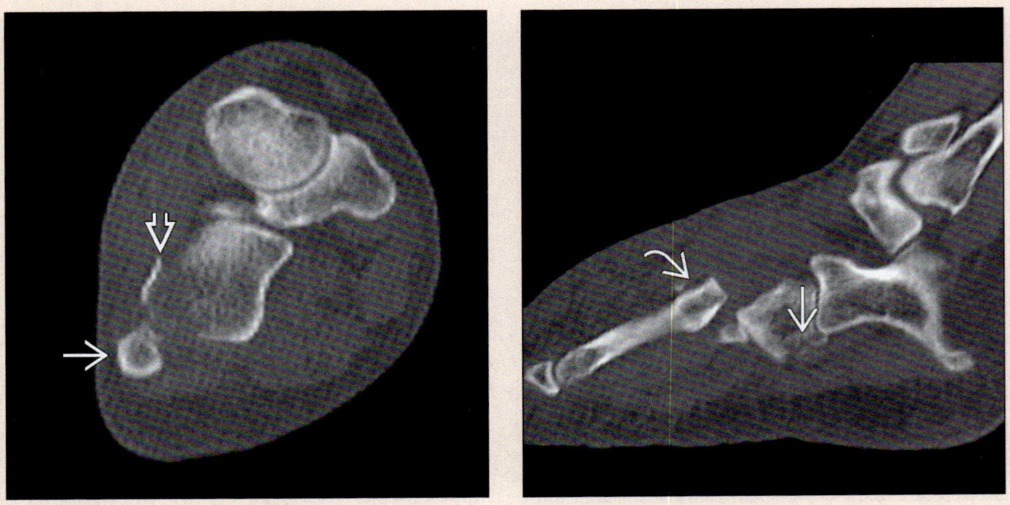

(Left) *Coronal bone CT shows a vertically oriented cuboid fracture. The 5th metatarsal is displaced proximally, indicating dislocation of TMT joint.* ***(Right)*** *Sagittal bone CT in the same patient shows dorsal dislocation of the 4th TMT and a small portion of the cuboid fracture.*

CUBOID FRACTURE

IMAGING

General Features

- Best diagnostic clue
  - Fracture: Deformity of cuboid, often without visible fracture line on radiographs
  - Dislocation: Loss of joint congruity, which may be subtle
- Location
  - Fracture: Proximal or distal rim or mid body
  - Dislocation: Chopart or Lisfranc joint or complete extrusion of cuboid

Radiographic Findings

- Distal rim: Sclerotic line parallels articular surface
  - Often see displacement of 4th &/or 5th tarsometatarsal (TMT) joint
- Proximal rim: Sclerotic line parallels articular surface
  - Often associated with dislocation of calcaneocuboid joint or avulsion of median eminence of navicular
- Mid body: Impaction, deformity of cuboid sulcus
- Stress fracture: Usually occult on radiographs
- Calcaneocuboid dislocation
  - Loss of parallelism of joint surfaces

CT Findings

- Usually see comminution not visible on radiographs

MR Findings

- Same as CT, also see bone marrow edema
- Useful for stress fractures

Imaging Recommendations

- Best imaging tool
  - CT scan
- Protocol advice
  - Sagittal, coronal reformations useful to see associated joint disruptions

DIFFERENTIAL DIAGNOSIS

Lisfranc Fracture-Dislocation

- More common than cuboid fracture
- May have dislocation of 4th-5th TMT joints with or without cuboid fx

Isolated Talonavicular Dislocation

- Usually dislocate both talonavicular and calcaneocuboid joints
- May dislocate talonavicular joint without calcaneocuboid joint, but not vice versa

Normal Mobility of Joint

- On non-weight-bearing views, cuboid articular surface extends slightly below calcaneus
- Joint margins remain parallel

PATHOLOGY

General Features

- Etiology
  - Different mechanisms of injury
    - Twisting injury
    - Axial load from toes toward heel
    - Shear force applied across midfoot
    - Stress fracture (rare)
- Associated abnormalities
  - Other foot fractures
  - Dislocation of Chopart &/or Lisfranc joints
  - Nutcracker injury
    - Avulsion of median eminence of navicular
    - Crush of calcaneocuboid joint, involving either cuboid or calcaneus

Relevant Anatomy

- Cuboid sulcus
  - Groove in mid body of cuboid, under which peroneus longus tendon passes
- Lisfranc joint: Tarsometatarsal joint
- Chopart joint: Midtarsal joint
  - Talonavicular + calcaneocuboid joint
  - Separate joint cavities that act together and are often injured together

CLINICAL ISSUES

Presentation

- Most common signs/symptoms
  - Trauma: Deformity is visible if dislocation present
    - Isolated fracture often overshadowed clinically by other associated injuries
  - Stress fracture: Vague midfoot pain, limp

Demographics

- Age
  - Toddlers may fracture either as stress or acute injury
  - Stress fractures seen in children or adult athletes
  - Trauma is most common in young adults

Natural History & Prognosis

- Prognosis depends on extent of associated joint injuries

DIAGNOSTIC CHECKLIST

Image Interpretation Pearls

- Fractures difficult to see on radiographs, look for deformity or sclerosis of cuboid
- Normal calcaneocuboid joint appears slightly incongruent on lateral non-weight-bearing radiographs

Reporting Tips

- Associated articular injuries of greater prognostic importance than cuboid fracture

SELECTED REFERENCES

1. Franco M et al: An uncommon cause of foot pain: the cuboid insufficiency stress fracture. Joint Bone Spine. 72(1):76-8, 2005
2. Simonian PT et al: Fracture of the cuboid in children. A source of leg symptoms. J Bone Joint Surg Br. 77(1):104-6, 1995
3. Blumberg K et al: The toddler's cuboid fracture. Radiology. 179(1):93-4, 1991
4. Sangeorzan BJ et al: Displaced fractures of the cuboid. J Bone Joint Surg Br. 72(3):376-8, 1990
5. Main BJ et al: Injuries of the midtarsal joint. J Bone Joint Surg Br. 57(1):89-97, 1975

(Left) AP radiograph shows Chopart fracture dislocation. There is a stellate cuboid fracture ➡. The cuboid and navicular are subluxated medially, and the head of the talus is impacted against the lateral margin of the navicular ➡. *(Right)* Lateral radiograph in the same patient shows the head of talus superimposed on the navicular ➡. The calcaneocuboid joint ➡ is obscured due to joint incongruity. The cuboid fracture is difficult to see, but the angular deformity of the cuboid sulcus ➡ is an important clue.

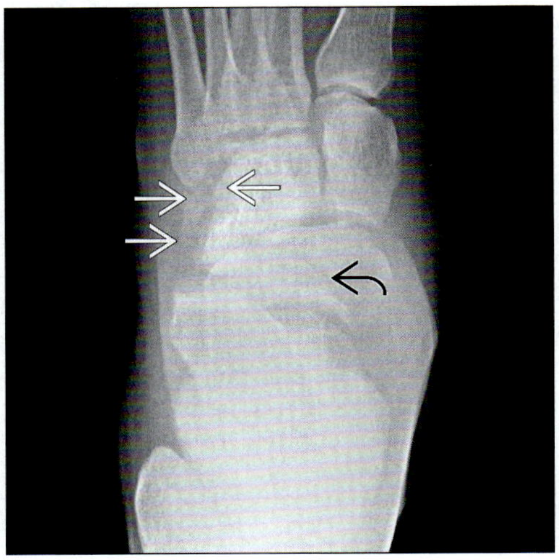

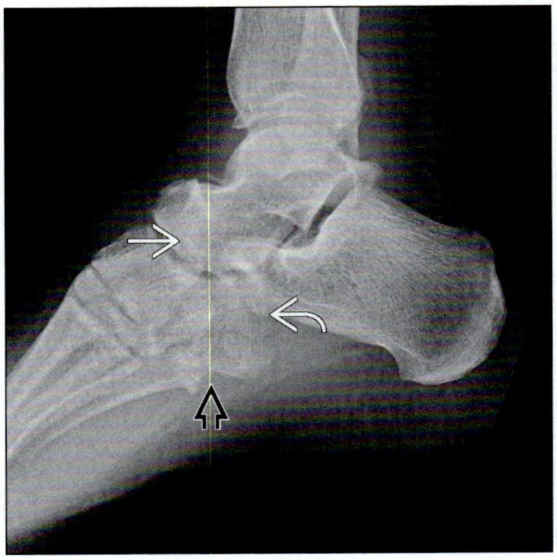

(Left) Sagittal bone CT in the same patient following closed reduction shows persistent talonavicular subluxation, fracture of the navicular ➡, and reduction of the calcaneocuboid joint. A small portion of the cuboid fracture ➡ is visible. *(Right)* Graphic shows simple dislocation of the Chopart joint, which is comprised of the talonavicular ➡ and calcaneocuboid ➡ joints. A simple dislocation without associated fracture is uncommon.

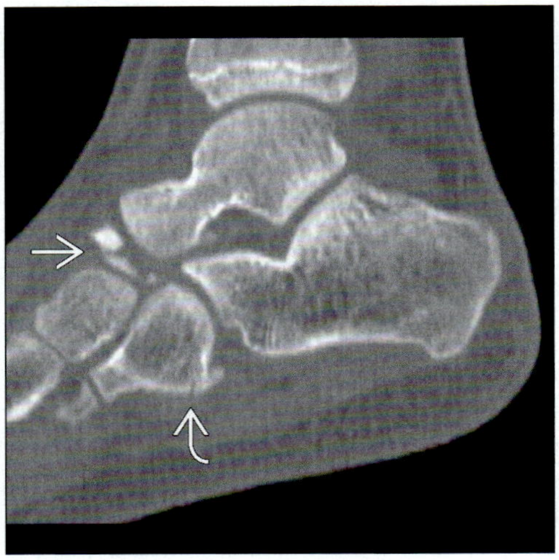

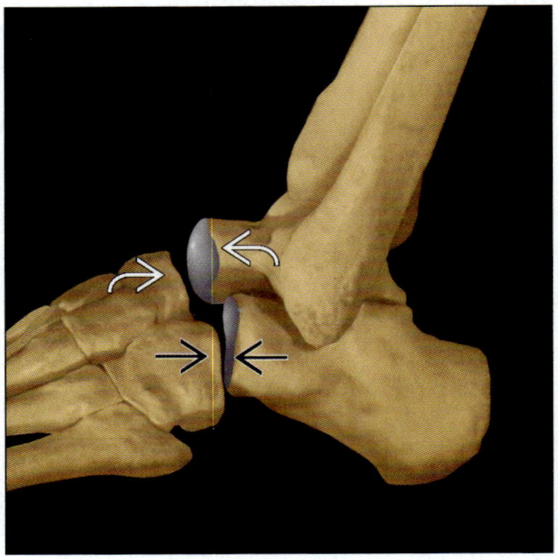

(Left) Oblique radiograph shows a sclerotic line indicating a mid body cuboid fracture ➡. Slight deformity of the cuboid sulcus ➡ is evident. *(Right)* Sagittal STIR MR shows an isolated cuboid fracture ➡ due to abduction injury. Note articular step-off ➡ at the TMT joint and contusion of the peroneus longus tendon ➡.

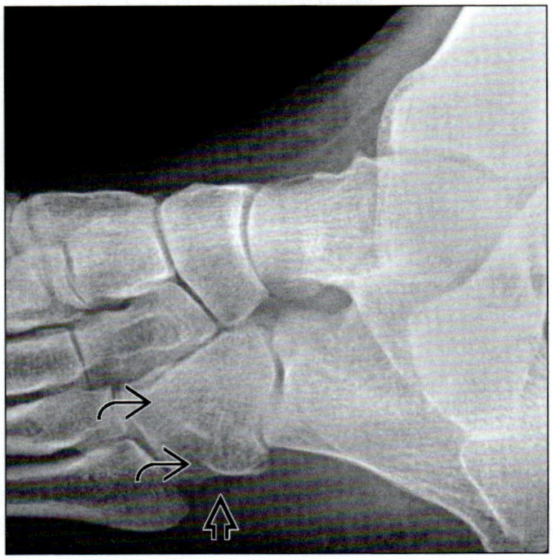

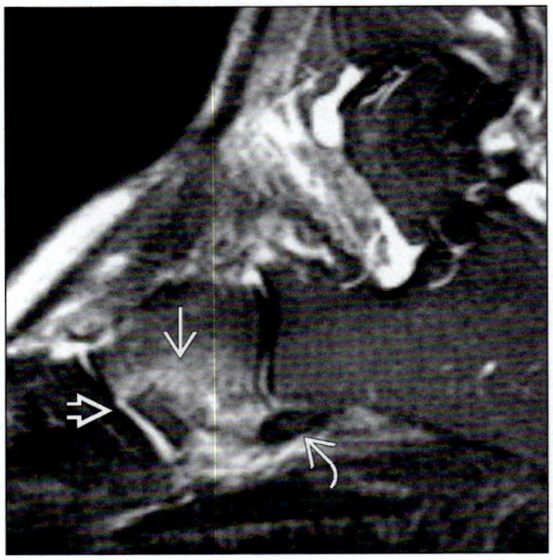

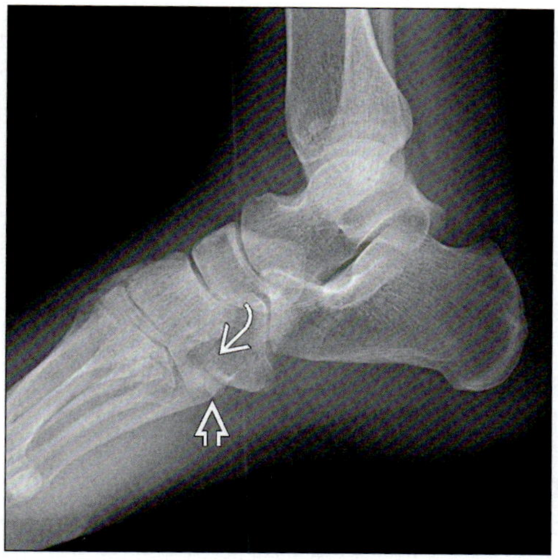

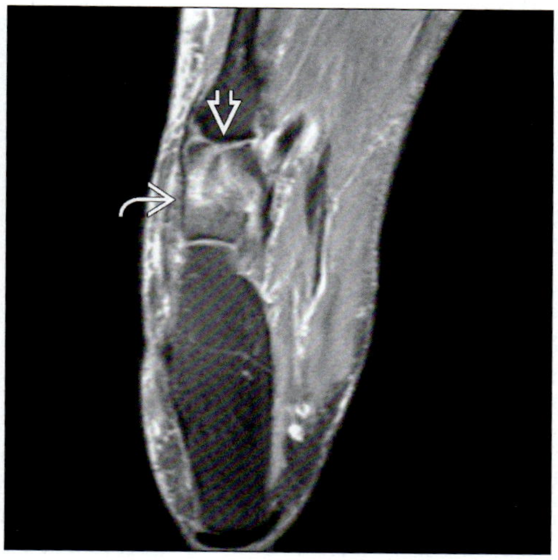

(Left) *Lateral radiograph shows a mid body cuboid fracture associated with dislocation of the 4th and 5th TMT joints.* **(Right)** *Axial PDWI FS MR shows a comminuted cuboid fracture, which was an isolated injury in a college football player. Note the impaction of the articular surface at the tarsometatarsal joint.*

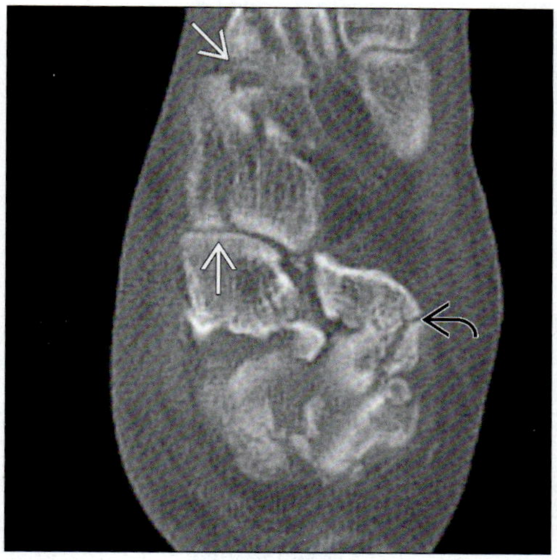

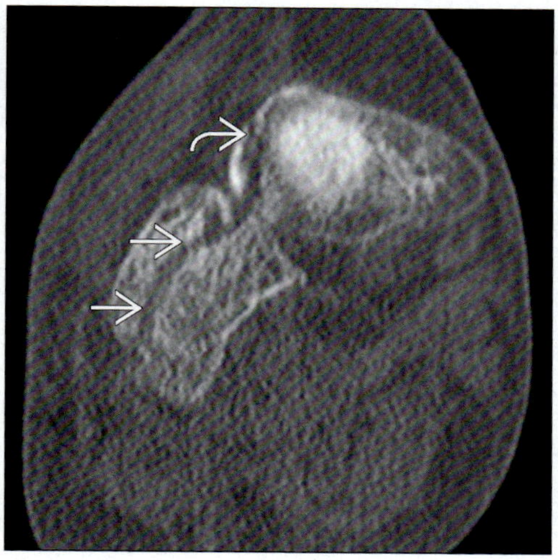

(Left) *Axial bone CT shows a longitudinally oriented fracture of the cuboid, which is associated with a compressive calcaneal fracture in a patient who fell from a height. The cuboid fracture disrupts both the calcaneocuboid and the Lisfranc joint.* **(Right)** *Coronal bone CT in the same patient shows a vertically oriented cuboid fracture, aligned with a navicular fracture. Cuboid fractures much more commonly occur in combination with other fractures than they do in isolation.*

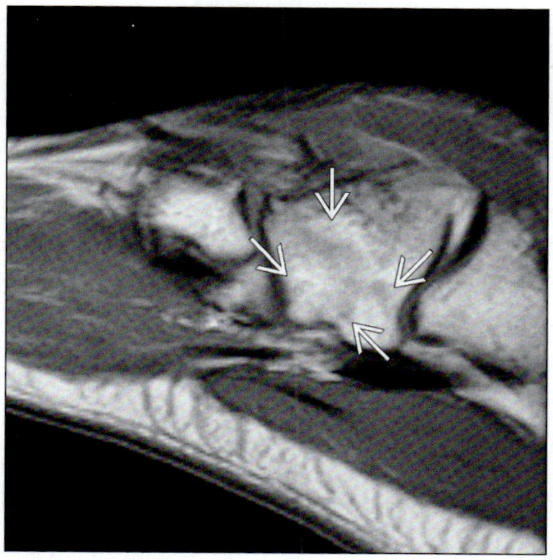

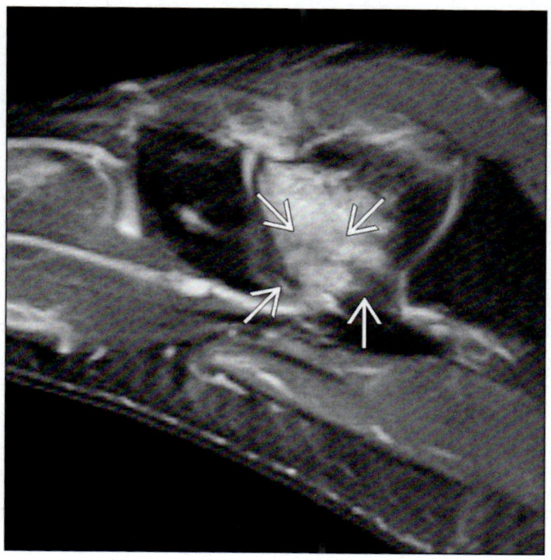

(Left) *Sagittal T1WI MR shows a cuboid stress fracture in an ice skater. As is typical of this fracture, multiple incomplete fracture lines are present in a stellate morphology.* **(Right)** *Sagittal STIR MR in the same patient shows extensive high signal intensity bone marrow edema involving the majority of the cuboid. Fracture lines are well seen, outlined by the marrow edema.*

LISFRANC LIGAMENT INJURY

Key Facts

Terminology

- Midfoot sprain: Isolated injury of Lisfranc ligament, without disruption of entire tarsometatarsal joint

Imaging

- Radiographs are more sensitive when weight-bearing; may see
  - Lateral &/or dorsal displacement of 2nd metatarsal base relative to 2nd cuneiform
  - Chip fracture at medial margin of 2nd metatarsal base
  - Flattening of longitudinal arch of foot
  - Radiographs may be false-negative even with stress
- MR is primary imaging method of diagnosis
  - Ligament disruption visible on axial, coronal images
  - 3 bands of Lisfranc ligament easily distinguishable

Top Differential Diagnoses

- Lisfranc dislocation: More extensive injury to Lisfranc joint, involving additional TMT articulations
- Neuropathic arthropathy
- 2nd metatarsal base fracture

Pathology

- 3 bands of ligament injured from dorsal to plantar
- If 2 of bands are torn, 2nd TMT is unstable

Clinical Issues

- Vague midfoot pain
- Swelling centered at 2nd TMT joint
- 3 bands of ligament injured from dorsal to plantar
- End result of injury: Flatfoot deformity and TMT osteoarthritis

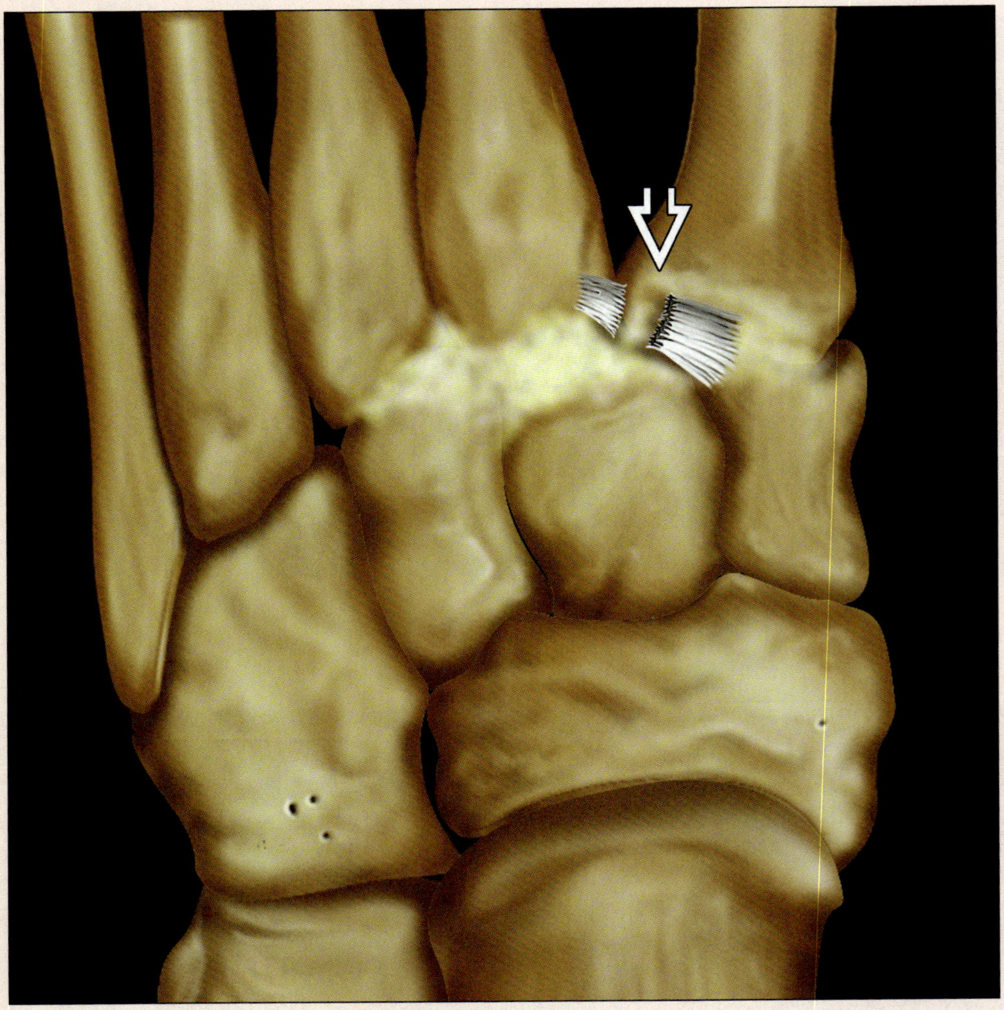

Graphic shows rupture of the dorsal band of Lisfranc ligament ⊳, which extends from the lateral margin of the 1st cuneiform to the medial margin of the 2nd metatarsal. The primary ligamentous stabilizer is the interosseous band of the ligament, which cannot be seen from a dorsal view. Since Lisfranc ligament injury, or midfoot sprain, occurs due to plantarflexion injury of the midfoot, the dorsal band of the ligament is the 1st injured, followed by the interosseous, and finally the plantar band.

LISFRANC LIGAMENT INJURY

TERMINOLOGY

Definitions

- Midfoot sprain: Isolated injury of Lisfranc ligament, without disruption of entire tarsometatarsal (TMT) joint

IMAGING

General Features

- Best diagnostic clue
  - Lateral displacement of base of 2nd metatarsal relative to 2nd cuneiform
    - May be evident only on weight-bearing or stress views
    - Radiographs may be false-negative even with stress
- Location
  - Between medial cuneiform and 2nd metatarsal base

Imaging Recommendations

- Best imaging tool
  - MR
- Protocol advice
  - Axial images must be aligned with axis of 2nd metatarsal
  - Coronal images perpendicular to 2nd metatarsal
  - Field of view should be centered on midfoot
  - Often useful to image foot with patient prone

Radiographic Findings

- Radiographs are more sensitive when weight-bearing or with eversion stress
- Lateral &/or dorsal displacement of 2nd metatarsal base relative to 2nd cuneiform
- Increased space between 1st and 2nd metatarsals
  - Finding is highly dependent on pronation or supination of foot, and is not reliable
- Chip fracture at medial margin of 2nd metatarsal base
  - Involves attachment of Lisfranc ligament
- Flattening of longitudinal arch of foot
  - Seen focally at tarsometatarsal joint
- Distance between cuneiforms not a sensitive sign of injury
  - Distance is highly dependent on supination/pronation of foot

CT Findings

- Since CT is not weight-bearing, displacement may be less apparent than on radiographs

MR Findings

- Ligament disruption visible on axial, coronal images
  - 3 bands of Lisfranc ligament easily distinguishable
  - Plantar and dorsal bands are thin, ribbon-like
  - Interosseous band is thick, contains distinct fascicles
- Bone marrow edema often seen in medial cuneiform &/or 2nd metatarsal

DIFFERENTIAL DIAGNOSIS

Lisfranc Dislocation

- More extensive injury to Lisfranc joint, involving additional TMT articulations
- Advanced imaging may change diagnosis from isolated Lisfranc ligament injury to true Lisfranc dislocation

Neuropathic Arthropathy

- Probably often begins with unrecognized Lisfranc ligament injury
  - Destabilizes joint and allows further injury

2nd Metatarsal Base Fracture

- Transverse fracture distal to Lisfranc ligament also destabilizes joint

PATHOLOGY

General Features

- Etiology
  - Plantar flexion of midfoot
    - Falling down stairs, tripping in sports or going downstairs

Staging, Grading, & Classification

- 3 bands of ligament injured from dorsal to plantar

Relevant Anatomy

- Lisfranc ligament: Major stabilizer of tarsometatarsal joint
  - Extends from medial cuneiform to base of 2nd metatarsal
  - 3 components: Dorsal, interosseous (most important), and plantar

CLINICAL ISSUES

Presentation

- Most common signs/symptoms
  - Vague midfoot pain
  - Swelling centered at 2nd TMT joint

Natural History & Prognosis

- When 2 or 3 bundles of Lisfranc ligament are ruptured, 2nd TMT is unstable
- End result of injury: Flatfoot deformity and TMT osteoarthritis

DIAGNOSTIC CHECKLIST

Image Interpretation Pearls

- Even slight lateral displacement of 2nd MT relative to middle cuneiform indicates ligament injury

SELECTED REFERENCES

1. Macmahon PJ et al: MRI of injuries to the first interosseous cuneometatarsal (Lisfranc) ligament. Skeletal Radiol. 38(3):255-60, 2009
2. Raikin SM et al: Prediction of midfoot instability in the subtle Lisfranc injury. Comparison of magnetic resonance imaging with intraoperative findings. J Bone Joint Surg Am. 91(4):892-9, 2009
3. Crim J: MR imaging evaluation of subtle Lisfranc injuries: the midfoot sprain. Magn Reson Imaging Clin N Am. 16(1):19-27, v, 2008
4. Nunley JA et al: Classification, investigation, and management of midfoot sprains: Lisfranc injuries in the athlete. Am J Sports Med. 30(6):871-8, 2002

*(**Left**) Axial T1WI MR shows a normal interosseous Lisfranc ligament extending from the lateral margin of the 1st cuneiform to the medial margin of the 2nd MT. Note that the 2nd TMT joint is recessed relative to the 1st and 3rd TMT joints, creating a mortise and tenon configuration which also stabilizes the joint. (**Right**) Axial PDWI FS MR at the corresponding level in an injured patient shows complete rupture of the interosseous Lisfranc ligament.*

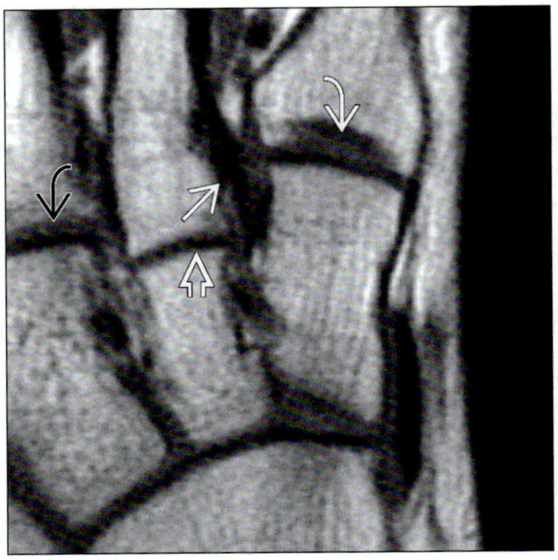

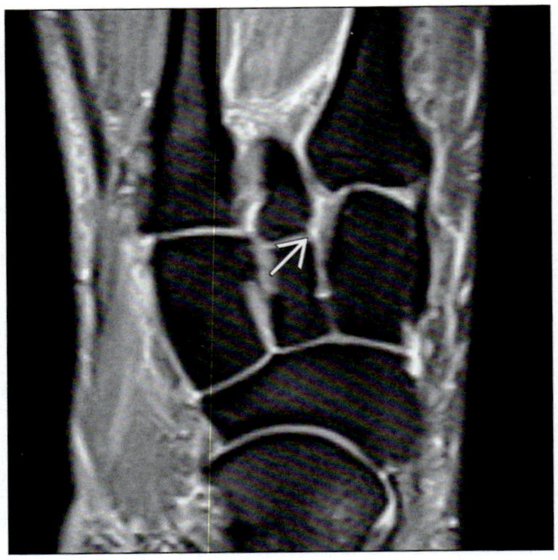

*(**Left**) Coronal T2WI FS MR shows the 3 distinct components of the Lisfranc ligament: Dorsal, interosseous, and plantar. Note that the interosseous ligament is rope-like and composed of multiple strands, while the dorsal and plantar ligaments are flat and ribbon-like. (**Right**) Coronal T2WI FS MR shows rupture of the dorsal, interosseous, and plantar Lisfranc ligaments. The coronal plane is useful for assessing the extent of Lisfranc ligament injury.*

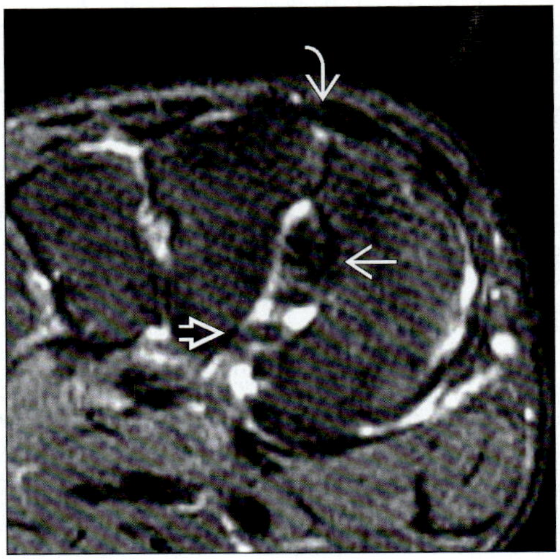

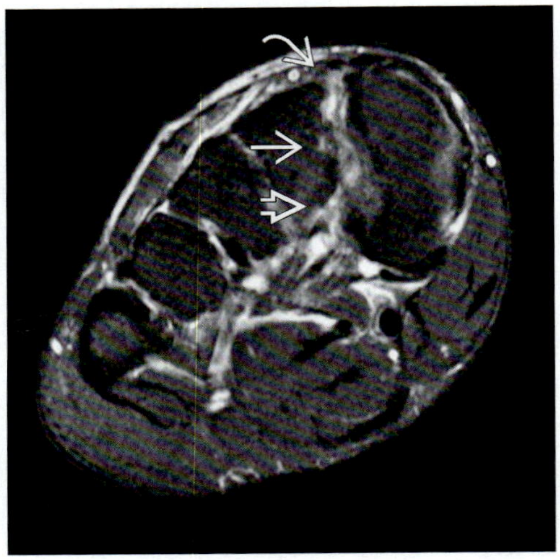

*(**Left**) AP radiograph shows normal relationships of the 2nd MT joint. The medial margin of the 2nd MT is aligned exactly with the medial margin of the 2nd cuneiform. Even slight offset of the joint indicates Lisfranc ligament injury. (**Right**) AP radiograph in a patient whose foot "folded under" while running downstairs shows lateral subluxation of the 2nd metatarsal base, indicating Lisfranc ligament injury. The joint was grossly unstable at surgery.*

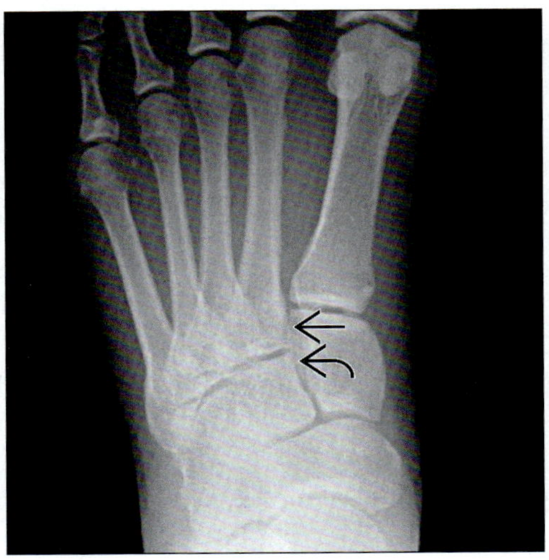

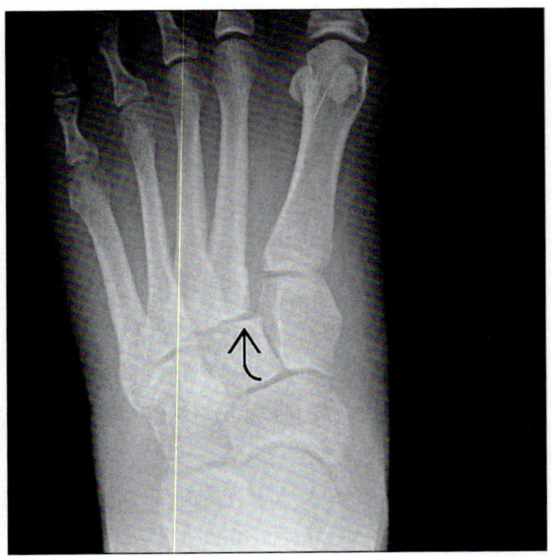

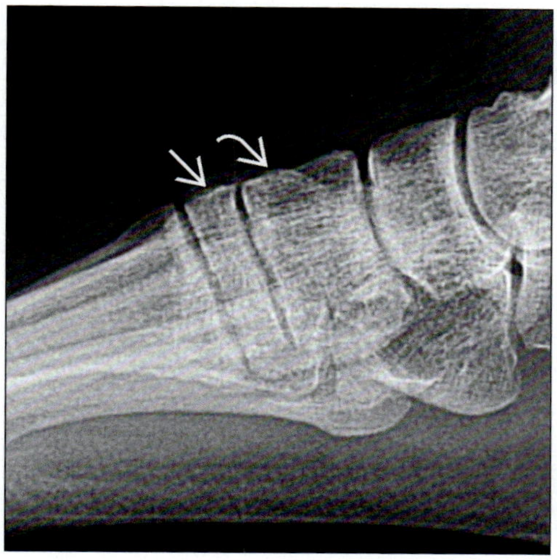

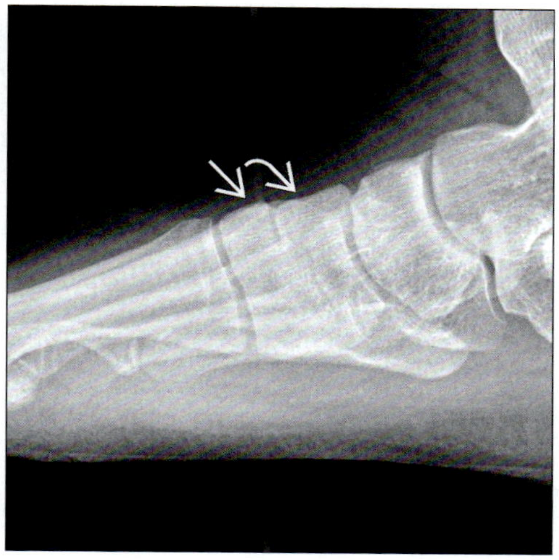

(Left) *Lateral radiograph shows normal TMT relationships. The 1st and 2nd metatarsal bases are superimposed but can be distinguished because the 1st MT is much thicker and the 2nd MT extends further proximally. The dorsal margin of 2nd MT is aligned with 2nd cuneiform.* ***(Right)*** *Lateral radiograph shows subluxation due to Lisfranc ligament injury. The 2nd MT base is 2 mm dorsal to the middle cuneiform. This subtle malalignment is an important sign of injury.*

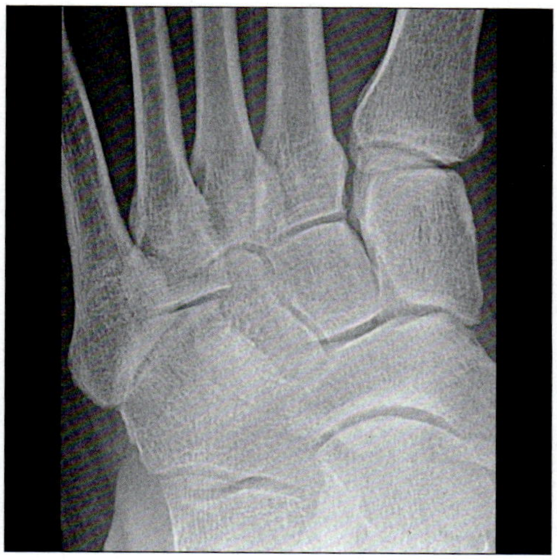

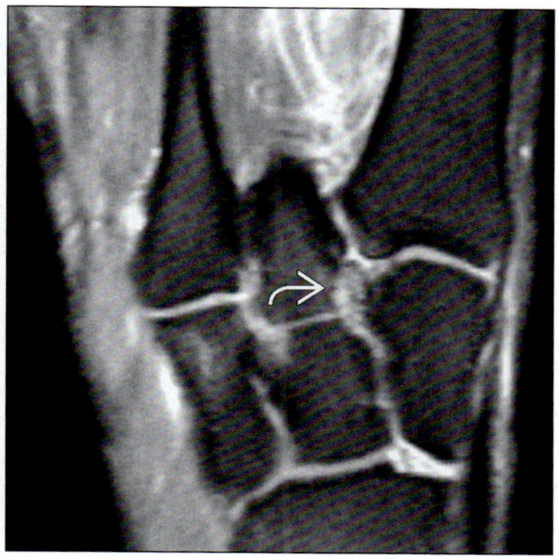

(Left) *AP radiograph shows normal relationships at the TMT joint and was false-negative for Lisfranc ligament injury. Due to pain, an MR was performed.* ***(Right)*** *Axial PDWI FS MR shows ruptured interosseous ligament. It is important to remember that while subluxation of the joint is specific for Lisfranc ligament injury, false-negative radiographs are not uncommon. If there is high clinical suspicion of Lisfranc ligament injury, an MR should be performed.*

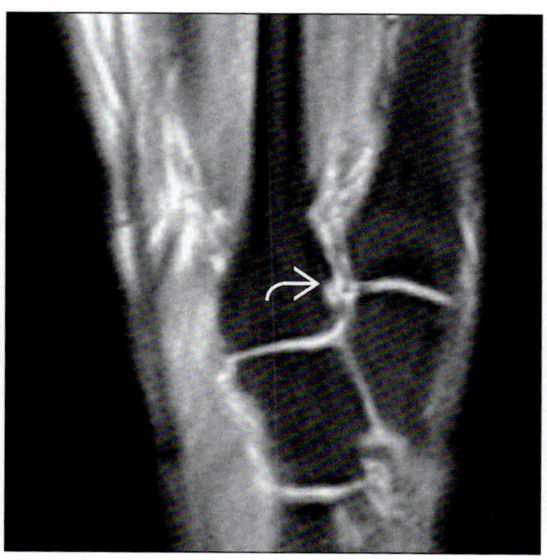

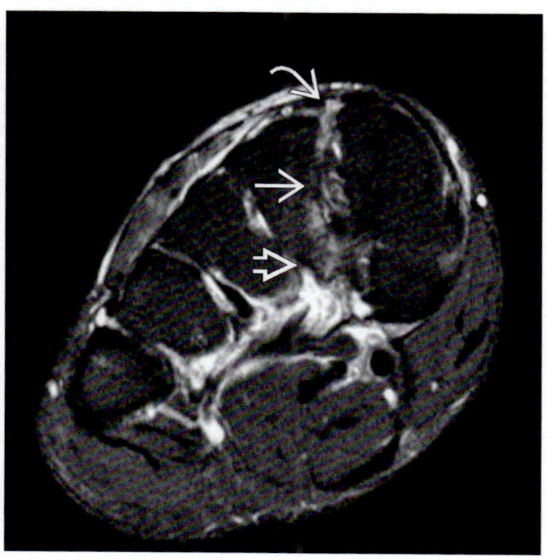

(Left) *Axial PDWI FS MR in the same patient shows rupture of the dorsal band of the Lisfranc ligament.* ***(Right)*** *Coronal T2WI FS MR in the same patient shows tears of dorsal and interosseous bands of Lisfranc ligament. The plantar band appears lax but continuous. The 2nd TMT joint was unstable under anesthesia.*

LISFRANC FRACTURE-DISLOCATION

Key Facts

Terminology
- Tarsometatarsal fracture-dislocation

Imaging
- Malalignment of MT relative to cuneiforms on AP radiographs
- Normal relationships on AP radiographs
  - 1st MT should be centered on 1st cuneiform
  - 2nd MT medial margin should be aligned with medial margin of 2nd cuneiform
  - 3rd MT lateral margin should be aligned with lateral margin of 3rd cuneiform
  - 5th MT styloid process should extend lateral to cuboid
- Dorsal subluxation of MT bases on lateral radiographs
- Fractures of MT, cuneiforms, cuboid are commonly present
- CT often shows fractures not detectable by radiograph

Top Differential Diagnoses
- Midfoot sprain
  - Isolated injury of Lisfranc ligament
- Chopart dislocation
- Neuropathic arthropathy (Charcot joint)

Pathology
- Majority are motor vehicle or industrial accidents
- Classified as homolateral or divergent
  - Homolateral: Displacement of all rays laterally
  - Divergent: 1st MT displaced medially and 2nd to 5th MT displaced laterally
  - Partial: Dislocation, which does not involve all TMT joints

*(**Left**) Graphic of foot shows dorsal view of a homolateral fracture-dislocation of the Lisfranc joint. There is a transverse fracture of the 2nd MT base ➡, and the Lisfranc ligament is ruptured ⮫. (**Right**) Graphic shows the dorsal view of a partial Lisfranc fracture-dislocation. The 1st MT ➡ is in a normal position while the 2nd-5th are dislocated laterally. Less commonly, the 1st TMT is dislocated medially while the 2nd-5th are dislocated laterally. The Lisfranc ligament is disrupted ⮫.*

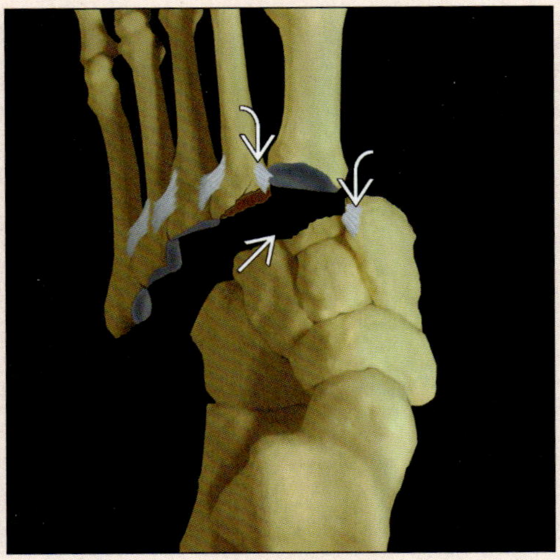

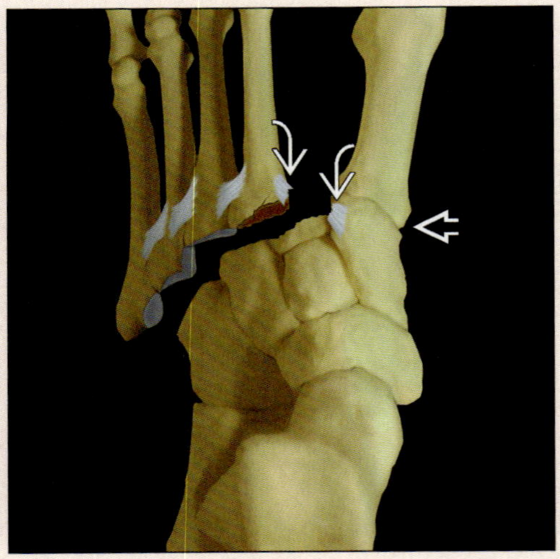

*(**Left**) AP radiograph shows a complete homolateral dislocation of the Lisfranc joint: All 5 tarsometatarsal joints are dislocated laterally. The 5th MT base is also displaced proximally almost to the level of the calcaneocuboid joint ➡. (**Right**) AP radiograph shows widening and lateral displacement of the 1st-3rd TMT joints, with apparently normal position of 4th and 5th metatarsals. Note the associated fracture of Lisfranc ligament attachment ➡.*

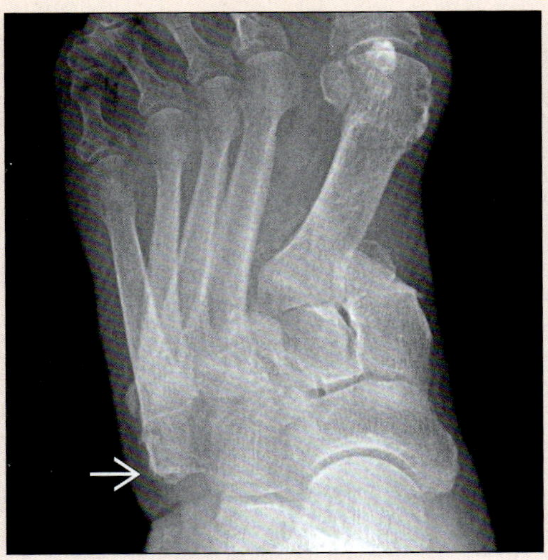

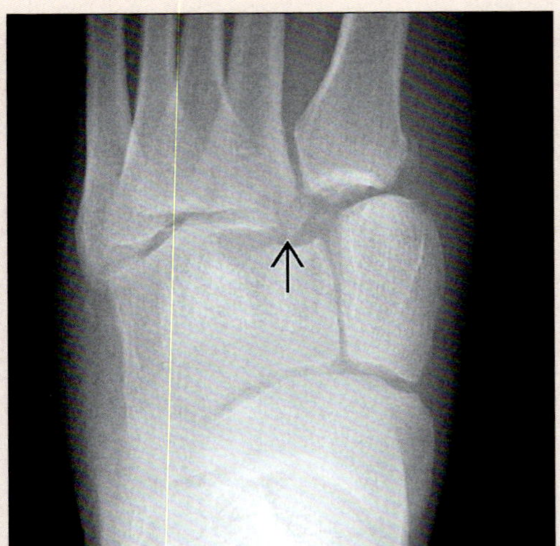

LISFRANC FRACTURE-DISLOCATION

TERMINOLOGY

Synonyms

- Tarsometatarsal (TMT) fracture-dislocation

IMAGING

General Features

- Best diagnostic clue
  - Malalignment of TMT joints
- Morphology
  - Usually gross disruption of joint

Radiographic Findings

- Malalignment of metatarsal (MT) relative to cuneiforms on AP radiographs
  - Normal alignment of TMT joint on AP weight-bearing radiographs
    - 1st MT should be centered on 1st cuneiform
    - 2nd MT medial margin should be aligned with medial margin of 2nd cuneiform
    - 3rd MT lateral margin should be aligned with lateral margin of 3rd cuneiform
    - 4th MT medial margin should be aligned with medial margin of cuboid
    - 5th MT styloid process should extend lateral to cuboid
- Dorsal or less commonly plantar displacement of MT bases on lateral radiographs
  - Dorsal margins of MT 1-2 and corresponding cuneiforms should align
  - 3rd-5th TMT joints difficult to see on lateral radiographs

CT Findings

- CT often shows fractures not detectable by radiograph

MR Findings

- Injury is difficult to fully assess on MR, since small fracture fragments are poorly seen

Imaging Recommendations

- Best imaging tool
  - Radiographs used for initial diagnosis
  - CT useful to evaluate extent of injury
- Protocol advice
  - Axial images parallel to axis of metatarsals
  - Sagittal images show dorsal displacement
  - Coronal images best for cuneiform fxs

DIFFERENTIAL DIAGNOSIS

Midfoot Sprain

- Isolated rupture of Lisfranc ligament
- Low-velocity injury (e.g., sports, stepping off curb)

Metatarsal Shaft Fracture

- Fractures of shafts often isolated, spare Lisfranc joint

Chopart Dislocation

- Presents with similar foot deformity

Charcot, Neuropathic

- Common at TMT joint
- Radiographs show joint destruction, fractures with bone debris, disorganization, dislocations

PATHOLOGY

General Features

- Etiology
  - Majority are motor vehicle or industrial accidents
  - Plantar-flexed foot + longitudinal force (as in equestrian injury with foot caught in stirrup)
  - Direct blow or crush
- Associated abnormalities
  - Tarsal bone fractures (50-90% of cases)
  - Fractures of MT bases, particularly 2nd
  - Dislocation of Chopart &/or naviculocuneiform joints

Staging, Grading, & Classification

- Homolateral: Displacement of all rays laterally
- Divergent: 1st MT displaced medially and 2nd to 5th MT displaced laterally
- Partial: Dislocation, which does not involve all TMT joints
- Isolated: Dislocation of single TMT joint

Relevant Anatomy

- Transverse ligaments unite bases of 2nd to 5th MT
  - Arrangement favors displacement of 2nd to 5th MT bases as unit
- Lisfranc ligament stabilizes 1st and 2nd digits
  - Ligament extends from lateral aspect of medial cuneiform to medial base of 2nd MT
- 2nd MT base is recessed between 1st and 3rd MT bases
  - Position likely reason 2nd MT base is most frequent site of MT fracture in Lisfranc injury

CLINICAL ISSUES

Presentation

- Clinical profile
  - High-energy injury
    - Motorcycle, ATV, horse, unrestrained motor vehicle accident

Treatment

- Percutaneous or open fixation

DIAGNOSTIC CHECKLIST

Image Interpretation Pearls

- Radiographs underestimate associated fractures

Reporting Tips

- Describe direction of dislocations & major fragments
- Do not get bogged down in minutiae of tiny fracture fragments

SELECTED REFERENCES

1. Desmond EA et al: Current concepts review: Lisfranc injuries. Foot Ankle Int. 27(8):653-60, 2006
2. Haapamaki V et al: Lisfranc fracture-dislocation in patients with multiple trauma: diagnosis with multidetector computed tomography. Foot Ankle Int. 25(9):614-9, 2004
3. Haapamaki VV et al: Ankle and foot injuries: analysis of MDCT findings. AJR Am J Roentgenol. 183(3):615-22, 2004
4. Buzzard BM et al: Surgical management of acute tarsometatarsal fracture dislocation in the adult. Clin Orthop Relat Res. (353):125-33, 1998

LISFRANC FRACTURE-DISLOCATION

*(**Left**) AP radiograph shows a Lisfranc joint injury. The 1st TMT is unaffected, the 2nd MT based overlies the 2nd cuneiform indicating dislocation, and the 3rd TMT is dislocated laterally. Cuboid impaction fracture is due to 4th and 5th TMT dislocation. There is also a naviculocuneiform fracture-dislocation. (**Right**) Axial bone CT 3D reformation in the same patient better shows the dorsal displacement of the 2nd through 5th TMT joints.*

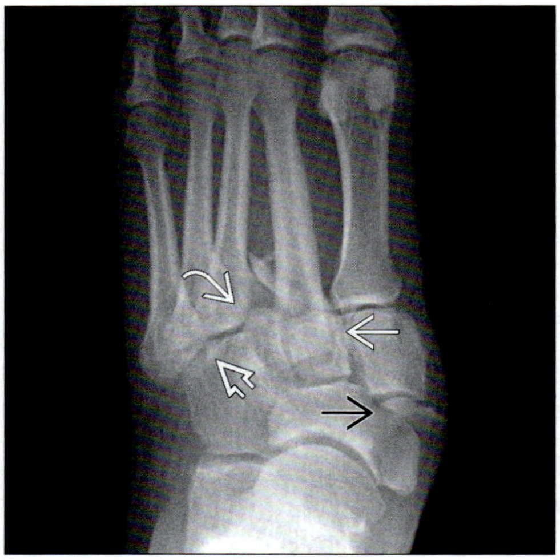

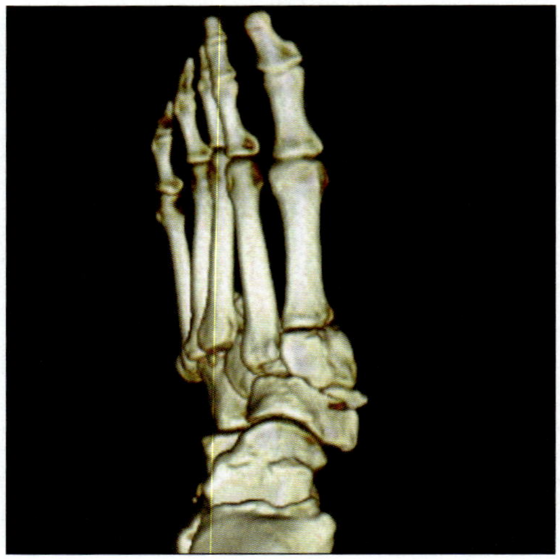

*(**Left**) AP radiograph shows an isolated 1st TMT fracture dislocation. There is an oblique fracture of the 1st MT base, subluxation of the 1st TMT joint, and a fracture of the medial cuneiform. (**Right**) AP radiograph shows what appears to be an isolated avulsion of the Lisfranc ligament from the base of the 2nd metatarsal. However, fragment displacement is more severe than in a typical midfoot sprain, and an MR was obtained to better assess other ligamentous structures.*

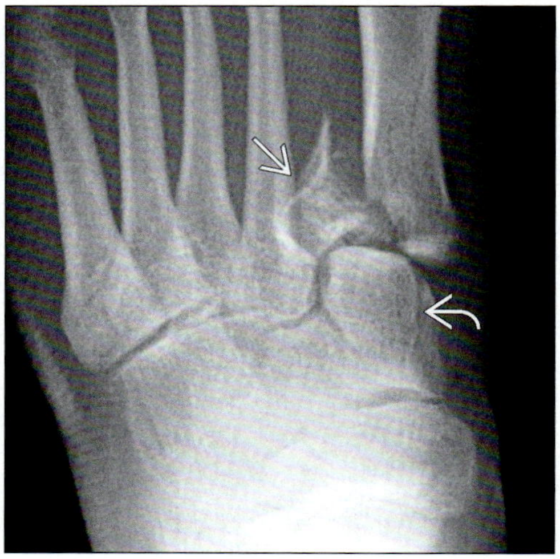

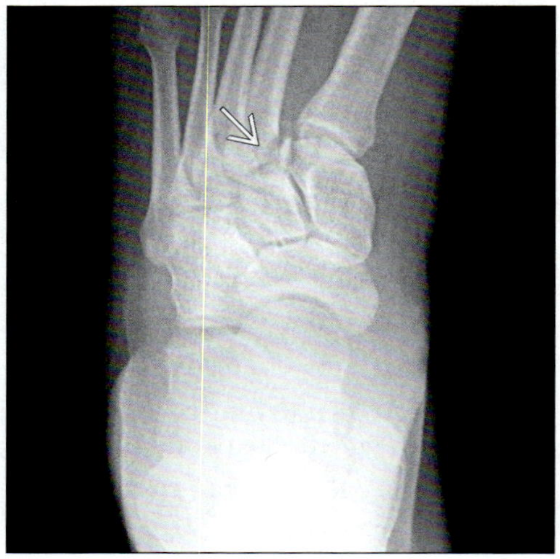

*(**Left**) PA radiograph demonstrates a divergent Lisfranc fracture-dislocation with medial dislocation of the 1st tarsometatarsal joint and lateral displacement of the 2nd and 3rd TMT joints. There are also 2 transverse fractures in the base of the 5th metatarsal. (**Right**) Lateral radiograph in the same patient shows dorsal displacement of the 1st metatarsal and uncovering of the medial cuneiform articular surface. The 5th metatarsal fractures are also seen.*

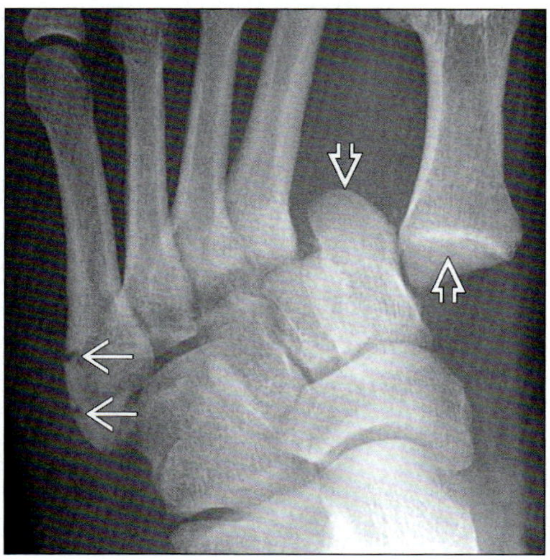

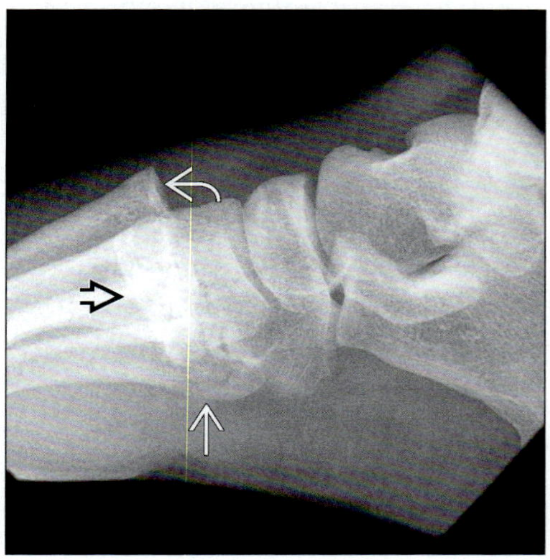

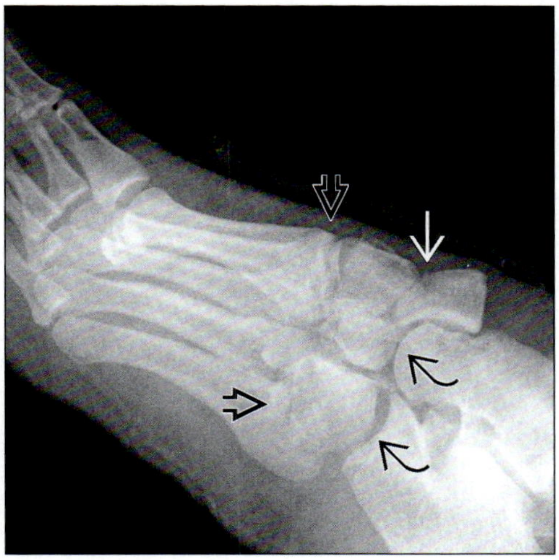

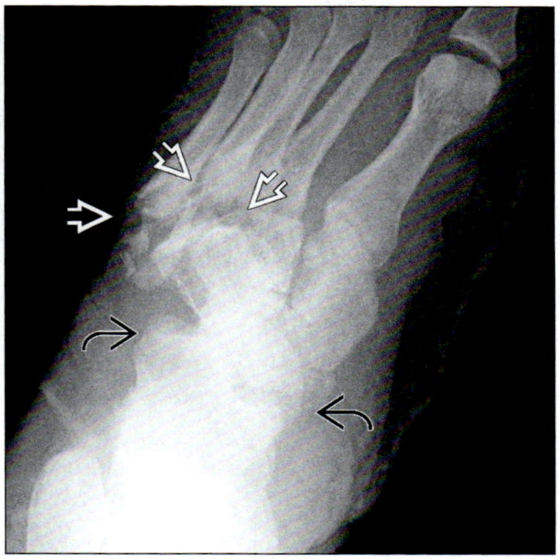

(Left) *Lateral radiograph shows dislocation of the 1st through 5th tarsometatarsal , naviculocuneiform , and Chopart joints from a motorcycle accident. There are multiple small bony fragments.* ***(Right)*** *AP radiograph in the same patient better shows metatarsal fractures and disruption of the Chopart joint . The 1st TMT appears normally aligned on this view, but dorsal widening of the joint on the lateral view indicates loss of joint stability.*

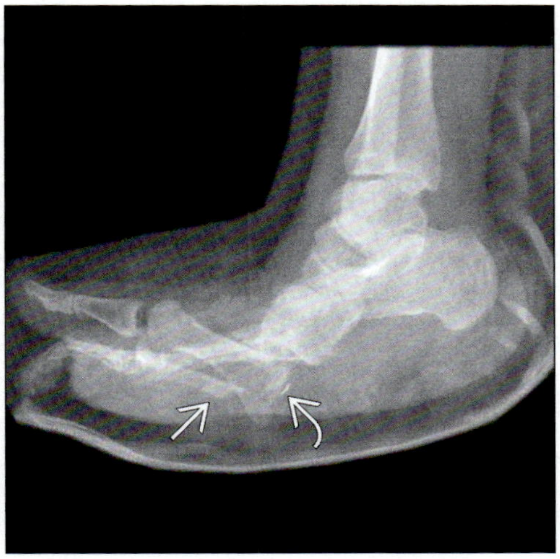

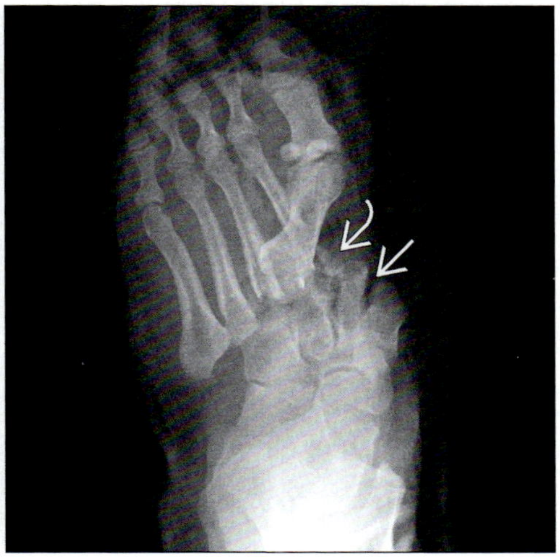

(Left) *Lateral radiograph shows an unusual pattern of Lisfranc joint dislocation, with complete plantar displacement of the 1st and 2nd MT. This was a parasailing injury.* ***(Right)*** *AP radiograph in the same patient shows a homolateral displacement of the metatarsal bases. The 1st cuneiform and the 2nd MT base are both fractured.*

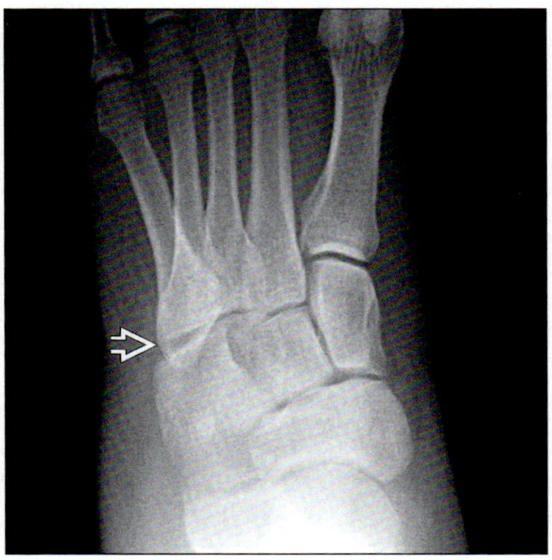

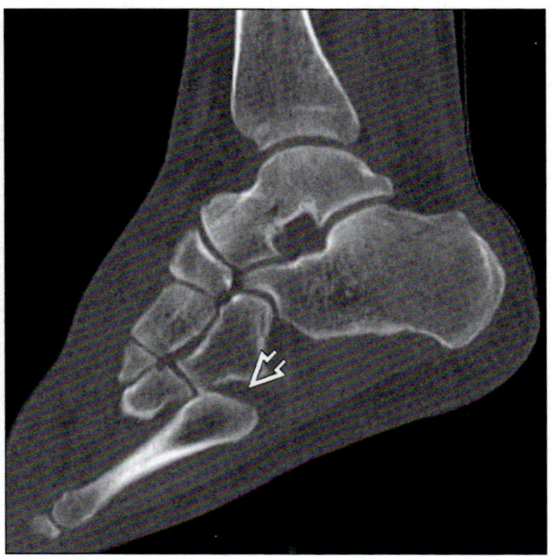

(Left) *AP radiograph shows an isolated medial dislocation of the 5th MT . Normally, the styloid process of the 5th MT extends laterally to the lateral margin of the cuboid. The other TMT joints appear intact, which was confirmed on the subsequent CT scan.* ***(Right)*** *Sagittal bone CT shows that the 5th MT articular surface is not articulating with the cuboid, but is displaced in a plantar direction. This unusual injury was due to a biking accident.*

METATARSAL FRACTURE

Key Facts

Imaging

- Common sites: 5th MT > 3rd > 2nd > 1st > 4th
- MT shaft > neck > > head
- Most fractures are transverse or short oblique
- Stress fractures are usually nondisplaced
  - May not be visible at time of initial pain/injury
  - Incomplete fracture line, which sometimes goes to completion
  - May be seen only as focal periosteal new bone 7-10 days post injury
- 5th MT: 3 types of fracture at proximal portion
  - Tuberosity avulsion fracture: Most proximal and generally extraarticular (1/10 intraarticular)
  - Jones fracture: At junction of diaphysis and metaphysis; always extraarticular
  - Stress fracture: Proximal diaphysis, usually distal to Jones
- MR will alter management in patients with MT pain and no visible fracture
  - T1WI best for visualization of fracture line
  - T2WI, STIR show marrow edema that may obscure fracture line

Top Differential Diagnoses

- Metatarsal tumor or infection
- Accessory centers of ossification
- Pseudoepiphysis

Pathology

- Associated injuries
  - Lisfranc joint injury
  - Cuneiform and other foot fractures
  - MTP dislocation

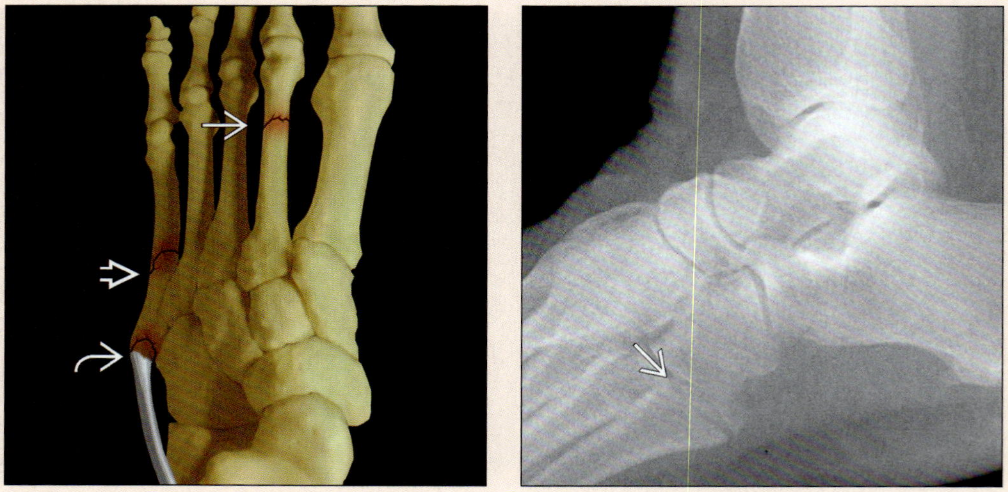

(Left) *Graphic of foot shows several common metatarsal fxs. The march fx → is a common fatigue fx usually of the 2nd or 3rd MT; traumatic fractures are also seen in this region. Fx of the 5th MT proximal shaft may be due to stress fx or a traumatic Jones fx →. The 5th MT tuberosity fx → is an avulsion injury due to traction by the peroneus brevis tendon and plantar aponeurosis.* ***(Right)*** *Lateral radiograph shows avulsion fx → of 5th MT base, due to inversion injury of the ankle.*

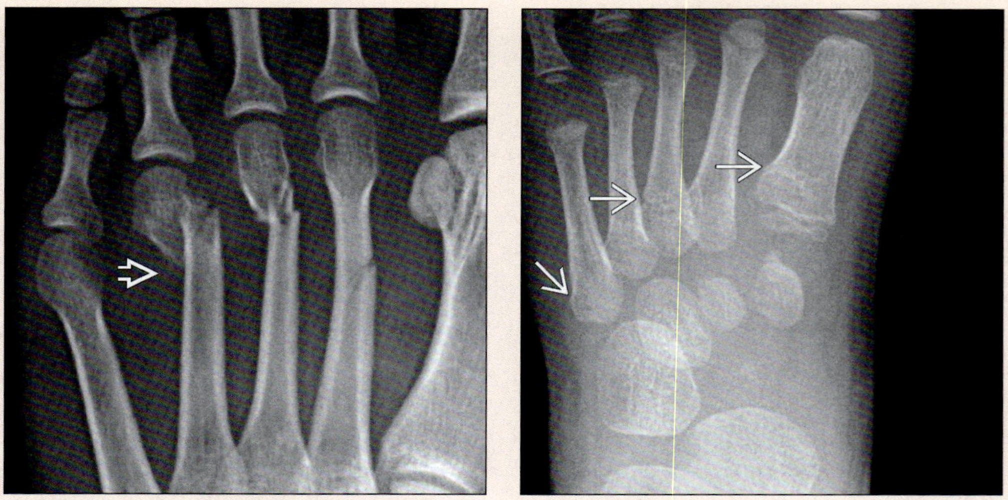

(Left) *AP radiograph shows fractures of the 2nd through 4th MT. There is complete lateral displacement of the 4th MT fracture →, with bayonet apposition of the fragments, while the 2nd MT fracture is nondisplaced. Traumatic MT fractures are commonly multiple.* ***(Right)*** *AP radiograph shows subtle fracture lines → at the base of the 5th, 3rd, and 1st MT in a child injured on a trampoline. The 5th MT fracture is sclerotic due to trabecular impaction.*

METATARSAL FRACTURE

TERMINOLOGY

Definitions

- Jones fracture: Traumatic transverse fx of 5th MT proximal shaft
- March fracture: Stress fracture (fx) of distal 2nd or 3rd metatarsal (MT) due to fatigue, typically in military recruits
- Stress fractures are due to repetitive injuries and divided into 2 types
  - Fatigue fx: Increased demand on normal bone
  - Insufficiency fx: Normal demand on weakened bone

IMAGING

General Features

- Location
  - 5th MT > 3rd > 2nd > 1st > 4th
  - Base 5th MT fractures: Tuberosity avulsion > Jones > stress fracture
  - Shaft, neck more commonly fractured than MT head
- Morphology
  - Most are transverse or short oblique
  - Fractures due to direct blow may have stellate morphology
  - 5th MT: 3 types of fracture at proximal portion
    - Tuberosity avulsion fracture: Most proximal and generally extraarticular (1/10 intraarticular)
    - Jones fracture: At junction of diaphysis and metaphysis; always extraarticular
    - Stress fracture: Proximal diaphysis, usually distal to Jones

Radiographic Findings

- Radiography
  - 3 views recommended
  - Fracture may be visible only on a single view

CT Findings

- Bone CT
  - Useful to evaluate joint involvement
  - Useful to evaluate for incomplete or nonunion
    - Incomplete healing of Jones fracture has high risk of developing nonunion

MR Findings

- T1WI best for visualization of fracture line
- T2WI, STIR show marrow edema that may obscure fracture line

Imaging Recommendations

- Best imaging tool
  - Radiograph is initial study, but nondisplaced fracture may be occult
  - MR used when finding stress fractures will change management
    - i.e., high-level athletes
- Protocol advice
  - Angle axial, sagittal images along MT shafts
  - Include full evaluation of Lisfranc ligament

DIFFERENTIAL DIAGNOSIS

Metatarsal Tumor or Infection

- On MR, marrow edema may obscure fracture line

Freiberg Infraction

- Osteonecrosis vs. OCD of 2nd-3rd MT head

Accessory Ossicles, Ankle and Foot

- Rounded contour, characteristic location

Pseudoepiphysis

- Occurs at opposite end of MT compared to expected location of epiphysis

PATHOLOGY

General Features

- Etiology
  - Multiple etiologies: Crush, blunt trauma, shear, twisting, stress
  - Stress fracture most common in runners
  - Stress fracture of 2nd MT base occurs in ballet dancers
- Associated abnormalities
  - Lisfranc joint injury
  - Cuneiform and other foot fractures
  - MTP dislocation

CLINICAL ISSUES

Presentation

- Most common signs/symptoms
  - Stress fracture: Sharp pain, patient can precisely localize
  - Trauma: Pain, soft tissue swelling

Demographics

- Epidemiology
  - 35% of foot fractures involve metatarsal

Natural History & Prognosis

- Most heal well
- Jones fracture is at particular risk for nonunion due to poor vascularity

Treatment

- Conservative: Flat sole boot, immobilization
- Pin fixation for significantly displaced fractures
- Screw or plate fixation for Jones fracture: Either acutely or for delayed union

SELECTED REFERENCES

1. Gehrmann RM et al: Current concepts review: Stress fractures of the foot. Foot Ankle Int. 27(9):750-7, 2006
2. Ashman CJ et al: Forefoot pain involving the metatarsal region: differential diagnosis with MR imaging. Radiographics. 21(6):1425-40, 2001
3. Harrington T et al: Overuse ballet injury of the base of the second metatarsal. A diagnostic problem. Am J Sports Med. 21(4):591-8, 1993
4. Lawrence SJ et al: Jones' fractures and related fractures of the proximal fifth metatarsal. Foot Ankle. 14(6):358-65, 1993

(Left) *AP radiograph shows an unusual longitudinal fracture ↷ of the 4th MT head associated with dislocation of the MTP joint. This patient fell off of her bicycle when cycling barefoot.* ***(Right)*** *Sagittal bone CT shows that the fracture bisects the articular surface ↷. Because of the articular injury, the patient was treated with ORIF.*

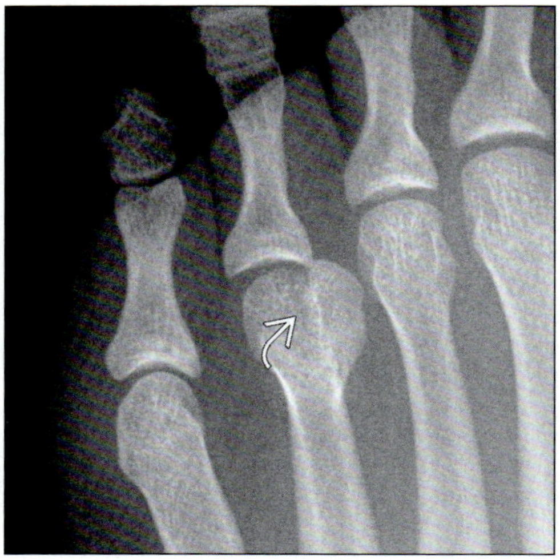

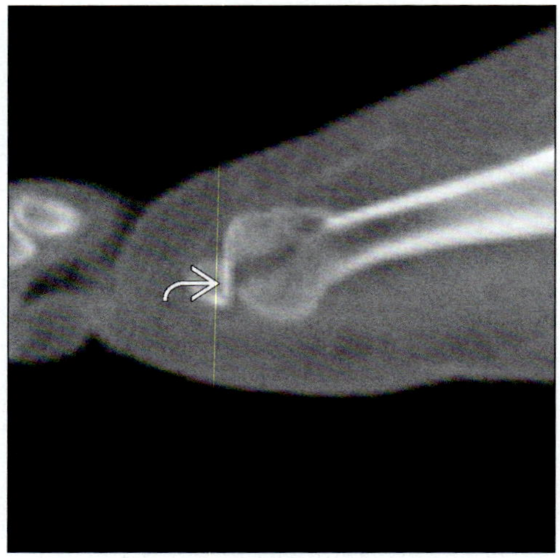

(Left) *Oblique radiograph shows a horizontal fracture line ➔ of the tuberosity of the 5th MT. This avulsion fracture is due to pull by peroneus brevis and lateral fibers of plantar aponeurosis.* ***(Right)*** *Oblique radiograph in a 12-year-old girl shows accessory center of ossification for tuberosity of the 5th metatarsal ⇨, distinguished by location, orientation, cortication, and rounded contours. The fracture of tuberosity ⇩, in distinction, is oriented transversely and noncorticated.*

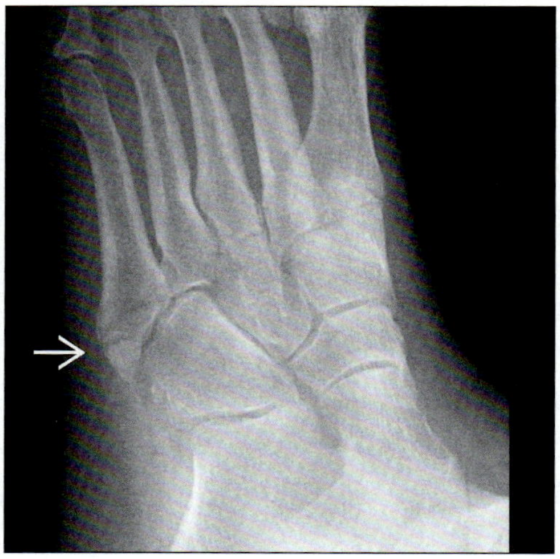

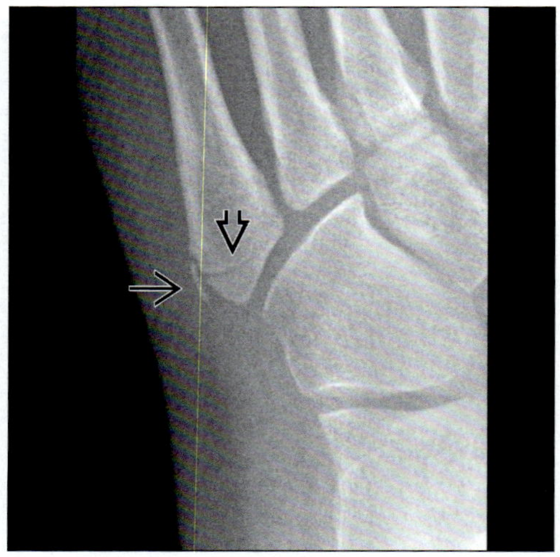

(Left) *Coronal NECT shows fracture ⇨ of the 2nd metatarsal base. A horizontal fracture such as this tends to occur in association with other midfoot fractures and dislocations due to high-energy mechanism.* ***(Right)*** *AP radiograph in a patient with disuse osteopenia shows slight cortical step-off and focal periosteal new bone/ peripheral callus ↷ at a 3rd MT stress fracture. The radiograph 2 weeks prior was normal.*

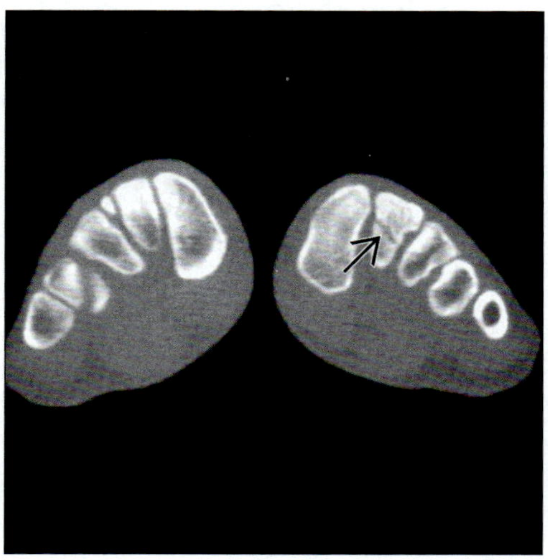

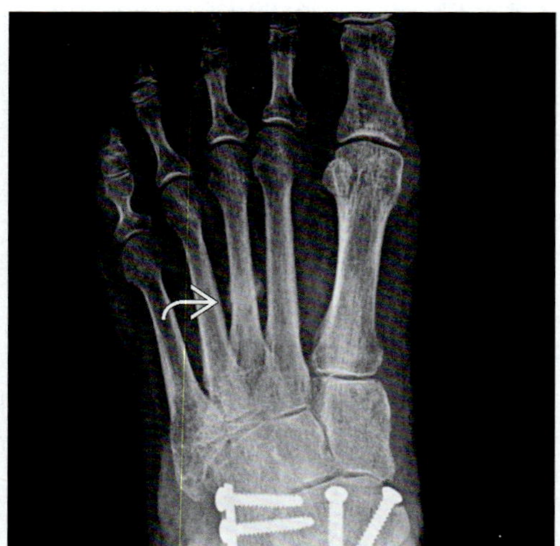

METATARSAL FRACTURE

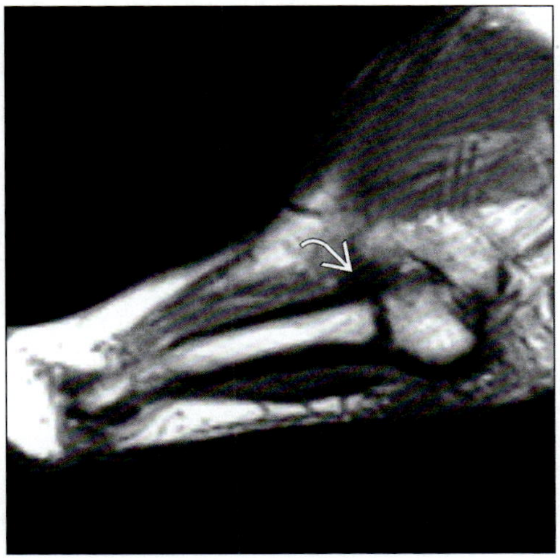

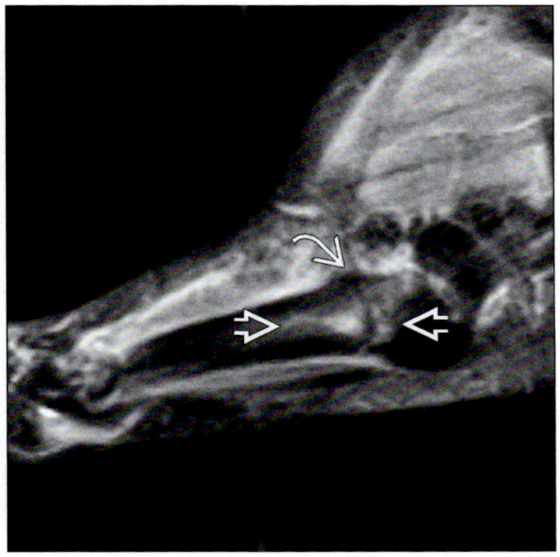

(Left) *Sagittal T1WI MR shows fracture of the 5th MT base in a football player complaining of vague lateral midfoot pain. This was a stress fracture; radiographs were equivocal.* ***(Right)*** *Sagittal STIR MR in the same patient shows a fracture line and surrounding marrow edema/blood. The band-like configuration of the marrow edema is characteristic of fracture. In contrast, tumor or infection show a rounded configuration of marrow edema.*

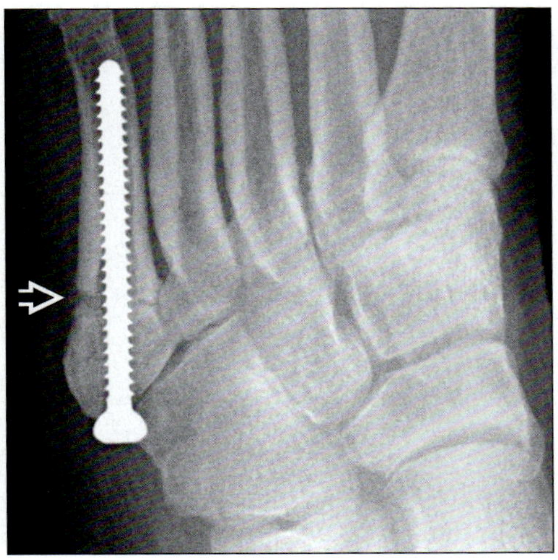

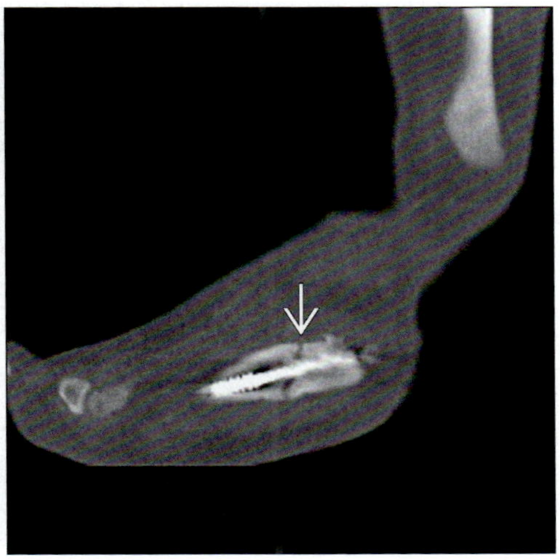

(Left) *Oblique radiograph in the same patient 1 year after initial presentation. A screw was placed after 4 months of failed conservative management, and the fracture margins appear blurred, but the patient had persistent pain.* ***(Right)*** *Sagittal bone CT in the same patient (same time as preceding radiograph) shows fracture nonunion with sclerotic margins and no bridging callus. CT is much more accurate than radiographs in the diagnosis of nonunion.*

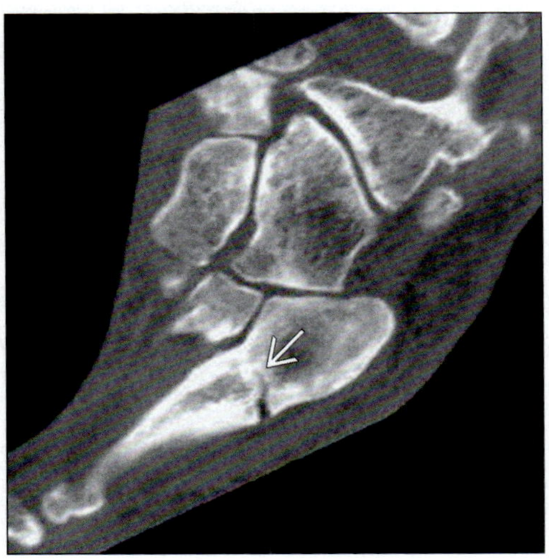

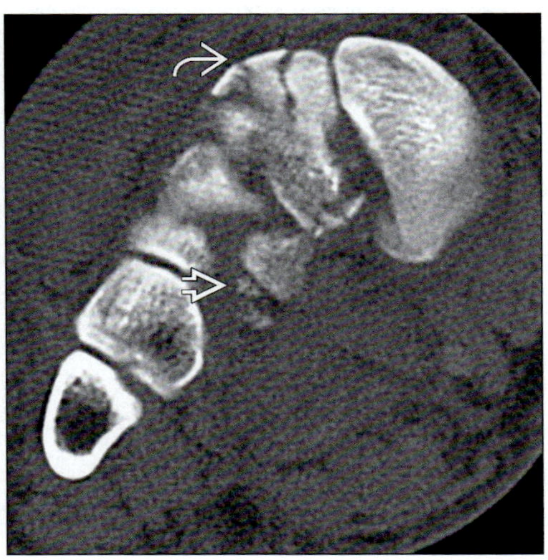

(Left) *Sagittal bone CT reformat shows an incompletely healed Jones fracture of the 5th MT base. Jones fracture carries a high risk for nonunion.* ***(Right)*** *Coronal bone CT shows highly comminuted fractures involving the 2nd cuneiform and base of the 3rd metatarsal, which were occult on radiograph. CT is valuable in such cases. This patient had a crush injury due to a snowmobile landing on the foot.*

PHALANGEAL FRACTURE AND DISLOCATION, FOOT

Key Facts

Imaging
- Usually radiographic diagnosis
- Often visible on only single view
- Optimize lateral film by centering on toes, coning image, using sling or pointer to move unaffected toes out of field of view
- CT or MR used for sesamoid fracture in athletes

Top Differential Diagnoses
- Bipartite or multipartite sesamoid
- Accessory epiphysis
- Osteomyelitis

Pathology
- Crush injury: Heavy object dropped on toe
- Stubbed toe: Tip of toe impacted against vertical surface
- Sesamoid fracture: Often stress fracture, may also occur as acute impaction or due to dorsiflexion injury

Clinical Issues
- Salter-Harris injury to 1st distal phalanx may be open injury; signs include
  - Laceration ± bleeding of nail fold

Diagnostic Checklist
- Be alert for osteomyelitis complicating Salter-Harris fractures of distal phalanges
- Toe fractures often missed because not included in search pattern for foot radiograph
- Fracture may be misdiagnosed as infection on MR
  - Bone marrow edema sometimes obscures fracture line

*(**Left**) Sagittal graphic shows a Salter-Harris type I fracture ➔ of the distal phalanx of the 1st toe. Salter-Harris fractures of the distal phalanx occur in children due to a "stubbed toe." Although the fracture appears innocuous, it is often an open fracture due to communication through a small laceration at the nail bed ➔, and patients are at risk for osteomyelitis. (**Right**) AP radiograph shows a comminuted Salter-Harris type IV fracture ➔ of the distal phalanx of the great toe.*

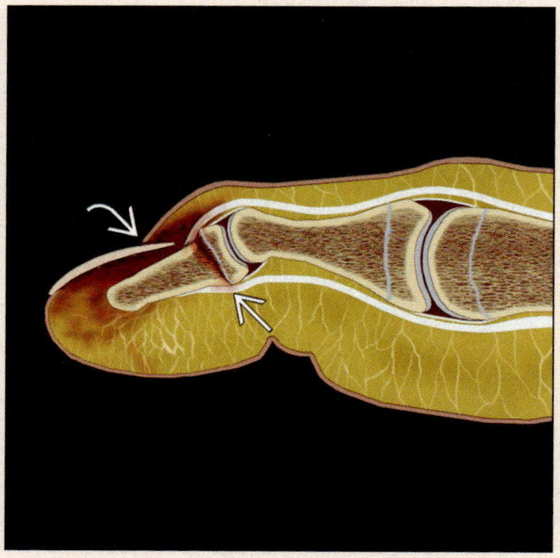

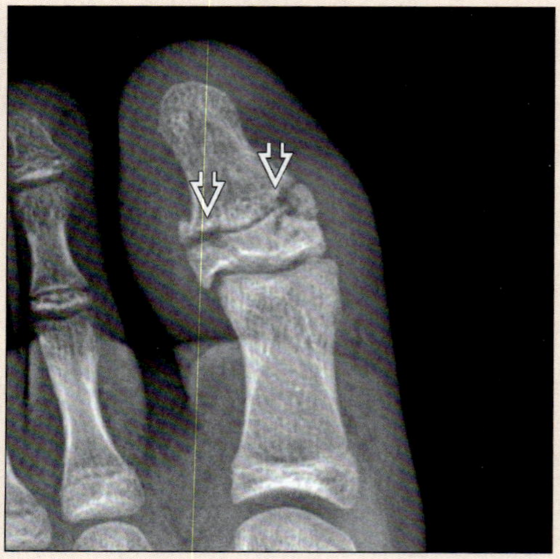

*(**Left**) Lateral radiograph in the same patient shows only slight displacement of the major fragments ➔. This was an open fracture. Note the severe soft tissue swelling at the nail bed ➔. (**Right**) AP radiograph shows lateral dislocation ➔ of the 2nd MTP. Dislocations are usually either lateral or dorsal. Lateral dislocations are almost always traumatic, while dorsal dislocations may be due to trauma or chronic stress on the plantar plate.*

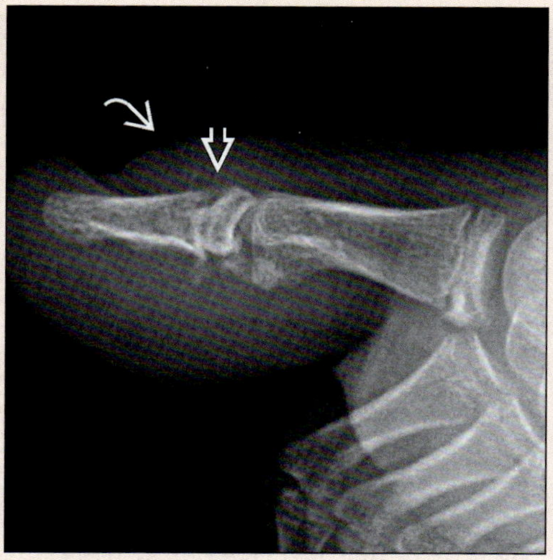

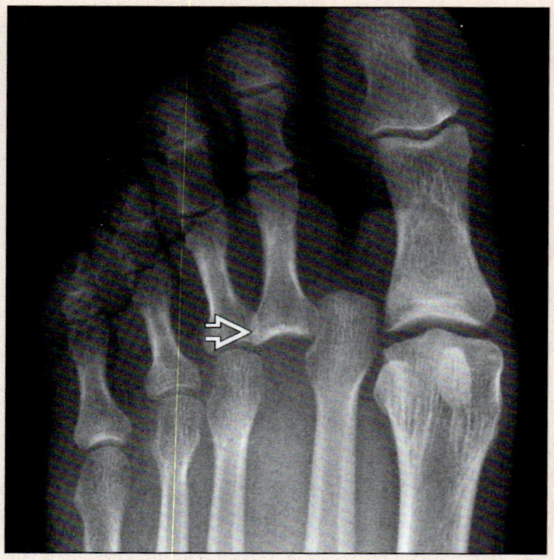

PHALANGEAL FRACTURE AND DISLOCATION, FOOT

IMAGING

General Features

- Morphology
  - Metatarsophalangeal joint dislocation
    - Dorsal: Trauma or chronic injury of plantar plate of MTP joint
    - Lateral: Trauma, usually see other injuries as well
  - Interphalangeal joint dislocation
    - Usually dorsal, due to trauma
  - Phalangeal fracture
    - Crush injury often stellate morphology
    - "Stubbed toe" usually transverse or oblique
    - Often involves growth plate in children
  - Sesamoid fracture
    - Single fracture line in coronal plane may mimic bipartite sesamoid
    - Fracture may have complex morphology, best seen on CT
    - Sesamoid may resorb after fracture

Radiographic Findings

- Fracture is often visible only on single view
- Optimize lateral film by centering on toes, coning image, using sling or pointer to move unaffected toes out of field of view
- Sesamoid fracture best seen on lateral view
  - Often not visible on sesamoid view

CT Findings

- Sesamoid fractures better seen than on radiographs
  - CT often shows unsuspected comminution

MR Findings

- Bone marrow edema high signal intensity on T2WI, STIR
- Fracture line may be obscured

Imaging Recommendations

- Best imaging tool
  - Usually radiographic diagnosis
  - CT or MR used for sesamoid fracture in athletes
- Protocol advice
  - CT of sesamoid fracture: Foot flat on gantry, sagittal and coronal reformations
  - MR of sesamoid fracture: FOV limited to forefoot, obtain all 3 planes

DIFFERENTIAL DIAGNOSIS

Bipartite or Multipartite Sesamoid

- Usually larger than fractured unipartite sesamoid
- Margins corticated
- May be unilateral (not present in other foot)

Accessory Epiphysis

- At opposite end of phalanx from epiphysis
- i.e., at distal end of 1st proximal phalanx, but at proximal end of remaining phalanges

Osteomyelitis

- If MR low-resolution, edema in phalanx may lead to erroneous diagnosis of osteomyelitis
- Use thin slices in sagittal plane to identify fracture line

PATHOLOGY

General Features

- Etiology
  - Crush injury: Heavy object dropped on toe
  - Stubbed toe: Tip of toe impacted against vertical surface
    - Distal phalanx forcibly plantarflexed
  - MTP dislocation, especially of 2nd toe, may occur chronically due to altered biomechanics
    - Short 1st MT, metatarsus primus varus, hallux valgus
  - Sesamoid fx: Often stress fracture, may also occur as acute impaction or due to dorsiflexion injury

CLINICAL ISSUES

Presentation

- Salter-Harris (SH) injury to 1st distal phalanx may be open injury
  - Laceration ± bleeding from nail fold

Natural History & Prognosis

- Most toe fractures heal with minimal deformity
- SH fracture of distal phalanx may be open injury
  - Communicates with germinal matrix of nail
  - Potential for osteomyelitis

Treatment

- Fracture usually treated by buddy taping to adjacent toe and wearing rigid sole shoe
- Dislocation may be fixed with K-wire
- Antibiotics for open SH fracture of distal phalanx

DIAGNOSTIC CHECKLIST

Consider

- Be alert for signs of osteomyelitis following SH fractures of distal phalanges

Image Interpretation Pearls

- Toe fractures often missed because not included in search pattern for foot radiograph
  - Always perform dedicated visual search of this region (magnify image on PACS)
- Fx may be misdiagnosed as infection on MR because bone marrow edema present, fx line difficult to see
  - Thin slice T1WI best for showing fracture line

SELECTED REFERENCES

1. Cohen BE: Hallux sesamoid disorders. Foot Ankle Clin. 14(1):91-104, 2009
2. Mittlmeier T et al: Sesamoid and toe fractures. Injury. 35 Suppl 2:SB87-97, 2004
3. Hatch RL et al: Evaluation and management of toe fractures. Am Fam Physician. 68(12):2413-8, 2003
4. Schenck RC Jr et al: Fractures and Dislocations of the Forefoot: Operative and Nonoperative Treatment. J Am Acad Orthop Surg. 3(2):70-78, 1995
5. Pinckney LE et al: The stubbed great toe: a cause of occult compound fracture and infection. Radiology. 138(2):375-7, 1981

PHALANGEAL FRACTURE AND DISLOCATION, FOOT

(Left) *Oblique radiograph shows a subtle, longitudinally oriented fracture → of the 5th toe. Toe fractures are easily overlooked on radiographs. Adjacent soft tissue swelling provides a useful hint to investigate closely for fracture.* ***(Right)*** *AP radiograph in the same patient better shows the fracture line →. Note the failure of segmentation of the middle and distal phalanges ↪, a common normal variant.*

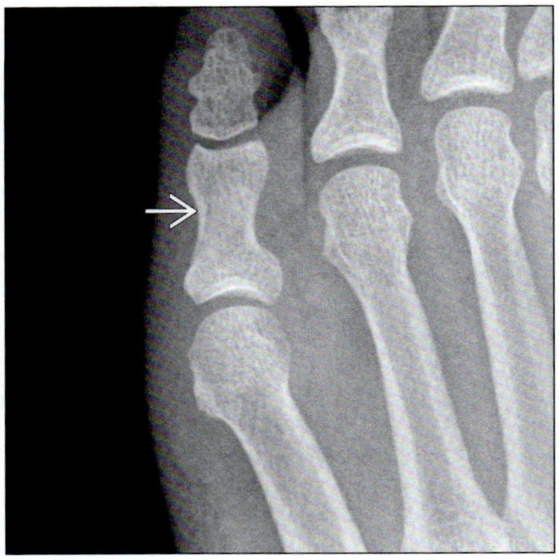

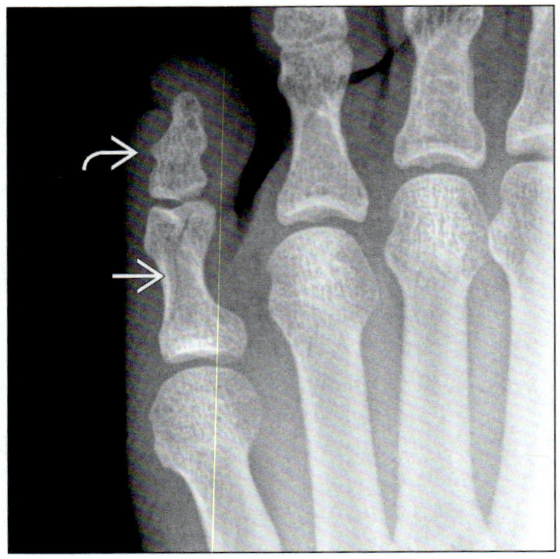

(Left) *AP radiograph shows a transverse fracture → of the 3rd proximal phalanx. This fracture is usually treated by "buddy taping" the toe to the adjacent toe and should heal without sequelae.* ***(Right)*** *Lateral radiograph shows a 5th toe fracture → that was not visible on other views. In order to reliably visualize fractures such as this, the image should be coned to the toes. A pointer ↪ is useful to isolate the affected toe on the lateral view.*

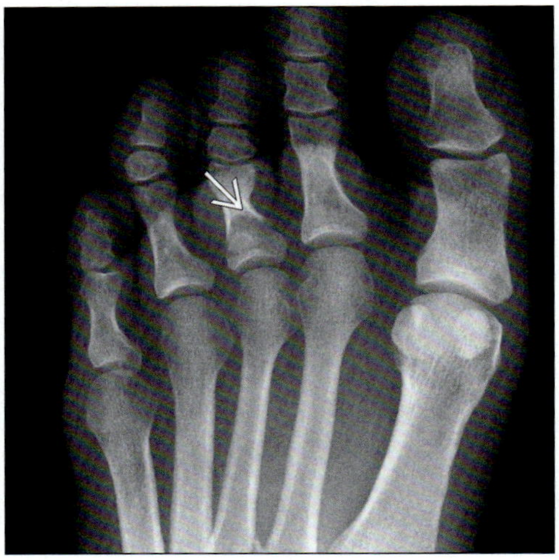

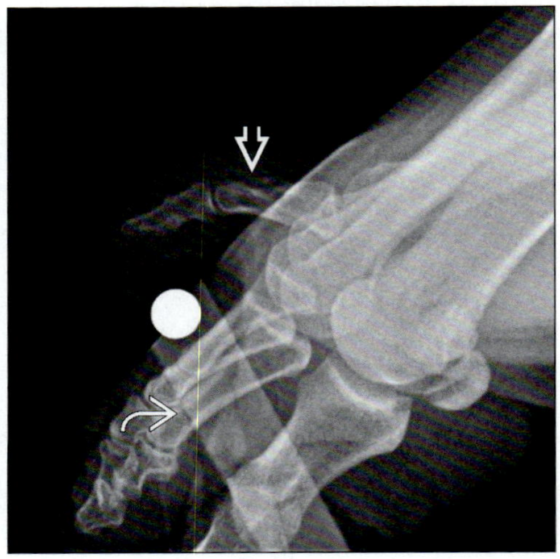

(Left) *AP radiograph shows a Salter-Harris type III fracture → involving the distal phalanx of the 1st toe. This is potentially an open fracture, with risk of osteomyelitis. The patient is close to skeletal maturity, but growth plates are still partly open.* ***(Right)*** *Lateral radiograph in the same patient shows the fracture → extending through the dorsal, unfused portion of the growth plate, and through the epiphysis. Note the proximity of the fracture to the nail bed ↪.*

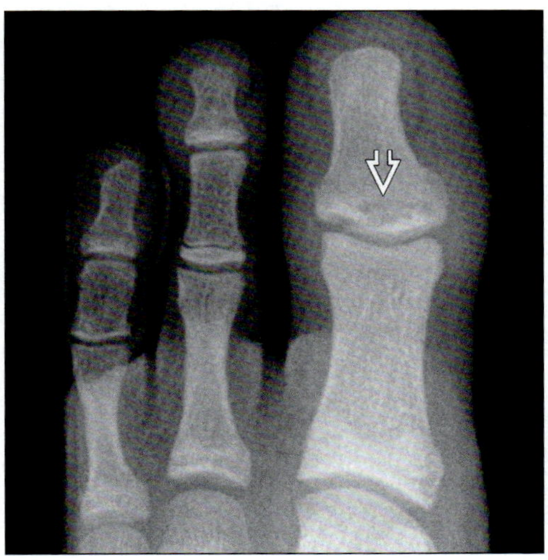

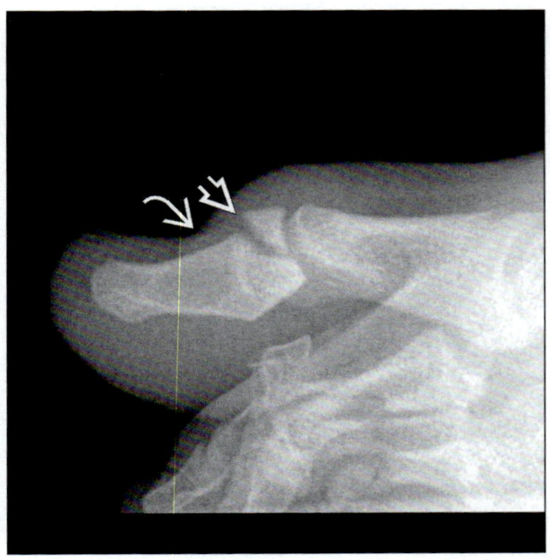

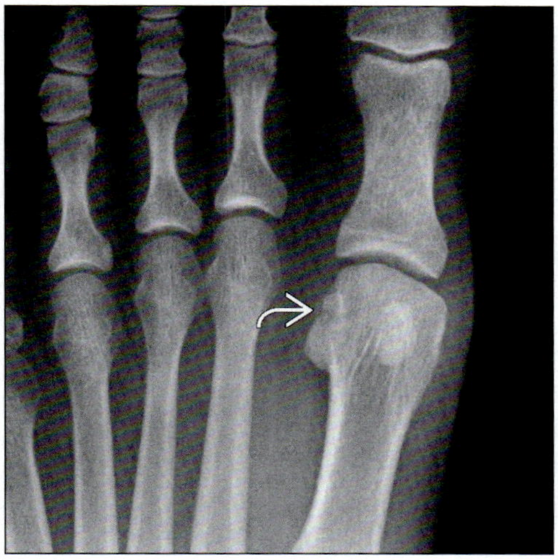
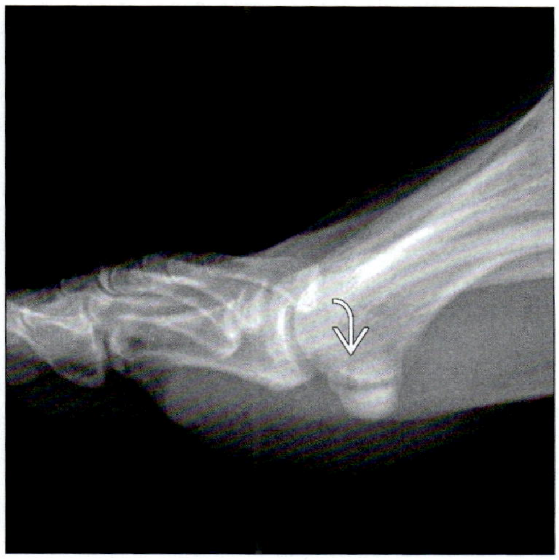

(Left) *AP radiograph shows an apparent bipartite lateral sesamoid of the 1st toe. Its margins are sclerotic, but the patient is tender over the sesamoid. This could be a nonunited fracture or an injured bipartite sesamoid.* ***(Right)*** *Lateral radiograph in the same patient also shows the separation of the sesamoid. Acute margins of the ossicles suggest a chronic fracture but are not a completely reliable sign.*

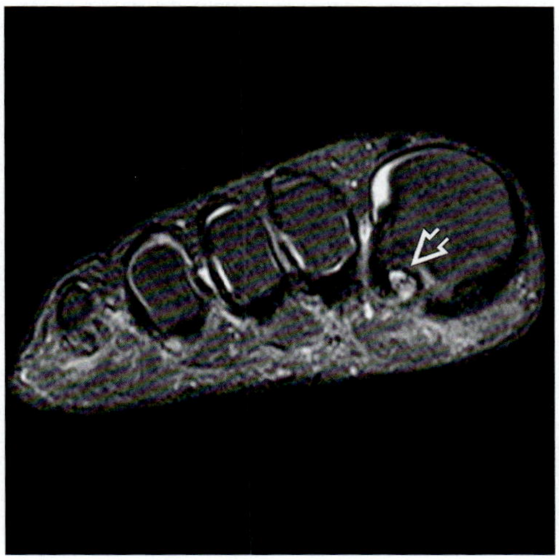
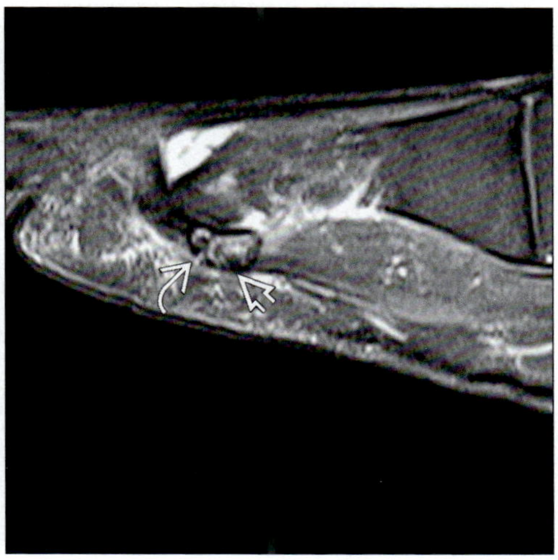

(Left) *Coronal T2WI FS MR in the same patient shows an irregular fracture line in the sagittal plane, which was not visible on radiographs. Surrounding edema indicates acute or subacute injury.* ***(Right)*** *Sagittal STIR MR in the same patient shows the sesamoid division seen on prior radiograph, as well as the acute fracture. In high-level athletes, it is sometimes worthwhile to perform MR of the toe to confirm an acute injury.*

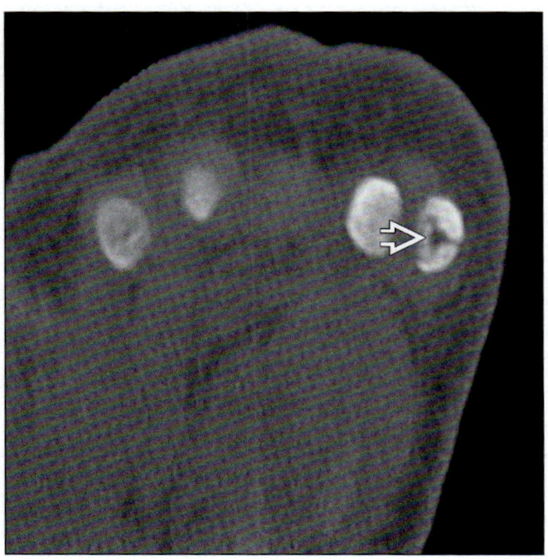
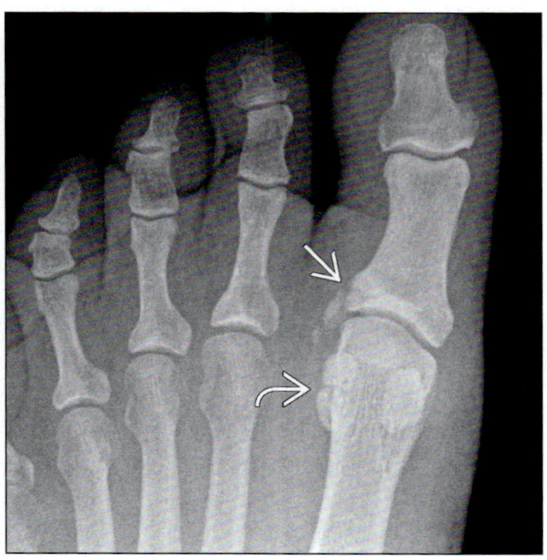

(Left) *Axial bone CT shows an acute sesamoid fracture. The margins are not sclerotic, and the fracture plane is jagged. The appearance mimics a cracked egg, unlike the smoother, more rounded appearance of a bipartite sesamoid.* ***(Right)*** *AP radiograph shows a subacute avulsion fracture of the adductor hallucis origin due to abduction injury (patient was barefoot). This has resulted in hallux varus deformity and 1st MTP instability. Lateral sesamoid is bipartite.*

STRESS AND INSUFFICIENCY FRACTURE, ANKLE AND FOOT

Key Facts

Terminology

- Chronic, repetitive injury that collectively leads to partial or complete fracture

Imaging

- Radiographs not sensitive to early fracture but still most common method of diagnosis
- MR used for cryptogenic pain or in athletes in whom definitive diagnosis will change management
- Radiographs and CT show sclerotic and jagged incomplete or complete fracture line
- Focal periosteal reaction may be only radiographic sign
- MR shows jagged fracture line with surrounding bone marrow edema
  - Fracture line may not be visible if correct plane and high resolution not used

Top Differential Diagnoses

- Traumatic fracture
- Osteomyelitis
- Stress response
- Inflammatory arthritis

Clinical Issues

- Most common in metatarsals, navicular, and calcaneus
- Also seen in medial malleolus, 1st toe sesamoids, cuboid, talus
- Pain increases with activity, improves with rest
- Most heal with conservative management
- Navicular stress fracture may progress to osteonecrosis
- 5th MT stress fracture has high nonunion rate

__(Left)__ Lateral graphic shows several common stress fractures in the foot: Posterior process of the calcaneus, cuboid, proximal 5th metatarsal shaft, and neck of the 2nd-4th metatarsals. Other common stress fractures (not shown) are talar neck, navicular, and 1st toe sesamoids. __(Right)__ Sagittal STIR MR shows an inferior calcaneal stress fracture. Stress fractures of the calcaneus occur in a variety of locations but are almost always perpendicular to the primary bony trabeculae.

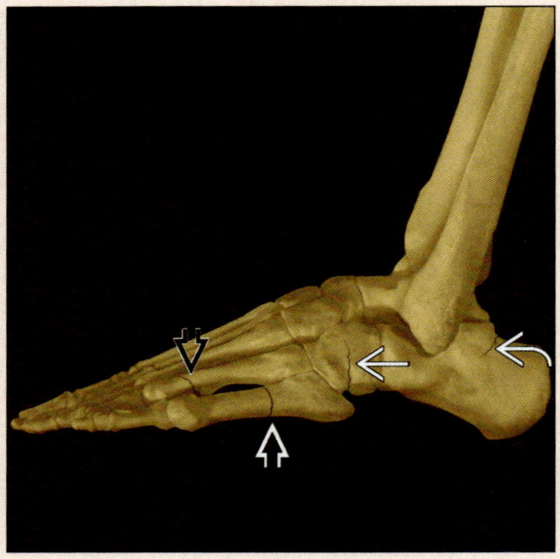

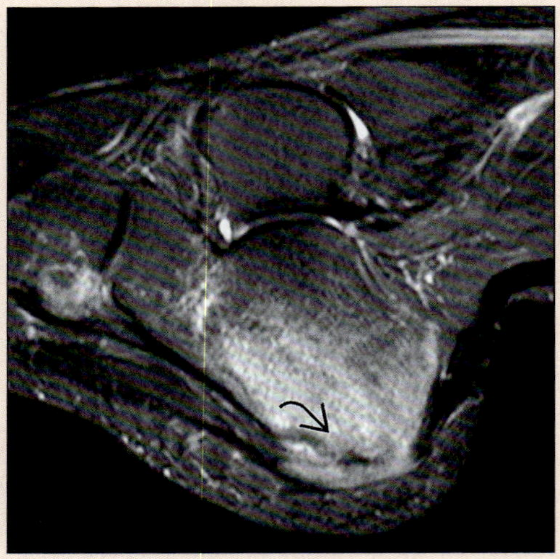

__(Left)__ Axial T2WI FS MR demonstrates stress fractures of the middle and lateral cuneiforms, as well as marrow edema in several other bones. This patient was a runner. __(Right)__ Axial PDWI FS MR shows a stress fracture of the 4th metatarsal that has progressed to a complete fracture. Periosteal reaction, which appears amorphous and of intermediate signal intensity, is seen centered at the fracture site. Bone marrow edema extends nearly the entire length of the metatarsal.

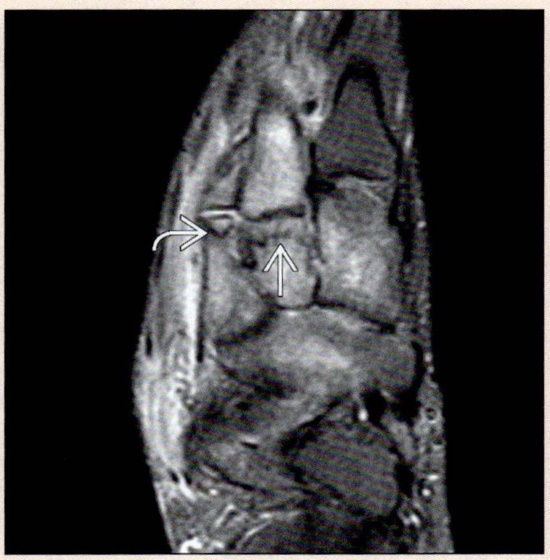

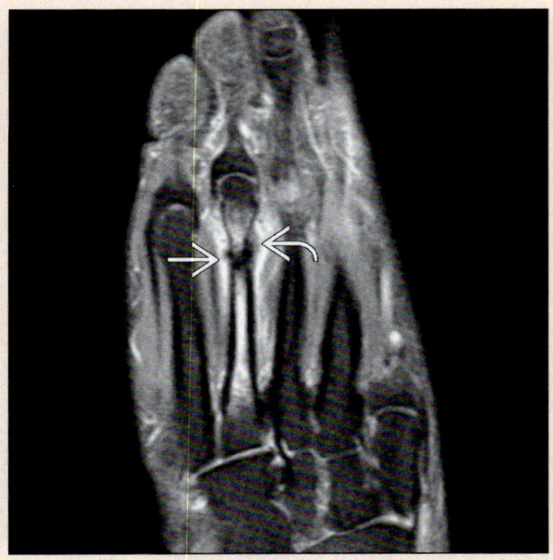

STRESS AND INSUFFICIENCY FRACTURE, ANKLE AND FOOT

TERMINOLOGY

Definitions

- Chronic, repetitive injury that leads to partial or complete macroscopic fracture
  - Stress fracture: Fracture due to abnormal stress on normal bone
  - Insufficiency fracture: Fracture due to normal stress on weak bone
  - Pathologic fracture: Fracture due to underlying tumor or infection
- In practice, often overlap between stress and insufficiency fractures

IMAGING

General Features

- Location
  - Most common in metatarsals, navicular, and calcaneus
  - Also seen in medial malleolus, 1st toe sesamoids, cuboid, or talus

Radiographic Findings

- Sclerotic line usually perpendicular to long axis of bone
  - Jagged contour
  - May be complete or incomplete
- Lucent band in bone reflects resorption at fracture site
- Focal periosteal reaction may be only radiographic sign
- Healing stress fracture often develops wide sclerotic band due to callus

CT Findings

- Sclerotic or lucent line, complete or incomplete
- More sensitive than radiographs, but less sensitive than MR

MR Findings

- Jagged and irregular incomplete or complete line through bone
- Surrounding bone marrow edema
- Edema may be quite extensive
  - In metatarsals, often involves nearly entire bone
- Fracture line may not be visible if correct plane and high resolution not used

Imaging Recommendations

- Best imaging tool
  - Radiographs not sensitive to early fracture but still most common method of diagnosis
  - MR used for cryptogenic pain or in athletes in whom definitive diagnosis will change management

Nuclear Medicine Findings

- Bone scan
  - Focal abnormal increased uptake in 1 or more bones of foot
    - May or may not be linear in morphology
    - May take up to 48 hours after injury to manifest
  - Pinhole collimation useful for better anatomic detail

DIFFERENTIAL DIAGNOSIS

Traumatic Fracture

- May be complete (traversing entire width of bone) or incomplete
- History is key in making diagnosis

Osteomyelitis

- Bone marrow edema without visible fracture line
- Surrounding soft tissue inflammation ± abscess

Stress Response

- Bone marrow edema due to chronic stress but without fracture line
- Radiates out from point of stress

Inflammatory Arthritis

- Bone marrow edema centered at joint
- Erosions usually visible on radiographs or MR

CLINICAL ISSUES

Presentation

- Most common signs/symptoms
  - Pain increases with activity, improves with rest
  - Hindfoot: Vague, deep pain
  - Metatarsals: Sharp, precisely localized, pinpoint pain
  - Sesamoids of 1st toe: Pain with running, dorsiflexion of toe

Natural History & Prognosis

- Most heal with conservative management
- Navicular stress fracture may progress to osteonecrosis
- 5th MT stress fracture has high nonunion rate

Treatment

- Rest, gradual return to activity usually adequate treatment
- Navicular, 5th MT stress fractures often require surgical fixation

DIAGNOSTIC CHECKLIST

Image Interpretation Pearls

- Early navicular stress fracture easily missed
  - Seen on most dorsal axial image through navicular
  - Seen at dorsal margin of navicular on coronal images

SELECTED REFERENCES

1. Brockwell J et al: Stress fractures of the foot and ankle. Sports Med Arthrosc. 17(3):149-59, 2009
2. Mann JA et al: Evaluation and treatment of navicular stress fractures, including nonunions, revision surgery, and persistent pain after treatment. Foot Ankle Clin. 14(2):187-204, 2009
3. Burne SG et al: Tarsal navicular stress injury: long-term outcome and clinicoradiological correlation using both computed tomography and magnetic resonance imaging. Am J Sports Med. 33(12):1875-81, 2005
4. Franco M et al: An uncommon cause of foot pain: the cuboid insufficiency stress fracture. Joint Bone Spine. 72(1):76-8, 2005

STRESS AND INSUFFICIENCY FRACTURE, ANKLE AND FOOT

(Left) *Lateral radiograph shows an advanced stress fracture of the calcaneal tuberosity that has progressed to involve cortex ➔, with mild cortical irregularity at the fracture site. Sclerosis more inferiorly ➔ is less well defined and, in isolation, would be more difficult to conclusively identify as stress fracture.* **(Right)** *Lateral radiograph shows a calcaneal stress fracture with a sclerotic line ⇨ adjacent to a lucent line ⇨ reflecting bone resorption as part of the healing process.*

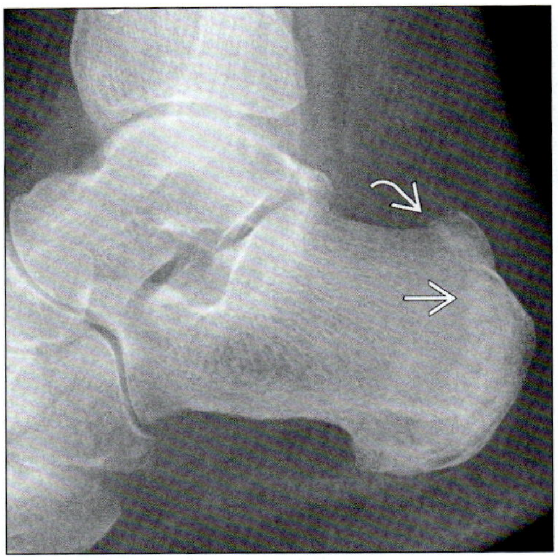

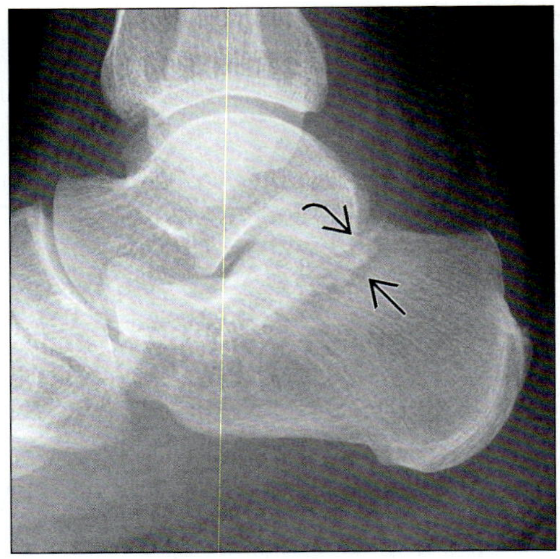

(Left) *Sagittal STIR MR shows a stress fracture ➔ surrounded by extensive bone marrow edema. The edema radiates out from the fracture, a pattern that can usually be distinguished from the edema pattern of tumor or infection, which is more ovoid in configuration.* **(Right)** *Sagittal T2WI FS MR from an 8-year-old girl with lateral foot pain related to gymnastics shows a vertical stress fracture ➔ in the cuboid surrounded by intense marrow edema. Note that the calcaneocuboid joint ➔ is normal.*

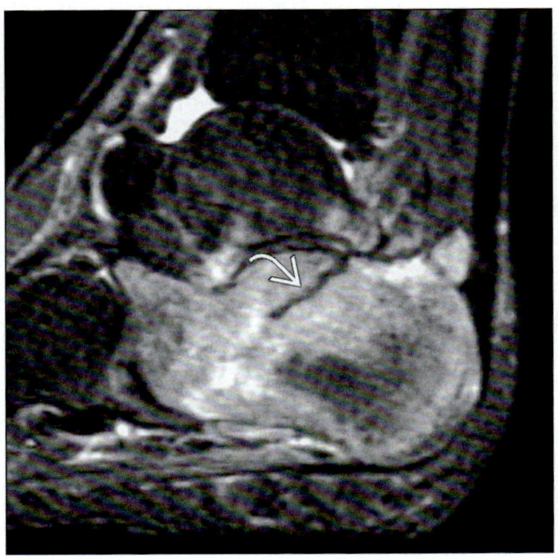

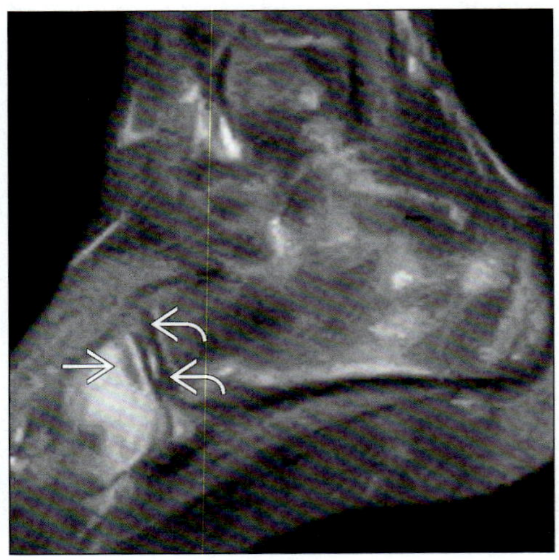

(Left) *AP radiograph from a runner with foot pain shows focal periosteal reaction at 2nd MT neck ➔, often the earliest radiographic sign of stress fracture. No fracture line is visible. This subtle finding may only be visible on a single radiographic projection and must be actively sought in the appropriate clinical setting.* **(Right)** *AP radiograph 6 weeks later from the same patient shows maturing periosteal reaction ⇨, but the fracture itself is still not visualized.*

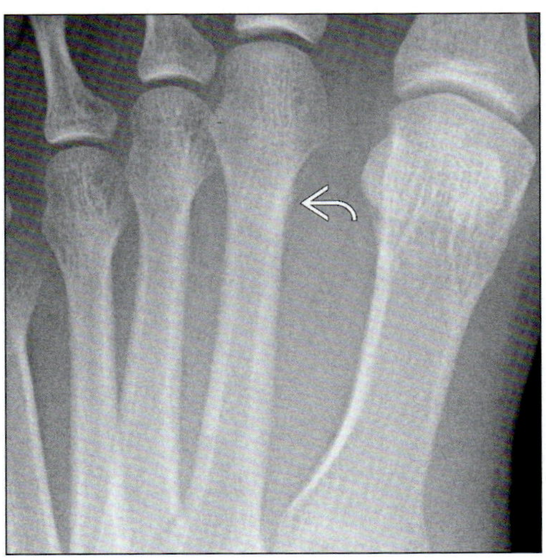

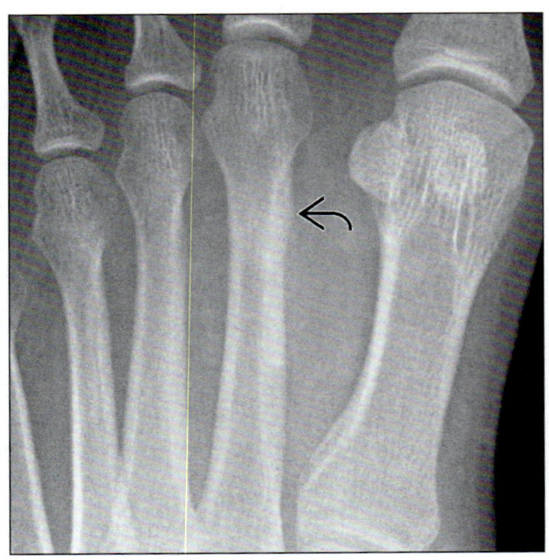

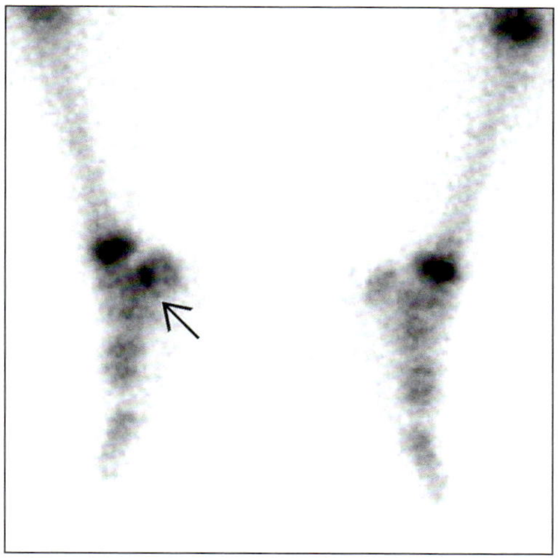

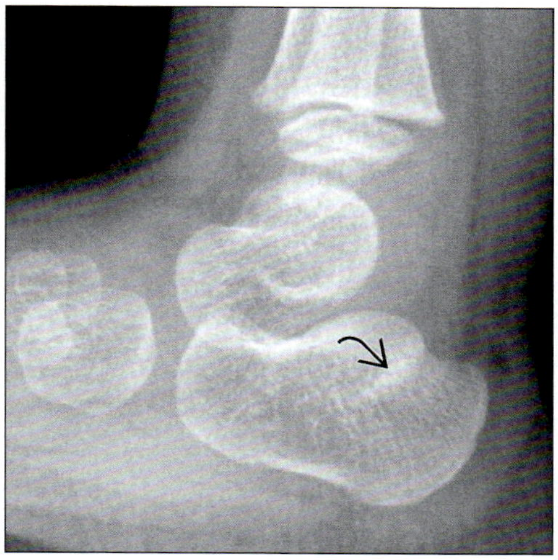

(Left) *Lateral bone scan shows focal uptake in the posterior process of the calcaneus in a young gymnast. Radiographs at this time were normal. The linear configuration of the abnormal uptake is characteristic of stress fracture, but due to the lack of specificity of bone scan, follow-up radiographs are useful to rule out other osteoblastic processes.* ***(Right)*** *Lateral radiograph from the same patient obtained 6 weeks later shows a sclerotic fracture line consistent with stress fracture.*

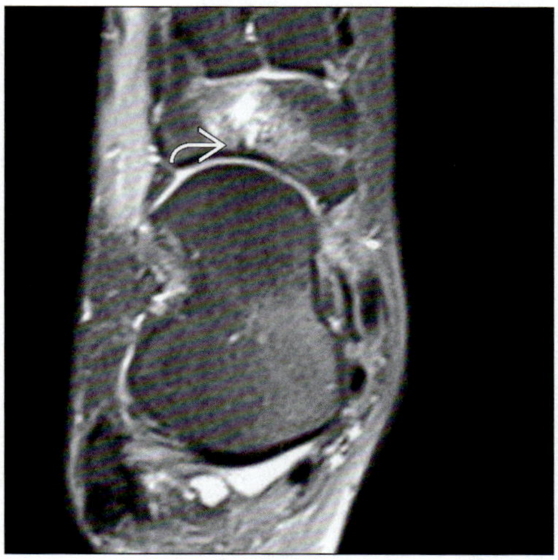

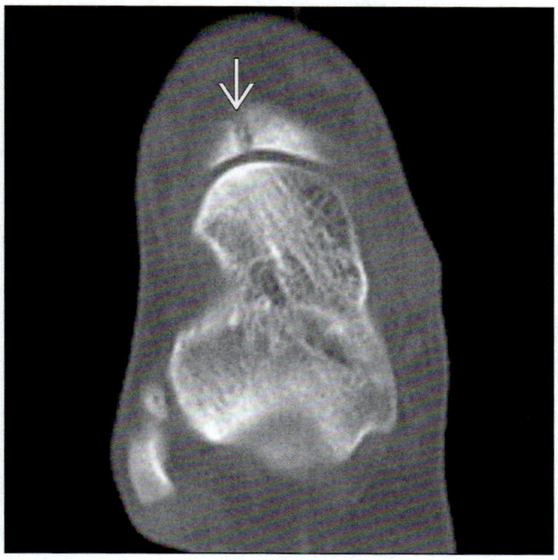

(Left) *Axial PDWI FS MR shows an early navicular stress fracture. Navicular stress fractures are almost always in this location. They originate at the dorsal, proximal cortex of the navicular and then progress distally and to the plantar surface. Osteonecrosis may develop as a long-term complication.* ***(Right)*** *Axial bone CT shows a more advanced navicular stress fracture in a runner. Injuries were bilateral in this patient.*

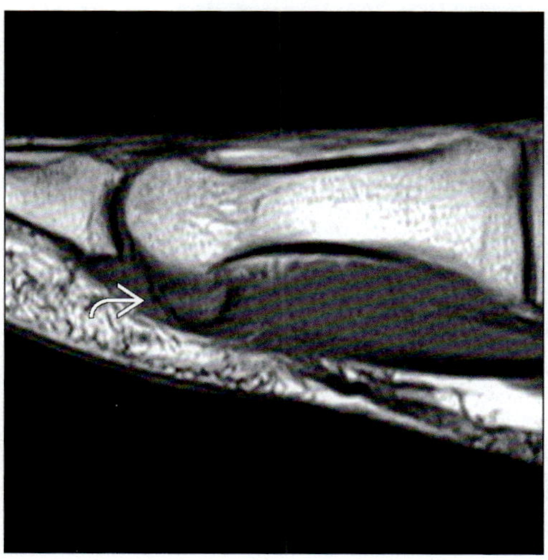

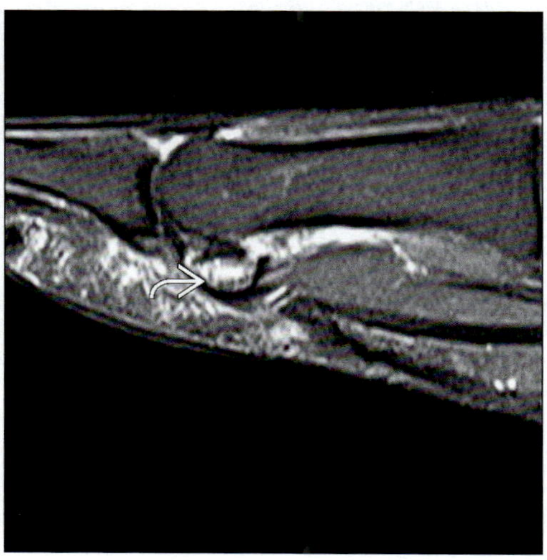

(Left) *Sagittal T1WI MR shows stress fracture of a 1st MTP joint sesamoid. Due to the small size of the sesamoid, the fracture line is difficult to see. These fractures can also be seen on CT, but other injuries presenting in a similar fashion on CT, such as ligament rupture, cannot be as well evaluated.* ***(Right)*** *Sagittal STIR MR from the same patient shows edema throughout the sesamoid. The horizontal fracture line is well seen against the background of edema.*

SALTER-HARRIS FRACTURE, ANKLE

Key Facts

Terminology

- Pediatric fracture involving unfused growth plate

Imaging

- SH-I fractures are often subtle; look for
  - Displacement of epiphysis
  - Widening of growth plate
- Triplane fracture of tibia (SH-IV)
  - Reflects complex pattern of fusion of distal tibial physis
  - Includes: Sagittal fracture through epiphysis, horizontal fracture through unfused portion of growth plate, coronal fracture through metaphysis
  - Characteristic spike of metaphyseal bone fragment seen on lateral radiograph
  - Multiplanar and 3D CT reformations very useful in understanding triplane fracture pattern
- Juvenile Tillaux fracture (SH-III)
  - Horizontal fracture line through anterolateral tibial growth plate (remainder of growth plate fused); fracture line extends sagittally out through epiphysis

Clinical Issues

- Distal tibia is 2nd most common site overall of SH fracture
- Fractures occur near skeletal maturity (~ ages 12-15)
- No significant growth disturbance
- SH-I of fibula probably present if skeletally immature patient with focal tenderness over distal fibular growth plate and no fracture visible

Diagnostic Checklist

- Describe number of fragments present for triplane fractures
- Failure of reduction suggests soft tissue interposition in fracture

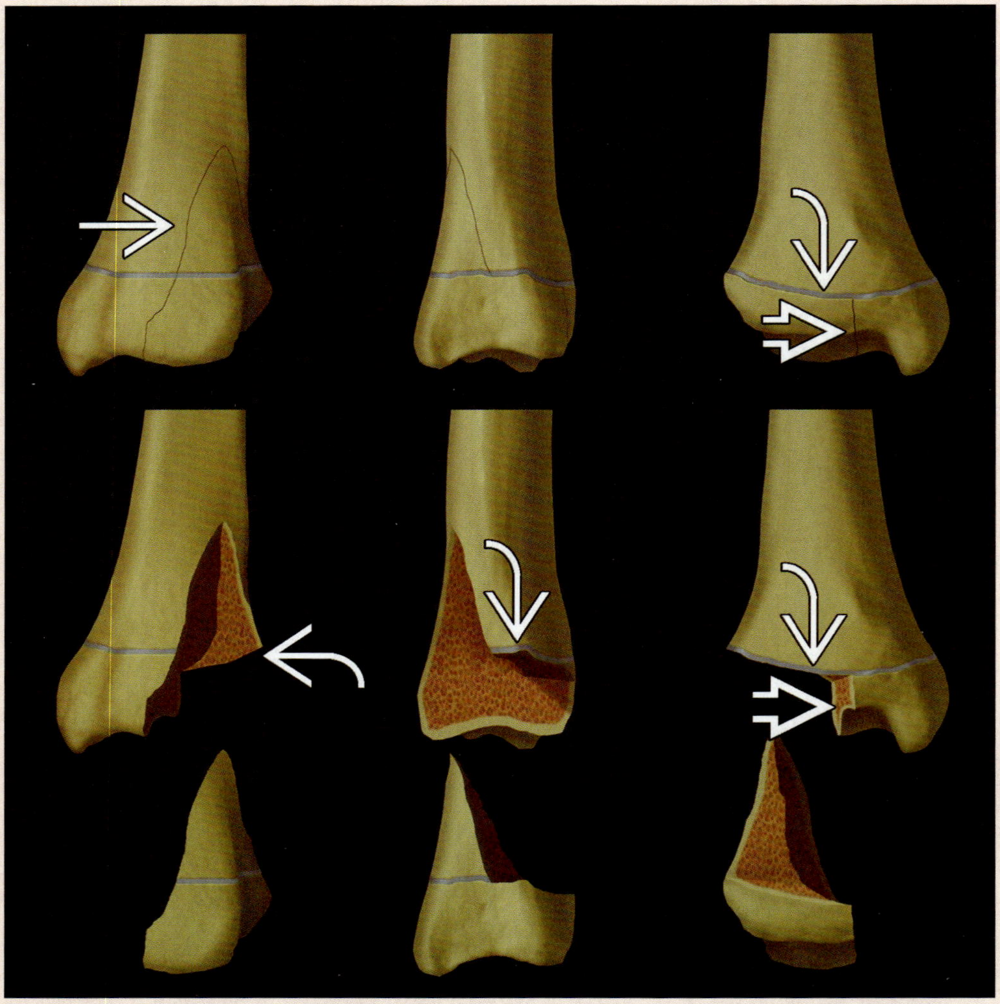

Graphic shows a 2-part triplane fracture of the tibia from posterior (left), lateral (middle), and anterior (right) viewpoints. The top row shows fragments in situ, while the bottom is an exploded view. The oblique coronal component of the fracture ➡ extends through both the epiphysis and metaphysis where they are fused, and through the metaphysis only where the growth plate is open. The horizontally oriented portion of the fracture ➡ runs along the unfused portion of the physis. The sagittally oriented portion of the fracture ➡ is through the epiphysis at the junction of the fused and unfused portions. Morphology of the triplane fracture will vary depending on the extent of physeal closure.

SALTER-HARRIS FRACTURE, ANKLE

TERMINOLOGY

Definitions

- Salter-Harris (SH) fx: Fx through unfused growth plate
- Kump bump: Superior undulation of distal tibial growth plate

IMAGING

General Features

- Best diagnostic clue
  - Displacement of epiphysis relative to metaphysis
- Morphology
  - Tibial growth plate has complex pattern of fusion
    - Fusion 1st at Kump bump in central portion of growth plate
    - Fusion proceeds medially (posterior → anterior)
    - Fusion then proceeds laterally along posterior portion
    - Anterolateral portion of growth plate last to fuse
  - Triplane fracture
    - SH-IV fracture
    - Includes
      - Sagittal fracture through epiphysis
      - Horizontal fracture through nonfused portion of growth plate
      - Coronal fracture through posterior portion of metaphysis and fused portion of epiphysis
      - May have 2 or 3 fracture fragments
  - Quadriplane fracture
    - Triplane fracture + additional oblique metaphyseal fracture
  - Juvenile Tillaux fracture (SH-III of distal tibia)

Radiographic Findings

- Displacement of epiphysis
- Widening of physis
- Fracture extension into metaphysis or epiphysis
- Triplane fracture
  - Lateral view shows bone spike through posterior metaphysis
  - Sagittal fracture through epiphysis difficult to see
- Juvenile Tillaux fracture (SH-III)
  - Sagittal fracture line through anterolateral margin of distal tibial epiphysis, extends horizontally across unfused physis

CT Findings

- Multiplanar and 3D reformations very useful in understanding triplane fracture pattern

MR Findings

- Useful to diagnose SH-V (crush injuries)
  - Bone marrow edema centered on physis
- Useful in evaluation of physeal bar

Imaging Recommendations

- Best imaging tool
  - Ankle radiographs, ± CT for complex injuries

DIFFERENTIAL DIAGNOSIS

Pilon Fracture

- Axial load injury causing splitting of tibial plafond

Ankle Sprain

- If growth plate is open, Salter-Harris fracture of fibula is more likely than ligament sprain

PATHOLOGY

Staging, Grading, & Classification

- Salter-Harris classification is standard for growth plate injuries
  - Type I: Physis only
  - Type II: Physis plus corner of metaphysis
  - Type III: Physis plus epiphysis
  - Type IV: Physis, metaphysis, and epiphysis
  - Type V: Crush injury to growth plate

CLINICAL ISSUES

Demographics

- Age
  - Occur near skeletal maturity (~ 12-15 years)
  - SH-I later than triplane
  - Juvenile Tillaux later than triplane
- Epidemiology
  - Distal tibia is 2nd most common site of SH fracture

Natural History & Prognosis

- Growth arrest may occur at fracture site due to premature closure of physis
  - Fractures occur close enough to skeletal maturity that there is no significant growth disturbance

DIAGNOSTIC CHECKLIST

Consider

- SH-I fracture may be occult on radiograph if nondisplaced

Image Interpretation Pearls

- Fusion of 1 part of growth plate does not preclude injury to another part
  - In appropriate patient age, look carefully at anterolateral tibia for SH injury
- Triplane fracture involves metaphysis; Tillaux does not

Reporting Tips

- Describe number of fragments present for triplane fractures
- Report incomplete reduction; may relate to soft tissue entrapment

SELECTED REFERENCES

1. Schnetzler KA et al: The pediatric triplane ankle fracture. J Am Acad Orthop Surg. 15(12):738-47, 2007
2. Brown SD et al: Analysis of 51 tibial triplane fractures using CT with multiplanar reconstruction. AJR Am J Roentgenol. 183(5):1489-95, 2004
3. Petit P et al: Acute fracture of the distal tibial physis: role of gradient-echo MR imaging versus plain film examination. AJR Am J Roentgenol. 166(5):1203-6, 1996
4. Cass JR et al: Salter-Harris Type-IV injuries of the distal tibial epiphyseal growth plate, with emphasis on those involving the medial malleolus. J Bone Joint Surg Am. 65(8):1059-70, 1983
5. Salter RB et al: Injuries involving the epiphyseal plate. J Bone Joint Surg Am. 45:587-622, 1963

SALTER-HARRIS FRACTURE, ANKLE

*(**Left**) Anteroposterior radiograph shows a Salter-Harris type I fracture of the fibula and a type III fracture of the medial malleolus due to supination-adduction type injury. (**Right**) AP radiograph shows a Salter-Harris type I fracture of the distal fibula, due to supination-adduction injury. Clues to diagnosis are lateral soft tissue swelling centered on the growth plate and mild widening of the growth plate relative to the lateral portion of distal tibial growth plate.*

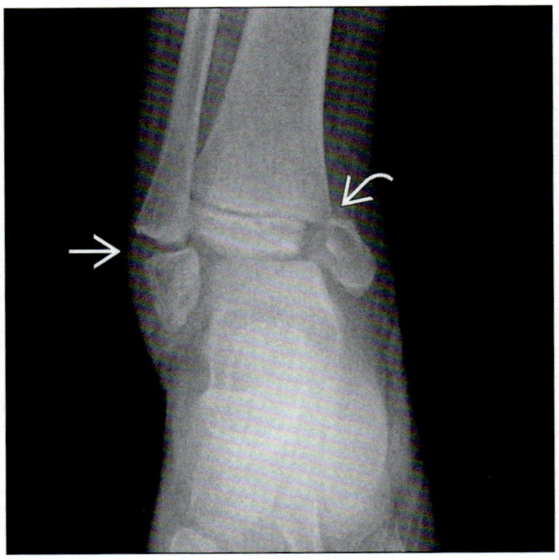

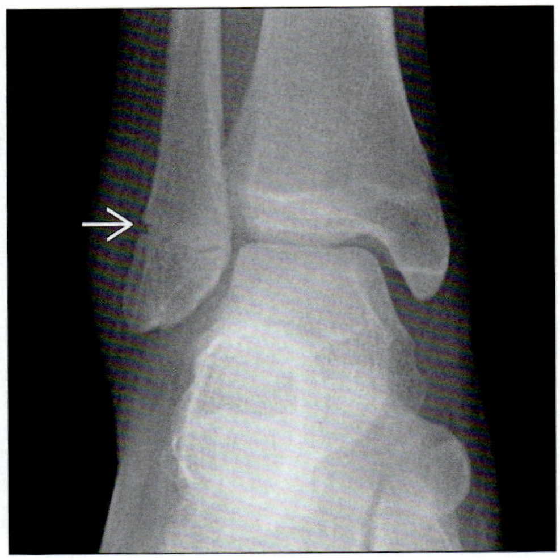

*(**Left**) Axial bone CT shows a juvenile Tillaux fracture, a Salter-Harris type IV fracture due to supination-external rotation injury. The anterolateral portion of the growth plate is the last part to fuse, and this corner is vulnerable to avulsion by the anterior inferior tibiofibular ligament. (**Right**) Coronal bone CT in the same patient shows the extension of the fracture along the growth plate and into the epiphysis.*

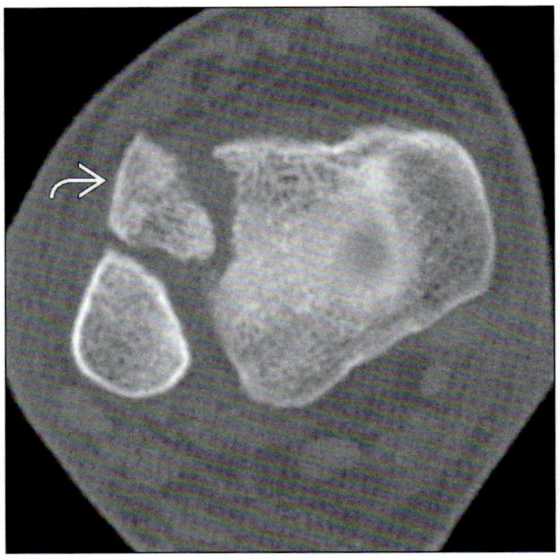

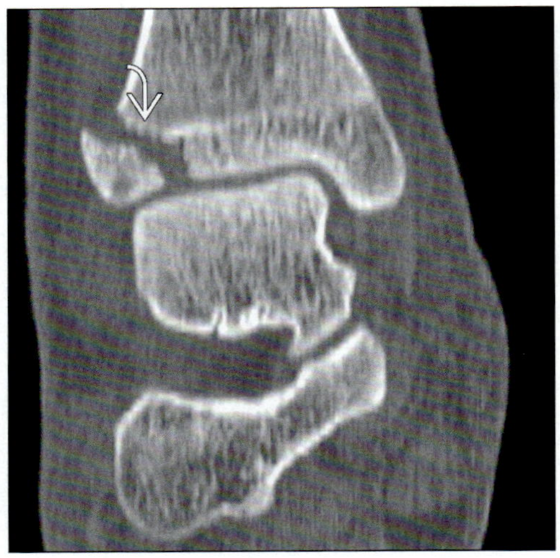

*(**Left**) Anteroposterior radiograph shows traumatic osteolysis, with widening and irregularity of the distal tibial physis 6 months after Salter I injury. Metaphyseal periosteal reaction is also present. This patient continued to walk on the leg after the initial injury; repetitive trauma resulted in osteolysis, which resolved when the patient stopped bearing weight. (**Right**) Lateral radiograph in the same patient shows widened, irregular growth plate from osteolysis.*

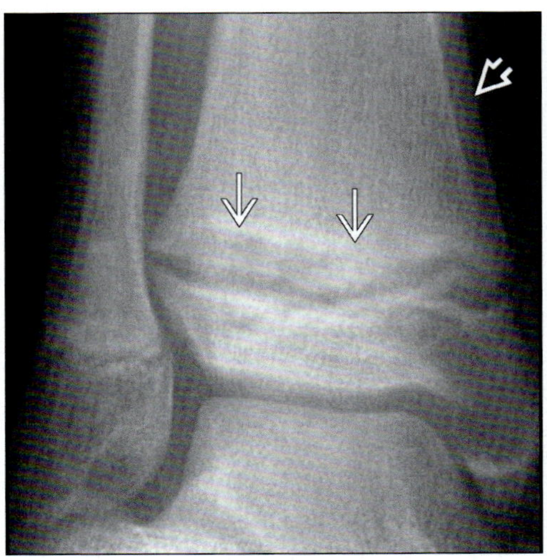

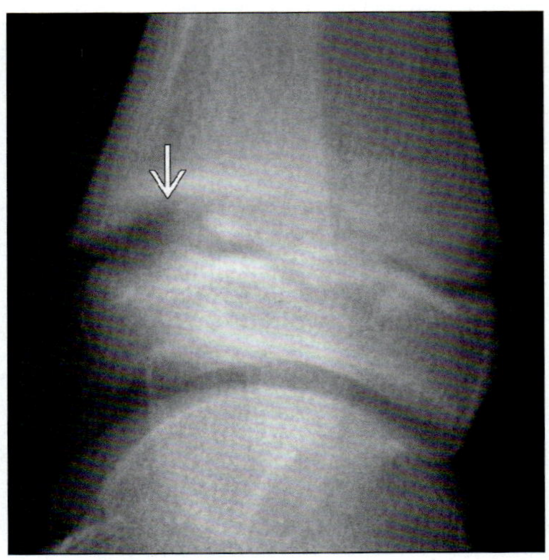

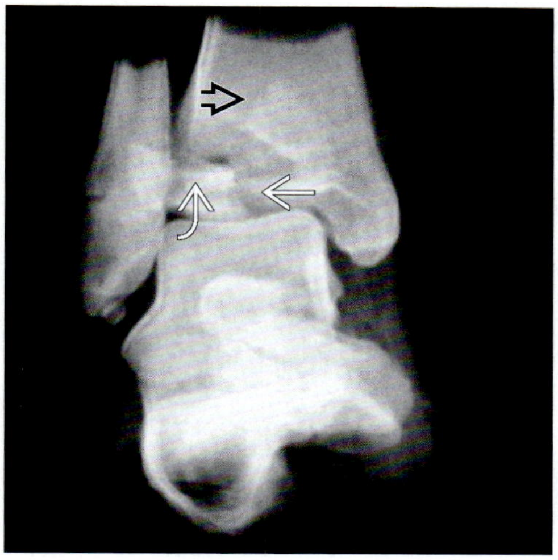

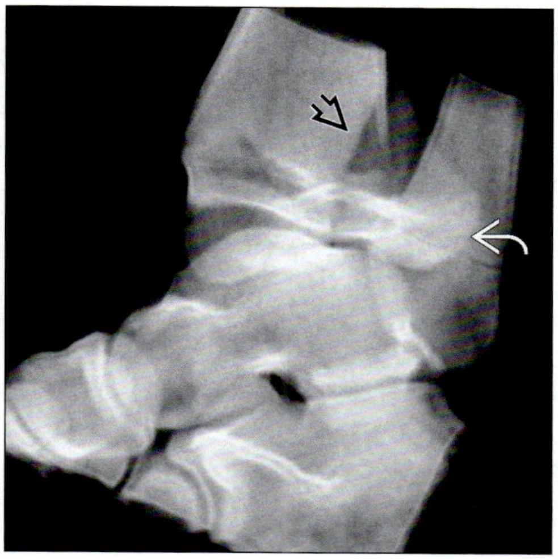

(Left) *AP bone CT 3D reformation shows a 2-part triplane fracture of the distal tibia. The sagittal fracture line is through the epiphysis, the axial fracture line is through unfused growth plate, & the coronal fracture line is through the metaphysis where growth plate is open.* ***(Right)*** *Lateral bone CT 3D reformation in the same patient better shows the coronal fracture line. The posteromedial fragment is displaced posteriorly. 3D reconstructions may aid in understanding complex fragment relationships.*

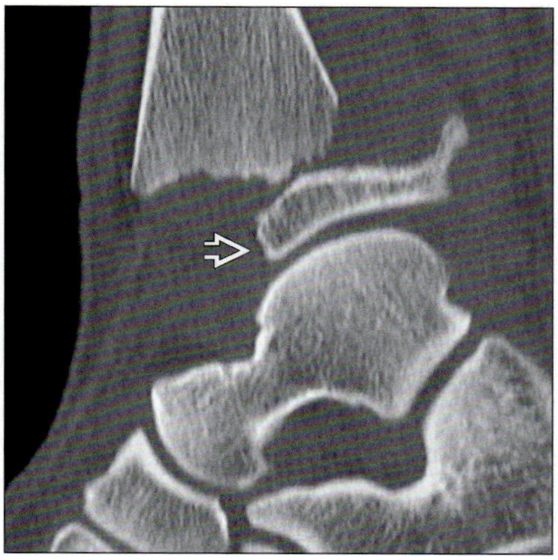

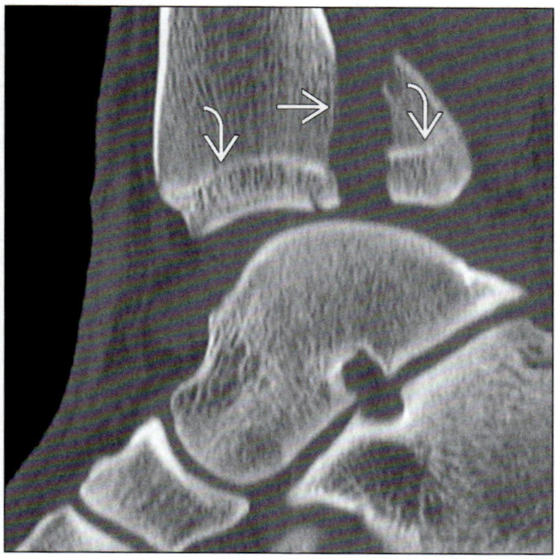

(Left) *Sagittal bone CT in the same patient through the lateral portion of fracture. In this region the growth plate is not yet fused, and the fracture is purely through the growth plate. There is posterior displacement of the epiphysis.* ***(Right)*** *Sagittal bone CT in the same patient through the medial portion of fracture shows the growth plate is fused and the coronal fracture line extends through both the metaphysis and the epiphysis. Planar reformats show the fused epiphysis well.*

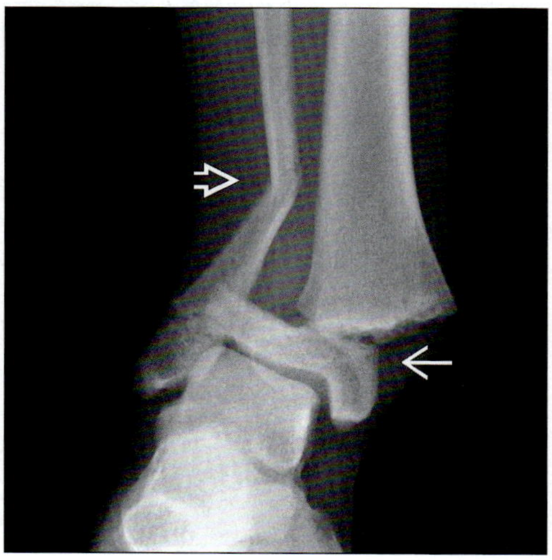

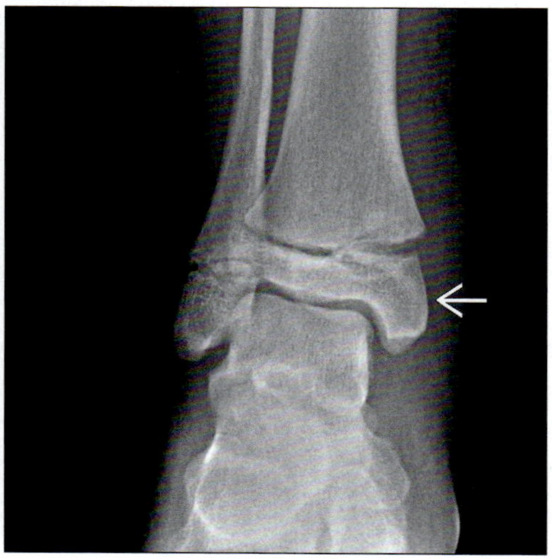

(Left) *AP radiograph shows a trampoline injury resulting in SH-I fracture of distal tibia with 50% lateral displacement of the epiphysis. There is an associated fracture of the distal fibular shaft.* ***(Right)*** *AP radiograph post closed reduction shows slight residual lateral displacement of epiphysis. This suggests soft tissue interposition along the growth plate. ORIF was performed and interposed periosteum was removed, allowing full reduction and internal fixation.*

FREIBERG INFRACTION

Key Facts

Terminology
- Osteochondral fracture vs. osteonecrosis of 2nd or 3rd metatarsal (MT) head

Imaging
- Metatarsal head contour abnormality
  - Flattening of metatarsal head is normal variant
  - Concavity of metatarsal head indicates advanced Freiberg infraction
  - May have "spikes" of bone at medial and lateral margins, with collapse of central portion of head
- Sclerosis ± subchondral fracture line
- Loose body (late finding)
- MR: Bone marrow edema, subchondral fracture line, articular depression, loose body

Top Differential Diagnoses
- Plantar plate rupture
- Morton neuroma
- Intermetatarsal bursitis
- Collateral ligament injury
- Fracture of metatarsal head or neck

Pathology
- Most likely due to chronic repetitive trauma

Clinical Issues
- Pain increased by weight-bearing, may awaken patient from sleep

Diagnostic Checklist
- Unstable 2nd MTP often misdiagnosed as Freiberg infraction
- Flattening of MT head in isolation is not diagnostic of Freiberg infraction

__(Left)__ AP radiograph shows subtle flattening and sclerosis → of 2nd metatarsal (MT) head, consistent with Freiberg infraction. The contour of a normal MT head may be mildly flat, and subchondral sclerosis or concavity of the articular surface is needed to make the radiographic diagnosis. __(Right)__ AP radiograph shows flattening of 3rd MT head → and subchondral sclerosis characteristic of advanced Freiberg infraction. A bowl-shaped, sclerotic line → demarcates the area of abnormality.

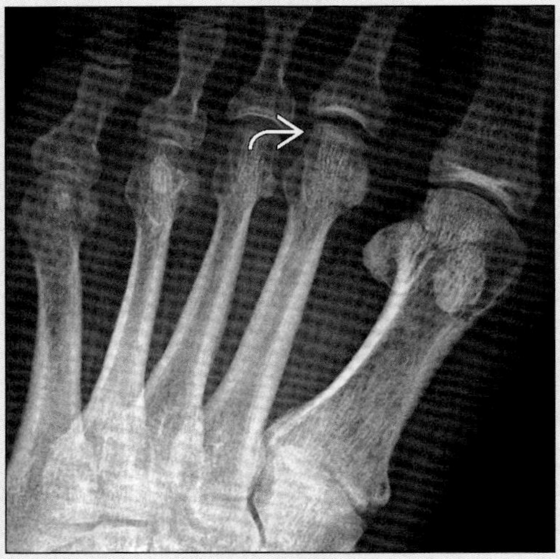

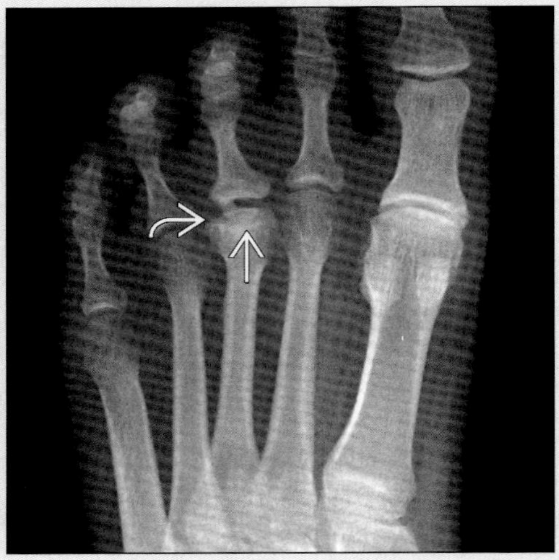

__(Left)__ Axial T1WI MR shows loss of normal fat signal → in the 2nd metatarsal head. Low signal intensity line around the area of abnormality → is MR equivalent to sclerosis seen on radiographs. __(Right)__ Sagittal T2WI FS MR shows advanced Freiberg infraction. A small free bone fragment → has developed in a dorsal location characteristic of Freiberg infraction. Joint fluid extends beneath the fragment →.

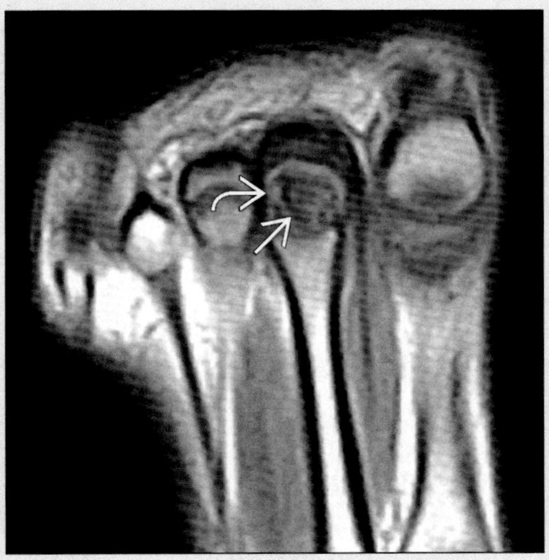

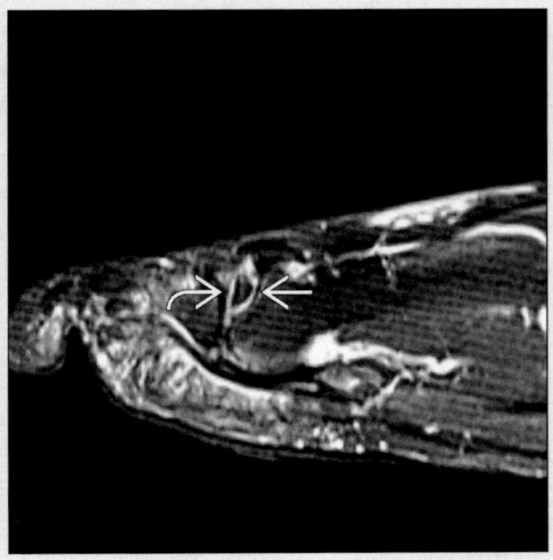

FREIBERG INFRACTION

TERMINOLOGY

Definitions

- Osteochondral fracture vs. osteonecrosis of 2nd or 3rd metatarsal (MT) head

IMAGING

Radiographic Findings

- Metatarsal head contour abnormality
  - Flattening of metatarsal head is normal variant, seen in 10% of asymptomatic population
  - Concavity of metatarsal head indicates advanced Freiberg infraction
    - May have "spikes" of bone at medial and lateral margins, with collapse of central portion of head
- Abnormal density in metatarsal head
  - Sclerosis &/or
  - Serpentine or bowl-shaped sclerotic line beneath articular surface
- Loose body (late finding)
  - Usually located dorsally, best seen on oblique radiograph

MR Findings

- Bone marrow edema
- Serpentine or bowl-shaped low signal intensity line beneath articular surface
- Concavity of articular surface of MT head

Imaging Recommendations

- Best imaging tool
  - MR

DIFFERENTIAL DIAGNOSIS

Plantar Plate Rupture

- May see dorsal MTP joint subluxation on radiographs
- MR shows discontinuity &/or redundancy of plantar plate
- Edema in subchondral bone often prominent on MR
  - Due to abnormal joint motion
  - Mimics Freiberg infraction on MR

Morton Neuroma

- MR shows enhancing mass between metatarsal heads

Intermetatarsal Bursitis

- MR shows mass between metatarsal heads
- May enhance if due to inflammatory arthropathy or if chronic synovitis present

Collateral Ligament Injury

- MR shows collateral ligament disruption

Fracture of Metatarsal Head or Neck

- Fracture line is proximal to location of Freiberg infraction

PATHOLOGY

General Features

- Etiology
  - Most likely due to chronic, repetitive trauma
  - Altered biomechanics increase stress on MT heads, ↑ risk of Freiberg infraction
    - Short 1st MT or metatarsus primus varus
    - High-heeled shoes with narrow toe box
  - Some authors believe it is a primary avascular necrosis
  - Sometimes associated with diabetes, systemic lupus erythematosus

Staging, Grading, & Classification

- Smillie classification
  - Stage I: Epiphyseal "fissure" surrounded by sclerotic bone
  - Stage II: Mild subchondral collapse
  - Stage III: Medial and lateral projections of MT head prominent due to subchondral collapse
  - Stage IV: MT head projections fracture, central area of collapse forms osteochondral loose body
  - Stage V: Extensive articular surface collapse, osteoarthritis

CLINICAL ISSUES

Presentation

- Most common signs/symptoms
  - Pain increased by weight-bearing, may awaken patient from sleep
- Other signs/symptoms
  - Soft tissue swelling at 2nd MTP, decreased range of motion

Demographics

- Most common in adolescent girls but can occur in adulthood

Treatment

- Off-loading of stress used for early stages
  - Metatarsal pads, wide toe box shoes
- Debridement of loose bodies, synovitis
- Surgery to improve biomechanics
  - Osteotomy of 2nd MT shaft or head
  - Realignment of metatarsus primus varus
- Excision arthroplasty of 2nd MTP joint

DIAGNOSTIC CHECKLIST

Consider

- Unstable 2nd MTP often misdiagnosed as Freiberg infraction

Image Interpretation Pearls

- Flattening of MT head in isolation is not diagnostic of Freiberg infraction

SELECTED REFERENCES

1. Carmont MR et al: Current concepts review: Freiberg's disease. Foot Ankle Int. 30(2):167-76, 2009
2. Nguyen VD et al: Freiberg's disease in diabetes mellitus. Skeletal Radiol. 20(6):425-8, 1991
3. Stanley D et al: Assessment of etiologic factors in the development of Freiberg's disease. J Foot Surg. 29(5):444-7, 1990
4. Young MC et al: Osteochondral disruption of the second metatarsal: a variant of Freiberg's infraction? Foot Ankle. 8(2):103-9, 1987

SYNDESMOSIS LIGAMENT INJURY, ANKLE

Key Facts

Terminology
- High ankle sprain

Imaging
- Radiographs
  - Lateral soft tissue swelling, superior to swelling seen with ankle sprain
  - AP view: < 1 cm overlap of tibia and fibula
  - Mortise view: < 1 mm overlap of tibia and fibula
  - AP or mortise: Lateral ankle gutter between talus and fibula > 4 mm
- Weight-bearing or eversion stress views may show abnormal joint widening when routine views are normal
- MR and CT
  - Fluid extending > 1 cm cephalad in syndesmotic ligament between tibia and fibula
  - Discontinuity or laxity of AITFL, PITFL
  - Widened tibiofibular syndesmosis

Top Differential Diagnoses
- Bassett ligament
- Ankle sprain

Pathology
- Associated abnormalities
  - Maisonneuve fx: Fx of proximal fibula
  - Malleolar fractures
  - Osteochondral injury of talar dome

Clinical Issues
- Chronic injury of AITFL may lead to anterolateral impingement

Diagnostic Checklist
- Isolated fractures of medial or posterior malleolus should raise suspicion for syndesmosis tear

*(**Left**) Graphic of ankle viewed from an anterior position, with foot plantarflexed, shows rupture of anterior inferior tibiofibular ligament (AITFL) and associated rupture of deltoid ligament. (**Right**) Mortise view radiograph shows mature but irregular periosteal new bone in the ankle of a 25-year-old man, related to remote syndesmosis injury. There is loss of the normal 1 mm overlap between tibia and fibula. Premature osteoarthritis has developed.*

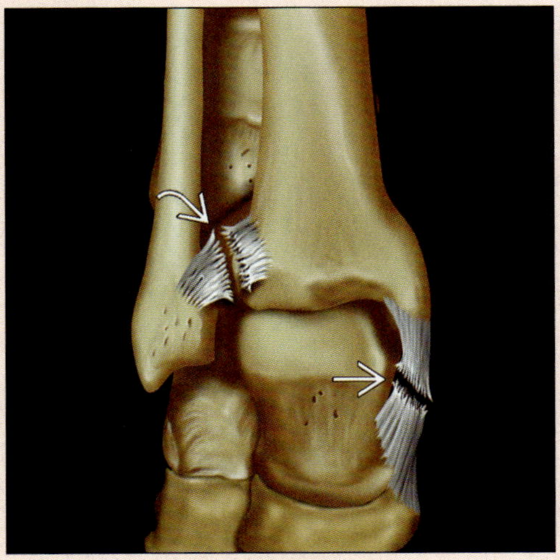

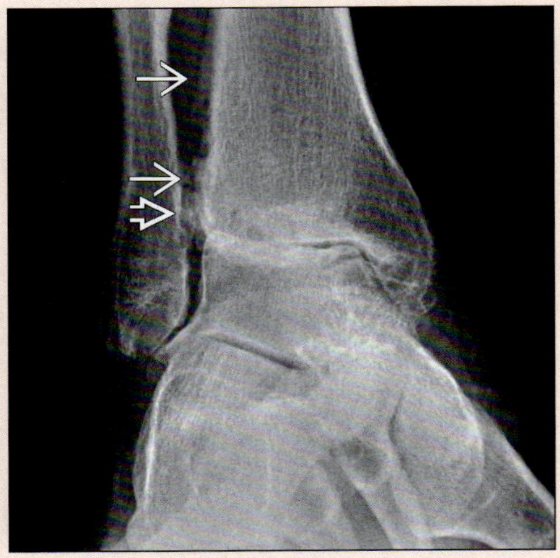

*(**Left**) Axial T2WI FS MR of acute syndesmosis injury shows a torn and irregular anterior inferior tibiofibular ligament and a fracture of the posterior malleolus at the posterior inferior tibiofibular ligament attachment. (**Right**) Axial MR arthrogram T1WI FS shows avulsed AITFL and widening of the anterior portion of the syndesmosis, with preserved PITFL. This pattern of injury occurs with ankle rotation and is often associated with injury to the anterior talofibular ligament.*

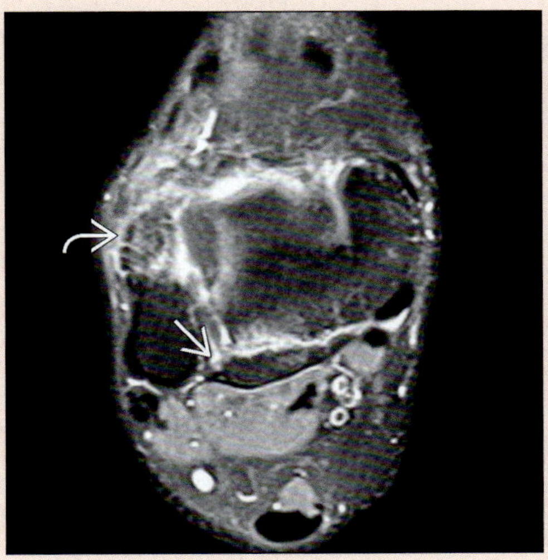

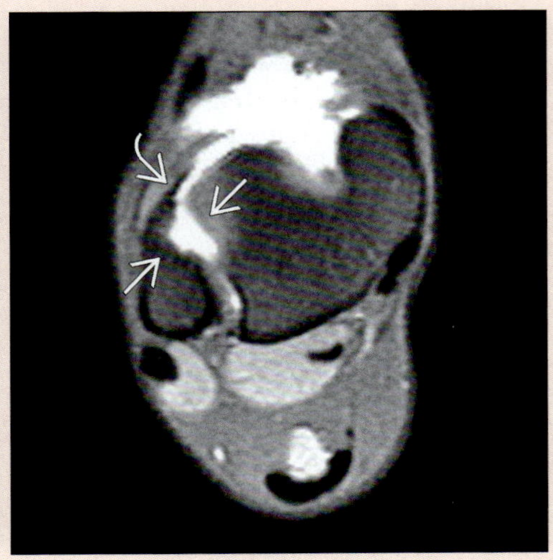

SYNDESMOSIS LIGAMENT INJURY, ANKLE

TERMINOLOGY

Synonyms

- High ankle sprain

IMAGING

General Features

- Best diagnostic clue
  - Radiographs: Widening of interval between tibia and fibula

Radiographic Findings

- Soft tissue swelling
  - Centered at plafond, higher than for ankle sprain
- Joint effusion
- Widening of tibiofibular syndesmosis
  - AP radiograph: 1 cm of overlap expected between tibia and fibula
  - Mortise radiograph
    - 1 mm of overlap expected between tibia and fibula
    - Widening of lateral ankle gutter
  - Weight-bearing or eversion stress views may show widening when routine views are normal
- Chronic injury: Periosteal reaction in syndesmosis

MR Findings

- Fluid extending > 1 cm cephalad in syndesmotic ligament between tibia and fibula
- May see isolated injury of anterior or posterior bands
- Anterior inferior tibiofibular ligament (AITFL) best seen on coronal images
  - Ligament extends superomedial to inferolateral
  - Avulsion of tibial attachment or midsubstance tear
- Normal AITFL often appears thick, heterogeneous on axial images
  - Due to obliquity of ligament, multiple fascicles
- Posterior inferior tibiofibular ligament (PITFL) best seen on axial images
  - Avulsion of tibial attachment or midsubstance tear
  - Bony avulsion more common than ligament tear
- Interosseous ligament best seen on coronal images
  - Fluid extending > 1 cm cephalad to plafond suggests tear

Imaging Recommendations

- Best imaging tool
  - Radiographs primary imaging method
  - MR used in chronic injury
- Protocol advice
  - Axial and coronal fluid-sensitive images needed

DIFFERENTIAL DIAGNOSIS

Bassett Ligament

- Accessory AITFL

Ankle Sprain

- Lateral collateral ligament injury, due to supination injury

Malleolar Fractures

- Nondisplaced malleolar fractures may be difficult to see on radiographs

PATHOLOGY

General Features

- Etiology
  - Pronation or external rotation
- Associated abnormalities
  - Pronation injury
    - Maisonneuve fx: Fx of proximal fibula
    - Posterior malleolar fracture
    - Medial malleolar fracture
    - Osteochondral injury of talar dome
  - External rotation injury
    - Lateral collateral ligament injury
    - Osteochondral injury of talar dome

Relevant Anatomy

- Tibiofibular syndesmosis consists of 3 ligaments
  - Anterior inferior tibiofibular ligament
    - Obliquely oriented from superomedial to inferolateral
    - Bassett ligament: Accessory ligament located inferior to AITFL, variably present
  - Posterior inferior tibiofibular ligament
    - Horizontal orientation
    - Slightly inferior to AITFL
    - Intermalleolar ligament (transverse ligament) is additional fascicle
      - Extends obliquely from posterior malleolus to tip of fibula
  - Interosseous ligament
    - Distal thickening of interosseous membrane
    - Ends about 1 cm above tibial plafond

CLINICAL ISSUES

Presentation

- Most common signs/symptoms
  - Pain, tenderness lateral ankle
- Other signs/symptoms
  - Ankle instability

Treatment

- Immobilization or ORIF

DIAGNOSTIC CHECKLIST

Consider

- Isolated fractures of medial or posterior malleolus should raise suspicion for syndesmosis ligament injury

SELECTED REFERENCES

1. Cheung Y et al: MRI of isolated distal fibular fractures with widened medial clear space on stressed radiographs: which ligaments are interrupted? AJR Am J Roentgenol. 192(1):W7-12, 2009
2. Han SH et al: Chronic tibiofibular syndesmosis injury: the diagnostic efficiency of magnetic resonance imaging and comparative analysis of operative treatment. Foot Ankle Int. 28(3):336-42, 2007
3. Brown KW et al: MRI findings associated with distal tibiofibular syndesmosis injury. AJR Am J Roentgenol. 182(1):131-6, 2004
4. Oae K et al: Injury of the tibiofibular syndesmosis: value of MR imaging for diagnosis. Radiology. 227(1):155-61, 2003

SYNDESMOSIS LIGAMENT INJURY, ANKLE

***(Left)** Mortise view radiograph shows syndesmosis sprain, with loss of the normal 1 mm overlap between the tibia and fibula. This sign may be mimicked if the ankle is overly internally rotated; however, if that were the case, the medial malleolus and talar body would overlap. **(Right)** Lateral radiograph in the same patient shows bony avulsion of posterior inferior tibiofibular ligament (PITFL). A joint effusion is visible anteriorly.*

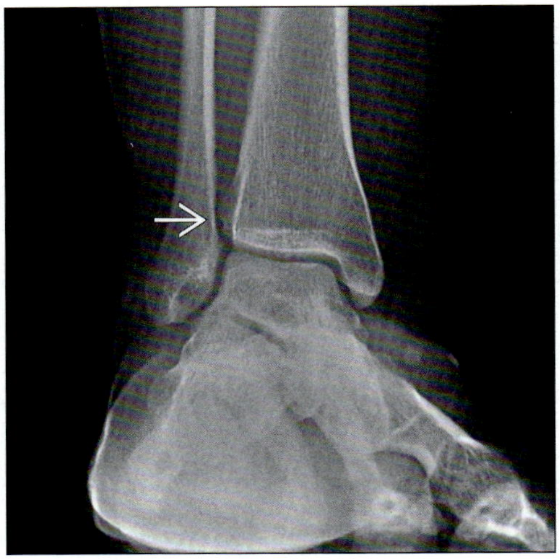

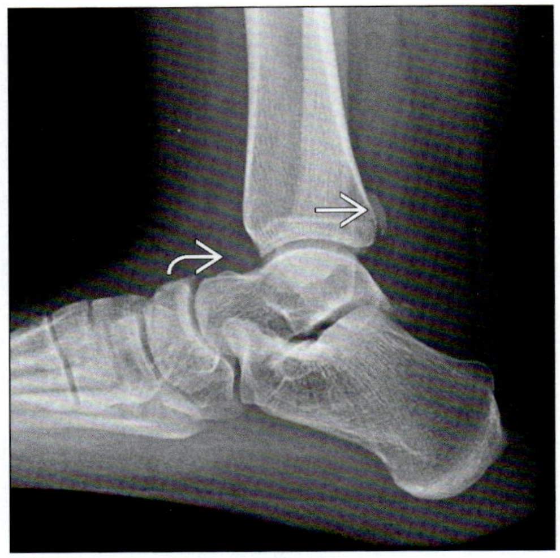

***(Left)** Coronal T2WI FS MR through posterior margin of the ankle joint shows partial tear of PITFL and discontinuity of intermalleolar ligament in a patient with instability following eversion injury. **(Right)** Coronal T2WI FS MR more anteriorly in the same patient shows a portion of AITFL flipped into syndesmosis. Fluid extends > 1 cm cephalad to plafond, indicating tear of syndesmotic ligament. Deltoid strain and medial malleolus marrow edema are also evident.*

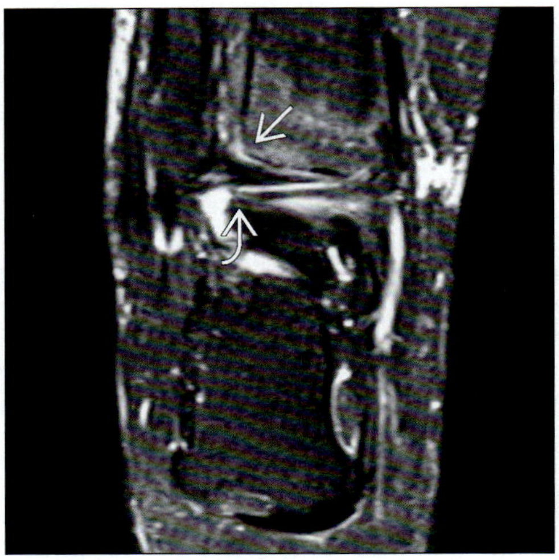

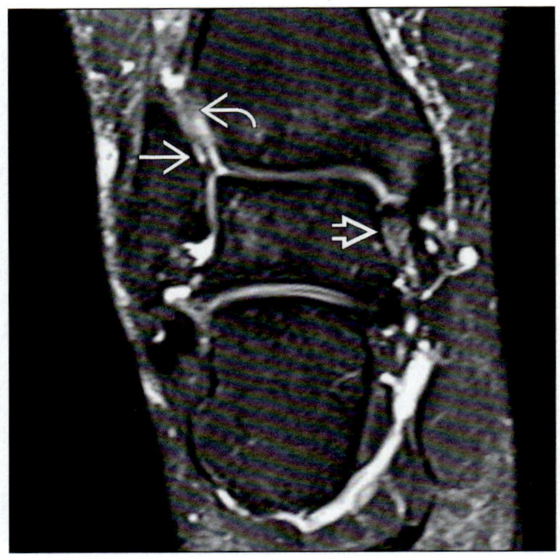

***(Left)** Coronal MR arthrogram T1WI FS shows tears of PITFL, as well as syndesmotic and intermalleolar ligaments. Tears are usually more easily seen with MR arthrography than with noncontrast imaging. **(Right)** Axial MR arthrogram T1WI FS in the same patient shows AITFL tear. The majority of the PITFL is located more inferiorly, but a small portion of the PITFL tear is visible on this image.*

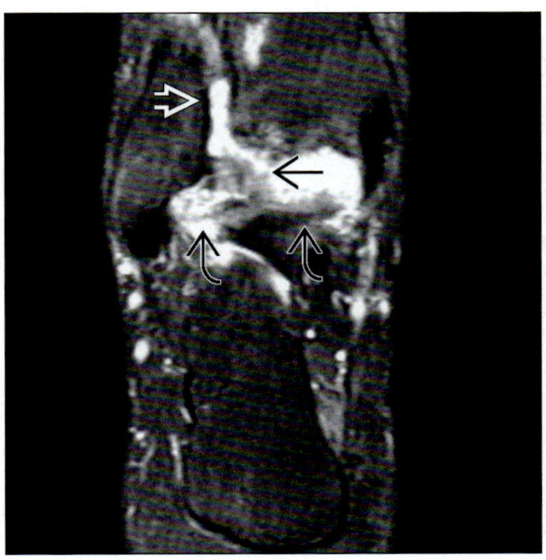

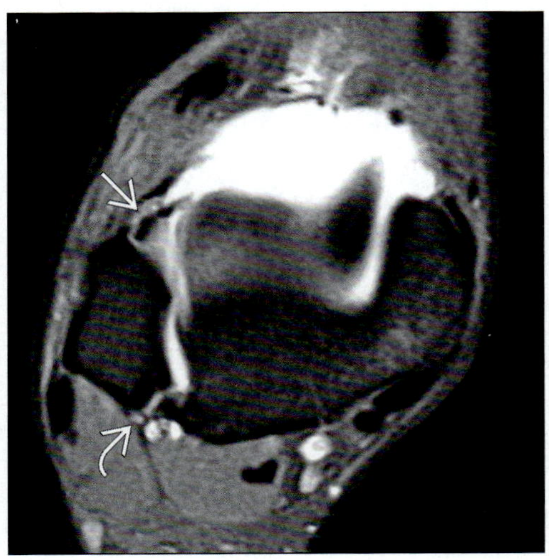

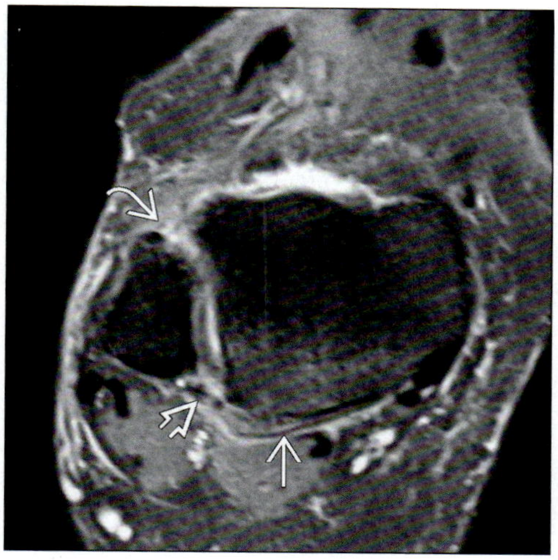
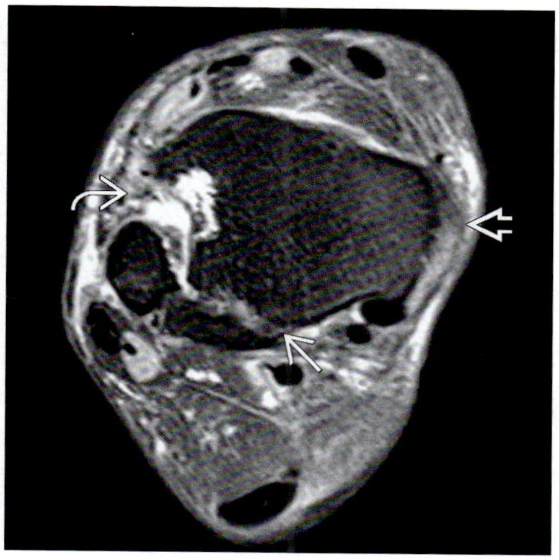

(Left) *Axial PDWI FS MR shows a chronic syndesmosis injury, with periosteal reaction → posteriorly. AITFL ligament ↷ is discontinuous. PITFL ⇨ is not discontinuous, but appears stretched and heterogeneous, consistent with ligament scarring.* ***(Right)*** *Axial PDWI FS MR shows an incompletely healed posterior malleolus fracture →. AITFL is torn ↷, and fibula is rotated laterally in fibular notch. Medial periosteal reaction ⇨ reflects deltoid ligament tear (not shown).*

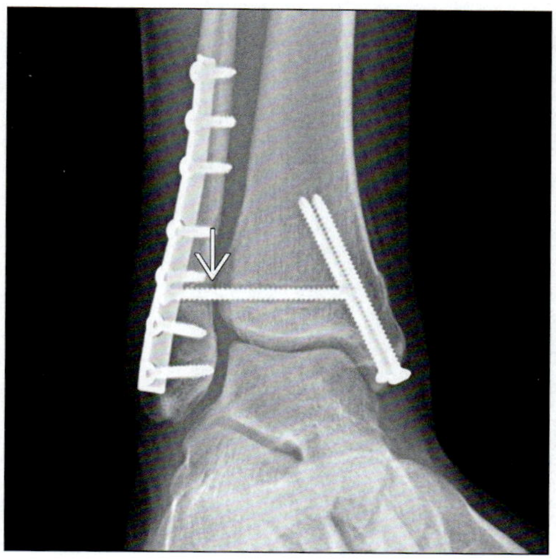
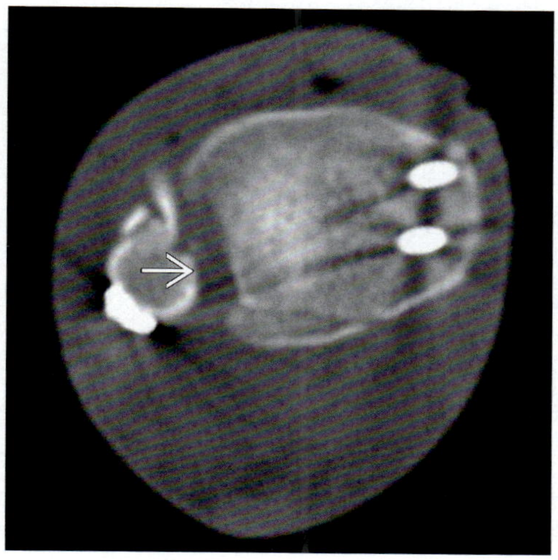

(Left) *Mortise view radiograph shows fixation of Weber C fibular fracture and medial malleolar fracture, as well as syndesmotic screw placement. Despite appropriate fixation, syndesmosis is widened, with < 1 mm of overlap →.* ***(Right)*** *Axial bone CT in the same patient shows a bone fragment → in the syndesmosis, which prevented adequate reduction. Patient was taken to OR for removal and repeat reduction. CT is useful in determining the cause of poor reduction or loss of reduction.*

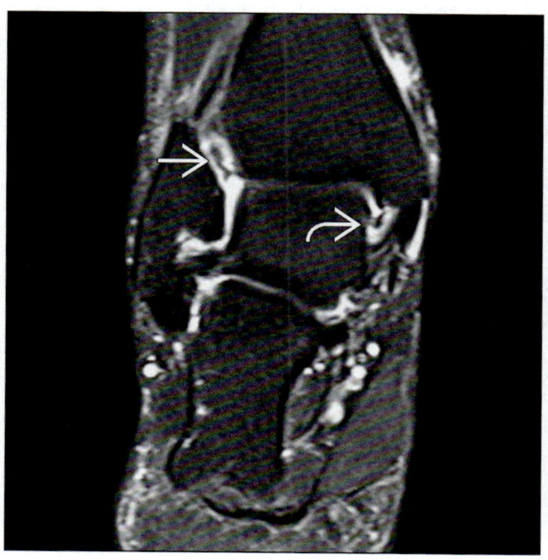
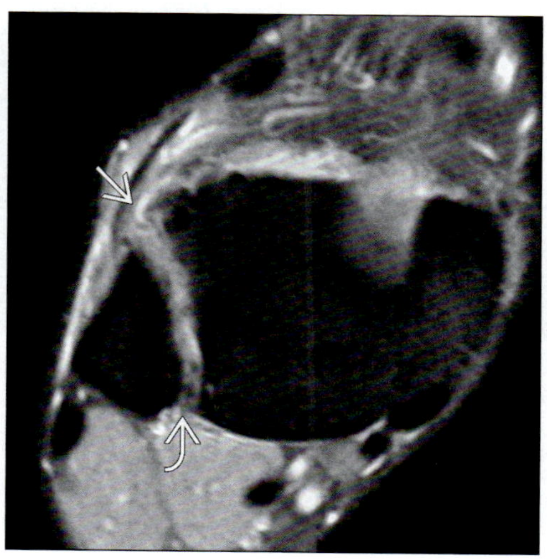

(Left) *Coronal T2WI FS MR demonstrates the anterior inferior tibiofibular ligament (AITFL) → torn and displaced superiorly. The syndesmosis is widened. The deep fibers of the deltoid ligament are partly torn ↷.* ***(Right)*** *Axial PDWI FS MR from the same patient demonstrates a lax, redundant anterior inferior tibiofibular ligament (AITFL) →. The posterior inferior tibiofibular ligament (PITFL) ↷ appears amorphous, likely representing a sprain.*

ANKLE SPRAIN

Key Facts

Terminology

- Ankle sprain: Injury of LCL without significant bony injury

Imaging

- Radiographs
  - Soft tissue swelling centered beneath tip of fibula
  - Medial soft tissues usually normal
  - ± joint effusion, bony avulsion of CFL
- MR
  - Reserved for cases of failed conservative management
  - Complete tear shows ligament discontinuity, extraarticular fluid
  - Partial tear shows lax ligament, which may be thick or thin

Top Differential Diagnoses

- Fracture of lateral process of talus
- Fracture of 5th metatarsal styloid process
- Malleolar fractures
- Syndesmosis sprain
- Salter-Harris I fracture of fibula

Pathology

- Anterior talofibular ligament (ATFL) torn 1st
- Calcaneofibular ligament (CFL) torn in more severe injury
- Posterior talofibular ligament (PTFL) rarely completely torn
- Associated injuries: Osteochondral injury of talar dome &/or tibial plafond, tear of peroneal tendons

Diagnostic Checklist

- If soft tissue swelling is centered at level of plafond consider syndesmosis sprain
- Be alert for fractures that may mimic ankle sprain, change treatment

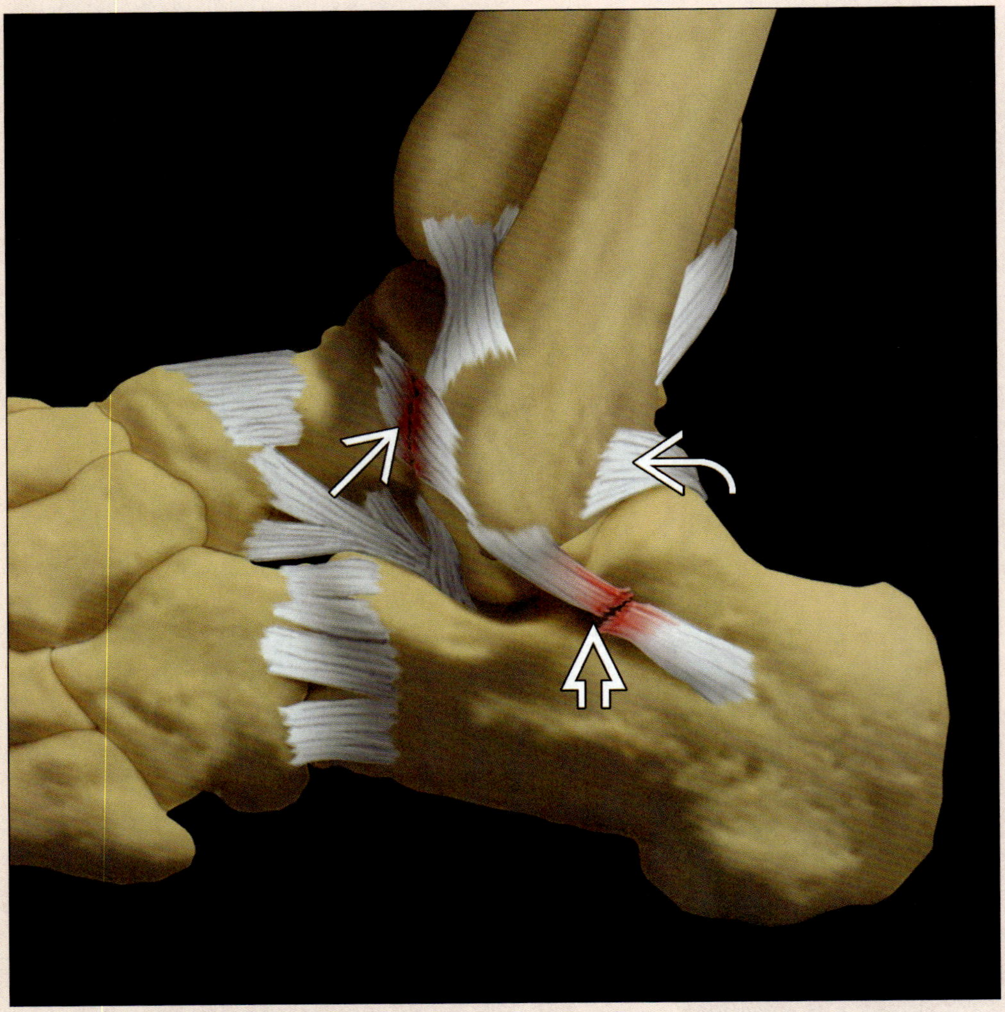

Graphic shows tears of the ATFL ➔ and CFL ⇨. The ATFL originates on the fibula 1 cm above its tip and inserts at the dorsal margin of the talar head/neck junction. The CFL originates at the tip of the fibula and inserts posteriorly on the calcaneus. Although the CFL is a discrete ligament, the ATFL is a band-like thickening of the anterior joint capsule and merges with the capsule inferiorly. The PTFL ➔ is intact; this ligament is rarely severely injured. It extends from the deep margin of the distal fibula to the posterior talus.

ANKLE SPRAIN

TERMINOLOGY

Abbreviations
- Lateral collateral ligament (LCL)

Definitions
- Injury of LCL without significant bony injury

IMAGING

Radiographic Findings
- Soft tissue swelling centered beneath tip of fibula
- ± joint effusion
- ± flake-like avulsion of tip of fibula

MR Findings
- Reserved for cases of failed conservative management
- Complete tear shows ligament discontinuity
  - Fluid often extends beyond confines of joint
- Partial tear shows lax ligament, which may be thick or thin
  - Often difficult to distinguish between partly and completely torn ligament on MR
- ATFL tear identified on axial images at level of dorsal talar neck
  - Capsule inferior to torn ligament may be thickened
  - This inferior capsular thickening should not be mistaken for ATFL
- CFL identified on coronal images at tip of fibula
  - And on axial images deep to peroneal tendons, extending posteriorly from fibular tip
- Posterior talofibular ligament identified on coronal or axial images in lateral ankle gutter
  - Normal ligament has striated appearance

Imaging Recommendations
- Best imaging tool
  - Radiographs used to exclude fracture
  - MR to assess degree of ligamentous injury

DIFFERENTIAL DIAGNOSIS

Fracture of Lateral Process of Talus
- Fracture seen across base of triangle on lateral radiograph

Fracture of 5th Metatarsal Styloid Process
- This region should be included on ankle series

Lateral Malleolar Fracture
- Best seen on mortise or lateral view

Syndesmosis Sprain
- Soft tissue swelling centered more superiorly
- Patients tender over syndesmosis more than fibular tip
- Medial soft tissue swelling often present

Salter-Harris I Fracture of Fibula
- Occurs instead of ankle sprain in children

PATHOLOGY

General Features
- Associated abnormalities
  - Osteochondral injury of talar dome &/or tibial plafond
  - Tear of peroneal tendons
  - Chip fracture of fibular tip

Staging, Grading, & Classification
- Ligaments tear in typical sequence: ATFL → CFL → PTFL

Relevant Anatomy
- LCL is composed of ATFL, CFL, and PTFL
  - Stabilizes against inversion injury
  - Also provides rotational and anteroposterior stability
- ATFL originates 1 cm above tip of fibula, inserts anteriorly at junction of talar body and neck
- CFL originates tip of fibula, inserts posteriorly on body of calcaneus
  - Deep to peroneal tendons
  - Forms floor of peroneal tendon sheath
- PTFL extends from deep margin of fibula to lateral margin of talus

CLINICAL ISSUES

Presentation
- Most common signs/symptoms
  - Inversion injury with lateral pain and soft tissue swelling

Natural History & Prognosis
- Most resolve, some develop chronic instability

Treatment
- Conservative management with immobilization
- Chronic instability may be treated surgically

DIAGNOSTIC CHECKLIST

Consider
- By Ottawa rules, radiographs are needed if
  - Pain is present at tip or posterior margin of fibula
  - Patient is unable to walk > 4 steps acutely and in emergency department

Image Interpretation Pearls
- If soft tissue swelling is centered at level of plafond consider syndesmosis sprain
  - Especially if medial soft tissue swelling ± medial malleolar fracture present

SELECTED REFERENCES

1. Oae K et al: Evaluation of anterior talofibular ligament injury with stress radiography, ultrasonography and MR imaging. Skeletal Radiol. 39(1):41-7, 2010
2. Bachmann LM et al: Accuracy of Ottawa ankle rules to exclude fractures of the ankle and mid-foot: systematic review. BMJ. 326(7386):417, 2003
3. Steinbach LS et al: Special focus session. MR arthrography. Radiographics. 22(5):1223-46, 2002
4. Cheung Y et al: MR imaging of ligamentous abnormalities of the ankle and foot. Magn Reson Imaging Clin N Am. 9(3):507-31, x, 2001
5. Kreitner KF et al: Injuries of the lateral collateral ligaments of the ankle: assessment with MR imaging. Eur Radiol. 9(3):519-24, 1999

(Left) *Mortise view radiograph shows lateral soft tissue swelling ➔ centered at the tip of the fibula, characteristic of ankle sprain. In contrast, soft tissue swelling from syndesmotic sprain is centered at the level of the plafond, and medial soft tissue swelling is usually present as well.* ***(Right)*** *Axial PDWI FS MR shows the normal appearance of the ATFL ➔ and PTFL ➔. ATFL is identified on the most superior cut where the talar neck is also visible. The PTFL is broad and has multiple bands.*

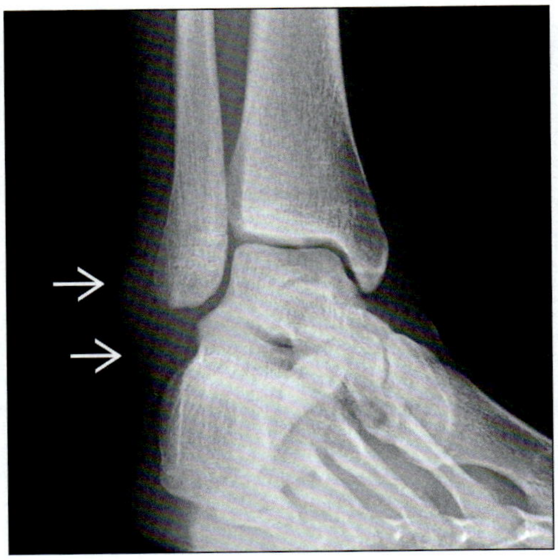

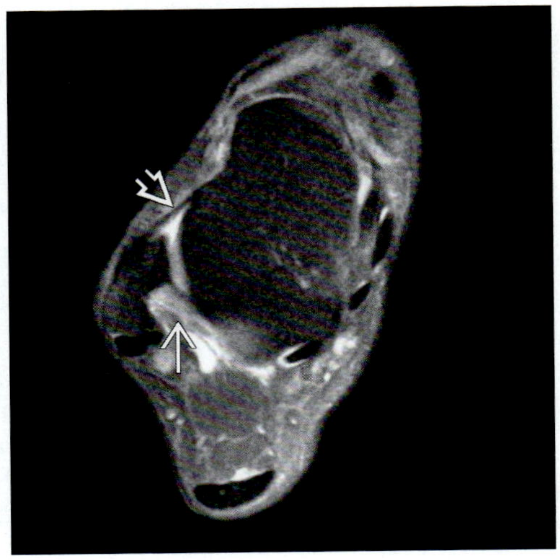

(Left) *Axial PDWI FS MR shows the ATFL ➔ detached from its attachment to the talus. A synovial cyst ➔ has developed along the anterior talar neck.* ***(Right)*** *Axial MR arthrogram T1WI FS shows extravasation of contrast anterolaterally ➔ through the expected location of the ATFL. Only a small stump of the ATFL is still visible ➔. Contrast in the peroneal tendon sheath ➔ indicates rupture of the CFL. Contrast in the flexor hallucis longus tendon sheath ➔ is a normal finding.*

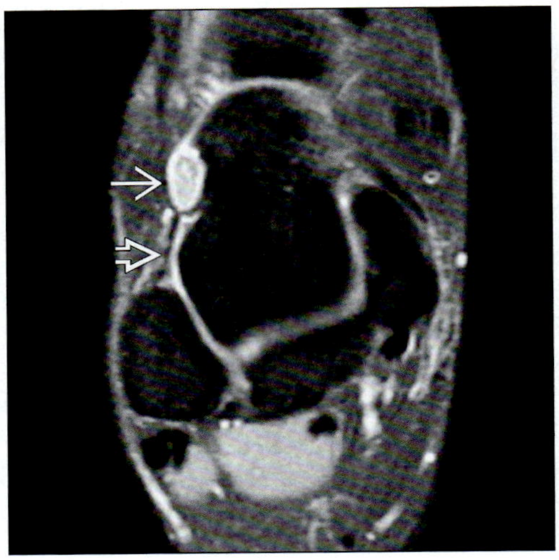

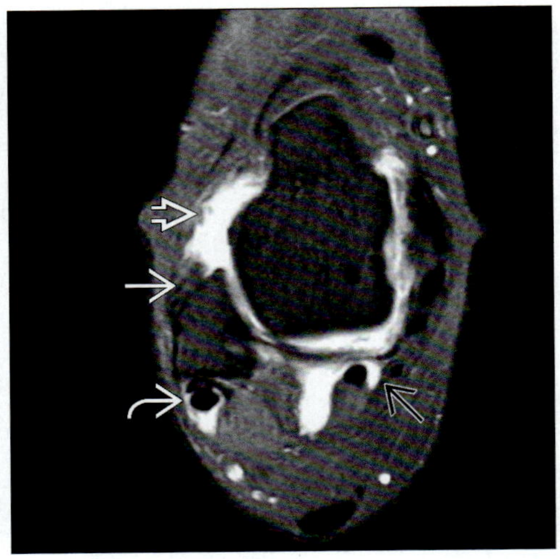

(Left) *Axial T2WI FS MR shows a small stump of ATFL ➔ at its fibular attachment and absence of ligament along its normal course to the junction of the talar body and neck ➔. Fluid lateral to the joint margins ➔ is a secondary sign of ligament tear.* ***(Right)*** *Axial arthrogram CT shows a defect ➔ in the ATFL. Contrast extends through the defect and lateral to the fibula. Thickened and retracted ligament stumps ➔ are visible adjacent to the defect.*

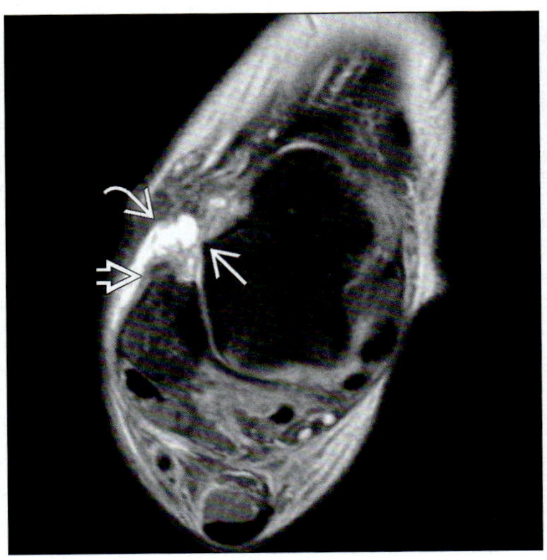

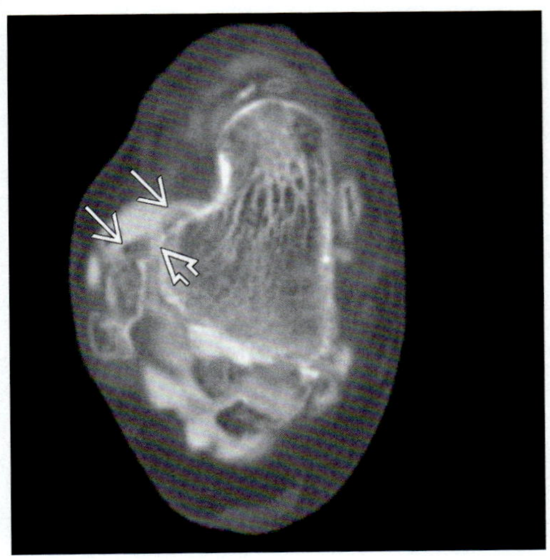

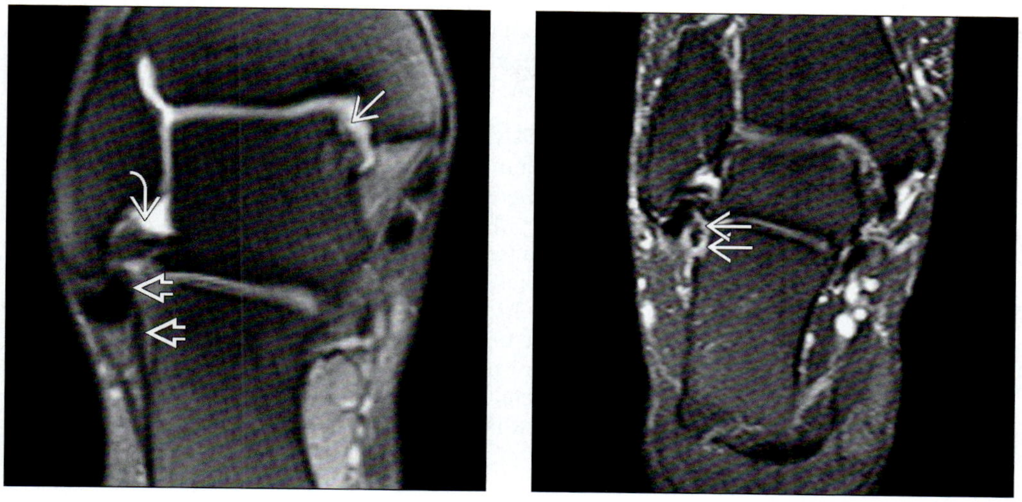

(Left) *Coronal MR arthrogram T1WI FS shows normal CFL ⇨ and PTFL ↪. The CFL lies deep to the peroneal tendons and forms the floor of the tendon sheath. It is normally seen on coronal images showing the PTFL. The CFL should appear thin and taut on MR. PTFL has multiple bands. Note the osteochondral lesion ➔ in the medial talar dome.* ***(Right)*** *Coronal T2WI FS MR shows a CFL tear ➔. Note that peroneal tendons have migrated slightly medially and superiorly through the ligament defect.*

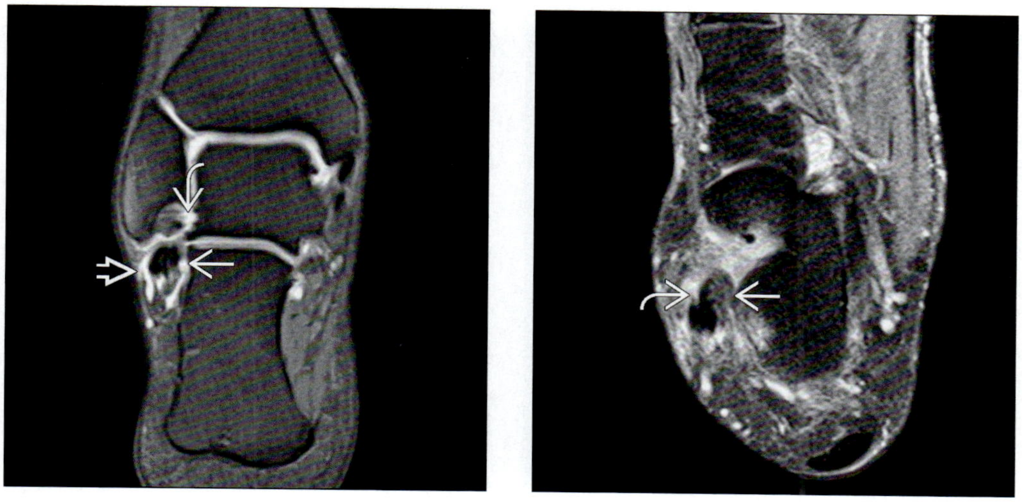

(Left) *Coronal MR arthrogram T1WI FS shows tears of both the CFL ➔ and the PTFL ↪. Contrast extends through the torn CFL into the peroneal tendon sheath ⇨.* ***(Right)*** *Axial T2WI FS MR shows a chronic tear of the CFL ➔; the ligament is markedly thickened and amorphous. Fluid is seen around the peroneal tendons, and splitting of the peroneus brevis tendon is also present ↪. Peroneal tendon injury is often associated with ankle sprain.*

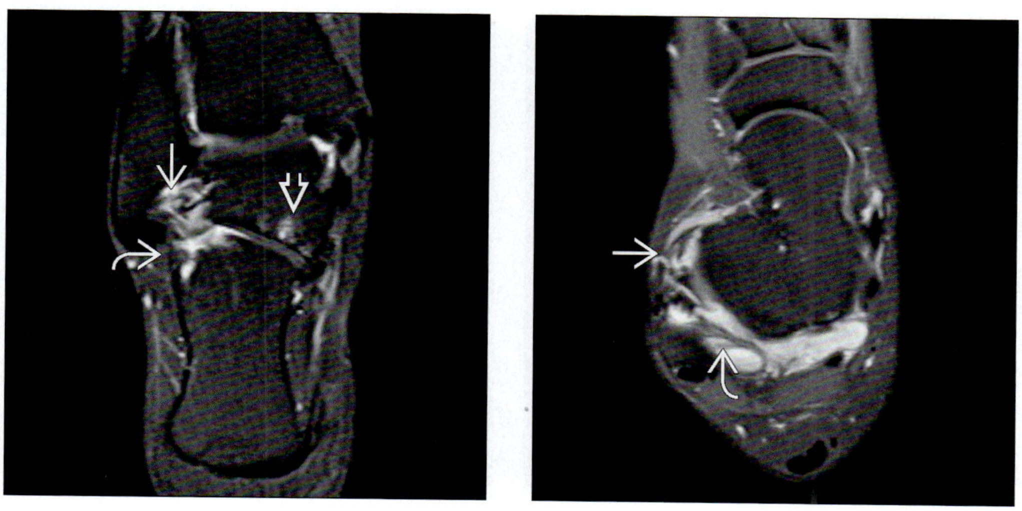

(Left) *Coronal T2WI FS MR shows an unusual pattern, with tear of the PTFL ➔ but relatively normal appearance of the CFL ↪. This is an unusual pattern suggesting a different injury mechanism in this case an eversion injury. Note osteochondral injury of the talus at the subtalar joint ⇨.* ***(Right)*** *Axial PDWI FS MR shows an ATFL tear ➔ and CFL tear ↪. This patient had chronic instability. It is still incompletely understood why some patients develop chronic instability after an ankle sprain.*

DELTOID LIGAMENT INJURY

Key Facts

Imaging

- Routine or stress radiographs used in acute setting
  - Widened medial clear space > 6 mm
- MR usually reserved for chronically unstable hindfoot
  - May see ligament discontinuity or thickened, lax fibers
- Deep deltoid best seen on coronal images
- Superficial deltoid origin best seen on axial images
- Tibiospring, tibiocalcaneal fibers best seen on coronal images

Top Differential Diagnoses

- Ankle sprain
- Fracture of medial malleolus
- Spring ligament injury
- Posterior tibial tendon tear & tendinopathy

Clinical Issues

- Pronation injury or chronic instability
- Medial ankle pain and tenderness
- Anteromedial impingement with chronic tear and scar formation

Diagnostic Checklist

- Term "deltoid ligament" often used to indicate deep component only, ignoring superficial component
  - To avoid confusion, make sure your reporting is in agreement with practice of your referring clinicians
- Deep deltoid ligament normally has multiple separate fascicles
  - Do not mistake striations for sprain
- Superficial deltoid fibers merge with spring ligament; include this entire region in MR search pattern

*(**Left**) Coronal graphic shows the most common pattern of deltoid ligament injury. The deep deltoid ligament is torn in its midsubstance, and there is an associated tear of the distal tibiofibular syndesmotic ligaments. This pattern occurs due to pronation injury. (**Right**) Graphic of the medial side of the ankle shows partial tear of the tibiocalcaneal and posterior tibiotalar bands of the superficial deltoid ligament. This pattern of partial tear is very unusual.*

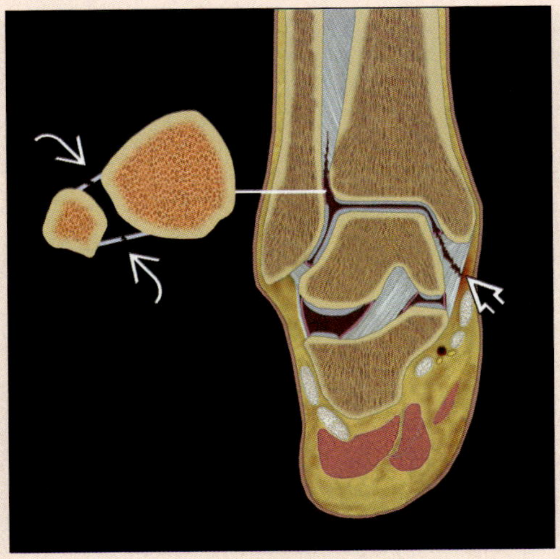

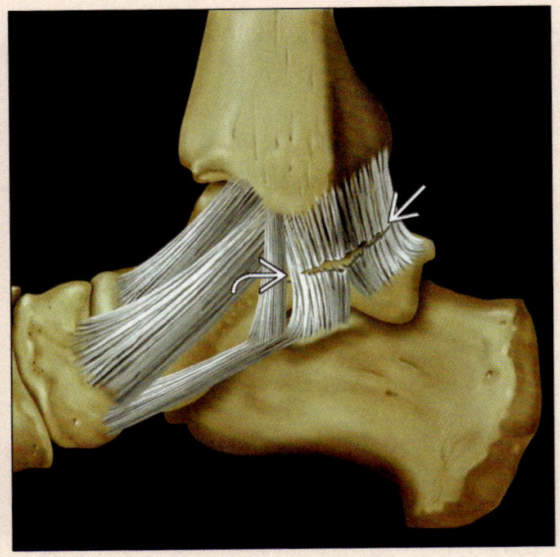

*(**Left**) Graphic shows partial tear of the superficial deltoid ligament, viewed posteriorly. The deep deltoid ligament, arising from the deep surface of the medial malleolus, is not shown. (**Right**) AP radiograph shows widening of the medial clear space of the ankle, which measures > 6 mm, indicating deep deltoid ligament tear. However, radiographs can be falsely negative. There is also a Weber type C fracture of the distal fibula.*

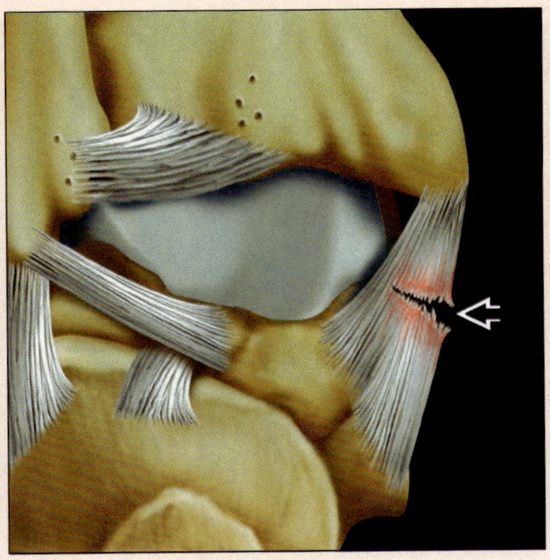

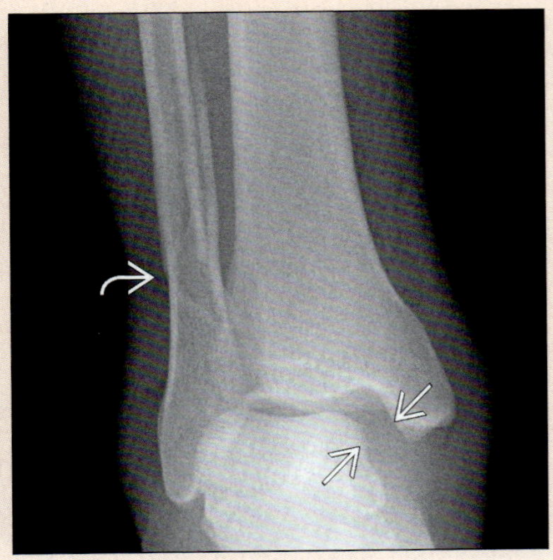

DELTOID LIGAMENT INJURY

TERMINOLOGY

Definitions

- Medial clear space: Radiographic interval between medial malleolus and medial talus

IMAGING

General Features

- Best diagnostic clue
  - Radiograph: Widened medial clear space > 6 mm on routine or stress views
  - MR: Coronal images show loss of normal fibers from medial talus to medial malleolus

Radiographic Findings

- Routine radiographs
  - Ankle joint effusion
  - Widening of medial clear space of ankle joint > 6 mm
  - Lateral shifting of talus relative to fibular notch of tibia
    - Medial, lateral, and superior ankle clear spaces should all be equal on mortise view
  - Valgus alignment of talus in mortise joint
  - Medial soft tissue swelling
    - Swelling may be present laterally as well, usually centered at syndesmosis
  - Presence of Weber C fracture raises suspicion for deltoid injury even if alignment is normal
- Valgus stress or weight-bearing radiography
  - Both methods increase sensitivity of radiographic signs of deltoid injury
  - May be false-negative
  - Valgus stress radiographs more reliable if patient is under anesthesia

CT Findings

- Same findings as for radiographs
- Hematoma in location of deltoid ligament

MR Findings

- Discontinuity of fibers
  - Deep component from deep margin of medial malleolus to concavity of medial talus
    - Midsubstance tear or avulsion from talus
  - Common origin on superficial surface of medial malleolus
    - May see fluid between medial malleolus and deltoid fibers
  - Distal superficial deltoid discontinuity
- Thickened, lax fibers
- Deep deltoid best seen on coronal images
- Tibiospring, tibiocalcaneal, posterior tibiotalar bands of superficial deltoid best seen on coronal images
- Tibionavicular, anterior tibiotalar bands of superficial deltoid best seen on axial images
- Common origin of anterior superficial deltoid best seen on axial images

Imaging Recommendations

- Best imaging tool
  - Radiographs and stress radiographs used in acute setting
  - MR usually reserved for chronically unstable hindfoot
- Protocol advice
  - MR: Coronal and axial fluid-sensitive sequences

DIFFERENTIAL DIAGNOSIS

Ankle Sprain

- Injury to lateral collateral ligaments of ankle
- Due to inversion injury

Fracture of Medial Malleolus

- More common than deltoid ligament injury

Spring Ligament Injury

- Pain more anterior, in midfoot

Posterior Tibial Tendon Tear & Tendinopathy

- Present with medial ankle pain, usually chronic

PATHOLOGY

General Features

- Etiology
  - Most commonly due to pronation injury
  - May also occur due to twisting injury
- Associated abnormalities
  - Syndesmotic ligament injury or medial malleolus fracture
    - Often seen with significant injury to deep deltoid
  - Posterior malleolus fracture
    - Reflects associated injury of syndesmosis
    - Due to avulsion of posterior inferior tibiofibular ligament
  - Distal fibular fracture
    - Weber type C or less commonly Weber type B
    - Not invariably seen, however, even with Weber type C fracture
    - Stress radiographs under anesthesia often performed during open reduction internal fixation (ORIF) of distal fibular fracture
      - In order to assess integrity of deltoid ligament
  - Maisonneuve fracture
    - Fracture of proximal fibula due to pronation injury
    - Pronation force ruptures deltoid ligament, separates tibia and fibula, ruptures syndesmotic ligaments
    - Force continues proximally to proximal fibula, gives oblique fracture
  - Posterior tibial tendinopathy/tear
    - May occur chronically, due to medial ankle instability from untreated deltoid ligament tear
  - Flexor retinaculum tear
    - Look for avulsion fracture on radiographs
    - Thin flake of bone in distal, medial tibial metaphysis
  - Spring ligament injury
    - Associated with tear of anterior, superficial deltoid ligament
  - Lateral collateral ligament injury
    - Most commonly associated with tears of superficial deltoid
      - Due to rotational component of injury

DELTOID LIGAMENT INJURY

- Also associated with deep deltoid ligament injury
  - Patients with repeated ankle injuries often have both supination and pronation episodes
- Ankle dislocation
  - If dislocation without medial malleolar fracture, deltoid is almost certainly torn

Relevant Anatomy

- Pronation: Complex foot motion consisting of eversion, abduction, and rotation, resulting in depression of medial arch
- Supination: Complex foot motion consisting of inversion, adduction, and rotation, resulting in elevation of medial arch
- Deltoid ligament divided into deep and superficial components
  - Deep deltoid
    - Stabilizes ankle from pronation and abduction forces
    - Considered most important portion of ligament
    - Multiple, separate well-defined bands
      - Similar to appearance of anterior cruciate ligament of knee
    - Medial attachment: Deep surface of anterior colliculus of medial malleolus
    - Lateral attachment: Concavity at medial border of talus
      - About 2/3 of concavity normally filled by deltoid fibers
      - Talar attachment slightly anterior, inferior to medial malleolar attachment
  - Superficial deltoid
    - Stabilizes ankle from rotational forces; importance is debated
    - Bands are named for their attachments
    - Common origin from medial malleolus
      - Should completely cover superficial margin of medial malleolus
    - Anterior insertions: Talar neck, navicular, sustentaculum tali, spring ligament
      - Tibiospring ligament merges with superomedial band of spring ligament
    - Posterior insertions: Posterior margin of talar body
- Medial clear space = medial ankle gutter
  - Distance from lateral margin of medial malleolus to medial margin of talus
    - Normal < 4 mm
    - Equivocal 4-6 mm
    - Abnormal > 6 mm
    - Measurement valid on AP or mortise view
    - All ankle spaces (medial, superior, and lateral) should be equal on mortise view

CLINICAL ISSUES

Presentation

- Most common signs/symptoms
  - Medial ankle pain and tenderness
- Other signs/symptoms
  - Often present late, with medial instability
  - Not uncommon for patients to be unstable both medially and laterally

Natural History & Prognosis

- Instability
- Osteoarthritis
- Anteromedial impingement
  - Due to scar from anterior, superficial deltoid fibers

Treatment

- Conservative management
  - Deltoid ligament injury per se usually treated with immobilization
  - Associated syndesmosis injury or malleolar fracture often treated with ORIF
- Surgical management
  - Deep deltoid rarely repaired, left to "scar in" while concomitant injuries may be repaired
  - Superficial deltoid may be repaired with sutures or anchors

DIAGNOSTIC CHECKLIST

Consider

- Term "deltoid ligament" often used to indicate deep component only, ignoring superficial component
  - To avoid confusion, make sure your reporting is in agreement with practice of your referring clinicians
- Associated syndesmosis injury

Image Interpretation Pearls

- Deep deltoid ligament normally has multiple separate fascicles giving striated appearance
  - Do not mistake striations for sprain
- Superficial deltoid fibers merge with spring ligament; include this entire region in MR search pattern

SELECTED REFERENCES

1. Cheung Y et al: MRI of isolated distal fibular fractures with widened medial clear space on stressed radiographs: which ligaments are interrupted? AJR Am J Roentgenol. 192(1):W7-12, 2009
2. Mengiardi B et al: Medial collateral ligament complex of the ankle: MR appearance in asymptomatic subjects. Radiology. 242(3):817-24, 2007
3. Hintermann B et al: Deltoid ligament injuries: diagnosis and management. Foot Ankle Clin. 11(3):625-37, 2006
4. Hintermann B et al: Medial ankle instability: an exploratory, prospective study of fifty-two cases. Am J Sports Med. 32(1):183-90, 2004
5. Hintermann B: Medial ankle instability. Foot Ankle Clin. 8(4):723-38, 2003
6. Miller CD et al: Deltoid and syndesmosis ligament injury of the ankle without fracture. Am J Sports Med. 23(6):746-50, 1995
7. Rasmussen O et al: Deltoid ligament. Functional analysis of the medial collateral ligamentous apparatus of the ankle joint. Acta Orthop Scand. 54(1):36-44, 1983
8. Beals TR et al: Deltoid ligament injuries in athletes: Techniques of repair and reconstruction. Operative Techniques in Sports Medicine, in press

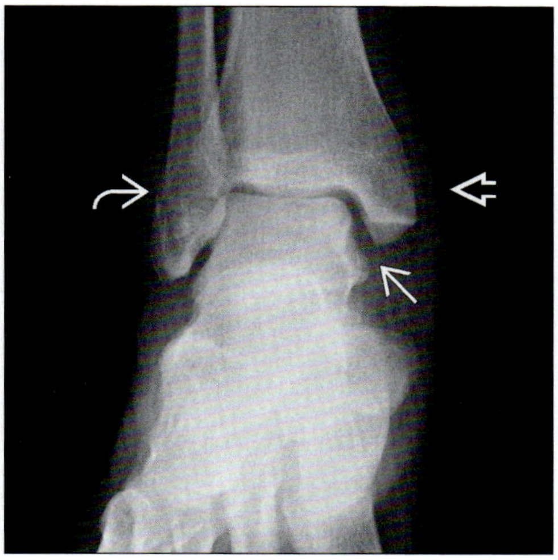

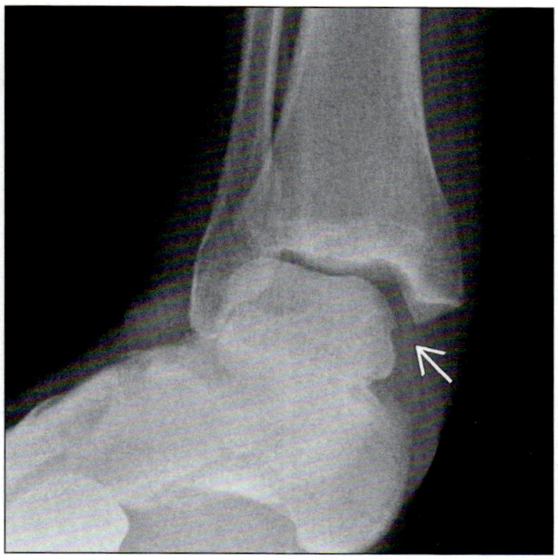

(Left) *AP radiograph shows Weber B fracture of fibula with more medial soft tissue swelling than expected, suggesting additional medial ligament injury. In addition, the medial ankle gutter is slightly wider inferiorly than superiorly, although overall mortise width is normal.* ***(Right)*** *AP radiograph in the same patient with eversion stress shows widening of medial gutter and external rotation of talus, confirming deltoid ligament injury.*

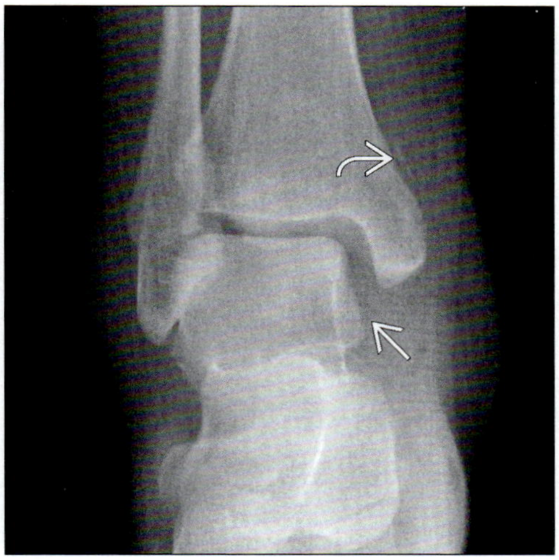

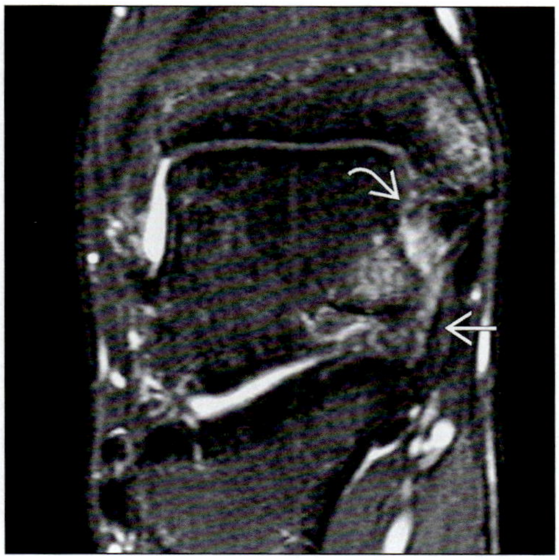

(Left) *AP radiograph shows medial gutter widening even on non-weight-bearing radiograph, indicating deltoid ligament rupture. Note severe medial soft tissue swelling and avulsion fracture of flexor retinaculum insertion.* ***(Right)*** *Coronal T2WI FS MR shows avulsion of deep deltoid ligament from its insertion on the concave portion of the medial talar body. The tibiocalcaneal band of the superficial component appears intact.*

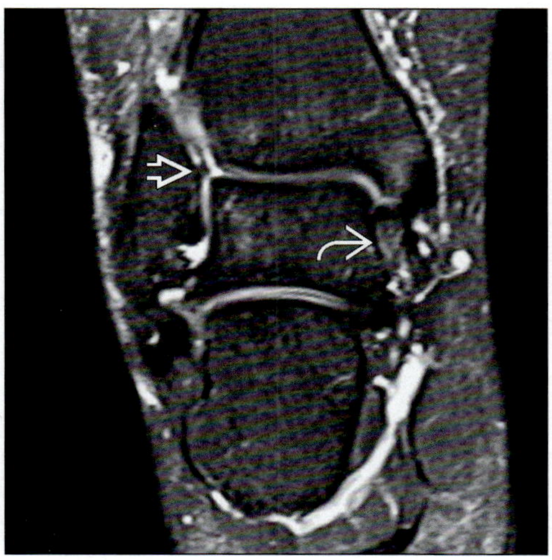

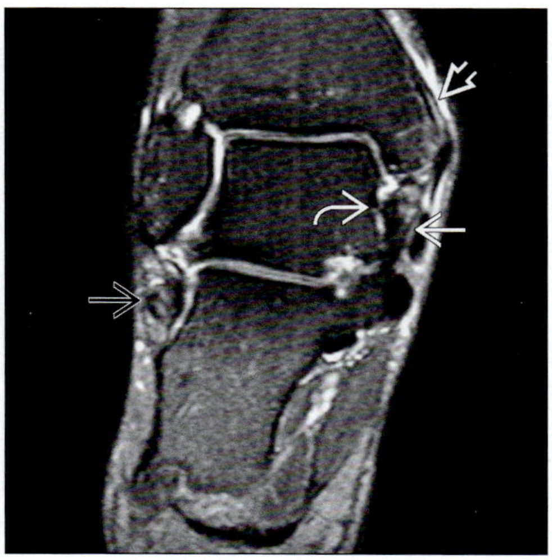

(Left) *Coronal T2WI FS MR shows sprain of deep deltoid fibers with loss of definition of the fiber bundles. There is an associated tear of the syndesmotic ligament.* ***(Right)*** *Coronal T2WI FS MR shows avulsion of the deep deltoid ligament from the talus, stripping of the origin of the superficial deltoid, and lax wavy tibiocalcaneal band of the superficial deltoid. Peroneal tendinopathy is also present, probably due to altered gait.*

DELTOID LIGAMENT INJURY

(Left) *Coronal T2WI FS MR shows acute deep deltoid ligament injury and adjacent bone bruise. There is partial detachment of the origin of superficial deltoid fibers.* *(Right)* *AP radiograph in the same patient, who returned 4 years later (next image) complaining of chronic pain and "giving way." There is interval development of medial ankle osteoarthritis. Heterotopic ossification is present at the site of superficial deltoid avulsion. Loose body is present laterally.*

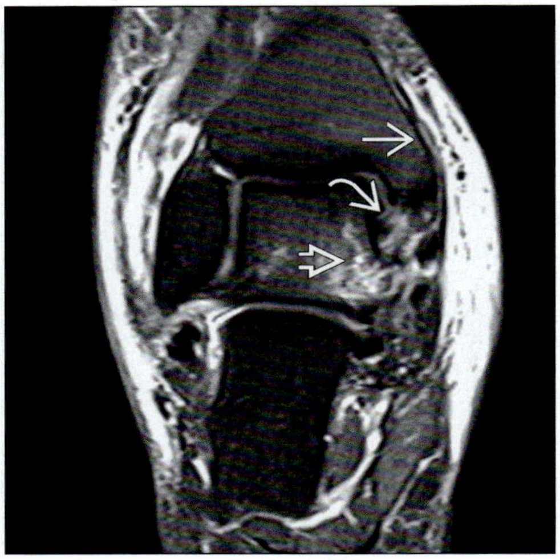

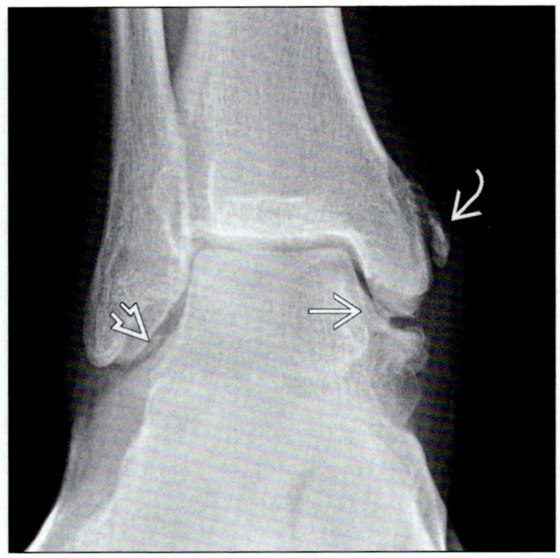

(Left) *Axial PDWI FS MR shows sprain of deep deltoid and anterior superficial deltoid. Because of the oblique orientation of the deep deltoid fibers, injuries are more reliably identified in the coronal than the axial plane on nonarthrographic MR.* *(Right)* *Axial PDWI FS MR shows complete avulsion of deep deltoid and thickened, amorphous anterior fibers of superficial deltoid.*

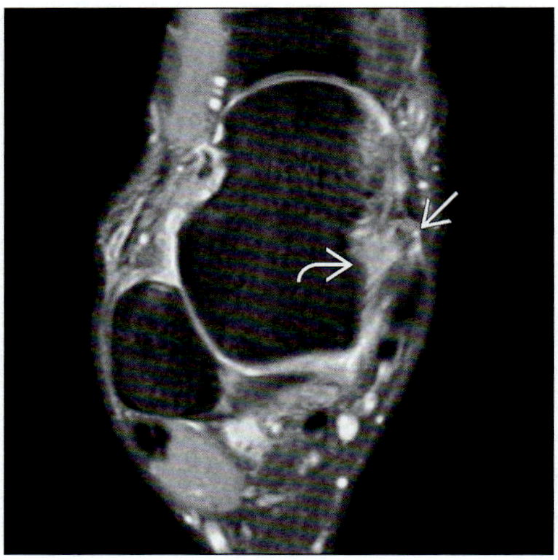

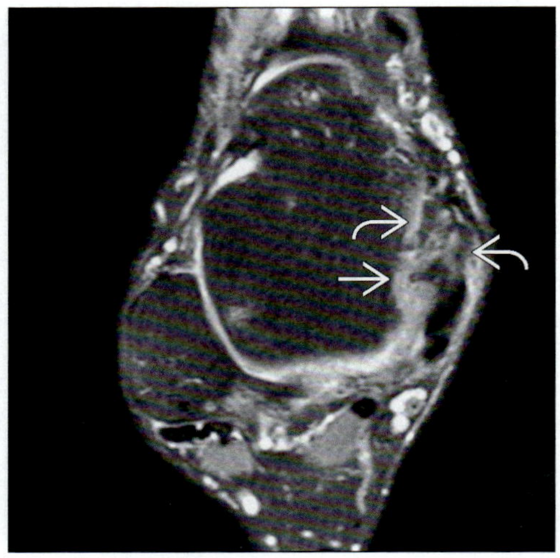

(Left) *Axial MR arthrogram T1WI FS shows tears of anterior fibers of superficial deltoid ligament as well as the deep deltoid. Contrast extends into the talonavicular joint due to tears of both the superficial deltoid and ATFL. Fluid around peroneal tendons indicates calcaneofibular ligament tear.* *(Right)* *Coronal MR arthrogram T1WI FS in the same patient shows thick, amorphous, torn anterior tibiotalar and talonavicular bands. Tibiospring band is intact.*

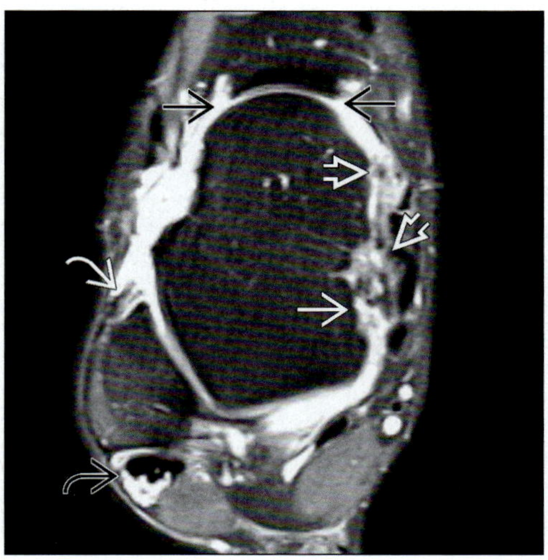

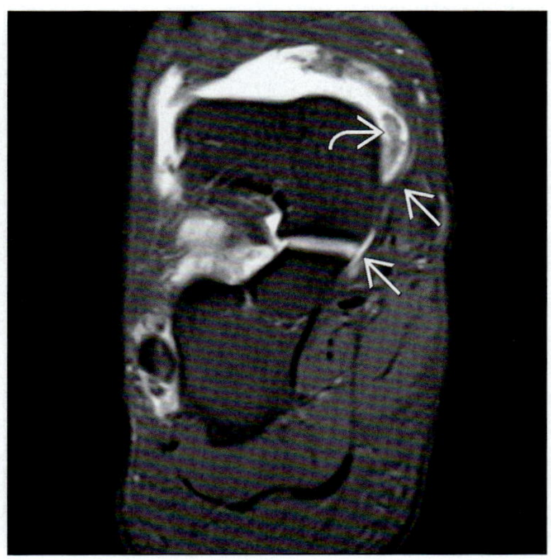

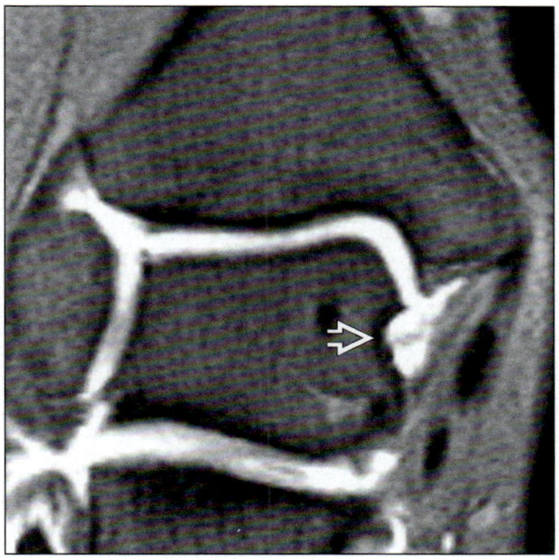

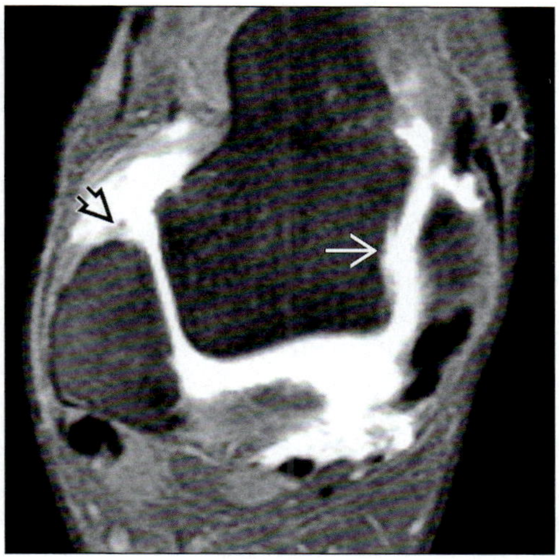

(Left) *Coronal MR arthrogram T1WI FS shows complete absence of deltoid fibers at normal attachment to medial talus. This patient had multiple ligament injuries and instability in both pronation and supination.* ***(Right)*** *Axial MR arthrogram T1WI FS in the same patient shows complete absence of deep deltoid attachment to the talus. This patient also has a rupture of the anterior talofibular ligament (ATFL).*

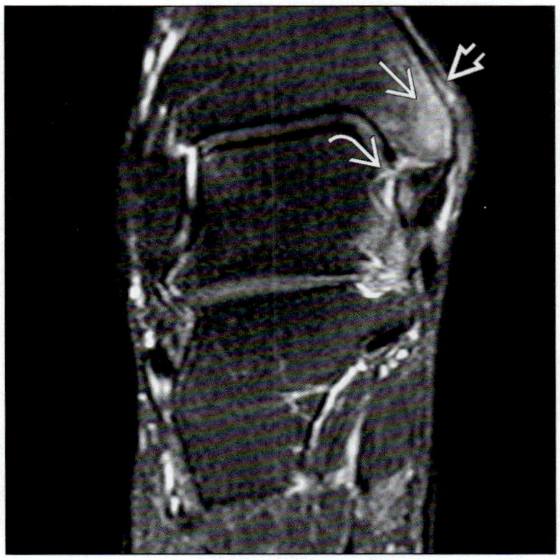

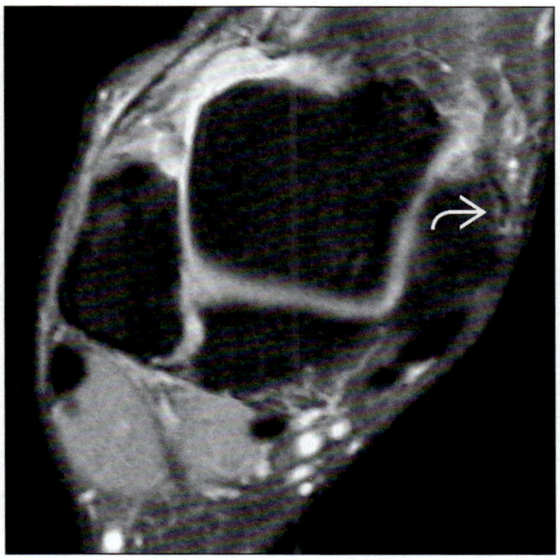

(Left) *Coronal T2WI FS MR shows torn deep deltoid fibers, bone marrow edema in medial malleolus, and avulsion of origin of superficial deltoid fibers.* ***(Right)*** *Axial T2WI FS MR shows avulsion of anterior, superficial deltoid origin in a patient whose deep deltoid fibers were intact. This pattern is often associated with rotational instability.*

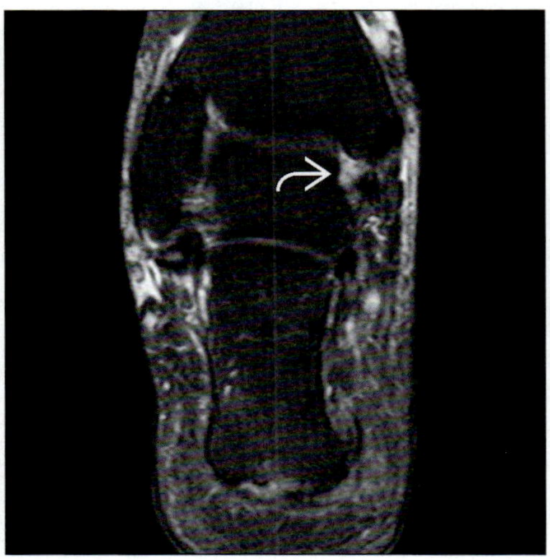

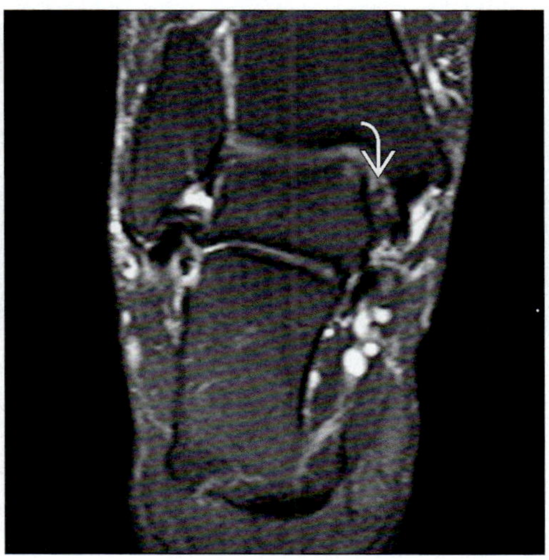

(Left) *Coronal T2WI FS MR shows complete rupture of posterior tibiotalar band of deltoid ligament. These fibers arise from the posterior colliculus of the medial malleolus.* ***(Right)*** *Coronal T2WI FS MR shows sprain of posterior tibiotalar fibers of deltoid ligament. Mild sprain such as this is probably not functionally significant.*

SPRING LIGAMENT INJURY

Key Facts

Terminology
- Synonym: Plantar calcaneonavicular ligament
- Spring ligament divided into 2 bands
  - Superomedial band: From sustentaculum tali to navicular
  - Inferior band: From body of calcaneus to navicular

Imaging
- Superomedial band more commonly affected, due to chronic stress and degeneration
  - Seen deep to posterior tibial tendon at level of sustentaculum tali insertion
  - Proximally merges with tibiospring ligament, which is often also torn
  - Torn ligament lax, wavy, &/or discontinuous
  - Best seen on coronal images
- Inferior band
  - Discontinuity best seen on sagittal images
  - Axial images difficult to interpret because of partial volume artifact

Top Differential Diagnoses
- Posterior tibial tendinopathy or tear
- Deltoid ligament sprain
- Navicular stress fracture

Clinical Issues
- Associated with posterior tibial tendinopathy or tear, deltoid ligament injury
- Usually degenerative phenomenon
- Rarely due to acute trauma
- Most common in overweight, middle-aged women with preexisting flatfoot deformity

Diagnostic Checklist
- High risk of spring ligament abnormality with chronic posterior tibial tendon tear

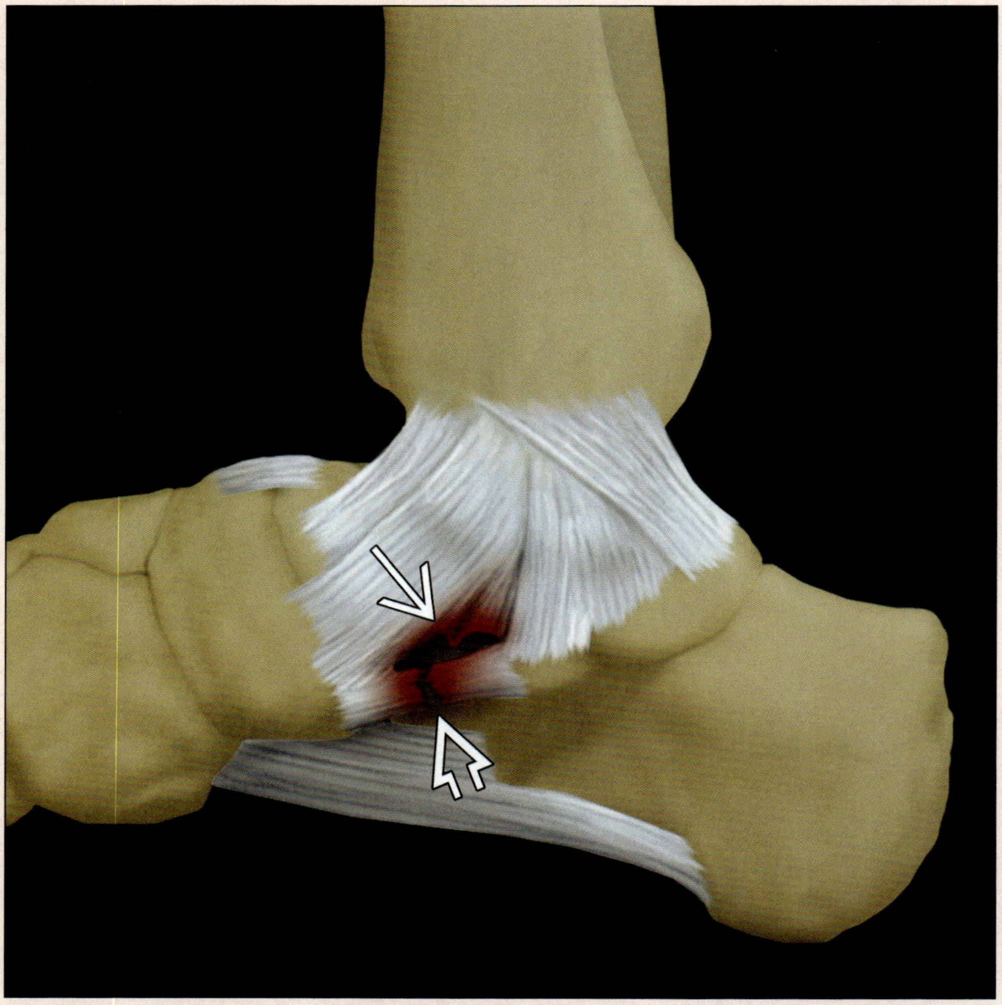

Graphic of the medial ankle shows a tear of the superomedial band of the spring ligament ➡ and the tibiospring band of the deltoid ligament ➡. The spring ligament merges superiorly with the tibiospring ligament, part of the superficial deltoid ligament complex. At MR, the tibiospring and spring ligaments appear continuous. The spring ligament lies deep to the posterior tibial tendon (not shown), and tears of the spring ligament are strongly associated with posterior tibial tendon abnormalities.

SPRING LIGAMENT INJURY

TERMINOLOGY

Synonyms

- Plantar calcaneonavicular ligament

IMAGING

General Features

- Best diagnostic clue
  - Ligament irregular, redundant, &/or discontinuous on MR
- Location
  - Superomedial band more commonly affected, due to chronic stress and degeneration
  - Inferior band may be torn with trauma

MR Findings

- Superomedial band
  - Seen deep to posterior tibial tendon at level of sustentaculum tali insertion
  - Proximally merges with tibiospring ligament, which is often also torn
  - Torn ligament appears lax, wavy, &/or discontinuous
- Medial plantar oblique, inferoplantar oblique
  - Striated appearance
  - Fluid between bands common, due to distension of talonavicular joint

Imaging Recommendations

- Best imaging tool
  - MR
- Protocol advice
  - FOV to include from tibiotalar joint to tarsometatarsal joints
    - Allows evaluation of commonly associated abnormalities
  - Axial, coronal, and sagittal fluid-sensitive sequences
  - Sagittal, coronal T1 or PD sequences for anatomy
  - Because of small size of ligament, need to carefully cross-check between imaging planes

DIFFERENTIAL DIAGNOSIS

Posterior Tibial Tendon (PTT) Tear & Tendinopathy

- Spring ligament laxity may occur secondary to advanced PTT dysfunction

Deltoid Ligament Sprain

- Also presents with medial ankle pain

Navicular Stress Fracture

- Vague, ill-defined midfoot pain

PATHOLOGY

General Features

- Associated abnormalities
  - Posterior tibial tendinopathy/tear
  - Superficial deltoid ligament injury
  - Flatfoot deformity
  - Obesity

Relevant Anatomy

- Spring ligament divided into 3 bands
  - Superomedial band: From medial surface of sustentaculum tali to navicular
    - Merges superiorly with tibiospring ligament (band of superficial deltoid ligament)
    - Deep to posterior tibial tendon near its insertion
  - Inferoplantar oblique (IPO): From body of calcaneus to navicular
  - Medioplantar oblique (MPO): From anterior margin of sustentaculum to navicular
  - Talonavicular joint recess between IPO and MPO

CLINICAL ISSUES

Presentation

- Most common signs/symptoms
  - Painful, flatfoot deformity

Demographics

- Most common in overweight, middle-aged women with preexisting flatfoot deformity

Natural History & Prognosis

- Progressive flatfoot deformity, osteoarthritis if not addressed

DIAGNOSTIC CHECKLIST

Consider

- High risk of spring ligament abnormality with posterior tibial tendon tear

Image Interpretation Pearls

- Tear of superomedial band usually best seen on coronal images, deep to posterior tibial tendon

SELECTED REFERENCES

1. Melão L et al: Ligaments of the transverse tarsal joint complex: MRI-anatomic correlation in cadavers. AJR Am J Roentgenol. 193(3):662-71, 2009
2. Kavanagh EC et al: MRI of rupture of the spring ligament complex with talo-cuboid impaction. Skeletal Radiol. 36(6):555-8, 2007
3. Subhas N et al: Diagnosis: isolated spring ligament tear demonstrated on magnetic resonance imaging. Orthopedics. 30(1):1, 70-2, 2007
4. Mengiardi B et al: Spring ligament complex: MR imaging-anatomic correlation and findings in asymptomatic subjects. Radiology. 237(1):242-9, 2005
5. Toye LR et al: MRI of spring ligament tears. AJR Am J Roentgenol. 184(5):1475-80, 2005
6. Balen PF et al: Association of posterior tibial tendon injury with spring ligament injury, sinus tarsi abnormality, and plantar fasciitis on MR imaging. AJR Am J Roentgenol. 176(5):1137-43, 2001
7. Yao L et al: MR imaging findings in spring ligament insufficiency. Skeletal Radiol. 28(5):245-50, 1999
8. Davis WH et al: Gross, histological, and microvascular anatomy and biomechanical testing of the spring ligament complex. Foot Ankle Int. 17(2):95-102, 1996
9. Rule J et al: Spring ligament of the ankle: normal MR anatomy. AJR Am J Roentgenol. 161(6):1241-4, 1993

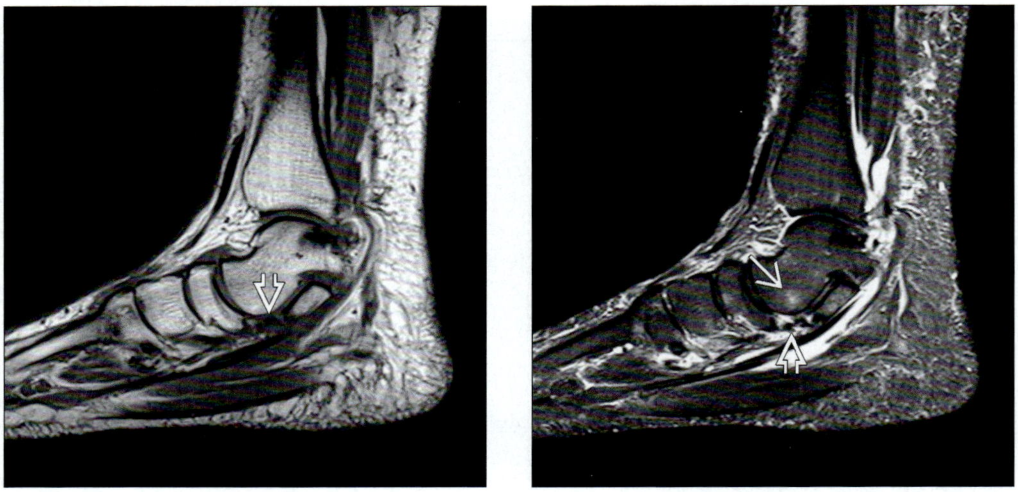

(Left) *Sagittal T1WI MR shows discontinuity of the superomedial band of spring ligament ⇨. The ligament should always be checked in all 3 planes, to exclude partial volume artifact.* ***(Right)*** *Sagittal STIR MR in the same patient confirms tear of the ligament ⇨. Ligament is lax and redundant. Bone marrow edema in the head of the talus ➡ is attributable to abnormal motion due to loss of support of the talar head by the spring ligament.*

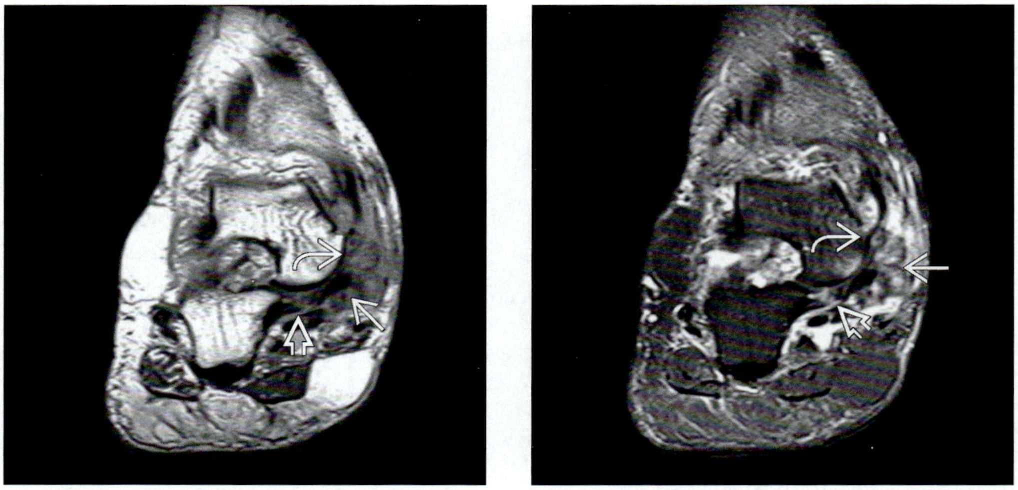

(Left) *Coronal PDWI MR in the same patient shows the thickened, heterogeneous medial band of spring ligament ⇨ merging with a poorly defined and torn tibiospring ligament ↪. The posterior tibial tendon ➡ is surrounded by amorphous soft tissue.* ***(Right)*** *Coronal T2WI FS MR in the same patient shows that the spring ligament ⇨ and tibiospring ligament ↪ are lax and discontinuous. Severe posterior tibial tenosynovitis ➡ is present.*

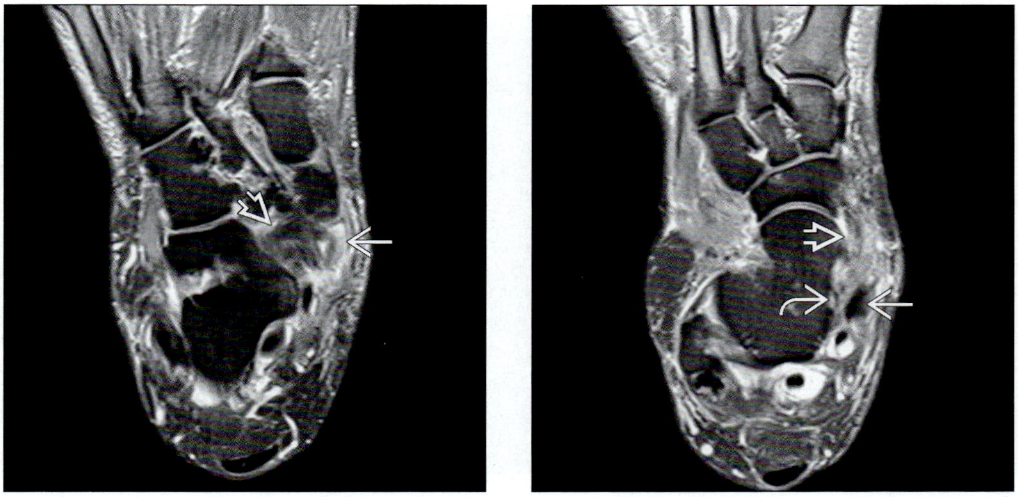

(Left) *Axial PDWI FS MR shows an intact inferoplantar oblique band ⇨ of spring ligament. Fluid ➡ is seen medially, and the supermedial band is not visible.* ***(Right)*** *Axial PDWI FS MR more superiorly in the same patient shows a tear of the superomedial band ⇨ of the spring ligament, tear of tibiospring ligament ↪, and posterior tibial tendinopathy ➡.*

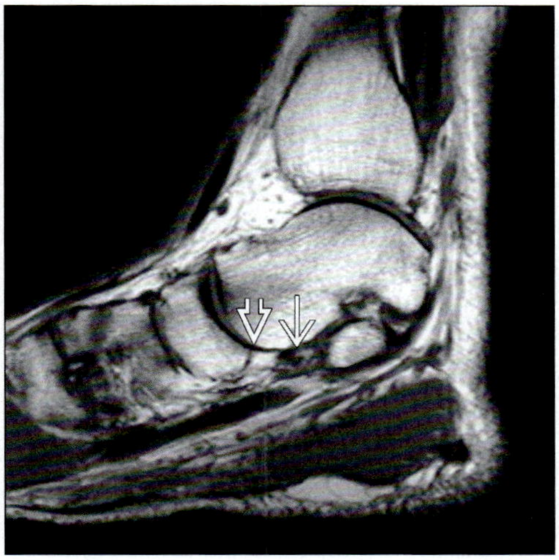

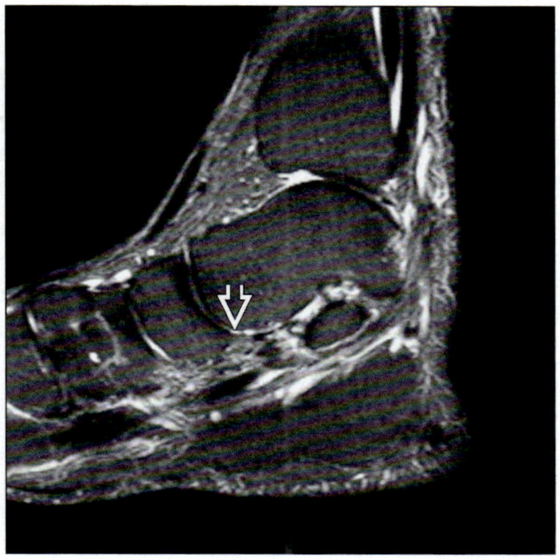

(Left) *Sagittal T1WI MR shows redundant, discontinuous spring ligament ➔. Expected site of ligament attachment to navicular is shown ➔.* ***(Right)*** *Sagittal STIR MR in the same patient shows amorphous, intermediate signal intensity material ➔ near the expected attachment of the spring ligament to the navicular. No normal ligament fibers are visible. Since spring ligament tears are usually chronic, scar tissue is commonly evident.*

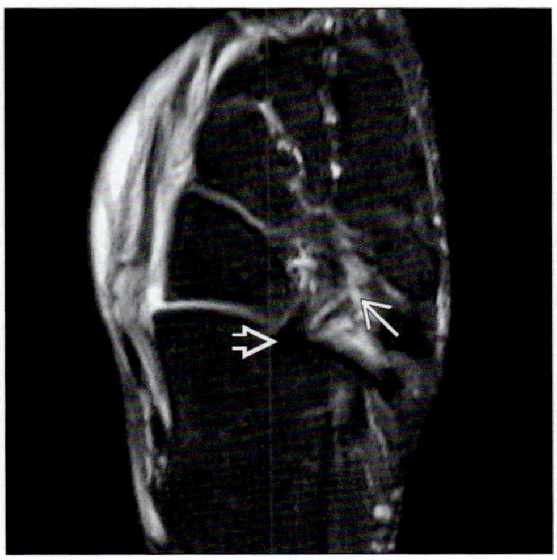

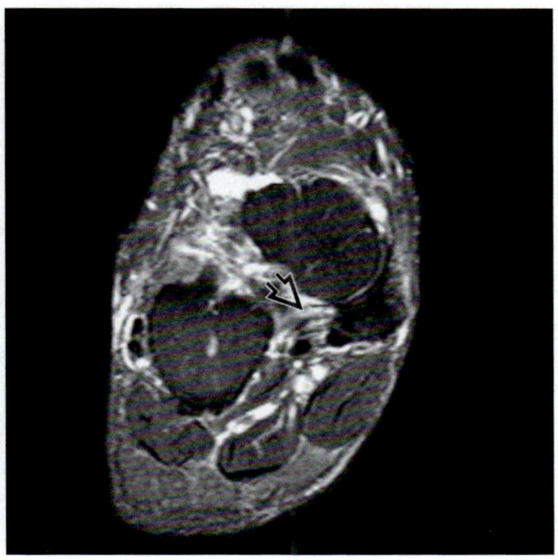

(Left) *Axial PDWI FS MR in a young patient with a traumatic tear of spring ligament shows lateral band fibers arising normally from the calcaneus ➔ but becoming discontinuous ➔ near their attachment to the navicular.* ***(Right)*** *Coronal T2WI FS MR in the same patient confirms discontinuity of fibers ➔ near their navicular attachment.*

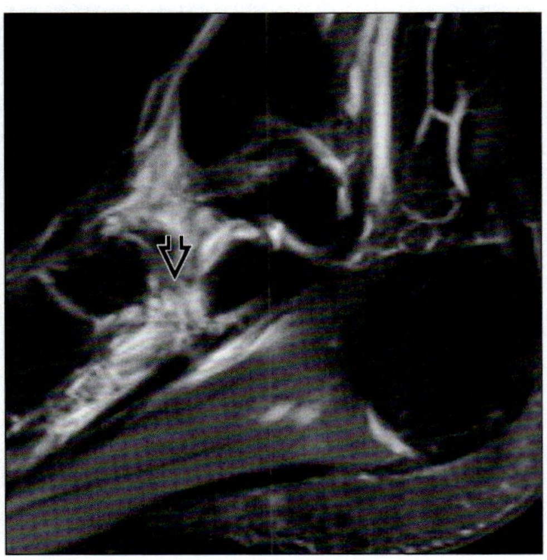

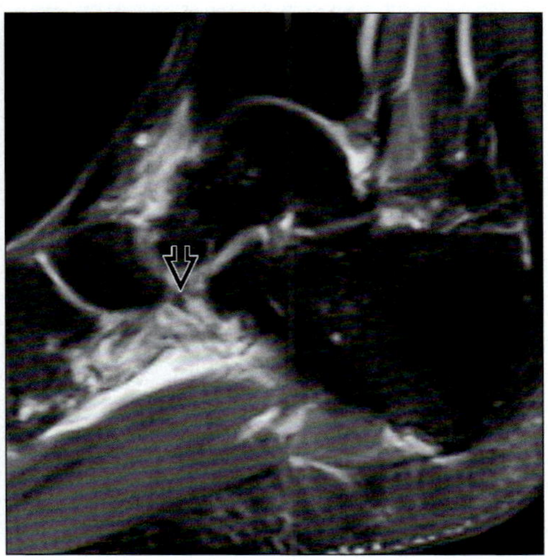

(Left) *Sagittal STIR MR in a young patient with traumatic injury to the spring ligament shows fluid in the expected location of the spring ligament ➔.* ***(Right)*** *Sagittal STIR MR more laterally in the same patient shows the tear of the inferoplantar oblique band ➔.*

SUPERIOR PERONEAL RETINACULUM AVULSION

Key Facts

Terminology

- Superior peroneal retinaculum: Fibrous band maintaining position of peroneal tendons posterior to fibula

Imaging

- Radiography
  - Cortical bone flake avulsion fracture at lateral margin of distal fibula
  - Soft tissue swelling centered above lateral malleolus
  - Periosteal reaction at lateral margin of distal fibula
- CT and MR
  - Detachment of retinaculum from fibula
  - Edema centered on retinaculum
  - Subluxation of peroneal tendon(s)
- Dynamic ultrasound shows tendon subluxation when foot is everted, dorsiflexed

Top Differential Diagnoses

- Peroneal tendon tear
- Peroneal tenosynovitis
- Lateral collateral ligament injury
- Lateral malleolus fracture

Pathology

- Associated abnormalities
  - Peroneal tendon subluxation, tenosynovitis
  - Hypoplastic fibular groove
  - Peroneus quartus
  - Fractures of calcaneus or medial malleolus

Clinical Issues

- Lateral ankle pain
- Subluxation of peroneal tendons
- Subjective instability on uneven ground

__(Left)__ Lateral view graphic with axial cross-section inset shows the peroneus longus → and peroneus brevis → tendons held in place by the superior peroneal retinaculum →. __(Right)__ AP radiograph shows avulsion of the fibular attachment of the superior peroneal retinaculum → associated with fracture of lateral process of talus →. This combination of fractures is unusual, but both can occur with a supination-external rotation-type mechanism.

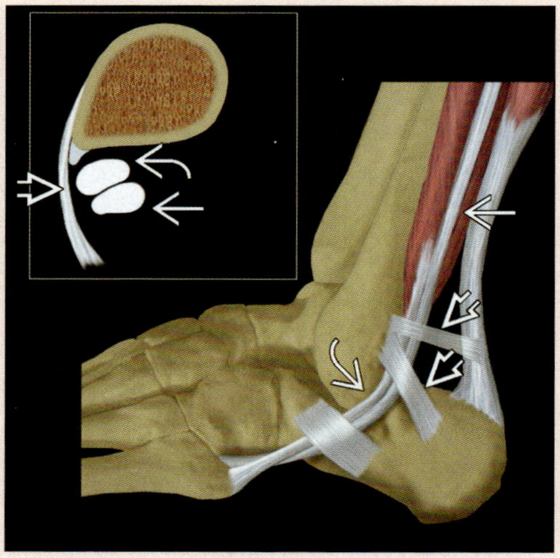

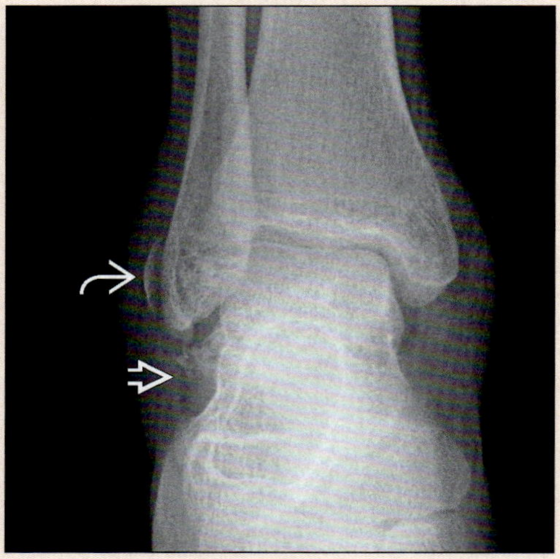

__(Left)__ Axial NECT obtained for evaluation of calcaneal fracture shows acute dislocation of the peroneus brevis →. Unfortunately, this was not recognized at the time of the injury, & the patient developed chronic peroneal tendon subluxation and tendinopathy. __(Right)__ Axial PDWI FS MR shows stripping of peroneal retinaculum →, which attaches anterior to its normal insertion. There is subluxation of the peroneus brevis tendon → around the lateral margin of the fibula.

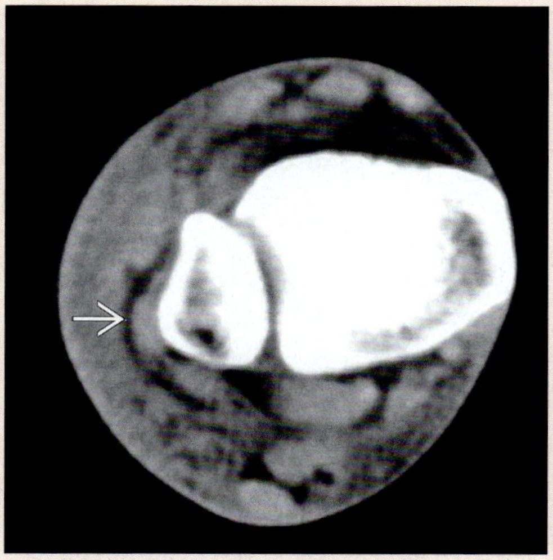

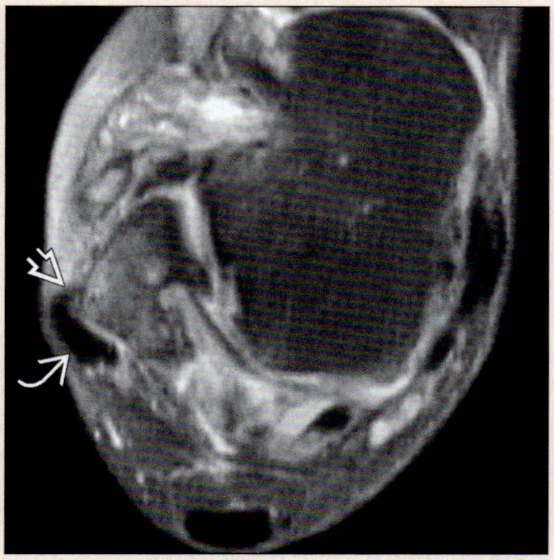

SUPERIOR PERONEAL RETINACULUM AVULSION

TERMINOLOGY

Definitions

- Superior peroneal retinaculum: Fibrous band maintaining position of peroneal tendons posterior to fibula

IMAGING

General Features

- Best diagnostic clue
  - Subluxation of tendons on MR or CT

Radiographic Findings

- Avulsion fracture of fibular attachment
  - Less common than purely soft tissue injury
  - Thin, vertically oriented flake of cortical bone
  - Posterolateral margin of distal fibular metaphysis
  - Better seen on mortise view than AP view
- Chronic injury
  - Periosteal reaction at lateral margin of distal fibula

CT Findings

- Insertion of retinaculum along lateral margin of fibula
  - Anterior to normal posterolateral attachment
- Poorly defined, thickened peroneal retinaculum
- Cortical avulsion fracture
  - ± entrapment of tendons between avulsion fracture and fibula
- Subluxation or dislocation of peroneal tendon(s)

MR Findings

- Abnormal position of retinaculum attachment
  - Attaches to lateral margin of fibula rather than posterolateral corner
  - Seen in partial tear or chronic tear with scar
  - This may be only sign of prior injury
- Complete detachment of retinaculum from fibula
  - Fluid visible between retinaculum and fibula
- Edema centered on retinaculum
- Thickened, poorly defined retinaculum
- Intermediate signal intensity in retinaculum
- Subluxation or dislocation of peroneal tendon(s)
  - Need not be present
  - Often transient, elicited by stress maneuvers
- Midsubstance tear of retinaculum (rare)
  - Discontinuity of retinaculum posterior to peroneal tendons
- Cortical fragments are difficult to see
  - Small, low-signal intensity cortical bone fragment blends with retinaculum
- Intermediate or high signal intensity periosteal reaction
- Fluid in tendon sheath (nonspecific)

Ultrasonographic Findings

- Dynamic ultrasound shows tendon subluxation when foot is everted & dorsiflexed

Imaging Recommendations

- Best imaging tool
  - MR
- Protocol advice
  - Seen best on axial fluid-sensitive sequences

DIFFERENTIAL DIAGNOSIS

Peroneal Tendon Tear

- Usually longitudinal split in tendon, rarely complete tear
- Seen in runners, other athletes
- May occur secondary to peroneal tendon subluxation

Peroneal Tenosynovitis

- Often associated with large peroneal tubercle of calcaneus
- May occur secondary to retinacular avulsion
  - Stenosing tenosynovitis: Extensive scarring of tendon sheath
- May occur secondary to inflammatory arthritis (e.g., rheumatoid)

Lateral Collateral Ligament Injury

- Avulsion fracture if present is at tip of fibula rather than lateral margin
- Due to avulsion of calcaneofibular ligament
- Soft tissue swelling is centered more inferiorly, at tip of fibula

Lateral Malleolus Fracture

- Also presents with twisting injury, lateral ankle pain
- Fracture traverses malleolus as opposed to avulsion of lateral margin occurring with retinacular injury

PATHOLOGY

General Features

- Associated abnormalities
  - Peroneal tendon subluxation
    - Often transient
  - Peroneal tendon tear
    - May occur secondary to tendon subluxation
    - Usually longitudinal split rather than full thickness tear
  - Convex margin of posterior fibula
    - Posterior surface of distal fibula normally flat
    - Convex posterior margin of fibula decreases stability of peroneal tendons
      - Increases stress on retinaculum, predisposes to injury
  - Peroneus quartus
    - Anomalous muscle posterior to peroneus longus and brevis predisposes to tendon subluxation
    - May increase stress on peroneal retinaculum
  - Calcaneus fracture
    - Retinaculum injured in association with intraarticular calcaneal fractures
    - Retinacular injury often unrecognized acutely, presents with chronic lateral pain
  - Lateral collateral ligament injury
    - Increases stress on peroneal retinaculum
  - Medial malleolar fracture
    - Twisting injury fractures medial malleolus, avulses peroneal retinaculum
  - Ankle dislocation
    - Retinacular tear may occur with dislocation in any direction

Staging, Grading, & Classification

- Oden classification

SUPERIOR PERONEAL RETINACULUM AVULSION

- Type I (most common)
  - Retinaculum partly stripped from lateral fibula
  - Lateral "pouch" between retinaculum and fibula
  - Tendons displace into pouch
- Type II
  - Retinaculum completely torn at fibula
- Type III
  - Small fibular fragment avulsed with retinaculum
  - Tendons may be entrapped between avulsed fragment and lateral margin of fibula
- Type IV (uncommon)
  - Retinaculum torn in midsubstance

Relevant Anatomy

- Peroneal retinaculum
  - Origin
    - Deep posterior aponeurosis (continuous with fascia of deep flexors)
    - Superficial posterior aponeurosis (continuous with fascia of Achilles tendon)
    - ± lateral margin of calcaneus
  - Insertion: Posterolateral margin of fibula
- Peroneal tendons at level of distal fibula
  - Peroneus longus
    - Tendinous, lies lateral to peroneus brevis
  - Peroneus brevis
    - Tendon and muscle both seen in axial cross section
  - Peroneus quartus variably present
- Distal fibula contour
  - Normally concave or flat
  - Convex contour may predispose to retinacular tear

CLINICAL ISSUES

Presentation

- Most common signs/symptoms
  - Lateral ankle pain
  - Sensation of popping or snapping at lateral ankle
  - Subluxation of peroneal tendons
    - Often transient
    - May be elicited on physical exam by
      - Dorsiflexion & eversion of foot
- Other signs/symptoms
  - Subjective instability on uneven ground
- Clinical profile
  - Often seen in runners as isolated injury

Natural History & Prognosis

- If untreated, high risk of chronic pain and tendon dysfunction

Treatment

- Conservative management with casting
  - High rate of treatment failure
  - Trend is to operative repair
- Surgical repair
  - Peroneal tendon symptoms may persist in chronic injury because of secondary tenosynovitis
  - Tendon sheath steroid injection a useful adjunct to surgery

DIAGNOSTIC CHECKLIST

Consider

- Be suspicious for retinacular tear in patients with subjective instability, peroneal tenosynovitis, or peroneal tendon tear

Image Interpretation Pearls

- Importance of fracture fragment on radiographs often overlooked
  - Located superior to avulsion of lateral collateral ligament
  - Prompt recognition of importance of fracture fragment changes treatment and outcomes
- Injury often overlooked on MR
  - Be alert to abnormal insertion of retinaculum along lateral margin of fibula

SELECTED REFERENCES

1. Kopp F: Traumatic peroneal dislocation with medial malleolus fracture: a case report. Foot Ankle Int. 29(7):756-8, 2008
2. Numkarunarunrote N et al: Retinacula of the foot and ankle: MRI with anatomic correlation in cadavers. AJR Am J Roentgenol. 188(4):W348-54, 2007
3. Ferran NA et al: Recurrent subluxation of the peroneal tendons. Sports Med. 36(10):839-46, 2006
4. Rosenberg ZS et al: MRI features of chronic injuries of the superior peroneal retinaculum. AJR Am J Roentgenol. 181(6):1551-7, 2003
5. Ho RT et al: Peroneal tendon dislocation: CT diagnosis and clinical importance. AJR Am J Roentgenol. 177(5):1193, 2001
6. Magnano GM et al: High-resolution US of non-traumatic recurrent dislocation of the peroneal tendons: a case report. Pediatr Radiol. 28(6):476-7, 1998
7. Murr S. Dislocation of the peroneal tendons with marginal fracture of the lateral malleolus. J Bone Joint Surg Br. 43:563-565, 1961

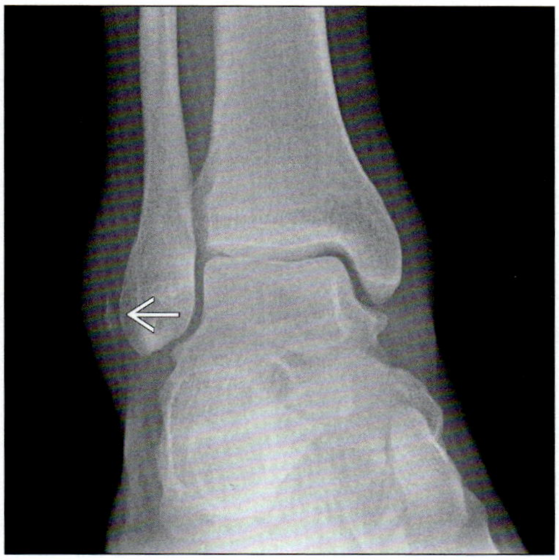

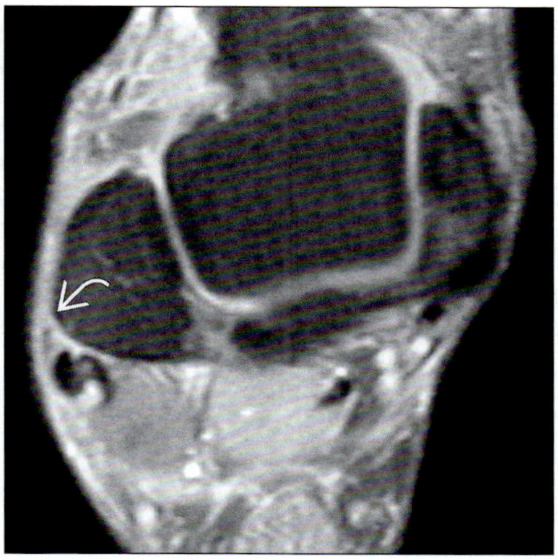

(Left) *Mortise view radiograph shows avulsion of the fibular attachment of the peroneal retinaculum and adjacent soft tissue swelling in an acute injury. The swelling is centered superior to the typical location seen in ankle sprains.* ***(Right)*** *Axial PDWI FS MR shows detachment of the peroneal retinaculum and edema superficial to the retinaculum. The patient complained of intermittent "snapping" at the lateral margin of the ankle.*

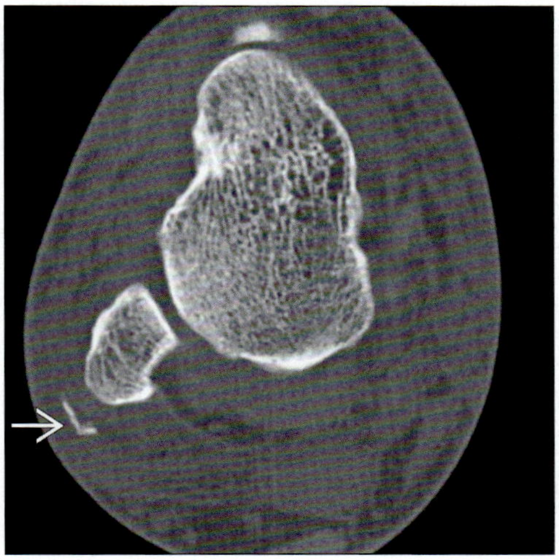

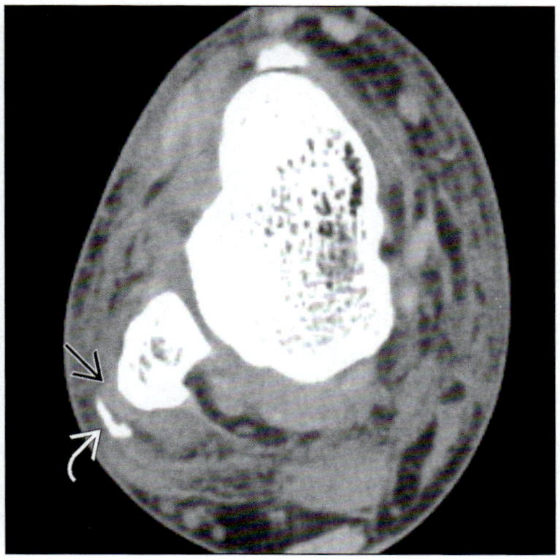

(Left) *Axial bone CT shows an avulsion fracture of the peroneal retinaculum. Prospectively, this was mistaken for avulsion of the calcaneofibular ligament, which attaches at the tip of the fibula.* ***(Right)*** *Axial soft tissue window at the same level in the same patient shows anterior subluxation of the peroneus brevis tendon, which is trapped between the avulsion fragment of the retinacular insertion and the fibula. This is an Oden type III injury.*

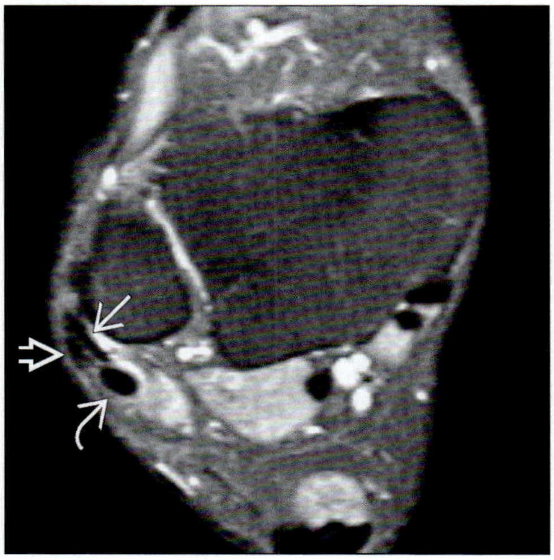

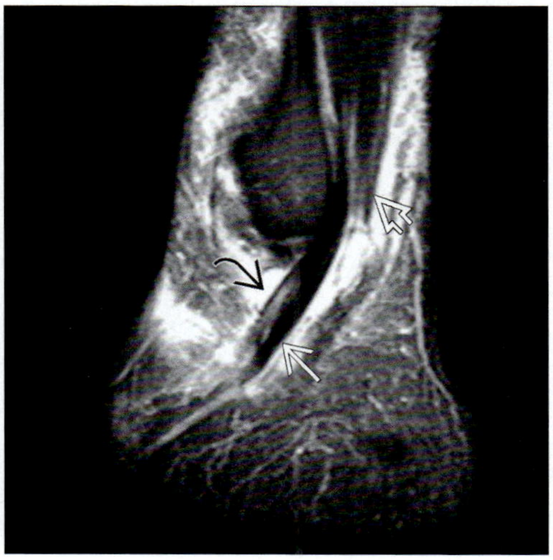

(Left) *Axial PDWI FS MR shows lateral subluxation of the peroneus brevis and peroneus longus. Peroneus quartus is also present. The presence of a peroneus quartus increases stress on flexor retinaculum and predisposes to subluxation.* ***(Right)*** *Sagittal STIR MR in the same patient shows peroneus quartus and subluxated peroneus longus and brevis. Longitudinal splitting of the brevis is attributable to the tendon instability.*

FLEXOR RETINACULUM AVULSION

Key Facts

Imaging

- Radiographs
  - Small flake of cortical bone avulsed from medial margin of distal tibial metaphysis (uncommon)
  - Focal periosteal reaction at medial margin of distal tibial metaphysis (chronic injury)
- CT, MR
  - Discontinuity of retinaculum at tibial attachment
  - Nonvisualization of retinaculum
  - Focal fluid adjacent to retinacular attachment
  - ± subluxation or dislocation of PTT
  - ± posterior tibial tendinopathy or tear

Top Differential Diagnoses

- Posterior tibial tendinopathy due to chronic stress
- Periostitis due to other causes
- Deltoid ligament injury

Pathology

- Usually isolated injury due to inversion, dorsiflexion
- Also seen with ankle dislocation or fracture

Clinical Issues

- Chronic injury: Patients present with medial ankle pain, PTT dysfunction

Diagnostic Checklist

- Acute injury may be missed on CT because of focus on bony injuries
- Retinaculum should always be visible on axial fluid-sensitive MR sequences and axial soft tissue algorithm CT
- PTT may be normally positioned or only slightly subluxated despite torn retinaculum

__(Left)__ Graphic shows ruptured flexor retinaculum. Retinacular injuries destabilize the posterior tibial tendon, leading to posterior tibial tendon dysfunction. The posterior tibial tendon should always lie lateral to the well-defined corner of the tibia where the retinaculum attaches; even slight medial displacement indicates retinacular injury. __(Right)__ Anteroposterior radiograph shows bony avulsion of flexor retinaculum attachment associated with distal fibular fracture.

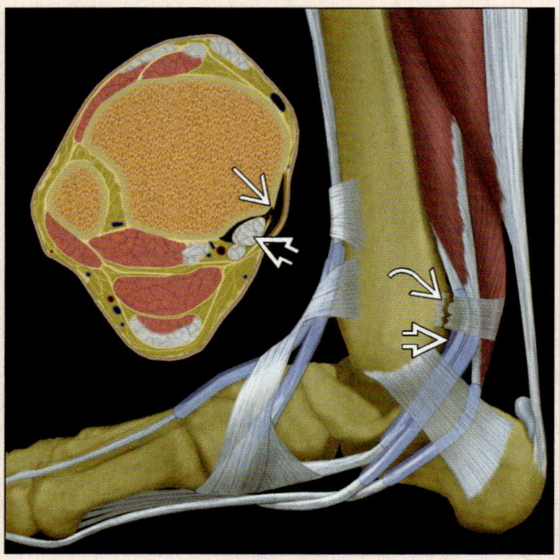

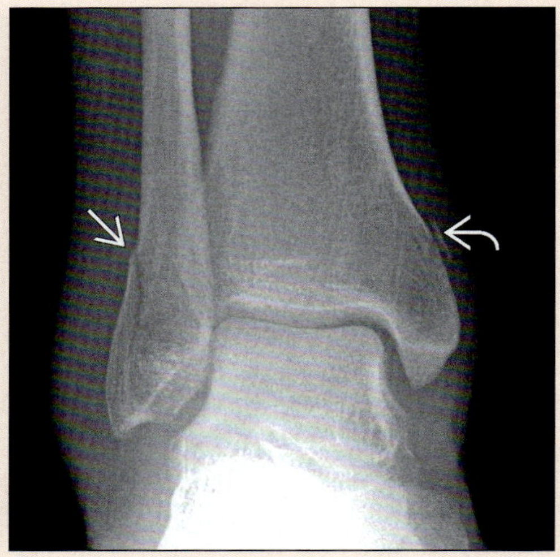

__(Left)__ Axial soft tissue algorithm CT shows acute dislocated posterior tibial tendon due to ankle dislocation. The flexor retinaculum is not visible; compare with the normal CT appearance of the peroneal retinaculum. __(Right)__ Axial MR arthrogram T1WI FS shows extension of contrast from the ankle joint around the posterior tibial tendon due to avulsion of flexor retinaculum and superficial deltoid fibers. The patient gave a history of vague medial ankle pain.

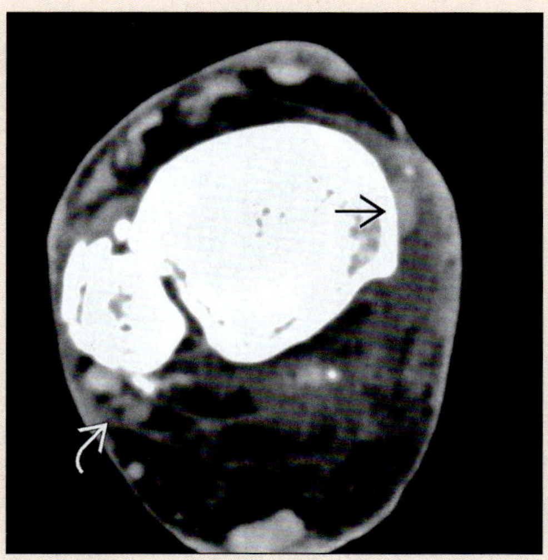

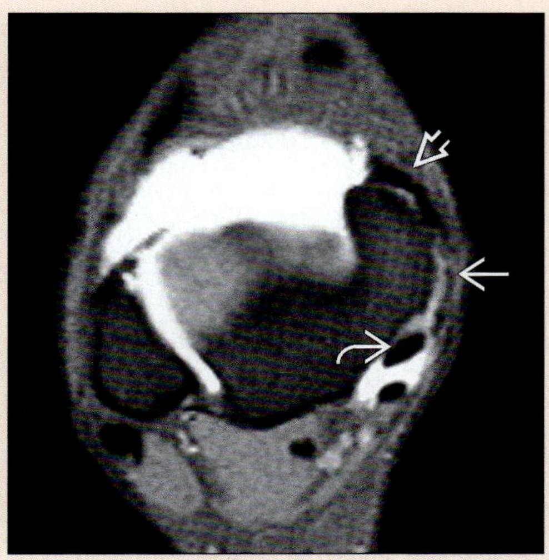

FLEXOR RETINACULUM AVULSION

IMAGING

General Features

- Best diagnostic clue
  - Medial subluxation of posterior tibial tendon (PTT) on axial MR

Radiographic Findings

- Avulsion fracture of medial tibial metaphysis
  - Thin flake of bone, vertically oriented
- Focal periosteal reaction at medial margin of distal tibial metaphysis

CT Findings

- Discontinuity of retinaculum at tibial attachment
- Nonvisualization of retinaculum
  - Normal retinaculum always visible on ST algorithm CT
- Avulsion fracture
- Subluxation or dislocation of posterior tibial tendon

MR Findings

- Discontinuity of retinaculum
- Low signal intensity periosteal reaction at medial malleolus adjacent to retinacular attachment
- Subluxation of PTT medially
- Dislocation of PTT anteriorly
  - Occurs in chronic injury

Ultrasonographic Findings

- If PTT normal in location, ask patient to perform active plantar flexion and inversion
  - Will cause tendon to subluxate if retinaculum torn

Imaging Recommendations

- Best imaging tool
  - MR
- Protocol advice
  - Best seen on axial fluid-sensitive sequences

DIFFERENTIAL DIAGNOSIS

Posterior Tibial Tendinopathy with Intact Retinaculum

- Patients with flat foot, often obese and middle-aged

Deltoid Ligament Sprain

- Pain at medial malleolus
- Often see widening of medial ankle gutter > 6 mm

Periostitis Due to Other Causes

- Chronic venous stasis, stress fracture, hypertrophic osteoarthropathy, infection, tumor

PATHOLOGY

General Features

- Etiology
  - Usually isolated injury due to inversion, dorsiflexion
  - May occur with other ankle injuries
  - Shallow sulcus at posteromedial tibia may predispose to injury
- Associated abnormalities
  - Malleolar or pilon fractures
  - Deltoid ligament sprain
  - Posterior tibial tendon dislocation, tendinopathy, and tear

Relevant anatomy

- Flexor retinaculum maintains position of flexor tendons posterior to medial malleolus
  - Originates on fascia of flexor hallucis longus muscle
  - Inserts on tubercle at posteromedial margin of tibia
  - Forms sling posterior to flexor digitorum longus and PTT

CLINICAL ISSUES

Presentation

- Most common signs/symptoms
  - May present as ankle "sprain" with pain over medial malleolus
  - Chronic injury: Patients present with medial ankle pain, PTT dysfunction
  - Dislocation of posterior tibial tendon
  - Intermittent subluxation, "snapping" of posterior tibial tendon

Treatment

- Usually requires surgical repair

DIAGNOSTIC CHECKLIST

Consider

- Acute injury may be missed on CT because of focus on bony injuries
- Nonvisualization of retinaculum on CT or MR indicates acute or chronic injury
- Posterior tibial tendon may be normally positioned despite torn retinaculum

Image Interpretation Pearls

- Subluxation of tendon is often slight
  - Posterior tibial tendon should never protrude medial to posterior margin of medial malleolus
- Complete, chronic dislocation of tendon may be missed on MR
  - Anteriorly displaced tendon may be overlooked
  - Flexor digitorum longus migrates into position vacated by PTT
  - Scarring seen without fluid signal intensity
  - Always check position and number of flexor tendons behind medial malleolus

SELECTED REFERENCES

1. Numkarunarunrote N et al: Retinacula of the foot and ankle: MRI with anatomic correlation in cadavers. AJR Am J Roentgenol. 188(4):W348-54, 2007
2. Miki T et al: Recurrent dislocation of tibialis posterior tendon. A report of two cases. Arch Orthop Trauma Surg. 118(1-2):96-8, 1998
3. Perlman MD et al: Traumatic dislocations of the tibialis posterior tendon: a review of the literature and two case reports. J Foot Surg. 29(3):253-9, 1990
4. Larsen E et al: Dislocation of the tibialis posterior tendon in two athletes. Am J Sports Med. 12(6):429-30, 1984

(Left) Anteroposterior radiograph shows acute avulsion of the retinaculum . Unfortunately, the importance of the small, flake-like avulsion fracture is often not recognized. Note also the widening of medial ankle gutter indicating associated injury of deltoid ligament. ***(Right)*** Axial PDWI FS MR shows flexor retinaculum stripped from its attachment to the medial malleolus . The posterior tibial tendon (PTT) is subluxated medially relative to posterior corner of the medial malleolus.

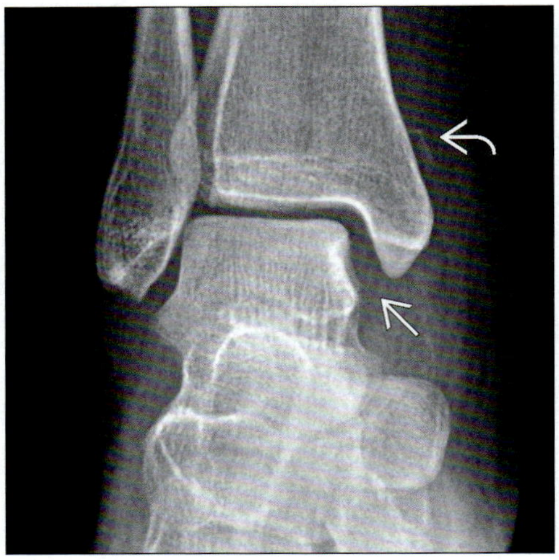

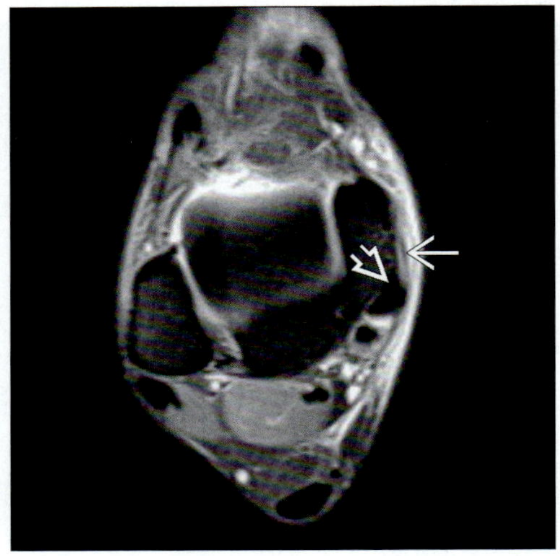

(Left) Sagittal STIR MR shows posterior tibial tenosynovitis and partial tear . This patient had associated flexor retinaculum injury, missed on sagittal images because of normal PTT position. In every patient with abnormal PTT, scrutinize the flexor retinaculum on axial images. ***(Right)*** Sagittal STIR MR shows PTT displaced anteriorly due to flexor retinacular tear. The flexor digitorum longus (FDL) has moved medially into the expected location of the PTT.

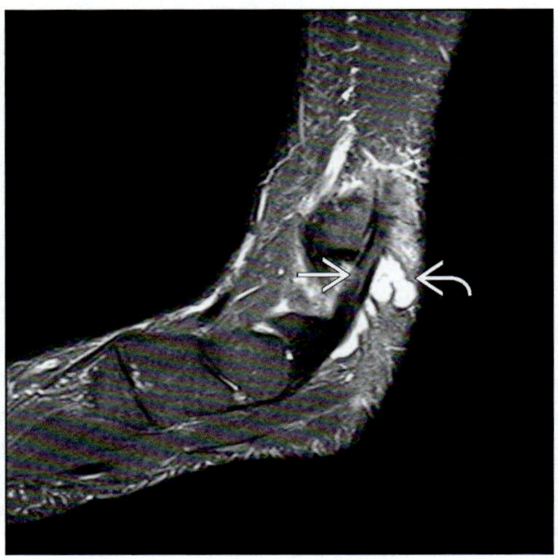

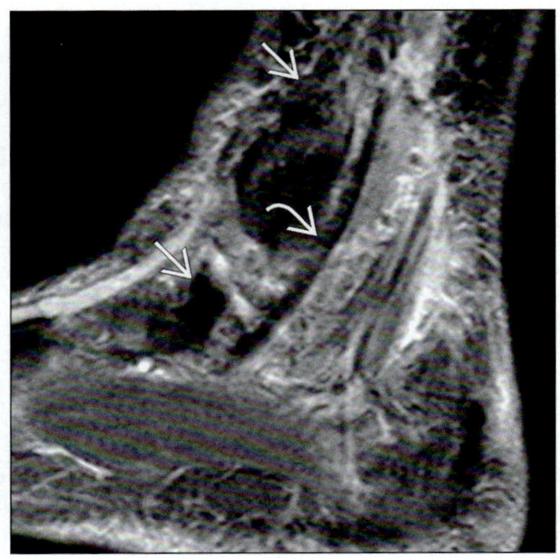

(Left) Axial PDWI FS MR shows chronic injury of flexor retinaculum, with focal discontinuity at its tibial attachment. Medial retinacular attachment to flexor hallucis longus muscle is intact. Periosteal reaction is seen adjacent to the avulsion site. PTT is normal in position despite injury to retinaculum. ***(Right)*** Coronal T2WI FS MR in the same patient shows periosteal reaction and extensive edema in the medial malleolus . Posterior tibial tendon appears normal.

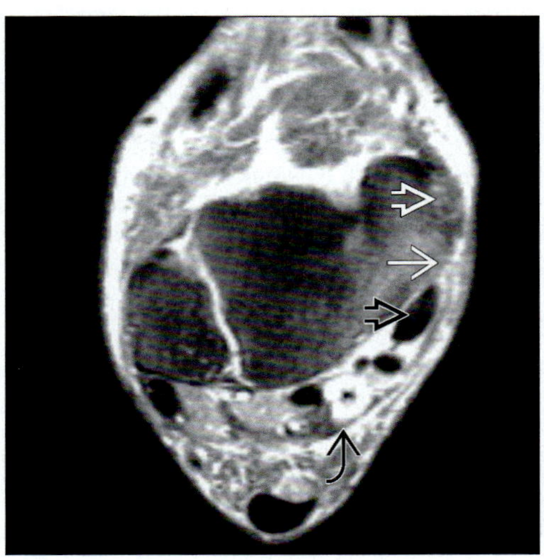

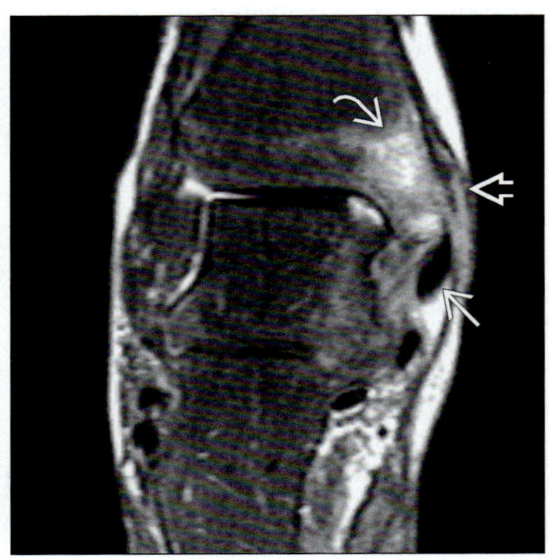

FLEXOR RETINACULUM AVULSION

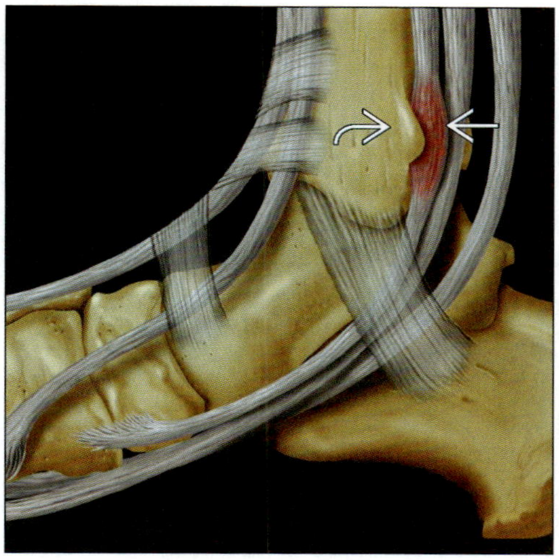

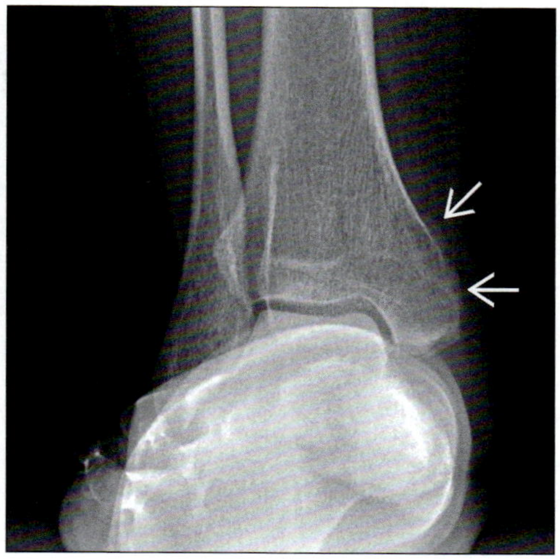

(Left) *Graphic of the medial aspect of the ankle shows chronic periosteal reaction* ↪ *at the insertion site of the flexor retinaculum. Focal posterior tibial tendinopathy* ➔ *has developed due to instability of the tendon.* ***(Right)*** *AP radiograph shows undulating periosteal reaction* ➔ *due to chronic tear of flexor retinaculum. A similar-appearing periosteal reaction may occur with cellulitis, venous stasis, osteomyelitis, or hypertrophic osteoarthropathy.*

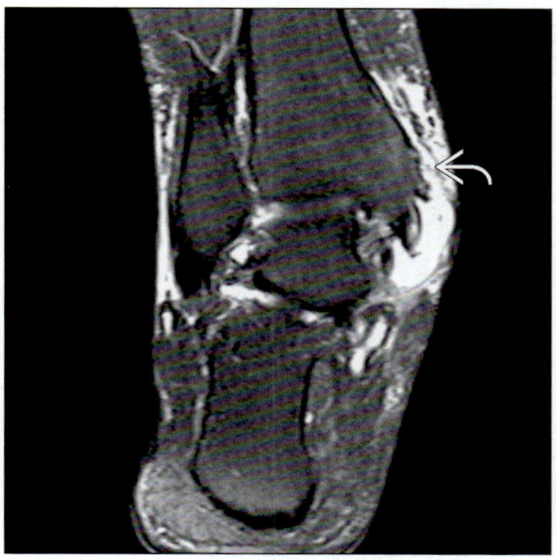

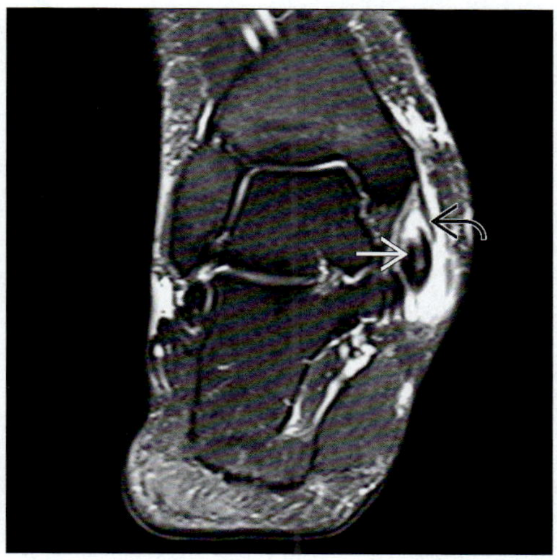

(Left) *Coronal T2WI FS MR of the same patient the periosteal reaction* ↩*, which has incorporated into the underlying bone, indicating chronic injury. Periosteal reaction may be more readily visible on radiographs than on CT.* ***(Right)*** *Coronal T2WI FS MR in the same patient further anteriorly shows a partial PTT tear* ➔ *and distended, irregular PTT tendon sheath* ↪*. The tendon sheath is continuous with flexor retinaculum.*

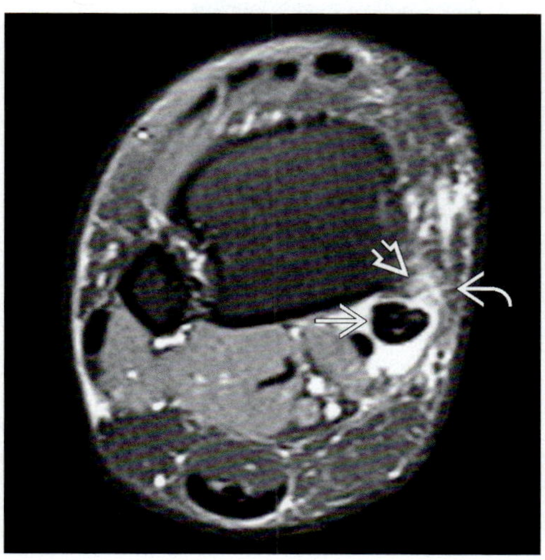

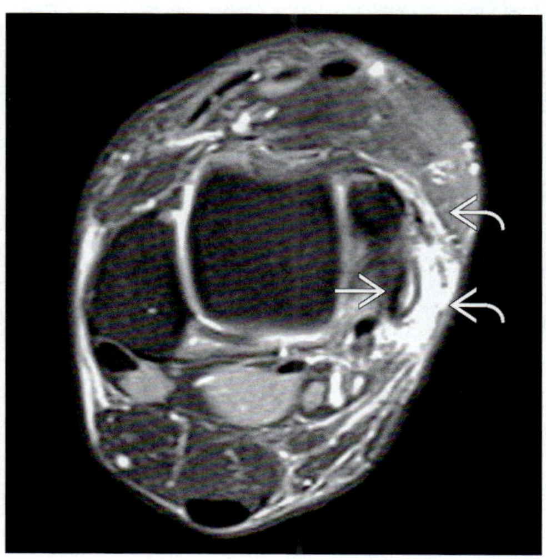

(Left) *Axial PDWI FS MR in the same patient shows the torn flexor retinaculum* ↩*. PTT* ➔ *is enlarged and contains intermediate signal intensity consistent with tenosynovitis. The tendon is slightly subluxated medially. PTT should not protrude medial to the posterior border of the tibia* ⇘*.* ***(Right)*** *Axial PDWI FS MR more inferiorly in the same patient shows a partial tear of the PTT* ➔ *and severe stripping of the flexor retinaculum* ↩ *from the medial malleolus.*

MTP LIGAMENT INJURY, DIGIT 1

Key Facts

Terminology

- Plantar plate: Plantar thickening of joint capsule
  - Includes attachments of sesamoids to proximal phalangeal bases
- Sesamophalangeal apparatus: Sesamoids + plantar plate
- Extensor expansion: Fibrous thickening of dorsal joint capsule, passes around extensor hallucis longus tendon
- Turf toe: Rupture of plantar plate due to hyperdorsiflexion
- Sand toe: Rupture of dorsal joint capsule due to plantar flexion
- Skimboarder's toe: Rupture of dorsal joint capsule due to hyperdorsiflexion
- Traumatic hallux valgus: Rupture of medial collateral ligament

Imaging

- Due to mobility and vulnerable position of 1st toe, injury may involve any quadrant of joint capsule
- MR best for diagnosis
  - Ligament may appear discontinuous, lax, or thickened and heterogeneous
  - Joint fluid may extravasate beyond joint capsule
  - ± bone marrow edema in sesamoids, 1st MT head, base proximal phalanx
  - Use small coil such as wrist or TMJ coil

Top Differential Diagnoses

- Sesamoid fracture
- Sesamoiditis
- Tendon rupture
- Metatarsal stress fracture

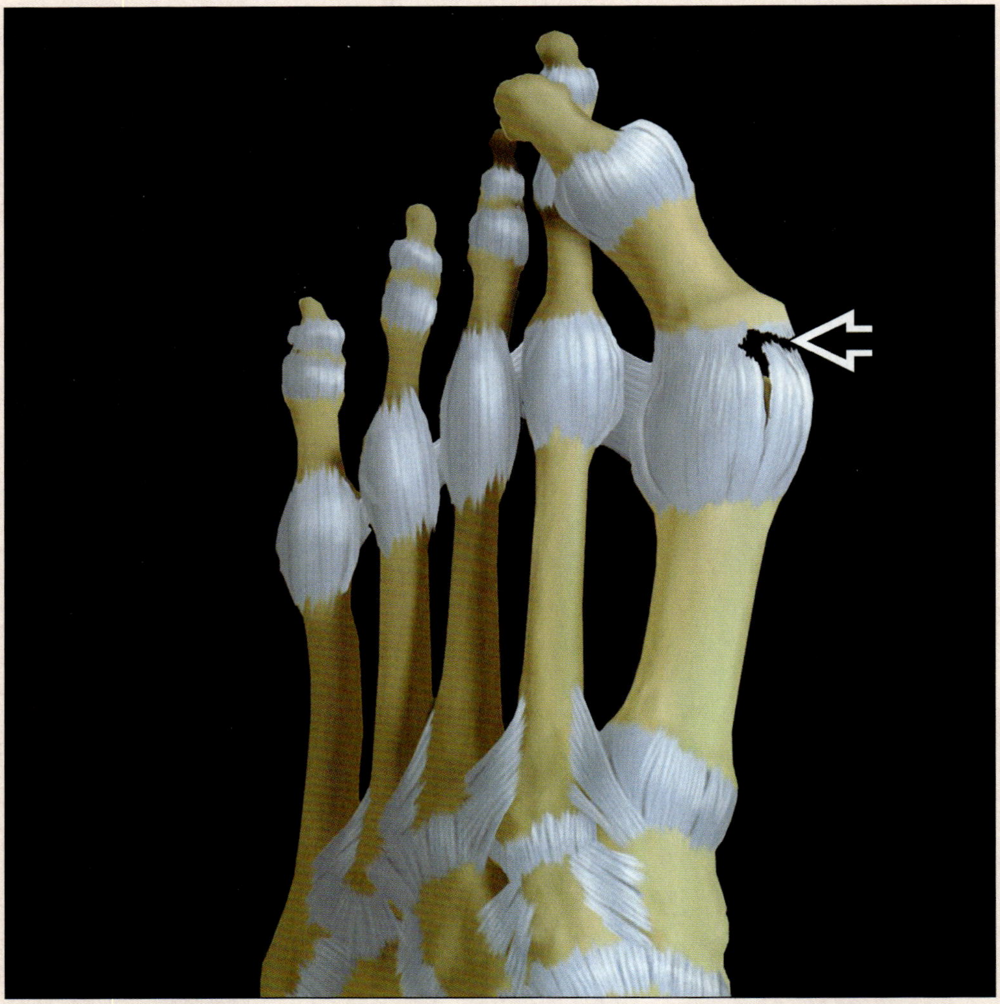

Graphic of the dorsum of the foot shows a traumatic hallux valgus. There is rupture ➡ of the medial collateral ligament of the 1st toe. Patients present with focal pain in this region and new onset of deformity. Treatment is conservative. The 1st toe is vulnerable to ligament injury due to its mobility and position; ligament injury may occur due to hyperdorsiflexion, plantarflexion, abduction, or adduction.

MTP LIGAMENT INJURY, DIGIT 1

TERMINOLOGY

Definitions

- Turf toe: Rupture of plantar plate due to hyperdorsiflexion
- Sand toe: Rupture of dorsal joint capsule due to plantar flexion
- Skimboarder's toe: Rupture of dorsal joint capsule due to hyperdorsiflexion
- Traumatic hallux valgus: Rupture of medial collateral ligament

IMAGING

General Features

- Location
  - Due to mobility and vulnerable position of 1st toe, injury may occur in any quadrant of joint capsule

Radiographic Findings

- Turf toe
  - Dorsal soft tissue swelling, widened joint space
  - Proximal displacement of sesamoids
  - Dorsal subluxation of proximal phalanx
- Traumatic hallux valgus
  - Medial soft tissue swelling, ± avulsion fracture
  - Acute abduction deformity of MTP joint
- Sand toe
  - Dorsal soft tissue swelling
- Lateral collateral ligament tear
  - Rarely see avulsion fracture or hallux varus

MR Findings

- Injured ligament may be discontinuous, lax, or thickened and heterogeneous in appearance
- Fluid often extravasates beyond joint capsule
- Bone bruises frequently present
- Turf toe: Injury plantar plate, either distal or proximal
  - ± interstitial tear of muscle attachments to sesamoids
  - ± tear of conjoined tendon attachments of sesamoids to proximal phalangeal base
  - ± bone marrow edema sesamoids, dorsal MTP joint
- Traumatic hallux valgus: Injury medial collateral ligament
  - ± bone marrow edema medially, laterally
- Lateral collateral ligament tear
  - ± medial bone marrow edema
- Sand toe, skimboarder's toe: Injury dorsal joint capsule
  - Bone marrow edema may be dorsal or plantar
  - ± displacement of extensor hallucis longus tendon
    - Due to loss of constraint by extensor expansion

Imaging Recommendations

- Best imaging tool
  - MR
- Protocol advice
  - Use small coil, such as wrist or TMJ coil, small field of view
  - Slice thickness 2.5-3.5 mm

DIFFERENTIAL DIAGNOSIS

Sesamoid Fracture

- Fracture line may be difficult to see on radiographs
- CT or MR show fracture in coronal or axial plane

Sesamoiditis

- Due to chronic, repetitive stress; no fracture line visible
- MR shows bone marrow edema

Tendon Rupture

- Tear of extensor or flexor hallucis longus will lead to joint malalignment

Metatarsal Stress Fracture

- Often not seen on radiographs until callus/periosteal reaction are visible

PATHOLOGY

Relevant Anatomy

- Plantar plate: Plantar thickening of joint capsule
  - Includes attachments of sesamoids to proximal phalangeal bases
    - Medial sesamoid: Conjoined tendon of medial short head of flexor hallucis brevis + abductor hallucis
    - Lateral sesamoid: Conjoined tendon of lateral short head of flexor hallucis brevis + adductor hallucis
- Sesamophalangeal apparatus: Sesamoids + plantar plate
  - Extensor expansion: Fibrous thickening of dorsal joint capsule, passes around extensor hallucis longus tendon

CLINICAL ISSUES

Presentation

- Most common signs/symptoms
  - History of forceful subluxation or dislocation of toe
  - Focal tenderness, bruising

Treatment

- Usually treated conservatively
- May be surgically repaired

DIAGNOSTIC CHECKLIST

Consider

- Dorsal capsule injury may occur due to either hyperdorsiflexion or to plantarflexion injury

SELECTED REFERENCES

1. Crain JM et al: MR imaging of turf toe. Magn Reson Imaging Clin N Am. 16(1):93-103, vi, 2008
2. Donnelly LF et al: Skimboarder's toe: findings on high-field MRI. AJR Am J Roentgenol. 184(5):1481-5, 2005
3. Ashman CJ et al: Forefoot pain involving the metatarsal region: differential diagnosis with MR imaging. Radiographics. 21(6):1425-40, 2001
4. Watson TS et al: Periarticular injuries to the hallux metatarsophalangeal joint in athletes. Foot Ankle Clin. 5(3):687-713, 2000

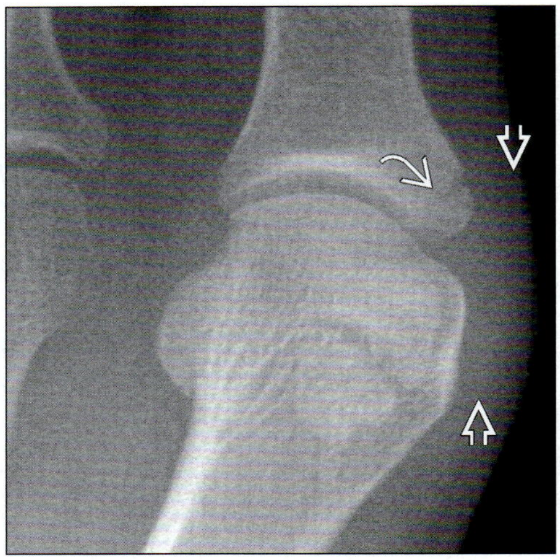

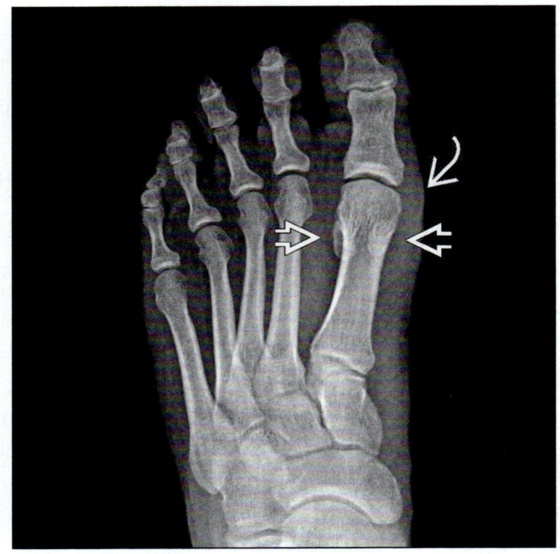

(Left) *AP radiograph shows findings of traumatic hallux valgus. There is an avulsion fracture at the attachment of the medial collateral ligament to the 1st MT base. Soft tissue swelling is present medially. There is abduction at the 1st MTP joint, but correlation to history is needed to determine if this is of new onset.* ***(Right)*** *AP radiograph of turf toe shows proximally displaced sesamoids due to rupture of the plantar plate. Soft tissue swelling is seen medially.*

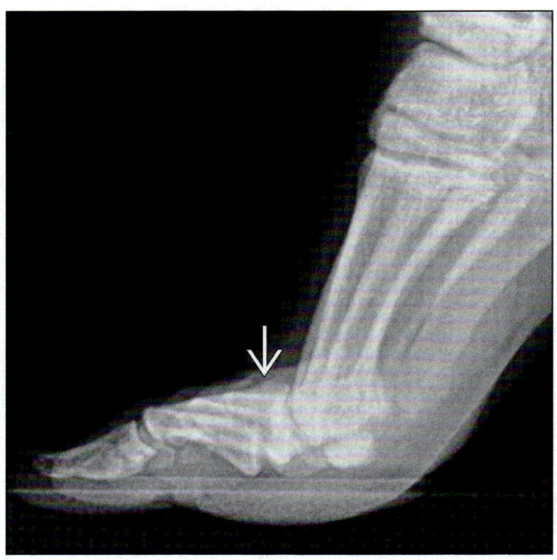

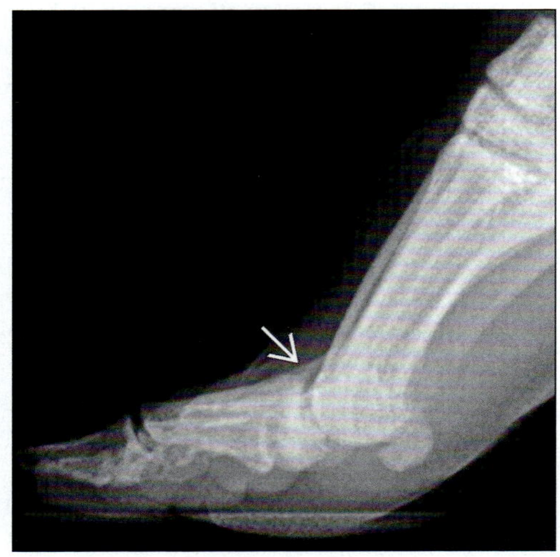

(Left) *Lateral radiograph obtained with patient standing on dorsiflexed toes shows hypermobility and dorsal subluxation of 1st MTP joint due to plantar plate rupture (turf toe). Stress views are difficult to interpret and must be compared to the uninjured contralateral foot.* ***(Right)*** *Lateral radiograph of contralateral normal foot shows less dorsiflexion of the 1st MTP joint because of restraint by the plantar plate.*

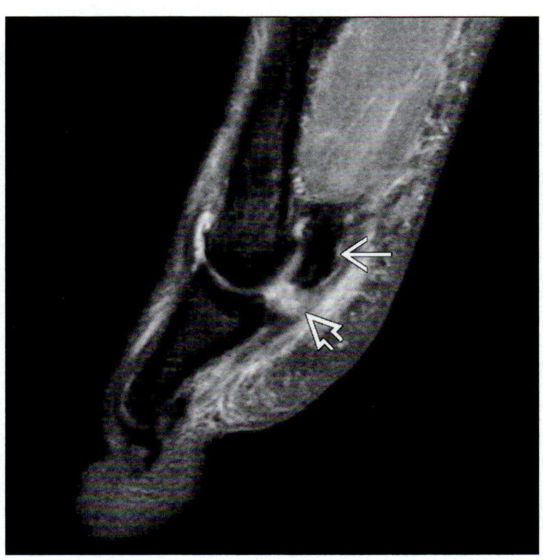

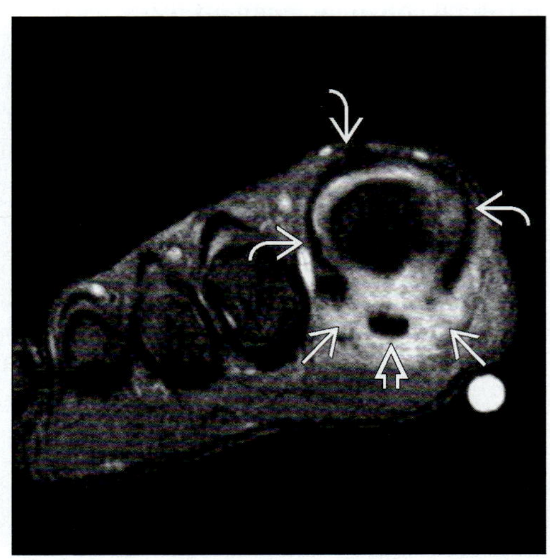

(Left) *Sagittal T2WI FS MR shows distal plantar plate rupture with proximal displacement of sesamoid.* ***(Right)*** *Coronal (short axis) T2WI FS MR in the same patient shows plantar plate rupture. The flexor hallucis longus tendon is almost always intact, as in this case. Medial, dorsal, and lateral collateral ligaments are intact.*

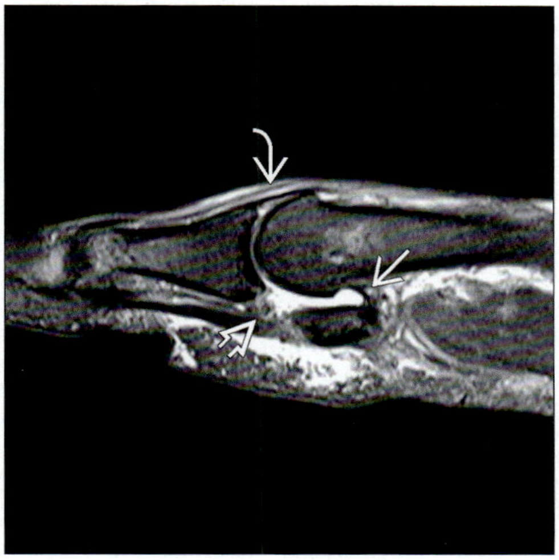

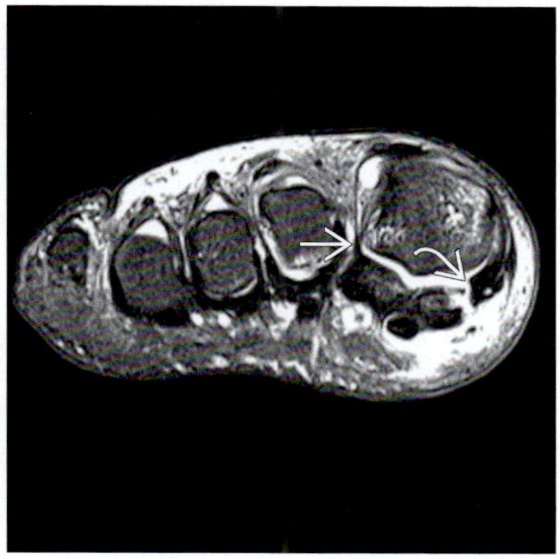

(Left) *Sagittal T2WI FS MR shows laxity of proximal plantar plate and distal tear. Note the normal dorsal joint capsule.* ***(Right)*** *Coronal (short axis) T2WI FS MR shows asymmetric turf toe. There is a tear of the medial portion of the plantar plate immediately distal to the medial sesamoid. The lateral portion is intact. This was a football injury, with a twisting force on a planted foot.*

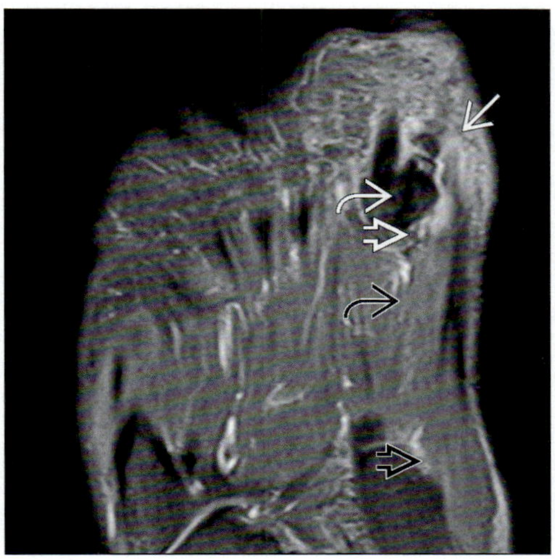

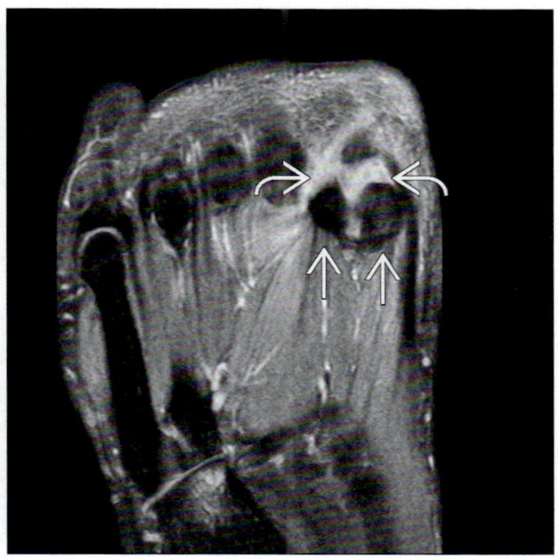

(Left) *Axial (long axis) PDWI FS MR shows a tear of the sesamophalangeal apparatus. The sesamoid is nearly completely detached from the flexor hallucis brevis. The abductor hallucis and medial head of the flexor hallucis brevis are well seen proximal to the tear.* ***(Right)*** *Axial (long axis) PDWI FS MR shows tear of plantar plate distal to sesamoids. The attachments of the intrinsic muscles to the sesamoids are well seen.*

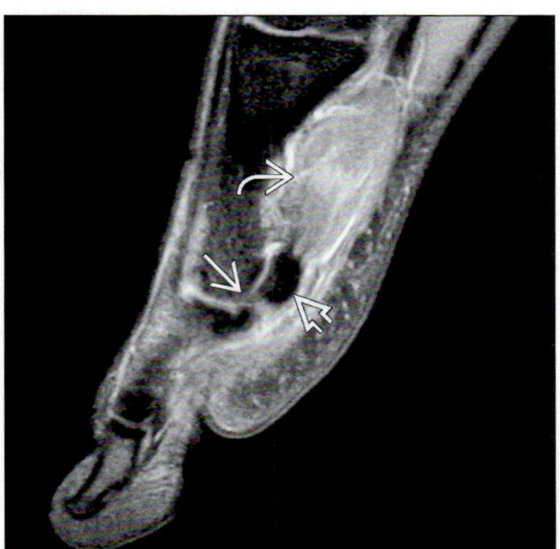

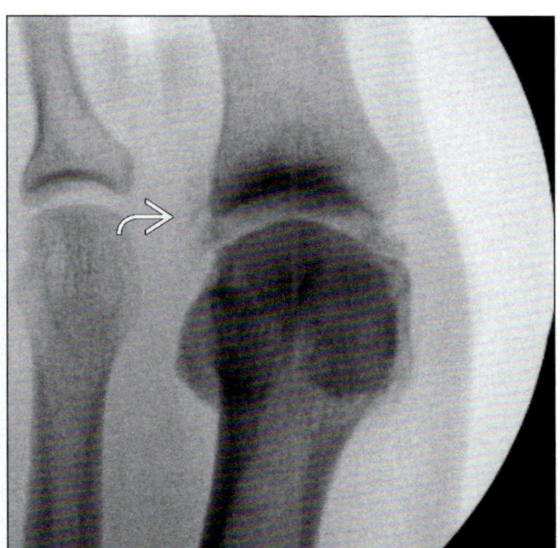

(Left) *Sagittal STIR MR shows lateral sesamoid pulled proximally by the intact flexor hallucis brevis muscle. Plantar plate is ruptured just distal to the sesamoid.* ***(Right)*** *AP arthrogram shows a tear of the lateral collateral ligament. Contrast was injected from a dorsal approach with a 25-gauge needle and has extravasated through the torn lateral capsule. This unusual injury occurs in barefoot patients, when the 1st toe is forcefully adducted from the remainder of the foot.*

MTP LIGAMENT INJURY, DIGITS 2-5

Key Facts

Imaging

- Radiographs sometimes show joint subluxation
- Diagnosis usually made on MR
  - Small FOV, thin slices critical in diagnosis
  - Synovial cysts common helpful sign
- Plantar plate injury
  - Dorsal subluxation of MTP joint
  - Focal discontinuity of plantar plate
  - Fluid extending proximally along MT
- Collateral ligament injury
  - Redundant, lax ligament
  - Abduction or adduction of joint
- Dorsal capsule injury
  - Redundant, lax capsule

Top Differential Diagnoses

- Morton neuroma
- Metatarsal stress fracture
- Freiberg infraction
- Rheumatoid arthritis
- Psoriatic arthritis
- Intermetatarsal bursitis

Pathology

- Associated with
  - Morton foot configuration
  - Hallux valgus

Clinical Issues

- Pain in push-off phase of gait
- Instability of MTP joint, often leading to chronic nontraumatic dislocation
- "Cock up" or "crossover" toe deformity

*(**Left**) Sagittal graphic shows joint capsule of a MTP joint. All of the digits except the 1st have the same appearance. The joint capsule is reinforced by dorsal, lateral, and plantar ligaments, which are thickenings of the capsule. The most commonly injured is the plantar plate →. (**Right**) Sagittal STIR MR shows a synovial cyst ↪ centered at the 2nd MTP. A thin neck → connects the cyst to the plantar plate region, but no ligament tear is visible on this image.*

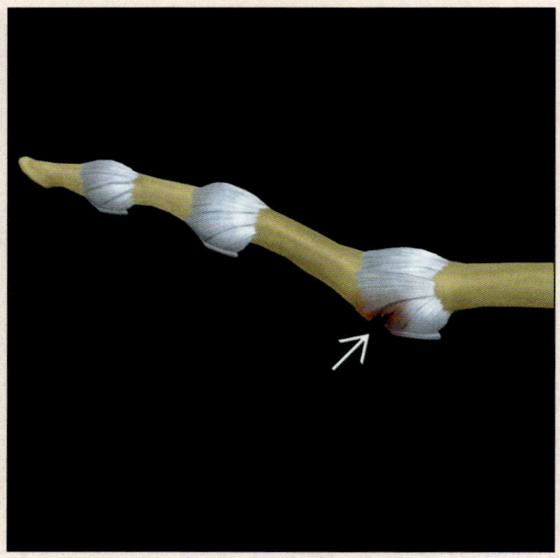

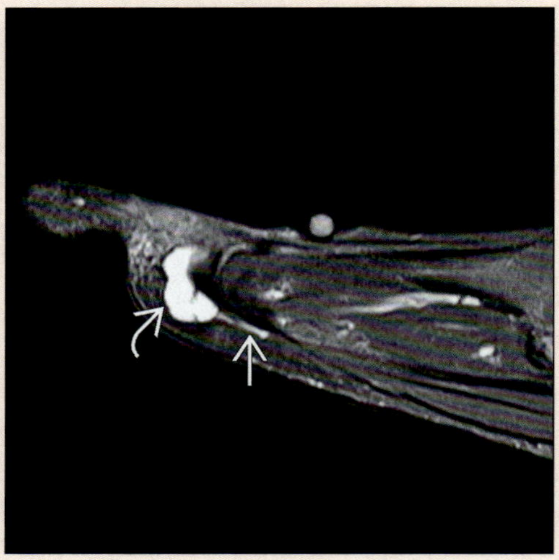

*(**Left**) Coronal T2WI FS MR in the same patient shows a tear in the plantar plate → of the 2nd MTP, as well as a tear of the 3rd ↪. Cysts adjacent to a joint should always raise concern for articular injury and can usually be traced to the injury site. (**Right**) Coronal T1WI C+ FS MR in the same patient best shows both the plantar plate rupture ↪ and the resultant synovial cyst →.*

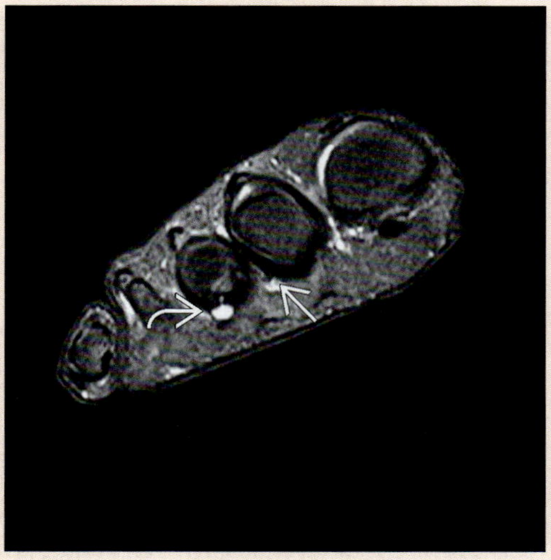

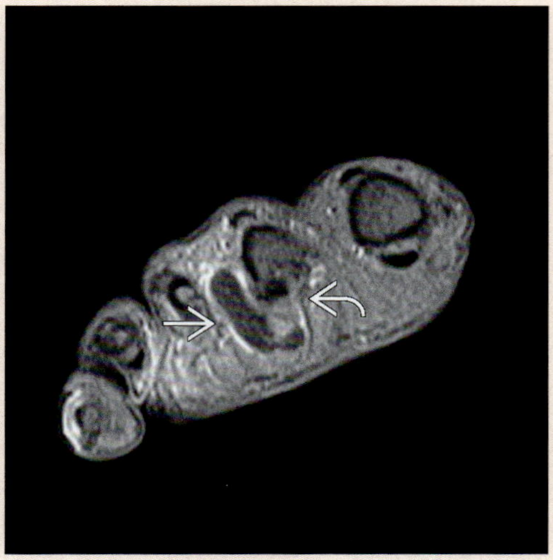

MTP LIGAMENT INJURY, DIGITS 2-5

IMAGING

General Features

- Best diagnostic clue
  - Radiographs: Subluxation of metatarsophalangeal (MTP) joint
  - MR: Fluid extending beyond normal joint contours
- Location
  - 2nd MTP most commonly injured

Radiographic Findings

- Subluxation of MTP joint (usually dorsal)

MR Findings

- Findings seen in any MTP ligament injury
  - Joint effusion
  - Bone marrow edema in head of MT due to abnormal motion
  - Synovial cyst
- Plantar plate injury
  - Focal discontinuity of plantar plate
  - Fluid at attachment of plantar plate
  - Extravasation of joint fluid
- Collateral ligament injury
  - Redundant, lax ligament
  - Abduction or adduction of joint
- Dorsal capsule injury
  - Redundant, lax capsule
  - Extravasation of joint fluid through capsular defect

Imaging Recommendations

- Best imaging tool
  - MR
- Protocol advice
  - FOV 12-14, centered at MTP, fluid-sensitive sequence in all 3 planes
  - Include MTs to evaluate for stress fracture
  - Plantar plate best seen on 3 mm sagittal images
  - Proton density with FS preferable to STIR because of better spatial resolution of very thin structures

DIFFERENTIAL DIAGNOSIS

Morton Neuroma

- Pain, enhancing mass between metatarsal heads

Metatarsal Stress Fracture

- MR useful to distinguish between stress fracture and plantar plate injury

Freiberg Infraction

- Collapse of 2nd or 3rd MT head
- Due to osteonecrosis or osteochondral injury (etiology debated)

Rheumatoid Arthritis

- Joint effusions, erosions
- More likely to affect 4th to 5th MTP joints early in course of disease

Psoriatic Arthritis

- Periostitis, joint effusion, erosions

Intermetatarsal Bursitis

- Bursa normally present between each of MT heads
- Bursitis develops due to inflammatory arthropathy or tight shoes

PATHOLOGY

General Features

- Associated abnormalities
  - Morton foot configuration
    - Short 1st MT relative to 2nd MT
  - Hallux valgus, metatarsus primus varus
  - Both conditions result in increased stress on 2nd and 3rd MT heads

Relevant Anatomy

- MTP joint capsule extends from MT neck to base of proximal phalanx
- Medial and lateral collateral ligaments merge with joint capsule
  - Dorsal and plantar interosseous muscles attach to proximal phalangeal joints adjacent to collateral ligaments
- Plantar plate: Condensation of joint capsule at plantar portion of joint capsule of MTP joints
  - Helps prevent dorsiflexion injury
  - Plantar plate is thin at origin from MT neck
    - Joint recess, curved and slightly redundant contour normally seen at MT neck origin
  - Plantar plate becomes thicker as extends to base of proximal phalanx
    - Attaches distally to phalangeal cartilage, which should not be mistaken for fluid
- Dorsally, capsule merges with extensor tendons to form extensor hood

CLINICAL ISSUES

Presentation

- Most common signs/symptoms
  - Pain in push-off phase of gait
  - Instability of MTP joint
    - Plantar plate rupture: Hyperdorsiflexion ("cock up") deformity, dislocation
    - Collateral ligament rupture: Abduction or adduction, "crossover" of adjacent toe

Treatment

- Metatarsal pads to relieve stress on toe
- Casting or pin fixation for complete tear

SELECTED REFERENCES

1. Smith BW et al: Disorders of the lesser toes. Sports Med Arthrosc. 17(3):167-74, 2009
2. Mohana-Borges AV et al: Lesser metatarsophalangeal joints: standard MR imaging, MR arthrography, and MR bursography--initial results in 48 cadaveric joints. Radiology. 227(1):175-82, 2003
3. Umans HR et al: The plantar plate of the lesser metatarsophalangeal joints: potential for injury and role of MR imaging. Magn Reson Imaging Clin N Am. 9(3):659-69, xii, 2001
4. Yao L et al: Plantar plate of the foot: findings on conventional arthrography and MR imaging. AJR Am J Roentgenol. 163(3):641-4, 1994
5. Thompson FM et al: Problems of the second metatarsophalangeal joint. Orthopedics. 10(1):83-9, 1987

MTP LIGAMENT INJURY, DIGITS 2-5

(Left) *AP radiograph of the foot shows slight medial subluxation and adduction of 2nd MTP, consistent with ligamentous laxity or injury. The patient sustained an injury while running barefoot and complained of persistent pain and instability at the 2nd MTP despite several months attempted immobilization.* ***(Right)*** *Axial PDWI FS MR in the same patient shows a torn MCL, with the proximal end curled distal to its normal origin at the sharply defined tubercle of the MT head.*

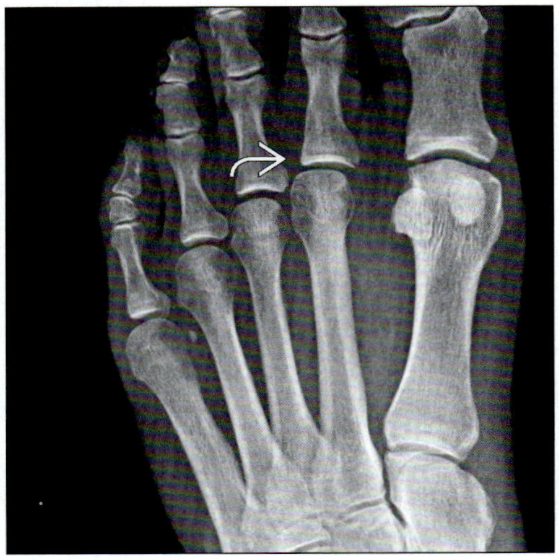

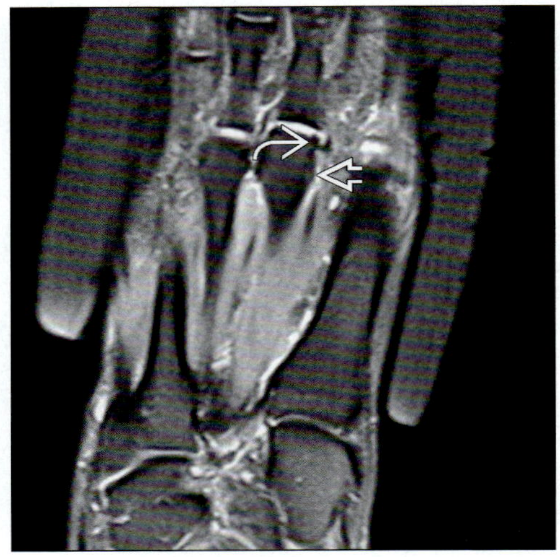

(Left) *Axial PDWI FS MR in the same patient more dorsally shows thick, lax lateral and medial collateral ligaments. Normal origin of the medial collateral ligament has no ligament attached.* ***(Right)*** *Coronal T2WI FS MR in the same patient shows nearly circumferential injury of the 2nd MTP joint capsule.*

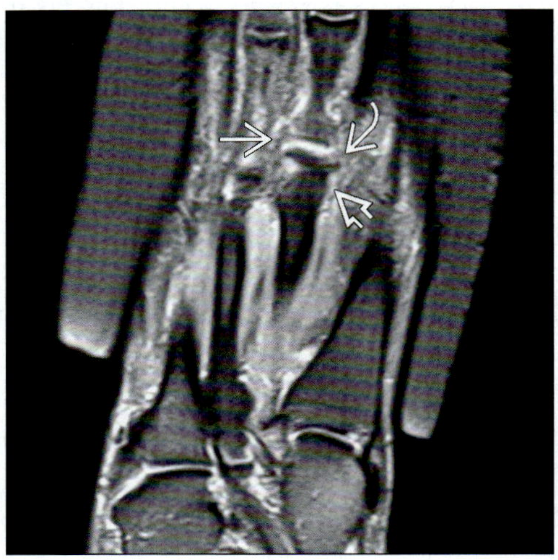

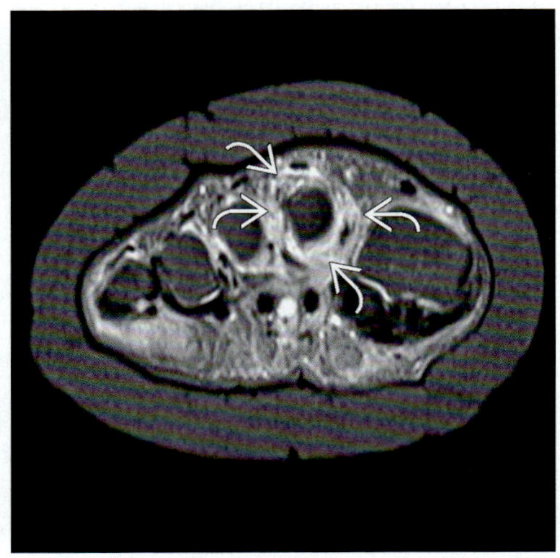

(Left) *Sagittal T1WI MR shows plantar plate rupture, with fluid plantar to the MTP joint, and poorly defined, thinned, lax plantar plate.* ***(Right)*** *Sagittal T2WI FS MR shows linear fluid in the plantar plate just proximal to its insertion. In addition, fluid extends further proximally than normal along the MT neck. Origin of the plantar plate is difficult to see unless images are high resolution and slices are thin (2.5-3 mm).*

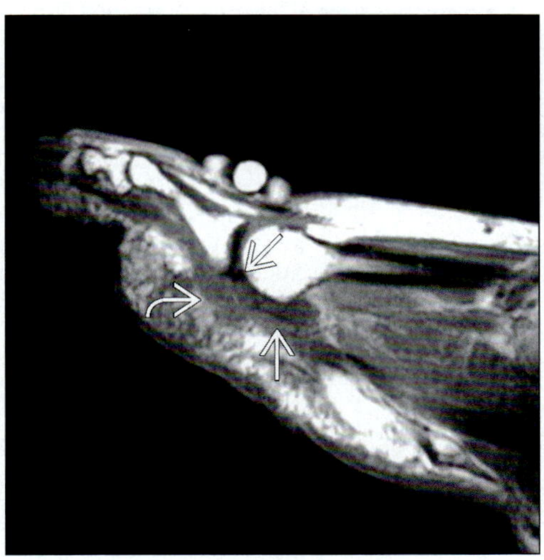

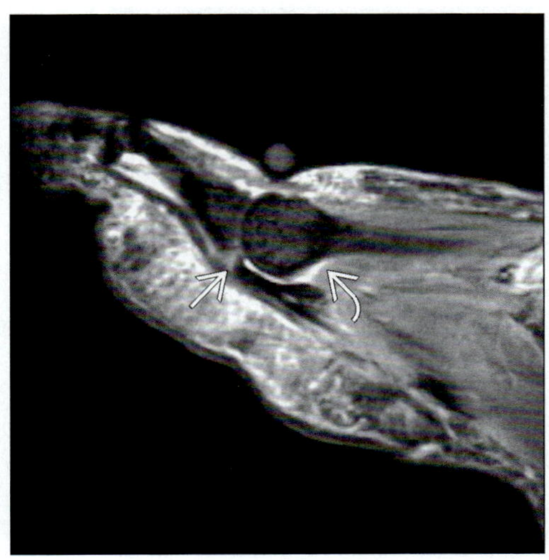

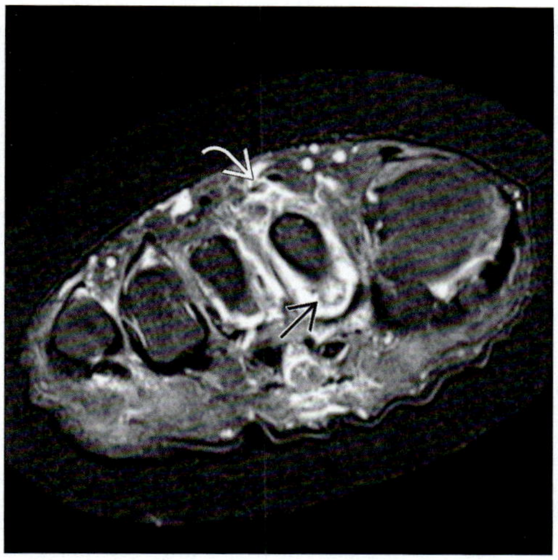

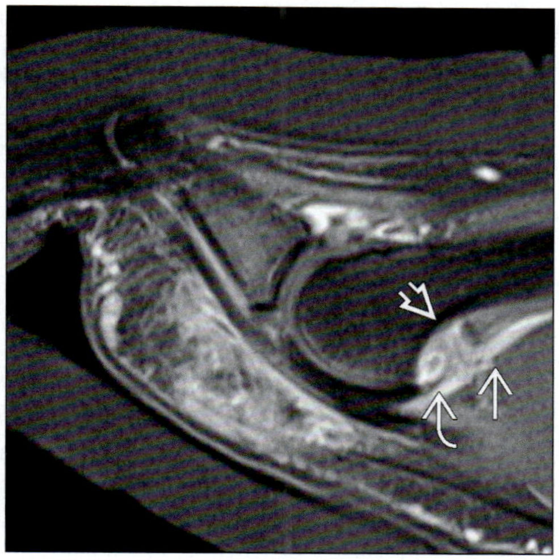

(Left) *Coronal T2WI FS MR in the same patient shows marked 2nd MTP joint distention and dorsal extravasation of fluid ➔. Curved structure in plantar fluid collection ➔ is a redundant portion of the torn plantar plate.* ***(Right)*** *Sagittal T2WI FS MR shows curved configuration of the proximal plantar plate ➔. Although the plate is well seen distally, the origin ➔ is avulsed. Fluid has extravasated through the avulsion of the ligament origin and extends along the MT shaft ➔.*

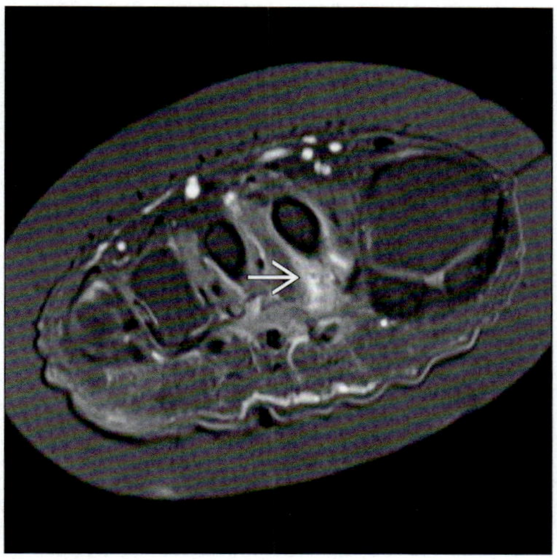

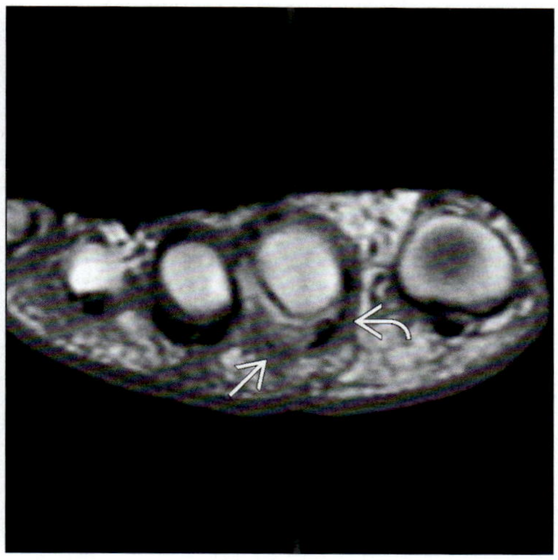

(Left) *Coronal T2WI FS MR more proximally in the same patient shows extravasated fluid ➔ along the MT neck from a plantar plate rupture.* ***(Right)*** *Coronal T1WI MR shows marked thinning of the lateral portion of the plantar plate ➔ of the 2nd MTP. Medial attachments ➔ are preserved.*

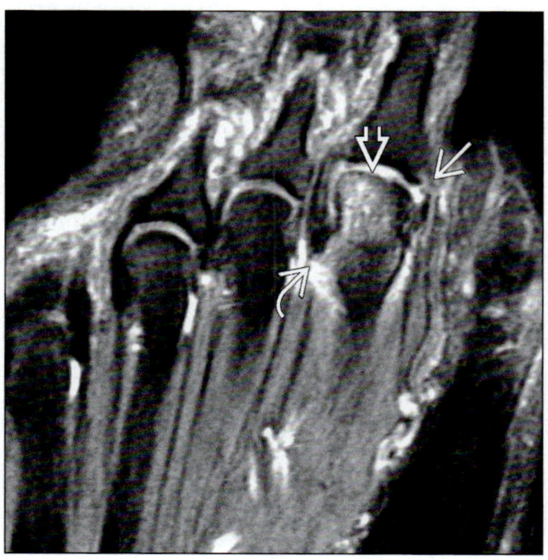

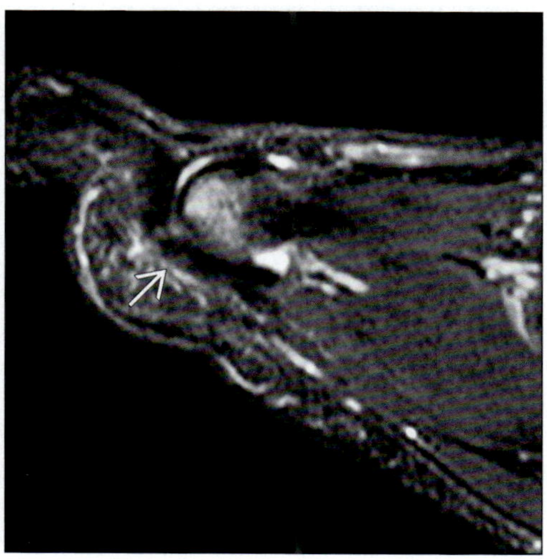

(Left) *Axial PDWI FS MR shows a tear of the proximal origin of the lateral collateral ligament ➔ and laxity of the medial collateral ligament ➔. Bone marrow edema in the 2nd MT head ➔ reflects abnormal motion but mimics Freiberg infraction.* ***(Right)*** *Sagittal STIR MR in the same patient shows a thickened, poorly defined, heterogeneous plantar plate ➔ without definite rupture. At surgery, the plantar plate was lax and scarred, and the collateral ligaments were completely torn.*

ACCESSORY MUSCLES, ANKLE AND FOOT

Key Facts

Terminology

- Morphologically normal muscle that is present in only a small percentage of the population
- Named for muscle with which it is closely associated

Imaging

- Fusiform mass equal in signal intensity to muscle
- If muscle is causing impingement, may see edema in fascia and around muscle
- Accessory soleus
  - Originates anterior margin of soleus or posterior cortex of tibia &/or fibula
  - Inserts on posterior process of calcaneus or on Achilles tendon
- Accessory flexor hallucis longus
  - Variable origin and insertion, often terminates in tarsal tunnel
  - Follows course of flexor hallucis longus
- Accessory flexor digitorum longus
  - Variable origin, inserts on quadratus plantae or flexor digitorum longus
- Peroneus quartus
  - Originates distal fibula, peroneus brevis or peroneus longus
  - Inserts on calcaneus, cuboid, peroneal retinaculum or 5th metatarsal

Clinical Issues

- Tarsal tunnel syndrome
- Subjective fullness
- Posterior impingement
- Achilles tendinopathy
- Ankle pain
- Incidental finding

(Left) *Graphic of the medial side of the ankle shows normal tendon anatomy. Accessory muscles may be associated with the Achilles tendon ➔, flexor hallucis longus tendon ➔, or flexor digitorum longus tendon ➔.* ***(Right)*** *Graphic of the lateral side of the ankle shows normal tendon anatomy. A peroneus quartus may occur adjacent to the peroneus brevis ➔.*

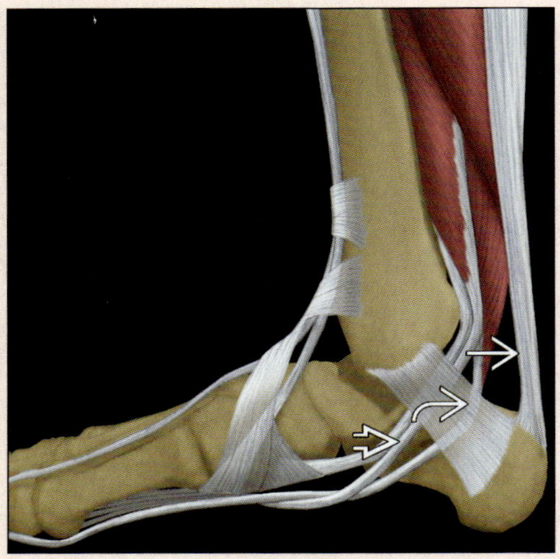

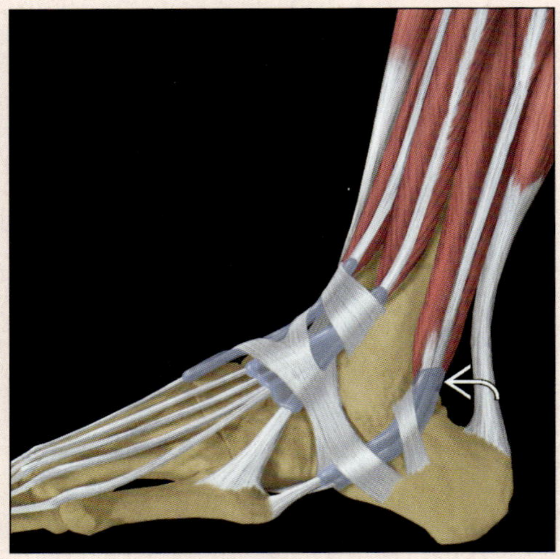

(Left) *Sagittal T1WI MR shows an accessory soleus muscle ➔ filling the pre-Achilles space and inserting on the calcaneus anterior to the normal Achilles tendon insertion.* ***(Right)*** *Sagittal T1WI MR shows an accessory FHL ➔ in a patient complaining of vague ankle pain after exercise. Accessory muscles may be an incidental finding, but they can cause symptoms due to increased bulk in a soft tissue compartment.*

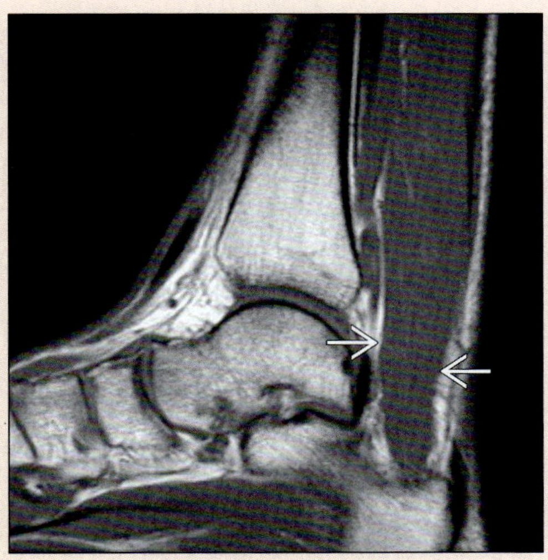

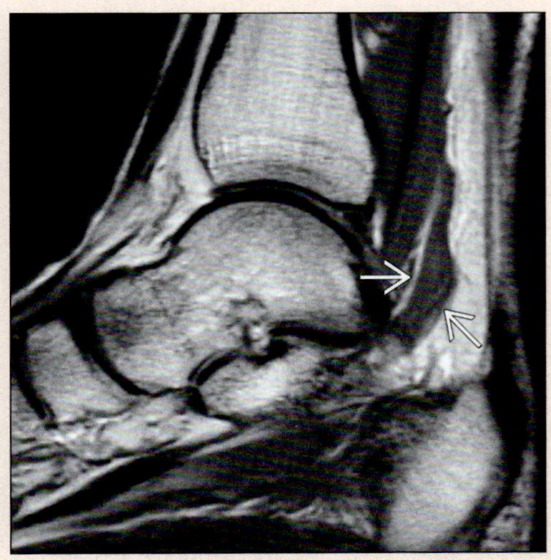

ACCESSORY MUSCLES, ANKLE AND FOOT

TERMINOLOGY

Definitions

- Morphologically normal muscle that is present in only a small percentage of the population
- Named for muscle with which it is closely associated

IMAGING

General Features

- Best diagnostic clue
  - Fusiform mass equal in CT attenuation or MR signal intensity to muscle
- Location
  - **Accessory soleus**
    - Originates on anterior margin of soleus or posterior cortex of tibia &/or fibula
    - Inserts on posterior process of calcaneus or on Achilles tendon
  - **Accessory flexor hallucis longus**
    - Variable origin and insertion, often terminates in tarsal tunnel
    - Follows course of flexor hallucis longus
  - **Accessory flexor digitorum longus**
    - Variable origin, inserts on quadratus plantae or flexor digitorum longus
  - **Peroneus quartus**
    - Originates on distal fibula, peroneus brevis, or peroneus longus
    - Inserts on 5th metatarsal, calcaneus, cuboid or peroneal retinaculum
- Morphology
  - Round, ovoid, or triangular in cross-section
    - May be mistaken for mass on imaging studies
  - Fusiform shape from superior to inferior
  - May attach to muscle, tendon, retinaculum, or bone

Radiographic Findings

- Soft tissue fullness beyond expected muscle contour

CT Findings

- Fusiform mass equal in density to muscle

MR Findings

- Fusiform mass equal in signal intensity to muscle
- If muscle is causing impingement, may see edema in fascia and around muscle
- If muscle is causing tarsal tunnel syndrome, may see denervation changes in intrinsic plantar muscles

Imaging Recommendations

- Best imaging tool
  - MR
- Protocol advice
  - Seen best on T1-weighted images
    - Blend with adjacent fat on T2WI FS, STIR
  - Most easily recognized on axial images

DIFFERENTIAL DIAGNOSIS

Intramuscular Neoplasm

- Usually has signal intensity distinct from muscle
- Usually enhances

Prior Surgery

- Many surgical procedures re-route tendons in foot
- Look for stigmata of surgery, check history

CLINICAL ISSUES

Presentation

- Most common signs/symptoms
  - Tarsal tunnel syndrome
    - Accessory flexor hallucis or digitorum longus may create mass effect in tarsal tunnel
    - Look for denervation changes in intrinsic muscles of foot
  - Subjective fullness
    - Patients with accessory soleus may complain of fullness behind ankle
  - Posterior impingement
    - Accessory flexor hallucis longus may limit extreme plantar flexion of foot
  - Achilles tendinopathy
    - Occurs in association with accessory soleus
  - Ankle pain
  - Incidental finding
    - Many patients are asymptomatic

Demographics

- Epidemiology
  - Prevalence
    - Peroneus quartus: 13-26%
    - Accessory flexor digitorum longus (FDL): 6-8%
    - Accessory soleus: 1-5.5%
    - Accessory flexor hallucis longus (FHL): 1%

Treatment

- Can be surgically excised if symptomatic
- Diagnostic injection of tendon sheath used to determine if accessory muscle is cause of patient's symptoms

DIAGNOSTIC CHECKLIST

Consider

- Can be overlooked if not performing specific checklist, including all extrinsic foot muscles and tendons

Image Interpretation Pearls

- Accessory muscles may be unilateral or bilateral
- Easily overlooked on fat-suppressed images

Reporting Tips

- Term "accessory flexor muscle" can be used if unsure whether muscle represents accessory FHL or FDL

SELECTED REFERENCES

1. Sookur PA et al: Accessory muscles: anatomy, symptoms, and radiologic evaluation. Radiographics. 28(2):481-99, 2008
2. Doda N et al: Symptomatic accessory soleus muscle: diagnosis and follow-up on magnetic resonance imaging. Br J Radiol. 79(946):e129-32, 2006
3. Cheung Y et al: MR imaging of the accessory muscles around the ankle. Magn Reson Imaging Clin N Am. 9(3):465-73, x, 2001
4. Cheung YY et al: Peroneus quartus muscle: MR imaging features. Radiology. 202(3):745-50, 1997

(Left) *Coronal T1WI MR shows an accessory soleus → inserting on the medial aspect of the calcaneus.* ***(Right)*** *Coronal T2WI FS MR in the same patient shows fluid → around the accessory soleus, suggesting peritendinitis. The patient complained of pain in this region associated with sustained exercise.*

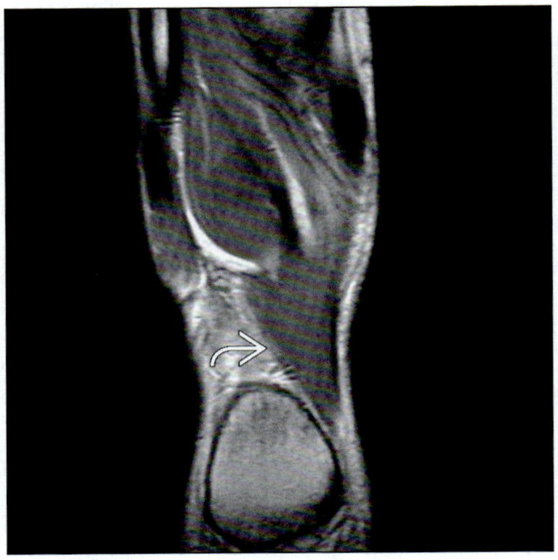

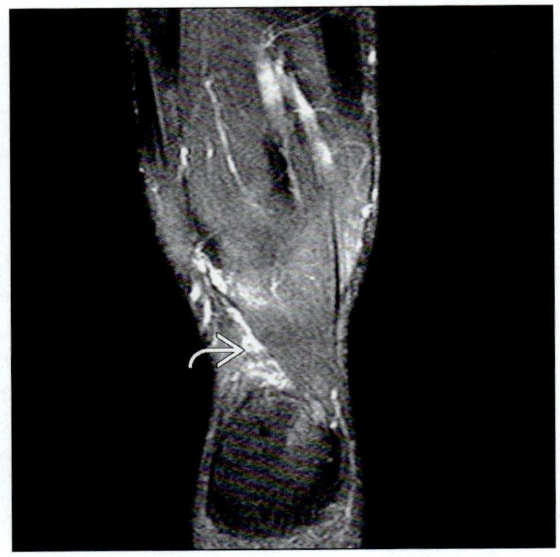

(Left) *Sagittal soft tissue algorithm CT shows an accessory flexor hallucis longus → resulting in tarsal tunnel syndrome. The fusiform shape of the muscle aids in distinguishing it from a neoplasm.* ***(Right)*** *Axial soft tissue algorithm CT in the same patient shows the accessory muscle → posterior to the flexor hallucis longus tendon →. This patient also has an os trigonum →.*

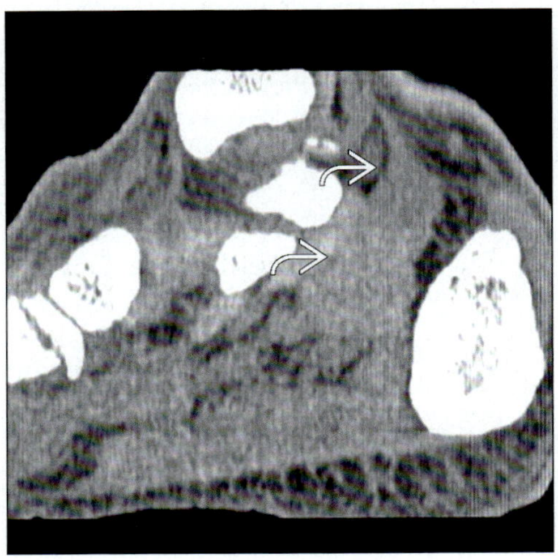

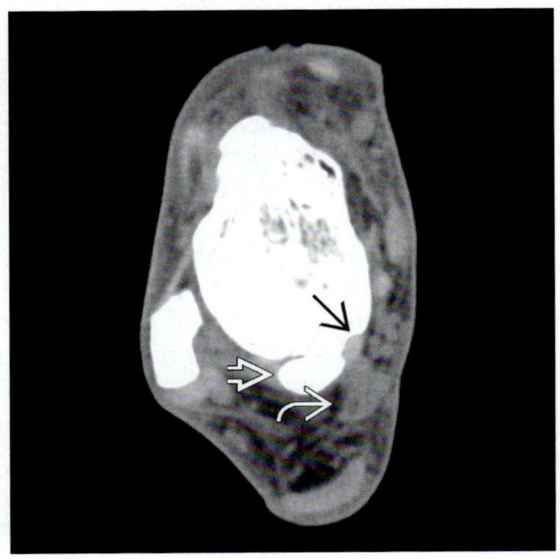

(Left) *Sagittal T1WI MR shows an accessory flexor hallucis longus muscle → in a ballet dancer who complained of pain when fully en pointe. Her pain resolved after resection of the muscle, and she felt she was better able to maintain the en pointe position.* ***(Right)*** *Sagittal STIR MR in the same patient shows that the accessory flexor hallucis muscle → is isointense to remaining musculature.*

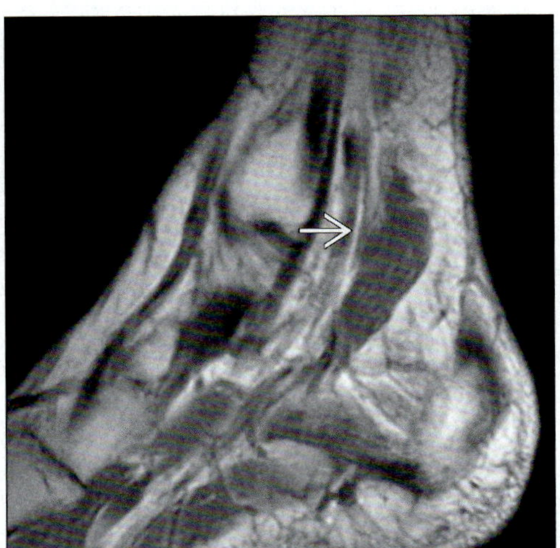

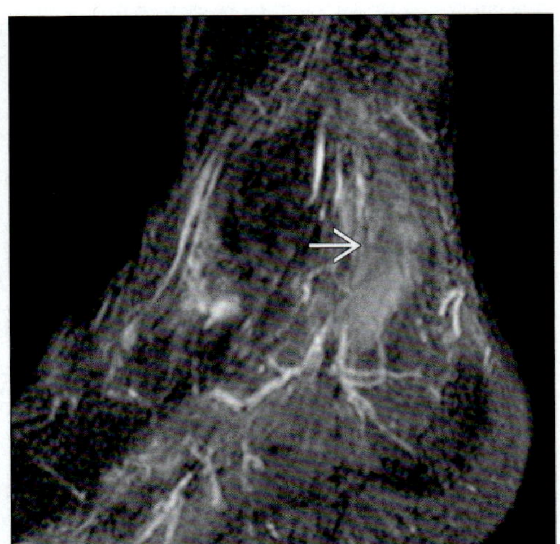

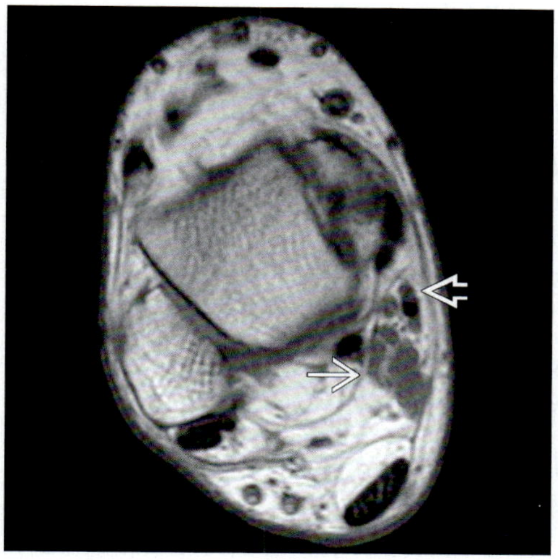

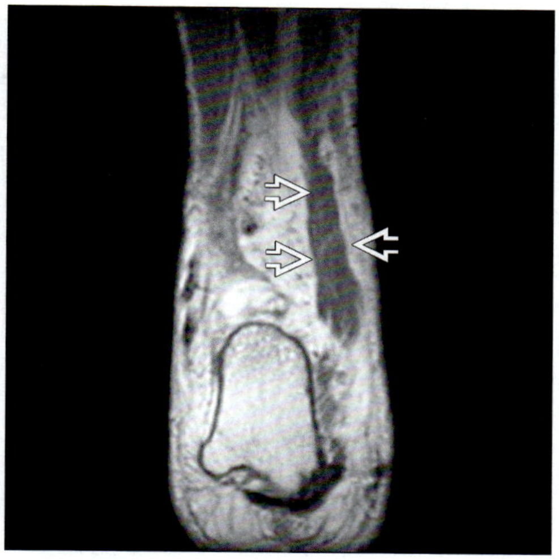

(Left) *Axial T1WI MR shows accessory flexor muscle most closely associated with the flexor digitorum longus tendon.* ***(Right)*** *Coronal T1WI MR in the same patient shows the characteristic fusiform shape of the accessory muscle.*

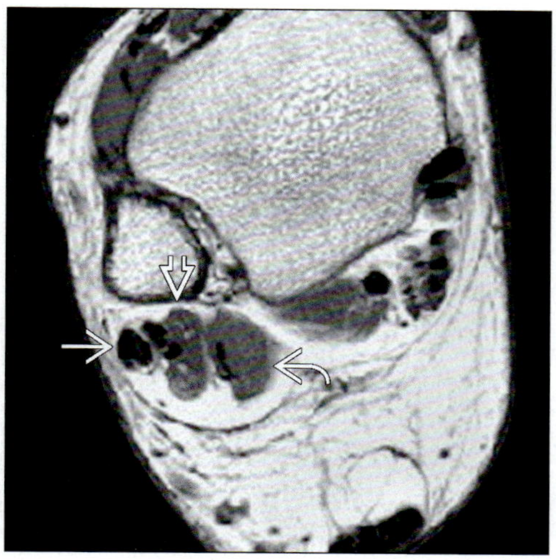

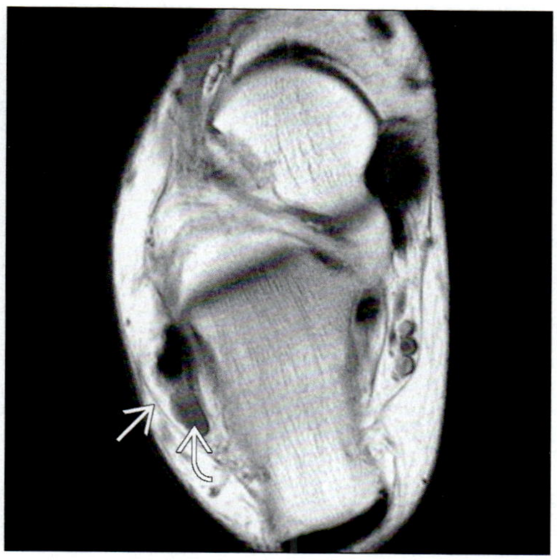

(Left) *Axial T1WI MR shows peroneus quartus adjacent to the peroneus longus and peroneus brevis tendons. There is still a small amount of peroneus brevis muscle visible at this level, which is a normal finding above the tibial plafond.* ***(Right)*** *Axial T1WI MR shows peroneus quartus muscle more distally. It lies within the inferior peroneal retinaculum.*

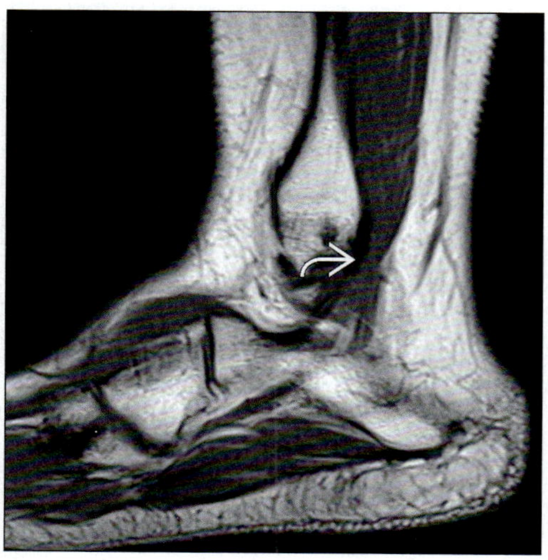

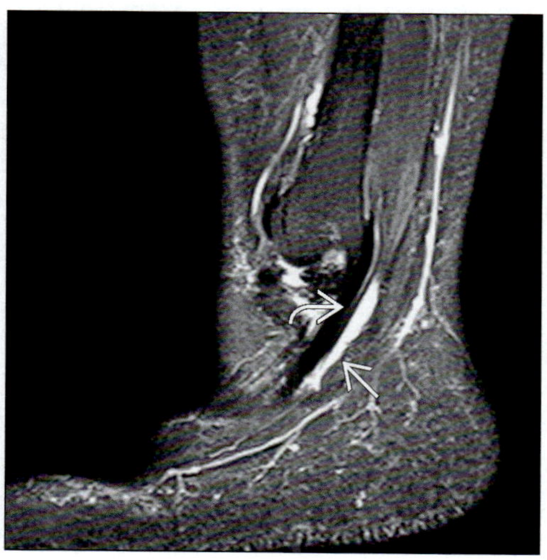

(Left) *Sagittal T1WI MR shows low-lying peroneus brevis muscle. This normal variant, like the peroneus quartus, increases pressure against the superior peroneal retinaculum and predisposes to retinacular injury and tendon dysfunction.* ***(Right)*** *Sagittal STIR MR shows low-lying peroneus brevis muscle in a patient presenting with clinical peroneal tenosynovitis. Note fluid in peroneal tendon sheath, a fairly reliable sign of clinical tenosynovitis.*

ACHILLES TENDON TEAR AND TENDINOPATHY

Key Facts

Terminology

- Tendinopathy: Tendon thickening ± mucoid degeneration, usually midsubstance of tendon
- Tendinitis: Inflammation at tendon insertion
- Paratenonitis: Inflammation in paratenon anterior to tendon

Imaging

- Paratenonitis: Normal tendon, fluid seen around anterior tendon margin
- Tendinopathy: Tendon > 8 mm AP dimension, convex anterior contour, ± abnormal T2 signal
  - Noninsertional tendinopathy (most common): Midsubstance of tendon
  - Insertional tendonitis: Near calcaneal insertion
- Tear: Usually midsubstance of tendon, sometimes musculotendinous junction
  - Interstitial tears: High signal parallel to long axis of tendon with intact surrounding fibers
  - Partial tears: Tendon thinning without complete interruption
  - Complete tear: Tendon discontinuity, ends usually retracted

Top Differential Diagnoses

- Tear medial head gastrocnemius or plantaris
- Haglund disease
- Diffuse idiopathic skeletal hyperostosis (DISH)
- Bursitis

Clinical Issues

- Tear: Sudden pop, patient falls, unable to plantar flex foot
- Paratenonitis: Painful, swollen tendon
- Tendinopathy: Clinically silent or pain and swelling

__(Left)__ Graphic shows a tear of the Achilles tendon at the most common location, in the midsubstance of the tendon. Note the pre-Achilles bursa between the posterior-superior margin of the calcaneal tuberosity and the Achilles tendon. The Achilles tendon inserts further distally on the calcaneal tuberosity. __(Right)__ Sagittal PDWI FS MR shows a partial thickness Achilles tendon tear, with fluid in torn anterior portion but preserved posterior fibers.

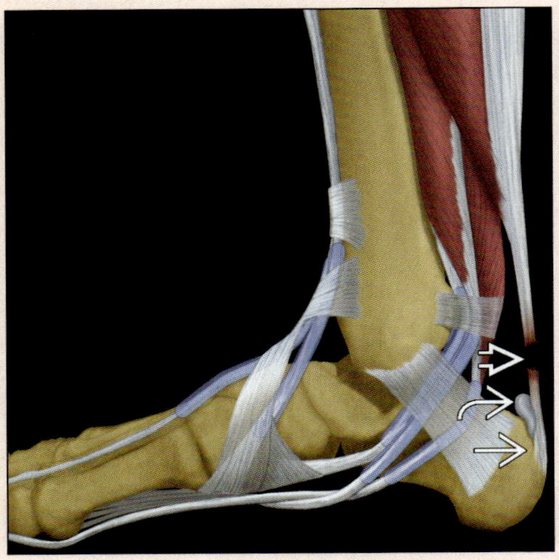

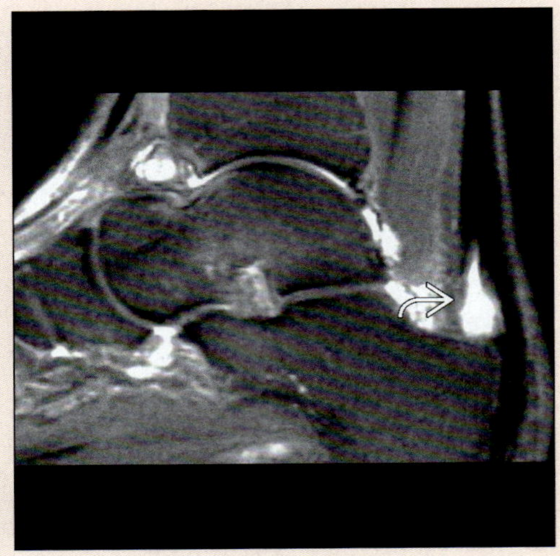

__(Left)__ Sagittal STIR MR shows an acute tear, with hematoma filling the gap between the torn tendon ends. The distance between the tendon ends is useful information for the surgeon and should be recorded in the radiology report. __(Right)__ Sagittal STIR MR shows a minimally retracted tear. Thickening of the tendon adjacent to the tear is due to underlying tendinopathy. Injury to the distal soleus muscle is also evident.

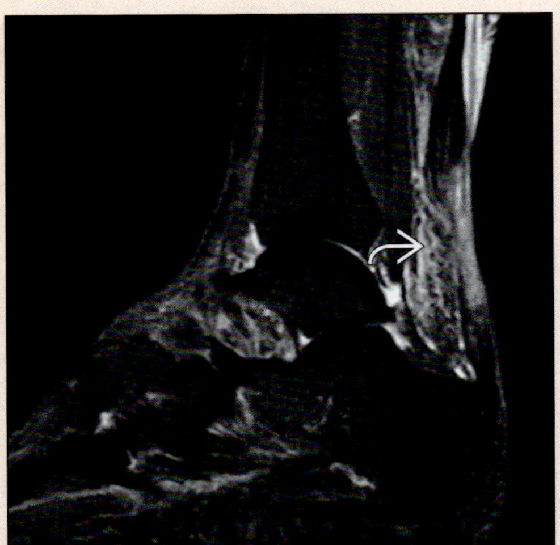

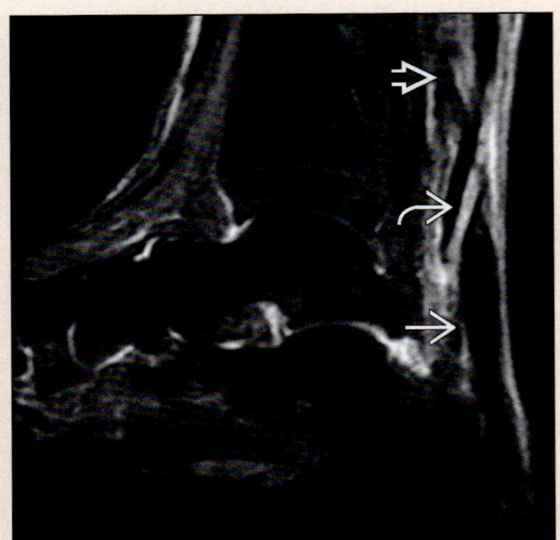

ACHILLES TENDON TEAR AND TENDINOPATHY

TERMINOLOGY

Definitions

- Paratenon: Loose connective tissue surrounding tendon
  - Found in tendons, such as Achilles tendon, which lack true tendon sheath
- Paratenonitis: Inflammation of paratenon
  - a.k.a. peritendinitis, paratendinitis, peritendinopathy
- Tendinitis: Inflammatory change in tendon
  - Occurs at Achilles tendon insertion on calcaneus
- Tendinopathy: Tendon degeneration without inflammation
  - Occurs in midsubstance of Achilles tendon

IMAGING

General Features

- Location
  - Noninsertional tendinopathy (most common): Midsubstance of tendon
  - Insertional tendonitis: Near calcaneal insertion
  - Tear: Usually midsubstance of tendon, sometimes musculotendinous junction
- Morphology
  - Tendinopathy: Tendon > 8 mm AP dimension, convex anterior contour
    - Normal < 6 mm AP dimension, concave anterior contour

Radiographic Findings

- Tendinopathy: Tendon thickening ± fluid in pre-Achilles (Kager) fat pad
  - Linear or globular calcifications frequently present
- Tear: Loss of tendon definition, fluid in Kager fat pad

MR Findings

- Paratenonitis: Normal tendon, fluid around anterior tendon margin, ± Kager fat pad
- Tendinopathy: Tendon thickening > 8 mm anteroposterior dimension
  - Tendon contour becomes convex at anterior margin (normally concave)
  - Tendon develops fusiform contour on sagittal images, thickest in midsubstance
  - Mucoid degeneration variably present
    - ↑ signal intensity < fluid on T2WI, STIR
- Insertional tendonitis: Achilles thickened distally, ill-defined longitudinal high signal at insertion site (may mimic partial tear)
- Tear: Classified as interstitial, partial, or complete
  - Interstitial tears: High signal parallel to long axis of tendon, with intact surrounding fibers
  - Partial tears: Heterogeneous high signal and tendon thinning without complete interruption
  - Complete tear: Usually see fluid-filled gap in tendon
- Postoperative findings
  - Tendon remains thick, heterogeneous on fluid sensitive sequences
  - Pre-Achilles fluid commonly present
  - Sutures visible at tendon repair site

Ultrasonographic Findings

- Achilles tendon tear
  - Defect in tendon, posterior acoustic shadowing
  - US can distinguish complete from partial tears with 92% accuracy

Imaging Recommendations

- Best imaging tool
  - MR
- Protocol advice
  - Requires both T1 (or PD) and fluid-sensitive (preferably fat-suppressed) sequences
  - Requires both sagittal and axial planes
  - Coronal plane can be used to confirm findings

DIFFERENTIAL DIAGNOSIS

Xanthoma

- Seen in hypercholesterolemia, especially inherited forms
- Mass containing lipid-laden macrophages in Achilles and other tendons (also subcutaneous)
- Causes lobulated thickening of tendon
- Intermediate signal intensity on all sequences

Tear of Medial Head Gastrocnemius or Plantaris

- In differential for "tennis leg"

Haglund Syndrome

- Retro-Achilles bursa compressed against posterior lateral calcaneal prominence
- Calcaneal tuberosity focally enlarged
- Fluid in retrocalcaneal bursa and retro-Achilles bursa

DISH

- Enthesophyte forms at tendon insertion
- Seen primarily in patients > 50 years old

Bursitis

- May have bursitis either in anatomic pre-Achilles bursa or in acquired retro-Achilles bursa

Accessory Soleus

- Causes pre-Achilles fullness, sometimes tenderness after exercise
- Small retrospective series suggests association with Achilles tendinopathy

Ossification of Achilles Tendon

- Usually secondary to prior surgery or trauma
- Occasionally seen in longstanding tendinopathy

PATHOLOGY

General Features

- Etiology
  - Paratenonitis occurs primarily in runners
  - Tendinopathy due to repetitive microtrauma and increasing tendon stiffness with age
    - 6-18% of runners
  - Tear occurs during sports with sudden starting and stopping (e.g., tennis and basketball)
  - Most tendinopathy and tears occur in relatively hypovascular portion (2-6 cm above insertion)
  - Preexisting conditions increase risk of tear

ACHILLES TENDON TEAR AND TENDINOPATHY

- Tendinopathy
- Hyperpronation/forefoot varus/cavus foot/ equinus deformities
- Systemic arthropathy: Rheumatoid, spondyloarthropathy
- Other systemic conditions predisposing to tendon weakness
  - Steroid use
  - Diabetes mellitus
  - Gout
  - Connective tissue disorders
- Fluoroquinolone treatment

- Associated abnormalities
  - Sural nerve injury: Sural nerve near lateral border of Achilles tendon
  - Plantaris tendon tear
  - Haglund deformity
  - Pre-Achilles bursitis
  - Accessory soleus muscle

Staging, Grading, & Classification

- Weinstabi classification
  - Type I: Inflammatory reaction
  - Type II: Degenerative changes
  - Type III: Partial tear
  - Type IV: Complete tear

Gross Pathologic & Surgical Features

- Tendon is composed of fibers arising from gastrocnemius and soleus muscles
- Tendinopathy: Intratendinous degeneration without significant inflammatory response
- Achilles paratenon may be affected by systemic inflammatory diseases, such as rheumatoid arthritis
- Watershed zone of tendon most commonly affected
  - Area 2-6 cm above calcaneal insertion

Microscopic Features

- Histologic evidence of tendinopathy found in 95% of torn Achilles tendons
  - Confirmation that tear is acute event superimposed on chronic process

CLINICAL ISSUES

Presentation

- Most common signs/symptoms
  - Paratenonitis: Painful, swollen tendon
  - Tendinopathy: Often clinically silent, may present with pain and swelling
  - Tear
    - Classic history: Felt sudden pop, fell, unable to plantar flex foot
    - Palpable tendon defect variably present
    - Soft tissue swelling may obscure tendon gap
    - Weak plantar flexion may be retained because plantaris tendon is intact
- Other signs/symptoms
  - Calf squeeze test: Circumferential compression to thickest portion of gastrocnemius muscle
    - Causes passive ankle flexion if Achilles is intact

Demographics

- Gender
  - M > > F
- Epidemiology
  - Tendinopathy and tear usually seen in "weekend warriors" in 3rd to 5th decades of life
  - Insertional tendinitis tends to occur in older, less active individuals
  - Tendon tear: 18 per 100,000
  - Most commonly injured ankle tendon; 3rd most frequent tendon rupture overall

Natural History & Prognosis

- Patients progress from tendinopathy to tear
- Tear
  - Nonsurgical rate of rerupture: 21%
  - Surgical rate of rerupture: 2-5%
  - Surgical repair favored in young, athletic patients with high activity levels
- Prolonged recovery from surgery: 2-3 months immobilization, restricted weight-bearing

Treatment

- Tendinopathy
  - Rest, stretching exercises, orthoses, anti-inflammatory medications
  - If severe, debridement of mucoid degeneration and oversewing of damaged fibers
- Partial tears: Surgery or cast immobilization
- Complete tears: Usually surgically repaired
  - May immobilize with cast acutely or if related to use of steroids
  - < 3 cm diastasis: End-to-end anastomosis
  - > 3 cm diastasis often requires tendon graft (usually flexor hallucis longus)

DIAGNOSTIC CHECKLIST

Consider

- Axial images: Measure tendon thickening, identify intact plantaris tendon

Image Interpretation Pearls

- Use both sagittal and axial images to distinguish partial from complete tear
- Check radiographs for calcific tendinopathy, which may mimic tear on MR

Reporting Tips

- Report length of gap between torn tendon ends

SELECTED REFERENCES

1. Courville XF et al: Current concepts review: noninsertional Achilles tendinopathy. Foot Ankle Int. 30(11):1132-42, 2009
2. Luck MD et al: High association between accessory soleus muscle and Achilles tendonopathy. Skeletal Radiol. 37(12):1129-33, 2008
3. Richards PJ et al: Achilles tendon ossification: pathology, imaging and aetiology. Disabil Rehabil. 30(20-22):1651-65, 2008
4. Sode J et al: Use of fluroquinolone and risk of Achilles tendon rupture: a population-based cohort study. Eur J Clin Pharmacol. 63(5):499-503, 2007
5. Chiodo CP et al: Current concepts review: acute ruptures of the achilles tendon. Foot Ankle Int. 27(4):305-13, 2006
6. Schweitzer ME et al: MR imaging of disorders of the Achilles tendon. AJR Am J Roentgenol. 175(3):613-25, 2000

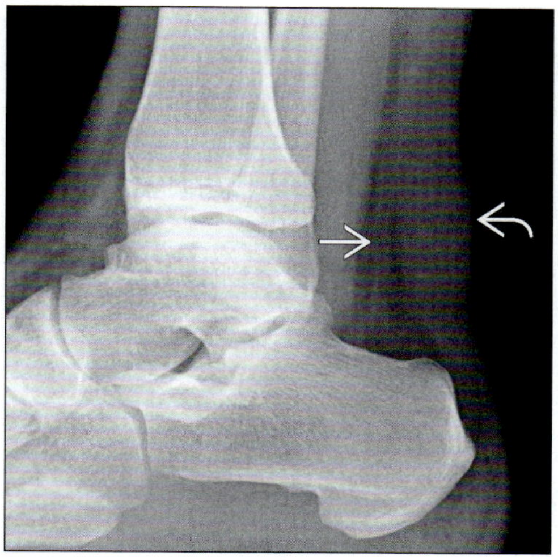

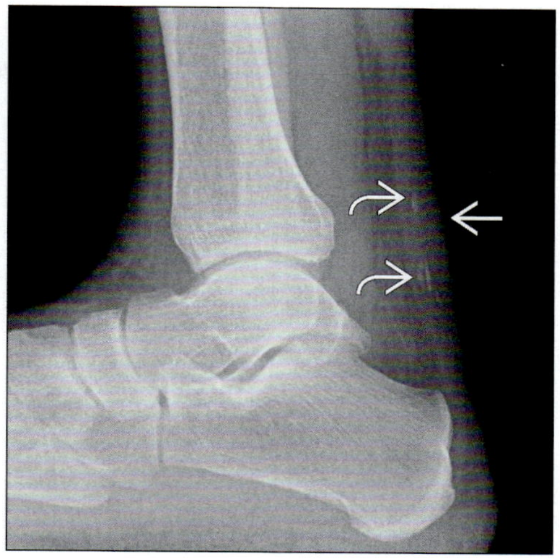

(Left) *Lateral radiograph shows radiographic appearance of tendinopathy. There is a markedly thickened Achilles tendon indenting the pre-Achilles (Kager) fat pad. Fluid stranding in Kager fat pad is typical of paratenonitis.* ***(Right)*** *Lateral radiograph of Achilles tendinopathy shows thickening of the Achilles tendon and tendon calcification. Calcification is a common finding in Achilles tendinopathy, and ossification may also occur.*

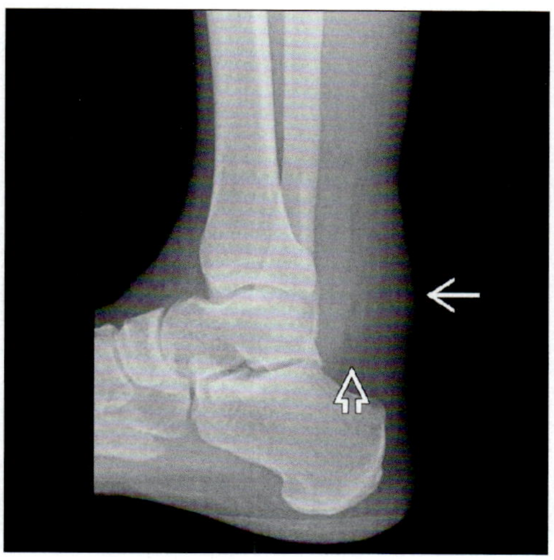

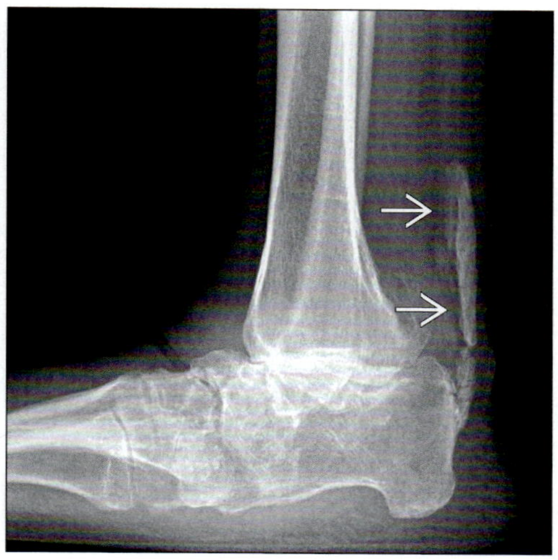

(Left) *Lateral radiograph shows radiographic findings of a tendon tear. There is severe soft tissue swelling in the region of the Achilles tendon and fluid nearly obliterating Kager fat. The changes in Kager fat are more extensive than those commonly seen in tendinopathy.* ***(Right)*** *Lateral radiograph shows extensive ossification of the Achilles tendon in a patient who underwent a tendon lengthening procedure in the remote past as part of correction of clubfoot deformity.*

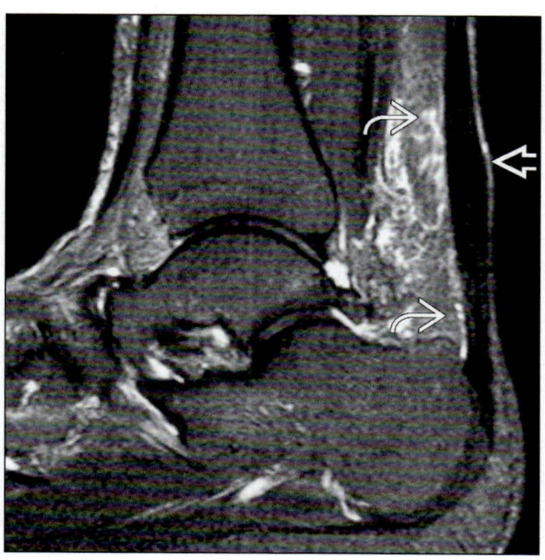

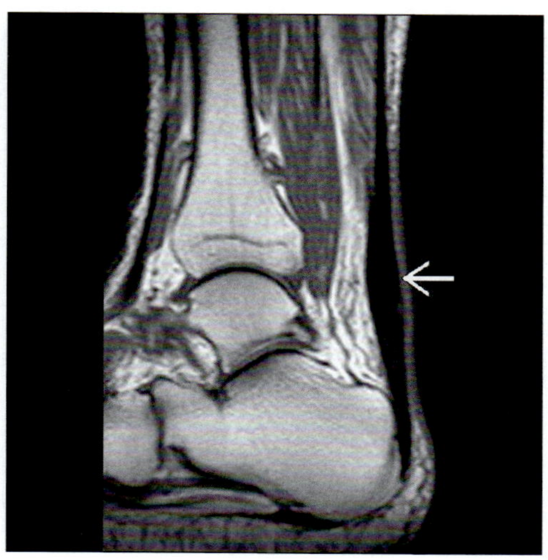

(Left) *Sagittal STIR MR shows fluid along the anterior margin of the Achilles tendon, indicating paratenonitis. Fusiform tendon thickening indicates tendinopathy, but in this case there is no abnormal signal to suggest mucoid degeneration.* ***(Right)*** *Sagittal PDWI MR shows fusiform enlargement within the Achilles tendon typical of tendinopathy. Changes are centered in the midportion of the tendon, where there is relative hypovascularity.*

ACHILLES TENDON TEAR AND TENDINOPATHY

(Left) *Xanthomatous changes are present diffusely in the markedly thickened Achilles tendon. The areas of xanthoma are slightly higher signal intensity than the tendon itself.* ***(Right)*** *Same patient as prior image shows multiple areas of increased signal intensity corresponding to xanthoma. The striations in the tendon are more regular, numerous, and extensive than those seen with tendinopathy and are lower signal intensity than partial thickness tears.*

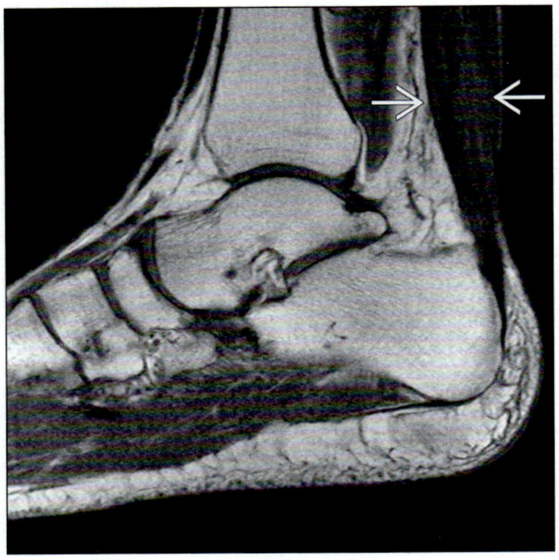

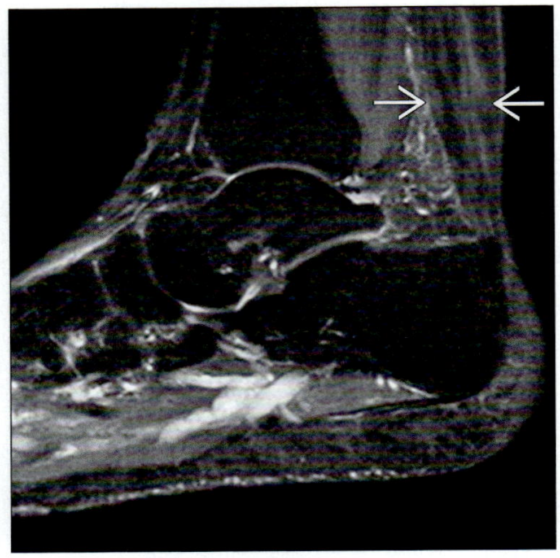

(Left) *Sagittal STIR MR shows severe tendinopathy with mucoid degeneration, insertional tendonitis, and paratenonitis.* ***(Right)*** *Sagittal graphic of insertional Achilles tendinitis. There is bony prominence of posterior process (Haglund deformity). The distal tendon is thickened and inflammatory changes are present in the tendon.*

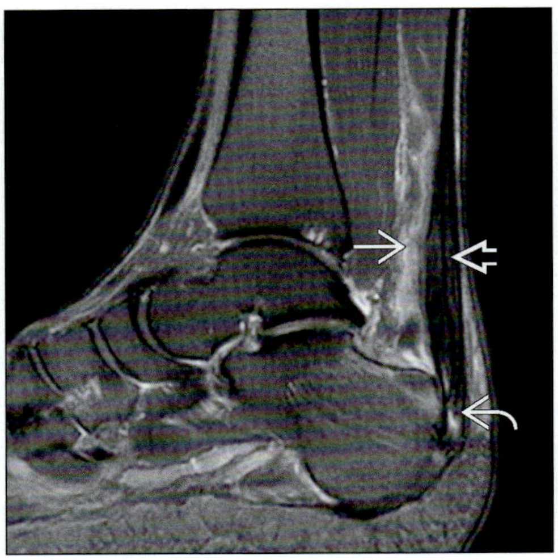

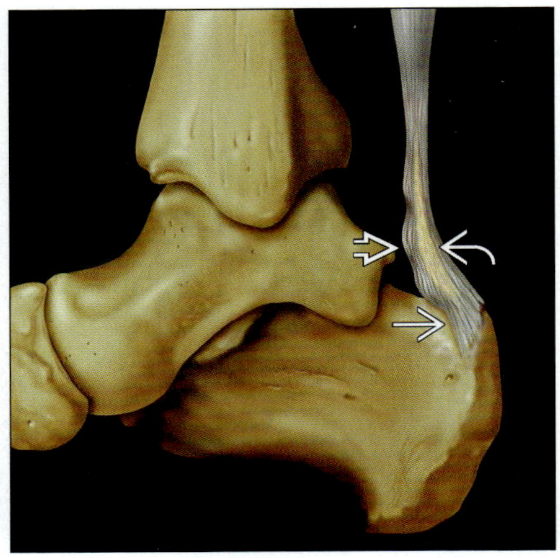

(Left) *Sagittal PD/intermediate FS MR shows early interstitial tear and paratenonitis.* ***(Right)*** *Sagittal PD/intermediate FS MR shows hyperintense linear signal in an insertional Achilles partial tear with reactive calcaneal marrow edema.*

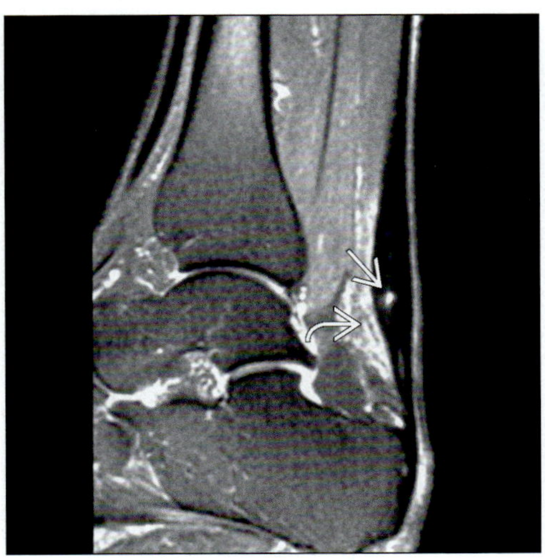

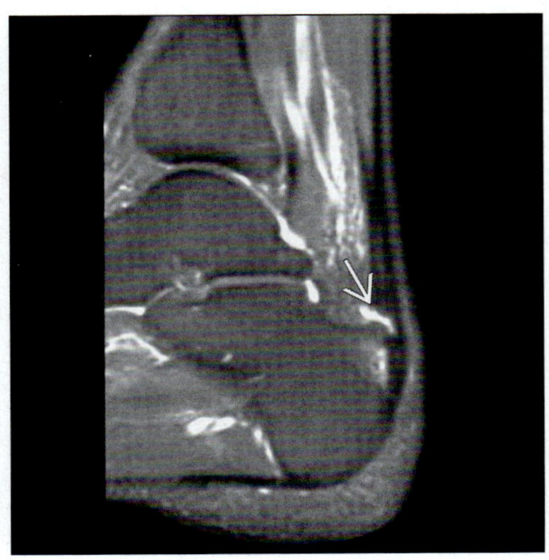

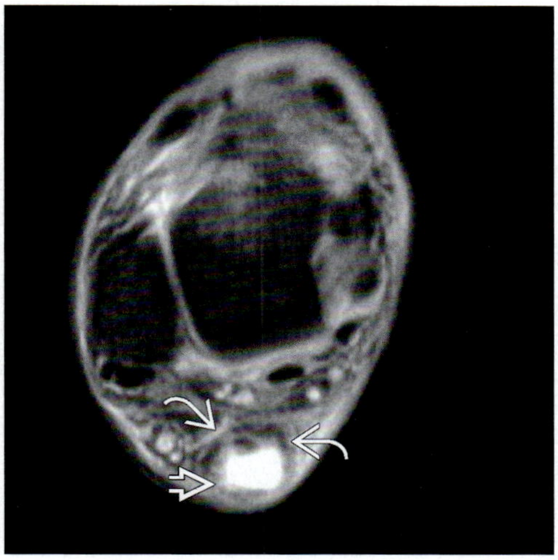

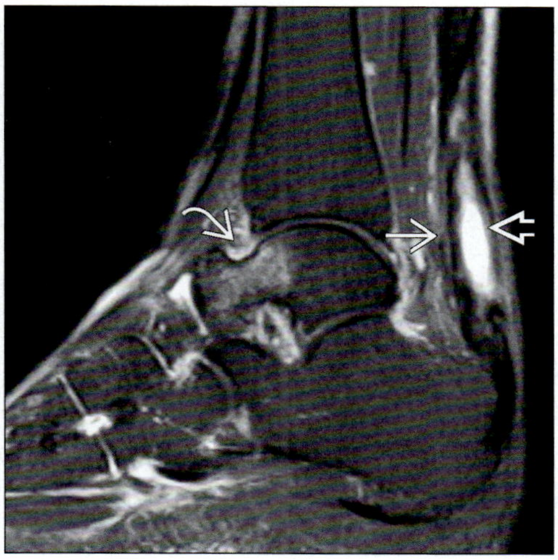

(Left) *Axial T2WI FS MR shows fluid filling the majority of the Achilles tendon, with a few low signal intensity tendon fibers still visible. On this single image, the tear could be complete or partial.* ***(Right)*** *Sagittal STIR MR in the same patient shows the high-grade partial thickness tear. Fluid fills a tendon defect, but taut, intact fibers are still present anteriorly. Bone marrow edema in the talus is also due to hyperdorsiflexion injury.*

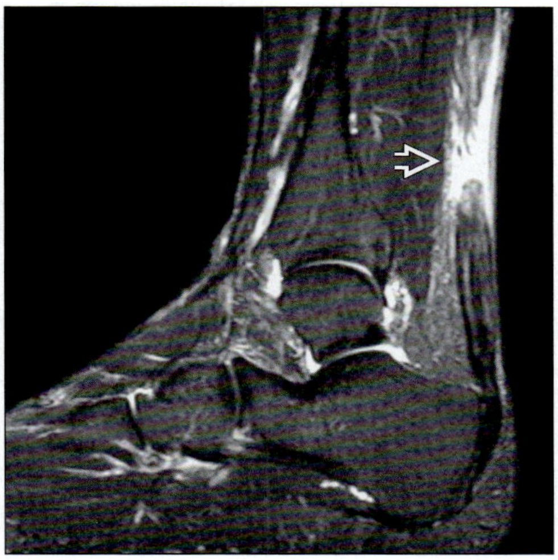

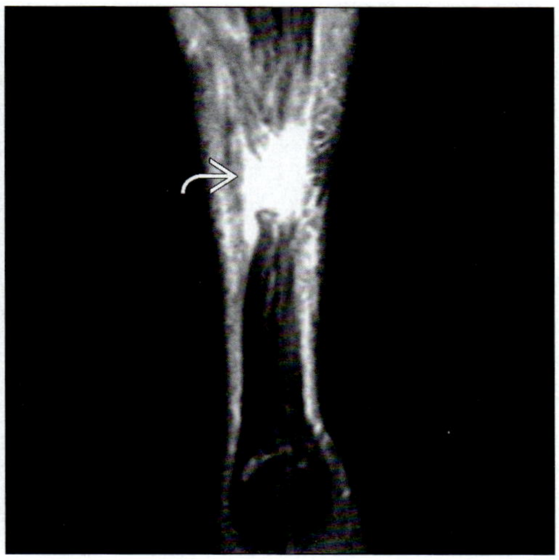

(Left) *Sagittal STIR MR shows a full thickness tendon tear, with retraction of tendon ends and fluid filling the tendon defect.* ***(Right)*** *Coronal T2WI FS MR is not a standard sequence for imaging an Achilles tendon but can be useful in distinguishing between high-grade partial tears and complete tears. This tear was complete in all planes.*

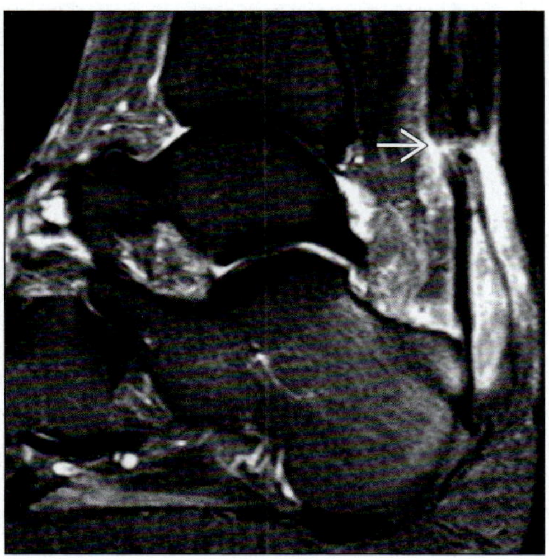

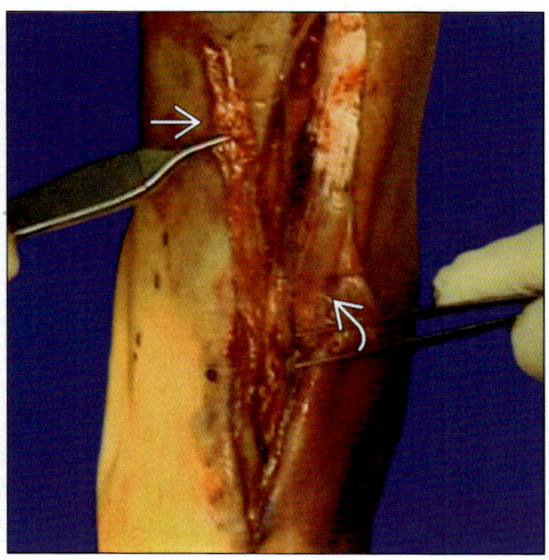

(Left) *Sagittal STIR MR shows what appears to be a complete, nondisplaced tear of the Achilles tendon. However, this was a partial thickness tear. The tautness of the tendon fibers is a clue that there are intact fibers not included on this image; this was confirmed on axial imaging.* ***(Right)*** *Posterior intraoperative photograph of the same patient shows torn Achilles fibers held in a clamp. Intact fibers are also visible.*

PLANTARIS TENDON INJURY

Key Facts

Terminology
- Tennis leg
- Partial or complete tear of plantaris myotendinous unit, usually around myotendinous junction

Imaging
- Injury may occur at myotendinous junction in proximal leg or in tendon near calcaneal insertion
- Proximal (myotendinous junction) rupture: Edema ± hematoma in proximal medial leg
  - Located between medial head of gastrocnemius and medial aspect of soleus
- Distal (tendinous) rupture: Discontinuity of plantaris tendon medial to distal Achilles tendon
  - Often seen in conjunction with partial or complete Achilles tendon tear

Top Differential Diagnoses
- Gastrocnemius or soleus muscle injury
  - May occur in conjunction with plantaris injury
- Achilles tendon injury
  - May occur in conjunction with plantaris tendon rupture

Pathology
- Eccentric contraction of plantaris muscle (muscle contracts while knee is extended and foot dorsiflexed)

Clinical Issues
- Acute onset of calf pain
- May also experience "pop"
- Plantaris muscle congenitally absent in 7-10% of population
- Usually self-limited symptoms
- Conservative therapy

*(**Left**) PA graphic shows the plantaris muscle arising from the lateral femoral condyle →. The myotendinous junction → is slightly below the knee joint. The tendon travels posterior to the soleus muscle → and inserts on the medial calcaneus → adjacent to the Achilles tendon. (**Right**) Coronal T2WI FS MR demonstrates rupture of the plantaris muscle in the proximal leg. There is edema and hemorrhage → in the space between the medial gastrocnemius → and the soleus → muscles.*

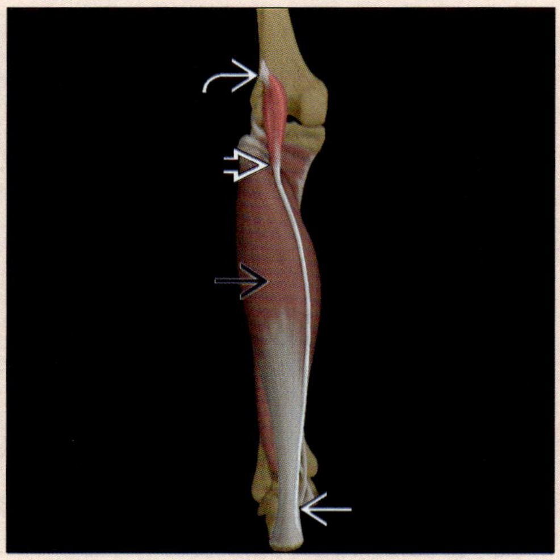

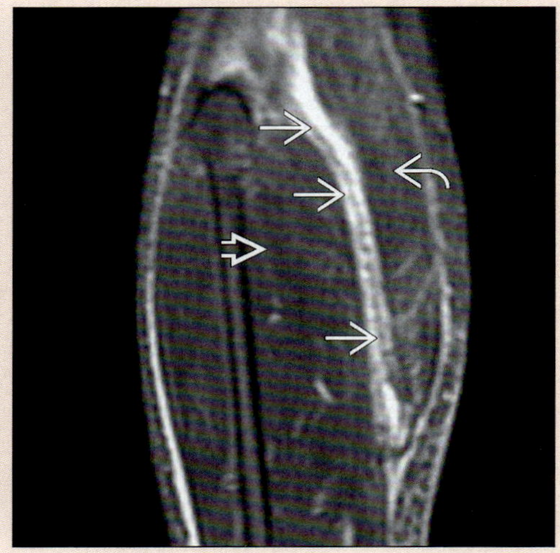

*(**Left**) Axial T2WI FS MR from the same patient shows a large hematoma → interposed between the medial head of the gastrocnemius → and the soleus muscle →, the space occupied by the plantaris muscle and its myotendinous junction. There is edema within the gastrocnemius as well, a common occurrence with plantaris rupture. (**Right**) Axial PDWI FS MR demonstrates rupture of the Achilles → and plantaris → tendons, with hemorrhagic fluid filling the space vacated in the distal leg.*

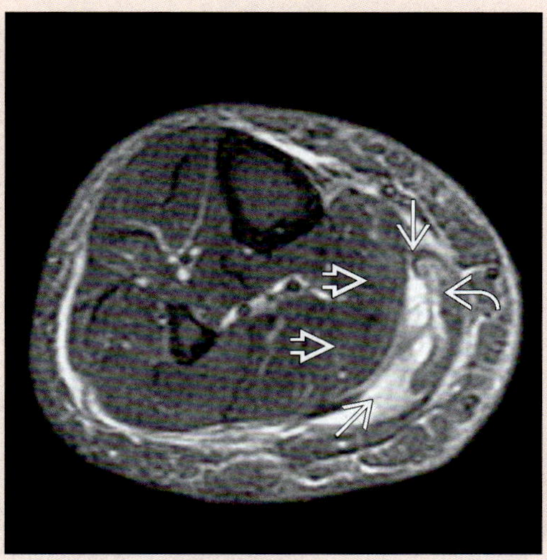

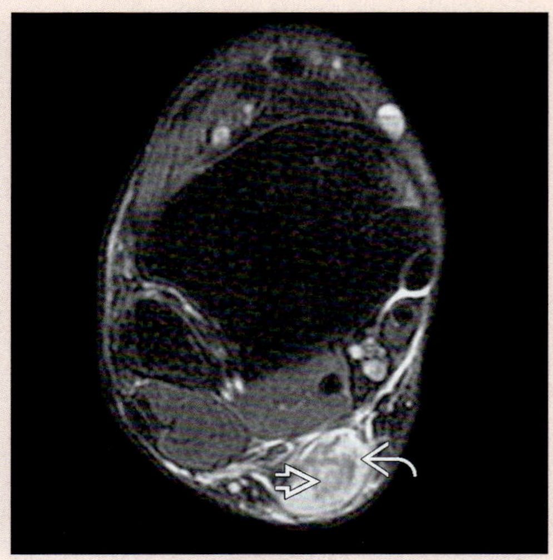

PLANTARIS TENDON INJURY

TERMINOLOGY

Synonyms
- Tennis leg

Definitions
- Partial or complete tear of plantaris myotendinous unit, usually around myotendinous junction

IMAGING

General Features
- Best diagnostic clue
  - Partial or complete disruption of plantaris myotendinous junction
- Location
  - Injury may occur at myotendinous junction in proximal leg or in tendon near calcaneal insertion

MR Findings
- Proximal (myotendinous junction) rupture
  - Edema ± hematoma in proximal medial leg
    - Located between medial head of gastrocnemius and medial aspect of soleus
    - Extends variable distance proximal-to-distal
    - May be associated with edema in gastrocnemius ± soleus muscles
  - Discontinuous plantaris muscle
- Distal (tendinous) rupture
  - Discontinuity of plantaris tendon medial to distal Achilles tendon
  - Often seen in conjunction with partial or complete Achilles tendon tear

Ultrasonographic Findings
- Discontinuity of plantaris myotendinous junction with hematoma formation
- Can be used to identify distal plantaris tendon rupture above calcaneus

Imaging Recommendations
- Best imaging tool
  - MR
- Protocol advice
  - Must include proximal leg as well as Achilles insertion in evaluation of plantaris injury

DIFFERENTIAL DIAGNOSIS

Gastrocnemius or Soleus Muscle Injury
- May occur in conjunction with plantaris injury

Achilles Tendon Injury
- May occur in conjunction with plantaris tendon rupture

PATHOLOGY

General Features
- Etiology
  - Eccentric contraction of plantaris muscle (muscle contracts while knee is extended and foot dorsiflexed)
- Associated abnormalities
  - Posterolateral corner of knee injury
  - Medial head of gastrocnemius tear

Gross Pathologic & Surgical Features
- Plantaris originates at lateral femoral condyle, superior and medial to lateral head of gastrocnemius
  - Also takes origin from oblique popliteal ligament
- Continues deep to lateral head of gastrocnemius
- Myotendinous junction about 10 cm below origin, at level of soleus origin
- Tendon travels between medial head of gastrocnemius and soleus muscles
- Extends medially and slightly anterior to Achilles tendon
- Inserts on medial calcaneus or occasionally on Achilles tendon

CLINICAL ISSUES

Presentation
- Most common signs/symptoms
  - Acute onset of calf pain
    - May also experience "pop"
    - Proximal calf pain if muscle belly or myotendinous junction injury
    - Distal calf/ankle pain if distal tendon injury
  - Calf swelling/edema

Demographics
- Epidemiology
  - Plantaris muscle congenitally absent in 7-10% of population

Natural History & Prognosis
- Usually self-limited symptoms

Treatment
- Conservative

DIAGNOSTIC CHECKLIST

Consider
- Plantaris rupture with posteromedial proximal leg pain and edema between medial gastroc and soleus

Image Interpretation Pearls
- Absence of distal plantaris tendon does not mean rupture; it may be congenitally absent

SELECTED REFERENCES

1. Boer R et al: Neglected Achilles tendon rupture with central insertional plantaris tendon hypertrophy: two cases. Strategies Trauma Limb Reconstr. 4(1):41-3, 2009
2. Das S: Plantaris tendon: cadaveric study with clinical implications. Surg Radiol Anat. 31(1):73-4, 2009
3. Lopez GJ et al: Plantaris rupture: A mimic of deep venous thrombosis. J Emerg Med. Epub ahead of print, 2009
4. Mozena J et al: Plantaris tendon ruptures. J Am Podiatr Med Assoc. 94(5):505-8, 2004
5. Saxena A et al: Magnetic resonance and cadaveric findings of the incidence of plantaris tendon. Foot Ankle Int. 21(7):570-2, 2000
6. Leekam RN et al: Using sonography to diagnose injury of plantaris muscles and tendons. AJR Am J Roentgenol. 172(1):185-9, 1999
7. Helms CA et al: Plantaris muscle injury: evaluation with MR imaging. Radiology. 195(1):201-3, 1995

FLEXOR HALLUCIS LONGUS TENDON TEAR, IMPINGEMENT, AND TENDINOPATHY

Key Facts

Terminology

- Os trigonum syndrome: Impingement of FHL by os trigonum
- FHL impingement syndrome: Os trigonum, large posterior process, or fracture of posterior process limits gliding of FHL through fibro-osseous tunnel at posterior process of talus

Imaging

- Fluid in tendon sheath normal finding because of communication with ankle joint
- Tenosynovitis: Tendon sheath is irregular or has lobulated contour
- MR can be false-negative
  - Tenosynography may reveal irregular, scarred tendon sheath despite normal MR
- FHL impingement
  - Large effusion in tendon sheath, with cut off at level of os trigonum or posterior process
- Tendon tear: May occur at musculotendinous junction or in forefoot

Top Differential Diagnoses

- Posterior tibial tendon tear & tendinopathy
- Achilles tendinopathy
- Plantar fasciitis

Pathology

- Associated with: Os trigonum, accessory FHL muscle, fracture of posterior process of talus

Clinical Issues

- Symptomatic os trigonum may be surgically excised
- Tenosynovitis may be treated with steroid injection under fluoroscopy or ultrasound

__(Left)__ Graphic of the medial side of the ankle shows tendinopathy of the flexor hallucis longus (FHL) ➾. Tendinopathy is usually centered in this region, where the FHL passes through a fibro-osseous tunnel between the medial and lateral tubercles of the posterior process of the talus. __(Right)__ Sagittal STIR MR shows irregular, lobulated fluid ➾ in proximal FHL tendon sheath, with a normal amount of fluid distally ➔. This pattern is characteristic of posterior tendon impingement.

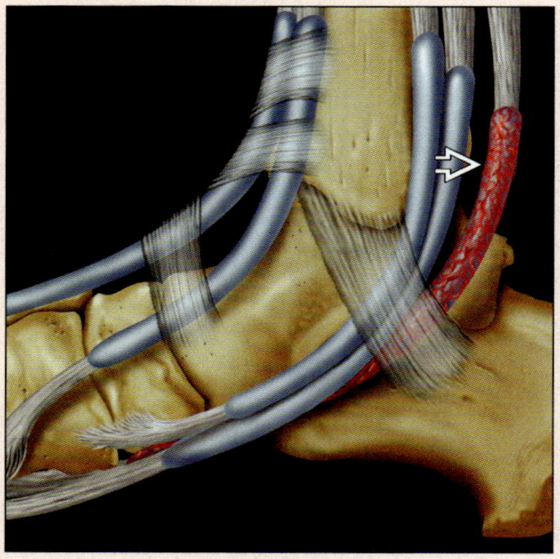

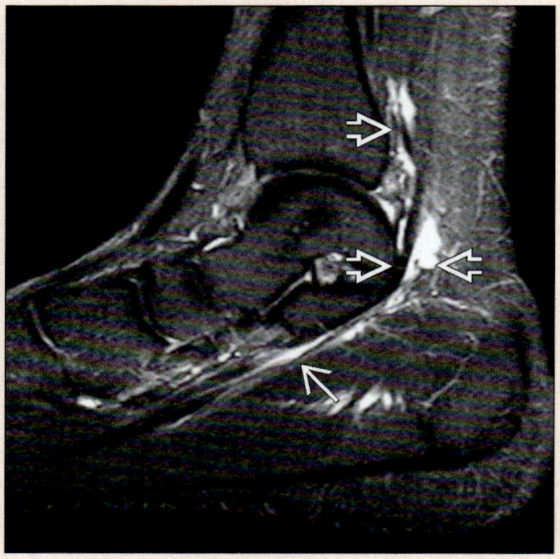

__(Left)__ Lateral therapeutic tenosynogram shows irregular dilation of tendon sheath ➾ with a cut off sign ➔ typical of posterior impingement. Corticosteroid injection under fluoroscopic guidance is an effective treatment of FHL tenosynovitis. __(Right)__ Sagittal T2WI FS MR shows 2 cartilaginous bodies ➔ in the flexor hallucis longus tendon sheath. Post-traumatic changes in the talar dome are present ➚.

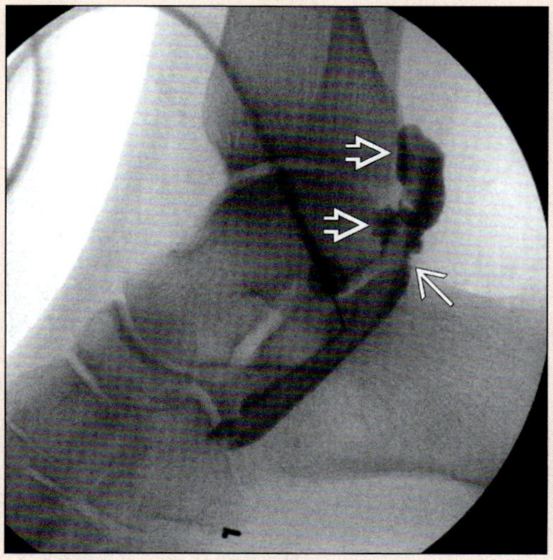

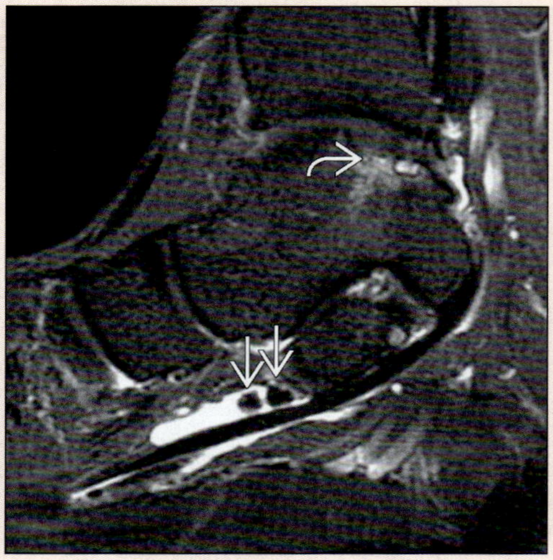

FLEXOR HALLUCIS LONGUS TENDON TEAR, IMPINGEMENT, AND TENDINOPATHY

TERMINOLOGY

Abbreviations
- Flexor hallucis longus (FHL)

Definitions
- Os trigonum: Accessory center of ossification at posterior process of talus
- Os trigonum syndrome: Impingement of FHL by os trigonum
- FHL impingement syndrome: Os trigonum, large posterior process, or fracture of posterior process limits gliding of FHL through fibro-osseous tunnel at posterior process of talus
- Check rein deformity: Tethering of FHL by flexor retinaculum
  - Due to scar from previous injury

IMAGING

MR Findings
- Tenosynovitis
  - Fluid around FHL tendon is normal finding
    - Due to communication of ankle joint with FHL tendon sheath
  - Diagnosis of tenosynovitis can be made when sheath is irregular or has lobulated contour
  - MR not 100% sensitive
- FHL impingement
  - Large effusion in tendon sheath with cut off at level of os trigonum or posterior process
- Tendon tear
  - Musculotendinous junction
  - Forefoot
    - Usually complete
    - Retracted, discontinuous tendon
- Intermittent tendon dislocation
  - Dynamic MR or ultrasound may diagnose this rare condition

Imaging Recommendations
- Best imaging tool
  - MR

Other Modality Findings
- Tenosynography may reveal irregular, scarred tendon sheath despite normal MR

DIFFERENTIAL DIAGNOSIS

Posterior Tibial Tendon Pathology
- Also presents with posteromedial ankle pain & swelling

Achilles Tendinopathy
- Posterior ankle pain with dorsiflexion

Plantar Fasciitis
- Medial heel pain

PATHOLOGY

General Features
- Associated abnormalities
  - Os trigonum
    - If mobile, predisposes to FHL impingement
    - Cysts at synchondrosis between os trigonum and talus suggest abnormal motion
  - Accessory FHL muscle
    - Increases stress on tendon as it passes through sulcus of posterior talar process
  - Inflammatory arthropathy
  - Fracture of posterior process of talus

Relevant Anatomy
- Flexor hallucis longus lies lateral to posterior tibial and flexor digitorum longus tendons
  - Tendon sheath communicates with ankle joint
- Runs through sulcus between medial and lateral tubercles of posterior process of talus
  - Fascia overlying tendon in sulcus creates fibro-osseous tunnel
- Distally, FHL runs between flexor hallucis brevis and sesamoids of 1st toe and inserts on distal phalanx

CLINICAL ISSUES

Presentation
- Other signs/symptoms
  - Tenosynovitis: Limited dorsiflexion of 1st toe when ankle is dorsiflexed
  - Tear: Inability to plantarflex 1st interphalangeal joint

Demographics
- Epidemiology
  - Most common in ballet dancers, basketball players

Treatment
- Symptomatic os trigonum may be surgically excised
- Tenosynovitis may be treated with steroid injection under fluoroscopy or ultrasound

DIAGNOSTIC CHECKLIST

Image Interpretation Pearls
- Fluid in FHL tendon sheath is normal, because sheath communicates with joint
  - Fluid probably abnormal if large effusion in sheath, irregular sheath contour, or abrupt cut off of fluid at posterior process/os trigonum

SELECTED REFERENCES

1. O'Loughlin P et al: Fracture of the medial tubercle of the posterior process of the talus: magnetic resonance imaging appearance with clinical follow-up. HSS J. 5(2):161-4, 2009
2. Michelson J et al: Tenosynovitis of the flexor hallucis longus: a clinical study of the spectrum of presentation and treatment. Foot Ankle Int. 26(4):291-303, 2005
3. Na JB et al: The flexor hallucis longus: tenographic technique and correlation of imaging findings with surgery in 39 ankles. Radiology. 236(3):974-82, 2005
4. Lo LD et al: MR imaging findings of entrapment of the flexor hallucis longus tendon. AJR Am J Roentgenol. 176(5):1145-8, 2001
5. Wredmark T et al: Os trigonum syndrome: a clinical entity in ballet dancers. Foot Ankle. 11(6):404-6, 1991

(Left) *Sagittal T2WI FS MR shows distended, lobulated flexor hallucis longus tendon sheath both above and below the level of the ankle.* ***(Right)*** *Sagittal STIR MR shows a focal fluid collection at FHL musculotendinous junction, as well as fluid in the posterior recess of the posterior subtalar joint.*

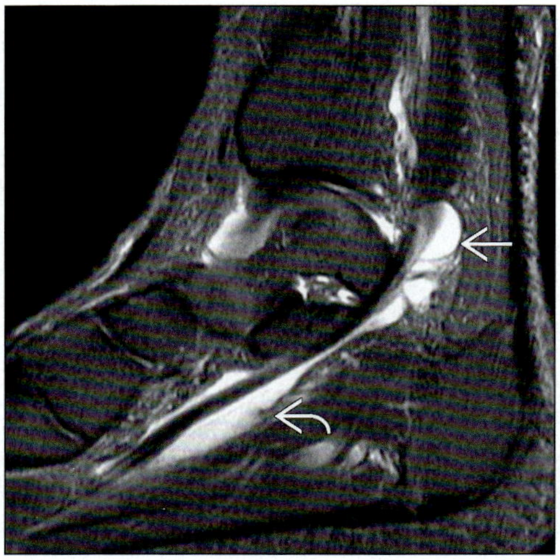

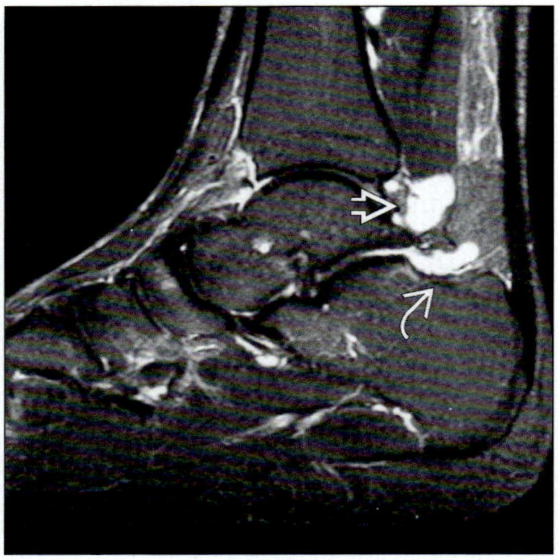

(Left) *Axial T2WI FS MR shows a partial tear at the FHL musculotendinous junction. This was an acute injury in a college football player.* ***(Right)*** *Axial PDWI FS MR shows small cyst in the FHL tendon sheath at the level of the musculotendinous junction, suggesting tenosynovitis. Symptoms resolved after therapeutic tenosynogram.*

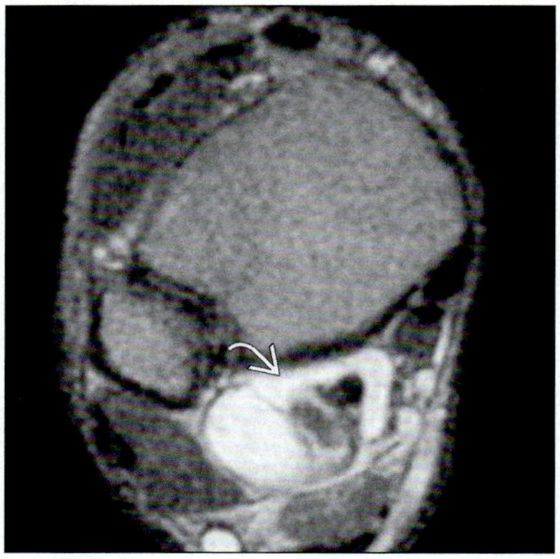

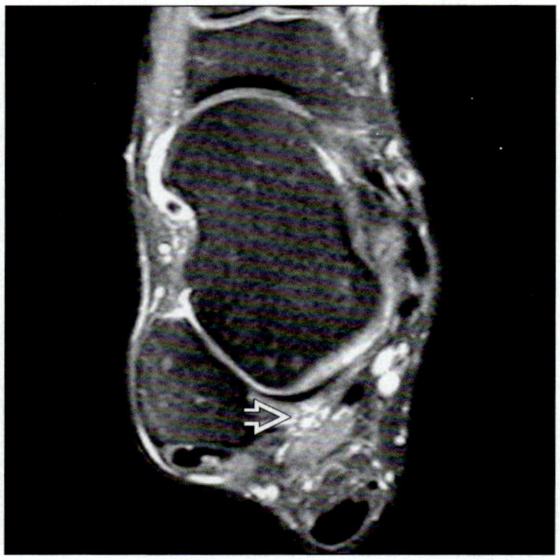

(Left) *Sagittal T1WI MR in a patient with chronic foot pain shows a retracted tear of the flexor hallucis longus tendon with moderate fat atrophy of the FHL muscle belly.* ***(Right)*** *Axial PDWI MR at the level of the ankle in the same patient shows scar in the expected location of the flexor hallucis longus. The posterior tibial and flexor digitorum longus tendons are intact.*

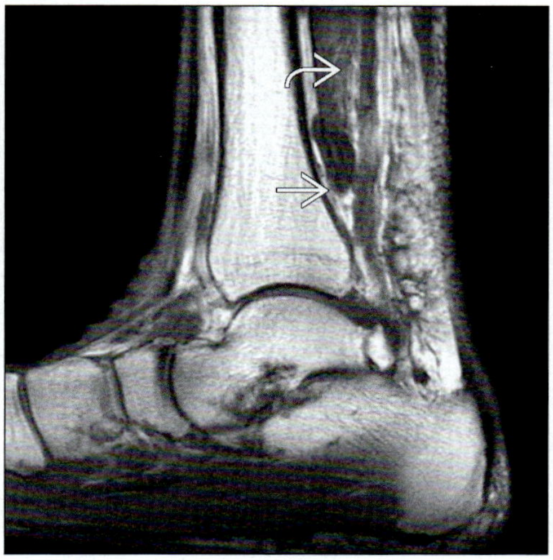

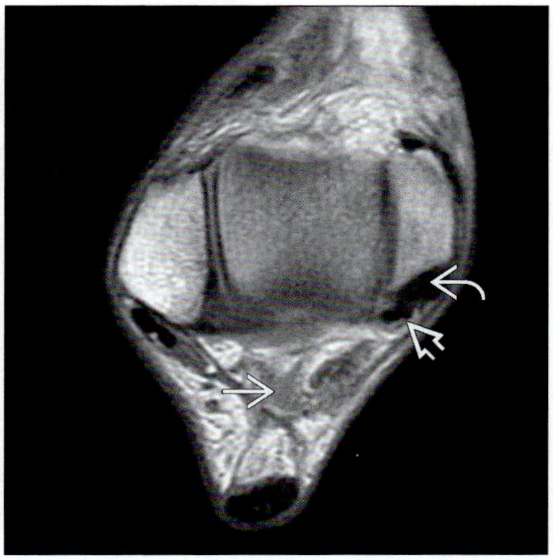

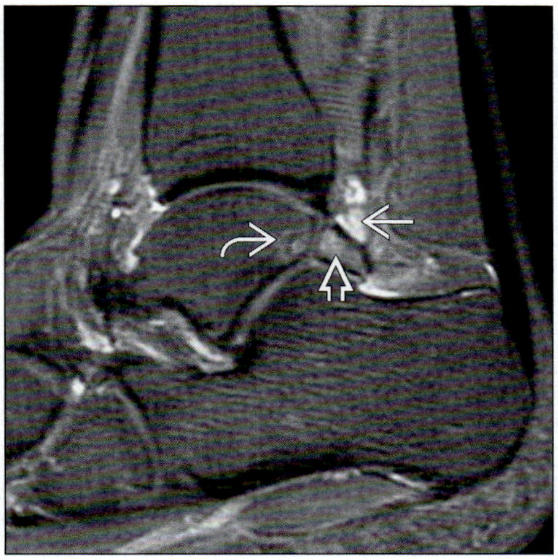

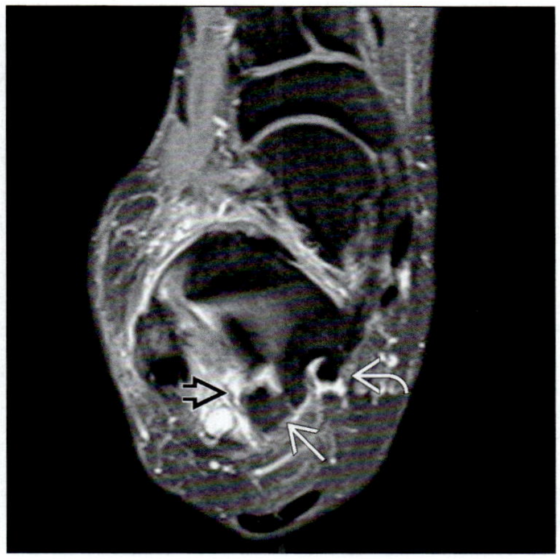

(Left) *Sagittal T2WI FS MR demonstrates a large os trigonum. There is reactive marrow edema in the os and in the adjacent talus. Fluid is present in the FHL tendon sheath.* ***(Right)*** *Axial PDWI FS MR shows os trigonum, with fluid in the synchondrosis between it and the talus. Findings are typical of os trigonum syndrome. An irregular fluid collection is also present around the FHL tendon.*

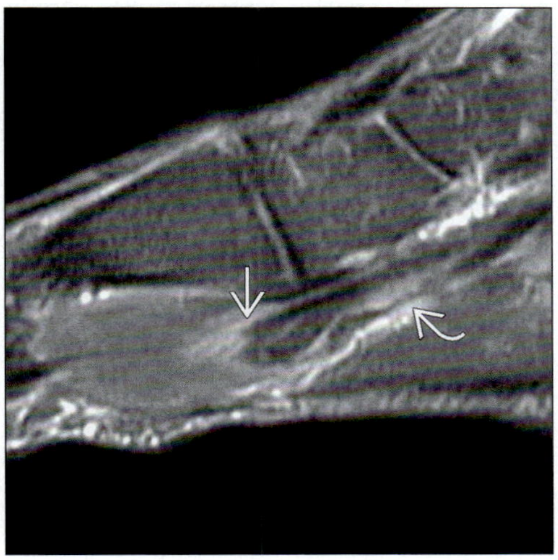

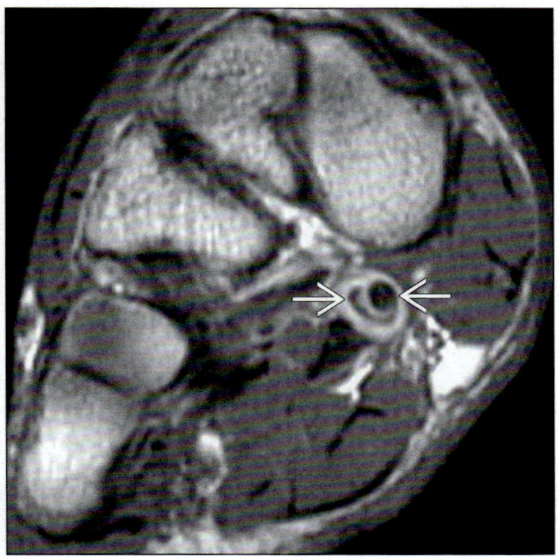

(Left) *Sagittal T2SI FS MR shows a longitudinal split tear of the flexor hallucis longus tendon in the arch of the foot. Note fluid in the tendon sheath proximal to tear.* ***(Right)*** *Coronal T1WI C+ MR in the same patient shows the split fibers of the FHL surrounded by enhancing synovium.*

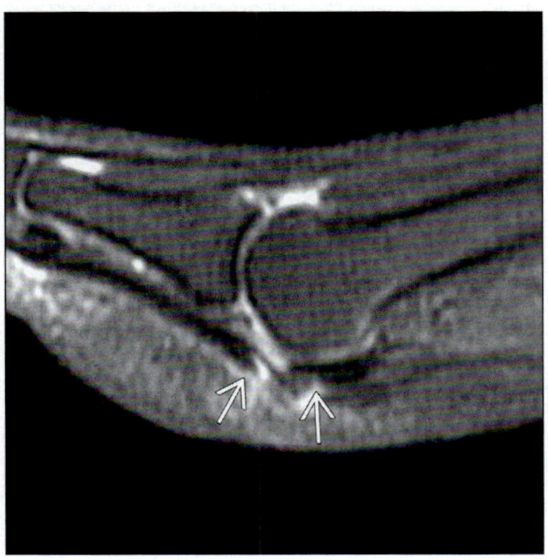

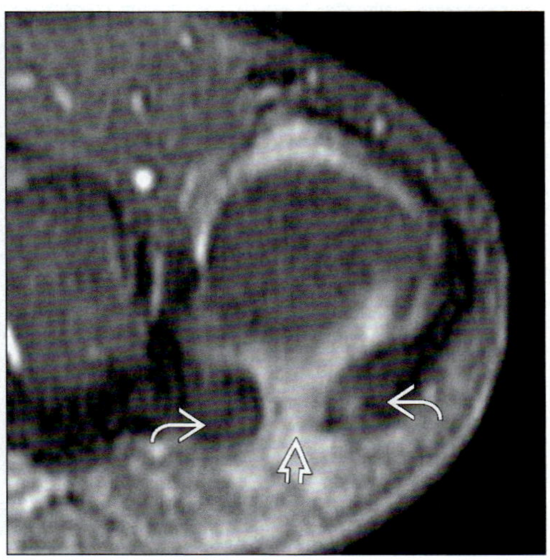

(Left) *Sagittal T2WI FS MR shows traumatic transection of the flexor hallucis longus tendon at the level of the 1st MTP joint. The torn edges of the FHL are separated by fluid and debris.* ***(Right)*** *Coronal T2WI FS MR in the same patient shows absence of tendon in its normal location between the sesamoids. The intersesamoid ligament is also torn, resulting in increased separation of the sesamoids.*

POSTERIOR TIBIAL TENDON TEAR AND TENDINOPATHY

Key Facts

Imaging

- Usually tears in horizontal portion, > 1 cm from navicular insertion
- Normal PTT is 2x size of flexor digitorum longus tendon
- Tendinopathy: Tendon enlarged, increased signal intensity but < fluid on fluid-sensitive sequences
- Tenosynovitis: Fluid within tendon sheath
  - Minimal amount of fluid is often seen in asymptomatic patients
- Tear
  - Grade I: Enlarged tendon, fluid between fibers
  - Grade II: Focal tendon thinning; torn fiber ends often identifiable
  - Grade III: Tendon discontinuity, ends often frayed or bulbous
- Injury to type II accessory navicular
  - Edema in accessory navicular, adjacent navicular
  - Fluid between accessory navicular and navicular
  - Proximal displacement of accessory navicular

Top Differential Diagnoses

- Flexor retinaculum avulsion
- Spring ligament injury

Pathology

- Associated conditions: Type II accessory navicular, rheumatoid arthritis, preexisting flatfoot deformity, obesity, spring ligament tear
- Sequelae of PTT tear: Flatfoot deformity, pronation of foot, osteoarthritis, lateral hindfoot impingement

Diagnostic Checklist

- Tendon normally increases slightly in signal intensity as it inserts on navicular

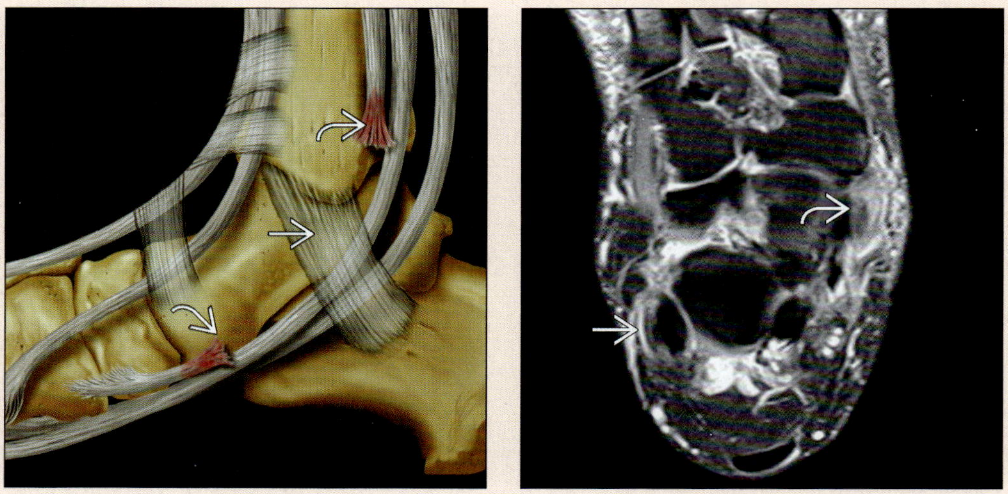

(Left) *Graphic of the medial ankle shows PTT tear with separation of the tendon ends ➔. This is the end stage of what is generally a chronic, degenerative process that begins with tendinopathy and progresses to partial and then to full thickness tear. Tendon passes beneath flexor retinaculum ➔.* ***(Right)*** *Axial PDWI FS MR shows complete PTT tear ➔. There is splitting of the peroneus brevis tendon ➔, which can occur in PTT dysfunction due to flatfoot deformity and subfibular impingement.*

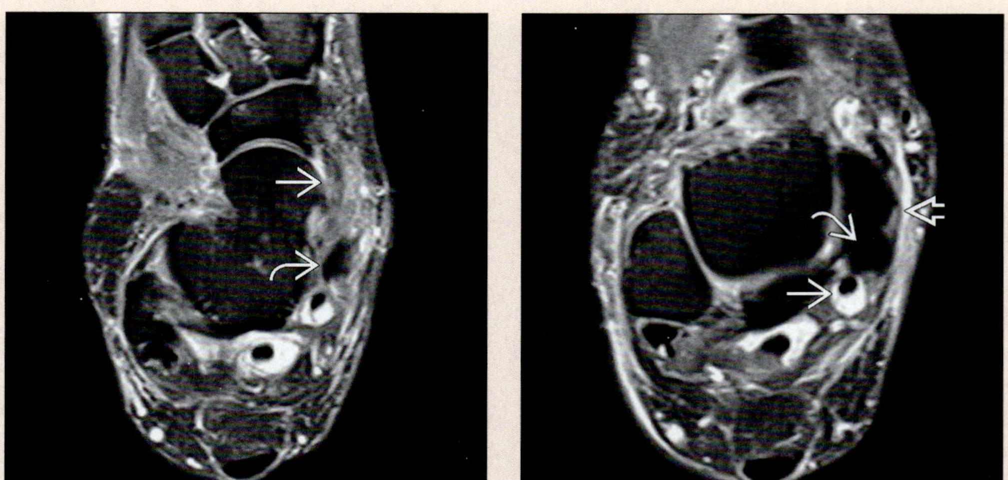

(Left) *Axial PDWI FS MR in the same patient shows that the PTT ➔ has an irregular contour proximal to the complete tear. Tibionavicular fibers of deltoid ligament ➔ are also torn.* ***(Right)*** *Axial T2WI FS MR shows enlarged PTT ➔ at the level of the ankle joint, due to tendon retraction. Normally, PTT is 2x the size of FDL ➔. Note stripping of flexor retinaculum ➔, which is commonly associated with PTT tear.*

POSTERIOR TIBIAL TENDON TEAR AND TENDINOPATHY

TERMINOLOGY

Abbreviations

- Posterior tibial tendon (PTT)

Synonyms

- Tibialis posterior tendon

Definitions

- Type I accessory navicular: Sesamoid in PTT, normally slightly proximal to navicular
- Type II accessory navicular: Accessory center of ossification united to parent navicular by synchondrosis
- Type III accessory navicular: Assimilated accessory navicular, resulting in enlarged median eminence of navicular

IMAGING

General Features

- Location
  - Tears usually occur in horizontal portion, > 1 cm from navicular insertion

Radiographic Findings

- Dorsolateral peritalar subluxation
  - Head of talus subluxates plantarly and medially
  - Abduction seen at talonavicular joint
- Flatfoot deformity
  - Nonspecific unless dorsolateral peritalar subluxation present
  - Predisposes to PTT abnormality
- Medial soft tissue swelling
  - Seen on AP radiograph of foot
  - Centered proximal to navicular
- Proximal displacement of type II accessory navicular
  - Type II is distinguishable from normal sesamoid in tendon (type I accessory navicular) by shape of ossicle
    - Type II accessory navicular has flat facet where it articulates with parent navicular
    - Type I accessory navicular is ovoid, normally seen proximal to location of type II ossicle
- Periosteal reaction along distal, medial tibial metaphysis
  - Suggests injury to flexor retinaculum
- Avulsion fracture of flexor retinaculum
  - Thin flake of bone along medial tibial metaphysis

CT Findings

- Tendinopathy: Enlarged tendon, slightly decreased attenuation
- Tenosynovitis: Low-attenuation fluid surrounds tendon within tendon sheath
- Tendon tear
  - Grade I: Thin stripes of low-attenuation fluid within tendon substance, tendon enlarged
  - Grade II: Tendon focally thinned
  - Grade III: Fluid attenuation fills tendon sheath

MR Findings

- Normal PTT is 2x size of flexor digitorum longus tendon
  - Increases slightly in signal intensity (SI) at its primary insertion on median eminence of navicular
- Tendinopathy: Tendon enlarged, increased SI but < fluid on fluid-sensitive sequences
- Tenosynovitis: Fluid within tendon sheath
  - Minimal amount of fluid is often seen in asymptomatic patients
  - Not all patients with tenosynovitis clinically have abnormal MR
- Tear
  - Grade I: Enlarged tendon, fluid between fibers
  - Grade II: Focal tendon thinning; torn fiber ends often identifiable
  - Grade III: Tendon discontinuity, ends often frayed or bulbous
    - In chronic tear, tendon retracted proximally into calf ("window shade" effect)
- Injury to type II accessory navicular
  - Bone marrow edema in accessory ossicle, adjacent navicular
  - Fluid between accessory ossicle and navicular
  - Proximal or dorsal displacement of ossicle

Ultrasonographic Findings

- Tenosynovitis: Fluid in tendon sheath
- Tendinopathy: Enlarged tendon with heterogeneous echotexture
- Tendon tear: Thickened, thinned, or discontinuous tendon
  - Often difficult to distinguish between tendinopathy and partial tear

Imaging Recommendations

- Best imaging tool
  - MR
- Protocol advice
  - Long and short axis of tendon, fluid-sensitive sequences in all 3 planes

DIFFERENTIAL DIAGNOSIS

Flexor Retinaculum Avulsion

- Usually an acute trauma
- Flake of bone may be visible from posteromedial margin of tibia

Spring Ligament Injury

- Results in flatfoot deformity

Plantar Fasciitis

- Medial hindfoot pain

Type I Accessory Navicular

- Sesamoid in PTT, normally separated from median eminence of parent navicular

PATHOLOGY

General Features

- Associated abnormalities
  - Preexisting flatfoot, obesity
    - Increase stress on tendon
  - Type II accessory navicular
    - Tendon attaches to accessory navicular

- Synchondrosis between accessory navicular and parent navicular vulnerable to pull from PTT
- Spring and tibiospring ligament tears
  - Spring ligament is static stabilizer of longitudinal arch of foot, while PTT is dynamic stabilizer
  - Tibiospring component of deltoid ligament forms sling beneath talar head, stressed with PTT injury
  - Both ligaments are stretched by PTT dysfunction, become lax and incompetent
  - Frank rupture of spring, tibiospring ligaments uncommon
- Lateral hindfoot (subfibular) impingement
  - Due to abnormal bony abutment from pronation of foot
  - Edema in lateral process talus, lateral margin of calcaneus, lateral malleolus
  - Secondary splitting of peroneal tendons
- Flexor retinaculum injury
  - Destabilizes tendon and allows it to displace medially and anteriorly
  - Small fracture fragment may be visible from medial tibial metaphysis
  - Periosteal reaction develops with chronic subluxation
- Rheumatoid arthritis
  - Rheumatoid involvement of synovium of tendon sheath **or**
  - Secondary to subtalar rheumatoid arthritis, resulting in hindfoot valgus and increased stress on tendon

Staging, Grading, & Classification

- Grade I: Intrasubstance splitting of tendon, tendon enlarged
- Grade II: Partial tear, tendon thinned
- Grade III: Complete tear

Relevant Anatomy

- PTT supinates and plantarflexes the foot
- Originates from tibia, interosseous membrane, fibula
- Most medial of extrinsic flexor tendons at ankle
- Position behind medial malleolus is maintained by flexor retinaculum
- Passes beneath medial malleolus, extends anteriorly in horizontal plane
- Main attachment is to median eminence of navicular (or type II accessory navicular when present)
  - Multiple distal slips attach to other tarsal bones and metatarsal bases
- Type II accessory navicular distinguishable from type I by its flat facet for articulation with navicular
  - Type I accessory navicular is not a risk factor for PTT dysfunction

CLINICAL ISSUES

Presentation

- Most common signs/symptoms
  - Flatfoot, with medial hindfoot pain
  - Difficulty standing on tiptoes
- Other signs/symptoms
  - Soft tissue swelling and tenderness
    - Along medial aspect of talar neck and at median eminence of navicular **or**
    - Posterior to medial malleolus
  - Lateral pain due to pronation, secondary lateral hindfoot impingement

Natural History & Prognosis

- Flatfoot deformity leads to midfoot osteoarthritis
- Chronic tear and resultant pronation may lead to lateral hindfoot impingement
  - Fluid in tarsal sinus
  - Tenosynovitis of peroneal tendons
  - Bone marrow edema at lateral margin calcaneus, talus, fibula

Treatment

- Tenosynovitis can be treated with fluoroscopic-guided steroid injection into tendon sheath
  - May approach tendon sheath either in horizontal portion or posterior to medial malleolus
    - Palpate tendon
    - Advance 25-gauge needle through tendon to bone
    - Withdraw needle slowly while injecting lidocaine
    - When injects easily, needle tip is in tendon sheath
    - Confirm location with contrast injection
    - Inject corticosteroid, long-acting anesthetic
    - Post-injection fluoro spot to confirm dispersal of contrast in sheath by medication
- Orthotics used to improve longitudinal arch of foot, decrease stress on midfoot
- Primary repair for acute tendon tear
- Surgical debridement for partial tear
- Cases with longstanding tear or severe deformity may be treated with arthrodesis

DIAGNOSTIC CHECKLIST

Consider

- Patients may have symptomatic tenosynovitis despite normal MR
  - Clinical response to tendon sheath injection confirms clinical diagnosis

Image Interpretation Pearls

- Tendon normally increases slightly in signal intensity as it inserts on navicular
- In chronic partial tears, signal intensity of torn region may be < fluid because of scar tissue formation

SELECTED REFERENCES

1. Donovan A et al: Extraarticular lateral hindfoot impingement with posterior tibial tendon tear: MRI correlation. AJR Am J Roentgenol. 193(3):672-8, 2009
2. Nallamshetty L et al: Evaluation of posterior tibial pathology: comparison of sonography and MR imaging. Skeletal Radiol. 34(7):375-80, 2005
3. Balen PF et al: Association of posterior tibial tendon injury with spring ligament injury, sinus tarsi abnormality, and plantar fasciitis on MR imaging. AJR Am J Roentgenol. 176(5):1137-43, 2001
4. Pomeroy GC et al: Acquired flatfoot in adults due to dysfunction of the posterior tibial tendon. J Bone Joint Surg Am. 81(8):1173-82, 1999
5. Khoury NJ et al: MR imaging of posterior tibial tendon dysfunction. AJR Am J Roentgenol. 167(3):675-82, 1996

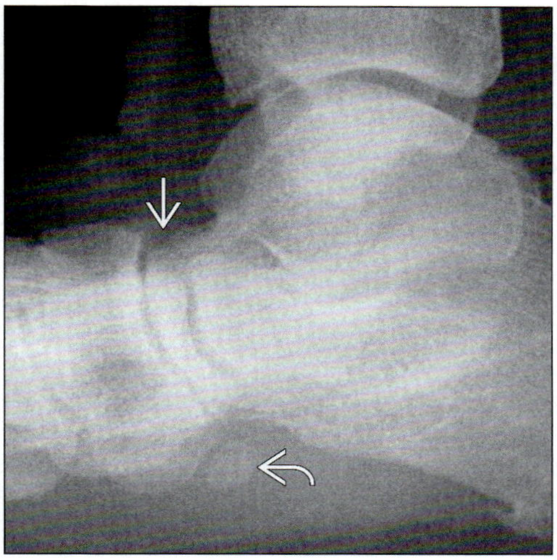

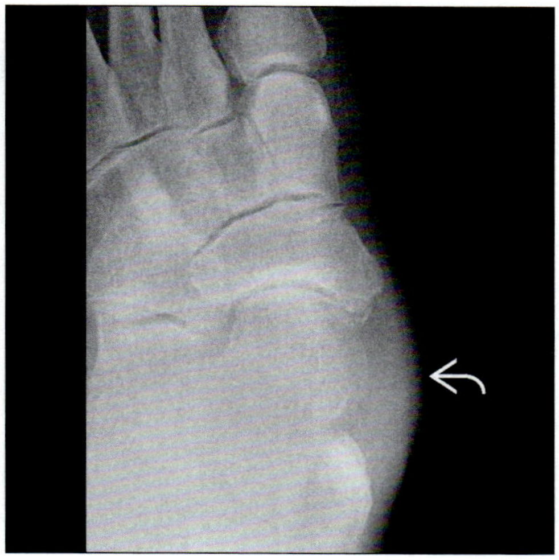

(Left) *Lateral radiograph shows inferior displacement of the head of the talus ➡ at the talonavicular joint. Note type II accessory navicular ↪, which is associated with PTT dysfunction.* ***(Right)*** *AP radiograph from the same patient shows abduction of talonavicular joint. In combination with the findings on the lateral radiograph, this is termed dorsolateral peritalar subluxation and constitutes radiographic evidence of PTT dysfunction. Note soft tissue swelling over course of PTT ↪.*

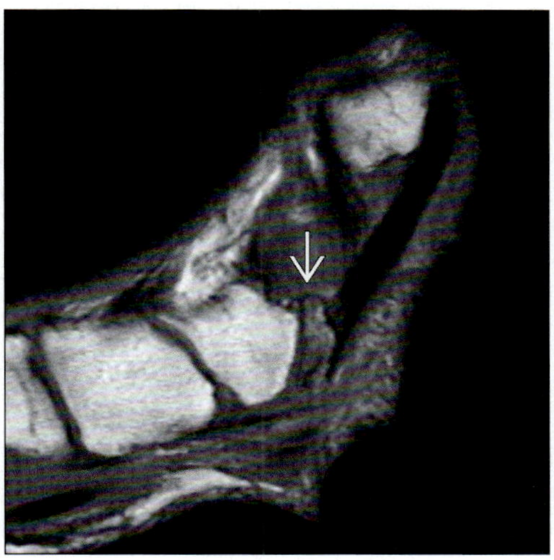

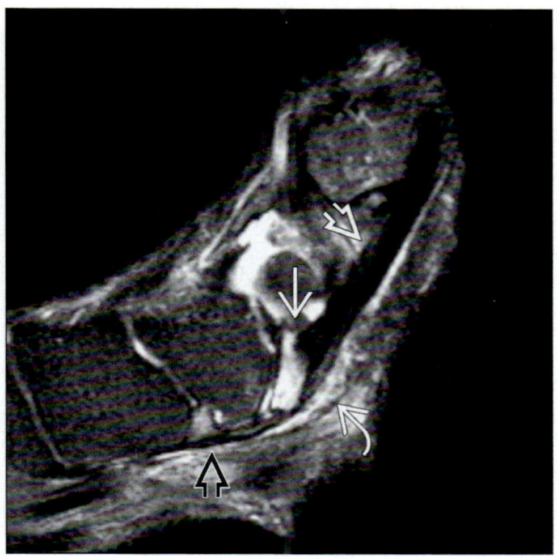

(Left) *Sagittal T1WI MR shows avulsion of type II accessory navicular. The ossicle ➡ is slightly displaced superiorly relative to the main navicular and contains low signal intensity marrow edema.* ***(Right)*** *Sagittal STIR MR from the same patient shows high signal intensity bone marrow edema in the ossicle ➡ and surrounding soft tissue hematoma ↪. PTT is intact ➡ proximal to the avulsion injury, and the distal tendon slip to the medial cuneiform ➡ remains intact as well.*

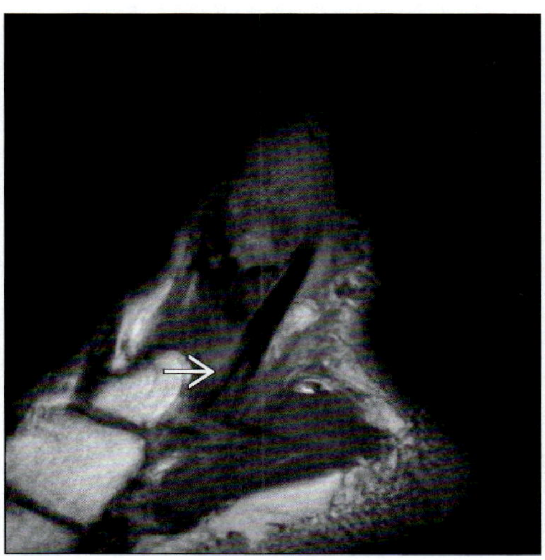

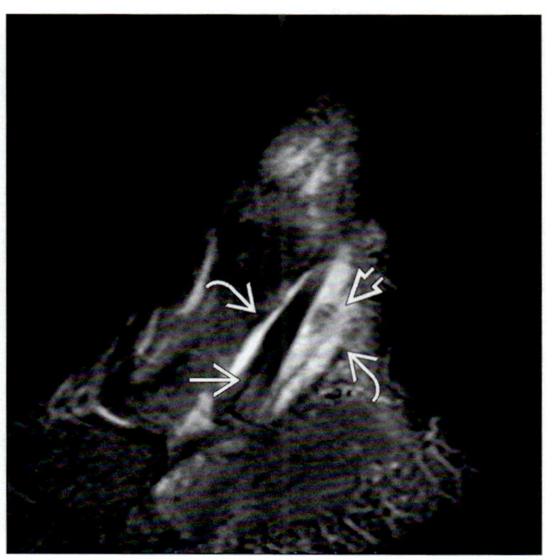

(Left) *Sagittal T1WI MR shows longitudinal splitting ➡ of the distal PTT, near its navicular insertion. Most PTT injuries occur slightly more proximally.* ***(Right)*** *Sagittal STIR MR from the same patient shows distal split ➡ and large amount of fluid in PTT sheath ↪. Amorphous material ➡ is visible in the tendon sheath. This patient had rheumatoid arthritis, and the amorphous material probably represents pannus.*

POSTERIOR TIBIAL TENDON TEAR AND TENDINOPATHY

*(**Left**) Axial soft tissue algorithm CT shows type II accessory navicular → in a normal position adjacent to navicular →. There is a large effusion in the tendon sheath →. Even if the accessory navicular is not displaced, it is associated with increased prevalence of PTT dysfunction. (**Right**) Sagittal soft tissue algorithm CT from the same patient shows fluid in tendon sheath → and stripes of low attenuation → in the substance of the tendon, indicating longitudinal splitting of the tendon.*

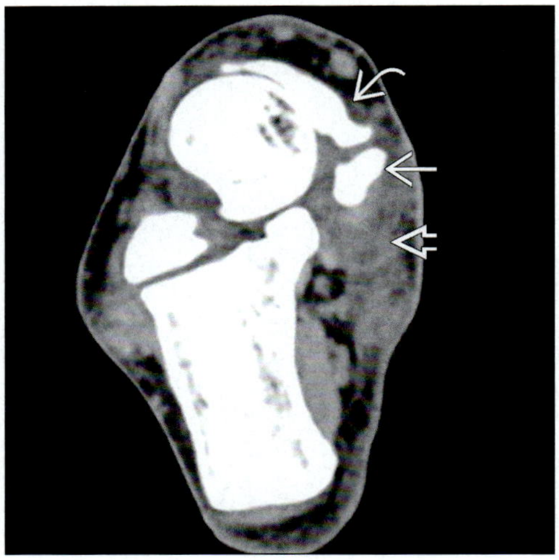

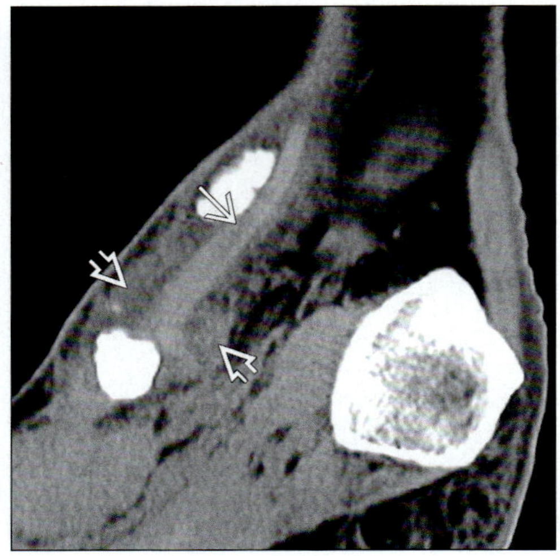

*(**Left**) Coronal T2WI FS MR shows effusion and adhesions → in the PTT sheath, reflecting chronic tenosynovitis. There is mildly increased signal intensity → centrally in the tendon, indicating tendinopathy. (**Right**) Axial PDWI FS MR from the same patient confirms severe tendinopathy → and tenosynovitis →. It is often difficult to distinguish between severe tendinopathy and a grade I tear.*

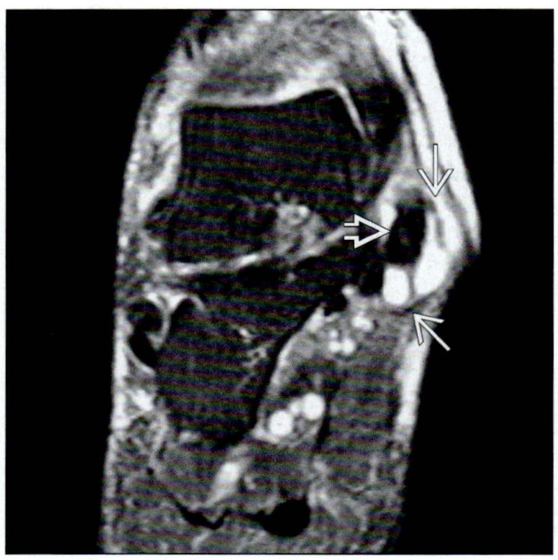

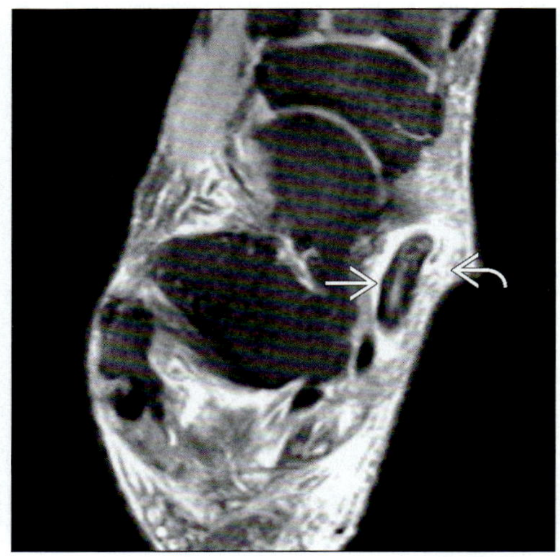

*(**Left**) Coronal T2WI FS MR shows longitudinal splitting of PTT → and tenosynovitis. This patient had longstanding PTT symptoms and has developed secondary impingement laterally, with associated soft tissue swelling →. (**Right**) Axial PDWI FS MR from the same patient confirms that the abnormal signal in the tendon → is due to severe longitudinal splitting, rather than tendinopathy. A few central discontinuous central fibers are still visible. Tear was nearly complete at time of surgery.*

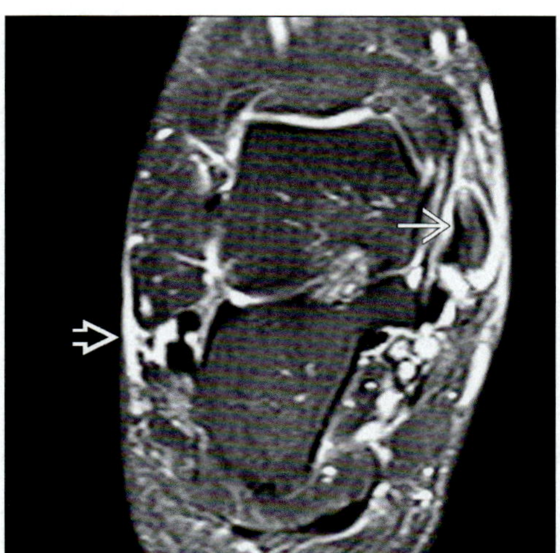

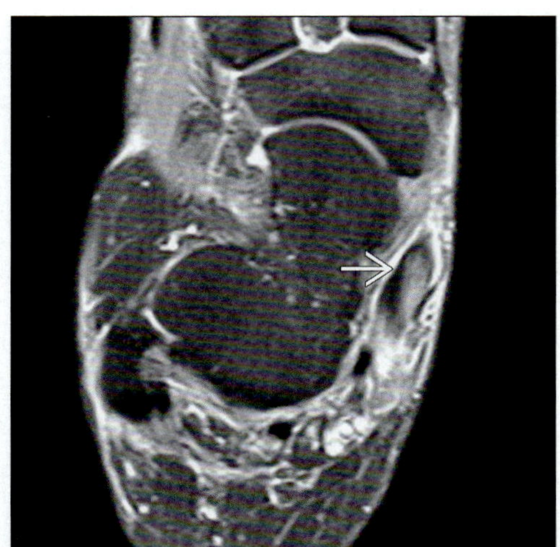

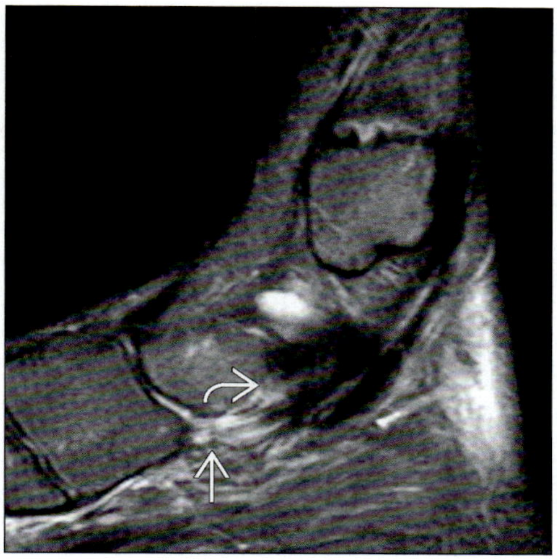

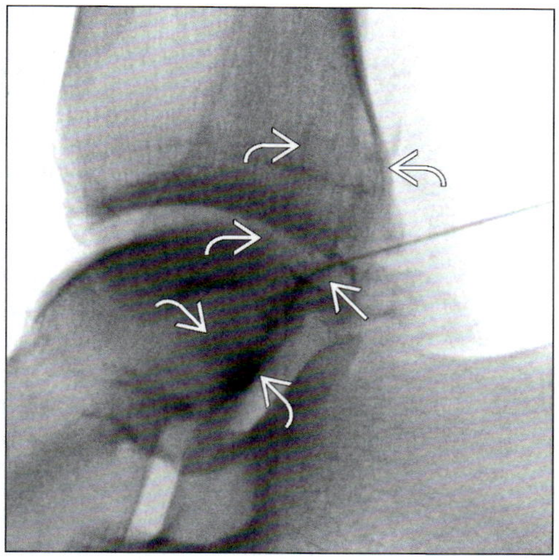

(Left) *Sagittal STIR MR shows traumatic avulsion of PTT in a teenaged patient. There is avulsion of PTT from navicular and 1st cuneiform attachments. This is an unusual pattern, more consistent with a traumatic episode rather than degeneration of the tendon.* ***(Right)*** *Lateral tenosynogram shows needle in posterior tibial tendon sheath, opacifying the sheath. Patient had preexisting flatfoot with worsening medial ankle pain. Symptoms completely resolved after steroid injection.*

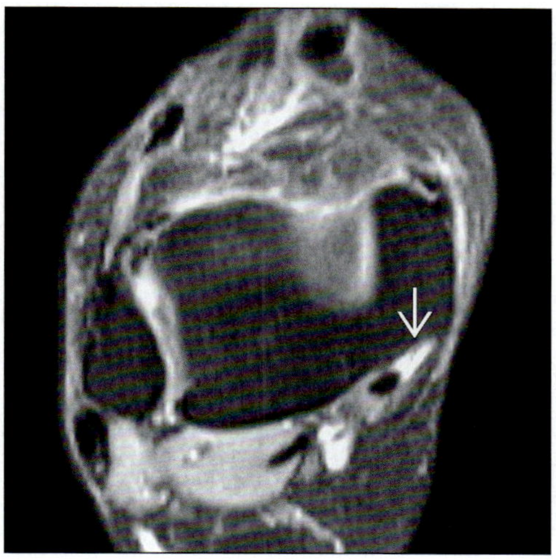

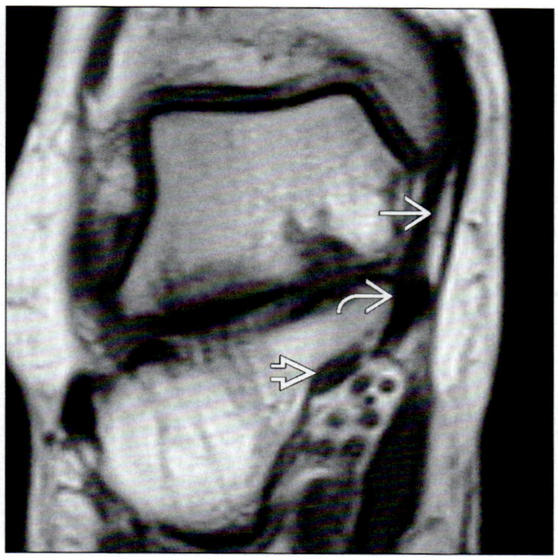

(Left) *Axial T2WI FS MR shows chronic complete tear of PTT. Tendon is absent from its normal position behind the medial malleolus. This finding is easily missed if a systematic checklist of normal structures is not performed.* ***(Right)*** *Coronal PDWI MR from the same patient shows absence of the posterior tibial tendon in the tendon sheath. Normal flexor digitorum longus and flexor hallucis longus tendons are seen.*

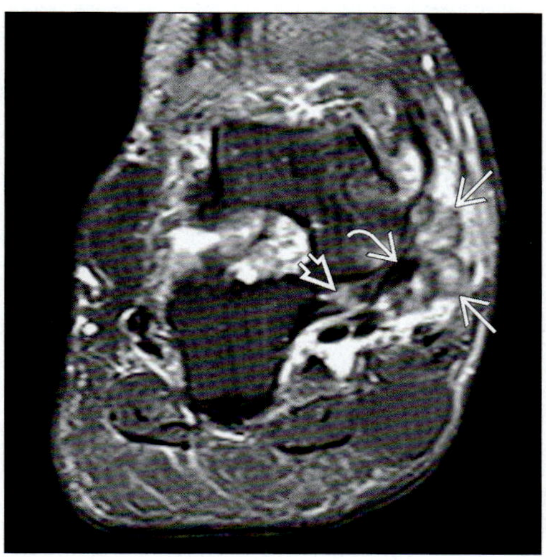

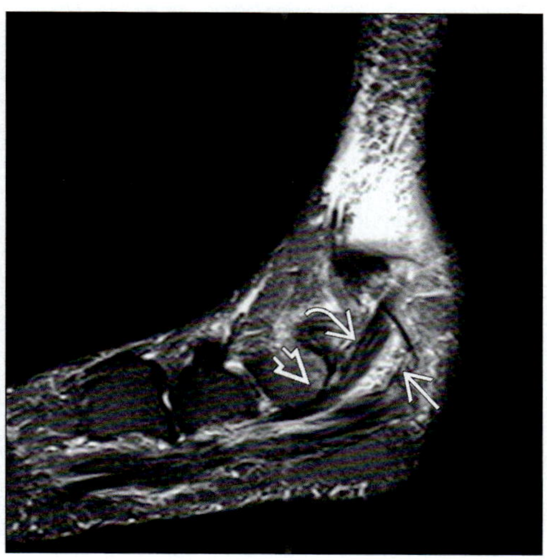

(Left) *Sagittal T2WI FS MR shows severe tendosynovitis and thinning of the posterior tibial tendon. A tear of the spring ligament is also present.* ***(Right)*** *Sagittal STIR MR from the same patient confirms thickening and heterogeneity of the posterior tibial tendon consistent with tendinopathy with normal attachment to the inferior margin of the navicular. Fluid and debris are present within the tendon sheath.*

EXTENSOR TENDON TEAR AND TENDINOPATHY

Key Facts

Imaging

- May be injured in distal leg, midfoot, or near insertion
- Tendinopathy: Tendon enlarged and amorphous with increased signal intensity (but less than fluid) on fluid-sensitive sequences
- Tenosynovitis: Fluid in tendon sheath
- Tears may be partial or complete
  - Grade 1: Tendon thickened, fluid signal intensity between fibers, focally discontinuous fibers
  - Grade 2: Tendon thinned
  - Grade 3: Discontinuity of tendon

Top Differential Diagnoses

- Bursitis
- Fracture
- Shin splints
- Other causes of foot drop

Clinical Issues

- Both tendinopathy and tear are rare
  - Usually occur in older males, present with foot drop
- Associated with systemic corticosteroid use, inflammatory arthropathy, gout

Diagnostic Checklist

- Easily overlooked because rarely injured; include extensor tendons in search pattern
- Use coronal images for horizontal portions of tendons, axial images for vertical portions
- Tendons are thin and oblique in course, usually cannot follow far on single sagittal slice
- Be alert for magic angle phenomenon as tendons change direction from ankle to foot

__(Left)__ Graphic of medial side of ankle shows a complete rupture of the anterior tibial tendon. The 3 extrinsic foot extensors are held against the dorsum of the foot by the superior and inferior extensor retinacula. __(Right)__ Sagittal T2WI FS MR shows complete disruption of the anterior tibial tendon with retraction of the tendon stump and fluid in the tendon sheath. Differentiating the tendons of the anterior compartment is best accomplished with axial imaging.

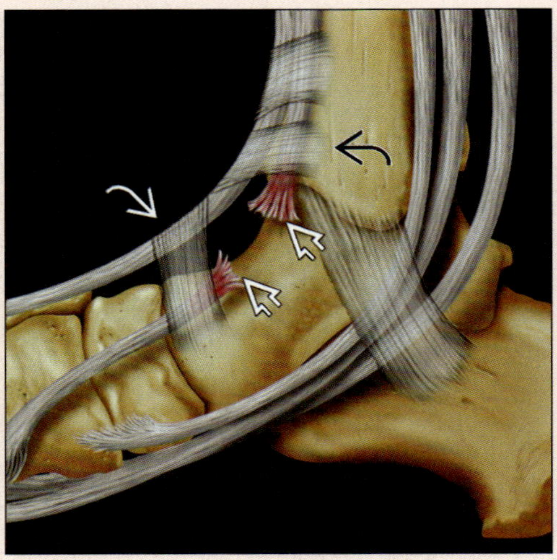

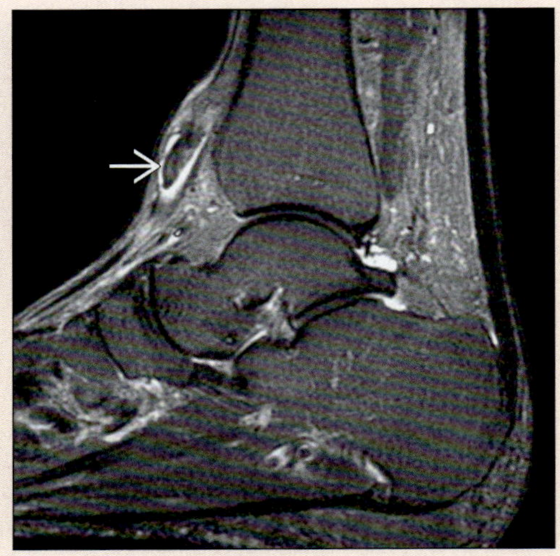

__(Left)__ Axial T2WI FS MR shows complete disruption of the anterior tibial tendon, with fluid in the empty tendon sheath. A skin marker has been placed over the palpable tendon defect. __(Right)__ Sagittal T2WI FS MR shows focal fusiform thickening of the extensor digitorum longus tendon, presenting as ankle pain on dorsiflexion of the toes. A small amount of fluid is tracking along the tendon sheath. A marker capsule was placed over the site of clinical swelling and tenderness.

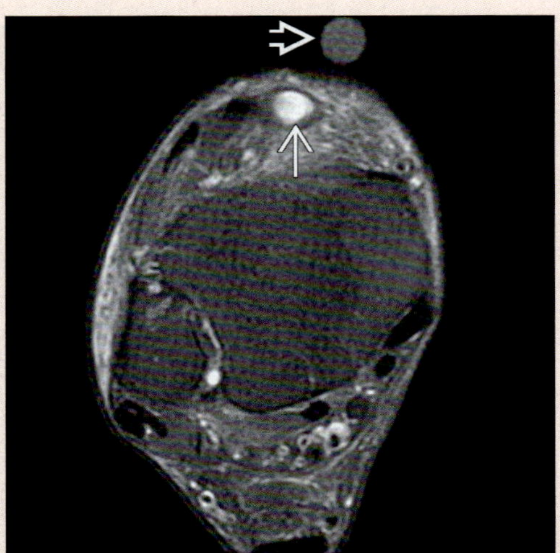

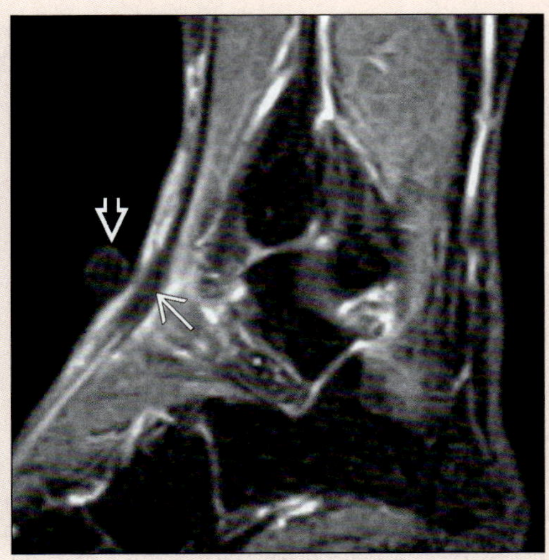

EXTENSOR TENDON TEAR AND TENDINOPATHY

IMAGING

General Features

- Location
  - May be injured in distal leg, midfoot, or near insertion

MR Findings

- Extensor tendons difficult to see as they change orientation from vertical in leg to horizontal in foot
  - Magic angle phenomenon may mimic tear
  - Confirm tendon integrity on sagittal images
- Tendinopathy: Tendon enlarged and amorphous with increased signal intensity (but less than fluid) on fluid-sensitive sequences
- Tenosynovitis: Fluid in tendon sheath
- Tears may be partial or complete
  - Grade 1: Tendon thickened, fluid signal intensity between fibers, focally discontinuous fibers
  - Grade 2: Tendon thinned
  - Grade 3: Discontinuity of tendon

Imaging Recommendations

- Best imaging tool
  - MR
- Protocol advice
  - Prone imaging with foot plantar flexed reduces magic angle effect on tendons

DIFFERENTIAL DIAGNOSIS

Bursitis

- Tibialis anterior bursa located near insertion of tendon on medial cuneiform
- May be irritated by poorly fitting shoes

Fracture

- Anterior margin of tibia, talus

Shin Splints

- Anterior tibial pain due to periostitis or inflammation of tibialis anterior muscle origin

Other Causes of Foot Drop

- e.g., peroneal nerve palsy or disc herniation

PATHOLOGY

General Features

- Etiology
  - Tendinopathy
    - Inflammatory arthropathy
    - Downhill running (anterior tibial tendon)
  - Tendon tear
    - Laceration by top of boot or by object falling on foot
    - Forced plantar flexion
    - Sports, e.g., fencing
    - Inflammatory arthropathy or gout
    - Hyperdorsiflexion of toe (skimboarder's toe)
- Associated abnormalities
  - Systemic corticosteroids
  - Diabetes mellitus
  - Inflammatory arthropathy and gout

Relevant Anatomy

- Extensor retinaculum holds extrinsic extensor tendons in place at ankle
- Tibialis anterior is most medial of 3 extrinsic extensors, inserts on medial cuneiform and 1st MT
- Extensor hallucis longus lies between tibialis anterior and extensor digitorum longus, inserts on distal phalanx of 1st toe
- Extensor digitorum longus has slips to 2nd through 5th toes
  - Slips insert on distal phalanges

CLINICAL ISSUES

Presentation

- Most common signs/symptoms
  - Pain with dorsiflexion
  - Foot drop
- Clinical profile
  - Spontaneous rupture may be seen in patients with diabetes, gout, or rheumatoid arthritis
  - Distal tibialis anterior tendinopathy seen in elderly, overweight females

Demographics

- Tendinopathy and tear are rare
- More common in older patients

Treatment

- Tendinopathy: Orthotics, rest, NSAIDs
- Tendon tear: Surgical repair

DIAGNOSTIC CHECKLIST

Image Interpretation Pearls

- Easily overlooked because rarely injured; include extensor tendons in search pattern
- Use coronal images for horizontal portions of tendons, axial images for vertical portions
- Tendons are thin and oblique in course, usually cannot follow far on single sagittal slice
- Be alert for magic angle phenomenon as tendons change direction from ankle to foot

SELECTED REFERENCES

1. Beischer AD et al: Distal tendinosis of the tibialis anterior tendon. Foot Ankle Int. 30(11):1053-9, 2009
2. Gwynne-Jones D et al: Closed tibialis anterior tendon rupture: a case series. Foot Ankle Int. 30(8):758-62, 2009
3. Jerome JT et al: Tibialis anterior tendon rupture in gout--case report and literature review. Foot Ankle Surg. 14(3):166-9, 2008
4. Negrine JP: Tibialis anterior rupture: acute and chronic. Foot Ankle Clin. 12(4):569-72, v, 2007
5. Lee MH et al: Tibialis anterior tendon and extensor retinaculum: imaging in cadavers and patients with tendon tear. AJR Am J Roentgenol. 187(2):W161-8, 2006
6. Mengiardi B et al: Magic angle effect in MR imaging of ankle tendons: influence of foot positioning on prevalence and site in asymptomatic subjects and cadaveric tendons. Eur Radiol. 16(10):2197-206, 2006
7. Donnelly LF et al: Skimboarder's toe: findings on high-field MRI. AJR Am J Roentgenol. 184(5):1481-5, 2005

EXTENSOR TENDON TEAR AND TENDINOPATHY

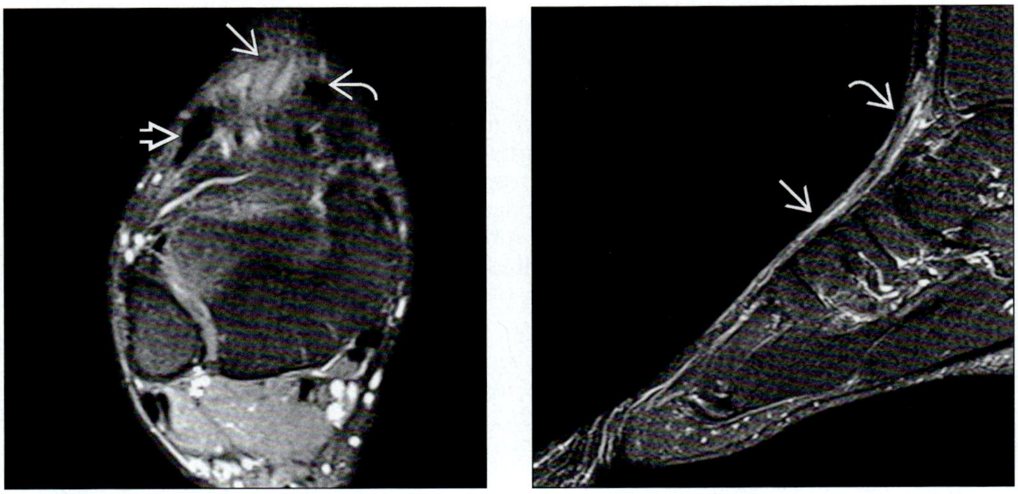

(Left) *Axial T2WI FS MR at level of ankle joint shows thin, split extensor hallucis longus (EHL) ➔, with fluid in the tendon sheath. Note the normal anterior tibial ↷ and extensor digitorum longus ⇨ tendons.* ***(Right)*** *Sagittal STIR MR in the same patient shows an area of complete rupture ➔. Proximal to the rupture, tendon is thin and heterogeneous in appearance ↷. On sagittal images, the thin EHL tendon is difficult to see; diagnosis should never be made from this plane alone.*

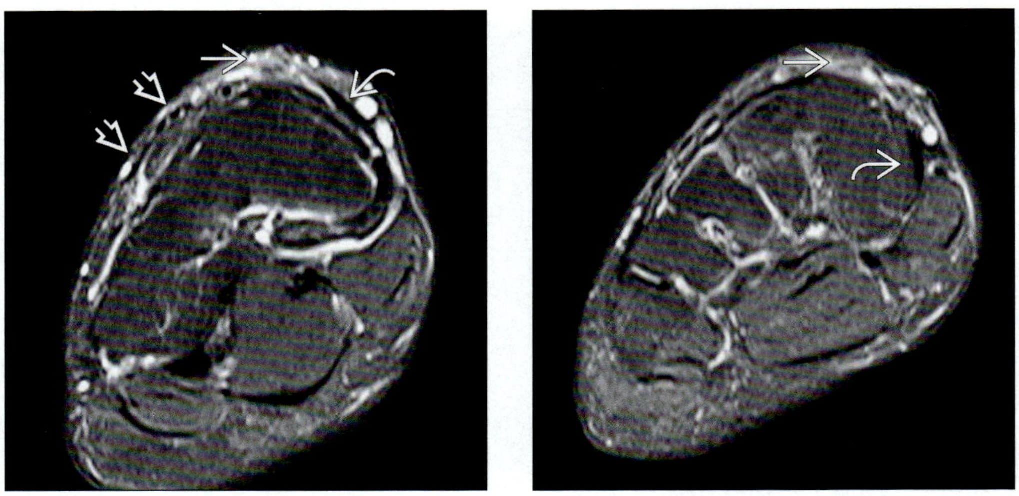

(Left) *Coronal T2WI FS MR at level of navicular in the same patient shows thin, split EHL tendon ➔. Anterior tibial tendon ↷ is obliquely oriented on this slice, as it extends medially to insert on the 1st cuneiform. Extensor digitorum longus tendon ⇨ has multiple slips.* ***(Right)*** *Coronal T2WI FS MR at level of metatarsal bases in the same patient shows only amorphous soft tissue ➔ in expected location of EHL tendon. Distal insertion of anterior tibial tendon on 1st MT base ↷ is visible.*

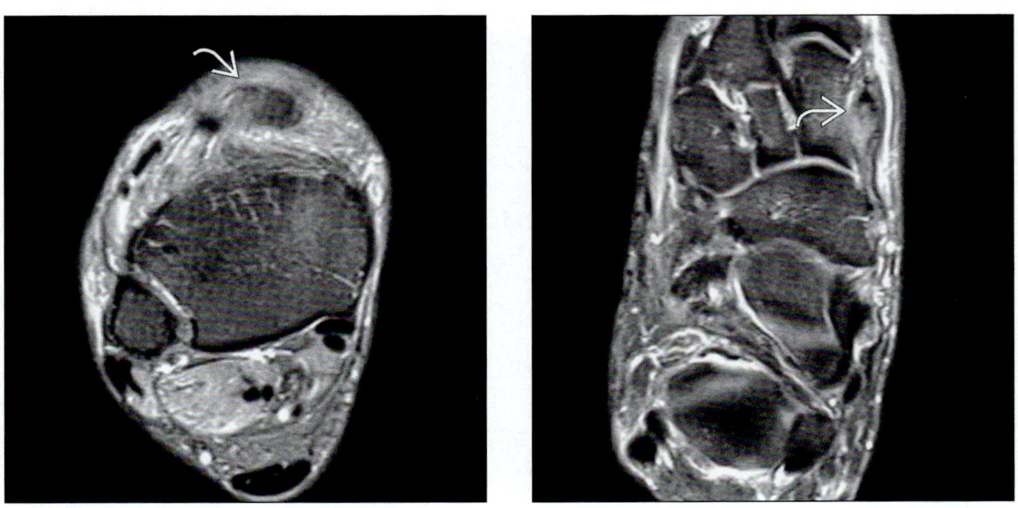

(Left) *Axial T2WI FS MR shows a thickened, heterogeneous anterior tibial ↷ tendon in a patient with chronic anterior ankle pain and tenderness. In this case, tendinopathy occurred due to impingement by a ski boot. Signal intensity in the tendon is less than fluid, excluding tear. Enlargement of tendon is useful to confirm that the abnormal signal is pathologic and not due to magic angle phenomenon.* ***(Right)*** *Axial PDWI FS MR shows severe tendinopathy of the anterior tibial tendon ↷ at its insertion.*

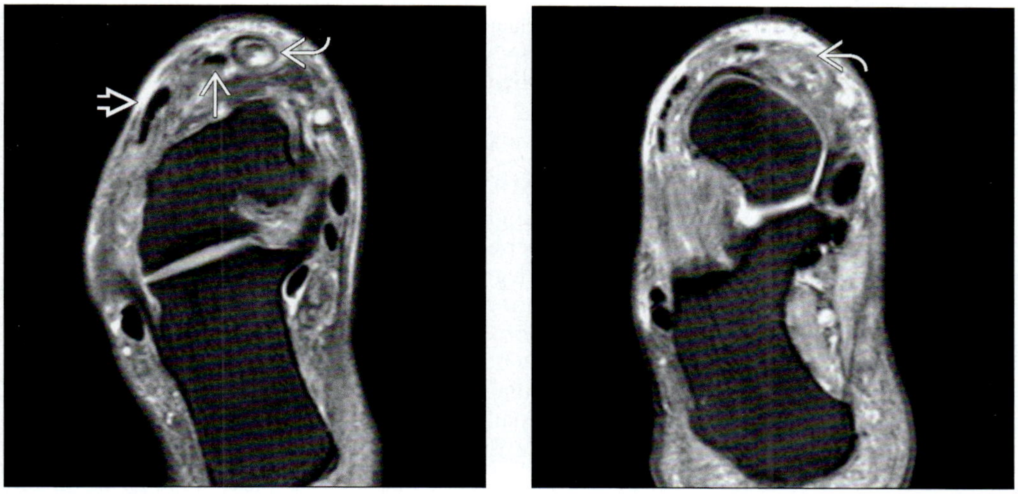

(**Left**) *Axial PDWI FS MR shows partial tear of the anterior tibial tendon. The image was obtained with the patient prone and foot plantar flexed, which diminishes magic angle phenomenon. Note the lack of magic angle phenomenon in the extensor hallucis longus and extensor digitorum longus tendons.* (**Right**) *Axial PDWI FS MR shows absence of anterior tibial tendon due to a complete retracted tendon tear.*

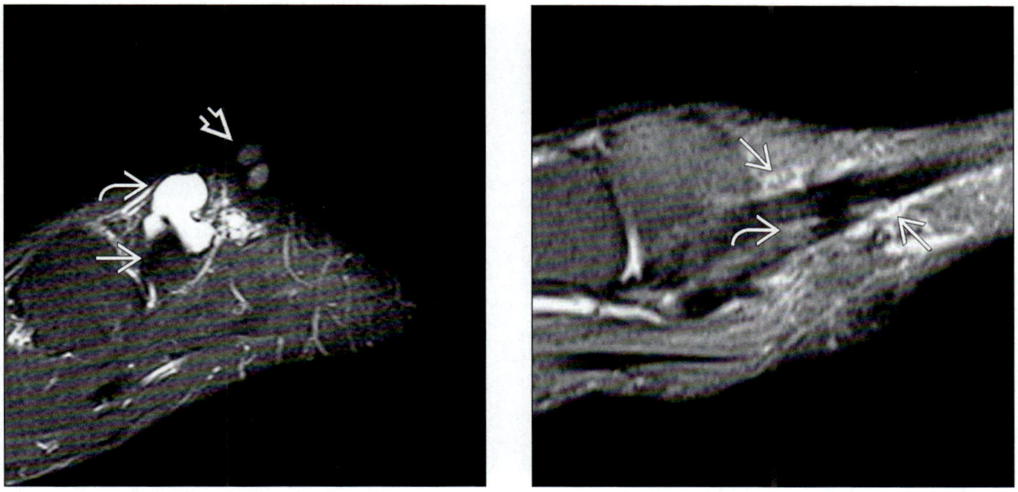

(**Left**) *Sagittal STIR MR shows a lobulated mass adjacent to the anterior tibial tendon insertion, consistent with anterior tibial tendon bursitis. This is not uncommon, usually presenting with pain with dorsiflexion and a palpable mass. Markers are at the proximal extent of the palpable mass.* (**Right**) *Sagittal STIR shows anterior tibial tendon partial tear, with fluid in distal tendon sheath and longitudinal splitting of the tendon.*

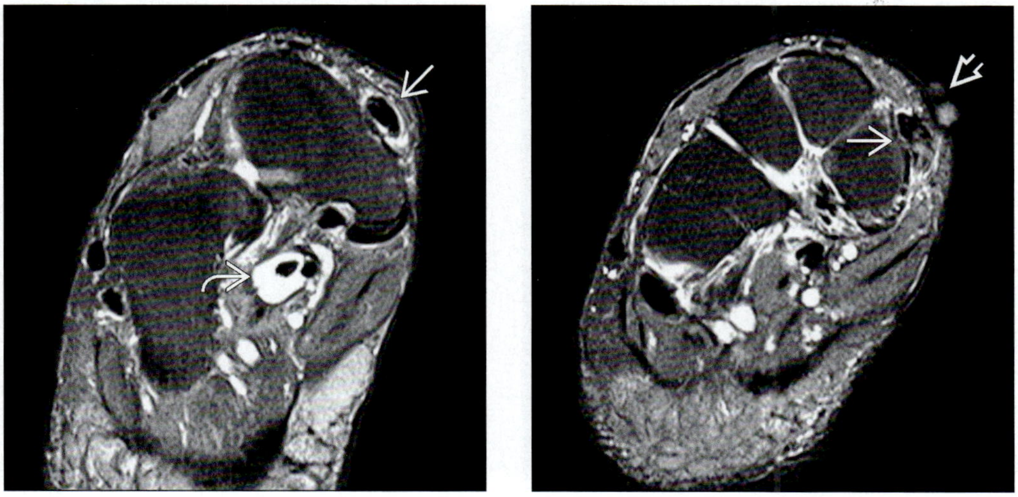

(**Left**) *Coronal T2WI FS MR shows fluid surrounding distal anterior tibial tendon. The tendon is intact. A small amount of fluid around the flexor hallucis longus and flexor digitorum longus tendons is a normal finding, owing to communication of the FHL tendon sheath with the ankle joint & communication of FHL & FDL in the midfoot.* (**Right**) *Coronal T2WI FS MR in the same patient shows partial thickness tear of distal anterior tibial tendon. Markers show the patient's site of pain.*

PERONEAL TENDON TEAR AND TENDINOPATHY

Key Facts

Imaging

- Splitting of tendons more common than full thickness tear
- Tendinopathy: Increased signal intensity in tendon, but < fluid signal on fluid-sensitive sequences
- Tenosynovitis: Fluid in peroneal tendon sheath
- Partial thickness tear
  - Longitudinally oriented fluid between tendon fibers
- Full thickness tear: Tendon discontinuity

Top Differential Diagnoses

- 5th metatarsal tuberosity fracture
- Cuboid stress fracture
- Superior peroneal retinaculum avulsion
- Ankle sprain

Pathology

- Associated abnormalities
  - Muscle anomalies: Peroneus quartus, low-lying PB muscle belly
  - Rheumatoid arthritis
  - Lateral collateral ligament tear
  - Bone marrow edema in peroneal tubercle

Diagnostic Checklist

- Magic angle phenomenon prominent as tendons curve beneath tip of fibula
- Minor degrees of peroneal tendon splitting are of uncertain clinical significance, and patients are often asymptomatic
- MR may be false-negative for tenosynovitis, which can be diagnosed and treated with tenosynogram
- PB normally has thin, flat contour near its insertion; do not mistake this for partial tear

__(Left)__ Graphic of lateral side of ankle shows inflammation of peroneal tendon sheath ➡. Peroneal tendons are held against the lateral margin of the calcaneus by the superior ➡ and inferior ➡ limbs of the inferior peroneal retinaculum. __(Right)__ Sagittal STIR MR shows that the peroneus brevis ➡ tendon is thin and irregular, while the peroneus longus ➡ is thickened. There is a large amount of fluid in the peroneal tendon sheath ➡.

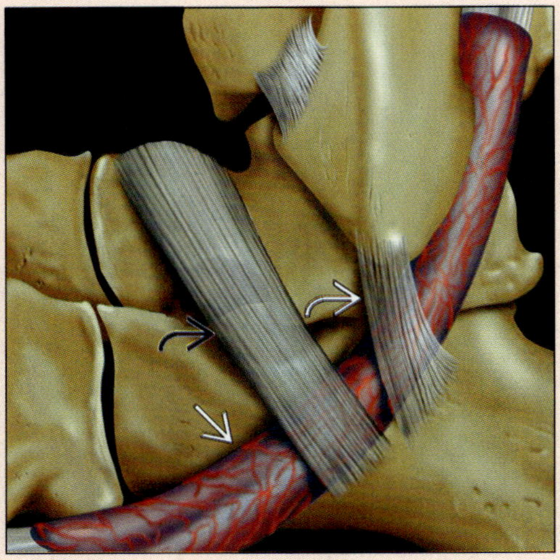

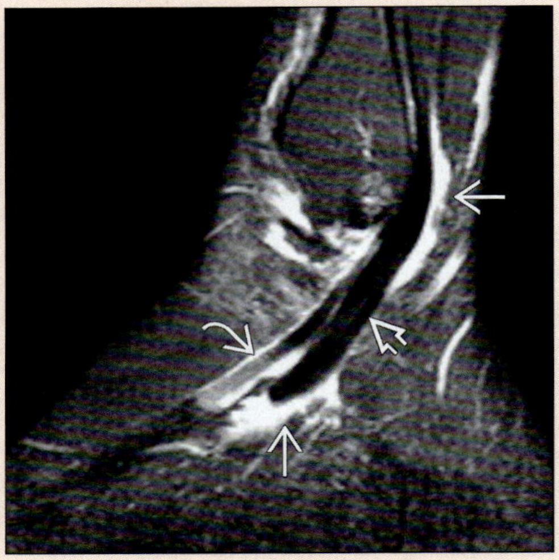

__(Left)__ Axial graphic shows a longitudinal split tear of the peroneus brevis tendon at the level of the lateral malleolus. Then tendon divides into medial ➡ and lateral ➡ limbs, and the peroneus longus tendon ➡ is displaced anteriorly between the 2 limbs. __(Right)__ Axial PDWI MR from a patient with a longitudinal split tear of the peroneus brevis shows the medial ➡ and lateral ➡ limbs of the tendon with the peroneus longus ➡ displaced anteriorly between them.

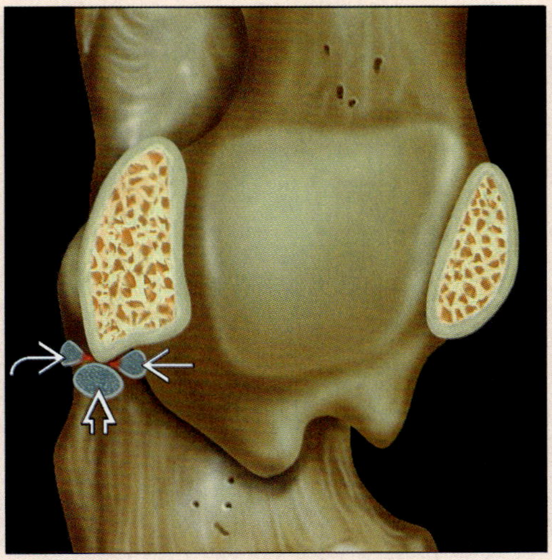

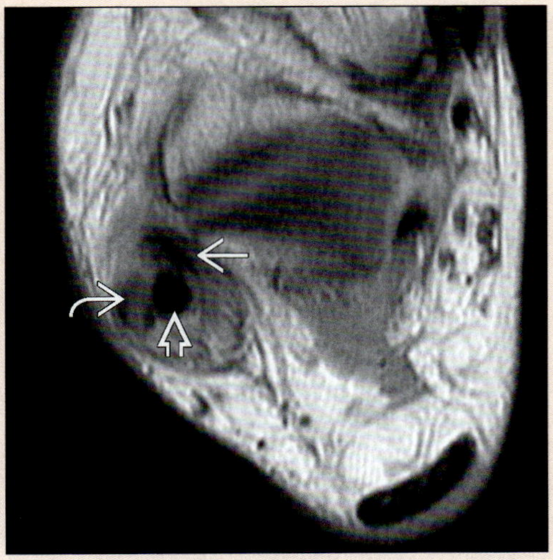

PERONEAL TENDON TEAR AND TENDINOPATHY

IMAGING

General Features

- Location
  - Usually in horizontal portion of tendons
  - Peroneus brevis (PB) more often affected than peroneus longus (PL)
- Morphology
  - Splitting of tendons more common than full thickness tear

Radiographic Findings

- Proximal displacement of os peroneum (sesamoid in PL) due to full thickness tear (rare)
- Enlarged peroneal tubercle of calcaneus sometimes associated with peroneal tendon abnormality

CT Findings

- Tenosynovitis: Low-attenuation fluid in sheath
- Partial thickness tear: Thinned or thickened tendon containing areas of low attenuation
- Full thickness tear: Low-attenuation fluid fills tendon sheath

MR Findings

- Magic angle artifact prominent as tendons pass beneath fibula
- Tendinopathy: Increased signal intensity in tendon, but < fluid signal on fluid-sensitive sequences
- Tenosynovitis: Fluid in peroneal tendon sheath
  - Contour of tendon sheath often abnormal if scarring present (stenosing tenosynovitis)
  - MR may be false-negative for tenosynovitis (no fluid)
  - MR may be false-positive for tenosynovitis
    - Peroneal tendon sheath communicates with ankle joint with tear of calcaneofibular ligament
- Tendon tear
  - Grade I: Separation of tendon fibers
    - Signal intensity may be less than fluid in chronic injuries
  - Grade II: Thinning of tendon
    - Often has "V" shape
      - PL often moves into center of "V" of split PB
  - Grade III: Tendon discontinuity

Imaging Recommendations

- Best imaging tool
  - MR
- Protocol advice
  - 3 planes orthogonal to axis of hindfoot
  - Fluid-sensitive sequences in all 3 planes
  - T1-weighted sequence in coronal &/or axial planes
    - Peroneus quartus, low-lying peroneus brevis muscle easily overlooked without T1-weighted images because nearly isointense to fat on fat-suppressed fluid-sensitive sequences

Nonvascular Interventions

- MR may be false-negative for tenosynovitis
- Tenosynography used for diagnosis
  - Tenosynovitis: Sheath enlarged, often has diverticula
  - Stenosing tenosynovitis: Sheath not distended overall but has irregular "beaded" contour
- Injection of corticosteroid under fluoroscopic guidance useful for treatment of tenosynovitis
  - Palpate tendons along lateral margin of hindfoot
    - Have patient evert foot to improve palpability
    - Usually best spot is about 1 fingerbreadth beneath and anterior to tip of fibula
  - Advance 25-gauge needle to bone
  - Withdraw needle slowly while injecting lidocaine
  - When lidocaine flows freely, needle tip is in tendon sheath
  - Confirm position with injection of iodinated contrast
  - Inject long-acting anesthetic, corticosteroid
  - Fluoro spot after injection shows dispersal of contrast through tendon sheath, confirming correct placement

DIFFERENTIAL DIAGNOSIS

5th Metatarsal Tuberosity Fracture

- Insertion of peroneus brevis tendon commonly avulsed

Cuboid Stress Fracture

- Marrow edema with underlying fracture line

Superior Peroneal Retinaculum Avulsion

- Causes peroneal tendon subluxation, lateral ankle pain

Ankle Sprain

- Injury to lateral collateral ligament leads to lateral swelling and pain
- Fluid in peroneal tendon sheath may be secondary to extravasation through torn calcaneofibular ligament

PATHOLOGY

General Features

- Associated abnormalities
  - **Superior peroneal retinaculum injury**
    - Permits lateral and anterior subluxation of tendons
    - Often intermittent, so tendons are in normal position on MR
    - May occur due to overuse or acute trauma
    - Sometimes see thin flake of cortical bone along lateral margin of fibula
  - **Peroneus quartus**
    - Accessory muscle arising above ankle, variable insertion
    - Allows less room for normal excursion of tendons
  - **Low-lying peroneus brevis muscle belly**
    - Muscle extends to tip of fibula or further
    - Decreases room beneath peroneal retinacula
  - **Lateral collateral ligament tear**
    - Also due to supination injury
  - **Rheumatoid arthritis or seronegative arthritis**
    - Causes tenosynovitis, longitudinal splitting but rarely complete rupture of tendon
  - **Peroneal tubercle**
    - Some authors believe tubercle is an incidental finding
    - Large tubercle or marrow edema in tubercle suggest tendon impingement

- Pronation of foot, especially due to chronic posterior tibial tendon dysfunction
  - Chronic pronation may impinge tendons as they pass beneath tip of fibula
- Convex posterior margin of fibula
  - Increases stress on tendons posterior to fibula
  - May predispose to peroneal retinaculum tear

Staging, Grading, & Classification

- Different staging systems used
- Most commonly used by foot surgeons
  - Grade I: Tendon thickened, splitting of fibers internally
  - Grade II: Tendon thinned by partial thickness tear
  - Grade III: Tendon completely torn
- Sobel system for peroneus brevis injury
  - Grade 1: Flattened tendon
  - Grade 2: Partial thickness split < 1 cm in length
  - Grade 3: Full thickness split < 2 cm in length
  - Grade 4: Full thickness split > 2 cm in length

Relevant Anatomy

- PL and PB originate from fibula in lateral compartment of leg
  - PL originates more proximally
- PL superficial to PB in leg, passes posterior to PB at ankle, and is inferior to PB in foot
- PB inserts on tuberosity at base of 5th MT
  - Has round contour in hindfoot, changes to ribbon-like contour as PB nears its insertion
- PL crosses under sulcus of cuboid, inserts on plantar aspect of midfoot
  - Multiple fascicles normally seen as PL fans out to its insertions
  - Inserts primarily on 1st cuneiform and metatarsal
  - May have slips to 5th MT base and posterior tibial tendon
- **Superior peroneal retinaculum**
  - Maintains position of tendons posterior to posterior malleolus
  - Attaches to fibula, lateral margin of calcaneus, and sometimes Achilles tendon
- **Inferior peroneal retinaculum**
  - Maintains position of tendons at lateral margin of calcaneus
  - Continuous with extensor retinaculum
- **Peroneal tubercle** of calcaneus variably present, may be quite large
  - PB courses above tubercle
  - PL courses below tubercle
- **Tendon sheaths**
  - Tendons share common sheath in hindfoot
    - Calcaneofibular ligament forms a portion of floor of tendon sheath
  - Common tendon sheath splits into separate sheaths as tendons diverge toward their insertions
  - Additional, separate tendon sheath may be present distally around PL

CLINICAL ISSUES

Presentation

- Most common signs/symptoms
  - Lateral ankle pain
- Other signs/symptoms
  - Pain and weakness with eversion of foot
  - "Snapping" of tendon around tip of fibula occurs when peroneal retinaculum injury is also present

Treatment

- Bracing, immobilization
- Physical therapy
- Corticosteroid injection under fluoroscopic guidance
- Tendon repair

DIAGNOSTIC CHECKLIST

Consider

- Minor degrees of peroneal tendon splitting are of uncertain clinical significance, and patients are often asymptomatic

Image Interpretation Pearls

- Tendons difficult to evaluate as they curve beneath the fibula; check all 3 planes
- MR may be false-negative for tenosynovitis, which can be diagnosed and treated with tenosynogram
- PB normally has thin, ribbon-like appearance near its insertion on 5th MT; do not mistake this for partial tear
- Use T1-weighted images to evaluate for peroneus quartus muscle and low-lying peroneus brevis muscle belly
- Always include evaluation of superior peroneal retinaculum together with tendons

SELECTED REFERENCES

1. Kijowski R et al: Magnetic resonance imaging findings in patients with peroneal tendinopathy and peroneal tenosynovitis. Skeletal Radiol. 36(2):105-14, 2007
2. Saupe N et al: Anatomic variants associated with peroneal tendon disorders: MR imaging findings in volunteers with asymptomatic ankles. Radiology. 242(2):509-17, 2007
3. Wang XT et al: Normal variants and diseases of the peroneal tendons and superior peroneal retinaculum: MR imaging features. Radiographics. 2005 May-Jun;25(3):587-602. Review. Erratum in: Radiographics. 2006 Mar-Apr;26(2):640. Radiographics. 25(5):1436, 2005
4. Rademaker J et al: Tear of the peroneus longus tendon: MR imaging features in nine patients. Radiology. 214(3):700-4, 2000
5. Mota J et al: Magnetic resonance imaging of the peroneal tendons. Top Magn Reson Imaging. 9(5):273-85, 1998
6. Rosenberg ZS et al: MR features of longitudinal tears of the peroneus brevis tendon. AJR Am J Roentgenol. 168(1):141-7, 1997
7. Sobel M et al: The dynamics of peroneus brevis tendon splits: a proposed mechanism, technique of diagnosis, and classification of injury. Foot Ankle. 13(7):413-22, 1992

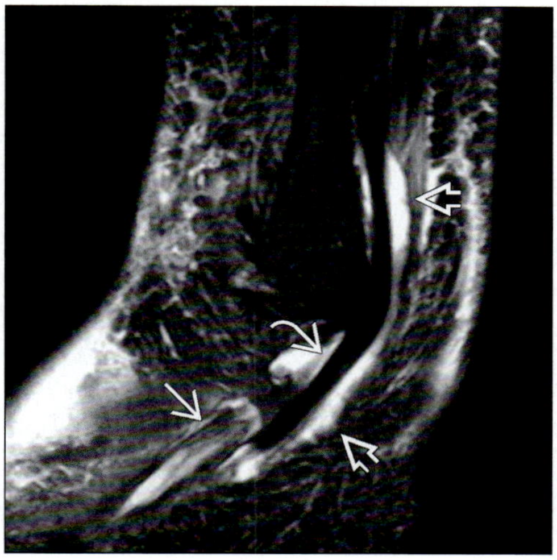

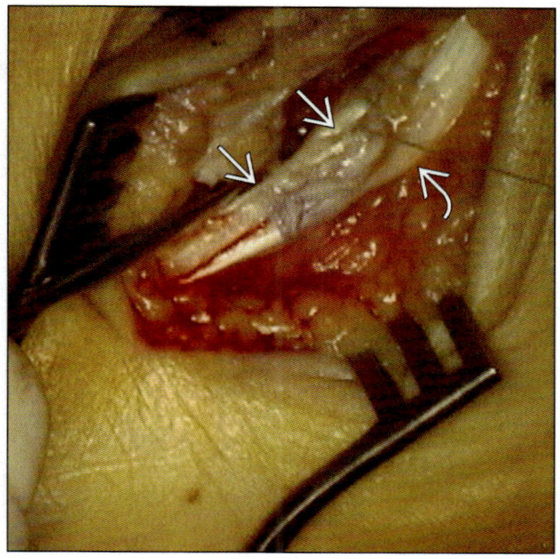

(Left) *Sagittal STIR MR shows extensive longitudinal splitting of peroneus brevis with intact peroneus longus. There is a large amount of fluid in the tendon sheath, and the sheath has an irregular contour, indicating tenosynovitis.* ***(Right)*** *Intraoperative photograph from the same patient shows longitudinal splits in peroneus brevis. Peroneus longus is intact.*

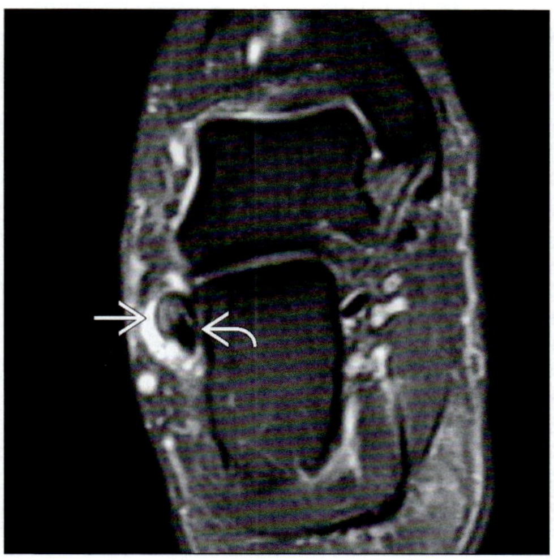

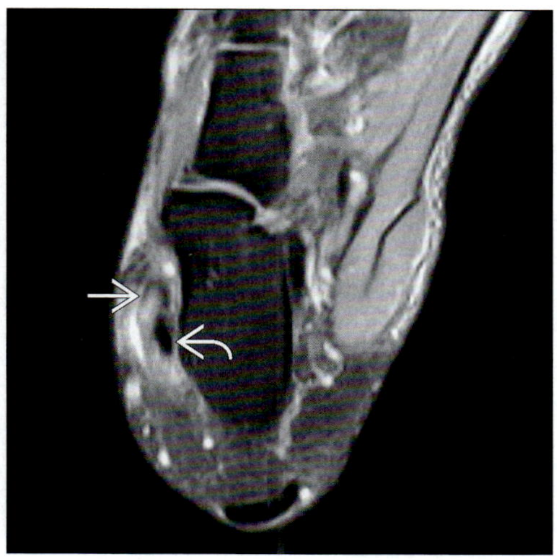

(Left) *Coronal T2WI FS MR shows split peroneus brevis tendon. The tendon has an inverted "V" shape in cross section, and peroneus longus is displaced superiorly into the defect in the peroneus brevis.* ***(Right)*** *Axial PDWI FS MR from the same patient shows severe splitting of peroneus brevis, which lies anterior to peroneus longus.*

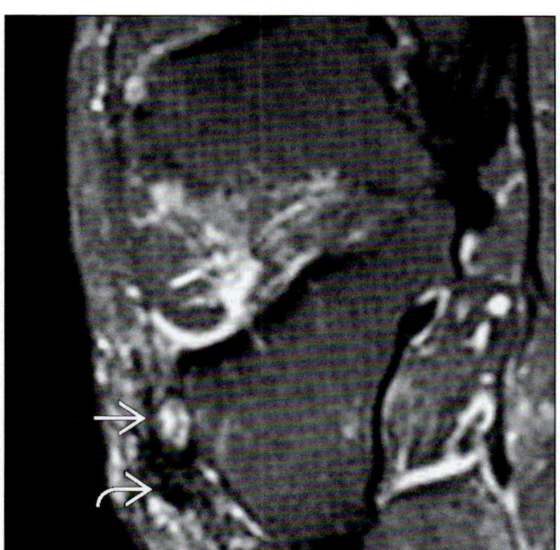

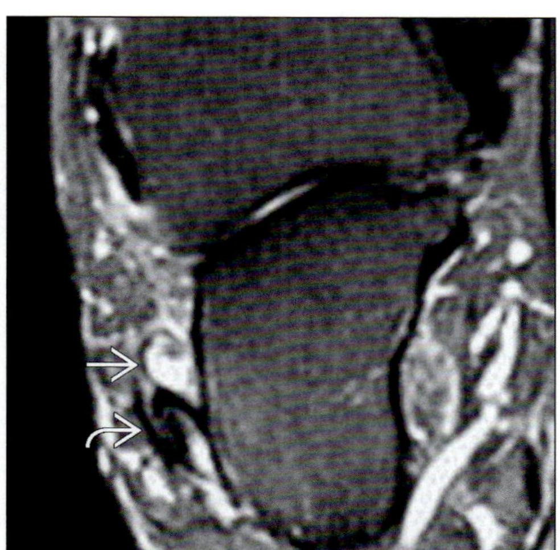

(Left) *Coronal T2WI FS MR shows complete absence of peroneus brevis and irregular, disorganized debris at site of peroneus longus. At time of surgery there was complete rupture of both tendons.* ***(Right)*** *Coronal T2WI FS MR shows absence of peroneus brevis. Peroneus longus has split into 2 V-shaped portions. The peroneus longus is less commonly torn than the peroneus brevis and rarely torn without concomitant injury of peroneus brevis.*

*(**Left**) Lateral radiograph shows fracture of os peroneum indicating peroneus longus injury. A fractured ossicle has irregular, uncorticated margins as seen in this example. In contrast, the commonly seen bipartite os peroneum has smooth, rounded, corticated margins. (**Right**) Sagittal STIR MR from the same patient shows fluid between fragments of fractured os peroneum. Fluid is also seen at the cuboid sulcus, where the peroneum longus normally passes.*

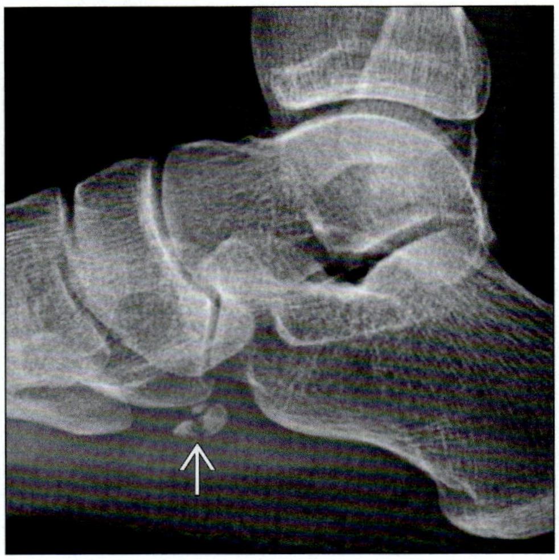

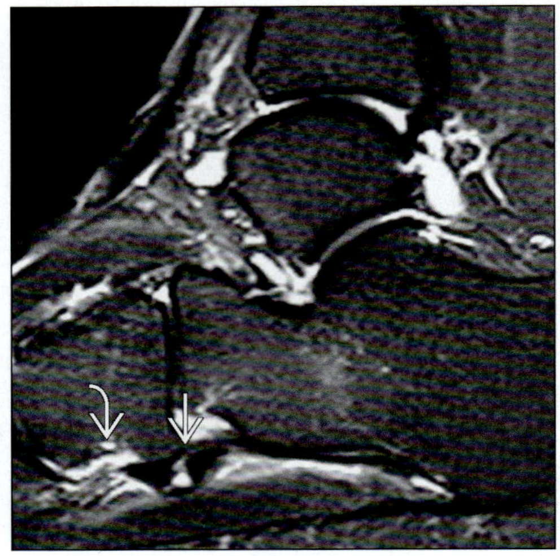

*(**Left**) Axial PDWI FS MR from the same patient shows fluid between fracture fragments and extensive surrounding edema. (**Right**) Coronal T2WI FS MR from the same patient shows thin, split peroneus longus tendon more distally. This patient was a runner who felt a "snap" while running on uneven ground.*

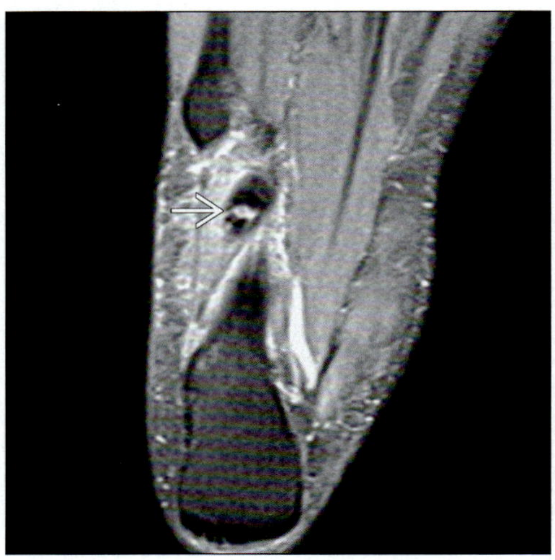

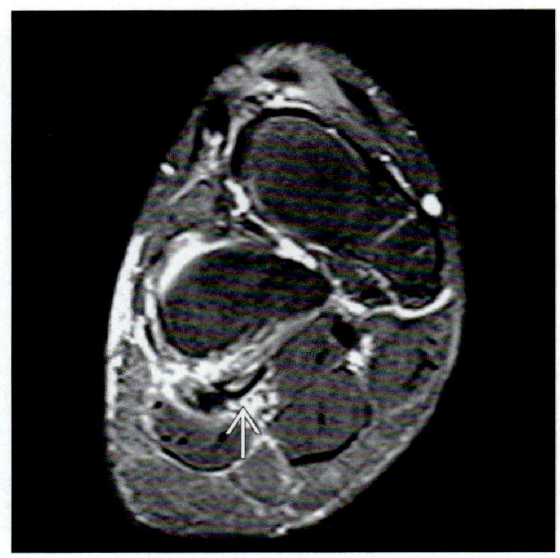

*(**Left**) Coronal T2WI FS MR shows a peroneus quartus associated with peroneal tendon dysfunction. Peroneus brevis is split, and peroneus longus has migrated into the peroneus brevis defect. (**Right**) Coronal T2WI FS MR shows splitting of both peroneus brevis and peroneus longus. There is bone marrow edema in the peroneal tubercle, suggesting abnormal stress on the tubercle.*

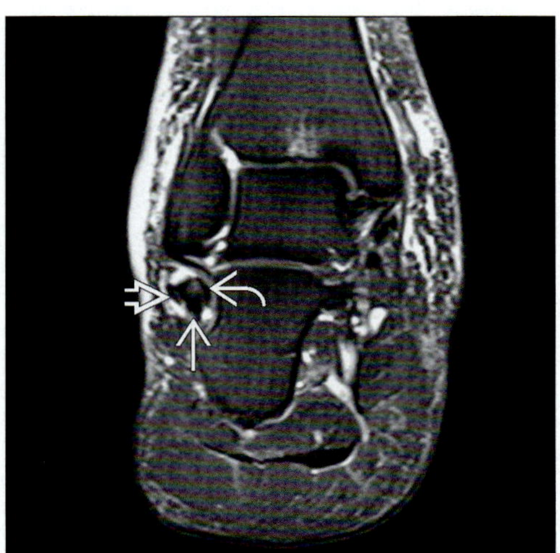

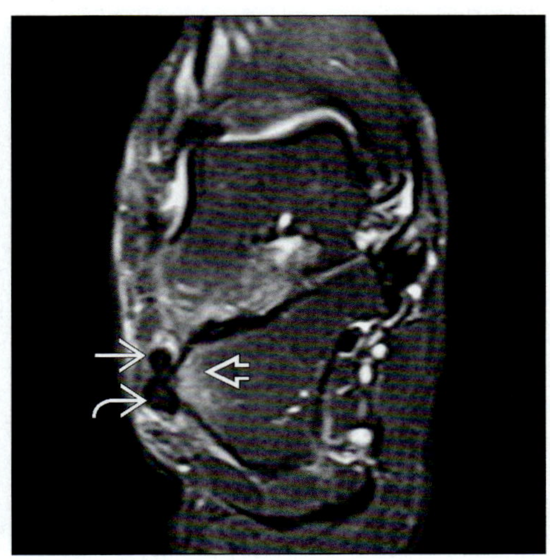

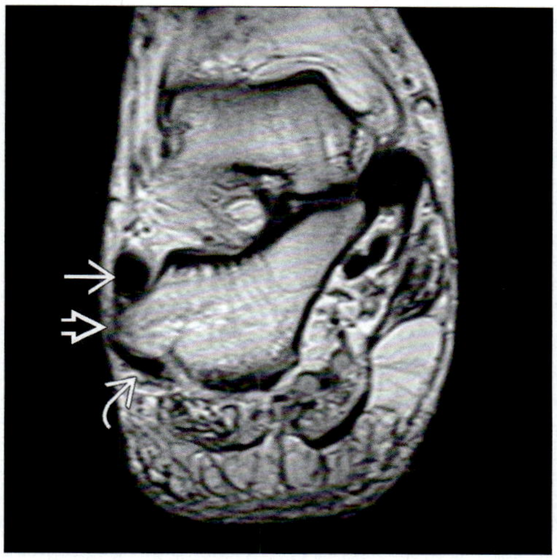
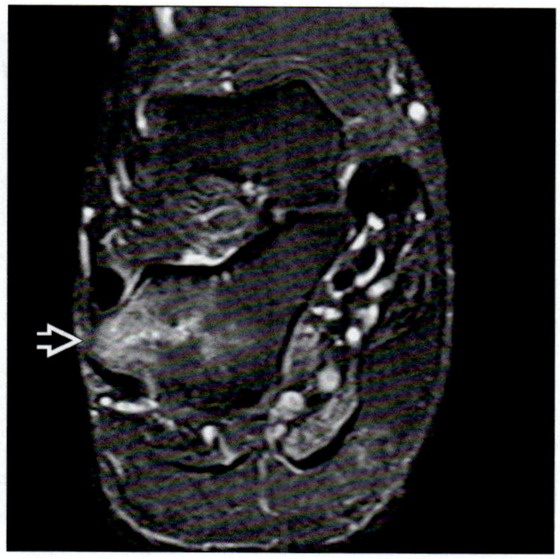

(Left) *Coronal PDWI MR shows a markedly enlarged peroneal tubercle of the calcaneus. Peroneus brevis is larger than normal, while peroneus longus is flattened.* ***(Right)*** *Coronal T2WI FS MR from the same patient shows bone marrow edema in the peroneal tubercle. The patient presented with symptoms of peroneal tenosynovitis. Despite lack of fluid around tendons, a tendon sheath injection was performed and relieved symptoms.*

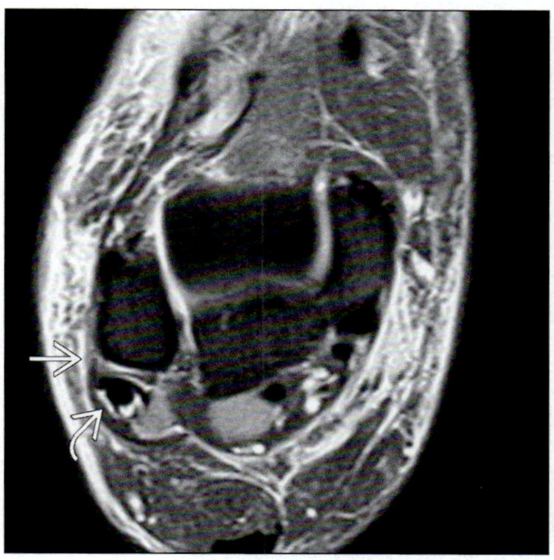
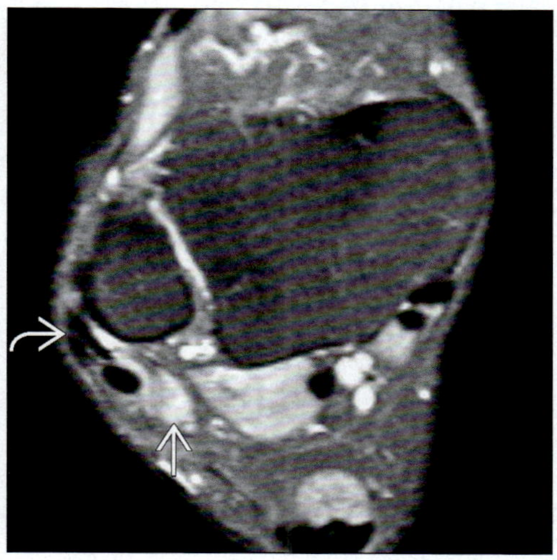

(Left) *Axial T2WI FS MR shows chronic injury of the peroneal retinaculum. There is fluid along the peroneal tendon sheath. Peroneal retinaculum injury is often unrecognized and is a major causative factor for peroneal tendinopathy.* ***(Right)*** *Axial PDWI FS MR shows anterolateral subluxation and splitting of the peroneus brevis tendon due to chronic tear of the peroneal retinaculum. Fluid is seen in the tendon sheath.*

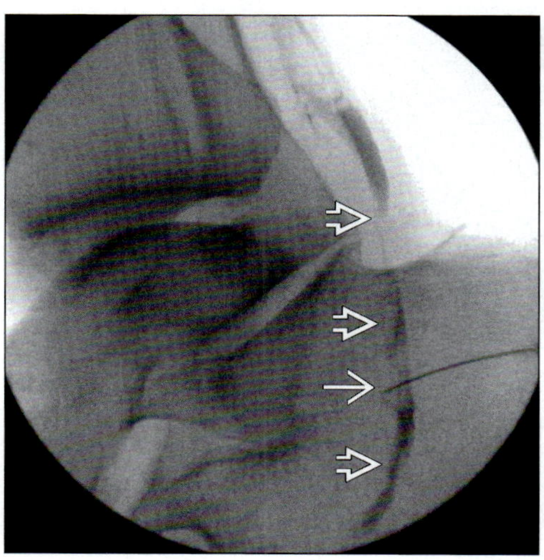
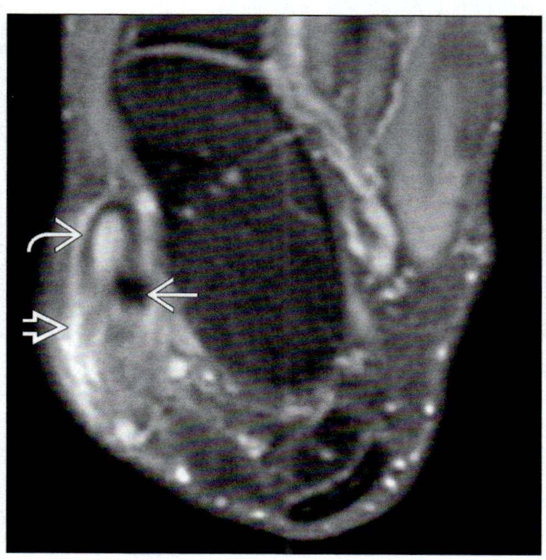

(Left) *Lateral tenosynogram shows needle tip in peroneal tendon sheath, and a beaded, irregular appearance of the tendon sheath. Patients with this finding have a high rate of pain relief from corticosteroid injection into the sheath.* ***(Right)*** *Axial PDWI FS MR shows splitting in the peroneus brevis and an intact peroneus longus. The inferior peroneal retinaculum is torn.*

TEARS OF INTRINSIC FOOT MUSCLE AND PLANTAR FASCIA

Key Facts

Imaging

- Location
  - Plantar fascia tears usually occur in hindfoot, most commonly involve central band
  - Intrinsic muscle tears usually in hindfoot but may occur distally
- Radiographs usually normal
- MR used for diagnosis
  - Fluid tracking between muscle fibers
  - Discontinuity of fascial or muscle fibers
  - Gap filled with fluid in acute injury
- May be partial or complete
- Discontinuity of fascia
- Fluid signal intensity in fascia
- Herniation of muscle through defect

Top Differential Diagnoses

- Plantar fibromatosis
- Plantar fascia release
- Foreign body

Pathology

- Usually occur in hindfoot
- Plantar fascia rupture often superimposed on fasciitis
- Plantar fascia rupture often associated with injury to underlying muscle
- Osseous avulsion of intrinsic muscle attachment more common than muscle tear

Clinical Issues

- Plantar fascia rupture may lead to flattening of longitudinal arch of foot
- Conservative treatment: Casting, protection from weight-bearing

Diagnostic Checklist

- Plantar fascia release MR may appear identical to tear

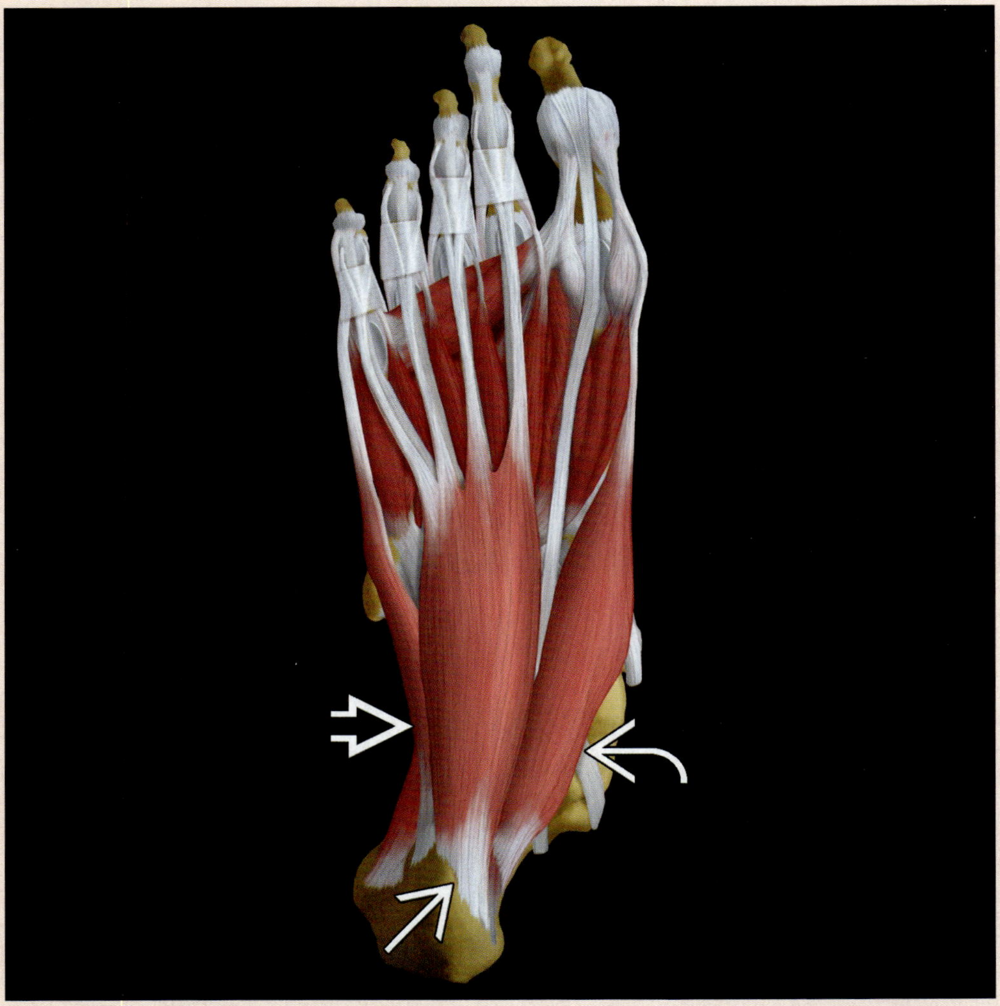

Graphic shows superficial layer of plantar foot musculature. Plantar fasciae overlie the 1st layer of muscles. The central band of the plantar fascia is the most frequently injured. It originates from the plantar and medial margin of the calcaneal tuberosity and overlies the flexor digitorum brevis ➡. The medial band originates from the medial margin of the calcaneal tuberosity and overlies the abductor hallucis ➡. The lateral band originates from the lateral margin of the calcaneal tuberosity and overlies the abductor digiti minimi ➡. Tears of the plantar fascia often involve the underlying muscles.

TEARS OF INTRINSIC FOOT MUSCLE AND PLANTAR FASCIA

IMAGING

General Features

- Location
  - Plantar fascia tears usually occur in hindfoot, most commonly involve central band
  - Intrinsic muscle tears usually in hindfoot but may occur distally

Radiographic Findings

- Radiographs usually normal
- Sometimes see osseous avulsion fracture of muscle attachment
  - Extensor digitorum brevis: Dorsal anterior calcaneus
  - Abductor hallucis brevis: Plantar medial calcaneal tuberosity
- Displacement of 1st toe sesamoids may be seen with tears of intrinsic muscles of 1st toe (rare)

MR Findings

- Muscle tear
  - Fluid tracking between muscle fibers
  - Discontinuity of fascial or muscle fibers
    - Gap filled with fluid in acute injury
- Plantar fascia tear
  - May be partial or complete
  - Discontinuity of fascia
  - Fluid signal intensity in fascia
  - Herniation of muscle through fascial defect
  - Tear or strain of underlying muscle

Imaging Recommendations

- Best imaging tool
  - MR

DIFFERENTIAL DIAGNOSIS

Plantar Fibromatosis

- Nodular or sheet-like thickening of plantar fascia
- Usually low signal intensity on all sequences
- May have moderate increase in signal intensity on STIR
- Enhances with gadolinium administration
  - Healing tear will also enhance

Plantar Fascia Release

- Performed for intractable plantar fasciitis
- Focal defect visible on MR
- Post-release MR may show adjacent fascia thickening (compared to preoperative MR)
  - Patients often asymptomatic, despite appearance

Foreign Body

- Intramuscular edema surrounding a foreign body may obscure the foreign body itself

PATHOLOGY

Gross Pathologic & Surgical Features

- Osseous avulsion of intrinsic muscle attachments is more common than muscle tear

Relevant Anatomy

- 3 bands of plantar fascia originate on posterior process of calcaneus and fan out to merge with distal fascia at level of MTP joints
  - Medial band: Superficial to abductor hallucis muscle
  - Central band: Superficial to flexor digitorum brevis muscle
  - Lateral band: Superficial to abductor digiti minimi muscle
- Some authors omit medial band, describe central band as medial band

CLINICAL ISSUES

Presentation

- Most common signs/symptoms
  - Acute onset of pain
  - Plantar fascia rupture often superimposed on fasciitis
- Intrinsic muscle tears are rare in isolation
  - Usually occur with fractures and dislocations of foot

Natural History & Prognosis

- Plantar fascia rupture may lead to flattening of longitudinal arch of foot

Treatment

- Conservative: Casting, protection from weightbearing

DIAGNOSTIC CHECKLIST

Image Interpretation Pearls

- Plantar fascia release will appear identical to tear on MR

SELECTED REFERENCES

1. Cheung JT et al: Consequences of partial and total plantar fascia release: a finite element study. Foot Ankle Int. 27(2):125-32, 2006
2. Saxena A et al: Plantar fascia ruptures in athletes. Am J Sports Med. 32(3):662-5, 2004
3. Theodorou DJ et al: Plantar fasciitis and fascial rupture: MR imaging findings in 26 patients supplemented with anatomic data in cadavers. Radiographics. 20 Spec No:S181-97, 2000
4. Yu JS et al: The plantar fasciotomy: MR imaging findings in asymptomatic volunteers. Skeletal Radiol. 28(8):447-52, 1999
5. Pelletier JP et al: Avulsion fracture of the calcaneus at the origin of the abductor hallucis muscle. J Foot Surg. 29(3):268-71, 1990

TEARS OF INTRINSIC FOOT MUSCLE AND PLANTAR FASCIA

__(Left)__ Axial PDWI FS MR shows tears of central and lateral bands of plantar fascia. The medial band is intact. Clinically, the central band is the most important, and it is also the most commonly injured. __(Right)__ Coronal T2WI FS MR shows absent central band of plantar fascia and hematoma in the underlying flexor digitorum brevis muscle. The patient was a college basketball player with an acute injury.

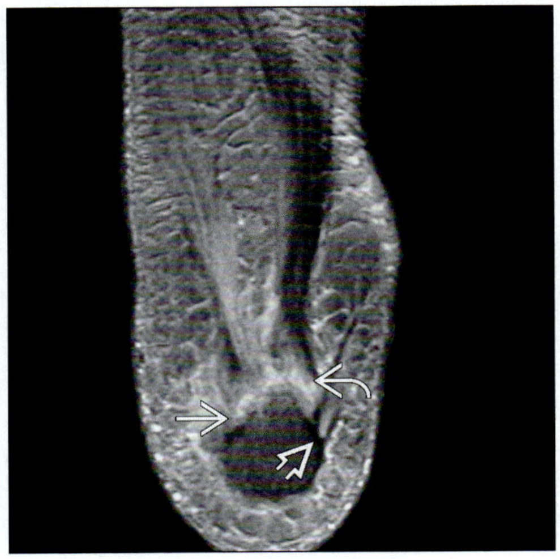

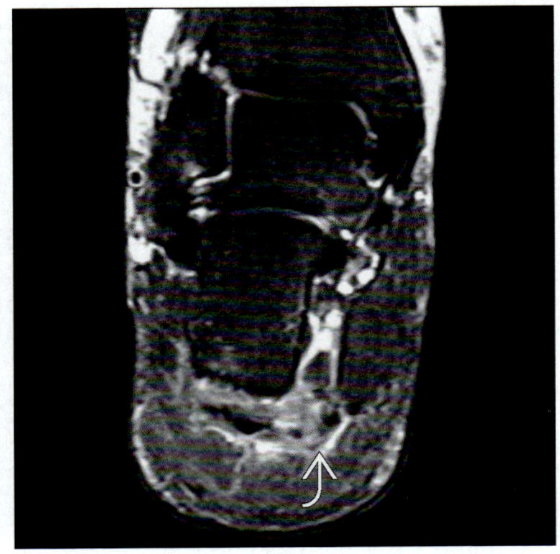

__(Left)__ Sagittal T1WI MR shows loss of low signal intensity fascia overlying flexor digitorum brevis. The patient complained of waxing and waning soft tissue fullness, which was thought to represent muscle herniation in this chronic tear. __(Right)__ Coronal T2WI FS MR in the same patient shows fluid along fascial defect. Chronic tears are often poorly defined and may enhance with gadolinium administration, mimicking fibromatosis.

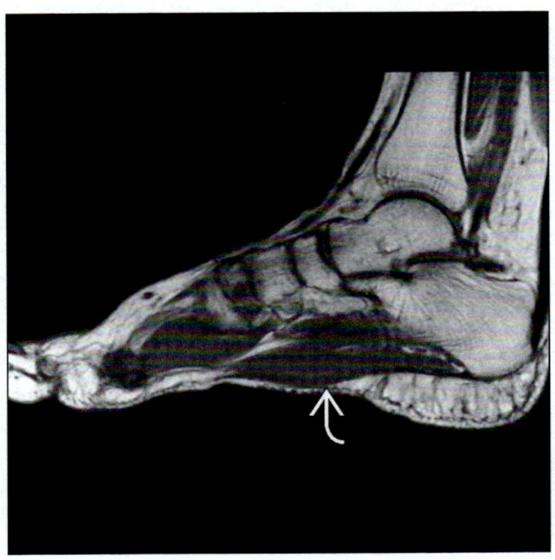

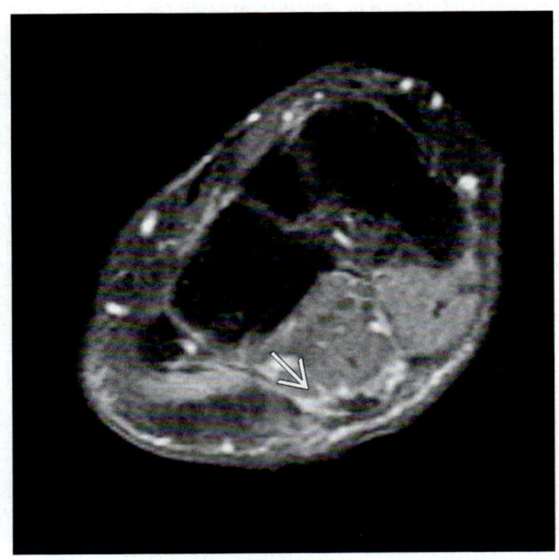

__(Left)__ Coronal T1WI MR shows plantar fat pad fibrosis presenting with nonspecific heel pain in this patient status post calcaneal osteotomy. The underlying fascia are intact. __(Right)__ AP radiograph shows distal displacement of medial sesamoid due to rupture of the medial head of the flexor hallus brevis. The patient was status post revision bunionectomy, with dorsiflexion deformity ("cockup deformity") of the 1st toe.

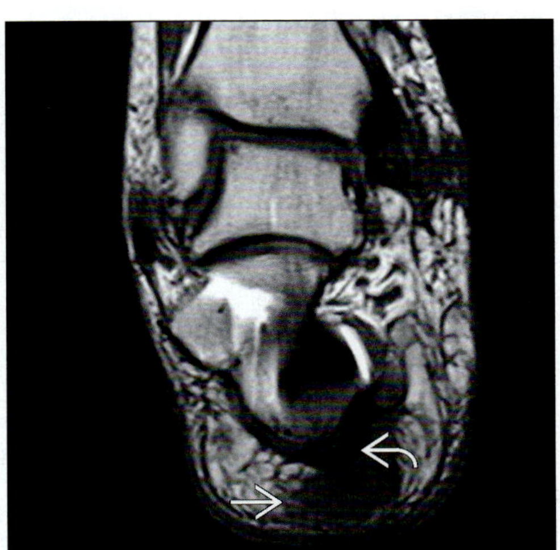

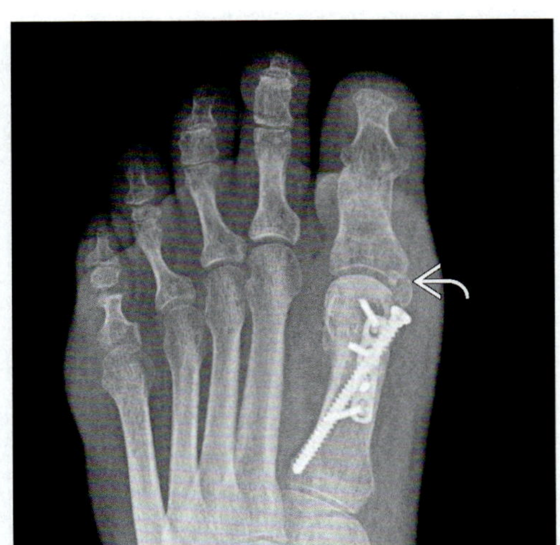

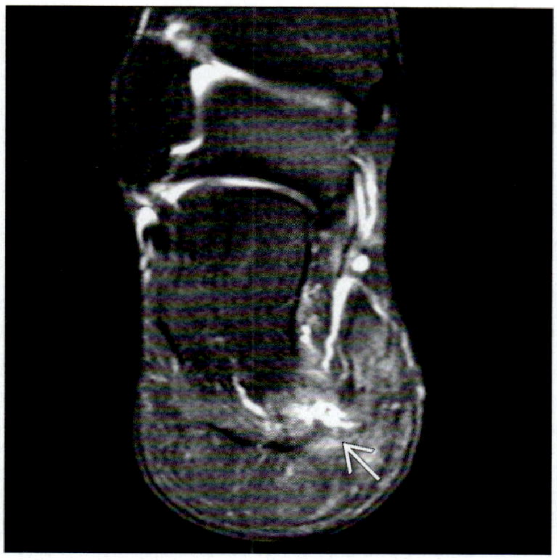

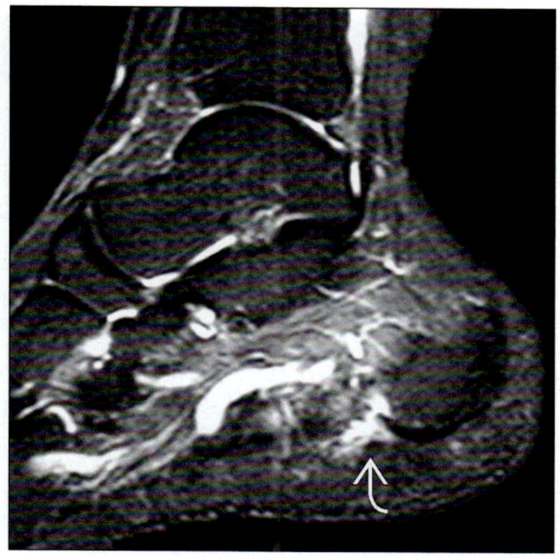

(Left) *Coronal T2WI FS MR shows focal defect in the plantar fascia → due to surgical release. Fascial release may be performed for severe fasciitis unresponsive to conservative treatment. This operation may result in flattening of medial arch and dorsolateral foot pain.* ***(Right)*** *Sagittal STIR MR shows plantar fascia release ↪. History is needed to distinguish this from spontaneous rupture.*

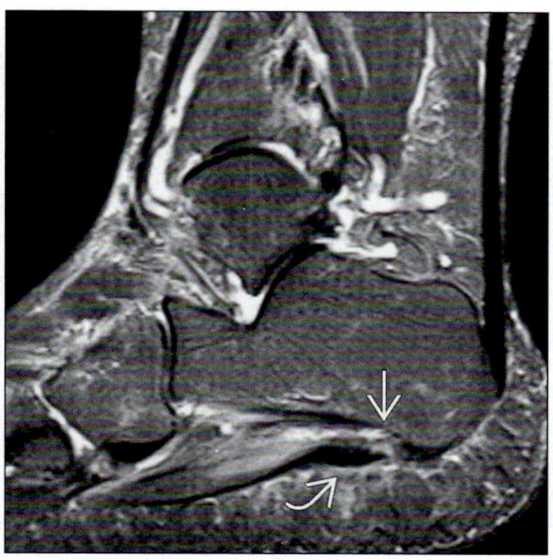

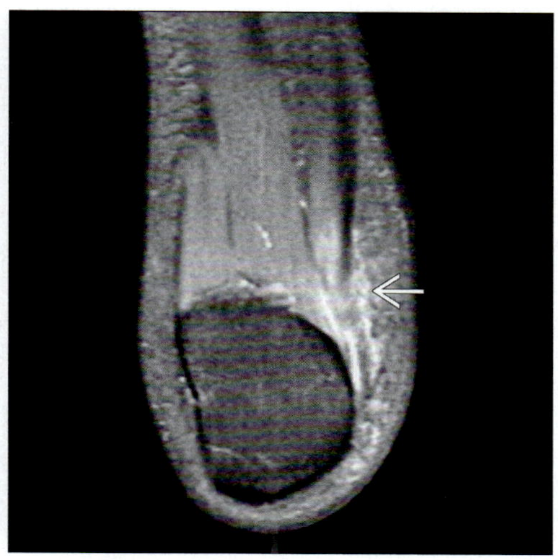

(Left) *Sagittal STIR MR shows classic plantar fascia tear → involving the origin of the central band. The retracted fibers ↪ are mildly thickened, a finding that should not be mistaken for plantar fibromatosis.* ***(Right)*** *Axial PDWI FS MR shows isolated tear of abductor hallucis muscle → in a gymnast. This is an unusual injury, and it was treated with rest and a boot.*

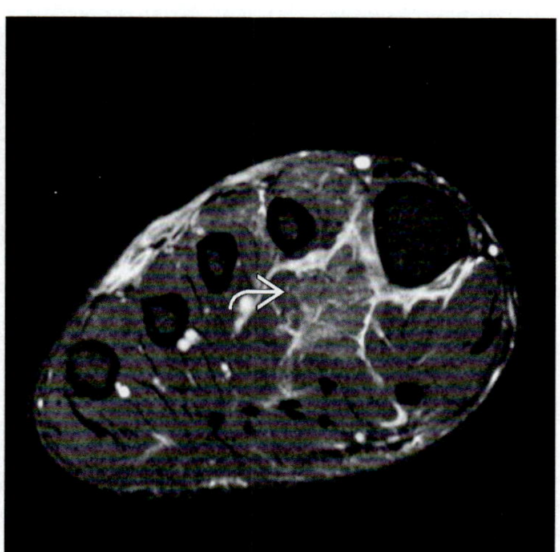

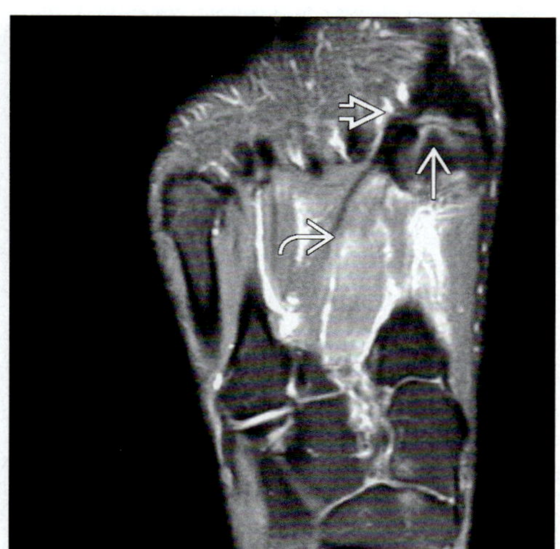

(Left) *Coronal T2WI FS MR shows strain of adductor hallucis muscle ↪ in a patient who suffered an abduction (i.e., away from 2nd toe) injury of the hallux. Injuries such as this are useful secondary signs of the mechanism of injury but in themselves are not treated.* ***(Right)*** *Axial PDWI FS MR in the same patient shows the adductor strain ↪. More importantly, there was injury to the 1st MT head → and the lateral joint capsule ⇨.*

ANTERIOR IMPINGEMENT SYNDROME, ANKLE

Key Facts

Terminology
- Mechanical limitation of dorsiflexion due to bony or soft tissue abnormality
- Divided into 3 types: Anterior, anteromedial, and anterolateral

Imaging
- Lateral radiograph may show osteophytes from anterior margin of tibial plafond or talar dome
- Diagnosis made primarily on MR
- Anteromedial impingement
  - Thickened, irregular soft tissues at anteromedial corner of ankle joint
  - From torn superficial deltoid ligament fibers
- Anterior impingement
  - Anterior osteophytes from tibial plafond or talar dome
  - Soft tissue impingement may occur from capsular scarring
- Anterolateral impingement
  - Meniscoid lesion at anterolateral aspect of ankle joint
  - Triangular, low signal intensity focus in expected location of ligament
  - From torn ATFL or AITFL
  - Bassett ligament (accessory AITFL) may also cause impingement

Top Differential Diagnoses
- Osteochondral injuries, ankle
- Tarsal coalition
- Talar ridge
- ATFL tear

Clinical Issues
- Pain, limitation of motion

*(**Left**) Graphic of the anterior ankle (foot plantarflexed) shows scar forming in the anterolateral ankle joint. This may occur due to chronic injury of the anterior talofibular ligament (ATFL) or anterior inferior tibiofibular ligament (AITFL). (**Right**) Axial PDWI FS MR shows a large meniscoid lesion due to scarring of the torn ATFL. Patient presented with vague anterolateral pain, focal tenderness, and limited dorsiflexion/eversion.*

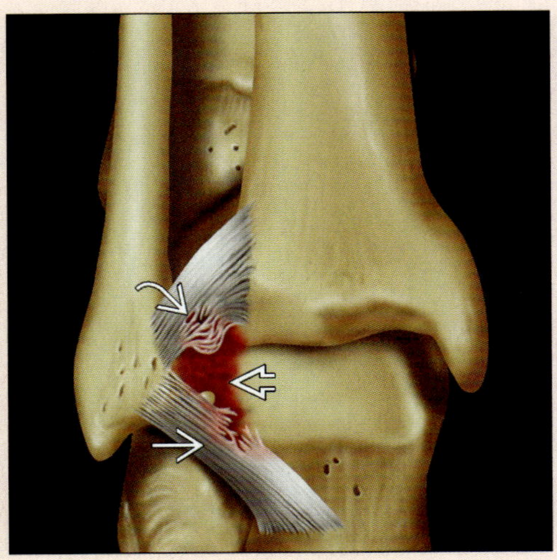

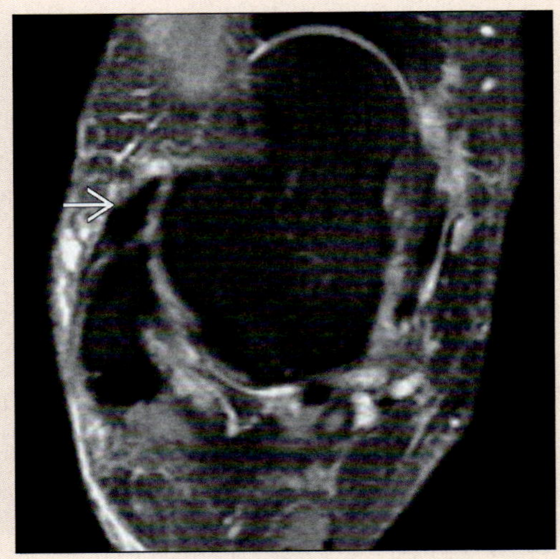

*(**Left**) Sagittal STIR MR shows anterior talar osteophyte causing anterior ankle impingement with severely limited dorsiflexion. There is cartilage loss and subchondral bone marrow edema in the anterior portion of the ankle joint. Cartilage more posteriorly is preserved. (**Right**) Axial PDWI FS MR shows a focal scar at the anteromedial ankle, resulting in impingement with dorsiflexion and inversion. This patient was a former gymnast.*

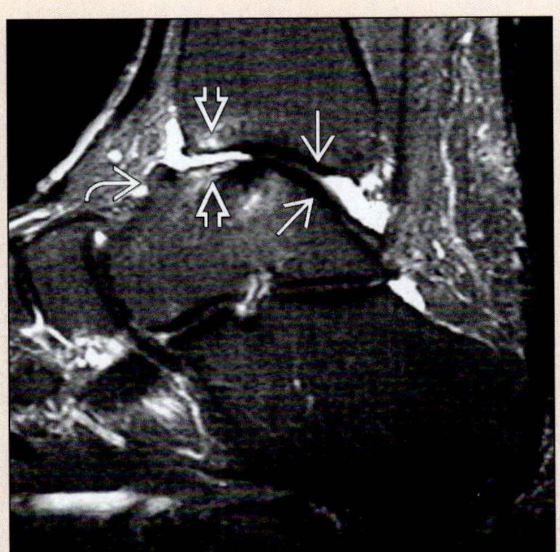

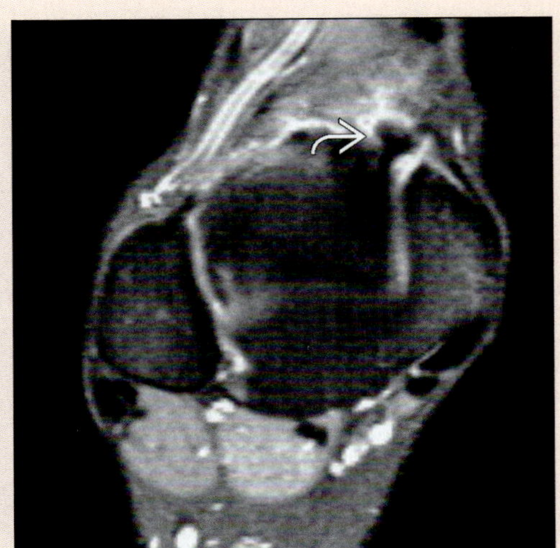

ANTERIOR IMPINGEMENT SYNDROME, ANKLE

TERMINOLOGY

Synonyms
- Anterolateral gutter syndrome

Definitions
- Mechanical limitation of dorsiflexion due to osseous or soft tissue impediment

IMAGING

General Features
- Location
  - 3 distinct locations: Anterior, anteromedial, and anterolateral

Radiographic Findings
- Lateral radiograph may show osteophytes from anterior margin of tibial plafond or talar dome

MR Findings
- Anteromedial impingement
  - Thickened, irregular soft tissues at anteromedial corner of ankle joint
- Anterior impingement
  - Anterior osteophytes from tibial plafond or talar dome
    - Bone marrow edema in osteophytes
  - Scar
    - Thickened band of tissue at anterior margin of ankle joint
- Anterolateral impingement
  - Meniscoid lesion at anterolateral aspect of ankle joint
    - Triangular, thick, low signal intensity structure between fibula and talus
    - Usually attached to fibula, discontinuous with talus
  - Thick Bassett ligament (accessory AITFL) may also cause impingement

Imaging Recommendations
- Best imaging tool
  - Lateral radiograph for anterior osteophytes
  - MR for marrow edema, cartilage loss, soft tissue impingement lesions

DIFFERENTIAL DIAGNOSIS

Osteochondral Injuries, Ankle
- Osteochondral fragment may dislodge into joint, limit motion

Tarsal Coalition
- Limited subtalar motion

Talar Ridge
- Normal projection from dorsal margin of talar neck
- Distal to talar dome osteophytes, which can result in anterior impingement

ATFL Tear
- Very common; does not result in impingement unless meniscoid lesion is also present

PATHOLOGY

General Features
- Etiology
  - Often related to prior sports trauma
    - Ligament tears with scar tissue formation
  - Osteoarthritis
  - Chronic synovitis with scar tissue formation

CLINICAL ISSUES

Presentation
- Most common signs/symptoms
  - Pain and limitation of motion
    - Pain reliably reproduced with appropriate maneuver
      - Anteromedial impingement: Dorsiflexion and inversion
      - Anterior impingement: Straight dorsiflexion
      - Anterolateral impingement: Dorsiflexion and eversion
- Other signs/symptoms
  - Focal tenderness at site of impingement
- Clinical profile
  - Often seen in athletes
    - Soccer players ("footballer's ankle")
    - Ballet dancers (especially with pliés)

Demographics
- Epidemiology
  - Anterior ankle osteophytes seen in 40-60% of professional athletes

Treatment
- Cheilectomy: Surgical resection of bone spur
- Arthroscopic debridement of soft tissue lesions

DIAGNOSTIC CHECKLIST

Consider
- Include evaluation of ankle gutters in routine checklist for ankle MR

SELECTED REFERENCES

1. Linklater J: MR imaging of ankle impingement lesions. Magn Reson Imaging Clin N Am. 17(4):775-800, vii-viii, 2009
2. Sanders TG et al: Impingement syndromes of the ankle. Magn Reson Imaging Clin N Am. 16(1):29-38, v, 2008
3. van den Bekerom MP et al: The distal fascicle of the anterior inferior tibiofibular ligament as a cause of tibiotalar impingement syndrome: a current concepts review. Knee Surg Sports Traumatol Arthrosc. 15(4):465-71, 2007
4. Haller J et al: MR-imaging of anterior tibiotalar impingement syndrome: agreement, sensitivity and specificity of MR-imaging and indirect MR-arthrography. Eur J Radiol. 58(3):450-60, 2006
5. Robinson P et al: Anteromedial impingement of the ankle: using MR arthrography to assess the anteromedial recess. AJR Am J Roentgenol. 178(3):601-4, 2002
6. Robinson P et al: Soft-tissue and osseous impingement syndromes of the ankle: role of imaging in diagnosis and management. Radiographics. 22(6):1457-69; discussion 1470-1, 2002

(Left) Axial MR arthrogram T1WI FS shows a large meniscoid lesion ➡ due to ATFL tear. Note enlargement of anterolateral recess ⇨ distal to the meniscoid lesion, with contrast extending through the incompetent ATFL. (Right) Coronal T2WI FS MR from the same patient also shows an anterolateral meniscoid lesion ➡. Axial images are the primary means of identifying meniscoid lesions, but they are also visible on coronal sequences.

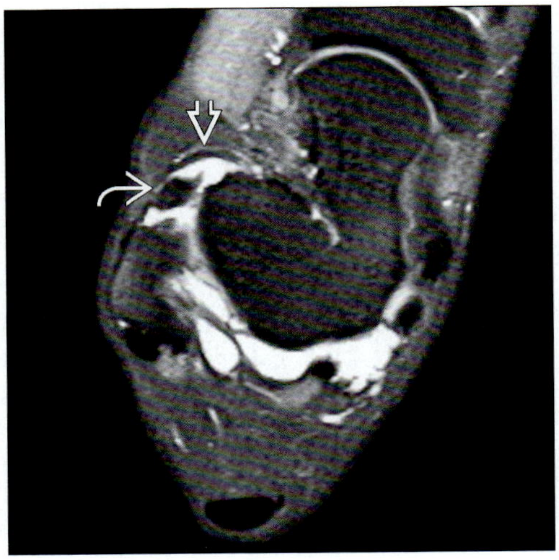

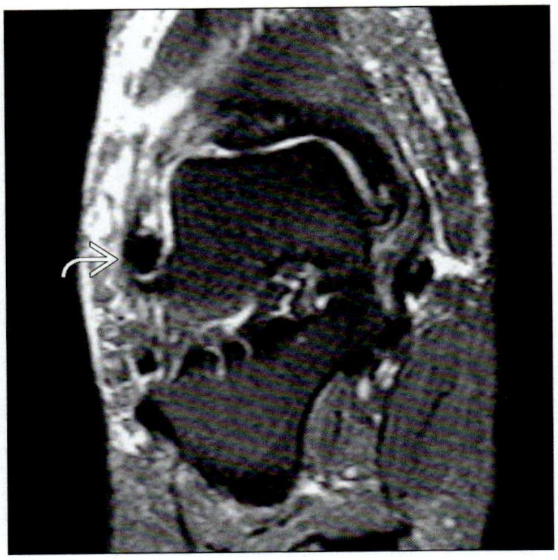

(Left) Axial PDWI FS MR shows scar ➡ from a torn AITFL. The AITFL is best visualized on axial images at the level of the tibial plafond. Meniscoid lesions in this region are smaller and less common than those from a torn ATFL. (Right) Axial bone CT arthrogram shows contrast outlining a triangular meniscoid lesion ➡ in the anterolateral gutter. If MR cannot be performed, CT arthrogram reliably demonstrates impingement lesions.

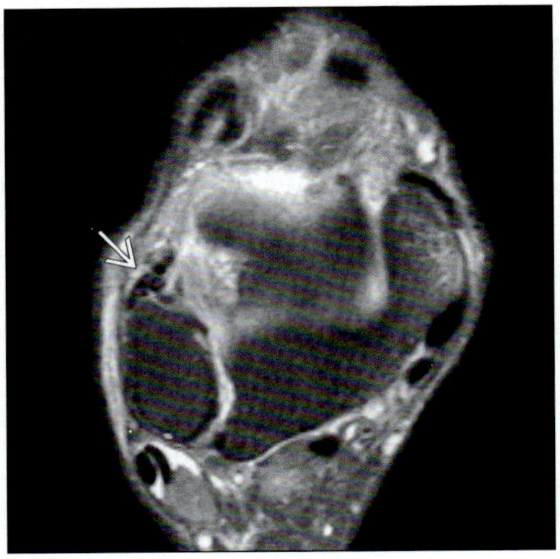

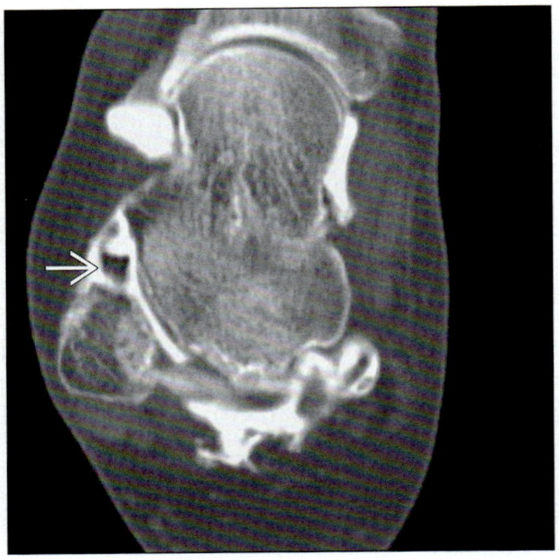

(Left) Axial PDWI FS MR shows scar ➡ in the anteromedial joint recess from tear of the superficial deltoid ligament. Patient presented with vague anteromedial pain and focal tenderness to palpation in the region of the scar. Note intact ATFL ➡. (Right) Axial T2WI FS MR shows bulky scar ➡ in the anteromedial recess in a patient with limited motion and medial ankle pain. There is also a small meniscoid lesion laterally ➡, which was asymptomatic.

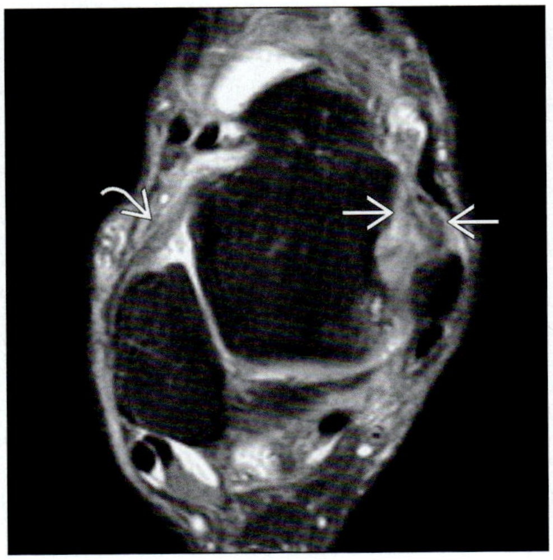

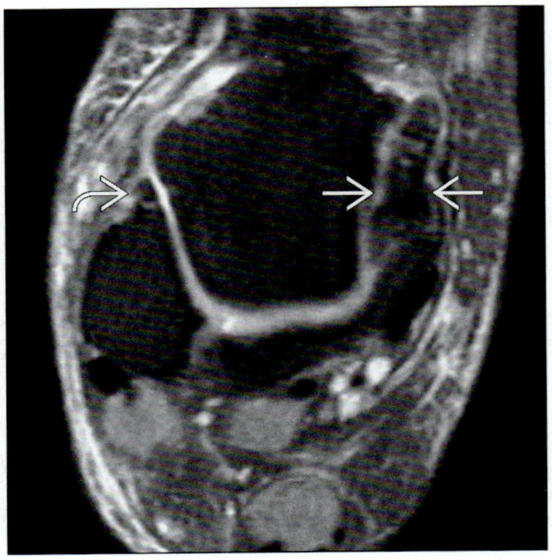

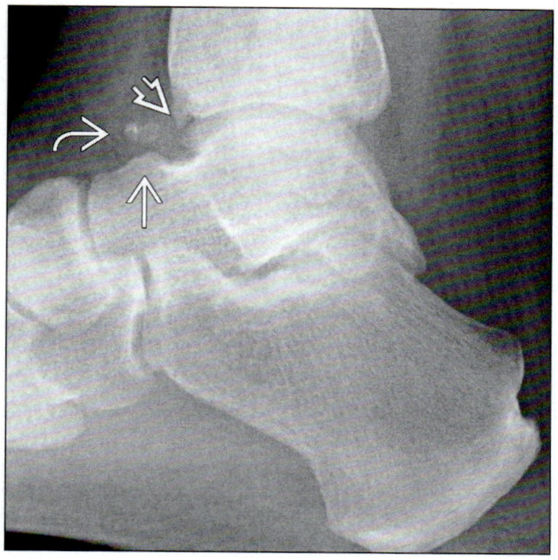

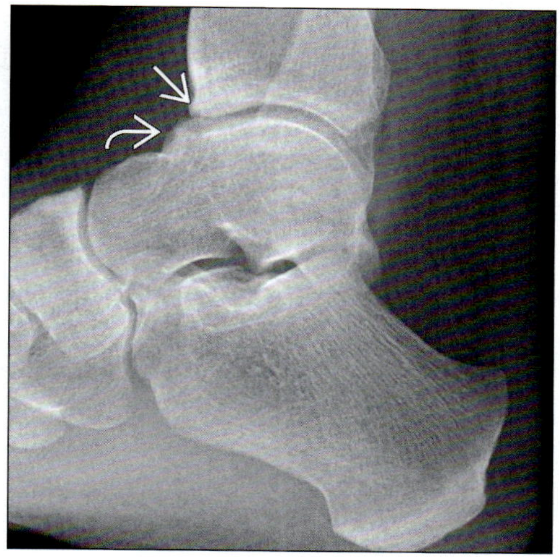

(Left) *Lateral radiograph shows anterior osteophytes limiting dorsiflexion. Note intraarticular bodies. The talar ridge (site of ankle capsule insertion) on the talar neck is a normal variant, which should not be mistaken for an osteophyte.* ***(Right)*** *Lateral radiograph shows spurs arising from the anterior margin of the talar dome and the anterior margin of the tibial plafond. Patient complained of ankle pain and limited dorsiflexion.*

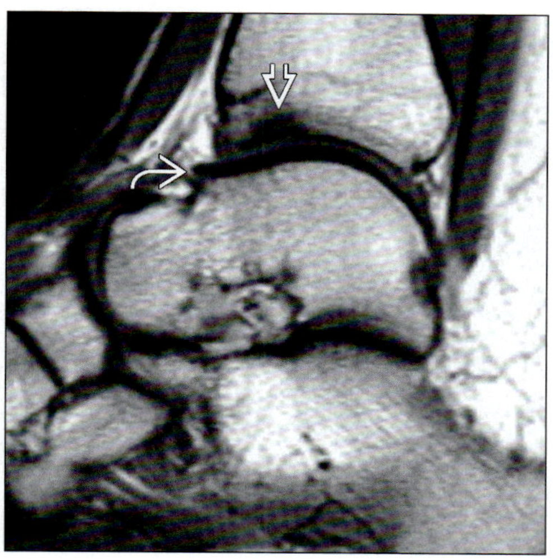

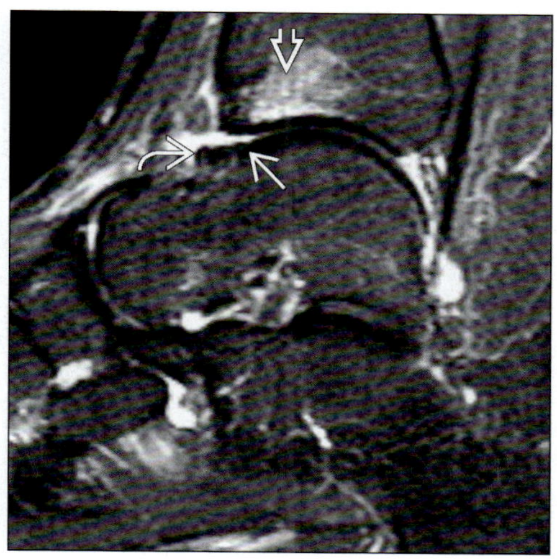

(Left) *Sagittal T1WI MR shows anterior talar dome osteophyte. There is focal bone marrow edema in the tibial plafond secondary to impingement. When ankle osteoarthritis is focal, as in this case, symptoms often respond well to cheilectomy.* ***(Right)*** *Sagittal STIR MR from the same patient shows osteophyte, loss of anterior articular cartilage, and tibial bone marrow edema. At time of arthroscopy, remainder of the joint appeared normal, and cheilectomy was performed.*

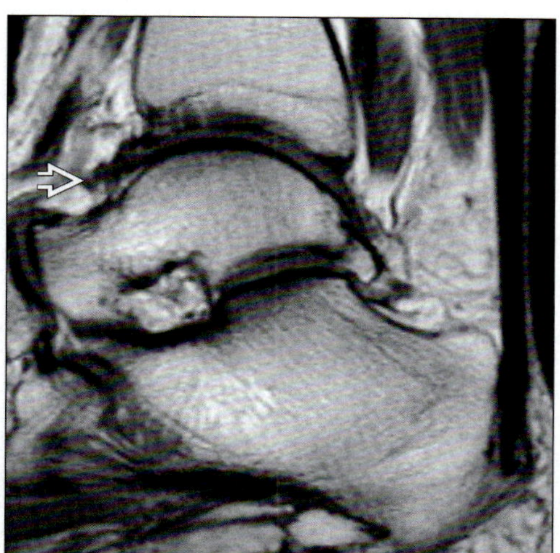

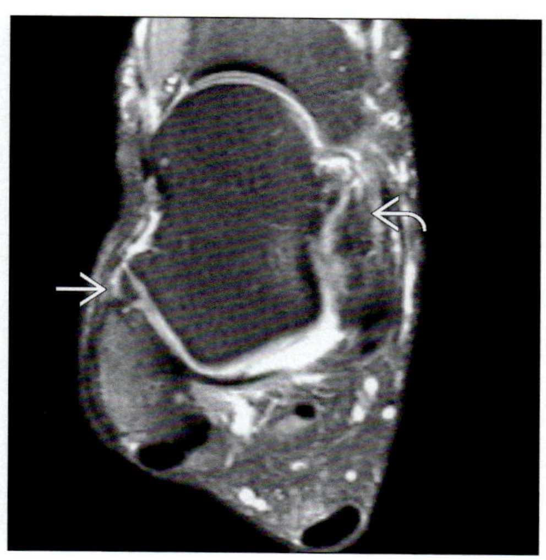

(Left) *Sagittal T1WI MR shows thick fibrous tissue in the anterior ankle joint causing anterior impingement. At arthroscopy, this corresponded to a band-like scar. Anterior impingement is usually due to osteophytes but can also reflect soft tissue lesions, as in this case.* ***(Right)*** *Axial PDWI FS MR shows a small lateral meniscoid lesion, which was clinically causing anterolateral impingement. Patient also had anteromedial impingement due to chronic injury of the superficial deltoid.*

POSTERIOR IMPINGEMENT, ANKLE

Key Facts

Terminology
- Focal abnormality limiting ankle plantarflexion and causing posterior ankle pain

Imaging
- Scar from posterior ligamentous tear
- Heterogeneous, thickened material posterior to ankle joint
- Posterior synovial cyst and synovitis
- Loose bodies in posterior joint recesses

Top Differential Diagnoses
- Posterior syndesmotic ligament tear
- Os trigonum syndrome
- Achilles tendinopathy
- Accessory flexor hallucis longus, digitorum longus, or soleus muscle

Pathology
- Categorized as posteromedial or syndesmotic

Clinical Issues
- Pain, limitation of motion with plantar flexion
- Pain often relieved with intraarticular steroid injection
- Arthroscopic or open debridement

Diagnostic Checklist
- Posterior ankle recess often distended in patients with chronic ankle pain, does not indicate impingement
- Look for irregular, thick, low signal intensity material in posterior ankle recesses

__(Left)__ Graphic shows torn intermalleolar ligament ➡ and posterior inferior tibiofibular ligament ⇨. Scar tissue due to tear of either of these ligaments may result in posterior ankle impingement. __(Right)__ Coronal T2WI FS MR shows torn intermalleolar ligament ➡ with displaced ligament ↷ causing posterior impingement. Posterior inferior tibiofibular ligament ⇨ is intact. Original injury was forced dorsiflexion. Isolated injury of intermalleolar ligament was confirmed at arthroscopy.

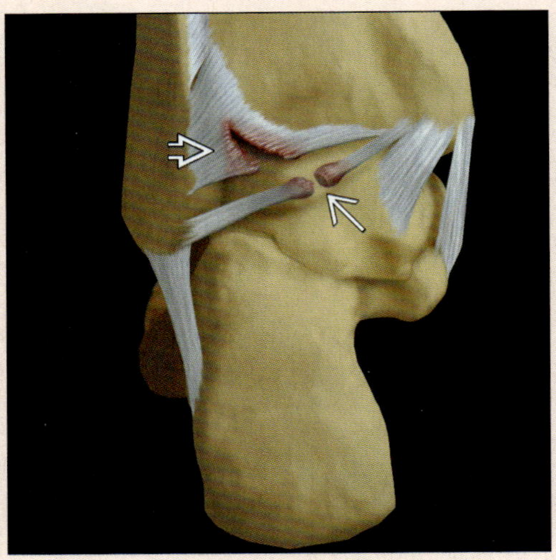

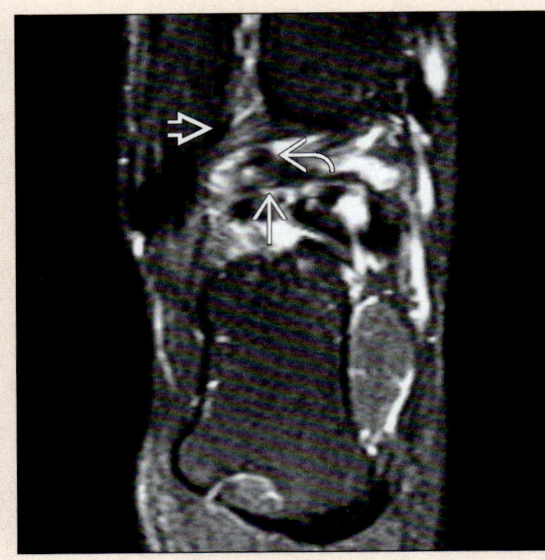

__(Left)__ Sagittal STIR MR shows thick, scarred intermalleolar ligament ➡ and posterior synovitis ↷ causing posterior impingement. __(Right)__ Coronal MR arthrogram T2WI FS shows synovial cyst ⇨ extending through chronically torn and scarred intermalleolar ligament ➡. Cyst fluid has slightly lower signal intensity than injected gadolinium contrast ↷. Posterior arthroscopy and debridement relieved pain, but symptoms recurred 1 year later and repeat MR showed recurrent scar.

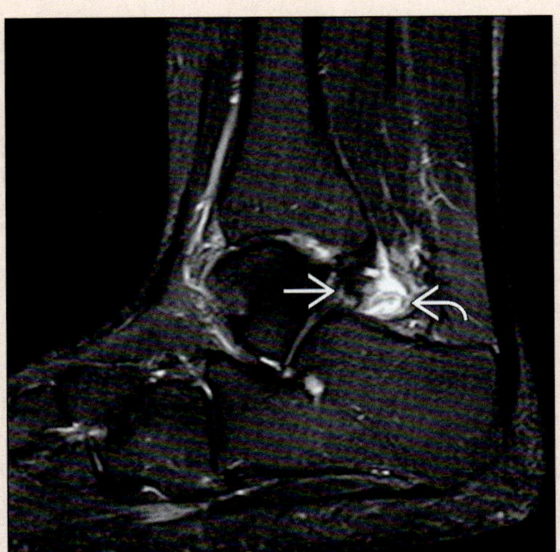

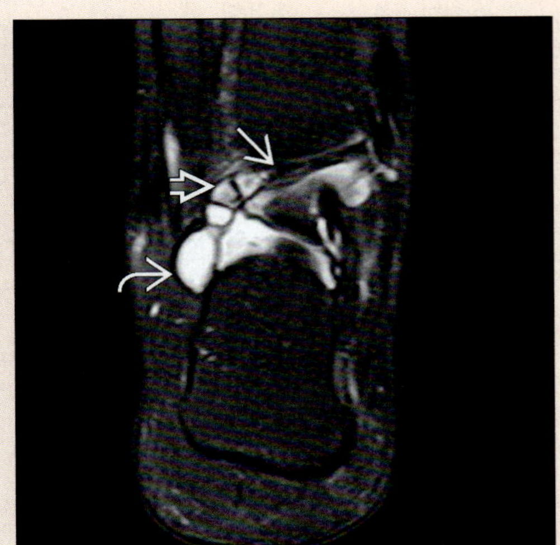

POSTERIOR IMPINGEMENT, ANKLE

TERMINOLOGY

Definitions

- Focal posterior ankle joint abnormality limiting ankle plantarflexion and causing posterior ankle pain

IMAGING

General Features

- Best diagnostic clue
  - Thick, fibrous material in posterior ankle joint

MR Findings

- Scar from posterior ligamentous tear
  - Discontinuity of posterior inferior tibiofibular ligament (PITFL) or intermalleolar ligament
    - Rarely, from torn posterior talofibular ligament
  - Heterogeneous, thickened material posterior to ankle joint
- Posterior synovial cyst and synovitis
  - Multiloculated fluid collection
  - Usually see connection to joint
  - Often will not fill with arthrographic contrast because of small cyst neck
- Loose bodies in posterior joint recesses
  - Easily missed on fluid-sensitive sequences

Imaging Recommendations

- Best imaging tool: MR

DIFFERENTIAL DIAGNOSIS

Posterior Syndesmotic Ligament Tear

- May have tear without clinical impingement

Os Trigonum Syndrome

- Large os trigonum causing impingement on flexor hallucis longus tendon, limited plantar flexion

Achilles Tendinopathy

- Fusiform thickening of Achilles tendon
- Increased signal intensity in Achilles tendon, < fluid

Posterior Osteochondral Lesion of Talar Dome

- Pain limits plantar flexion, although no impingement lesion present

Accessory Flexor Hallucis Longus, Digitorum Longus, or Soleus Muscle

- May limit extreme plantar flexion
- Fusiform in long axis, ovoid in cross section
- Usually same signal intensity as normal muscle
  - May show mild edema when impingement present

PATHOLOGY

General Features

- Etiology
  - May be caused by synovial cysts and synovitis, scarred ligaments, or loose bodies

Gross Pathologic & Surgical Features

- Not visible from routine anterior ankle arthroscopy portals
  - Posterior arthroscopy must be performed

Relevant Anatomy

- Posterior syndesmotic ligament divided into 2 components
  - Posterior inferior tibiofibular ligament (PITFL)
    - Triangular shape, wider laterally
    - Originates from posterolateral corner of tibia
    - Inserts slightly inferiorly on posterior margin of fibula
  - Intermalleolar ligament (a.k.a. transverse ligament)
    - Thin, discrete band inferior to PITFL
    - By early reports, only visible in small percentage of ankles but almost always visible on high-resolution MR
    - Originates from posterior margin of posterior malleolus
    - Courses inferiorly and laterally
    - Inserts on posterior margin of fibular tip
- Ankle and subtalar joints communicate in 15% of patients

CLINICAL ISSUES

Presentation

- Most common signs/symptoms
  - Pain with plantar flexion
  - Limited plantar flexion
  - Focal tenderness at posterior ankle joint either medially or laterally

Natural History & Prognosis

- Untreated, limits athletic activity
- Good results reported from debridement

Treatment

- Pain often relieved with intraarticular steroid injection
  - Useful when pain caused by synovitis or synovial cyst
- Arthroscopic or open debridement

DIAGNOSTIC CHECKLIST

Image Interpretation Pearls

- Posterior ankle recess often distended in patients with chronic ankle pain, does not indicate impingement
- Look for irregular, thick, low signal intensity material in posterior ankle recesses

SELECTED REFERENCES

1. Calder JD et al: Return to training and playing after posterior ankle arthroscopy for posterior impingement in elite professional soccer. Am J Sports Med. 38(1):120-4, 2010
2. Lee JC et al: Posterior impingement syndromes of the ankle. Semin Musculoskelet Radiol. 12(2):154-69, 2008
3. Sanders TG et al: Impingement syndromes of the ankle. Magn Reson Imaging Clin N Am. 16(1):29-38, v, 2008
4. Robinson P et al: Soft-tissue and osseous impingement syndromes of the ankle: role of imaging in diagnosis and management. Radiographics. 22(6):1457-69; discussion 1470-1, 2002
5. Fiorella D et al: The MR imaging features of the posterior intermalleolar ligament in patients with posterior impingement syndrome of the ankle. Skeletal Radiol. 28(10):573-6, 1999

OS TRIGONUM SYNDROME

Key Facts

Terminology
- Os trigonum (OT): Secondary ossification center at posterior process of talus
- OT syndrome: Posterior ankle impingement secondary to OT

Imaging
- Fluid in synchondrosis between OT & lateral tubercle
- Bone marrow edema in OT & posterior talus
- Isolated FHL tenosynovitis
- Posterior ankle or subtalar synovitis

Top Differential Diagnoses
- Fracture of trigonal process (Shepherd fracture)
- Other posterior ankle impingement syndromes
- FHL tenosynovitis

Pathology
- Disruption of synchondrosis
- Acute or stress fracture of OT
- Soft tissue impingement related to OT

Clinical Issues
- Related to sports involving forceful plantar flexion or push-off maneuvers
    - Ballet most common
- Posterior ankle pain with plantar flexion

Diagnostic Checklist
- Bone marrow edema in OT strongly suggests OT syndrome
- Do not assume OT syndrome based simply on presence of OT

*(**Left**) Lateral graphic of the ankle shows an unfused ossicle ➔ adjacent to the posterior process of the talus ➔. While often incidental, such ossicles can cause impingement and posterior ankle pain. (**Right**) Lateral radiograph shows a large os trigonum ➔ with cysts and osteophytes indicating osteoarthritis. There is increased density in the surrounding soft tissues, suggesting inflammation. Additional imaging is needed to confirm that the os trigonum is the cause of pain.*

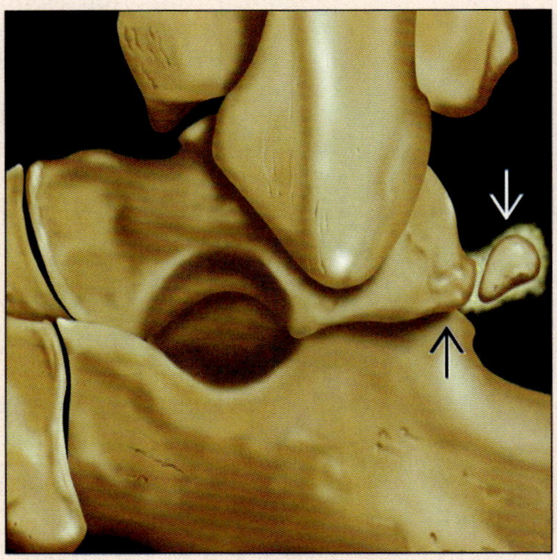

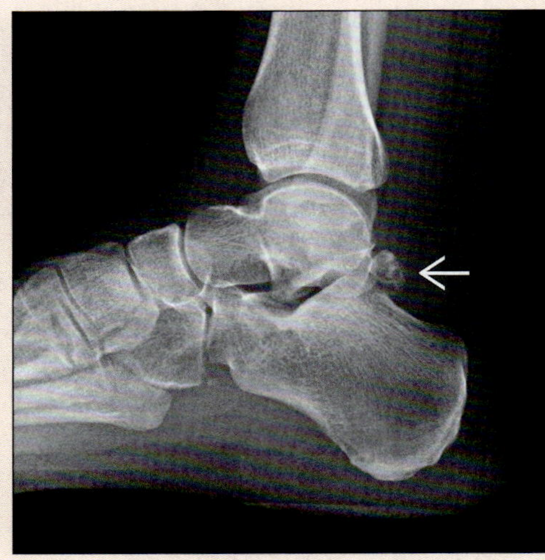

*(**Left**) Sagittal T1WI FS from a subtalar MR arthrogram in the same patient shows contrast ➔ between the os and the posterior talus indicating disruption of the synchondrosis in the clinical setting of associated pain. (**Right**) Sagittal T2WI FS MR in the same patient shows edema throughout the os trigonum ➔ and milder edema on the talar side of the synchondrosis ➔. Findings are consistent with acute or subacute disruption of the synchondrosis and contusion of the os trigonum.*

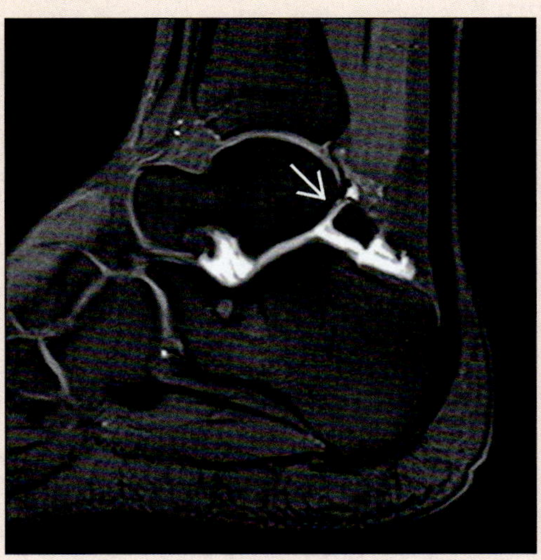

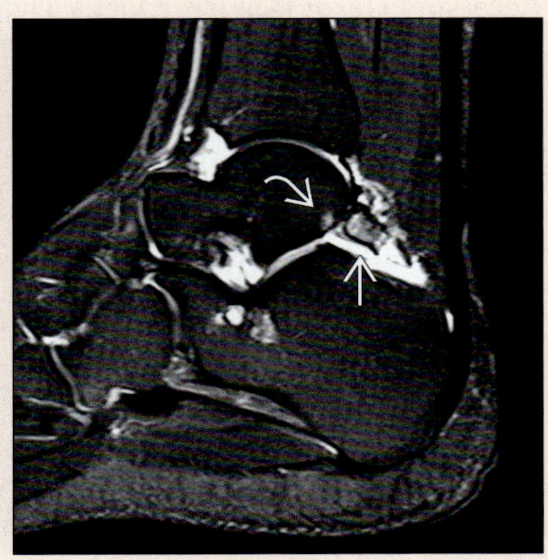

OS TRIGONUM SYNDROME

TERMINOLOGY

Definitions

- Os trigonum (OT): Secondary ossification center at posterior process of talus
- OT syndrome: Posterior ankle impingement secondary to OT
  - Disruption of synchondrosis
  - Acute or stress fracture of OT
  - Soft tissue impingement related to OT

IMAGING

General Features

- Best diagnostic clue
  - Fluid in synchondrosis between OT & lateral tubercle
  - Bone marrow edema in OT & posterior talus
  - Isolated flexor hallucis longus (FHL) tenosynovitis
  - Posterior ankle or subtalar synovitis
- Size
  - OT may enlarge with time
- Morphology
  - Variable shape
  - Occasionally fragments

Radiographic Findings

- Os trigonum
  - Additional imaging needed to confirm as cause of symptoms
- Osteoarthritis of synchondrosis and between calcaneus and OT
  - Subchondral cysts, osteophytes
- Soft tissue fullness along posterior margin of OT

Nuclear Medicine Findings

- Negative bone scan useful in excluding OT syndrome
- Positive bone scan nonspecific

CT Findings

- Better evaluation of osseous anatomy
- May differentiate OT from fracture of posterior process
  - Acute fracture has nonsclerotic, irregular margins

MR Findings

- Excessive fluid within synchondrosis
  - No data found regarding this, but authors believe a completely separate os may have a small amount of fluid between os and talus
  - Presence of fluid should be interpreted within the context of other findings, including clinical presentation
- Widening of synchondrosis
  - Rarely, OT may displace if synchondrosis disrupted
- Osteoarthritis of synchondrosis and between calcaneus and OT
  - Reactive sclerosis or cystic change, osteophytes
- Edema in OT, lateral tubercle, or both
  - May represent contusion or occult fractures
- Edema in soft tissues surrounding OT
- Synovitis in posterior recesses of tibiotalar joint, posterior subtalar joint, and FHL
  - May be focal, involving posterior ligaments (intermalleolar or posterior talofibular)
  - IV contrast may improve detection of small areas of synovitis
- Isolated tenosynovitis of FHL
  - Fluid around proximal FHL
  - Abrupt tapering of fluid at level of OT relatively specific for impingement
- Posterior capsular or ligament thickening
- Evaluate integrity of ligaments, look for loose bodies

Ultrasonographic Findings

- Nodular, hyperechoic thickening of posterolateral joint capsule
- May be used to guide therapeutic injection

Arthrography

- Local anesthetic/steroid injection may confirm as source of pain

Imaging Recommendations

- Best imaging tool
  - MR ± intraarticular contrast

DIFFERENTIAL DIAGNOSIS

Fracture of Trigonal Process (Shepherd Fracture)

- May be difficult to differentiate from OT at imaging
  - Acute fracture should have irregular, nonsclerotic margins
  - Acute fracture may be more anterior, at junction of talar body and posterior process
- Similar symptoms and treatment

Posterior Ankle Impingement Syndromes

- OT syndrome is most common form of osseous impingement
- Others bony causes include
  - Prominent trigonal process (Stieda process)
  - Prominent posterior tibial osteophytes
  - Intraarticular loose bodies
  - Periarticular calcification
- Nonosseous causes
  - Ganglion cyst
  - Low-lying FHL muscle belly
  - Other anomalous muscles

FHL Tenosynovitis

- Causes posterior ankle pain radiating to arch
- OT syndrome often associated with FHL tenosynovitis
- FHL tenosynovitis may be unrelated to OT, but suspect OT syndrome when tenosynovitis isolated to FHL

Normal Fluid in FHL Tendon Sheath

- FHL tendon sheath normally communicates with tibiotalar joint in 20%
  - Smooth, uniform distension of tendon sheath
  - Tenosynovitis causes septations, irregular contour, and abrupt narrowing

PATHOLOGY

General Features

- Etiology
  - Maximal plantar flexion → compression of OT between tibia and calcaneus
    - Described as nutcracker phenomenon

OS TRIGONUM SYNDROME

- Related to sports involving forceful plantar flexion or push-off maneuvers
  - Ballet most common
    - Ankle frequently in extreme plantar flexion, held there for long periods
  - Soccer
    - Ball kicking → repeated sudden forced plantar flexion
  - Also seen in downhill running, basketball, football, volleyball, gymnastics, ice skating, others

Staging, Grading, & Classification

- Anatomic variants of posterolateral talus
  - Type I: Normal tubercle
  - Type II: Enlarged lateral tubercle of posterior process (Stieda process)
  - Type III: Unfused separate ossicle
  - Type IV: OT fused to posterior process by synchondrosis or syndesmosis

Gross Pathologic & Surgical Features

- Synchondrosis fissuring or disruption
- Osteoarthritis between os and calcaneus and across synchondrosis
- Acute or stress fracture of OT or lateral process
- Entrapment of synovial and capsular tissues between tibia and OT with thickening & fibrosis
  - Posterior capsule and posterior talofibular, posterior tibiofibular, & intermalleolar ligaments at risk
- FHL tendon sheath thickening or tenosynovitis
  - Scarring of FHL sheath → tethering of FHL, limited plantar & dorsiflexion of great toe

Relevant Anatomy

- Secondary ossification center forms between ages of 8-10 years in girls and 11-13 years in boys
- Osseous fusion usually occurs within 1 year
  - Forms lateral tubercle of posterior process (trigonal process)
- May remain completely unfused or, more commonly, form synchondrosis or syndesmosis with posterior talus
- OT articulates with lateral tubercle and superior aspect of calcaneus
- FHL runs between lateral and medial tubercles of posterior process of talus
- Posterior talofibular ligament inserts on lateral tubercle or posterior facet of OT
- Medial talocalcaneal and posterior tibiotalar ligaments insert on medial tubercle

CLINICAL ISSUES

Presentation

- Most common signs/symptoms
  - Posterior ankle pain with plantar flexion
- Other signs/symptoms
  - Pain may also be exacerbated by dorsiflexion and weight-bearing
  - Posterior tenderness between Achilles and peroneal tendons
  - Occasional palpable soft tissue thickening
  - Decreased subtalar range of motion
  - Peroneal spasm
  - Presentation may be acute or chronic
    - Usually arises insidiously in predisposed athletes (repetitive stress)
      - Repetitive forced plantar flexion → chronic injury to posterior osseous and soft tissues
      - Often present when significant soft tissue component develops
      - May present with stress fracture of OT (or trigonal process)
    - May present acutely or subacutely following single traumatic episode
      - Acute injury to OT or synchondrosis
      - Inversion injury with strain on calcaneofibular and posterior talofibular ligaments
      - Acute ligamentous injury elsewhere in ankle → posterior instability and posterior symptoms

Demographics

- Gender
  - OT syndrome more common in females
    - Due to high association with ballet
- Epidemiology
  - OT present in 7-14% of population
    - Bilaterality reported in 1.4-50%

Natural History & Prognosis

- Conservative treatment often adequate

Treatment

- Diagnosis confirmed by relief of symptoms following injection of local anesthetic
- Conservative treatment is 1st line
  - Rest, splinting, anti-inflammatory agents, steroid injections, occasional casting
- Surgery, frequently endoscopic
  - Resection of os trigonum
  - Resection of impinging osteophytes
  - Resection of hypertrophic or fibrotic soft tissue
  - Resection of hypertrophic or scarred FHL sheath

DIAGNOSTIC CHECKLIST

Consider

- Evaluate other internal derangements of ankle that often coexist and may precipitate presentation

Image Interpretation Pearls

- Marrow edema in OT strongly suggests OT syndrome
- Do not assume OT syndrome based on presence of normal-appearing OT

SELECTED REFERENCES

1. Rathur S et al: Posterior ankle impingement: os trigonum syndrome. Am J Orthop. 38(5):252-3, 2009
2. Hopper MA et al: Ankle impingement syndromes. Radiol Clin North Am. 46(6):957-71, v, 2008
3. Scholten PE et al: Hindfoot endoscopy for posterior ankle impingement. J Bone Joint Surg Am. 90(12):2665-72, 2008
4. Maquirriain J: Posterior ankle impingement syndrome. J Am Acad Orthop Surg. 13(6):365-71, 2005
5. Robinson P et al: Soft-tissue and osseous impingement syndromes of the ankle: role of imaging in diagnosis and management. Radiographics. 22(6):1457-69; discussion 1470-1, 2002
6. Bureau NJ et al: Posterior ankle impingement syndrome: MR imaging findings in seven patients. Radiology. 215(2):497-503, 2000

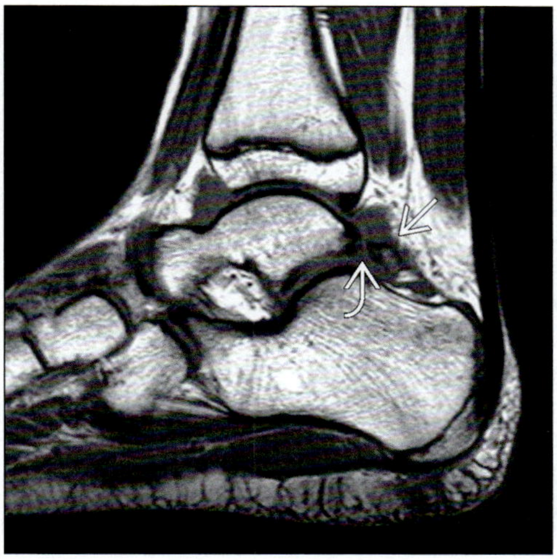

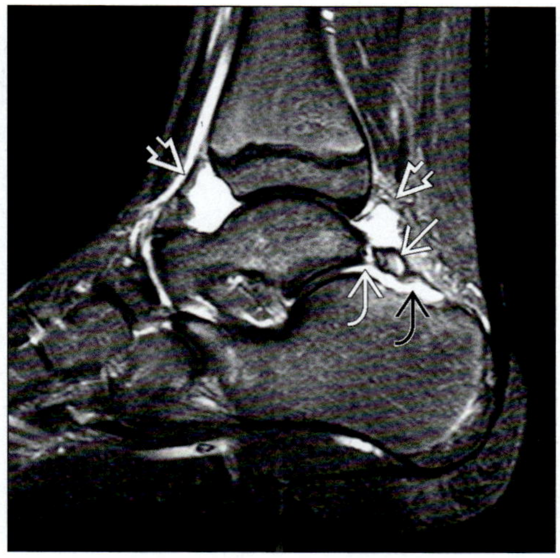

(Left) *Sagittal T1WI MR shows a gap between the os trigonum (OT) or lateral tubercle and posterior talus. The osseous margins are irregular, suggesting arthritis or fracture.* ***(Right)*** *Sagittal STIR MR in the same patient shows tibiotalar and subtalar effusions and fluid between the OT and posterior talus. There is edema in the OT and surrounding soft tissues, confirming OT abnormality; the syndrome can be associated with posterior synovitis of the tibiotalar or subtalar joints (not shown).*

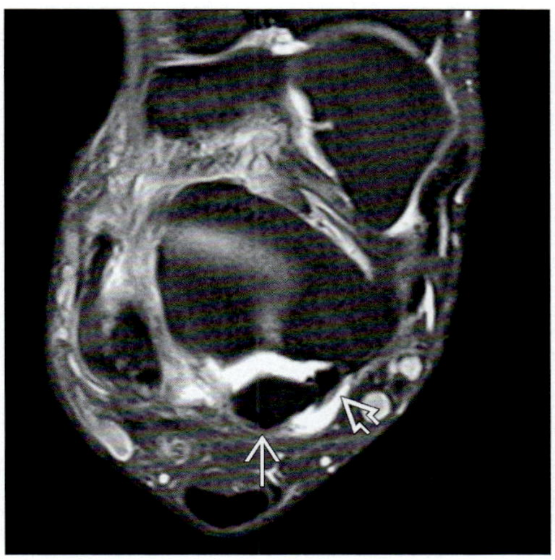

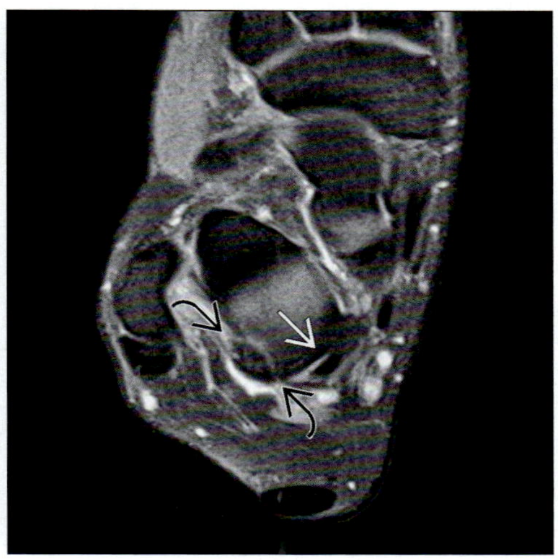

(Left) *Axial PDWI FS MR shows a displaced os trigonum and impingement of the flexor hallucis longus (FHL) between the os and posterior talus. This may result in FHL tendinopathy, stenosing tenosynovitis, or occasionally tethering of FHL. Treatment is resection of the os.* ***(Right)*** *Axial PDWI FS MR in a different patient shows a fracture of the talus involving the lateral tubercle of the posterior process, sometimes called Shepherd fracture. Note the adjacent FHL.*

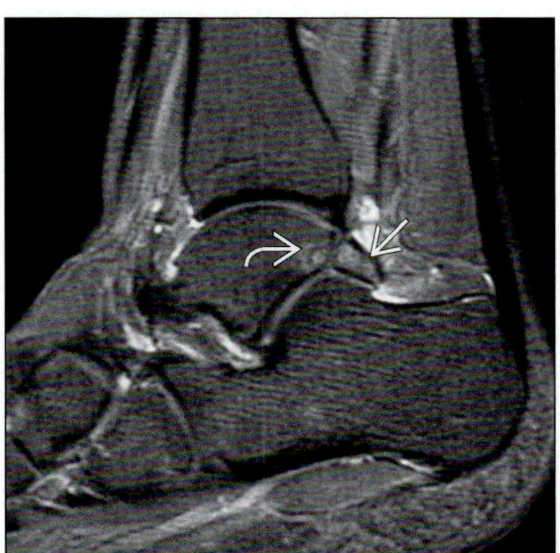

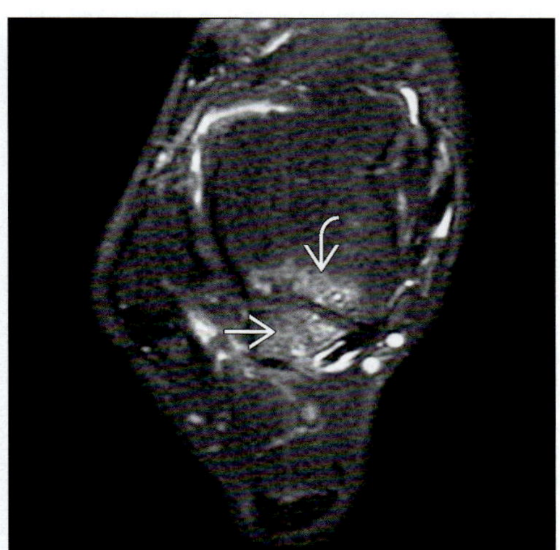

(Left) *Sagittal T2WI FS MR demonstrates marrow edema in an os trigonum and in the adjacent posterior talus. There is no fluid in the intervening synchondrosis. Findings are consistent with os trigonum syndrome without fracture.* ***(Right)*** *Axial T2WI FS MR in the same patient shows marrow edema, as well as small degenerative cysts in both the os trigonum and the talus. Chronic micromotion at the synchondrosis is the likely cause.*

HAGLUND SYNDROME

Key Facts

Terminology

- Haglund deformity: Enlarged superior margin of calcaneal posterior process
- Haglund triad (Haglund syndrome): Haglund deformity + insertional tendinopathy + pre-Achilles bursitis

Imaging

- Initial diagnosis suggested on radiographs, confirmed on MR
- Radiographic findings
  - Soft tissue fullness in pre-Achilles bursa region
  - Triangular osseous prominence at posterosuperior margin calcaneal posterior process
- MR findings
  - Osseous prominence at posterosuperior margin posterior calcaneal process
  - Marrow edema radiates out from osseous prominence
  - Pre-Achilles or retro-Achilles bursitis
  - Achilles peritendinitis, insertional tendinopathy

Top Differential Diagnoses

- Achilles insertional tendinopathy
- Inflammatory arthritis
- Accessory soleus muscle
- Stress fracture

Clinical Issues

- Focal pain at posterior calcaneal tuberosity
- Palpable soft tissue and osseous prominence ("pump bump")
- Pain with plantar flexion of foot
- Treated conservatively or with surgical excision of Haglund deformity

__(Left)__ Graphic of medial side of ankle shows inflammation in the pre-Achilles bursa ➔ and at the Achilles tendon insertion ➔ due to prominence of the posterior superior margin of the calcaneus (Haglund deformity). __(Right)__ Lateral radiograph shows typical Haglund deformity ➔. There is soft tissue density ➔ in Kager fat consistent with edema from impingement by the Haglund deformity. Calcific Achilles insertional tendinitis ➔ is also present.

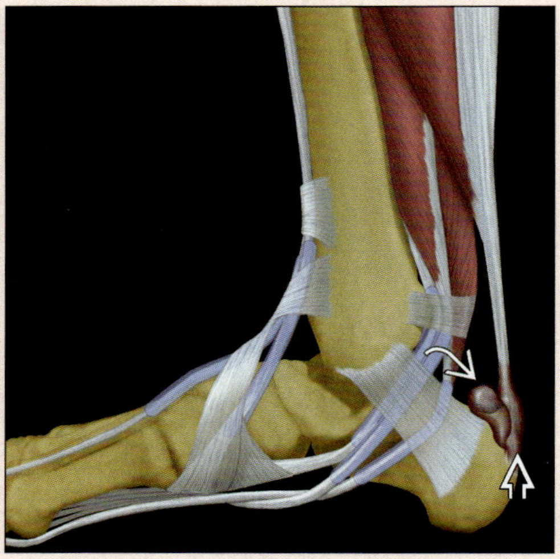

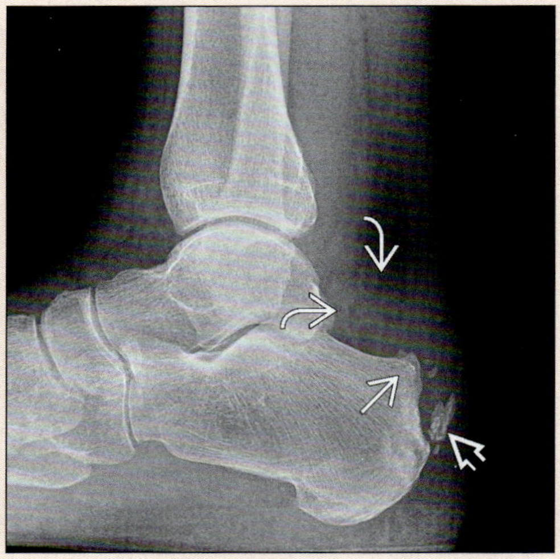

__(Left)__ Sagittal T1WI MR shows extensive soft tissue reaction to Haglund deformity ➔. This includes fluid in the pre-Achilles ➔ bursa, inflammatory changes ➔ in the retro-Achilles and pre-Achilles fat, and insertional Achilles tendinopathy ➔. __(Right)__ Sagittal STIR MR from the same patient shows extensive calcaneal marrow edema ➔. Pre-Achilles bursitis ➔ is more readily apparent. Insertional tendinopathy shows streaks ➔ of increased signal intensity not as bright as fluid.

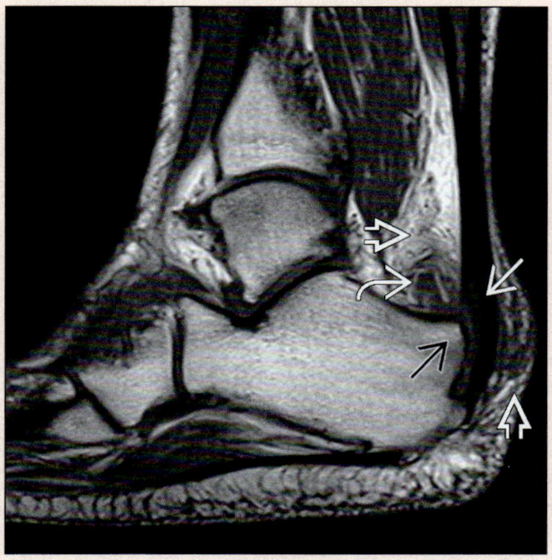

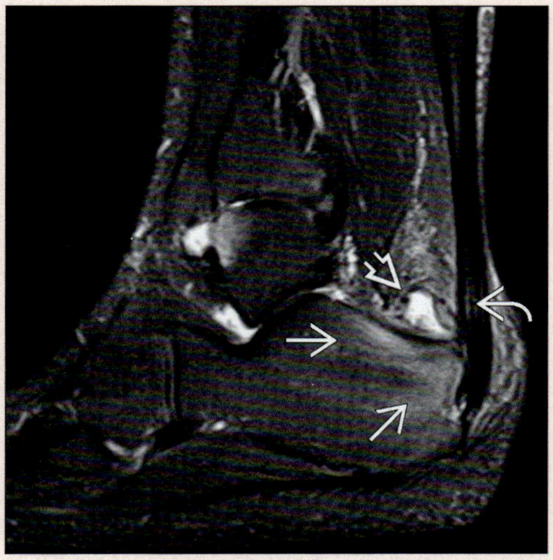

HAGLUND SYNDROME

TERMINOLOGY

Synonyms

- "Pump bump": Osseous and soft tissue prominence caused by Haglund deformity
- Pre-Achilles bursa = retrocalcaneal bursa

Definitions

- Haglund deformity: Enlarged superior margin of calcaneal posterior process
- Haglund triad (Haglund syndrome): Haglund deformity + insertional Achilles tendinopathy + pre-Achilles bursitis

IMAGING

Radiographic Findings

- Osseous prominence of posterosuperior margin of calcaneus
  - No specific radiographic criteria for what makes this prominence too large
- Soft tissue fullness in pre-Achilles bursa region

MR Findings

- Osseous prominence of posterosuperior margin of calcaneus
  - Bone marrow edema often striking, radiates out from deformity
- Pre-Achilles or retro-Achilles bursitis
- Edema in Kager (pre-Achilles) fat pad
- Achilles peritendinitis
- Achilles insertional tendinopathy
  - Streaky areas of increased signal intensity within thickened tendon

Ultrasonographic Findings

- Thickening, heterogeneity of distal Achilles tendon near its insertion
- Fluid in pre-Achilles or retro-Achilles bursae

DIFFERENTIAL DIAGNOSIS

Achilles Insertional Tendinopathy

- Degeneration of Achilles tendon near its calcaneal attachment
- Often produces calcification
- Often but not always associated with Haglund deformity

Rheumatoid Arthritis or Seronegative Spondyloarthropathy

- Often involves pre-Achilles bursa
- Heel pain often overshadowed by multiple other painful foci
- Erosions at posterior process of calcaneus

Accessory Soleus Muscle

- Creates soft tissue density in pre-Achilles fat
- May cause local discomfort with exercise

Stress Fracture

- Often involves posterior calcaneal process
- Produces similar bone marrow edema pattern
- Low signal intensity fracture line with surrounding edema

PATHOLOGY

General Features

- Etiology
  - Overuse, tight-fitting shoes
    - Especially high heels

Relevant Anatomy

- Pre-Achilles bursa (a.k.a. retrocalcaneal bursa)
  - Between superior margin of posterior process and Achilles tendon
  - Normal anatomic structure, may contain small amount of fluid
- Retro-Achilles bursa
  - Adventitious bursa
  - Develops in response to pressure
- Pre-Achilles fat (a.k.a. Kager fat)
  - Triangular fat pad between Achilles tendon, deep flexor tendons, and superior margin of calcaneal posterior process

CLINICAL ISSUES

Presentation

- Most common signs/symptoms
  - Focal pain at posterior calcaneal tuberosity
  - Palpable soft tissue and osseous prominence ("pump bump")
- Other signs/symptoms
  - Pain with plantar flexion of foot

Treatment

- Modification of shoes to relieve stress
- Steroid injection into pre-Achilles bursa
- Surgical excision of Haglund deformity

DIAGNOSTIC CHECKLIST

Image Interpretation Pearls

- Pre-Achilles bursa may contain small amount of fluid in asymptomatic patients

SELECTED REFERENCES

1. Den Hartog BD: Insertional achilles tendinosis: pathogenesis and treatment. Foot Ankle Clin. 14(4):639-50, 2009
2. Erdem CZ et al: MR imaging features of foot involvement in patients with psoriasis. Eur J Radiol. 67(3):521-5, 2008
3. Ortmann FW et al: Endoscopic bony and soft-tissue decompression of the retrocalcaneal space for the treatment of Haglund deformity and retrocalcaneal bursitis. Foot Ankle Int. 28(2):149-53, 2007
4. Yodlowski ML et al: Surgical treatment of Achilles tendinitis by decompression of the retrocalcaneal bursa and the superior calcaneal tuberosity. Am J Sports Med. 30(3):318-21, 2002
5. Narváez JA et al: Painful heel: MR imaging findings. Radiographics. 20(2):333-52, 2000
6. Bottger BA et al: MR imaging of the normal and abnormal retrocalcaneal bursae. AJR Am J Roentgenol. 170(5):1239-41, 1998

TARSAL SINUS SYNDROME

Key Facts

Imaging

- MR best imaging tool, findings depend on etiology
- Post-traumatic (most common)
  - Discontinuity or laxity of cervical ligament
  - Poorly defined ligament due to adjacent scarring
  - Heterotopic ossification along course of ligament
  - Normal fat signal in tarsal sinus lost, replaced by fluid or fibrous tissue
  - Bone marrow edema adjacent to cervical ligament attachments
- Ganglion cyst
  - Well-defined fluid collection, usually elongated
  - May be associated with arthritis or tarsal coalition
- Inflammatory arthritis
  - Fluid in tarsal sinus
  - Synovitis, erosion in subtalar joint

Top Differential Diagnoses

- Joint effusion from subtalar joint
- Bifurcate ligament tear
- Talar neck stress fracture

Pathology

- Most cases due to trauma
  - Inversion injury affecting subtalar joint
- May be nontraumatic
  - Tarsal coalition, ganglion cyst, synovitis, neoplasm

Clinical Issues

- Lateral hindfoot pain and focal tenderness
- Subjective instability
- Often treated with subtalar arthroscopy and debridement

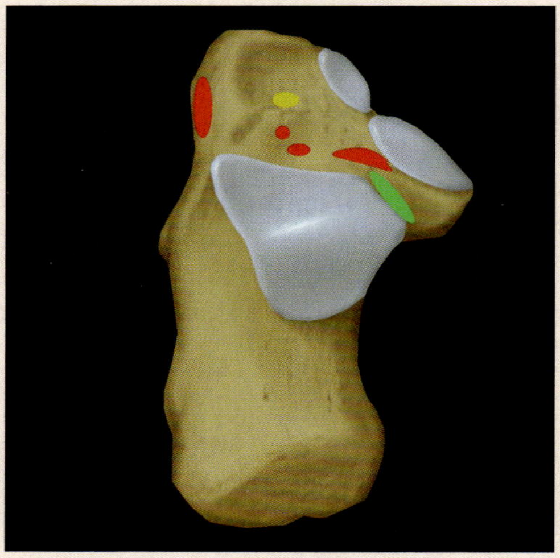

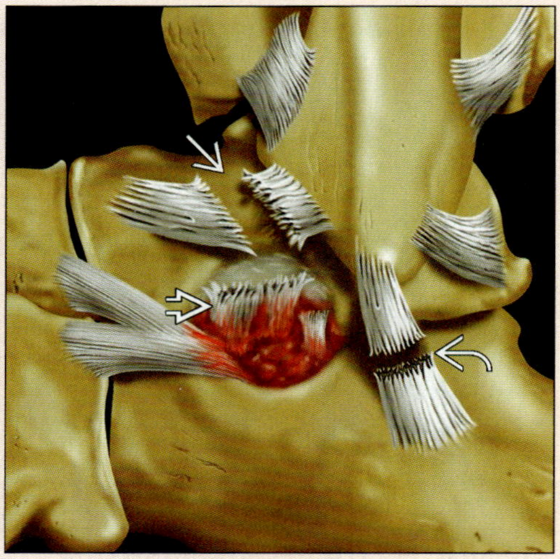

(Left) *Graphic of dorsal surface of calcaneus shows ligamentous attachments in tarsal sinus. The inferior extensor retinaculum (red) has multiple bands extending into the tarsal sinus adjacent to the cervical ligament (yellow). Interosseous ligament (green) is located posteromedial to the sinus in the tarsal canal.* ***(Right)*** *Graphic shows characteristic tarsal sinus syndrome, with inflammation of the sinus and torn anterior talofibular ➔ and calcaneofibular ➔ ligaments in addition to ligaments in the sinus ➔.*

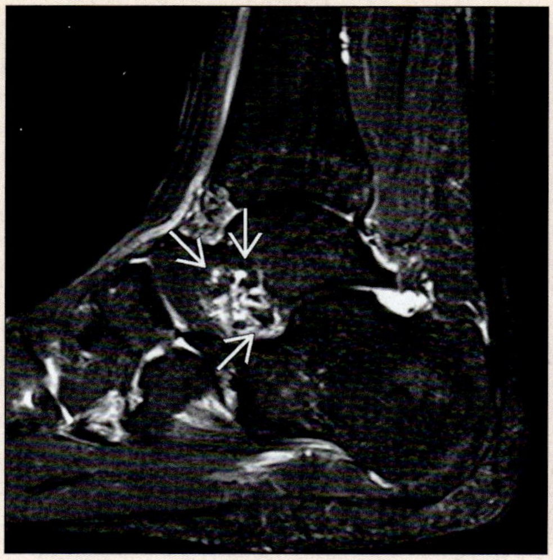

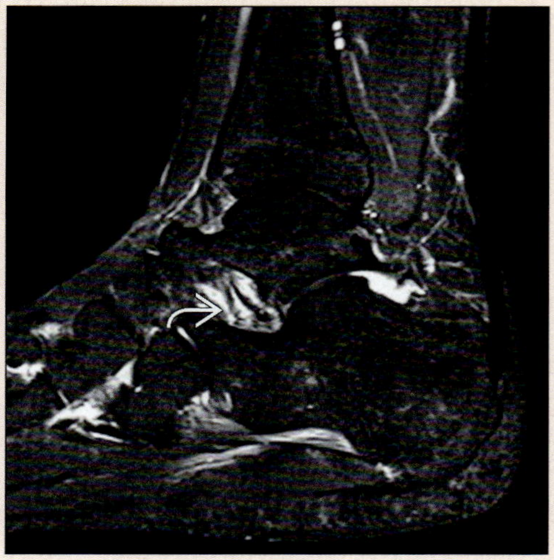

(Left) *Sagittal STIR MR shows fluid and low signal intensity scar ➔ replacing normal fat in tarsal sinus. The cervical ligament is not visible and could not be seen on adjacent sagittal or coronal images.* ***(Right)*** *Sagittal STIR MR further laterally in the same patient shows intact fibers of inferior extensor retinaculum ➔. Close attention is required to distinguish between retinacular fibers originating outside the sinus and cervical ligament fibers.*

TARSAL SINUS SYNDROME

TERMINOLOGY

Synonyms

- Sinus tarsi syndrome

IMAGING

General Features

- Best diagnostic clue
  - Soft tissue filling tarsal sinus

MR Findings

- Tarsal sinus syndrome due to trauma
  - Cervical ligament
    - Discontinuity or laxity
    - Poorly defined ligament due to adjacent scarring
    - Heterotopic ossification along course of ligament
  - Tarsal sinus fat
    - Normal fat signal lost, replaced by fluid or fibrous tissue
  - Bone marrow
    - Edema adjacent to talar and calcaneal attachments of cervical ligament
    - Less commonly, edema adjacent to interosseous ligament attachments
  - Inferior extensor retinaculum
    - Discontinuity, laxity variably present
  - Interosseous ligament
    - Less commonly torn than cervical ligament
- Tarsal sinus syndrome due to ganglion cyst
  - Well-defined fluid collection, usually multilocular and elongated
  - May arise from coalition or arthritis
- Tarsal sinus syndrome due to inflammatory arthritis
  - Synovitis extending into sinus, joint space narrowing, erosions, tenosynovitis of multiple tendons
- Rarely, soft tissue neoplasm
  - May mimic cyst
  - Enhances with gadolinium administration

Imaging Recommendations

- Best imaging tool
  - MR
- Protocol advice
  - Coronal and sagittal planes most useful

DIFFERENTIAL DIAGNOSIS

Joint Effusion from Subtalar Joint

- Anterior recess of subtalar joint extends into tarsal sinus
- Joint effusion enlarges this recess but is not usually a cause of tarsal sinus symptoms

Bifurcate Ligament Tear

- Ligament originates from anterior process of calcaneus
- Inserts on navicular and cuboid
- Tear causes edema at anterolateral margin of tarsal sinus but not within sinus

Talar Neck Stress Fracture

- Bone marrow edema in talar neck
- Incomplete or complete fracture line oriented perpendicular to talar neck
- May have edema in tarsal sinus

PATHOLOGY

General Features

- Etiology
  - Trauma
    - Inversion injury affecting subtalar joint
    - Rupture of cervical ligament
    - Scar tissue causes pain, subjective impingement and instability
    - Heterotopic ossification may also develop
  - Neoplasm
  - Tarsal coalition
    - Osseous overgrowth may extend into tarsal sinus
    - Ganglion cyst at nonosseous coalition site may also extend into tarsal sinus
    - Abnormal motion may cause fibrosis in tarsal sinus
  - Inflammatory arthropathy
    - Synovitis related to rheumatoid or psoriatic arthritis
- Associated abnormalities
  - Lateral collateral ligament injury
    - Injury to calcaneofibular ligament usually present when there is injury to cervical ligament

Relevant Anatomy

- Tarsal sinus lies between talar neck and calcaneus and contains
  - Fat
  - Cervical ligament
    - Between neck of talus and dorsal margin of calcaneus
  - Insertions of extensor retinaculum
    - 3 bands, inserting at lateral, mid, and medial portions of sinus

CLINICAL ISSUES

Presentation

- Most common signs/symptoms
  - Lateral hindfoot pain and focal tenderness over tarsal sinus
  - Subjective instability

Treatment

- Corticosteroid injection
- Subtalar arthroscopy with debridement

SELECTED REFERENCES

1. Lee KB et al: Subtalar arthroscopy for sinus Tarsi syndrome: arthroscopic findings and clinical outcomes of 33 consecutive cases. Arthroscopy. 24(10):1130-4, 2008
2. Lektrakul N et al: Tarsal sinus: arthrographic, MR imaging, MR arthrographic, and pathologic findings in cadavers and retrospective study data in patients with sinus tarsi syndrome. Radiology. 219(3):802-10, 2001
3. Oloff LM et al: Subtalar joint arthroscopy for sinus tarsi syndrome: a review of 29 cases. J Foot Ankle Surg. 40(3):152-7, 2001
4. Beltran J: Sinus tarsi syndrome. Magn Reson Imaging Clin N Am. 2(1):59-65, 1994

TARSAL SINUS SYNDROME

(Left) *Coronal PDWI MR shows soft tissue filling the tarsal sinus. Bone marrow edema in the talar neck suggests abnormal stress.* ***(Right)*** *Coronal T2WI FS MR in the same patient shows lax cervical ligament, which is avulsed from the talar neck. There is heterogeneous scar lateral to the ligament. Bands of extensor retinaculum entering the tarsal sinus are also visible. Bone marrow edema in the talar neck is a clue to the diagnosis, but it may also occur with stress fracture.*

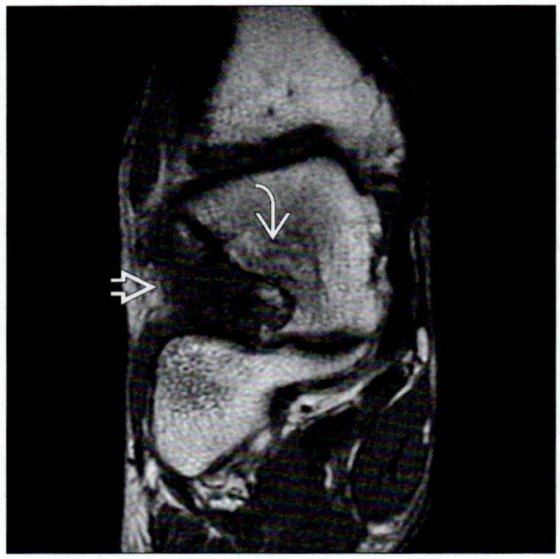

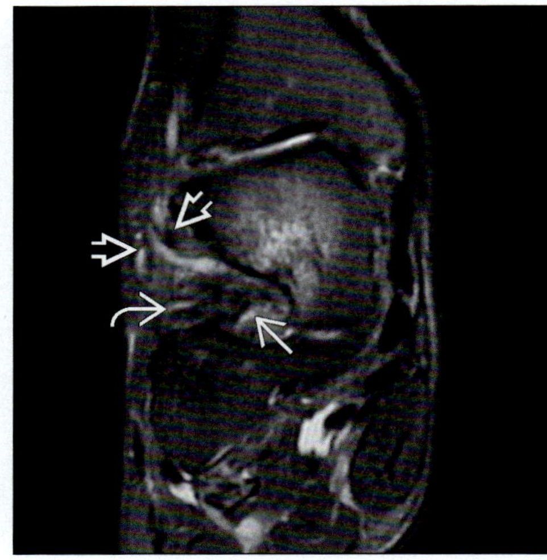

(Left) *Coronal PDWI MR shows amorphous soft tissue filling the majority of the tarsal sinus. The cervical ligament is not visible.* ***(Right)*** *Coronal T2WI FS MR in the same patient allows identification of the wavy, torn fibers of the cervical ligament. This patient had a history of subtalar dislocation, which frequently results in cervical ligament rupture. He presented 1 year later with tenderness in the tarsal sinus and subjective instability.*

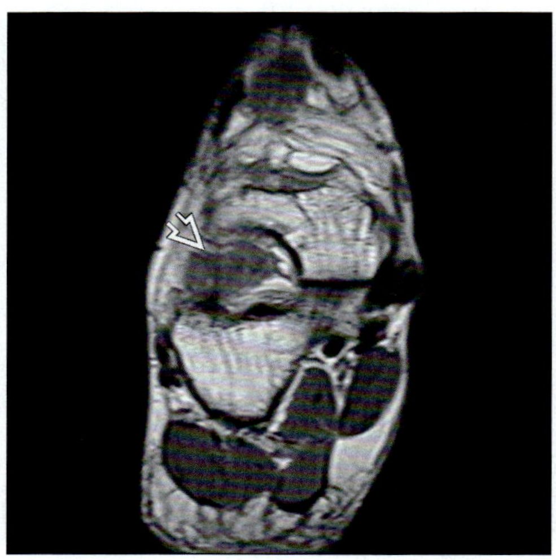

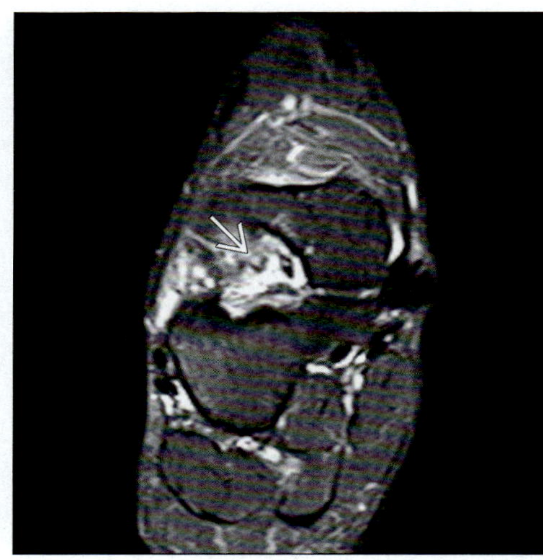

(Left) *Sagittal T1WI MR shows soft tissue in tarsal sinus in a patient with no history of trauma, but vague lateral hindfoot pain. A small lipoma with central cystic degeneration is incidentally noted.* ***(Right)*** *Sagittal STIR MR slightly more medially in the same patient confirms that the cervical ligament is intact, but there is bone marrow edema in the talar neck, centered at the ligament attachment.*

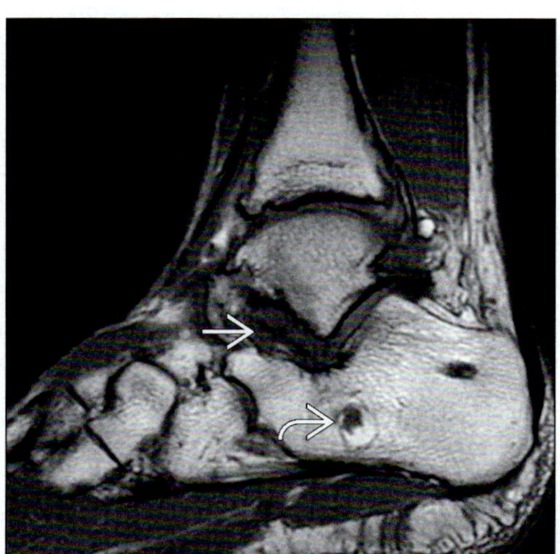

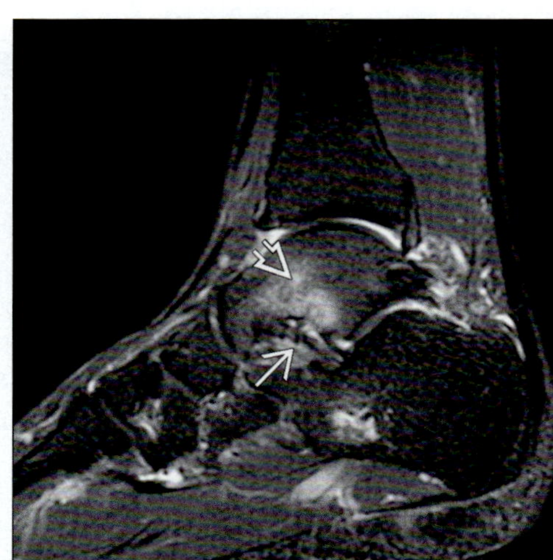

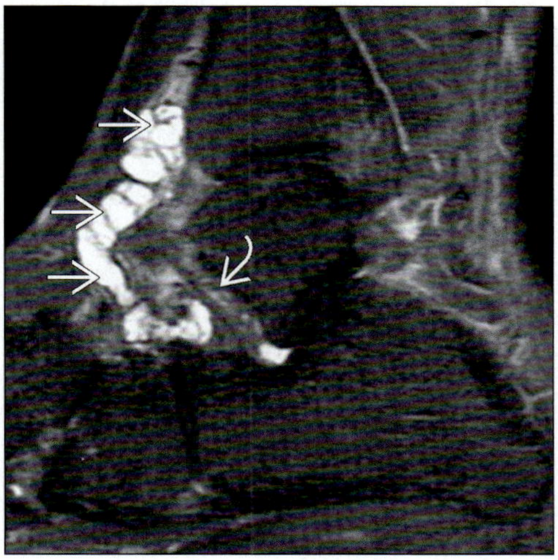
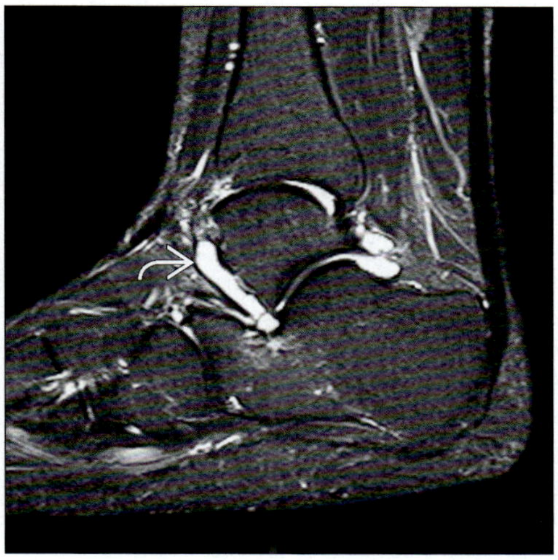

*(**Left**) Sagittal STIR MR shows a ganglion cyst ➡ extending into the tarsal sinus and causing impingement. A normal lateral band of inferior extensor retinaculum ↪ is visible. (**Right**) Sagittal STIR MR shows a synovial cyst ↪ arising from the posterior subtalar joint and causing tarsal sinus pain. The normal anterior recess of the joint, in contrast, is limited to the posterior-inferior margin of the sinus. Subsequent subtalar arthrogram showed filling of the cyst.*

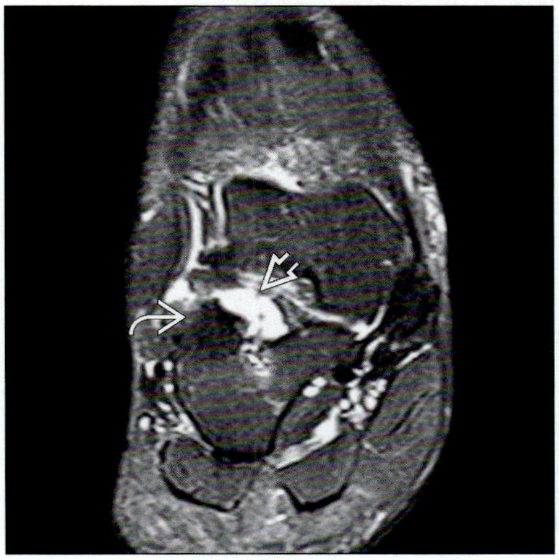
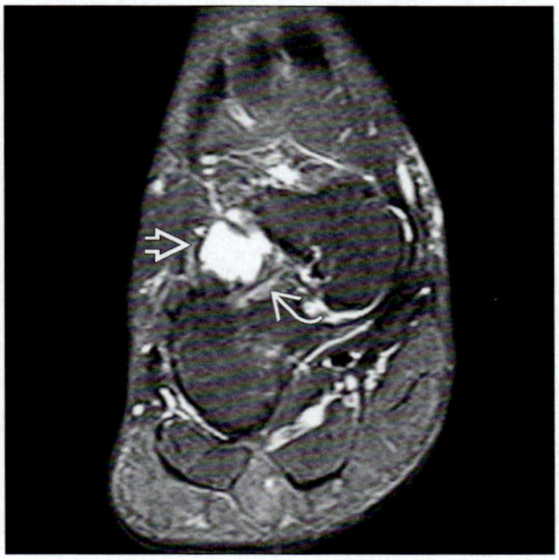

*(**Left**) Coronal T2WI FS MR shows a post-traumatic exostosis ↪ of the calcaneus extending into the tarsal sinus. A focal fluid collection ⇨ has developed, probably as a result of the exostosis, and causes tarsal sinus pain. (**Right**) Coronal T2WI FS MR slightly further anterior in the same patient shows the fluid collection ⇨. Note that the cervical ligament ↪ is intact. The patient had pain but no evidence of instability.*

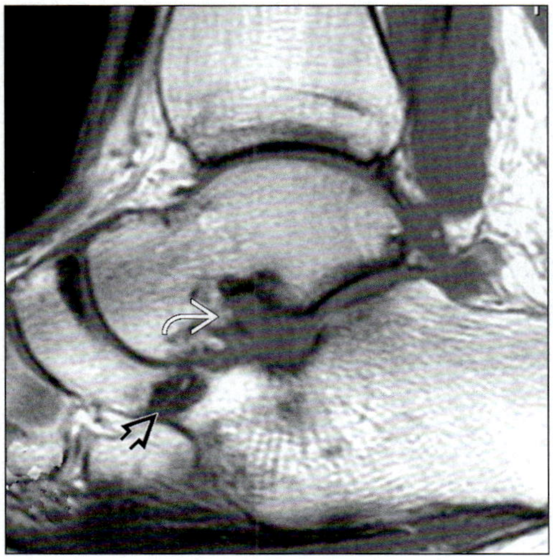
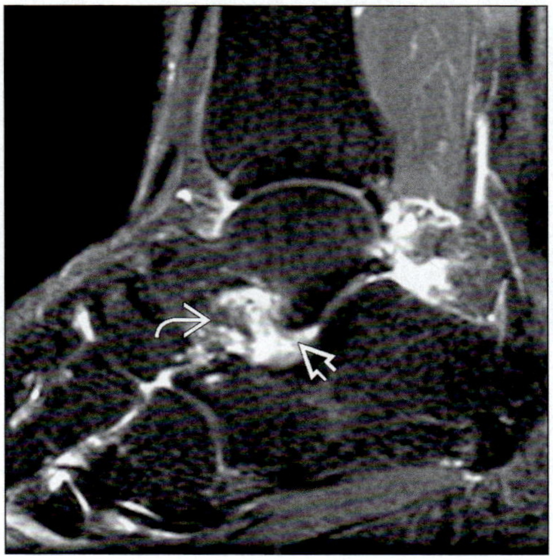

*(**Left**) Sagittal T1WI MR shows a nonosseous calcaneonavicular coalition ⇨ and amorphous material in the tarsal sinus ↪. Coalition alters hindfoot motion and can result in tarsal sinus syndrome. (**Right**) Sagittal STIR MR in the same patient shows a redundant, irregular cervical ligament ↪. A blunted configuration of the tip of the talar lateral process ⇨ is evident and is a helpful secondary sign of tarsal coalition.*

TARSAL TUNNEL SYNDROME

Key Facts

Terminology

- Entrapment neuropathy of tibial nerve in tarsal tunnel

Imaging

- Denervation of plantar intrinsic muscles
  - Uniform high signal intensity on fluid-sensitive sequences
- Mass in tarsal tunnel
  - Ganglion cyst: Follows fluid signal intensity, homogeneous, ± thin rim of enhancement with gadolinium
  - Nerve sheath tumor: Round or ovoid shape, often shows "tail" where it arises from nerve
  - Venous varicosities: Serpentine, enlarged vessels in tarsal tunnel, can be followed beyond tunnel
- Scar: Amorphous material around tibial nerve, often shows gadolinium enhancement
- Osseous impingement: Fracture malunion, subtalar coalition

Top Differential Diagnoses

- Plantar fasciitis
- Diabetic neuropathy

Clinical Issues

- Burning, tingling pain, numbness at plantar aspect of foot

Diagnostic Checklist

- Fat atrophy of plantar intrinsic muscles can be incidental finding, more common in older patients
- Intramuscular edema is most reliable sign of muscle denervation

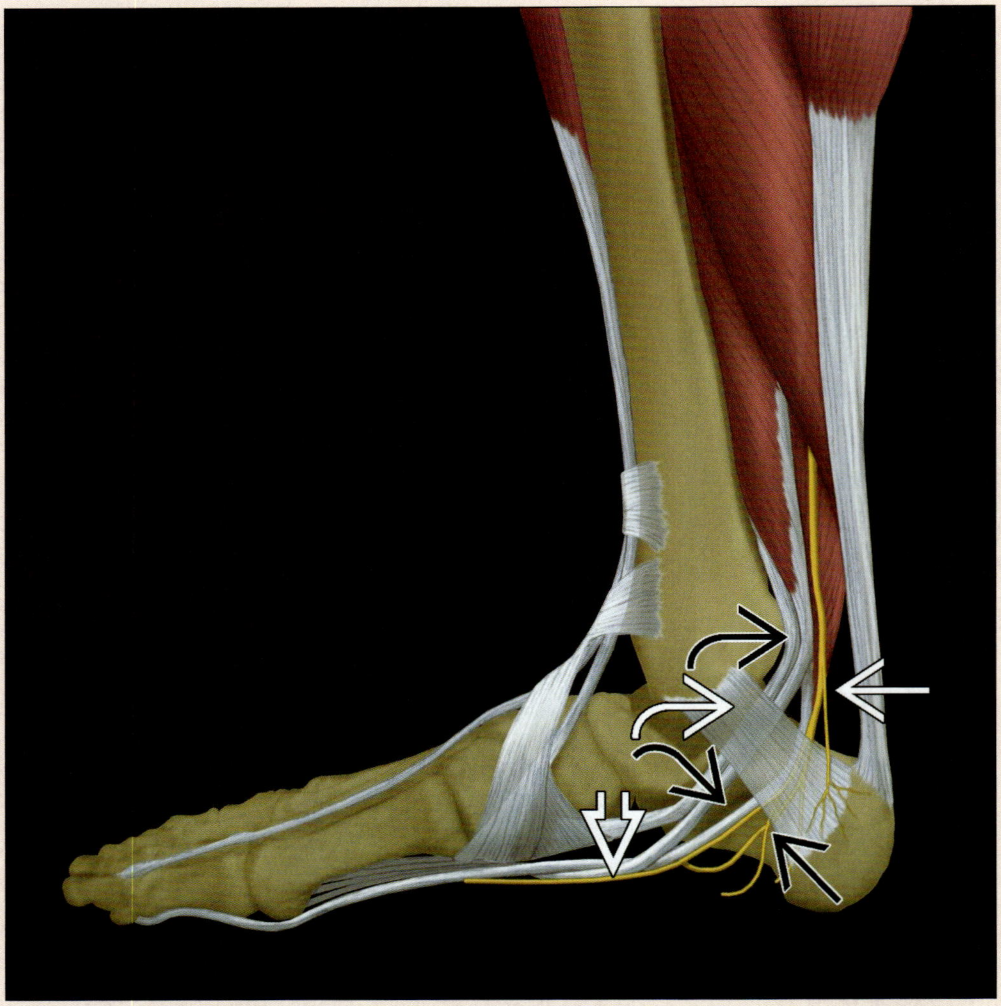

Sagittal graphic of the medial side of the ankle shows the borders and contents of the tarsal tunnel. The inferior flexor retinaculum ➔ forms the superficial border of the tarsal tunnel, and the floor is composed of the adjacent talus and calcaneus. Tendons ➔ of the posterior tibialis, flexor digitorum longus, and flexor hallucis longus also occupy the tunnel. The tibial nerve is vulnerable to compression throughout its course from behind the medial malleolus to the midfoot. The lateral calcaneal nerve ➔ arises in the proximal tarsal tunnel. More distally, the tibial nerve divides into medial ➔ and lateral ➔ plantar nerves.

TARSAL TUNNEL SYNDROME

TERMINOLOGY

Definitions

- Entrapment neuropathy of tibial nerve in tarsal tunnel

IMAGING

General Features

- Location
  - Fibroosseous tunnel deep to inferior flexor retinaculum
- Tibial nerve bifurcates at tibial tunnel into 2 branches
  - Medial plantar nerve
    - Motor branches to abductor hallucis, flexor digitorum and hallucis brevis, 1st lumbrical
    - Plantar digital nerves to 1st-3rd toes, medial aspect of 4th toe
  - Lateral plantar nerve
    - Motor branches to flexor digiti minimi brevis, lumbricals, interossei, adductor hallucis
    - Plantar digital nerves to 5th toe and lateral aspect of 4th toe

MR Findings

- Mass in tarsal tunnel
  - Ganglion cyst: Follows fluid signal intensity, homogeneous, ± thin rim of enhancement with gadolinium
  - Nerve sheath tumor: Round or ovoid shape, often shows "tail" where it arises from nerve
    - Mildly heterogeneous high signal intensity on T2WI, diffuse gadolinium enhancement
  - Venous varicosities: Serpentine, enlarged vessels in tarsal tunnel, can be followed beyond tunnel
  - Accessory muscle: Fusiform, elongated, follows muscle signal intensity on all sequences
- Scar: Amorphous material around tibial nerve, often shows gadolinium enhancement
- Altered osseous morphology of tarsal tunnel
  - Malunited fracture
  - Subtalar coalition with osseous overgrowth
- Enlargement of tibial nerve: Not a reliable finding
- Denervation of plantar intrinsic muscles
  - High signal intensity on fluid-sensitive sequences
    - Uniform appearance throughout affected muscle(s)
    - No disruption of muscle fibers
  - Fat atrophy of muscles is a late finding
  - May involve all intrinsic foot muscles innervated by tibial nerve
  - May involve only muscles innervated by individual nerve branches
    - Depending on location where nerve impingement arises
- Post-contrast enhancement of nerve

Imaging Recommendations

- Best imaging tool
  - MR

DIFFERENTIAL DIAGNOSIS

Plantar Fasciitis

- Edema & thickening of plantar fascia

Diabetic Neuropathy

- Diffuse denervation of all plantar intrinsic muscles

Isolated Fat Atrophy of Plantar Muscles

- Seen in asymptomatic patients, increases with age

PATHOLOGY

General Features

- Etiology
  - Up to 50% of cases are idiopathic
  - Osseous impingement on tunnel: Due to trauma or subtalar coalition
  - Space-occupying lesion: Ganglion cyst, nerve sheath tumor, venous varicosities, accessory muscle, lipoma, synovitis
  - Scar tissue: Post-traumatic or postsurgical
  - Biomechanical: Valgus heel + pronated forefoot
- Relevant anatomy: Tarsal tunnel
  - Roof: Inferior flexor retinaculum
  - Floor: Medial borders of talus and calcaneus
  - Contains extrinsic flexor tendons and neurovascular bundle
  - Posterior tibial nerve gives off lateral calcaneal branch, then divides into medial and lateral plantar nerves

CLINICAL ISSUES

Presentation

- Most common signs/symptoms
  - Burning, tingling pain, numbness at plantar aspect of foot
- Other signs/symptoms
  - Positive Tinel sign: Percussion of tibial nerve at tarsal tunnel reproduces symptoms

Treatment

- Conservative: NSAIDs, orthotics, physical therapy
- Surgical: Fasciotomy, excision of space-occupying lesion

DIAGNOSTIC CHECKLIST

Consider

- Fat atrophy of plantar intrinsic muscles can be incidental finding, more common in older patients
- Intramuscular edema is most reliable sign of muscle denervation

SELECTED REFERENCES

1. Schmid DT et al: Fatty muscle atrophy: prevalence in the hindfoot muscles on MR images of asymptomatic volunteers and patients with foot pain. Radiology. 253(1):160-6, 2009
2. Kinoshita M et al: Tarsal tunnel syndrome in athletes. Am J Sports Med. 34(8):1307-12, 2006
3. Hochman MG et al: Nerves in a pinch: imaging of nerve compression syndromes. Radiol Clin North Am. 42(1):221-45, 2004
4. Delfaut EM et al: Imaging of foot and ankle nerve entrapment syndromes: from well-demonstrated to unfamiliar sites. Radiographics. 23(3):613-23, 2003

(Left) Sagittal STIR MR shows a synovial cyst ➔ in the tarsal tunnel deviating the neurovascular bundle ➩ posteriorly and inferiorly. *(Right)* Axial T1WI C+ FS MR from the same patient shows that the mass ➔ has homogeneously low signal intensity, without any enhancement, confirming its cystic nature. Enhancing blood vessels are present adjacent to the mass ➔. There is mild enhancement of the abductor hallucis muscle ➔ related to denervation.

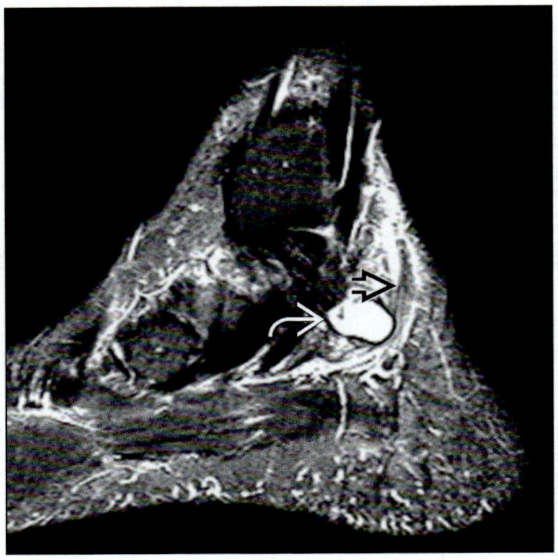

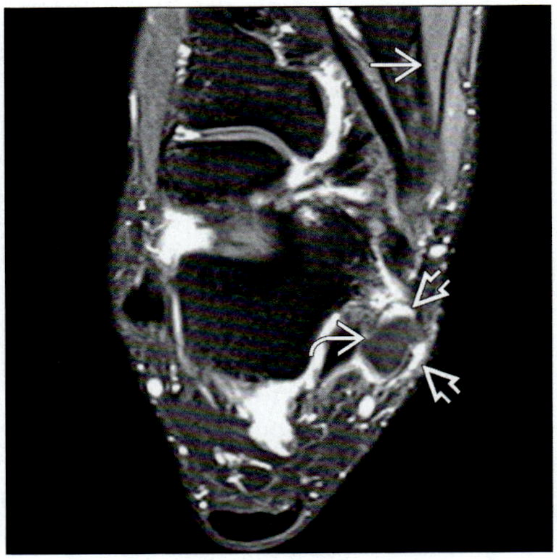

(Left) Sagittal STIR MR shows a synovial cyst ➔ arising from the subtalar joint and entering the posterior portion of the tarsal tunnel, posterior to the sustentaculum tali. The tibial nerve ➔ appears enlarged, a finding sometimes seen in tarsal tunnel syndrome. However, nerve enlargement does not reliably correlate with patient symptoms. *(Right)* Coronal T2WI FS MR from the same patient confirms the multilocular cyst ➔ adjacent to the enlarged tibial nerve ➔.

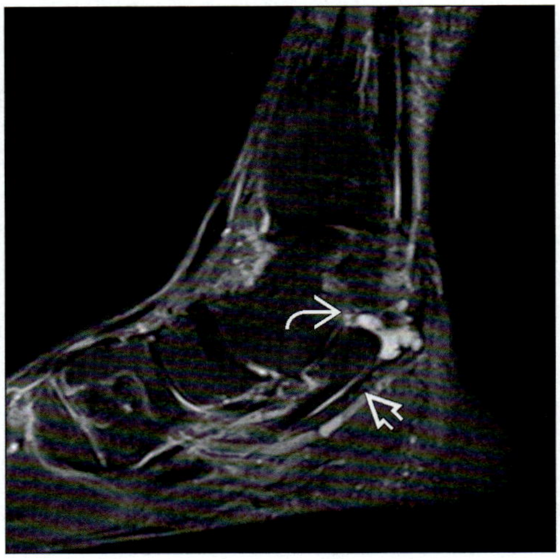

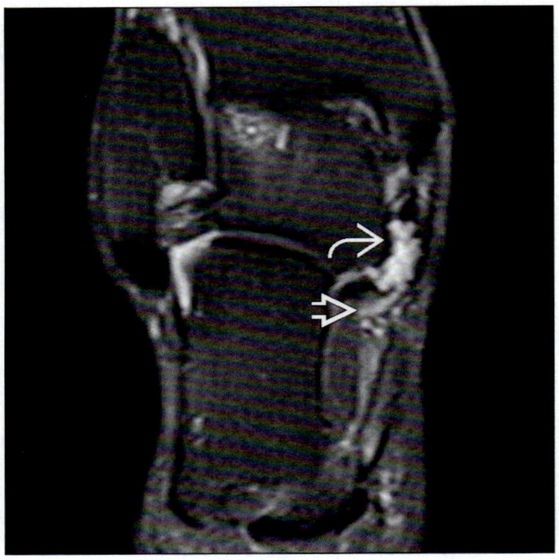

(Left) Coronal T2WI FS MR from a patient with tarsal tunnel syndrome shows severe denervation edema ➔ of muscles innervated by the medial plantar nerve. Note that the abductor digiti minimi ➔ and quadratus plantae ➔ (innervated by lateral plantar nerve) are not affected. *(Right)* Coronal T2WI FS MR shows isolated mild abductor hallucis muscle denervation ➔ due to distal tarsal tunnel syndrome.

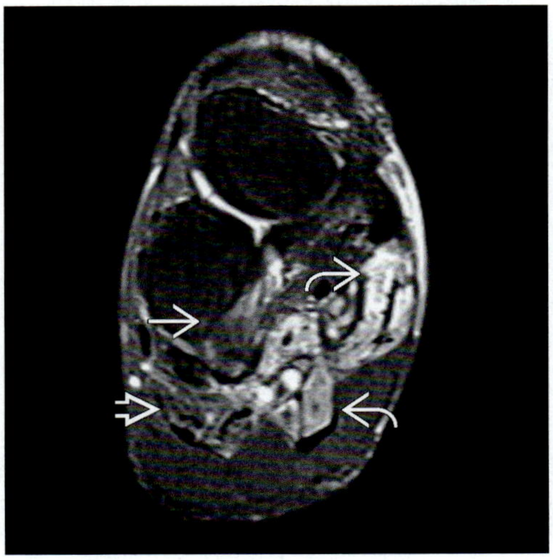

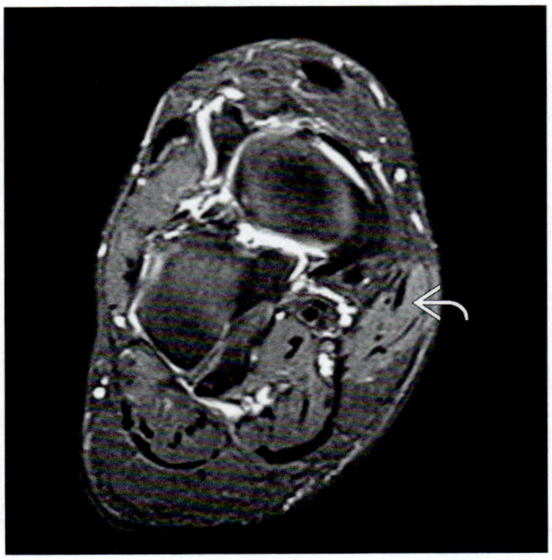

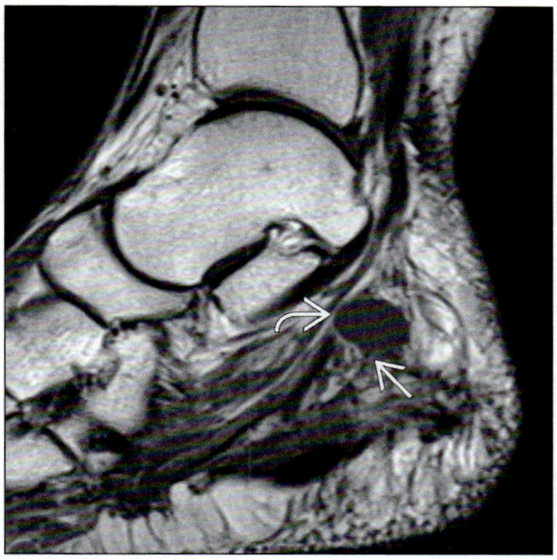

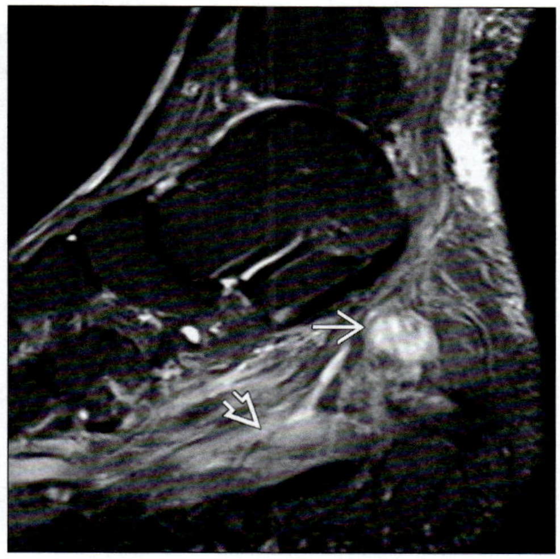

(Left) *Sagittal T1WI MR shows a low signal intensity mass in the tarsal tunnel. The mass has a small "tail" characteristic of nerve sheath tumors. At surgery this was found to be a malignant peripheral nerve sheath tumor.* ***(Right)*** *Sagittal STIR MR from the same patient shows that the mass is heterogeneously high in signal intensity. The degree of heterogeneity is greater than expected for a synovial cyst. There is denervation edema in the abductor hallucis muscle.*

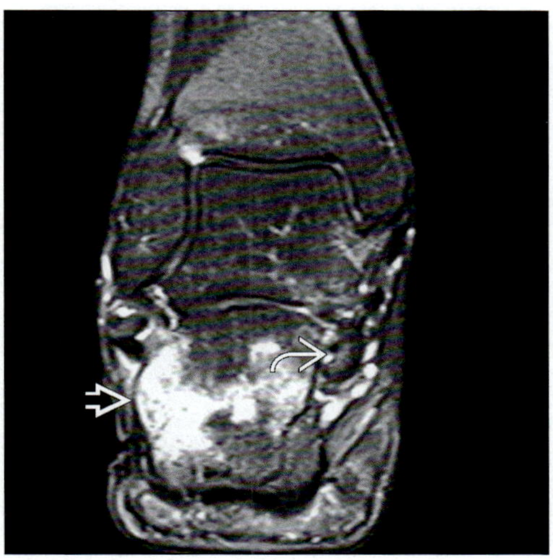

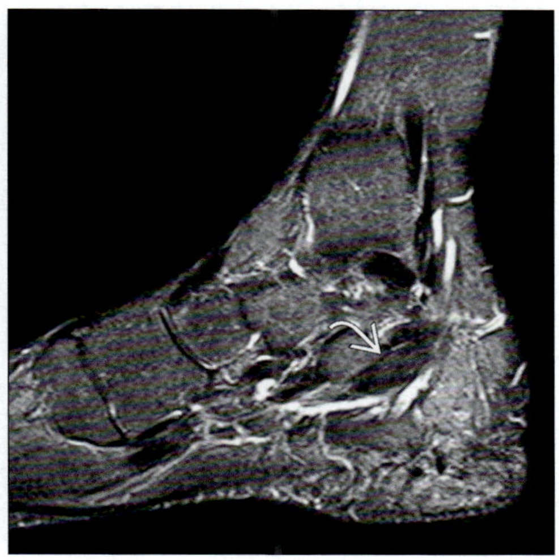

(Left) *Coronal T2WI FS MR shows post-traumatic deformity of the calcaneus due to subacute fracture. Tibial nerve is enlarged due to osseous impingement in this patient who presented with tarsal tunnel syndrome.* ***(Right)*** *Sagittal STIR MR from the same patient shows low to intermediate signal intensity scar around the tibial nerve. Scar in this location is easily overlooked at MR. Normally, nerve should be surrounded only by vessels and fat.*

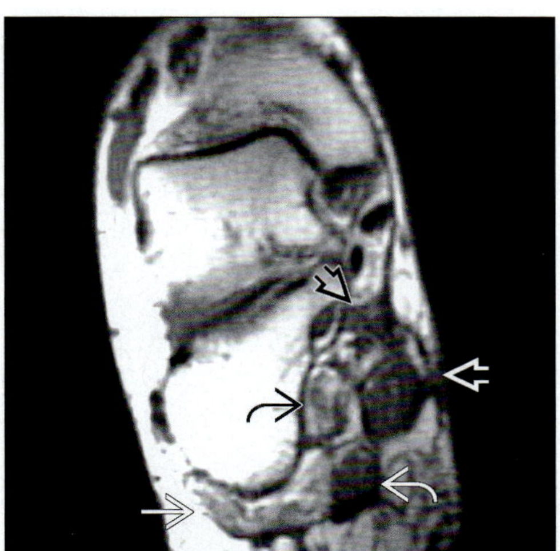

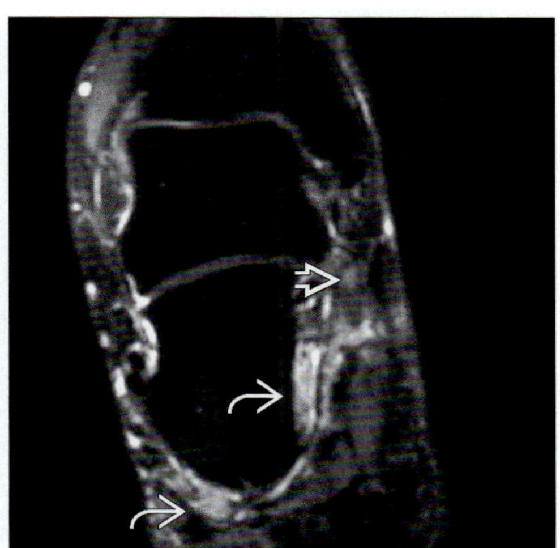

(Left) *Coronal PDWI MR from a patient with tarsal tunnel syndrome shows atrophy of quadratus plantae and abductor digiti minimi muscles, innervated by the lateral plantar nerve. Muscles innervated by the medial plantar nerve (abductor hallucis and flexor digitorum brevis) are spared. Tibial nerve is surrounded by scar.* ***(Right)*** *Coronal T2WI FS MR from the same patient shows denervation edema of the affected muscles. Tibial nerve appears normal.*

PLANTAR FASCIITIS

Key Facts

Imaging
- Almost always involves central band of plantar fascia
  - Superficial to flexor digitorum brevis muscle
- Fusiform thickening of proximal portion of plantar fascia
- Increased signal intensity in plantar fascia
- ± edema in subcutaneous fat &/or flexor digitorum brevis muscle
- ± edema in subjacent calcaneus
- Fibrosis of plantar heel pad

Top Differential Diagnoses
- Plantar fibromatosis
- Plantar fascia rupture
- Stress fracture of calcaneus
- Calcaneal posterior process impaction fracture
- Impingement of lateral calcaneal nerve
- Spondyloarthropathy

Pathology
- Due to chronic, repetitive trauma
- Predisposing factors: Obesity, pronated foot

Clinical Issues
- Pain at inferior margin of heel with weight-bearing
- Worst in morning
- Tenderness over inferior margin of calcaneal posterior process
- Treatment usually conservative
  - Rest, orthotics, heel cup, stretching

Diagnostic Checklist
- Heel pain is often assumed to be due to plantar fasciitis
- Exclude other causes of heel pain on MR evaluation

*(**Left**) Graphic shows thickening and inflammation ➡ of the central band of the plantar fascia. A plantar calcaneal spur ➡ is present here but is not reliably associated with plantar fasciitis. In fact, many plantar calcaneal spurs are not associated anatomically with the plantar fascia. (**Right**) Sagittal STIR MR shows mild plantar fasciitis ➡. There is only minimal thickening of the fascia, but edema is present superficial and deep to the fascia.*

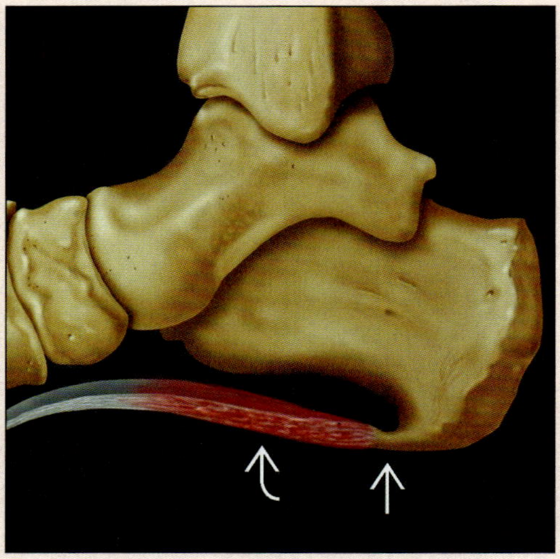

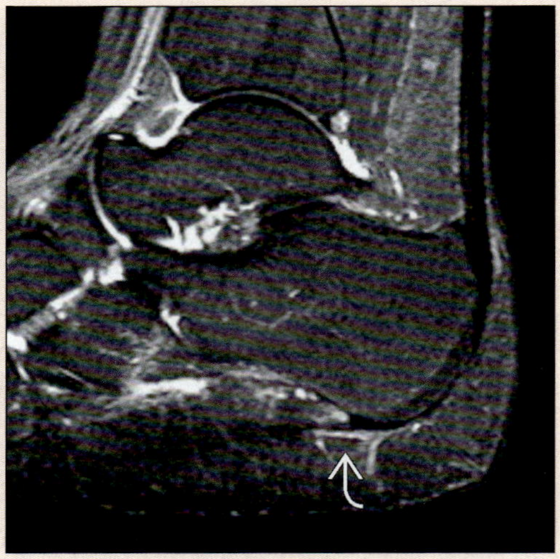

*(**Left**) Sagittal STIR MR shows thickened fascia ➡ with surrounding edema, and edema in the posterior process of the calcaneus ➡. The absence of cortical erosion distinguishes this case from enthesopathy due to spondyloarthropathy. (**Right**) Coronal T2WI FS MR from the same patient shows characteristic appearance of plantar fasciitis, with thickening of central band of plantar fascia ➡, high signal intensity ➡ deep and superficial to the thickened fascia, and calcaneal marrow edema ➡.*

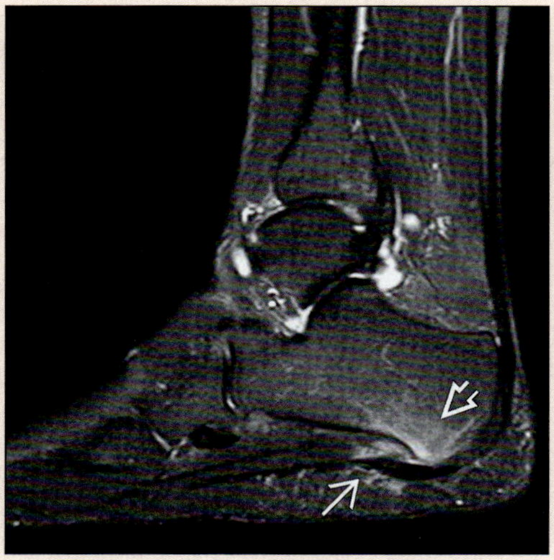

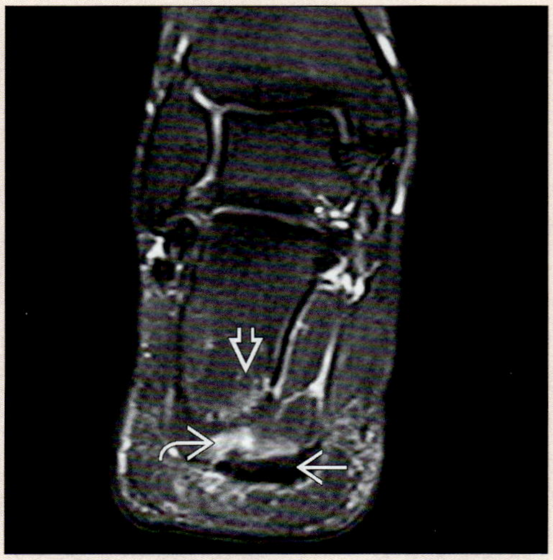

PLANTAR FASCIITIS

TERMINOLOGY

Definitions

- Low-grade tendon injury at origin of short flexors of foot due to chronic repetitive injury

IMAGING

General Features

- Location
  - Almost always involves central band of plantar fascia
    - Superficial to flexor digitorum brevis muscle

Radiographic Findings

- Plantar calcaneal spur shows little correlation to plantar fasciitis
- Many spurs do not originate in plantar fascia

MR Findings

- Fusiform thickening of proximal portion of plantar fascia
- Increased signal intensity in plantar fascia
- ± edema in subcutaneous fat &/or flexor digitorum brevis muscle
- ± edema in subjacent calcaneus

Ultrasonographic Findings

- Thickening of plantar fascia > 4 mm

Imaging Recommendations

- Best imaging tool
  - MR
- Protocol advice
  - Sagittal & coronal fluid-sensitive sequences most useful

DIFFERENTIAL DIAGNOSIS

Plantar Fibromatosis

- Nodular thickening of plantar fascia
- Usually more distally located

Plantar Fascia Rupture

- MR shows discontinuity of fascia
- Due to athletic injury or steroid injection for plantar fasciitis

Stress Fracture of Calcaneus

- Radiographs: Band of sclerosis in posterior process
- MR: Band of low signal intensity on T1WI, high signal on fluid-sensitive sequences

Calcaneal Posterior Process Impaction Fracture

- Due to fall, landing on heel
- Focal fracture line, often with angular deformity of cortex
- Best seen on Harris (axial) view or MR

Impingement of Lateral Calcaneal Nerve

- At level of tarsal tunnel, causes focal heel pain

Spondyloarthropathy

- Presenting symptom of reactive arthropathy may be heel pain
- Also seen with psoriasis or ankylosing spondylitis
- Radiographs or MR show erosions of plantar fascia origin

Fibrosis of Plantar Heel Pad

- Occurs with aging, diabetes mellitus

PATHOLOGY

General Features

- Etiology
  - Predisposing factors: Obesity, pronated foot
  - Chronic, repetitive trauma

Relevant Anatomy

- Plantar fascia
  - Originates from inferior margin of posterior process of calcaneus
  - Fans out at level of MTP joints, merges with MTP fascia
  - 3 bands
    - Medial: Superficial to abductor hallucis m.
    - Central: Superficial to flexor digitorum brevis m.
    - Lateral: Superficial to abductor digiti minimi m.
  - Some authors omit medial band, describe central band as medial band

CLINICAL ISSUES

Presentation

- Most common signs/symptoms
  - Pain at inferior margin of heel with weight-bearing
    - Worst in morning
  - Tenderness over inferior margin of calcaneal posterior process

Treatment

- Rest, orthotics, heel cup, stretching
  - May require 6-12 months to respond to treatment
- Extracorporeal shock wave therapy
- Corticosteroid injection
  - Risk of plantar fascia rupture
- Surgical release of plantar fascia

DIAGNOSTIC CHECKLIST

Consider

- Heel pain is often assumed to be due to plantar fasciitis
  - Exclude other causes of heel pain on MR evaluation

SELECTED REFERENCES

1. League AC: Current concepts review: plantar fasciitis. Foot Ankle Int. 29(3):358-66, 2008
2. Neufeld SK et al: Plantar fasciitis: evaluation and treatment. J Am Acad Orthop Surg. 16(6):338-46, 2008
3. Vohra PK et al: Ultrasonographic evaluation of plantar fascia bands. A retrospective study of 211 symptomatic feet. J Am Podiatr Med Assoc. 92(8):444-9, 2002
4. Theodorou DJ et al: Plantar fasciitis and fascial rupture: MR imaging findings in 26 patients supplemented with anatomic data in cadavers. Radiographics. 20 Spec No:S181-97, 2000
5. Grasel RP et al: MR imaging of plantar fasciitis: edema, tears, and occult marrow abnormalities correlated with outcome. AJR Am J Roentgenol. 173(3):699-701, 1999

ANKLE AND SUBTALAR INSTABILITY

Key Facts

Imaging

- Lateral (inversion) ankle instability
  - Varus tilt of talus with weight-bearing or inversion stress
  - Anterior translation of talus ("clown foot")
  - Disruption of ATFL and usually CFL
  - 40% of lateral instability patients reported to also have deltoid ligament insufficiency
- Medial (eversion) ankle instability
  - Widening of medial ankle gutter
  - Valgus tilt of talus
  - Deltoid ligament, syndesmosis injuries
- Subtalar instability
  - Hindfoot valgus, pes planus
  - Lateral subluxation of calcaneus
  - Calcaneofibular ligament injury
  - Less commonly, cervical ligament and interosseous ligament injury

Pathology

- Usually 2 or more ligaments disrupted to cause instability
- 10-20% of ankle sprains (lateral collateral ligament injury) develop chronic ankle instability
- 40% of unstable ankles have abnormality of deltoid ligament
- Instability may also occur due to prior surgery or rheumatoid arthritis

Clinical Issues

- Sensation of ankle "giving way" especially on uneven ground

Diagnostic Checklist

- MR often shows ligament abnormalities greater than suspected clinically
  - Helps direct surgical planning

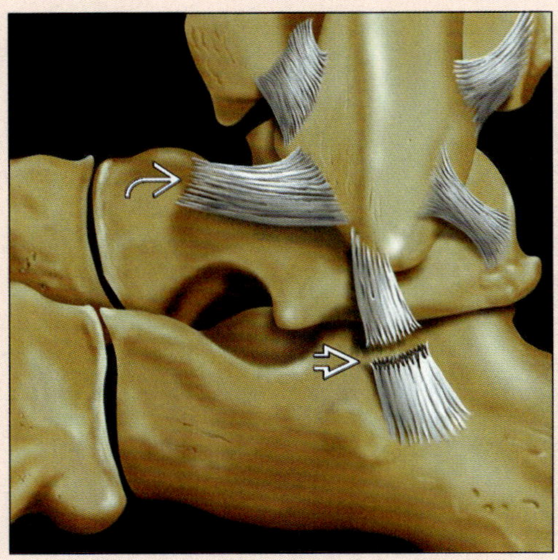

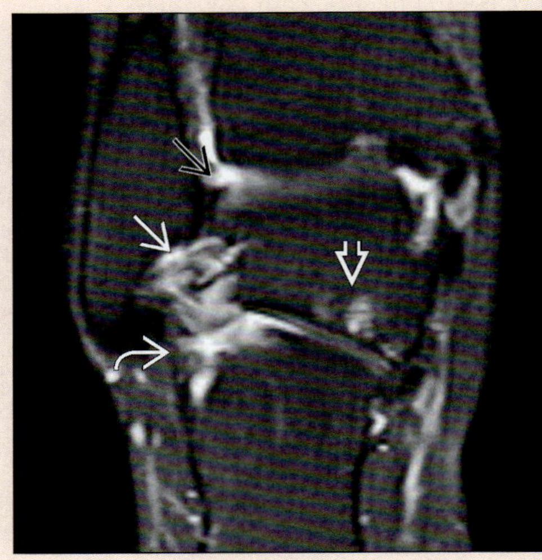

__(Left)__ Graphic of lateral ankle shows tears of anterior talofibular from the talar insertion ➔ and midsubstance calcaneofibular ➔ ligaments. Patients with chronic ankle instability usually have at least 2 torn ligaments. __(Right)__ Coronal T2WI FS MR of chronic ankle and subtalar instability shows torn posterior talofibular ligament (PTFL) ➔, an uncommon finding. In this case, the syndesmotic ligaments ➔, CFL ➔, and ATFL (not shown) were also torn. Bone marrow edema ➔ reflects abnormal subtalar motion.

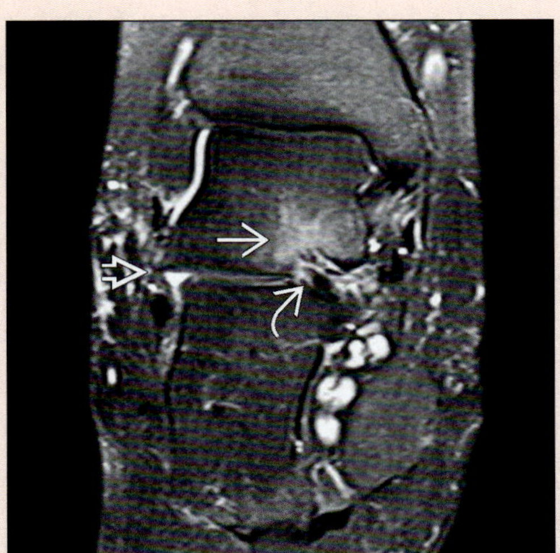

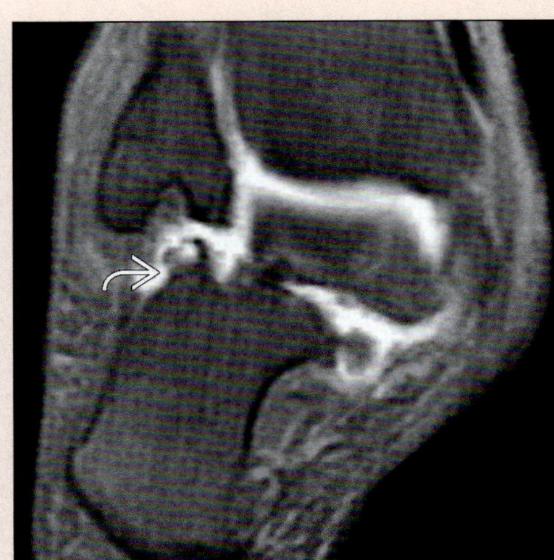

__(Left)__ Coronal T2WI FS MR shows rupture of interosseous ligament ➔ in a patient with subtalar instability. Bone marrow edema is present in the talus ➔ adjacent to expected attachment of interosseous ligament. There is subluxation of the subtalar joint ➔. __(Right)__ Coronal MR arthrogram T1WI FS shows gross subtalar instability, with lateral subluxation ➔ of the calcaneus and valgus hindfoot deformity. There is free communication of contrast from ankle to subtalar joint, a normal variant.

ANKLE AND SUBTALAR INSTABILITY

TERMINOLOGY

Definitions

- Functional instability: Patient has sensation of ankle "giving way"
- Mechanical instability: Instability can be demonstrated on physical exam or stress radiography
- Instability may be present with inversion, eversion, or rotation of ankle

IMAGING

General Features

- Best diagnostic clue
  - Malalignment on weight-bearing ankle series

Radiographic Findings

- Findings best seen on weight-bearing or stress views
- Hindfoot alignment view useful to evaluate weight-bearing axis
  - Line drawn down center of tibial metaphysis and articular surface
  - Lowest weight-bearing portion of calcaneus should be < 8 mm medial or lateral to this line
- Lateral (inversion) ankle instability
  - Loss of parallelism of joint surfaces
  - Varus tilt of talus with inversion stress
    - Stress radiography may be false-positive or false-negative for instability
      - More reliable when patient is under anesthesia
    - Some authors state > 15° varus indicates lateral instability
      - Unreliable; wide variability in "normal" (5-23°)
    - Some authors state > 5° difference in varus tilt between ankles indicates lateral instability
      - Unreliable; instability often bilateral so comparison not valid
  - Varus tilt of talus with patient standing
  - Osteoarthritis centered at medial corner of ankle joint
  - Anterior translation of talus ("clown foot")
- Medial (eversion) ankle instability
  - Loss of parallelism of joint surfaces
  - Widening of medial ankle gutter
  - Valgus tilt of talus
  - Osteoarthritis of lateral corner of ankle joint
- Rotational instability
  - Difficult to see on radiographs and difficult to quantify on physical exam
  - Talar neck may appear more prominent on AP view, shorter on lateral view due to rotation of talus
- Subtalar instability
  - Hindfoot valgus, pes planus
  - Lateral subluxation of calcaneus
    - Lateral margin under tip of fibula

MR Findings

- Lateral ankle instability
  - Usually > 1 ligament is abnormal
  - Anterior talofibular ligament (ATFL) tear in isolation usually not sufficient to cause clinical lateral instability
  - 40% of lateral instability patients reported to also have deltoid ligament insufficiency
  - Bone marrow edema at medial portion of joint due to abnormal osseous abutment
- Medial ankle instability
  - Deltoid ligament injury
  - Tibiofibular syndesmosis injury
  - Bone marrow edema at lateral portion of joint due to abnormal osseous abutment
- Rotational instability
  - Usually injury to both medial and lateral ligaments
- Subtalar instability
  - Calcaneofibular ligament (CFL) injury
  - Less commonly, cervical ligament and interosseous ligament injury

Imaging Recommendations

- Best imaging tool
  - Weight-bearing &/or stress radiographs to evaluate alignment
  - MR to show extent of ligament injuries & aid in surgical planning
- Protocol advice
  - Due to multiplicity of ligaments at risk, need fluid-sensitive sequences with fat saturation in all 3 planes

DIFFERENTIAL DIAGNOSIS

Osteoarthritis

- Osteophytes, joint space narrowing, ± subluxation
- Joint mobility decreased

Charcot, Neuropathic

- Florid joint destruction, bone debris, disorganization, dislocation, joint distension, preserved bone density

PATHOLOGY

General Features

- Etiology
  - 10-20% of ankle sprains (lateral collateral ligament injury) develop chronic inversion ankle instability
    - Usually 2 or more ligaments disrupted to cause instability
    - Instability still controversial in terms of etiology
  - Risk of instability higher in cavus foot
  - 40% of unstable ankles have abnormality of deltoid ligament
  - Syndesmotic ligament injury associated with eversion instability
  - Injury to cervical, interosseous ligaments associated with subtalar instability
  - Instability secondary to prior surgery
    - Fusion of ankle joint increases stress on subtalar joint and vice versa
    - Unfused joint may become unstable or merely osteoarthritic
  - Rheumatoid arthritis
    - Involvement of ankle, subtalar joints common
    - May lead to instability, subluxation, hindfoot valgus
- Associated abnormalities
  - Osteochondral lesions of talar dome and gutters &/ or tibial plafond
  - Cartilage loss, osteophytes, loose bodies

ANKLE AND SUBTALAR INSTABILITY

- Peroneal tendon injury
- Superior peroneal retinaculum injury
- Flexor retinaculum injury
- Synovitis
- Ankle soft tissue impingement lesions
  - e.g., ankle may be unstable laterally and impinge medially due to scar from ligament injury
- Subtalar instability often associated with ankle instability

Staging, Grading, & Classification

- Classified as inversion, eversion, or rotational

Relevant Anatomy

- Calcaneofibular ligament (CFL) key to subtalar and ankle stability
  - Extends from tip of fibula posteriorly and inferiorly to calcaneus
  - Deep to peroneal tendons
  - Stabilizes from inversion stress
- Anterior talofibular ligament (ATFL) provides rotational stability to ankle
  - Extends from fibula 1 cm above tip to junction of talar body and neck
- Superficial deltoid ligament provides rotational stability to both joints
  - Fibers fan from superficial surface of medial malleolus to talus, sustentaculum tali, navicular, spring ligament
- Deep deltoid ligament stabilizes ankle from eversion stress
  - Extends from deep surface of medial malleolus (primarily posterior colliculus) to concavity at medial border of talar body
- Syndesmotic ligaments stabilize ankle from eversion stress
  - Extend from tibia to fibula at level of plafond
- Cervical, interosseous ligaments provide subtalar stability
  - Cervical ligament: In tarsal sinus, extends from talar neck laterally and inferiorly to superior border of calcaneus
  - Interosseous ligament: In tarsal canal, at posteromedial margin of tarsal sinus
- Peroneal tendons provide dynamic stability
- Inferior extensor retinaculum provides some stability to subtalar joint
  - 3 bands inserting on lateral, mid, and medial dorsal surface of calcaneus in tarsal sinus

CLINICAL ISSUES

Presentation

- Most common signs/symptoms
  - History of recurrent ankle sprains
  - Sensation of ankle "giving way" especially on uneven ground
  - Laxity of ankle on physical exam
- Other signs/symptoms
  - Pain and tenderness at lateral ankle (lateral instability) or medial ankle (medial instability)
  - Difficult to detect subtalar instability on physical exam

Natural History & Prognosis

- Limitation of athletic activity
- Impingement lesions
- Premature osteoarthritis

Treatment

- Physical therapy, bracing, lace-up boots
- Reconstruction of ankle ligaments; many procedures described
  - Anatomic reconstructions usually preferred to reconstruction using peroneal tendons

DIAGNOSTIC CHECKLIST

Consider

- MR often shows ligament abnormalities greater than suspected clinically
  - Helps direct surgical planning

Image Interpretation Pearls

- Be alert for "satisfaction of search"
  - Ligament injuries are usually multiple

Reporting Tips

- Diagnosis of instability is based on clinical findings, not imaging
  - Report abnormalities but do not attempt to determine ankle stability

SELECTED REFERENCES

1. Weindel S et al: Subtalar instability: a biomechanical cadaver study. Arch Orthop Trauma Surg. 130(3):313-9, 2010
2. Ferran NA et al: Ankle instability. Sports Med Arthrosc. 17(2):139-45, 2009
3. Alparslan L et al: Lateral ankle instability: MR imaging of associated injuries and surgical treatment procedures. Semin Musculoskelet Radiol. 12(4):346-58, 2008
4. Kirby AB et al: Magnetic resonance imaging findings of chronic lateral ankle instability. Curr Probl Diagn Radiol. 34(5):196-203, 2005
5. Hintermann B et al: Medial ankle instability: an exploratory, prospective study of fifty-two cases. Am J Sports Med. 32(1):183-90, 2004
6. Michelson J et al: The effect of ankle injury on subtalar motion. Foot Ankle Int. 25(9):639-46, 2004
7. Hintermann B et al: Arthroscopic findings in patients with chronic ankle instability. Am J Sports Med. 30(3):402-9, 2002
8. Keefe DT et al: Subtalar instability. Etiology, diagnosis, and management. Foot Ankle Clin. 7(3):577-609, 2002
9. Larsen E et al: Association of ankle instability and foot deformity. Acta Orthop Scand. 61(2):136-9, 1990

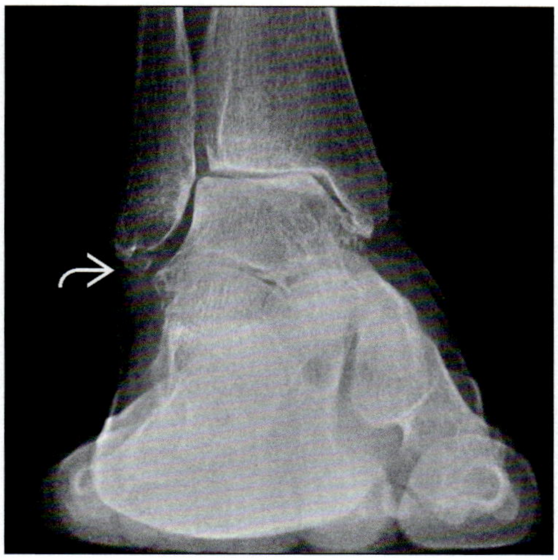

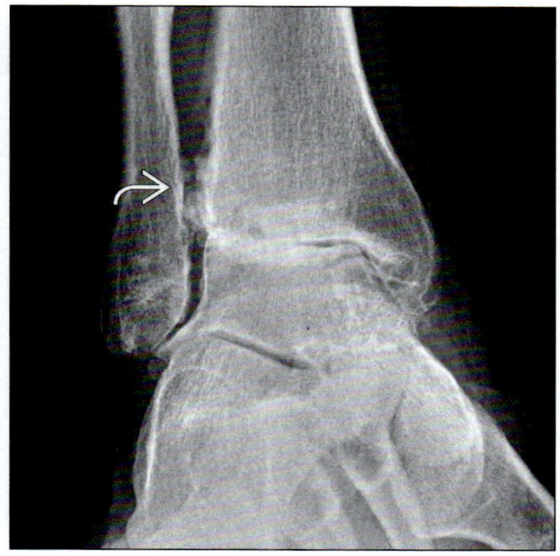

(Left) *AP radiograph shows lateral ankle instability. There is varus tilt of the talus and loss of normal parallelism of the joint. Secondary osteoarthritis is present medially. Lateral ossicle is old avulsion of calcaneofibular ligament.* ***(Right)*** *AP radiograph shows osteoarthritis, which has developed due to instability from chronic syndesmotic tear. Note the widened syndesmosis and new bone formation at site of ligament injury.*

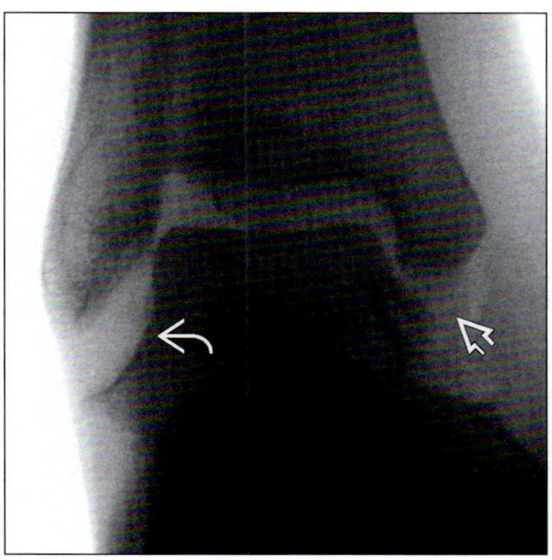

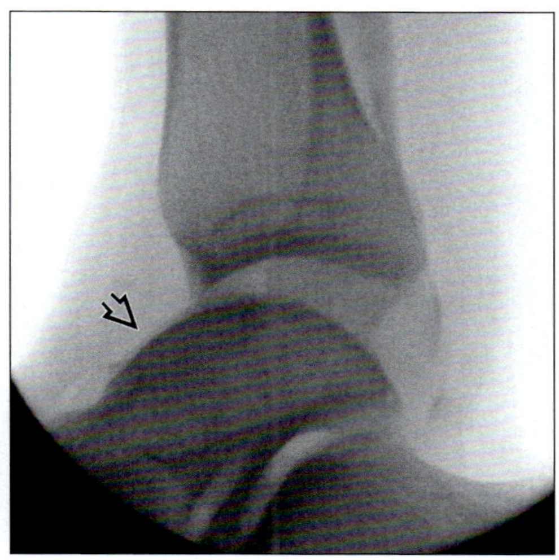

(Left) *AP fluoroscopic spot radiograph with inversion stress under anesthesia shows marked widening of lateral ankle gutter and loss of parallel contours of medial gutter, indicating tears of both lateral collateral and deltoid ligaments. Weight-bearing radiographs showed normal alignment.* ***(Right)*** *Lateral fluoroscopic radiograph with anterior stress under anesthesia in the same patient shows 50% anterior subluxation of talar dome relative to tibial plafond.*

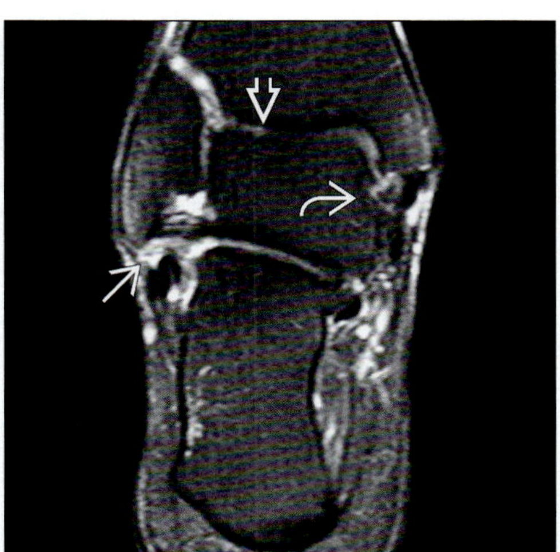

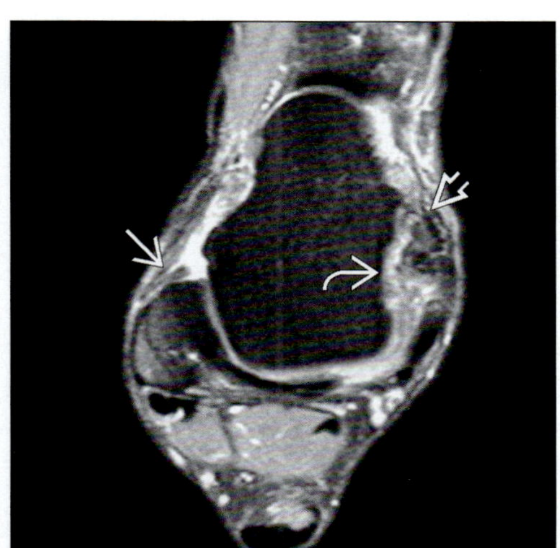

(Left) *Coronal T2WI FS MR in the same patient shows tears of the calcaneofibular and deep deltoid ligaments. Cartilage loss is evident in the talar dome.* ***(Right)*** *Axial PDWI FS MR in the same patient shows torn anterior talofibular ligament with only a small stump remaining. Deep deltoid is torn, and extensive scar formation is present anteromedially. Medial and lateral ligament reconstructions were performed, with successful restoration of ankle stability.*

*(**Left**) Axial PDWI FS MR shows acute superimposed on chronic injury in a patient with chronic lateral instability and recent sprain. ATFL and deep deltoid ligaments are torn. There is also a tear of the flexor retinaculum. (**Right**) Coronal T2WI FS MR in the same patient shows deltoid sprain with edema at talar attachment of the ligament. The calcaneofibular ligament is completely torn.*

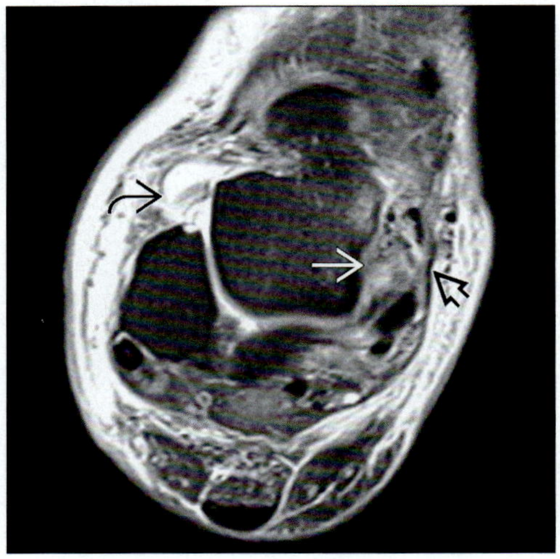

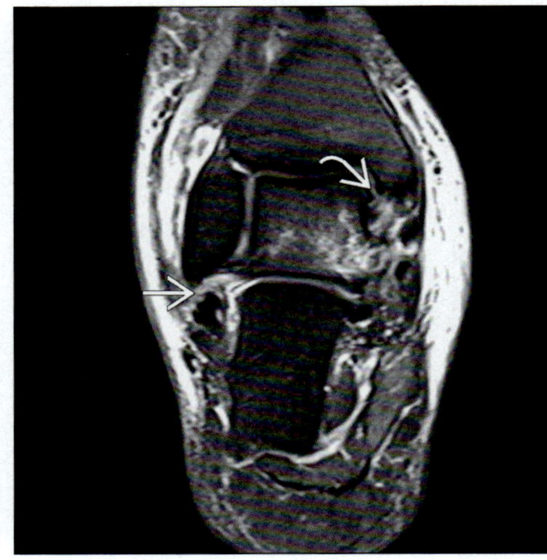

*(**Left**) Mortise view radiograph shows medial instability with valgus tilt of talus and loss of joint congruity. Medial ossicle is from prior deltoid ligament avulsion. (**Right**) Coronal T2WI FS MR shows tears of CFL, anterior inferior tibiofibular ligament, anterior deep deltoid ligament, and superficial deltoid origin. The patient had both medial and lateral instability symptoms. Interosseous ligament is intact.*

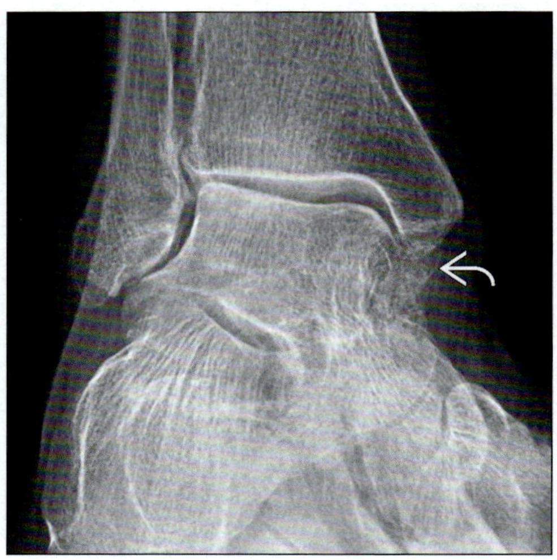

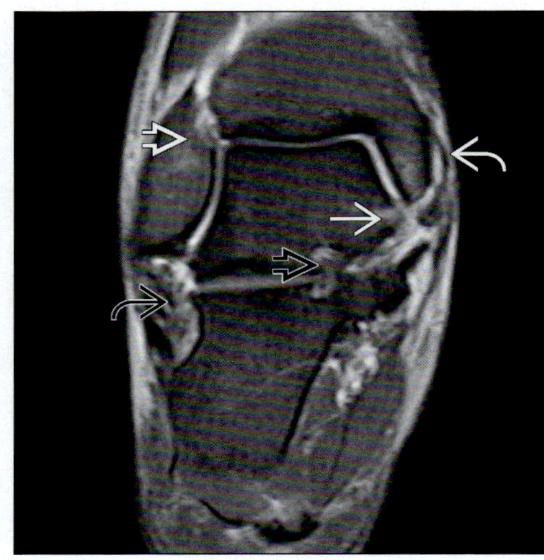

*(**Left**) Sagittal T1WI C+ FS MR shows ankle instability due to rheumatoid arthritis. There is anterior subluxation of the joint resulting in "clown foot" deformity. Synovitis and bone marrow edema are present. (**Right**) Lateral radiograph shows subtalar instability due to neuropathic arthropathy. Characteristic findings include the "5 Ds": Preserved bone density, bone debris, joint destruction and distension, and disorganization.*

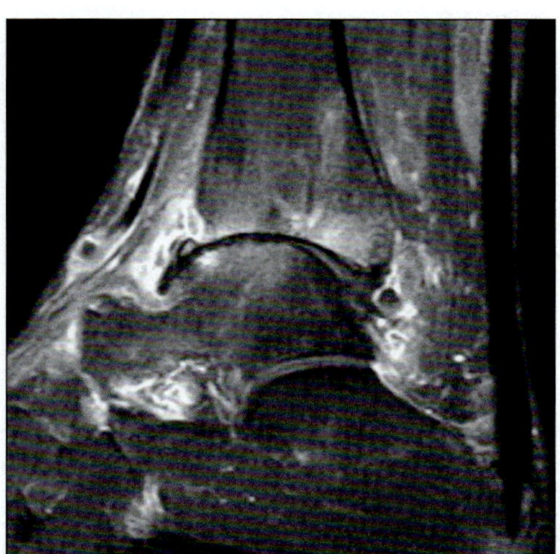

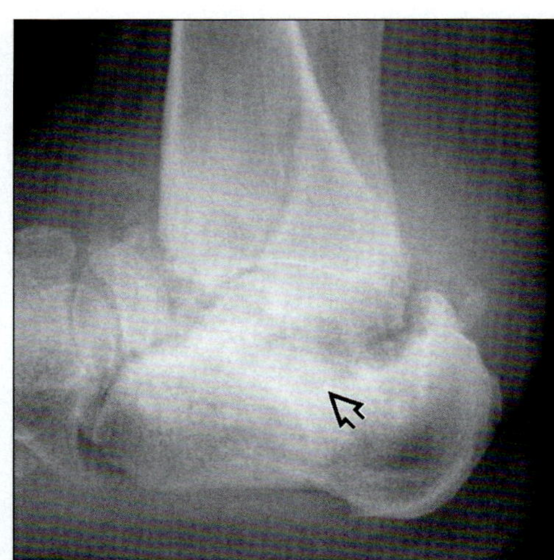

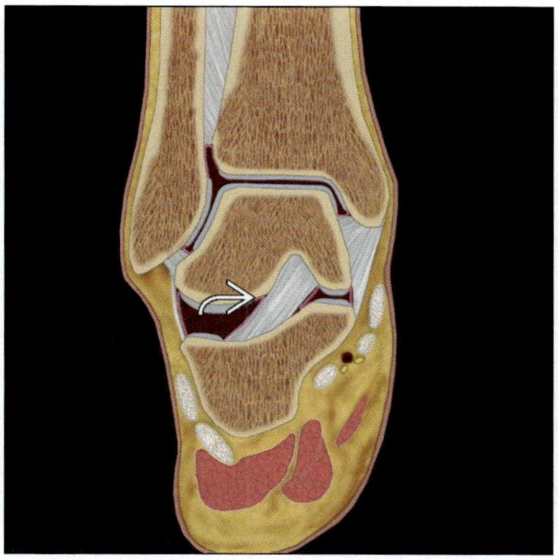

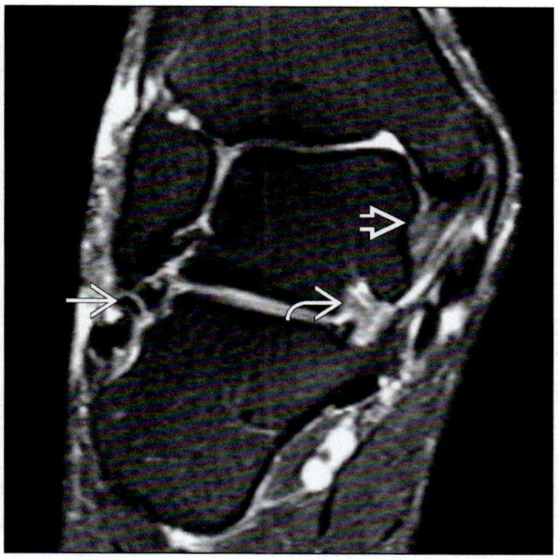

(Left) *Coronal graphic shows normal interosseous ligament. Ligament is nearly vertical in orientation and lies in tarsal canal. The interosseous ligament is less commonly injured than the calcaneofibular or cervical ligaments.* ***(Right)*** *Coronal T2WI FS MR shows torn interosseous ligament as well as torn calcaneofibular ligament. Deep deltoid fibers are intact.*

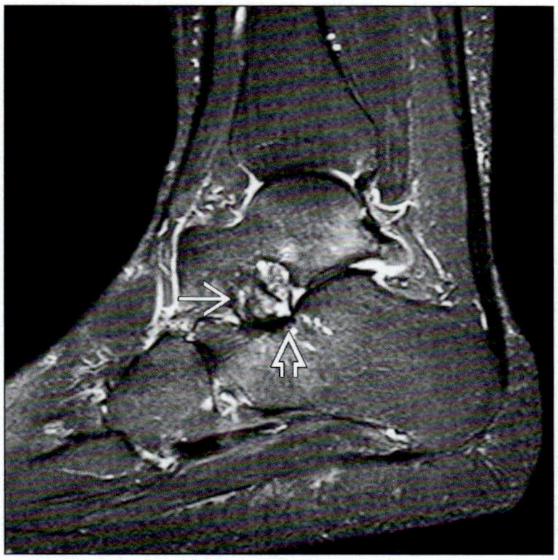

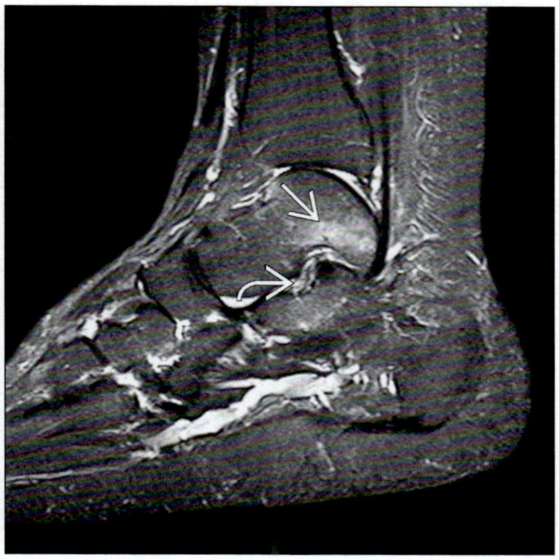

(Left) *Sagittal STIR MR shows thickened, scarred cervical ligament and subluxation of the posterior subtalar joint. Normally, the anterior margin of the talar articular surface should align with the angle of Gissane, but in this case the calcaneus is displaced anteriorly.* ***(Right)*** *Sagittal STIR MR in the same patient shows disruption of the interosseous ligament and bone marrow edema in the talus. Ankle instability was not demonstrable on physical exam in this patient.*

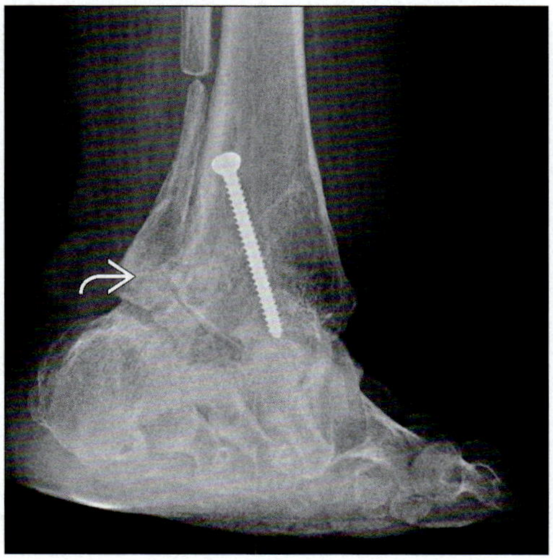

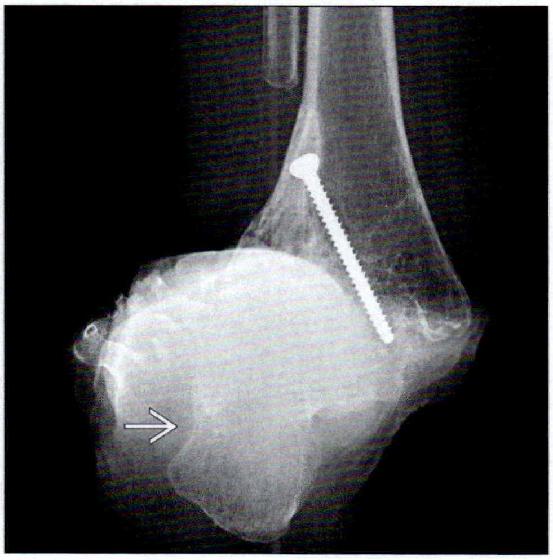

(Left) *Mortise view radiograph shows lateral subluxation of the calcaneus due to subtalar instability developing after ankle fusion. Ankle fusion increases stress to the subtalar joint and can lead to instability &/or osteoarthritis of the subtalar joint.* ***(Right)*** *Hindfoot alignment radiograph in the same patient better shows the lateral displacement of the calcaneus. The most inferior portion of the calcaneal tuberosity should be < 8 mm from a line drawn through the long axis of the tibia.*

SEVER DISEASE

Key Facts

Imaging

- Radiographs
  - Fragmented, sclerotic apophysis of calcaneal posterior process
  - These findings are often also seen in asymptomatic children
- MR
  - Bone marrow edema in and adjacent to posterior process apophysis

Top Differential Diagnoses

- Stress fracture
  - MR often shows stress fracture adjacent to apophysis in cases of clinical Sever disease
  - Some literature suggests Sever disease is actually unrecognized stress fracture
- Plantar fasciitis
- Juvenile idiopathic arthritis
- Tarsal tunnel syndrome
- Haglund syndrome

Pathology

- Posterior calcaneal apophysis appears at age 7 years, fuses at age 15 years
- Frequently multipartite & sclerotic

Clinical Issues

- Posterior process pain in skeletally immature individual
- Self-limited, treated with rest

Diagnostic Checklist

- MR often shows that children with clinical presentation of Sever disease have a different cause of heel pain
- Key to diagnosis: Combination of imaging findings and clinical symptoms

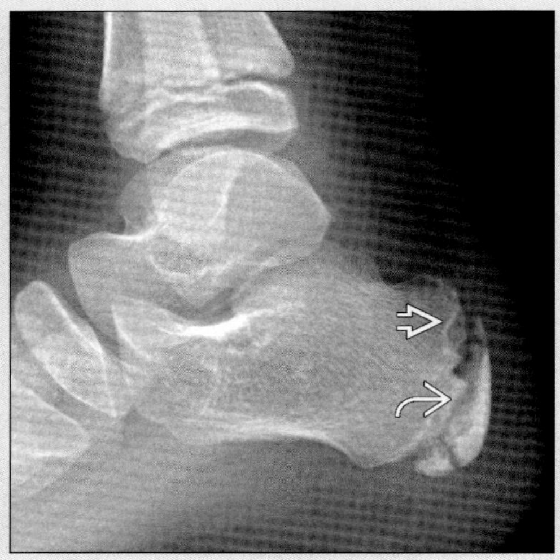
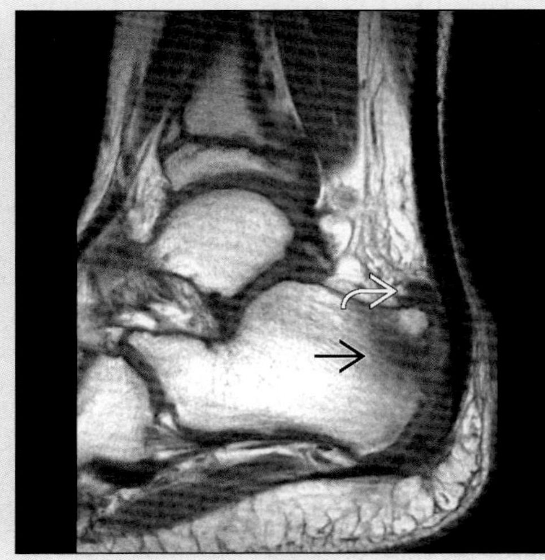

(Left) *Lateral radiograph in a 12-year-old boy with clinical Sever disease shows a sclerotic, fragmented calcaneal apophysis and serrated metaphyseal contour, often seen in asymptomatic children. Sever disease may reflect Achilles insertional tendonitis, stress response of the apophysis to tugging from the Achilles tendon, or stress fracture of the calcaneus.* ***(Right)*** *Sagittal T1WI MR shows bone marrow edema in the calcaneus and pre-Achilles bursitis in a patient with clinical Sever disease.*

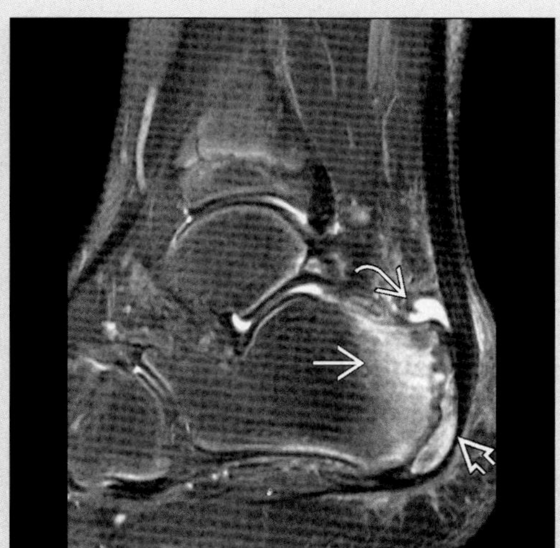
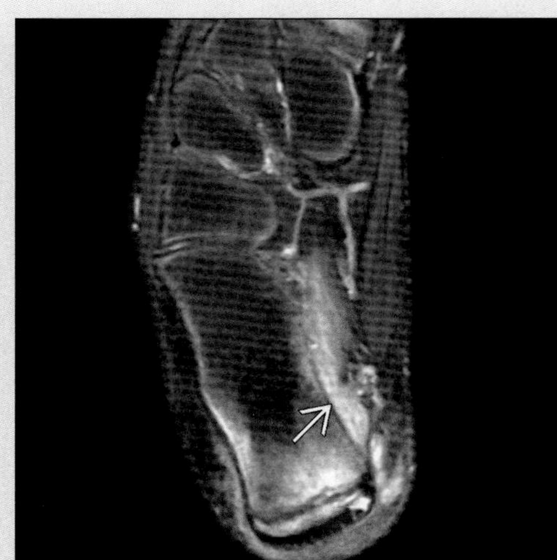

(Left) *Sagittal T2WI FS MR in the same patient shows that bone marrow edema involves not only the metaphysis but also the apophysis. There is fluid in the pre-Achilles bursa and less well-defined edema posteriorly.* ***(Right)*** *Axial T2WI FS MR shows that edema involves the quadratus plantae muscle as well as the calcaneus. Whether or not the name Sever disease is warranted, stress injuries to the posterior calcaneus are common in children.*

SEVER DISEASE

TERMINOLOGY

Synonyms

- Osteochondrosis of heel

Definitions

- Described as "apophysitis" of skeletally immature posterior calcaneal apophysis
- Apophysis: Secondary growth center of a bone, usually not contributing to longitudinal growth

IMAGING

Imaging Recommendations

- Best imaging tool
  - MR

Radiographic Findings

- Fragmented, sclerotic apophysis of calcaneal posterior process not a reliable sign
  - These findings are also seen in asymptomatic children

MR Findings

- Bone marrow edema in and adjacent to posterior process apophysis
- Pre-Achilles (retrocalcaneal) bursitis

DIFFERENTIAL DIAGNOSIS

Stress Fracture

- MR often shows stress fracture adjacent to apophysis in cases of clinical Sever disease
- Some literature suggests Sever disease is actually unrecognized stress fracture

Plantar Fasciitis

- Thickening of plantar fascia
- Edema in and surrounding fascia

Juvenile Idiopathic Arthritis

- Inflammation of Achilles &/or plantar fascia origin
- Uncommon in children but may have juvenile onset of spondyloarthropathy

Tarsal Tunnel Syndrome

- Impingement on lateral calcaneal branch of tibial nerve causes heel pain

Haglund Syndrome

- Bony prominence leads to bursitis & heel pain

Achilles Tendinopathy

- Due to chronic, repetitive stress on tendon
- MR shows tendon thickening ± mild increase in intratendinous signal

PATHOLOGY

General Features

- Etiology
  - Associated with athletic activity, especially jumping
  - Due to pull by Achilles tendon on apophysis
- Associated abnormalities
  - Pre-Achilles bursitis

Relevant Anatomy

- Posterior calcaneal apophysis appears at age 7 years, fuses at age 15 years
- Frequently multipartite and sclerotic
- Adjacent metaphysis often has scalloped contour that is not completely congruent with apophysis

CLINICAL ISSUES

Presentation

- Most common signs/symptoms
  - Posterior process pain in skeletally immature individual
  - Pain exacerbated by exercise, relieved by rest

Demographics

- Age
  - Usually ages 8-14, slightly younger in girls than boys
- Gender
  - M > F

Natural History & Prognosis

- Self-limited

Treatment

- Rest

DIAGNOSTIC CHECKLIST

Consider

- MR often shows that children with clinical presentation of Sever disease have a different cause of heel pain

Image Interpretation Pearls

- Key to diagnosis: Combination of imaging findings and clinical symptoms

SELECTED REFERENCES

1. Kose O: Do we really need radiographic assessment for the diagnosis of non-specific heel pain (calcaneal apophysitis) in children? Skeletal Radiol. 39(4):359-61, 2010
2. Scharfbillig RW et al: Sever's disease: what does the literature really tell us? J Am Podiatr Med Assoc. 98(3):212-23, 2008
3. Ogden JA et al: Sever's injury: a stress fracture of the immature calcaneal metaphysis. J Pediatr Orthop. 24(5):488-92, 2004
4. Volpon JB et al: Calcaneal apophysitis: a quantitative radiographic evaluation of the secondary ossification center. Arch Orthop Trauma Surg. 122(6):338-41, 2002

LATERAL LIGAMENT RECONSTRUCTION, ANKLE

Key Facts

Terminology

- Gould modification of Broström procedure
  - Direct repair of ATFL ± CFL, augmented by inferior extensor retinaculum
  - Surgery currently most recommended
- Peroneus brevis (PB) tendon used for a variety of nonanatomic repairs
  - Tendon ligated, rerouted to improve lateral stability

Imaging

- Broström and modified Broström procedure normal findings
  - Repair should appear taut
  - Repair tissue often thick, poorly defined margins
- Signs of Broström repair failure
  - Lax, distended lateral joint capsule
  - Native or arthrographic fluid extending beyond joint capsule
- Peroneal tendon rerouting procedures
  - Tendon should be visible as low signal intensity structure in osseous tunnels
- Should be able to follow rerouted tendon to anastomoses

Diagnostic Checklist

- Possible causes of persistent pain after lateral ankle ligament repair
  - Failure of surgical repair
  - Medial ankle instability due to deltoid ligament tear
  - Osteochondral lesion
  - Anterolateral impingement due to scar tissue
  - Peroneal tendon tear &/or retinaculum injury

__(Left)__ Axial PDWI FS MR shows an intact Gould modification of Broström procedure. Extensor retinaculum → merges with adjacent ATFL →. The repair is taut, and the anterolateral joint recess → is normal in size. __(Right)__ Axial MR arthrogram T1WI of failed Broström repair shows repaired tendon no longer attaching to the talus → and ballooning of the anterolateral joint recess → with contrast extending anterior to normal talar ATFL insertion. Artifact → is seen at fibular anchor.

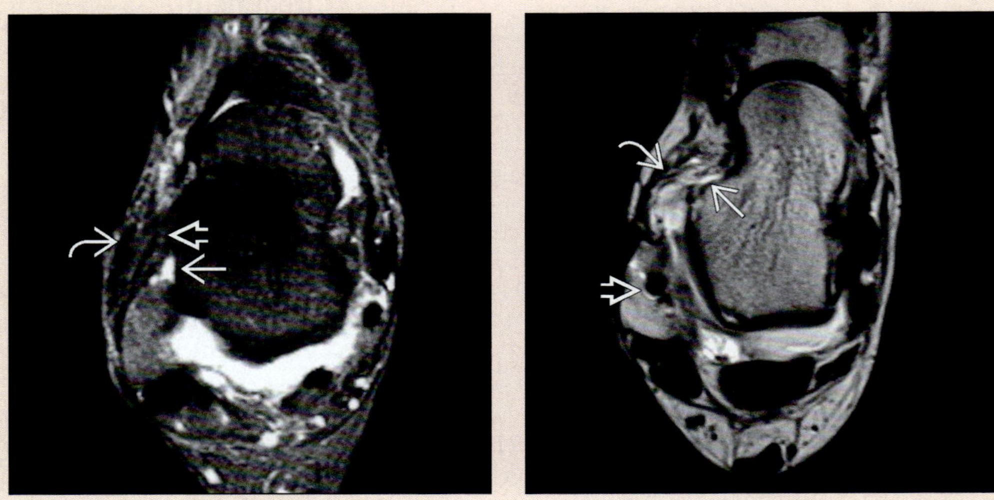

__(Left)__ Sagittal T1WI MR shows peroneus brevis loop. Peroneus brevis (PB) has been cut, and distal portion looped → through a tunnel in the fibula, then re-anastomosed to the distal PB →. PB inserts normally on the 5th metatarsal base →. Proximal end of the cut PB was anastomosed to the peroneus longus (not shown). __(Right)__ Axial PDWI FS MR of PB loop shows intact PB tendon → in fibular tunnel. The proximal stump of the PB has been sutured to the peroneus longus →.

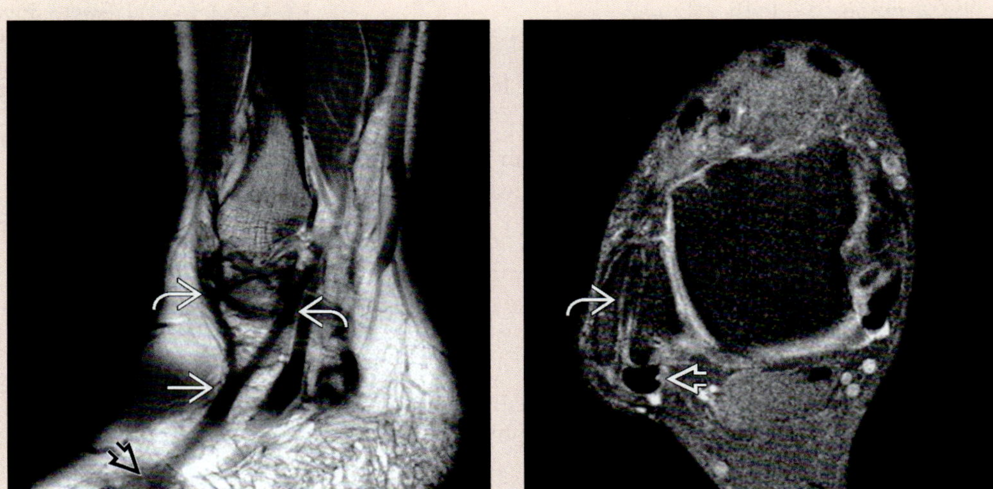

LATERAL LIGAMENT RECONSTRUCTION, ANKLE

TERMINOLOGY

Definitions

- Broström procedure
  - Direct repair and plication of lateral collateral ligament (LCL)
  - Always includes anterior talofibular ligament (ATFL)
  - May also include calcaneofibular ligament (CFL)
- Gould modification of Broström procedure
  - Repair of ATFL ± CFL, augmented by inferior extensor retinaculum
  - Surgery currently most recommended
- Peroneus brevis (PB) tendon used for a variety of nonanatomic repairs
  - PB tendon loop
    - PB is cut, distal portion looped through horizontal fibular tunnel, reattached to distal peroneus brevis
  - Watson-Jones procedure
    - PB is cut, distal portion looped through fibula, talar neck, back through fibula, reattached to distal peroneus brevis
  - Chrisman-Snook procedure
    - PB is split, anterior portion looped through fibula, attached to periosteum of talus and calcaneus and reattached to distal peroneus brevis
  - Evans procedure
    - PB is cut, routed through vertical fibular tunnel, reattached to peroneus longus distally

IMAGING

Radiographic Findings

- Suture anchors often visible
  - Anchor displacement is useful sign of repair failure
- May see cylindrical bone tunnels if peroneal tendon rerouted
  - Round, sharply demarcated lucency if seen end-on

MR Findings

- Broström procedure
  - Normal findings
    - Repair should appear taut
      - Repair tissue often appears thick, margins poorly defined
      - Heterogeneous signal intensity in repaired tissue
      - If Gould modification performed, extensor retinaculum merges with ATFL
    - Anterolateral joint recess should be small & triangular
    - Metal artifact at lateral malleolus, talar neck ± calcaneus
  - Signs of repair failure
    - Lax, distended lateral joint capsule
    - Native or arthrographic fluid extending beyond joint capsule
    - Bone marrow edema persisting more than a few months postoperatively
      - Due to abnormal motion of ankle joint
- Peroneal tendon rerouting procedures
  - Normal findings
    - Tendon should be visible as low signal intensity structure in osseous tunnels
      - Nonvisualization of tendon in tunnel implies tear
    - Should be able to follow rerouted tendon to anastomoses
      - Helpful to know what type of surgery performed in order to evaluate anastomoses
      - If tendon splitting performed, tendon will have ribbon-like configuration
    - Fluid usually seen around tendons, does not imply failure of repair
- Signs of repair failure
  - Discontinuity of repair tissue or tendon
  - Lax, distended lateral joint capsule

Imaging Recommendations

- Best imaging tool
  - MR
- Protocol advice
  - MR arthrography often improves visualization of repair

CLINICAL ISSUES

Presentation

- Most common signs/symptoms
  - Recurrent pain ± sensation of giving way may reflect failure of repair

Natural History & Prognosis

- 80-90% success rate of reconstructive surgery
- Multiple possible causes for persistent pain after lateral ankle ligament repair
  - Failure of surgical repair
  - Medial ankle instability due to deltoid ligament tear
  - Osteochondral lesion
  - Anterolateral impingement due to scar tissue
  - Peroneal tendon tear &/or retinaculum injury

DIAGNOSTIC CHECKLIST

Reporting Tips

- Knowledge of type of previous surgery makes MR report more meaningful to surgeon

SELECTED REFERENCES

1. Ferran NA et al: Ankle instability. Sports Med Arthrosc. 17(2):139-45, 2009
2. Li X et al: Anatomical reconstruction for chronic lateral ankle instability in the high-demand athlete: functional outcomes after the modified Broström repair using suture anchors. Am J Sports Med. 37(3):488-94, 2009
3. Chien AJ et al: Imaging appearances of lateral ankle ligament reconstruction. Radiographics. 24(4):999-1008, 2004
4. Colville MR: Surgical treatment of the unstable ankle. J Am Acad Orthop Surg. 6(6):368-77, 1998
5. Liu SH et al: Comparison of lateral ankle ligamentous reconstruction procedures. Am J Sports Med. 22(3):313-7, 1994

TENDON REPAIR, ANKLE

Key Facts

Terminology

- Tendon repair: Tendon defect debrided, tendon ends re-anastomosed
  - Graft may be used for large defects
- Tendon transfer: Rerouting of intact tendon in order for it to take on function of torn or denervated tendon
- Tendon lengthening: Z-shaped partial cut of tendon

Imaging

- Radiographs
  - Bone tunnels and soft tissue anchors may be visible
  - Corrective osteotomy may be seen in addition to tendon repair
- MR
  - Repaired tendon may appear thick or thin
  - Fluid-sensitive sequences: Heterogeneous signal intensity, but less SI than fluid in intact repair
  - Low signal intensity scar may be seen around tendon
  - Repair normally enhances with gadolinium
- Tendon transfers show relatively normal tendon in new position
  - For Achilles tear: Flexor hallucis longus tendon transferred to posterior process of calcaneus
  - For posterior tibial tendon tear: Anterior tibial tendon or flexor digitorum brevis tendon transferred to navicular
- Stenosing tenosynovitis can be a complication of tendon surgery
  - Tendon sheath thickened, ± fluid in sheath
  - MR may appear normal
  - Diagnosis and treatment may be performed with tenosynography

__(Left)__ Sagittal T1WI MR shows what appears to be a thickened Achilles tendon ➮ but actually represents the coalescence of repaired Achilles tendon and the transferred flexor hallucis longus (FHL). Note that the FHL ➙ muscle is positioned posterior to its usual location immediately posterior to the tibia. __(Right)__ Axial T1WI MR in the same patient shows the posterior position of the FHL ➙ and the thickened, repaired Achilles tendon ➮.

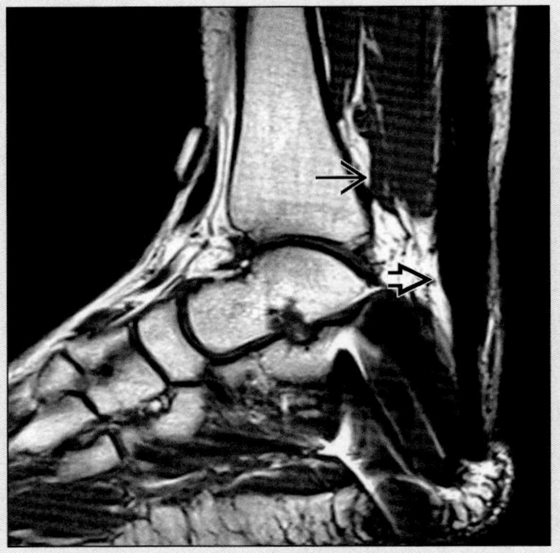

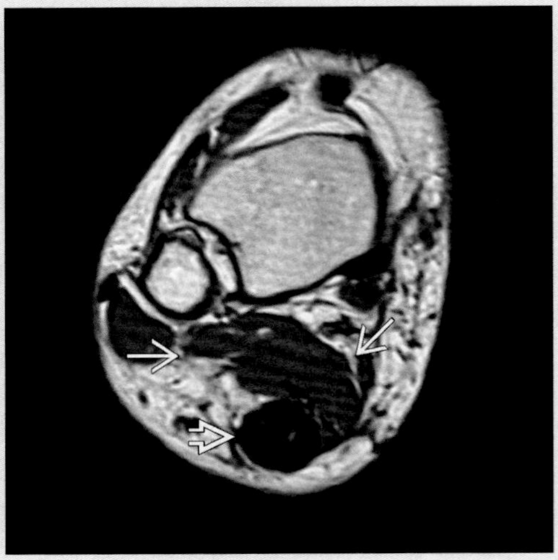

__(Left)__ Lateral radiograph in the same patient shows attachment tunnel ➙ for FHL. The combined FHL/Achilles tendon outline ➮ is thickened. Calcaneal osteotomies were also performed for realignment. __(Right)__ Axial soft tissue algorithm CT shows calcification ➙ at repair of peroneus brevis tendon. The tendon is poorly defined and surrounded by scar. Tendon integrity is difficult to assess on CT after surgical repair, because scar and tendon have a similar appearance.

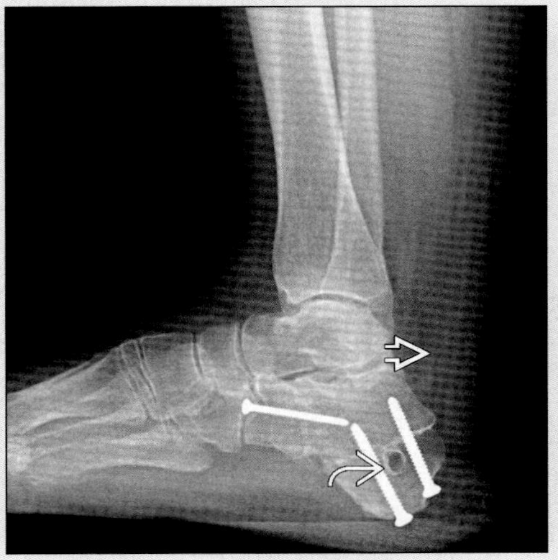

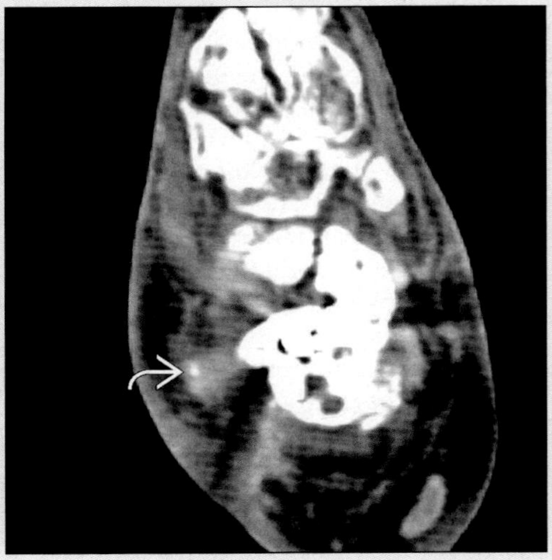

TENDON REPAIR, ANKLE

TERMINOLOGY

Definitions

- Tendon repair: Tendon defect debrided, tendon ends re-anastomosed
  - Graft may be used for large defects
- Tendon transfer: Rerouting of intact tendon in order for it to take on function of torn or denervated tendon
- Tendon lengthening: Z-shaped partial cut of tendon

IMAGING

Radiographic Findings

- Soft tissue fullness may be seen where tendon has been rerouted
- Bone tunnels and soft tissue anchors may be visible
- Corrective osteotomy may be performed in addition to tendon repair
  - Most commonly calcaneal realignment in conjunction with posterior tibial tendon repair

CT Findings

- Tendon may appear thick or thinned
- Scar tissue often obscures tendon outlines
- Calcification/ossification may develop in tendon

MR Findings

- Primary repair
  - Repaired tendon may appear thick or thinned
  - Fluid-sensitive sequences: Heterogeneous signal intensity, but less SI than fluid in intact repair
    - Heterogeneous signal does not imply failure
  - Metal artifact may be seen at repair site
  - Scar may be seen around tendon
    - Low signal intensity on all sequences
  - Repair normally enhances with gadolinium
- Tendon transfer: Rerouting of a tendon to take function of torn tendon
  - Achilles tendon tear
    - Flexor hallucis longus (FHL) tendon transferred to posterior process of calcaneus
    - FHL deviates posteriorly from its normal course & merges with Achilles tendon
  - Posterior tibial tendon tear
    - Anterior tibial tendon transferred to navicular
      - Tunnel placed from plantar to dorsal
    - Flexor digitorum longus tendon transferred to navicular
      - Suture anchor seen at medial eminence of navicular
- Achilles tendon lengthening
  - Tendon appears thin & heterogeneous
  - Scar often present in pre-Achilles fat
- Stenosing tenosynovitis
  - Complication of tendon repair
  - Tendon sheath thickened, ± fluid in sheath
  - MR may appear normal
  - Diagnosis and treatment may be performed with tenosynography

Imaging Recommendations

- Best imaging tool
  - MR to evaluate for recurrent tear

CLINICAL ISSUES

Presentation

- Most common signs/symptoms
  - Failure of repair presents with sudden "pop" & pain
  - Chronic pain & catching may be due to adhesions

Natural History & Prognosis

- Adhesions at repair site may result in painful adhesive tenosynovitis

DIAGNOSTIC CHECKLIST

Consider

- Tendon rerouting may also be performed for neurologic deficits
  - Example: Posterior tibial tendon to extensor tendons for drop-foot

Image Interpretation Pearls

- Repaired tendon may look irregular and contain heterogeneous signal in asymptomatic patient

SELECTED REFERENCES

1. Heckman DS et al: Tendon disorders of the foot and ankle, part 2: achilles tendon disorders. Am J Sports Med. 37(6):1223-34, 2009
2. Reddy SS et al: Surgical treatment for chronic disease and disorders of the achilles tendon. J Am Acad Orthop Surg. 17(1):3-14, 2009
3. DiPaola M et al: Tendon transfers and realignment osteotomies for treatment of stage II posterior tibial tendon dysfunction. Foot Ankle Clin. 12(2):273-85, vi, 2007
4. Myerson MS et al: Treatment of stage II posterior tibial tendon deficiency with flexor digitorum longus tendon transfer and calcaneal osteotomy. Foot Ankle Int. 25(7):445-50, 2004

AMPUTATION, FOOT/ANKLE

Key Facts

Terminology

- Many different amputations performed
- Toes: Distal amputation for osteomyelitis or resection arthroplasty of deformity
- Transmetatarsal: Through MT shafts
- Lisfranc joint: Usually disarticulation of tarsometatarsal joints, sometimes 2nd &/or 1st MT base left in situ
- Chopart joint: Disarticulation of talonavicular and calcaneocuboid joints
- Syme amputation: Disarticulation of ankle joint + resection of malleoli
- Calcanectomy: Excision of calcaneus only

Imaging

- Radiographs
  - Stump often develops sclerosis due to stress
  - Amputation site should develop well-defined rim of cortication
  - Loss of cortication at stump site is indicative of infection
  - Small amount of unilaminar periosteal reaction at amputation site is normal
  - Altered biomechanics may cause abnormalities of adjacent bones
- MR
  - Bone marrow usually homogeneously fatty after amputation
  - Small foci of mildly increased signal on STIR often seen, due to fibrosis or infarct
  - Confluent areas of low T1, high T2 or STIR probably due to osteomyelitis

Diagnostic Checklist

- MR important part of assessment of need for amputation

*(**Left**) AP radiograph in a diabetic patient shows disarticulation of the 1st MTP joint, amputation of the 4th toe at the base of the proximal phalanx, and amputation of the 5th ray at the metatarsal shaft. Erosions of the 4th metatarsal head ➔ reflect septic arthritis. (**Right**) Oblique radiograph shows 1st TMT disarticulation. This has altered biomechanics, resulting in subluxation of the 2nd TMT joint ➔, dislocation of the 2nd MTP joint ➔, and Freiberg infraction of the 3rd MT head ➔.*

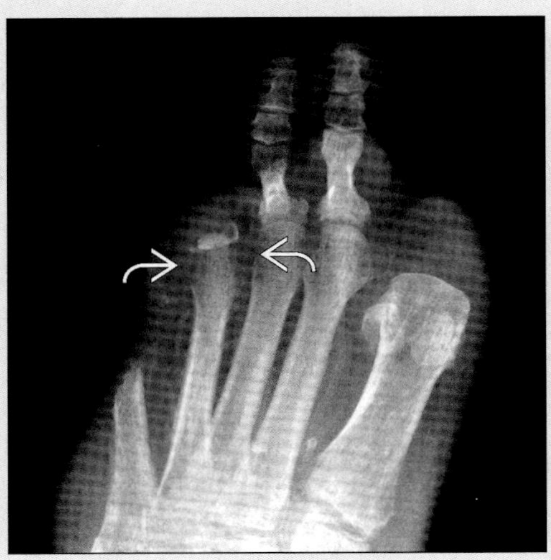

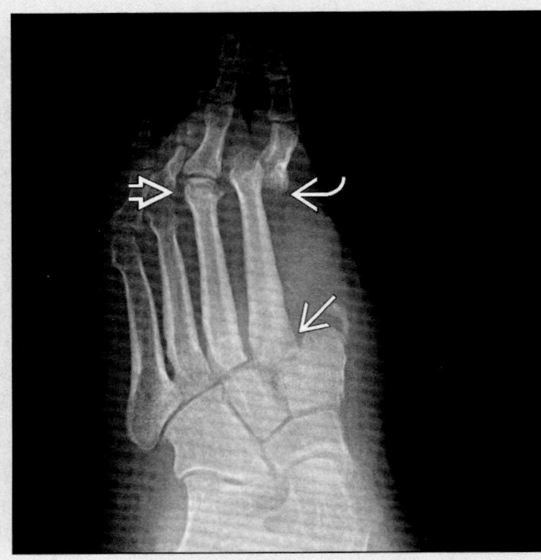

*(**Left**) Lateral radiograph shows Chopart amputation, with additional resection of the anterior portion of the calcaneus ➔. Bones are osteopenic due to disuse, but there is no evidence of infection. (**Right**) AP radiograph shows below the knee amputation (BKA). Amputation sites are well corticated ➔; loss of cortication is an important sign of osteomyelitis. A small amount of unilaminar periosteal new bone is present ➔ and within normal limits. Bony stump is covered by a soft tissue flap.*

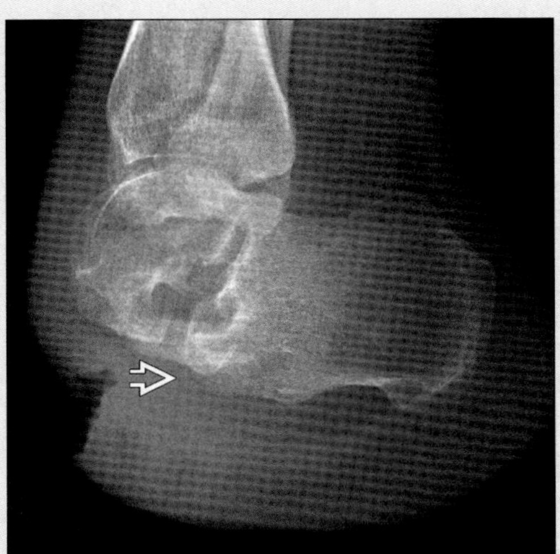

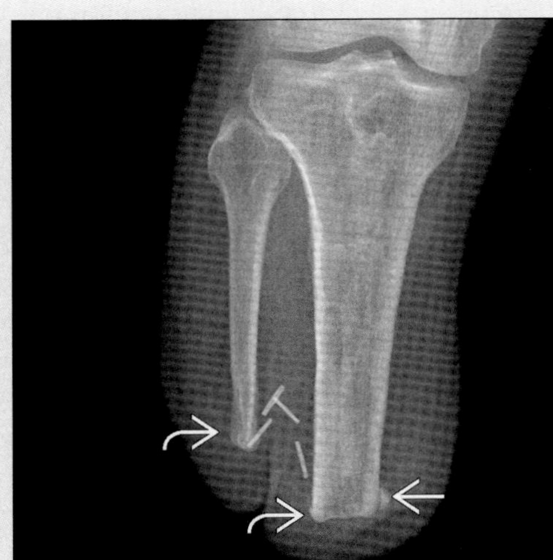

AMPUTATION, FOOT/ANKLE

TERMINOLOGY

Definitions

- Disarticulation: Amputation at level of joint, leaving proximal joint surface intact
- Resection arthroplasty: Amputation of phalangeal or metatarsal (MT) head, leaving distal osseous structures in place
  - Performed for toe deformities

IMAGING

General Features

- Location
  - Toes: Distal amputation for osteomyelitis or resection arthroplasty for deformity
  - Transmetatarsal: Through MT shafts
  - Lisfranc joint: Usually a disarticulation of tarsometatarsal joints, sometimes 2nd &/or 1st MT base left in situ
  - Chopart joint: Disarticulation of talonavicular and calcaneocuboid joints
  - Syme amputation: Disarticulation of ankle joint + resection of malleoli
    - Heel pad used to cushion amputation site
  - Calcanectomy: Excision of calcaneus only
  - Below the knee amputation (BKA): Amputation at level of tibial and fibular shafts

Radiographic Findings

- Stump often develops sclerosis due to stress
  - Amputation site should develop well-defined rim of cortication
    - Loss of cortication at stump site is indicative of infection
- Small amount of unilaminar periosteal reaction at amputation site is normal
  - Usually well-defined, matures in early months after amputation
- Extensive, fluffy, or multilaminar periosteal reaction suggests infection or insufficiency fracture

MR Findings

- Bone marrow usually homogeneously fatty after amputation
- Small foci of mildly increased signal on STIR often seen
  - Probably due to fibrosis or infarcts
- Confluent areas of low T1, high T2 or STIR signal probably due to osteomyelitis
- Sinus tract extending to bone is almost always associated with osteomyelitis

DIFFERENTIAL DIAGNOSIS

Acroosteolysis

- Loss of terminal tufts of digits
- Due to thermal injury, vasculitis, scleroderma, etc.

Amniotic Band Syndrome

- Congenital amputation due to in utero constriction of extremity by strands of amnion

Fibular Hemimelia

- Variable hypoplasia or aplasia of fibula and lateral bones of foot
- Most common lower limb skeletal anomaly, usually unilateral
- Associated with proximal focal femoral deficiency

PATHOLOGY

Indications for Amputation

- Osteomyelitis
- Trauma
- Charcot joint complicated by nonhealing soft tissue ulcer
- Tumor
- Fibular hemimelia

CLINICAL ISSUES

Natural History & Prognosis

- Syme amputation requires prosthesis designed to transfer weightbearing from amputation site
- Many amputations alter foot biomechanics
  - May lead to fractures, dislocations, ulcers in remaining portions of foot
- Significant failure rate of amputation for osteomyelitis in diabetic patients
  - Most significantly linked to control of blood glucose levels
  - Also linked to smoking, duration of ulcer
- Chopart amputation falls into equinus due to muscle imbalance
  - Tendon transfers performed in attempt to prevent equinus
    - Tibialis anterior transferred to insert on talar neck
    - Extensor tendons transferred to sole of foot
- Patients may suffer phantom limb pain
- Traumatic neuroma may develop and cause pain
  - MR shows fusiform nerve thickening at amputation
  - Neuroma enhances with gadolinium administration

DIAGNOSTIC CHECKLIST

Consider

- MR important part of assessment of need for amputation
  - Osteomyelitis that improves on MR over 3-month interval can often be treated medically rather than with amputation

SELECTED REFERENCES

1. Valabhji J et al: Conservative management of diabetic forefoot ulceration complicated by underlying osteomyelitis: the benefits of magnetic resonance imaging. Diabet Med. 26(11):1127-34, 2009
2. Younger AS et al: Risk factors for failure of transmetatarsal amputation in diabetic patients: a cohort study. Foot Ankle Int. 30(12):1177-82, 2009
3. Anthony T et al: Transmetatarsal amputation: assessment of current selection criteria. Am J Surg. 192(5):e8-11, 2006
4. Baumhauer JF et al: Total calcanectomy for the treatment of chronic calcaneal osteomyelitis. Foot Ankle Int. 19(12):849-55, 1998

INDEX

A

INDEX

INDEX

INDEX

INDEX

C

INDEX

INDEX

D

INDEX

INDEX

INDEX

INDEX

G

INDEX

H

INDEX

INDEX

INDEX

L

INDEX

M

INDEX

INDEX

INDEX

N

O

INDEX

P

INDEX

INDEX

Q

R

INDEX

INDEX

S

INDEX

INDEX

INDEX

T

INDEX

INDEX

U

V

W

X